The chemistry of

organic germanium, tin and lead compounds

THE CHEMISTRY OF FUNCTIONAL GROUPS

A series of advanced treatises founded by Professor Saul Patai and under the general editorship of Professor Zvi Rappoport

The chemistry of alkenes (2 volumes)
The chemistry of the carbonyl group (2 volumes)
The chemistry of the ether linkage
The chemistry of the amino group
The chemistry of the nitro and nitroso groups (2 parts)
The chemistry of carboxylic acids and esters
The chemistry of the carbon–nitrogen double bond
The chemistry of amides
The chemistry of the cyano group
The chemistry of the hydroxyl group (2 parts)
The chemistry of the azido group
The chemistry of acyl halides
The chemistry of the carbon–halogen bond (2 parts)
The chemistry of the quinonoid compounds (2 volumes, 4 parts)
The chemistry of the thiol group (2 parts)
The chemistry of the hydrazo, azo and azoxy groups (2 volumes, 3 parts)
The chemistry of amidines and imidates (2 volumes)
The chemistry of cyanates and their thio derivatives (2 parts)
The chemistry of diazonium and diazo groups (2 parts)
The chemistry of the carbon–carbon triple bond (2 parts)
The chemistry of ketenes, allenes and related compounds (2 parts)
The chemistry of the sulphonium group (2 parts)
Supplement A: The chemistry of double-bonded functional groups (3 volumes, 6 parts)
Supplement B: The chemistry of acid derivatives (2 volumes, 4 parts)
Supplement C: The chemistry of triple-bonded functional groups (2 volumes, 3 parts)
Supplement D: The chemistry of halides, pseudo-halides and azides (2 volumes, 4 parts)
Supplement E: The chemistry of ethers, crown ethers, hydroxyl groups and their sulphur analogues (2 volumes, 3 parts)
Supplement F: The chemistry of amino, nitroso and nitro compounds and their derivatives (2 volumes, 4 parts)
The chemistry of the metal–carbon bond (5 volumes)
The chemistry of peroxides
The chemistry of organic selenium and tellurium compounds (2 volumes)
The chemistry of the cyclopropyl group (2 volumes, 3 parts)
The chemistry of sulphones and sulphoxides
The chemistry of organic silicon compounds (3 volumes, 6 parts)
The chemistry of enones (2 parts)
The chemistry of sulphinic acids, esters and their derivatives
The chemistry of sulphenic acids and their derivatives
The chemistry of enols
The chemistry of organophosphorus compounds (4 volumes)
The chemistry of sulphonic acids, esters and their derivatives
The chemistry of alkanes and cycloalkanes
Supplement S: The chemistry of sulphur-containing functional groups
The chemistry of organic arsenic, antimony and bismuth compounds
The chemistry of enamines (2 parts)
The chemistry of organic germanium, tin and lead compounds (2 volumes, 3 parts)
The chemistry of dienes and polyenes (2 volumes)
The chemistry of organic derivatives of gold and silver

UPDATES

The chemistry of α-haloketones, α-haloaldehydes and α-haloimines
Nitrones, nitronates and nitroxides
Crown ethers and analogs
Cyclopropane derived reactive intermediates
Synthesis of carboxylic acids, esters and their derivatives
The silicon–heteroatom bond
Synthesis of lactones and lactams
Syntheses of sulphones, sulphoxides and cyclic sulphides

Patai's 1992 guide to the chemistry of functional groups — *Saul Patai*

C–Ge, C–Sn, C–Pb

The chemistry of
organic germanium, tin and lead compounds

Volume 2

Part 1

Edited by

ZVI RAPPOPORT

The Hebrew University, Jerusalem

2002

JOHN WILEY & SONS, LTD

An Interscience® Publication

Telephone (+44) 1243 779777

Email (for orders and customer service enquiries): cs-books@wiley.co.uk
Visit our Home Page on www.wileyeurope.com or www.wiley.com

Other Wiley Editorial Offices

John Wiley & Sons Inc., 111 River Street, Hoboken, NJ 07030, USA

Jossey-Bass, 989 Market Street, San Francisco, CA 94103-1741, USA

Wiley-VCH Verlag GmbH, Boschstr. 12, D-69469 Weinheim, Germany

John Wiley & Sons Australia Ltd, 33 Park Road, Milton, Queensland 4064, Australia

John Wiley & Sons (Asia) Pte Ltd, 2 Clementi Loop #02-01, Jin Xing Distripark, Singapore 129809

John Wiley & Sons Canada Ltd, 22 Worcester Road, Etobicoke, Ontario, Canada M9W 1L1

Library of Congress Cataloging-in-Publication Data

The chemistry of organo-germanium, tin, and lead compounds / edited by Zvi Rappoport and Yitzhak Apeloig.
p. cm.—(Chemistry of functional groups)
Includes bibliographical references and index.
ISBN 0-471-49738-X (v. 2 : alk. paper)
1. Organogermanium compounds. 2. Organotin compounds. 3. Organolead compounds. I. Rappoport, Zvi. II. Apeloig, Yitzhak. III. Series.

QD412.G5 C49 2001
547′.05684–dc21

2001026197

British Library Cataloguing in Publication Data

A catalogue record for this book is available from the British Library

ISBN 0-471-49738-X

Typeset in 9/10pt Times by Laserwords Private Limited, Chennai, India
Printed and bound in Great Britain by Biddles Ltd, Guildford, Surrey
This book is printed on acid-free paper responsibly manufactured from sustainable forestry in which at least two trees are planted for each one used for paper production.

Dedicated to
the memory of

Nahum

and

Zeez

Contributing authors

Klavdiya A. Abzaeva	A. E. Favorsky Institute of Chemistry, Siberian Branch of the Russian Academy of Sciences, 1 Favorsky Str., 664033 Irkutsk, Russia
Yuri I. Baukov	Department of General and Bioorganic Chemistry, Russian State Medical University, 1 Ostrovityanov St, 117997 Moscow, Russia
Sergey E. Boganov	N. D. Zelinsky Institute of Organic Chemistry of the Russian Academy of Sciences, Leninsky prospect, 47, 119991 Moscow, Russian Federation
Michael W. Carland	School of Chemistry, The University of Melbourne, Victoria, Australia, 3010
Annie Castel	Laboratoire d'Hétérochimie Fondamentale et Appliquée, UMR 5069 du CNRS, Université Paul Sabatier, 31062 Toulouse cedex, France
Marvin Charton	Chemistry Department, School of Liberal Arts and Sciences, Pratt Institute, Brooklyn, New York 11205, USA
Alexey N. Egorochkin	G. A. Razuvaev Institute of Metallorganic Chemistry of the Russian Academy of Sciences, 49 Tropinin Str., 603950 Nizhny Novgorod, Russia
Mikhail P. Egorov	N. D. Zelinsky Institute of Organic Chemistry of the Russian Academy of Sciences, Leninsky prospect, 47, 119991 Moscow, Russian Federation
Valery I. Faustov	N. D. Zelinsky Institute of Organic Chemistry of the Russian Academy of Sciences, Leninsky prospect, 47, 119991 Moscow, Russian Federation
Eric Fouquet	Laboratoire de Chimie Organique et Organométallique, Université Bordeaux I, 351, Cours de la Liberation, 33405 Talence Cedex, France
Gernot Frenking	Fachbereich Chemie, Philipps-Universität Marburg, Hans-Meerwein-Strasse, D-35032 Marburg, Germany
Inga Ganzer	Fachbereich Chemie, Philipps-Universität Marburg, Hans-Meerwein-Strasse, D-35032 Marburg, Germany
Ionel Haiduc	Department of Chemistry, University of Texas at El Paso, El Paso, Texas 79968, USA

Michael Hartmann	Fachbereich Chemie, Philipps-Universität Marburg, Hans-Meerwein-Strasse, D-35032 Marburg, Germany
Luba Ignatovich	Latvian Institute of Organic Synthesis, Aizkraukles 21, Riga, LV-1006 Latvia
Klaus Jurkschat	Lehrstuhl für Anorganische Chemie II der Universität Dortmund, D-44221 Dortmund, Germany
Thomas M. Klapötke	Department of Chemistry, Ludwig-Maximilians-University Munich, Butenandtstr. 5-13 (Building D), D-81377 Munich, Germany
Karl W. Klinkhammer	Institute for Inorganic Chemistry, University of Stuttgart, Pfaffenwaldring 55, D-70569 Stuttgart, Germany
Stanislav Kolesnikov	N. D. Zelinsky Institute of Organic Chemistry, Russian Academy of Sciences, 47 Leninsky prospect, 119991 Moscow, Russian Federation
Alexander I. Kruppa	Institute of Chemical Kinetics and Combustion, Novosibirsk-90, 630090 Russia
Vladimir Ya. Lee	Department of Chemistry, University of Tsukuba, Tsukuba, Ibaraki 305-8571, Japan
Tatyana V. Leshina	Institute of Chemical Kinetics and Combustion, Novosibirsk-90, 630090 Russia
Conor Long	School of Chemical Sciences, Dublin City University, Dublin 9, Ireland
Edmunds Lukevics	Latvian Institute of Organic Synthesis, Aizkraukles 21, Riga, LV-1006, Latvia
Heinrich Chr. Marsmann	Universität Paderborn, Fachbereich Chemie, Anorganische Chemie, Warburger Straße 100, D-30095 Paderborn, Germany
Michael Mehring	Lehrstuhl für Anorganische Chemie II der Universität Dortmund, D-44221 Dortmund, Germany
Josef Michl	Department of Chemistry and Biochemistry, University of Colorado, Boulder, CO 80309-0215, USA
Oleg M. Nefedov	N. D. Zelinsky Institute of Organic Chemistry, Russian Academy of Sciences, 47 Leninsky prospect, 119991 Moscow, Russian Federation
Renji Okazaki	Department of Chemical and Biological Sciences, Faculty of Science, Japan Women's University, 2-8-1 Mejirodai, Bunkyo-ku, Tokyo 112–8681, Japan
Keith H. Pannell	Department of Chemistry, University of Texas at El Paso, El Paso, Texas 79968, USA
Mary T. Pryce	School of Chemical Sciences, Dublin City University, Dublin 9, Ireland
Olga Pudova	Latvian Institute of Organic Synthesis, Aizkraukles 21, Riga, LV-1006, Latvia
Claudia M. Rienäcker	Department of Chemistry, Ludwig-Maximilians-University Munich, Butenandtstr. 5-13 (Building D), D-81377 Munich, Germany

José M. Riveros	Institute of Chemistry, University of São Paulo, Caixa Postal 26077, São Paulo, Brazil, CEP 05513-970
Pierre Riviere	Laboratoire d'Hétérochimie Fondamentale et Appliquée, UMR 5069 du CNRS, Université Paul Sabatier, 31062 Toulouse cedex, France
Monique Riviere-Baudet	Laboratoire d'Hétérochimie Fondamentale et Appliquée, UMR 5069 du CNRS, Université Paul Sabatier, 31062 Toulouse cedex, France
Carl H. Schiesser	School of Chemistry, The University of Melbourne, Victoria, Australia, 3010
Akira Sekiguchi	Department of Chemistry, University of Tsukuba, Tsukuba, Ibaraki 305-8571, Japan
Hemant K. Sharma	Department of Chemistry, University of Texas at El Paso, El Paso, Texas 79968, USA
Keiko Takashima	Department of Chemistry, University of Londrina, Caixa Postal 6001, Londrina, PR, Brazil, CEP 86051-970
Stanislav N. Tandura	N. D. Zelinsky Institute of Organic Chemistry, Russian Academy of Sciences, 47 Leninsky prospect, 119991 Moscow, Russian Federation
Marc B. Taraban	Institute of Chemical Kinetics and Combustion, Novosibirsk-90, 630090 Russia
Norihiro Tokitoh	Institute for Chemical Research, Kyoto University, Gokasho, Uji, Kyoto 611-0011, Japan
Frank Uhlig	Universität Dortmund, Fachbereich Chemie, Anorganische Chemie II, Otto-Hahn-Str. 6, D-44221 Dortmund, Germany
Olga S. Volkova	Institute of Chemical Kinetics and Combustion, Novosibirsk-90, 630090 Russia
Mikhail G. Voronkov	A. E. Favorsky Institute of Chemistry, Siberian Branch of the Russian Academy of Sciences, 1 Favorsky Str., 664033 Irkutsk, Russia
Ilya Zharov	Department of Chemistry and Biochemistry, University of Colorado, Boulder, CO 80309-0215, USA

Foreword

The preceding volume on *The Chemistry of Organic Germanium, Tin and Lead Compounds* in 'The Chemistry of Functional Groups' series (S. Patai, Ed.) appeared in 1995. The appearance of the present two-part volume seven years later reflects the rapid growth of the field.

The book covers two types of chapters. The majority are new chapters on topics which were not covered in the previous volume. These include chapters on reaction mechanisms involving the title organic derivatives, on reactive intermediates derived from them, like cations and carbene analogs, on NMR spectra, and on gas phase and mass spectrometry of organic germanium, tin and lead derivatives. There are chapters on their alkaline and alkaline earth metal compounds, on highly reactive multiply-bonded derivatives involving the title elements and on their hypervalent compounds, their synthetic applications, biological activities, polymers, cage compounds, unsaturated three membered ring derivatives and a new germanium superacid.

The second group of chapters are updates or extensions of material included in previous chapters. These include chapters on theory, on comparison of the derivatives of the three metals, on new advances in structural and photochemistry and in substituent effects and acidity, basicity and complex formation.

The volume opens with a new historical chapter on the genesis and evolution of organic compounds of the three elements, written by one of the pioneers in the field. We hope that such a historical background adds perspectives to those working both in the field and outside it.

The contributing authors to the book come from nine countries including some from Russia and Latvia who contributed several chapters. Part of the work in the field in these countries was covered by articles in Russian which were frequently not easily available to non-Russian readers. We now have many references including *Chemical Abstract* citations which will facilitate access to these articles.

The literature coverage in the book is mostly up to mid- or late-2001.

One originally planned chapter on radical reactions was not delivered, but part of the material can be found in another, more mechanistically oriented chapter.

This and the preceding volume should be regarded as part of a larger collection of books which appeared in recent years in 'The Chemistry of Functional Groups' series and deal with the chemistry of organic derivatives of the group 14 elements (excluding carbon). These also include four parts on the chemistry of organic silicon compounds (Z. Rappoport and Y. Apeloig, Eds., Vol. 2, parts 1–3, 1998 and Vol. 3, 2001) which follow two earlier volumes (S. Patai and Z. Rappoport, Eds., 1989) and an update volume, *The Silicon-Heteroatom Bond* (1991). The 136 chapters in the ten volumes cover extensively the main aspects of the chemistry of this group in the periodic table. Some comparisons of the derivatives of these groups appear both in the present and in earlier volumes.

This book was planned to be coedited by Prof. Y. Apeloig from the Technion in Haifa, Israel, but he was elected to the presidency of his institute and was unable to proceed with the editing beyond its early stage. I want to thank him for the effort that he invested and for his generous advice. I also want to thank the authors for their contributions.

I will be grateful to readers who draw my attention to mistakes in the present volume, or mention omissions and new topics which deserve to be included in a future volume on the chemistry of germanium, tin and lead compounds.

Jerusalem
April 2002

ZVI RAPPOPORT

The Chemistry of Functional Groups Preface to the series

The series 'The Chemistry of Functional Groups' was originally planned to cover in each volume all aspects of the chemistry of one of the important functional groups in organic chemistry. The emphasis is laid on the preparation, properties and reactions of the functional group treated and on the effects which it exerts both in the immediate vicinity of the group in question and in the whole molecule.

A voluntary restriction on the treatment of the various functional groups in these volumes is that material included in easily and generally available secondary or tertiary sources, such as Chemical Reviews, Quarterly Reviews, Organic Reactions, various 'Advances' and 'Progress' series and in textbooks (i.e. in books which are usually found in the chemical libraries of most universities and research institutes), should not, as a rule, be repeated in detail, unless it is necessary for the balanced treatment of the topic. Therefore each of the authors is asked not to give an encyclopaedic coverage of his subject, but to concentrate on the most important recent developments and mainly on material that has not been adequately covered by reviews or other secondary sources by the time of writing of the chapter, and to address himself to a reader who is assumed to be at a fairly advanced postgraduate level.

It is realized that no plan can be devised for a volume that would give a complete coverage of the field with no overlap between chapters, while at the same time preserving the readability of the text. The Editors set themselves the goal of attaining reasonable coverage with moderate overlap, with a minimum of cross-references between the chapters. In this manner, sufficient freedom is given to the authors to produce readable quasi-monographic chapters.

The general plan of each volume includes the following main sections:

(a) An introductory chapter deals with the general and theoretical aspects of the group.

(b) Chapters discuss the characterization and characteristics of the functional groups, i.e. qualitative and quantitative methods of determination including chemical and physical methods, MS, UV, IR, NMR, ESR and PES — as well as activating and directive effects exerted by the group, and its basicity, acidity and complex-forming ability.

(c) One or more chapters deal with the formation of the functional group in question, either from other groups already present in the molecule or by introducing the new group directly or indirectly. This is usually followed by a description of the synthetic uses of the group, including its reactions, transformations and rearrangements.

(d) Additional chapters deal with special topics such as electrochemistry, photochemistry, radiation chemistry, thermochemistry, syntheses and uses of isotopically labelled compounds, as well as with biochemistry, pharmacology and toxicology. Whenever applicable, unique chapters relevant only to single functional groups are also included (e.g. 'Polyethers', 'Tetraaminoethylenes' or 'Siloxanes').

This plan entails that the breadth, depth and thought-provoking nature of each chapter will differ with the views and inclinations of the authors and the presentation will necessarily be somewhat uneven. Moreover, a serious problem is caused by authors who deliver their manuscript late or not at all. In order to overcome this problem at least to some extent, some volumes may be published without giving consideration to the originally planned logical order of the chapters.

Since the beginning of the Series in 1964, two main developments have occurred. The first of these is the publication of supplementary volumes which contain material relating to several kindred functional groups (Supplements A, B, C, D, E, F and S). The second ramification is the publication of a series of 'Updates', which contain in each volume selected and related chapters, reprinted in the original form in which they were published, together with an extensive updating of the subjects, if possible, by the authors of the original chapters. A complete list of all above mentioned volumes published to date will be found on the page opposite the inner title page of this book. Unfortunately, the publication of the 'Updates' has been discontinued for economic reasons.

Advice or criticism regarding the plan and execution of this series will be welcomed by the Editors.

The publication of this series would never have been started, let alone continued, without the support of many persons in Israel and overseas, including colleagues, friends and family. The efficient and patient co-operation of staff-members of the publisher also rendered us invaluable aid. Our sincere thanks are due to all of them.

The Hebrew University
Jerusalem, Israel

SAUL PATAI
ZVI RAPPOPORT

Sadly, Saul Patai who founded 'The Chemistry of Functional Groups' series died in 1998, just after we started to work on the 100th volume of the series. As a long-term collaborator and co-editor of many volumes of the series, I undertook the editorship and I plan to continue editing the series along the same lines that served for the preceeding volumes. I hope that the continuing series will be a living memorial to its founder.

The Hebrew University
Jerusalem, Israel
June 2002

ZVI RAPPOPORT

Contents

List of abbreviations used

Ac	acetyl (MeCO)
acac	acetylacetone
Ad	adamantyl
AIBN	azoisobutyronitrile
Alk	alkyl
All	allyl
An	anisyl
Ar	aryl
Bn	benzyl
Bz	benzoyl (C_6H_5CO)
Bu	butyl (also *t*-Bu or Bu^t)
CD	circular dichroism
CI	chemical ionization
CIDNP	chemically induced dynamic nuclear polarization
CNDO	complete neglect of differential overlap
Cp	η^5-cyclopentadienyl
Cp*	η^5-pentamethylcyclopentadienyl
DABCO	1,4-diazabicyclo[2.2.2]octane
DBN	1,5-diazabicyclo[4.3.0]non-5-ene
DBU	1,8-diazabicyclo[5.4.0]undec-7-ene
DIBAH	diisobutylaluminium hydride
DME	1,2-dimethoxyethane
DMF	*N,N*-dimethylformamide
DMSO	dimethyl sulphoxide
ee	enantiomeric excess
EI	electron impact
ESCA	electron spectroscopy for chemical analysis
ESR	electron spin resonance
Et	ethyl
eV	electron volt

Fc	ferrocenyl
FD	field desorption
FI	field ionization
FT	Fourier transform
Fu	furyl(OC_4H_3)
GLC	gas liquid chromatography
Hex	hexyl(C_6H_{13})
c-Hex	cyclohexyl(c-C_6H_{11})
HMPA	hexamethylphosphortriamide
HOMO	highest occupied molecular orbital
HPLC	high performance liquid chromatography
i-	iso
Ip	ionization potential
IR	infrared
ICR	ion cyclotron resonance
LAH	lithium aluminium hydride
LCAO	linear combination of atomic orbitals
LDA	lithium diisopropylamide
LUMO	lowest unoccupied molecular orbital
M	metal
M	parent molecule
MCPBA	*m*-chloroperbenzoic acid
Me	methyl
MNDO	modified neglect of diatomic overlap
MS	mass spectrum
n	normal
Naph	naphthyl
NBS	*N*-bromosuccinimide
NCS	*N*-chlorosuccinimide
NMR	nuclear magnetic resonance
Pc	phthalocyanine
Pen	pentyl(C_5H_{11})
Pip	piperidyl($C_5H_{10}N$)
Ph	phenyl
ppm	parts per million
Pr	propyl (also *i*-Pr or Pr^i)
PTC	phase transfer catalysis or phase transfer conditions
Py, Pyr	pyridyl (C_5H_4N)

R	any radical
RT	room temperature
s-	secondary
SET	single electron transfer
SOMO	singly occupied molecular orbital
t-	tertiary
TCNE	tetracyanoethylene
TFA	trifluoroacetic acid
THF	tetrahydrofuran
Thi	thienyl(SC_4H_3)
TLC	thin layer chromatography
TMEDA	tetramethylethylene diamine
TMS	trimethylsilyl or tetramethylsilane
Tol	tolyl(MeC_6H_4)
Tos or Ts	tosyl(*p*-toluenesulphonyl)
Trityl	triphenylmethyl(Ph_3C)
Xyl	xylyl($Me_2C_6H_3$)

In addition, entries in the 'List of Radical Names' in *IUPAC Nomenclature of Organic Chemistry*, 1979 Edition, Pergamon Press, Oxford, 1979, p. 305–322, will also be used in their unabbreviated forms, both in the text and in formulae instead of explicitly drawn structures.

CHAPTER 1

Genesis and evolution in the chemistry of organogermanium, organotin and organolead compounds

MIKHAIL G. VORONKOV and KLAVDIYA A. ABZAEVA

A. E. Favorsky Institute of Chemistry, Siberian Branch of the Russian Academy of Sciences, 1 Favorsky Str., 664033 Irkutsk, Russia
e-mail: voronkov@irioch.irk.ru

The task of science is to induce the future from the past

Heinrich Herz

The chemistry of organic germanium, tin and lead compounds — Vol. 2
Edited by Z. Rappoport

I. INTRODUCTION

Germanium, tin and lead are members of one family, called the silicon subgroup. Sometimes these elements are called mesoids as well, due both to their central position in the short version of Mendeleev's Periodic Table and to their valence shells, which occupy an intermediate place among the I–VII Group elements[1]. They can also be called the heavy elements of Group 14 of the Periodic Table.

The history of the silicon prototype of this family and its organic derivatives is elucidated in detail in the literature[2–5]. In contrast, we could not find any special accounts dealing with the history of organic germanium, tin and lead compounds. The only exception is a very brief sketch on the early history of the chemistry of organotin compounds[6]. Some scattered information on the organic compounds of germanium, tin and lead can be found in some monographs and surveys. In this chapter we try to fill the gaps in this field.

Humanity first encountered the heavy elements of Group 14 at different times; with germanium, it happened quite unusually in the middle of the 19th century. As with the discovery of the planet Neptune[7], which was first predicted by astronomers and almost immediately discovered, Mendeleev, who predicted the existence of three hither to unknown elements, reported at the Russian Chemical Society session on December 10, 1870 on the discovery of one of these elements as follows: '...to my mind, the most interesting among undoubtedly missing metals will be one that belongs to Group IV and the third row of the Periodic Table, an analog of carbon. It will be a metal, following silicon, so we call it '*eca*-silicon'[8]. Moreover, Mendeleev even predicted the physical and chemical properties of the virtual element[9–12]. Having no conclusive proof of the existence of *eca*-silicon, Mendeleev himself began experimental investigations aimed at finding it in different minerals[13]. It is noteworthy that as early as 1864 Newlands[14] and Meyer[15] suggested the possible existence of an element like *eca*-silicon and predicted its atomic weight. However, Mendeleev was the first to predict properties of the element in detail.

Fifteen years later the German chemist Winkler[16,17], working at the Freiberg Academy of Mines, was able to isolate during the investigation of a recently discovered mineral argirodit (Ag_6GeS_5) a new element in its free state. Initially, Winkler wanted to name the new element neptunium, after the newly discovered planet Neptune. However, this name seemed to be given for another falsely discovered element, so he called the new element germanium in honor of his motherland[18–21]. At the time several scientists sharply objected to this name. For example, one of them indicated that the name sounded like that of the flower Geranium while another proposed for fun to call the new element Angularium, i.e. angular (causing debates). Nevertheless, in a letter to Winkler, Mendeleev encouraged the use of the name germanium. It took same time until the identity of *eca*-silicon and germanium was established[18–22]. Polemics, as to which element germanium is analogous flared up ardently. At first, Winkler thought that the newly discovered element filled the gap between antimony and bismuth. Having learned about Winkler's discovery, almost simultaneously in 1886 Richter (on February 25, 1886) and Meyer (on February 27, 1886) wrote him that the discovered element appeared to be *eca*-silicon. Mendeleev first suggested that germanium is *eca*-cadmium, the analog of cadmium. He was surprised by the origin of the new element, since he thought that *eca*-silicon would be found in titanium–zirconium ores. However, very soon, he rejected his own suggestion and on March 2, 1886, he wired Winkler about the identity of germanium and *eca*-silicon. Apparently, this information raised doubts in Winkler's mind about the position of germanium in the Periodic Table. In his reply to Mendeleev's congratulation he wrote: '. . .at first I was of the opinion that the element had to fill up the gap between antimony and bismuth and coincide with *eca*-stibium in accordance with your wonderful, perfectly developed Periodic Table. Nevertheless, everything showed us we dealt with a perfectly well developed Periodic Table. But everything implied that we are dealing with *eca*-silicon[23]. The letter was read at the Russian Physical and Chemical Society section on March 7. Winkler reported that the properties of the element and its common derivatives corresponded closely to those predicted for *eca*-silicon. A second letter by Winkler was read in a Chemical Section meeting of the Russian Physical and Chemical Society on May 1,1886. Winkler reported that the properties of germanium and its simpler derivatives were surprisingly very similar to those predicted for *eca*-silicon[22,24]. This is reported in Winkler's paper in the Journal of the Russian Physical and Chemical Society entitled 'New metalloid Germanium', translated into Russian at the author's request[25,26].

An inspection of Table 1 impresses one by the precise way in which Mendeleev predicted the properties of germanium and its elementary derivatives.

In 1966, Rochow[27] somewhat criticized the accuracy of Mendeleev's predictions of the properties of *eca*-silicon (germanium). He stated: 'Mendeleev predicted that *eca*-silicon would decompose steam with difficulty, whereas germanium does not decompose it at

TABLE 1. The properties of *eca*-silicon (Es) and its derivatives predicted by Mendeleev[9–12,19,20] in comparison with the properties of germanium and several germanium derivatives[24–30]

Properties	M = Es	M = Ge
Atomic weight	72.0	72.3
Specific weight	5.5	5.469
Atomic volume	13.0	13.2
Specific weight of MO_2	4.7	4.703
B.p. of MCl_4	*ca* 90°	88°
Specific weight of MCl_4	1.9	1.887
B.p. of $M(C_2H_5)_4$	*ca* 160°	160°
Specific weight of $M(C_2H_5)_4$	0.96	1.0

all. This is to say that germanium is less metallic than was predicted. Mendeleev also said that acids would have a slight action on the element, but they have none; again it is a more negative element than was predicted. There are many more chemical facts[31] which point in the same direction: germanium is more electronegative than was expected by interpolation, and it actually behaves a great deal like arsenic'. Rochow was right to some extent. It is known[32,33] that in accordance with Mendeleev's predictions germanium has more metallic characteristics than silicon; in a thin layer or under high temperatures germanium reacts with steam, and it reacts very slowly with concentrated H_2SO_4, HNO_3, HF and Aqua Regia. In relation to the Allred and Rochow electronegativity scale[34,35] the electronegativity of germanium is higher than that of silicon. However, according to other scales[36–39] and to Chapter 2 of this book, the electronegativity of germanium is lower or approximately the same as that for silicon. As illustrated in Table 1 Mendeleev predicted not only the possibility of existence, but also the properties of the simple organogermanium derivative Et_4Ge.

It is noteworthy that Winkler synthesized Et_4Ge in 1887[23,29]. Its properties were consistent with those predicted by Mendeleev. Organogermanium chemistry was born at this time.

In contrast to germanium the exposure of mankind to tin and lead was much earlier and not so dramatic[18–21,28]. These two elements belong to the seven main elements known to ancient man[40]. Up to the seventeenth century, tin and lead were often confused, as is witnessed by their Latin names, i.e. Plumbum album, Plumbum candidum (Sn) and Plumbum nigrum (Pb). Tin was known in countries of the Near East at least from the middle of the third millennium BC. Lead became known to the Egyptians at the same time as iron and silver, and very probably earlier than tin[19,28].

Many of Mendeleev's predecessors (Pettenkofer, Dumas, Cooke, Graham and others) assumed that tin and lead cannot belong to the same group as silicon[12] and Mendeleev was the first to include them in the same group of his Periodic Table with silicon and *eca*-silicon. He made this courageous prediction based on the assumption that the unknown element *eca*-silicon should have properties intermediate between metals and nonmetals and that all these elements, including carbon, should belong to one group.

The forefather of the chemistry of organic compounds of tin and lead was the Swiss chemist Carl Löwig. In the middle of the nineteenth century in the Zurich University laboratory (which was not set up to handle toxic compounds), he developed for the first time several methods for the synthesis of common organic derivatives of these two elements and described their properties[41–44].

Following Edward Frankland, who paid attention to organotin compounds as early as 1853[45], Löwig became one of the founders of organometallic chemistry but, unfortunately, historians of chemistry have forgotten this. In spite of his work with rather toxic organotin and organolead compounds during a period of several years in the absence of safety precautions, Löwig lived a long life and died only in 1890 due to an accident.

It is necessary to outline the nomenclature that we use before starting to develop the genesis and evolution of the chemistry of organic derivatives of heavy elements of Group 14. From the moment of their appearance and to some extent up to now, the names of organic derivatives of tin and lead were based on the name of the corresponding metals. It should be mentioned that tin and lead are called quite differently in English, German, French and Russian — Tin, Zinn, Etein, Олово, and Lead, Blei, Plomb, Свинец, respectively. In addition, archaic names of these compounds (such as trimethyltin oxide and alkylgermanium acid) are incompatible with the modern nomenclature of organosilicon compounds, which are the prototypes of this mesoid group. In this chapter we use the nomenclature of organic compounds of germanium, tin and lead approved by IUPAC[46] in analogy with the nomenclature of organosilicon compounds, based on their Latin names (Germanium,

Stannum, Plumbum). It is not the central metallic atom that is named, but only its hydride MH_4 (germane, stannane, plumbane) and the substituents which replace hydrogen atoms in the hydride molecule. Compounds in which the metal atom valence is either higher or lower than 4 are named in analogy to the nomenclature of organosilicon compounds.

In this chapter, we have tried to gain some insight into the genesis and development of the chemistry of organic germanium, tin and lead compounds up to the end of the 20th century. We have also paid attention to the work of the early researchers which was sometimes forgotten in spite of their tedious work under more difficult conditions than in the present time, which laid the fundamental laws of the chemistry of organic germanium tin and lead compounds. The organic chemistry of the heavy elements (Ge, Sn, Pb) of the silicon sub-group has been previously reviewed extensively either in reviews devoted to organic derivatives of all these elements[1,47–73] or in separate reviews on organogermanium[74–86], organotin[87–106] and organolead compounds[107–112].

Valuable information can be also found in chapters devoted to organometallic compounds[113–123] and in many surveys[124–138]. Excellent bibliographical information on reviews devoted to organogermanium (369 references)[79], organotin (709 references)[100] and organolead compounds (380 references)[112] have been published in Russia. Unfortunately, all the literature cited did not review the historical aspect, so our attempt to extract from that vast body of information the chronological order of the genesis and development of the organic chemistry of germanium tin, and lead compounds was not an easy task. It forces us to re-study numerous original publications, in particular those published in the 19th century. Nevertheless, the references presented in chronological order still do not shed light on the evolution of this chemistry, but they have important bibliographic value.

II. ORGANOGERMANIUM COMPOUNDS

A. Re-flowering after Half a Century of Oblivion

Up to the middle of the 20th century organogermanium derivatives were the least understood among the analogous compounds of the silicon subgroup elements. As mentioned above[23,29] the first organogermanium compound, i.e. tetraethylgermane, was synthesized for the first time by Winkler in 1887 by the reaction of tetrachlorogermane and diethylzinc[23,29], i.e. a quarter century later than the first organic compounds of silicon, tin and lead were obtained.

The synthesis of Et_4Ge proved unequivocally that the germanium discovered by Winkler belong to Group IV of the Periodic Table and that it was identical to Mendeleev's *eca*-silicon. Consequently, Winkler was the forefather of both the new germanium element and also the chemistry of its organic derivatives, whereas Mendeleev was their Nostradamus.

During the period between 1887 and 1925 no new organogermanium compound was reported. The forty years of the dry season resulted mainly from the scarcity and high prices of germanium and its simplest inorganic derivatives. This reflected the low natural reserves of argirodit, the only mineral source of germanium known at that time. The picture changed dramatically when in 1922 new sources of germanium were discovered. In particular, 0.1–0.2% of Ge were found in a residue of American zinc ore after zinc removal[139,140]. Dennis developed a method for the isolation of tetrachlorogermane from the ore[141]. In 1924, 5.1% of Ge was found in germanite, a mineral from southwestern Africa. Rhenierite, a mineral from the Belgian Congo, containing 6–8% of Ge[142], became another source of germanium. In 1930–1940, processing wastes of coal ashes and sulfide ores became the main sources of germanium[34,141,143,144]. These developments allowed American, English and German chemists to start in 1925 to carry out fundamental investigations of organogermanium compounds, in spite of the fact that germanium was still very expensive[145–150].

Thus, the chemistry of organogermanium compounds actually started to develop in the second quarter of the twentieth century. Its founders were L. M. Dennis, C. A. Kraus, R. Schwartz and H. Bayer, whose results were published in 1925–1936. A period of low activity then followed in this field and was resumed only in the middle of the century by leaders such as E. Rochow, H. Gilman, H. H. Anderson, O. H. Johnson, R. West and D. Seyferth. Organogermanium chemistry started to flourish in the sixties when many new investigators joined the field. These included the French chemists M. Lesbre, J. Satge and P. Mazerolles, the German chemists M. Schmidt, H. Schmidbaur, M. Wieber, H. Schumann and J. Ruidisch, the English chemists F. Glockling and C. Eaborn, the Russian chemists V. F. Mironov, T. K. Gar, A. D. Petrov, V. A. Ponomarenko, O. M. Nefedov, S. P. Kolesnikov, G. A. Razuvaev, M. G. Voronkov and N. S. Vyazankin, the Dutch chemist F. Rijkens the American chemist J. S. Thayer and others.

Activity was stimulated by the intensive development of the chemistry of organometallic compounds, particularly of the silicon and tin derivatives. The chemistry of organogermanes was significantly developed as well due to the essential role of germanium itself and its organic derivatives in electronics[151,152], together with the discovery of their biological activities (including anticancer, hypotensive, immunomodulating and other kinds of physiological action)[80,81,86,153]. In addition, a progressive decrease in the prices of elemental germanium and its derivatives expanded their production and helped their growth. The rapid expansion of organogermanium chemistry is clearly evident due to the increase in the number of publications in this field.

From 1888 till 1924 there were no publications and prior to 1934 just 26 publications were devoted to organogermanes[154]. Only 25 references on organogermanium compounds were listed in an excellent monograph by Krause and Grosse published in 1937[155]; 60 publications appeared before 1947[156], 99 before 1950[157] and 237 during the period 1950–1960[48,78] By 1967 the number of publications was over 1800 and by 1971 it exceeded 3000[36,37]. By 1970 about 100 publications had appeared annually[36,79] and by this time 370 reviews dealing with organogermanium compounds had appeared[79].

In 1951 already 230 organogermanium compounds were known[157], in 1961 there were 260[158] and in 1963 there were more than 700[159].

As the chemistry of organogermanium compounds is three-quarters of a century younger than the organic chemistry of tin and lead, it is reasonable to consider in this chapter the most important references published before 1967, when two classical monographs were published[36,37,78]. Due to space limitation we will avoid, where possible, citing reaction equations in the hope that they will be clear to the readers.

B. Organometallic Approaches to a C−Ge and Ge−Ge Bond

Thirty-eight years after Winkler developed the organozinc method for the synthesis of tetraethylgermane, Dennis and Hance[160] reproduced it, but this method for synthesis of aliphatic germanium derivative was not used later. However, in the years 1927–1935 arylzinc halides were used for the synthesis of tetraarylgermanes[23,161–165].

Application of Grignard reagents in organometallic synthesis led to the synthesis of common aliphatic, aromatic and alicyclic germanium derivatives during the years 1925–1932. Dennis and Hance[160] were the first to produce in 1925 tetraalkylgermanes R_4Ge (R = Me, Et, Pr, Bu, Am)[145,160,166–169] from Grignard reagents. Kraus and Flood[148] used organomagnesium reagents for the synthesis of tetraalkylgermanes. In 1925 Morgan and Drew[149], and later Kraus and Foster[161] synthesized tetraphenylgermane, the first compound having a Ph–Ge bond, from $GeCl_4$ and PhMgBr. The maximum (70–75%) yield was reached at a $GeCl_4$: PhMgBr ratio of 1 : 5[170,171].

In 1934 Bauer and Burschkies[172], and only later other researchers[173–176] showed for the first time that a reaction of $GeCl_4$ and Grignard reagents results in hexaorganyldigermanes R_3GeGeR_3 (R = 4-MeC_6H_4 and $PhCH_2$). In 1950, Johnson and Harris[173] noted the formation of hexaphenyldigermane in the reaction of $GeCl_4$ with an excess of PhMgBr. Glocking and Hooton[177,178] later found that if the above reaction was carried out in the presence of magnesium metal, hexaphenyldigermane $Ph_3GeGePh_3$ was produced in a higher yield along with Ph_4Ge. Seyferth[176] and Glockling and Hooton[178] concluded that the intermediate product in the reaction of $GeGl_4$ and ArMgBr leading to $Ar_3GeGeAr_3$ was $Ar_3GeMgBr$.

In line with this assumption Gilman and Zeuech[179] found in 1961 that Ph_3GeH reacted with several Grignard reagents (such as CH_2=$CHCH_2MgX$ or ArMgBr) to give Ph_3GeMgX (X = Cl, Br). The latter has cleaved THF, since a product of the reaction followed by hydrolysis seemed to be $Ph_3Ge(CH_2)_4OH$. Mendelsohn and coworkers[180] indicated the possibility of the formation of R_3GeMgX in the reaction of $GeCl_4$ and Grignard reagents.

In the period 1931–1950 the organomagnesium syntheses became the laboratory practice for preparing tetraorganylgermanes.

Tetraalkyl- and tetraarylgermanes containing bulky organic substituents could be synthesized only with difficulty, if at all, using Grignard reagents. In this case the reaction resulted in triorganylhalogermane[181–183].

Organylhalogermanes $R_{4-n}GeX_n (n = 1–3)$ were prepared for the first time in 1925 by Morgan and Drew[149], who isolated phenylbromogermanes $Ph_{4-n}GeBr_n$ (n = 1, 3) together with tetraphenylgermane from the reaction of $GeBr_4$ and PhMgBr. However, the organomagnesium synthesis of organylhalogermanes has not found much use due to the simultaneous production of other compounds and the difficulty of separating them. The only exceptions were R_3GeX products having bulky R substituents[172,181,183,184].

In the reaction of $HGeCl_3$ and MeMgBr, Nefedov and Kolesnikov[185] obtained a mixture of both liquid and solid permethyloligogermanes $Me(Me_2Ge)_nMe$.

In 1932, Krause and Renwanz[186] synthesized the first heterocyclic organogermanium compound, tetra-2-thienylgermane, from the corresponding Grignard reagent. In the same year Schwarz and Reinhardt[150] synthesized by the same method the first germacycloalkanes (1,1-dichloro- and 1,1-diethyl-1-germacyclohexanes). They also synthesized tetra-N-pyrrolylgermane by the reaction of $GeCl_4$ and potassium pyrrole.

Since 1926 the organomagnesium synthesis was also used for preparing more complex tetraorganylgermanes[145,162,163,169,172,187–190] such as R_3GeR', $R_2GeR'_2$ and $R_2GeR'R''$.

The first unsaturated organogermanium compounds having α,β- or β,γ-alkynyl groups at the Ge atom were synthesized in 1956–1957 by Petrov, Mironov and Dolgy[191,192] and by Seyferth[176,193,194] using Grignard or Norman reagents.

In 1925, the Dennis group used along with the organozinc and organomagnesium synthesis of tetraorganylgermanes, also the Wurtz–Fittig reaction (i.e. the reaction of aryl halides with sodium metal and tetrahalogermanes[168,187,195]). The Wurtz–Fittig reaction was extensively employed for the synthesis of organogermanium compounds having Ge–Ge bonds such as R_3GeGeR_3. The first representative of the $Ph_3GeGePh_3$ series was synthesized in 1925 by Morgan and Drew[149], and subsequently by Kraus and coworkers[161,196], using the reaction of triphenylbromogermane and sodium metal in boiling xylene. Analogously, Bauer and Burschkies[172] produced in 1934 R_3GeGeR_3, R = 4-MeC_6H_4 and $PhCH_2$. In addition, they found that the reaction of $GeCl_4$, Na and RBr (R = 4-MeC_6H_4) led to R_3GeGeR_3 in good yield together with R_4Ge. In 1932,

Kraus and Flood[148] found that hexaethyldigermane was not formed in the reaction of triethylbromogermane and sodium metal in boiling xylene. However, they produced hexaethyldigermane by heating Et_3GeBr and Na in a sealed tube at 210–270 °C without solvent or by the reaction of Et_3GeBr and Na in liquid ammonia.

The possibility of producing diphenylgermylene alkali metal derivatives like Ph_2GeM_2 (M = Li, Na) was shown in 1952 by Smyth and Kraus[197] when they obtained Ph_2GeNa_2 by cleavage of Ph_4Ge with concentrated solution of sodium in liquid ammonia. In 1930, Kraus and Brown[198] produced a mixture of perphenyloligocyclogermanes $(Ph_2Ge)_n$ by the reaction of sodium metal with diphenyldichlorogermane in boiling xylene. However, only in 1963 did Neumann and Kühlein[199] show that the main crystalline product of the reaction is octaphenylcyclotetragermane $(Ph_2Ge)_4$. Cleavage of $(Ph_2Ge)_n$ with sodium in liquid ammonia resulted in Ph_2GeNa_2. Reaction of $(PhGe)_4$ with iodine which resulted in cleavage of the Ge–Ge bond, allowed the authors[199] to synthesize the first organotetragermanes involving three Ge–Ge bonds $X[Ph_2Ge]_4X$ (X = I, Me, Ph). By the reaction of diphenyldichlorogermane and lithium (or sodium naphthalene) Neumann and Kuhlein[175,199,200] isolated higher perphenylcyclogermanes with $n = 5$ (37%) and $n = 6$ (17%). It is particularly remarkable that, unlike their homologs with $n = 4$, these compounds could not be cleaved with iodine.

In 1962–1965 Nefedov, Kolesnikov and coworkers[201–205] investigated the reaction of Me_2GeCl_2 with lithium metal in THF. The main products were $(Me_2Ge)_6$ (80% yield) at 20–45 °C and the polymer $(Me_2Ge)_n$ (50% yield) at 0 °C.

In 1966 Shorygin, Nefedov, Kolesnikov and coworkers[206] were the first to investigate and interpret the UV spectra of permethyloligogermanes $Me(Me_2Ge)_nMe$($n = 1$–5). The reaction of Et_2GeCl_2 with Li in THF led mostly to polydiethylgermane $(Et_2Ge)_n$[207]. At the same time Mironov and coworkers[208,209] obtained dodecamethylcyclohexagermane $(Me_2Ge)_6$ by the same procedure.

In 1969, Bulten and Noltes[210] synthesized the perethyloligogermanes $Et(Et_2Ge)_nEt$ ($n = 2$–6) by the organolithium method. The oligomer with $n = 6$ was thermally stable and heating at 250 °C for 8 hours resulted in only 20% decomposition.

By a reaction of Li amalgam with Ph_2GeBr_2, Metlesics and Zeiss[211] produced 1,2-dibromotetraphenyldigermane instead of the cyclic oligomers obtained previously in a similar reaction with Li metal. A reaction of Li amalgam with $PhGeBr_3$ gave $PhBr_2GeGeBr_2Ph$, the thermolysis of which resulted in $PhGeBr_3$.

Curiously, the reaction of phenyltrichlorogermane with sodium or potassium produced a compound $(PhGe)_n$, which Schwarz and Lewinsohn[187] mistook for hexaphenylhexagermabenzene Ph_6Ge_6. Five years later Schwartz and Schmeisser[212] found that the action of potassium metal on $PhGeCl_3$ yielded a product, assigned by them to be a linear hexamer having terminal Ge(III) atoms i.e. a biradical of a structure $^\bullet$(PhGe=GePh)$^\bullet$. They thought that this structure could be confirmed by addition reactions with bromine, iodine and oxygen, which indeed took place. However, HI and HBr were not involved in the addition reactions.

Two dozen years later Metlesics and Zeiss[213] obtained the same product by the reaction of $PhGeCl_3$ with Li amalgam. They found that the product was a polymer consisting of $(PhGe)_n$, $(Ph_2Ge)_n$ and $(PhGeO)_n$ chains.

In 1950–1960 it was found that triarylgermyl derivatives of alkali metals could be obtained by cleavage of Ge–H[214,215], C–Ge[174,195,216,217], Ge–Ge[218–221] and Ge–Hal[222] bonds by Li, Na or K in the appropriate solvents.

In 1950, Glarum and Kraus[214] investigated the reaction of alkylgermanes $R_{4-n}GeH_n$ ($n = 1$–3) and sodium metal in liquid ammonia. They found that alkylgermanes $RGeH_3$ reacted with Na to give $RGeH_2Na$.

As early as in 1927, Kraus and Foster[161] produced for the first time triphenylgermylsodium as its ammonia complex $Ph_3Ge(NH_3)_3Na$. They also found that the reaction of Ph_3GeNa with H_2O or NH_4Br in liquid ammonia led quantitatively to Ph_3GeH. The reaction of Ph_3GeNa and Ph_3GeF in liquid ammonia resulted in $Ph_3GeGePh_3$.

In 1957–1959, Gilman and coworkers[220,222] found that $Ph_3GeGePh_3$ was cleaved by sodium in THF solution in the presence of PhBr and Ph_4Ge to give Ph_3GeNa.

In 1932 it was found that the reaction of Ph_3GeNa with organic halides RX gave Ph_3GeR[196], whereas when R = Ph, Ph_3Ge[196,198,223,224] was isolated. The reaction of Ph_3GeNa with oxygen led to Ph_3GeONa[161]. In the years 1950–1952, Kraus and coworkers further developed this chemistry by studying the reactions of Ph_3GeNa with organic mono- and dihalides of different structure, such as $HCCl_3$, CCl_4[197], BCl_3[224] or $HSiCl_3$[225]. The product of the latter reaction was $(Ph_3Ge)_3SiH$.

In 1930, Kraus and Brown[198,226] prepared octaphenyltrigermane by the reaction of Ph_3GeNa and Ph_2GeCl_2. It was the first organogermanium compound with more than one Ge–Ge bond. The two Ge–Ge bonds could readily be cleaved by bromine. Kraus and Scherman[224] synthesized in 1933 the first unsymmetrical hexaorganyldigermane $Ph_3GeGeEt_3$ by the reaction of Ph_3GeNa and Et_3GeBr.

In 1932, Kraus and Flood[148] prepared the first compound having a Ge–Sn bond ($Ph_3GeSnMe_3$) by the reaction of Ph_3GeNa and Me_3SnBr. In 1934, Kraus and Nelson[227] synthesized $Ph_3GeSiEt_3$ by the reaction of Ph_3GeNa and Et_3SiBr.

The reaction of hexaethyldigermane and potassium in ethylamine solution led Kraus and Flood[148] to the first synthesis of triethylgermylpotassium. Its reaction with ethyl bromide resulted in Et_4Ge. However, attempts to cleave hexamethyldigermane either by potassium or by its alloy with sodium were unsuccessful[228].

The action of potassium metal with Me_3GeBr without solvent resulted in $Me_3GeGeMe_3$[228]. Gilman and coworkers[217–220,229] synthesized Ph_3GeK by cleavage of $Ph_3GeGePh_3$ with a sodium and potassium alloy in THF in the presence of an initiator (PhBr or Ph_4Ge). Triphenylgermylpotassium was produced in a 26% yield during a slow cleavage of Ph_3CGePh_3 by the same alloy[195]. The development of a method for the synthesis of Ph_3GeK opened a route to carry out its addition to double bonds, such as 1,1-diphenylethylene, which resulted in $Ph_2CH_2CH_2GePh_3$[220], or to activated conjugated bonds[220].

Lithium metal has been used for organogermanium synthesis since 1932, but organolithium compounds were used only since 1949[173,230]. Lithium and its organic derivatives were used in three approaches: (1) reactions of lithium and organogermanium compounds; (2) reactions of organolithium compounds with organic and inorganic germanium compounds; (3) synthesis based on compounds having a Ge–Li bond.

Although fundamental research in this field was undertaken in Gilman's laboratory, Kraus and Flood[148] were the pioneers in using lithium for the synthesis of organogermanium compounds. In 1932, they discovered that the reaction of Et_3GeX (X = Cl, Br) and lithium in ethylamine resulted in $Et_3GeGeEt_3$. With excess lithium, the Ge–Ge bond of hexaethyldigermane was cleaved to give Et_3GeLi. When the latter was treated with NH_3 or NH_4Br in an ethylamine solution, Et_3GeH was formed.

In 1950, Glarum and Kraus[214] developed a very convenient method for the synthesis of alkylgermyllithium compounds ($RGeH_2Li$) by the reaction of $RGeH_3$ with lithium in ethylamine solution. An analogous reaction of R_2GeH_2 and lithium led to R_2GeHLi. Later, Vyazankin, Razuvaev and coworkers[231–234] synthesized Et_3GeLi in $>90\%$ yield by the reaction of lithium and $(Et_3Ge)_2Hg$ or $(Et_3Ge)_2Tl$.

Gilman and coworkers[216,222] obtained Ph_3GeLi by a simpler method. The reaction of Ph_3GeBr with Li in THF gave the compound, although in a lower (52%) yield. In 1956,

Gilman and Gerow[229,235] synthesized Ph_3GeLi in 70% yield by the cleavage of Ph_4Ge with lithium metal in a diglyme solution. They later showed that aryl groups were cleaved from the Ge atom in the same solvent much more easily than alkyl or phenyl groups.

Tamborski and coworkers[236] found that the reaction of Ph_3GeCl and lithium metal in THF involved the intermediate formation of $Ph_3GeGePh_3$ and resulted in Ph_3GeLi.

Gross and Glockling[237] developed in 1964 a very effective method for the synthesis of $(Ph_2CH_2)_3GeLi$ based on the cleavage of $(PhCH_2)_4Ge$ by lithium in diglyme. Gross and Glockling[237,238] found that, when tetrabenzylgermane is treated with lithium, two $PhCH_2$ groups were cleaved, and $(PhCH_2)_2GeLi_2$ was probably formed.

The organolithium synthesis proved to be the simplest and most convenient route to organogermanium compounds, including those carrying bulky substitutes on the Ge atom. The method was first used in 1930 by Kraus and Brown[198] and found many applications shortly after.

In 1949, Johnson and Nebergall[230] showed that the use of RLi for R_4Ge production resulted in higher yields than that for RMgX. Ten years later Gilman and coworkers[174] found that the reaction of $GeBr_4$ and EtLi led to Et_4Ge and $Et_3GeGeEt_3$. In 1953, Summers[239] discovered that reaction of PhLi with GeI_2 gave a polymer $(Ph_2Ge)_n$. In contrast, the reaction of GeI_2 with Bu_2Hg produced 1,2-diiodotetrabutyldigermane[240].

Developed by Nefedov, Kolesnikov and coworkers[185,203], the reaction of RLi with $HGeCl_3$ resulted in linear and cyclic oligomers and polymers consisting of alternate Ge−Ge bonds.

The Ge−H bonds in triarylgermanes were cleaved as well by organolithium compounds to form Ar_3GeLi[235]. Together with the latter Ar_3GeR and $Ar_3GeGeAr_3$ were also formed[173,235]. Johnson and Harris[173] investigated the reaction of PhLi and Ph_3GeH and found that, depending on the mixing sequence of the reagents, the product could be either Ph_4Ge or $Ph_3GeGePh_3$. Trialkylgermanes reacted less readily than triarylgermanes with RLi (R = Bu, Ph)[241].

In 1956, Gilman and Gerow[229,235] and then Brook and Peddle[242] developed an effective, nearly quantitative method for the synthesis of Ph_3GeLi by the reaction of Ph_3GeH and BuLi.

Gilman and coworkers[220,229,235,243] found that Ph_3GeLi could be added to 1,1-diphenylethylene, 1-octadecene and benzalacetophenone (but not to 1-octene, cyclohexene and E-stilbene). The reaction of Ph_3GeLi with enolizable ketones followed equation 1[244].

$$Ph_3GeLi + CH_3COPh \longrightarrow Ph_3GeH + LiCH_2COPh \qquad (1)$$

On the other hand, addition of Ph_3GeLi to benzophenone gave $Ph_2(Ph_3Ge)COH$[244]. An analogous addition of Ph_3GeLi to formaldehyde and benzaldehyde led to Ph_3GeCH_2OH[244] and $Ph(Ph_3Ge)CHOH$[242,245], respectively. Triphenylgermyllithium adds to 1,4-benzalacetone (equation 2)[218] and reacts as a metal-active reagent with CH acids such as fluorene[195,246].

$$Ph_3GeLi + PhCH{=}CHCOPh \longrightarrow Ph(Ph_3Ge)CHCH_2COPh \qquad (2)$$

Chalcogens E (E = O, S, Se, Te) readily insert into the Ge−Li bond. For example, reaction of E with PhGeLi yields Ph_3GeELi (E = O, S, Se, Te)[247], Brook and Gilman found that triphenylgermyllithium was oxidized to Ph_3GeOLi, and carbon dioxide could easily be inserted into the molecule to give $Ph_3GeCOOLi$[235]. Thermal decomposition of $Ph_3GeCOOH$ led to Ph_3GeOH[195]. Triphenylgermyllithium cleaved the oxirane ring with ring opening to give $Ph_3GeCH_2CH_2OLi$[248].

The reactions of GeI_2 with organic lithium, manganese, aluminum and mercury derivatives[185,201,239,240,249] were widely investigated as a possible route for producing diorganylgermylene R_2Ge. However, the reaction proceeds in a complex manner and has no preparative application. However, Glocking and Hooton[249] discovered later that the reactions of GeI_2 and phenyllithium or mesitylmagnium bromide led to the corresponding products Ar_3GeLi or $Ar_3GeMgBr$ whose hydrolysis resulted in Ar_3GeH. The first bulky oligogermane, i.e. $(Ph_3Ge)_3GeH$, was obtained in 1963 by this reaction[249]. A year later Vyazankin and coworkers[250] synthesized methyl-tris(triphenylgermyl)germane $(Ph_3Ge)_3GeMe$.

C. Nonorganometallic Approaches to a C−Ge Bond

E. G. Rochow, whose name became famous due to his discovery of the direct synthesis of organohalosilanes from elementary silicon[2,4,5], tried to develop an analogous method for the synthesis of organohalogermanes. In 1947 he showed that the methylhalogermanes $MeGeX_3$ and Me_2GeX_2 were formed in the reaction of methyl chloride or methyl bromide and elementary germanium in the presence of copper or silver metals at 300–400 °C[213]. Later, he added EtCl, PrCl and PhCl[251–255] to the reaction. Generally, a mixture of alkylhalogermanes $R_{4-n}GeX_n$ ($n = 2$, 3) was obtained in the process. The product ratios were dependent on the temperature and the catalyst structure. When MeCl and EtCl were used a mixture of R_2GeCl_2 and $RGeCl_3$, R = Me, Et, was formed in a ratio very close to 2 : 1. The yields of metyltrichlorogermane were increased on increasing the temperature and were dependent on the copper content in the contact mass, as well as on the addition of Sb, As and $ZnCl_2$[66,191,256] to the reaction mixture.

In 1956–1958, this direct organylhalogermanes synthesis was thoroughly investigated at the Petrov, Mironov and Ponomarenko laboratory[66,191,257,258]. A variety of halides, such as allyl and methallyl chloride, allyl bromide[258] and CH_2Cl_2[259] (but, not vinyl chloride), were found to react. With the latter, $MeGeCl_3$ (27%), $Cl_3GeCH_2GeCl_3$ (23%) and $(CH_2GeCl_2)_3$ (19%) were produced. Alkyltribromogermanes $RGeBr_3$ (R = Pr, Bu) were synthesized by the reaction of the corresponding alkyl bromides with sponged germanium at 300–340 °C.

Alkyliodogermanes were produced by direct synthesis only in 1963–1966[260–264]. It is noteworthy that no compounds having Ge−H bonds (such as $RGeHCl_2$ or R_2GeHCl) were formed during the direct synthesis of alkylchlorogermanes, in contrast with the direct synthesis of alkylchlorosilanes.

A hydrogermylation reaction (the term was first introduced by Lukevics and Voronkov[52,53,77], i.e. the addition of organic and inorganic germanium derivatives having Ge−H bonds to unsaturated compounds) was first performed by Fischer, West and Rochow[265] in 1954. They isolated hexyltrichlorogermane (in 22% yield) after refluxing for 35 hours a mixture of trichlorogermane and 1-hexene in the presence of a benzoyl peroxide initiator. Two years later, the reaction of $HGeCl_3$ and other alkenes in the presence of the same initiator was carried out at 70–85 °C to give the appropriate alkyltrichlorogermanes in low yields (9–24%)[266] as well. In 1957, Gilman and coworkers added $HGeCl_3$ to 1-octene[267], 1-octadecene[217], cyclohexene[267], allyltriphenylsilane[268] and -germane[217] in the presence of benzoyl peroxide or under UV radiation.

In 1958 Ponomarenko and coworkers[269] found that $HGeCl_3$ was exothermally added to ethylene at 40 atm pressure in the presence of H_2PtCl_6 to give $EtGeCl_3$ in 25% yield. In the same year Mironov and Dzhurinskaya in Petrov's laboratory unexpectedly discovered that the reaction of $HGeCl_3$ and diverse unsaturated compounds proceeded exothermally at room temperature and without either catalyst or initiator[270–272]. On the contrary, the presence of either a catalyst or an initiator actually decreased the yield of the hydrogermylation products[270–272].

A noncatalytic hydrogermylation reaction was carried out at 85 °C in a sealed tube in 1956[266]. Furthermore, $HGeBr_3$[273], $HGeI_3$[274], R_2GeHCl (at 100–150 °C)[275], R_2GeHBr (at 150 °C)[275], R_2GeH_2 (at 140–150 °C)[276] and R_3GeH (at 50–200 °C)[276–278] were reacted in the noncatalytic hydrogermylation process. However, addition of R_3GeH to unsaturated compounds proceeded more easily in the presence of H_2PtCl_6[52,53].

In 1962, Satge and Lesbre[279,280] carried out for the first time hydrogemylation of the carbonyl group of aldehydes and ketones.

The best method for the synthesis of aryltrihalogermanes based on the reaction of aryl iodides and GeX_4 (X = Cl, Br) in the presence of copper powder was discovered by Mironov and Fedotov[281,282] in 1964. Bauer and Burschkies[181] discovered in 1932 an unusual way of Ge–C bond formation by condensation of $GeCl_4$ and aromatic amines according to equation 3. The reaction products were isolated as the corresponding substituted phenylgermsesquioxanes.

$$R_2NC_6H_5 + GeCl_4 \longrightarrow Cl_3GeC_6H_4NR_2 \cdot HCl \xrightarrow{H_2O} 1/n(R_2NC_6H_4GeO_{1.5})_n \quad (3)$$

In 1955, Seyferth and Rochow[283] developed a nontrivial method of Ge–Ge bond formation based on the insertion of a carbene (H_2C: formed from diazomethane) into a Ge–Cl bond of $GeCl_4$ to form $ClCH_2GeCl_3$. Later, Seyferth and coworkers[284,285] extended this approach to the formation of the $GeCH_2X$ (X = Cl, Br) group by the reaction of dihalocarbenes (generated from $PhHgCX_2Br$) with Ge–H bonds.

Kramer and Wright[286,287] and Satge and Riviére[288] demonstrated the possibility of carbene (formed from diazomethane) insertion into the Ge–H bond to give a Ge–CH_3 moiety. However, this reaction is of no practical application. It was more interesting to insert substituted carbenes (generated from diazo derivatives such as ethyl diazoacetate, diazoacetone and diazoacetophenone) into Ge–H bonds in the presence of copper powder. In this case a Ge–CH_2X group was formed, where X was the corresponding functional group[276,277,289].

In 1958, Nesmeyanov and coworkers[290] found that decomposition of aryldiazonium tetrafluoroborates with zinc dust in the presence of $GeCl_4$ resulted in formation of aryltrichlorogermanes in <30% yield, isolated as the corresponding arylgermsesquioxanes.

In 1960, Volpin, Kursanov and coworkers[291–293] showed that dihalogermylenes add to multiple bonds by reacting GeI_2 with tolan (PhC≡CPh) at 220–230 °C[292]. The main product of the reaction was assigned to 1,1-diiodo-2,3-diphenyl-1-germa-2-cyclopropene, which the authors considered to be a new three-membered heterocyclic aromatic system[291,292,294]. When this substance was allowed to react with RMgX (R = Me, Et), the iodine atoms were replaced by alkyl substituents, whereas upon the action of NaOH they were substituted by OH groups. The OH groups of the hydroxy derivative obtained were replaced by halogen[291] on reaction with HCl or HBr. However, it was established later that the isolated adduct was actually 1,1,4,4-tetraiodo-2,3,5,6-tetraphenyl-1,4-digerma-2,5-cyclohexadiene[295–298].

Reaction of GeI_2 and acetylene at 130–140 °C and 10 atm[299] gave 44% yield of an adduct whose structure was assigned to 1,1-diiodo-1-germa-2-cyclopropene (i.e. 1,1-diiododigermyrene)[299]. Its iodine atoms were replaced by OH and Cl atoms[299] and by Me groups using known reactions. However, X-ray analysis established the structure of the isolated chlorinated compound as 1,1,4,4-tetracloro-1,4-digerma-2,5-cyclohexadiene. Hydrogenation of 1,1,4,4-tetramethyl derivative synthesized from the latter afforded the 1,1,4,4-tetramethyl-1,4-digermacyclohexane, whereas its bromination led to $Me_2Ge(CH{=}CHBr)Br$[297]. Simultaneously, a polymer $(-I_2Ge-CH{=}CH-)_n$[299] with average molecular weight of 4300 (after removal of lower molecular weight fractions) was formed in a 56% yield. Probably the low molecular weight polymer fractions had macrocyclic structures resembling their silicon analog $(-R_2SiCH{=}CH-)_n$[300]. The reaction of acetylene with $GeBr_2$ leads to analogous polymers.

In 1960, Russian chemists found that GeI_2 acts easily with diarylmercuranes Ar_2Hg to give $Ar_{4-n}GeI_n$ ($n = 1, 2$) in good yield[301], together with ArHgI and Hg. In contrast, dialkyl mercury derivatives reduced GeI_2 to Ge metal, but did not form dialkyldiiodogermanes (one of the products was $I_2RGeGeRI_2$)[240].

In 1963, Mironov and Gar[302] showed that $GeCl_2$ and $GeBr_2$[273,303] (generated from $HGeX_3$) add to 1,3-butadiene to give the corresponding 1,1-dihalo-1-germa-3-cyclopentene. Analogously[304,305], GeI_2 adds to 2-methyl- and 2,3-dimethylbutadiene.

Another approach to the formation of a C−Ge bond resulting in organyltrihalogermanes was based on the reaction of dihalogermylenes (GeX_2) with organic halides. For this purpose, the more stable and easily available GeI_2 was usually used. In 1933, Flood and coworkers[306,307] discovered that the reaction of GeI_2 with alkyl iodides proceeds smoothly to give alkyltriiodogermanes. Pope[308] and Pfeiffer[309] and their coworkers performed analogous synthesis of $RSnI_3$ from SnI_2 as early as 1903. This reaction can be regarded as an insertion of diodogermylene into the C−I bond. F_3GeGeI_3[310], ICH_2GeI_3, $PhGeI_3$, $MeOCH_2GeI_3$ and $EtOCH_2GeI_3$ were also similarly synthesized at 110–290 °C in sealed ampoules.

In 1965, Mironov and Gar[273,303] found that allyl bromide adds easily to $GeBr_2$ to form allyltribromogermane in a 65% yield. In 1935, Tchakirian and Lewinsohn[311] used a complex of $GeCl_2$ and CsCl, i.e. cesium trichlorogermane ($CsGeCl_3$), to synthesize $RGeCl_3$. Heating $CsGeCl_3$ with PhI at 250 °C afforded phenyltrichlorogermane in 80% yield. Alkyl iodides also reacted similarly under similar conditions[312,313]. However, this method did not find any application.

D. C−Ge Bond Cleavage. Organylhalogermanes

The C−Ge bond is less stable toward heterolytic and homolytic cleavage reactions than the C−Si bond, but it is more stable than the C−Sn and C−Pb bonds. This is consistent with the bond energies of these bonds (see Chapter 2).

The first example of heterolytic cleavage of the C−Ge bond was the cleavage of tetraorganylgermanes (and later of organylhalogermanes) by halogens or hydrogen halides (mainly Br_2 and HBr). A synthetic method of organylhalogermanes ($R_{4-n}GeX_n$, $n = 1$–3) based on this reaction has been widely used. It was first used in 1927 in the laboratories of Kraus[161] and Dennis[145] and afterwords by many chemists[162,163,165,172,173,187,314].

In 1927, Kraus and Foster[161] showed that refluxing tetraphenylgermane with a bromine solution in CCl_4 for 7 hours gave triphenylbromogermane. In the same year, Orndorff, Tabern and Dennis[145] discovered that by using 1,2-dibromoethane as a solvent, the reaction was completed within a few minutes. The second phenyl group could be also cleaved, but with difficulty. However, with excess bromine, or by adding $AlBr_3$, catalyst more Ar_2GeBr_2 was obtained in satisfactory yields. In 1931, Schwarz and Lewinsohn[187] cleaved the Ar−Ge bond in many tetraarylgermanes by bromine.

In 1932, Kraus and Flood[148] obtained Et_3GeBr in 82% yield during bromination of tetraethylgermanium in an EtBr media. R_3GeBr derivatives (R = Pr[315], Bu[314,316]) were then synthesized by the same method. The feasibility of cleavage of substituents attached to the Ge atom by reaction with bromine decreases in the following order: $4\text{-}PhC_6H_4 > Ph > CH_2{=}CHCH_2 > Bu > i\text{-}Pr > Pr > Et > Me$[77].

In the early 1950s Anderson[315,317] used bromine, or bromine and iodine halides in the presence of iron powder (i.e. FeX_3 formed *in situ*) to cleave the C−Ge bond.

In a number of cases, cleavage of R_4Ge with bromine gave mixtures of $R_{4-n}GeBr_n$ (n = 1–3) which were difficult to separate. Fuchs and Gilman[318] suggested separating such mixtures by hydrolysis to the corresponding oxygen derivatives followed by their retransformation to halides. Organyliodogermanes were obtained by C−Ge bond cleavage with iodine and AlI_3 catalyst. $EtGeI_3$ was obtained from Et_2GeI_2 by this method[319].

Organyliodogermanes and organylfluorogermanes were prepared by the reaction of isostructural organylhalogermanes (chlorides and bromides) with NaI in acetone or with SbF_3[184], respectively.

In 1930, Dennis and Patnode[167] used HBr for the first time to cleave the C−Ge bond. In each case, the reaction did not continue beyond the stage of forming R_3GeBr[162,163,167,320]. By this approach, they obtained Me_3GeBr from Me_4Ge. Five years later Simons[163] showed that the rate of the C−Ge bond cleavage by HBr decreased in the following order of the Ge substituents: 4-MeC_6H_4 > 3-MeC_6H_4 > Ph > $PhCH_2$.

R_4Ge (R = Me, Et) cleavage by HF was carried out by Gladstein and coworkers[321] in 1959. R_4Ge reacted with HCl or HI only in the presence of aluminum halides[322].

It is noteworthy that under the action of sulfuric acid the C−Ge bond of $(PhCH_2)_4Ge$, was not cleaved, and $(HSO_3C_6H_4CH_2)_4Ge$ was formed[145]. In the early 1960s it was shown that the C−Ge bond could be cleaved by $AlCl_3$[323] and particularly easily by $GaCl_3$ and $InCl_3$.

In 1963 Razuvaev, Vyazankin and coworkers[324,325] found that alkyl halides in the presence of $AlCl_3$ cleaved the C−Ge bond in tetraalkylgermanes to give trialkylhalogermanes in good yield. This reaction was later used by other investigators[326,327].

In 1931, Schwarz and Lewinsohn[187] first obtained $PhGeCl_3$ in 75% yield by the cleavage of Ph_4Ge with tetrachlorogermane in an autoclave at 350 °C during 36 hours.

The cleavage reactions of the Ge−halogen bond leading to the formation of germanium−pnicogen and germanium−chalcogen bonds are considered in Sections II.F and II.C, respectively. Hence, we only indicate that in 1955 Rochow and Allred[328] found that Me_2GeCl_2 dissociates to Me_2Ge^{2+} and $2Cl^-$ ions in dilute aqueous solutions.

E. Compounds having a Ge−H Bond

The first representative of organogermanium hydrides $R_{4-n}GeH_n$ (n = 1–3) was triphenylgermane. Kraus and Foster[161] obtained it in 1927 by reaction of NH_4Br and triphenylgermylsodium in liquid ammonia. Five years later Kraus and Flood[148] similarly synthesized triethylgermane.

In 1950 the first alkyl germanes $RGeH_3$ (R = Me, Et, Pr, *i*-Am) were obtained by Kraus and coworkers[214,329] by the reaction of $NaGeH_3$ and alkyl bromides or chlorides (the same method was also used later[330,331]). They also synthesized the first dialkylgermane *i*-$AmEtGeH_2$ from *i*-AmBr and $EtGeH_2Li$ in an ethylamine media (References 330 and 331). Analogously, the reaction of *i*-AmEtGeHLi and EtI led to *i*-$AmEt_2GeH$[214,329].

It is remarkable that according to Kraus[332,333] the reaction of $NaGeH_3$ and PhBr in liquid ammonia gave benzene and the monomeric germylene GeH_2. Onyszchuk[331] added H_3GeBr, Me_3GeBr, Me_3SiCl, Me_2SiCl_2 and MeI to $NaGeH_3$ and obtained the corresponding substituted compounds containing Ge−Ge and Ge−Si bonds.

In 1953, West[334] succeeded in obtaining Ph_3GeH and Me_2GeH_2 by reducing Ph_3GeBr and $(Me_2GeS)_n$ with zinc amalgam and hydrochloric acid. However, $MeGeCl_3$ was not reduced by this method.

The most accessible synthesis of organohydrogermanes was based on the reduction of the corresponding organohalogermanes ($R_{4-n}GeX_n$, $n = 1-3$) with complex hydrides such as $LiAlH_4$[173,183,230,237,318,335–342], $NaBH_4$[276,343], and $LiAlH(OBu\text{-}t)_3$[322,344]. The less reactive lithium hydride and deuteride have been also recommended for this reduction[270,345], and sodium hydride in the presence of boron or aluminum derivatives was also used.

The Ge−Cl bonds in $(c\text{-}C_6H_{11})_3GeX$ (X = Cl, Br) were first reduced with the Ge−H bonds with $LiAlH_4$ in 1947 by Finholt and coworkers[336]. Two years later this method of organohydrogermane synthesis was implemented by Johnson and Nebergall[230]. In particular, Johnson and Harris[173] obtained in this way the first diarylgermane Ph_2GeH_2. Johnson and Nebergall[230] succeeded in reducing the Ge−O bond of $(c\text{-}C_6H_{11})_3GeOH$ and $Ph_3GeOGePh_3$ by $LiAlH_4$ to $(c\text{-}C_6H_{11})_3GeH$ and Ph_3GeH, respectively.

Lesbre and Satge obtained trialkylgermanes by reducing trialkylalkoxygermanes[280], trialkyl(alkylthio)germanes[346] and triethyl(diphenylphosphinyl)germane[347] with $LiAlH_4$. In 1963, the reduction of the corresponding halides with $LiAlH_4$ gave the optically active organogermanes $RPh(1\text{-}C_{10}H_7)GeH$ (R = Me[348], Et[349]), which were resolved to the optically active enantiomers.

Triorganylgermanes were also formed by the reaction of $GeCl_4$ and organylmagnesium halides having bulky substituents such as *i*-Pr, $2\text{-}MeC_6H_4$ and $c\text{-}C_6H_{11}$[178,180,319]. The intermediates of this reaction seem to be triorganylgermylmagnesium halides R_3GeMgX, whose hydrolysis gave R_3GeH[178]. Triethylgermane was formed by cleavage of the Ge−M bonds of Et_3GeM (M = Li, Cd, Hg, Bi) with water, alcohols or acetic acid[231,249,350].

In 1961, Satge and Lesbre[276,351] used trialkylgermanes in the presence of AlX_3 for partial reduction of R_2GeX_2 (X = Cl, Br) to R_2GeHX. An analogous reaction was performed four years earlier in organosilicon chemistry[352].

The same authors[276,351] also synthesized dialkylhalogermanes R_2GeHX (X = Br, I) by the reaction of R_2GeH_2 and haloalkanes in the presence of AlX_3. Again, the analogous organosilicon reaction was reported four years earlier[352].

Mironov and Kravchenko[353] suggested an original synthesis of alkyldichlorogermanes $RGeHCl_2$ based on alkylation of the $Et_2O \cdot HGeCl_3$ complex with tetraalkylstannanes and tetraalkyl-plumbanes. The reaction with Me_4Sn resulted in 80% yield of $MeGeHCl_2$. The reaction with higher tetraalkylstannanes was complicated with by-processes.

In 1950, Johnson and Harris[173] found that thermal decomposition of Ph_3GeH gave Ph_2GeH_2 and Ph_4Ge. The diphenyldigermane product was also unstable and decomposed slowly even at room temperature, forming tetraphenylgermane as one of the products. Phenylgermane decomposed to Ph_2GeH_2 and GeH_4[354] at 200 °C. The reaction proceeded instantly in the presence of $AlCl_3$ even at room temperature. In contrast, the alkylgermanes $R_{4-n}GeH_n$ were more stable and their stability toward thermolysis increased on decreasing the value of n[276]. At 400–450 °C tricyclohexylgermane decomposed to elementary germanium, cyclohexene and hydrogen, and at *ca* 360 °C cyclohexane, benzene and polycondensed compounds having $c\text{-}C_6H_{11}$ groups were formed[355]. In contrast, the thermal decomposition of $(c\text{-}C_6H_{11})_3SiH$ proceeded at 600–650 °C. Since 1949, it was established that the first products of Ge−H bond oxidation, e.g. of R_3GeH, were triorganylgermanoles R_3GeOH, which then condensed to give digermoxanes $R_3GeOGeR$[230,337,356].

Kraus, Flood and Foster[148,161], and much later other research chemists[173,183,230,276,357], discovered that organic germanium hydrides $R_{4-n}GeH_n$ ($n = 1-3$) reacted extremely readily with halogens to form corresponding halides $R_{4-n}GeX_n$ (X = Cl, Br, I).

Even in 1927, Kraus and Foster[161] showed that triphenylgermane reacted with HCl to give the triphenylchlorogermane. Thirty years later Anderson[337] conducted an analogous reaction of trialkylgermane, e.g. triethylgermane and hydrochloric acid. HCl and HBr reacted with $RGeH_3$ and R_2GeH_2 only in the presence of $AlCl_3$ or $AlBr_3$[276,358,359].

In 1953, Anderson[337] found that the reaction of concentrated H_2SO_4 with trialkylgermanes gave hydrogen and bis(trialkylgermyl) sulfates $(R_3GeO)_2SO_2$. According to Satge[276], the reaction of Et_3GeH and benzenesulfonic acid leads similarly to Et_3GeOSO_2Ph.

$(Et_3GeO)_3B$ was obtained by the reaction of Et_3GeH and H_3BO_3 in the presence of copper powder[360]. An analogous reaction of Et_3SiH and H_3BO_3 in the presence of colloidal nickel was reported four years earlier[361]. Bu_3GeH reacts quantitatively with acetic acid[360] in the presence of copper. Perfluoroalkanecarboxylic acids reacted smoothly with Et_3GeH without any catalyst to form triethylperfluoroacyloxygermanes[337]. In contrast, Cl_3CCOOH, Br_3CCOOH and ICH_2COOH were reduced to CH_3COOH by Et_3GeH[337]. Anderson also conducted the reaction of R_2GeH_2 with H_2SO_4[337].

In 1962, Lesbre and Satge[360,362] found that R_3GeH condensed with water or with alcohols, glycols and phenols (R′OH) in the presence of copper powder to form hydrogen and R_3GeOH or R_3GeOR', respectively. The reaction of Bu_2GeH_2 and 1,4-butanediol led to 2,2-dibutyl-1,3-dioxa-2-germacyclopentane[360].

Unlike Si−H and especially Sn−H bonds, the Ge−H bond is rather stable to alkaline hydrolysis or alcoholysis[363]. For example, $R_{4-n}GeH_n$ (R = alkyl; $n = 1-3$) did not react with a 20% NaOH solution. According to Fuchs and Gilman[318] trihexylgermane did not react with aqueous-alcoholic KOH solution, whereas Ph_3GeH reacted easily with a similar solution[318], and $HexGeH_3$ and R_2GeH_2 reacted very slowly at 80 °C[276].

Organogermanium hydrides are very good reducing agents. In 1957, Anderson[337] showed that Et_3GeH reduced transition metal salts to their lower valence state (Cu^{II} to Cu^{I}, Ti^{IV} to Ti^{III} or Ti^{II}, V^{IV} to V^{III}, Cr^{IV} to Cr^{III}) or to the free metals (Au, Hg, Pd, Pt).

In 1961, Satge[276] found out that Et_3GeH reduced $GeCl_4$ first to $GeCl_2$ and then to Ge^0. Nametkin and coworkers used an analogous reaction to reduce $TiCl_4$ to $TiCl_2$[364]. In ether, the reaction gave a $2Et_2O \cdot HGeCl_3$ complex[356].

In 1961, it was found that organogermanium hydrides $R_{4-n}GeH_n$ reduced organic halogen derivatives in the absence of catalysts to the corresponding hydrocarbons[276,351]. The reaction is easier the higher the value of n and the atomic number of the halogen.

Bu_2GeH_2 reduces iodobenzene with greater difficulty than it reduces aliphatic monohalides[276]. At 220 °C, Bu_3GeH reduces CCl_4 to $HCCl_3$ almost quantitatively[276].

Triorganylgermanes readily reduce acyl chlorides[173,276] and chloromethyl ether, preferably in the presence of traces of $AlCl_3$[276,288,354,356]. In 1964, it was found that organogermanium hydrides also readily reduced N-halosuccinimides[354,365].

In 1962, Lesbre and Satge[360] pointed out that trialkyl(alkylthio)germanes R_3GeSR' were formed by condensation of trialkylgermanes and thioles in the presence of asbestos platinum. Reduced nickel proved later to be the best catalyst for the reaction[346].

In 1966, Vyazankin and Bochkarev[366–369] found that, depending on the reaction conditions, heating of triethylgermane and elementary sulfur, selenium and tellurium gave the respective triethylgermylchalcogenols Et_3GeEH (E=S, Se) or bis(triethylgermyl)chalcogenides $(Et_3Ge)_2E$ (E = S, Se, Te). The latter were also formed when diethylselenide[366,367] and diethyltelluride[368] were used instead of Se and Te. The reaction of Et_3GeEH and Et_3SnH afforded unsymmetrical chalcogenides $Et_3GeESnEt_3$ (E = S, Se)[367,368]. Vyazankin and coworkers determined that the M−H bond reactivity with chalcogens increased considerably in the following order for M: Si < Ge < Sn.

F. Organogermanium Chalcogen Derivatives

Organogermanium compounds in which the Ge is bonded to a Group 16 element (chalcogen) were first encountered in 1925, concerning this field of organometallic chemistry.

The first compounds having germoxane Ge–O bonds were Ph_3GeOH, $Ph_3GeOGePh_3$ and $(Ph_2GeO)_4$. In 1925, Morgan and Drew[149] synthesized hexaphenyldigermoxane in quantitative yield by the reaction of aqueous-alcoholic $AgNO_3$ with Ph_3GeBr. The germoxane quantitatively generated Ph_3GeBr by reaction with concentrated HBr. In 1930, Kraus and Wooster[370] obtained $Ph_3GeOGePh_3$ by hydrolysis of Ph_3GeNH_2. They discovered that the digermoxane was cleaved to Ph_3GeONa and Ph_3GeNa by Na in liquid ammonia.

In 1933, Simons and coworkers[162] showed that hexaaryldigermoxanes $Ar_3GeOGeAr_3$ (Ar = 3-MeC_6H_4, 4-MeC_6H_4) were formed not only by the reaction of aqueous-alcoholic $AgNO_3$ with Ar_3GeBr, but also by a 0.5N NaOH solution. $(2\text{-}MeC_6H_4)_3GeCl$ and aqueous-alcoholic $AgNO_3$ gave $(2\text{-}MeC_6H_4)_3GeOH$. In 1934, Bauer and Burschkies[172] obtained $(PhCH_2)_3GeOGe(CH_2Ph)_3$ by the same method. When concentrated HHal was added to the latter, the corresponding tribenzylhalogermanes were isolated.

In 1930, Dennis and Patnode[167] first reported that self-condensation of trimethylgermanol Me_3GeOH under anhydrous conditions led to $Me_3GeOGeMe_3$ which, however, was neither characterized nor examined. In 1961, Schmidt and Ruidisch[371,372], Griffiths and Onyszchuk[373] and others in 1966[374,375] simultaneously synthesized hexamethyldigermoxane by the reaction of Me_3GeX (X = Cl, Br) and Ag_2CO_3. In 1932, Kraus and Flood[148] obtained hexaethyldigermoxane $Et_3GeOGeEt_3$ nearly quantitatively by hydrolysis of Et_3GeBr with aqueous KOH or NaOH solutions. It was transformed to the corresponding triethylhalogermanes by reaction with concentrated HCl or HBr. The reaction of $Et_3GeOGeEt_3$ and Li gave an equimolar mixture of Et_3GeOLi and Et_3GeLi.

In 1951, Anderson[376] obtained hexaethyldigermoxane by reacting Et_3GeBr with Ag_2CO_3 and studied its cleavage by HNCS. Later, he obtained $R_3GeOGeR_3$ with R = Pr[184,315,376], *i*-Pr[184], Bu[314,376] and investigated their cleavage by organic[377] and inorganic[337,378,379] acids. $Me_3GeOGeMe_3$ was even cleaved with such an exotic reagent as $Me(PO)F_2$[374]. Hexaorganyldigermoxanes carrying bulky substituents could not generally be obtained by hydrolysis of the corresponding triorganylhalogermanes. However, they were produced by other methods. For example, in 1953, Anderson[184] synthesized $R_3GeOGeR_3$, R = *i*-Pr by the reaction of *i*-Pr_3GeBr and Ag_2CO_3. The cleavage of hexaisopropyldigermoxanes with inorganic acids HX resulted in *i*-Pr_3GeX (X = F, Cl, Br, I, NCS).

Triphenylgermanol was the first organogermanium compound containing the Ge–OH group. Contrary to expectations, attempts by Morgan and Drew[149] and Kraus and Foster[161] to obtain Ph_3GeOH by hydrolysis of Ph_3GeBr had failed and $Ph_3GeOGePh_3$ was always the only reaction product. Nevertheless, in 1954, Brook and Gilman[195] obtained high yield of Ph_3GeOH by the reaction of Ph_3GeBr in aqueous-alcoholic KOH. However, Kraus and Foster[161] synthesized triphenylgermanol for the first time in 1927 by hydrolysis of Ph_3GeONa or by treating the latter with NH_4Br in liquid ammonia. The Ph_3GeONa was prepared by oxidation of Ph_3GeNa in the same solvent. In 1966, the synthesis of Ph_3GeOH by a slow hydrolysis of Ph_3GeH[380] was reported.

Dennis and Patnode[167] assumed the existence of trimethylgermanol, but neither they nor Schmidt and Ruidisch[371] succeeded in isolating it. Schmidt and Ruidisch used titrimetric and cryoscopic methods to show that Me_3GeCl was hydrolyzed by water to Me_3GeOH, but its attempted isolation from the aqueous solution failed and only $Me_3GeOGeMe_3$ was isolated. However, lithium trimethylgermanolate Me_3GeOLi was obtained by cleavage of $Me_3GeOGeMe_3$ with methyllithium[372]. Et_3GeBr hydrolysis had not resulted in triethylgermanol[148] and hexaethyldigermoxane was always formed instead.

It was not possible to isolate trialkylgermanols R_3GeOH with R = Me, Et, Pr, Bu until 1970, since they turned out to be considerably less stable than the isostructural trialkylsilanols and trialkylstannanols. Nevertheless, when the germanium atom was bonded to bulky substituents such as *i*-Pr, *c*-C_6H_{11}, 2-MeC_6H_4 and 1-$C_{10}H_7$, the corresponding rather stable triorganylgermanoles were isolated. Thus, in 1932, Bauer and Burschkies[181] synthesized tricyclohexylgermanol by the reaction of $(c\text{-}C_6H_{11})_3GeBr$ with aqueous-alcoholic $AgNO_3$. Johnson and Nebergall[230] repeated this reaction after 17 years. Simons and coworkers[162] similarly obtained $(2\text{-}MeC_6H_4)_3GeOH$ from the appropriate chloride.

In 1952, West[183] successfully used for the first time the reaction of R_3GeX for the synthesis of $(1\text{-}C_{10}H_7)_3GeOH$. The latter was so stable that it was transformed slowly and partially to the corresponding digermoxane only at 175 °C during 24 hours.

Triisopropylgermanol was first synthesized by Anderson in 1954[184,381] by hydrolysis of $i\text{-}Pr_3GeBr$ in aqueous 6N NaOH solution. Later, he used $i\text{-}Pr_3GeCl$[379] for obtaining the same product which he obtained by alkali hydrolysis of the reaction products of $GeBr_4$ with excess *i*-PrMgBr (i.e. $i\text{-}Pr_3GeBr$)[184,337]. The $i\text{-}Pr_3GeOH$ was then converted to $i\text{-}Pr_3GeX$ by reaction with HX (X = F, Cl, I)[184].

Compounds with R = Ph were the first representatives of perorganylcyclogermoxanes $(R_2GeO)_n$ and the corresponding linear polymers $HO(R_2GeO)_nH$. In 1925, Morgan and Drew[149] isolated two products from hydrolysis of Ph_2GeBr_2 which were described as $HO(Ph_2GeO)_4H$ and $(Ph_2GeO)_4$ and named according to Kipping's nomenclature 'trianhydrotetrakisdiphenylgermanediol' and 'tetraanhydrotetrakisdiphenylgermanediol', respectively.

Five years later Kraus and Brown[226] found out that the solid products of hydrolysis of Ph_2GeBr_2 with concentrated aqueous ammonia have the $(Ph_2GeO)_n$ structure.

In 1960, Metlesics and Zeiss[213] investigated the thermal decomposition of $(Ph_2GeO)_4$ and $(Ph_2GeO)_n$ in vacuum, which resulted in $(Ph_2GeO)_3$. In the same year Brown and Rochow[382] similarly obtained $(Me_2GeO)_3$ from thermolysis of the products of the hydrolysis of Me_2GeCl_2.

In 1932, Flood[188] isolated two products with a composition of Et_2GeO from the aqueous NaOH hydrolysis of Et_2GeBr_2. A liquid was identified as hexaethylcyclotrigermoxane $(Et_2GeO)_3$, where the other, an insoluble solid, was ascribed to the dimer. In 1950, Anderson[378] reproduced the experiment, and suggested that the latter was octaethylcyclotetragermoxane $(Et_2GeO)_4$.

In 1948, Rochow[252,383] discovered that the hydrolysis product of Me_2GeCl_2 was easily dissolved in water in contrast to the hydrolysis product of Me_2SiCl_2. The solution was evaporated without leaving any residue, indicating the formation of volatile hydrolysis products. This was also observed in the reaction of Me_2GeCl_2 with aqueous ammonia. This led Rochow to the conclusion that the hydrolysis reaction of Me_2GeCl_2 was reversible.

In 1948, Trautman and Ambrose[384] patented a method for producing $(Et_2GeO)_3$. In 1953, Anderson[184] synthesized $(i\text{-}Pr_2GeO)_3$, the first cyclogermoxane having rather bulky substituents at the Ge atom, by hydrolysis of the reaction products of $GeBr_4$ and *i*-PrMgBr with aqueous NaOH. Hexaisopropylcyclotrigermoxane cleavage by appropriate acids gave $i\text{-}Pr_2GeX_2$ (X = F, Cl, Br, I).

According to Mazerolles[385], the oxidation of germacycloalkanes $(CH_2)_nGeH_2$, $n = 4$ gave the corresponding cyclotrigermoxane $[(CH_2)_4GeO]_3$ but, when $n = 5$, a mixture of $[(CH_2)_5GeO]_4$ and $(CH_2)_5Ge(H)OH$ was formed.

The pioneers of organogermanium chemistry, Morgan and Drew[149], were the first to synthesize polyorganylgermoxanol and polyorganylgermsesquioxane, which for a long time were termed organyl germanoic acid and its anhydride, respectively. The amorphous polymer soluble in alkalis, obtained by hydrolysis of $PhGeBr_3$, had a composition varying from $PhGeO_2H$ to $PhGeO_{1.5}$, depending on the reaction conditions. The authors were sure that the product had a structure intermediate between those of phenylgermanoic acid PhGeOOH and its anhydride $(PhGeO)_2O$.

In 1927, Orndorff, Tabern and Dennis[145] synthesized the aforementioned anhydride, i.e. polyphenylgermsesquioxane $(PhGeO_{1.5})_n$, by treatment of $PhGeCl_3$ with a dilute aqueous ammonia solution. The anhydride had high solubility in alkalis and could be re-precipitated from the alkali solution by carbon dioxide. Other polyorganylgermsesquioxanes $(RGeO_{1.5})_n$ with $R = PhCH_2$, $4\text{-}MeC_6H_4$ and $Me_2NC_6H_4$ were produced analogously. Five years later Bauer and Burschkies[181] described a few more polyorganylgermsesquioxanes.

The first polyalkylgermsesquioxane $(EtGeO_{1.5})_n$ was obtained by Flood[188] as a by-product of a reaction that he investigated. A year later he synthesized it by the reaction of $EtGeI_3$ and Ag_2O or by hydrolysis of $(EtGeN)_n$, the product of ammonolysis of $EtGeI_3$[306]. In 1939, Tchakirian[386] obtained 'alkylgermanium acids' RGeOOH (R = Me, Et) by hydrolysis of $RGeCl_3$. Analogously, 'germanomalonic acid' $CH_2(GeOOH)_2$ was synthesized from $CH_2(GeCl_3)_2$.

Organoxy and acyloxy derivatives having Ge−OR and Ge−OOCR groups as well as a heterogermoxane Ge−OM group (M = a metal or a nonmetal atom) belong to organogermanium compounds with germoxane bonds. In 1949, Anderson[387] reported the formation of alkylalkoxygermanes $Et_{4-n}Ge(OR)_n$ (R = Me, Et, Bu; $n = 1, 2$) during the reaction of $Et_{4-n}Ge(NCO)_n$ and the appropriate alcohols; he did not isolate or characterize the compounds. In 1954, West and coworkers[388] described the synthesis of all the methylmethoxygermanes by the reaction of Me_3GeI, Me_2GeCl_2 and $MeGeCl_3$ with sodium methoxide. In 1956, Anderson[389] also synthesized the first trialkylaryloxygermanes $Et_3GeOC_6H_4R$ (R = 3-Me, 2-NH_2)[389]. As early as in 1962, Lesbre and Satge[360] produced the Et_3GeOPh, the simplest representative of this series. Et_3GeOCH_2Ph[280] was synthesized at the same time[280]. In 1961, Griffiths and Onyszchuk[373] similarly obtained Me_3GeOMe. In 1954, Brook and Gilman[195] pointed out that one of the first arylalkoxygermanes Ph_3GeOMe was the thermal decomposition product of $Ph_3GeCOOMe$ at 250 °C with CO elimination. For comparison, triphenylmethoxygermane was synthesized from Ph_3GeBr and MeONa. In 1968, Peddle and Ward[390] discovered the rearrangement of Ph_3GeCH_2OH to Ph_3GeOMe.

In 1961, Griffiths and Onyszchuk[373] found that the reaction of $MeGeH_2Br$ and MeONa at −80 °C gave $MeGeH_2OMe$, which slowly decomposed to form the polymer $(MeGeH)_n$ and MeOH.

In 1962, two new approaches to trialkylalkoxygermane were introduced at the Satge laboratory[36]. The first was based on the dehydrocondensation of trialkylgermanes and alcohols or glycols. Using R_2GeH_2 led not only to $R_2GeOR'_2$, but also to R_2GeHOR'. Dehydrocondensation of Bu_2GeH_2 with $HO(CH_2)_4OH$ resulted in 2,2-dibutyl-1,3-dioxa-2-germacycloheptane. A year later Wieber and Schmidt[391] synthesized one of the simplest heterocyclic systems, 2,2-dimethyl-1,3-dioxa-2-germacyclopentane, by the reaction of Me_2GeCl_2 and ethylene glycol in the presence of Et_3N. They also produced the benzyl derivatives of 2,2-dimethyl-1,3-dioxa-2-germacyclopentane and 2,2-dimethyl-1,4-dioxa-2-germacyclohexane[392]. The second approach[280] to compounds R_3GeOR' was the addition reaction of R_2GeH to carbonyl compounds in the presence of copper powder.

In 1964, Satge[362] used a re-alkoxylation reaction, with alcohols having boiling points higher than those of MeOH or EtOH, to replace the alkoxy group in R_3GeOR' ($R' = Me$, Et) by another alkoxy group. This method was used later by other investigators[390,393–395].

Mehrotra and Mathur[396] in 1966 investigated extensively the cleavage reaction of $Bu_3GeOGeBu_3$, $(Bu_2GeO)_n$ and $(Ph_2GeO)_n$ by alcohols, glycols, mono-, di-, and triethanolamine and acetylacetone (which apparently reacted via its enol form). Diverse noncyclic and cyclic compounds having Ge−O−C groups were produced. Voronkov and coworkers[397,398] first obtained 1-organylgermatranes $RGe(OCH_2CH_2)_3N$ (R = Alk, Ar) by cleavage of $(RGeO_{1.5})_n$ with triethanolamine.

In 1962, Lesbre and Satge[280] demonstrated that the Ge−O bond in alkylalkoxygermanes was cleaved with HI, RCOOH, Ac_2O, PhCOCl, $PhSO_2OH$ and $LiAlH_4$ much more easily than that of the Ge−O−Ge group.

Organogermanium peroxides $Pr_3GeOOCMe_3$, $Pr_3GeOOC_{10}H_{17}$-1, $Pr_2Ge(OOC_{10}H_{17}$ 1$)_2$ and $Pr_3GeOOGePr_3$ ($1\text{-}C_{10}H_{17}$ = 1-decalyl) having the germperoxane Ge−O−O group were first synthesized by Davies and Hall[399,400]. They were produced by the reaction of the appropriate hydroperoxide with Pr_3GeCl or with Pr_2GeCl_2 in the presence of a tertiary amine. A year later Rieche and Dahlmann[401] synthesized Ph_3GeOOR and $Ph_3GeOOGePh_3$ from Ph_3GeBr, NH_3 and ROOH with $R = CPh_{3-n}Me_n$ ($n = 1–3$).

Johnson and Nebergall[230] discovered the series of organylacyloxygermanes; they synthesized $(c\text{-}C_6H_{11})_3GeOOCCH_3$ by the reaction of tricyclohexylgermanol and acetic anhydride.

In 1950, Anderson[378] obtained alkylacyloxygermanes $Et_{4-n}Ge(OOCR)_n$ (R = H, Me, CH_2SGeEt_3; $n = 1, 2$) by cleavage of $Et_3GeOGeEt_3$ or $(Et_2GeO)_n$ by carboxylic acids or by their anhydrides. Within a year he used $Pr_3GeOGePr_3$[402] in these reactions. However, his attempt to cleave germoxanes $R_3GeOGeR_3$ and $(R_2GeO)_3$ with $R = i\text{-}Pr$ by carboxylic acids was unsuccessful[184]. However, he later succeeded in obtaining the expected products $i\text{-}Pr_2Ge(OOCR)_2$ with R = Me, Et, Pr, Bu by the reaction of $i\text{-}Pr_2GeX_2$ and the silver salts of the corresponding carboxylic acids[379]. He used a similar reaction of Et_3GeBr and ArCOOAg for the synthesis of $Et_3GeOOCAr$ (Ar = Ph, $2\text{-}H_2NC_6H_4$)[389]. Tributylacyloxygermanes could also be prepared by the reaction of Bu_3GeI and RCOOAg[314].

However, silver chloroacetate and benzoate did not react with Et_3GeCl. Anderson[381] recommended the reaction of $i\text{-}Pr_3GeCl$ with RCOOAg as the best method for synthesis of $i\text{-}Pr_2GeOOCR$. During the synthesis of trialkylacyloxygermanes he discovered that a thermal decomposition of $i\text{-}Pr_3GeOOCCH_2CH_2Cl$ gave $i\text{-}Pr_3GeCl$ and $CH_2{=}CHCOOH$[381]. For $Et_3GeOOCH$ synthesis he used lead formate[317].

In 1956, Anderson[379] studied the reaction of MeCOOAg with a series of Et_3GeX (X = I, Br, Cl, H, SR_3, CN, NCS, NCO, $OGeR_3$). He found that the reactivity of the Ge−X bond in R_3GeX (R = Et, i-Pr) with respect to silver salts decreased in the following order of X : I > $SGeR_3$ > Br > CN > Cl > NCS > NCO > $OGeR_3$ $\geqslant$ OCOR $\gg$ F ('the Anderson row').

In 1951, Anderson[402] first carried out the re-esterification reaction of organylacyloxygermanes with carboxylic acids. He also studied esterification of R_3GeOH with carboxylic acids (R′OOH) leading to $R_3GeOOCR'$ in the presence of anhydrous Na_2SO_4 or with H_2SO_4[381].

In 1957, Anderson found that perfluoroalkanoic acids dehydrocondense with Et_3GeH without catalyst to give $Et_3GeOOCR$ ($R = CF_3$, C_2F_5, C_3F_7), whereas the reaction did not occur with acetic acid. In the reaction with Et_3GeH the RCOOH (R = Cl_3C, Br_3C, ICH_2) behaved in a quite different manner and were reduced to CH_3COOH[337].

In the period of 1951 till 1957, Anderson enriched the acyloxygermane chemistry by sixty-six compounds $R_{4-n}Ge(OOCR')_n$ with R = Et, Pr, *i*-Pr, Bu, *c*-C_6H_{11}; R′ = H, Alk, Ar, haloalkyl; $n = 1-3$[184,314,317,377–379,381,402,403].

In 1962, Lesbre and Satge[360] discovered that the dehydrocondensation reaction of trialkylgermane and carboxylic acids could be catalyzed by copper powder. For instance, the reaction of Bu_3GeH and MeCOOH gave $Bu_3GeOOCMe$ in 60% yield.

In 1954, Brook and Gilman synthesized $Ph_3GeOOCGePh_3$ by the reaction of Ph_3GeBr and $Ph_3GeCOONa$[195].

A year later Brook[404] discovered that under short heating to 200 °C triphenylgermanecarboxylic acid $Ph_3GeCOOH$ eliminated CO and H_2O and transformed to $Ph_3GeOOCGePh_3$. Further heating of the latter afforded $Ph_3GeOGePh_3$[405]. Evidently, Ph_3GeOH was an intermediate product in the initial stage of the thermolysis.

The first organogermanium compounds having a metal-germoxane Ge−O−M group were alkali metal triorganylgermanolates R_3GeOM (R = Ph, Et; M = Li, Na, K) produced in 1925–1932 by Morgan and Drew[149], as well as by Kraus and coworkers[148,161]. Later, they and many other investigators obtained and used Li and Na germanolates and $R_3GeOMgX$ for synthetic purposes.

Among heterogermoxanes in which the germanium atom was bonded via oxygen to a nonmetal (metalloid) atom, bis(trialkylgermyl)sulfates $(R_3GeO)_2SO_2$ with R = Et, Pr, *i*-Pr[337,378,381,402] and cyclic dialkylgermylene sulfates $(R_2GeOSO_2O)_2$ with R = Me, Et, Pr, *i*-Pr[317,379,406] were the first to be synthesized by Anderson in 1950–1956. For the synthesis of compounds with R = Et, the reactions of H_2SO_4 with $Et_3GeOGeEt_3$ and $(Et_2GeO)_4$[378] were first used. Later, $(R_3GeO)_2SO_2$ and $(R_2GeOSO_2O)_2$ with R = Me, Et, Pr, *i*-Pr were obtained by the reaction of H_2SO_4 with $R_3GeOOCMe$, *i*-Pr_3GeOH and $R_2Ge(OOCMe)_2$[379,389,402,406]. Anderson[337] used the reactions of Et_3GeH and H_2SO_4 or $HgSO_4$ for synthesis of bis(triethylgermyl) compounds. In 1951, he also produced the first organogermanium compound having a Ge−O−N group, i.e. Et_3GeONO_2 by the reaction of Et_3GeBr and $AgNO_3$[376]. In 1955, Rochow and Allred[328] obtained Me_2GeCrO_4, isostructural to $(R_2GeSO_4)_2$ by the reaction of Me_2GeCl_2 and K_2CrO_4 in aqueous media.

In 1961, Satge[276] carried out the dehydrocondensation of Et_3GeH and $PhSO_3H$, which resulted in Et_3GeOSO_2Ph.

In 1950–1967 Schmidt, Schmidbaur and Ruidisch synthesized a large series of heterogermoxanes having Ge−O−M groups with M = B, Al[407], Ga[408], In[408], Si[405,409], N[410,411], P[410,412], As[413], S[405], Se[412,413], Cl[414], V[412], Cr[409] and Re[412]. Those were mostly the trimethylgermyl esters of the corresponding inorganic acids such as $(Me_3GeO)_nY$ with Y = NO_2, ClO_3, ReO_3 ($n = 1$); SO_2, SeO_2, CrO_2 ($n = 2$); B, PO, AsO, VO ($n = 3$). They were produced by hexamethyldigermoxane cleavage with anhydrides of inorganic acids: P_2O_5[410], As_2O_5[413], SO_3[405], SeO_3[412,413], V_2O_5[412], CrO_3[409,412] and Re_2O_7[412]. The same types of compounds were synthesized by reaction of Me_3GeCl with the silver salts of the corresponding acids[410,412,413]. Similarly, Srivastava and Tandon prepared $(Ph_3GeO)_2Y$ with Y = SO_2 and SeO_2[415].

In 1961–1964 Schmidbaur, Schmidt and coworkers[405,416], synthesized a series of compounds $R_{4-n}Ge(OSiR'_3)_n$ with R, R′ = Me, Et; $n = 1-3$. For example, trimethyl(trimethylsiloxy)germane $Me_3GeOSiMe_3$ and dimethylbis(trimethylsiloxy)germane $Me_2Ge(OSiMe_3)_2$ were obtained by the reaction of alkali metal trimethylsilanolates Me_3SiOM and Me_3GeCl or Me_2GeCl_2. Schmidbaur and Schmidt[409] studied the cleavage of trimethyl(trimethylsiloxy)germane with sulfuric and chromic anhydrides (SO_3 and CrO_3), which gave $Me_3GeOSO_2OSiMe_3$ and $Me_3GeOCrO_2OSiMe_3$, respectively. When

$Me_3GeOSiMe_3$ reacts with $AlCl_3$[407] and $POCl_3$[412], only the Si−O bond cleaves, thus leading to $Me_3GeOAlCl_2$ and $Me_3GeOPOCl_2$, respectively.

By dehydrocondensation of $B(OH)_3$ with Et_3GeH in 1962 Lesbre and Satge[360] obtained tris(triethylgermyl)borate $(Et_3GeO)_3B$.

Cleavages of the Ge−O bond in Ge−O−Ge and Ge−O−Si groups are much easier than that for Si−O−Si groups[417,418]. This indicates that the Ge−O bond is highly reactive. However, the heterolytic cleavage of Sn−O and Pb−O bonds was much easier (see Sections III and IV) than that for the Ge−O bond.

In 1967 Armer and Schmidbaur[408] obtained metallogermoxanes $(Me_3GeOMMe_3)_2$ with M = Al, Ga, In as well as $(Me_3GeOGaPh_2)_2$, $(Ph_3GeOGaMe_2)_2$ and $(Ph_3GeOGaPh_3)_2$. All these compounds seemed to be dimers. Subsequently, Davies and coworkers[419] synthesized the similar tin and lead derivatives $Ph_3GeOSnEt_3$ and $Ph_3GeOPbPh_3$.

Organogermanes possessing a germthiane Ge−S bond were first prepared at the Dennis laboratory in 1927[145]. These were three-dimensional polyarylgermosesquithianes $(RGeS_{1.5})_n$ with R = Ph, 4-MeC_6H_4, $Et_2NC_6H_4$, produced by the action of H_2S on the corresponding $(RGeO_{1.5})_n$. At the time the compounds were considered to be the sulfur analogs of anhydrides of arylgermanoic acids, $(RGe{=}S)_2S$.

Five years later a series of organylgermsesquithianes $(RGeS_{1.5})_n$ (R = Ph, 4-MeC_6H_4, 1-$C_{10}H_7$, $Me_2NC_6H_4$, $Et_2NC_6H_4$) was synthesized by the same method by Bauer and Burschkies[181] and later by an easier method by Reichle[420].

In 1967, when studying the reaction of $MeGeBr_3$ and H_2S in the presence of Et_3N, Moedritzer[421] prepared oligomeric $(MeGeS_{1,5})_4$, apparently of tetrahedral structure.

Cyclic perorganylcyclogermthiane oligomers $(R_2GeS)_n$ became known much later. In 1948–1950 Rochow[251,252] obtained the first $(Me_2GeS)_n$ by the reaction of H_2S and Me_2GeCl_2 in a 6N H_2SO_4 solution. The crystalline product which has a specific pepper and onion smell was slowly hydrolyzed to H_2S when exposed to atmospheric moisture, and also in boiling water. In dilute acids the hydrolysis is much faster. This was patented later[422,423], Brown and Rochow[424] found subsequently that the compound was a trimer, i.e. hexamethylcyclotrigermthiane, $(Me_2GeS)_3$, having the same structure in solution and in the gas phase. In 1963, Ruidisch and Schmidt[425] also produced $(Me_2GeS)_3$ by reaction of H_2S with Me_2GeCl_2 in the presence of Et_3N.

In 1956, Anderson[379] synthesized the first four-membered tetraisopropylcyclodigermdithiane $(i\text{-}Pr_2GeS)_2$ by the reaction of $i\text{-}Pr_2GeI_2$ with Ag_2S. In 1965 its analog $(Bu_2GeS)_2$ was obtained by passing gaseous H_2S through a solution of $Bu_2Ge(OR)_2$ (R = Bu, *i*-Bu) in the corresponding alcohol[426]. When R = *t*-Bu the reaction occurred only in the presence of $PhSO_3H$.

In 1963, Schmidt and Schumann[427] found that heating Bu_4Ge with sulfur at 250 °C gave $(Bu_2GeS)_3$ and Bu_2S. $Bu_{4-n}GeSBu_n$, $n = 1, 2$ were the intermediate products. A similar reaction of sulfur and Ph_4Ge at 270 °C gave elementary germanium and Ph_2S, the final thermolysis products of the intermediate $(Ph_2GeS)_3$[427]. The latter was first obtained by Reichle[420] in 1961 by the reaction of $(Ph_2GeO)_3$ with H_2S *in statu nascendi* in aqueous media. In 1963, Henry and Davidson[428] obtained $(Ph_2GeS)_n$ with $n = 2, 3$ by the reaction of $Ph_2Ge(SNa)_2$ and PhCOCl.

Investigations of the chemical transformations of $(R_2GeS)_n$ started in 1953. West[334] succeeded in reducing $(Me_2GeS)_3$ to Me_2GeH_2 by reaction with zinc amalgam and HCl in an alcoholic media. In 1956, Anderson[379] described the reactions of $(i\text{-}Pr_2GeS)_2$ with silver bromide, cyanide and acetate. Moedritzer and van Wazer[429] investigated the exchange reactions of $(Me_2GeS)_3$ with Me_2GeX_2 (X = Cl, Br, I), and with $(Me_2SiS)_3$[430].

Monomeric organogermanium compounds having digermthiane (Ge−S−Ge) groups, i.e. hexaorganyldigermthianes $R_3GeSGeR_3$, R = Et, Ph, 4-MeC_6H_4, 4-PhC_6H_4 and

$PhCH_2$, were produced for the first time by Burschkies[431] in 1936 via the reaction of the corresponding R_3GeBr with aqueous or alcoholic Na_2S solutions. The reaction of $(c\text{-}C_6H_{11})_3GeBr$ and Na_2S resulted in hexacyclohexyldigermperthiane $(c\text{-}C_6H_{11})_3GeSSGe(C_6H_{11}\text{-}c)_3$.

In 1956, Anderson[379] synthesized $(i\text{-}Pr)_3GeSGe(i\text{-}Pri)_3$ by reacting Ag_2S with $i\text{-}Pr_3GeI$. In 1965–1966, Satge and Lesbre[346] and Cumper and coworkers[432] produced hexaalkyldigermthianes $R_3GeSGeR_3$, R = Et, Bu in high yield by the same method.

In 1966 Abel, Brady and Armitage[433], and in 1968 Wieber and Swarzmann[434] used the reaction of triorganylhalogermanes with H_2S in the presence of nitrogen bases for the synthesis of hexaorganyldigermthianes in analogy to the widely used method in organisilicon chemistry.

In 1963, Ruidisch and Schmidt[425] discovered that the thermal decomposition of lithium trimethylgermanthiolate afforded $Me_3GeSGeMe_3$ and Li_2S. They found that $Me_3GeSSiMe_3$ thermally disproportionated to $Me_3GeSGeMe_3$ and $Me_3SiSSiMe_3$. In 1966, Vayzankin and coworkers[367,368] decomposed Et_3GeSH at 130 °C to H_2S and $Et_3GeSGeEt_3$. The latter product also produced in the reaction of Et_3GeSH with Et_2Hg[369]. Finally, hexaorganyldigerthianes $R_3GeSGeR_3$ were obtained by the reaction of R_3GeSLi with R_3GeX (R = Me, Ph; X = Cl, Br)[247,425].

In 1962, Henry and Davidson[435] synthesized octaphenyltrigermdithian $Ph_3GeSGePh_2SGePh_3$, one of the first perorganyloligogermdithianes. They also obtained hexaphenyldigermperthiane $Ph_3GeSSGePh_3$ having a Ge–S–S–Ge group by the oxidation of Ph_3GeSH[428,435] with iodine. We note that the first compound of this type $c\text{-}Hex_3GeSSGeHex\text{-}c_3$ was described by Burschkies[431] as early as 1936.

The trialkylgermylthio derivatives of Group 14 elements R_3GeSMR_3 (M = Si, Sn, Pb) are analogs of hexaalkyldigermthiane. They were first synthesized in the Schmidt laboratory[247,425,436]. For example, R_3GeSMR_3 (M = Si, Sn, Pb; R = Me, Ph) was obtained by the reaction of R_3GeSLi with R_3MCl. Unsymmetrical compounds $R_3GeSMR'_3$ were obtained by the reaction of R_3GeSLi with R'_3MX (M = Si, Sn)[436]. In 1966, Vayzankin and coworkers[368] obtained $Et_3GeSSnEt_3$ by dehydrocondensation of Et_3GeSH and Et_3SnH.

Triorganyl(organylthio)germanes R_3GeSR' should be considered as organogermanium compounds having a Ge–S–M group, when M = C. Anderson[389] was the first to synthesize nine representatives of the Et_3GeSR series by heterofunctional condensation of triethylacetoxygermanes with aliphatic and aromatic thiols RSH, R = Alk, Ar(C_6–C_{12}).

In 1956, Anderson[389] obtained triethyl(organylthio)germanes by cleavage of $Et_3GeOGeEt_3$ with aromatic and aliphatic thiols. Later, Satge and Lesbre[346] used this reaction for synthesis of Et_3GeSBu. In 1966, Abel and coworkers[433] employed the reaction for the preparation of $Me_3GeSCMe_3$ from $Me_3GeOGeMe_3$ and Me_3CSH. They also demonstrated that the reaction of Me_3GeOEt and PhSH resulted in Me_3GeSPh. Satge and Lesbre[346] obtained Bu_3GeSPh, $Bu_3GeSC_8H_{17}\text{-}n$ and $Et_2Ge(SPh)_2$ by the reaction of PhSH or $n\text{-}C_8H_{17}SH$ with Bu_3GeOMe and $Et_2Ge(OMe)_2$. They also cleaved $(Et_3GeO)_n$ with thiophenol to form $Et_2Ge(SPh)_2$. Similar transformations of a Si–O bond to a Si–S bond did not occur with organosilicon compounds. Anderson[377,389] discovered re-thiylation of trialkyl(organylthio)germanes by higher alkanethiols and arenethiols. This process occurred smoothly only on heating >170 °C and when a sufficiently wide range exists between the boiling points of the starting and the resultant thiols[346]. Following Anderson[377,389], other researchers[346,347,375] used this reaction. Ph_3GeSH[428] was also used in the reaction. The re-thiylation reaction resembles the reaction of PhSH with $Et_3GeSGeEt_3$ at 180–190 °C which gives Et_3GeSPh and H_2S[346].

Satge and Lesbre[346] used the reaction of thiols with R_3GeNMe_2 for the synthesis of triorganyl(alkylthio)germanes R_3GeSR' (R = *i*-Pr, *t*-Bu, Octyl; R′ = Bu). A year after, Abel, Armitage and Brady[375] employed the reaction of Me_3GeNEt_2 and BuSH to produce Me_3GeSBu.

In 1962, Davidson, Hills and Henry[437] obtained triphenyl(organylthio)germanes Ph_3GeSR by the reaction of Ph_3GeSNa with organic halides (R = Me, Bu, CH_2Ph, COPh, CH_2SMe), and by reaction of the latter halides with Ph_3GeSH in the presence of pyridine. For the synthesis of organyl(organylthio)germanes $R_{4-n}Ge(SR')_n$ ($n = 1$, 2) the reactions of organogermanium halides with mercaptanes or with sodium mercaptides RSNa[369,375] in the presence of organic bases[369,438] were used. The reaction of Me_3GeSLi and Me_3CSH resulting in $Me_3GeSCMe_3$ was also described. Lead alkanethiolates $Pb(SR')_2$ with R′ = Et, Bu[346,369,429] were also employed for the synthesis of $R_{4-n}Ge(SR')_n$ ($n = 1$, 2) from $R_{4-n}GeX_n$ (X = Cl, Br). Abel, Armitage and Brady[375] succeeded in substituting the bromine atom in Me_3GeBr by an alkylthio group by the action of Me_3SiSR (R = Et, *i*-Pr).

By using dimethyldichlorogermane and aliphatic or aromatic dithiols, Wieber and Schmidt[391,392,439] designed new heterocyclic systems in 1963–1964. In 1968, the reaction of Me_2GeCl_2 with $HS(CH_2)_nSH$ ($n = 2$, 3) in the presence of Et_3N enabled them to obtain the first of 2,2-dialkyl-1,3-dithia-2-germacycloalkanes[392,439]. By the reaction of Me_2GeCl_2 or $MeCH_2(Cl)GeCl_2$ with 4-methylbenzene-1,2-dithiol in the presence of Et_3N they produced the two isomeric ring-methyl derivatives 2,2-dimethyl-4,5-benza-1,3-dithia-2-germacyclopentane and 2-chloro-2-dimethyl-5,6-benza-1,4-dithia-2-germacyclohexane, respectively[392]. Similarly, 2,2-dimethyl-1-oxa-3-thia-2-germacyclopentane[391] was obtained from 2-mercaptoethanol.

In 1962, Lesbre and Satge[360] first carried out the dehydrocondensation of organogermanium compounds having the Ge–H bond with thiols, by reacting Ph_3GeH with BuSH in the presence of a platinum catalyst to give Ph_3GeSBu. They later used this reaction in the presence of nickel catalyst[346]. Thus, $Et_3GeSCH_2CH_2SGeEt_3$ was formed by the dehydrocondensation reaction of Et_3GeH with $HSCH_2CH_2SH$. In addition, Satge and Lesbre[346] discovered that triethylgermane cleaved MeSSMe to give Et_3GeSMe and MeSH.

Several reactions of triorganyl(alkylthio)germanes were investigated in 1962–1965. The Ge–S bond in these compounds was found to be chemically more stable than an Si–S bond, but much more reactive than Sn–S and Pb–S bonds. According to Satge and Lesbre[346] and Hooton and Allred[440], long exposure of triorganyl(alkylthio)germanes (Et_3GeSBu[346], Me_3GeSMe and Ph_3GeSMe[440]) to water either caused no change or only a slight hydrolysis (for Et_3GeSMe)[346]. The alcoholysis of R_3MSR' (M = Ge) was much more difficult than that for M = Si[346,440]. Compounds R_3GeSR' are easily oxidized by hydrogen peroxide up to $R_3GeOGeR_3$[440]; $LiAlH_4$ reduces them to R_3GeH[346] and aniline does not react with them. The Ge–S bond of Ph_3GeSMe was so reactive that it was cleaved with methyl iodide to Ph_3GeI and $Me_3S^+I^-$[440]. Similarly, dimethyl sulfate transforms Me_3GeSMe to Me_3GeOSO_2OMe[440] and $Me_3S^+[MeSO_4]^-$. When organolithium or organomagnium compounds R′M (R′ = alkyl; M = Li[435], MgX[346]) reacted with R_3GeSMe (R = Et, Ph), the SMe group was replaced by alkyl groups giving R_3GeR' derivatives.

Triorganylgermanethiols R_3GeSH were latecomers in organogermanium chemistry. The first representative of this class, i.e. Ph_3GeSH, was produced only in 1963 by Henry and Davidson[428] by the reaction of Ph_3GeBr with H_2S in the presence of pyridine.

In 1966, Vyazankin and coworkers[367,368] found that triethylgermanethiol Et_3GeSH was formed by heating Et_3GeH with sulfur at 140 °C. Attempts of Henry and Davidson[428] to obtain diphenylgermanedithiol from Ph_2GeBr_2 had failed, although they suggested that a rather labile $Ph_2Ge(SH)_2$ could exist in the reaction mixture.

The first alkali metal triorganylgermanethiolate R_3GeSM was Ph_3GeSNa, synthesized by Henry, Davidson and coworkers[435,437] by the reaction of Ph_3GeBr with excess Na_2S in an alcoholic solution.

Ruidisch and Schmidt[425,441] developed a new synthesis of lithium trimethylgermanethiolate Me_3GeSLi in quantitative yield by the reaction of $(Me_2GeS)_3$ with MeLi. Later, Vyazankin and coworkers[369] produced the analog Et_3GeSLi by the reaction of Li with Et_3GeSH in THF. Ph_3GeSLi was synthesized by the reaction of Ph_3GeLi with sulfur in THF[247,436,442,443]. Unlike the labile $Ph_2Ge(SH)_2$, its di-sodium salt, which was isolated by Henry and Davidson[428] as the trihydrate $R_2Ge(SNa)_2 \cdot 3H_2O$ from the reaction of Ph_2GeBr_2 with Na_2S, turned out to be rather stable.

The chemical transformations of Et_3GeSH and Ph_3GeSH have been extensively investigated by Vyazankin and coworkers[367,368] and by Henry and Davidson[428]. The latter authors showed that the reaction of Ph_3GeSH with PhCOCl and $(SCN)_2$ resulted in $Ph_3GeSCOPh$ and Ph_3GeSCN, respectively. They failed in an attempted addition of Ph_3GeSH to an activated double bond.

Organogermanium compounds with Ge−Se bond were prepared much later than their sulfur analogs. All of them were obtained in Schmidt's and the Vyazankin's laboratories. The first compound was hexamethylcyclotrigermselenane $(Me_2GeSe)_3$, which Schmidt and Ruf[444] obtained by the reaction of Me_2GeCl_2 with Na_2Se in 1961, together with higher cyclogermselenanes $(Me_2GeSe)_n$ and a minor amount of the linear polymer $Cl(Me_2GeSe)_nCl$. Two years later $(Me_2GeSe)_3$ was synthesized again in Schmidt's[445,446] laboratory. Its analogs and homologs $(R_2GeSe)_n$ as well as all the organylgermsesquiselenanes $(RGeSe_{1.5})_n$ were not described until 1970.

In 1963, Ruidisch and Schmidt[445] generated the first hexaalkyldigermselenane $R_3GeSeGeR_3$, R = Me, together with Li_2Se by thermal decomposition at >65 °C of $Me_3GeSeLi$. The precursor $Me_3GeSeLi$ was quantitatively produced by cleavage of $(Me_2GeSe)_n$ with methyllithium[445] or by the action of selenium on Me_3GeLi[445]. $Ph_3GeSeLi$[247,436,442,443] was obtained similarly. $Me_3GeSeGeMe_3$ was also synthesized from $Me_3GeSeLi$ and Me_3GeCl[445]. Lithium triethylgermaneselenolate was prepared by the reaction of Et_3GeSeH with Li in THF[369] whereas the reaction of MeMgI upon Et_3GeSeH resulted in $Et_3GeSeMgI$[369]. The reaction of the latter with Et_3GeBr gave $Et_3GeSeGeEt_3$[369].

In 1965, $Ph_3GeSeGePh_3$ was synthesized in the same laboratory by reaction of $Ph_3GeSeLi$ with Ph_3GeBr[247]. In 1966, Vyazankin and coworkers[366–369] found that $Et_3GeSeGeEt_3$ was obtained in 22% yield upon heating Et_3GeH and Se at 200 °C. It was suggested that Et_3GeSeH was an intermediate in the reaction and, indeed, it was obtained in 63% yield at 200 °C[367,368]. Heating Et_3GeSeH at 130 °C for a long time gave 37% of $Et_3GeSeGeEt_3$[367]. When trialkylgermanes R_3GeH reacted with Se at 200 °C, R_3GeSeH (R = *i*-Pr, *c*-C_6H_{11}) were obtained in 67% and 31% yield, respectively[447] together with the corresponding hexaorganyldigermselenanes $R_3GeSeGeR_3$. A more effective synthesis of $Et_3GeSeGeEt_3$ in 45% yield was the thermal (200 °C) reaction of Et_3GeSeH with Et_2Se[366]. A convenient synthesis of hexaethyldigermselenane was the reaction of $Et_3GeSeLi$ and Et_3GeBr[369]. The reaction of Et_3GeSeH with Et_2Hg at 20 °C afforded $Et_3GeSeGeEt_3$[369].

Triethyl(organylseleno)germanes Et_3GeSeR with R = Bu[369], CH_2Ph[369], CH_2CH_2Ph[448] and CH_2CH_2COOEt[448] became known in 1967–1969. Compounds with R = Bu and CH_2Ph were produced by the reaction of $Et_3GeSeLi$ with BuBr and $PhCH_2Cl$. Unexpectedly, the reaction of $Et_3GeSeLi$ and 1,2-dibromoethane gave $Et_3GeSeGeEt_3$ and $CH_2{=}CH_2$[369]. Other compounds were prepared by hydroselenation (i.e. by photochemical addition of Et_3GeSeH to styrene and ethyl acrylate[448]). $Et_3GeSeBu$ was also synthesized by the reaction of $(Et_3Ge)_2Hg$ and BuSeH[449]. Compounds having a Ge−Se−M group (M = Si, Sn, Pb) were first obtained in Schmidt's[247,442,445,450], laboratory in 1963–1965. These were $R_3GeSeMR_3$, with M = Si[445], R = Me; M = Sn[247,442,450], Pb[247], R = Ph, and were produced by the reaction of the corresponding $R_3GeSeLi$ and R_3MX (X = Cl, Br)[247]. $Ph_3GeSeSnPh_3$ was also synthesized, but with the 'opposite' reagents Ph_3GeBr with $Ph_3SnSeLi$[442,450]. Finally, the Vyazankin group obtained $Et_3GeSeGeEt_3$ by the condensation of Et_3GeSeH with Et_3SnH[367] or of $Et_3GeSeLi$ with Et_3SnCl[369].

In 1968, Mazerolles and coworkers[451] found that selenium inserted into the C−Ge bond of octaorganylgermacyclobutanes $R_2Ge(CR_2)_3$ gave octaorganyl-1-seleno-2-germcyclopentane.

The Vyazankin group studied some cleavage reactions of the Ge−Se bond. The reaction of $Et_3GeSeGeEt_3$ with bromine resulted in Et_3GeBr and Se, that with HCl led to Et_3GeCl and H_2Se[449] and, with sulfur, $Et_3GeSGeEt_3$[452] was formed.

Organogermanium compounds having Ge−Te bonds were also first prepared in Schmidt's and Vyazankin's laboratories in 1965–1967. Seven compounds [$R_3GeTeGeR_3$ (R = Et, Ph, c-C_6H_{11}), R_3GeTeR (R = Et) and $R_3GeTeMR_3$ (M = Si, Sn, Pb; R = Et, Ph)] were prepared in which the germanium atom was bound to the Group 14 element by the tellurium atom. $Ph_3GeTeLi$ was synthesized along with these compounds by the reaction of Ph_3GeLi with tellurium in THF[247,436,442,443]. The reaction of $Ph_3GeTeLi$ with Ph_3GeBr, Ph_3SnCl and Ph_3PbCl gave the corresponding $Ph_3GeTeMPh_3$ (M = Ge, Sn, Pb)[247].

Hexaethyldigermtellurane was obtained by heating Et_3GeH either with tellurium at 190–210 °C[447] or with diethyltelluride at 140 °C[368,447] in 75% and 58% yields, respectively. It was synthesized by the reaction of Et_3GeH with $(Et_3Si)_2Te$[452]. $Et_3GeTeEt$ was obtained for the first time (in 28–39% yield) by heating Et_3GeH with Et_2Te at 140 °C[368,447]. The reaction of $Et_3GeTeEt$ with Et_3MH (M = Si, Ge, Sn) at 20 °C resulted in 60% $Et_3GeTeMEt_3$[368,447]. When $Et_3GeTeGeEt_3$ reacted with Et_3SnH at 170 °C, Et_3GeH and $(Et_3Sn)_2Te$ were produced[368].

Vyazankin and coworkers[452] found that in the reaction of elementary S and Se with $Et_3GeTeGeEt_3$ the tellurium atom was replaced by the other chalcogen.

G. Organogermanium Pnicogen Derivatives

Among organogermanium derivatives in which the Ge atom is bound to Group 15 elements (pnicogens), the compounds having Ge−N bonds were the first to be studied.

The first compound of this family was tris(triphenylgermyl)amine $(Ph_3Ge)_3N$, prepared by Kraus and Foster[161] in 1927 by the reaction of Ph_3GeBr and liquid ammonia. In Kraus's laboratory[197,198,224,370] all the triphenylgermylamines of the $(Ph_3Ge)_nNH_{3-n}$ series, namely Ph_3GeNH_2, $(Ph_3Ge)_2NH$ and $(Ph_3Ge)_3N$, were synthesized. The hydrolytically very unstable Ph_3GeNH_2 was produced by the reaction of gaseous ammonia and Ph_3GeBr in an inert solvent[370]. It was also synthesized by reaction of Ph_3GeBr and KNH_2. With excess of KNH_2 the product was Ph_3GeNHK[370], which could be converted back to Ph_3GeNH_2 with NH_4Br. Kraus and coworkers found that Ph_3GeNH_2 was formed as a side product of the reaction of Ph_3GeNa with aryl halides[197] or methylene dihalides[196,329] in liquid ammonia. They pointed out that by eliminating ammonia, Ph_3GeNH_2 could be condensed to the first representative of hexaorganyldigermazanes i.e. $Ph_3GeNHGePh_3$[370]. When heating to 200 °C, Ph_3GeNH_2 was entirely converted to $(Ph_3Ge)_3N$[370].

In 1930 Kraus and Brown[226] synthesized $(Ph_2GeNH)_n$, $n = 3$ or 4 (although they considered the product to be 'diphenylgermanium imine' $Ph_2Ge = NH$), by the reaction of Ph_2GeCl_2 and liquid NH_3. The compound was hydrolytically unstable.

The first hexaalkyldigermazane $Et_3GeNHGeEt_3$ was obtained in 1932 by Kraus and Flood[148] by reaction of Et_3GeBr with Na in liquid ammonia. Its hydrolysis gave hexaethyldigermoxane $Et_3GeOGeEt_3$. Ammonolysis of Et_2GeBr_2 gave $(Et_2GeNH)_3$, which was hydrolyzed extremely easily to $(Et_2GeO)_n$ ($n = 3, 4$)[188]. Flood[188] in 1932, and much later Rijkens and van der Kerk[76,77], obtained $(R_2GeNH)_3$, R = Et, Bu by the reaction of Na in liquid ammonia with Et_2GeBr_2 and Bu_2GeCl_2, respectively. In 1933, Flood[306] found that during ammonolysis of $EtGeX_3$ (X = I, Br) a solid product corresponding to EtGeN, 'ethylgermanium nitride', was formed. Its hydrolysis resulted in polyethylgermsesquioxane $(Et_3GeO_{1.5})_n$, 'ethylgermanoic anhydride'. Therefore, Flood had prepared the first three-dimensional polyethylgermazane.

In 1931, Thomas and Southwood[453] obtained pseudo-organic organyl- and diorganylamine derivatives of two- and four-valence germanium such as $Ge(NHR)_2$ (R = Et, Ph), $Ge(NEt_2)_2$, $Ge(NHPh)_4$ and $Ge(NC_5H_9\text{-}c)_4$.

Laubengayer and Reggel[454], in 1943, synthesized Me_3GeNMe_2, the first organogermanium compound having a Ge$-$NR$_2$ group, by reacting Me_3GeCl and $LiNMe_2$. Analogous compounds with R = $SiMe_3$ were produced much later from Me_3GeCl or Me_2GeCl_2 and $NaN(SiMe_3)_2$[455,456].

In 1952, Anderson[457] synthesized a series of ethyl(dialkylamino)germanes $EtGe(NR_2)_3$ (R = Me, Et) by the direct reaction of $EtGeCl_3$ with dialkylamines. In 1949–1951 he discovered a new class of organogermanium compounds having Ge$-$N bonds, the alkylisocyanatogermanes $R_{4-n}Ge(NCO)_n$ (R = Et, Pr, *i*-Pr, Bu; $n = 1-3$). They were obtained from $R_{4-n}GeCl_n$ and AgNCO[314,315,387]. Rochow was an invisible participant in the work, since he gave Anderson Et_2GeCl_2 and $EtGeCl_3$[387]. In 1956, Anderson[379] obtained $i\text{-}Pr_2Ge(NCO)_2$ from the reaction of $i\text{-}Pr_2GeCl_2$ and AgNCO, Anderson hydrolyzed the alkylisocyanatogermanes, and their hydrolysis rates appeared to be the faster for the compounds with higher values of n. The cleavage of the Ge$-$N bond in ethylisocyanatogermanes with alcohols R$'$OH resulted in $R_{4-n}Ge(OR')_n$ (R = Me, Et, Bu; R$'$ = Me, Et; $n = 1, 2$) and formation of H_2NCOOR[387]. At the same time Anderson synthesized the first alkylisothiocyanatogermanes $R_{4-n}Ge(NCS)_n$ (R = Et, Pr, Bu; $n = 2$, 3)[184,376,379,458]. Compounds such as R_3GeNCS and $R_2Ge(NCS)_2$, R = Et, Pr, Bu were obtained in 1951 by cleavage of $R_3GeOGeR_3$ and $(R_2GeO)_3$ with HNCS generated *in situ*[376]. Analogously, $i\text{-}Pr_2Ge(NCS)_2$ was obtained from $(i\text{-}Pr_2GeO)_3$. Exchange processes have also been studied, such as those of Et_3GeNCS with AgNCO and of Et_3GeCN with AgNCS[376].

It is remarkable that the rather intensive investigations on nitrogen-containing organogermanium compounds during a quarter of a century were followed by reduced activity. From 1952 till 1963 they were mentioned only in seven publications[184,328,337,381,458–460], five of which devoted to compounds having the Ge$-$NCY bond (Y = O, S). No new compounds having the digermazane group (Ge$-$N$-$Ge) have been reported during this period.

The activity in the field was then resumed. In 1963, Onyszchuk[331] carried out the reaction of Me_3GeBr, Et_2GeCl_2 and Ph_2GeCl_2 with liquid ammonia at $-78\,^{\circ}C$ which gave the 1 : 1 adducts. On raising the temperature the products were converted to the corresponding ammonolysis products $[Me_3GeNH_3]^+Br^-$ and $(R_2GeNH)_n$ ($n = 2, 3$).

In 1964, Ruidisch and Schmidt[461] synthesized hexamethyldigermazane by the reaction of Me_3GeCl and gaseous NH_3 in diethyl ether. At $-60\,^{\circ}C$ a considerable quantity of $(Me_3Ge)_3N$[462] was formed. In the same year the authors also obtained organogermanium azides $Me_{4-n}Ge(N_3)_n$ ($n = 1, 2$) by reacting $Me_{4-n}GeCl_n$ and NaN_3[463]. At the same time Thayer and West[464] as well as Reichle[465] synthesized Ph_3GeN_3 from Ph_3GeBr.

In 1964, Rijkens and van der Kerk[77] obtained hexabutylcyclotrigermazane $(Bu_2GeNH)_3$ by the reaction of Bu_2GeCl_2 with a Na solution in liquid NH_3.

In 1964–1966, Satge and coworkers[36,37] used reactions of alkylhalogermanes with amino lithium and organomagnesium derivatives to generate Ge–N bonds.

Satge and Baudet[462] synthesized in 1966 hexaethyldigermazane by the reaction of Et_3GeCl and $LiNH_2$ in THF. The extremely unstable Et_3GeNH_2 was a probable intermediate in the reaction.

At the same year Massol and Satge[356] discovered that the ammonolysis of $Et_{3-n}GeH_nCl$ ($n = 1, 2$) led to the corresponding trigermylamines $(Et_{3-n}GeH_n)_3N$ (when $n = 1$, $Et_2GeHNHGeHEt_2$ was also formed). By contrast, Et_3GeCl ($n = 0$) did not give $(Et_3Ge)_3N$ on reaction with ammonia. That indicated a steric effect of the $R_{3-n}GeH_n$ group on the chlorides during ammonolysis. Accordingly Me_3GeNMe_2, which has less bulky substituents than Et_3Ge, underwent ammonolysis to give $(Me_3Ge)_3N$. The latter was formed also by the reaction of Me_3GeCl and $LiN(GeMe_3)_2$ or LiN_3[462] as well as by the reaction of MeLi and $(ClMe_2Ge)_3N$[461].

According to Wieber and Schwarzmann[434], ammonolysis of $ClCH_2Me_2GeCl$ resulted in $ClCH_2Me_2GeNHGeMe_2CH_2Cl$, the first carbon functionalized hexaalkyldigermazane derivative.

In 1964–1965, Rijkens and coworkers synthesized a series of nitrogen heterocycles (pyrrole, pyrazole, imidazole, triazole, succinimide, phthalimide), N-triorganylgermyl derivatives, and studied their properties[76,466].

In 1969, Highsmith and Sisler[467] attempted to repeat the reaction of Ph_3GeBr and ammonia, described by Kraus and Foster[161], but they obtained only $Ph_3GeNHGePh_3$ instead of $(Ph_3Ge)_3N$.

Since the first synthesis of organogermanium nitrogen derivatives it was found that the Ge–N bonds display high reactivity, especially an easy protolysis with water, alcohols, phenols, carboxylic acids, hydrohalic acids, SH-, NH-, PH- and CH-acids, etc[36,37,77,346,347]. All these reactions were initiated by electrophilic attack of the reactant proton on the nitrogen atom[76,468,469].

In particular, Anderson[457] in 1952 found out that the Ge–N bond in $EtGe(NMe_2)_3$ was cleaved by HI to give $EtGeI_3$. From 1964 ammonolysis[76,347], aminolysis[347], amidolysis[347] and hydrazinolysis[470] reactions of trialkyl(dimethylamino)germanes R_3GeNMe_2 (R = Me, Et) resulting in $R_3GeNHGeR_3$, R_3GeNHR', $R_3GeNHCOR'$ and $R_3GeNHNHR'$, respectively, were discovered.

Under strict reaction conditions (sometimes in the presence of $(NH_4)_2SO_4$) peralkylgermazanes $(R_3Ge)_nNH_{3-n}$[76,462] ($n = 2, 3$) were cleaved.

Schmidt and Ruidisch[471] in 1964 were the first to cleave the Ge–N–Ge group with organometallic reagents in the reaction of $(Me_2GeNMe)_3$ and MeLi, which gave $Me_3GeN(Li)Me$.

In 1964–1969, cleavage reactions of the Ge–N bond by anhydrides, carboxylic acids chloroanhydrides[347], chloramine[472], metal halides[473], and trimethylchlorometalanes Me_3MCl (M = Si, Ge, Sn, Pb) were described.

The addition reactions of trialkyl(dimethylamino)germanes to activated double and triple bonds were discovered in 1967–1968[474,475].

The first investigations of Satge and coworkers on the introduction of organic and inorganic compounds having M=Y groups[470] (CO_2, CS_2[347], PhNCO, PhNCS[476], F_3CCOCF_3[477]) into the Ge–N bond are of particular interest. Glockling and Hooton[478] were the first to obtain in 1963 organogermanium compounds having Ge–P bonds (e.g. Et_3GePPh_2) by reaction of Et_3GeBr with Ph_2PLi. A year later Satge and coworkers[347,462] synthesized the same compound by cleavage of Et_3GeNMe_2 with diphenylphosphine.

In 1969, Schumann-Ruidisch and Kuhlmey[479] carried out analogous reactions of Me_3GeNMe_2 with RPH_2 (R = Me, Ph), which resulted in $(Me_3Ge)_2PR$ and $Me_3GePHPh$.

Norman[480] proposed a new approach to the synthesis of compounds R_3GePH_2 by reaction of R_3GeCl with $LiAl(PH_2)_4$.

In 1965, Brooks and coworkers[481] discovered that the reaction of $Ph_{4-n}GeBr_n$ (n = 2, 3) and Ph_2PLi afforded $Ph_{4-n}Ge(PPh_2)_n$. Satge and Couret[474] similarly synthesized Et_3GePEt_2 from Et_3GeCl and Et_2PLi.

In 1965–1966, Schumann and coworkers[482,483] carried out the condensation of Ph_3GeCl with PH_3 and $PhPH_2$, which led to $(Ph_3Ge)_3P$ and $(Ph_3Ge)_2PPh$, respectively. Ph_3GeOH[483] was formed by cleavage of these compounds by an alcoholic KOH solution. $(Me_3Ge)_3P$[484] was synthesized by the reaction of Me_3GeNMe_2 with PH_3.

The Ge–P bond turned out to be extremely active. For example, $R_3GePR'_2$ ($R = R'$ = Et, Ph) was easily cleaved by water, alcohols, carboxylic acids, HCl, HBr, thiols, aniline and ammonia[474,481,485]. Oxidation of Et_3GePPh_2 by oxygen involved insertion into the Ge–P bond as well and resulted in $Et_3GeOPOPh_2$. The latter was also produced in the reaction of $Et_3GeOGeEt_3$ and Ph_2POOH[481]. When Et_3GePPh_2 reacted with bromine, Et_3GeBr and Ph_2PBr were formed. Butyllithium cleavage of Et_3GePPh_2 led to Et_3GeBu and Ph_2PLi. It is noteworthy that the Ge–P bond in Et_3GePPh_2 was cleaved even by methyl iodine to give Et_3GeI and Ph_2PMe. The reaction of Ph_2PMe and excess of MeI gave $[Me_2Ph_2P]^+I^-$[347,481]. When Et_3GePPh_2 and AgI were added to the reaction mixture, the complex $[Et_3GePPh_2 \cdot AgI]_4$[481] was produced.

CS_2, PhNCS, PhNCO, PrCHO, PhCHO, $CH_2{=}C{=}O$, $CH_2{=}CHCN$ and $PhC{\equiv}CH$ [474,486,487] insert into the Ge–P bond of Et_3GePR_2 (R = Et, Ph) similarly to their insertion into the Ge–N bond. Et_3GePEt_2 added to α,β-unsaturated aldehydes at the 1,4-positions[486].

There was only one report before 1970 on organogermanium arsenic derivatives. In 1966, Schumann and Blass[484] prepared $(Me_3Ge)_3As$ by the reaction of Me_3GeNMe_2 with AsH_3 and described some of its properties.

H. Compounds having a Hypovalent and Hypervalent Germanium Atom

The formation of inorganic compounds of hypovalent (divalent) germanium such as dihalo germanium GeX_2 (i.e. dihalogermylenes) was already noted by Winkler[16,23] in the 19th century. He reported the existence of $GeCl_2$ in HCl solution and of GeF_2 as the reduction product of K_2GeF_6 by hydrogen. However, only in the beginning of the 20th century did fundamental investigations of dihalo germanium, including monomeric GeX_2, start[77,488–491].

In 1926–1934, some methods for the gas-phase generation of monomeric inorganic derivatives of divalent germanium such as H_2Ge[332,492], F_2Ge[493,494], Cl_2Ge[495,496] and Br_2Ge[497] were developed. Dennis and Hance[498] obtained solid GeI_2 for the first time in 1922. It turned out not to be a monomer, since the germanium atom was surrounded octahedrally with six iodine atoms[499] in its crystal lattice. However, at a high temperature GeI_2 dissociated to form the monomeric molecules.

Interesting complexes of GeI_2 and CH_3NH_2[500] or Me_4NI have been described. The reaction of GeI_2 with NH_3 gave germanium(II) imide Ge=NH, which could be hydrolyzed to $Ge(OH)_2$, i.e. $(H_2O \cdot GeO)$ and NH_3[501]. Complexes of GeF_2 with Et_2O and with Me_2SO[502] were described in 1960–1962 and series of complexes of GeF_2[503], $GeCl_2$[504] and $GeBr_2$[505,506] were obtained as well. Thus, the inorganic chemistry of germylenes was born almost simultaneously with their organogermanium chemistry.

Organogermanium derivatives R_2Ge, which are often regarded as monomers, proved to be cyclic oligomers or linear polymers. The first attempt to synthesize monomeric diorganylgermylenes was made by Kraus and Brown[198]. In 1930 they tried to obtain diphenylgermylene by reduction of Ph_2GeCl_2 with sodium metal in boiling xylene, but the product was a mixture of cyclic oligomers $(Ph_2Ge)_n$. Only in 1963 did Neumann and Kühlein[199] determine that the main product of the reaction was octaphenylcyclotetragermane $(Ph_2Ge)_4$, i.e. a tetravalent germanium derivative.

In the 1960s, Nefedov, and his coworkers, Kolesnikov[201–207,507], Neumann and coworkers[175,199,200,508,509], Glockling and Hooton[249] and other investigators[36,37,69,77,185,210,239,240,490,510,511] started to study the generation of diorganylgermylenes. However, the reduction reactions of diorganyldihalogermanes by alkali metals, as well as the reaction of dihalogermanes with organometallic compounds (cf. Section II.B.) always resulted in the formation of cyclic oligomers, linear polymers or the insertion products of the R_2Ge moiety into bonds of the solvents or the reagents. For example, in 1954 Jacobs[240] tried to produce dialkylgermylenes by reaction of GeI_2 with a series of organometallic compounds (EtLi, BuLi, Bu_2Zn, Et_2Hg, Bu_2Hg). However, the only organogermanium compound that he was able to isolate was $IBu_2GeGeBu_2I$. The latter was also formed along with metallic mercury in the reaction of GeI_2 and Bu_2Hg.

In spite of these failures, all the authors had no doubts that diorganylgermylenes were the intermediates in the reactions studied.

Nefedov and coworkers[204,207,507] confirmed the generation of dimethylgermylene Me_2Ge in the reaction of Me_2GeCl_2 and Li based on the fact that its addition product to ethylene was formed. According to Vyazankin and coworkers[512] diethylgermylene was evidently an intermediate in the thermal (200 °C) decomposition of $Et_3GeGeEt_3$ with $AlCl_3$ catalyst which resulted in $(Et_2Ge)_n$ and Et_4Ge. In 1966, Bulten and Noltes[513] observed an analogous decomposition of $ClEt_2GeGeEt_2Cl$ and $Et_3GeGeEt_2Cl$. In both cases one of the products obtained was $(Et_2Ge)_n$, formed along with Et_2GeCl_2 or Et_3GeCl, respectively. The intermediate generation of Et_2Ge was confirmed by its insertion into the Ge–Cl bond of the precursor chloride with the formation of oligomers such as $Et_3Ge(Et_2Ge)_nCl$ ($n = 1, 2$).

Neumann and Kühlein[175,199,509] found in 1963 another precursor of diorganylgermylenes, the organomercurygermanium polymer $(-Ph_2Ge-Hg-)_n$, which was synthesized by the reaction of Ph_2GeH_2 and Et_2Hg. Unfortunately, in the early 1960s the Neumann laboratory did not have available spectroscopic techniques for the identification of the highly reactive short-lived diorganylgermylenes and other labile intermediates.

Nefedov and coworkers[202] had proven in their first publication that Me_2Ge: was the intermediate formed in the reaction of Me_2GeCl_2 and Li, since when the reaction was conducted in the presence of styrene, 1,1-dimethyl-3,4-diphenyl-1-germacyclopentane was formed in 40% yield.

The possibility of thermal generation of diorganylgermylene was established for the first time at the Nefedov laboratory in 1964–1965[203,205,207,507]. Thermolysis of $(Me_2Ge)_n$ where n is *ca* 55 at 350–400 °C led to Me_2Ge which was identified by its addition products to tolan and ethylene, together with its dimeric and polymeric biradicals[207] and to $(Me_2Ge)_n$ ($n = 6$, 5 and 4) as well. Shorigin, Nefedov and coworkers[206] were the first to obtain the UV spectra of polydiorganylgermylenes $(Me_2Ge)_n$.

The publications of Glockling and Hooton[249] and of Summers[239] were of special interest, because they reported the formation of diphenylgermylene from $Ph_2GeHOMe$ by α-elimination of methanol. This led to the conclusion that the intermediate products in the formation of R_2Ge from R_2GeX_2 or GeX_2(X = halogen) were $R_2Ge(X)M$ (M = Li, Na, K, MgX), which further decomposed to MX by an α-elimination process.

It was surprising that Neumann did not investigate the thermal and photochemical reactions of the decomposition of $(Ph_2Ge)_n$.

Cited as follows, publications of Nefedov and Kolesnikov[201–207,507], Neumann[175,199,200,508,509] and their coworkers can be regarded as the beginning of the chemistry of diorganylgermylenes.

Entrapping and subsequent investigations of diorganylgermylenes in hydrocarbon or argon matrices were carried out only in the 1980s[491,508]. Metlesics and Zeiss[211] showed the possible existence of organylhalogermylenes. They considered PhClGe: to be the intermediate formed in the reaction of $PhGeCl_3$ and Li amalgam.

Kinetically stable diorganylgermylenes $[(Me_3Si)_2CH]_2Ge$ and $[2,4,6\text{-}(Me_3C)_3C_6H_8\text{-}c]_2Ge$[491,514,515] were obtained and described in the last quarter of the 20th century. Different transformations of diorganylgermylenes (especially their insertion and dimerization reactions[69,508,510,516]) were studied in the 1980s. New precursors of diorganylgermylenes such as 7,7-diorganyl-7-germabenzonorbornadienes, $Ar_2Ge(SiMe_3)_2$, $Me_2Ge(N_3)_2$, and some heterocyclic compounds having endocyclic Ge−Ge[508] bonds were discovered at that very time, but we cannot dwell on these investigations in more detail.

Short-lived Ge-centered free radicals of $R_3Ge^•$ belong to the hypovalent (trivalent) germanium derivatives. In 1953–1957, Gilman and coworkers[195,217,219,229,517,518], based on the dissociation of hexaphenylethane to free $Ph_3C^•$ radicals, tried to obtain the $Ph_3Ge^•$ radical by dissociation of Ph_3GeMPh_3 (M = C, Si, Ge, Sn).

The stability of the Ge−Ge bond in $Ph_3GeGePh_3$ to homolytic cleavage to $Ph_3Ge^•$ radicals was evidenced from the fact that the compound melted at 336 °C without decomposition[249]. Hexaethyldigermane was thermally stable as well and could be distilled under atmospheric pressure at 265 °C[148].

The results of thermal decomposition of polydimethylgermylenes $(Me_2Ge)_n$[205] ($n \geqslant 2$) provide evidence in favor of the formation of Ge-centered biradicals $^•(Me_2Ge)_n{}^•$. It is suggested that the initial step of the thermal decomposition (>400 °C) of tetraalkylgermanes R_4Ge (R = Me, Et), which are widely used in producing germanium films[151,152], involves the formation of free radicals $R_3Ge^•$. We note that Gaddes and Mack[519] in 1930 were the first to carry out thermal cleavage of Et_4Ge starting at *ca* 420 °C. The final cleavage products seemed to be Ge and C_4H_{10}. It is very likely that the data on Ge-centered organogermanium free radicals reviewed in the period under discussion are limited to what was reported in the references mentioned above.

Compounds $R_2Ge{=}Y$ (Y = GeR_2, CR_2, NR′, O, S), in which the Ge atom is three-coordinated and is bonded by a π(p−p) bond with a Ge atom or with another element, can be considered as hypovalent germanium derivatives. The simplest concept of germanium atoms binding in the $R_2Ge{=}GeR_2$ molecule can be presented in the following way: $R_2\dot{G}e{-}\dot{G}eR_2$[514,516].

Information about compounds having Ge=Y bonds were published much later than the period considered above of organogermanium chemistry evolution. We only refer to some pertinent reviews[491,520–523].

Hypervalent germanium derivatives are compounds having penta-, hexa- and sometimes heptacoordinate germanium atom. Numerous publications are devoted to inorganic and pseudo-organic (with no C−Ge bonds) derivatives of this type[159,488,518,524–528].

Of particular interest are pseudo-organic compounds of hypervalent germanium such as germanium tetrahalide complexes with amines, complexes of GeX_4 with β-diketones[159,527], polyatomic alcohols and phenols[159,488,524], phthalocyanines[159], and others. The first labile hypervalent organogermanium compound $Ph_3Ge(NH_3)_3Na$ was obtained by Kraus and Foster[161] by cleavage of $Ph_3GeGePh_3$ with sodium in liquid ammonia.

The formation of organic derivatives of penta- and hexacoordinate germanium is due to a later time when the reactions of organohalogermanes were studied with ammonia and amines (see Section II.F). When these reactions were conducted at low temperatures, 1 : 1 and 1 : 2 adducts were formed. When heated >0 °C, the complexes of organohalogermanes with ammonia, primary and secondary amines decomposed to give compounds with a Ge—N bond and quaternary ammonium salts[331]. Such reactions were described for the first time in 1926–1933[148,188,453,457,529–533]. However, Kraus and Flood[148] found that the only reaction product of Et_3GeBr and liquid ammonia was the monoadduct $[Et_3GeNH_3]^+Br^-$ [75]. Organohalogermanes and tertiary amines formed rather stable 1 : 1 or 1 : 2 complexes, which were unstable toward hydrolysis.

Sowa and Kenny[534] in 1952 patented the unusual complex compounds $[R_{4-n}Ge(N^+R'_3)_n]X^-{}_n$ ($n = 1–4$) obtained from $R_{4-n}GeX_n$ and tertiary amines.

The first stable intramolecular complexes of pentacoordinate organogermanium derivatives (1-organylgermatranes $RGe(OCH_2CH_2)_3N$) having a transannular Ge ← N bond were synthesized in the Voronkov laboratory in 1965[397]. Their synthesis was based on the direct use of $RGeCl_3$ and $(RGeO_{1.5})_n$[397,398]. Their molecular and crystalline structure[83,535,536], UV spectra[537], ^1H[538], ^{13}C[539] and ^{15}N[540] NMR spectra as well as their biological activity[398,541–543] have been investigated. Mironov, Gar and coworkers[82,84] later contributed to the investigations of germatranes and their biological activity.

Beginning from 1989, another interesting series of intramolecular organogermane complexes, such as Ge-substituted N-germylmethyllactames, were investigated extensively by Baukov, Pestunovich, Voronkov, Struchkov and others[544,545].

I. Biological Activity

The biological activity of germanium compounds and their influence on the biosphere have been considered in detail in an excellent monograph of Latvian and Russian chemists published in 1990 (in which 767 references are cited[86]) as well as in earlier reviews by the same authors[82,546–548].

Investigations of the effect of inorganic germanium compounds on living organisms began in 1922 when it was discovered that germanium dioxide stimulated erythropoesis (production of red blood cells). In the same year the toxicity of GeO_2 was determined for the first time[549–553]. The results of germanium dioxide toxicological studies were published in 1931–1944[553–556]. The growing interest in the chemistry of germanium, especially in the middle of the 20th century, led to numerous investigations of the biological activity of inorganic compounds of this element (GeO_2, RGeOOH and its salts, metal hexafluorogermanates, GeH_4, $GeCl_4$, GeF_4, GeS_2), which were undertaken mostly after 1953[48,75,86].

Even in the first half of the last century it was already established that many organogermanium compounds did not suppress *Trypanosoma, Spirochaeta, Pneumococcus, Streptococcus*[557,558] and test rat sarcoma[559]. Moreover, in 1935 Carpenter and coworkers[560] found that $(Me_2GeO)_n$ stimulated the growth of many kinds of microorganisms. Much later, Rochow and Sindler[561] found that $(Me_2GeO)_4$ did not show either toxic or irritating action on mammals (hamsters, rabbits). However, this oligomer exerted a teratogenic effect on chicken embryos and was more toxic to them than acetone[476,562]. The toxicity of $(R_2GeO)_n$ (R = Me, Et, Bu; $n = 3, 4$) was determined by Rijkens and van der Kerk[76] and by Caujiolle and coworkers[563] in 1964–1966.

In 1936, Rothermundt and Burschkies[557] tried to establish the possibility of chemotherapeutic use of organogermanium compounds. They determined the toxicity of many types of substances such as R_4Ge, R_3GeX, R_3GeGeR_3, $(RGeO_{1,5})_n$ and $(ReGeS_{1,5})_n$, where R = alkyl, cyclohexyl, aryl or benzyl. The conclusion reached was that organogermanes are of moderate therapeutic use because of their total low toxicity. In another article Burschkies[431] has reported that these compounds are of no chemotherapeutic use. Nevertheless, Rijkens and coworkers[75,564] thought that this statement was premature.

In 1962, Kaars[564] first investigated the fungicidal activity of trialkyl(acetoxy)germanes. In contrast to analogous tin and lead derivatives, they were inactive. Triethyl(acetoxy)germane appeared to be considerably less toxic to rats (LD_{50}125–250 mg kg^{-1} *per-os*) than isostructural tin and lead compounds. Its homologs, $R_3GeOOCMe$ (R = Pr, Bu), did not show any toxic action[565]. In general, no specific biological activity of compounds of type $R_3GeOOCMe$ has been found. The toxicity of alkylhalogermanes $Bu_{4-n}GeCl_n$ ($n = 0$–3) or $RGeI_3$ (R = Me, Et, Pr) was within a range of 50–1300 mg kg^{-1}[566] on intraperitoneal administration.

The toxicity of hexaalkyldigermoxanes $R_3GeOGeR_3$ (R = Me–Hex)[76,567] was determined in 1963–1964.

Italian pharmacologists in 1963–1966 studied extensively the toxicity of tetraalkylgermanes. All the compounds were practically nontoxic (LD_{50}2300–8100 mg kg^{-1}), except i-Pr_4Ge (LD_{50} 620 mg kg^{-1}). It is noteworthy that the toxicity of $Et_3GeCH_2CH{=}CH_2$ (LD_{50} 114 mg kg^{-1}) was 40 times lower than that of its saturated analog Et_3GePr.

In 1969, diphenyl(iminodiacetoxy)germane was recommended for use as an insecticide[568]. The lower toxicity of organic germanium compounds compared to that of isostructural silicon compounds was reasonably confirmed by Voronkov and coworkers[398,569] in 1968; they found that 1-phenylgermatrane was 100 times less toxic than 1-phenylsilatrane (LD_{50} 0.3–0.4 and 40 mg kg^{-1}, respectively), although it showed an analogous physiological action.

Nevertheless, $PhGe(OCH_2CH_2)_3N$ was not the most toxic organogermaniun derivative. Toxicological investigations in 1979 with other 1-organylgermatranes $RGe(OCH_2CH_2)_3N$ [84,86] showed that most of them had low toxicity (LD_{50}1300–10000 mg kg^{-1}). Compounds with R = H and $BrCH_2$ showed LD_{50} of 320 and 355mg kg^{-1}, respectively[86]. The most toxic compounds were 1-(2-thienyl)germatrane and 1-(5-bromo-2-thienyl)germatrane (LD_{50} 16.5 and 21 mg kg^{-1}, respectively). Nevertheless, their toxicity was 10–12 times lower than that of 1-(2-thienyl)silatrane (LD_{50} 1.7 mg kg^{-1})[570]. It is remarkable that 1-(3-thienyl)germatrane was several times less toxic (LD_{50} 89 mg kg^{-1}) than that of its isomer mentioned above.

The discovery of a wide spectrum of biological activity of the organogermanium drug Ge-132 has stimulated extensive investigations in the field of synthesis and pharmacology of carbofunctional polyorganylgermsesquioxanes $(RGeO_{1.5})_n$. For this purpose Asai established a special Germanium Research Institute and a clinic[81,571] in Tokyo. It should be mentioned that a cytotoxic antitumor drug 2-(3-dimethylaminopropyl)-8,8-diethyl-2-aza-8-germaspiro[4,5]decane, 'spirogermanyl'[572,573], was developed in 1974.

Further events in bio-organogermanium chemistry, which was born soon after bio-organosilicon chemistry[547,548], have been described in a monograph[86].

The practical application of organogermanium compounds has been developed since the last quarter of the 20th century. They were used in medicine and agriculture as drugs and biostimulants[86,574] as well as in the microelectronic industry to produce thin films of elementary germanium[151,152].

III. ORGANOTIN COMPOUNDS

A. How it All Began

The chemistry of organotin compounds was born in the middle of the 19th century almost simultaneously with the birth of the chemistry of organolead compounds. Organic derivatives of these two Group 14 elements started to develop three quarters of a century earlier than those of germanium, their neighbor in the Periodic Table. Due to this large age difference, the review of the evolution of organotin compounds will cover a period of only 110 years, up to the beginning of the 1960s.

Carl Jacob Löwig (1803–1890), a professor at Zürich University, laid the foundation for the chemistry of organotin compounds. He is honored by the synthesis of the first organic compounds of tin in 1852[41]. Polydiethylstannylene $(Et_2Sn)_n$ was obtained in his unpretentious laboratory before other organotin compounds by the reaction of ethyl iodide with an alloy of tin containing 14% of sodium (he found that the optimal Sn:Na ratio is 6 : 1). Triethyliodostannane and hexaethyldistannane were formed together with it. At a later date it was discovered that another reaction product was tetraethylstannane[575,576]. Consequently, Löwing[41,577–582] became the founder of the direct synthesis of organotin compounds. During his investigations he observed that the polydiethylstannylene obtained was easily oxidized in air to a white precipitate, which by modern concepts is a mixture of perethyloligocyclostannoxanes $(Et_2SnO)_n$. The latter was prepared by the reaction of Et_2SnI_2 with Ag_2O or with aqueous ammonia. Löwig found that the action of alcoholic HCl solution on $(Et_2SnO)_n$ led to Et_2SnCl_2. By the reaction of a solution of KOH saturated with hydrogen sulfide with Et_2SnCl_2, Löwig obtained oligodiethylcyclostannathianes $(Et_2SnS)_n$ as an amorphous precipitate having a penetrating foul smell. However, all the other compounds obtained had not quite a sweet smell, and they irritated the eyes and mucous membranes.

The reaction of Et_2Sn with bromine and chlorine (with iodine, a fire was created) resulted in the corresponding Et_2SnX_2 (X = Cl, Br). When Et_2Sn reacted with HCl, Et_2SnCl_2 was also formed[41].

Triethyliodostannane was converted to hexaethyldistannoxane by treatment with aqueous ammonia, and hexaethyldistannoxane was converted to triethylchlorostannane by reaction with HCl.

Löwig[41] and then Cahours[583] obtained diethylstannyldinitrate $Et_2Sn(ONO_2)_2$ and triethylstannylnitrate Et_3SnONO_2[41,584], by the reaction of HNO_3 with Et_2SnO and $Et_3SnOSnEt_3$, respectively. The reaction of diethyliodostannane with Ag_2SO_4 gave diethylstannylenesulfate $Et_2SnO_2SO_2$[41,583,585].

In spite of the rather tedious investigations of Löwig which were conducted at the level of 19th century chemistry, they resulted both in syntheses and the study of reactivities of the first organotin compounds. He interpreted his results by the then predominant theory of radicals and used the obsolete values of 6 and 59, respectively, for the carbon and tin atomic weights.

Though a very experienced detective is required to investigate the Löwig publications, it is clear that Löwig had in his hands the first representatives of the main classes of the organotin compounds, i.e. $(R_2Sn)_n$, R_3SnSnR_3, $(R_2SnO)_n$, $R_3SnOSnR_3$, R_3SnX, R_2SnX_2, as their ethyl derivatives. It is rather interesting to compare the Löwig formulas and names for his organotin compounds with the modern ones (Table 2).

It is regretful that the Löwig papers devoted to organotin compounds were published only during one year[41,577–579,586]. He then stopped the investigations in this field. Evidently, this was caused by his leaving Zürich for Breslau, where he was invited to take Bunsen's position.

Bunsen, who accepted the chair in Heidelberg University, left his new laboratory in Breslau to his successor. Löwig's termination of his organotin investigations possibly reflects his unwillingness to impose severe hazards upon himself and the people surrounding him by the poisoning and irritating vapors of the organotin and organolead compounds to which he was exposed in Zürich. Nevertheless, Löwig did not forget the organic tin and lead compounds and his publications[43,587], where he did not fail to mention his priority and which are part of the history of organometallic compounds[579–582], bear witness to this fact.

TABLE 2. Names and formulas of organotin compounds synthesized by Löwig

Modern formula	Löwig's formula	Löwig's name (*in German*)
$Et_3SnSnEt_3$	$Sn_4(C_4H_5)_3$	Acetstannäthyl
Et_4Sn	$Sn_4(C_4H_5)_5$	Äethstannäthyl
Et_3SnOEt		Acetstannäthyl-oxyd
Et_2SnCl_2		Chlor-Elaylstannäthyl
Et_2SnBr_2		Brom-Elaylstannäthyl
Et_2SnI_2		Iod-Elaylstannäthyl
$(Et_2SnO)_n$	Et_2SnO	Elaylstannäthyl-oxyd
$Et_3SnOSnEe_3$		Methylenstannäthyl-oxyd
Me_3SnONO_2	$Sn_2(C_4H_5)_3O,NO_5$	Salpetersäure Methstannäthyl-oxyd
Et_3SnONO_2	$Sn_4(C_4H_5)_4O,NO_5$	Salpetersäures Elaylstannäthyl-oxyd
$(Et_3SnO)_2SO_2$	$Sn_2(C_4H_5)_3O,SO_3$	Schwefelsäure Methstannäthyl-oxyd
Et_3SnONO_2		Salpetersäure Acetstannäthyl-oxyd

In spite of Löwig's outstanding research, which laid the foundation of organotin chemistry, it should be noted that he shared the laurels of the discoverer with two other founders of organometallic chemistry: Edward Frankland (1825–1899), a professor of the Royal Chemical College in London, and August Cahours (1813–1891), a professor of the Ecole Centrale in Paris. It is generally believed that Frankland's first article devoted to the organotin synthesis appeared in 1853[45]. Actually, the results of his pioneer research were published a year earlier in a journal that was of little interest to chemists[588]. There is no reference to this article in monographs and reviews dealing with organotin compounds. Frankland reported in this article that he used his earlier discovered organozinc method for the syntheses of organotin compounds[588]. By the reaction of $SnCl_2$ with diethylzinc, he first synthesized tetraethylstannane Et_4Sn and studied some of its reactions. In the course of his investigations Frankland[588] together with Lawrence[589] first discovered the cleavage reactions of the C–Sn bond.

In the reaction of Et_4Sn with sulfur dioxide in the presence of air oxygen he obtained ethyl triethylstannylsulfonate Et_3SnOSO_2Et ('stantriethylic ethylsulfonate'). The action of H_2SO_4 on the latter led to bis(triethylstannyl) sulfate $(Et_3SnO)_2SO_2$ ('stantriethylic sulphate'). Finally, Frankland prepared polydiethylstannylene $(Et_2Sn)_n$ by the reduction of Et_2SnI_2 with zinc in hydrochloric acid[45]. All these data were reproduced and extended in his later publications[45,589–591]. In 1853, he isolated crystals of Et_2SnI_2 and some amount of Et_3SnI[45] by heating a tin foil with ethyl iodide at 180 °C in a sealed tube. He observed this reaction also under sunlight, i.e. he reported the first photochemical process in organometallic chemistry. In 1879, Frankland and Lawrence[589] demonstrated that the action of R_2Zn (R = Me, Et) on Et_2SnI_2 resulted in Et_2SnMe_2 and Et_4Sn, respectively. He found that HCl cleaved Et_2SnMe_2, but the cleavage products were not identified.

Frankland's[45,588–591] research is also a corner stone of organotin chemistry. He favored the valence ideas and the use of modern graphic formulas of organotin compounds. Moreover, his research destroyed the border between inorganic and organic chemistry.

In 1852, simultaneously Cahours[575,583,592–602] together with Löwig and Frankland became interested in organotin compounds. Together with Löwig and Frankland he belongs to the great pioneers of organotin chemistry in the 19th century and he made an essential contribution to its development. His first organotin investigation concerned the synthesis of diethylstannylene (almost simultaneously with those described by Löwig[41] and Frankland[45]) and other reactions followed[575,592,593]. In 1853, Cahours showed that MeI reacted with Sn at 150–180 °C to give Me_4Sn and Me_3SnI[593]. The hydrolysis of Me_3SnI gave $Me_3SnOSnMe_3$, whose cleavage with aqueous acids (HX, H_nY) resulted in Me_3SnX

(X = Cl, Br, I, S, OOCMe, NO_3) and $(Me_3Sn)_nY$ (Y = S, CO_3, $(OOC)_2$, $n = 2$; PO_4, $n = 3$)[583]. By hydrolysis of Me_2SnI_2, $(Me_2SnO)_n$ was obtained, and it was cleaved with the corresponding acids to Me_2SnX_2[583,593]. Cahours[596] was the first to demonstrate the possibility of replacing the halogens in alkylhalostannanes (Et_3SnI, Et_2SnI_2) by the anions of the corresponding silver salts (AgCN, AgNCO, AgSCN) using their reactions with Et_3SnI as an example. He also obtained hexaethyldistannathiane $Et_3SnSSnEt_3$ by the reaction of Et_3SnCl with H_2S in alcoholic media[596].

Following these three fathers of organotin chemistry, other luminaries of the chemical science of the 19th century such as Buckton[603,604], Ladenburg[605–607], and then at the turn of the century Pope and Peachey[308,608–612] and Pfeiffer and coworkers[309,613–620], were engaged in the development of organotin chemistry. In the 20th century this development is associated with the names of well-known scientists such as Krause, Schmidt and Neumann (in Germany), Kraus, Druce, Bullard, Gilman, Rochow, Anderson, Seyferth and West (in the USA), Kocheshkov, Nesmeyanov, Razuvaev, Nefedov, Koton, Kolesnikov and Manulkin (in the USSR), van der Kerk (in the Netherlands), Lesbre (in France) and Nagai and Harada (in Japan) and their numerous colleagues. Together they synthesized about 1800 organotin compounds up to 1960, in *ca.* 950 publications.

We shall now follow systematically these developments.

B. Direct Synthesis

The reactions of metals, their intermetal derivatives or alloys (often in the presence of a catalyst or a promoter) with organic halides and with some other organic compounds such as lower alcohols and alkylamines can be regarded as a direct synthesis of the organometallic compounds. As mentioned in Section III.A, Löwig and, to a lesser extent, Frankland were the originators of the direct synthesis of organotin compounds. Since 1860, they were followed by Cahours[575,583,598,600,602], who used the reaction of alkyl iodides with a tin–sodium alloy (10–20%) in a sealed tube at 100–200 °C. Cahours[575] obtained Et_2SnI_2 by the reaction of EtI with tin metal at 140–150 °C as well as at 100 °C under sunlight irradiation (according to Frankland). He also established that by increasing the Na content (from 5 to 20%) in the Sn–Na alloy, the reaction of EtI with the alloys led to the formation of Et_3SnI and then to Et_4Sn. Cahours synthesized a series of trialkyliodostannanes R_3SnX[575,583,598–602] and tetralkylstannanes R_4Sn (R = Me, Et, Pr)[575,583,592,598] by the reaction of alkyl iodides with the Sn–Na alloy. Among other products he observed the formation of the corresponding dialkyldiiodostannanes and hexaalkyldistannoxanes[598,600,602]. He found that increasing the sodium content of the alloy led predominantly to tetralkylstannanes, and decreasing its content led to dialkyldiiodostannanes. He also demonstrated that EtBr reacted analogously to alkyl iodides to give Et_3SnBr. Cahours prepared a series of tetraalkylstannanes $(C_nH_{2n+1})_4Sn$ (n = 1[583,594]; 2[575]; 3[598]; 4, 5[600]) by heating the corresponding alkyl iodides and bromides with the Sn–Na alloy in a sealed tube. A simultaneous formation of the corresponding trialkyliodostannanes[575,583,594,598–602] and trialkylbromostannanes[583] was observed. Other researchers[585,589,607,621–625] then synthesized tetraalkyl- and dialkyldihalostannanes by the reaction of alkyl halides with an Sn–Na alloy. Neiman and Shushunov[626,627] investigated the kinetics of the reaction of tin alloys containing 8.8 and 18.2% Na with EtBr at a wide range of temperatures and pressures in 1948. Depending on the alloy compositions and reaction conditions, the products were Et_4Sn or Et_2SnBr_2. When using this process, they were the first to discover a topochemical reaction with a longer induction period at higher rather than lower temperatures. Following Cahours, they also found that using an alloy with a high Na content led to Et_4Sn. When the sodium content in the alloy corresponded to a $NaSn_4$ composition, Et_4Sn was formed at a temperature <60 °C, but at >60–160 °C the main product was Et_2SnBr_2.

Pure tin was also used for the direct synthesis of organotin compounds. In 1948, unlike previous investigations, Harada obtained sodium stannite Na_2Sn not by the metal fusion, but by the reaction of tin with sodium in liquid ammonia[628]. The reaction of Na_2Sn with EtBr led to $(Et_2Sn)_n$.

Following Frankland[45], Cahours[575,583,598,600,602] established that alkyl halides reacted with melted tin to give dialkyldihalostannanes. Consequently, both authors became the founders of the direct synthesis of organylhalostannanes from metallic tin. Nevertheless, the first attempts to use alkyl bromides in the reaction with tin were unsuccessful[629,630]. In 1911, Emmert and Eller[622] first obtained carbofunctional organotin compounds $(EtCOOCH_2)_2SnI_2$ by the reaction of metallic tin with ethyl iodoacetate. In 1928–1929, Kocheshkov[631–633] discovered that dibromomethane and dichloromethane reacted with tin at 180–220 °C to give almost quantitative yields of $MeSnX_3$ (X = Cl, Br) according to equation 4.

$$3CH_2X_2 + 2Sn \longrightarrow 2MeSnX_3 + C \qquad (4)$$

He suggested that $CH_2{=}SnX_2$ was an intermediate product in the process and that $MeSnX_3$ was formed by addition of HX to the intermediate, which, in turn, was the insertion product of Sn into CH_2X_2. The reaction of CH_2I_2 with Sn at 170–180 °C led only to carbon and SnI_4. It is noteworthy that benzyl chloride acted with tin powder in water or in alcohol under mild conditions to give $(PhCH_2)_3SnCl$[634] in 85% yield.

In 1958, Kocheshkov and coworkers[635] showed that alkylbromostannane can be prepared from tin and alkyl bromides under ionizing irradiation[603].

The development of the direct syntheses of organotins involves a mysterious and even detective story, as told by Letts and Collie[636–638]. They wanted to prepare diethylzinc according to Frankland by heating ethyl iodide with zinc metal. To their great surprise, tetraethylstannane was the main reaction product[636]. They could not guess that so much tin was present in the commercial zinc that they purchased. Their further experiments with mixtures of tin and zinc led to the same result. They also found that heating tin powder with EtZnI at 150 °C resulted in Et_4Sn[636–638]. Consequently, Letts and Collie proposed the following scheme (equations 5 and 6) for the reaction.

$$Et_2Zn + Sn \longrightarrow Et_2Sn + Zn \qquad (5)$$

$$2\ Et_2Sn \longrightarrow Et_4Sn + Sn \qquad (6)$$

Anyway, it is doubtful whether Letts and Collie thought about zinc as the catalyst of the reaction of tin with alkyl halides (in spite of their demonstration) since this fact was established considerably later. The authors also found that the reaction of EtI with a Sn–Zn alloy (33–50%) containing 5% of Cu gave a maximum yield of Et_4Sn. Thus, long before Rochows's finding the catalytic influence of copper in the direct synthesis of organometallic compounds was observed[636].

Since 1927, Harada[639–642] studied the influence of addition of zinc to the Sn–Na alloy in its reaction with haloalkanes (MeI[640], EtI[639], PrI[642], EtBr[641,642] and others). Among other factors, he found that boiling ethyl bromide with an Sn alloy containing 14% of Na and 12–22% of Zn resulted in remarkable Et_4Sn yields. The promotion by zinc during the direct synthesis was further studied by other researchers[643–648]. In particular, it was shown that Bu_3SnCl[649] was the product of the reaction of BuCl and tin–sodium alloy containing 2% of Zn. In 1957, Zietz and coworkers[649] found that the reaction of higher alkyl chlorides with an Sn–Na alloy containing 2% of Zn at 150–180 °C led to a mixture of R_4Sn and R_3SnCl (R = Pr, Bu, Am) with a high tin conversion. Under milder conditions, in the same reaction with exactly the same Sn−Na alloy, the product $(R_2Sn)_n$ with R = Et, Bu[649], was formed. Cu, Cd, Al[650] were suggested in 1958 as activators of the alloys of the compositions Na_4Sn (43.5% Na) and Na_2Sn (28% Na). In the presence of these metals, even higher alkyl chlorides (C_8−C_{12}) also reacted with the alloys.

From the end of the 19th century, alkyl chlorides and bromides (often under pressure)[628,651] successfully reacted with melted tin, preferably in the presence of catalytic amounts of copper or zinc[652,653]. These data are mostly presented in patents[643,644,650,654,655].

In 1953, Smith[656] patented the reaction of MeCl with Sn, which led to Me_2SnCl_2 at 300 °C. However, already in 1949–1951 Smith and Rochow thoroughly investigated the reaction of gaseous MeCl with melted tin under ordinary pressure, but they did not publish the results though they were presented in a thesis submitted by Smith to Harvard University. The existence of the above patent[656] induced them to report their result in 1953. Smith and Rochow[652] studied the influence of added 25 elements to the reaction of methyl chloride with melted tin at 300–350 °C. The best catalysts found were Cu, Ag and Au.

Naturally, copper was further used as a catalyst. Under appropriate conditions, the main reaction product was Me_2SnCl_2 but small quantities of $MeSnCl_3$ and Me_3SnCl were also formed. The yield of Me_3SnCl was increased by the addition of sodium to tin[652]. Methyl bromide reacted with liquid tin at 300–400 °C to form Me_2SnBr_2[657], whereas in these conditions (385 °C) methyl iodide was completely decomposed. The products of the thermolysis were gaseous hydrocarbons and iodine. The iodine reacts with Sn to give SnI_2 and the reaction of SnI_2 with MeI gave $MeSnI_3$[657]. In the same article Smith and Rochow[657] reported that under conditions analogous to those used for the direct synthesis of Me_2SnCl_2, tin reacted very slowly with EtCl and BuCl underwent a complete thermal decomposition. They also found that MeX (X = Cl, Br, I) reacted with tin monoxide containing 10% Cu at 300 °C to form Me_3SnX[657].

In 1954, van der Kerk and Luijten[658] found that in the direct synthesis of tetraorganylstannanes the tin — sodium alloy can be replaced with a tin–magnesium alloy[659–661]. A mercury catalyst (Hg or $HgCl_2$) was required for this variant and the process was conducted at 160 °C under pressure.

In further investigations it was possible to conduct this reaction at atmospheric pressure in a solvent capable of influencing the ratio of the reaction products R_4Sn and R_3SnX. The method of the alloy preparation played an important role, with the content of magnesium being at most 21–29% (Mg_2Sn). In the absence of the catalyst (mercury salts or amines) alkyl chlorides did not react with these alloys[650]. In the reaction of a Sn−Na alloy (containing 4–5% Cu) with alkyl bromides or iodides in solution, up to 60% of dialkyldihalostannanes R_2SnX_2 as well as R_4Sn and R_3SnX[650] were formed.

The direct synthesis of aromatic tin compounds was realized for the first time in the 19th century. In 1889 Polis[651] and in 1926 Chambers and Scherer[662] obtained tetraphenylstannane by a longtime boiling of bromobenzene with an Sn−Na alloy in the presence of the initiator ethyl acetate[651] or without it[662]. In the reaction of PhBr with an alloy of Li_4Sn composition the yield of Ph_4Sn was only 13%[663]. Aryl halides did not react with tin alone at temperatures <200 °C[664]. In 1938, Nad' and Kocheshkov[665] obtained tetraphenylstannane by heating Ph_3SnCl with an Sn−Na alloy. The formation of tetraarylstannanes in the reaction of aryl halides with Sn−Na alloy was probably preceded by arylation of the tin with sodium aryls, which were the intermediates of this process. This mechanism was confirmed by the alkylation of tin with phenylmagnesium bromide[650]. Consequently, organometallic compounds were actually used to synthesize tetraarylstannanes from metallic tin or its alloys with sodium. Thus, in 1938, Talalaeva and Kocheshkov[666,667] obtained tetraarylstannanes in a reasonable yield by boiling lithium aryls with tin powder or its amalgam. Nad' and Kocheshkov[665] carried out the reaction of PhHgCl with an Sn−Na alloy in boiling xylene with better results (50% yield of Ph_4Sn). They found that the reaction between PhHgCl and Na_2Sn involved the intermediate formation of $(Ph_2Sn)_n$ and $Ph_3SnSnPh_3$. The latter disproportionated to Ph_4Sn and Ph_2Sn. According to their data, the reaction of PhHgCl with Sn gave Ph_2SnCl_2, which was disproportionated to Ph_3SnCl and $SnCl_4$[665]. The data on the direct synthesis of organotin compounds are summarized in a monograph[650].

C. Organometallic Synthesis from Inorganic and Organic Tin Halides

Frankland[588,589,591] was the first to synthesize in 1852 organotin compounds using the reaction of Et_2Zn with $SnCl_2$ to give Et_4Sn (Section II.A). One year later Cahours[575] obtained Et_4Sn by the reaction of Et_3SnI with Et_2Zn. He synthesized the first mixed tetraalkylstannane Me_3SnEt[583] by the reaction of Me_3SnI with Et_2Zn and in 1862 he analogously prepared Et_3SnMe[596]. Under Butlerov's guidance Morgunov[668] obtained Me_2SnEt_2 by the reaction of Me_2SnI_2 with Et_2Zn, although they did not succeed in synthesizing it from Et_2SnI_2 and Me_2Zn in the pure form according to Frankland. In 1900, Pope and Peachey[608] also used the organozinc method to prepare Me_3SnEt. During 12 years they obtained the first organotin compound containing asymmetric tin atom (MeEtPrSnI) using the appropriate dialkyl zinc. The asymmetric iodide was converted into an optically active salt with $[\alpha]_D = +95°$ by the reaction with silver d-camphorsulfonate.

Buckton in 1859 was the first to use tin tetrachloride to synthesize organotin compounds[603]. At that time, the reaction of $SnCl_4$ with Et_2Zn was the common route to Et_4Sn. Pope and Peachey[608] used this method only after four decades. In 1926, Chambers and Scherer[662] obtained Ph_4Sn by the organozinc method. Kocheshkov, Nesmeyanov and Potrosov[669] synthesized $(4\text{-}ClC_6H_4)_4Sn$ in the same way in 1934. However, at the beginning of the 20th century the organozinc method of organotin compounds synthesis lost its importance.

The use of Grignard reagents led to revolutionary developments in the synthesis of organotin compounds. It started in 1903 when Pope and Peachey[611] obtained R_4Sn, R = Et, Ph in a good yield from $SnCl_4$ and RMgBr. Just one year later this method was used by Pfeiffer and Schnurmann to synthesize Et_4Sn, Ph_4Sn and $(PhCH_2)_3SnCl$[670]. In 1904, Pfeiffer and Heller[615] reacted SnI_4 with the Grignard reagent MeMgI to obtain Me_3SnI. In 1954, Edgell and Ward[671] used Et_2O and, in 1957, Seyferth[672] and Stone[673] used THF as the solvent in this reaction and that improved the yield of R_4Sn.

From 1914, the Grignard method of synthesis completely displaced the organozinc method and was widely used[674]. Up to 1960, fifty publications reporting the use of this method appeared[125,675].

In 1927, Kraus and Callis[643] patented the method of preparing tetraorganylstannanes by the reaction of Grignard reagents with tin tetrahalides. In 1926, Law[676] obtained mixed tetraorganylstannanes, such as $Et(PhCH_2)_2SnBu$ and $Et(PhCH_2)SnBu_2$ from $Et(PhCH_2)_2SnI$ and $Et(PhCH_2)SnI_2$, by the Grignard method. In 1923, Böeseken and Rutgers[677] demonstrated that a Grignard reagent was able to cleave the Sn−Sn bond: the reaction of PhMgBr with $(Ph_2Sn)_n$ led to Ph_4Sn, $Ph_3SnSnPh_3$ and $Ph_{12}Sn_5$ (the first linear perorganylpolystannane).

It is noteworthy that in 1912 Smith and Kipping[678] applied the Barbier synthesis, i.e. the addition of organic halide to a mixture of Mg and $SnCl_4$ in ether (without preliminary preparation of the Grignard reagent) to obtain organylchlorostannanes in a good yield[678,679].

Organolithium synthesis of organotin compounds, in particular $(4\text{-}MeC_6H_4)_4Sn$ from $SnCl_4$, was first described by Austin[680] in 1932. In 1942, Talalaeva and Kocheshckov[667] used this method to obtain Ar_4Sn (e.g. Ar = 4-PhC_6H_4) when Grignard reagents failed to react. Bähr and Gelius[681] used the appropriate aryllithiums to synthesize tetra(9-phenanthryl)- and tetra(1-naphthyl)stannane from $SnCl_4$. Organolithium compounds were also used to synthesize 1,1-diorganylstannacycloalkanes[682,683].

An interesting spirocyclic system was created by the reaction of $SnCl_4$ with 1,2-bis(2′-lithiumphenyl)ethane by Kuivila and Beumel[682] in 1958. Spirocyclic compounds were also obtained in the reaction of $SnCl_4$ with 1,4-dilithium-1,2,3,4-tetraphenylbutadiene[683,684] or with ethyl bis(2-lithiumphenyl)amine[685].

In some cases the organolithium compounds cleaved the C−Sn bond[686–689]. However, these obstacles were successfully overcome by converting the organolithium compounds to the Grignard reagent by adding a magnesium halide[646,686,688,690]. The organolithium synthesis was also extensively used, especially for attaching vinyl and aryl groups[686–688,691–702] to the tin atom. It should be noted that in 1955, Gilman and Wu[694] obtained 4-$Ph_3SnC_6H_4NMe_2$ by the reaction of Ph_3SnCl with 4-$Me_2NC_6H_4Li$. In 1958, Bähr and Gelius[703] prepared $(4\text{-}PhC_6H_4)_3SnBr$ by reacting 4-PhC_6H_4Li with $SnBr_4$ in a 3.5 : 1 molar ratio.

Both organolithium and organomagnesium syntheses of tetra(*tert*-butyl)stannane had failed[704]. Up to 1960 organolithium compounds were seldom used to synthesize aliphatic tin derivatives[705]. In 1951, the reaction of PhLi with $SnCl_2$ allowed Wittig and coworkers[706] to obtain $(Ph_2Sn)_n$ in a good yield. With excess PhLi, Ph_3SnLi was also formed. Immediately after Wittig's work in 1953, Gilman and Rosenberg[707] developed a method to synthesize Ar_3SnLi from ArLi and $SnCl_2$. The reaction of Ar_3SnLi with appropriate aryl halides gave Ar_4Sn. This method was also successfully used to synthesize tetraalkylstannanes[708]. In 1956, Fischer and Grübert[709] obtained for the first time dicyclopentadienyltin by the reaction of cyclopentadienyllithium with $SnCl_2$.

Ladenburg[605,606] first used organosodium synthesis (i.e. the Würtz reaction) of organotin compounds in 1871. He synthesized Et_3SnPh by the reaction of Et_3SnI with Na and PhBr in ether medium. That was the first aromatic tin compound. He obtained $EtPhSnCl_2$[605] in the same way. In 1889, Polis[651] found that the reaction of $SnCl_4$ with Na and PhCl in boiling toluene did not result in Ph_4Sn. Nevertheless, when a 25% Na — 75% Sn alloy reacted with PhBr, using the MeCOOEt as an initiator, he obtained Ph_4Sn. However, during the following century this method was forgotten. Nevertheless, Dennis and coworkers[168] and Lesbre and Roues[691] used the reaction of aryl bromides and Na with $SnCl_4$ in ether, benzene or toluene to prepare tetraarylstannanes and the method was even patented[710,711]. Only in 1954–1958 was the Würtz reaction used to synthesize tetraalkylstannanes[629,712–716]. $SnCl_4$ could be replaced with alkylchlorostannanes, and Bu_4Sn was obtained in 88% yield[629] by the reaction of Bu_2SnCl_2 with BuCl and Na in petroleum ether. Organosodium compounds were used for the synthesis of organotin derivatives in 1954 by Zimmer and Sparmann[693], who obtained tetra(1-indenyl)stannane from 1-indenylsodium with $SnCl_4$. Five years later Hartmann and coworkers[701] synthesized $(PhC{\equiv}C)_4Sn$ by the reaction of $SnCl_4$ with $PhC{\equiv}CNa$, and by the reaction of $NaC{\equiv}CNa$ with R_3SnX (R = Ar, $PhCH_2$) they prepared $R_3SnC{\equiv}CSnR_3$[717,718].

In 1926, Chambers and Scherer[662] obtained the first organotin compound containing an Sn−Na bond. By reacting Ph_3SnBr with Na in liquid ammonia, they synthesized Ph_3SnNa and investigated its transformations. For example, the reactions of Ph_3SnNa with aryl halides resulted in Ph_3SnAr, with $ClCH_2COONa$, Ph_3SnCH_2COONa was formed, and PhHgI gave $Ph_3SnHgPh$[662]. They were the first to cleave the C−Sn bond by metallic Na, demonstrating that Ph_4Sn reacted with Na in liquid ammonia to form consecutively

Ph_3SnNa and Ph_2SnNa_2. The reaction of the latter with Ph_2SnBr_2 in liquid ammonia gave the polymeric substance $(Ph_2Sn)_n$[662].

The Würtz-type reaction was applied in the syntheses of organotin compounds, containing an Sn−Sn bond. Law prepared hexabenzyldistannane $(PhCH_2)_3SnSn(CH_2Ph)_3$ for the first time in 1926 by the reaction of Na with $(PhCH_2)_3SnCl$ in toluene[676].

Just a few reactions of $R_{4-n}SnCl_n$ ($n = 1-4$) with silver[719,720], mercury, aluminum, thallium and lead were described[576,721]. As early as 1878, Aronheim[722] found that prolonged heating of Ph_2Hg with $SnCl_4$ resulted in Ph_2SnCl_2 (33% yield). Only three quarters of a century later $PhSnCl_3$[723,724] was similarly obtained. In 1930 and 1931, Nesmeyanov and Kocheshkov[725–727] demonstrated that tin dihalides can react with organomercury compounds. In the reaction of SnX_2 (X = Cl, Br) with Ar_2Hg in ethanol or acetone they obtained diaryldihalostannanes Ar_2SnX_2. In 1922, Goddard and coworkers[728,729] found that the reaction of $SnCl_4$ with Ph_4Pb led to Ph_2SnCl_2 and Ph_2PbCl_2. By the reaction of Ph_3SnCl with $AgC{\equiv}CCH(OEt)_2$, Johnson and Holum[730] obtained $Ph_3SnC{\equiv}CCH(OEt)_2$ in 1958. Finally, in 1957 and 1959, Zakharkin and Okhlobystin[720,731] found that the reaction of $SnCl_4$ with R_3Al[720,731] could be employed to synthesize tetraalkylstannanes.

In conclusion it should be noted that organometallic synthetic methods of organylhalostannanes were not as widely used as in the synthesis of the isostructural compounds of silicon and germanium. Section III.E explains the reason for this.

D. Organotin Hydrides

In 1922, Kraus and Greer[732] synthesized trimethylstannane Me_3SnH, the first organotin compound containing an Sn−H bond, by the reaction of sodium trimethylstannane with ammonium bromide in liquid ammonia. In 1926, Chambers and Scherer[662] used this method for the synthesis of triphenylstannane R_3SnH and diphenylstannane R_2SnH_2. In 1943, Malatesta and Pizzotti[733] obtained Et_3SnH and Ph_3SnH by the same method. In 1951, Wittig and coworkers[706] used Ph_3SnLi in this reaction. The chemistry of organotin hydrides started to develop extensively when, in 1947, Finholt, Schlesinger and coworkers[336,648], who developed the reduction method of organometallic halides by $LiAlH_4$, used this method for the synthesis of trimethyl-, dimethyl- and methylstannane from $Me_{4-n}SnCl_n$ ($n = 1-3$). This method was widely applied later to obtain organotin hydrides[48,125,675,734]. Thus, in 1955–1958, Et_3SnI[735], Ph_3SnCl[736], Et_2SnCl_2[737] and Pr_2SnCl_2[738] were reduced to the appropriate hydrides by $LiAlH_4$. In 1953, West[334] failed to reduce triphenylhalostannanes with zinc in hydrochloric acid, unlike the reduction of triphenylbromogermane. In 1957–1958, Kerk and coworkers[738,739] developed the reduction method of R_3SnCl to R_3SnH (R = Et, Pr, Bu, Ph) with amalgamated aluminum in aqueous medium. As a result of this research 20 organotin hydrides $R_{4-n}SnH_n$ ($n = 1-3$) became known up to 1960.

Beginning from 1929, Ipatiev and his nearest coworkers Razuvaev and Koton tried to hydrogenate Ph_4Sn under drastic conditions (60 atm, 220 °C)[740–744], but neither formation of the compounds containing an Sn−H bond nor hydrogenation of the aromatic cycle was observed. Instead, hydrogenolysis of the C−Sn bond with formation of metallic tin and benzene took place. In this respect we note that, when in 1989 Khudobin and Voronkov[745] tried to reduce Bu_2SnCl_2 by R_3SiH in the presence of colloidal nickel, the products were metallic tin, butane and R_3SiCl. R_3SiH reduced tetrachlorostannane to $SnCl_2$ to give R_3SiCl and H_2. Organotin hydrides $R_{4-n}SnH_n$ are not among the stable organotin compounds. Their stability increases (i.e. their reactivity decreases) on decreasing the number n[737] of hydrogens at the tin atom. Even the early researchers observed that many organotin hydrides $R_{4-n}SnH_n$ (especially with R = Me,

Et and $n = 2$, 3) were slowly decomposed at room temperature and easily oxidized by air oxygen[336,732,735–737,746]. However, Me_3SnH and Me_2SnH_2 are little changed when stored in a sealed ampoule at room temperature during 3 months and 3 weeks, respectively[336]. $MeSnH_3$ decomposed under these conditions less than 2%[747] during 16 days. Distillation of butylstannane under atmospheric pressure at *ca* 100 °C failed because of its complete decomposition. However, at 170 °C and 0.5 mm the high-boiling triphenylstannane was so stable that its distillation succeeded[736,738] but it decomposed under sunlight exposure. In 1926, Chambers and Scherer[662] found that diphenylstannane Ph_2SnH_2 decomposed to Ph_2Sn at > -33 °C. In contrast, van der Kerk and coworkers[748] found that Ph_2SnH_2 decomposed to Ph_4Sn and metallic tin only on heating >100 °C in vacuum. Apparently, this process involves the intermediate formation of Ph_2Sn[662] which further disproportionated. Et_2SnH_2 was decomposed with explosion in contact with oxygen. In 1926–1929 it was shown that oxidation of trialkylstannanes and triphenylstannanes gave different products under different conditions. Bullard and coworkers[749,750] and later Anderson[735] obtained trialkylstannanols. According to Chambers and Scherer[662], Ph_3SnH gave hexaphenyldistannane. The latter is the product of reaction of Ph_3SnH with amines, as Noltes and van der Kerk[751] had found. Diphenylstannane was dehydrocondensed into the yellow modification of $(Ph_2Sn)_n$ in the presence of amines. In contrast, the reaction of Ph_3SnH with thiols gave hexaphenyldistannathiane $Ph_3SnSSnPh_3$[751]. In 1950, Indian researchers[752] found that the reaction of Pr_3SnH with aqueous-alcoholic NaOH solution gave Pr_3SnOH. In 1922, Kraus and Greer[732] found that the reaction of Me_3SnH with concentrated HCl led to Me_3SnCl. In 1951, Wittig and coworkers[706] converted Ph_3SnH to Ph_3SnCl by the same method. Noltes showed in his dissertation (1958) that triorganylstannanes reacted analogously with carboxylic acids to form $R_3SnOOCR'$ and that organotin hydrides reacted vigorously with halogens to give the corresponding halides. In 1955, Gilman and coworkers[267,736] found that Ph_3SnH in the presence of benzoyl peroxide formed Ph_4Sn without precipitation of metallic tin. However, in the presence of excess $(PhCOO)_2$ the product was $Ph_3SnOCOPh$. According to Kraus and Greer, and to Chambers and Scherer, R_3SnH (R = Me[732], Ph[662]) reacted with Na in liquid NH_3 to give R_3SnNa. In 1949, Gilman and Melvin[753] pointed out that Ph_4Sn and LiH were formed in the reaction of PhLi on Ph_3SnH. In contrast, Wittig and coworkers[706] found that the reaction of Ph_3SnH with MeLi led to Ph_3SnLi and CH_4. Nevertheless, in 1953, Gilman and Rosenberg[754] found that this reaction resulted in Ph_3SnMe and LiH. Lesbre and Buisson[755] developed the reaction of trialkylstannanes with diazo compounds ($R'CHN_2$), which gave R_3SnCH_2R' (R = Pr, Bu; R′ = H, COOEt, COMe, COPh, CN) along with nitrogen.

The hydrostannylation reaction[756] is of great importance in organotin chemistry. This term was proposed by Voronkov and Lukevics[52,53] in 1964. The reaction is based on the addition of organotin compounds, containing at least one Sn−H bond to multiple bonds (C=C, C≡C, C=O etc.)[52,53,77,265]. It is of special interest for the synthesis of carbofunctional organotin compounds. This reaction was first carried out by van der Kerk, Noltes and coworkers[748,751,757,758] in 1956. They found that trialkylstannanes R_3SnH (R = Pr, Bu) were easily added to the double bonds in $CH_2{=}CHR'$ (R′ = Ph, CN, COOH, COOMe, CH_2CN, $CH(OEt)_2$), to give the adducts $R_3SnCH_2CH_2R'$ in 95% yield. Hydrostannylation proceeded easily in the absence of catalysts by heating mixtures of both reagents at 80–100 °C for several hours. In 1958, monosubstituted ethylene derivatives with R′ = $CONH_2$, CH_2OH, COCMe, CH_2OOCMe, 4-C_5H_4N, OPh, Hex, $C_6H_4CH{=}CH_2$ were involved in the reaction with Ph_3SnH. It was found that Ph_3SnH was involved in the hydrostannylation process more easily than trialkylstannanes R_3SnH with R = Pr, Bu[759]. For example, the attempted addition of R_3SnH to $CH_2{=}CHCH_2OH$

had failed, while Ph_3SnH was easily added to allyl alcohol[756]. In 1959, van der Kerk and Noltes[758] carried out the first hydrostannylation of dienes. The addition of dialkylstannanes to dienes and acetylenes gave polymers in some cases[759]. However, in 1959, Noltes and van der Kerk obtained the cyclic diadduct 1,1,2,4,4,5-hexaphenyl-1,4-distannacyclohexane by the addition of Ph_2SnH_2 to $PhC{\equiv}CH$[758]. He also hydrostannylated $Ph_3GeCH{=}CH_2$ and $Ph_2Si(CH{=}CH_2)_2$ with triphenyl- and diphenylstannane. The reaction of Me_3SnH with $HC{\equiv}CPh$ led to $Me_3SnCH{=}CHPh$[759]. Only the *trans*-adduct[758] was isolated by hydrostannylation of phenylacetylene by triphenylstannane, but its addition to propargyl alcohol gave a mixture of *cis*- and *trans*-adducts. The hydrostannylation of alkynes proceeded more easily than that of alkenes, as confirmed by the lack of reactivity of Ph_3SnH with $HexCH{=}CH_2$, whereas it easily added to $BuC{\equiv}CH$. Nevertheless, in the reaction of R_3SnH with acetylenic hydrocarbons the diadducts[748,758,759] could also be obtained. Dialkylstannanes R_2SnH_2 (R = Pr[759], Bu[760]) were first used as hydrostannylating agents in 1958, and Ph_2SnH_2 in 1959[758,761]. The addition of R_2SnH_2 to the monosubstituted ethylenes $CH_2{=}CHR'$ at 60–80 °C resulted in the diadducts $R_2Sn(CH_2CH_2R')_2$[758,760,761] and addition of Ph_2SnH_2 to $F_2C{=}CF_2$[758,760] at 80 °C proceeded similarly. Analogously, organylstannanes $RSnH_3$ were added to three molecules of unsaturated compounds[758]. Unlike the hydrosilylation reaction, neither Pt nor H_2PtCl_6 catalyzed the hydrostannylation reactions. Addition of hydroquinone did not inhibit this reaction, thus arguing against a free radical mechanism. Dutch researchers[757,758,760] concluded that the hydrosilylation is an ionic process.

Since 1957 the triorganylstannanes Bu_3SnH and Ph_3SnH attracted scientists' attention as effective reducing reagents. They easily reduced alkyl-[759], alkynyl-[751] and aryl halides[748,762], amines[748,751] and mercaptans[751] to the corresponding hydrocarbons, but reduced ketones to the corresponding alcohols[751,763]. Hydrostannylation of the carboxylic group was not observed, distinguishing it from the hydrosilylation. However, Neumann found that in the presence of radical reaction initiators triorganylstannanes were added easily to aldehydes $R'CH_2O$ with the formation of the R_3SnOCH_2R' adducts (R = Alk; R′ = Alk, Ar)[756,764]. Kuivila and Beumel[765,766] established that the ability of organotin hydrides to reduce aldehydes and ketones was decreased along the series: $Ph_2SnH_2 > Bu_2SnH_2 > BuSnH_3 > Ph_3SnH > Bu_3SnH$. In 1957, Dutch chemists[748] showed that benzoyl chloride was reduced to benzaldehyde with Ph_3SnH, and Anderson[735] discovered that Et_3SnH reduced halides and oxides of Group 13 elements to their lowest oxidation state or even to the free metals. Noltes reported that Pr_3SnH reduced BF_3 and $AlCl_3$ to the free elements[756]. At the end of the last century, organotin hydrides were widely applied for the reduction of different organic, organometallic and inorganic compounds[721,767].

E. Organylhalostannanes. The C–Sn Bond Cleavage

Among the first organotin compounds of special importance are the organylhalostannanes $R_{4-n}SnX_n$. We would like to review here their development and approach to their synthesis in the absence of metallic tin, other metals or organometallic compounds, which have not been considered in the previous sections. Their properties will be considered as well.

Historically, the first and basic nonorganometallic method for the synthesis of organylhalostannanes was the C–Sn bond cleavage reaction by halogens and inorganic halides. As reported in section III.A, Frankland[591] and Cahours[595–597] first observed the C–Sn bond cleavage of tetraalkylstannanes by halogens in 1859 and 1860–1862, respectively. In 1867, following Frankland, Morgunov[668] demonstrated that the reaction of iodine with Me_2SnEt_2 resulted in Et_2SnI_2. In 1871, Ladenburg[768] found that, depending on the reagent ratio (1 : 1, 1 : 2, 1 : 3), the reaction of Et_4Sn with I_2 resulted in Et_3SnI, Et_2SnI_2 and $IEt_2SnSnEt_2I$, respectively. The cleavage of Et_3SnPh by iodine led to Et_3SnI and PhI[606]. Thus, he was the first to show that the Sn−Ar bond is weaker than the Sn-alkyl bond. In 1872, he cleaved Me_4Sn by iodine and obtained Me_3SnI[607]. Ladenburg[768] also found that, contrary to Frankland[591], the reaction between iodine and Me_2SnEt_2 led to $MeEt_2SnI$ and Et_2SnI_2. He demonstrated the Sn−Sn bond cleavage by alkyl iodides, e.g. by the reaction of $Et_3SnSnEt_3$ with EtI at 220 °C, which gave Et_3SnI and C_4H_{10}[768].

In 1900 Pope and Peachey[608], who intended to synthesize a mixed trialkyliodostannane having an asymmetric tin atom, cleaved Me_3SnEt and $Me_2SnEtPr$ by iodine and obtained Me_2SnEtI and MeEtPrSnI, respectively. In 1912, Smith and Kipping[678] demonstrated that it was easier to cleave a $PhCH_2$−Sn than an Et−Sn bond and that cleavage of a Ph−Sn was easier in the reaction of R_3SnCH_2Ph (R = Ph, Et) with iodine, which led to Et_3SnI and $R_2(PhCH_2)_2SnI$, respectively. Sixteen years later Kipping[679] found the following decreasing cleavage ability of the R−Sn bond: 2-MeC_6H_4 > 4-MeC_6H_4 > Ph > $PhCH_2$. He obtained four organotin compounds containing the asymmetric tin atoms: Ph(4-MeC_6H_4)($PhCH_2$)SnI, Ph(4-MeC_6H_4)($PhCH_2$)SnOH, BuPh($PhCH_2$)SnI and EtBu($PhCH_2$)SnI during the multistage process of the C−Sn bond cleavage by iodine followed by a new C−Sn bond formation with a Grignard reagent. Unfortunately, he failed to isolate them as pure optically active isomers. In 1924, Krause and Pohland[769] showed that one of the phenyl groups was cleaved in the reaction of iodine with triphenylhexylstannane. In 1889, Polis[651] found that iodine did not cleave the Ph−Sn bond in Ph_4Sn and this was confirmed by Bost and Borgstrom[770]. Steric factors were evidently predominant, i.e. the Ar−Sn bond in $ArSnR_3$ (R = Me, Et) was easier to cleave by iodine than the Sn−R bond.

Manulkin[771] together with Naumov[772] extensively studied for the first time the cleavage of alkyl radicals at the tin atom by iodine. They found that their cleavage from R_4Sn (R = Me, Et, Pr, Bu, *i*-Am) to form R_3SnI became more difficult (i.e. required higher temperature) on increasing their length. The same was demonstrated for the homologous series of tetraalkylstannanes $(C_nH_{2n+1})_4$ Sn with $n = 1-7$, and for the mixed series R_3SnR', where R, R′ were alkyl groups of various length[773,774]. Contrary to Cahours[595,597], Manulkin showed that the reaction of iodine with Me_3SnR (R = Et, Bu, Am, *i*-Bu, *i*-Am) led to Me_2RSnI and that Et_3SnBu was transformed to Et_2BuSnI[775]. He was also able to cleave two or even all four R groups from the tin atom in R_4Sn (R = Me, Et) under more drastic conditions (160–170 °C). Thus, he was the first to find that in the reaction of halogens X_2 with R_4Sn, one or two R substituents were first cleaved in consequent steps whereas the remaining two groups were cleaved simultaneously with the formation of SnX_4. He was unsuccessful in stopping the process at the $RSnX_3$ formation. In 1957, Koton and Kiseleva[776] were the first to demonstrate that the allyl group was easily cleaved from tin atom by iodine: the reaction of iodine with $CH_2{=}CHCH_2SnPh_3$ led to Ph_3SnI and $CH_2{=}CHCH_2I$.

Only few publications[651,777] were devoted to the use of chlorine to obtain organylchlorostannanes. In 1870, Ladenburg[778] obtained Et_2SnCl_2 by chlorination of hexaethyldistannane $Et_3SnSnEt_3$, i.e. a cleavage of both the C−Sn and Sn−Sn bond took place. In the reaction of hexaethyldistannane with chloroacetic acid at 250 °C he obtained Et_2SnCl_2 as well as C_2H_6 and C_4H_{10}[778].

In the 20th century the cleavage reaction of R_4Sn by halogens (mainly by bromine and iodine) was widely used for the syntheses of R_3SnX and R_2SnX_2, at yields which were dependent on the reaction conditions and the ratios of the reagents. Thus, the first syntheses of organylhalostannanes by the cleavage of R_4Sn and R_3SnX were carried out in

1900–1925[609,610,630,678,732,769,777,779–782]. Sixty publications in the period before 1960, reporting the use of the cleavage of R_4Sn for organylbromo- and organyliodonstannane syntheses by halogens (mainly bromine and iodine), were reviewed[125,675].

Bromine was mostly used to easily cleave aryl substituents from the tin atom. In 1899 Polis[651] synthesized Ph_2SnBr_2 by the reaction of bromine and Ph_4Sn and in 1918 Krause[779] obtained Ph_3SnBr by the same reaction. The cleavage of Ph_4Sn by bromine and chlorine to form Ph_3SnX was carried out by Bost and Borgstrom[770] in 1929. Unlike iodine, ICl reacted extremely easily with Ph_4Sn to give Ph_3SnCl and PhI. In 1931 Bullard and Holden[783], and in 1941–1946 Manulkin[771,773–775] began to investigate in detail the hydrocarbon radical cleavage from the tin atom. The Manulkin studies showed that the tin–alkyl bond became more difficult to cleave as the alkyl group length increased (and it was more difficult when its tail was branched). Secondary alkyl groups (e.g. Me_2CH) were cleaved more easily from the tin atom than primary ones[773]. These investigations enabled one to arrange the substituents according to the ease of their cleavage by halogens from the tin atom as follows: All > Ph > $PhCH_2$ > CH_2=CH > Me > Et > Pr > *i*-Bu > Bu > *i*-Am > Am > Hex ⩾ Heptyl > Octyl.

Following Frankland[591], Buckton[604,784] in 1859 demonstrated the possibility of the C–Sn bond cleavage in tetraalkylstannanes by hydrohalic acids. In 1870, Ladenburg[778] found that HCl cleaved the C–Sn and Sn–Sn bonds in $Et_3SnSnEt_3$ with the formation of Et_3SnCl and that HCl cleaved the phenyl group from $PhSnEt_3$ with the formation of Et_3SnCl[605]. In 1878, Aronheim[722] showed that HCl could cleave two phenyl groups of Ph_2SnCl_2 with the formation of $SnCl_4$. He also reported that the reaction of Ph_2SnCl_2 with the gaseous HBr and HI was not accompanied by the Ph–Sn bond cleavage, but was an exchange reaction, which resulted in Ph_2SnClX (X = Br, I). The reaction between Ph_2SnX_2 (X = Br, I) and HBr and HI led to SnX_4.

In 1927, Bullard and Robinson[785] studied the cleavage reaction of Ph_2SnMe_2 by hydrogen chloride, which resulted in Me_2SnCl_2. Four years later Bullard and Holden[783] isolated $MeEtSnCl_2$ from the reaction of HCl with Me_2SnEt_2. This result showed that both ethyl and methyl groups were cleaved. Under the action of HCl on Et_2SnR_2, R = Pr, Ph the products $EtPr_2SnCl$ and Et_2SnCl_2[783] were obtained, respectively. The facility of alkyl group cleavage from the tin atom with hydrogen halides decreased in line with the above-mentioned substituent order with the halogens. However, the order may be different in cleavage by HCl and by iodine. For example, in 1928 Kipping[679] found that HCl the reaction with $(PhCH_2)_3SnEt$ cleaved the ethyl group, but the reaction of halogens led to the benzyl group cleavage. He also demonstrated that in the reaction of concentrated HCl with tetraarylstannanes two aryl groups might be cleaved[679]. During the action of hydrogen halides on the silicon organotin derivatives $R_2Sn(CH_2SiMe_3)_2$, R = Me, Bu the (trimethylsilyl)methyl group was the first to be cleaved[672]. In contrast, halogens cleaved preferentially the R–Sn bond of these compounds. In 1938, Babashinskaya and Kocheshkov[696] studied the facility of the reaction of HCl with $Ar_2SnAr'_2$ and found that the C–Sn bond cleavage by hydrogen chloride became more difficult in the following order (the 'electronegative row' of substituents): 2-thienyl > 4-$MeOC_6H_4$ > 1-$C_{10}H_7$ > Ph > *c*-C_6H_{11}. In 1946, Manulkin[774] showed that $Me_2EtSnCl$ was formed in the reaction of HCl with Me_3SnEt. In 1958, Bähr and Gelius[703] cleaved by HCl all the three isomers of $(PhC_6H_4)_4Sn$ to $(PhC_6H_4)_2SnCl_2$. Finally, in 1957, Koton and Kiseleva[776] demonstrated for the first time that an allyl group easily cleaves from the tin atom under the action of alcoholic HCl solution. The ease of the cleavage followed the order of compounds: $(CH_2=CHCH_2)_4Sn$ > $(CH_2=CHCH_2)_2SnPh_2$ > $CH_2=CHCH_2SnPh_3$. The cleavage of tetraalkylstannane with HCl at room temperature to $SnCl_4$ was especially easy in this series. Further, the high reactivity in homolytic processes of the C–Sn bond in the $CH_2{=}CHCH_2Sn$ moiety was extensively used in synthesis[786–788]. In 1957–1958, Seyferth[672,789,790] demonstrated that, under the action of hydrogen halides, a vinyl group was cleaved more easily from the tin atom than an alkyl one, less easily than the phenyl group. In this process an addition of HX to the double bond was not observed. In the 20th century the application of the C–Sn bond cleavage by hydrogen halides was limited. From 1928 to 1948 it was used only in 7 laboratories[774,783,785,791–794].

Developed in 1859 by Buckton[604] and then studied by Cahours[596] in 1862, by Ladenburg[605,606] in 1871, by Pope and Peachey[308] in 1903 and by Goddard and Goddard[729] in 1922, the cleavage reaction of tetraorganylstannane by tin tetrahalides became the most important method for the synthesis of organylhalostannanes. Neumann[90] named it the co-proportionation reaction (originally '*komproportionierung*'). In general, it may be presented by equation 7.

$$R_4Sn + \frac{n}{4-n}SnX_4 \longrightarrow \frac{4}{4-n}R_{4-n}SnX_n (R = \text{Alk, Ar; } X = \text{Cl, Br, I}; n = 1\text{–}3) \quad (7)$$

The first stage of this process is the cleavage of one organic substituent R with the formation of R_3SnX and $RSnX_3$. A further reaction of the latter led to R_2SnX_2 and an excess of SnX_4 led to $RSnX_3$[653]. In 1871, Ladenburg[606] was the first to show that the presence of both aryl and alkyl groups at the tin atom in the reaction with $SnCl_4$ led to the reaction described in equation 8.

$$Et_3SnPh + SnCl_4 \longrightarrow Et_2SnCl_2 + EtPhSnCl_2 \quad (8)$$

Unlike the synthesis of organylhalostannanes based on dealkylation by halogens, hydrohalic acids and other inorganic and organic halides of R_4Sn and R_3SnX, the co-proportionation reaction enabled one to keep all the organic substituents in the products, i.e. the number of R–Sn bonds is the same in the precursor and in the products. In 1929–1945, this reaction was studied extensively by Kocheshkov and his coworkers[633,727,795–806]. In particular, by the reaction of tetraarylstannanes and diarylhalostannanes with $SnCl_4$ under severe conditions (150–220 °C) they obtained aryltrihalostannanes for the first time. In 1938, Kocheshkov and coworkers synthesized $(4\text{-}PhC_6H_4)_2SnBr_2$ by the reaction of $(4\text{-}PhC_6H_4)_4Sn$ with $SnBr_4$ at 160–210 °C. According to Zimmer and Sparmann[693] (1954) the reaction of $SnBr_4$ with Ph_4Sn at 220 °C led to Ph_2SnBr_2. The reaction of diaryldibromostannanes with $SnBr_4$ at 150 °C enabled Kocheshkov[633,795] to obtain a number of aryltribromostannanes in 1929. Two years later he showed that both $SnCl_4$ and $SnBr_4$ could be widely used to synthesize $ArSnX_3$ (X = Cl, Br)[806]. In 1933 he reacted R_4Sn, R_3SnX, R_2SnX_2 (R = Me, Et, Pr) with SnX_4 (X = Cl, Br)[796]. In 1950, Razuvaev[807] first conducted the photochemical reaction of $SnCl_4$ with Ph_4Sn and obtained Ph_2SnCl_2 almost quantitatively. This allowed the temperature of the reaction to be reduced to 200 °C[633,795] and it also showed that the process proceeded via a free-radical mechanism. Unfortunately, these data remained unknown to the general circle of researchers. During the first 60 years of the 20th century the co-proportionation reaction had been referred to in 50 publications[631,653,798]. In 1957, Rosenberg and Gibbons[808] used tetravinylstannane in the reaction with SnX_4 at 30 °C which led to $(CH_2{=}CH)_2SnCl_2$.

For the first time tetraiodostannane was used in the reaction with tetraalkylstannanes by Pope and Peachey[308,611] in 1903. They demonstrated that heating Me_4Sn with SnI_4 at $>$100 °C led to Me_3SnI and $MeSnI_3$. Ph_4Sn did not react with SnI_4 even at 240 °C.

In 1871, Ladenburg[606] was the first to study the reaction of $SnCl_4$ with a nonsymmetric tetraorganylstannane. As a result Et_2SnCl_2 and $EtPhSnCl_2$ were obtained from Et_3SnPh.

In 1945, Pavlovskaya and Kocheshkov[798] showed that in the reaction of $SnCl_4$ with triarylalkylstannanes Ar_3SnR, $ArSnCl_3$ and $RSnCl_3$ were easily formed. In 1933 Kocheshkov[796], and in 1963 Neumann and Burkhardt[809] as well as Seyferth and Cohen[810] found that dialkyldihalostannanes[662,796,797,800] R_2SnX_2 (X = Cl, Br) and alkyltrihalostannanes[809] reacted with $SnCl_4$ analogously to tetraalkylstannanes and trialkylhalostannanes, but at a higher temperature (200–215 °C).

In 1878, Aronheim[722] was able to disproportionate (i.e. '*retrokomproportionierung*') organylhalostannanes when Ph_2SnCl_2 was transformed to Ph_3SnCl and $SnCl_4$, as well as the catalytic influence of NH_3 and sodium amalgam[811] on this reaction. He also showed that the reaction of Ph_2SnCl_2 with NaOH led to Ph_3SnCl and SnO_2, and that the reaction of $NaNO_3$ with Ph_2SnCl_2 in acetic acid solution resulted in Ph_3SnCl[812].

During the first half of the 20th century it was found that the C–Sn bond in tetraorganylstannanes could be cleaved by the halides of mercury[644,703,771,813–815], aluminum[816,817], phosphorus[629], arsenic[818], bismuth[818,819] and iron[817] with formation of the corresponding organylhalostannanes. In this case tetraorganylstannanes acted as alkylating and arylating agents and could be used for preparative purposes.

In 1936 Kocheshkov, Nesmeyanov and Puzyreva[820] found that $HgCl_2$ cleaved the Sn–Sn bond in both R_3SnSnR_3 and $(R_2Sn)_n$ with the formation of R_3SnCl and R_2SnCl_2, respectively.

In 1903 and 1904 Pfeiffer and Heller[309,615] developed a new synthetic approach to organyltrihalostannanes. By conducting the reaction of SnI_2 with MeI in a sealed tube at 160 °C, they obtained $MeSnI_3$. In 1911, Pfeiffer[613] decided to replace in the reaction the SnI_2 by Et_2Sn, which he probably regarded as a monomer. Indeed, heating of $(Et_2Sn)_n$ with EtI at 150 °C led to Et_3SnI. In 1936, Lesbre and coworkers modified Pfeiffer's reaction[821]. They replaced tin dihalides with the double salts with the halides of heavy alkaline metals $MSnX_3$ (M = K, Rb, Cs; X = Cl, Br), which enabled them to obtain organyltrichloro- and -tribromostannanes. The reaction of $KSnCl_3$ with excess of RI at 110 °C led to $RSnI_3$, R = Me, Et, Pr with 44, 37 and 25% yields, respectively. In 1953, Smith and Rochow[822] found that the reaction of $SnCl_2$ with MeCl led to $MeSnCl_3$.

In 1935, Nesmeyanov, Kocheshkov and Klimova[823] found that the decomposition of the double salts of aryldiazonium chlorides and $SnCl_4$, i.e. $[(ArN_2Cl)_2SnCl_4]$ (more exactly, $[(ArN_2)_2{}^+[SnCl_6]^{2-})$, by tin powder gave Ar_2SnCl_2. Sometimes the reaction product turned out to be $ArSnCl_3$. Two years later Waters[824] simplified this method by allowing the tin powder to act directly on phenyldiazonium chloride. Later, he found that the reaction proceeded via decomposition of PhN_2Cl into $Ph^•$ and $Cl^•$ radicals, whose interaction with tin led to Ph_2SnCl_2[825]. In 1957–1959 Reutov and coworkers[826,827] found that the decomposition of double chloronium, bromonium and iodonium salts of tin dichloride $Ar_2XCl \cdot SnCl_2$ by the tin powder led to Ar_2SnCl_2, ArX (X = Cl, Br, I) and $SnCl_2$. This reaction was simplified by decomposing the mixture of Ar_2XCl and $SnCl_2$ by the tin powder. In 1959 Nesmeyanov, Reutov and coworkers[828] obtained diphenyldichlorostannane by decomposition of complexes of diphenylhalonium dichlorides and $SnCl_4$, i.e. $[Ph_2Y]_2{}^{2+}SnCl_6{}^{2-}$ (Y = Cl, Br), by the tin powder.

The attempt of Aronheim[722] to obtain $PhSnCl_3$ by the thermal reaction of $SnCl_4$ with benzene (analogously to the Michaelis synthesis of $PhPCl_2$) had failed in 1878. The reaction products at 500 °C were biphenyl, $SnCl_2$ and HCl. The chemical properties of organylhalostannanes began to be studied extensively after their synthesis. The first property was their ability to be hydrolyzed by water, especially in the presence of bases. As early as 1852–1860 Löwig, Frankland and Cahours obtained $(R_2SnO)_n$ with R = Me[583], Et[41,45]; $Et_3SnSnEt_3$[583] and R_3SnOH (R = Me, Et)[583] by the reaction of alkylhalostannanes with aqueous-alcoholic alkaline solution.

In 1862, Cahours[596] first showed that the halogen in organylhalostannanes could be easily substituted by a pseudohalide group in the reactions with silver pseudohalides, such as the reaction of Et_3SnI and Et_2SnI_2 with AgCN, AgNCO and AgSCN. In 1878 Aronheim[722] substituted the chlorine atom in Ph_2SnCl_2 by the action of HI, H_2O, NH_3 and EtONa which resulted in Ph_2SnClI, $Ph_2Sn(OH)Cl$, $Ph_2Sn(NH_2)Cl$ and $Ph_2Sn(OEt)_2$, respectively. The products $Ph_2Sn(OH)Cl$ and $Ph_2Sn(NH_2)Cl$ were of special interest, since no stable isostructural silicon and germanium analogs had been known. The $Ph_2Sn(OH)Cl$ was also obtained by the hydrolysis of $Ph_2Sn(NH_2)Cl$ and it was transformed to Ph_2SnCl_2 by the action of HCl. The stability of $Ph_2Sn(OH)Cl$ is amazing, as it does not undergo intramolecular dehydrochlorination. It is more amazing that, according to Aronheim's data, the intermolecular heterofunctional condensation of Ph_3SnCl with Ph_3SnOH resulted in $Ph_3SnOSnPh_3$[722]. It might be assumed that compounds $R_2Sn(X)Cl$, X = OH, NH_2 were either dimeric or bimolecular complexes $R_2SnCl_2 \cdot R_2SnX_2$. In 1879, Aronheim[812] continued to study exchange reactions of Ph_3SnCl. In this respect the interesting investigations of the Russian chemist Gustavson[829], who developed the exchange reactions of $SnCl_4$ with mono-, di- and triiodomethanes, should be mentioned. The mixtures of $SnCl_4$ with CH_3I, CH_2I_2 and CHI_3 were stored in the dark in sealed ampoules at room temperature for 7 years. No reaction was observed with CHI_3, but in the mixture of CH_2I_2 with $SnCl_4$ 0.7–1.2% of the chlorine was displaced with iodine, while in the mixture of $SnCl_4$ with CH_3I 33–34% of the chlorines were displaced. These data should be added to the Guinness Book of Records.

The substitution of alkylhalostannanes by the reaction with silver salts was first realized by Cahours[592,596]. In 1852 he found that the reaction of Et_2SnI_2 with $AgNO_3$ and $AgSO_4$ resulted in $Et_2Sn(NO_3)_2$ and Et_2SnSO_4, respectively[592]. Ten years later he synthesized $Et_2Sn(SCN)_2$ and $Et_2Sn(CN)I$ in the same way, and Et_3SnCN, Et_3SnSCN and Et_3SnNCO by the reaction of Et_3SnI with AgCN, AgSCN and AgNCO, respectively[596]. In 1860, Kulmiz[584] used a similar reaction with silver salts for the synthesis of a series of Et_3SnX derivatives (cyanide, carbonate, cyanate, nitrate, phosphate, arsenate, sulfate).

In 1954, Anderson and Vasta[830] studied the exchange reactions of Et_3SnX with silver salts AgY. They showed that the substitution ability of X by Y is decreased in the following order of Y ('the Anderson row'): SMe > $SSnEt_3$ > I > Br > CN > SNC > Cl > $OSnEt_3$ > NCO > OCOMe > F. None of these groups could replace the F atom in Et_3SnF. In contrast, the SMe group in Et_3SnSMe can be replaced by any group Y in this series. The simplest synthesis of organylfluoro- and iodostannanes was by the exchange reaction of an appropriate organylchloro- and bromostannanes with alkali metal halides (KF, NaI etc.). The exchange of the halogen atoms of nonfluoro organylhalostannanes for fluorine, i.e. the preparation of organylfluorostannanes, was first realized by Krause and coworkers[769,781,831–833], in the reaction of KF and Ph_3SnCl in aqueous-alcohol medium. A number of researchers used the exchange reactions of organylchlorostannanes with the sodium salts of organic and inorganic acids. For example, Kocheshkov and coworkers[633,795,804] and jointly with Nesmeyanov[727,806], and only more than 20 years later Seyferth[834], obtained Ar_2SnI_2 by the reaction of Ar_2SnX_2 (X = Cl, Br) with NaI in acetone or ethanol. In 1929, Kocheshkov[633,795] found that $PhSnCl_3$ (which could be easy hydrolyzed by boiling water) reacted with HX (X = Br, I) in water to give $PhSnX_3$.

In the first half of the 20th century it was shown that the C–Sn bond in organotin compounds, especially in tetraorganylstannanes, was easily cleaved by both heterolytic and homolytic mechanisms. This fact makes the C–Sn bond quite different (regarding its thermal and chemical stability) from the C–Si and C–Ge bonds and brought it close to the C–Pb bond. In 1945, Waring and Horton[835] studied the kinetics of the thermal decomposition of tetramethylstannane at 440–493 °C, or at 185 °C at a low pressure

(5 mm). Metallic tin, methane and some amounts of ethylene and hydrogen turned out to be the prevalent products of the thermolysis reaction. Indian researchers[752] revised their data and concluded that the reaction is of a kinetic order of 1.5 and proceeds by a free-radical mechanism. Long[836] investigated the mechanism of tetramethylstannane thermolysis in more detail.

In 1958, Prince and Trotman-Dickenson[837] studied the thermal decomposition of Me_2SnCl_2 at 555–688 °C in the presence of toluene as the radical carrier. The process proceeded homolytically according to equation 9.

$$Me_2SnCl_2 \longrightarrow 2Me^{\bullet} + SnCl_2 \qquad (9)$$

In 1956 and 1959, Dutch researchers[757,838] first observed the thermal cleavage of $Ph_3SnCH_2CH_2COOH$, which led to C_6H_6 and $Ph_2Sn^+CH_2CH_2COO^-$. The latter was the first zwitterionic organotin compound.

F. Compounds Containing an Sn−O Bond

As reported in Section III.A, oxygen-containing organotin compounds with the stannoxane Sn−O bond, such as $(R_2SnO)_n$[41,45,583,600,602,722], $R_3SnOSnR_3$[583,598], R_3SnOH[583,598–602] and $[R(HO)SnO]_n$[839], became known in the second half of the 19th century. They appeared first in the laboratories of Löwig (1852), Frankland (1853), Cahours (1860), Aronheim (1878) and Meyer (1883)[839]. The main synthetic method of compounds of the $(R_2SnO)_n$ and $R_3SnOSnR_3$ type was alkaline hydrolysis of diorganyldihalostannanes and triorganylhalostannanes. In 1913, Smith and Kipping[780] were the first to report that the so-called diorganyl tin oxides R_2SnO were not monomers, as previously considered. This is the reason why their archaic name has to be taken out of use. They concluded that these compounds were formed in a dehydrocondensation process of the primary hydrolysis products of R_2SnX_2 and were typical polymers, i.e. polydiorganylstannoxane-α,ω-diols $HO(R_2SnO)_nH$, which are solids mostly insoluble in water and organic solvents. The authors succeeded in isolating a low molecular weight oligomeric intermediate, i.e. hexabenzyltristannoxane-1,5-diol $HOR_2SnOSnR_2OSnR_2OH$ ($R = PhCH_2$), from the dehydrocondensation of $R_2Sn(OH)_2$. According to Kipping's nomenclature, it was named 'di-unhydro-tri-(dibenzyltin)-dihydroxide'.

In 1951, Solerio[840] reported that compounds with the R_2SnO formula could be monomeric as well, when the tin atom carries bulky substituents, such as diorganylstannanones $R_2Sn{=}O$, $R = C_{12}H_{25}$. When the substituents R are less bulky, the substrates are still polymers. Thus, Solerio can be considered as the founder of the chemistry of diorganylstannanones $R_2Sn{=}O$, the first organotin compounds of three-coordinated tin, bonded to one of its substituents by a double bond.

Many years after Löwig's initial study of the oxidation of diethylstannylene to Et_2SnO by air oxygen, the reaction was studied properly in the 20th century by Pfeiffer[613], Krause and Becker[781] and Chambers and Scherer[662].

In 1952, Nesmeyanov and Makarova[841] developed the synthetic method for 'diaryltin oxides' Ar_2SnO by the reaction of $SnCl_2$ with $[ArN_2]^+ \cdot [BF_4]^-$ and with zinc powder in acetone, followed by aqueous hydrolysis with ammonia. The yields of $(Ar_2SnO)_n$ never exceeded 41%. Along with it small amounts of triarylstannanols and arylstannane acids were isolated. In 1957, Reutov and coworkers[826] succeeded in significantly increasing the yields of $(Ar_2SnO)_n$ up to 80% using the Harada reaction[628,641,842]. In 1939–1949, Harada[842,843] described a series of compounds with a composition of $R_2SnO \cdot R_2SnX_2$ whose molecular structure has not yet been determined.

The Sn−O bond in $(R_2SnO)_n$ and in $R_3SnOSnR_3$ was very reactive. It was hydrolyzed by alkalis, and decomposed by alcohols, glycols[844] and inorganic and organic acids[662,722].

In 1860, Cahours[583] began to study nonprotolytic, heterolytic cleavage reactions of the Sn−O−Sn group and showed that polydiethylstannoxane reacted with PCl_5 to give diethyldichlorostannane. The cleavage reactions of this group by $SnCl_4$[845], $SiBr_4$[845], $HgCl_2$[804,846], I_2 and H_2S[631,798,847] were studied only in the 20th century.

During the period 1920–1940, studies of thermal reactions of organotin compounds having Sn−O bonds[641,662,848–851] had started. All the reactions proceeded with a C−Sn bond cleavage followed by a disproportionation process. In 1926, Chambers and Scherer[662] found that thermolysis of Ph_3SnOH gave $(Ph_2SnO)_n$, Ph_4Sn and H_2O. According to Schmitz-DuMont[852] the product of the dehydrocondensation, i.e. $Ph_3SnOSnPh_3$, was also formed. In 1929, Kraus and Bullard[848] observed an analogous thermal destruction of Me_3SnOH. According to Harada[641,842,849] (1939–1940) thermolysis of triethylstannanol occurred in another way (equation 10).

$$3Et_3SnOH \xrightarrow[-C_2H_6]{} 3Et_2SnO \longrightarrow Et_3SnOSnEt_3 + SnO_2 \quad (10)$$

Kraus and Bullard[848] found that $Me_3SnOSnMe_3$ thermolysis led to Me_4Sn and $(Me_2SnO)_n$[848]. They also showed that thermal decomposition of $(Me_2SnO)_n$ gave Me_4Sn, C_2H_6, SnO_2 and SnO. Unlike this, the thermolysis of $(Et_2SnO)_n$ led to $Et_3SnOSnEt_3$ and SnO_2[641]. According to Druce[850,851] (1920–1921) the thermal destruction of $[Me(OH)SnO]_n$ resulted in CH_4, SnO_2, CO_2 and H_2O. The thermolysis of $[Et(OH)SnO]_n$ proceeded in two simultaneous directions to give C_2H_6, SnO_2 or EtOH and SnO[670,851].

The first trialkylstannanols R_3SnOH, R = Me, Et were synthesized by Frankland[45] (in 1853), and Cahours and coworkers[583,600,602] (in 1860) by the action of alkaline aqueous solutions on the corresponding trialkylhalostannanes. In 1928, Kipping[679] used aqueous ammonia solution for this purpose. Ladenburg[606] (1871), Aronheim[722] (1878), Hjortdahl[853] (1879), Werner and Pfeiffer[585] (1898) similarly obtained triorganylstannanols. Aronheim[722] synthesized triphenylstannanol Ph_3SnOH in 1878. The first trialkylstannanol containing bulky substituents at the Sn atom, $(t\text{-}Bu)_3SnOH$, was synthesized by Krause and Weinberg[647] in 1930. During the period from 1903 to 1960 trialkylstannanols were mentioned in 50 publications[675,789,854,855]. In some cases the Sn−OH bond was also formed by hydrolytic cleavage of the XCH_2−Sn bond, when X was an electronegative substituent (N≡C, EtOOC).

Trialkylstannanols turned out to be rather stable compounds and this was their main difference from their isostructural silicon and germanium compounds[583,598–602,614,617,647,665,722,856]. They could be dehydrated to hexaalkyldistannoxanes only in the presence of dehydrating agents. For example, Harada[576] obtained hexamethyldistannoxane from trimethylstannanol only when it was distilled from sodium[576]. Unlike R_3SiOH, the R_3SnOH (R = Alk) are strong bases[857,858]. Nevertheless, triphenylstannanol, as well as its silicon analogs are still weak acids[859]. According to the ebullioscopy data, the compounds R_3SnOH (R = Me, Et, $PhCH_2$) were associated to some extent[641,780,848] in boiling benzene. Trialkylstannanols were not converted to stannolates even by the action of Na metal. According to Harada[639,640] (1927, 1929), the reaction of Me_3SnOH with Na in liquid ammonia did not give Me_3SnNa, but $Me_3SnSnMe_3$. Kraus and Neal[860] also found that the latter was obtained in the reaction of Me_3SnOPh with Na in the same solvent. The reaction of R_3SnOH with inorganic acids (e.g. HCl, HBr, HI, H_2SO_4) enabled an easy replacements of the hydroxyl group by the anions of the acids[647].

The first attempts to obtain dialkylstannandiols $R_2Sn(OH)_2$ by the hydrolysis of dialkylhalostannanes were unsuccessful. These compounds turned out to be extremely unstable and they dehydrated immediately to amorphous polyperorganylstannoxane-α,ω-diols

$HO(R_2SnO)_nH$. However, in the first half of the 20th century diorganylstannandiols containing bulky substituents (R = *c*-Hex[769], *t*-Bu, *t*-Am[647]) were synthesized.

In 1954, Anderson[845] concluded that the basicity of organotin compounds having Sn−O bonds decreases on increasing the number of oxygen atoms surrounding the Sn atom, i.e. in the series: $(R_3Sn)_2O > (R_2SnO)_n > (R_2SnO_{1.5})_n > SnO_2$.

In the Krause[647,769] laboratory it was established in 1924 and 1930 that the reaction of $R_2Sn(OH)_2$ with HCl or HBr resulted in R_2SnX_2 (X = Cl, Br). Simultaneously, an interesting disproportionation reaction was discovered according to equation 11.

$$2[R(MO)SnO]_n \longrightarrow (R_2SnO)_n + nM_2SnO_3 (M = Na, K). \qquad (11)$$

In 1878, first Aronheim[722] and then Kipping (1928)[679] and Krause and Weinberg (1930)[647] synthesized stable diorganylhalostannanols $R_2Sn(OH)X$, which are stable crystalline substances[640,650,679]. Organotin compounds $R(OH)_2SnOSn(OH)ClR$[861], $RSn(OH)_2Cl$[862] and $[R(Cl)Sn(O)]_n$[862] as well as compounds containing the >Sn(OH)Cl group were obtained only in the 1960s.

Silicon compounds having the >Si(OH)Cl group have not yet been identified. They immediately undergo disproportionation into hydrohalic acid and a short-lived highly reactive diorganylsilanones $R_2Si{=}O$, which in turn quickly oligomerize or are inserted into the bond of a trapping reagent[863–866]. The higher stability of diorganylhalostannanols in comparison with their organosilicon analogs can be ascribed to two factors: (1) a longer distance between halogen and oxygen atoms, and (2) a higher stability of the O−H bond due to the higher basicity of the ≡SnOH group. It is more likely that these compounds are cyclic dimers $[R_2Sn(OH)Cl]_2$ or $[R_2Sn(OH)_2 \cdot R_2SnCl_2]$, or even high oligomers.

Organylstannantriols $RSn(OH)_3$ have not yet been isolated. Consequently, organotin compounds $R_2Sn(OH)_2$ and $RSn(OH)_3$ are less stable than their isostructural compounds of silicon and germanium, which in turn are not highly stable. However, their formation as intermediate compounds in hydrolysis reactions of R_2SnX_2 and $RSnX_3$ seems likely.

The attempted synthesis of organylstannantriols, which was begun by Pope and Peachey[308] in 1903 and continued by Kocheshkov and coworkers[795,801–804,806], always resulted in obtaining their dehydration products, which were assigned the structure of '*organylstannone acids*' RSnOOH. We use this term although it does not correspond to their structure. In 1883, Meyer[839] obtained these compounds, for the first time after developing a simple and efficient method for their preparation, although not in high yields. He found that the action of methyl iodide with aqueous alcoholic solution of sodium stannite (formed from $SnCl_2$ and NaOH) gives a white crystalline powder which corresponds to the MeSnOOH formula. The latter was easily soluble in hydrochloric acid with the formation of $MeSnCl_3$.

Meyer's reaction can be described by equation 12.

$$SnCl_2 + 3MOH \xrightarrow[-2MCl,\ -H_2O]{} Sn(OH)OM \xrightarrow{+RX} R(X)Sn(OH)OM \longrightarrow \qquad (12)$$

$$\xrightarrow[-HX]{} RSnOOM \xrightarrow{+CO_2,\ H_2O} RSnOOH + MHCO_3 \quad (M = Na, K; X = I, Br)$$

At the same time Meyer also isolated 'pyro acid' of $(MeSn)_4O_7H_2$ composition. It appeared to be a cross-linked polymer, corresponding to the formula $HO(MeSnO_{1.5})_4H$.

At the beginning of the 20th century Pfeiffer and coworkers[309,614,615,617,867] and Pope and Peachey[308], then Druce[850,851,868–870] and Lambourne[871] improved the Meyer[839] method and synthesized a series of 'alkylstannone acids' and studied their properties. Unfortunately, the Meyer reaction was hardly suitable for the synthesis of arenestannone acids[872].

In 1903, Pfeiffer and Lehnardt[309,617] and Pope[308], and others[850,851,868,869,871,873,874] suggested another method for the synthesis of organylstannone acids (as their Na and K salts). It was based on the reaction of alkyltrihalostannanes with aqueous alcoholic alkaline solutions. In 1929 following Pope, Kocheshkov and coworkers[795,801–804] developed a method for the synthesis of arylstannone acids, based on hydrolysis of $ArSnCl_3$. It is interesting to note that according to Kocheshkov[633,795] the hydrolysis of $ArSnBr_3$ was more difficult than that of $ArSnCl_3$. During the hydrolysis of $ArSnX_3$ by alkali solutions the arylstannates salts were formed, but not the free acids, and then the free acids were isolated by the action of CO_2. In 1957, Koton and Kiseleva obtained the first unsaturated allylstannone acid by heating tetraallylstannane with water in a sealed ampoule at 170 °C[776].

Zhukov[875] following Pfeiffer and Lehnardt[309] and Pope and Peachey (1903)[308], Druce[850,868] and then Kocheshkov and Nad'[799] and Solerio[876] showed that alkylstannone acids were easily decomposed by hydrohalic acids to $RSnX_3$. This reaction was used extensively for the synthesis of pure organyltrihalostannanes. In 1938, in Kocheshkov laboratory[804] it was found that the reaction of RSnOOH with HX proceeded with R–Sn bond cleavage to give RH and SnX_4 under severe conditions. According to Pope and Peachey[308], MeSnOOH was transformed in boiling aqueous alkali to a mixture of $(Me_2SnO)_n$ and Me_3SnOH with simultaneous formation of CH_4. In 1934, Lesbre and Glotz[873] found that the transformation of alkylstannone acids RSnOOH to $(R_2SnO)_n$ became easier with the decrease in the size of the alkyl radical R. Arenestannone acids did not undergo this reaction.

The so-called organylstannone acids are polymers, which could be assigned the structure of polyorganyl(hydroxy)stannoxanes $[RSn(OH)O]_n$ or $HO[RSn(OH)O]_nH$. It is interesting that they were not hydrolyzed on heating and were not converted to polyorganylstannsesquioxanes $(RSnO_{1.5})_n$. The properties of their hydrolysis products were strikingly different from those of their isostructural organyltrichlorosilanes with regard to solubility in water and methanol and the high reactivity. They were easily decomposed by acids, alkalis, hydrogen sulfide or mercaptans.

Lambourne[846,871] showed that the action of carboxylic acids RCOOH on 'methylstannone acid' gave 1,3,5-trimethylpentaacyloxytristannoxanes $Me(RCOO)_2SnOSn(OCOR)MeOSn(OCOR)_2Me$, which hydrolyzed to the cyclic trimers 1,3,5-trimethyltriacyloxycyclotristannoxanes $[Me(RCOO)SnO]_3$. The data obtained led him to conclude that methylstannone acid was the cyclic trimer $[Me(HO)SnO]_3$.

The first representative of hexaalkyldistannoxanes $R_3SnOSnR_3$ with R = Et (incorrectly named earlier 'trialkyltin oxides') was obtained by Cahours[583] and Kulmiz[584] in 1860. Unlike their isostructural silicon and germanium analogs, the preparation of lower hexalkyldistannoxanes with R = Me, Et by hydrolysis of trialkylhalostannanes failed even in the presence of alkalis. This was caused by the fact that the Sn-O-Sn group in these compounds was extremely easily cleaved by water, so that the equilibrium of the trialkylstannanol dehydration with their primary alkaline hydrolysis products (equation 13) was almost completely to the left. Probably, this was the reason that, after the Cahours[583,598] and Kulmiz[584] reports, hexaethyldistannoxane appeared in chemical publications again only in 1939[641,849]. Hexamethyldistannoxane was first synthesized in the Kraus[782,848] laboratory in 1925–1929 (and then by Bähr[794]), where it was obtained by $Me_3SnSnMe_3$ oxidation. In 1940, Harada[576,842] synthesized $Me_3SnOSnMe_3$ by the reaction of Me_3SnOH with metallic sodium. Krause and Pohland[769] synthesized $Ph_3SnOSnPh_3$, the first representative of hexaaryldistannoxanes in 1924. In the last century, the higher hexaalkyldistannoxanes began to be obtained by dehydration of corresponding trialkylstannanols in the presence of dehydrating agents (P_2O_5, $CaCl_2$) or even at high temperature (preferably in vacuum[642,877–879]). The higher hexaalkyldistannoxanes (beginning from R = Bu) were synthesized by the reaction of the corresponding trialkylhalostannanes with aqueous[662] or alcoholic alkaline[880] solutions. Anderson and Vasta[830] obtained $Et_3SnOSnEt_3$ by the reaction of Ag_2O with Et_3SnX or with $Et_3SnSSnEt_3$.

$$2R_3SnX + 2MOH \xrightarrow[-2MX]{} 2R_3SnOH \rightleftharpoons R_3SnOSnR_3 + H_2O$$

$$(M = Na, K;\ X = Cl, Br;\ R = Me, Et) \qquad (13)$$

The properties of hexaalkyldistannoxanes[583,584,598,641,769,782] were very different in comparison with those of their silicon isostructural analogs $R_3SiOSiR_3$. The ability of the Sn−O−Sn group to be decomposed by water, alcohols, phenols, diols[844,881], organic and inorganic acids, SH acids (H_2S, RSH)[584,882], organic and inorganic halides and pseudohalides is consistent with later investigations, which demonstrated the cleavage of the Sn−O bond in $R_3SnOSnR_3$, $(R_2SnO)_n$, R_3SnOR' and R_3SnOH by NH acids ($RCONH_2$, $(RCO)_2NH$[759], pyrrole, pyrazole, imidazole, benzotriazole)[883–885] and by CH acids (RC≡CH[886–893]; $CH_2(CN)_2$, $CH_2(COOMe)_2$[894], fluorene), as well as by H_2O_2[895], CO_2[896,897] and RCOCl[898]. The majority of reactions showed a significant difference between the Sn−O−Sn and Si−O−Si groups. The latter was not decomposed by SH, NH and CH acids, and usually reacted with weak OH acids only in the presence of catalysts.

Anderson[845] showed that $Et_3SnOSnEt_3$ was decomposed by many halides and pseudohalides of B, Si, Ge and Sn, i.e. $EtOBCl_2$, $Me_3SiOCOMe$, $MeSi(OCOCF_3)_3$, $Me_2Si(OCOCF_3)_2$, Ph_2SiF_2, $MeOSi(NCO)_3$, $SiBr_4$, Pr_3GeF, *i*-PrGeOH, $GeCl_4$, $SnCl_4$, $SnBr_4$, $SnCl_2$, Et_2SnCl_2, PCl_3, $AsCl_3$ and $SbCl_3$[845].

Some classes of compounds having the Sn−O−M moiety (with M = C, metalloid or metal) can be combined to give a wide range of organotin compounds. The first class, having the Sn−O−C group, include organic compounds of tin with alkoxy, aryloxy or acyloxy groups at the Sn atom. Organylalkoxystannanes $R_{4-n}Sn(OR')_n$, namely Me_3SnOEt, Et_3SnOEt and $Ph_2Sn(OEt)_2$, were first obtained by Ladenburg[899,900] and Aronheim[722] in the 1870s. However, the basic investigations of compounds containing the Sn−O−R group were carried out in the 20th century[41,599,604,653]. They were obtained from the corresponding organylhalostannanes with sodium alcoholates[860,901–903] or phenolates or by the reaction of organylstannanols, polydialkylstannoxanes, organylacetoxystannanes and organylhalostannanes with alcohols[881,904] or phenols[607,714,722,882,900,905,906]. In 1956, Koton[907] showed the possibility of the Sn−C bond cleavage by alcohols. The studied reaction of $(H_2C{=}CHCH_2)_4Sn$ led to the cleavage of all four Sn−C bonds to give $Sn(OEt)_4$.

Three years later, D'Ans and Gold[901] found that triorganylaryloxystannanes with electron-withdrawing substituents in the aromatic ring (halogen, NO_2) could be obtained only by the reaction of the corresponding phenols with organylhalostannanes in the presence of sodium hydride in THF. Finally, R. and G. Sasin[855] succeeded in cleaving the C−Sn bond of Et_4Sn by phenol to obtain Et_3SnOPh. Organylalkoxystannanes were interesting synthons due to their high reactivity. Yakubovich and coworkers[908] first showed the possibility of transforming organylalkoxystannanes to the corresponding organylhalostannanes by reaction with acyl halides in 1958. The reaction of $(Et_2SnO)_2$ with MeCOF consequently led to $Et_2Sn(OEt)F$ and Et_2SnF_2 together with MeCOOEt.

The first organylacyloxystannanes $R_{4-n}Sn(OCOR')_n$ were obtained by Cahours[583] (1860), Kulmiz[584] (1860) and Frankland and Lawrence[589] (1879). They were synthesized by the reaction of carboxylic acids or their anhydrides with $(R_2SnO)_n$, R_3SnOH or $R_3SnOSnR_3$. Cahours[583,598,600] obtained 30 $R_2Sn(OCOR')_2$ and $R_3SnOCOR'$ type compounds with R = Me, Et, Pr, Bu, *i*-Bu, *i*-Am; $R' = C_nH_{2n+1}$; $n = 0-11$, as well as the corresponding derivatives of hydroxycarboxylic acids (citrates and tartrates) by using this method. Kulmiz[584] synthesized triethylacyloxystannanes $Et_3SnOCOR'$, R' = H, Me, Pr, Ph and triethylstannyl esters of oxalic and tartaric acids, as well. He also used the reaction of $(Et_3Sn)_2SO_4$ with $Ba(OCOR')_2$ and of $(Et_3Sn)_2CO_3$ with RCOOH for the synthesis of these compounds. Frankland and Lawrence[589] were less 'pretentious' and had made only triethylacetoxystannane. Further, organylacyloxystannanes were obtained by the Sn−O bond cleavage with carboxylic acids by Quintin[909] (1930), Kocheshkov and coworkers[820] (1936), Smyth[910] (1941), Anderson[911] (1957), Shostakovskii and coworkers[881] (1958). By this method the two latter authors obtained trimethylacryloxystannane, which was used further for the synthesis of organotin polymers. Anderson[911] synthesized 12 triethylacyloxystannanes by the cleavage of hexaethyldistannoxane with the corresponding halogen-substituted alkanecarboxylic acids.

Another approach to the synthesis of organylacyloxystannanes based on the reaction of organylhalostannanes with salts of carboxylic acids, including silver salts[830], was first offered by Pope and Peachey[608,609] in 1900, then used by Pfeiffer, Lehnard and coworkers[617] in 1910. In 1955−1958 the re-esterification reaction[905,912] started to be used for the synthesis of organylacyloxystannanes[905,912]. Anderson[845,911] (1954, 1957) found that dialkyldiacyloxystannanes were formed in the reaction of $(R_2SnO)_n$ with esters. For the first time the ability of carboxylic acids to cleave the C−Sn bond of R_4Sn was explored by Lesbre and Dupont[913] (1953), by R. and G. Sasin[855] and then by Koton and Kiseleva[776,907] (1957), Seyferth and coworkers[672,789,790,818] (1957−1958) and Rosenberg and coworkers[879] (1959).

In the second half of the last century a strong interest was developed in alkylacyloxystannanes due to the discovery of the high fungicide activity of $R_3SnOCOR'$ and the possibility that $R_2Sn(OCOR')_2$ could be applied as polyvinyl chloride stabilizers (see Sections III.J and III.K).

Organylcyanatostannanes $R_{4-n}Sn(OCN)_n$ belong to compounds containing the Sn−O−C group. A series of such compounds with $n = 1$ were synthesized by Zimmer and Lübke[914] (1952) and Anderson and Vasta[830] (1954).

Some derivatives of oxygen-containing inorganic acids (such as H_3BO_3, HNO_3, H_3PO_4 and H_2SO_4) can be also classified as belonging to organotin compounds, having the Sn−O−M group, where M is a metalloid. Unlike the isostructural organosilicon compounds, trialkystannyl and dialkylstannyl derivatives of strong inorganic acids have an ionic structure, so they can be referred to as organotin salts. As early as 1898 Werner and Pfeiffer[585] showed that diethylstannylenesulfate Et_2SnSO_4 (which has no monomeric organosilicon analog) and many other similar compounds were dissociated in water into Et_2Sn^{2+} and $SO_4{}^{2-}$ ions. In the 19th century the first organotin salts of this kind were obtained by Löwig[41] : Et_2SnSO_4, $Et_2Sn(NO_3)_2$, Et_3SnNO_3, $(Et_3Sn)_2SO_4$; by Cahours[583,599] : Me_2SnSO_4, $(Me_3Sn)_2SO_4$, $(Et_3Sn)_2SO_4$, Et_2SnSO_4, $(i\text{-}Bu_3Sn)_2SO_4$, $Et_2Sn(NO_3)_2$; by Buckton[604] : $(Et_3Sn)_2SO_4$; by Kulmiz[584] : $(Et_3Sn)_2CO_3$, Et_3SnNO_3, $(Et_3Sn)_2SO_4$, $(Et_3Sn)_3PO_4$, $(Et_3Sn)_3AsO_4$; by Frankland[589] : $(Et_3Sn)_2SO_4$; and by Hjortdahl[621,853] : $(Me_3Sn)_2SO_4$, $(Et_3Sn)_2SO_4$, $(Et_3Sn)_2SeO_4$. In 1898 Werner and Pfeiffer[585] obtained Et_2SnHPO_4 and Et_2SnSO_4. Six years later Pfeiffer and Schnurmann[670] described $(Et_3Sn)_2CO_3$ again.

After these investigations, organotin salts did not attract attention until almost the middle of the 20th century. In the second half of the 20th century interest in these compounds increased sharply owing to the discovery of some useful properties of organotin compounds having the Sn−O−M group. During these years numerous organotin salts with M = B[915,916], N[917], P[790,918–921], As[922], S[623,673,782,905,923], Se and I[915,924] were synthesized (only the periodical publications are cited here). A large number of patents cited in a review[631] were devoted to these salts.

Na and Li stanolates belong to these compounds, since they have Sn−O−M groups (M = metal). Unlike the isostructural compounds of silicon and germanium, the preparation of R_3SnOM (M = Na, Li) by the direct reaction of sodium and lithium with the appropriate stannanols had failed. Compounds of these type were synthesized by Chambers and Scherer[662] in 1926, and later by Harada[641] via the oxidation of R_3SnNa in 1939. In 1963 Schmidbaur and Hussek[925] obtained R_3SnOLi by the cleavage of hexaorganyldistannoxanes with organolithium compounds. Me_3SnOLi turned out to be a hexamer. The attempt of Harrison[926] to obtain Bu_3SnOLi by cleavage of $(Bu_2SnO)_n$ with butyllithium resulted in the formation of Bu_4Sn.

Dimethylstannylene salts of inorganic acids, which came to light in Rochow's laboratory in 1952–1953[915,924], could be assigned to organotin compounds, having the Sn−O−M group with M = Sb, V, Mo, W. They were obtained by the reaction of Me_2SnCl_2 with the corresponding acids and their salts in the aqueous medium. Rochow attributed the ease of such reactions to the complete dissociation of Me_2SnCl_2 in water to the Me_2Sn^{2+} and Cl^- ions.

In 1959, Wittenberg and Gilman[927] obtained dimethylstannylene salts of phosphorus, arsenic, molybdenic and tungsten acids by the reaction of Me_2SnCl_2 with the corresponding acids. In 1950–1960, many compounds containing the Sn−O−Si group were synthesized by the reaction of triorganylsilanolates of alkaline metals with organylhalostannanes[928–936]. In 1952, $Ph_3SnOSiPh_3$ and $(Me_3SiO)_2Sn$[937] were synthesized by the reaction of Ph_3SiONa and Me_3SiONa with Ph_3SnBr and $SnCl_2$, respectively. In 1957, Papetti and Post[928] obtained $Ph_3SnOSiPh_3$ by reacting Ph_3SiONa with Ph_3SnCl. In 1961, Okawara and Sugita[938] synthesized triethyl(trimethylsilyloxy)stannane $Et_3SnOSiMe_3$ and found that its reaction with CO_2 gave $(Et_3Sn)_2CO_3$. Okawara and coworkers[939–941] (1950, 1961) obtained tetraalkyl-1,3-bis(trimethylsilyloxy)distannoxanes $R_2(Me_3SiO)SnOSn(OSiMe_3)R_2$ (R = Me, Et, Pr, Bu), which turned out to be dimeric, by co-hydrolysis of R_2SnCl_2 with Me_3SiCl in aqueous ammonia. These compounds were recently shown to be centrosymmetric tricyclic ladder dimers in which all the tin atoms were pentacoordinated[942]. The syntheses of these compounds by co-hydrolysis of Me_3SiCl with $ClR_2SnOSnR_2Cl$ were carried out in order to confirm their structures. Labile dialkylbis(trimethylsilyloxy)stannanes $R_2Sn(OSiMe_3)_2$ were obtained similarly. All these compounds tended to disproportionate to form α,ω-bis(trimethylsiloxy)polydialkylstannoxanes. These investigations founded the basis for the chemistry of stannosiloxanes[640,910,926,943,944] and their practical use.

In the middle of the 20th century synthetic methods started to develop, and the properties[926] were studied of metal−stannoxane monomers and polymers having a Sn−O−M group, where M = Ge, Pb, Ti, P, as well as their analogs, containing SnEM (E = S, Se, Te, NR) chains.

G. Compounds Containing an Sn−E Bond (E = S, Se, N, P)

Unlike silicon and germanium, tin and lead belong to the family of chalcophile elements (according to the Goldshmidt geochemical classification), which have a high affinity to sulfur. In this connection the stability of the stannathiane Sn−S bond (in the Sn−S−Sn group) and the ease of its formation differ strongly from the high reactivity of the Si−S and the Ge−S bonds. The Sn−S bond can be compared with the siloxane bond in the Si−O−Si group. Consequently, the distannathiane Sn−S−Sn group has a special place in organotin chemistry[88,125,675,757,945] just like the disiloxane Si−O−Si group, which played a most important role in organosilicon chemistry.

The first reaction, which showed the easy conversion of the Sn−O to the Sn−S bond, is due to Kulmiz[584,946]. In 1860, he found that triethylstannanol could be converted to hexaethyldistannathiane $Et_3SnSSnEt_3$ by reaction with hydrogen sulfide. In 1953, Sasin showed that hydrogen sulfide easily cleaved the distannoxane group in hexaalkyldistannoxanes[882] with the formation of hexaalkyldistannathianes. Analogously, trialkylalkoxystannanes[88] reacted with H_2S. Hydrogen sulfide also cleaved the Sn−O bonds in the oligomers $(R_2SnO)_n$ and polymers $[R(OH)SnO]_n$. After Kulmiz's investigations, organotin compounds containing the Sn−S bond did not attract the attention of chemists until the end of the 19th century, probably because of their low reactivity and the reluctance to work with hydrogen sulfide and its derivatives. However, in the first half of the last century the incredible ease of the Sn−S bond formation was supported again by the easy cleavage of the Sn−O bond and other Sn−X bonds (X = halogen, H, Sn and even C[789,818]) by H_2S. In 1903, Pfeiffer and Lehnardt[309,614] found that the action of H_2S on methyltrihalostannanes gave an unknown polymethylstannasesquithiane $(MeSnS_{1.5})_n$, which was assigned the $(MeSn{=}S)_2S$ structure. Analogously, in 1931, Nesmeyanov and Kocheshkov[727,806] obtained the first polyarylstannasesquithianes $(ArSnS_{1.5})_n$ by the reaction of H_2S with aryltrihalostannanes.

Pfeiffer and coworkers[616] (1910), and then Kocheshkov[796,947] and Nesmeyanov[727,806,815] (1931–1933) carried out the easy hydrothiolysis of organotin halides $R_{4-n}SnX_n$, $n = 1$–3 (equations 14–16).

$$2R_3SnX + H_2S \longrightarrow R_3SnSSnR_3 + 2HX \tag{14}$$

$$R_2SnX_2 + H_2S \longrightarrow \frac{1}{n}(R_2SnS)_n + 2HX \tag{15}$$

$$RSnX_3 + 1.5H_2S \longrightarrow \frac{1}{n}(RSnS_{1.5})_n + 3HX, \text{ where X = Cl, Br, I)} \tag{16}$$

With organosilicon halides, the same reactions proceeded only in the presence of an acceptor of hydrogen halide. In the first half of the last century the monomeric structures were assigned to $R_2Sn{=}S$ and $(RSn{=}S)_2S$, obtained in the reactions mentioned above. In 1942, Harada[948,949] and later other investigators[924,950,951] found that the compounds of the composition R_2SnS (R = Ph) were cyclic trimers $(Ph_2SnS)_3$, i.e. hexaphenylcyclotristannathianes. It is remarkable that the reactions of organylhalostannanes with alkali metals or ammonium sulfides and hydrosulfides as well as with H_2S proceed smoothly even in aqueous medium[727,815]. This method for the synthesis of $(Ar_2SnS)_3$ was first proposed by Kocheshkov[802] and Nesmeyanov and Kocheshov[727,806] in 1931 and used later by them[801,804,952] and by Harada[642,948,949], Seyferth[672] and Edgar and Teer[953] for the syntheses of compounds of the $R_3SnSSnR_3$ and $(R_2SnS)_3$[804] series. In 1938, first Nad' and Kocheshkov[665] and then Pang and Becker[954] obtained hexaaryldistannathianes. The first representatives of the hexaalkyldistannathiane series $R_3SnSSnR_3$, R = Me, Et, Pr were obtained by Harada[948,949] in 1942. Organyl(organylthio)stannanes $R_{4-n}Sn(SR')_n$ containing the Sn−S−C group were first obtained in the 1950s. It was found that the Sn−O bonds in the Sn−O−Sn and Sn−OH groups were easily cleaved by mercaptans like hydrogen sulfide. That was evidently proved by the reaction studied by Stefl and Best[955] (1957) and Ramsden and coworkers[956] (1954) (equation 17).

$$\frac{1}{n}[R(OH)SnO]_n + 3\ HSR' \xrightarrow{125-150\,^\circ C} RSn(SR')_3 + 2\ H_2O \qquad (17)$$

Cycloalkyldistannoxanes $(R_2SnO)_2$ were also cleaved by thiols R′SH to give $R_2Sn(SR')_2$[956,957]. As is evident by the numerous patent data, not only alkane- and alkenethiols, but also their carbofunctional derivatives such as mercaptoalcohols, mercaptoacids and their ethers and esters, were applied in the reaction with organotin compounds containing an Sn–O bond. Pang and Becker[954] obtained the first triorganyl(organylthio)stannane Ph_3SnSPh in 1948. In 1953–1958, Sasin and coworkers[882,958] synthesized a series of trialkyl(organylthio)stannanes R_3SnSR' (R = Et, Pr; R′ = Alk, Bn, Ar). In 1957–1961, compounds of this series[959], including Ph_3SnSPh[960], were obtained by the reaction of sodium thiolates with organotin halides. The first patents dealing with the methods of obtaining organyl(organylthio)stannanes by the reaction of the corresponding halides and mercaptans in the presence of an HHal acceptor were issued in 1953–1956[961–963].

Whereas aliphatic and aromatic thiols cleaved the C–Sn bond in tetraalkylstannanes to give trialkylorganylthiostannanes[768,855], the analogous reaction in organosilicon chemistry is absolutely unusual. According to Seyferth[789,818] (1957), the vinyl group was especially easy to cleave from tin atom by mercaptans.

In 1933, Bost and Baker[964] first carried out the C–Sn bond cleavage by elemental sulfur. They recommended the reaction of Ar_4Sn and S as a method for the synthesis of Ar_3SnSAr. Furthermore, in 1962–1963 Schmidt, Bersin and Schumann[965,966] (for a review, see Reference[967]) studied the cleavage of Bu_4Sn, Ph_4Sn and Ph_3SnCl by sulfur. In spite of the high stability of the Sn–S–Sn group in comparison with the Sn–O–Sn group, in 1954 Anderson[845] was able to cleave it by the action of n-$C_{12}H_{25}SiI_3$, $SiBr_4$, $GeCl_4$, $SnCl_4$, $SnCl_2$, PCl_3 and $AsCl_3$ on $Et_3SnSSnEt_3$ with the formation of Et_3SnX (X = Cl, Br, I) together with the corresponding inorganic sulfides.

In 1950, Tchakirian and Berillard[874] obtained for the first time organotin compounds containing the Sn–Se bond. Those were polyalkylstannasesquiselenanes $(RSnSe_{1.5})_n$ which were formed by the reaction of $[R(OH)SnO]_n$ with H_2Se. Monomeric compounds containing the Sn–Se bonds were synthesized in the 1960s. The cleavage reactions of the Sn–O bonds by H_2Se were unprecedented.

The majority of organotin compounds containing the stannazane Sn–N bond appeared only in the early 1960s. Their late appearance was probably caused by the fact that the reaction of organylhalostannanes with ammonia, primary and secondary amines did not result in the corresponding amino derivatives (as for the isostructural Si and Ge derivatives), but in stable complexes containing a hypervalent tin atom (Section III.I). The first compound containing the Sn–N bond, triethylstannylisocyanide Et_3SnNC, was synthesized by Kulmiz[584] by the reaction of Et_3SnI with AgCN in 1860. He also obtained *N*-triethylstannylcarbamide and this synthesis was no longer reproduced. In 1927, Bullard and Robinson[785] obtained a mixture of $(Me_3Sn)_3N$ and Me_3SnPh by the reaction of Me_3SnNa with PhBr in liquid NH_3, but they failed to isolate tris(trimethylstannyl)amine. Nevertheless, they can be considered as the founders of modern synthetic methods of organotin compounds having Sn–N bonds. In 1930, Kraus and Neal[968] reported success in obtaining amino(trimethyl)stannane Me_3SnNH_2 by the reaction of hexamethyldistannane or trimethylstannane with sodium amide in liquid ammonia. However, they could neither isolate it nor describe its properties. Between 1930 and 1960 only organotin sulfonamide derivatives[88,714,969–971], trialkylstannylisocyanates R_3SnNCO[830] and isocyanides R_3SnNC[714,795,830,924,972] were synthesized, but many organotin complexes containing the N $\rightarrow$ Sn bonds were obtained (Section III.I). Kettle[973] (1959) pointed out the formation of aminodimethylstannylsodium $Me_2Sn(NH_2)Na$ by the reaction of dimethylstannane with sodium in liquid ammonia. Up to 1960 no compound of the type $R_{4-n}Sn(NR^1R^2)_n$ had been synthesized.

A revolutionary breakthrough, which marks the birth of the most important compounds containing the Sn−N bond, was made by Wiberg and Rieger[974]. They patented the preparation method of trialkyl(alkylamino)stannanes by the reaction of trialkylchlorostannanes with lithium alkylamides. In 1962, this method was improved by Abel and coworkers[975] and Jones and Lappert[976] and was further widely practiced. Jones and Lappert[976] synthesized 23 compounds by this method and studied their numerous addition and insertion reactions. In the same period Sisido and Kozima developed a method for obtaining trialkyl(dialkylamino)stannanes based on the reaction of trialkylchlorostannane with dialkylaminomagnesium bromides[977]. In 1962, Abel and coworkers[975] also developed an original exchange method to synthesize trialkyl(alkylamino)stannanes by Si−N bond cleavage using trialkylbromostannanes according to equation 18.

$$Me_3SnBr + Me_3SiNHEt \xrightarrow{\Delta} Me_3SnNHEt + Me_3SiBr \qquad (18)$$

The intermediate of this process was a complex of the precursor reagents, which decomposed to the final reaction products.

This pioneer research marked the start of vigorous development of the chemistry of organotin compounds containing Sn−N bonds[862,973]. Numerous publications appeared in reviews[737,738,926,949,973,978,979] as well as in parts 18 and 19 of Gmelin's Handbook[87].

The first compound containing the Sn−P bond was synthesized in 1947 by B. Arbuzov and Pudovik[980], who applied the A. Arbuzov reaction to organotin halides by demonstrating that R_3SnX reacted with $P(OR')_3$ at 105 °C with the formation of $R_3SnPO(OR')_2$ ($R = R' =$ Me, Et). The Sn−P bond in these compounds was easily cleaved by Cl_2, HCl, MeCOCl and aqueous KOH. The reaction of Et_3SnI with $NaPO(OEt)_2$ in EtOH gave $Et_3SnSnEt_3$. In 1947 Arbuzov and Grechkin showed that R_2SnX_2 reacted with $P(OMe)_3$ with the formation of $R_2Sn[PO(OMe)_2]_2$[918]. The reaction of $MeSnI_3$ and $P(OMe)_3$ resulted in $MeSn[PO(OMe)_2]_3$. The reaction of Et_2SnI_2 with $NaPO(OEt)_2$ proceeded in two directions with the formation of $Et_2Sn[PO(OEt)_2]_2$ and Et_2Sn. The latter was oxidized to Et_2SnO[980] by the air oxygen. In 1959, Kuchen and Buchwald[981] obtained R_3SnPPh_2 by the reaction of R_3SnBr with Ph_2PNa.

Organotin compounds having Sn−As and Sn−Sb bonds were mentioned briefly in a patent[982] issued in 1935.

Since 1963, a number of organotin compounds in which the tin atom was bonded to B, P, As and Sb atoms were synthesized. However, these studies are beyond the period of history covered in this chapter.

H. Compounds Containing Sn−Sn or Sn−M Bond

Compounds containing the Sn−Sn bonds corresponding in general to the R_3SnSnR_3 and $(R_2Sn)_n$* formulas appeared in the early days of organotin chemistry. Until the middle of the 20th century, these compounds were considered as the three-valent (R_3Sn) and two-valent (R_2Sn) tin derivatives[334].

As described in Section III.A, the first compound of this type was polydiethylstannane $(Et_2Sn)_n$, which was synthesized by Löwig[41] in 1852 as one of the products of the reaction

* Hereafter, oligomers and polymers $(R_2Sn)_n$ will be denoted as R_2Sn unless otherwise noted, and monomers, i.e. (diorganylstannylenes), as R_2Sn.

of ethyl iodide and a tin–sodium alloy, and by Frankland in 1853 by reducing Et_2SnI_2 with zinc in the HCl[45]. In 1859–1860, Buckton[604] and Cahours[575] synthesized the same compound. Already in 1911, Pfeiffer[613] obtained R_2Sn, R = Me, Et by reducing R_2SnCl_2 with sodium amalgam in ether. In 1925, Kraus and Greer[983] synthesized Me_2Sn by the reaction of Me_2SnBr_2 and Na in liquid ammonia. Excess of Na gave Me_2SnNa_2[641,983]. Harada[628,641] used this method to obtain Et_2Sn. In 1925, Kraus and Greer[983] synthesized Me_2Sn by the reaction of Me_2SnNa_2 and Me_2SnBr_2. In 1959, Kettle[973] synthesized Me_2Sn by the reaction of metallic sodium and Me_2SnH_2 followed by decomposition of Me_2SnNa_2 obtained by ammonium bromide.

In 1920, Krause and Becker[781] for the first time prepared Ph_2Sn by the reaction of $SnCl_2$ and PhMgBr. In 1923, Böeseken and Rutgers[677] observed the formation of Ph_2Sn when Ph_2SnNa_2 reacted with Ph_2SnBr_2 in liquid ammonia. In 1926, Chambers and Scherer[662] reacted Ph_2SnBr_2 and Na in liquid ammonia to synthesize Ph_2Sn.

In 1939–1959 Me_2Sn[973,983] Et_2Sn[628,641,649,842], Ph_2Sn[706,984–986] and their analogs were obtained by the methods described above.

It should be mentioned that the compounds of structure R_2Sn, which were once considered as monomers and later proved to be oligomers or polymers $(R_2Sn)_n$, did not always correspond to this formula. In 1964, Neumann and König[987] pointed out that when Ph_2Sn was synthesized by the reaction of alkali metals and Ph_2SnX_2, not only a Sn–Sn bond but also C–Sn bonds were created, making the structure of the formed polymers more complicated. The latter polymers were assigned the $R_3Sn(R_2Sn)SnR_3$ and $R(R_3Sn)_2Sn(R_2Sn)_nSnR_3$ (R = Ph)[806,987,988] structures. However, at the same time the cyclic oligomer dodecaphenylcyclohexastannane $(Ph_2Sn)_6$ was isolated in the reaction of $SnCl_2$ and PhMgBr together with the higher oligomers and polymers[987]. In 1961, Kuivila and coworkers[989] showed that Ph_2SnH_2 in the presence of amines underwent the dehydrocondensation to perphenylcyclostannanes $(Ph_2Sn)_n$. Neumann and König[987] obtained a series of dodecaphenylcyclohexastannanes in a high yield from the corresponding Ar_2SnH_2 in the presence of pyridine. In the dehydrocondensation of Ph_2SnH_2 in DMF they succeeded in obtaining $(Ph_2Sn)_5$. Consequently, four-, five-, six- and nine-membered peralkyl-, perbenzyl- and percyclohexylcyclostannanes $(R_2Sn)_n$[988,990,991] with R = *t*-Bu, $PhCH_2$ ($n = 4$); *c*-Hex ($n = 5$); Et, Bu, *i*-Bu ($n = 6, 9$) were synthesized using this method. Thus, the investigations of Neumann and König[987,990,991] clarified the structures of the compounds corresponding to the $(R_2Sn)_n$ composition.

The first representative of hexaorganyldistannanes R_3SnSnR_3 with R = Et was obtained in 1860 by Cahours[575] and then in 1869–1872 by Ladenburg[607,768,900,992]. Cahours isolated $Et_3SnSnEt_3$ from the reaction products of EtI with a tin–sodium alloy and Ladenburg synthesized it from the reaction of Et_3SnI and metallic Na. Ladenburg determined the molecular weight of the product of Et_3SnI with Na by its vapor elasticity. This enabled him to assign the $Et_3SnSnEt_3$ formula to the product, instead of Et_3Sn, as was considered before and even some time later. In 1908, Rügheimer[993] repeated this synthesis and carried out a precise measurement of the molecular weight (MW) of $Et_3SnSnEt_3$ in ether by the ebullioscopic method. He found that the MM value decreased on dilution. When the solvent to substance ratio was 5.55 : 1, the MW was 235, and when it was 38.7 : 1, MW = 368 (for Et_6Sn_2, MM = 411). These data apparently indicated that hexaethyldistannane was dissociated to the free radicals $Et_3Sn^{\bullet}$[993] in the dilute solutions. Rügheimer followed Ladenburg by pointing out that this compound was the derivative of four-valent tin and contained an Sn–Sn bond. In 1917, Grüttner[994] (the same method was used later by Kraus and Eatough[995]) obtained R_3SnSnR_3 (R = Et, Pr, *i*-Bu) from R_3SnCl and Na at 120 °C by a similar method and corroborated Ladenburg's data when he determined the molecular mass of hexaethyldistannane by cryoscopy in benzene. He also synthesized the mixed hexaalkyldistannanes of $REt_2SnSnEt_2R$ (R = Pr, *i*-Bu)[994] from REt_2SnBr and Na.

Only in 1925 did Kraus and Bullard[996] and Kraus and Session[782] obtain hexamethyldistannane by the reaction of Me_3SnBr with Na solution in liquid ammonia. In 1929, Harada[640] described the preparation of $Me_3SnSnMe_3$ by the reaction of Me_3SnOH and sodium in liquid ammonia. In 1924, Krause and Poland[769] obtained hexacyclohexyldistannane by the reaction of $SnCl_4$ and *c*-HexMgBr. In 1926, Law synthesized $(PhCH_2)_3SnSn(CH_2Ph)_3$ by the reaction of Na with $(PhCH_2)_3SnCl$[676]. In 1937, Riccoboni[997] developed a synthesis of R_3SnSnR_3 by the electrochemical reduction of R_3SnCl in methanol.

Krause and Becker[781] in 1920 synthesized the first representative hexaaryldistannane $Ar_3SnSnAr_3$ (Ar = Ph) by the reaction of triarylbromostannane and sodium in liquid ammonia. Krause and Weinberg[832] synthesized a series of other hexaaryldistannanes in 1929. According to Nad' and Kocheshkov[665] hexaaryldistannanes were among the reaction products of arylmercurochlorides and tin–sodium alloy. In 1920, Krause and Becker[781] (and later Bonner and coworkers[998]) established that when the reaction of $SnCl_4$ with PhMgBr was carried out under defined conditions, it lead to $Ph_3SnSnPh_3$. Hexaaryldistannane with R = 2-PhC_6H_4[999] was obtained analogously.

The attempt of Kraus and coworkers[782,1000,1001] to obtain compounds $(R_3Sn)_4C$ by the reaction of CCl_4 and R_3SnNa (R = Me, Et, Ph) gave instead R_3SnSnR_3. In 1951, Razuvaev and Fedotova[985] found that R_3SnSnR_3 could be prepared by the reaction of $(R_2Sn)_n$ and $Ph_3CN{=}NPh$. Wittig and his coworkers[706] (1951) and Gilman and Rosenberg[767,1002] (1952) offered a convenient method for the synthesis of $Ph_3SnSnPh_3$ by the reaction of Ph_3SnLi with Ph_3SnX (X = Cl, Br). In 1953, Gilman and Rosenberg[707] also found that the main reaction product of $(2\text{-}MeC_6H_4)_3SnLi$ with 2-iodotoluene was R_3SnSnR_3 (R = 2-MeC_6H_4).

In contrast, in the reaction of Ph_3SnLi with EtI or $PhCH_2Cl$ they obtained Ph_4Sn and Ph_3SnR[767] (R = Et, CH_2Ph). The interaction of Ph_3SnNa with O_2, CO_2, SO_2, PhCOCl and PhSSPh[960] gave $Ph_3SnSnPh_3$.

Finally, $Ph_3SnSnPh_3$ was formed slowly in the reaction of Ph_3SnM (M = Li, Na) with Et_2O, THF, EtOH and BuOH. Wittig and coworkers[706] showed that the action of lithium on Ph_3SnBr in liquid ammonia followed by treatment with NH_4Br led to a mixture of $Ph_3SnSnPh_3$ and Ph_3SnH. As reported in Section III.D, hexaphenyldistannane was formed by the dehydrocondensation of Ph_3SnH in the presence of aliphatic amines[751]. $Ph_3SnSnPh_3$ was also formed on reduction of carbonyl compounds by triphenylstannane. It is remarkable that according to Johnson and coworkers[1003,1004] 1,2-dihalotetraalkyldistannanes were formed in the reaction of R_2SnCl_2 with EtONa or with highly basic amines in ethanol. Finally, it must be remembered that dodecaorganylpentastannanes $R(R_2Sn)_nSnR_3$ containing 5 tin atoms ($n = 4$) in a linear chain were first synthesized by Böeseken and Rutgers[677] in 1923 (with R = Ph) and by Kraus and Greer[983] in 1925 (with R = Me). The latter authors also described $EtMe_2SnMe_2SnMe_2Et$, which is unstable in air. In 1932, Kraus and Neal[860] obtained dodecamethyltetrastannane (R = Me, $n = 3$). Individual linear peroorganylpolystannanes containing more than five Sn atoms in the chain were unknown until 1956[1005]. Böeseken and Rutgers[677] synthesized the first bulky perorganyloligostannane, i.e. tetrakis(triphenylstannyl)stannane $(Ph_3Sn)_4Sn$, by the reaction of Ph_3SnNa and $SnCl_4$. One cannot but mention that macrocyclic perethylcyclostannanes $(Et_2Sn)_n$ with $n = 8$, 9[1006], 10[1007] were synthesized in 1963. Thus, the possibility that ten tin atoms can be bonded to each other in a closed chain was shown.

Peroorganylstannylmetals $R_3SnMR'_3$ (M = Si, Ge) were first obtained in the laboratory of Kraus[161,782,996] by the reaction of R_3SnX with R'_3MNa (formed by the action of Na on R'_3MX). In 1933, Kraus and Eatough[995] obtained $Ph_3SnSiMe_3$ by the reaction of Ph_3SnLi with Me_3SiCl. Afterward, Gilman and Rosenberg[767] synthesized $Ph_3SnSiPh_3$ by the reaction of Ph_3SnLi with Ph_3SiCl in 1952. In 1961, Blake and coworkers[960] obtained $Bu_3SnSiMe_3$[960] by the reaction of Bu_3SnLi and Me_3SiCl. Analogously, Gilman and Gerow[217] synthesized $Ph_3SnGePh_3$ by the reaction of Ph_3SnCl with Ph_3GeK in 1957. The attempt of Buckton[604] to obtain a compound containing a Sn−Pb bond failed in 1859, but three quarters of a century later the synthesis of $Me_3SnPbPh_3$ was patented[982]. In all the studies mentioned above it was demonstrated that the Sn−Sn bond is more reactive than the Si−Si and the Ge−Ge bonds and is closer in reactivity to the Pb−Pb bond.

Chemical transformations of organotin compounds containing the Sn−Sn bond started to develop in the 19th century by Cahours[575] (1860) and Ladenburg[900] (1870). They found that halogens cleaved this bond easily in $(R_2Sn)_n$ and R_3SnSnR_3 to form R_2SnX_2 and R_3SnX, respectively. In the last century the reaction of halogens with R_2Sn and R_3SnSnR_3 was carried out in the laboratories of Krause[647,769,781,831] (1918–1929), Böeseken[677] (1923), Kraus[161,782,995] (1925, 1927, 1933), Law[676] (1926), Kocheshkov[665] (1938) and Gilman[217] (1957).

Kraus and Session[782] (1925), Kraus and Bullard[996] (1926), Harada[641] (1939) and Brown and Fowbes[1008] (1958) observed that R_3SnSnR_3 is slowly oxidized by the air oxygen to $R_3SnOSnR_3$.

In 1925, Kraus and Session[782] showed that elementary sulfur reacted easily with the Sn−Sn bond of hexaalkyldistannanes to form hexaalkyldistannathianes.

In 1908, Rügheimer[993] observed that hexaalkyldistannane were slowly oxidized by air to $R_3SnOSnR_3$ under exposure to air. In 1870, Ladenburg[778] showed that $Et_3SnSnEt_3$ was cleaved by H_2SO_4 with the formation of inflammable gas and an oil-like product, which crystallized on cooling and was probably Et_2SnSO_4. When the latter was recrystallized from the hot HCl, Et_2SnCl_2 was isolated. Ladenburg obtained Et_3SnCl and Sn by the reaction of $Et_3SnSnEt_3$ with $SnCl_4$. Therefore, he was the first to discover that hexaalkyldistannanes possess reductive properties. In 1871, Ladenburg was also the first to cleave the Sn−Sn bond by organic halides. The reaction of $Et_3SnSnEt_3$ with EtI at 220 °C led to Et_3SnI and butane, but with $ClCH_2COOH$ it led to Et_2SnCl_2, C_2H_6, C_4H_{10} and CO_2[768]. Continuing this research, Ladenburg[606] cleaved $Et_3SnSnEt_3$ by MeI (at 220 °C) and $ClCH_2COOEt$. In 1917, Grüttner[994] followed these studies and showed that hexaalkyldistannanes were cleaved by EtI at 180 °C to R_3SnI and $EtSnR_3$ (R = Et, Pr, *i*-Bu). He also reported that hexaalkyldistannanes were slowly oxidized by air to $R_3SnOSnR_3$. Krause and Pohland[769] and Kraus and Bullard[996] found an unusual reaction of hexamethyldistannane with $CaCl_2$ in the presence of air, which led to trimethylchlorostannane. In 1917, Grüttner[994] first showed that hexaethyldistannane was cleaved by $HgCl_2$ to give Et_3SnCl and mercury. Then Kraus and Session[782] (1925) and Kocheshkov, Nesmeyanov and Puzyreva[1009] (1937) carried out an analogous reaction. In 1936, the latter authors[820] found that the Sn−Sn bond in hexaethyldistannane was cleaved by aromatic organomercury compounds Ar_2Hg and ArHgCl, with the formation of Et_3SnAr and metallic mercury (with ArHgCl, Et_3SnCl was also formed). The products Et_2SnAr_2 and Hg were obtained in the reaction of Ar_2Hg with Et_2Sn. The reactions of $HgCl_2$ with $Et_3SnSnEt_3$ and Et_2Sn gave Hg as well as Et_3SnCl and Et_2SnCl_2, respectively.

During 1917–1957 it was found that the Sn−Sn bond was also cleaved by $AgNO_3$[641,781,831], $BiBr_3$[994], sodium amide[968] and organolithium compounds[217,707].

The ability of the Sn–Sn bond to be cleaved by alkali metals was first established by Kraus and Session[782] in 1925. They found that hexaorganyldistannanes were cleaved by sodium in liquid ammonia to give triorganylstannylsodium[148,227,782]. Subsequently, Gilman and Marrs[246] showed that lithium could also cleave hexaphenyldistannane in THF. The Sn–Sn bond in R_2Sn was as reactive as that in R_3SnSnR_3 and was similarly cleaved by the reagents mentioned above. For example, in 1911 Pfeiffer and coworkers[618] reported that reactions of Et_2Sn with oxygen and halogens resulted in the formation of Et_2SnO and Et_2SnX_2 (X = Cl, Br, I), respectively. In 1958, Bähr and Gelius discovered an unusual reaction of R_2Sn (R = 2-, 3- and 4-PhC_6H_4) with $SnCl_2$, which led to Sn and R_3SnSnR_3[999]. The latter product with R = 2-PhC_6H_4 was isolated in two crystal modifications. The precursor (2-$PhC_6H_4)_2Sn$ was the cyclic trimer, since its molecular weight determined in 1,2-dibromoethane after careful purification is 1300[1010].

I. Compounds of Nontetracoordinated Tin

Already in 1862 Cahours[596] obtained adducts with the composition $R_3SnI \cdot 2B$ (R = Me, Et; B = NH_3, *i*-$C_5H_{11}NH_2$, $PhNH_2$) and for the first time he drew attention to the tendency of organotin compounds to complex with organic bases and ammonia. One quarter of a century later, Werner and Pfeiffer[585] reproduced these data and obtained complexes $Et_3SnI \cdot 2B$. They also obtained complexes with the composition $Et_2SnX_2 \cdot 2B$ (X = Cl, Br, I; B = NH_3, Py) and considered their structure according to Werner's coordination theory[1011]. Richardson and Adams[1012] reported adducts with the composition $SnX_4 \cdot 4[PhNH_2 \cdot HX]$ (X = Cl, Br). Werner assigned to the latter complex the $[SnX_2(HX \cdot PhNH_2)_4]X_2$ structure, and the Richardson complexes were probably mixtures of $[PhNH_3]_2{}^+[SnX_6]^{2-}$ and $2PhNH_2 \cdot HX$. Twenty-seven years later Pfeiffer and coworkers[309,618,620] continued the research of his teacher. He obtained and studied many complexes of organotin halides with amines which contained a hexacoordinated Sn atom, namely $Me_2SnI_2 \cdot 2Py$ (1903)[309]; $R_2SnX_2 \cdot 2Py$ (R = Me, Et, Pr, Bu, Ph, 4-MeC_6H_4; X = Cl, Br, I); $MeSnX_3 \cdot 2Py$ (X = Cl, Br) (1911)[618]; $R_2SnX_2 \cdot 2B$ (R = Me, Et, Pr; X = Cl, Br, I; B = $PhNH_2$, Py, quinoline) (1924)[620]. At that time he and then others described the adducts $Pr_{4-n}SnX_n \cdot 2B$ (B = Py; $n = 1-3$) containing hexacoordinated tin atom[616–618,1013–1015]. In 1911 he obtained the $RSnX_3 \cdot 2Py$[618] complexes. Pfeiffer and coworkers[309,618] also synthesized the first complexes of the type $R_{4-n}SnX_n \cdot 2B \cdot 2HX$ (R = Me, Et, Pr, Ph; X = Cl, Br; B = Py, quinoline, PhNHMe, $PhNH_2$; $n = 1-3$). It may be assumed that the structure of these complexes corresponded to the $[BH^+]_2[R_{4-n}SnX_{n+2}]^{2-}$ formula with a hexacoordinated Sn atom. Ten years later Druce[850,851,868,1016] obtained the similar adducts $RSnX_3 \cdot 2B \cdot 2HX$. It can now be stated that these complexes corresponded to the general formula $[BH]_2{}^+[RSnCl_5]^{2-}$ (R = Me, Et, *i*-Pr; B = Py, $PhNH_2$, PhMeNH), as well as to $\{[PyH]_2{}^+[i\text{-}PrSnBr_5]^{2-}\}$[851,868,1016].

In 1923, Kraus and Greer[777] obtained the 1 : 2 complexes of Me_3SnX (X = Cl, I) and Me_2SnCl_2 with NH_3, $PhNH_2$ and pyridine. Within a year Krause and Pohland[769] obtained the adduct (*c*-$C_6H_{11})_3SnCl \cdot NH_3$, which was prepared despite the presence of three bulky substituents at the tin atom. In 1937, Karantassis and Basillides[624] described a series of complexes of the composition $Me_2SnI_2 \cdot 2B$ (B = Py, $PhNH_2$, 2-$MeC_6H_4NH_2$, $PhNEt_2$, 4-MePy, 2-methylquinoline). In 1934, Kocheshkov[1014] obtained the adducts of $R_{4-n}SnX_n \cdot 2Py$ (R = Me, Et; X = Cl, Br) and, in 1936, the $Et_3SnBr \cdot NH_3$ adduct[820]. One such complex, namely ($Me_2SnCl_2 \cdot 2Py$), was obtained in the Rochow laboratory in 1957[1015]. In 1934–1935, Kocheshkov and coworkers[799,803,1014] prepared the complexes of Py and aryltrihalostannanes $ArSnX_3 \cdot 2Py$. In 1958, Reutov and coworkers[1017] synthesized a series of coordinated compounds of the composition $[ArN_2]_2{}^+[MeSnCl_5]^{2-}$ and $[ArN_2]_2{}^+[Et_2SnCl_4]^{2-}$. In 1959, Nesmeyanov, Reutov and coworkers[828] obtained the unusual complexes $[Ph_2I^+]_2[SnCl_4Y_2]^{2-}$, containing hexacoordinated tin atom, by the reaction of diphenylhaloiodides Ph_2IY with $SnCl_4$. Reaction of these complexes with tin powder gave Ph_2SnCl_2.

Almost all these complexes had a hexacoordinated tin atom in octahedral environment. It is remarkable that only five intermolecular coordination compounds $Me_3SnX \cdot B$ (Cl, Br, I; B = NH_3[777,782,1001,1018], Py[777], $PhNH_2$[777]) containing pentacoordinated tin atom were obtained in 1923–1934. In 1901, Kehrmann[1019] described an unusual 1 : 1 complex of triphenylchloromethane with tetrachlorostannane. The formula $[Ph_3C]^+[SnCl_5]^-$ could describe its structure. Unlike it and according to Gustavson[829], methyl iodide (as well as H_2CI_2 and HCI_3) did not form the addition products in the reaction with $SnCl_4$, but the reactants were involved in a very slow exchange reaction between the iodine and chlorine atoms. The tin atom in the 1 : 1 complexes is placed in the center of a trigonal bipyramid[1020–1022] or have the ionic tetrahedral structure $[Me_3SnB]^+X^-$. The simple statistics mentioned above indicate that the complexes of the pentacoordinated tin atom were less stable thermodynamically, and consequently they were easily converted into their analogs having the hexacoordinated Sn atom.

In 1924, Pfeiffer and coworkers[620] reported the existence of the sole adduct $Et_2SnCl_2 \cdot 3NH_3$, in which the tin atom was heptacoordinated. In the fourteen years 1910–1924, Pfeiffer had shown that intermolecular complexes, having octacoordinated tin atom, are not rarities. He obtained $Ph_2SnBr_2 \cdot 4Py$[616], $MeSnI_3 \cdot 4Py$[618], $Ph_2SnCl_2 \cdot 4Py$[618], $Me_2SnI_2 \cdot 4NH_3$, $Et_2SnCl_2 \cdot 4NH_3$ and $Et_2SnBr_2 \cdot 4NH_3$[620]. It could not be stated unequivocally that the tin atom in these complexes was octacoordinated. However, it is probable that their structure corresponded to the formula. $[R_{4-n}SnB_4]^{n+}nX^-$ in which the Sn atom was hexacoordinated and the complexes are salts or ion pairs. As a whole, according to Gol'dshtein and coworkers[1023,1024] the tendency of organylchlorostannanes for complexing decreased in the following series: $PhSnCl_3 > Bu_2SnCl_2 > Ph_2SnCl_2 > Ph_3SnCl > Bu_3SnCl$, i.e. with the decreasing number of chlorine atoms at the central Sn atom.

Alkylhalostannanes form stable complexes with oxygen-containing ligands. First, the aliphatic organotin bases having an Sn–O bond in which the oxygen atom has a strong nucleophilic reactivity belong to such ligands. The first complex of such a type, $Et_2SnO \cdot Et_2SnI_2$, was obtained by Strecher[1025] in 1858. From 1914[1026] complexes of the compositions $(R_2SnO)_n \cdot R_2SnX_2$[641,1026,1027] ($n = 1, 2$); $HO(R_2SnO)_3H \cdot R'_2SnX_2$[628,641,1027–1029]; $(R_3Sn)_2O \cdot R'_3SnX$[1030]; $R_3SnOH \cdot R'_3SnX$[641,1031,1032] (R, R′ = Alk; X = Cl, Br, I) were described. This series can be supplemented with the $Me_2SnO \cdot Me_2Sn(OH)I$[589] adduct. Such complexes often appeared as by-products when the syntheses of organotin compounds having a Sn–X or Sn–O bond were carried out. It is noteworthy that many solid organotin compounds having the stannoxane bond were found to be coordinated polymers rather than monomers, as hitherto considered, because their molecules were bonded by donor–acceptor (bridge) Sn–O → Sn bonds.[50]

Hypervalent intramolecular organotin compounds are stannatranes $RSn(OCH_2CH_2)_3N$ (N→Sn) and dragonoides $Y(CH_2)_3SnX_3$ (Y→Sn) (Y = an atom having at least one lone electron pair, e.g. N, O, Cl), which appeared only in the 1970s and hence do not yet belong to history.

The first attempt of Zelinskii and Krapivin[921,944] to prove the ionization of alkylhalostannanes was undertaken in 1896. By electroconductive investigations of methanolic solutions of Et_3SnI and Et_2SnI_2 they established that these compounds behaved similarly to weak electrolytes in the aqueous medium, i.e. according to the dilution law. From the second quarter of the 20th century the possible existence of organotin cations

R_3Sn^+ and R_2Sn^{2+} in solutions was raised. In the Kraus[1013,1033,1034] (1923–1924) and Rochow[915,924,1035] (1952–1957) laboratories the dissociation of alkylchlorostannanes in water and in organic solvents was studied extensively. The solutions of ethyl- and methylhalostannanes in water, in lower alcohols, in acetone and in pyridine displayed comparatively efficient electrolytic conduction, but their conductivity in ether, nitrobenzene and nitromethane was insignificant[1013,1033,1034,1036]. The ionization constant of Me_3SnCl in EtOH was 10^{-5} at 25 °C[1034]. Rochow and coworkers[915,924] found that Me_2SnCl_2 was dissociated in water into Me_2Sn^{2+} and Cl^-. The solutions were acidic, indicating that partial hydrolysis (*ca* 10% in very dilute solutions)[1037] took place. According to the conductometric titration data of solutions of the organylhalostannanes Me_3SnCl, Me_2SnCl_2, Ph_3SnCl, Ph_3SnF (as well as Ph_3MCl, where M = Si, Ge, Pb) they did not dissociate into ions[1035] in such an aprotic solvent as pure DMF.

Dissociation of Me_2SnCl_2 and Me_3SnCl and their analogs in H_2O enable the displacement of the halogen atoms in organylhalostannanes by other atoms and groups present in the aqueous medium.

The Rügheimer's[993] ebullioscopic molecular weight measurements (1908) of $Et_3SnSnEt_3$ in ether indicated the possibility of the generation of a free organotin radical $R_3Sn^•$ (Section III.H). These measurements showed that the apparent molecular weight of hexaethyldistannane decreased on decreasing its concentration in ether solvent. In 1925, Kraus and Session[782] achieved similar results when they found that $Me_3SnSnMe_3$ was almost completely dissociated in dilute solutions into free $Me_3Sn^•$ radicals. Bullard, in his doctoral dissertation (1925), found that $Me_3SnSnEt_3$ was formed from an equimolecular mixture of $Me_3SnSnMe_3$ and $Et_3SnSnEt_3$ in boiling benzene solution. In his opinion this indicated the intermediacy of $Me_3Sn^•$ and $Et_3Sn^•$ free radicals. Stable free radicals $R_3Sn^•$ with $R = CH(Me_3Si)_2$ were first obtained by Lappert[514,1038] by photochemical disproportionation of $[(Me_3Si)_2CH]_2Sn$ at the end of the last century.

The history of organic hypovalent (divalent) tin derivatives R_2Sn seemed as old as the rest of the chemistry of organotin compounds, but this is not so because almost all compounds with the structure R_2Sn synthesized for over 100 years were the cyclic oligomers or polymers of tetravalent tin (see Section III.H), but not the monomers as originally thought. Old arguments which supported the monomeric structures of these compounds, such as the facile addition reactions of halogens, hydrohalic acids, oxygen and sulfur to R_2Sn, were in fact due to Sn–Sn bond cleavage. Nevertheless, many investigators in the past encountered monomeric diorganylstannylenes R_2Sn, which were the intermediates in reactions developed by them. Organotin compounds $R_2Sn{=}Y$ ($Y = SnR_2$[516], CR_2, OS, Se[523] etc.), having three-coordinated Sn atom, bonded by a double bond atom or to another Sn element, can be considered as the hypovalent tin derivatives. However, they also appeared only at the end of the past century and their historical development lies beyond the scope of this chapter. In 1926 Chambers and Scherer[662], and then Schmitz-DuMont and Bungard[1039] observed the formation of the first representative of these labile compounds, i.e. diphenylstannylene Ph_2Sn, in the thermal dissociation of diphenylstannane Ph_2SnH_2. However, Krause and Becker[781] were the first who had Ph_2Sn in their hands in 1920. In 1943, Jensen and Clauson-Kass[984] confirmed this fact when they showed that freshly prepared (according to the Krause) diphenylstannylene was monomeric and that it slowly polymerized on storing to give the pentamer $(Ph_2Sn)_5$, hexamer and higher oligomers. Diphenylstannylene, which was diamagnetic in all the polymerization stages, maintained a constant value of its dipole moment (1.0 D). This gave rise to the suggestion that, when formed as intermediates, the oligomers were biradicals obtained according to equation 19.

$$Ph_2Sn: + Ph_2Sn \longrightarrow Ph_2\dot{S}n - \dot{S}nPh_2 \xrightarrow{+Ph_2Sn:} Ph_2\dot{S}n - (Ph_2Sn) - \dot{S}nPh_2 \text{ and so on} \quad (19)$$

The synthesis of stable diorganylstannylenes, the true divalent organotin derivatives, was carried out only in the second half of the 20th century[68,97,105,508,514,1038,1040–1043]. The first stable diorganylstannylene (named 'homoleptic'[1040]) $[(Me_3Si)_2CH]_2Sn$ appeared in the Lappert laboratory[1044,1045] after 1975. However, these investigations[1040] and those about stable free radicals $R_3Sn^{\bullet}$[58,516] lie beyond the scope of this chapter.

J. Biological Activity

Among investigations into the biological activities of organometallic compounds, those of organotin derivatives are highly important[91,99,1046–1049], being comparable only with those on the biological activity of mercury and lead compounds. The majority of investigations into the organotin compounds were related to their toxicity that influenced the working process and the experimentalist's health. In 1951 there were four cases of poisoning with Me_4Sn and Et_4Sn which were reported to result from careless treatment of these substances in the laboratory[1050].

Already in 1853, Frankland[45] paid attention to the toxicity of organotin compounds. But the first experimental investigations of their toxicity were conducted by White[1051,1052] in 1881 and 1886. He found that, unlike inorganic tin salts, triethylacetoxystannane was highly toxic for frogs, rabbits and dogs. In 1886, Ungar and Bodländer[1053] studied the toxicity of some organotin compounds on mammalians.

Only forty years later did the investigations on toxicity and biological activity of organotin compounds restart in 1926 due to Hunt[1054]. Collier[1055] found in 1929 that the toxicity of aromatic organotin compounds increased in the order: $Ph_4Sn < Ph_3SnSnPh_3 < Ph_3SnPr < Ph_3SnBr$. According to Lesbre and coworkers[1056] aliphatic tin derivatives are more toxic than the aromatic ones, and R_3SnX is more toxic than R_2SnX_2 and R_4Sn (R = alkyl). In 1954–1955 the toxic action of the organotin compounds on warm-blooded animals was determined[1056,1057]. It was found that in the $Et_{4-n}SnX_n$ series the toxicity of the compounds with $n = 1$, i.e. Et_3SnX ($LD_{50} = 5–10\ mg\,kg^{-1}$), was the highest. The Et_2SnX_2 poisoning was reduced by 2,3-dimercaptopropanol (dimercaptol-BAL), demonstrating an equal toxic action of R_2SnX_2 to that of organic mercury and lead compounds. At the same time, antagonists for Et_3SnX were not found. Seifter[1058] (1939), Gilman[1059] (1942), Glass and coworkers[1060] (1942) and McCombie and Saunders[971] (1947) were involved in the search of organotin compounds as war poisoning substances during World War II. As a result of their investigations the structure–toxicity relationship of organotin compounds was established. Trialkylstannane derivatives R_3SnX, R = Me, Et stimulated progress of the momentum reversible paralysis and retarded encephalopathy. These compounds and also R_2SnX_2 derivatives, R = Me, Et, Pr, Bu possessed dermato-vesical, lachrymatorial and skin-irritating influence. However, none of them was employed as war poisoning agents. In 1940–1942, toxicological studies of the organotin compounds started at the Medical Research Council in Great Britain and in Toulouse University. In 1955 Stoner, Barnes and Duff[1057] studied the toxicity and biological activity of the $R_{4-n}SnX_n$ (R = Alk; $n = 0–3$) series. They found that the influence of Et_3SnX and Et_2SnX_2 was quite different and that only the toxic effect of the latter compound was suppressed by 2,3-dimercaptopropanol. The results of investigations of the influence of tetraalkylstannane under intravenous, intramuscular, oral and intraperitoneal infusion[125] were summarized in

Meynier's doctoral dissertation (1955) and published in 1956[1061]. In 1955–1959 a series of physiological investigations of organotin compounds and a mechanistic study of their influence on laboratory animals were carried out[923,1061–1066].

In 1950 in Utrecht intensive investigations of toxicity and fungicidal activities of organotin compounds begun under van der Kerk supervision[88,713,1067]. First, the fungicide activities of $Et_{4-n}SnX_n$ ($n = 0$–4) were investigated and it was found that Et_3SnCl ($n = 1$) possessed the maximal fungicide action. Compounds R_4Sn, R_2SnX_2 and $RSnX_3$, having different R and X substituents, were less active in comparison with R_3SnX.

Further studies of the fungicide activities of compounds R_3SnX showed that the activity was almost independent of the type of the substituent X. This led to the conclusion that toxicity of R_3SnX was conditioned by the R_3Sn^+ ion or probably by the undissociated R_3SnOH. In contrast, the R substituents affected strongly the fungicidal properties of the R_3SnX (X = OCOMe) series. This effect was maximal at R = Pr, Bu. Further investigation of $R'R_2SnX$ (R = Me, Et; $R' = C_nH_{2n+1}$; $n = 2$–12) derivatives showed that the fungicide activity was dependent only on the total number of carbon atoms in the three-alkyl groups bonded to the tin atom. The maximal activity was achieved when this number was 9–12.

In 1938 it was found that organotin derivatives of proteins and nucleoproteins and the products of their hydrolysis could be used to treat some skin and blood diseases[1068,1069].

Kerr[1070] and Walde[1071] found that $Bu_2Sn(OCOC_{11}H_{23})_2$ was a very effective medication against some intestinal worm infections of chickens. Later, this medication was patented.

Physicians called attention to the effective antimicrobial action of organotin compounds in the middle of the last century. In 1958, 'Stalinon', a medical preparation consisting of diethyldiiodostannane and isolinolenic acid esters[1072], was produced in France for the treatment of staphylococcus infections. Hexabutyldistannoxane in combination with formaldehyde[1046] was used as a remedy against *Staphylococcus aureus*.

The insecticide action of organotin compounds attracted attention in the first half of the 20th century. In 1929 and 1930 a great number of compounds $R_{4-n}SnX_n$ were patented as a remedy against moth[1073–1075]. In 1952 a patent on the application of trialkylchlorostannanes as insecticides was issued[1076]. In 1946, some organotin compounds were patented as the active components of anti-overgrowing coatings[1077]. Somewhat later the stable bioprotective organotin coatings were developed on the basis of monomers of the $R_3SnS(CH_2)_nSi(OR')_3$ type[1078].

K. Practical Use

In the second half of the 20th century organotin compounds found extensive applications in different technological fields[50,125,151] and in agriculture[125,1046,1079]. In 1980 the annual world production of organotin compounds was 35,000 tons[152] and 28,000 tons of metallic tin[1080] were used as the precursor.

The practical application of the organotin compounds started with the fundamental investigations of Yngve, who found that several organotin compounds were excellent photo- and thermo-stabilizers of polyvinyl chloride and other chlorinated polymers, and received a patent in 1940[1081]. For the next several years, many other organotin compounds were patented[1082–1085]. Compounds of the types $Bu_2Sn(OCOR)_2$ and $Bu_2Sn(OCO)_2R'$ (R' = divalent organic radical, preferably unsaturated) were found to be the best stabilizers. Just up to 1960, 82 patents, cited in reviews[125,675] and in several articles[1086–1090], were devoted to PVC organotin stabilizers. In 1953, Kenyon[1091] first started to investigate the mechanism governing the influence of the organotin stabilizers. In 1953–1958 organotin compounds were offered as stabilizers for liquid chlorinated dielectrics[1092,1093], chloro-containing dye stuffs for rubber[1094] and polystyrene[1095,1096] and as inhibitors of corrosion[1097]. In 1949, based on Hart's[1098] investigations, a patent for the use of tribenzylalkylstannanes $(PhCH_2)_3SnR$ as antioxidants for protecting rubbers from cracking was issued. In 1954–1959 a series of different R_2SnX_2 and R_3SnX compounds which were identical to already known polyvinyl chloride stabilizers[1064] were patented for similar use.

Patents dealing with possible practical use of organotin compounds as components for catalytic systems for polymerization of olefins[125,675] appeared during the same period of time. The investigations of chemists and biologists from Utrecht[713] proposed the practical use of organotin compounds as biocides (fungicides[1099], insecticides[1100]) and biocide coatings and impregnations[1101,1102]. Compounds Et_3SnX (X = OH, OCOMe) were found to effectively suppress ordinary types of fungus which damaged wood. Consequently, they recommended these compounds for practical use, for example[1047,1103], to protect timber in mines from biodegradation, and against fabric damage (cotton, jute) by insects and fungus. Further, the R_3SnX compounds were proposed as highly effective means against plant diseases (pesticides)[1099], for the bioprotection of hemp, sisal ropes[1104], and paper[1047], as insecticides[1105] and as fungistatic agents for dyes[1106].

IV. ORGANOLEAD COMPOUNDS

A. Introduction

Organolead compounds came into the world in 1852–1853, i.e. at the same time as organotin chemistry was born. The Swedish chemist Löwig, mentioned extensively in section III as one of the founders of organotin chemistry, is also the father of organolead compounds. He had in his hands for the first time simple representatives of organolead compounds such as $Et_3PbPbEt_3$ and the triethylplumbane derivatives Et_3PbX (X = I, Br, Cl, OH, NO_3, $0.5SO_4$)[42,43,1107]. In the 19th century and at the beginning of the 20th century, the development of the chemistry of organolead compounds was not intensive, although its basis was founded at this period. Only in the years 1915–1925 did organolead chemistry start to develop more quickly due to the efforts of Grüttner and Krause[674,1108–1119] and Krause and coworkers[186,781,814,833,1120–1127]. The systematic investigations showed that organolead compounds could be divided into two main classes: derivatives of tetravalent lead ($R_{4-n}PbX_n$) and divalent lead (R_2Pb). Gilman's investigations carried out in 1937–1952 (for reviews see References 1128 and 1129) contributed significantly to the chemistry of organolead compounds. His investigations led to the development of metalloorganic lead derivatives, such as Ph_3PbLi, which turned out to be an important synthon for the synthesis of different organolead compounds. The studies of Kocheshkov and coworkers[54,156,1130–1134], who were the first to study the possible existence of aryltriacyloxyplumbanes $RPb(OCOR')_3$, were unknown until 1950.

What had Löwig done and what would follow from his work? Among the organic derivatives of the heavy elements of Group 14 only the organolead compounds attracted the least attention. This is evident by the number of publications in this field, which totaled 350[741,1135,1136] up to the middle of the 20th century (of which only 20 appeared in the 19th century) and 420 publications appeared up to 1963[109,1129]. Due to their high toxicity, low thermal and chemical stability, and the similarity of the methods for their synthesis and chemical properties with those of their isostructural tin compounds, there was in general less interest in the organolead compounds. In addition, there was a dominant opinion that the fundamental investigations of organolead compounds could not lead to new developments in comparison with organotins. Nevertheless, the chemistry of tetraalkylplumbanes led to two important discoveries in the 1920s, namely the thermal generation of free radicals by Paneth and Lautsch in 1929–1931[1137–1139] and the discovery of antiknock additives for motor fuels by Midgley and coworkers in 1923[1140].

B. Synthesis from Metallic Lead and its Alloys

In 1852, Löwig[1107] obtained hexaethyldiplumbane $Et_3PbPbEt_3$, initially confused with Et_4Pb, by heating ethyl iodide with a lead–sodium alloy. In the 20th century this became the predecessor of the industrial synthesis of Et_4Pb. Following Löwig, Polis[1141] (1887), Ghira[1142,1143] (1893, 1894), Tafel[1144] (1911), Calingaert[1145] (1925), Fichter and Stein[1146] (1931), Goldach[1147] (1931) and others[47,54,110,1129] studied the reaction of lead–sodium alloy with organic halides. Tafel[1144] and Ghira[1142,1143] established the correct structure of the substance obtained by Löwig[42,43,1107]. Kraus and Callis[643] found the optimal conditions for the industrial production of Et_4Pb, namely the reaction of Pb–Na alloy with EtCl, which was cheaper and more readily available than EtI, which proceeded according to equation 20.

$$4NaPb + 4EtCl \longrightarrow Et_4Pb + 3Pb + 4NaCl \qquad (20)$$

Consequently, 3/4 of the lead was recovered and could be used further. In the laboratory this method had only limited use. In particular, it was used by Calingaert and coworkers[1148] and Saunders and coworkers[1149,1150] to obtain R_4Pb with R = Me, Et, Pr, *i*-Bu in 1948 and 1949, respectively. From 1927[1151,1152] many dozens of patents appeared following the investigations of Kraus and Callis, to protect the method for preparing the R_4Pb (R = Me, Et) by the reaction of Pb–Na alloy (sometimes with addition of K, Li, Mg, Ca) with RCl, RBr and $(EtO)_2SO_2$ under different conditions[47,109,110]. It is impossible to demonstrate all of them here, but we point out that one of the first patents for the preparation of Et_4Pb was granted to Kraus in 1928[1153]. Even in 1950 patents for the preparation of Et_4Pb from lead alloys with Mg[1154,1155] and Ca[1156] were published. Hence the reaction of alkyl halides with lead–sodium alloy, discovered by Löwig[42,43,1107] opened the way for the industrial production of tetraethyllead. A total of 166,000 tons (1/6 of the US lead production) was used to produce tetraethyllead[1157].

Already in 1887, Polis[1141] obtained tetraphenylplumbane Ph_4Pb by the reaction of PhBr with Pb–Na alloy in the presence of ethyl acetate. Calingaert[1145] found that the reaction of alkyl halides with Pb–Na alloy was promoted also by water and by the other compounds, and hydrogen was formed in the reaction with the alloy. We note that in 1853 Cahours[1158] found that metallic lead reacted at a low rate with EtI on heating to give unidentified organolead compounds. Although this was not of any practical interest, metallic lead, but not its alloy, was used successfully for the synthesis of R_4Pb. In 1911, Tafel[1144] showed that the electrochemical reduction of acetone on lead cathode in sulfuric acid solution led to formation of *i*-Pr_4Pb. In 1925, the electrochemical synthesis of tetraalkylplumbanes from alkyl bromides and iodides by using a lead cathode was patented[1159,1160]. The intermediate in this process was dialkylplumbylene, which was rapidly transformed into tetraalkylplumbane at the high temperature of the cathode electrolyte. In 1942, Nad' and Kocheshkov[1161] found that the reaction of Ar_2PbCl_2 with metallic lead or Pb–Na alloy in boiling xylene led to Ar_3PbCl and $PbCl_2$.

C. Metalloorganic Approaches to Organolead Compounds

Historically, the first metalloorganic method for the synthesis of organolead compounds was based on the use of zinc dialkyls. It was not surprising that the method was first used by Frankland and Lawrence[591,1162], who used zinc dialkyl in other reactions. In 1859 they synthesized R_2PbCl_2 by the reaction of $PbCl_2$ with R_2Zn (R = Me, Et). In 1859 Buckton[603,604,1163] obtained R_4Pb by the reaction of $PbCl_2$ with R_2Zn (R = Me, Et). The effort of Tafel[1144] to synthesize $(i\text{-Pr})_4Pb$ in 1911 failed. In 1925, Meyer[1164] described the organozinc synthesis of Et_4Pb from $PbCl_2$.

The reactions of lead dihalides with organomagnesium compounds then became widely used and a convenient laboratory method. Unlike organic derivatives of silicon, germanium and tin, which were usually prepared from $MHal_4$ (M = Si, Ge, Sn) according to the Grignard method, the lead tetrahalides $PbHal_4$ could not be used for this purpose because of their extraordinary instability. The organomagnesium synthesis of organolead compounds was first applied by Pfeiffer and Trüskier[1165] in 1904 and then by Möller and Pfeiffer in 1916[1166]. They obtained both a tetraalkyl- and a tetraarylstannane R_4Pb (R = Et, Ph) by the reaction of organylmagnesium halides RMgX with $PbCl_2$. Metallic lead was the by-product of the reaction. Later, Ph_4Pb was similarly synthesized in the laboratories of Krause[1124] (1925), Gilman[1167,1168] (1927 and then 1939), Kocheshkov[1169] (1937) and others. This method was not suitable for the preparation of some other tetraarylplumbanes; e.g. $Ar_3PbPbAr_3$[1120,1170,1171] was the main product of the reaction of ArMgX with $PbCl_2$. Krause and Reissaus[1122] managed to carry out the reaction of PhMgBr with $PbCl_2$ in such a way that the main reaction product was $Ph_3PbPbPh_3$. In 1914, Grüttner and Krause[674,1121] succeeded in obtaining tetraacyclohexylplumbane according to the Grignard method. In 1916–1918, Grüttner and Krause[1110,1114,1115,1117] used Grignard reagents for the synthesis of R_4Pb with R = Me, Et, Pr, *i*-Pr, *i*-Bu, *i*-Am. In 1916, Möller and Pfeiffer[1166] were the first to use organylhaloplumbanes in the Grignard reaction. They obtained Ph_2PbEt_2 by the reaction of Ph_2PbBr_2 with EtMgBr. In 1919, Krause and Schmitz[814] synthesized mixed tetraorganylplumbanes $(1\text{-}C_{10}H_7)_2PbR_2$ (R = Et, Ph) by the Grignard method. It was also found that in the reaction of $2,5\text{-}Me_2C_6H_3MgBr$ with $PbCl_2$ only $(2,5\text{-}Me_2C_6H_3)_3PbPb(C_6H_3Me_2\text{-}2,5\ C_6H_3\text{-}2,3\ Me_2)_3$ was obtained, but not tetra-*p*-xylylplumbane. The reaction of $PbCl_2$ with $2\text{-}MeC_6H_4MgBr$ proceeded analogously[1122]. There were no doubts that such a result was due to the steric hindrances. In 1928 the organomagnesium method of the synthesis of Et_4Pb was patented[1172–1175]. The use of the Grignard reagent enabled one to obtain compounds of the types R_3PbPbR_3, R_2Pb, $R_{4-n}PbR'_n$ ($n = 0\text{–}3$) and $R_2R'R''Pb$[1129] from PbX_2. Compounds such as 1,1-diorganylplumbacycloalkanes belong to this group, and the first representative 1,1-diethylplumbacyclohexane $Et_2Pb(CH_2)_5\text{-}c$ was obtained by the reaction of Et_2PbCl_2 with $BrMg(CH_2)_5MgBr$ by Grüttner and Krause[1112] in 1916. Tetrabenzylplumbane, which is extremely easily oxidized by air (sometimes with inflammation), was first synthesized by Hardtmann and Backes[710] by the Grignard method and two years later by Krause and Schlöttig[833] and then by Lesbre[1176]. Unstable tetravinylplumbane was first synthesized by the action of the Norman reagent $CH_2{=}CHMgX$ (X = Cl, Br) on $PbCl_2$ or on $Pb(OCOMe)_2$ by Juenge and Cook[1177] in 1959. As early as in 1916, Grüttner and Krause[1110] and Möller and Pfeiffer[1166] observed that in the reaction of the Grignard reagent with $PbCl_2$ the reaction mixture became red. This was explained by the intermediate formation of colored diorganylplumbylenes R_2Pb. However, all attempts to isolate dialkylplumbylenes from the solutions failed[1145]. In contrast, several publications[984,1122,1178,1179] were devoted to diarylplumbylenes Ar_2Pb, before 1961. In 1922, Krause and Reissaus[1122] isolated the red powder-like Ar_2Pb together with $Ar_3PbPbAr_3$ from the reaction products of $PbCl_2$ with ArMgBr (Ar = Ph, $4\text{-}MeC_6H_4$) at 0 °C.

In 1932, Austin[680] used for the first time organolithium compounds for the synthesis of organoleads. He reported that the reaction of ArLi with $PbCl_2$ led to Ar_2Pb. Further heating of the latter led to products such as Ar_4Pb, $Ar_3PbPbAr_3$ and Pb[680,1180,1181].

However, in 1941 Bindschadler and Gilman[1182] concluded that the reaction proceeded in another way. The reaction mixture of $PbCl_2$ with PhLi at $-5\,^{\circ}C$ was not red colored due to Ph_2Pb, and free lead was not isolated. In addition, boiling of $Ph_3PbPbPh_3$ in an ether–toluene mixture did not result in the formation of Ph_4Pb. Based on these facts they concluded that $Ph_3PbPbPh_3$, Ph_3PbLi and, finally, Ph_4Pb[1182] were consequently formed in the reaction. Austin[680,1180] obtained R_3PbAr and R_2PbAr_2 (R = Ar) by the reaction of ArLi with R_3PbCl and R_2PbCl_2, respectively. In 1940, Gilman and Moore[1183] used the reaction of ArLi with $R_{4-n}PbX_n$ ($n = 1, 2$) for the synthesis of R_3PbAr and R_2PbAr_2. Austin[680,1180] in 1932 obtained optically active $PrPh(2\text{-}MeC_6H_4)Pb(C_6H_4OOct\text{-}i)$ by the reaction of $(i\text{-}OctOC_6H_4)Li$ with optically active $PrPh(2\text{-}MeC_6H_4)PbX$. Talalaeva and Kocheshkov[666,667] were the first to describe the reaction of PhLi with lead powder which resulted in low yield of Ph_4Pb and metallic Li. Replacement of lead with its amalgam increased the output of the products and reduced the time of the reaction[667]. In 1950, Gilman and Jones[1184] found that the reaction of MeLi with PbI_2 and MeI resulted in Me_4Pb formation. Metallic lead and Me_2PbI_2 were the intermediate products of the reaction. The reaction of $PbCl_2$ with ArLi and with the appropriate aryl iodide was carried out analogously and led to Ar_4Pb (Ar = Ph, $4\text{-}Me_2NC_6H_4$)[239,1185].

In 1941, in the Gilman laboratory, triphenylplumbyllithium was first synthesized by the addition of excess PhLi to $PbCl_2$ in ether at $-10\,^{\circ}C$[1182]. In 1951, Gilman and Leeper[316] developed another synthesis of triphenylplumbyllithium Ph_3PbLi by the reaction of $Ph_3PbPbPh_3$ with metallic Li. In 1917, Schlenk and Holtz[1186] and later Hein and Nebe[1187] (1942) found that metallic Na cleaved R_4Pb in ether solvent. In 1938, Calingaert and Soroos[1188] found that alkylhaloplumbanes reacted with a stoichiometric amount of Na in liquid ammonia to give hexaalkyldiplumbanes R_3PbPbR_3 (R = Me, Et). Gilman and Bailie[791,1170], Foster and coworkers[1189] and Bindschadler[1190] observed that R_3PbPbR_3 was formed by the reaction of Na with R_3PbX (R = Alk, Ph; X = Cl, Br) in ammonia, and that the dark-red solution of R_3PbNa was formed. In 1941, Bindschadler[1190] succeeded in obtaining R_3PbNa by the cleavage of R_4Pb by sodium in liquid ammonia. The ease of the R–Pb bond cleavage was found to decrease in the following order for R : $CH_2CH{=}CH_2 > i\text{-}Bu > Bu > Et > Me > Ph > 4\text{-}Me_2NC_6H_4$. Thus, for example, $Et_2PhPbNa$[1190] was formed from the reaction of sodium with Et_3PbPh in the liquid ammonia. However, the best way for obtaining Et_3PbNa became the cleavage of Et_4Pb by sodium in liquid ammonia. Ph_3PbNa was prepared similarly from $Ph_3PbPbPh_3$[1190]. In 1951, Gilman and Leeper[316] found that $Ph_3PbPbPh_3$ was cleaved by K, Rb, Ca, Sr, Ba in liquid ammonia. In 1926, Hardmann and Backes[710] patented the method of tetraalkylplumbane preparation by the reaction of $PbCl_2$ and RX with Na in toluene.

The transformations of compounds Ph_3PbM (M = Li, Na) and their possible use for synthetic purposes started to develop in 1939, but the basic investigations in this field were carried out after 1960.

In 1939, Gilman and Bailie[791,1170] demonstrated that the reaction of Ar_3PbNa with $PhCH_2Cl$ or Ph_3CCl led to Ar_3PbR (R = CH_2Ph, CPh_3). In 1950 in Gilman's laboratory[1191] Et_3PbNa, which turned out to be more reactive than Ph_3PbNa, was introduced as a reagent in the reaction with organic halides. The reaction of Et_3PbNa with $PhCH_2Cl$ was 'abnormal' and led mainly to formation of stilbene. In 1959, Et_3PbNa was used for the synthesis of $Et_3PbCH{=}CHPh$ by Glockling and Kingston[1192].

Triphenylplumbyllithium was introduced into synthetic practice by Gilman and Summers[239,1193] only in 1952. In 1952, D'Ans and coworkers[986] used Ph_3PbLi to obtain fluorenyllithium. In 1932, Shurov and Razuvaev[1194] studied the transfer of phenyl radicals, formed by the thermolysis of Ph_nM (M = metal) to another metal atom, which formed more thermally stable phenyl derivatives. They found that the reaction of Ph_4Pb with Sn led to the formation of Ph_4Sn and Pb at 300–375 °C.

Shurov and Razuvaev[1194] tried, but failed to prepare phenyl derivatives of lead by the reaction of metallic lead with Ph_2Hg, as well as with Ph_3Bi. Aromatic mercury compounds were first used for the synthesis of organolead compounds in 1932 when Austin[1136] obtained Ph_3PbCl by the reaction of Ph_2Hg with Ph_2PbCl_2, but he failed when synthesizing Ph_4Pb and $(p\text{-}MeC_6H_4)_2PbCl_2$ by this method. In 1934, Nesmeyanov and Kocheshkov[813] reported that the reaction of Ph_4Pb with $HgCl_2$ led to Ph_3PbCl or Ph_2PbCl_2 along with PhHgCl. In 1942, Nad' and Kocheshkov[1161] found that the reaction of Ar_2Hg with tetraacetoxyplumbane $Pb(OCOMe)_4$ proceeds easily at room temperature in $CHCl_3$, to give $Ar_2Pb(OCOMe)_2$. The same reaction with Et_2Hg took three months. These authors first used this reagent for the synthesis of organolead compounds. This reaction enabled one to obtain otherwise almost inaccessible compounds, like Ar_2PbX_2 having reactive substituents in the aromatic ring. The reaction of tetraacetoxyplumbane with $(ClCH{=}CH)_2Hg$ was used by Nesmeyanov and coworkers[1195,1196] for the preparation of $(ClCH{=}CH)_2Pb(OCOMe)_2$ in 1948. In 1956–1964, the reaction of Ar_2Hg with $Pb(OCOR)_4$ was used extensively for the synthesis of $ArPb(OCOR)_3$ in Kocheshkov's laboratory[1197–1200].

Hein and Klein[1201] obtained hexaethyldiplumbane by the reaction of an alkaline solution of Et_3PbCl with aluminum powder. In 1959, Razuvaev, Vyazankin and coworkers[1202,1203] showed that Et_2Pb was formed in this reaction along with $Et_3PbPbEt_3$. This reaction was a usual reduction process and organoaluminum compounds were not its intermediate products. The use of the reaction for the synthesis of organolead compounds began only in 1957. Its use was complicated by the fact that both aluminum alkyls and $AlCl_3$, which are obtained by the reaction of the organoaluminum compounds with $PbCl_2$, cleaved the C–Pb bond in the formed organolead compounds[731]. Therefore, the reaction of R_3Al with $PbCl_2$ had to be carried out in the presence of alkali metals halides, which reacted with $AlCl_3$ or when $PbCl_2$ was replaced by $Pb(OCOMe)_2$[731] or PbF_2 (when the inert AlF_3 was formed). In 1957, Jenker[1204] used this method. In 1957–1958, the methods for the preparation of tetraalkylplumbanes by the reaction of $PbCl_2$ with $LiAlEt_4$[1205] or with equivalent amounts of R_3Al and RI[1206] were patented.

D. Nonorganometallic Approaches to the Formation of a C–Pb Bond

The Nesmeyanov reaction based on a decomposition of double aryldiazonium salts by the powdered metals had little importance for the synthesis of organolead compounds because of the low yields of the products[1207]. In 1936 Kocheshkov, Nesmeyanov and Gipp prepared Ph_3PbCl by the decomposition of $PhN_2Cl \cdot PbCl_2$ with zinc powder in ether medium[1208]. Ph_2PbCl_2 was prepared when copper powder and acetone were used in the reaction. In both cases the yields of phenylchloroplumbanes were small. In 1945 Nesmeyanov, Kocheshkov and Nad'[1209] succeeded in obtaining Ph_4Pb in 16.5% yield by the decomposition of PhN_2BF_4 by powdered pure lead at 6 °C. When the alloy of lead with 10% Na was used instead, the yield of Ph_4Pb increased to 30%[1110]. Tetra-*p*-xylylplumbane $(4\text{-}MeC_6H_4)_4Pb$ was synthesized analogously in 18% yield. Aliphatic diazo compounds were originally used for the synthesis of organotin compounds by Yakubovich[1210,1211] in his laboratory in 1950 and 1952. He showed that Et_3PbCl and Et_2PbCl_2 reacted with diazomethane in the presence of powdered bronze to give Et_3PbCH_2Cl and $Et_2Pb(CH_2Cl)Cl$ or $Et_2Pb(CH_2Cl)_2$, respectively.

In 1960, Becker and Cook[1212] found that the reaction of trialkylplumbanes R_3PbH (R = Me, Et) with diazoethane at −80 °C in ether led to R_3PbEt in a low yield.

The hydroplumbylation reaction (addition of organolead hydrides to multiple bonds)[53] was first carried out by Becker and Cook[1212]. They showed that Me_3PbH added to ethylene in diglyme at 0 °C under pressure of 17–35 atm to give Me_3PbEt in 92% yield. Further investigations were performed by Neumann and Kühlein[1213] and by Leusink and van der Kerk[1214] in 1965. The addition of R_3PbOH or $R_3PbOCOR'$ to ketene, which was studied only in 1965, was of specific interest[1215].

In 1958, Panov and Kocheshkov[1216] found another route to the formation of the C−Pb bond, namely the interaction of tetraacyloxyplumbanes with aromatic and heteroaromatic compounds (the plumbylation reaction). They showed that the reaction of thiophene with $Pb(OCOPr\text{-}i)_4$ at room temperature during 10 days led to unstable $RPb(OCOR')_3$ (R = 2-thienyl; R′ = *i*-Pr), which was disproportionated to $R_2Pb(OCOR')_2$ and $Pb(OCOR')_4$.

Alkylhaloplumbanes Et_3PbX (X = Cl, Br, I) were synthesized by Löwig[42,43,1107] in 1852–1853. He found that the evaporation of an alcoholic solution of $Et_3PbPbEt_3$ (formed from EtI and a Pb−Na alloy) resulted in the formation of bis(triethylplumbyl)carbonate $(Et_3Pb)_2CO_3$ and Et_3PbOH. Treatment of the products with hydrohalic acids gave Et_3PbX, X = Cl, Br, I. Analogously, the treatment of the above products with HNO_3 and H_2SO_4 resulted in the formation of Et_3PbNO_3 and $(Et_3Pb)_2SO_4$, respectively.

In 1860, Klippel[1217,1218] obtained a series of triethylacyloxyplumbanes $Et_3PbOCOR$ with R = H, Me, Pr, Ph, as well as the corresponding oxalates, tartrates, cyanides and cyanates.

E. Cleavage of the C−Pb and Pb−Pb Bond

Among the C−Pb bond cleavage reactions, thermo- and photo-induced homolytic cleavage is of special theoretical and practical interest.

As early as 1887 Polis[1141] observed that Ph_4Pb decomposed at 300 °C to free metallic lead. In 1927, Zechmeister and Csabay[1219] showed that the reaction occurred even at 270 °C to give biphenyl. Thermal decomposition of Ph_4Pb was studied thoroughly by Razuvaev, Bogdanov and Koton in 1929–1934[1220–1224]. It was also shown that the thermolysis of tetraphenylplumbane at 200 °C under normal pressure or at 175 °C in ethanol under autogenic pressure resulted in metallic lead and biphenyl. The process was catalyzed by metals, which decreased the initial decomposition temperature to 150 °C. The catalysis by the metal decreased in the order: Pd > Au > Ag > Ni. Dull and Simons[1225] (1933) showed that thermolysis of Ph_4Pb gave benzene, biphenyl and terphenyl. The ratio of the products was temperature-dependent. In 1933, Dull and Simons[1226] found that the thermolysis of Ph_4Pb in the presence of evaporated mercury involved the formation of Ph_2 and Ph_2Hg, indicating the intermediate formation of phenyl radicals. Krause and Schmitz in 1919 found that the thermal decomposition of Ph_3PbEt gave lead at 235 °C i.e. at a lower temperature than that for Ph_4Pb[814]. The data indicated that replacement of the aryl with an alkyl substituent decreased the thermolysis temperature of tetraorganylplumbanes.

From 1929, the Paneth[1137–1139,1227] discovery, was published, that the thermal decomposition of lower tetraalkylplumbanes R_4Pb (R = Me, Et) at *ca* 400 °C led to metallic lead and free $CH_3^{\bullet}$ or $C_2H_5^{\bullet}$ radicals, respectively. These free radicals transformed the smooth surface of the metals Pb, Zn, Cd, As and Sb into the corresponding metal alkyls. This prominent discovery corroborated the existence of the free radicals and made a name for Paneth. Later, Calingaert[1228] (1925), Taylor and Jones[1229] (1930), Simons, McNamee, and Hurd[1230] (1932), Meinert[1231] (1933), Cramer[1232] (1934) and Garzuly[1233] (1935) studied the thermal decomposition of tetraalkylplumbanes. Taylor and Jones[1229] found that the thermal decomposition of Et_4Pb at 250–300 °C led to metallic lead and a mixture of gaseous and liquid hydrocarbons (C_2H_4, C_2H_6, C_4H_8, C_6H_{12}), formed by the ethyl radicals generated in this process. According to Calingaert[1228], the thermolysis of tetraethylplumbane over pumice gave a mixture of butane (40%), ethane and ethylene. Simons, McNamee and Hurd[1230] identified the gaseous hydrocarbons HC≡CH, CH_2=CH_2, MeCH=CH_2, Me_2C=CH_2, CH_4, C_2H_6, and small amounts of liquid hydrocarbons as well as H_2 among the products of Me_4Pb thermolysis. Razuvaev, Vyazankin and Vyshinskii[1234] (1959) showed that the thermal decomposition of Et_4Pb was a multiple chain process involving the consequent cleavage of $Et_3Pb^•$ and the intermediate formation of $Et_3PbPbEt_3$ and Et_2Pb which terminated with lead precipitation. A year later these authors studied the kinetics of the thermolysis of Et_4Pb and its mixtures with $Et_3PbPbEt_3$[1203]. The catalytic effect of the formed metallic lead on this process was also established. The investigations of Razuvaev and coworkers demonstrated for the first time that during the homolytic cleavage of the C−Pb bonds in R_4Pb an intermediate formation of a Pb−Pb bond took place. The easy decomposition of the intermediates R_3PbPbR_3 and R_2Pb resulted finally in metallic lead. As a consequence of the homolytic C−Pb and Pb−Pb bond cleavages we deal with their reaction in this section in spite of the fact that Section IV. J is devoted to organolead compounds containing Pb−Pb bonds. The dissociation of tetraalkylplumbanes into free radicals was carried out photochemically under UV irradiation. In 1936, Leighton and Mortensen[1235] showed that the photolysis of gaseous Me_4Pb resulted in lead and ethane. Photolytic decomposition of Ph_4Pb in aromatic hydrocarbons was investigated in McDonald's[1236] (1959) and Razuvaev's[1237] (1963) laboratories. The formation of metallic lead and biphenyl in benzene solution[1237] as well as the formation of 2- or 3-isopropylbiphenyl in cumene medium[1236] was observed. The use of a ^{14}C-labelled benzene and cumene solvents showed that, on photolysis of Ph_4Pb, the formed phenyl radicals reacted with the solvent. Hexaphenyldiplumbane $Ph_3PbPbPh_3$ was apparently the intermediate decomposition product. It confirmed that Pb-centered free radicals $R_3Pb^•$ were the first products of the R_4Pb thermolysis.

In 1918, Grüttner[1118] was the first who called attention to the thermal decomposition of organylhaloplumbanes and found that Ph_3PbBr was decomposed to give $PbBr_2$ even at its melting point (166 °C). In 1925, Calingaert[1145] started to investigate in detail the thermolysis of organylhaloplumbanes. He found that during thermal decomposition of Et_3PbX (X = Cl, Br), Et_4Pb and Et_2PbX_2 were formed. This observation initiated a study of the thermal disproportionation (dismutation) reactions of organylhaloplumbanes. Twenty-three years later Calingaert and coworkers[1148] found that Et_3PbBr was spontaneously decomposed at room temperature with formation of Et_2PbBr_2 within 50 hours. In 1932, Austin[1136] showed that Ph_3PbCl was transformed to Ph_4Pb and Ph_2PbCl_2 in boiling butanol. The products of the disproportionation reaction of Et_3PbCl were Et_4Pb and $PbCl_2$. In 1938, Evans[1238] pointed out that Bu_4PbCl, $PbCl_2$ and BuCl were the products of the thermal decomposition of Bu_2PbCl_2. In 1939, Gilman and Apperson[1239] found that the thermolysis of Et_2PbCl_2 behaved analogously. In 1948, Calingaert and coworkers[1148] studied the hydrothermal decomposition of Et_3PbX and Et_2PbX_2 (X = Cl, Br) during steam distillation: Et_3PbX was transformed to Et_2PbX_2 and Et_4Pb and Et_2PbX_2 to Et_3PbX, PbX_2 and C_4H_{10}, respectively. The authors assumed that the extremely unstable $EtPbX_3$ was the intermediate product of this reaction. As a summary: the decomposition products of Et_3PbX and Et_2PbX_2 were identical, but their ratios were different. Hydrothermal decomposition of Et_2PbBr_2 occurred instantly, and for Et_2PbCl_2 it happened over a period of two minutes. In contrast, Et_3PbX rather slowly decomposed by steam, but Et_3PbBr decomposed faster than Et_3PbCl. The thermolysis of Et_3PbOH and $Et_2Pb(OH)_2$[1148,1239,1240] and organylacyloxyplumbanes $R_{4-n}Pb(OCOR')_n$[1241] was also studied in 1939–1962 (see Section IV.F).

The hydrogenolysis of the C−Pb bond in R_4Pb (R = Me, Et, Ph) was first studied in the Ipatiev[741–743,1220–1224] laboratory. Since 1929, his coworkers Razuvaev and

Bogdanov[1220–1222] as well as Koton[1222,1224] illustrated that Ph_4Pb was decomposed under a pressure of 60 atm hydrogen at 175–225 °C to metallic lead and benzene. Tetraalkylplumbanes R_4Pb (R = Me, Et) under such conditions precipitated a metallic lead even at 125 °C and 100 °C, respectively[1220,1221]. In 1930–1932, Adkins and coworkers[1242–1245] followed the Russian scientists in studying the hydrogenation of tetraorganylplumbanes. They found that R_4Pb (R = Alk) was cleaved by hydrogen with formation of the corresponding alkanes RH and Pb[1242]. Hydrogenolysis of tetraarylplumbanes Ar_4Pb (Ar = Ph, 4-MeC_6H_4) led to a quantitative formation of the corresponding diaryls and metallic lead at 200 °C under H_2 pressure of 125 atm. Tetraheptylplumbane under these conditions was transformed to tetradecane in only 62% yield. In 1931, Adkins and Covert[1243] found that Ni catalyzed the cleavage of tetraalkyl- and tetraarylplumbanes. In 1932, Zartmann and Adkins[1245] found that catalytically active Ni significantly decreased the thermolysis temperature of R_4Pb (R = Alk, Ar) to 200 °C under H_2 pressure. The hydrocarbons R–R were formed in a high yield as the recombination products of the R radicals. In the absence of Ni the precursor Ph_4Pb did not change under the experimental conditions, and under nitrogen pressure at 200 °C it did not change with or without Ni. These data contradicted the results gained by Ipatiev and his coworkers[1220,1223]. In 1933, Razuvaev and Koton[743,1222] studied a catalytic effect of Cu, Ag, Au, Ni and Pd on the destruction of Ph_4Pb by hydrogen under pressure. In the presence of these metals (except Pd) its decomposition proceeded at low temperatures and led to Pb and C_6H_6. Palladium catalyzed only the thermal decomposition of Ph_4Pb (but not the hydrogenolysis process) to form biphenyl, but not benzene. It cannot be believed that Ipatiev remembered in the twilight of his life the investigations on hydrogenolysis of metalloorganic compounds carried out during his Soviet period. In the article of Gershbein and Ipatiev[744] published already after Ipatiev's death, the hydrogenolysis results of Ph_4M (M = Pb, Sn) obtained at the Ipatiev laboratory in the USSR were confirmed without using new experiments. It was reported that, at 200 °C and under an initial H_2 pressure of 60 atm, Ph_4Pb was decomposed to Pb, C_6H_6 and a trace amount of Ph_2 (i.e. nothing new). The composition of the products remained unchanged when copper powder was added to the reaction (as was known earlier). The appearance of this article was unfortunate.

The heterolytic cleavage of the C–Pb bond was especially easy in a series of organometallic compounds of the silicon subgroup. In 1887, Polis[1246] was the first to find C–Pb bond cleavage in tetraalkylplumbanes with halogens. He demonstrated that by bubbling chlorine through a CS_2 solution of Ph_4Pb, the Ph_2PbCl_2 was the product formed. Similarly, Ph_4Pb with bromine in CS_2 or in $CHCl_3$ media was transformed to Ph_2PbBr_2. A year later Polis[1247] synthesized $(4\text{-}MeC_6H_4)_2PbX_2$ (X = Cl, Br, I) by the action of chlorine, bromine and iodine on $(4\text{-}MeC_6H_4)_4Pb$. In 1904, Pfeiffer and Trüskier[1165] prepared Et_3PbCl by chlorination of Et_4Pb with strong cooling. Following him in 1916, Grüttner and Krause[1110] showed that halogens cleaved only one of the R–Pb bonds in tetraalkylplumbanes with formation of R_3PbX only at low temperatures (−70 °C). The reaction of gaseous chlorine with R_4Pb (R = Me, Et) at −70 °C in ethyl acetate solution led to R_3PbCl in a quantitative yield. The chlorination of Me_3PbCl at −10 °C also resulted in a quantitative formation of Me_2PbCl_2. Later, R_3PbX or R_2PbX_2 were synthesized similarly by the reaction of chlorine or bromine with R_4Pb (R = Me, Et, Pr, *i*-Bu, *i*-Am, *c*-C_6H_{11}) at an appropriate temperature[1108,1110,1114,1188]. In 1921, Grüttner and Krause[1108] succeeded in synthesizing $(c\text{-}C_6H_{11})_3PbI$ and $(c\text{-}C_6H_{11})_2PbI_2$ by cleavage of $(c\text{-}C_6H_{11})_4Pb$ with iodine. Only in 1938, by the reaction of iodine with Me_4Pb in ether at 60 °C, did Calingaert and Soroos[1188] prepare Me_3PbI in 60% yield. The realization of the reaction of iodine with Et_4Pb at −65 °C allowed Juenge and Cook[1177] (1959) to synthesize Et_3PbI (in 73% yield). The reaction of halogens with Ar_4Pb even at

$-75\,^{\circ}C$ resulted in cleavage of two aryl groups with the formation of Ar_2PbX_2. Gerchard and Gertruda Grüttner[1119] succeeded in obtaining Ar_3PbBr by the reaction of bromine in pyridine solution with Ar_4Pb at $-15\,^{\circ}C$, i.e. with the $Py \cdot Br_2$ complex. In 1939, Gilman and Bailie[1170] used this method to synthesize $(3\text{-}MeC_6H_4)_3PbBr$. They also obtained 88% of Ph_3PbI by the reaction of iodine with Ph_4Pb in $CHCl_3$ at room temperature. Investigations of Grüttner and Krause[1114] (1917) and later Calingaert and Soroos[1188] (1938) demonstrated that, during the action of halogens on mixed tetraalkylplumbanes, the smaller alkyl group could be eliminated more easily. A phenyl group[1115,1124] still cleaved easily from a Pb atom and a cyclohexyl group[1121] was eliminated with more difficulty. When Me_3PbEt was brominated at $-70\,^{\circ}C$, $Me_2EtPbBr$ was formed, and at $-10\,^{\circ}C$, $MeEtPbBr_2$ was the product. By the reaction of bromine with i-Am(Pr)$PbMe_2$, i-Am(Pr)MePbBr (the latter compound with an assymetric lead atom) and $AmPrPbBr_2$ were subsequently obtained.

A series of organolead dihalides $RR'PbX_2$ ($R =$ Et, Pr, Bu; $R' =$ Bu, i-Bu, i-Am; $X =$ Cl, Br)[1114,1128] was prepared by the detachment of the low alkyl radicals from mixed tetraalkylplumbanes with bromine or chlorine. Juenge and Cook[1177] prepared $(CH_2{=}CH)_2PbCl_2$ by chlorination of $(CH_2{=}CH)_4Pb$ in acetic acid solution at room temperature in 1959. It was remarkable that chlorine cleaved the C$-$Pb bond more easily than it was added to the double bond. In 1921, Krause[1121] demonstrated that $(c\text{-}C_6H_{11})_{4-n}PbX_n$ ($X =$ Br, I; $n = 1, 2$) was obtained preferably by the $(c\text{-}C_6H_{11})_3PbPb(C_6H_{11}\text{-}c)_3$ cleavage with bromine or iodine. In 1917, Grüttner and Krause[1114] found that cleavage of $(i\text{-}Bu)_3PbCl$ by bromine gave $(i\text{-}Bu)_2PbClBr$ and i-BuBr. When Flood and Horvitz[856] (1933) studied the cleavage of R_3MX ($M =$ Si, Ge, Sn, Pb; $X =$ Hal) with halogens, they found that Ph_3PbX ($X =$ Cl, I) reacted with iodine in CCl_4 to form PhI, Ph_2PbClI and Ph_2PbI_2, respectively.

The ability of the C$-$Pb bond to be cleaved with proton acids was shown in the 19th century. In 1859, Buckton[604] was the first to introduce the cleavage reaction of alkyl radical from the Pb atom by the action of gaseous HCl on Et_4Pb with the formation of Et_3PbCl. Others[1165,1248] followed the procedure. Cahours[596] (1862) and Pfeiffer and Trüskier[1165] (1904) repeated the reaction. Pfeiffer and coworkers[1166,1249] obtained organylhaloplumbanes by bubbling dry HCl or HBr through an ethereal R_4Pb solution.

Browne and Reid[1250] (1927), Gilman and Robinson[1251] (1930) and Catlin[1252] (1935) found that the reaction of saturated HCl solution with Et_4Pb led to Et_3PbCl. Gilman and Robinson[1251] showed that the reaction of HCl with Et_4Pb could lead to Et_3PbCl and Et_2PbCl_2, depending on the reaction conditions. In 1939, Gilman and Bailie[1170] obtained Et_3PbBr when gaseous HBr reacted with Et_4Pb. Austin[1253] (1931), Gilman and coworkers[1170,1184,1254] (1939, 1950), Bähr[1255] (1947) and Juenge and Cook[1177] (1959) also described the Ar_4Pb cleavage by HCl. Möller and Pfeiffer[1166] (1916), Hurd and Austin[1180,1256] (1931, 1933), Gilman and coworkers[1257] (1933), Calingaert, Soroos and Shapiro[1258] (1940), Stuckwisch[1259] (1943), Calingaert and coworkers[1260] (1945), Heap and coworkers[1241] (1951) and Koton and coworkers[1261] (1960) studied the relative order of elimination of organic substituents from the lead atom in mixed tetraorganylplumbanes. In 1931, Austin[1262] showed that the reaction of gaseous HCl with $PhPbEt_3$ led to Et_3PbCl and C_6H_6. Two year later[1180] he found that the more electronegative group (according to the 'Kharasch row'[1263]) was the first to cleave when HCl acted on mixed tetraarylplumbanes. Thus, for example[1264,1265], Ph_3PbCl[1180] was formed from $4\text{-}MeC_6H_4PbPh_3$, $PrPh_2PbCl$ from $PrPbPh_3$ and $PrPh(2\text{-}MeC_6H_4)PbCl$[1180] from $Pr(2\text{-}MeC_6H_4)_2PbPh$. According to Gilman and coworkers[1264,1265] (1932, 1936) and other investigators mentioned above, the ease of eliminating the substituents from lead atom decreased in the following order: 2-Thi > 2-Fu > 1-$C_{10}H_7$ > All > CH=CHPh. Alkyl groups, as well as CH_2Ph, $CH_2CH_2CH=CH_2$ and 4-$MeOC_6H_4$ were bonded more strongly to the lead atom than Ph[1254,1257]. Delhaye and coworkers[1266] found a second order kinetics for the cleavage of Me_3PbPh with HCl in methanol. In 1935, Yakubovich and Petrov[1267] obtained both Et_3PbCl and Et_2PbCl_2 by the reaction of gaseous HCl with Et_4Pb. In the second quarter of the 20th century (1945) numerous methods for the preparation of organolead compounds were used in the Calingaert laboratory[1260]. It was found that R_3PbBr and R_2PbCl_2 prepared by the reaction of R_4Pb (R = Alk) with HBr and HCl in ether were contaminated with $PbBr_2$ and $PbCl_2$. However, pure R_3PbCl was obtained in a high yield by bubbling HCl through a 5–10% solution of R_4Pb in hexane. This method surpassed the methods described previously for the preparation of R_3PbCl, which used concentrated hydrochloric acid[596,1250]. In 1945, Calingaert and coworkers[1260] prepared Me_3PbCl when cleaving Me_4Pb with hydrogen chloride. Later, R_3PbCl with R = Pr, *i*-Bu, CH_2=CH[1149,1150,1177] and Pr_2PbCl_2[1150] were prepared analogously. In 1951, Saunders and coworkers[1241] prepared Et_2PbCl_2 in 80% yield when bubbling dry HCl through an Et_4Pb solution in toluene at 90 °C. Under long-time boiling, all ethyl groups were cleaved off to give $PbCl_2$. Earlier, in 1949, they considered the reaction of R_4Pb with saturated HCl in ether solution to be the best method for the synthesis of trialkylchloroplumbanes[1149].

In 1887–1888, Polis[1246,1247] showed that the C–Pb bonds in Ar_4Pb (Ar = Ph, 4-MeC_6H_4) were cleaved by inorganic and organic acids (HNO_3, HCOOH, MeCOOH) with the formation of appropriate salts Ar_2PbX_2 (X = NO_3, OOCH, OOCMe). In the following century the reaction of tetraarylplumbane with organic acids was carried out (see Section IV.F) by Goddard and coworkers[728] (1922), Gilman and Robinson[1251] (1930) and Koton[1268,1269] (1939, 1941). In 1916, Möller and Pfeiffer[1166] found that aryl groups were cleaved off more easily than alkyls from lead atom of Ph_2PbEt_2 with inorganic acids. In 1925, Krause and Schlöttig[1124] reached the same conclusion when they cleaved Ph_2PbR_2 (R = Me, Et, *c*-C_6H_{11}), while Calingaert[1145] (1925) and Hurd and Austin[1256] (1931) concluded likewise when conducting the cleavage of $PhPbEt_3$. In 1931–1932, Austin[680,1262] demonstrated that the (2-MeC_6H_4)–Pb and (4-MeC_6H_4)–Pb bonds cleaved more easily than the Ph–Pb bond. According to Austin[1262] (1931) and McCleary and Degering[1270] (1938), two ethyl groups are usually cleaved off from Et_4Pb in the reaction with nitric acid with the formation of $Et_2Pb(NO_3)_2$. The reaction of Et_4Pb with H_2SO_4 proceeds in the same way. Jones and coworkers[1271] carried out the reaction of HNO_3, H_2SO_4 and HCl with R_4Pb (R = Pr, Bu, Am) which led to R_2PbX_2.

In 1930, Gilman and Robinson[1248] showed that HSO_3Ph cleaved Et_4Pb to form Et_3PbSO_3Ph. Gilman and Robinson (1929) obtained selectively Ph_3PbCl or Ph_2PbCl_2[1272] by the reaction of gaseous HCl with Ph_4Pb.

Remarkably, according to Krause and Schlöttig[1124] (1925) even NH_4Cl cleaved at 170–180 °C the C–Pb bond of Et_4Pb with formation of Et_3PbCl. Analogously, in 1948, Koton[1273] prepared Ph_3PbCl by heating a mixture of Ph_4Pb and $Me_3N{\cdot}HCl$ at 130 °C. In the early part of the last century it was established that tetraorganylplumbanes R_4Pb (R = Alk, Ar) cleaved by some metal and nonmetal halides with the formation of R_3PbX and R_2PbX_2. So triethylchloroplumbane, as well as the products of ethylation of the corresponding element chlorides were formed during the reaction of Et_4Pb with $HgCl_2$[774,1274,1275], $AlCl_3$[817,1239,1276], $SiCl_4$[1250,1277], $TiCl_4$[1278], PCl_5, $BiCl_3$[819] and $FeCl_3$[1239] and also RCOCl (R = Me, Ph)[1250]. In particular, Gilman and Apperson[1239] found in 1939 that the first reaction product of Et_4Pb with $AlCl_3$ was Et_2PbCl_2. The further process could be described by equations 21a and 21b.

$$2Et_2PbCl_2 \longrightarrow Et_3PbCl + PbCl_2 + EtCl \qquad (21a)$$

$$Et_2PbCl_2 \longrightarrow PbCl_2 + C_4H_{10} \quad (21b)$$

In 1934 Kocheshkov and Nesmeyanov[813,1279] and in 1949 Hein and Schneiter[1280] carried out the dearylation reaction of Ph_4Pb by mercury dihalides, which led to Ph_3PbX and Ph_2PbX_2 ($X = Cl$, Br). According to Panov and Kocheshkov[1281–1283] (1952, 1955) $Hg(OOCR)_2$ smoothly cleaved off phenyl groups from Ph_4Pb in the corresponding carboxylic acid medium to consequently form $Ph_{4-n}Pb(OCOR)_n$ with $n = 1–4$. In the case of tetraalkylplumbane R'_4Pb ($R' = Alk$) such reaction resulted in $R'_2Pb(OCOR)_2$ formation. In 1949–1959 the possible dearylation process of Ph_4Pb with $TlCl_3$[728], PCl_3[1284–1286], $AsCl_3$[1284,1285], $SbCl_3$ and $SbCl_5$[1285] was shown to result in the formation of Ph_2PbCl_2 and Ph_2TlCl, Ph_2PCl, Ph_2AsCl, Ph_2SbCl and Ph_2SbCl_3, respectively. The results mentioned above showed that tetraorganylplumbanes could be used as specific alkylating and arylating agents. In 1919, Krause and Schmitz[814] showed the possibility of C–Pb bond cleavage by silver nitrate in the case of Ph_4Pb. The reaction products were $Ph_2Pb(NO_3)_2$ and metastable PhAg, which easily decomposed to Ph_2 and Ag. The R_4Pb ($R = Alk$, Ar) cleavage by silver nitrate was further used by many investigators[54], as shown by the 13 publications devoted to the reaction.

The coproportionation reaction ('*komproportionierung*'), which was so well developed in organotin chemistry, did not attract attention in organolead chemistry for a long time. This was because $PbCl_4$, which should be used in this reaction, was both unstable and has a chlorination action. In 1932, Austin[1136] showed that the interaction of Ph_4Pb and Ph_2PbCl_2 led to Ph_3PbX. In 1968, Willemsens and van der Kerk[1287] replaced $PbCl_4$ with the more stable $Pb(OOCMe)_4$ in the presence of catalytic amounts of mercury diacetate (Section IV.F). The processes of radical rearrangement in a mixture of two tetraorganylplumbanes (which could be attributed to coproportionation) in the presence of Lewis acids (such as BF_3, $AlCl_3$, $SnCl_4$, EtPbX) as catalysts were studied in detail by Calingaert and coworkers[1276,1288–1291]. When carried out at relatively low temperatures[1276,1288–1291] these processes led to a mixture of tetraorganylplumbanes including all possible combinations of substituents present in the starting reagents. However, isomerization of alkyl groups was not observed. For example, during the coproportionation of an equimolecular mixture of Me_4Pb and Et_4Pb (mol%): Me_3PbEt (25%), Et_3PbMe (25%) and Me_2PbEt_2 (37.5%) were formed together with only about 6.25% of the unreacted precursors Me_4Pb and Et_4Pb. Calingaert and coworkers[1258,1260] (1940–1945) called attention to the dealkylation reaction of nonsymmetric tetraalkylplumbanes with HCl, which was often accompanied by disproportionation of the formed trialkylchloroplumbanes that led to several reaction products.

Further disproportionation reaction is important in organolead chemistry. As reported in Section III. C, tetraorganylplumbanes were obtained by reacting $PbCl_2$ with organometallic compounds via the intermediates $:PbR_2$. The processes were accompanied by cleavage and formation of C–Pb and Pb–Pb bonds as described by equations 22 and 23.

$$3R_2Pb \longrightarrow R_3PbPbR_3 + Pb \quad (22)$$

$$R_3PbPbR_3 \longrightarrow R_4Pb + Pb \quad (23)$$

The results of these reactions depended essentially on the nature of the substituent (mainly on steric factors). The first reaction of aliphatic organometallic compounds with $PbCl_2$ was so fast that it was impossible to stop it at the stage of R_2Pb formation. However, $Ar_2PbAr = Ph$, $4\text{-}MeC_6H_4$ was proved to be rather stable and it was possible to synthesize it by the organomagnesium method at a low temperature. Even at 20 °C the reaction resulted in $Ar_3PbPbAr_3$, and at the temperature of boiling ether it led to Ar_4Pb. It might be emphasized that these reactions depended considerably on the nature of the substituent at the lead atom. The studies of Krause and Reissaus[1122,1292] (1921, 1922), Austin[1262] (1931), Calingaert and Soroos[1188] (1938) and Gilman and Bailie[1170] (1939) clearly demonstrated that the steric factor in the disproportionation reaction of R_3PbPbR_3 played a very essential role. When R = Ph and 4-MeC_6H_4 the reaction led easily to R_4Pb[1122,1170,1292]. If R = 2-MeC_6H_4 the disproportionation became difficult[1122,1170,1262,1292] and when R = 2,4,6-$Me_3C_6H_2$, 2,4-$Me_2C_6H_3$ and *c*-C_6H_{11} the process did not proceed. According to Calingaert and Soroos[1188] and Gilman and Bailie[1170] the tendency to disproportionate increased in the following order: 2,4,6-$Me_3C_6H_2$ < *c*-C_6H_{11} < 1-$C_{10}H_7$ < 2-ROC_6H_4 < 2-MeC_6H_4 < 4-ROC_6H_4 < 4-MeC_6H_4 < 3-MeC_6H_4 < Ph < Et < Me. Calingaert[1145] (1925) was the first to observe the disproportionation of alkylchloroplumbanes. Later, together with coworkers he found that a mixture of $Me_{4-n}PbEt_n$ ($n = 0-4$) as well as of $MeEt_2PbCl$ and Et_3PbCl[1258] was formed by boiling $EtMe_2PbCl$. In 1932, Austin[1136] reported the transformation of Ph_3PbCl into Ph_4Pb and Ph_2PbCl_2. In 1960, Razuvaev and coworkers[1293] found out that thermal decomposition of Et_3PbBr at 70 °C led to Et_4Pb and Et_2PbBr_2. Reducing agents[1161,1294] promoted the disproportionation of organylhaloplumbanes, and Gilman and Barnett[1294] showed that Ph_3PbCl was transformed into Ph_4Pb in 70% yield in the presence of hydrazine[1294]. In analogous conditions Ph_4Pb was also obtained from Ph_2PbCl_2. In 1942, Nad' and Kocheshkov[1161] observed the transformation of Ar_2PbCl_2 (Ar = Ph, 2-MeC_6H_4) into Ar_3PbCl in the presence of metallic lead powder or its alloys with Na. In 1959–1961 investigations, carried out in the Razuvaev[1202,1293,1295–1297] laboratory, showed that the disproportionation reactions of organolead compounds should be divided into thermal and catalytic reactions. It was established that $Et_3PbPbEt_3$, which was usually stable in the absence of air at room temperature, was easily disproportionated with the formation of Et_4Pb and Pb[1293,1295,1296] in the presence of a catalytic amount of HgX_2, EtHgX (X = Cl, Br), AlX_3 (X = Cl, Br), Et_3SnCl, Et_3PbBr[1296] or Et_2PbBr_2 and $BrCH_2CH_2Br$[1293]. All these catalytic reactions were not accompanied by evolution of gaseous products. According to the patent literature, silica or activated carbon[1298,1299] could be used for the catalytic disproportionation. Free $Et^\bullet$ radicals stimulated the formation of gaseous products and were generated along with the formation of Et_4Pb and Pb in the thermal disproportionation of $Et_3PbPbEt_3$. The intermediate product of this process was $PbEt_2$[1202,1203,1297,1300].

Thermal disproportionation of Et_3PbOH at 150 °C and its kinetics were studied by Alexandrov and coworkers in 1959[1240]. The thermolysis reaction products were Et_4Pb, $Et_2Pb(OH)_2$ as well as H_2O, C_2H_4, C_2H_6 and C_4H_{10}. In 1961, Alexandrov and Makeeva[1301] showed that $Et_2Pb(OOCMe)_2$ disproportionated into $Et_3PbOOCMe$ and $EtPb(OOCMe)_3$, but the latter immediately decomposed to $Pb(OOCMe)_2$ and MeCOOEt. Analogously, $Et_2Pb(OOCCH_2Cl)_2$ disproportionated[971,1241].

F. Compounds having a Pb−O Bond

The majority of organolead compounds having the Pb−O bond have the following formulas: $R_{4-n}Pb(OH)_n$ ($n = 1-3$), $R_3PbOPbR_3$, $(R_2PbO)_n$, $(RPbOOH)_n$, $R_{4-n}Pb(OR')_n$ ($n = 1, 2$) and $R_{4-n}Pb(OOCR')_n$ ($n = 1-3$). They were studied less intensively than their organogermanium and organotin analogs. Nevertheless, the number of known organolead compounds with a Pb−O bond reached 200 by 1953. In 1853, Löwig[42] obtained the first representative of trialkylplumbanols Et_3PbOH by the reaction of Et_3PbI or $(Et_3Pb)_2CO_3$ with moist silver oxide or with aqueous alkali in ether medium. He showed that the compound was a typical base, which was neutralized by inorganic acids HX (X = Cl, Br, I, NO_3, $0.5SO_4$) with the formation of the corresponding salts Et_3PbX. In 1860,

Klippel[1217,1218], following Löwig[42] synthesized Et_3PbOH (which he considered to be a monohydrate of hexaethyldiplumboxane) by the reaction of Et_3PbI with moist Ag_2O, followed by water treatment. He found also that Et_3PbOH was formed in the reaction of Et_3PbNO_3 with alcoholic KOH solution. However, Klippel[1217,1218] found this method less convenient. He synthesized a series of triethylacyloxyplumbanes $Et_3PbOOCR$ (R = H, Me, Pr, Ph) as well as triethylplumbyl derivatives of oxalic, tartaric, hydrocyanic and cyanic acids by the neutralization of Et_3PbOH with the corresponding acids. In the 19th century Buckton[604] (1859) and Cahours[596] (1862) also synthesized trialkylplumbanols. In the 20th century Pfeiffer and Trüskier[1249] (1916), Krause and Pohland[1123] (1922), Calingaert and coworkers[1140] (1923), Browne and Reid[1250] (1927), Bähr[1255] (1947) and Saunders and Stacey[1150] (1949) used the methods mentioned above for the synthesis of R_3PbOH. Calingaert and coworkers[1260] (1945) found that the reaction of Et_3PbX (X = Cl, Br, I) with aqueous alkali in ether did not lead to pure Et_3PbOH due to contamination by the starting Et_3PbX. They showed that pure Et_3PbOH could be obtained by modification of two methods described earlier. The ether was replaced by benzene during the alkaline hydrolysis of Et_3PbX, and an aqueous solution of Et_3PbCl was used during the Ag_2O hydrolysis. The yield of triethylplumbanol then reached 93%. It was also established that the reaction of an aqueous solution of Et_3PbOH with CO_2 led to $(Et_3Pb)_2CO_3$, and with excess of CO_2 to $Et_3Pb(HCO_3)$, a compound which was previously unknown. A hydrolytic method for the synthesis of Ar_3PbOH (mainly Ph_3PbOH) from Ar_3PbX was described by Grüttner[1118] (1918) and Krause and Pohland[1123,1302] (1922, 1938). In 1921, Krause[1121] obtained the first tricyclohexylplumbanol by the reaction of $(c\text{-}C_6H_{11})_3PbI$ with 30% KOH.

Another method for the preparation of R_3PbOH was based on the oxidation of R_3PbPbR_3 by potassium permanganate in acetone. Austin[1262] (1931) and Bähr[1255] (1947) obtained Ar_3PbOH (Ar = Ph, 2,4-$Me_2C_6H_3$) in the same way. In 1959, Razuvaev and coworkers[1303] isolated Et_3PbOH when $Et_3PbPbEt_3$ was oxidized by organic peroxides.

Jones and coworkers[1271] (1935), Schmidt[1304] (1938), Calingaert and coworkers[1148,1260] (1945, 1948), Saunders and coworkers[1241] (1951) and Alexandrov and coworkers[1240] (1959) synthesized diorganylplumbanediols $R_2Pb(OH)_2$ (R = Alk, Ar). In 1935 and 1940, Lesbre[1176,1305] reported the synthesis of organylplumbanetriols by the reaction of alkyl iodides with an alkaline solution of lead oxide (i.e. $NaPb(OH)_3$) at 5 °C. These compounds were regarded as hydrated alkylplumbane acids.

Trialkylplumbanols as well as triorganylstannanols have no tendency to undergo the reaction of anhydrocondensation and that is their main difference from R_3MOH with M = Si, Ge. Only in 1960–1962 did Brilkina and coworkers[1306,1307] succeed in transforming R_3PbOH to $R_3PbOPbR_3$ by the action of metallic sodium, which did not form R_3PbONa. Up to 1964[109] only three hexaorganyldiplumboxanes $R_3PbOPbR_3$ with R = Et[43,1217,1218,1306–1311], *i*-Am[1217,1218] and Ph[1118,1254,1256,1306] appeared in the literature. Löwig[43] (1853) reported the first representative of hexaalkyldiplumboxanes $Et_3PbOPbEt_3$, which was obtained by alkaline hydrolysis of Et_3PbI. In 1860, Klippel[1217,1218] synthesized $R_3PbOPbR_3$ with R = *i*-Am by the reaction of *i*-Am_3PbI with moist silver oxide, followed by water treatment.

Et_3PbOH ('*methplumbäthyloxydhydrat*') was obtained analogously from Et_3PbI. The reaction of the latter with CO_2 led to $(Et_3Pb)_2CO_3$. Although Löwig[43] and other authors reported that they had obtained $Et_3PbOPbEt_3$ by different methods involving water or even air moisture, it could not be true because this compound is extremely unstable hydrolytically. Apparently they dealt with Et_3PbOH. In 1918, Grüttner[1118] mentioned for the first time hexaphenyldiplumboxane $Ph_3PbOPbPh_3$. He assumed that it was obtained by the reaction of Ph_3PbBr with hot alcoholic KOH or NaOH solution, followed by treatment with water or by shaking of Ph_3PbBr with 10% aqueous alkali in the cold. Actually, it was Ph_3PbOH. Up to 1960 hexaorganyldiplumboxanes were neither isolated nor characterized. The compounds with R = Et, Ph were hardly formed because their syntheses were conducted in aqueous or water–alcohol media, in which they were very easily hydrolyzed with the formation of R_3PbOH. Austin[1253,1262] (1931) and Bähr[1255] (1947) assumed that the labile $Ph_3PbOPbPh_3$, the isolation and characterization of which had failed, was apparently the intermediate in the oxidation reaction of $Ph_3PbPbPh_3$ which led to Ph_3PbOH. At the beginning of the 1960s Russian chemists[1306,1307] developed the most convenient preparative method of hexaorganyldiplumboxane. It was based on the reaction of triorganylplumbanols with dispersed Na in benzene. Compounds $R_3PbOPbR_3$ with R = Et[1307,1309,1310] and Ph[1306] were obtained by this method and characterized.

In 1856, Klippel[1217,1218] obtained and then published in 1860 the data which indicated the ease of Pb–O–Pb group protolysis by water and acids. Particularly, he showed that during the synthesis of $Et_3PbOPbEt_3$ its monohydrate, i.e. Et_3PbOH, was formed upon contact with water. He also cleaved $R_3PbOPbR_3$ with R = *i*-Am by hydrochloric and sulfuric acids. In 1960–1961, Alexandrov and coworkers[1307,1309] showed that $R_3PbOPbR_3$ with R = Et, Ph was easily protolyzed not only by water with formation of R_3PbOH (especially in aqueous methanol or dioxane), but also by alcohols already at $-10\,^{\circ}C$. By the way, Et_3PbOH (in 95–100% yield) and Et_3PbOR[1309] (R = Me, Et, CH_2Ph, CMe_2Ph) were formed from $Et_3PbOPbEt_3$. Analogously, $Et_3PbOPbEt_3$ was cleaved by organic hydroperoxides ROOH with the formation of Et_3PbOH in 95–100% yield and Et_3PbOOR (R = Me_3C, Me_2PhC). Hexaethyldiplumboxane decomposed with formation of Et_4Pb, $(Et_2PbO)_n$, ethylene and ethane[1309] even at $70–90\,^{\circ}C$. Hexaphenyldiplumboxane disproportionated with the formation of Ph_4Pb and $(Ph_2PbO)_n$ in xylene at $100\,^{\circ}C$[1306].

The first dialkylplumbanediols $R_2Pb(OH)_2$ were synthesized only in the middle of the 20th century. All were synthesized from R_2PbX_2 by alkaline hydrolysis or by the reaction with moist silver oxide[1148,1240,1241,1260,1271,1304]. The first $R_2Pb(OH)_2$ with R = Bu, Am were prepared by Jones and coworkers[1271] in 1935. Later, $Et_2Pb(OH)_2$ was synthesized in the laboratory of Calingaert[1260] by the reaction of Et_2PbCl_2 with Ag_2O in water. $Et_2Pb(OH)_2$ was isolated as hexahydrate, which transformed into polymeric $[Et_2PbO]_n$, losing water even at room temperature. It was shown that $Et_2Pb(OH)_2$ was a rather weak base, like NH_4OH. Its aqueous solutions were neutralized by strong acids (HX) with the formation of the corresponding salts Et_2PbX_2, and by saturating with CO_2 it led to Et_2PbCO_3. Calingaert and coworkers studied the decomposition of Et_3PbOH and $Et_2Pb(OH)_2$ during their contact with water steam at $100\,^{\circ}C$[1260]. It was found that $Et_2Pb(OH)_2$ was more stable than Et_3PbOH. The initial products of the hydrothermal disproportionation of the latter were Et_4Pb and $Et_2Pb(OH)_2$, which in turn decomposed into $Pb(OH)_2$ and gaseous hydrocarbons. In 1951, Heap and coworkers[1241] found also that $Et_2Pb(OH)_2$ was easily dehydrated in vacuum at room temperature, and the $(Et_2PbO)_n$ formed slowly decomposed with isolation of PbO at $100\,^{\circ}C$. Shushunov, Brilkina and Alexandrov[1312] (1959) found that high yield of $Et_2Pb(OH)_2$ and insignificant yield of Et_3PbOH were formed as intermediate products during the oxidation of Et_4Pb by oxygen in nonane or in trichlorobenzene.

In 1959, Alexandrov and coworkers[1240] reported that $Et_2Pb(OH)_2$ decomposed on heating with explosion. The thermal decomposition of both $Et_2Pb(OH)_2$ and Et_3PbOH was studied in nonane medium at $40–120\,^{\circ}C$ and PbO, Et_4Pb, ethylene, ethane and butane were isolated. The intermediate decomposition product of Et_3PbOH under mild conditions was $Et_2Pb(OH)_2$, and thermal decomposition of the latter led back to Et_3PbOH. In 1938, Schmidt[1304] reported the formation $Ar_2Pb(OH)_2$. Unlike triarylplumbanols, these compounds were extremely unstable and easily transformed into polydiarylplumboxanes $(Ar_2PbO)_n$.

As early as in the 19th century the polymeric diorganylplumboxanes $(Et_2PbO)_n$ were first synthesized. Already in 1853, Löwig[43] was the first to obtain polydialkylplumboxane $(Et_2PbO)_n$ in the reaction of alkali with Et_2PbI_2. In 1916, Grüttner and Krause[1110] synthesized first $(Me_2PbO)_n$. In 1887, Polis[1246] obtained $(Ph_2PbO)_n$ by the reaction of alkali with Ph_2PbI_2. In 1927, Zechmeister and Csabay[1219] had reproduced this synthesis. In 1955, Kocheshkov and Panov[1313] demonstrated that $(Ar_2PbO)_n$ with Ar = 4-MeC_6H_4 could be prepared by the reaction of $Ar_2Pb(NO_3)_2$ with KOH. According to them, diaryldiacyloxyplumbanes were hydrolyzed with formation of $(Ar_2PbO)_n$ much more easily than the corresponding diaryldihaloplumbanes. In 1943, Hein and coworkers[1314] obtained the first polydicyclohexylplumboxane. Polydiorganylplumboxanes did not receive the special attention of investigators and the number of publications dealing with them did not exceed 10 until 1960. The polymers, corresponding to the RPbOOH formula, i.e. the so-called organylplumbane acids, were described in more detail. Such compounds with R = Me, Et, Pr, *i*-Pr, Bu, $CH_2{=}CHCH_2$ and $PhCH_2$ were first obtained by Lesbre[1305] (1935) by the reaction of the corresponding organic iodides RI with alkaline PbO solution at 5 °C[1176,1305] according to equation 24.

$$RI + NaPb(OH)_3 \xrightarrow[-NaI]{} RPb(OH)_3 \longrightarrow RPbOOH + H_2O \qquad (24)$$

Lesbre assumed that organylplumbanetriols were intermediates of this reaction. Arylplumbane acids ArPbOOH were first obtained at the Koshechkov laboratory[1198,1283,1313,1315] by the hydrolysis of $RPb(OCOR')_3$ with aqueous alcoholic ammonia solution. These polymeric compounds ('acids') turned out to be bases which were easily dissolved in mineral and organic acids. They could not be neutralized by aqueous Na_2CO_3 or NH_3 solution but dissolved with difficulty only in 15–20% KOH[1315]. On long-time drying ArPbOOH converted into polyarylplumbsesquioxanes $(ArPbO_{1.5})_n$[1198,1315].

Organylacyloxyplumbanes $R_{4-n}Pb(OOCR')_n$, organolead carbonates $(R_3Pb)_2CO_3$, R_2PbCO_3 and organylorganoxyplumbanes $R_{4-n}Pb(OR')_n$ are classified as organoleads containing the Pb−O−C group. The latter were unknown until the second half of the last century. For the first time they appeared in Gilman and coworkers'[1308] article. In 1962, the formation of Et_3PbOR by the reaction of $Et_3PbPbEt_3$ with ROH was reported[1311]. In 1964, Rieche and Dahlmann[1316] developed three methods for the synthesis of organolead peroxides described in equations 25– 27.

$$R_3PbX + NaOOR' \longrightarrow R_3PbOOR' + NaX \qquad (25)$$

$$Ph_3PbBr + HOOR' + NaNH_2 \longrightarrow Ph_3PbOOR' + NaBr + NH_3 \qquad (26)$$

$$R_3PbOR'' + HOOR' \longrightarrow R_3PbOOR' + R''OH \qquad (27)$$

$$(X = Cl, Br;\ R = Alk, Ar;\ R' = Alk, ArAlk;\ R'' = Alk)$$

The triorganyl(organylperoxy)plumbanes proved to be hydrolytically very unstable and were easily transformed into the corresponding triorganylplumbanols even under the action of air moisture. Only in 1963–1967 was a simple method for the synthesis of R_3PbOR' found: by the reaction of R_3PbX (X = Cl, Br) with R'ONa[1310,1311,1316–1319] under conditions which completely excluded any contact with air moisture. Trialkylalkoxyplumbanes R_3PbOR' attained importance only in 1966, when Davies and Puddephatt[1318] studied their reactions with RNCO, PhNCS, CS_2, $Cl_3CCH{=}O$, $(Cl_3C)_2C{=}O$, $Cl_3CC{\equiv}N$ and other compounds.

The first organylacyloxylplumbanes were synthesized in the 19th century. Klippel[1217,1218] (1860) synthesized triethylacyloxyplumbanes $Et_3PbOOCR$ with R = H,

Me, Pr, Ph by the reaction of the corresponding acids with Et_3PbOH (he thought that they were monohydrates of hexaethyldiplumboxane) or with $(Et_3Pb)_2CO_3$ (the product of Et_3PbH with CO_2). Browne and Reid[1250] applied this method for the synthesis of triethylacyloxyplumbane in 1927. In 1952, Panov and Kocheshkov[1281] used the cleavage reaction of $(Ar_2PbO)_n$ by carboxylic RCOOH (R = Me, i-Pr) acids for synthesis of $Ar_2Pb(OOCR)_2$. Polis[1246,1247] prepared in 1887 $Ar_2Pb(OOCR)_2$ (Ar = Ph, 4-MeC_6H_4; R = H, Me) by heating Ar_4Pb with RCOOH. In addition, he demonstrated that diaryldiacyloxyplumbanes were involved in an exchange reaction with NH_4SCN, $K_2Cr_2O_7$ and H_2S. In 1927, Browne and Reid[1250] used for the first time the cleavage reaction of Et_4Pb by eight different carboxylic acids (from acetic to pelargonic) in the presence of silica as catalyst for the synthesis of trialkylacyloxyplumbanes. Analogously, five diethyl(haloacetoxy)plumbanes $Et_2PbOOCCH_{3-n}X_n$ with X = Cl, Br; $n = 1-3$ were synthesized. By the same method he obtained $Et_2Pb(OOCMe)_2$, i.e. he showed the possibility of the cleavage of two ethyl groups from Et_4Pb by acetic acid. An attempt at synthesis of $Pb(OOCMe)_4$ by the same method was unsuccessful. Browne and Reid[1250] also found that on heating Et_4Pb with acetic acid at over 90 °C, $Et_2Pb(OOCMe)_2$ was formed. Later, other experiments confirmed these data[971,1150,1308]. For instance, on heating Et_4Pb with PhCOOH at 100 °C, $Et_2Pb(OOCPh)_2$[1241] was prepared. In 1922, Goddard, Ashley and Evans[728] found that on heating Ph_4Pb with aliphatic or aromatic carboxylic acids, two phenyl groups were easily eliminated with the formation of $Ph_2Pb(OOCR)_2$. This method for synthesis of diaryldiacyloxyplumbanes was used later by Koton[1268,1269] (1939, 1941) and by Panov and Kocheshkov[1282,1283,1313] (1952, 1955). These experiments had established that the reaction rate of the acidolysis of tetraalkylplumbanes decreased as the lengths of the alkyl radicals increased.

Goddard, Ashley and Evans[728] (1922), Gilman and Robinson[1248] (1930), Koton[1268,1269] (1939, 1941) and Calingaert and coworkers[1260] (1945) also used this method to prepare triethylacyloxyplumbane. The latter authors[1260] showed that the use of silica for the $Me_3PbOOCMe$ synthesis was optional. Browne and Reid[1250] (1927) developed another synthesis of triethylacyloxyplumbanes based on the reaction of $Et_3PbOOCMe$ with RCOOK (R = Bu, Ph) in aqueous media. They carried out a similar reaction with KCN which resulted in Et_3PbCN[1250]. In 1930 and 1953 Gilman and Robinson[1248,1308] used this method. Thereafter, Calingaert and coworkers[1260] (1945), Saunders and coworkers[971,1149,1150,1241,1320] (1947–1951) and Gilman and coworkers[1308] (1953) obtained $R_3PbOOCR$ from R_4Pb.

In 1934, Kocheshkov and Alexandrov[1321] found a method for the preparation of triphenylacyloxyplumbanes based on the reaction of Ph_3PbCl with potassium salts of carboxylic acids. They first synthesized $Ph_3PbOOCCH_2COOEt$ by this method. Thermal decomposition of the latter at 160–165 °C resulted in Ph_3PbCH_2COOEt and CO_2. Analogously, $Ph_3PbOOCCH(Ph)COOEt$ was obtained and its thermolysis led to $Ph_3PbCH(Ph)COOEt$. Another method for the preparation of triorganylacyloxyplumbanes, based on the neutralization reaction of triorganylplumbanols by carboxylic acids, was used by Gilman and coworkers[1308] in 1953. They observed that sometimes the reaction of triethylplumbanol with some carboxylic acids was accompanied by cleavage of one ethyl group that led to diethyldiacyloxyplumbanes. The reaction of carboxylic acids with triarylplumbanols, developed by Koton[1322,1323], was of special synthetic interest.

Nad' and Kocheshkov[1161] first established the possibility of reacylation of organylacyloxyplumbanes by high carboxylic acids in 1942. This reaction was used at the laboratories of Kocheshkov[1197–1199,1216,1282,1283,1315], Nesmeyanov[1195] (1948) and Saunders[1241] (1951). In 1947, McCombie and Saunders[971] showed that trialkylacyloxyplumbanes could be obtained by reacylation of triethylplumbylcarbonate by carboxylic acids.

In 1953, Gilman and coworkers[1308] proposed an unusual method for reacylation of triethylacetoxyplumbane. They found that insoluble $Et_3PbOOCR$ was immediately precipitated when an aqueous solution of triethylacetoxylplumbane was mixed with the sodium salts of higher carboxylic acids RCOONa.

In 1952, Panov and Kocheshkov[1281] employed the reaction of trialkylacyloxyplumbane cleavage by mercury salts of carboxylic acids $Hg(OOCR')_2$ for the synthesis of $R_2Pb(OOCR')_2$. Triethylacetoxyplumbane was also obtained by Razuvaev and coworkers[903] using $Et_3PbPbEt_3$ cleavage of $Pb(OOCMe)_4$ in benzene media.

In 1942, Nad' and Kocheshkov[1161] studied in detail the reaction of $Pb(OOCMe)_4$ with diarylmercury in $CHCl_3$ at room temperature. This appeared to be a useful method for the preparation of diaryldiacetoxyplumbane. In 1948, Nesmeyanov and coworkers[1195,1196] used it for the synthesis of $(ClCH{=}CH)_2Pb(OOCMe)_2$.

Organolead compounds of the $RPbX_3$ series (R = organic substituent) were unknown up to 1952. However, in 1935–1940, Lesbre[1176,1305,1324,1325] reported the synthesis of alkyltriiodoplumbanes $RPbI_3$ (but their physical constants were not given) by the reaction of alkyl iodides with $CsPbCl_3$. However, Capinjola at the Calingaert laboratory[1148] could not reproduce Lesbre's data. In accordance with that, Druce in 1922[1326] and Gilman and Apperson in 1939[1239] pointed out the high instability of $RPbX_3$ (X = halogen). The first stable representatives of organolead compounds of type $RPbX_3$ turned out to be arylacyloxyplumbanes $ArPb(OOCR)_3$, which were obtained in a high yield by Kocheshkov, Panov and Lodochnikova[1197–1199,1281,1283] by the reaction of $Hg(OOCR)_2$ with $Ar_2Pb(OOCR)_2$[1281] in RCOOH media or by the reaction of Ar_2Hg with $Pb(OOCR)_4$ (R = Me, Et, *i*-Pr) in $CHCl_3$ in 1956–1959. Aryltriacyloxyplumbanes were transformed into $Ar_2Pb(OOCR)_2$[1199] by the action of Ar_2Hg. In 1952, Panov and Kocheshkov[1281] first prepared arylplumbane acids $(ArPbOOH)_n$ by the reaction of $ArPb(OOCR)_3$ with weak alkali solution or aqueous ammonia. They also carried out re-esterification of aryltriacyloxyplumbanes with carboxylic acids having higher boiling temperatures than MeCOOH (e.g. PhCOOH or $CH_2{=}CMeCOOH$). By the same method the labile $Et_2Pb(OOCMe)_2$ was transformed into the more stable $Et_2Pb(OOCCH_2Cl)_2$.

In 1930 the first organolead sulfonates were obtained by Gilman and Robinson[1248] by heating Et_4Pb with 4-$MeC_6H_4SO_2OH$ in the presence of silica to form $Et_3PbOSO_2C_6H_4Me$-4. In 1953, Gilman and coworkers[1308] synthesized triethylplumbylsulfonates and sulfinates Et_3PbOSO_2R, $Et_3PbOSOR$ from $Et_3PbOOCMe$ and the sodium salts of the corresponding acids. In 1936, Schmidt[1304] prepared them by the reaction of Ph_2PbO with sodium pyrocatecholdisulfonate.

Triorganylplumbane and diorganylplumbane esters of oxygen-containing inorganic acids R_3PbX and R_2PbX_2, where X were the acid anions, could be considered as organolead compounds formally having the plumboxane bond. However, not all of them had a Pb–O–M group with a covalent Pb–O bond and so they were properly salts. Particularly, this concerns the derivatives of oxygen-containing strong inorganic acids in which the M atom is highly electronegative (Cl, S, N etc.). For example, organolead ethers of H_2SO_4 and HNO_3 were typical salts. The compounds of this type, i.e. Et_3PbNO_3[43,1217,1218], $(Et_3Pb)_2SO_4$[43,1163,1217,1218] and $[(i\text{-}Am)_3Pb]_2SO_4$[1217,1218], were first obtained by Löwig[43] (1853), Buckton[1163] (1859) and Klippel[1217,1218] (1860). In 1887, Polis[1246,1247] first obtained $Ph_2Pb(NO_3)_2$ by the reaction of Ph_2PbCl_2 with $AgNO_3$. Compounds such as $Ph_2Pb[(OH)CO_2]_2$, $(Ph_2Pb)_3(PO_4)_2$, Ph_2PbCrO_4, $Ph_2Pb(OH)CN$ and Ph_2PbBr_2 were synthesized by the exchange reactions of $Ph_2Pb(NO_3)_2$ with Na_2CO_3, Na_3PO_4, $K_2Cr_2O_7$, KCN and KBr, respectively. In 1930, Gilman and Robinson[1248] synthesized $Et_3PbOPO(OH)_2$ by heating Et_4Pb with H_3PO_4. The reaction of aqueous or alcoholic solution of R_3PbOH or $R_2Pb(OH)_2$ with the corresponding acids was the basic method for synthesis of R_3PbA and R_2PbA_2 (A = acid anion). In the past century Tafel[1144] (1911), Pfeiffer and Trüskier[1249] (1916), Goddard and coworkers[728] (1922), Vorlander[1327] (1925), Buck and Kumro[1328] (1930), Austin and Hurd[1256,1262] (1931), Challenger and Rothstein[1329] (1934), Jones and coworkers[1271] (1935), Gilman, Woods and Leeper[316,1330] (1943, 1951), McCombie and Saunders[971] (1947), Nesmeyanov and coworkers[1195] (1948), and Saunders and coworkers[1241] (1951) synthesized a series of R_3PbA and R_2PbA_2 by this method. In addition, arylsulfonates[1149] and iodates[728] were among the anions in the series given above along with sulfates and nitrates.

G. Compounds having a Pb−S, Pb−Se and Pb−Te Bond

As indicated in Section III. G, according to the Goldschmidt geochemical classification lead as well as tin belong to the chalcofile elements, i.e. they have high affinity to sulfur. Consequently, numerous organolead compounds possessing the plumbathiane Pb−S bond have been easily formed in many reactions involving hydrogen sulfide, alkaline metal sulfides, sulfur and some other sulfurizing agents with various organolead derivatives.

The main organolead derivatives of this type have the following general formulas: $R_3PbSPbR_3$, $(R_2PbS)_n$, $(RPbS_{1.5})_n$, $R_{4-n}Pb(SR')_n$ ($n = 1, 2$). Compounds containing the Pb−S−H bond do not appear in this list, due to their extreme instability. In contrast with isostructural compounds of tin, organolead compounds containing the Pb−S bond attracted only a little attention of researchers and industrial chemists. The number of known compounds of this type, which was less than 50[1331] by 1967, bears witness to this fact. On the one hand this was due to their unacceptability to be used as synthons and reagents, and, on the other hand, due to the seemingly absence of any future practical use. Only a few patents dealing with the application of Me_3PbSMe[1332,1333], $R_3PbSCH_2CONH_2$[1334] and $Me_3PbSPbMe_3$[1332] as potential motor engine antiknocks and the use of compounds $R_3PbSCH_2CONH_2$[1334] and Me_3PbSCH_2COOMe[1335] as potential pesticides were issued.

Organolead compounds containing sulfur appeared in chemical circles in the 19th century. The first one was hexaethyldiplumbathiane $Et_3PbSPbEt_3$, which was prepared by Klippel[1217,1218] employing the reaction of Et_3PbCl with an aqueous solution of Na_2S in 1860. Significantly later, in 1945 the above reaction was repeated at the Calingaert laboratory at 0 °C[1260]. It was found during the reaction that $Et_3PbSPbEt_3$ was slowly oxidized by air oxygen to $(Et_3Pb)_2SO_4$. In 1887, Polis[1246] synthesized $(Ar_2PbS)_3$, Ar = Ph, 4-MeC_6H_4 by the reaction of $Ar_2Pb(OOCMe)_2$ with H_2S. Only in the second half of the 20th century[1336–1338] was $(Ph_2PbS)_3$ synthesized again, and it was proved that it was a trimer. Other compounds of the series of $R_3PbSPbR_3$ were obtained in 1911–1917. In 1911, Tafel[1144] synthesized its representative with R = i-Pr, and its analogs with R = c-C_6H_{11} and Me were synthesized by Grüttner and Krause[674,1110]. In 1917, Grüttner and Krause[1113] obtained MeEtPbS, Pr(i-Bu)PbS and Pr(i-Am)PbS. At last, in 1918, Grüttner synthesized $Ph_3PbSPbPh_3$[1119] for the first time. After this pioneer research no organolead compound having the Pb−S−Pb group was obtained up to 1945. Henry and Krebs[443,1337] (1963) found that the reaction of Ph_3PbCl with Na_2S proceeded in a different direction with formation of Ph_3PbSNa. The latter interacted with RI (R = Me, Et) to give Ph_3PbSR. In 1965, Davidson and coworkers[437] obtained $Ph_3PbSPbPh_3$ in a quantitative yield by the reaction of Ph_3PbX (X = Cl, Br) with H_2S in the presence of Et_3N or pyridine. Organyl(organylthio)plumbanes $R_{4-n}Pb(SR')_n$ were prepared by the reaction of the corresponding organylhaloplumbanes with mercaptides or thiophenolates of alkali metals or of silver or lead. Saunders and coworkers[971,1149,1150] first described this type of compound (Et_3PbSEt, Et_3PbSPh) in 1947 and 1949. They were synthesized by the reaction of Et_3PbOH with RSH or Et_3PbCl with NaSR. These compounds slowly hydrolyze by water and they turned out to be effective sternutators[1304]. A convenient method for the synthesis of Ph_3PbSR (R = Me, Et, Pr, Bu, Ph, CH_2Ph, COMe, COPh) based on the use of $Pb(SR)_2$ was elaborated by Krebs and Henry[1337] (1963) and later applied by Davidson and coworkers[437] (1965). This method proved to be unsuitable for the preparation of $Ph_2Pb(SR)_2$. These authors[437] also attempted to obtain Ph_3PbSR by the cleavage of R_4Pb by thiols, but they were unsuccessful. Compounds of the Ph_3PbSR series turned out to be hydrolytically stable, but it was impossible to distill them without decomposition. This research[437] demonstrated that the thermal stability of the M−SR bond in R_3MSR (M = Si, Ge, Sn, Pb; R = Alk, Ar) is diminished on increasing the atomic number of M, but their hydrolytic stability had increased. Abel and Brady[1339] obtained Me_3PbSEt (in 53% yield) by the reaction of Me_3PbCl with EtSH in aqueous NaOH solution in 1965. In 1951, Saunders and coworkers[1241] illustrated that the reaction of Et_4Pb with MeCOSH resulted in $Et_3PbSCOMe$ and in the presence of silica, in $Et_2Pb(SCOMe)_2$, indicating that thiols were capable of cleaving the C−Pb bond.

Davidson and coworkers[437] synthesized diphenyl(diorganylthio)plumbanes $Ph_2Pb(SR)_2$ by condensation of Ph_2PbX_2 (X = Cl, Br) with RSH (R = Alk, Ar) in benzene in the presence of Et_3N or Py as an acceptor of the hydrohalic acids. These compounds appeared to be unstable and decomposed by heating according to equation 28.

$$3Ph_2Pb(SPh)_2 \longrightarrow 2Ph_3PbSPh + Pb(SPh)_2 + PhSSPh \qquad (28)$$

By the reaction of Ph_2PbCl_2 with $HSCH_2CH_2SH$ in the presence of Et_3N, 2,2-diphenyl-1,3-dithio-2-plumbacyclopentane[437] was obtained. Finally, Davidson and coworkers[437] synthesized a series of carbofunctional triphenyl(organylthio)plumbanes $Ph_3PbS(CH_2)_nX$, where X = COOMe, $CONH_2$ ($n = 1$); OH, NH_2 ($n = 2$); $Ph_3PbSC_6H_4X$-4 (X = Cl, NH_2, NO_2) and $Ph_3PbSC_6F_5$. They also prepared the first organolead derivative of a natural hormone, i.e. Ph_3PbS-17-β-mercaptotestosterone.

The Pb−S bond in trialkylthiocyanatoplumbanes R_3PbSCN was definitely ionic. Klippel[1217,1218] obtained the first compound of the Et_3PbSCN series by the reaction of Et_3PbCl with AgSCN in alcoholic media as early as 1860. However, Saunders and coworkers[971,1149] (1947, 1949) could not repeat this reaction. They synthesized the same compound by the reaction of Et_3PbCl with KSCN in alcohol, and Gilman and coworkers[1308] obtained it by the reaction of $Et_3PbOOCMe$ with KSCN in 1953. Ethyl(thioacetoxy)plumbanes $Et_{4-n}Pb(SCOOMe)_n$ with $n = 1$, 2 were described by Saunders and coworkers[1241] in 1951.

Organolead compounds having the Pb−Se and Pb−Te bonds became known only in 1962–1965[1340–1342]; they were $Me_3PbSeMe$[1341,1342], $Me_3PbSePh$[1342], $Ph_3PbSeLi$, $Ph_3PbSePbPh_3$, $Ph_3PbTeLi$ and $Ph_3PbTePbPh_3$[1340]. The salt-like triethyl(selenocyanato)plumbane was synthesized by Heap and Saunders[1149] (1949) by the reaction of Et_3PbCl with KSeCN, and by Gilman and coworkers[1308] (1953) by the exchange reaction of $Et_3PbOOCMe$ with KSeCN; it could also be regarded as a compound containing a Pb−Se bond.

In 1961, $R_3PbSPbR_3$ (R = Alk) were proposed as motor fuel antiknocks[1343].

H. Compounds having a Pb−N Bond

Compounds with a Pb−N bond are the least studied in organolead chemistry. By 1953 there were only less than 10 of them[109]. The syntheses of the first representatives of this class were published by McCombie and Saunders[971,1149] in 1947–1950, but preliminary reports on their syntheses were given in 1940. These were *N*-trialkylplumbylarene sulfonamides $R_3PbNR'SO_2Ar$, -phthalimides $R_3PbN(CO)_2C_6H_4$-*o* (R = Et, Pr) and -phthalhydrazides $R_3PbNHN(CO)_2C_6H_4$-*o*. They were synthesized by the reaction of the corresponding sodium derivatives with R_3PbCl or by the reaction of the free acids or phthalimide with R_3PbOH. In 1953, Gilman and his coworkers[1308] synthesized analogous compounds by reaction of organolead bases R_3PbOH or $(R_3Pb)_2O$ with NH acids such as sulfonamides or imides.

Willemsens and van der Kerk[110] applied this method for the synthesis of *N*-trialkylplumbyl-substituted heterocycles containing the endocyclic NH group in 1965. In some cases trialkylhaloplumbanes were also used to synthesize organolead compounds containing the Pb−N bond using reagents containing an N−H bond. In this process an excess of a nitrogen base served as an acceptor of the hydrogen halide[456,471].

In the 1950s, some patents were granted for the use of *N*-trialkylplumbyl phthalimide and -phthalhydrazide as fungicides[1344,1345] and of $Et_3PbNHCHMeEt$ as a herbicide[1346]. The latter was obtained by the reaction of Et_3PbCl with NaNHCHMeEt. It is remarkable that the preparation of triorganyl(dialkylamino)plumbanes $R_3PbNR'_2$ (R = Me, Et) (as well as that of their organotin analogs)[1347] from R_3PbX (X = Cl, Br) was successful only when lithium dialkylamides $LiNR'_2$ were used as the aminating agents. This was caused by the ability of triorganylhaloplumbanes (as well as R_3SnX) to form adducts (less stable than the corresponding tin complexes) during the reaction with ammonia and amines but not to substitute the halogen atom by an amino group. Neumann and Kühlein[1348,1349] first used this method of synthesis in their laboratory in 1966.

The Pb−N bond turned out to be rather active. For example, trialkyl(dialkylamino)plumbanes hydrolyzed extremely rapidly by water, whereas *N*-trialkylplumbyl derivatives of sulfonamides, amides and imides of carboxylic acids as well as their nitrogen heterocycles were hydrolytically rather stable. In the 1960s, the cleavage reactions of the Pb−N bond in $R_3PbNR'_2$ by inorganic and organic acids, alcohols and NH acids (e.g. re-amination by organometallic hydrides) were developed[1350]. We shall not consider them here since this period is not yet regarded as historical.

In the 1960s at the Schmidt laboratory[456,471], organometallic compounds containing Pb−N−M (M = Si, Ge) bonds were synthesized[1350]. In 1964, Sherer and Schmidt[456] obtained trimethylbis(trialkylsilylamino)plumbanes $Me_3PbN(SiR_3)_2$, R = Me, Et by the reaction of Me_3PbCl with $NaN(SiR_3)_2$[456,1351]. Schmidt and Ruidish[471] (1964) prepared analogously $Me_3PbN(GeMe_3)_2$ from $LiN(GeMe_3)_2$.

One year later Sherer and Schmidt[1352] carried out the reaction of Me_3PbCl with $LiN(SiMe_3)Me$, which led to $Me_3PbN(SiMe_3)Me$. In 1961 and 1963 Lieber and coworkers[1353,1354] synthesized phenylazidoplumbanes $Ph_{4-n}Pb(N_3)_n$ ($n = 1, 2$) from $Ph_{4-n}Pb(OH)_n$ with HN_3. In 1964, Reichle[465] reported that, contrary to expectations, Ph_3PbN_3 proved to be rather thermostable and decomposed with formation of Ph_4Pb and N_2 under thermolysis.

The number of organolead compounds having a Pb−N bond reached fifty by 1968[1350].

Pfeiffer and Trüskier[1249] obtained the first organic compounds of hypervalent lead, having coordinated N→Pb bonds in 1916. They were isolated during recrystallization of Ph_2PbX_2 (X = Br, Cl, NO_3) from pyridine and corresponded to the formula $Ph_2PbX_2 \cdot 4Py$. These complexes were stable only under pyridine atmosphere. In the absence of the latter they lost two molecules of pyridine and transformed into hexacoordinated lead complexes $Ph_2PbX_2 \cdot 2Py$. In ammonia atmosphere Ph_2PbBr_2 formed the unstable complex $Ph_2PbBr_2 \cdot 2NH_3$, which easily lost ammonia[1355]. Even these limited data showed that the complexes of organylhaloplumbanes Ph_3PbX and Ph_2PbX_2 with nitrogen bases were unstable and they were not studied further. The preparation of complexes of amines with $RPbX_3$ and PbX_4 failed, apparently due to their redox reaction with the addend. Nevertheless, the stable complexes $[Me_3PbPy]^+ClO_4^-$, and $[Me_2PbPy_2]_2{}^{2+}2ClO_4^-$ and $[Me_3Pb(OP(NMe_2)_3)_2]^+ClO_4^-$ were described in 1966. However, they had an ammonium structure, i.e. the lead atom was tetracoordinated but not hypervalent[1356]. One should note that, with respect to DMSO, organylhaloplumbanes served as rather strong Lewis acids. In 1966 the stable complexes $Ph_2PbX_2 \cdot 2OSMe_2$ (X = Cl, Br) containing hexacoordinated lead atom because of the presence of two O→Pb bonds[1357] were synthesized. They were so stable that they could be synthesized even in aqueous medium. The melting point of $Ph_2PbCl_2 \cdot 2OSMe_2$, 171 °C, witnessed its thermal stability. In 1964, Matviyov and Drago[1358] prepared the complexes of R_3PbCl (R = Me, Et) with tetramethylenesulfoxide (B) of compositions $R_3PbCl \cdot B$ and $R_3PbCl \cdot 2B$, $Me_2PbCl_2 \cdot 2B$ and $R_3PbCl_2 \cdot 4B$. $Me_3PbCl \cdot B$ had a trigonal-bipyramidal structure, i.e. its lead atom was pentacoordinated. The second and third complexes were apparently of octahedral structure and in $Me_2PbCl_2 \cdot 4B$ the lead atom was octacoordinated. Later, the analogous complex $[Me_2Pb(OSMe_2)_4](ClO_4)_2$ was obtained. In 1961, Duffy and Holliday[1359] showed that the reaction of Me_3PbCl with KBH_4 in liquid NH_3 at −70 °C led to an adduct of $Me_3Pb(BH_4) \cdot nNH_3$ composition with $n \geqslant 2$ (probably, it was a mixture of $Me_3PbH \cdot NH_3$ and $H_3N \cdot BH_3$). The product obtained at −5 to +20 °C decomposed with the formation of Me_3PbH, NH_3 and $H_3N \cdot BH_3$. The Me_3PbH obtained reacted instantly with liquid ammonia at −78 °C with the formation of an unstable green adduct, which evidently was $Me_3PbH \cdot NH_3$. Based on the ^1H NMR data, the authors ascribed to the product the very unlikely structure of ammonium trimethylplumbate Me_3PbNH_4 containing the Me_3Pb^- anion. This complex evolved CH_4 and NH_3 at −78 °C and was slowly transformed into $Me_3PbPbHMe_2 \cdot NH_3$, which in the authors' opinion provided the red color of the reaction mixture. However, it was most probably Me_2Pb:, which provided the red color according to equations 29 and 30.

$$Me_3PbH \cdot NH_3 \longrightarrow Me_2Pb\colon + CH_4 + NH_3 \quad (29)$$

$$Me_3PbH \cdot NH_3 + Me_2Pb \longrightarrow Me_3PbPbHMe_2 \cdot NH_3 \quad (30)$$

Pentamethyldiplumbane ammoniate decomposed to Me_4Pb, $Me_3PbPbMe_3$, CH_4, H_2 and Pb at −45 °C. The solution of Me_3PbH in Me_3N was less stable than its solution in liquid ammonia.

I. Organolead Hydrides

The first investigations of organolead hydrides $R_{4-n}PbH_n$ ($n = 1, 2$) were conducted only in the 1960s. The reason for their late appearance was their extreme instability due to the presence of the Pb−H bond. Early attempts to obtain organolead hydrides R_3PbH by the reaction of R_3PbNa (R = Et, Ph) with NH_4Br in liquid ammonia[1170,1360] or by catalytic hydrogenation of $Ph_3PbPbPh_3$ had failed. In 1958, Holliday and Jeffers[1361] were the first to report the preparation of Me_3PbH, when they found that it was formed by decomposition of Me_3PbBH_4 in liquid ammonia. Later, Duffy and Holliday[1359,1362,1363] used the reduction reaction of R_3PbCl by KBH_4 in liquid ammonia to prepare R_3PbH (R = Me, Et). An intermediate of this reaction was R_3PbBH_4, which eliminated R_3PbH at −5 °C. In 1960, Amberger[1364] synthesized R_3PbH and R_2PbH_2 (R = Me, Et) by the reduction of the appropriate organolead chlorides by $LiAlH_4$ in Me_2O. Becker and Cook[1212] (1960) used for this purpose the reaction of R_3PbX (X = Cl, Br) with KBH_4 in liquid ammonia or with $LiAlH_4$ in Me_2O at −78 °C. Dickson and West[1365] succeeded in obtaining some amount of Et_3PbH by the decomposition of Et_3PbNa by ammonium bromide in liquid ammonia in 1962.

Neumann and Kühlein[1213,1366] used the reduction of R_3PbCl by $LiAlH_4$ for the synthesis of R_3PbH, R = Pr, Bu, *i*-Bu, *c*-C_6H_{11} in 1965. They also synthesized Bu_2PbH_2 from Bu_2PbCl_2. Such solvents as Me_2O, Et_2O, THF or diglyme, which interacted with the $AlCl_3$ formed, were used for this purpose since the $AlCl_3$ caused decomposition of the R_3PbH[1212,1317,1366]. In 1966, Amberger and Hönigschmid-Grossich[1367] demonstrated that trialkylmethoxyplumbanes R_3PbOMe reacted with B_2H_6 at $-35\,^{\circ}C$ to form R_3PbBH_4. Further treatment at $-78\,^{\circ}C$ by MeOH resulted in R_3PbH with R = Me, Et, Pr, Bu. Even without methanolysis, Me_3PbBH_4 slowly decomposed in ether with formation of Me_3PbH at $-78\,^{\circ}C$[1362]. In 1965, Neumann and Kühlein[1213,1366] showed that Et_3PbCl could be reduced by Bu_3PbH to Et_3PbH, which was removed from the reaction mixture by distillation. High-boiling organotin hydrides R_3SnH and R_2SnHCl (R = Bu, Ph) were employed as reductants of Et_3PbX. Thus, within the period from 1960 till 1965, 10 organolead hydrides were synthesized. The low organolead hydrides $R_{4-n}PbH_n$ (R = Me, Et; $n = 1, 2$) were liquids, which decomposed at temperatures below $0\,^{\circ}C$[1212,1367]. Duffy and coworkers[1363] (1962) identified methane as a gaseous product of the Me_3PbH decomposition. According to Amberger and Hönigschmid-Grossich[1367] high trialkylplumbanes started to decompose to R_4Pb, R_3PbPbR_3, Pb and H_2[1367] without air in vacuum at -30 to $-20\,^{\circ}C$. Neumann and Kühlein[1213,1366] showed in 1965 that Pr_3PbH was completely decomposed (disproportionated) to Pr_4Pb, $Pr_3PbPbPr_3$, Pb, C_3H_8 and H_2[1366] within 24 hours. Propane appeared in the product of the hydrogen atom cleavage from Pr_3PbH.

Becker and Cook[1212] (1960) proposed a rather complicated scheme for the homolytic decomposition of R_3PbH (Scheme 1). It was possible that this process was simpler, involving the intermediate formation of PbR_2.

$$2R_3PbH \xrightarrow{h\nu} 2R_3Pb^{\bullet} + H_2$$

$$2R_3Pb^{\bullet} \longrightarrow R_3PbPbR_3$$

$$R_3Pb^{\bullet} + R_3PbPbR_3 \longrightarrow R_4Pb + R_3PbPbR_2^{\bullet}$$

$$R_3PbPbR_2^{\bullet} \longrightarrow R_4Pb + Pb + R^{\bullet}$$

$$R^{\bullet} + R_3PbH \longrightarrow R_3Pb^{\bullet} + RH$$

$$R^{\bullet} + R_3PbPbR_3 \longrightarrow R_4Pb + R_3Pb^{\bullet}$$

SCHEME 1

In 1960, Becker and Cook[1212] were the first to succeed in carrying out the reaction of hydroplumbylation (a term first suggested by Voronkov in 1964[53]). They demonstrated that Me_3PbH was added to ethylene in diglyme at 35 atm and $0\,^{\circ}C$ with the formation of Me_3PbEt in 92% yield. Unlike that, Duffy and coworkers[1363] found that trialkylplumbanes did not add to ethylene in Me_2O media or without solvent at normal pressure. Neumann's[1366] attempts to carry out the hydroplumbylation reaction of $CH_2{=}CHR$ (R = C_6H_{13}, CH_2OH, CH_2OOCMe) with Bu_3PbH at $0\,^{\circ}C$ or $20\,^{\circ}C$ were unsuccessful as well. Nevertheless, Blitzer and coworkers[1368] patented a method of addition of organolead hydrides to terminal olefins and cyclohexene in 1964. In 1965, Neumann and Kühlein[1213,1366] found that Bu_3PbH was added to compounds having terminal activated double bonds $CH_2{=}CHR$ (R = CN, COOMe, Ph) at $0\,^{\circ}C$. In 1965, Leusink and van der Kerk[1214] showed that Me_3PbH added easily to $HC{\equiv}C{-}CN$ and $HC{\equiv}C{-}COOMe$.

The *cis*-adduct was the first product of the hydroplumbylation of cyanoacetylene and it was consequently converted into the *trans*-isomer at temperatures from −78 °C to 0 °C. At about the same time Neumann and Kühlein[1213] carried out a similar reaction of Bu_3PbH with HC≡CPh that led to the *trans*-adduct. They also showed that Bu_3PbH did not add to the C=O bonds of aldehydes and ketones. In contrast, they showed that in the reaction of Bu_3PbH with PhN=C=S the hydroplumbylation of thiocarbonyl group proceeded with the formation of Bu_3PbS−CH=NPh. They also found that Bu_3PbN(CH=O)Ph, the product of the N=C bond hydroplumbylation, was formed in the reaction of Bu_3PbH with PhN=C=O at −70 °C. Phenylisocyanurate $(PhNC{=}O)_3$ and $Bu_3PbPbBu_3$ were the final products of the reaction. In 1968, Neumann and Kühlein[1369] investigated the mechanism of the hydroplumbylation reaction, which was found to proceed by both radical and ionic processes.

In 1960, Becker and Cook[1212] pointed out that $R_{4-n}PbH_n$ (R = Me, Et; $n = 1, 2$) reacted with diazoalkanes $R'CHN_2$ (R′ = H, Me) with the formation of both the products of hydrides disproportionation and the insertion of the R′CH group into the Pb−H bond from −80 to −0 °C.

Duffy and coworkers[1359,1363] found that R_3PbH (R = Me, Bu) were decomposed under ammonia and amines action. Trimethylplumbane reacted with liquid ammonia to give green and then red solutions (evidently connected with an intermediate formation of $Me_3Pb^\bullet$ and Me_2Pb) and Me_4Pb, Pb and CH_4 were formed. Addition of $PbCl_2$ to an Me_3PbH solution in NH_3 led to $Me_3PbPbMe_3$ in almost a quantitative yield[1359]. Organolead hydrides were extremely easily oxidized by air oxygen (Me_3PbH was oxidized with an explosion)[1212] and they turned out to be strong reductants (more effective than organic hydrides of Ge and Sn). In 1960, Neumann[1370] found that trialkylplumbanes reacted with ethyl iodide even at temperatures from −60 up to −40 °C with the formation of ethane. Holliday and coworkers[1363] (1962) found that Me_3PbH reacted with HCl to give Me_3PbCl and H_2 as well as some amount of $Me_3PbPbMe_3$ at −112 °C. Along with them Me_2PbCl_2 and CH_4 were identified at −78 °C.

In 1965, Neumann and Kühlein[1213,1366] reduced aliphatic halogens, and carbonyl, nitro and nitroso compounds, and Et_3SnCl as well, by tributylplumbane at 0 °C and 20 °C. A higher temperature was found unacceptable due to the decomposition of Bu_3PbH.

J. Compounds Containing a Pb−Pb Bond

Almost all the known compounds having a Pb−Pb bond are hexaorganyldiplumbanes R_3PbPbR_3 and only a few of them do not correspond to this formula. Hexaorganyldiplumbanes have been regarded for a long time as trivalent lead derivatives and it is a wonder that even such leaders of metalloorganic chemistry as Gilman (with Towne) in 1939[1254] and Kocheshkov in 1947[156] and with Panov even in 1955[1313] gave the R_3Pb formula to these compounds. Some historical aspects and data concerning the synthesis and transformations of hexaorganyldiplumbanes were given in Sections IV.B, IV.C and IV.E because they were connected with the quoted data. Herein we consider the historical developments of the investigations of compounds having the Pb−Pb bond in more detail.

As reported in Section IV.B, Löwig[43] obtained hexaethyldiplumbane, the first organolead compound having the Pb−Pb bond, in 1953. It was difficult to decide whether this compound was Et_6Pb_2 (Et_3Pb radical by Löwig) or Et_4Pb according to his data, which were based on the atomic weights known at that time. In 1859, Buckton[1371] reported that the compound described by Löwig was apparently Et_4Pb and this was confirmed by Ghira[1372] in 1894. Moreover, he stated: ‘At the present time no lead compounds of the type PbX_3 or Pb_2X_6 have ever been reported, studied or isolated.’

In 1919, Krause and Schmitz[1120] obtained for the first time hexaaryldiplumbane $Ar_3PbPbAr_3 (Ar = 2,5\text{-}Me_2C_6H_3)$ by the reaction of $2,5\text{-}Me_2C_6H_3MgBr$ with $PbCl_2$. In 1921, Krause[1121] synthesized hexacyclohexyldiplumbane analogously. He wrote that tetracyclohexylplumbane which was obtained by Grüttner[674] in 1914 was apparently nonpure. The synthesis of R_3PbPbR_3 from RMgX and $PbCl_2$ was further used by Krause and Reissaus[1122,1292] (1921, 1922), Calingaert and coworkers[1188,1373] (1938, 1942) and Gilman and Bailie[1170] (1939). It was established that R_2Pb were the labile intermediates of this reaction, which is described by equations 31 and 32.

$$2RMgX + PbX_2 \longrightarrow R_2Pb + 2MgX_2 \quad (31)$$

$$3R_2Pb \longrightarrow R_6Pb_2 + Pb \quad (32)$$

Hexaethyldiplumbane, whose chemical composition and structure were unequivocally proved, was obtained by electrolysis of Et_3PbOH with lead cathode in alcoholic medium by Calingaert and coworkers[1140] only in 1923. The electrochemical method for the R_3PbPbR_3 synthesis was further developed by the Calingaert group[1188,1373] in 1938–1942 and by Italian chemists[1374] in 1960. In 1960, Vyazankin and coworkers[1203] found that during the electrochemical synthesis of $Et_3PbPbEt_3$ a new product, identified as Et_2Pb, was formed along with it. Hein and Klein[1201] found that compounds R_3PbPbR_3 (R = Me, Et) were easily formed by the reduction of R_3PbCl by Al, Zn or Pb in alkaline solution. In the years 1938 and 1939, the method for R_3PbPbR_3 synthesis based on the reaction of R_3PbX (R = Alk, Ar; X = Cl, Br, I) with Na in liquid ammonia[791,1170,1188,1189] began to develop. This fact was more surprising since even in 1947 Kocheshkov related to the formation of R_3PbPbR_3 from a reduction of R_3PbX to R_3Pb[156]. He referred to the magnetochemical evidence of this fact given by Preckel and Selwood in 1940[1375].

Bright red tetrakis(triphenylplumbyl)plumbane $(Ph_3Pb)_4Pb$ obtained by the simultaneous hydrolysis and oxidation of Ph_3PbLi or Ph_2Pb by H_2O_2 at low temperature by Willemsens and van der Kerk[109,1376,1377] turned out to be the first organolead compound having several Pb–Pb bonds. Tetrakis(triphenylplumbyl)plumbane was an unstable compound which decomposed into $Ph_3PbPbPh_3$ and Pb at storage. This indicated that the Pb–Pb–Pb bond system was quite unstable.

Gilman and Woods[1330] and Leeper[1378] in 1943 and Gilman and Leeper in 1951[316] described the condensation of diorganyldihaloplumbanes with lithium and calcium. Foster and coworkers[1189] (1939) carried out the reaction of Ph_3PbCl with $[Na_4Pb_9]$.

For the synthesis of hexaaryldiplumbanes Gilman and coworkers[1170,1185] (1939, 1952) and Podall and coworkers[1379] (1959) used lithium aryls. In 1941, Bindschadler[1190] obtained hexaphenyldiplumbane by the reaction of Ph_3PbNa with $BrCH_2CH_2Br$. Hein and Nebe[1187] synthesized hexacyclohexyldiplumbane by the reaction of $(c\text{-}C_6H_{11})_3PbNa$ with mercury. In 1931, Goldach[1147] found that hexaisopropyldiplumbane was formed by the reaction of acetone with an Na–Pb alloy in sulfuric acid. Hexamethyldiplumbane was isolated by the reaction of Me_3PbCl with the adduct $Me_3PbH \cdot NH_3$ in liquid ammonia by Duffy and Holliday[1359]. In 1962, the same authors[1363] observed that $Me_3PbPbMe_3$ was the product of the thermal dehydrocondensation of Me_3PbH. In the first half of the 20th century, twenty hexaorganyldiplumbanes were synthesized by the methods described above.

All hexaalkyldiplumbanes described in the literature turned out to be thermally unstable liquids which decomposed on distillation. In 1923, Calingaert and coworkers reported that $Et_3PbPbEt_3$ dissociated into the $Et_3Pb^{\bullet}$ radicals in dilute solutions[1140]. However, in concentrated solutions $Et_3PbPbEt_3$ was the main species. The molecular weights found for R_3PbPbR_3 with R = Ph[1336,1380,1381], $2,4,6\text{-}Me_3C_6H_2$[1382] and $c\text{-}C_6H_{11}$[1380] showed that all the compounds corresponded to the formula given above. In particular, the thermolysis data of $Et_3PbPbEt_3$ obtained by Razuvaev and coworkers[1202,1295,1300,1383] and other investigations[984,1239,1384] corroborated the structure.

In contrast, hexaaryldiplumbanes were crystalline substances and were successfully purified by recrystallization.

All the R_3PbPbR_3 disproportionated with the formation of R_4Pb in up to 90% yields[1170] and to Pb during the thermolysis. The starting temperature for this process depended on the nature of R. As for hexaalkyldiplumbanes, Calingaert and coworkers[1373] (1942) reported that they similarly decomposed even on distillation. According to Krause and Reissaus[1122] (1922), hexaaryldiplumbanes decomposed around their melting points of 117 °C ($R = 3\text{-}MeC_6H_4$) and 255 °C ($R = 1\text{-}C_{10}H_7$, $2,4,6\text{-}Me_3C_6H_2$). Gilman and Bailie[1170] (1939) found that the thermal stability of R_3PbPbR_3 increased in the following order for R: Me < Et < Ph < $3\text{-}MeC_6H_4$ < $4\text{-}MeC_6H_4$ < $4\text{-}MeOC_6H_4$ < $4\text{-}EtOC_6H_4$ $\ll$ $2\text{-}MeC_6H_4$ < $2\text{-}MeOC_6H_4$ < $2\text{-}EtOC_6H_4$ < $c\text{-}C_6H_{11}$, $2,4,6\text{-}Me_3C_6H_2$ < $1\text{-}C_{10}H_7$.

In 1951–1963, a number of investigations established that the thermolysis of hexaorganyldiplumbanes is catalyzed by silica[1298] (1951), activated charcoal[1299] (1956), $AlCl_3$[1295] (1960), as well as by lead, which is formed during the thermolysis process[1300] (autocatalysis) and also by UV irradiation[1237] (1963).

In 1960, Razuvaev, Vyazankin and Chshepetkova[903] found that $Et_3PbPbEt_3$ decomposed with a Pb–Pb bond cleavage in the presence of a catalytic amount of free-radical initiators such as benzoyl peroxide or tetraacetoxyplumbane at room temperature.

In 1942, Calingaert and coworkers[1373] showed that the 1 : 1 $Me_3PbPbMe_3$–$Et_3PbPbEt_3$ system gave at 100 °C a mixture of tetraalkylplumbanes of the following composition (%): Me_4Pb (18), Me_3PbEt (15), Me_2PbEt_2 (23), $MePbEt_3$ (31), Et_4Pb (13). The yield of lead was 5% of the theoretical calculated value. These data indicated that during the thermolysis of hexaalkyldiplumbanes, alkyl radicals, the corresponding Pb-centered free radicals as well as dialkylplumbylenes Alk_2Pb were formed. Indeed, in 1959 Razuvaev and coworkers[1202] established that the thermal decomposition of hexaethyldiplumbane proceeded in accordance with equations 33 and 34.

$$Et_3PbPbEt_3 \longrightarrow Et_4Pb + Et_2Pb \qquad (33)$$

$$Et_2Pb \longrightarrow 2Et^{\bullet} + Pb \qquad (34)$$

One year later[1203] they also studied the kinetics of the thermolysis of mixtures of $Et_3PbPbEt_3$ with Et_4Pb or with Et_2Pb at 135 °C. The data confirmed that the process proceeded according to equations 33 and 34. As a result of their investigations they concluded that the thermal decomposition of $Et_3PbPbEt_3$ was different from its disproportionation reaction, which occurred in the presence of catalysts.

In 1962, Razuvaev and coworkers[1383] studied the decomposition of $Ph_3PbPbPh_3$ in solutions and in the presence of metal salts. Krebs and Henry[1337] studied the same reaction in boiling MeCOOH. Belluco and Belluco[1385] used a radiochemical method to show that the intermediate of the thermolysis was diphenylplumbylene Ph_2Pb. As early as 1860 Klippel[1217,1218] observed the photochemical decomposition of hexaorganyldiplumbanes. He found that $Et_3PbPbEt_3$ decomposed under light and isolated metallic lead.

In 1919, Krause and Schmitz[1120] observed that the yellow color of the solution of R_3PbPbR_3 ($R = 2,4\text{-}Me_2C_6H_3$) quickly disappeared under sunlight to give a white precipitate. They concluded that the compound obtained decomposed under light irradiation. Two years later Krause[1121] reported that hexacyclohexyldiplumbane was also decomposed by light, but it was absolutely stable in the dark. According to Krause and Reissaus[1122,1292] its molecular weight was decreased when it was diluted in benzene. Analogously, the molecular weights of R_3PbPbR_3 with R = Ph and 4-MeC_6H_4 depended on the concentration of their solutions. In 1923, Calingaert and coworkers[1140] reached the same conclusion. Lesbre and coworkers[1171] determined cryoscopically the molecular weight of hexamesityldiplumbane. However, EPR data indicated that this compound did not dissociate into free radicals $R_3Pb^\bullet$[1382] in benzene. An EPR study of R_3PbPbR_3 in the crystal and in solutions in C_6H_6 and $CHCl_3$ also did not detect any dissociation into free radicals[1385]. Willemsens[109] tried to ascribe the difference between the EPR and the cryoscopic data to the imperfection of the latter method. However, this explanation does not stand up to criticism because an analogous decrease of the molecular weight in dilute solutions of hexaorganyldistannanes R_3SnSnR_3 was established as well by ebullioscopy (see Section III.H). It must be assumed that the decrease of the molecular weight of hexaorganyldiplumbanes in dilute solutions was not caused by their dissociation into free radicals $R_3Pb^\bullet$, but was caused by their decomposition into R_4Pb and R_2Pb. In accordance with that, Razuvaev and coworkers[1202] observed that the concentration of Et_4Pb, which was usually presented in $Et_3PbPbEt_3$ increased with time.

A pale yellow or pink color[109] indicated the presence of R_2Pb in the solution of R_3PbPbR_3 in organic solvents.

In 1943, Hein and coworkers[1314] studied the auto-oxidation process of hexacyclohexyldiplumbanes and found that it took place only under ultraviolet irradiation. Obviously, this observation allowed Peters[1386] to patent the use of this compound for the preparation of photosensitive films in 1961. In 1961–1963, Aleksandrov and coworkers[1310,1387,1388] investigated in detail the oxidation of $Et_3PbPbEt_3$ by oxygen at low temperatures. The final products of this reaction were Et_3PbOH, C_2H_6, C_2H_4 and PbO, and $Et_3PbOPbEt_3$ was the intermediate. Aleksandrov and coworkers[1303] (1959) studied the oxidation of $Et_3PbPbEt_3$ by α-hydroperoxoisopropylbenzene $HOOCMe_2Ph$ and 1,4-bis(α-hydroperoxoisopropyl)benzene 1,4-$(HOOCMe_2)_2C_6H_4$, which led to Et_3PbOH formation. In the first case $Et_3PbOOCMe_2Ph$ and in the second $(Et_3PbOOCMe_2)_2C_6H_4$ were formed. The two compounds were the first organolead peroxides. The reaction of $Et_3PbPbEt_3$ with $Et_3PbOOCMe_2Ph$ led to $Et_3PbOPbEt_3$ and $Et_3PbOCMe_2Ph$. The oxidation product of hexaethyldiplumbane by benzoyl peroxide was $Et_3PbOOCPh$. In 1960, Razuvaev and coworkers[903] found that the Pb–Pb bond in $Et_3PbPbEt_3$ was cleaved by MeCOOH to give $Et_3PbOOCMe$.

The reactions studied above were nonradical because they could not be initiated by AIBN. This suggested that a concerted cleavage of the Pb–Pb bond took place in the cyclic intermediate as shown in structure **1**. According to Austin[1253] (1931) and Bähr[1255] (1947), the R_3PbPbR_3 oxidation by $KMnO_4$ led to R_3PbOH. In 1959, Podall and coworkers[1379] established that the hydrogenolysis of $Ph_3PbPbPh_3$ led to metallic lead formation, as well as to Ph_4Pb or Ph_2 (depending on the reaction conditions and the catalyst used).

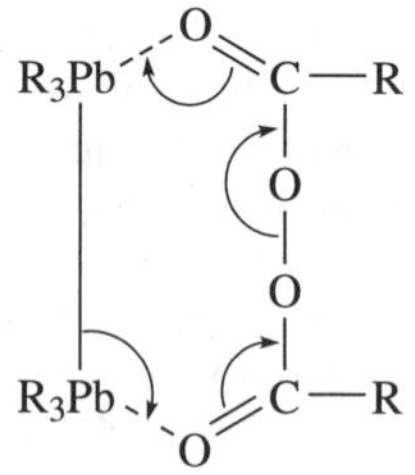

(1)

As early as in 1856, Klippel[1217,1218] carried out the Pb−Pb bond cleavage by halogens. He found that hexaethyl- and hexaisoamyldiplumbane reacted easily with iodine in ether, to form R_3PbI (R = Et, *i*-Am). In 1919, Krause and Schmitz[1120], by reacting R_3PbPbR_3 (R = 2,5-$Me_2C_6H_3$) with bromine, confirmed that hexaorganyldiplumbanes decomposed by halogens. When pyridine was used as the solvent, R_3PbBr was formed, but when chloroform was used the product was R_2PbBr_2. After 2–3 years, in the Krause laboratory, the cleavages of R_3PbPbR_3 by bromine or iodine when R = *c*-C_6H_{11}[1121], Ph or 4-MeC_6H_4[1122] were studied and the corresponding R_3PbX (X = Br, I) were obtained in good yield. In the period 1931–1961, some reports had appeared about the halogenation of the R_3PbPbR_3 series with R = Ar[791,1170,1389], $PhCH_2CH_2$[833], *c*-C_6H_{11}[791,1187]. Depending on the reaction conditions R_3PbX, R_2PbX_2 and PbX_2 were prepared in different ratios.

In 1964, Willemsens and van der Kerk[1377] found that reaction of $(Ph_3Pb)_4Pb$ with iodine led to Ph_3PbI and PbI_2, thus confirming the branched structure of the compound. Remarkably, even in 1947 Kocheshkov[156] considered the Pb−Pb bond cleavage in R_3PbPbR_3 as an oxidation reaction of trivalent lead (R_3Pb) which gave the tetravalent R_3PbX derivatives.

In 1923, Calingaert and coworkers[1140] showed that $Et_3PbPbEt_3$ was cleaved by HCl with the formation of Et_3PbCl, $PbCl_2$ and C_2H_6. In 1931, Austin[1262] obtained $(2\text{-}MeC_6H_4)_3PbBr$ by the cleavage of hexa-*ortho*-tolyldiplumbane by HBr.

In 1939, the R_3PbPbR_3 cleavage by hydrohalic acids was frequently used to form R_3PbX[791,1170,1254]. Belluco and coworkers[1390] (1962) as well as Krebs and Henry[1337] (1963) concluded that the reaction of R_3PbPbR_3 with hydrohalic acid was not a single-stage process because R_3PbH was not formed. In their opinion, the process was more complicated and could be described by Scheme 2 (for X = Cl). The general equation of the process was equation 35.

$$R_3PbPbR_3 \rightleftharpoons [R_3Pb] \longrightarrow R_4Pb + [R_2Pb]$$

$$R_4Pb + HX \longrightarrow R_3PbX + RH$$

$$[R_2Pb] + 2HX \longrightarrow R_2PbX_2 + 2RH$$

SCHEME 2

$$R_6Pb_2 + 3HX \longrightarrow R_3PbX + PbX_2 + 3RH \qquad (35)$$

In 1964, Emeleus and Evans[1391] found that the C−Pb bond was the first to be cleaved and the Pb−Pb bond was cleaved next in the reactions of HCl with R_3PbPbR_3. The process of formation of $PbCl_2$ was unclear and hence the reaction mechanism was represented by the two equations 36 and 37.

$$R_3PbPbR_3 + 2HCl \longrightarrow ClR_2PbPbR_2Cl \longrightarrow R_4Pb + PbCl_2 \qquad (36)$$

$$R_3PbPbR_3 + 3HCl \longrightarrow R_3PbPbCl_3 \longrightarrow R_3PbCl + PbCl_2 \qquad (37)$$

The data of Gilman and Apperson[1239] (1939) served as proof of the intermediate formation of ClR_2PbPbR_2Cl, so they proposed that the reaction of R_3PbPbR_3 with $AlCl_3$ could be described by equation 38.

$$R_3PbPbR_3 + AlCl_3 \longrightarrow ClR_2PbPbR_2Cl \longrightarrow R_4Pb + PbCl_2 \qquad (38)$$

According to a later point of view of Gilman and coworkers[1129], the mechanism of the reaction of hexaorganyldiplumbanes with aluminum chloride can be represented by Scheme 3, which is summarized by equation 39.

$$R_3PbPbR_3 \longrightarrow R_4Pb + R_2Pb$$

$$R_2Pb + 2AlCl_3 \longrightarrow PbCl_2 + 2RAlCl_2$$

$$R_4Pb + AlCl_3 \longrightarrow R_3PbCl + RAlCl_2$$

SCHEME 3

$$R_3PbPbR_3 + 3AlCl_3 \longrightarrow R_3PbCl + PbCl_2 + 3RAlCl_2 \quad (39)$$

Scheme 3 did not require the initial cleavage of the C−Pb bond by $AlCl_3$ as well as the intermediate formation of ClR_2PbPbR_2Cl, which has not yet been identified.

In 1952, Kocheshkov and Panov[1281] found that $Ar_3PbPbAr_3$ (Ar = 4-MeC_6H_4) was cleaved by HNO_3 to form Ar_3PbNO_3. An excess of HNO_3 led to $Ar_2Pb(NO_3)_2$.

Razuvaev and coworkers[903] showed that $Pb(OOCMe)_4$ cleaved $Et_3PbPbEt_3$ in benzene media with a formation of $Et_3PbOOCMe$ in 1960. In 1963, Krebs and Henry[1337] found that the Pb−Pb bond in R_3PbPbR_3 was cleaved by the reaction of MeCOOH, MeCOSH, S and $BrCH_2CH_2Br$. In 1943, Hein and coworkers[1314] studied the reaction of hexacyclohexyldiplumbane with polyhalomethanes. The organolead products of this reaction were R_3PbX, R_2PbX_2 (R = *c*-C_6H_{11}) and PbX_2.

Krohn and Shapiro[1392] (1951) patented the cleavage reaction of R_3PbPbR_3 by alkyl halides as a method for the preparation of R_4Pb (R = Alk) in a high yield from R_3PbPbR_3 and RX (X = Br, I) at 20–100 °C. In 1960, Razuvaev, Vyazankin and their coworkers[1293,1389] investigated thoroughly the reaction of hexaethyldiplumbane with organobromides. They found that the reaction of $Et_3PbPbEt_3$ with EtBr, $BrCH_2CH_2Br$ and $BrCH_2CHBrCH_3$ led to Et_4Pb as well as to Et_3PbBr, Et_2PbBr_2, $PbBr_2$ and Pb in heptane media at 40–70 °C. When used in catalytic amounts, the bromides initiated the disproportionation of $Et_3PbPbEt_3$ into Et_4Pb and Pb.

In 1860, Klippel[1217,1218] found that the reaction of $Et_3PbPbEt_3$ with $AgNO_3$ in alcoholic media led to Et_3PbNO_3 and metallic silver. According to Krause and Grosse[155], during the reaction of hexaorganyldiplumbanes with $AgNO_3$ in alcohols at low temperature the reaction mixture became green colored, which was attributed to the formation of R_3PbAg. In 1960–1961, Belluco and coworkers[1374] and Duffy and Holliday[1359] studied the reaction of $Et_3PbPbEt_3$ with an alcoholic solution of $AgNO_3$ at room temperature, from which triethylplumbyl nitrate and metallic silver were isolated. Thus, they reproduced the results of Klippel[1217,1218] one hundred years later.

In 1931–1962 the reactions of the R_3PbPbR_3 cleavage by chlorides of Cu[1393,1394], Au[1394], Hg[1394], Al[1239,1395], Ti[1176] and Fe[1256,1394] were studied. In 1939, Gilman and Bailie[1170] found that sterically hindered R_3PbPbR_3 with R = 2-MeC_6H_4, 2,4,6-$Me_3C_6H_2$ and *c*-C_6H_{11} were cleaved with MgI_2 (Mg itself did not apparently exhibit any effect) giving R_3PbI. Unlike this, the reaction of R_3PbPbR_3 having no bulky substituents with a MgI_2−Mg system led to R_4Pb, Pb and RMgI. Probably, it proceeded through an intermediate formation of R_3PbI and R_2Pb. In 1963, Belluco and coworkers[1396] studied the cleavage of $Et_3PbPbEt_3$ by chlorides and oxychlorides of sulfur. It was found that the yield of Et_3PbCl was lower as the nucleophilicity of the sulfur atom increased, i.e. in the order: $SO_2Cl_2 > SOCl_2 > SCl_2 > S_2Cl_2$.

An unexpected addition of R_3PbPbR_3 to multiple bonds was reported by Gilman and Leeper[316] in 1951. They suggested that the reaction of $Ph_3PbPbPh_3$ with maleic anhydride led to 2,3-bis(triphenylplumbyl)succinic anhydride. However, in 1964, Willemsens[109] noted that the product of the reaction was apparently diphenylplumbylen maleate formed from an admixture of maleic acid, which was present in its anhydride. This conclusion was corroborated by the absence of any reaction of $Ph_3PbPbPh_3$ with pure maleic anhydride. The formation of diphenylplumbylenmaleate (along with Ph_4Pb) was assumed to result from decomposition of an intermediate product bis(triphenylplumbyl) maleate.

In 1941, Bindschadler and Gilman[1182] showed that PhLi cleaved $Ph_3PbPbPh_3$ with formation of Ph_3PbLi and Ph_4Pb. Gilman and Bailie[791,1170] (1939) and Foster and coworkers[1189] (1939) found that the reaction of $Ar_3PbPbAr_3$ with Na in liquid ammonia led to Ar_3PbNa, whose solution was dark-red colored. It was found in 1941–1953 that hexaphenyldiplumbane was similarly cleaved by alkali and alkali earth metals (Li, K, Rb, Ca, Sr, Ba) in liquid ammonia at the Gilman[316,1182,1378] laboratory. Hein and coworkers[118,1397] (1942, 1947) found that sodium in ether media cleaved hexacyclohexyldiplumbane. In 1962, Tamborski and coworkers[1398] showed that $Ph_3PbPbPh_3$ was cleaved by Li in THF to form Ph_3PbLi in a high yield.

In 1922, Krause and Reissaus[1122] succeeded in isolating two monomers of diarylplumbylenes Ar_2Pb (Ar = Ph, 2-MeC_6H_4) in about 4% yield by the reaction of $PbCl_2$ with ArMgBr at 2 °C. For a long period they were the only representatives of organic compounds of two-valent Pb. Unlike analogous compounds of the other elements of the silicon subgroup R_2M (M = Si, Ge, Sn), diarylplumbylenes could not be transformed into oligomers or polymers of the $(R_2M)_n$ type, but they easily disproportionated into Ar_4Pb and Pb at about 2 °C. These data became additional proof of the inability of lead to form chains longer than Pb–Pb–Pb.

K. Biological Activity and Application of Organolead Compounds

Even the first investigators of organolead compounds encountered its harmful physiological action. Thus, for example, in 1860 Klippel[1217,1218] reported that the vapors of hexaethyldiplumbane affected the mucous membranes and respiratory tract and caused a lachrymatory action and prolonged cold. Similarly, the hexaisoamyldiplumbane vapors irritated the mucous membranes. Klippel even tasted this substance and found that it caused a long-time scratching irritation of his tongue and even of his throat. It must be assumed that trialkylplumbanols, which were formed in the reaction of R_3PbPbR_3 with moisture from the air and CO_2, caused all these symptoms. Krause and Pohland[1123] (1922) felt the irritation action of the R_3PbX (R = Alk) dust. Browne and Reid[1250] (1927) and Gilman and coworkers[1248,1399] (1930, 1931) found that the organolead compounds of the Et_3PbX type showed sternutatory and irritating actions and caused rhinitis symptoms.

In the end of the 1940s McCombie and Saunders synthesized large amounts of Et_3PbCl and felt the symptoms of a severe attack of influenza, which, however, disappeared at night and returned by day[971]. High toxicity was the main effect of organolead compounds on living organisms. Obviously, the first researchers in the field felt this. It is noteworthy that the organolead derivatives turned out to be more toxic than inorganic lead compounds and even pure lead. From 1925 the toxicity of tetraethyllead started to be studied thoroughly because of its wide application as an antiknock of motor fuels[1400,1401]. The toxic and physiological action of Et_4Pb and other organolead compounds was summarized in several monographs and reviews[109,130,154,1402–1405]. The majority of these investigations were carried out in the second half of the 20th century.

Already in the first half of the 20th century, it was established that the first symptoms of Et_4Pb poisoning were a drop in body temperature, a marked decrease in blood pressure, sleeplessness, headaches, nightmares and hallucinations. Higher doses of tetraethyllead caused insanity. The indicated emotional and nervous deviations indicated that the lipid-soluble Et_4Pb was absorbed rapidly by the soft and nervous tissues and concentrated in the latter. In 1925, Norris and Gettler[1406] found that a high concentration of lead occurred in brain, liver and kidney tissues. It was also established that tetraethyllead was able to penetrate human or animals through the integuments or by breathing its vapors. Extra large doses of Et_4Pb (in comparison with other highly toxic substances) caused a lethal outcome. Tetraethyllead was used as a poison in the mystery novel of Ellery Queen 'The Roman Hat Mystery'. The chronic effect of small doses of tetraethyllead due, for instance, to long respiration of its vapors or a lasting contact with its solutions in motor fuel (ethylated gasoline) resulted in serious poisoning. Removal of tetraethyllead and its metabolites from the body occurred very slowly owing to the resistance of Et_4Pb to hydrolysis and the insolubility of the resulting inorganic lead compounds in tissue liquids. Like tetraalkylstannanes, the toxicity of Et_4Pb depended on the cleavage of one C−Pb bond *in vivo* which resulted in the formation of the highly toxic cation Et_3Pb^+[1407].

An international arms race started shortly after World War II and was concerned with the creation of new types of chemical weapons, which inspired many prominent scientists in the USA, England, USSR and other countries to conduct investigations in this field. Organolead compounds were also involved in such studies and a search of their suppressing effect on human disturbances was started. In 1939–1941, Saunders in England carried out secret and extensive investigations for the Ministry of Supply with the aim of creating chemical weapons based on organolead compounds, having sternutatory and irritation action. Detailed data about these investigations were published[1149,1150,1320,1408–1410] in 1946–1950. They synthesized many compounds of the R_3PbX and R_2PbX_2 series. Remarkably, the authors and their coworker-volunteers tested the effects on themselves. They entered a special room, where an alcoholic solution of a tested compound in several concentrations was dispersed. The activity of the compound was determined by the time that the investigators could stay in the room. It was established that the derivatives of the R_3PbX (R = Alk; X = Hal, OH, OR′, OOCR′, SR′, $NHSO_2R'$, OCN, CN, SCN, $N(CO)_2C_6H_4$-*o* etc.) type were sternutators and irritating agents. The influence rate of the alkyl substituents R and X on the irritating effect of R_3PbX[971] was also studied. On the whole, the activity of these compounds was increased for the following R substituents in the order: Me < Et < Bu < Pr. Hence, the Pr_3PbX compounds turned out to be the most active. Their representatives with X = $OOCCH{=}CH_2$, OOCCH=CHMe, $OOCCH_2CH_2Cl$, $N(CO)_2C_6H_4$-*o*, $NHSO_2C_6H_4Me$-*p* and $NHSO_2Me$ were the most effective among the compounds mentioned above, and their unbearable concentration in air was lower than 1 ppm. The most powerful sternutators were $Pr_3PbNHSO_2R$ with R = $CH{=}CH_2$ and Ph, with an unbearable concentration of 0.1 ppm. All the investigators ran out of the room after 40 seconds when the compound with R = Ph in the mentioned concentration was spread. Compounds of Ar_3PbX and R_2PbX_2 type had no effect at all or a little sternutatory action. It is noteworthy that the investigations of McCombie and Saunders, which had doubtless nonhumane but pragmatic aims, made a valuable contribution to the chemistry of organolead compounds. Their work resulted in the synthesis of many new substances of this class and led to new developments or improvements of their preparative methods. Analogous investigations were carried out in Gilman's laboratory on the other side of the Atlantic Ocean. The results were published in an article by Gilman, Spatz and Kolbezen[1308] only in 1953.

In 1928–1929, Evans and coworkers[1411] and Krause[1126] started to investigate the possible use of organolead compounds as medicines, mainly against cancer.

In 1938, Schmidt[1304] examined their application against cancer from a historical aspect. He obtained many complex lead compounds of different types which did not have the C–Pb bond. Along with them a series of organolead compounds Me_2PbCl_2, $Me_2Pb(OH)_2$, Me_3PbCl, Ph_4Pb, $(2,5\text{-}Me_2C_6H_3)_3PbPb(C_6H_3Me_2\text{-}2,5)_3$, $Ph_2Pb(OOCMe)_2$, $Ph_2Pb(OH)_2$, $(p\text{-}O_2NC_6H_4)_2Pb(OH)_2$ and $(p\text{-}H_2NC_6H_4)_2Pb(OH)_2$ was synthesized. These compounds were transformed into the corresponding water-soluble Na-aryllead pyrocatecholdisulfonates by the reaction with Na-pyrocatecholdisulfonate. Carcinogenic activities of the above seventeen synthesized compounds mentioned above were studied on mice carcinoma and partially on Brown–Pearce tumors. From all the compounds studied only the above-mentioned diarylsulfonatoplumbanes had a definite carcinogenic action. Testing radioactive lead compounds did not confirm the expected high activity. However, comparatively insufficient investigations in this field as well as studies of the effect of organolead compounds on plants and the possibility of using them in plant cultivation, as well as their use as components of antifouling paints, appeared only after 1970[1412]. Nevertheless, even in 1952–1953 *N*-triethylplumbyl derivatives of phthalimide and phthalohydrazide were patented as fungicides[1344,1345]. In 1959, a patent for the application of triethyl(diisobutylamino)plumbane as a herbicide[1346] was granted.

Practical use of organolead compounds will be hardly extended due to their high toxicity and the possibility of sustainable pollution of the environment by the lead compounds. In this connection it must be indicated that the production of tetraethyllead, which achieved 270,000 tons by 1964 only in the USA, started to be reduced at the end of the 20th century.

In the second half of the past century, there were numerous patents dealing with the application of organolead compounds as polymerization catalysts or as pesticides[109]. However, they did not find any practical application. Regarding the same is true of Me_4Pb, which began to be used as an antiknock additive along with Et_4Pb in the 1960s.

V. CONCLUSION

The concepts and development of the chemistry of organic compounds of Group 14 of the Periodic Table heavy elements, i.e. germanium, tin and lead, are presented in a historical sequence in the earlier sections of this chapter. We have tried to tell the reader not only about the achievements of researchers in this field of organometallic chemistry, but also to give the names of pioneer researchers and their close successors. The development of organolead and organotin chemistry proceeded almost simultaneously and their study was actually synchronous in the middle of the 19th century. The investigations of organogermanium compounds were started in 1925.

The research interests in organic compounds of the elements above were not the same throughout the historical development of organometallic chemistry. The tin derivatives turned out to be the focus of interest in comparison with organogermaniums, which were less attractive, while organolead compounds attracted the least attention of scientists. Table 3 demonstrates these facts. The number of publications devoted to organic compounds of the elements of the silicon subgroup (mezoids) published in 1966 and in 1969[47] are presented. In these years the main fields of practical application of organic compounds of the silicon subgroup were determined.

It is not difficult to see that the number of published works generally corresponds to the importance of the elements in various fields of human activity. It is remarkable that the chemistry of organotin compounds was the most intensively developed in these years. In the 1960s, the rate of development of organosilicon chemistry was lower than that of the chemistry of organogermanium compounds. The dynamics of the research and progress in the field of organolead compounds both in the previous and subsequent years was relatively minimal. At the same time organolead compounds, the first of them being tetraethyllead, found practical application. There was a time when the industrial production of this antiknock additive of motor fuels exceeded the total output production

TABLE 3. The number of investigations devoted to organic compounds of the elements of the silicon subgroup, carried out in 1966 and in 1969

	Number of articles		Relative increase in the number of publications (%)
	1966	1969	
Si	615	823	34
Ge	148	208	40
Sn	207	537	159
Pb	71	82	15

of all the organotin and organogermanium compounds. At the end of the 20th century, organogermanium compounds found practical application as biologically active products.

Laboratory research on organic compounds of the silicon subgroup elements showed that they ought to be divided into two subgroups (dyads) in accordance with their similarity in chemical properties and biological activity. Silicon and germanium derivatives were placed in the first one while the tin and lead derivatives belong to the second.

Unlike this chapter, the history of organogermanium, organotin and organolead compounds has no end and will probably never have one. The initiation of various new research tendencies in this field of metalloorganic chemistry, which took shape at the end of the 20th and beginning of the 21st centuries, is a witness to this. Some of them are mentioned in Chapter 2. Nevertheless, it must be acknowledged with sorrow that the number of publications devoted to organic compounds of the elements reviewed in this chapter among the organometallic papers is decreasing more and more due to the rapidly growing interest in the transition metal organic derivatives and their complexes.

While working on this chapter, the first author recollected with pleasure, pride and a slight sadness his close acquaintance and friendly connections with many of the heroes of this narration whom he had met not only at international forums or in laboratories throughout the world, but also at home and in other everyday situations. They include H. Gilman, E. Rochow, R. West, D. Seyferth and A. MacDiarmid (USA); M. Schmidt, W. Neumann and H. Schmidbaur (Germany); K. A. Kocheshkov, A. N. Nesmeyanov, G. A. Razuvayev, O. M. Nefedov, V. F. Mironov, M. M. Koton, S. P. Kolesnikov and N. S. Vyazankin (Russia). At the same time, these reminiscences caused some sorrow in that the age of the author has become historical.

The authors cordially thank Dr. Andrey Fedorin for his extensive and valuable assistance in correcting and preparing the manuscript for publication.

This chapter is dedicated to friends and colleagues whose contribution to the organometallic chemistry of the last century was outstanding.

VI. REFERENCES

1. A. N. Egorochkin and M. G. Voronkov, *Electronic Structure of Si, Ge and Sn Compounds*, SB RAS Publishing House, Novosibirsk, 2000.
2. K. Rumpf, *Gmelin Handbook of Inorganic Chemistry*, 8th Edition, *Silicon*, Part A1, *History*, Springer-Verlag, Berlin, 1984, p. 168.
3. M. G. Voronkov, *Organosilicon Chemistry in Papers of Russian Scientists*, Leningrad University Publishing House, Leningrad, 1952, p. 103.
4. E. G. Rochow, in *Progress in Organosilicon Chemistry* (Eds. B. Marciniec and J. Chojnowski), Gordon and Breach Publishers, Amsterdam, 1995, pp. 3–15.
5. J. Y. Corey, in *The Chemistry of Organic Silicon Compounds* (Eds. S. Patai and Z. Rappoport), Part 1, Wiley, Chichester, 1989, pp. 1–56.
6. J. W. Nicholson, *J. Chem. Educ.*, **66**, 621 (1989).

7. E. A. Grebennikov and Yu. A. Ryabov, *Searches and Discoveries of the Planets*, Science, Moscow, 1975, pp. 157–175.
8. D. I. Mendeleev, *J. Russ. Phys. Chem. Soc.*, **3**, 25 (1871).
9. D. I. Mendeleev, *Ann. Suppl.*, **8**, 133 (1871).
10. D. I. Mendeleev, *Fundamentals of the Chemistry*, 11 Edn., Vol. 2, GosKhimTechIzdat, Leningrad, 1932, p. 63, 370; *Chem. Abstr.*, **27**, 1786 (1933).
11. D. I. Mendeleev, *Total Transactions Collection*, Vol. 2, GosKhimTechIzdat, Leningrad, 1934, pp. 140–215.
12. R. B. Dobrotin, *Leningrad Univ. Bull. (USSR)*, **10**, 55 (1956).
13. B. M. Kedrov, *Khim. Redkikh Elementov, Akad. Nauk SSSR*, **1**, 7 (1954).
14. J. A. R. Newlands, *Chem. News*, **10**, 59 (1864).
15. L. Meyer, *Die Modernen Theorien der Chemie und ihre Bedeutung für die chemische Statik*, 1864.
16. C. A. Winkler, *J. Prakt. Chem.*, **34**, 182 (1886).
17. C. A. Winkler, *Chem. Ber.*, **19**, 210 (1886).
18. N. A. Figurovsky, *Discovery of the Elements and the Origin of their Names*, Science, Moscow, 1970, pp. 65–66; *Chem. Abstr.*, **71**, B119983r (1971).
19. D. N. Trifonov and V. D. Trifonov, *How the Chemical Elements were Discovered*, Education, Moscow, 1980, pp. 134–136.
20. F. J. Moore, *A History of Chemistry*, State Publishing House, Moscow, 1925, p. 291.
21. F. J. Moore, *A History of Chemistry*, McGraw-Hill, New York, 1939, p. 447.
22. Records of the Russian Physical Chemical Society Session in *Zh. Rus. Fiz. Khim. Obshch.*, **18**, 179 (1886).
23. C. A. Winkler, *J. Prakt. Chem.*, **36**, 177 (1887).
24. Records of the Russian Physical Chemical Society Session in *Zh. Rus. Fiz. Khim. Obshch.*, **18**, 317 (1886).
25. D. I. Mendeleev, *The Periodic Law*, Classics Series (Ed. B. N. Menschutkin), Natural Science, Moscow, 1926, p. 180.
26. C. A. Winkler, *Zh. Rus. Fiz. Khim. Obshch.*, **18**, 4, 185 (1886).
27. E. G. Rochow, in *Comprehensive Inorganic Chemistry*, (Eds. J. S. Hailar, J. H. Emeleus, R. Nyholm and A. F. Trotman-Dickenson) Vol. 2, Pergamon Press, Oxford, 1973, p. 1–41.
28. I. R. Selimkhanov and V. V. Ivanov, in *Beginning and Development of Chemistry from the Ancient Times to the 17th Century*, Science, Moscow, 1983, p. 52–54, 57–60; *Chem. Abstr.*, **100**, 173867 (1984).
29. C. A. Winkler, *Chem. Ber.*, **20**, 677 (1887).
30. C. A. Winkler, *Chem. Ber.*, **30**, 6 (1897).
31. A. G. Brook, *Adv. Organomet. Chem.*, **7**, 95 (1968/69).
32. E. G. Rochow, *J. Appl. Phys.*, **9**, 664 (1938).
33. *Gmelins Handbuch der Anorganischen Chemie*, 8 Auflage, Germanium, Erg. Band, 45, Verlag Chemie, Weinheim, 1958.
34. A. L. Allred and E. G. Rochow, *J. Am. Chem. Soc.*, **79**, 5361 (1957).
35. A. L. Allred and E. G. Rochow, *J. Inorg. Nucl. Chem.*, **5**, 269 (1958).
36. M. Lesbre, P. Mazerolles and J. Satge, *The Chemistry of Germanium*, Wiley, London, 1971, p. 701.
37. M. Lesbre, P. Mazerolles and J. Satge, *Organic Compounds of Germanium*, World, Moscow, 1974, p. 472.
38. M. G. Voronkov and I. F. Kovalev, *Latvijas PSR Zinatny Akademijas Vêstis, Kimijas sërija*, **2**, 158 (1965); *Chem. Abstr.*, **64**, 37b (1966).
39. R. S. Drago, *J. Inorg. Nucl. Chem.*, **15**, 237 (1960).
40. W. H. Brock, *The Fontana History of Chemistry*, Harper Collins Publ., London, 1992, p. 9.
41. C. Löwig, *Ann. Chem.*, **84**, 308 (1852).
42. C. Löwig, *Ann. Chem.*, **85**, 318 (1853).
43. C. Löwig, *J. Prakt. Chem.*, **60**, 304 (1853).
44. H. Landolt, *Chem. Ber.*, **23**, 1013 (1890).
45. E. Frankland, *Ann. Chem.*, **85**, 329 (1853).
46. IUPAC, *Nomenclature of Inorganic Chemistry*, Second Edn., London, 1971.
47. Yu. Shmidt, *Metalloorganic Compounds*, Onti-KhimTeoret, Leningrad, 1937, p. 377.

48. R. Ingham, S. Rosenberg, G. Gilman and F. Rijkens, *Organotin and Organogermanium Compounds*, Foreign Literature, Moscow, 1962, p. 265.
49. P. Pascal, *Nouveau Traite de Chemie Minerale*, Vol. 8, *Germanium*, Etain, Plomb, Masson, Paris, 1963.
50. J. H. Harwood, *Industrial Applications of the Organometallic Compounds. A. Structure Survey*, Chapman and Hall, London, 1963.
51. H. D. Kaesz and F. G. Stone *Organometallic Chemistry* (Ed. H. Zeiss), World, Moscow, 1964.
52. E. Y. Lukevics and M. G. Voronkov, *Hydrosilylation, Hydrogermylation, Hydrostannylation*, Acad. Sci. Latv. SSR, Riga, 1964, p. 371; *Chem. Abstr.*, **63**, 1472a (1965).
53. E. Y. Lukevics and M. G. Voronkov, *Organic Insertion Reactions of the Group IV Elements*, Consultants Bureau, New York, 1966, p. 413; *Chem. Abstr.*, **65**, 16802f (1966).
54. K. A. Kocheshkov, N. N. Zemlyansky, N. I. Sheverdina and E. M. Panov, *Methods of Elementoorganic Chemistry. Germanium, Tin and Lead*, Science, Moscow, 1968, p. 704.
55. D. A. Kochkin and I. N. Azerbaev, *Tin- and Lead-organic Monomers and Polymers*, Science, Alma-Ata, 1968; *Chem. Abstr.*, **69**, B20211f (1969).
56. P. Poson, *Metalloorganic Compounds*, Moscow, World, 1970.
57. J. H. Harwood, *Industrial Applications of the Organometallic Compounds*, Chemistry, Leningrad, 1970.
58. R. A. Jackson, *Silicon, Germanium, Tin and Lead Radicals*, The Chemical Society, London, 1970.
59. N. A. Chumaevsky, *Vibrational Spectra of the Elementoorganic Compounds of the Group IV Elements*, Science, Moscow, 1971; *Chem. Abstr.*, **76**, B66255n (1972).
60. D. A. Armitage, *Inorganic Rings and Cages*, E. Arnold (Publishers) Ltd., London, 1972, pp. 152–266.
61. I. N. Azerbaev and D. A. Kochkin, *Organic Compounds of Tin and Lead*, Knowledge, Moscow, 1972, p. 31; *Chem. Abstr.*, **78**, B4353a (1973).
62. D. S. Matteson, *Organometallic Reaction Mechanisms of the Nontransition Elements*, Academic Press, New York, 1974, p. 353.
63. E. G. Rochow and E. W. Abel, *The Chemistry of Germanium. Tin and Lead*, Pergamon, Oxford, 1975, pp. 1–146.
64. M. J. Taylor, *Metal-to-Metal Bonded States of the Main Group Elements*, Academic Press, New York, 1975, p. 212.
65. E. A. Chernyshev, M. V. Reshetova and A. D. Volynskikh, *Silicon-, Germanium- and Tin-containing Derivatives of the Transition Elements Compounds*, NIITEKhIM, Moscow, 1975.
66. V. I. Shyraev, E. M. Stepina, T. K. Gar and V. F. Mironov, *Direct Synthesis of the Organic Compounds of Tin and Germanium*, Chemical Industry Ministry, Moscow, 1976, p. 118.
67. A. G. MacDiarmid, *Organometallic Compounds of the Group IV Elements*, Vol. 1–2, M. Dekker, New York, 1968.
68. W. P. Neumann, in *The Organometallic and Coordination Chemistry of Ge, Sn and Pb* (Eds. M. Gielen and P. Harrison), Freund Publ., Tel Aviv, 1978, p. 51.
69. G. Wilkinson, F. G. Stone and E. W. Abel (Eds.), *Comprehensive Organometallic Chemistry*, Vol. 2, Chap. 11, Pergamon, Oxford, 1982, p. 399.
70. P. G. Harrison, *Organometallic Compounds of Germanium. Tin and Lead*, Chapman & Hall, London, 1985, p. 192.
71. E. A. Chernyshev, T. K. Gar and V. F. Mironov, *Investigations in the Chemistry of the Adamantane Structures of Silicon, Germanium and Tin*, NIITEKhIM, Moscow, 1989, p. 73.
72. E. Lukevics and L. Ignatovich (Eds.), *Frontiers of Organogermanium, -Tin and -Lead Chemistry*, Latvian Institute of Organic Synthesis, Riga, 1993, p. 346.
73. S. Patai (Ed.), *The Chemistry of Organic Germanium. Tin and Lead Compounds*, Wiley, Chichester, 1995, p. 997.
74. *Gmelin Handbook of Inorganic Chemistry*, 8th Edn., *Organogermanium Compounds*, Part 1–7, Springer-Verlag, Berlin, 1988–1997.
75. J. Eisch, *The Chemistry of Organometallic Compounds: The Main Group Elements*, McMillan, New York, 1967.
76. F. Rijkens, *Organogermanium Compounds. A Survey of the Literature*, Institute of Organic Chemistry, Utrecht, 1960.

77. F. Rijkens and G. J. van der Kerk, *Investigations in the Field of Organogermanium Chemistry*, Germanium Research Committee, TNO, Utrecht, 1964.
78. V. F. Mironov and T. K. Gar, *Organogermanium Compounds*, Science, Moscow, 1967; *Chem. Abstr.*, **68**, 114732z (1968).
79. T. K. Gar, V. F. Mironov and K. V. Praven'ko, *Bibliographic Index of Survey Literature on the Organogermanium Compounds*, MKhP, Moscow, 1977.
80. F. Glockling, *The Chemistry of Germanium*, Academic Press, London, 1969.
81. K. Asai, *Miracle Cure, Organic Germanium*, 3rd. Edn., Japan Publications Inc., Tokyo, 1980.
82. T. K. Gar and V. F. Mironov, *Biologically Active Germanium Compounds*, NIITEKhIM, Moscow, 1982; *Chem. Abstr.*, **110**, 57708f (1989).
83. S. N. Tandura, S. N. Gurkova, A. I. Gusev and N. V. Alekseev, *The Structure of the Biologically Active Germanium Compounds with Extended Coordination Sphere*, NIITEKhIM, Moscow, 1983; *Chem. Abstr.*, **104**, 95841c (1986).
84. N. Yu. Khromova, T. K. Gar and V. F. Mironov, *Germatranes and their Analogs*, NIITEKhIM, Moscow, 1985.
85. E. Ya. Lukevics and L. M. Ignatovich, *Synthesis and Reactions of Aryl- and Getarylgermatranes*, Latvian Institute of Organic Synthesis, Riga, 1986.
86. E. Ya. Lukevics, T. K. Gar, L. M. Ignatovich and V. F. Mironov, *Biological Germanium Compounds*, Zinatne, Riga, 1990.
87. *Gmelin Handbook of Inorganic Chemistry*, 8th Edn., *Organotin Compounds*, Part 1–25, Springer-Verlag, Berlin, 1975–1997.
88. J. G. Luijten and G. J. van der Kerk, *Investigation in the Field of Organotin Chemistry*, Tin Research Institute, Greenford, Middlesex, England, 1955.
89. A. J. Leusink, *Hydrostannation*, Schotanus & Jens Utrecht, Utrecht, 1966.
90. W. P. Neumann, *Die Organische Chemie des Zinns*, F. Enke Verlag, Stuttgart, 1967.
91. R. C. Poller, *The Chemistry of Organotin Compounds*, Academic Press, New York, 1970.
92. W. P. Neumann, *The Organic Chemistry of Tin*, Wiley, London, 1970.
93. A. K. Sawyer, *Organotin Compounds*, Vol. 1–3, Dekker, New York, 1970–1972.
94. D. A. Kochkin, *Syntheses and Properties of Organotin Compounds*, Kalinin State University, 1975.
95. K. D. Bos, *Organic and Organometallic Chemistry of Divalent Tin*, Drukkerij B. V. Elinkwijk, Utrecht, 1976.
96. J. Zuckerman (Ed.), *Organotin Compounds. New Chemistry and Applications*, American Chemical Society, Washington, 1976.
97. V. I. Shiryaev, V. F. Mironov and V. P. Kochergin, *Two-valent Tin Compounds in Synthesis of the Organotin Compounds*, NIITEKhIM, Moscow, 1977, p. 69.
98. V. I. Shiryaev and V. P. Kochergin, *Stannylenes as the Electrodonor Ligands in Complexes*, NIITEKhIM, Moscow, 1978, p. 30.
99. A. L. Klyatsh'itskaya, V. T. Mazaev and A. M. Parshina, *Toxic Properties of Organotin Compounds*, GNIIKhTEOS, Moscow, 1978, p. 47.
100. V. I. Shiryaev, L. V. Papevina and K. V. Praven'ko, *Bibliographic Index of the Survey Literature on the Organotin Compounds*, GNIIKhTEOS, Moscow, 1980, p. 161.
101. S. J. Blunden, P. A. Cusack and R. Hill, *The Industrial Uses of the Tin Chemicals*, Royal Society of Chemistry, London, 1985.
102. A. Rahm, J. Quintard and M. Pereype, *Tin in Organic Synthesis*, Butterworths, Stoneham, 1986, p. 304.
103. V. I. Shiryaev and E. M. Stepina, *Status and Usage Perspective of the Organotin Compounds*, NIITEKhIM, Moscow, 1988, p. 65.
104. G. Harrison, *Chemistry of Tin*, Chapman & Hall, New York, 1989.
105. J. D. Donaldson and S. M. Grimers, in *Frontiers of Organogermanium, -Tin and -Lead Chemistry* (Eds. E. Lukevics and L. Ignatovich), Latvian Institute of Organic Synthesis, Riga, 1993, pp. 29–40.
106. P. J. Smith (Ed.), *Organometallic Compounds of Bivalent Tin*, Vol. 1, Blackie Academic & Professional, New York, 1998.
107. *Gmelin Handbook of Inorganic Chemistry*, 8th Edn., *Organolead Compounds*, Part 1–5, Springer-Verlag, Berlin, 1987–1996.
108. V. V. Korshak and G. S. Kolesnikov, *Tetraethyllead*, GosKhimIzdat, Moscow, 1946.

109. L. C. Willemsens, *Organolead Chemistry*, International Lead Zinc Research Organization, New York, 1964.
110. L. C. Willemsens and G. J. van der Kerk, *Investigations in the Field of Organolead Chemistry*, International Lead Zinc Research Organization, New York, 1965.
111. H. Shapiro and F. W. Frey, *The Organic Compounds of Lead*, Interscience, New York, 1968.
112. L. V. Papevina, K. V. Praven'ko and V. F. Mironov, *Bibliographic Index of the Survey Literature on Organolead Compounds*, MinKhimProm, Moscow, 1978.
113. P. F. Runge, in *Organometallverbindungen*, Teil 1, Wissenschaftliche Verlaggeschaft, Stuttgart, 1932, pp. 654–662.
114. K. Kakimoto, K. Miyao and M. Akiba, in *Frontiers of Organogermanium, -Tin and -Lead Chemistry* (Eds. E. Lukevics and L. Ignatovich), Latvian Institute of Organic Synthesis, Riga, 1993, pp. 319–329.
115. P. M. Treichel and F. G. A. Stone, in *Advances in Organometallic Chemistry*, Vol. 1, Academic Press, New York, 1964, pp. 143–220.
116. F. J. Bajer, in *Progress in Infrared Spectroscopy*, Vol. 2, Plenum Press, New York, 1964, pp. 151–176.
117. R. Ingham and H. Gilman, in *Inorganic Polymers*, World, Moscow, 1965.
118. T. G. Brilkina and V. A. Shushunov, in *Reactions of the Metalloorganic Compounds with Oxygen and Peroxides*, Science, Moscow, 1966; *Chem. Abstr.*, **71**, 30583a (1969).
119. H. Shapiro and F. W. Frey, in *Kirk-Othmer Encyclopedia of Chemical Technology*, 2nd Edn., Vol. 12, Interscience Publishers, New York, 1967, pp. 282–299.
120. B. Aylett, in *Progress in Stereochemistry*, Vol. 4 (Eds. W. Klyne and P.B.D. de la Mare), Butterworth, London, 1969, pp. 213–371.
121. C. H. Yoder and J. J. Zuckerman, in *Preparative Inorganic Reactions*, Vol. 6 (Ed. W. L. Jolly), Wiley Interscience, New York, 1971, pp. 81–155.
122. E. W. Abel and D. A. Armitage, in *Advances in Organometallic Chemistry*, Vol. 5, Acad. Press, New York, 1967, pp. 1–92.
123. P. D. Lickiss, in *Organometallic Compounds of Bivalent Tin*, Vol. 1 (Ed. P. J. Smith), Blackie Academic & Professional, New York, 1998, pp. 176–202.
124. E. C. Baughan, *Quart. Rev.*, **7**, 10 (1953).
125. R. K. Ingham, S. D. Rosenberg and H. Gilman, *Chem. Rev.*, **5**, 459 (1960).
126. H. Schumann and M. Schmidt, *Angew. Chem.*, **77**, 1049 (1965); *Angew Chem., Int. Ed. Engl.*, **4**, 1009 (1965).
127. R. S. Drago, *Rec. Chem. Prog.*, **26**, 3, 157 (1965).
128. K. Moedritzer, *Organomet. Chem. Rev.*, **1**, 2, 179 (1966).
129. W. E. Davidsohn and M. C. Henry, *Chem. Rev.*, **67**, 73 (1967).
130. J. M. Barnes and L. Magos, *Organomet. Chem. Rev.*, **A3**, 137 (1968).
131. Yu. A. Alexandrov, *Organomet. Chem. Rev.*, **A6**, 209 (1970).
132. D. Seyferth, *Pure. Appl. Chem.*, **23**, 391 (1970).
133. C. F. Shaw and A. L. Allred, *Organomet. Chem. Rev.*, **A5**, 95 (1970).
134. T. Tanaka, *Organomet. Chem. Rev.*, **A5**, 1 (1970).
135. E. C. Pant, *J. Organomet. Chem.*, **66**, 321 (1974).
136. K. P. Butin, V. N. Shishkin, I. P. Beletskaya and O. A. Reutov, *J. Organomet. Chem.*, **93**, 139 (1975).
137. P. J. Davidson, M. F. Lappert and R. Pearce, *Chem. Rev.*, **76**, 219 (1976).
138. P. G. Harrison, *Coord. Chem. Rev.*, **20**, 1 (1976).
139. U. S. Bureau of Mines, *Minerals Yearbook*, U. S. Department of the Interior, Washington, 1949, p. 1311.
140. J. A. O'Connor, *Chem. Eng.*, **59**, 158 (1952).
141. L. M. Dennis and J. Papish, *J. Am. Chem. Soc.*, **43**, 2131 (1921).
142. J. F. Vaes, *Ann. Soc., Geol. Belg. Bull.*, **72**, 19 (1948).
143. L. McCabe, *Ind. Eng. Chem.*, **44**, 113A (1951).
144. H. Lundin, *Trans. Am. Electrochem. Soc.*, **63**, 149 (1933).
145. W. R. Orndorff, D. L. Tabern and L. M. Dennis, *J. Am. Chem. Soc.*, **49**, 2512 (1927).
146. K. Burschkies, *Angew. Chem.*, **48**, 478 (1935).
147. L. R. Gadders and E. Mack, *J. Am. Chem. Soc.*, **52**, 4372 (1930).
148. C. A. Kraus and E. A. Flood, *J. Am. Chem. Soc.*, **54**, 1635 (1932).
149. G. T. Morgan and H. D. Drew, *J. Chem. Soc.*, **127**, 1760 (1925).

150. R. Schwarz and W. Reinhardt, *Chem. Ber.*, **65**, 1743 (1932).
151. G. A. Razuvaev, B. G. Gribov, G. A. Domrachev and B. A. Salamatin, *Organometallic Compounds in Electronics*, Science, Moscow, 1972; *Chem. Abstr.*, **78**, B141643j (1973).
152. B. G. Gribov, G. A. Domrachev, B. V. Zhuk, B. S. Kaverin, B. I. Kozyrkin, V. V. Mel'nikov and O. N. Suvorova, *Precipitation of Films and Covers by Decomposition of Metalloorganic Compounds*, Science, Moscow, 1981, p. 322; *Chem. Abstr.*, **96**, B147644d (1982).
153. S. G. Ward and R. C. Taylor, in *Metal-Based Anti-Tumor Drugs* (Ed. M. F. Gielen), Freund Publ., London, 1988.
154. J. Schmidt, *Organometallverbindungen*, II Teil, Wissenschaftliche Verlaggesellschaft, 1934, p. 376.
155. E. Krause and A. von Grosse, *Die Chemie der Metalloorganischen Verbindungen*, Börnträger, Berlin, 1937, pp. 372–429.
156. A. Kocheshkov, *Synthetic Methods in the Field of the Metalloorganic Compounds of the Group IV Elements*, Russ. Acad. Sci., Moscow, 1947; *Chem. Abstr.*, **47**, 6434g (1953).
157. O. H. Johnson, *Chem. Rev.*, **48**, 259 (1951).
158. M. Dub, *Organometallverbindungen*, Vol. 2, Springer-Verlag, Berlin, 1961.
159. D. Quane and R. S. Bottei, *Chem. Rev.*, **63**, 4, 403 (1963).
160. L. M. Dennis and F. E. Hance, *J. Am. Chem. Soc.*, **47**, 370 (1925).
161. C. A. Kraus and L. S. Foster, *J. Am. Chem. Soc.*, **49**, 457 (1927).
162. J. K. Simons, E. C. Wagner and J. H. Müller, *J. Am. Chem. Soc.*, **55**, 3705 (1933).
163. J. K. Simons, *J. Am. Chem. Soc.*, **57**, 1299 (1935).
164. I. H. Lengel and V. H. Dibeler, *J. Am. Chem. Soc.*, **74**, 2683 (1952).
165. M. Dub, in *Organometallic Compounds, Methods of Synthesis, Physical Constants and Chemical Reactions*, 2nd Edn. (Ed. W. Weise), Vol. 2, Springer-Verlag, Berlin, 1967, pp. 1–157.
166. L. M. Dennis, *Z. Anorg. Chem.*, **174**, 97 (1928).
167. L. M. Dennis and W. J. Patnode, *J. Am. Chem. Soc.*, **52**, 2779 (1930).
168. W. R. Orndorff, D. L. Tabern and L. M. Dennis, *J. Am. Chem. Soc.*, **47**, 2039 (1925).
169. L. M. Dennis and F. E. Hance, *J. Phys. Chem.*, **30**, 1055 (1926).
170. D. M. Harris, W. H. Nebergall and O. H. Johnson, *Inorg. Synth.*, **5**, 72 (1957).
171. Z. M. Manulkin, A. B. Kuchkarev and S. A. Sarankina, *Dokl. Akad. Nauk SSSR*, **149**, 318 (1963); *Chem. Abstr.*, **59**, 5186c (1963).
172. H. Bauer and B. Burschkies, *Chem. Ber.*, **67**, 1041 (1934).
173. O. H. Johnson and D. M. Harris, *J. Am. Chem. Soc.*, **72**, 5554, 5566 (1950).
174. H. Gilman, B. Hughes and C. W. Gerow, *J. Org. Chem.*, **24**, 352 (1959).
175. W. P. Neumann and K. Kühlein, *Ann. Chem.*, **683**, 1 (1965).
176. D. Seyferth, *J. Am. Chem. Soc.*, **79**, 2738 (1957).
177. F. Glockling and K. A. Hooton, *Inorg. Synth.*, **8**, 31 (1966).
178. F. Glockling and K. A. Hooton, *J. Chem. Soc.*, 3509 (1962).
179. H. Gilmar and E. A. Zeuech, *J. Org. Chem.*, **26**, 3035 (1961).
180. J. C. Mendelsohn, F. Metras and J. Valade, *C. R. Acad. Sci., Paris*, **261**, 756 (1965).
181. H. Bauer and B. Burschkies, *Chem. Ber.*, **65**, 955 (1932).
182. E. Worral, *J. Am. Chem. Soc.*, **62**, 3267 (1940).
183. R. West, *J. Am. Chem. Soc.*, **74**, 4363 (1952).
184. H. H. Anderson, *J. Am. Chem. Soc.*, **75**, 814 (1953).
185. O. M. Nefedov and S. P. Kolesnikov, *Izv. Akad. Nauk SSSR, Ser. Khim.*, 773 (1964); *Chem. Abstr.*, **61**, 3136e (1964).
186. E. Krause and G. Renwanz, *Chem. Ber.*, **65**, 777 (1932).
187. R. Schwarz and M. Lewinsohn, *Chem. Ber.*, **64**, 2352 (1931).
188. E. A. Flood, *J. Am. Chem. Soc.*, **54**, 1663 (1932).
189. H. Bauer and B. Burschkies, *Chem. Ber.*, **66**, 1156 (1933).
190. O. H. Johnson and W. H. Nebergall, *J. Am. Chem. Soc.*, **70**, 1706 (1948).
191. A. D. Petrov, V. F. Mironov and I. E. Dolgy, *Izv. Akad. Nauk SSSR, Ser. Khim.*, 1146 (1956); *Chem. Abstr.*, **51**, 4938 (1957).
192. A. D. Petrov, V. F. Mironov and I. E. Dolgy, *Izv. Akad. Nauk SSSR, Ser. Khim.*, 1491 (1957); *Chem. Abstr.*, **52**, 7136 (1958).
193. D. Seyferth, in *Progress in Inorganic Chemistry*, Vol. 3, Interscience, New York, 1962, pp. 129–280.
194. D. Seyferth, *Rec. Chem. Prog.*, **26**, 87 (1965).

195. A. G. Brook and H. Gilman, *J. Am. Chem. Soc.*, **76**, 77 (1954).
196. C. A. Kraus and H. S. Nutting, *J. Am. Chem. Soc.*, **54**, 1622 (1932).
197. F. Smyth and C. A. Kraus, *J. Am. Chem. Soc.*, **74**, 1418 (1952).
198. C. A. Kraus and C. L. Brown, *J. Am. Chem. Soc.*, **52**, 4031 (1930).
199. W. P. Neumann and K. Kühlein, *Tetrahedron Lett.*, 1541 (1963).
200. W. P. Neumann and K. Kühlein, *Ann. Chem.*, **702**, 13 (1967).
201. O. M. Nefedov and A. I. Ioffe, *Zh. Rus. Fiz. Khim. Obshch.*, **19**, 305 (1974); *Chem. Abstr.*, **81**, 62542 (1974).
202. O. M. Nefedov, M. N. Manakov and A. D. Petrov, *Zh. Rus. Fig. Khim. Obshch.*, **147**, 1376 (1962); *Chem. Abstr.*, **59**, 5185 (1963).
203. O. M. Nefedov, S. P. Kolesnikov and V. I. Schejtschenko, *Angew. Chem.*, **76**, 498 (1964).
204. O. M. Nefedov, M. N. Manakov and A. D. Petrov, *Plaste und Kautschuk*, **10**, 721 (1963); *Chem. Abstr.*, **60**, 12366d (1964).
205. O. M. Nefedov, G. Garzo, T. Székely and W. I. Schirjaew, *Dokl. Akad. Nauk SSSR*, **164**, 822 (1965); *Chem. Abstr.*, **64**, 2178a (1996).
206. P. P. Shorigin, W. A. Petukhov, O. M. Nefedov, S. P. Kolesnikov and V. I. Shiryaev, *Teor. Exp. Chem., Acad. Sci. Ukr.*, **2**, 190 (1966); *Chem. Abstr.*, **65**, 14660g (1966).
207. O. M. Nefedov and T. Székely, in *International Symposium on Organosilicon Chemistry Science*, Commun. Prag., 1965, p. 65.
208. V. F. Mironov, A. L. Kravchenko and A. D. Petrov, *Dokl. Akad. Nauk SSSR*, **155**, 843 (1964); *Chem. Abstr.*, **60**, 15899f (1964).
209. V. F. Mironov, A. L. Kravchenko and L. A. Leytes, *Izv. Akad. Nauk SSSR, Ser. Khim.*, 1177 (1966); *Chem. Abstr.*, **65**, 16997e (1966).
210. E. J. Bulten and J. G. Noltes, *J. Organomet. Chem.*, **16**, 8 (1969).
211. W. Metlesics and H. Zeiss, *J. Am. Chem. Soc.*, **82**, 3321 (1960).
212. R. Schwarz and E. Schmeisser, *Chem. Ber.*, **69**, 579 (1936).
213. W. Metlesics and H. Zeiss, *J. Am. Chem. Soc.*, **82**, 3324 (1960).
214. S. N. Glarum and C. A. Kraus, *J. Am. Chem. Soc.*, **72**, 5398 (1950).
215. C. A. Kraus, *J. Chem. Educ.*, **26**, 45 (1949).
216. H. Gilman and C. W. Gerow, *J. Am. Chem. Soc.*, **77**, 4675 (1955).
217. H. Gilman and C. W. Gerow, *J. Org. Chem.*, **22**, 334 (1957).
218. H. Gilman and C. W. Gerow, *J. Am. Chem. Soc.*, **77**, 5740 (1955).
219. H. Gilman and C. W. Gerow, *J. Am. Chem. Soc.*, **77**, 5509 (1955).
220. H. Gilman and C. W. Gerow, *J. Am. Chem. Soc.*, **79**, 342 (1957).
221. C. W. Gerow, *Iowa State Coll. J. Sci.*, **31**, 418 (1957).
222. J. George, J. Peterson and H. Gilman, *J. Am. Chem. Soc.*, **82**, 403 (1960).
223. J. G. Milligan and C. A. Kraus, *J. Am. Chem. Soc.*, **72**, 5297 (1950).
224. C. A. Kraus and C. S. Scherman, *J. Am. Chem. Soc.*, **55**, 4694 (1933).
225. R. B. Booth and C. A. Kraus, *J. Am. Chem. Soc.*, **74**, 1418 (1952).
226. C. A. Kraus and C. L. Brown, *J. Am. Chem. Soc.*, **52**, 3690 (1930).
227. C. A. Kraus and W. K. Nelson, *J. Am. Chem. Soc.*, **56**, 195 (1934).
228. M. P. Brown and G. W. Fowles, *J. Chem. Soc.*, 2811 (1958).
229. H. Gilman and C. W. Gerow, *J. Am. Chem. Soc.*, **78**, 5823 (1956).
230. O. H. Johnson and W. H. Nebergall, *J. Am. Chem. Soc.*, **71**, 1720 (1949).
231. N. S. Vyazankin and O. A. Kruglaya, *Usp. Khim.*, **35**, 8, 1388 (1966); *Chem. Abstr.*, **67**, 100168n (1967).
232. N. S. Vyazankin, E. N. Glagyshev, G. A. Razuvaev and S. P. Korneeva, *Zh. Obshch. Khim.*, **36**, 952 (1966); *Chem. Abstr.*, **65**, 8955f (1966).
233. N. S. Vyazankin, G. A. Razuvaev, E. N. Glagyshev and S. P. Korneeva, *J. Organomet. Chem.*, **7**, 357 (1966).
234. N. S. Vyazankin, G. A. Razuvaev, V. T. Bychkov and V. L. Zvezdin, *Izv. Akad. Nauk SSSR, Ser. Khim.*, 562 (1966); *Chem. Abstr.*, **65**, 5483b (1966).
235. H. Gilman and C. W. Gerow, *J. Am. Chem. Soc.*, **78**, 5435 (1956).
236. C. Tamborski, F. E. Ford, W. L. Lehn, G. J. Moore and E. Soloski, *J. Org. Chem.*, **27**, 619 (1962).
237. J. Gross and F. Glockling, *J. Chem. Soc.*, 4125 (1964).
238. F. Glockling, *Quart. Rev.*, **20**, 45 (1966).
239. L. Summers, *Iowa State Coll. J. Sci.*, **26**, 292 (1952); *Chem. Abstr.*, **47**, 8673 (1953).

240. G. Jacobs, *C. R. Acad. Sci., Paris*, **238**, 1825 (1954).
241. M. B. Hughes, *Diss. Abstr.*, **19**, 1921 (1958); *Chem. Abstr.*, **53**, 9025a (1959).
242. A. G. Brook and G. J. Peddle, *J. Am. Chem. Soc.*, **85**, 2338 (1963).
243. G. Tai, H. Mook and H. Gilman, *J. Am. Chem. Soc.*, **77**, 649 (1955).
244. D. A. Nicholson and A. L. Allred, *Inorg. Chem.*, **4**, 1747 (1965).
245. A. G. Brook, M. A. Quigley, G. J. Peddle, N. V. Schwartz and C. M. Wagner, *J. Am. Chem. Soc.*, **82**, 5102 (1960).
246. H. Gilman, O. L. Marrs, W. J. Trepka and J. W. Diehl, *J. Org. Chem.*, **27**, 1260 (1962).
247. H. Schumann, K. F. Thom and M. Schmidt, *J. Organomet. Chem.*, **4**, 22 (1965).
248. H. Gilman, D. Adke and D. Wittenberg, *J. Am. Chem. Soc.*, **81**, 1107 (1959).
249. F. J. Glockling and K. A. Hooton, *J. Chem. Soc.*, 1849 (1963).
250. N. S. Vyazankin, G. A. Razuvaev, S. P. Korneeva and R. F. Galiullina, *Dokl. Akad. Nauk SSSR*, **155**, 839 (1964).
251. E. G. Rochow, *J. Am. Chem. Soc.*, **72**, 198 (1950).
252. E. G. Rochow, *J. Am. Chem. Soc.*, **70**, 1801 (1948).
253. E. G. Rochow, *US Patent* 2444270 (1948); *Chem. Abstr.*, **42**, 7318b (1948).
254. E. G. Rochow, *US Patent* 2451871 (1948); *Chem. Abstr.*, **43**, 2631e (1949).
255. E. G. Rochow, R. Didchenko and R. C. West, *J. Am. Chem. Soc.*, **73**, 5486 (1951).
256. G. Ya. Zueva, A. G. Pogorelov, V. I. Pisarenko, A. D. Snegova and V. A. Ponomarenko, *Izv. Akad. Nauk SSSR, Neorg. Mat.*, 1359 (1966); *Chem. Abstr.*, **66**, 2625w (1967).
257. V. A. Ponomarenko and G. Ya. Vzenkova, *Izv. Akad. Nauk SSSR, Ser. Khim.*, 994 (1957); *Chem. Abstr.*, **52**, 4473c (1958).
258. V. F. Mironov, N. G. Dzhurinskaya and A. D. Petrov, *Izv. Akad. Nauk SSSR, Ser. Khim.*, 2095 (1961); *Chem. Abstr.*, **56**, 10176f (1962).
259. V. F. Mironov and T. K. Gar, *Izv. Akad. Nauk SSSR, Ser. Khim.*, 1887 (1964); *Chem. Abstr.*, **62**, 2787h (1965).
260. *Netherlands Patent* 106446 (1963).
261. R. Zablotna, K. Akerman and A. Szuchnik, *Bull. Acad. Polon Sci. Ser. Sci. Chim.*, **12**, 695 (1964); *Chem. Abstr.*, **62**, 13167a (1965).
262. R. Zablotna, K. Akerman and A. Szuchnik, *Bull. Acad. Polon Sci. Ser. Sci. Chim.*, **13**, 527 (1965); *Chem. Abstr.*, **64**, 5131b (1966).
263. R. Zablotna, K. Akerman and A. Szuchnik, *Bull. Acad. Polon Sci., Ser. Sci. Chim.*, **14**, 731 (1966); *Chem. Abstr.*, **66**, 76108 (1967).
264. R. Zablotna, K. Akerman and A. Szuchnik, *Poland Patent* 51899; *Chem. Abstr.*, **67**, 108762h (1967).
265. A. K. Fischer, R. C. West and E. G. Rochow, *J. Am. Chem. Soc.*, **76**, 5878 (1954).
266. R. Riemschneider, K. Menge and P. Z. Klang, *Z. Naturforsch.*, **11b**, 115 (1956).
267. R. Fuchs and H. Gilman, *J. Org. Chem.*, **22**, 1009 (1957).
268. R. H. Menn and H. Gilman, *J. Org. Chem.*, **22**, 684 (1957).
269. V. A. Ponomarenko, G. Ya. Vzenkova and Yu. P. Egorov, *Dokl. Akad. Nauk SSSR*, **122**, 405 (1958); *Chem. Abstr.*, **53**, 112e (1959).
270. A. D. Petrov, V. F. Mironov and N. G. Dzhurinskaya, *Dokl. Akad. Nauk SSSR*, **128**, 302 (1959); *Chem. Abstr.*, **54**, 7546g (1960).
271. V. F. Mironov, N. G. Dzhurinskaya and A. D. Petrov, *Dokl. Akad. Nauk SSSR*, **131**, 98 (1960); *Chem. Abstr.*, **54**, 11977h (1960).
272. N. G. Dzhurinskaya, A. D. Petrov and V. F. Mironov, *Dokl. Akad. Nauk SSSR*, **138**, 1107 (1961); *Chem. Abstr.*, **55**, 24544d (1961).
273. V. F. Mironov and T. K. Gar, *Izv. Akad. Nauk SSSR, Ser. Khim.*, 855 (1965); *Chem. Abstr.*, **63**, 5666g (1965).
274. G. Manuel and P. Mazerolles, *Bull. Soc. Chim. France*, 2715 (1966).
275. M. Lesbre, J. Satge and M. Massol, *C. R. Acad. Sci., Paris*, **256**, 1548 (1963).
276. J. Satge, *Ann. Chim. (France)*, **6**, 519 (1961).
277. M. Lesbre and J. Satge, *C. R. Acad. Sci., Paris*, **247**, 471 (1958).
278. M. C. Henry and M. F. Downey, *J. Org. Chem.*, **26**, 2299 (1961).
279. J. Satge and M. Lesbre, *Bull. Soc. Chim. France*, 703 (1962).
280. M. Lesbre and J. Satge, *C. R. Acad. Sci., Paris*, **254**, 1453 (1962).
281. V. F. Mironov and N. S. Fedotov, *Zh. Obshch. Khim.*, **34**, 4122 (1964); *Chem. Abstr.*, **62**, 9136 (1965).

282. V. F. Mironov and N. S. Fedotov, *Zh. Obshch. Khim.*, **36**, 556 (1966); *Chem. Abstr.*, **65**, 743 (1966).
283. D. Seyferth and E. G. Rochow, *J. Am. Chem. Soc.*, **77**, 907 (1955).
284. D. Seyferth and J. M. Burliton, *J. Am. Chem. Soc.*, **85**, 2667 (1963).
285. D. Seyferth, J. M. Burliton, H. Dertouzos and H. D. Simmons, *J. Organomet. Chem.*, **7**, 405 (1967).
286. K. Kramer and A. N. Wright, *J. Chem. Soc.*, 3604 (1963).
287. K. Kramer and A. N. Wright, *Angew. Chem.*, **74**, 468 (1962).
288. J. Satge and P. Riviere, *Bull. Soc. Chim. France*, 1773 (1966).
289. F. Rijkens, M. J. Janssen, W. Drenth and G. J. van der Kerk, *J. Organomet. Chem.*, **2**, 347 (1964).
290. A. N. Nesmeyanov, L. I. Emel'yanova and L. G. Makarova, *Dokl. Akad. Nauk SSSR*, **122**, 403 (1958).
291. M. E. Vol'pin and D. N. Kursanov, *Zh. Obschch. Khim.*, **32**, 1137 (1962); *Chem. Abstr.*, **58**, 1332c (1963).
292. M. E. Vol'pin and D. N. Kursanov, *Izv. Akad. Nauk SSSR, Ser. Khim.*, 1903 (1960); *Chem. Abstr.*, **55**, 14419e (1961).
293. M. E. Vol'pin, Yu. D. Koreshkov, V. T. Dulova and D. N. Kursanov, *Tetrahedron*, **18**, 107 (1962).
294. M. E. Vol'pin, Yu. D. Koreshkov, V. T. Dulova and D. N. Kursanov, *Zh. Obshch. Khim.*, **32**, 1137 (1962); *Chem. Abstr.*, **58**, 9111b (1963).
295. F. Johnson and R. S. Gohlke, *Tetrahedron Lett.*, 1291 (1962).
296. F. Johnson, R. S. Gohlke and W. A. Nasutavicus, *J. Organomet. Chem.*, **3**, 233 (1965).
297. M. E. Vol'pin, Yu. T. Struchkov, L. V. Vilkov, V. S. Mastryukov, V. T. Dulova and D. N. Kursanov, *Izv. Akad. Nauk SSSR, Ser. Khim.*, 2067 (1963); *Chem. Abstr.*, **60**, 5532h (1964).
298. N. G. Bokii and Yu. T. Struchkov, *Zh. Strukt. Khim.*, **7**, 133 (1966); *Chem. Abstr.*, **66**, 122 (1966).
299. M. E. Vol'pin, V. T. Dulova and D. N. Kursanov, *Izv. Akad. Nauk SSSR, Ser. Khim.*, **3**, 727 (1963); *Chem. Abstr.*, **59**, 10104e (1963).
300. M. G. Voronkov, O. G. Yarosh, G. Yu. Turkina and A. I. Albanov, *J. Organometal. Chem.*, **491**, 216 (1995); *Chem. Abstr.*, **123**, 33160p (1995); M. G. Voronkov, O. G. Yarosh, G. Yu. Turkina and A. I. Albanov, *Zh. Obshch. Khim.*, **64**, 2, 435 (1994); *Chem. Abstr.*, **121**, 205467w (1994).
301. L. I. Emel'yanova and L. G. Makarova, *Izv. Akad. Nauk SSSR, Ser. Khim.*, 2067 (1960); *Chem. Abstr.*, **55**, 13347c (1961).
302. V. F. Mironov and T. K. Gar, *Izv. Akad. Nauk SSSR, Ser. Khim.*, 587 (1963); *Chem. Abstr.*, **59**, 3941h (1963).
303. V. F. Mironov and T. K. Gar, *Izv. Akad. Nauk SSSR, Ser. Khim.*, 755 (1965); *Chem. Abstr.*, **63**, 2993g (1965).
304. G. Manuel and P. Mazerolles, *Bull. Soc. Chim. France*, 2447 (1965).
305. P. Maze, P. Mazerolles and G. Manuel, *Bull. Soc. Chim. France*, 327 (1966).
306. E. A. Flood, *J. Am. Chem. Soc.*, **55**, 4935 (1933).
307. E. A. Flood, K. L. Godfrey and L. S. Foster, *Inorg. Synth.*, **2**, 64 (1950).
308. W. J. Pope and S. J. Peachey, *Proc. Roy. Soc. (London)*, **72**, 7 (1903).
309. P. Pfeiffer and R. Lehnardt, *Chem. Ber.*, **36**, 1054 (1903).
310. H. C. Clark and C. J. Willis, *Proc. Chem. Soc.*, 282 (1960).
311. A. Tchakirian and M. Lewinsohn, *C. R. Acad. Sci., Paris*, **201**, 835 (1935).
312. A. Tchakirian, *Ann. Phys. (Paris)*, **12**, 415 (1939).
313. P. S. Poskosin, *J. Organomet. Chem.*, **12**, 115 (1968).
314. H. H. Anderson, *J. Am. Chem. Soc.*, **73**, 5800 (1951).
315. H. H. Anderson, *J. Am. Chem. Soc.*, **73**, 5440 (1951).
316. H. Gilman and R. W. Leeper, *J. Org. Chem.*, **16**, 466 (1951).
317. H. H. Anderson, *J. Am. Chem. Soc.*, **74**, 2371 (1952).
318. R. Fuchs and H. Gilman, *J. Org. Chem.*, **23**, 911 (1958).
319. P. Mazerolles, *Diss. Doct. Sci. Phys.*, Univ. Toulouse (1959).
320. P. Mazerolles and M. Lesbre, *C. R. Acad. Sci., Paris*, **248**, 2018 (1959).
321. B. M. Gladstein, V. V. Rode and L. Z. Soborovskii, *Zh. Obshch. Khim.*, **29**, 2155 (1959); *Chem. Abstr.*, **54**, 9736c (1960).

322. D. F. van de Vondel, *J. Organomet. Chem.*, **3**, 400 (1965).
323. L. A. Leytes, Yu. P. Egorov, G. Ya. Zueva and V. A. Ponomarenko, *Izv. Akad. Nauk SSSR, Ser. Khim.*, 2132 (1961); *Chem. Abstr.*, **58**, 2029a (1963).
324. N. S. Vyazankin, G. A. Razuvaev and O. S. D'yachkovskaya, *Zh. Obshch. Khim.*, **33**, 613 (1963); *Chem. Abstr.*, **59**, 1670e (1963).
325. N. S. Vyazankin, G. A. Razuvaev and E. N. Gladyshev, *Dokl. Akad. Nauk SSSR*, **155**, 830 (1964); *Chem. Abstr.*, **60**, 15901e (1964).
326. V. F. Mironov and A. L. Kravchenko, *Izv. Akad. Nauk SSSR, Ser. Khim.*, 1026 (1965); *Chem. Abstr.*, **63**, 8392e (1965).
327. H. Sakurai, K. Tominaga, T. Watanabe and M. Kumada, *Tetrahedron Lett.*, 5493 (1966).
328. E. G. Rochow and A. L. Allred, *J. Am. Chem. Soc.*, **77**, 4489 (1955).
329. G. K. Teal and C. A. Kraus, *J. Am. Chem. Soc.*, **72**, 4706 (1950).
330. V. W. Laurie, *J. Chem. Phys.*, **30**, 1210 (1959).
331. M. Onyszchuk, *Angew. Chem.*, **75**, 577 (1963).
332. C. A. Kraus and E. Carney, *J. Am. Chem. Soc.*, **56**, 765 (1934).
333. C. A. Kraus, *J. Chem. Educ.*, **29**, 8, 417 (1952).
334. R. West, *J. Am. Chem. Soc.*, **75**, 6080 (1953).
335. O. H. Johnson and L. V. Jones, *J. Org. Chem.*, **17**, 1172 (1952).
336. A. E. Finholt, A. C. Bond, K. E. Wilzbach and H. T. Schlesinger, *J. Am. Chem. Soc.*, **69**, 2692 (1947).
337. H. H. Anderson, *J. Am. Chem. Soc.*, **79**, 326 (1957).
338. O. H. Johnson, W. H. Nebergall and D. M. Harris, *Inorg. Synth.*, **5**, 76 (1957).
339. M. Lesbre and J. Satge, *Bull. Soc. Chim. France*, 789 (1959).
340. J. Satge, R. Mathis-Noel and M. Lesbre, *C. R. Acad. Sci., Paris*, **249**, 131 (1959).
341. O. H. Johnson and D. M. Harris, *Inorg. Synth.*, **5**, 74 (1957).
342. F. E. Brinckman and F. G. A. Stone, *J. Inorg. Nucl. Chem.*, **11**, 24 (1959).
343. J. E. Griffiths, *Inorg. Chem.*, **2**, 375 (1963).
344. S. Sujishi and J. Keith, *J. Am. Chem. Soc.*, **80**, 4138 (1958).
345. V. F. Mironov, V. A. Ponomarenko, G. Ya. Vsenkova, I. E. Dolgy and A. D. Petrov, in *Chemistry and Practical Application of Organosilicon Compounds*, Vol. 1 (Eds. M. G. Voronkov and S. N. Borisov), CBTI, Leningrad, 1958, p. 192.
346. J. Satge and M. Lesbre, *Bull. Soc. Chim. France*, 2578 (1965).
347. J. Satge, M. Lesbre and M. Baudet, *C. R. Acad. Sci., Paris*, **259**, 4733 (1964).
348. A. G. Brook and G. J. Peddle, *J. Am. Chem. Soc.*, **85**, 1869 (1963).
349. R. W. Bott, C. Eaborn and I. D. Varma, *Chem. Ind.*, 614 (1963).
350. N. S. Vyazankin, G. A. Razuvaev and O. S. Kruglaya, *Organomet. Chem. Rev.*, **A3**, 323 (1968).
351. M. Lesbre and J. Satge, *C. R. Acad. Sci., Paris*, **252**, 1976 (1961).
352. B. N. Dolgov, S. N. Borisov and M. G. Voronkov, *Zh. Obshch. Khim.*, **27**, 716 (1957); *Chem. Abstr.*, **51**, 16282b (1957).
353. V. F. Mironov and A. L. Kravchenko, *Dokl. Akad. Nauk SSSR*, **158**, 656 (1964).
354. P. Riviere and J. Satge, *Bull. Soc. Chim. France*, 4039 (1967).
355. G. G. Petukhov, S. S. Svireteva and O. N. Druzhkov, *Zh. Obshch. Khim.*, **36**, 914 (1966); *Chem. Abstr.*, **65**, 10460d (1966).
356. M. Massol and J. Satge, *Bull. Soc. Chim. France*, 2737 (1966).
357. H. H. Anderson, *J. Am. Chem. Soc.*, **82**, 3016 (1960).
358. E. Amberger and H. Boeters, *Angew. Chem.*, **73**, 114 (1961).
359. H. H. Anderson, *J. Am. Chem. Soc.*, **79**, 4913 (1957).
360. M. Lesbre and J. Satge, *C. R. Acad. Sci., Paris*, **254**, 4051 (1962).
361. N. F. Orlov, B. N. Dolgov and M. G. Voronkov, in *Chemistry and Practical Use of the Organosilicon Element Compounds*, Vol. 6, Rus. Acad. Sci., Leningrad, 1961, pp. 123–126; *Chem. Abstr.*, **55**, 9316a (1961).
362. J. Satge, *Bull. Soc. Chim. France*, 630 (1964).
363. V. G. Schoff and C. Z. Harzdorf, *Z. Anorg. Allg. Chem.*, **307**, 105 (1960).
364. G. V. Sorokin, M. V. Pozdnyakova, N. I. Ter-Asaturova, V. N. Perchenko and N. S. Nametkin, *Dokl. Akad. Nauk SSSR*, **174**, 376 (1967); *Chem. Abstr.*, **67**, 117351q (1967).
365. W. Gee, R. A. Shaw and B. C. Smith, *J. Chem. Soc.*, 2845 (1964).

366. N. S. Vyazankin, M. N. Bochkarev and L. P. Sanina, *Zh. Obshch. Khim.*, **36**, 166 (1966); *Chem. Abstr.*, **64**, 14212d (1966).
367. N. S. Vyazankin, M. N. Bochkarev and L. P. Sanina, *Zh. Obshch. Khim.*, **36**, 1961 (1966); *Chem. Abstr.*, **66**, 76114y (1966).
368. N. S. Vyazankin, M. N. Bochkarev and L. P. Sanina, *Zh. Obshch. Khim.*, **36**, 1154 (1966); *Chem. Abstr.*, **66**, 14212g (1966).
369. N. S. Vyazankin, M. N. Bochkarev, L. P. Sanina, A. N. Egorochkin and S. Ya. Khoroshev, *Zh. Obshch. Khim.*, **37**, 2576 (1967); *Chem. Abstr.*, **68**, 87365y (1968).
370. C. A. Kraus and C. B. Wooster, *J. Am. Chem. Soc.*, **52**, 372 (1930).
371. M. Schmidt and I. Ruidisch, *Z. Anorg. Allg. Chem.*, **311**, 331 (1961).
372. J. Ruidisch and M. Schmidt, *Chem. Ber.*, **96**, 821 (1963).
373. J. E. Griffiths and M. Onyszchuk, *Can. J. Chem.*, **39**, 339 (1961).
374. B. M. Gladstein, I. P. Kulyukin and L. Z. Soborovsky, *Zh. Obshch. Khim.*, **36**, 488 (1966); *Chem. Abstr.*, **72**, 99033s (1972).
375. E. W. Abel, D. A. Armitage and D. B. Brady, *J. Organomet. Chem.*, **5**, 130 (1966).
376. H. H. Anderson, *J. Am. Chem. Soc.*, **73**, 5439 (1951).
377. H. H. Anderson, *J. Am. Chem. Soc.*, **72**, 2089 (1950).
378. H. H. Anderson, *J. Am. Chem. Soc.*, **72**, 194 (1950).
379. H. H. Anderson, *J. Am. Chem. Soc.*, **78**, 1692 (1956).
380. E. Amberger, W. Stoeger and R. Hönigschmid-Grossich, *Angew. Chem.*, **78**, 459 (1966); *Int. Ed. Engl.*, **5**, 522 (1966).
381. H. H. Anderson, *J. Org. Chem.*, **20**, 536 (1955).
382. M. P. Brown and E. G. Rochow, *J. Am. Chem. Soc.*, **82**, 4166 (1960).
383. E. G. Rochow, *Organometallic Chemistry*, Chapman and Hall, London, 1965.
384. C. E. Trautman and H. H. Ambrose, *US Patent*, 2416360; *Chem. Abstr.*, **42**, 2760 (1948).
385. P. Mazerolles, *Bull. Soc. Chim. France*, 1907 (1962).
386. A. Tchakirian, *Ann. Chim.*, **12**, 415 (1939).
387. H. H. Anderson, *J. Am. Chem. Soc.*, **71**, 1799 (1949).
388. R. West, H. R. Hunt and R. O. Whipple, *J. Am. Chem. Soc.*, **76**, 310 (1954).
389. H. H. Anderson, *J. Org. Chem.*, **21**, 869 (1956).
390. G. J. Peddle and J. E. Ward, *J. Organomet. Chem.*, **14**, 131 (1968).
391. M. Wieber and M. Schmidt, *Z. Naturforsch.*, **18b**, 847 (1963).
392. M. Wieber and M. Schmidt, *Angew. Chem.*, **76**, 615 (1964).
393. S. Mathur, G. Chandra, A. K. Rai and R. C. Mehrotra, *J. Organomet. Chem.*, **4**, 371 (1965).
394. R. C. Mehrotra and S. Mathur, *J. Organomet. Chem.*, **7**, 233 (1967).
395. R. C. Mehrotra and S. Mathur, *J. Organomet. Chem.*, **6**, 425 (1966).
396. R. C. Mehrotra and S. Mathur, *J. Organomet. Chem.*, **6**, 11 (1966).
397. M. G. Voronkov, G. I. Zelchan and V. F. Mironov, *USSR Patent* 190897 (1967); *Chem. Abstr.*, **68**, 691 (1968).
398. M. G. Voronkov, G. I. Zelchan and V. F. Mironov, *Khim. Geterotsikl. Soedin.*, 227 (1968); *Chem. Abstr.*, **68**, 69128u (1968).
399. A. G. Davies and C. D. Hall, *Chem. Ind. (London)*, 1695 (1958).
400. A. G. Davies and C. D. Hall, *J. Chem. Soc.*, 3835 (1959).
401. A. Rieche and J. Dahlmann, *Angew. Chem.*, **71**, 194 (1959).
402. H. H. Anderson, *J. Am. Chem. Soc.*, **73**, 5798 (1951).
403. H. H. Anderson, *J. Chem. Soc.*, 900 (1955).
404. A. G. Brook, *J. Am. Chem. Soc.*, **77**, 4827 (1955).
405. H. Schmidbaur and M. Schmidt, *Chem. Ber.*, **94**, 1138 (1961).
406. H. H. Anderson, *J. Am. Chem. Soc.*, **74**, 2370 (1952).
407. H. Schmidbaur and M. Schmidt, *Chem. Ber.*, **94**, 1349 (1961).
408. B. Armer and H. Schmidbaur, *Chem. Ber.*, **100**, 1521 (1967).
409. H. Schmidbaur and M. Schmidt, *Chem. Ber.*, **94**, 2137 (1961).
410. M. Schmidt and I. Ruidisch, *Chem. Ber.*, **95**, 1434 (1962).
411. M. Schmidt and I. Ruidisch, German Patent, 1179550 (1950); *Chem. Abstr.*, **62**, 1688 (1965).
412. M. Schmidt, H. Schmidbaur and I. Ruidisch, *Angew. Chem.*, **73**, 408 (1961).
413. M. Schmidt, I. Ruidisch and H. Schmidbaur, *Chem. Ber.*, **94**, 2451 (1961).
414. I. Ruidisch and M. Schmidt, *Z. Naturforsch.*, **18b**, 508 (1963).
415. T. N. Srivastava and S. K. Tandon, *Z. Anorg. Allgem. Chem.*, **353**, 87 (1967).

416. H. Schmidbaur and H. Hussek, *J. Organomet. Chem.*, **1**, 235 (1964).
417. M. G. Voronkov, V. P. Mileshkevich and Yu. A. Yuzhelevskii, *The Siloxane Bond*, Consultants Bureau, New York, 1978.
418. M. G. Voronkov and S. V. Basenko, *Heterolytic Cleavage Reactions of the Siloxane Bond*, Soviet Scientific Reviews (Ed. M. E. Vol'pin), Vol. 15, Part 1, Harwood Academic Publishers, London, 1990, pp. 1–83.
419. A. G. Davies, P. G. Harrison and T. A. Silk, *Chem. Ind. (London)*, 949 (1968).
420. W. T. Reichle, *J. Org. Chem.*, **26**, 4634 (1961).
421. K. Moedritzer, *Inorg. Chem.*, **6**, 1248 (1967).
422. British Thomson-Houston Co. Ltd. *UK Patent* 654571 (1951); *Chem. Abstr.*, **46**, 4561b (1952).
423. E. G. Rochow, *US Patent* 2506386 (1950); *Chem. Abstr.*, **44**, 7344 (1950).
424. C. L. Brown and E. G. Rochow, *J. Am. Chem. Soc.*, **82**, 4166 (1960).
425. I. Ruidisch and M. Schmidt, *Chem. Ber.*, **96**, 1424 (1963).
426. S. Mathur, G. Chandra, A. K. Rai and R. C. Mehrotra, *J. Organomet. Chem.*, **4**, 294 (1965).
427. M. Schmidt and H. Schumann, *Z. Anorg. Allgem. Chem.*, **325**, 130 (1963).
428. M. Henry and W. E. Davidson, *Can. J. Chem.*, **41**, 1276 (1963).
429. K. Moedritzer and J. R. van Wazer, *J. Am. Chem. Soc.*, **87**, 2360 (1965).
430. K. Moedritzer and J. R. van Wazer, *Inorg. Chim. Acta*, **1**, 152 (1967).
431. B. Burschkies, *Chem. Ber.*, **69**, 1143 (1936).
432. C. W. Cumper, A. Melnikoff and A. I. Vogel, *J. Am. Chem. Soc.*, **88**, 242 (1966).
433. E. W. Abel, D. A. Armitage and D. B. Brady, *Trans. Faraday Soc.*, **62**, 3459 (1966).
434. M. Wieber and G. Schwarzmann, *Monatsch. Chem.*, **99**, 255 (1968).
435. M. C. Henry and W. E. Davidson, *J. Org. Chem.*, **27**, 2252 (1962).
436. H. Schumann, K. F. Thom and M. Schmidt, *J. Organomet. Chem.*, **1**, 167 (1963).
437. W. E. Davidson, K. Hills and M. C. Henry, *J. Organomet. Chem.*, **3**, 285 (1965).
438. E. V. van der Berghe, D. F. van der Vondel and G. P. van der Kelen, *Inorg. Chim. Acta*, **1**, 97 (1967).
439. M. Wieber and M. Schmidt, *J. Organomet. Chem.*, **1**, 336 (1964).
440. K. A. Hooton and A. L. Allred, *Inorg. Chem.*, **4**, 671 (1965); *Chem. Abstr.*, **62**, 16290a (1965).
441. M. Schmidt and I. Ruidisch, *German Patent* 1190462 (1965); *Chem. Abstr.*, **63**, 631g (1965).
442. H. Schumann, K. F. Thom and M. Schmidt, *J. Organomet. Chem.*, **2**, 361 (1964).
443. M. Henry, *J. Org. Chem.*, **28**, 225 (1963).
444. M. Schmidt and F. Ruf, *Angew. Chem.*, **73**, 64 (1961).
445. I. Ruidisch and M. Schmidt, *J. Organomet. Chem.*, **1**, 160 (1963).
446. M. Schmidt and F. Ruf, *J. Inorg. Nucl. Chem.*, **25**, 557 (1963).
447. N. S. Vyazankin, M. N. Bochkarev and L. P. Sanina, *Zh. Obshch. Khim.*, **37**, 1037 (1967); *Chem. Abstr.*, **68**, 13099f (1968).
448. N. S. Vyazankin, M. N. Bochkarev and L. P. Mayorova, *Zh. Obshch. Khim.*, **39**, 468 (1969); *Chem. Abstr.*, **70**, 115222m (1969).
449. M. N. Bochkarev, L. P. Sanina and N. S. Vyazankin, *Zh. Obshch. Khim.*, **39**, 135 (1969); *Chem. Abstr.*, **70**, 96876j (1969).
450. H. Schumann, K. F. Thom and M. Schmidt, *Angew. Chem.*, **75**, 138 (1963); *Angew. Chem., Int. Ed. Engl.*, **2**, 99 (1963).
451. P. Mazerolles, J. Dubac and M. Lesbre, *J. Organomet. Chem.*, **12**, 143 (1968).
452. N. S. Vyazankin, M. N. Bochkarev and L. P. Sanina, *Zh. Obshch. Khim.*, **38**, 414 (1968); *Chem. Abstr.*, **69**, 96844 (1968).
453. J. S. Thomas and W. W. Southwood, *J. Chem. Soc.*, 2083 (1931).
454. A. W. Laubengayer and L. Reggel, *J. Am. Chem. Soc.*, **65**, 1783 (1943).
455. O. J. Scherer and M. Schmidt, *Angew. Chem.*, **75**, 642 (1963).
456. O. J. Scherer and M. Schmidt, *J. Organomet. Chem.*, **1**, 490 (1964).
457. H. H. Anderson, *J. Am. Chem. Soc.*, **74**, 1421 (1952).
458. H. H. Anderson, *J. Am. Chem. Soc.*, **83**, 547 (1961).
459. A. I. Barchukov and A. M. Prokhorov, *Optika I Spektroskopiya*, **4**, 799 (1958); *Chem. Abstr.*, **52**, 16875e (1958).
460. M. V. George, P. B. Talukdar, C. W. Gerow and H. Gilman, *J. Am. Chem. Soc.*, **82**, 4562 (1960).

461. I. Ruidisch and M. Schmidt, *Angew. Chem.*, **76**, 229 (1964).
462. J. Satge and M. Baudet, *C. R. Acad. Sci., Paris*, **263C**, 435 (1966).
463. I. Ruidisch and M. Schmidt, *J. Organomet. Chem.*, **1**, 493 (1964).
464. J. S. Thayer and R. West, *Inorg. Chem.*, **3**, 406 (1964).
465. W. T. Reichle, *Inorg. Chem.*, **3**, 402 (1964).
466. F. Rijkens, M. J. Janssen and G. J. van der Kerk, *Recl. Trav. Chim. Pays-Bas*, **84**, 1597 (1965).
467. R. E. Highsmith and H. H. Sisler, *Inorg. Chem.*, **8**, 996 (1969).
468. J. G. Luijten, F. Rijkens and G. J. van der Kerk, in *Advances in Organometallic Chemistry*, Vol. 3, Academic Press, New York, 1965, pp. 397–446.
469. J. G. Luijten and G. J. van der Kerk, in *Organometallic Compounds of the Group IV Elements*, Vol. 1, Part II (Ed. A. G. MacDiarmid), Dekker, New York, 1968, pp. 92–172.
470. J. Satge, M. Baudet and M. Lesbre, *Bull. Soc. Chim. France*, 2133 (1966).
471. M. Schmidt and I. Ruidisch, *Angew. Chem.*, **76**, 686 (1964).
472. R. E. Highsmith and H. H. Sisler, *Inorg. Chem.*, **8**, 1029 (1969).
473. O. J. Scherer, *Angew. Chem.*, **81**, 871 (1969).
474. J. Satge and C. Couret, *C. R. Acad. Sci., Paris*, **264**, 2169 (1967).
475. T. A. George and M. F. Lappert, *J. Organomet. Chem.*, **14**, 327 (1968).
476. F. Caujiolle, R. Huron, F. Moulas and S. Cros, *An. Pharm. Fr.*, **24**, 23 (1966).
477. E. W. Abel and J. P. Crow, *J. Chem. Soc. (A)*, 1361 (1968).
478. F. Glockling and K. A. Hooton, *Proc. Chem. Soc.*, 146 (1963).
479. I. Schumann-Ruidisch and J. Kuhlmey, *J. Organomet. Chem.*, **16**, 26p (1969).
480. A. D. Norman, *Chem. Commun.*, 812 (1968).
481. E. H. Brooks, F. Glockling and K. A. Hooton, *J. Chem. Soc.*, 4283 (1965).
482. H. Schumann and M. Schmidt, *Inorg. Nucl. Chem. Lett.*, **1**, 1 (1965).
483. H. Schumann, P. Schwabe and M. Schmidt, *Inorg. Nucl. Chem. Lett.*, **2**, 309 (1966).
484. H. Schumann and H. Blass, *Z. Naturforsch*, **21b**, 1105 (1966).
485. S. Gradock, E. A. Ebsworth, C. Davidson and L. A. Woodward, *J. Chem. Soc. (A)*, 1229 (1967).
486. J. Satge and C. Couret, *C. R. Acad. Sci., Paris*, **267**, 173 (1968).
487. J. Satge and C. Couret, *Bull. Soc. Chim. France*, 333 (1969).
488. I. V. Tanaev and M. Ya. Shpirt, *The Chemistry of Germanium*, Chemistry, Moscow, 1967.
489. O. H. Johnson, *Chem. Rev.*, **48**, 259 (1951).
490. J. Satge, M. Massol and P. Riviere, *J. Organomet. Chem.*, **56**, 1 (1973).
491. J. Satge, in *Frontiers of Organogermanium, -Tin and-Lead Chemistry* (Eds. E. Lukevics and L. Ignatovich), Latvian Institute of Organic Synthesis, Riga, 1993, pp. 55–69.
492. P. Royen and R. Schwarz, *Z. Anorg. Allg. Chem.*, **215**, 288 (1933).
493. J. Bardet and A. Tchakirian, *C. R. Acad. Sci., Paris*, **186**, 637 (1928).
494. L. M. Dennis and A. W. Laubengayer, *Z. Phys. Chem.*, **130**, 520 (1927).
495. L. M. Dennis and H. Hunter, *J. Am. Chem. Soc.*, **51**, 1151 (1929).
496. L. M. Dennis, W. R. Orndorff and D. L. Tabern, *J. Phys. Chem.*, **30**, 1049 (1926).
497. F. Brewer and L. M. Dennis, *J. Phys. Chem.*, **31**, 1101, 1530 (1927).
498. L. M. Dennis and F. E. Hance, *J. Am. Chem. Soc.*, **44**, 2854 (1922).
499. W. Jolly and W. Latimer, *J. Am. Chem. Soc.*, **74**, 5752 (1952).
500. D. A. Everest, *J. Chem. Soc.*, 1670 (1952).
501. W. C. Johnson, G. H. Morey and A. E. Kott, *J. Am. Chem. Soc.*, **54**, 4278 (1932).
502. E. L. Muetterties, *Inorg. Chem.*, **1**, 342 (1962).
503. J. L. Margrave, K. G. Sharp and P. W. Wilson, *Fortschr. Chem. Forsch.*, **26**, 1 (1972).
504. A. Tchakirian, *C. R. Acad. Sci., Paris*, **192**, 233 (1931).
505. T. Karantassis and L. Capatos, *C. R. Acad. Sci., Paris*, **199**, 64 (1934).
506. T. Karantassis and L. Capatos, *C. R. Acad. Sci., Paris*, **201**, 74 (1935).
507. O. M. Nefedov and M. N. Manakov, *Angew. Chem.*, **76**, 270 (1964); *Angew. Chem., Int. Ed. Engl.*, **3**, 226 (1964).
508. W. P. Neumann, *Chem. Rev.*, **91**, 311 (1991).
509. W. P. Neumann, *Angew. Chem., Int. Ed. Engl.*, **2**, 555 (1963).
510. J. Satge, *Pure Appl. Chem.*, **56**, 137 (1984).
511. J. Satge, P. Riviere and J. Barrau, *J. Organomet. Chem.*, **22**, 599 (1970).

512. N. S. Vyazankin, E. N. Gladyshev, S. P. Korneva and G. A. Razuvaev, *Zh. Obshch. Khim.*, **34**, 1645 (1964); *Chem. Abstr.*, **66**, 76050z (1967).
513. E. J. Bulten and J. G. Noltes, *Tetrahedron Lett.*, **29**, 3471 (1966).
514. M. F. Lappert, *Silicon, Germanium, Tin and Lead Compounds*, **9**, 129 (1986).
515. P. J. Davidson, D. H. Harris and M. F. Lappert, *J. Chem. Soc., Dalton Trans.*, 2268 (1976).
516. T. Tsumuraya, S. A. Batcheller and S. Masamune, *Angew. Chem.*, **103**, 916 (1991); *Int. Ed. Engl.*, **30**, 902 (1991).
517. T. C. Wu and H. Gilman, *J. Am. Chem. Soc.*, **75**, 3762 (1953).
518. A. G. Brook, H. Gilman and L. S. Miller, *J. Am. Chem. Soc.*, **75**, 4759 (1953).
519. F. Gaddes and G. Mack, *J. Am. Chem. Soc.*, **52**, 4372 (1930).
520. C. J. Attridge, *Organomet. Chem. Rev.*, **A5**, 323 (1970).
521. P. Jutzi, *Angew. Chem., Int. Ed. Engl.*, **14**, 232 (1975).
522. P. Jutzi, in *Frontiers of Organogermanium, -Tin and -Lead Chemistry* (Eds. E. Lukevics and L. Ignatovich), Latvian Institute of Organic Synthesis, Riga, 1993, pp. 147–158.
523. J. N. Tokitoh, Y. Matsuhashi, T. Matsumoto, H. Suzuki, M. Saito, K. Manmaru and R. Okazaki, in *Frontiers of Organogermanium, -Tin and -Lead Chemistry* (Eds. E. Lukevics and L. Ignatovich), Latvian Institute of Organic Synthesis, Riga, 1993, pp. 71–82.
524. V. A. Nazarenko and A. M. Andrianov, *Usp. Khim.*, **34**, 1313 (1965); *Chem. Abstr.*, **63**, 15832a (1965).
525. I. R. Beattie, *Quart. Rev.*, **17**, 382 (1963).
526. M. Gielen, C. Dehouck, H. Mokhton-Jamat and J. Topart, in *Review of Silicon, Germanium, Tin and Lead Compounds*, **1**, 9 (1972). Pt. 1, Sci. Publ. Div., Freund Publ. House Ltd, Tel Aviv, 1972.
527. R. C. Mehrotra, R. Bohra and D. P. Gaur, *Metal β-Diketonates and Allied Derivatives*, Academic Press, London, 1978, p. 382.
528. M. Grosjean, M. Gielen and J. Nasielski, *Ind. Chim. Belge*, **28**, 721 (1963).
529. W. Pugh and J. S. Thomas, *J. Chem. Soc.*, 1051 (1926).
530. J. S. Thomas and W. Pugh, *J. Chem. Soc.*, 60 (1931).
531. R. Schwarz and P. W. Schenk, *Chem. Ber.*, **63**, 296 (1930).
532. T. Karantassis and L. Capatos, *C. R. Acad. Sci., Paris*, **193**, 1187 (1931).
533. T. Karantassis and L. Capatos, *Bull. Soc. Chim. France*, **53**, 115 (1933).
534. F. J. Sowa and E. J. Kenny, *US Patent* (1952), 2580473; *Chem. Abstr.*, **46**, 4823d (1952).
535. Ya. Ya. Bleidelis, A. A. Kemme, G. I. Zelchan and M. G. Voronkov, *Khim. Geterotsikl. Soed.*, 617 (1973); *Chem. Abstr.*, **79**, 97945d (1973).
536. M. G. Voronkov, R. G. Mirskov, A. L. Kunetsov and V. Yu. Vitkovskij, *Izv. Akad. Nauk SSSR, Ser Khim.*, 1846 (1979); *Chem. Abstr.*, **91**, 20636u (1979).
537. V. A. Petukhov, L. P. Gudovich, G. I. Zelchan and M. G. Voronkov, *Khim. Geterotsikl. Soedin.*, **6**, 968 (1969); *Chem. Abstr.*, **70**, 105542f (1970).
538. M. G. Voronkov, S. N. Tandura, B. Z. Shterenberg, A. L. Kuznetsov, R. G. Mirskov, G. I. Zelchan, N. Yu. Khromova and T. K. Gar, *Dokl. Akad. Nauk SSSR*, **248**, 134 (1979); *Chem. Abstr.*, **92**, 94518z (1980).
539. V. I. Glukhikh, M. G. Voronkov, O. G. Yarosh, S. N. Tandura, N. V. Alekseev, N. Yu. Khromova and T. K. Gar, *Dokl. Akad. Nauk SSSR*, **258**, 387 (1981); *Chem. Abstr.*, **95**, 114291n (1981).
540. V. A. Pestunovich, B. Z. Shterenberg, S. N. Tandura, V. P. Baryshok, M. G. Voronkov, N. V. Alekseev, N. Yu. Khromova and T. K. Gar, *Izv. Akad. Nauk SSSR, Ser. Khim.*, 2179 (1980); *Chem. Abstr.*, **94**, 46234 (1981).
541. V. B. Kazimirovskaya, M. G. Voronkov, A. T. Platonova, L. N. Kholdeeva, Yu. B. Pisarskii, G. M. Barenboim, T. B. Dmitrievskaya, T. M. Gavrilova, T. K. Gar and V. P. Baryshok, in *Biologically Active Compounds of Silicon, Germanium, Tin and Lead*, Irkutsk Institute of Organic Chemistry SB RAS, Irkutsk, 1980 pp. 89–90.
542. M. G. Voronkov, J. P. Romadan and I. B. Masheika, *Z. Chem.*, **8**, 252 (1968).
543. M. G. Voronkov and V. P. Baryshok, *J. Organomet. Chem.*, **239**, 199 (1982).
544. Yu. Baukov and V. Pestunovich, in *Frontiers of Organogermanium, -Tin and -Lead Chemistry* (Eds. E. Lukevics and L. Ignatovich), Latvian Institute of Organic Synthesis, Riga, 1993, pp. 159–170.
545. I. D. Kalikhman, A. I. Albanov, O. B. Bannikova, L. I. Belousova, S. V. Pestunovich, M. G. Voronkov, A. A. Macharashvili, V. E. Shklover, Yu. T. Struchkov, T. I. Khaustova, G. Ya.

Zueva, E. P. Kramarova, A. G. Shipov, G. I. Oleneva and Yu. I. Baukov, *Zh. Metalloorg. Khim.*, **2**, 637 (1989); *Chem. Abstr.*, **112**, 139209z (1990).
546. E. Y. Lukevics and L. M. Ignatovich, *Zh. Metalloorg. Khim.*, **2**, 184 (1989); *Chem. Abstr.*, **113**, 230371c (1990).
547. M. G. Voronkov and I. G. Kuznetsov, *Silicon in Living Nature*, Science, Novosibirsk, 1984, p. 157; *Chem. Abstr.*, **103**, 100676p (1985).
548. M. G. Voronkov and I. G. Kuznetsov, *Silicon in Living Nature*, Japanese-Soviet Interrelation Company, Wakayama, 1988.
549. F. S. Hammet, J. E. Nowrey and J. H. Mueller, *J. Exp. Med.*, **35**, 173 (1922).
550. J. H. Mueller and M. S. Iszard, *Am. J. Med. Sci.*, **163**, 364 (1922).
551. J. H. Mueller and M. S. Iszard, *J. Metabol. Res.*, **3**, 181 (1923).
552. L. Kast, H. Croll and H. Schmidt, *J. Lab. Clin. Med.*, **7**, 643 (1922).
553. W. C. Heuper, *Am. J. Med. Sci.*, **181**, 820 (1931).
554. J. H. Mueller, *J. Pharmacol. Exp. Ther.*, **42**, 277 (1931).
555. F. S. Hammet, J. E. Nowrey and J. H. Mueller, *J. Exp. Med.*, **35**, 507 (1922).
556. G. C. Harrold, S. F. Meek and C. P. McCord, *Ind. Med.*, **13**, 233 (1944).
557. M. Rothermundt and K. Burschkies, *Z. Immunitätsforsch*, **87**, 445 (1936).
558. R. Schwarz and H. Schols, *Chem. Ber.*, **74**, 1676 (1941).
559. H. Oikawa and K. Kikuyo, *Japan Patent* 46002498 (1971); *Chem. Abstr.*, **75**, 6108 (1971).
560. P. L. Carpenter, M. Fulton and C. A. Stuart, *J. Bacteriol.*, **29**, 18 (1935).
561. E. G. Rochow and B. M. Sindler, *J. Am. Chem. Soc.*, **72**, 1218 (1950).
562. F. Caujiolle, D. Caujiolle, S. Cros, O. Dao-Huy-Giao, F. Moulas, Y. Tollon and J. Caylas, *Bull. Trav. Soc. Pharm.*, **9**, 221 (1965).
563. D. Caujiolle, O. Dao-Huy-Giao, J. L. Foulquier and M.-C. Voisin, *Ann. Biol. Clin. (Paris)*, **24**, 479 (1966).
564. S. Kaars, F. Rijkens, G. J. van der Kerk and A. Manten, *Nature*, **201**, 736 (1967).
565. J. E. Cremer and W. N. Aldridge, *Br. J. Ind. Med.*, **21**, 214 (1964).
566. F. Caujiolle, D. Caujiolle, O. Dao-Huy-Giao, J. L. Foulquier and E. Maurel, *C. R. Acad. Sci.*, **262**, 1302 (1966).
567. H. Bouisson, F. Caujiolle, D. Caujiolle and M.-C. Voisin, *C. R. Acad. Sci.*, **259**, 3408 (1964).
568. H. Langer and G. Horst, *US Patent* 3442922 (1969); *Chem. Abstr.*, **72**, 12880 (1970).
569. M. Voronkov, G. Zeltschan, A. Lapsina and W. Pestunovich, *Z. Chem.*, **8**, 6, 214 (1968); *Chem. Abstr.*, **68**, 43301r (1968).
570. M. G. Voronkov, G. I. Zelchan, V. I. Savushkina, B. M. Tabenko and E. A. Chernyshev, *Khim. Geterotsikl. Soedin.*, 772 (1976); *Chem. Abstr.*, **85**, 143181a (1976).
571. K. Asai, *Organic Germanium, A Medical Godsend*, Kogakusha Ltd., Tokyo 1977.
572. L. Rice, J. W. Wheeler and C. F. Geschicter, *J. Heterocycl. Chem.*, **11**, 1041 (1974).
573. M. C. Henry, E. Rosen, C. D. Port and B. S. Levine, *Cancer Treatment Rep.*, **64**, 1207 (1980).
574. E. J. Bulten and G. J. M. van der Kerk, in *New Uses for Germanium* (Ed. F. I. Mets), Midwest Research Institute, Kansas City, 1974, pp. 51–62.
575. A. Cahours, *Ann. Chem.*, **114**, 227 (1860).
576. T. Harada, *Sci. Papers Inst. Phys.-Chem. Res. (Tokyo)*, **38**, 115 (1940); *Chem. Abstr.*, **35**, 1027 (1941).
577. C. Löwig, *Z. Prakt. Chem.*, **57**, 385 (1852).
578. C. Löwig, *Zürich Mittheil*, **2**, 556 (1850–1852); *Chem. Zbl.*, **23**, 849, 865, 889 (1852).
579. C. Löwig, *Schles. Gesell. Ubersicht (Breslau)*, 26 (1853).
580. C. Löwig, *Z. Prakt. Chem.*, **60**, 348 (1853).
581. C. Löwig, *Z. Prakt. Chem.*, **65**, 355 (1855).
582. C. Löwig, *Ann. Chem.*, **102**, 376 (1857).
583. A. Cahours, *Ann. Chem.*, **114**, 354 (1860).
584. P. Kulmiz, *Z. Prakt. Chem.*, **80**, 60 (1860).
585. A. Werner and P. Pfeiffer, *Z. Anorg. Chem.*, **17**, 82 (1898).
586. C. Löwig, *Z. Gesammt. Naturw.*, **1**, 34 (1853).
587. C. Löwig, *Schles. Gesell. Ubersicht (Breslau)*, 31 (1853).
588. E. Frankland, *Phil. Trans.*, **142**, 418 (1852).
589. E. Frankland and A. Lawrence, *J. Chem. Soc.*, **35**, 130 (1879).
590. E. Frankland, *J. Chem. Soc.*, **6**, 57 (1854).

591. E. Frankland, *Ann. Chem.*, **111**, 44 (1859).
592. A. Cahours, A. Riche, *C. R. Acad. Sci., Paris*, **35**, 91 (1852).
593. A. Cahours, *Ann. Chem.*, **88**, 316 (1853).
594. A. Cahours, *Ann. Chem.*, **114**, 367 (1860).
595. A. Cahours, *Ann. Chem.*, **114**, 372 (1860).
596. A. Cahours, *Ann. Chem.*, **122**, 48 (1862).
597. A. Cahours, *Ann. Chem.*, **122**, 60 (1862).
598. A. Cahours, *C. R. Acad. Sci., Paris*, **76**, 133 (1873).
599. A. Cahours, *C. R. Acad. Sci., Paris*, **77**, 1403 (1873).
600. A. Cahours and E. Demarcay, *Compt. Rend.*, **89**, 68 (1879).
601. A. Cahours, *C. R. Acad. Sci., Paris*, **88**, 725 (1879).
602. A. Cahours and E. Demarcay, *C. R. Acad. Sci., Paris*, **88**, 1112 (1879).
603. G. B. Buckton, *Ann. Chem.*, **109**, 218 (1859).
604. G. B. Buckton, *Ann. Chem.*, **112**, 220 (1859).
605. A. Ladenburg, *Ann. Chem.*, **159**, 251 (1871).
606. A. Ladenburg, *Chem. Ber.*, **4**, 17 (1871).
607. A. Ladenburg, *Ann. Supl.*, **8**, 55 (1872).
608. W. J. Pope and S. J. Peachey, *Proc. Chem. Soc.*, **16**, 42 (1900).
609. W. J. Pope and S. J. Peachey, *Proc. Chem. Soc.*, **16**, 116 (1900).
610. W. J. Pope and S. J. Peachey, *Proc. Chem. Soc.*, **19**, 290 (1903).
611. W. J. Pope and S. J. Peachey, *Chem. News*, **87**, 253 (1903).
612. W. J. Pope and S. J. Peachey, *Proc. Chem. Soc.*, **16**, 2365 (1928).
613. P. Pfeiffer, *Chem. Ber.*, **44**, 1269 (1911).
614. P. Pfeiffer and R. Lehnardt, *Chem. Ber.*, **36**, 3027 (1903).
615. P. Pfeiffer and I. Heller, *Chem. Ber.*, **37**, 4618 (1904).
616. P. Pfeiffer, B. Friedmann and H. Rekate, *Ann. Chem.*, **376**, 310 (1910).
617. P. Pfeiffer, R. Lehnardt, H. Luftensteiner, R. Prade, K. Schnurmann and P. Truskier, *Z. Anorg. Chem.*, **68**, 102 (1910).
618. P. Pfeiffer, B. Friedmann, R. Lehnardt, H. Luftensteiner, R. Prade and K. Schnurmann, *Z. Anorg. Allg. Chem.*, **71**, 97 (1911).
619. P. Pfeiffer, B. Friedmann, R. Lehnardt, H. Luftensteiner, R. Prade and K. Schnurmann, *Z. Anorg. Allg. Chem.*, **87**, 229 (1914).
620. P. Pfeiffer, B. Friedmann, R. Lehnardt, H. Luftensteiner, R. Prade and K. Schnurmann, *Z. Anorg. Allg. Chem.*, **133**, 91 (1924).
621. M. Hjortdahl, *Z. Krist. Mineral.*, **4**, 286 (1879).
622. B. Emmert and W. Eller, *Chem. Ber.*, **44**, 2328 (1911).
623. A. Grimm, *Z. Prakt. Chem.*, **62**, 385 (1954).
624. T. Karantassis and K. Basillides, *C. R. Acad. Sci., Paris*, **205**, 460 (1937).
625. T. Karantassis and K. Basillides, *C. R. Acad. Sci., Paris*, **206**, 842 (1938).
626. M. B. Neuman and V. A. Shushunov, *Zh. Fiz. Khim.*, **22**, 145 (1948); *Chem. Abstr.*, **42**, 5315h (1948).
627. M. B. Neuman and V. A. Shushunov, *Dokl. Akad. Nauk SSSR*, **60**, 1347 (1948); *Chem. Abstr.*, **42**, 5315h (1948).
628. T. Harada, *J. Sci. Res. Inst. (Tokyo)*, **43**, 31 (1948); *Chem. Abstr.*, **43**, 4632 (1949).
629. G. J. M. van der Kerk and J. G. A. Luijten, *J. Appl. Chem.*, **4**, 301 (1954).
630. G. Grüttner, E. Krause and M. Wiernik, *Chem. Ber.*, **50**, 1549 (1917).
631. K. A. Kocheshkov, *Chem. Ber.*, **61**, 1659 (1928).
632. K. A. Kocheshkov, *Zh. Rus. Fiz. Khim. Obshch.*, **60**, 1191 (1928); *Chem. Abstr.*, **23**, 2931 (1929).
633. K. A. Kocheshkov, *Zh. Rus. Fiz. Khim. Obshch.*, **61**, 1385 (1929); *Chem. Abstr.*, **24**, 1360 (1930).
634. K. Sisido and J. Kinukawa, *Japan Patent* 6626 (1953); *Chem. Abstr.*, **49**, 9690 (1955).
635. L. V. Abramova, N. I. Sheverdina and K. A. Kocheshkov, *Dokl. Akad. Nauk SSSR*, **123**, 681 (1958); *Chem. Abstr.*, **53**, 4928e (1959).
636. E. A. Letts and N. Collie, *Phil. Mag.*, **22**, 41 (1886).
637. E. A. Letts and N. Collie, *Proc. Chem. Soc.*, **2**, 166 (1886).
638. E. A. Letts and N. Collie, *Jahresber.*, 1601 (1886).
639. T. Harada, *Bull. Chem. Soc. Japan*, **2**, 105 (1927).

640. T. Harada, *Bull. Chem. Soc. Japan*, **4**, 266 (1929).
641. T. Harada, *Sci. Papers Inst. Phys.-Chem. Res. (Tokyo)*, **35**, 290 (1939).
642. T. Harada, *Rep. Sci. Res. Inst. (Japan)*, **24**, 177 (1948); *Chem. Abstr.*, **45**, 2356 (1951).
643. C. A. Kraus and C. C. Callis, *USA Patent* 16399447 (1927); *Chem. Abstr.*, **21**, 3180 (1927).
644. H. Polkinhorne and C. G. Tapley, *UK Patent* 736822 (1955); *Chem. Abstr.*, **50**, 8725 (1956).
645. C. E. Arntzen, *Iowa State Coll. J. Sci.*, **18**, 6 (1943); *Chem. Abstr.*, **38**, 61 (1943).
646. H. Gilman and C. E. Arntzen, *J. Org. Chem.*, **15**, 994 (1950).
647. E. Krause and K. Weinberg, *Chem. Ber.*, **63**, 381 (1930).
648. A. E. Finholt, A. C. Bond and H. I. Schlesinger, *J. Am. Chem. Soc.*, **69**, 1199 (1947).
649. J. R. Zietz, S. M. Blitzer, H. E. Redman and G. C. Robinson, *J. Org. Chem.*, **22**, 60 (1957).
650. S. M. Blitzer and J. R. Zietz, *US Patent* 2852543 (1958); *Chem. Abstr.*, **53**, 4135 (1959).
651. A. Polis, *Chem. Ber.*, **22**, 2915 (1889).
652. A. C. Smith and E. G. Rochow, *J. Am. Chem. Soc.*, **75**, 4103 (1953).
653. E. G. Rochow, *US Patent* 2679506 (1959); *Chem. Abstr.*, **49**, 4705 (1955).
654. C. A. Kraus and C. C. Callis, *Canadian Patent* 226140 (1926); *Chem. Abstr.*, **21**, 917 (1927).
655. E. I. Du Pont, de Nemours and Co., *UK Patent* 469518 (1936); *Chem. Abstr.*, **32**, 592 (1938).
656. F. A. Smith, *US Patent* 2625559 (1953); *Chem. Abstr.*, **47**, 11224b (1953).
657. A. C. Smith and E. G. Rochow, *J. Am. Chem. Soc.*, **75**, 4105 (1953).
658. G. J. M. van der Kerk and J. G. A. Luijten, *J. Appl. Chem.*, **4**, 307 (1954).
659. J. Ireland, *UK Patent* 713727 (1954); *Chem. Abstr.*, **49**, 12530 (1955).
660. G. J. M. van der Kerk, *German Patent* 946447 (1956); *Chem. Abstr.*, **53**, 7014 (1959).
661. H. Polkinhorne, C. G. Tapley, *UK Patent* 761357 (1956); *Chem. Abstr.*, **51**, 11382 (1957).
662. R. F. Chambers and P. C. Scherer, *J. Am. Chem. Soc.*, **48**, 1054 (1926).
663. R. F. Chambers and P. C. Scherer, *UK Patent* 854776 (1960); *Chem. Abstr.*, **55**, 11362 (1961).
664. K. Sisido, S. Kozima and T. Tusi, *J. Organomet. Chem.*, **9**, 109 (1967).
665. M. M. Nad' and K. A. Kocheshkov, *Zh. Obshch. Chim.*, **8**, 42 (1938); *Chem. Abstr.*, **32**, 5387 (1938).
666. T. V. Talalaeva and K. A. Kocheshkov, *Zh. Obshch. Khim.*, **8**, 1831 (1938); *Chem. Abstr.*, **33**, 5819 (1939).
667. T. V. Talalaeva and K. A. Kocheshkov, *Zh. Obshch. Khim.*, **12**, 403 (1942); *Chem. Abstr.*, **37**, 3068 (1943).
668. N. Morgunov, *Ann. Chem.*, **144**, 157 (1867).
669. K. A. Kocheshkov, A. N. Nesmeyanov and V. I. Potrosov, *Chem. Ber.*, **67**, 1138 (1934).
670. P. Pfeiffer and K. Schnurmann, *Chem. Ber.*, **37**, 319 (1904).
671. W. E. Edgell and C. H. Ward, *J. Am. Chem. Soc.*, **76**, 1169 (1954).
672. D. Seyferth, *J. Am. Chem. Soc.*, **79**, 5881 (1957).
673. D. Seyferth and F. G. A. Stone, *J. Am. Chem. Soc.*, **79**, 515 (1957).
674. G. Grüttner, *Chem. Ber.*, **47**, 3257 (1914).
675. A. N. Nesmeyanov and K. A. Kocheshkov, *Uch. Zap. Mosk. Univ.*, **3**, 283 (1934); *Chem. Abstr.*, **28**, 2285 (1934).
676. K. K. Law, *J. Chem. Soc.*, 3243 (1926).
677. J. Böeseken and J. J. Rutgers, *Recl. Trav. Chim. Pays-Bas*, **42**, 1017 (1923).
678. T. A. Smith and F. S. Kipping, *J. Chem. Soc.*, **101**, 2553 (1912).
679. F. S. Kipping, *J. Chem. Soc.*, **132**, 2365 (1928).
680. P. R. Austin, *J. Am. Chem. Soc.*, **54**, 3726 (1932).
681. G. Bähr and R. Gelius, *Chem. Ber.*, **91**, 818 (1958).
682. H. G. Kuivila and O. F. Beumel, *J. Am. Chem. Soc.*, **80**, 3250 (1958).
683. F. C. Leavitt, T. A. Manuel and F. Johnson, *J. Am. Chem. Soc.*, **81**, 3163 (1959).
684. F. C. Leavitt, T. A. Manuel, F. Johnson, L. I. Matternas and D. S. Lehmann, *J. Am. Chem. Soc.*, **82**, 5099 (1960).
685. H. Gilman and E. A. Zuech, *J. Am. Chem. Soc.*, **82**, 2522 (1960).
686. H. Gilman and L. A. Gist, *J. Org. Chem.*, **22**, 368 (1957).
687. H. Gilman, F. W. Moore and R. G. Jones, *J. Am. Chem. Soc.*, **63**, 2482 (1941).
688. H. Gilman and T. N. Gorlan, *J. Org. Chem.*, **17**, 1470 (1952).
689. L. A. Woods and H. Gilman, *Proc. Iowa Acad. Sci.*, **48**, 251 (1941); *Chem. Abstr.*, **36**, 3492 (1942).
690. H. Gilman and C. E. Arntzen, *J. Am. Chem. Soc.*, **72**, 3823 (1950).

691. M. Lesbre and J. Roues, *Bull. Soc. Chim. France*, **18**, 490 (1951).
692. H. Zimmer and H. W. Sparmann, *Naturwissenschaften*, **40**, 220 (1953).
693. H. Zimmer and H. W. Sparmann, *Chem. Ber.*, **87**, 645 (1954).
694. H. Gilman and T. C. Wu, *J. Am. Chem. Soc.*, **77**, 328 (1955).
695. R. H. Meen and H. Gilman, *J. Org. Chem.*, **22**, 564 (1957).
696. S. S. Babashinskaya and K. A. Kocheshkov, *Zh. Obshch. Khim.*, **8**, 1850 (1938); *Chem. Abstr.*, **33**, 5820 (1939).
697. G. B. Bachman, C. L. Carlson and M. Robinson, *J. Am. Chem. Soc.*, **73**, 1964 (1951).
698. H. Zimmer and H. G. Mosle, *Chem. Ber.*, **87**, 1255 (1954).
699. H. Zimmer and H. Gold, *Chem. Ber.*, **89**, 712 (1956).
700. D. Seyferth and M. A. Weiner, *Chem. Ind. (London)*, 402 (1959).
701. H. Hartmann, H. Niemöller, W. Reiss and B. Karbstein, *Naturwissenschaften*, **46**, 321 (1959).
702. H. Gilman and L. A. Gist, *J. Org. Chem.*, **22**, 250 (1957).
703. G. Bähr and R. Gelius, *Chem. Ber.*, **91**, 812 (1958).
704. R. H. Prince, *J. Chem. Soc.*, 1759 (1959).
705. A. N. Borisov and N. V. Novikova, *Izv. Akad. Nauk SSSR, Ser. Khim.*, 1370 (1959); *Chem. Abstr.*, **54**, 8608i (1960).
706. G. Wittig, F. J. Meyer and G. Lange, *Ann. Chem.*, **571**, 167 (1951).
707. H. Gilman and S. D. Rosenberg, *J. Org. Chem.*, **18**, 1554 (1953).
708. H. Gilman and S. D. Rosenberg, *J. Am. Chem. Soc.*, **75**, 2507 (1953).
709. E. O. Fischer and H. Grübert, *Z. Naturforsch.*, **11b**, 423 (1956).
710. M. Hardmann and P. Backes, *German Patent* 508667 (1926); 1930; *Chem. Abstr.*, **25**, 713 (1931).
711. J. O. Harris, *US Patent* 2431038 (1947); *Chem. Abstr.*, **42**, 1606 (1948).
712. J. G. A. Luijten and G. J. M. van der Kerk, *Tin Research Inst. Publ., Apr.*, (1952), *Chem. Abstr.*, **49**, 11544 (1955).
713. G. J. M. van der Kerk and J. G. A. Luijten, *Ind. Chim. Belge*, **21**, 567 (1956).
714. G. J. M. van der Kerk and J. G. A. Luijten, *J. Appl. Chem.*, **6**, 49 (1956).
715. G. J. M. van der Kerk and J. G. A. Luijten, *J. Appl. Chem.*, **7**, 369 (1957).
716. C. R. Gloskey, *US Patent* 2805234 (1957); *Chem. Abstr.*, **52**, 2050 (1958).
717. C. Beermann and H. Hartmann, *Z. Anorg. Allg. Chem.*, **276**, 54 (1954).
718. H. Hartmann and H. Honig, *Angew. Chem.*, **69**, 614 (1957).
719. Karl-Chemie Akt.-Ges., *UK Patent* 802796 (1958); *Chem. Abstr.*, **53**, 9061 (1959).
720. L. I. Zakharkin and O. Y. Okhlobystin, *Dokl. Akad. Nauk SSSR*, **116**, 236 (1957); *Chem. Abstr.*, **52**, 6167 (1958).
721. A. N. Nesmeyanov, A. N. Borisov and N. V. Novikova, *Izv. Akad. Nauk SSSR, Ser. Khim.*, 644 (1959); *Chem. Abstr.*, **53**, 21626 (1959).
722. B. Aronheim, *Ann. Chem.*, **194**, 145 (1878).
723. L. K. Stareley, H. P. Paget, B. B. Goalby and J. B. Warren, *J. Chem. Soc.*, 2290 (1950).
724. G. Bähr and G. Zoche, *Chem. Ber.*, **88**, 1450 (1955).
725. K. A. Kocheshkov and A. N. Nesmeyanov, *Zh. Rus. Fiz. Khim. Obshch.*, **62**, 1795 (1930); *Chem. Abstr.*, **25**, 3975 (1931).
726. A. N. Nesmeyanov and K. A. Kocheshkov, *Chem. Ber.*, **63**, 2496 (1930).
727. A. N. Nesmeyanov and K. A. Kocheshkov, *Zh. Obshch. Khim.*, **1**, 219 (1931); *Chem. Abstr.*, **25**, 927 (1931).
728. A. E. Goddard, J. N. Ashley and R. B. Evans, *J. Chem. Soc.*, **121**, 978 (1922).
729. D. Goddard and A. E. Goddard, *J. Chem. Soc.*, **121**, 256 (1922).
730. O. H. Johnson and J. R. Holum., *J. Org. Chem.*, **23**, 738 (1958).
731. L. I. Zakharkin and O. Y. Okhlobystin, *Izv. Akad. Nauk SSSR, Ser. Khim.*, 1942 (1959); *Chem. Abstr.*, **53**, 15958f (1959).
732. C. A. Kraus and W. N. Greer, *J. Am. Chem. Soc.*, **44**, 2629 (1922).
733. L. Malatesta and R. Pizzotti, *Gazz. Chim. Ital.*, **73**, 344 (1943).
734. F. G. A. Stone, *Hydrogen Compounds of the Group IV Elements*, Prentice-Hall International, London, 1962.
735. H. H. Anderson, *J. Am. Chem. Soc.*, **79**, 4913 (1957).
736. H. Gilman and J. Eisch, *J. Org. Chem.*, **20**, 763 (1955).

737. C. R. Dillard, E. H. McNeill, D. E. Simmons and J. D. Veldell, *J. Am. Chem. Soc.*, **80**, 3607 (1958).
738. G. J. M. van der Kerk, J. G. Noltes and J. G. A. Luijten, *J. Appl. Chem.*, **7**, 366 (1957).
739. G. J. M. van der Kerk, J. G. Noltes and J. G. A. Luijten, *Chem. Ind. (London)*, 1290 (1958).
740. V. N. Ipatiev, G. A. Razuvaev and I. F. Bogdanov, *Izv. Akad. Nauk SSSR, Ser. Khim.*, 155 (1929); *Chem. Zbl.*, **1**, 1263 (1930).
741. M. M. Koton, *Zh. Obshch. Khim.*, **2**, 345 (1932); *Chem. Abstr.*, **27**, 275 (1933).
742. M. M. Koton, *Chem. Ber.*, **66**, 1213 (1933).
743. M. M. Koton, *Zh. Obshch. Khim.*, **4**, 653 (1934); *Chem. Abstr.*, **29**, 3662 (1935).
744. L. L. Gershbein and V. N. Ipatiev, *J. Am. Chem. Soc.*, **74**, 1540 (1952).
745. Yu. I. Khudobin and M. G. Voronkov, *Zh. Metalloorg. Khim.*, **2**, 1305 (1989); *Chem. Abstr.*, **113**, 64284 (1990).
746. H. J. Emeleus and S. F. A. Kettle, *J. Chem. Soc.*, 2444 (1958).
747. D. R. Lide, *J. Chem. Phys.*, **29**, 1605 (1951).
748. G. J. M. van der Kerk, J. G. Noltes and J. G. A. Luijten, *J. App. Chem.*, **7**, 356 (1957).
749. R. H. Bullard and R. A. Vingee, *J. Am. Chem. Soc.*, **51**, 892 (1929).
750. C. A. Kraus and R. H. Bullard, *J. Am. Chem. Soc.*, **52**, 4056 (1930).
751. J. G. Noltes and G. J. M. van der Kerk, *Chem. Ind. (London)*, 294 (1959).
752. T. V. Sathyamurthy, S. Swaminathan and L. M. Yeddanapalli, *J. Indian Chem. Soc.*, **27**, 509 (1950).
753. H. Gilman and H. W. Melvin, *J. Am. Chem. Soc.*, **71**, 4050 (1949).
754. H. Gilman and S. D. Rosenberg, *J. Am. Chem. Soc.*, **75**, 3592 (1953).
755. M. Lesbre and R. Buisson, *Bull. Soc. Chim. France*, 1204 (1957).
756. G. J. M. van der Kerk and J. G. Noltes, *Ann. Chem., Acad. Sci.*, New York, 1965.
757. J. G. Noltes and G. J. M. van der Kerk, *Functionally Substituted Organotin Compounds*, Tin Research Inst., Greenford, Middlesex, England, 1958.
758. G. J. M. van der Kerk and J. G. Noltes, *J. App. Chem.*, **9**, 106 (1959).
759. G. J. M. van der Kerk, J. G. A. Luijten and J. G. Noltes, *Angew. Chem.*, **70**, 298 (1958).
760. C. G. Krespan and V. A. Engelhardt, *J. Org. Chem.*, **23**, 1565 (1958).
761. E. M. Pearce, *J. Polym. Sci.*, **40**, 136; 272 (1959).
762. L. A. Rothman and E. I. Becker, *J. Org. Chem.*, **24**, 294 (1959).
763. H. G. Kuivila and O. F. Beumel, *J. Am. Chem. Soc.*, **80**, 3798 (1958).
764. W. P. Neumann, *Angew. Chem.*, **76**, 849 (1964).
765. H. G. Kuivila and O. F. Beumel, *J. Am. Chem. Soc.*, **83**, 1246 (1961).
766. H. G. Kuivila, in *Advances in Organometallic Chemistry*, Vol. 1, Academic Press, New York, 1964 pp. 47–89.
767. H. Gilman and S. D. Rosenberg, *J. Am. Chem. Soc.*, **74**, 531 (1952).
768. A. Ladenburg, *Chem. Ber.*, **4**, 19 (1871).
769. E. Krause and R. Pohland, *Chem. Ber.*, **57**, 532 (1924).
770. R. W. Bost and P. Borgstrom, *J. Am. Chem. Soc.*, **51**, 1922 (1929).
771. Z. M. Manulkin, *Zh. Obshch. Khim.*, **11**, 386 (1941); *Chem. Abstr.*, **35**, 5854 (1941).
772. S. N. Naumov and Z. M. Manulkin, *Zh. Obshch. Khim.*, **5**, 281 (1935); *Chem. Abstr.*, **29**, 5071 (1935).
773. Z. M. Manulkin, *Zh. Obshch. Khim.*, **14**, 1047 (1944); *Chem. Abstr.*, **41**, 89g (1947).
774. Z. M. Manulkin, *Zh. Obshch. Khim.*, **16**, 235 (1946); *Chem. Abstr.*, **42**, 6742f (1948).
775. Z. M. Manulkin, *Zh. Obshch. Khim.*, **13**, 42 (1943); *Chem. Abstr.*, **38**, 331 (1944).
776. M. M. Koton and T. M. Kiseleva, *Zh. Obshch. Khim.*, **27**, 2553 (1957); *Chem. Abstr.*, **52**, 7136g (1958).
777. C. A. Kraus and W. N. Greer, *J. Am. Chem. Soc.*, **45**, 3078 (1923).
778. A. Ladenburg, *Chem. Ber.*, **3**, 647 (1870).
779. E. Krause, *Chem. Ber.*, **51**, 912 (1918).
780. T. A. Smith and F. S. Kipping, *J. Chem. Soc.*, **103**, 2034 (1913).
781. E. Krause and R. Becker, *Chem. Ber.*, **53**, 173 (1920).
782. C. A. Kraus and N. V. Sessions, *J. Am. Chem. Soc.*, **47**, 2361 (1925).
783. R. H. Bullard and F. R. Holden, *J. Am. Chem. Soc.*, **53**, 3150 (1931).
784. G. B. Buckton, *Ann. Chem.*, **109**, 22 (1859).
785. R. H. Bullard and W. R. Robinson, *J. Am. Chem. Soc.*, **49**, 138 (1927).

786. M. G. Voronkov, V. I. Rakhlin and R. G. Mirskov, in *Advances in Organometallic Chemistry*, Mir, Moscow, 1985, pp. 196–216.
787. M. G. Voronkov, V. I. Rakhlin and R. G. Mirskov, *Khim. Mashinostroenie*, **30**, 35 (1986); *Chem. Abstr.*, **107**, 778494u (1987).
788. V. I. Rakhlin, R. G. Mirskov and M. G. Voronkov, *Zh. Org. Khim.*, **32**, 807 (1996); *Chem. Abstr.*, **126**, 171078f (1997).
789. D. Seyferth, *J. Am. Chem. Soc.*, **79**, 2133 (1957).
790. A. Saitow, E. G. Rochow and D. Seyferth, *J. Org. Chem.*, **23**, 116 (1958).
791. J. C. Bailie, *Iowa State Coll. J. Sci.*, **14**, 8 (1939); *Chem. Abstr.*, **34**, 6241 (1940).
792. A. P. Skoldinov and K. A. Kolesnikov, *Zh. Obshch. Khim.*, **12**, 398 (1942); *Chem. Abstr.*, **37**, 3064[2] (1943).
793. R. N. Meals, *J. Org. Chem.*, **9**, 211 (1944).
794. G. Bähr, *Z. Anorg. Chem.*, **256**, 107 (1948).
795. K. A. Kocheshkov, *Chem. Ber.*, **62**, 996 (1929).
796. K. A. Kocheshkov, *Chem. Ber.*, **66**, 1661 (1933).
797. K. A. Kocheshkov, *Zh. Obshch. Khim.*, **4**, 1359 (1934).
798. M. E. Pavlovskaya and K. A. Kocheshkov, *Dokl. Akad. Nauk SSSR*, **46**, 263 (1945); *Chem. Abstr.*, **40**, 5697[1] (1946).
799. K. A. Kocheshkov, M. M. Nad' and A. P. Aleksandrov, *Zh. Obshch. Khim.*, **6**, 1672 (1936); *Chem. Abstr.*, **31**, 2590 (1937).
800. K. A. Kocheshkov, *Zh. Obshch. Khim.*, **5**, 211 (1935); *Chem. Abstr.*, **29**, 5071 (1935).
801. K. A. Kocheshkov, *Zh. Obshch. Khim.*, **4**, 1434 (1934); *Chem. Abstr.*, **29**, 3660 (1935).
802. K. A. Kocheshkov, *Chem. Ber.*, **67**, 713 (1934).
803. K. A. Kocheshkov, *Zh. Obshch. Khim.*, **5**, 1158 (1935); *Chem. Abstr.*, **30**, 1036 (1936).
804. E. I. Pikina, T. V. Talalaeva and K. A. Kocheshkov, *Zh. Obshch. Khim.*, **8**, 1844 (1938).
805. K. A. Kocheshkov, *Rus. Chem. Rev.*, **3**, 83 (1934); *Chem. Abstr.*, **51**, 10556e (1957).
806. A. N. Nesmeyanov and K. A. Kocheshkov, *Chem. Ber.*, **64**, 628 (1931).
807. G. A. Razuvaev, in *Syntheses of Organic Compounds* (Eds. A. N. Nesmeyanov and P. A. Bobrov), Vol. 1, Academic Publishers, Moscow, 1950, p. 41.
808. S. D. Rosenberg and A. J. Gibbons, *J. Am. Chem. Soc.*, **79**, 2138 (1957).
809. W. P. Neumann and G. Burkhardt, *Ann. Chem.*, **663**, 11 (1963).
810. D. Seyferth and H. M. Cohen, *Inorg. Chem.*, **2**, 652 (1963).
811. B. Aronheim, *Ann. Chem.*, **194**, 171 (1878).
812. B. Aronheim, *Chem. Ber.*, **12**, 509 (1879).
813. A. N. Nesmeyanov and K. A. Kocheshkov, *Chem. Ber.*, **67**, 317 (1934).
814. E. Krause and M. Schmitz, *Chem. Ber.*, **52**, 2150 (1919).
815. I. T. Eskin, A. N. Nesmeyanov and K. A. Kocheshkov, *Zh. Obshch. Khim.*, **8**, 35 (1938); *Chem. Abstr.*, **32**, 53867 (1938).
816. G. Calingaert, H. Soroos and V. Hnizda, *J. Am. Chem. Soc.*, **62**, 1107 (1940).
817. Z. M. Manulkin, *Zh. Obshch. Khim.*, **18**, 299 (1948); *Chem. Abstr.*, **42**, 67421 (1948).
818. D. Seyferth, *Naturwissenschaften*, **44**, 34 (1957).
819. Z. M. Manulkin, *Zh. Obshch. Khim.*, **20**, 2004 (1950); *Chem. Abstr.*, **45**, 5611i (1951).
820. K. A. Kocheshkov, A. N. Nesmeyanov and B. P. Puzyreva, *Chem. Ber.*, **69**, 1639 (1936).
821. A. Tchakirian, M. Lesbre and M. Lewinsohn, *C. R. Acad. Sci., Paris*, **202**, 138 (1936).
822. A. C. Smith and E. G. Rochow, *J. Am. Chem. Soc.*, **75**, 4103 (1953).
823. A. N. Nesmeyanov, K. A. Kocheshkov and V. A. Klimova, *Chem. Ber.*, **68**, 1877 (1935).
824. W. A. Waters, *J. Chem. Soc.*, 2007 (1937).
825. W. A. Waters, *J. Chem. Soc.*, 864 (1939).
826. O. A. Ptitsina, O. A. Reutov and M. F. Turchinskii, *Dokl. Akad. Nauk SSSR*, **114**, 110 (1957); *Chem. Abstr.*, **52**, 1090a (1958).
827. O. A. Ptitsina, O. A. Reutov and M. F. Turchinskii, *Nauchn. Dokl. Vysshei Shkoly, Khim. i Tekhnol.*, **1**, 138 (1959); *Chem. Abstr.*, **53**, 17030i (1959).
828. A. N. Nesmeyanov, O. A. Reutov, T. P. Tolstaya, O. A. Ptitsina, L. S. Isaeva, M. F. Turchinskii and G. P. Bochkareva, *Dokl. Akad. Nauk SSSR*, **125**, 1265 (1959); *Chem. Abstr.*, **53**, 21757d (1959); *Chem. Abstr.*, **53**, 17030k (1959).
829. G. Gustavson, *Zh. Rus. Fiz. Khim. Obshch.*, **23**, 253 (1894).
830. H. H. Anderson and J. A. Vasta, *J. Org. Chem.*, **19**, 1300 (1954).
831. E. Krause, *Chem. Ber.*, **51**, 1447 (1918).

832. E. Krause and K. Weinberg, *Chem. Ber.*, **62**, 2235 (1929).
833. E. Krause and O. Schlöttig, *Chem. Ber.*, **63**, 1381 (1930).
834. D. Seyferth, *J. Am. Chem. Soc.*, **77**, 1302 (1955).
835. C. E. Waring and W. S. Horton, *J. Am. Chem. Soc.*, **67**, 540 (1945).
836. L. H. Long, *J. Chem. Soc.*, 3410 (1956).
837. S. J. M. Prince and A. F. Trotman-Dickenson, *Trans. Faraday Soc.*, **54**, 1630 (1958).
838. G. J. M. van der Kerk, J. G. Noltes and J. G. A. Luijten, *J. Appl. Chem.*, **9**, 113 (1959).
839. G. Meyer, *Chem. Ber.*, **16**, 1439 (1883).
840. A. Solerio, *Gazz. Chim. Ital.*, **81**, 664 (1951).
841. A. N. Nesmeyanov and L. G. Makarova, *Dokl. Akad. Nauk SSSR*, **87**, 421 (1952); *Chem. Abstr.*, **48**, 623 (1954).
842. T. Harada, *Sci. Papers Inst. Phys.-Chem. Res. (Tokyo)*, **38**, 146 (1940); *Chem. Abstr.*, **35**, 2470 (1941).
843. T. Harada, *Bull. Chem. Soc. Japan*, **15**, 481 (1940).
844. M. G. Voronkov and Yu. P. Romodan, *Khim. Geterotsikl. Soedin.*, 892 (1966); *Chem. Abstr.*, **65**, 8943 (1966).
845. H. H. Anderson, *J. Org. Chem.*, **19**, 1766 (1954).
846. H. Lambourne, *J. Chem. Soc.*, **125**, 2013 (1924).
847. A. N. Nesmeyanov and K. A. Kocheshkov, *Methods of Organic Elemental Chemistry*, 2nd Edn., Science, Moscow, 1968.
848. C. A. Kraus and R. H. Bullard, *J. Am. Chem. Soc.*, **51**, 3605 (1929).
849. T. Harada, *Sci. Papers Inst. Phys.-Chem. Res. (Tokyo)*, **36**, 501 (1939); *Chem. Abstr.*, **34**, 3671 (1940).
850. J. G. F. Druce, *Chem. News*, **120**, 229 (1920).
851. J. G. F. Druce, *J. Chem. Soc.*, **119**, 758 (1921).
852. O. Schmitz-DuMont, *Z. Anorg. Allg. Chem.*, **248**, 289 (1941).
853. M. Hjortdahl, *C. R. Acad. Sci., Paris*, **88**, 584 (1879).
854. M. P. Brown, E. Cartwell and G. W. A. Fowles, *J. Chem. Soc.*, 506 (1960).
855. R. Sasin and G. S. Sasin, *J. Org. Chem.*, **20**, 770 (1955).
856. E. A. Flood and L. Horvitz, *J. Am. Chem. Soc.*, **55**, 2534 (1933).
857. G. Z. Bredig, *Z. Physik. Chem.*, **13**, 288 (1894).
858. F. Hein and H. Meininger, *Z. Anorg. Allg. Chem.*, **145**, 95 (1925).
859. R. H. Prince, *J. Chem. Soc.*, 1783 (1959).
860. C. A. Kraus and A. M. Neal, *J. Am. Chem. Soc.*, **54**, 2403 (1932).
861. J. G. A. Luijten, *Recl. Trav. Chim. Pays-Bas*, **85**, 873 (1966).
862. H. H. Anderson, *Inorg. Chem.*, **2**, 912 (1964).
863. M. G. Voronkov, *J. Organomet. Chem.*, **557**, 143 (1998).
864. M. G. Voronkov, *Main Group Metal Chem.*, **2**, 235 (1998).
865. M. G. Voronkov, *Russ. Chem. Bull., Chem. Ser.*, **5**, 824 (1998); *Chem. Abstr.*, **129**, 175985j (1998).
866. M. G. Voronkov, *Zh. Obshch. Khim.*, **68**, 950 (1998); *Chem. Abstr.*, **130**, 25119m (1999).
867. P. Pfeiffer, *Chem. Ber.*, **35**, 3303 (1902).
868. J. G. F. Druce, *J. Chem. Soc.*, **121**, 1859 (1922).
869. L. G. F. Druce, *Chem. News*, **127**, 306 (1923).
870. J. G. F. Druce, *Chemist and Druggist*, **112**, 643 (1930).
871. H. Lambourne, *J. Chem. Soc.*, **121**, 2533 (1922).
872. M. Lesbre, *C. R. Acad. Sci., Paris*, **202**, 136 (1936).
873. M. Lesbre and G. Glotz, *C. R. Acad. Sci., Paris*, **198**, 1426 (1934).
874. A. Tchakirian and P. Berillard, *Bull. Soc. Chim. France*, 1300 (1950).
875. I. Zhukov, *Chem. Ber.*, **38**, 2691 (1905).
876. A. Solerio, *Gazz. Chim. Ital.*, **85**, 61 (1955).
877. G. J. M. van der Kerk and J. G. A. Luijten, *J. Appl. Chem.*, **6**, 56 (1956).
878. Metal & Thermit Corporation, *UK Patent* 797976 (1958); *Chem. Abstr.*, **53**, 3061 (1959).
879. S. D. Rosenberg, E. Debreczeni and E. L. Weinberg, *J. Am. Chem. Soc.*, **81**, 972 (1959).
880. Wacker-Chemie GMBH, *German Patent* 957483; *Chem. Abstr.*, **53**, 18865 (1959).
881. M. F. Shostakovskii, V. N. Kotrelev, D. A. Kochkin, G. I. Kuznetsova, S. P. Kalinina and V. V. Borisenko, *Zh. Prikl. Khim.*, **31**, 1434 (1958); *Chem. Abstr.*, **53**, 3040d (1959).
882. G. S. Sasin, *J. Org. Chem.*, **18**, 1142 (1953).

883. R. Okawara and M. Ohara, *J. Organomet. Chem.*, **1**, 360 (1964).
884. J. G. A. Luijten, M. J. Janssen and G. J. M. van der Kerk, *Recl. Trav. Chim. Pays-Bas*, **81**, 202 (1962).
885. G. J. M. van der Kerk, J. G. A. Luijten and M. J. Janssen, *Chimia*, **16**, 10 (1962).
886. W. P. Neumann and F. G. Kleiner, *Tetrahedron Lett.*, 3779 (1964).
887. M. F. Shostakovskii, V. M. Vlasov and R. G. Mirskov, *Zh. Obshch. Khim.*, **34**, 1354, 2843, 3178 (1964); *Chem. Abstr.*, **61**, 6771 (1964).
888. M. F. Shostakovskii, V. M. Vlasov and R. G. Mirskov, *Dokl. Akad. Nauk SSSR*, **159**, 869 (1964); *Chem. Abstr.*, **62**, 7788d (1965).
889. M. F. Shostakovskii, V. M. Vlasov, R. G. Mirskov and I. E. Loginova, *Zh. Obshch. Khim.*, **33**, 2843, 3178 (1964); *Chem. Abstr.*, **62**, 4046h (1965).
890. M. F. Shostakovskii, N. V. Komarov, L. S. Guseva, V. I. Misyunas, A. M. Sklyanova and T. D. Burnashova, *Zh. Obshch. Khim.*, **163**, 390 (1965); *Chem. Abstr.*, **63**, 11601d (1965).
891. M. F. Shostakovskii, N. V. Komarov, L. S. Guseva and V. I. Misyunas, *Zh. Obshch. Khim.*, **34**, 401 (1964); *Chem. Abstr.*, **62**, 2788 (1965).
892. I. F. Lutsenko, S. P. Ponomarev and O. P. Petrii, *Zh. Obshch. Khim.*, **32**, 896 (1962); *Chem. Abstr.*, **58**, 3455 (1963).
893. M. F. Shostakovskii, V. M. Vlasov, R. G. Mirskov and I. M. Korotaeva, *Zh. Obshch. Khim.*, **35**, 401 (1965); *Chem. Abstr.*, **62**, 13167 (1965).
894. A. J. Leusink, J. W. Marsman, H. A. Budding, J. G. Noltes and G. J. M. van der Kerk, *Recl. Trav. Chim. Pays-Bas*, **84**, 567 (1965).
895. R. L. Dannley and W. Aue, *J. Org. Chem.*, **30**, 3845 (1965).
896. K. Sisido and S. Kozima, *J. Org. Chem.*, **27**, 4051 (1962).
897. K. Jones and M. F. Lappert, *J. Organomet. Chem.*, **3**, 295 (1965).
898. G. A. Razuvaev, O. A. Chshepetkova and N. S. Vyazankin, *Zh. Obshch. Khim.*, **32**, 2152 (1962); *Chem. Abstr.*, **58**, 12589 (1963).
899. A. Ladenburg, *Ann. Supl.*, **8**, 79 (1872).
900. A. Ladenburg, *Chem. Ber.*, **3**, 353 (1870).
901. J. D'Ans and H. Gold, *Chem. Ber.*, **92**, 3076 (1959).
902. A. N. Nesmeyanov, I. F. Lutsenko and C. V. Ponomarev, *Dokl. Akad. Nauk SSSR*, **124**, 1073 (1959); *Chem. Abstr.*, **53**, 14984a (1959).
903. G. A. Razuvaev, N. S. Vyazankin and O. A. Chshepetkova, *Zh. Obshch. Khim.*, **30**, 2498 (1960); *Chem. Abstr.*, **55**, 14290f (1961).
904. D. A. Kochkin, V. N. Kotrelev, S. P. Kalinina, G. I. Kuznetsova, G. I. Laǐne, L. V. Chervova, A. I. Borisova and V. V. Borisenko, *Vysokomol. Soed.*, **1**, 1507 (1959); *Chem. Abstr.*, **54**, 14107h (1960).
905. G. S. Sasin and R. Sasin, *J. Org. Chem.*, **20**, 387 (1955).
906. D. A. Kochkin, V. N. Kotrelev, M. F. Shostakovskii, S. P. Kalinina, G. I. Kuznetsova and V. V. Borisenko, *Vysokomol. Soed.*, **1**, 482 (1959); *Chem. Abstr.*, **54**, 5150g (1960).
907. M. M. Koton, *Zh. Obshch. Khim.*, **26**, 3212 (1956); *Chem. Abstr.*, **51**, 8031c (1957).
908. A. A. Yakubovich, S. P. Makarov and V. A. Ginzburg, *Zh. Obshch. Khim.*, **28**, 1036 (1958); *Chem. Abstr.*, **52**, 17094a (1958).
909. C. Quintin, *Ing. Chim.*, **14**, 205 (1930); *Chem. Abstr.*, **26**, 2182 (1932).
910. C. P. Smyth, *J. Am. Chem. Soc.*, **63**, 57 (1941).
911. H. H. Anderson, *J. Org. Chem.*, **22**, 147 (1957).
912. T. M. Andrews, F. A. Bower, B. R. Laliberge and J. C. Montermoso, *J. Am. Chem. Soc.*, **80**, 4102 (1958).
913. M. Lesbre and R. Dupont, *C. R. Acad. Sci., Paris*, **237**, 1700 (1953).
914. H. Zimmer and H. Lübke, *Chem. Ber.*, **85**, 1119 (1952).
915. E. G. Rochow, D. Seyferth and A. C. Smith, *J. Am. Chem. Soc.*, **75**, 3099 (1953).
916. O. Danek, *Collect. Czech. Chem. Commun.*, **26**, 2035 (1961).
917. K. Yasuda and R. Okawara, *J. Organomet. Chem.*, **3**, 76 (1965).
918. B. A. Arbuzov and N. P. Grechkin, *Zh. Obshch. Khim.*, **17**, 2166 (1947); *Chem. Abstr.*, **44**, 5832c (1950).
919. B. A. Arbuzov and N. P. Grechkin, *Izv. Akad. Nauk SSSR, Ser. Khim.*, 440 (1956); *Chem. Abstr.*, **50**, 16661h (1956).
920. B. A. Arbuzov and N. P. Grechkin, *Zh. Obshch. Khim.*, **17**, 2158 (1947); *Chem. Abstr.*, **42**, 4522h (1948).

921. N. Zelinskii and S. Krapivin, *Zh. Rus. Khim. Obshch.*, 579 (1896).
922. A. W. Walde, H. E. van Essen and T. W. Zbornik, *US Patent* 2762821; *Chem. Abstr.*, **51**, 4424 (1957).
923. W. N. Aldridge and J. E. Cremer, *J. Biochem.*, **61**, 406 (1955).
924. K. Gingold, E. G. Rochow, D. Seyferth, A. C. Smith and R. C. West, *J. Am. Chem. Soc.*, **74**, 6306 (1952).
925. H. Schmidbauer and H. Hussek, *Angew. Chem., Int. Ed. Engl.*, **2**, 328 (1963).
926. P. G. Harrison, *Organomet. Chem. Rev.*, **A4**, 379 (1969).
927. D. Wittenberg and H. Gilman, *Quart. Rev.*, **13**, 116 (1959).
928. S. Papetti and H. W. Post, *J. Org. Chem.*, **22**, 526 (1957).
929. K. A. Andrianov and A. A. Zhdanov, *Izv. Akad. Nauk SSSR, Ser. Khim.*, 779 (1958); *Chem. Abstr.*, **52**, 19916i (1958).
930. M. M. Chamberlein, *Tech. Rep. 1*, Western Reserve Univ., 1960.
931. M. M. Chamberlein, G. Kern, G. A. Jabs, D. Germanas, A. Grene, K. Brain and B. Wayland, *US Dept. Com. Office Tech. Serv., PB Rep.* 152086 (1960); *Chem. Abstr.*, **58**, 2508 (1963).
932. V. Gutmann and A. Meller, *Monatsch.*, **91**, 519 (1960).
933. H. Schmidbaur and M. Schmidt, *J. Am. Chem. Soc.*, **83**, 2963 (1961).
934. H. H. Takimoto and J. B. Rust, *J. Org. Chem.*, **26**, 2467 (1961).
935. D. Seyferth and D. L. Alleston, *Inorg. Chem.*, **2**, 418 (1963).
936. H. Schmidbauer and H. Hussek, *Angew. Chem., Int. Ed.*, **1**, 244 (1963).
937. W. S. Tatlock and E. G. Rochow, *J. Org. Chem.*, **17**, 1555 (1952).
938. R. Okawara and K. Sugita, *J. Am. Chem. Soc.*, **83**, 4480 (1961).
939. R. P. Okawara, *Angew. Chem.*, **62**, 231 (1950).
940. R. Okawara, *Proc. Chem. Soc.*, 383 (1961).
941. R. Okawara, D. G. White, K. Fujitani and H. Sato, *J. Am. Chem. Soc.*, **83**, 1342 (1961).
942. J. Beckmann, K. Jurkschat, U. Kaltenbrunner, S. Rabe, M. Schürmann, D. Dakternieks, A. Duthie and O. Möller, *Organometallics*, **19**, 4887 (2000).
943. J. H. Gladstone, *J. Chem. Soc.*, **59**, 290 (1891).
944. N. Zelinskii and S. Krapivin, *Z. Phys. Chem.*, **21**, 35 (1896).
945. Anon, *Chem. Eng. News*, **36**, 34, 40 (1958).
946. P. Kulmiz, *Jaresber.*, 377 (1860).
947. K. A. Kocheshkov and A. N. Nesmeyanov, *Chem. Ber.*, **64**, 628 (1931).
948. T. Harada, *Bull. Chem. Soc. Japan*, **17**, 281 (1942).
949. T. Harada, *Bull. Chem. Soc. Japan*, **17**, 283 (1942).
950. M. Antler, *Ing. Eng. Chem.*, **51**, 753 (1959).
951. M. P. Brown, R. Okawara and E. G. Rochow, *Spectrochim. Acta*, **16**, 595 (1960).
952. T. V. Talalaeva, N. A. Zaitseva and K. A. Kocheshkov, *Zh. Obshch. Khim.*, **16**, 901 (1946); *Chem. Abstr.*, **41**, 2014 (1947).
953. S. A. Edgar and P. A. Teer, *Poultry Sci.*, **36**, 329 (1957).
954. M. Pang and E. I. Becker, *J. Org. Chem.*, **29**, 1948 (1964).
955. E. P. Stefl and C. E. Best, *US Patent* 2731482 (1956); *Chem. Abstr.*, **51**, 13918 (1957).
956. H. E. Ramsden, H. W. Buchanan, J. M. Church and E. W. Johnson, *UK Patent* 719421 (1954); *Chem. Abstr.*, **50**, 397 (1956).
957. Firestone Tire & Rubber Co. *UK Patent* 728953 (1955); *Chem. Abstr.*, **49**, 13693 (1955).
958. G. S. Sasin, A. L. Barror and R. Sasin, *J. Org. Chem.*, **23**, 1366 (1958).
959. Badische Anilin- & Soda-Fabrik Akt.-Ges. *UK Patent* 781905 (1957); *Chem. Abstr.*, **52**, 2049 (1958).
960. D. Blake, G. E. Coates and J. M. Tate, *J. Chem. Soc.*, 618 (1961).
961. E. L. Weinberg and E. W. Johnson, *US Patent* 2648650 (1953); *Chem. Abstr.*, **48**, 10056 (1954).
962. W. E. Leistner and W. E. Setzler, *US Patent* 2752325 (1956); *Chem. Abstr.*, **51**, 1264 (1957).
963. H. E. Ramsden, *UK Patent* 759382 (1956); *Chem. Abstr.*, **51**, 13918 (1957).
964. R. W. Bost and H. R. Baker, *J. Am. Chem. Soc.*, **55**, 1112 (1933).
965. M. Schmidt, H. J. Dersin and H. Schumann, *Chem. Ber.*, **95**, 1428 (1962).
966. H. Schumann and M. Schmidt, *Chem. Ber.*, **96**, 3017 (1963).
967. M. G. Voronkov, N. S. Vyazankin, E. N. Deryagina, A. S. Nakhmanovich and V. A. Usov, *Reactions of Sulfur with Organic Compounds*, Consultant Bureau, New York, 1987.
968. C. A. Kraus and A. M. Neal, *J. Am. Chem. Soc.*, **52**, 695 (1930).

969. G. P. Mack and E. Parker, *US Patent* 2618625 (1952); *Chem. Abstr.*, **47**, 1977 (1953).
970. G. P. Mack and E. Parker, *US Patent* 2634281 (1953); *Chem. Abstr.*, **48**, 1420 (1954).
971. H. McCombie and B. C. Saunders, *Nature*, **159**, 491 (1947).
972. D. Seyferth and N. Kahlem, *J. Org. Chem.*, **25**, 809 (1960).
973. S. F. A. Kettle, *J. Chem. Soc.*, 2936 (1959).
974. E. Wiberg and R. Rieger, *German Patent* 1121050 (1960); *Chem. Abstr.*, **56**, 14328 (1962).
975. E. W. Abel, D. Brady and B. R. Lerwill, *Chem. Ind.*, 1333 (1962).
976. K. Jones and M. F. Lappert, *Proc. Chem. Soc.*, 358 (1962).
977. K. Sisido and S. Kozima, *J. Org. Chem.*, **29**, 907 (1964).
978. N. G. Pai, *Proc. Roy. Soc. (London), Ser.* A, **149**, 29 (1935).
979. C. A. Kraus and W. H. Kahler, *J. Am. Chem. Soc.*, **55**, 3537 (1933).
980. B. A. Arbuzov and A. N. Pudovik, *Zh. Obshch. Khim.*, **17**, 2158 (1947); *Chem. Abstr.*, **42**, 4522a (1948).
981. W. Kuchen and H. Buchwald, *Chem. Ber.*, **92**, 227 (1959).
982. Standard Oil Development Company, *UK Patent* 445813 (1935); *Chem. Abstr.*, **30**, 6936 (1936).
983. C. A. Kraus and W. N. Greer, *J. Am. Chem. Soc.*, **47**, 2568 (1925).
984. K. A. Jensen and N. Clauson-Kass, *Z. Anorg. Allg. Chem.*, **250**, 277 (1943).
985. G. A. Razuvaev and E. I. Fedotova, *Zh. Obshch. Khim.*, **21**, 1118 (1951); *Chem. Abstr.*, **46**, 5006g (1952).
986. J. D'Ans, H. Zimmer, E. Endrulat and K. Lübke, *Naturwissenschaften*, **39**, 450 (1952).
987. W. P. Neumann and K. König, *Ann. Chem.*, **677**, 1 (1964).
988. L. Riccoboni and P. Popov, *Atti ist. Veneto Sci., Lettere ed Arti, Classe Sci. Mat. Nat.*, **107**, II, 123 (1949); *Chem. Abstr.*, **44**, 6752 (1950).
989. H. G. Kuivila, A. K. Sawyer and A. G. Armour, *J. Org. Chem.*, **26**, 1426 (1961).
990. W. P. Neumann, *Angew. Chem.*, **74**, 122 (1962).
991. W. P. Neumann, *Angew. Chem.*, **74**, 215 (1962).
992. A. Ladenburg, *Ann. Chem. (Supl)*, **8**, 67 (1869).
993. L. Rügheimer, *Ann. Chem.*, **364**, 51 (1908).
994. G. Grüttner, *Chem. Ber.*, **50**, 1808 (1917).
995. C. A. Kraus and H. Eatough, *J. Am. Chem. Soc.*, **55**, 5008 (1933).
996. C. A. Kraus and R. H. Bullard, *J. Am. Chem. Soc.*, **48**, 2131 (1926).
997. L. Riccoboni, *Atti ist. Veneto Sci., Lettere ed Arti, Classe Sci. Mat. Nat.*, **96**, II, 183 (1937); *Chem. Abstr.*, **33**, 7207 (1939).
998. T. G. Bonner, J. M. Clayton and G. Williams, *J. Chem. Soc.*, 1705 (1958).
999. G. Bähr and R. Gelius, *Chem. Ber.*, **91**, 825 (1958).
1000. C. A. Kraus and H. Eatough, *J. Am. Chem. Soc.*, **55**, 5014 (1933).
1001. C. A. Kraus and A. M. Neal, *J. Am. Chem. Soc.*, **52**, 4426 (1930).
1002. H. Gilman and S. D. Rosenberg, *J. Org. Chem.*, **18**, 680 (1953).
1003. O. H. Johnson and H. E. Fritz, *J. Org. Chem.*, **19**, 74 (1954).
1004. O. H. Johnson, H. E. Fritz, D. O. Halvorson and R. L. Evans, *J. Am. Chem. Soc.*, **77**, 5857 (1955).
1005. M. K. Zaikina, *Uch. Zap. Kazan. Universitet*, **116**, 129 (1956).
1006. W. P. Neumann, *Angew. Chem.*, **75**, 679 (1963).
1007. N. N. Zemlyanskii, E. M. Panov and K. A. Kocheshkov, *Dokl. Akad. Nauk SSSR*, **146**, 1335 (1962); *Chem. Abstr.*, **58**, 9110 (1963).
1008. M. P. Brown and W. A. Fowbes, *J. Chem. Soc.*, 2811 (1958).
1009. A. N. Nesmeyanov, K. A. Kocheshkov and V. P. Puzyreva, *Dokl. Akad. Nauk SSSR*, **7**, 118 (1937); *Chem. Abstr.*, **31**, 4290 (1937).
1010. G. Bähr and R. Gelius, *Chem. Ber.*, **91**, 829 (1958).
1011. A. Werner, *Neuere Anschanungen auf dem Gebiete der Anorganischen Chemie*, F. Vieweg & Sohn Akt.-Ges., Braunschweig, 1923.
1012. G. M. Richardson and M. Adams, *J. Am. Chem. Soc.*, **22**, 446 (1900); *Chem. Zbl.*, **1**, 282 (1900).
1013. C. A. Kraus and W. N. Greer, *J. Am. Chem. Soc.*, **45**, 2946 (1923).
1014. K. A. Kocheshkov, *Uch. Zap. Mosk. Univ.*, **3**, 297 (1934); *Chem. Abstr.*, **30**, 8184 (1936).
1015. A. B. Thomas and E. G. Rochow, *J. Inorg. Nucl. Chem.*, **4**, 205 (1957).
1016. L. G. F. Druce, *J. Chem. Soc.*, **113**, 715 (1918).

1017. O. A. Reutov, O. A. Ptitsina and N. D. Patrina, *Zh. Obshch. Khim.*, **28**, 588 (1958); *Chem. Abstr.*, **52**, 17151a (1958).
1018. C. A. Kraus and F. C. Schmidt, *J. Am. Chem. Soc.*, **56**, 2297 (1934).
1019. F. Kehrmann, *Chem. Ber.*, **34**, 3818 (1901).
1020. I. R. Beattie and G. P. McQuillan, *J. Chem. Soc.*, 1519 (1963).
1021. I. R. Beattie, G. P. McQuillan and R. Hulme, *Chem. Ind.*, 1429 (1962).
1022. R. Hulme, *J. Chem. Soc.*, 1524 (1963).
1023. I. P. Gol'dstein, E. N. Gurianova, E. D. Delinskaya and K. A. Kocheshkov, *J. Russ. Chem. Soc.*, **136**, 1079 (1960).
1024. I. P. Gol'dshtein, N. A. Faizi, N. A. Slovokhotova, E. N. Gur'yanova, I. M. Viktorova and K. A. Kocheshkov, *Dokl. Akad. Nauk SSSR*, **138**, 839 (1961); *Chem. Abstr.*, **55**, 27208c (1961).
1025. A. Strecher, *Ann. Chem.*, **105**, 306 (1858).
1026. P. Pfeiffer and O. Brach, *Z. Anorg. Chem.*, **87**, 229 (1914).
1027. T. Harada, *Sci. Papers Inst. Phys.-Chem. Res. (Tokyo)*, **42**, 57, 59, 62, 64 (1947); *Chem. Abstr.*, **43**, 7900 (1949).
1028. T. Harada, *Sci. Papers Inst. Phys.-Chem. Res. (Tokyo)*, **35**, 497 (1939); *Chem. Abstr.*, **34**, 3674 (1940).
1029. G. P. Mack and E. Parker, *German Patent* 838212; *Chem. Zbl.*, 6771 (1952).
1030. T. Harada, *Bull. Chem. Soc. Japan*, **15**, 455 (1940).
1031. C. A. Kraus and T. Harada, *J. Am. Chem. Soc.*, **47**, 2416 (1925).
1032. T. Harada, *Sci. Papers Inst. Phys.-Chem. Res. (Tokyo)*, **36**, 504 (1939); *Chem. Abstr.*, **34**, 3675 (1940).
1033. C. A. Kraus and C. C. Callis, *J. Am. Chem. Soc.*, **45**, 2624 (1923).
1034. C. A. Kraus, *J. Am. Chem. Soc.*, **46**, 2196 (1924).
1035. A. B. Thomas and E. G. Rochow, *J. Am. Chem. Soc.*, **79**, 1843 (1957).
1036. J. Teply and J. Maly, *Chem. Zvesti*, **7**, 463 (1953); *Chem. Abstr.*, **48**, 10427 (1954).
1037. E. G. Rochow, D. Seyferth and A. C. Smith, *J. Am. Chem. Soc.*, **75**, 2877 (1953).
1038. M. F. Lappert, in *Frontiers of Organogermanium, -Tin and -Lead Chemistry* (Eds. E. Lukevics and L. Ignatovich), Latvian Institute of Organic Synthesis, Riga, 1993, pp. 9–27.
1039. O. Schmitz-DuMont and G. Bungard, *Chem. Ber.*, **92**, 2399 (1959).
1040. M. Mehring, C. Löw, M. Schürmann, F. Uhlig and K. Jurkschat, *Organometallics*, **19**, 4613 (2000).
1041. R. W. Leeper, L. Summers and H. Gilman, *Chem. Rev.*, **54**, 101 (1954).
1042. O. M. Nefedov and M. N. Manakow, *Angew. Chem.*, **78**, 1039 (1966).
1043. W. P. Neumann, *Silicon, Germanium, Tin and Lead*, 81 (1978).
1044. D. E. Goldberg, P. B. Hitchcock, M. F. Lappert and K. M. Thomas, *J. Chem. Soc., Chem. Commun.*, 261 (1976).
1045. T. Fjeldberg, A. Haaland, B. E. R. Schilling, M. F. Lappert and A. J. Thorne, *J. Chem. Soc., Dalton Trans.*, 1551 (1986).
1046. J. G. A. Luijten, in *Organotin Compounds*, Vols. 1–3, (Ed. A. K. Sawayer), Dekker, New York, 1970–1972.
1047. W. J. Connolly, *Paper Trade J.*, **141**, 31, 46 (1957); *Chem. Abstr.*, **52**, 2404 (1958).
1048. A. K. Sÿpesteyn, J. G. A. Luijten and G. J. M. van der Kerk, in *Fungicides: An Advance Treatise* (Ed. D. S. Torgeson), Academic Press., New York, 1969.
1049. J. M. Barnes and H. B. Stoner, *Pharmacol. Rev.*, **11**, 211 (1959).
1050. O. Zeman, E. Gadermann and K. Hardebeck, *Deutsche Arch. Klin. Med.*, **198**, 713 (1951); *Chem. Abstr.*, **46**, 6265 (1952).
1051. T. P. White, *Arch. Experim. Pathol. Pharmakol.*, **13**, 53 (1881).
1052. T. P. White, *J. Pharmacol.*, **17**, 166 (1886).
1053. E. Ungar and G. Bodländer, *Z. Hyg. Infektionskrank.*, **2**, 241 (1886); *Chem. Zbl.*, **85**, 644 (1887).
1054. R. Hunt, *J. Pharmacol.*, **28**, 367 (1926).
1055. W. A. Collier, *Z. Hyg. Infektionskrank.*, **110**, 169 (1929); *Chem. Zbl.*, **I**, 2554 (1929).
1056. F. M. Caujolle, M. Lesbre and D. Meynier, *C. R. Acad. Sci., Paris*, **239**, 1091 (1954).
1057. H. B. Stoner, J. M. Barnes and J. I. Duff, *Br. J. Pharmacol.*, **10**, 6 (1955).
1058. J. Seifter, *J. Pharmacol. Exper. Ther.*, **66**, 32 (1939).
1059. H. Gilman, *Organotin Compounds*, PB. 60004 OSRD Report, 1942.

1060. H. G. Glass, J. M. Coon, C. C. Lushbaugh and J. Lust, *Toxicity and Vesicant Action of Various Organic Tin Compounds*, PB 50814, Univ. of Chicago Tox. Lab., 1942, p. 27.
1061. F. Caujolle, M. Lesbre and D. Meynier, *Ann.. Pharm. Franc.*, **14**, 88 (1956); *Chem. Abstr.*, **50**, 13288 (1956).
1062. P. N. Magee, H. B. Stoner and J. M. Barnes, *J. Pathol. Bacteriol.*, **73**, 107 (1957); *Chem. Abstr.*, **51**, 1264 (1957).
1063. J. M. Barnes and P. N. Magee, *J. Pathol. Bacteriol.*, **75**, 267 (1958).
1064. G. B. Fahlstrom, *Tin and Its Uses (Quarterly J. of the Tin Research Institute)*, **46**, 4 (1959).
1065. W. N. Aldridge, *J. Biochem.*, **69**, 367 (1958).
1066. H. B. Stoner and C. J. Threlfall, *Biochem. J.*, **69**, 376 (1958).
1067. A. Manten, H. L. Klopping and G. J. M. van der Kerk, *Ant. v. Leeuwenhock*, **16**, 282 (1950).
1068. S. L. Ruskin, *US Patent* 2115751 (1938); *Chem. Abstr.*, **32**, 4725 (1938).
1069. C. S. Wright, *Med. Rec.*, **147**, 453 (1938); *Chem. Zbl.*, **11**, 1444 (1938).
1070. K. B. Kerr, *Poultry Sci.*, **31**, 328 (1952); *Chem. Abstr.*, **46**, 11557 (1952).
1071. K. B. Kerr and A. W. Walde, *US Patent* 2702775, 2702776, 2702777, 2702778 (1955); *Chem. Abstr.*, **49**, 7816 (1955).
1072. H. P. *Br. Med. J.*, **1**, 515 (1958).
1073. J. G. Farbenindustrie Akt.-Ges., *UK Patent* 303092 (1927); *Chem. Abstr.*, **23**, 45841 (1929).
1074. E. Hardtmann, M. Hardtmann, P. Kümmel, *US Patent* 1744633 (1930); *Chem. Abstr.*, **24**, 1524 (1930).
1075. E. Hardtmann, P. Kümmel and M. Hardtmann, *German Patent* 485646; *Chem. Abstr.*, **24**, 1230 (1930).
1076. N. V. de Bataafsche, Petroleum Maatschappij, *Dutch Patent* 68578 (1951); *Chem. Abstr.*, **46**, 5781 (1952).
1077. E. I. Du Pont de Nemours and Co, W. H. Tisdale, *UK Patent* 578312 (1946).
1078. M. G. Voronkov, N. F. Chernov, A. A. Baigozhin, *Zh. Prikl. Khim.*, **69**, 1594 (1996); *Chem. Abstr.*, **126**, 119078s (1997).
1079. M. S. Blum and F. A. Bower, *J. Econ. Entomol.*, **50**, 84 (1957); *Chem. Abstr.*, **51**, 10823 (1957).
1080. F. E. Brinckman and J. M. Bellama (Eds.), *Organometals and Organometalloids—Occurrence and Fate in the Environment, Am. Chem. Soc.*, 1978.
1081. V. Yngve, *US Patent* 2219463 (1940); *Chem. Abstr.*, **35**, 1145 (1941).
1082. V. Yngve, *US Patent* 2267777, 2267778, 2267779 (1991); *Chem. Abstr.*, **36**, 2647 (1942).
1083. V. Yngve, *US Patent* 2307092 (1943); *Chem. Abstr.*, **37**, 3532 (1943).
1084. W. M. Quattlebaum and C. A. Noffsinger, *US Patent* 2307157; *Chem. Abstr.*, **37**, 3533 (1943).
1085. E. W. Rugeley and W. M. Jr. Quattlebaum, *US Patent* 2344002 (1944); *Chem. Abstr.*, **38**, 3392 (1944).
1086. G. P. Mack, *Modern Plastics Encyclopedia*, **37**, 333 (1959).
1087. D. E. Winkler, *Ind. Eng. Chem.*, **50**, 863 (1958).
1088. V. W. Fox, J. G. Hendricks and H. J. Ratti, *Ind. Eng. Chem.*, **41**, 1774 (1949).
1089. G. P. Mack, *Modern Plastics Encyclopedia*, **35**, 377 (1957).
1090. H. V. Schmith, *Plastics*, **17**, 264 (1952).
1091. A. S. Kenyon, *Nat. Bur. Stand. (U. S.), Circ.*, **525**, 81 (1953); *Chem. Abstr.*, **48**, 7338 (1954).
1092. E. L. Raab and F. C. Gorsline, *US Patent* 2734926 (1956); *Chem. Abstr.*, **50**, 8945 (1956).
1093. E. S. Hedges, *Tin and Its Uses (Quarterly J. of the Tin Research Institute)*, **38**, 8 (1957).
1094. P. H. Zwalneppel, *Tin and Its Uses (Quarterly J. of the Tin Research Institute)*, **45**, 7 (1958).
1095. J. W. Churchill, *US Patent* 2643242 (1953); *Chem. Abstr.*, **47**, 10904 (1953).
1096. J. W. Churchill, *Canada Patent* 506310 (1954); *Chem. Abstr.*, **48**, 10904a (1954).
1097. W. T. Rossiter and C. C. Currie, *US Patent* 2742368 (1956); *Chem. Abstr.*, **50**, 11713 (1956).
1098. E. J. Hart, *US Patent* 2476661; *Chem. Abstr.*, **43**, 9519 (1949).
1099. O. R. Klimmer, *Arzneim-Forsch.*, **19**, 934 (1969).
1100. K. R. S. Ascher and S. Nissim, *World Rev. Post Control*, **3**, 7 (1964); **3**, 188 (1964).
1101. C. J. Evans and P. J. Smith, *J. Oil Colour Chem. Assoc.*, **58**, 5, 160 (1975).
1102. H. Plum, *Chim. Peint*, **35**, 4, 127 (1972); *Chem. Abstr.*, **77**, 84251h (1972).
1103. H. C. Stecker, *Tin and Its Uses (Quarterly J. of the Tin Research Institute)*, **41**, 13 (1957).
1104. S. C. Britton, *Tin and Its Uses (Quarterly J. of the Tin Research Institute)*, **36**, 10 (1956); **38**, 6 (1957).

1105. M. S. Beum and F. A. Bower, *J. Econ. Entomol.*, **50**, 84 (1957); *Chem. Abstr.*, **51**, 10823 (1957).
1106. M. H. M. Arnold, *J. Oil Colour Chem. Assoc.*, **39**, 900 (1956).
1107. C. Löwig, *Chem. Zentr.*, 575 (1852).
1108. G. Grüttner and E. Krause, *Chem. Ber.*, **54**, 2065 (1921).
1109. G. Grüttner and E. Krause, *Chem. Ber.*, **49**, 1125 (1916).
1110. G. Grüttner and E. Krause, *Chem. Ber.*, **49**, 1415 (1916).
1111. G. Grüttner and E. Krause, *Chem. Ber.*, **49**, 1546 (1916).
1112. G. Grüttner and E. Krause, *Chem. Ber.*, **49**, 2666 (1916).
1113. G. Grüttner and E. Krause, *Chem. Ber.*, **50**, 202 (1917).
1114. G. Grüttner and E. Krause, *Chem. Ber.*, **50**, 278 (1917).
1115. G. Grüttner and E. Krause, *Chem. Ber.*, **50**, 574 (1917).
1116. G. Grüttner and E. Krause, *Chem. Ber.*, **50**, 1559 (1917).
1117. G. Grüttner, *Ann. Chem.*, **415**, 338 (1918).
1118. G. Grüttner, *Chem. Ber.*, **51**, 1298 (1918).
1119. G. Grutter and G. Grüttner, *Chem. Ber.*, **51**, 1293 (1918).
1120. E. Krause and N. Schmitz, *Chem. Ber.*, **52**, 2165 (1919).
1121. E. Krause, *Chem. Ber.*, **54**, 2060 (1921).
1122. E. Krause and G. G. Reissaus, *Chem. Ber.*, **55**, 888 (1922).
1123. E. Krause and E. Pohland, *Chem. Ber.*, **52**, 1282 (1919).
1124. E. Krause and O. Schlöttig, *Chem. Ber.*, **58**, 427 (1925).
1125. E. Krause and G. Renwanz, *Chem. Ber.*, **60**, 1582 (1927).
1126. E. Krause, *Chem. Ber.*, **62**, 135 (1929).
1127. E. Krause, *Chem. Ber.*, **62**, 1877 (1929).
1128. H. Gilman, in *Organic Chemistry. An Advanced Treatise*, 2nd Edn., Vol. 1, Chap. 5, Wiley, New York, 1943, pp. 489–580.
1129. H. Gilman, W. H. Atwell and F. K. Cartledge, in *Adv. Organomet. Chem.*, Vol. 4, Academic Press, New York, 1966, pp. 63–94.
1130. K. A. Kocheshkov and A. P. Aleksandrov, *Chem. Ber.*, **67**, 527 (1934).
1131. A. P. Aleksandrov and A. N. Nesmeyanov, *Zh. Obshch. Khim.*, **4**, 1102 (1934); *Chem. Abstr.*, **29**, 3993 (1935).
1132. K. A. Kocheshkov and A. N. Nesmeyanov, *Zh. Obshch. Khim.*, **6**, 172 (1936); *Chem. Abstr.*, **30**, 4834 (1936).
1133. K. A. Kocheshkov and A. P. Aleksandrov, *Zh. Obshch. Khim.*, **7**, 93 (1937); *Chem. Abstr.*, **31**, 4291 (1937).
1134. K. A. Kocheshkov and G. M. Borodina, *Izv. Akad. Nauk SSSR, Ser. Khim.*, 569 (1937); *Chem. Abstr.*, **32**, 2095 (1938).
1135. E. G. Rochow, D. T. Hurd and R. N. Lewis, *The Chemistry of Organolead Compounds*, Wiley, New York, 1957, 344 pp.
1136. P. R. Austin, *J. Am. Chem. Soc.*, **54**, 3287 (1932).
1137. F. Paneth and W. Lautsch, *Chem. Ber.*, **64**, 2702, 2708 (1931).
1138. F. Paneth and W. Lautsch, *Naturwissenschaften*, **18**, 307 (1930).
1139. F. Paneth and W. Hofeditz, *Chem. Ber.*, **62**, 1335 (1929).
1140. T. Midgley Jr., C. A. Hochwalt and G. Calingaert, *J. Am. Chem. Soc.*, **45**, 1821 (1923).
1141. A. Polis, *Chem. Ber.*, **20**, 716 (1887).
1142. A. Ghira, *Atti Rep. Acc. Dei Linc. Roma*, **2**, 216 (1893).
1143. A. Ghira, *Gazz. Chim. Ital.*, **24**, 42 (1894).
1144. J. Tafel, *Chem. Ber.*, **44**, 323 (1911).
1145. G. Calingaert, *Chem. Rev.*, **2**, 43 (1925).
1146. F. Fichter and I. Stein, *Helv. Chim. Acta*, **14**, 1205 (1931).
1147. A. Goldach, *Helv. Chim. Acta*, **14**, 1436 (1931).
1148. G. Calingaert, H. Shapiro, F. J. Dykstra and L. Hess, *J. Am. Chem. Soc.*, **70**, 3902 (1948).
1149. R. Heap and B. C. Saunders, *J. Chem. Soc.*, 2983 (1949).
1150. B. S. Saunders and G. J. Stacey, *J. Chem. Soc.*, 919 (1949).
1151. G. Calingaert, *US Patent* 1622233 (1927); *Chem. Abstr.*, **21**, 1546h (1927).
1152. W. S. Calcott, A. E. Parmelee and F. R. Lorriman, *UK Patent* 280169 (1927); *Chem. Abstr.*, **22**, 3042f (1928).
1153. C. A. Kraus, *US Patent* 1694268 (1928); *Chem. Abstr.*, **23**, 970 (1929).

1154. G. Calingaert and H. Shapiro, *US Patent* 2535190, 2535191, 2535192 (1950); *Chem. Abstr.*, **45**, 3864 (1951).
1155. H. Shapiro, *US Patent* 2535235, 2535236, 2535237 (1950); *Chem. Abstr.*, **45**, 3865 (1951).
1156. I. T. Krohn and H. Shapiro, *US Patent* 2594183, 2594225 (1952); *Chem. Abstr.*, **47**, 145i, 146a (1953).
1157. S. W. Turnrer and B. A. Fader, *Ind. Eng. Chem.*, **54**, 4, 52 (1962).
1158. A. Cahours, *C. R. Acad. Sci., Paris*, **36**, 1001 (1853).
1159. G. Calingaert, *US Patent* 1539297 (1925); *Chem. Abstr.*, **19**, 2210f (1925).
1160. B. Mead, *US Patent* 1567159 (1925); *Chem. Abstr.*, **20**, 607 (1926).
1161. M. M. Nad' and K. A. Kocheshkov, *Zh. Obshch. Khim.*, **12**, 409 (1942); *Chem. Abstr.*, **37**, 3068 (1943).
1162. E. Frankland and A. Lawrence, *J. Chem. Soc.*, **35**, 244 (1879).
1163. G. B. Buckton, *Chem. Gazette*, 276 (1859).
1164. M. Meyer, *Chem. News*, **131**, 1 (1925); *Chem. Abstr.*, **19**, 2637 (1925).
1165. P. Pfeiffer and P. Trüskier, *Chem. Ber.*, **37**, 1125 (1904).
1166. S. Möller and P. Pfeiffer, *Chem. Ber.*, **49**, 2441 (1916).
1167. H. Gilman and J. Robinson, *J. Am. Chem. Soc.*, **49**, 2315 (1927).
1168. W. C. Setzer, R. W. Leeper and H. Gilman, *J. Am. Chem. Soc.*, **61**, 1609 (1939).
1169. K. A. Kocheshkov and T. M. Borodina, *Izv. Akad. Nauk SSSR, Ser. Khim.*, 569 (1937); *Chem. Abstr.*, **32**, 2095 (1938).
1170. H. Gilman and J. C. Bailie, *J. Am. Chem. Soc.*, **61**, 731 (1939).
1171. M. Lesbre, J. Satge and D. Voigt, *C. R. Acad. Sci., Paris*, **246**, 594 (1958).
1172. C. A. Kraus and C. C. Callis *US Patent* 1690075 (1928); *Chem. Abstr.*, **23**, 245 (1929).
1173. H. W. Daudt, *French Patent* 642120 (1928); *Chem. Abstr.*, **23**, 1143i (1929).
1174. H. W. Daudt, *UK Patent* 283913 (1928); *Chem. Abstr.*, **22**, 4134 (1928).
1175. H. W. Daudt, *UK Patent* 279106 (1928); *Chem. Abstr.*, **22**, 2836d (1928).
1176. M. Lesbre, *C. R. Acad. Sci., Paris*, **210**, 535 (1940).
1177. E. C. Juenge and C. E. Cook, *J. Am. Chem. Soc.*, **81**, 3578 (1959).
1178. L. D. Apperson, *Iowa State Coll. J. Sci.*, **16**, 7 (1941); *Chem. Abstr.*, **36**, 4476 (1942).
1179. F. Glockling, K. Hooton and D. Kingston, *J. Chem. Soc.*, 4405 (1961).
1180. P. R. Austin, *J. Am. Chem. Soc.*, **55**, 2948 (1933).
1181. P. R. Austin, H. Gilman and J. C. Bailie, *J. Am. Chem. Soc.*, **61**, 731 (1939).
1182. E. Bindschadler and H. Gilman, *Proc. Iowa Acad. Sci.*, **48**, 273 (1941); *Chem. Abstr.*, **36**, 1595 (1942).
1183. H. Gilman and F. N. Moore, *J. Am. Chem. Soc.*, **62**, 1843 (1940).
1184. H. Gilman and R. G. Jones, *J. Am. Chem. Soc.*, **72**, 1760 (1950).
1185. H. Gilman, L. Summers and R. W. Leeper, *J. Org. Chem.*, **17**, 630 (1952).
1186. W. Schlenk and J. Holtz, *Chem. Ber.*, **50**, 262 (1917).
1187. F. Hein and E. Nebe, *Chem. Ber.*, **75**, 1744 (1942).
1188. G. Calingaert and H. Soroos, *J. Org. Chem.*, **34**, 535 (1938).
1189. L. S. Foster, W. M. Dix and I. J. Grumtfest, *J. Am. Chem. Soc.*, **61**, 1685 (1939).
1190. E. Bindschadler, *Iowa State Coll. J. Sci.*, **16**, 33 (1941); *Chem. Abstr.*, **36**, 4476 (1942).
1191. H. Gilman and D. S. Melstrom, *J. Am. Chem. Soc.*, **72**, 2953 (1950).
1192. F. Glockling and D. Kingston, *J. Chem. Soc.*, 3001 (1959).
1193. H. Gilman and L. Summers, *J. Am. Chem. Soc.*, **74**, 5924 (1952).
1194. A. E. Shurov and G. A. Razuvaev, *Chem. Ber.*, **65**, 1507 (1932).
1195. A. N. Nesmeyanov, R. K. Fredlina and A. Kochetkov, *Izv. Akad. Nauk SSSR, Ser. Khim.*, 127 (1948); *Chem. Abstr.*, **43**, 1716 (1949).
1196. A. N. Nesmeyanov and A. E. Borisov, *Dokl. Akad. Nauk SSSR*, **60**, 67 (1948); *Chem. Abstr.*, **43**, 560 (1949).
1197. E. M. Panov, V. I. Lodochnikova and K. A. Kocheshkov, *Dokl. Akad. Nauk SSSR*, **111**, 1042 (1956); *Chem. Abstr.*, **51**, 9512b (1957).
1198. V. I. Lodochnikova, E. M. Panov and K. A. Kocheshkov, *Izv. Akad. Nauk SSSR, Ser. Khim.*, 1484 (1957); *Chem. Abstr.*, **52**, 7245 (1958).
1199. V. I. Lodochnikova, E. M. Panov and K. A. Kocheshkov, *Zh. Obshch. Khim.*, **29**, 2253 (1959); *Chem. Abstr.*, **54**, 10967f (1960).
1200. V. I. Lodochnikova, E. M. Panov and K. A. Kocheshkov, *Zh. Obshch. Khim.*, **34**, 4022 (1964); *Chem. Abstr.*, **62**, 9164c (1965).

1201. F. Hein and A. Klein, *Chem. Ber.*, **71**, 2381 (1938).
1202. G. A. Razuvaev, N. S. Vyazankin and N. N. Vyshinskii, *Zh. Obshch. Khim.*, **29**, 3662 (1959); *Chem. Abstr.*, **54**, 17015 (1960).
1203. G. A. Razuvaev, N. S. Vyazankin and N. N. Vyshinskii, *Zh. Obshch. Khim.*, **30**, 967 (1960); *Chem. Abstr.*, **54**, 23646 (1960).
1204. H. Jenker, *Z. Naturforsch.*, **12b**, 909 (1957).
1205. S. M. Blitzer and T. H. Pearson, *US Patent* 2859227 (1958); *Chem. Abstr.*, **53**, 9149d (1959).
1206. E. H. Dobratz, *US Patent* 2816123 (1957); *Chem. Abstr.*, **52**, 7344g (1958).
1207. L. G. Makarova, *Reactions and Investigations Methods of Organic Compounds*, Vol. 3, GNTIKhL, Moscow, 1954; *Chem. Abstr.*, **49**, 9535h (1955).
1208. K. A. Kocheshkov, A. N. Nesmeyanov and N. K. Gipp, *Zh. Obshch. Khim.*, **6**, 172 (1936); *Chem. Abstr.*, **30**, 4834 (1936).
1209. A. N. Nesmeyanov, K. A. Kocheshkov and M. M. Nad', *Izv. Akad. Nauk SSSR, Ser. Khim.*, 522 (1945); *Chem. Abstr.*, **42**, 5870 (1948).
1210. A. Ya. Yakubovich, S. P. Makarova, V. A. Ginsburg, G. I. Gavrilov and E. N. Merkulova, *Dokl. Akad. Nauk SSSR*, **72**, 69 (1950); *Chem. Abstr.*, **45**, 2856 (1951).
1211. A. Ya. Yakubovich, E. N. Merkulova, S. P. Makarova and G. I. Gavrilov, *Zh. Obshch. Khim.*, **22**, 2060 (1952); *Chem. Abstr.*, **47**, 9257i (1953).
1212. W. E. Becker and S. E. Cook, *J. Am. Chem. Soc.*, **82**, 6264 (1960).
1213. W. P. Neumann and K. Kühlein, *Angew. Chem.*, **77**, 808 (1965).
1214. A. J. Leusink and G. J. M. van der Kerk, *Recl. Trav. Chim. Pays-Bas*, **84**, 1617 (1965).
1215. L. C. Willemsens and G. J. M. van der Kerk, *J. Organomet. Chem.*, **4**, 241 (1965).
1216. E. M. Panov and K. A. Kocheshkov, *Dokl. Akad. Nauk SSSR*, **123**, 295 (1958); *Chem. Abstr.*, **53**, 7133b (1959).
1217. J. Klippel, *Jahresber.*, **380**, 383 (1860).
1218. J. Klippel, *Z. Prakt. Chem.*, **81**, 287 (1860).
1219. L. Zechmeister and J. Csabay, *Chem. Ber.*, **60**, 1617 (1927).
1220. G. A. Razuvaev and I. F. Bogdanov, *Zh. Rus. Fiz. Khim. Obshch.*, **61**, 1791 (1929).
1221. G. A. Razuvaev and I. F. Bogdanov, *Dokl. Akad. Nauk SSSR*, 159 (1929); *Chem. Abstr.*, **24**, 2660 (1930).
1222. G. A. Razuvaev and M. M. Koton, *Chem. Ber.*, **66**, 854 (1933).
1223. V. N. Ipatiev, G. A. Rasuviev and I. F. Bogdanov, *Chem. Ber.*, **63**, 335 (1930).
1224. M. M. Koton, *J. Am. Chem. Soc.*, **56**, 1118 (1934).
1225. M. F. Dull and J. H. Simons, *J. Am. Chem. Soc.*, **55**, 4328 (1933).
1226. M. F. Dull and J. H. Simons, *J. Am. Chem. Soc.*, **55**, 3898 (1933).
1227. F. Paneth and K. Herzfeld, *L. Elektrochem.*, **37**, 577 (1931).
1228. G. Calingaert, *US Patent* 1539297 (1925); *Chem. Abstr.*, **19**, 2210 (1925).
1229. H. L. Taylor and W. N. Jones, *J. Am. Chem. Soc.*, **52**, 1111 (1930).
1230. J. H. Simons, R. W. McNamee and C. D. Hurd, *J. Phys. Chem.*, **36**, 939 (1932).
1231. R. N. Meinert, *J. Am. Chem. Soc.*, **55**, 979 (1933).
1232. P. L. Cramer, *J. Am. Chem. Soc.*, **56**, 1234 (1934).
1233. R. Garzuly, *Z. Prakt. Chem.*, **142**, 141 (1935).
1234. G. A. Razuvaev, N. S. Vyazankin and N. N. Vyshinskii, *Zh. Obshch. Khim.*, **29**, 3662 (1959); *Chem. Abstr.*, **54**, 17015 (1960).
1235. P. A. Leighton and R. A. Mortensen, *J. Am. Chem. Soc.*, **58**, 448 (1936).
1236. B. J. McDonald, D. Bryce-Smith and B. Pendilly, *J. Chem. Soc.*, 2174 (1959).
1237. G. A. Razuvaev, G. G. Petukhov and Yu. A. Kaplin, *Zh. Obshch. Khim.*, **33**, 2394 (1963); *Chem. Abstr.*, **59**, 14014 (1964).
1238. D. P. Evans, *J. Chem. Soc.*, 1466 (1938).
1239. H. Gilman and L. D. Apperson, *J. Org. Chem.*, **4**, 162 (1939).
1240. Yu. A. Aleksandrov, T. G. Brilkina and V. A. Shushunov, *Tr. Khim. i Khim. Tekhnol.*, **3**, 623 (1959); *Chem. Abstr.*, **56**, 14314c (1962).
1241. R. Heap, B. S. Saunders and G. J. Stacey, *J. Chem. Soc.*, 658 (1951).
1242. H. Adkins, *Chem. Ber.*, **63**, 335 (1930).
1243. H. Adkins and L. W. Covert, *Z. Phys. Chem.*, **35**, 1684 (1931).
1244. H. Adkins and R. Connor, *J. Am. Chem. Soc.*, **53**, 1091 (1931).
1245. W. H. Zartmann and H. Adkins, *J. Am. Chem. Soc.*, **54**, 3398 (1932).
1246. A. Polis, *Chem. Ber.*, **20**, 3331 (1887).

1247. A. Polis, *Chem. Ber.*, **21**, 3424 (1888).
1248. H. Gilman and J. Robinson, *J. Am. Chem. Soc.*, **52**, 1975 (1930).
1249. P. Pfeiffer, P. Trüskier and P. Disselkampf, *Chem. Ber.*, **49**, 2445 (1916).
1250. O. H. Browne and E. E. Reid, *J. Am. Chem. Soc.*, **49**, 830 (1927).
1251. H. Gilman and J. Robinson, *Recl. Trav. Chim. Pays-Bas*, **49**, 766 (1930).
1252. W. E. Catlin, *Iowa State Coll. J. Sci.*, **10**, 65 (1935); *Chem. Abstr.*, **30**, 935 (1936).
1253. P. R. Austin, *J. Am. Chem. Soc.*, **53**, 3514 (1931).
1254. H. Gilman and E. B. Towne, *J. Am. Chem. Soc.*, **61**, 739 (1939).
1255. G. Bähr, *Z. Anorg. Allg. Chem.*, **253**, 330 (1947).
1256. C. D. Hurd and P. R. Austin, *J. Am. Chem. Soc.*, **53**, 1543 (1931).
1257. H. Gilman, E. B. Towne and H. L. Jones, *J. Am. Chem. Soc.*, **55**, 4689 (1933).
1258. G. Calingaert, H. Soroos and H. Shapiro, *J. Am. Chem. Soc.*, **62**, 1104 (1940).
1259. C. G. Stuckwisch, *Iowa State Coll. J. Sci.*, **18**, 92 (1943); *Chem. Abstr.*, **38**, 728 (1944).
1260. G. Calingaert, F. J. Dykstra and H. Shapiro, *J. Am. Chem. Soc.*, **67**, 190 (1945).
1261. M. M. Koton, T. M. Kiseleva and N. P. Zapevalova, *Zh. Obshch. Khim.*, **30**, 186 (1960); *Chem. Abstr.*, **54**, 22436 (1960).
1262. P. R. Austin, *J. Am. Chem. Soc.*, **53**, 1548 (1931).
1263. M. S. Kharasch, O. Reinmuth and F. R. Mayo, *J. Chem. Educ.*, **5**, 404 (1928); **8**, 1903 (1931); **11**, 82 (1934); **13**, 7 (1936).
1264. H. Gilman and E. B. Towne, *Recl. Trav. Chim. Pays-Bas*, **51**, 1054 (1932).
1265. H. Gilman and M. Lichtenwalter, *Recl. Trav. Chim. Pays-Bas*, **55**, 588 (1936).
1266. A. Delhaye, J. Nasielski and M. Planchon, *Bull. Soc. Chim. Belg.*, **69**, 134 (1960).
1267. A. I. Yakubovich and I. Petrov, *Z. Prakt. Chem.*, **144**, 676 (1935).
1268. M. M. Koton, *Zh. Obshch. Khim.*, **9**, 2283 (1939); *Chem. Abstr.*, **34**, 5049 (1940).
1269. M. M. Koton, *Zh. Obshch. Khim.*, **11**, 376 (1941); *Chem. Abstr.*, **35**, 5870 (1941).
1270. R. F. McCleary and E. F. Degering, *Ind. Eng. Chem.*, **30**, 64 (1938).
1271. W. J. Jones, D. P. Evans, T. Gulwell and D. C. Griffiths, *J. Chem. Soc.*, 39 (1935).
1272. H. Gilman and J. Robinson, *J. Am. Chem. Soc.*, **51**, 3112 (1929).
1273. M. M. Koton, *Zh. Obshch. Khim.*, **18**, 936 (1948); *Chem. Abstr.*, **43**, 559 (1949).
1274. M. S. Kharasch, *US Patent* 1987685 (1925); *Chem. Abstr.*, **29**, 1436 (1935).
1275. *UK Patent* 331494 (1929); *Chem. Zbl.*, **I**, 2688 (1930).
1276. G. Calingaert, H. Beatty and H. Soroos, *J. Am. Chem. Soc.*, **62**, 1099 (1940).
1277. M. Manulkin, *Uzb. Khim. Zh.*, **2**, 41 (1958); *Chem. Abstr.*, **53**, 9112a (1959).
1278. C. E. Bawn and J. Gladstone, *Proc. Chem. Soc.*, 227 (1959).
1279. K. A. Kocheshkov and A. N. Nesmeyanov, *Zh. Obshch. Khim.*, **4**, 1102 (1934); *Chem. Abstr.*, **29**, 3993 (1935).
1280. F. Hein and H. Schneiter, *Z. Anorg. Allg. Chem.*, **259**, 183 (1949).
1281. E. M. Panov and K. A. Kocheshkov, *Dokl. Akad. Nauk SSSR*, **85**, 1037 (1952); *Chem. Abstr.*, **47**, 6365 (1953).
1282. E. M. Panov and K. A. Kocheskov, *Zh. Obshch. Khim.*, **25**, 489 (1955); *Chem. Abstr.*, **50**, 3271a (1956).
1283. K. A. Kocheshkov and E. M. Panov, *Izv. Akad. Nauk SSSR, Ser. Khim.*, 711 (1955); *Chem. Abstr.*, **50**, 7075 (1956).
1284. M. S. Kharasch, E. V. Jensen and S. Weinhouse, *J. Org. Chem.*, **14**, 429 (1949).
1285. L. Maier, *Tetrahedron Lett.*, **6**, 11 (1959).
1286. B. Bartocha and M. Y. Gray, *Z. Naturforsch.*, **14b**, 350 (1959).
1287. L. C. Willemsens and G. J. M. van der Kerk, *J. Organomet. Chem.*, **13**, 357 (1968).
1288. G. Calingaert and H. Beatty, *J. Am. Chem. Soc.*, **61**, 2748 (1939).
1289. G. Calingaert, H. Beatty and N. R. Neal, *J. Am. Chem. Soc.*, **61**, 2755 (1939).
1290. G. Calingaert and H. Soroos, *J. Am. Chem. Soc.*, **61**, 2758 (1939).
1291. G. Calingaert, H. A. Beatty and L. Hess, *J. Am. Chem. Soc.*, **61**, 3300 (1939).
1292. E. Krause and G. G. Reissaus, *Chem. Ber.*, **54**, 2060 (1921).
1293. G. A. Razuvaev, N. S. Vyazankin and Yu. I. Dergunov, *Zh. Obshch. Khim.*, **30**, 1310 (1960); *Chem. Abstr.*, **55**, 362b (1961).
1294. H. Gilman and M. M. Barnett, *Recl. Trav. Chim. Pays-Bas*, **55**, 563 (1936).
1295. G. A. Razuvaev, N. S. Vyazankin, Yu. I. Dergunov and O. S. D'yachkovskaya, *Dokl. Akad Nauk SSSR*, **132**, 364 (1960); *Chem. Abstr.*, **54**, 20937g (1960).

1296. G. A. Razuvaev, Yu. I. Dergunov and N. S. Vyazankin, *Zh. Obshch. Khim.*, **31**, 998 (1961); *Chem. Abstr.*, **55**, 23321 (1961).
1297. G. A. Razuvaev, N. S. Vyazankin, Yu. I. Dergunov and N. N. Vyshinskii, *Zh. Obshch. Khim.*, **31**, 1712 (1961); *Chem. Abstr.*, **55**, 24546h (1961).
1298. W. Me Dyer and R. D. Closson, *US Patent* 2571987 (1951); *Chem. Abstr.*, **46**, 3556 (1952).
1299. T. W. Gittins and E. J. Mattison, *US Patent* 2763673 (1956); *Chem. Abstr.*, **51**, 4414c (1957).
1300. G. A. Razuvaev, N. S. Vyazankin and N. N. Vyshinskii, *Zh. Obshch. Khim.*, **30**, 4099 (1960); *Chem. Abstr.*, **55**, 24546f (1960).
1301. Yu. A. Aleksandrov and T. I. Makeeva, *Tr. Khim. i Khim Tekhnol.*, **4**, 365 (1961); *Chem. Abstr.*, **56**, 493b (1962).
1302. E. Krause and E. Pohland, *Chem. Ber.*, **55**, 1282 (1922).
1303. Yu. A. Aleksandrov, T. G. Brilkina, A. A. Kvasova, G. A. Razuvaev and V. A. Shushunov, *Dokl. Akad. Nauk SSSR*, **129**, 321 (1959); *Chem. Abstr.*, **54**, 7608 (1960).
1304. H. Schmidt, in *Medicine in its Clinical Aspects*, V. III, Bayer, Leverkusen, 1938, pp. 394–404.
1305. M. Lesbre, *C. R. Acad. Sci., Paris*, **200**, 559 (1935).
1306. T. G. Brilkina, M. K. Safonova and V. A. Shushunov, *Zh. Obshch. Khim.*, **32**, 2684 (1962); *Chem. Abstr.*, **58**, 9112b (1963).
1307. Yu. A. Aleksandrov, T. G. Brilkina and V. A. Shushunov, *Tr. Khim. i Khim. Tekhnol.*, **3**, 381 (1960); *Chem. Abstr.*, **55**, 27023 (1961).
1308. H. Gilman, S. M. Spatz and M. J. Kolbezen, *J. Org. Chem.*, **18**, 1341 (1953).
1309. Yu. A. Aleksandrov, T. G. Brilkina and V. A. Shushunov, *Dokl. Akad. Nauk SSSR*, **136**, 89 (1961); *Chem. Abstr.*, **55**, 27027 (1961).
1310. N. N. Vyshinskii and N. K. Rudnevskii, *Opt. i Spektrosk.*, **10**, 797 (1961); *Chem. Abstr.*, **58**, 4059a (1963).
1311. N. N. Vyshinskii, Yu. A. Aleksandrov and N. K. Rudnevskii, *Dokl. Akad. Nauk SSSR, Ser. Khim.*, **26**, 1285 (1962); *Chem. Abstr.*, **58**, 7517e (1963).
1312. V. A. Shushunov, T. G. Brilkina and Yu. A. Aleksandrov, *Tr. Khim. Khim. Tekhnol.*, **2**, 329 (1959); *Chem. Abstr.*, **55**, 27027 (1959).
1313. K. A. Kocheshkov and E. M. Panov, *Izv. Akad. Nauk SSSR, Ser. Khim.*, 718 (1955); *Chem. Abstr.*, **50**, 7076d (1956).
1314. F. Hein, E. Nebe and W. Reimann, *Z. Anorg. Allg. Chem.*, **251**, 1251 (1943).
1315. E. M. Panov and K. A. Kocheshkov, *Dokl. Akad. Nauk SSSR*, **85**, 1293 (1952); *Chem. Abstr.*, **47**, 6887 (1953).
1316. A. Rieche and J. Dahlmann, *Ann. Chem.*, **675**, 19 (1964).
1317. E. Amberger and R. Hönigschmid-Grossich, *Chem. Ber.*, **98**, 3795 (1965).
1318. A. G. Davies and R. J. Puddephatt, *J. Organomet. Chem.*, **5**, 590 (1966).
1319. A. G. Davies and R. J. Puddephatt, *J. Chem. Soc.*, 2663 (1967).
1320. B. S. Saunders and G. J. Stacey, *J. Chem. Soc.*, 1773 (1948).
1321. K. A. Kocheshkov and A. P. Aleksandrov, *Chem. Ber.*, **67**, 527 (1934).
1322. M. M. Koton and T. M. Kiseleva, *Izv. Akad. Nauk SSSR, Ser. Khim.*, 1783 (1961); *Chem. Abstr.*, **56**, 8741a (1962).
1323. M. M. Koton and F. S. Florinskii, *Zh. Obshch. Khim.*, **32**, 3057 (1962); *Chem. Abstr.*, **58**, 9111 (1963).
1324. M. Lesbre, *C. R. Acad. Sci., Paris*, **204**, 1822 (1937).
1325. M. Lesbre, *C. R. Acad. Sci., Paris*, **206**, 1481 (1938).
1326. I. Druce, *Chem. News*, **124**, 215 (1922).
1327. D. Vorlander, *Chem. Ber.*, **58**, 1893 (1925).
1328. J. S. Buck and D. M. Kumro, *J. Pharmacol.*, **38**, 161 (1930).
1329. F. Challenger and E. Rothstein, *J. Chem. Soc.*, 1258 (1934).
1330. H. Gilman and L. A. Woods, *J. Am. Chem. Soc.*, **65**, 435 (1943).
1331. E. W. Abel and D. A. Armitage, in *Adv. Organomet. Chem.*, Vol. 5, Academic Press, New York, 1967, pp. 1–83.
1332. P. Ballinger, *US Patent* 3073853 (1963); *Chem. Abstr.*, **58**, 12599 (1963).
1333. P. Ballinger, *US Patent* 3116127 (1963); *Chem. Abstr.*, **60**, 6684 (1964).
1334. P. Ballinger, *US Patent* 3081325 (1963); *Chem. Abstr.*, **59**, 6440 (1963).
1335. P. Ballinger, *US Patent* 3073854 (1963); *Chem. Abstr.*, **58**, 12599 (1963).

1336. W. Libe and R. C. Menzies, *J. Chem. Soc.*, 617 (1950).
1337. A. W. Krebs and M. C. Henry, *J. Org. Chem.*, **28**, 1911 (1963).
1338. W. T. Reichle, *Inorg. Chem.*, **1**, 650 (1963).
1339. E. N. Abel and D. B. Brady, *J. Chem. Soc.*, 1192 (1965).
1340. H. Schumann, K. F. Thom and M. Schmidt, *J. Organomet. Chem.*, **4**, 28 (1965).
1341. W. L. Richardson, *US Patent* 3010980 (1961); *Chem. Abstr.*, **56**, 11620 (1962).
1342. W. L. Richardson, *US Patent* 3116126 (1963); *Chem. Abstr.*, **60**, 6686 (1964).
1343. W. L. Richardson, M. R. Barusch, G. J. Kautsky and R. E. Steinke, *J. Chem. Eng. Data*, **6**, 305 (1961); *Ind. Eng. Chem.*, **53**, 305 (1961).
1344. W. B. Ligett, R. D. Closson and C. N. Wolf, *US Patent* 2595798 (1952); *Chem. Abstr.*, **46**, 7701 (1952).
1345. W. B. Ligett, R. D. Closson and C. N. Wolf, *US Patent* 2640006 (1953); *Chem. Abstr.*, **47**, 8307 (1953).
1346. D. O. De Pree, *US Patent* 2893857 (1959); *Chem. Abstr.*, **53**, 18372 (1959).
1347. K. Jones and M. F. Lappert, *J. Chem. Soc.*, 1944 (1965).
1348. K. Kuhlein, Dissertation, Univ. Giessen (1966).
1349. W. P. Neumann and K. Kühlein, *Tetrahedron Lett.*, 3419 (1966).
1350. J. G. A. Luijten and G. G. M. van der Kerk, in *Adv. Organomet. Chem.*, Vol. 7, Academic Press, New York, 1967, pp. 92–172.
1351. O. J. Scherer and M. Schmidt, *Z. Naturforsch.*, **18b**, 415 (1963).
1352. O. J. Scherer and M. Schmidt, *J. Organomet. Chem.*, **3**, 156 (1965).
1353. E. Lieber and F. M. Keane, *Chem. Ind.*, 747 (1961).
1354. E. Lieber, C. N. R. Rao and F. M. Keane, *J. Inorg. Nucl. Chem.*, **25**, 631 (1963).
1355. M. Kumada, K. Naka and M. Ishikawa, *J. Organomet. Chem.*, **2**, 136 (1964).
1356. G. D. Shier, S. Russell and R. S. Drago, *J. Organomet. Chem.*, **6**, 359 (1966).
1357. H. G. Langer and A. H. Blut, *J. Organomet. Chem.*, **5**, 288 (1966).
1358. N. A. Matviyov and R. S. Drago, *Inorg. Chem.*, **3**, 337 (1964).
1359. R. Duffy and A. K. Holliday, *J. Chem. Soc.*, 1679 (1961).
1360. A. E. Finholt, A. C. Bond, K. E. Wilzbach and H. I. Schlesinger, *J. Am. Chem. Soc.*, **69**, 2693 (1947).
1361. A. K. Holliday and W. Jeffers, *J. Inorg. Nucl. Chem.*, **6**, 134 (1958).
1362. R. Duffy and A. K. Holliday, *Proc. Chem. Soc.*, 124 (1959).
1363. R. Duffy, J. Feeney and A. K. Holliday, *J. Chem. Soc.*, 1144 (1962).
1364. E. Amberger, *Angew. Chem.*, **72**, 494 (1960).
1365. R. S. Dickson and B. O. West, *Aust. J. Chem.*, **15**, 710 (1962).
1366. W. P. Neumann, *Chem. Eng. News*, **43**, 38, 49 (1965).
1367. E. Amberger and R. Hönigschmid-Grossich, *Chem. Ber.*, **99**, 1673 (1966).
1368. S. M. Blitzer, M. W. Farrar, T. H. Pearson and J. R. Zietz, *US Patent* 3136795 (1964); *Chem. Abstr.*, **61**, 5691 (1964).
1369. W. P. Neumann and K. Kühlein, in *Adv. Organomet. Chem.*, Vol. 7, Academic Press, New York, 1967, pp. 241–312.
1370. W. P. Neumann, *Ann. Chem.*, **629**, 23 (1960).
1371. G. B. Buckton, *Proc. Chem. Soc.*, **9**, 685 (1859).
1372. A. Ghira, *Gazz. Chim. Ital.*, **24**, 142 (1894).
1373. G. Calingaert, H. Soroos and H. Shapiro, *J. Chem. Soc.*, **64**, 462 (1942).
1374. U. Belluco, G. Tagliavini and R. Barbieri, *Ric. Sci.*, **30**, 1675 (1960); *Chem. Abstr.*, **55**, 14175 (1961).
1375. R. Preckel and W. Selwood, *J. Am. Chem. Soc.*, **62**, 2765 (1940).
1376. L. C. Willemsens and G. J. M. van der Kerk, *J. Organomet. Chem.*, **2**, 260 (1964).
1377. L. C. Willemsens and G. J. M. van der Kerk, *J. Organomet. Chem.*, **2**, 275 (1964).
1378. R. W. Leeper, *Iowa State Coll. J. Sci.*, **18**, 57 (1943); *Chem. Abstr.*, **38**, 726 (1944).
1379. H. E. Podall, H. E. Petree and J. R. Zietz, *J. Org. Chem.*, **24**, 1222 (1959).
1380. L. Malatesta, *Gazz. Chim. Ital.*, **73**, 176 (1943); *Chem. Abstr.*, **38**, 5128 (1944).
1381. G. L. Lewis, P. F. Oesper and C. P. Smyth, *J. Am. Chem. Soc.*, **62**, 3243 (1940).
1382. E. Müller, F. Günter, K. Scheffler and H. Fettel, *Chem. Ber.*, **91**, 2888 (1958).
1383. G. A. Razuvaev, Yu. I. Dergunov and N. S. Vyazankin, *Zh. Obshch. Khim.*, **32**, 2515 (1962); *Chem. Abstr.*, **57**, 15136a (1962).
1384. G. Bähr and G. Zoche, *Chem. Ber.*, **88**, 542 (1955).

1385. U. Belluco and G. Belluco, *Ric. Sci., Rend. Suppl.*, **32**, 76, 110 (1962); *Chem. Abstr.*, **57**, 13786 (1962).
1386. R. A. Peters, *US Patent* 2967105 (1961); *Chem. Abstr.*, **55**, 9129 (1961).
1387. Yu. A. Aleksandrov and N. N. Vyshinskii, *Tr. Khim. i Khim. Technol.*, **4**, 656 (1962); *Chem. Abstr.*, **58**, 3453 (1963).
1388. Yu. A. Aleksandrov, T. G. Brilkina and V. A. Shushunov, *Tr. Khim. i Khim. Tekhnol.*, **4**, 3 (1961); *Chem. Abstr.*, **56**, 492 (1962).
1389. N. S. Vyazankin, G. A. Razuvaev and Yu. I. Dergunov, *Tr. Khim. i Khim. Tekhnol.*, **4**, 652 (1961); *Chem. Abstr.*, **58**, 543 (1963).
1390. U. Belluco, A. Peloso, L. Cattalini and G. Tagliavini, *Ric. Sci., Rend.*, **2**, 269 (1962); *Chem. Abstr.*, **59**, 1667 (1963).
1391. H. J. Emeleus and P. R. Evans, *J. Chem. Soc.*, 511 (1964).
1392. I. T. Krohn and H. Shapiro, *US Patent* 2555891 (1951); *Chem. Abstr.*, **46**, 523 (1952).
1393. G. A. Razuvaev, M. S. Fedotov and T. B. Zavarova, *Tr. Khim. i Khim. Tekhnol.*, **4**, 622 (1961); *Chem. Abstr.*, **58**, 543 (1963).
1394. U. Belluco and G. Belluco, *Ric. Sci., Rend. Suppl.*, **32**, 102 (1962); *Chem. Abstr.*, **57**, 13786 (1962).
1395. S. B. Wiczer, *US Patent* 2447926 (1948); *Chem. Abstr.*, **42**, 7975 (1948).
1396. U. Belluco, L. Cattalini, A. Peloso and G. Tagliavini, *Ric. Sci., Rend.*, **3**, 1107 (1963); *Chem. Abstr.*, **61**, 677 (1964).
1397. F. Hein and E. Heuser, *Z. Anorg. Allg. Chem.*, **254**, 138 (1947).
1398. C. Tamborski, F. E. Ford, W. L. Lehn, G. J. Moore and E. J. Soloski, *J. Org. Chem.*, **27**, 619 (1962).
1399. H. Gilman and O. M. Gruhzit, *Pharmacol. J.*, **41**, 1 (1931).
1400. T. Midgley, *J. Ind. Eng. Chem.*, **17**, 827 (1925).
1401. T. Midgley and P. R. Boyd, *J. Ind. Eng. Chem.*, **14**, 896 (1922).
1402. S. K. Hall, *Environ. Sci. Technol.*, **6**, 36 (1972).
1403. W. Bolanowska, *Med. Pracy*, **16**, 476 (1965); *Chem. Abstr.*, **64**, 14841e (1966).
1404. H. J. Thomas, *Air, Qual. Monogr.*, **7**, 53 (1969).
1405. A. Browder, M. Joselow and D. Louria, *Medicine (Baltimore)*, **52**, 121 (1973).
1406. C. Norris and A. O. Gettler, *J. Am. Med. Assoc.*, **85**, 818 (1925).
1407. W. Bolanowska and J. M. Wisniewskaknypl, *Biochem. Pharmacol.*, **21**, 2018 (1972).
1408. B. C. Saunders, G. J. Stacey, F. Wild and I. G. E. Wilding, *J. Chem. Soc.*, 699 (1948).
1409. B. C. Saunders, *J. Chem. Soc.*, 684 (1950).
1410. H. McCombie and B. C. Saunders, *Nature*, **158**, 382 (1946).
1411. F. Bischoff, L. C. Maxwell, R. D. Evans and F. R. Nuzum, *J. Pharmacol.*, **34**, 85 (1928).
1412. J. S. Thayer, *J. Organomet. Chem.*, **76**, 265 (1974).

CHAPTER 2

Similarities and differences of organic compounds of germanium, tin and lead

MIKHAIL G. VORONKOV

A. E. Favorsky Institute of Chemistry, Siberian Branch of the Russian Academy of Sciences, 1 Favorsky Str., 664033 Irkutsk, Russia
email:voronkov@irioch.irk.ru

and

ALEXEY N. EGOROCHKIN

G. A. Razuvaev Institute of Metallorganic Chemistry of the Russian Academy of Sciences, 49 Tropinin Str., 603950 Nizhny Novgorod, Russia
email: lopatin@imoc.sinn.ru

The chemistry of organic germanium, tin and lead compounds — Vol. 2
Edited by Z. Rappoport

I. INTRODUCTION

The problem of the similarity and difference of organic compounds of the heavy silicon subgroup elements — germanium, tin and lead — has many aspects. The four aspects listed below seem to us to be especially important and interesting.

1. Similarities and differences between Ge, Sn, Pb (M) atoms and their M−Y bonds (Y = inorganic, organic or organometallic substituents). In some cases the specific character of the M atom and the M−Y bonds is more clearly seen when compared with properties and bonds of carbon and especially silicon atoms — the lighter elements of the group 14 column of the periodic table. The peculiarities of M atoms and M−Y bonds determine to a considerable extent the similarity and difference of the physical properties and chemical behaviour of organic compounds of the elements in question.

2. Classical and recent quantum-chemical concepts about the electronic structure of M element compounds. The dependence of intramolecular electronic, steric and stereoelectronic interactions on the nature of the M atom and the M−Y bonds.

3. Similarities and differences of electronic effects of the R_3M groups (R = organic substituent) in compounds of the R_3MX (X = an atom with at least one unshared electron pair — halogen, O−, S−, N< etc.) and R_3MR_π (R_π = an aromatic or α,β-unsaturated group) type. The intramolecular electronic interactions in these compounds are determined by the magnitude and sign of charge which is induced on the reaction centre R_π or X by physical or chemical influence. Such dependencies allow one to verify the existing theoretical concepts about the mechanism of these intramolecular electronic interactions. Therefore, the physical properties of R_3MR_π and R_3MX substances as well as of their molecular complexes allow one to model heterolytic reactions of these compounds.

4. Similarities and differences of chemical reactions of organic M element derivatives. The specific character of these processes becomes apparent in comparison with the analogous reactions of compounds of the lighter elements of group 14 — carbon and silicon. Unfortunately, so far there are no systematic investigations of the reactivity of R_3MR_π and R_3MX type compounds as a function of M (C, Si, Ge, Sn, Pb) based on a representative sample of R, R_π and X groups. Only such an approach will enable one to establish the validity of existing theoretical concepts of the electronic structure of silicon subgroup element compounds for a detailed description of their reactivity and comparison of chemical and physical properties. Until now only isolated fragments of this complete idealized picture have been published, some of which we consider in this chapter.

We shall take a brief look at some characteristics of M (Ge, Sn and Pb) atoms in comparison with Si and C. In the ground state the electron configurations of C, Si, Ge, Sn and Pb atoms are [He] $2s^22p^2$, [Ne] $3s^23p^2$, [Ar]$3d^{10}4s^24p^2$, [Kr]$4d^{10}5s^25p^2$ and [Xe]$5d^{10}6s^26p^2$, respectively.

The atomic covalent radius (one half of the M−M [1]distance) has been used for a long time for estimates of the nature of chemical bonds. Its magnitude correlates with the M−M bond energy. The notion of the van der Waals radius of an atom is ambiguous[3]. The sum of van der Waals radii of two atoms is defined in crystallography as the minimum distance at which they can approach each other.

The covalent and van der Waals radii of M atoms increase as their atomic number rises (Table 1). When going from C to Si the covalent atomic radius increases sharply, and it increases gradually from Si to Pb. The value of the van der Waals radius of lead (2.02 Å)

TABLE 1. Selected properties of C, Si, Ge, Sn and Pb atoms

Property	C	Si	Ge	Sn	Pb	Reference
Atomic number	6	14	32	50	82	
Atomic weight	12.011	28.0855	72.59	118.69	207.2	2
Covalent radius (Å)	0.77	1.17	1.22	1.42	1.48	4
	0.77	1.18	1.22	1.40	1.47	2
van der Waals radius (Å)	1.70	2.10	2.15	2.17	2.02	5
Ionization potential (eV)	11.26	8.15	7.88	7.34	7.42	2
Electronegativity	2.60	1.90	2.00	1.93	2.45	6
	2.46	1.89	1.71	1.59	—	7
	2.6	1.9	2.0	2.0	2.1	8
	2.746	2.138	2.618	2.298	2.291	4
Atomic refraction (cm^3 mol^{-1})	2.5	7.5	9.7	13.3	34.3	9
	4	9	12	16	34	8
Covalent refraction (cm^3 mol^{-1})	2.1	9	11	16.5	19	8
Polarizability ($10^{-24} cm^3$ mol^{-1})	0.82	3.59	4.39	6.34	7.10	2

stands out in the general trend. Owing to their large radii, the atoms of Ge, Sn and Pb are sterically accessible for nucleophilic interaction and it favours their transition to the hypervalent state.

The first ionization potential Ip (the energy required for removal of one electron from the neutral atom) of the M (Ge, Sn, Pb) atoms is less than that of Si and, the more so, of C. Therefore, the M element compounds are characterized by greater electrophilicity in comparison with isostructural derivatives of C and Si.

According to Pauling, the electronegativity of an atom in the molecule is defined by its capacity to pull electrons from substituents bonded to it[1]. It was established later that the electronegativity depends also on the valence state (the hybridization) of the atom[4,6,8]. Carbon is the most electronegative among the group 14 elements (Table 1). As the atomic number increases, the electronegativity of the group 14 elements changes non-monotonically (see, however, Reference 1). The electronegativity values and series listed in Table 1 can change sharply under the influence of intramolecular resonance interactions, e.g. by varying the substituents bonded to C, Si and M atoms[10].

The refraction R and polarizability $\alpha = (3/4\pi N_A)R$, where N_A is the Avogadro number, characterize the deformability of the atom's electron shell under external influence (charge, electric field). The atomic and covalent refraction relate to isolated atoms and to atoms involved in a covalent bond, respectively. The R and α values grow as the atomic number of M increases. The high polarizability of M atoms and of M−X bonds is an important property of compounds of heavier group 14 elements. The sensitivity of conjugation in the molecule to changes in the charge on the reaction centre grows as the polarizability of the central M atom increases; see, for example, Reference 11.

We shall now consider some properties of M−X bonds (M = Ge, Sn, Pb) in comparison with Si−X and C−X. As the atomic number of M increases, these bond distances d (Table 2) become longer. It is caused by the increase in the covalent radius of the group 14 element as its atomic number rises. The d values of the Me_3M−Me and Me_3M−MMe_3 bonds coincide to within 0.05 Å with the sum of covalent radii of the atoms forming this bond. The Si−Cl bond distances in $SiCl_4$ are 0.15 Å shorter than the sum of covalent radii of Si and Cl atoms. As the atomic number of M increases, the difference between the experimental d values in MCl_4 molecules and the expected ones (based on the sum of the

TABLE 2. Selected bond distance (Å) data

Bond	M				
	C	Si	Ge	Sn	Pb
$D_3M{-}H$[a]	1.092	1.480	1.532	1.701	—
$Me_3M{-}CH_3$[b]	1.539	1.875	1.980	2.143	2.238
$Me_3M{-}(C{\equiv}CH)$[b]	1.498	1.826	1.896[c]	2.082	—
M−N[d]	1.451[b]	1.734[e]	1.836[b]	2.038[f]	—
M−O[g]	1.416[b]	1.631[e]	1.766[b]	1.940[b]	—
$Me_3M{-}Cl$[b]	1.803	2.022	2.170	2.37	2.706
$Me_3M{-}Br$[b]	1.94	2.235	2.323	2.49	2.852
$I_3M{-}I$[a]	2.15	2.43	2.50	2.64	—
$Me_3M{-}MMe_3$	1.582[b]	2.340[e]	2.403[h]	2.776[i]	2.88[j]

[a] Reference 12.
[b] Reference 13.
[c] Measured for $H_3Ge{-}(C{\equiv}CH)$.
[d] Measured for $(H_3M)_3N$ (M = C, Si, Ge) and $(Me_3Sn)_3N$ compounds.
[e] Reference 14.
[f] Reference 15.
[g] Measured for $(H_3M)_2O$ (M = C, Si, Ge) and $(Me_3Sn)_2O$ compounds.
[h] Measured for H_3GeGeH_3 (Reference 16).
[i] Reference 17.
[j] Reference 18.

covalent radii) diminishes (0.12, 0.09 and *ca* 0.04 Å for M = Ge, Sn and Pb, respectively). The shortened interatomic distances in silicon halogenides led to the hypothesis of d–n conjugation between the 3d-orbitals of Si and the unshared electron pairs of the halogen atom[1,19]. Modern ideas of conjugation in compounds of the silicon subgroup elements are discussed in Section II.

The electric dipole moment, $\vec{\mu}$, of an electroneutral molecule, $\vec{\mu} = q\vec{1}$ (where $\vec{1}$ is the radius vector directed from the centre of gravity of negative charges to the centre of gravity of positive charges and q is the absolute value of each charge), reflects its polarity. The quantity μ may be represented as a vector sum of dipole moments of separate bonds: $\vec{\mu} = \Sigma_{i=1}^{n}\mu_i$. The μ_i values consist of several components. Therefore, the calculation of atom charges on the basis of μ and μ_i values is a complicated problem. The μ_i values also depend on the electronic effects of the substituents attached to the bond in question. The dipole moments of five series of isostructural compounds of group 14 elements are listed in Table 3. These compounds may be divided into two types.

For the compounds of the first type (Me_3MSPh and Me_3MSMMe_3) the μ value increases as the atomic number of M increases. It correlates well with one of the electronegativity scales of the M atoms[7]. For the compounds of the second type (Me_3MOMMe_3, Me_3MCl, Me_3MBr) the μ value decreases or does not virtually vary as M changes from C to Si. It is contradictory to the electronegativity values of the C and Si atoms. Based on the minimum electronegativity of Si among all of the group 14 elements, the Si compounds should have the highest μ values. However, the dipole moments of Si compounds are in fact the smallest. When M changes further from Si to Ge, Sn, Pb, the μ value grows. This does not correlate with some of the electronegativity scales[6,8]. The traditional explanation of these anomalies in the μ values is based on the hypothesis of d–n conjugation[20–22].

TABLE 3. Dipole moments (D) of some compounds

Series	M					Reference
	C	Si	Ge	Sn	Pb	
Me_3MOMMe_3	1.20	0.78	1.41	1.60	—	20
Me_3MSPh	1.49	1.76	2.18	2.58	3.20	21
Me_3MSMMe_3	1.52	1.75	2.18	2.62	3.08	21
Me_3MCl	2.18	2.13	2.91	3.57	4.82	22
Me_3MBr	2.24	2.28	2.98	3.61	4.46[a]	22

[a]Measured for Et_3PbBr (Reference 23).

TABLE 4. Polarity of M–X bonds in MX_4 molecules[a]

X	M				
	C	Si	Ge	Sn	Pb
F	27	44	42	42	39
Cl	6	22	19	19	17
Br	3	17	14	14	12
I	0	10	9	9	6

[a]In percent, relative to a hypothetical pure ionic bond. From Reference 8.

According to Pauling[1], the polarity p of a chemical bond is the measure of its ionicity. It is related to dipole moment μ_i by the equation: $p = \mu_i/d$, where d is the interatomic distance. The polarity of M–X bonds in MX_4 molecules[8] is illustrated in Table 4. The polarity of a specific M–X bond increases significantly as M changes from C to Si, and it diminishes slightly on going from Si to Ge, Sn and Pb. At the same time, the polarity of a specific M–X bond decreases sharply as the atomic number of the halogen X increases.

The refraction (R_D) and polarizability (α) are important properties of M–X bonds. The R_D values are not universal, because the refraction magnitude depends on the type of substituents attached to a given bond. As a rule, R_D and α values of M–X bonds rise as the atomic number of the M element increases. Thus, R_D values of $M{-}C_{Alk}$ bonds for M = C, Si, Ge, Sn and Pb are 1.30, 2.47, 3.05, 4.17 and 5.25 $cm^3\,mol^{-1}$, respectively[8]. The R_D values of M–F bonds (M = C, Si, Ge) change non-monotonically as the atomic number of M atom increases. This may be due to the polarizability anisotropy of these bonds[24]. The polarizability anisotropy (γ) of a σ-bond is the difference of polarizabilities along and perpendicular to the bond axis. γ values, except for M–F bonds, increase as the atomic number of M increases. For example, the γ values of M–Cl bonds in MCl_4 for M = C, Si, Ge, Sn and Pb are 2.40, 3.05, 4.24, 4.96 and 6.22 $Å^3$, respectively[25]. Thus, M–X bonds of heavy elements of group 14 are characterized by higher polarizability and polarizability anisotropy in comparison with the corresponding Si–X and C–X bonds.

It appears from the above that attempts to consider the nature of M–X bonds (M = Si, Ge, Sn, Pb) in analogy to C–X bonds on the basis of simple electronegativity usually fail. A deeper understanding of the observed trends is possible only on the basis of modern concepts on the electronic structure of compounds of the silicon subgroup elements.

II. ELECTRONIC STRUCTURE

In the second half of the 20th century, numerous physical and chemical investigations were conducted. They suggest that the electronic structure of isostructural compounds of carbon on the one hand and of silicon subgroup elements on the other differ fundamentally (see, for example, References 26–29). This difference is more clearly manifested in various conjugation effects which are discussed below. R_3M substituents in compounds of the type R_3MR_π and R_3MX (M = Si, Ge, Sn, Pb; for the R, R_π and X symbols, see Section I), unlike in $R_3MCH_2R_\pi$, have two resonance effects with respect to reaction centres R_π and X—both as donor and as acceptor. Depending on the atomic number of the M element, these effects change in the following way[10,28,30–34].

1. Resonance acceptor effect of R_3M substituents towards R_π (d–π conjugation) and X (d–n conjugation) decreases as M changes in the order: Si > Ge > Sn > Pb. The question whether the R = Pb group still has resonance acceptor properties is under debate.

2. The resonance donor effect of R_3M substituents towards R_π (σ–π conjugation), shown schematically in **1**, increases as M changes in the order: C < Si < Ge < Sn < Pb. One can observe a weak σ–π conjugation effect even in organic compounds.

$$R_3M-CH{=}CH_2$$

(1)

3. The resonance donor effect of R_3M substituents towards X (σ–n conjugation), shown schematically in **2**, increases as M changes in the order: Si < Ge < Sn < Pb. The σ–n conjugation effect (unlike σ–π conjugation) was disregarded for a long time, though there were indications of its existence[30]. The importance of this effect was demonstrated recently[31].

$$R_3M-\ddot{S}-$$

(2)

4. The R_3MCH_2 substituent has only a donor effect in $R_3MCH_2R_\pi$ molecules towards R_π (σ–π conjugation), shown schematically in **3**. It is enhanced as M changes in the series: C < Si < Ge < Sn < Pb[32].

$$R_3M-CH_2-CH{=}CH_2$$

(3)

5. The σ–π conjugation effect in R_3MR_π molecules is enhanced when there is a positive charge on the reaction centre R_π. The higher the atomic number of the M element, the stronger the conjugation effect. The resonance acceptor properties of the R_3M substituents towards R_π are almost the same for M = Ge and Sn, if there is no partial positive charge on R_π. Therefore, the change of sign of the overall resonance effect under the influence of the positive charge on R_π is more typical for R_3GeR_π and R_3SnR_π compounds[32,33].

Thus, the dominance of d–π or d–n conjugation acceptor effects in R_3MR_π and R_3MX compounds is more typical for M = Si and probably Ge, and the σ–π or σ–n conjugation donor effect dominates for M = Pb.

The overall resonance effect of the R_3M substituents depends on the nature of R_π and X and also on the value of the effective positive charge on the M atom and reaction centre R_π.

The existence of $\sigma-\pi$ or $\sigma-n$ conjugation effects is by now widely accepted. However, opinions on the mechanisms of $d-\pi$ or $d-n$ conjugation differ. We shall consider briefly the evolution of theoretical views on the acceptor resonance effects mentioned above.

The first stage was based on the hypothesis that unoccupied nd-orbitals of silicon subgroup elements in R_3MR_π and R_3MX participate in conjugation with π- or n-orbitals of the reaction centre bonded to them. It was established that the π-acceptor properties of the M atom weaken with increase of its atomic number and that there is a close analogy between $d-\pi$ or $d-n$ interaction and the conjugation effects in organic molecules[35]. This hypothesis agrees qualitatively with general quantum-chemical ideas. But quantitative information concerning the participation of nd-orbitals of M, which was obtained by semi-empirical quantum-chemical methods, proved to be ambiguous and often led to opposing conclusions[35,36].

In the second stage an erroneous interpretation of hyperconjugation as an alternative to the $d-\pi$ conjugation appeared[37] (for more details see Reference 33). However, it led to the realization of the important role of hyperconjugation in organic molecules.

In the third stage of the evolution of views on conjugation effects, the acceptor properties of M atoms of the silicon subgroup towards R_π and X were explained not by the presence of unoccupied nd-orbitals, but by the participation of antibonding σ^* orbitals of the M−Y bonds in molecules of the Y_3MR_π and Y_3MX type (Y = inorganic or organic substituent)[38]. The important role of σ^* orbitals in the hyperconjugation was first pointed out as long ago as 1973 by Pitt[30].

The consideration of the participation of antibonding σ^* orbitals of M−Y bonds (the $n-\sigma^*$ conjugation effect) as an alternative to the nd-orbital hypothesis was shown afterwards by non-empirical quantum-chemical methods to be wrong[39,40]. This effect takes place, for example, in compounds of nitrogen and carbon (M = N, C), for which the nd-orbital contributions are known to be absent.

The fourth stage of evolution was the unifying concept, which is considered in detail elsewhere[30,33]. It was based[30] on two fundamental principles of physical organic chemistry: linear free-energy relationships as well as independence and additivity of the influence of inductive, resonance and steric effects on free-energy change[33]. According to Pitt[30], the energy of frontier π and π^* orbitals of hydrocarbons HR_π (R_π = Ph, HC=CH_2 etc.) changes in the following way when Me_3M (M = a silicon subgroup element) is substituted for the H atom (Figure 1). The substituents Me_3M have a positive inductive +I effect, so that their influence leads to an increase in the energy (destabilization) of the π and π^* orbitals. Further destabilization occurs owing to hyperconjugation, i.e. mixing of the σ(M−C) orbitals of Me_3M with the π and π^* orbitals. The degree of mixing of σ, π is higher than that of σ, π^*. Therefore, the HOMO of Me_3MR_n is more destabilized than the LUMO.

We shall now consider the acceptor component of the overall resonance effect in Me_3MR_π molecules. The most important factor here is the proximity of the energies of the nd-orbitals of M and of the antibonding σ^*-orbitals of the M−C fragment of Me_3M. Thus, the energy of nd-orbitals of Si, Ge, Sn and Pb atoms is −1.97, −1.8, −1.93 and −1.78 eV[41], respectively, and the energy of the σ^*(M−C) orbitals in Me_4M (M = Si, Ge and Sn) is 3.8, 3.4 and 2.89 eV[42], respectively. The mixing of nd and σ^*(M−C) results in the formation of two new orbitals. One of them (designated as d,σ^*) has a lower energy than the initial d-orbitals.

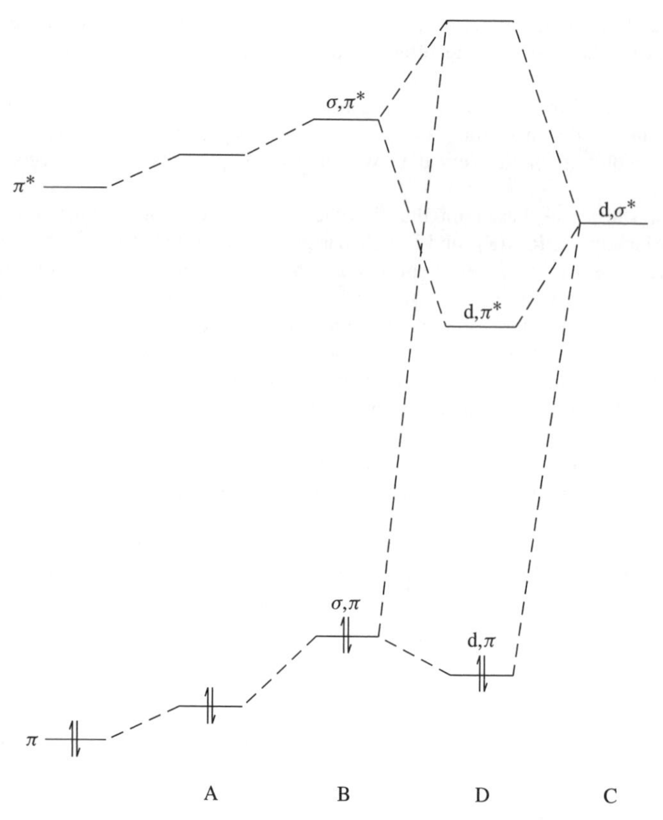

FIGURE 1. The influence of the substituents Me_3M on the energies of the frontier orbitals π and π^* of Me_3MR_π compounds: (A) inductive +I effect of Me_3M; (B) $\sigma-\pi$ and $\sigma-\pi^*$ conjugation; (C) mixing of σ^*(M−C) orbitals of the Me_3M substituent with the nd-orbitals of the M atoms (a mixed d,σ^*-orbital with a lower energy is illustrated); (D) mixing of d,σ^*-orbitals with σ,π^* and σ,π; the formation of the HOMO (d,π) and LUMO (d,π^*) of the molecule

The σ,π- and σ,π^*-orbitals also interact with the d,σ^*-orbital. This results in the stabilization (decrease in energy) of the σ,π- and σ,π^*-orbitals, involving the formation of two new orbitals with lower energy, namely d,π and d,π^*. The designations d,π and d,π^* represent resonance acceptor properties of Me_3M substituents towards the frontier π- and π^*-orbitals of Me_3MR_π (see also References 30 and 33).

Thus, in the fourth stage of evolution of views on conjugation[30] the fundamental idea that the acceptor resonance effect is a joint influence of unoccupied nd- and antibonding σ^*-orbitals was formed.

More recently, another explanation of the acceptor effect of M (Si) towards X (F, O, N) in M−X bonds (M = Si) was proposed[14,43]. The high bond energy of Si−X bonds was attributed to an increase in the contribution of ionic structures of the type $H_3Si^+X^-$. However, the polarities of all M−X bonds (M = Si, Ge, Sn, Pb) are quite similar (Table 4), whereas according to all available data the d−n conjugation weakens when the atomic number of M increases[26,28,31−35]. Therefore, these ideas cannot be considered as an alternative to d−n conjugation.

Later on the intramolecular interactions in R_3MR_π and R_3MX were studied by non-empirical quantum-chemical calculations. This provided important information on the existence of two effects of opposite direction — a donor effect (hyperconjugation of $\sigma-\pi$ or $\sigma-n$ type) and an acceptor effect (d–π or d–n conjugation). We discuss below several of the important studies on this question.

First we shall consider the allylalkylchlorostannanes $R_{3-n}Cl_nSnCH_2CH{=}CH_2$ (R = Me, n-Bu; $n = 0–2$), in which the $\sigma-\pi$ conjugation is the only operative resonance effect[44]. It weakens when n increases. This seems to be connected with a decrease in the number of bonding σ(Sn–C) orbitals. According to Cauletti and coworkers[44], the steric effect of the butyl group, which hinders the optimal mutual orientation of the interacting σ(Sn–C) and π(C=C) orbitals for $\sigma-\pi$ conjugation, plays a certain role when R is changed from Me to Bu.

The acceptor properties of the central M atom are well studied by quantum-chemical methods for M = Si[36]. These properties weaken[45] if M = Ge, for example, changing from c-$PrSiH_3$ to c-$PrGeH_3$. Starting with M = Sn, the donor component of the overall resonance effect of the R_3M groups becomes more dominant. Thus strong mixing of the σ(Sn–C) and π(C≡C) orbitals has been established for $Me_3SnC{\equiv}CX$ (X = H, Me, $SnMe_3$)[46]. The strength of the $\sigma-\pi$ conjugation effect depends on the energy of the initial orbitals. Therefore, this effect is enhanced in compounds $R_3SnC{\equiv}CH$ when R is changed from Me to Et[46]. The HOMO energy of molecules $Sn(C{\equiv}CX)_4$ depends on the type of X. As X changes from Me to $SiMe_3$, stabilization (energy decrease) of the HOMO occurs. This is caused by the acceptor properties (d–π conjugation) of the silicon atom[47].

The metallamines $(H_3M)_3N$ (M = Si, Ge) have planar structures due to the rehybridization of the central nitrogen atom, which results from electrostatic repulsion of the M···M and H···H atoms as well as from the resonance acceptor effect of the M atoms[48,49]. The calculated values of the HOMO energy, which are 9.11, 9.61 and 8.91 eV for M = C, Si, Ge, respectively, agree satisfactorily with the experimental values (8.44, 9.7 and 9.2 eV[48,49], respectively). These data confirm that d–n conjugation is more important for M = Si than for M = Ge. In comparison with the d–n conjugation, the $\sigma-n$ conjugation effect (the interaction of the $2p_z$ orbital of the nitrogen atom with the σ(M–H) orbitals) exerts less influence on the HOMO and decreases as M changes in the series: C > Si > Ge[48]. In $Me_nSn(NCS)_{4-n}$ ($n = 2, 3$), the HOMO is localized mainly on sulphur atom and the influence of the electron-accepting NCS groups on the tin atom is very small[50].

Quantum-chemical calculations of MH_4 and MCl_4 (M = C, Si, Ge, Sn, Pb) demonstrated that it is necessary to take into account the relativistic effects, which are proportional to the square of the atomic number of M and therefore essential when M = Ge, Sn, Pb[51]. This was considered in calculations of the $Me_{4-n}SnCl_n$ ($n = 0–4$) series[50]. The mixing of orbitals of the unshared electron pairs of the chlorine atoms with the σ(Sn–C) orbitals ($\sigma-n$ conjugation) intensifies with the rise of the number of methyl groups. On the contrary, increase in the number of chlorine atoms is accompanied by an increase in the population of the 5d-orbitals of the tin atom due to the d–n conjugation[50]. The calculated HOMO energies[50] and NMR chemical shifts of $Me_{4-n}SnCl_n$[52] conform satisfactorily with experimental values.

The HOMO of $(Me_3Sn)_2S$ and $(Me_2SnS)_3$ is localized mainly on the unshared electron pairs of the sulphur atoms[53], but the orbitals of methyl groups also contribute due to $\sigma-n$ conjugation. This contribution exceeds 20% for $(Me_3Sn)_2S$.

Among derivatives of group 14 elements, the lead compounds are characterized by the highest degree of hyperconjugation. Some important features of the $\sigma-n$ conjugation in the halo lead compounds $R_{4-n}PbX_n$ (R = H, Me; X = F, Cl; $n = 0–4$) were established by quantum-chemical methods[54]. The σ–n conjugation effect is enhanced on going from $H_{4-n}PbCl_n$ to $H_{4-n}PbF_n$ and further to $Me_{4-n}PbF_n$. The aptitude for σ–n conjugation of σ(Pb–C) bonds is stronger, and that of the σ(Pb–F) bonds is weaker, than that of σ(Pb–H) bonds. Therefore, the σ–n conjugation effect changes non-additively when n in $H_{4-n}PbF_n$ decreases[54].

The interaction of σ-orbitals of M–C and M′–C bonds (σ–σ conjugation) in the dimetalla compounds, $R_3MM'R_3$, was evaluated by quantum-chemical calculations. Thus, the HOMO of $Me_3SnSnMe_3$ is localized mainly on the Sn atoms but it also includes contributions of C and H atoms, whose magnitude exceeds 30%[55]. Calculations of $H_3MM'H_3$ (M and M′ = C, Si, Ge, Sn, Pb) show the necessity to take into account both the bonding and antibonding orbitals of M–H, M′–H and M–M′[16]. Orbital interactions are enhanced as the atomic numbers of M and M′ increase[55].

It is usual to consider the inductive interaction as a universal electron effect[56]. This means that the inductive influence of the R_3M substituent on the reaction (indicator) centre R_π and X does not depend on the reaction centre type. This seems to be true for M = C, but for M = Si, Ge, Sn, Pb some comments, given below, are required.

The inductive effect of an inorganic, organic or organometallic substituent, Y, is characterized quantitatively by several widespread constants, such as σ_I and others (σ^*, σ', F etc.)[34,57]. The relationship $\sigma_I(Y) = \sigma'(Y) = F(Y) = 0.45\sigma^*(CH_2Y)$ is well known. The inductive constant σ_I of a certain Y substituent (Table 5) reflects several electronic interactions. The most important independent mechanisms of polarization of bonds between the substituent Y and the reaction centre are the field and the inductive interactions. The contribution of each mechanism depends on the type of substituent Y[29].

We shall consider the most reliable[34,37] values of σ_I for three series of isostructural substituents containing M (Table 5). In the Me_3M series the highest σ_I value is observed for M = C. This corresponds to the highest electronegativity of the carbon atom among all group 14 elements (Table 1). As mentioned above, the existing electronegativity series of M (M = Si, Ge, Sn, Pb) are inadequate. It is caused not only by differences in the methods for determining electronegativity, but also by features of the electron structure of investigated objects[34]. The most important is the presence of hyperconjugation contributions (the σ–σ, σ–π and σ–n conjugation effects) which complicate the determination of electronegativity and inductive constants σ_I.

Unfortunately, the methods for determining the electronegativity of M and the σ_I constants of R_3M substituents (in particular, Me_3M) do not yet provide a rigorous separation of the inductive and hyperconjugation effects.

Negative σ_I values of substituents Me_3M and Me_3MCH_2 for M = Si, Ge, Sn, Pb are higher in absolute magnitude than those for M = C (Table 5). This is in accord with the higher electronegativity of C.

TABLE 5. The σ_I constants of isostructural substituents (Y) containing a group 14 element (M)[a]

Y	M				
	C	Si	Ge	Sn	Pb
Me_3M	−0.07	−0.15	−0.11	−0.13	−0.12
Me_3MCH_2	−0.03	−0.05	−0.04	−0.05	−0.04
Cl_3M	+0.37[b]	+0.39	+0.63	+0.80	—

[a]From Reference 34 unless otherwise specified.
[b]Reference 57.

The donor inductive effect of the Me_3M groups (M = Ge, Sn, Pb) is approximately equal ($\sigma_I = -0.12 \pm 0.01$), and somewhat higher in the case of Me_3Si ($\sigma_I = -0.15$), pointing to the anomalous position of Si among the group 14 elements. The positive σ_I values of Cl_3M groups rise with increasing atomic number of M (Table 5) and this points to their electron-accepting character. It is caused by the presence of three high electronegative chlorine atoms at M and probably by the field effect[29]. The −I effect of the Cl_3M groups increases in direct proportion to the magnitude of the dipole moment μ of the M−Cl bond[56]. For M = C, Si and Ge the μ values are *ca* 1.3, 2.1 and 3 D, respectively[29], i.e. they change in parallel to the σ_I values of Cl_3M substituents.

III. SPATIAL STRUCTURE

In the discussion of the spatial structure of organic compounds of the heavy silicon subgroup elements it is necessary to dwell on their molecular structure, structure of condensed state and complexes of donor−acceptor type.

When studying the molecular structure of group 14 elements, such as R_4M, it is important to take into account the presence of the non-bonded interactions $R \cdots R$ between substituents caused by repulsion of their electron shells, along with the valence bonds in the MR fragments. Non-bonded interactions can decrease when the interatomic distances M−R increase or when the R_4M molecule is arranged in a conformation with minimum $R \cdots R$ interactions. Several examples will illustrate these points.

The M−O−M fragments in $(PhCH_2)_3MOM(CH_2Ph)_3$ (M = Si, Ge, Sn) have a linear structure[58]. In contrast, the M−O−M angles in Ph_3MOMPh_3 are 127.9, 180, 135.2 and 137.3° for M = C, Si, Ge and Sn, respectively[59]. For M = C, the C−O−C angle is the smallest in the series, in spite of the possibility of non-bonded interactions between the phenyl groups which are located at the opposite carbon atoms. The Ge−O−Ge and Sn−O−Sn angles in Ph_3MOMPh_3 are somewhat larger, while the Si−O−Si fragment has a linear structure. All this indicates that the magnitude of M−O−M angles is determined mainly by electronic effects.

Steric hindrance for free rotation around the M−C bonds in n-Pr_3MCl decreases along the series: Si > Ge > Sn due to successive decrease of the non-bonded interactions between the propyl groups[60].

The interatomic distances d(M−M) in R_3MMR_3 (M = Ge, Sn) depend not only on the electronic effects, but also on the bulk of the R substituents. Thus the Ge−Ge bond distances in the series: H_3GeGeH_3, $Ph_3GeGePh_3$ and $(Me_3C)_3GeGe(CMe_3)_3$ are 2.403, 2.437 and 2.710 Å, respectively[16,61,62].

For R = Me_3C, the d(Ge−Ge) value exceeds noticeably the double atomic radius of germanium (2.44 Å) clearly demonstrating the influence of non-bonded interactions between the substituents.

Non-bonded interactions in distannane derivatives R_3SnSnR_3 are weaker owing to larger covalent radius of tin relative to germanium. Thus, the d(Sn−Sn) distances in $Ph_3SnSnPh_3$, $(Me_3C)_3SnSn(CMe_3)_3$, $(Me_3C)_3SnSn(CMe_3)_2Sn(CMe_3)_3$ and $(2,6\text{-}Et_2C_6H_3)_3SnSn(2,6\text{-}Et_2C_6H_3)_3$ are 2.77, 2.894, 2.966 and 3.052Å, respectively[63–65]. The d(Sn−Sn) distances in the last three molecules exceed the double covalent radius of the tin atom (2.82 Å). However, the observed lengthening of the Sn−Sn bond is less than that of the Ge−Ge bond in similar digermane derivatives. There is no evidence of significant non-bonded interactions in hexaphenyldiplumbane $Ph_3PbPbPh_3$ and its heteroanalogues Ph_3PbMPh_3 (M = Ge, Sn). The experimental values of d(Pb−M), 2.623, 2.829 and 2.844 Å for M = Ge, Sn and Pb, respectively[66,67], are all less than the sums of the covalent radii of Pb and M atoms (2.69, 2.88 and 2.94 Å, respectively). The non-bonded interactions of substituents in hexasubstituted benzene derivatives $C_6(MMe_3)_6$

cause deformation of the planar benzene ring which distorts into a chair form, i.e. such substitution causes dearomatization of the central benzene ring[68]. It is noteworthy that the distortion of the ring is more significant for M = Si than for M = Ge. This indicates that d–π conjugation of M with the aromatic ring might be important.

In some cases the molecular structure of organic compounds of heavy group 14 elements is determined by a stereoelectronic effect. According to quantum-chemical calculations, the neutral $C_6H_5CH_2MH_3$ and their protonated forms (carbocations) 4-$HC_6{}^+H_5CH_2MH_3$ (M = Si, Ge, Sn, Pb) adopt a *gauche* conformation (the C–C–M angle is 90°)[69]. The σ–π conjugation effect attains a maximum in these conformers. The σ–π conjugation is enhanced when changing from the neutral molecules to the carbocations (i.e. substituted phenonium ions) and when M becomes heavier.

The structure of organic compounds of heavy group 14 elements in the condensed state, for example R_4M, depends not only on intramolecular R···R (R = Alk, Ar, H or other nucleophobic substituent) non-bonded interactions, but also on intermolecular R···R and R···M interactions. If the M atom is greatly shielded by bulky R substituents, the intra- and inter-molecular R···R interactions prevail. If the M atom is weakly shielded by R substituents, the possibility of intermolecular R···M non-bonded interactions as well as M···M (if R = H) interactions become more important. This favours the association of R_4M molecules[70]. As the atomic number of M increases, the shielding of the central M atom in isostructural molecules R_4M diminishes. This is the reason why intra-molecular non-bonded interactions are more typical for M = Ge, and inter-molecular interactions are more typical for M = Sn and Pb.

Non-tetrahedral structures of organic derivatives of the silicon subgroup elements are often caused by inter- or intra-molecular coordination interaction X → M. This takes place in compounds where there is a nucleophilic substituent at the central M atom. An electronegative X atom, which has at least one unshared electron pair (X = N, P, O, S, halogen) and is directly bonded to M, can be such a substituent. Compounds of this kind tend to be involved in inter-molecular coordination. There can be also a heteroatom X as part of the organic substituent at M; in this case an intra-molecular coordination usually occurs[71–73]. Such compounds, which contain five- or six-membered coordination rings, include, for example, draconoides (**4**)[74,75], their analogues (**5**)[76,77], metalloatranes (**6**)[78] and others. The stability of a coordination bond X → M increases with the atomic number of M: Si < Ge < Sn < Pb.

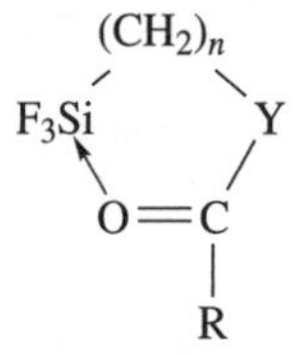

(**4**) (**5**) (**6**)

Y = O, S, NCH_3, CH_2

$n = 1, 2$

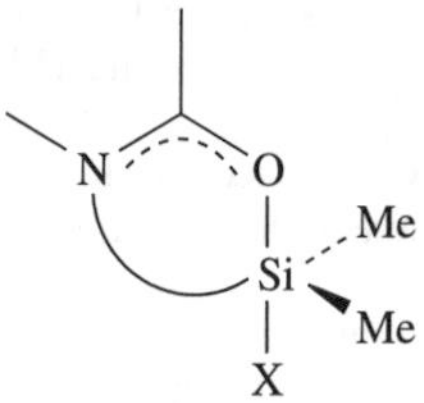

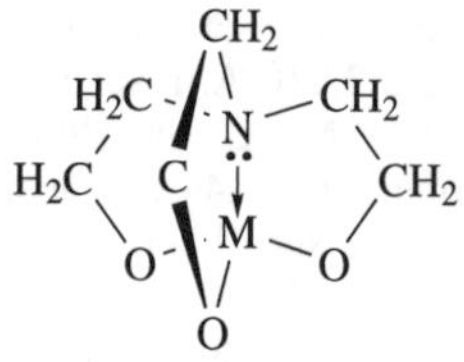

The donor–acceptor interaction X → M results in a rise of the coordination number of the M atom. For M = Si, Ge it amounts to a coordination number of 5, 6 and 7 (very seldom). For M = Sn, Pb, the coordination number can be as high as 8. The higher coordination number reflects mainly the increase in the atomic number of M and therefore its steric accessibility.

The M atom is tetravalent in many of its organic compounds and the substituents are placed at the vertices of a tetrahedron. If the coordination number of M atom is 5, 6, 7 and 8, then the central M atom has a configuration of a trigonal bipyramid, octahedron, pentagonal bipyramid and distorted cube, respectively.

Steric factors play a marked role in inter- and intra-molecular coordination of compounds of heavy elements of group 14. The complexation requires an approach of the donor and acceptor centres to an optimal distance. If these centres are shielded by bulky substituents, the complexation becomes difficult or impossible. Thus, for example, the tributylalkoxystannanes Bu_3SnOR are monomeric for any R. At the same time the dibutyldialkoxystannanes $Bu_2Sn(OR)_2$ are monomeric only when containing bulky R substituents such as CH_2CHMe_2 or CMe_3. The butylalkoxystannanes $BuSn(OR)_3$ are monomeric only when the alkyl substituents R are not smaller alkyl radicals (Me, Et, Pr)[79].

Lead compounds with very large bulky groups such as $Ph_3PbOSiPh_3$[80] and [2,4,6-$(Me_3Si)_2CHC_6H_2]_2PbBr_2$[81] are monomeric due to steric shielding of the Pb atom by the bulky aryl substituents.

Unlike many associated compounds of tin and lead, the germanium compounds of type $R_{4-n}GeX_n$ ($n = 1-3$) are usually monomeric[82,83].

The inter-molecular complexes of compounds of the type Ph_nMR_{4-n} (M = Si, Ge, Sn) with molecular O_2, which have been studied by the method of low-temperature luminescence quenching[84a], are of special interest. Towards O_2 these compounds can possess both π-donor (owing to the phenyl groups) and v-acceptor[84b] (owing to the M atom) properties. The v-acceptor properties of the M atom increase along the series: Si < Ge < Sn. When M = Si, the v-acceptor properties are not revealed and only the π-donor ability of the aromatic rings is realized, on which the steric effects of the SiR_{4-n} substituents have no influence. For M = Ge, these compounds exhibit both π-donor and v-acceptor properties. On going from $PhGeH_3$ to Ph_4Ge, the equilibrium constant for complexation with oxygen (K_C) diminishes by a factor of about 20 due to a sharp increase in the steric shielding of the Ge atom (the π-donor properties of these two compounds are almost equal). The K_C values of the complexation reactions of SnR_4 with O_2 decrease as the steric bulk of the alkyl radicals R increases in the series: Me < Bu < pentyl[84a].

The complexes of tetraalkylstannanes SnR_4 with tetracyanoethylene (TCNE)[85] are more stable than those with oxygen. The Sn−C bond serves as a σ-donor in this case. The K_C values of such complexes are also defined predominantly by the degree of shielding of the donor centre. With change of R these constants diminish in the series: Me >> Et > i-Pr > i-Bu > t-Bu. The steric hindrance to the complexation of $Me_{4-n}PbEt_n$ with TCNE is smaller in comparison with the isostructural tin compounds as expected on the basis of the larger Pb interatomic Pb−C distance[85].

In complexes of $R_3SnC{\equiv}CSnR_3$ with iodine, the C≡C bond is the donor centre. If R = Me, the interaction with I_2 occurs at a very high rate, and if R = Me_3C, complex formation with I_2 is not observed[86].

Finally, we shall briefly discuss the compounds of the silicon subgroup elements in their divalent state (germylenes, stannylenes and plumbylenes) R_2M: (R = alkyl, aryl). The chemical bonds in R_2M: are formed by the p_x and p_y orbitals of M. The p_z orbital is unoccupied and there is an unshared electron pair in an s-orbital of M. Therefore, the R_2M: compounds have both electrophilic and nucleophilic properties. The valence angle R−M−R diminishes as the atomic number of M increases. The distances of M−R bonds in R_2M: are less than in the corresponding tetracoordinate R_4M derivatives[87]. The spatial structure of R_2M: compounds, the shielding of the reaction centre (M:) as well as the possibility for R_2M: molecules to transit from a singlet state into a triplet state depend on the steric bulk of the R substituents. If the R substituents are bulky, some R_2M:

compounds can form the $R_2M{=}MR_2$ dimers. The stability of the M=M bond decreases as the atomic number of M becomes larger. Thus, for R = 2-*t*-Bu-4, 5,6-Me_3C_6H, the M=M-type dimers are formed only for M = Ge and Sn, while the corresponding plumbylene exists as a monomer R_2Pb:[88].

IV. PHYSICAL PROPERTIES

A great number of investigations was dedicated to studying the physical properties of organic compounds of germanium, tin and lead. We discuss here only those that include a comparative study of these compounds as well as studies which verify or develop theoretical concepts of their electronic structure.

A. Vibrational Spectra

Table 6 presents a list of the force constants k of some M−X bonds which are calculated by means of normal coordinate analysis, for example[89,90]. The k values characterize the curvature of the potential well close to the equilibrium internuclear distance. It is widely believed that the higher the value of k, the stronger the corresponding chemical bond. Such a relation is found in narrow series of isostructural compounds. It follows particularly from the k values of $H_{4-n}MX_n$, where M = C, Si, Ge; X = F, Cl, Br, I; $n = 1-3$[90].

It follows from Table 6 that the values of force constants of M−C and M−X bonds diminish for all given series of isostructural compounds as the atomic number of M increases. The decrease in the strength of these bonds on going from M = Si to M = Ge, Sn and Pb is confirmed by the parallel behaviour of the values of k and of the bond dissociation energy (D). Thus the D values in the series $Me_3M{-}H$ for M = Si, Ge, Sn and Pb are respectively 378, 343, 309 and 259 kJ mol^{-1}[98,99], and in series $Me_3M{-}MMe_3$ 354, 259, 234 and 228 kJ mol^{-1}[98,100]. When the number of X atoms (n) is increased, the k values of the M−X (X = Cl, Br, I) bonds increase for a fixed M atom in all series $Me_{4-n}MX_n (n = 1-4)$. A similar dependence between the values of k(M−C) and n in the same series $Me_{4-n}MX_n$ exists as a general trend only. In the same series for fixed M and n values the k(M−X) values diminish as the atomic number of the halogen (X) increases. So the D values of $Cl_3Ge{-}Cl$, $Br_3Ge{-}Br$ and $I_3Ge{-}I$ bonds are 338, 270 and 213 kJ mol^{-1}[101,102], respectively. The change from Me_4M to Me_3MMMe_3 (i.e. substitution of one methyl group by a more electron-donating substituent MMe_3) for a fixed M atom leads to a decrease in k(M−C). If methyl groups in Me_4M are replaced by hydrogen atoms (change to Me_3MH and $MeMH_3$), the k(M−C) as well as k(M−H) values increase. Unfortunately, the interrelation of the force constants and donor−acceptor properties of the substituents at M are poorly explored as yet.

The influence of the R and X substituents on the stretching modes of the M−H bond in IR spectra of the series $R_{3-n}X_nMH$ (M = Si, Ge, Sn) has been investigated in some detail. The form of these vibrations is highly characteristic. Therefore, their frequency ν(M−H) and intensity A(M−H), which characterize the M−H bond strength and polarity respectively, depend only on the electronic effects of the R and X substituents. If the R substituents at the M have only inductive influence upon ν and A, the correlations given in equations 1 and 2 are observed[35] (for M = Si and Ge).

$$\nu_{\mathrm{ind}}(\mathrm{M{-}H}) = a + 23\sum\sigma^* \qquad (1)$$

$$A^{1/2}{}_{\mathrm{ind}}(\mathrm{M{-}H}) = b - 20\sum\sigma^* \qquad (2)$$

If one, two or three X substituents form d,n or d,π bonds with the M, the experimental values of ν_{exp} and $A^{1/2}{}_{\mathrm{exp}}$ differ from those calculated according to equations 1 and 2.

TABLE 6. Force constant data (mdyn Å^{-1}) for M–X, M–C and M–M stretching modes[a]

Compound	k	M			
		Si	Ge	Sn	Pb
$MeMH_3$	M–H	3.14	2.97	2.41	—
	M–C	3.08	2.96	2.27	—
Me_3MH	M–H	2.77	2.44[b]	1.99[c]	—
	M–C	2.94	4.74[b]	2.63[c]	—
Me_4M	M–C	2.94	2.73[d]	2.19[e]	1.88[f]
Me_3MCl	M–Cl	2.54	2.22[c]	1.92[e]	1.41[f]
	M–C	3.23	2.86[d]	2.41[e]	—
Me_2MCl_2	M–Cl	2.70	2.27[d]	—	—
	M–C	3.34	2.94[d]	—	—
$MeMCl_3$	M–Cl	2.88	2.50[d]	—	—
	M–C	3.36	3.06[d]	—	—
MCl_4	M–Cl	3.47	3.11	2.76	2.31
Me_3MBr	M–Br	2.44	1.78	1.56	1.24[f]
	M–C	3.24	3.10	2.50	—
Me_2MBr_2	M–Br	2.54	2.07	1.95	—
	M–C	3.39	3.08	2.47	—
$MeMBr_3$	M–Br	2.64	2.28	2.06	—
	M–C	3.37	3.12	2.53	—
MBr_4	M–Br	2.61	2.39	2.20	—
Me_3MI	M–I	—	1.13[g]	1.16[e]	1.07[f]
	M–C	—	2.87[g]	2.34[e]	—
MI_4	M–I	1.91	1.74	1.62	—
Me_3MMMe_3[h]	M–M	1.70	1.54	1.39	0.98
	M–C	2.49	2.47	2.08	1.74

[a] From Reference 89, unless otherwise specified.
[b] Reference 91.
[c] Reference 92.
[d] Reference 93.
[e] Reference 94.
[f] Reference 95.
[g] Reference 96.
[h] Reference 97.

The differences from the theoretical values, $\Delta\nu = \nu_{ind} - \nu_{exp}$ and $\Delta A^{1/2} = A^{1/2}{}_{exp} - A^{1/2}{}_{ind}$, constitute a quantitative characteristic of the d–n or d–π conjugation. The correlations given in equations 3[103] and 4[104]

$$\Delta\nu_{Si} = 1.77\Delta\nu_{Ge} - 4 \tag{3}$$

$$\Delta A^{1/2}{}_{Si} = 1.11\Delta A^{1/2}{}_{Ge} + 0.07 \tag{4}$$

are valid for isostructural compounds of silicon and germanium. These correlations suggest that the d–n and d–π conjugation is weaker for M = Ge than for M = Si. As the effective positive charge on M increases (it may be characterized by the sum of the σ_p constants of R and X), the d–n and d–π conjugation is enhanced, more so for M = Si than for

M = Ge. This follows from the correlations in equations 5[104] and 6[105].

$$\Delta A^{1/2}{}_{\mathrm{Si}} = 1.21 \sum \sigma_{\mathrm{p}} + 0.47 \tag{5}$$

$$\Delta A^{1/2}{}_{\mathrm{Ge}} = 1.09 \sum \sigma_{\mathrm{p}} + 0.36 \tag{6}$$

The correlations 1 and 2 fail for alkylstannanes $R_{4-n}SnH_n$ (n = 1–3)[106]. This is caused by the enhancement of σ–σ conjugation between the C–H and M–H bonds (shown in **7**) with increase in the atomic number of M. According to this and other data[54], the validity of correlations 1 and 2 for organic lead compounds containing the Pb–H bond is even less probable.

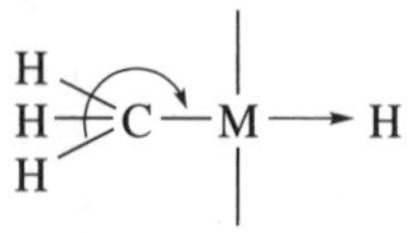

(7)

The valence vibrations of the C≡C bond in IR spectra of $Me_3MC{\equiv}CR$ (M = Si, Ge, Sn) are characteristic. The intensity A of the ν(C≡C) bands is found to be related linearly (equation 7) to the resonance constants $\sigma_R{}^0$ of the R substituent.

$$A^{1/2} = a + b\sigma_{\mathrm{R}}{}^{0} \tag{7}$$

The coefficient a depends on the atomic number of M while b is almost independent of M[107]. The $A^{1/2}$ values are also linearly correlated with the Δq_π values (calculated by the *ab initio* method), which denote the π-electron exchange between R and the triple bond in HC≡CR[108]. According to these correlations the $\sigma_R{}^0$ and Δq_π (the magnitude of the π-electron transfer from the π-system to Me_3M) values for $Me_3MC{\equiv}CR$ have been calculated to be: Si (+0.12, 0.028e), Ge (+0.06, 0.022e) and Sn (+0.04, 0.016e). The positive values of $\sigma_R{}^0$ indicate the acceptor character of the Me_3M groups (the d–π conjugation effect), which weakens in the series Si > Ge > Sn[107].

The electronic effects of the Me_3M substituents (M = C to Pb), bonded to n- or π-donor centres, were studied by the method of hydrogen bond IR spectroscopy. This method is demonstrated in equation 8.

$$\mathrm{Y}_i\mathrm{A{-}H} + \mathrm{BX}_i \rightleftharpoons \mathrm{Y}_i\mathrm{A}^{\delta-}{-}\mathrm{H}\cdots\mathrm{B}^{\delta+}\mathrm{X}_i \tag{8}$$

If three of the four variables (acceptor A–H, donor centre B, substituents X_i and Y_i) are fixed (Y_i, A and B), a series of H-bonded complexes are formed, such as the phenol complexes shown in equation 9.

$$\mathrm{PhO{-}H} + \mathrm{O(X}_i)_2 \rightleftharpoons \mathrm{PhO{-}H}\cdots\mathrm{O(X}_i)_2 \tag{9}$$

The frequency shift $\Delta\nu = \nu(\mathrm{OH}) - \nu(\mathrm{OH}\cdots\mathrm{O})$, where ν(OH) is the frequency of the stretching mode of the O–H bond of the isolated phenol and $\nu(\mathrm{OH}\cdots\mathrm{O})$ in the presence of the electron donor $O(X_i)_2$, is very informative for this series. The quantity $\Delta\nu$ is linearly correlated with the change of enthalpy (energy of donor–acceptor bond in the H-complex) and free energy (stability of the H-complex)[109], as well as with the value of effective charge q on the donor centre B, which was calculated by quantum-chemical methods[108]. The second correlation shows the dominant role of electrostatic interactions in the formation of the hydrogen bond. The correlations given in equations 10 and 11 are also valid for the H-complexes.

$$\Delta\nu = a \sum \sigma_p + b \tag{10}$$

$$\Delta\nu = c \sum \sigma_I + d \sum \sigma_R + k \tag{11}$$

They allow one to calculate the resonance constants σ_R of the Me_3M substituents and to estimate the effective charge component of the donor centre, resulting from conjugation[31,109–115].

The $\Delta\nu$ values for H-complexes formed by the electron donors of π-type (Table 7) and n-type (Table 8) obey a general rule. The $\Delta\nu$ values decrease in all 8 series of $Me_3M(CH_2)_nR_\pi$ and Me_3MX molecules when changing from $M = C$ to Si and then increase in the order Si < Ge < Sn < Pb. Such dependence of $\Delta\nu$ on the atomic number of M cannot be explained by a simple correlation with the electronegativity of M. According to the minimum electronegativity of the Si atom (Table 1) the $\Delta\nu$ values of its compounds should have been the highest.

In contrast, the experimental values of $\Delta\nu$ for M = Si are the lowest. This strongly supports the maximum acceptor ability of silicon towards donor centres R_π and X as a

TABLE 7. $\Delta\nu$ values (cm^{-1}) measured in the IR spectra of H-complexes of $Me_3M(CH_2)_nR_\pi$ with PhOH and calculated σ_R and q_π values

M	n		R_π			
			Ph[a]	$H_2C{=}CH$[b]	2-Fu[c]	$Me_3CC{\equiv}C$[d]
		$\Delta\nu$	61	65	72	140
C	0	σ_R	−0.13	−0.13	−0.13	−0.13
		q_π	—	−0.012	−0.015	−0.014
		$\Delta\nu$	55	56	67	131
Si	0	σ_R	0.05	0.05	−0.02	0.00
		q_π	0.004	0.018	0.002	0.010
		$\Delta\nu$	57	64	70	153
Ge	0	σ_R	0.01	−0.05	−0.09	−0.18
		q_π	−0.003	0.002	−0.008	−0.022
		$\Delta\nu$	—	67	79	170
Sn	0	σ_R	0.01	−0.06	−0.18	−0.24
		q_π	—	−0.002	−0.029	−0.034
		$\Delta\nu$	58	80	—	144
Si	1	σ_R	−0.20	−0.24	—	−0.24
		q_π	−0.036	−0.033	—	−0.032
		$\Delta\nu$	65	85	—	148
Ge	1	σ_R	−0.23	−0.29	—	−0.24
		q_π	−0.039	−0.042	—	−0.032
		$\Delta\nu$	—	—	—	152
Sn	1	σ_R	−0.24	—	—	−0.30
		q_π	—	—	—	−0.041

[a]Reference 111.
[b]Reference 112.
[c]Reference 113.
[d]Reference 114.

result of the d–π and d–n conjugation effect. As the atomic number of M increases, its resonance acceptor effect diminishes. The σ_R parameters, which characterize the overall resonance effect (both acceptor and donor) of the Me_3M substituents towards the π-system and reflect its magnitude and sign, diminish as well (Table 7). However, these parameters are not universal. For example, the σ_R values of $SnMe_3$ for R_π = Ph and for R = $Me_3CC{\equiv}C$ are respectively +0.01 and −0.24. The first value points to a balance between acceptor and donor effects of $SnMe_3$ while the second value indicates that in $Me_3CC{\equiv}CSnMe_3$, the Me_3Sn substituent is a donor (σ–π conjugation). The σ–π conjugation effect in Me_3MR_π and $Me_3MCH_2R_\pi$ increases as M becomes heavier (i.e. Sn > Ge > Si).

The q_π parameter characterizes the resonance effects of substituents in isolated neutral molecules[108], and the $\sigma_R{}^0$ parameter in molecules having formed a hydrogen bond (at the same time a partial positive charge δ^+ appears on the donor centre B (equation 8)[111–115]). Therefore the correlation between the q_π and $\sigma_R{}^0$ parameters is rigorous[108] while between the q_π and σ_R parameters it is only approximate. The q_π values (Table 7) calculated from $\Delta\nu$ are approximate as well. The δ^+ charge on R_π is higher, the stronger the σ–π conjugation in Me_3MR_π and $Me_3MCH_2R_\pi$ and the higher the atomic number of M. This becomes apparent in the differences between the $\sigma_R{}^0$ and σ_R parameters of the Me_3M substituents. The $\sigma_R{}^0$ [107] and σ_R (in parentheses) values in $Me_3MC{\equiv}CCMe_3$ are +0.12 (0.00), +0.06 (−0.18) and +0.04 (−0.24) for M = Si, Ge and Sn, respectively. Thus, even relatively weak perturbations of the electronic structure of the donor molecules Me_3MR_π due to hydrogen bonding can cause reversal of the sign of the resonance effect of $GeMe_3$ and $SnMe_3$ substituents.

The $\Delta\nu_R = \Delta\nu_{ind} - \Delta\nu$ parameters are given in the Table 8. The $\Delta\nu_{ind}$ values characterize the inductive effect of a substituent only approximately[31,115,116]. Hence, also the $\Delta\nu_R$ values may be used only for rough estimates of the inductive effect. The $\Delta\nu_R$ values are positive in the series Me_3MNMe_2 for M = Si, Ge, Sn. Lower experimental values of $\Delta\nu$ in comparison with the expected ones based only on the inductive effect of the Me_3M substituents ($\Delta\nu_{ind}$) can be taken as evidence of the reduced n-donor properties of

TABLE 8. $\Delta\nu$ values (cm^{-1}) measured in the IR spectra of H-complexes of Me_3MX with $CDCl_3$

M		X			
		$NMe_2{}^a$	$OMMe_3{}^{b,c}$	$SMMe_3{}^{b,c}$	Cl^d
C	$\Delta\nu$	96	33	40	74
	$\Delta\nu_R$	0	0	0	0
Si	$\Delta\nu$	61	13	29	55
	$\Delta\nu_R$	58	263	62	29
Ge	$\Delta\nu$	76	55	38	90
	$\Delta\nu_R$	36	−126	−37	−14
Sn	$\Delta\nu$	108	84	43	113
	$\Delta\nu_R$	8	−324	−57	−34
Pb	$\Delta\nu$	—	—	51	—
	$\Delta\nu_R$	—	—	−127	—

[a] From References 110 and 116.
[b] From Reference 117.
[c] $\Delta\nu_R$ values from Reference 31 (acceptor = PhOH).
[d] From Reference 115 (acceptor = PhOH).

the nitrogen atom due to the d–n conjugation. This effect weakens as the atomic number of the M element increases. The d–n conjugation dominates also in the series Me_3MX ($X = OMMe_3, SMMe_3, Cl$) for M = Si. However, starting with M = Ge the resonance donor effect of the σ–n conjugation of the Me_3M group prevails, and it increases along the series Ge < Sn < Pb. This is clearly indicated by the negative $\Delta\nu_R$ values.

B. Nuclear Magnetic Resonance (NMR) Spectra

The main parameters of NMR spectroscopy are the chemical shift δ_A (the difference of the magnetic shielding of a reference (σ_r) and studied (σ_A) nucleus) and the spin–spin coupling constant J between nuclei. The shielding constant of a nucleus A (σ_A) includes diamagnetic ${\sigma_A}^d$ and paramagnetic ${\sigma_A}^p$ components, as well as a $\Sigma\sigma_{AB}$ term (magnetic anisotropy) characterizing the shielding of the nucleus A by the electrons of the other nuclei (B); see equation 12[118].

$$\sigma_A = {\sigma_A}^d + {\sigma_A}^p + \Sigma\sigma_{AB} \tag{12}$$

The ${\sigma_A}^d$ component prevails for ^1H nuclei. According to some data[119] the ${\sigma_A}^p$ component prevails for the ^{13}C, ^{19}F, ^{29}Si, ^{73}Ge, ^{117}Sn, ^{119}Sn and ^{207}Pb nuclei, but according to others[120] both components (${\sigma_A}^d$ and ${\sigma_A}^p$) are significant. When estimating the $\Sigma\sigma_{AB}$ term in equation 12, appreciable difficulties arise. Calculations of $\Sigma\sigma_{AB}$ taking the inductive effect of X into account have been carried out for simple systems R_3MX (R = Me, Et; X = F, Cl, Br, OMe)[121]. According to these studies the d–n conjugation has been established to decrease along the series Si > Ge, F > Cl > Br and F > OMe.

The main efforts were aimed, however, at searching systems where the chemical shifts δ and coupling constants J are connected with the electronic effects of the substituents by simple dependencies. The primary idea was to move the variable substituents R_3M as far as possible from the indicator centre A (i.e. minimizing the $\Sigma\sigma_{AB}$ term in equation 12). At the same time, the sensitivity of δ_A to the effect of R_3M must not be lost[122,123]. Table 9 presents 5 examples of such systems. The ^{13}C chemical shifts depend on the inductive and resonance effects of the Me_3M and Me_3MCH_2 groups. The series 1–3 illustrate the effect of Me_3M substituents in Me_3MR_π on the ^{13}C chemical shifts in position 4 of benzene and naphthalene rings, as well as of the β-carbon atom of the vinyl group. Depending on M the δ values change in the following sequence: C < Si > Ge >

TABLE 9. Carbon-13 chemical shifts δ^a of Me_3MR_π and $Me_3MCH_2R_\pi$

No.	Series	Position	M				
		^{13}C	C	Si	Ge	Sn	Pb
1	Me_3MPh^b	δ (C-4)	122.5	128.8	128.3	128.2	127.5
2	$Me_3MNaph\text{-}1^b$	δ (C-4)	127.4	129.7	129.1	128.8	128.1
3	$p\text{-}Me_3M{-}C_6H_4{-}CH{=}CH_2{}^c$	$\delta(C_\beta)$	111.84	113.45	113.22	113.22	113.00
4	$Me_3MCH_2Ph^d$	δ (C-4)	125.8	123.9	123.8	123.0	—
5	$p\text{-}Me_3MCH_2{-}C_6H_4{-}CH{=}CH_2{}^c$	$\delta(C_\beta)$	112.43	111.62	111.54	111.22	111.21

^{a}In ppm vs Me_4Si.
^{b}Reference 122.
^{c}Reference 123.
^{d}Reference 124. The $^1J(^{13}C_\alpha{-}^{13}C_1)$ values for M = C, Si, Ge and Sn are 36.0, 40.9, 42.5 and 42.8 Hz, respectively.

Sn > Pb. This trend agrees with a maximal resonance acceptor effect (d–π conjugation) for M = Si mentioned above repeatedly, which weakens as the atomic number of M increases.

In series 4 and 5, $Me_3MCH_2R_\pi$, the δ values decrease as M changes along the series C > Si > Ge > Sn > Pb. This is yet another confirmation that σ–π conjugation becomes stronger when M becomes heavier.

The values of coupling constants ^1J (^{13}C–^{13}C) in organic compounds are proportional to the C–C bond order[125]. In series 4 this constant characterizes the interaction of nuclei of aliphatic (C_α) and aromatic (C_1) carbon atoms. The $^1J(^{13}C_\alpha-^{13}C_1)$ value increases when M changes from M = C to M = Sn, corresponding to an increase of the C_{ar}–CH_2 bond order, i.e. to the enhancement of the σ–π conjugation as the atomic number of M increases.

^{13}C and ^{19}F NMR spectroscopy are classical methods for determining the resonance $\sigma_R{}^0$ and inductive σ_I constants of X substituents bonded to an aromatic ring[34,57]. This method is based on correlations of the type shown in equation 13 for chemical shifts δ of ^{13}C and ^{19}F atoms in the *para*- and *meta*-positions to X in the spectra of C_6H_5X, as well as of *p*-FC_6H_4X and *m*-FC_6H_4X.

$$\delta = a\sigma_I + b\sigma_R{}^0 + c \tag{13}$$

The $\sigma_R{}^0$ values characterize in addition (see above) the similarities and differences of the resonance interactions of Ge-, Sn- and Pb-containing substituents with a benzene ring. The principal similarity between the R_3M groups lies in a smaller d–π conjugation and a larger σ–π conjugation in R_3MPh and R_3MCH_2Ph for M = Ge, Sn, Pb than for M = Si. The main difference between the R_3M groups results from the fact that as the atomic number of M increases from Ge to Pb, the d–π conjugation weakens and the σ–π conjugation becomes stronger.

The chemical shifts of heavy group 14 nuclei are interconnected by the correlations[126,127] shown in equations 14–17, where n is the number of data points and r is the correlation coefficient.

$$\delta(^{29}Si) = 0.787\delta(^{13}C) - 61.7 \quad r = 0.825 \quad n = 13 \tag{14}$$

$$\delta(^{119}Sn) = 5.119\delta(^{29}Si) - 18.5 \quad r = 0.990 \quad n = 48 \tag{15}$$

$$\delta(^{119}Sn) = 2.2\delta(^{73}Ge) - 11.3 \quad r = 0.984 \quad n = 14 \tag{16}$$

$$\delta(^{207}Pb) = 2.424\delta(^{119}Sn) + 74.8 \quad r = 0.975 \quad n = 35 \tag{17}$$

The high sensitivity of the chemical shifts of ^{73}Ge, ^{119}Sn and ^{207}Pb to substituent effects calls for a detailed study of the resonance interactions in these organometallic compounds.

C. Photoelectron Spectra and UV Spectra of Charge Transfer (CT) Complexes

Important information on the similarities and differences of germanium, tin and lead compounds was obtained using two mutually complementary types of spectroscopy. Photoelectron spectroscopy is widely used to determine the first (Ip) and subsequent ionization potentials of molecules. According to Koopmans' theorem, the Ip is equated with the HOMO energy (equation 18)[128].

$$\text{Ip} = -E_{\text{HOMO}}. \tag{18}$$

The electronic absorption spectroscopy of charge transfer (CT) complexes of donor molecules of π-, n- and σ-type (DX) with π- and σ-acceptors (A = TCNE, I_2 etc.) allows one to study the influence of the X substituents bonded to a donor centre, D, on the energies of charge transfer bands, $h\nu_{CT}$[129]. The $h\nu_{CT}$ and Ip parameters are connected by a linear dependence given in equation 19.

$$h\nu_{CT} = a\text{Ip} + b \tag{19}$$

If steric effects of substituents do not affect the formation of the complex and A (e.g. TCNE) is constant, coefficients a and b in equation 19 depend on D only[130]. In approximation 18, $h\nu_{CT}$ and Ip depend on the inductive and resonance effects of the X substituents for constant D[128,129]. As the donor properties of X become stronger, the HOMO energy increases and the $h\nu_{CT}$ and Ip values decrease.

In Table 10 we bring data for 9 series of group 14 compounds. As M changes from C to Si, the Ip and $h\nu_{CT}$ values (in the UV spectra of the CT complexes with TCNE) of compounds with the Me_3M group (series 1, 2 and 4–9) increase. However, these values decrease along the series: Si > Ge > Sn > Pb. The Ip values in series 1 for M = Si and Ge, as well as in series 8 for M = C and Si, are close. The given sequence of the Ip and $h\nu_{CT}$ change does not agree with the electronegativity scales of group 14 elements (Table 1). This is additional evidence that the maximal d–π and d–n conjugations are reached for M = Si, which weakens gradually as the atomic number of M increases. On the contrary, the σ–π conjugation effect increases when changing consecutively from M = C to M = Pb. Therefore, the Ip and $h\nu_{CT}$ values decrease smoothly in series 3 as the atomic number of M increases.

TABLE 10. First vertical ionization potentials Ip (eV) and energies of charge transfer bands $h\nu_{CT}$ (eV) in spectra of organic compounds of group 14 elements (M)

No.	Series	Ip/($h\nu_{CT}$)				
		C	Si	Ge	Sn	Pb
1	Me_3MPh[a,b]	8.74 (2.80)	8.94 (2.93)	8.95 (2.90)	8.75 (2.78)	8.54 (2.60)
2	p-$Me_3MC_6H_4MMe_3$[c]	8.40 (2.44)	8.98 (2.73)	8.60 (2.68)	8.50 (2.59)	8.25 (2.48)
3	Me_3MCH_2Ph[b,d]	8.77 (2.83)	8.42 (2.50)	8.40 (2.40)	8.21 (2.19)	— (2.01)
4	Me_3MFu-3[e]	(2.53)	(2.63)	(2.58)	(2.53)	(2.48)
5	Me_3MThi-2[a,e]	8.32 (2.46)	8.64 (2.54)	8.52 (2.50)	8.49 (2.48)	8.46 (2.47)
6	Me_3MSMMe_3[f]	8.18	8.74	8.40	8.22	7.78
7	Me_3MSMe[f]	8.38	8.69	8.50	8.37	8.13
8	Me_3MCl[g]	10.76	10.76	10.35	10.16	9.70
9	Me_3MBr[g]	10.05	10.23	9.78	9.60	9.30

[a] Reference 130.
[b] References 131 and 132.
[c] References 128 and 133.
[d] Reference 134.
[e] References 135 and 136.
[f] Reference 137.
[g] Reference 138.

The above views regarding the dependence of the Ip and $h\nu_{CT}$ values on the atomic number of M were recently shown to be rather simplified and the approximation in equation 18 to be rough[130]. The donor component $D^{+\bullet}X$ of the compact radical ion-pair which is formed in the excited state of a CT complex in solution according to equation 20

$$A + DX \leftrightarrow [A, DX] \xleftrightarrow{h\nu_{CT}} [A^{-\bullet}, D^{+\bullet}X] \quad (20)$$

and the radical cation generated by photoionization of individual DX molecules in the gas phase (equation 21)

$$DX \xrightarrow{h\nu} D^{+\bullet}X + e^{-} \quad (21)$$

are very similar in their electronic structure.

Therefore, the Ip and $h\nu_{CT}$ values obey, to a high degree of accuracy, equations of the type given in equation 22:

$$\text{Ip (or } h\nu_{CT}) = k + l\sigma_I + m\sigma_R^{+} + n\sigma_\alpha. \quad (22)$$

The quantity σ_I in equation 22 is an inductive X substituent constant and σ_R^{+} is an electrophilic resonance constant. The constant σ_R^{+} accounts for the influence of X on the reaction centre $D^{+\bullet}$, which has large positive charge arising in the processes shown in equations 20 and 21. The constant σ_α denotes an electrostatic attraction between the positive charge of the radical-cation and the dipole moment induced by this charge in the X substituent. The coefficients k, l, m and n depend on the type of donor centre D. When correlating the Ip and $h\nu_{CT}$ values according to equation 22, these coefficients are equal for $D = \text{const}$[130].

Equation 22 allows one to calculate parameters σ_R^{+} on the basis of the Ip or $h\nu_{CT}$, σ_I and σ_α values. Characteristic examples of the σ_R^{+} values of the Me_3M and Me_3MCH_2 groups in 6 series of organic compounds of the silicon subgroup elements calculated in such a way are given in Table 11.

The σ_R^{+} values of the Me_3M (series 1–4) and Me_3MCH_2 (series 5 and 6) groups depend essentially on the nature of the radical-cation centre. It was already pointed out in Section IV.A that the σ–π conjugation increases under the influence of a partial positive

TABLE 11. Parameters σ_R^{+} and σ_R[a] of Me_3M and Me_3MCH_2 groups

No.	Series	$\sigma_R^{+}/(\sigma_R)$				References
		Si	Ge	Sn	Pb	
1	Me_3MPh	0.02 (0.05)	−0.10 (0.01)	−0.21 (0.01)	−0.26 —	34, 139
2	Me_3MThi-2	0.25	0.01	−0.01	−0.05	140
3	$Me_3MC{\equiv}CH$	0.00 (0.00)	−0.22 (−0.18)	−0.36 (−0.24)	— —	34, 139
4	Me_3MSMMe_3	0.15	−0.10	−0.15	−0.31	141
5	Me_3MCH_2Ph	−0.49 (−0.20)	−0.59 (−0.23)	−0.76 (−0.24)	−0.99 —	34, 139
6	$Me_3MCH_2CH{=}CH_2$	−0.65 (−0.24)	−0.83 (−0.29)	−1.00 —	— —	142

[a]Values of σ_R (in parentheses) from Table 6.

charge developing on the donor centre when an H-complex is formed. This is illustrated in equation 8. With CT complex formation and photoionization taking place according to equations 20 and 21, the positive charge on the donor centre rises sharply (in comparison with that in H-complexes). Therefore, the $\sigma-\pi$ conjugation increase is even greater than upon formation of a hydrogen bond. This shows up in smaller $\sigma_R{}^+$ values in comparison with the σ_R values. The absolute value of the difference $|\sigma_R{}^+ - \sigma_R|$ increases as the atomic number of M increases (Table 11).

The $\sigma_R{}^+ - \sigma_R$ differences allow one to support the presence of both d$-\pi$ and $\sigma-\pi$ conjugation in phenyl derivatives of the silicon subgroup elements[11]. It doing so it was taken into consideration that $\sigma_R{}^+ - \sigma_R = \sigma_p{}^+ - \sigma_p$, because $\sigma_p{}^+ = \sigma_R{}^+ + \sigma_I$ and $\sigma_p = \sigma_R + \sigma_I$. For compounds of the general formula $(Me_3M)_mX_nR_{3\text{-}m\text{-}n}CPh$ (M = Si, Ge, Sn, Pb; X = inorganic or organic substituent; R = H, alk; $m, n = 0$–3), the correlation given in equation 23 (where ΣR_D is the sum of the refractions of its bonds) is valid.

$$\sigma_p{}^+ - \sigma_p = -0.117\Sigma R_D + 0.53 \quad r = 0.968 \quad n = 19 \tag{23}$$

For all benzene substituents which are resonance donors, $\sigma_p > \sigma_p{}^+$. It follows from equation 23 that the differences $\sigma_p{}^+ - \sigma_p$ characterizing the strengthening of the $\sigma-\pi$ conjugation increase with the enhancement of the polarizability of all the bonds within the substituent bonded to the aromatic ring. The quantitative characteristic of the overall substituent polarizability is the sum of the refractions of its bonds, ΣR_D (see Section I). The values of $\sigma_p{}^+$ and σ_p are approximately equal for organic substituents which are resonance acceptors. If the $R_{3\text{-}n}X_nM$ substituents had only a resonance acceptor effect (the d$-\pi$ conjugation), the correlation in equation 23 would fail for compounds $R_{3\text{-}n}X_nMPh$. In fact, equation 24

$$\sigma_p{}^+ - \sigma_p = -0.025\Sigma R_D + 0.23 \quad r = 0.985 \quad n = 30 \tag{24}$$

similar to equation 23, is valid for these compounds. Hence the $R_{3\text{-}n}X_nM$ substituents possess two resonance effects towards a phenyl group—they act both as a donor (the $\sigma-\pi$ conjugation) and an acceptor (the d$-\pi$ conjugation). If the two effects operate in opposite directions, the sensitivity of the value of $\sigma_p{}^+ - \sigma_p$ to the parameter ΣR_D is reduced. The smaller slope of equation 24 in comparison with that of equation 23 points it out[11].

It also follows from these equations that the polarizability of M and of the M$-$R bonds is the most important factor determining the enhancement of the $\sigma-\pi$ conjugation in $R_3MCH_2R_\pi$ and R_3MR_π molecules when a positive charge develops on the reaction (indicator) centre R_π. As the atomic number of M increases, the polarizability of M and of the M$-$R bonds increases (see Section I). Therefore, the $\sigma-\pi$ conjugation in an isostructural series increases under the influence of a positive charge on R_π both when going from M = C, Si to M = Ge, Sn, Pb and when M changes along the series Ge < Sn < Pb.

The electronic structure and physical properties of organic compounds of germanium, tin and lead (in comparison with isostructural derivatives of silicon and carbon) discussed in the previous sections lead to the following main conclusions regarding their similarities and differences.

Points of similarity as a function of M in compounds of the types R_3MR_π, R_3MX and $R_3MCH_2R_\pi$ (R = organic substituent; M = Ge, Sn, Pb; R_π = aromatic or α,β-unsaturated group; X = halogen, N<, O$-$, S$-$ or other atoms having at least one unshared electron pair) are:

1. Large atomic radius of M and accessibility of M atoms to nucleophilic attack.
2. High polarizability of the electron shell of M atoms as well as of M$-$C and M$-$X bonds.

3. Relatively low ionization potentials of M atoms and their compounds.

4. Lower electronegativity of M in comparison with carbon.

5. Virtually equal negative σ_I values (-0.12 ± 0.01) of all the Me_3M groups (M = Ge, Sn, Pb).

6. The presence of two components, acceptor and donor, in the total resonance effect of R_3M substituents towards a reaction (indicator) centre R_π or towards X.

7. A complex mechanism of the resonance acceptor effect (the d–π conjugation in R_3MR_π and the d–n conjugation in R_3MX, for M = Ge, Sn, Pb). This effect is absent when M = C. The acceptor effect includes the participation of unoccupied nd-orbitals of M and of antibonding σ^*-orbitals of the M–R bonds of the R_3M fragments (R = alkyl).

8. The important role of the resonance donor effect (hyperconjugation) in R_3MR_π [the σ–π conjugation; mixing of σ(M–R) orbitals with the π-orbitals of the R_π group] and R_3MX molecules [the σ–n conjugation; mixing of σ(M–R) orbitals with the n-orbitals of an X fragment].

9. The presence of a strong resonance donor σ–π conjugation effect in $R_3MCH_2R_\pi$ molecules.

10. Enhancement of the d–π and d–n conjugation when the effective positive charge on M increases.

11. Enhancement of the σ–π conjugation when increasing the effective positive charge on the reaction (indicator) centre R_π. In some cases there is a reversal of the donor–acceptor properties of R_3M substituents towards R_π when the charge of R_π is changed.

12. The tendency to expansion of the coordination sphere of M, which increases when going from Ge to Sn, Pb.

13. Change of interatomic distances and valence angles under the influence of non-bonded interactions (steric effect).

Points of difference of R_3MR_π, R_3MX and $R_3MCH_2R_\pi$ compounds (M = Ge, Sn, Pb) as a function of M are:

1. Increase of the atomic radius of M and of its accessibility to a nucleophilic attack when increasing the atomic number of M.

2. Enhancement of the polarizability of the M atom as well as of M–C and M–X bonds when increasing the atomic number of M.

3. Decrease of the ionization potentials of organic compounds of Ge, Sn, Pb when increasing the atomic number of M.

4. Absence of the commonly accepted united scale of electronegativity for M.

5. Complicated mechanism of the inductive effect of R_3M groups involving increase in the field effect contribution with increase in the atomic number of M.

6. Different ratio of the acceptor and donor resonance effects of R_3M for M = Ge, Sn, Pb.

7. Weakening of the d–π and d–n conjugation effects with increase in the atomic number of M.

8. Enhancement of the σ–π and σ–n conjugation effects with increase in the atomic number of M.

9. An increased role of σ–π conjugation in $R_3MCH_2R_\pi$ molecules in comparison with R_3MR_π, where this effect competes with the d–π conjugation. There is an enhancement of the σ–π conjugation in $R_3M(CH_2)_nR_\pi$ ($n = 0, 1$) when going consecutively from M = Ge to Pb.

10. The enhancement of the d–π and d–n conjugation effects when increasing the positive charge on M differs for M = Ge, Sn, Pb and becomes smaller as the atomic number of M increases.

11. The sensitivity of the σ–π conjugation in R_3MR_π molecules to the influence of the effective positive charge of R_π rises as the atomic number of M increases. Therefore,

the reversal in the donor–acceptor properties of R_3M substituents (i.e. transformation of resonance acceptors to donors) is most probable for M = Pb.

12. The enhancement of the tendency to increase the coordination number of M when going consecutively from M = Ge to Pb.

13. Weakening of the importance of the steric effects of substituents on the spatial structure around the M atom with increase in the atomic number of M.

In general, it seems that, on the basis of physical, chemical and biological properties, one should divide the compounds of the silicon subgroup elements into two groups, one with M = Si, Ge and the other with M = Sn, Pb.

V. CHEMICAL PROPERTIES

Differences in the chemical behaviour of organic compounds of the silicon subgroup elements are caused mainly by the increase in the covalent atomic radius as well as in bond distances and polarity (and therefore in steric accessablilty of M atom) and the decrease in the bond dissociation energy when the atomic number of M increases. In the same order, i.e. Pb > Sn > Ge > Si, the electrophilicity of M and its tendency for complexation[26] increases and its electronegativity (according to the spectroscopic hydride scale which is based on the M–H bond stretching modes in X_3MH and MH_4), which excludes the influence of conjugation and association effects[7], gradually diminishes. The stability of compounds having a low-valent M atom (e.g. metallenes R_2M: and free radicals R_3M•) also rises as the atomic number of M increases. The chemical properties of organic compounds of the silicon subgroup elements were first compared in 1934 and next in 1947 by Kocheshkov[143,144], and 20 years later almost simultaneously and independently by Schmidt[145] and Glockling[146]. They were the first to have taken note of the fact that Mendeleyev's assertion[147–149], that properties of isostructural compounds of silicon, germanium and tin must change regularly in accordance with the position of these elements in the periodic table, is not always valid. The Allred and Rochow's[6] electronegativity scale of C, Si, Ge, Sn, Pb is a clear verification of this point. According to this scale the electronegativity of lead is close to that of carbon, and the electronegativity of silicon is less than that of germanium.

Numerous literature data mentioned in the previous chapter, including the data in References 144–146 and 150, allow one to conclude that the chemical properties of organic compounds of germanium, tin and lead have much in common, though there are essential differences in some cases. In general, organic compounds of germanium are closer in their properties to isostructural derivatives of silicon than to those of tin and lead. This is reflected in particular by a higher thermal and chemical stability of C–Si, Si–O, C–Ge and Ge–O bonds in comparison with corresponding bonds of Sn and Pb. At the same time, Sn–S and Pb–S bonds are much more stable than Si–S and Ge–S bonds. All in all the reactivity of bonds of organic derivatives of the silicon subgroup elements (M) as well as of compounds containing non-quadrivalent M atoms increases in the series Si < Ge < Sn < Pb.

A. C–M Bonds

The thermal and chemical stability of C–M bonds (M = Si, Ge, Sn, Pb), and therefore of all organic compounds of the silicon subgroup elements, decreases, both in homolytic and in heterolytic processes, when the atomic number of the M element increases. For example, the thermal stability of tetraalkyl derivatives R_4M diminishes essentially when M is changed consecutively from Si to Pb[151,152]. The ease of oxidation of R_4M compounds and the ease of cleavage of C–M bonds by halogens, protic and aprotic acids, etc., increases in the same order. When halogens and hydrohalogens react with R_4M (M = Ge) only one R−Ge bond is cleaved, and only in the presence of $AlCl_3$ are two M−R bonds cleaved. At the same time, if M = Sn and Pb, two or all four R−M bonds decompose easily under mild conditions (for M = Pb even at temperatures below 0 °C). R−M bonds (R = alkyl, aryl) are not cleaved by nucleophilic reagents unless reacted under severe conditions[153–156]. An acid cleavage rate of the M−Ph bond in R_3MPh (R = alkyl), which occurs as an electrophilic aromatic substitution process, increases in the following order: Si < Ge << Sn << Pb. The sharp distinction between the reactivity of the isostructural compounds with M = Si, Ge on the one hand and Sn, Pb on the other is explained by a change in the reaction mechanism by which they react. The C(sp^2)−M bond in vinyl derivatives is cleaved by electrophilic reagents faster, the heavier is M[157,158]. The stability of the C(sp)−M bond in ethynyl derivatives of the silicon subgroup elements like $R_3MC{\equiv}CR$ decreases as the atomic number of M increases. For M = Sn, Pb it is easily cleaved even by water and alcohols.

B. M−H Bonds

In accordance with the dissociation energy of M−H bonds, the reactivity of organometallic hydrides $R_{4-n}MH_n$ increases sharply as the atomic number of M and the number of hydrogen atoms (n) increase. The thermal stability of compounds of this series decreases in the same direction (especially for M = Pb). Trialkylplumbanes R_3PbH (especially with short-chain R) begin to decompose even below −78 °C. Unlike the Si−H bond, which is easily decomposed by water solutions of inorganic bases, the M−H bonds for M = Ge, Sn, Pb fail to react with them and their hydrolysis is more difficult. When R_3SiH reacts with organolithium reagents, the hydrogen atom is substituted by an organic function. In contrast, the reaction of R_3MH, where M = Ge, Sn, leads to R_3MLi (in the case of M = Sn, both reactions, alkylation and lithiation, can occur). Hydrosilylation of unsaturated compounds in the absence of catalysts or free radical initiators occurs only at a temperature above 250 °C. In contrast, hydrogermylation occurs often under mild temperature conditions.

Hydrostannylation occurs without the need for a catalyst and it occurs at temperatures below 100 °C (usually 20–80 °C, in some cases below 0 °C). It is noteworthy that the hydrosilylation catalysts do not affect the rate of this reaction. Hydroplumbylation, a poorly studied reaction, is carried out in the absence of both catalysts and initiators.[159]

As the atomic number of M increases, the possibility of reducing R_3MH rises sharply as well. For example, the reduction of organohalides by trialkylsilanes (M = Si) requires the use of catalysts ($AlCl_3$, Ni, Pt and others), but the heavier R_3MH (M = Ge, Sn, Pb) react in the absence of catalysts. The reactivity of the M−H bond in R_3MH towards elemental chalcogens increases as the atomic number of M increases.

Catalytic reactions of triorganylsilanes with ammonia and amines lead to the formation of the Si−N bond. In contrast, reaction of triphenylstannane with primary amines gives an Sn−Sn bond, i.e. $Ph_3SnSnPh_3$.

It is remarkable that triorganylgermanes R_3GeH can be synthesized by the reduction of germyl halides R_3GeX by zinc amalgam in hydrochloric acid. An analogous reaction of R_3MH for M = Si, Sn fails.

C. M−M Bonds

As the atomic number of M increases, the thermal stability of the M−M bond decreases. The ease of cleavage of the M−M bond by halogens, trifluoroiodomethane, phenyllithium, potassium and sodium alkoxides, alkali metals, sulphur etc. increases in the same order.

The M−M bonds in R_3MMR_3, $R_3M(R_2M)_nMR_3$ or $(R_2M)_n$ for M = Sn, Pb are cleaved by water, alcoholic solution of $AgNO_3$, alkyl iodides, mercury and bismuth halides and are oxidized by air. In contrast, these compounds with M = Si, Ge are stable under similar reaction conditions. The stability of the highly reactive $R_2M{=}MR_2$, as well as the tendency of R_2M: to dimerize, diminish as the atomic number of M increases[160–162].

Compounds of a RM≡MR pattern are unknown till now, but quantum-chemical calculations point out the possibility that M≡M bonds between Si and Ge atoms with appropriate substituents may exist[163,164]. The synthesis of the organolead ArPbPbAr, where Ar = 2, 6-(2, 4, 6-i-$Pr_3C_6H_2)_2C_6H_3$, was recently reported[164]. However, the X-ray data showed that the Pb−Pb bond is probably a single bond, i.e. the Pb−Pb bond (3.19 Å) is a little shorter than in metallic lead (3.49 Å) and the Pb−Pb−C angle is 94.3°.

D. M−X bonds (X = halogen)

M−X bond ionicity rises as the atomic number of M increases. Organylfluorosilanes and organylfluorogermanes $R_{4-n}MF_n$ (M = Si, Ge) are monomeric, but organylfluorostannanes (M = Sn) are polymeric due to an intermolecular F → Sn coordination. The lead compounds of the RMX_3 series (X = halogen) are extremely unstable (as well as PbX_4) and virtually unknown.

The M−X bonds of M = Ge, Sn, Pb are hydrolysed by basic water solutions (for M = Si and Ge, just by water), forming ≡M−OH or ≡M−O−M≡ groups.

Halides R_3MX (M = Si, Ge) react easily with water to produce first R_3MOH. If M = Sn or Pb, the ethers R_3MOMR_3 are formed instantaneously. Hydrolysis of R_2MX_2 results usually in the formation of oligomeric or polymeric metalloxanes, while for M = Pb, monomeric dialkyllead dihydroxides $R_2Pb(OH)_2$ are formed.

Depending on the hydrolysis conditions, R_2SiX_2 can lead to the formation of both $R_2Si(OH)_2$ and $(R_2SiO)_n$. Hydrolysis of RMX_3 for M = Si, Ge leads as a result to polymers $(R_3MO_{1.5})_n$, and for M = Sn, Pb to $[RM(OH)O]_n$ and $HO[RM(OH)O]_nH$. Thus, the hydrolysis of $R_{4-n}MX_n$ is facilitated when the atomic number of M decreases and that of the X halogen increases.

E. M−O Bonds

As the atomic number of M increases, the strength of the M−O bonds decreases and the ease of cleavage of the ≡M−O−M≡ group by water, alcohols and acids rises[165]. The stability towards protolysis of oligomers and polymers $R_nMO_{2-0.5n}$ is enhanced as the number of organic substituents at M (n) decreases. Polyorganylsilsesquioxanes $(RSiO_{1.5})_m$ are stable towards protic acids. In contrast, even $(RGeO_{1.5})_m$ and $[RSn(OH)O]_m$ react easily with organic and mineral acids; the substitution of oxygen atoms by the anion of the acid. Sn−O−Sn and Sn−O−C groups are decomposed easily by −SH, −NH and even CH acids. Unlike this, Si−O−Si groups are not cleaved by −SH, −NH and CH acids and they react with weak OH acids only in the presence of catalysts.

While linear polydiorganylsiloxanes, such as $X(R_2MO)_nMR_2X$ (M = Si), are easily formed and are stable, their structural analogues for M = Ge are poorly studied. At the same time, the analogous polymers for M = Sn and Pb are well known, but the metalloxane chains in these polymers are associated.

R_3MOH (R = alkyl; M = Si) are weak acids, and for M = Ge, Sn, Pb they are quite strong bases. While R_3MOH (M = Si, Ge) condense easily to give R_3MOMR_3, R_3SnOH and R_3PbOH condense only under the action of dehydrating agents. R_3MOH, M = Sn and Pb, also do not form metal derivatives R_3MONa when interacting with metal sodium, in contrast to R_3MOH (R = Si, Ge).

The stability of nitrates Me_3MONO_2 and the ionicity of the M−O bond in these molecules rises sharply as the atomic number of M increases. The compounds of this type with M = Sn, Pb are salt-like.

F. M−S Bonds

The M−S bonds with M = Si, Ge on the one hand and with M = Sn, Pb on the other are very different in their stability and reactivity. With M = Si, Ge these compounds are decomposed easily by water, alcohols and protic acids. In the case of M = Sn, Pb, they are converted into Sn−O and Pb−O bonds under the action of water solutions of strong bases. The reaction of R_3MCl with H_2S for M = Sn, Pb leads smoothly to R_3MSMR_3. If M = Si, Ge, the reaction occurs only in the presence of nitrogen bases.

Affinity of the silicon subgroup elements towards a thiocarbonyl group (C=S) decreases in the order Sn > Pb > Ge >> Si[166], but even silicon forms coordination bonds of the type C=S → Si (see, for instance, Reference 167).

G. M−N Bonds

Ammonolysis and aminolysis of M−X bonds (X = Cl, Br, I) for M = Si, Ge leads to the formation of $M{-}NR_2$ (R = H, alkyl) or M−N(R)−M groups, respectively, and for M = Sn, Pb the complexes are formed in which M possesses a coordination number of 5, 6, 7 and even 8 (for NH_3).

In the case that M = Ge, such adducts of 1:1 and 1:2 composition appear only at low temperature and are intermediates in the reaction of Ge−N bond formation. Organic compounds of tin and lead having M−N bonds are less accessible than those of silicon and germanium. They can be obtained by the interaction of the corresponding halides with metal derivatives of ammonia and amines. The M−N bonds in $Me_3MNHMMe_3$ and Me_3MNMe_2 (M = Si, Ge, Sn, Pb) are decomposed by HX acids (OH, SH, NH, PH and even CH acids). The hydrolysis is easier, the larger the atomic number of M. The basicity of the nitrogen atom in the M−N bonds increases in the same order.

H. Compounds Containing a Hypervalent M Atom

The ability of the heavier group 14 elements (M) to increase their coordination number rises with the increase of: (a) the atomic number of M, (b) the number of halogen atoms or other electronegative substituents bonded to M. At the same time, many complexes of quadrivalent Pb compounds are unstable due to redox interactions of the central Pb atom with the ligands.

Silicon and germanium compounds of the $R_{4-n}MX_n$ (X = halogen; $n = 2, 4$) series interact with bases (B) forming complexes of 1:1 and 1:2 composition (the latter are more stable) in which the coordination number of M is 5 or 6, respectively. However, there exist several intramolecular complexes of Si and Ge in which the coordination number of M is 7.

Organylhalostannanes form inter- and intra-molecular complexes in which the coordination number of Sn atom is 5, 6, 7 and even 8.

The range of hypervalent tin derivatives is the broadest. Among these compounds there are, for example, stable coordination compounds such as $R_2SnX_2 \cdot H_2NPh$, $R_2SnX_2 \cdot 2H_2NPh$, $R_2SnX_2 \cdot nNH_3$ ($n = 2, 3, 4$), $[R_2SnO]_n \cdot R_2SnX_2$ ($n = 1, 2$), $HO(R_2SnO)_3H \cdot R_2SnX_2$, $(R_3Sn)_2O \cdot R_3SnX$ and $R_3SnOH \cdot R_3SnX$. Similar isostructural complexes of organylhalosilanes and -germanes exist only at very low temperature (if they exist at all).

In contrast to $R_{4-n}MX_n$ (M = Si, Ge), the compounds with M = Sn can form complexes having the $R_{4-n}SnX_n \cdot 2B \cdot 2HX$ composition (X = Cl, Br; B = pyridine, quinoline, PhMeNH, $PhNH_2$; $n = 1$–4). These complexes correspond to structure $[BH^+]_2 \cdot [R_{4-n}SnX_{n+2}]^{2-}$ with a hexacoordinated Sn atom.

While compounds of R_3MX type with M = Si, Ge are stable as monomers, the isostructural analogues with M = Sn, Pb form stable complexes of 1:1 composition with dipolar aprotic solvents (DMF, DMSO, sulpholane)[168–170]. 1-Organylmetallatranes $RM(OCH_2CH_2)_3N$[78,171,172] are another striking example of the different coordination ability of organic compounds of the silicon subgroup elements. Compounds of this type are always monomeric both in the crystal state and in solution, for M = Si, Ge. The M atom in these compounds does not adopt a hexacoordinated structure under the action of nucleophilic agents. On the contrary, 1-organylstannatranes, $RM(OCH_2CH_2)_3N$, M = Sn, which have no bulky substituents at the tin atom, are associated in the crystal state and even in solvents (except water), i.e. the Sn atom is hexacoordinated[78]. This is evident from NMR and Mössbauer spectra. Organylplumbatranes $RPb(OCH_2CH_2)_3N$ remain unknown.

Unlike Si and Ge acetylacetonates (acac), in which the central M atom is hexacoordinated, the coordination number of the Sn atom in $RSn(acac)_3$ is 7[173]. Hexacoordinated Si, Ge, Sn, Pb complexes with tropolone (Tp) are known. Tp_4M molecules, where the metal atom is octacoordinated for M = Sn, Pb or hexacoordinated for M = Si, Ge, are of particular interest. The latter compounds correspond to the structure $Tp_3M^+Tp^-$ and exist as salts or ion pairs[174].

I. Compounds Containing a Low-valent M Atom

The majority of organic compounds of group 14 elements, corresponding to the formula R_2M, for which a bivalent state was assigned in the past to the M atom, are in fact cyclic oligomers or linear polymers, in which the M atoms are quadrivalent. In recent years labile monomeric R_2M: compounds (metallenes) have attracted considerable interest[160,175–177]. These are six-electron derivatives of divalent C, Si, Sn, Pb—carbenes[178–180], silylenes[181–186], germylenes[82,83,177,187,188], stannylenes[177,189–194] and plumbylenes[195,196]. The carbenes (M = C) differ essentially from the silylenes (M = Si). The reactivity of silylenes also differs from that of the germylenes (M = Ge). Diorganylsilylenes are so unstable that it is possible to isolate them only in argon matrix at very low temperature[197]. Diorganylgermylenes are monomeric in the gas phase. Some of them can exist under ordinary conditions as dimers, such as $R_2Ge{=}GeR_2$[87]. Diorganylstannylenes are more stable[160,189,192,198–200]. Metallenes are thermochromic. R_2M: (M = Ge, Sn, Pb) are practically colourless at very low temperature. However, at ordinary temperatures R_2Ge, R_2Sn and R_2Pb have canary, terracotta and purple colours, respectively. Their colour is caused by a metal-centred p $\rightarrow$ p electron transition as a result of sp-mixing.

Diorganylmetallenes have both electron-donating (nucleophilic) and weak electron-accepting (electrophilic) properties. They are good π-acceptors. Diorganylmetallenes exhibit Lewis base properties, because they can coordinate with unoccupied orbitals of electrophilic molecules, most often with those of transition metal derivatives[177,200–202]. In this case weaker ligands can be displaced by diorganylmetallenes from their coordination compounds (for example, carbonyls).

The first ionization potential of R_2M: is rather low. Its values for $[(Me_3Si)_2CH]_2M$ (M = Ge, Sn, Pb) are 7.75, 7.42 and 7.25 eV, respectively. In contrast to the numerous known structures of germylene and especially stannylene[200] complexes, the structures of only a few silylene complexes are known.

Singlet diorganylmetallenes add easily electrophilic agents such as hydrohalogens, halocarbons, acyl halogenides and halogens, according to the general equation 25,

$$R_2M{:} + R^1X \longrightarrow R_2R^1MX \tag{25}$$

where R^1 = H, alkyl, acyl, halogen; X = halogen. This reaction can be also considered as an insertion reaction of R_2M into the $R^1{-}X$ bond[176,188].

R_2M: complexes with strong Lewis acids (such as BX_3 and AlX_3) are more stable. R_2M: can also be added to the 1,4-position of 1,3-dienes, forming the corresponding heterocycles (cyclometallation reaction).

The donor ability (nucleophilicity) of R_2M:, including the stability of complexes being formed, increases as M becomes heavier. The acceptor properties of M in R_2M: (such as the ability to form the adducts with Lewis bases, for example, with pyridine and piperidine at $-30\,^{\circ}C$) are determined by the low-lying unoccupied atomic d- and p_z-orbitals[160]. Stable free radicals $R_3M^{\bullet}$ (M = Ge, Sn) are obtained by a photochemical disproportionation reaction of R_2M: in a hydrocarbon solvent medium[160]; see equation 26, R = $(Me_3Si)_2CH$.

$$2R_2M{:} \xrightarrow{h\nu} R_3M^{\bullet} + 1/n(RM)_n \tag{26}$$

The very high stability (10 years at room temperature) of $R_3M^{\bullet}$ radicals is due to the very bulky R substituents at M hindering recombination, as well as the comparatively low values of the M−H bond energy. Therefore, hydrogen atom abstraction from the hydrocarbon solvent turns out to be unfavourable.

In contrast, the half-life of $R_3Si^{\bullet}$ radicals is 10 min at ordinary temperature. It is noteworthy that a photochemical decomposition of $[(Me_3Si)_2CH]_2Pb$ fails to proceed according to equation 26, but leads instead to homolysis of both C−Pb bonds and to the formation of a lead mirror deposition (equation 27).

$$2R_2Pb \xrightarrow{h\nu} 2R^{\bullet} + Pb \tag{27}$$

This result supports the data presented in Section V.A, that of all C−M bonds, the C−Pb bonds can undergo homolytic cleavage, forming carbon-centred free radicals most easily.

VI. BIOLOGICAL ACTIVITY

The biological activity of organometallic compounds is determined by the nature of the metal atom, the molecular structure and their chemical properties. Biological action of organic compounds of the silicon subgroup can be divided into two groups, Si and Ge derivatives on the one hand and Sn and Pb derivatives on the other, according to certain differences in their chemical properties and molecular structure. The biological activity of these two groups differs sharply.

Silicon is a microbiogenous element and it plays an important role in vital activities of living matter of our planet. Some of its organic compounds are drugs, pesticides and biostimulators[203–206]. Most organosilicon compounds are slightly toxic, although there are surprising exceptions (1-arylsilatranes are highly toxic[207–209]). The role of germanium as a microbiogenous element has not been proved yet, although its presence is established in fungi, plants and animals, as well as in coal formed from plant remains[210–212]. The similarity of the biogeochemical history of germanium and silicon is striking[205–206].

The metabolism of silicon and germanium compounds in living organisms is closely related and mutually balanced. It is connected to a high extent with isomorphism of silicon and germanium, i.e. with the ability of germanium to substitute silicon in biological systems[205,206]. Organic compounds of germanium are close in their biological action as well as in their chemical properties to isostructural compounds of silicon and differ sharply from toxic organic derivatives of tin and lead[211,213].

A high-toxic compound has not been found as yet among an abundance of organogermanium compounds. Differences in the toxic action of some isostructural compounds of silicon and germanium can be illustrated by one example: 1-phenylgermatrane ($LD_{50} = 48$ mg/kg)[214] is almost 150 times less toxic than 1-phenylsilatrane ($LD_{50} = 0.3$ mg/kg)[207–209]. However, substitution of silicon in organisms of plants and hydrocole by germanium, accompanied by its excess in the environment, can lead to their death.

A large number of organic germanium compounds characterized by different kinds of biological activity have been synthesized and studied[215,216]. Some of them have already found application in medicine and agriculture as drugs[212,216] and biostimulators[212].

Almost all organic compounds of tin and lead have toxic action[211,213,217–220]. Therefore, their appearance in the environment where they are accumulated[221–225] is highly dangerous for all living organisms[221,222,226–229]. Three biochemical mechanisms regarding the influence of organic compounds of tin and lead on the cellular function and viability have been established. They include a cellular lipid metabolism, disruptions of cytosolic calcium homeostasis and a destruction of cytoskeletal components. Man or animals having been poisoned with organic compounds of tin and lead are affected mainly in the central nervous system[219,230–232]. They are especially dangerous for man, because they are accumulated in the brain and cause degeneration of primitive brain cells, inhibit growth of neurons, disrupt cytoskeletal structures and change lipid metabolism of human cells and cell morphology[220].

Cytotoxicity of organic derivatives of tin and lead depends mainly on their topological characteristics and to a lesser extent on the nature of the metal atom.

R_3MX are the most poisonous in the $R_{4-n}MX_n$ (M = Sn, Pb) series; this is also observed for the isostructural compounds of silicon[203]. The toxic action of compounds of tin and lead is similar. Derivatives of R_3MX inhibit oxidative phosphorylation, whereas R_2MX_2 binds thiol enzymes groups[211]. Fungicidal activity of trialkylstannane R_3MX derivatives is maximal when the number of carbon atoms is 9 in all three R substituents. On the whole, the toxicity of organic compounds of tin and lead decreases as the bulk of the substituents about the central metal atom increases.

The toxicity of R_4M (M = Sn, Pb) is caused by decomposition *in vivo* of one R–M bond with formation of R_3M^+. This biochemical reaction, proceeding much more easily for M = Pb than for M = Sn, explains the higher toxicity of isostructural lead compounds. Change in the nature of the R substituent at the Pb atom affects the toxicity to a lesser extent than for M = Sn.

In spite of its toxicity (and sometimes owing to it) organic compounds of tin and lead have found wide use during the last century as effective fungicides (R_3SnX), insecticides, antihelmintic and medicinal agents. Other industrial uses of organotin and organolead compounds are as polyvinyl chloride stabilizers, antiknock agents for vehicle fuels (Et_4Pb), polymeric materials for radiation shielding etc.[233–235].

VII. REFERENCES

1. L. Pauling, *The Nature of the Chemical Bond*, 3rd edn., Cornell University Press, New York, 1960.
2. K. Saito, *Chemistry and Periodic Table*, Iwanami Shoten, 1979 (Japanese).

3. U. Burkert and N. L. Allinger, *Molecular Mechanics*, ACS Monograph 177, American Chemical Society, Washington, D.C., 1982.
4. R. T. Sanderson, *J. Am. Chem. Soc.*, **105**, 2269 (1983).
5. A. Bondi, *J. Phys. Chem.*, **68**, 441 (1964).
6. A. L. Allred and E. G. Rochow, *J. Inorg. Nucl. Chem.*, **5**, 269 (1958).
7. M. G. Voronkov and I. F. Kovalev, *Izv. Akad. Nauk Latv. SSR, Ser. Khim.*, 158 (1965); *Chem. Abstr.*, **64**, 37b (1966).
8. S. S. Batsanov, *Structural Refractometry*, Vysshaya Shkola, Moscow, 1976; *Chem. Abstr.*, **86**, 163485b (1977).
9. E. Lippincott and J. Stutman, *J. Phys. Chem.*, **68**, 2926 (1964).
10. A. N. Egorochkin, in *Organometallic Compounds and Radicals* (Ed. M. I. Kabachnik), Nauka, Moscow, 1985, pp. 265–275; *Chem. Abstr.*, **104**, 109944c (1986).
11. A. N. Egorochkin, G. A. Razuvaev and M. A. Lopatin, *J. Organomet. Chem.*, **344**, 49 (1988).
12. E. A. V. Ebsworth, in *Organometallic Compounds of the Group IV Elements*, Vol. 1 (Ed. A. G. MacDiarmid), Dekker, New York, 1968, pp. 1–104.
13. L. V. Vilkov, V. S. Mastryukov and N. I. Sadova, *Physical Methods for Studying Organic Compounds: Determination of the Geometric Structure of Free Molecules*, Khimiya, Leningrad, 1978; *Chem. Abstr.*, **91**, 210816a (1979).
14. W. S. Sheldrick, in *The Chemistry of Organic Silicon Compounds*, Part 1 (Eds. S. Patai and Z. Rappoport), Chap. 3, Wiley, Chichester, 1989, pp. 227–303.
15. L. S. Khaikin, A. V. Belyakov, G. S. Koptev, A. V. Golubinskii, L. V. Vilkov, N. V. Girbasova, E. T. Bogoradovskii and V. S. Zavgorodnii, *J. Mol. Struct.*, **66**, 191 (1980).
16. P. v. R. Schleyer, M. Kaupp, F. Hampel, M. Bremer and K. Mislow, *J. Am. Chem. Soc.*, **114**, 6791 (1992).
17. A. Haaland, A. Hammel and H. Thomassen, *Z. Naturforsch., B*, **45**, 1143 (1990).
18. P. G. Harrison, in *Comprehensive Organometallic Chemistry*, Vol. 2 (Eds. G. Wilkinson, G. A. Stone and E. W. Abel), Pergamon Press, Oxford, 1982, pp. 629–680.
19. L. O. Brockway and F. T. Wall, *J. Am. Chem. Soc.*, **56**, 2373 (1934).
20. C. W. H. Cumper, A. Melnikoff and A. I. Vogel, *J. Chem. Soc. (A)*, 246 (1966).
21. S. Sorriso, A. Foffani, A. Ricci and R. Danieli, *J. Organomet. Chem.*, **67**, 369 (1974).
22. S. Sorriso, A. Ricci and R. Danieli, *J. Organomet. Chem.*, **87**, 61 (1975).
23. O. A. Osipov, V. I. Minkin and A. D. Garnovsky, *Handbook of Dipole Moments*, Vysshaya Shkola, Moscow, 1971; *Chem. Abstr.*, **76**, 118929g (1972).
24. R. S. Armstrong and R. J. H. Clark, *J. Chem. Soc., Faraday Trans. 2*, **72**, 11 (1976).
25. V. S. Dernova, I. F. Kovalev and M. G. Voronkov, *Dokl. Akad. Nauk SSSR*, **202**, 624 (1972); *Chem. Abstr.*, **76**, 105797n (1972).
26. H. Kwart and K. King, *d-Orbitals in the Chemistry of Silicon, Phosphorus and Sulfur*, Springer, Berlin, 1977.
27. C. J. Attridge, *Organomet. Chem. Rev. A*, **5**, 323 (1970).
28. A. N. Egorochkin, N. S. Vyazankin and S. Ya. Khorshev, *Usp. Khim.*, **41**, 828 (1972); *Chem. Abstr.*, **77**, 74338c (1972).
29. A. N. Egorochkin and M. G. Voronkov, *Electronic Structure of Organic Compounds of Silicon, Germanium and Tin*, Publishing House of the Siberian Branch of Russian Academy of Sciences, Novosibirsk, 2000 (Russian).
30. C. G. Pitt, *J. Organomet. Chem.*, **61**, 49 (1973).
31. A. N. Egorochkin, S. E. Skobeleva and V. L. Tsvetkova, *Metalloorg. Khim.*, **3**, 570, 576 (1990); *Chem. Abstr.*, **113**, 97733d, 97734e (1990).
32. A. N. Egorochkin, *Russ. Chem. Rev.*, **53**, 445 (1984).
33. A. N. Egorochkin, *Russ. Chem. Rev.*, **61**, 600 (1992).
34. A. N. Egorochkin and G. A. Razuvaev, *Russ. Chem. Rev.*, **56**, 846 (1987).
35. A. N. Egorochkin and S. Ya. Khorshev, *Usp. Khim.*, **49**, 1687 (1980); *Chem. Abstr.*, **94**, 3323j (1981).
36. Y. Apeloig, in *The Chemistry of Organic Silicon Compounds*, Part 1 (Eds. S. Patai and Z. Rappoport), Chap. 2, Wiley, Chichester, 1989, pp. 57–225.
37. C. G. Pitt, *J. Chem. Soc., Chem. Commun.*, 816 (1971).
38. A. E. Reed and P. v. R. Schleyer, *Inorg. Chem.*, **27**, 3969 (1988).
39. A. E. Reed, L. A. Curtiss and F. Weinhold, *Chem. Rev.*, **88**, 889 (1988).
40. R. P. Arshinova, *Metalloorg. Khim.*, **3**, 1127 (1990); *Chem. Abstr.*, **114**, 42829e (1991).

41. O. P. Charkin, *Stability and Structure of Gaseous Inorganic Molecules, Radicals and Ions*, Nauka, Moscow, 1980; *Chem. Abstr.*, **94**, 90792c (1981).
42. A. Modelli, D. Jones, L. Favaretto and G. Distefano, *Organometallics*, **15**, 380 (1996).
43. B. T. Luke, J. A. Pople, M.-B. Krogh-Jesperson, Y. Apeloig, J. Changrasekhar and P. v. R. Schleyer, *J. Am. Chem. Soc.*, **108**, 260 (1986).
44. C. Cauletti, C. Furlani, F. Grandinetti and D. Marton, *J. Organomet. Chem.*, **315**, 287 (1986).
45. M. Dakkouri, *J. Am. Chem. Soc.*, **113**, 7109 (1991).
46. C. Cauletti, C. Furlani, G. Granozzi, A. Sebald and B. Wrackmeyer, *Organometallics*, **4**, 290 (1985).
47. M. V. Andreocci, M. Bossa, C. Cauletti, S. Stranges, K. Horchler and B. Wrackmeyer, *J. Mol. Struct., THEOCHEM*, **254**, 171 (1992).
48. L. Noodleman and N. L. Paddock, *Inorg. Chem.*, **18**, 354 (1979).
49. (a) P. Livant, M. L. McKee and S. D. Worley, *Inorg. Chem.*, **22**, 895 (1983).
(b) Y. Mo, Y. Zhang and J. Gao, *J. Am. Chem. Soc.*, **121**, 5737 (1999).
50. R. Bertoncello, J. P. Daudey, G. Granozzi and U. Russo, *Organometallics*, **5**, 1866 (1986).
51. S. G. Wang and W. H. E. Schwarz, *J. Mol. Struct., THEOCHEM*, **338**, 347 (1995).
52. S. Berger, W. Bock, G. Frenking, V. Jonas and F. Muller, *J. Am. Chem. Soc.*, **117**, 3820 (1995).
53. C. Cauletti, F. Grandinetti, G. Granozzi, A. Sebald and B. Wrackmeyer, *Organometallics*, **7**, 262 (1988).
54. M. Kaupp and P. v. R. Schleyer, *J. Am. Chem. Soc.*, **115**, 1061 (1993).
55. G. Granozzi, R. Bertoncello and E. Tondello, *J. Electron Spectrosc. Relat. Phenom.*, **36**, 207 (1985).
56. (a) A. N. Vereshchagin, *Inductive Effect*, Nauka, Moscow, 1987; *Chem. Abstr.*, **108**, 230511v (1988).
(b) M. Charton, *Prog. Phys. Org. Chem.*, **13**, 120 (1981).
57. C. Hansch, A. Leo and R. W. Taft, *Chem. Rev.*, **91**, 165 (1991).
58. C. Glidewell and D. C. Liles, *Acta Crystallogr.*, **B 35**, 1689 (1979).
59. C. Glidewell and D. C. Liles, *J. Organomet. Chem.*, **212**, 291 (1981).
60. C. J. Attridge and J. Struthers, *J. Organomet. Chem.*, **25**, C17 (1970).
61. M. Drager and L. Ross, *Z. Anorg. Allg. Chem.*, **460**, 207 (1980).
62. M. Weidenbruch, F.-T. Grimm, M. Herrndorf, A. Schafer, K. Peters and H. G. von Schnering, *J. Organomet. Chem.*, **341**, 335 (1988).
63. H. Preut, H.-J. Haupt and F. Huber, *Z. Anorg. Allg. Chem.*, **396**, 81 (1973).
64. H. Puff, B. Breuer, G. Gehrke-Brinkmann, P. Kind, H. Reuter, W. Schuh, W. Wald and G. Weidenbruck, *J. Organomet. Chem.*, **363**, 265 (1989).
65. L. R. Sita and R. D. Bickerstaff, *J. Am. Chem. Soc.*, **111**, 6454 (1989).
66. H. Preut and F. Huber, *Z. Anorg. Allg. Chem.*, **419**, 92 (1976).
67. H.-J. Koglin, M. Drager, N. Kleiner and C. Schneider-Koglin, *Abstracts of VIIth International Conference on Organometallics and Coordination Chemistry of Germanium, Tin and Lead*, Riga, 1992, p. 96.
68. W. Weissensteiner, I. I. Schuster, J. F. Blount and K. Mislow, *J. Am. Chem. Soc.*, **108**, 6664 (1986).
69. K. Ya. Burshtein and P. P. Shorygin, *Dokl. Akad. Nauk SSSR*, **296**, 903 (1987); *Chem. Abstr.*, **108**, 166707j (1988).
70. L. N. Zakharov, G. A. Domrachev and Yu. N. Safyanov, *Dokl. Akad. Nauk SSSR*, **293**, 108 (1987); *Chem. Abstr.*, **108**, 131871f (1988).
71. P. G. Harrison, T. J. King and M. A. Healy, *J. Organomet. Chem.*, **182**, 17 (1979).
72. V. P. Feshin, L. S. Romanenko and M. G. Voronkov, *Usp. Khim.*, **50**, 461 (1981); *Chem. Abstr.*, **95**, 41599g (1981).
73. J. J. Park, D. M. Collins and J. L. Hoard, *J. Am. Chem. Soc.*, **92**, 3636 (1970).
74. M. G. Voronkov and V. A. Pestunovich, *Abstracts of IXth International Symposium on Organosilicon Chemistry*, Edinburgh, 1990, p. A13.
75. M. G. Voronkov, *Izv. Akad. Nauk SSSR, Ser. Khim.*, 2664 (1991); *Bull. Acad. Sci. USSR, Div. Chem. Sci. (Engl. Transl.)*, **40**, 2319 (1991).
76. M. G. Voronkov, V. A. Pestunovich and Yu. I. Baukov, *Metalloorg. Khim.*, **4**, 1210 (1991); *Organomet. Chem. USSR (Engl. Transl.)*, **4**, 593 (1991).
77. V. V. Negrebetsky and Yu. I. Baukov, *Russ. Chem. Bull.*, **46**, 1912 (1997).

78. M. G. Voronkov and V. P. Baryshok, *J. Organomet. Chem.*, **239**, 199 (1982).
79. J. D. Kennedy, W. McFarlane, P. J. Smith, R. F. M. White and L. Smith, *J. Chem. Soc., Perkin Trans. 2*, 1785 (1973).
80. P. G. Harrison, T. J. King, J. A. Richards and R. C. Phillips, *J. Organomet. Chem.*, **116**, 307 (1976).
81. N. Kano, N. Tokitoh and R. Okazaki, *Organometallics*, **16**, 2748 (1997).
82. M. Lesbre, P. Mazerolles and J. Satge, *The Organic Compounds of Germanium*, Wiley, London, 1971.
83. V. F. Mironov and T. K. Gar, *Organic Compounds of Germanium*, Nauka, Moscow, 1967; *Chem. Abstr.*, **68**, 114732z (1968).
84. (a) A. N. Egorochkin, *Russ. Chem. Rev.*, **54**, 786 (1985).
(b) R. S. Mulliken and W. B. Person, *Molecular Complexes*, Wiley, New York, 1969.
85. J. K. Kochi, *Pure Appl. Chem.*, **52**, 571 (1980).
86. L. N. Zakharov, A. N. Egorochkin, M. A. Lopatin, I. A. Litvinov, O. N. Kataeva, V. A. Naumov, N. V. Girbasova, A. V. Belyakov, E. T. Bogoradovsky and V. S. Zavgorodny, *Metalloorg. Khim.*, **1**, 809 (1988); *Chem. Abstr.*, **111**, 194919k (1989).
87. S. N. Tandura, S. N. Gurkova and A. I. Gusev, *Zh. Struct. Khim.*, **31**, 154 (1990); *Chem. Abstr.*, **113**, 65418p (1990).
88. M. Sturmann, M. Weidenbruch, K. W. Klinkhammer, F. Lissner and H. Marsmann, *Organometallics*, **17**, 4425 (1998).
89. V. S. Dernova and I. F. Kovalev, *Vibrational Spectra of Organic Compounds of IVB Group Elements*, Saratov University Press, Saratov, 1979 (Russian).
90. J. Aron, J. Bunnell, T. A. Ford, N. Mercau, R. Aroca and E. A. Robinson, *J. Mol. Struct., THEOCHEM*, **110**, 361 (1984).
91. Y. Imai and K. Aida, *Bull. Chem. Soc. Jpn.*, **54**, 3323 (1981).
92. A. V. Belyakov, V. S. Nikitin and M. V. Polyakova, *Zh. Obshch. Khim.*, **65**, 81 (1995); *Chem. Abstr.*, **123**, 83402q (1995).
93. S. V. Markova, *Trudy Fiz. Inst., Akad. Nauk SSSR*, **35**, 150 (1966); *Chem. Abstr.*, **66**, 109813f (1967).
94. V. S. Nikitin, M. V. Polyakova and A. V. Belyakov, *Zh. Obshch. Khim.*, **63**, 1785 (1993); *Russ. J. Gen. Chem. (Engl. Transl.)*, **63**, 1246 (1993).
95. R. J. H. Clark, A. G. Davies and R. J. Ruddephatt, *J. Am. Chem. Soc.*, **90**, 6923 (1968).
96. J. W. Anderson, G. K. Barker, J. E. Drake and R. T. Hemmings, *Can. J. Chem.*, **49**, 2931 (1971).
97. B. Foutal and T. G. Spiro, *Inorg. Chem.*, **10**, 9 (1971).
98. R. A. Jackson, *J. Organomet. Chem.*, **166**, 17 (1979).
99. R. Walsh, *Acc. Chem. Res.*, **14**, 246 (1981).
100. I. M. T. Davidson and A. V. Howard, *J. Chem. Soc., Faraday Trans. 1*, **71**, 69 (1975).
101. F. Glockling, *The Chemistry of Germanium*, Academic Press, New York, 1969.
102. E. G. Rochow and E. W. Abel, *The Chemistry of Germanium, Tin and Lead*, Pergamon Press, Oxford, 1975.
103. A. N. Egorochkin, S. Ya. Khorshev, N. S. Ostasheva, J. Satge, P. Riviere, J. Barrau and M. Massol, *J. Organomet. Chem.*, **76**, 29 (1974).
104. E. I. Sevastyanova, S. Ya. Khorshev and A. N. Egorochkin, *Dokl. Akad. Nauk SSSR*, **258**, 627 (1981); *Chem. Abstr.*, **95**, 131836m (1981).
105. A. N. Egorochkin, E. I. Sevastyanova, S. Ya. Khorshev, S. Kh. Ratushnaya, J. Satge, P. Riviere, J. Barrau and S. Richelme, *J. Organomet. Chem.*, **162**, 25 (1978).
106. Y. Kawasaki, K. Kawakami and T. Tanaka, *Bull. Chem. Soc. Jpn.*, **38**, 1102 (1965).
107. A. N. Egorochkin, S. E. Skobeleva, T. G. Mushtina and E. T. Bogoradovsky, *Russ. Chem. Bull.*, **47**, 1526 (1998).
108. S. Marriott and R. D. Topsom, *J. Mol. Struct.*, **106**, 277 (1984).
109. M. D. Joesten and L. I. Schaad, *Hydrogen Bonding*, Dekker, New York, 1974.
110. J. Mack and C. H. Yoder, *Inorg. Chem.*, **8**, 278 (1969).
111. A. N. Egorochkin and S. E. Skobeleva, *Russ. Chem. Bull.*, **43**, 2043 (1994).
112. A. N. Egorochkin, S. E. Skobeleva and V. L. Tsvetkova, *Russ. Chem. Bull.*, **42**, 1316 (1993).
113. A. N. Egorochkin, S. E. Skobeleva and T. G. Mushtina, *Russ. Chem. Bull.*, **44**, 280 (1995).

114. A. N. Egorochkin, S. E. Skobeleva, V. L. Tsvetkova, E. T. Bogoradovsky and V. S. Zavgorodny, *Metalloorg. Khim.*, **5**, 1342 (1992); *Organomet. Chem. USSR (Engl. Transl.)*, **5**, 658 (1992).
115. A. N. Egorochkin, S. E. Skobeleva and V. L. Tsvetkova, *Metalloorg. Khim.*, **3**, 656 (1990); *Chem. Abstr.*, **113**, 97735f (1990).
116. S. E. Skobeleva, A. N. Egorochkin, E. T. Bogoradovsky and A. A. Petrov, *Izv. Akad. Nauk SSSR, Ser. Khim.*, 1294 (1982); *Chem. Abstr.*, **97**, 163150r (1982).
117. E. W. Abel, D. A. Armitage and D. B. Brady, *Trans. Faraday Soc.*, **62**, 3459 (1966).
118. M. Karplus and J. A. Pople, *J. Chem. Phys.*, **38**, 2803 (1963).
119. G. Engelhardt, R. Radeglia, H. Jancke, E. Lippmaa and M. Magi, *Org. Magn. Reson.*, **5**, 561 (1973).
120. W. H. Flygare, *Chem. Rev.*, **74**, 653 (1974).
121. A. N. Egorochkin, N. S. Vyazankin, A. I. Burov and S. Ya. Khorshev, *Izv. Akad. Nauk SSSR, Ser. Khim.*, 1279 (1970); *Chem. Abstr.*, **73**, 130364q (1970).
122. M. Bullpitt, W. Kitching, W. Adcock and D. Doddrell, *J. Organomet. Chem.*, **116**, 161 (1976).
123. W. F. Reynolds, G. K. Hamer and A. R. Bassindale, *J. Chem. Soc., Perkin Trans. 2*, 971 (1977).
124. J. B. Lambert and R. A. Singer, *J. Am. Chem. Soc.*, **114**, 10246 (1992).
125. L. B. Krivdin and G. A. Kalabin, *Prog. NMR Spectrosc.*, **21**, 293 (1989).
126. T. N. Mitchell, *J. Organomet. Chem.*, **255**, 279 (1983).
127. E. Liepins, I. Zicmane, L. M. Ignatovich and E. Lukevics, *J. Organomet. Chem.*, **389**, 23 (1990).
128. H. Bock and B. Solouki, in *The Chemistry of Organic Silicon Compounds*, Part 1 (Eds. S. Patai and Z. Rappoport), Chap. 9, Wiley, Chichester, 1989, pp. 555–653.
129. R. S. Mulliken and W. B. Person, *Molecular Complexes*, Wiley, New York, 1969.
130. A. N. Egorochkin, O. V. Zderenova and S. E. Skobeleva, *Russ. Chem. Bull.*, **49**, 997 (2000).
131. H. Bock and H. Alt, *J. Am. Chem. Soc.*, **92**, 1569 (1970).
132. H. Bock, W. Kaim and H. Tesmann, *Z. Naturforsch., B*, **33**, 1223 (1978).
133. W. Kaim, H. Tesmann and H. Bock, *Chem. Ber.*, **113**, 3221 (1980).
134. P. K. Bischof, M. J. S. Dewar, D. W. Goodman and J. B. Jones, *J. Organomet. Chem.*, **82**, 89 (1974).
135. M. A. Lopatin, V. A. Kuznetsov, A. N. Egorochkin, O. A. Pudova, N. P. Erchak and E. Ya. Lukevics, *Dokl. Akad. Nauk SSSR*, **246**, 379 (1979); *Chem. Abstr.*, **91**, 174320y (1979).
136. A. N. Egorochkin, V. A. Kuznetsov, M. A. Lopatin, N. P. Erchak and E. Ya. Lukevics, *Dokl. Akad. Nauk SSSR*, **258**, 391 (1981); *Chem. Abstr.*, **95**, 114292p (1981).
137. G. Distefano, A. Ricci, F. P. Colonna, D. Pietropaolo and S. Pignataro, *J. Organomet. Chem.*, **78**, 93 (1974).
138. A. Flamini, E. Seprini, F. Stefani, S. Sorriso and G. Cardaci, *J. Chem. Soc., Dalton Trans.*, 731 (1976).
139. A. N. Egorochkin, S. E. Skobeleva and T. G. Mushtina, *Russ. Chem. Bull.*, **47**, 1436 (1998).
140. A. N. Egorochkin, S. E. Skobeleva and T. G. Mushtina, *Russ. Chem. Bull.*, **47**, 2352 (1998).
141. A. N. Egorochkin, M. G. Voronkov, S. E. Skobeleva, T. G. Mushtina and O. V. Zderenova, *Russ. Chem. Bull.*, **49**, 26 (2000).
142. A. N. Egorochkin, S. E. Skobeleva and T. G. Mushtina, *Russ. Chem. Bull.*, **46**, 1549 (1997).
143. K. A. Kocheshkov, *Usp. Khim.*, **3**, 1 (1934); *Russ. Chem. Rev.*, **3**, 1 (1934).
144. K. A. Kocheshkov, *Synthetic Methods in the Field of Metalloorganic Compounds of Group IV Elements*, AN SSSR, Moscow, 1947 (Russian).
145. M. Schmidt, *Pure Appl. Chem.*, **13**, 15 (1966).
146. F. Glockling, *Quart. Rev.*, **20**, 45 (1966).
147. D. I. Mendeleev, *Zh. Russ. Phiz. Khim. Obshch.*, **3**, 25 (1871).
148. D. I. Mendeleev, *Ann. Chem. Pharm.*, **8**, 133 (1872).
149. D. I. Mendeleev, *Complete Works*, Vol. 2, Goschimizdat, Leningrad, 1934 (Russian).
150. K. A. Kocheshkov, N. N. Zemlyansky, N. I. Sheverdina and E. M. Panov, *Methods in Organo Elemental Chemistry*, Nauka, Moskow, 1968 (Russian).
151. G. A. Rasuvaev, B. G. Gribov, G. A. Domrachev and B. A. Salamatin, *Organometallic Compounds in Electronics*, Nauka, Moscow, 1972; *Chem. Abstr.*, **78**, 141643j (1973).

152. B. G. Gribov, G. A. Domrachev, B. V. Zhuk, B. S. Kaverin, B. I. Kozyrkin, V. V. Melnikov and O. N. Suvorova, *Deposition of Films and Coatings by Decomposition of Organometallic Compounds*, Nauka, Moscow, 1981; *Chem. Abstr.*, **96**, 147644d (1982).
153. C. Eaborn and R. W. Bott, in *Organometallic Compounds of the Group IV Elements*, Vol. 1 (Ed. A. G. MacDiarmid), Dekker, New York, 1968, pp. 405–536.
154. F. Glockling and K. A. Hooton, in *Organometallic Compounds of the Group IV Elements*, Vol. 1 (Ed. A. G. MacDiarmid), Dekker, New York, 1968, pp. 2–90.
155. J. G. A. Luijten and G. J. M. Van der Kerk, in *Organometallic Compounds of the Group IV Elements*, Vol. 1 (Ed. A. G. MacDiarmid), Dekker, New York, 1968, pp. 91–189.
156. L. S. Willemsens and G. J. M. Van der Kerk, in *Organometallic Compounds of the Group IV Elements*, Vol. 1 (Ed. A. G. MacDiarmid), Dekker, New York, 1968, pp. 191–229.
157. H. D. Kaesz and F. G. A. Stone, in *Organometallic Chemistry* (Ed. H. Zeiss), Chapman and Hall, London, 1960, pp. 115–180.
158. D. Seyferth, in *Progress in Inorganic Chemistry*, Vol. 3 (Ed. A. Cotton), Interscience, New York, 1962, pp. 129–280.
159. E. Y. Lukevits and M. G. Voronkov, *Organic Insertion Reactions of Group IV Elements*, Consultants Bureau, New York, 1966.
160. M. F. Lappert, *Silicon, Germanium, Tin and Lead Compounds*, **9**, 129 (1986).
161. M. Weidenbruch, *Eur. J. Inorg. Chem.*, 373 (1999).
162. P. P. Power, *Chem. Rev.*, **99**, 3463 (1999).
163. S. Nagase, K. Kobayashi and N. Takagi, *J. Organomet. Chem.*, **611**, 264 (2000).
164. L. Pu, B. Twamley and P. P. Power, *J. Am. Chem. Soc.*, **122**, 3524 (2000).
165. P. G. Harrison, *Organomet. Chem. Rev., A*, **4**, 379 (1969).
166. K. Tani, N. Kitaoka, K. Yamada and H. Mifune, *J. Organomet. Chem.*, **611**, 190 (2000).
167. M. S. Sorokin, S. G. Shevchenko, N. N. Chipanina, Yu. L. Frolov, M. F. Larin and M. G. Voronkov, *Metalloorg. Khim.*, **3**, 419 (1990); *Chem. Abstr.*, **113**, 78476f (1990).
168. N. A. Matwiyoff and R. S. Drago, *Inorg. Chem.*, **3**, 337 (1964).
169. K. Hills and M. G. Henry, *J. Organomet. Chem.*, **3**, 159 (1965).
170. H. G. Langer and A. H. Blut, *J. Organomet. Chem.*, **6**, 288 (1966).
171. M. G. Voronkov, V. M. Dyakov and S. V. Kirpichenko, *J. Organomet. Chem.*, **233**, 1 (1982).
172. V. Pestunovich, S. Kirpichenko and M. Voronkov, in *The Chemistry of Organic Silicon Compounds*, Vol. 2, Part 2 (Eds. Z. Rappoport and Y. Apeloig), Wiley, Chichester, 1998, pp. 1447–1537.
173. R. C. Mehrotra, R. Bohra and D. P. Gaur, *Metal β-Diketonates and Allied Derivatives*, Academic Press, London, 1978.
174. E. L. Muetterties and C. M. Wright, *J. Am. Chem. Soc.*, **86**, 5123 (1964); *J. Am. Chem. Soc.*, **87**, 4706 (1965); *J. Am. Chem. Soc.*, **88**, 4856 (1966).
175. O. M. Nefedow and M. N. Manakow, *Angew. Chem.*, **78**, 1039 (1966).
176. O. M. Nefedov, S. P. Kolesnikov and A. I. Ioffe, *Organomet. Chem. Rev.*, **5**, 181 (1977).
177. W. P. Neuman, *Chem. Rev.*, **91**, 311 (1991).
178. J. Nine, *Divalent Carbon*, Ronald Press, New York, 1964.
179. W. Kirmse, *Carbene Chemistry*, Academic Press, New York, 1971.
180. R. A. Moss and M. Jones, *React. Intermed.*, **3**, 45 (1985).
181. W. H. Atwell and D. R. Wegenberg, *Angew. Chem., Int. Engl. Ed.*, **8**, 469 (1969).
182. J. L. Margrave and P. W. Wilson, *Acc. Chem. Res.*, **4**, 145 (1971).
183. W. H. Atwell and D. R. Wegenberg, *Intra-Sci. Chem. Rep.*, **7**, 139 (1973).
184. P. L. Timms, *Acc. Chem. Res.*, **6**, 118 (1973).
185. E. A. Chernyshev, N. G. Komalenkova and S. A. Bashkirova, *Usp. Khim.*, **45**, 1782 (1976); *Chem. Abstr.*, **85**, 94442a (1976).
186. E. M. Arnett, T. C. Hofelich and G. W. Schriver, *React. Intermed.*, **3**, 189 (1985).
187. K. L. Bobbitt, D. Lei, V. M. Maloney, B. S. Parker, J. M. Raible and P. P. Gaspar, in *Frontiers of Organogermanium, -tin and -lead Chemistry* (Eds. E. Lukevics and L. Ignatovich), Latvian Institute of Organic Synthesis, Riga, 1993, pp. 29–40.
188. J. Satge, in *Frontiers of Organogermanium, -tin and -lead Chemistry* (Eds. E. Lukevics and L. Ignatovich), Latvian Institute of Organic Synthesis, Riga, 1993, pp. 55–70.
189. K. D. Bos, *Organic and Organometallic Chemistry of Divalent Tin*, Drukkerij B. V. Elinkwijk, Utrecht, 1976.
190. W. P. Neumann, *The Organic Chemistry of Tin*, Wiley, London, 1970.

191. A. G. Davies and P. J. Smith, in *Comprehensive Organometallic Chemistry*, Vol. 2 (Eds. G. Wilkinson, G. A. Stone and E. W. Abel), Pergamon Press, Oxford, 1982, pp. 519–627.
192. V. I. Shiryaev, V. F. Mironov and V. P. Kochergin, *Compounds of Divalent Tin and Synthesis of Organotin Compounds*, NIITEKhIM, Moscow, 1977 (Russian).
193. W. P. Neumann, in *The Organometallic and Coordination Chemistry of Ge, Sn, and Pb* (Eds. M. Gielen and P. Harrison), Freund Publ., Tel Aviv, 1978, p. 51.
194. W. P. Neumann, *Nachr. Chem. Tech. Lab.*, **30**, 190 (1982).
195. R. W. Leeper, L. Summer and H. Gilman, *Chem. Rev.*, **54**, 101 (1954).
196. L. C. Willemsen, *Organolead Chemistry*, International Lead, Zinc Research Organization, New York, 1964.
197. C.-S. Liu and T.-L. Hwang, *Adv. Inorg. Chem. Radiochem.*, **29**, 1 (1985).
198. J. W. Connolly and C. Hoff, *Adv. Organomet. Chem.*, **19**, 123 (1981).
199. S.-W. Ng and J. J. Zuckerman, *Adv. Inorg. Chem. Radiochem.*, **29**, 297 (1985).
200. V. I. Shiryaev, V. P. Kochergin and V. F. Mironov, *Stannylenes as Electron-donating Ligands in Complexes*, NIITEKhIM, Moscow, 1977 (Russian).
201. E. A. Chernyshev, M. D. Reshetova and A. D. Volynskikh, *Silicon- and Tin-containing Derivatives of Compounds of Transition Elements*, NIITEKhIM, Moscow, 1975 (Russian).
202. W. Petz, *Chem. Rev.*, **86**, 1019 (1986).
203. M. G. Voronkov, G. I. Zelchan and E. Ya. Lukevits, *Silicon and Life*, Zinatne, Riga, 1978; *Chem. Abstr.*, **90**, 34092e (1979).
204. M. G. Voronkov, G. I. Zelchan and E. Lukevitz, *Silicium und Leben*, Akademie-Verlag, Berlin, 1975.
205. M. G. Voronkov and I. G. Kuznetsov, *Silicon and Natural Living*, Nauka, Novosibirsk, 1984; *Chem. Abstr.*, **103**, 100676p (1985).
206. M. G. Voronkov and I. G. Kuznetsov, *Silicon in Living Nature*, Japanese–Soviet Interrelation Company, Wakayama, 1988.
207. M. G. Voronkov, *Top. Curr. Chem.*, **84**, 77 (1979).
208. M. G. Voronkov and V. M. Dyakov, *Silatranes*, Nauka, Novosibirsk, 1978; *Chem. Abstr.*, **92**, 76655n (1980).
209. M. G. Voronkov, in *Biochemistry of Silicon and Related Problems* (Eds. G. Bendz and I. Lindguist), Plenum Press, New York, 1978, pp. 393–433.
210. K. Asai, *Miracle Cure Organic Germanium*, Japan Publ., Tokyo, 1980.
211. J. S. Thayer, *J. Organomet. Chem.*, **76**, 265 (1974).
212. E. Ya. Lukevics, T. K. Gar, L. M. Ignatovich and V. F. Mironov, *Biological Activity of Germanium Compounds*, Zinatne, Riga, 1990; *Chem. Abstr.*, **114**, 159437s (1991).
213. J. M. L. Mages, *Organomet. Chem. Rev., A*, **3**, 137 (1968).
214. M. G. Voronkov, G. I. Zelchan, V. F. Mironov, Ya. Ya. Blejdelis and A. A. Kemme, *Khim. Geterotsikl. Soed.*, **2**, 227 (1968); *Chem. Abstr.*, **69**, 87129v (1968).
215. S. G. Ward and R. C. Taylor, in *Metal-Based Anti-Tumour Drugs* (Ed. M. F. Gielen), Freund Publ., London, 1988, pp. 1–54.
216. J. S. Thayer, *Appl. Organomet. Chem.*, **1**, 227 (1987).
217. J. M. Barnes and H. B. Stoner, *Pharmacol. Rev.*, **11**, 211 (1959).
218. R. K. Ingam, S. D. Rosenberg and H. Gilman, *Chem. Rev.*, **60**, 439 (1960).
219. A. L. Klyashitskaya, V. T. Mazaev and A. M. Parshina, *Toxic Properties of Organotin Compounds*, GNIIKhTEOS, Moscow, 1978 (Russian).
220. H. F. Krug, A. Kofer and H. Duterich, in *Frontiers of Organogermanium, -tin and -lead Chemistry* (Eds. E. Lukevics and L. Ignatovich), Latvian Institute of Organic Synthesis, 1993, pp. 273–281.
221. P. Grandjean, in *Lead versus Health* (Eds. M. Rutter and R. R. Jones), Wiley, London, 1983, pp. 179–203.
222. J. Boyer, *Toxicology*, **55**, 253 (1989).
223. L. Friberg and N. K. Mottet, *Biol. Trace Elem. Res.*, **21**, 201 (1989).
224. G. Erg, E. J. Tierney, G. J. Olson, F. E. Brinckman and J. M. Bellama, *Appl. Organomet. Chem.*, **5**, 33 (1991).
225. E. Luedke, E. Lucero and G. Erg, *Main Group Metal Chem.*, **14**, 59 (1991).
226. P. J. Craig, in *Comprehensive Organometallic Chemistry*, Vol. 2 (Eds. G. Wilkinson, G. A. Stone and E. W. Abel), Pergamon Press, Oxford, 1982, pp. 979–1020.

227. S. J. Blunden and A. H. Chapman, in *Organometallic Compounds in the Environment—Principles and Reactions* (Ed. P. J. Craig), Longman Group Ltd, Harlow, 1986, pp. 111–159.
228. C. N. Hewitt, in *Organometallic Compounds in the Environment—Principles and Reactions* (Ed. P. J. Craig), Longman Group Ltd, Harlow, 1986, pp. 160–205.
229. N. J. Snoij, A. H. Penninks and W. Seinen, *Environ. Res.*, **44**, 335 (1987).
230. W. Bolanowska, *Med. Pracy*, **16**, 476 (1965).
231. D. Bryce-Smith, *Chem. Br.*, **8**, 240 (1972).
232. A. Browder, M. Joselow and D. Louria, *Medicine*, **52**, 121 (1973).
233. J. Harwood, *Industrial Application of the Organometallic Compounds*, Chapman and Hall, London, 1963.
234. A. L. Prokhorova, in *Preparation and Uses of Organolead Compounds*, NIITEKhIM, Moscow, 1972, pp. 3–23 (Russian).
235. V. I. Shiryaev and E. M. Stepina, *The State and Prospects of the Use of Organotin Compounds*, NIITEKhIM, Moscow, 1988 (Russian).

CHAPTER 3

Theoretical studies of organic germanium, tin and lead compounds

INGA GANZER, MICHAEL HARTMANN and GERNOT FRENKING

Fachbereich Chemie, Philipps-Universität Marburg, Hans-Meerwein-Strasse, D-35032 Marburg, Germany
Fax: +49-6421-282-5566; E-mail: frenking@chemie.uni-marburg.de

The chemistry of organic germanium, tin and lead compounds — Vol. 2
Edited by Z. Rappoport

I. LIST OF ABBREVIATIONS

AIMP	*Ab Initio* Model Potential
AO	Atomic Orbital
ASE	Aromatic Stabilization Energy
BDE	Bond Dissociation Energy
BO	Born–Oppenheimer
B3LYP	Becke's 3-parameter fit using the correlation functional by Lee, Yang and Parr
CASSCF	Complete Active Space Self-Consistent Field
CC	Coupled-Cluster Theory
CCSD(T)	Coupled-Cluster Theory with Singles, Doubles and Noniterative Approximation of Triples
CI	Configuration Interaction
CIPSI	Configuration Interaction by Perturbation with Multiconfigurational Zero-Order Wave Function Selected by Iterative Process
CISD	Configuration Interaction with Singles and Doubles
DFT	Density Functional Theory
ECP	Effective Core Potential
GGA	Generalized Gradient Approximation
HF	Hartree–Fock
HOMO	Highest Occupied Molecular Orbital
KS	Kohn–Sham
LANL	Los Alamos National Laboratory
LCAO	Linear Combination of Atomic Orbitals
LDA	Local Density Approximation
LUMO	Lowest Unoccupied Molecular Orbital
MNDO	Modified Neglect of Diatomic Differential Overlap
MCSCF	Multiconfiguration Self-Consistent Field
MO	Molecular Orbital
MP	Møller–Plesset Perturbation Theory
NICS	Nuclear Independent Chemical Shift
PE	Photoelectron
PES	Potential Energy Surface
QCISD	Quadratic Configuration Interaction with Singles and Doubles
QCISD(T)	Quadratic Configuration Interaction with Singles, Doubles, and Noniterative Approximation of Triples
SAC-CI	Symmetry Adapted Cluster–Configuration Interaction
SVP	Split-Valence basis set plus Polarisation functions
ZORA	Zero-Order Regular Approximation
ZPE	Zero-Point Energy

II. INTRODUCTION

The enormous progress in the development of theoretical methods and the dramatic increase in computer power have made it possible for quantum chemical investigations of heavy-atom molecules to become a standard research tool in chemistry in the last decade. While the 1980s can be considered as the age where application of *ab initio* methods to classical organic molecules which contain elements of the first and second full rows of the periodic system were routinely done in organic chemistry[1], the 1990s saw the conquest of inorganic compounds with all elements of the periodic system including transition metals[2] by accurate quantum chemical methods. Numerous theoretical studies have been reported in the last 10 years about compounds which contain atoms that were once considered to be elusive for reliable theoretical calculations.

Two methods are mainly responsible for the breakthrough in the application of quantum chemical methods to heavy atom molecules. One method consists of pseudopotentials, which are also called effective core potentials (ECPs). Although ECPs have been known for a long time[3], their application was not widespread in the theoretical community which focused more on all-electron methods. Two reviews which appeared in 1996 showed that well-defined ECPs with standard valence basis sets give results whose accuracy is hardly hampered by the replacement of the core electrons with parameterized mathematical functions[4,5]. ECPs not only significantly reduce the computer time of the calculations compared with all-electron methods, they also make it possible to treat relativistic effects in an approximate way which turned out to be sufficiently accurate for most chemical studies. Thus, ECPs are a very powerful and effective method to handle both theoretical problems which are posed by heavy atoms, i.e. the large number of electrons and relativistic effects.

The second method which dramatically changed the paradigm of the dominant computational method in chemistry is density functional theory (DFT)[6]. The introduction of gradient corrected functionals into quantum chemistry altered the view that DFT is not reliable enough for the calculation of molecules. After initial resistance by the theoretical establishment it is now generally accepted that DFT is the most cost-effective quantum chemical method which usually gives reliable results for molecules in the electronic ground state. DFT has specific problems like any approximate method which should be known. Some of them are different from problems in *ab initio* theory and some of them are the same. In the next section we will shortly address the most important features of the present DFT methods and their strength and weakness.

The goal of this chapter is to give a summary of the most important results of quantum chemical studies in the field of organic germanium, tin and lead compounds which have been reported since 1990. The first volume in this series, published in 1989, covered the chemistry of organic germanium, tin and lead compounds, and included a chapter by Basch and Hoz (BH) which covered the theoretical work in this field up to 1989[7]. Therefore, we have focused on more recent work in order to show the progress which has been made in the last decade. However, our chapter is written in a different way than the previous one by Basch and Hoz, who discussed the nature of the E−C bond (E = Ge, Sn, Pb) while presenting results for compounds with the formula $H_3E{-}Y$ where Y is a ligand. Theoretical research in the field has been greatly extended recently to compounds which have different molecular connectivities than $H_3E{-}Y$. In particular, molecules which have multiple bonds of Ge to Pb have been reported. The review is an attempt to cover all theoretical work published in the field since 1990. This includes publications about combined experimental and theoretical studies. We wish to draw the attention of the reader to a recently published related review about theoretical aspects of compounds containing Si, Ge, Sn and Pb by Karni et al.[8]. A review about experimental and theoretical studies of main group analogues of carbenes, olefines and small ring compounds was published by Driess and Grützmacher[9].

The great diversity of the published work in the field made it difficult to find a simple ordering scheme for presenting the results. We decided to present first a summary of theoretical studies about relevant parent compounds of group-14 elements which makes it possible to compare the structures and the nature of the bonding of the elements Ge, Sn and Pb with C and Si. We have chosen the divalent carbene analogues EH_2 and EX_2 (X = halogen) and the tetravalent compounds EH_4, EX_4 and $H_3E{-}EH_3$. We also discuss recent work about the ethylene analogues E_2H_4 and E_2X_4 and the protonated species $E_2H_5^+$.

The section about parent compounds includes also theoretical work about the group-14 homologues of acetylene E_2H_2, cyclopropenium cation $E_3H_3^+$ and benzene E_6H_6. Although these molecules do not strictly belong to the class of organometallic compounds, they are important for an understanding of the molecules which have E−C bonds.

The section about organometallic compounds of Ge, Sn and Pb was divided into neutral closed-shell molecules, which comprise the largest part of the chapter, cations and anions and finally radicals. The section about closed-shell molecules is further divided into papers which report about structures and properties of molecules with multiple and those with single bonds of Ge, Sn and Pb, and work that concentrates on the elucidation of reaction mechanisms. We included all studies which report about compounds of the heavier elements Ge to Pb even when the main focus of the work was on the lighter atoms C and Si or other elements. Thus, some work which is discussed here will also be found in the chapter which focuses on Si compounds[8].

III. THEORETICAL METHODOLOGY

Because this chapter is a follow-up of previous work in the field[7] it is not necessary to repeat the basics of *ab initio* methods. This has been done in detail by Basch and Hoz, who also discuss the most important atomic properties of Ge, Sn and Pb. We also recommend the theoretical section in the chapter by Apeloig[10] about organosilicon compounds in this series who gave an excellent overview about the most important aspects of *ab initio*, semiempirical and force-field methods. The reader will find there an explanation of the most common standard methods which will be mentioned in this review without further explanation. We will focus in the following on those theoretical and computational aspects of methods which are particularly important for heavy-atom molecules that have been advanced in the last decade, i.e. ECPs and DFT. We also briefly discuss relativistic effects. We point out that semiempirical methods[11] and force field parameters[12] are available for the elements Ge, Sn and Pb. However, the application of the two methods has not gained much popularity and not many papers have been published in the field. Most reports are restricted to special problems[13].

The following is a very short outline of the basic ideas of the relevant theoretical methods and aims at giving experimental chemists an understanding of the underlying principles. For those readers who wish to learn more about present methods in computational chemistry, we recommend the textbook *Introduction to Computational Chemistry* by Jensen[14]. An excellent book about the theory and application of DFT given from a chemist's point of view is *A Chemist's Guide to Density Functional Theory* by Koch and Holthausen[15]. Two reviews are available which discuss the application of ECPs to heavy atom molecules[4,5]. We also mention the *Encyclopedia of Computational Chemistry* which contains a large number of reviews written by experts about nearly all aspects of the field[16].

A. Density Functional Theory

It is perhaps helpful to introduce the fundamental concepts of DFT by comparing it with *ab initio* methods which are based on Hartree−Fock theory. The basic idea of the latter is that the many-electron wave function of a molecule is approximated by a set of one-electron functions which give the energy of a single electron in the field of (i) fixed nuclei and (ii) the average of the remaining electrons. The former approximation is known as the Born−Oppenheimer (BO) approximation. It holds also in DFT. The error caused by the BO model is negligible in most cases for molecules in their electronic ground state. The second approximation introduces correlation energy. It is the difference between the nonrelativistic electron–electron interactions in HF theory which are calculated using an average field of the electronic charge distribution and the sum of the exact, i.e. individual electron–electron interactions. There are several methods in *ab initio* theory which can be used to calculate correlation energy. The most popular ones are Møller–Plesset perturbation theory (MP), coupled-cluster theory (CC) and configuration interaction (CI).

The central equations in *ab initio* theory are the Hartree–Fock equations:

$$F\varphi_i = \varepsilon_i \varphi_i \tag{1}$$

The Fock operator F contains terms for the kinetic energy, the nuclei–electron attraction and the averaged electron–electron repulsion. The values ε_i are the energies of the electrons which, after proper addition, give the Hartree–Fock (HF) total energy E^{HF} of the molecule. The one-electron functions φ_i (molecular orbitals) give in a simplified view the 3-dimensional distribution of the electron i in space. The total wave function of all electrons can be constructed from the φ_i via the so-called Slater determinant. Each φ_i is expressed in terms of a linear combination of atomic orbitals (LCAO). The size of the basis set determines the calculated energy E^{HF} of the molecule. The correlation energy E^{corr} is estimated in a separate calculation after the HF equations have been solved. The nuclear repulsion energy $E^{\mathrm{N-N}}$ is calculated classically using the fixed positions of the nuclei. The total energy of the molecule is then given by the sum of three terms:

$$E^{\mathrm{tot}} = E^{\mathrm{HF}} + E^{\mathrm{corr}} + E^{\mathrm{N-N}} \tag{2}$$

Thus, the accuracy of a (nonrelativistic) *ab initio* calculation is determined by two factors, i.e. the size and quality of the basis set and the method by which correlation energy has been calculated. The accuracy can systematically be improved by choosing better basis sets and better methods for calculating correlation energy.

The working equations in DFT look very similar to the Hartree–Fock equations. They are called Kohn–Sham (KS) equations[6,15,17]:

$$F^{\mathrm{KS}}\phi_i = e_i \phi_i \tag{3}$$

The difference between equations 1 and 3 is the form of the operator F. The Kohn–Sham operator F^{KS} is constructed with the goal that the resulting one-electron functions ϕ_i yield a total ground-state electron density distribution of the molecule which is correct. Hohenberg and Kohn showed that the total energy of a molecule in the electronic ground state is uniquely determined by the electron density[6a]. It follows that the electron density which is calculated via equations 3 gives directly the correct *total* energy E^{tot} via proper summation of the one-electron energies ϕs_i, unlike the HF equations 1 which give the HF energy E^{HF}. Thus, the operator F^{KS} contains a term for the correlation energy.

The problem with the conceptually simple form of the KS equations is the choice of the functionals which determine the operator F^{KS} in equations 3. While the HF operator F is known but yields only approximate energies, the operator F^{KS} gives the correct energy but the functional is not known. The form of F^{KS} can only be guessed! The popularity of DFT in molecular quantum chemistry which came in the last decade is because the guesses have been very successful. New types of functionals have been suggested for F^{KS} which yield a much higher accuracy than earlier functionals.

Before we discuss in brief the new functionals, we wish to comment on the one-electron functions ϕ_i in equations 3 and the basis sets used in DFT. The functions ϕ_i have originally been calculated with the purpose of describing the total electron density in terms of one-electron functions. It was recently suggested that the Kohn–Sham orbitals ϕ_i can also be used for qualitative MO models in the same way as HF orbitals were employed in the past[18]. Thus, orbital interaction diagrams can be constructed and frontier orbital analyses of chemical reactivity can be carried out using Kohn–Sham orbitals which are calculated by DFT methods[19]. KS orbitals have the advantage over HF orbitals in that their energies are an approximation to the total energy of the molecule, while the latter give only the HF energy. The shape of the KS orbitals ϕ_i is very similar to the shape of HF orbitals φ_i, but the occupied orbitals ϕ_i are higher in energy than the occupied φ_i and the unoccupied ϕ_i are lower in energy than the vacant φ_i.

A fortunate finding of test calculations was that the same basis sets of Gaussian-type functions which are used as standard basis sets in *ab initio* calculations can be used for DFT calculations. It was also found that the same ECPs which have been optimized for *ab initio* methods can be employed for DFT methods[20]. Users of the program package Gaussian may, e.g., simply choose DFT/6-31G(d) instead of HF/6-31G(d) or MP2/6-31G(d). The only choice which one has to make is the DFT functional.

Mathematical expressions for the functionals which are found in the Kohn–Sham operator F^{KS} are usually derived either from the model of a uniform electron gas or from a fitting procedure to calculated electron densities of noble gas atoms[15]. Two different functionals are then derived. One is the exchange functional F_X and the other the correlation functional F_C, which are related to the exchange and correlation energies in *ab initio* theory. We point out, however, that the definition of the two terms in DFT is slightly different from *ab initio* theory, which means that the corresponding energies cannot be directly compared between the two methods.

In order to run a DFT calculation the user has to choose a combination of F_X and F_C, which together define the DFT method to be employed. Mathematical expressions for F_X and F_C were first derived as a function of the electron density $\rho(r)$. This is called the Local Density Approximation (LDA). A significant improvement in the accuracy of the calculated results was achieved when not only the electron density $\rho(r)$ but also its gradient $\nabla\rho(r)$ was used for deriving mathematical expressions for F_X and F_C. This is called the Generalized Gradient Approximation (GGA), which gives gradient corrected functionals. They are sometimes called nonlocal functionals which is a misnomer, because the gradient $\nabla\rho(r)$ is also a local function.

The situation at present is that the nonlocal exchange functional suggested by Becke (B) in 1988[21] has been established as a standard expression in DFT calculations. The choice of the best correlation functional is less obvious than the choice of the exchange functional. The presently most popular correlation functionals are those of Perdew (P86)[22], Lee et al. (LYP)[23], Perdew and Wang (PW91)[24] and Vosko et al. (VWN)[25].

The situation in choosing proper combinations of exchange and correlation functionals became a bit confusing in the early 1990s when different functionals were combined and the resulting energy expression was given by a multiparameter fit of the functionals to a set of well-established experimental values, i.e. the so-called G2 set[26]. The most commonly used functional combination of this type is the 3-parameter fit of Becke (B3)[27]. A widely used variant of the B3 hybrid functional termed B3LYP[28], which is slightly different from the original formulation of Becke, employs the LYP expression for the nonlocal correlation functional F_C. It seems that the B3LYP hybrid functional is at present the most popular DFT method for calculating molecules. Other widely used combinations of functionals are BP86, which gives particularly good results for vibrational frequencies[29], BPW91 and BLYP. It should be noted that the development of new functionals is presently an area of active research. New DFT procedures may soon come and replace the above functionals as standard methods. The present state of development in the field has been the topic of a special issue of the *Journal of Computational Chemistry*[30]. It is a wise idea to estimate the accuracy of a functional for a particular problem at the beginning of a research project, by running some test calculations before the final choice of the DFT is made. The disadvantage of DFT compared with conventional *ab initio* methods is that the DFT calculations cannot systematically be improved toward better results by going to higher levels of theory.

1. Basis sets, effective core potentials and relativistic effects

As said above it is possible to use the same Gaussian-type standard basis sets of *ab initio* theory for DFT calculations. Concerning the quality of the basis set which is necessary to obtain reliable results, it is advisable to use for Ge at least a split-valence basis set which should be augmented by a d-type polarization function such as 6-31G(d). Better basis sets of triple-zeta quality with more polarization functions up to 6-311G(3df) have been developed for Ge which belong to the standard basis sets in Gaussian 98[31]. Other basis sets for Ge are available, e.g., from the compilations of Huzinaga et al.[32] and Poitier et al.[33] and from the work of Ahlrichs et al.[34].

All-electron basis sets are also available for the heavier atoms Sn and Pb, but relativistic effects become so important for these elements that they must be considered in the theoretical method. This is the reason why most workers choose for Sn and Pb quasi-relativistic ECPs with valence-only basis sets which significantly reduce the computational costs while at the same time the most important relativistic effects are considered. The error introduced by the approximate treatment of relativity and replacement of the core electrons by a pseudopotential is for most chemical questions negligible compared with the truncation of the basis set and the approximate calculation of correlation energy[4,5]. ECPs may actually be used also for Ge and even for Si and C. It is uncommon to employ ECPs for the lighter elements carbon and silicon, but calculations of germanium compounds frequently use ECPs rather than all-electron basis sets. Relativistic effects may be neglected for the calculation of energies and geometries of germanium compounds if chemical accuracy is sufficient.

The program package Gaussian 98[31] has three different ECPs as standard for heavy atom calculations. One popular set of ECPs has been developed by Hay and Wadt (HW) and includes parameters for Si to Pb[35]. Standard versions of the HW ECPs in Gaussian 98[31] have minimal basis sets (LANL2MB) and double-zeta basis sets (LANL2DZ). Additional d-type polarization functions for the HW ECPs have been published in the literature[36] but they must be added by hand to the basis sets. The user should know, however, that the HW ECPs for Si and Ge are nonrelativistic, while only the ECPs for Sn and Pb include scalar-relativistic effects[35]. Another important point is that the valence basis sets of HW have only three Gaussian functions for the s and p orbitals, written as (3/3). The LANL2DZ keyword splits them into (21/21), which has the same quality as the 3-21G all-electron basis set for the valence electrons. Four Gaussian functions are needed for the valence s and p orbitals in order to achieve the same quality as 6-31G after contraction to (31/31). This is provided by the ECPs of Stevens et al.[37,38] and the ECPs of the Stuttgart/Dresden group[39,40]. The ECPs of the two groups are also available as standard options in Gaussian 98[31]. The Stuttgart/Dresden group is particularly active in developing ECPs for all elements of the periodic system. An overview of the pseudopotentials which are available from the group is given on their web site[41]. Other ECPs for main group elements which include functions for the heavier group-14 elements have been published by the group of Christiansen[42–45], by Gomez and Pacios[46] and by Bouteiller et al.[47]. The former group also developed ECPs for a larger valence space which explicitly includes outermost core electrons. An overview of the most widely used ECPs is given in Table 1. It should be noted that it is possible to use nonstandard ECPs in Gaussian 98[31] through the general basis set option. Finally, we mention the so-called *ab initio* model potentials (AIMP) which have been developed and advocated by Huzinaga and coworkers[48,49]. The main difference between ECPs and AIMPs is that the latter have the correct nodal structure of the valence orbitals, while the ECPs are smoothed out in the core region. This means, e.g., that an ECP 4s valence orbital of Ge is nodeless, while the 4s orbital of an AIMP has three nodes. Test calculations have shown that the results of the two methods are very close to each other if a similar size basis set is used[50]. For a detailed discussion of ECPs and AIMPs see References 4 and 5.

Relativistic effects may be also considered by other methods than pseudopotentials. It is possible to carry out relativistic all-electron quantum chemical calculations of molecules. This is achieved by various approximations to the Dirac equation, which is the relativistic analogue to the nonrelativistic Schrödinger equation. We do not want to discuss the mathematical details of this rather complicated topic, which is an area where much progress has been made in recent years and where the development of new methods is a field of active research. Interested readers may consult published reviews[51–53]. A method which has gained some popularity in recent years is the so-called Zero-Order Regular Approximation (ZORA) which gives rather accurate results[54]. It is probably fair to say that

TABLE 1. Overview of common pseudopotentials for main group elements

Authors	Reference	Atoms	Method[a]	Type[b]	Valence basis set[c]
Hay/Wadt	35	Na–K	ECP	NR	[3/3]
Hay/Wadt	35	Rb–B	ECP	R	[3/3]
Stevens et al.	37	Li–Ar	ECP	NR	[4/4]
Stevens et al.	38	K–Rd	ECP	R	[5/5][d]
Stuttgart/Dresden	39	B–I [p]	ECP	R	[4/4][e]
Stuttgart/Dresden	40	Tl–Rn	ECP	R	[4/4][e]
Christiansen and coworkers	42	Li–Ar	ECP	R	[4/4]
Christiansen and coworkers	43	K–Kr	ECP	R	[3/3][f], [3/3/4][g]
Christiansen and coworkers	44	Rb–Xe	ECP	R	[3/3][h], [3/3/4][g]
Christiansen and coworkers	45	Cs–Rd	ECP	R	[3/3][i], [3/3/4][g], [55/5/4][j]
Christiansen and coworkers	46	B–Kr [q]	ECP	R	[311/311][k]
Bouteiller and coworkers	47	Li–Kr	ECP	NR	[4/4]
Huzinaga	49	Li–Xe	AIMP	NR	[5/5][l], [7/6][m], [9/8][n], [11/10][o]

[a]ECP = effective core potential, AIMP = *ab initio* model potential.
[b]NR = nonrelativistic, R = relativistic.
[c]Number of valence s and p Gaussian functions [s/p].
[d]The valence basis set for group 1 and group 2 elements is [4/4].
[e]The valence basis set for group 16 and group 17 elements is [4/5].
[f]For group 1 and 2 the valence basis set is [5/4].
[g]Valence basis set for group 13–18 elements which has the $(n-1)$ d electrons in the valence shell.
[h]Valence basis set for group 1 and 2 elements is [5/5].
[i]Valence basis set for Tl– Rn.
[j]Valence basis set for Cs and Ba including the outermost 5s, 5p and 5d electrons.
[k]The s and p valence basis sets were optimized for the contraction scheme 311.
[l]Valence basis set for Li–Ne.
[m]Valence basis set for Na–Ar.
[n]Valence basis set for K–Kr.
[o]Valence basis set for Rb–Xe.
[p]Only main group elements of groups 13–17.
[q]Only main group elements of groups 13–18.

approximate relativistic all-electron calculations of molecules achieve a similar accuracy to relativistic ECPs but at higher computational costs. The advantage of approximate all-electron methods is that the results may in principle become improved by going to higher levels of theory.

IV. QUANTUM CHEMICAL STUDIES OF GROUP-14 COMPOUNDS

In the following we review quantum chemical work about organic germanium, tin and lead compounds which has been published since 1990. The presentation of the results is organized as follows. First, we discuss in brief some relevant theoretical work about parent compounds of Ge to Pb. Then we summarize calculations of organometallic compounds. The latter section is divided into studies of neutral closed-shell molecules, charged species and radicals. E is used for any of the group-14 elements C to Pb.

A. Parent Compounds

Although the chapter focuses on calculations of organometallic compounds of group-14 elements Ge, Sn and Pb, we also review recent theoretical work about parent compounds of the elements which show the differences between the structures and energies of the carbon and silicon compounds and the heavier analogues of germanium, tin and lead. This part covers papers which report about theoretical studies of EH_2, EX_2, EH_4, EX_4, E_2H_6, E_2H_4, $E_2H_5^+$, E_2X_4, E_2H_2, $E_3H_3^+$ and E_6H_6 (X = halogen) that have been published since 1990.

1. EH_2 and EX_2 (X = halogen)

Table 2 shows calculated geometries and theoretically predicted energy differences between the ($^1A_1$) singlet state and ($^3B_1$) triplet state of the divalent group-14 compounds EH_2, EF_2 and ECl_2. CH_2 is the only EX_2 species which has a triplet ground state. The energy difference between the triplet and singlet state in favor of the latter shows the order C << Si < Ge ~ Sn < Pb. Chlorine and particularly fluorine strongly favor the singlet state over the triplet state of EX_2.

Benavides-Garcia and Balasubramanian studied also the dibromides and diiodes of Ge, Sn and Pb[58,59]. The latter work gives also results for the monohalogen systems GeHCl, GeHBr and GeHI. The geometries and singlet–triplet excitation energies are intermediate between the values of the germanium dihydrides and dihalogens[59]. The authors calculated also the positive ions ECl_2^+, EBr_2^+ and EI_2^+ (E = Sn to Pb) in the $^2A_1$ ground state and $^2B_1$ excited state[58].

The electronic ground states of EH_2 for E = Si to Pb and the hydrogenation energy yielding EH_4 have been calculated by Barandiarán and Seijo[60] and by Dyall[61]. Table 3 shows the theoretically predicted reaction energies for the reaction $EH_4 \rightarrow EH_2 + H_2$. The calculations predict that the reaction becomes less endothermic from Si to Pb in intervals of *ca* 20 kcal mol^{-1}. The hydrogenation of PbH_2 is nearly thermoneutral. The reaction energies of H_2 loss from EH_4 and $MeEH_3$ have been calculated by Hein et al.[62]. The results are discussed in the next section (see Table 6).

The spectroscopic constants of the dihydrides SiH_2, GeH_2 and SnH_2 and their cations and anions have been calculated by Mineva et al.[63]. The neutral dihydrides and their donor–acceptor complexes with various AH_3 and AH_2 species have been the subject of a theoretical work by Schöller and Schneider[64]. Table 4 shows the calculated bond energies of the complexes. It becomes obvious that the bond strength of the EH_2 complexes has the order $SiH_2 > GeH_2 > SnH_2$. The donor–acceptor complex of GeH_2 with water has also been calculated by Nowek and Leszczynski[65a].

TABLE 2. Calculated geometries and relative energies (kcal mol^{-1}) of the $^1A_1$ and $^3B_1$ states of EX_2(X = H, F, Cl); distances R in (Å), angles A in (deg)

EH_2[a]					
	C	Si	Ge	Sn	Pb
			Singlet ($^1A_1$)		
R (E–X)	1.150	1.555	1.620	1.793	1.880
A (X–E–X)	100.1	93.4	92.3	92.4	91.5
			Triplet ($^3B_1$)		
R (E–X)	1.116	1.511	1.559	1.734	1.827
A (X–E–X)	129.8	118.3	118.9	116.8	115.2
ΔE^b_{S-T}	10.6	−16.8	−24.1	−23.7	−39.1
EF_2					
	C	Si	Ge	Sn	Pb
			Singlet ($^1A_1$)		
R (E–X)	1.279	1.598	1.742	1.889	2.091
A (X–E–X)	104.7	99.5	97.0	95.7	98.2
			Triplet ($^3B_1$)		
R (E–X)	1.303	1.596	1.732	1.878	2.060
A (X–E–X)	118.3	114.0	112.8	111.8	118.9
ΔE^b_{S-T}	−46.3	−71.0	−75.3	−73.7	−88.4
ECl_2					
	C[b]	Si[c]	Ge[d]	Sn[e]	Pb[e]
			Singlet ($^1A_1$)		
R (E–X)	1.756	2.073	2.191	2.363	2.542
A (X–E–X)	109.4	101.7	100.5	98.4	100.8
			Triplet ($^3B_1$)		
R (E–X)	1.730	2.049	2.040	2.336	2.599
A (X–E–X)	125.5	118.2	118.6	116.0	139.9
ΔE^b_{S-T}	−20.5	−55.2	−60.3	−60.0	−69.7

[a] Reference 55.
[b] Reference 58. Negative values indicate that the singlet state is more stable than the triplet state.
[c] Reference 56.
[d] Reference 57.
[e] Reference 59.

TABLE 3. Reaction energies ΔE (kcal mol^{-1}) without ZPE corrections for $XH_4 \rightarrow XH_2 + H_2$[a]

	DHF[b]	PT[c]	QR-AIMP[d]
X=Si	62.3 (−0.6)	62.6 (−0.3)	
X=Ge	42.4 (−3.1)	42.6 (−2.9)	41.7 (−2.1)
X=Sn	23.2 (−7.7)	26.6 (−4.3)	21.6 (−6.6)
X=Pb	−6.2 (−27.4)	7.1 (−13.5)	−2.2 (−18.7)

[a] Numbers in parentheses give the relativistic effects. See also Table 6.
[b] Dirac–Hartree–Fock calculations, Reference 61.
[c] Relativistic correction included by perturbation theory, Reference 61.
[d] Quasi-relativistic *ab initio* model potential calculations, Reference 60.

TABLE 4. Binding energies (kcal mol^{-1}) for $H_2E{-}AH_n$ adduct formation of silylene, germylene and stannylene with various AH_3 and AH_2 units[a]

AH_n	$EH_2 = SiH_2$	GeH_2	SnH_2
NH_3	24.6	20.9	19.4
PH_3	20.9	16.3	12.5
AsH_3	16.3	13.0	10.2
SbH_3	17.1	14.0	10.8
BiH_3	11.3	9.5	7.4
OH_2	13.2	11.5	11.6
SH_2	12.4	10.3	9.0
SeH_2	12.5	10.6	9.3
TeH_2	14.7	12.3	10.4

[a]Reference 64.

The $^1A_1$ state of EH_2 with E = C to Pb has been calculated in the context of theoretical studies by Trinquier which focused on the structures and isomers of the heavier analogues of the ethyl cation $E_2H_5^+$[66]. The author reports calculated proton affinities of ($^1A_1$) EH_2 and E_2H_4 which are discussed below in Section IV.A.3.

2. EH_4, EX_4 (X = halogen) and E_2H_6

The structures and properties of group-14 tetrahydrides EH_4 have been the subject of several comparative theoretical studies in the last decade[61,67–69]. Table 5 shows calculated values of the E−H bond distances R_e, total bond dissociation energies of the four E−H bonds D_o and the force constants of the totally symmetric stretching mode k_e which were published by Wang and Schwarz[67]. The authors investigated also the tetrachlorides ECl_4 and these results are also shown in Table 5. The bond energies and force constants of the EH_4 species follow the trend C > Si > Ge > Sn > Pb. The D_o and k_e values of the ECl_4 molecules show that the heavier tetrahalides $SiCl_4$ to $PbCl_4$ have a similar bond strength to the tetrahydrides, while CCl_4 has a significantly weaker bond than CH_4.

The stability and vibrational spectra of EH_4 and the methyl-substituted analogues $MeEH_3$ have been the subject of a high-level theoretical study at the CCSD(T) level

TABLE 5. Calculated bond distances R_e(Å), total bond dissociation energies D_0 (kcal mol^{-1}) which include ZPE corrections and force constants of the totally symmetric mode k_e (N cm^{-1}) of the molecules EH_4 and ECl_4, using relativistic gradient-corrected DFT[a]

	CH_4	SiH_4	GeH_4	SnH_4	PbH_4
R_e	1.10	1.49	1.55	1.73	1.78
D_0	418.56	321.47	281.81	245.14	202.71
k_e	20.40	11.00	9.81	7.96	7.22
	CCl_4	$SiCl_4$	$GeCl_4$	$SnCl_4$	$PbCl_4$
R_e	1.84	2.08	2.20	2.39	2.48
D_0	262.20	321.01	265.66	248.83	198.33
k_e	12.80	11.10	8.80	7.52	6.10

[a]Reference 67.

by Hein et al.[62]. Table 6 shows the theoretically predicted reaction energies for H_2 loss. Note that the calculated reaction energies which include ZPE effects clearly predict that PbH_4 is unstable with regard to $PbH_2 + H_2$. However, the reaction has a large activation barrier. Figure 1 shows the calculated reaction profiles for the reactions $EH_4 \rightarrow EH_2$. All EH_4 molecules have large activation energies for H_2 loss. Table 6 shows that methyl substitution lowers the relative stability of the tetravalent molecules $MeEH_3$ compared with $EH_2 + CH_4$. The thermodynamic stability of EH_4 toward loss of H_2 calculated by different theoretical methods was already shown in Table 3. The values there and in Table 6 are very similar.

TABLE 6. Calculated reaction energies and ZPE corrections (kcal mol^{-1}) for hydrogen loss from EH_4 and H_3ECH_3[a]

E	$EH_4 \rightarrow EH_2+H_2$				$H_3ECH_3 \rightarrow EH_2+CH_4$			
	SCF	MP2	CCSD(T)	ZPE	SCF	MP2	CCSD(T)	ZPE
Si	68.3	68.6	64.9	−7.2	56.1	61.1	57.2	−3.2
Ge	44.9	45.8	42.5	−6.4	30.4	36.1	32.8	−2.5
Sn	23.9	23.0	20.2	−4.8	8.7	13.2	10.3	−1.0
Pb	−2.2	−1.6	−3.2	−4.5	−18.9	−12.7	−14.5	−0.8

[a]Reference 62.

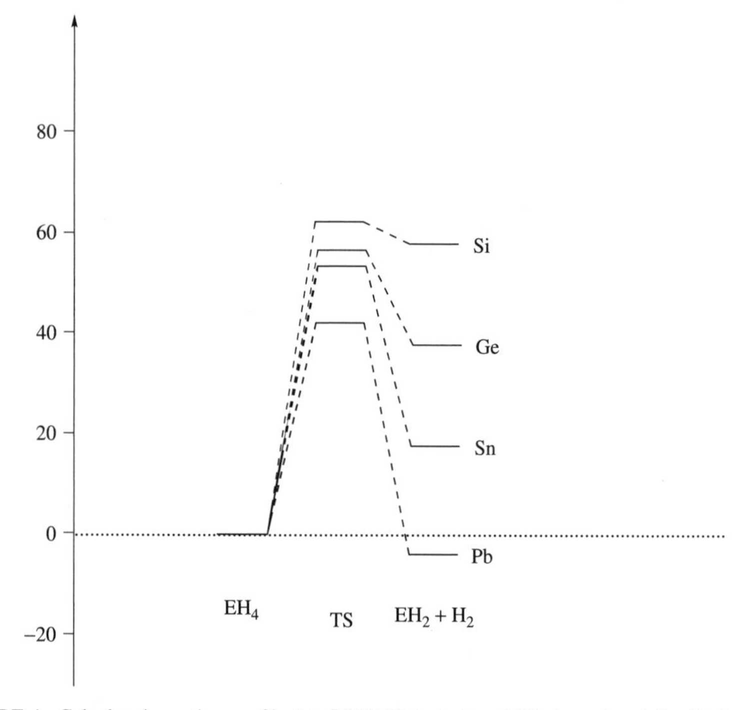

FIGURE 1. Calculated reaction profile (at CCSD(T) including ZPE corrections) for H_2 loss from EH_4 molecules. Reprinted with permission from Reference 62. Copyright 1993 American Chemical Society

Several theoretical studies compared properties of EH_4 molecules for different group-14 elements E. Oliveira et al. calculated the IR intensities, polar tensors and core electron energies of EH_4 and the deuteriated species ED_4 with E = C to Sn[70]. The work includes also results for the fluorosilanes and fluorogermanes. NMR spin–spin coupling constants of EH_4 with E = C to Sn have been calculated by different groups[71–73]. Theoretically predicted magnetic shielding constants of SnH_4 and SnX_4 (X = Cl to I) have been reported by Kaneko et al.[74].

The homonuclear and heteronuclear homologues of ethane $H_3E{-}E'H_3$ (E, E′ = C to Pb) have been studied in an extensive investigation by Schleyer et al.[75]. The focus of the paper is the size and origin of the energy barrier for rotation about the E–E′ bond. Table 7 shows the calculated E–E′ distances of the staggered energy minimum conformations and of the eclipsed conformations and the theoretically predicted rotational barriers. The energy barrier for rotation about the E–E′ bond becomes smaller when the atom E or E′ becomes heavier, but it does not vanish even for E, E′ = Pb. The analysis of the different factors which contribute to the energy barrier led the authors to conclude that the origin of the barrier is *not* steric repulsion between the hydrogen atoms. The lower energy of the staggered conformation is rather due to the stabilizing hyperconjugative interaction between the filled E–H σ-orbital and the vacant E′–H σ^*-orbital[75].

The equilibrium geometries of $H_3E{-}EH_3$ for E = C to Ge have been calculated at very high levels of theory by Leszczynski et al.[76]. Trinquier investigated in a detailed theoretical study the stabilities of the methylene–methane type complexes $H_2E{-}H{-}EH_3$ with respect to isomerization to the more stable $H_3E{-}EH_3$ isomers and to dissociation into $EH_2 + EH_4$[77]. Table 8 shows the calculated relative energies of the molecules and fragments. Figure 2 exhibits the calculated reaction profiles for the rearrangement of the isomers. It becomes obvious that the $H_2E{-}H{-}EH_3$ isomers of E = C, Si are very shallow minima on the potential energy surface which will be very difficult to observe

TABLE 7. E-E′ distances R(Å), and rotational barriers ΔE_{rot} (kcal mol^{-1}) for $H_3E{-}E'H_3$ molecules (E, E′ = C, Si, Ge, Sn, Pb)[a,b]

		All-electron calculations			Pseudopotential calculations		
E	E′	R_{st}	R_{ec}	ΔE_{rot}	R_{st}	R_{ec}	ΔE_{rot}
C	C	1.542	1.556	2.751	1.526	1.539	2.776
C	Si	1.883	1.893	1.422	1.883	1.893	1.388
C	Ge	1.990	1.999	1.104	1.996	2.004	0.986
C	Sn	2.188	2.193	0.498	2.178	2.184	0.520
C	Pb	2.275	2.278	0.204	2.242	2.246	0.321
Si	Si	2.342	2.355	0.949	2.355	2.364	0.823
Si	Ge	2.409	2.420	0.613	2.425	2.433	0.682
Si	Sn	2.610	2.617	0.581	2.610	2.616	0.476
Si	Pb	2.695	2.701	0.486	2.640	2.645	0.358
Ge	Ge	2.499	2.513	0.664	2.499	2.506	0.528
Ge	Sn	2.662	2.667	0.445	2.669	2.675	0.408
Ge	Pb	2.741	2.745	0.395	2.705	2.709	0.315
Sn	Sn	2.850	2.855	0.412	2.843	2.847	0.350
Sn	Pb	2.928	2.930	0.309	2.869	2.873	0.286
Pb	Pb	3.012	3.015	0.214	2.897	2.900	0.230

[a] Reference 75.
[b] R_{st} and R_{ec} are the E–E′ distances for the staggered and eclipsed conformations, respectively.

TABLE 8. Calculated relative energies (kcal mol^{-1}) of E_2H_6 species[a]

	C	Si	Ge	Sn	Pb
$2EH_3$[b]	93.5	69.6	64.2	58.5	50.8
$EH_4 + EH_2$	110.9	53.3	41.4	33.6	17.9
$H_3E{-}H{-}EH_2$	109.8	48.5	36.3	23.8	8.8
$H_3E{-}EH_3$	0.0	0.0	0.0	0.0	0.0

[a] At the MP4 level. From Reference 77.
[b] The energy of two EH_3 radicals with respect to $H_3E{-}EH_3$ at the MP2 level.

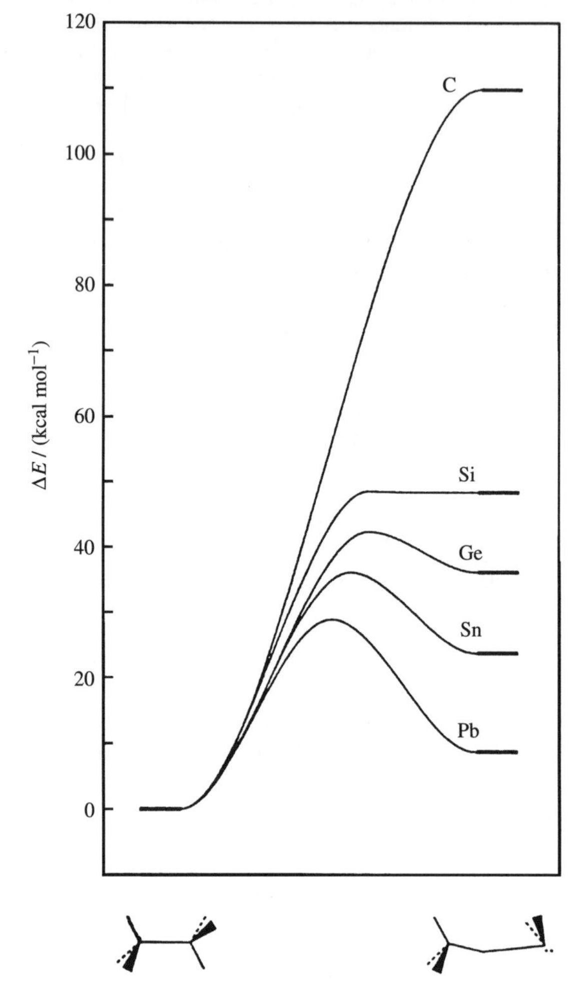

FIGURE 2. Calculated reaction profile at MP4 for the rearrangement of $H_3E{-}EH_3$ to $H_3E{-}H{-}EH_2$ complexes. Reproduced by permission of The Royal Society of Chemistry from Reference 77

experimentally. The barriers for rearrangments of the complexes to the more stable classical isomers become larger for the heavier elements Ge to Pb.

3. E_2H_4, $E_2H_5^+$ and E_2X_4 (X = halogen)

It is well known that compounds with multiple bonds between the heavier group-14 elements are much less stable than their carbon homologues. Although numerous olefin analogous compounds of the general formula R_2EER_2 could be isolated in the recent past, they usually require substituents which sterically or electronically protect the E−E bond against nucleophilic and electrophilic attack[78]. Theoretical work has been published which addresses the question about the structure and bonding situation in the parent compounds $H_2E{=}E'H_2$ and explains the different behavior of the heavier analogues where E and/or $E' =$ Si to Pb[79].

Jacobsen and Ziegler calculated the hetero- and homonuclear analogous olefins $H_2E{=}CH_2$ and $H_2E{=}EH_2$ (E = C to Pb) at the DFT level and they analyzed the bonding situation of the molecules with an energy decomposition method[80]. The calculations predict that the heteronuclear compounds $H_2E{=}CH_2$ have a planar geometry, while the heavier homonuclear compounds $H_2E{=}EH_2$ with E = Si to Pb have a *trans*-bent geometry as shown in Figure 3. Table 9 shows relevant results of the work. The bending angle Φ increases and the energy difference between the planar (D_{2h}) structure and the bent (C_{2h}) equilibrium structure becomes higher along the sequence Si < Ge < Sn < Pb. The E=E and E=C bond dissociation energies are much lower than the C=C BDE.

Jacobsen and Ziegler concluded from the analysis of the bonding situation that the $H_2E{=}CH_2$ and $H_2E{=}EH_2$ species have double bonds, because the intrinsic π-bonding, which in case of the *trans*-bent species arises from the $b_u(\pi)$ orbitals, makes an important contribution to the overall bond strength[80]. The authors suggest that the weaker bonds of the heavier analogues come from two factors. One is the high excitation energy of the ($^1A_1$) ground state of the EH_2 fragments to the $^3B_1$ excited state which is the electronic reference state for the double bonding in H_2EEH_2. The second important factor is the high intra-atomic and interatomic Pauli repulsion which is said to be mainly responsible for the change in the geometry from planar to *trans*-bent form[80].

The π-bond strength of $H_2E{=}E'H_2$ (E = Ge, Sn; E' = C to Sn) was the subject of a theoretical investigation at the MP2 and MCSCF+CI level by Windus and Gordon[81]. These authors estimated the strength of the π-bonding by evaluating the rotational barriers and by investigating thermochemical cycles. The results are shown in Table 10. Both methods give nearly the same results for the strength of the π-bonds, which have the energy order C > Si $\sim$ Ge > Sn. Note that according to the data given in Table 10, Sn has about the same π-bond strength in $H_2Sn{=}EH_2$ independent of the other group-14 element E.

The bonding situation in homonuclear systems E_2H_4 with a linkage $H_2E{-}EH_2$ and isomeric forms where the EH_2 fragments are bonded to each other through one or two

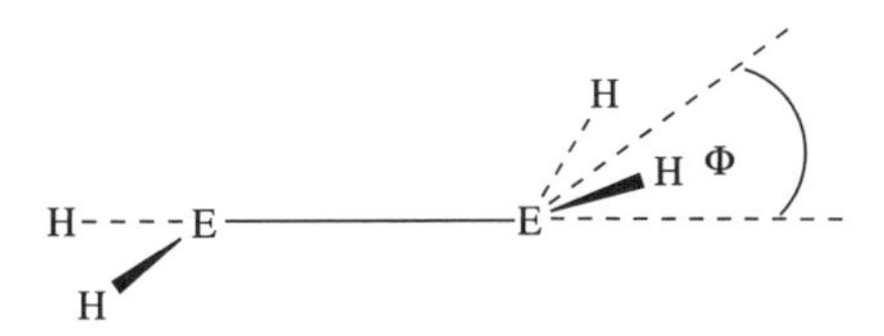

FIGURE 3. Definition of the bending angle Φ of the *trans*-bent forms of $H_2E{=}EH_2$ with E = Si to Pb. See Table 9

TABLE 9. Calculated geometries, bond dissociation energies D_e (kcal mol^{-1}) and energy differences E_{rel} between the planar and *trans*-bent structures of $H_2E{=}E'H_2$ molecules at BP86[a]

$H_2E{=}E'H_2$	Symmetry	Distances			Angles					Energies	
		E=E′	E–H	E′–H	H–E–H	H–E′–H	H–E=E′	H–E′=E	Φ^b	E_{rel}	D_e $H_2E{=}E'H_2$
$H_2C{=}CH_2$	D_{2h}	1.323	1.093		116.5		121.7		0.0		176.6
$H_2Si{=}SiH_2$	C_{2h}	2.150	1.483		112.4		119.1		36.1	0.0	59.8
$H_2Si{=}SiH_2$	D_{2h}	2.127	1.478		115.6		122.2		0.0	1.7	
$H_2Si{=}CH_2$	C_{2v}	1.687	1.480	1.092	114.7	115.5	122.6	122.2	0.0		110.4
$H_2Ge{=}GeH_2$	C_{2h}	2.245	1.538		109.5		117.0		47.3	0.0	43.02
$H_2Ge{=}GeH_2$	D_{2h}	2.205	1.521		116.9		121.5		0.0	5.5	
$H_2Ge{=}CH_2$	C_{2v}	1.770	1.526	1.090	115.4	117.4	123.3	121.5	0.0		85.3
$H_2Sn{=}SnH_2$	C_{2h}	2.569	1.727		105.8		117.9		51.0	0.0	28.9
$H_2Sn{=}SnH_2$	D_{2h}	2.501	1.698		113.0		123.5		0.0	7.4	
$H_2Sn{=}CH_2$	C_{2v}	1.945	1.698	1.089	109.6	117.0	125.2	121.5	0.0		66.9
$H_2Pb{=}PbH_2$	C_{2h}	2.819	1.794		107.7		115.2		53.6	0.0	10.0
$H_2Pb{=}PbH_2$	D_{2h}	2.693	1.771		125.6		117.2		0.0	3.2	
$H_2Pb{=}CH_2$	C_{2v}	2.045	1.774	1.090	114.3	116.2	122.8	121.9	0.0		45.7

[a]Bond distances in Å, bond angles in deg. From Reference 80.
[b]Bond angle; for definition see Figure 3.

TABLE 10. Theoretical π-bond strengths (kcal mol^{-1}) of $H_2E{-}E'H_2$ molecules in their equilibrium geometry[a]

E	E′	Thermocycle	Rotation
Ge	C	33	32.2
Ge	Si	26	25.7
Ge	Ge	28	25.4
Sn	C	21	20.9
Sn	Si	21	21.5
Sn	Ge	23	21.6
Sn	Sn	20	19.7

[a]Estimated by thermodynamical cycles and by calculations of the rotational barrier. From Reference 81.

TABLE 11. Nature of the E_2H_4 stationary points on the potential energy surface[a,b]

E	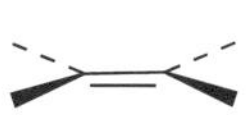				
C	G Min		TS	SP 2	TS
Si	G Min		Min	Min	Min
Ge	TS	G Min	Min	Min	Min
Sn	TS	Min	G Min	Min	Min
Pb	TS	TS	G Min	Min	Min

[a]Reference 65b.
[b]G Min, global energy minimum; Min, energy minimum; TS, transition state; SP 2, second-order saddle point.

TABLE 12. Calculated relative energies (in kcal mol^{-1}) of E_2H_4 units[a]

Structure	Symmetry	C_2H_4	Si_2H_4	Ge_2H_4	Sn_2H_4	Pb_2H_4
$2\ EH_2(^1A_1)$		192.0	53.7	35.9	33.2	28.7
$H_3E{-}EH\ (^1A')$	C_s	79.1 65.3[b]	9.8	2.4	7.0	17.5
$HE(H)_2EH$ (H-bridged)	C_{2v}, *cis*	140.3	25.2	11.6	2.3	2.0
	C_{2h}, *trans*	164.7	22.5	9.0	0	0
$H_2E{=}EH_2$	C_{2h}, *trans*-bent			0	9.1	23.9
$H_2E{=}EH_2$	D_{2h}, planar	0	0	3.2	18.5	43.7

[a]At the CIPSI level. From Reference 65b.
[b]$(^3A')$ triplet state.

E−H−E bridges have been studied in a series of papers by Trinquier and Malrieu[65b,82–85]. Table 11 shows the nature of the stationary points which have been found on the Hartree–Fock potential energy surfaces of E_2H_4[65b]. Table 12 gives the relative energies at the CIPSI correlated level using the HF-optimized geometries of the different isomers. Because Si_2H_4 is found to be planar at the HF level, the calculated energy at the correlated level gives the planar form as the lowest energy structure. Geometry optimization of Si_2H_4 at correlated levels gives the *trans*-bent form as lowest-energy species which has a small barrier for becoming planar[10]. The *trans*-bent form is the energetically lowest-lying isomer of Ge_2H_4 while the most stable structures of Sn_2H_4 and Pb_2H_4 are the *trans*-bridged forms. The *trans*-bent form of Sn_2H_4 is a higher-lying isomer while the *trans*-bent form of Pb_2H_4 is a transition state and not a minimum on the PES. A later study by Trinquier revealed that yet another low-lying isomer which has a direct E–E linkage and one E–H–E bridge exists as minimum on the Sn_2H_4 and Pb_2H_4 potential energy surface. The latter isomers were calculated to lie 8 and 15 kcal mol^{-1} higher than the doubly-bridged forms of Sn_2H_4 and Pb_2H_4[82].

Figure 4 shows a qualitative model which was suggested for the chemical bonding in the planar and *trans*-bent species H_2EEH_2. The σ/π bonding in the planar structures arises according to this model from the interactions of the EH_2 fragments in the $^3B_1$ triplet state, while the bonding in the *trans*-bent form are explained in terms of donor–acceptor interactions between the EH_2 moieties in the $^1A_1$ state. Trinquier and Malrieu proposed a criterion which links the singlet–triplet energy difference ΔE_{ST} of EH_2 with the occurrence of the *trans*-bent form and the doubly bridged form[85]. They suggested that a *trans*-bent form should become more stable than the planar structure if the sum of the ΔE_{ST} values is higher than half of the total ($\sigma + \pi$) EE bond energy, i.e. if the following condition is met:

$$\Sigma\Delta E_{ST} \geqslant \tfrac{1}{2}E_{\sigma+\pi}$$

The doubly bridged form was predicted to exist only as long as the following condition holds:

$$\Delta E_{ST} \geqslant \tfrac{1}{2}E_{\sigma+\pi}$$

Table 13 shows the estimated energies and the expected structures of the E_2H_4 systems[65b]. The predictions made by the qualitative model are generally in agreement with the calculated energy minima on the E_2H_4 potential energy surfaces[65b,82].

A related theoretical study by Trinquier focused on trends in electron-deficient bridged compounds of A_2H_6 (A = B, Ga), E_2H_4 (E = C to Pb) and $E_2H_6^{2+}$ (E = C, Si)[83]. Bridged and unbridged forms of the ethyl cation and its heavier analogues $E_2H_5^+$ (E = Si

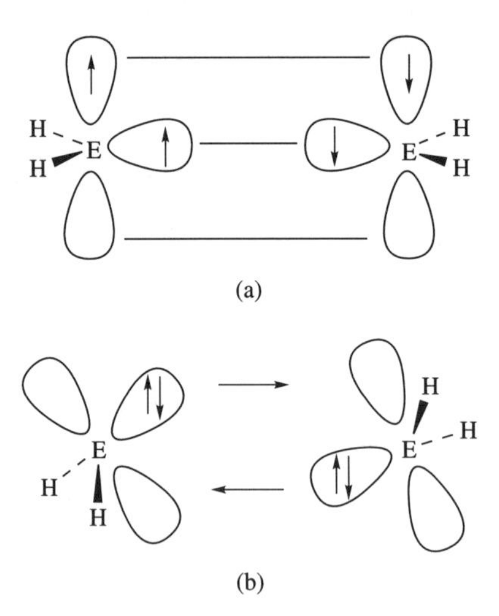

FIGURE 4. Qualitative molecular orbital model for the E–E bonding in (a) planar and (b) *trans*-bent $H_2E{=}EH_2$

TABLE 13. Estimated singlet-triplet separations ΔE_{ST} in EH_2 versus E=E bond energies[a]

	ΔE_{ST}	$1/4E_{\sigma+\pi}$	$1/2E_{\sigma+\pi}$	Expected structure
$H_2C{=}CH_2$	−9	43	86	planar
$H_2Si{=}SiH_2$	18–21	18–19	35–38	*trans*-bent (borderline)
$H_2Ge{=}GeH_2$	22	15–17	30–35	*trans*-bent
$H_2Sn{=}SnH_2$	23	15–17	30–35	*trans*-bent
$H_2Pb{=}PbH_2$	41	12–15	25–30	doubly bridged

[a] All values in kcal mol^{-1}. From Reference 65b.

to Pb) have also been studied by the same author[66]. Figure 5 exhibits the investigated structures. Table 14 shows the theoretically predicted relative energies of the stationary points on the $E_2H_5^+$ potential energy surface. The nonclassical form which has a $H_2E{-}EH_2$ moiety bridged by H^+ is the global minimum on the $C_2H_5^+$ and $Si_2H_5^+$ PES. The classical form $H_3Si{-}SiH_2^+$ is nearly degenerate with the nonclassical isomer, however. The classical form is the lowest lying isomer of $Ge_2H_5^+$. The most stable structures of $Sn_2H_5^+$ and $Pb_2H_5^+$ have a geometry where an EH^+ moiety is bonded to EH_4 through two hydrogen atoms of the latter species. The same paper reports also calculated proton affinities of EH_2 and E_2H_4. Table 15 shows the theoretical results at the MP4 level. The proton affinities of EH_2 show a regular trend C > Si > Ge > Sn > Pb, while the values of the E_2H_4 species first increase but then decrease. Note that the values for E_2H_4 refer to the energetically lowest lying forms of the molecules[66].

Trinquier and Barthelat calculated also the structures and energies of different isomers of E_2F_4 with E = C to Pb[56]. The planar π-bonded structure $F_2E{=}EF_2$ was found to be

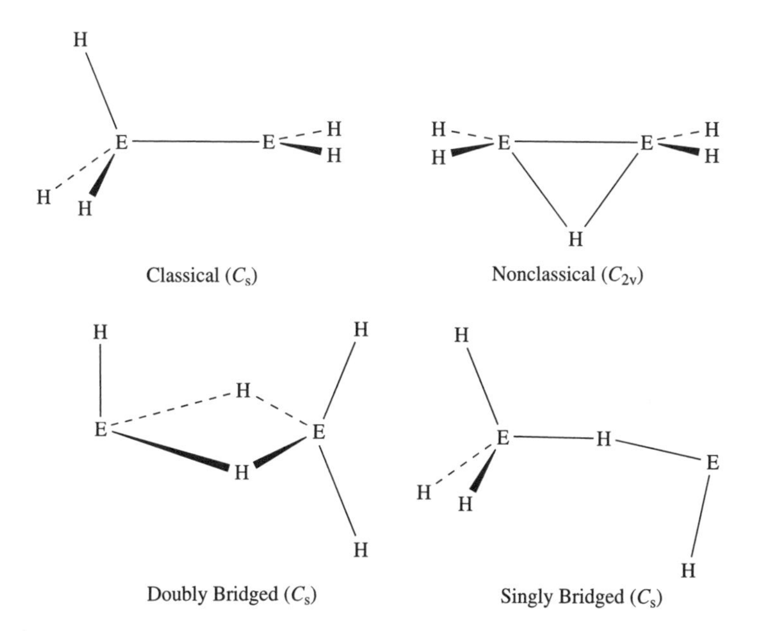

FIGURE 5. Calculated $E_2H_5^+$ species by Trinquier[66]. Relative energies are given in Table 14

TABLE 14. Calculated relative energies (kcal mol^{-1}) of $E_2H_5^+$ systems shown in Figure 5[a]

Molecules[b]	C	Si	Ge	Sn	Pb
$E_2H_4 + H^+$	172.3	208.7	207.3	204.9	190.8
$EH_3^+ + XH_2$	150.7	65.4	55.3	55.5	50.6
$EH^+ + XH_4$	151.3	41.4	32.3	29.7	26.1
Doubly bridged C_s	102.9	12.0	5.5	0	0
Singly bridged C_s	117.2	16.6	7.8	3.7	1.5
Classical C_s	7.6[c]	0.1	0	7.3	15.6
Nonclassical C_{2v}	0	0	5.3	13.2	26.5

[a] At the MP4 level. From Reference 66.
[b] The fragments correspond to the preferred isomer for E_2H_4, the $^1A_1$ state for EH_2 and the $^1\Sigma^+$ state for EH^+.
[c] Eclipsed conformation.

TABLE 15. Calculated proton affinities (kcal mol^{-1}) of EH_2 and E_2H_4[a]

E	$EH_2(^1A_1)$	E_2H_4
C	220	166
Si	204	203
Ge	194	202
Sn	193	200
Pb	179	185

[a] At MP4. From Reference 66.

TABLE 16. Calculated relative energies (kcal mol^{-1}) of E_2F_4 species[a]

		C	Si	Ge	Sn	Pb
$2EF_2$ ($^1A_1$)		56.3	6.3	23.8	50.9	62.3
$F_2E{-}EF_2(^3B)$		44.0	38.8	74.6	102.4	134.4
$F_2E{=}EF_2$		0	50.5[b]	97.3[b]	133.1[b]	212.2[c]
$F_3E{-}EF$		38.9	0	32.1	60.0	88.7[b]
FE(μ-F)$_2$EF	C_{2v}	118.9[b]	4.2	1.9	2.6	4.8
	C_{2h}	118.8[b]	3.7	0	0	0

[a] At CIPSI using HF optimized geometries. From Reference 55.
[b] Not a true minimum.
[c] First excited singlet state.

a true energy minimum for E = C, a saddle point for E = Si, Ge, Sn and an energy minimum on the first excited singlet state for E = Pb. The energetically lowest lying forms of the heavier analogues E_2F_4 are $F_3Si{-}SF$ for the silicon species, while doubly fluorine bridged structures were predicted to be the lowest lying forms of Ge_2F_4, Sn_2F_4 and Pb_2F_4[56]. Table 16 shows the relative energies of the stationary point on the PES which were obtained at the CIPSI correlated level using HF optimized geometries. The calculated values of the fluorinated species may be compared with the energies of the analogous hydrogen system E_2H_4 which are given in Table 12. Note that SnF_2 and PbF_2 are much more stabilized by forming the most stable dimer than SnH_2 and PbH_2, respectively, while

TABLE 17. Calculated reaction energies ($kcal\,mol^{-1}$) of E_2F_4 and EF_2 species[a,b]

E	$2EF_2 \rightarrow EF_4 + E$	$EF_4 + E \rightarrow E_2F_4$	$EF_2 + F_2 \rightarrow EF_4$
C	+4.0	−53.5	−189.6
Si	+1.6	−12.9	−242.5
Ge	+37.6	−66.0	−167.7
Sn	+51.5	−108.0	−154.1
Pb	+105.8	−173.8	−70.9

[a] At the HF level. From Reference 55.
[b] Each species is in its ground state: E, ^3P; EF_2, $^1A_1$, C_2F_4, π-bonded form; Si_2F_4, F_3Si-SiF; other E_2F_4: *trans*-bridged form.

CF_2, SiF_2 and GeF_2 are less stabilized by dimerization than their dihydrides. Table 17 gives some relevant reaction energies of E_2F_4 and EF_2 species which were predicted by Trinquier and Barthelat[56]. The authors report also the energies for addition of F_2 to EF_2 which may be compared with the reactions energies for H_2 addition to EH_2 given in Tables 3 and 6. It is obvious that F_2 addition to EF_2 is much more exothermic than H_2 addition to EH_2 for all elements E.

The homo- and heteroatomic ethylene homologues $H_2EE'H_2$ (E, E′ = C, Si, Ge) have been calculated with quantum chemical methods in a theoretical paper by Horner et al. which focused on the strain energies of the three-membered cyclic compounds cyc-$EE'E''H_6$ (E, E′, E″ = C, Si, Ge)[86]. The authors estimated the strain energies of the rings by calculating the reaction enthalpies of homodesmotic reactions at the CCSD level using HF optimized geometries. They also give the theoretically predicted reaction enthalpies of the addition of $E''H_2 + H_2EE'H_2 \rightarrow$ cyc-$EE'E''H_6$[86]. The latter quantity was also estimated from a bond additivity scheme using standard energy values. Table 18 shows the calculated results. It becomes obvious that the ring strain of the three-membered rings E_2H_6 increases with $E = C < Si < Ge$. Of the ten rings studied, cyclogermirane cyc-GeC_2H_6 is by far the least stable molecule with respect to dissociation, being only about 20 $kcal\,mol^{-1}$ more stable than GeH_2 + ethylene.

The reaction course for addition of the singlet EH_2 and EF_2 species with E = C to Sn to ethylene has been investigated with quantum chemical methods by Sakai[87]. The author calculated also the transition states for the addition reactions. A related work by Boatz et al. reported about the reactions of EH_2 and EF_2 (E = C to Sn) with acetylene yielding metallacyclopropenes[88]. The results of both studies are discussed below in the section on ‘reaction mechanisms’.

The ethylene and formaldehyde analogues H_2EEH_2 and H_2EO (E = Si to Pb), respectively, have been the subject of a more recent theoretical work by Kapp et al. which was carried out at the DFT level[89]. The calculated results were in general agreement with the work by Trinquier except for H_2PbPbH_2. In contrast to the latter author[65b,82] it was found that the *trans*-bent form of H_2PbPbH_2 is a true energy minimum on the PES[89]. Table 19 gives theoretically predicted reaction energies of H_2EEH_2 and H_2EO. The $H_2E{-}EH_2$ bond energies are somewhat lower than those which are shown in Table 9 but the trend is the same. Note that the inclusion of relativistic effects has a strong influence on the results of the lead compounds.

The structures and relative energies of germasilene (H_2GeSiH_2) and its isomers silylgermylene (H_3SiGeH) and germylsilylene (H_3GeSiH) in the singlet and triplet states have been calculated in a theoretical work by Grev et al.[90]. The geometries were optimized at the TCSCF level while the energies were calculated at the CISD level. The

TABLE 18. Energetics ΔE of the decomposition reactions cyc-$XH_2YH_2ZH_2 \rightarrow XH_2$(singlet) + $H_2Y\ ZH_2$ and strain enthalpies of group 14 cyclotrimetallanes[a]

Reaction	ΔE (kcal mol^{-1})		Strain enthalpy (kcal mol^{-1})
	ab initio	estimated	CCSD/DZ+d//HF/DZ+d
cyc-$C_3H_6 \rightarrow CH_2 + C_2H_4$	98.6 (84.9)	101	28.1 (26.1)
cyc-$Si_3H_6 \rightarrow SiH_2 + Si_2H_4$	62.3 (58.7)	68	34.5 (36.8)
cyc-$Ge_3H_6 \rightarrow GeH_2 + Ge_2H_4$	47.1 (47.5)	43	37.3 (39.3)
cyc-$SiC_2H_6 \rightarrow SiH_2 + C_2H_4$	43.2 (33.9)	57	35.2 (36.7)
cyc-$SiC_2H_6 \rightarrow CH_2 + CSiH_4$	112.8 (101.6)	119	35.2 (36.7)
cyc-$GeC_2H_6 \rightarrow GeH_2 + C_2H_4$	18.3 (9.3)	30	35.8 (38.1)
cyc-$GeC_2H_6 \rightarrow CH_2 + CGeH_4$	107.3 (96.9)	116	35.8 (38.1)
cyc-$CSi_2H_6 \rightarrow SiH_2 + CSiH_4$	60.3 (56.4)	67	37.0 (39.7)
cyc-$CSi_2H_6 \rightarrow CH_2 + Si_2H_4$	120.3 (109.5)	129	37.0 (39.7)
cyc-$CGe_2H_6 \rightarrow GeH_2 + CGeH_4$	44.5 (43.3)	48	39.2 (41.7)
cyc-$CGe_2H_6 \rightarrow CH_2 + Ge_2H_4$	100.1 (91.0)	107	39.2 (41.7)
cyc-$CSiGeH_6 \rightarrow GeH_2 + CSiH_4$	42.9 (40.2)	46	38.2 (40.8)
cyc-$CSiGeH_6 \rightarrow SiH_2 + CGeH_4$	62.3 (59.9)	70	38.2 (40.8)
cyc-$CSiGeH_6 \rightarrow CH_2 + SiGeH_4$	110.2 (110.7)	118	38.2 (40.8)
cyc-$GeSi_2H_6 \rightarrow GeH_2 + Si_2H_4$	52.5 (51.4)	54	35.6 (37.8)
cyc-$GeSi_2H_6 \rightarrow SiH_2 + SiGeH_4$	59.8 (58.8)	64	35.6 (37.8)
cyc-$SiGe_2H_6 \rightarrow GeH_2 + SiGeH_4$	49.7 (49.7)	49	36.3 (38.6)
cyc-$SiGe_2H_6 \rightarrow SiH_2 + Ge_2H_4$	57.3 (56.7)	59	36.3 (38.6)

[a] At CCSD//HF. Energy values in parentheses are HF/DZ+d results. From Reference 86.

TABLE 19. Calculated B3LYP reaction energies (kcal mol^{-1})[a] of reactions of E_2H_4 and EH_2 species[b]

E	$H_2E{=}EH_2 + O_2 \rightarrow$ cyclo-$(H_2EO)_2$[c,d]	$2H_2E + O_2 \rightarrow 2H_2E{=}O$[d]	$H_2E{=}EH_2 \rightarrow 2H_2E$[c]	$2H_2E{=}O \rightarrow$ cyclo-$(H_2EO)_2$
C	−50.6	−222.6	160.2[e]	11.78
Si	−193.3	−149.8	53.4	−96.9
Ge	−117.0	−68.0	32.7	−81.7
Sn	−101.0	−28.0	22.5	−95.5
n-Pb[f]	−104.7	−25.1	24.5	−104.1
Pb	−38.8	25.2	10.5	−74.5

[a] All energies include ZPE corrections.
[b] Reference 89.
[c] Calculated with the C_{2h} minimum structures of the Si, Ge, Sn, *n*-Pb and Pb compounds (planarization barriers: 1.18, 6.54, 10.13, 7.17 and 22.79 kcal mol^{-1}, respectively).
[d] Reactions with singlet oxygen are 38.6 kcal mol^{-1} more exothermic.
[e] Calculation with triplet CH_2.
[f] n-Pb: lead computed with a nonrelativistic pseudopotential.

π-bond energy of germasilene is predicted to be 25 kcal mol^{-1}, essentially identical with those of H_2SiSiH_2 and H_2GeGeH_2. The bond dissociation energy, however, was found to decrease in the order Si=Si > Si=Ge > Ge=Ge and in each case was smaller than the bond energy of the corresponding single bonds in the saturated systems disilane, germylsilane and digermane[90]. Table 20 shows theoretical energy differences between isomers of the compounds which have been calculated or which were estimated by taking the differences between the π-bond energies D_π and the bond energies D(E–H). The two sets of data agree nicely with each other.

TABLE 20. Comparison of predicted and theoretically determined isomeric 0 K enthalpy differences (kcal mol^{-1})[a]

Reaction	$\Delta H_{predict} = D_\pi + E_2(M'-H) - E_1(M-H)$	$\Delta H_{ab\ initio}$
$H_2Ge{=}SiH_2 \rightarrow H_3Ge{-}SiH$	$25 + 68.5 - 84.0 = 9.5$	9.4
$H_2Ge{=}SiH_2 \rightarrow HGe{-}SiH_3$	$25 + 57.4 - 90.6 = -8.2$	−6.3
$H_2Ge{=}GeH_2 \rightarrow H_3Ge{-}GeH$	$25 + 57.4 - 84.0 = -1.6$	−2.0
$H_2Si{=}SiH_2 \rightarrow H_3Si{-}SiH$	$25 + 68.5 - 90.6 = 2.9$	6.4
$H_2Si{=}CH_2 \rightarrow H_3Si{-}CH$	$38 + 111.2 - 90.6 = 58.6$	45.2
$H_2Si{=}CH_2 \rightarrow HSi{-}CH_3$	$38 + 68.5 - 104.6 = 1.9$	3.6
$H_2Ge{=}CH_2 \rightarrow HGe{-}CH_3$	$31 + 57.4 - 104.6 = -16.2$	−17.6

[a] Reference 90.

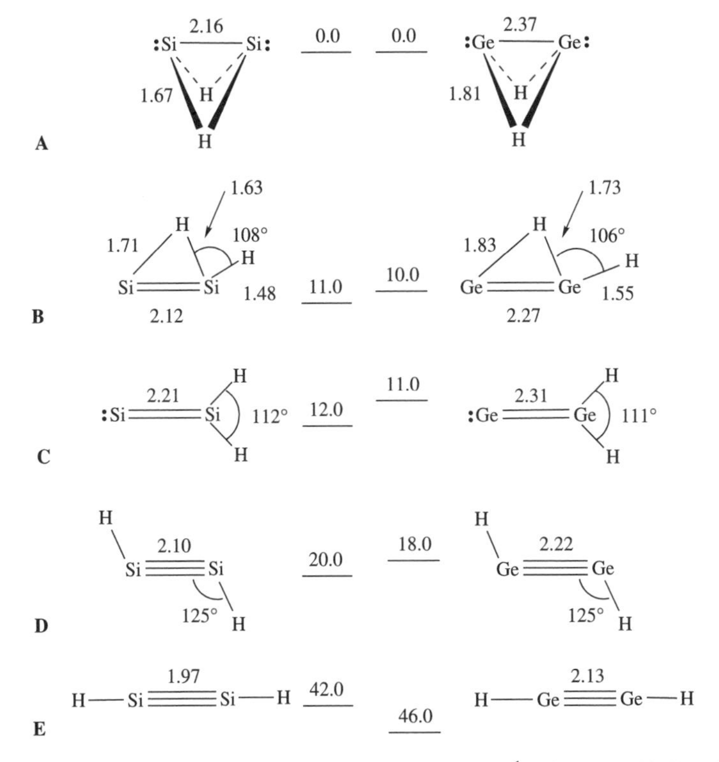

FIGURE 6. Calculated geometries and relative energies (kcal mol^{-1}) of relevant Si_2H_2 and Ge_2H_2 isomers **A**–**E**. The linear forms **E** are not minima on the PES. Distances in Å, angles in deg. Data are from Reference 91 except for structure **B** of Ge_2H_2, which have been reproduced from Reference 92 by permission of Academic Press

4. E_2H_2, $E_3H_3^+$ and E_6H_6

Quantum chemical investigations of the heavier group-14 analogues of acetylene aroused considerable interest because the calculations predicted that the energetically lowest lying Si_2H_2 and Ge_2H_2 isomers are very different from what is known about C_2H_2. Figure 6 shows the theoretically predicted stationary points on the E_2H_2 (E = Si, Ge) PES.

The results for the silicon and germanium species are very similar[91–93]. The doubly bridged structure **A** is the global energy minimum on the PES. The singly bridged isomer **B** and the energetically nearly degenerate vinylidene analogue **C** are 10–12 kcal mol^{-1} higher in energy than **A**. The isomer with a H–E–E–H linkage **D** has a *trans*-bent geometry which is 18–20 kcal mol^{-1} above **A**. The linear form **E** is not a minimum on the PES of Si_2H_2 and Ge_2H_2. **E** is much higher (40–42 kcal mol^{-1}) in energy than **A**.

The homonuclear species Ge_2H_2 and the heteronuclear systems $GeCH_2$ and $GeSiH_2$ have been calculated in an extensive study by Boone and coworkers[94]. Figure 7 shows schematically the structural isomers which were investigated. Table 21 shows the calculated results. The doubly bridged isomer **A** is also the global energy minimum on the $GeSiH_2$ PES. The vinylidene form $H_2C{=}Ge$ is clearly the lowest lying isomer on the $GeCH_2$ PES. Hydrogen-bridged structures of $GeCH_2$ were not found as energy minima on the PES[94]. More recent calculations of $GeCH_2$ isomers by Stogner and Grev agree with the previous results[95].

Theoretical investigations of the heavier analogues Sn_2H_2 and Pb_2H_2 have recently been published[96–98]. Figure 8 shows the calculated potential energy surfaces of E_2H_2 for E = Si to Pb at the MP2 level by Nagase and coworkers[96]. The qualitative features of the four systems are the same. The doubly bridged isomer **A** is also predicted as the global energy minimum of Sn_2H_2 and Pb_2H_2. The energy difference between the linear form HEEH and the most stable form **A** is particularly large for E = Pb. The latter isomer was also calculated by Han et al.[97].

The potential energy surface of Pb_2H_2 has recently been the subject of a detailed theoretical study at the DFT and MP2 level by Frenking and coworkers[98]. The geometries and relative energies of the calculated isomers are shown in Figure 9. The global energy minimum is the doubly-bridged isomer. The singly-bridged form **B** is a transition state at the DFT level, while it is an energy minimum at MP2[98]. The most important aspect of the work by Frenking and coworkers concerns the relative energies of the *trans*-bent isomers **D1** and **D2**. The former structure is the analogue of the *trans*-bent form **D** which is shown in Figure 6. However, **D1** lies energetically higher than another *trans*-bent form

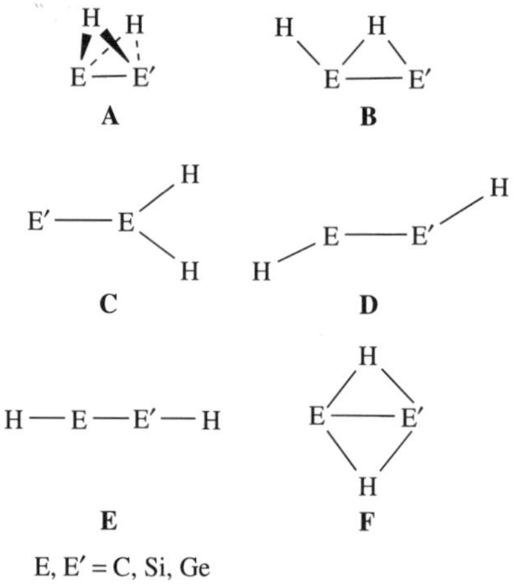

FIGURE 7. Structures **A**–**F** of Ge_2H_2, $GeSiH_2$ and $GeCH_2$ which have been investigated in Reference 94. The calculated bond lengths and angles and the relative energies are given in Table 21

TABLE 21. Calculated energies (kcal mol^{-1}) and equilibrium geometries of the structural isomers of Ge_2H_2, $GeSiH_2$ and $GeCH_2$ shown in figure 7[a]

	A	**B**	**B′**	**C**	**C′**	**D**	**E**[b]	**F**[b]
				Ge_2H_2				
E_{rel}	0.00	8.69		13.67		17.67	42.36	8.36
R(Ge−Ge)	2.365	2.234		2.298		2.188	2.067	2.543
R(Ge−H)[c]	1.745	1.789 (H_B)		1.528		1.535	1.492	1.678
R(Ge−H)[c]	1.745	1.534 (H_T)		1.528		1.535	1.492	1.678
∠(H−Ge−Ge)[c]	47.34	48.77 (H_B)		124.19		125.54	180	40.72
∠(Ge−Ge−H)[c]		110.70 (H_T)					180	
∠(H−Ge−Ge−H)	105.83	180		180		180	180	180
				$GeSiH_2$[d]				
E_{rel}	0.00	5.98	11.50	8.42	19.67	17.56	40.04	10.28
R(Ge−Si)	2.291	2.187	2.172	2.263	2.242	2.144	2.028	2.461
R(Ge−H)[c]	1.766	1.821 (H_B)	1.532(H_T)		1.526	1.533	1.492	1.677
R(Si−H)[c]	1.657	1.484 (H_T)	1.695(H_B)	1.479		1.485	1.456	1.599
∠(Si−Ge−H)[c]	45.99	46.64 (H_B)	158.85 (H_T)		124.01	127.60	180	40.12
∠(Ge−Si−H)[c]	50.03	107.82 (H_T)	51.11 (H_B)	123.55		123.32	180	42.52
∠(H−Ge−Si−H)	104.44	180	180	180	180	180	180	180
				$GeCH_2$[e]				
E_{rel}		106.91	60.17	0.00	104.30	42.40	46.20	
R(Ge−C)		1.819	1.735	1.796	1.851	1.698	1.668	
R(Ge−H)[c]		1.513 (H_T)	1.564 (H_B)		1.518	1.499	1.482	
R(C−H)[c]		2.261 (H_B)	1.081 (H_T)	1.087		1.079	1.069	
∠(C−Ge−H)[c]		155.43 (H_T)	86.75 (H_B)		121.58	148.66	180	
∠(Ge−C−H)[c]		42.89 (H_B)	165.44 (H_T)	122.72		139.94	180	

[a] At the MP2/TZV (2df, 2p) level of theory. From Reference 94. Bond distances R in Å and bond angles in deg.
[b] Transition state.
[c] H_B denotes bridging hydrogen, H_T denotes terminal hydrogen.
[d] Structure **B** has a terminal Si−H bond and **B′** has a terminal Ge−H bond. Structure **C** has a GeH_2 moiety and **C′** has a SiH_2 moiety.
[e] Structure **B** has a terminal Ge−H bond and **B′** has a terminal C−H bond. Structure **C** has a GeH_2 moiety and **C′** has a CH_2 moiety.

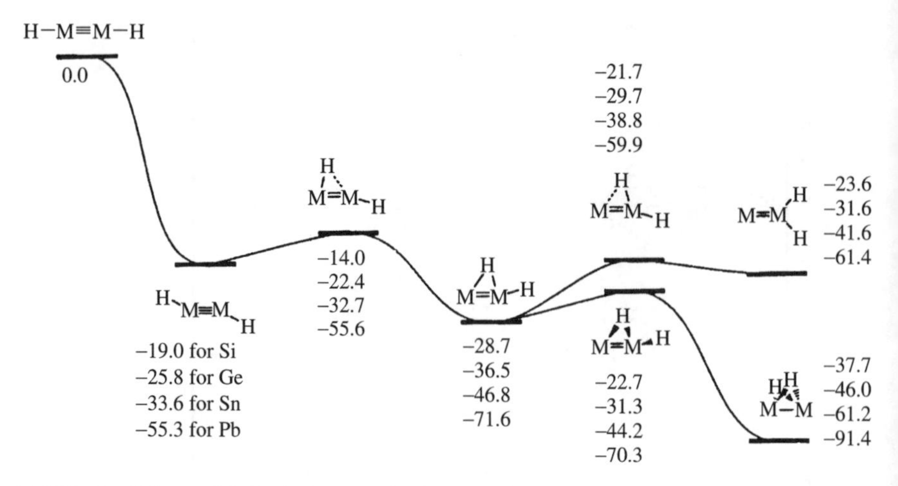

FIGURE 8. Calculated PES of M_2H_2 species for M = Si to Pb at the MP2 level. Relative energies are given in kcal mol^{-1}. Reprinted from Reference 96 with permission from Elsevier Science

D2, which has an acute bond angle of *ca* 90° (Figure 9). The analysis of the electronic structure of **D1** and **D2** showed that the HOMO of the former isomer is the Pb−Pb π-orbital, while the LUMO is a lead lone-pair type orbital. Structure **D2** has the frontier orbitals reversed, i.e. the lead lone-pair orbital is the HOMO and the Pb−Pb π-orbital is the LUMO. This is shown in Figure 10. It means that lead has a valence shell with an electron sextet. Nevertheless structure **D2** is lower in energy than **D1**. However, both structures **D1** and **D2** are not minima on the PES. The results were important, though, because the calculations showed that bulky substituents stabilize **D2** so much that it not only becomes an energy minimum but even the global minimum on the PES[98].

The heavier analogues of the cyclopropenium cation $E_3H_3^+$ with E = Si to Pb and other structural isomers of $E_3H_3^+$ have been calculated by Jemmis et al. at the DFT (B3LYP) level[99]. Figure 11 shows the calculated isomers and the relative energies and nature of the stationary points on the PES. The classical cyclopropenium form **1(Si)** is the most stable isomer of $Si_3H_3^+$. A second isomer which has a nonplanar C_{3v} symmetry **2(Si)** is 23.7 kcal mol^{-1} higher in energy than **1(Si)**. The former isomer becomes clearly more stable than the cyclopropenium form in case of the germanium analogues. Structure **2(Ge)** is 17.4 kcal mol^{-1} lower in energy than **1(Ge)**. The bridged structures **2(Sn)** and **2(Pb)** are predicted to be the only minimum on the respective PES. The cyclopropenium forms of $Sn_3H_3^+$ and $Pb_3H_3^+$ are not minima on the PES[99].

The structures, stabilities and properties of the heavier analogues of benzene and other structural C_6H_6 isomers have been calculated for Si_6H_6 and Ge_6H_6 but not for Sn_6H_6 and Pb_6H_6. Matsunaga et al. optimized at the HF level the geometries of benzene and the planar analogues $Si_3C_3H_6$, $Ge_3C_3H_6$, Si_6H_6, $Ge_3Si_3H_6$ and Ge_6H_6 among other heteroatomic benzene analogues[100]. Inspection of the vibrational frequencies showed that only the planar (D_{3h}) forms of $Si_3C_3H_6$ and $Ge_3C_3H_6$ are minima on the PES while the D_{6h} geometries of Si_6H_6 and Ge_6H_6 and the D_{3h} form of $Ge_3Si_3H_6$ are transition states of higher-order saddle points. A following paper by Matsunaga and Gordon reported about other isomeric forms of the molecules[101]. The authors calculated the equilibrium geometries of the chair and boat forms of the $EE'H_6$ molecules (E, E′ = C, Si, Ge and

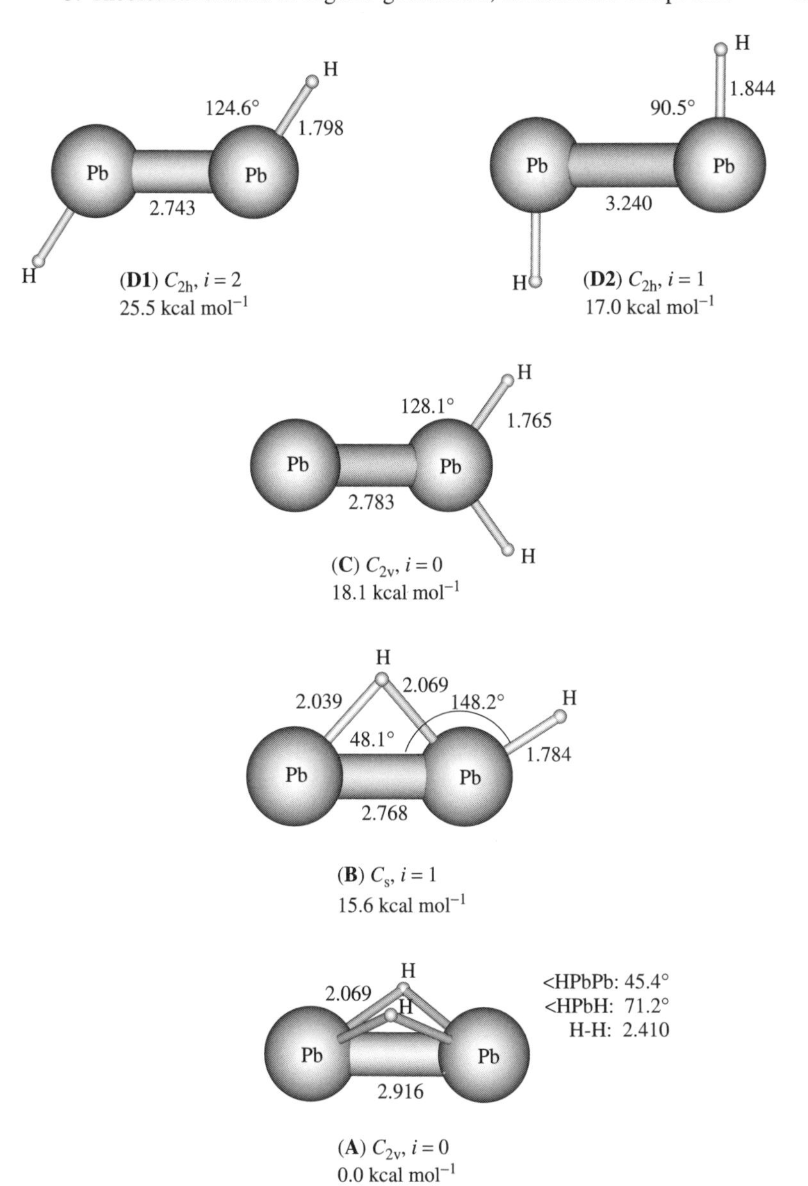

FIGURE 9. Calculated geometries and relative energies of Pb_2H_2 species at B3LYP. Bond lengths are given in Å, angles in deg. The number of imaginary frequencies i indicates if the structure is an energy minimum ($i = 0$), a transition state ($i = 1$) or a higher-order saddle point ($i = 2$). Reproduced by permission of Wiley-VCH from Reference 98

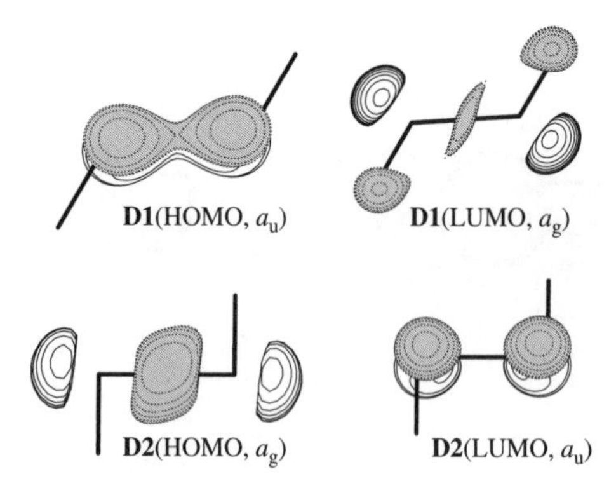

FIGURE 10. Plots of the frontier orbitals of structures **D1** and **D2** of Pb_2H_2 (see Figure 9 for the notation). Reproduced by permission of Wiley-VCH from Reference 98

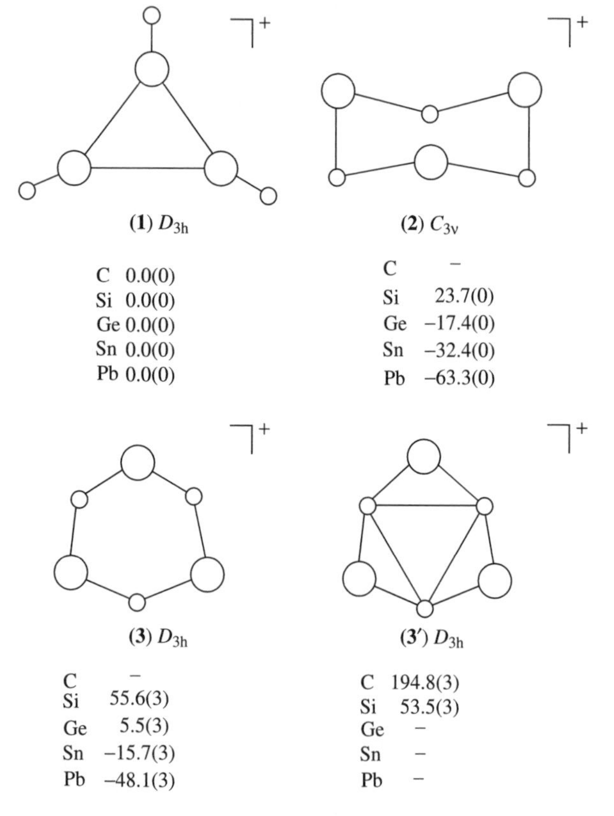

FIGURE 11. Calculated structures and relative energies (kcal mol^{-1}) of $E_3H_3^+$ isomers (E = C to Pb). The number of imaginary frequencies i is given in parentheses. Reprinted with permission from Reference 99. Copyright 1995 American Chemical Society

group13/15 elements) besides the planar structures. They also investigated the geometries and relative energies of the prismane analogue. Figure 12 shows the calculated isomers. Table 22 gives the relative energies of the stationary points on the PES. The prismane isomers of C_6H_6, $Si_3C_3H_6$ and $Ge_3C_3H_6$ are higher in energy than the planar isomers which are the global minima on the PES. The prismane analogues of Si_6H_6, $Si_3Ge_3H_6$ and Ge_6H_6 become the energetically lowest lying structures which are slightly more stable than the chair conformation of the 6-membered rings. The paper reports also about other heteroatomic C_6H_6 isomers with group13/group15 atoms in the ring[101].

The aromaticity of the planar (D_{6h}) and chair (D_{3d}) forms of Si_6H_6 and Ge_6H_6 and other heteroatom benzene analogues has been the subject of a theoretical study by Schleyer and coworkers[102]. The authors used three different criteria for estimating the aromatic character of the molecules. They are: (i) the calculated NICS (Nuclear Independent Chemical Shift)[103] values at the center of the ring, (ii) the ring size adjusted aromaticity index ρ which is based on the calculated magnetic susceptibility[104] and

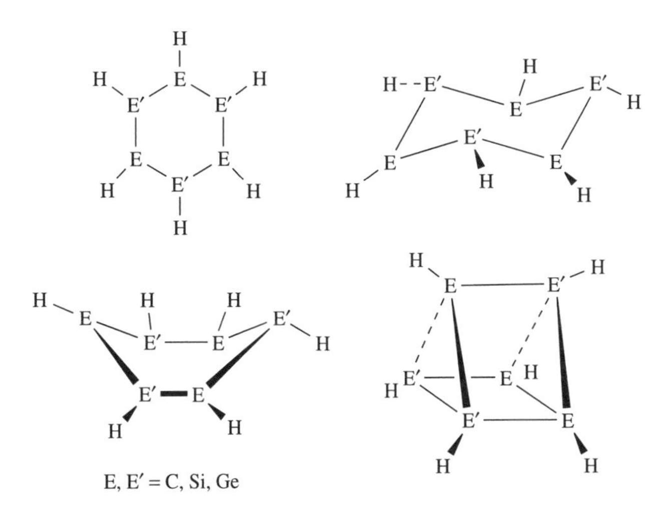

FIGURE 12. Calculated $E_nE'_{6-n}H_6$ isomers (E, E′ = C, Si, Ge) with planar, chair, boat and prismane structure. Relative energies are given in Table 22. The boat form is not an energy minimum on the PES. Reprinted with permission from Reference 101. Copyright 1994 American Chemical Society

TABLE 22. Calculated relative energies (kcal mol^{-1}) of $E_3E'_3H_6$ isomers shown in Figure 12[a]

	Planar	Chair	Prismane
C_6H_6	0.0	—	113.9
$Si_3C_3H_6$	0.0	—	35.3
$Ge_3C_3H_6$	0.0	—	27.2
Si_6H_6	0.0[b]	−3.0	−13.6
$Si_3Ge_3H_6$	0.0[b]	−4.8	−6.4
Ge_6H_6	0.0[b]	−10.7	−11.1

[a] At MP2 using HF optimized geometries. From Reference 101.
[b] Not a minimum on the potential energy surface.

(iii) the calculated ASE (Aromatic Stabilization Energy) values which are based on the energies of homodesmotic reactions[105]. The authors found that benzene is more aromatic than its sila and germa homologues on the basis of these three criteria even in D_{6h} symmetry, but Si_6H_6 and Ge_6H_6 still have significant aromatic character even in their D_{3d} equilibrium geometries[102].

B. Substituted Compounds

The previous sections discussed theoretical work of parent compounds of heavier group-14 elements which are homologues of important carbon reference compounds. In the next sections we will discuss in brief theoretical studies of organometallic molecules of germanium, tin and lead which were published since 1990.

1. Neutral closed-shell molecules

Most theoretical work on group-14 organometallic compounds was devoted to neutral closed-shell species. We divided the presentation of the papers into two parts. The first part focuses on theoretical studies of geometries and properties of molecules. The second part describes work where reaction mechanisms of chemical reactions of organometallic Ge-, Sn- and Pb-organic compounds have been investigated theoretically.

a. Structures and properties. i. Compounds with multiple bonds of Ge, Sn, Pb. Several theoretical papers which investigate compounds with multiple bonds of heavier group-14 elements Ge to Pb have been published recently. Most of them focus on germanium compounds. A theoretical study of Cotton et al. about multiple bonding between main group elements gave the experimental and calculated geometry using DFT (B3PW91) methods of $Me_2Ge{=}GeMe_2$[106]. Figure 13 shows the theoretically predicted and experimentally observed bond lengths and angles. The agreement is quite good.

Two papers by Khabashesku et al. report about theoretical and experimental work of various germenes $R^1R^2Ge{=}CH_2$[107,108]. One paper is a joint experimental/theoretical work about the gas-phase pyrolytic generation of $Me_2Ge{=}CH_2$ from four-membered cyclic precursors[107]. The authors give the calculated geometries of dimethylgermene $Me_2Ge{=}CH_2$ (Table 23) and methylgermylene $H_3C{-}GeH$ (Table 24) at different levels of theory. Comparison of the calculated vibrational spectra of the molecules and the experimental spectrum suggests that both species are formed in the pyrolysis[107]. The second paper

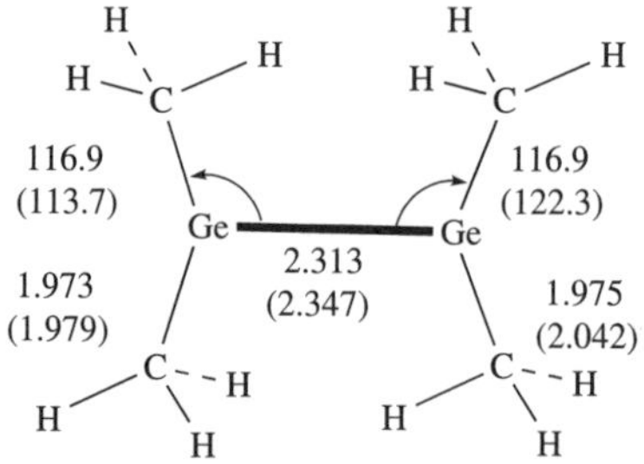

FIGURE 13. Optimized geometry of $Me_2Ge{=}GeMe_2$ at B3PW91. Experimental values are given in parentheses. Bond lengths in Å, angles in deg. Reprinted with permission from Reference 106. Copyright 1998 American Chemical Society

TABLE 23. Optimized geometry of 1,1-dimethyl-1-germene[a]

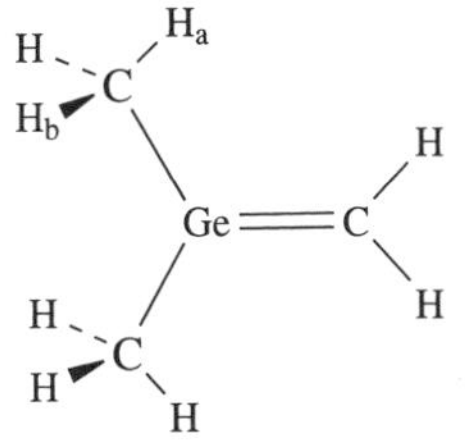

	Bond length (Å)					Angle (deg)			
	Ge=C	Ge−C	C−H	C−H_a	C−H_b	CGeC	HCGe	H_aCGe	H_bCGe
HF/DZ+d	1.761	1.947	1.085	1.091	1.093	123.0	122.1	110.3	110.7
HF/6-311G(d,p)	1.765	1.954	1.076	1.083	1.085	123.2	121.9	110.3	110.4
MP2/DZ+d	1.772	1.937	1.100	1.105	1.106	122.8	121.6	110.0	110.6
B3LYP/6-311G(d,p)	1.780	1.962	1.083	1.090	1.092	122.7	121.5	109.8	110.5

[a]Reference 107.

TABLE 24. Optimized geometry of methylgermylene[a]

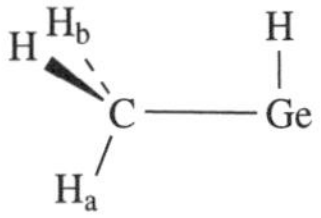

	Bond length (Å)				Angle (deg)		
	Ge−C	Ge−H	C−H_a	C−H_b	HGeC	H_aCGe	H_bCGe
HF/DZ+d	1.991	1.596	1.092	1.097	94.6	112.4	110.1
MP2/DZ+d	1.980	1.595	1.105	1.111	93.6	112.8	109.6
B3LYP/6-311G(d,p)	2.012	1.604	1.091	1.096	93.2	112.5	109.6

[a]Reference 107.

by Khabashesku et al. deals with five substituted germenes[108]. Table 25 shows the optimized geometries, atomic partial charges and calculated dipole moments of $H_2Ge{=}CH_2$, $MeHGe{=}CH_2$, $Me_2Ge{=}CH_2$, $FHGe{=}CH_2$ and $H_2Ge{=}CHF$. The authors also calculated the reaction profiles of the head-to-head and head-to-tail dimerization of the germenes. The results will be discussed below in the section about reaction mechanisms.

Another combined experimental paper of Khabashesku et al. reported about pyrolytic generation of dimethylgermanone $Me_2Ge{=}O$ and its dimer[109]. The calculated geometries of the two compounds and the parent molecule $H_2Ge{=}O$ are shown in Figure 14. The authors calculated the vibrational spectra of the organogermanium compounds. Comparison with the observed FTIR spectrum showed that both molecules are generated by pyrolysis from three different precursors. The observed frequency and the calculated force

TABLE 25. Optimized parameters of germene monomers at B3LYP/6–311G(d)[a]

Germene	Bond distances (Å)					Valence angles (deg)				Total net charge		Dipole moment (D)
	Ge=C	Ge–H	Ge–R	C–H	C–R	HGeC	RGeC	HCGe	RCGe	Ge	C	
$H_2Ge=CH_2$	1.778	1.525		1.082		122.5		121.4		+0.371	−0.496	0.591
$MeHGe=CH_2$	1.779	1.531	1.958	1.082	1.083	121.0	124.2	121.3	121.7	+0.472	−0.524	1.462
$Me_2Ge=CH_2$	1.780		1.962	1.083			122.7	121.5		+0.615	−0.551	1.800
$FHGe=CH_2$	1.764	1.518	1.755	1.081	1.080	133.2	120.0	119.3	121.5	+0.753	−0.519	1.998
$H_2Ge=CHF$	1.794	1.518	1.518	1.085	1.353	117.9	122.3	124.3	123.4	+0.172	+0.036	1.624

[a] At $B3LY_p$/6-311G(d). From Reference 108.

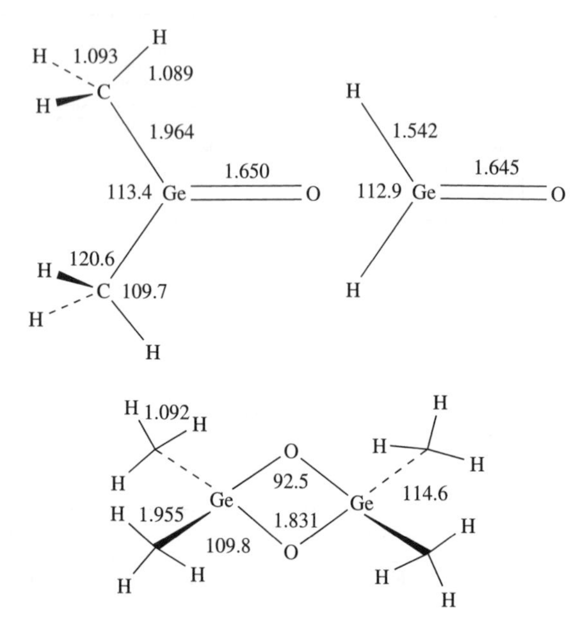

FIGURE 14. Calculated geometries of $Me_2Ge=O$, $H_2Ge=O$ and $(Me_2GeO)_2$. Bond distances in Å, angles in deg. Reprinted with permission from Reference 109. Copyright 1998. American Chemical Society

constant and bond order of the Ge=O moiety in $Me_2Ge=O$ was found to be lower than those in the parent germanone $H_2Ge=O$ and in $F_2Ge=O$[109].

The geometries of the whole series of group-14 homologues of acetone $Me_2E=O$ and the parent system $H_2E=O$ (E = C to Pb) have been calculated at different levels of theory by Kapp et al.[110]. The authors investigated also the relative energies of other isomers of the molecules and the structures and energies of the trimers $(H_2EO)_3$ as well as the hydrated species H_3EOH. Figure 15 shows the calculated species. The relative energies are given in Table 26. The theoretically predicted values show that the heavier monomeric H_2EO species with E = Si to Pb are higher in energy than the carbene-like isomers HEOH. The energy differences between the two isomeric forms of the silicon compound are rather small but they are much higher for E = Ge to Pb. It is interesting to note that the H_2EO

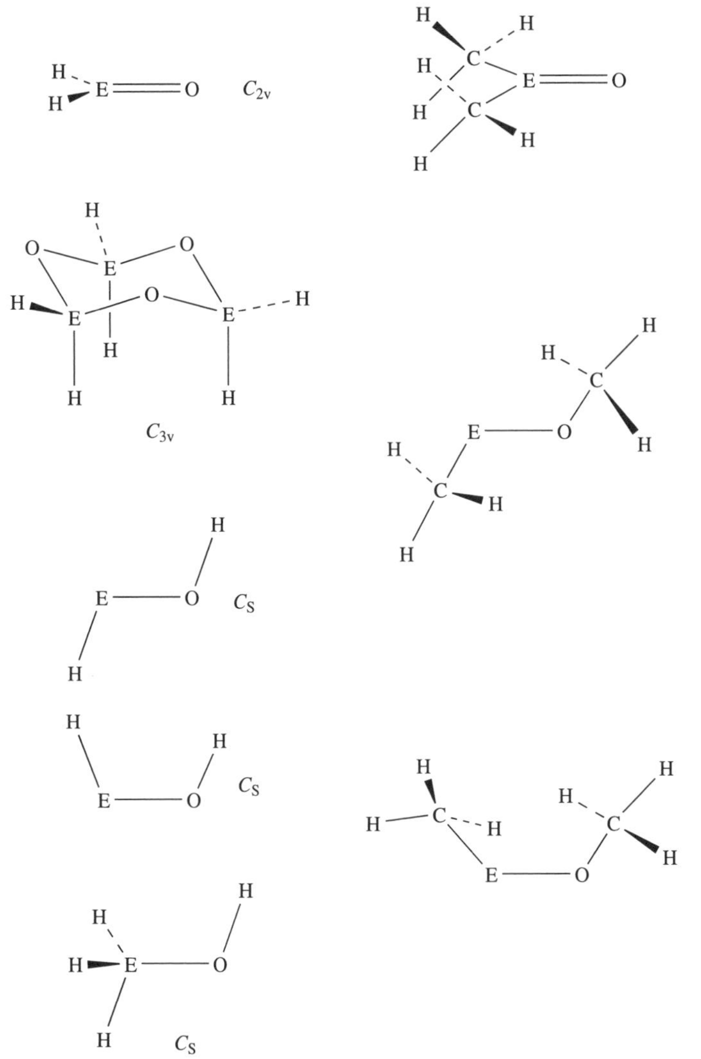

FIGURE 15. Calculated isomers of H_2EO, Me_2EO, $(H_2EO)_3$ and H_3EOH with E = C to Pb. The relative energies are given in Table 26. Reprinted with permission from Reference 110. Copyright 1996 American Chemical Society

molecules are unstable toward cyclotrimerization and hydration yielding H_3EOH for all elements E except Pb. Substitution of hydrogen by methyl groups enhances the keto form Me_2EO relative to the carbene-like isomers, but the latter remains more stable than the former for E = Ge to Pb[110].

The effect of bulky organic groups on the structures and relative stabilities of R_2Pb_2 isomers has been the subject of a theoretical study by Frenking and coworkers and was already mentioned above in the context of discussing Pb_2H_2 isomers[98]. The authors

TABLE 26. Calculated relative energies of H_2EO and Me_2EO isomers shown in Figure 15 at various levels of theory and hydration energies of H_2EO (kcal mol^{-1})[a]

	$H_2E{=}O$	*cis*-HEOH	*trans*-HEOH	*cyclo*-$(H_2EO)_3$[b]	H_3EOH[c]	$Me_2E{=}O$	*cis*-MeEOMe	*trans*-MeEOMe
				C				
MP4	0.0	59.8	54.0			0.0	75.8	66.9
CCSD(T)	0.0	58.1	52.5			0.0	74.1	65.4
B3LYP	0.0	57.5	52.2	−9.7	−25.8	0.0	72.2	63.7
				Si				
MP4	0.0	0.6	0.5			0.0	24.6	22.8
CCSD(T)	0.0	−2.6	−2.7			0.0	21.9	20.3
B3LYP	0.0	−2.1	−1.8	−66.9	−46.9	0.0	22.4	21.1
				Ge				
MP4	0.0	−26.3	−26.2			0.0	−4.8	−6.6
CCSD(T)	0.0	−30.9	−30.7			0.0	−9.1	−10.7
B3LYP	0.0	−31.1	−30.7	−53.7	−51.4	0.0	−10.2	−11.5
				Sn				
MP4	0.0	−42.1	−41.6			0.0	−22.6	−23.3
CCSD(T)	0.0	−49.7	−48.5			0.0	−29.3	−29.9
B3LYP	0.0	−49.7	−49.0	−61.7	−58.1	0.0	−31.0	−31.5
				Pb				
MP4	0.0	−64.6	−63.7			0.0	−46.7	−47.0
CCSD(T)	0.0	−71.2	−70.1			0.0	−54.1	−54.1
B3LYP	0.0	−70.7	−69.6	−46.9	−56.1	0.0	−55.3	−55.3

[a] Reference 110.
[b] Trimerisation energy per H_2EO molecule.
[c] Hydration energy.

calculated the phenyl-substituted systems Ph_2Pb_2 at the DFT (B3LYP) and MP2 levels of theory. Figure 16 shows the B3LYP optimized geometries and relative energies of stationary points on the PES. The doubly bridged isomer **A(Ph)** (see Figure 6 for the notation) remains the global energy minimum form. The DFT calculations give an asymmetrically bridged geometry which is shown in Figure 16. MP2 calculations give a symmetrically bridged structure with Pb−C distances of 2.579 Å[98]. The *trans*-bent forms **D1(Ph)** and **D2(Ph)** are still higher in energy than **A(Ph)** (compare the results for Pb_2H_2 which are shown in Figure 9). The former two species are not minima on the PES. Geometry optimization at B3LYP of the sterically much more crowded molecule Ar*PbPbAr* (Ar* = 2,6-$Ph_2C_6H_3$) gave the *trans*-bent form **D2(Ar*)** shown in Figure 16 as an energy minimum structure. The calculated bond lengths and angles of **D2(Ar*)** are in very good agreement with the values of the X-ray structure analysis of 2,6-$Tip_2H_3C_6PbPbC_6H_3$-2,6-Tip_2 (Tip = 2,4,6-$(i\text{-Pr})_3C_6H_2$)[111].

The conformational profile of 2,3-digermabutadiene $H_2C{=}GeH{-}HGe{=}CH_2$ has been investigated by Jouany et al. and compared with the parent butadiene molecule at the MP4 level using HF optimized geometries[112]. The authors found that the conjugation about the central Ge−Ge bond is significantly decreased. The relatively long Ge−Ge bond reduces the steric hindrance which is causing the s-*cis* isomer of butadiene to distort into a nonplanar *gauche* form. Therefore, 2,3-digermabutadiene has two planar energy minimum conformations. Figure 17 shows the calculated conformational pathways of 2,3-digermabutadiene and butadiene. Energy minima of 2,3-digermabutadiene are predicted when the rotational angle Φ is either 0° (s-*trans*) or 180° (s-*cis*). The calculated energies at MP4 including ZPE corrections predict that the s-*cis* form of 2,3-digermabutadiene

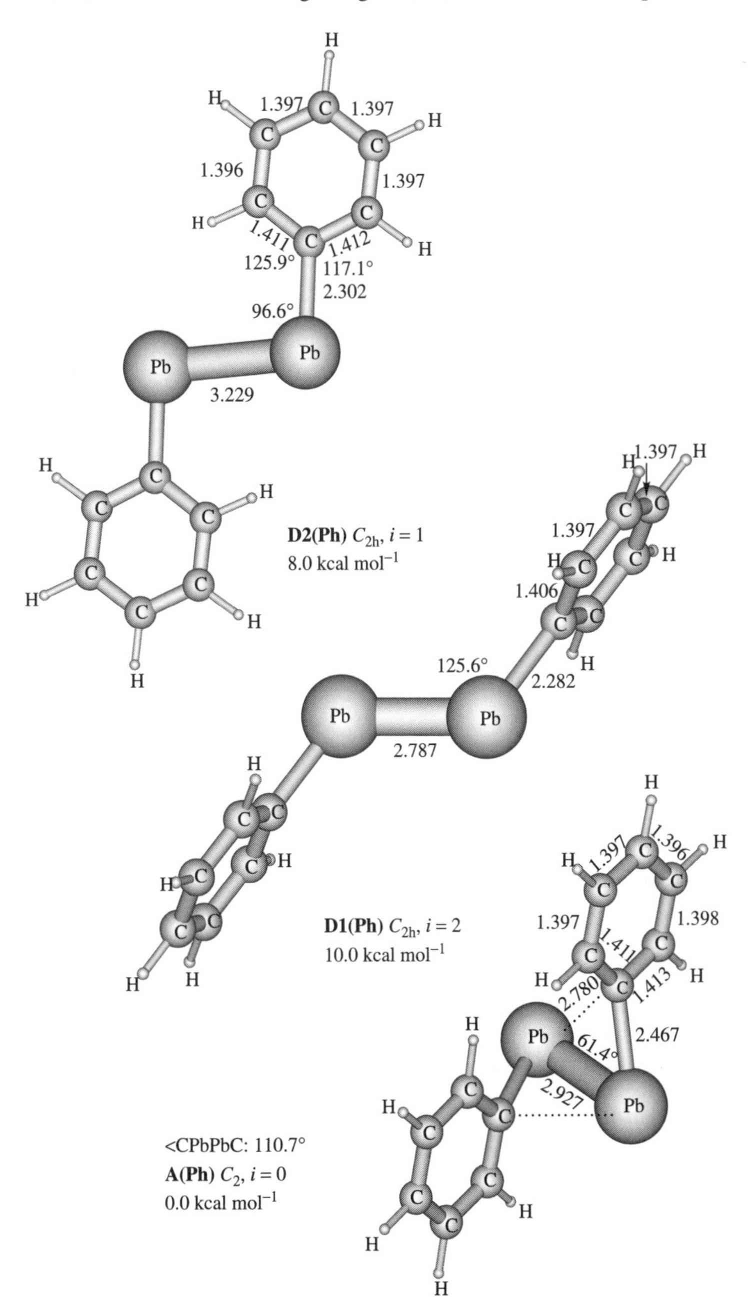

FIGURE 16. Optimized structures of Pb_2Ph_2 and $Pb_2Ph^*_2$ at B3LYP. Bond lengths in Å, angles in deg. Relative energies of Pb_2Ph_2 are given in kcal mol^{-1}. The number of imaginary frequencies i is given in *italics*. Reproduced by permission of Wiley-VCH from Reference 98

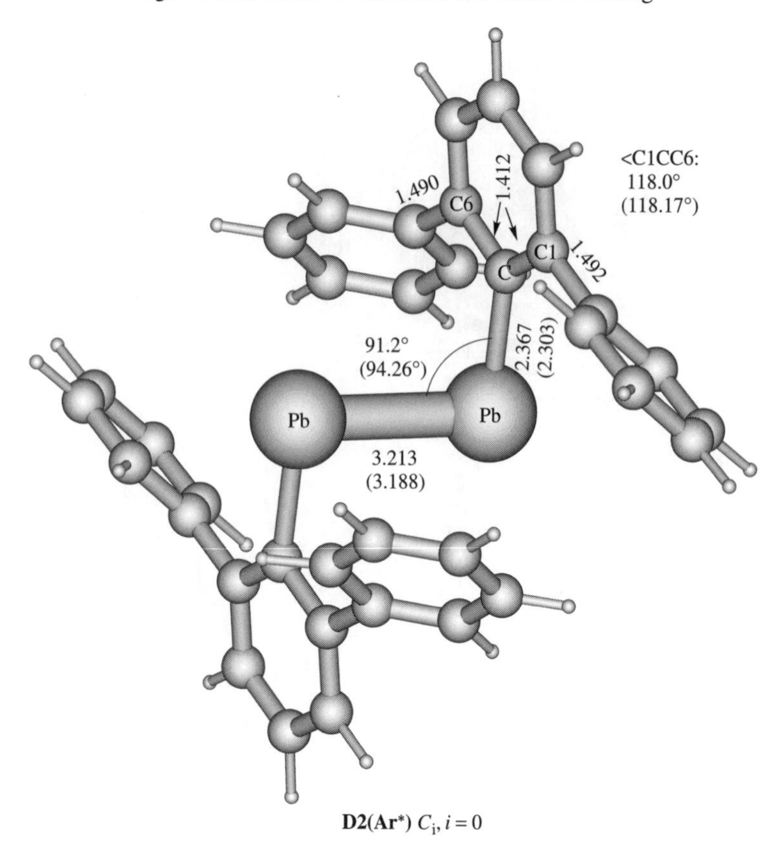

D2(Ar*) C_i, $i = 0$

FIGURE 16. (*continued*)

is 0.3 kcal mol^{-1} less stable than the s-*trans* form. The rotational barrier relative to the global energy minimum is 1.6 kcal mol^{-1} [112].

A combined experimental/theoretical study of germacyclopentadienes which includes species with Ge=C double bonds has been reported by Khabashesku et al.[113]. The authors photolyzed matrix-isolated 1,1-diazido-1-germacyclopent-3-ene and deuteriated analogues. They identified the reaction products by comparing the experimental IR spectra with calculated vibrational frequencies and IR intensities of molecules which might be formed during the reaction. Figure 18 shows the optimized geometries at the HF level of those compounds which could be identified as reaction products[113]. The authors observed during irradiation at selected wavelengths a photoconversion of **3** into **5** and a reversible interconversion of **4** and **6**, which provide experimental evidence for a germylene-to-germene rearrangement.

Comparison of theoretical data with experimental spectra led also to the identification of compounds which have germanium–nitrogen double and triple bonds. Foucat et al. report about the results of flash vacuum thermolysis of substituted germacyclopentenes and DFT (B3LYP) calculations of model compounds of possible reaction products[114]. The authors took the experimental photoelectron (PE) spectra and compared them with

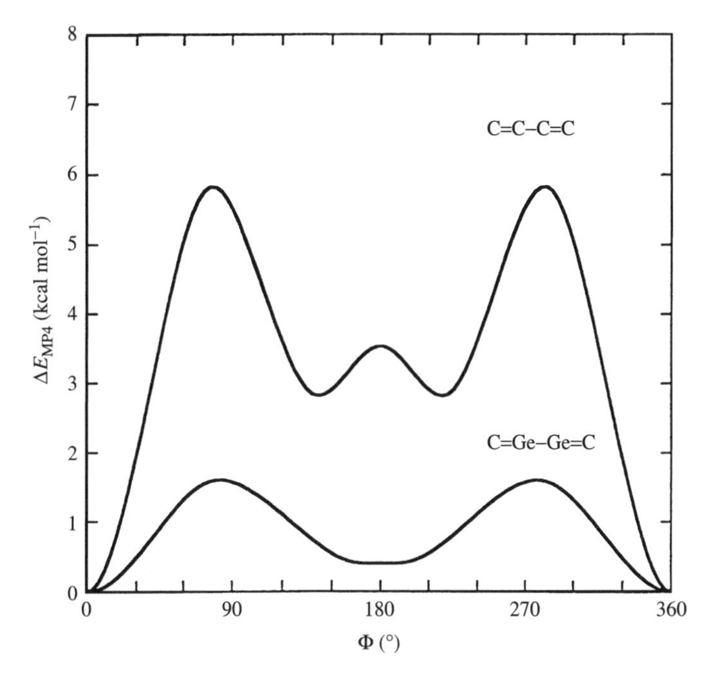

FIGURE 17. Calculated rotational profiles of butadiene and 2,3-digermabutadiene at MP4. Reprinted with permission from Reference 112. Copyright 1994 American Chemical Society

the ionization energies of model compounds (Figure 19) which were calculated by taking the energy difference between the neutral molecule and the cation. They identified the compounds **7**, **8** and **9** which have $SiMe_3$ groups instead of the less bulky substituents of the model compounds **7M**, **8M** and **9M**, respectively (Figure 19). These authors have also identified **10** which has a *t*-butyl group instead of CH_3 as in **10M** and the parent molecule GeNH (**11**). The calculated geometries of the molecules shown in Figure 19 are given in the paper[114].

Theoretical studies of group-14 organometallic compounds with multiply bonded tin and lead compounds are rare. Márquez et al. calculated at the HF level the structures and relative energies of vinylstannane $H_3Sn{-}CH{=}CH_2$ and its isomers $H_2C{=}SnH{-}CH_3$ and $H_2Sn{=}CH{-}CH_3$[115]. The former molecule is predicted to be 16.4 and 26.5 kcal mol^{-1} lower in energy than the latter two isomers, respectively. The $H_2E{=}E'H{-}E$ moieties of the three isomers are planar. The authors calculated also the neutral, cation and anion of the allyl-like species $H_2SnCHCH_2$ and $H_2CSnHCH_2$ at the CASSCF level. The cations were found to be planar, whereas the neutral radical and the anion prefer distorted geometries[115]. The cations and anions of the allylic system $CH_2CHEH_2{}^{+/-}$ with E = C to Pb) have been the subject of a theoretical study at the HF and MP2 level by Gobbi and Frenking[116]. The results of the latter work will be discussed in the section about cations and anions.

Compounds with multiple bonds between a transition metal and group-14 elements as the heavier analogues of carbenes — transition metal complexes of the type $(CO)_5Cr{-}EH_2$ (E = C to Sn) — have been calculated at the DFT level by Jacobsen and Ziegler[117]. The authors report the equilibrium geometries and the transition metal–E bond dissociation

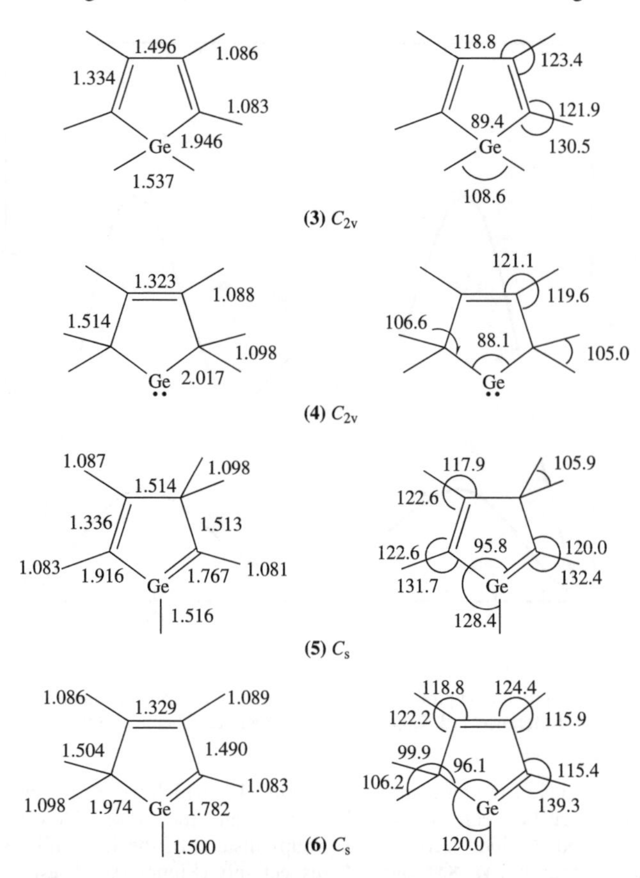

FIGURE 18. Optimized geometries of methyl-substituted germacyclopentene and germacyclopentadiene isomers. Bond lengths in Å, angles in deg. Reprinted with permission from Reference 113. Copyright 1996 American Chemical Society

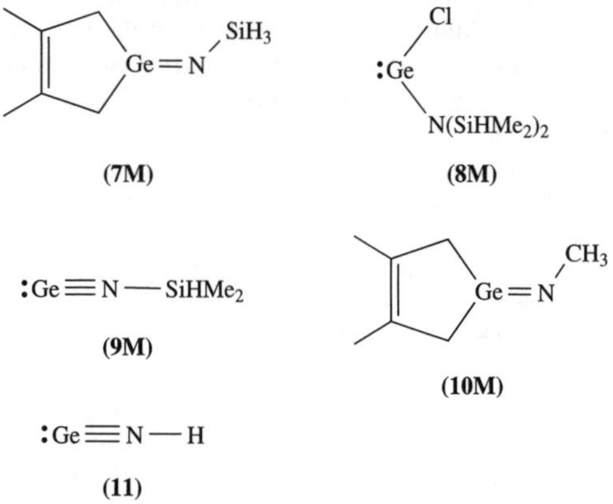

Ge=N—SiH_3

(7M)

Cl

:Ge

$N(SiHMe_2)_2$

(8M)

:Ge≡N—$SiHMe_2$

(9M)

CH_3

Ge=N

(10M)

:Ge≡N—H

(11)

FIGURE 19. Germanium–nitrogen compounds which have been investigated in Reference 114

energies. The theoretically predicted $(CO)_5Cr{-}EH_2$ bond energies decrease for the heavier group-14 elements with the trend E = C(67.2 kcal mol^{-1}) > Si(57.6 kcal mol^{-1}) > Ge(44.5 kcal mol^{-1}) > Sn(40.0 kcal mol^{-1}). The transition metal–ligand bonding interactions are analyzed with an energy decomposition scheme which shows that the intrinsic π-bonding of the Cr–E bonds has the trend E = C(48.3 kcal mol^{-1}) > Si(19.6 kcal mol^{-1}) > Ge(17.2 kcal mol^{-1}) > Sn(12.2 kcal mol^{-1})[117].

ii. Compounds with single bonds of Ge, Sn, Pb. Numerous theoretical studies of methyl derivatives of EH_4 (E = Ge to Pb) have been published in the past. Almlöf and Faegri investigated the relativistic effects on the molecular structure of EH_4 and the tetramethyl compounds EMe_4 (E = C to Pb) using the Breit-Pauli Hamiltonian and first order perturbation theory[118]. They found that relativistic effects shorten the Pb–H and Pb–C distances by 0.05 Å. The authors report also the calculated force constants for the breathing mode. The electronic structure of EMe_4 molecules for all elements E was probed in a combined experimental/theoretical investigation by Aoyama et al.[119]. The authors report also the experimental ionization spectra of the molecules. The observed bands are assigned to three orbital groups, σ_{EC}, σ_{CH} and ns(E), by comparing the experimental values with MO calculations at the HF level. With increasing size of the central atom, the relative band intensity of the spectral lines for the σ_{EC} and ns(E) orbitals decreases. This is interpreted by diminishing contributions of the electron distribution on the methyl groups for these orbitals on going from E = C to E = Pb[119].

The double ionization energies of EMe_4 (E = Si to Pb) to triplet electronic states of the dications have been calculated by Phillips et al. using a modified MSX_α method[120]. The results are used to interpret the peak positions in experimental spectra obtained by double-charge-transfer spectroscopy. The experimental spectra correlate quite well with the calculated average double-ionization energies. The geometry and vibrational spectrum of $SnMe_4$ have been calculated by Papakondylis et al. at the HF, MP2 and MP4 levels of theory[121]. The results were found to be in good agreement with experiment. The electronic excitation spectra of SnH_4 and $SnMe_4$ was the subject of a theoretical study at the SAC-CI level by Yasuda et al.[122]. The calculated results led to new assignments of the bands in the higher energy region up to the first ionization potential. DFT studies of the force constants and electrical properties of methylsilane and methylgermane have been published in a very detailed combined experimental/theoretical work by Mathews et al. which reports also the experimental IR spectrum of $^{13}CD_3GeH_3$[123].

The geometries of Me_2EX_2 with E = Si, Ge and X = F, Cl have been optimized at the CISD level of theory by Vacek et al.[124]. The authors compare the theoretical values with the results of electron diffraction data. The calculated data shown in Table 27 agree with experiment in that the C–E–C angles are significantly larger than 109°28′. On the basis of the calculated data the authors question the 124° experimental C–Ge–C angle of Me_2GeBr_2. The paper gives also the harmonic vibrational analysis at the HF level of the four compounds[124].

The calculated equilibrium geometries of the molecules Me_2ECl_2 for all elements E = C to Pb have been reported by Jonas et al.[125]. Table 28 shows generally good agreement between theory and experiment. The experimental values for the bond angles of the tin compound have large error bars. The theoretical values of the latter compound are probably more accurate than the experimental data. The authors analyzed the hybridization of the E–C and E–Cl bonds in order to find out if there is a correlation between the hybridization of the bond orbitals and the bond angles. Such a correlation is suggested by Bent's rule which states that 'Atomic s character concentrates in orbitals directed toward electropositive substituents'[126]. Because chlorine is more electronegative than carbon, it follows that the E–Cl bonds should have a lower %s character than the E–C

TABLE 27. Calculated bond lengths (Å) and angles (deg) of Me_2EX_2[a]

E	X	Theory	r(E−C)	r(E−X)	α(C−E−C)	α(X−E−X)	α(E−C−H′)	α(E−C−H″)
Si	F	CISD/DZd	1.8474	1.6001	115.9	105.7	111.0	111.5
Si	Cl	CISD/DZd	1.8537	2.0545	114.3	108.4	111.2	110.9
Ge	F	CISD/DZd	1.9246	1.7344	120.9	103.7	109.5	110.6
Ge	Cl	CISD/DZd	1.9333	2.1647	117.6	106.9	109.8	110.2

[a]Reference 124. All structures have C_{2v} symmetry.

TABLE 28. Calculated and experimental bond lengths (Å) and angles (deg) of Me_2ECl_2[a]

Structure	Method[b]	X−Cl	X−C	C−X−C	Cl−X−Cl
Me_2CCl_2	HF/II	1.798	1.521	113.0	108.3
	MP2/II	1.793	1.516	113.1	108.7
	expt	1.799	1.523	113.0(±0.4)	108.3(±0.3)
Me_2SiCl_2	HF/II	2.069	1.867	114.5	107.8
	MP2/II	2.061	1.860	114.2	108.2
	expt	2.055	1.845	114.7(±0.3)	107.2(±0.3)
Me_2GeCl_2	HF/II	2.184	1.949	118.6	106.2
	MP2/II	2.183	1.954	118.3	106.6
	expt	2.155	1.926	121.7(±1.4)	106.1(±0.6)
Me_2SnCl_2	HF/II	2.379	2.159	122.1	105.4
	MP2/II	2.380	2.161	122.0	105.9
	expt	2.327	2.109	110.1(±9.1)	107.5(±3.9)

[a]Reference 125.
[b]Basis set II which is described in Reference 4 uses ECPs for the heavy atoms and valence basis sets of DZP quality.

bonds. The calculated and observed C−E−C and Cl−E−Cl angles can be explained with Bent's rule because bond angles of sp-hybridized bonds become larger when the %s character increases. Table 29 shows that the calculated hybridization of the E−C and E−Cl bonds indeed show that the former bonds have a higher %s(E) character than the latter[125].

A recent theoretical study at the MP2 level of theory by Boyd and Boyd reported about the effects of protonation and deprotonation on the bond dissociation energies of compounds of third-row elements[127]. The authors give calculated energies of the homolytic Ge−C and C−C bond cleavage of $RGeH_3$ and $RGeH_2^-$ where R = CH_3, C_2H_5, C_2H_3, C_2H. They also give theoretically predicted bond energies of the heterolytic bond cleavages of R−GeH_3 yielding $R^+ + GeH_3^-$ and $R^- + GeH_3^+$ and for R−GeH_2^- yielding $R^+ + GeH_2^{2-}$ and $R^- + GeH_2$, respectively[127].

Several investigations have been published where the geometries of heavier group-14 molecules were determined by gas-phase electron diffraction and where the results were compared with theoretical calculations. Smart et al. report about the molecular structure of tin(II) acetate $Sn(O_2CMe)_3$ which is shown in Figure 20[128]. The calculated bond lengths

TABLE 29. Results of the NBO analysis of Me_2ECl_2 at MP2/II[a,b,c]

	E−C				E−Cl			
	% E	% s(E)	% p(E)	% d(E)	% E	% s(E)	% p(E)	% d(E)
Me_2CCl_2	52.5	31.4	68.5	0.1	46.1	18.6	81.1	0.2
Me_2SiCl_2	26.4	29.3	69.2	1.5	22.8	20.7	76.8	2.6
Me_2GeCl_2	29.1	30.7	68.8	0.5	22.3	19.3	79.4	1.3
Me_2PbCl_2	31.2	31.8	68.2	0.0	18.6	18.2	81.6	0.2

[a]Reference 125.
[b]% E gives the central atom part of the E−C and E−Cl bonds; % s(E), % p(E) and % d(E) give the hybridization of the E−C and E−Cl bonds at the central atom E.
[c]Basis set II which is described in Reference 4 uses ECPs for the heavy atoms and valence basis sets of DZP quality.

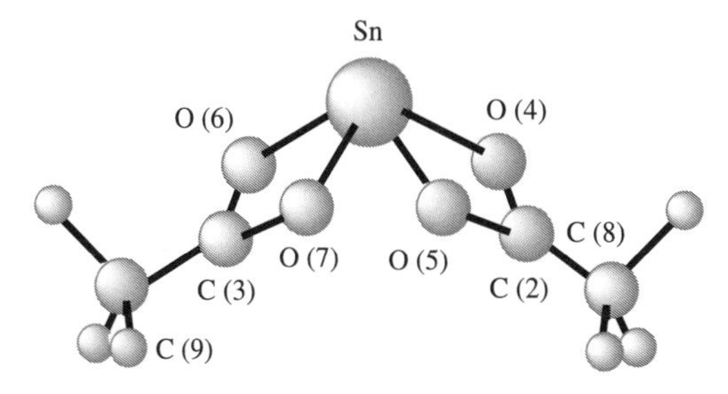

FIGURE 20. Optimized structure of tin(II) acetate reported in Reference 128. The calculated bond lengths and angles are given in Table 30

and angles agree very well with the experimental results (Table 30). Even the HF/3−21G* geometry optimization gives a rather accurate geometry.

The molecular structure of $C(GeBr_3)_4$ has also been determined by gas-phase electron diffration and by DFT calculations by Haaland et al.[129]. Table 31 shows the experimental and theoretical bond lengths and angles of the molecule which has T symmetry. The calculated values are in very good agreement with experiment. The geometries of tetragermylmethane parent compound $C(GeH_3)_4$ and trigermylmethane $HC(GeH_3)_3$ have more recently been determined by gas-phase electron diffraction and by DFT calculations by Kouvetakis et al.[130]. Figure 21 shows the calculated and experimental bond lengths and angles.

A paper by Dakkouri which reported about the molecular structure of (trifluorosilylmethyl)cyclopropane gave also the results of HF calculations of the conformational profiles of a series of cyclopropyl-CH_2-EH_3 (E = C to Ge) and R-EX_3 (R = methyl, cyclopropyl, cyclopropyl-CH_2-; E = C to Ge; X = H, F)[131]. The geometry of a methyl-substituted derivative of germacyclobutane determined by gas-phase electron diffraction and HF calculations was reported by Haaland et al.[132]. Figure 22 shows some relevant bond lengths and angles. The theoretical and experimental values are in very good agreement.

The compounds Me_3EONMe_2 with E = Si, Ge have been synthesized and the geometries of the molecules were determined by gas-phase electron diffraction and by MP2/6-31G(d) calculations by Mitzel et al.[133]. The structures of the molecules are interesting because weak attractive interactions by the nitrogen atom of the $ONMe_2$

TABLE 30. Calculated and experimental bond distances (Å) and bond angles (deg) of tin(II) acetate shown in Figure 20[a]

Parameter	HF/3–21G*	HF/DZ(P)	MP2/ DZ(P)	Expt[b]
Sn–O(4)	2.312	2.359	2.352	2.337 (12)
Sn–O(5)	2.119	2.145	2.184	2.192 (8)
C(2)–O(4)	1.254	1.236	1.271	1.245 (5)
C(2)–O(5)	1.296	1.272	1.301	1.275 (5)
C(2)–C(8)	1.489	1.500	1.503	1.510 (5)
C(8)–H (mean)	1.081	1.083	1.094	1.121 (10)
C(2)–Sn–C(3)	94.6	97.3	96.3	95.1 (13)
O(4)–Sn–O(5)	58.0	56.8	58.3	58.1 (2)
O(4)–Sn–O(6)	123.3	124.4	123.2	121.0 (4)
O(4)–Sn–O(7)	81.3	83.6	81.8	80.0 (4)
O(5)–Sn–O(7)	88.4	90.0	90.8	90.0 (3)
Sn–O(4)–C(2)	89.5	88.0	88.0	86.2 (6)
Sn–O(5)–C(2)	97.3	97.2	94.8	93.5 (4)
O(4)–C(2)–O(5)	115.3	118.0	118.8	122.0 (4)
O(4)–C(2)–C(8)	124.5	122.9	122.3	120.0 (3)
O(5)–C(2)–C(8)	120.3	119.1	118.9	117.0 (3)
C(2)–C(8)–H (mean)	109.4	109.8	109.3	111.6 (11)
$\tau(O_2CCH_3)$[c]	20.2	20.0	18.5	16.8 (11)
O(4)–Sn–C(2)–O(5)	179.6	178.6	178.2	176.3 (16)

[a]Reference 128.
[b]Gas-phase electron diffraction.
[c]Bending angle of the methyl group given by the angle between the midpoint of the O^4–O^5 axis, C^2 and C^8 (see Figure 20).

TABLE 31. Calculated and experimentally interatomic distances (Å), bond and torsion angles (deg) of $C(GeBr_3)_4$[a]

	Expt[b]	DFT
C–Ge	2.042(8)	2.051
Ge–Br	2.282(3)	2.297
Ge···Ge	3.33 (1)	3.348
C···Br	3.61 (1)	3.627
Br···Br[c]	3.64 (1)	3.663
Ge···Br	3.77 (2)	3.873
Ge···Br	4.58 (2)	4.490
Ge···Br	5.32 (1)	5.386
Br···Br	3.48 (6)	3.926
Br···Br	4.11 (4)	3.909
Br···Br	6.03 (2)	6.170
Br···Br	6.12 (2)	6.203
Br···Br	6.33 (4)	6.103
Br···Br	7.20 (2)	7.252
∠(C–Ge–Br)	112.9 (5)	112.9
∠(Br–Ge–Br)	105.9 (5)	105.8
τ(Ge–C–Ge–Br)[d]	31.4 (9)	39.6

[a]Reference 129.
[b]Gas-phase electron diffraction.
[c]Within a CBr_3 group.
[d]Torsion angle.

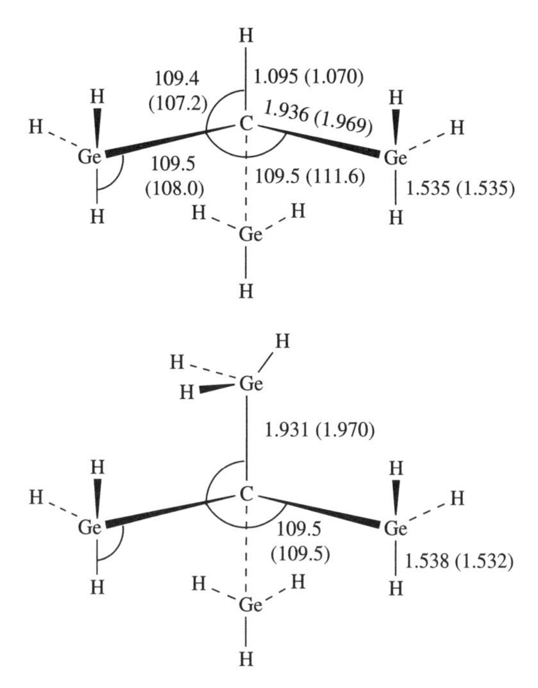

FIGURE 21. Calculated bond lengths and angles of $HC(GeH_3)_3$ and $C(GeH_3)_4$ at the DFT (B3PW91) level. Experimental values obtained from gas-phase electron diffraction are given in parentheses. Bond distances are in Å, angles in deg. Reprinted with permission from Reference 130. Copyright 1998. American Chemical Society

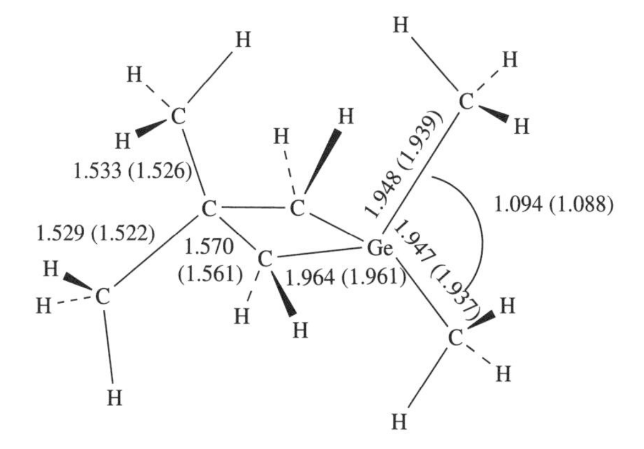

FIGURE 22. Calculated bond lengths and angles of 1,1,3,3-tetramethylgermacyclobutane at the HF level. Experimental values obtained from gas-phase electron diffraction are given in parentheses. Bond distances are in Å, angles in deg. Reprinted from Reference 132 with permission from Elsevier Science

substituent lead to pentacoordinated atoms E. Figure 23 shows the relevant calculated and experimental values of the bond lengths and angles. The calculated Si−N distance of the silicon compound is in good agreement with experiment, but the theoretical Ge−N distance of the Ge compound is *ca* 0.1 Å too long. The authors discuss the difference between theory and experiment. They point out that earlier calculations showed that the 6-31G(d) basis set is not large enough to give good geometries of molecules which have β donor interactions as in Me_3EONMe_2[133]. Calculations with the larger basis set 6-311G(d) were not possible because of the size of the molecules.

The insufficient size of the 6-31G(d) basis set for calculating β donor interactions came also to the fore in a paper by Feshin and Feshina who optimized the geometry of $Cl_3GeCH_2CH_2C(O)NH_2$ at the HF/6-31G(d) and B3LYP/6-31G(d) levels of theory[134]. Figure 24 shows the relevant calculated (B3LYP/6-31G(d)) and experimental bond lengths. It becomes obvious that the calculated Ge−O distance of the pentacoordinated Ge compound is much too long. The HF/6-31G(d) calculations gave nearly the same value (2.607 Å) as B3LYP/6-31G(d).

A paper by Campbell et al. which reported about a combined experimental/theoretical study of cyclotrigallazane gave also calculated geometries of cyclotriborazane, cyclotrialumazane and 1,3,5-trigermacyclohexane $(H_2GeCH_2)_3$ which is isoelectronic to cyclotrigallazane[135]. The geometry optimizations were carried out for the chair and the twist-boat conformations. The calculations at the MP2 level predict that the chair conformation of $(H_2GeCH_2)_3$ is 1.5 kcal mol^{-1} lower in energy than the twist-boat conformation. The theoretically predicted and experimentally observed bond distances

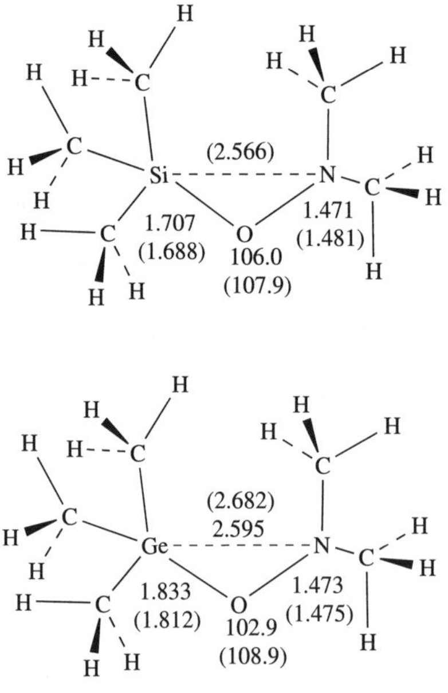

FIGURE 23. Calculated bond lengths of Me_3EONMe_2 with E = Si (top) and E = Ge (bottom) at the MP2 level. Experimental values obtained from gas-phase electron diffraction are given in parentheses. Bond distances are in Å, angles in deg. Reprinted with permission from Reference 133. Copyright 1999 American Chemical Society

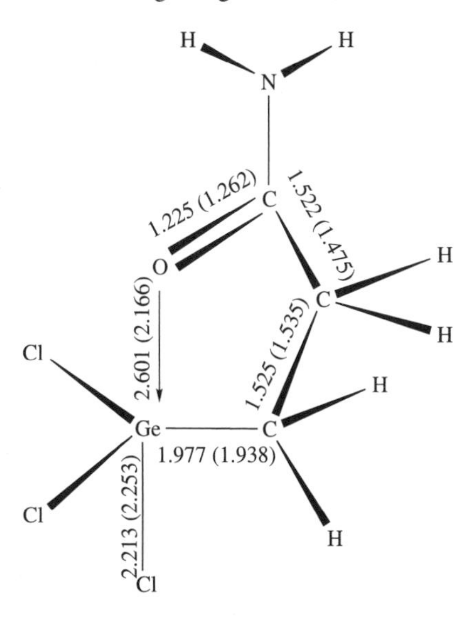

FIGURE 24. Calculated bond lengths of $Cl_3GeCH_2CH_2C(O)NH_2$ at the B3LYP/6-31G(d) level. Experimental values are given in parentheses. Bond distances are in Å. Adapted from Reference 134

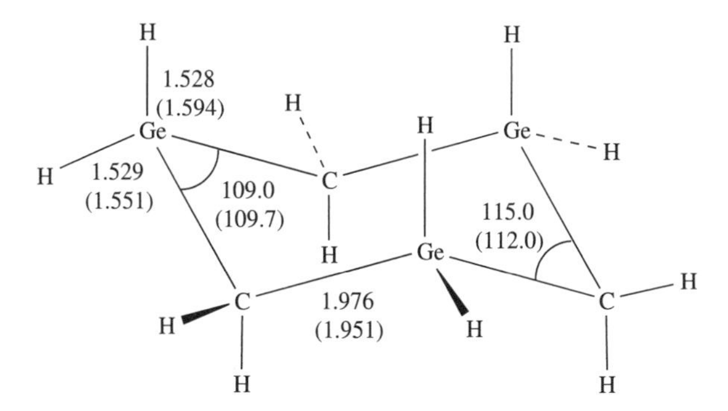

FIGURE 25. Calculated bond lengths of 1,3,5-trigermacyclohexane $(H_2GeCH_2)_3$ at the MP2 level. Experimental values obtained from X-ray structure analysis are given in parentheses. Bond distances are in Å, angles in deg. Reprinted with permission from Reference 135. Copyright 1998 American Chemical Society

and angles of the chair conformation are shown in Figure 25. The agreement between theoretical and experimental values which come from X-ray structure analysis is quite good. The larger differences of the Ge−H bond lengths are due to the experimental difficulty of locating the position of hydrogen atoms.

The influence of crystal packing forces on the molecular geometry of some diorganotin compounds has been investigated in a combined crystallographic and theoretical study by Buntine et al.[136]. The authors report about calculated geometries at the HF and DFT (B3LYP and BLYP) levels of theory of the model compounds $MeSnCl_3$ and Me_3SnCl which were taken as reference systems in order to estimate the accuracy of the theoretical values. The newly investigated tin compounds were $Ph_2Sn(S_2CNEt_2)_2$, t-$Bu_2Sn(S_2CN(c$-$Hex_2))Cl$, $Vin_2SnCl_2(bipy)\cdot 0.5C_6H_6$, $MePhSnCl_2(bipy)\cdot 0.25CHCl_3$ and $Me_2SnCl_2(phen)$ (c-Hex = cyclohexyl, bipy = 2, 2′-bipyridyl and phen = phenanthroline) where the tin atom is penta- and hexacoordinated. The authors found that the geometries of the isolated molecules are more symmetric than in the solid state and that the Sn–ligand distances tend to be shorter in the solid state than in the gas phase[136]. It has previously been shown that the latter conclusion is valid for other kinds of donor–acceptor bonds[137].

The electronic structures and geometries of six-coordinated $SnCl_2(trop)_2$ and $SnMe_2(trop)_2$ (trop = tropolone) have been calculated at the HF level by Bruno et al.[138]. The calculations predict in agreement with X-ray structure analysis that the *cis* arrangement of the tropolone ligand is more stable than the *trans* arrangement. The gas-phase UV spectra were assigned using the calculated molecular orbitals[138]. The structure of some 'paddle-wheel' tin and lead complexes $(\eta^5$-$Cp_3)E^-$ with various counterions X^+ were the topic of a combined experimental and theoretical work by Armstrong et al.[139]. Figure 26 shows schematically the calculated model compounds with the relevant bond lengths and angles. The analysis of the electronic structure shows that the naked species $(\eta^5$-$Cp_3)E^-$ (E = Sn, Pb) **13(Sn)** and **13(Pb)** are best formulated as triorganometal anions, while the unsolvated $(\eta^5$-$Cp_2)E(\eta$-$Cp)Na$ complexes **12(Sn)** and **12(Pb)** are loose-contact complexes of Cp_2E and CpNa[139]. The effect of the NH_3 molecule in **12′(Sn)** which mimics the solvent is to lengthen the Na–Cp distance and to move the bridging Cp ligand into closer contact with Sn.

Adducts of stannocene and plumbocene Cp_2E (E = Sn, Pb) with the bidentate Lewis bases TMEDA (tetramethylethylenediamine) and 4,4′-Me_2bipyl (4,4′-dimethyl-2,2′-bipyridyl) have also been studied theoretically and experimentally by Armstrong et al.[140]. Figure 27 shows the HF optimized geometries and the most important bond lengths and atomic partial charges. The authors found that the association of the metallocenes with TMEDA is energetically more favorable than with 4,4′-Me_2bipyl despite the presence of longer E–N bonds in the solid state of the TMEDA adducts. This finding was explained with the greater reorganization energy of the former Lewis base compared with the latter[140].

Another combined experimental/theoretical paper by Armstrong et al. reported about the observation of a Pb–Li bond in the complex Ph_3Pb–Li(pmdeta) (pmdeta = $(Me_2NCH_2CH_2)_2NMe$) and HF calculations of the model compounds Ph_3ELi (E = Sn, Pb)[141]. Figure 28 shows the optimized geometries and the relevant atomic partial charges of the molecules. The analysis of the E–Li bonds showed that the s and p orbitals of E are involved in the bonding interactions. Model calculations on solvated $Ph_3SnLi(NH_3)$ showed that the effect of the NH_3 ligand is a lengthening and weakening of the Sn–Li bond[141].

The heavier analogues of the Arduengo carbene imidazol-2-ylidene **14** (Figure 29) with E = Si, Ge have been the topic of several theoretical papers in the last decade. Arduengo et al. reported about photoelectron spectra and DFT (BP86) calculations of **14C**, **14Si** and **14Ge** with R = t-butyl[142]. The assignment of the PE spectra with the help of the DFT calculations showed that the first band of **14C** arises from the in-plane lone-pair orbital of the carbene carbon atom. The first bands of the silylene and germylene compounds **14Si** and **14Ge**, however, come from the π orbital of the C=C double bond. The authors analyzed the bonding situation in these compounds. They suggested that the π-bonding in the heavier homologues **14Si** and **14Ge** contributes little to the E–N interactions because the contour line diagrams of the π-valence electron density 0.7 Å above the molecular plane shows vanishing contributions by the p(π) electrons of E[142]. The authors suggested

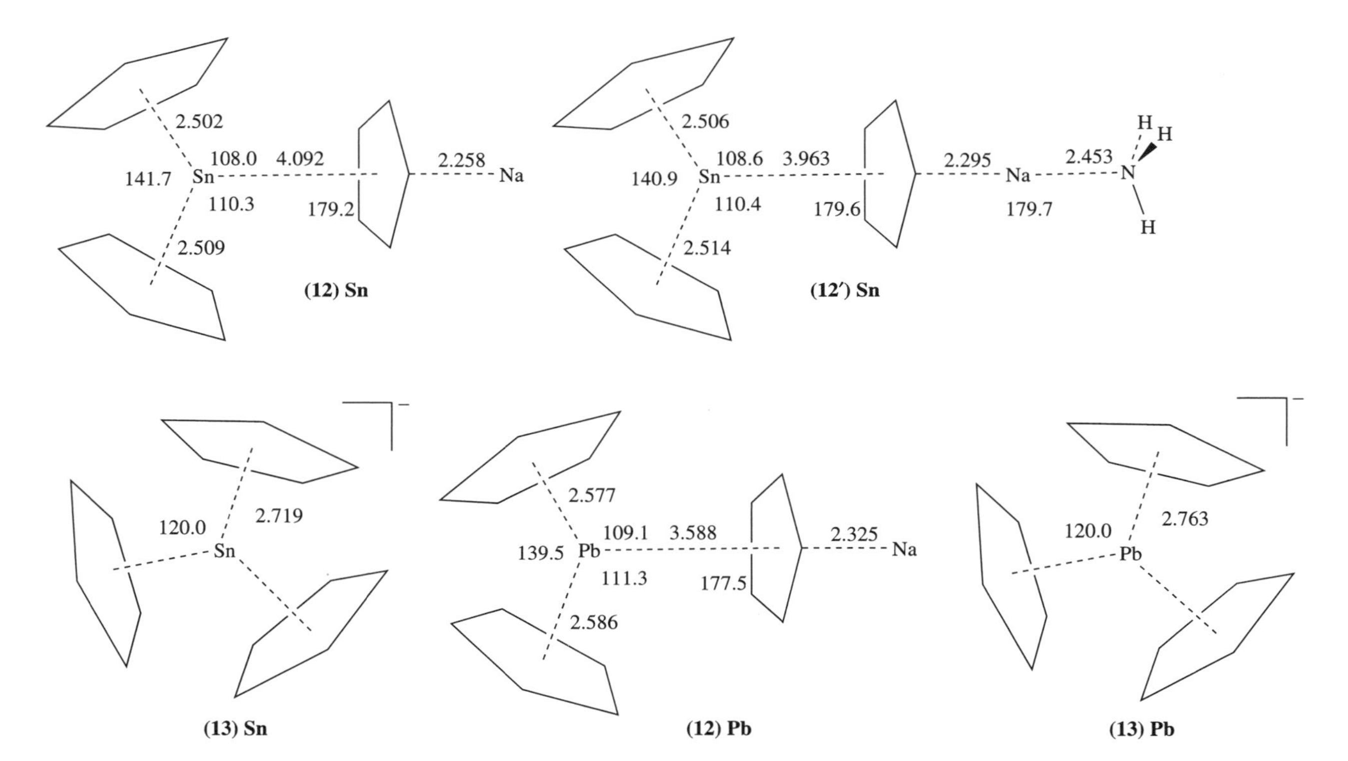

FIGURE 26. Calculated bond lengths and bond angles of Cp_3E^-, Cp_3ENa (E = Sn, Pb) and $Cp_3Na(NH_3)$ at the HF level. Bond distances are in Å, angles in deg. Reproduced by permission of The Royal Society of Chemistry from Reference 138

a 'chelated atom' bonding model for **14E** (E = Si, Ge) which is shown schematically in Figure 30.

The bonding model of Arduengo et al.[142] was later criticized in two theoretical studies of **14E** and the saturated analogues **15E** (Figure 29) with R = H at the MP4 level using MP2 optimized geometries by Apeloig, Schwarz and coworkers[143a] for E = C, Si and by Boehme and Frenking[143b] for E = C, Si, Ge. The latter authors showed that the method of electron density mapping suggested by Arduengo et al. applied to pyridine gives no significant π-electron density distribution between nitrogen and carbon which would lead to the conclusion that there is no cyclic π-delocalization in the pyridine ring. The bonding analysis by Boehme and Frenking led them to conclude that the π-delocalization becomes smaller with the trend **14C** > **14Si** > **14Ge** and **15C** > **15Si** > **15Ge** and that the unsaturated series **14E** has a more delocalized π-character than **15E**. The same conclusion was reached for the carbon and silicon species which were analyzed in the theoretical work of Apeloig, Schwarz and coworkers[143a]. Boehme and Frenking[143b] found that even the saturated cyclic molecule **15Ge** has a significantly populated germanium p(π) orbital and thus a strong N→Ge π-donation. They also optimized the geometries of numerous five-membered heterocyclic compounds, among them **14Ge**, **15Ge** and the Ge-hydrogenated compound **14GeH$_2$**[143b]. The calculated geometries are shown in Figure 31. Note that the calculated Ge−N bond length of **14Ge** is clearly longer than the experimental value. This means that the calculated p(π) charge of **14Ge** underestimates the degree of π-delocalization. Boehme and Frenking calculated also the heats of hydrogenation of the unsaturated compounds **14E** at element E yielding the tetravalent compounds **14EH$_2$**. The calculated values are -20.8 kcal mol^{-1} for E = C, -9.8 kcal mol^{-1} for

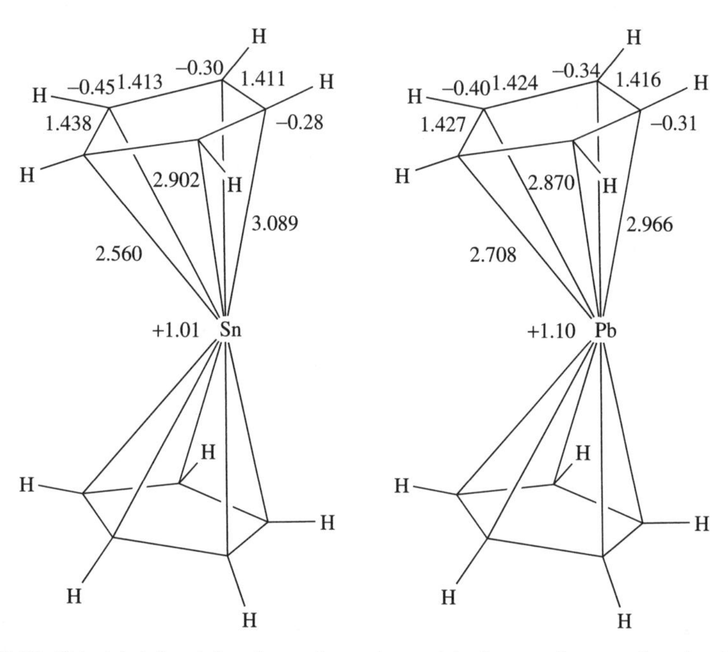

FIGURE 27. Calculated bond lengths and atomic partial charges (in parentheses) of Cp_2E, Cp_2E(TMEDA) and Cp_2E(4,4′-Me_2bipyridyl) (E = Sn, Pb) at the HF level. Bond distances are in Å. Reprinted with permission from Reference 140. Copyright 1998 American Chemical Society

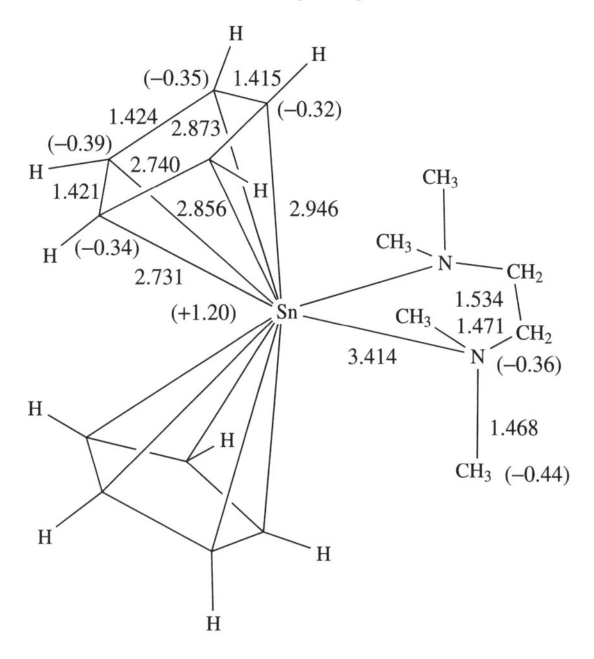

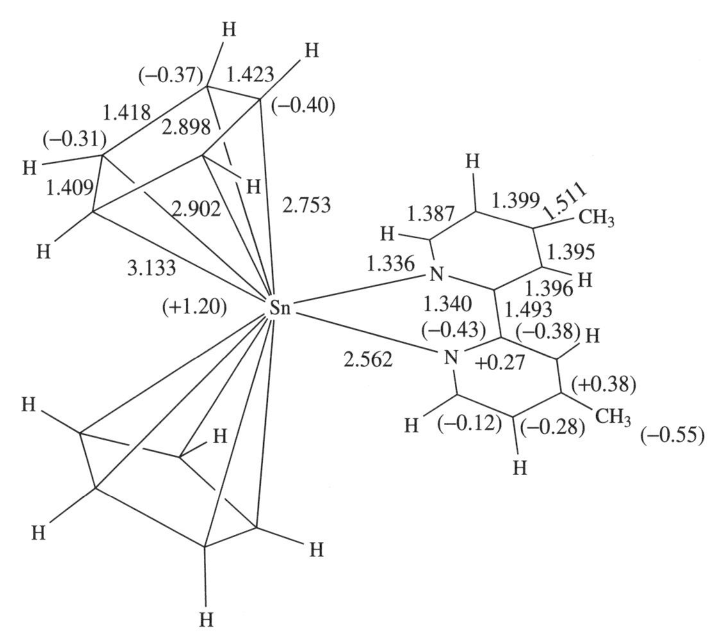

FIGURE 27. (*Continued*)

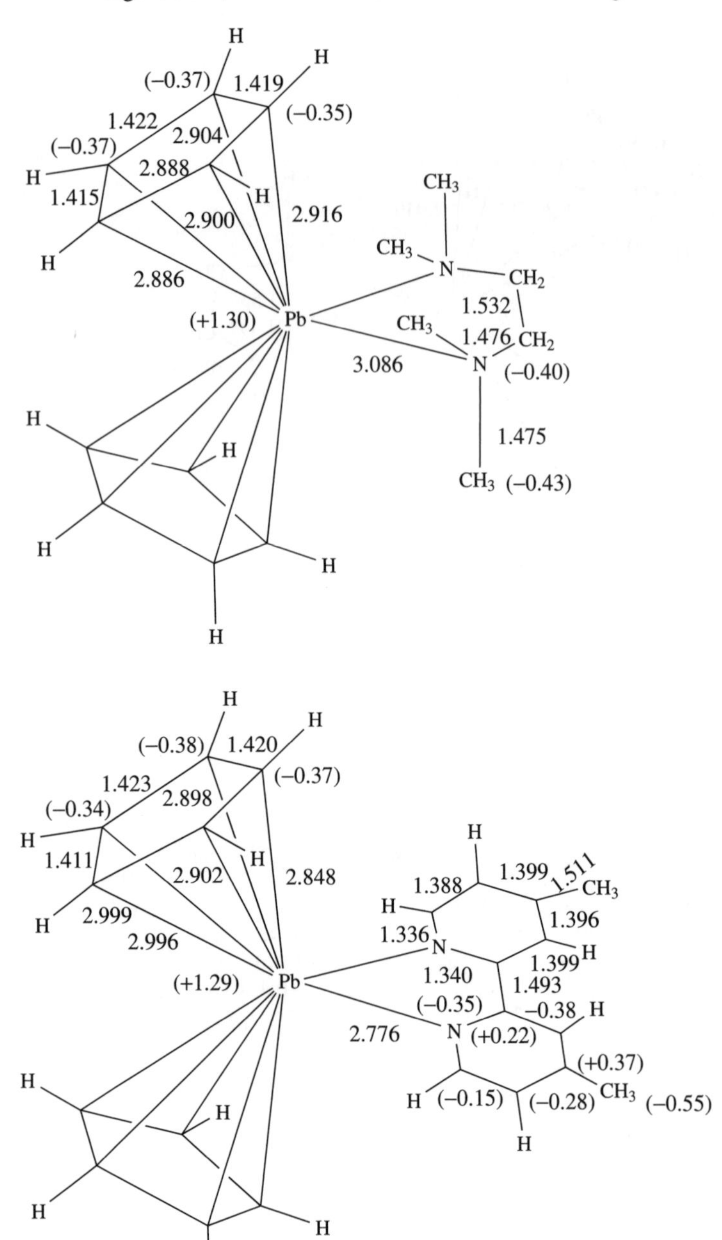

FIGURE 27. (*Continued*)

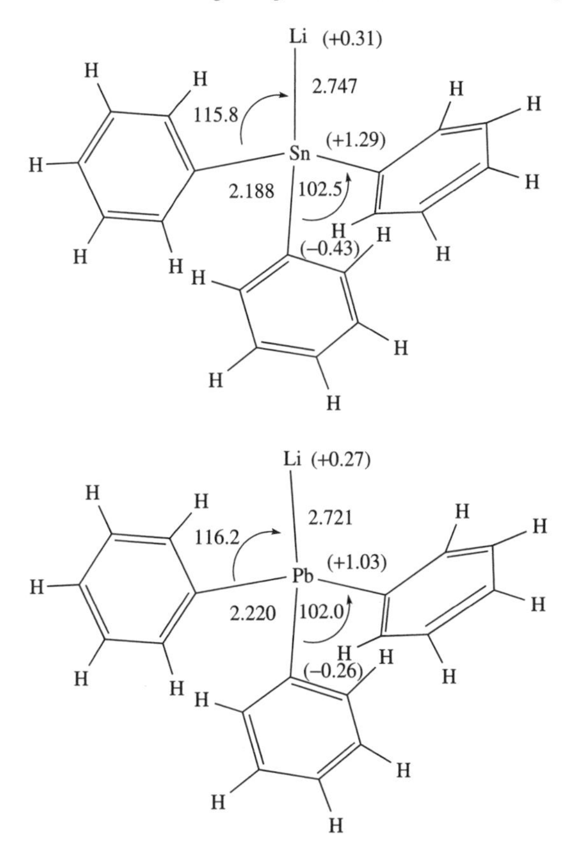

FIGURE 28. Calculated bond lengths and atomic partial charges (in parentheses) of Ph_3ELi (E = Sn, Pb) at the HF level. Bond distances are in Å, angles in deg. Reproduced by permission of The Royal Society of Chemistry from Reference 141

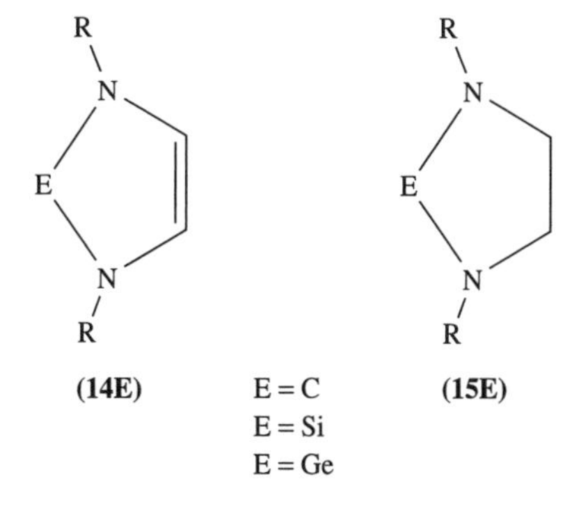

FIGURE 29. Arduengo-type carbenes with divalent atoms E = C, Si, Ge

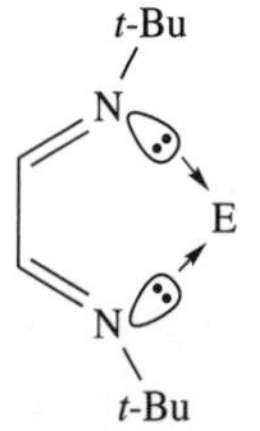

t-Bu

N

E

N

t-Bu

(14) E = Si, Ge

FIGURE 30. Bonding model of the interactions between the nitrogen atoms and the heavier elements E = Si, Ge in the imidazol-2-ylidenes, suggested by Arduengo et al. Reprinted with permission from Reference 142. Copyright 1994 American Chemical Society

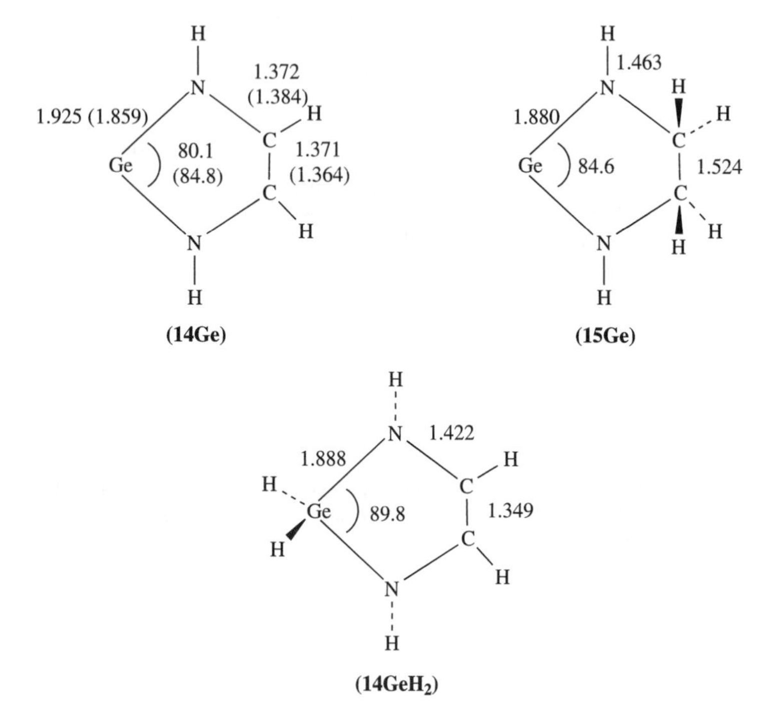

FIGURE 31. Calculated bond lengths and bond angles at the MP2 level of germaimidazol-2-ylidene **14Ge** and the hydrogenated compounds **15Ge** and **14GeH$_2$**. Bond distances are in Å, angles in deg. Reprinted with permission from Reference 143. Copyright 1996 American Chemical Society

E = Si and +23.2 kcal mol^{-1} for E = Ge[143b]. The geometries and bonding situations of stable carbenes **14C**, silylenes **14Si**, germylenes **14Ge** and **15Ge** with various substituents R have been calculated with MNDO and *ab initio* methods by Heinemann et al.[144]. The analysis of the bonding situation in the above molecules and some germanium model compounds with Ge−N bonds led the authors conclude that 'Electronic stabilization via $p\pi-p\pi$ delocalization is an important bonding feature in amino-substituted silylenes and germylenes'.

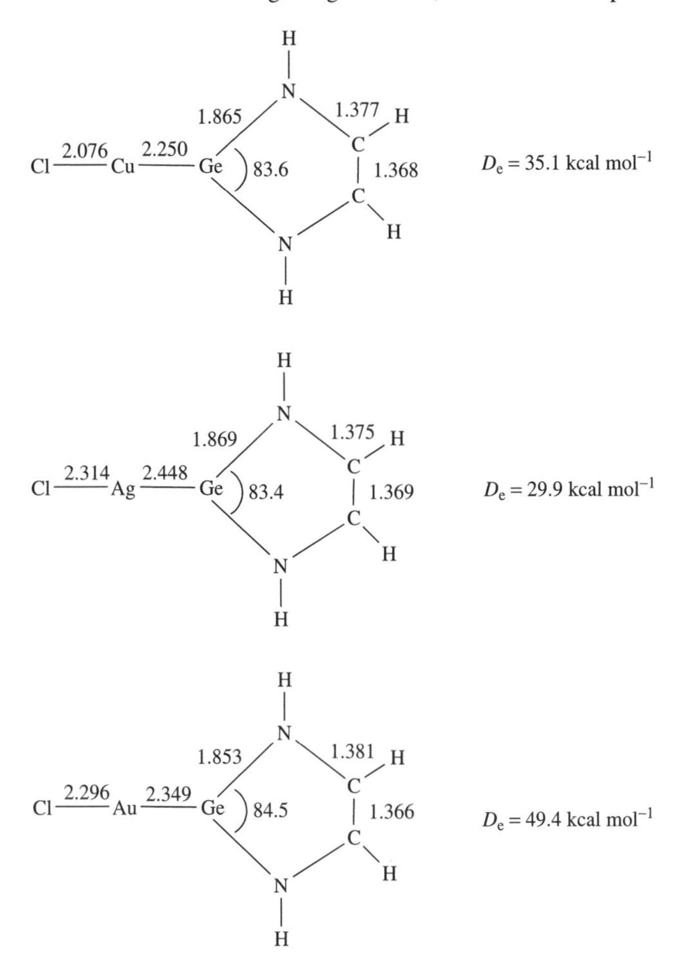

$D_e = 35.1$ kcal mol^{-1}

$D_e = 29.9$ kcal mol^{-1}

$D_e = 49.4$ kcal mol^{-1}

FIGURE 32. Calculated geometries at MP2 and transition metal (TM)–Ge bond dissociation energies D_e [at CCSD(T)] of the ClTM complexes (TM = Cu, Ag, Au) with the ligand germaimidazol-2-ylidene. Bond distances are in Å, angles in deg. Reprinted with permission from Reference 145. Copyright 1998 American Chemical Society

Donor–acceptor complexes of compounds **14E** (Figure 29) with CuCl, AgCl and AuCl as ligands have also been investigated by Boehme and Frenking[145]. The geometries were optimized at MP2 and the metal–ligand BDEs were calculated at CCSD(T). Figure 32 shows the optimized geometries and the theoretically predicted bond energies (D_e) of the germanium compounds. The strongest bond is calculated for the gold complex and the weakest bond is predicted for the silver complex. The same trend was found for the analogous carbene and silylene complexes[145]. The analysis of the metal–ligand bond showed that there is mainly ligand→metal σ-donation and very little metal→ligand π-back-donation. The authors investigated also the degree of aromaticity in the free ligands **14E** and in the ClCu-**14E** complexes using the NICS (Nuclear Independent Chemical

Shift) method suggested by Schleyer et al.[103]. The calculated NICS values indicate that the molecules **14E** have a significant aromatic character which becomes slightly enhanced in the CuCl complexes[145].

Divalent compounds of group-14 elements Si to Pb where the elements E are stabilized via intramolecular mono- and bidentate chelation, shown schematically in Figure 33, have been the subject of an extensive theoretical study by Schöller et al.[146]. The complexes are experimentally known for E = Si, Ge, Sn and L = P, but not yet for the other elements which were calculated. Table 32 gives relevant calculated bond lengths and angles. The analysis of the bonding situation shows that the central element E is weakly coordinated by the axial E–L bonds which become somewhat stronger when

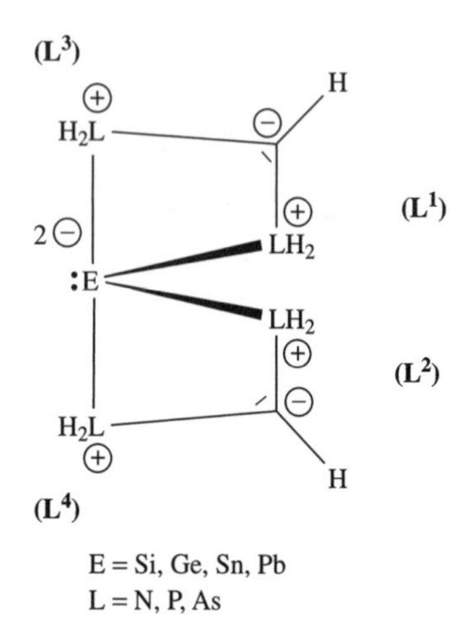

FIGURE 33. Schematic representation of the divalent group-14 compounds studied by Schöller et al. The calculated bond lengths and angles are given in Table 32. Reproduced by permission of Wiley-VCH from Reference 146

TABLE 32. Optimized bond lengths (Å) and angles (deg) of divalent group-14 compounds shown in Figure 33[a]

E	L	Symmetry	E–L^1	E–L^2	E–L^3	E–L^4	∠ L^{1}EL^2	∠ L^{3}EL^4
Si	N	C_2	1.956	1.956	2.165	2.165	96.0	142.0
	P	C_1	2.331	2.338	2.401	3.245	93.5	138.2
	As	C_1	2.435	2.459	2.502	3.514	90.6	135.2
Ge	N	C_2	2.079	2.079	2.284	2.284	94.8	137.8
	P	C_2	2.432	2.432	2.749	2.749	92.8	140.9
	As	C_1	2.525	2.548	2.664	3.200	89.6	139.3
Sn	N	C_2	2.301	2.301	2.417	2.417	89.9	130.4
	P	C_2	2.639	2.639	2.900	2.900	90.5	132.7
	As	C_2	2.731	2.731	3.008	3.008	87.8	133.8
Pb	N[b]	C_2	2.420	2.420	2.462	3.462	85.1	131.8
	P	C_2	2.691	2.691	2.929	2.929	90.6	131.6
	As	C_2	2.776	2.776	3.021	3.021	89.1	134.1

[a]Reference 146.
[b]Not an energy minimum structure.

$L = N$. The authors calculated also the barriers for the degenerate rearrangement of the complexes[146].

The stability of divalent group-14 diyl compounds $E(PH_2)_2$ with respect to isomerization to the systems with E=P double bonds $(PH_2)HE{=}PH$ (E = Si to Pb) has been reported in a combined experimental/theoretical work about diphosphanyl and diarsanyl substituted carbene homologues by Driess et al.[147]. Table 33 shows the calculated energies of the adducts, transition states and products of the rearrangement of the model compounds. It becomes obvious that the stability of the diyl form $E(PH_2)_2$ **F** relative to **G** increases with Si < Ge < Sn < Pb.

Several authors investigated also the electronic structure of group-14 organometallic compounds. Day et al. reported calculations using DFT, Hartree–Fock and semiempirical (PM3) methods of the structures and absorption spectra of metal phtalocyanine complexes of copper, tin and lead in the gas phase and in solution[148]. The solvent effect was treated with the COSMO model[149]. The electronic spectra were calculated with the ZINDO method. The optical spectrum of a nickel porphyrazine complex which has four bulky $Sn(t\text{-}Bu)_2$ substituents coordinated at the *meso*-nitrogen atoms was calculated by Liang et al. using local DFT[150]. The theoretical optical spectra including oscillator strengths were found to be in good agreement with experimental absorption. The electronic structure of tin acetylides $Sn(C{\equiv}CMe)_4$ and $Sn(C{\equiv}CSiMe_3)_4$ was probed in a combined theoretical/experimental work by HF calculations and gas-phase UV PE spectroscopy by Andreocci et al.[151]. The calculated MO energy levels of the valence electrons were used to assign the experimentally observed bands. The same procedure was used by Aoyama et al. who calculated the compounds Me_3EPh (E = C to Pb) at the HF level in order to assign the bands which were observed in ionization electron spectra measurements of the compounds[152].

The geometries and electronic structures of group-14 metalloles from silole to stannole, together with the parent cyclopentadiene, having two thienyl groups at the 2,5-positions have been investigated in a combined experimental/theoretical study by Yamaguchi et al.[153]. The authors give the experimental and theoretical geometries at the DFT (B3LYP) level and the UV-Vis and fluorescence spectra of the molecules. Figure 34 shows the calculated energy levels of the HOMO and LUMO. The authors found that the central

TABLE 33. Calculated energies (kcal mol^{-1}) of the diyls **F** and the transition states **G** for its rearrangement to the doubly-bonded isomers **H**[a]

$:E(PH_2)_2$ → $[H_2P{-}E{=}P{-}H \cdots H]^{\ddagger}$ → $(H)(H_2P)E{=}PH$

F **G** **H**

E	F	G(TS)	H
Si	0.0	29.2	−18.2
Ge	0.0	40.4	−2.1
Sn	0.0	54.6	14.8
Pb	0.0	70.6	32.4

[a] Reference 147.

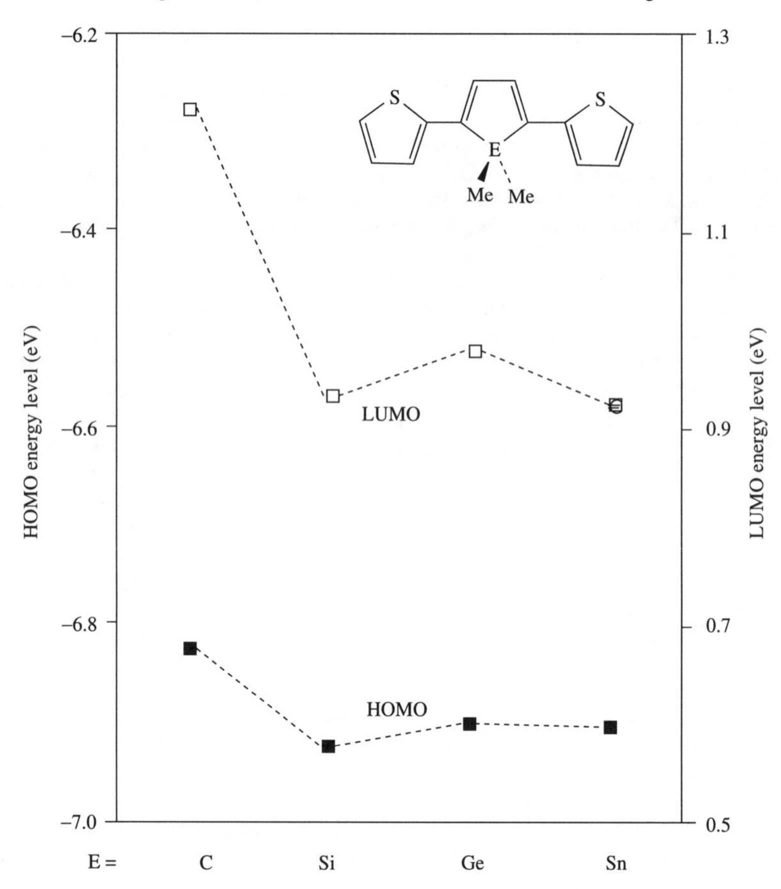

FIGURE 34. Calculated energy levels at B3LYP of the HOMO and LUMO of group-14 metalloles which have two thienyl substituents at the 2,5 position with E = C to Sn. Reprinted with permission from Reference 153. Copyright 1998 American Chemical Society

group-14 elements Si, Ge and Sn affect the LUMO energy levels of the π-electron system to almost the same extent through σ^*–π^* conjugation[153]. As a consequence, the systems with E = Si to Sn have comparable absorption maxima in the UV-Vis spectra, while the absorption maximum of the Cp parent system lies at much shorter wavelengths.

Theoretical studies have been undertaken in order to investigate the aromatic character of analogous compounds of organic molecules where carbon is substituted by heavier group-14 elements Si to Pb. Baldridge and Gordon published in 1988 a theoretical study at the HF level of potentially aromatic metallocycles[154]. A more recent work by Goldfuss and Schleyer at the DFT (B3LYP) level focused on the structures and the bonding situation in the neutral and positively and negatively charged group-14 metalloles which are derived from cyclopentadiene and the heavier analogues[155]. The authors used the NICS method[103] and structural and energetic criteria for analyzing the degree of aromaticity in the cyclic compounds $C_4H_4EH_2$, $C_4H_4EH^-$, $C_4H_4EH^+$, C_4H_4EHLi, $C_4H_4ELi_2$ and the singlet and triplet state of C_4H_4E (E = C to Pb). Figure 35 shows the optimized

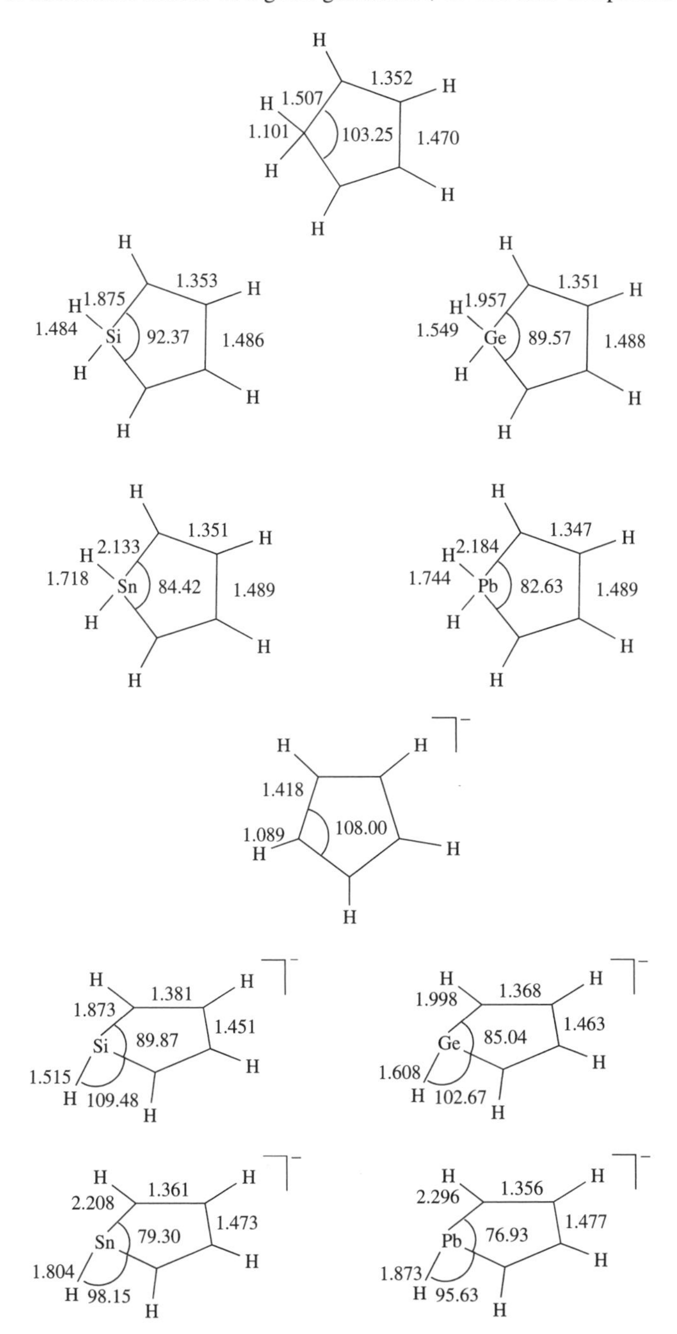

FIGURE 35. Calculated bond lengths and bond angles at B3LYP of the group-14 metalloles $C_4H_4EH_2$, $C_4H_4EH^-$ and $C_4H_4EH^+$. Bond distances are in Å, angles in deg. Reprinted with permission from Reference 155. Copyright 1997 American Chemical Society

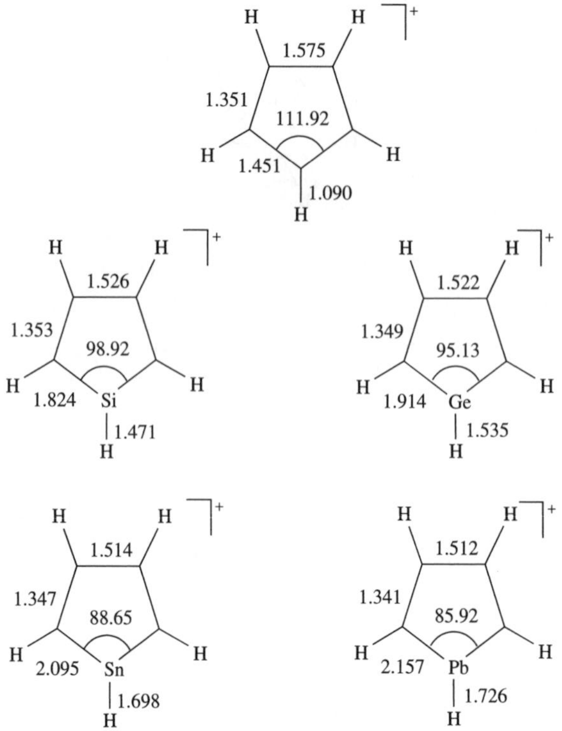

FIGURE 35. (*Continued*)

geometries of $C_4H_4EH_2$, $C_4H_4EH^-$, $C_4H_4EH^+$. The neutral parent systems and the cations have a planar C_4E skeleton, but the heavier $C_4H_4EH^-$ anions have a pyramidal environment at the heteroatom E = Si to Pb[155]. The increasing pyramidality at E down group-14 elements results in strongly decreased aromaticity of metallolyl anions $C_4H_4EH^-$. In contrast, calculations of the anions with enforced planar geometries are significantly more aromatic. The antiaromatic character of $C_5H_5{}^+$ becomes much less in the heavier analogues $C_4H_4EH^+$. The C_4H_4E species in the singlet state exhibit nearly as localized geometries as the $C_4H_4EH^+$ cations, but the C_4H_4E triplets are more delocalized. The geometries of the lithiated species C_4H_4EHLi and $C_4H_4ELi_2$ exhibit $C_4H_4EH^-$ (in case of C_4H_4EHLi) or $C_4H_4E^{2-}$ (in case of $C_4H_4ELi_2$) cyclic moieties which are capped by one or two Li^+, respectively[155].

The structures and aromatic character of tria- and pentafulvenes and their exocyclic silicon, germanium and tin derivatives (Figure 36) has been the subject of a quantum chemical study at *ab initio* levels of theory using various correlated methods and DFT (B3LYP) by Saebø et al.[156]. The calculations predict that the triafulvenes with E = Si to Sn have nonplanar geometries and that the equilibrium structures have *trans*-bent conformations as shown in Table 34. The pentafulvenes have planar geometries. The authors take the bond alternation and the charge distribution in the compounds as criterion for assigning the molecule as more or less aromatic. They suggest that there is some contribution from 2-π-aromatic resonance forms in the heterosystems of the triafulvenes

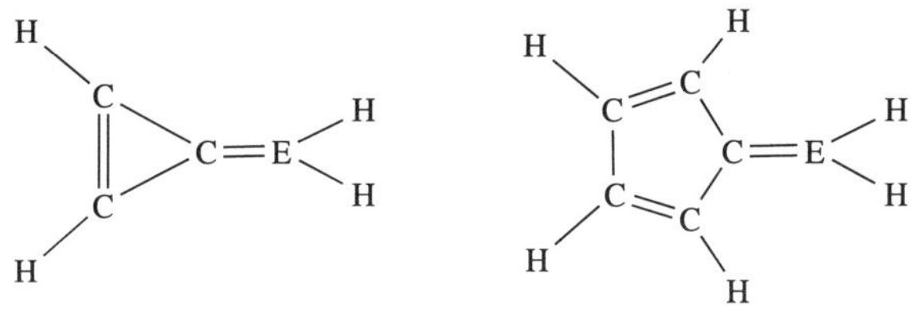

FIGURE 36. Sketch of the calculated tria- and pentafulvenes with E = C to Sn[156]. Table 34 shows the optimized bond lengths and angles at B3LYP

TABLE 34. Schematic representation of the *trans*-bent conformations of the 4-heterosubstituted triafulvenes shown in Figure 36 calculated at the MP2/TZ+2P level[a]

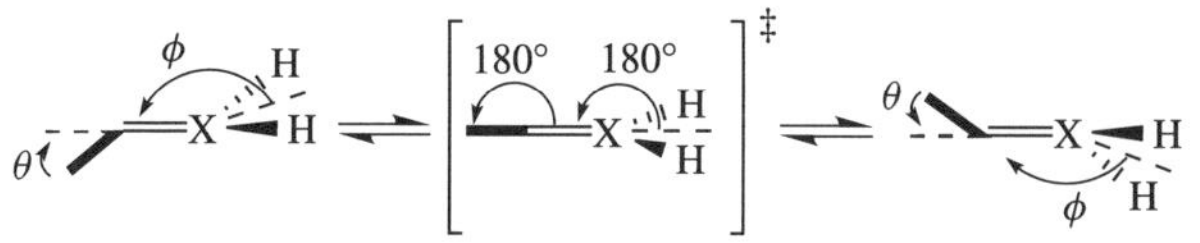

	θ	ϕ	$\Delta E^{\ddagger}$ (kcal mol^{-1})
X = Si	11.2	142.6	1.6
X = Ge	16.5	124.7	5.1
X = Sn	14.1	97.7	10.0

[a]Reference 156.

which becomes enhanced as the heteroatom becomes more pyramidal. The pentafulvene series, however, exhibits evidence for a relatively small contribution from aromatic-like resonance structures[156].

The peculiar tendency of lead to prefer the oxidation state Pb(II) in inorganic compounds while the oxidation state Pb(IV) is prevalant in organolead compounds has been the topic of a theoretical study by Kaupp and Schleyer at highly correlated MP4 and QCISD(T) levels of theory[157]. The authors calculated the structures and energies of a series of halogenated lead hydrides and methyllead compounds R_nPbX_{4-n} (R = H, Me; X = F, Cl; $n = 0$–4) and R_nPbX_{2-n} ($n = 0$–2). The relative stabilities of Pb(II) and Pb(IV) compounds were estimated by calculations of model reactions. Table 35 shows one set of reactions between tetravalent and divalent lead compounds where the products have a higher number of electronegative substituents in the tetravalent species. All reactions are endothermic. The endothermicity is particularly high for the formation of PbF_4. A simple bonding model is proposed to explain the thermodynamic observations. The increase of the positive metal charge upon halogen substitution results in greater contraction of the 6s orbtials than the 6p orbitals of Pb. Hence, the 6p orbitals are less effective in sp^n hybridization, and electronegatively substituted Pb(IV) compounds become destabilized. The proposed concept emphasizes the size difference between the s and p valence orbitals, in contrast to the traditional term 'inert pair effect' which implies that the 6s orbital is too low in energy to hybridize with the 6p orbital. Geometrical aspects and the influence of relativistic effects are also discussed[157].

The progress in calculating NMR chemical shifts of heavier nuclei made it possible to calculate ^{73}Ge and ^{119}Sn NMR chemical shifts. Figure 37 shows the comparison of

TABLE 35. Calculated energies (kcal mol^{-1}) for isodesmic reactions between divalent and tetravalent lead methyl fluorides[a,b]

	HF	MP4[c]
$(CH_3)_4Pb + PbF_2 \rightarrow (CH_3)_3PbF + CH_3PbF$	8.9	10.3
$(CH_3)_4Pb + CH_3PbF \rightarrow (CH_3)_3PbF + (CH_3)_2Pb$	9.8	10.6
$(CH_3)_3PbF + PbF_2 \rightarrow (CH_3)_2PbF_2 + CH_3PbF$	15.1	16.5
$(CH_3)_3PbF + CH_3PbF \rightarrow (CH_3)_2PbF_2 + (CH_3)_2Pb$	16.0	16.3
$(CH_3)_2PbF_2 + PbF_2 \rightarrow CH_3PbF_3 + CH_3PbF$	29.1	29.9
$(CH_3)_2PbF_2 + CH_3PbF \rightarrow CH_3PbF_3 + (CH_3)_2Pb$	29.9	30.0
$CH_3PbF_3 + PbF_2 \rightarrow PbF_4 + CH_3PbF$	48.4	47.4
$CH_3PbF_3 + CH_3PbF \rightarrow PbF_4 + (CH_3)_2Pb$	49.3	47.5
$(CH_3)_4Pb + PbF_2 \rightarrow (CH_3)_2PbF_2 + (CH_3)_2Pb$	24.9	26.4
$(CH_3)_3PbF + PbF_2 \rightarrow CH_3PbF_3 + (CH_3)_2Pb$	45.0	46.3
$(CH_3)_2PbF_2 + PbF_2 \rightarrow PbF_4 + (CH_3)_2Pb$	78.3	77.3
$(CH_3)_4Pb + 2PbF_2 \rightarrow PbF_4 + 2Pb(CH_3)_2$	103.2	103.7

[a]Reference 157b.
[b]Quasi-relativistic Pb pseudopotential used.
[c]Reactions involving $(CH_3)_4Pb$ have been treated at the MP4SDQ level, all others at MP4SDTQ.

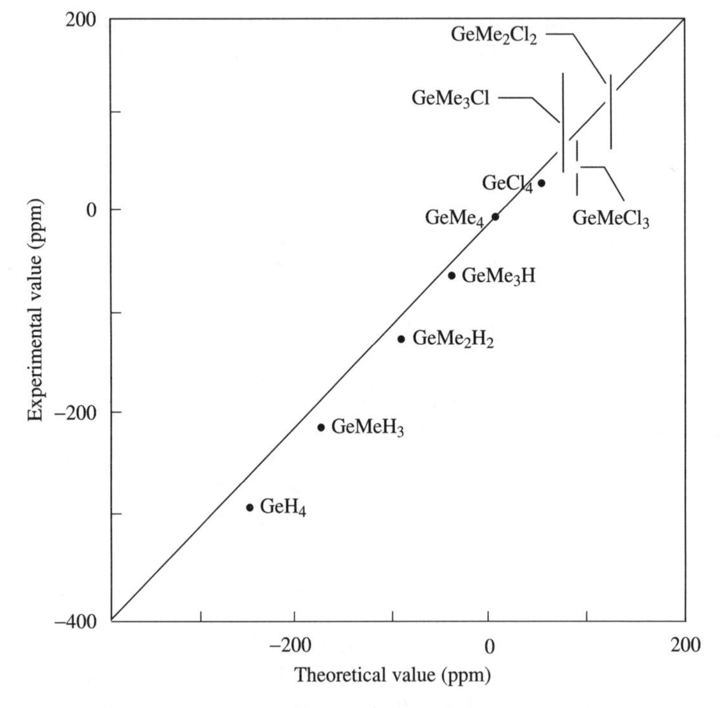

FIGURE 37. Comparison of calculated and experimental ^{73}Ge NMR chemical shifts of some simple germanium compounds. The experimental values of $GeMe_{4-n}Cl_n$ ($n = 1-3$) were not available. They have been estimated from the analogues silicon compounds using empirical correlation factors. Reproduced by permission of John Wiley & Sons, Inc. from Reference 158

theoretically predicted (using *ab initio* finite perturbation theory) ^{73}Ge chemical shifts of simple germanium compounds with experimental values[158]. It becomes obvious that the agreement between theory and experiment is quite good. The authors investigated the diamagnetic and paramagnetic contributions to the NMR chemical shifts. Nakatsuji et al. reported also about calculated ^{119}Sn chemical shifts of some simple Sn(IV) compounds[159]. Figure 38 shows a comparison of the theoretical and experimental values. The agreement is less satisfactory than for the germanium compounds. This may be due to relativistic effects which become much more important in calculating the resonances of the heavier Sn atom. Relativistic effects were neglected in the paper by Nakatsuji et al.[159].

Finally we want to mention a recent theoretical study of a hexanuclear tin cluster (Figure 39) which has been calculated at the CIS level of theory by Arnold et al[160]. The geometry of a model compound with hydrogens instead of methyls (Figure 39) was optimized at the CIS level of theory in the lowest energy ground ($^1A_1$) and excited ($^1B_1$) state. The calculations suggest that the structural distortion which is observed for the methyl substituted system is probably caused by a first order Jahn-Teller effect and not by tin-tin bonding as previously assumed[160].

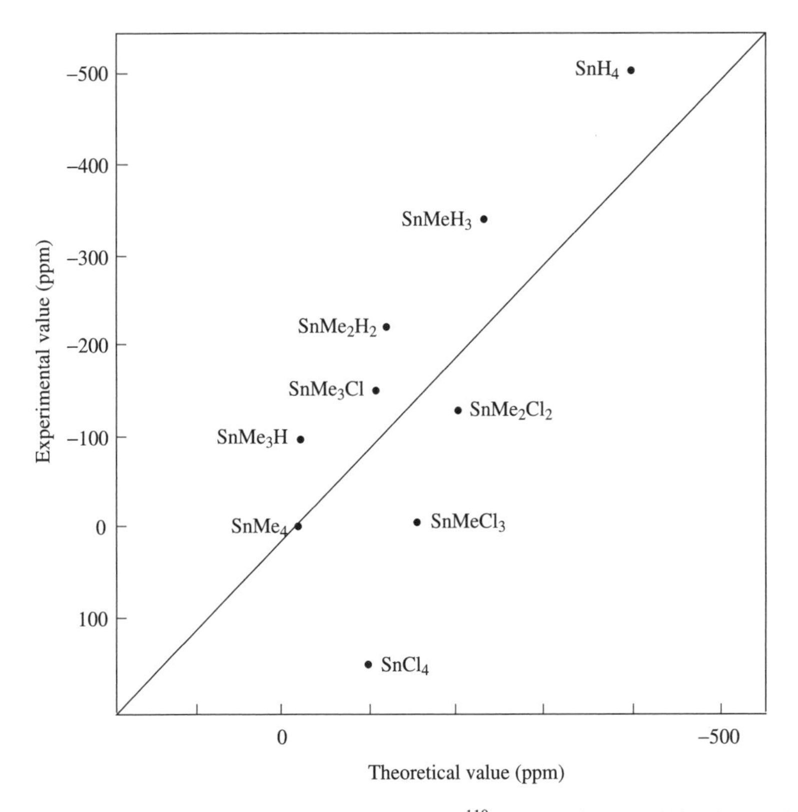

FIGURE 38. Comparison of calculated and experimental ^{119}Sn NMR chemical shifts of some simple tin compounds. Reprinted with permission of Reference 159. Copyright 1992 American Chemical Society

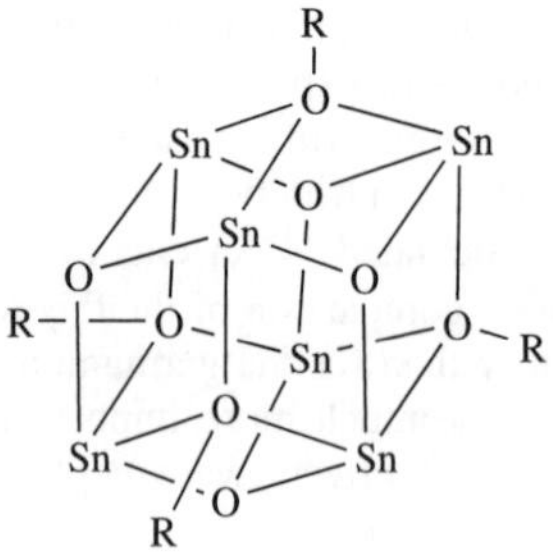

R = Me
R = H

FIGURE 39. Schematic representation of the hexanuclear tin cluster calculated in Reference 160

b. Reaction mechanisms. Organometallic compounds of germanium and tin have become important agents in many reactions and thus have been the topic of several theoretical studies. Organolead compounds play a less prominent role. This may be the reason why we could not find any theoretical work which reports about reaction mechanisms of organolead compounds.

Kudin et al. reported HF and DFT (B3LYP) calculations of the dimerization of simple germenes $H_2Ge{=}CH_2$, $MeHGe{=}CH_2$, $Me_2Ge{=}CH_2FHGe{=}CH_2$ and $H_2Ge{=}CHF$[108]. The authors report about the theoretically predicted transition states of the head-to-head and head-to-tail reactions which lead to 1,2- and 1,3-digermacyclobutane, respectively. The reaction pathways which lead to *cis* and *trans* isomers of the asymmetrically substituted germenes $MeHGe{=}CH_2$, $FHGe{=}CH_2$ and $H_2Ge{=}CHF$ have also been investigated. The calculations predict that the formation of the 1,3-digermacyclobutanes (head-to-tail reaction) has lower activation barriers and is more exothermic than the formation of the 1,2-digermacyclobutanes (head-to-head reaction), except for $H_2Ge{=}CHF$. The calculated activation parameters ($\Delta E^{\ddagger}$, and ZPE corrected activation enthalpies $\Delta H^{\ddagger}$) and the reaction energies ΔE and enthalpies ΔH are given in Table 36. The calculations predict that the head-to-head product of dimerization of $H_2Ge{=}CHF$ is more stable than the head-to-tail product. The calculations are not very conclusive about the height of the activation barriers of the two reactions. The HF calculations give a higher barrier for the head-to-head addition, but the B3LYP optimization did not give a transition state for this reaction. Figure 40 shows three types of reaction profiles which were suggested for the head-to-tail dimerization of the germenes. The calculations predict that dimerization of $H_2Ge{=}CH_2$ and $MeHGe{=}CH_2$ toward the 1,3-isomer proceeds along the A type profile, while $FHGe{=}CH_2$ should dimerize without a barrier as shown in reaction profile C. The head-to-tail dimerization process of $Me_2Ge{=}CH_2$ and $H_2Ge{=}CHF$ is predicted to proceed either along reaction path A or path B, depending on the level of theory[108].

The protoype Ge−H insertion reaction of GeH_2 with GeH_4 yielding Ge_2H_6 was studied in a combined experimental/theoretical work by Becerra et al.[161]. The calculations at the MP2 and G2 levels of theory predict that the reaction proceeds via initial formation of a weakly bonded donor–acceptor complex which may exist in two different conformations. The following rearrangement to digermane takes place with a very low (<3 kcal mol^{-1}) activation barrier. The authors give also the calculated heat of formation of GeH_2 (60.2 kcal mol^{-1}) which is in good agreement with the experimental value of $\Delta H_f^{o} = 56.6 \pm 2.7$ kcal mol^{-1}.

TABLE 36. Summary of transition state energies ($\Delta E^{\ddagger}$ and $\Delta H^{\ddagger}$) and dimerization energies (ΔE and ΔH) (kcal mol^{-1}) of the dimerization of germenes[a]

Germene	Theretical level	Head-to-tail				Head-to-head			
		$\Delta E^{\ddagger}$	$\Delta H^{\ddagger}$	ΔE	ΔH	$\Delta E^{\ddagger}$	$\Delta H^{\ddagger}$	ΔE	ΔH
$H_2Ge{=}CH_2$	(1)	4.9	6.0	−98.5	−95.1	32.5	32.9	−79.8	−75.3
	(2)	5.3	6.4	−78.7	−75.3	34.0	34.4	−76.1	−71.6
	(3)	2.5	3.6	−68.7	−65.3	20.3	20.7	−68.5	−64.0
$MeHGe{=}CH_2$	(1)	0.4	0.9	−98.9	−96.3	33.4	33.5	−79.1	−75.3
(*trans*)	(2)	2.6	3.1	−78.3	−75.7	36.7	36.8	−74.2	−70.4
	(3)	0.4	0.9	−68.5	−65.9	21.6	21.7	−67.2	−63.4
$MeHGe{=}CH_2$	(1)	0.5	1.0	−98.9	−96.3	33.5	33.6	−79.3	−75.5
(*cis*)	(2)	2.7	3.2	−78.2	−75.6	36.8	36.9	−73.8	−70.0
	(3)	0.6	1.1	−68.4	−65.8	21.7	21.8	−66.8	−63.0
$Me_2Ge{=}CH_2$	(1)	−3.8	−3.6	−99.4	−97.4	33.9	33.7	−79.0	−75.7
	(2)	0.4	0.6	−77.9	−75.9	39.3	39.1	−71.8	−68.5
	(3)	−1.2	−1.0	−68.4	−66.4	22.8	22.6	−65.3	−62.0
$FHGe{=}CH_2$	(1)	no TS		−123.0	−120.0	26.6	26.4	−93.3	−89.1
(*trans*)	(2)	no TS		−94.0	−91.0	27.9	27.7	−86.1	−81.9
	(3)	no TS		−82.8	−79.8	13.8	13.6	−78.9	−74.7
$FHGe{=}CH_2$	(1)	no TS		−122.5	−119.5	26.8	26.6	−92.0	−87.8
(*cis*)	(2)	no TS		−93.3	−90.3	28.8	28.6	−84.9	−80.7
	(3)	no TS		−82.2	−79.2	14.6	14.4	−77.7	−73.5
$H_2Ge{=}CHF$	(1)	−3.7	−3.1	−95.9	−93.1	18.7	18.9	−97.8	−93.9
(*trans*)	(2)	11.6	12.2	−77.1	−74.3	20.2	20.4	−91.6	−87.7
	(3)	<5.8[b]	<6.4[b]	−64.9	−62.1	not found		−77.6	−73.7
$H_2Ge{=}CHF$	(1)	11.7	12.3	−94.7	−91.9	13.9	14.1	−93.1	−89.1
(*cis*)	(2)	11.7	12.3	−76.4	−73.6	21.8	22.0	−90.8	−86.8
	(3)	<6.2[b]	<6.8[b]	−64.3	−61.5	not found		−77.3	−73.3

[a] At RHF/3-21G (1), RHF/6-311G(d,p) (2), and B3LYP/6-311G(d,p) (3). From Reference 108.
[b] Upper limits of the activation energy and enthalpy. The optimized structure has two imaginary frequencies. V. N. Khabashesku, personal communication to G. F.

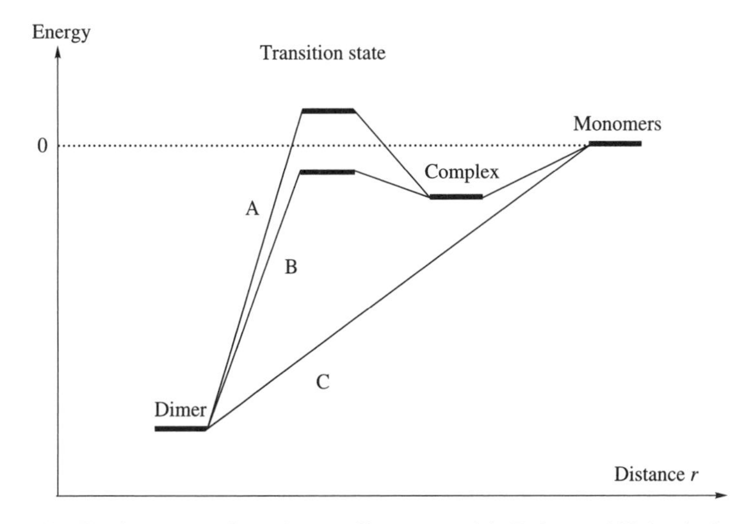

FIGURE 40. The three types of reactions profiles suggested in Reference 108 for the head-to-tail dimerization of simple germenes

The oxygen-to-carbon migration of a germyl substituent in the free anion $H_2C{-}O{-}GeH_3^-$ was investigated in a theoretical study at the CASSCF and MP4 levels of theory by Antoniotti and Tonachini[162a]. The authors calculated in a later work the same process in the presence of a lithium counterion[162b]. Figure 41 shows the theoretically predicted reaction profile for rearrangement of the GeH_3 group in the free anion and a comparison with the analogous reaction of the SiH_3 migration. The strongly exothermic rearrangement of $H_2C{-}O{-}GeH_3^-$ to $(GeH_3)H_2C{-}O^-$ proceeds with a small activation barrier of 2.1 kcal mol^{-1} which involves the rotation of the CH_2 group (the alternative pathway with CH_2 inversion has a barrier of 7.0 kcal mol^{-1}) via a cyclic structure which is, however, not an energy minimum structure. Thus, the overall reaction of the germyl anion has a very low activation barrier. The mechanisms of the analogous silyl migration involve the formation of a cyclic intermediate and is predicted to be significantly different from its germanium analogue. The authors calculated also the dissociative pathway which involves cleavage of the germanium–oxygen bond. The bond energy was found to be *ca* 10 kcal mol^{-1} which indicates that the dissociative pathway is not competitive with the direct 1,2-shift[162a].

The calculated reaction profile for the germyl migration in the presence of a lithium counterion is significantly different from that of the free anion. The reaction is less exothermic (−24 kcal mol^{-1}) than in case of the free anion (−30 kcal mol^{-1}) and the nondissociative pathway which proceeds without an intermediate has a barrier of 16.7 kcal mol^{-1} [162b]. However, the dissociative pathway has still a higher activation energy of 38.1 kcal mol^{-1}. The authors calculated also other structures on the PES and found an electrostatically bound complex of $H_2CO{-}LiGeH_3$ which is 10.1 kcal mol^{-1} lower in energy than the reactant molecule $(Li)H_2COGeH_3$ in the most stable form which has the lithium in a bridging position between carbon and oxygen[162b].

The reaction profiles of the [1+2] addition of EH_2 and EF_2 (E = C, Si, Ge, Sn) in the ($^1A_1$) singlet state to ethylene yielding the cyclopropanes cyclic-$C_2H_4EH_2$ and cyclic-$C_2H_4EF_2$, respectively, have been calculated by Sakai at the MP2 and MP4 levels of theory[87]. Figure 42 shows stationary points which were found for the reaction EH_2 +

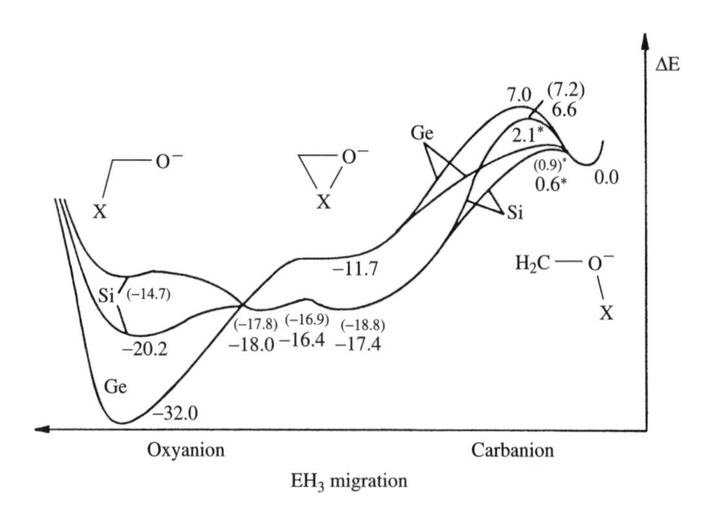

FIGURE 41. Calculated reaction profiles for rearrangement of GeH_3 and SiH_3 in the anions $CH_2OEH_3^-$. The energy values are given in kcal mol^{-1}. Reprinted with permission from Reference 162a. Copyright 1996 American Chemical Society

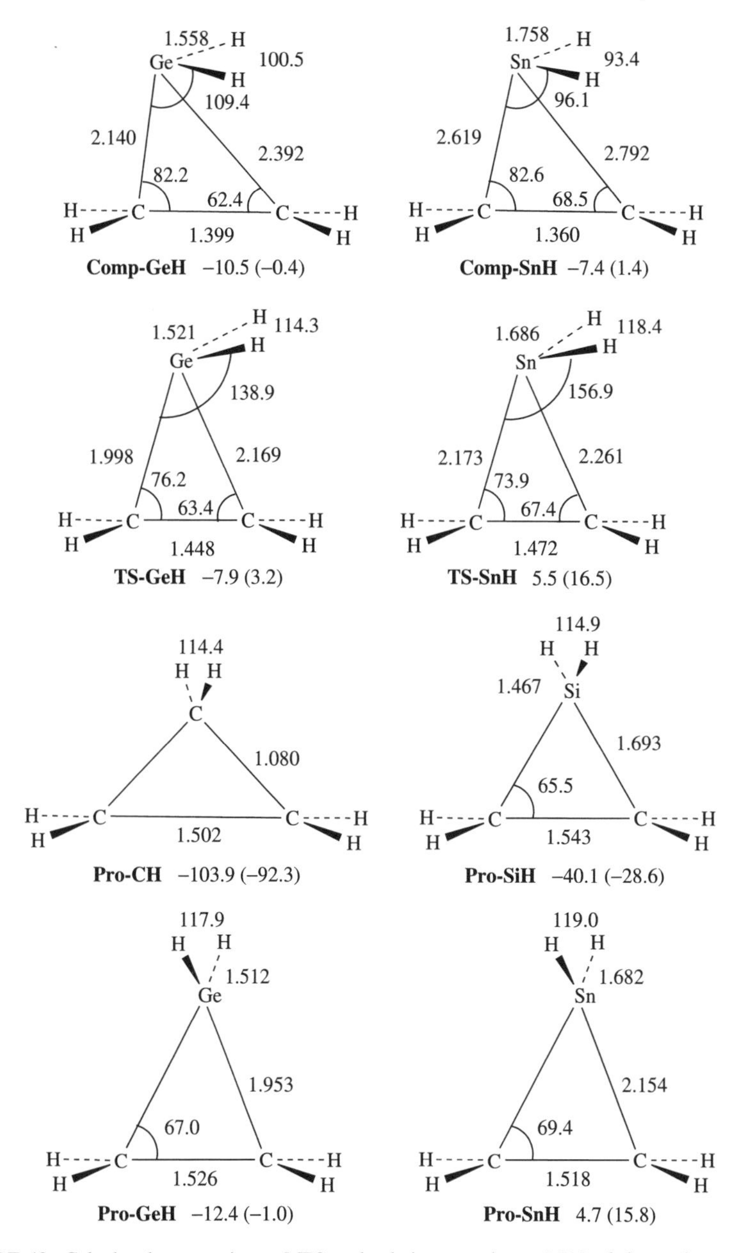

FIGURE 42. Calculated geometries at MP2 and relative energies at MP4 of the stationary points of the reaction path for the addition of ($^1A_1$) EH_2 to ethylene. Precursor complexes **Comp-EH**, transition states **TS-EH** and products **Pro-EH**. Bond distances are in Å, bond angles in deg. The energy values are ΔH° values and they are in kcal mol^{-1}. The numbers in parentheses give the ΔH^{298} values. Reproduced by permission of John Wiley & Sons, Inc. from Reference 87

C_2H_4. The addition of CH_2 and SiH_2 takes place without a barrier yielding the product molecules in strongly exothermic reactions. The reactions of the heavier analogues GeH_2 and SnH_2 lead first to side-on bonded complexes. The latter are stable at 0 K but the calculated Gibbs free energy at 298.15 K shows that they disappear at higher temperature. The [1+2] addition of GeH_2 at room temperature has a small barrier which becomes higher for reaction of SnH_2. The calculations predict that cyclic-$C_2H_4GeH_2$ should only exist at low temperatures while cyclic-$C_2H_4SnH_2$ is thermodynamically unstable at all temperatures[87].

Figure 43 shows the stationary points for the reaction of EF_2 with C_2H_4. Weakly bonded complexes for all reactions are predicted as minima on the PES which are unstable at

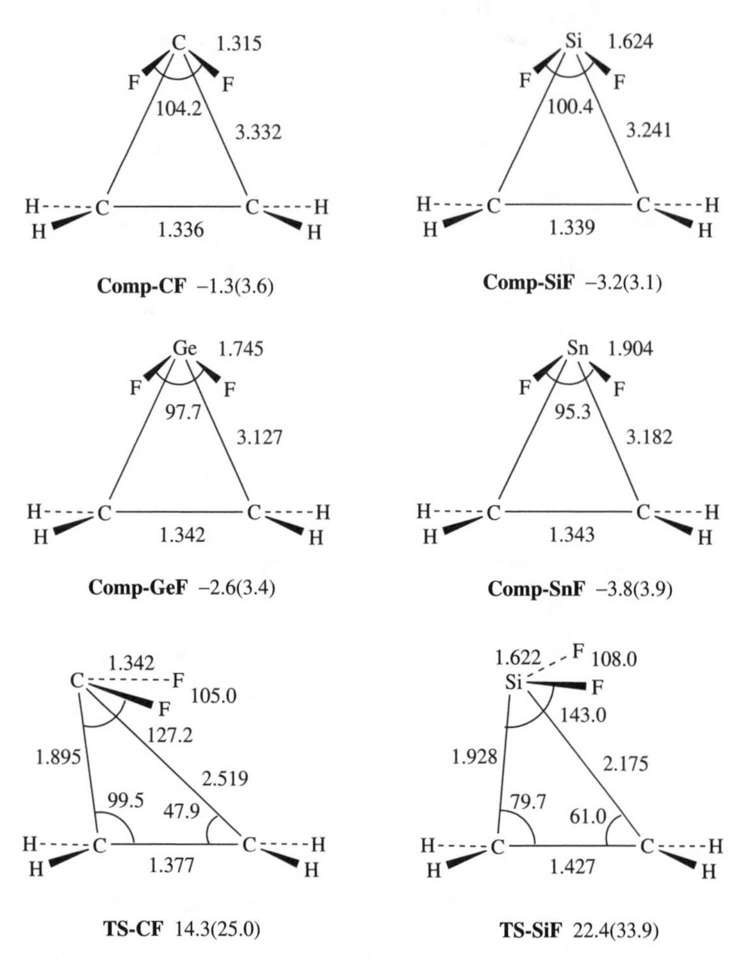

FIGURE 43. Calculated geometries at MP2 and relative energies at MP4 of the stationary points of the reaction path for the addition of ($^1A_1$) EF_2 to ethylene. Precursor complexes **Comp-EF$_2$**, transition states **TS-EF$_2$** and products **Pro-EF$_2$**. Bond distances are in Å, bond angles in deg. The energy values are ΔH° values and they are in kcal mol^{-1}. The numbers in parentheses give the ΔH^{298} values. Reproduced by permission of John Wiley & Sons, Inc. from Reference 87

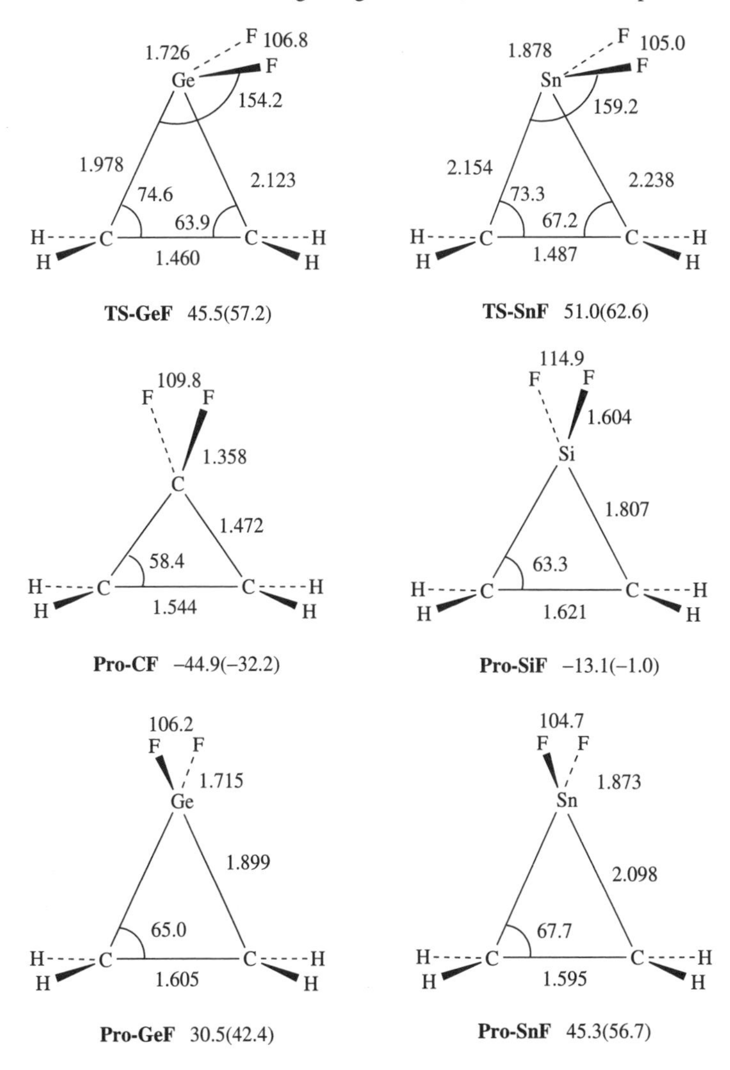

FIGURE 43. (*Continued*)

298.15 K. Note that the initial complexes of EF_2 with ethylene have a different geometrical shape than the EH_2 complexes which are shown in Figure 42. The C–EF_2 distances of **Comp-EF** are the same and the fluorine atoms are located on the same side of the three-membered ring while the hydrogen atoms of **Comp-EH** are on opposite sides of the ring plane. The author explained the differences between **Comp-EF** and **Comp-EH** with the repulsive interactions between the π-electrons of ethylene and fluorine[87]. The [1+2] addition of CF_2 has a significant barrier and is much less exothermic than the addition of CH_2. The barrier for addition of SiF_2 is even higher than for CF_2, but cyclic-$C_2H_4SiF_2$ is only kinetically stable at room temperature. The cyclic compounds cyclic-$C_2H_4GeF_2$ and cyclic-$C_2H_4SnF_2$ are thermodynamically unstable, but there is an energy barrier of *ca* 15 kcal mol^{-1} for EF_2 loss from the former compound and a barrier of *ca* 6 kcal mol^{-1} for the latter molecule. The author reports also about IRC calculations of the $EF_2 + C_2H_4$ addition reaction[87].

The analogous [1+2] addition of EH_2 and EF_2 to acetylene yielding the metallacyclopropenes cyclic-$C_2H_2EH_2$ and cyclic-$C_2H_2EF_2$ with E = C, Si, Ge, Sn has been calculated by Boatz et al. at the HF and MP2 levels and also at MP4 for the carbon and silicon systems[88]. As for the addition to ethylene, the cycloaddition of EH_2 to acetylene is predicted at the correlated level to proceed without a barrier. Calculated transition states at the HF level disappear at higher levels of theory. The formation of the metallacyclic compounds cyclic-$C_2H_2EH_2$ is exothermic but the reaction energies depend strongly on the metal E. The calculated enthalpies of formation for the reaction ($^1A_1$) $EH_2 + C_2H_2 \rightarrow$ cyclic-$C_2H_2EH_2$ at MP2/3-21G(d) corrected to 298 K (ΔH^{298}) are -101.5 kcal mol^{-1} for E = C, -56.8 kcal mol^{-1} for E = Si, -20.0 kcal mol^{-1} for E = Ge and -12.5 kcal mol^{-1} for E = Sn. The [1+2] addition of EF_2 to C_2H_2 yielding cyclic-$C_2H_2EF_2$ has significant activation barriers and is much less exothermic than in case of the EH_2 addition or is even endothermic. The calculated reaction barriers $\Delta H^{\ddagger}$ (reaction enthalpies ΔH^{298} are given in parentheses) are $\Delta H^{\ddagger} = 14.1$ kcal mol^{-1} ($\Delta H^{298} = -47.2$ kcal mol^{-1}) for E = C, $\Delta H^{\ddagger} = 14.3$ kcal mol^{-1} ($\Delta H^{298} = -40.9$ kcal mol^{-1}) for E = Si, $\Delta H^{\ddagger} = 38.4$ kcal mol^{-1} ($\Delta H^{298} = 14.4$ kcal mol^{-1}) for E = Ge and $\Delta H^{\ddagger} = 27.5$ kcal mol^{-1} ($\Delta H^{298} = 16.5$ kcal mol^{-1}) for E = Sn[87].

A theoretical study of the degenerate 1,3-allyl migration of EH_3 substituents with E = C, Si, Ge, Sn was published by Takahasi and Kira[163]. The authors found that there are two transition states TS_{ret} and TS_{inv} which yield retention (TS_{ret} or an inversion TS_{inv}) of the migrating EH_3 group. The geometry of TS_{inv} has a square pyramidal form while TS_{ret} is a trigonal bipyramid (Figure 44). Table 37 gives the calculated activation energies for the two transition states at the Hartree–Fock level. The most imporant result is that the pathway with retention of the CH_3 group has a higher energy than the reaction which proceeds with inversion of the methyl group, while the heavier EH_3 groups migrate with retention. The trend of the activation energies for the group-14 elements E is: C > Si > Ge > Sn[163].

The influence of carbon group substituents ER_3 (E = C to Sn; R = H, Me, *t*-Bu) on the energy barrier of bond shift and electrochemical reduction of substituted cyclooctatetraenes (COT-ER_3) has been studied with experimental and theoretical methods by Staley et al.[164]. The ring inversion transition state (Figure 45) was taken as a model for the steric interactions in the bond shift transition state which could not be calculated

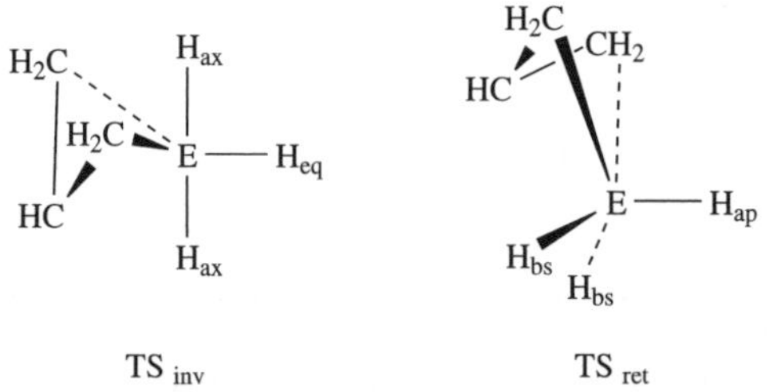

FIGURE 44. Schematic representation of the transition states of the degenerate 1,3-allyl migration of EH_3 (E = C to Sn) with inversion (TS_{inv}) and retention (TS_{ret}) of the EH_3 group. Reprinted with permission from Reference 163. Copyright 1997 American Chemical Society

TABLE 37. Comparison of activation energies ΔE_a (kcal mol^{-1}) for 1,3-migration in $CH_2{=}CHCH_2EH_3$ (E = C, Si, Ge and Sn)[a]

	C	Si		Ge	Sn
	HF/6-31G(d)	HF/6-31G(d)	HF/Lanel1DZ	HF/Lanel1DZ	HF/Lanl1DZ
E_a(ret)	133.6	64.0	79.6	72.9	55.4
E_a(inv)	116.9	75.1	84.6	77.5	62.3
ΔE_a^b	−16.7	11.1	5.0	4.6	6.9

[a]Reference 163.
[b]$\Delta E_a = E_a(\text{inv}) - E_a(\text{ret})$.

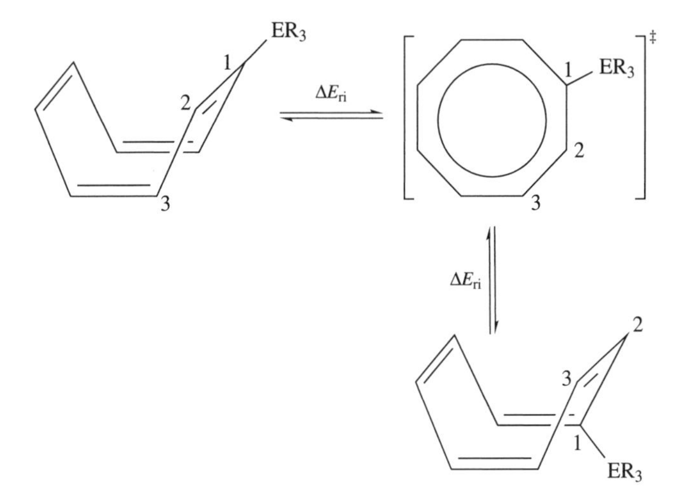

FIGURE 45. Schematic representation of the ring inversion (ri) transition state of cyclooctatetraene. The calculated energies are given in Table 38. Reprinted with permission from Reference 164. Copyright 1998 American Chemical Society

directly because of the size of the molecules and the multiconfigurational character of the transition state. The validity of the model was supported by a correlation between the calculated activation energies of the ring inversion of COT-R with R = H, Me, $SiMe_3$ and *t*-Bu, with the experimental values of the free activation enthalpies of the bond shift reaction. Table 38 shows the calculated activation energies of the ring inversion of COT and COT-ER_3 with E = C to Sn and R = H, Me. In spite of the rather low level of theory (HF/3-21G) it turns out that the ΔE_{ri} values for the systems with EMe_3 substituents are in fairly good agreement with the experimental results of the bond shift reaction. The latter process has $\Delta G^{\ddagger}$ (298 K) values of 18.1, 16.4, 16.2, and 16.2 kcal mol^{-1} for CMe_3, $SiMe_3$, $GeMe_3$ and $SnMe_3$, respectively[164]. Note that the difference in activation energy of the ER_3 substituent is the highest when one goes from CH_3 to CMe_3. It follows that steric effects play a crucial role in the relative activation barriers.

The influence of substituents X of 5-substituted 1,3-cyclopentadienes on the diastereoselectivity of the Diels–Alder addition with various nucleophiles has been studied at the HF/6-31G(d) level by Xidos et al.[165]. The calculations with ethylene as nucleophile were also carried out with the substituents X = EH_3 (E = C, Si, Ge, Sn). Figure 46 shows

TABLE 38. Calculated energies of the ring inversion transition state (ΔE_{ri}) shown in Figure 45 for substituted cyclooctatetraenes. [a]

Substituent	ΔE_{ri}(kcal mol^{-1})		
	HF/3-21G	HF/3-21G + ZPE	HF/ 6-31G(d) + ZPE
H	15.9	16.7	13.9
$C(CH_3)_3$	20.6	21.6	
$Si(CH_3)_3$	19.1	20.0	
$Ge(CH_3)_3$	18.1		
$Sn(CH_3)_3$	19.6		
CH_3	17.6	18.5	15.7
SiH_3	19.1	19.9	16.6
GeH_3	18.2	19.2	
SnH_3	19.8	20.6	

[a] Reference 164.

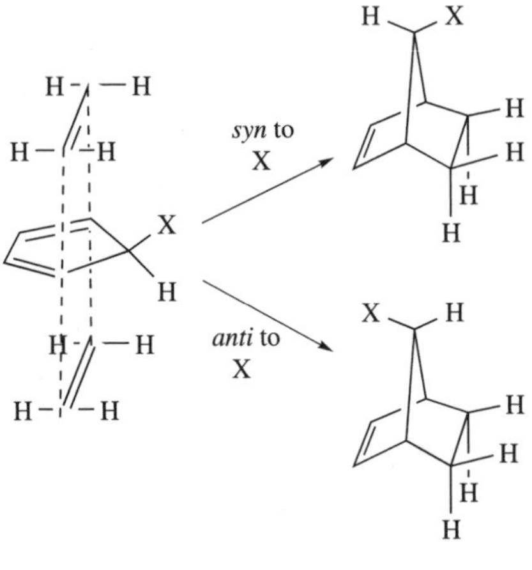

FIGURE 46. Schematic representation of the Diels–Alder reaction of 1,3-cyclopentadienes with ethylene yielding *syn* and *anti* products

the investigated reactions. The authors optimized the transition states for the *syn* and *anti* attack of ethylene with respect to X. The energy differences between the transition states were compared with the experimentally observed diastereoselectivity of substituted cyclopentadienes. The calculated facial stereoselectivities are in excellent agreement with experimental data. For the EH_3 substituted cyclopentadienes it was found that CH_3 favors the *anti* addition by 0.83 kcal mol^{-1}. The calculations predict that the heavier analogues SiH_3, GeH_3 and SnH_3 lead to higher preferences for *anti* addition by 6.4, 6.7 and 9.2 kcal mol^{-1}, respectively[165].

The role of $SnCl_4$ as catalyst in [2+2] cycloaddition reactions of olefines which are activated by selenophenyl and silyl groups in the 1,1 position with vinyl ketones has been examined in a combined experimental/theoretical study by Yamazaki et al.[166]. Calculations at the HF level showed that the formation of a chelate complex where the selenium atom of the olefin and the oxygen atom of the keto group are bonded as ligands to the

Lewis acid $SnCl_4$ is unlikely because of the large Sn−Se separation which was found in a zwitterionic intermediate. The latter structure has a Sn−O donor–acceptor bond but no Sn−Se bond. The calculations showed that a second $SnCl_4$ may perhaps bind to Se during the reaction[166].

Yamazaki et al. investigated in a combined experimental/theoretical work also the role of $SnCl_4$ in the formal [2+1] cycloaddition of 1-seleno-2-silylethenes to various vinylketones (Figure 47a) which give cyclopropane compounds rather than cyclobutanes that would be the result of a [2+2] cycloaddition[167]. Calculations of various possible intermediates led the authors to suggest that the formation of the $SnCl_4$ stabilized intermediate **I(*trans*)** which is shown in Figure 47b is responsible for the formation of the cyclopropane. The authors give the geometries and energies calculated at the HF level of various model compounds[167].

A related study by Yamazaki et al. investigated the reaction profile of the $SnCl_4$-catalyzed [2+1] cycloaddition of 1-seleno-2-silylethenes to 2-phosphonoacrylates[168]. Calculations at the HF level suggest that the crucial intermediate for this reaction is the chelate complex which is shown in Figure 48. The authors give the optimized geometries, energies and charge distribution of the complex and other possible intermediates of the reactions[168].

Hobson et al. investigated in a combined experimental/theoretical study the origin of 1,5-induction in Sn(IV)halide-promoted reactions of 4-alkoxyalk-2-enylstannanes **16** with aldehydes to give product **17**[169]. Figure 49 shows the investigated reaction course and the postulated intermediates **18** and transition state **19**. Figure 50 displays the optimized B3LYP geometry of transition state **19** for *cis* and *trans* addition of the aldehyde. The former structure is much lower in energy than the latter, which is in agreement with the experimentally observed formation of *cis* alkenes[169].

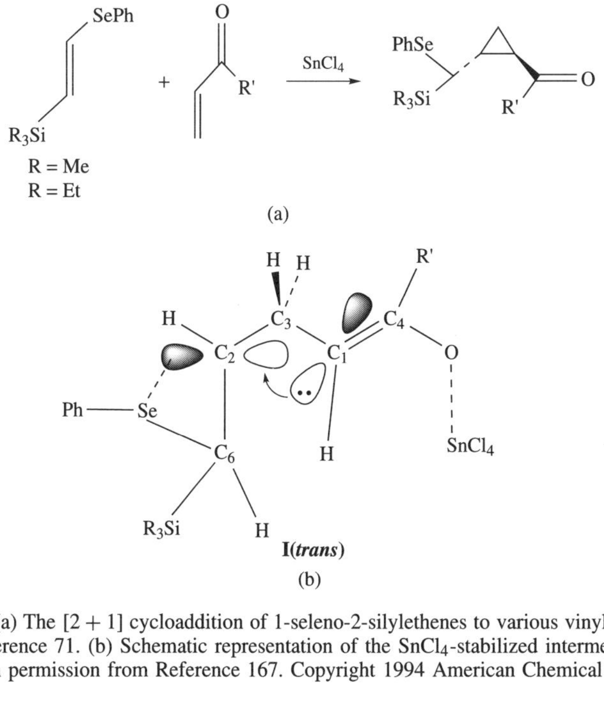

FIGURE 47. (a) The [2 + 1] cycloaddition of 1-seleno-2-silylethenes to various vinylketones investigated in Reference 71. (b) Schematic representation of the $SnCl_4$-stabilized intermediate **I(*trans*)**. Reprinted with permission from Reference 167. Copyright 1994 American Chemical Society

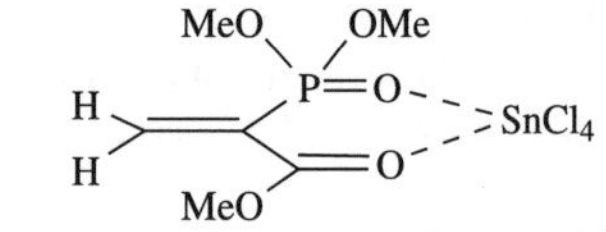

MeO
OMe
P=O
SnCl4
H
H
=O
MeO

FIGURE 48. Schematic drawing of the $SnCl_4$ chelate complex which was calculated as the crucial intermediate of the $SnCl_4$-catalyzed [2+1] cycloaddition of 1-seleno-2-silylethenes to 2-phosphonoacrylates. Reprinted with permission from Reference 168. Copyright 1998 American Chemical Society

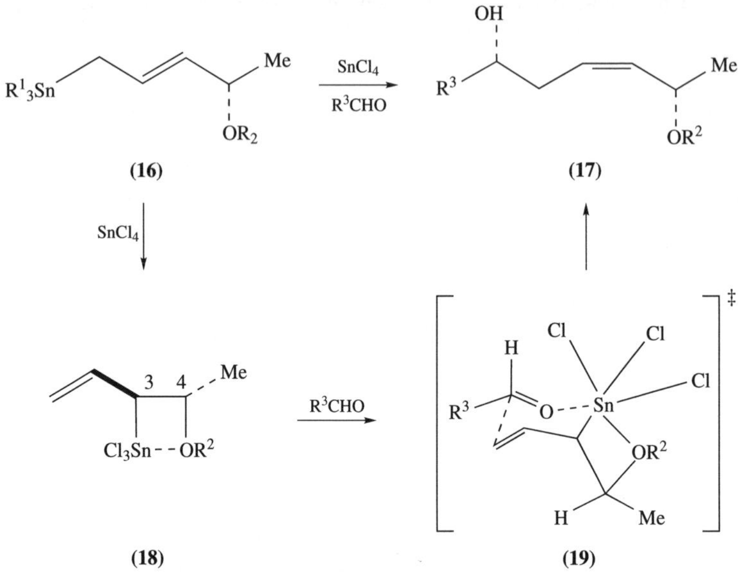

R1 3Sn
Me
OR2
SnCl4
R3CHO
OH
R3
Me
OR2
(16)
(17)
SnCl4
3
4
Me
Cl3Sn--OR2
R3CHO
H
Cl
Cl
Cl
R3
O
Sn
OR2
H
Me
‡
(18)
(19)

FIGURE 49. Postulated intermediates and transition state (**19**) of the Sn(IV)halide-promoted reactions of 4-alkoxyalk-2-enylstannanes with aldehydes[169]

Cl
Cl
Cl
Sn
2.222
1.267
2.380
2.197
C
O
2.370
C2
C3
O
C1
1.376
1.434
Me
C4
‡
(19) cis
ΔE = 1.9 kcal mol−1
(a)
Cl
Cl
Cl
1.284
2.182
Sn
3.297
C
O
2.374
O
Me
2.046
C2
C3
C4
C1
1.388
1.429
‡
(19) trans
ΔE = 12.0 kcal mol−1
(b)

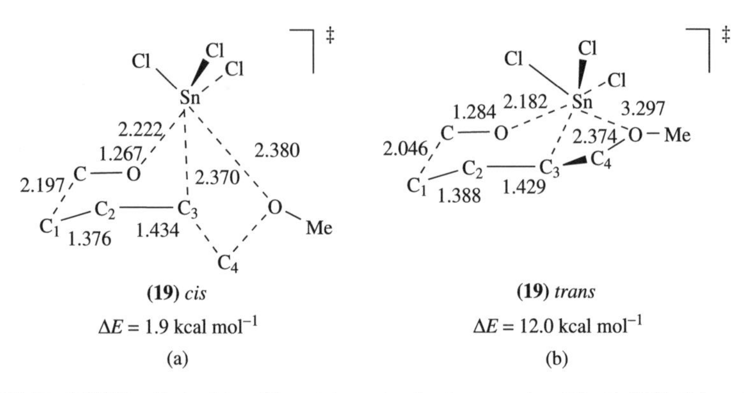

FIGURE 50. B3LYP optimized transition states and activation energies of the Sn(IV)halide-catalyzed *cis*- and *trans*-addition of formaldehyde to 4-alkoxyalk-2-enylstannanes. Bond lengths are in Å, energies in kcal mol^{-1}. Reproduced by permission of The Royal Society of Chemistry from Reference 169

The atomic charge distribution and the polarity of the LUMO of HC≡C$-$$SnF_3$ and HC≡C$-$$SnH_3$ have been calculated at the HF level by Yamaguchi et al.[170]. The authors investigated experimentally the reaction of substituted phenols with acetylene in the presence of $SnCl_4$ yielding *ortho*-vinyl phenols as main products. They argued that the electronic nature of the tin reagent which is strongly influenced by the substituents is crucial for the reaction path. The authors speculated that a nucleophilic attack of phenoxytin at the electrophilic β-carbon atom of stannylacetylene takes place, but they could not present a transition state for the reaction[170].

An *ab initio* investigation at the HF level of the reaction pathways of organotin enolate addition to benzaldehye and bromoethane have recently been reported by Yasuda et al.[171]. The investigated reactions are shown in Figure 51. Figure 52 gives the optimized geometries and atomic charge distribution of calculated tin compounds which are representative of triorganotin enolates, triorganotin alkoxides and triorganotin bromides. It also shows the negatively charged complexes which are formed when a Br^- is bonded to Sn. The calculated complexation energies are rather high. The reaction pathways of the two reactions shown in Figure 51 have been calculated with and without Br^- as ligand which is attached to tin. Figure 53 shows the optimized stationary points which were found for the addition of $Me_3SnOC(Me){=}CH_2$ (**20**) to benzaldehyde (reaction 1 in Figure 51). Figure 54 gives the theoretically predicted energy profile for this reaction. It becomes obvious that the activation barrier without the Br^- ligand is lower than the reaction barrier of the five-coordinated tin compound. The authors give also the calculated intermediates and the reaction profile for reaction 2 of Figure 51. The energy difference between the transition states of reaction 2 with and without the Br^- ligand are even higher in favor of the latter than for reaction 1[171].

The influence of the $SnCl_3$ substituent on the olefin insertion reaction into the Pt$-$H bond of a platinum model compound has been investigated at the MP2 (for the geometries) and MP4 (for the energies) levels of theory by Rocha and De Almeida[172]. Figure 55 shows the calculated reaction profile of the ethylene insertion into the Pt$-$H bond of $PtH(PH_3)_2X$, where X is Cl or $SnCl_3$. The calculations reveal that the $SnCl_3$ substituent stabilizes the pentacoordinated intermediates much more than Cl. The reaction proceeds through a rate-determining four-center transition state where the hydrogen atom migrates to the β-carbon atom of ethylene. The activation barrier for $X = SnCl_3$ is much lower than for $X = Cl$. The authors give the geometries and energies of the intermediates and the transition states. They also discuss the nature of the metal$-$ligand bonding in the relevant intermediates[172].

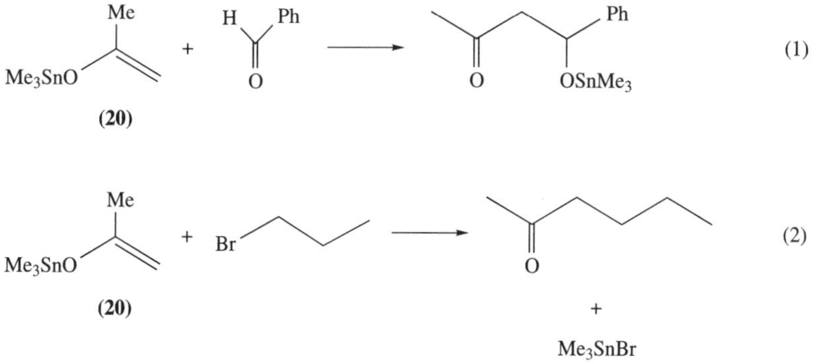

Me
Me_3SnO (**20**) + H Ph O ⟶ O Ph OSnMe$_3$ (1)

Me
Me_3SnO (**20**) + Br ⟶ O + Me_3SnBr (2)

FIGURE 51. Reactions which have been calculated in Reference 171

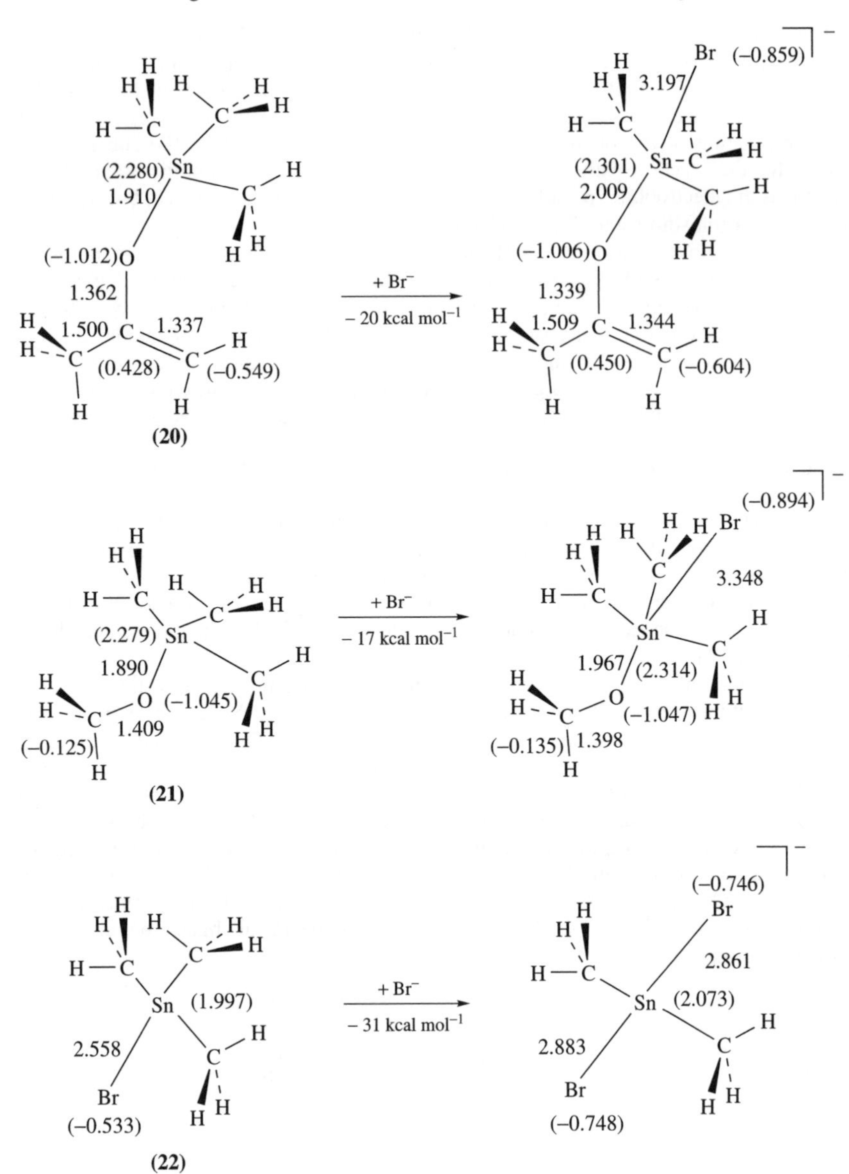

FIGURE 52. Optimized geometries and energies at the HF level for Br^- addition to tin compounds **20**–**22** taken from Reference 171. Bond distances are in Å. The atomic partial charges are given in parentheses. Reprinted with permission from Reference 171. Copyright 2000 American Chemical Society

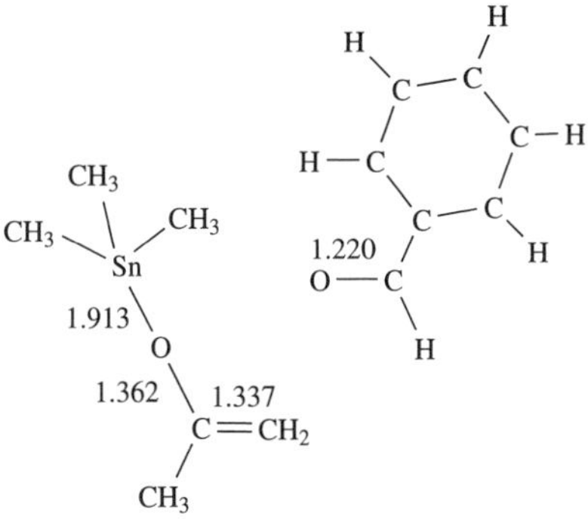

FIGURE 53. Calculated stationary points at the HF level for the addition of $Me_3SnO\text{-}C(Me){=}CH_2$ to benzaldehyde with and without Br^-. Precursor complexes **23** and **23(Br^-)**, transition states **24** and **24(Br^-)** and products **25** and **25(Br^-)**. Bond distances are in Å. See Figure 54 for the corresponding energy profiles. Reprinted with permission from Reference 171. Copyright 2000 American Chemical Society

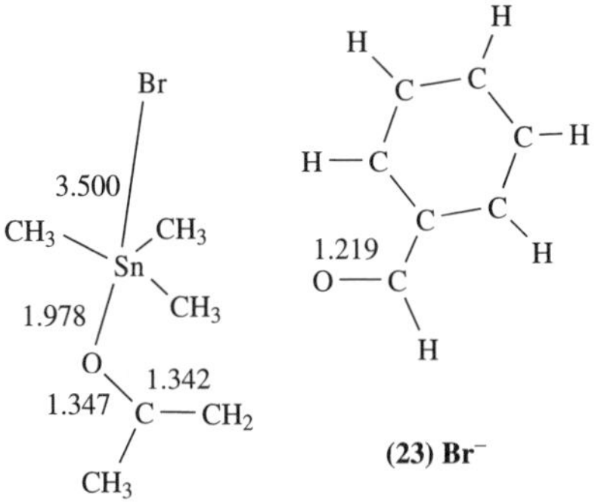

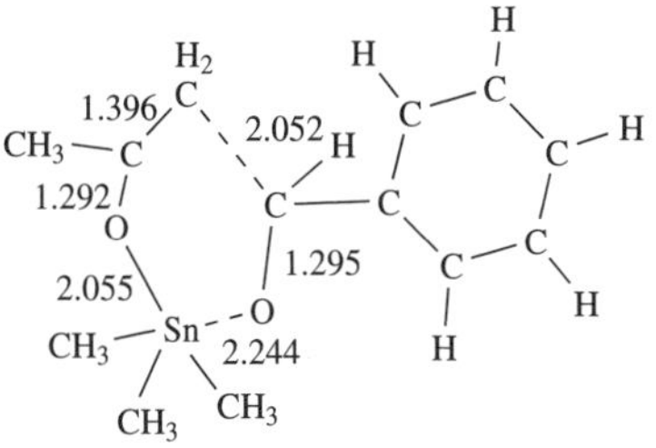

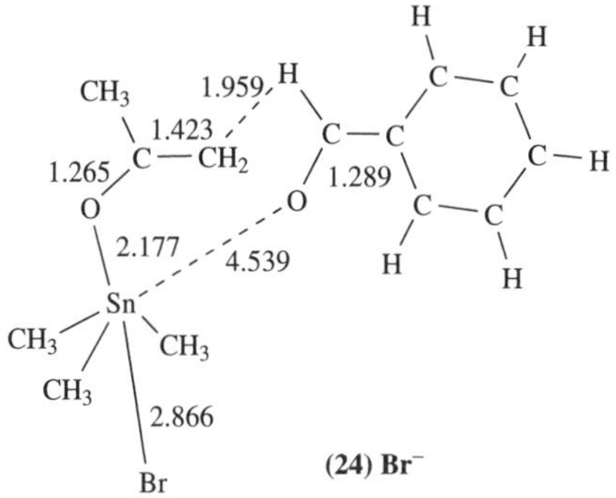

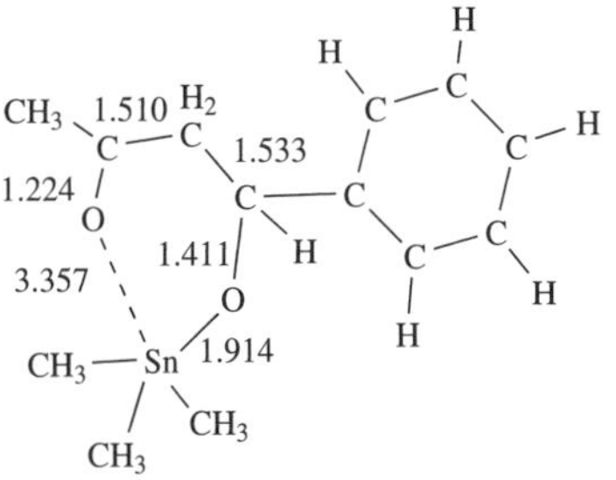

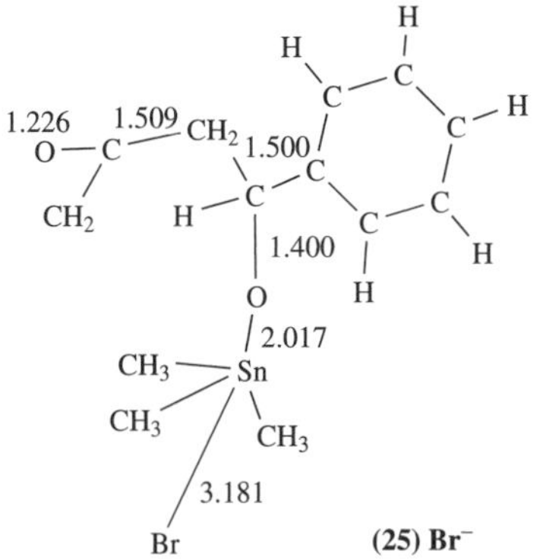

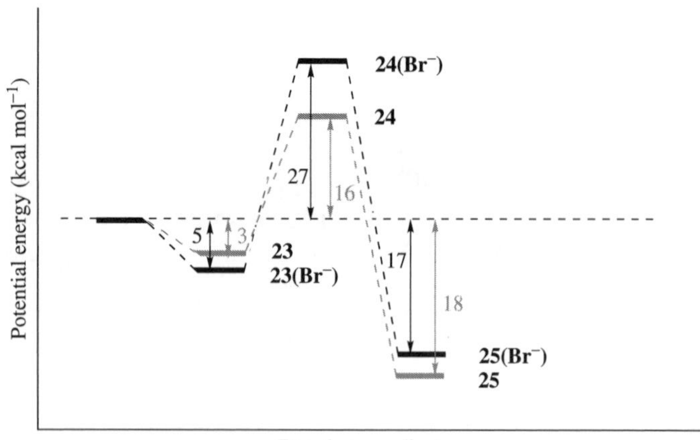

FIGURE 54. Calculated reaction profiles at the HF level for the addition of Me_3SnO-C(Me)=CH_2 to benzaldehyde with and without Br^-. Precursor complexes **23** and **23(Br^-)**, transition states **24** and **24(Br^-)** and products **25** and **25(Br^-)**. See Figure 53 for the calculated geometries of these structures. Reprinted with permission from Reference 171. Copyright 2000 American Chemical Society

A theoretical study at the HF level on the reaction mechanisms of the regioselective silastannation of acetylenes with a model palladium catalyst has been published by Hada et al.[173]. Figure 56 shows the calculated reaction profile for the addition of $H_3Si-SnH_3$ to RC≡CH (R = Me) in the presence of the model catalyst $Pd(PH_3)_2$. The transition states TS3 and TS4 could not be localized on the PES. The given energies are upper limits of the ligand exchange reactions. The authors give also the energy profiles for the insertion step of RC≡CH with R = H, CN and OCH_3 and they analyze the electronic structure of the intermediates[173].

2. Cations and anions

Theoretical studies have been published which investigated the changes in the geometries and bonding situations in carbocations when carbon is substituted by a heavier group-14 atom Si to Pb. Gobbi and Frenking calculated the structures and analyzed the bonding situation in the allyl cations and anions $CH_2CHEH_2{}^{+/-}$ for E = C to Pb at the HF and MP2 levels[116]. The allyl cations are predicted to have a planar geometry. All allyl cations are stabilized by π-conjugative interactions. The strength of the resonance interactions as measured by the rotational barrier decreases from 37.8 kcal mol^{-1} (E = C) to 14.1 kcal mol^{-1} (E = Si), 12.0 kcal mol^{-1} (E = Ge), 7.2 kcal mol^{-1} (E = Sn) and 6.1 kcal mol^{-1} (E = Pb). The allyl cations are additionally stabilized by σ-bonding and through-space charge interactions, which have the same magnitude as the resonance stabilization. The equilibrium geometries of the heavy-atom allyl anions have strongly pyramidal EH_2 groups. The planar forms are much higher in energy. The calculations suggest that there is no resonance stabilization in the allyl anions, except in the parent anion $CH_2CHCH_2{}^-$. The electronic structure of the molecules was investigated using the Laplacian of the electron density distribution[116].

The trend of the π-donor ability of the halogens X = F to I in the cations $EX_3{}^+$ and EH_2X^+ (E = C to Pb) and in the isoelectronic neutral compounds AX_3 and AH_2X

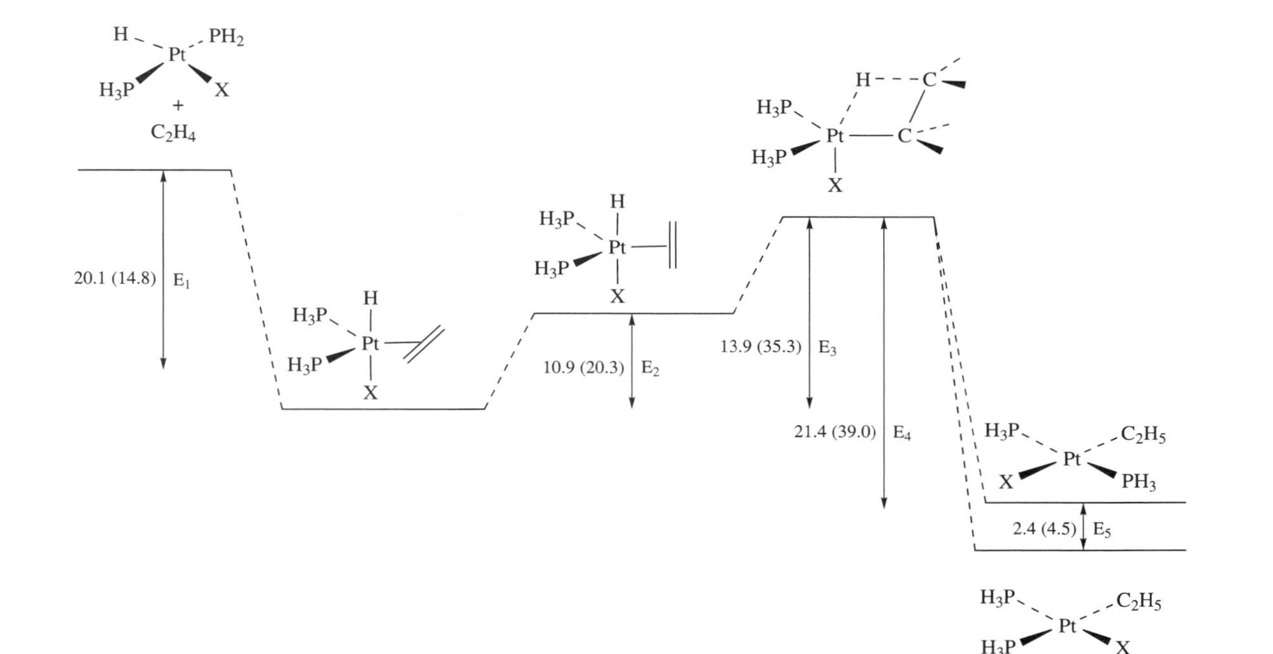

FIGURE 55. Calculated reaction profile at the MP4//MP2 level of ethylene insertion into the Pt−H bond of $PtH(PH_3)_2X$. The calculated energies ($kcal\,mol^{-1}$) refer to $X = SnCl_3$; the numbers in parentheses refer to X = Cl. Reprinted with permission from Reference 172. Copyright 1998 American Chemical Society

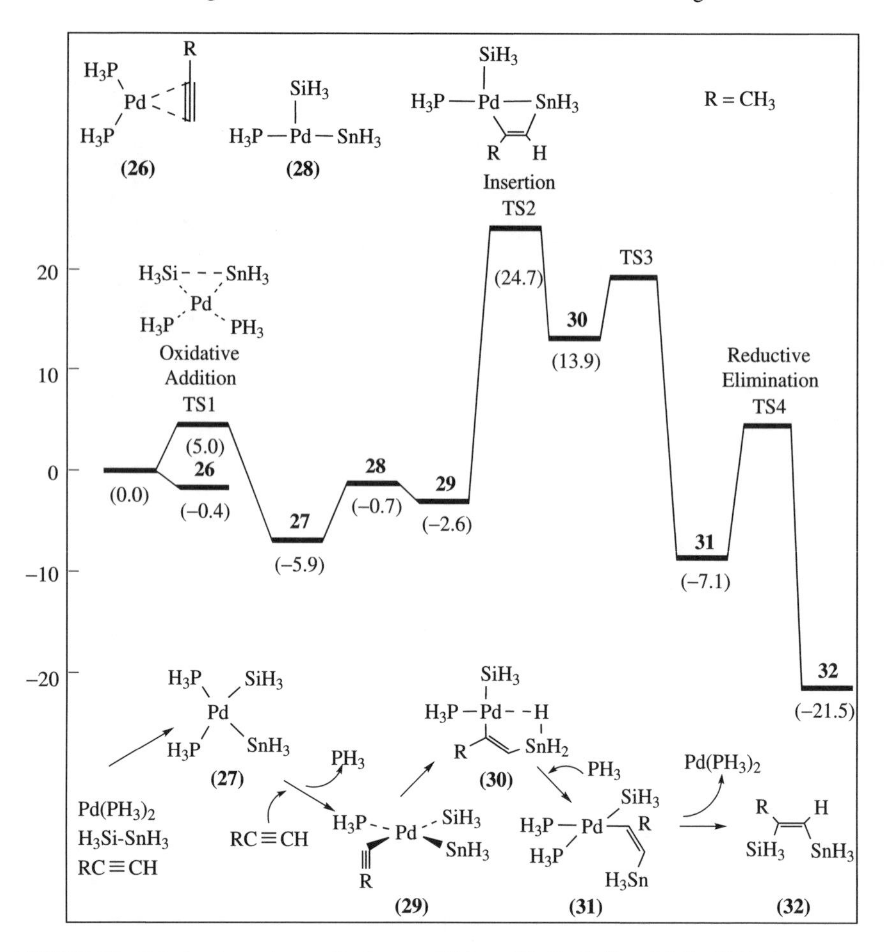

FIGURE 56. Calculated reaction profile for the addition of $H_3Si{-}SnH_3$ to MeC≡CH in the presence of the model catalyst $Pd(PH_3)_2$. Energies are in kcal mol^{-1}. The energies of transition states TS3 and TS4 are only approximate. Reprinted with permission from Reference 173. Copyright 1994 American Chemical Society

(A = B, Al, Ga, In, Tl) was the focus of a theoretical study at the MP2 level by Frenking and coworkers[174]. The strength of the π-donation was probed by the population of the p(π) AO of the central atoms E and A, by calculating the reaction energy of isodesmic reactions and by calculation of the complexation energies with H_2O. All three criteria suggest that the π-donor strength has the trend F < Cl < Br < I. Figure 57 shows a diagram of the theoretically predicted BDEs of the complexes $X_3E^+{-}OH_2$. It becomes obvious that carbon plays a special role among the group-14 elements. However, all $EX_3{}^+{-}OH_2$ species exhibit a trend of the BDEs X = F > Cl > Br > I which indicates that iodine stabilizes the cation $EX_3{}^+$ the most and fluorine the least[174].

The potential energy surface of $C_2GeH_5{}^+$ was investigated by Antoniotti et al. at correlated levels up to QCISD(T) using HF optimized geometries[175]. Figure 58 shows the

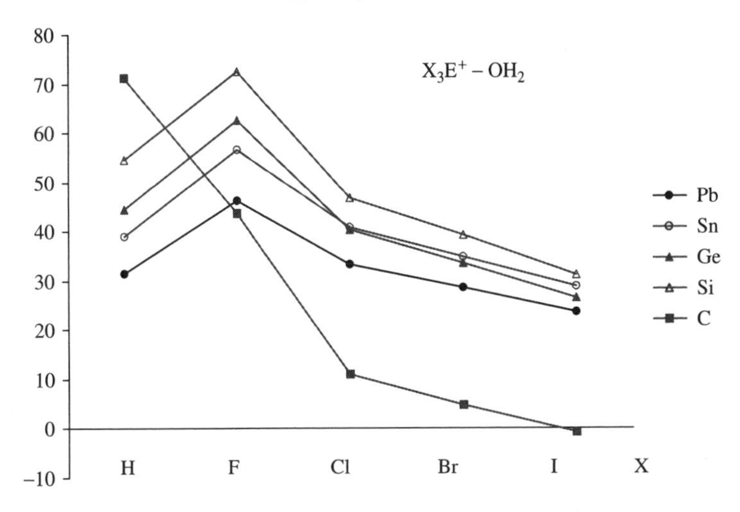

FIGURE 57. Calculated complexation energies at the MP2 level of $X_3E^+{-}OH_2$ with E = C to Pb and X = H, F to I. Bond energies are in kcal mol^{-1}. Reprinted with permission from Reference 174. Copyright 1997 American Chemical Society

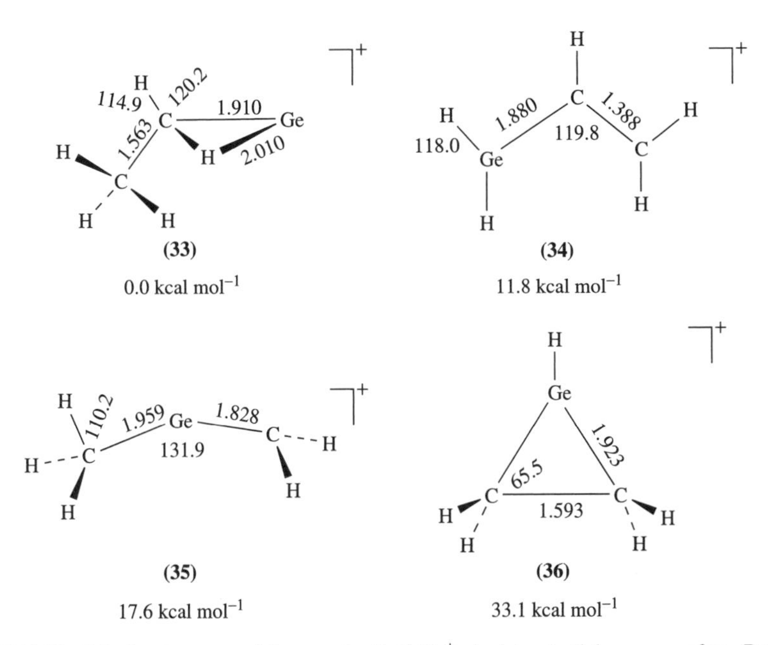

FIGURE 58. Calculated energy minima on the $GeC_2H_5^+$ singlet potential energy surface. Relative energies have been calculated at QCISD(T) using HF optimized geometries. Bond distances are in Å, angles in deg. Reprinted with permission from Reference 175. Copyright 1993 American Chemical Society

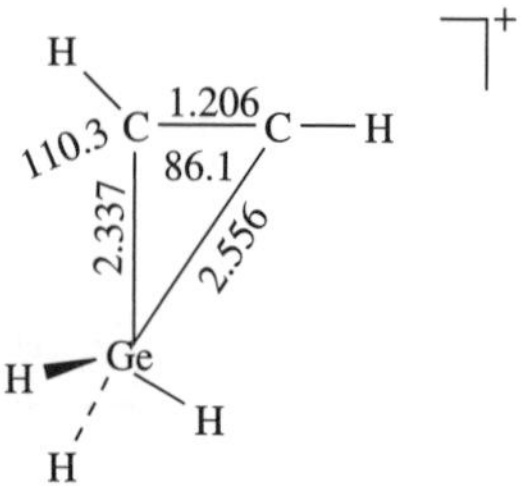

(37)

35.7 kcal mol^{-1}

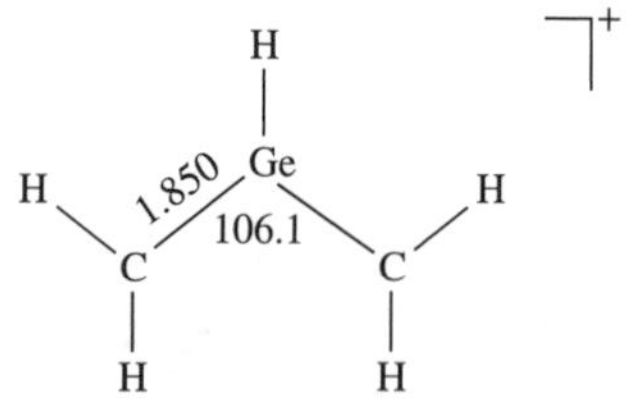

(38)

47.3 kcal mol^{-1}

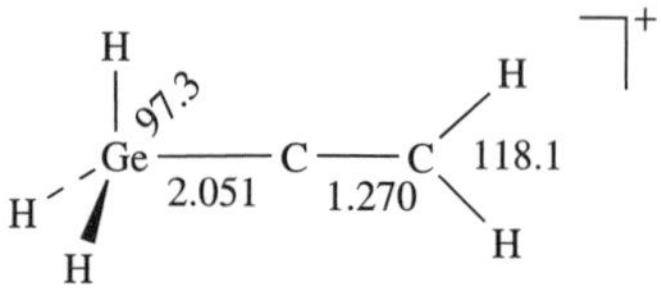

(39)

48.4 kcal mol^{-1}

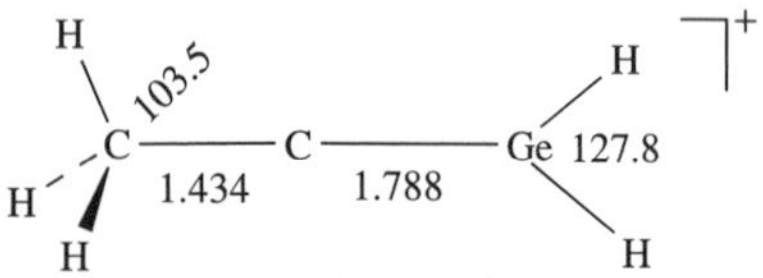

(40)

66.3 kcal mol^{-1}

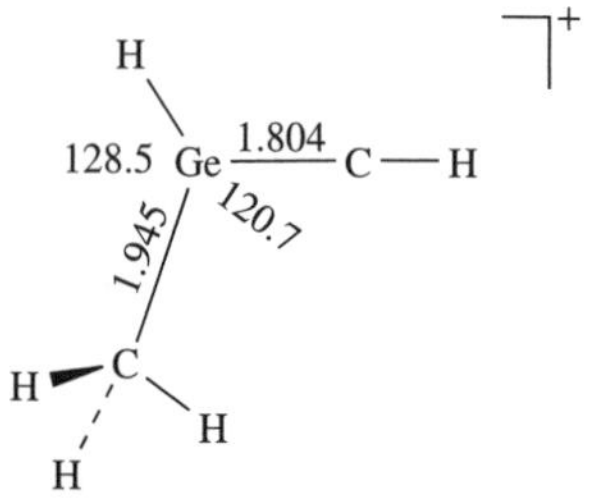

(41)

73.1 kcal mol^{-1}

FIGURE 58. (*Continued*)

optimized structures and the relative energies of the energy minima **33**–**41** which have been found. The global energy minimum structure is the hydrogen-bridged nonclassical cation **33**, which is 11.8 kcal mol^{-1} more stable than the 1-germaallyl cation **34**, the second most stable isomer on the PES. There are seven other energy minima which were found on the PES. The authors give also the structures and energies of some $C_2GeH_5^+$ transition states[175]. The same group investigated also the $C_2GeH_7^+$ PES[176]. The geometries were optimized at the MP2 level and the energies were calculated at QCISD(T) using MP2 optimized structures. Figure 59 gives the geometries and the relative energies of four structures which were found to be energy minima on the $C_2GeH_7^+$ PES. The global energy minimum structure is the classical 2-germapropyl-2 cation **42** which has C_{2v} symmetry. The energetically nearly degenerate two rotamers **43** and **44** of the 1-germapropyl-1 cation are predicted to be *ca* 20 kcal mol^{-1} less stable than **42**. The nonclassical form **45** is much higher in energy. The authors give also the structures and

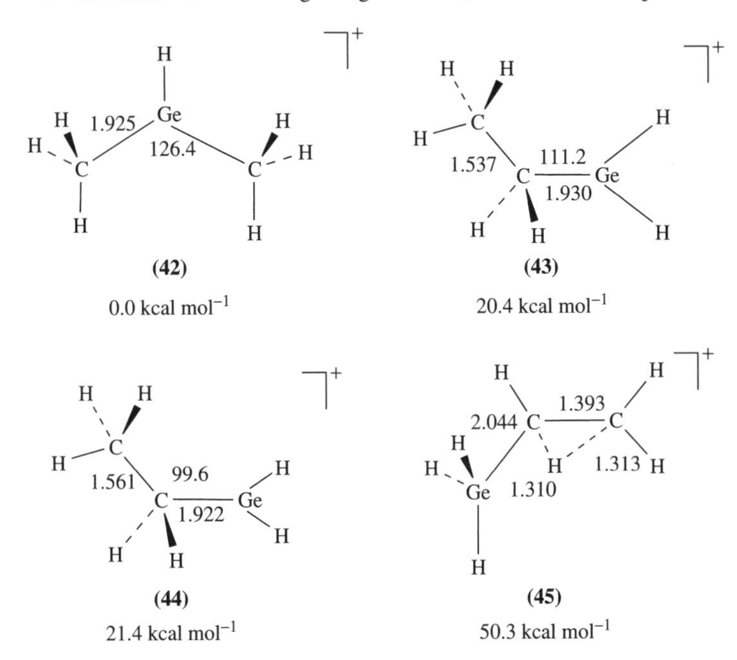

FIGURE 59. Calculated energy minima on the $GeC_2H_7{}^+$ singlet potential energy surface. Relative energies (kcal mol^{-1}) have been calculated at QCISD(T) using MP2 optimized geometries. Bond distances are in Å, angles in deg. Reprinted with permission from Reference 176. Copyright 1995 American Chemical Society

energies of some transition states and higher-order saddle points on the $C_2GeH_7{}^+$ PES[176]. Some of the $C_2GeH_7{}^+$ isomers were previously calculated at the HF level by Nguyen et al.[177].

The relative energies of classical and nonclassical isomers of tropylium, silatropylium and germatropylium cations were calculated by Nicolaides and Radom at the G2 level of theory using MP2 optimized geometries[178]. Figure 60 shows the structures **46–53** which were found as energy minima on the $C_6H_7E^+$ (E = C, Si, Ge) PES. Table 39 gives the relative energies. It becomes obvious that the classical isomers **46** and **47** (which is identical to **48** when E = C) are the most stable $C_7H_7{}^+$ carbocations, while for E = Si and Ge the nonclassical form **50Si** and **50Ge** becomes the global minimum on the respective PES.

The substituent effect of group-14 substituents on the stability of the bicyclic carbocations $\mathbf{J}^+$ and $\mathbf{K}^+$ (Figure 61) has been the subject of a theoretical study at the MP2 level by Hrovat and Borden[179] and a combined experimental/theoretical work by Adcock et al.[180]. The former workers calculated $\mathbf{J}^+$ and the parent system **JH** with X = H, SiH_3, SnH_3. They found that the cations are stabilized by SiH_3 and even more by SnH_3 via hyperconjugation[179]. Adcock et al. calculated the 4X-1-norbonyl cation $\mathbf{K}^+$ and the parent compound **KH** with X = H, SiH_3, Me, $SiMe_3$ and $SnMe_3$[180]. Table 40 shows the theoretically predicted relative hydride affinities of the cations. It becomes obvious that $SiMe_3$ and particularly $SnMe_3$ have a large stabilizing effect on the 1-norbonyl cation. The calculated energies also show that the hyperconjugative stabilization of the $SiMe_3$

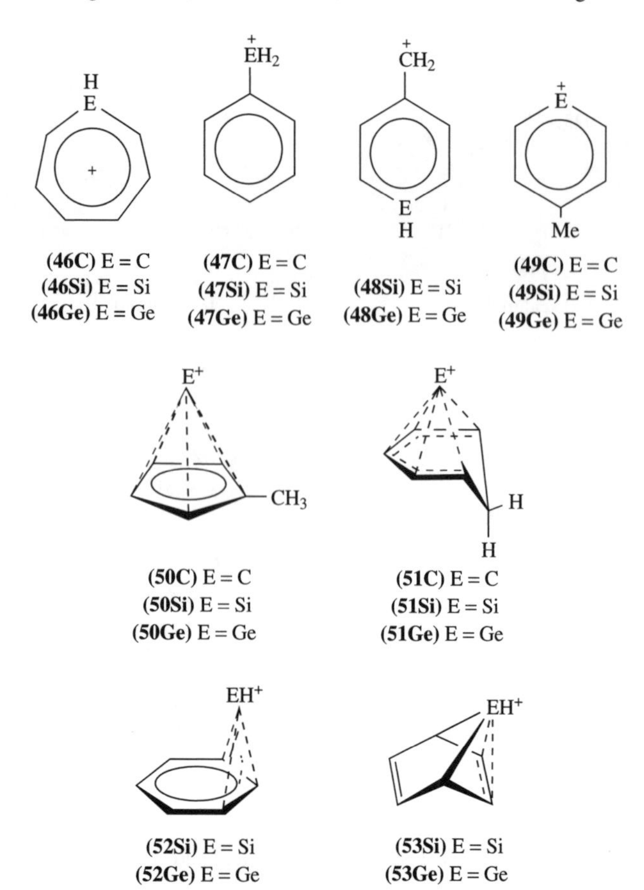

FIGURE 60. Schematic representation of the energy minima which were found on the $EC_6H_7^+$ singlet potential energy surface. Relative energies are given in Table 39. Reprinted with permission from Reference 178. Copyright 1997 American Chemical Society

TABLE 39. Relative G2(MP2, SVP)[a] isomer energies (kcal mol^{-1}) at 298 K of the cations shown in Figure 60[b]

E	46E	47E	48E	49E	50E	51E	52E	53E
C	0	6.9	6.9	50.2	66.2	97.0	—[c]	36.8
Si	34.2	25.1	31.8	67.9	0	30.4	34.9	36.8
Ge	55.9	39.9	52.8	78.9	0	28.0	32.3	—[c]

[a] G2(MP2, SVP) is an approximation scheme for obtaining correct total energies using QCISD(T) and MP2 calculations. For details see Reference 178.
[b] Reference 178.
[c] No energy minimum found.

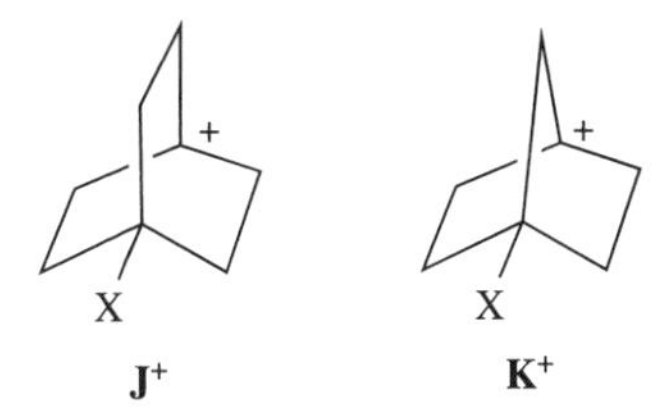

FIGURE 61. Bridged carbocations with group-14 substituents X which have been calculated in References 179 and 180. For the calculated energies see Table 40

TABLE 40. MP2/6-31G(d) calculated energies (kcal mol^{-1}) of substituted norbornyl cation $\mathbf{K}^+$ (Figure 61) and **KH**[a]

Structure	$E(\mathbf{K}^+) - E(\mathbf{KH})^b$
$\mathbf{K}^+$(X = H)	0
KH (X = H)	
$\mathbf{K}^+$(X = SiH_3)	−3.0
KH (X = SiH_3)	
$\mathbf{K}^+$(X = Me)	−3.5
KH (X = Me)	
$\mathbf{K}^+$(X = $SiMe_3$)	−11.1
KH (X = $SiMe_3$)	
$\mathbf{K}^+$(X = $SnMe_3$)	−15.5
KH (X = $SnMe_3$)	

[a]Reference 180.
[b]Energies relative to $E(\mathbf{K}^+$, X = H) − E(**KH**, X = H), i.e. relative hydride affinities with respect to the parent ion.

group is not well reproduced by the SiH_3 substituent. This is an important result because SiR_3 groups are often modeled in theoretical studies by SiH_3. The authors also discuss the different geometries and the electronic structure of the $\mathbf{K}^+$ and **KH** systems[180].

The geometries and relative energies of isomers of the (3-oxopropyl) trimethylstannane radical cation $Me_3Sn{-}CH_2CH_2CHO^{+\cdot}$ have been calculated at *ab initio* and DFT (B3LYP) levels of theory by Yoshida and Izawa[181]. The calculations were part of a combined experimental and theoretical study of the intramolecular assistance by carbonyl groups in electron transfer driven cleavage of C−Sn bonds. Figure 62 shows the structures of three energy minima on the PES. The cyclic isomers **54** and **55** are clearly lower in energy than structure **56**.

The β- and γ-effects of group-14 elements were probed by Sugawara and Yoshida using intramolecular competition between γ-elimination of tin and β-elimination of Si, Ge and Sn in reactions of α-acetoxy(arylmethyl)stannanes with allylmetals where the metal is Si, Ge or Sn[182]. In order to gain information about the stability of carbocations substituted by group-14 elements the authors carried out *ab initio* calculations at the MP2 level of theory. Figure 63 shows the optimized structures and relative energies of the cations $H_3SnCH_2CH_2CHCH_2EH_3^+$ (E = Si, Ge, Sn) which have been calculated. Two structures **57(E)** and **58(E)** were found for E = Sn, while three energy minima **57(E)**, **58(E)** and **59(E)** were found for E = Si, Ge. The energetically lowest lying form in all cases is structure **57(E)**. The three-membered cyclic structure **59(E)** is a low-lying energy minimum for E = Si and an energetically rather high-lying isomer for E = Ge, while it is not an energy minimum for E = Sn[182].

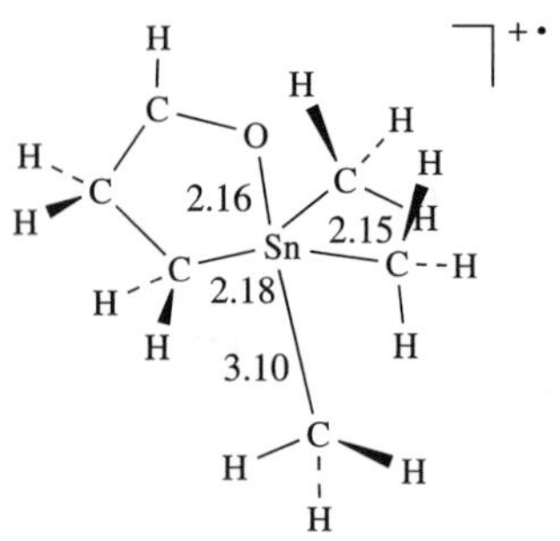

(54)
0.0 kcal mol^{-1}

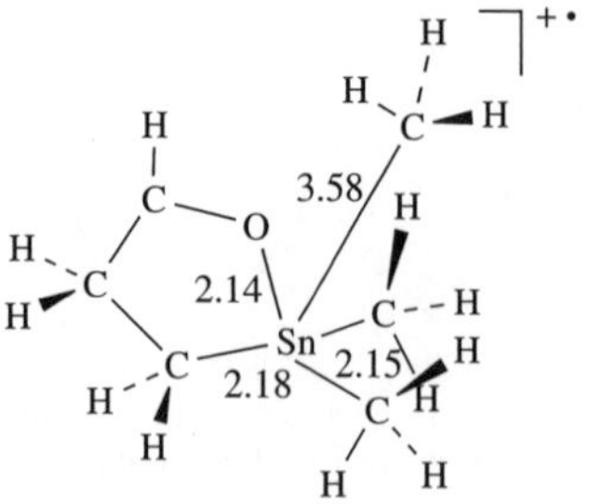

(55)
4.4 (4.8) kcal mol^{-1}

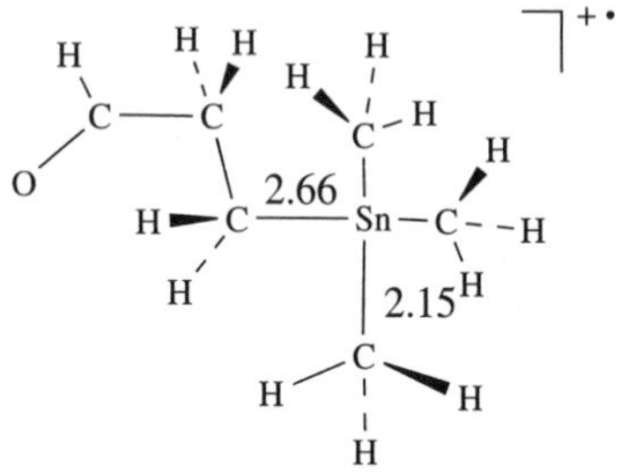

(56)
30.9 (23.8) kcal mol^{-1}

FIGURE 62. Calculated energy minimum structures of the radical cation $Me_3Sn{-}CH_2CH_2CHO^{+\bullet}$. Relative energies have been calculated at MP2/3-21G(d) and at B3LYP/3-21G(d) (in parentheses). Bond distances are given in Å, angles in deg. Reprinted with permission from Reference 181. Copyright 1997 American Chemical Society

Relativistic effects on the metal–carbon bond strengths of Me_2M, where M = Au^-, Hg, Tl^+ and Pb^{2+}, were studied by Schwerdtfeger[183]. The author found that in Me_2Pb^{2+} the BDEs of the Pb^{2+}–Me bonds increase by *ca* 15% and the force constant of the symmetric Me–Pb^{2+}–Me stretching mode by *ca* 20% when relativistic effects are included in the calculations.

The structures and stabilization energies of methyl anions X-CH_2^- with main group substituents X from the first five periods have been investigated at correlated *ab initio* levels by El-Nahas and Schleyer[184]. The work includes calculations of the anions $H_3Ge{-}CH_2^-$ and $H_3Sn{-}CH_2^-$ and the neutral parent compounds. The optimized geometries of the energy minima are shown in Figure 64. The structures have a pyramidal XCH_2^- geometry. The inversion barrier of the CH_2 group is predicted to be 3.2 kcal mol^{-1} for X = GeH_3 and 3.0 kcal mol^{-1} for X = SnH_3. The EH_3 substituents stabilize the methyl anion relative to CH_3^- by 19.7 kcal mol^{-1} (E = Ge) and by 23.7 kcal mol^{-1} (E = Sn)[184].

Negatively charged species of group-14 compounds have also been calculated by Anane et al. at the G2 level[185]. The authors optimized the donor–acceptor complexes of the

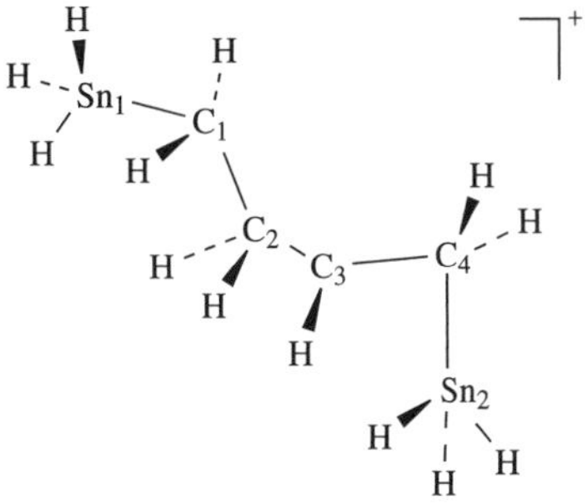

(57) Sn

0.0 kcal mol^{-1}

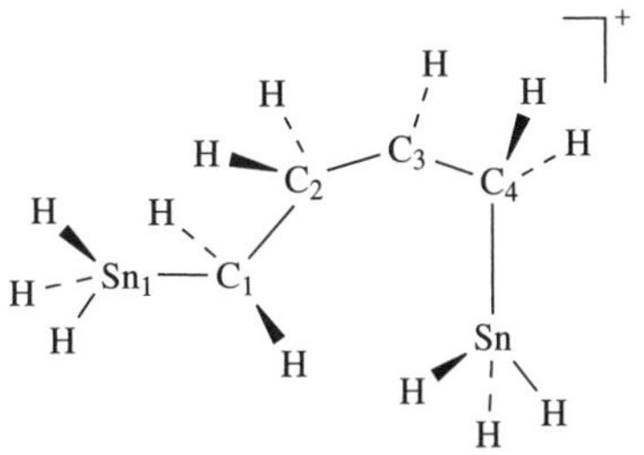

(58) Sn

1.5 kcal mol^{-1}

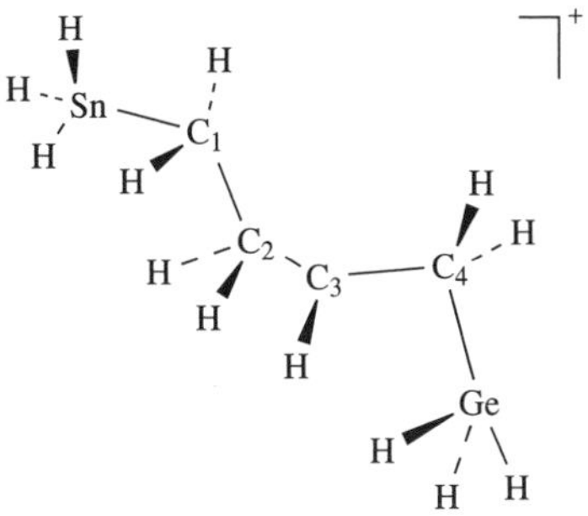

(57) Ge

0.0 kcal mol^{-1}

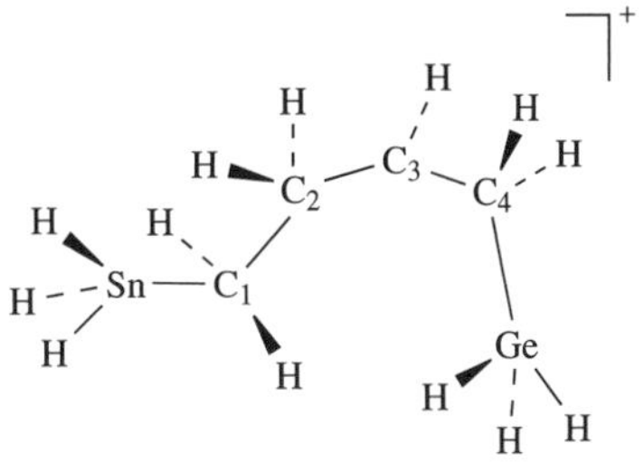

(58) Ge

2.0 kcal mol^{-1}

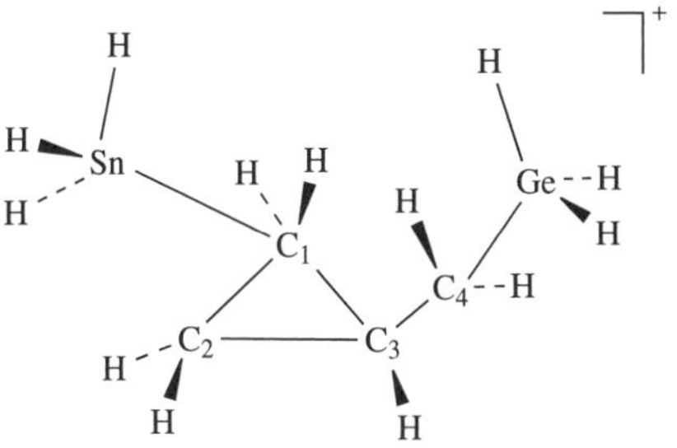

(59) Ge

7.1 kcal mol^{-1}

FIGURE 63. Calculated energy minimum structures and relative energies at MP2 of the cations $H_3SnCH_2CH_2CHCH_2EH_3{}^+$ (E = Si, Ge, Sn). Bond distances are in Å, angles in deg. Reprinted with permission from Reference 182. Copyright 2000 American Chemical Society

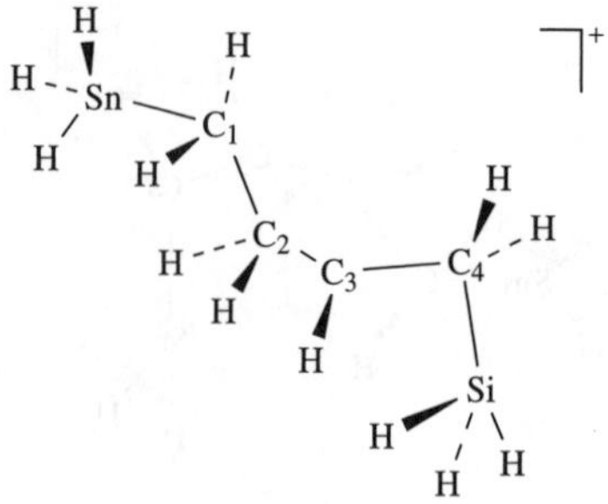

(57) Si
0.0 kcal mol^{-1}

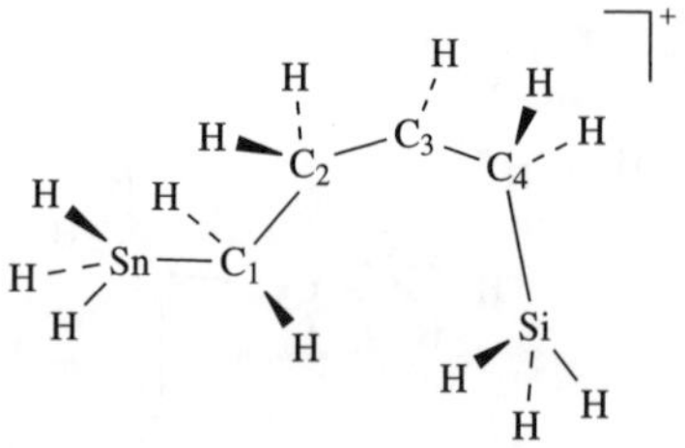

(58) Si
2.9 kcal mol^{-1}

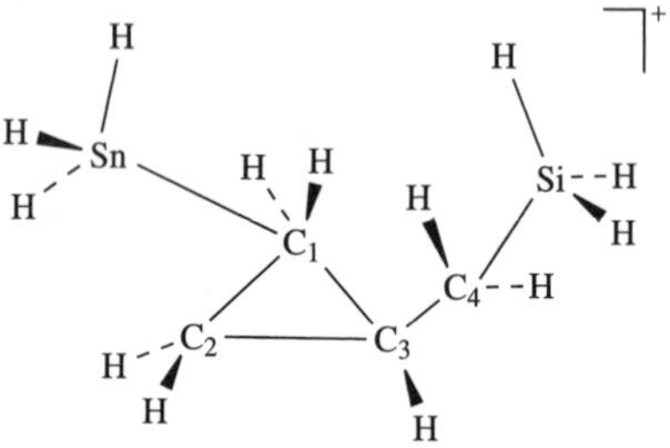

(59) Si
3.3 kcal mol^{-1}

FIGURE 63. (*Continued*)

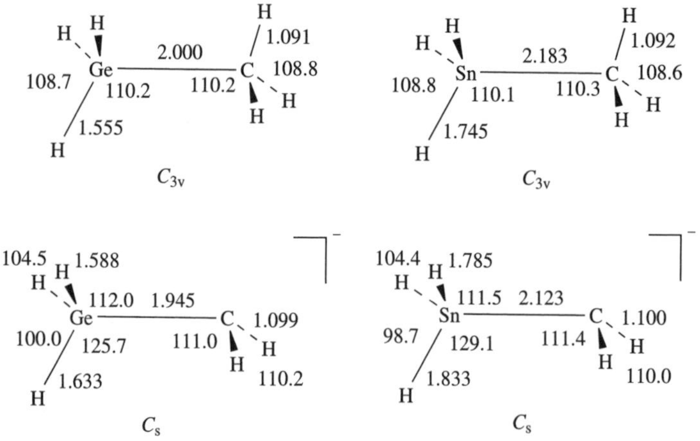

C_{3v} C_{3v}

C_s C_s

FIGURE 64. Calculated energy minimum structures of H_3ECH_3 and $H_3ECH_2^-$ at MP2. Bond distances are in Å, angles in deg. Reproduced by permission of John Wiley & Sons, Inc. from Reference 184

Lewis acids AlH_3 with the anionic Lewis bases EH_3^- (E = C, Si, Ge) and with neutral species AH_3 (A = N, P, As). The $H_3Al{-}EH_3^-$ BDEs (D_e) at the G2 level are predicted to be 84.7 kcal mol^{-1} (E = C), 54.4 kcal mol^{-1} (E = Si) and 49.9 kcal mol^{-1} (E = Ge)[185].

The geometries of the cyclopentadienyl anions $C_4H_4ESiH_3^-$ with E = C, Si, Ge, Sn have been optimized at the HF level by Freeman et al.[186]. The calculated molecules were used as models for the methyl-substituted systems $C_4Me_4ESiMe_3^-$. The X-ray structure analysis of the latter compound with E = Si was also reported. Table 41 gives the theoretically predicted relevant bond lengths and angles of $C_4H_4ESiH_3^-$ and the experimental data for $C_4Me_4GeSiMe_3^-$. The authors conclude that the heavier analogues of the Cp^- ring with E = Si, Ge, Sn are clearly not aromatic. This becomes obvious by the large alteration of the C−C bond lengths and the pyramidal arrangement of the substituents at the atom E[186].

3. Radicals

It is well known that molecules of the heavier main-group elements with unpaired electrons are usually more stable than the respective radicals of the first-row elements. Theoretical studies have been carried out which investigate the structures and energies of neutral molecules of germanium, tin and lead which have unpaired electrons. The group of Schiesser has been particularly active in the field. They published several papers which report about quantum chemical investigations of reactions of free radicals containing the heavier group-14 elements[187–194].

The homolytic substitution reaction at sulfur, selenium and tellurium carrying EH_3 groups (E = C to Sn) was calculated by Schiesser and Smart at the QCISD//MP2 level[187]. Figure 65 shows the reactions which were studied. Figure 66 gives the optimized transition states for substitution of EH_3 by a methyl group and the calculated activation energies. The $\Delta E_1^{\ddagger}$ values are the activation barriers with respect to $CH_3YH + EH_3$ (Y = S, Se, Te) and the $\Delta E_2^{\ddagger}$ values are the barriers with respect to $EH_3YH + CH_3$. The trend of the activation energies with respect to atom E is Si > Ge > Sn and the trend with respect to

TABLE 41. Calculated bond lengths (Å) and angles (deg) of $C_4H_4ESiH_3^-$ [a]

	E			
	C	Si	Ge [b]	Sn
α [c]	177.3	104.5	99.7 (100.1)	94.7
$d(C_1-C_2) = d(C_3-C_4)$	1.408	1.367	1.361(1.36)	1.358
$d(C_2\text{-}C_3)$	1.428	1.477	1.484(1.46)	1.493

[a] At the RHF level. From Reference 186.
[b] Experimental values are given in parentheses.
[c] Out-of-plane bonding angle of SiH_3.

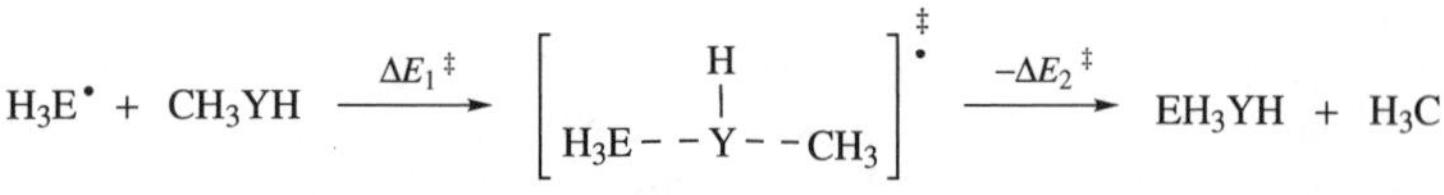

$$H_3E^{\bullet} + CH_3YH \xrightarrow{\Delta E_1^{\ddagger}} \left[\begin{array}{c} H \\ | \\ H_3E--Y--CH_3 \end{array} \right]^{\ddagger\bullet} \xrightarrow{-\Delta E_2^{\ddagger}} EH_3YH + H_3C^{\bullet}$$

E = Si, Ge, Sn
Y = S, Se, Te

FIGURE 65. Homolytic substitution reactions which were investigated in Reference 187. The calculated transition states are shown in Figure 66

Y is S > Se > Te; i.e. the activation barrier becomes lower when atom E or Y becomes heavier[187].

Schiesser et al. also calculated at the HF and correlated (MP2 and QCISD) levels the reaction profiles of some free-radical homolytic substitution reactions at silicon, germanium and tin centers[188]. The reactions which were studied are the methyl and hydrogen substitution of EH_3 and EMe_3 by hydrogen and methyl radicals, respectively. Figure 67 gives the optimized structures of the transition states and the calculated activation energies $\Delta E^{\ddagger}$. The authors found shallow energy minima $Me{-}EH_3{-}Me$ at correlated levels with pentacoordinated atoms E which are slightly (<1 kcal mol^{-1}) lower in energy than the transition states $Me{-}EH_3{-}Me^{\ddagger}$. The pentacoardinated structures become higher in energy than the transition states when ZPE corrections are included[188]. The calculated barriers for substituting hydrogen by methyl show the trend Ge > Si > Sn; i.e. they do *not* follow the size of the atoms. The activation barrier for breaking the E−Me bonds shows the regular trend Si > Ge > Sn. The differences between the values for Si and Ge are, however, much less than between Ge and Sn.

Schiesser and Skidmore reported a combined experimental/theoretical study of free-radical substitution reactions of aryltellurides with stannyl, germyl and silyl radicals[189]. The authors calculated at the QCISD/MP2 level the adducts, products and transition states of the reactions which are shown in Figure 68. The optimized geometries of the transition states and the activation barriers for the forward ($\Delta E_1^{\ddagger}$) and reverse ($\Delta E_2^{\ddagger}$) reactions are shown in Figure 69. The related substitution reactions with methyl radical were studied before (see Figure 66). The calculated activation barriers for the forward and reverse reactions show that the trend for the group-14 groups EH_3 is Si > Ge > Sn. The trend for the alkyl groups is Me < Et < *i*-Pr for the forward reactions while the opposite trend Me < Et < *i*-Pr is predicted for the reverse reaction[189].

Schiesser and Styles investigated in another theoretical study at the QCISD//MP2 level the reaction course of the 1,2-migration of silyl, germyl and stannyl substituents of radicals $H_3E{-}CH_2X^{\bullet}$ to $H_3EX{-}CH_2^{\bullet}$ ($X = CH_2$, NH, O; E = Si, Ge, Sn)[190]. Figure 70 gives the optimized geometries of the transition states and the calculated reaction barriers. The degenerate rearrangement of $H_3ECH_2CH_2^{\bullet}$ has nearly the same activation energy for E = Si and E = Ge while the reaction barrier for the stannyl group is lower. The forward reactions of the aminyl and oxyl radicals $H_3E{-}CH_2NH^{\bullet}$ and $H_3E{-}CH_2O^{\bullet}$ yielding $H_3ENH{-}CH_2^{\bullet}$ and $H_3EO{-}CH_2^{\bullet}$, respectively, have lower barriers for rearrangements than the methylene analogues. Thus, the activation barrier of the forward reactions for rearrangment of $H_3EX{-}CH_2^{\bullet}$ show the trend E = Ge > Si > Sn and $X = CH_2$ > NH > O[190]. Note that the geometry and the energy of the transition structure of the stannyloxo

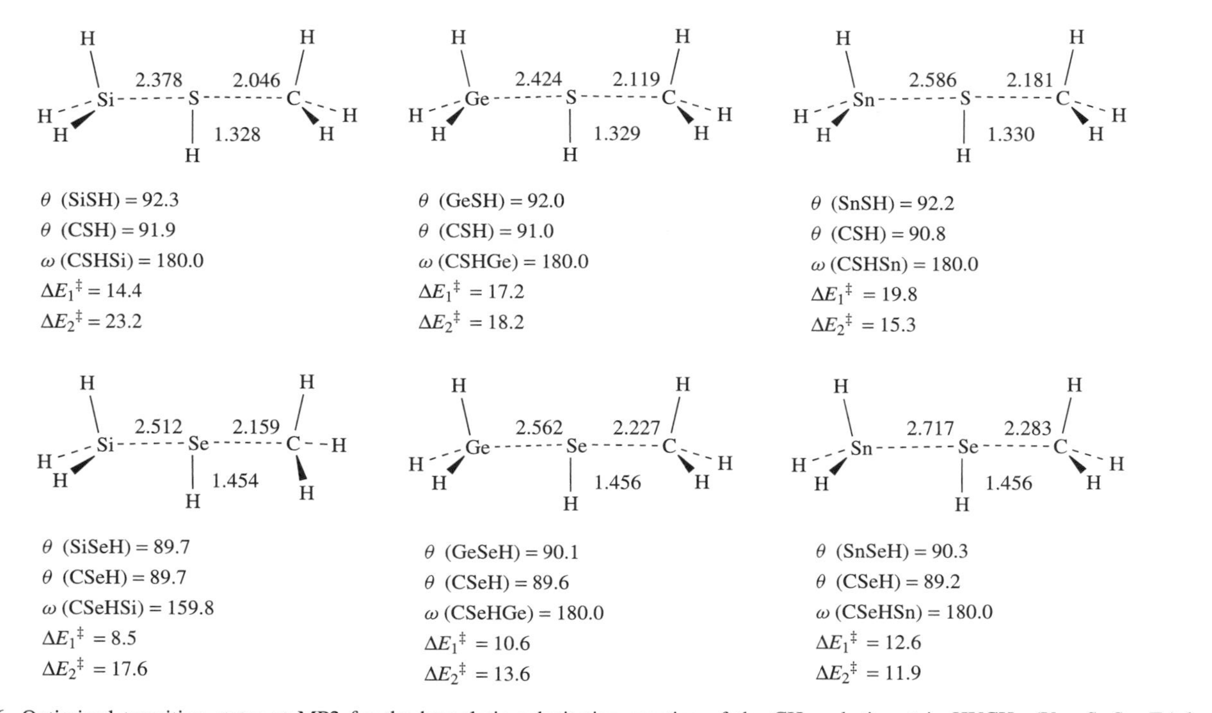

FIGURE 66. Optimized transition states at MP2 for the homolytic substitution reaction of the CH_3 substituent in $HYCH_3$ (Y = S, Se, Te) by $H_3E^\bullet$ (E = Si, Ge, Sn). Bond distances are in Å, angles in deg. The calculated activation energies (kcal mol^{-1}) at QCISD//MP2 refer to the forward reaction ($\Delta E_1^\ddagger$) and the reverse reaction ($\Delta E_2^\ddagger$), respectively (see Figure 65). Reprinted from Reference 187 with permission from Elsevier Science

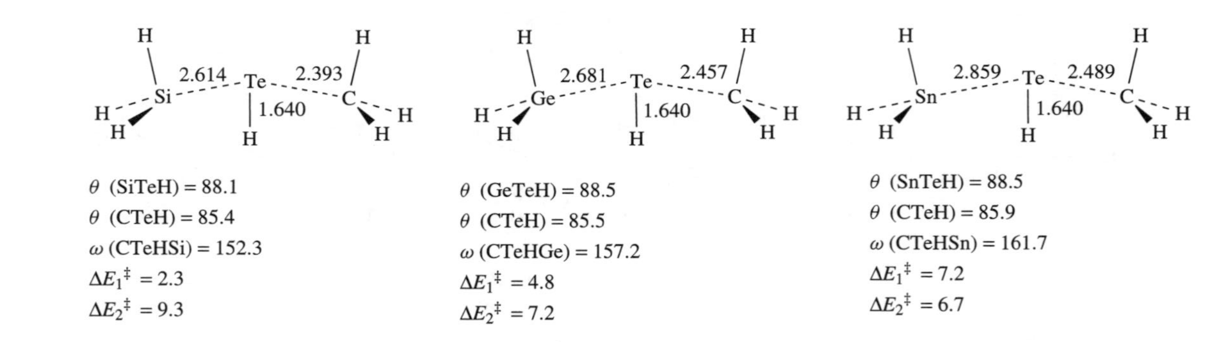

FIGURE 66. (*Continued*)

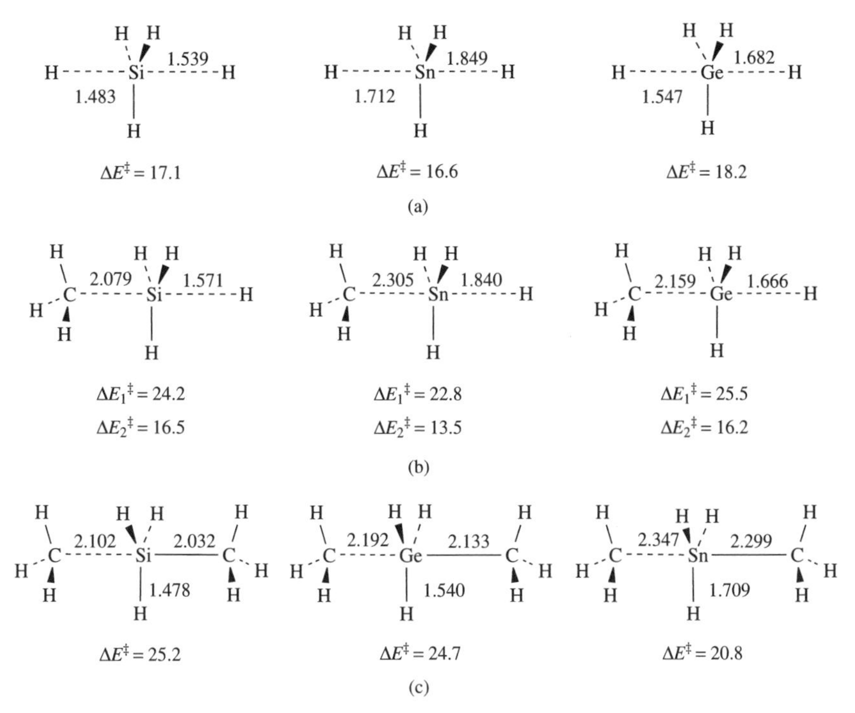

FIGURE 67. Optimized transition states at MP2 for the homolytic methyl and hydrogen substitution reaction of EH_4 and EH_3Me by $H^\bullet$ and $Me^\bullet$ (E = Si, Ge, Sn). Bond distances are in Å. The calculated activation energies (kcal mol^{-1}) at QCISD//MP2 refer in the case of (b) to the forward reaction ($\Delta E_1{}^\ddagger$) and the reverse reaction ($\Delta E_2{}^\ddagger$), respectively. Reproduced by permission of The Royal Society of Chemistry from Reference 188

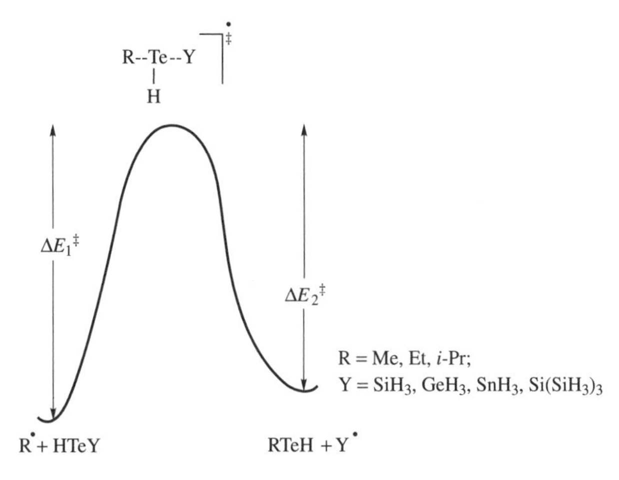

FIGURE 68. Free-radical substitution reactions of hydridotellurides $HTeEH_3$ (E = Si, Ge, Sn) and $HTeSi(SiH_3)_3$ with alkyl groups $R^\bullet$. The optimized transition states and calculated activation energies are given in Figure 69

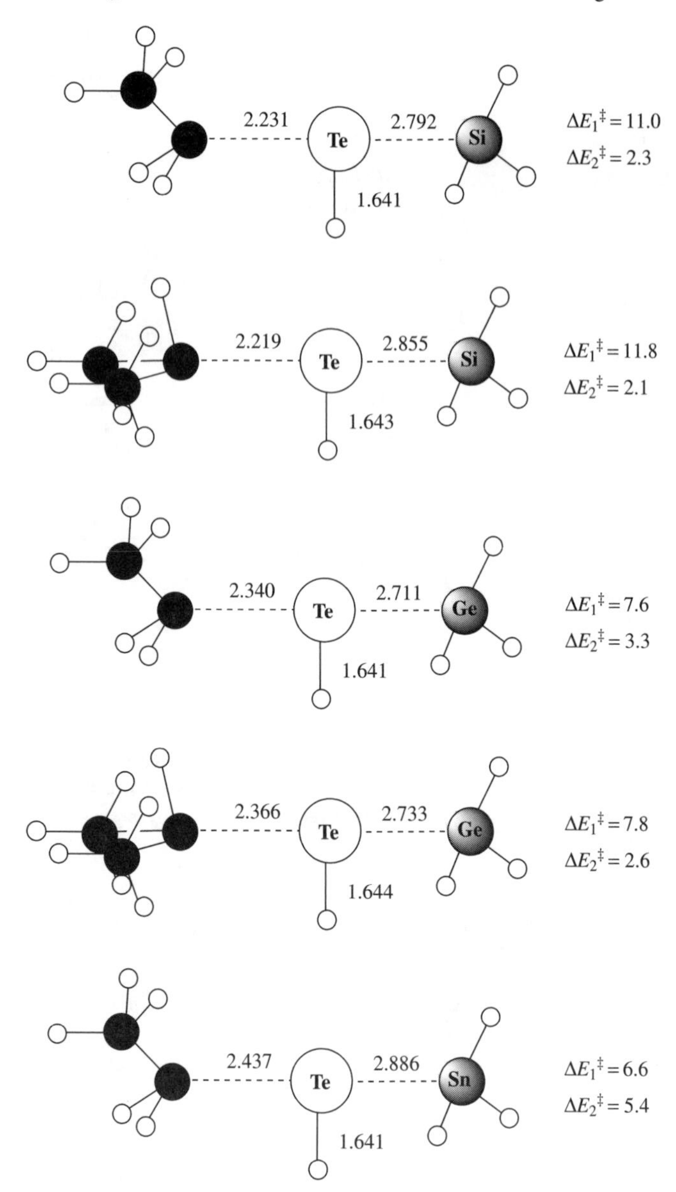

FIGURE 69. Optimized transition states at MP2 for the homolytic free-radical substitution reactions shown in Figure 68 of hydridotellurides $HTeEH_3$ (E = Si, Ge, Sn) and $HTeSi(SiH_3)_3$ with alkyl groups. Bond distances are in Å. The calculated activation energies (kcal mol^{-1}) at QCISD//MP2 refer to the forward reaction ($\Delta E_1^{\ddagger}$) and the reverse reaction ($\Delta E_2^{\ddagger}$), respectively. Reprinted from Reference 189 with permission from Elsevier Science

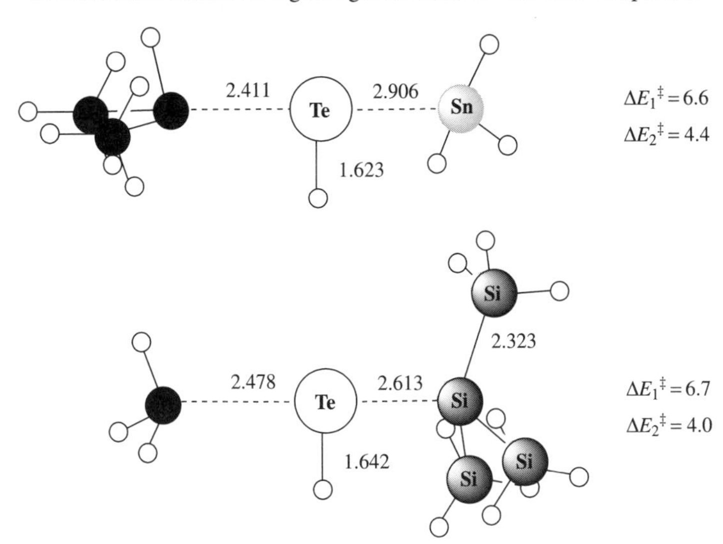

FIGURE 69. (*Continued*)

species $H_3SnCH_2O\bullet$ is given only at the HF level. The results of the other species show that the inclusion of correlation effects at QCISD lowers the barriers of the stannyl radicals by 5–8 kcal mol^{-1}.

Another paper from the Schiesser group by Dakternieks et al. reported about *ab initio* and semiempirical calculations of the hydrogen abstraction from R_3SnH by alkyl radicals $R^\bullet$ [191]. Figure 71 shows the investigated reactions. Figure 72 gives the optimized transition states and the activation barriers for the forward and reverse reactions at the QCISD//MP2 level of theory. It becomes obvious that larger alkyl groups R decrease the activation barrier for hydrogen abstraction from R_3SnH with the order R = Me > Et > *i*-Pr > *t*-Bu[191].

Hydrogen abstraction reactions from trialkylsilanes and trialkylgermanes by hydrogen atom or alkyl radicals are the topic of another lengthy theoretical study at *ab initio* and semiempirical levels by Dakternieks et al.[192]. The alkyl groups which were considered are methyl, ethyl, isopropyl and *tert*-butyl. The calculated activation barriers for hydrogen abstraction from silicon were found to be higher than from germanium. Table 42 gives the theoretically predicted activation energies for the forward and reverse hydrogen abstraction reactions of R_3GeH with $R'^\bullet$. The forward reactions are strongly exothermic and have lower barriers than the reverse reactions because the R′–H bonds are stronger than the Ge–H bonds. Methyl substituents at germanium have little influence on the activation barriers of the forward reaction, and their effect on the reverse reaction is also rather small. The trend of the activation barriers for the forward reaction with different radicals is Me > Et > H > *i*-Pr > *t*-Bu[192].

A third theoretical paper by Dakternieks et al. about hydrogen abstraction reactions calculated the equilibrium structures and transition states for hydrogen transfer between silyl, germyl and stannyl radicals and their hydrides[193]. Figure 73 shows the transition states for the homonuclear and heteronuclear hydrogen transfer from EH_4 to $EH_3^\bullet$ (E = Si, Ge, Sn). The energy barrier of the homonuclear hydrogen exchange shows the expected trend: Si > Ge > Sn. The barriers of the heteronuclear reactions between EH_4 and $E'H_3^\bullet$

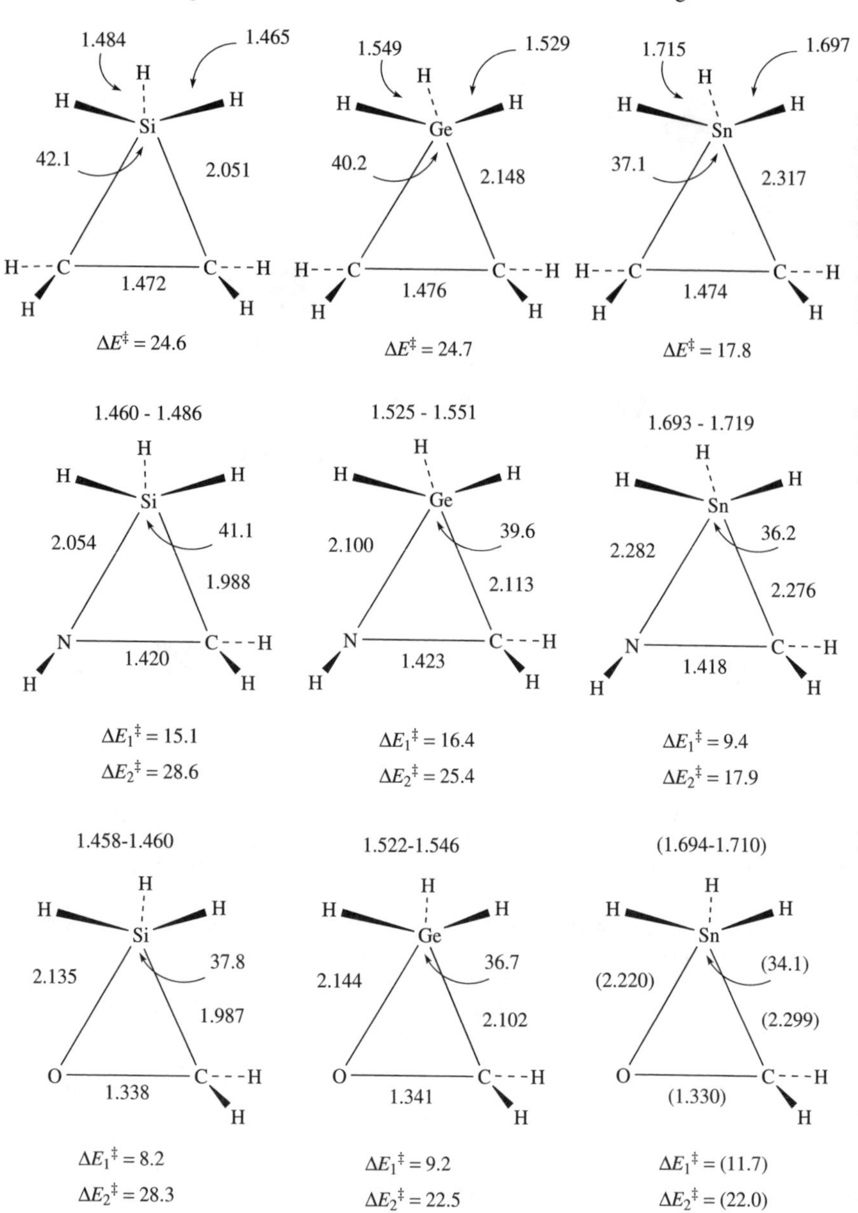

FIGURE 70. Calculated transition states at MP2 of the 1,2-migration of silyl, germyl and stannyl substituents of radicals $H_3E\text{-}CH_2X^{\bullet}$ to $H_3EX\text{-}CH_2^{\bullet}$ ($X = CH_2$, NH, O; $E = Si$, Ge, Sn). Bond distances are in Å, angles in deg. The calculated activation energies (kcal mol^{-1}) at QCISD//MP2 refer to the forward reaction ($\Delta E_1{}^{\ddagger}$) and the reverse reaction ($\Delta E_2{}^{\ddagger}$), respectively. Reproduced by permission of The Royal Society of Chemistry from Reference 190

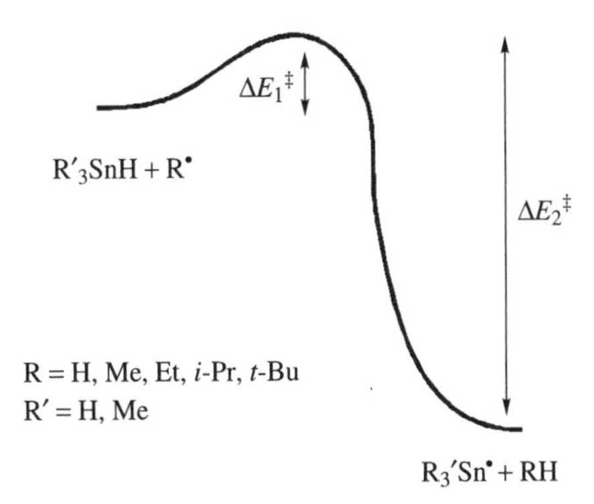

FIGURE 71. Schematic representation of the reaction profile for some hydrogen abstraction reactions of trialkylstannane with hydrogen and alkyl radicals[191]. The optimized transition states and calculated activation energies are given in Figure 72

are higher than those of the homonuclear reactions of elements E and E′ if the reaction is endothermic ($\Delta E_1{}^{\ddagger}$ in Figure 73) and they are lower if the reaction is exothermic ($\Delta E_2{}^{\ddagger}$ in Figure 73). Figure 74 shows the transition state structures for the homonuclear and heteronuclear hydrogen transfer between Me_3EH and $H_3E^{\bullet}$ (forward reaction) and between EH_4 and $Me_3E^{\bullet}$ (reverse reaction). The calculated activation barriers for the forward reaction $\Delta E_1{}^{\ddagger}$ and for the reverse reaction $\Delta E_2{}^{\ddagger}$ are also given. Note that the energies of the methyl-substituted systems are only given at MP2 while the parent systems were also calculated at QCISD. The trend of the theoretically predicted activation barriers is Si > Ge > Sn. The authors discuss also the geometries of the transition states[193].

The latest paper about radical reactions of group-14 compounds by the Schiesser group reports *ab initio* and DFT calculations of the frontside and backside radical substitution reactions of $H_3EE'H_3$ with $E''H_3^{\bullet}$ (E, E′, E″ = Si, Ge, Sn)[194]. Figure 75 shows schematically the structures of the transition states for the backside attack (transition state **L**) and for the frontside attack (transition state **M**) which have been calculated at different levels of theory. Table 43 gives the theoretically predicted activation energies. The first three entries give the calculated values for the degenerate substitution reaction of the systems with E = E′ = E″. At the highest level of theory (CCSD(T)/DZP//MP2/DZP) it is found that the silicon and germanium compounds favor the backside attack by 2–3 kcal mol^{-1} over the frontside attack, while the frontside and backside attack of the stannyl compounds are nearly degenerate. Calculations of the silicon systems at CCSD(T) with larger basis sets up to cc-pVDZ do not change the results significantly[194]. Note that the B3LYP data for the activation energies in Table 43 are always too low.

The next six entries in Table 43 give the activation barriers of the degenerate substitution reactions where E = E″ with different central atoms E′. It is found that the backside attack is more favored than the frontside attack by 3–4 kcal mol^{-1} (CCSD(T)//MP2) except when E = E″ = Sn. The remaining 9 entries give the barriers for the frontside and backside nondegenerate substitution of EH_3 groups by $E''H_3$. Table 43 gives for **L** and **M** the activation barriers for the forward reaction ($\Delta E_1{}^{\ddagger}$ with respect to $H_3EE'H_3 + E''H_3^{\bullet}$) and for the reverse reaction ($\Delta E_2{}^{\ddagger}$ with respect to $H_3E''E'H_3 + EH_3^{\bullet}$). The backside attack is in most cases favored over the frontside attack. In a few cases involving stannyl

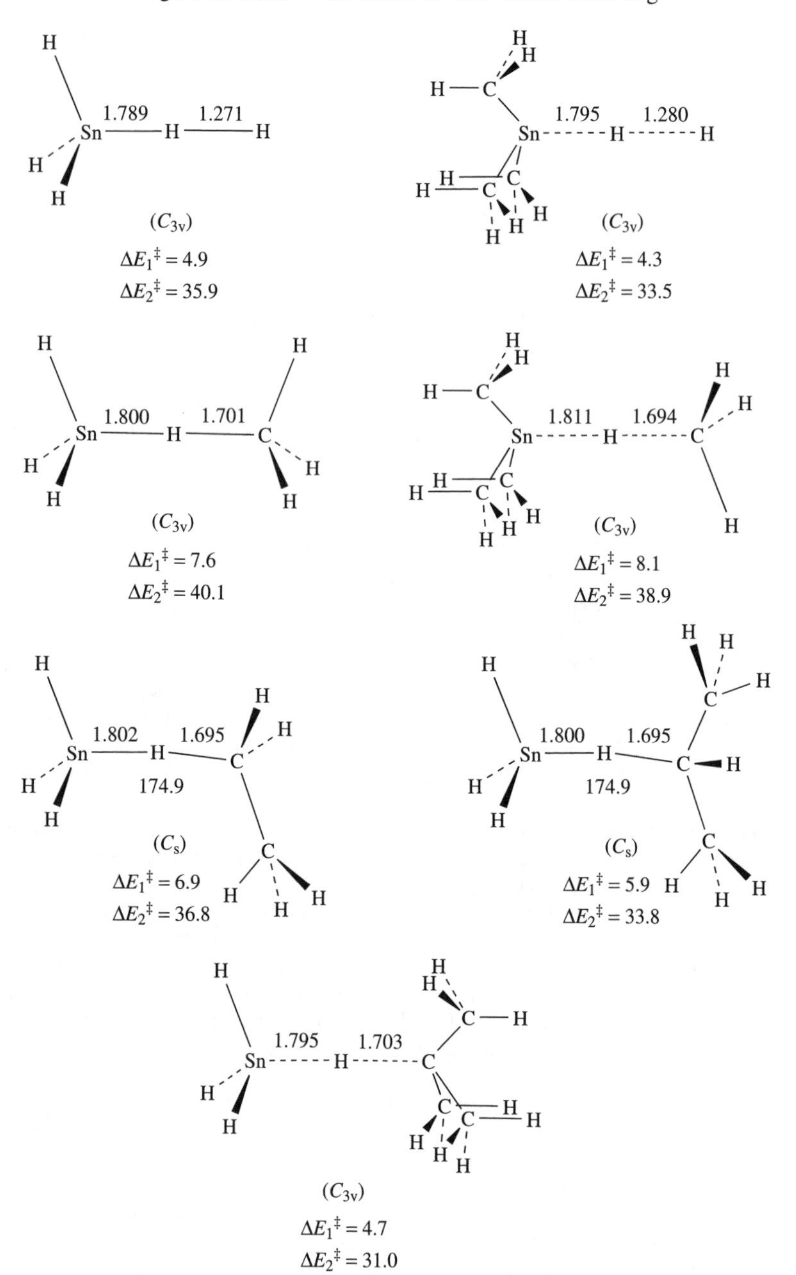

FIGURE 72. Calculated transition states at MP2 of the hydrogen abstraction reactions of trialkylstannane with hydrogen and alkyl radicals which are shown in Figure 71. Bond distances are in Å, angles in deg. The calculated activation energies (kcal mol^{-1}) at QCISD//MP2 (MP2 values in parentheses) refer to the forward reaction ($\Delta E_1^{\ddagger}$) and the reverse reaction ($\Delta E_2^{\ddagger}$), respectively. Reproduced by permission of The Royal Society of Chemistry from Reference 191

TABLE 42. Calculated energy barriers (kcal mol^{-1}) at QCISD//MP2 for the forward ($\Delta E_1{}^{\ddagger}$) and reverse ($\Delta E_2{}^{\ddagger}$) hydrogen atom abstraction reactions of germanes by hydrogen atoms and various alkyl radicals R[a]

Germane	R	Method	$\Delta E_1{}^{\ddagger}$	$\Delta E_2{}^{\ddagger}$
$H_3Ge{-}H$	H	MP2/DZP	8.7	27.7
		QCISD/DZP	6.8	28.3
$H_3Ge{-}H$	Me	MP2/DZP	9.8	34.7
		QCISD/DZP	10.0	33.1
$H_3Ge{-}H$	Et	MP2/DZP	8.8	31.5
		QCISD/DZP	9.8	30.4
$H_3Ge{-}H$	*i*-Pr	MP2/DZP	7.6	28.5
		QCISD/DZP	8.7	27.2
$H_3Ge{-}H$	*t*-Bu	MP2/DZP	5.9	25.5
		QCISD/DZP	7.6	24.5
$MeH_2Ge{-}H$	H	MP2/DZP	8.4	26.9
		QCISD/DZP	6.6	27.5
$MeH_2Ge{-}H$	Me	MP2/DZP	10.0	34.3
		QCISD/DZP	10.3	32.7
$MeH_2Ge{-}H$	Et	MP2/DZP	9.2	31.2
		QCISD/DZP	9.9	29.9
$MeH_2Ge{-}H$	*i*-Pr	MP2/DZP	7.9	28.2
		QCISD/DZP	9.2	27.1
$MeH_2Ge{-}H$	*t*-Bu	MP2/DZP	6.3	25.3
$Me_2HGe{-}H$	H	MP2/DZP	8.2	25.9
		QCISD/DZP	6.4	26.6
$Me_2HGe{-}H$	Me	MP2/DZP	10.1	33.7
$Me_2HGe{-}H$	Et	MP2/DZP	9.3	30.6
$Me_2HGe{-}H$	*i*-Pr	MP2/DZP	8.3	27.9
$Me_3Ge{-}H$	H	MP2/DZP	7.9	25.0
		QCISD/DZP	6.1	25.6
$Me_3Ge{-}H$	Me	MP2/DZP	10.1	33.1
		QCISD/DZP	10.5	31.7
$Me_3Ge{-}H$	Et	MP2/DZP	9.4	30.1

[a]Reference 192.

groups, however, the frontside attack becomes competitive and is even predicted to be slightly more favorable than the backside attack. The authors tried also to locate transition states for the frontside attack of the systems involving the methyl radical. However, except for reactions of methylstannane they were unable to locate transition states for frontside attack at correlated levels of theory[194].

Small germyl radicals in the ground state and first excited state have been calculated using DFT (B3PW91) and *ab initio* methods by BelBruno[195]. The author reports the optimized geometries of the doublet ground states and quartet excited states of GeH, $GeCH_3$ and GeC_2H_5. An experimental and theoretical study of gaseous products in the radiolysis of GeH_4/C_2H_4 mixtures was reported by Antoniotti et al.[196]. The authors carried out *ab initio* calculations at the QCISD(T)/6-311G(3df,2p) level using MP2/DZP

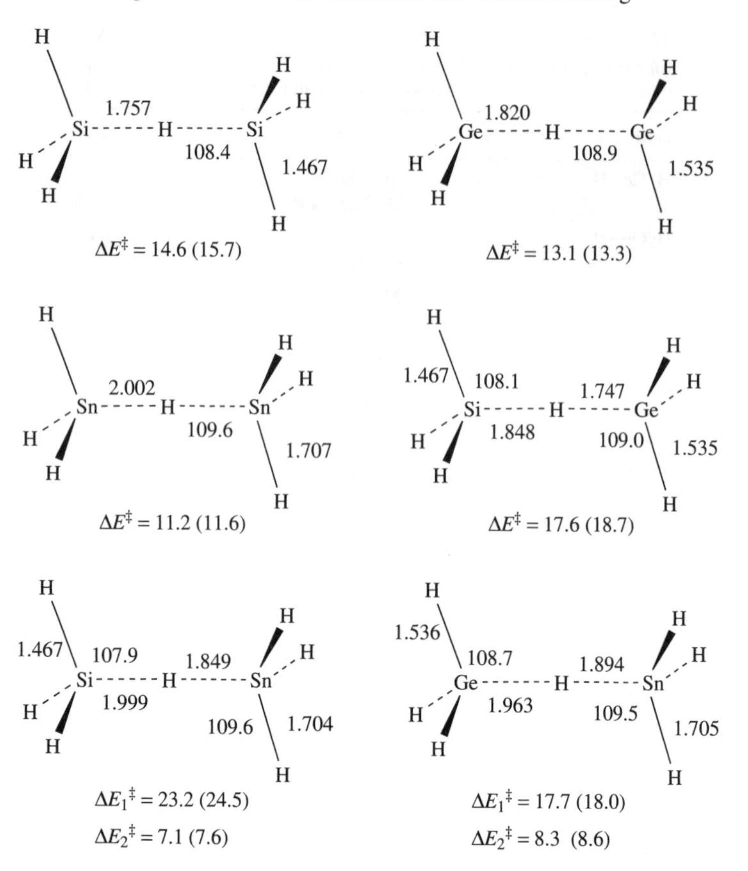

FIGURE 73. Calculated transition states at MP2 of the hydrogen abstraction reactions of EH_4 by $EH_3^{\bullet}$ radicals (E = Si, Ge, Sn). Bond distances are in Å, angles in deg. The calculated activation energies (kcal mol^{-1}) at QCISD/MP2 (MP2 values in parentheses) refer to the forward reaction ($\Delta E_1{}^{\ddagger}$) and the reverse reaction ($\Delta E_2{}^{\ddagger}$), respectively. Reprinted with permission from Reference 193. Copyright 1998 American Chemical Society

optimized geometries of GeC_2H_n ($n = 4$–7) molecules. Figures 76–79 show the geometries and relative energies of the stationary points on the PES. Three energy minima were found for GeC_2H_4 (Figure 76). The singlet state of germylidenecyclopropane **60** is predicted to be the global energy minimum followed by the triplet state **61** which is 17.6 kcal mol^{-1} higher in energy. Structure **62** in Figure 76 is the side-on bonded complex of germylene with acetylene and is 38.4 kcal mol^{-1} less stable than **60**. The authors do not report about germacyclopropene[196]. The earlier work by Boatz et al. which was discussed above gave at the MP2 level only the C_{2v} symmetric structure of germacyclopropane

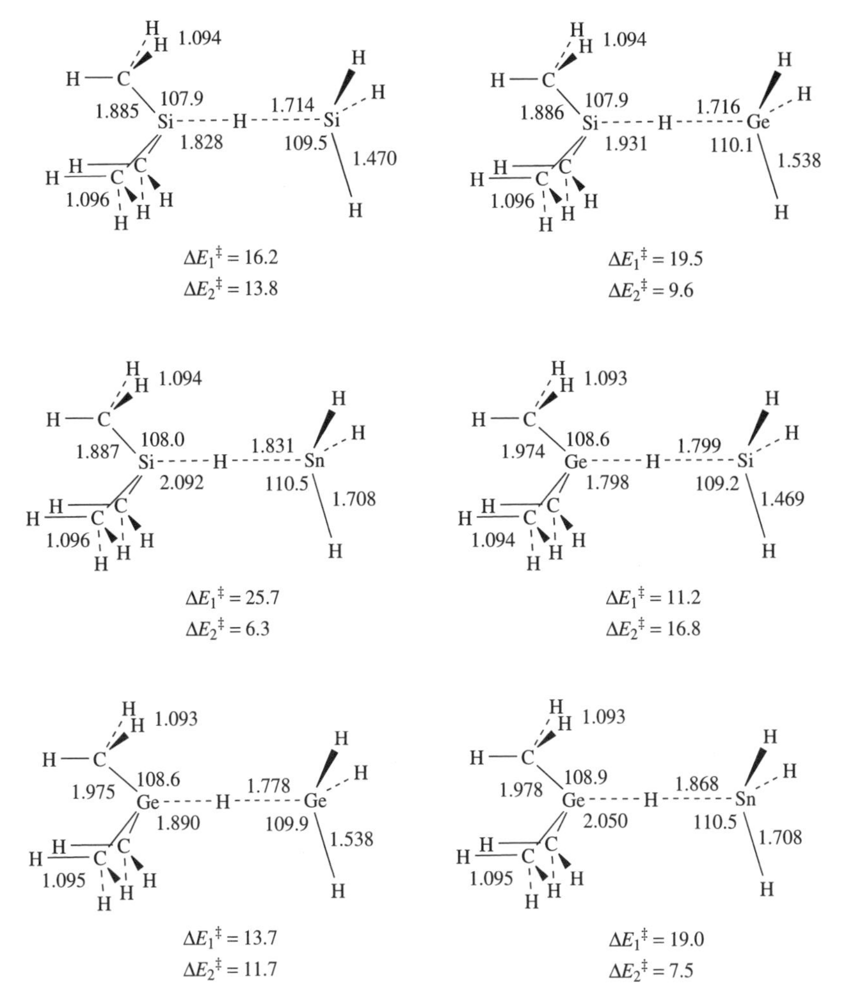

FIGURE 74. Calculated transition states at MP2 of the hydrogen abstraction reactions of Me_3EH by $EH_3^\bullet$ radicals (E = Si, Ge, Sn). Bond distances are in Å, angles in deg. The calculated activation energies (kcal mol^{-1}) at MP2 refer to the forward reaction ($\Delta E_1^\ddagger$) and the reverse reaction ($\Delta E_2^\ddagger$), respectively. Reprinted with permission from Reference 193. Copyright 1998 American Chemical Society

as an energy minimum while the complex **62** was not found as an energy minimum structure[88].

Five energy minima and one transition state which are shown in Figure 77 were found on the GeC_2H_5 doublet PES[195]. Structures **63** and **64** are rotational isomers with the

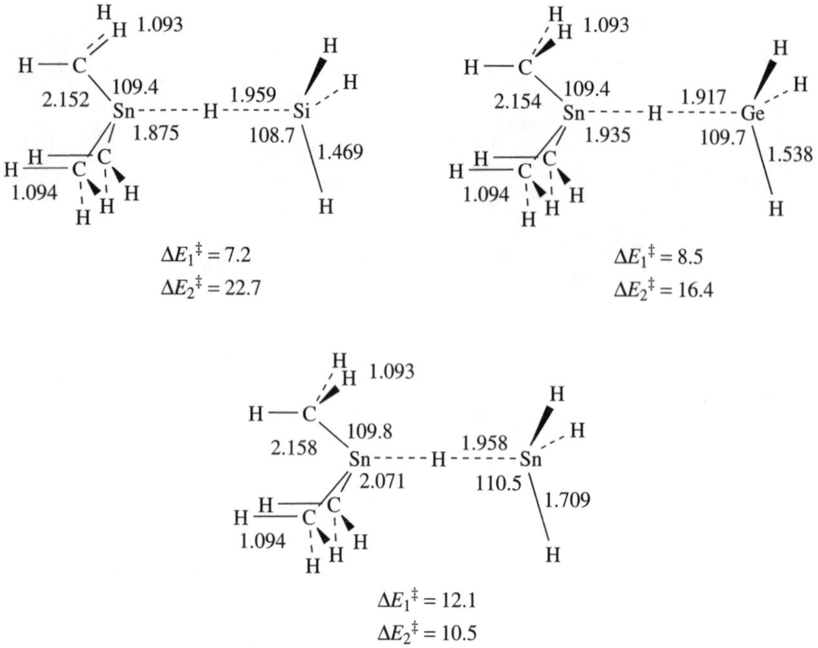

$\Delta E_1^{\ddagger} = 7.2$
$\Delta E_2^{\ddagger} = 22.7$

$\Delta E_1^{\ddagger} = 8.5$
$\Delta E_2^{\ddagger} = 16.4$

$\Delta E_1^{\ddagger} = 12.1$
$\Delta E_2^{\ddagger} = 10.5$

FIGURE 74. (*Continued*)

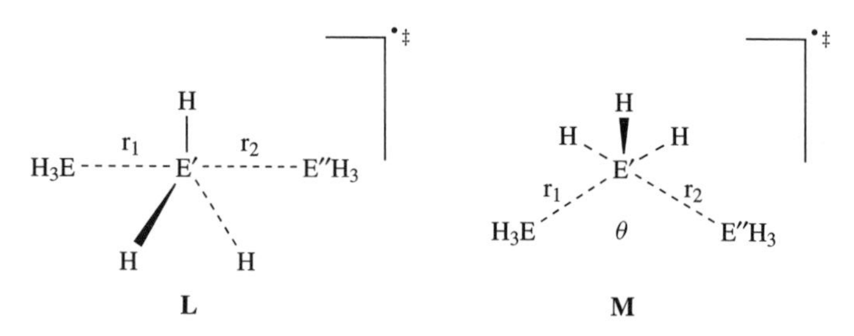

FIGURE 75. Schematic representation of the transition states of EH_3 substitution in $H_3EE'H_3$ by $E''H_3^{\bullet}$ (E = Si, Ge, Sn) for the backside attack **L** and frontside attack **M**[194]. The calculated activation barriers are given in Table 43

connectivity $H_2Ge{-}CH{-}CH_2$ which are energetically nearly degenerate. Structure **65** is the transition state for the interconvertion of **63** and **64**. The other energy minima **66**, **67** and **68** are 21–27 kcal mol^{-1} higher in energy than **63**.

Five singlets and four triplets have been found as stationary points on the GeC_2H_6 PES (Figure 78). All singlets are lower in energy than the triplets. Germylethene **69** is the global energy minimum. Structure **70** is the transition state for rotation about the Ge−C bond. Germyleneethane **71** is only 4.3 kcal mol^{-1} higher in energy than **69**. Germacyclopropane **72** and 1-germapropene **73** are 10.0 and 14.7 kcal mol^{-1} less stable than **69**. The triplet states **74**, **75** and **77** are high-lying energy minima on the GeC_2H_6 PES. The triplet structure **76** is a transition state.

TABLE 43. Calculated energy barriers ($\Delta E^{\ddagger}$) (kcal mol^{-1}) for the degenerate and nondegenerate homolytic substitution of silyl, germyl and stannyl radicals of silylgermane, silylstannane and germylstannane by $E''H_3$ (E'' = Si, Ge, Sn) radicals[a]

E	E′	E″	method	L[b]		M[b]	
				$\Delta E_1^{\ddagger}$	$\Delta E_2^{\ddagger}$	$\Delta E_1^{\ddagger}$	$\Delta E_2^{\ddagger}$
Si	Si	Si	HF/DZP	29.8	29.8	31.4	31.4
			MP2/DZP	16.9	16.9	19.7	19.7
			QCISD/DZP//MP2/DZP	17.0	17.0	19.5	19.5
			CCSD(T)/DZP//MP2/DZP	15.6	15.6	18.1	18.1
			B3LYP/DZP	12.2	12.2	15.6	15.6
Ge	Ge	Ge	HF/DZP	28.3	28.3	30.2	30.2
			MP2/DZP	16.9	16.9	20.0	20.0
			QCISD/DZP//MP2/DZP	16.9	16.9	19.8	19.8
			CCSD(T)/DZP//MP2/DZP	15.6	15.6	18.3	18.3
			B3LYP/DZP	12.0	12.0	16.0	16.0
Sn	Sn	Sn	HF/DZP	25.3	25.3	23.9	23.9
			MP2/DZP	15.6	15.6	15.8	15.8
			QCISD/DZP//MP2/DZP	15.1	15.1	15.1	15.1
			CCSD(T)/DZP//MP2/DZP	14.0	14.0	14.1	14.1
			B3LYP/DZP	10.1	10.1	11.8	11.8
Ge	Si	Ge	HF/DZP	29.7	29.7	30.9	30.9
			MP2/DZP	17.4	17.4	19.9	19.9
			QCISD/DZP//MP2/DZP	17.3	17.3	19.5	19.5
			CCSD(T)/DZP//MP2/DZP	15.8	15.8	18.1	18.1
			B3LYP/DZP	12.7	12.7	15.8	15.8
Sn	Si	Sn	HF/DZP	26.7	26.7	24.2	24.2
			MP2/DZP	15.4	15.4	15.1	15.1
			QCISD/DZP//MP2/DZP	14.7	14.7	14.3	14.3
			CCSD(T)/DZP//MP2/DZP	13.2	13.2	13.1	13.1
			B3LYP/DZP	9.3	9.3	10.7	10.7
Si	Ge	Si	HF/DZP	28.7	28.7	31.2	31.2
			MP2/DZP	16.7	16.7	20.3	20.3
			QCISD/DZP//MP2/DZP	17.0	17.0	20.2	20.2
			CCSD(T)/DZP//MP2/DZP	15.6	15.6	18.9	18.9
			B3LYP/DZP	11.8	11.8	16.3	16.3
Sn	Ge	Sn	HF/DZP	25.9	25.9	23.8	23.8
			MP2/DZP	15.3	15.3	15.1	15.1
			QCISD/DZP//MP2/DZP	14.8	14.8	14.5	14.5
			CCSD(T)/DZP//MP2/DZP	13.4	13.4	13.3	13.3
			B3LYP/DZP	9.2	9.2	10.9	10.9
Si	Sn	Si	HF/DZP	27.8	27.8	31.1	31.1
			MP2/DZP	17.4	17.4	21.4	21.4
			QCISD/DZP//MP2/DZP	17.5	17.5	21.2	21.2
			CCSD(T)/DZP//MP2/DZP	16.5	16.5	20.1	20.1
			B3LYP/DZP	13.8	13.8	17.4	17.4
Ge	Sn	Ge	HF/DZP	27.3	27.3	30.0	30.0
			MP2/DZP	17.3	17.3	20.8	20.8
			QCISD/DZP//MP2/DZP	17.2	17.2	20.3	20.3
			CCSD(T)/DZP//MP2/DZP	16.1	16.1	19.3	19.3
			B3LYP/DZP	12.6	12.6	16.9	16.9

(continued overleaf)

TABLE 43. (*Continued*)

E	E′	E″	method	L[b]		M[b]	
				$\Delta E_1{}^{\ddagger}$	$\Delta E_2{}^{\ddagger}$	$\Delta E_1{}^{\ddagger}$	$\Delta E_2{}^{\ddagger}$
Si	Si	Ge	HF/DZP	27.1	31.5	29.0	33.4
			MP2/DZP	14.5	19.6	17.4	22.4
			QCISD/DZP	14.5	19.7	17.2	22.4
			CCSD(T)/DZP	13.1	18.4	15.8	21.1
			B3LYP/DZP	10.0	14.6	13.4	18.2
Si	Si	Sn	HF/DZP	24.1	34.4	26.0	36.2
			MP2/DZP	12.5	24.2	14.5	26.2
			QCISD/DZP	12.2	23.9	14.4	26.1
			CCSD(T)/DZP	11.0	22.9	13.1	25.0
			B3LYP/DZP	7.8	19.0	10.7	21.9
Si	Ge	Ge	HF/DZP	27.1	31.0	28.5	32.4
			MP2/DZP	15.1	19.6	17.6	22.2
			QCISD/DZP	14.9	19.5	17.2	21.9
			CCSD(T)/DZP	13.5	18.1	15.9	20.5
			B3LYP/DZP	10.3	14.6	13.7	18.0
Si	Ge	Sn	HF/DZP	24.4	33.1	26.0	34.6
			MP2/DZP	13.2	23.2	15.2	25.2
			QCISD/DZP	12.7	22.7	14.9	24.9
			CCSD(T)/DZP	11.4	21.6	13.5	23.8
			B3LYP/DZP	8.4	18.1	11.4	21.0
Si	Sn	Ge	HF/DZP	25.0	27.7	22.5	25.3
			MP2/DZP	13.7	17.2	13.3	16.7
			QCISD/DZP	13.0	16.6	12.5	16.1
			CCSD(T)/DZP	11.6	15.2	11.3	14.9
			B3LYP/DZP	7.7	11.0	9.0	12.3
Si	Sn	Sn	HF/DZP	23.1	29.3	20.7	26.9
			MP2/DZP	12.1	20.0	11.3	19.2
			QCISD/DZP	11.2	19.1	10.7	18.5
			CCSD(T)/DZP	10.0	18.0	9.4	17.6
			B3LYP/DZP	6.3	13.9	7.2	14.8
Ge	Si	Sn	HF/DZP	25.5	31.4	28.1	33.7
			MP2/DZP	14.2	20.8	17.3	23.9
			QCISD/DZP	14.4	20.9	17.3	23.8
			CCSD(T)/DZP	13.0	20.3	16.0	22.7
			B3LYP/DZP	9.4	15.7	13.4	19.8
Ge	Ge	Sn	HF/DZP	25.5	30.3	27.6	32.3
			MP2/DZP	14.6	20.2	17.5	23.0
			QCISD/DZP	14.5	19.9	17.2	22.6
			CCSD(T)/DZP	13.2	18.8	15.6	21.2
			B3LYP/DZP	9.8	15.1	13.6	18.9
Ge	Sn	Sn	HF/DZP	24.0	27.3	22.0	25.4
			MP2/DZP	13.4	17.8	13.1	17.5
			QCISD/DZP	12.9	17.2	12.6	16.9
			CCSD(T)/DZP	11.5	16.1	11.4	15.9
			B3LYP/DZP	7.6	12.6	9.1	13.3

[a]Reference 194.
[b]Values for **L** refer to backside attack, values for **M** refer to frontside attack (see Figure 75).

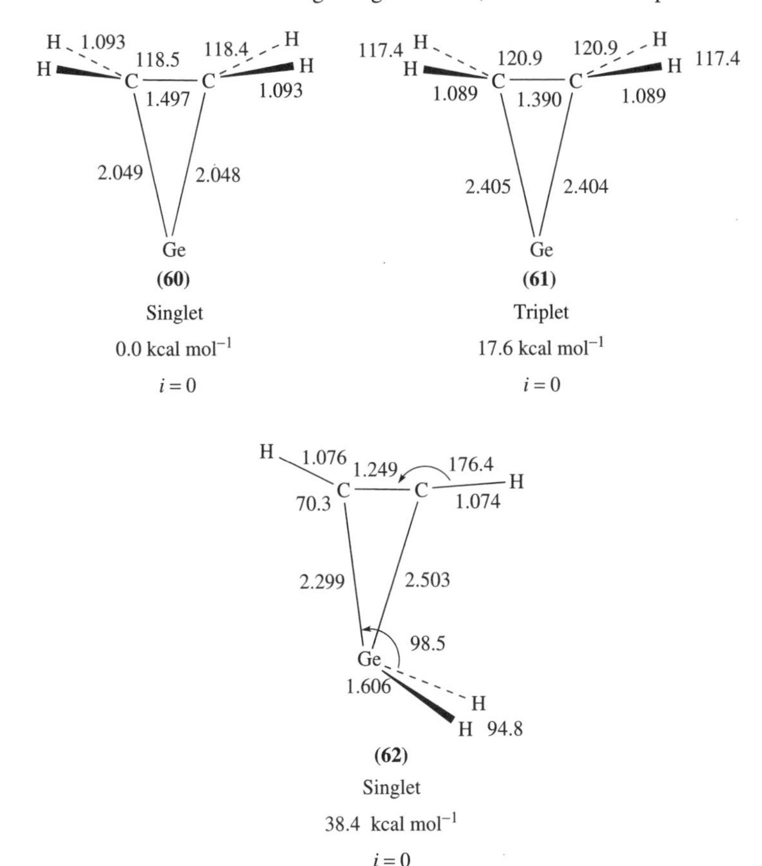

FIGURE 76. Optimized energy minima ($i = 0$) at MP2 on the singlet and triplet PES of GeC_2H_4 which were reported in Reference 195. Calculated relative energies at QCISD(T). Bond distances are in Å, angles in deg. Reproduced by permission of The Royal Society of Chemistry from Reference 195

Three energy minima and one transition state were located on the GeC_2H_7 doublet PES (Figure 79). The germyleneethane radical **78** is clearly the lowest lying form while **79** and **80** are 16–17 kcal mol^{-1} higher in energy. Structure **81** is a transition state. The authors calculated also the reaction energies of possible reactions of GeH_n ($n = 0–3$) with ethylene. The results are shown in Table 44.

V. CLOSING REMARKS

This review of the theoretical literature published since 1990 about organic germanium, tin and lead compounds demonstrates how important quantum chemical methods have become in modern chemical research. Prior to 1990, molecules containing heavier atoms than third-row elements were only rarely treated with the goal of obtaining accurate data about the geometry and energy of the compounds. The general acceptance of pseudopotentials and DFT as reliable quantum chemical methods, together with

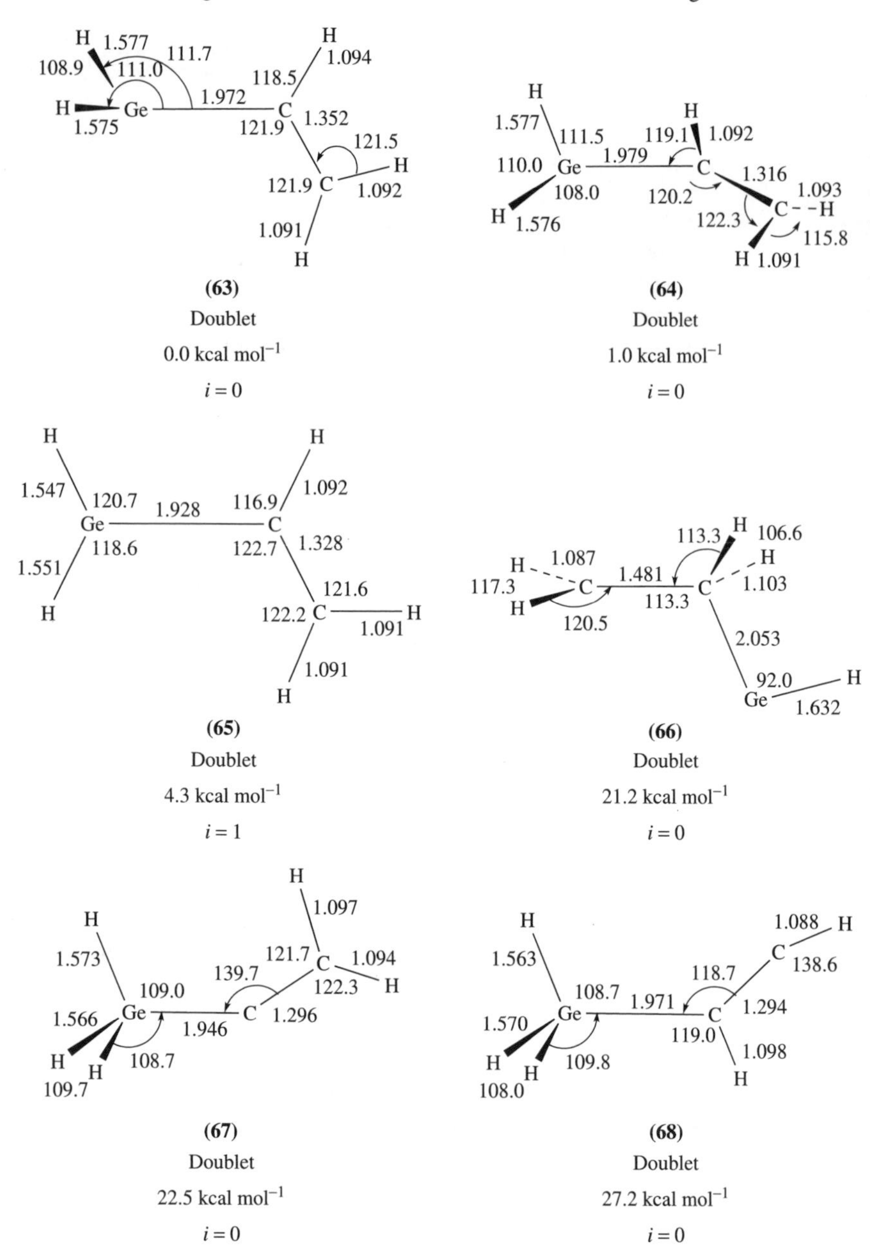

FIGURE 77. Optimized energy minima ($i = 0$) and transition states ($i = 1$) at MP2 on the doublet PES of GeC_2H_5. Calculated relative energies at QCISD(T). Bond distances are in Å, angles in deg. Reproduced by permission of The Royal Society of Chemistry from Reference 195

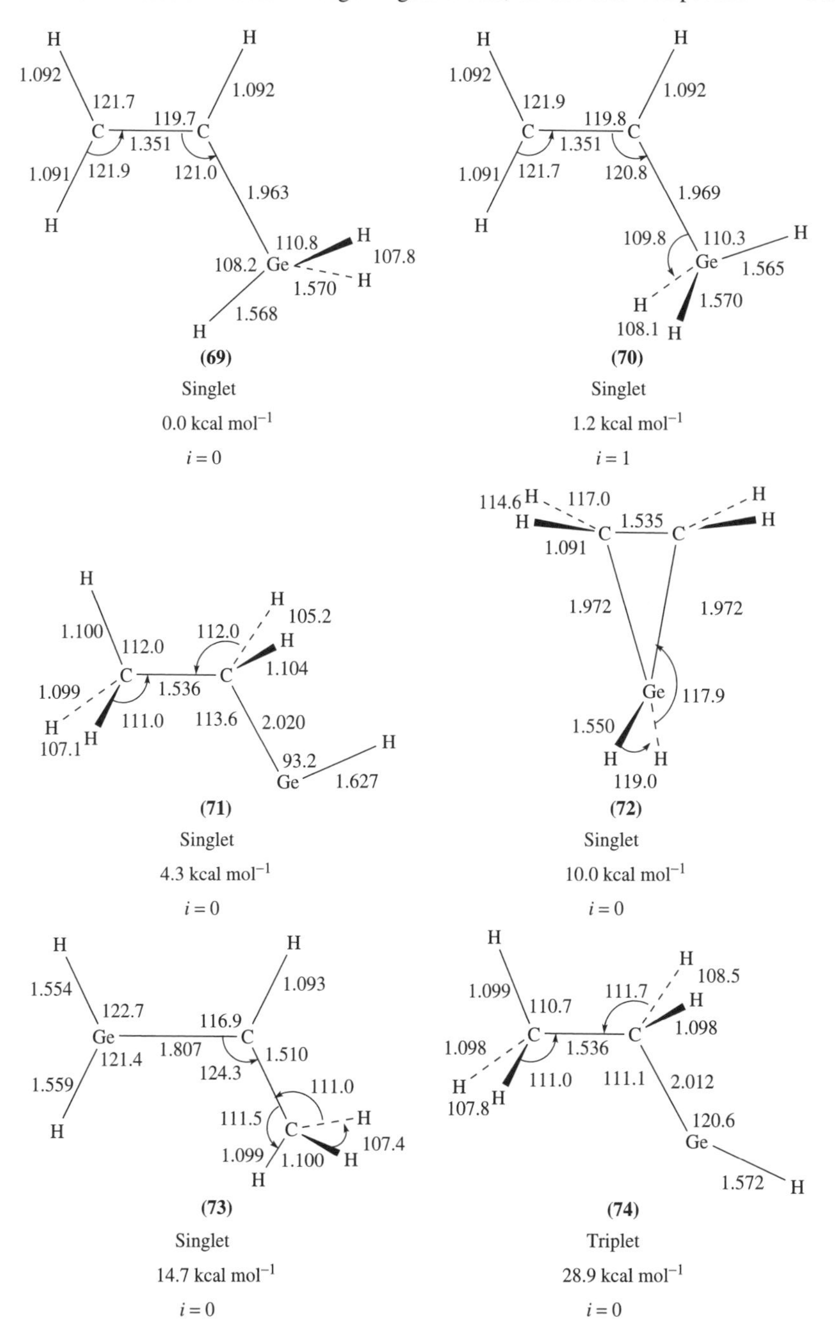

FIGURE 78. Optimized energy minima ($i = 0$) and transition states ($i = 1$) at MP2 on the singlet and triplet PES of GeC_2H_6. Calculated relative energies at QCISD(T). Bond distances are in Å, angles in deg. Reproduced by permission of The Royal Society of Chemistry from Reference 195

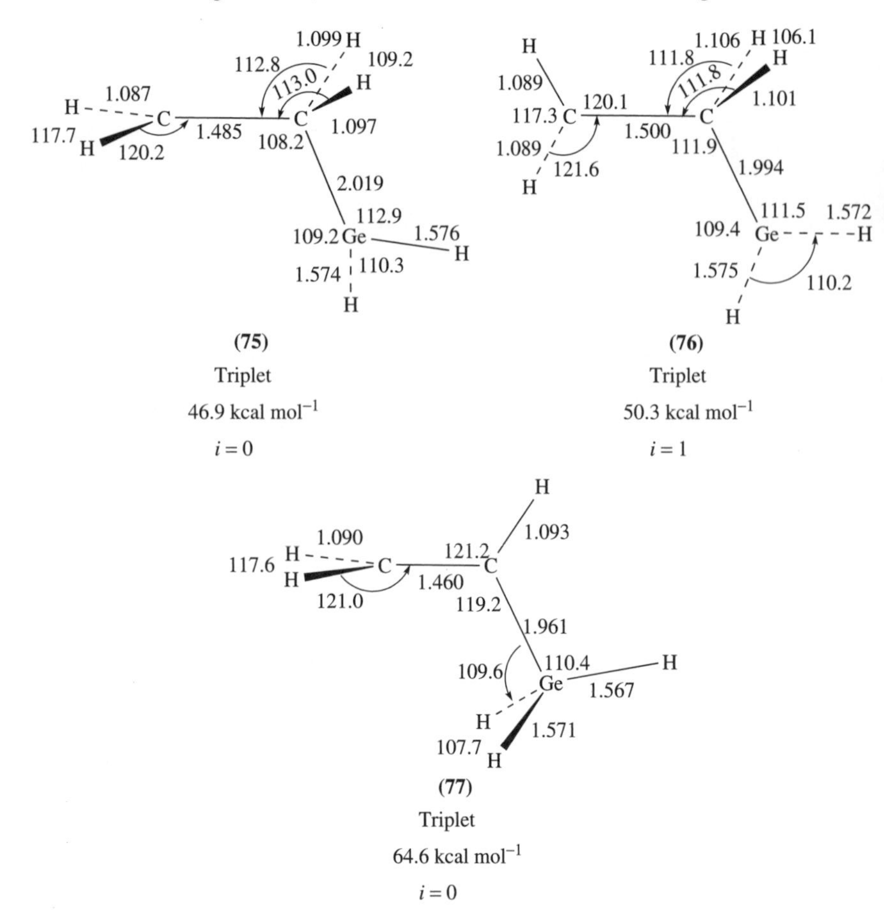

FIGURE 78. (*Continued*)

ongoing developments of faster algorithms and computers, made it possible that information about heavier group-14 compounds produced by quantum chemical calculations is almost routinely used as complementary to experimental research. This is particularly evident in those numerous studies where both experimental and theoretical methods are used for the investigations. Relativistic effects and the large number of core electrons no longer present an insuperable obstacle when calculating molecular structures or reactions pathways of heavy-atom molecules. It is to be expected that the number of theoretical studies of organometallic compounds of Ge, Sn and Pb will further increase in the next decade.

VI. ACKNOWLEDGMENTS

This work was financially supported by the Deutsche Forschungsgemeinschaft and the Fonds der Chemischen Industrie.

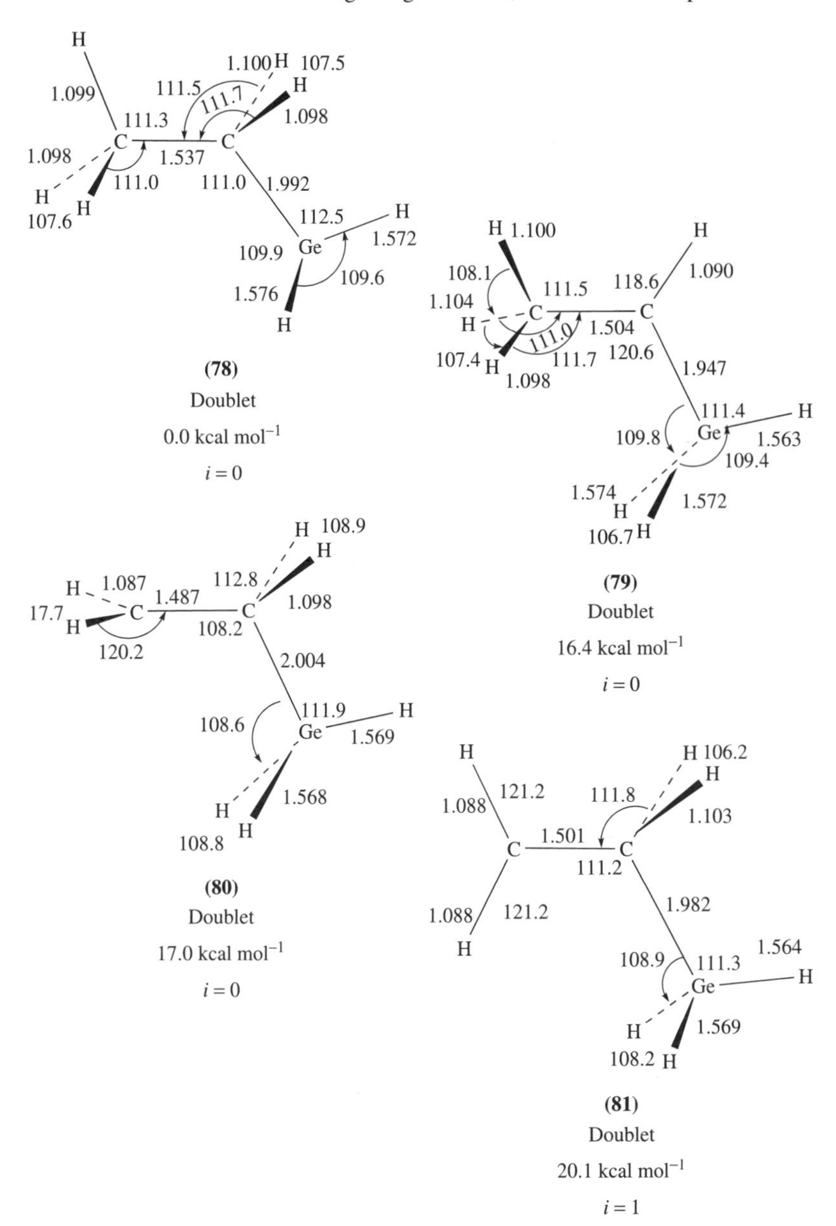

FIGURE 79. Optimized energy minima ($i = 0$) and transition states ($i = 1$) at MP2 on the doublet PES of GeC_2H_7. Calculated relative energies at QCISD(T). Bond distances are in Å, angles in deg. Reproduced by permission of The Royal Society of Chemistry from Reference 195

TABLE 44. Calculated enthalpies of reactions of GeH_n with C_2H_4 leading to the most stable products shown in Figures 76–79[a]

Reaction	ΔH° (kcal mol^{-1})
$Ge + C_2H_4 \rightarrow GeC_2H_4$	−31.5
$GeH_2 + C_2H_4 \rightarrow GeC_2H_4 + H_2$	−3.0
$GeH + C_2H_4 \rightarrow GeC_2H_5$	−19.4
$GeH_2 + C_2H_4 \rightarrow GeC_2H_5 + H$	49.0
$GeH_3 + C_2H_4 \rightarrow GeC_2H_5 + H_2$	5.8
$GeH_2 + C_2H_4 \rightarrow GeC_2H_6$	−35.2
$GeH_3 + C_2H_4 \rightarrow GeC_2H_6 + H$	23.6
$GeH_3 + C_2H_4 \rightarrow GeC_2H_7$	−25.2

[a]Reference 196.

VII. REFERENCES

1. W. J. Hehre, L. Radom, P. v. R. Schleyer and J. A. Pople, *Ab Initio Molecular Orbital Theory*, Wiley, New York, 1986.
2. See the special thematic issue about computational transition metal chemistry edited by E. R. Davidson, *Chem. Rev.*, **100** (2000).
3. (a) H. Hellmann, *J. Chem. Phys.*, **3**, 61 (1935).
   (b) L. Szasz, *Pseudopotential Theory of Atoms and Molecules*, Wiley, New York, 1986.
   (c) M. Krauss and W. J. Stevens, *Annu. Rev. Phys. Chem.*, **35**, 357 (1984).
   (d) M. C. Zerner, *Mol. Phys.*, **23**, 963 (1972).
   (e) P. A. Christiansen, W. C. Ermler and K. S. Pitzer, *Annu. Rev. Phys. Chem.*, **36**, 407 (1985).
   (f) K. Balasubramanian and K. S. Pitzer, *Adv. Chem. Phys.*, **67**, 287 (1987).
   (g) P. Durand and J. P. Malrieu, *Adv. Chem. Phys.*, **67**, 321 (1987).
4. G. Frenking, I. Antes, M. Boehme, S. Dapprich, A. W. Ehlers, V. Jonas, A. Neuhaus, M. Otto, R. Stegmann, A. Veldkamp and S. F. Vyboishchikov, in K. B. Lipkowitz and D. B. Boyd (Eds.), *Reviews in Computational Chemistry*, Vol. 8, VCH, New York, 1996, pp. 63–144.
5. T. R. Cundari, M. T. Benson, M. L. Lutz and S. O. Sommerer, in K. B. Lipkowitz and D. B. Boyd (Eds.), *Reviews in Computational Chemistry*, Vol. 8, VCH, New York, 1996, pp. 145–202.
6. (a) P. Hohenberg and W. Kohn, *Phys. Rev. B*, **136**, 864 (1964).
   (b) J. Labanowski and J. Andzelm (Eds.), *Density Functional Methods in Chemistry*, Springer-Verlag, Heidelberg, 1991.
   (c) L. Levy, *Proc. Natl. Acad. Sci. USA*, **76**, 6052 (1979).
   (d) W. Kohn and L. J. Sham, *Phys. Rev. A*, **140**, 1133 (1965).
   (e) R. G. Parr and W. Yang (Eds.), *Density Functional Theory of Atoms and Molecules*, Oxford University Press, New York, 1988.
   (f) R. F. Nalewajski (Ed.), *Topics in Current Chemistry*, **180–183**, Springer-Verlag, Berlin, 1996.
7. H. Basch and T. Hoz, in S. Patai (Ed.), *The Chemistry of Organic Germanium, Tin and Lead Compounds*, Chap. 1, Wiley, Chichester, 1995.
8. M. Karni, J. Kapp, P. v. R. Schleyer and Y. Apeloig, in Z. Rappoport and Y. Apeloig (Eds.), *The Chemistry of Organic Silicon Compounds*, Vol. 3, Chap. 1, Wiley, Chichester, 2001.
9. M. Driess and H. Grützmacher, *Angew. Chem.*, **108**, 900 (1996); *Angew. Chem., Int. Ed. Engl.*, **35**, 828 (1996).
10. Y. Apeloig, in S. Patai and Z. Rappoport (Eds.), *The Chemistry of Organic Silicon Compounds*, Chap. 2, Wiley, Chichester, 1989, pp. 57–275.
11. (a) Ge parameters for MNDO: M. J. S. Dewar, G. L. Grady and E. F. Healy, *Organometallics*, **5**, 186 (1986).
    (b) Sn parameters for MNDO: M. J. S. Dewar, G. L. Grady and J. J. P. Stewart, *J. Am. Chem. Soc.*, **106**, 6771 (1984).

(c) Pb parameters for MNDO: M. J. S. Dewar, M. K. Holloway, G. L. Grady and J. J. P. Stewart, *Organometallics*, **4**, 1973 (1985).
(d) Ge parameters for AM1: M. J. S. Dewar and C. Jie, *Organometallics*, **8**, 1544 (1989).
(e) Sn parameters for AM1: M. J. S. Dewar, E. F. Healy, D. R. Kuhn and A. J. Holder, *Organometallics*, **10**, 431 (1991).
(f) Ge, Sn, Pb parameters for PM3: J. J. P. Stewart, *J. Comput. Chem.*, **12**, 320 (1991).
(g) Ge parameters for SINDO(Ge): K. Jug, G. Geudtner and T. Homann, *J. Comput. Chem.*, **21**, 974 (2000).

12. (a) P. Comba and T. W. Hambley, *Molecular Modeling of Inorganic Compounds*, VCH, Weinheim, 1995.
(b) N. L. Allinger, M. I. Quinn, K. Chen, B. Thompson and M. R. Frierson, *J. Mol. Struct.*, **194**, 1 (1989).
13. Selective examples are:
(a) R. D. Hancock, *Prog. Inorg. Chem.*, **28**, 187 (1989).
(b) R. D. Hancock, J. S. Weaving and H. M. Marques, *J. Chem. Soc., Chem. Commun.*, 1176 (1989).
(c) M. G. B. Drew and D. G. Nicholson, *J. Chem. Soc., Dalton Trans.*, 1543 (1986).
14. F. Jensen, *Introduction to Computational Chemistry*, Wiley, New York, 1999.
15. W. Koch and M. Holthausen, *A Chemist's Guide to Density Functional Theory*, Wiley, New York, 2000.
16. P. v. R. Schleyer, N. L. Allinger, T. Clark, J. Gasteiger, P. A. Kollman, H. F. Schaefer III and P. R. Schreiner (Eds.), *Encyclopedia of Computational Chemistry*, Vols. I–VI, Wiley-VCH, New York, 1998.
17. W. Kohn and L. J. Sham, *Phys. Rev. A*, **140**, 1144 (1965). The KS equations can only be solved iteratively like the HF equations because the operator F^{KS} contains terms for the electron–electron interactions.
18. F. M. Bickelhaupt and E. J. Barends, *Rev. Comput. Chem.*, **15**, 1 (2000).
19. (a) A. Diefenbach, F. M. Bickelhaupt and G. Frenking. *J. Am. Chem. Soc.*, **122**, 6449 (2000).
(b) J. Uddin and G. Frenking, *J. Am. Chem. Soc.*, **123**, 1683 (2001).
(c) Y. Chen and G. Frenking, *J. Chem. Soc., Dalton Trans.*, 434 (2001).
20. T. V. Russo, R. L. Martin and P. J. Hay, *J. Phys. Chem.*, **99**, 17085 (1995).
21. A. D. Becke, *Phys. Rev. A*, **38**, 3098 (1988).
22. J. P. Perdew, *Phys. Rev. B*, **33**, 8822 (1986).
23. C. Lee, W. Yang and R. G. Parr, *Phys. Rev. B*, **37**, 785 (1988).
24. (a) J. P. Perdew and Y. Wang, in P. Ziesche and H. Schrig (Eds.), *Electronic Structure of Solids '91*, Akademie-Verlag, Berlin, 1991.
(b) J. P. Perdew and Y. Wang, *Phys. Rev.*, **845**, 13244 (1992).
25. S. H. Vosko, L. Wilk and M. Nuisar, *Can. J. Phys.*, **58**, 1200 (1980).
26. L. A. Curtiss, K. Raghavachari, G. W. Trucks and J. A. Pople, *J. Chem. Phys.*, **94**, 7221 (1991).
27. (a) A. D. Becke, *J. Chem. Phys.*, **98**, 1372 (1993).
(b) A. D. Becke, *J. Chem. Phys.*, **98**, 5648 (1993).
28. P. J. Stephens, F. J. Devlin, C. F. Chabalowski and M. J. Frisch, *J. Phys. Chem.*, **98**, 11623 (1994).
29. (a) V. Jonas and W. Thiel, *Organometallics*, **17**, 353 (1998).
(b) V. Jonas and W. Thiel, *J. Chem. Phys.*, **102**, 1 (1995).
30. Special thematic issue about DFT: *J. Comput. Chem.*, **20** (1999), issue 1.
31. Gaussian 98: M. J. Frisch, G. W. Trucks, H. B. Schlegel, G. E. Scuseria, M. A. Robb, J. R. Cheeseman, V. G. Zakrzewski, J. A. Montgomery, R. E. Stratmann, J. C. Burant, S. Dapprich, J. M. Milliam, A. D. Daniels, K. N. Kudin, M. C. Strain, O. Farkas, J. Tomasi, V. Barone, M. Cossi, R. Cammi, B. Mennucci, C. Pomelli, C. Adamo, S. Clifford, J. Ochterski, G. A. Petersson, P. Y. Ayala, Q. Cui, K. Morokuma, D. K. Malick, A. D. Rabuck, K. Raghavachari, J. B. Foresman, J. Cioslowski, J. V. Ortiz, B. B. Stefanov, G. Liu, A. Liashenko, P. Piskorz, I. Komaromi, R. Gomberts, R. L. Martin, D. J. Fox, T. A. Keith, M. A. Al-Laham, C. Y. Peng, A. Nanayakkara, C. Gonzalez, M. Challacombe, P. M. W. Gill, B. G. Johnson, W. Chen, M. W. Wong, J. L. Andres, M. Head-Gordon, E. S. Replogle and J. A. Pople, Gaussian Inc., Pittsburgh, PA, 1998.

32. S. Huzinaga, J. Andzelm, M. Klobukowski, E. Radzio-Andzelm, Y. Sakai and H. Tatekawi, *Gaussian Basis Sets for Molecular Calculations*, Elsevier, Amsterdam, 1984.
33. R. Poirier, R. Kari and I. G. Csizmadia, *Handbook of Gaussian Basis Sets*, Elsevier, Amsterdam, 1985.
34. (a) A. Schäfer, H. Horn and R. Ahlrichs, *J. Chem. Phys.*, **97**, 2571 (1992).
    (b) A. Schäfer, C. Huber and R. Ahlrichs, *J. Chem. Phys.*, **100**, 5829 (1994).
35. W. W. Wadt and P. J. Hay, *J. Chem. Phys.*, **82**, 284 (1985).
36. A. Höllwarth, M. Böhme, S. Dapprich, A. W. Ehlers, A. Gobbi, V. Jonas, K. F. Köhler, R. Stegmann, A. Veldkamp and G. Frenking, *Chem. Phys. Lett.*, **208**, 237 (1993).
37. W. J. Stevens, H. Basch and M. Krauss, *J. Chem. Phys.*, **81**, 6026 (1984).
38. W. J. Stevens, M. Krauss, H. Basch and P. G. Jasien, *Can. J. Chem.*, **70**, 612 (1992).
39. A. Bergner, M. Dolg, W. Küchle, H. Stoll and H. Preuss, *Mol. Phys.*, **80**, 1431 (1993).
40. W. Küchle, M. Dolg, H. Stoll and H. Preuss, *Mol. Phys.*, **74**, 1245 (1991).
41. http://www.theochem.uni-stuttgart.de/pseudopotentiale/
42. L. F. Pacios and P. A. Christiansen, *J. Chem. Phys.*, **82**, 2664 (1985).
43. M. M. Hurley, L. F. Pacios, P. A. Christiansen, R. B. Ross and W. C. Ermler, *J. Chem. Phys.*, **84**, 6840 (1986).
44. L. A. LaJohn, P. A. Christiansen, R. B. Ross, T. Atashroo and W. C. Ermler, *J. Chem. Phys.*, **87**, 2812 (1987).
45. R. B. Ross, J. M. Powers, T. Atashroo, W. C. Ermler, L. A. LaJohn and P. A. Christiansen, *J. Chem. Phys.*, **93**, 6654 (1990).
46. L. F. Pacios and P. C. Gómez, *Int. J. Quantum Chem.*, **49**, 817 (1994).
47. Y. Bouteiller, C. Mijoule, M. Nizam, J. C. Barthelat, J. P. Daudey, M. Pelissier and B. Silvi, *Mol. Phys.*, **65**, 295 (1988).
48. Y. Sakai and S. Huzinaga, *J. Chem. Phys.*, **76**, 2537 (1982).
49. S. Huzinaga, L. Seijo and Z. Barandiarán, *J. Chem. Phys.*, **86**, 2132 (1987).
50. M. Klobukowski, *Theor. Chim. Acta*, **83**, 239 (1992).
51. P. Pyykkö, *Chem. Rev.*, **88**, 563 (1988).
52. B. Hess, in P. v. R. Schleyer, N. L. Allinger, T. Clark, J. Gasteiger, P. A. Kollman, H. F. Schaefer III and P. R. Schreiner (Eds.), *Encyclopedia of Computational Chemistry*, Vol. IV, Wiley-VCH, New York, 1998, p. 2499.
53. C. van Wüllen, *J. Comput. Chem.*, **20**, 51 (1999).
54. (a) E. van Lenthe, E. J. Baerends and J. G. Snijders, *J. Chem. Phys.*, **99**, 4597 (1993).
    (b) E. van Lenthe, E. J. Baerends and J. G. Snijders, *J. Chem. Phys.*, **101**, 9783 (1994).
55. N. Matsunaga, S. Koseki and M. S. Gordon, *J. Chem. Phys.*, **104**, 7988 (1996).
56. G. Trinquier and J.-C. Barthelat, *J. Am. Chem. Soc.*, **112**, 9121 (1990).
57. S. K. Shin, W. A. Goddard III and J. L. Beauchamp, *J. Phys. Chem.*, **94**, 6963 (1990).
58. M. Benavides-Garcia and K. Balasubramian, *J. Chem. Phys.*, **100**, 2821 (1994).
59. M. Benavides-Garcia and K. Balasubramian, *J. Chem. Phys.*, **97**, 7537 (1992).
60. Z. Barandiarán and L. Seijo, *J. Chem. Phys.*, **101**, 4049 (1994).
61. K. G. Dyall, *J. Chem. Phys.*, **96**, 1210 (1992).
62. T. A. Hein, W. Thiel and T. J. Lee, *J. Phys. Chem.*, **97**, 4381 (1993).
63. T. Mineva, N. Russo, E. Sicilia and M. Toscano, *Int. J. Quantum Chem.*, **56**, 669 (1995).
64. W. W. Schoeller and R. Schneider, *Chem. Ber.*, **130**, 1013 (1997).
65. (a) A. Nowek and J. Leszczyński, *J. Phys. Chem. A*, **101**, 3788 (1997).
    (b) G. Trinquier, *J. Am. Chem. Soc.*, **112**, 2130 (1990).
66. G. Trinquier, *J. Am. Chem. Soc.*, **114**, 6807 (1992).
67. S. G. Wang and W. H. E. Schwarz, *J. Mol. Struct.*, **338**, 347 (1995).
68. K. G. Dyall, P. R. Taylor, K. Faegri, Jr. and H. Partridge, *J. Chem. Phys.*, **95**, 2583 (1991).
69. O. Visser, L. Visscher, P. J. C. Aerts and W. C. Nieuwpoort, *Theor. Chim. Acta*, **81**, 405 (1992).
70. A. E. de Oliveira, P. H. Guadagnini, R. Custódio and R. E. Burns, *J. Phys. Chem. A*, **102**, 4615 (1998).
71. O. L. Malkina, D. R. Salahub and V. G. Malkin, *J. Chem. Phys.*, **105**, 8793 (1996).
72. S. Kirpekar, T. Enevoldsen, J. Oddershede and W. T. Raynes, *Mol. Phys.*, **91**, 897 (1997).
73. S. Kirpekar, H. J. A. Jensen and J. Oddershede, *Theor. Chim. Acta*, **95**, 35 (1997).

74. H. Kaneko, M. Hada, T. Nakajima and H. Nakatsuji, *Chem. Phys. Lett.*, **261**, 1 (1996).
75. P. v. R. Schleyer, M. Kaupp, F. Hampel, M. Bremer and K. Mislow, *J. Am. Chem. Soc.*, **114**, 6791 (1992).
76. J. Leszczyński, J. Q. Huang, P. R. Schreiner, G. Vacek, J. Kapp, P. v. R. Schleyer and H. F. Schaefer III, *Chem. Phys. Lett.*, **244**, 252 (1995).
77. G. Trinquier, *J. Chem. Soc., Faraday Trans.*, **89**, 775 (1993).
78. (a) R. Okazaki and R. West, *Adv. Organomet. Chem.*, **39**, 232 (1996).
(b) P. B. Hitchcock, M. F. Lappert, S. J. Miles and A. J. Thorne, *J. Chem. Soc., Chem. Commun.*, 480 (1984).
(c) K. M. Baines and W. G. Stibbs, *Adv. Organomet. Chem.*, **39**, 275 (1996).
(d) D. E. Goldberg, D. H. Harris, M. F. Lappert and K. M. Thomas, *J. Chem. Soc., Chem. Commun.*, 261 (1976).
(e) M. Stürmann, M. Weidenbruch, K. W. Klinkhammer, F. Lissner and H. Marsmann, *Organometallics*, **17**, 4425 (1998).
(f) M. Weidenbruch, *Eur. J. Inorg. Chem.*, 373 (1999).
(g) P. P. Power, *J. Chem. Soc., Dalton Trans.*, 2939 (1998).
79. An important review about the chemical bonding of heavier main group elements was published by: W. Kutzelnigg, *Angew. Chem.*, **96**, 262 (1984); *Angew. Chem., Int. Ed. Engl.*, **23**, 272 (1984).
80. H. Jacobsen and T. Ziegler, *J. Am. Chem. Soc.*, **116**, 3667 (1994).
81. T. L. Windus and M. S. Gordon, *J. Am. Chem. Soc.*, **114**, 9559 (1992).
82. G. Trinquier, *J. Am. Chem. Soc.*, **113**, 144 (1991).
83. G. Trinquier and J.-P. Malrieu, *J. Am. Chem. Soc.*, **113**, 8634 (1991).
84. G. Trinquier and J.-P. Malrieu, *J. Am. Chem. Soc.*, **109**, 5303 (1987).
85. J.-P. Malrieu and G. Trinquier, *J. Am. Chem. Soc.*, **111**, 5916 (1989).
86. D. A. Horner, R. S. Grev and H. F. Schaefer III, *J. Am. Chem. Soc.*, **114**, 2093 (1992).
87. S. Sakai, *Int. J. Quantum Chem.*, **70**, 291 (1998).
88. J. A. Boatz, M. S. Gordon and L. R. Sita, *J. Phys. Chem.*, **94**, 5488 (1990).
89. J. Kapp, M. Remko and P. v. R. Schleyer, *Inorg. Chem.*, **36**, 4241 (1997).
90. R. S. Grev, H. F. Schaefer III and K. M. Baines, *J. Am. Chem. Soc.*, **112**, 9467 (1990).
91. R. S. Grev, *Adv. Organomet. Chem.*, **33**, 125 (1991).
92. Z. Palágyi, H. F. Schaefer III and E. Kapuy, *J. Am. Chem. Soc.*, **115**, 6901 (1993).
93. R. S. Grev, B. J. Deleeuw and H. F. Schaefer III, *Chem. Phys. Lett.*, **165**, 257 (1990).
94. A. J. Boone, D. H. Magers and J. Leszczynski, *Int. J. Quantum Chem.*, **70**, 925 (1998).
95. S. M. Stogner and R. S. Grev, *J. Chem. Phys.*, **108**, 5458 (1998).
96. S. Nagase, K. Kobayashi and N. Takagi, *J. Organomet. Chem.*, **611**, 264 (2000).
97. Y.-K. Han, C. Bae, Y. S. Lee and S. Y. Lee, *J. Comput. Chem.*, **19**, 1526 (1998).
98. Y. Chen, M. Hartmann, M. Diedenhofen and G. Frenking, *Angew. Chem.*, **113**, 2107 (2001); *Angew. Chem., Int. Ed.*, **40**, 2051 (2001).
99. E. D. Jemmis, G. N. Srinivas, J. Leszczynski, J. Kapp, A. A. Korkin and P. v. R. Schleyer, *J. Am. Chem. Soc.*, **117**, 11361 (1995).
100. N. Matsunaga, T. R. Cundari, M. W. Schmidt and M. S. Gordon, *Theor. Chim. Acta*, **83**, 57 (1992).
101. N. Matsunaga and M. S. Gordon, *J. Am. Chem. Soc.*, **116**, 11407 (1994).
102. P. v. R. Schleyer, H. Jiao, N. J. R. van Eikema Hommes, V. G. Malkin and O. L. Malkina, *J. Am. Chem. Soc.*, **119**, 12669 (1997).
103. P. v. R. Schleyer, C. Maerker, A. Dransfeld, H. Jiao and N. J. R. van Eikema Hommes, *J. Am. Chem., Soc.*, **118**, 6317 (1996).
104. B. Maoche, J. Gayoso and O. Ouamerali, *Rev. Roum. Chem.*, **26**, 613 (1984); *Chem. Abstr.*, **102**, 113558 (1984).
105. W. H. Fink and J. C. Richards, *J. Am. Chem. Soc.*, **113**, 3393 (1991).
106. F. A. Cotton, A. H. Cowley and X. Feng, *J. Am. Chem. Soc.*, **120**, 1795 (1998).
107. V. N. Khabashesku, K. N. Kudin, J. Tamás, S. E. Boganov, J. L. Margrave and O. M. Nefedov, *J. Am. Chem. Soc.*, **120**, 5005 (1998).
108. K. N. Kudin, J. L. Margrave and V. N. Khabashesku, *J. Phys. Chem. A*, **102**, 744 (1998).
109. V. N. Khabashesku, S. E. Boganov, K. N. Kudin, J. L. Margrave and O. M. Nefedov, *Organometallics*, **17**, 5041 (1998).

110. J. Kapp, M. Remko and P. v. R. Schleyer, *J. Am. Chem. Soc.*, **118**, 5745 (1996).
111. L. Pu, B. Twamley and P. P. Power, *J. Am. Chem. Soc.*, **122**, 3524 (2000).
112. C. Jouany, S. Mathieu, M.-A. Chaubon-Deredempt and G. Trinquier, *J. Am. Chem. Soc.*, **116**, 3973 (1994).
113. V. N. Khabashesku, S. E. Boganov, D. Antic, O. M. Nefedov and J. Michl, *Organometallics*, **15**, 4714 (1996).
114. S. Foucat, T. Pigot, G. Pfister-Guillouzo, H. Lavayssière and S. Mazières, *Organometallics*, **18**, 5322 (1999).
115. A. Márquez, J. Anguiano, G. González and J. F. Sanz, *J. Organomet. Chem.*, **486**, 45 (1995).
116. A. Gobbi and G. Frenking, *J. Am. Chem. Soc.*, **116**, 9287 (1994).
117. H. Jacobsen and T. Ziegler, *Inorg. Chem.*, **35**, 775 (1996).
118. J. Almlöf and K. Faegri, Jr., *Theor. Chim. Acta*, **69**, 437 (1986).
119. M. Aoyama, S. Masuda, K. Ohno, Y. Harada, M. C. Yew, H. H. Hua and L. S. Yong, *J. Phys. Chem.*, **93**, 1800 (1989).
120. C. Phillips, M. Vairamani, F. M. Harris, C. P. Morley, S. R. Andrews and D. E. Parry, *J. Chem. Soc., Faraday Trans.*, **91**, 1901 (1995).
121. A. Papakondylis, A. Mavridis and B. Bigot, *J. Phys. Chem.*, **98**, 8906 (1994).
122. K. Yasuda, N. Kishimoto and H. Nakatsuji, *J. Phys. Chem.*, **99**, 12501 (1995).
123. S. Mathews, J. L. Duncan, D. C. McKean and B. A. Smart, *J. Mol. Struct.*, **413–414**, 553 (1997).
124. G. Vacek, V. S. Mastryukov and H. F. Schaefer III, *J. Phys. Chem.*, **98**, 11337 (1994).
125. V. Jonas, C. Boehme and G. Frenking, *Inorg. Chem.*, **35**, 2097 (1996).
126. H. A. Bent, *Chem. Rev.*, **61**, 275 (1961).
127. S. L. Boyd and R. J. Boyd, *J. Phys. Chem. A*, **103**, 7087 (1999).
128. B. A. Smart, L. E. Griffiths, C. R. Pulham, H. E. Robertson, N. W. Mitzel and D. W. H. Rankin, *J. Chem. Soc., Dalton Trans.*, 1565 (1997).
129. A. Haaland, D. J. Shorokhov, T. G. Strand, J. Kouvetakis and M. O'Keeffe, *Inorg. Chem.*, **36**, 5198 (1997).
130. J. Kouvetakis, A. Haaland, D. J. Shorokhov, H. V. Volden, G. V. Girichev, V. I. Sokolov and P. Matsunaga, *J. Am. Chem. Soc.*, **120**, 6738 (1998).
131. M. Dakkouri, *J. Mol. Struct.*, **413–414**, 133 (1997).
132. A. Haaland, S. Samdal, T. G. Strand, M. A. Tafipolsky, H. V. Volden, B. J. J. van de Heisteeg, O. S. Akkerman and F. Bickelhaupt, *J. Organomet. Chem.*, **536–537**, 217 (1997).
133. N. W. Mitzel, U. Losehand and A. D. Richardson, *Inorg. Chem.*, **38**, 5323 (1999).
134. V. P. Feshin and E. V. Feshina, *Main Group Metal Chemistry*, **22**, 351 (1999).
135. J. P. Campbell, J.-W. Hwang, V. G. Young, Jr., R. B. von Dreele, C. J. Cramer and W. L. Gladfelter, *J. Am. Chem. Soc.*, **120**, 521 (1998).
136. M. A. Buntine, V. J. Hall, F. J. Kosovel and E. R. T. Tiekink, *J. Phys. Chem. A*, **102**, 2472 (1998).
137. V. Jonas, G. Frenking and M. T. Reetz, *J. Am. Chem. Soc.*, **116**, 8741 (1994).
138. G. Bruno, G. Lanza, G. Malandrino, and I. Fragalà, *J. Chem. Soc., Dalton Trans.*, 965 (1997).
139. D. R. Armstrong, M. J. Duer, M. G. Davidson, D. Moncrieff, C. A. Russell, C. Stourton, A. Steiner, D. Stalke and D. S. Wright, *Organometallics*, **16**, 3340 (1997).
140. D. R. Armstrong, M. A. Beswick, N. L. Cromhout, C. N. Harmer, D. Moncrieff, C. A. Russell, P. R. Raithby, A. Steiner, A. E. H. Wheatley and D. S. Wright, *Organometallics*, **17**, 3176 (1998).
141. D. R. Armstrong, M. G. Davidson, D. Moncrieff, D. Stalke and D. S. Wright, *J. Chem. Soc., Chem. Commun.*, 1413 (1992).
142. A. J. Arduengo III, H. Bock, H. Chen, M. Denk, D. A. Dixon, J. C. Green, W. A. Herrmann, N. L. Jones, M. Wagner and R. West, *J. Am. Chem. Soc.*, **116**, 6641 (1994).
143. (a) C. Heinemann, T. Müller, Y. Apeloig and H. Schwarz, *J. Am. Chem. Soc.*, **118**, 2023 (1996).
(b) C. Boehme and G. Frenking, *J. Am. Chem. Soc.*, **118**, 2039 (1996).
144. C. Heinemann, W. A. Herrmann and W. Thiel, *J. Organomet. Chem.*, **475**, 73 (1994).
145. C. Boehme and G. Frenking, *Organometallics*, **17**, 5801 (1998).
146. W. W. Schöller, A. Sundermann, M. Reiher and A. Rozhenko, *Eur. J. Inorg. Chem.*, 1155 (1999).

147. M. Drieß, R. Janoschek, H. Pritzkow, S. Rell and U. Winkler, *Angew. Chem.*, **107**, 1746 (1995); *Angew. Chem., Int. Ed. Engl.*, **34**, 1614 (1995).
148. P. N. Day, Z. Wang and R. Pachter, *J. Mol. Struct.*, **455**, 33 (1998).
149. A. Klamt and G. Schuurmann, *J. Chem. Soc., Perkin Trans. 2*, 799 (1993).
150. X. L. Liang, D. E. Ellis, O. V. Gubanova, B. M. Hoffman and R. L. Musselman, *Int. J. Quantum Chem.*, **52**, 657 (1994).
151. M. V. Andreocci, M. Bossa, C. Cauletti, S. Stranges, K. Horchler and B. Wrackmeyer, *J. Mol. Struct.*, **254**, 171 (1992).
152. M. Aoyama, S. Masuda, K. Ohno, Y. Harada, M. C. Yew, H. H. Hua and L. S. Yong, *J. Phys. Chem.*, **93**, 5414 (1989).
153. S. Yamaguchi, Y. Itami and K. Tamao, *Organometallics*, **17**, 4910 (1998).
154. K. K. Baldridge and M. S. Gordon, *J. Am. Chem. Soc.*, **110**, 4204 (1988).
155. B. Goldfuss and P. v. R. Schleyer, *Organometallics*, **16**, 1543 (1997).
156. S. Saebø, S. Stroble, W. Collier, R. Ethridge, Z. Wilson, M. Tahai and C. U. Pittman, Jr., *J. Org. Chem.*, **64**, 1311 (1999).
157. (a) M. Kaupp and P. v. R. Schleyer, *Angew. Chem.*, **104**, 1240 (1992).
(b) M. Kaupp and P. v. R. Schleyer, *J. Am. Chem. Soc.*, **115**, 1061 (1993).
158. H. Nakatsuji and T. Nakao, *Int. J. Quantum Chem.*, **49**, 279 (1994).
159. H. Nakatsuji, T. Inoue and T. Nakao, *J. Phys. Chem.*, **96**, 7953 (1992).
160. F. P. Arnold, J. K. Burdett and L. R. Sita, *J. Am. Chem. Soc.*, **120**, 1637 (1998).
161. R. Becerra, S. E. Boganov, M. P. Egorov, V. I. Faustov, O. M. Nefedov and R. Walsh, *J. Am. Chem. Soc.*, **120**, 12657 (1998).
162. (a) P. Antoniotti and G. Tonachini, *Organometallics*, **15**, 1307 (1996).
(b) P. Antoniotti and G. Tonachini, *Organometallics*, **18**, 4538 (1999).
163. M. Takahashi and M. Kira, *J. Am. Chem. Soc.*, **119**, 1948 (1997).
164. S. W. Staley, R. A. Grimm and R. A. Sablosky, *J. Am. Chem. Soc.*, **120**, 3671 (1998).
165. J. D. Xidos, R. A. Poirier, C. C. Pye and D. J. Burnell, *J. Org. Chem.*, **63**, 105 (1998).
166. S. Yamazaki, H. Fujitsuka, S. Yamabe and H. Tamura, *J. Org. Chem.*, **57**, 5610 (1992).
167. S. Yamazaki, M. Tanaka, A. Yamaguchi and S. Yamabe, *J. Am. Chem. Soc.*, **116**, 2356 (1994).
168. S. Yamazaki, T. Takada, T. Imanishi, Y. Moriguchi and S. Yamabe, *J. Org. Chem.*, **63,** 5919 (1998).
169. L. A. Hobson, M. A. Vincent, E. J. Thomas and I. H. Hillier, *Chem. Commun.*, 899 (1998).
170. M. Yamaguchi, M. Arisawa, K. Omata, K. Kabuto, M. Hirama and T. Uchimaru, *J. Org. Chem.*, **63**, 7298 (1998).
171. M. Yasuda, K. Chiba and A. Baba, *J. Am. Chem. Soc.*, **122**, 7549 (2000).
172. W. R. Rocha and W. B. De Alemeida, *Organometallics*, **17**, 1961 (1998).
173. M. Hada, Y. Tanaka, M. Ito, M. Murakami, H. Amii, Y. Ito and H. Nakatsuji, *J. Am. Chem. Soc.*, **116**, 8754 (1994).
174. G. Frenking, S. Fau, C. M. Marchand and H. Grützmacher, *J. Am. Chem. Soc.*, **119**, 6648 (1997).
175. P. Antoniotti, P. Benzi, F. Grandinetti and P. Volpe, *J. Phys. Chem.*, **97**, 4945 (1993).
176. P. Antoniotti, F. Grandinetti and P. Volpe, *J. Phys. Chem.*, **99**, 17724 (1995).
177. K. A. Nguyen, M. S. Gordon, G. Wang and J. B. Lambert, *Organometallics*, **10**, 2798 (1991).
178. A. Nicolaides and L. Radom, *J. Am. Chem. Soc.*, **119**, 11933 (1997).
179. D. A. Hrovat and W. T. Borden, *J. Org. Chem.*, **57**, 2519 (1992).
180. W. Adcock, C. I. Clark and C. H. Schiesser, *J. Am. Chem. Soc.*, **118**, 11541 (1996).
181. J. Yoshida and M. Izawa, *J. Am. Chem. Soc.*, **119**, 9361 (1997).
182. M. Sugawara and J. Yoshida, *J. Org. Chem.*, **65**, 3135 (2000).
183. P. Schwerdtfeger, *J. Am. Chem. Soc.*, **112**, 2818 (1990).
184. A. M. El-Nahas and P. v. R. Schleyer, *J. Comput. Chem.*, **15**, 596 (1994).
185. H. Anane, A. Jarid and A. Boutalib, *J. Phys. Chem. A*, **103**, 9847 (1999).
186. W. P. Freeman, T. D. Tilley, F. P. Arnold, A. L. Rheingold and P. K. Gantzel, *Angew. Chem.*, **107**, 2029 (1995); *Angew. Chem., Int. Ed. Engl.*, **34**, 1887 (1995).
187. C. H. Schiesser and B. A. Smart, *Tetrahedron*, **51**, 6051 (1995).
188. C. H. Schiesser, M. L. Styles and L. M. Wild, *J. Chem. Soc., Perkin Trans. 2*, 2257 (1996).
189. C. H. Schiesser and M. A. Skidmore, *J. Organomet. Chem.*, **552**, 145 (1998).
190. C. H. Schiesser and M. L. Styles, *J. Chem. Soc., Perkin Trans. 2*, 2335 (1997).
191. D. Dakternieks, D. J. Henry and C. H. Schiesser, *J. Chem. Soc., Perkin Trans. 2*, 1665 (1997).

192. D. Dakternieks, D. J. Henry and C. H. Schiesser, *J. Chem. Soc., Perkin Trans. 2*, 591 (1998).
193. D. Dakternieks, D. J. Henry and C. H. Schiesser, *Organometallics*, **17**, 1079 (1998).
194. S. M. Horvat, C. H. Schiesser and L. M. Wild, *Organometallics*, **19**, 1239 (2000).
195. J. J. BelBruno, *J. Chem. Soc., Faraday Trans.*, **94**, 1555 (1998).
196. P. Antoniotti, P. Benzi, M. Castiglioni and P. Volpe, *Eur. J. Inorg. Chem.*, 323 (1999).

CHAPTER 4

Recent advances in structural chemistry of organic germanium, tin and lead compounds

KARL W. KLINKHAMMER

Institute for Inorganic Chemistry, University of Stuttgart, Pfaffenwaldring 55, D-70569 Stuttgart, Germany
Fax: (49)711-6854241; e-mail: Klink@uni-mainz.de

The chemistry of organic germanium, tin and lead compounds — Vol. 2
Edited by Z. Rappoport

I. INTRODUCTION

A. List of Abbreviations

Most of the chemistry reported here depends on the presence of bulky ligands which will be denoted by the following abbreviations:

R =	Abbreviation	R =	Abbreviation	R =	Abbreviation
$CHMe_2$	*i*-Pr	$CH(SiMe_3)_2$	Bsi	2-*t*-Bu-4,5,6-Me_3C_6H	Bmp
CMe_3	*t*-Bu	$C(SiMe_3)_3$	Tsi	2,6-$[P(O)(OEt)_2]_2$-4-*t*-BuC_6H_2	R^P
$SiMe_3$	Tms			$C_{10}H_{15}$ (adamantyl)	Ad
$Si(SiMe_3)_3$	Hyp	$Si(Bu\text{-}t)_3$	Sup		

2,6-Di-substituted and 2,4,6-tri-substituted aryl groups:

$2,6\text{-}R_2C_6H_3$		$2,4,6\text{-}R_3C_6H_2$	
R =	Abbreviation	R =	Abbreviation
Et	Det	Me	Mes
i-Pr	Dip	Et	Tret
Mes	Btm	*i*-Pr	Tip
Trip	Btp	*t*-Bu	Mes*
2′-naphthyl	Btn	CF_3	Ar^F
2′-*i*-PrC_6H_4	Bip	$CH(SiMe_3)_2$	Tbt
CH_2NMe_2	Dmdm		
NMe_2	Ar^N		

GED denotes electron diffraction (gas phase).

B. General Comments

In the first edition of this book Mackay gave an excellent overview about most types of compounds containing E−C bonds, E being a heavier member of Group 14 (tetrels): germanium, tin and lead[1]. In his introduction he stated that most of the differences between C and the heavier tetrels derive from the ability of carbon (and the disability of its higher congeners) to form

(i) strong element−element bonds giving homonuclear chains and rings, and
(ii) π-overlap of p-orbitals and hence alkenes, alkynes, aromatic compounds and the like.

Secondly, he stressed

(iii) the existence of a distinct two-valent state already for Ge and its increasing stability through Sn to Pb, and

(iv) the ability of the heavier elements to show higher coordination numbers than 4, resulting in different structures for compounds of the same stoichiometry and providing lower-energy intermediates.

During the last five years only statement (ii) has had to be revised. It is now well established that it is *not* the insufficiency of the heavier group 14 elements of π-bonding that prevents them from giving multiply-bonded species — π-bonds may in fact be very strong, even or especially for Pb[2a,b]. Instead, an imbalance between increasing promotion energies of the constituting tetrylene (tetrelandiyl) ER_2 or tetrylyne (tetrelantriyl) ER fragments from their singlet or doublet ground-state to the triplet or quartet valence-state, respectively, on the one hand, and generally decreasing bond energies going from lighter to heavier elements on the other hand are responsible for this fact[2,3]. Additionally, the *kinetic lability* of such compounds, partially related to (iv), surely prevented many unusual multiple-bonded systems from being isolated under ambient conditions. The electronic structure of carbene homologues R_2E, the major building blocks of such systems, has been investigated by many groups using quantum-mechanical methods[2–4]. Their chemistry and the nature of bonding within their dimers, i.e. alkene homologues $R_2E{=}ER_2$, and their cyclic oligomers $c\text{-}E_nR_{2n}$ is now understood more deeply. On the other hand, only little really novel information is available about the title elements in their 'normal' oxidation state. Thus, this chapter will deal mostly with *low valent* germanium, tin and lead derivatives such as carbene homologues, mono- and oligocyclic systems and species containing homonuclear and heteronuclear multiple bonds between Ge, Sn and Pb. Numerous reviews have been published, especially on multiply bonded compounds of group 14[5] and, in line with the title of this chapter, I will concentrate on the *structural* aspects of these species.

Some of the novel systems are known only with silyl, germyl or other inorganometallic substituents, but not with simple organyl substituents. Since no principal differences from the related organyl derivatives are expected, these compounds are also included in the present discussions. Excluded are compounds in which the tetrel exclusively bears halido, amido and alkoxo substituents or other hetero-element atoms with lone-pairs.

II. TETRYLENES (TETRELANEDIYLS) R_2E AND THEIR DERIVATIVES

A. Tetrylenes with Two-coordinate Tetrel Atom

Numerous compounds with divalent two-coordinate tetrel atoms have been synthesized during the last twenty years and many of them have been characterized by absolute structure methods[5c]. In this chapter I will discuss the structural features of such derivatives which bear at least one σ-bonded organyl group or related homologous inorganometallic substituents with a heavier tetrel bonded directly to the divalent atom. Most of the cyclopentadienyl compounds will be excluded from this discussion, since in most cases these are coordinated in a multi-hapto π-fashion and so the bonding becomes more ionic. Moreover, the structural chemistry of such derivatives has grown to such an extent during recent years[6] and many of them form complicated 3-dimensional network structures or polynuclear oligomeres, and so they should be treated in a different chapter.

Before going into a detailed discussion of the structural peculiarities of homoleptic and heteroleptic tetrylenes (Tables 1–7), two features common to all such species should be briefly mentioned:

(a) small C—E—C bond angles which, in most cases, are significantly smaller than the ideal tetrahedral angle of about 109.5°;

(b) C—E bond lengths which are markedly longer than the appropriate parameters in the corresponding tetravalent species.

Both features are based on the fact that all known (isolable) heavier tetrylenes ER_2 exist in a singlet ground-state, the lone-pair on E residing in an almost pure s-type orbital. As a consequence almost pure p-orbitals are utilized for bonding to the substituents, leading to small C—E—C angles and, since pure p-orbitals are less directed than sp^n-hybrides, to longer E—C bonds. Only in cases where extreme sterical congestion or significant additional interactions such as Lewis-acid/Lewis-base interactions are present, widening of the C—E—C angles or shortening of the C—E bonds may be observed.

1. Homoleptic species — diorganyltetrylenes R_2E

In this chapter the focus is laid on species comprising two E—C bonds, including however those rare examples where two different organyl groups are bonded to the two-coordinate tetrel atom (Table 1) and which are not homoleptic within a stricter definition.

a. Germylenes (germandiyls). Many germylenes are stable in the gas phase or in dilute solutions only, and they tend to dimerise to digermenes in concentrated solutions and in the solid-state. Hence Bsi_2Ge **(1)** is monomeric in the gas phase[7], in solution an equilibrium with its dimer $Bsi_2Ge{=}GeBsi_2$ is observed and in the solid-state digermene molecules with a typical *trans*-bent conformation (see Section III) has been found[8]. This delicate equilibrium between germylene and digermene — which is discussed in more detail in Section III — may be, however, influenced by small structural (and electronic) changes as can be seen by the example of the cyclic analogue **(2)**, which is strictly monomeric in solution and even in the solid-state (C—Ge 202 pm; C—Ge—C 90.98°)[9]; the same holds, besides, for the analogous tin derivative **3** (C—Sn, 213 pm; C—Sn—C, 86.7°)[10].

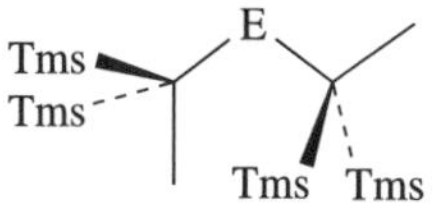

Tms Tms E Tms Tms

(1) E = Ge
(*syn, anti*)

Tms Tms E Tms Tms

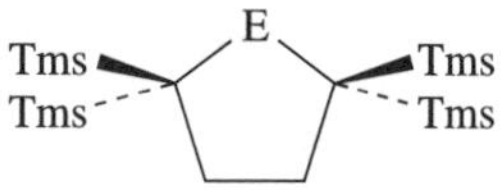

(2) E = Ge
(3) E = Sn
(4) E = Si
(*syn, syn*)

The stability of the monomer can be rationalized by the pronounced rigidity of the ligand system and by the fact that in the case of the cyclic system all Me_3Si groups must be orientated towards the Ge atom (*syn, syn*-conformer) whereas in Bsi_2Ge **(1)** two of them may turn away (*syn, anti*-conformer) and are indeed turned away in the solid-state structure of the respective digermene; in the gas-phase Bsi_2Ge adopts the *syn, syn*-conformation, however. This rigid orientation in the cyclic compounds leads to an increased sterical congestion (destabilization of the dimer) and to additional hyperconjugative stabilization of monomer **2** (Figure 1), since all four Me_3Si groups are in the right orientation for interaction of the Si—C σ-bond orbitals with the empty p-orbital on Ge; the latter is in turn crucial for interaction with external electron-pair donors such as a second germylene entity. Indeed, the C—Si bonds to the $SiMe_3$ groups are slightly elongated in the germylene **2** (190 pm) compared with other derivatives of the substituent where no similar hyperconjugation is possible. Notably, the same cyclic substituent was very recently utilized to synthesize the first isolable dialkylsilylene **(4)** (C—Si, 190.4 pm; C—Si—C, 93.87°); here the C—$SiMe_3$ bond (190.6 pm) is even more elongated[11]. The acyclic derivative Bsi_2Si and its dimer $Bsi_2Si{=}SiBsi_2$ are still unknown.

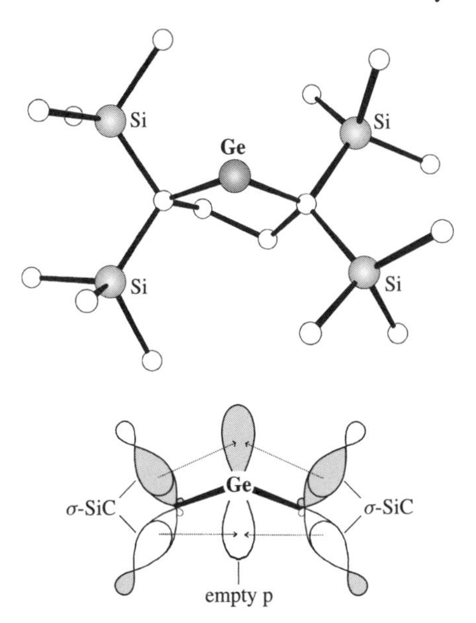

FIGURE 1. Molecular structure of cyclic germylene **2** (top) and scheme for the proposed hyperconjugative σ(SiC) → p(Ge) interaction (bottom)

TABLE 1. Parameters (bond length in pm, angle in deg) of germylenes R^1R^2Ge with two-coordinate tetrel atom

R^1	R^2	Ge−R^1	Ge−R^2	R^1−Ge−R^2	Reference	Remarks
Homoleptic						
	I(C, C)[a]	201	202	91.0	9	
Bis	Bis	204	204	107	7	GED
Bis	Tris	201	207	111.3	12	
Btm	Btm	203	203	114.4	13	
Btn	Btn	203	204	102.7	14	
Ar^F	Ar^F	207	208	100.0	15	Ge−F contacts
Ar^N	Ar^N	202	202	105.1	16	Ge−N contacts 239.4; 272.2
Mes*	Mes*	204	205	108.0	17	
Heteroleptic						
Btp	Cl	199	220.3	101.3	18	
Btm	$BtmGe^-$	202	242.2	111.3		
	II (Si, N)[a]	246.1 (Si)	183 (N)	97.5	19	
Btp	$Cp(CO)_3Cr$	199	259.0	117.8	20a	
Btp	$Cp(CO)_3W$	199	268.1	114.7	20a	
For comparison						
F	F	173.2	173.2	97	21	GED
Cl	Cl	218.3	218.3	100	22	GED
Br	Br	233.7	233.7	101	23	
$N(SiMe_3)_2$	$N(SiMe_3)_2$	187	189	107.1	24	

[a] For the definition of substituents *I* and *II*, see Chart 1 atomic symbols in parentheses denote the type of atom binding to Sn.

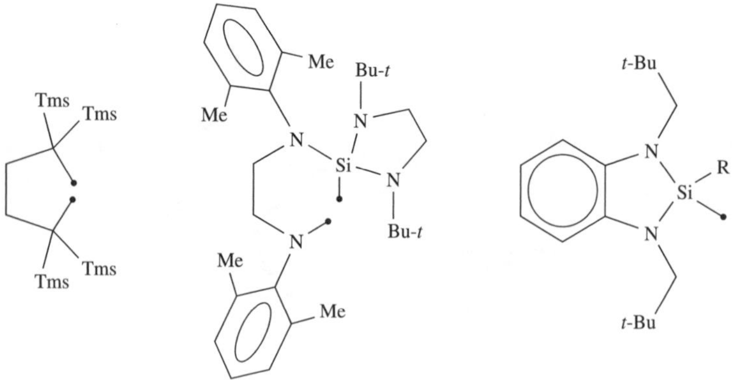

I II III

R = (a) $N(Tms)_2$; R = (b)Ar^N

CHART 1. Abbreviations used in tables. Heavy dots indicate the connection point of the substituent

Factors that are definitely known to contribute to the stabilization of germylene monomers are (a) substituents even bulkier than Bsi or (b) electron-withdrawing ligands. Apart from cyclopentadienyl ligands, especially bulky aryl substituents such as 2,4,6-tri-*tert*-butylphenyl (supermesityl; Mes*) and *o,o'*-disubstituted terphenyls 2,6-(2',4',6'-$R_3C_6H_2)_2C_6H_3$ (e.g., R = Me, *i*-Pr) have been used to stabilize not only germylenes, but also many other sensitive compounds which tend to dimerise otherwise.

The extreme steric congestion present in Mes^*_2Ge (**5**)[17] and to a somewhat lesser extent in the homologous stannylene Mes^*_2Sn (**6**)[25] (since the Sn-C bond is markedly longer than the Ge−C bond) can be easily recognized by a strong distortion of the phenyl rings and the out-of-plane positioning of the tetrel atom. In the germylene **5** the two aryl rings are distorted to a quite different extent (Table 2a); the Ge atom is lying within the best plane through the only slightly distorted ring, but 158 pm (!) above the best plane through the other ring (Figure 2). In the stannylene **6**, the aryl rings are both substantially distorted and the Sn atom lies out-of-plane by 105 and 112 pm with respect to both (Table 2b). The corresponding plumbylene Mes^*_2Pb (**7**), because of its low kinetic stability, rearranges at about −30 °C to the isomeric alkyl(aryl)plumbylene $Mes^*PbCH_2CMe_2$(3,5-*t*-$Bu_2C_6H_3$) (**8**)[26].

Utilizing extremely bulky terphenyl substituents Power and coworkers recently reported the synthesis of another strictly monomeric germylene: Btm_2Ge (**9**)[13]. Despite the huge 2,6-dimesitylphenyl (Btm) substituents the distortion around Ge observed for **9** is much smaller than for the supermesityl derivative **5**. Thus Power and coworkers claimed that these terphenyl ligands are different from other bulky groups by protecting the space surrounding the central atom, whereas most of the other bulky ligands such as Mes* occupy the space in the immediate proximity of the central atom. This is even more true for the 2,6-bis(1'-naphthyl)phenyl (Btn) substituent **10** which was used very recently by Schmidbaur and coworkers in preparing the 'ligand-protected strain-free' diarylgermylene Btn_2Ge[14]; herein short Ge−C bonds (203 pm) and a very small C−Ge−C angle (102.7°) have been determined. Therefore he claimed these values to be the 'natural' bond parameters of diarylgermylenes.

In the case of the derivative Ar^F_2Ge (**11**)[15] the introduction of the electron-withdrawing Ar^F-substituent (2,4,6-$(F_3C)_3C_6H_2$) destabilizes the hypothetical dimer for electronic

TABLE 2. Deviation from planarity measured by the distance (pm) to the best plane through the respective aromatic ring in Mes^*_2Ge (**5**) and in Mes^*_2Sn (**6**)[a]

Ring I				Ring II			
			(a) Mes^*_2Ge (**5**)				
Ge	+158.0			Ge	−0.9		
C11	+11.5			C21	−3.3		
C12	−6.5	C121	−39.8	C22	+3.1	C221	+16.2
C13	−3.4			C23	−0.8		
C14	+8.3	C141	+30.0	C24	−1.3	C241	−7.4
C15	−3.2			C25	+1.1		
C16	−6.7	C161	−45.7	C26	+1.2	C261	+18.0
			(b) Mes^*_2Sn (**6**)				
Sn	−105.1			Sn	+118.4		
C11	+9.8			C21	+10.6		
C12	−5.8	C121	−39.6	C22	+5.5	C221	+38.1
C13	−2.6			C23	+4.1		
C14	+7.2	C141	+23.3	C24	−8.3	C241	−26.9
C15	−3.1			C25	+3.0		
C16	−5.4	C161	−36.9	C26	+6.4	C261	+42.5

[a]For atom numbering see Figure 2.

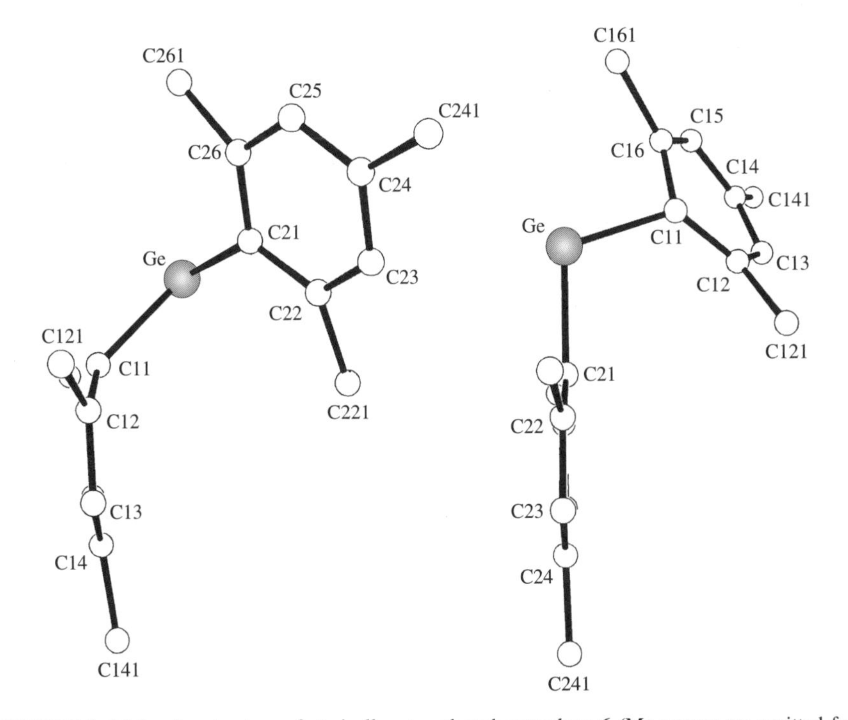

FIGURE 2. Molecular structure of sterically encumbered germylene **6** (Me groups are omitted for clarity) illustrating the distortion present: (left) view within the best plane through ring I; (right) view within the best plane through ring II (see Table 2a)

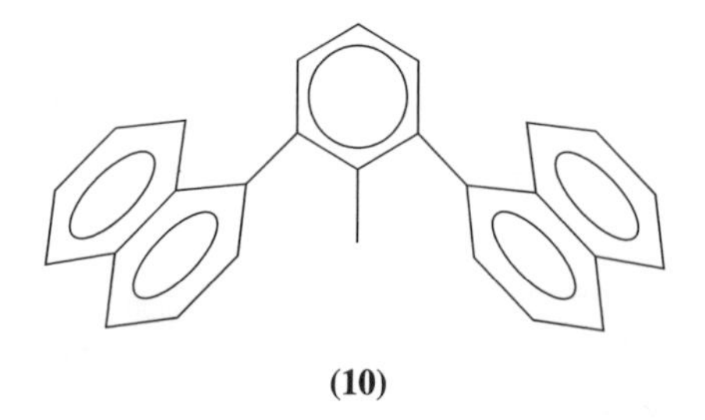

(10)

reasons (see Section III) and simultaneously stabilizes the monomer by intramolecular $F \rightarrow Ge$ interactions. A particular orientation of the *ortho*-CF_3 groups is observed (Figure 3); one group for each aryl substituent is orientated in a way that short $Ge \cdots F$ distances are formed (255.2 and 256.3 pm), indicating an interaction of a lone-pair on F with the empty p-orbital on Ge. The two remaining *ortho*-CF_3 groups show orientations less effective for interaction and hence lead to significantly longer $Ge \cdots F$ distances (278.3 and 278.9 pm). The monomers of the homologues Sn and Pb derivatives of **11**, Ar^F_2Sn **(12)** and Ar^F_2Pb **(13)**[27,28] are stabilized by analogous interactions, the discrimination between the short and long $F \rightarrow E$ distances being somewhat smaller (see below).

Finally, I would like to make some remarks on related germylenes bearing silyl or germyl substituents instead of organyl groups. In sharp contrast to bis(silyl)stannylenes $(R_3Si)_2Sn$ and plumbylenes $(R_3Si)_2Pb$, there is no report of an isolable bis(silyl)germylene $(R_3Si)_2Ge$ in the literature. If bulky tris(alkyl)silyl substituents are employed, dimers, i.e. digermenes $(R_3Si)_2Ge{=}Ge(SiR_3)_2$, are obtained (see Section III). If the hypersilyl group: $Si(SiMe_3)_3$ (Hyp)—a tris(silyl)silyl group—is introduced, neither germylenes nor digermenes are isolated. Instead, rearrangements to cyclic products are observed. Heine and Stalke, for instance, reported the almost quantitative formation of hexakis(trimethylsilyl)disilagermirane **(14)**[29] when reacting $(thf)_3LiSi(SiMe_3)_3$ with $GeCl_2 \cdot$ dioxane. In an analogous reaction using $(thf)_3LiGe(SiMe_3)_3$ instead, Geanangel and coworkers observed the formation of the cyclotetragermane $(HypGeCl)_4$—a tetramer of the mono-substitution product—and hexakis(trimethylsilyl)cyclotrigermane **(15)**[30]. In our laboratory, we investigated the reactions of $GeCl_2 \cdot$ dioxane and $HSiCl_3$ with unsolvated lithium hypersilanide $LiSi(SiMe_3)_3$. Again, products which could be isolated in high yields were **14** and the homologous trisilirane **16**, respectively[31]. Most probably, germylenes and silylene of type **17** are intermediates in the formation of these cyclic products, which then undergo a

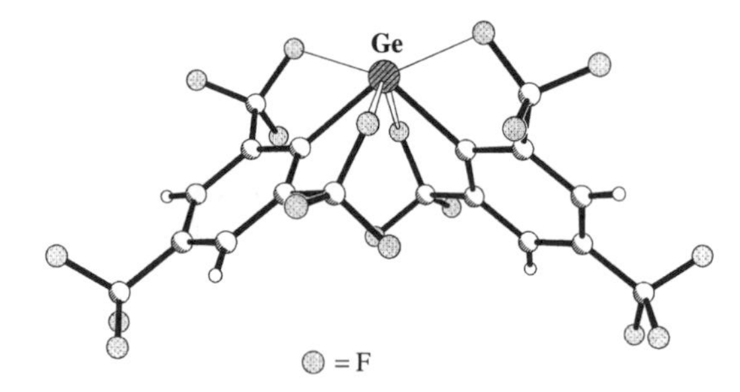

FIGURE 3. Molecular structure of germylene **11**. The stronger $F \rightarrow Ge$ interactions are represented as wide (=), the weaker ones as thin (−) lines

rapid diatropic rearrangement by the shift of two Me_3Si groups from the hypersilyl or the respective tris(trimethylsilyl)germyl substituent to the electron-deficient Ge(II) or Si(II) centre (Scheme 1). Further very probable intermediates, in fact the first ones which could be postulated in the course of these reactions, are silagermenes, germenes or silenes of type **18**. A related compound, the silene **(19)**, has been unambiguously identified as the only product from heating the cyclic silylene **(4)** (equation 1)[11]. It is noteworthy that the respective stannylene and plumbylene with two hypersilyl substituents Hyp_2Sn **20** and Hyp_2Pb **21**[32] show no tendency for rearrangement and may be isolated as dimer and monomer, respectively. The reason for this different behaviour probably lies in the different energy balance of broken and new-formed bonds: while along the rearrangement of **17** to **14, 15** and **16** two relatively strong bonds (Si−Si or Ge−Si) are broken, but three nearly equally strong bonds (Ge−Si, Si−Si or Ge−Ge) are formed, in the case of **20** and **21** two strong bonds are to break, but only one new strong Si−Si and two weak Sn−Si or Pb−Si bonds would form.

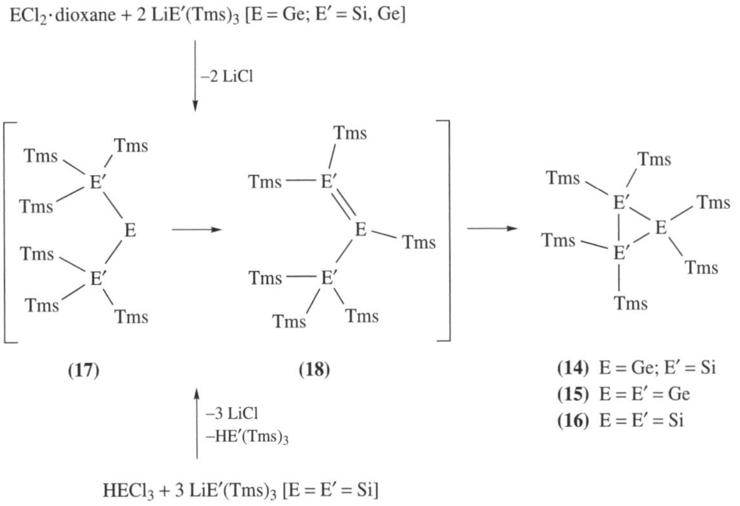

ECl_2·dioxane + 2 $LiE'(Tms)_3$ [E = Ge; E′ = Si, Ge]

−2 LiCl

(17) **(18)**

(14) E = Ge; E′ = Si
(15) E = E′ = Ge
(16) E = E′ = Si

−3 LiCl
$-HE'(Tms)_3$

$HECl_3$ + 3 $LiE'(Tms)_3$ [E = E′ = Si]

SCHEME 1. Putative rearrangement paths of germylenes and silylenes bearing silyl or germyl substituents. Proposed intermediates are given in brackets

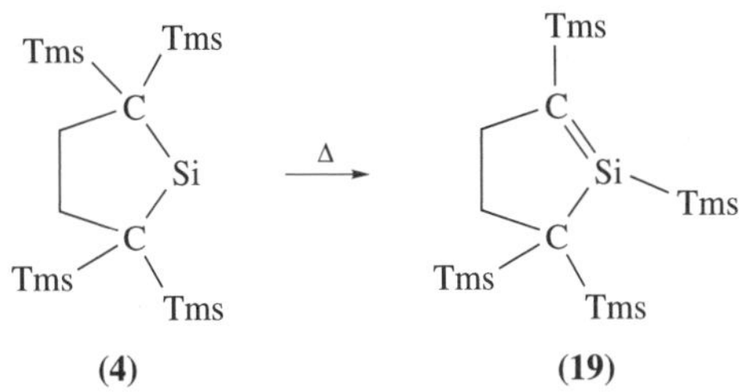

(4) —Δ→ **(19)** (1)

b. Stannylenes (stannandiyls). Due to the larger singlet–triplet gap (see Section III), the tendency of stannylenes for dimerisation or oligomerization is significantly lower than for the lighter congeners, hence all known dimers readily dissociate in solution or in the gas phase (see, for example, $SnBsi_2$[7,42]).

TABLE 3. Parameters (bond length in pm, angle in deg) of stannylenes R^1R^2Sn with two-coordinate tetrel atom

R^1	R^2	$Sn-R^1$	$Sn-R^2$	R^1-Sn-R^2	Reference	Remarks
Homoleptic						
Bis	Bis	222	222	97	7	GED
Ar^F	Ar^F	228	228	98.2	27a	four Sn···F contacts: 266; 268; 281; 283
Ar^F	Ar^F	228	229	95.1	27b	loose dimer: Sn-Sn: 363.9; four Sn···F contacts: 269; 271; 280; 282
$I(C, C)^a$		222	222	86.7	10	
Btm	Btm	223	223	114.7	13	
Ar^N	Ar^N	222	221	105.6	16, 33	Sn-N contacts: 261; 267
Mes*	Mes*	226	227	103.6	25	
	compound **22** (*C, C*)	231 230	231 232	98.7 99.2	34	two N-Sn-N units present: Sn-N: 216–226; N-Sn-N: 103.0–104.0
IIIa (Si)^a	*IIIa* (Si)^a	271.2	271.2	106.8	35	
Heteroleptic						
Ar^F	compound **23** (*C*)	225 228	212 215	108.4 108.7	36	two independent molecules two Sn···F contacts: 273, 273
Btp	I	221	276.6	102.7	18	
Btp	$N(Tms)_2$	223	209	108.4	18	
Ar^N	*IIIb (Si)*^{a}	221	263.7	107.0	37	two Sn-N contacts: 257; 258
Btp	$Btp(Me)_2Sn$	223	289.1	101.2	38	
For comparison						
Cl	Cl	233.5	233.5	99	39	GED
Br	Br	250.1	250.1	100	40	GED
I	I	268.8	268.8	105	39	GED
$N(Tms)_2$	$N(Tms)_2$	209	210	105	41	GED

^{a}For the definition of substituents *I* and *III*, see Chart 1 atomic symbols in parentheses denote the type of atom binding to Sn.

Less than ten homoleptic stannylenes are known which are strictly monomeric in the solid state and which are not stabilized by substantial further coordination at the divalent Sn (Table 3). Additionally, one monomeric homoleptic bis(silyl)stannylene **24**[35] had been reported, formed unexpectedly by the reaction of a bis(amido)silylene **25** and a bis(amino)stannylene **26** alongside a formal insertion of the electron-deficient Si atom of **25** into a Sn−N bond of **26**. A most probable reaction path, starting with subsequent addition and rearrangement steps, is depicted in Scheme 2. The second known bis(silyl)stannylene, Hyp_2Sn **(20)**[32], is monomeric only in dilute solution and forms dimers with very unusual structural features in the solid state (Section III). Very recently, Power and coworkers reported the first example of triorganylstannyl-substituted stannylene **27** with two-coordinate Sn[43]. Such compounds are discussed as being intermediates in rearrangement reactions of distannenes, since they are thought to be easily accessible isomers on the hypersurface of the latter.

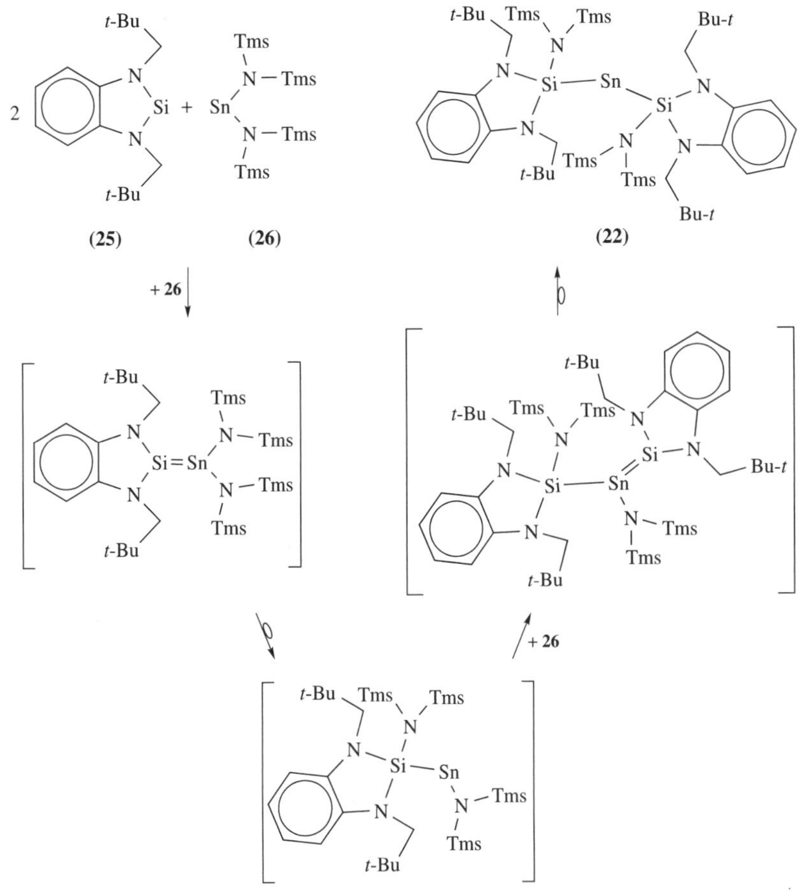

(25) (26) (22)

SCHEME 2. Probable mechanism for the formation of stannylene **22** from stannylene **26** and silylene **25**

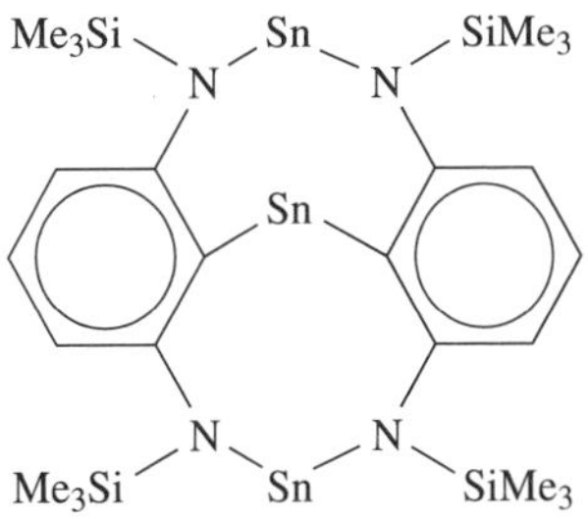

(22)

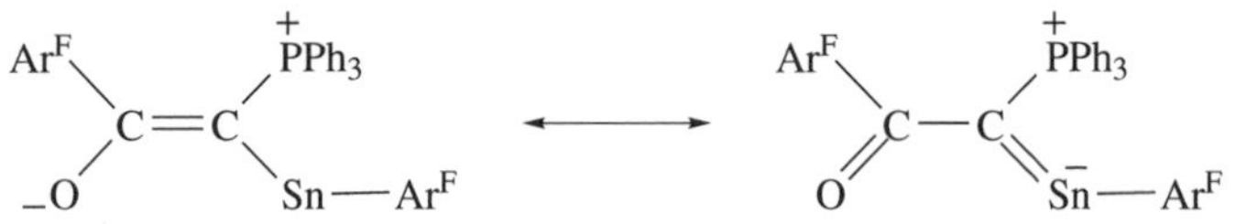

(23)

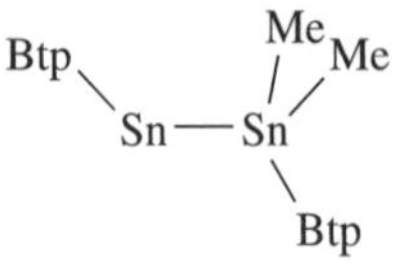

(27)

The crystal structures of three bis(organyl)stannylenes reveal weak coordinative intramolecular interactions between Sn and electron-donating groups and one of them shows additional weak intermolecular interaction to adjacent stannylene species in the crystal.

The tris(trifluoromethyl)phenyl derivative (2,4,6-$(F_3C)_3C_6H_2)_2Sn$ (Ar^F_2Sn) **12** crystallizes in two polymorphs. The yellow polymorph[27a], exhibiting no notable intermolecular interactions, displays two sets of relatively tight intramolecular contacts from *ortho*-CF_3 groups to the divalent tin, a shorter set: Sn−F = 266.3, 268.1 pm, and a longer set: 280.7, 283.3 pm (cf **11**). In the second red polymorph[27b] these contacts are still present; the shorter set, however, is significantly lengthened by 5 pm (av.), whereas the longer set is nearly unchanged. The reason probably lies in competing weak *inter*molecular Sn$\cdots$Sn interactions between two stannylene moieties with a distorted *trans*-bent arrangement of the substituents. Similar conformations are found for 'real' distannenes, though the Sn−Sn distances in the latter are much shorter than the value of 363.9 pm found for $(\mathbf{12})_2$ which is significantly shorter than twice the van der Waals radius of Sn. Similar Sn$\cdots$F interactions are also observed in two heteroleptic stannylenes: Ar^F(Hyp)Sn **(28)**[44] and Ar^F−C(O)=C(PPh_3)−SnAr^F **(23)**[36]; here, one Ar^F group is replaced by a hypersilyl ($(Me_3Si)_3Si$) substituent or an oxoethenyl group, respectively. In the dimer of compound **28** the Sn−Sn bond is strengthened by the strongly electron-releasing silyl group, leading to a Sn−Sn bond length of 283.3 pm. This strengthening of the Sn−Sn bond — by competitive usage of the empty p-orbital on Sn by the second stannylene moiety — leads to dramatically reduced F$\cdots$Sn interactions with Sn−F distances of 294.9 and 295.6 pm.

The remaining stannylene with weak intramolecular stabilization by a Lewis base is the bis[2,6-(*N,N*-dimethylamino)phenyl]stannylene $[2,6\text{-}(Me_2N)_2C_6H_3]_2Sn$ (Ar^N_2Sn) **(29)**[33]. Two Me_2N groups show close Sn$\cdots$N contacts of 259.9 and 266.3 pm, the proposed lone-pair being directed to the space above and below the C−Sn−C plane where the empty p-orbital of Sn is assumed (Figure 4). Similar interactions are found in the related heteroleptic aryl(silyl)stannylene **42** mentioned in Section II. A.2.b[37].

One further unique diarylstannylene which is worth discussing in more detail is the polycyclic derivative **(22)**[34]. This particular molecule possesses three divalent Sn atoms, two of them being coordinated by two amido ligands and the remaining one by two aryl substituents. The Sn atom (Sn_A) of the bis(aryl)stannylene fragment shows no unusual tight contacts to any other atom — there are close intramolecular contacts to the other Sn atoms, but they are most probably superimposed by the geometric needs of a bicyclic

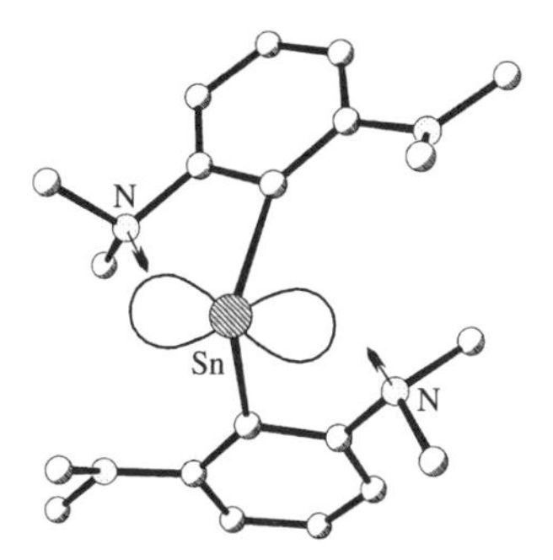

FIGURE 4. Molecular structure of stannylene **29**, showing weak Lewis acid/Lewis base interactions of lone-pairs on N with empty p-orbital of Sn

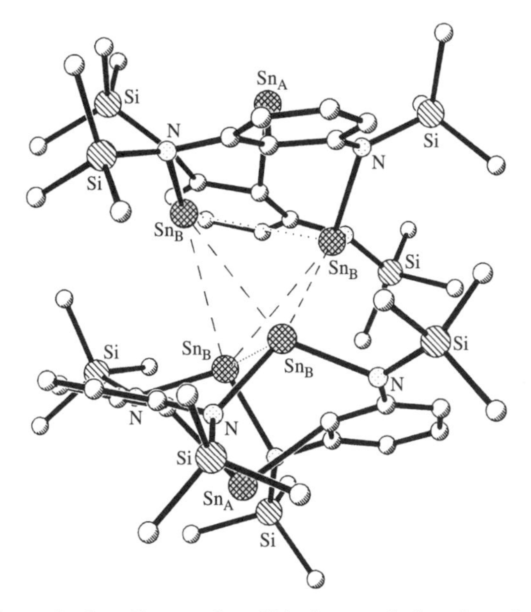

FIGURE 5. The dimer of polycyclic stannylene **23** in the crystal; short Sn $\cdots$ Sn contacts are given as dashed lines (- -)

system. The two bis(amino)stannylene (Sn_B) fragments, however, do display unusual short contacts (Figure 5) which are not forced, since they are *intermolecular*. Within the tetrahedral array $Sn_B \cdots Sn_B$ distances between 367 and 388 pm are observed, thus lying between typical covalent and van der Waals interactions. There are many examples of similar unsupported contacts in heavy element chemistry, the nature of which is not understood in most cases and is still under controversial discussion. In main-group element chemistry some related electron-deficient species such as some In(I) and many T1(I) derivatives exhibit a similar type of short intermolecular contacts in the solid state. Various closed-shell species[45] such as stibanes, bismutanes, selenans and tellanes, and related compounds tend also to associate in the condensed phase.

c. Plumbylenes (plumbandiyls). In contrast to the lighter congeners, until very recently, plumbylenes were thought to be strictly monomeric even in the solid state, though weakly bonded dimers and even cyclo-oligomers had been postulated by quantum-chemical calculations — at least for the parent compound PbH_2. Unlike carbenes, their higher homologues should have different possibilities of dimerisation, thereby not only and not

TABLE 4. Parameters (bond length in pm, angle in deg) of plumbylenes R^1R^2Pb with two-coordinate tetrel atom[a]

R^1	R^2	$Pb-R^1$	$Pb-R^2$	R^1-Pb-R^2	Reference	Remarks
Homoleptic						
Bis	Bis	230	230	103.6	46	GED
Bis	Bis	231	232	93.6	47	loose dimer Pb–Pb: 412.9
Bmp	Bmp	236	238	103.0	26	
Ar^F	Ar^F	236	237	94.5	28	
Btm	Btm	233	233	114.5	13	
Tbt	Tbt	233	233	116.3	48	
IV (C, C)		240	241	117.1	49	
Tret	Tret	235	234	99.4	50	$MgBr_2(thf)_4$ adduct
Mes	Mes	231	232	97.4	51	$MgBr_2(thf)_4$ adduct loose dimer Pb–Pb 335.5
Btp	Me	227	227	101.4	52	
Btp	*t*-Bu	229	233	100.5	52	
Btp	Ph	232	226	95.6	52	
V (C)	*V (C)*	230	230	95.2	53	weak Pb···N contacts 266.1; 270.8
V (C)	*V (C)*	228	231	94.8	53	weak Pb···N contacts 268.3; 272.7
Hyp (Si)	Hyp (Si)	270.0	270.4	113.6–115.7	32	4 independent molecules
Heteroleptic						
compound **8**	*(C, C)*	234	249	94.7	26	
Btm	Hyp (Si)	229	271.2	109.2	54	
Ar^F	Hyp (Si)	237	270.6	96.6	44	loose dimer Pb–Pb 353.7 Pb···F contacts: 276.6; 278.3
Bmp	Hyp (Si)	236	271.1	106.0	26	loose dimer Pb–Pb: 336.9
Btp	Br	233	278.9	95.4	52	loose dimer
		230.6	278.4	98.0		Pb–Br–Pb–Br ring
Btp	$Cp(CO)_3Cr$	229	290.9	113.6	55	
Btp	$Cp(CO)_3Mo$	229	298.5	110.0	55	
Btp	$Cp(CO)_3W$	228	298.1	108.6	55	
Btp	$Cp(CO)_3W$	228	300.6	109.4	55	
For comparison						
F	F	203.6	203.6	96	40	GED
Cl	Cl	244.4	244.4	98.0	40	GED
Br	Br	259.7	259.7	100	40	GED
I	I	280.4	280.4	100	40	GED
$N(Tms)_2$	$N(Tms)_2$	222.2	226.0	103.6	41	

[a]For the definition of *IV* and *V*, see Chart 2; atomic symbols in parentheses denote the type of atom binding to Pb.

even preferably yielding alkene homologues E_2R_4, but also by forming three-centre bonds with substituents R in a bridging position. This topic will be discussed in detail later (Section III). Here we will mainly discuss the structural features of the well-characterized monomers (Table 4).

The first bis(organyl)plumbylene isolated and structurally characterized (at first in the gas phase only) was the dark blue bis[bis(trimethylsilyl)methyl]plumbylene Bsi_2Pb **30**[46]. Recently we were able to determine the crystal structure of the low-melting solid and, to our surprise, found (very) weakly associated dimers (Figure 6)[49]. Despite a Pb···Pb distance *not much* shorter than twice the van der Waals radius of Pb, the dimer adopts a *trans*-bent conformation typical for the heavier congeners of the alkenes. In spite of the same conformation (*syn, syn*) in the gas phase and in the solid state, the structure displays markedly differing C−Pb−C angles of 103.6° (**30**) and 93.6°((**30**)$_2$), respectively.

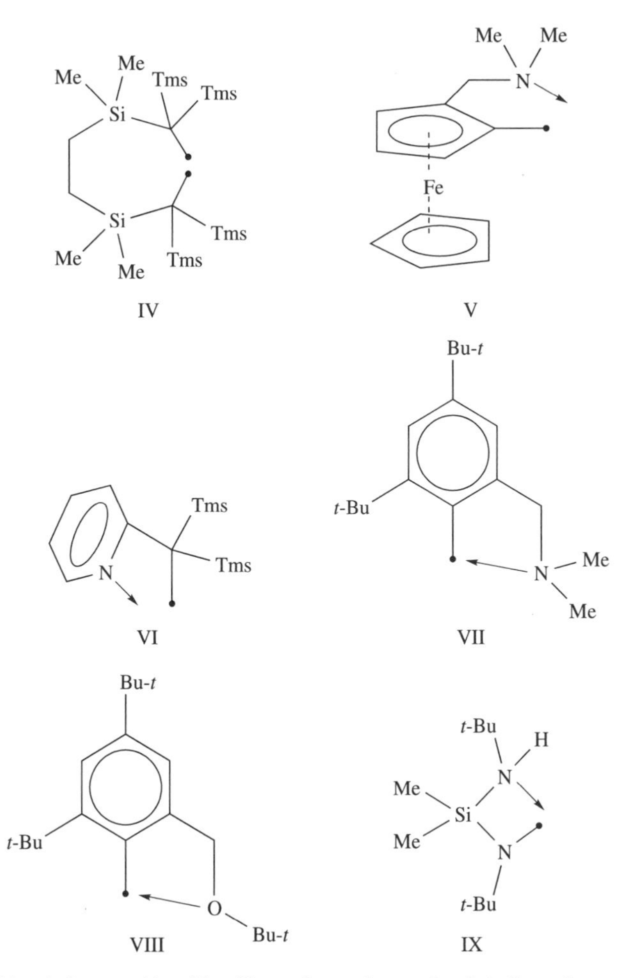

CHART 2. Abbreviations used in tables. Heavy dots and arrow heads indicate the connection point of the substituent

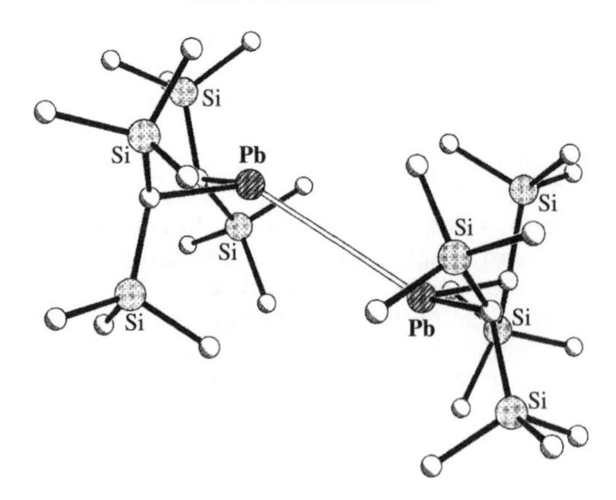

FIGURE 6. Molecular structure of plumbylene **30** in the crystal: formation of loose dimers

There are only two other homoleptic plumbylenes which form dimers in the solid state and which will be discussed in more detail later, i.e. bis(2,4,6-triisopropylphenyl)plumbylene (Tip_2Pb) and dimesitylplumbylene (Mes_2Pb). All remaining derivatives are monomers for steric and/or electronic reasons. Pure steric congestion had been achieved by substituents such as 2,6-$Mes_2C_6H_3$ (Btm)[13], 2-*t*-Bu-4,5,6-Me_3CH (Bmp)[26] or the bidentate ligand $[H_2CSi(Me)_2C(SiMe_3)_2]_2$ (in **31**)[49].

Surprisingly, in compound **31** two relatively short Pb$\cdots$Si contacts to adjacent Me_3Si groups are observed[49] which are only about 20% longer than the covalent Pb(II)−Si bonds in heteroleptic aryl(silyl)plumbylenes.

(31)

$[2,4,6\text{-}(F_3C)_3C_6H_2]_2Pb$ (Ar^F_2Pb) **13**[28] as its lighter congeners **11**[15] and **12**[27] comprises electronegative substituents and short intramolecular Pb$\cdots$F contacts (278.4; 279.3; 284.0; 296.7 pm), thus preventing the compound from dimerisation. Replacing one Ar^F substituent by the electron-releasing hypersilyl group yielding Ar^F(Hyp)Pb **(32)**[44] again favours the formation of a Pb−Pb bond. Since a Pb−Pb interaction is markedly weaker than an analogous Sn−Sn interaction (this especially holds for *trans*-bent double bonds), no lengthening of Pb$\cdots$F contacts is observed (as it is for the pair **12** and **28**), but rather an even slight shortening, reflecting perhaps that the number of F $\rightarrow$ Pb contacts is reduced from four to two.

An unprecedented type of intramolecular coordination is found in the alkyl(aryl)plumbylene **8**[26]. This particular plumbylene is formed via a rearrangement from homoleptic bis(supermesityl)plumbylene **7** (equation 2) which, in contrast to the respective derivatives of Ge and Sn (**5** and **6**), could not be structurally characterized so far due to its low stability. In the course of the rearrangement process a C−H bond of one *tert*-butyl group may interact with the electron-deficient two-coordinate Pb atom and subsequently undergo an addition reaction to the Pb−C bond. There is one structural feature remaining in the resulting heteroleptic species **8** which may serve as a possible model for this proposed $H_3C\cdots Pb$ interaction: two *tert*-butyl substituents of the remaining aryl substituent indeed show unusual short contacts of one methyl group each with the divalent Pb atom (Figure 7). These particular methyl groups are positioned in the voids above and below the C−Pb−C plane where the empty p-orbital of the Pb atom is thought to be located. This orientation leads to $Pb\cdots C$ distances of 280 and 283 pm, which are only about 15% or 20% longer than the covalent $Pb-C_{alkyl}$ and $Pb-C_{aryl}$ bonds in **8**. Unfortunately, but not unexpectedly, the positions of the hydrogen atoms at these particular methyl groups could not be located from the diffraction data. Thus no detailed picture of the interaction may be derived from the present structural data. However, a donor/acceptor interaction with C−H bonds as Lewis base and the electron-deficient Pb atom as Lewis acid seems to be possible.

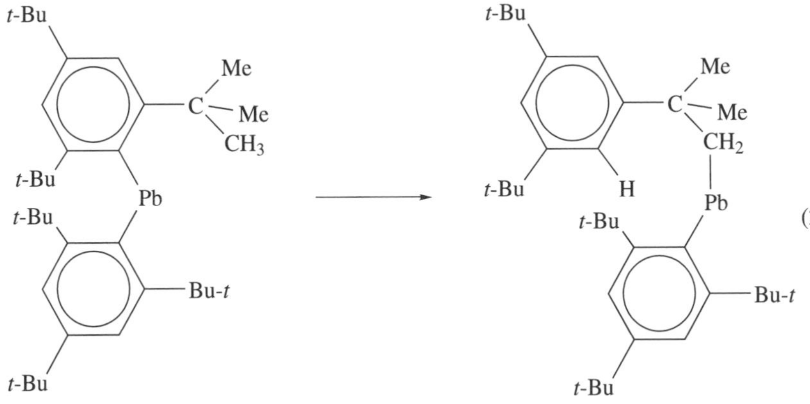

(2)

(7) **(8)**

Very recently, Power and coworkers reported the synthesis and structural characterization of a series of alkyl(aryl) and bis(aryl) substituted plumbylenes **33a–c** with one extremely bulky terphenyl ligand Btp = 2,6-$Tip_2C_6H_3$(Tip = 2,4,6-(i-Pr)$_3C_6H_2$) and one simple alkyl or aryl group such as methyl, *tert*-butyl and phenyl, by reacting the heteroleptic bromo(aryl)plumbylene 2,6-BtpPbBr with the appropriate Grignard or lithium organyl agent[52]. Despite the extreme bulk of the Btp substituent, the observed Pb−C bond lengths as well as the C−Pb−C angles are in the lower part of the range found for Pb(II) derivatives.

2. *Heteroleptic species* R−E−X

Heteroleptic carbene homologues with two-coordinate tetrel atoms still bearing one organyl group are relatively rare, since in most cases the second substituent is an amido, alkoxy or halido group which possesses one or more free electron pairs and thus tend to

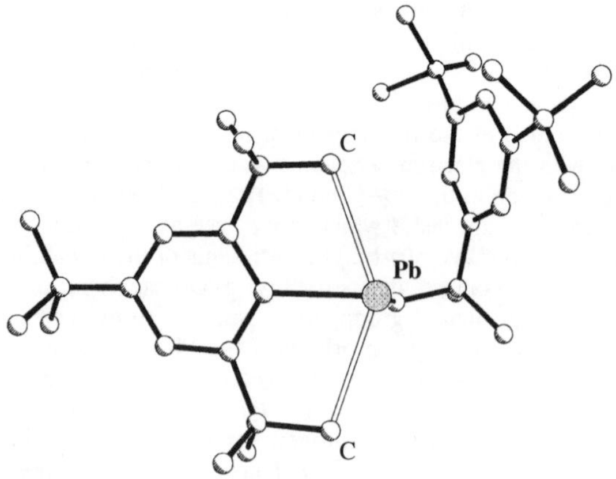

FIGURE 7. Molecular structure of plumbylene **8** in the crystal: short $H_3C\cdots Pb$ contacts are depicted as wide (=) sticks

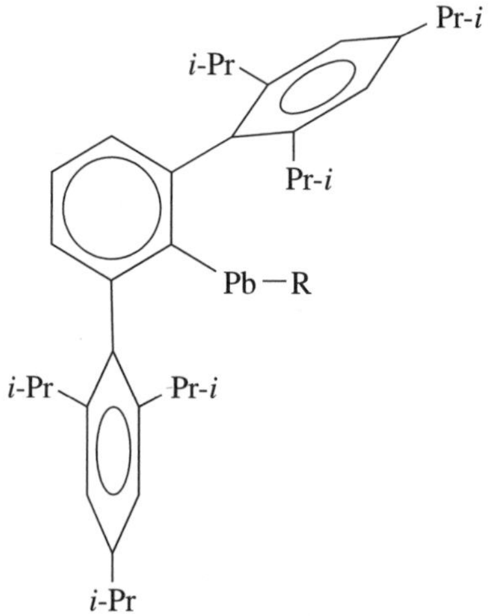

(33a) R = Me
(33b) R = *t*-Bu
(33c) R = Ph

give rise to the formation of bridged oligomers. Only if the second substituent has almost no Lewis basic properties, such as a silyl group, stannyl group or an appropriate transition metal fragment, and in cases where oligomerization is prevented by steric congestion, monomers with two-coordinate tetrel atoms may be observed.

a. Germylenes (germandiyls). There are only few papers about structure elucidation of heteroleptic germylenes with two-coordinate Ge bearing at least one organyl ligand. Power and coworkers reported the synthesis of the synthetically very valuable aryl(chloro)germylene BtpGeCl (**34**)[18]. Its existence as a monomeric species is due to the extremely bulky Btp substituent. Compound **35** with the related, but smaller, Btm = 2,6-$Mes_2C_6H_3$ ligand dimerises in the solid state to a unique digermene (see below)[56]. The crystal structure of **34** shows Ge−C = 198.9 pm, Ge−Cl = 220.3 pm and C−Ge−Cl = 101.3°; hence the Ge−C bond in **34** is markedly shorter than in all structurally characterized dialkyl and diarylgermylenes ranging from 201 to 208 pm, and the GeCl distance is somewhat longer than in the gas-phase structure of $GeCl_2$[22] [GED: Ge−Cl = 218.3(4), Cl−Ge−Cl = 100.3(4)°].

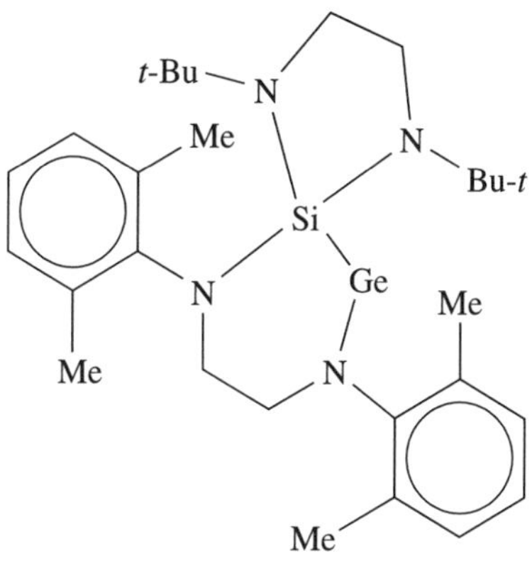

t-Bu
N
Me
N
Bu-t
Si
Ge
N
Me
Me
N
Me

(**36**)

Apart from the silyl(amino)germylene **36**, which is obtained from the reaction of a cyclic silylene with a cyclic bis(amino)germylene[19] (cf **24**), the remaining heteroleptic germylenes with two-coordinate Ge which had been structurally characterized are the metallogermylenes **37a,c**[20a]. As discussed in the literature[20a,b], these compounds may be regarded in a good approximation as germylenes. In other species comprising two-coordinate Ge atoms, a varying degree of multiple-bond character may be present, depending on the nature of the bonded metal fragment, i.e. its π-accepting and π-donating ability. We will not go into detail here, since it is beyond the scope of this chapter. Nevertheless, the two species, (**37a,c**) must be addressed as metallogermylenes, since the bonded metal fragment as a 17-electron moiety is neither a good π-acceptor nor a good π-donor. As expected, the M−Ge−C unit (M = Cr, W) is strongly bent, with M−Ge−C = 117.8°(M = Cr) and 114.7°(M = W); the Ge−M bond with 259 and 268 pm, respectively, is definitely a single bond. The compounds were synthesized by metathesis of Na[M(η^5-C_5H_5)$(CO)_3$] · 2DME and the heteroleptic aryl(chloro)germylene **34** at low temperatures. Heating the resulting germylenes under reflux in toluene or irradiating with UV light leads to CO loss and formation of the related metallogermylynes (**38a,c**) (equation 3). The Mo analogue **37b** could not be isolated at all; it looses CO at temperatures below −20 °C and yields directly the metallogermylyne **38b** under the given reaction conditions.

Btp–Ge–Cl (34) + $[(\eta^5\text{-}C_5H_5)M(CO)_3]^-$ Na^+ ⟶ Btp–Ge–$M(\eta^5\text{-}C_5H_5)(CO)_3$ (37) + Cl^-

(37) —($h\nu$ or Δ)→ Btp–Ge≡$M(\eta^5\text{-}C_5H_5)(CO)_2$ (38) + CO (3)

(a) M = Cr
(b) M = Mo
(c) M = W

b. Stannylenes (stannandiyls). Heteroleptic stannylenes are known with a greater variety of substituents, although the total number is also small. In solution, even a hydrido species BtpSnH was characterized very recently[38]. However, it dimerises in the solid state and will be discussed below. The same bulky aryl ligand allowed also the isolation of the monomeric aryl(iodo) and aryl(amido)stannylenes BtpSnI **(39)** and $BtpSn[N(Tms)_2]$ **(40)**[18]. Both resemble the related aryl(chloro)germylene **34** discussed above. Whereas the Sn–I bond in **39** with 276.6 pm is again markedly lengthened compared with SnI_2 (268.8 pm), no change in Sn–N bond lengths is observed going from $Sn[N(Tms)_2]_2$ **(26)**[41] to **40**.

A remarkable compound, comprising divalent and tetravalent tin atoms within one molecule, is the already mentioned triorganylstannylstannylene **27**[38]; the observed Sn(II)–Sn(IV) bond distance of 286.9 pm is much longer than the Sn(IV)–Sn(IV) bonds in distannanes; responsible are the larger covalent radius of Sn(II) compared with Sn(IV) and the extremely bulky Btp group present. It should be noted that a closely related donor-stabilized stannylene **41**[57,58] is known which exhibits the same feature, but comprises a three-coordinate Sn(II) centre with intramolecular Sn ← N coordination (Sn–N: 228.8 pm). In spite of the higher coordination number at Sn(II), but expected from the presence of less demanding substituents, a somewhat shorter Sn–Sn bond (286.9 pm) is found. Whereas the heteroleptic stannylenes bearing one extremely bulky Btp group are accessible from Sn(II) halides and successive treatment with appropriate nucleophiles, the remaining heteroleptic species had been synthesized from diarylstannylenes with other electronically or coordinatively unsaturated compounds.

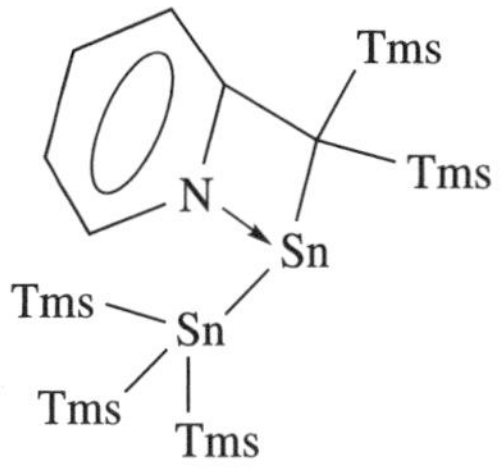

(41)

The unique oxoethenyl substituted stannylene **23** is obtained via the reaction of $Ar^{F}_{2}Sn$ with a ketene[36]. In spite of two close F···Sn contacts (274 pm), a very short Sn−C bond (212.2 pm) to the oxoethenyl moiety is found. This short Sn−C bond as well as C−O and C−C bond lengths of 129 and 141 pm, respectively, and the virtually coplanar O−C−C−Sn arrangement, indicates π-delocalization and Sn−C multiple bonding.

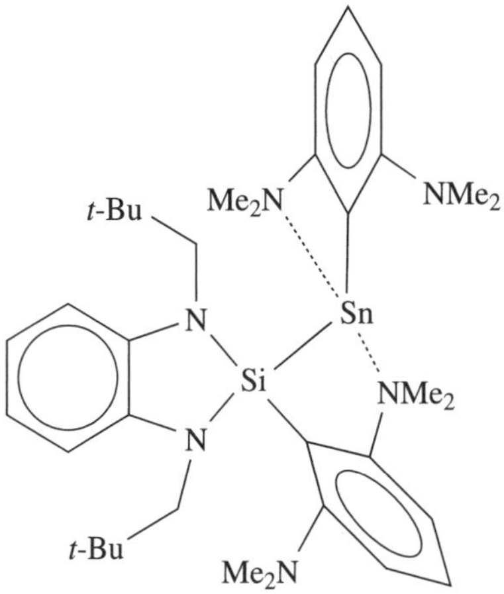

(42)

All aryl(silyl)stannylenes known are monomers in dilute solution. One of them, the stannylene **42**[37], stays as a monomer even in the solid state, most probably for steric reasons. It is formed alongside an unusual insertion reaction of silylene **25** into one tin carbon bond of the diarylstannylene $Ar^{N}_{2}Sn$ (cf **24** and **36**). The others bearing at least one hypersilyl group (Hyp = $Si(SiMe_3)_3$) all form dimers (distannenes) in the solid state[54,59,60]. They are synthesized via ligand exchange between the respective homoleptic stannylenes or between Hyp_2Sn **(20)** and CuMes. The proposed reaction scheme for the ligand exchange between stannylenes[59] is similar to the one given for the formation of **24**, **36** and **42** (see above), since in all cases the first step is a formation of doubly-bonded mixed dimers, which then rearrange by migration of a substituent from one of the doubly-bonded atoms to the other, the first step having almost no activation barrier and the second step in most cases having a very small barrier.

c. Plumbylenes (plumbandiyls). Heteroleptic plumbylenes with two-coordinate Pb are known with very bulky substituents only. Apart from the already discussed alkyl(aryl)plumbylene **8**[26], either a hypersilyl or a terphenyl substituent (Btm or Btp) or even both are present. Whereas most terphenyl-substituted plumbylenes are monomers both in solution and in the solid state, most of the hypersilyl derivatives — according to the low electronegativity of the Hyp group — give dimers in the solid state. Despite the long Pb–Pb bonds, the latter may be addressed as diplumbenes $R_2Pb{=}PbR_2$ and will be discussed in detail in Section III. Terphenyl-substituted plumbylenes never form dimers with a Pb–Pb bond for steric reasons, even if the second substituent is as small as methyl (see **33a–c**)[52]. Only if the second substituent is a halogen, such as Br, may dimers be observed (see below)[52].

We recently synthesized a new class of heteroleptic plumbylenes which, in spite of the presence of Lewis-basic centres, form no oligomers but only monomers with a three-coordinate Pb. The reaction of azides (RN_3; R = $SiMe_3$; 1-adamantyl) with Hyp_2Pb **(21)** did not lead to iminoplumbanes $Hyp_2Pb{=}NR$ as intended; instead, the formation of hypersilyl(triazenyl)plumbylenes **43a**,**b** (equation 4) had been observed[61], putatively formed via formation of an azide adduct and subsequent migration of one hypersilyl substituent from the lead to the terminal nitrogen atom. The crystal structure of **43a** had been determined (Figure 8) and revealed the presence of a pyramidally coordinated Pb and an η^2-coordinated triazenyl moiety. The latter shows two markedly different N–N bonds of 129 and 135 pm, the longer one directed towards a slightly pyramidally coordinated and the shorter one to a planar coordinated nitrogen atom, indicating at least a partial localization of the double bond.

$$Hyp_2Pb\ \mathbf{(21)} + N{=}N{=}N{-}R \longrightarrow \mathbf{(43)}\ [R{-}N{=}N{-}N(Hyp)\text{ chelating }Pb{-}Hyp] \tag{4}$$

(a) R = 1-Ad
(b) R = Tms

A further interesting class of heteroleptic plumbylenes are the metalloplumbylenes $BtpPb[M(CO)_3Cp]$ (**44**, M = Cr, Mo, W)[55] — the homologues of the metallogermylenes **37** mentioned above — which were synthesized only very recently by reaction of BtpPbBr with $(thf)_2Na[M(CO)_3Cp]$(M = Cr, Mo, W). The nature of the metal fragment ($17e^-$) as well as long M–E bonds and small C–Pb–M angles between 108.6 pm (W) and 113.6 pm (Cr) clearly indicate the absence of notable M $\rightarrow$ Pb backbonding, i.e. of M–Pb multiple bonding.

B. Tetrylenes with Higher-coordinate Tetrel Atoms

There is a lot of structural information available on intramolecular and intermolecular Lewis acid/Lewis base complexes of tetrylenes. In this section I will concentrate on basic structural principles which are valid throughout the series of the tetrylenes.

Whereas amino groups are used most frequently as *intra*molecular Lewis bases (Tables 5–7), stable unsupported intermolecular complexes are typically accomplished by carbene-type donors, such as carbenes, isonitriles and ylides (Table 8). To my knowledge, no unsupported intermolecular adduct of ethers, amines or phosphanes have been structurally characterized so far (except for the more ionic cyclopentadienyl derivatives[62]).

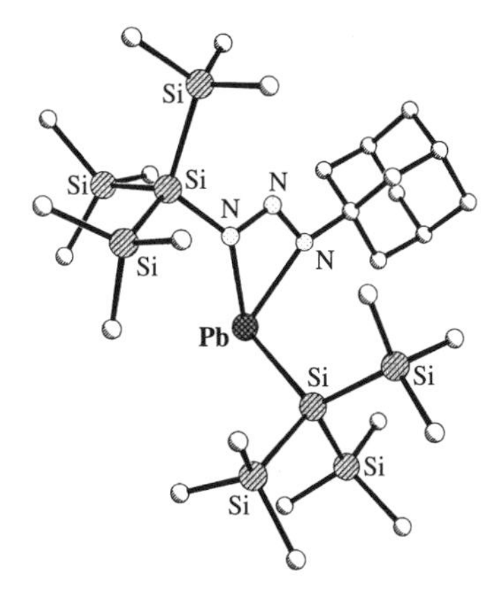

FIGURE 8. Molecular structure of the heteroleptic plumbylene **43a** in the crystal

Most probably, it is the softness of the E(II) centre that is mainly responsible for this fact. Nevertheless, *intra*molecular E $\leftarrow$ N bonds may be relatively short and may lead to significant differences in reactivity compared to tetrylenes missing such interactions, since the acceptor orbital at the tetrel is (partially) blocked towards interactions with external Lewis bases (see discussion in Section III). Substantial structural changes within the R—Sn—R skeleton (apart from those caused by steric interactions) are not observed and are not expected for most species, however, since bonding of the Lewis-basic centres is accomplished by the LUMO of the R—E—R fragment which is an empty p-orbital of the respective tetrel E. Contributions of antibonding R—E orbitals — for symmetry reasons — should only play a significant role if E—R—π—bonds were present. Thus, the Ge—Cl and Sn—Cl bond lengths in the carbene complexes of $GeCl_2$ and $SnCl_2$, **45a–c**, are by about 12 pm longer than in gaseous $GeCl_2$ and $SnCl_2$ (218 pm[22] and 234 pm[39], respectively, some Sn—Hal π—bonding)[22], whereas the Sn—C bond in the isonitrile complex Ar^F_2Sn $\leftarrow$ CNMes is only by 3 pm longer than in the parent stannylene (228 pm; no Sn-C π-bonding)[63]. The length of the E $\leftarrow$ D bond in turn may also be influenced by different Lewis-acidities of the tetrel in different tetrylenes.

Thus, in the series of the stannylenes **41** and **46a–e**[57,58,64,65] the shortest intramolecular Sn $\leftarrow$ N contact of 226 pm is observed for **46a** which is still about 20 pm longer than typical covalent and Sn(II)—N bonds. A similar observation was very recently made for two germylenes (**47a,b**)[66] with intramolecular Ge $\leftarrow$ O contacts: again the aryl(chloro)germylene **47a** exhibits a by far shorter Ge $\leftarrow$ O bond of 207 pm than the respective diaryl derivative **47b** (219 pm). If the acidity of the tetrel is enhanced by coordination of an external Lewis acid, as can be seen by the only two examples of main-group Lewis-acid adducts **48a,b**[16] to tetrylenes (see Table 9), the shortening in the E $\leftarrow$ D bond is dramatic. Whereas the parent germylene Ar^N_2Ge and stannylene **29** display only weak Ge $\leftarrow$ N and Sn $\leftarrow$ N interactions (Ge-N: 239; Sn-N: 261, 267 pm, respectively), the complexation with BH_3 results in a shortening of these contacts by 28 pm and 18 pm, respectively.

In the unique adduct **49**[67] (Figure 9), which is built from four stannylene sub-units, the interplay between different Lewis bases bonded to the same Lewis-acidic stannylene can be demonstrated impressively. The compound comprises three chloro(dimethylamino)-3-methylbut-2-yl stannylene fragments, two of which (A and A′) act as Lewis acid and

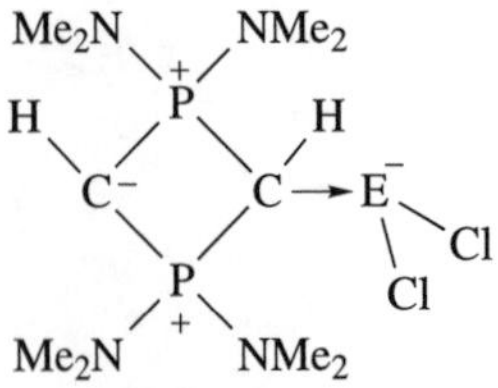

(**45a**) E = Ge
(**45b**) E = Sn

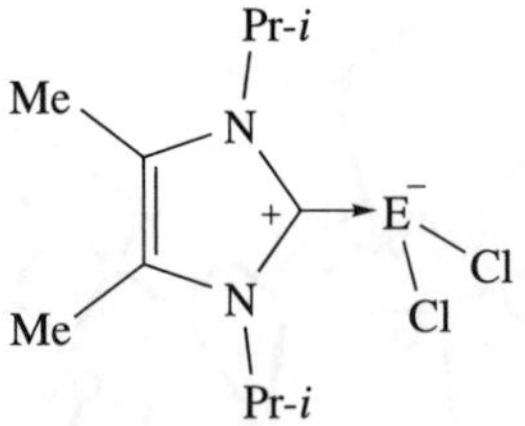

(**45c**) E = Sn

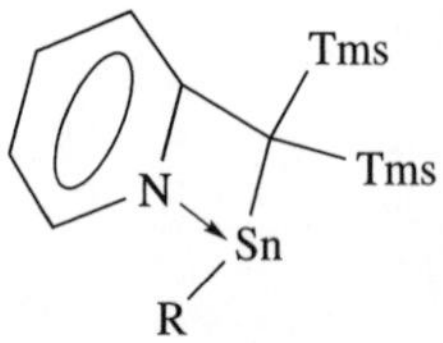

(**46**)

(**a**) R = Cl
(**b**) R = $N(SiMe_3)_2$
(**c**) R = Tip
(**d**) R = $CH(PPh_2)_2$
(**e**) R = Hyp

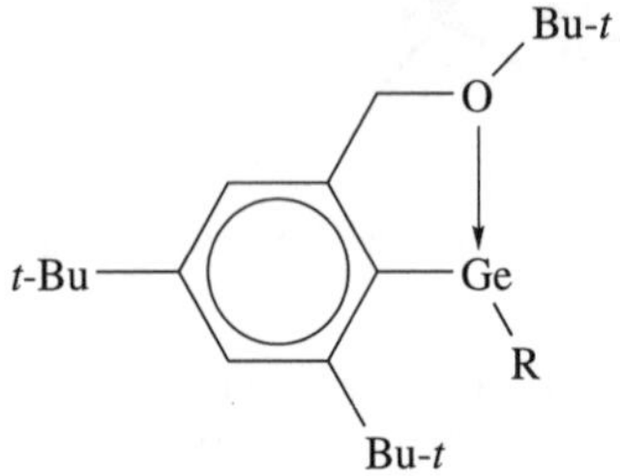

(**47**)

(**a**) R = Cl
(**b**) R = t-$Bu_2(t$-$BuO)C_6H_2$

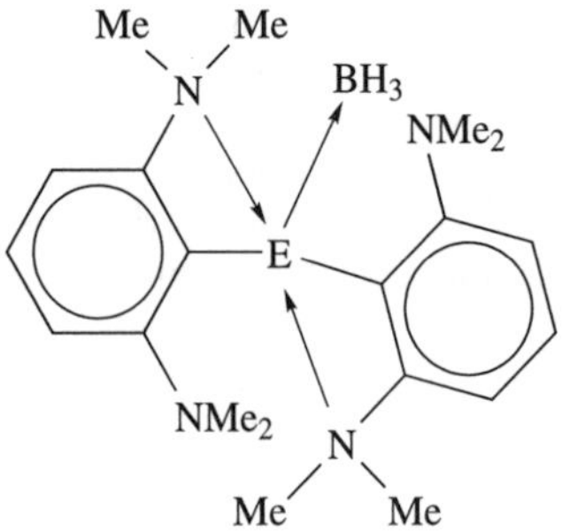

(**48**)

(**a**) E = Ge
(**b**) E = Sn

as Lewis base towards a bridging chloride anion and a $SnCl^+$-cation, respectively, and one such fragment (B) which acts only as base towards the $SnCl^+$-cation (C). Since no competing donor is present, the observed $N_B \rightarrow Sn_C$ bond in fragment B is by far shorter (221 pm) than the respective bond in the fragments A (N_A-Sn_C: 243 and $N_{A'}-Sn_C$: 247 pm), where the bridging chloride anion Cl_D is competing for the same acceptor orbital on the Sn_A atom. Similar reasons lead to the markedly different Sn–Sn bonds: the longer Sn_B-Sn_C bond (315.6 pm) is formed by fragment B, since it competes with the terminal

TABLE 5. Parameters (bond length in pm, angle in deg) of germylenes R^1R^2Ge with higher coordinated tetrel atom (donor atom D)

R^{1a}	R^2	D	$Ge-R^1$	$Ge-R^2$	Ge–D	R^1-Ge-R^2	R^1-Ge-D	R^2-Ge-D	Reference	Remarks
VI (C)	Cl	N	214 214	229.5 229.6	208; 208	101.9 102.5	67.2 66.9	91.9 93.0	57	
VI (C)	$CH(PPh_2)_2$	N	214	210	209	105.8	66.6	98.7	57	
VII (C)	Tip	N	201	208	216	104.2	80.3	93.4	68	
VII (C)	Cl	N	202	232.7	209	94.1	83.1	93.1	68	
VIII (C)	Cl	O	201	233.3	207	96.3	82.0	91.2	66	
VIII (C)	*VIII (C)*	O	204	206	219	109.1	80.3	91.5	66	
Bsi	MCp^b	C	204	224; 225		106.5; 104.3		—	69	η^2-coordination (Mcp)
Tsi	Mcp^b	C	213	225; 230		115.6; 119.0		—	70	η^2-coordination (Mcp)
Mes*	Mcp^b	C	209	231; 232		100.8; 100.0		—	70	
For comparison										
Cl	MCp^b	C	238.4	221; 222		96.8; 100.6		—	71	η^2-coordination (Mcp)
BF_4	MCp^b	F	293.7 (F)	225–226		—		—	71	η^5-coordination (Mcp)
Cl	$Me_4C_5(C_2H_4NMe_2)$	C N	236.8	218; 240 229		—			72	

[a] For the definition of substituents *VI, VII* and *VIII*, see Chart 2; atomic symbols in parentheses denote the type of atom binding to Ge.
[b] Mcp = pentamethylcyclopentadienyl.

TABLE 6. Parameters (bond length in pm, angle in deg) of stannylenes R^1R^2Sn with higher coordinated tetrel atom (donor atom D)

R^{1a}	R^2	D	$Sn-R^1$	$Sn-R^2$	Sn–D	R^1-Sn-R^2	R^1-Sn-D	R^2-Sn-D	Reference	Remarks
VI	Cl	N	232.9	244.0	227	101.4	61.1	91.7	65	two independent molecules
			232.4	244.6	226	101.3	61.7	89.0		
VI	$N(SiMe_3)_2$	N	235.6	214	230	105.4	61.1	97.3	65	
VI	Tip	N	237.2	225	235	111.9	60.5	107.6	64	
VI	$CH(PPh_2)_2$	N	235.9	231	230	100.8	61.1	93.0	57	
VI	$Sn(SiMe_3)_3$	N	230.4	286.9 (*Sn*)	229	108.3	61.3	89.7	57, 58	
VI	Hyp	N	234.5	272.4 (*Si*)	234	113.5	61.3	93.4	57	
IX	$C_5H_5(\pi)$	N	247.4	212	247	102.1	88.8	67.6	73	
IX	$C_9H_7(\sigma)$	N	234.3	212	231	98.2	92.5	70.7	73	
R^P	$SiPh_3$	O	222.9	275.1 (*Si*)	245	96.4	76.7	89.2	74	
					254		74.9	88.4		
Mes*	Mes^*CS_2	S	222.3	265.3 (*S*)	266.2	91.7	99.2	119.3	75	two independent molecules
			223.0	263.7 (*S*)	266.6	92.7	94.5	119.5		
$CH(PPh_2)_2$ (*C*)	$CH(PPh_2)_2$ (*P*)	P	228.2	265.9 (*P*)	267.8	90.3	99.4	63.4	76	

[a]For the definition of substituents *VI* and *IX*, see Chart 2 atomic symbols in parentheses denote the type of atom bonding to Sn.

TABLE 7. Parameters (bond length in pm, angle in deg) of plumbylenes R^1R^2Pb with higher coordinated tetrel atom (donor atom D)

R^{1a}	R^2	D	$Pb-R^1$	$Pb-R^2$	Pb–D	R^1-Pb-R^2	R^1-Pb-D	R^2-Pb-D	Reference
$CH(PPh_2)_2$ (*C*)	$CH(PPh_2)_2$ (*P*)	P	237	275.8 (P)	278.2	88.7	96.6	61.5	76
Hyp	$HypN_3Ad$-1	N	273.7	234	237	100.1	105.8	54.2	61

[a]atomic symbols in parentheses denote the type of atom binding to Pb.

TABLE 8. Parameters (bond length in pm, angle in deg) of unsupported Lewis-base adducts to tetrylenes L → ER_2(E = Ge, Sn, Pb) with donor atoms D

R	Base	D	Ge–R[1]	Ge–R[2]	Ge–D	R[1]–Ge–R[2]	R[1]–Ge–D	R[2]–Ge–D	Reference
Germylenes									
I	*Xa*[a]	C	263.9	268.1	210	99.4	100.2	95.4	79
Cl	$HC[P(NMe_2)_2]_2CH$	C	229.9	232.9	207	97.7	91.5	91.4	80
$N(Tms)_2$	c-$C[C(NPr_2^{-i})]_2$	C	197	199	209	105.7	98.6	98.6	81
Stannylenes									
R	Base	D	Sn–R[1]	Sn–R[2]	Sn–D	R[1]–Sn–R[2]	R[1]–Sn–D	R[2]–Sn–D	
Ar^F	CNMes	C	231	232	240	102.6	104.9	83.4	63
Hyp	CNBu-*t*	C	265.0	267.9	228	115.4	90.6	91.7	76
Hyp	CNHex-*c*	C	265.2	267.8	226	115.0	91.5	86.4	76
Tip	*Xb*[a]	C	230.8	232.0	238	106.7	92.6	109.5	82
Cl	*Xb*[a]	C	245.6	245.8	229	95.9	92.5	93.6	83
Cl	$HC[P(NMe_2)_2]_2C(H)$	C	248.0	248.0	227	95.4	86.6	91.2	80
$Me_2Si[N(Bu^{-t})]_2$	CH_2PPh_3	C	210.4	213	240	72.8	92.2	90.7	84
$Me_2Si[N(Bu^{-t})]_2$	CH_2PPh_3	C	211.3	212	244	72.7	99.5	91.8	85
$N(Tms)_2$	c-$C[C(NPr_2^{-i})]_2$	C	215.6	221	230	110.6	94.3	95.2	81
Plumbylenes									
R	Base	D	Pb–R[1]	Pb–R[2]	Pb–D	R[1]–Pb–R[2]	R[1]–Pb–D	R[2]–Pb–D	Reference
Tret	$BrMg(thf)_4Br$	Br	234	235	296.4	99.4	88.6	100.9	50
Trip	*Xb*[a]	C	237	238	254	105.2	90.1	108.8	86
Hyp (Si)	CNBu-*t*	C	273	275	250	114.0	89.7	90.8	76
$N(Tms)_2$	c-$C[C(NPr_2^{-i})]_2$	C	230	231	242	110.2	95.0	91.9	81

[a] For the definition of substituents *Xa* and *Xb*, see Chart 3.

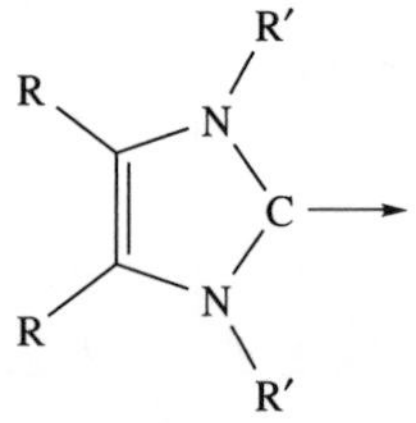

X

(a) R = H, R′ = Mes
(b) R = Me, R′ = *i*-Pr

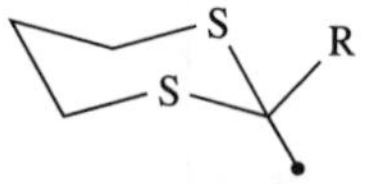

XI

(a) R = H
(b) R = Tms

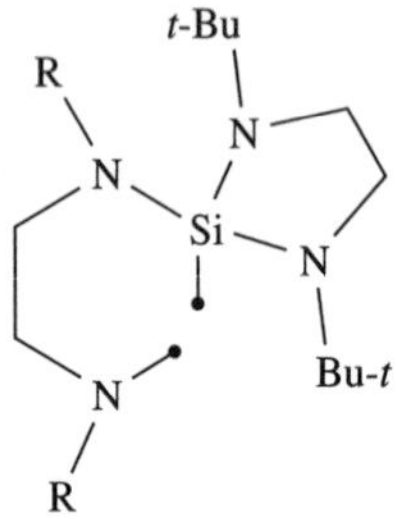

XII

(a) R = *t*-Bu
(b) R = *i*-Pr

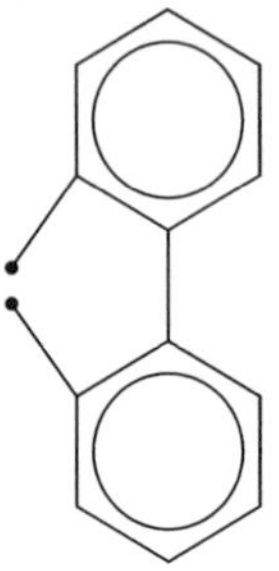

XIII

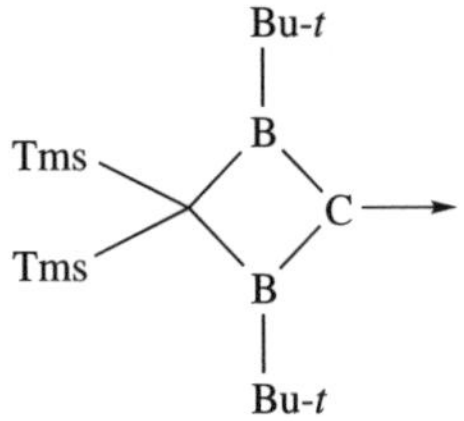

XIV

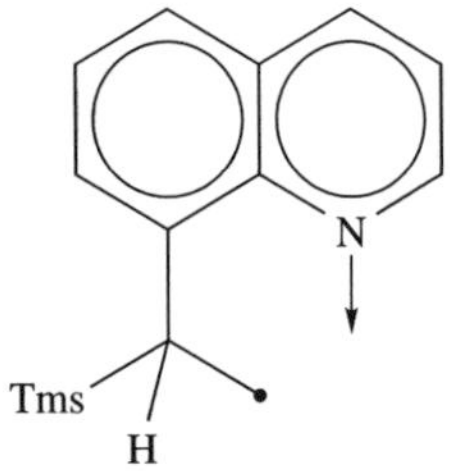

XV

CHART 3. Abbreviations used in tables. Heavy dots and arrow heads indicate the connection point of the substituent

chloride Cl_C for the same acceptor orbital on Sn_C ($Sn_B-Sn_C-Cl_C$: 157.4°), whereas the two stannylene units A and A′ may use two different acceptor orbitals ($Sn_A-Sn_C-Sn_{A'}$: 84.3°) and hence form the shorter bonds (287.3 and 288.2 pm, respectively).

Apart from the heteroleptic plumbylenes **43a,b**, one homoleptic species is known to form intramolecular N → Pb contacts: bis-1-[2-(*N*,*N*′-dimethylaminoethyl)ferrocenyl] plumbylene (**50**)[53], the Pb−N distances in **50** with 266.1 and 270.8 pm being again much longer than covalent $Pb^{II}-N$ bonds (*ca* 220 pm) or the Pb−N bonds in the plumbylene **43a** (234.4; 232.2 pm).

TABLE 9. Parameters (bond length in pm, angle in deg) of Lewis-acid adducts to tetrylenes $A \leftarrow ER_2$ with acceptor atom A

Germylenes										
R	Acid	A	$Ge-R^1$	$Ge-R^2$	Ge–A	R^1-Ge-R^2	R^1-Ge-A	R^2-Ge-A	Reference	Remarks
Ar^N	BH_3	B	195.9	196.2	204	112.7	117.2	125.7	16	one Ge–N contact: 211.0
Stannylenes										
R	Acid	A	$Sn-R^1$	$Sn-R^2$	Sn–A	R^1-Sn-R^2	R^1-Sn-A	R^2-Sn-A		
Ar^N	BH_3	B	217.0	217.0	226.2	119.0	120.5	120.5	16	two Sn–N contacts: 245.6; 245.6

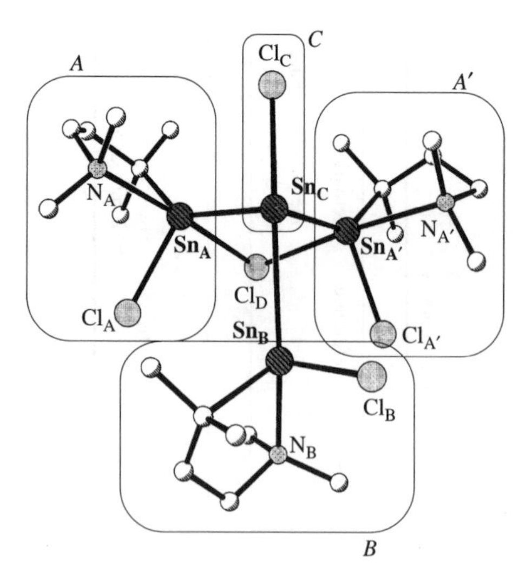

FIGURE 9. Molecular structure of adduct **49** in the crystal, showing its four constituting stannylene sub-units

(50) **(51)**

Apart from the unique adduct **51**[50], in which two plumbylene moieties are bridged by a $MgBr_2(thf)_4$ molecule via Br → Pb interactions [Pb−Br 296.4(2) pm], and few cyclopentadienyl derivatives where chelating amines interact with the two-valent tetrel atom[62], unsupported adducts of Lewis bases with tetrylenes have been restricted to carbene-type bases so far. However, only donor carbenes, such as Arduengo carbenes, isonitriles or ylids, form simple adducts with dative C → E interactions (Table 8). Carbenes or analogous species which also exhibit a substantial acceptor ability form multiple bonds instead (see Section III).

The adducts of Lewis bases known so far often show re-dissociation in solution or under low pressure. If the acceptor ability of the tetrylene is enhanced, stable compounds with shorter (stronger) C → E bonds are obtained, however. One possibility of enhancing the acceptor ability of the tetrylene is by the introduction of electropositive substituents such as silyl groups. As shown by theoretical calculations, such substituents will simultaneously raise the energy of the HOMO (lone-pair on E) and lower the energy of the LUMO (empty p-orbital on E) and therefore lead to both an enhanced donor and an enhanced acceptor ability of the tetrylene. As will be discussed in Section III, such substitution also enhances the tendency to form dimers (ditetrenes E_2R_4) or cyclo-oligomers. Thus dihypersilylstannylene Hyp_2Sn **(20)** and Hyp_2Pb **(21)** both form adducts with isonitriles: **52, 53** (**53a**: Figure 10)[61] which show no tendency of re-dissociation in solution or if stored under vacuum, whereas an analogous adduct of Ar^F_2Sn **(12)**, i.e. **54** easily dissociates[63] (weak Sn $\leftarrow$ F interactions of the CF_3 substituents are present and may further decrease the acceptor ability of **12**). Consequently, the C $\rightarrow$ E distances observed for the Hyp_2E adducts **52** and **53** are significantly smaller than those of the known diaryltetrylenes by about 10 pm (Sn) or 5 pm (Pb). It is noteworthy that the adducts **52–54** have no cumulated double bonds, thus they are not the analogues of ketenimines $R_2C{=}C{=}NR'$: all compounds comprise pyramidal tetrel atoms E with R−E−C angles of about 90°.

$$E \leftarrow C \equiv N - R' \quad (\text{E bonded to R, R})$$

(52) (E = Sn; R = Hyp; R′ = *t*-Bu, *c*-Hex)

(53) (E = Pb; R = Hyp; R′ = *t*-Bu, *c*-Hex)

(54) (E = Sn; R = Ar^F; R′ = Mes)

Apart from BH_3 and some certain stannylenes such as $SnCl_2$ (see Section III) which are typical soft acids, no further main-group Lewis acid has been found so far that forms stable adducts with the heavier tetrylenes. Such adducts are known, however, with electron-deficient (Lewis-acidic) transition metal fragments. Adducts have been characterized where the tetrylene acts as a Lewis base towards one, two or even three Lewis-acidic fragments. It is small wonder that, apart from peculiarities derived from steric congestion, such adducts show similar structural features to those of carbon monoxide, since tetrylenes and CO are isolobal, i.e. both have similar frontier orbitals. It is beyond the scope of this chapter to give an overview of that rapidly growing area. The interested reader is referred to additional literature[77,78].

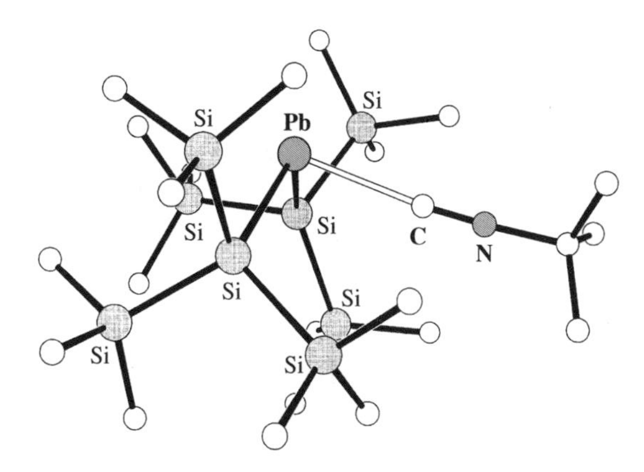

FIGURE 10. Molecular structure of the pivalonitrile adduct **53**, R′ = t-Bu of dihypersilylplumbylene

C. Oligomers

Tetrylenes comprising small organyl substituents (including hydrogen) or Lewis-basic groups X with free electron pairs are usually not stable under ambient conditions and are prone to oligomerize (Ge, Sn) or to disproportionate (Pb). The oligomerization in principle can lead (a) to alkene-homologous dimers (discussed in Section III), (b) to cyclo-oligomers with E—E bonds, (c) to cyclo-oligomers with E—X—E bridges (Table 10) or (d) to infinite polymers, for which detailed structural information is not available to date and which are not treated in this chapter.

There are few open-chain oligomers which are Lewis-acid/Lewis-base adducts of different tetrylenes and which are discussed together with ditetrenes in Section III.

1. *Cyclo-oligomers with* E—E *bonds*

Cyclo-oligomers *c*-$(R_2E)_x$ $(x > 2)$ with E—E bonds are known for Ge and Sn only[87]; cycloplumbanes *c*-$(R_2Pb)_x$ had never been detected, although according to quantum-chemical calculations they should form exothermically from appropriate plumbylenes and may adopt very unusual structures[88].

Only little new information on Ge and Sn homocycles was obtained since the earlier volume of this book was published. Thus, only few remarks on recent results will be made. Geanangel and coworkers reported several hypersilyl- or tris(trimethylsilyl)germyl-substituted stannacycles and germacycles, all synthesized from E(II) halides and appropriate alkali metal silyls or germyls using different solvents. The cyclotetratetrelanes $E_4[E'(SiMe_3)_3]_4Cl_4$ [E = Ge (**55**), Sn (**56**); E′ = Si (**a**), Ge (**b**)][89] — being tetramers of the respective heteroleptic chloro(silyl) and chloro(germyl)tetrylenes — all consist of puckered four-membered rings [like the related $Ge_4(Bu\text{-}t)_4Cl_4$[90]] with the four chlorine atoms in all-*trans* orientation. Only the hypersilyl-substituted compounds (**55a, 56a**) (co-crystallizing with pentane or benzene) gave well-ordered crystal structures, whereas the germyl-substituted species show severe disordering. The Ge—Ge and Sn—Sn bond lengths within the $E_4Hyp_4Cl_4$ (**55a, 56a**) species could be determined reliably to range from 250.6 to 255.8 and 281.1 to 291.5 pm, respectively. Surprisingly, in cyclotetrastannane **56a** (Figure 11) alternating long and short bonds are observed. The synthetic procedure leading to $Ge_4Hyp_4Cl_4$ (**55a**) yields also the unexpected cyclotrisilane $Ge_3(SiMe_3)_6$ (**15**) as already mentioned in an earlier section. Due to the low electronegativity of the silyl groups, which diminishes the singlet—triplet gap for the parent germylene fragments $Ge(SiMe_3)_2$ (see the next section), unusually short Ge—Ge bonds of 246.0 pm are observed for **15**, which is the shortest yet reported value for cyclotrigermanes.

A cyclic compound which is not related to germylenes but to germylynes, i.e. $Ge_4(Si(Bu\text{-}t)_3)_4$ (**57**), had been reported by Wiberg and coworkers[91]. This is the first tetragermatetrahedrane (Figure 12). It had been obtained in low yields by the reaction of Sup—$GeCl_2$—$GeCl_2$—Sup or $GeCl_2$ dioxane with NaSup. In this species even shorter Ge—Ge bonds ranging from 243.1 to 244.7 pm are observed. Attempts to synthesize its tin analogue $Sn_4(SiBu\text{-}t)_4$ led to a hexastannaprismane instead (see Section III).

2. *Cyclo-oligomers with* E—X—E *bridges*

Typical candidates prone to oligomerize via formation of E—X—E bridges are tetrylenes bearing at least one substituent with free electron pairs, such as halide, alkoxide and amide or, alternatively, bearing substituents that may form strong three-centre two-electron bonds such as hydride. Many tetrylenes with two such groups (EX_2) have been known for a long time and in many cases form cyclo-oligomers, but are beyond the scope of this

TABLE 10. Parameters (bond length in pm, angle in deg) of E–X–E bridged oligomers of tetrylenes [R–E–X] (E = Ge, Sn, Pb)[a]

Germylenes R	X	Ge–R	Ge–X	Ge–X′	R–Ge–X	R–Ge–X′	Ge–X–Ge′ X–Ge–X′	References	Remarks
Mcp[b]	Br	220 229	270.6	313.3	98.2 95.2	98.9 135.8	93.8 85.5	71	dimer η^2-coordination (Mcp)
Stannylenes R	**X**	**Sn–R**	**Sn–X**	**Sn–X′**	**R–Sn–X**	**R–Sn–X′**	**Sn–X–Sn′ X–Sn–X′**		
Btp	H	221	189(3)	195(3)	92	94	109 71	38	dimer
Btm	Cl	222	260.1	268.5	92.4	102.1	98.3 81.7	56	dimer
$C(SiMe_2Ph)_3$	Cl	230	259.5	277.9	99.1	111.1	101.6 78.4	92	dimer
$C(SiMe_3)_2(SiMe_2OMe)$	Cl	229	253.8	295.2	103.1	99.5	99.6 80.4	92	dimer; Sn–O 241.6 O–Sn–R 158.8
c-C_4S_2(*XIa*)[c]	$N{=}C(NMe_2)_2$	229	218	219	93.8	94.3	102.3 77.7	93	dimer
c-$C_4S_2(SiMe_3)$(*XIb*)[c]	$N{=}C(NMe_2)_2$	238	218	218	93.1	95.7	102.8 77.2	93	dimer
Plumbylenes R	**X**	**Pb–R**	**Pb–X**	**Pb–X′**	**R–Pb–X**	**R–Pb–X′**	**Pb–X–Pb′ X–Pb–X′**	**Reference**	**Remarks**
$C(SiMe_2Ph)_3$	Cl	242	272.4	283.5	99.0	110.8	100.6 79.4	92	dimer orthorhombic
$C(SiMe_2Ph)_3$	Cl	244	272.9	296.3	98.5	112.0	92.9 87.1	94	dimer monoclinic
$C(SiMe_3)_2(SiMe_2OMe)$	Cl	237	268.1	286.8	103.4	100.9	97.7 82.3	92	dimer; Pb–O 259.8 O–Sn–R 157.4
Tsi	Cl	235(mean)	271–274	—	99–104	—	90.3 (t); 115.7; 112.0 92.6 (t); 87.7; 91.2	92	trimer (boat conformation)
Btp	Br	233 231	278.9 278.4	301.6 299.0	95.4 98.0	—	94.8; 94.4 85.1; 85.6	52	dimer

[a]The symbols Ge′ and X′ indicate atoms of the REX moieties linked to the present one by X or Ge bridges, respectively.
[b]Mcp = methylcyclopentadienyl.
[c]For the structure of substituents *XI a, b*, see Chart 3.

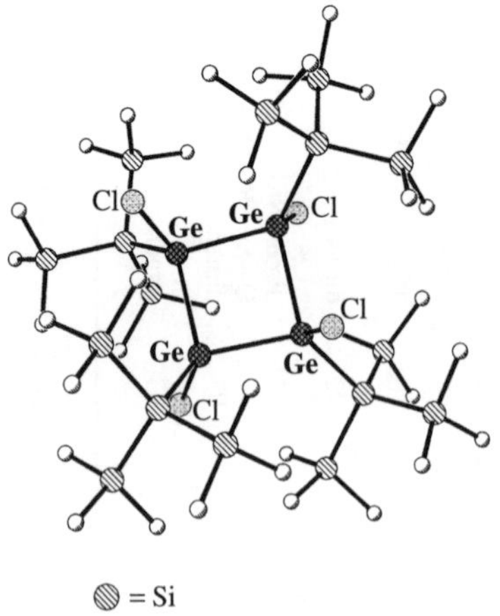

FIGURE 11. Molecular structure of cyclotetragermane **56a**

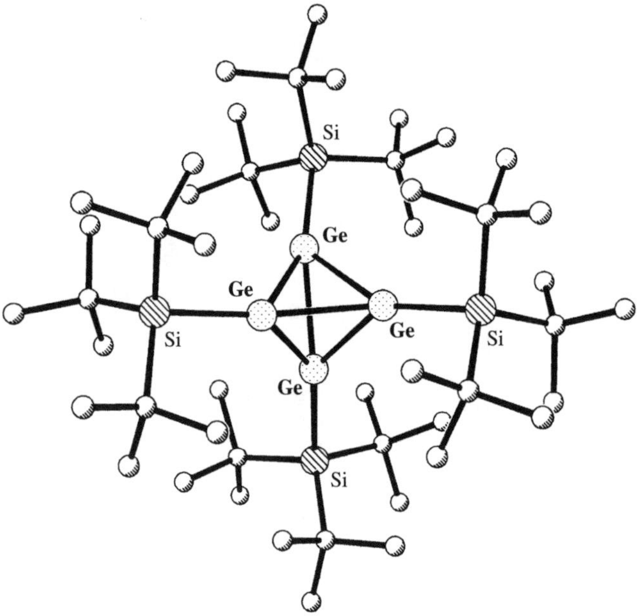

FIGURE 12. Molecular structure of tetragermatetrahedrane **57**

chapter since they have no E—C bond. There are only a dozen heteroleptic species with at least one hydrocarbyl substituent which meet the mentioned requirement; nine of them are bridged by halide, two by imido groups, and very recently one species was reported comprising bridging hydride.

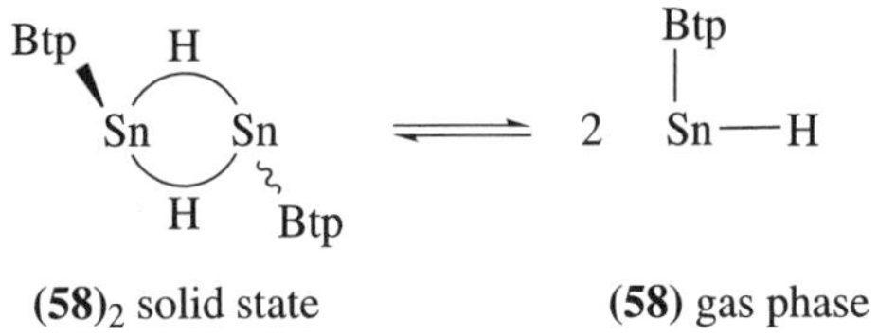

$$(\mathbf{58})_2 \text{ solid state} \rightleftharpoons 2\ \text{BtpSn–H}\ (\mathbf{58}) \text{ gas phase} \tag{5}$$

This unique hydride of two-valent tin, BtpSnH (**58**)[38], had been synthesized by Power's group via the reaction of BtpSnCl with $LiAlH_4$. While the analogous lead derivative BtpPbBr (**59**) is reduced by $LiAlH_4$ and yields a plumbylyne dimer BtpPbPbBtp[95] (Section III), BtpSnCl undergoes a substitution reaction, and a dark blue solution of the monomeric hydride **58** is formed. By crystallization from benzene orange crystals of the dimer, $(\mathbf{58})_2$ are obtained (equation 5). The X-ray crystal data were of sufficient quality to locate the bridging hydride. Within twice the standard deviation all Sn–H bonds are of equal length (189(3) and 195(3) pm, respectively). The terphenyl groups occupy *trans* positions at the planar ring, with tilt angles of 93.3°. All remaining dimers adopt similar *trans*-bridged structures (Table 10). While all halides have almost planar rings with asymmetric bridges (approximate C_i symmetry), i.e. tetrylene entities may still be recognized, the two imido-substituted species comprise planar rings with symmetric bridges (approximate C_{2h} symmetry); see Scheme 3.

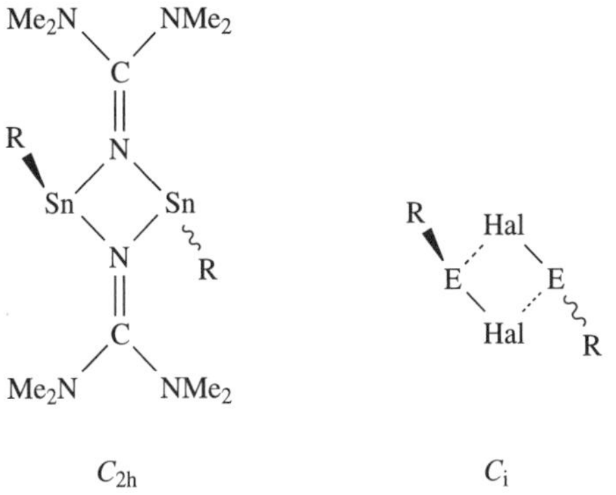

C_{2h} C_i

SCHEME 3. Different symmetries of the dihalo and di-imido dimers

For all asymmetric cases the constituting tetrylene moieties within the dimers are not only recognized by the shorter E–X bond, but also by the smaller C–Sn–X angle. Whether oligomers are formed at all, and which kind of oligomer is more stable should be mainly governed by the interplay of the bulk of the hydrocarbyl group and the length of the E–X bond, whereas the question whether symmetric or asymmetric bridges are formed should depend on both steric and electronic effects. Thus, while BtpSnI **39**[18] is monomeric, the related chloride with the smaller Btm group BtmSnCl (**59**) forms dimers in the solid state[56], as does BtpPbBr **60** in which a longer E–C bond and a smaller halide again allow for interactions of the parent tetrylenes[52]. A very unsymmetrical bridge (Pb–Br 278.7 and 300.3 pm, both mean values) is found for $(\mathbf{60})_2$, however. Very unsymmetrical bridges are also found for the dimers of the tetrylenes $(Me_3Si)_2[(MeO)Me_2Si]C–E–Cl$ (**61**), (**a**) E = Sn, (**b**) E = Pb[92], where competing O → E intramolecular interactions are present. As expected from the relative strength of the O → E bonds and the E–Cl bond, a by far looser dimer is found for the stannylene **61a** (Figure 13, left): the E–X bond lying approximately in line with the O → E interaction is longer by 41.4 ((**61a**)$_2$) and 18.7 pm ((**61b**)$_2$), respectively, than the other E–X bond. The differences in the E–X bond lengths are less pronounced, 18.4 and 11.1 pm, respectively, for the related stannylene dimer $[(PhMe_2Si)_3CSnCl]_2$(**62a**)$_2$[92] and one polymorph of $[(PhMe_2Si)_3CPbCl]_2$(**62b**)$_2$ with no donor functionality within the substituent. The second polymorph (**62b**)$'_2$[94] exhibits again more strongly differing Pb–Cl bonds (23.4 pm), however. Perhaps interactions with a phenyl group of the hydrocarbyl substituent are responsible for this fact, since in (**62b**)$'_2$—in contrast to (**62b**)$_2$—a phenyl group would be in the right orientation, i.e. opposite to one (the longer) Pb–Cl bond of the Pb_2Cl_2 ring (Figure 13, right).

Cl
Btm~Ge=Ge~Cl
Btm

((**59**)$_2$)

There is only one R–E–X trimer known so far: $[(Me_3Si)_3CPbCl]_3$(**63**)$_3$[92]. Whereas most trimers known from other fields of chemistry comprise six-membered rings in either a planar or a chair conformation, the Pb_3Cl_3 ring in (**63**)$_3$ adopts a very unusual boat-type structure with approximate C_s symmetry as depicted in Figure 14. All Pb–Cl bond lengths are the same within twice their standard deviation, 271(2) to 274(2) pm, thus lying in between the values found for the shorter and longer bonds within RPbCl dimers. Different are the angles around the Cl atoms, however: whereas those for the two Cl_B atoms at the 'bottom' of the boat are large (112° and 116°), the one at the 'top' Cl_T is much smaller (90°). No reasons are obvious, neither for the boat conformation itself nor for the differing environments of the Cl atoms.

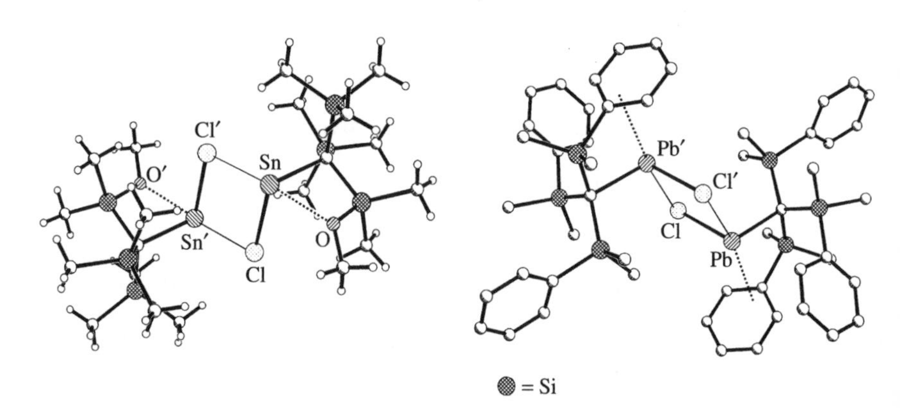

FIGURE 13. Molecular structure of the chloro(alkyl)stannylene **61a** (left) and of chloro(alkyl)plumbylene **62b**′ (right) in the crystal, both forming asymmetric halogen-bridged dimers due to competitive intramolecular Lewis acid/Lewis base interactions. These are depicted as dashed lines (- - -) and the longer E–Cl bonds as thin solid lines

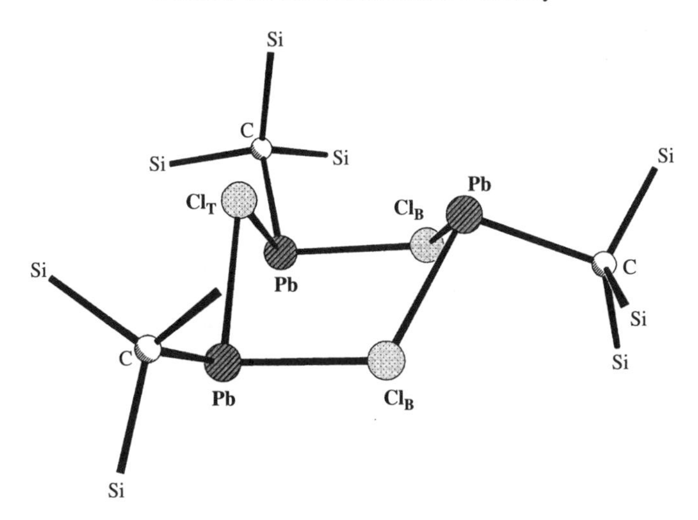

FIGURE 14. Plot of the central part of the molecular structure of trimeric chloro(alkyl)plumbylene **(63)**$_3$ in the crystal, showing its unusual boat conformation

III. MULTIPLE-BONDED SYSTEMS

A. Introduction

Since 1996 when the earlier volume of this book was published, the field of multiple-bonded systems containing heavier tetrel atoms has been developed intensely. The main focus has shifted from Si to the heavier elements Ge, Sn and Pb. Even for the heaviest element of this series, Pb, several species have been synthesized meantime, comprising (at least formal) double and even triple bonds. The structural details revealed by experiment and quantum chemistry, however, challenge long-established models of bonding, and especially the well-established relationships of bond order, bond strength and bond length. It showed that multiple bonds may be much longer and weaker than single bonds. Numerous published theoretical papers deal with this topic[2–4,96,97], and it is beyond the scope of the present chapter to deal with the details of these theoretical analyses, especially since several excellent reviews have been published recently[5b–g]. Only the most important results will be summarized briefly.

Alkenes and alkynes represent the prototypic models for doubly and triply bonded species, respectively. The C−C double and triple bond within these systems is much shorter and stronger than the corresponding C−C single bond in alkanes. A very simple relationship between bond order (n) and bond length d_n (equation 6) had been given by Pauling and may be also extended to fractional bond orders in conjugated systems[98]. Consequently, the C−C double (triple) bond as well as other classical homonuclear double (triple) bonds are by 21 pm (34 pm) shorter than the related single bond.

$$d_n = d_1 - 71\ \text{pm} \cdot \log(n)\ (d_1 = \text{bond order for } n = 1) \tag{6}$$

The inspection of bond lengths determined by absolute structure methods or *ab initio* calculations for compounds with homonuclear multiple bonds reveals that a similar relationship is valid only for boron and the heavier members of group 15 and 16. Nitrogen and oxygen display a much stronger shortening of about 30 and 27 pm, respectively, going from single to double bonds, since in N–N or O–O single-bonded species crowding of lone-pairs on these relatively small atoms causes strong repulsive interactions, which are significantly diminished if the number of substituents or lone-pairs is reduced by going from single- to double-bonded species. For the heavier tetrels Ge and Sn, on the other hand, the shortening is much less pronounced than for carbon; for Sn and Pb even a *lengthening* for the E–E bond may be observed going from singly- to doubly-bonded systems. Moreover, most digermenes and distannenes and all diplumbenes are in equilibrium with their carbene homologous fragments (tetrylenes) ER_2 in solution. The seemingly paradoxical observation that a higher bond order does not necessarily imply a shorter and stronger bond is traced back to the different electronic ground states of the parent tetrylenes compared to the parent carbenes. While carbenes typically have triplet ground states or at least a very low lying triplet state, all known higher homologues possess a singlet ground state, due to energetically and spatially stronger separation of valence s- and p-orbitals[96]. Only triplet tetrylenes can directly form classical double bonds by making two covalent interactions with their singly occupied valence orbitals, whereas the analogous interaction of singlet species (in the same orientation) would lead to strong Pauli repulsion between doubly occupied orbitals (Scheme 4). Therefore, a promotion step is necessary in all those cases where the tetrylene fragment possesses a singlet ground state. However, the more energy is required for the initial promotion ($2 \cdot \Delta E_{ST}$ where ΔE_{ST} is the singlet–triplet energy difference, since *two* tetrylene fragments must be promoted), the more the overall bond energy E_{total} is reduced. A critical point is reached if the double-bond formation from these triplet species (snapping process[2a]) gains less energy ($E_{\sigma+\pi}$) than the promotion costs; then, for obvious reasons, no classical double bond may form (equation 7).

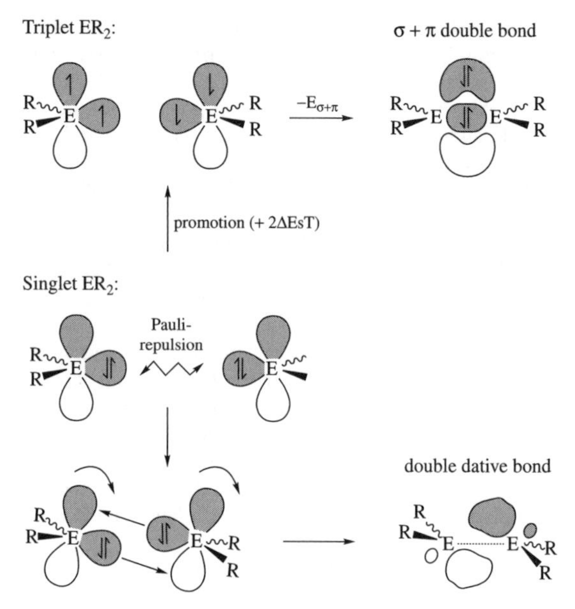

SCHEME 4. Formation of classical $\sigma + \pi$ double bond and double dative bond from triplet and singlet tetrylene moieties, respectively

$$E_{\sigma+\pi} > 2 \cdot \Delta E_{ST} \Longleftrightarrow \Delta E_{ST} < 1/2 E_{\sigma+\pi} \qquad (7)$$

However, even in cases where equation 7 does not hold, i.e. $E_{\sigma+\pi} < 2 \cdot \Delta E_{ST}$, a bonding interaction between two tetrylene fragments is possible, and a double bond may be formed: no classical double bond, of course, but a double donor–acceptor bond (double dative bond). This idea was already proposed by Lappert and coworkers in 1976 when they succeeded in the synthesis of the first stable distannene $Bsi_2Sn{=}SnBsi_2$[42] and found that the compound has no planar, but a *trans*-bent distorted $C_2Sn{=}SnC_2$ skeleton. A double donor–acceptor bond may form, if the tetrylene fragments, prior to interaction, are tilted in an appropriate way, such that the lone-pair of one fragment may interact with the empty p-orbital of the other (Scheme 4). The resulting bond energy depends — as it always does for dative bonding — on the difference in orbital energies between the interacting orbitals; the larger the difference, the weaker the bond. A model widely accepted and used nowadays to predict quantitatively the limits for the stability of both types of double-bonded systems, classical and dative ones, is the CGMT model[3]. As a simple quantum-chemical model it not only provides equation 7 for the existence of classical double bonds, but also equation 8 for the existence of double donor–acceptor bonds.

$$E_{\sigma+\pi} > \Delta E_{ST} \Longleftrightarrow \Delta E_{ST} < E_{\sigma+\pi} \qquad (8)$$

Since, in general, the promotion energy ΔE_{ST} of the two tetrylene fragments of the double-bonded system increases going down group 14, and since the expected E–E($\sigma + \pi$) bond strength decreases in the same direction, it is obvious that the tendency to form classical double bonds will be reduced going from C to Pb. According to the calculated or estimated values for ΔE_{ST} or $E(\sigma + \pi)$ of the parent hydrido derivatives (EH_2 and E_2H_4), one would expect that of the heavier group 14 elements, only Si (and perhaps Ge) should be able to form classical double bonds. Double donor–acceptor bonds should be observed for Ge and Sn, whereas plumbylenes PbR_2, finally, should remain monomers. Calculations at high *ab initio* levels reveal, however, some degree of distortion from planarity already for Si_2H_4, although the potential surface is very shallow in the questionable region. Moreover, these calculations predict that even diplumbene Pb_2H_4 is a stable dimer, but with a very low dissociation energy of about 25 to 40 kJ mol^{-1}, depending on the employed quantum-chemical method. (Note that the comparison of dissociation energies for compounds with double donor–acceptor bonds with the values of related singly-bonded species is very problematic, since in the former case a bond is broken *heterolytically* to relatively low-energy closed-shell monomers, whereas in the latter case a *homolytic* bond breakage occurs giving usually high-energy radical species.) It should be noted at this point that there are several other minima on the hypersurfaces of the heavier tetrylene dimers (Scheme 5); for three of them (B–D) it is not (or not exclusively) the lone-pair of one tetrylene unit that is serving as Lewis base to the other unit, but a E–H bond, finally leading to hydrogen-bridged species with three-centre two-electron E$\cdots$H$\cdots$E bonds. Isomer E is a tetryltetrylene, a mixed valence species (cf **27**) and F is no minimum at all for the parent E_2H_4 derivatives; it may be observed, however, with substituents other than hydrogen (see below). The relative energies of the isomers are dependent on the element E and, if other substituents than H are introduced, on the nature of the employed substituent.

All synthetically accessible alkene homologues bear other than hydrogen substituents, therefore additional influences on the double bond are present by the nature of the substituent pattern. It was shown that the promotion energy E_{ST} may be strongly influenced by the chosen substituent on E: the lower the electronegativity, the smaller E_{ST}, and vice versa[88,99]. Finally, one should consider that for synthetically accessible tetrylenes, steric effects (large substituents) may also lead to changes in ΔE_{ST} — by altering bond lengths and the valence angle at E — and may reduce or enhance the dissociation energy of $R_2E{=}ER_2$ by additional van der Waals interactions within the periphery of the molecule.

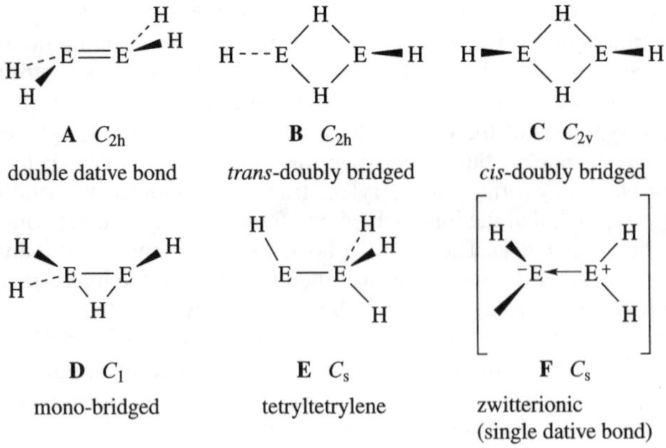

SCHEME 5. Local minima A to E on the hypersurface of tetrylenes E_2H_4 and a zwitterionic form F in hydrocarbyl-substituted species

The CGMT model was intended to describe (non-cyclic) homonuclear and heteronuclear double-bonded systems. However, in the meantime other multiple-bonded species have also been synthesized, as will be seen in the following. They also often exhibit structural features which are not familiar from the analogous carbon derivatives, and are not yet fully understood. Some of them are, however, related to the phenomena discussed above and may be understood on a similar base. I will not go into details in these cases, but will refer to recent literature when available.

In the following sections alkene homologues will be discussed first, i.e. species exhibiting a double bond between two heavier tetrel atoms. Then we will switch to molecules comprising double bonds between heavier tetrels and carbon, and finally relatives of ketones and imines will be discussed. Structural information about all structurally characterized compounds are found in following tables: digermenes (Table 11), distannenes (Table 12), diplumbenes (Table 13), cyclic species (Table 14), germenes and stannenes (Table 15) and heteroketones as well as heteroimines (Table 16).

B. Alkene Homologues (Ditetrenes)

According to the CGMT model the skeleton of ditetrenes $R_2E{=}ER_2$ bearing organyl or related non-donor substituents may exhibit at least two possible ideal conformations:

(I) planar conformation (D_{2h} symmetry) $\rightarrow$ a classical double bond.
(II) *trans*-bent conformation (C_{2v} symmetry) $\rightarrow$ a double dative bond (characterized by a *trans*-bent angle κ as defined in Scheme 6).

Depending on the nature of R, various structural changes or distortions are observed:

(a) lengthening of the E=E bond,
(b) torsion about the E=E bond (characterized by the twist angle τ defined in Scheme 6),
(c) formation of a zwitterionic form (single dative bond) $R_2Sn \rightarrow SnR'_2$.

For diplumbenes and distannenes — in accordance with the presence of the weakest E=E bond of group 14 and a shallow potential curve for E–E stretching — a distortion of type (a) is predominantly found due to steric strain or electronic destabilization, the lengthening of the E=E bond being more pronounced for E = Pb. For few cases, for instance if very sterically demanding silyl substituents are present which strengthen the

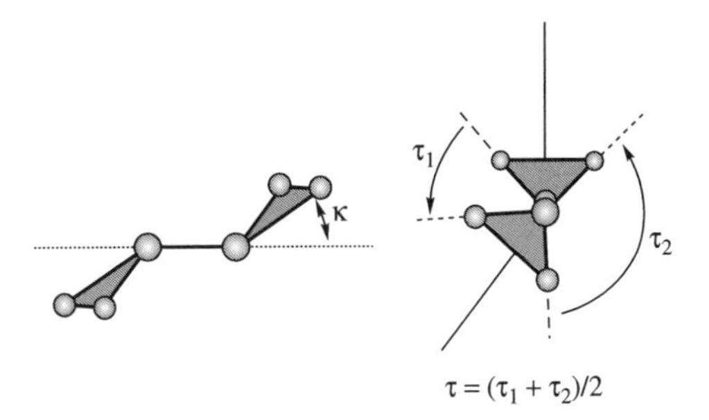

SCHEME 6. Definition of two characteristic parameters describing the conformation of ditetrenes. κ is the tilt angle of the R_2E plane towards the E-E bond; τ is the twist angle, defined as arithmetic mean of the (signed) torsion angles τ_1 and τ_2, thus giving the distortion from an ideal *trans*-bent conformation ($\tau_1 = -\tau_2$)

Sn=Sn bond electronically, a relatively short bond and a large torsion angle τ have been observed. Finally, three distannenes exhibit the zwitterionic conformation which has not been observed for other ditetrenes to date. The digermenes possessing the strongest E=E bond among the discussed ditetrenes frequently display distortion mode (b), whereas lengthening of the bond is not as pronounced as for the heavier congeners.

1. Digermenes

The first ever structurally characterized digermene, Lappert's $Bsi_2Ge{=}GeBsi_2$ (**64**), is the only tetraalkyldigermene known to date[100]. It exhibits the expected undistorted *trans*-bent conformation ($\kappa = 32°$; $\tau = 0$) and the longest Ge=Ge bond (234.7 pm) among tetraorganyldigermenes, being, however, still markedly shorter than Ge−Ge single bonds in digermanes (Ge_2H_6: 240.3 pm[101a]; Ge_2Ph_6: 243.7 pm[101b]). Three out of four tetraaryldigermenes (Table 11) exhibit slightly distorted *trans*-bent structures and significantly shorter Ge=Ge bonds (221.3–230.1 pm) than **64**. The *trans*-bent angles κ for the two derivatives with the shortest Ge=Ge bond Dep_4Ge_2 (**65**)[102] and Tip_4Ge_2 (**66**)[103] with 12° and 12.3°, respectively, are markedly smaller than that for **64**. Thus bond lengths and conformation are close to those values expected for a classical, planar Ge=Ge bond, indicating that digermenes are near the borderline to classical double-bond systems, as also derived from the CGMT model. Both the other two tetraaryldigermenes, Bmp_4Ge_2 (**67**)[104] and Mes(Dip)Ge=Ge(Dip)Mes (**68**)[105], show some unexpected features. (**67**), bearing the most sterically demanding substituents among this series, is (not unexpectedly) strongly twisted about the Ge=Ge bond ($\tau = 21.2°$), but, in spite of a longer Ge=Ge bond (225.2 pm), it seemingly adopts a planar conformation. The large thermal parameters of the Ge atoms indicate dynamic or static disordering, however, and so perhaps the presence of the expected *trans*-bent conformation. The remaining species, the only unsymmetrically substituted tetraaryldigermene **68**, was obtained as the unexpected *Z*-isomer and exhibits an only slightly twisted Ge=Ge double bond ($\tau = 3.4°$).

In spite of sterically very demanding substituents, both structurally characterized tetrasilyldigermenes[106] (Table 11) show nearly planar conformations with *trans*-bent angles κ of 6–7° only. The observed E=E bond lengths (226.7; 229.8 pm) are, however, significantly longer than those found for **65** and **66**, thus leading to the conclusion that the extent of *trans*-bent distortion is mainly governed by the electronic effects of the substituents: electropositive substituents, such as silyl groups, diminish the promotion energy E_{ST} and thus favour small *trans*-bent angles or even the planar form;

TABLE 11. Parameters (bond length in pm, angle in deg) of digermenes $R^1R^2Ge{=}GeR^1R^2$

R^1	R^2	E=E	E−R	R−E−R	κ	τ	Reference	Remarks
Det	Det	221.2	196; 196	115.4	12.0	10.8	102	
Tip	Tip	221.3	196; 196;	117.0	12.3	1.1	103	
			196; 197	111.4	13.7			
Bmp	Bmp	225.2	201; 201;	128.0	0.6	21.2	104	large thermal parameter at Ge
			201; 202	128.5	0.4			
Mes	Dip	230.1	197	109.9	35.4	3.4	105	*Z*-isomer
			199					
Bsi	Bsi	234.6	198; 204	112.5	32.2	0	100	
XIIa (Si, N)[a]		245.4	243.8 (Si); 186 (N)	102.9	41.3	36.8	107	*Z*-isomer
			244.5 (Si); 186 (N)	102.0				
XIIb (Si, N)[a]		245.3	243.2 (Si); 184 (N)	101.4	47.3	0	19	two independent molecules; *E*-isomer
		246.0	244.9 (Si); 185 (N)	101.1				
i-Pr_2MeSi (Si)	i-Pr_2MeSi (Si)	226.7	240.0; 240.6	117.0	7.1	0	106	two independent molecules
		227.0	239.9; 240.1	118.3	5.9			
i-Pr_3Si (Si)	i-Pr_3Si (Si)	229.8	242.7; 244.3	115.2	16.4	0	106	
Btp	Cl	244.3	200 (C)	109.1	49.0	0	56	Ge disordered
			212 (Cl)					
Btp	Na	239.4	207 (C)		—	—	108	doubly bridged R-E=E : 102.4°
			307.4; 312.1 (Na)					
Tetragermabutadiene		234.4	Ge^1: 199; Ge^2: 200	107.8; 109.6	33.2; 31.1	4.8	110	*cis*-conformation
Tip_6Ge_4		235.7	Ge^3: 200; Ge^4: 199	109.4; 108.9	35.3; 31.1	6.4		$Ge^1{-}Ge^2{-}Ge^3{-}Ge^4$: 22.5°

[a]For the definition of substituents *XII a,b* see Chart 3; atomic symbols in parentheses denote the type of atom binding to Ge.

electronegative substituents enhance the promotion energy and thus favour large *trans*-bent angles. The bond length, however, depends on at least two factors: (a) the type of substituent — electropositive groups will shorten and electronegative groups will lengthen the bond — and (b) the steric demand of the substituent — small groups will allow for a short bond while large groups will stretch the bond and may cause twisting. This proposal, which must be corroborated by more experimental evidence, of course, is supported by the observation that alkenes may be twisted or suffer bond stretching, but never show any *trans*-bent distortion, even if very large substituents are present.

The structures of three heteroleptic digermenes, **69–71**, have been reported. In **69** each Ge bears a terphenyl group and a Cl atom[56]; in **70**[19] and **71**[107] tris(amino)silyl and amino substituents are present at each Ge. While **69** and **70** are *E*-isomers, **71** adopts the unusual *Z*-form. The electronegative chloro or amino groups cause large *trans*-bent angles between 39° and 47.3° and long Ge=Ge bonds between 244.3 and 246.0 pm.

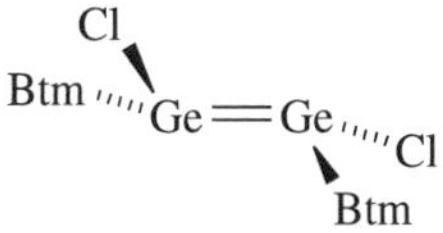

(69)

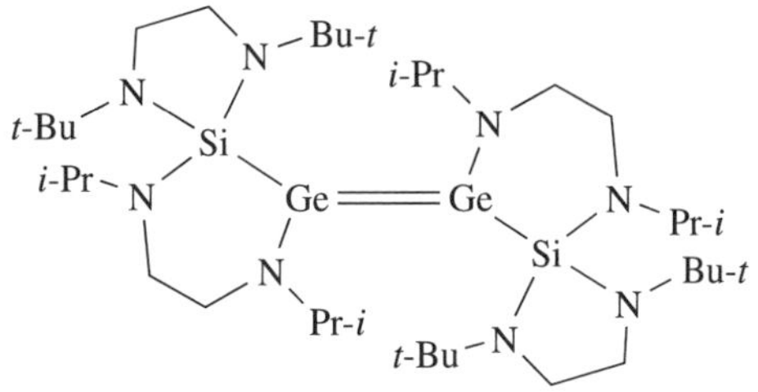

(70)

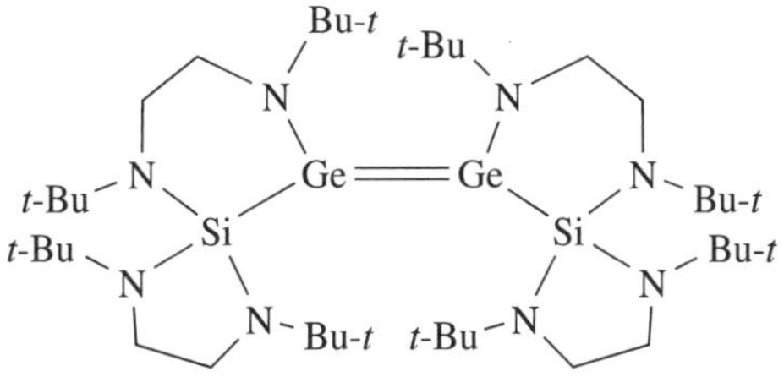

(71)

The dianionic digermene $[BtpGe{=}GeBtp]^{2-}$ **(72a)** and the respective distannene $[BtpSn{=}SnBtp]^{2-}$ **(72b)** were synthesized by Power and coworkers by reduction of the appropriate aryl(chloro)germylene BtpGeCl **(34)** and aryl(chloro)stannylene BtpSnCl with an excess alkali metal in benzene[108], respectively. Doubly metal-bridged $[BtpE{=}EBtp]^{2-}$ units are observed in the solid state (**72b**: Figure 15). The cations are coordinated by the aromatic rings of the Tip fragments of the utilized terphenyl ligand and the π-electrons of the dianion. Since the formation of the double bond in these RE=ER anions needs no s-p promotion, **72a** and **72b** can be regarded as systems with almost classical double bonds, as are the isoelectronic neutral dipnictenes RY=YR(Y = As, Sb). The Ge=Ge and Sn=Sn bonds (239.4 pm and 277.6 pm, resp.) in **72a** and **72b** are about 10 pm longer than the respective As=As and Sb=Sb bond distances observed for dipnictenes[109]. Despite similar covalent radii for Ge and As as well as for Sn and Sb, this can be traced back to electrostatic repulsion between adjacent negative charges in **72a** and **72b**.

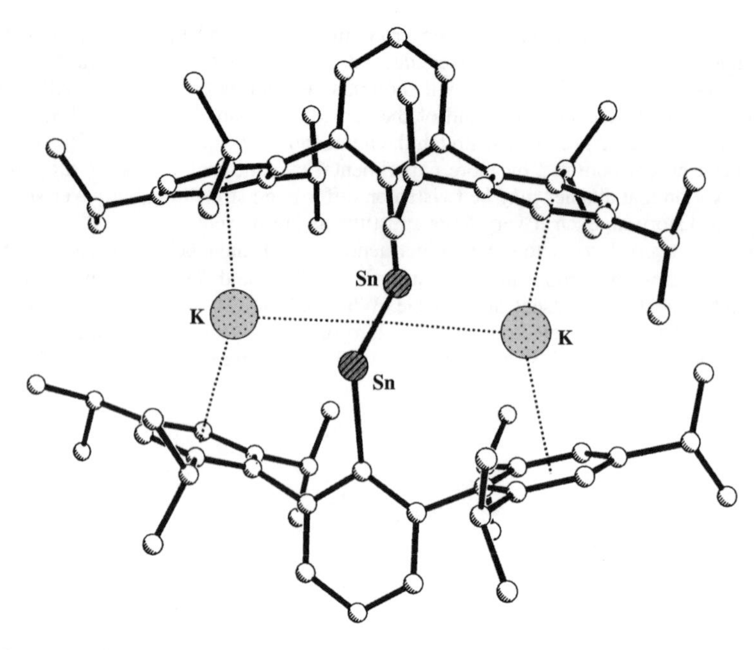

FIGURE 15. Molecular structure of the dipotassium salt of dianionic distannene **72b**

Very interestingly, if equimolar amounts of alkali metal are used for the reduction of the aryl(chloro)stannylene, radical anions (which will be discussed later) are formed.

2. Distannenes

The distannenes show the greatest variety of structures among the ditetrenes (Table 12) although, in accordance with the CGTM model, and apart from the dianion **72b**[108] for which this model does not apply, no distannene with a planar skeleton is known. Most organyl- and silyl-substituted distannenes adopt undistorted or only slightly distorted *trans*-bent structures; $Hyp_2Sn{=}SnHyp_2$ **73**[32], which is the only tetrasilyldistannene known, is strongly twisted, however. Since an Sn=Sn bond is generally weaker than a respective Ge=Ge bond, a greater variation in bond lengths is observed among the *trans*-bent structures for Sn=Sn, ranging from 270.2 pm for Mes(Hyp)Sn=Sn(Hyp)Mes **(74)**[54] to 363.9 pm for $Ar^F_2Sn{=}SnAr^F_2$ **(75)**[27b]. The influence of the electronegativity of the substituent can be impressively demonstrated by comparing the structural parameters of tetraaryldistannenes with those of the respective heteroleptic diaryldihypersilyldistannenes (Table 12). The substitution of two aryl groups for two hypersilyl groups generally results in much shorter Sn=Sn bonds and smaller tilt angles κ. Having the most electronegative substituents among the series, Ar^F_2Sn **(12)** forms only a loose dimer **(75)** with an Sn−Sn distance of 363.9 pm. The substitution of two Ar^F in **75** for two strongly electropositive hypersilyl groups shortens the Sn=Sn bond by about 80 pm (!) to give 283.3 pm in the respective heteroleptic derivative $Hyp(Ar^F)Sn{=}Sn(Ar^F)Hyp$ **(76)**[59]. Similar differences are found for the couple $Bmp_2Sn{=}SnBmp_2$ **(77)**[111]/Hyp(Bmp)Sn=Sn(Bmp)Hyp **(78)**[60] (**77** has no simple *trans*-bent structure; see below). Finally **74**, the species bearing the smallest aryl group among all diaryl-dihypersilyldistannenes, exhibits the shortest Sn=Sn

TABLE 12. Parameters (bond length in pm, angle in deg) of distannenes $R^1R^2Sn{=}SnR^1R^2$

R^1	R^2	Compound	Sn=Sn	Sn–R[b]	R–Sn–R	κ	τ	Reference	Remarks
Bsi	Bsi	**79**	276.8	221; 223	109.2	41	0	100	
Bmp	Bmp	**77**	291.0	227; 227	114.4	64.4	10.7	111	$Bmp_2Sn \rightarrow SnBmp_2$
				219; 222	114.6	21.3			zwitterionic
Ar^F	Ar^F	**75**	363.9	228; 229	95.1	45	0	27b	
Ar^F	Hyp	**76**	283.3	226 (*C*)	103.8	41.5	0	59	
				262.4 (*Si*)					
Mes	Hyp	**74**	270.2	218 (*C*)	105.5	39.4	0	54	
				260.2 (*Si*)					
Hyp	Hyp	**73**	282.5	266.7 (*Si*); 267.8 (Si)	120.5	28.6	62.7	32	
Bmp	Hyp	**78**	279.1	222 (*C*)	109.0	44.9	0	60	*E*-isomer
				263.5 (*Si*)					
$C_2Sn \rightarrow SnC'_2$		**80**[a]	300.9	221 (*C*, donator, mean)	89.5	1.4	14.0	115	zwitterionic
				223 (*C′*, acceptor, mean)	90.0	81.0			λ^4, λ^3 Sn atoms
				227 ($N \rightarrow$ Sn)					
$C_2Sn \rightarrow SnCl_2$		**81**[a]	296.1	220 (*C*, donator, mean)	100.1	0.1	87.0	112	zwitterionic
				245 (*Cl*, acceptor, mean)	94.6	83.8			λ^5, λ^3 Sn atoms
				241 ($N \rightarrow$ Sn)					
$C_2Sn \rightarrow SnN_2$		**82**[a]	304.9	218 (*C*, donator, mean)	110.4	18.1	12.0	113	zwitterionic
				208 (*N*, acceptor, mean)	88.1	81.3			λ^5, λ^3 Sn atoms
				254 ($N \rightarrow$)					2 independent
			308.7	219 (*C*, donator, mean)	110.7	16.8	6.5		molecules
				209 (*N*, acceptor, mean)	86.9	81.5			
				255 ($N \rightarrow$)					
Btp	K	**72b**[a]	277.6	227 (*C*)	—	—	—	108	doubly bridged
				357.9; 359.1 (K)					R–E=E : 107.5

[a] See text; [b] atomic symbols in parentheses denote the type of atom binding to Sn.

bond (270.2 pm) found so far, even shorter than in Lappert's $Bsi_2Sn{=}SnBsi_2$ **(79)**[8]. The substitution of aryl for hypersilyl also allowed the synthesis of the first diplumbenes, as shown below. Nevertheless, tetrahypersilyldistannene **(73)**, in spite of comprising four electropositive groups, has no extraordinary short Sn=Sn bond (282.5 pm). However, due to the enormous steric demand of the four hypersilyl groups it adopts a distorted *trans*-bent structure with a small tilt angle ($\kappa = 28.6°$), but very large torsion angle τ of 62.7° (Figure 16).

The other four known distannenes all adopt a novel zwitterionic structure with a single dative bond, i.e. one stannylene fragment serves as electron-pair donor (Lewis base), the other as acceptor (Lewis acid): $R_2Sn \rightarrow SnR'_2$, hence they are not distannenes within a stricter definition.

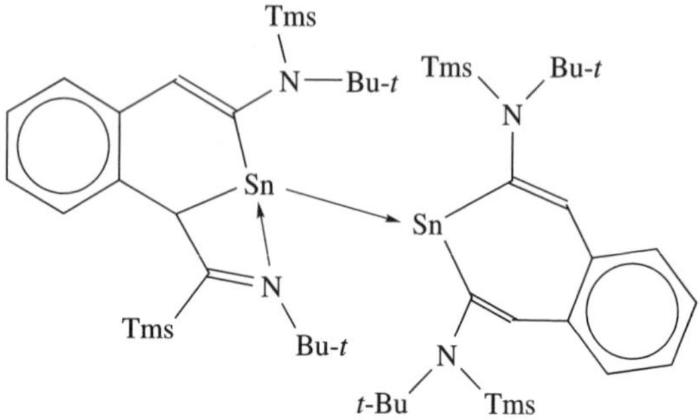

(80)

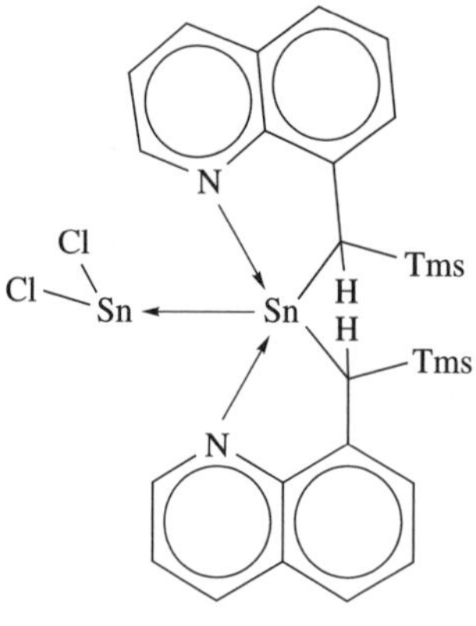

(81)

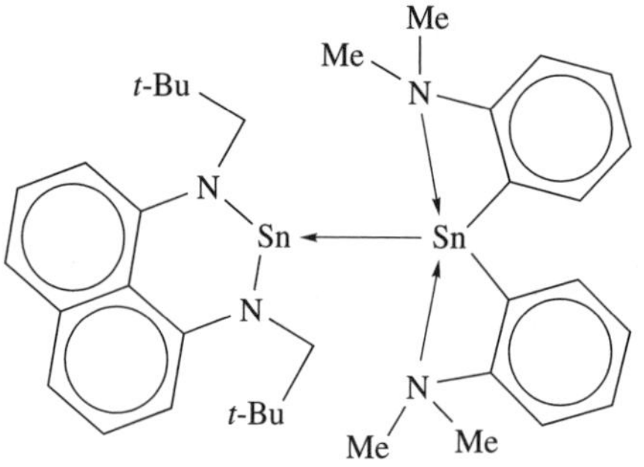

(82)

For three of these compounds **(80–82)**[112,113] this particular structure type may be expected, since herein two different stannylene moieties ($R \neq R'$) having different acceptor abilities interact. The remaining example is the tetra(aryl)distannene $Bmp_2Sn{=}SnBmp_2$ **(77)**[111] (Figure 17) with four identical substituents ($R = R'$)[114]. It is not understood to date why **77** adopts this particular conformation, although steric reasons may be responsible. According to the different 'dative bond orders', the Sn–Sn bond distances for the zwitterionic compounds are significantly longer than for those with

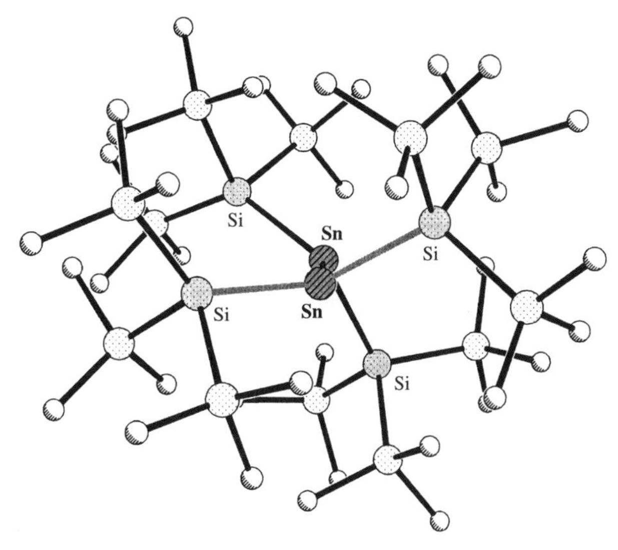

FIGURE 16. Molecular structure of distannene **73** in the crystal. Projection along the Sn=Sn bond illustrating the large twist angle $\tau = 62.7°$

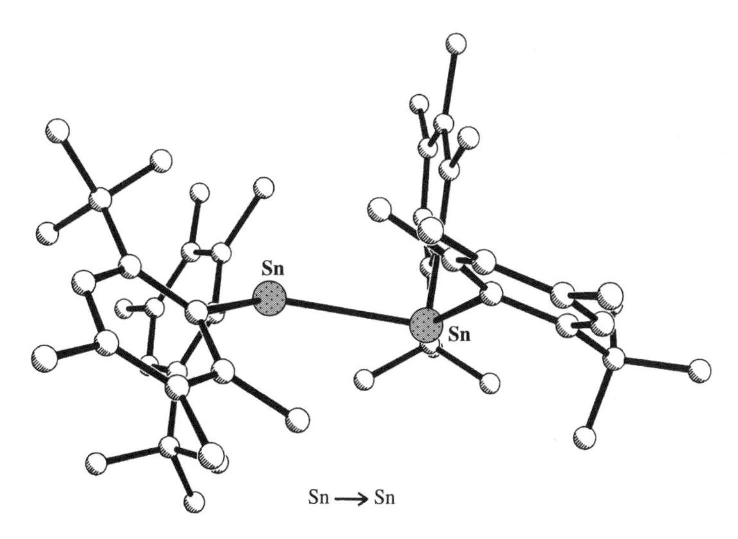

FIGURE 17. Molecular structure of zwitterionic distannene **77**

trans-bent conformation, the values ranging from 291.0 pm **(77)** to 308.7 pm **(82)**[113]. In compounds **80–82** one or two additional intramolecular nitrogen donors block the acceptor orbital of one Sn atom, thus making the respective stannylene unit a pure electron donor. This is demonstrated in Figure 18 for **81**. Hence, these distannenes are closely related to the adducts of Lewis bases such as carbenes or ylids to stannylenes, as described in Section II. The acceptor stannylenes comprise SnC_2 **(80)**[115], $SnCl_2$ **(81)**[112] or SnN_2

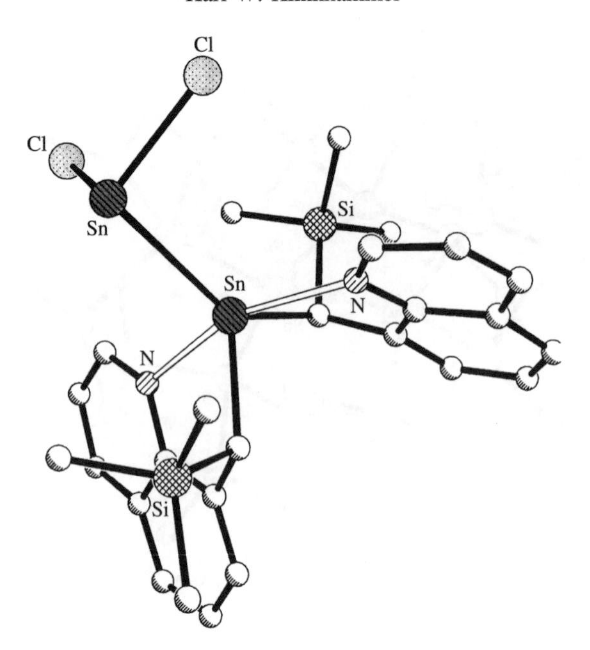

FIGURE 18. Molecular structure of zwitterionic distannene **81**. Intramolecular N $\rightarrow$ Sn interactions are depicted as open lines (=)

(82)[113] central units. The acceptor orbital of these fragments is oriented perpendicular to the SnR'_2 planes, whereas the donor orbital of the Lewis-basic tetrylene R_2Sn should lie approximately within the plane, formed by the tin atoms and the α-atom of the R groups. Therefore, the respective tilt angles κ should be near $90°(ER'_2)$ and $0°(ER_2)$, respectively. The observed deviations from these ideal values may be due to the steric demands of the R and R′ groups.

3. Diplumbenes

For a long time it was thought that plumbylenes, in contrast with their lighter congeners, would show no tendency to dimerise to doubly-bonded species. The first *ab initio* calculations inspecting the hypersurface of Pb_2H_4 also seemed to exclude the possibility of even a *trans*-bent diplumbene A (Scheme 5); $H_2Pb{=}PbH_2$ was calculated to be actually a saddle point[2c]. Instead, a doubly hydrogen-bridged isomer B was calculated to be the global minimum, whereas several other isomers were calculated as at least local minima (Scheme 5). Later, calculations at higher levels confirmed structure B as a global minimum, but they revealed that *trans*-bent diplumbene A is also a local minimum on the hypersurface. After all, the calculated dissociation energy for $H_2Pb{=}PbH_2$ is very small (about 20–40 $kJ\,mol^{-1}$)[2a,b,59].

It was not until 1998 that the first plumbylene dimer had been isolated and structurally characterized: the heteroleptic diaryl-dihypersilyl-substituted *trans*-bent species $[Hyp(Ar^F)Pb]_2$ **(83a)**[59]. It is generated via an at first unexpected ligand exchange reaction between the homoleptic species Hyp_2Pb **(21)** and $Ar_2^F Sn$ **(12)**, the originally intended mixed ditetrene $Hyp_2Pb{=}SnAr_2^F$ being a probable intermediate of the reaction. It was demonstrated later that such exchange reactions generally occur when Hyp_2Sn (**20**) or Hyp_2Pb (**21**) react with other diaryl- or dialkyltetrylenes[51,60,116]. The presence of electropositive substituents, which leads to significant shortening of Sn=Sn bonds for the respective distannenes, seems here to favour the formation of plumbylene dimers. The observed Pb−Pb distance of 353.7 pm for **83a** is, however, much longer than the calculated double bond length for the parent Pb_2H_4 [281.9 (DFT)[2a] — 295.0 pm(CCSD)[59]]. Two reasons — both related to the employed aryl group — become obvious when the structure is analysed in detail. (a) The strongly electronegative Ar^F group, by its electronic influence, strongly disfavours the formation of tetrylene dimers, and (b) the lone-pairs of the fluorine atoms of the CF_3 groups compete with the only weak Lewis-basic plumbylene lone-pair for the acceptor orbital of the second plumbylene moiety (short F···Pb contacts are formed; see Section II).

TABLE 13. Parameters (bond length in pm, angle in deg) of diplumbenes RR′Pb=PbRR′[a]

R,R′	Pb=Pb	Pb−R or Pb−R′	R−Pb−R′	κ	τ	Reference	Remarks
Tip, Tip	305.2	229; 229 231; 231	97.3 102.3	43.9 51.3	16.0	117	
Mes, Mes	335.5	230; 232	97.4	58.5	0	51	$Mg(THF)_4Br_2$ adduct Pb···Br 315.7
Mes, Hyp	290.3	231 (*C*) 268.1 (*Si*)	102.5	46.0	0	54	
Tip, Hyp	299.0	230 (*C*) 271.7 (*Si*)	108.9	42.7	0	51	
Ar^F, Hyp	353.7	237 (*C*) 270.5 (*Si*)	96.6	40.8	0	59	
Bmp, Hyp	337.0	236 (*C*) 271.1 (*Si*)	106.0	46.5	0	116	

[a] Atomic symbols in parantheses denote the type of atom binding to Pb.

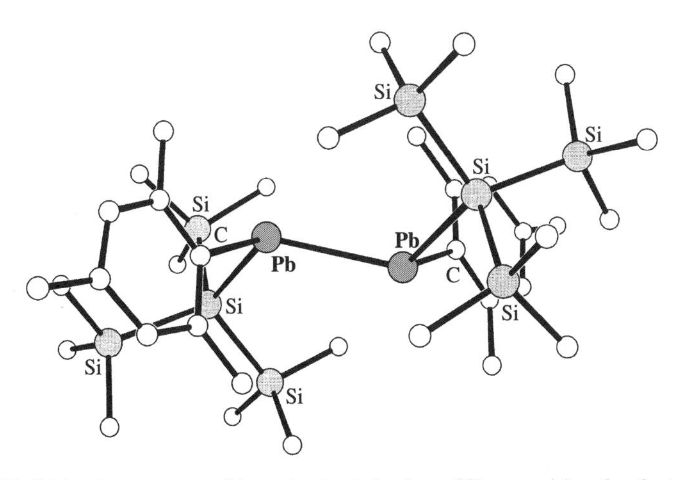

FIGURE 19. Molecular structure of heteroleptic diplumbene **83b** comprising the shortest Pb=Pb double bond observed so far

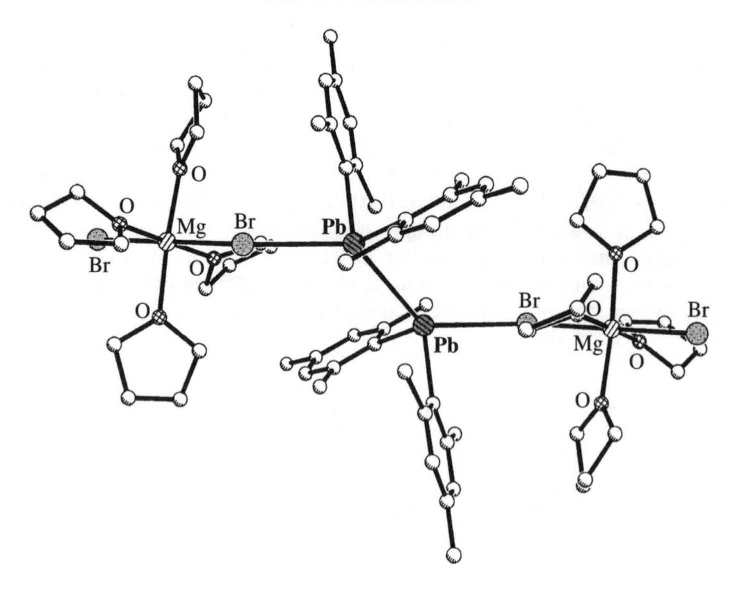

FIGURE 20. Molecular structure of the adduct of diplumbene **85** with two $Mg(thf)_4Br_2$ molecules

Thus, the next logical step was the replacement of Ar^F by other groups which are less electronegative and bear no donor groups. To date, the synthesis of three further heteroleptic diplumbenes with *trans*-bent geometry (Table 13)[51,54,116] had been reported; the species with the shortest Pb=Pb bond so far — with 290.3 pm now in the range of the calculated values for Pb_2H_4 — is the mesityl derivative Mes(Hyp)Pb=Pb(Hyp)Mes **83b** (Figure 19)[54]. Again ligand exchange led to the formation of **83b**, but here mesitylcopper CuMes was used as reaction partner for $PbHyp_2$ **(21)**. The synthesis of a first homoleptic *trans*-bent diplumbene $Tip_2Pb{=}PbTip_2$ **(84)**[117] by reaction of TipMgBr with $PbCl_2$ showed, however, that the existence of diplumbenes is not restricted to species comprising silyl groups. As can be seen again by comparing the observed Pb=Pb bond length (305.2 pm) with that of the appropriate heteroleptic species Tip(Hyp)Pb=Pb(Hyp)Tip **83c** (299.0 pm), the formation of diplumbenes is at least favoured by the utilization of such electropositive substituents. **84** is the only diplumbene known which comprises a twisted Pb=Pb double bond ($\tau = 16°$). If instead MesMgBr is reacted with $PbCl_2$, again the expected diplumbene $Mes_2Pb{=}PbMes_2$ **(85)** is formed, but from the obtained solution it was isolated as a unique $Mg(thf)_4Br_2$ adduct (Figure 20)[51]. Herein the Pb=Pb bond is markedly lengthened to 335.5 pm by competing interaction of the p-orbitals on both Pb atoms with bromide anions. Replacing mesityl for 2,4,6-triethylphenyl finally leads to a complex **(51)** of two separated plumbylene fragments which are bridged by one $Mg(thf)_4Br_2$ molecule (see above)[50].

The data for all these diplumbenes are given in Table 13.

4. Cyclic ditetrenes and their derivatives

When the earlier edition of this book was in preparation, no cyclic digermene or distannene was known. Meanwhile, several cyclotrigermenes and mixed Si/Ge heterocycles with Si=Si or Si=Ge bond have been synthesized by Sekiguchi's group. The only cyclotristannene c-Sup$_4$Sn$_3$ **(86)** was obtained by isomerization of a tristannaallene[118] and will be discussed together with this unique cumulated compound in the next section.

$$\text{(86)}: c\text{-}[(t\text{-Bu})_3\text{Si}]_2\text{Sn}\text{–}\text{Sn}[\text{Si(Bu-}t)_3]{=}\text{Sn}[\text{Si(Bu-}t)_3]$$

(86)

$$\text{(87), (88)}: c\text{-}[\text{R}][\text{Si(Bu-}t)_3]\text{Ge}\text{–}\text{Ge}[\text{Si(Bu-}t)_3]{=}\text{Ge}[\text{Si(Bu-}t)_3]$$

(87) (**a**) R = (t-Bu)$_3$Si
(**b**) R = (t-Bu)$_3$Ge

(88) R = Hyp (**a**), (t-Bu)$_3$Ge, Tm$_3$Ge, Mes

The two symmetrically substituted cyclotrigermenes ((t-Bu)$_3$E)$_4$Ge$_3$ **87a** and **87b** are formed from GeCl$_2$·dioxane and the appropriate alkali metal tetryl (t-Bu)$_3$EM (M = Li, Na)[119]. Unsymmetrically substituted derivatives **88** were obtained by addition of different alkali metal silyls or germyls to salts of the cyclotrigermenium cation **89**[120] (of equation 10 below). The heteronuclear disilagermirenes **90a** and **90b** were finally prepared by reduction of a (t-Bu)$_2$MeSiSiBr$_3$/((t-Bu)$_2$MeSi)GeCl$_2$ mixture[121]. At first the 1-disilagermirene **90a** is isolated from the reaction mixture and may then be photochemically isomerized to **90b** (equation 9).

$$\text{(90a)}: c\text{-}[(t\text{-Bu})_2\text{MeSi}]_2\text{Si}{=}\text{Si}[\text{Si(Bu-}t)_2\text{Me}]\text{–}\text{Ge}[\text{Si(Bu-}t)_2\text{Me}] \xrightarrow{h\nu} \text{(90b)}: c\text{-}[(t\text{-Bu})_2\text{MeSi}]_2\text{Si}\text{–}\text{Ge}[\text{Si(Bu-}t)_2\text{Me}]{=}\text{Si}[\text{Si(Bu-}t)_2\text{Me}] \quad (9)$$

(90a) **(90b)**

The structural parameters from the crystal structures of the two cyclotrigermenes **87a** and **87b** are very similar, the Ge=Ge bond length of 224.1 and 226.0 pm, respectively, being somewhat shorter than in silyl-substituted acyclic digermenes (227.0–229.8 pm). The endocyclic Ge–Ge bonds with 250.6 and 252.2 pm are by far longer than those in the acyclic derivatives (239.9–244.3 pm). Since the Ge$_3$ ring lies on a crystallographic mirror plane, both Ge atoms of the Ge=Ge bond have a planar coordination (in agreement with a short Ge=Ge bond). In sharp contrast to this finding, the double-bonded germanium atoms in the asymmetrically substituted cyclotrigermene **88a** show pyramidal coordination (*cis*-bent) and a somewhat longer Ge=Ge bond of 226.4 pm. A pronounced *trans*-bent conformation is found, however, for 2-disilagermirene **90b**. Apart from the torsion angle including the double-bonded atoms and the connected Si atoms of 40.3(5)°, which is a little larger than for the 1-sila-isomer (37.0°), no reliable data can be given for **90b**, due to disordering of Si and Ge. Isomer **90a** is well ordered in the crystal, the Si=Si bond being 214.6 pm and the Si–Ge bonds being 241.5 and 242.0 pm.

$(t\text{-Bu})_3Si$ $Si(Bu\text{-}t)_3$
Ge
Ge=Ge
$(t\text{-Bu})_3Si$ $Si(Bu\text{-}t)_3$

(87a)

$[Ph_3C]^+[BAr_4]^-$ (10)

$(t\text{-Bu})_3Si$
Ge
(2π)
Ge—Ge
$(t\text{-Bu})_3Si$ $Si(Bu\text{-}t)_3$

(89) $[BAr_4]^-$

Treatment of the cyclotrigermene **87a** with trityl tetraarylborates yields salts containing the corresponding cyclotrigermenium cation **89**[122,123]. Its structural parameters are very similar among the different salts. The Ge–Ge bond lengths fall into a narrow range between 232.6 and 233.5 pm, thus lying in between typical single and double bond values, being indicative of a delocalized 2π-aromatic system. The Ge atoms exhibit almost planar coordination, although, depending on the counter-anion present, the out-of-plane positioning of the connected silicon atoms may reach up to 30 pm.

Power reported the synthesis of a cyclogermenyl radical Btm_3Ge_3 **(91)**[124] by dehalogenation of the heteroleptic terphenyl(chloro)germylene BtmGeCl **(35)** (see above) by KC_8. Unfortunately, the X-ray crystal structure analysis revealed severe disordering for the Ge atoms, therefore no reliable structural data are available to date. Nevertheless, the ESR spectrum of the compound indicate the localization of the unpaired electron on one Ge atom

Crystallographic data on the cyclic compounds are given in Table 14.

C. Other Systems with Multiple Bonds between Heavier Tetrel Atoms

In recent years a couple of other compounds featuring multiple bonds between tetrel atoms have been synthesized, perhaps showing and defining one of the future courses of tetrel chemistry. Among them are the first examples of cumulated and conjugated double bonds and the very first E–E triple bond, or, strictly speaking, a *formal* analogue of alkynes[97,125], comprising a REER skeleton. Anionic and radical species with no stable analogue in carbon chemistry were also found to be (kinetically) stable for its heavier congeners.

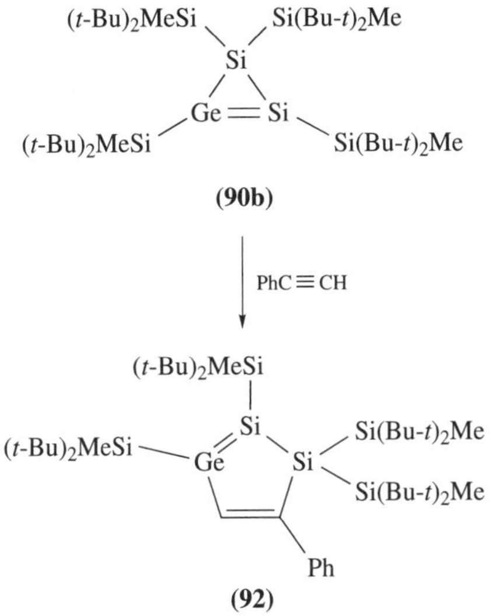

(**90b**)

$\xrightarrow{PhC\equiv CH}$ (11)

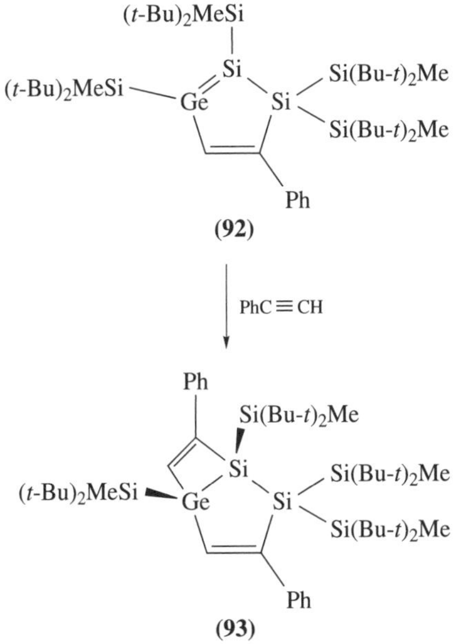

(**92**)

(**92**)

$\xrightarrow{PhC\equiv CH}$ (12)

(**93**)

The first structurally authenticated heteronuclear double bond between heavier tetrels was prepared by Sekiguchi's group via cycloaddition of phenylacetylene across the Ge=Si double bond of the appropriate disilagermirene **90b** (equation 11). The bicyclic intermediate rearranges quickly to form the 1,2,3-germadisilol **92**, comprising formally conjugated C=C and Ge=Si bonds. Despite the almost planar geometry of the ring, conjugation was

TABLE 14. Parameters (bond length in pm, angle in deg) of cyclotrigermenes, cyclotristannenes and cyclotrigermenium cations[a]

Cyclotrigermenes

R^1	R^2	Ge–R^1	Ge–R^2	Ge=Ge	Ge–Ge	Ge=Ge–Ge	Ge–Ge–Ge	Reference	Remarks
$(t\text{-Bu})_3Si$	$(t\text{-Bu})_3Si$	244.7	262.9	224.1	252.2	63.6	52.8	119	
$(t\text{-Bu})_3Ge$	$(t\text{-Bu})_3Ge$	249.0	264.7	224.0	250.6	63.4	53.1	119	
$(t\text{-Bu})_3Si$	$(t\text{-Bu})_3Si$ Hyp	251.0	251.0	226.4	249.8 250.7	62.9; 63.3	53.8	120	

Cyclotristannene

R^1	R^2	Sn–R^1	Sn–R^2	Sn=Sn	Sn–Sn	Sn=Sn–Sn	Sn–Sn–Sn	Reference	Remarks
$(t\text{-Bu})_3Si$	$(t\text{-Bu})_3Si$	not reported		258 260	285, 286 284, 285	63.1 63.1	54.2 54.0	118	severely disordered substituents space-group uncertain; 2 independent molecules

Cyclotrigermenium cations

R	R–Ge	Ge–Ge	Reference	Remarks
$(t\text{-Bu})_3Si$	243.9–244.5	232.6–233.3	122	$[BPh_4]$ salt
$(t\text{-Bu})_3Si$	242.8–244.8	232.9–234.3	123a,b	$[B(3,5\text{-}(CF_3)_2C_6H_3]$ salt
$(t\text{-Bu})_3Si$	242.0–246.0	233.1–233.5	123b	$B(2,3,5,6\text{-}F_4\text{-}4\text{-}(SiMe_2Bu\text{-}t)C_6]$ salt

[a] R^1 is attached to –E=E and R^2 to –E–E.

neither detected by spectroscopic means nor supported by the structural data, however[126]. The Ge=Si bond length of 225.0 pm meets the expectation[127], being 17 pm shorter than the Ge–Si single bond in the same molecule. The [2 + 2] cycloaddition of a second phenylacetylene moiety furnishes finally a novel bicyclic system **93** (equation 12) with a Ge–Si single bond (Ge–Si: 243.2 pm); note that for a conjugated system, a [4 + 2] cycloaddition would have been expected rather than a [2 + 2] cycloaddition. Other novel bicyclic systems have been obtained by analogous reaction of the unsymmetrically substituted cyclotrigermenes **88**[128].

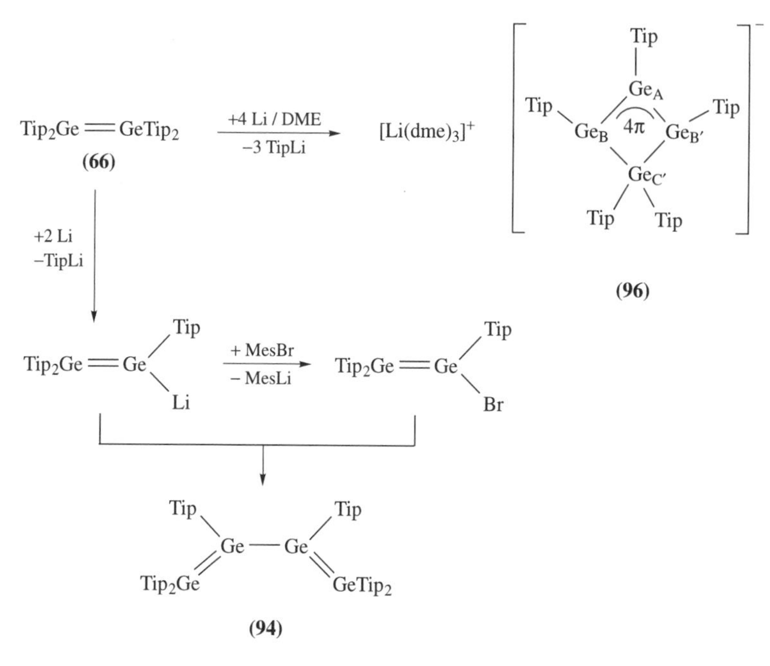

SCHEME 7. Synthesis of **94** and **96**

Conjugated bonds are indeed found within the tetragermabutadiene Tip_6Ge_4 **(94)** synthesized by Weidenbruch's group[110]. It was obtained by an analogous route to that of the respective tetrasilabutadiene Tip_6Si_4 **(95)** (Scheme 7)[129]. Like is lighter congener **95** the tetragermabutadiene **94** has 2,3-ditipyl *cis*-configuration. For steric reasons the Ge_4 skeleton is not planar, but shows a dihedral angle of 22.5° — much smaller than that observed for **95** (51°). The Ge=Ge distances are 235.7 and 234.4 pm and thus markedly longer than in the parent digermene **66** (221.3 pm), but still lying in the range for double bonds, while the central Ge–Ge bond with 245.8 pm lies in the range of normal single bonds. The main evidence for conjugative interactions between the two Ge=Ge bonds comes from the large bathochromic shift of the long-wave visible absorption to 560 nm, giving a deep blue solution; ordinary digermenes are yellow or orange, showing much lower shifts than 500 nm[106,130]. Both digermene fragments of **94** exhibit a pronounced *trans*-bent conformation with respective κ-values of 31.1 and 35.4°, values much larger than for other known digermenes (<22.3°).

Slight changes in the reaction conditions furnished via an elusive reaction path a salt with a very unusual cyclic C_s symmetric tetragermaallyl anion **96** (Scheme 7)[110]. Two different Ge−Ge bond lengths are found within the planar ring system, one of them (Ge_A-Ge_B and $Ge_A-Ge_{B'}$) with 236.8 pm clearly indicating multiple-bond character, whereas the other (Ge_B-Ge_C and $Ge_{B'}-Ge_C$) with 251.2 pm is typical for an elongated single bond. A markedly longer Ge−Ge multiple bond of 242.2 pm is found for the open-chain trigermaallyl-type anion $[Btm_3Ge_3]^-$ **97**, being a reduction product from the respective cyclogermyl radical **91**[124]. The anion **97** with two two-coordinated Ge atoms may also be addressed as bisgermylene, one terphenyl group and two germylidyne moieties serving as substituents to a germanide anion. Whereas the terminal Ge atoms exhibit C−Ge−Ge angles of 111.3°, typical for sterically congested germylenes, the Ge_3C_3 skeleton is almost planar and a very wide Ge−Ge−Ge angle (159.2°) is observed, implying delocalization of the negative charge by π-bonding to the adjacent Ge atoms. Consequently, the observed Ge−Ge distances (242.2 pm) are somewhat shorter than expected for Ge−Ge single bonds in congested digermanes[131].

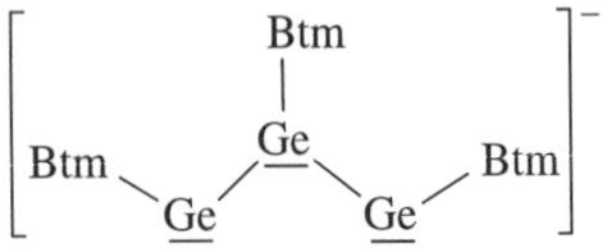

$$\left[\text{Btm}-\underline{\text{Ge}}-\text{Ge}(\text{Btm})=\underline{\text{Ge}}-\text{Btm} \right]^-$$

(97)

Terphenyl(chloro) and terphenyl(bromo)tetrylenes Ar−E−X (Ar = Btp; X = Cl, Br) have been used by Power's group to synthesize a series of multiply-bonded dinuclear species with a varying amount of multiple-bond character[95,132–134]. They originally intended to synthesize alkyne homologues ArGe≡GeR and ArSn≡SnAr by intermolecular reductive elimination of alkali metal halides from BtpGeCl **(34)** and BtpSnCl, respectively. Presumably such compounds may have formed at first, but due to the presence of low-lying empty p-orbitals (or π^*-orbitals) on E, they came up with reduced anionic species. In the case of E = Ge only a doubly reduced compound with Ge=Ge double bond **(72a)** was isolated (see above). The analogous Sn derivative **72b** is obtained when using an excess of an alkali metal.

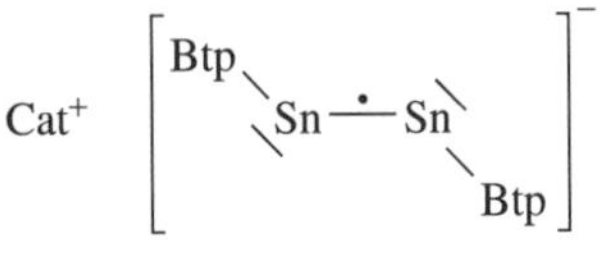

$$\text{Cat}^+ \left[\text{Btp}-\text{Sn}\overset{\bullet}{-}\text{Sn}-\text{Btp} \right]^-$$

Cat =

(98) (**a**) $[K(thf)_6]$

(**b**) $[K(thf)_3(db18\text{-}cr\text{-}6)]$

(**c**) $[Na(thf)_3]$ (db18-cr-6 = dibenzo-18-crown-6)

If a stoichiometric amount of alkali metal is used instead, in the case of E = Sn singly reduced species, the radical anion $[Btp_2Sn_2]^{\cdot -}$ **(98)** is formed as the predominant product. Depending on the employed alkali metal and donor-solvent present, three different crystalline compounds were obtained and structurally characterized: **98a** and **98b** built from isolated ions[132] and **98c** comprising contact ion pairs with a Na−Sn bond[133]. The

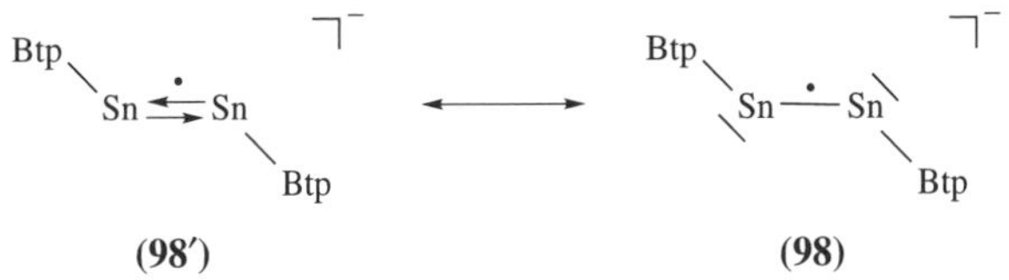

Btp–Sn⇄Sn–Btp ⟷ Btp–Sn–Sn–Btp

(98′) **(98)**

SCHEME 8. Structure of **98**

structural parameters of the solvates are, however, very similar, only the Sn−Sn−C angles are somewhat larger for **98c**, probably for steric reasons. Thus the direct interaction of the anion and the cation seems to have only little influence on the structure of **98c** (Figure 21). The observed Sn−Sn distances of 278.2 to 282.4 are in the same region as for sterically encumbered distannenes and are shorter than in encumbered distannanes[134]. While the Sn−Sn distances match both possible bonding schemes: **98′** having a dative double bond augmented by a one-electron π-bond or **98** having a single bond augmented by a one-electron π-bond, the small Sn−Sn−C angles of 95.2–98° show a major contribution of hybrid **98** (Scheme 8).

Switching from alkali metal to $LiAlH_4$ as reducing agent for Btp−E−X the first alkyne analogue could be finally isolated and structurally characterized, not for E = Ge and Sn, however, but for E = Pb. While in the Sn case the novel Sn(II) hydride BtpSnH **(58)** was isolated instead (see Section II), the putative Pb(II) hydride is not stable and dehydrogenates, finally yielding BtpPb≡PbBtp **(99)**[95].

The structural features of the dark green diplumbyne **99**, a very long Pb−Pb bond of 318.8 pm and a very small C−Pb−Pb angle of 94.3°, favour more the description

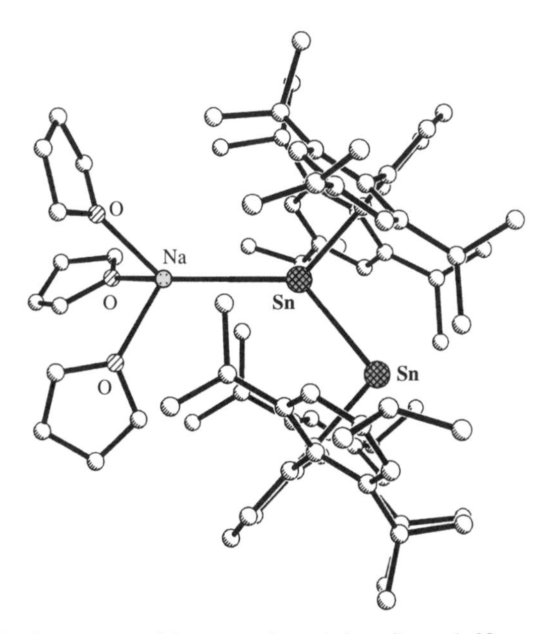

FIGURE 21. Molecular structure of the contact ion-pair in sodium salt **98c** comprising the radical anion $[Btp_2Sn_2]^-$

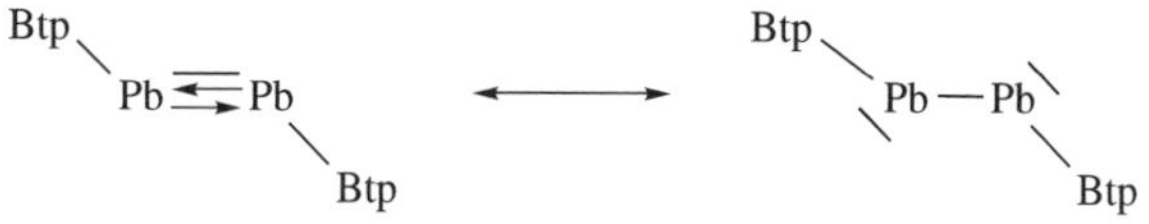

Btp–Pb⇌Pb–Btp (100) ⟷ Btp–Pb—Pb–Btp (100′)

SCHEME 9. Structure of **99**

as a diplumbylene with a lone-pair on each Pb atom (Scheme 9, hybride **100′**) than as a diplumbyne with a Pb≡Pb triple bond (hybride **100**). The double dative bonds, which may be used for the description of ditetrynes[4,97,125], are expected to be weakest for E = Pb; therefore, one can look forward with eager expectation to what future experiments will reveal for the other tetrels, especially for Si and Ge. However, even Pb is capable of forming relatively strong and short triple bonds, as was shown by the structure of a Pb_2 dianion, recently synthesized and structurally characterized by Rutsch and Huttner as the transition metal complex $[Ph_4P]_2[Pb_2[W(CO)_5]_4]$ **(101)**[135]. In this complex a $[Pb_2]^-$ unit is found which is isoelectronic to Bi_2. Other than for diplumbynes, but similar to the dianions $[BtpE{=}EBtp]^-$ **72a** and **72b** (see above), the bond energy is not reduced by a preceding promotion step, since the low-energy s-orbital may stay doubly occupied and thus is not involved in bonding. Consequently, the observed Pb≡Pb distance of 281 pm in **101**, despite putative coloumb repulsion, is very short.

Four interesting compounds derive from the reaction of $Sn(NTms_2)_2$ **(26)** with supersilylsodium NaSi(Bu-*t*)$_3$ (NaSup): Wiberg and coworkers reported that, depending on the solvent and the reaction conditions, a heterocumulene, a cyclotristannene or two striking cage compounds are obtained. At a low temperature in pentane tetra(supersilyl)tristannaallene $Sup_2Sn{=}Sn{=}SnSup_2$ **(102)** is formed in about 20% yield[118]. It could be isolated as dark blue crystals at −25 °C. If allowed to stay for a longer period at room temperature, it rearranges quantitatively to the isomeric dark red-brown tetrasupersilylcyclotristannene **(86)** [half-life: 9.8 h at 25 °C(C_6D_6)].

In contrast to allenes $R_2C{=}C{=}CR_2$, which for most cases are almost linear, the tristannaallene **102** adopts a bent structure with an Sn−Sn−Sn angle of 155.9° (Figure 22). Both terminal Sn atoms display pyramidal coordination with tilt angles κ of 48° and 42.9° for Sn1 and Sn3, respectively. The two stannylene moieties are twisted with respect to each other by 66.7°. Both Sn=Sn double bonds with 268.4(1) and 267.5(1) pm are shorter than the shortest bond found so far for distannenes in Hyp(Mes)Sn=Sn(Mes)Hyp **(74)** (270.2 pm)[54], probably owing to the presence of electropositive substituents and small steric repulsion across the double bond. The repulsion between adjacent supersilyl groups is large, however, leading to Si−Sn−Si angles of about 134°.

An even shorter Sn=Sn double bond is found for the cyclotristannene **86**[118]: though the quality of the diffraction data is poor, the Sn=Sn distance could be determined with sufficient accuracy to be 259 pm (mean value). Moreover, the Sn atoms have an almost planar surrounding, thus based on these parameters cyclotristannene **86** is the only example of classical double bonding within an $R_2Sn{=}SnR_2$ fragment. Due to steric reasons the Sn−Sn single bonds with 286 pm (mean value) are relatively long.

If cylclotristannene **86** is heated to 100 °C for several days or if the solvent of the initial reaction mixture is replaced by *t*-BuOMe, hexasupersilyl hexastannaprismane Sup_6Sn_6 **103**, the first hexastannaprismane is obtained[136]. The Sn−Sn bonds range from 291 to 294 pm and match those Sn−Sn bonds in encumbered distannanes. Very striking is the orientation of the supersilyl groups towards the Sn−Sn−bonds within the triangular faces of the prismane: the projection along the 3-fold axis of the prismane (Figure 23) reveals an

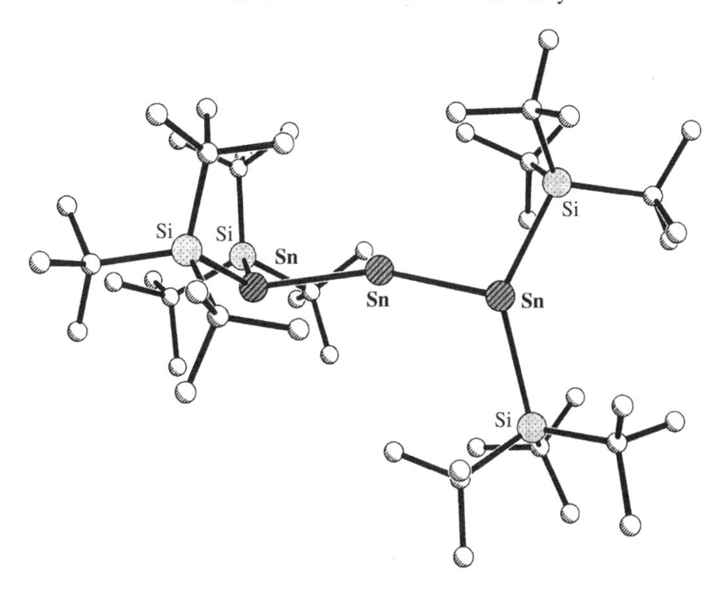

FIGURE 22. Molecular structure of tristannaallene **102** with non-linear Sn_3 backbone

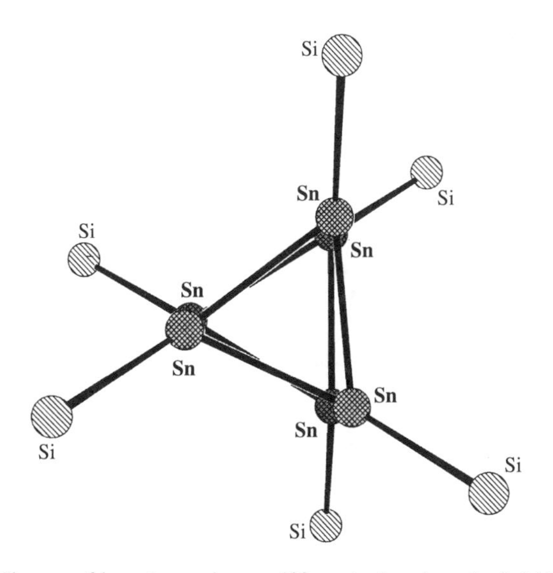

FIGURE 23. Sn_6Si_6 core of hexastannaprismane **103**; projection along the 3-fold axis

orientation of the three substituents at each face that resembles the proposed arrangements of the substituents in cyclotriplumbanes[88] or in one triangular face of the tetrameric thallium(I) alkyl $(TsiTl)_4$ **(104)**[137]. At this point, it is an open question whether a similar electronic structure is responsible for these distortions in **103** and **104**.

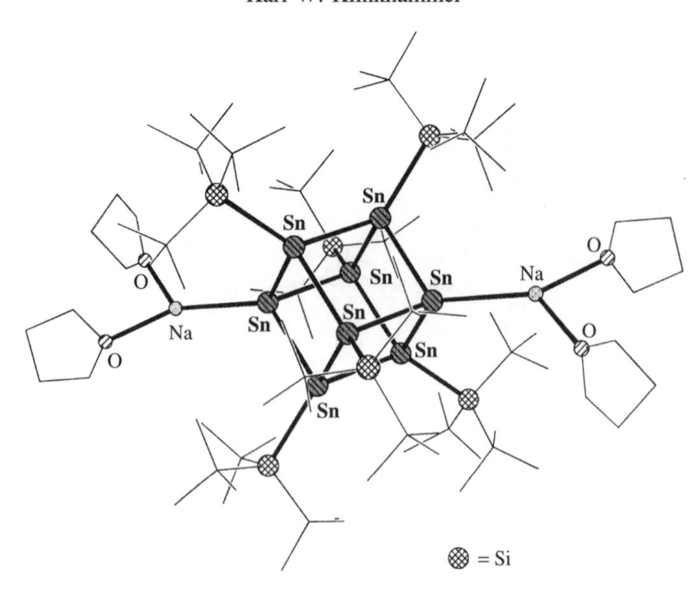

FIGURE 24. Molecular structure of the disodium salt **105** of an octastannacubane

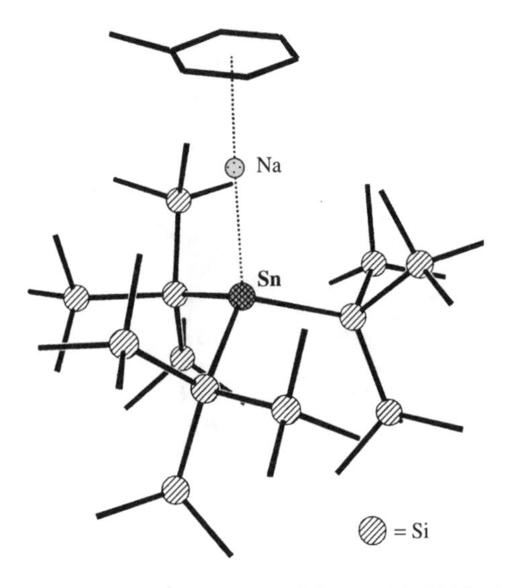

FIGURE 25. Molecular structure of sodium trihypersilylstannanide **106** in the crystal

If the thf-solvate of NaSup is employed in the reaction with $Sn(NTms_2)_2$ **(26)**, a further cage compound is obtained: the salt $[(thf)_2Na]_2[Sn_8Sup_6]$ **(105)**[138]. In the solid state a contact triple ion is observed, consisting of a cubic Sn_8 cage where all, but two opposite corners, are substituted by supersilyl groups (Figure 24). The free corners bind the two sodium ions; the observed Na–Sn distance of 310 pm is only slightly longer than that found for $Hyp_3Sn-Na(\eta^6$-toluene) **106**[139] (Figure 25; Na–Sn: 307 pm), but markedly shorter than that within the ion pair $[(thf)_3Na][Sn_2Btp_2]$ **(98c)** (324 pm). The Sn–Sn bonds in stannacubane **105** range from 287 to 292 pm and are somewhat shorter than in prismane **103**.

D. Multiple Bonds to Carbon: Germenes, Stannenes and Plumbenes

While the chemistry of silenes $R_2Si{=}CR_2$ has been well developed, structural information about germenes $R_2Ge{=}CR_2$ and stannenes $R_2Sn{=}CR_2$ is still scarce, and no plumbene $R_2Pb{=}CR_2$ had been isolated to date.

Although several germenes have been described[148], only one simple germene, i.e. a derivative bearing non-functionalized organyl substituents, the fluorenylidene derivative $Mes_2Ge{=}C_{13}H_8$ **(107)**, with mesityl groups on Ge, has been structurally characterized so far[140]. The Ge=C bond is by about 18 pm shorter than the standard Ge–C single bond, which is close to 198 pm, but about 21 pm shorter than in the corresponding sterically crowded germane $Mes_2Ge(H){-}C_{13}H_9$ **(108)** (201 pm)[140]. Thus the shortening is almost the same as for the couple C–C/C=C and the one derived from Pauling's logarithmic estimation (equation 6).

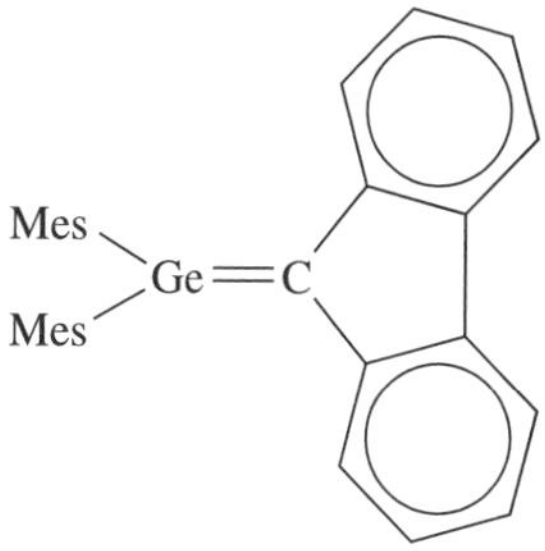

(107) 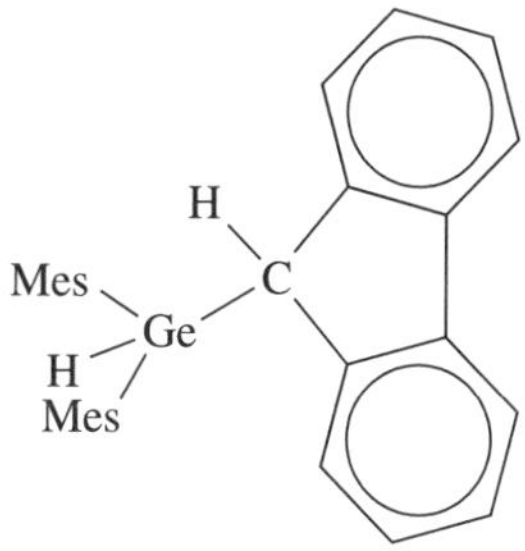 **(108)**

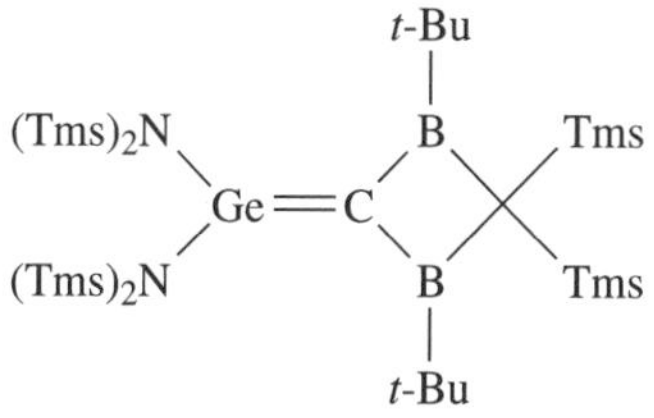

(109)

A somewhat longer and twisted bond ($\tau = 35.7°$) is found for the cryptodiborylcarbene adduct **109** to bis[bis(trimethylsilyl)amino]germylene[141]. The value of 182.7 pm is still in accordance with the presence of a Ge=C double bond, but sharply contrasting the much larger values found for the Ge–C bonds in the carbene adducts **110a–c**[79–81] which are even longer than the Ge–C single bond in the crowded germane **108**. Moreover, the Ge atoms within these adducts are strongly pyramidalized, whereas the Ge atom in **109** is almost planar (κ: Ge: 1.7°; C: 4.7°). The bonds in **110a–c** should therefore be addressed as dative C → Ge interactions (the respective carbenes are obviously pure donor carbenes).

TABLE 15. Parameters (bond length in pm, angle in deg) of compounds with heteronuclear double bonds to carbon $R^1R^2E{=}CR^3_2$ (E=Ge, Sn)

R^1	R^2	$R^{3\,a}$	E	E=Y	E–R^1 E–R^2	R^1–E–R^2	R^1–E=C R^2–E=C	κ	τ	Reference	Remarks
Mes	Mes	*XIII*	Ge	180	193	115.0	122.7	3.1 (Ge); 2.1 (C)	3.6	140	two independent molecules
					194		121.6;				
				181	193	115.9	122.1	7.5 (Ge); 0.1 (C)	3.6		
					194		122.4				
$(Tms)_2N$	$(Tms)_2N$	*XIV*	Ge	183	182	110.5	125.3	1.7 (Ge); 4.7 (C)	35.7	141	cryptocarbene adduct
					184		124.2				
Tip[b]	Tbt[b]	CS_2Ge(Tbt)Tip[b]	Ge	177	196	112.9	116.4	5.3 (Ge); 1.1 (C)	0	142	benzene solvate
					192		130.4				
Tip[c]	Tip[c]	C=C(Tms)Ph[b]	Ge	178	195	114.0	111.2	32.6 (Ge)	—	143	germaallene
					195		123.2				C=C 131.4; Ge=C=C 159.2
Bmp[d]	Bmp[d]		Ge	182	199	130.7	113.7	2.8 (Ge); 1.9 (C)	2.9	144	conjugated system with two
					199		115.5				Ge=C and one C≡C bond
Bmp	Bmp	*XIV*	Ge	177	202	112.5	123.8		34	145	Two independent molecules
				185	202	113.1	122.9		33		cryptocarbene adduct
Bsi	Bsi	*XIV*	Sn	203	215	104.8	129.2	5.0 (Sn); 15.4 (C)	60.9	146	cryptocarbene adduct
					217		125.7				
Bmp	Bmp	*XIV*	Sn	203	217	118.5	122.2	4.4 (Sn); 4.5 (C)	36.5	147	cryptocarbene adduct
					219		119.1				
Bmp	Hyp	*XIV*	Sn	203	218	105.4	120.0	13.2 (Sn); 9.7 (C)	11.9	145	cryptocarbene adduct
					262.2						

[a] For the definition of substituent *XIII* and *XIV*, see chart 3.
[b] Compound **111**.
[c] Compound **112**.
[d] Compound **113a**.

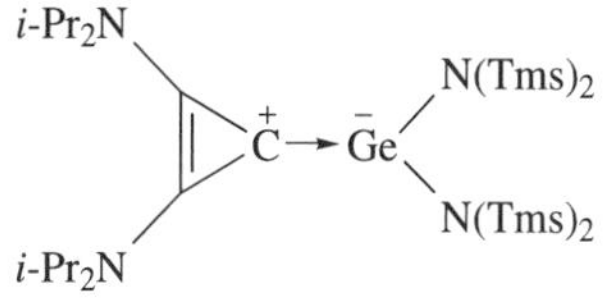

(a) 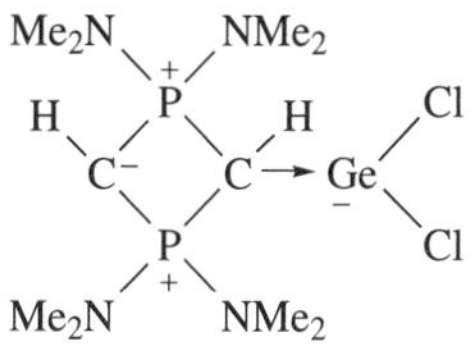 (b) 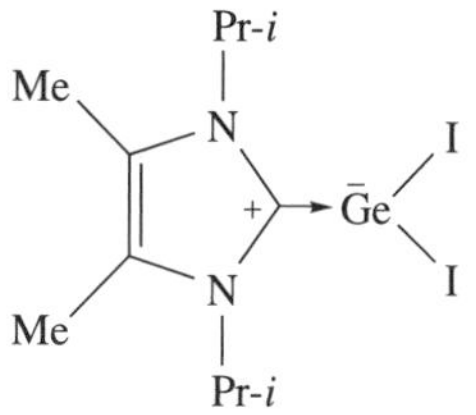 (c)

(110)

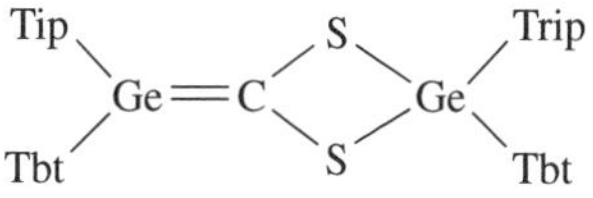

(111) 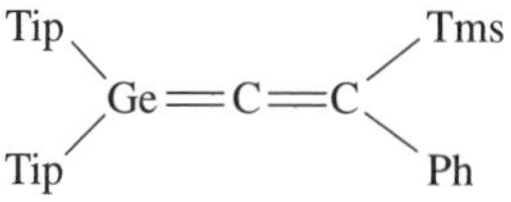 **(112)**

A further germene structurally characterized is **111**[142], the reaction product of CS_2 and the germylene Tbt(Tip)Ge, comprising a cyclic 1,2,3-dithiagermetandiyl unit as carbene fragment. It exhibits the shortest Ge=C bond (177.0 pm) of all molecules with a Ge=C double bond, even shorter than the corresponding bond in the recently characterized 1-germaallene **112**[143], a germene with a Ge=C bond to a sp-hybridized carbon atom. The Ge=C=C fragment of this cumulated multiply-bonded system is bent with a Ge−C−C angle of 159.2°, and the GeC_2 plane of the germylene fragment is markedly tilted towards the Ge=C bond ($\kappa = 32.6°$) (Figure 26), both observations indicating a somewhat different bonding mode compared to homonuclear allenes $R_2C{=}C{=}CR_2$. The central Si−C−C plane of the C(Tms)Ph carbene unit of **112** is orientated almost perpendicular to the GeC_2 plane of the germylene unit, as it is found for most allenes, however. Very recently two 1,6-digermahexadienynes **113a** and **113b**, comprising two germene units linked by a acetylene bridge, were synthesized[144]. The conjugation of the germene moieties could be proven by UV−Vis spectroscopy. The structure determination of **113a** indeed revealed an almost coplanar Ge_2C_4 backbone (Figure 27) and short Ge=C bonds of 181.9 pm (although somewhat longer than for the fluorenylidene germane **107**), however the main structural evidence for conjugation being the very short C(sp)−C(sp^2) single bond (140.7 pm).

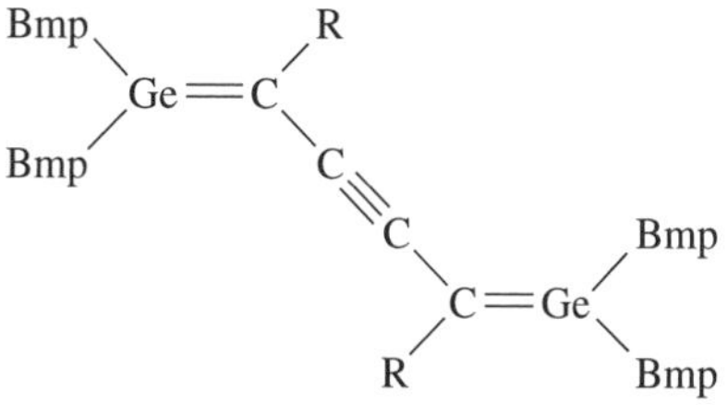

(113)

(a) R = Bu
(b) R = Ph

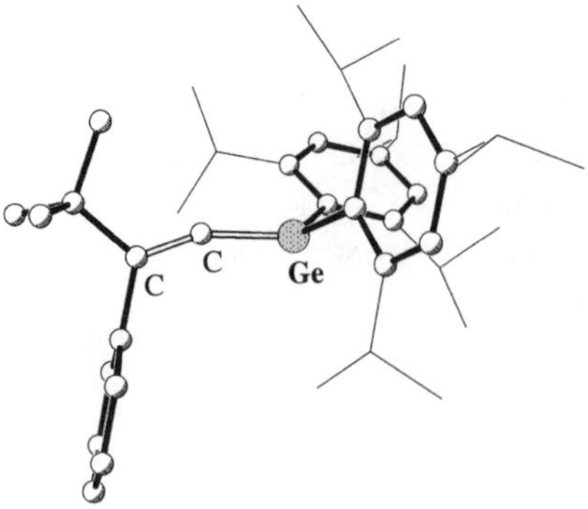

FIGURE 26. Molecular structure of germaallene **112** with a non-linear GeC_2 backbone and *trans*-bent Ge=C unit

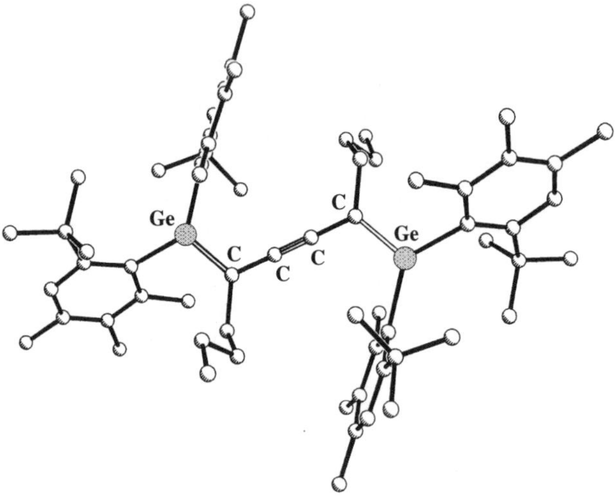

FIGURE 27. Molecular structure of 1,6-digermahexadienyne **113a** with a planar Ge_2C_4 core

Although simple stannenes such as the violet ditipyl fluorenylidenstannane **114** had been synthesized[149], the only structurally characterized compounds which exhibit proper structural features for stannenes are the three cryptodiborylcarbene adducts **115a–c**[145–147]. In spite of large to medium twist angles (60.9°, 36.5° and 11.9°, respectively) between the carbene and the stannylene moieties, short Sn–C bonds of 203–204 pm are observed, being substantially shorter than Sn–C single bonds (*ca* 215 pm) or C → Sn dative bonds in donor carbene adducts (>230 pm). Quantum-chemical calculations lead to similar or even shorter bond lengths for the parent species $H_2Sn{=}CH_2$ depending on the method used: 206.3 pm (MCSCF)[2b] or 194.5 pm (DFT)[2a] with slightly *trans*-bent or planar skeleton, respectively. According to these calculations even plumbenes should be accessible; DFT methods predict a planar structure for $H_2Pb{=}CH_2$ with a Pb=C bond of 204.5 pm, thus again having a substantially shorter bond than Pb–C single-bonded species.

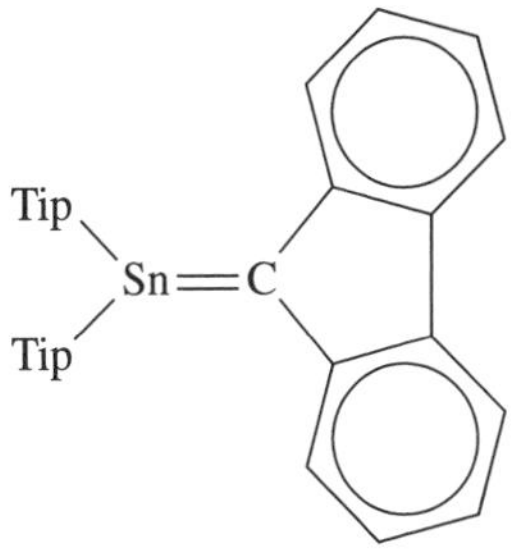

(114)

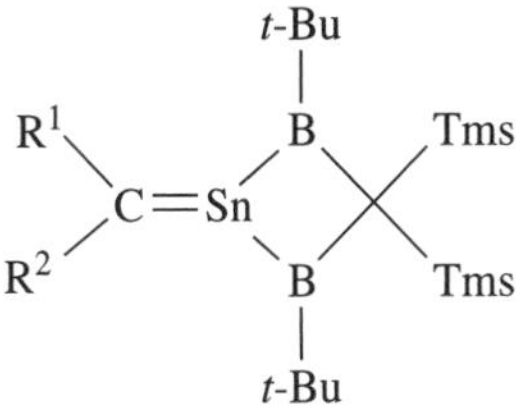

(115) (**a**) $R^1 = R^2 = Bsi$
(**b**) $R^1 = R^2 = Bmp$
(**c**) $R^1 = Bmp$; $R^2 = Hyp$

E. Heavier Homologues of Ketones $R_2E{=}Y$ (E = Ge–Pb; Y = O–Te)

Whereas aldehydes and ketones are one of the most important classes of organic compounds containing multiple bonds, no species with a E=O double bond with E = Si, Ge, Sn, Pb could be isolated as a pure compound in the condensed phase[148b,161]. The germanone Tbt(Tip)Ge=O **116** prepared by Tokitoh's group could be detected in solution only. At room temperature, however, by insertion of the Ge=O bond into a C–Si bond of an *ortho*-Bsi substituent of the employed Tbt ligand, it rearranges quickly to a mixture of diastereomeric benzogermacyclobutanes (equation 13)[162].

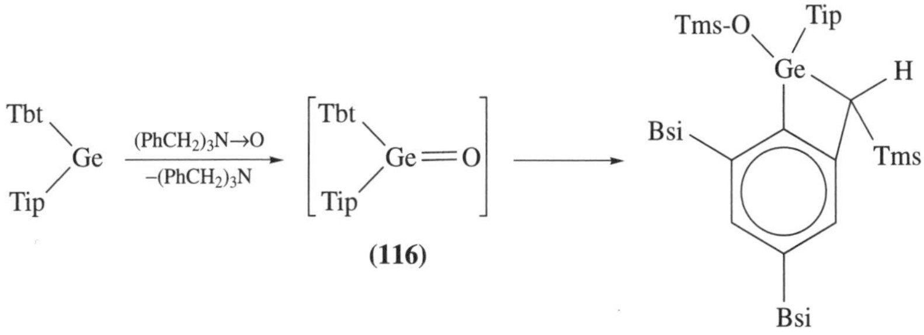

$$Tbt(Tip)Ge \xrightarrow[-(PhCH_2)_3N]{(PhCH_2)_3N\rightarrow O} [Tbt(Tip)Ge{=}O] \longrightarrow \text{benzogermacyclobutane} \quad (13)$$

(116)

$$\textbf{117} \xrightarrow{+S} \textbf{118} \quad (14)$$

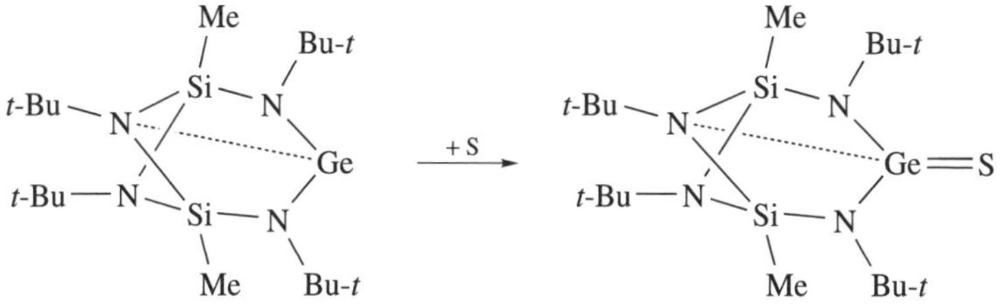

(117) **(118)**

The first structural characterization of homologues of ketones (strictly speaking, of ureas), where E and Y are elements of the later rows of the periodic table, were reported in 1989 by Veith and coworkers. By sulfurization of the bis(amino)germylene **117**, the germanethione **118** (Ge–Se: 206.3 pm) that is stabilized by intramolecular $N \rightarrow Ge$ complexation was obtained (equation 14)[163]. Several analogous or similar

TABLE 16. Structural parameters (bond length in pm, angle in deg) for compounds with heteronuclear double bonds to main-group elements other than carbon $R^1R^2E{=}YR^3$[a]

R^1	R^2	E	Y	E–R^1 E–R^2	R^1–E–R^2	E=Y	E=Y–R^3	Reference	Remarks
Bsi	Bsi	Ge	$(N)_2SiMes_2$	197 197	116.3	168.1	137.3	150	two Ge=N units C_2 symmetry
Bsi	Bsi	Ge	NSi(Bu-*t*)$_2$N$_3$	195 196	122.9	170.4	136.0	150, 151	
Bsi	Bsi	Sn	NSi(Bu-*t*)$_2$N$_3$	215 216	124.8	190.5	130.6	151	
Mes	Mes	Ge	PMes*	194 196	112.9	213.8	107.5	152	
Mes	*t*-Bu	Ge	PMes*	196 201	110.6	214.4	103.1	153	
Tbt	Tip	Ge	S	192 195	118.4	204.9	—	154, 155	
Tbt	Tip	Ge	Se	193 194	119.1	218.0	—	155, 156	
Tbt	Bsi	Ge	Se	199 192	119.7	217.3	—	155	
VI	*VI*	Ge	Se	205 206	121.2	224.7	—	157	two Ge···N contacts Ge···N: 216.2; 217.7
Tbt	Tip	Ge	Te	193 195	117.6	239.8	—	155, 158	
Tbt	Bsi	Ge	Te	198 193	117.9	238.4	—	158	
VI	*VI*	Ge	Te	206 206	120.5	248.0	—	157	two Ge···N contacts Ge···N: 217.1; 217.1
Tbt	Bip	Sn	Se	223 220	122.5	237.5	—	159	
XV	*XV*	Sn	Se	218 219	107.3	239.8	—	160	two Sn···N contacts Sn···N: 236.2; 239.0
XV	*XV*	Sn	Te	217 217	105.5	261.8	—	160	two Sn···N contacts Sn···N: 238.1; 238.1

[a] For the definition of substituents *VI* and *XV*, see charts 2 and 3, respectively.

species comprising the structural fragment $N_2Ge{=}Y$ were synthesized afterwards and structurally characterized[5b,148b]. Real ketone homologues $R_2E{=}Y$ (R = hydrocarbyl), however, with no further stabilization by Lewis bases, were reported by Okazaki and Tokitoh only in 1993[161]. They used the extremely sterically demanding 2,4,6-tris[bis(trimethylsilyl)methyl]phenyl (Tbt) group for protecting the reactive E=Y bond from dimerisation; one should note that the chalcogen atom itself bears no substituent, thus it must be protected by the substituents bonded to the tetrel.

(a) [Tbt(Ar)E] $\xrightarrow{1/n\,Y_n}$ Tbt(Ar)E=Y

E = Ge; Y = S, Se, Te

(b) Tbt(Ar)EY₄ $\xrightarrow[-3\,R_3PY]{3\,R_3P}$ Tbt(Ar)E=Y   (15)

R = Ph, NMe_2

E = Ge, Sn; Y = S, Se

Two different routes were developed: (a) chalcogenation of appropriate (transient) tetrylenes Tbt(Ar)E and (b) dechalcogenation of chalcogenametallolanes $Tbt(R)EY_4$ (equation 15). Thus, several species with Ge=S, Ge=Se, Ge=Te or Sn=Se bonds were synthesized and structurally characterized so far (Table 16). All compounds have similar structures [see, for example, Tbt(Tip)Ge=Te (**119**)[115,158] in Figure 28] with E=Y multiple bonds much shorter (by *ca* 20 pm) than the respective single bonds; the values from diffraction experiments match very well the calculated values (B3LYP) for the parent species $H_2E{=}Y$[164].

(120)

E = Ge; Y = Se, Te

(121)

E = Sn; Y = Se, Te

Meller[157] and Leung[160] and their coworkers succeeded in preparing heteroketones stabilized by intramolecular interactions to Lewis-basic centres, such as **120** and **121**,

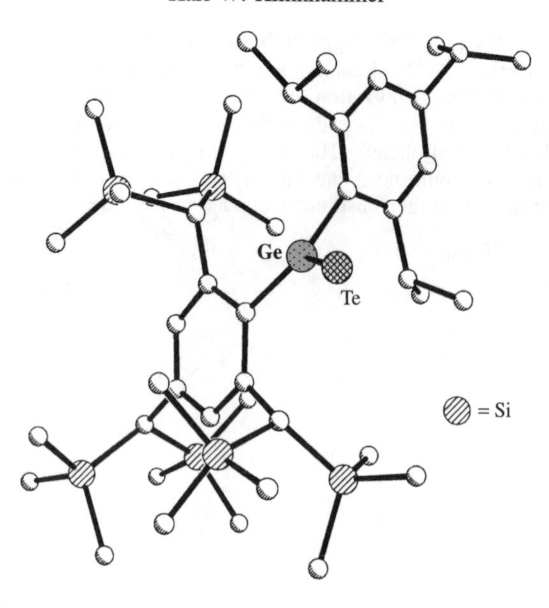

FIGURE 28. Molecular structure of Tbt(Tip)Ge=Te **(119)** with planar coordinated Ge

by chalcogenation of appropriate tetrylenes. In this manner even a compound with Sn=Te double bond (**122**[160]; Figure 29) could be characterized. The observed E=Y(E = Ge, Sn; Y = Se, Te) bonds are significantly elongated compared with the calculated values, since the N $\rightarrow$ E dative interactions weaken the E$-$Yπ-bond by competing with the chalcogen for the empty p-orbital of the tetrylene fragment. As could be expected by the intrinsic strength of these N $\rightarrow$ E dative interactions, this lengthening is more pronounced for Ge than for Sn.

For E = Pb, neither base-stabilized nor base-free derivatives have been isolated so far. The plumbanethione **123** could be indeed prepared by desulfurization of the respective tetrathiaplumbolane at $-78\,^{\circ}$C and trapped with several reagents, but has not been isolated to date. Instead, at ambient conditions the plumbylene **124** and the head-to-tail dimer, the 1,3,2,4-dithiadiplumbetane **125**, is obtained (equation 16)[165].

Tbt(Tip)Pb(S–S–S–S) $\xrightarrow[-3\,R_3PS]{3\,R_3P}$ [Tbt(Tip)Pb=S] **(123)** $\rightarrow$ Tbt–Pb–S–Tbt **(124)** and $[Tip_2Pb(\mu\text{-}S)]_2$ **(125)**

R = Ph, NMe_2

(16)

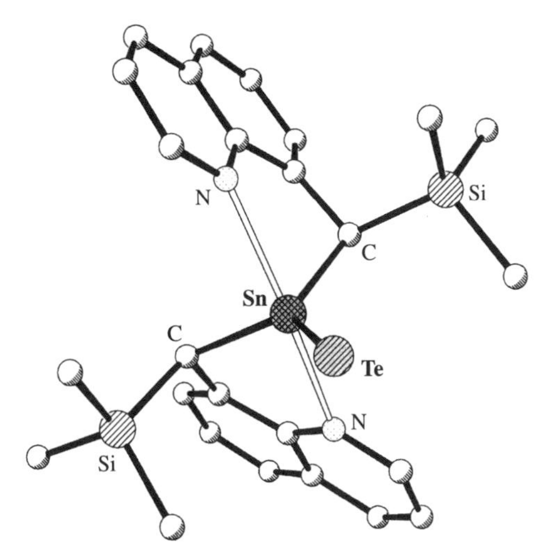

FIGURE 29. Molecular structure of heteroketone **122** comprising intramolecular N $\rightarrow$ Sn interactions

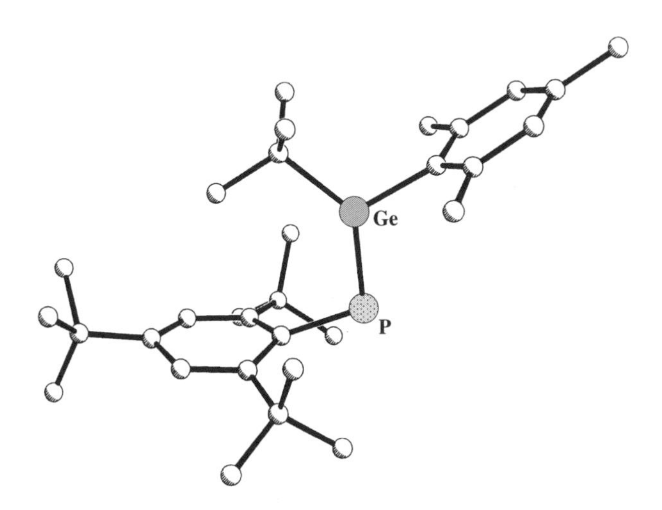

FIGURE 30. Molecular structure of phosphagermene **126b** with planar coordinated Ge.

F. Other Multiple-bonded Compounds

Examples of compounds with heavier tetrels multiply-bonded to other main-group elements are still scarce. Only a handful of species have been structurally characterized with double bonds between Ge or Sn and pnictogenes $R^1R^2E{=}XR^3$ and even less with hydrocarbyl substituents on E (Table 16). No such derivative is known for E = Pb and no example with X = As, Sb and Bi. Iminogermanes $R^1R^2GeNR^3$ and iminostannanes are usually prepared by reaction of an azide R^3N_3 with the appropriate tetrylene R_2E or the corresponding dimer, while phosphagermenes $R^2R^1GePR^3$ **(126a, 126b)**[152,153],

(126a) R^1 = Mes
(126b) R^1 = *t*-Bu

(127)

the only species structurally characterized with both E and X being the heavier elements of their respective groups, have been prepared by salt elimination reactions. All known compounds feature almost planar coordination geometry at E (see, for example, **126b**[153]: Figure 30), bent geometry at X and short E=X bonds, again about 20 pm shorter than typical single-bonded species. Using a bis(azide), even a bis(iminogermane) **(127)**[150] could be prepared; its structural parameters match those of other iminogermanes.

IV. REFERENCES

1. K. M. Mackay, in *The Chemistry of Organic Germanium, Tin and Lead Compounds* (Ed. S. Patai), Wiley, Chichester, 1995, p. 97.
2. (a) H. Jacobsen and T. Ziegler, *J. Am. Chem. Soc.*, **116**, 3667 (1994).
   (b) T. L. Windus and M. S. Gordon, *J. Am. Chem. Soc.*, **114**, 9559 (1992).
   (c) G. Trinquier and J.-P. Malrieu, *J. Am. Chem. Soc.*, **112**, 2130 (1990).
3. E. A. Carter and W. A. Goddard, *J. Phys. Chem.*, **90**, 998 (1986); G. Trinquier and J.-P. Malrieu, *J. Am. Chem. Soc.*, **109**, 5303 (1987); G. Trinquier and J.-P. Malrieu, *J. Am. Chem. Soc.*, **111**, 5916 (1989); G. Trinquier and J.-P. Malrieu, in *The Chemistry of Double Bonded Functional Groups* (Ed. S. Patai), Vol. 2, Part 1, Wiley, Chichester, 1989, p. 1.
4. H. Grützmacher and T. F. Fässler, *Chem. Eur. J.*, **6**, 2317 (2000).
5. (a) R. Okazaki and N. Tokitoh, *Acc. Chem. Res.*, **33**, 625 (2000).
   (b) P. P. Power, *Chem. Rev.*, **99**, 3463 (1999).
   (c) M. Weidenbruch, *Eur. J. Inorg. Chem.*, 373 (1999).
   (d) P. P. Power, *J. Chem. Soc., Dalton Trans.*, 2939 (1998).
   (e) R. Okazaki and R. West, in *Multiply-Bonded Metals and Metalloids* (Eds. R. West and F. G. A. Stone), Academic Press, San Diego, 1996, p. 232.
   (f) K. M. Baines and W. G. Stibbs, in *Multiply-Bonded Metals and Metalloids* (Eds. R. West and F. G. A. Stone), Academic Press, San Diego, 1996, p. 275.
   (g) M. Driess and H. Grützmacher, *Angew. Chem., Int. Ed. Engl.*, **35**, 827 (1996).
6. For some recent examples, see: J. S. Overby, T. P. Hanusa and V. G. Young Jr., *Inorg. Chem.*, **37**, 163 (1998); D. R. Armstrong, M. J. Duer, M. G. Davidson, D. Moncrieff, C. A. Russell, C. Stourton, A. Steiner, D. Stalke and D. S. Wright, *Organometallics*, **16**, 3340 (1997); M. A. Beswick, H. Gornitzka, J. Karcher, M. E. G. Mosquera, J. S. Palmer, P. R. Raithby, C. A. Russell, D. Stalke, A. Steiner and D. S. Wright, *Organometallics*, **18**, 1148 (1999); S. P. Constantine, P. B. Hitchcock and G. A. Lawless, *Organometallics*, **15**, 3905 (1996); S. P. Constantine, G. M. De Lima, P. B. Hitchcock, J. M. Keates and G. A. Lawless; *J. Chem. Soc., Chem. Commun.*, 2337 (1996); J. S. Overby, T. P. Hanusa and P. D. Boyle; *Angew. Chem., Int. Ed. Engl.*, **36**, 2378 (1997); M. A. Beswick, C. Lopez-Casideo, M. A. Payer, P. R. Raithby, C. A. Russell, A. Steiner and D. S. Wright; *J. Chem. Soc., Chem. Commun.*, 109 (1997); S. P. Constantine, P. B. Hitchcock, G. A. Lawless and G. M. De Lima, *J. Chem. Soc., Chem. Commun.*, 1101 (1996); M. Veith, C. Mathur and V. Huch, *Organometallics*, **15**, 2858 (1996); M. Veith, C. Mathur, S. Mathur and V. Huch, *Organometallics*, **16**, 1292 (1997); S. P. Constantine, G. M. De Lima, P. B. Hitchcock, J. M. Keates, G. A. Lawless and I. Marziano, *Organometallics*, **16**, 793 (1997); M. Veith, M. Olbrich, W. Shihua and V. Huch, *J. Chem. Soc., Dalton Trans.*, 161 (1996); H. Sitzmann, R. Boese and P. Stellberg, *Z. Anorg. Allg. Chem.*, **622**, 751 (1996).

7. T. Fjeldberg, A. Haaland, B. E. R. Schilling, P. B. Hitchcock, M. F. Lappert and A. J. Thorne, *J. Chem. Soc., Dalton Trans.*, 1551 (1986).
8. P. B. Hitchcock, M. F. Lappert, S. J. Miles and A. J. Thorne, *J. Chem. Soc., Chem. Commun.*, 480 (1984); D. E. Goldberg, P. B. Hitchcock, M. F. Lappert, K. M. Thomas, A. J. Thorne, T. Fjeldberg, A. Haaland and B. E. R. Schilling, *J. Chem. Soc., Dalton Trans.*, 2387 (1986).
9. M. Kira, S. Ishida, T. Iwamoto, M. Ichinohe, C. Kabuto, L. Ignatovich and H. Sakurai, *Chem. Lett.*, 263 (1999).
10. M. Kira, R. Yauchibara, R. Hirano, C. Kabuto and H. Sakurai, *J. Am. Chem. Soc.*, **113**, 7785 (1991).
11. M. Kira, S. Ishida, T. Iwamoto and C. Kabuto, *J. Am. Chem. Soc.*, **121**, 9722 (1999).
12. P. Jutzi, A. Becker, H. G. Stammler and B. Neumann, *Organometallics*, **10**, 1647 (1991).
13. R. S. Simons, L. Pu, M. M. Olmstead and P. P. Power, *Organometallics*, **16**, 1920 (1997).
14. G. L. Wegner, R. J. F. Berger, A. Schier and H. Schmidbaur, *Organometallics*, **20**, 418 (2001).
15. J. E. Bender IV, M. M. B. Holl and J. W. Kampf, *Organometallics*, **16**, 2743 (1997).
16. C. Drost, P. B. Hitchcock and M. F. Lappert, *Organometallics*, **17**, 3838 (1998).
17. P. Jutzi, H. Schmidt, B. Neumann and H.-G. Stammler, *Organometallics*, **15**, 741 (1996).
18. L. Pu, M. M. Olmstead, P. P. Power and B. Schiemenz, *Organometallics*, **17**, 5602 (1998).
19. A. Schäfer, W. Saak and M. Weidenbruch, *Z. Anorg. Allg. Chem.*, **624**, 1405 (1998).
20. (a) L. Pu, B. Twamley, S. T. Haubrich, M. M. Olmstead, B. V. Mork, R. S. Simons and P. P. Power, *J. Am. Chem. Soc.*, **122**, 650 (2000).
(b) A. C. Filippou, A. I. Philippopoulos, P. Portius and D. U. Neumann, *Angew. Chem., Int. Ed. Engl.*, **39**, 2778 (2000).
21. H. Takeo and R. F. Curl, *J. Mol. Spectrosc.*, **43**, 21 (1972).
22. G. Schultz, J. Tremmel, I. Hargittai, I. Berecz, S. Bohátka, N. D. Kagramanov, A. K. Maltsev and O. M. Nefedov, *J. Mol. Struct.*, **55**, 207 (1979).
23. G. Schultz, J. Tremmel, I. Hargittai, N. D. Kagramanov, A. K. Maltsev and O. M. Nefedov, *J. Mol. Struct.*, **82**, 107 (1982).
24. R. W. Chorley, P. B. Hitchcock, M. F. Lappert, W.-P. Leung, P. P. Power and M. M. Olmstead, *Inorg. Chim. Acta*, **198**, 203 (1992).
25. M. Weidenbruch, J. Schlaefke, A. Schäfer, K. Peters, H. G. von Schnering and H. Marsmann, *Angew. Chem., Int. Ed. Engl.*, **33**, 1846 (1994).
26. M. Stürmann, M. Weidenbruch, K. W. Klinkhammer, F. Lissner and H. Marsmann, *Organometallics*, **17**, 4425 (1998).
27. (a) H. Grützmacher, H. Pritzkow and F. T. Edelmann, *Organometallics*, **10**, 23 (1991).
(b) U. Lay, H. Pritzkow and H. Grützmacher, *J. Chem. Soc., Chem. Commun.*, 260 (1992).
28. S. Brooker, J.-K. Buijink and F. T. Edelmann, *Organometallics*, **10**, 25 (1991).
29. A. Heine and D. Stalke, *Angew. Chem., Int. Ed. Engl.*, **33**, 113 (1994).
30. S. P. Mallela, S. Hill and R. A. Geanangel, *Inorg. Chem.*, **36**, 6247 (1997).
31. K. W. Klinkhammer, *Silylderivate der schweren Alkalimetalle in der Synthese niedervalenter Hauptgruppenelementverbindungen* (Habilitation-Thesis in German), Verlag Ulrich Grauer, Stuttgart, 1998.
32. K. W. Klinkhammer and W. Schwarz, *Angew. Chem., Int. Ed. Engl.*, **34**, 1334 (1995).
33. C. Drost, P. B. Hitchcock, M. F. Lappert and J.-M. Pierssens, *J. Chem. Soc., Chem. Commun.*, 1141 (1997).
34. H. Braunschweig, C. Drost, P. B. Hitchcock, M. F. Lappert and J.-M. Pierssens, *Angew. Chem., Int. Ed. Engl.*, **36**, 261 (1997).
35. B. Gehrhus, P. B. Hitchcock and M. F. Lappert, *Angew. Chem., Int. Ed. Engl.*, **36**, 2514 (1997).
36. H. Grützmacher, W. Deck, H. Pritzkow and M. Sander, *Angew. Chem., Int. Ed. Engl.*, **33**, 456 (1994).
37. C. Drost, B. Gehrhus, P. B. Hitchcock and M. F. Lappert, *J. Chem. Soc., Chem. Commun.*, 1845 (1997).
38. B. E. Eichler and P. P. Power, *J. Am. Chem. Soc.*, **122**, 8785 (2000).
39. K. V. Ermakov, B. S. Butayev and V. P. Spiridonov, *J. Mol. Struct.*, **248**, 143 (1991).
40. A. G. Gershikov, E. Z. Zasorin, A. V. Demidov and V. P. Spiridonov, *Russ. J. Struct. Chem.*, **27**, 375 (1986); A. V. Demidov, A. G. Gershikov, E. Z. Zasorin, V. P. Spiridonov and A. A. Ivanov, *Russ. J. Struct. Chem.*, (*Engl. Transl.*), **24**, 7 (1983).
41. T. Fjeldberg, H. Hope, M. F. Lappert, P. P. Power and A. J. Thorne, *J. Chem. Soc., Chem. Commun.*, 639 (1983).

42. D. E. Goldberg, D. H. Harris, M. F. Lappert and K. M. Thomas, *J. Chem. Soc., Chem. Commun.*, 21 (1976); P. J. Davidson, D. H. Harris and M. F. Lappert, *J. Chem. Soc., Dalton Trans.*, 2268 (1976); J. D. Cotton, P. J. Davidson, M. F. Lappert, J. D. Donaldson and J. Silver, *J. Chem. Soc., Dalton Trans.*, 2286 (1976).
43. B. E. Eichler and P. P. Power, *J. Am. Chem. Soc.*, **122**, 5444 (2000).
44. K. W. Klinkhammer, T. F. Fässler and H. Grützmacher, *Angew. Chem., Int. Ed. Engl.*, **37**, 124 (1998).
45. P. Pyykkö, *Chem. Rev.*, **97**, 597 (1997).
46. A. Haaland and M. F. Lappert, unpublished results, cited in: M. F. Lappert, *Main Group Metal Chem.*, **17**, 183 (1994).
47. M. F. Lappert and K. W. Klinkhammer, unpublished results.
48. N. Kano, K. Shibata, N. Tokitoh and R. Okazaki, *Organometallics*, **18**, 2999 (1999).
49. C. Eaborn, T. Ganicz, P. B. Hitchcock, J. D. Smith and S. E. Sozerli, *Organometallics*, **16**, 5621 (1997).
50. M. Stürmann, W. Saak and M. Weidenbruch, *Z. Anorg. Allg. Chem.*, **625**, 705 (1999).
51. M. Stürmann, W. Saak, M. Weidenbruch and K. W. Klinkhammer, *Eur. J. Inorg. Chem.*, 579 (1999).
52. L. Pu, B. Twamley and P. P. Power, *Organometallics*, **19**, 2874 (2000).
53. N. Seidel, K. Jacob, A. A. H. van der Zeijden, H. Menge, K. Merzweiler and C. Wagner, *Organometallics*, **19**, 1438 (2000).
54. K. W. Klinkhammer, M. Niemeyer and J. Klett, *Chem. Eur. J.*, **5**, 2531 (1999).
55. L. Pu, P. P. Power, I. Boltes and R. Herbst-Irmer, *Organometallics*, **19**, 352 (2000).
56. R. S. Simons, L. Pu, M. M. Olmstead and P. P. Power, *Organometallics*, **16**, 1920 (1997).
57. S. Benet, C. J. Cardin, D. J. Cardin, S. P. Constantine, P. Heath, H. Rashid, S. Teixeira, J. H. Thorpe and A. K. Todd, *Organometallics*, **18**, 389 (1999).
58. C. J. Cardin, D. J. Cardin, S. P. Constantine, A. K. Todd, S. J. Teat and S. Coles, *Organometallics*, **17**, 2144 (1998).
59. K. W. Klinkhammer, T. F. Fässler and H. Grützmacher, *Angew. Chem., Int. Ed. Engl.*, **37**, 124 (1998).
60. M. Stürmann, W. Saak, K. W. Klinkhammer and M. Weidenbruch, *Z. Anorg. Allg. Chem.*, **625**, 1955 (1999).
61. K. W. Klinkhammer, to be published.
62. D. R. Armstrong, M. A. Beswick, N. L. Cromhout, C. N. Harmer, D. Moncrieff, C. A. Russell, P. R. Raithby, A. Steiner, A. E. H. Wheatley and D. S. Wright, *Organometallics*, **17**, 3176 (1998); D. R. Armstrong, M. A. Beswick, N. L. Cromhout, C. N. Harmer, D. Moncrieff, C. A. Russell, P. R. Raithby, A. Steiner, A. E. H. Wheatley and D. S. Wright, *Organometallics*, **17**, 3176 (1998); M. A. Beswick, N. L. Cromhout, C. N. Harmer, P. R. Raithby, C. A. Russell, J. S. B. Smith, A. Steiner and D. S. Wright, *J. Chem. Soc., Chem. Commun.*, 1977 (1996).
63. H. Grützmacher, S. Freitag, R. Herbst-Irmer and G. S. Sheldrick, *Angew. Chem., Int. Ed. Engl.*, **31**, 437 (1992).
64. C. C. Cardin, D. J. Cardin, S. P. Constantine, M. G. B. Drew, H. Rashid, M. A. Convery and D. Fenske, *J. Chem. Soc., Dalton Trans.*, 2749 (1998).
65. L. M. Engelhardt, B. S. Jolly, M. F. Lappert, C. L. Raston and A. H. White, *J. Chem. Soc., Chem. Commun.*, 336 (1988); B. S. Jolly, M. F. Lappert, L. M. Engelhardt, A. H. White and C. L. Raston, *J. Chem. Soc., Chem. Commun.*, 2653 (1993).
66. P. Jutzi, S. Keitemeyer, B. Neumann, A. Stammler and H.-G. Stammler, *Organometallics*, **20**, 42 (2001).
67. K. Jurkschat, C. Klaus, M. Dargatz, A. Tschach, J. Meunier-Piret and B. Mahieu, *Z. Anorg. Allg. Chem.*, **577**, 122 (1989).
68. H. Schmidt, S. Keitemeyer, B. Neumann, H.-G. Stammler, W. W. Schoeller and P. Jutzi, *Organometallics*, **17**, 2149 (1998).
69. P. Jutzi, B. Hampel, M. B. Hursthouse and A. J. Howes, *Organometallics*, **5**, 1944 (1986).
70. P. Jutzi, A. Becker, C. Leue, H. G. Stammler, B. Neumann, M. B. Hursthouse and A. Karaulov, *Organometallics*, **10**, 3838 (1991).
71. J. G. Winter, P. Portius, G. Kociok-Kohn, R. Steck and A. C. Filippou, *Organometallics*, **17**, 4176 (1998).
72. P. Jutzi, H. Schmidt, B. Neumann and H.-G. Stammler, *J. Organomet. Chem.*, **499**, 7 (1995).

73. M. Veith, M. Olbrich, W. Shihua and V. Huch, *J. Chem. Soc., Dalton Trans.*, 161 (1996).
74. M. Mehring, C. Löw, M. Schürmann, F. Uhlig, K. Jurkschat and B. Mahieu, *Organometallics*, **19**, 4613 (2000).
75. M. Weidenbruch, U. Grobecker, W. Saak, E.-M. Peters and K. Peters, *Organometallics*, **17**, 5206 (1998).
76. A. L. Balch and D. E. Oram, *Organometallics*, **5**, 2159 (1986).
77. Reviews: M. F. Lappert and R. S. Rowe, *Coord. Chem. Rev.*, **100**, 267 (1990); (Ge only) H. Ogino and H. Tobita, *Adv. Organomet. Chem.*, **42**, 223 (1998).
78. Recent papers: (R_2Ge) J. S. McIndoe and B. K. Nicholson, *J. Organomet. Chem.*, **577**, 181 (1999); W. K. Leong, F. W. B. Einstein and R. K. Pomeroy, *Organometallics*, **15**, 1589 (1996); (R_2Sn) P. McArdle, L. O'Neill and D. Cunningham, *Inorg. Chim. Acta*, **291**, 252 (1999); J. J. Schneider, J. Hagen, D. Bläser, R. Boese, F. F. Biani, P. Zanelli and C. Krüger, *Eur. J. Inorg. Chem.*, 1987 (1999); C. J. Cardin, D. J. Cardin, M. A. Convery, Z. Dauter, D. Fenske, M. M. Devereux and M. B. Power, *J. Chem. Soc., Dalton Trans.*, 1133 (1996); J. J. Schneider, N. Czap, D. Bläser and R. Boese, *J. Am. Chem. Soc.*, **121**, 1409 (1999); J. J. Schneider, J. Hagen, D. Spickermann, D. Bläser, R. Boese, F. F. de Biani, F. Laschi and P. Zanelli, *Chem. Eur. J.*, **6**, 237 (2000); Reference 54; (R_2Pb) N. C. Burton, C. J. Cardin, D. J. Cardin, B. Twamley and Y. Zubavichus, *Organometallics*, **14**, 5708 (1995); W. K. Leong, F. W. B. Einstein and R. K. Pomeroy, *J. Cluster Sci.*, **7**, 121 (1996).
79. A. J. Arduengo III, H. V. R. Dias, J. C. Calabrese and F. Davidson, *Inorg. Chem.*, **32**, 1541 (1993).
80. E. Fluck, M. Spahn, G. Heckmann and H. Borrmann, *Z. Anorg. Allg. Chem.*, **612**, 56 (1992).
81. H. Schumann, M. Glanz, F. Girgsdies, F. E. Hahn, M. Tamm and A. Grzegorzewski, *Angew. Chem., Int. Ed. Engl.*, **36**, 2232 (1997).
82. A. Schäfer, M. Weidenbruch, W. Saak and S. Pohl, *J. Chem. Soc., Chem. Commun.*, 1157 (1995).
83. N. Kuhn, T. Kratz, D. Bläser and R. Boese, *Chem. Ber.*, **128**, 245 (1995).
84. M. Veith and V. Huch, *J. Organomet. Chem.*, **293**, 161 (1985).
85. M. Veith and V. Huch, *J. Organomet. Chem.*, **308**, 263 (1986).
86. F. Stabenow, W. Saak and M. Weidenbruch, *J. Chem. Soc., Chem. Commun.*, 1131 (1999).
87. T. Tsumuraya, S. A. Batcheller and S. Masamune, *Angew. Chem., Int. Ed. Engl.*, **30**, 902 (1991); M. Weidenbruch, *Chem. Rev.*, **95**, 1479 (1995); E. Hengge and R. Janoschek, *Chem. Rev.*, **95**, 1495 (1995).
88. S. Nagase, K. Kobayashi and M. Nagashima, *J. Chem. Soc., Chem. Commun.*, 1302 (1992); S. Nagase, *Polyhedron*, **10**, 1299 (1991).
89. S. P. Mallela, Y. Saar, S. Hill and R. A. Geanangel, *Inorg. Chem.*, **38**, 2957 (1999); S. P. Mallela, W.-P. Su, Y.-S. Chen, J. D. Korp and R. A. Geanangel, *Main Group Chem.*, **2**, 315 (1998); S. P. Mallela, S. Hill and R. A. Geanangel, *Inorg. Chem.*, **36**, 6247 (1997); S. P. Mallela and R. A. Geanangel, *Inorg. Chem.*, **33**, 1115 (1994).
90. A. Sekiguchi, T. Yatabe, H. Naito, C. Kabuto and H. Sakurai, *Chem. Lett.*, 1697 (1992).
91. N. Wiberg, W. Hochmuth, H. Nöth, A. Appel and M. Schmidt-Amelunxen, *Angew. Chem., Int. Ed. Engl.*, **35**, 1333 (1996).
92. C. Eaborn, P. B. Hitchcock, J. D. Smith and S. E. Sozerli, *Organometallics*, **16**, 5653 (1997).
93. A. J. Edwards, M. A. Paver, P. R. Raithby, M.-A. Rennie, C. A. Russell and D. S. Wright, *J. Chem. Soc., Dalton Trans.*, 1587 (1995).
94. C. Eaborn, K. Izod, P. B. Hitchcock, S. E. Sozerli and J. D. Smith, *J. Chem. Soc., Chem. Commun.*, 1829 (1995).
95. L. Pu, B. Twamley and P. P. Power, *J. Am. Chem. Soc.*, **122**, 3524 (2000).
96. W. Kutzelnigg, *Angew. Chem., Int. Ed. Engl.*, **23**, 272 (1984).
97. K. W. Klinkhammer, *Angew. Chem., Int. Ed. Engl.*, **36**, 2320 (1997).
98. L. Pauling, *Die Natur der Chemischen Bindung* (in German), VCH, Weinheim, 1968, p. 64.
99. Y. Apeloig and M. Karni, *J. Chem. Soc., Chem. Commun.*, 1048 (1985); S. G. Bott, P. Marshall, P. E. Wagenseller, Y. Wong and R. T. Conlin, *J. Organomet. Chem.*, **499**, 11 (1995).
100. P. B. Hitchcock, M. F. Lappert, S. J. Miles and A. J. Thorne, *J. Chem. Soc., Chem. Commun.*, 480 (1984); D. E. Goldberg, P. B. Hitchcock, M. F. Lappert, K. M. Thomas, A. J. Thorne, T. Fjeldberg, A. Haaland and B. E. R. Schilling, *J. Chem. Soc., Dalton Trans.*, 2387 (1986).
101. (a) L. Pauling, A. W. Laubergayer and J. L. Hoard, *J. Am. Chem. Soc.*, **60**, 605 (1938).

(b) M. Dräger and L. Ross, *Z. Anorg. Allg. Chem.*, **460**, 207 (1980).
102. J. T. Snow, S. Murakami, S. Masamune and D. J. Williams, *Tetrahedron Lett.*, **25**, 4191 (1984).
103. H. Schäfer, W. Saak and M. Weidenbruch, *Organometallics*, **18**, 3159 (1999).
104. M. Weidenbruch, M. Stürmann, H. Kilian, S. Pohl and W. Saak, *Chem. Ber.*, **130**, 735 (1997).
105. S. A. Batcheller, T. Tsumuraya, O. Tempkin, W. M. Davis and S. Masamune, *J. Am. Chem. Soc.*, **112**, 9394 (1990).
106. M. Kira, T. Iwamoto, T. Maruyama, C. Kabuto and H. Sakurai, *Organometallics*, **15**, 3767 (1996).
107. A. Schäfer, W. Saak, M. Weidenbruch, H. Marsmann and G. Henkel, *Chem. Ber. Recl.*, **130**, 1733 (1997).
108. L. Pu, M. O. Senge, M. M. Olmstead and P. P. Power, *J. Am. Chem. Soc.*, **120**, 12682 (1998).
109. A. H. Cowley, J. G. Lasch, N. C. Norman and M. Pakulski, *J. Am. Chem. Soc.*, **105**, 5506 (1983); A. H. Cowley, N. C. Norman and M. Pakulski, *J. Chem. Soc., Dalton Trans.*, 383 (1985); C. Couret, J. Escudié, Y. Madaule, H. Ranaivonjatovo and J.-G. Wolf, *Tetrahedron Lett.*, **24**, 2769 (1993); N. Tokitoh, T. Arai, T. Sasamori, R. Okazaki, S. Nagase, H. Uekusa and Y. Ohashi, *J. Am. Chem. Soc.*, **120**, 433 (1998); B. Twamley, C. D. Sofield, M. M. Olmstead and P. P. Power, *J. Am. Chem. Soc.*, **121**, 3357 (1999).
110. H. Schäfer, W. Saak and M. Weidenbruch, *Angew. Chem., Int. Ed. Engl.*, **39**, 3703 (2000).
111. M. Weidenbruch, H. Kilian, K. Peters, H. G. von Schnering and H. Marsmann, *Chem. Ber.*, **128**, 983 (1995).
112. W. P. Leung, W.-H. Kwok, F. Xue and T. C. W. Mak, *J. Am. Chem. Soc.*, **119**, 1145 (1997).
113. C. Drost, P. B. Hitchcock and M. F. Lappert, *Angew. Chem., Int. Ed. Engl.*, **38**, 1113 (1999).
114. M. Weidenbruch, H. Kilian, K. Peters, H. G. von Schnering and H. Marsmann, *Chem. Ber.*, **128**, 983 (1995).
115. W.-P. Leung, H. Cheng, R.-B. Huang, Q.-C. Yang and T. C. W. Mak, *J. Chem. Soc., Chem. Commun.*, 451 (2000).
116. M. Stürmann, M. Weidenbruch, K. W. Klinkhammer, F. Lissner and H. Marsmann, *Organometallics*, **17**, 4425 (1998).
117. M. Stürmann, W. Saak, H. Marsmann and M. Weidenbruch, *Angew. Chem., Int. Ed. Engl.*, **38**, 187 (1999).
118. N. Wiberg, H.-W. Lerner, S.-K. Vasisht, S. Wagner, K. Karaghiosoff, H. Nöth and W. Ponikwar, *Eur. J. Inorg. Chem.*, 1211 (1999).
119. A. Sekiguchi, H. Yamazaki, C. Kabuto and H. Sakurai, *J. Am. Chem. Soc.*, **117**, 8025 (1995).
120. A. Sekiguchi, N. Fukaya, M. Ichinohe, N. Takagi and S. Nagase, *J. Am. Chem. Soc.*, **121**, 11587 (1999).
121. V. Y. Lee, M. Ichinohe and A. Sekiguchi, *J. Am. Chem. Soc.*, **122**, 9034 (2000).
122. A. Sekiguchi, M. Tsukamoto and M. Ichinohe, *Science*, **275**, 60 (1996).
123. (a) M. Ichinohe, N. Fukaya and A. Sekiguchi, *Chem. Lett.*, 1045 (1998).
(b) A. Sekiguchi, N. Fukaya, M. Ichinohe and Y. Ishida, *Eur. J. Inorg. Chem.*, 1155 (2000).
124. M. M. Olmstead, L. Pu, R. S. Simons and P. P. Power, *J. Chem. Soc., Chem. Commun.*, 1595 (1997).
125. For debates on this topic, dealing mainly with the isoelectronic triple-bonded dianion [RGaGaR], see: J. Su, X.-W. Crittendon and G. H. Robinson, *J. Am. Chem. Soc.*, **119**, 5471 (1997); Y. Xie, R. S. Grev, J. Gu, H. F. Schaefer, P. v. R. Schleyer, J. Su, X.-W. Li and G. H. Robinson, *J. Am. Chem. Soc.*, **120**, 5471 (1998) and A. J. Downs, *Coord. Chem. Rev.*, **189**, 59 (1999); F. A. Cotton, A. H. Cowley and X. Feng, *J. Am. Chem. Soc.*, **120**, 1795 (1998); I. Bytheway and Z. Lin, *J. Am. Chem. Soc.*, **120**, 12133 (1998); F. A. Cotton and X. Feng, *Organometallics*, **17**, 128 (1998); T. L. Allen, P. P. Power and W. H. Funk, *J. Chem. Soc., Dalton Trans.*, 407 (2000).
126. V. Y. Lee, M. Ichinohe and A. Sekiguchi, *J. Am. Chem. Soc.*, **122**, 12604 (2000).
127. V. Y. Lee, M. Ichinohe, A. Sekiguchi, N. Tagaki and S. Nagase, *J. Am. Chem. Soc.*, **122**, 9034 (2000).
128. N. Fukaya, M. Ichinohe and A. Sekiguchi, *Angew. Chem., Int. Ed. Engl.*, **39**, 3881 (2000).
129. M. Weidenbruch, S. Willms, W. Saak and G. Henkel, *Angew. Chem., Int. Ed. Engl.*, **36**, 2503 (1997).
130. K. M. Baines and W. G. Stibbs, *Adv. Organomet. Chem.*, **39**, 275 (1996).
131. M. Weidenbruch, F.-T. Grimm, M. Herrndorf, A. Schäfer, K. Peters and H. G. von Schnering, *J. Organomet. Chem.*, **341**, 335 (1988); H. Ohgaki, Y. Kabe and W. Ando,

Organometallics, **14**, 2139 (1995); R. I. Bochkova, Y. N. Drozkov, E. A. Kuz'min, L. N. Bochkarev and M. N. Bochkarev, *Koord. Khim.* **13**, 1126 (1987).
132. M. Olmstead, R. S. Simons and P. P. Power, *J. Am. Chem. Soc.*, **119**, 11705 (1997).
133. L. Pu, S. T. Haubrich and P. P. Power, *J. Organomet. Chem.*, **582**, 100 (1999).
134. C. Schneider-Koglin, K. Behrends and M. Dräger, *J. Organomet. Chem.*, **448**, 29 (1993); H. Puff, B. Breuer, G. Gehrke-Brinkmann, P. Kind, H. Reuter, W. Schuh, W. Wald and G. Weidenbruck, *J. Organomet. Chem.*, **363**, 265 (1989).
135. P. Rutsch and G. Huttner, *Angew. Chem., Int. Ed. Engl.*, **39**, 3697 (2000).
136. N. Wiberg, H.-W. Lerner, H. Nöth and W. Ponikwar, *Angew. Chem., Int. Ed. Engl.*, **38**, 1103 (1999).
137. W. Uhl, S. U. Keimling, K. W. Klinkhammer and W. Schwarz, *Angew. Chem., Int. Ed. Engl.*, **36**, 64 (1997).
138. N. Wiberg, H.-W. Lerner, S. Wagner, H. Nöth and T. Seifert, *Z. Naturforsch.*, **B54**, 877 (1999).
139. K. W. Klinkhammer, *Chem. Eur. J.*, **3**, 1418 (1997).
140. M. Lazraq, J. Escudie, C. Couret, J. Satgé, M. Dräger and R. Dammel, *Angew. Chem., Int. Ed. Engl.*, **27**, 828 (1988).
141. H. Meyer, G. Baum, W. Massa and A. Berndt, *Angew. Chem., Int. Ed. Engl.*, **26**, 798 (1987).
142. N. Tokitoh, K. Kishikawa and R. Okazaki, *J. Chem. Soc., Chem. Commun.*, 1425 (1995).
143. B. E. Eichler, D. R. Powell and R. West, *Organometallics*, **17**, 2147 (1998).
144. F. Meiners, W. Saak and M. Weidenbruch, *Organometallics*, **19**, 2835 (2000).
145. M. Stürmann, W. Saak, M. Weidenbruch, A. Berndt and D. Sclesclkewitz, *Heteroatom Chem.*, **10**, 554 (1999).
146. H. Meyer, G. Baum, W. Massa, S. Berger and A. Berndt, *Angew. Chem., Int. Ed. Engl.*, **26**, 5546 (1997).
147. M. Weidenbruch, H. Kilian, M. Stürmann, S. Pohl, W. Saak, H. Marsmann, D. Steiner and A. Berndt, *J. Organomet. Chem.*, **530**, 255 (1997).
148. For recent reviews, see:
(a) J. Escudié, C. Couret and H. Ranaivonjatovo, *Coord. Chem. Rev.*, **178–180**, 562 (1998).
(b) J. Barrau and G. Rima, *Coord. Chem. Rev.*, **178–180**, 593 (1998).
149. G. Anselme, H. Ranaivonjatovo, J. Escudié, C. Couret and S. Satgé, *Organometallics*, **11**, 2748 (1992).
150. W. Ando, T. Ohtaki and Y. Kabe, *Organometallics*, **13**, 434 (1994).
151. T. Ohtaki, Y. Kabe and W. Ando, *Heteroatom Chem.*, **5**, 313 (1994).
152. M. Dräger, J. Escudie, C. Couret, H. Ranaivonjatovo and J. Satgé, *Organometallics*, **7**, 1010 (1988).
153. H. Ranaivonjatovo, J. Escudie, C. Couret, J. Satgé and M. Dräger, *New J. Chem.*, **13**, 389 (1989).
154. N. Tokitoh, T. Matsumoto, K. Manmaru and R. Okazaki, *J. Am. Chem. Soc.*, **115**, 8855 (1993).
155. T. Matsumoto, N. Tokitoh and R. Okazaki, *J. Am. Chem. Soc.*, **121**, 8811 (1999).
156. T. Matsumoto, N. Tokitoh and R. Okazaki, *Angew. Chem., Int. Ed. Engl.*, **33**, 2316 (1994).
157. G. Ossig, A. Meller, C. Brönneke, O. Müller, M. Schäfer and R. Herbst-Irmer, *Organometallics*, **16**, 2116 (1997).
158. N. Tokitoh, Y. Matsuhashi, K. Shibata, T. Matsumoto, H. Suzuki, M. Saito, K. Manmaru and R. Okazaki, *Main Group Metal Chem.*, **17**, 55 (1994).
159. M. Saito, N. Tokitoh and R. Okazaki, *J. Am. Chem. Soc.*, **119**, 11124 (1997).
160. W.-P. Leung, W.-H. Kwok, T. C. Low, Z. Y. Zhou and T. C. W. Mak, *J. Chem. Soc., Chem. Commun.*, 505 (1996).
161. For a recent review, see: R. Okazaki and N. Tokitoh, *Acc. Chem. Res.*, **33**, 625 (2000).
162. N. Tokitoh, T. Matsumoto and R. Okazaki, *J. Am. Chem. Soc.*, **119**, 2337 (1997).
163. M. Veith, S. Becker and V. Huch, *Angew. Chem., Int. Ed. Engl.*, **28**, 1237 (1989).
164. H. Suzuki, N. Tokitoh, R. Okazaki, S. Nagase and M. Goto, *J. Am. Chem. Soc.*, **120**, 11096 (1998).
165. N. Kano, *A Study on Stable Divalent Organolead Compounds*, PhD Thesis, The University of Tokyo, 1998.

CHAPTER 5

Gas-phase chemistry and mass spectrometry of Ge-, Sn- and Pb-containing compounds

JOSÉ M. RIVEROS

Institute of Chemistry, University of São Paulo, Caixa Postal 26077, São Paulo, Brazil, CEP 05513-970
Fax: 55-11-3818-388; e-mail: jmrnigra@quim.iq.usp.br

and

KEIKO TAKASHIMA

Department of Chemistry, University of Londrina, Caixa Postal 6001, Londrina, PR, Brazil, CEP 86051-970
Phone/Fax: 55-43-371-4286; e-mail: keikotak@onda.com.br

The chemistry of organic germanium, tin and lead compounds — Vol. 2
Edited by Z. Rappoport

I. INTRODUCTION

Unlike the lighter elements of group 14, namely C and Si, considerably less is known about the gas-phase chemistry and mass spectrometry of Ge-, Sn- and Pb-containing compounds. The poor volatility of the majority of Ge, Sn and Pb compounds has been a major drawback for studies at or near room temperature. In the case of mass spectrometry, the additional problems associated with thermal stability and the complex isotopic patterns displayed by these elements have traditionally restricted the routine use of mass spectra as analytical tools for identifying typical organometallic derivatives of Ge, Sn and Pb. However, the variety of ionization techniques presently available, such as electron ionization (EI), chemical ionization (CI), fast atom bombardment (FAB), matrix assisted laser desorption ionization (MALDI) and electrospray (ESI), among others, coupled with the high resolution and high mass capabilities of modern instruments can nowadays circumvent many of the early problems. Thus, mass spectrometry can increasingly provide a complementary approach for the characterization of organometallic compounds containing elements of group 14 of the periodic table[1,2]. This can be exemplified by germenes and stannanes, an area that has enjoyed phenomenal growth and interest in recent years, because of their potential applications as synthons in organometallic and organic chemistry. While many germenes have been characterized by a combination of spectroscopic techniques such as NMR and IR[3], new advances in this area are increasingly taking full advantage of the versatility of modern mass spectrometry for more complete structural identification[4].

The small number of fundamental mass spectrometric studies on Ge, Sn and Pb derivatives also accounts for our poor knowledge of their thermochemistry. Heats of formation, ionization energies, bond energies and electron affinities of even simple Ge, Sn and Pb species are still scarce and subject to considerable uncertainty, as illustrated in the most recent NIST database[5].

In spite of the limitations imposed by volatility, studies related to the gas-phase chemistry of simple organogermanes have grown in number in the last two decades because of their implications regarding deposition of Ge through chemical vapor deposition processes (CVD) aimed at surface modifications and film formation. Thus, the characterization of the elementary reactions responsible for the mechanism of these processes has become an important area of research. Likewise, the characterization of Ge clusters by gas-phase techniques has also become a growing field because of the relevance of clusters to semiconductors[6]. Moreover, the detailed outcome of gas-phase reactions involving organostannanes and Pb compounds are extremely useful in mapping out the environmental effects of these elements[7].

Three main topics relevant to the gas-phase chemistry of Ge, Sn and Pb derivatives are discussed in the present chapter: (a) the mass spectrometry related to organometallic compounds of group 14 with particular emphasis on the more general aspects; (b) the gas-phase ion chemistry comprising the thermochemistry, structure and reactivity of ions; and (c) gas-phase reactions involving neutral species.

II. MASS SPECTROMETRY OF Ge, Sn AND Pb DERIVATIVES

The principles of mass spectrometry as applied to organic compounds have been extensively investigated and are well described according to different functional groups in some of the classical literature of the field[8,9]. By comparison, there are few systematic approaches towards the interpretation of mass spectral fragmentations for the organometallic derivatives of Ge, Sn and Pb. As a rule, the mass spectra of these organometallic compounds display in high abundance fragment ions that retain the metal element. This is a consequence of the much weaker E−H, E−C or E-halogen bond energies (E = Ge, Sn, Pb) when compared with the carbon analogs, and the lower ionization energy of the corresponding fragment containing Ge, Sn or Pb. Some of the early reviews[10,11] on the subject were based mostly on the mass spectra of the trialkyl or tetraalkyl derivatives of these elements. Unfortunately, most of the early studies lacked the information that can be obtained from more recent experimental techniques such as tandem mass spectrometry that are essential for the full understanding of fragmentation processes and ion structures. Nevertheless, the mass spectrometry of organostannanes is probably the best studied among these compounds because of the increasing use of Sn reagents in organic synthesis.

As indicated above, the mass spectra of Ge, Sn and Pb organometallic compounds are uniquely characterized by fragment ions bearing the metal element. These ions are recognized by the distinct isotopic composition of the higher elements of group 14. Several isotopes of Ge, Sn and Pb are known to occur with significant natural abundances, as shown in Tables 1–3. Thus, peaks corresponding to ions containing any of these elements are responsible for characteristic isotope patterns such as those displayed in Figures 1–3.

TABLE 1. Isotopic composition and relative abundance of Ge isotopes

Isotope	70	72	73	74	76
Natural abundance (%)	20.5	27.4	7.8	36.5	7.8

TABLE 2. Isotopic composition and relative abundance of the most common Sn isotopes

Isotope	116	117	118	119	120	122	124
Natural abundance (%)	14.7	7.7	24.3	8.6	32.4	4.6	5.6

TABLE 3. Isotopic composition and relative abundance of Pb isotopes

Isotope	204	206	207	208
Natural abundance (%)	1.4	24.1	22.1	52.4

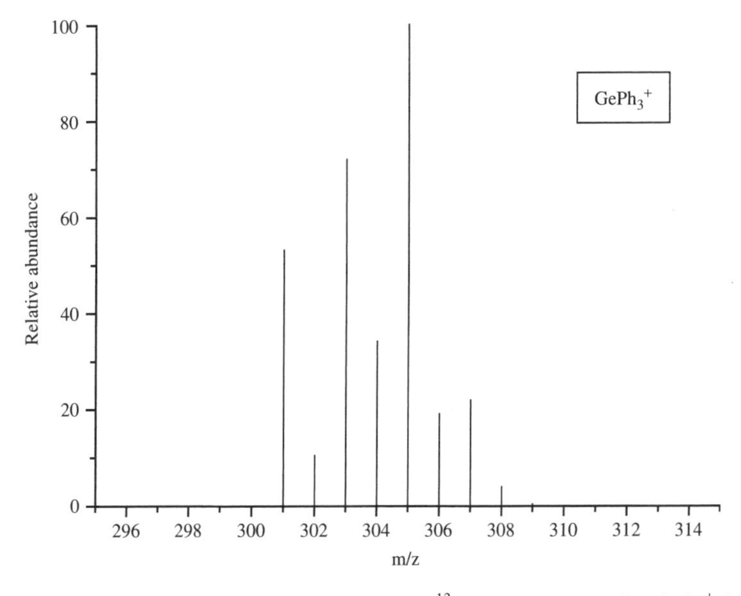

FIGURE 1. Typical isotopic pattern (including that of the ^{13}C) encountered for the Ph_3Ge^+ fragment ion in the mass spectra of a typical Ph_3GeX compound

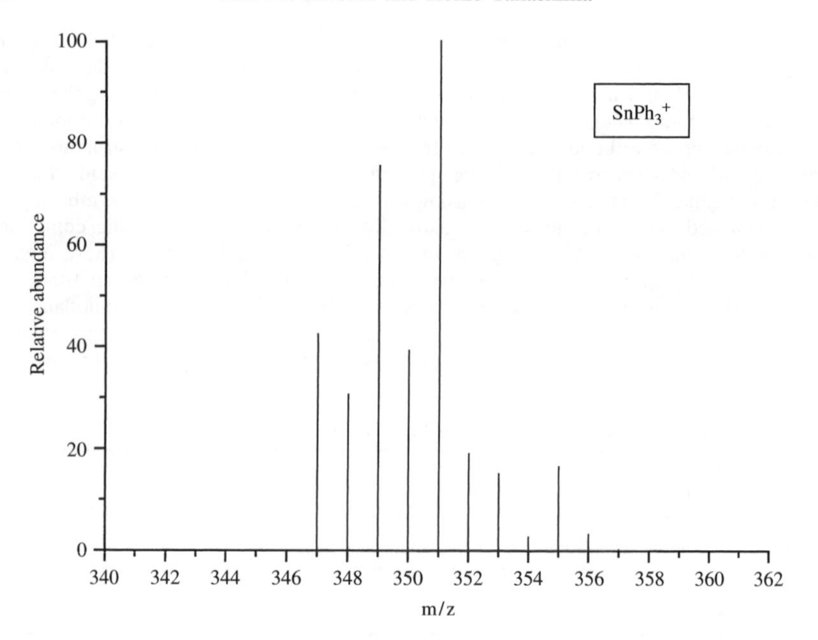

FIGURE 2. Typical isotopic pattern (including that of the ^{13}C) encountered for the Ph_3Sn^+ fragment ion in the mass spectra of a typical Ph_3SnX compound

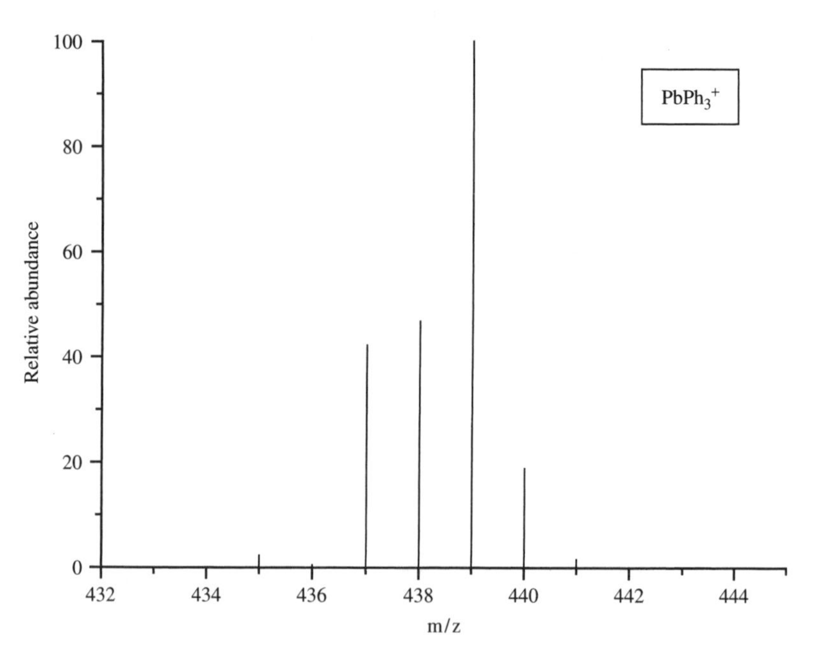

FIGURE 3. Typical isotopic pattern (including that of the ^{13}C) encountered for the Ph_3Pb^+ fragment ion in the mass spectra of a typical Ph_3PbX compound

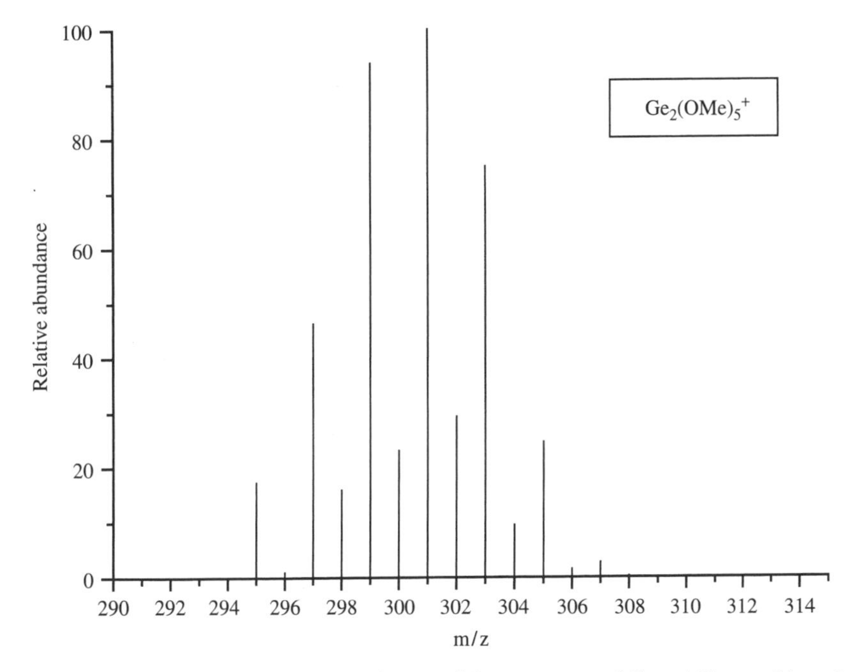

FIGURE 4. Typical isotopic pattern for an ion containing two atoms of Ge and illustrated here for the $Ge_2(OMe)_5^+$ ion

These patterns become considerably more complex for species containing more than one atom of Ge, Sn or Pb, or a mixture of these metals, leading to potentially very dense isotopic distribution as illustrated in Figure 4 for the $Ge_2(OMe)_5^+$ ion containing two atoms of Ge. An even more complex pattern often results from ions differing in their chemical composition by one or two hydrogens, as is often the case for compounds containing alkyl groups bonded directly to the metal element. In these cases, the use of high resolution mass spectrometry and careful analysis of relative intensities becomes essential for the full elucidation of the fragmentation patterns.

A second important characteristic of the mass spectra of Ge-, Sn- and Pb-containing compounds is the tendency to display abundant peaks corresponding to the M(IV) and M(II) oxidation states of these elements.

A. Mass Spectra of Simple Hydrides of Ge, Sn and Pb

The first mass spectrometric studies of the simple hydrides[12] date back to the 1950s, and were directed towards the elucidation of important parameters such as ionization and bond energies of GeH_4, SnH_4 and PbH_4. However, a complete analysis and characterization of the mass spectra of germane, stannane and plumbane emerged from the work of Saalfeld and Svec[13,14]. For all three hydrides, spectra obtained in a sector type mass spectrometer by electron ionization at 60 eV, or higher, revealed negligible peaks corresponding to the molecular ions, $EH_4^{+\bullet}$. The observed fragmentation patterns were similar to those encountered in silane with the EH_3^+ and $EH_2^{+\bullet}$ fragments as the most abundant species in the mass spectra. The relative abundance of the fragment ions in the mass spectra obtained at 60 eV follows the order:

$$GeH_4 \qquad GeH_2^{+\bullet} \geqslant GeH_3^+ > Ge^{+\bullet} > GeH^+$$

$$SnH_4 \qquad SnH_3^+ > Sn^{+\bullet} > SnH_2^{+\bullet} \geqslant SnH^+$$

$$PbH_4 \qquad PbH_2^{+\bullet} \geqslant PbH_3^+ > Pb^{+\bullet} > PbH^+$$

While the actual relative abundances are often dependent on the type of mass spectrometer on which the spectrum is recorded because of the time window that is used to sample the ions, it is important to emphasize the significant contribution of the EH^+ fragment and of the neat metal cation in the recorded spectra. In fact, the presence of the bare metal cation and of EH^+ among the fragment ions is a common feature in the mass spectra of organogermanes and organostannanes.

The mass spectra of GeH_4 and SnH_4 obtained by field ionization[15] are simpler with the EH_3^+ fragments as the base peaks. Since this is a soft ionization method, fragmentation is considerably reduced. For example, no metal cation fragment is observed and the abundance of the EH^+ is also considerably less than that observed by electron ionization. This technique also allows for the observation of GeH_5^+ and SnH_5^+, a fact that is rare and has prevented the characterization of the proton affinity of these substrates (see below).

The determination of accurate and reliable values for the ionization energy (IE) of these simple hydrides has proved a difficult experimental task. Yields of the corresponding molecular ions $XH_4^{+\bullet}$ are very low and their ground state geometries are expected to differ considerably from the neutral species because of Jahn–Teller distortions[16,17] raising doubts about the true adiabatic ionization energies obtained from experiments. For GeH_4, the most comprehensive photoionization study[18] yields an IE $\leqslant$ 10.53 eV but the actual value may be as low as 10.44 eV. This value is well below the 11.3 eV threshold value determined from photoelectron spectroscopy[19–21]. For SnH_4, photoelectron spectroscopy[19] yields a threshold value of 10.75 eV that is probably more representative of the vertical ionization energy, and thus higher than the adiabatic IE. No values are available for PbH_4.

The mass spectra of some higher hydrides such as Ge_2H_6, Ge_3H_8 and Sn_2H_6 have also been the subject of early investigations[22–24]. Unlike GeH_4, the mass spectrum of digermane (Ge_2H_6) exhibits a reasonably intense molecular ion. For Ge_2H_6 and Sn_2H_6, the base peak corresponds to $X_2H_2^{+\bullet}$ resulting from elimination of two hydrogen molecules from the molecular ion. However, very extensive fragmentation is observed in both cases, particularly for Ge_2H_6. The relative abundance of the important fragment ions follows the order:

$$Ge_2H_6 \qquad Ge_2H_2^{+\bullet} > Ge_2H_4^{+\bullet} > Ge_2^{+\bullet} \geqslant Ge_2H^+ \gg Ge_2H_6^{+\bullet}$$

$$Sn_2H_6 \qquad Sn_2H_2^{+\bullet} > Sn_2^{+\bullet} \sim Sn_2H^+ \gg Sn_2H_3^+$$

A somewhat similar extensive fragmentation is observed for Ge_3H_8 with negligible formation of the molecular ion. The base peak in this case corresponds to Ge_3^+ (relative abundance 100) while other important fragment ions have been identified as Ge_3H^+ (90.1), $Ge_2H_2^{+\bullet}$(69.5), $Ge_2H_4^{+\bullet}$ (58) and $Ge_2^{+\bullet}$ (30.1). These results clearly reveal the tendency for the mass spectra of the higher hydrides to be dominated by ions with high metal content resulting from successive elimination of hydrogen molecules.

B. Mass Spectra of Simple Alkyl Derivatives of Ge, Sn and Pb

The most studied compounds by mass spectrometry of Ge, Sn and Pb have been the tri- and tetraalkyl derivatives. Detailed analyses of the results obtained with the simple tetraalkyl systems provide considerable insight on the general behavior observed in the mass spectra of Ge, Sn and Pb organometallics.

The initial mass spectrometric studies of Me_4Ge, Me_4Sn and Me_4Pb established the Me_3E^+ (E = Ge, Sn, Pb) ions as the most important fragment peaks[25–28]. Appearance energies and estimates for the heats of formation of the parent neutral were used to estimate the heats of formation of the different Me_nE^+ ($n = 0–3$) ions. However, all of the thermochemistry involved in these processes remains unsettled including the heat of formation of the neutrals. Nevertheless, the fragmentation patterns of the tetramethyl and tetraethyl derivatives were thoroughly studied by assignment of the metastable transitions. These metastable transitions helped to establish some general rules regarding the mass spectra of organometallic compounds of Ge, Sn and Pb[29–31].

For the EMe_4 series, the Me_3E^+ ion is the base peak followed by the MeE^+ fragment. This latter fragment becomes increasingly important from Ge to Pb, in agreement with the progressive importance of the +2 oxidation state in going from Ge to Pb. A much smaller intensity is observed for the $Me_2E^{+\bullet}$ fragment while the peak corresponding to the molecular ion is negligible. Figure 5 displays the mass spectrum of Me_4Ge obtained at 70 eV.

The metastable transitions suggest that the fragmentation pattern of the tetramethyl derivatives follows Scheme 1, where $HEMe_2{}^+$ and H_2EMe^+ are very minor fragments in the spectra.

Scheme 1 has also been confirmed by studies carried out by unimolecular and collision induced dissociation of Me_3E^+ ions[32]. In these experiments, Me_3Ge^+ was found to undergo two low-energy processes: (a) loss of methyl, and (b) loss of ethylene presumably through the mechanism shown in Scheme 2.

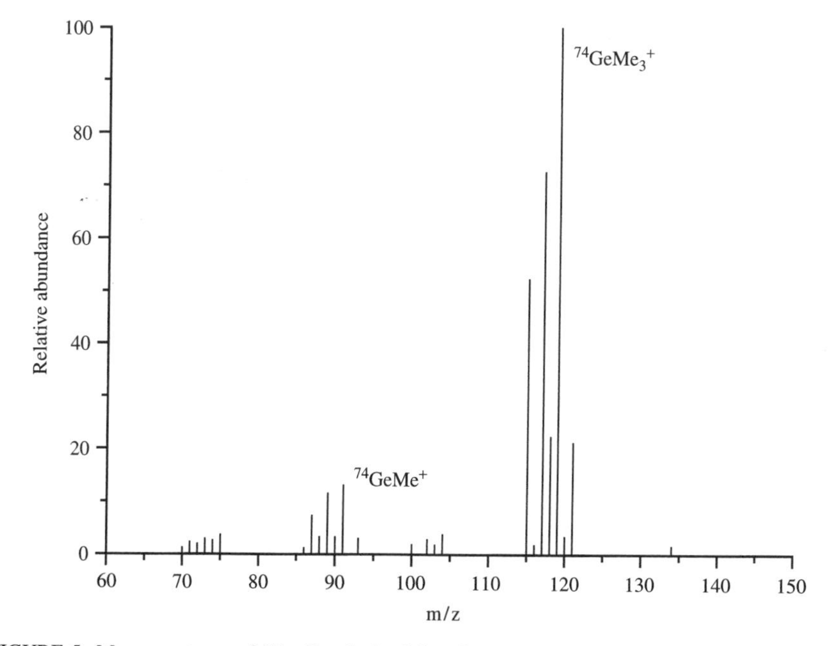

FIGURE 5. Mass spectrum of Me_4Ge obtained by electron ionization at 70 eV in the authors' homemade FTMS instrument at the University of São Paulo

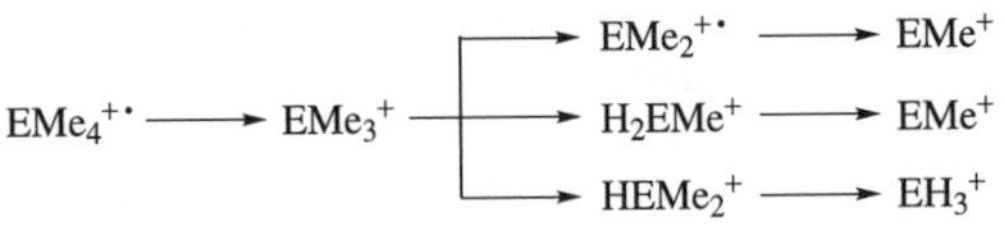

$$EMe_4^{+\bullet} \longrightarrow EMe_3^{+} \longrightarrow \begin{cases} EMe_2^{+\bullet} \longrightarrow EMe^{+} \\ H_2EMe^{+} \longrightarrow EMe^{+} \\ HEMe_2^{+} \longrightarrow EH_3^{+} \end{cases}$$

SCHEME 1

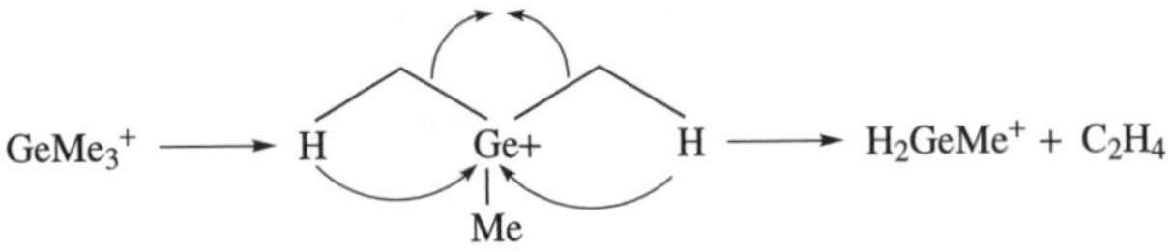

$$GeMe_3^{+} \longrightarrow H \quad Ge+ \quad H \longrightarrow H_2GeMe^{+} + C_2H_4$$

SCHEME 2

By comparison, Me_3Sn^+ and Me_3Pb^+ only undergo successive losses of methyl groups but no loss of ethylene as shown in Scheme 2. This behavior presumably reflects the decreasing carbon–metal bond strength. On the other hand, a different low energy process is observed for the Me_3E^+ ions (with E = Sn, Pb) resulting in the elimination of $:CH_2$, a rare process in mass spectra. The proposed mechanism for this process is shown in Scheme 3. Loss of $:CH_2$ is not enhanced in the collisional activation spectra. Similar conclusions were obtained from experiments designed to identify the neutral fragments generated from ionic fragmentation of the tetramethyl compounds[33]. All of them show loss of $^{\bullet}CH_3$ to be the most important neutral fragment while no C_2H_4 loss is detected for the case of Pb.

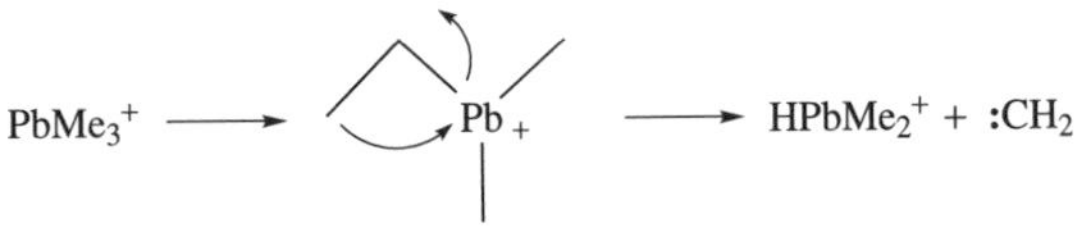

$$PbMe_3^{+} \longrightarrow Pb_{+} \longrightarrow HPbMe_2^{+} + :CH_2$$

SCHEME 3

The ionization energies for the EMe_4 series follow the expected trend, decreasing from Ge to Pb, although there is some discrepancy regarding the exact value for $PbMe_4$. Values obtained from electron impact (EI)[34] and by photoelectron spectroscopy (PES)[35,36] are compared in Table 4.

The mass spectra of the corresponding tetraethyl derivatives are characterized by a strong base peak corresponding to the Et_3E^+ fragment ion (see Figure 6) and low abundances of the molecular ion. For Ge and Sn, successive loss of neutral ethylene[37] starting with Et_3E^+ yields relatively intense peaks corresponding to Et_2EH^+ and $EtEH_2^+$. In the

TABLE 4. Trend in ionization energies (in eV) for the Me_4E (E = Ge, Sn and Pb) as obtained by electron (El) impact and by photoelectron spectroscopy (PES)

Method	GeH_4	SnH_4	PbH_4
El (Reference 34)	9.29 ± 0.14	9.33 ± 0.04	9.38 ± 0.10
PES (Reference 35)	8.76 ± 0.12	8.93 ± 0.04	8.85 ± 0.10
PES (Reference 36)	8.26 ± 0.17	8.50 ± 0.04 eV	8.83 ± 0.10 eV

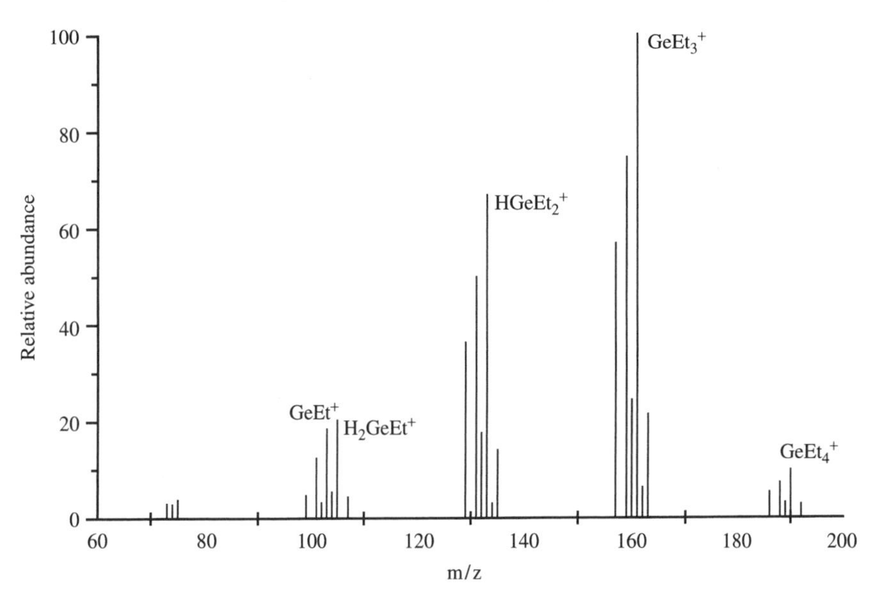

FIGURE 6. Mass spectrum of Et_4Ge obtained by electron ionization at 70 eV in the authors' home-made FTMS instrument at the University of São Paulo

case of Pb, the preferred fragmentation pathway is successive losses of Et that are rationalized in terms of the weak Pb−C bond. The identification of the metastable transitions reveals that Scheme 4 adequately represents these mechanisms. Nevertheless, it should be emphasized that there are no detailed studies on the actual structure of the different ions and alternative connectivities are always possible for gas-phase positive ions.

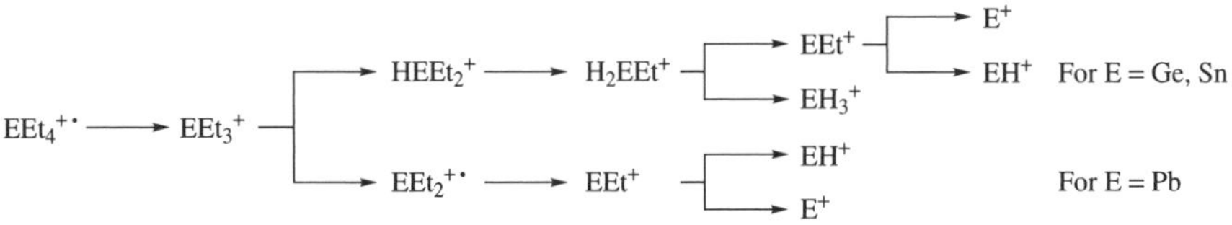

$EEt_4^{+\bullet} \rightarrow EEt_3^+ \rightarrow HEEt_2^+ \rightarrow H_2EEt^+ \rightarrow EEt^+$ ($\rightarrow E^+$, $\rightarrow EH^+$), EH_3^+ For E = Ge, Sn

$EEt_3^+ \rightarrow EEt_2^{+\bullet} \rightarrow EEt^+ \rightarrow EH^+$, E^+ For E = Pb

SCHEME 4

Higher symmetrical tetraalkyl derivatives of Ge and Sn have also been extensively studied by mass spectrometry[30,38–42]. For saturated alkyl groups, some general rules can be established:

(a) The molecular ion peak is generally negligible under electron ionization.

(b) Ions containing only carbon and hydrogen, or ions resulting from a carbon−carbon bond fission, are rare but their abundance tends to increase with larger alkyl groups.

(c) The abundance of the R_3E^+ fragments decreases in going from R = Pr to R = Hex, but subsequent fragmentation by loss of a neutral olefin to yield abundant HER_2^+ and H_2ER^+ is essentially the only pathway as confirmed by the study of metastable transitions. While no studies have been carried out with deuteriated derivatives, it is assumed that the

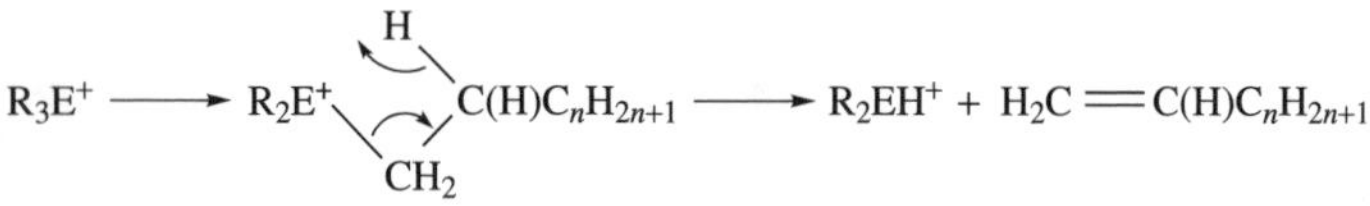

$$R_3E^+ \longrightarrow R_2E^+ \cdots \underset{CH_2}{\overset{H}{}} C(H)C_nH_{2n+1} \longrightarrow R_2EH^+ + H_2C{=}C(H)C_nH_{2n+1}$$

SCHEME 5

mechanism for olefin elimination for alkyl groups with β hydrogens proceeds as shown in Scheme 5.

Unlike the spectra obtained by conventional electron ionization, the mass spectra of a number of tetraalkyl and trialkyl derivatives of Sn under field ionization conditions reveal almost exclusively formation of the molecular ions, i.e. $ER_4^{+\bullet}$, and fragmentation by loss of a single alkyl group[43].

The mass spectra of Ph_4Ge, Ph_4Sn and Ph_4Pb are of considerable interest as prototype systems for the widely used triphenyl derivatives (particularly those of Sn)[10,39,44,45]. As in the case of simple alkyl groups, the most important fragment ion is Ph_3E^+. For Ge, the relative abundance of the other fragment ions follows the order $Ph_2E^{+\bullet} > PhE^+$, but this trend is completely reversed for Sn and Pb ($Ph_2E^{+\bullet}$ is negligible for the Pb compound). The detailed mechanism for the fragmentation processes in Ph_4Ge and Ph_4Sn has been characterized recently by tandem mass spectrometry[46–48]. In the meantime, the study of metastable transitions has established the importance of processes involving the elimination of biphenyl (presumably) as shown in equations 1 and 2.

$$Ph_4Sn^{+\bullet} \longrightarrow Ph_2Sn^{+\bullet} + Ph_2 \quad (1)$$

$$Ph_2Sn^{+\bullet} \longrightarrow Sn^{+\bullet} + Ph_2 \quad (2)$$

Interestingly enough, both the Sn and Pb derivatives reveal very strong bare metal ion peaks. Other less important metastable transitions have also been well characterized in Reference 45.

The relative abundance of the Ge and Sn ions originating from the primary loss of $C_{12}H_{10}$ (equation 1), measured in the source and in the first and second field-free regions of a multi-sector mass spectrometer, suggests that equation 1 is a fast process. This observation implies that the elimination of biphenyl is not a concerted process but that it is likely to be mediated by the type of structure shown in Scheme 6. On the other hand, the kinetic energy release during the loss of the second phenyl group suggests a second mechanism by which $Ph_2Ge^{+\bullet}$ ions are also formed through a sequential process.

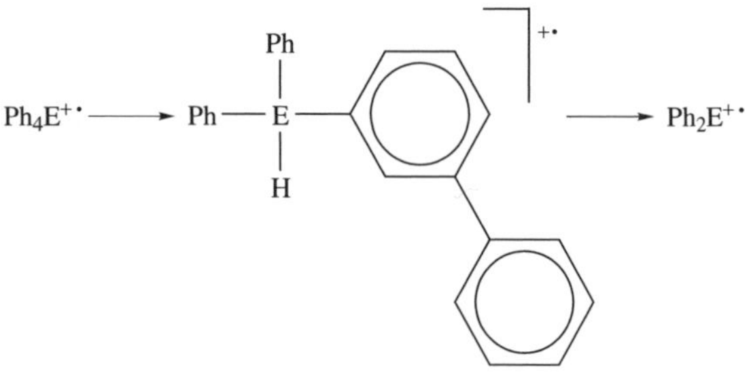

$Ph_4E^{+\bullet}$ → [Ph–E(Ph)(H)–C$_6$H$_4$–C$_6$H$_5$]$^{+\bullet}$ → $Ph_2E^{+\bullet}$

SCHEME 6

The FAB spectra of the Ph_4E compounds show no molecular ions, or molecular ions coordinated to a molecule of the liquid matrix, and do not differ considerably from EI spectra except for the appearance of the Ph_3E^+ and PhE^+ ions coordinated to a molecule of the matrix[49,50].

The effect of substituents on the mass spectral disintegration pattern of a large number of Ar_4Sn derivatives has been investigated both by EI and by FAB[51,52]. In all cases, the Ar_3Sn^+, $Ar_2Sn^{+\bullet}$ and $ArSn^+$ species dominate the EI mass spectra with no molecular ion peaks. By comparison, the FAB mass spectra display the Ar_3Sn^+ and $ArSn^+$ species, and ions coordinated to one molecule of the liquid matrix (Mat) in the form of $Ar_3Sn^+(Mat)$. Collisional activation shows the processes given by equations 3 and 4 to be very characteristic of the FAB-MS of these compounds.

$$Ar_3Sn(Mat)^+ \longrightarrow ArSn(Mat)^+ + Ar_2 \quad (3)$$

$$ArSn(Mat)^+ \longrightarrow Sn(Mat)^{+\bullet} + Ar^\bullet \quad (4)$$

The EI spectra of the Ar_4Sn compounds reveal some further interesting features[51–54]. For *m*- and *p*-substituted halophenyl (F, Cl) aromatics, there are noticeable peaks corresponding to $Sn(halogen)^+$ that have been identified as originating from a halide migration in the fragmentation process as shown in equation 5.

$$(p\text{- or } m\text{-}FC_6H_4)Sn^+ \longrightarrow SnF^+ + C_6H_4 \quad (5)$$

The appearance of SnF^+ ions is also observed in the mass spectra of (*p*- or *m*-$CF_3C_6H_4)_4Sn$ and it becomes the major ion in the spectra of the perfluorinated $(C_6F_5)_4Sn$ species. This fluorine migration has been rationalized in terms of hard/soft acid base theory (HSAB) by considering Sn^{2+} as a hard acid that favors fluoride transfer[55].

Another interesting rearrangement observed in the spectra of the Ar_4Sn species occurs in the case of Ar = *o*-Tol for which a prominent peak corresponding to the $ArSnC_7H_6^+$ ion has been identified. This fragment ion is not apparent in the spectra of the *m*- or *p*-tolyl system. This position effect has been claimed to be a typical example of the '*ortho* effect' that is commonly encountered in organic mass spectrometry[56].

C. Mass Spectra of Mixed Alkyl and Aryl Derivatives of Ge and Sn

A wide variety of mass spectrometric studies has been carried out for mixed hydrides, and mixed alkyl and aryl derivatives[30,38,41,44,45,57–64]. While many of the main features regarding E–C or E–H bond cleavages, and olefin elimination, are similar to those discussed in the previous section, there are some specific aspects that need to be considered explicitly.

The mass spectrum[30] of Me_3GeH yields as a base peak the Me_2GeH^+ ion (relative abundance 100) reflecting the fact that the Ge–H bond dissociation energy is larger than that of the corresponding Ge–C bond energy. However, important fragments are also observed (in decreasing order of abundance): $MeGe^+$ (35), $Me_2G^{+\bullet}$ (28) and Me_3Ge^+ (22). For the corresponding Et_3GeH, the base peak is the $EtGeH_2^+$ ion resulting from elimination of ethylene from the second most important fragment Et_2GeH^+. A detailed analysis of the mass spectra of a number of R_3EH compounds (R = Me, Et, Pr, Bu and E = Ge, Sn) shows that the observed fragmentation patterns are consistent with the predictions of the quasi-equilibrium theory (QET) of mass spectra[65]. Similar conclusions were obtained in a study involving Me_2GeH_2 and Me_2SnH_2 including the combined use of photoelectron spectroscopy and mass spectrometry[66].

The presence of phenyl groups changes the situation considerably. For Ph_3GeH, elimination of benzene from the molecular ion results in the base peak $Ph_2Ge^{+\bullet}$. The fragment $PhGe^+$ is also important and its formation is attributed to elimination of benzene from Ph_2GeH^+, and to elimination of biphenyl from Ph_3Ge^+. For Ph_3GeR (R = Me, Et, CH_2Ph), the Ph_3Ge^+ fragment is the predominant peak in the spectra with negligible amounts of the mixed fragment Ph_2GeR^+. For these cases, mass spectrometry is not an effective means to distinguish among different triphenyl compounds as all the compounds yield low abundances of fragment ions containing the distinct R group.

For mixed alkyl stannanes[41] like $RSnMe_3$ and $RSnEt_3$ (with R = Me, Et, Pr, *i*-Pr, Bu, *t*-Bu, *c*-Hex, *p*-$MeC_6H_4CH_2$, *m*-$MeC_6H_4CH_2$ and *o*-$MeC_6H_4CH_2$) the most notorious fragment ions are Me_3Sn^+ and Et_3Sn^+, respectively, although several other fragment ions can also be observed as a result of different bond scissions and elimination of olefins. In all cases, the important fragment ions are those corresponding to Sn(IV) or Sn(II) species that are reached by loss of an alkyl radical from the next higher unfavorable valence state, or by loss of an alkane or alkene molecule from a favorable valence state. A somewhat more complex pattern is observed in the mass spectra[61] of *s*-$BuRMe_2Sn$ compounds for which several fragment ions of the IV oxidation state of Sn are possible.

A very extensive set of data is available for a variety of (*p*-YC_6H_4)EMe_3 compounds (Y = NO_2, CF_3, Br, F, Me, OMe, OH, $OSiMe_3$) for E = Ge, Sn. For this series[67], the $(M\text{-}15)^+$ ion is consistently the base peak except for the case of Y = F, where the EMe^+ becomes the most abundant species. As in many previous examples, the molecular ion appears with negligible abundance or is simply not detected. An interesting observation has been the positional dependence of the fragmentation processes for $(MeOC_6H_4)SnMe_3$. For the three possible isomers, the $(M\text{-}Me)^+$ is the base peak but the *ortho* isomer reveals two important differences with respect to the *meta*- and *para*-isomers[68]: (1) the spectrum of the *ortho* isomer exhibits an $(M\text{-}C_2H_6)^{+\bullet}$ fragment with significant abundance that has been assumed to proceed through the mechanism shown Scheme 7. This mechanism is consistent with the results obtained by deuterium substitution. (2) $PhSn^+$ originating by

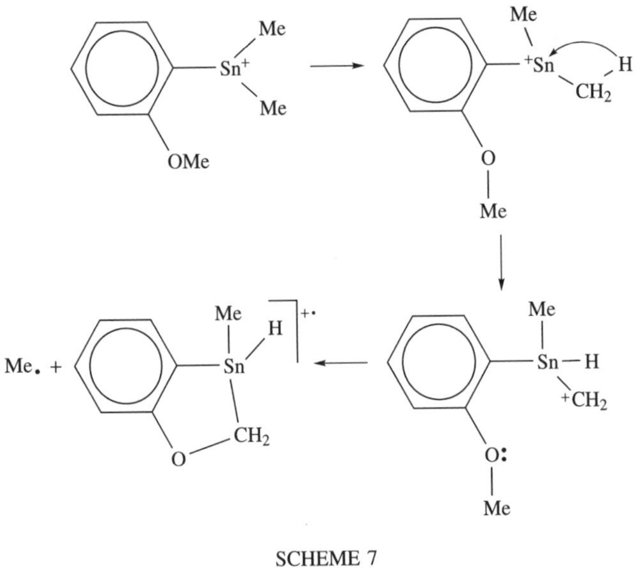

Me Sn+ Me OMe → Me +Sn H CH2 O Me ↓ Me Sn—H +CH2 O: Me ← Me• + Me H Sn CH2 O +•

SCHEME 7

loss of CH_2O (presumably as shown in Scheme 8) is two times more abundant for the *ortho* isomer than for the other two isomers.

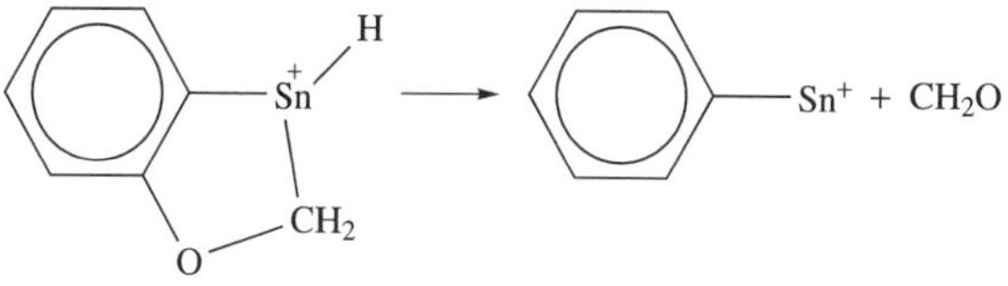

SCHEME 8

For the case of aryl-substituted stannanes[45,69], the most important fragment ions are those with the highest number of phenyl groups attached to Sn. In fact, cleavage of the alkyl-tin bond in the molecular ion is always the preferred fragmentation route for $RSnPh_3$ compounds rather than cleavage of the phenyl-tin bond[70]. As in the case of the equivalent germanes, mass spectrometry is not well suited for analytical applications for these compounds.

Closely associated sets of compounds that have also been investigated by mass spectrometry are species like R_nEX_{4-n} (E = Ge, Sn and X = halogen). The early investigations on organogermanes[30] reveal that for Me_3GeCl, Et_3GeCl and Et_3GeBr, the most abundant ions result from loss of an alkyl radical ($R^\bullet$) by the molecular ion to yield R_2EX^+. Multiple chlorine substitution on simple organogermanes along the series Me_3GeCl, Me_2GeCl_2 and $MeGeCl_3$ has been shown to favor loss of a $Me^\bullet$ for the first two compounds to yield $(M\text{-}Me)^+$ as the most abundant ion. In the meantime, loss of $Cl^\bullet$ is responsible for the base peak[71] in the mass spectrum of $MeGeCl_3$. This trend becomes somewhat different for Ph_3GeX derivatives[30,49] reflecting the differences between Ge−X bond strengths and the stability of the Ph_3Ge^+ ion. The resulting spectra show that the relative abundance of the first cleavage of the molecular ions follows the order:

$$Ph_2GeCl^+ > Ph_3Ge^+ \text{ (for } Ph_3GeCl\text{)}$$

$$Ph_3Ge^+ > Ph_2GeBr^+ \text{ (for } Ph_3GeBr\text{)}$$

$$Ph_3Ge^+ \gg Ph_2GeI^+ \text{ (for } Ph_3GeI\text{)}$$

However, the base peak for all three compounds results from the fragmentation shown in equation 6.

$$Ph_3GeX^{+\bullet} \longrightarrow Ph_2^{+\bullet} + PhGeX \qquad (6)$$

This is one of the rare cases where a non-containing Ge ion is the base peak. The effect of multiple chlorine substitution on the mass spectra of the phenyl derivatives[46] is shown in Table 5.

TABLE 5. Relative abundance of the most important ions in the mass spectra for $Ph_{3-n}GeCl_n$[a]

Compound	Relative abundance
Ph_3GeCl	$Ph_2^{+\bullet} > Ph^+ > Ph_3Ge^+ \sim Ph_2GeCl^+$
Ph_2GeCl_2	$Ph_2^{+\bullet} > Ph_2GeCl^+ > Ph^+$
$PhGeCl_3$	$PhCl^{+\bullet} > Ph^+ \sim PhGeCl_2^+ > PhGeCl_3^{+\bullet}$

[a]The full spectra with the appropriate relative intensities have been reported in Reference 46.

A somewhat similar situation is observed for the tin derivatives[61,72,73]. For a large number of R_3SnX (R = Me, Et, Pr, Bu, *i*-Bu and X = F, Cl, Br, I) the base peak of the spectra corresponds to the R_2SnX^+ ion except for the case of Pr_3SnI, where the base peak is R_2SnH^+. Loss of the alkyl group has also been demonstrated in the mass spectra of (1-, 2- and 3- butenyl)$_3$GeBr, where R_2SnBr^+ and $SnBr^+$ are the major fragment ions[74]. For the Ph_3SnX series[45,49], the base peak corresponds to the Ph_2SnX^+ ion for C = F, Cl and Br, and to Ph_3Sn^+ for X = I without any appreciable formation of $Ph_2^{+\bullet}$. The corresponding Pb compounds display a similar trend[49].

Finally, the mass spectra of a series of PhMeSnRR′ compounds[75] have been compared for R = Me, Et, $PhCH_2$ and R′ = $PhCH_2$. Loss of $PhCH_2^{\bullet}$ is found to give rise to the base peak ion, $MePhSnR^+$, for all cases.

D. Further Examples of Mass Spectra of Ge, Sn and Pb Compounds

The previous sections covered a systematic approach of simple families of Ge, Sn and Pb compounds that have been among the best characterized over the years. However, there are several isolated examples in the literature that describe in detail important characteristics of the mass spectrometry of relevant Ge, Sn and Pb compounds.

Germacyclopentanes, **1**, and germacyclopentenes, **2**, are good examples of Ge-containing cyclic compounds. Their mass spectra[76] exhibit a noticeable molecular ion but the individual fragmentations depend on the substituents. For **1**, the base peak is $Ge^{+\bullet}$ for R = simple alkyl, and there is a general propensity towards the loss of the groups attached to the Ge. The relative abundance of $Ge^{+\bullet}$ is greatly reduced for R = Ph, and $PhGe^+$ becomes the base peak. Another characteristic feature of these spectra is the elimination of ethylene (as shown in Scheme 9), that has been identified as occurring through expulsion of a moiety containing C2 and C3. Germacyclopentenes, **2**, do not display an ion corresponding to ethylene loss and the base peak corresponds to $RR'Ge^+$. The mechanism for this process is shown in Scheme 10.

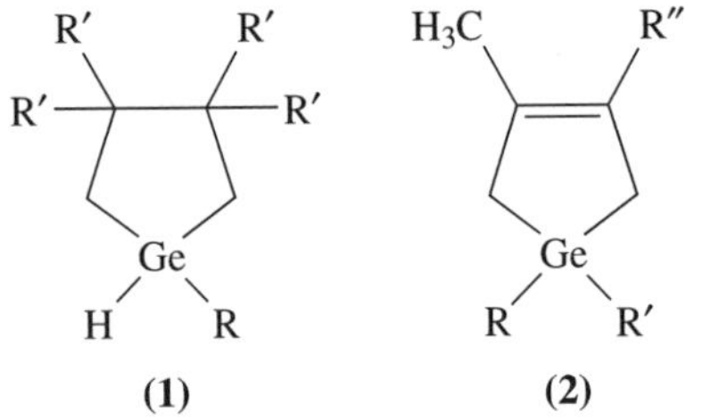

R′ R′ R′ R′ Ge H R
(1)

H_3C R″ Ge R R′
(2)

D D D D Ge H n-C_4H_9 → D D Ge H n-C_4H_9 + CH_2CD_2

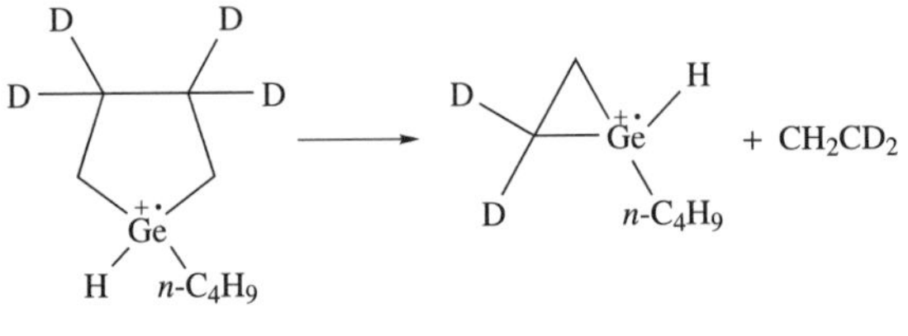

SCHEME 9

Other germacyclopentenes and germacyclopentanols have also been characterized by mass spectrometry with particular emphasis on the elimination processes and migrations

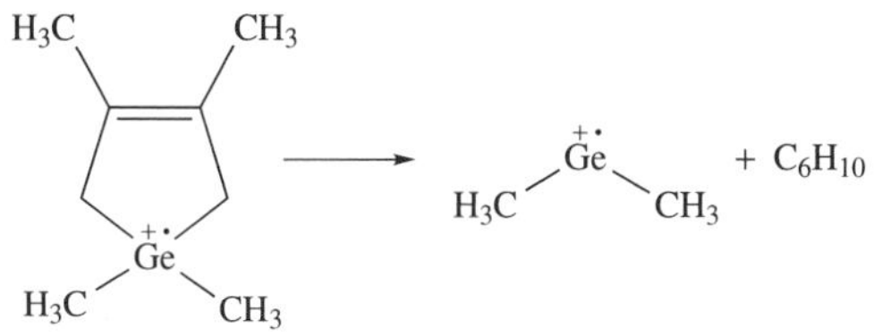

SCHEME 10

leading to the major fragment ions[77,78] Another example of a cyclic compound is found in the reported mass spectrum[79] of 5-plumbaspiro[4,4]-nonane (**3**) for which the base peak corresponds to Pb^+.

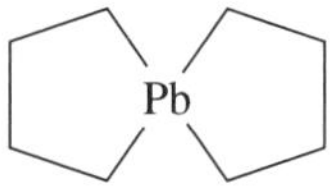

(**3**)

A somewhat different behavior is observed in the mass spectrum of the alkoxygermanes. For example, the mass spectrum of $Ge(OMe)_4$ reveals two important cleavage pathways[80,81] (Scheme 11): (a) loss of formaldehyde to yield the molecular ion of simpler alkoxygermanes and (b) Ge–O cleavage.

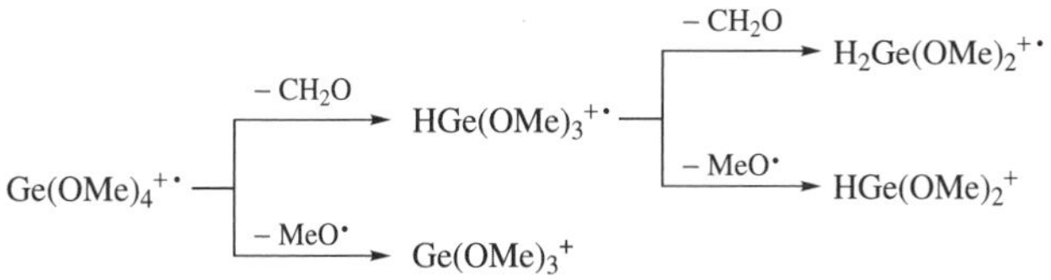

SCHEME 11

Skeletal rearrangements are less common in the mass spectra of Ge, Sn and Pb compounds but some good examples have been illustrated for the case of ketoorganotins[82] and hydroxyorganotins[83]. For example, the mass spectra of a number of $RCO(CH_2)_nSnMe_3$ species reveal that for $n = 3$, very strong peaks are observed that correspond to loss of ethylene. Scheme 12 illustrates the proposed rearrangement that accounts for the formation of the $RCO(CH_2)SnMe_3^+$ ion. A very similar mechanism has been invoked for the observation of very abundant fragment ions, Me_2SnOH^+, in the mass spectra of the stannyl alcohols $Me_2Sn(CH_2)_nOH$.

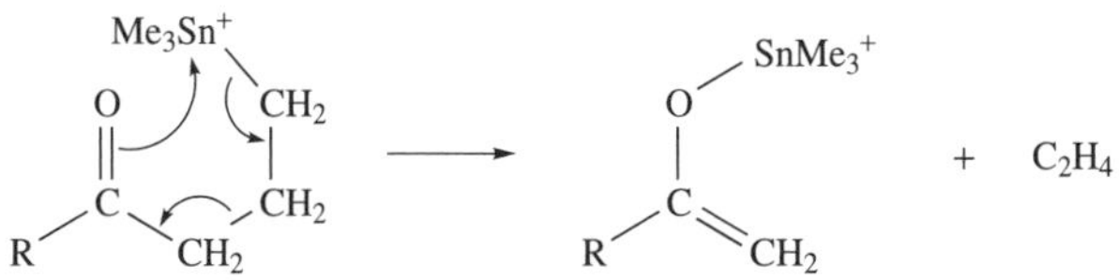

SCHEME 12

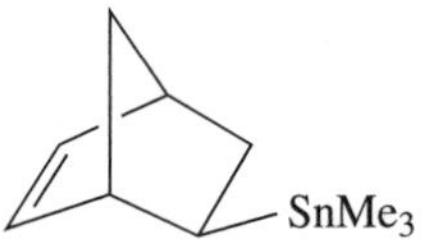

$SnMe_3$ $SnMe_3$

(4) 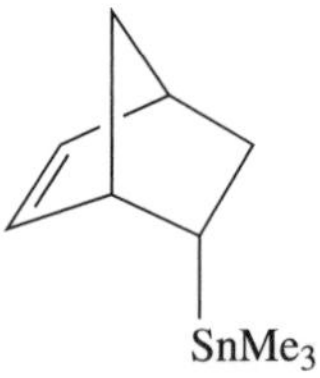 **(5)**

Me_3Sn $SnMe_3$

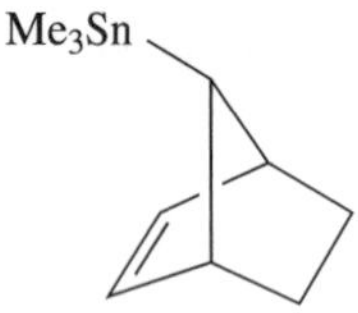

(6) 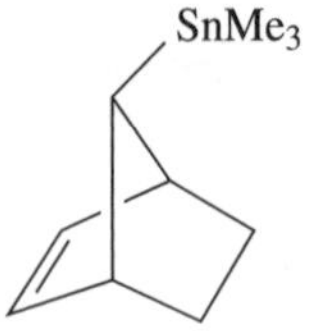 **(7)**

The effect of a norbornene group on the fragmentation pattern of organotin compounds has been reported for a number of trimethylstannylnorbornene isomers[84]. Comparison of the spectra of the *exo*- and *endo*-5-trimethylstannylnorborn-2-enes, **4** and **5,** reveal similar fragmentation but some differences in the relative abundances of ions. On the other hand, the mass spectra of the *syn*- (**6**) and *anti*-7-trimethylstannylnorborn-2-ene (**7**) exhibit major differences. At 70 eV, the base peak of **7** is a non-metal-containing ion, namely $C_7H_9{}^+$, a rare situation for these organometallic compounds.

Mass spectrometry has also been used to characterize organostannoxanes of the type $Me_2Sn(Cl)OC(O)CF_3$ and their products of hydrolysis[85]. The base peak in all cases has been tentatively assigned to the CF_3OSn^+ ion but the mechanism leading to this rearrangement has not been explored in detail.

Additional examples of the application of mass spectrometry have been described for less common compounds such as furyl and thienyl germanes[86], germatranes[87], acetylene derivatives of Ge[88] and alkyl and aryltin oxinates[89,90].

E. Mass Spectra of Compounds Containing Metal–Metal Bonds or More Than One Element of Group 14

Glocklin and coworkers[91], starting with Me_6Ge_2, initially reported the mass spectra of several alkylpolygermanes. The base peak for this compound corresponds to Me_3Ge^+ with a strong fragment ion corresponding to $Me_5Ge_2{}^+$ resulting from a Ge–Me bond cleavage. Similar results were obtained for the other hexa-substituted dimetals[92,93]. Me_3Ge^+ is also the base peak for Me_8Ge_3, but in the case of Et_8Ge_3 the mass spectrum yields trigermanium ions in high abundance due to successive eliminations of ethylene. The primary process for higher oligomers of Ge has been illustrated to be cleavage of a germanium–alkyl bond. By analogy, early studies on simple polytin compounds showed a similar behavior[94]. For example, trimethylstannyl methanes, $(Me_3Sn)_nCH_{4-n}$ ($n = 1$–4), exhibit a base peak corresponding to loss of a methyl group by cleavage of a Sn–C bond. The fragments Me_3Sn^+ and $MeSn^+$ are also present in large abundances for $n = 2$ and 3. However, without additional mass spectrometric information such as collisional activation spectra (CAD), it is difficult to establish clearly the fragmentation pathways due to the intricate isotopic pattern of these spectra.

A rather surprising finding is the fact that Ge−C and Sn−C bond cleavage *are* the major fragmentation pathways over metal−metal bond cleavage in compounds such as $Bu_3GeGeBu_3$ and $Bu_3SnSnBu_3$, resulting in $Bu_3EEBu_2{}^+$ as the base peak. This behavior, unlike what is observed in Si, parallels the trend observed in photoinduced electron transfer reactions[95]. This argument also follows the proposed similarity between electron transfer reactions and fragmentation pathways in the mass spectrum of organometallic compounds[96].

An interesting case is the cyclic phenyl derivatives $(Ph_2E)_n$ ($n = 4$–6) of Ge and Sn for which a small abundance of the molecular ions was reported followed by a series of fragmentations and rearrangements leading to Ph_3Ge^+ and $PhGe^+$, or Ph_3Sn^+ and $PhSn^+$, as the most abundant fragment ions[97]. However, there is considerable doubt about the actual mechanisms for the proposed processes as poor thermal stability of these compounds may actually be responsible for some of the experimental observations[98]. Furthermore, analysis of the metastable transitions in this case is complicated by the isotopic complexity of the spectra.

The tendency for the R_3E^+ ions to be the base peak in the mass spectra of compounds containing more than one Sn atom is well illustrated for the case of sulfides[99] such as $(Me_3Sn)_2S$ and $(Ph_3Sn)_2S$. For cyclic polysulfides, *cyc*-$(Me_2SnS)_3$, *cyc*-$(Bu_2SnS)_3$ and *cyc*-$(Ph_2SnS)_3$, the mass spectra show significant abundance of $R_3Sn_2S_2{}^+$ (the base peak for R = Bu), R_3SnS^+ (the base peak for R = Me) and R_2SnS^+ (the base peak for R = Ph) besides the R_3Sn^+ fragment.

Compounds containing a metal−metal bond between different elements of group 14 represent an interesting case for mass spectrometry. For example, the mass spectrum[24] of H_3GeSiH_3 shows $GeSiH_4{}^+$ and $GeSiH_2{}^+$ as the most abundant peaks while the abundance of $GeH_3{}^+$ is higher than that of $SiH_3{}^+$, as might be predicted by the expected trend in ionization energies of the corresponding radicals. However, the most extensive study was carried out for a series of $R_3EE'R'_3$ compounds involving Ge and Si, and Sn and Ge[100]. In general, bond cleavage of the molecular ion proceeds primarily in the direction of the weakest bond of the molecule but product ions originating from cleavage of all four main bonds are encountered in the mass spectrum. While the resulting spectra are quite complex, identification of the metastable transitions has made it possible to identify different processes. These can be exemplified for the case of $Ph_3GeSnMe_3$ with their respective relative abundances:

(a) bond cleavage from the molecular ion giving rise to $Ph_3GeSnMe_2{}^+$ (58), Ph_3Ge^+ (100);

(b) molecule elimination (i.e. biphenyl) by cleavage of two bonds as in $Ph_3Ge^+ \longrightarrow PhGe^+$ (42);

(c) rearrangement of ions by group transfer to yield ions such as Ph_2GeMe^+ (48) and $PhSn^+$ (45).

While these results show great preference for the formation of Ge-containing ions, the stabilizing effect of the phenyl groups plays an enormous role. By comparison, the most abundant ions in the mass spectrum of the corresponding $Ph_3SnGeMe_3$ follow the rules above with the difference that the resulting fragments are Sn-containing ions.

An even wider set of $R_3EE'R'_3$ compounds [E = Si, E′ = Ge; and R or R′ = Ph, Me, $(\eta^5\text{-}C_5H_4)Fe(C_5H_5)$, $(\eta^5\text{-}C_5H_4)Fe(CO)_2$ and $(\eta^5\text{-}C_9H_7)Fe(CO)_2$] has been recently investigated by mass spectrometry[101]. Several general conclusions stem from this study: (a) significant ligand exchange occurs in the ions for non-metal-substituted Si−Ge isomers as shown above; (b) Fe−Si−Ge complexes undergo preferential Si−Ge cleavage while the Fe−Ge−Si undergo preferential Fe−Ge bond cleavage; (c) in the case of the symmetrical complex $(\eta^5\text{-}C_5H_4)Fe(C_5H_5)Si(Me)_2Ge(Me)_2(\eta^5\text{-}C_5H_4)Fe(C_5H_5)$, Si−Ge bond cleavage is the major fragmentation, and $(\eta^5\text{-}C_5H_4)Fe(C_5H_5)Si(Me)_2{}^+$ is observed as the most abundant ion.

The mass spectrum of $PhSn(Me)_2Ge(Ph)_2CH_2CH_2CH_2COCH_3$ is one example where cleavage of the metal–metal bond is the major fragmentation pathway resulting in $(Ph)_2Ge(CH_2CH_2CH_2COCH_3)^+$ (assumed structure) as the base peak[102]. This is also the case when the keto group is replaced by dioxolane. However, in both cases Ge- and Sn-containing ions such as $PhSn^+$ and $PhGe^+$ are important fragments.

It is therefore possible to conclude that in the case of compounds containing metal–metal bonds of elements of group 14, cleavage of this bond is an important process in mass spectrometry but not necessarily the preferred fragmentation pathway. Furthermore, the most abundant fragment ions from the metal–metal bond cleavage are ions containing the lighter element of the group.

III. GAS-PHASE ION CHEMISTRY

The fundamental aspects related to the thermochemistry, structure and reactivity of gas-phase ions are usually considered the domain of gas-phase ion chemistry. By extension, some of these same properties are often obtained for simple neutrals and radicals from methods used in gas-phase ion chemistry. A wide range of experimental techniques can be used for this purpose, and instrumental developments have contributed a great deal to our knowledge of gas-phase ions. Theoretical calculations have also played an important role and gas-phase ion chemistry has witnessed a very lively interplay between experiment and theory in recent years.

This section discusses these different aspects of positive and negative Ge, Sn and Pb ions. Negative ions have not been discussed in the previous section as they have rarely been used to identify Ge, Sn and Pb compounds[103], even though negative ion chemical ionization based on Cl^- attachment can often be a useful technique for compounds that display low abundance of the molecular ion in conventional mass spectrometry[104].

A. Thermochemistry, Structure and Reactivity Related to the Gas-phase Positive Ion Chemistry of Ge, Sn and Pb Compounds

Photoelectron spectroscopy has been extensively used for Ge, Sn and Pb compounds to understand the bonding of these systems and to determine ionization energies[19–21,105–119]. Mass spectrometry has also been used in this respect although in general with a lower energy resolution. The results are reviewed in Reference 5 and in many cases the main uncertainty in the reported ionization energies is related to the ability in determining the true adiabatic IE.

Trends in ionization energies have been illustrated previously but other examples are of particular interest. The halides of Ge, Sn and Pb can be singled out because they have been extensively characterized for both the (IV) and (II) oxidation states. As in the case of C and Si, the tetrafluoro derivatives display very high ionization energies, and the observed trends in Ge demonstrate the decrease in IE in going from F to I, and the decrease in going from the (IV) to the (II) oxidation state, i.e.

GeF_4 (15.69 eV)[105,120] > $GeCl_4$ (11.88 eV)[105] > $GeBr_4$ (10.62 eV)[121] > GeI_4 (9.84 eV)[122]

GeF_2 (11.65 eV)[123] > $GeCl_2$ (10.55 eV)[123] > $GeBr_2$ (10.02 eV)[123] > GeI_2 (9.08 eV)[123]

By comparison, determination of the ionization energies and thermodynamic data for the Sn and Pb dihalides represent classical benchmarks in high temperature photoelectron spectroscopy and mass spectrometry[124–127]. The combined results for the Pb series shows a similar trend to that shown for the IE of the GeX_2 compounds.

$$PbF_2\ (11.5\ eV) > PbCl_2\ (10.2\ eV) > PbBr_2\ (9.6\ eV) > PbI_2\ (8.86\ eV)$$

Thermochemistry for some Ge-, Sn- and Pb-fragment ions has been obtained from appearance energies in mass spectrometry. This is a traditional method that can be used for determining bond energies, the ionization energy of simple radicals and heats of formation of ions. A typical example of this approach for characterizing neutral and ionic Me_3Sn and Sn−X bond energies was used in an extensive study by Yergey and Lampe[58]. However, values obtained by this method rely on knowledge of the thermochemistry of the stable neutrals[34,128]. The experimental difficulties in determining threshold appearance energies in mass spectrometry are well known and most of the values derived from the early literature must be viewed with great caution. This is particularly true because of the revised values for the neutral thermochemistry of organometallics[5].

The question of bond dissociation energies for the elements considered here has also been an area of lively discussion. The accurate determination of bond dissociation energies from mass spectrometric, photoionization or gas kinetics experiments is often subject to considerable uncertainty. The most recent critical review dealing with experimental methods recommends a value of 343 ± 8 KJ mol^{-1} for the $H_3Ge{-}H$ bond energy[129] derived from the photoionization of monogermane[18]. Other bond energies for these elements are less certain and values obtained by high-level theoretical calculations[130] are probably very good estimates for bond energies of Ge, Sn and Pb.

Considerable insight about fundamental aspects of the behavior of simple Ge, Sn and Pb species can be obtained from studies aimed at characterizing the reactivity of gas-phase ions. Reactions of the primary ions obtained by electron ionization of GeH_4 with the neutral monogermane precursor have been characterized both by low- and high-pressure mass spectrometric techniques[131,132], and more recently by ion trap techniques (ITMS)[133]. The ability to select a particular isotopic species (usually the ^{70}Ge-containing ion) in low pressure studies carried out by Fourier Transform Mass Spectrometry (FTMS) has been essential in understanding the mechanism of these processes. The main results can be summarized as follows:

(i) The primary ions GeH^+ and $GeH_2{}^+$ react readily with GeH_4 to yield $GeH_3{}^+$ with the resulting ion retaining the original isotope (equations 7 and 8),

$$^{70}GeH^+ + GeH_4 \longrightarrow {}^{70}GeH_3{}^+ + GeH_2 \qquad (7)$$

$$^{70}GeH_2{}^+ + GeH_4 \longrightarrow {}^{70}GeH_3{}^+ + GeH_3 \qquad (8)$$

(ii) $^{70}Ge^+$ ions in turn react with GeH_4 by hydrogen atom abstraction to yield $^{70}GeH^+$, a reaction similar to that shown in equation 8.

(iii) GeH_3^+, both a primary and secondary ion in these processes, reacts with GeH_4 by hydride abstraction as verified by the isotope scrambling shown in equation 9,

$$^{70}GeH_3{}^+ + GeH_4 \longrightarrow GeH_3{}^+ + {}^{70}GeH_4 \qquad (9)$$

(iv) $GeH_2{}^+$ and Ge^+ also undergo slow condensation reactions followed by elimination of molecular hydrogen as illustrated in equations 10 and 11,

$$GeH_2{}^+ + GeH_4 \longrightarrow Ge_2H_4{}^+ + H_2 \qquad (10a)$$

$$\longrightarrow Ge_2H_2^+ + 2H_2 \quad (10b)$$

$$Ge^+ + GeH_4 \longrightarrow Ge_2H_2^+ + H_2 \quad (11)$$

(v) By comparison, GeH_3^+ undergoes very slow condensation-type reactions to yield $Ge_2H_5^+$ and $Ge_2H_3^+$. These reactions are not observed in the time regime of the ITMS experiments[133].

(vi) Protonated germane, GeH_5^+, is not observed as a product ion in these reactions, and this fact is further discussed below.

(v) Digermanium species, $Ge_2H_n^+$ ($n = 2–7$), become more important ionic products in experiments carried out by high-pressure mass spectrometry (*ca* 0.1 Torr) where termolecular processes become favorable.

One of the most important thermochemical parameters in ion chemistry is the proton affinity (PA) (equation 12), and considerable experimental and theoretical work has been carried out in the last 30 years to determine this property accurately.

$$M(g) + H^+(g) \longrightarrow MH^+(g) \quad PA(M) = -\Delta H^\circ \quad (12)$$

Yet, the experimental determination of the proton affinity of GeH_4 has been a considerable problem due to the difficulty in observing the GeH_5^+ ion. An early ion beam experiment[134] concluded that the proton affinity of germane was higher than that of acetylene, C_2H_2, and that proton transfer from H_3S^+ to GeH_4 was endothermic. In the most recent update of gas-phase proton affinities[135], the recommended value for the PA of germane at 298 K amounts to 713.4 kJ mol^{-1}. However, this value is probably too high since recent high level calculations[136] placed the proton affinity of GeH_4 at 0 K at 654.4 kJ mol^{-1} while at 298 K a value 673 kJ mol^{-1} is predicted by G2 calculations[137].

The difficulty encountered in observing GeH_5^+ in the gas phase accounts for the uncertainty in the experimental value for the proton affinity of GeH_4. *Ab initio* calculations[136–138] reveal that the structure of the GeH_5^+ ion corresponds to that of a germyl cation weakly attached to molecular hydrogen, $GeH_3^+(H_2)$. The binding energy of such a moiety has been estimated[138,139] to be in the range of 30 to 35 kJ mol^{-1}, a value that is indicative of the poor stability and ease of dissociation of the GeH_5^+ ion at room temperature.

A considerably more complex reactivity pattern has been observed for the ions generated from CH_3GeH_3 reacting with the parent neutral[140,141]. The minor fragments GeH_m^+ ($m = 0, 2, 3$) react rapidly with CH_3GeH_3 to yield ions CH_3Ge^+ (of unknown atom connectivity) and $CH_3GeH_2^+$ (assumed structure). The CH_3Ge^+ and $CH_3GeH_2^+$ ions, obtained both as main fragments by electron ionization and by ion/molecule reactions, undergo a variety of condensation reactions that yield ions with the generic composition $Ge_2CH_n^+$ ($n = 4, 5, 6, 7$ and 9) as well as $GeC_2H_7^+$. The structure of these ions has not been elucidated and it is presently unclear as to what are the most stable isomers for these different ions.

The structure of $[GeCH_2]^+$ and $[GeCH_3]^+$ ions has been examined by collisional activation (CA) and neutralization–reionization (NR) mass spectrometry[142]. Regardless of how these ions are formed, the prevalent connectivity for these ions is $GeCH_n^+$ ($n = 2$, 3). The modeling of the cationic and neutral surfaces for the $[GeCH_2]$ species suggests the unlikelihood that significant amounts of neutral or cationic germaacetylene or germavinylidene can be generated in these experiments. Similar experiments with the $[GeCH]^+$ ion also suggest that $GeCH^+$ is the preferred connectivity and theoretical calculations support the conclusion that this structure is substantially more stable than the $HGeC^+$ isomer[143].

The proton affinity of CH_3GeH_3 is also unknown. An *ab initio* calculation[144] at the Hartree–Fock level predicts that the structure of protonated methylgermane corresponds to a $CH_3GeH_2^+(H_2)$ complex and that the proton affinity of CH_3GeH_3 is 22 kJ mol^{-1} higher than that of germane.

Interest in the fundamental processes involved in formation of amorphous Ge-containing compounds by chemical vapor deposition processes assisted by radiolysis has been responsible for a number of studies dedicated to the reactions of Ge ions with O_2, NH_3, PH_3, SiH_4 and simple alkanes and alkenes[131,145–150].

These studies have revealed a number of interesting observations:

(i) O_2 reacts with $Ge_2H_2^{+\bullet}$, a secondary product ion formed through ion/molecule reactions of germane (see reactions 10b and 11), to yield $GeOH^+$ through the reaction given in equation 13.

$$Ge_2H_2^{+\bullet} + O_2 \longrightarrow GeOH^+ + (HGeO)^\bullet \qquad (13)$$

This unusual reaction has been proposed to occur via a four-center mechanism[131]. While no structural information was derived from the original experiments, calculations at the HF level predict that a linear structure corresponding to protonated germanium oxide, $GeOH^+$, is considerably more stable than the alternative linear $HGeO^+$ structure[151].

(ii) The primary hydrogen-containing ions obtained from GeH_4 react with NH_3 primarily by proton transfer to form NH_4^+ while condensation-type reactions are much slower. The notable exception is GeH_2^+, which undergoes significant reaction to yield both $GeNH_4^+$ and $GeNH_3^+$ (structures unknown) by hydrogen elimination[131,149]. The secondary product $Ge_2H_2^{+\bullet}$, as in the above reaction, also reacts with NH_3 by elimination of hydrogen through the reaction in equation 14.

$$Ge_2H_2^{+\bullet} + NH_3 \longrightarrow Ge_2NH_3^{+\bullet} + H_2 \qquad (14)$$

(iii) Interest in the mechanism for the formation of ions containing Ge–C bonds has motivated a number of studies involving GeH_4 with simple hydrocarbons[146], ethylene[145], allene[146,150] and alkynes. Primary ions of germane undergo very slow reaction with methane and ethane and experiments carried out under methane or ethane chemical ionization conditions reveal the formation of small amounts of $GeCH_5^+$, $GeCH_7^+$ and $GeC_2H_9^+$ as the result of tertiary reactions[146]. Other ions like $GeC_2H_5^+$ and $GeC_2H_7^+$ are observed as minor product ions. By comparison, reaction with alkynes like C_2H_2 and propyne, C_3H_4, gives rise to abundant amounts of $GeC_2H_3^+$ (in the case of acetylene) and $GeC_3H_n^+$ ($n = 3, 4, 5$) in the case of propyne under chemical ionization conditions through a sequential set of ion/molecule reactions. Likewise, chemical ionization of GeH_4 and C_2H_4 mixtures produce significant amounts of $GeC_2H_5^+$ and $GeC_2H_7^+$ ions. Since reactions observed under chemical ionization conditions are promoted both by ions originating from GeH_4 and those from the corresponding carbon compound, the overall mechanism for these reactions is quite complex. On the other hand, the reaction of GeH_3^+ with C_2H_4 has been shown to yield $GeC_2H_5^+$ in a tandem mass spectrometer[152] via initial formation of a $GeC_2H_7^+$ adduct.

The structure of the newly formed ions containing Ge–C bonds is a matter of considerable interest. While there is no experimental evidence at present, theoretical calculations[153] predict that for the $GeC_2H_5^+$ ion the lowest energy structures correspond to $CH_3CH(H)Ge^+$, with one hydrogen bridging the middle carbon and germanium, and to $H_3GeCHCH_2^+$, reminiscent of an allyl cation. On the other hand, theoretical calculations[154] predict that for the $GeC_2H_7^+$ ion, the most stable structure corresponds to that of a dimethyl germyl ion, $(CH_3GeHCH_3)^+$.

(iv) A mixture of GeH_4 and CO yields $GeCO^+$ as the most important cross reaction product under chemical ionization conditions[145]. The structure of this ion is an interesting problem with respect to the question of model systems for main group carbonyls. A recent study by collisional activation spectroscopy[155] strongly supports the idea that this ion retains a $Ge^+{-}CO$ connectivity. At lower pressures, experiments carried out by FTMS reveal that the main reaction (equation 15a) leading to a cross ionic product is promoted by a secondary Ge-containing ion[145]. In this case, the reaction in equation 15b is a minor channel.

$$Ge_2H_5^+ + CO \longrightarrow GeH_5O^+ + GeC \tag{15a}$$

$$\longrightarrow GeH_3O^+ + (GeCH_2) \tag{15b}$$

In both cases, the resulting ions involve the formation of a Ge−O covalent bond. A very similar set of reactions is observed with CO_2 with formation of GeH_5O^+ and GeH_3O^+.

(v) Ionic reactions between fragment and secondary ions of PH_3 and GeH_4 give rise to a number of interesting ions from a bonding point of view[148]. The following reactions, among others, have been identified by ion trap mass spectrometry to proceed rapidly and to lead to the formation of ions with Ge−P bonds (equations 16–18):

$$P_2^+ + GeH_4 \longrightarrow GePH_2^+ + PH_2 \tag{16a}$$

$$\longrightarrow GeP_2^+ + 2H_2 \tag{16b}$$

$$P_2H^+ + GeH_4 \longrightarrow GePH_2^+ + PH_3 \tag{17a}$$

$$\longrightarrow GeP_2H_3^+ + H_2 \tag{17b}$$

$$GeH_2^+ + PH_3 \longrightarrow GePH_3^+ + H_2 \tag{18}$$

(vi) Reactions between ionic fragments in GeH_4/SiH_4 and GeH_4/CH_3SiH_3 mixtures yield $GeSiH_5^+$ as the most important Ge−Si reaction product[147,156].

Very similar studies to those described above have also been carried out for methylgermane and the same substrates discussed above[140,141,149,156,157]. While the number of reaction channels is even greater in these cases, these reactions reveal the pathways for Ge−C, Ge−Si, Ge−O, Ge−N and Ge−P bond formation.

Very different and distinct ion chemistry has been observed in the reaction between the fragment ions obtained by electron ionization of tetramethoxygermane, $Ge(OMe)_4$, and the parent neutral[81]. Reactions in this system proceed by nucleophilic addition followed by elimination of formaldehyde and/or elimination of methanol. An overview of the reactions of the different ions with $Ge(OMe)_4$ is shown in Scheme 13 for the even electron ions, and in Scheme 14 for the radical ions originating from tetramethoxygermane. In these schemes, the neutral reagent of the ion/molecule reactions, $Ge(OMe)_4$, is not shown for the sake of simplicity but the schemes include the neutral products that are eliminated upon addition of the reagent ion to the parent neutral molecule.

While the actual structure of the product ions has not been clearly established, it is likely that the product ions $Ge_2(OMe)_6^{+\bullet}$, $Ge_2(OMe)_4^{+\bullet}$, $Ge_2(OMe)_5^+$ and $Ge_2(OMe)_3^+$ are species where Ge−Ge bonds have been formed in the process of elimination of MeOH and H_2CO. This implies that while the reaction of germyl-type cations like $Ge(OMe)_3^+$ and $HGe(OMe)_2^+$ are probably initiated by attachment of a Ge to an oxygen lone pair in $Ge(OMe)_4$, the reaction must proceed through a rearrangement that allows for an incipient Ge−Ge bond formation.

Additional examples of the ionic reactivity of germanium and tin systems in the gas phase deserve particular attention. For example, tritiated methyl cations have been used

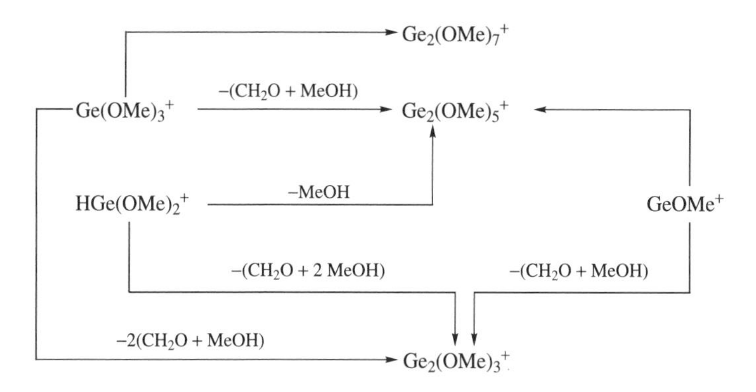

SCHEME 13

$$HGe(OMe)_3^{+\bullet} \xrightarrow{-MeOH} Ge(OMe)_6^{+\bullet}$$

$$HGe(OMe)_3^{+\bullet} \xrightarrow{-(CH_2O + 2\,MeOH)} Ge_2(OMe)_4^{+\bullet}$$

$$H_2Ge(OMe)_2^{+\bullet} \xrightarrow{-2\,MeOH} Ge_2(OMe)_4^{+\bullet}$$

SCHEME 14

to promote a methide abstraction from tetramethylgermane and tetramethylstannane[158] (equation 19).

$$CT_3^+ + Me_4E \longrightarrow MeCT_3 + Me_3E^+ \quad (E = Ge, Sn) \tag{19}$$

This same kind of reaction was shown to occur with Me_3Si^+ as the reagent ion[159] and the same substrates. The mechanism for such unusual reactions has not been elucidated, but a proposal has been made for the participation of an intermediate containing a Me^+ group bridging the C, or Si, and Ge, or Sn, center through a two-electron three-center bond. A somewhat similar reaction has been reported in radiolytical and chemical ionization experiments with trimethylgermylbenzene (equation 20), where the actual mechanism here involves an *ipso* substitution presumably through a sigma-type complex[160,161].

$$Me_3Si^+ + PhGeMe_3 \longrightarrow Me_3Ge^+ + PhSiMe_3 \tag{20}$$

The R_3E^+ ions (E = Si, Ge, Sn) are strong electrophiles that display great propensity to form adducts with n-donor bases (B)[162,163]. This has been shown to occur with simple aliphatic neutrals containing N, O or S as shown in equation 21.

$$Me_3Ge^+ + {:}B \longrightarrow [MeGe_3B]^+ \tag{21}$$

The relative binding strength of the adducts follows the qualitative trend of the gas-phase basicities of the donor base, i.e. t-BuOH > i-PrOH > EtOH > MeOH > H_2O. The binding to alcohols can be sufficiently selective to the point where 1,2-cyclopentanediols isomers have been distinguished by their reactivity towards Me_3Ge^+ in tandem mass spectrometry[164]. The actual binding energy has been measured for H_2O (equation 22)[165] and it reveals the reasonable stability of such adducts.

$$Me_3Ge^+ + H_2O \longrightarrow [Me_3GeOH_2]^+ \quad \Delta H^\circ = -119.7 \pm 2.0 \text{ kJ mol}^{-1} \quad (22)$$
$$\Delta G^\circ(300 \text{ K}) = -76.5 \pm 3.0 \text{ kJ mol}^{-1}$$

The Me_3Ge^+ cation can also form adducts with arenes[166] and quantitative gas-phase equilibrium measurements show that the stability of the adducts ($-\Delta G^\circ$) obeys the order 1,3-$Me_2C_6H_4 > H_2O >$ MePh. The thermochemistry and reactivity of the $[Me_3Ge^+.\text{arene}]$ adducts suggest that the most likely structure is that of a sigma complex.

Similar kind of data is available for the gas-phase adducts of Me_3Sn^+ with alcohols, amines, H_2O, ketones, esters and simple arenes[167]. Using methanol as a standard, the values for H_2O are shown in equation 23.

$$Me_3Sn^+ + H_2O \longrightarrow [Me_3SnOH_2]^+ \quad \Delta H^\circ = -107 \pm 4 \text{ kJ mol}^{-1} \quad (23)$$
$$\Delta G^\circ(300 \text{ K}) = -76 \pm 6 \text{ kJ mol}^{-1}$$

The relative trends for the stability of the adducts of Me_3Sn^+ also parallel very closely the relative order of proton affinity of the different neutral bases.

While similar data are not available for the corresponding Me_3Pb^+ ion, the thermochemistry of the gas-phase association of Pb^+ with several molecules of NH_3 and H_2O, and with one molecule of MeOH, $MeNH_2$ and benzene has been thoroughly characterized[168–171]. By comparison with the above examples, the association of Pb^+ with H_2O is weaker than that of the trimethylgermanium and trimethyltin ions (see equation 24)[169].

$$Pb^+ + H_2O \longrightarrow Pb^+(H_2O) \quad \Delta H^\circ = -93.7 \text{ kJ mol}^{-1} \quad (24)$$

In summary, the gas-phase ion chemistry of Ge, Sn and Pb positive ions reveals some very unusual reactivity and will probably witness important advances in the near future regarding structural aspects of these ions.

B. Thermochemistry, Structure and Reactivity of Negative Ions of Ge, Sn and Pb

Negative ions are less common species in mass spectrometry for a variety of reasons: (a) Stable negative ions are only detected for molecules or radicals with positive electron affinities (equation 25).

$$M(g) + e^- \longrightarrow M^-(g) \qquad EA(M) = -\Delta E \quad (25)$$

(b) Direct formation of negative ions by electron impact generally occurs through a resonance process over a narrow range of electron energy that coincides with capture of the electron to yield either a long-lived anion, or undergoes dissociation to yield a fragment negative ion. These mechanisms have been reviewed in the literature[172]. On the other hand, stable negative ions for which no convenient precursor is available for direct formation can often be generated indirectly by ion/molecule reactions. Electrospray ionization is an alternative method for obtaining negative ions since ions are introduced in the mass spectrometer directly from solution.

Negative ions containing Ge, Sn or Pb have been obtained in mass spectrometry from the halides of these elements. For example, GeF_4 has been shown to yield F^-, $GeF_3{}^-$ and $GeF_4{}^-$ and traces of $GeF_2{}^-$ by electron impact[173]. Formation of these ions is typically a resonance process that is observed at electron energies between 8 and 11 eV. In principle, the appearance energy (AE) of a species like GeF_3^- can be used to determine important thermochemical parameters (equation 26).

$$GeF_4 + e^- \longrightarrow GeF_3^- + F \quad (26a)$$

$$AE(GeF_3^-) = D(F_3Ge-F) + EA(F_3Ge^\bullet) + E^* \quad (26b)$$

In equation 26b, E* represents the excess energy (vibrational, translational and eventually electronic) with which the fragments of equation 26a are formed. By assuming a $F_3Ge{-}F$ bond energy of 518 $kJ\,mol^{-1}$ obtained from the positive ion data and measuring the translational energy of GeF_3^- formed at the onset, a value for the electron affinity of $F_3Ge^\bullet$ was thus obtained. The uncertainty in this approach is well illustrated by the fact that three different experiments[174–176] yield very different values for the electron affinity of $F_3Ge^\bullet$, namely 3.1 eV, 1.6 eV and 1.1 eV. Thus, this method suffers serious limitations for determining electron affinities. As a general guideline, recent high-level calculations[137,177] estimate the EA of $F_3Ge^\bullet$ to be in the range of 3.5 to 3.7 eV.

Similar mass spectrometric experiments[178–180] were carried out for $GeCl_4$ and $GeBr_4$, and for SnF_4, $SnCl_4$, $SnBr_4$ and SnI_4. For these compounds, the observed negative ions are EX_3^-, EX_2^-, X_2^- and X^- (E = Ge, Sn; X = Cl, Br, I). Appearance energy measurements coupled with measurements of the translational energies of the ions were again used to obtain estimates of the electron affinities for the different EX_2 and $^\bullet EX_3$ species. However, the electron affinities estimated for $GeCl_3$ and $SnCl_3$ from these experiments are substantially lower than those obtained from the energy threshold for reactions studied by atom beam techniques[181,182] (equation 27).

$$M + ECl_4 \longrightarrow M^+ + Cl_3E^- + Cl \quad (M = Cs, K; E = Ge, Sn) \quad (27)$$

The only other family of compounds that have been characterized by negative ion mass spectrometry has been the organo-tin compounds $RSnCl_3$ (R = Me, *n*-Bu, Octyl, Dodecyl and Ph), R_2SnCl_2 (R = Me, Et, *n*-Bu, Octyl, Dodecyl and Ph) and R_3SnCl (R = Me, *n*-Pr, *n*-Bu, *c*-Hex, Ph). For all of these compounds, primary negative ions are observed that correspond to loss of an alkyl group generating the $[R_{n-1}SnCl_{4-n}]^-$ ($n = 1-3$) ions and chloride ions[103]. These observations clearly indicate the well-known ease for halides to yield negative ions by electron impact in the gas phase.

Accurate measurements of the electron affinity of simple Ge and Sn radicals have been obtained by threshold photodetachment experiments carried out in ion cyclotron resonance experiments[183]. In these experiments, measurement of the threshold frequency for removing the electron from the anion yields an upper limit for the electron affinity of the species, as shown for GeH_3^- in equation 28.

$$GeH_3^- + h\nu_{th} \longrightarrow {}^\bullet GeH_3 + e^- \quad EA({}^\bullet GeH_3) \leqslant h\nu_{th} \quad (28)$$

These experiments can provide the true electron affinity in cases where the geometry of the anion and the neutral are reasonably similar or where favorable Franck–Condon factors allow observation of the adiabatic transition. Using this approach[184,185], the electron affinities of $^\bullet GeH_3$, $^\bullet GeMe_3$ and $^\bullet SnMe_3$ have been obtained and are listed in Table 6.

For the case of $^\bullet GeH_3$, recent high-level theoretical calculations[137,186] suggest that the actual value for the electron affinity is somewhat lower than the upper limit suggested by experiment. The calculations estimate values in the range of 1.55 to 1.60 eV.

Electron affinities can be determined with even higher accuracy by photoelectron spectroscopy of negative ions. This technique has been used to determine the electron affinity

TABLE 6. Electron affinity and gas-phase acidity of some simple Ge and Sn systems

Radical	EA	Compound	ΔH°_{acid} (kJ mol^{-1})
$^{\bullet}GeH_3$	$\leqslant$ 1.74 (Reference 184)	GeH_4	1502.0 ± 5.1 (Reference 190)
$^{\bullet}GeMeH_2$	—	$MeGeH_3$	1536.6 ± 5.0 (Reference 190)
$^{\bullet}GeMe_3$	1.38 ± 0.03 (Reference 185)	Me_3GeH	1512 ± 12 (Reference 185)
$^{\bullet}SnMe_3$	1.70 ± 0.06 (Reference 185)	Me_3SnH	1460 ± 8 (Reference 185)

of Ge (1.233 ± 0.003 eV) and Sn (1.113 ± 0.020 eV)[187] and of gas-phase germanium clusters[188,189].

The gas-phase acidity of some of the simplest organogermanes and of trimethylstannane have been determined by a combination of gas-phase equilibrium measurements and proton-transfer bracketing experiments using FTMS[185,190]. The experimental values of ΔH°_{acid} are shown in Table 6 and refer to the enthalpy change associated with reaction 29.

$$AH(g) \longrightarrow A^{-}(g) + H^{+}(g) \quad \Delta H^{\circ}_{acid}(AH) = \Delta H^{\circ} \tag{29}$$

Methyl substitution decreases the gas-phase acidity of the Ge system similarly to what has been found in methylsilane[191]. On the other hand, the increase observed in trimethylgermane is more difficult to analyze as relative variations of ΔH°_{acid} along a given family reflect changes in electron affinity and in Ge−H bond dissociation energies.

Very few studies are available on the reactivity of negative ions in Ge, Sn and Pb compounds. However, some interesting results have emerged from the study of gas-phase reactions between $Ge(OMe)_4$ and simple nucleophiles (F^-, MeO^-) using FTMS techniques[192]. The low-pressure reaction (10^{-8} Torr range) of F^- reveals that the reaction proceeds by addition to yield a pentacoordinated Ge anion that can undergo selective elimination as shown in Scheme 15.

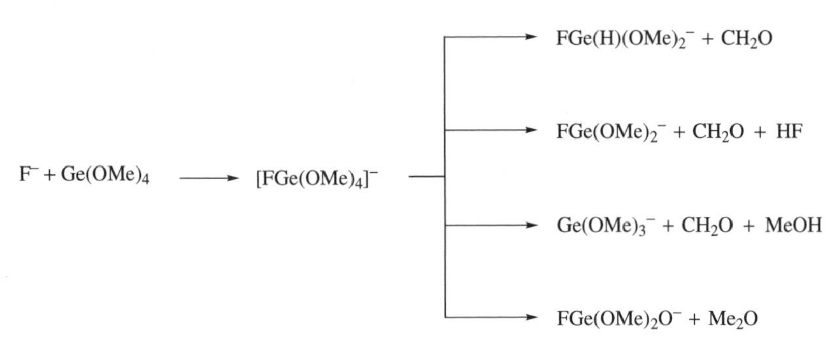

SCHEME 15

While the $FGe(OMe)_4{}^-$ adduct is the main reaction product, it is interesting to note that the most favorable elimination process involves loss of MeOH and CH_2O to yield $Ge(OMe)_3{}^-$. This behavior is very reminiscent of what is observed in the positive ion chemistry of tetramethoxygermane[81]. An additional interesting observation in this system is the fact that the pentacoordinated anion can undergo successive exchange reactions with fluorine-containing neutrals (BF_3, SO_2F_2) (equation 30).

$$FGe(OMe)_4{}^- + BF_3 \longrightarrow F_2Ge(OMe)_3{}^- + (MeO)BF_2 \tag{30}$$

This type of exchange reaction is similar to what has been observed in the corresponding Si ions[193].

Theoretical calculations[192] indicate that the most stable structure for the $FGe(OMe)_4^-$ ion corresponds to a trigonal bipyramid with the F occupying an apical position.

IV. GAS-PHASE CHEMISTRY OF NEUTRAL Ge, Sn AND Pb SPECIES

Gas-phase reactions of some of the simplest Ge, Sn and Pb compounds have been the subject of a number of investigations. For example, the detailed mechanism responsible for the thermal and photochemical decomposition of germanes, as well as some of the simple radical- and atom-molecule reactions of germanes, have attracted considerable attention in recent years because of the interest in Ge deposition processes. In the case of Sn and Pb compounds, the main interest has been related to environmental problems.

Monogermane was first shown to undergo mercury photosensitized decomposition[194] at low pressures and at 298 K through the overall equation (31).

$$Hg(6\,^3P_1) + GeH_4 \longrightarrow Hg(6\,^1S_0) + Ge + 2H_2 \tag{31}$$

Later studies[195] revealed that Ge_2H_6, and small amounts of Ge_3H_8 and Ge_4H_{10}, are also produced in this process. The significant yield of HD obtained in the Hg sensitized photolysis of GeH_4 and GeD_4 mixtures led to the suggestion that reaction 32 is the primary photochemical quenching process.

$$Hg(6\,^3P_1) + GeH_4 \longrightarrow Hg(6\,^1S_0) + {}^{\bullet}GeH_3 + H^{\bullet} \tag{32}$$

The same type of decomposition in the presence of NO yields substantial amounts of GeH_3OGeH_3, lending further support to the idea that reaction 32 is the main process[196]. While the mercury photosensitized reaction is generally accepted to proceed primarily via Ge−H bond cleavage[197], a different process (equation 33) involving the initial formation of germylene, $:GeH_2$, becomes the most important primary reaction under direct photochemical and thermal decomposition.

$$GeH_4 \longrightarrow :GeH_2 + H_2 \tag{33}$$

The direct photodecomposition of GeH_4 in argon-GeH_4 matrices by vacuum-UV radiation[198] in the temperature range of 4–25 K was shown by spectroscopic measurements to generate $:GeH_2$ and ${}^{\bullet}GeH_3$. A more detailed and recent study has shown that the 147-nm gas-phase photolysis[199] of GeH_4 proceeds by both primary channels with relative quantum yields of $\Phi_{33}(:GeH_2) = 0.66$ and $\Phi_{32}({}^{\bullet}GeH_3) = 0.34$. Based on these primary processes a number of possible radical reactions have been proposed to account for the final products, including formation of the higher germanes.

The gas-phase thermal decomposition of GeH_4 proved to be a more complex process. In fact, pyrolysis of GeH_4 at $T \sim 600$ K revealed a heterogeneous reaction sensitive to surface effects[200]. The use of shock tube techniques[201] over an effective temperature range of 962–1063 K allowed for the study of the reaction free of heterogeneous effects. Under these conditions, the activation energy for the thermal decomposition of GeH_4 amounts to 212.3 ± 14.9 kJ mol^{-1}. This activation energy is considerably less than that of the Ge−H bond energy (*ca* 343 kJ mol^{-1}), indicating that the primary process in the mechanism involves a collision-assisted three-center elimination of hydrogen and formation of germylene (equation 34).

$$M + GeH_4 \longrightarrow :GeH_2 + H_2 + M \tag{34}$$

Initial formation of germylene in the thermal process has also been inferred for the decomposition process sensitized by multiphoton vibrational excitation of SiF_4 using a pulsed CO_2 laser[202]. Thus, it is clear that decomposition of GeH_4 through its ground electronic state proceeds by initial formation of $:GeH_2$ while decomposition from excited electronic states can proceed both by formation of $:GeH_2$ and $^{\bullet}GeH_3$.

The homogeneous gas-phase thermal decomposition of methylgermane is considerably more complicated since several primary processes are possible in principle[203]. Three primary processes have been proposed based on the following considerations: (i) the activation energy of the reaction measured from the temperature dependence of the rate; (ii) the question of the source of methyl radicals in the overall mechanism; and (iii) measurements of the yields of HD and D_2 in the decomposition of CH_3GeD_3. These processes can then be identified as: (a) a three-center elimination of H_2 accounting for 40% of the overall reaction (equation 35a); (b) a three-center elimination of CH_4 accounting for about 30% of the overall reaction (equation 35b); and (c) a four-center elimination of hydrogen (HD in the case of $MeGeD_3$) accounting for about 30% of the reaction (equation 35c).

$$CH_3GeH_3 \longrightarrow CH_3GeH + H_2 \quad (35a)$$

$$\longrightarrow CH_4 + :GeH_2 \quad (35b)$$

$$\longrightarrow H_2C = GeH_2 + H_2 \quad (35c)$$

Formation of methyl radicals in this case was considered to be the result of secondary reactions arising from further decomposition of methylgermylene, CH_3GeH.

Higher germanes have also been studied for the purpose of investigating routes for chemical vapor deposition (CVD) of germanium. For example, the results obtained from the sensitized thermal decomposition of Me_4Ge promoted by multiphoton vibrational excitation of SF_6 using a pulsed CO_2 laser[204,205] are consistent with the pyrolytic decomposition[206–208] that points out to a Ge–C bond cleavage as the primary process (equation 36).

$$Me_4Ge \longrightarrow Me^{\bullet} + {}^{\bullet}GeMe_3 \quad (36)$$

By comparison, the photolysis of Me_4Ge with an argon fluoride laser at 193 nm yields trimethylgermane, ethane, methane, ethylene and layers of Ge as the end products[209]. These results have been interpreted as indicative of electronically excited Me_4Ge undergoing preferential molecular elimination (equations 37a and 37b) over the dissociation shown in equation 36.

$$(Me_4Ge)^* \longrightarrow Me_2Ge: + C_2H_6 \quad (37a)$$

$$\longrightarrow CH_4 + Me_2Ge = CH_2 \quad (37b)$$

Ethylgermanes, Et_nGeH_{4-n} ($n = 1$–4), yield primarily ethane and ethylene as the main end product of 193 nm laser photolysis in ratios that vary from 2 to 5 depending on the number of Et groups present in the original molecule[210]. Very small quantities of ethylgermanes with a lower number of Et groups in the molecule are also observed as minor products. Experiments carried out in the presence of GeD_4 suggest that ethane and ethylene are probably formed in a primary process which results in their elimination (equations 38a and 38b).

$$Et_nGeH_{4-n}{}^* \longrightarrow Et_{n-1}GeH_{3-n} + C_2H_6 \quad (38a)$$

$$\longrightarrow C_2H_4 + Et_{n-1}GeH_{5-n} \quad (38b)$$

By comparison, the thermal dissociation of the ethylgermanes has been proposed to proceed by cleavage of the Ge−C bond (equation 48) followed by hydrogen abstraction by the nascent ethyl radical[211,212].

$$Et_nGeH_{4-n}^{*} \longrightarrow Et_{n-1}GeH_{4-n} + {}^{\bullet}Et \quad (39)$$

Unfortunately, the unequivocal elucidation of these mechanisms requires real time measurements of the nascent products and an analysis of the kinetics of the propagation steps. Some studies on individual elementary reactions are discussed below.

The chemical vapor deposition processes obtained from the laser induced multiphoton decomposition of neat organogermanes or sensitized by SF_6 have also been characterized for several systems. For example, the decomposition of $Ge(OMe)_4$ leads to the formation of organoxogermanium polymers[213], while $EtOGeMe_3$ leads to materials rich in Ge and containing small amounts of oxygen and carbon[214]. In the latter case, two primary processes have been proposed to be responsible for the chain reactions leading to the final products (equations 40a and 40b)[215].

$$EtOGeMe_3 \longrightarrow Me^{\bullet} + Me_2Ge(OEt)^{\bullet} \quad (40a)$$

$$\longrightarrow EtO^{\bullet} + Me_3Ge^{\bullet} \quad (40b)$$

However, the ethylene and acetaldehyde appearing among the final products may also originate through non-radical mechanisms, namely β elimination processes that are prevalent for these alkoxygermanes (equations 41 and 42).

$$EtOGeMe_3 \longrightarrow Me_3GeH + CH_3CHO \quad (41)$$

$$EtOGeMe_3 \longrightarrow Me_3GeOH + C_2H_4 \quad (42)$$

A few gas-phase atom-GeH_4 elementary reactions have been studied in detail because of their role in the general mechanism of thermal deposition processes. The most widely studied reaction is that involving H atoms and GeH_4 for which hydrogen abstraction (equation 43) is considered the only pathway.

$$H + GeH_4 \longrightarrow H_2 + {}^{\bullet}GeH_3 \quad (43)$$

Recent measurements[216,217] for this reaction in the range of $T = 293-473$ K have settled the question of the rate constant that was in serious doubt from the early experiments[218,219]. Reaction 43 is characterized by a very low activation energy of 7.3 ± 0.2 kJ mol^{-1}, and an isotope effect $k_H/k_D = 2.0 \pm 0.4$ at 300 K obtained by comparison with the reaction of H atom with fully deuteriated monogermane[220]. The rate constant for the prototype reaction 43 increases with progressive methyl substitution on the germane. This fact has been used to suggest a lowering of the Ge−H bond dissociation energy upon methyl substitution[221].

The dynamics of hydrogen abstraction reactions promoted by F, O, OH and OD with monogermane have been studied as a function of the vibrational and rotational state by infrared chemiluminescence[222–224]. While this technique provides enormous insight on the energy disposal in a reaction, it also led to a value of 326 ± 4 kJ mol^{-1} for the H_3Ge−H bond dissociation energy at 0 K. This value is somewhat lower than the value of 346 ± 10 kJ mol^{-1} obtained from a kinetic study of reaction 44 and its thermochemistry[225,226].

$$I_2 + GeH_4 \longrightarrow GeH_3I + HI \quad (44)$$

This same approach was used by Doncaster and Walsh to study the iodine reaction with Me_3GeH and to derive a bond dissociation energy of 340 ± 10 kJ mol^{-1} for $Me_3Ge{-}H$[227]. These results suggest that methyl substitution has no appreciable effect on the Ge−H bond energy within experimental error. On the other hand, iodine substitution as in GeH_3I was shown[228] to decrease the Ge−H bond energy by 14 kJ mol^{-1}.

Hydrogen atom abstraction reactions are usually characterized by very low activation energies (E_a). This is the case for the reaction of the t-BuO$^\bullet$ radical with germane[229] for which E_a was determined to be 7.9 kJ mol^{-1}. A somewhat higher activation energy has been measured for the hydrogen abstraction by O (^{3}P) from methyl groups attached to germanium[230]. For example, an activation energy of 22.3 kJ mol^{-1} was obtained for Me_4Ge, 13.5 kJ mol^{-1} for Et_4Ge and 16.6 kJ mol^{-1} for $(MeO)_4Ge$. These higher activation energies reflect the fact that the C−H bond energies are higher than the Ge−H bond energy in monogermane.

A new and exciting development in the gas-phase chemistry of Ge has been the ability to characterize the behavior of germylenes, RR′Ge:[231]. While the chemistry of germylenes has been extensively explored in solution to promote a series of reactions[232], it is only recently that GeH_2 has been detected in the gas phase by laser-induced fluorescence as a result of the 193 nm photolysis of phenylgermane[233]. This has opened the possibility to investigate some of the elementary reactions involved in CVD processes using germanes.

The rate constants for the reaction of GeH_2, generated by laser flash photolysis from either $PhGeH_3$ or 3,4-dimethylgermacyclopentane, with O_2, C_2H_2, i-C_4H_8, 1,3-C_4H_6, C_3H_8 and Me_3SiH has been recently determined by time-resolved laser induced fluorescence[234]. Analysis of the end products for the reaction with 2,3-dimethyl-1,3-butadiene reveals addition of GeH_2 onto the double bond to yield 3,4-dimethylgermacyclopentane, as shown in Scheme 16.

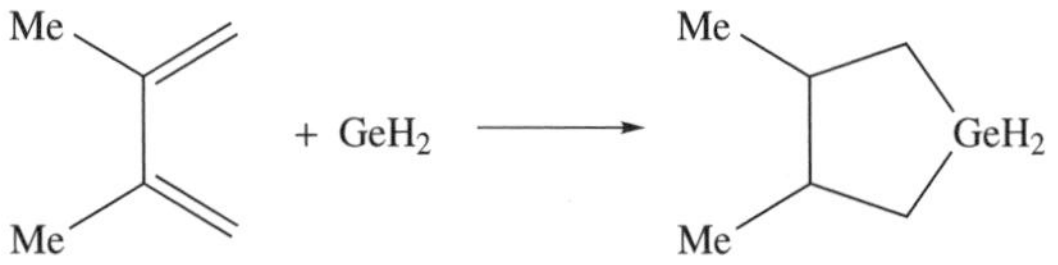

SCHEME 16

From these measurements it has been possible to conclude that GeH_2 can insert readily into Si−H bonds but not into C−H bonds, and can undergo addition to π bonds.

More detailed studies on the insertion of GeH_2 have been obtained from the reaction with Me_3SiH and GeH_4 (equations 45–47)[235,236].

$$GeH_2 + Me_3SiH \longrightarrow Me_3SiGeH_3 \qquad (45)$$

$$GeH_2 + GeH_4 \longrightarrow Ge_2H_6 \qquad (46)$$

Both of these reactions proceed with very high rate constants at room temperature (within a factor of 5 of the collision rate for GeH_4) and insertion in the Ge−H bond is more facile than for the Si−H bond. Reaction 46 has been found to be pressure-dependent, so that formation of Ge_2H_6 is best described as a third body assisted association. The analogous reaction to 46, but using SiH_2, reveals that silylene insertion in the Ge−H bond is faster than that of the corresponding germylene[237].

These fast insertion reactions display a negative activation energy[238], and high level *ab initio* calculations[236] confirm the suggestion that these reactions proceed through a transition state where a hydrogen bridge between the two germanium atoms mediates formation of the final product in reaction 46.

By comparison with the reactivity of GeH_2, dimethylgermylene ($GeMe_2$) has been found to be unreactive towards C−H, Si−H and Ge−H insertion but reacts rapidly with π-bonded systems[239].

A very recent study[240] reveals that an activation energy of at least 19 ± 6 kJ mol^{-1} can be estimated from the upper limit for the rate constant of the insertion of germylene into molecular hydrogen (equation 47).

$$GeH_2 + D_2 \longrightarrow GeH_2D_2 \tag{47}$$

Theoretical calculations predict a significant barrier of 58 kJ mol^{-1} for this reaction.

Much less information is available on the gas-phase chemistry of Sn derivatives. The thermal decomposition of SnH_4 (equation 48) was characterized as a heterogeneous process where Sn was found to promote the reaction autocatalytically[241,242].

$$SnH_4 \longrightarrow Sn + 2H_2 \tag{48}$$

The 193 nm laser induced photodecomposition of SnH_4 suggested that stannylene, SnH_2, is the main intermediate of this process in analogy with what is observed for silane and germane[243]. A similar photolysis study at 147 nm results in the final products as shown in reaction 48 plus a small amount of Sn_2H_6. However, the quantum yield of hydrogen, $\Phi(H_2)$, at 0.20 Torr of SnH_4 was measured to be 11.4 ± 0.6, indicating a somewhat more complex mechanism than that predicted by a simple initial formation of SnH_2 and hydrogen[244]. The quantum yield decreases with increasing pressure, but extrapolation to zero pressure yields a value similar to that obtained at 0.20 Torr. At the 147 nm wavelength, it is possible to consider two primary pathways that are energetically feasible (equations 49 and 50).

$$SnH_4 \longrightarrow SnH_2 + 2H \tag{49}$$

$$\longrightarrow SnH + 3H \tag{50}$$

The ensuing propagation steps promoted by the hydrogen atoms can lead to quantum yields as high as 12 for hydrogen if reaction 50 is the exclusive channel.

Distannane, Sn_2H_6, is known to have extremely poor thermal stability leading to Sn and hydrogen[245]. Recent measurements[246] show that the activation energy for thermal decomposition of Sn_2H_6 is only 5.3 ± 0.6 kJ mol^{-1} but the low pre-exponential factor obtained for the rate constant suggests that the decomposition is a heterogeneous process.

The trend for heterogeneous reactions in these systems is well illustrated in studies carried out for gas-phase auto-oxidation of simple tin compounds at higher temperatures (200–450 °C)[247]. These reactions are of environmental concern and serve as complementary information to what is known about lead-containing systems. For Et_4Sn, auto-oxidation is believed to be initiated by the initial decomposition shown in equation 51 followed by reaction with O_2 (equation 52), by analogy to the mechanism for auto-oxidation of organic compounds.

$$Et_4Sn \longrightarrow Et_3Sn^{\bullet} + Et^{\bullet} \tag{51}$$

$$Et_3Sn^{\bullet} + O_2 \longrightarrow EtSn(OH)_2{}^{\bullet} + C_4H_8 \tag{52}$$

Generation of OH radicals can then lead to formation of triethyltin hydroxide, Et_3SnOH, which undergoes heterogeneous dehydration and oxidation.

However, for environmental purposes the most important gas-phase chemistry is that related to the tetraalkyl lead derivatives. Me_4Pb and Et_4Pb have been used extensively as a gasoline additive to prevent pre-ignition, but considerable concern has been raised over the last 30 years regarding its long-lasting effects on the environment and as a health hazard. While these compounds tend to yield a complex mixture of lead salts upon combustion[248], incomplete combustion accounts for a non-negligible amount of lead-containing fuel to enter the atmosphere.

A classical study on the fate of tetraalkyl compounds in the atmosphere[249] shows the effect of photochemical oxidation on the breakdown of R_4Pb. Direct ozonolysis of Me_4Pb (equation 53) presumably proceeds by analogy to the process in tetraethyltin[250].

$$Me_4Pb + O_3 \longrightarrow Me_3PbOOH + CH_2O \qquad (53)$$

However, reactions promoted by O (from photolysis of O_3) are much faster and more important in the atmosphere. Reaction 54 leads to formation of OH, and the ensuing reactions are responsible for the degradation of Me_4Pb in the atmosphere.

$$Me_4Pb + O \longrightarrow Me_3PbCH_2 + OH \qquad (54)$$

The rate constants for the reaction of OH with Me_4Pb and Et_4Pb reveal that consumption of the tetraalkyl lead is very rapid[251]. The fact that these reactions are much faster than for the corresponding carbon analogs has led to the suggestion that hydrogen abstraction cannot be the dominant reaction channel. The fact that Et_4Pb reacts much faster than Me_4Pb is consistent with the proposal that the initial step in the reaction is addition of OH to yield an R_4PbOH intermediate from which either H_2O or an alkyl radical can be eliminated. From these rate constant measurements and considering typical tropospheric concentrations of OH, a lifetime of *ca* 50 hours was estimated for Me_4Pb and of 4 hours for Et_4Pb in ambient air.

A different type of gas-phase chemistry has also been explored for simple germanium, tin and lead compounds, namely that induced by radiolysis. While germanium compounds have been the main targets and modeling of CVD processes the ultimate goal, radiolysis involves reactions of transient neutral and ionic species for which the overall mechanism is not always clear. On the other hand, analysis of the ultimate solid products[252] obtained from radiolysis of different mixtures provides an exciting approach towards the synthesis of polymers and thin films.

V. ACKNOWLEDGMENTS

The authors thank the support of the São Paulo Science Foundation (FAPESP), the Brazilian Research Council (CNPq) and the library facilities of the Institute of Chemistry of the University of São Paulo. We particularly thank Jair J. Menegon for his continuous assistance with matters pertaining to managing our databases.

VI. REFERENCES

1. S. P. Constantine, D. J. Cardin and B. G. Bollen, *Rapid Commun. Mass Spectrom.*, **14**, 329 (2000).
2. G. Lawson, R. H. Dahm, N. Ostah and E. D. Woodland, *Appl. Organomet. Chem.*, **10**, 125 (1996).
3. See, for example:
   (a) M. Lazraq, C. Couret, J. Escudie, J. Satge and M. Soufiaoui, *Polyhedron,* **10**, 1153 (1991).
   (b) J. Escudie, C. Couret and H. Ranaivonjatovo, *Coord. Chem. Rev.,* **180**, 565 (1998).

(c) A. Schafer, W. Saak, M. Weidenbruch, H. Marsmann and G. Henkel, *Chem. Ber.-Recueil,* **130**, 1733 (1997).
(d) N. P. Toltl and W. J. Leigh, *J. Am. Chem. Soc.,* **120**, 1172 (1998).
4. See, for example:
(a) J. Barrau, G. Rima and T. El Amraoui, *J. Organomet. Chem.,* **570**, 163 (1998).
(b) M. Weidenbruch, M. Sturmann, H. Kilian, S. Pohl and W. Saak, *Chem. Ber.-Recueil,* **130**, 735 (1997).
(c) C. E. Dixon, M. R. Netherton and K. M. Baines, *J. Am. Chem. Soc.,* **120**, 10365 (1998).
(d) V. N. Khabashesku, K. N. Kudin, J. Tamas, S. E. Boganov, J. L. Margrave and O. M. Nefedov, *J. Am. Chem. Soc.,* **120**, 5005 (1998).
5. NIST Standard Reference Database Number 69, National Institute of Standards and Technology, Gaithersburg, MD, 2000 available at http://webbook.nist.gov/chemistry/
6. K. A. Gingerich, R. W. Schmude, M. S. Baba and G. Meloni, *J. Chem. Phys.,* **112**, 7443 (2000).
7. S. M. Colby, M. Stewart and J. P. Reilly, *Anal. Chem.,* **62**, 2400 (1990).
8. H. Budzikiewicz, C. Djerassi and D. H. Williams, *Mass Spectrometry of Organic Compounds*, Holden-Day, San Francisco, 1967.
9. F. W. McLafferty, *Interpretation of Mass Spectra*, 3rd edn., University Science Books, Mill Valley, 1980.
10. D. B. Chambers, F. Glockling and J. R. Light, *Quart. Rev.,* **22**, 317 (1968).
11. V. Yu. Orlov, *Russ. Chem. Rev.,* **42**, 529 (1973).
12. H. Neurt and H. Clasen, *Z. Naturforsch.,* **7A**, 410 (1952).
13. F. E. Saalfeld and H. J. Svec, *J. Inorg. Nucl. Chem.,* **18**, 98 (1961).
14. F. E. Saalfeld and H. J. Svec, *Inorg. Chem.,* **2**, 46 (1963).
15. G. G. Devyatyk, I. L. Agafonov and V. I. Faerman, *Russ. J. Inorg. Chem.,* **16**, 1689 (1971).
16. T. Kudo and S. Nagase, *Chem. Phys. Lett.,* **148**, 73 (1988).
17. T. Kudo and S. Nagase, *Chem. Phys. Lett.,* **156**, 289 (1989).
18. B. Ruscic, M. Schwarz and J. Berkowitz, *J. Chem. Phys.,* **92**, 1865 (1990).
19. A. W. Potts and W. C. Price, *Proc. R. Soc. London, Ser. A,* 165 (1972).
20. B. P. Pullen, T. A. Carlson, W. E. Moddeman, G. K. Schweitz, W. E. Bull and F. A. Grimm, *J. Chem. Phys.,* **53**, 768 (1970).
21. S. Cradock, *Chem. Phys. Lett.,* **10**, 291 (1971).
22. G. P. van der Kelen and D. F. van de Vondel, *Bull. Soc. Chim. Belg.,* **69**, 504 (1960).
23. F. E. Saalfeld and H. J. Svec, *Inorg. Chem.,* **2**, 50 (1963).
24. F. E. Saalfeld and H. J. Svec, *J. Phys. Chem.,* **70**, 1753 (1966).
25. V. H. Dibeler and F. L. Mohler, *J. Res. Natl. Bur. Stand.,* **47**, 337 (1951).
26. V. H. Dibeler, *J. Res. Natl. Bur. Stand.,* **49**, 235 (1952).
27. B. G. Hobrock and R. W. Kiser, *J. Phys. Chem.* **65**, 2186 (1961).
28. B. G. Hobrock and R. W. Kiser, *J. Phys. Chem.,* **66**, 155 (1962).
29. J. J. de Ridder and G. Dijkstra, *Recl. Trav. Chim. Pays-Bas*, **86**, 737 (1967).
30. F. Glockling and J. R. C. Light, *J. Chem. Soc., A,* 717 (1968).
31. K. G. Heumann, K. Bachmann, E. Kubasser and K. H. Lieser, *Z. Naturforsch. B,* **28**, 108 (1973).
32. G. S. Groenewold, M. L. Gross, M. M. Bursey and P. R. Jones, *J. Organomet. Chem.,* **235**, 165 (1982).
33. G. D. Flesch and H. J. Svec, *Int. J. Mass Spectrom. Ion Processes,* **38**, 361 (1981).
34. M. F. Lappert, J. B. Pedley, J. Simpson and T. R. Spalding, *J. Organomet. Chem.,* **29**, 195 (1971).
35. A. E. Jonas, G. K. Schweitzer, F. A. Grimm and T. A. Carlson, *J. Electron Spectrosc. Relat. Phenom.,* **1**, 29 (1973).
36. S. Evans, J. C. Green, J. P. Maier, D. W. Turner, P. J. Joachim and A. F. Orchard, *J. Chem. Soc., Faraday Trans. 2,* **68**, 905 (1972).
37. J. T. Bursey, M. M. Bursey and D. G. Kingston, *Chem. Rev.,* **73**, 191 (1973).
38. A. Carrick and F. Glockling, *J. Chem. Soc., A,* 623 (1966).
39. J. L. Occolowitz, *Tetrahedron Lett.,* 5291 (1966).
40. J. J. de Ridder, G. Vankoten and G. Dijkstra, *Recl. Trav. Chim. Pays-Bas,* **86**, 1325 (1967).
41. S. Boue, M. Gielen and J. Nasielski, *Bull. Soc. Chim. Belg.,* **77**, 43 (1968).
42. M. Gielen, B. De Poorter, M. T. Sciot and J. Topart, *Bull. Soc. Chim. Belg.,* **82**, 271 (1973).

43. R. Weber, F. Visel and K. Levsen, *Anal. Chem.*, **52**, 2299 (1980).
44. D. B. Chambers, F. Glockling, J. R. C. Light and M. Weston, *J. Chem. Soc., Chem. Commun.*, 281 (1966).
45. D. B. Chambers, F. Glockling and M. Weston, *J. Chem. Soc., A,* 1759 (1967).
46. N. Ostah and G. Lawson, *Appl. Organomet. Chem.*, **9**, 609 (1995).
47. N. Ostah and G. Lawson, *Appl. Organomet. Chem.* **14**, 383 (2000).
48. B. Pelli, A. Sturaro, P. Traldi, F. Ossola, M. Porchia, G. Rossetto and P. Zanella, *J. Organomet. Chem.*, **353**, 1 (1988).
49. J. M. Miller and A. Fulcher, *Can. J. Chem.*, **63**, 2308 (1985).
50. J. M. Miller, H. Mondal, I. Wharf and M. Onyszchuk, *J. Organomet. Chem.*, **306**, 193 (1986).
51. J. M. Miller, Y. C. Luo and I. Wharf, *J. Organomet. Chem.*, **542**, 89 (1997).
52. J. M. Miller and I. Wharf, *Can. J. Spectrosc.*, **32**, 1 (1987).
53. T. Chivers, G. F. Lanthier and J. M. Miller, *J. Chem. Soc., A,* 2556 (1971).
54. J. M. Miller, *Can. J. Chem.*, **47**, 1613 (1969).
55. J. M. Miller, T. R. B. Jones and G. B. Deacon, *Inorg. Chim. Acta,* **32**, 75 (1979).
56. H. Schwarz, *Top. Curr. Chem.*, **73**, 231 (1978).
57. A. L. Yergey and F. W. Lampe, *J. Am. Chem. Soc.*, **87**, 4204 (1965).
58. A. L. Yergey and F. W. Lampe, *J. Organomet. Chem.*, **15**, 339 (1968).
59. M. Gielen and J. Nasielski, *Bull. Soc. Chim. Belg.*, **77**, 5 (1968).
60. M. Gielen, M. R. Barthels, M. Declercq and J. Nasielski, *Bull. Soc. Chim. Belg.*, **80**, 189 (1971).
61. M. Gielen and G. Mayence, *J. Organomet. Chem.*, **46**, 281 (1972).
62. J. K. Khandelwal and J. W. Pinson, *Spectrosc. Lett.*, **6**, 41 (1973).
63. J. W. Pinson and J. K. Khandelwal, *Spectrosc. Lett.*, **6**, 745 (1973).
64. M. Aoyama, S. Masuda, K. Ohno, Y. Harada, M. C. Yew, H. H. Hua and L. S. Yong, *J. Phys. Chem.*, **93**, 1800 (1989).
65. K. Hottmann, *J. Prakt. Chem.*, **323**, 399 (1981).
66. J. Lango, L. Szepes, P. Csaszar and G. Innorta, *J. Organomet. Chem.*, **269**, 133 (1984).
67. S. M. Moerlein, *J. Organomet. Chem.*, **319**, 29 (1987).
68. O. Curcurutto, P. Traldi, G. Capozzi and S. Menichetti, *Org. Mass Spectrom.*, **26**, 119 (1991).
69. M. Gielen and K. Jurkschat, *Org. Mass Spectrom.*, **18**, 224 (1983).
70. M. Gielen, B. De Poorter, R. Liberton and M. T. Paelinck, *Bull. Soc. Chim. Belg.*, **82**, 277 (1973).
71. J. Tamas, A. K. Maltsev, O. M. Nefedov and G. Czira, *J. Organomet. Chem.*, **40**, 311 (1972).
72. M. Gielen and G. Mayence, *J. Organomet. Chem.*, **12**, 363 (1968).
73. K. Licht, H. Geissler, P. Koehler, K. Hottmann, H. Schnorr and H. Kriegsmann, *Z. Anorg. Allg. Chem.*, **385**, 271 (1971).
74. C. A. Dooley and J. P. Testa, *Org. Mass Spectrom.*, **24**, 343 (1989).
75. M. Gielen, *Org. Mass Spectrom.*, **18**, 453 (1983).
76. A. M. Duffield, C. Djerassi, P. Mazerolles, J. Dubac and G. Manuel, *J. Organomet. Chem.*, **12**, 123 (1968).
77. C. Lageot, J. C. Maire, P. Riviere, M. Massol and J. Barrau, *J. Organomet. Chem.*, **66**, 49 (1974).
78. C. Lageot, J. C. Maire, G. Manuel and P. Mazerolles, *J. Organomet. Chem.*, **54**, 131 (1973).
79. K. C. Williams, *J. Organomet. Chem.*, **19**, 210 (1969).
80. G. Dube, *Z. Chem.*, **9**, 316 (1969).
81. L. A. Xavier and J. M. Riveros, *Int. J. Mass Spectrom.*, **180**, 223 (1998).
82. H. G. Kuivila, K.-H. Tsai and D. G. I. Kingston, *J. Organomet. Chem.*, **23**, 129 (1970).
83. D. G. I. Kingston, H. P. Tannenbaum and H. G. Kuivila, *Org. Mass Spectrom.*, **9**, 31 (1974).
84. J. D. Kennedy and H. G. Kuivila, *J. Chem. Soc., Perkin Trans. 2,* 1812 (1972).
85. C. S. C. Wang and J. M. Shreeve, *J. Organomet. Chem.*, **49**, 417 (1973).
86. I. Mazeika, S. Grinberga, M. Gavars, A. P. Gaukhman, N. P. Erchak and E. Lukevics, *Org. Mass Spectrom.*, **28**, 1309 (1993).
87. S. Rozite, I. Mazeika, A. Gaukhman, N. P. Erchak, L. M. Ignatovich and E. Lukevics, *J. Organomet. Chem.*, **384**, 257 (1990).

88. M. L. Bazinet, W. G. Yeomans and C. Merritt, *Int. J. Mass Spectrom. Ion Processes,* **16**, 405 (1975).
89. C. D. Barsode, P. Umapathy and D. N. Sen, *J. Indian Chem. Soc.,* **52**, 942 (1975).
90. P. Umapathy, S. N. Bhide, K. D. Ghuge and D. N. Sen, *J. Indian Chem. Soc.,* **58**, 33 (1981).
91. F. Glockling, J. R. C. Light and R. G. Strafford, *J. Chem. Soc., A,* 426 (1970).
92. J. J. de Ridder and G. Dijkstra, *Org. Mass Spectrom.,* **1**, 647 (1968).
93. M. Gielen, J. Nasielski and G. Vandendunghen, *Bull. Soc. Chim. Belg.,* **80**, 175 (1971).
94. D. R. Dimmel, C. A. Wilkie and P. J. Lamothe, *Org. Mass Spectrom.,* **10**, 18 (1975).
95. S. Kyushin, S. Otani, Y. Nakadaira and M. Ohashi, *Chem. Lett.,* 29 (1995).
96. J. K. Kochi, *Pure Appl. Chem.,* **52**, 571 (1980).
97. K. Kuhlein and W. P. Neumann, *J. Organomet. Chem.,* **14**, 317 (1968).
98. J. J. de Ridder and J. G. Noltes, *J. Organomet. Chem.,* **20**, 287 (1969).
99. P. G. Harrison and S. R. Stobart, *J. Organomet. Chem.,* **47**, 89 (1973).
100. D. B. Chambers and F. Glockling, *J. Chem. Soc., A,* 735 (1968).
101. A. Guerrero, J. Cervantes, L. Velasco, J. Gomez-Lara, S. Sharma, E. Delgado and K. Pannell, *J. Organomet. Chem.,* **464**, 47 (1994).
102. M. Gielen, S. Simon, M. van de Steen and T. W. -Aosc, *Org. Mass Spectrom.,* **18**, 451 (1983).
103. R. H. Dahm, G. Lawson and N. Ostah, *Appl. Organomet. Chem.,* **9**, 141 (1995).
104. R. C. Dougherty, *Anal. Chem.,* **53**, 625A (1981).
105. P. J. Bassett and D. R. Lloyd, *J. Chem. Soc., A,* 641 (1971).
106. G. Beltram, T. P. Fehlner, K. Mochida and J. K. Kochi, *J. Electron Spectrosc. Related Phenom.,* **18**, 153 (1980).
107. R. Boschi, M. F. Lappert, J. B. Pedley, W. Schmidt and B. T. Wilkins, *J. Organomet. Chem.,* **50**, 69 (1973).
108. C. Cauletti, F. Grandinetti, A. Sebald and B. Wrackmeyer, *Inorg. Chim. Acta,* **117**, L37 (1986).
109. A. Chrostowska, V. Metail, G. Pfister-Guillouzo and J. C. Guillemin, *J. Organomet. Chem.,* **570**, 175 (1998).
110. S. Cradock and R. A. Whiteford, *Trans. Faraday Soc.,* **67**, 3425 (1971).
111. S. Cradock, R. A. Whiteford, W. J. Savage and E. A. Ebsworth, *J. Chem. Soc., Faraday Trans. 2,* **68**, 934 (1972).
112. S. Cradock, E. A. Ebsworth and A. Robertson, *J. Chem. Soc., Dalton Trans.,* 22 (1973).
113. J. E. Drake, B. M. Glavincevski and K. Gorzelska, *J. Electron Spectrosc. Relat. Phenom.,* **17**, 73 (1979).
114. J. E. Drake, B. M. Glavincevski and K. Gorzelska, *J. Electron Spectrosc. Relat. Phenom.,* **16**, 331 (1979).
115. J. E. Drake and K. Gorzelska, *J. Electron Spectrosc. Relat. Phenom.,* **21**, 365 (1981).
116. J. E. Drake, K. Gorzelska and R. Eujen, *J. Electron Spectrosc. Relat. Phenom.,* **49**, 311 (1989).
117. J. E. Drake, K. Gorzelska, G. S. White and R. Eujen, *J. Electron Spectrosc. Relat. Phenom.,* **26**, 1 (1982).
118. S. Foucat, T. Pigot, G. Pfister-Guillouzo, H. Lavayssiere and S. Mazieres, *Organometallics,* **18**, 5322 (1999).
119. G. Lespes, A. Dargelos and J. Fernandez-Sanz, *J. Organomet. Chem.,* **379**, 41 (1989).
120. D. R. Lloyd and P. J. Roberts, *J. Electron Spectros. Related. Phenom.,* **7**, 325 (1975).
121. J. C. Creasey, I. R. Lambert, R. P. Tuckett, K. Codling, J. F. Leszek, P. A. Hatherly and M. Stankiewicz, *J. Chem. Soc., Faraday Trans.,* **87**, 3717 (1991).
122. P. Burroughs, S. Evans, A. Hamnett, A. F. Orchard and N. V. Richardson, *J. Chem. Soc., Faraday Trans. 2,* **70**, 1895 (1974).
123. Vertical values, see G. Jonkers, S. M. Van Der Kerk, R. Mooyman and C. A. De Lange, *Chem. Phys. Lett.,* **90**, 252 (1982).
124. I. Novak and A. W. Potts, *J. Electron Spectrosc. Relat. Phenom.,* **33**, 1 (1984).
125. J. W. Hastie, H. Bloom and J. D. Morrison, *J. Chem. Phys.,* **47**, 1580 (1967).
126. S. Evans and A. F. Orchard, *J. Electron Spectrosc. Relat. Phenom.,* **6**, 207 (1975).
127. J. Berkowitz, *Adv. High Temp. Chem.,* **3**, 123 (1971).
128. J. C. Baldwin, M. F. Lappert, J. B. Pedley and J. S. Poland, *J. Chem. Soc., Dalton Trans.,* 1943 (1972).

129. J. Berkowitz, G. B. Ellison and D. Gutman, *J. Phys. Chem.,* **98**, 2744 (1994).
130. H. Basch, *Inorg. Chim. Acta,* **252**, 265 (1996).
131. P. Benzi, L. Operti, G. A. Vaglio, P. Volpe, M. Speranza and R. Gabrielli, *J. Organomet. Chem.,* **354**, 39 (1988).
132. J. K. Northrop and F. W. Lampe, *J. Phys. Chem.,* **77**, 30 (1973).
133. L. Operti, M. Splendore, G. A. Vaglio, A. M. Franklin and J. F. J. Todd, *Int. J. Mass Spectrom. Ion Processes,* **136**, 25 (1994).
134. S. N. Senzer, R. N. Abernathy and F. W. Lampe, *J. Phys. Chem.,* **84**, 3066 (1980).
135. E. P. L. Hunter and S. G. Lias, *J. Phys. Chem. Ref. Data,* **27**, 413 (1998).
136. P. R. Schreiner, H. F. Schaefer III and P. v. R. Schleyer, *J. Chem. Phys.,* **101**, 2141 (1994).
137. N. H. Morgon and J. M. Riveros, *J. Phys. Chem., A,* **102**, 10399 (1998).
138. E. F. Archibong and J. Leszczynski, *J. Phys. Chem.,* **98**, 10084 (1994).
139. S. Roszak, P. Babinec and J. Leszczynski, *Chem. Phys.,* **256**, 177 (2000).
140. L. Operti, M. Splendore, G. A. Vaglio, P. Volpe, M. Speranza and G. Occhiucci, *J. Organomet. Chem.,* **433**, 35 (1992).
141. L. Operti, M. Splendore, G. A. Vaglio and P. Volpe, *Organometallics,* **12**, 4509 (1993).
142. P. Jackson, R. Srinivas, N. Langermann, M. Diefenbach, D. Schroder and H. Schwarz, *Int. J. Mass Spectrom.,* **201**, 23 (2000).
143. P. Jackson, M. Diefenbach, D. Schroder and H. Schwarz, *Eur. J. Inorg. Chem.,* 1203 (1999).
144. S. Kohda-Sudoh, S. Katagiri, S. Ikuta and O. Nomura, *J. Mol. Struct.* (THEOCHEM), **31**, 113 (1986).
145. P. Benzi, L. Operti, G. A. Vaglio, P. Volpe, M. Speranza and R. Gabrielli, *J. Organomet. Chem.,* **373**, 289 (1989).
146. P. Benzi, L. Operti, G. A. Vaglio, P. Volpe, M. Speranza and R. Gabrielli, *Int. J. Mass Spectrom. Ion Processes,* **100**, 647 (1990).
147. L. Operti, M. Splendore, G. A. Vaglio and P. Volpe, *Spectrochim. Acta, Part A,* **49**, 1213 (1993).
148. P. Benzi, L. Operti, R. Rabezzana, M. Splendore and P. Volpe, *Int. J. Mass Spectrom. Ion Processes,* **152**, 61 (1996).
149. P. Antoniotti, L. Operti, R. Rabezzana and G. A. Vaglio. *Int. J. Mass Spectrom.,* **183**, 63 (1999).
150. P. Benzi, L. Operti and R. Rabezzana, *Eur. J. Inorg. Chem.,* 505 (2000).
151. P. Antoniotti and F. Grandinetti, *Gazz. Chim. Ital.,* **120**, 701 (1990).
152. K. P. Lim and F. W. Lampe, *Org. Mass Spectrom.,* **28**, 349 (1993).
153. P. Antoniotti, P. Benzi, F. Grandinetti and P. Volpe, *J. Phys. Chem.,* **97**, 4945 (1993).
154. P. Antoniotti, F. Grandinetti and P. Volpe, *J. Phys. Chem.,* **99**, 17724 (1995).
155. P. Jackson, R. Srinivas, S. J. Blanksby, D. Schroder and H. Schwarz, *Chem. Eur. J.,* **6**, 1236 (2000).
156. L. Operti, M. Splendore, G. A. Vaglio and P. Volpe, *Organometallics,* **12**, 4516 (1993).
157. M. Castiglioni, L. Operti, R. Rabezzana, G. A. Vaglio and P. Volpe, *Int. J. Mass Spectrom.,* **180**, 277 (1998).
158. N. A. Gomzina, T. A. Kochina, V. D. Nefedov, E. N. Sinotova and D. V. Vrazhnov, *Russ. J. Gen. Chem.,* **64**, 403 (1994).
159. A. C. M. Wojtyniak, X. P. Li and J. A. Stone, *Can. J. Chem.,* **65**, 2849 (1987).
160. B. Chiavarino, M. E. Crestoni and S. Fornarini, *Organometallics,* **14**, 2624 (1995).
161. B. Chiavarino, M. E. Crestoni and S. Fornarini, *J. Organomet. Chem.,* **545/546**, 45 (1997).
162. V. C. Trenerry and J. H. Bowie, *Org. Mass Spectrom.,* **16**, 344 (1981).
163. V. C. Trenerry, G. Klass, J. H. Bowie and I. A. Blair, *J. Chem. Res. (S),* 386 (1980).
164. W. D. Meyerhoffer and M. M. Bursey, *J. Organomet. Chem.,* **373**, 143 (1989).
165. J. A. Stone and W. J. Wytenburg, *Can. J. Chem.,* **65**, 2146 (1987).
166. B. Chiavarino, M. E. Crestoni and S. Fornarini, *J. Organomet. Chem.,* **545/546**, 53 (1997).
167. J. A. Stone and D. E. Splinter, *Int. J. Mass Spectrom. Ion Processes,* **59**, 169 (1984).
168. B. C. Guo and A. W. Castleman, *Zeit. Phys. D.,* **19**, 397 (1991).
169. I. N. Tang and A. W. Castleman, *J. Chem. Phys.,* **57**, 3638 (1972).

170. B. C. Guo and A. W. Castleman, *Int. J. Mass Spectrom. Ion Processes,* **100**, 665 (1990).
171. B. C. Guo, J. W. Purnell and A. W. Castleman, *Chem. Phys. Lett.,* **168**, 155 (1990).
172. J. G. Dillard, *Chem. Rev.,* **73**, 589 (1973).
173. P. W. Harland, S. Cradock and J. C. J. Thynne, *Int. J. Mass Spectrom. Ion Phys.,* **10**, 169 (1972).
174. P. W. Harland, S. Cradock and J. C. J. Thynne, *Inorg. Nucl. Chem. Lett.,* **9**, 53 (1973).
175. J. L. Franklin, J. L.-F. Wang, S. L. Bennett, P. W. Harland and J. L. Margrave, *Adv. Mass Spectrom.,* **6**, 319 (1974).
176. J. L.-F. Wang, J. L. Margrave, and J. L. Franklin, *J. Chem. Phys.,* **60**, 2158 (1974).
177. Q. Li, G. Li, W. Xu, Y. Xie and H. F. Schaefer III, *J. Chem. Phys.,* **111**, 7945 (1999).
178. R. E. Pabst, J. L. Margrave and J. L. Franklin, *Int. J. Mass Spectrom. Ion Phys.,* **25**, 361 (1977).
179. S. L. Bennett, J. L.-F. Wang, J. L. Margrave and J. L. Franklin, *High Temp. Sci.,* **7**, 142 (1975).
180. R. E. Pabst, D. L. Perry, J. L. Margrave and J. L. Franklin, *Int. J. Mass Spectrom. Ion Phys.,* **25**, 323 (1977).
181. B. P. Mathur, E. W. Rothe and G. P. Reck, *Int. J. Mass Spectrom. Ion Phys.,* **31**, 77 (1979).
182. K. Lacmann, M. J. P. Maneira, A. M. C. Moutinho and U. Weigmann, *J. Chem. Phys.,* **78**, 1767 (1983).
183. D. M. Wetzel and J. I. Brauman, *Chem. Rev.,* **87**, 607 (1987).
184. K. J. Reed and J. I. Brauman, *J. Chem. Phys.,* **61**, 4830 (1974).
185. E. A. Brinkman, K. Salomon, W. Tumas and J. I. Brauman, *J. Am. Chem. Soc.,* **117**, 4905 (1995).
186. P. M. Mayer, J. F. Gal and L. Radom, *Int. J. Mass Spectrom. Ion Processes,* **167**, 689 (1997).
187. T. M. Miller, A. E. S. Miller and W. C. Lineberger, *Phys. Rev. A,* **33**, 3558 (1986).
188. G. R. Burton, C. S. Xu, C. C. Arnold and D. M. Neumark, *J. Chem. Phys.,* **104**, 2757 (1996).
189. Y. Negishi, H. Kawamata, T. Hayase, M. Gomei, R. Kishi, F. Hayakawa, A. Nakajima and K. Kaya, *Chem. Phys. Lett.,* **269**, 199 (1997).
190. M. Decouzon, J.-F. Gal, J. Gayraud, P. C. Maria, G.-A. Vaglio and P. Volpe, *J. Am. Soc. Mass Spectrom.,* **4**, 54 (1993).
191. D. M. Wetzel, K. E. Salomon, S. Berger and J. I. Brauman, *J. Am. Chem. Soc.,* **111**, 3835 (1989).
192. N. H. Morgon, L. A. Xavier and J. M. Riveros, *Int. J. Mass Spectrom.,* **196**, 363 (2000).
193. N. H. Morgon, A. B. Argenton, M. L. P. Silva and J. M. Riveros, *J. Am. Chem. Soc.,* **119**, 1708 (1997).
194. H. Romeyn and W. A. Noyes Jr., *J. Am. Chem. Soc.,* **54**, 4143 (1932).
195. Y. Rousseau and G. J. Mains, *J. Phys. Chem.,* **70**, 3158 (1966).
196. R. Varma, K. R. Ramaprasad, A. J. Sidorelli and B. K. Sahay, *J. Inorg. Nucl. Chem.,* **37**, 563 (1975).
197. R. J. Cvetanovic, in *Progress in Reaction Kinetics* (Ed. G. Porter), Vol. 2, Chap. 2, Pergamon, London, 1964, pp. 39–130.
198. G. R. Smith and W. A. Guillory, *J. Chem. Phys.,* **56**, 1423 (1972).
199. M. V. Piserchio and F. W. Lampe, *J. Photochem. Photobiol. A: Chem.,* **60**, 11 (1991).
200. K. Tamaru, M. Boudart and H. Taylor, *J. Phys. Chem.,* **59**, 806 (1955).
201. C. G. Newman, J. Dzarnoski, M. A. Ring and H. E. O'Neal, *Int. J. Chem. Kinet.,* **12**, 661 (1980).
202. J. Blazejowski and F. W. Lampe, *J. Phys. Chem.,* **93**, 8038 (1989).
203. J. Dzarnoski, H. E. O'Neal and M. A. Ring, *J. Am. Chem. Soc.,* **103**, 5740 (1981).
204. A. E. Stanley, *J. Photochem. Photobiol. A: Chem.,* **99**, 1 (1996).
205. M. Jakoubkova, Z. Bastl. P. Fiedler and J. Pola, *Infrared Phys. Technol.,* **35**, 633 (1994).
206. J. E. Taylor and T. S. Milazzo, *J. Phys. Chem.,* **82**, 847 (1978).
207. R. L. Geddes and E. Marks, *J. Am. Chem. Soc.,* **52**, 4372 (1930).
208. G. P. Smith and R. Patrick, *Int. J. Chem. Kinet.,* **15**, 167 (1983).
209. J. Pola and R. Taylor, *J. Organomet. Chem.,* **437**, 271 (1992).

210. J. Pola, J. P. Parsons and R. Taylor, *J. Chem. Soc., Faraday Trans.*, **88**, 1637 (1992).
211. A. E. Stanley, R. A. Johnson, J. B. Turner and A. H. Roberts, *Appl. Spectrosc.*, **40**, 374 (1986).
212. T. A. Sladkova, O. O. Berezhanskaya, B. M. Zolotarev and G. A. Razuvaev, *Bull. Acad. Sci. USSR, Div. Chem. Sci.*, 144 (1978).
213. J. Pola, R. Fajgar, Z. Bastl and L. Diaz, *J. Meter. Chem.*, **2**, 961 (1992).
214. R. Fajgar, M. Jakoubkova, Z. Bastl and J. Pola, *Appl. Surf. Sci.*, **86**, 530 (1995).
215. R. Fajgar, Z. Bastl, J. Tlaskal and J. Pola, *Appl. Organomet. Chem.*, **9**, 667 (1995).
216. D. E. Nava, W. A. Payne, G. Marston and L. J. Stief, *J. Geophys. Res.*, **98**, 5531 (1993).
217. N. A. Arthur and I. A. Cooper, *J. Chem. Soc., Faraday Trans.*, **91**, 3367 (1995).
218. K. Y. Choo, P. P. Gaspar and A. P. Wolf, *J. Phys. Chem.*, **79**, 1752 (1975).
219. E. R. Austin and F. W. Lampe, *J. Phys. Chem.*, **81**, 1134 (1977).
220. N. A. Arthur, I. A. Cooper and L. A. Miles, *Int. J. Chem. Kinet.*, **29**, 237 (1997).
221. E. R. Austin and F. W. Lampe, *J. Phys. Chem.*, **81**, 1546 (1977).
222. K. C. Kim, D. W. Setser and C. M. Bogan, *J. Chem. Phys.*, **60**, 1837 (1974).
223. B. S. Agrawalla and D. W. Setser, *J. Chem. Phys.*, **86**, 5421 (1987).
224. N. I. Butkovskaya and D. W. Setser, *J. Chem. Phys.*, **106**, 5028 (1997).
225. M. J. Almond, A. M. Doncaster, P. N. Noble and R. Walsh, *J. Am. Chem. Soc.*, **104**, 4717 (1982).
226. P. N. Noble and R. Walsh, *Int. J. Chem. Kinet.*, **15**, 547 (1983).
227. (a) A. M. Doncaster and R. Walsh, *J. Phys. Chem.*, **83**, 578 (1979).
(b) A. M. Doncaster and R. Walsh, *J. Chem. Soc., Chem. Commun.*, 446 (1977).
228. P. N. Noble and R. Walsh, *Int. J. Chem. Kinet.*, **15**, 561 (1983).
229. Y.-E. Lee and K. Y. Choo, *Int. J. Chem. Kinet.*, **18**, 267 (1986).
230. M. Eichholtz, A. Schneider, J. T. Vollmer and H. G. Wagner, *Z. Phys. Chem.-Int. J. Res. Phys. Chem. Chem. Phys.*, **199**, 267 (1997).
231. D. Q. Lei, M. E. Lee and P. P. Gaspar, *Tetrahedron*, **53**, 10179 (1997).
232. W. P. Neumann, *Chem. Rev.*, **91**, 311 (1991).
233. (a) K. Saito and K. Obi, *Chem. Phys. Lett.*, **215**, 193 (1993).
(b) K. Saito and K. Obi, *Chem. Phys.*, **187**, 381 (1994).
234. R. Becerra, S. E. Boganov, M. P. Egorov, O. M. Nefedov and R. Walsh, *Chem. Phys. Lett.*, **260**, 433 (1996).
235. R. Becerra and R. Walsh, *Phys. Chem. Chem. Phys.*, **1**, 5301 (1999).
236. R. Becerra, S. E. Boganov, M. P. Egorov, V. I. Faustov, O. M. Nefedov and R. Walsh, *J. Am. Chem. Soc.*, **120**, 12657 (1998).
237. R. Becerra, S. Boganov and R. Walsh, *J. Chem. Soc., Faraday Trans.*, **94**, 3569 (1998).
238. R. Becerra, S. E. Boganov, M. P. Egorov, O. M. Nefedov and R. Walsh, *Mendeleev Commun.*, 87 (1997).
239. R. Becerra, S. E. Boganov, M. P. Egorov, V. Y. Lee, O. M. Nefedov and R. Walsh, *Chem. Phys. Lett.*, **250**, 111 (1996).
240. R. Becerra, S. E. Boganov, M. P. Egorov, V. I. Faustov, O. M. Nefedov and R. Walsh, *Can. J. Chem.* **78**, 1428 (2000).
241. K. Tamaru, *J. Phys. Chem.*, **60**, 610 (1956).
242. S. F. Kettle, *J. Chem. Soc.*, 2569 (1961).
243. R. F. Meads, D. C. J. Mardsen, J. A. Harrison and L. F. Phillips, *Chem. Phys. Lett.*, **160**, 342 (1989).
244. D. J. Aaserud and F. W. Lampe, *J. Phys. Chem.*, **100**, 10215 (1996).
245. (a) W. L. Jolly, *Angew. Chem.*, **72**, 268 (1960).
(b) W. L. Jolly, *J. Am. Chem. Soc.*, **83**, 335 (1961).
246. D. J. Aaserud and F. W. Lampe, *J. Phys. Chem. A*, **101**, 4114 (1997).
247. Y. Y. Baryshnikov, I. L. Zakharov, T. I. Lazareva and M. V. Trushina, *Kinet. Catal.*, **32**, 709 (1991).
248. P. D. E. Biggins and R. M. Harrison, *Environ. Sci. Technol.*, **13**, 558 (1979).
249. R. M. Harrison and D. P. H. Laxen, *Environ. Sci. Technol.*, **12**, 1384 (1978).
250. Y. A. Aleksandrov, N. G. Sheyanov and V. A. Shushunov, *J. Gen. Chem. USSR (Engl. Transl.)*, **39**, 957 (1969).
251. O. J. Nielsen, D. J. O'Farrell, J. J. Treacy and H. W. Sidebottom, *Environ. Sci. Technol.*, **25**, 1098 (1991).

252. For some typical radiolysis studies, see:
(a) P. Antoniotti, P. Benzi, M. Castiglioni, L. Operti and P. Volpe, *Chem. Mater.*, **4**, 717 (1992).
(b) P. Benzi, M. Castiglioni and P. Volpe, *J. Matar. Chem.*, **4**, 1067 (1994).
(c) P. Benzi, M. Castiglioni, E. Trufa and P. Volpe, *J. Matar. Chem.*, **6**, 1507 (1996).
(d) P. Antoniotti, P. Benzi, M. Castiglioni and P. Volpe, *Eur. J. Inorg. Chem.*, 323 (1999).

CHAPTER 6

Further advances in germanium, tin and lead NMR

HEINRICH CHR. MARSMANN

Universität Paderborn, Fachbereich Chemie, Anorganische Chemie, Warburger Straße 100, D-30095 Paderborn, Germany
Fax: +49-5251-60 3485; E-mail: hcm@ac16.uni-paderborn.de

and

FRANK UHLIG

Universität Dortmund, Fachbereich Chemie, Anorganische Chemie II, Otto-Hahn-Str. 6, D-44221 Dortmund, Germany
Fax: +49-231-755 5048; E-mail: fuhl@platon.chemie.uni-dortmund.de

The chemistry of organic germanium, tin and lead compounds — Vol. 2
Edited by Z. Rappoport

I. INTRODUCTION

Despite their environmental problems organic tin and lead containing compounds are still in the focus of interest of both academic as well as industrial research. One of the most powerful tools for the characterization of derivatives in this field of chemistry is the application of NMR experiments. The properties of the isotopes of group 14 elements used for NMR experiments are shown in Table 1.

Therefore, it is no surprise that the group 14 elements, with the exception of germanium, are well investigated by NMR spectroscopy. This survey covers the development of NMR observations for ^{73}Ge, ^{119}Sn and ^{207}Pb between 1995 and 2000 for organometallic compounds. In these years *Chemical Abstracts* lists more than 500 entries for ^{119}Sn NMR results and nearly 50 entries for ^{207}Pb NMR. In contrast, only a limited number of germanium resonances is found in the literature. Consequently, it will not be possible to give tin all the attention it deserves. However, there are data collections giving a much more detailed view of results obtained so far.

II. GERMANIUM NMR

A. History and Technical Details

In comparison with the homologous elements of germanium, for example silicon and tin, only little is known about germanium NMR. Germanium NMR had a 'boom' during the 1980s but research is proceeding relatively leisurely in recent years. The limitations in the observations of germanium resonance are mainly due to the nature of the element. Germanium has only one NMR active isotope, ^{73}Ge, with a low gyromagnetic

TABLE 1. Properties of isotopes of group 14 elements

	Natural abundance (%)	Nuclear spin	Magnetic moment μ^a	Sensitivity		Receptivity relative to ^{13}C
				rel.^{b}	abs.^{c}	
^{13}C	1.108	$\frac{1}{2}$	0.7022	1.59×10^{-2}	1.76×10^{-4}	1
^{29}Si	4.7	$\frac{1}{2}$	−0.5548	7.84×10^{-3}	3.69×10^{-4}	2.1
^{73}Ge	7.76	$\frac{9}{2}^d$	−0.8768	1.4×10^{-3}	1.08×10^{-4}	0.61
^{115}Sn	0.35	$\frac{1}{2}$	−0.9132	3.5×10^{-2}	1.22×10^{-4}	0.69
^{117}Sn	7.61	$\frac{1}{2}$	−0.9949	4.52×10^{-2}	3.44×10^{-3}	19.5
^{119}Sn	8.58	$\frac{1}{2}$	−1.0409	5.18×10^{-3}	4.44×10^{-3}	25.2
^{207}Pb	22.6	$\frac{1}{2}$	0.5843	9.16×10^{-3}	2.07×10^{-3}	11.8

^{a}In multiples of the nuclear magneton[1].
^{b}For equal number of nuclei at constant field.
^{c}Product of the relative sensitivity and the natural abundance.
^{d}Quadrupole moment, -0.18×10^{-28} m.

ratio (-0.9332 rad s^{-1} T^{-1} $\times 10^7$), a large nuclear spin ($\frac{9}{2}$)' and is observed at low resonance frequencies. Most of the ^{73}Ge chemical shifts, known from the literature, were measured with respect to tetramethylgermane (Me_4Ge: $\delta = 0$ ppm) or $GeCl_4$ [converted to Me_4Ge; $\delta(Me_4Ge) = \delta(GeCl_4) - 30.9$]. Cited negative values mean that the resonance is at a lower frequency than Me_4Ge. Absolute frequencies, relative to the ^{1}H signal of Me_4Si = 100 MHz, were 3.488423 MHz $\pm$ 10 Hz for $GeCl_4$ and 3.488418 $\pm$ 10 Hz for Me_4Ge[2,3].

The first pioneering work of high resolution ^{73}Ge nuclear magnetic resonance[3,4] was undertaken by the groups of Kaufmann and Spinney at the beginning of the 1970s. More far-reaching progress is linked with the research groups of Lukevics[5], Mackay[6] and Takeuchi[7] in the early 1980s.

B. Problems in ^{73}Ge NMR Spectroscopy

In recent years a number of ^{73}Ge chemical shifts were published. However, the method has still a vast number of limitations and therefore only a few hundred compounds are characterized by ^{73}Ge NMR. Earlier reviews concerning ^{73}Ge NMR chemical shifts are given in the literature[8–12].

Especially, organometallic chemists might encounter a number of difficulties in attempting to use ^{73}Ge NMR for the characterization of samples. Although a wide variety of classes of germanium containing compounds has been observed, including germanium hydrides, alkyl- and polyalkylgermanes, alkyl derivatives, tetra-alkoxides, tetraarylgermanes and mixed tetrahalides, as far as we are aware no chemical shifts have been given in the literature for mixed alkyl/aryl halides of germanium[13]. For example, in the ^{73}Ge NMR spectra of 1-chloro-1,1-dimethyldigermane ($H_3Ge-GeMe_2Cl$) only the signal for the GeH_3 unit was observed[13]. Similarly, the homoleptic $Ge(OR)_4$ species were observable but mixed compounds of type $R_nGe(OR')_{4-n}$ were not, with one exception. This exception, i.e. $Me_3GeONMe_2$, is remarkable from another point of view as well. The ^{73}Ge chemical shift has a value of +522 ppm[14], which is the only known chemical shift in this low field region. All other described germanium resonances are in a range between +100 and −1200 ppm.

The germanium NMR data fit very well the linear relationships found between the chemical shifts of the group 14 elements[15]. There are linear ^{73}Ge/^{29}Si and ^{73}Ge/^{119}Sn chemical shift correlations between derivatives of silicon, germanium and tin[13].

C. Current Progress in ^{73}Ge NMR

One of the major problems of working with ^{73}Ge is the baseline roll as a result of the acoustic ringing, which creates difficulties with the expected broader lines for many germanium derivatives. A solution to this problem might be the use of pulse techniques such as RIDE, PHASE or EXSPEC[13,16]. The first example for the application of such techniques was in solving the behavior of several exchange processes[13].

The use of such pulse techniques and higher processing capacities of computers should allow some progress in this field in the near future, for example in getting a better understanding of the effects of substituents of the field gradients and therefore on the line widths in the germanium spectra.

D. ^{73}Ge Chemical Shifts of Organometallic Compounds

^{73}Ge chemical shifts of selected organometallic compounds are given in Table 2.

TABLE 2. ^{73}Ge Chemical Shifts of Organometallic Compounds

Structural formula		Chemical shift (ppm)	Linewidth (Hz) (solvent)	Reference
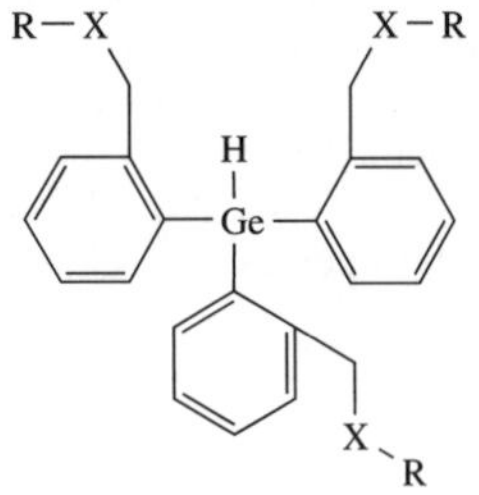 Tris[2-(R–X–CH$_2$)C$_6$H$_4$]GeH	R = Me; X = O	−85.4	350 ($CDCl_3$)	17
	R = Et; X = O	−85.4	350 ($CDCl_3$)	
	R = *t*-Bu; X = O	−84.0	350 ($CDCl_3$)	
	R = Me; X = S	−92.7	270 ($CDCl_3$)	
	R = Me; X = NMe	−89	900 ($CDCl_3$)	
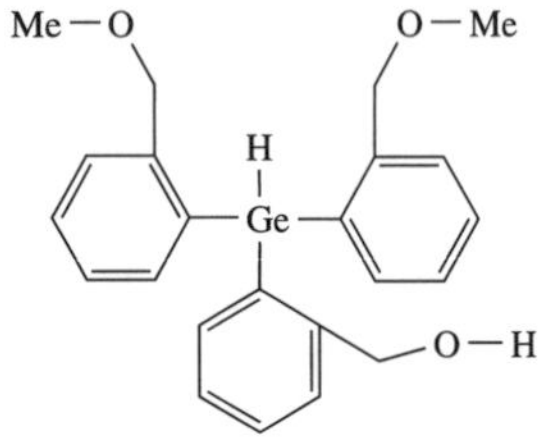 Bis[2-(MeOCH$_2$)C$_6$H$_4$][2-(HOCH$_2$)C$_6$H$_4$]GeH		−84.7	250 ($CDCl_3$)	17
$(C_6H_5)_3GeH$		−57.0	20 ($CDCl_3$)	17
		−55.4[a]	— (C_6D_6)	22
$(2\text{-}MeC_6H_4)_3GeH$		−84.0	20 ($CDCl_3$)	17
$(2\text{-}MeC_6H_4)_4Ge$		−33.2	80 ($CDCl_3$)	17
$(Me_3Sn)_4Ge$		−515.2	— (C_6D_6)	18
$(Me_3Sn)_2GeMe_2$		−169.0	— (C_6D_6)	18
$Me_3Sn{-}GeMe_2{-}GeMe_2{-}SnMe_3$		−133.2	— (C_6D_6)	18
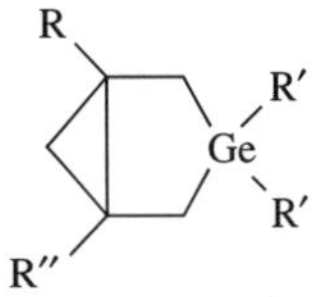 3-Germabicyclo[4.1.0]heptane with R, R″ on the bridgehead carbons and 3,3-R′$_2$ on Ge	R = H; R′ = Me; R″ = H	57	— ($CDCl_3$)	19
	R = H; R′ = Et; R″ = H	76	— ($CDCl_3$)	
	R = Me; R′ = Et; R″ = H	82	— ($CDCl_3$)	
	R = R″ = Me; R′ = Et	46	— ($CDCl_3$)	
Ph_4Ge		−32.9	— ($CDCl_3$)	20 2 21
$[Me_2N(CH_2)_3]_3GePh{\cdot}HCl$		−25.1	—	20
$(4\text{-}MeC_6H_4)GePh_3$		−32.5	— ($CDCl_3$)	21
$(4\text{-}MeC_6H_4)_2GePh_2$		−31.7	— ($CDCl_3$)	21
$(4\text{-}MeC_6H_4)_3GePh$		−30.8	— ($CDCl_3$)	21
$(4\text{-}MeC_6H_4)_4Ge$		−31.2	— ($CDCl_3$)	21
$Me_3GeONMe_2$		522.0	— (C_6D_6)	14
$4\text{-}MeOC_6H_4GeH_3$		−189.9[b]	— (C_6D_6)	22

TABLE 2. (*continued*)

Structural formula	Chemical shift (ppm)	Linewidth (Hz) (solvent)	Reference
$(4\text{-}MeOC_6H_4)_2GeH_2$	−112.0	— (C_6D_6)	22
$4\text{-}MeC_6H_4GeH_3$	-190.6^c	— (C_6D_6)	22
$MesGeH_3$	-234.3^d	— (C_6D_6)	22
$C_6H_5GeH_3$	-190.0^a	— (C_6D_6)	22
$(C_6H_5)_2GeH_2$	-108.8^e	— (C_6D_6)	22

^{a} $^1J_{GeH}$: 98 Hz;
^{b} $^1J_{GeH}$: 97 Hz;
^{c} $^1J_{GeH}$: 96 Hz;
^{d} $^1J_{GeH}$: 95 Hz;
^{e} $^1J_{GeH}$: 94 Hz.

III. TIN NMR

A. Introduction

As mentioned above, this review can not discuss all the ^{119}Sn NMR results published in the recent 5 years[23–389]. A large number of them are compiled in other data collections[23–25]. Elemental tin consist of ten stable isotopes. Three of them have a spin of $\frac{1}{2}$ and therefore a magnetic moment (see Table 1). Because of its high abundance and its higher magnetic moment leading to a sensitivity of 25.2 compared to that of carbon-13, tin-119 is mostly used for NMR purposes, although in some studies the isotopes ^{115}Sn[26] or ^{117}Sn[278–288] are used instead. Referencing is always done directly or indirectly to $SnMe_4$. Positive shifts mean shifts to a higher frequency or to a lower field.

B. Technical Details

The sensitivity of the ^{119}Sn experiment makes tin NMR spectroscopy a standard procedure for every tin containing compound. Most techniques of observing tin NMR were developed prior to the period covered in this review, but a few examples are presented here. Single resonance experiments and simple proton–tin decoupling do not present too many challenges except for the negative sign of the magnetic moments of the active tin isotopes, which could lead to a decrease in signal strength or even to a null signal, if an unfavorable mix of relaxation paths occurs. Reverse gated decoupling is then the best choice if quantitative results are required. If coupling constants to protons are known, pulse sequences such as ^1H–^{119}Sn INEPT[136] are used frequently. For special purposes such as the determination of coupling constants, variations of this method such as HEED INEPT[27,136,146] are of advantage because the central line, which can be rather broad and sometimes masking the satellites is suppressed. One cause of this could be the relaxation by chemical shift anisotropy[25].

For the determination of the sign of coupling constants involving tin, 2D correlation spectra in the form of ^1H–^{119}Sn HETCOR[27–29,136], HMQC[30,31] or HMBC[43,265] are employed. Due to the good sensitivity, Sn–Sn COSY experiments can be conducted in a reasonable time[32].

Solid state NMR can be performed for crystalline or amorphous samples. Tetracyclohexyltin, $(c\text{-}C_6H_{11})_4Sn$ ($\delta = -97.0$ ppm), is often used as a secondary standard[46]. The spinning side bands are very extensive due to the often sizeable chemical shift anisotropy which is especially pronounced for higher coordination numbers[41]. The line representing the isotropic shift is not found easily. Very often it is one of smaller signals[47]. Careful analysis is then needed to obtain the chemical shift tensors around the tin atom and the central line giving the isotropic shift. Extreme care should also be exercised in interpreting MAS chemical shift data without knowing the crystal structure of the compounds. A recent example is provided by Clayden and Pugh[48]. To overcome the often high chemical shift anisotropies, high rotation frequencies must be used. To diminish the problem of unstable cross-polarization behavior at high spinning speeds, variable-amplitude cross-polarization pulse programs (VACP) have been used[44]. High rotational speeds lead to high sample temperatures because of the friction between the rotor and the bearing gas. Use of the ^{119}Sn chemical shift of $Sm_2Sn_2O_7$ to determine the sample temperature has been proposed[45].

Besides the opportunity to measure chemical shift anisotropies, MAS NMR also offers the possibility to measure the scalar couplings to other isotopes with magnetic moments. Examples to determine the spatial components of these couplings can be found in the literature[48,49]. Even couplings to quadrupolar nuclei not observable in solution can be deduced from the MAS spectra[48]. 2D techniques such as HETCOR are also applicable for solids[49].

The $^{x}J(^{119}Sn{-}X)$ coupling constants of tin derivatives were discussed recently in a number of reviews and need not be repeated in this chapter[24,25,383].

C. Chemical Shifts

1. General remarks

Tin has a rather complicated chemistry. First, it must be distinguished between its two valency states Sn(II) and Sn(IV). Second, in both valency states tin may act as a Lewis acid connecting the tin with one or more donor atoms such as oxygen, nitrogen and other donors. Consequently, besides the so-called 'normal' coordination numbers of two for Sn(II) and four for Sn(IV), higher numbers up to eight are frequently encountered. The number and kind of donor atoms attached to tin change very easily, e.g. by using different solvents or between the solid and the dissolved state, making the assignment of chemical shifts to different types of tin compounds far from straightforward. In some studies of the biological effects of tin, the samples have many donor groups in the organic part and the environment of the tin is not very well defined. A proposal to classify tin chemical shifts by statistical methods has been advanced by Bartel and John[53].

2. Tin(II) compounds

A selection of chemical shifts of tin(II) compounds with direct tin-group 14 element or tin-group 15 element bonds encountered at the time of the report are collected in Table 3. More organometallic tin(II) derivatives are found in the literature[54–71]. The tin-119 chemical shifts for tin(II) compounds cover a range from nearly 4000 to −3000 ppm. Due to this wide range it is impossible to recognize simple correlations between the coordination number at the tin center or the substituents directly bonded to tin(II) and the chemical shift value. What one can suggest are predictions within only one class of compounds, e.g. for the derivatives shown below with a given substitution pattern when the R group is changed selectively (see also Table 3).

R = $CH(PPh_2)_2$	397	R = Cl	793.4	R = Cl	−100
R = Tip	474	R = *t*-Bu	1904.4	R = $CH(SiMe_3)_2$	258
R = $Si(SiMe_3)_3$	876	R = $N(SiMe_3)_2$	1196.8	R = $SiPh_3$	192
R = $Sn_B(SiMe_3)_3$	897	R = Sn_BMe_2R''	2856.9	R = Sn_BMe_3	217
Tip = 2,4,6-(i-Pr)$_3C_6H_2$		R′ = Tip			
		R″ = 2,6-Tip_2 C_6H_3			

This difficulty arises also from another problem in those compounds: the fact that equilibria exist between stannylenes in their true state as R_2Sn and their dimeric counterparts, the distannenes $R_2Sn{=}SnR_2$. For this reason, the separation of compounds belonging to Tables 3 and 4 is probably open to debate. Recent reviews about the chemical behavior of stannylenes, distannenes and their relatives in group 14 are available[123,125,127–130].

A rough correlation also exists between the shift of tin in the oxidation state +2 in transition metal complexes and the coordination number, so that additional ligands also cause additional shielding. Although this is not exactly the topic of this review, some Sn(II) compounds with a surrounding of three transition metal atoms constitute the low shielding limit of tin so far. The shift of 3924 ppm for Sn(II) in **1**, being in the center of a triangle of Cr atoms, is thought to be the result of low-lying π^*-orbitals[70]. If the triangle is formed by Mn atoms in **2**, a still lower shielding of 3301 ppm is observed[71].

$[\{(CO)_5Cr\}_3Sn]^{2-}$ **(1)** 3924 ppm

$[\{Cp^*(CO)_2Mn\}_3Sn]$ **(2)** 3301 pm

The increase of the coordination number by exchanging one metal center by two oxygen or nitrogen donor atoms in compounds **3** leaves the shift range between 600 and 1500 ppm.

$(TM)_2SnD_2$ **(3)** $(TM)SnD_3$ **(4)**

TM = transition metal fragment; D = halogen, organic amine or alkoxy group

One transition metal atom and three other donor atoms around the tin in **4** result in resonances of tin(II) between 170 and 344 ppm. For transition metal tin(II) compounds with similar substituents, the ^{119}Sn chemical shifts for the tin compounds are significantly lower (high field shifted) than for the chromium or molybdenum derivatives[55–57]. The order of shifts below is in good agreement with the results for tin(IV) compounds.

$$\delta[\{(Z)_5Cr\}_2Sn\ (bipy)] \approx \delta[\{(Z)_5Mo\}_2Sn\ (bipy)] > \delta[\{(Z)_5W\}_2Sn\ (bipy)]$$

$$Z = CO,\ cyclopentadienyl$$

The high coordination numbers typical for the cyclopentadienyl ligand lead to strong shielding as exemplified in Table 3. The low coordination of the central tin atom in the tristannaallene $(t\text{-}Bu_3Si)_2Sn{=}Sn{=}Sn(Si(Bu\text{-}t)_3)_2$ also results in a strong deshielding. The mesomeric structures of this compound lead to an interpretation in which the terminal tin is regarded as tin(II)[97].

TABLE 3. ^{119}Sn NMR data of tin(II) compounds

Compound	Chemical shift (ppm)	Solvent	Reference
$[(2,4,6\text{-}i\text{-}Pr_3C_6H_2)Sn(2,4,6\text{-}R_3C_6H_2)]_2$; R = $CH(SiMe_3)_2$	2208	$C_6D_5CD_3$	72
$[(2\text{-}t\text{-}Bu\text{-}4,5,6\text{-}Me_3C_6H)_2Sn]_2$	1329 1331 at 298 K 1401 at 373 K	$C_6D_5CD_3$ $C_6D_5CD_3$ $C_6D_5CD_3$	73 74
$2,6\text{-}R_2C_6H_3$–Sn-X; R = Mes; X = $2,6\text{-}Mes_2C_6H_3$ Cl (dimer)	635 562	C_6D_6 C_6D_6	75
$2,4,6\text{-}(CF_3)_3C_6H_2$–Sn–$Si(SiMe_3)_3$	168	C_6D_6	76

TABLE 3. (*continued*)

Compound		Chemical shift (ppm)	Solvent	Reference
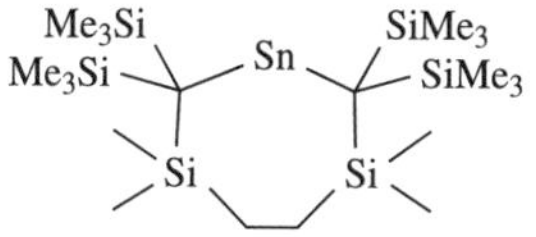 Cyclic dialkylstannylene (Sn bonded to two $C(SiMe_3)_2$ groups, bridged by $-SiMe_2CH_2CH_2SiMe_2-$)		2299	C_6D_6	77
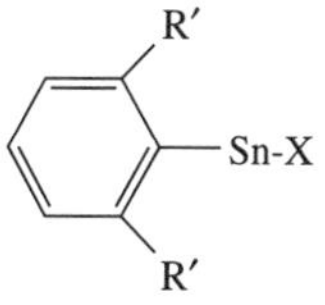 2,6-$R'_2C_6H_3$–Sn-X	X = H	698.7[a]	$C_6D_5CD_3$	80
	Cl	793.4		81
	L	1140		82
	t-Bu	1904.4	C_6D_6	83
	$Cr(\eta^5\text{-}C_5H_5)(CO)_3$	2297.7		84
	$N(SiMe_3)_2$	1196.8		82
	Sn_BMe_2R''	2856.9 Sn_B: 257.4		83
	$[Sn(\mu\text{-}Cl)R'']_2$	625.2		85
R' = 2,4,6-(i-Pr)$_3C_6H_2$ (Tip);	R'' = C_6H_3-2,6-Trip$_2$			
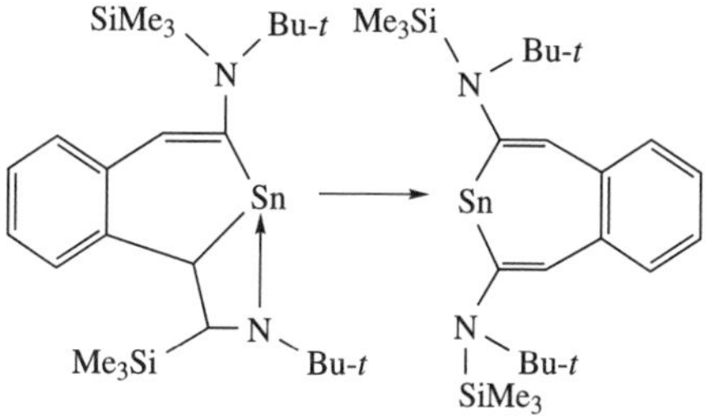 Benzostannacycle with $N(SiMe_3)(Bu\text{-}t)$, Me_3Si and $N\to Sn$ ($Bu\text{-}t$) groups → stannepine with two $N(SiMe_3)(Bu\text{-}t)$ groups		333.04	C_6D_6/C_5D_5N	78
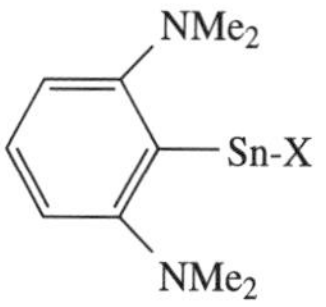 2,6-$(NMe_2)_2C_6H_3$–Sn-X	X = Cl	360	C_6D_6	79
	$N(SiMe_3)_2$	422		
	$CH(SiMe_3)_2$	758		
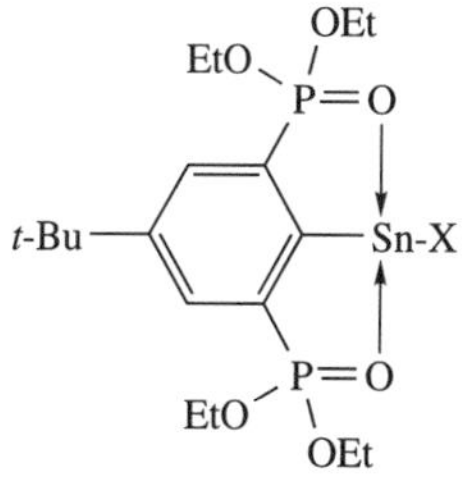 4-t-Bu-2,6-[$(EtO)_2P=O$]$_2C_6H_2$–Sn-X ($P=O\to Sn$)	X = Cl	−100	$C_6D_5CD_3$	86
	Br	−68		
	SPh	2		
	$CH(SiMe_3)_2$	258	$C_6H_5CH_3/D_2O$	
	$SiPh_3$	192		
	Sn_BPh_3	109 Sn_B: −43	$C_6D_5CD_3$	
	Sn_BMe_3	217 Sn_B: 11	THF/D_2O	
$[(PhMe_2Si)_3CSnCl]_2$		777	C_6D_6	87
$[(MeOMe_2Si)(Me_3Si)_2CSnCl]_2$		469	C_6D_6	87

(*continued overleaf*)

TABLE 3. (*continued*)

Compound		Chemical shift (ppm)	Solvent	Reference
$CH_2Bu\text{-}t$, N, R, Si, R, N, Sn, $CH_2Bu\text{-}t$	$R = N(SiMe_3)_2$	620.94	C_6D_6	88
	$2,6\text{-}(Me_2N)_2C_6H_3$	412		
Me_3Si, R, Me_3Si, Sn, N	$R = CH(PPh_2)_2$	397	C_6D_6	89
	$2,4,6\text{-}(i\text{-}Pr)_3C_6H_2$	474		
	$Si(SiMe_3)_3$	876	C_6D_6	90
	$Sn_B(SiMe_3)_3$	897		
		Sn_B: −502		
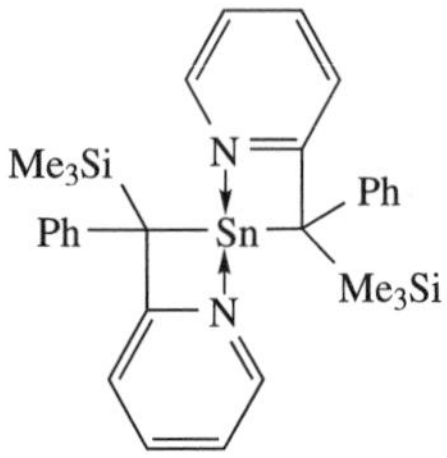 Me_3Si, N, Ph, Ph, Sn, Me_3Si, N		156.18	C_6D_6	91
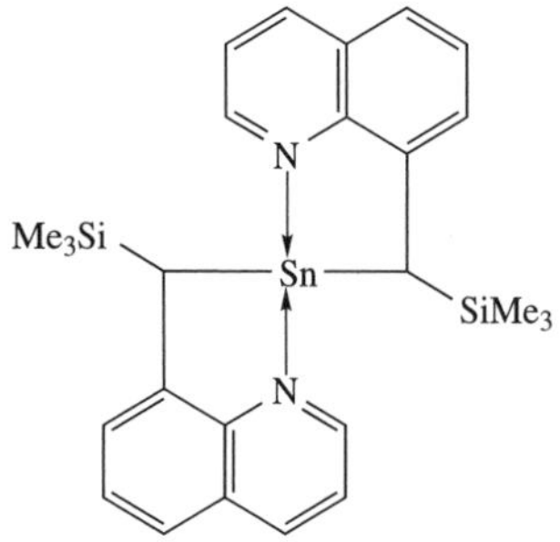 N, Me_3Si, Sn, $SiMe_3$, N		141.73	C_6D_6	91
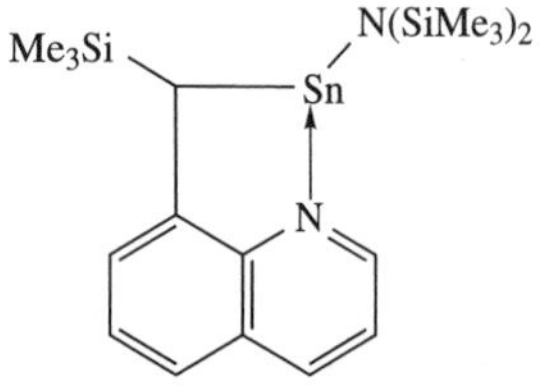 $N(SiMe_3)_2$, Me_3Si, Sn, N		141.73	C_6D_6	92
$(i\text{-}Pr_5Cp)_2Sn$		−2262	—[b]	93
$[(\eta^5\text{-}C_5H_5)_2Sn(\mu\text{-}C_5H_5)]$		−2312		94

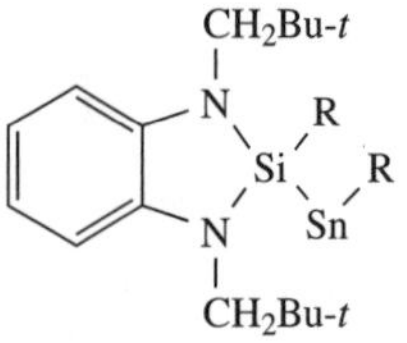

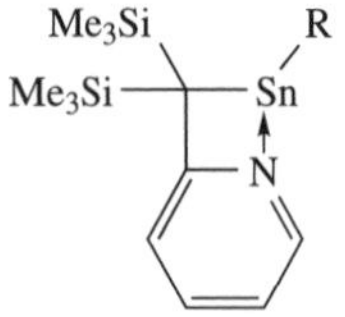

TABLE 3. *(continued)*

Compound		Chemical shift (ppm)	Solvent	Reference
$[\{(\eta^3\text{-}C_5H_5)_3Sn\}_2]^{2-}$		−1730		94
$[\{(\eta^5\text{-}Me_4(Me_3Si)Cp)_2Sn\}_2]$		−2171	C_6D_6	121
$[\{(\eta^5\text{-}Me_4(t\text{-}BuMe_2Si)Cp)_2Sn\}_2]$		−2236	C_6D_6	121
$[\{(\eta^5\text{-}(PhC(=O))C_5H_4)_2Sn\}_2]$		−2137.4	THF-d_8	95
$[(t\text{-}Bu_3Si)_2Sn]_2$		412.6	C_6D_6	95
$(t\text{-}Bu_3Si)_2Sn = Sn_A = Sn_B(Si(Bu\text{-}t)_3)_2$ [a]		Sn_A: 2233 Sn_B: 503	$C_6D_5CD_3$	97
$t\text{-}Bu_2Sn$ ring with $t\text{-}BuSn{=}SnBu\text{-}t$ (cyclotristannene)		t-BuSn: −694 $t\text{-}Bu_2Sn$: 412	C_6D_6	97
$CpSn(\mu\text{-}OBu\text{-}t)_2M\text{-}OBu\text{-}t$	M = Sn Pb	−79.56/−362.9 −103.3	C_6D_6	98
$RSn(\mu\text{-}NBu\text{-}t)_2SiMe_2$	R = Cp Indenyl	257.1 120.8	C_6D_6 $C_6D_5CD_3$	99
$(Me_3Si)_2NSn(N\text{-}c\text{-}C_6H_{11})_2C\text{—}R$	R = Me t-Bu	17.6 10.3	C_6D_6	100
$(Me_3Si)_2N-Sn(\mu\text{-}O)_2Sn-OSiMe_3$		Sn−N: 45.7 Sn−O: 205.7	$C_6D_5CD_3$	101
$[((Me_3Si)_3Si)Me_3SiN]_2Sn$		549.61	C_6D_6	102
1,8-naphthalenediyl $(t\text{-}BuCH_2N)_2Sn$	complex with: $Ge(C_6H_3(NMe_2)_2)_2\text{-}2,6$ $Sn_B(C_6H_3(NMe_2)_2)_2\text{-}2,6$	183 −58 Sn_B: −30/275	C_6D_6	103
2,3-naphthalenediyl $(t\text{-}BuCH_2N)_2Sn$		288	C_6D_6	104

(continued overleaf)

TABLE 3. (*continued*)

Compound	Chemical shift (ppm)	Solvent	Reference
[structure: tridentate amide Sn(N(R)CH$_2$CH$_2$)$_2$NMe] $R = i$-Pr	129	C_6D_6	105
Me_3Si	158		
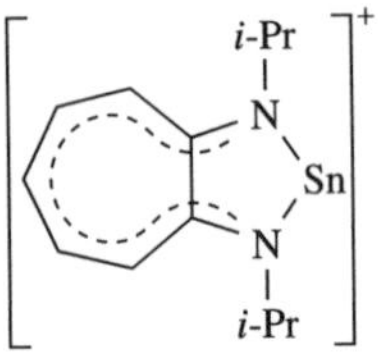 [structure: aminotroponiminate Sn cation, N-*i*-Pr]$^+$ $[Cp_2Zr_2Cl_7]^-$	734	$CDCl_3$	106
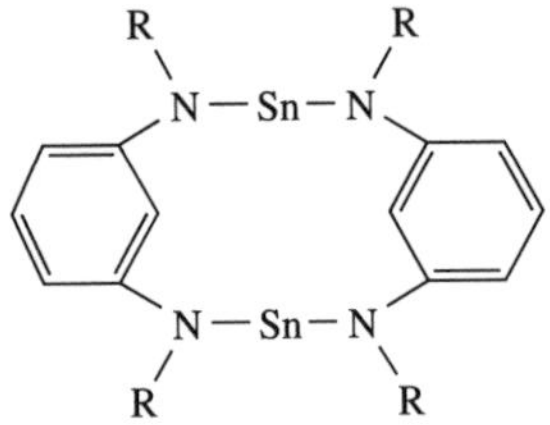 [structure: bis(N-Sn-N) macrocycle with two phenylene rings] $R = SiMe_3$	505	C_6D_6	107
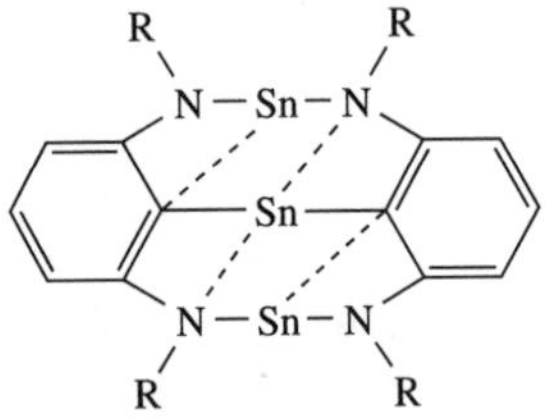 [structure: tri-tin cluster with two phenylene rings] $R = SiMe_3$	−888	C_6D_6	107
$[(PhMe_2Si)_2N]_2Sn$	501	C_6D_6	108
$[2,6\text{-}(i\text{-}Pr)_2C_6H_3(Me_3Si)N]_2Sn$	440	C_6D_6	108
$[(Me_3Si)_2NSn(\mu\text{-}Cl)]_2$	39	$C_6D_5CD_3$	109
$[(Me_3Si)_2NSn(\mu\text{-}OSO_2CF_3)]_2$	−270	THF-d_8	109
$(Me_3Si)_2N(ArO)Sn$	−52	C_6D_6	110
[structure: Sn_4N_4 cubane with $SiR'R_2$ groups] $R = R' = Me$	782	C_6D_6/ $C_6D_5CD_3$	233
Et	753		
Ph	565		
$R = Me$; $R' = t$-Bu	741		

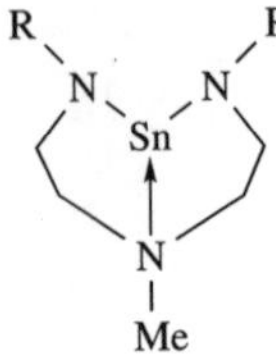

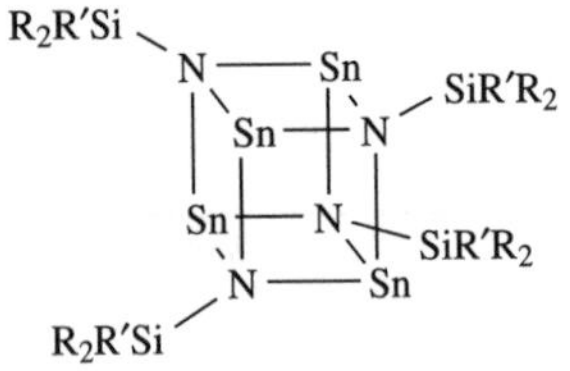

TABLE 3. (*continued*)

Compound		Chemical shift (ppm)	Solvent	Reference
Et_3Si, N, Sn, $SiEt_3$, Sn, N, $(OC)_4Fe$, Sn, N, $SiEt_3$, N, Sn, Et_3Si, $Fe(CO)_4$		489/520	C_6D_6	382
$[(Tip_2Si(F))_2P]_2Sn$[b]		1551		261
$[((Me_3Si)_2As)_2Sn]_2$		475 (*cis*)[c] 671 (*trans*)[c]	C_6D_6	111
t-Bu_3Si, P, Sn, $Si(Bu$-$t)_3$, X, P, Sn, P, $Si(Bu$-$t)_3$, P, Sn, t-Bu_3Si	X = Ba Sn	743.9[d] 1234.3	C_6D_6	387
NPr-*i*, $SnCl_2$, NPr-*i*		−59.4	C_6D_6	112
N, $Sn\overset{-}{T}ip_2$, $\overset{+}{N}$		710	C_6D_6	114
$[Me_4N][Sn(SOCPh)_3]$		−227	CH_3CN (0.03 M)	122
OEt, EtO, P=O, M, Sn, Cl, P=O, EtO, OEt	M = $W(CO)_5$ $Cr(CO)_5$ $Fe(CO)_4$	−74 131 54	$C_6H_5CH_3/D_2O$	86

(*continued overleaf*)

TABLE 3. (*continued*)

Compound		Chemical shift (ppm)	Solvent	Reference
$(OC)_nM = SnRR'$	M = $W(CO)_5$	799	C_6D_6	115
	$Mo(CO)_5$	928.5		116
	$Cr(CO)_5$	1001		
	$Fe(CO)_3$	889		117
	$Ni(CO)_3$	956		
	free stannylene	960		
$RR'Sn = Mn(CO)_4 = SnRR'$ R = 2,4,6-$(t\text{-Bu})_3C_6H_2$ R′ = 2,6-$(t\text{-Bu})_2$-4-$EtMe_2CC_6H_2$		1006	C_6D_6	118
$[Cp(OC)Fe(=SnR_2)(P(NMe)_2C_2H_4)]^+$ TfO^-	R = Me	495.8	$CDCl_3$	119
	n-Bu	589.06	$CDCl_3$	
$R_3P-Pt(Ph_2PCH_2PPh_2)Fe(CO)_3$, μ-Sn[N(Bu-*t*)]$_2SiMe_2$	R = Ph	615.1	$CDCl_3$	120
	R = 4-MeC_6H_4	608.2	$CDCl_3$	
$Me_3P-M(Ph_2PCH_2PPh_2)Fe(CO)_3$; *t*-Bu—N—Sn, Me_2Si—N(*t*-Bu); Si—OMe, OMe, O	M = Pt	128.6	$CDCl_3$	120
	M = Pd	81.2		

[a] At 253 K;
[b] No signal observed in solution, MAS spectra only;
[c] Broad signal;
[d] AMM′X type.

3. Tin(IV) compounds

a. Tin with a coordination number of four. Despite the importance of tin in a coordination sphere greater than four, a number of tin compounds are at least thought to be connected to four substituents only. The wealth of data is organized into 10 tables. Table 5 comprises trimethylstannyl derivatives. Other publications of stannanes with Me_3Sn units are found in the literature[149–156]. A number of vinyl-substituted stannanes with and without the trimethylstannyl moiety is summarized elsewhere[159–167,269–274]. Table 6 is devoted to aliphatic triorganostannyl derivatives, while Table 7 contains an assortment of compounds with substituents, such as H, Si or the halogenes. Aryl-substituted tin compounds can be found in Tables 8–10. Table 11 lists compounds with at least one transition metal bond to tin. Compounds containing two tin–carbon bonds are found in Table 12 while Table 13 presents compounds with tin–tin bonds. The limited space of this review,

TABLE 4. ^{119}Sn NMR data of stanna-alkenes, germanes and phosphenes

Structural formula		Chemical Shift [ppm]	Solvent	Reference
$Tip_2Sn{=}GeMes_2$[a]		360[b]	ether-toluene	124 125 125
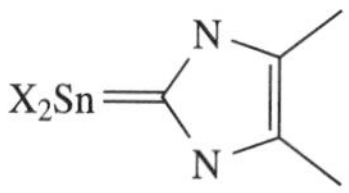 $X_2Sn{=}$ (N, N)	$X = 2,4,6\text{-}(t\text{-}Bu)_3C_6H_2$	710		114 125
	$X = Cl$	−59.4		112
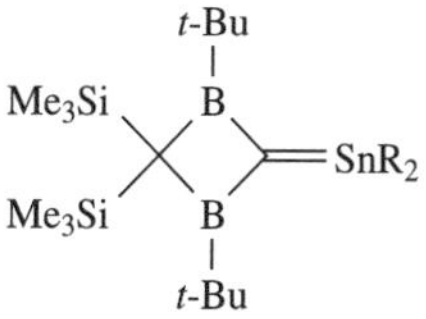 $(Me_3Si)_2$, B(t-Bu), B(t-Bu) $=SnR_2$	$R = 2\text{-}t\text{-}Bu\text{-}4,5,6\text{-}Me_3C_6H$	374	C_6D_6	113
$(Me_3Si)_2CHP{=}SnTip_2$[a]		606.0	—	123 125

[a] Tip = $2,4,6\text{-}(i\text{-}Pr)_3C_6H_2$;
[b] at 253 K.

however, does not allow us to present all these compounds in this way; the remainder can be found elsewhere[196–224].

Most tin NMR data are just used for characterizing the substances. There are some series of compounds measured to obtain a better understanding of tin chemical shifts. Examples are 3-X-bicyclo[1.1.1]pent-1-yltrimethylstannes (**5**), with different X substituents (cf. Table 5) which show that the tin chemical shifts are influenced by rear lobe interactions of the orbitals of the bridgehead carbons[134].

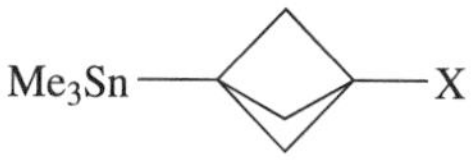

Me_3Sn—[bicyclo[1.1.1]pentane]—X

(**5**)

For aryl-substituted stannanes, some relationships between the electronic properties of the aryl groups and the chemical shifts have been observed. For instance, there is a dependence of shifts and the resonance parameters $\sigma_R{}^o$ and σ_R^{51}. The effect is larger for the *p*-substituted aryl groups XC_6H_4(X = Me, F, Cl, Br) than for the corresponding *m*-substituted derivatives and there are also linear correlations with the ^{207}Pb chemical shifts of the corresponding lead compounds[191]. Dräger and coworkers describe the ^{119}Sn chemical shifts of tetraarylstannanes to charge transfers between π- and σ^*-orbitals[181]. A collection of such data is found in Tables 9 and 10.

A wide variety of compounds are known which contain at least one transition metal–tin bond. These derivatives undergo different types of reactions, such as substitution of ligands at the tin or the metal center, photochemical reactions and so on. Selected tin derivatives of such transition metal complexes are shown in Table 11.

Compounds of the type R_2SnX_2, where X = H, Si, N, P, O, S or a substituted carbon, are listed in Table 12. Cyclic compounds of this type are summarized in Table 13.

Selected derivatives with at least one tin-tin bond are given in Table 14.

TABLE 5. Selected ^{119}Sn NMR data of trimethyltin substituted compounds

Compound			Chemical shift (ppm)	Solvent	Reference
$[Me_3Sn]_4C$			49.3	C_6D_6	131
$[Me_3Sn]_2CMe_2$			−30.2	C_6D_6	131
$[Me_3Sn{-}C{\equiv}C{-}SiMe_2]_2(C{\equiv}C)$			−74.5	$C_6D_5CD_3$	132
$[Me_3Sn{-}C{\equiv}C{-}SiMe_2{-}C{\equiv}C]_2SiMe_2$			−70.9	$CDCl_3$	133
$Me_3Sn{-}(C{\equiv}C{-}SiMe_2)_3{-}C{\equiv}C{-}SiMe_3$			−74.2	$C_6D_5CD_3$	133
CH(R)OH; Me_3Sn, $SnMe_3$	R = H		1.0 12.6	$CDCl_3$	275
	Me		−1.0 6.3		
$SnMe_3$; Me_3Sn, $SnMe_3$			−32	C_6D_6	389
Me_3Sn, $SnMe_3$; Me_3Sn, $SnMe_3$			−34.7	C_6D_6	389
R, $SnMe_3$; R′, $SnMe_3$	R = H	R′ = Me	−32.2 −32.9	$CDCl_3$	276
	MeO	MeO	−28.3		
$SnMe_3$; $SnMe_3$			−28.9	$CDCl_3$	276
$(H_2C)_n$; $SnMe_3$, $SnMe_3$	$n = 5$		−57.2	$CDCl_3$	276
	6		−59.1		
	7		−50.7		
	8		−54.5		
$(H_2C)_n$; $SnMe_3$, $SnMe_3$	$n = 5$		−7.5	$CDCl_3$	276
	6		−13.7		
	7		−5.8		
	8		−3.0		
X, Y; $SnMe_3$, $SnMe_3$	X = O	Y = C	−42.2 −52.6	$CDCl_3$	276
	S	C	−32.4 −42.2		
	N	N	−52.2 −53.5		

TABLE 5. (*continued*)

Compound		Chemical shift (ppm)	Solvent	Reference
Me_3Sn–(1-sila-cyclopentadienyl, $SiMe_2$, 3-BEt_2, 4-Et)–$SiMe_2$–(1-sila-cyclopentadienyl, $SiMe_2$, 4-Et, 3-BEt_2)–$SnMe_3$		−50.0	C_6D_6	133
Me_3Sn–(bicyclo[1.1.1]pentyl)–X	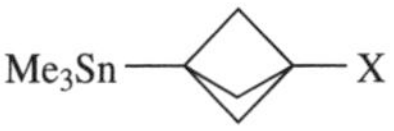 X = H	−46.9	$CDCl_3$	134
	CN	−32.5		
	CF_3	−31.0		
	COOMe	−34.1		
	$CONMe_2$	−35.4		
	COOH	−33.8		
	OMe	−14.7		
	Ph	−36.1		
	4-FC_6H_4	−35.6		
	t-Bu	−37.4		
	Me	−41.2		
	$SnMe_3$	−69.7		
$Me_3SnC{\equiv}COEt$		−61.5	$C_6D_5CD_3$	135
$[Me_3Sn]_2C{=}C{=}O$		32.7	$C_6D_5CD_3$	135
$[Me_3Sn]_2CHCOOEt$		23.8	$C_6D_5CD_3$	135
$Me_3SnCH_2(2\text{-}C_5H_4N)$		8.6	C_6D_6	136
$Me_3SnCH_2PPh_2$		3.6	C_6D_6	137
Me_3SnCH_2PHPh		5.8	C_6D_6	137
$[Me_3SnCH_2]_2PPh$		4.1	C_6D_6	137
$(CO)_3Co{-}Co(CHO)_3$ bridged alkyne complex of $Me_3SnC{\equiv}CMe$		15.6	C_6D_6	138
$[(\eta^5\text{-}Me_4(Me_3Sn)Cp)_2Fe]$		−4.2	C_6D_6	139
1,3-di-R-1,3,2-diazaborolidine, 2-$SnMe_3$	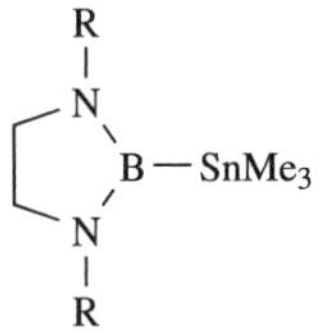 R = *t*-Bu	152	C_6D_6	140
	2,6-$Me_2C_6H_3$	146	C_6D_6	140
Diborapolyhedrane: B–Et, B–$SnMe_3$; C substituents Et, Et(Me), Me, Me(Et)		−117.1	C_6D_6	141
$[Me_3Sn]_4Si$		−34.1	C_6D_6	131

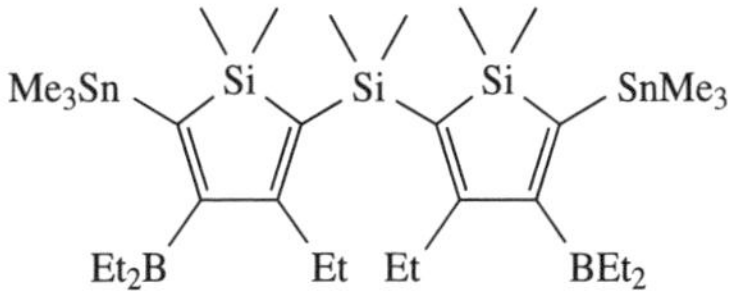

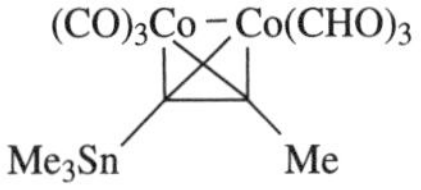

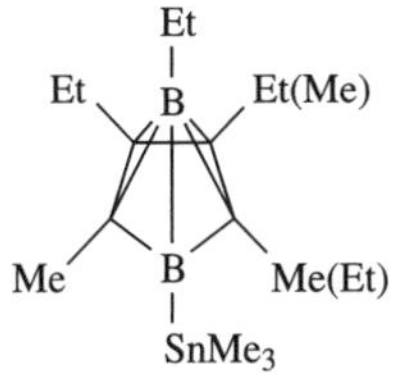

(*continued overleaf*)

TABLE 5. (*continued*)

Compound			Chemical shift (ppm)	Solvent	Reference
$Me_3Sn{-}(SiMe_2)_n{-}SnMe_3$	$n = 1$		−99.0	C_6D_6	131
			−97.6	$CDCl_3$	142
	2		−106.4	$CDCl_3$	
	3		−104.7	$CDCl_3$	
	4		−103.4	C_6D_6	
	6		−104.8	$CDCl_3$	
$Me_3SnSiCl_3$			−70	C_6D_6	143
$[Me_3Sn]_2Si(SiCl_3)_2$			−55.2	C_6D_6	143
$[Me_3Sn]_4Ge$			−25.1	C_6D_6	131
			−25.2	C_6D_6	144
$[Me_3Sn]_2GeMe_2$			−79.5	C_6D_6	131
$Me_3SnPbEt_3$			−36.2	C_6D_6	145
$Me_3SnPb(Pr\text{-}i)_3$			−7.5	C_6D_6	145
$[Me_3Sn]_3N$			86.3	$C_6D_5CD_3$	136
$[Me_3Sn]_2NPh$			63.0	$C_6D_6CD_3$	136
B—N($SnMe_3$)($SnMe_3$)			48.6	$C_6D_5CD_3$	136
B—N($SnMe_3$)—B			35.6	$C_6D_5CD_3$	136
N—$SnMe_3$			72.9	C_6D_6	146
R, N, $SnMe_3$	R = H		73.1	C_6D_6	146
	Me		64.3		
N, $SnMe_3$			69.0	C_6D_6	146
$Me_3SnNH(2\text{-}C_5H_4N)$			26.8	C_6D_6	146
$Me_3SnNHPh$			46.4	C_6D_6	146
$Me_3SnN(R)C(=O)R'$	R	R′			
	H	CF_3	77.0	$CDCl_3$	157
	H	Me	39.8		
	Me	Me	41.1		
	Me	Ph	49.7		

TABLE 5. (*continued*)

Compound			Chemical shift (ppm)	Solvent	Reference
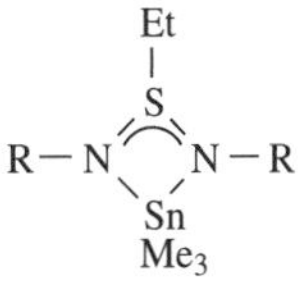 [4-membered ring: Me_3Sn, two N–R, S–Et]	R = *t*-Bu		−48.6	$C_6D_5CD_3$	158
	$SiMe_3$		−1.6		
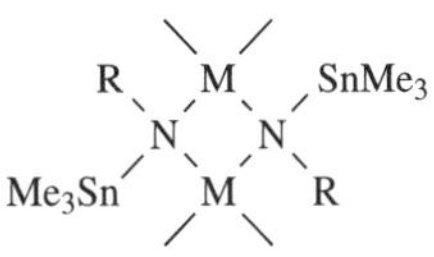 [$(R)(Me_3Sn)N$–M ring dimer]	M	R			
	Al	Pr	78.5	C_6D_6	147
	Al	*i*-Pr	54.0		
	Al	Bu	68.5		
	Ga	Me	94.2		
	Ga	Pr	80.5		
	Ga	*i*-Pr	56.5		
	Ga	*i*-Bu	70.0		
	In	Me	102.9		
			103.9		
	In	Pr	90.0		
	In	*i*-Pr	66.8		
	In	*i*-Bu	81.1		
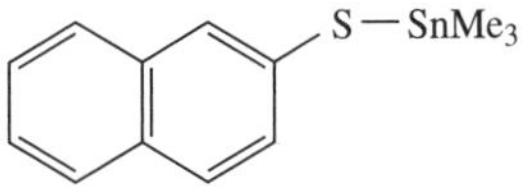 2-Naphthyl–S–$SnMe_3$			96.7	$CDCl_3$	148

TABLE 6. Triorganotin compounds R_3SnX (R = aliphatic substituents)

Compound	Chemical shift (ppm)	Solvent	Reference
R = Ethyl			
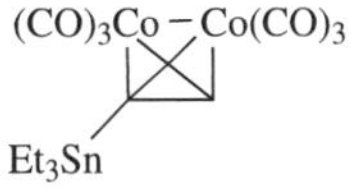 $(CO)_3Co$–$Co(CO)_3$ alkyne complex with Et_3Sn	10.6	C_6D_6	138
$Et_3SnC_6F_5$	−6.8	$CDCl_3$	168
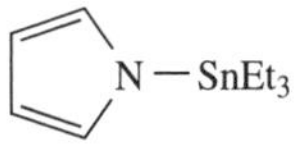 Pyrrol-1-yl–$SnEt_3$ (N–$SnEt_3$)	−48.0	C_6D_6	146
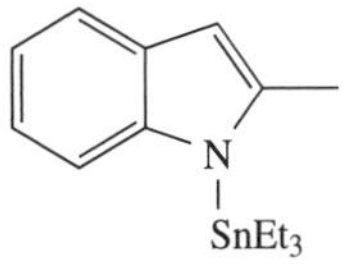 2-Methylindol-1-yl–$SnEt_3$ (N–$SnEt_3$)	53.9	C_6D_6	146
Et_3Sn–$SiCl_3$	−59	C_6D_6	143
$(Et_3Sn)_2Si(SiCl_3)_2$	−45.0	C_6D_6	143
Et_3Sn–$Pb(Bu\text{-}t)_3$	31.9	C_6D_6	145

(*continued overleaf*)

TABLE 6. (*continued*)

Compound	Chemical shift (ppm)	Solvent	Reference
R = *t*-Butyl			
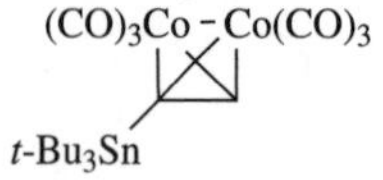 $(CO)_3Co-Co(CO)_3$ / *t*-Bu_3Sn	23.6	C_6D_6	138
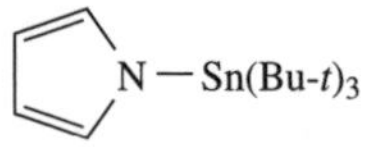 N—$Sn(Bu\text{-}t)_3$	−106.9	C_6D_6	146
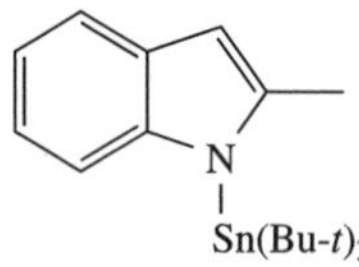 N / $Sn(Bu\text{-}t)_3$	−38.3	C_6D_6	146
t-$Bu_3Sn-Pb(Bu\text{-}t)_3$	5.8	C_6D_6	135
R = *n*-Butyl			
$(Bu_3Sn)_2CHCHMeCO_2Me$	10.1	CD_2Cl_2	169
$(Bu_3Sn)_2CHCH(CH_2CH_2CH=CH_2)CO_2Me$	4.3 0.9	CD_2Cl_2	169
$(Bu_3Sn)_2CHCH(CH_2Ph)CO_2Me$	4.6 0.6	CD_2Cl_2	169
$(Bu_3Sn)_2CHCH(CH_2NMe_2)CO_2Me$	5.0 1.8	CD_2Cl_2	169
Bu_3Sn-α-D-glucuronate	139.7	$CDCl_3$	170
Bu_3Sn-α-D-glucuronate	142.2	$CDCl_3$	170
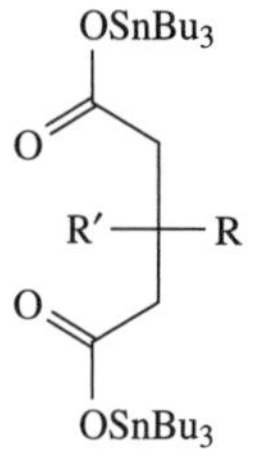 $OSnBu_3$, O, R′—R, O, $OSnBu_3$: R = CO_2SnBu_3, R′ = OH	121.0 114.4	$CDCl_3$	171
R′ = H	106.0 106.7		
$CO_2[H_2N(C_6H_{11}\text{-}c)_2]$, R′ = OH	109.7		
R′ = H	105.6		
$CO_2[H_3NC_6H_{11}\text{-}c]$, R′ = OH	110.9		
$CO_2[H_2N(CH_2)_5]$, R′ = OH	105.3		
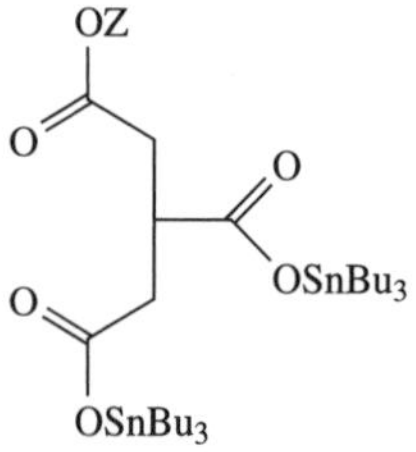 OZ, O, O, $OSnBu_3$, O, $OSnBu_3$: Z = H	136.3 121.2	$CDCl_3$	171
$[H_2N(C_6H_{11}\text{-}c)_2]$	101.8		

TABLE 6. (*continued*)

Compound		Chemical shift (ppm)	Solvent	Reference
Bu_3Sn–CH_2–(3-methylisoxazol-5-yl)		125.4	$CDCl_3$	172
Bu_3Sn–(3-methylisoxazol-5-yl)		116.9	$CDCl_3$	172
$Bu_3Sn-SiCl_3$		−72	C_6D_6	143
$Bu_3Sn-Si(SiCl_3)_3$		−46.8	C_6D_6	143
$(Bu_3Sn)_2Si(SiCl_3)_2$		−52.7	C_6D_6	143
$Bu_3Sn-Pb(Bu\text{-}t)_3$		20.5	C_6D_6	145
Bu_3SnO–C(=O)–(3,5-dimethylisoxazol-4-yl)		116.9	$CDCl_3$	172
Bu_3SnO–C(=O)–CH_2–(3-methylisoxazol-5-yl)		125.5	$CDCl_3$	172
2-naphthyl–S–$SnBu_3$		84.0	$CDCl_3$	148
$Bu_3Sn-SePh$		57.8	—	173
$Bu_3Sn-Se(C_6H_4F\text{-}4)$		59.6	—	173
$Bu_3Sn-TePh$		−1.3	—	173
$Bu_3Sn-Te(C_6H_4F\text{-}4)$		0.4	—	173
R = various aliphatic substituents				
$(Me_2N(CH_2)_3)_3Sn\text{-}R$	R = Ph	−41.7	$CDCl_3$	174
	Cl	2.4		
$(Ph-C{\equiv}C)_3SnR$	R = Me	−239.0	$CDCl_3$	175
	Bu	−242.0		
	i-Pr	−235.5		
	t-Bu	−232.9		
	$(CH_2)_2CO_2Me$	−257.4	CD_2Cl_2	
$(PhCH_2)_3SnPh$		−63,2	$CDCl_3$	168
((−)-Menthyl)$_3$SnZ	Z = H	−102.9	C_6D_6	385
	Cl	93.6		

(*continued overleaf*)

TABLE 6. (*continued*)

Compound		Chemical shift (ppm)	Solvent	Reference
$C_4H_4N-Sn(CH_2Ph)_3$ (pyrrol-1-yl)		−37.8	C_6D_6	146
2-naphthyl-$S-SnR_3$	$R = CH_2Ph$	18.0	$CDCl_3$	148
	c-C_6H_{11}	10.1	$CDCl_3$	

TABLE 7. Compounds with X_3Sn moieties (X = H, Si, halogen)

Compound		Chemical shift (ppm)	Solvent	Reference
$H_3Sn-CH=CH_2$		−361	C_6D_6	259
$H_3Sn-CH=C=CH_2$		−338.4	C_6D_6	260
$H_3Sn-C\equiv CH$		−320.6	C_6D_6	260
$((t\text{-}BuNCH)_2SiCl)_3SnCl$		−222.6	C_6D_6	176
$[\eta^5\text{-}(X_3Sn)C_5H_4]_2Fe$	X = H	−330.6	C_6D_6	139
	Cl	−23.2		
$Cl_3Sn-CHPh-CHMe-C(O)Ph$		−151.7 −151.6	$CDCl_3$	250
$Cl_3Sn-(CHPh)_2-C(O)Ph$		−153.8	$CDCl_3$	250
$Cl_3Sn-(CH_2)_n-OH$	$n = 3$	−235 (−112)	$(CD_3)_2C=O$	287
	4	−252 (−147)	C_6D_6	384
	5	−179 (−80)		
$Cl_3Sn-(CH_2)_n-OAc$	$n = 3$	−38	C_6D_6	384
	4	−47		
	5	−67		
Br_3SnPh		−289.4	$CDCl_3$	168
$Br_3Sn-CH_2C\equiv CH$		−202.7	$CDCl_3$	177
$Br_3Sn-CH=C=CH_2$		−251.0	$CDCl_3$	177
Carborane cage: B(Et), Et, Et(Me), Me, Me(Et), B–SnX_3	X = Cl	44.0	C_6D_6	141
	Br	96.1		

TABLE 8. Triaryltin compounds (R_3SnX)

Compound		Chemical shift (ppm)	Solvent	Reference
R_3SnR' (R' = aliphatic substituent)				
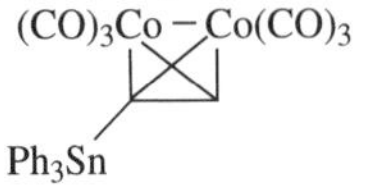 $(CO)_3Co-Co(CO)_3$ … Ph_3Sn		−114.6	$CDCl_3$	138
Ph_3SnCH_2Br		−129.4	CD_2Cl_2	178
$[Ph_3Sn^*CH_2]_2SnPh_2$		−79.0	$CDCl_3$	178
$[Ph_3Sn^*CH_2SnPh_2]_2$		−79.9	$CDCl_3$	178
$[Ph_3Sn(CH_2)_2]_2$		−100.2	$CDCl_3$	179
$Ph_3Sn(CH_2)_3NMe_2$		−101.9	$CDCl_3$	174
$Ph_3Sn(CH_2)_3NMe_2{\bullet}HCl$		−103.0	$CDCl_3$	174
$Ph_3Sn-C{\equiv}C-C{\equiv}C-SnPh_3$		−170.1 −165.2 −164.6	$CDCl_3$	180
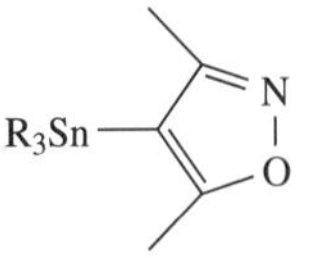 R_3Sn (isoxazole, N, O)	R = Ph 4-MeC_6H_4 4-ClC_6H_4	−142.2 −138.8 −134.1	$CDCl_3$	172
$(C_6F_5)_3SnEt$		−136.3	$CDCl_3$	168
$(C_6F_5)_3SnCH_2CH=CH_2$		−153.7	$CDCl_3$	168
R_3SnR' (R' = aromatic)				
$Ph_3Sn(C_6H_4Me\text{-}2)$		−126.7 −123.1[a]	$CDCl_3$	181
$Ph_3Sn(C_6H_4Me\text{-}3)$		−130.2 −120.2[a]	$CDCl_3$	181
$Ph_3Sn(C_6H_4Me\text{-}4)$		−129.1 −119.9[a]	$CDCl_3$	181
$PhSn(C_6H_4Me\text{-}2)_3$		−121.7 −123.6[a]	$CDCl_3$	181
$PhSn(C_6H_4Me\text{-}3)_3$		−130.0 −111.4[a]	$CDCl_3$	181
$PhSn(C_6H_4Me\text{-}2)_3$		−126.1 −118.8[a]	$CDCl_3$	181
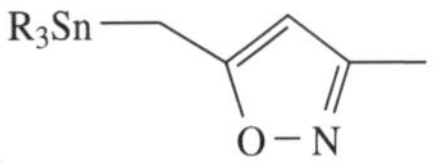 R_3Sn (isoxazole, O−N)	R = Ph 4-MeC_6H_4 4-ClC_6H_4	−114.0 −110.6 −107.1	$CDCl_3$	172
$Ph_3SnC_6F_5$		−139.3	$CDCl_3$	168
$(C_6F_5)_3SnPh$		−186.9	$CDCl_3$	168
R_3SnOR'				
3β-(Ph_3Sn)cholest-5-ene		−114.8 −117.1	$CDCl_3$	182
$Ph_3Sn(CH_2)_3OC(O)Me$		−100.2	$CDCl_3$	182
$Ph_3Sn(CH_2)_3OCH_2Ph$		−100.3	$CDCl_3$	182

(*continued overleaf*)

TABLE 8. (*continued*)

Compound		Chemical shift (ppm)	Solvent	Reference
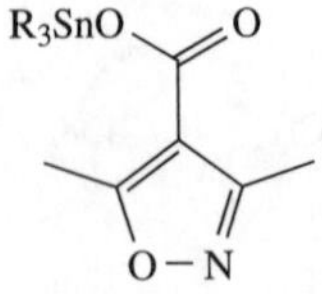 R_3SnO(C=O)-(3,5-dimethylisoxazol-4-yl)	R = Ph	−106.4	$CDCl_3$	172
	4-MeC_6H_4	−94.5		
3-(Ph_3Sn)propyl 2,3:5,6-di-*O*-isopropylidene-α-mannofuranoside		−100.1	$CDCl_3$	183
3-(Ph_3Sn)propyl 2,3:5,6-di-*O*-isopropylidene-β-mannofuranoside		−99.5		183
3-(Ph_3Sn)propyl-α-D-mannopyranoside		−100.1		183
1,2:5,6-di-*O*-isopropylidene-3-*O*-3-(Ph_3Sn)propyl-α-D-glucofuranose		−99.7		183
1,2:3,4-di-*O*-isopropylidene-6-*O*-(Ph_3Sn)methyl-α-D-galactopyranose		−146.2	$CDCl_3$	175
3-(Ph_3Sn)propyl hepta-*O*-acetyl-β-D-lactoside		−99.6		183
3β-[(Ph_3Sn)methoxy]cholest-5-ene		−144.6	$CDCl_3$	184
3β-[(Ph_3Sn)propoxy]cholest-5-ene		−100.3	$CDCl_3$	184
R_3SnN				
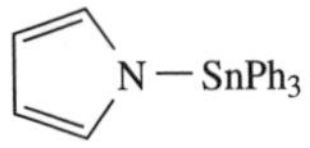 N-$SnPh_3$ pyrrole		−106.2	C_6D_6	146
Ph_3SnN_3		−86.2	$CDCl_3$	185
		−85.6	C_6D_6	
R_3SnB				
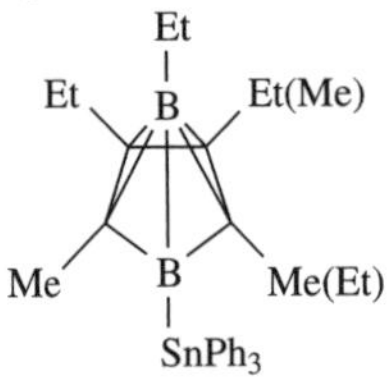 Carborane (B-Et; Et, Et(Me), Me, Me(Et); B-$SnPh_3$)		−137.8	C_6D_6	141
R_3SnZ (Z = Si, Pb)				
$Ph_3Sn-(SiMe_2)_n-SnPh_3$	$n = 1$	−155.5	$CDCl_3$	142
	2	−157.5		
	3	−155.0		
	4	−156.4		
	6	−156.6		
$Ph_3Sn-SiMe_2-CH_2CH_2-PPh_2$		−137.2	C_6D_6	256
$Ph_3Sn-Pb(Bu\text{-}t)_3$		−330	C_6D_6	145

[a] Solid state NMR

TABLE 9. Triaryl substituted halotin compounds

Compound	Chemical shift (ppm)	Solvent	Reference
Ph_3SnCl	−44.8	$CDCl_3$	186
$(3\text{-}MeC_6H_4))_3SnCl$	−42.3	$CDCl_3$	186
$(3,5\text{-}Me_2C_6H_3)_3SnCl$	−39.7	$CDCl_3$	186
$(3\text{-}MeOC_6H_4)_3SnCl$	−44.0	$CDCl_3$	186
$(2\text{-}MeC_6H_4))_3SnCl$	−32.3	$CDCl_3$	186
Mes_3SnCl	−84.4	$CDCl_3$	186
$(2\text{-}MeOC_6H_4)_3SnCl$	−56.7	$CDCl_3$	186
$(C_6F_5)_3SnCl$	−123.9	$CDCl_3$	168
Ph_3SnBr	−60.0	$CDCl_3$	186
$(3\text{-}MeC_6H_4)_3SnBr$	−56.9	$CDCl_3$	186
$(3,5\text{-}Me_2C_6H_3)_3SnBr$	−53.5	$CDCl_3$	186
$(3\text{-}MeOC_6H_4)_3SnBr$	−58.5	$CDCl_3$	186
$(2\text{-}MeC_6H_4)_3SnBr$	−54.0	$CDCl_3$	186
Mes_3SnBr	−74.3	$CDCl_3$	186
$(2\text{-}MeOC_6H_4)_3SnBr$	−74.3	$CDCl_3$	186
$(3\text{-}ClC_6H_4)_3SnBr$	−67.6	$CDCl_3$	186
$(3\text{-}FC_6H_4)_3SnBr$	−67.3	$CDCl_3$	186
$(3\text{-}F_3CC_6H_4)_3SnBr$	−67.8	$CDCl_3$	186
$(C_6F_5)_3SnBr$	−198.3	$CDCl_3$	168
Ph_3SnI	−114.5	$CDCl_3$	186
$(4\text{-}MeC_6H_4)_3SnI$	−10.8	$CDCl_3$	186
$(3\text{-}MeC_6H_4)_3SnI$	−108.5	$CDCl_3$	186
$(3,5\text{-}Me_2C_6H_3)_3SnI$	−103.7	$CDCl_3$	186
$(3\text{-}MeOC_6H_4)_3SnI$	−110.5	$CDCl_3$	186
$(2\text{-}MeC_6H_4)_3SnI$	−121.8	$CDCl_3$	186
Mes_3SnI	−217.0	$CDCl_3$	186
$(2\text{-}MeOC_6H_4)_3SnI$	−135.6	$CDCl_3$	186
$(4\text{-}ClC_6H_4)_3SnI$	−11.8	$CDCl_3$	186

TABLE 10. Tetraaryl compounds of the type R_4Sn

Compound	Chemical shift (ppm)	Solvent	Reference
Ph_4Sn	−128.1	$CDCl_3$	181
	−121.1[a]		
$(4\text{-}MeC_6H_4)_4Sn$	−124.6	$CDCl_3$	181
	−118.8[a]		
$(3\text{-}MeC_6H_4)_4Sn$	−128.0	$CDCl_3$	186
	−129.8		159
	−107.5		
$(2\text{-}MeC_6H_4)_4Sn$	−122.6	$CDCl_3$	186
	−124.5		181
	−127.4[a]		
$(3,5\text{-}Me_2C_6H_3)_4Sn$	−127.5	$CDCl_3$	181
$(2\text{-}MeOC_6H_4)_4Sn$	−136.3	$CDCl_3$	186
$(3\text{-}MeOC_6H_4)_4Sn$	−125.1	$CDCl_3$	186
$(3\text{-}ClC_6H_4)_4Sn$	−126.3	$CDCl_3$	186
$(3,5\text{-}Cl_2C_6H_3)_4Sn$	−122.6	$CDCl_3$	186

(*continued overleaf*)

TABLE 10. (*continued*)

Compound	Chemical shift (ppm)	Solvent	Reference
$(3\text{-}FC_6H_4)_4Sn$	−126.7	$CDCl_3$	186
$(3,5\text{-}F_2C_6H_3)_4Sn$	−119.6	$CDCl_3$	186
$(4\text{-}F_3CC_6H_4)_4Sn$	−134.0	$CDCl_3$	186
$(3\text{-}F_3CC_6H_4)_4Sn$	−126.3	$CDCl_3$	186

[a] Solid state NMR

TABLE 11. Transition metal–tin complexes

Compound		Chemical shift (ppm)	Solvent	Reference
$[(Ph_3P)_2Cl(CO)(Bu_3Sn)Ru]$		187.6	$CDCl_3$	191
$[(Ph_3P)_2Cl(CO)_2(Me_3Sn)Os]$		−103.74	$CDCl_3$	191
$[(Ph_3P)_2Cl(CO)(Bu_3Sn)(4\text{-}MeC_6H_4CN)Ru]$		−37.6	$CDCl_3$	191
$[(Ph_3P)_2Cl(CO)(Me_3Sn)(4\text{-}MeC_6H_4CN)Os]$		−128.7	$CDCl_3$	191
$[(Ph_3P)_2(CO)(Bu_3Sn)(\eta^2\text{-}S_2CNMe_2)Ru]$		6.6	$CDCl_3$	191
$[(Ph_3Sn)Me(CO)_2(i\text{-Pr-DAB})Ru]$[a]		−31.0	C_6D_6	187
$[(Ph_3Sn)[Mn(CO)_5](CO)_2(i\text{-Pr-DAB})Ru]$[a]		−49.6	C_6D_6	188
$[(Ph_3Sn)[Co(CO)_5](CO)_2(i\text{-Pr-DAB})Ru]$[a]		−52	THF-d_8	188
$[(COD)(SPh)_2Cl(Ph_3Sn)Pt]$[b]		−50.8	$CDCl_3$	189
$[(Ph_3P{=}N{=}PPh_3)(Cl_3Sn)(Cl)Pt]$		−2.2	$CDCl_3$	190
$[(i\text{-Pr-DAB})(CO)_2(Ph_3Sn)_2Ru]$[a]		−53	C_6D_6	192
$[K(15\text{-}C\text{-}5)_2]_2[Zr(CO)_4(SnMe_3)_4]$		49.5[c]	THF-d_8	193
$[K(15\text{-}C\text{-}5)_2]_2[Zr(CO)_4(SnMe_3)_2]$		16.4[c]	THF-d_8	194
$[Cp(CO)_2Fe]_2Sn(OSiPh_2)_2O$		358.6	C_6D_6	246
$[Cp(CO)_2W]_2Sn(OSiPh_2)_2O$		114.8	C_6D_6	246
$[Cp(CO)_2W]t\text{-}BuSn(OSiPh_2)_2O$		71.5	CH_2Cl_2	246
$[PtMe_2(Me_2SnZ)_2(Bu_2bipy)]$	Z = S	138.1	CD_2Cl_2	249
		51.9		
	Se	−60.3		
	Te	−168.1		
		−340.3		
$[PtMe_2\{(Me_2SnZ)(Ph_2SnZ)\}(Bu_2bipy)]$	Z = S		CD_2Cl_2	249
		−51.7		
		−32.7		
	Se	−57.4		
		−72.5		
	Te	−334.5		
		−160.5		
$[PtMe_2\{(Me_2SnZ)(Me_2SnZ')\}(Bu_2bipy)]$	Z = S		CD_2Cl_2	249
	Z′ = Se	−41.3		
		−51.6		
	S Te	−50.5		
		−128.4		
	Se Te	−86.0		
		−193.2		
$[PtMe_2\{(Ph_2SnS)(Ph_2SnSe)\}(Bu_2bipy)]$		−173.1	CD_2Cl_2	249
		−22.3		

TABLE 11. *(continued)*

Compound		Chemical shift (ppm)	Solvent	Reference
$[PtMe_2\{(Me_2SnSe)(Ph_2SnS)\}(Bu_2bipy)]$		−188.6 23.6	CD_2Cl_2	249
$[Cp(CO)_2W]Sn(Bu\text{-}t)X_2$	X = Ph Cl	−11.2 211.8	C_6D_6	246

[a]DAB = N,N′-diisopropyl-1,4-diaza-1,3-butadiene;
[b]COD = cyclooctadiene;
[c]99% ^{13}CO enriched product, 20 °C.

TABLE 12. Compounds of the type R_2SnX_2

Compound		Chemical shift (ppm)	Solvent	Reference
$\{[t\text{-}Bu_2SnO][Ph_2SiO]_2\}_n$		−167.1[a] −119.5[b]	$CDCl_3$	234
3α-$[Ph_2Sn(L)]$ cholest-5-ene		−52.4	$CDCl_3$	182
3β-$[Ph_2Sn(L)]$ cholest-5-ene		−50.9	$CDCl_3$	182
$[Ph_2ISn(CH_2)_2]_2$		−55.1	$CDCl_3$	179
$[t\text{-}Bu_3Pb\text{-}SnMe_2\text{-}(\eta^5\text{-}C_5H_4)]_2Fe$		7.8	C_6D_6	139
$[ClSnMe_2\text{-}(\eta^5\text{-}C_5H_4)]_2Fe$		125.5	C_6D_6	235
$[HSnMe_2\text{-}(\eta^5\text{-}C_5H_4)]_2Fe$		−102.4	C_6D_6	235
$X[SnMe_2\text{-}(\eta^5\text{-}C_5H_4)]_2Fe$	X = O S Se Te	64.6 66.5 28.1 −70.7	C_6D_6	235
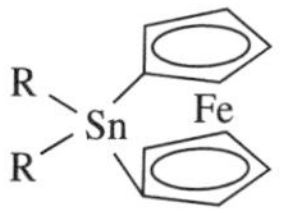 R, R, Sn, Fe	R = 2,4,6-$(i\text{-}Pr)_3C_6H_2$	−129.5 −129.3[a]	C_6D_6	238
$RP(Me_2Sn(\eta^5\text{-}C_5H_4))Fe$	R = Me t-Bu c-C_6H_{11} Ph	−47.6 −5.3 1.4 1.3	C_6D_6	248
$ZPhP(MeSn(\eta^5\text{-}C_5H_4)_2)Fe$	Z = $Cr(CO)_5$ $Mo(CO)_5$ $W(CO)_5$	11.1 11.6 13.1	C_6D_6	248
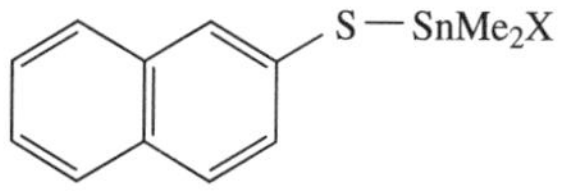 $S\text{—}SnMe_2X$	X = Cl S-Naph-2	53.0 126.1	$CDCl_3$	148
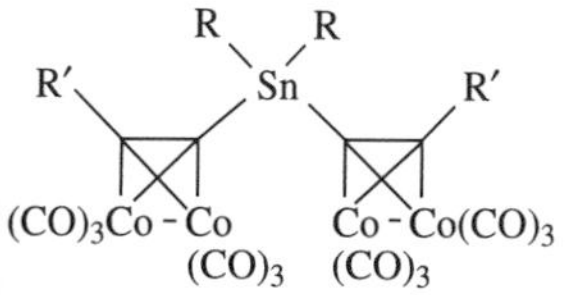 R, R, Sn, R′, R′, $(CO)_3Co\text{-}Co$, $Co\text{-}Co(CO)_3$, $(CO)_3$, $(CO)_3$	R = Me R′ = Me Me H Et H t-Bu H Ph H	−6.0 −4.9 −20.5 −71.3 −114.6	$CDCl_3$ C_6D_6 $CDCl_3$	138
$[t\text{-}Bu_2ClSnO]_2Si(Bu\text{-}t)_2$		−67.9	$CDCl_3$	237

(continued overleaf)

TABLE 12. *(continued)*

Compound		Chemical shift (ppm)	Solvent	Reference
t-$Bu_2ClSnOSi(Bu$-$t)_2Cl$		−58.5	$CDCl_3$	237
t-$Bu_2ClSnOSi(Bu$-$t)_2OH$		−56.8	$CDCl_3$	237
t-$Bu_2Sn(OSiH(Bu$-$t)_2)_2$		−161.4 −160.2	$CDCl_3$	237
t-$Bu_2Sn(OSiF(Bu$-$t)_2)_2$		−164.7 −162.1	$CDCl_3$	237
t-$Bu_2ClSnOSiH(Bu$-$t)_2$		−58.4	$CDCl_3$	237
t-$Bu_2Sn(OSiFEt_2)_2$		−161.9	$CDCl_3$	237
t-$Bu_2Sn(OSiF(Pr$-$i)_2)_2$		−159.4	$CDCl_3$	237
t-$Bu_2Sn(OSiFPh_2)_2$		−152.6	$CDCl_3$	237
t-$Bu_2Sn(OSiCl_2(Bu$-$t))_2$		−160.1	$CDCl_3$	237
t-$Bu_2ClSnOSiCl_2(Bu$-$t)$		−53.2	$CDCl_3$	237
t-$Bu_2Sn(OSiF_2(Bu$-$t))_2$		−164.7	$CDCl_3$	237
$[(Me_3Si)_2CH]_2Sn(NCO)_2$		−4.7	$C_6D_5CD_3$	58
$[(Me_3Si)_2CH]_2SnI_2$		−21.8	$C_6D_5CD_3$	58
$[Me_2N(CH_2)_3]_2SnPh_2$		−103.0	$CDCl_3$	174
$[Me_2N(CH_2)_3]_2SnCl_2$		−50.0	$CDCl_3$	174
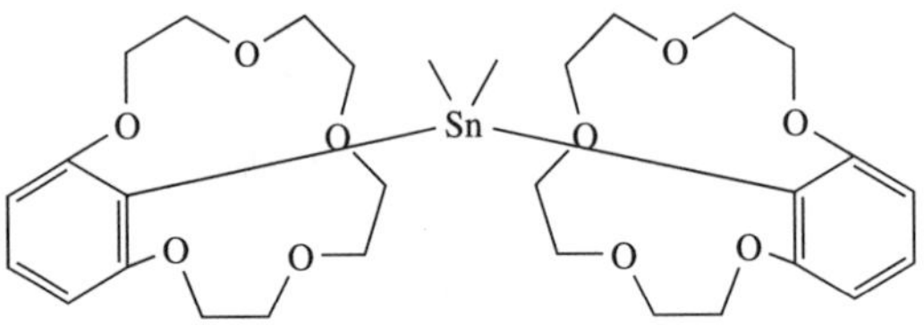		−99.7	$CDCl_3$	388
$PrR_2Sn(H)$	R = Bu	−89.0	C_6D_6	386
	Ph	−137		
	c-C_6H_{11}	−87.9		
i-Pr_3Si−$Sn(H)(C_6H_{11}$−$c)_2$		−196.6	C_6D_6	386
t-$Bu_2Sn(H)$−$(SiMe_2)_n$−$Sn(H)(Bu$-$t)_2$	n = 1	−111.3	C_6D_6	142
	2	−123.6		
	3	−120.4		
	4	−122.2		
	5	−120.1		241
	6	−121.7		142
t-$Bu_2Sn(X)$−$(SiMe_2)_n$−$Sn(X)(Bu$-$t)_2$	X = Cl n = 2	104.1	D_2O-cap[c]	142
	3	100.8	$CDCl_3$	
	4	98.1		
	5	101.8	D_2O-cap[c]	241
	6	100.3	$CDCl_3$	142
	X = Br n = 2	102.9	D_2O-cap[c]	142
	3	97.4		
	4	97.7	$CDCl_3$	
	5	93.1	D_2O-cap[c]	241
	6	98.3		142

TABLE 12. (*continued*)

Compound		Chemical shift (ppm)	Solvent	Reference
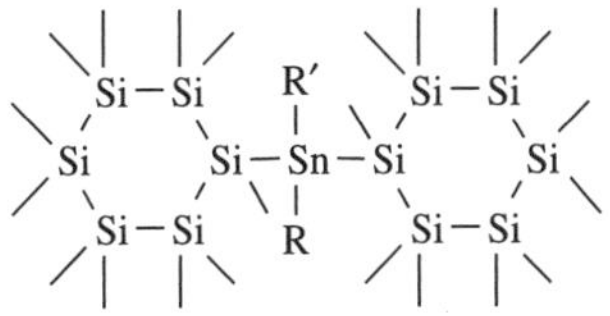	R R′			
	Me Me	−206.3	$CDCl_3$	242
	Me Ph	−191.2		
	Ph Ph	−192.3		
	4-MeC_6H_4 4-MeC_6H_4	−192.3		
$(C_6F_5)_2SnR_2$	R = $CH_2CH=CH_2$	−101.6	$CDCl_3$	168
	Cl	−88.9		
	Br	−236.5		
	CH_2Ph	−102.0		
	Ph	−158.2		
	Et	−61.5		
$(CH_2=CH)_2SnH_2$		−263.3	C_6D_6	259
$(CH_2=CH)_2Sn(H)Cl$		−88.9	C_6D_6	259
$HC\equiv C-CH_2Sn(Cl)_2CH=CH_2$		−13.2	$CDCl_3$	177
$H_2C=C=CSn(Cl)_2CH=CH_2$		−35.7	$CDCl_3$	177
$(Me(X)YSn\text{-}\eta^5\text{-}C_5H_4)_2Fe$	X Y			
	Me Cl	125.5	C_6D_6	139
	Cl Cl	100.4		
	Me H	−102.4		
	H H	−210.5		
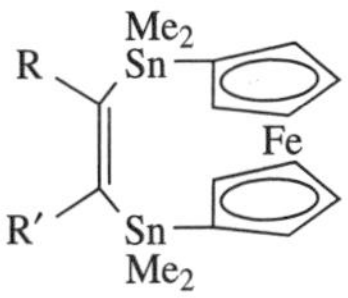	R = H; R′ = H	−71.4	C_6D_6	244
	Me	−73.4		
		−61.5		
	Me′[d]	−40.1		
		−17.9		
	Pr	−72.7		
		−62.4		
	Bu	−72.5		
		−62.4		
	n-C_5H_{11}	−72.5		
		−62.5		
	Ph	−66.7		
		−56.6		
	CH(OH)Et	−66.1		
		−64.4		
	R′ = R = MeOOC	−35.7		
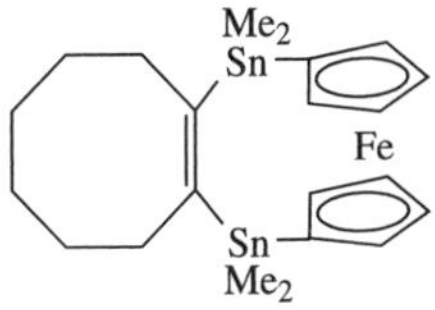		−63.3	C_6D_6	244
(−)-MenSnMe$_2$(CHPh−CHPhCOOMe)[e,f]		1.7	$CDCl_3$	245
		−1.1		
(−)-MenSnMe$_2$(CHPh−CHPhCOOMe)[e]		4.9	$CDCl_3$	245
(−)-MenSnMe$_2$(CHPh−CHPhCOO(−)-Men)[e,f]		−2.8	$CDCl_3$	245

(*continued overleaf*)

TABLE 12. *(continued)*

Compound		Chemical shift (ppm)	Solvent	Reference
(−)-MenSnMe$_2$(CHPh−CHPhCOO(−)-Men)[e]		4.0 6.6	$CDCl_3$	245
(−)-MenSnMe$_2$(CHPh−CHPhCN)[e]		0.5	$CDCl_3$	245
$Me_2Sn(C{\equiv}C{-}SiMe_2{-}C{\equiv}CR)_2$	R = H	−164.5	C_6D_6	133
	Bu	−166.4		
	t-Bu	−165.4		
	i-Pent	−166.0		
	Ph	−164.8		
	$SiMe_3$	−164.1		
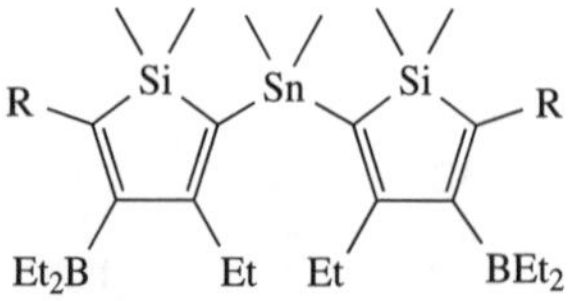 R, Si, Sn, Si, R, Et_2B, Et, Et, BEt_2	R = H	−103.4	C_6D_6	133
	Bu	−105.0		
	t-Bu	−106.2		
	3-Me-Bu	−104.8		
	Ph	−104.3		
	$SiMe_3$	−102.6		

[a]MAS spectra.
[b]equilibrium between 6-membered ring (in solution) and polymer (solid state).
[c]Measured in THF with a D_2O-capillary.
[d]Me in a second isomer.
[e]Men = Menthyl.
[f]Differs from the following compound in the conformation of the Ph group.

TABLE 13. Ring systems containing tin in the ring skeleton

Compound		Chemical shift (ppm)	Solvent	Reference
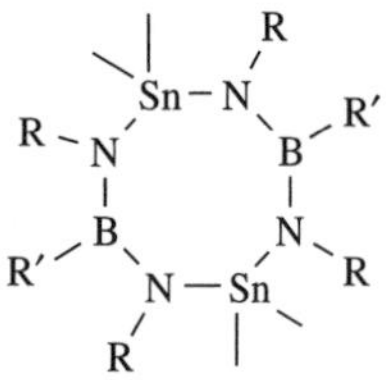 R, Sn−N, R′, R−N, B, B, N, R′, N−Sn, R, R	R = 2,4,6-(i-Pr)$_3C_6H_2$		$CDCl_3$	236
	R′ = Me	40.5		
	Ph	42.5		
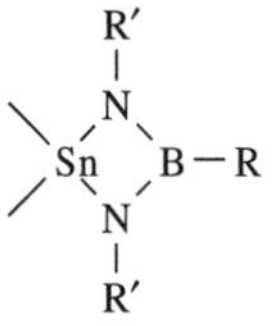 R′, N, Sn, B−R, N, R′	R = 2,4,6-(i-Pr)$_3C_6H_2$		$CDCl_3$	236
	R′ = i-Pr	135.2		
	t-Bu	121.9		
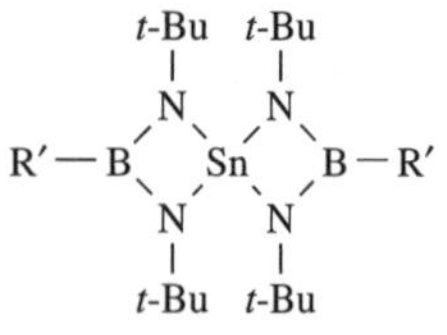 *t*-Bu, *t*-Bu, N, N, R′−B, Sn, B−R′, N, N, *t*-Bu, *t*-Bu	R′ = 2,4,6-(i-Pr)$_3C_6H_2$	−117.8	$CDCl_3$	236

TABLE 13. (*continued*)

Compound		Chemical shift (ppm)	Solvent	Reference
$[t\text{-}Bu_2SnO]_3$		−83.5 −84.3	$CDCl_3$	237
$[t\text{-}Bu_2SnO]_2Si(Bu\text{-}t)_2$		−107.2 −106.6[a]	$CDCl_3$	237
$[t\text{-}Bu_2SnOSi(Bu\text{-}t)_2]_2$		−178.5 −178.1[a]	$CDCl_3$	237
$[t\text{-}Bu_2SnOSiPh_2]_2$		−149.5 −145.6[a]	$CDCl_3$	237
$t\text{-}Bu_2Sn[OSi(Bu\text{-}t)_2]_3$		−153.1 −150.2[a]	$CDCl_3$	237
trans-$[t\text{-}Bu_2SnOSiCl(Bu\text{-}t)]_2$ *cis*-$[t\text{-}Bu_2SnOSiCl(Bu\text{-}t)]_2$		−166.6 −167.7	$CDCl_3$	237
trans-$[t\text{-}Bu_2SnOSiF(Bu\text{-}t)]_2$ *cis*-$[t\text{-}Bu_2SnOSiF(Bu\text{-}t)]_2$		−161.5 −163.1 −161.2[a] −164.9[a]	$CDCl_3$	237
$[t\text{-}Bu_2SnO]_2SiF(Bu\text{-}t)$		−100.8	$CDCl_3$	237
$[t\text{-}Bu_2SnO]_2SiFPh$		−98.6	$CDCl_3$	237
$S[t\text{-}Bu_2SnO]_2Si(Bu\text{-}t)_2$		−3.6	$CDCl_3$	237
$S[t\text{-}Bu_2SnO]_2SiF(Bu\text{-}t)$		8.8	$CDCl_3$	237
$S[t\text{-}Bu_2SnO]_2SiPh_2$		10.7	$CDCl_3$	237
$O-SiPh_2$, Ph_2Si, O, O, $Sn(Bu\text{-}t)_2$, $t\text{-}Bu_2Sn-O$		−125.7 −123.0[a] −132.3[a]	$CDCl_3$	237
$t\text{-}Bu_2Sn[OSiMe_2CH_2]_2$		−135.7	$CDCl_3$	237
$t\text{-}Bu_2Si[(OBu\text{-}t)_2SnO]_2SiF(Bu\text{-}t)$		−169.2	$CDCl_3$	237
R, Et, N, Sn, B, Et, Et, Et	R = Et R = $SnMe_3$	124.1 131.8 109.4($SnMe_3$)	$C_6D_5CD_3$	28
$SnMe_3$, N, Sn, B−Et, Me_3Sn, Et, Et		15.7(Me_3Sn-N) 142.5(Me_2Sn) 50.8(Me_3Sn-C)	C_6D_6	28
R, R, P, Sn, B, Et, Et, Et	R = $SiMe_3$ Ph	29.8 19.3	$C_6D_5CD_3$ $CDCl_3$	28
$(t\text{-}Bu)_2Sn(OGePh_2)_2O$		−92.4 −92.3[a]	$CDCl_3$	246

(*continued overleaf*)

TABLE 13. (*continued*)

Compound		Chemical shift (ppm)	Solvent	Reference
$(R_2SnTe)_3$	$R = CH_2Ph$	−148.5	$CDCl_3$	247
	Ph	−206	CD_2Cl_2	249
$(Me_2SnNR)_3$	R = Me	90.1	C_6D_6	147
	Et	79.9		
	Pr	79.1		
	i-Pr	56.5		
	Bu	78.3		
	t-Bu	110.2		
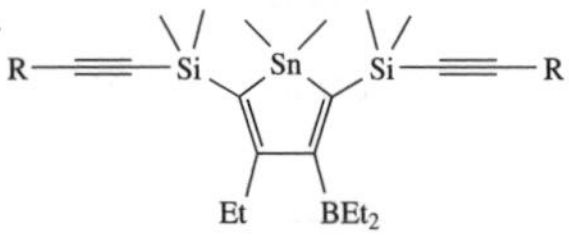 R Si Sn Si R Et BEt_2	R = H	136.5	C_6D_6	147
	Bu	132.6		
	t-Bu	133.6		
	n-C_5H_{11}	132.4		
	Ph	134.8		
	$SiMe_3$	134.4		
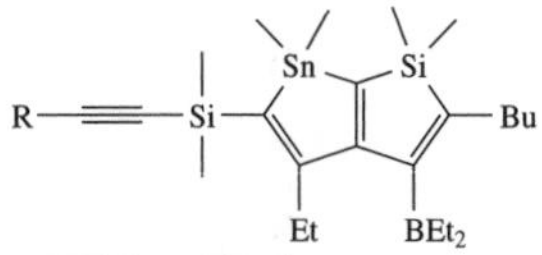 R Si Sn Si Bu Et BEt_2	R = Bu	42.3	C_6D_6	147
	n-C_5H_{11}	42.1		
	$SiMe_3$	37.4		
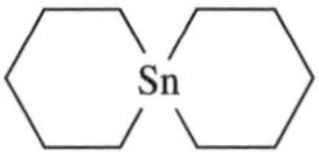 Sn		−82.8	$CDCl_3$	252
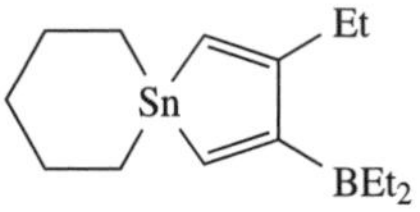 Sn Et BEt_2		−21.7	$C_6D_5CD_3$	252
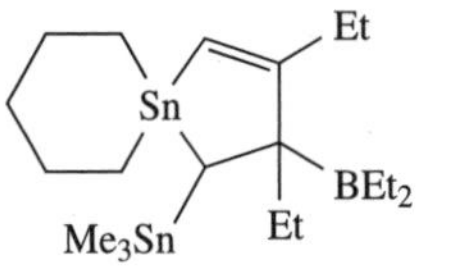 Sn Et BEt_2 Me_3Sn Et		45.8	$C_6D_5CD_3$	252
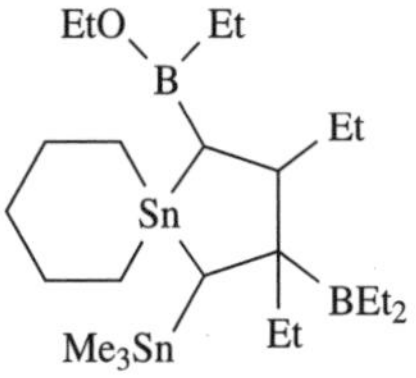 EtO Et B Sn Et BEt_2 Me_3Sn Et		34.8	$C_6D_5CD_3$	252
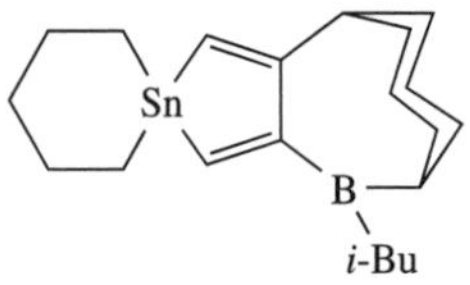 Sn B *i*-Bu		−38.6	$C_6D_5CD_3$	252

TABLE 13. (*continued*)

Compound			Chemical shift (ppm)	Solvent	Reference
[structure: Sn-spiro bicyclic compound with Me_3Sn, *i*-Bu, B–OBu]			18.1	$C_6D_5CD_3$	252
[structure: R_2Sn in six-membered ring with five $SiMe_2$]	R = Me		−223.1	$CDCl_3$	239
	Ph		−242.6		
[structure: R_2Sn in seven-membered ring with six $SiMe_2$]	R = Me		−224.1	$CDCl_3$	239
	Ph		−243.6		
[structure: six-membered ring with two $t\text{-}Bu_2Sn$ and four $SiMe_2$]			−152.6	THF/D_2O-cap.	240
[structure: five-membered ring with two $t\text{-}Bu_2Sn$, SiR_2 and two $SiMe_2$]	R = Me		−147.6	$CDCl_3$	240
	Ph		−152.3		
[structure: spiro Sn with N-*t*-Bu groups, $(Me_2Si)_n$ and X]	X = Se	*n* = 1	−128.4	C_6D_6	251
	X = $SiMe_2$	*n* = 1	−154.0		
	X = Se	*n* = 2	−128.8		
[structure: cage with two Sn–Cl and P-*t*-Bu groups]	P = *t*-Bu		21.19[b]	CD_2Cl_2	253

[a]MAS spectra;

[b]higher-order spin systems.

TABLE 14. Compounds with at least one tin–tin moiety

Compound		Chemical shift (ppm)	Solvent	Reference
$[Me_3Sn^*]_4Sn$[a]		−81.5	C_6D_6	131
$[Me_3Sn^*]_2SnMe_2$[a]		−99.5	C_6D_6	131
$[Ph_3Sn^*]_4Sn$[a]		−135.5	C_6D_6	194
$(Me_3Si)_3Si$ Cl Cl−Sn−Sn−$Si(SiMe_3)_3$ $(Me_3Si)_3Si$−Sn−Sn−Cl Cl $Si(SiMe_3)_3$		192.2	$CDCl_3$	255
t-Bu_2Sn—Sn(Bu-t)$_2$ Me_2Si—$SiMe_2$		−19.4	C_6D_6	243
t-Bu_2Sn—Sn(Bu-t)$_2$ Me_2Si $SiMe_2$ $[SiMe_2]_n$	$n = 1$	−98.5	C_6D_6	243
	2	−99.7		
X-Bu_2Sn−(t-Bu)$_2Sn_A$ −Bu_2Sn_B-X	X = CH_2CH_2OEt	Sn_A: −86.9 Sn_B: −82.6	C_6D_6 or	254
X-Bu_2Sn−Bu_2Sn−(t-Bu)$_2Sn_A$−Bu_2Sn_B−Sn_CBu_2-X		Sn_A: −35.8 Sn_B: −205.8 Sn_C: −91.1	$C_6D_5CD_3$	
X-$(Bu_2Sn)_3$−(t-Bu)$_2Sn_A$−Bu_2Sn_B−Sn_CBu_2−Sn_DBu_2-X		Sn_A: −27.7 Sn_B: −192.5 Sn_C: −206.5 Sn_D: −84.7		
X-Bu_2Sn−Bu_2Sn−Bu_2Sn_A−Bu_2Sn_B−Sn_CBu_2-X		Sn_A: −197.1 Sn_B: −206.2 Sn_C: −83.0		
$(R_2Sn)_x$	R = n-Bu	−189.6	C_6D_6	257
		−178.9	$CDCl_3$	258
	n-Hex	−190.9	C_6D_6	259
	n-Oct	−190.7		
t-Bu t-Bu t-Bu t-Bu N Se N Me_2Si(N)(N)Sn—Sn(N)(N)$SiMe_2$ t-Bu t-Bu		−121.3[b] −116.0[c]	C_6D_6 $C_6D_5CD_3$/ CH_2Cl_2	251
Me_2Sn Fe $(Me_2Sn)_n$	$n = 1$	−43.4	C_6D_6	235
	$n = 2$	−102.5 (Sn−Cp) −249.3		262

TABLE 14. (*continued*)

Compound			Chemical shift (ppm)	Solvent	Reference
$Me_3Sn_A{-}Me_2Sn_B{-}SiR_2R'$	R = Me	R′ = Me	Sn_A: −111.7 Sn_B: −263.6	$CDCl_3$	156
	Me	*i*-Pr	Sn_A: −110.8 Sn_B: −274.6		
	Ph	*i*-Pr	Sn_A: −147.7 Sn_B: −241.1		
$Me_3Sn_A{-}Me_2Sn_B$ $-CR=CH\text{-}SiMe_3$	R = Ph		Sn_A: −98.4 Sn_B: −140.5	$CDCl_3$	156
	CH_2OMe		Sn_A: −97.3 Sn_B: −150.4		

[a] Sn* is the Sn atom displaying the shift given.
[b] At 298 K.
[c] At 223 K.

Compounds containing one or two silyl substituents at one tin atom display tin-119 chemical shifts which are similar to those of the corresponding tin hydrides (±30 ppm) as summarized below.

$$\delta^{119}Sn\text{: } R_3SnH \approx R_3Sn{-}Si$$

$$\delta^{119}Sn\text{: } R_2SnH_2 \approx Si{-}R_2Sn{-}Si$$

Examples for such derivatives are given in Tables 5, 8, 12 and 13. Bulky substituents (*t*-Bu, *i*-Pr) at the tin and/or silicon atoms may cause slightly higher deviations. Exceptions are compounds with a relatively high ring strain and therefore smaller bond angles such as, for example, four-membered Si−Sn rings (Chart 1). The ^{119}Sn resonance is shifted significantly to lower field values and this effect is in good agreement with results shown for rings containing carbon[24] instead of the silyl moieties.

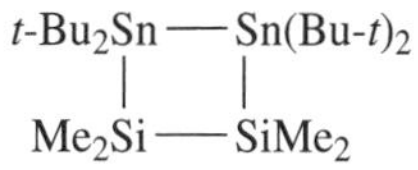

t-Bu₂Sn — Sn(Bu-t)₂ / Me₂Si — SiMe₂ (four-membered ring): −19.6

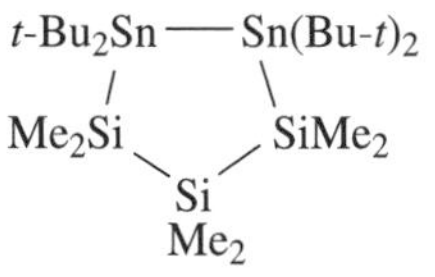

t-Bu₂Sn — Sn(Bu-t)₂ / Me₂Si, SiMe₂ / Si Me₂ (five-membered ring): −98.5

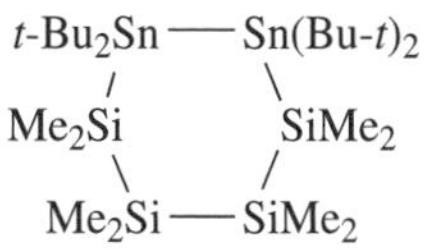

t-Bu₂Sn — Sn(Bu-t)₂ / Me₂Si, SiMe₂ / Me₂Si — SiMe₂ (six-membered ring): −99.7

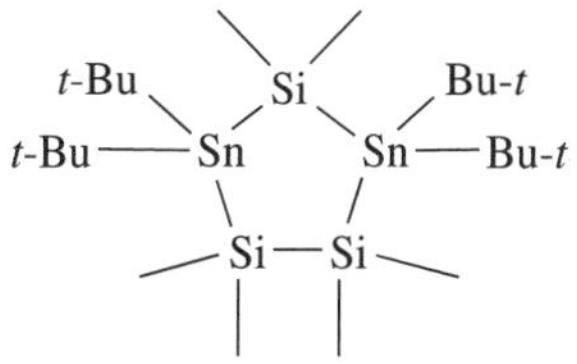

t-Bu, t-Bu — Sn, Si, Sn — Bu-t, Bu-t / Si — Si (five-membered ring): −147.6

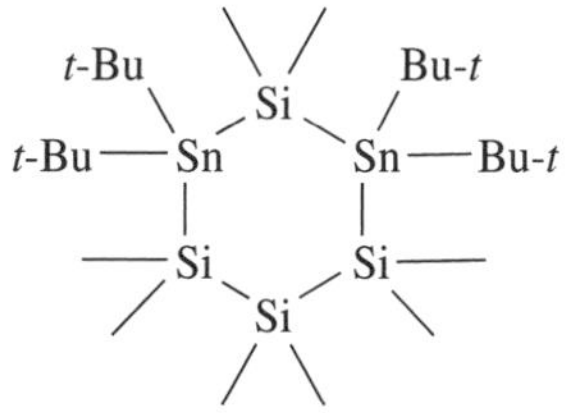

t-Bu, t-Bu — Sn, Si, Sn — Bu-t, Bu-t / Si, Si / Si (six-membered ring): −152.6

CHART 1. ^{119}Sn NMR chemical shift of tin-modified cyclosilanes[240,243]

In contrast to carbon–tin rings the differences between the ^{119}Sn chemical shifts of five- and six-membered Si–Sn rings are much smaller, possibly due to the larger Si–Si and Si–Sn bond lengths, atomic radii and therefore to the larger bond angles.

b. Tin(IV) compounds with coordination number higher than four. In the presence of Lewis bases tin can act as a Lewis acid. Higher coordination numbers than four are then possible and are very common. Coordination numbers of five[299–314] or six[310–363] are common but arrangements of seven and eight atoms surrounding the tin are rare. Especially for octacoordinated tin atoms only a few examples are known and they are from pure inorganic chemistry (e.g. $Sn(NO_3)_4$). ^{119}Sn chemical shifts for some selected compounds having tin surrounded by seven ligands are found in Table 15. The geometry around the tin atom can change easily. A good demonstration of this point are the carboxylato derivatives of the tetraorganodistannoxanes. The three structures proposed for this type of compound are sketched in Chart 2. The structures consist of a central four-membered ring with two endocyclic tin atoms connected via oxygen bridges to the two exocyclic tin atoms. The carboxylato ligands O_2C–R (R = Me, *t*-Bu, *p*-Tol, C_6F_5) bond with one (X) or two oxygen atoms (Y) with the tin atom. The solution spectra consist of two lines at −140 and −220 ppm. One of the two signals is broader than the other. Two groups tried to assign these resonances. The main tool for studying such compounds was 2D ^{1}H–^{119}Sn correlation NMR spectroscopy[30,32]. Both groups agree that in solution, the first structure of chart 1 is present, but disagree in the assignment of signals to the endo- and exocyclic tin atoms. The ^{119}Sn CP-MAS spectra for their compounds give three signals of equal intensity for the dibutyl but four for the dimethyl derivative (X, Y = MeCOO). This is interpreted by Willem and coworkers by the presence of the second structure in Chart 2.

Extensive spinning side bands are found and analyzed for the dimethyl derivative whereas the dibutyldistannaoxane does not show any such bands.

Solid state NMR indicates the third type structure of Chart 2, when R = Me, X = Y = *t*-BuCOO.

Similar cluster compounds are described in References 225–228 and 265. Another interesting cluster is $\{(BuSn)_{12}O_{14}(OH)_6\}^{2+}$ displaying two resonances at *ca* −282 and *ca* −460 ppm due to penta- and hexacoordinated tin[344,348,353].

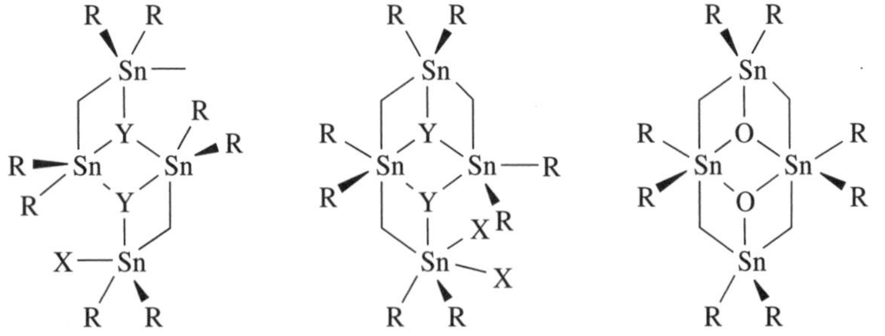

CHART 2. Sketches of tetramer diorganostannanes X = OCOR′, Y = O_2CR' (R′ = Me, *t*-Bu, *p*-Tol, C_6F_5, R = Me, Bu)

Compound			Chemical shift (ppm)	Solvent	Reference
$RSnR'_3$	R = Me	R′ 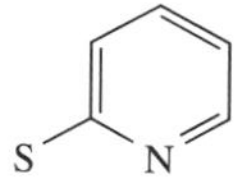 = (S, N)	−398	$CDCl_3$	263
			−469	DMSO-d_6	264
	Bu		−387	$CDCl_3$	
	Ph		−448		
			−504	DMSO-d_6	
	4-ClC_6H_4		−445	$CDCl_3$	
	2-ClC_6H_4		−420		
	4-MeC_6H_4		−446		
	R = Me	R′ 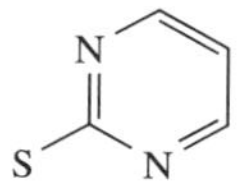 = (N, S, N)	−267	$CDCl_3$	
			−372	DMSO-d_6	
	Bu		−280	$CDCl_3$	
	Ph		−402		
			−491		
$RSnR'_3$					264
	R = 4-MeC_6H_4	R′ 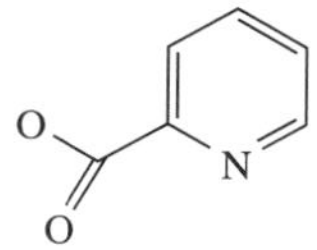 = (O, O, N)	−620	$CDCl_3$	
	R = 4-MeC_6H_4	R′ 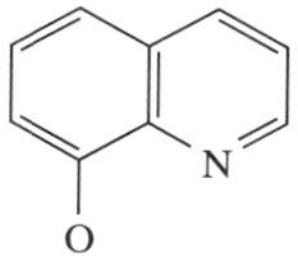 = (O, N)	−611	$CDCl_3$	
$[(c\text{-}C_6H_{11})NH_2]_2[Bu_2SnR_2]$	R 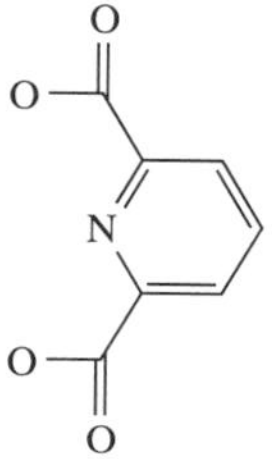 = (O, O, N, O, O)		−392.0 −424.9[a]	$CDCl_3$	266

(continued overleaf)

TABLE 15. (*continued*)

Compound			Chemical shift (ppm)	Solvent	Reference
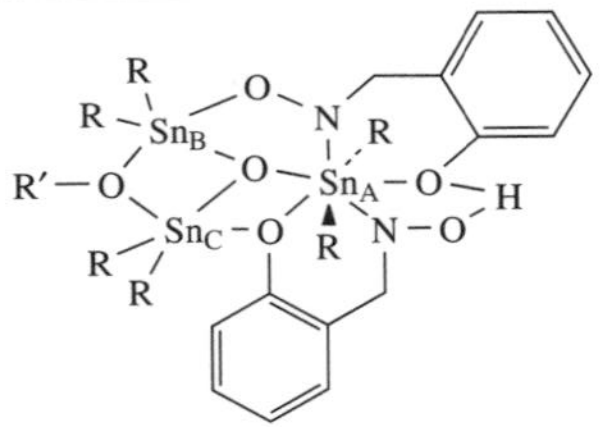 R, R–Sn_B–O–N, R′–O, Sn_A–O–H, Sn_C–O, N–O, R, R, R					
R	R′				
Bu	H		Sn_A: −444.6 Sn_B: −120.3 Sn_C: −105.7	C_6D_6	265
Me	4-MeC_6H_4		Sn_A: −452.1 Sn_B: −138.1 Sn_C: −123.6	$CDCl_3$	267
	3,5-$Me_2C_6H_3$		Sn_A: −452.2 Sn_B: −138.1 Sn_C: −125.0		
	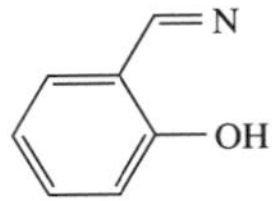 N, OH		Sn_A: −451.4 Sn_B: −122.9 Sn_C: −109.1	C_6D_6	268
RO, RO, P, S, R′, Sn, O, S, P, OR, OR, R′, O, R_2Sn, SnR_2, O, R′, RO, RO, P, O, Sn, O, P, OR, OR, S, S, R′	R	R′			
	Pr	Me	Sn_A: −124 Sn_B: −146	C_6D_6	277
	Pr	Pr	Sn_A: −147 Sn_B: −177		

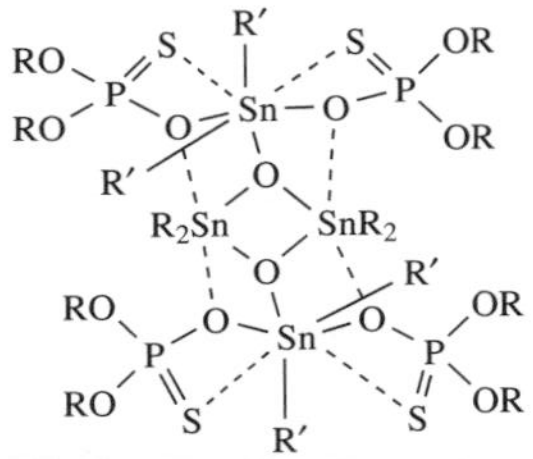

Many ^{119}Sn NMR data exist for tin complexes with ligands of biological importance and activity such as *in vitro* antitumor activity or antimicrobial activity[229–232,365–381]. The exact nature of the coordination sphere around the tin is not always given.

Triorganotin hydrides are common reductants in a large number of organic and organometallic syntheses. In many cases these reductions proceed by a normal radical reaction pathway. However, ionic reductions using tin hydrides also occur, but they were only sparsely investigated. Studies of ionic reduction processes with tin hydrides by using ^{119}Sn NMR are found elsewhere[298,310].

IV. LEAD NMR

A. History and Technical Details

Since the first investigations about lead NMR in the late 1950s[390] ^{207}Pb NMR was, among other methods such as X-ray crystallography, one of the major tools for the determination of lead containing compounds in solution and in the solid state. In line with our own experience, solution state ^{207}Pb NMR measurements are relatively simple to generate. Lead-207 chemical shifts were measured with respect to Me_4Pb or 1 M $Pb(NO_3)_2$ in water [$\delta(Me_4Pb) = \delta(Pb(NO_3)_2) + 2961$ppm]. Absolute frequencies, relative to the ^{1}H signal of Me_4Si = 100 MHz, were 20.920597 MHz for Me_4Pb (80% in toluene)[391].

A number of reviews dealing with ^{207}Pb NMR, sometimes in addition to other nuclei, were published over the years[392–395]. A summary of most NMR parameters and also chemical shifts was given by Wrackmeyer and Horchler in 1989[394].

The ^{207}Pb chemical shifts cover a range from +11 000 ppm to nearly −6500 ppm for lead(II) cyclopentadienyl derivatives. Analogous to ^{29}Si/^{73}Ge, ^{29}Si/^{119}Sn or ^{73}Ge/^{119}Sn a linear relationships between ^{119}Sn and ^{207}Pb were found[394,396–398].

B. Recent Progress in ^{207}Pb NMR Shifts

Selected ^{207}Pb NMR data of organometallic lead compounds are given in Table 16. In over 50 publications ^{207}Pb NMR chemical shifts have been reported since 1995. One of the major topics in recent years was investigations dealing with high resolution solid state ^{207}Pb NMR. Besides pure inorganic compounds and theoretical studies[399–402] a wide variety of ^{207}Pb NMR of organometallic derivatives with Pb−X(X = C, N, P) and Pb−O−C[403–418] bonds were described both in solution and in the solid state.

^{207}Lead NMR deals mainly with compounds containing lead in the oxidation state IV. Only a limited, but in recent years increasing number of lead(II) derivatives were determined by ^{207}Pb NMR[394,401,419–423]. An overview of compounds with direct lead−carbon, lead−nitrogen or−phosphorus bonds published since 1995 is given in Table 16[419–443].

Due to the wide range of ^{207}Pb chemical shifts, especially for lead(II) compounds, it is (in analogy to the tin compounds discussed above) impossible to recognize simple correlations between the coordination number at the lead atoms or the substituents directly bonded to lead(II) and the chemical shift value. What one can only do is make predictions within one class of compounds.

Lead-207 NMR will never be such an important spectroscopic tool like tin-119 NMR; however, it is very useful for the solution of many new and old problems in lead chemistry and will attract the interest of both the preparative as well as analytical chemist in the years to come.

TABLE 16. ^{207}Pb chemical shifts of organometallic compounds

Compound		Chemical shift (ppm) ($^{1}J_{Pb-X}$, Hz)	Solvent	Reference
Et_4Pb		74.8	$CDCl_3$	425
Et_2PbPh_2		−38	$CDCl_3$	425
Et_2PbMe_2		53	$CDCl_3$	425
Ph_4Pb		−179.0	$CDCl_3$	424
		−178.0	$CDCl_3$	425
		−146.7[a]		428
$(C_6F_5)_4Pb$		−391	$CDCl_3$	437
R-$PbPh_3$	R = 4-MeC_6H_4	−179.5/−145.9[a]	$CDCl_3$	424
	3-MeC_6H_4	−181.5/−147.8[a]		428
	2-MeC_6H_4	−170.4/−146.9[a]		428
(4-$MeC_6H_4)_2PbPh_2$	R = 4-MeC_6H_4	−176.3/−142.0[a]	$CDCl_3$	424
	3-MeC_6H_4	−180.9/−146.4[a]		428
	2-MeC_6H_4	−161.2/−151.1[a]		428
(4-$MeC_6H_4)_3PbPh$	R = 4-MeC_6H_4	−174.0/−150.0[a]	$CDCl_3$	424
	3-MeC_6H_4	−180.3/−128.0[a]		428
	2-MeC_6H_4	−152.9/−150.2[a]		428
(4-$MeC_6H_4)_4Pb$		−171.3/−148.8[a]	$CDCl_3$	424
				428
(3-$MeC_6H_4)_4Pb$		−179.5/−119.3[a]	$CDCl_3$	428
(2-$MeC_6H_4)_4Pb$		−166.7/−159.5[a]	$CDCl_3$	428
Tip_2PbMe_2[b]		−234	$CDCl_3$	420
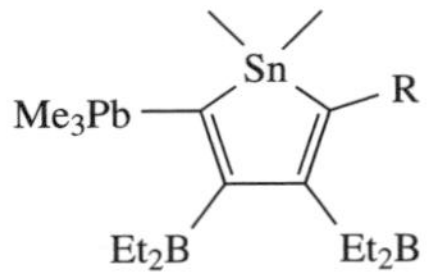 Me_3Pb-stannole (Me_2Sn ring, 3,4-bis(Et_2B), 5-R)	R = NMe_2	−38.0	$C_6D_5CD_3$	435
	Me	−47.2	C_6D_6	

	NHC_6H_5	156.3		
t-Bu_3Pb-R	R = Me	101.1	C_6D_6	438
	Et	40.1		
	n-Pr	41.1		
	n-Bu	41.5		
	CH_2Ph	7.9		
	$CH(SiMe_3)_2$	124.5		
t-Bu_3Pb-R^c	R = $SiMe_3$	−47.4(−207.6)	C_6D_6	438
	$SiMe_2Bu$-t	−32.0(102.5)		
	$SiMe_2SiMe_3$	18.7 (57.9)		
	$SiMe_2Ph$	65.1 (151.4)		
	$SiPh_3$	67.2		
t-Bu_3Pb−$GeMe_3$		21	C_6D_6	438
t-Bu_3Pb-R^d	R = $SnMe_3$	110.5 (1637)	C_6D_6	438
	$SnEt_3$	131.7 (2504)		
	$SnBu_3$	120.9 (2441)		
	$Sn(Bu$-$t)_3$	158.0 (6685)		
	$SnPh_3$	135.7 (3581)		
i-Pr_3Pb−$SnMe_3{}^d$		12.9 (303)	C_6D_6	438
Et_3Pb_A-R	R = $Pb_B(Pr$-$i)_3$	166.0/61.0 (−6836)	C_6D_6	438
	$SnMe_3$	135.2 $(-1398)^d$		
t-Bu_3Pb_A-R	R = Pb_BMe_3	213.6/0.4 (−7380)	C_6D_6	438
	Pb_BEt_3	261.2/83.0 (−8126)		
	$Pb_B(Pr$-$i)_3$	309.1/164.8 (−9114)		
	$Pb_B(Bu$-$t)_3$	335.1		
	$Pb_B(C_6H_{13})_3$	302.3/75.4 (−9200)		
	$Pb_B(C_6H_{11}$-$c)_3$	132.6/23.6 (−8911)		
$Me_2(R)Si$-$(t$-$Bu_2)$ Pb−$Si(R)Me_2$	R = Me	−264.8(180)	C_6D_6	438
	t-Bu	−252.7(80.6)		
t-Bu_2MePb−$PbMe(Bu$-$t)_2$		201	C_6D_6	438

(*continued overleaf*)

TABLE 16. (*continued*)

Compound		Chemical shift (ppm) ($^{1}J_{Pb-X}$, Hz)	Solvent	Reference
t-Bu_3Pb–NHR	R = H	115.9	C_6D_6	426
	Ph	44.8		
	$SiMe_3$	87.1		
Me_3Pb–NHR	R = Ph	156.3	C_6D_6	426
	$SiMe_2Bu$-t	155.2		
$PbMe_3$, N, Me, Me (2,5-dimethylpyrrol-1-yl)		170.3	C_6D_6	431
N, N–Pb(Bu-t)$_3$ (pyrazol-1-yl)		68.1	C_6D_6	426
Me, Sn–Pb–(Bu-t)$_3$, Me, Fe, (t-Bu)$_3$–Pb–Sn		105.0 (1985)[d]	C_6D_6	438
Pb, R, R (plumbolane)	R = Me	205.3	$CDCl_3$	425
	Ph	149.3		
Pb (spiro-biplumbolane)		416.3	$CDCl_3$	425

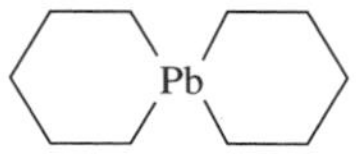

Compound				δ	Solvent	Ref.
(Pb spiro-bicyclic structure)				−199	$CDCl_3$	425
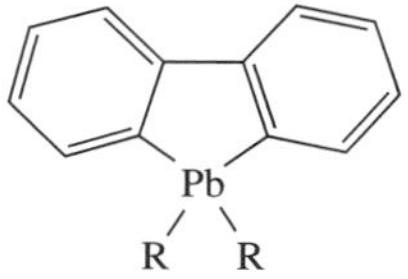 (dibenzoplumbole structure: Pb, R, R)	R = Me			0.3	$CDCl_3$	425
	Ph			−95.4		
(spiro-bis(dibenzoplumbole) structure: Pb)				10.7	$CDCl_3$	425
(tricyclic structure: E, Pb, R, R, X, X)						
E = CH_2	R = Me	X = H		−201.4	$CDCl_3$	425
CH_2	Ph	H		−284.1		
NMe	Me	H		−238.0		
NMe	Ph	H		−334.7		
NMe	Me	Br		−219.9		
NMe	Ph	Br		−333.4		
O	Me	H		−41.5		
O	Ph	H		−363.4		
Mes_3PbBr				−97	$CDCl_3$	401
Mes_2PbBr_2				−149	$CDCl_3$	401

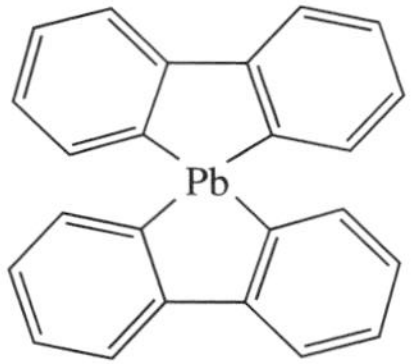

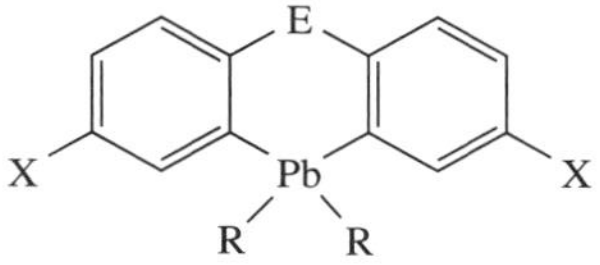

(*continued overleaf*)

TABLE 16. (*continued*)

Compound		Chemical shift (ppm) ($^1J_{Pb-X}$, Hz)	Solvent	Reference
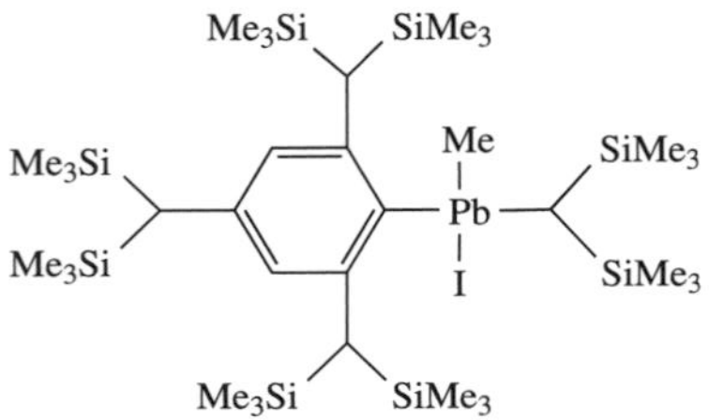		−10	$CDCl_3$	420
	R = $CH(SiMe_3)_2$	−191	$CDCl_3$	420
	CH_2SiMe_3	−176		
	$CHMe_2$	−195		
	R = 2,4,6-$(Me_3SiCH_2)_3C_6H_2$	−137	$CDCl_3$	420
	Tip	−143		
	$(Me_3Si)_3CH$	83		
$RPb(OOCCH_3)_3$	R = Ph	−839	$CDCl_3$	439
	4-MeC_6H_4	−812		

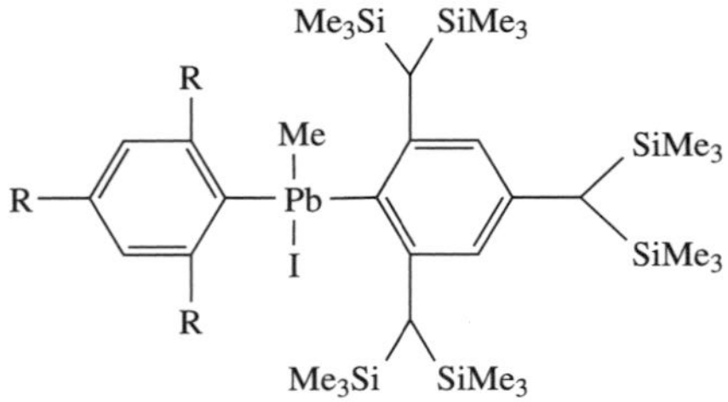

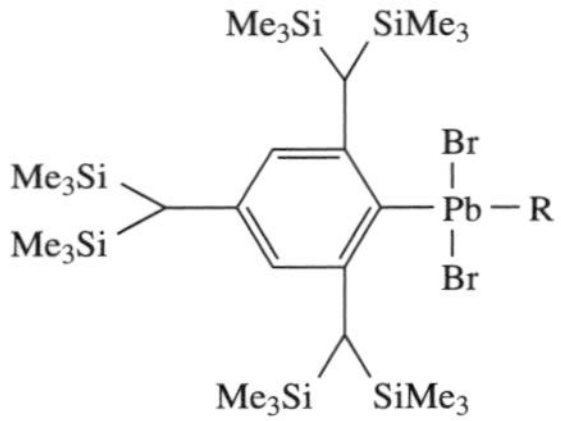

$R_2Pb(OOCCH_3)_2$	R = Ph	−587	$CDCl_3$	439
	$4-MeC_6H_4$	−609		
	$2-MeC_6H_4$	−534		
	$4-ClC_6H_4$	−695		
	$2-ClC_6H_4$	−570		
$RPb(OOCPy-z)_3$	R = Ph	−904	$CDCl_3$	441
	$4-MeC_6H_4$	−885		
	$2-MeC_6H_4$	−901		
	$2-ClC_6H_4$	−1038		
$4-MeOC_6H_4Pb(OOCCH_3)_3$		−791	$CDCl_3$	427
$4-MeOC_6H_4Pb(OOCCH_3)_2(OOCR)$	R = $4-MeC_6H_4$	−802	$CDCl_3$	427
$4-MeOC_6H_4Pb(OOCCH_3)(OOCR)_2$	R = $4-MeC_6H_4$	−808	$CDCl_3$	427
$4-MeOC_6H_4Pb(OOCR)_3$	R = $4-MeC_6H_4$	−813	$CDCl_3$	427
$Ph_3Pb-O-SiPh_3$		−107.3	C_6H_6[e]	440
$Ph_3Pb-O-GePh_3$		−89.6	C_6H_6[e]	440
$Ph_3Pb-O-SnPh_3$		−76.2	C_6H_6[e]	440
$Ph_3Pb-O-PbPh_3$		−63.6	C_6H_6[e]	440
$Tip_2Pb(SPh)_2$[b]		69	$CDCl_3$	420
$Tip_2Pb(SePh)_2$[b]		−261	$CDCl_3$	420
S—S, S, S, Pb, R′, R	R = R′ = Tip[b]	288.6	$CDCl_3$	420
	R = $2,4,6-((Me_3Si)_2CH)_3C_6H_2$ R′ = $2,4,6-(Me_3SiCH_2)_3C_6H_2$	312.3		
	R = Tip[b] R′ = $(Me_3Si)_2CH$	314.5		

(*continued overleaf*)

TABLE 16. (*continued*)

Compound		Chemical shift (ppm) ($^1J_{Pb-X}$, Hz)	Solvent	Reference
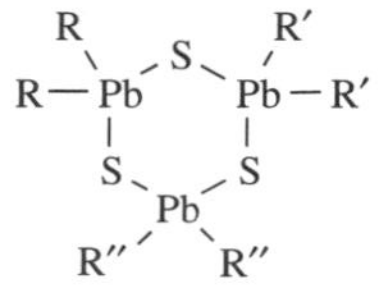 [cyclic $(R_2PbS)_3$: R, R on Pb; R′, R′ on Pb; R″, R″ on Pb]	R = R′ = R″ = Ph R = R′ = R″ = 2-MeC_6H_4 R = R′ = R″ = 4-MeC_6H_4 37 compounds with different substituents R, R′ and R″ are also given	172.7–173.5 180.9–181.3 172.6–173.3	$CDCl_3$	432
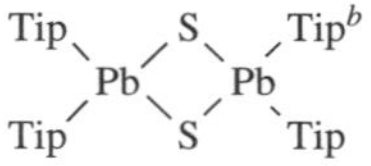 [$Tip_2Pb(\mu$-S$)_2PbTip_2$][b]		81.6	$CDCl_3$	420
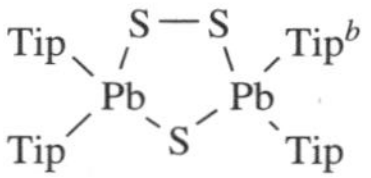 [Tip_2Pb(S–S)(S)$PbTip_2$ ring][b]		130	$CDCl_3$	420
$Pb(N(SiMe_3)_2)_2$		4916	C_6D_6	419
$Pb(CH(SiMe_3)_2)_2$		10050	C_6D_6	420
2,4,6-$[(SiMe_3)_2CH]_3C_6H_2Pb[CH(SiMe_3)_2]$		8884	$C_6D_5CD_3$	420
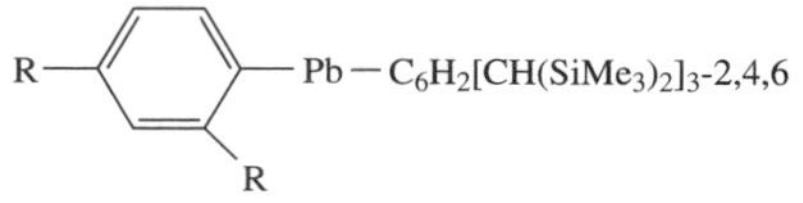 2,4-$R_2C_6H_3$–Pb–$C_6H_2[CH(SiMe_3)_2]_3$-2,4,6	R = $CH(SiMe_3)_2$ CH_2SiMe_3 $CHMe_2$	9751 8873 8888	$C_6D_5CD_3$	433
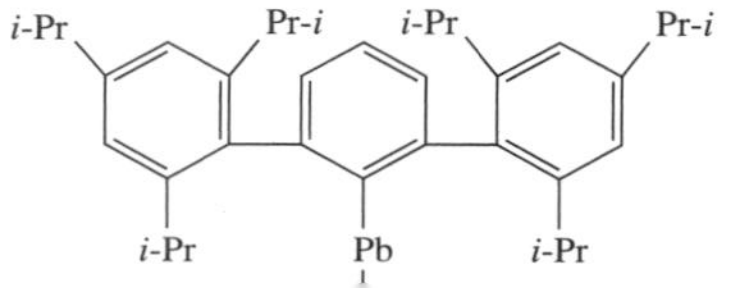 [2,6-bis(2,4,6-tri-*i*-Pr-phenyl)phenyl–Pb]	R = Me *t*-Bu Ph	7420 7853 6657	C_6D_6	423

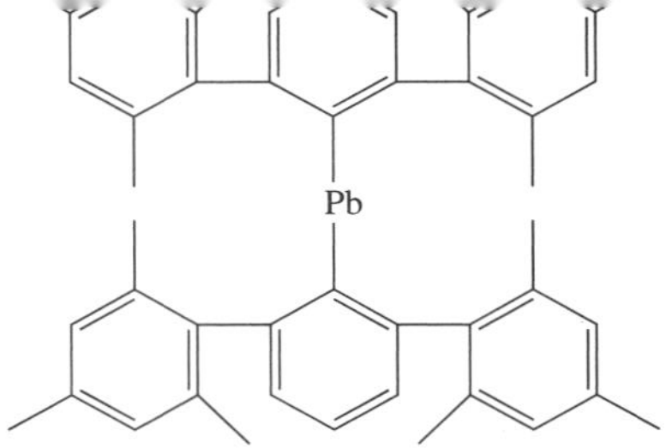 Pb	3870	C_6D_6	442
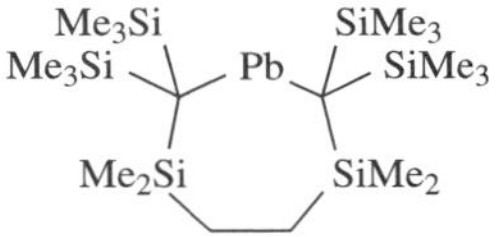 Me3Si, Me3Si, Pb, SiMe3, SiMe3, Me2Si, SiMe2	9112	$C_6D_5CD_3$	433
$(2,3,4\text{-}Me_3\text{-}6\text{-}t\text{-}BuC_6H)_2Pb$	6927	C_6D_6	443
$2,3,4\text{-}Me_3\text{-}6\text{-}t\text{-}BuC_6H{-}Pb{-}Si(SiMe_3)_3$	7545	C_6D_6	443
$2,4,6\text{-}(t\text{-}Bu)_3C_6H_2{-}Pb{-}CH_2C_6H_3(Bu\text{-}t)_2\text{-}3,5$	5067	C_6D_6	44
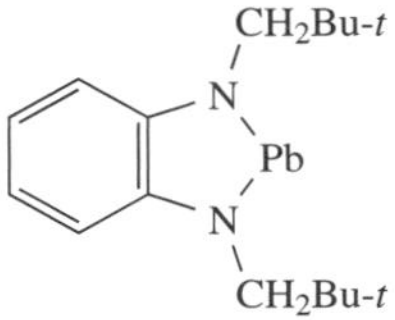 CH2Bu-t, N, Pb, N, CH2Bu-t	3300	$C_6D_5CD_3$	421
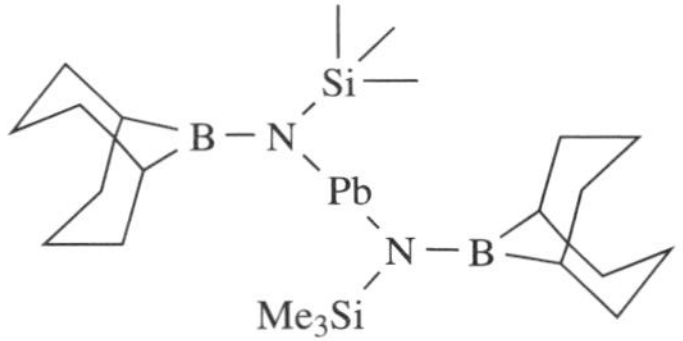 Si, B, N, Pb, N, B, Me3Si	5170	C_6D_6	419

(*continued overleaf*)

TABLE 16. (*continued*)

Compound	Chemical shift (ppm) ($^{1}J_{Pb-X}$, Hz)	Solvent	Reference
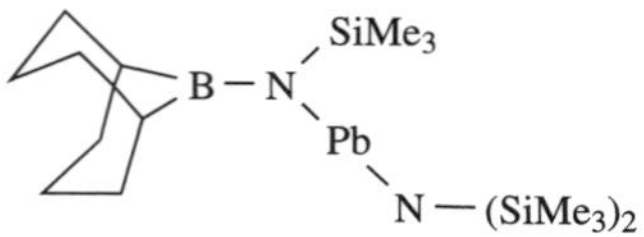 B—N(SiMe$_3$)—Pb—N—(SiMe$_3$)$_2$	4968	C_6D_6	419
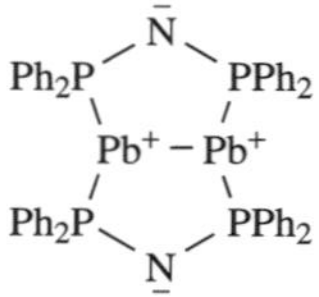 Ph$_2$P, N̄, PPh$_2$, Pb$^+$—Pb$^+$, Ph$_2$P, N̲, PPh$_2$	−608	C_6D_6	429
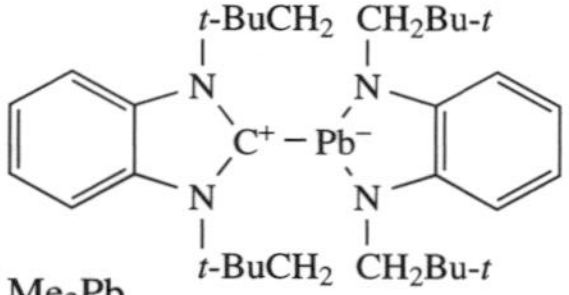 t-BuCH$_2$, CH$_2$Bu-t, N, C$^+$—Pb$^-$, N, t-BuCH$_2$, CH$_2$Bu-t	3000[f] 1800[g]	$C_6D_5CD_3$	421
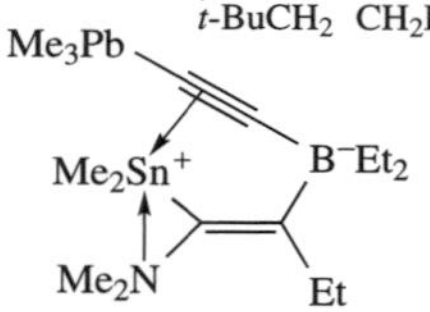 Me$_3$Pb, Me$_2$Sn$^+$, B$^-$Et$_2$, Me$_2$N, Et	−27.6	$CDCl_3$	435

$[Me_3Pb]_4[Fe(CN)_6]^{\bullet}H_2O$		190/200[a]		430
$[Me_3Pb]_4[Ru(CN)_6]^{\bullet}H_2O$		201[a]		430
$[Me_3Pb]_4[Fe(CN)_6]$		139[a]		430
$[Me_3Pb]_4[Ru(CN)_6]$		130[a]		430
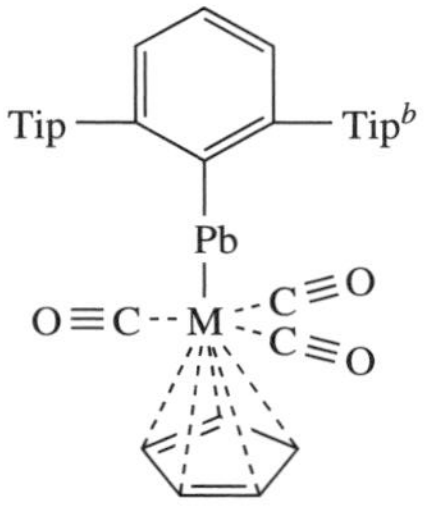 (2,6-Tip₂C₆H₃)Pb–M(CO)₃(η⁵-C₅H₅) (structure; Tip[b])	M = Cr	9563	C_6D_6	422
	Mo	9659		
	W	9374		

[a] In the solid state;
[b] Tip = 2,4,6-(i-Pr)$_3C_6H_2$;
[c] $^1J(^{207}Pb-^{29}Si)$;
[d] $^1J(^{207}Pb-^{119}Sn)$;
[e] External D_2O;
[f] At 358 K;
[g] At 198 K.

V. REFERENCES

1. R. C. Weast and M. J. Astle, Eds., *CRC Handbook of Chemistry and Physics*, CRC Press Boca Raton, 1981.
2. P. J. Watkinson and K. M. Mackay, *J. Organomet. Chem.*, **275**, 39 (1984).
3. J. Kaufmann, W. Sahn and A. Schwenk, *Z. Naturforsch.*, **26A**, 1384 (1971).
4. R. G. Kidd and H. G. Spinney, *J. Am. Chem. Soc.*, **95**, 88 (1973).
5. I. Zicmane, E. Liepins, E. Lukevics and T. K. Gar, *J. Gen. Chem. USSR (Engl. Transl.)*, **52** 780 (1982).
6. K. M. Mackay, P. J. Watkinson and A. L. Wilkins, *J. Chem. Soc., Dalton Trans.*, 133 (1984)
7. Y. Takeuchi, T. Harazano and N. Kakimoto, *Inorg. Chem.*, **23**, 3835 (1984).
8. Y. Takeuchi, I. Zicmane, G. Manuel and R. Boukheroub, *Bull. Chem. Soc. Jpn.*, **66**, 1732 (1993).
9. R. K. Harris, J. D. Kennedy and W. MacFarlane, in *NMR and the Periodic Table* (Eds. R. K Harris and B. E. Mann), Academic Press, London, 1978, p. 340.
10. J. D. Kennedy and W. MacFarlane, in *Multinuclear NMR* (Ed. J. Mason), Plenum Press, New York, 1987, p. 310.
11. K. M. Mackay and R. A. Thomson, *Main Group Metal Chem.*, **10**, 83 (1987).
12. E. Liepins, M. V. Petrova, E. T. Bogoradovskii and V. S. Zavgorodnii, *J. Organomet. Chem.* **410**, 287 (1991).
13. R. A. Thompson, A. L. Wilkins and K. M. Mackay, *Phosphorus, Sulfur, and Silicon*, **150–151** 319 (1999).
14. N. W. Mitzel, U. Losehand and A. D. Richardson, *Inorg. Chem.*, **38**, 5323 (1999).
15. T. N. Mitchell, *J. Organomet. Chem.*, **255**, 279 (1983).
16. A. L. Wilkins, R. A. Thompson and K. M. Mackay, *Main Group Metal Chem.*, **13**, 219 (1990)
17. Y. Takeuchi, H. Yamamoto, K. Tanaka, K. Ogawa, J. Harada, T. Iwamoto and H. Yuge *Tetrahedron*, **54**, 9811 (1998).
18. B. Wrackmeyer and P. Bernatowicz, *J. Organomet. Chem.*, **579**, 133 (1999).
19. Y. Takeuchi, K. Ogawa, G. Manuel, R. Boukherroub and I. Zicmane, *Main Group Meta. Chem.*, **17**, 121 (1994).
20. A. Zickgraf, M. Beuter, U. Kolb, M. Dräger, R. Tozer, D. Dakternieks and K. Jurkschat *Inorg. Chim. Acta*, **275–276**, 203 (1998).
21. M. Charisse, A. Zickgraf, H. Stenger, E. Bräu, C. Desmarquet, M. Dräger, S. Gerstmann, D Dakternieks and J. Hook, *Polyhedron*, **17**, 4497 (1998).
22. F. Riedmüller, G. L. Wegner, A. Jockisch and H. Schmidbaur, *Organometallics*, **18**, 4317 (1999).
23. P. J. Smith and A. P. Tupciauskas, *Annu. Rep. NMR Spectrosc.*, **8**, 291 (1978).
24. B. Wrackmeyer, *Annu. Rep. NMR Spectrosc.*, **16**, 73 (1985).
25. B. Wrackmeyer, *Annu. Rep. NMR Spectrosc.*, **38**, 203 (1999).
26. J.-C. Meurice, M. Vallier, M. Ratier, J.-G. Duboudin and M. Petraud, *J. Chem. Soc., Perkin Trans. 2*, 1311 (1996).
27. B. Wrackmeyer, G. Kehr, H. E. Maisel and H. Zhou, *Magn. Reson. Chem.*, **36**, 39 (1998).
28. B. Wrackmeyer, S. Kerschl and H. E. Maisel, *Main Group Metal Chem.*, **21**, 89 (1998).
29. B. Wrackmeyer and P. Bernatowicz, *J. Organomet. Chem.*, **579**, 133 (1999).
30. F. Ribot, C. Sanchez, A. Meddour, M. Gielen, E. R. T. Tiekin, M. Biesemans and R. Willem *J. Organomet. Chem.*, **552**, 177 (1998).
31. J.-C. Meurice, J.-G. Duboudin and M. Ratier, *Organometallics*, **18**, 1699 (1999).
32. O. Primel, M.-F. Llauro, R. Petiaud and A. Michel, *J. Organomet. Chem.*, **558**, 19 (1998).
33. A. I. Kruppa, M. B. Taraban, N. V. Shokhirev, S. A. Svarovsky and T. V. Leshina, *Chem. Phys. Lett.*, **258**, 316 (1996).
34. G. A. Aucar, E. Botek, S. Gomez, E. Sproviero and R. H. Contreras, *J. Organomet. Chem.* **524**, 1 (1996).
35. R. Challoner and A. Sebald, *J. Magn. Reson.*, **A122**, 85 (1996).
36. T. N. Mitchell and B. Kowall, *Magn. Reson. Chem.*, **33**, 325 (1995).
37. F. Fouquet, T. Roulet, R. Willen and I. Pianet, *J. Organomet. Chem.*, **524**, 103 (1996).
38. B. Wrackmeyer, G. Kehr, H. Zhou and S. Ali, *Magn. Reson. Chem.*, **34**, 921 (1996).
39. F. Kayser, M. Biesemans, F. Fu, H. Pan, M. Gielen and R. Willem, *J. Organomet. Chem.* **486**, 263 (1995).

40. Y. K. Grishin, I. F. Leshcheva and T. I. Voevodskaya, *Vestn. Mosk. Univ., Ser. 2: Khim.*, **37**, 387 (1996); *Chem. Abstr.*, **126**, 165, 605 (1997).
41. R. K. Harris, S. E. Lawrence, S.-W. Oh and V. G. Kumar Das, *J. Mol. Struct.*, **347**, 309 (1995).
42. H. Kaneko, M. Hada, T. Nakajima and H. Nakatsuji, *Chem. Phys. Lett.*, **258**, 261 (1996).
43. J. C. Martins, F. Kayser, P. Verheyden, M. Gielen, R. Willem and M. Biesemans, *J. Magn. Reson.*, **124**, 218 (1997).
44. H. Ahari, Ö. Dag, S. Petrov, G. A. Ozin and R. L. Bedard, *J. Phys. Chem. B*, **102**, 2356 (1998).
45. B. Langer, I. Schnell, H. W. Spiess and A.-R. Grimmer, *J. Magn. Reson.*, **138**, 182 (1999).
46. R. K. Harris and A. Sebald, *Magn. Reson. Chem.*, **25**, 1058 (1987).
47. J. Li and H. Kessler, *Microporous Mesoporous Mater.*, **27**, 57 (1999).
48. N. J. Clayden and L. Pugh, *J. Mater. Sci. Lett.*, **17**, 1563 (1998).
49. D. Christendat, I. Wharf, F. G. Morin, I. S. Butler and D. F. R. Gilson, *J. Magn. Reson.*, **131**, 1 (1998).
50. J. C. Cherryman and R. K. Harris, *J. Magn. Reson.*, **128**, 21 (1997).
51. C. Hansch and A. Leo, *Substituent Constants for Correlation Analysis in Chemistry and Biology*, Wiley, New York, 1979.
52. M. Arshadi, D. Johnels and U. Edlund, *Chem. Commun.*, 1279 (1996).
53. H.-G. Bartel and P. E. John, *Z. Phys. Chem.*, **209**, 141 (1999).
54. P. Kircher, G. Huttner, B. Schiemenz, K. Heinze, L. Zsolnai, O. Walter, A. Jacobi and A. Driess, *Chem. Ber.*, **130**, 687 (1997).
55. P. Kircher, G. Huttner, K. Heinze, B. Schiemenz, L. Zsolnai, M. Büchner and A. Driess, *Eur. J. Inorg. Chem.*, 703 (1998).
56. P. Kircher, G. Huttner, K. Heinze and L. Zsolnay, *Eur. J. Inorg. Chem.*, 1057 (1998).
57. M. Veith, C. Mathur, S. Mathur and V. Huch, *Organometallics*, **16**, 1292 (1997).
58. P. B. Hitchcock, M. F. Lappert and L. J.-M. Pierssens, *Organometallics*, **17**, 2686 (1998).
59. M. Veith, C. Mathur and V. Huch, *J. Chem. Soc., Dalton Trans.*, 995 (1997).
60. B. Wrackmeyer and J. Weidinger, *Z. Naturforsch.*, **52b**, 947 (1997).
61. A. K. Varshney, S. Varshney and H. L. Singh, *Synth. React. Inorg. Met.-Org. Chem.*, **29**, 245 (1999).
62. M. Alanfandy, R. Willem, B. Mahieu, M. Alturky, M. Gielen, M. Biesemans, F. Legros, F. Camu and J.-M. Kauffmann, *Inorg. Chim. Acta*, **255**, 175 (1997).
63. Y. Zhang, S. Xu, G. Tian, W. Zhang and X. Zhou, *J. Organomet. Chem.*, **544**, 43 (1997).
64. J. Barron, G. Rima and T. El-Amraoui, *Organometallics*, **17**, 607 (1998).
65. D. J. Teff, C. D. Mineor, D. V. Baxter and K. G. Coulton, *Inorg. Chem.*, **37**, 2547 (1998).
66. J. Barron, G. Rima and T. El-Amraoui, *Inorg. Chim. Acta*, **241**, 9 (1996).
67. D. Agustin, G. Rima, H. Gornitzka and J. Barron, *Eur. J. Inorg. Chem.*, 693 (2000).
68. D. Agustin, G. Rima, H. Gornitzka and J. Barron, *Inorg. Chem.*, **39**, 5492 (2000).
69. D. Agustin, G. Rima and J. Barron, *Main Group Metal Chem.*, **20**, 791 (1997).
70. P. Kircher, G. Huttner and K. Heinze, *J. Organomet. Chem.*, **562**, 217 (1998).
71. B. Schiemenz, G. Huttner, I. Zsolnay, P. Kircher and T. Dierks, *Chem. Ber.*, **128**, 187 (1995).
72. M. Saito, N. Tokitoh and R. Okazaki, *Organometallics*, **15**, 4531 (1996).
73. M. A. Della Bona, M. C. Cassoni, J. M. Keats, G. A. Lawless, M. F. Lappert, M. Stürmann and M. Weidenbruch, *J. Chem. Soc., Dalton Trans.*, 1187 (1998).
74. M. Weidenbruch, H. Marsmann, H. Kilian and H. G. von Schnering, *Chem. Ber.*, **128**, 983 (1995).
75. R. S. Simons, L. Pu, M. M. Olmstead and P. P. Power, *Organometallics*, **16**, 1920 (1997).
76. K. W. Klinkhammer, T. F. Fässler and H. Grützmacher, *Angew. Chem.*, **110**, 114 (1998); *Angew. Chem., Int. Ed. Engl.*, **37**, 124 (1998).
77. C. Eaborn, M. S. Hill, P. B. Hitchcock, D. Patel, J. D. Smith and S. Zhang, *Organometallics*, **19**, 49 (2000).
78. W.-P. Leung, H. Cheng, Q.-C. Yang and T. C. W. Mak, *Chem. Commun.*, 451 (2000).
79. C. Drost, P. B. Hitchcock, M. F. Lappert and L. J.-M. Pierssens, *Chem. Commun.*, 1141 (1997).
80. B. E. Eichler and P. P. Power, *J. Am. Chem. Soc.*, **122**, 8785 (2000).
81. R. S. Simons, L. Pu, M. M. Olmstead and P. P. Power, *J. Am. Chem. Soc.*, **119**, 11705 (1997).
82. L. Pu, M. M. Olmstead and P. P. Powers, *Organometallics*, **17**, 5602 (1998).

83. B. E. Eichler and P. P. Power, *Inorg. Chem.*, **39**, 5444 (2000).
84. B. E. Eichler, B. L. Phillips, P. P. Power and M. P. Augustine, *Inorg. Chem.*, **39**, 5453 (2000).
85. L. Pu, M. O. Senge, M. M. Olmstead and P. P. Power, *J. Am. Chem. Soc.*, **120**, 12682 (1998).
86. M. Mehring, C. Löw, M. Schürmann, F. Uhlig, K. Jurkschat and B. Mahieau, *Organometallics*, **19**, 4613 (2000).
87. C. Eaborn, P. B. Hitchcock, J. D. Smith and S. E. Sözerli, *Organometallics*, **16**, 5653 (1997).
88. C. Drost, B. Gehrhus, P. B. Hitchcock and M. F. Lappert, *Chem. Commun.*, 1845 (1997).
89. C. J. Cardin, D. J. Cardin, S. P. Constantine, M. G. B. Drew, D. Fenske and H. J. Rashid, *J. Chem. Soc., Dalton Trans.*, 2749 (1998).
90. S. Benet, C. J. Cardin, D. J. Cardin, S. P. Constantine, P. Heath, H. Rashid, S. Teixeira, J. H. Thorpe and A. K. Todd, *Organometallics*, **18**, 389 (1999).
91. W.-P. Leung, W.-H. Kwok, L.-H. Weng, L. T. C. Law, Z. Y. Zhou and T. C. W. Mak, *J. Chem. Soc., Dalton Trans.*, 4301 (1997).
92. W.-P. Leung, W.-H. Kwok, L. T. C. Low, Z.-Y. Zhou and T. C. W. Mak, *Chem. Commun.*, 505 (1996).
93. H. Sitzmann, R. Boese and P. Stellberg, *Z. Anorg. Allg. Chem.*, **622**, 751 (1996).
94. D. R. Armstrong, M. J. Duer, M. G. Davidson, D. Moncrieff, C. A. Russel, C. Stourton, A. Steiner, D. Stalke and D. S. Wright, *Organometallics*, **16**, 3340 (1997).
95. M. Westerhausen, M. Hartmann, N. Makropolus, B. Wienecke, M. Wienecke, W. Schwarz and D. Stalke, *Z. Naturforsch.*, **53b**, 117 (1998).
96. M. Weidenbruch, A. Stilter, H. Marsmann, K. Peters and H. G. von Schnering, *Eur. J. Inorg. Chem.*, 1333 (1998).
97. N. Wiberg, H.-W. Lerner, S.-K. Vasisht, S. Wagner, K. Karaghiosoff, H. Nöth and W. Ponikwar, *Eur. J. Inorg. Chem.*, 1211 (1999).
98. M. Veith, C. Mathur and V. Huch, *Organometallics*, **15**, 2858 (1996).
99. M. Veith, M. Olbrich, W. Shihua and V. Huch, *J. Chem. Soc., Dalton Trans.*, 191 (1996).
100. S. R. Foley, G. P. A. Yap and D. S. Richardson, *Organometallics*, **18**, 4700 (1999).
101. R. Xi and L. R. Sita, *Inorg. Chim. Acta*, **270**, 118 (1998).
102. M. Westerhausen, J. Greul, H.-D. Hansen and W. Schwarz, *Z. Anorg. Allg. Chem.*, **622**, 1295 (1996).
103. C. Drost, P. B. Hitchcock and M. F. Lappert, *Angew. Chem.*, **111**, 1185 (1999); *Angew. Chem., Int. Ed. Engl.*, **38**, 1113 (1999).
104. J. Heinicke and A. Oprea, *Heteroatom Chem.*, **9**, 439 (1998).
105. J.-L. Foure, H. Gornitzka, R. Reau, D. Stalke and G. Bertrand, *Eur. J. Inorg. Chem.*, 2295 (1999).
106. H. V. Rasika Diaz and W. Jin, *J. Am. Chem. Soc.*, **118**, 9123 (1996).
107. H. Braunschweig, C. Drost, P. B. Hitchcock, M. F. Lappert and L.J.-M. Pierssens, *Angew. Chem.*, **109**, 285 (1997); *Angew. Chem., Int. Ed. Engl.*, **36**, 261 (1997).
108. J. R. Babcock, L. Liable-Sands, A. L. Rheingold and L. R. Sita, *Organometallics*, **18**, 4437 (1999).
109. P. D. Hitchcock, M. F. Lappert, G. A. Lawless, G. M. de Lima and L. J.-M. Pierssens, *J. Organomet. Chem.*, **601**, 142 (2000).
110. J. Barrau, G. Rima and T. El-Amraoui, *J. Organomet. Chem.*, **561**, 167 (1998).
111. M. Westerhausen, M. M. Enzelberger and W. Schwarz, *J. Organomet. Chem.*, **491**, 83 (1995).
112. N. Kuhn, T. Kratz, D. Bläser and R. Boese, *Chem. Ber.*, **128**, 245 (1995).
113. M. Weidenbruch, H. Kilian, M. Stürmann, S. Pohl, W. Saak, H. Marsmann, D. Steiner and A. Berndt, *J. Organomet. Chem.*, **530**, 255 (1997).
114. A. Schäfer, M. Weidenbruch, W. Saak and S. Pohl, *J. Chem. Soc., Chem. Commun.*, 1157 (1995).
115. M. Weidenbruch, A. Stilter, J. Schlaefke, K. Peters and H. G. von Schnering, *J. Organomet. Chem.*, **501**, 67 (1995).
116. M. Weidenbruch, A. Stilter, K. Peters and H. G. von Schnering, *Z. Anorg. Allg. Chem.*, **622**, 534 (1996).
117. M. Weidenbruch, A. Stilter, K. Peters and H. G. von Schnering, *Chem. Ber.*, **129**, 1565 (1996).
118. M. Weidenbruch, A. Stilter, W. Saak, K. Peters and H. G. von Schnering, *J. Organomet. Chem.*, **560**, 125 (1998).
119. H. Nakazawa, Y. Yamaguchi, K. Kawamura and K. Miyoshi, *Organometallics*, **16**, 4626 (1997).

120. M. Knorr, E. Hallauer, V. Huch, M. Veith and P. Braunstein, *Organometallics*, **15**, 3868 (1996).
121. S. P. Constantine, H. Cox, P. B. Hitchcock and G. A. Lawless, *Organometallics*, **19**, 317 (2000).
122. J. J. Vittal and P. A. W. Dean, *Acta Crystallogr., Sect. C.*, **C52**, 1180 (1996).
123. A. Kandri Rodi, H. Ranaivonjatovo, J. Escudie and A. Kerbal, *Main Group Metal Chem.*, **19**, 199 (1996).
124. M.-A. Chaubon, J. Escudie, H. Ranaivonjatovo and J. Satge, *Chem. Commun.*, 2621 (1996).
125. M.-A. Chaubon, J. Escudie, H. Ranaivonjatovo and J. Satge, *Main Group Metal Chem.*, **19**, 145 (1996).
126. M. Weidenbruch, H. Kilian, K. Peters, H. G. von Schnering and H. Marsmann, *Chem Ber.*, **128**, 983 (1995).
127. M. Weidenbruch, *Eur. J. Inorg. Chem.*, 373 (1999).
128. P. P. Power, *Chem. Rev.*, **99**, 3463 (1999).
129. H. Grützmacher and T. F. Fässler, *Chem. Eur. J.*, **6**, 2317 (2000).
130. M. F. Lappert, *J. Organomet. Chem.*, **600**, 144 (2000).
131. B. Wrackmeyer and P. Bernatowicz, *J. Organomet. Chem.*, **579**, 133 (1999).
132. B. Wrackmeyer, G. Kehr, J. Süß and E. Molla, *J. Organomet. Chem.*, **562**, 207 (1999).
133. B. Wrackmeyer, G. Kehr, J. Süß and E. Molla, *J. Organomet. Chem.*, **577**, 82 (1999).
134. W. Adcock and A. R. Krstic, *Magn. Reson. Chem.*, **35**, 663 (1997).
135. B. Wrackmeyer and S. V. Ponomarev, *Z. Naturforsch.*, **54b**, 705 (1999).
136. B. Wrackmeyer, J. Weidinger, H. Nöth, W. Storch, T. Seifert and M. Vosteen, *Z. Naturforsch.*, **53b**, 1494 (1998).
137. B. Kowall and J. Heinecke, *Main Group Metal Chem.*, **20**, 379 (1997).
138. B. Wrackmeyer, H. E. Maisel, G. Kehr and H. Nöth, *J. Organomet. Chem.*, **532**, 201 (1997).
139. M. Herberhold, W. Millius, U. Steffl, K. Vitzthum, B. Wrackmeyer, R. H. Herber, M. Fontani and P. Zanello, *Eur. J. Inorg. Chem.*, 145 (1999).
140. L. Weber, E. Dobbert, H.-G. Stammler, B. Neumann, R. Boese and D. Bläser, *Eur. J. Inorg. Chem.*, 491 (1999).
141. B. Wrackmeyer and A. Glöckle, *Main Group Metal Chem.*, **20**, 181 (1997).
142. F. Uhlig, C. Kayser, R. Klassen, U. Hermann, L. Brecker, M. Schürmann, K. Ruhlandt-Senge and U. Englich, *Z. Naturforsch.*, **54b**, 278 (1999).
143. W.-W. du Mont, L. Müller, R. Martens, P. M. Papathomas, B. A. Smart, H. E. Robertson and D. W. H. Rankin, *Eur. J. Inorg. Chem.*, 1381 (1999).
144. B. Wrackmeyer and P. Bernatowicz, *Magn. Reson. Chem.*, **37**, 418 (1999).
145. M. Herberhold, V. Tröbs and B. Wrackmeyer, *J. Organomet. Chem.*, **541**, 391 (1997).
146. B. Wrackmeyer, G. Kehr, H. E. Maisel and H. Zhou, *Magn. Reson. Chem.*, **36**, 39 (1998).
147. K. Schmid, H.-D. Hausen, K.-W. Klinkhammer and J. Weidlein, *Z. Anorg. Allg. Chem.*, **625**, 945 (1999).
148. A. Kalsoom, M. Mazhar, A. Saqib, M. F. Mahon, K. C. Malloy and M. Iqbal Chaudry, *Appl. Organomet. Chem.*, **11**, 47 (1997).
149. B. Wrackmeyer, G. Kehr, H. Zhou and A. Saqib, *Magn. Reson. Chem.*, **34**, 921 (1996).
150. B. Wrackmeyer, S. Kerschl and H. E. Maisel, *Main Group Metal Chem.*, **21**, 89 (1998).
151. B. Wrackmeyer, H. E. Maisel, B. Schwarze, M. Milius and R. Köster, *J. Organomet. Chem.*, **541**, 97 (1997).
152. B. Wrackmeyer and J. Weidinger, *Z. Naturforsch.*, **52b**, 947 (1997).
153. A. D. Ayala, A. B. Chopa, N. N. Giagante, L. C. Köll, S. D. Mandolesi and J. C. Podesta, *An. Asoc. Quim. Argent.*, **86**, 139 (1998).
154. Yu. K. Gun'ko, L. Nagy, W. Brüser, V. Lorenz, A. Fischer, S. Gießmann, F. T. Edelmann and K. Jacob, *Monatsh. Chem.*, **130**, 45 (1999).
155. W. Adcock, C. I. Clark, A. Houmam, A. R. Krstic and J.-M. Saveant, *J. Org. Chem.*, **61**, 2891 (1996).
156. P. Bleckmann, U. Englich, U. Hermann, I. Prass, K. Ruhlandt-Senge, M. Schürmann, C. Schwittek and F. Uhlig, *Z. Naturforsch.*, **54b**, 1188 (1999).
157. S. Geetha, M. Ye and J. G. Verkade, *Inorg. Chem.*, **34**, 6158 (1995).
158. B. Wrackmeyer, U. Klaus and W. Milius, *Inorg. Chim. Acta*, **250**, 327 (1996).
159. B. Wrackmeyer, H. E. Maisel and W. Milius, *Z. Naturforsch.*, **50b**, 809 (1995).
160. N. Asao, J.-X. Liu, T. Sudoh and Y. Yamamoto, *J. Org. Chem.*, **61**, 4568 (1996).

161. T. Janati, J.-C. Guillemin and M. Soufiaoui, *J. Organomet. Chem.*, **486**, 57 (1995).
162. B. Wrackmeyer, G. Kehr and J. Süß, *Main Group Metal Chem.*, **18**, 127 (1995).
163. B. Wrackmeyer, U. Dörfler, W. Milius and M. Herberhold, *Z. Naturforsch.*, **51b**, 851 (1996).
164. B. Wrackmeyer, S. Kerschl, H. E. Maisel and W. Milius, *J. Organomet. Chem.*, **490**, 197 (1995).
165. B. Wrackmeyer, K. Horchler and S. Kundler, *J. Organomet. Chem.*, **503**, 289 (1995).
166. B. Wrackmeyer, U. Klaus and W. Milius, *Chem. Ber.*, **128**, 679 (1995).
167. B. Wrackmeyer and U. Klaus, *J. Organomet. Chem.*, **520**, 211 (1996).
168. Jian-Xie Chen, K. Sakamoto, A. Orita and J. Otera, *J. Organomet. Chem.*, **574**, 58 (1999).
169. J.-C. Meurice, J.-G. Duboudin, M. Ratier, M. Petraud, R. Willem and M. Biesemans, *Organometallics*, **18**, 1699 (1999).
170. A. Lycka, J. Holocek and D. Micak, *Collect. Czech. Chem. Commun.*, **62**, 1169 (1997).
171. J. Holecek, A. Lycka, D. Micak, L. Nagy, G. Vanko, J. Brus, S. Shanunga, S. Raj, H. K. Fun and S. W. Ng, *Collect. Czech. Chem. Commun.*, **64**, 1028 (1999).
172. S. Selvaratnam, S. W. Ng, N. W. Ahmad and V. G. Kumar Das, *Main Group Metal Chem.*, **22**, 321 (1999).
173. C. H. Schiesser and M. A. Skidmore, *J. Organomet. Chem.*, **552**, 145 (1998).
174. A. Zickgraf, M. Beuter, U. Kolb, M. Dräger, R. Tozer, D. Dakternieks and K. Jurkschat, *Inorg. Chim. Acta*, **275–276**, 203 (1998).
175. R. Willem, M. Biesemans, P. Jaumier and B. Jousseaume, *J. Organomet. Chem.*, **572**, 233 (1999).
176. M. K. Denk, K. Hatano and A. J. Lough, *Eur. J. Inorg. Chem.*, 1067 (1998).
177. J.-C. Guillemin and K. Malagu, *Organometallics*, **18**, 5259 (1999).
178. R. Altmann, K. Jurkschat, M. Schürmann, D. Dakternieks and A. Duthie, *Organometallics*, **16**, 5716 (1997).
179. S. M. S. V. Doidge-Harrison, R. A. Howie, J. N. Low and J. L. Wardell, *J. Chem. Crystallogr.*, **27**, 291 (1997).
180. F. Carre, S. G. Dutremez, C. Guerin, B. J. L. Henner, A. Jolivet and V. Tomberli, *Organometallics*, **18**, 770 (1999).
181. M. Charisse, A. Zickgraf, H. Stenger, E. Bräu, C. Desmarquet, M. Dräger, S. Gerstmann, D. Dakternieks and J. Hook, *Polyhedron*, **17**, 4497 (1998).
182. H. J. Buchanan, P. J. Cox, S. M. S. V. Doidge-Harrison, R. A. Howie, M. Jaspars and J. L. Wardell, *J. Chem. Soc., Perkin Trans. 1*, 3657 (1997).
183. S. J. Garden and J. L. Wardell, *Main Group Metal Chem.*, **20**, 711 (1997).
184. H. J. Buchanan, P. J. Cox and J. L. Wardell, *Main Group Metal Chem.*, **21**, 751 (1998).
185. I. Wharf, R. Wojtowski, C. Bowes, A.-M. Lebuis and M. Onyszchuk, *Can. J. Chem.*, **76**, 1827 (1998).
186. I. Wharf and M. G. Simard, *J. Organomet. Chem.*, **532**, 1 (1997).
187. M. P. Aarnts, D. J. Stufkens, A. Oskam, J. Fraanje and K. Gubitz, *Inorg. Chim. Acta*, **256**, 93 (1997).
188. M. P. Aarnts, A. Oskam, D. J. Stufkens, J. Fraanje, K. Goubitz, N. Veldman and A. L. Spek, *J. Organomet. Chem.*, **531**, 191 (1997).
189. R. H. Vaz, A. Abras and R. M. Silva, *J. Braz. Chem. Soc.*, **9**, 57 (1998).
190. L. Kollar, S. Gladiali, M. J. Tenorio and W. Weissensteiner, *J. Cluster Sci.*, **9**, 321 (1998).
191. P. R. Craig, K. R. Flower, W. R. Roper and L. J. Wright, *Inorg. Chim. Acta*, **240**, 385 (1995).
192. M. P. Aarnts, M. P. Wilms, K. Peelen, J. Fraanje, K. Goubitz, F. Hartl, D. J. Stufkens, E. V. Baerends and A. Vlcek Jr., *Inorg. Chem.*, **35**, 5468 (1996).
193. J. E. Ellis, P. Yuen and M. Jang, *J. Organomet. Chem.*, **507**, 283 (1996).
194. K. Ruhlandt-Senge, U. Englich and F. Uhlig, *J. Organomet. Chem.*, **613**, 139 (2000).
195. P. J. Cox, O. A. Melvin, S. J. Garden and J. L. Wardell, *J. Chem. Crystallogr.*, **25**, 469 (1995).
196. J. C. Podesta, A. B. Chopa, G. E. Radivoy and C. A. Vitale, *J. Organomet. Chem.*, **494**, 11 (1995).
197. C. A. Vitale and J. C. Podesta, *J. Chem. Soc., Perkin Trans. 1*, 2407 (1996).
198. P. J. Cox, O. A. Melvin, S. J. Garden and J. L. Wardell, *J. Chem. Crystallogr.*, **25**, 469 (1995).
199. W. Adcock, C. I. Clark, A. Houmam, A. R. Krstic and J.-M. Saveant, *J. Org. Chem.*, **61**, 2891 (1996).
200. D. Crich, X.-Y. Jiao, Q. Yao and J. S. Harwood, *J. Org. Chem.*, **61**, 2368 (1996).

201. D. Marton, U. Russo, D. Stivanello, G. Tagliavini, P. Ganis and G. C. Valle, *Organometallics*, **15**, 1645 (1996).
202. L. Barton, H. Fang, D. K. Srivastava, T. A. Schweitzer and N. P. Rath, *Appl. Organomet. Chem.*, **10**, 183 (1996).
203. C. H. Schiesser and M. A. Skidmore, *Chem. Commun.*, 1419 (1996).
204. L. R. Allain, C. A. L. Filgueiras, A. Abras and A. G. Ferreira, *J. Braz. Chem. Soc.*, **7**, 247 (1996).
205. M. Charisse, B. Mathiasch, M. Dräger and U. Russo, *Polyhedron*, **14**, 2429 (1995).
206. L. A. Uzal and J. L. Wardell, *Quim. Anal.*, **14**, 158 (1995).
207. P. B. Hitchcock, E. Jang and M. F. Lappert, *J. Chem. Soc., Dalton Trans.*, 3179 (1995).
208. M. Herberhold, S. Gerstmann and B. Wrackmeyer, *Phosphorus, Sulfur and Silicon*, **113**, 89 (1996).
209. D. Hänssgen, T. Oster and M. Nieger, *J. Organomet. Chem.*, **526**, 59 (1996).
210. D. N. Kravtsov, A. S. Peregudov and V. M. Pachevskaya, *Russ. Chem. Bull.*, **45**, 441 (1996).
211. G. Ossig, A. Meller, S. Freitag, O. Müller, H. Gornitzka and R. Herbst-Irmer, *Organometallics*, **15**, 408 (1996).
212. I. Wharf, H. Lamparski and R. Reeleder, *Appl. Organomet. Chem.*, **11**, 969 (1997).
213. R. Willem, H. Dalil, P. Broekaert, M. Biesemans, L. Ghys, K. Nooter, D. de Vos, F. Ribot and M. Gielen, *Main Group Metal Chem.*, **20**, 535 (1997).
214. J. Beckmann, M. Biesemans, K. Hassler, K. Jurkschat, J. C. Martins, M. Schürmann and R. Willem, *Inorg. Chem.*, **37**, 4891 (1998).
215. Liu Hua, Xie Qing-Ian, Wang Ji-Tao and M. Mazhar, *Heteroatom Chem.*, **9**, 298 (1998).
216. G.-Y. Yeap, N. Ishizawa and Y. Nakamura, *J. Coord. Chem.*, **44**, 325 (1998).
217. T. S. B. Baul, S. Dhar, N. Kharbani, S. M. Pyke, R. Butcher and F. E. Smith, *Main Group Metal Chem.*, **22**, 413 (1999).
218. M. Bhagat, A. Singh and R. C. Mehrota, *Indian J. Chem., Sect. A*, **37A**, 820 (1999).
219. P. Boudjuk, M. P. Remington, D. G. Grier, W. Triebold and B. R. Jarabek, *Organometallics*, **18**, 4534 (1999).
220. F. Carre, S. G. Dutremez, C. Guerin, B. J. L. Henner, A. Jolivet and V. Tomberli, *Organometallics*, **18**, 770 (1999).
221. Jian-xie Chen, K. Sakamoto, A. Orita and J. Otera, *J. Organomet. Chem.*, **574**, 58 (1999).
222. M. B. Faraoni, L. C. Koll, C. A. Vitale, A. B. Chopa and J. C. Podesta, *Main Group Metal Chem.*, **22**, 289 (1999).
223. D. H. Hunter and C. McRoberts, *Organometallics*, **18**, 5577 (1999).
224. J. C. Martins, R. Willem, F. A. G. Mercier, M. Gielen and M. Biesemans, *J. Am. Chem. Soc.*, **121**, 3284 (1999).
225. Jiaxun Tao, Feng Pan, Qian Gu, Wenjing Xiao, Wenguan Lu, Zhesheng Ma, Nicheng Shi and Ruji Wang, *Main Group Metal Chem.*, **22**, 489 (1999).
226. J. Gimenez, A. Michel, R. Petiaud and M.-F. Llauro, *J. Organomet. Chem.*, **575**, 286 (1999).
227. J. Holecek, M. Nadvornik, K. Handlik, V. Pejchal, R. Vitek and A. Lycka, *Collect. Czech. Chem. Commun.*, **62**, 279 (1997).
228. M. Danish, S. Ali, A. Badash, M. Mazhar, H. Masood, A. Malik and G. Kehr, *Synth. React. Inorg. Met.-Org. Chem.*, **27**, 863 (1997).
229. E. Balestrieri, L. Bellugi, A. Boicelli, M. Giomini, A. M. Giulani, M. Giustini, L. Marciani and P. J. Saadler, *J. Chem. Soc., Dalton Trans.*, 4099 (1997).
230. S. Belwal and R. V. Singh, *Bol. Soc. Chil. Quim.*, **42**, 363 (1997).
231. S. Belwal, H. Taneja, A. Dandia and R. V. Singh, *Phosphorus, Sulfur, Silicon Relat. Elem.*, **127**, 49 (1997).
232. A. Bacchi, A. Bonardi, M. Carcelli, P. Mazza, P. Pelagatti, C. Pelizzi, O. Pelizzi, C. Solinas and F. Zani, *J. Inorg. Biochem.*, **69**, 101 (1998).
233. M. Veith, M. Opsölder, M. Zimmer and V. Huch, *Eur. J. Inorg. Chem.*, **109**, 2328 (1997).
234. J. Beckmann, K. Jurkschat, D. Schollmeyer and M. Schürmann, *J. Organomet. Chem.*, **543**, 229 (1997).
235. M. Herberhold, U. Steffl, M. Milius and B. Wrackmeyer, *J. Organomet. Chem.*, **533**, 109 (1997).
236. T. Albrecht, G. Elter, M. Noltemeier and A. Meller, *Z. Anorg. Allg. Chem.*, **624**, 1514 (1998).
237. J. Beckmann, B. Mahieu, W. Nigge, D. Schollmeyer, M. Schürmann and K. Jurkschat, *Organometallics*, **17**, 5697 (1998).

238. H. K. Sharma, F. Cervantes-Lee, J. S. Mahmoud and K. H. Pannell, *Organometallics*, **18**, 399 (1998).
239. F. Uhlig, C. Kayser, R. Klassen and M. Schürmann, *J. Organomet. Chem.*, **556**, 165 (1998).
240. F. Uhlig, U. Hermann, G. Reeske and M. Schürmann, *Z. Anorg. Allg. Chem.*, **627**, 543 (2001).
241. F. Uhlig and U. Hermann, unpublished results.
242. F. Uhlig, U. Englich, U. Hermann, C. Marschner, E. Hengge and K. Ruhlandt-Senge, *J. Organomet. Chem.*, **605**, 22 (2000).
243. U. Hermann, M. Schürmann and F. Uhlig, *J. Organomet. Chem.*, **585**, 211 (1999).
244. M. Herberhold, U. Steffl and B. Wrackmeyer, *J. Organomet. Chem.*, **577**, 76 (1999).
245. S. D. Mandolesi, K. C. Koll and J. C. Podesta, *J. Organomet. Chem.*, **587**, 74 (1999).
246. J. Beckmann, K. Jurkschat, U. Kaltenbrunner, N. Pieper and M. Schürmann, *Organometallics*, **18**, 1586 (1999).
247. P. Boudjouk, M. P. Remington Jr., D. G. Grier, W. Tribold and B. R. Jarabek, *Organometallics*, **18**, 4534 (1999).
248. M. Herberhold, U. Steffl and B. Wrackmeyer, *Z. Naturforsch.*, **54b**, 57 (1999).
249. M. C. Janzen, H. A. Jenkins, L. M. Rendina, J. J. Vittal and R. J. Puddephat, *Inorg. Chem.*, **38**, 2123 (1999).
250. A. B. Chopa and A. P. Murray, *Main Group Metal Chem.*, **21**, 347 (1998).
251. B. Wrackmeyer, C. Köhler, W. Milius and M. Herberhold, *Z. Anorg. Allg. Chem.*, **621**, 1625 (1995).
252. B. Wrackmeyer, U. Klaus, W. Milius, E. Klaus and T. Schaller, *J. Organomet. Chem.*, **517**, 235 (1996).
253. D. Bongert, G. Heckmann, H.-D. Hausen, W. Schwarz and H. Binder, *Z. Anorg. Allg. Chem.*, **622**, 1793 (1996).
254. K. Shibata, C. S. Weinert and L. R. Sita, *Organometallics*, **17**, 2241 (1998).
255. S. P. Mallela, Y. Saar, S. Hill and R. A. Geanangel, *Inorg. Chem.*, **38**, 2957 (1999).
256. H. Gilges, G. Kickelbick and U. Schubert, *J. Organomet. Chem.*, **548**, 57 (1997).
257. T. Imori, V. Lu, H. Cai and T. D. Tilley, *J. Am. Chem. Soc.*, **117**, 9931 (1995).
258. N. Devylder, M. Hill, K. C. Molloy and G. J. Price, *Chem. Commun.*, 711 (1996).
259. T. Janati, J.-C. Guillemin and M. Soufiaoui, *J. Organomet. Chem.*, **486**, 57 (1995).
260. L. Lasalle, T. Janati and J.-C. Guillemin, *J. Chem. Soc., Chem. Commun.*, 669 (1995).
261. M. Drieß, R. Janoschek, H. Pritzkow and U. Winkler, *Angew. Chem.*, **107**, 1746 (1995); *Angew. Chem., Int. Ed. Engl.*, **34**, 1614 (1995).
262. M. Herberhold, U. Steffl, W. Milius and B. Wrackmeyer, *Z. Anorg. Allg. Chem.*, **624**, 386 (1998).
263. F. Huber, R. Schmiedgen, M. Schürmann, R. Barbieri, G. Ruisi and A. Silvetri, *Appl. Organomet. Chem.*, **11**, 869 (1997).
264. M. Schürmann, R. Schmiedgen, F. Huber, A. Silvestri, G. Ruisi, A. Barbieri Paulsen and R. Barbieri, *J. Organomet. Chem.*, **584**, 103 (1999).
265. A. Meddour, F. Mercier, J. C. Martins, M. Gielen, M. Biesemans and R. Willem, *Inorg. Chem.*, **36**, 5712 (1997).
266. S. W. Ng, V. G. Kumar Das, J. Holecek, A. Lycka, M. Gielen and M. G. B. Drew, *Appl. Organomet. Chem.*, **11**, 39 (1997).
267. R. Willem, A. Bouhdid, A. Meddour, C. Camacho-Camacho, F. Mercier, M. Gielen, C. Sanchez and E. R. T. Tiekink, *Organometallics*, **16**, 4377 (1997).
268. R. Willem, A. Bouhdid, F. Kayser, A. Delmotte, M. Gielen, J. C. Martins, M. Biesemans, B. Mahieu and E. R. T. Tiekink, *Organometallics*, **15**, 1920 (1996).
269. T. N. Mitchell and M. Schütze, *Tetrahedron*, **55**, 1285 (1999).
270. T. N. Mitchell, M. Schütze and F. Gießelmann, *Synlett*, 183 (1997).
271. M. Niestroj, W. P. Neumann and T. N. Mitchell, *J. Organomet. Chem.*, **519**, 45 (1996).
272. T. N. Mitchell and F. Gießelmann, *Synlett*, 475 (1996).
273. T. N. Mitchell and F. Gießelmann, *Synlett*, 333 (1995).
274. T. N. Mitchell, F. Gießelmann and K. Kwetkat, *J. Organomet. Chem.*, **492**, 191 (1995).
275. T. N. Mitchell and B. Kowall, *J. Organomet. Chem.*, **490**, 239 (1995).
276. T. N. Mitchell, K. Böttcher, P. Bleckmann, B. Costisella, C. Schwittek and C. Nettelbeck, *Eur. J. Org. Chem.*, 2413 (1999).
277. V. B. Mokal and V. M. Jain, *Bull. Chem. Soc. Jpn.*, **68**, 1149 (1995).

278. M. Kemmer, H. Dalil, M. Biesemans, J. C. Martins, B. Mahieu, E. Horn, D. De Vos, E. R. T. Tiekink, R. Willem and M. Gielen, *J. Organomet. Chem.*, **608**, 63 (2000).
279. C. C. Camacho, D. De Vos, B. Mahieu, M. Gielen, M. Kemmer, M. Biesemans and R. Willem, *Main Group Metal Chem.*, **23**, 381 (2000).
280. D. Kumar Dey, S. Kumar, M. Gielen, M. Kemmer, M. Biesemans, R. Willem, V. Gramlich and S. Mitra, *J. Organomet. Chem.*, **590**, 88 (1999).
281. M. Kemmer, L. Ghys, M. Gielen, M. Biesemans, E. R. T. Tiekink and R. Willem, *J. Organomet. Chem.*, **582**, 195 (1999).
282. J. K. Tsagatakis, N. A. Chaniotakis, K. Jurkschat, S. Damoun, P. Geerlings, A. Bouhdid, M. Gielen, I. Verbruggen, M. Biesemans, J. C. Martins and R. Willem, *Helv. Chim. Acta*, **82**, 531 (1999).
283. M. Nath, R. Yadav, G. Eng, T.-T. Nguyen and A. Kumar, *J. Organomet. Chem.*, **577**, 1 (1999).
284. H. Dalil, M. Biesemans, R. Willem and M. Gielen, *Main Group Metal Chem.*, **21**, 741 (1998).
285. R. Willem, I. Verbruggen, M. Gielen, M. Biesemans, B. Mahieu, T. S. B. Baul and E. R. T. Tiekink, *Organometallics*, **17**, 5758 (1998).
286. M. Kemmer, M. Gielen, M. Biesemans, D. De Vos and R. Willem, *Metal-Based Drugs*, **5**, 189 (1998).
287. M. Biesemans, R. Willem, S. Damoun, P. Geerlings, E. R. T. Tiekink, P. Jaumier, M. Lahcini and B. Jousseaume, *Organometallics*, **17**, 90 (1998).
288. R. Willem, A. Bouhdid, M. Biesemans, J. C. Martins, D. De Vos, E. R. T. Tiekink and M. Gielen, *J. Organomet. Chem.*, **514**, 203 (1996).
289. J. Klein, B. Schulze, R. Borsdorf and S. Weng Ng, *J. Prakt. Chem.*, **337**, 242 (1995).
290. M. D. Couce, G. Faraglia, U. Russo, L. Sindellari and G. Valle, *J. Organomet. Chem.*, **513**, 77 (1996).
291. R. Willem, A. Bouhdid, M. Biesemann, J. C. Martins, D. de Vos, E. R. T. Tiekink and M. Gielen, *J. Organomet. Chem.*, **514**, 203 (1996).
292. F. Marchetti, C. Pettinari, A. Cingolani, G. G. Lobbia, A. Cassetta and L. Barba, *J. Organomet. Chem.*, **517**, 141 (1996).
293. D. Dakternieks, K. Jurkschat and H. Zhu, *Organometallics*, **14**, 2512 (1995).
294. Q. Xie, Z. Yang and L. Jiang, *Main Group Metal Chem.*, **19**, 509 (1996).
295. F. Fu, H. Li, D. Zhu, Q. Fang, H. Pan, E. R. T. Tiekink, F. Kayser, M. Biesemans, I. Verbruggen, R. Willem and M. Gielen, *J. Organomet. Chem.*, **490**, 163 (1995).
296. R. J. Rao and H. B. Wankhade, *Main Group Metal Chem.*, **18**, 675 (1995).
297. J. Sharma, Y. Singh and A. Kumar Rai, *Phosphorus, Sulfur and Silicon*, **112**, 19 (1996).
298. T. Kawakami, I. Shibata and A. Baba, *J. Org. Chem.*, **61**, 82 (1996).
299. M. Herberhold, S. Gerstmann, W. Milius and B. Wrackmeyer, *Phosphorus, Sulfur and Silicon*, **112**, 261 (1996).
300. T. Kawakami, T. Sugimoto, I. Shibata, A. Baba, H. Matsuda and N. Sonoda, *J. Org. Chem.*, **60**, 2677 (1995).
301. M. Hill, M. F. Mahon and K. C. Molloy, *J. Chem. Soc., Dalton Trans.*, 1857 (1996).
302. A. S. Sall, L. Diop and U. Russo, *Main Group Metal Chem.*, **18**, 243 (1995).
303. M. D. Couce, V. Cherchi, G. Faraglia, U. Russo, L. Sindellari, G. Valle and N. Zancan, *Appl. Organomet. Chem.*, **10**, 35 (1996).
304. S. P. Narula, S. Kaur, R. Shankar, S. K. Bharadwaj and R. K. Chadha, *J. Organomet. Chem.*, **506**, 181 (1996).
305. H. Fang, D. Zhao, N. P. Rath, L. Brammer and L. Barton, *Organometallics*, **14**, 1700 (1995).
306. T. N. Mitchell and B. Godry, *J. Organomet. Chem.*, **516**, 133 (1996).
307. T. N. Mitchell and B. Godry, *J. Organomet. Chem.*, **490**, 45 (1995).
308. M. Beuter, U. Kolb, A. Zickgraf, E. Bräu, M. Bletz and M. Dräger, *Polyhedron*, **16**, 4005 (1997).
309. C. Pettinary, F. Marchetti, A. Gregori, A. Cingolani, J. Tanski, M. Rossi and F. Caruso, *Inorg. Chim. Acta*, **257**, 37 (1997).
310. L. Carlton, R. Weber and D. C. Levendis, *Inorg. Chem.*, **37**, 1264 (1998).
311. R. Schmiedgen, F. Huber, A. Silvestri, G. Ruisi, M. Rossi and R. Barbieri, *Appl. Organomet. Chem.*, **12**, 861 (1998).
312. M. Nath and S. Goyal, *Main Group Metal Chem.*, **19**, 75 (1996).
313. R. J. Rao and H. B. Wankhade, *Main Group Metal Chem.*, **19**, 239 (1996).
314. R. Kapoor, V. Sood and P. Kapoor, *Polyhedron*, **14**, 489 (1995).

315. M. Danish, S. Ali, M. Mazhar, A. Badshah, M. I. Choudhary, H. G. Alt and G. Kehr, *Polyhedron*, **14**, 3115 (1995).
316. J. S. Casas, E. E. Castellano, F. J. Garcia Barros, A. Sanchez, A. Sanchez Gonzalez, J. Sordo and J. Zukerman-Schpector, *J. Organomet. Chem.*, **519**, 209 (1996).
317. F. Caruso, D. Leonesi, F. Marchetti, E. Rivalora, M. Rossi, V. Tomov and C. Pettinari, *J. Organomet. Chem.*, **519**, 29 (1996).
318. C. Pettinari, *Main Group Metal Chem.*, **19**, 489 (1996).
319. J. S. Casas, E. Garcia Martinez, A. Sanchez Gonzalez, J. Sordo, U. Casellato, R. Graziani and U. Russo, *J. Organomet. Chem.*, **493**, 107 (1996).
320. G. G. Lobbia, P. Cecchi, C. Santini, S. Calogero, G. Valle and F. E. Wagner, *J. Organomet. Chem.*, **513**, 139 (1996).
321. T. Tavridou, U. Russo, D. Marton, G. Valle and D. Kovala-Demertzi, *Inorg. Chim. Acta*, **231**, 139 (1995).
322. E. V. Grigoriev, N. S. Yashina, A. A. Prischenko, M. V. Livantsov, V. S. Petrosyan, W. Massa, K. Harms, S. Wocadlo and L. Pellerito, *Appl. Organomet. Chem.*, **9**, 11 (1995).
323. G. G. Lobbia, P. Cecchi, R. Spagna, M. Colapietro, A. Pifferie and C. Pettinari, *J. Organomet. Chem.*, **485**, 45 (1995).
324. D. P. Arnold and E. R. T. Tiekink, *Polyhedron*, **14**, 1785 (1995).
325. A. Lorenzotti, G. Sclavi, A. Cingolani, E. Rivarla, M. Colapietro, A. Cassetta and C. Pettinari, *J. Organomet. Chem.*, **496**, 69 (1995).
326. S. Yu. Bylikin, A. G. Shipov, V. N. Negrebetsky, L. S. Smirnova, Yu. I. Baukov, Yu. E. Ovchinnikov and Yu. T. Struchkov, *Russ. Chem. Bull.*, **45**, 2627 (1995).
327. S. Nagabrahmanandachari and K. C. Kumara Swamy, *Indian J. Chem.*, **34A**, 658 (1995).
328. A. Chaturvedi, R. K. Sharma, P. N. Nagar and A. Kumar Rai, *Phosphorus, Sulfur and Silicon*, **112**, 179 (1996).
329. G. G. Lobbia, G. Valle, S. Calogero, P. Cecchi, C. Santini and F. Marchetti, *J. Chem. Soc., Dalton Trans.*, 2475 (1996).
330. R. G. Ramirez, R. A. Toscano, C. Silvestru and I. Haiduc, *Polyhedron*, **15**, 3857 (1996).
331. D. A. Lewis, D. J. Williams, A. M. Slain and J. D. Woollins, *Polyhedron*, **15**, 555 (1996).
332. J. S. Casas, A. Catineiras, E. Garcia Martinez, A. Sanchez Gonzalez, J. Sordo and E. M. V. Lopez, *Polyhedron*, **15**, 891 (1996).
333. T. V. Drovetskaja, N. S. Yashina, T. V. Leonova, V. S. Petrosyan, J. Lorberth, S. Wocadlo, W. Massa and J. Pebler, *J. Organomet. Chem.*, **507**, 201 (1996).
334. F. Caruso, M. Giomini, A. M. Guilini and E. Rivarola, *J. Organomet. Chem.*, **506**, 67 (1996).
335. J. Sharma, Y. Singh, R. Bohra and A. Kumar Rai, *Polyhedron*, **15**, 1097 (1996).
336. C. Pettinari, F. Marchetti and A. Cingolani, *Polyhedron*, **15**, 1263 (1996).
337. M. Nath, S. Goyal and D. Whalen, *Bull. Chem. Soc. Jpn.*, **69**, 605 (1996).
338. N. Chopra, L. C. Damunde, P. A. W. Dean and J. J. Vittal, *Can. J. Chem.*, **74**, 2095 (1996).
339. G. G. Lobbia, P. Cecchi, G. Valle, S. Calogero and C. Santini, *Main Group Metal Chem.*, **19**, 571 (1996).
340. G. G. Lobbia, P. Cecchi, S. Calogero, G. Valle, M. Chiarini and L. Stievano, *J. Organomet. Chem.*, **503**, 297 (1995).
341. M. Mehring, C. Löw, M. Schürmann and K. Jurkschat, *Eur. J. Inorg. Chem.*, 887 (1999).
342. M. Mehring, M. Schürmann and K. Jurkschat, *Organometallics*, **17**, 1227 (1998).
343. S. E. Dann, A. R. J. Genge, W. Levason and G. Reid, *J. Chem. Soc., Dalton Trans.*, 2207 (1997).
344. F. Ribot, F. Banse, C. Sanchez, M. Lahcini and B. Jouseaume, *J. Sol-Gel Sci. Tech.*, **8**, 529 (1997).
345. J. Beckmann, K. Jurkschat, B. Mahieu and M. Schürmann, *Main Group Metal Chem.*, **21**, 113 (1998).
346. R. Altmann, K. Jurkschat, M. Schürmann, D. Dakternieks and A. Duthie, *Organometallics*, **17**, 5858 (1998).
347. L. Carlton, R. Weber and D. C. Levendis, *Inorg. Chem.*, **37**, 1264 (1998).
348. F. Ribot, C. Sanchez, R. Willem, J. C. Martins and M. Biesemans, *Inorg. Chem.*, **37**, 911 (1998).
349. C. Pettinari, F. Marchetti, A. Cingolani, D. Leonesi, E. Mundorff, M. Rossi and F. Caruso, *J. Organomet. Chem.*, **557**, 187 (1998).

350. N. Pieper, C. Klaus-Mrestani, M. Schürmann and K. Jurkschat, *Organometallics*, **16**, 1043 (1997).
351. J. Susperregui, M. Bayle, J. M. Leger, G. Deleris, M. Biesemans, R. Willem, M. Kemmer and M. Gielen, *J. Organomet. Chem.*, **545–56**, 559 (1997).
352. S.-G. Teoh, S.-H. Ang and J.-P. Declerq, *Polyhedron*, **16**, 3729 (1997).
353. C. Eychenne-Baron, F. Ribot and C. Sanchez, *J. Organomet. Chem.*, **567**, 137 (1998).
354. C. Pettinary, M. Pellei, M. Miliani, A. Cingolani, A. Cassetta, L. Barba, A. Pifferi and E. Rivarola, *J. Organomet. Chem.*, **553**, 345 (1998).
355. R. P. Singh, *J. Coord. Chem.*, **44**, 101 (1998).
356. M. Veith, S. Mathur, C. Mathur and V. Huch, *Organometallics*, **17**, 1044 (1998).
357. S. Calogero, G. Valle, G. G. Lobbia, C. Santini, P. Cecchi and L. Stiefano, *J. Organomet. Chem.*, **526**, 269 (1996).
358. A. Bansal, N. Fahmi and R. V. Singh, *Main Group Metal Chem.*, **22**, 111 (1999).
359. F. Caruso, M. Rossi, F. Marchetti and C. Pettinary, *Organometallics*, **18**, 2398 (1999).
360. D. K. Dey, M. K. Das and H. Nöth, *Z. Naturforsch.*, **54b**, 145 (1999).
361. C. Eychenne-Baron, F. Ribot, N. Steunou and C. Sanchez, *Organometallics*, **19**, 1940 (2000).
362. A. Gamard, B. Jousseaume, T. Toupance and G. Campet, *Inorg. Chem.*, **38**, 4671 (1999).
363. A. R. J. Genge, W. Levason and G. Reid, *Inorg. Chim. Acta*, **288**, 142 (1999).
364. W. M. Teles, N. G. Fernandes, A. Abras and C. A. L. Filgueiras, *Transition Met. Chem.*, **24**, 321 (1999).
365. S. lanelli, P. Mazza, M. Orchestrerie, C. Pelizzi, G. Pelizzi and F. Zani, *J. Inorg. Biochem.*, **60**, 89 (1995).
366. J. S. Casas, M. V. Castano, M. S. Garcia-Tasende, T. Perez-Alvarez, A. Sanchez and J. Sordo, *J. Inorg. Biochem.*, **61**, 97 (1996).
367. P. J. Cox, S. J. Garden, R. A. Howie, O. A. Melvin and J. L. Wardell, *J. Organomet. Chem.*, **516**, 213 (1996).
368. D. Kovala-Demertzi, P. Tauridou, U. Russo and M. Gielen, *Inorg. Chim. Acta*, **239**, 177 (1995).
369. F. Kayser, M. Biesemans, A. Delmotte, R. Willem and M. Gielen, *Bull. Soc. Chim. Belg.*, **104**, 27 (1995).
370. M. Nath and S. Goyal, *Met.-Based Drugs*, **2**, 297 (1995).
371. L. Jiang, Z.-Q. Yang, Q.-L. Xie and S.-X. Shan, *Huaxue Xuebao*, **53**, 1034 (1995); *Chem. Abstr.*, **124**, 146328 (1996).
372. A. Kalsoom, M. Mazhar, S. Ali, M. I. Choudhary and K. C. Molloy, *J. Chem. Soc. Pak.*, **18**, 320 (1996); *Chem. Abstr.*, **126**, 343618 (1997).
373. M. Gielen, F. Kayser, O. B. Zhidkova, V. T. Kampel, V. I. Bregadze, D. de Vos, B. Mahieu and R. Willem, *Met.-Based Drugs*, **2**, 37 (1995).
374. M. Nath, S. Goyal, C. L. Sharma, G. Eng and D. Whalen, *Synth. React. Inorg. Met.-Org. Chem.*, **25**, 821 (1995).
375. M. Gielen, E. R. T. Tiekink, A. Bouhdid, D. de Vos, M. Biesemans, I. Verbruggen and R. Willem, *Appl. Organomet. Chem.*, **9**, 639 (1995).
376. M. Gielen, A. Bouhdid, F. Kayser, M. Biesemans, D. de Vos, B. Mahieu and R. Willem, *Appl. Organomet. Chem.*, **9**, 251 (1995).
377. Z. Yang. T. Bakas, A. Sanchez-Diaz, C. Charalampopoulos, J. Tsangaris and N. Hadjiliadis, *J. Inorg. Biochem.*, **72**, 133 (1998).
378. J. S. Casas, M. C. Rodriguez-Argüelles, U. Russo, A. Sanchez, J. Sordo, A. Vazquez-Lopez, S. Pinelli, P. Lunghi, A. Bomati and R. Albertini, *J. Inorg. Biochem.*, **69**, 383 (1998).
379. M. Nath, S. Goyal, S. Goyal, G. Eng and N. Ogwuru, *Synth. React. Inorg. Met.-Org. Chem.*, **28**, 1619 (1998).
380. R. Parkash and S. Singh, *Synth. React. Inorg. Met.-Org. Chem.*, **29**, 1091 (1999).
381. X. H. Wang, H. C. Dai and J. F. Liu, *Polyhedron*, **18**, 2293 (1999).
382. M. Veith, M. Opsölder, M. Zimmer and V. Huch, *Eur. J. Inorg. Chem.*, 1143 (2000).
383. B. Wrackmeyer, in *Physical Organometallic Chemistry—Advanced Applications of NMR to Organometallic Chemistry*, Vol. 1 (Eds. M. Gielen, R. Willem and B. Wrackmeyer), Wiley, London, 1996, p. 87.
384. B. Jousseaume, M. Lahcini, M.-C. Rascle, F. Ribot and C. Sanchez, *Organometallics*, **14**, 685 (1995).

385. C. Lucas, C. C. Santini, M. Prinz, M.-A. Cordonnier, J.-M. Basset and M.-F. Connil, *J. Organomet. Chem.*, **520**, 101 (1996).
386. M.-F. Connil, B. Jousseaume and M. Pereyre, *Organometallics*, **15**, 4469 (1996).
387. M. Westerhausen, M. Krofta, N. Wiberg, J. Knizek, H. Nöth and A. Pfitzner, *Z. Naturforsch.*, **53b**, 1489 (1998).
388. I. D. Kostas, G.-J. M. Gruter, O. S. Ackermann, F. Bickelhaupt, H. Kooijman, W. J. J. Smeets and A. L. Spek, *Organometallics*, **15**, 4450 (1996).
389. N. Rot, F. J. J. de Kanter, F. Bickelhaupt, W. J. J. Smeets and A. L. Spek, *J. Organomet. Chem.*, **593–594**, 369 (2000).
390. L. H. Piette and H. E. Waever, *J. Chem. Phys.*, **28**, 735 (1958).
391. W. MacFarlane, *Proc. R. Soc. London*, **A306**, 185 (1968).
392. J. D. Kennedy and W. MacFarlane, in *NMR and the Periodic Table* (Eds. R. K. Harris and B. E. Mann), Academic Press, London, 1978, p. 366.
393. J. D. Kennedy and W. MacFarlene, in *Multinuclear NMR* (Ed. J. Mason), Plenum Press, London, 1987, p. 305.
394. B. Wrackmeyer and K. Horchler, *Annu. Rep. NMR Spectrosc.*, **22**, 249 (1989).
395. A. Sebald, *NMR (Solid State NMR II, Inorganic Matter)*, **31**, 91 (1994).
396. J. D. Kennedy, W. MacFarlane and G. S. Pyne, *J. Chem. Soc., Dalton Trans.*, 2332 (1977).
397. T. N. Mitchell, *J. Organomet. Chem.*, **255**, 279 (1983).
398. B. Wrackmeyer, K. Horchler and C. Stader, *J. Magn. Reson.*, **83**, 601 (1989).
399. J. A. Gonzales, A. G. Ancor, M. C. Ruiz de Azua and R. H. Contreros, *J. Quantum Chem.*, **61**, 823 (1997).
400. A. Rodriguez-Fortea, P. Alemany and T. Ziegler, *J. Phys. Chem.*, **103**, 8288 (1999).
401. a. T. M. Klapötke, J. Knizek, H. Nöth, B. Krumm and C. M. Rienäcker, *Polyhedron*, **18**, 839 (1999).
b. T. M. Klapötke, J. Knizek, H. Nöth, B. Krumm and C. M. Rienäcker, **18**, 1687 (1999).
402. E. Faynon, I. Farnan, C. Bessada, J. Coutures, D. Massiot and J. P. Coutures, *J. Am. Chem. Soc.*, **119**, 6837 (1997).
403. S. Rupprecht, S. J. Franklin and K. N. Raymond, *Inorg. Chim. Acta*, **235**, 185 (1995).
404. M. G. Maloney, D. R. Paul and R. T. Thompson, *Main Group Metal Chem.*, **18**, 295 (1995).
405. J. Caruso, M. J. Hampden-Smith and E. N. Duesler, *J. Chem. Soc., Chem. Commun.*, 1041 (1995).
406. G. Liliane, S. Jagner and M. Hakansson, *Inorg. Chem.*, **34**, 628 (1995).
407. J. E. H. Buston, T. D. W. Claridge and M. G. Maloney, *J. Chem. Soc., Perkin Trans. 2*, 639 (1995).
408. D. E. Fenton, R. W. Matthews, M. McPartlin, B. P. Murphy, I. J. Scowen and P. A. Tasker, *J. Chem. Soc. Dalton Trans.*, 3421 (1996).
409. D. J. Teff, J. G. Huffmann and K. G. Caulton, *J. Am. Chem. Soc.*, **118**, 4030 (1996).
410. S. Rupprecht, K. Langemann, T. Lügger, J. M. McCormick and K. N. Raymond, *Inorg. Chim. Acta*, **243**, 79 (1996).
411. G. D. Fallon, L. Spiccia, B. O. West and Q. Zhang, *Polyhedron*, **16**, 19 (1997).
412. T. D. W. Claridge, E. J. Netleton and M. G. Maloney, *Magn. Reson. Chem.*, **35**, 159 (1997).
413. M. Veith, C. Mathur and V. Huch, *J. Chem. Soc., Dalton Trans.*, 995 (1997).
414. M. Veith, C. Mathur, S. Mathur and V. Huch, *Organometallics*, **16**, 1292 (1997).
415. L. G. Hubert-Pfalzgraf, S. Daniele, R. Papiernik, M.-C. Massiani, B. Septe, J. Vaissermann and J.-C. Daran, *J. Mater. Chem.*, **7**, 753 (1997).
416. L. G. Hubert-Pfalzgraf, S. Suoad, R. Papiernik, B. Septe, J. Vaissermann and J.-C. Daran, *J. Mater. Chem.*, **7**, 2053 (1997).
417. Y.-S. Kye, S. Connolly, B. Herreros and G. S. Harbison, *Main Group Metal Chem.*, **22**, 373 (1999).
418. G. D. Fallon, L. Spiccia, B. O. West and Q. Zhang, *J. Sol-Gel Sci. Technol.*, **16**, 119 (1999).
419. B. Wrackmeyer and J. Weidinger, *Z. Naturforsch.*, **52b**, 947 (1997).
420. N. Kano, K. Shibata, N. Tokitoh and R. Okazaki, *Organometallics*, **18**, 2999 (1999).
421. B. Gehrhus, P. B. Hitchcock and M. F. Lappert, *J. Chem. Soc., Dalton Trans.*, 3094 (2000).
422. L. Pu, P. P. Power, I. Boltes and R. Herbst-Irmer, *Organometallics*, **19**, 352 (2000).
423. L. Pu, P. Twamley and P. P. Power, *Organometallics*, **19**, 2874 (2000).
424. M. Charisse, B. Mathiasch, M. Dräger and U. Russo, *Polyhedron*, **14**, 2429 (1995).

425. D. C. Van Beelen, J. Van Rijn, K. D. Heringa, J. Wolters and D. de Vos, *Main Group Metal Chem.*, **20**, 37 (1997).
426. M. Herberhold, V. Tröbs, H. Zhou and B. Wrackmeyer, *Z. Naturforsch.*, **52b**, 1181 (1997).
427. J. E. H. Buston, T. D. W. Claridge, R. G. Compton and M. G. Maloney, *Magn. Reson. Chem.*, **36**, 140 (1998).
428. M. Charisse, A. Zickgraf, H. Stenger, E. Bräu, C. Desmarquet, M. Dräger, S. Gerstmann D. Dakternieks and J. Hook, *Polyhedron*, **17**, 4497 (1998).
429. A. Winkler, W. Walter, F. W. Heinemann, V. Garcia-Montavlo, M. Moll and J. Ellermann, *Eur. J. Inorg. Chem.*, 437 (1998).
430. A. K. Brimah, P. Schwarz, R. D. Fischer, N. A. Davies and R. K. Harris, *J. Organomet. Chem.*, **568**, 1 (1998).
431. B. Wrackmeyer, G. Kehr, H. E. Meisl and H. Zhou, *Magn. Reson. Chem.*, **36**, 39 (1998).
432. H. Stenger, B. S. Schmidt and M. Dräger, *Organometallics*, **14**, 4374 (1995).
433. C. Eaborn, T. Ganicz, P. B. Hitchcock, J. D. Smith and S. E. Sözerli, *Organometallics*, **16**, 5621 (1997).
434. M. C. Kuchta, J. M. Hahn and G. Parkin, *J. Chem. Soc., Dalton Trans.*, 3559 (1999).
435. B. Wrackmeyer, K. Horchler and S. Kundler, *J. Organomet. Chem.*, **503**, 289 (1995).
436. B. Wrackmeyer, G. Kejr, H. Zhou and S. Ali, *Magn. Reson. Chem.*, **34**, 921 (1996).
437. T. M. Klapötke, B. Krumm, M. Niemetz, K. Polborn and C. M. Rienäcker, *J. Fluorine Chem.*, **104**, 129 (2000).
438. M. Herberhold, V. Tröbs and B. Wrackmeyer, *J. Organomet. Chem.*, **541**, 391 (1997).
439. M. Schürmann and F. Huber, *J. Organomet. Chem.*, **530**, 121 (1997).
440. D. N. Kravtsov, A. S. Peregudov and V. M. Pachevskaya, *Russ. Chem. Bull.*, **45**, 441 (1996).
441. M. Schürmann, Ph. D. Thesis, Shaker Verlag Aachen, Germany, 1995.
442. R. S. Simons, L. Pu, M. M. Olmstead and P. P. Power, *Organometallics*, **16**, 1920 (1997).
443. M. Stürmann, M. Weidenbruch, K. W. Klinkhammer, F. Lissner and H. Marsmann, *Organometallics*, **17**, 4425 (1998).

CHAPTER 7

Recent advances in acidity, complexing, basicity and H-bonding of organo germanium, tin and lead compounds*

CLAUDIA M. RIENÄCKER and THOMAS M. KLAPÖTKE

Department of Chemistry, Ludwig-Maximilians-University Munich, Butenandtstr. 5-13 (Building D), D-81377 Munich, Germany
Fax: +49 89 21807492; e-mail: tmk@cup.uni-muenchen.de

I. ABBREVIATIONS

The following abbreviations are used in addition to the well-known abbreviations which are listed in each volume.

232tet	3,7-diazanonane-1,9-diamine
323tet	4,7-diazadecane-1,10-diamine

* In this chapter, full lines are used both for covalent chemical bonds as well as for partial bonds and for coordination.

The chemistry of organic germanium, tin and lead compounds — Vol. 2
Edited by Z. Rappoport

ac	acetate
amp	2-aminomethylpyridine
bmimt	bis(1-methyl-2-imidazolylthio)methane
bpy	2,2′-bipyridine
CVD	chemical vapor deposition
cyclam	1,4,8,11-tetraazacyclotetradecane
dien	3-azapentane-1,5-diamine (diethylenetriamine, '22')
dmphen	2,9-dimethyl-1,10-phenanthroline
dpa	bis(2-pyridyl)amine
en	ethane-1,2-diamine (ethylenediamine)
fc	ferrocene
Fp	$(\eta^5\text{-}C_5H_5)Fe(CO)_2$
HL^4	2-cyanaminofluoren-9-one
HL^{12}	2-dimethylaminoethanol
H_2L^6	N-(2-hydroxyphenyl)-2-hydroxy-1-naphthylaldimine
HMPA	hexamethylphosphorotriamide
IFp	$(\eta^5\text{-}C_9H_7)Fe(CO)_2 = (\eta^5\text{-indenyl})Fe(CO)_2$
L^1	1,4,7-tris(4-*tert*-butyl-2-mercaptobenzyl)-1,4,7- triazacyclononane
L^2	1,4,7-trimethyl-1,4,7-triazacyclononane
L^3	1,3,5-trimethyl-1,3,5-triazacyclohexane
L^5H	bis(2-methoxy-3-*tert*-butyl-5-methylphenyl)methane, $(2\text{-MeO-3-}t\text{-Bu-5-Me-}C_6H_2)_2CH_2$
L^{11}	3,10,17,24-tetraaza-29,30-dioxapentacyclo[24.2.1.1^{12,15}.0^{4,9}0^{18,23}]-triaconta-1(28),2,4,6,8,10,12,14,16,18,20,22,24,26-tetradecaene
l.s.	low-spin
ma	2-MeOC_6H_4
Me_8taa	octamethyldibenzotetraaza[14]annulene dianion
Mes	$2,4,6\text{-Me}_3C_6H_2$, mesityl
phen	1,10-phenanthroline
pn	propane-1,2-diamine
py	pyridine
tacn	1,4,7-triazacyclononane
tctscH	thiophene-2-carboxaldehyde thiosemicarbazone
tet-b	(7R*,14R*)-5,5,7,12,12,14-hexamethyl-1,4,8,11-tetraazacyclotetradecane
tpy	2,2′ : 6′,2″ -terpyridine
trien	3,6-diazaoctane-1,8-diamine, 222tet
trz	2,4,6-tris(2-pyridyl)-1,3,5-triazine

II. OUTLINE

The aim of this review is to focus on the hydrogen bonding, the acidity and basicity and complexing chemistry concerning the organo-element chemistry of germanium, tin and lead. This chapter is not exhaustive in scope, but contains the most recent work of the last five to six years, since another review in this series was published[1]. This chapter emphasizes the synthesis, reactions and molecular structures of the class of compounds outlined above (less attention is paid to mechanism, spectroscopic properties and applications which can be found in other specialized chapters). Especially, the single-crystal X-ray diffraction technique has elucidated many novel and unusual structures of molecules and of the solid state in general. Not unexpectedly, certain organo-element compounds present problems concerning their classification as n-coordinated species, since it is sometimes difficult to distinguish between a weak long-range interaction in the solid state and the fact that two atoms can be forced a little bit closer together by crystal lattice effects.

Since organo-element chemistry is the discipline dealing with compounds containing at least one direct element–carbon bond, in this chapter we discuss germanium, tin and lead species in which at least one organic group is attached through a carbon, and as an extension also compounds where the organic group is connected through a nitrogen or an oxygen atom.

III. INTRODUCTION

Considering the chemical reactivity and group trends of germanium, tin and lead, it can be stated that germanium is somewhat more reactive and more electropositive than silicon. Alkyl halides react with heated Ge (as with Si) to give the corresponding organogermanium halides. Tin, however, is notably more reactive and electropositive than Ge and Pb powder is pyrophoric whereas the reactivity of the metal is usually greatly diminished by the formation of a thin, coherent protective oxide layer. The steady trend towards increasing stability of M^{II} rather than M^{IV} compounds in general in the sequence Ge, Sn, Pb is an example for the 'inert-pair effect', which is well established for the heavier main group metals. Table 1 summarizes the physical properties of the group 14 elements Ge, Sn and Pb[1].

IV. COMPLEXING, ACIDITY, BASICITY AND H-BONDING

A. Introduction

The concept of second- or outer-sphere coordination, originally introduced by Werner[3], has played a major role in the subsequent development of the theory of bonding in metal complexes and has recently re-emerged as a means of describing higher-order

TABLE 1. Physical properties of Group 14 elements Ge, Sn and Pb

	^{32}Ge	^{50}Sn	^{82}Pb
Electron configuration	$[Ar]3d^{10}4s^24p^2$	$[Kr]4d^{10}5s^25p^2$	$[Xe]4f^{14}5d^{10}6s^26p^2$
Atomic weight ($g\,mol^{-1}$)	72.61	118.71	207.20
Electonegativity:			
Pauling	2.01	1.96	2.33
Allred–Rochow	2.02	1.72	1.55
Sanderson	2.31	2.02	2.0
Ionization potential (eV)[2]	(1) 7.899	7.344	7.416
	(2) 15.934	14.632	15.032
	(3) 34.220	30.502	31.937
	(4) 45.710	40.734	42.32
Relative electron density	17.4	17.8	15.3
B.E.(E-E) ($kcal\,mol^{-1}$)^a	45	36	—
B.E.(E-C) ($kcal\,mol^{-1}$)^a	61	54	31
B.E.(E-H) ($kcal\,mol^{-1}$)^a	69	60	49
B.E.(E-Cl) ($kcal\,mol^{-1}$)^a	82	77	58
Covalent bond radius (Å)	1.225	1.405	(1.750)
van der Waals radius (Å)	2.10	2.17	2.02

^{a}B.E. = Bond energy

bonding interactions in complexes with crown ether ligands and in systems involving supramolecular or host–guest interactions[4]. In molecular compounds, the preference for inner- over outer-sphere coordination may be expected to be dependent primarily on (i) the size of the central atom, (ii) the symmetries and energies of the available unoccupied orbitals, (iii) the electronegativity differences and (iv) special structural features of the ligating groups. Accordingly, with certain metals and ligands, complexes with unusual coordination numbers and geometries were obtained, but because of the manifold nature of the metal–ligand interactions, predictions as to the behaviour of a given metal or ligand are not generally possible.

We divided Section IV into three parts according to the three elements of Group 14 germanium, tin and lead. The main interest is the complexing chemistry of these elements.

B. Reactions

1. Germanium

Reports on the design, synthesis and modulation of the redox and photochemical function of germanium(IV) porphyrin-based, 'axial-bonding'-type hybrid trimers are of interest, because photochemically active supramolecular arrays are investigated concerning their ability to transport charge, ions or energy[5]. In Scheme 1 the synthesis of germanium compounds is shown.

Analysis of the UV-visible, ESR and redox potential data suggests the absence of any exciton coupling between the porphyrin rings and these trimers. Energies of the singlet and charge transfer states are shown in Figure 1. H_2 means H_2L^a, Ge means $[(L^b)$ Ge $(OH)_2]$ and Zn means ML^a M^{Zn}.

The compound $[L^1FeGeFeL^1]$ $[PF_6]_2$ (L^1 = 1,4,7-(4-*t*-butyl-2-mercaptobenzyl)-1,4,7-triazacyclononane) (Figure 2) can be synthesized from GeI_2 in a reaction with $[L^1Fe^{III}]$ and the addition of $NaPF_6$. This complex can be one-electron oxidized by $[Ni^{III}(tacn)_2]$ $[ClO_4]_3$ (tacn = 1,4,7-triazacyclononane) to form $[L^1FeGeFeL^1]$ $[ClO_4]_3$[6]. The oxidation state in the dication cannot be described as l.s.$Fe^{III}Ge^{II}$l.s.Fe^{III} species (l.s. = low-spin) but rather as a mixed valent species with a formal oxidation state distribution l.s.$Fe^{2.4}Ge^{3.2}Fe^{2.4}$. One-electron oxidation of the dication yields the trication and affects only the iron centres and not the respective central main group ion which possesses formally a +III oxidation state. An oxidation state distribution of l.s.$Fe^{2.9}Ge^{3.3}$l.s.$Fe^{2.9}$ can be envisaged in the trications yielding an $S_t = \frac{1}{2}$ ground state.

Compounds which contain a Si–Ge bond and similar complexes are of interest for mass spectrometry because of their fragmentation. Experiments were carried out on the following compounds: $Me_3SiGePh_3$, $Ph_3SiGeMe_3$, $FpSiMe_2GeMe_3$ (**a**), $FpGeMe_2SiMe_3$ (**b**), (Fp = $(\eta^5-C_5H_5)Fe(CO)_2$), $IFpSiMe_2GeMe_3$ (**c**), $IFpGeMe_2SiMe_3$ (**d**), $IFpSiMe_2$-$GePh_3$ (**e**), $IFpGeMe_2SiPh_3$ (**f**), (IFp = $(\eta^5$-$C_9H_7)Fe(CO)_2$) and $fcSiMe_2GeMe_2fc$ (fc = ferrocenyl)[7]. For the non-iron-containing compounds $R_3SiGeR'_3$ an exchange of R groups was observed, which is shown in the measured $[R_{3-n}R'_nSi]^+$ and $[R'_{3-n}R_nGe]^+$ ions. For the metal-substituted complexes containing the grouping Fe–Si–Ge, fragmentation occurs predominantly via Si–Ge bond cleavage with formation of ions containing the silylene ligand $[Fe{=}SiR_2]^+$. For some of the above-mentioned compounds, Scheme 2 shows the suggested fragmentation patterns.

The products formed during a reaction of an allenic stannane with a Lewis acidic element halide have been less investigated. One example is the propargyltrichlorogermane ($HC{\equiv}CCH_2GeCl_3$), which can be prepared from the starting material $GeCl_4$ and $H_2C{=}C{=}CHSnBu_3$ at 40 °C. Even if the latter is absolutely pure, a mixture of products

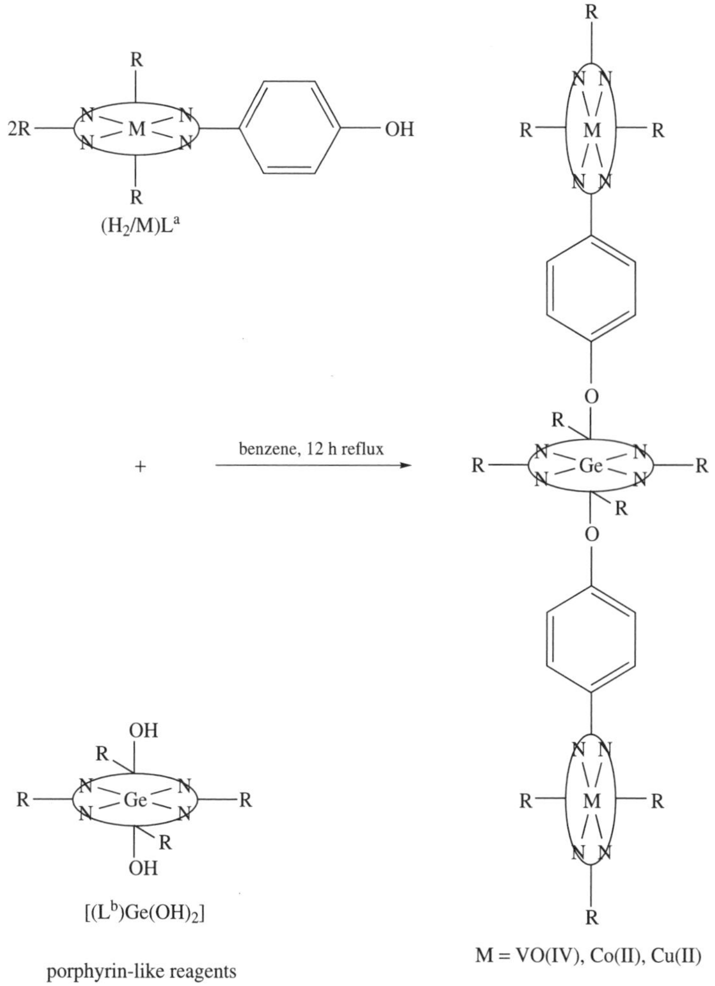

$(H_2/M)L^a$

$[(L^b)Ge(OH)_2]$

porphyrin-like reagents

benzene, 12 h reflux

M = VO(IV), Co(II), Cu(II)

SCHEME 1. Synthesis of a germanium 'axial-bonding'-type hybrid trimer. Reprinted with permission from Reference 5. Copyright 1999 American Chemical Society

was obtained ($HC{\equiv}CCH_2GeCl_3$ 48%, $H_2C{=}C{=}CHGeCl_3$ 1%). As the temperature of the reaction is substantially lower than the temperature of isomerization, it is possible to conclude that the formation of the propargylic compound occurs via an allenyl–propargyl transposition with more than 99% selectivity.

Heating up the propargyltrichlorogermane to 120 °C for a few hours led only to very small amounts of allenyltrichlorogermane ($H_2C{=}C{=}CHGeCl_3$), which can be synthesized from $HC{\equiv}CCH_2SnPh_3$ and $GeCl_4$ at 50 °C[8].

The calix[*n*]arenes are a class of 'chalice-like' macrocyclic molecules, which are useful ligands for divalent Ge. The divalent Ge complexes [*t*-Bu $calix_2^{(TMS)}$]Ge are observed by a reaction of [*t*-Bu $calix_2^{(TMS)}$]H_2 with $Ge[N(SiMe_3)_2]_2$ (Scheme 3)[9]. For this Ge system

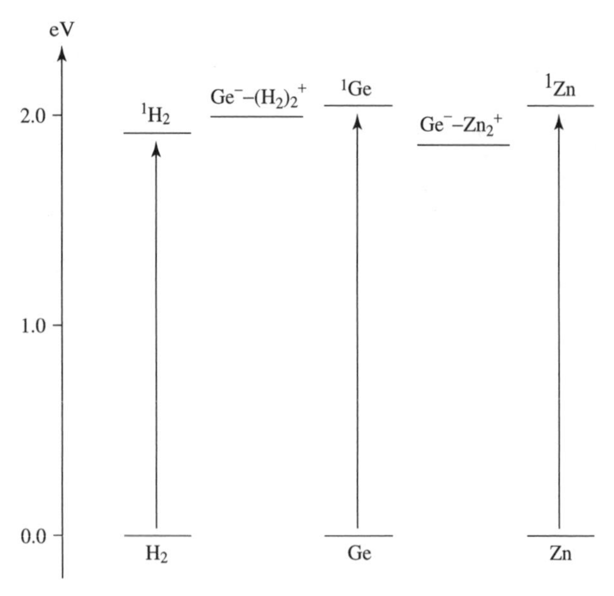

FIGURE 1. Energies of the singlet and charge transfer states pertaining to the photoactive array. Reprinted with permission from reference 5. Copyright 1999 American Chemical Society

two isomeres (*exo*: Figure 3 and *endo*: Figure 4) could be isolated, where the location of Ge with the respect to the calixarene cavity differs. This is the first pair of *exo* and *endo* isomers to be structurally characterized. For a schematic diagram clarifying the *exo/endo* coordination at Ge see Scheme 3.

The nature of multiple bonding between germanium and the heavier chalcogens in the complexes (η^4-Me_8taa)GeE (E = Se, Te) is best described as an intermediate between the $Ge^+–E^-$ and Ge=E resonance structures. The preparation of these complexes involves the addition of the elemental chalcogen to (η^4-Me_8taa)Ge, which is synthesized by the metathesis of $GeCl_2$(1,4-dioxane) and $Li_2[Me_8taa]$ (Me_8taa = octamethyldibenzotetraaza[14]annulene dianion). The molecular structures of both complexes are shown in Figures 5 and 6[10].

In the reaction of lithiated 2-hydroxybiphenyl and 2'-hydroxy-*m*-terphenyl with germanium dichloride–dioxane complex and $GeCl_4$, respectively, spirocyclic germabiphenanthrene compounds (Scheme 4) and 1-oxa-10-germaphenanthrene (Scheme 5) were formed[11].

The first crystallographic evidence for a neutral and, according to the authors, hypervalent germanium(IV) complex (*N.B.* these complexes are definitely hypercoordinated) with sulphur-induced hexacoordination of germanium in a spirocyclic complex with two sterically hindered eight-membered 12*H*-dibenzo[*d*,*g*][1,2,3,6,2]dioxathiagermocin rings is shown in Figure 7. Scheme 6 presents the synthesis of the complex[12].

The apparent hypercoordination of the germanium atom in complexes is an interesting research field. A reaction between *t*-butyltrichlorgermane (*t*-$BuGeCl_3$) and mercaptoacetic acid ($HSCH_2CO_2H$) affords a novel type of pentacoordinate germanium compound via *t*-$BuGe(SCH_2CO_2H)_3$, which loses one mole of SCH_2CO_2H. Figure 8 shows the molecular structure of the formed product 2-(2-*t*-butyl-5-oxo-1,2,3-oxathiagermolan-2-ylthio)acetic

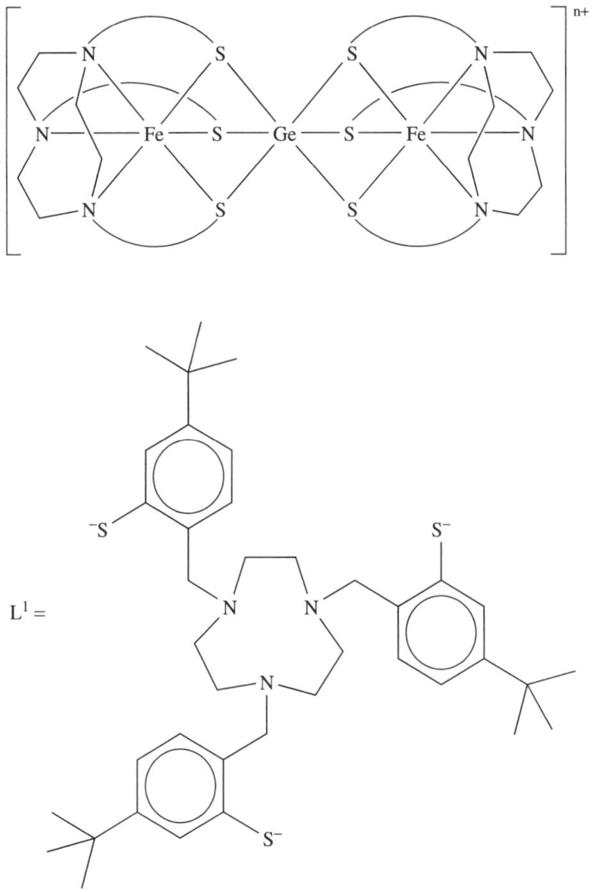

FIGURE 2. Structure of $[L^1FeGeFeL^1]^{n+}$. Reprinted with permission from Reference 6. Copyright 1999 American Chemical Society

acid[13]. For this compound some characteristic points could be observed: (a) The germanium atom is pentacoordinate with trigonal bipyramidal structure. (b) The Ge1−S1, Ge1−S2 and Ge1−C1 bonds are equatorial while the Ge1−O1 and Ge1−O3 bonds are apical with equal lengths (ca 2.04 Å, a little longer than the standard Ge−O covalent bond length (ca 1.7–1.8 Å). (c) Four atoms, Ge1, S1, S2 and C1, are coplanar with S1, S2 and C1 in a trigonal planar arrangement around Ge. (d) The O1−Ge1−O3 hypercoordinate bond is nearly perpendicular to the S1-S2-C1 plane, though the angle (166.7°) slightly deviates from the ideal trigonal bipyramidal structure.

The summary of all these characteristic points to a pentacoordinated organogermanium compound where no steric enforcement is involved to enhance the hypercoordination. The rotation about the $S-CH_2$ bond, which should be feasible, would move the CO_2H group far apart from germanium and should be noted. This in turn appears to suggest that the Ge−O interaction is strong enough to lead to hypercoordination.

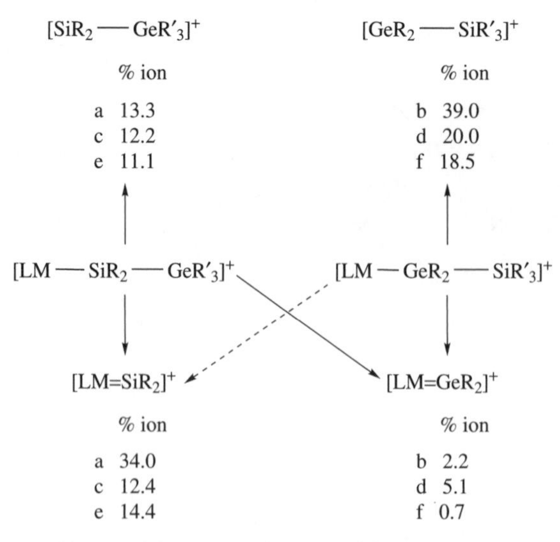

SCHEME 2. Suggested fragmentation patterns for some of the above-mentioned compounds. Reproduced by permission of Elsevier Science from Reference 7

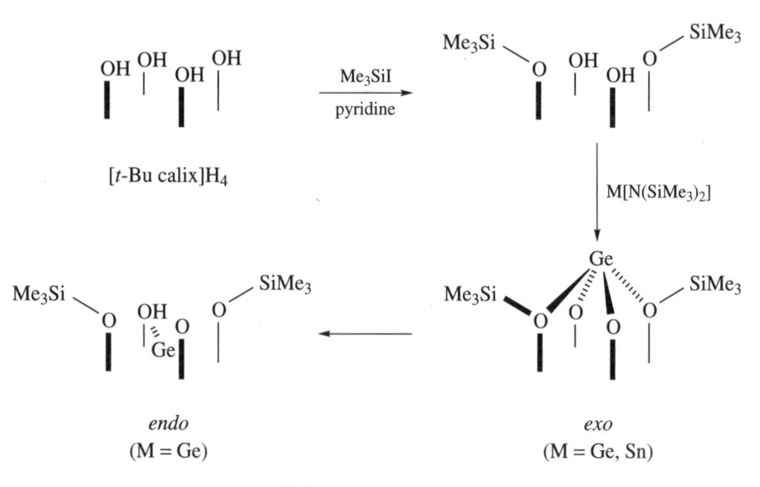

SCHEME 3. Synthesis of [t-Bu calix$_2^{(TMS)}$]Ge. Reproduced by permission of the Royal Society of Chemistry from Reference 9

Compounds of the type Ph_3GeCH_2SR were obtained by the reactions shown in equations 1–3[14].

$$Ph_3GeBr + LiCH_2SMe \longrightarrow Ph_3GeCH_2SMe \quad (1)$$

$$(Ph_3GeBr + Li) \longrightarrow Ph_3GeLi \xrightarrow{ClCH_2SPh} Ph_3GeCH_2SPh \quad (2)$$

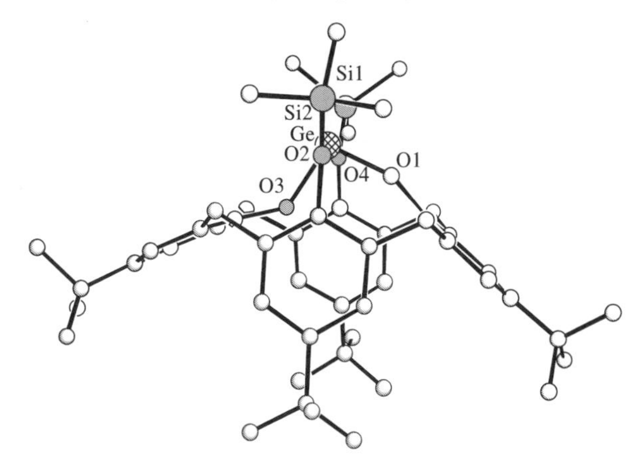

FIGURE 3. The molecular structure of *exo*-[*t*-Bu $calix_2^{(TMS)}$]Ge. Reproduced by permission of the Royal Society of Chemistry from Reference 9

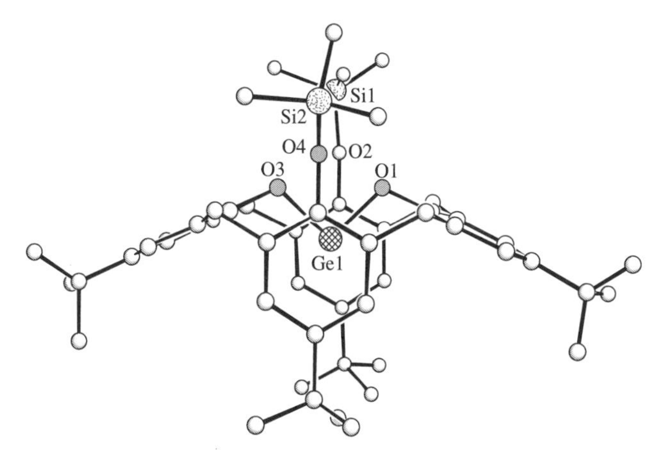

FIGURE 4. The molecular structure of *endo*-[*t*Bu $calix_2^{(TMS)}$]Ge. Reproduced by permission of the Royal Society of Chemistry from Reference 9

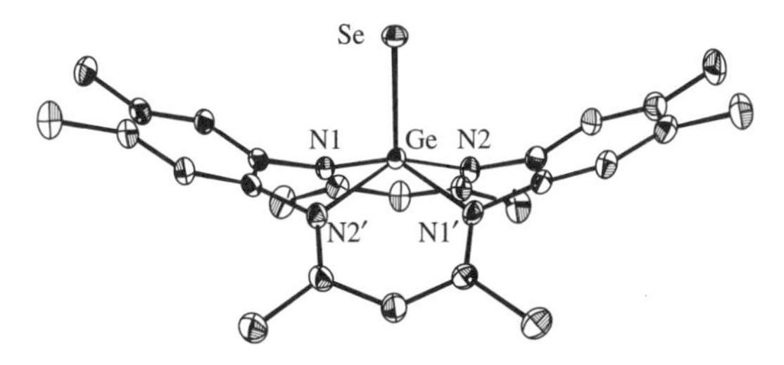

FIGURE 5. Molecular structure of (η^4-Me_8taa)GeSe. Reproduced by permission of the Royal Society of Chemistry from Reference 10

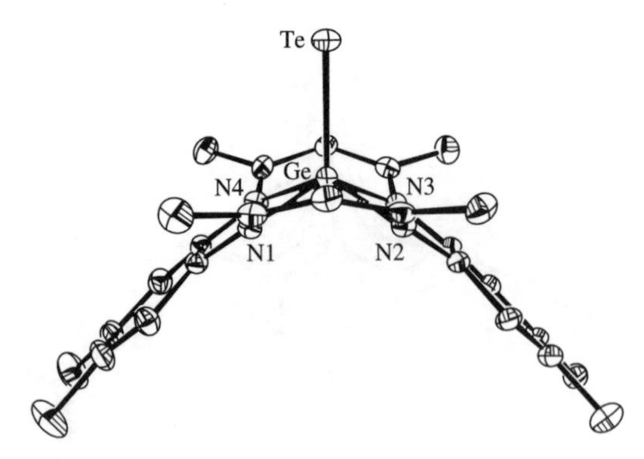

FIGURE 6. Molecular structure of (η^4-Me_8taa)GeTe. Reproduced by permission of the Royal Society of Chemistry from Reference 10

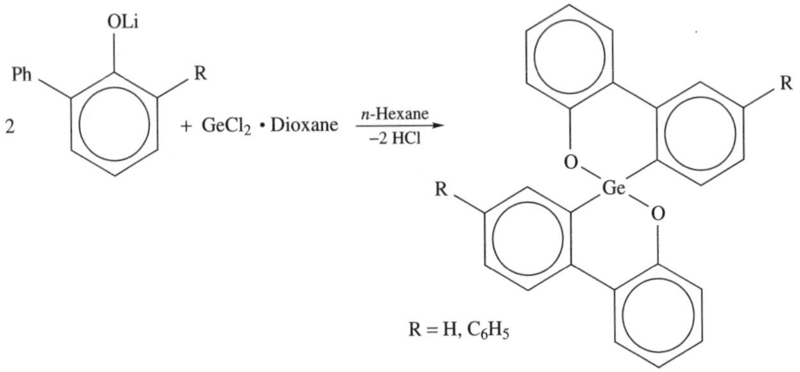

OLi
Ph R
2 + $GeCl_2$ • Dioxane $\xrightarrow[-2\ HCl]{n\text{-Hexane}}$
R O Ge O R
R = H, C_6H_5

SCHEME 4. Synthesis of spirocyclic germabiphenanthrene compounds from Reference 11. Permission granted by Gordon and Breach publishers, copyright OPA (Overseas Publishers Association)

OLi
Ph
3 + $GeCl_4$ $\xrightarrow[-3\ LiCl,\ -HCl]{n\text{-Hexane}}$
Ph Ph
O O
Ge
O

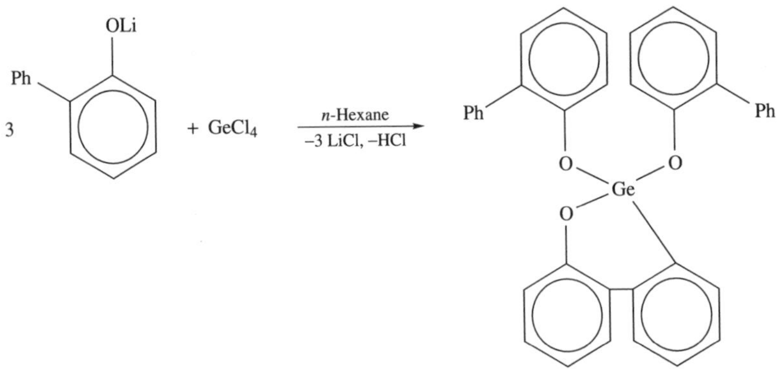

SCHEME 5. Synthesis of an 1-oxa-10-germaphenanthrene from Reference 11. Permission granted by Gordon and Breach publishers, copyright OPA (Overseas Publishers Association)

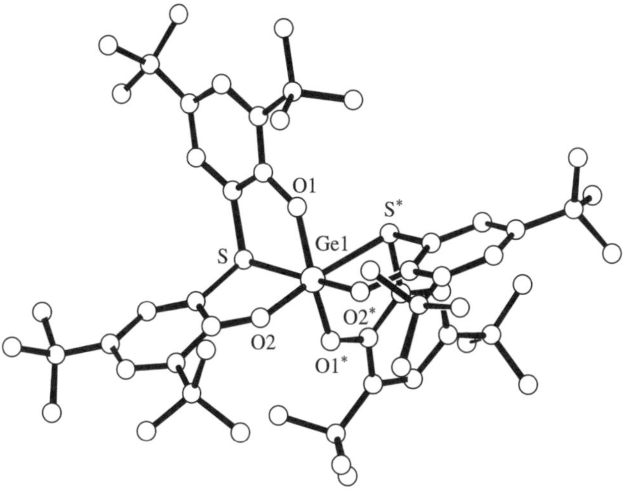

FIGURE 7. Molecular structure of a neutral hypervalent germanium(IV) complex. Reprinted with permission from Reference 12. Copyright 1997 American Chemical Society

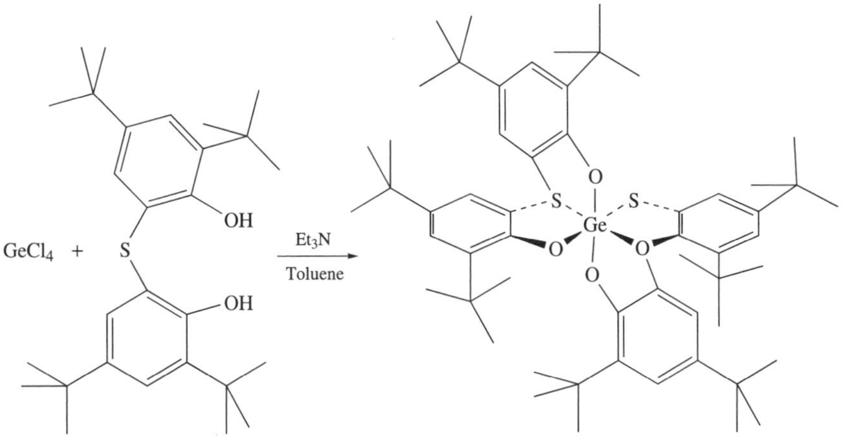

SCHEME 6. Synthesis of a neutral hypervalent germanium(IV) complex. Reprinted with permission from Reference 12. Copyright 1997 American Chemical Society

$$Ph_3GeCH_2Br + NaSR \longrightarrow Ph_3GeCH_2SR \quad (3)$$

$$R = pt\text{-}BuC_6H_4$$

A mild and effective reagent for the formation of alkyl- and arylsulphinylmethylgermanium compounds (R = Me, p-$R^1C_6H_4$, R^1 = H, Me or t-Bu) from Ph_3GeCH_2SR is m-chloroperbenzoic acid (m-$ClC_6H_4CO_3H$). No cleavage of the phenyl–germanium bonds occurred during these reactions. The structure of the sulphoxide is shown in

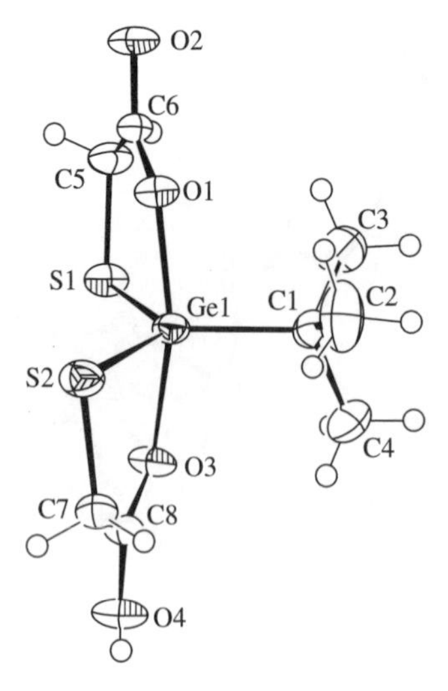

FIGURE 8. Molecular structure of product 2-(2-*t*-butyl-5-oxo-1,2,3-oxathiagermolan-2-ylthio)acetic acid. Reproduced by permission of the Royal Society of Chemistry from Reference 13

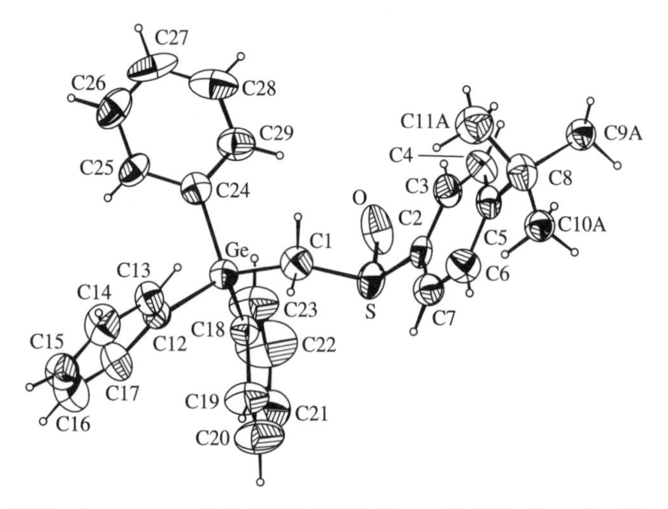

FIGURE 9. Molecular structure of $Ph_3GeCH_2S(O)R$, R = *pt*-BuC_6H_4. Reproduced by permission of Elsevier Science from Reference 14

Figure 9. Further oxidation of the sulphoxides ($Ph_3GeCH_2S(O)R$) to the sulphones $Ph_3GeCH_2S(O)_2R$ also occurs smoothly using *m*-chloroperbenzoic acid.

Compounds like 1,4,7-trimethyl-1,4,7-triazacyclononane (L^2) and 1,3,5-trimethyl-1,3,5-triazacyclohexane (L^3) are useful ligands in germanium chemistry. Reactions of L^2 with $GeCl_4$ and L^3 with $GeBr_4$ both in the ratio 1 : 1 in acetonitrile solution form ionic compounds with GeX_3^+ cations: $[GeCl_3(L^2)]Cl_3^-$•MeCN and $[GeBr_3(L^2)]Br_3^-$•MeCN. Acetonitrile is trapped in the lattice as solvate molecules[15].

The ligand binding of L^2 and L^3 to Ge(IV) are for both compounds terdentate N-donor (η^3). The resulting GeX_3^+ cations are effectively stabilized by (η^3) azamacrocyclic chelation to give six-coordinate species that show the anticipated *fac*-octahedral metal geometry. The structures of the cations are given in Figures 10 and 11.

Interestingly, for these compounds simple halide ions constitute the counter anion in salt formation. The compactness of the ring, especially in the case of L^3, does impose a severe steric constriction at the metal centre. In the case of uncomplexed L^2 the preferred endodentate conformation identifies it as an (almost) ideal ligand for occupation of three metal coordination sites (*fac*-octahedral)[16–20]. For free L^3 there are four chair conformers

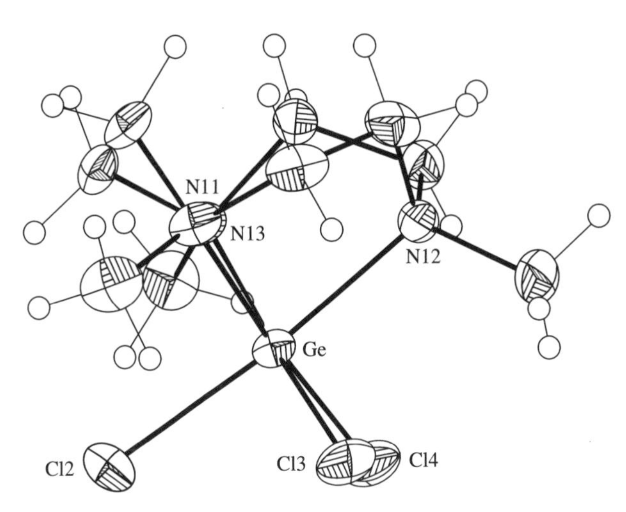

FIGURE 10. Molecular structure of the $[GeCl_3(L^2)]^+$ cation. Reproduced by permission of the Royal Society of Chemistry from Reference 15

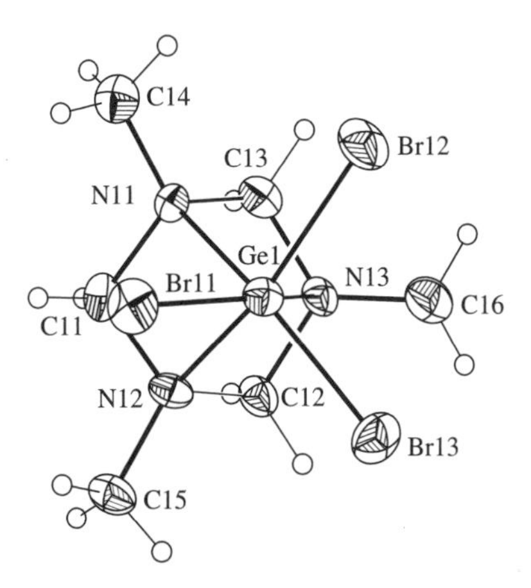

FIGURE 11. Molecular structure of the $[GeBr_3(L^3)]^+$ cation. Reproduced by permission of the Royal Society of Chemistry from Reference 15

possible, of which the *aee* arrangement with two methyl groups in equatorial positions and the remaining one in an axial location is preferred[21–23]. Formation of an η^3 complex is associated with rearrangement of all three methyl groups to equatorial sites, thereby facilitating maximal lone pair interactions with a metal ion acceptor.

2. Tin

Cyanamides are pseudo-halide nitrogen ligands that are readily coordinated to metals. A novel compound is 2-cyanaminofluoren-9-one (HL^4)[24]. Its thallium salt $Tl^+(L^4)^-$ (Scheme 7) is useful as a transmetallating agent in a reaction with trimethyltin chloride to produce the corresponding tin cyanamide complex [$SnMe_3L^4$] (Scheme 8).

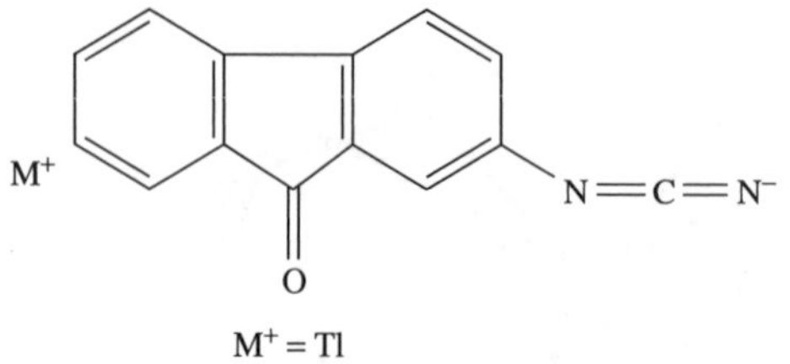

M^+ N=C=N$^-$

O

$M^+ = Tl$

SCHEME 7. $Tl^+(L^4)^-$. Reproduced by permission of the Royal Society of Chemistry from Reference 24

Cl—$SnMe_3$

$Tl^+ (L^4)^-$ | CH_2Cl_2, r.t., overnight

N=C=N—$SnMe_3$

O

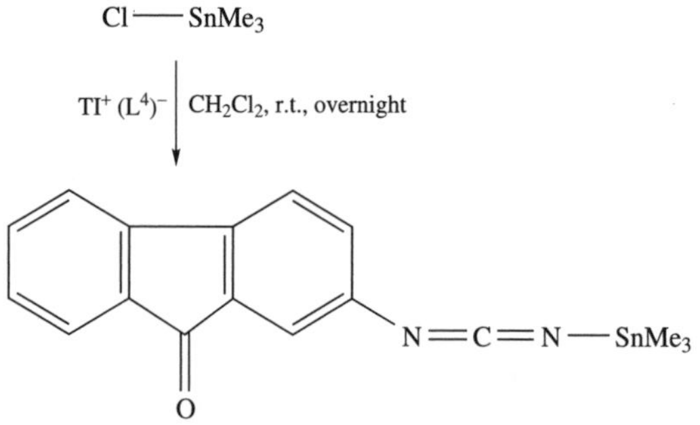

SCHEME 8. [$SnMe_3L^4$]. Reproduced by permission of the Royal Society of Chemistry from Reference 24

The synthesis of [PtCl($SnMe_2Cl$)(dmphen)(*E*-MeO_2CCH=$CHCO_2Me$)] (dmphen = 2,9-Me_2-1,10-phenanthroline) is given in equation 4. In this reaction a platinum(0) nucleophile [Pt(N,N′-chelat)(olefin)] and an organometal electrophile R_mSnX_n (X = Cl, Br, I; R = hydroxycarbyl group) are involved[25]. It has been found that the adduct is stabilized by the presence of electron-donor olefins on the platinum and by electron-withdrawing groups on the electrophilic metal. It has also been found that the influence of the halide moving onto the platinum can be rationalized in terms of the relative softness of the two metals involved in the equilibrium. Figure 12 shows the structure of [PtCl($SnMe_2Cl$)(dmphen)*E*-MeO_2CCH=$CHCO_2Me$)]. The coordination geometry is the expected one, i.e. the anionic ligands Cl and $SnMe_2Cl$ define the axial sites of the trigonal bipyramid [Cl–Pt–Sn angle 177.5(2)°], while the phenanthroline and the fumarate double

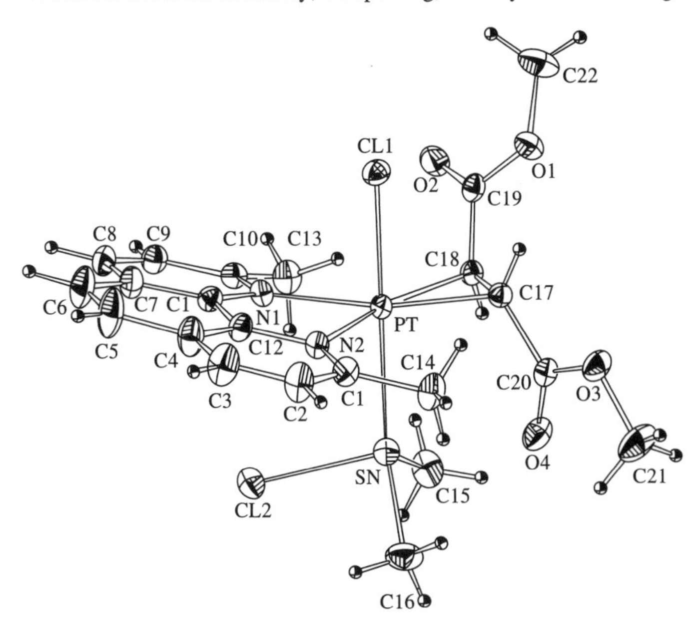

FIGURE 12. ORTEP drawing of [PtCl(SnMe$_2$Cl)(dmphen)(E-MeO$_2$CCH=CHCO$_2$Me)] with 30% probability displacement ellipsoids for atoms and arbitrarily small radii for hydrogen atoms. Reprinted with permission from Reference 25. Copyright 1996 American Chemical Society

bond occupy the equatorial coordination sites. The phenanthroline plane is not coincident with the coordination plane [Pt, N(1), N(2)], and it is tilted by 13° towards the chloride ligand in order to optimize the contacts of the methyl groups with the axial ligands. The molecule is chiral because of the prochiral nature of the fumarate ligand and is asymmetric because the conformations of the CO$_2$Me and SnMe$_2$Cl groups do not conform to any regularity.

$$[\mathrm{Pt(dmphen)}(E\text{-}\mathrm{MeO_2CCH{=}CHCO_2Me}) + \mathrm{Me_2SnCl_2} \rightleftharpoons [\mathrm{PtCl(SnMe_2Cl)(dmphen)}(E\text{-}\mathrm{MeO_2CCH{=}CHCO_2Me})] \quad (4)$$

The use of a polynuclear dimethylamido compound [{Sn(NMes)$_2$}{Sn(μ-NMe$_2$)}$_2$], Mes = 2,4,6-Me$_3$C$_6$H$_2$, as a reagent in reactions with [RNHLi] allows the formation of imido Sn(II) anions. One example of this reaction type is the deprotonation reaction of primary amido and phosphido lithium complexes ([REHLi]; E = N, P) with the cubane [SnNBu-t]$_4$, which give heterometallic complexes containing Sn(II) imido and phosphinidene anions, e.g. [{Sn(μ-PCy)}$_2${(μ-PCy}]$_2$(Li•THF)$_4$ containing a metallacyclic [{Sn(μ-PCy)}$_2${μ-PCy}]$_2^{4-}$ moiety[26]. The reaction shown in Scheme 9 of the polynuclear dimethylamido Sn(II) reagent [{Sn(NMes)$_2$}{Sn(μ-NMe$_2$)}$_2$] with [LiN(H)ma](ma = 2-MeOC$_6$H$_4$) in a 1 : 2 ratio follows an unexpected pathway in which elimination of [Sn(NMe$_2$)$_2$], rather than facial deprotonation, gives rise to the novel heterobimetallic ladder complex [[MesNHSn(μ-Nma)]$_2$(Li•THF)$_4$] (the molecular structure is shown in Figure 13), containing the first example of a dinuclear Sn(II) imido dianion, [(MesNH)Sn(μ-Nma)]$_2^{2-}$.

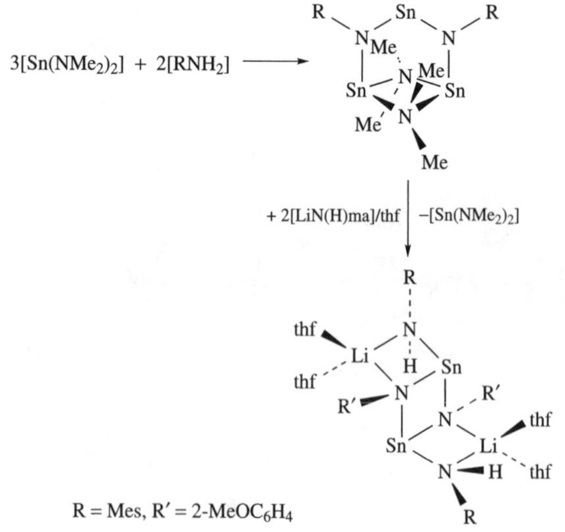

$3[Sn(NMe_2)_2] + 2[RNH_2] \longrightarrow$

$+ 2[LiN(H)ma]/thf$ $-[Sn(NMe_2)_2]$

$R = Mes, R' = 2\text{-}MeOC_6H_4$

SCHEME 9. Synthesis of $[[MesNHSn(\mu\text{-}Nma)]_2(Li{\bullet}THF)_4]$. Reprinted with permission from Reference 26. Copyright 1998 American Chemical Society

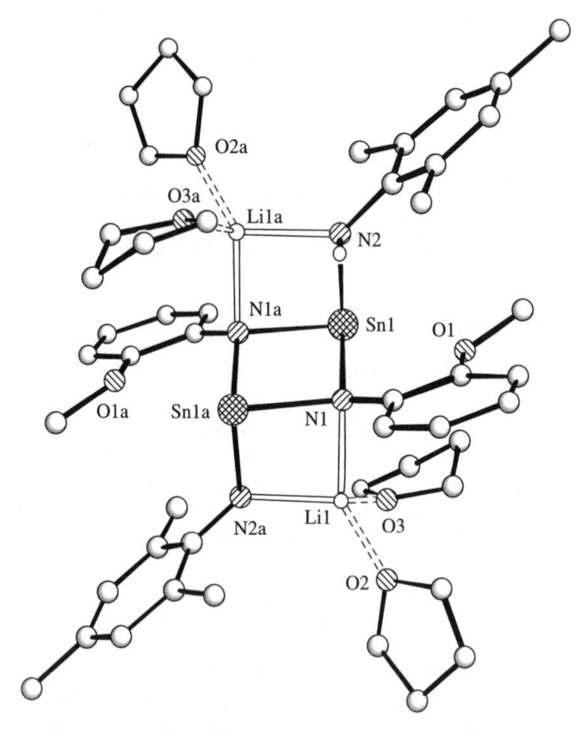

FIGURE 13. Molecular structure of $[[MesNHSn(\mu\text{-}Nma)]_2(Li{\bullet}THF)_4]$. H atoms, except those attached to N which have been omitted for clarity. Copyright 1998 American Chemical Society

Covalent fitting of Lewis acidic centres such as tin in suitable organic molecular structures results in multidentate Lewis acidic host molecules, which were shown to be efficient in coordinating anions and neutral Lewis bases[27–40]. The guest selectivity and stability of the host–guest complexes of these compounds strongly depend on the preorganization of the host molecule, i.e. the more rigid the host, the better the selectivity to be expected. Electrochemical studies show the potential ability of organotin(IV) halides to act as carriers in phosphate-selective electrodes[41–44]. Because of their potential use in electrochemical sensing, interest in the synthesis of redox-active host molecules with Lewis acidic centres is high[45].

For this reason, tin containing ferrocenophanes are being investigated by Altmann and coworkers[45]. Their synthesis is given in Scheme 10[45] and the structures of two of them are given in Figures 14 and 15.

Tin(IV) complexes are models for chemical investigations about biologically relevant ions in oxidation state four like vanadium, molybdenum and manganese. Examples are in Figure 16 and X-ray structure of one of them is given in Figure 17. Useful ligands are all diacidic tridentate chelate compounds[46,47]. Tin(IV) chelates have a good solvolytic and redox stability, so it is easy to investigate their structural behaviour. The metal chelate compounds are made via a ligand exchange reaction with bis(acetylacetonato)dichlorotin(IV) as well as via the reaction between the ligands and tin(II) chloride. In this case air oxygen oxidizes tin(II) and the tin(IV) chelate compounds are formed[48].

Compounds of the type $[SnR_2X_2(L\text{-}L)]$, where L-L is a bidentate *N*,*N*′-donor ligand, have been extensively studied with regard to antitumour activity[49]. The mechanism of this activity is still under discussion. However, studies of the structure–activity relationship show that the Sn–N distance is important, the average Sn–N bond length being $\geqslant 2.39$ Å among the active complexes and <2.39 Å among the inactive complexes. Sn–N distances which suggest possible antitumour activity are present in the eight-membered $CS_2C_2N_2Sn$ rings of $[SnEt_2Br_2(bmimt)]$ (Figure 18) and $[SnBu_2Cl_2(bmimt)]$ (Figure 19), with bmimt = bis(1-methyl-2-imidazolylthio)methane (Figure 20)[50].

Hydrolysis reactions of organotin(IV) halides have been established as viable sources of oligomeric tin clusters[51]. Given that tin–oxygen oligomers of predetermined size may have use in industrial applications[52], the rational synthesis of such species remains an important synthetic target. Intramolecular donors may allow specific control over the geometry of tin sites from which additional linkages can form during hydrolysis. For this reason the polyfunctional ligand system L^5H, bis(2-methoxy-3-*t*-butyl-5-methylphenyl)methane and tin complexes with this ligand were synthesized (equations 5–8, Figures 21–24)[53].

$$L^5H + n\text{-BuLi} \xrightarrow{\text{THF}/-20\,^\circ\text{C}} L^5Li \xrightarrow{\text{THF}/Ph_3SnCl} L^5\text{-}SnPh_3 \qquad (5)$$

$$L^5SnPh_3 + 2HgCl_2 \xrightarrow{\text{acetone}/0\,^\circ\text{C}} PhCl_2SnL^5 + 2PhHgCl \qquad (6)$$

$$L^5SnPh_3 + 2X_2 \xrightarrow{CH_2Cl_2 \text{ or MeOH}/25 \text{ or } 0\,^\circ\text{C}} PhX_2SnL^5 + 2PhX \qquad (7)$$

$$X = Br, I$$

$$PhBr_2SnL^5 \xrightarrow{2PhSNa/MeOH} Ph(SPh)_2SnL^5 + 2NaBr \qquad (8)$$

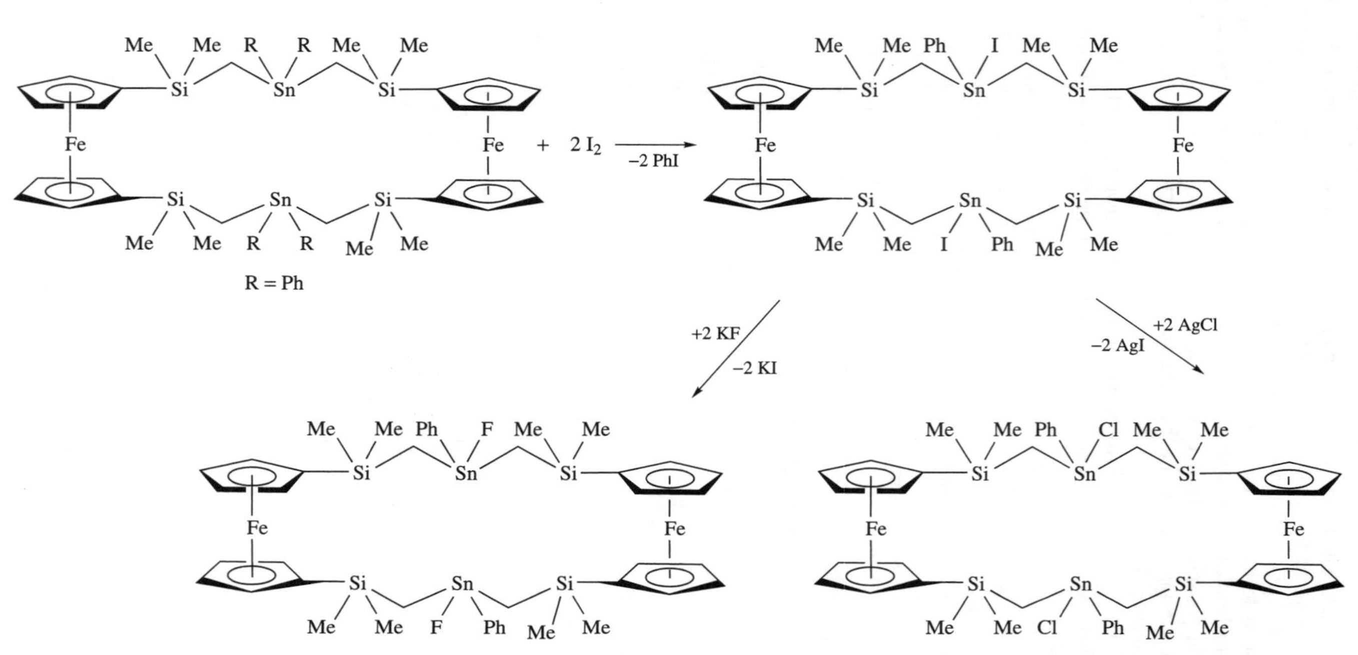

SCHEME 10. Synthesis of different tin containing ferrocenophanes. Reprinted with permission from Reference 45. Copyright 2000 American Chemical Society

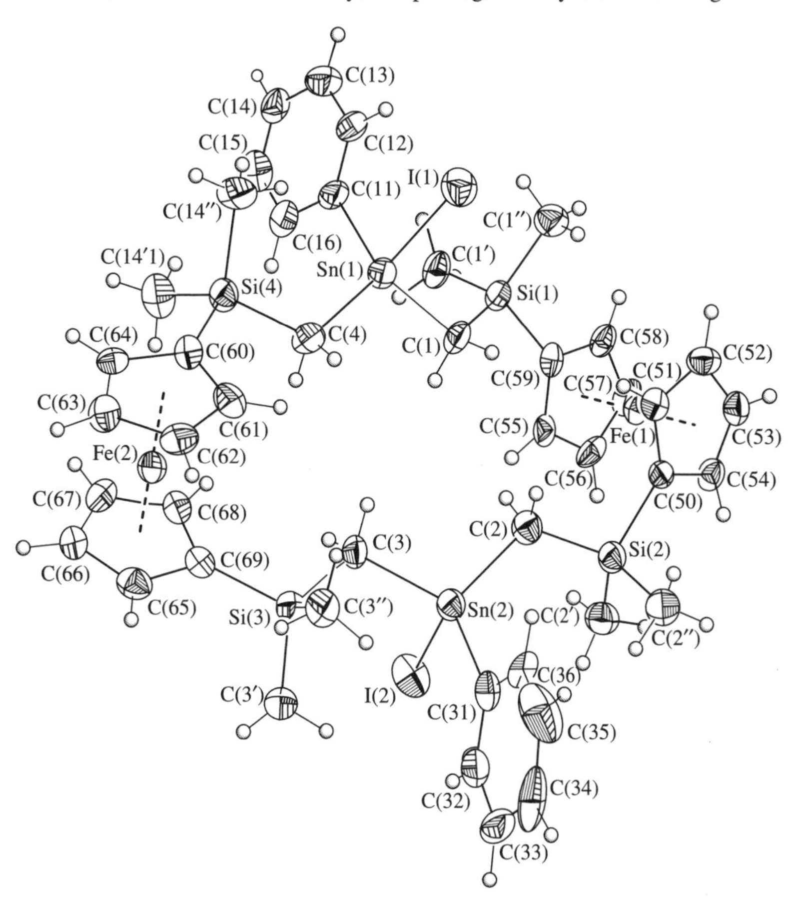

FIGURE 14. Molecular structure of [fc($SiMe_2CH_2SnPhICH_2SiMe_2$)$_2$fc]. Reprinted with permission from Reference 45. Copyright 2000 American Chemical Society

The preparation of materials with high chemical reactivity, especially η^6-arene lability, is an interesting field of research. The lability of the arene ring facilitates its displacement by other ligands, which could be an initiative step in some catalytic reactions. For this reason, the synthesis of additional stannyl complexes like (η^6-arene)Cr(CO)$_2$($HSnR_3$) and (η^6-arene)Cr(CO)$_2$(SnR_3)$_2$ were investigated[54]. Figures 25 and 26 show the molecular structures of two stannyl complexes, where the arene is 1,4-$C_6H_4(OCH_3)_2$ and R = Ph. Both compounds are prepared in a reaction between (η^6-1,4-$C_6H_4(OCH_3)_2Cr(CO)_3$) and $HSnPh_3$ in a hexane/toluene mixture. The separation of the two products was done on a silica gel column with hexane/toluene as the eluting solvent. The bis-stannyl compound was eluted first.

The chemotherapeutic properties, especially the antitumour activities, of diorganotin compounds continue to be the focus of many reports[55,56]. An interesting compound is

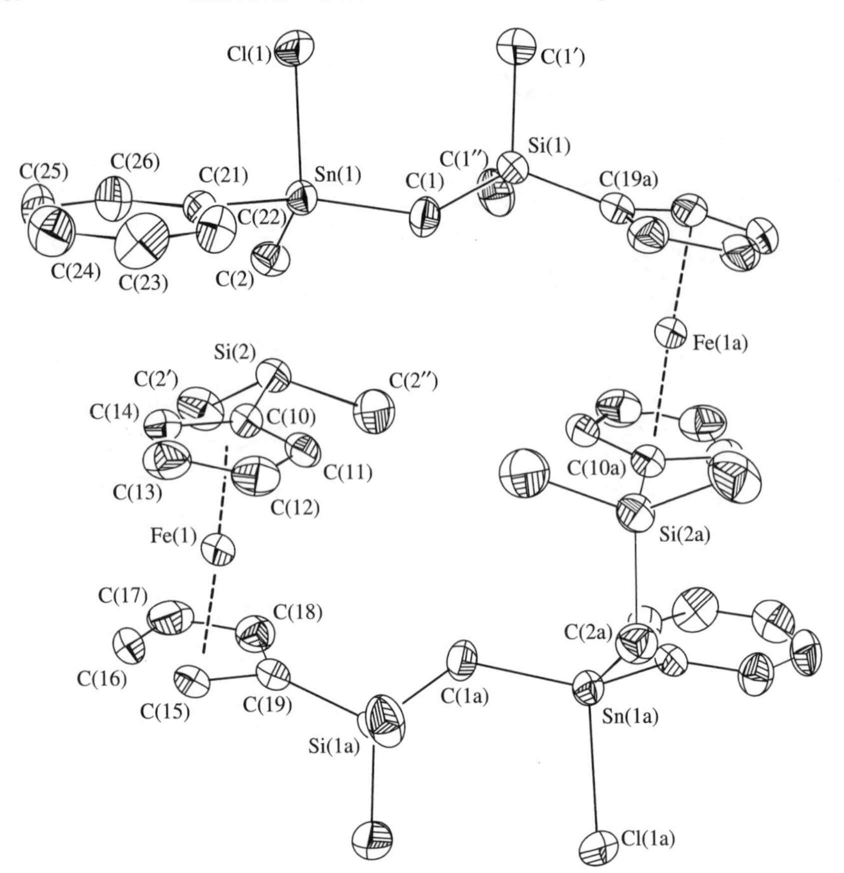

FIGURE 15. Molecular structure of [fc($SiMe_2CH_2SnPhClCH_2SiMe_2$)$_2$fc]. Reprinted with permission from Reference 45. Copyright 2000 American Chemical Society

[$Ph_2Sn(2\text{-}OC_{10}H_6CH{=}NCH_2COO)$]$SnPh_2Cl_2$[57], which is prepared in a reaction between diphenyltin dichloride and $Ph_2Sn(2\text{-}OC_{10}H_6CH{=}NCH_2COO)$ in refluxing benzene. The crystal structure of the product shows a monomeric 1 : 1 donor–acceptor dinuclear tin complex. Each of the two tin atoms, Sn(1) and Sn(2), has a five-coordination geometry in a distorted trigonal bipyramidal arrangement (Figure 27).

Complexes that exhibit terminal multiple bonds to the heavier chalcogens are subjects of considerable attention. Terminal chalcogenido complexes (η^4-Me_8taa)SnE (E = S, Se; Me_8taa = octamethyldibenzotetraaza[14]annulene dianion) react with MeI to give the corresponding methylchalcogenolate derivatives, [(η^4-Me_8taa)Sn(EMe)]I[58]. The molecular structure of the methylseleno derivative is shown in Figure 28.

An interesting part of chemistry comprises the divalent complexes of the Group 14 elements, germanium, tin and lead, supported by tetradentate nitrogen and oxygen donor ligands. The [Salen$^{R,R'}$] ligand system, obtained by condensation of a salicylaldehyde

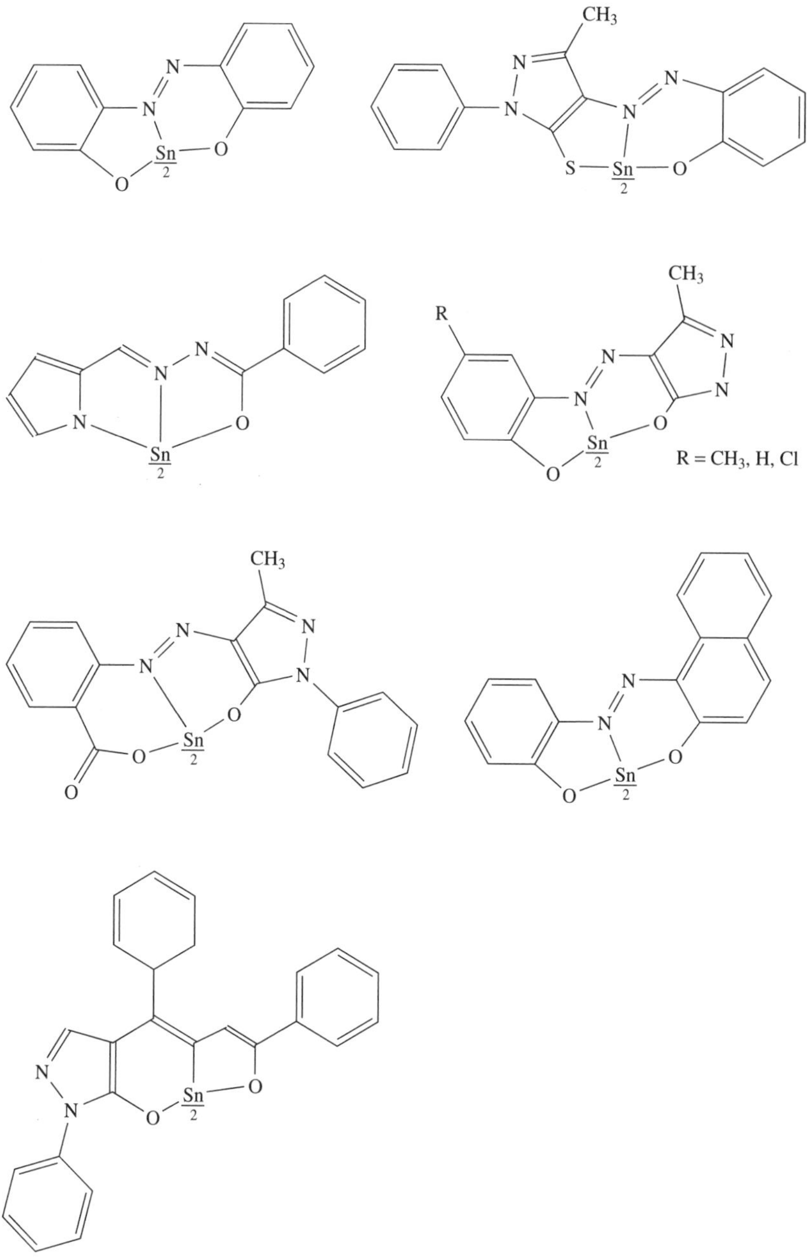

R = CH_3, H, Cl

FIGURE 16. Some tin(IV) chelate complexes, which are used as models for biologically relevant ions. Reproduced by permission of Verlag der Zeitschrift für Naturforschung from Reference 46

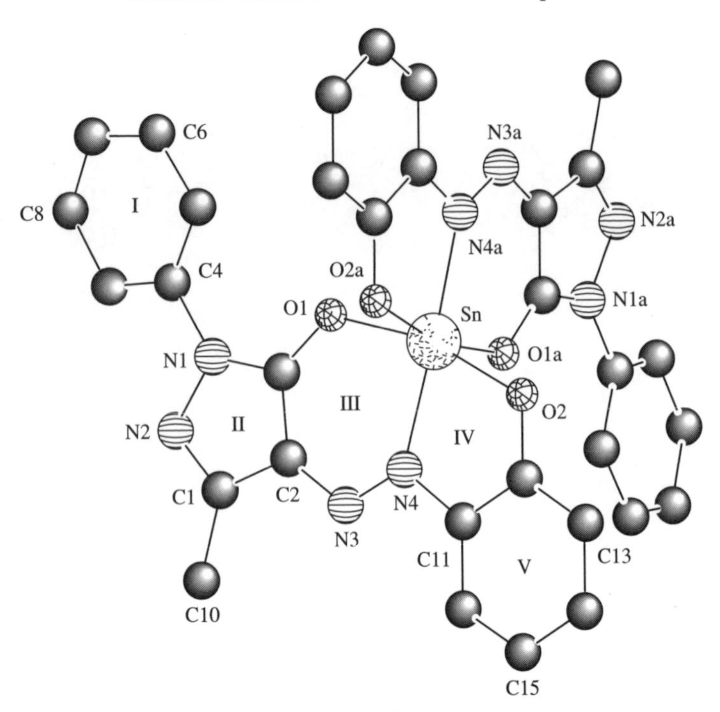

FIGURE 17. Molecular structure of bis[4-(2′-hydroxyphenylazo)-3-methyl-1-phenylpyrazol-5-onato(2-)]tin(IV). Reproduced by permission of Verlag der Zeitschrift für Naturforschung from Reference 46

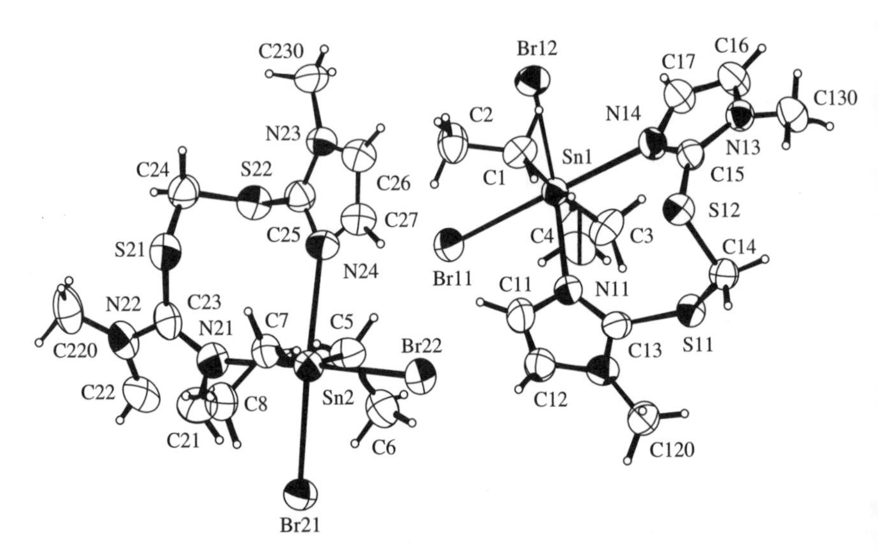

FIGURE 18. Molecular structure of [$SnEt_2Br_2$(bmimt)]. Reproduced by permission of John Wiley & Sons, Inc. from Reference 50

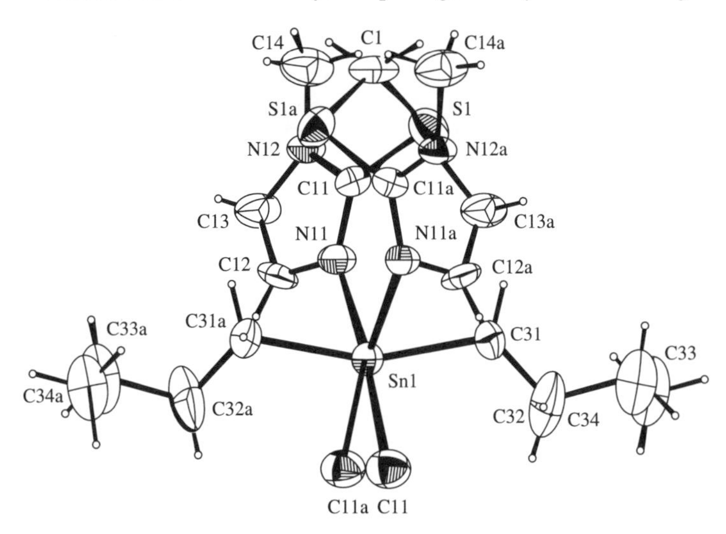

FIGURE 19. Molecular structure of [$SnBu_2Cl_2$(bmimt)]. Reproduced by permission of John Wiley & Sons, Inc. from Reference 50

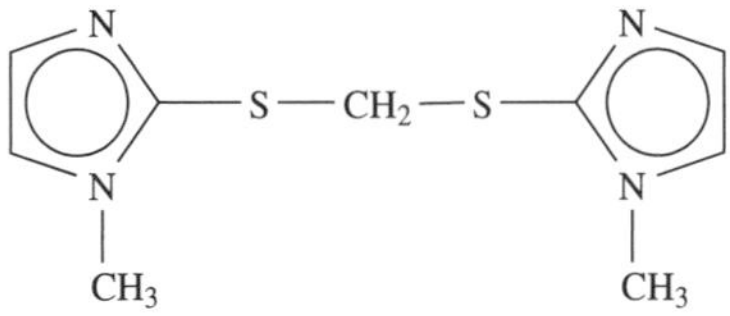

FIGURE 20. bmimt = bis(1-methyl-2-imidazolylthio)methane. Reproduced by permission of John Wiley & Sons, Inc. from Reference 50

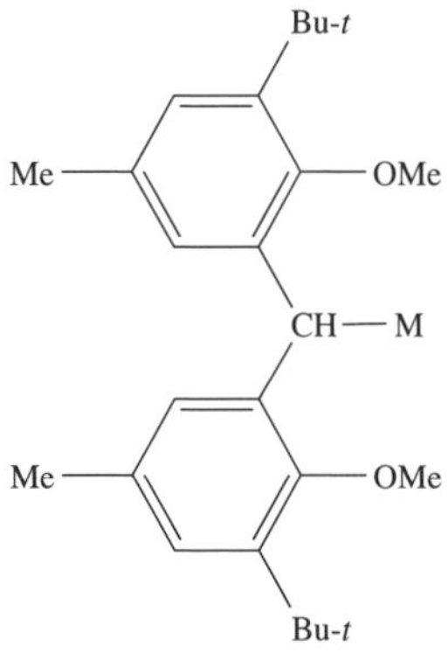

M = $SnPh_3$, $SnPhCl_2$, $SnPhI_2$,
$SnPhBr_2$, $SnPh(SPh)_2$

FIGURE 21. Tin complexes with the L^5H ligand. Reprinted with permission from Reference 53. Copyright 1997 American Chemical Society

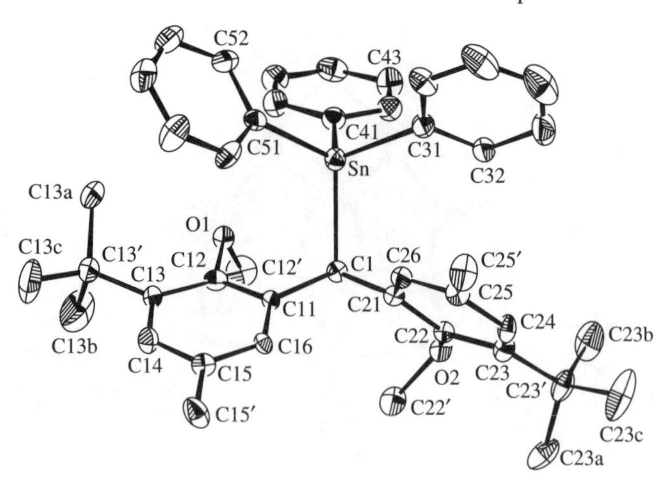

FIGURE 22. Molecular structure of L^5-$SnPh_3$. Reprinted with permission from Reference 53. Copyright 1997 American Chemical Society

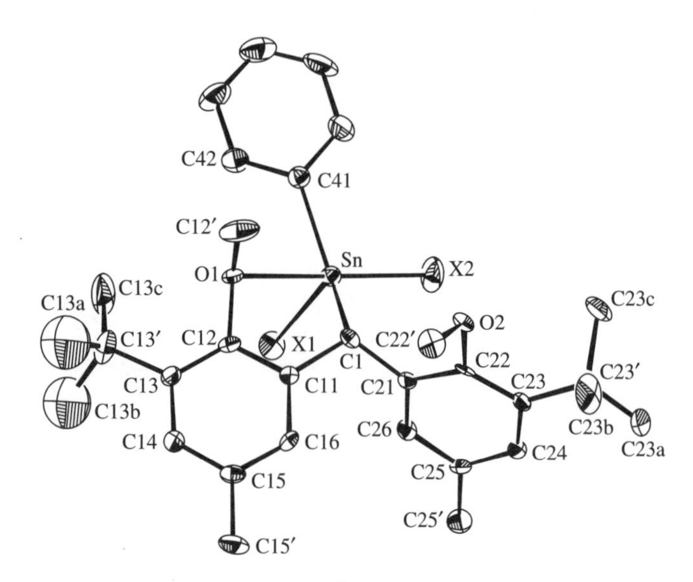

FIGURE 23. Molecular structure of PhX_2SnL^5, X = Cl, Br, I. Reprinted with permission from Reference 53. Copyright 1997 American Chemical Society

derivative with 1,2-ethylenediamine, has proved to be very useful in coordination chemistry[59]. The divalent tin complex, $[Salen^{t-Bu,Me}]Sn$ is readily prepared by the reaction of $[(Me_3Si)_2N]_2Sn$ with $[Salen^{t-Bu,Me}]H_2$ (Scheme 11). It could also be prepared by the reaction of $SnCl_2$ with $[Salen^{t-Bu,Me}]H_2$ in the presence of Et_3N[60]. A notable feature of the structures of $[Salen^{t-Bu,Me}]Sn$ is that the $[Salen^{t-Bu,Me}]$ ligand is not flat

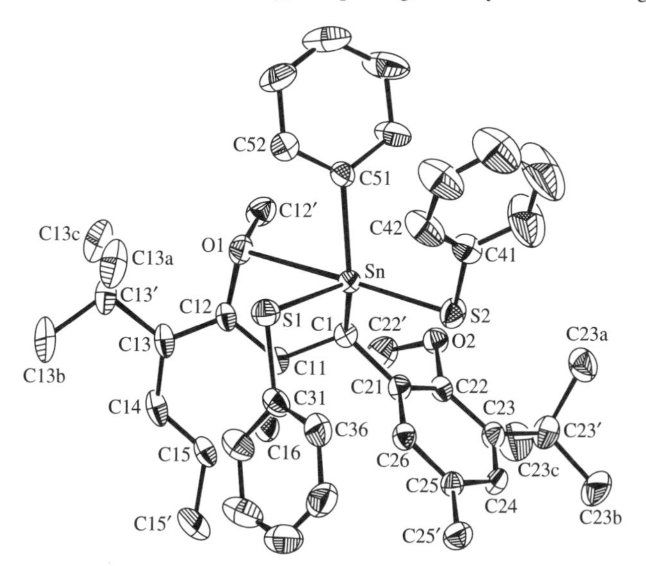

FIGURE 24. Molecular structure of $Ph(SPh)_2SnL^5$. Reprinted with permission from Reference 53. Copyright 1997 American Chemical Society

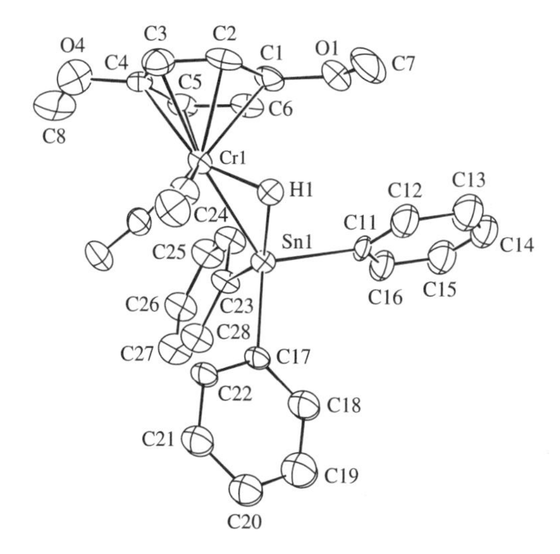

FIGURE 25. Molecular structure of $(\eta^6\text{-}1,4\text{-}C_6H_4(OCH_3)_2Cr(CO)_2)(HSnPh_3)$. Reproduced by permission of Elsevier Science from Reference 54

but adopts a slightly twisted conformation, such that the four coordinating atoms of the $[N_2O_2]$ core are not coplanar but deviate significantly from their mean plane (Figure 29).

Tin(II) compounds are known to have both acid and base properties; the metal centre can either react with electrophiles or act as a Lewis acid and thus be susceptible to

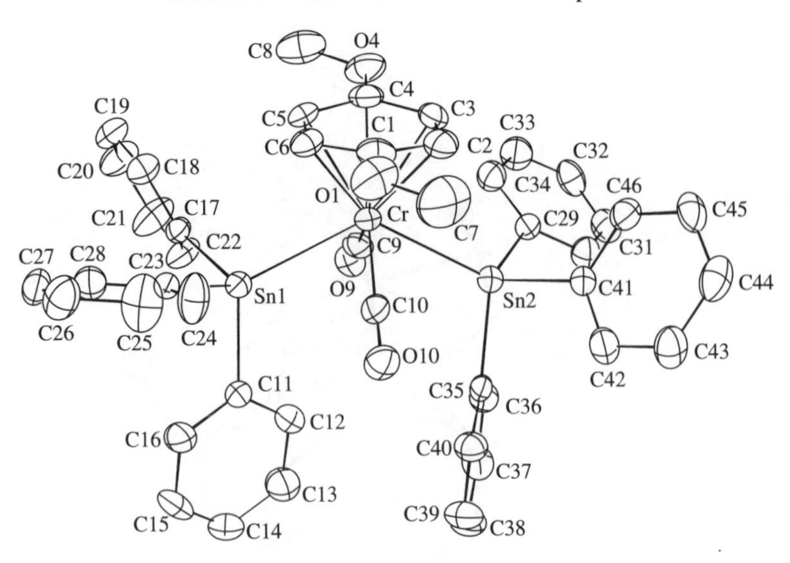

FIGURE 26. Molecular structure of (η^6-1,4-$C_6H_4(OCH_3)_2Cr(CO)_2)(SnPh_3)$. Reproduced by permission of Elsevier Science from Reference 54

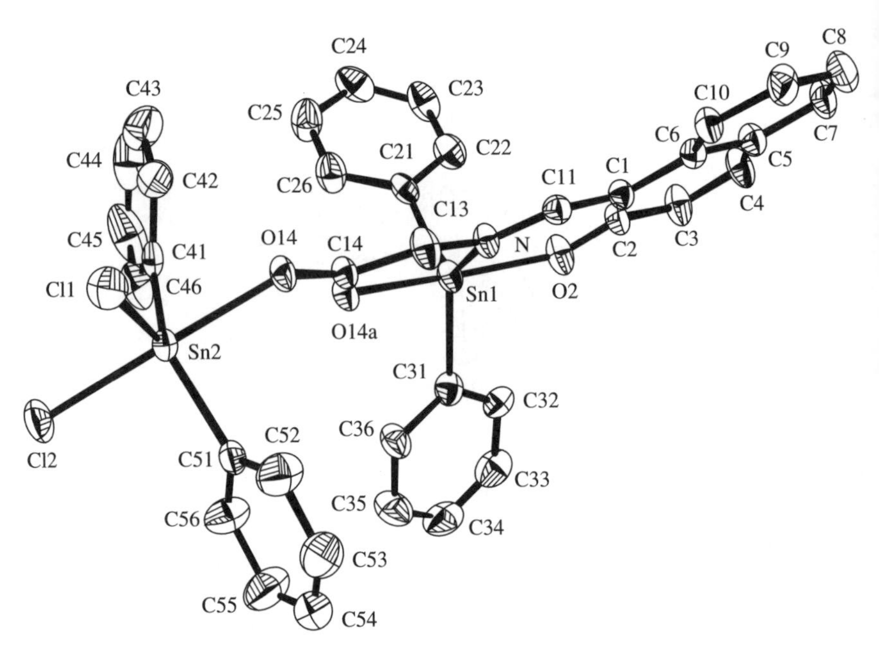

FIGURE 27. Molecular structure of $[Ph_2Sn(2\text{-}OC_{10}H_6CH{=}NCH_2COO)]SnPh_2Cl_2$. Reproduced by permission of Elsevier Science from Reference 57

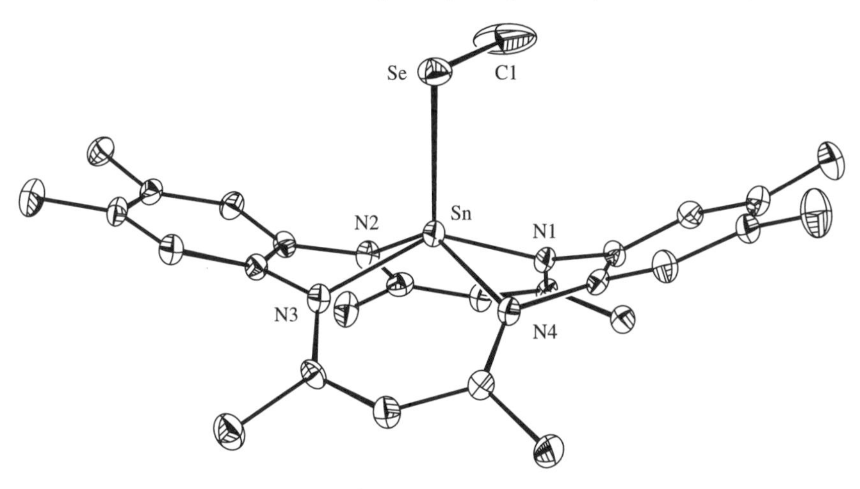

FIGURE 28. Molecular structure of $[(\eta^4\text{-Me}_8\text{taa})\text{Sn(SeMe)}]^+$. Reproduced by permission of the Royal Society of Chemistry from Reference 58

nucleophilic attack. It has been shown that nucleophilic addition of organolithium reagents to organotin(II) compounds formed triorganostannate compounds. Alkylation using 1 equivalent of $[\text{LiR}^{\text{N}}(\text{TMEDA})]$ with $SnCl_2$ afforded the binuclear chlorotin(II) alkyl $[\text{R}^{\text{N}}\text{SnCl}]_2$ in 75% yield. In similar reactions using 3 equivalents of $[\text{LiR}^{\text{N}}(\text{TMEDA})]$ $[\{(\text{SnR}^{\text{N}}_3)\text{Li}\}(\mu^3\text{-Cl})\{\text{Li(TMEDA)}_2\}(\mu^2\text{-Cl})]$ (for structure, see Figure 30) and with 3.5 equivalents of $[\text{LiR}^{\text{N}}(\text{TMEDA})]$, $[\{(\text{SnR}^{\text{N}}_3)\text{Li}\}(\mu^3\text{-Cl})\text{Li(TMEDA)}]_2$ (for structure, see Figure 31) are formed[61] $[\text{R}^{\text{N}} = \text{CH(Si(}t\text{-Bu)Me}_2)\text{C}_5\text{H}_4\text{N-2}$, TMEDA = N,N,N',N'-tetramethylethylene diamine]. $[\{(\text{SnR}^{\text{N}}_3)\text{Li}\}(\mu^3\text{-Cl})\{\text{Li(TMEDA)}_2\}(\mu^2\text{-Cl})]$ is a binuclear molecule with $(\text{R}^{\text{N}})^-$ acting as a bidentate C,N-bridging mode between two tin(II) atoms forming an eight-membered ring in a 'boat' conformation.

Triphenyltin chloride forms a dimeric hydrated complex with 1,10-phenanthroline in which the coordinated water molecule is linked by hydrogen bonds ($\text{O}\cdots\text{N} = 2.96$ and 3.02 Å) to two 1,10-phenanthroline bases[62]. Addition of four methyl substituents to the 1,10-phenanthroline ligand increases the basicity of its N atoms, enhancing the propensity for hydrogen bonding with water of aquachlorotriphenyltin molecules. In the complex shown in Figure 32 (1 : 1 adduct of aquachlorotriphenyltin with 3,4,7,8-tetramethyl-1,10-phenanthroline)[63] the ligand forms much shorter hydrogen bonds (2.661 and 2.767 Å) with aquachlorotriphenyltin than does 1,10-phenanthroline. In the adduct, $[\text{SnCl(C}_6\text{H}_5)_3(\text{H}_2\text{O})]\bullet\text{C}_{16}\text{H}_{16}\text{N}_2$, the aquaorganotin moiety is linked by hydrogen bonds through its axial bonded water molecule to the substituted 1,10-phenanthroline moiety. The Sn atom has *trans*-trigonal bipyramidal coordination, with aqua and chloro ligands in the axial positions.

The formation of $\text{SnR}_2[(\text{OPPh}_2)(\text{SPPh}_2)\text{N}]_2$, with R = Me, Ph, is done by metathesis reactions between SnR_2Cl_2 and $\text{K}[(\text{OPPh}_2)(\text{SPPh}_2)\text{N}]$ in toluene[64]. The molecular structures of both compounds are shown in Figures 33 and 34. The central tin atom occupies a centre of inversion. The SnC_2 unit is linear (C−Sn−C 180°). The two asymmetric ligand moieties are monometallic biconnective, thus resulting in a distorted-octahedral

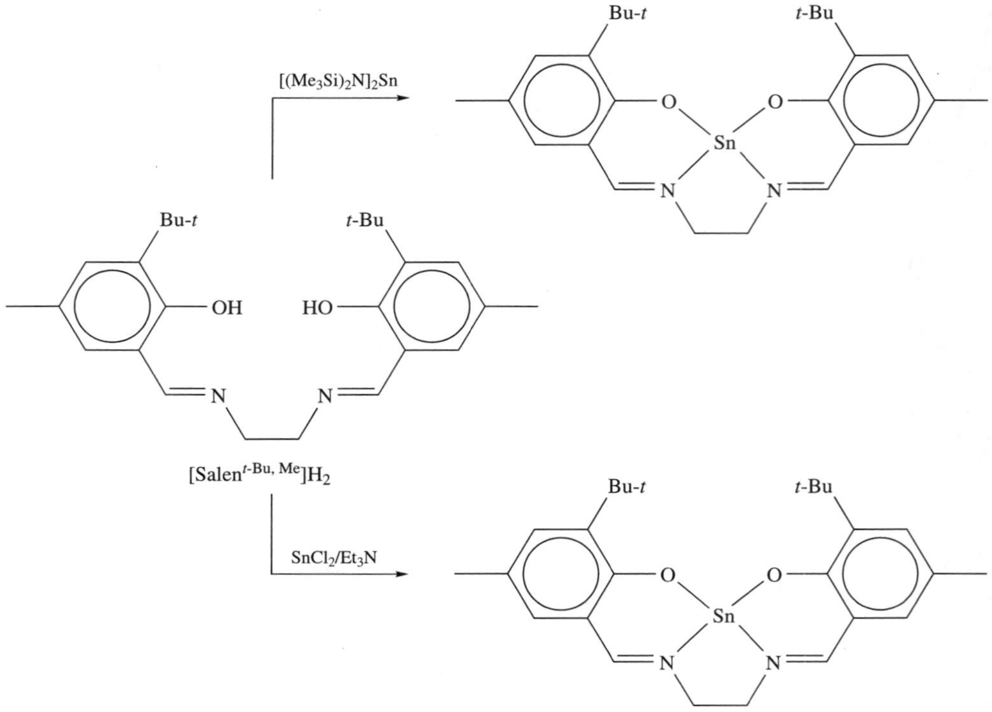

$[(Me_3Si)_2N]_2Sn$

$SnCl_2/Et_3N$

$[Salen^{t\text{-}Bu,\,Me}]H_2$

SCHEME 11. Synthesis route for $[Salen^{t-Bu,Me}]Sn$. Reproduced by permission of the Royal Society of Chemistry from Reference 59

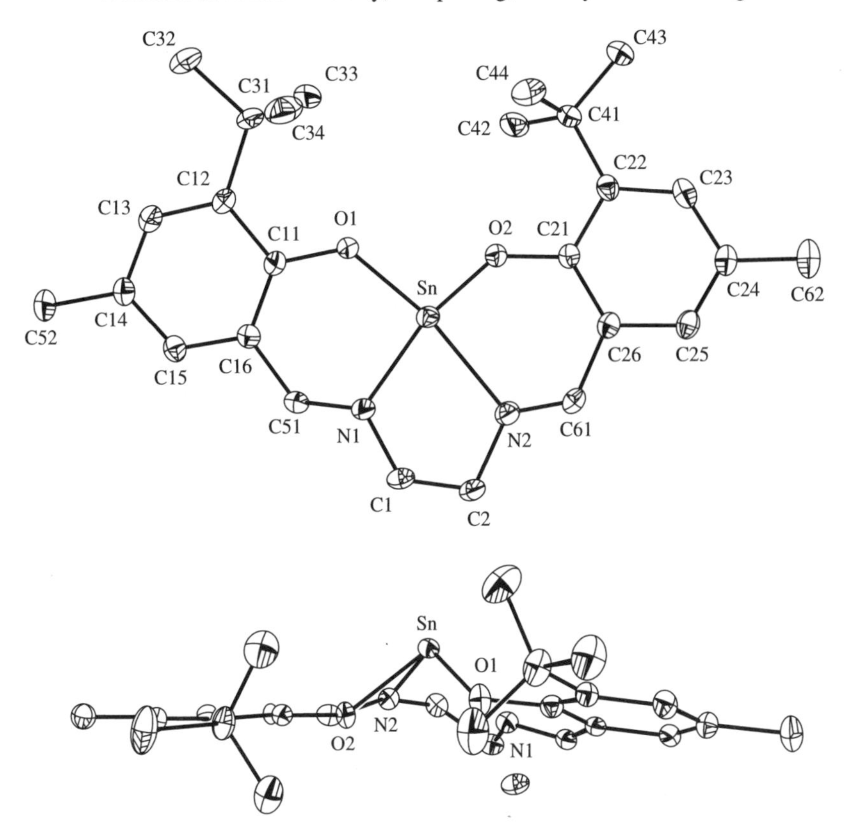

FIGURE 29. Two views of the molecular structure of [Salen$^{t-\text{Bu,Me}}$]Sn. Reproduced by permission of the Royal Society of Chemistry from Reference 59

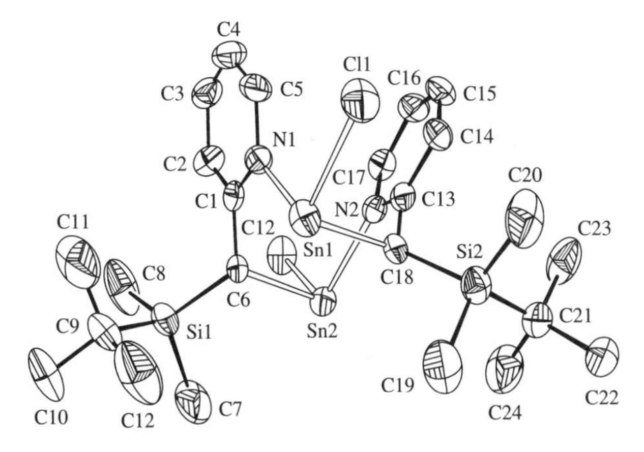

FIGURE 30. Molecular structure of [{(SnR$_3^N$)Li}(μ^3-Cl){Li(TMEDA)$_2$}(μ^2-Cl)]. Reprinted with permission from Reference 61. Copyright 1999 American Chemical Society

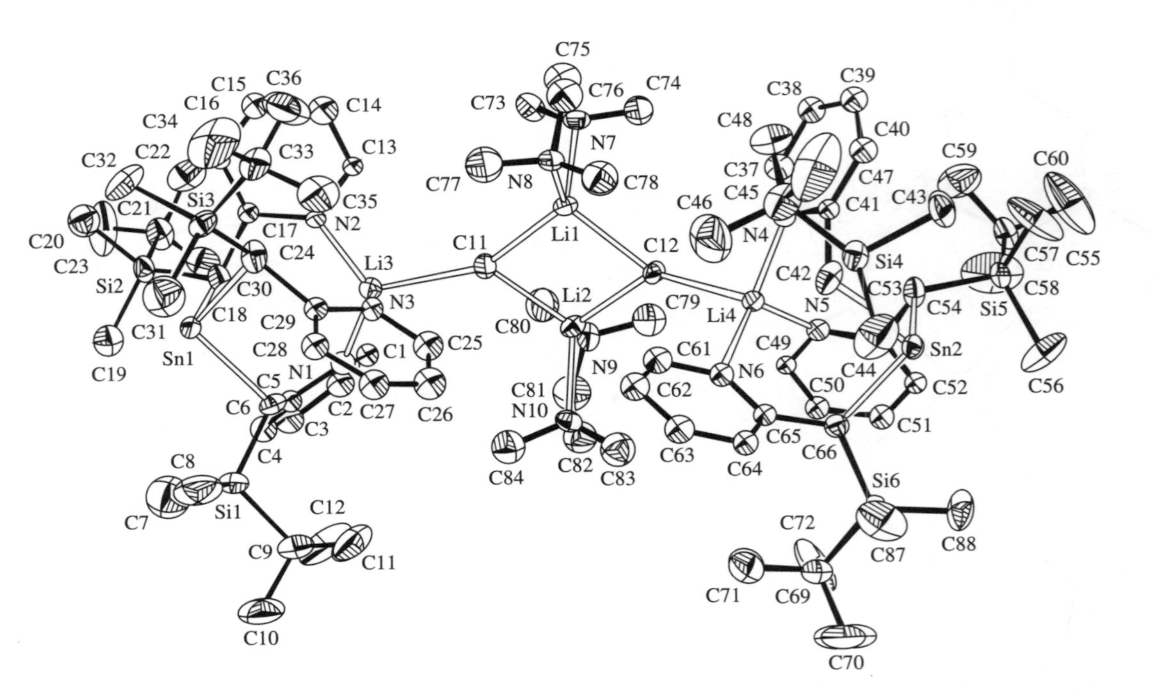

FIGURE 31. Molecular structure of $[\{(SnR_3^N)Li\}(\mu^3\text{-}Cl)Li(TMEDA)]_2$. Reprinted with permission from Reference 61. Copyright 1999 American Chemical Society

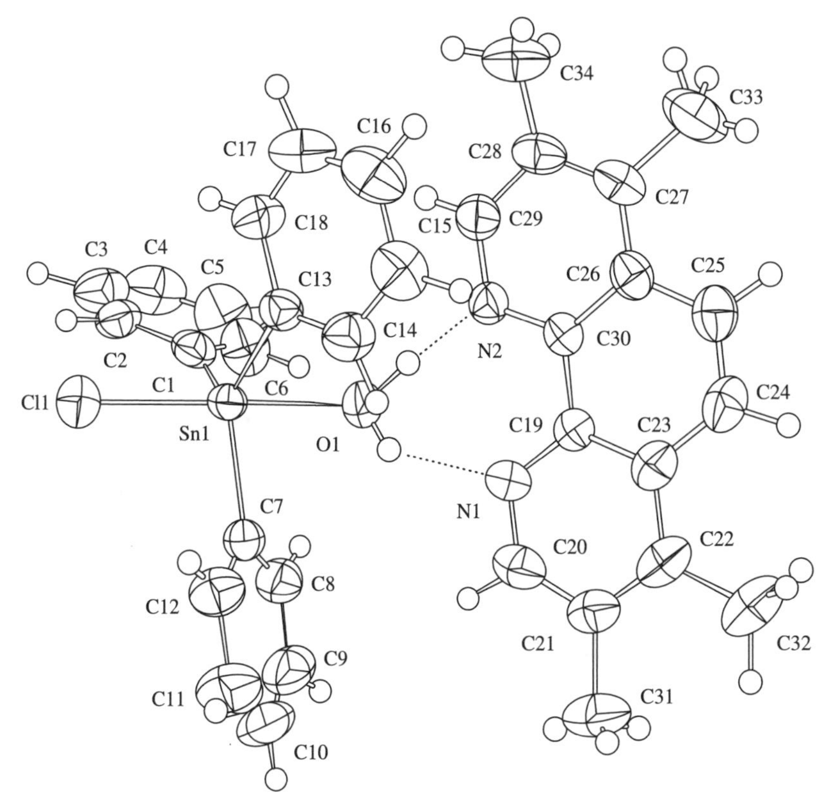

FIGURE 32. Molecular structure of $[SnCl(C_6H_5)_3(H_2O)]{\bullet}C_{16}H_{16}N_2$. Reproduced by permission of the International Union of Crystallography, Nunskgaard International Publishers from Reference 63

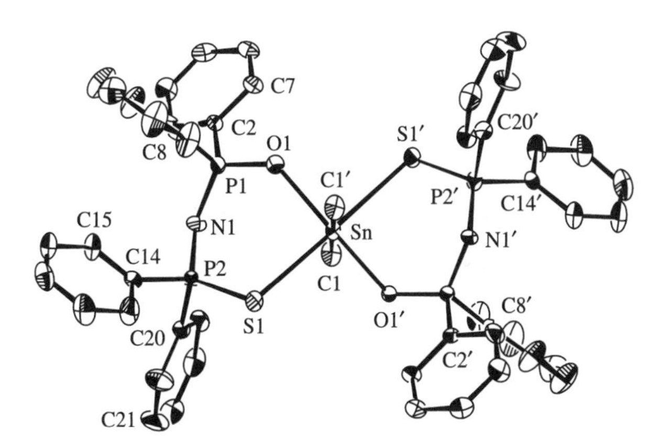

FIGURE 33. Molecular structure of $SnMe_2[(OPPh_2)(SPPh_2)N]_2$. Reproduced by permission of the Royal Society of Chemistry from Reference 64

coordination around tin, with the carbon atoms of the organic groups in axial positions. The equatorial SnO_2S_2 system is planar, with the *trans* positions occupied by pairs of the same donor chalcogen atoms (O−Sn−O and S−Sn−S angles 180°). The bidentate nature of the monothio ligand units leads to an inorganic bicyclic system, $NP_2SOSnOSP_2N$, with the metal as spiro atom. Although some delocalization of the π electrons over the OPNPS systems is suggested by the magnitude of the bonds, the $SnOSP_2N$ rings are not planar but exhibit a twisted-boat conformation (Figure 35).

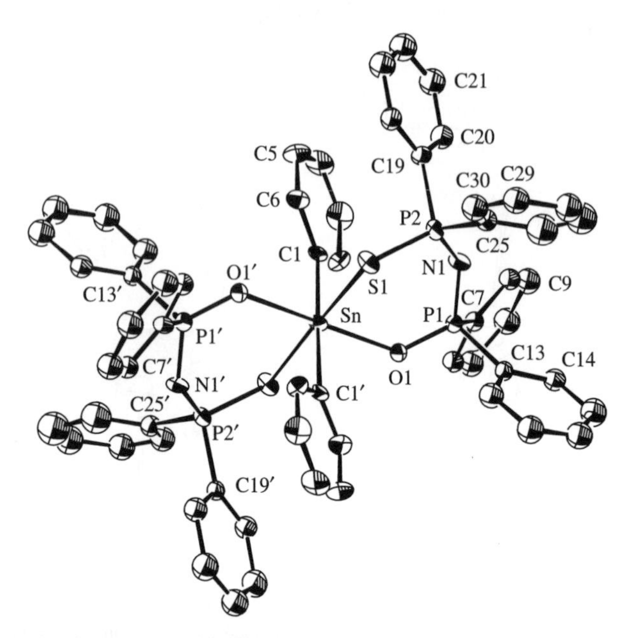

FIGURE 34. Molecular structure of $SnPh_2[(OPPh_2)(SPPh_2)N]_2$. Reproduced by permission of the Royal Society of Chemistry from Reference 64

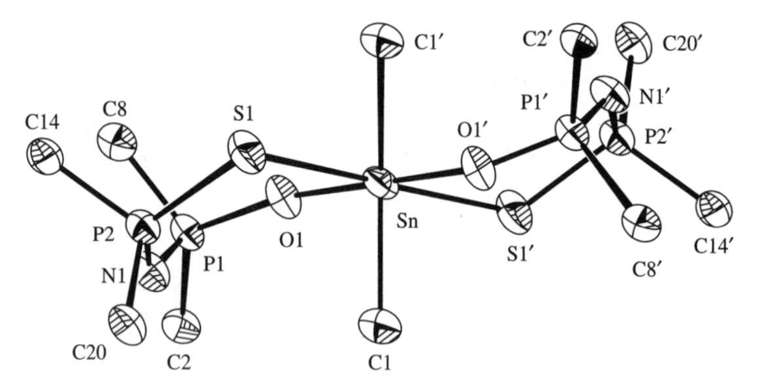

FIGURE 35. Conformation of the inorganic $SnOSP_2N$ chelate rings in $SnMe_2[(OPPh_2)(SPPh_2)N]_2$ (only *ipso*-carbon atoms of the phenyl groups are shown for clarity). Reproduced by permission of the Royal Society of Chemistry from Reference 64

Fluorine-doped tin oxide thin films deposited by chemical vapour deposition (CVD) techniques are used as transparent conductors in various applications[65–69]. For this reason the goal of this research field is to synthesize thin-film precursors that have the potential to deposite tin oxide or fluorine-doped tin oxide. The ligand hexafluoroisopropoxide, OR_f with $R_f = CH(CF_3)_2$, is used because alkoxide complexes are known to be viable oxide film precursors[70–72] and metal hexafluoroisopropoxide complexes are reported to decompose to metal fluorides under certain conditions[73]. $Sn(OR_f)_2$ and the amine adducts $Sn(OR_f)_2L$, $L = HNMe_2$ or pyridine, are prepared in high yield from bis(amido)tin(II) compounds (Scheme 12)[74]. $Sn(OR_f)_2$ is proposed to be a dimer with bridged alkoxide ligands (Figure 36). The crystal structure of $Sn(OR_f)_2(HNMe_2)$ (Figure 37) shows it to have a trigonal pyramidal geometry. The compounds are volatile solids, an important attribute if they are used as conventional CVD tin oxide precursors.

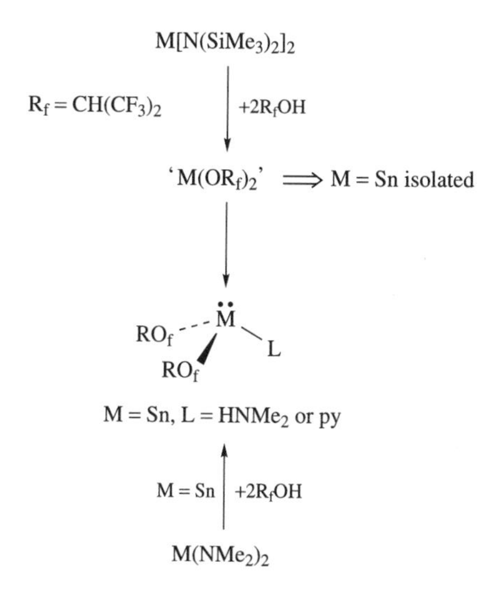

SCHEME 12. Synthesis route of $Sn(OR_f)_2$ and $Sn(OR_f)_2L$, $L = HNMe_2$ or pyridine, $R_f = CH(CF_3)_2$. Reprinted with permission from Reference 74. Copyright 1996 American Chemical Society

A useful ligand in tin chemistry is thiophene-2-carboxaldehyde thiosemicarbazone, 2-C_4H_3S-CH=N-NH-C(S)NH_2 (tctscH). It is formed during a reaction of thiophene-2-carboxaldehyde with thiosemicarbazide. The complex $SnPh_2Cl$(tctsc) is obtained in a reaction of $SnPh_2Cl_2$ and this ligand in a ratio 1 : 1 and the compound $SnCl_2(tctsc)_2$ is the product of $SnPhCl_3$ and the ligand in a ratio 1 : 2[75]. In both complexes, the tctsc ligand functions as a bidentate anion, coordinates to the central Sn atom through the thiol-S

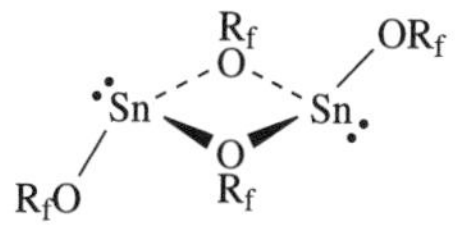

FIGURE 36. Proposed structure of '$Sn(OR_f)_2$'. Reprinted with permission from Reference 74. Copyright 1996 American Chemical Society

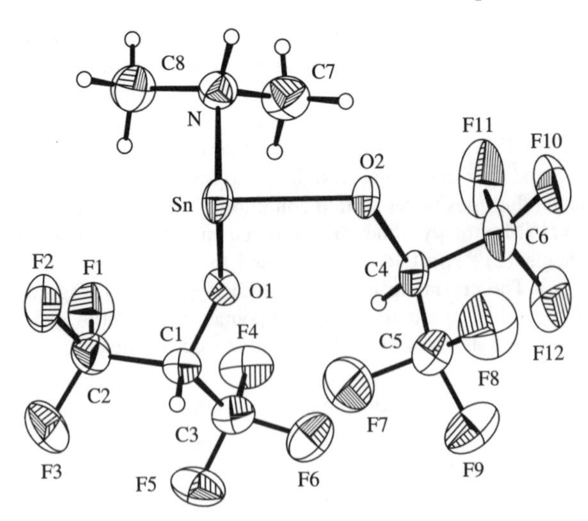

FIGURE 37. Molecular structure of $Sn(OR_f)_2(HNMe_2)$, $R_f = CH(CF_3)_2$. Reprinted with permission from Reference 74. Copyright 1996 American Chemical Society

atom and the azomethine-N atom, yielding a five-membered chelate ring after the enolization and deprotonation of the thiol proton. The occurrence of the enolization process is supported by the shortening of the C(1)—N(2) bond (increase in bond order). During the formation of these complexes, a conformational change of the ligand from *trans* to *cis* configuration [refer to S(1) and N(3) atoms] occurs so as to enable it to coordinate in a bidentate manner[76,77]. $SnPh_2Cl(tctsc)$ (Figure 38) exists in a distorted trigonal bipyramidal geometry about the tin atom, where the Cl and azomethine-N atoms, which are most electronegative, occupy the axial position. In the formation of the complex $SnCl_2(tctsc)_2$ (Figure 39), dephenylation has taken place, where the phenyl group is released from the coordination to the tin atom, so that an idealized geometry in the product is achieved. Complex $SnCl_2(tctsc)_2$ exists in a distorted octahedral geometry.

2-Alkoxycarbonylpropyltin trichlorides, $ROCOCH(CH_3)CH_2SnCl_3$, have attracted considerable attention[78] ever since their syntheses were first reported because of the variety of coordination geometries about the tin atom and also due to their applicability as PVC stabilizers with low mammalian toxicities[79–86]. The $ROCOCH(CH_3)CH_2$ moiety acts as a C, O chelating ligand in the solid state and non-coordinating solvents[80,81], but the intramolecular coordination by the carbonyl oxygen of the ester group can be broken by additional donor molecules[81–84]. In this research field some complexes of $ROCOCH(CH_3)CH_2SnCl_3$ with hexamethylphosphoramide (HMPA) and *N*-(2-hydroxyphenyl)-2-hydroxy-1-naphthylaldimine (H_2L^6) were prepared (equations 9–12). The structures of the complexes are displayed in Figure 40, and the X-ray diffraction of **1a** is shown in Figure 41. These complexes are air-stable and soluble in benzene and common polar organic solvents, such as methanol, chloroform, acetone and nitrobenzene, but insoluble in saturated hydrocarbons such as hexane and petroleum ether.

$$ROCOCH(CH_3)CH_2SnCl_3 + HMPA \longrightarrow \underset{\textbf{(1a)–(1c)}}{ROCOCH(CH_3)CH_2SnCl_3{\bullet}HMPA} \quad (9)$$

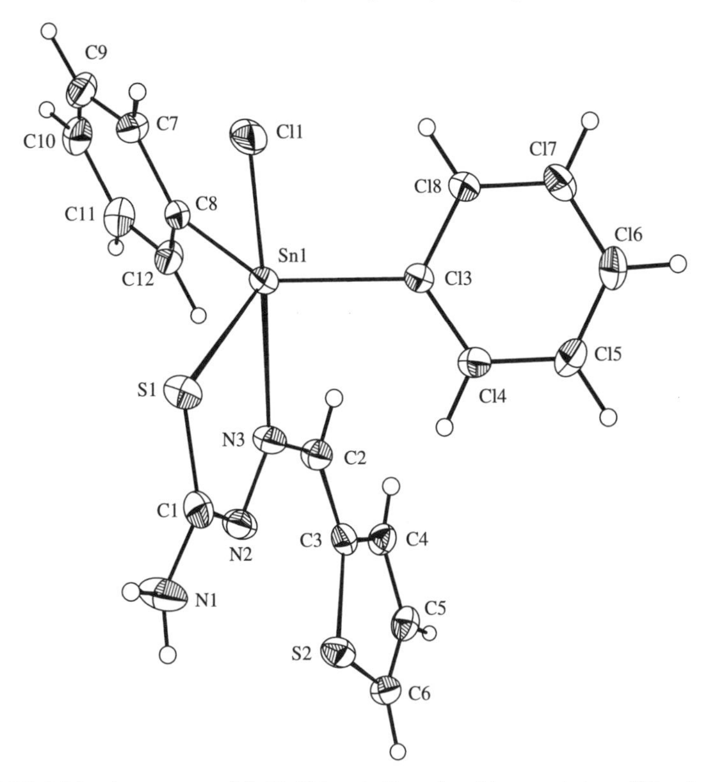

FIGURE 38. Molecular structure of $SnPh_2Cl(tctsc)$. Reproduced by permission of Elsevier Science from Reference 75

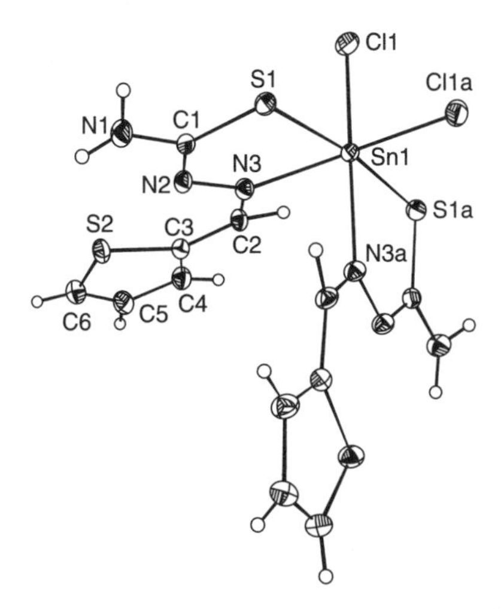

FIGURE 39. Molecular structure of $SnCl_2(tctsc)_2$. Reproduced by permission of Elsevier Science from Reference 75

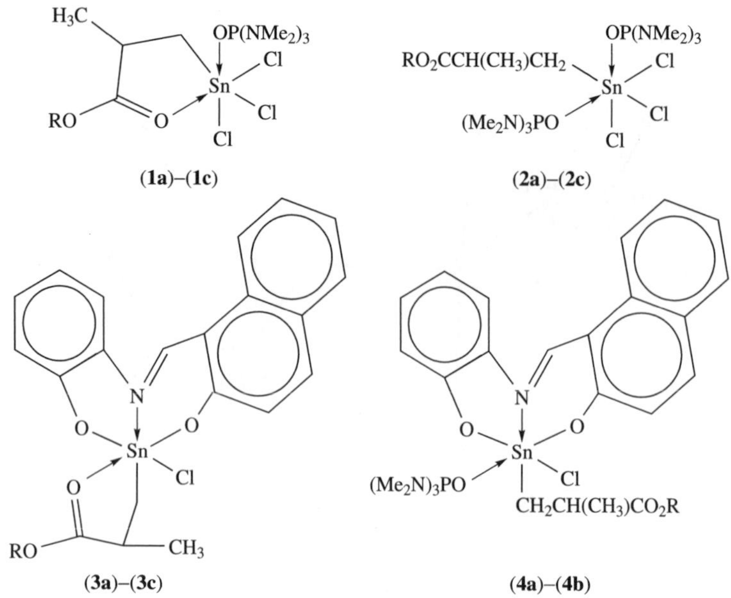

(1a)–(1c) (2a)–(2c)

(3a)–(3c) (4a)–(4b)

FIGURE 40. Suggested structures of the complexes. From Reference 78, copyright Marcel Dekker

$$\mathbf{1a{-}1c} + \mathrm{HMPA} \longrightarrow \underset{\mathbf{(2a){-}(2c)}}{\mathrm{ROCOCH(CH_3)CH_2SnCl_3{\bullet}2HMPA}} \qquad (10)$$

$$\mathrm{ROCOCH(CH_3)CH_2SnCl_3 + H_2L^6 + Et_3N} \longrightarrow \mathrm{ROCOCH(CH_3)CH_2SnClL^6} + \underset{\mathbf{(3a){-}(3c)}}{\mathrm{2Et_3N{\bullet}HCl}} \qquad (11)$$

$$\mathbf{3a} \text{ or } \mathbf{3b} + \mathrm{HMPA} \longrightarrow \underset{\mathbf{(4a),\ (4b)}}{\mathrm{ROCOCH(CH_3)CH_2SnClL^6{\bullet}HMPA}} \qquad (12)$$

(a) R = Me, **(b)** R = Et, **(c)** R = $CH_2{=}CHCH_2$

Organotin complexes show a spectrum of biological effects[87]. Their chemotherapeutic properties, including antitumor activity, have been extensively investigated. The chelation between the organotin(IV) and a Schiff based ligand enhances the α-CH acidity of the amino acid fragment in such complexes. The carbanions formed from the organotin(IV) complexes e.g., **5**–**7** (Scheme 13) are stabilized by resonance. Since the pK_a value of complex **5** in DMSO is 17.43[88], its α-CH acidity is greater than that of fluorene (p$K_a = 22.6$)[89] and the Ni^{2+} chelate with the Schiff base derived from glycine (p$K_a = 18.8$)[90], and it approaches that of mononitro compounds RNO_2, R = Me, Et (p$K_a = 17$)[89]. Because of the relatively high thermodynamic acidity and great kinetic stability, the carbanions formed from complexes **5–7** are able to condense, not only under strongly basic conditions, such as the use of 1.5 N MeONa giving **8–15** (Scheme 13), but also in the presence of the weak base Et_3N (Scheme 14).

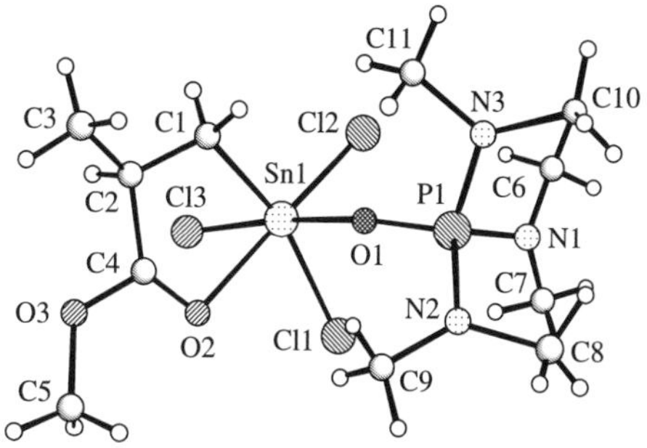

FIGURE 41. Molecular structure of **1a**, $MeOCOCH(CH_3)CH_2SnCl_3$•HMPA. From Reference 78, copyright Marcel Dekker

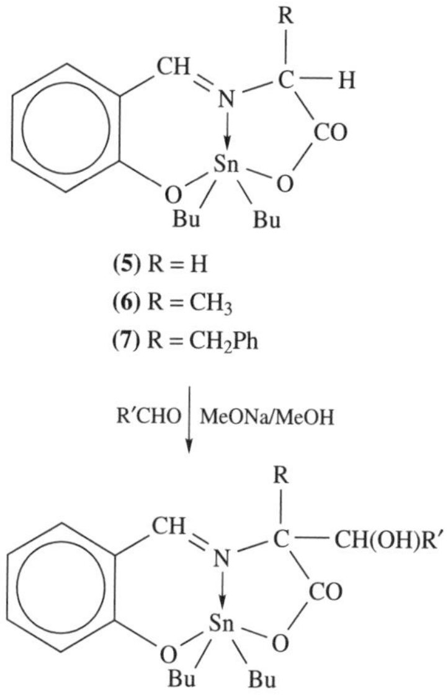

(**5**) R = H
(**6**) R = CH_3
(**7**) R = CH_2Ph

R′CHO | MeONa/MeOH

(**8**) R=CH_2OH, R′=H
(**9**) R=CH_3, R′=H
(**10**) R=CH_2Ph, R′=H
(**11**) R=H, R′=Ph
(**12**) R=H, R′=2-$CH_3OC_6H_4$
(**13**) R=H, R′=4-$CH_3OC_6H_4$
(**14**) R=H, R′=2-$CH_3C_6H_4$
(**15**) R=H, R′=4-$CH_3C_6H_4$

SCHEME 13. Reproduced by permission of John Wiley & Sons, Inc. from Reference 88

Hetero-atom-substituted stannylenes, which are electron-deficient compounds, are known in a larger number[91]. This deficiency is compared with the interaction between the 5pπ orbital of the tin atom and the free electron pairs at the hetero atom. In contrast, alkyl- and aryl-substituted compounds are not very well investigated[92–95]. The first-mentioned donor-free diarylstannylene is bis[(2,4,6-tris(tert-butyl)phenyl]tin[96]. In solution, this compound isomerized to a less bulky alkylarylstannylene (Scheme 15).

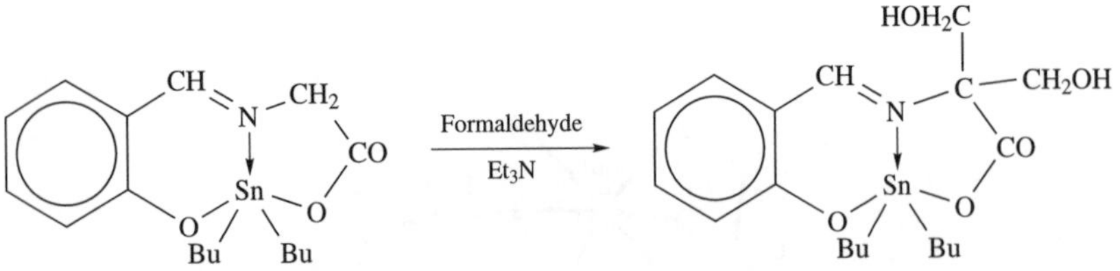

SCHEME 14. Reproduced by permission of John Wiley & Sons, Inc. from Reference 88

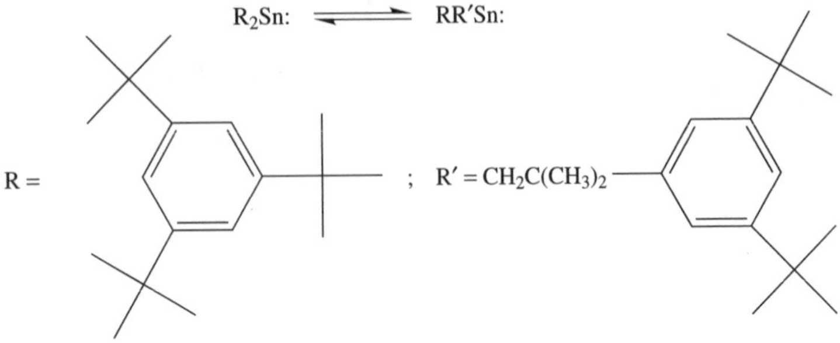

SCHEME 15. Reproduced by permission of Wiley-VCH from Reference 98

$$[(OC)_5Cr(THF)] + SnRR' \xrightarrow[-THF]{} [(OC)_5Cr{=}SnRR']$$

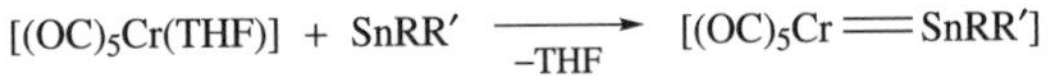

SCHEME 16. Reproduced by permission of Wiley-VCH from Reference 98

The rearrangement could be corroborated via addition and cycloaddition reactions and also by the synthesis of the $[(OC)_5W{=}SnRR']$ complex[97]. The analogous chromium complex is formed via a reaction of chromium hexacarbonyl in THF (Scheme 16)[98]. The molecular structure is shown in Figure 42. The molybdenum complex is formed in the same way (Scheme 16, Figure 43)[98].

3. Lead

Among toxic metals, lead is one of the principal poisoning metals, the environmental occurrence of which is mainly due to inorganic industrial derivatives and organic compounds from antiknock agents in petroleum[1,10]. Macrocyclic ligands can be used as effective sequestering agents for this toxic metal[99]. Phenanthrolinophane is a useful ligand for this purpose (Figure 44). It is rigid, and provides two aromatic nitrogens whose unshared electron pairs are beautifully placed to act cooperatively in binding cations[100]. An interesting compound is the $[(PbL^7Br)_2(\mu\text{-Br})][PbL^7Br_2]Br]{\bullet}5H_2O$ complex (Figure 45)[101]. It crystallizes in a triclinic crystal structure. The asymmetric unit contains two independent complexes, (a) $[(PbL^7Br)_2(\mu\text{-Br})]^+$ and (b) $[PbL^7Br_2]$, a bromine as a counter ion and five water solvent molecules. Complex (a) consists of two PbL^7 units bridged by one bromine anion. The overall conformations of L^7 and coordination geometry for the metal atoms are very similar in the two PbL^7 units. Atom Pb(1) is coordinated by the five nitrogens N(1)–N(5) and two bromide anions, one of them weakly interacting. The resulting arrangement for the seven donor atoms around the lead ion is rather asymmetric.

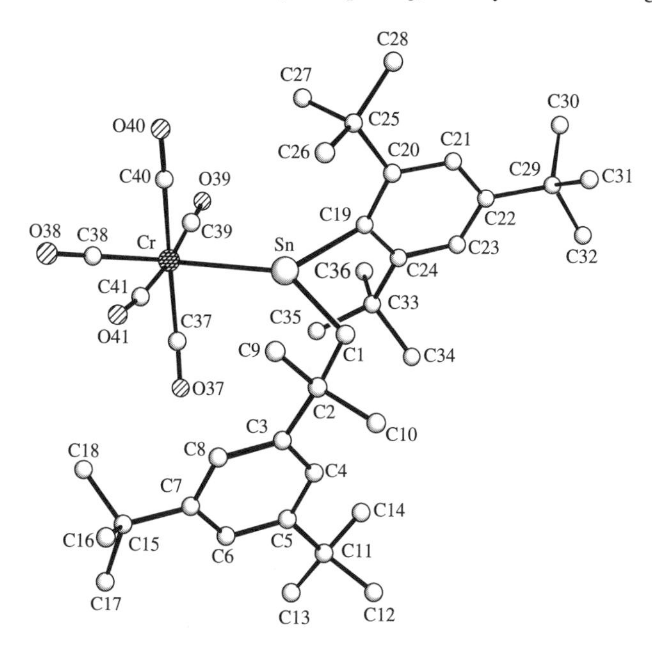

FIGURE 42. Molecular structure of [$(OC)_5Cr{=}SnRR'$]. Reproduced by permission of Wiley-VCH from Reference 98

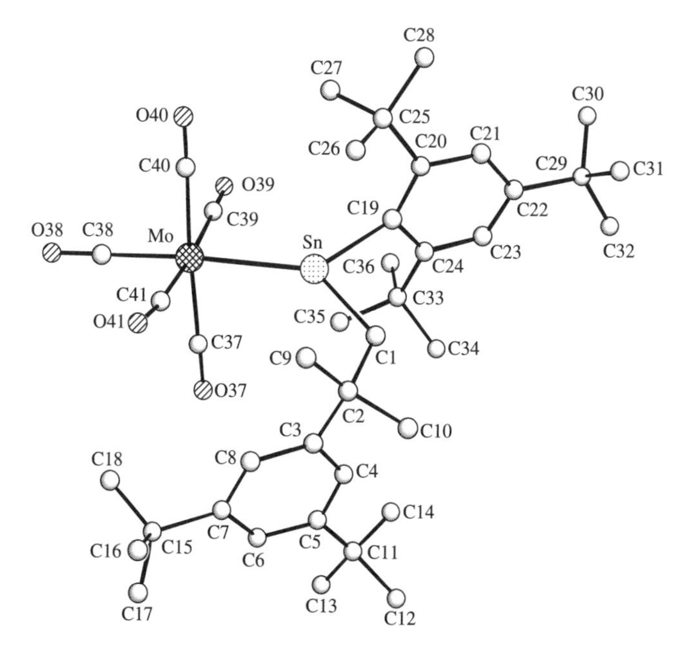

FIGURE 43. Molecular structure of [$(OC)_5Mo{=}SnRR'$]. Reproduced by permission of Wiley-VCH from Reference 98

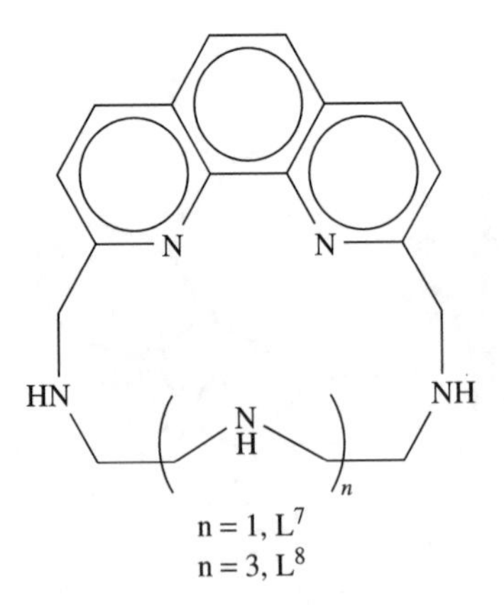

FIGURE 44. Ligands with a phenanthroline moiety. Reproduced by permission of the Royal Society of Chemistry from Reference 101

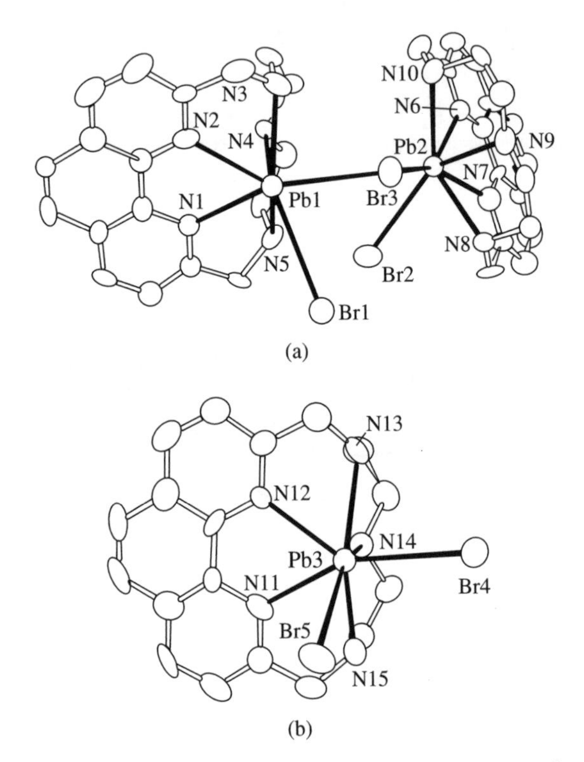

FIGURE 45. Molecular structures of (a) $[(PbL^7Br)_2(\mu\text{-}Br)]^+$ and of (b) $[PbL^7Br_2]$. Labels of the carbon atoms are omitted for clarity. Reproduced by permission of the Royal Society of Chemistry from Reference 101

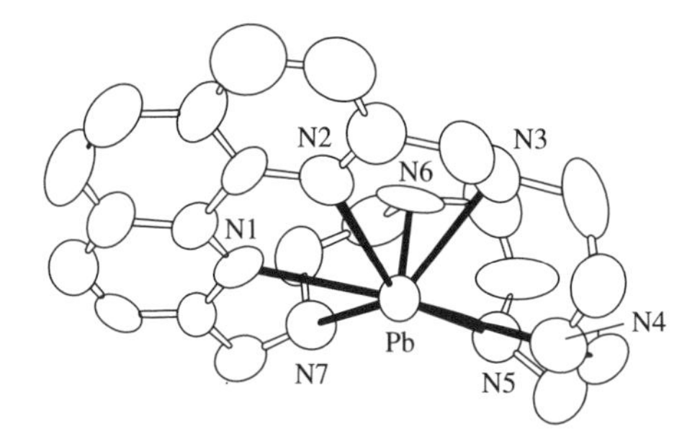

FIGURE 46. Molecular structure of $[PbL^8]^{2+}$. Labels of the carbon atoms are omitted for clarity. Reproduced by permission of the Royal Society of Chemistry from Reference 101

In the case of complex (b), $[PbL^7Br_2]$, the coordination geometry of the metal and the conformation of the macrocycle are almost equal to that found in complex (a).

Another interesting compound in this area is $[PbL^8][ClO_4][BPh_4]$ (Figure 46). It consists of the $[PbL^8]^{2+}$ cation and the two anions $[ClO_4]^-$ and $[BPh_4]^-$. The metal atom is seven-coordinated by the nitrogen atoms of the macrocycle. The resulting arrangement for the seven donor atoms around the lead ion is rather asymmetric, leaving a zone free from coordinated donor atoms, which is occupied by the lone pair of Pb^{2+}.

The 1 : 1 adducts of different lead(II) (pseudo-)halides with 1,10-phenanthroline (phen), (phen)PbX_2 (X = Cl, Br, I) and 2,2′-bipyridine (bpy), (bpy)PbX_2 (X = Cl, I, SCN), take the form of a one-dimensional polymer disposed along the *c* axis of the assigned cell. In the halides, the lead atom is six-coordinate, the *N*,*N*′-bidentate ligand being necessarily *cis* in the coordination sphere and the polymer being generated by a succession of $Pb(\mu\text{-}X)_2Pb$ rhombus. In the thiocyanate, the environment, although derivative, is more complex by virtue of more elaborate bridging behaviour of the thiocyanate group[102]. The structure of (bpy)PbI_2 is shown in Figures 47 and 48.

3,6-diformylpyridazine (Figure 49) is a building block for different macrocycles[103]. Interest in these macrocycles and their metal complexes is based on their potential relevance as structural models for metalloproteins[104] and, in particular, on the ability of pyrazine to mediate magnetic exchange[105–122], a property reminiscent of analogous phenol-bridged complexes[123–125]. When lead(II) ions are used as templates, two different macrocycle sizes can be isolated depending on the reaction conditions employed. Specifically, a 1 : 1 : 1 ratio of 3,6-diformylpyridazine:1,3-diaminopropane:lead(II) perchlorate resulted in the formation of $Pb_2L^9(ClO_4)_4$ (for the structure of ligand L^9, see Figure 50) whereas a 2 : 2 : 1 ratio gave $[Pb_2L^{10}][ClO_4]_4$ (for the structure of ligand L^{10}, see Figure 51). The $[Pb_2L^{10}]^{4+}$ cation is shown in Figure 52. The ability of the same metal to template two different macrocycle ring sizes efficiently, in this case lead(II) perchlorate and L^9 *vs.* L^{10} macrocycles, simply by employing different reagent ratios, is first mentioned by Brooker and coworkers[103]. In previous studies, a given metal salt templated the formation of only one specific macrocycle ring size, so that to obtain a macrocycle of different size quite different reaction conditions had to be employed, for example a different template ion or a different anion.

Some other Lewis base adducts of lead(II) compounds are the 2 : 1 adducts of the *N*,*N*′-bidentate aromatic base 1,10-phenanthroline (phen) with lead(II) nitrate and perchlorate[126].

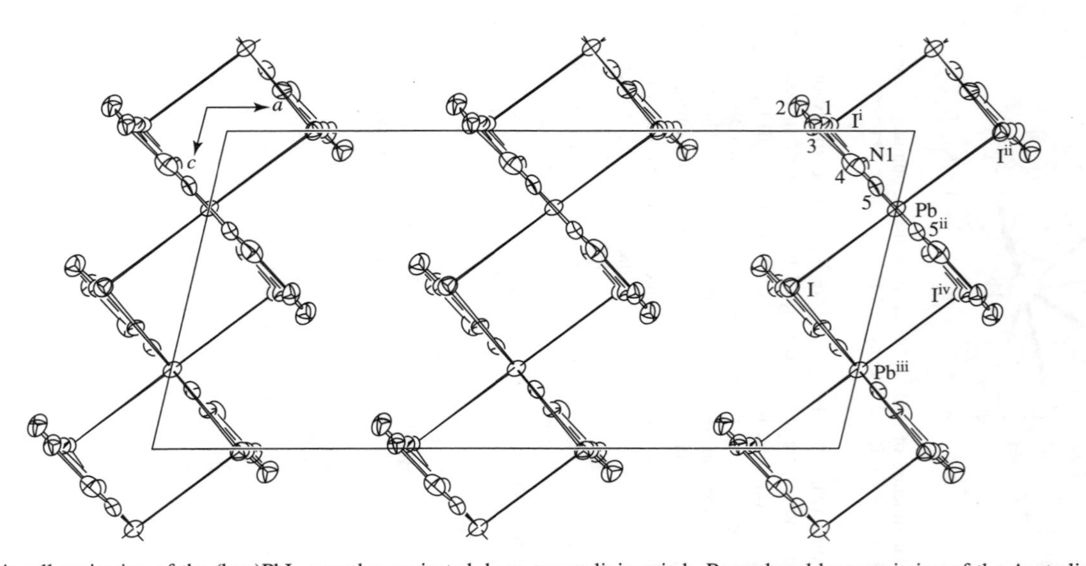

FIGURE 47. Unit cell projection of the $(bpy)PbI_2$ complex projected down monoclinic axis b. Reproduced by permission of the Australian Academy of Science and CSIRO from Reference 102

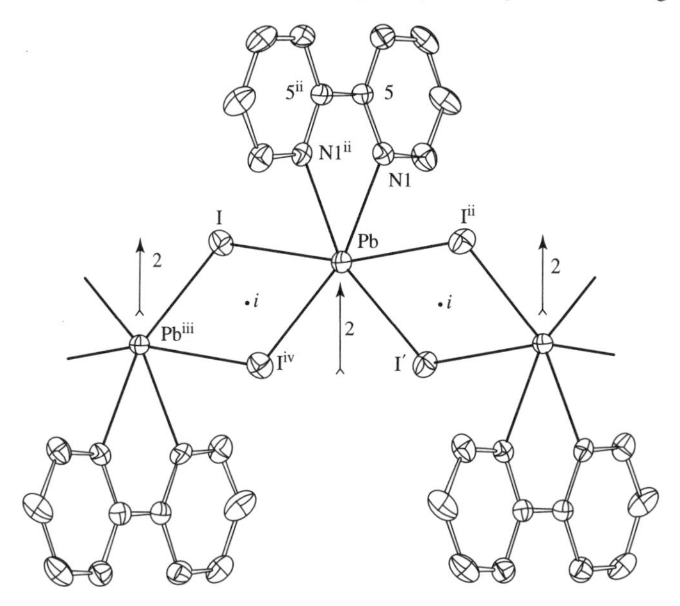

FIGURE 48. A strand of the (bpy)PbI_2 polymer projected normal to the plane containing the 2 *axis* and the polymer axis (i.e. down a^*). Reproduced by permission of the Australian Academy of Science and CSIRO from Reference 102

FIGURE 49. 3,6-diformylpyridazine. Reproduced by permission of the Royal Society of Chemistry from Reference 103

L^9

FIGURE 50. Structure of ligand L^9. Reproduced by permission of the Royal Society of Chemistry from Reference 103

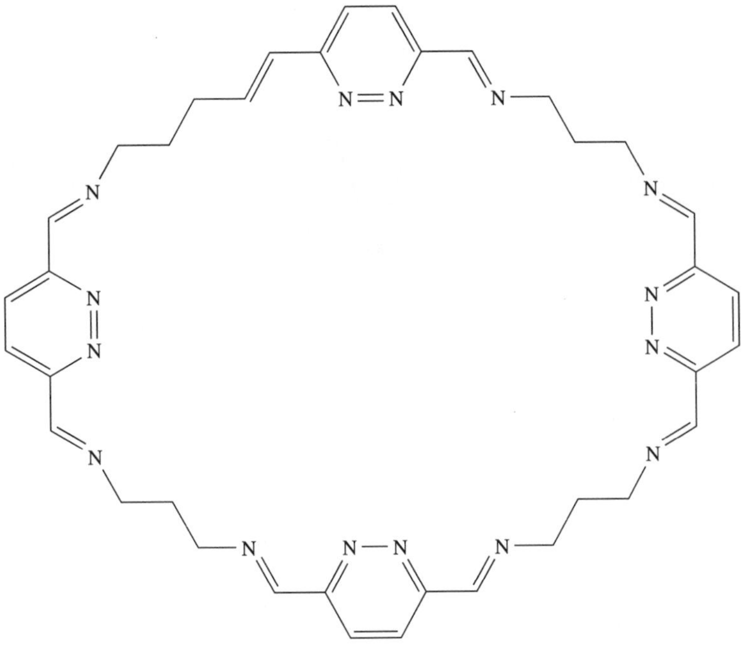

L^{10}

FIGURE 51. Structure of ligand L^{10}. Reproduced by permission of the Royal Society of Chemistry from Reference 103

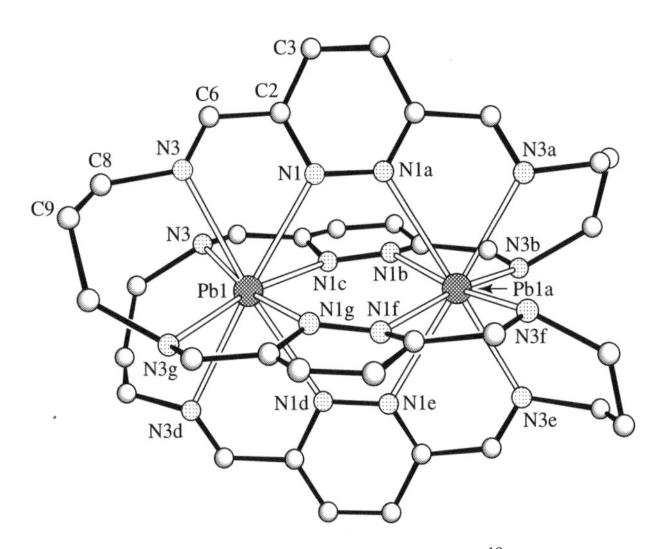

FIGURE 52. Perspective view of the cation of the complex $[Pb_2L^{10}][ClO_4]_4$. Reproduced by permission of the Royal Society of Chemistry from Reference 103

$[(phen)_2Pb(NO_3)_2]$ is monoclinic whereas $[(phen)_2Pb(ClO_4)_2]$ is triclinic. Both systems are mononuclear with eight-coordinate PbN_4O_4 coordination environments incorporating a pair of O,O'-bidentate anions. The complexes are essentially of the type $ML_2L'_2$ where L, here phen, and L′, here the oxoanion, both act as bidentate ligands. In projection normal to the *quasi-2 axis* (Figures 53 and 54) the two pairs of ligand type display the feature of being compressed towards each other from either pole, more so at the oxoanion end, so that if sterically active lone pairs are to be postulated, they are most likely to be found between the oxoanions and directed along the *quasi-2 axis*.

If 2,2′-bipyridine (bpy) is used instead of phen, the situation is different[127]. $[(bpy)_2Pb(NO_3)_2]$ (Figures 55 and 56) is triclinic and $[(bpy)_2Pb(ClO_4)_2]$ (Figures 57 and 58) is monoclinic. The bpy systems are centrosymmetric dimers; in each case the

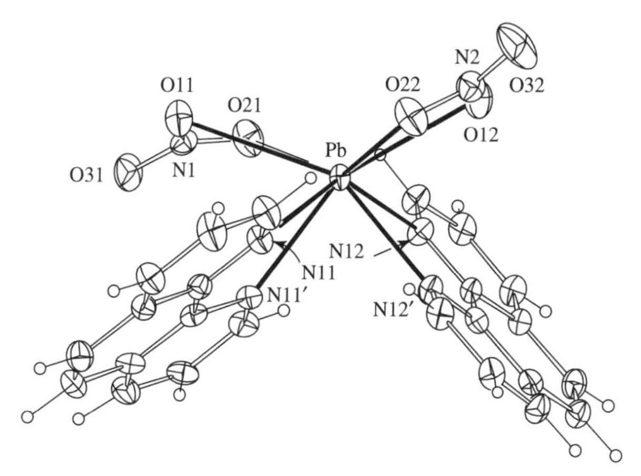

FIGURE 53. Projection of $(phen)_2Pb(NO_3)_2$ normal to the *quasi-2 axis*. Reproduced by permission of the Australian Academy of Science and CSIRO from Reference 126

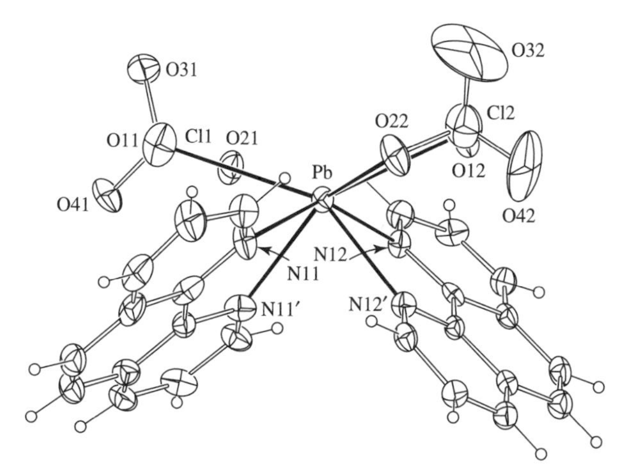

FIGURE 54. Projection of $(phen)_2Pb(ClO_4)_2$ normal to the *quasi-2 axis*. Reproduced by permission of the Australian Academy of Science and CSIRO from Reference 126

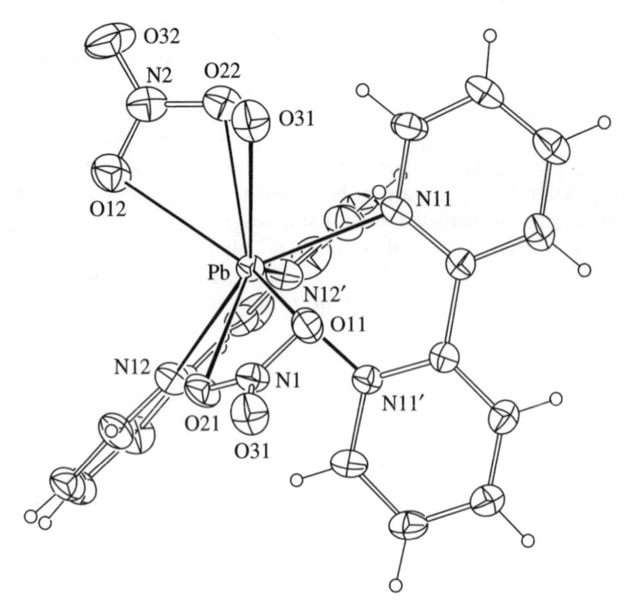

FIGURE 55. Projection of the asymmetric unit of the $(bpy)_2Pb(NO_3)_2$ down the Pb $\cdots$ Pb vector. Reproduced by permission of the Australian Academy of Science and CSIRO from Reference 127

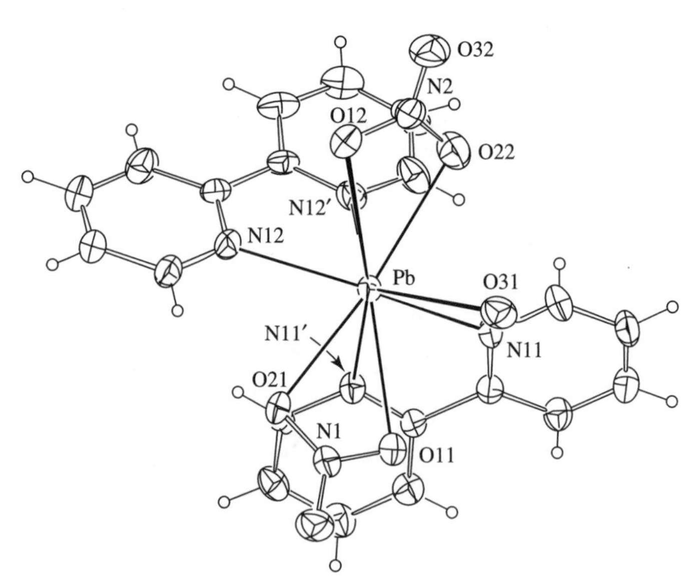

FIGURE 56. Projection of the asymmetric unit of the $(bpy)_2Pb(NO_3)_2$ down the *quasi-2 axis*. Reproduced by permission of the Australian Academy of Science and CSIRO from Reference 127

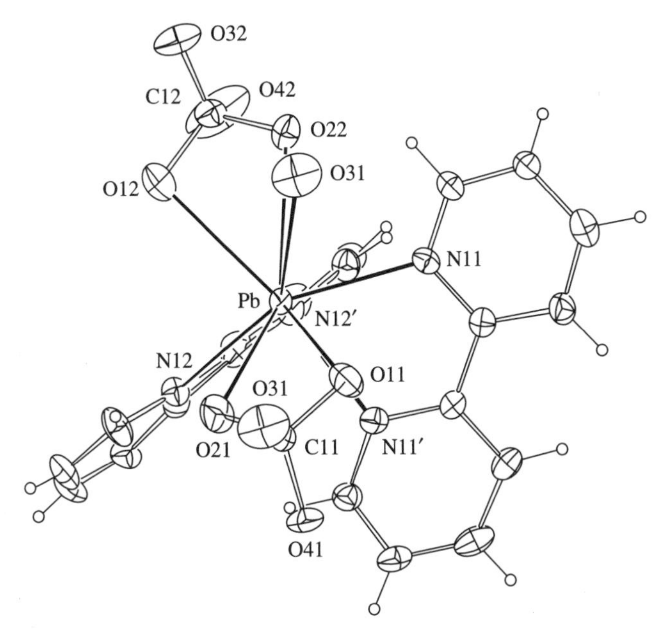

FIGURE 57. Projection of the asymmetric unit of the $(bpy)_2Pb(ClO_4)_2$ down the Pb $\cdots$ Pb vector. Reproduced by permission of the Australian Academy of Science and CSIRO from Reference 127

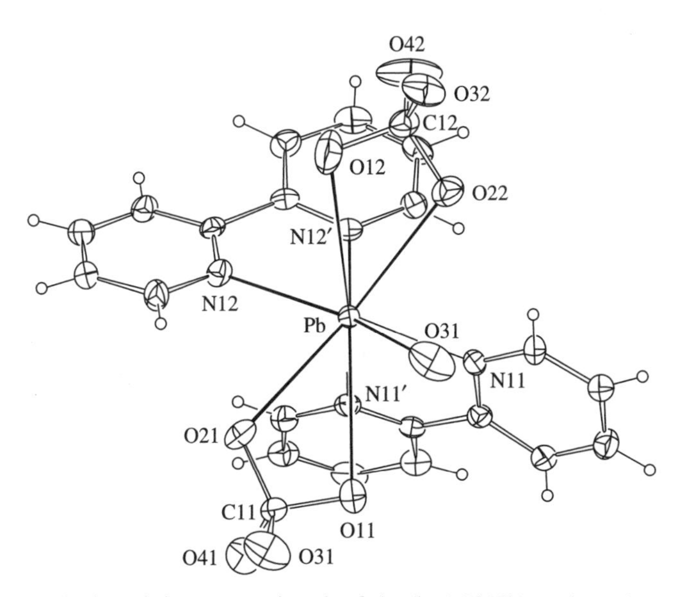

FIGURE 58. Projection of the asymmetric unit of the $(bpy)_2Pb(ClO_4)_2$ down the *quasi-2 axis*. Reproduced by permission of the Australian Academy of Science and CSIRO from Reference 127

coordination environment incorporates a pair of N,N'-bidentate aromatic bases and an O,O'-bidentate anion, but the other anion is not only O,O'-bidentate, but also bridges by a third oxygen atom the other lead atom so that it is nine-coordinate PbN_4O_5. In another synthesis a mixture of lead(II) perchlorate and lead(II) acetate (ac) in 1 : 1 stoichiometry were used in a reaction with an overall 1 : 2 lead(II) salt/2,2′-bipyridine ratio. The crystalline product shows a discrete mononuclear complex of $[(bpy)_2Pb(ClO_4)(ac)]$ (Figures 59 and 60).

The complexes $(tpy)Pb(oxoanion)_2$, oxoanion = ClO_4^-, NO_3^-, NO_2^-, tpy = 2,2′ : 6′,2″-terpyridine, all have a monoclinic structure[128]. One-half of the $[(tpy)Pb(oxyanion)_2]$ (H_2O) formula unit comprises the asymmetric unit of the structure, the lead atom lying on a crystallographic *2-axis* which also passes through the axis of the central ring of the tpy ligand, defining its polarity and relating the two halves of that ligand, and also relating the associated anionic components of the coordination sphere, one anion only being crystallographically independent. The lead environment comprises the N_3-tridentate ligand at one pole of the symmetry element, a pair of symmetry-related O,O'-chelating anions lying more or less equatorial, and a final pair or monodentate oxygen or nitrogen atoms bridging from anions associated with adjacent lead atoms about the other pole of the symmetry axis (Figures 61–63).

In this series of complexes of lead(II) nitrate and nitrogen containing ligands in a ratio 1 : 1, many different ligands are used. Some examples are the multidentate aliphatic

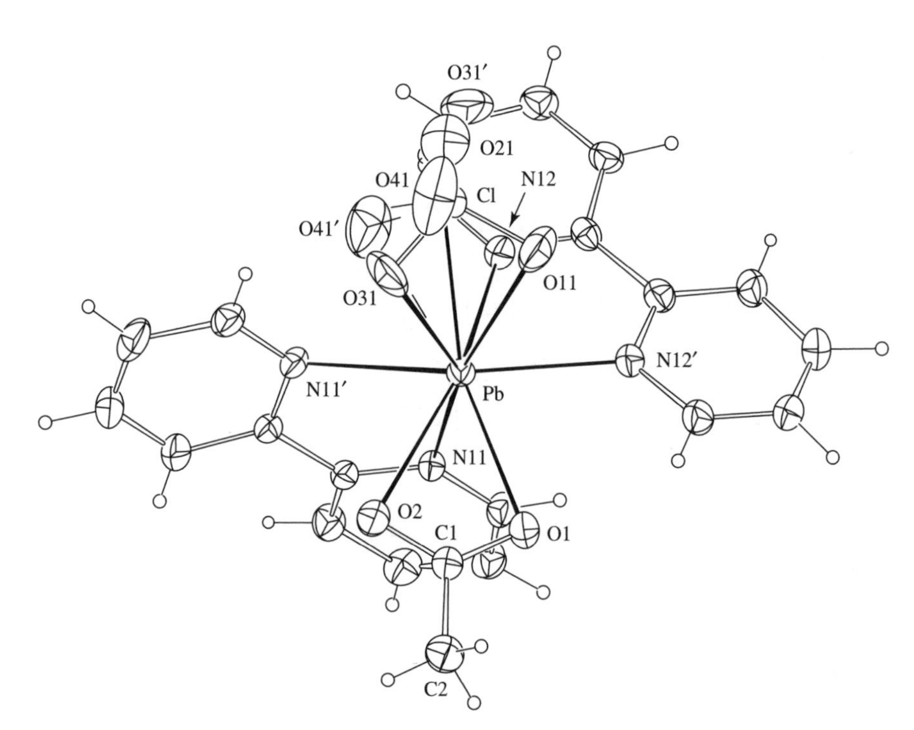

FIGURE 59. $(bpy)_2Pb(ac)(ClO_4)$ in projection down its incipient twofold axis. Reproduced by permission of the Australian Academy of Science and CSIRO from Reference 127

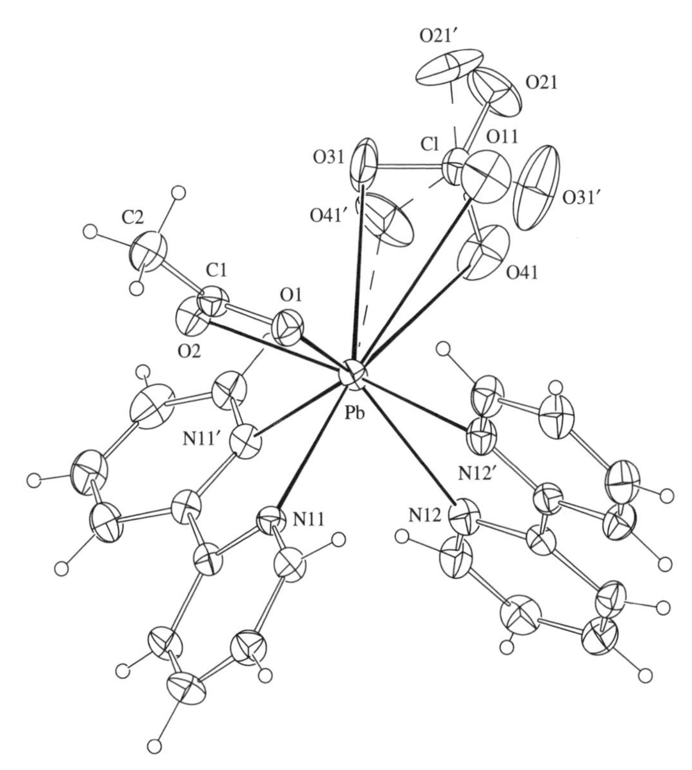

FIGURE 60. A projection of $(bpy)_2Pb(ac)(ClO_4)$ normal to the twofold axis. Reproduced by permission of the Australian Academy of Science and CSIRO from Reference 127

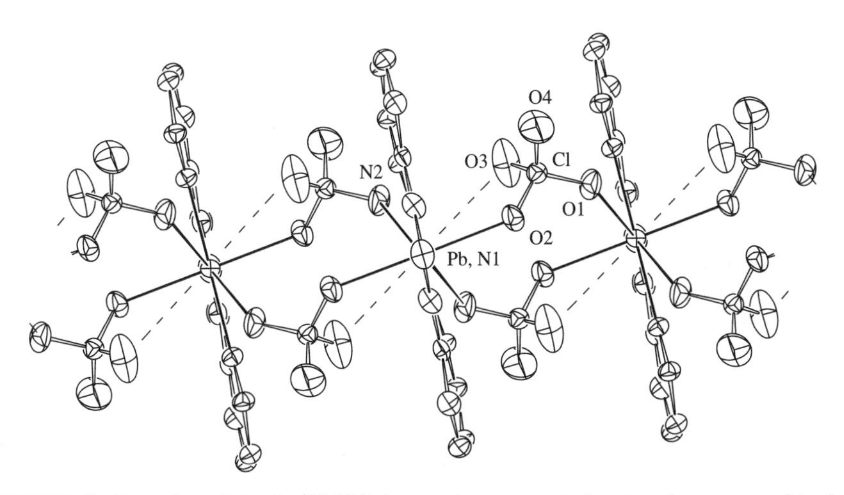

FIGURE 61. Projection of the $(tpy)Pb(ClO_4)_2$ complex: a strand of each polymer, normal to the plane of *b* and the *ac* cell diagonal. Reproduced by permission of the Australian Academy of Science and CSIRO from Reference 128

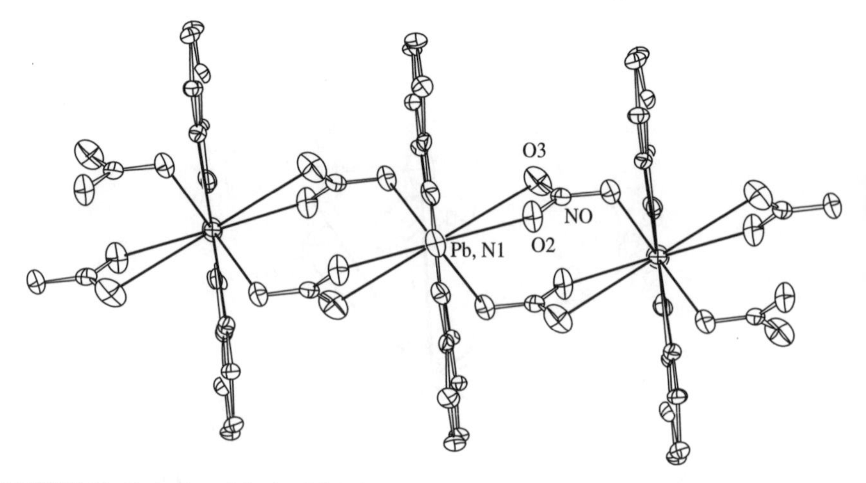

FIGURE 62. Projection of the $(tpy)Pb(NO_3)_2$ complex: a strand of each polymer, normal to the plane of *b* and the *ac* cell diagonal. Reproduced by permission of the Australian Academy of Science and CSIRO from Reference 128

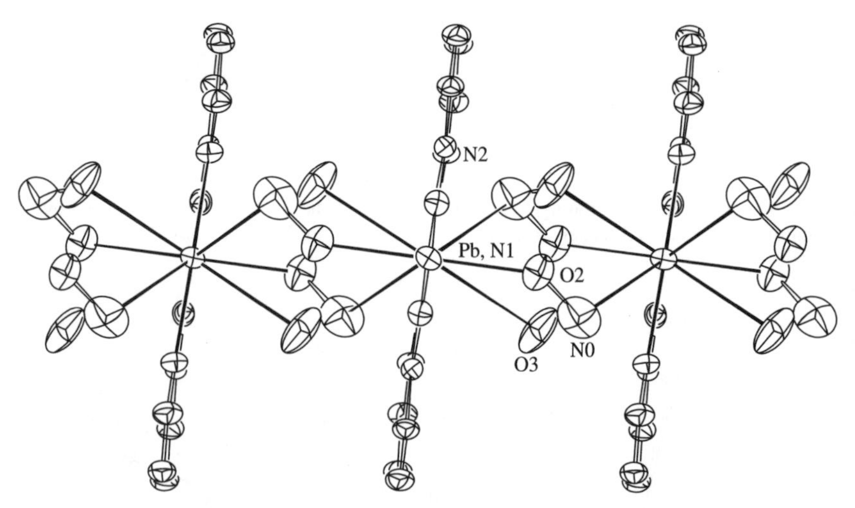

FIGURE 63. Projection of the $(tpy)Pb(NO_2)_2$ complex: a strand of each polymer, normal to the plane of *b* and the *ac* cell diagonal. Reproduced by permission of the Australian Academy of Science and CSIRO from Reference 128

nitrogen bases of increasing denticity (the number of active coordination sites a ligand has) and chain length; ethane-1,2-diamine = ethylenediamine = en, 3-azapentane-1, 5-diamine = diethylenetriamine = '22' = dien, 3,6-diazaoctane-1,8-diamine = 222tet = trien, 3,7-diazanonane-1,9-diamine = 232tet, 4,7-diazadecane-1,10-diamine = 323tet[129]. $(en)Pb(NO_3)_2$ (Figure 64) has an interesting structure. It is a double-stranded one-dimensional polymer, lying parallel to axis a, with the axis of the polymer disposed across the *bc* cell diagonal; the lead atoms of the two strands are linked by $Pb_2(O(12))_2$

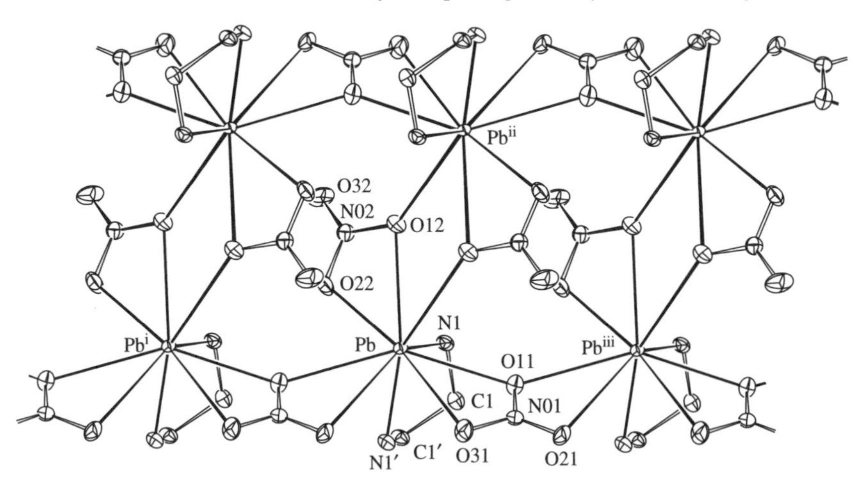

FIGURE 64. The (en)$Pb(NO_3)_2$ structure projected normal to the 'plane' of the polymer, showing relevant transformations of the asymmetric unit. Reproduced by permission of the Australian Academy of Science and CSIRO from Reference 129

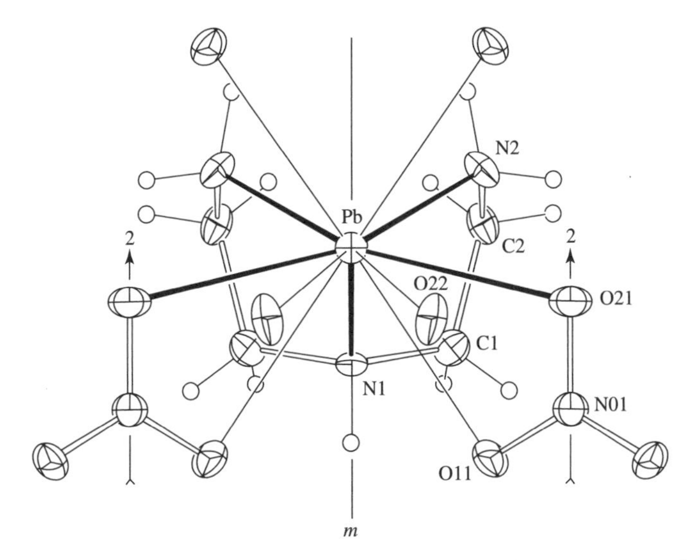

FIGURE 65. The (dien)$Pb(NO_3)_2$ structure projected down *c*, the lead environment showing the approach of the oxygen atoms O(22) of nitrate 2. Reproduced by permission of the Australian Academy of Science and CSIRO from Reference 129

centrosymmetric, necessarily planar rhombus, separated by the *a* translation. The lead environment is nine-coordinate PbN_2O_7, with four of the oxygen atoms being bridging, and the array essentially being comprised of three bidentate and one tridentate ligands. (dien)$Pb(NO_3)_2$ (Figure 65), the principal motifs in the array, which is polymeric, are the lead atom and the saturated triamine; each ligand is associated with one metal atom,

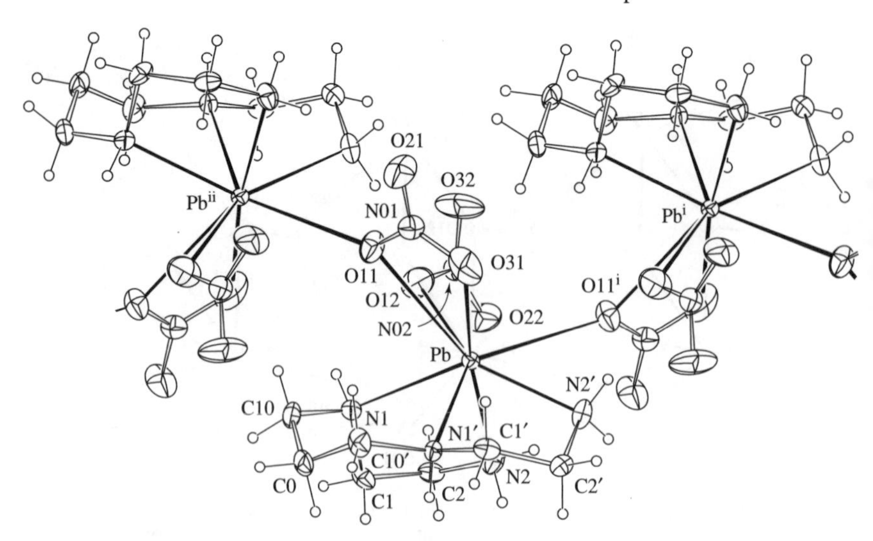

FIGURE 66. The structure of (232tet)Pb$(NO_3)_2$ projected normal to the polymer string. Reproduced by permission of the Australian Academy of Science and CSIRO from Reference 129

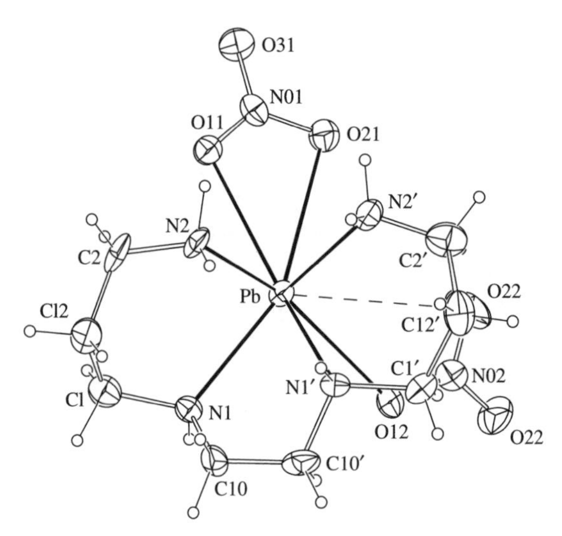

FIGURE 67. The 'molecular' unit of (323tet)Pb$(NO_3)_2$; the dotted 'contact' O(22) is very long at 3.72(2) Å. Reproduced by permission of the Australian Academy of Science and CSIRO from Reference 129

and adopts a *quasi-facial* coordination mode with the fused chelate rings in mirror image conformations. The structure of (232tet)Pb$(NO_3)_2$ is shown in Figure 66, and that of (323tet)Pb$(NO_3)_2$ in Figure 67.

Adducts of lead(II) bromide with ethane-1,2-diamine (en) and propane-1,2-diamine (pn) are of 2 : 1 and 1 : 1 stoichiometry, respectively[130]. $[(en)_2PbBr_2]_{(\infty|\infty)}$ (Figure 68) is a

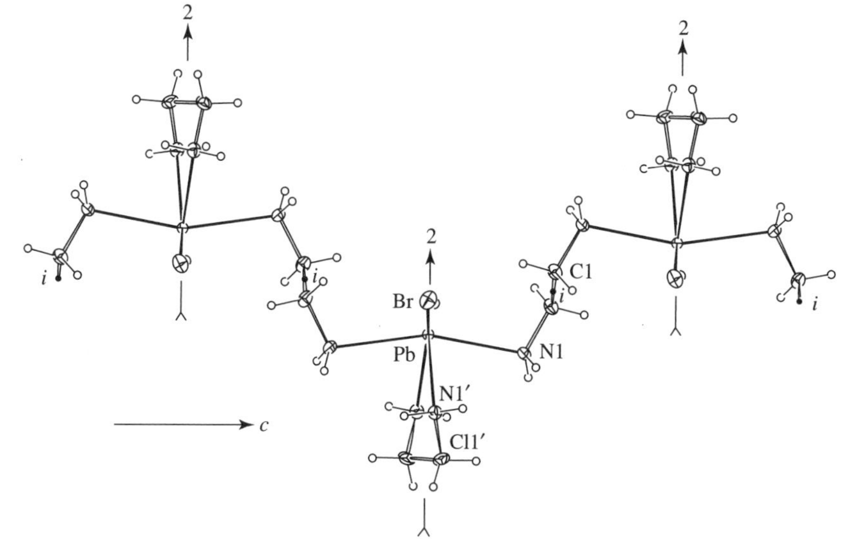

FIGURE 68. Projection of a strand of the polymer of $(en)_2PbBr_2$ along a^* (i.e. normal to the *bc* plane). Reproduced by permission of the Australian Academy of Science and CSIRO from Reference 130

single-stranded linear polymer parallel to *c*; the lead atom, lying on a crystallographic twofold axis, is six-coordinate PbN_4Br_2, with the bromine atoms *cis* in the coordination sphere and lying opposite to a bidentate en ligand. The *trans* sites are linked into a chain by bridging en ligands. $[(pn)PbBr_2]_{(\infty|\infty)}$ (Figure 69) is a more complex two-dimensional polymer; the pn is bidentate, but the lead atoms are now eight-coordinate with doubly and quadruply bridging bromines linking them into a polymeric sheet. The adduct with 323tet (4,7-diazadecane-1,10-diamine) is of a $(323tet)_2(PbBr_6)_3$ stoichiometry, best represented as $\{[Pb(323tet)]_2[PbBr_6]\}_{(\infty|\infty)}$ (Figure 70). About one lead atom type, the ligand is quadridentate, lying on one face of the coordination sphere; the region opposite is occupied by four triply bridging bromine atoms from $[PbBr_6]$ anionic units (which contain quasi-octahedral, centrosymmetric lead, with a pair of *trans*-bromine atoms terminal and the others bridging) linking the array into a two-dimensional polymer.

Adducts of 1,4,8,11-tetraazacyclotetradecane (cyclam) with lead(II) perchlorate and $(7R^*,14R^*)$-5,5,7,12,12,14-hexamethyl-1,4,8,11-tetraazacyclotetradecane (tet-b) with lead(II) nitrate, perchlorate and acetate (ac) have the ratio 1 : 1[131]. $[(cyclam)Pb(ClO_4)]$ is orthorhombic (Figure 71), $[(tet\text{-}b)Pb(NO_3)_2]_2$ and $[(tet\text{-}b)Pb(ClO_4)_2]_2{\bullet}H_2O$ are monoclinic (Figures 72 and 73) and $[(tet\text{-}b)Pb(ac)_2]_2{\bullet}2H_2O$ is triclinic (Figure 74). In all complexes, the macrocycle-N_4 ligand occupies one side of the coordination sphere of the lead atom, with anionic oxygens opposed; the cyclam/perchlorate complex is, like the nitrate, mononuclear with seven-coordinate $(N_4)PbO_3$ with a bidentate O,O'-and a unidentate O-perchlorate. In the tet-b acetate, the anionic oxygen atoms are surprisingly sparse, comprising simply a bidentate acetate, in a mononuclear $(N_4)PbO_2$ environment with the other (lattice) acetate bonded to the macrocycle axial NH hydrogens. The nitrate and perchlorate complexes involve bridging anions: in the nitrate, a central centrosymmetric PbO_2Pb array is found, the lead atoms being bridged by one oxygen of a bidentate nitrate, the other nitrate being unidentate and the coordination sphere $(N_4)PbO_4$; in the perchlorate, again a centrosymmetric dimer is found, the lead atoms being linked by

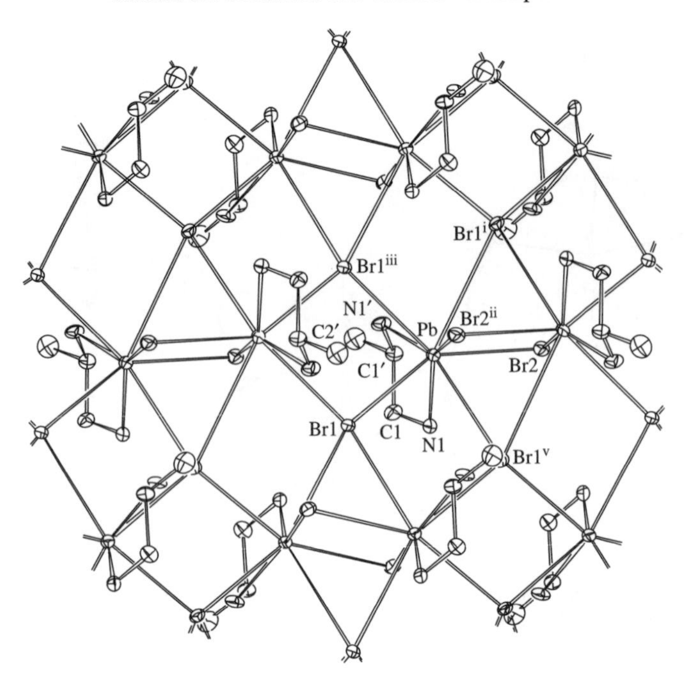

FIGURE 69. Projection of a sheet of the polymer of (pn)$PbBr_2$ down a. Reproduced by permission of the Australian Academy of Science and CSIRO from Reference 130

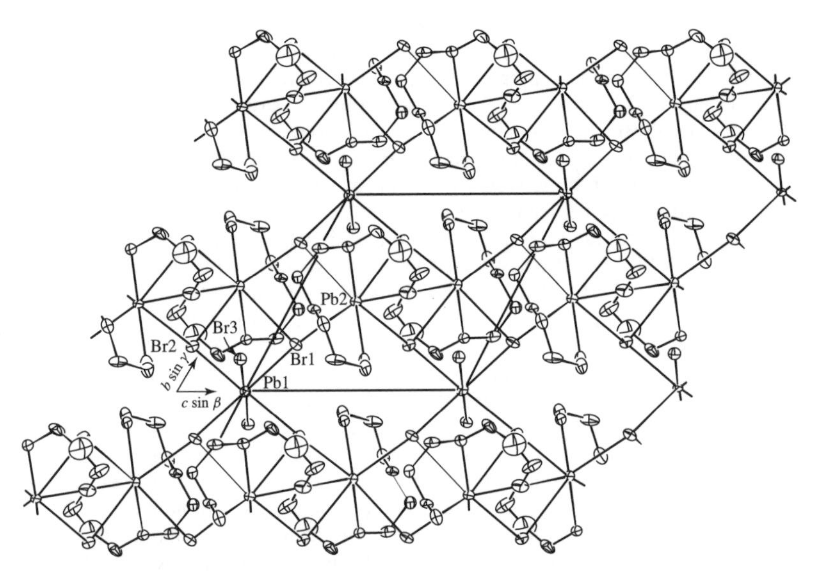

FIGURE 70. Unit cell contents of $(323tet)_2(PbBr_2)_3$ down a, showing the polymeric sheet; note also the quasi-anti-square-prismatic environment of 'cationic' Pb(2). Reproduced by permission of the Australian Academy of Science and CSIRO from Reference 130

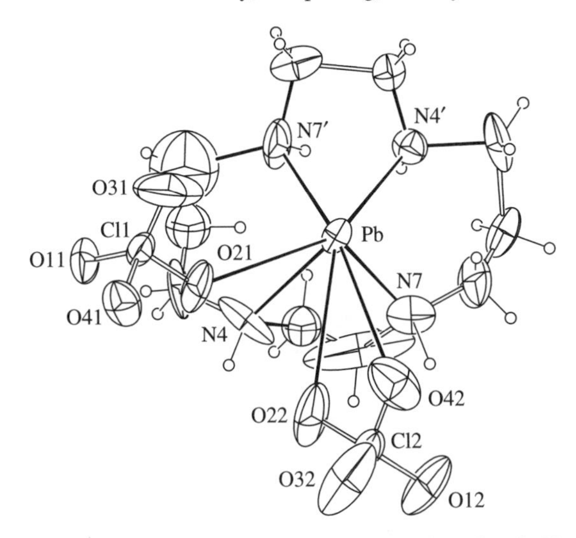

FIGURE 71. Projection of the 'molecule' of (cyclam)Pb$(ClO_4)_2$ normal to the N_4 plane, with ellipsoids showing the large vibrational amplitudes. Reproduced by permission of the Australian Academy of Science and CSIRO from Reference 131

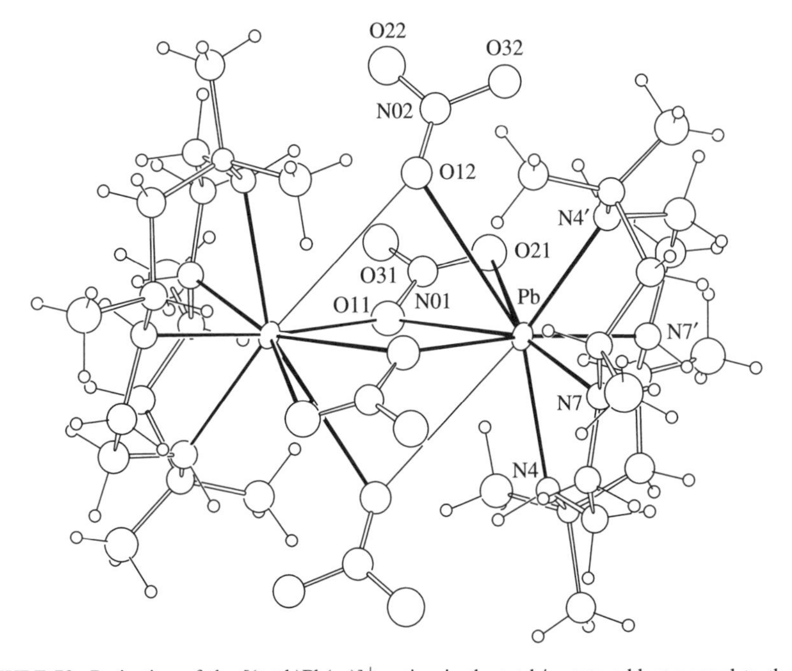

FIGURE 72. Projection of the $[(\text{tet-b})\text{Pb(ac)}]^+$ cation in the tet-b/acetate adduct normal to the N_4 plane. Reproduced by permission of the Australian Academy of Science and CSIRO from Reference 131

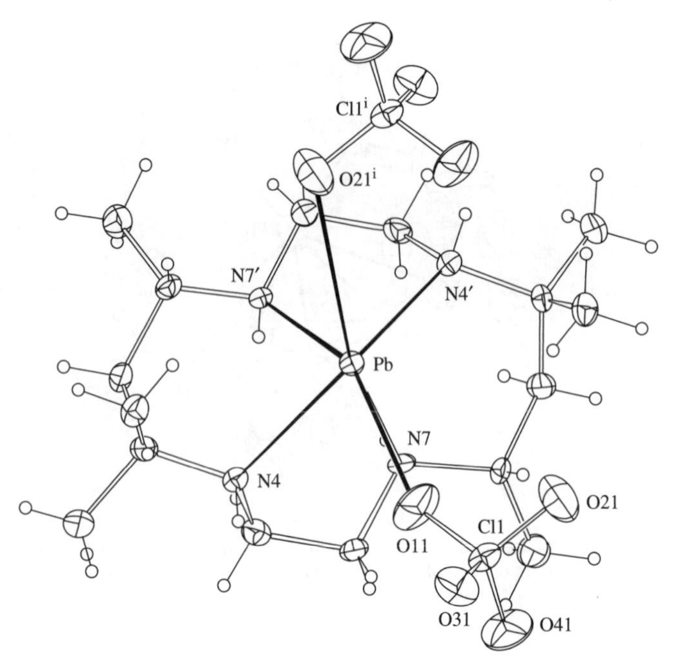

FIGURE 73. Projection of (tet-b)Pb$(ClO_4)_2$ showing the lead environment, normal to the ligand plane. Reproduced by permission of the Australian Academy of Science and CSIRO from Reference 131

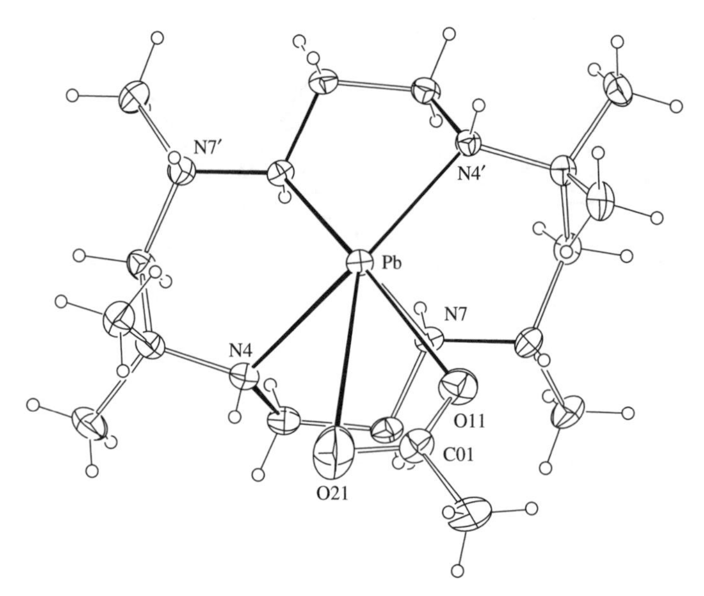

FIGURE 74. Projection of the (tet-b)Pb$(NO_3)_2$ dimer normal to the Pb(O(21))$_2$Pb plane. Reproduced by permission of the Australian Academy of Science and CSIRO from Reference 131

O,O'-bridging perchlorates [again with $(N_4)PbO_2$ coordination environment], the complex being essentially $[(\text{tet-b})Pb(OClO_2O)_2Pb(\text{tet-b})](ClO_4)_2 \cdot 2H_2O$.

Some other 1 : 1 adducts with tet-b are known. For example, $[(\text{tet-b})PbCl_2]$ (Figure 75) and $[(\text{tet-b})PbI_2]$ (Figure 76) are monoclinic, $[(\text{tet-b})Pb(NCS)_2]$ (Figure 77) is orthorhombic[132]. All are discrete mononuclear $[(\text{tet-b})PbX_2]$ entities in which the macrocyclic N_4 ligand occupies one 'face' of the N_4PbX_2 coordination sphere. The thiocyanate ligands being N-bonded, interesting hydrogen-bonding interactions are found, columns of molecules being formed by way of hydrogen bonding between the coordinated (pseudo-)halides and the NH hydrogen atoms which project to the 'rear' face of the ligand of the next molecule, opposite the metal. The bromine analogue complex is monoclinic and is best formulated as $[(\text{tet-b})PbBr]Br$ (Figure 78), only one of the bromine entities being

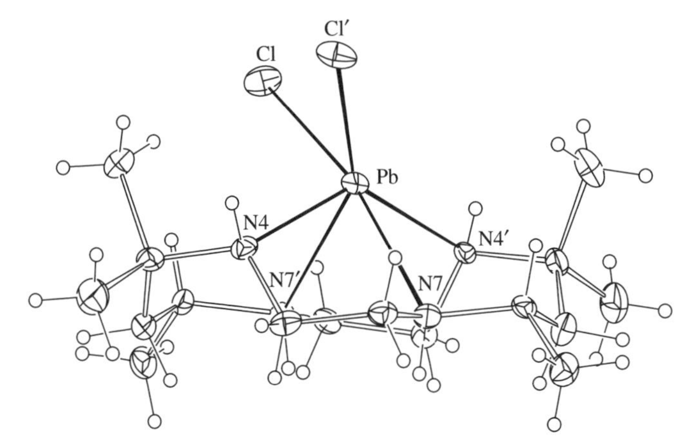

FIGURE 75. Molecule of $[(\text{tet-b})PbCl_2]$ normal to the macrocycle axis. Reproduced by permission of the Australian Academy of Science and CSIRO from Reference 132

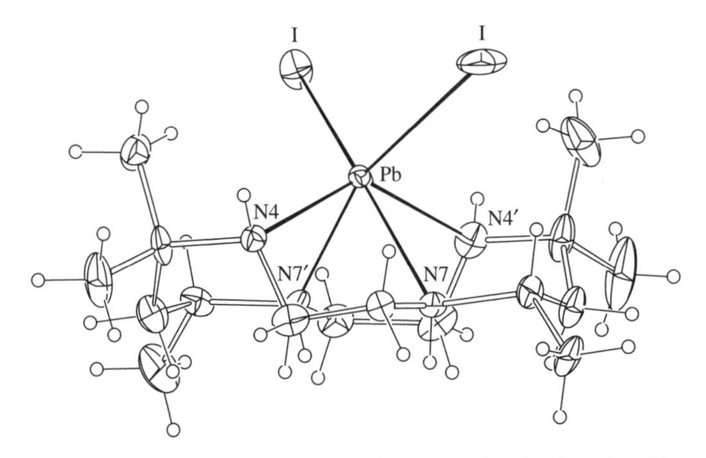

FIGURE 76. Molecule of $[(\text{tet-b})PbI_2]$ normal to the macrocycle axis. Reproduced by permission of the Australian Academy of Science and CSIRO from Reference 132

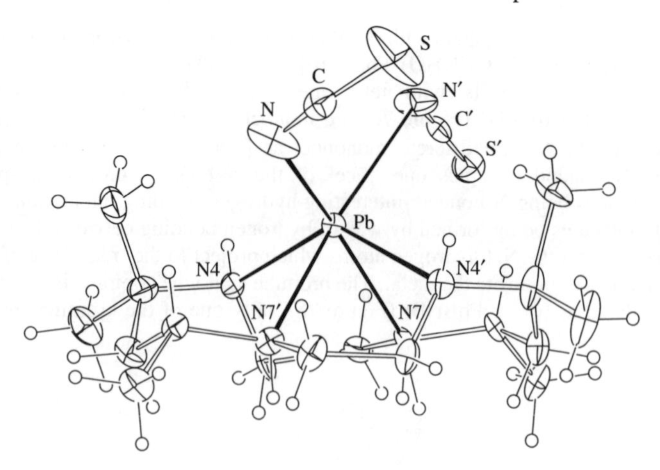

FIGURE 77. Molecule of [(tet-b)Pb(NCS)$_2$] normal to the macrocycle axis. Reproduced by permission of the Australian Academy of Science and CSIRO from Reference 132

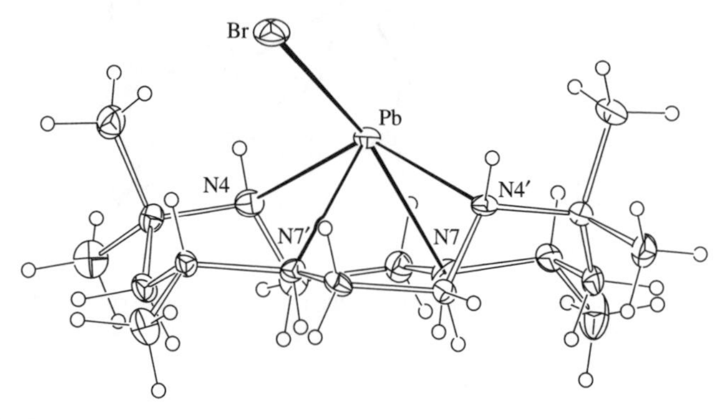

FIGURE 78. One of the two molecules of (tet-b)PbBr$_2$ viewed normal to the macrocycle axis. Reproduced by permission of the Australian Academy of Science and CSIRO from Reference 132

bound to the lead, the other being fully dissociated by hydrogen bonding/ion pairing to the 'rear' side of adjacent ligands, forming hydrogen-bonded sheets rather than columns.

The bpy (2,2′-bipyridine) and phen (1,10-phenanthroline) ligands are useful compounds in lead(II) complex chemistry. [(bpy)Pb(NO$_3$)$_2$]$_{(\infty|\infty)}$•H$_2$O (Figure 79) and [(phen)Pb(NO$_3$)$_2$]$_{(\infty|\infty)}$•H$_2$O are monoclinic, [(phen)Pb(ac)$_2$]•4H$_2$O is triclinic (Figure 80)[133]. The nitrates are one-dimensional polymers along *b*, successive lead atoms being linked by one oxygen of one of the nitrate groups, each of the other oxygen atoms completing a chelate to either side; the lead environment is completed as N$_2$PbO$_7$ by the other nitrate (as a chelate), the bidentate base and the water molecule. In the acetate which is a centrosymmetric dimer (manifested in two distinct independent dimers), the two lead atoms are linked by bridging oxygen atoms derived from pairs of chelating acetate moieties about each lead atom. The coordination sphere of each lead atom is completed by the bidentate aromatic base.

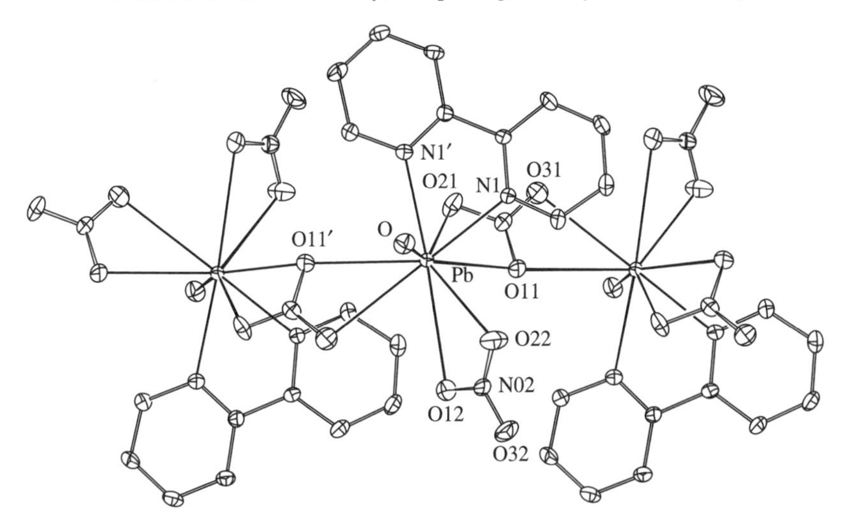

FIGURE 79. View of a single strand of the polymer of $[(bpy)Pb(NO_3)_2]_{(\infty|\infty)}\bullet H_2O$ down *a*. Reproduced by permission of the Australian Academy of Science and CSIRO from Reference 133

Another ligand is dpa (bis(2-pyridyl)amine). It also forms an adduct with lead(II) salts, for example $[(dpa)PbBr_2]_{(\infty|\infty)}$ (Figure 81) and $[(dpa)PbI_2]_{(\infty|\infty)}$ are both monoclinic[134]. These two complexes are linear polymers with six-coordinate (cis-N_2)Pb(μ-X)$_4$ environments linked in infinite $\cdots(\mu\text{-X})_2\text{Pb}(\mu\text{-X})_2\cdots$ one-dimensional chains, and with dpa being bidentate. A 2 : 1 adduct of dpa is $[(dpa)_2Pb(ac)_2]_{(\infty|\infty)}$ (Figure 82), which is monoclinic[134]. The complex is a linear polymer along *c*; the lead atom lies on a crystallographic twofold axis with a coordination environment comprising a pair of symmetry-related dpa ligands, in this case unidentate, and a pair of symmetry-related bidentate acetate ligands with the first oxygen atom performing an additional bridging function to adjacent symmetry-related lead atoms.

2 : 1 adducts of 2-aminomethylpyridine (amp) with lead(II) nitrate and thiocyanate are $[(amp)_2Pb(NO_3)_2]_2$ (Figure 83) and $[(amp)_2Pb(SCN)_2]_2$ (Figure 84)[135]. Both complexes are centrosymmetric dimers; the coordination environment is made up in each case of a pair of N,N'-bidentate bases, one terminally bound anion (O,O'-chelating nitrate or S-bonded thiocyanate) and bridging anions. In the case of the thiocyanates these bridge end-on, so that the lead(II) environment is seven-coordinate PbN_5S_2; in the nitrate, the anion chelates through two of its oxygen atoms, bridging via the third, so that the lead(II) environment is nine-coordinate PbN_4O_5.

The abbreviation trz denotes 2,4,6-tris(2-pyridyl)-1,3,5-triazine. In a reaction with lead(II) nitrate in 1 : 1 and 1 : 2 stoichiometry two complexes are formed, $[(trz)Pb(NO_3)_2]_{(\infty|\infty)}$ (Figure 85) and $[(trz)_2Pb(NO_3)_2]$ (Figure 86), respectively[136]. $[(trz)Pb(NO_3)_2]_{(\infty|\infty)}$ is an infinite polymer; the plane of the tridentate trz ligand lies normal to the polymer axis with unsymmetrically bidentate nitrate groups to either side. The third oxygen of each nitrate group bridges to the next lead atom in the polymer chain. $[(trz)_2Pb(NO_3)_2]$ is a methanol monosolvate. The complex species is mononuclear, with the lead atom located on a crystallographic *2-axis* and 10-coordinated by pairs of symmetric-related tridentate trz and bidentate nitrate ligands. From this complex a hexahydrate compound is also known.

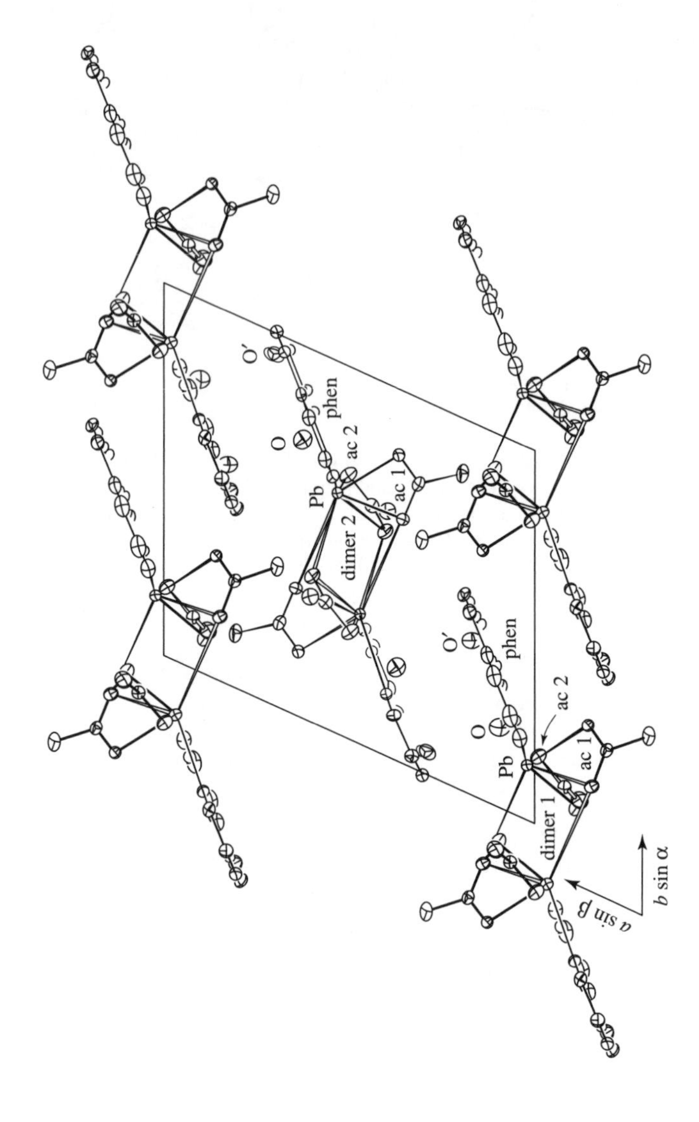

FIGURE 80. Unit cell contents of $(phen)Pb(ac)_2 \bullet 2H_2O$. Reproduced by permission of the Australian Academy of Science and CSIRO from Reference 133

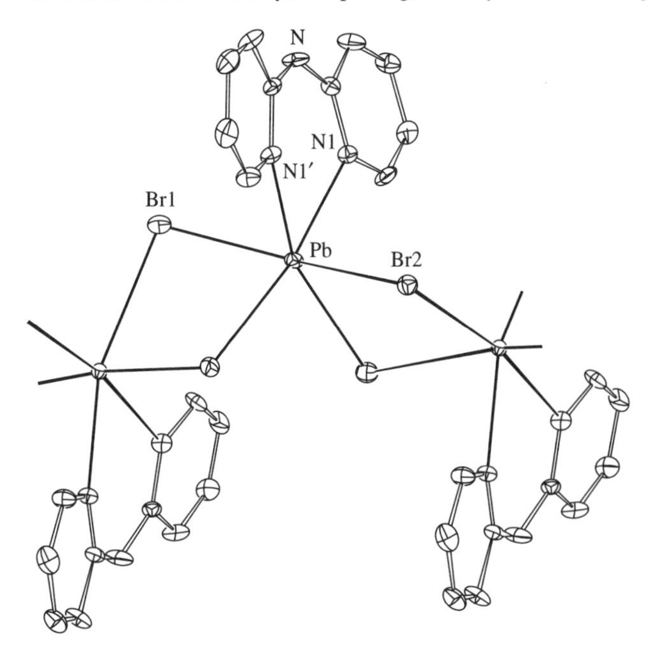

FIGURE 81. The polymer strand of $(dpa)PbBr_2$, normal to its axis. Reproduced by permission of the Australian Academy of Science and CSIRO from Reference 134

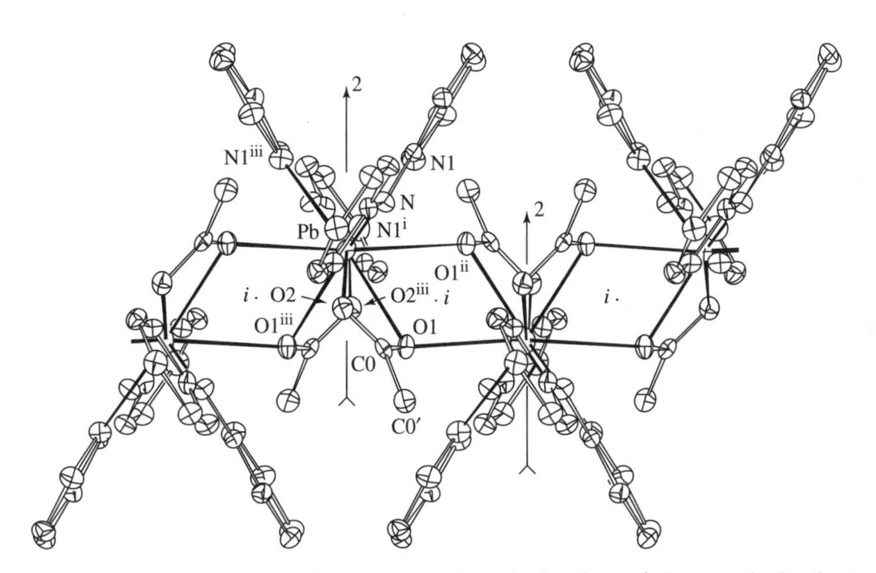

FIGURE 82. The polymer $(dpa)_2Pb(ac)_2$ as shown in projection down a^* (i.e. onto the *bc* plane). Reproduced by permission of the Australian Academy of Science and CSIRO from Reference 134

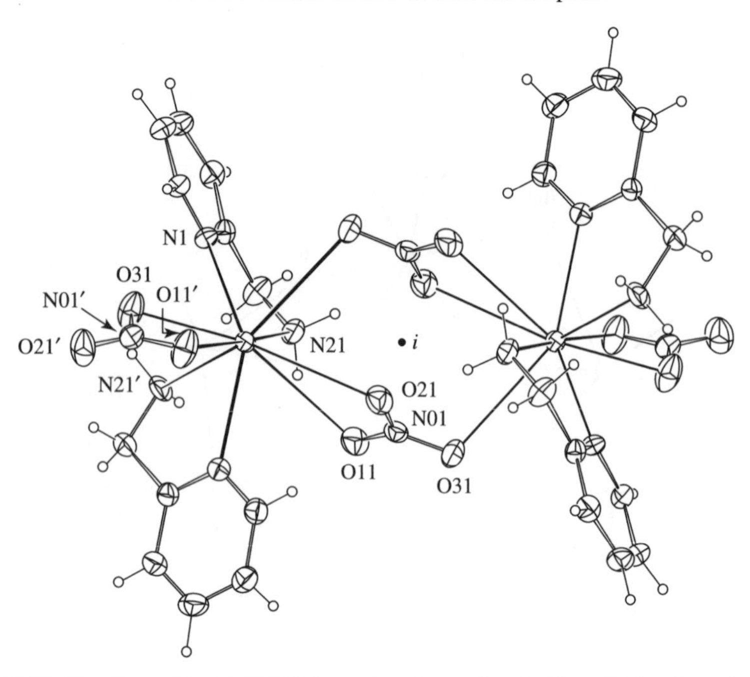

FIGURE 83. The dimer of $(amp)_2Pb(NO_3)_2$, projected normal to the Pb $\cdots$ Pb line. Reproduced by permission of the Australian Academy of Science and CSIRO from Reference 135

trz forms also 1 : 1 adducts with lead(II) chloride, bromide, iodide and thiocyanate, the chloride and bromide being methanol monosolvates[137]. $[(trz)PbCl_2]_{(\infty|\infty)}$•MeOH is monoclinic (Figure 87); the bromide is related (Figure 88), having a derivative triclinic cell. $[(trz)PbI_2]_{(\infty|\infty)}$ (Figure 89), also solvated, is triclinic. In all three compounds an infinite PbX_2PbX_2Pb polymer is found with the plane of tidentate trz lying quasi-normal to the polymer axis. The thiocyanate (monoclinic) (Figure 90) is a column of dimeric units stacked up *b*, successive lead atoms being bridged by thiocyanate sulphur atoms packed in between them and quasi-parallel to *a*, and by thiocyanates parallel to *b* which link pairs of lead atoms in each dimer with the same *b* coordinate by pairs of bridging nitrogens and bridging sulphurs from the adjacent pair.

One of the smallest and simple ligands is pyridine (py). In a reaction with lead(II) thiocyanate it forms a 1 : 1 adduct[138]. $[(py)Pb(SCN)_2]_{(\infty|\infty)}$ is triclinic (Figure 91). The structure is a two-dimensional polymer in the *bc* plane with eight-coordination (py-N)PbN_3S_4 linked by the familar four-membered Pb_2S_2 and eight-membered $Pb_2(SCN)_2$ motifs by way of bridging thiocyanate groups; one of the latter, unusually, has a bifurcating bridging nitrogen atom leading to the introduction of Pb_2N_2 motifs.

The chemistry of tetraimino macrocyclic complexes is of considerable interest because of their applications for modeling bioinorganic systems, catalysis and analytical practice[139]. The lead(II) complex $[Pb(L^{11})_2](BPh_4)_2$•$2CH_3CN$ (L^{11} = 3,10,17,24-tetraaza-29,30-dioxapentacyclo$[24.2.1.1^{12,15}.0^{4,9}0^{18,23}]$-triaconta-1(28),2,4,6,8,10,12,14, 16,18,20,22,24,26-tetradecaene, Figure 92; see Figures 93 and 94) is prepared by a metathesis reaction of $[Pb(L^{11})](ClO_4)_2$•H_2O complex in methanol solution with

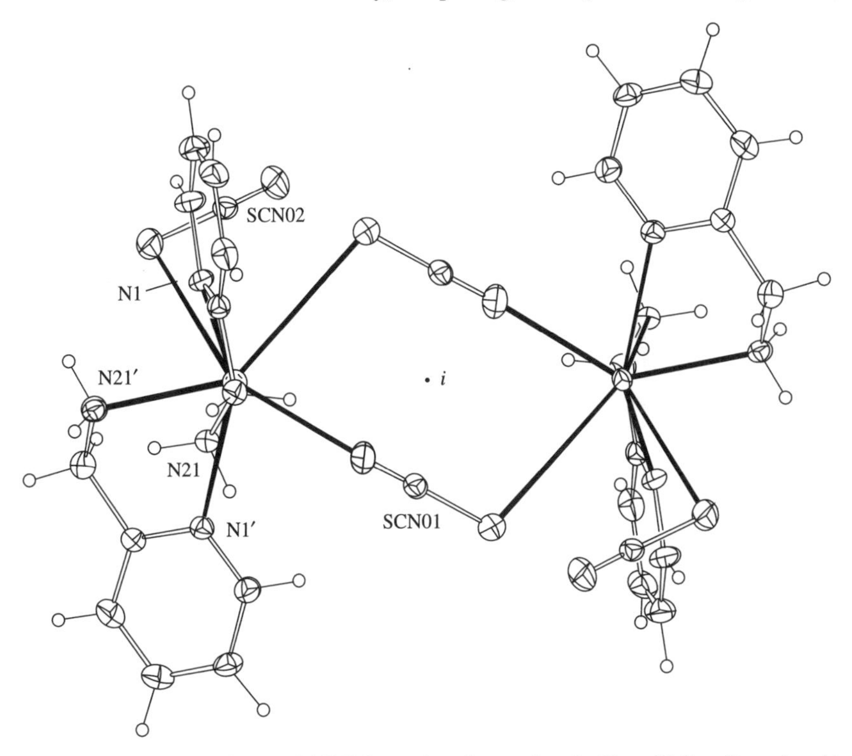

FIGURE 84. The dimer of $(amp)_2Pb(SCN)_2$, projected normal to the Pb ⋯ Pb line. Reproduced by permission of the Australian Academy of Science and CSIRO from Reference 135

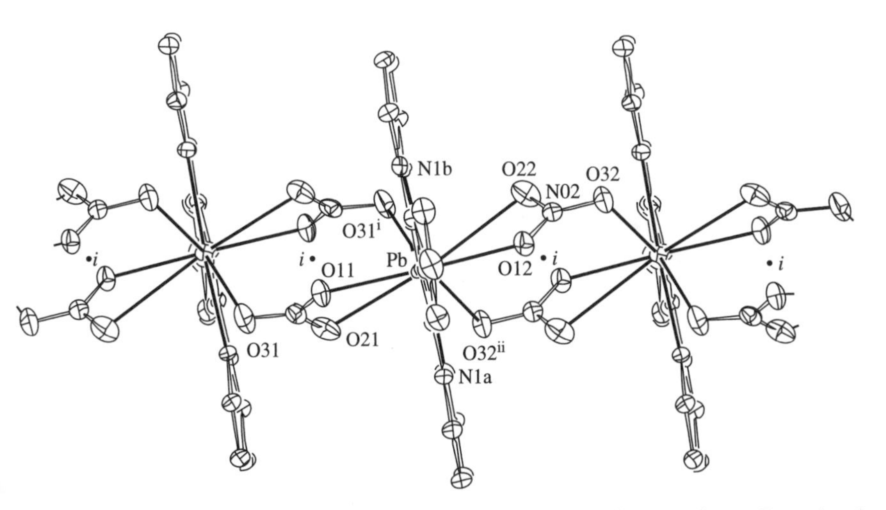

FIGURE 85. A polymer strand of $(trz)Pb(NO_3)_2$ is projected down the *quasi-2-axis*. Reproduced by permission of the Australian Academy of Science and CSIRO from Reference 136

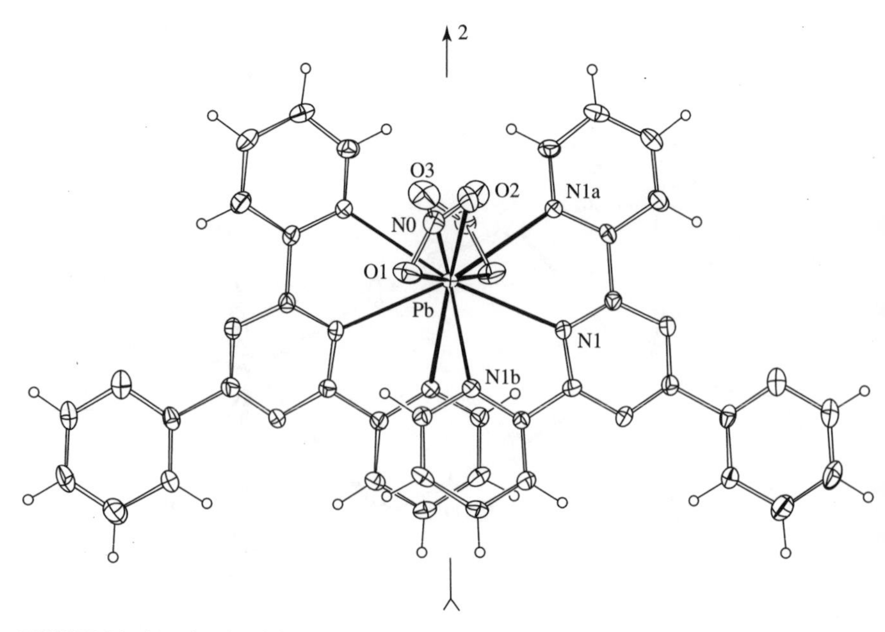

FIGURE 86. A molecule of the methanol-solvated $(trz)_2Pb(NO_3)_2$ adduct projected normal to the *2-axis*. Reproduced by permission of the Australian Academy of Science and CSIRO from Reference 136

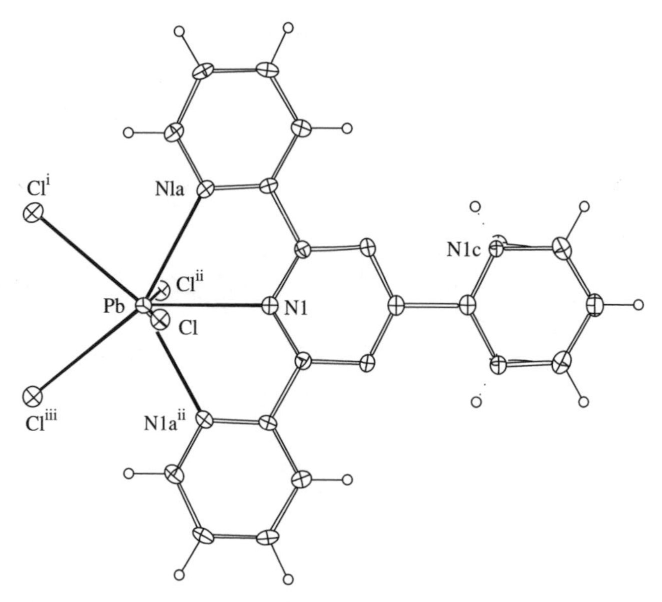

FIGURE 87. Structure of $(trz)PbCl_2$ projected normal to the ligand 'plane', showing the lead coordination environment. Reproduced by permission of the Australian Academy of Science and CSIRO from Reference 137

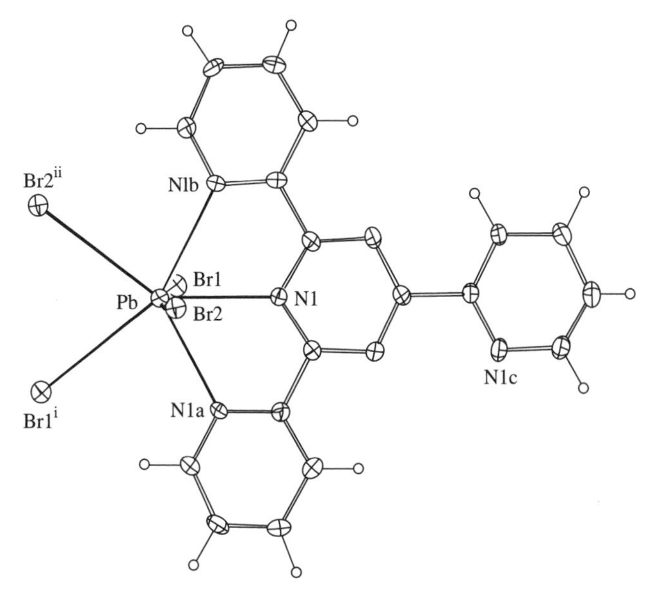

FIGURE 88. Structure of (trz)$PbBr_2$ projected normal to the ligand 'plane', showing the lead coordination environment. Reproduced by permission of the Australian Academy of Science and CSIRO from Reference 137

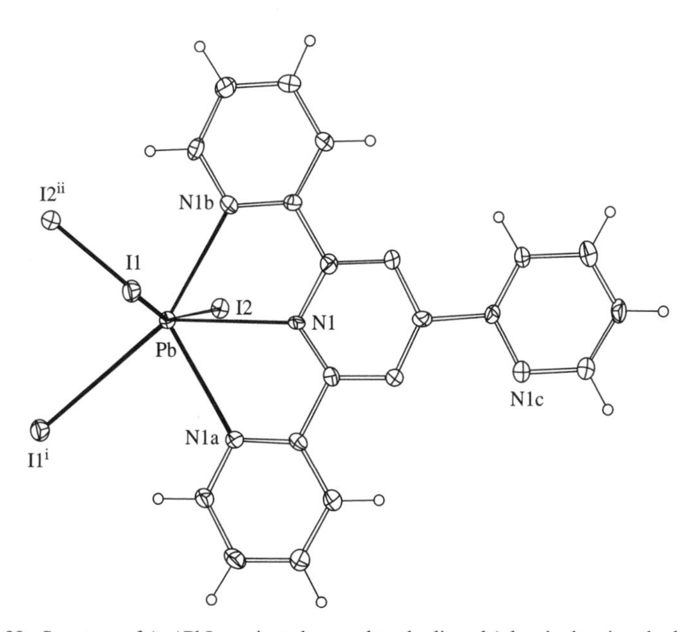

FIGURE 89. Structure of (trz)PbI_2 projected normal to the ligand 'plane', showing the lead coordination environment. Reproduced by permission of the Australian Academy of Science and CSIRO from Reference 137

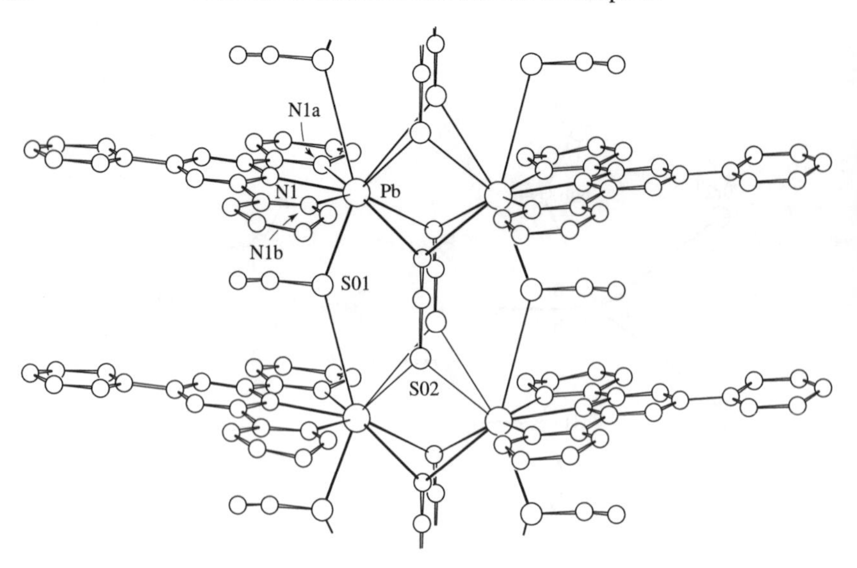

FIGURE 90. The $(trz)Pb(SCN)_2$ polymer projected normal to *b*. Reproduced by permission of the Australian Academy of Science and CSIRO from Reference 137

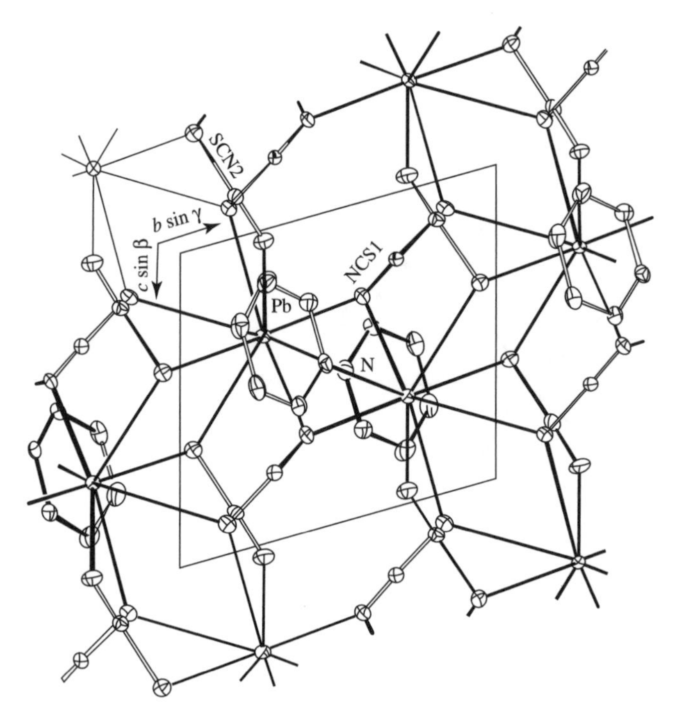

FIGURE 91. The unit cell of $(py)Pb(SCN)_2$ projected down *a*, normal to the polymer plane. Reproduced by permission of the Australian Academy of Science and CSIRO from Reference 138

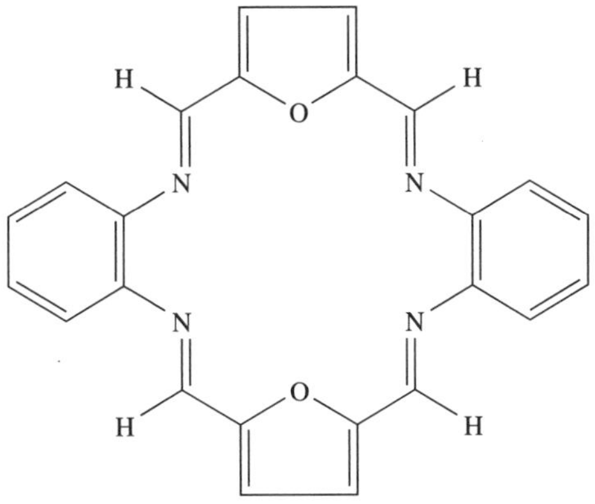

L^{11}

FIGURE 92. Structure of L^{11} = 3,10,17,24-tetraaza-29,30-dioxapentacyclo[24,2,1,$1^{12,15}$,$0^{4,9}$$0^{18,23}$]-triaconta-1(28),2,4,6,8,10,12,14,16,18,20,22,24,26-tetradecaene. Reproduced by permission of Elsevier Science from Reference 140

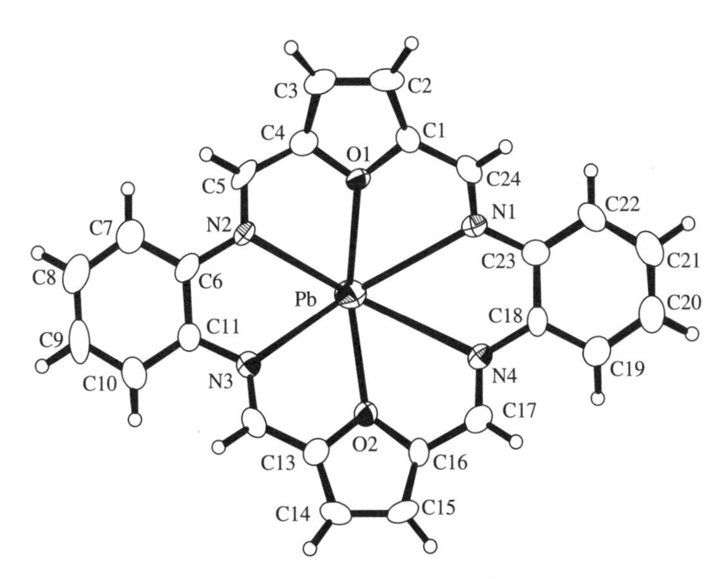

FIGURE 93. ORTEP drawings of the lead(II) complex $[Pb(L^{11})_2](BPh_4)_2$, showing the Pb atom and one of the two macrocycles. Reproduced by permission of Elsevier Science from Reference 140

$NaBPh_4$[140]. This complex has a sandwich-type structure and the macrocycles show a folded conformation, in which two oxygen atoms lie on the opposite side of the 'N_4' plane from the lead atom (Figure 93). The perchlorate complex has a perchlorate anion as a ligand. If the perchlorate anions are replaced by non-coordinating tetraphenylborate anions, the coordination sphere of the Pb(II) ion becomes unsaturated; thus the Pb^{2+} ion may prefer to coordinate two macrocyclic molecules.

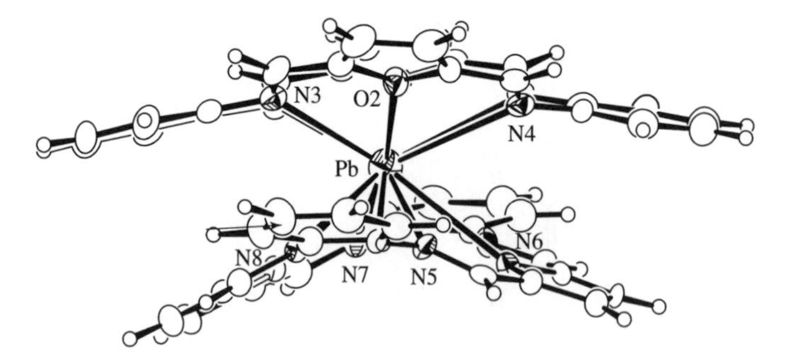

FIGURE 94. A side view of a cation of the sandwich-type $[Pb(L^{11})_2]^{2+}$ complex. Reproduced by permission of Elsevier Science from Reference 140

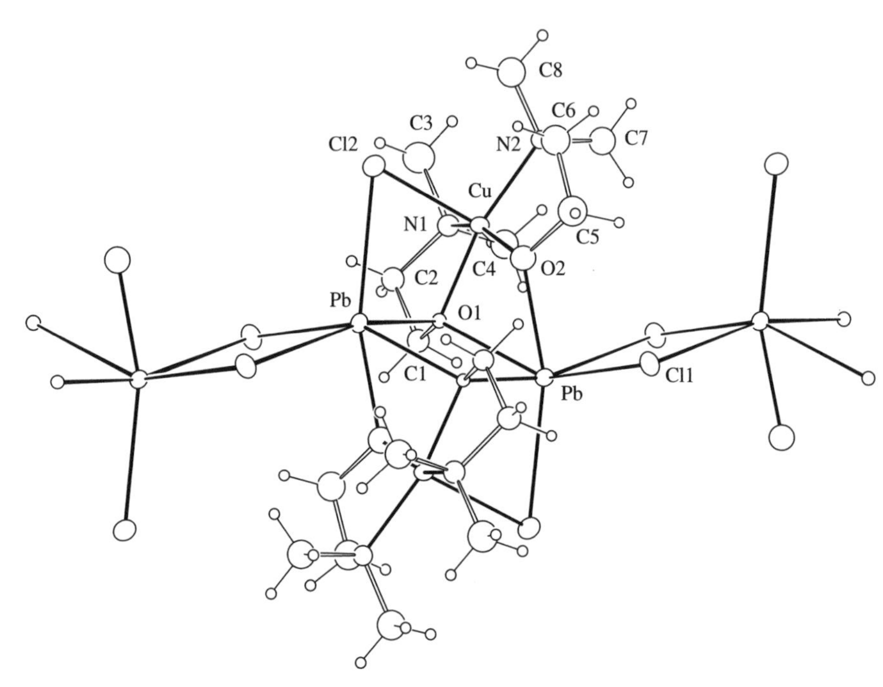

FIGURE 95. Fragment of the polymeric chain present in the crystal structure of $[CuPbCl_2(L^{12})_2]_n \bullet n/2H_2O$. Reproduced by permission of Elsevier Science from Reference 143

The interest in mixed-metal complexes may be attributed to increased recognition of the importance of polynuclear centres in biological catalytic processes and the potential application of expected metal–metal interactions to synthetic systems possessing useful magnetic and electrochemical properties[141,142]. For this reason two mixed-metal complexes of copper(II) and lead(II) with the ligand HL^{12} (2-dimethylaminoethanol) are investigated, $[CuPbCl_2(L^{12})_2]_n \bullet n/2\,H_2O$ (Figure 95) and $[CuPbI_2(L^{12})_2]_2$ (Figure 96)[143].

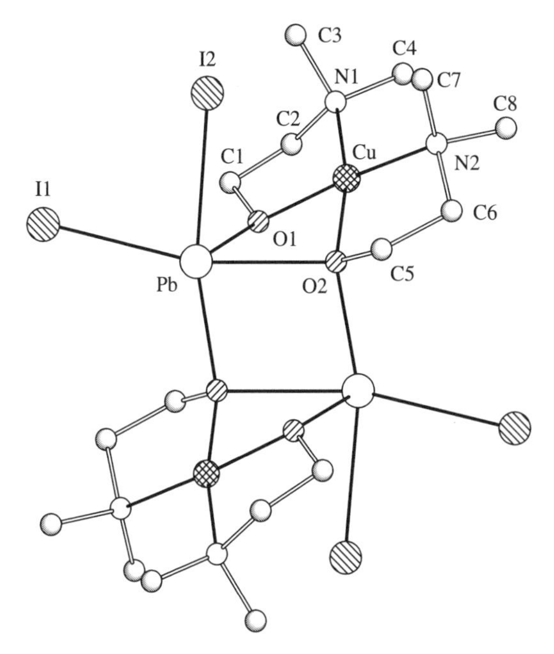

FIGURE 96. Molecular structure of $[CuPbI_2(L^{12})_2]_2$ (hydrogen atoms are omitted for clarity). Reproduced by permission of Elsevier Science from Reference 143

The metal complex motif of both compounds is a centrosymmetric tetranuclear dimer with square-pyramidal coordination geometry around the Cu atom and highly distorted octahedral coordination to the Pb atom. The Cu and Pb atoms are bridged by alkoxide oxygens from L^{12} to form a $Pb_2Cu_2O_4$ core which displays a flattened chair conformation. In the chloride complex tetranuclear units are connected successively via μ-chloro bridging between two Pb sites forming polymeric chains. The Pb atom is six-coordinate, being surrounded by three oxygen atoms from the three L^{12} groups and three chloride atoms with substantial departure from an ideal octahedral geometry. In the iodide complex the I atom bonded to the Pb atom bridges to Cu and Pb atoms of adjacent complex molecules to produce a layer parallel to the *bc* plane. The one-pot synthesis of these complexes (reaction of copper powder with lead salt in non-aqueous solution of 2-dimethylaminoethanol, HL^{12}) has the merit of mild reaction conditions and short reaction time, good yield and its versatility. It is possible to produce predictable mixed-metal complexes by reacting different metal powders and metal salts in solutions of different aminoalcohols or other complexing agents.

The mesityl group (Mes = 2,4,6-$(CH_3)_3C_6H_2$), being a relatively bulky substituent, has been widely used in the chemistry of silicon and germanium in order to stabilize and isolate new types of compounds, such as disilenes and digermenes[144–150]. In contrast, there are a few reports of mesityl derivatives of lead, such as Mes_4Pb[151], Mes_3PbCl and Mes_3PbI[152,153]. The reaction of mesityllithium (prepared from mesityl bromide) with lead(II) chloride in THF results in the unexpected formation of trimesityllead bromide and dimesityllead dibromide as a side product[154]. Their structures are given in Figures 97 and 98. The desired product mesityllead(II) chloride was not detected. It is believed

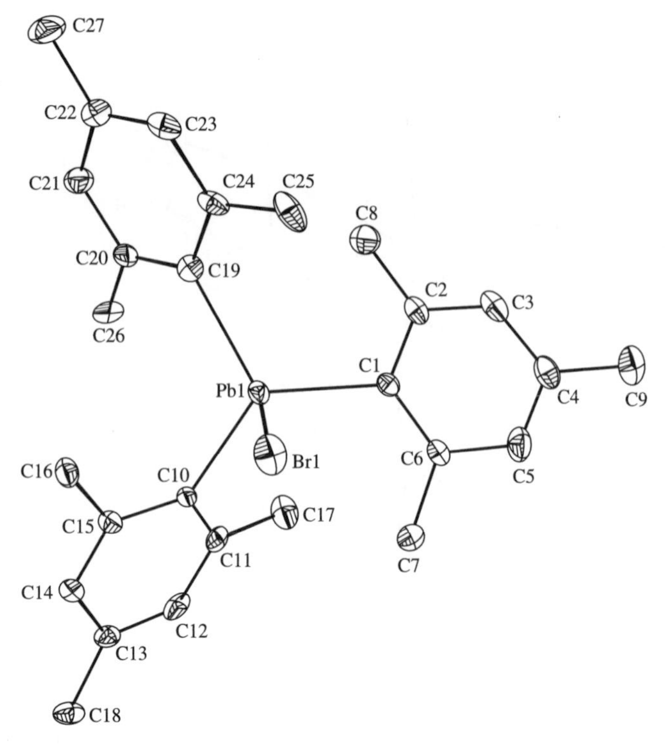

FIGURE 97. ORTEP view of Mes_3PbBr (25% probability, H atoms are omitted for clarity). Reproduced by permission of Elsevier Science from Reference 154

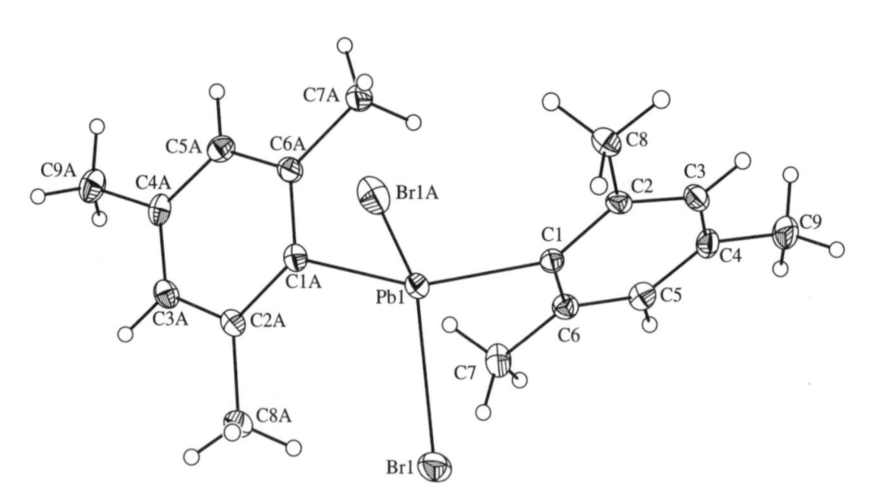

FIGURE 98. ORTEP view of Mes_2PbBr_2 (25% probability). Reproduced by permission of Elsevier Science from Reference 154

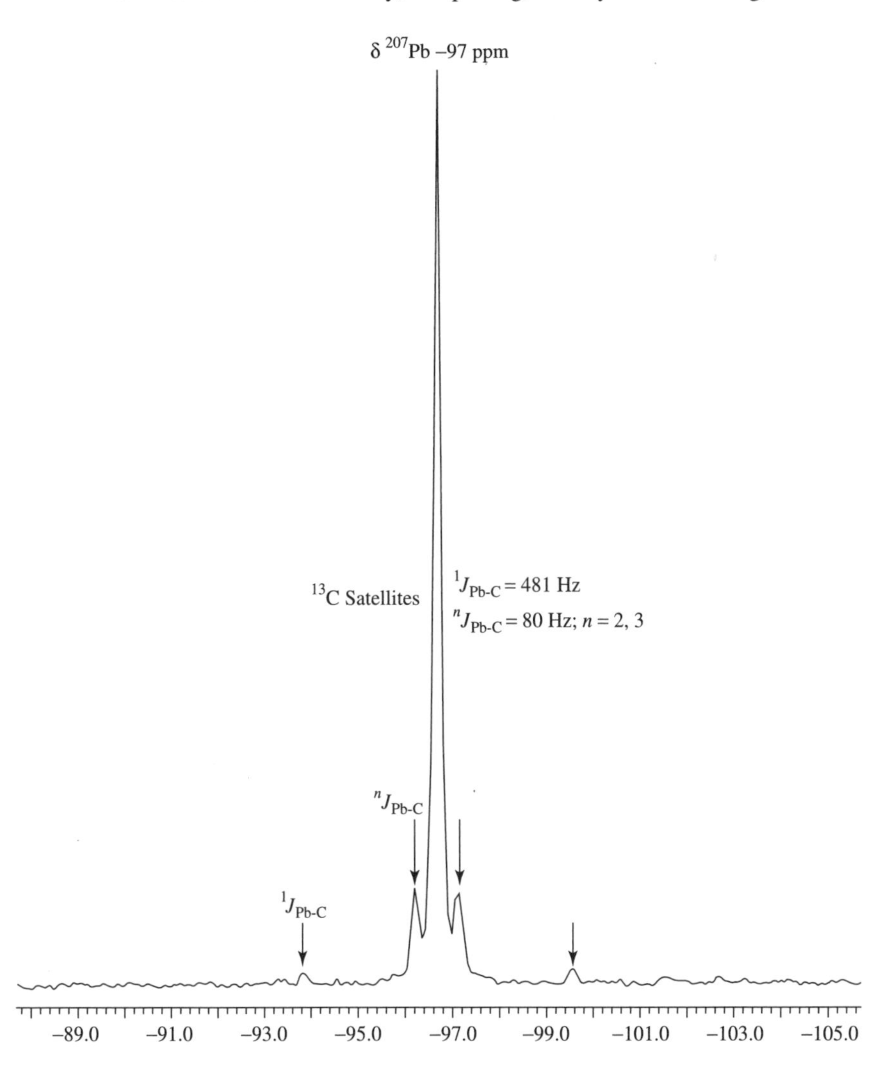

FIGURE 99. $^{207}Pb\{^1H\}$ NMR of Mes_3PbBr in $CDCl_3$. Reproduced by permission of Elsevier Science from Reference 154

that the lead–bromine bond was formed via exchange of the Cl for Br (arising from the precursor mesityl bromide). Initially formed plumbylenes, Mes_2Pb, MesPbCl or MesPbBr, which can be postulated as being intermediates, could then react with mesityl bromide to produce Mes_3PbBr and Mes_2PbBr_2[155]. The $^{207}Pb\{^1H\}$ NMR spectrum of Mes_3PbBr in $CDCl_3$ (Figure 99) shows the ^{13}C satellites[156]. For C-1, the ^1J coupling is visible; the smaller couplings, ^2J and ^3J, consisting of the satellites C-2, C-3 and 2,6-CH_3, are also present but cannot be resolved.

V. ACKNOWLEDGEMENTS

The authors thank Prof. Zvi Rappoport for suggesting the problem to us. We also thank the Fond der Chemischen Industrie for financial support of this work and Mrs C. Nowak for her help with the many diagrams.

VI. REFERENCES

1. A. Schulz and T. M. Klapötke, Chap. 12 in *The Chemistry of Organic Germanium, Tin and Lead Compounds* (Ed. S. Patai), Wiley, Chichester, 1995, p. 537.
2. I. Omae, *J. Organomet. Chem. Library*, **21**, 355 (1989).
3. A. Werner, *Ber. Dtsch. Chem. Ges.*, **45**, 121 (1912).
4. J. F. Stoddard and R. Zarzycki, *Recl. Trav. Chim. Pays-Bas*, **107**, 515 (1988) and references cited therein.
5. L. Giribabu, T. A. Rao and B. G. Maiya, *Inorg. Chem.*, **38**, 4971 (1999).
6. T. Glaser, E. Bill, T. Weyhermüller, W. Meyer-Klaucke and K. Wieghardt, *Inorg. Chem.*, **38**, 2632 (1999).
7. A. Guerrero, J. Cervantes, L. Velasco, J. Gomez-Lara, S. Sharma, E. Delgado and K. Pannell, *J. Organomet. Chem.*, **464**, 47 (1994).
8. J.-C. Guillemin and K. Malagu, *Organometallics*, **18**, 5259 (1999).
9. T. Hascall, A. L. Rheingold, I. Guzei and G. Parkin, *J. Chem. Soc., Chem. Commun.*, 101 (1998).
10. M. C. Kuchta and G. Parkin, *J. Chem. Soc., Chem. Commun.*, 1351 (1994).
11. W. Maringgele and A. Meller, *Phosphorus, Sulphur, Silicon*, **90**, 235 (1994).
12. S. D. Pastor, V. Huang, D. NabiRahni, S. A. Koch and H.-F. Hsu, *Inorg. Chem.*, **36**, 5966 (1997).
13. Y. Takeuchi, K. Tanaka, K. Tanaka, M. Ohnishi-Kameyama, A. Kalmán and L. Párkányi, *J. Chem. Soc., Chem. Commun.*, 2289 (1998).
14. J. L. Wardell and P. J. Cox, *J. Organomet. Chem.*, **515**, 253 (1996).
15. G. R. Willey, T. J. Woodman, U. Somasundaram, D. R. Aris and W. Errington, *J. Chem. Soc., Dalton Trans.*, 2573 (1998).
16. P. Chaudhuri and K. Wieghardt, *Prog. Inorg. Chem.*, **35**, 329 (1987).
17. G. Backes-Dahmann, W. Hermann, K. Wieghardt and J. Weiss, *Inorg. Chem.*, **24**, 485 (1985).
18. P. Chaudhuri, D. Ventur, K. Wieghardt, E.-M. Peters and A. Simon, *Angew. Chem., Int. Ed. Engl.*, **24**, 57 (1985).
19. P. J. Dearochers, K. W. Nebesny, M. J. La Barre, M. A. Bruck, G. F. Neilson, R. P. Sperline, J. H. Enemark, G. Backes and K. Wieghardt, *Inorg. Chem.*, **33**, 15 (1994).
20. P. Jeske, K. Wieghardt and B. Nubur, *Inorg. Chem.*, **33**, 47 (1994).
21. C. H. Bushweller, M. Z. Lourandos and J. A. Brunelle, *J. Am. Chem. Soc.*, **96**, 1591 (1974).
22. R. A. Y. Jones, A. R. Katritzky and M. Snarey, *J. Chem. Soc., B*, 135 (1970).
23. V. J. Baker, I. J. Ferguson, A. R. Katritzky, P. C. Patel and S. Rahimi-Rastgoo, *J. Chem. Soc., Perkin Trans. 2*, 377 (1978).
24. C. J. Adams, *J. Chem. Soc., Dalton Trans.*, 2059 (1999).
25. V. G. Albano, C. Castellari, M. Monari, V. De Felice, A. Panunzi and F. Ruffo, *Organometallics*, **15**, 4012 (1996).
26. R. E. Allan, M. A. Beswick, N. Feeder, M. Kranz, M. E. G. Mosquera, P. R. Raithby, A. E. H. Wheatley and D. S. Wright, *Inorg. Chem.*, **37**, 2602 (1998).
27. T. J. Karol, J. P. Hyde Jr., H. G. Kuivila and J. A. Zubieta, *Organometallics*, **2**, 103 (1983).
28. H. G. Kuivila, T. J. Karol and K. Swami, *Organometallics*, **2**, 909 (1983).
29. K. Swami, J. P. Hutchinson, H. G. Kuivila and J. A. Zubieta, *Organometallics*, **3**, 1687 (1984).
30. M. Austin, K. Gebreyes, H. G. Kuivila, K. Swami and J. A. Zubieta, *Organometallics*, **6**, 834 (1987).
31. M. T. Blanda and M. Newcomb, *Tetrahedron Lett.*, **30**, 3501 (1989).
32. M. Newcomb, J. H. Horner, M. T. Blada and P. J. Squatritto, *J. Am. Chem. Soc.*, **111**, 6294 (1989).
33. J. H. Horner, P. J. Squatritto, N. McGuire, J. P. Riebenspies and M. Newcomb, *Organometallics*, **10**, 1741 (1991).

34. M. Gielen, K. Jurkschat, J. Meunier-Piret and M. Van Meerssche, *Bull. Soc. Chim. Belg.*, **93**, 379 (1984).
35. K. Jurkschat, H. G. Kuivila, S. Liu and J. A. Zubieta, *Organometallics*, **8**, 2755 (1989).
36. K. Jurkschat, A. Rühlemann and A. Tzschach, *J. Organomet. Chem.*, **381**, C53 (1990).
37. K. Jurkschat, F. Hesselbarth, M. Dargath, J. Lehmann, E. Kleinpeter, A. Tzschach and J. Meunier-Piret, *J. Organomet. Chem.*, **388**, 259 (1990).
38. D. Dakternieks, K. Jurkschat, H. Zhu and E. R. T. Tiekink, *Organometallics*, **14**, 2512 (1995).
39. R. Altmann, K. Jurkschat, M. Schürmann, D. Dakternieks and A. Duthie, *Organometallics*, **16**, 5716 (1997).
40. R. Altmann, K. Jurkschat, M. Schürmann, D. Dakternieks and A. Duthie, *Organometallics*, **17**, 5858 (1998).
41. N. A. Chaniotakis, K. Jurkschat and A. Rühlemann, *Anal. Chim. Acta*, **282**, 345 (1993).
42. J. K. Tsagatakis, N. A. Chaniotakis and K. Jurkschat, *Helv. Chim. Acta*, **77**, 2191 (1994).
43. N. A. Chaniotakis, J. K. Tsagatakis, K. Jurkschat and R. Willem, *React. Funct. Polym.*, **34**, 183 (1997).
44. J. K. Tsagatakis, N. A. Chaniotakis and K. Jurkschat, *Quim. Anal.*, **16**, 105 (1997).
45. R. Altmann, O. Gausset, D. Horn, K. Jurkschat and M. Schürmann, *Organometallics*, **19**, 430 (2000).
46. W. Banße, E. Ludwig, E. Uhlemann, H. Mehner, F. Weller and K. Dehnicke, *Z. Anorg. Allg. Chem.*, **607**, 177 (1992).
47. W. Banße, E. Ludwig, E. Uhlemann, H. Mehner and D. Zeigan, *Z. Anorg. Allg. Chem.*, **620**, 2099 (1994).
48. W. Banße, N. Jäger, E. Ludwig, U. Schilde, E. Uhlemann, A. Lehmann and H. Mehner, *Z. Naturforsch.*, **52b**, 237 (1997).
49. A. J. Crowe, in *Metal Complexes in Cancer Chemotherapy* (Ed. B. K. Keppler), VCH, Weinheim, 1993, p. 369 and references cited therein.
50. J. S. Casas, A. Castiñeiras, E. García Martínez, P. Rodríguez Rodríguez, U. Russo, A. Sánchez, A. Sánchez González and J. Sordo, *Appl. Organomet. Chem.*, **13**, 69 (1999).
51. P. G. Harrison (Ed.), *Chemistry of Tin*, Chap. 2, Blakie, London, 1989.
52. T. J. Pinnavaia, *Science*, **220**, 4595 (1983).
53. D. Dakternieks, K. Jurkschat, R. Tozer, J. Hook and E. R. T. Tiekink, *Organometallics*, **16**, 3696 (1997).
54. A. Khaleel, K. J. Klabunde and A. Johnson, *J. Organomet. Chem.*, **572**, 11 (1999).
55. M. Gielen, *Metal-Based Drugs*, **1**, 213 (1994).
56. A. J. Crowe, *Drugs of the Future*, **12**, 40 (1987).
57. L. E. Khoo, Y. Xu, N. K. Goh, L. S. Chia and L. L. Koh, *Polyhedron*, **16**, 573 (1997).
58. M. C. Kuchta and G. Parkin, *J. Chem. Soc., Chem. Commun.*, 1669 (1996).
59. M. C. Kuchta, J. M. Hahn and G. Parkin, *J. Chem. Soc., Dalton Trans.*, 3559 (1999).
60. A. M. van den Bergen, J. D. Cashion, G. D. Fallon and B. O. West, *Aust. J. Chem.*, **43**, 1559 (1990).
61. W.-P. Leung, L.-H. Weng, W.-H. Kwok, Z.-Y. Zhou, Z.-Y. Zhang and T. C. W. Mak, *Organometallics*, **18**, 1482 (1999).
62. E. J. Gabe, F. L. Lee and F. E. Smith, *Inorg. Chim. Acta*, **90**, L11 (1984).
63. S. W. Ng, *Acta Crystallogr., Sect. C*, **C52**, 354 (1996).
64. R. Rösler, J. E. Drake, C. Silvestru, J. Yang and I. Haiduc, *J. Chem. Soc., Dalton Trans.*, 391 (1996).
65. U. V. S. Rao, J. S. Kumar and K. N. Reddy, *Prog. Cryst. Growth Charact.*, **15**, 187 (1987).
66. J. Proscia and R. G. Gordon, *Thin Solid Films*, **214**, 175 (1992).
67. R. G. Gordon, J. Proscia, F. B. Ellis (Jr.) and A. E. Delahoy, *Solar Energy Mater.*, **18**, 263 (1989).
68. G. K. Bhagavat and K. B. Sundaram, *Thin Solid Films*, **63**, 197 (1979).
69. M. Mizuhashi, Y. Gotoh and K. Adachi, *Jpn. J. Appl. Phys.*, **27**, 2053 (1988).
70. J. A. Aboaf, *J. Electrochem. Soc.*, **114**, 948 (1967).
71. M. Adachi, K. Okuyama, N. Tohge, M. Shimada, J. Sato and M. Muroyama, *Jpn. J. Appl. Phys.*, **32**, L748 (1993).
72. G. A. Battiston, R. Gerbasi, M. Pochia and A. Marigo, *Thin Solid Films*, **239**, 186 (1994).
73. J. A. Samuels, W.-C. Chiang, C.-P. Yu, E. Apen, D. C. Smith, D. V. Baxter and K. G. Caulton, *Chem. Mater.*, **6**, 1684 (1994).

74. S. Suh and D. M. Hoffman, *Inorg. Chem.*, **35**, 6164 (1996).
75. S.-G. Teoh, S.-H. Ang, H.-K. Fun and C.-W. Ong, *J. Organomet. Chem.*, **580**, 17 (1999).
76. M. Mathew and G. J. Palenik, *Acta Crystallogr., Sect. B*, **B27**, 59 (1971).
77. M. J. M. Campbell, *Coord. Chem. Rev.*, **15**, 279 (1975).
78. L. Tian, Z. Zhou, B. Zhao, H. Sun, W. Yu and P. Yang, *Synth. React. Inorg. Met.-Org. Chem.*, **30**, 307 (2000).
79. R. E. Hutton and V. Oakes, *Adv. Chem. Ser.*, **157**, 123 (1976).
80. R. E. Hutton, J. W. Burley and V. Oakes, *J. Organomet. Chem.*, **156**, 369 (1978).
81. R. M. Haigh, A. G. Davies and M. W. Tse, *J. Organomet. Chem.*, **174**, 163 (1979).
82. D. K. Deb and A. K. Ghosh, *Z. Anorg. Allg. Chem.*, **539**, 229 (1986).
83. D. K. Deb and A. K. Ghosh, *Polyhedron*, **5**, 863 (1986).
84. L. Tian, B. Zhao and F. Fu, *Synth. React. Inorg. Met.-Org. Chem.*, **28**, 175 (1998).
85. F. Fu, Z. Li, L. Tiang, H. Pan, F. Kayser, R. Willem and M. Gielen, *Bull. Soc. Chim. Belg.*, **101**, 279 (1992).
86. M. Gielen, H. Pan and E. N. T. Tiekink, *Bull. Soc. Chim. Belg.*, **102**, 447 (1993).
87. A. K. Eaxens, *Appl. Organomet. Chem.*, **1**, 39 (1987).
88. J. Wang and X. Yang, *Heteroatom Chem.*, **7**, 211 (1996).
89. W. S. Matthews, J. E. Baras, J. E. Bartmess, F. G. Bordwell, F. J. Corno, G. E. Drucker, Z. Margolin, R. S. McCallum, G. J. McCallum and N. R. Vanier, *J. Am. Chem. Soc.*, **97**, 7006 (1975).
90. Y. N. Belokon, V. I. Bakhmutov, N. I. Chernoglazova, K. A. Kochetkov, S. V. Vitt, N. S. Garbalinskaya and V. M. Belikov, *J. Chem. Soc., Perkin Trans., 1*, 305 (1988).
91. M. Veith and O. Recktenwald, *Top. Curr. Chem.*, **104**, 1 (1982).
92. P. J. Davidson and M. F. Lappert, *J. Chem. Soc., Chem. Commun.*, 317 (1973).
93. T. Fjiedberg, A. Haaland, B. E. R. Schilling, M. F. Lappert and A. J. Thorne, *J. Chem. Soc., Dalton Trans.*, 1551 (1986).
94. M. Kira, R. Yauchibura, R. Hirano, C. Kabuto and H. Sakurai, *J. Am. Chem. Soc.*, **113**, 7785 (1991).
95. H. Grützmacher, H. Pritzkow and F. T. Edelmann, *Organometallics*, **10**, 23 (1991).
96. M. Weidenbruch, J. Schlaefke, A. Schäfer, K. Peters, H. G. von Schnering and H. Marsmann, *Angew. Chem., Int. Ed. Engl.*, **33**, 1846 (1994).
97. M. Weidenbruch, A. Stilter, J. Schlaefke, K. Peters and H. G. von Schnering, *J. Organomet. Chem.*, **501**, 67 (1995).
98. M. Weidenbruch, A. Stilter, K. Peters and H. G. von Schnering, *Z. Anorg. Allg. Chem.*, **622**, 534 (1996).
99. M. N. Hughes, *The Inorganic Chemistry of Biological Processes*, Wiley, New York, 1981.
100. P. G. Sammes and G. Yahioglu, *Chem. Soc. Rev.*, 328 (1994).
101. C. Bazzicalupi, A. Bencini, V. Fusi, C. Giogi, P. Paoletti and B. Valtancoli, *J. Chem. Soc., Dalton Trans.*, 393 (1999).
102. G. A. Bowmaker, J. M. Harrowfield, H. Miyamae, T. M. Shand, B. W. Skelton, A. A. Soudi and A. H. White, *Aust. J. Chem.*, **49**, 1089 (1996).
103. S. Brooker and R. J. Kelly, *J. Chem. Soc., Dalton Trans.*, 2117 (1996).
104. P. Hubberstey and C. E. Russell, *J. Chem. Soc., Chem. Commun.*, 959 (1995).
105. F. Abraham, M. Lagrenee, S. Sueur, B. Mernari and C. Bremard, *J. Chem. Soc., Dalton Trans.*, 1443 (1991).
106. J. E. Andrew, P. W. Ball and A. B. Blake, *J. Chem. Soc., Chem. Commun.*, 143 (1969).
107. P. Dapporto, G. De Munno, A. Sega and C. Mealli, *Inorg. Chim. Acta*, **83**, 171 (1984).
108. M. Ghedini, F. Neve, F. Morazzoni and C. Oliva, *Polyhedron*, **4**, 497 (1985).
109. L. K. Thompson, S. K. Mandal, E. J. Gabe, F. L. Lee and A. W. Addison, *Inorg. Chem.*, **26**, 657 (1987).
110. P. J. Steel, *Coord. Chem. Rev.*, **106**, 227 (1990).
111. M. P. Gamasa, J. Gimeno, E. Lastra, J. M. Rubio Gonzalez and S. Garcia-Granda, *Polyhedron*, **9**, 2603 (1990).
112. S. S. Tandon, L. K. Thompson, M. E. Manuel and J. N. Bridson, *Inorg. Chem.*, **33**, 5555 (1994).
113. J. E. Andrew, A. B. Blake and L. F. Fraser, *J. Chem. Soc., Dalton Trans.*, 800 (1975).
114. P. W. Ball and A. B. Blake, *J. Chem. Soc., A*, 1415 (1969).

115. L. K. Thompson, V. T. Chacko, J. A. Elvidge, A. B. P. Lever and R. V. Parish, *Can. J. Chem.*, **47**, 4141 (1969).
116. D. A. Sullivan and G. J. Palenik, *Inorg. Chem.*, **16**, 1127 (1977).
117. D. Attanasio, G. Dessy and V. Fare, *Inorg. Chim. Acta*, **104**, 99 (1985).
118. T. C. Woon, R. McDonald, S. K. Mandal, L. K. Thompson, S. P. Connors and A. W. Addison, *J. Chem. Soc., Dalton Trans.*, 2381 (1986).
119. L. Chen, L. K. Thompson and J. N. Bridson, *Inorg. Chem.*, **32**, 2938 (1993).
120. L. Chen, L. K. Thompson and J. N. Bridson, *Can. J. Chem.*, **71**, 1086 (1993).
121. L. Chen, L. K. Thompson, S. S. Tandon and J. N. Bridson, *Inorg. Chem.*, **32**, 4063 (1993).
122. S. S. Tandon, L. K. Thompson, J. N. Bridson and M. Bubenik, *Inorg. Chem.*, **32**, 4621 (1993).
123. P. Guerriero, P. A. Vigato, D. E. Fenton and P. C. Hellier, *Acta Chem. Scand.*, **46**, 1025 (1992).
124. K. K. Nanda, L. K. Thompson, J. N. Bridson and K. Nag, *J. Chem. Soc., Chem. Commun.*, 1337 (1994).
125. S. S. Tandon, L. K. Thompson, J. N. Bridson and C. Benelli, *Inorg. Chem.*, **34**, 5507 (1995).
126. I. Bytheway, L. M. Engelhardt, J. M. Harrowfield, D. L. Kepert, H. Miyamae, J. M. Patrick, B. W. Skelton, A. A. Soudi and A. H. White, *Aust. J. Chem.*, **49**, 1099 (1996).
127. L. M. Engelhardt, J. M. Harrowfield, H. Miyamae, J. M. Patrick, B. W. Skelton, A. A. Soudi and A. H. White, *Aust. J. Chem.*, **49**, 1111 (1996).
128. L. M. Engelhardt, J. M. Harrowfield, H. Miyamae, J. M. Patrick, B. W. Skelton, A. A. Soudi and A. H. White, *Aust. J. Chem.*, **49**, 1135 (1996).
129. J. M. Harrowfield, H. Miyamae, B. W. Skelton, A. A. Soudi and A. H. White, *Aust. J. Chem.*, **49**, 1029 (1996).
130. J. M. Harrowfield, H. Miyamae, T. M. Shand, B. W. Skelton, A. A. Soudi and A. H. White, *Aust. J. Chem.*, **49**, 1043 (1996).
131. J. M. Harrowfield, H. Miyamae, T. M. Shand, B. W. Skelton, A. A. Soudi and A. H. White, *Aust. J. Chem.*, **49**, 1051 (1996).
132. J. M. Harrowfield, H. Miyamae, B. W. Skelton, A. A. Soudi and A. H. White, *Aust. J. Chem.*, **49**, 1067 (1996).
133. J. Harrowfield, H. Miyamae, B. W. Skelton, A. A. Soudi and A. H. White, *Aust. J. Chem.*, **49**, 1081 (1996).
134. J. M. Harrowfield, H. Miyamae, B. W. Skelton, A. A. Soudi and A. H. White, *Aust. J. Chem.*, **49**, 1121 (1996).
135. J. M. Harrowfield, H. Miyamae, B. W. Skelton, A. A. Soudi and A. H. White, *Aust. J. Chem.*, **49**, 1127 (1996).
136. J. M. Harrowfield, D. L. Kepert, H. Miyamae, B. W. Skelton, A. A. Soudi and A. H. White, *Aust. J. Chem.*, **49**, 1147 (1996).
137. J. M. Harrowfield, H. Miyamae, B. W. Skelton, A. A. Soudi and A. H. White, *Aust. J. Chem.*, **49**, 1157 (1996).
138. J. M. Harrowfield, H. Miyamae, B. W. Skelton, A. A. Soudi and A. H. White, *Aust. J. Chem.*, **49**, 1165 (1996).
139. P. Guerriero, S. Tamburini and P. A. Vigato, *Coord. Chem. Rev.*, **139**, 17 (1995).
140. T. Tsubomura, M. Ito and K. Sakai, *Inorg. Chim. Acta*, **284**, 149 (1999).
141. J.-M. Lehn, *Angew. Chem., Int. Ed. Engl.*, **27**, 89 (1988).
142. O. Kahn, *Molecular Magnetism.*, VCH Publishers, New York, 1993.
143. O. Y. Vassilyeva, L. A. Kovbasyuk, V. N. Kokozay, B. W. Skelton and W. Linert, *Polyhedron*, **17**, 85 (1998).
144. T. Tsumuraya, S. A. Batcheller and S. Masamune, *Angew. Chem., Int. Ed. Engl.*, **30**, 902 (1991).
145. R. West, *Angew. Chem., Int. Ed. Engl.*, **26**, 1201 (1987).
146. R. West, M. J. Fink and J. Michl, *Science*, **214**, 1343 (1981).
147. S. Collin, S. Murakami, H. Tobita and D. J. Williams, *J. Am. Chem. Soc.*, **105**, 7776 (1983).
148. K. M. Baines and J. A. Cooks, *Organometallics*, **10**, 3419 (1991).
149. M. Riviere-Baudet, A. Morere, J. F. Britten and M. Onyszchuk, *J. Organomet. Chem.*, **423**, C5 (1992).
150. S. Masamune, in E. R. Corey, J. Y. Corey and P. P. Gaspar (Eds.), *Silicon Chemistry*, Ellis Horwood, Chichester, 1988, p. 257.
151. H. Gilman and J. Bailie, *J. Am. Chem. Soc.*, **61**, 731 (1939).

152. H. K. Sharma, R. J. Villazana, F. Cervantes-Lee, L. Parkanyi and K. H. Pannell, *Phosphorus, Sulphur, Silicon*, **87**, 257 (1994).
153. B. C. Pant and W. E. Davidson, *J. Organomet. Chem.*, **39**, 295 (1972).
154. T. M. Klapötke, J. Knizek, B. Krumm, H. Nöth and C. M. Rienäcker, *Polyhedron*, **18**, 839 (1999).
155. F. Glocking, K. Hooton and D. Kingston, *J. Chem. Soc., A*, 4405 (1961).
156. T. M. Klapötke, J. Knizek, B. Krumm, H. Nöth and C. M. Rienäcker, *Polyhedron*, **18**, 1687 (1999).

CHAPTER **8**

Structural effects on germanium, tin and lead compounds

MARVIN CHARTON

Chemistry Department, School of Liberal Arts and Sciences, Pratt Institute, Brooklyn, New York 11205, USA
Fax: 718-722-7706; e-mail: mcharton@pratt.edu

The chemistry of organic germanium, tin and lead compounds — Vol. 2
Edited by Z. Rappoport

I. THE NATURE OF STRUCTURAL EFFECTS

A. Introduction

This article is a supplement to the chapter entitled 'Substituent effects of germanium, tin, and lead groups'[1]. Included is work published after the appearance of the original chapter, topics which were not discussed previously and topics for which further examples would be useful. Readers should consult the glossary in the original chapter for the definitions of terms and variables. Those terms and variables which are new will be defined in Appendix I of this supplement. A list of abbreviations used is given below. The objective of both the original chapter and this supplement is to describe methods and parameters for the quantitative description of structural effects on chemical reactivity, chemical and physical properties, and biological activities of germanium, tin and lead compounds.

ABBREVIATIONS

Ak	alkyl	*i*-Bu	isobutyl	1-Vn	vinylidene	2-Fr	2-furyl
c-Ak	cycloalkyl	*t*-Bu	*tert*-butyl	2-Vn	vinylene	3-Fr	3-furyl
Me	methyl	Pe	pentyl	Ph	phenyl	2-Tp	2-thienyl
Et	ethyl	*i*-Pe	isopentyl	Pn	phenylene	3-Tp	3-thienyl
Pr	propyl	Har	heterocycle	1-Nh	1-naphthyl	Py	pyridyl
i-Pr	isopropyl	Hx	hexyl	2-Nh	2-naphthyl	Ac	acetyl
c-Pr	cyclopropyl	*c*-Hx	cyclohexyl	C_2	ethynylene	Hl	halogen
Bu	butyl	Vi	vinyl				

B. Structure–Property Quantitative Relationships (SPQR)

Structural variations in a chemical species (molecule, ion, radical, carbene, benzyne etc.) generally result in changes in some measured property of the species. The property measured may be a chemical reactivity (rate or equilibrium constant, oxidation potential etc.), chemical property (resulting from a difference in intermolecular forces between an initial and a final state), a physical property (either of the ground state or of an excited state) or a biological activity. The change in the measured property that results from a structural variation is a structural effect. Structural effects within a set of related species can be modeled by the correlation of the measured properties with appropriate parameters using statistical methods. The resulting equation is called a structure–property quantitative relationship (SPQR). The parameters required for modeling structural effects may be obtained from physicochemical reference data sets, quantum chemical calculations, topological methods, comparative molecular field analysis (COMFA) or molecular mechanics (restricted to steric effects). An alternative to statistical methods is the use of neural networks.

SPQR have three functions:

1. They are predictive. Once the SPQR has been determined, the value of the property can be calculated for any chemical species for which the structural effect parameters are available. This makes possible the design of chemical species with specific chemical, physical or biological properties.
2. They are explicative. SPQR can be used to explain structural effects on a measured property. In the case of chemical reactivity they can provide information useful in determining reaction mechanisms.
3. They are archival. Information regarding structural effects on measurable properties can easily and concisely be stored in this way.

It must be noted that in order to be explicative, SPQR must be obtained either by using pure parameters or by using composite parameters of known composition. A pure parameter is a parameter which represents a single structural effect. A composite parameter is a parameter that represents two or more structural effects.

Data sets are of three types. The most frequently encountered type has the form XGY in which X is a variable substituent, Y is an active site (an atom or group of atoms responsible for the observed phenomenon) and G is a skeletal group to which X and Y are bonded. A second type has the form XY; the substituent X is directly bonded to the active site Y. In the third type, designated X_Y, the entire chemical species is both active site and variable substituent.

Structural effects are of three types: electrical effects, steric effects and intermolecular force effects.

II. ELECTRICAL EFFECTS

It has long been known that a substituent X in an XGY system can exert an electrical effect on an active site Y. It is also well known that the electrical effect which results when X is bonded to an sp^3 hybridized carbon atom differs from that observed when X is bonded to an sp^2 or an sp hybridized carbon atom. As electron delocalization is minimal, in the first case, it has been chosen as the reference system. The electrical effect observed in systems of this type is a universal electrical effect which occurs in all systems. In the second type of system, a second effect (resonance effect) occurs due to delocalization, which is dependent both on the inherent capacity for delocalization and on the electronic demand of the active site. In systems of the second type the overall (total) electrical effect is assumed to be a combination of the universal and the delocalized electrical effects. For many years an argument has sometimes raged (and at other times whimpered) concerning the mode of transmission of the universal electrical effect. Two models were proposed originally by Derick[2], a through bond model (the inductive effect) and a through space model (the field effect). These proposals were developed into the classical inductive effect (CIE)[3] and the classical field effect (CFE)[4] models. As the CIE model could not account for the observed dependence of the electrical effect on path number, a modified version was introduced (the MIE model)[5]. The matter has recently been treated in some detail[6,7]. The dependence on molecular geometry is in best agreement with a modified field effect (MFE) model[8].

Electrical effects are conveniently described by the triparametric (three independent variables) LDR equation (equation 1):

$$Q_X = L\sigma_{lX} + D\sigma_{dX} + R\sigma_{eX} + h \tag{1}$$

or relationships derived from it. The parameters are described below.

σ_l is the localized (field) electrical effect parameter; it is identical to σ_I and σ_F. Though other localized electrical effect parameters such as σ_I^q have been proposed, there is no advantage to their use. The σ^* parameter and the F parameter have sometimes been used as localized electrical effect parameters; such use is generally incorrect as both of these parameters contain a small but significant delocalized effect contribution. As was noted above, the available evidence is strongly in favor of an electric field model for transmission of the localized effect.

σ_d is the intrinsic delocalized (resonance) electrical effect parameter; it represents the delocalized electrical effect in a system with no electronic demand.

σ_e is the electronic demand sensitivity parameter; it adjusts the delocalized effect of a group to meet the electronic demand of the system.

The electrical effect is characterized by two quantities derived from equation 1:

The electronic demand, η, is a property of a system or of a composite electrical effect parameter that is itself a function of both σ_d and σ_e. It is defined as R/D where R and D are the coefficients of σ_e and σ_d, respectively.

The percent delocalized effect, P_D, is defined by equation 2:

$$P_D = \frac{100D}{L + D} \tag{2}$$

Diparametric equations can be obtained from equation 1 in two ways. One alternative is to combine σ_l and σ_d to form a composite parameter with a fixed value of P_D. The other is to combine σ_d and σ_e to form a composite parameter with a fixed value of η. These composite substituent constants are designated $\sigma_{Ck'}$ (where k' is P_D) and σ_D, respectively. A monoparametric equation results when a composite electrical effect parameter is obtained by combining all three pure electrical effect parameters with fixed values of both P_D and η. The Hammett substituent constants are of this type. The choice of electrical effect parameterization depends on the number of data points in the data set to be modeled. When using linear regression analysis, the number of degrees of freedom, N_{DF}, is equal to the number of data points, N_{dp}, minus the number of independent variables, N_{iv}, minus one. When modeling physicochemical data, N_{DF}/N_{iv} should be at least 2, and preferably 3 or more. As the experimental error in the data increases, N_{DF}/N_{iv} should also increase. Values of electrical effect substituent constants used in Section V are given in Table 1.

III. STERIC EFFECTS

A. Introduction

A short review of the origins and early development of steric effects is given elsewhere[1]. Steric effects are proximity effects that result from and are related to substituent size.

TABLE 1. Electrical effect substituent constants used in applications[a]

	σ_l	σ_d	σ_e	$\sigma_{c14.3}$	$\sigma_{c16.7}$	σ_{c50}	σ_{c60}
Ak, *c*-Ak							
Me	−0.01	−0.14	−0.030	−0.03	−0.04	−0.15	−0.22
Et	−0.01	−0.12	−0.036	−0.03	−0.03	−0.13	−0.19
c-Pr	0.01	−0.17	−0.069	−0.02	−0.02	−0.16	−0.25
Pr	−0.01	−0.15	−0.036	−0.04	−0.04	−0.16	−0.24
i-Pr	0.01	−0.16	−0.040	−0.02	−0.02	−0.15	−0.22
Bu	−0.01	−0.15	−0.036	−0.04	−0.04	−0.16	−0.24
i-Bu	−0.01	−0.14	−0.036	−0.03	−0.04	−0.15	−0.22
t-Bu	−0.01	−0.15	−0.036	−0.04	−0.04	−0.16	−0.24
Pe	−0.01	−0.14	−0.036	−0.03	−0.04	−0.15	−0.22
CH_2Bu-*t*	0.00	−0.16	−0.040	−0.03	−0.03	−0.16	−0.24
c-Hx	0.00	−0.14	−0.036	−0.02	−0.03	−0.14	−0.21
CH_2Z							
CH_2Br	0.20	−0.08	−0.026	0.19	0.18	0.12	0.08
CH_2OH	0.11	−0.10	−0.025	0.09	0.09	0.01	−0.04
CH_2Cl	0.17	−0.06	−0.024	0.16	0.16	0.11	−0.08
CH_2CN	0.20	−0.01	−0.011	0.20	0.20	0.19	0.18
CH_2OMe	0.11	−0.10	−0.041	0.09	0.09	0.01	−0.04
CH_2CH_2CN	0.09	−0.11	−0.024	0.07	0.07	−0.02	−0.08
CH_2Vi	0.02	−0.16	−0.039	−0.01	−0.01	−0.14	−0.22
CH_2OEt	0.11	−0.10	−0.041	0.09	0.09	0.01	−0.04
CH_2GeMe_3	−0.02	−0.31	−0.028	−0.07	−0.08	−0.29	−0.49
CH_2SiMe_3	−0.03	−0.30	−0.029	−0.08	−0.09	−0.27	−0.48
CH_2SnMe_3	−0.03	−0.16	−0.028	−0.06	−0.06	−0.19	−0.29
CH_2-2-Tp	0.06	−0.12	−0.028	0.04	0.04	−0.06	−0.12
CH_2-2-Fr	0.05	−0.12	−0.028	0.03	0.03	−0.07	−0.13
$CH_2CH_2CO_2Et$	0.08	−0.12	−0.027	0.06	0.06	−0.04	−0.10
$CH_2CHMeCO_2Me$	0.07	−0.12	−0.027	0.05	0.05	−0.05	−0.11
CH_2NEt_2	0.03	−0.12	−0.038	0.01	0.01	−0.09	−0.15
CH_2Ph	0.03	−0.13	−0.057	0.01	0.00	−0.10	−0.17
CH_2CH_2Py-2	0.02	−0.13	−0.03	0.00	−0.01	−0.11	−0.18
CH_2CH_2Py-4	0.02	−0.13	−0.03	0.00	−0.01	−0.11	−0.18
CZ_3							
CF_3	0.40	0.13	−0.026	0.42	0.43	0.53	0.60
CCl_3	0.36	0.10	−0.018	0.38	0.38	0.46	0.51
$C(SiMe_3)_3$	−0.09	−0.21	−0.028	−0.13	−0.13	−0.30	−0.41
VnX							
Vi	0.11	−0.08	−0.12	0.10	0.09	0.03	−0.01
2-VnVi	0.12	−0.37	−0.12	0.06	0.05	−0.25	−0.44
C_2Z							
C_2H	0.29	−0.02	−0.10	0.29	0.29	0.27	0.26
Ar							
C_6F_5	0.31	0.08	−0.068	0.32	0.33	0.39	0.43
Ph	0.12	−0.12	−0.12	0.10	0.10	0.00	−0.06
PnZ							
$4-PnNMe_2$	0.09	−0.32	−0.12	0.04	0.03	−0.23	−0.39
$4-PnNEt_2$	0.08	−0.27	−0.12	0.04	0.03	−0.19	−0.34

continued overleaf

TABLE 1. (*continued*)

	σ_l	σ_d	σ_e	$\sigma_{c14.3}$	$\sigma_{c16.7}$	σ_{c50}	σ_{c60}
Har							
2-Fr	0.17	−0.18	−0.13	0.14	0.13	−0.01	−0.10
3-Fr	0.09	−0.13	−0.12	0.07	0.06	−0.04	−0.11
2-Tp	0.19	−0.20	−0.11	0.16	0.15	−0.01	−0.11
3-Tp	0.10	−0.15	−0.11	0.07	0.07	−0.05	−0.13
(CO)Z, CN							
CHO	0.30	0.27	−0.10	0.35	0.35	0.57	0.71
CO_2H	0.30	0.17	−0.041	0.33	0.33	0.47	0.56
Ac	0.30	0.25	−0.095	0.34	0.35	0.55	0.68
CO_2Me	0.32	0.16	−0.070	0.35	0.35	0.48	0.56
CO_2Et	0.30	0.18	−0.064	0.33	0.34	0.48	0.57
CN	0.57	0.12	−0.055	0.59	0.59	0.69	0.75
Si							
$SiBr_3$	0.36	0.07	0.018	0.37	0.37	0.42	0.47
$SiCl_3$	0.36	0.10	−0.017	0.38	0.38	0.46	0.51
SiF_3	0.41	0.14	−0.004	0.43	0.44	0.58	0.62
SiH_3	0	0.13	−0.038	0.02	0.03	0.13	0.20
$SiMeBr_2$	0.21	0.09	−0.035	0.23	0.23	0.30	0.35
$SiMeCl_2$	0.21	0.23	−0.072	0.25	0.26	0.44	0.56
$SiMeF_2$	0.24	−0.02	−0.015	0.24	0.24	0.22	0.21
$SiMe_2Cl$	0.06	0.10	−0.030	0.08	0.08	0.16	0.21
$SiMe_3$	−0.11	0.13	−0.046	−0.09	−0.08	0.02	0.09
$SiMe_2Et$	−0.11	0.13	−0.046	−0.09	−0.08	0.02	0.09
$SiEt_3$	−0.11	0.13	−0.046	−0.09	−0.08	0.02	0.09
$SiMe_2Ph$	−0.07	0.17	−0.043	−0.04	−0.04	0.10	0.19
$SiPh_3$	−0.04	0.33	−0.055	0.02	0.03	0.29	0.46
Ge							
$GeMe_3$	−0.08	0.11	−0.050	−0.06	−0.06	0.03	0.09
$GeEt_3$	−0.08	0.11	−0.050	−0.06	−0.06	0.03	0.09
$GePh_2Br$	0.11	0.28	−0.065	0.16	0.17	0.39	0.53
$GePh_3$	−0.05	0.24	−0.053	−0.01	0.00	0.19	0.31
Sn							
$SnMe_3$	−0.09	0.12	−0.051	−0.07	−0.07	0.03	0.09
$SnEt_3$	−0.09	0.12	−0.051	−0.07	−0.07	0.03	0.09
$SnPh_2Cl$	0.09	0.35	−0.084	0.15	0.16	0.44	0.62
$SnPh_3$	−0.04	0.33	−0.055	0.02	0.03	0.29	0.46
Pb							
$PbPh_3$	−0.04	0.39	−0.055	0.06	0.04	0.35	0.55
N							
N_3	0.43	−0.27	−0.12	0.38	0.38	0.16	0.02
NH_2	0.17	−0.68	−0.13	0.06	0.03	−0.51	−0.85
NHMe	0.13	−0.67	−0.18	0.02	0.00	−0.54	−0.88
NMe_2	0.17	−0.66	−0.24	0.06	0.04	−0.49	−0.82
NEt_2	0.15	−0.65	−0.18	0.04	0.02	−0.50	−0.83
NO	0.37	0.31	−0.056	0.42	0.43	0.68	0.84
NO_2	0.67	0.18	−0.077	0.70	0.71	0.85	0.94

TABLE 1. (*continued*)

	σ_l	σ_d	σ_e	$\sigma_{c14.3}$	$\sigma_{c16.7}$	σ_{c50}	σ_{c60}
O							
OH	0.35	−0.57	−0.044	0.25	0.24	−0.22	−0.51
OMe	0.30	−0.55	−0.064	0.21	0.19	−0.25	−0.53
OAc	0.38	−0.24	−0.005	0.34	0.33	0.14	0.02
OEt	0.28	−0.55	−0.070	0.19	0.17	−0.27	−0.55
OPr-*i*	0.27	−0.55	−0.067	0.18	0.16	−0.28	−0.56
OBu	0.28	−0.55	−0.067	0.19	0.17	−0.27	−0.55
$OSiMe_3$	0.25	−0.44	−0.053	0.18	0.16	−0.19	−0.41
OPh	0.40	−0.51	−0.083	0.31	0.30	−0.11	−0.37
S							
SH	0.27	−0.40	−0.098	0.20	0.19	−0.13	−0.33
SMe	0.30	−0.38	−0.13	0.24	0.22	−0.08	−0.27
SEt	0.26	−0.39	−0.12	0.19	0.18	−0.13	−0.33
SPh	0.31	−0.34	−0.17	0.25	0.24	−0.03	−0.20
SO_2							
SO_2Me	0.59	0.13	−0.052	0.31	0.62	0.72	0.79
SO_2Ph	0.56	0.08	−0.082	0.57	0.58	0.64	0.68
Other							
H	0	0	0	0	0	0	0
Br	0.47	−0.27	−0.028	0.42	0.42	0.20	0.06
Cl	0.47	−0.28	−0.011	0.42	0.41	0.19	0.05
F	0.54	−0.48	0.041	0.46	0.44	−0.06	−0.18
I	0.40	−0.20	−0.057	0.37	0.36	−0.20	0.10

[a] For abbreviations see Section I.A. In the $\sigma_{c,k'}$ values $k' = P_D$. Thus $\sigma_{c14.3}$ had 14.3% delocalized effect.

B. The Nature of Steric Effects

1. Primary steric effects

These effects are due to repulsions between electrons in valence orbitals on adjacent atoms which are not bonded to each other. They supposedly result from the interpenetration of occupied orbitals on one atom by electrons on the other, resulting in a violation of the Pauli exclusion principle. *All primary steric interactions raise the energy of the system in which they occur.* Their effect on chemical reactivity is to either decrease or increase a rate or equilibrium constant, depending on whether steric repulsions are greater in the reactant or in the product (equilibria) or transition state (rate).

2. Secondary steric effects

These effects on chemical reactivity can result from the shielding of an active site from the attack of a reagent, from solvation, or both. They may also be due to a steric effect that determines the concentration of the reacting conformation of a chemical species. The secondary steric effect of a nonsymmetric group will also depend on its conformation.

3. Direct steric effects

Direct steric effects can occur when the active site at which a measurable phenomenon occurs is in close proximity to the substituent. Among the many skeletal groups exhibiting direct steric effects are vicinally (1,2) substituted skeletal groups such as *ortho*-substituted benzenes, **1**, *cis*-substituted ethylenes, **2**, the *ortho* (1,2-, 2,1- and 2,3-) naphthalenes, **3, 4** and **5**, respectively, and *peri* skeletal groups such as 1,8-substituted naphthalenes, **6**. Other vicinal examples are *cis*-1,2-disubstituted cyclopropanes, *cis*-2,3-disubstituted norbornanes and *cis*-2,3-disubstituted [2.2.2]bicyclooctanes, **7, 8** and **9**, respectively. Some skeletal groups do not always show steric effects. 2,3-Disubstituted five-membered ring heteroarenes such as thiophenes and selenophenes are generally free of steric effects. This is probably due to the larger XCC angle in these systems as compared with ethene and benzene systems. Geminally substituted (1,1) skeletal groups such as disubstituted methanes, **10**, and 1,1-disubstituted ethenes, **11**, are also generally free of steric effects.

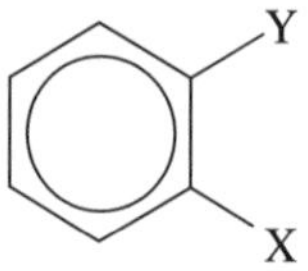

Y X **(1)**

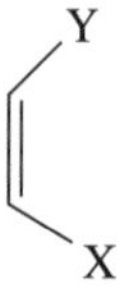

Y X **(2)**

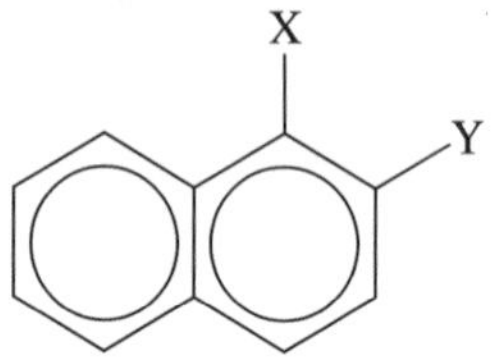

X Y **(3)**

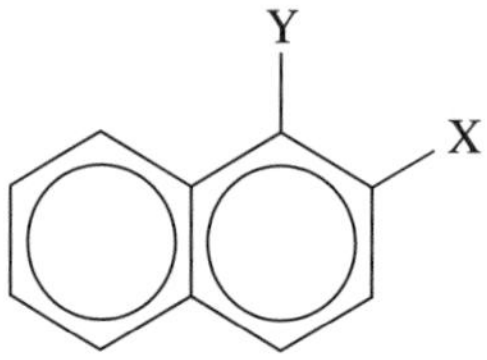

Y X **(4)**

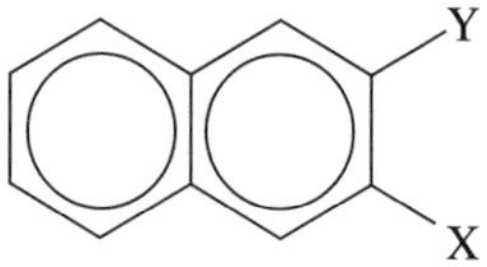

Y X **(5)**

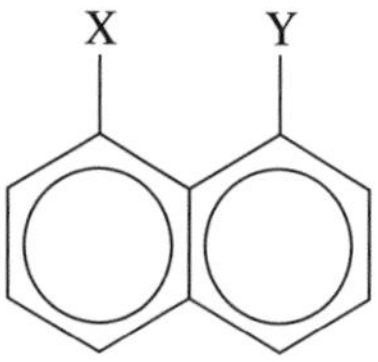

X Y **(6)**

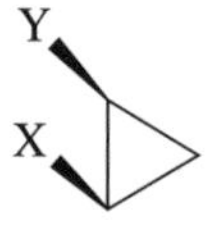

Y X **(7)**

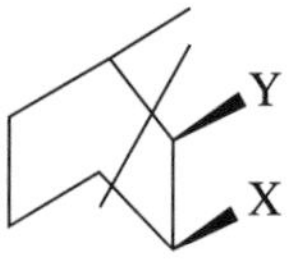

Y X **(8)**

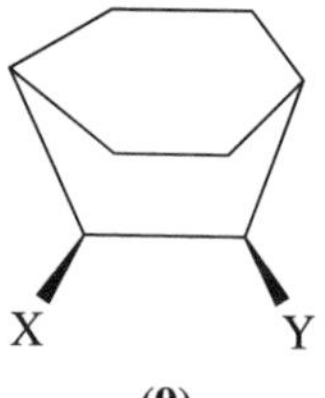

X Y **(9)**

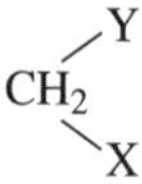

CH_2 Y X **(10)**

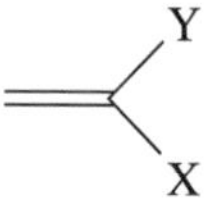

Y X **(11)**

4. *Indirect steric effects*

These effects are observed when the steric effect of the variable substituent is relayed by a constant substituent between it and the active site, as in **12**, where Y is the active site, Z is the constant substituent and X is the variable substituent. This is a type of buttressing effect.

Y Z X

(12)

5. *The directed nature of steric effects*

This is easily shown by considering, for example, the ratio r of the steric parameter ν for any five carbon alkyl group to that for 1-pentyl. Values of r are: 1-Pe, 1; 2-Pe, 1.54; 3-Pe, 2.22; CH_2Bu-*s*, 1.47; CH_2Bu-*i*; 1.00; CH_2Bu-*t*, 1.97; CMe_2Pr, 2.40; CHMePr-*i*, 1.90. All of these groups have the same volume and therefore the same bulk, but they differ in their steric effect[9]. In order to account for this it is necessary to consider what happens when a nonsymmetric substituent is in contact with an active site. We consider the simple case of a spherical active site Y in contact with a nonsymmetric substituent $CZ^LZ^MZ^S$, where the superscripts L, M and S represent the largest, the medium-sized and the smallest Z groups, respectively. The C−G bond and the Y−G bond are of comparable length. There are three possible conformations of this system (Figure 1). As all steric repulsions raise the energy of the system, the preferred conformation will be the one that results in the lowest energy increase. This is the conformation which presents the smallest face to the active site, conformation (a). From this observation we have the minimum steric interaction (MSI) principle which states: *a nonsymmetric substituent prefers that conformation which minimizes steric interactions*. The directed nature of steric effects results in a conclusion of vital importance: that in general *the volume of a substituent is not an acceptable measure of its steric effect*[10]. There are still some workers who are unable to grasp this point. It is nevertheless true that group volumes are not useful as steric parameters except in the case of substituents that are roughly spherical, and not always then. They are actually measures of group polarizability. In short, for a range of different substituent shapes in a data set *steric effects are not directly related to bulk, polarizability is.*

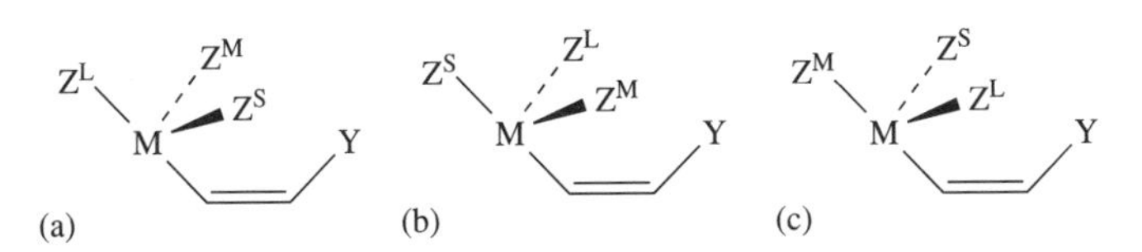

FIGURE 1. Possible conformations of a *cis*-1,2-substituted ethene having a spherical reaction site in contact with a tetrahedral substituent consisting of a central atom M bearing large (Z^L), medium (Z^M) and small (Z^S) sized groups. The energies of the conformations are (a) lowest and (c) highest. The same types of confirmation occur in other 1,2- and 1,3-disubstituted systems in which substituent and reaction site are in contact

C. The Monoparametric Model of Steric Effects

van der Waals radii, r_v, have long been held to be an effective measure of atomic size[11]. Charton proposed the use of the van der Waals radius as a steric parameter[12] and developed a method for the calculation of group van der Waals radii for tetracoordinate symmetric top substituents MZ_3 such as the methyl and trifluoromethyl groups[13a]. In later work the hydrogen atom was chosen as the reference substituent and the steric parameter ν was defined by equation 3:

$$\upsilon_X \equiv r_{VX} - r_{VH} = r_{VX} - 1.20 \quad (3)$$

where r_{VX} and r_{VH} are the van der Waals radii of the X and H groups in Angstrom units[13b].

Expressing r_V in these units is preferable to the use of picometers, because the coefficient of the steric parameter is then comparable in magnitude to the coefficients of the electrical effect parameters. Whenever possible, ν parameters are obtained directly from van der Waals radii or calculated from them. An equation has been derived which makes possible the calculation of ν values for nonsymmetric tetrahedral groups of the types $MZ_2^SZ^L$ and $MZ^SZ^MZ^L$ in which the Z groups are symmetric[14]. These are considered to be primary values. For the greater number of substituents, however, ν parameters must be calculated from the regression equations obtained for correlations of rate constants with primary values. The values obtained in this manner are considered to be secondary ν values. All other measures of atomic size are a linear function of van der Waals radii[10b]. There is therefore no reason for preferring one measure of atomic size over another. As values of ν were developed for a wide range of substituent types with central atoms including oxygen, nitrogen, sulfur and phosphorus as well as carbon, these parameters provide the widest structural range of substituents for which a measure of the steric effect is available.

1. Steric classification of substituents

Such classification is useful in understanding the way in which different types exert steric effects[9]. Substituents may be divided into three categories based on the degree of conformational dependence of their steric effects:

1. No conformational dependence (NCD). Groups of this type include monoatomic substituents such as hydrogen and the halogens, $M^a \equiv M^b$ substituents such as ethynyl and cyano and MZ_3 groups.
2. Minimal conformational dependence (MCD). Among these groups are: (a) Nonsymmetric substituents with the structure $MH_n(lp)_{3-n}$, such as the hydroxyl and amino groups (lp is a lone pair). (b) Nonsymmetric substituents with the structure $MZ_2^SZ^L$, where S stands for small and L for large.
3. Strong conformational dependence (SCD). These groups have the structures: (a) $MZ_2^LZ^S$ and $MZ^LZ^MZ^S$, where the superscript M indicates medium. (b) Planar π-bonded groups MZ^LZ^S where M and either or both Zs are sp^2 hybridized, such as phenyl, acetyl, nitro ($X_{p\pi}$ groups). (c) Quasi-planar π-bonded groups such as dimethylamino and cyclopropyl.

The steric parameter for NCD groups can be obtained directly from van der Waals radii or calculated from them. The values for MCD groups are often obtainable from van der Waals radii, although in some cases they must be derived as secondary values

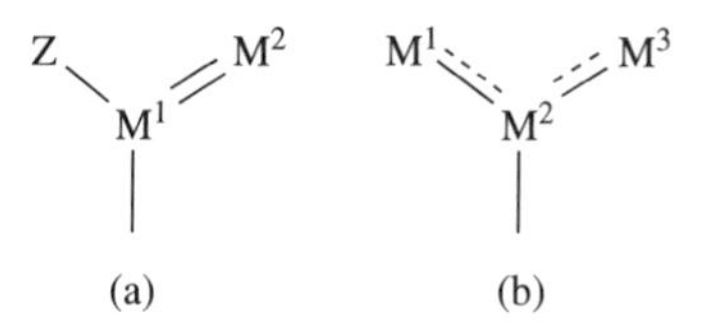

FIGURE 2. Examples of types of planar-bonded groups. (a) Doubly bonded groups such as ZC=O, $Z^1C{=}NZ^2$, ZC=S, ZN=O, N=N, C=C etc. (b) Aryl, heteroaryl, nitro, carboxylate etc

from regression equations obtained by correlating rate constants with known values of the steric parameter. Steric parameters for SCD groups of the nonsymmetric type are usually obtainable only from regression equations.

2. Planar π-bonded groups

In the case of planar π-bonded groups, the maximum and minimum values of the steric parameter are available from the van der Waals radii (Figure 2). A more detailed discussion is provided elsewhere[1]

D. Multiparametric Models of Steric Effects

In some cases a simple monoparametric model of the steric effect is insufficient. Examples are when the active site is itself large and nonsymmetric, or alternatively when the phenomenon studied is some form of bioactivity in which binding to a receptor determines the activity. The failure of the monoparametric model is due to the fact that a single steric parameter cannot account for the variation of the steric effect at various points in the substituent. The use of a multiparametric model of steric effects that can represent the steric effect at different segments of the substituent is required. Five multiparametric models are available: that of Verloop and coworkers[15], the simple branching model, the expanded branching model, the segmental model and the composite model. The Verloop model will not be discussed[1].

1. The branching equations

The simple branching model[10b] for the steric effect is given by equation 4:

$$S\psi = \sum_{i=1}^{m} a_i n_i + a_b n_b \tag{4}$$

where $S\psi$ represents the steric effect parameterization, a_i and a_b are coefficients, n_i is the number of branches attached to the i-th atom, and n_b is the number of bonds between the first and last atoms of the group skeleton. It follows that n_b is a measure of group length. Unfortunately, it is frequently highly collinear in group polarizability, which greatly limits its utility. For saturated cyclic substituents it is necessary to determine values of n_i from an appropriate regression equation. For planar π-bonded groups n_i is taken to be 1 for each atom in the group skeleton. For other groups n_i is obtained simply by counting branches. The model makes the assumption that all of the branches attached to a skeleton atom are equivalent. This is at best only a rough approximation. Distinguishing between branches results in an improved model called the expanded branching equation (equation 5):

$$S\psi = \sum_{i=1}^{m}\sum_{j=1}^{3} a_{ij}n_{ij} + a_b n_b \tag{5}$$

which allows for the difference in steric effect that results from the order of branching[10b]. This difference follows from the MSI principle. The first branch has the smallest steric effect, because a conformation in which it is rotated out of the way of the active site is preferred. In this conformation the active site is in contact with two hydrogen atoms. The preferred conformation in the case of a second branch has the larger of the two branches directed out of the way. The smaller branch and a hydrogen atom are in contact with the active site. When there are three branches, the largest will be directed out of the way and the other two will be in contact with the active site. The problem with the expanded branching method is that it requires a large number of parameters. Data sets large enough to permit its use are seldom seen.

2. *The segmental model*[10b]

This model is often the simplest and most effective of the multiparametric models. In this model each atom of the group skeleton together with the atoms attached to it constitutes a segment of the substituent. Applying the MSI principle, the segment is considered to have that conformation which presents its smallest face to the active site. The segment is assigned the ν value of the group which it most resembles. Values of the segmental steric parameters ν_i, where i designates the segment number, are given in Table 2. Numbering starts from the first atom of the group skeleton which is the atom that is attached to the rest of the system. The segmental model is given by equation 6:

$$S\psi = \sum_{i=1}^{m} S_i \upsilon_i \tag{6}$$

When only steric effects are present, equation 7 applies:

$$Q_X = S\psi_X \tag{7}$$

In the general case, electrical effects are also present and the general form of the LDRS equation (equation 8):

$$Q_X = L\sigma_{lX} + D\sigma_{dX} + R\sigma_{eX} + S\psi_X + h \tag{8}$$

is required.

3. *The composite model*

This model is a combination of the monoparametric ν model with the simple branching model. This method has proven useful in modelling amino acid, peptide and protein properties[10b]. It is an improvement over the simple branching model and requires only one additional parameter.

E. Bond Length Difference as a Factor in Steric Effects

The steric effect exerted by some group X is a function of the lengths of the substituent-skeletal group (X–G) and active site-skeletal group (Y–G) bonds[16]. The steric parameters

TABLE 2. Steric effect parameters used in applications[a]

	υ	υ_1	υ_2	n_1	n_2
Ak, *c*-Ak					
Me	0.52	0.52	0	0	0
Et	0.56	0.52	0.52	1	0
Pr	0.68	0.52	0.52	1	1
i-Pr	0.78	0.78	0	2	0
Bu	0.68	0.52	0.52	1	1
i-Bu	0.98	0.52	0.78	1	2
t-Bu	1.24	1.24	0.52	3	0
Pe	0.68	0.52	0.52	1	1
i-Pe	0.68	0.52	0.52	1	1
c-Hx	0.87			1.5	0.74
Hx	0.73	0.52	0.52	1	1
CH_2Z					
CH_2Br	0.64	0.52	0.65		
CH_2Cl	0.60	0.52	0.55		
CH_2OMe	0.63	0.52	0.32		
CH_2CH_2CN	0.68	0.52	0.52		
CH_2GeMe_3	1.53				
CH_2SiMe_3	1.46				
CH_2-2-Tp	0.70	0.52	0.57		
CH_2-2-Fr	0.70	0.52	0.57		
$CH_2CH_2CO_2Et$	0.68	0.52	0.52		
$CH_2CHMeCO_2Me$		0.52	0.78		
CH_2NEt_2		0.52	0.63		
CH_2CH_2Py-2	0.68	0.52	0.52		
CH_2CH_2Py-4	0.68	0.52	0.52		
Vn					
Vi	0.57	0.57	0.57		
Ar					
Ph	0.57	0.57	0.57		
PnZ					
4-$PnNMe_2$	0.57	0.57	0.57		
4-$PnNEt_2$	0.57	0.57	0.57		
Har					
2-Fr	0.57	0.57	0.57		
3-Fr	0.57	0.57	0.57		
2-Tp	0.57	0.57	0.57		
3-Tp	0.57	0.57	0.57		
Si					
$SiMe_3$	1.40	1.40	0.52		
$SiMe_2Et$	1.40	1.40	0.52		
Ge					
$GeMe_3$	1.44	1.44	0.52		

continued overleaf

TABLE 2. (*continued*)

	υ	υ_1	υ_2	n_1	n_2
O					
OH	0.32	0.32	0		
$OSiMe_3$		0.32	1.40		
Other					
H	0	0	0	0	0
M	l_{CM}[b]				
C	1.537				
Si	1.87				
Ge	1.946				
Sn	2.143				
Pb	2.3				

[a] For abbreviations, see section I.A.
[b] The bond length of the CM bond.

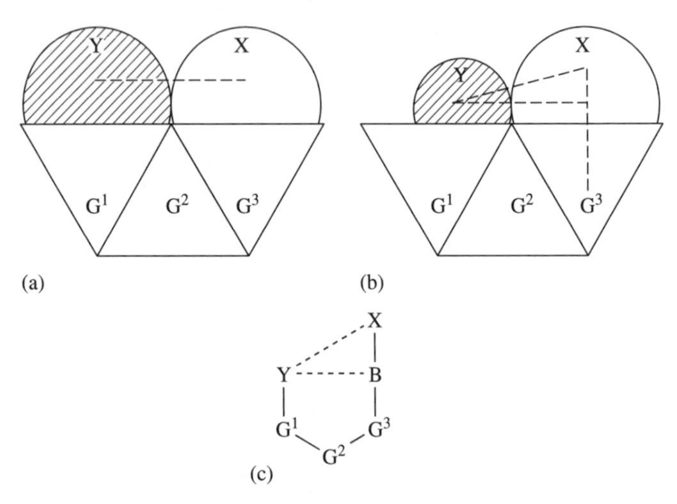

FIGURE 3. The effect of bond length on steric effects. (a) A system in which substituent and X and reaction site Y have comparable X–G and Y–G bond lengths; G^1, G^2 and G^3 are atoms of the skeletal group. (b), (c) A system in which X–G and Y–G bond lengths are significantly different

described above function best when they are of comparable length. In that case the contact between X and Y is that shown in Figure 3a. If the YG bond is much shorter than the XG bond, the contact is as shown in Figure 3b. In that case, the distance from Y to the X–G bond is less in Figure 3b than it is in Figure 3a although the XY distance in both Figures 3a and 3b is the sum of the van der Waals radii, r_{VX} and r_{VY}. The effective size of the van der Waals radius of X is reduced. Steric parameters were originally derived for systems like that in Figure 3a. In a system like Figure 1b, corrected steric parameters are needed. An approximate value of the effective van der Waals radius of X, $r^{c}{}_{VX}$, can be calculated for the case in which the X–G and Y–G bonds are parallel to each other from a consideration of Figure 3c and Scheme 1, where l_{XG} and l_{YG} are the lengths of the X–G and Y–G bonds, respectively.

$$\overline{XY} = r_{VX} + r_{VY}$$

$$\overline{XD} = l_{XG} - l_{YG}$$

$$\overline{DY} = [(\overline{XY})^2 - (\overline{XD})^2]^{1/2}$$

$$\overline{DY} = [(r_{VX} + r_{VY})^2 - (l_{XG} - l_{YG})^2]^{1/2}$$

$$\overline{DY} = r_{VY} + r_{VX}^c$$

$$r_{VX}^c = [(r_{VX} + r_{VY})^2 - (l_{XG} - l_{YG})^2]^{1/2} - r_{VY}$$

SCHEME 1

Values of steric effect substituent constants used in the applications are given in Table 2.

IV. INTERMOLECULAR FORCES

A. Introduction

Inter- and intramolecular forces (imf) are of major importance in the quantitative description of structural effects on bioactivities and chemical properties. They may make a significant contribution to chemical reactivities and some physical properties as well. Common types of intermolecular forces and their parameterization are given in Table 7 of Reference 1.

B. Parameterization of Intermolecular Forces

1. Hydrogen bonding

Two parameters are required for the description of hydrogen bonding. One is required to account for the hydrogen atom donating capacity of a substituent and another to account for its hydrogen atom accepting. The simplest approach is the use of n_H, the number of OH and/or NH bonds in the substituent, and n_n, the number of lone pairs on oxygen and/or nitrogen atoms as parameters[17]. The use of these parameters is based on the argument that if one of the phases involved in the phenomenon studied includes a protonic solvent, particularly water, then hydrogen bonding will be maximized. For such a system, hydrogen bond parameters defined from equilibria in highly dilute solution in an 'inert' solvent are unlikely to be a suitable model. This type of parameterization accounts only for the number of hydrogen donor and hydrogen acceptor sites in a group. It does not take into account differences in hydrogen bond energy. An improved parameterization would result from the use of the hydrogen bond energy for each type of hydrogen bond formed[18]. For each substituent, the parameter E_{hbX} would be given by equation 9:

$$E_{hbX} = \sum_{i=1}^{m} n_{hbi} E_{hbi} \tag{9}$$

where E_{hbX} is the hydrogen bonding parameter, E_{hbi} is the energy of the i-th type of hydrogen bond formed by the substituent X and n_{hbi} is the number of such hydrogen bonds. The validity of this parameterization is as yet untested. In any event, the site number parameterization suffers from the fact that, though it accounts for the number of hydrogen bonds formed, it does not differentiate between their energies and can therefore be only an approximation. A recent definition of a scale of hydrogen bond acceptor values from 1-octanol–water partition coefficients of substituted alkanes shows that the site number method strongly overestimates the hydrogen acceptor capability of the nitro group and seriously underestimates that of the methylsulfoxy group[19]. Much remains to be done in properly parameterizing hydrogen bonding.

2. van der Waals interactions

These interactions (dipole–dipole, dd; dipole–induced dipole, di; and induced dipole–induced dipole, ii) are a function of dipole moment (μ) and polarizability. It has been shown that the dipole moment cannot always be replaced entirely by the use of electrical effect substituent constants as parameters[17,18]. This is because the dipole moment has no sign. Either an overall electron donor group or an overall electron acceptor group may have the same value of μ. It has also been shown that the bond moment rather than the molecular dipole moment is the parameter of choice. The dipole moments of MeX and PhX were taken as measures of the bond moments of substituents bonded to sp^3 and sp^2 hybridized carbon atoms, respectively, of a skeletal group. Application to substituents bonded to sp hybridized carbon atoms should require a set of dipole moments for substituted ethynes.

The polarizability parameter used here, α, is given by equation 10:

$$\alpha \equiv \frac{MR_X - MR_H}{100} = \frac{MR_X}{100} - 0.0103 \tag{10}$$

where MR_X and MR_H are the group molar refractivities of X and H, respectively[18]. The factor 1/100 is introduced to scale the α parameter so that its coefficients in the regression equation are roughly comparable to those obtained for the other parameters used. Many other polarizability parameters have been proposed, including parachor, group molar volumes of various kinds, van der Waals volumes and accessible surface areas. Any of these will serve as they are all highly collinear in each other[20,21]. The advantage of α is that it is easily estimated either by additivity from the values for fragments or from group molar refractivities calculated from equation 11:

$$MR_X = 0.320n_c + 0.682n_b - 0.0825n_n + 0.991 \tag{11}$$

where n_c, n_b and n_n are the numbers of core, bonding and nonbonding electrons, respectively, in the group X[20].

3. Charge transfer interactions

These interactions can be roughly parameterized by the indicator variables n_A and n_D, where n_A takes the value 1 when the substituent is a charge transfer acceptor and 0 when it is not, and n_D takes the value 1 when the substituent is a charge transfer donor and 0 when it is not. An alternative parameterization makes use of the first ionization potential of MeX (ip_{MeX}) as the electron donor parameter and the electron affinity of MeX as the electron acceptor parameter. Usually, the indicator variables n_A and n_D are sufficient. This parameterization accounts for charge transfer interactions directly involving the substituent. If the substituent is attached to a π-bonded skeletal group, then the skeletal group is capable of charge transfer interaction the extent of which is modified by the substituent. This is accounted for by the electrical effect parameters of the substituent.

TABLE 3. Intermolecular force substituent constants used in applications[a]

	α	$\mu(sp^2)$	$\mu(sp^3)$	n_H	n_n
Ak, *c*-Ak					
Me	0.046	0.37	0	0	0
Et	0.093	0.37	0	0	0
c-Pr	0.125	0.48		0	0
Pr	0.139	0.37	0	0	0
i-Pr	0.140	0.40	0	0	0
Bu	0.186	0.37	0	0	0
i-Bu	0.186			0	0
t-Bu	0.186	0.52	0	0	0
Pe	0.232			0	0
CH_2Bu-*t*	0.232			0	0
c-Hx	0.257			0	0
Hx	0.278			0	0
CH_2Z					
CH_2Br	0.124	1.87	2.069	0	0
CH_2Cl	0.095	1.83	1.895	0	0
CH_2OH	0.062	1.71	1.58	1	2
CH_2CN	0.091	3.43	3.53	0	0
CH_2OMe	0.114			0	2
CH_2CH_2CN	0.145	3.92		0	0
CH_2Vi	0.135	0.364	0.438	0	0
CH_2OEt	0.160			0	2
CH_2GeMe_3	0.300			0	0
CH_2SiMe_3	0.285	0.68		0	0
CH_2SnMe_3	0.353			0	0
CH_2-2-Tp	0.276		0.81	0	0
CH_2-2-Fr	0.215		0.65	0	2
$CH_2CH_2CO_2Et$	0.256		1.84	0	4
$CH_2CHMeCO_2Me$	0.256		1.84	0	4
CH_2NEt_2	0.278		0.612	0	1
CH_2Ph	0.290	0.22	0.37	0	0
CH_2CH_2Py-2	0.212			0	1
CH_2CH_2Py-4	0.212			0	1
CZ_3					
CF_3	0.040	2.86	2.321	0	0
CCl_3	0.191		1.95	0	0
$C(SiMe_3)_3$	0.760			0	0
VnX					
Vi	0.100	0.13	0.364	0	0
2-VnVi	0.190			0	0
C_2Z					
C_2H	0.085	0.70	0.7809	1	0
Ar					
C_6F_5	0.230	1.99	1.73	0	0
Ph	0.243	0	0.37	0	0
PnZ					
4-$PnNMe_2$	0.388		1.60	0	1
4-$PnNEt_2$	0.475		1.81	0	1

continued overleaf

TABLE 3. (*continued*)

	α	$\mu(sp^2)$	$\mu(sp^3)$	n_H	n_n
Har					
2-Fr	0.169		0.65	0	1
3-Fr	0.169		1.03	0	1
2-Tp	0.230		0.674	0	0
3-Tp	0.230		0.81	0	0
(CO)Z, CN					
CHO	0.059	2.92	2.69	0	2
CO_2H	0.059	1.86	1.70	1	4
Ac	0.102	2.88	2.93	0	2
CO_2Me	0.118	1.92	1.706	0	4
CO_2Et	0.164	1.846	1.84	0	4
CN	0.053	4.14	3.9185	0	0
Si					
$SiBr_3$	0.338			0	0
$SiCl_3$	0.251			0	0
SiF_3	0.098			0	0
SiH_3	0.101			0	0
$SiMeBr_2$	0.305			0	0
$SiMeCl_2$	0.247			0	0
$SiMeF_2$	0.145			0	0
$SiMe_2Cl$	0.243			0	0
$SiMe_3$	0.239	0.42		0	0
$SiMe_2Et$	0.285			0	0
$SiEt_3$	0.380			0	0
$SiMe_2Ph$	0.436			0	0
$SiPh_3$	0.830			0	0
Ge					
$GeMe_3$	0.254			0	0
$GeEt_3$	0.392			0	0
$GePh_2Br$	0.681			0	0
$GePh_3$	0.845			0	0
Sn					
$SnMe_3$	0.307			0	0
$SnEt_3$	0.380			0	0
$SnPh_2Cl$	0.705			0	0
$SnPh_3$	0.898			0	0
Pb					
$PbPh_3$	0.915			0	0
N					
N_3	0.092	1.56	2.17	0	1
NH_2	0.044	1.49	1.296	2	1
NHMe	0.093	1.77	1.01	1	1
NMe_2	0.145	1.60	0.612	0	1
NEt_2	0.232	1.81		0	1
NO					
NO_2	0.063	4.28	3.56	0	4

TABLE 3. (*continued*)

	α	$\mu(sp^2)$	$\mu(sp^3)$	n_H	n_n
O					
OH	0.018	1.40	1.77	1	2
OMe	0.068	1.36	1.31	0	2
OAc	0.114	1.69	1.706	0	4
OEt	0.114	1.38	1.22	0	2
OPr-*i*	0.160			0	2
$OSiMe_3$	0.259		1.18	0	2
OBu	0.206			0	2
OPh	0.267	1.13	1.36	0	2
S					
SH	0.082	1.21	1.52	0	0
SMe	0.128	1.29	1.06	0	0
SEt	0.174			0	0
SPh	0.333	1.37	1.50	0	0
$\mathbf{SO_2}$					
SO_2Me	0.125	4.73		0	4
SO_2Ph	0.322	5.00	4.73	0	4
Other					
H	0	0	0	0	0
Br	0.079	1.70	1.84	0	0
Cl	0.050	1.70	1.895	0	0
F	−0.001	1.66	1.8549	0	0
I	0.129	1.71	1.618	0	0
Solvents	α	μ			
CCl_4	0.241	0			
CH_2Cl_2	0.146	1.59			
PhCN	0.296	3.99			
MeCN	0.099	3.51			
Silanes	α	μ			
SiH_4	0.111	0			
$SiCl_4$	0.301	0			
$SiBr_4$	0.417	0			
SiI_4	0.587	0			
$SiMe_4$	0.285	0			
SiH_3Br	0.180	1.32			
SiH_3Cl	0.151	1.292			
$SiHBr_3$	0.398	0.79			
SiH_2Cl_2	0.191	1.181			
$SiHCl_3$	0.231	0.855			
$SiClF_3$	0.148	0.49			
SiH_3I	0.230	1.62			
$SiHF_3$	0.077	1.26			
$SiMeCl_3$	0.296	1.91			
$SiMe_2H_2$	0.183	0.75			
$SiMeH_3$	0.147	0.73			
Germanes	α	μ			
GeH_4	0.126	0			
$GeCl_4$	0.316	0			

continued overleaf

TABLE 3. (*continued*)

	α	$\mu(sp^2)$	$\mu(sp^3)$	n_H	n_n
$GeBr_4$	0.432	0			
GeI_4	0.632	0			
$GeMe_4$	0.300	0			
GeH_3Br	0.205	1.31			
GeH_3Cl	0.176	2.124			
GeH_2Cl_2	0.213	2.22			
Stannanes	α	μ			
SnH_4	0.179	0			
$SnCl_4$	0.369	0			
$SnBr_4$	0.485	0			
SnI_4	0.685	0			
$SnMe_4$	0.353	0			
Plumbanes	α	μ			
$PbCl_4$	0.386	0			
$PbMe_4$	0.370	0			

[a] For abbreviations, see Section I. A.

4. Intermolecular force equation

The intermolecular force (IMF) equation is a general relationship for the quantitative description of intermolecular forces. It is written as equation 12:

$$Q_X = L\sigma_{lX} + D\sigma_{dX} + R\sigma_{eX} + M\mu_X + A\alpha_X + H_1 n_{HX} + H_2 n_{nX} + Ii_X + B_{DX}n_{DX} + B_{AX}n_{AX} + S\psi_X + B^o \quad (12)$$

Values of intermolecular force substituent constants used for the substituent applications are set forth in Table 3.

V. APPLICATIONS

A. Chemical Reactivities (QSRR)

We now consider the application of the methods and parameters described above to substituents and/or active sites containing Si, Ge, Sn and Pb.

Eaborn and Singh[22] have reported rate constants for H-T exchange in tri(4-substituted phenyl)-tritio-germanes, $(4-XPn)_3GeT$, with methoxide ion in methanol at 20, 30 and 40 °C (set **CR1**, Table 4). The data were correlated with the LDRT equation which has the form of equation 13:

$$Q_X = L\sum\sigma_{lX} + D\sum\sigma_{dX} + R\sum\sigma_{eX} + T\tau + h \quad (13)$$

where τ is defined as 100 divided by T_K, the temperature in degrees Kelvin. This model was chosen to permit the inclusion of every available rate constant. All of the regression equations presented in this section are the best obtained. In this case the regression equation is equation 14:

$$\log k_{X/T} = 3.19(\pm 0.268)\sigma_{lX} + 2.98(\pm 0.247)\sigma_{dX} - 62.3(13.5)\tau + 23.5(\pm 4.45) \quad (14)$$

$100R^2$, 95.54; $A\ 100R^2$, 94.73; F, 71.44; S_{est}, 0.363; S^o, 0.250; N_{dp}, 14;

P_D, 48.2(±4.92), η, 0; r_{ld}, 0.487; $r_{l\tau}$, 0.000; $r_{d\tau}$, 0.000.

TABLE 4. QSRR data sets[a]

CR1. 10^6k_s ($l\,mol^{-1}\,s^{-1}$), 4-X^1Pn-4-X^2Pn-4-X^3PnGeT + MeO^- in MeOH. X^1, X^2, X^3, T (°C), 10^6k_s: H, H, H, 20, 148; H, H, H, 30, 530; H, H, H, 40, 1520; Cl, Cl, Cl, 30, 37300; Me, Me, Me, 20, 14.2; Me, Me, Me, 30, 41.5; Me, Me, Me, 40, 161; MeO, MeO, MeO, 30, 13.2; Ph, Ph, Ph, 20, 156; Ph, Ph, Ph, 30, 570; Ph, Ph, Ph, 40, 17500; NO_2. H, H, 30, 680000; CN, H, H, 30, 360000; F, H, H, 30, 600.

CR2.[b] 10^4k_2 ($dm^3\,mol^{-1}\,s^{-1}$). $Me_3SnOP_n - X - Y + MeSO_2Cl$ in various solvents s_v at 130 °C Sv, X, 10^4k_2: CCl_4, OMe, 5.37; CCl_4, Me, 3.99; CCl_4, H, 3.74; CCl_4, Cl, 2.81; CCl_4, NO_2, 0.417; CH_2Cl_2, OMe, 5.13; CH_2Cl_2, Me, 3.23; CH_2Cl_2, H, 1.20; CH_2Cl_2, Cl, 0.466; CH_2Cl_2, NO_2, 0.174; PhCN, OMe, 17.3; PhCN, Me, 9.67; PhCN, H, 7.56; PhCN, Cl, 6.18; PhCN, NO_2, 7.58; MeCN, OMe, 5.04; MeCN, Me, 6.28; MeCN, H, 8.58; MeCN, Cl, 13.4; MeCN, NO_2, 56.8.

CR3. k_{rel}, $(MeO)_3SiOCH_2X$ methanolysis catalyzed by Et_2NH. X, k_{rel},: Me, 1.00; Et, 0.60; Pr, 0.53; Bu, 0.54; *t*-Bu, 0.049[c]; $GeMe_3$, 0.30; $SiMe_3$, 0.15; $SiMe_2Et$, 0.11; CH_2SiMe_3, 0.49; CH_2GeMe_3, 0.54; CH_2OMe, 2.10; CH_2Cl, 3.23; CH_2Ph, 1.48; CH_2Vi, 3.27.

CR4. k_{rel}, $(MeO)_3SiOCH_2CH_2X$ methanolysis catalyzed by Et_2NH. X, k_{rel},: H, 1.00; Me, 0.60; Et, 0.53; Pr, 0.54; *t*-Bu, 0.52; CH_2Bu-*t*, 0.52; $GeMe_3$, 0.54; $SiMe_3$, 0.49; CH_2SiMe_3, 0.50; CH_2GeMe_3, 0.52; OMe, 2.10; Cl, 3.23; CH_2Cl, 0.82; CH_2OMe, 0.87.

CR5. k_2 ($l\,mol^{-1}\,s^{-1}$), $Ak^1SnAk^2{}_3 + I_2$ in MeOH at 20 °C. Ak^1, Ak^2, k_2: Me, Me, 1.77; Et, Me, 0.256; Pr. Me, 0.056; Bu, Me, 0.132; *i*-Pr, Me, 0.01; Me, Et, 3.58; Et, Et, 0.22; Pr, Et, 0.065; Bu, Et, 0.060; *i*-Pr, Et, 0.004.

CR6. k_2 ($l\,mol^{-1}\,s^{-1}$), $AkSnMe_3 + Hl_2$ in AcOH at 20 °C. Ak, Hl, k_2: Me, I, 0.0610; Et, I, 0.00950; Pr, I, 0.00166; Bu, I, 0.00317; *i*-Pr, I, 0.00046; *t*-Bu, I, 0.00005; Me, Cl, 2.92; Et, Cl, 1.21; Pr, Cl, 0.36; Bu, Cl, 0.35; *i*-Pr, Cl; 0.03.

CR7. 10^2k_2 ($l\,mol^{-1}\,s^{-1}$), $Ak_4Sn + HgCl_2$ in aq. MeOH at 298 K. Ak, ϕ_{MeOH}[d], 10^2k_2: Me, 0.999, 155; Et, 0.999, 0.333, Pr, 0.999, 0.0628; Bu, 0.999, 0.0615; *i*-Bu, 0.999, 0.00800; Me, 0.914, 259; Et, 0.914, 0.630; Pr, 0.914, 0.113; Bu, 0.914, 0.104; *i*-Bu, 0.914, 0.0138; Me, 0.714, 730; Et, 0.714, 2.44; Pr, 0.714, 0.392; Bu, 0.714, 0.323; *i*-Bu, 0.714, 0.0393.

CR8. 10^5k_2 ($l\,mol^{-1}\,s^{-1}$), Ak_4Pb + AcOH in AcOH at various *T*. Ak, *T* °C, 10^5k_2: Me, 24.9, 1.15; Et, 24.9, 0.80; Pr, 24.9, 0.225; Bu, 24.9, 0.31; *i*-Pe, 24.9, 0.30; Me, 49.8, 17.1; Et, 49.8, 10.9; Pr, 49.8, 3.1; Bu, 49.8, 4.3; *i*-Pe, 49.8, 4.2; Me, 60.0, 41.2; Et, 60.0, 28.2; Pr, 60.0, 8.2; Bu, 60.0, 10.6; *i*-Pe, 60.0, 12.2.

[a] For abbreviations, see Section I.A.
[b] Sets **CR2a**, **CR2b**, **CR2c** and **CR2d** are rate constants in CCl_4, CH_2Cl_2, PhCN and MeCN, respectively.
[c] Excluded from the best regression equation.
[d] ϕ is the mole fraction.

There is no dependence on σ_e, thus the electronic demand is zero, contrary to the conclusions of Eaborn and Singh.

Kozuka and coworkers[23] have reported second-order rate constants for the reaction

$$Me_3SnOPnX\text{-}4 + MeSO_2Cl \rightarrow Me_3SnCl + MeSO_2OPnX\text{-}4$$

in the solvents tetrachloromethane, dichloromethane, benzonitrile and acetonitrile. The number of data points in each solvent is only five though at least they are well chosen (sets **CR2a** to **CR2d** in Table 4). Plots of $\log k$ against σ_{pX} show that the reactions in tetrachloromethane and dichloromethane go by a different mechanism from that in acetonitrile. Furthermore, 3-nitrophenoxytrimethylstannanes in benzonitrile reacts by a different mechanism from that which characterizes the other members of set **CR2c**. Sets **CR2a, C2b, CR2d** and, on exclusion of the nitro compound, set **CR2c** were all correlated separately with the CR equation 15:

$$Q_X = C\sigma_{cX} + R\sigma_{eX} + h \tag{15}$$

where σ_{cX} is a composite substituent constant obtained by combining σ_l and σ_d so as to have a particular P_D value. The regression equations obtained for sets **CR2a** through **CR2d**, respectively, are equations 16–19:

$$\log k_{2X} = -0.725(\pm 0.0265)\sigma_{c60X} + 3.22(\pm 0.433)\sigma_{eX} + 0.546(\pm 0.0197) \tag{16}$$

$$100R^2, 99.81; A\ 100R^2, 99.75; F, 530.3; S_{est}, 0.0273; S^{\circ}, 0.0686; N_{dp}, 5;$$

$$P_D, 60; \eta, -2.96(\pm 0.384); r_{ce}, 0.324.$$

$$\log k_{2X} = -1.48(\pm 0.163)\sigma_{c50X} + 5.94(\pm 2.14)\sigma_{eX} + 0.0151(\pm 0.0925) \tag{17}$$

$$100R^2, 97.67; A\ 100R^2, 96.89; F, 41.89; S_{est}, 0.130; S^{\circ}, 0.241; N_{dp}, 5;$$

$$P_D, 50; \eta, 4.01(\pm 1.37); r_{ce}, 0.409.$$

$$\log k_{2X} = -1.04(\pm 0.247)\sigma_{c60X} + 3.02(\pm 2.32)\sigma_{eX} + 0.847(\pm 0.0553) \tag{18}$$

$$100R^2, 99.37; A\ 100R^2, 99.04; F, 78.38; S_{est}, 0.0268; S^{\circ}, 0.159; N_{dp}, 4;$$

$$P_D, 60; \eta, -1.94(\pm 1.42); r_{ce}, 0.972.$$

$$\log k_{2X} = 1.14(\pm 0.163)\sigma_{c50X} - 1.17(\pm 0.343)\sigma_{eX} + 0.921(\pm 0.0153) \tag{19}$$

$$100R^2, 99.87; A\ 100R^2, 99.82; F, 752.1; S_{est}, 0.0215; S^{\circ}, 0.0576; N_{dp}, 5;$$

$$P_D, 50; \eta, -1.03(\pm 0.301); r_{ce}, 0.331.$$

The results are uncertain due to the small number of data points in each set. It seems likely, however, that sets **CR2a** through **CR2c**, all of which have a negative value of C and about the same P_D value, possess the same mechanism. Set **CR2d**, which has a positive value of C, seems to have a different mechanism. None of the four solvents studied is a hydrogen donor in hydrogen bonding; none of the substituents studied has this capacity. Then the solvent effect should be dependent only on dipole moment and polarizability. The data for sets **CR2a, CR2b** and **CR2c** excluding nitrophenoxytrimethylstannanes were combined into a single data set and correlated with equation 20:

$$Q_{X/S\nu} = C\sigma_{cX} + R\sigma_{eX} + A\alpha_{S\nu} + M\mu_{S\nu} + b^o \tag{20}$$

The best regression equation was obtained on exclusion of the value for Cl in dichloromethane; it is equation 21:

$$\log k_{2,X/S\nu} = -0.890(\pm 0.0666)\sigma_{c60X} + 3.52(\pm 0.556)\alpha_{S\nu} + 0.0468(\pm 0.0202)\mu_{S\nu} - 0.412(\pm 0.121) \tag{21}$$

$$100R^2, 97.15; A\ 100R^2, 96.58; F, 102.3; S_{est}, 0.108; S^{\circ}, 0.203; N_{dp}, 13;$$

$$P_D, 60; \eta, 0; r_{c\alpha}, 0.168; r_{c\mu}, 0.205; r_{\alpha\mu}, 0.428; C_{\sigma c}, 49.1; C_{\alpha}, 46.8; C_{\mu}, 4.10.$$

None of the zeroth-order partial correlation coefficients involving σ_e was significant. It seems fairly certain that the electronic demand in the reaction in set **CR2abc** is 0. The rate of the reaction is accelerated by electron donor substituents and decelerated by electron acceptor groups. This situation is reversed in set **CR2d**. The best explanation of these observations is that the mechanism in the most polar solvent, acetonitrile, is different from that in the least polar solvents, tetrachloromethane and dichloromethane, while in benzonitrile which is intermediate in polarity a change in mechanism is occurring with nitrophenoxytrimethylstannanes. A mechanism in accord with these results is shown in Scheme 2. If, in less polar media, S−O bond formation is more advanced than Sn−O bond breaking, there will be a partial positive charge on oxygen and the reaction will be favored by electron donor groups. If, in the more polar medium, this situation is reversed, then there will be a partial negative charge on oxygen and the reaction will be favored by electron acceptor substituents.

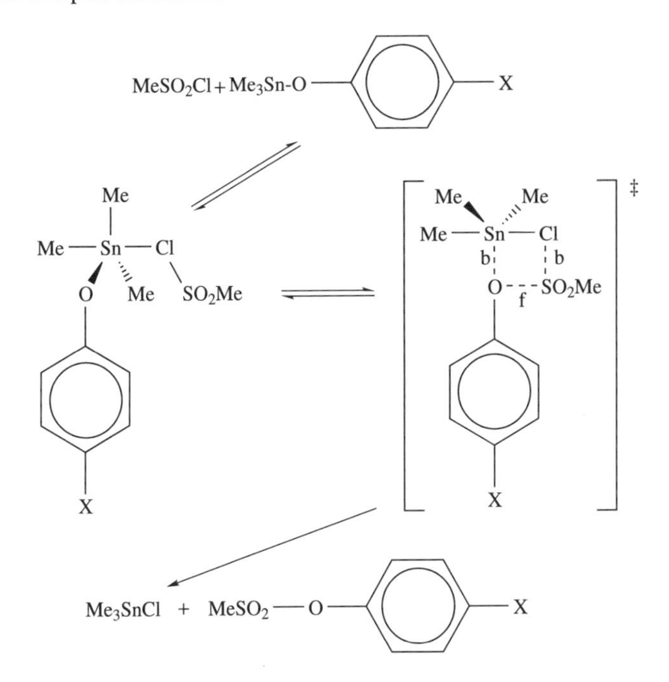

SCHEME 2

Pola, Bellama and Chvalovsky[24] have reported relative rates for the methanolysis in methanol catalyzed by ethylamine of $X(CH_2)_nOSi(OMe)_3$ ($n = 1$, set **CR3**; $n = 2$, set **CR4**; Table 4). Correlation with the LDRS equation is in the form of equation 22:

$$Q_X = L\sigma_{lX} + D\sigma_{dX} + R\sigma_{eX} + S\upsilon_X + h \qquad (22)$$

The regression equation for $n = 1$, obtained on the exclusion of the value for X = t-Bu, is equation 23:

$$\log k_{rel,X} = 4.85(\pm 0.449)\sigma_{lX} - 0.209(\pm 0.0393) \qquad (23)$$

$100r^2$, 91.39; F, 116.8; S_{est}, 0.141; S^o, 0.319; N_{dp}, 13, P_D, 0; η, 0.

The regression equation for $n = 2$, obtained on the exclusion of the value for X = H, is equation 24:

$$\log k_{rel,X} = 1.54(\pm 0.112)\sigma_{lX} - 0.230(\pm 0.0190) \qquad (24)$$

$100r^2$, 94.48; F, 186.4; S_{est}, 0.0641; S^o, 0.255; N_{dp}, 13, P_D, 0; η, 0.

We now consider structural effects in the halodealkylation of tetralalkylstannanes[25] (sets **CR5** and **CR6**), their reaction with $HgCl_2$[26] (set **CR7**) and the reaction of tetraalkylplumbanes with acetic acid[27] (set **CR8**). The data sets are given in Table 4. As only alkyl groups vary in these data sets and the electrical effects of alkyl groups are constant, only steric effects and polarizability need to be considered for parameterization. In order to provide a sufficiently large data set for analysis, rate constants for bromo- and iododealkylation were combined into a single data set by means of the Zeta method. In this method the parameterization needed to combine data sets is obtained by choosing a data point from each set to be combined for which all the other parameters are the same. The quantity ζ was defined as $\log k_2$ for the methyl group and the simple branching equation (equation 4) was used to account for the steric effect. Only branching at the first and second carbon atoms of the alkyl group was parameterized. The correlation equation is equation 25:

$$Q_{Ak/Hl} = a_1 n_1 + a_2 n_2 + Z\zeta + a^o \tag{25}$$

The regression equation is equation 26:

$$\log k_{2,Ak/Hl} = -1.03(\pm 0.0592)n_1 - 0.287(\pm 0.152)n_2 + 1.19(\pm 0.0879)\zeta + 0.213(\pm 0.146) \tag{26}$$

$$100R^2, 98.23; A\ 100R^2, 97.79; F, 129.7; S_{est}, 0.239; S^o, 0.167; N_{dp}, 11;$$

$$r_{12}, 0.165; r_{1\zeta}, 0.199; r_{2\zeta}, 0.069; C_{n1}, 59.0; C_{n2}, 15.4; C_{\zeta}, 29.6; B_1/B_2, 3.59.$$

There is a clear dependence on steric effects with branching at C^1 having a much greater effect than branching at C^2.

The same method has been applied to the iododealkylation of $Ak^1SnAk^2_3$ with Ak^2 equal to Me or Et. Again, ζ was defined as $\log k_2$ for the methyl group. Correlation of the data set with equation 25 gave regression equation 27:

$$\log k_{2,Ak^1/Ak^2} = -1.30(\pm 0.107)n_1 - 0.327(\pm 0.138)n_2 + 0.492(\pm 0.138) \tag{27}$$

$$100R^2, 95.67; A\ 100R^2, 95.13; F, 77.30; S_{est}, 0.213; S^o, 0.248; N_{dp}, 10;$$

$$r_{12}, 0.000; C_{n1}, 79.1; C_{n2}, 12.1; B_1/B_2, 3.98.$$

Steric effects account completely for the observed reactivity; there is no dependence on the nature of Ak^2. Again, the effect of branching at C^1 predominates over that at C^2.

Second-order rate constants for the reaction of mercury(II) chloride with tetraalkylstannanes according to equation 28:

$$Ak_4Sn + HgCl_2 \rightarrow Ak_3SnCl + AkHgCl \tag{28}$$

in aqueous methanol of varying concentrations were correlated with equation 29:

$$Q_{Ak/Sv} = a_1 n_1 + a_2 n_2 + F\phi + a^o \tag{29}$$

where ϕ is the mole fraction of methanol and F is its coefficient. The regression equation is equation 30:

$$\log k_{Ak/Sv} = -2.59(\pm 0.320)n_1 - 1.08(\pm 0.261)n_2 - 2.63(\pm 0.853)\phi + 4.794(\pm 0.781) \tag{30}$$

$$100R^2, 94.96; A\ 100R^2, 94.12; F, 69.11; S_{est}, 0.392; S^o, 0.262; N_{dp}, 15;$$

$$r_{12}, 0.612; r_{1\phi}, 0.000; r_{2\phi}, 0.000; C_{nl}, 42.9; C_{n2}, 17.9; C_{\phi}, 39.2; B_1/B_2, 2.40.$$

Steric effects account for the result of structural variation. Though the steric effect of branching at C^1 is again predominant, the extent is significantly less.

First-order rate constants for the acetolysis of tetraalkylplumbanes in acetic acid at various temperatures according to equation 31:

$$Ak_4Pb + AcOH \rightarrow Ak_3PbOAc + AkH \tag{31}$$

were correlated with equation 32:

$$Q_{Ak/T} = a_1n_1 + a_2n_2 + T\tau + a^o \tag{32}$$

The regression equation is equation 33:

$$\log k_{Ak/T} = -0.174(\pm 0.0495)n_1 - 0.454(\pm 0.0404)n_2 - 43.5(\pm 1.03)\tau + 14.7(\pm 0.327) \tag{33}$$

$$100R^2, 99.47; A\ 100R^2, 94.12; F, 69.11; S_{est}, 0.392; S^o, 0.262; N_{dp}, 15;$$

$$r_{12}, 0.612; r_{1\phi}, 0.000; r_{2\phi}, 0.000; C_{n1}, 42.9; C_{n2}, 17.9; C_{\phi}, 39.2; B_1/B_2, 0.383.$$

The structural effect is again accounted for by steric effects. What is unusual is that, contrary to the results obtained for the three previous data sets, steric effects are predominantly due to branching at C^2.

B. Chemical Properties (QSCR)

1. Phase change properties

a. Melting points (T_m). The melting points of $MX^1X^2X^3X^4$, where M is a group 14 element other than carbon[28], can be modeled by a relationship derived from the IMF equation. The melting points of the subset MX_4, where X is H, Cl, Br, I or Me (set **CP1a**, Table 5), is a particularly simple case as the dipole moment is zero for these compounds and hydrogen bonding is not possible. As all of these compounds have similar molecular geometry they are likely to have similar crystal lattices; this packing effect should be constant. Then only induced dipole–induced dipole interactions are possible and the correlation equation is equation 34:

$$Q_X = A\alpha + a_o \tag{34}$$

Correlation of the data subset with equation 34 gives equation 35:

$$T_m = 613(\pm 19.9)\alpha + 17.3(\pm 8.05) \tag{35}$$

$$100r^2, 98.44; F, 946.5; S_{est}, 12.9; S^o, 0.133; N_{dp}, 17.$$

Clearly, the assumptions made in the choice of equation 34 were justified. In order to include the remainder of the data set, it is necessary to introduce a term in the dipole moment μ. Another necessary term results from the fact that while all four surfaces of the tetrahedral molecule MX_4 are equivalent, those of related molecules are not. Thus, $MX^1_2X^2_2$ has two sets of two equivalent faces while $MX^1_3X^2$ has a set of three equivalent

TABLE 5. QSCR data sets[a]

CP1. T_m, K, $MX^1X^2X^3X^4$. $MX^1X^2X^3X^4$, T_m: SiH_4, 88.5; $SiCl_4$, 203; $SiBr_4$, 279; SiI_4, 394; $SiMe_4$, 174; GeH_4, 107; $GeCl_4$, 224; $GeBr_4$, 299; GeI_4, 407; $GeMe_4$, 185; SnH_4, 123; $SnCl_4$, 239; $SnBr_4$, 303; SnI_4, 418; $SnMe_4$, 218; $PbCl_4$, 258; $PbMe_4$, 246; SiH_3Br, 179; SiH_3Cl, 155; SiH_3I, 216; SiH_3Me, 117; $SiCl_3Me$, 195; $SiCl_3H$, 147; $SiBr_3H$, 200; SiF_3H, 142; SiF_3Cl, 135; SiH_2Cl_2, 151; $SiMe_2H_2$, 123; GeH_3Br, 241; GeH_3Cl, 221; GeH_2Cl_2, 205.

CP1a. This set consists of those members of set **CP1** that have the formula MX_4.
CP1b. This set consists of all other members of set **CP1**.

CP2. T_b, K, $MX^1X^2X^3X^4$. $MX^1X^2X^3X^4$, T_b: SiH_4, 161; SiF_4, 187; $SiCl_4$, 331; $SiBr_4$, 427; SiI_4, 576; $SiMe_4$, 300; GeH_4, 185; $GeCl_4$, 356; $GeBr_4$, 460; GeI_4, 621; $GeMe_4$, 317; SnH_4, 221; $SnCl_4$, 388; $SnBr_4$, 480; SnI_4, 638; $SnMe_4$, 351; $PbMe_4$, 383; SiH_3Br, 275; SiH_3Cl, 243; SiH_3I, 319; SiH_3Me, 216; $SiCl_3Me$, 340; SiF_3H, 368; SiF_3Cl, 203; SiH_2Cl_2, 281; $SiCl_3H$, 306; $SiBr_3H$, 382; $SiMe_2H_2$, 253; GeH_3Br, 325; GeH_3Cl, 301; GeH_2Cl_2, 343.

CP2a. This set consists of those members of set **CP2** that have the formula MX_4.
CP2b. This set consists of all other members of set **CP2**.

CP3. $\Delta\nu_{OH}$ (cm^{-1}), $XC_2SiMe_3 + PhOH$ in CCl_4. X, $\Delta\nu_{OH}$: H, 99; Me, 122; *t*-Bu, 131; Ph, 106; C_6F_5, 49; SMe, 107; SEt, 109; Cl, 70; Br, 75; I, 79; CH_2NEt_2, 125; CH_2Cl, 80; CH_2Br, 84; $SiMe_3$, 121; $GeMe_3$, 145; $GeEt_3$, 146; $SnMe_3$, 158; $SnEt_3$, 162; CH_2SiMe_3, 145.

CP4. $\Delta\nu_{OH}$ (cm^{-1}), $XC_2SnMe_3 + PhOH$ in CCl_4. X, $\Delta\nu_{OH}$: H, 136; Me, 154; Et, 160; *t*-Bu, 170; Ph, 140; C_6F_5, 74; SMe, 135; SEt, 136; Cl, 90; Br, 99; CH_2NEt_2, 160; CH_2OEt, 140; CH_2Cl, 111; CH_2Br, 110; $SiMe_3$, 158; $SiEt_3$, 163; $GeMe_3$, 181; $GeEt_3$, 183; $SnMe_3$, 194; CH_2SiMe_3, 181.

CP5. $\Delta\nu_{CH}$ (cm^{-1}), XC_2H in dimethylformamide. X, $\Delta\nu_{CH}$: *i*-Pr, 66; Bu, 64; *i*-Bu, 64; *t*-Bu, 67; OPh, 75; Ph, 80; CH_2OMe, 80; SEt, 77; CO_2Et, 100; C_6F_5, 107; CH_2Cl, 86; CH_2Br, 84; $SiMe_3$, 71; $SiEt_3$, 70; $GeMe_3$, 67; $SnMe_3$, 69; $SnEt_3$, 66; CH_2SiMe_3, 68; CH_2GeMe_3, 68.

CP6. $\Delta\nu_{CH}$ (cm^{-1}), XC_2H in tetrahydrofuran. X, $\Delta\nu_{CH}$: *i*-Bu, 56; *t*-Bu, 58; CH_2Ph, 62; Ph, 73; CH_2OMe, 67; SEt, 67; CH_2Cl, 78; CH_2Br, 77; C_6F_5, 100; $SiMe_3$, 66; $SiEt_3$, 63; $GeMe_3$, 61; $SnMe_3$, 56; $SnEt_3$, 54; CH_2SiMe_3, 58; CH_2GeMe_3, 59.

CP7. *A* ($KJ\,mol^{-1}$) values for *c*-HxX[b]. X, *A*: CMe_3, 21; $SiMe_3$, 10.5; $GeMe_3$, 8.8; $SnMe_3$, 3.9; $PbMe_3$, 2.8.

[a] For abbreviations, see Section I.A.
[b] C. H. Bushweller, in *Conformational Behavior of Six Membered Rings* (Ed. E. Juaristi), VCH Press, New York, 1995, pp. 25–58.

faces and a set of one face. We assume that the face with the largest value of $\Sigma\alpha_X$ will preferentially bind to the crystal surface. Then we assign values of the probability ω as 1 for all MX_4, 0.75 to $MX^1_3X^2$ when the set of three equivalent faces has the higher value of $\Sigma\alpha_X$, 0.25 when the reverse is the case and 0.50 to $MX^1{}_2X^2{}_2$. Thus, the correlation equation becomes equation 36:

$$T_m = A\alpha + M\mu + P\omega + a_o \tag{36}$$

Correlation of the entire data set (sets **CP1a** and **CP1b**) with equation 36 gave, on the exclusion of the values for $SiCl_3Me$ and GeH_3Br, the regression equation 37:

$$T_m = 594(\pm 18.7)\alpha + 52.3(\pm 5.63)\mu + 97.6(\pm 18.1)\omega - 72.2(\pm 18.5) \tag{37}$$

$$100R^2, 97.97; A\ 100R^2, 97.80; F, 385.4; S_{est}, 13.5; S^\circ, 0.154; N_{dp}, 28;$$

$$r_{\alpha\mu}, 0.504; r_{\alpha\omega}, 0.402; r_{\mu\omega}, 0.726; C_\alpha, 44.2; C_\mu, 19.5; C_\omega, 36.3.$$

As a hypothetical reference compound for calculating C, we have chosen that for which $\alpha = 0.2$, $\mu = 1$ and $\omega = 1$. Though α and μ are highly significant, the zeroth-order partial correlation coefficient of μ with ω indicates collinearity and casts some doubt on the validity of a dependence on ω. That dependence is likely but uncertain. Polarizability makes the greatest contribution. The coefficients of α in equations 35 and 37 do not differ significantly. There is, of course, a considerable difference in the intercepts.

b. Boiling points(T_b). The T_b values for MX_4 compounds[28] (set **CP2a**, Table 5) were correlated with equation 34 to give the regression equation 38:

$$T_b = 836(\pm 20.0)\alpha + 78.0(\pm 7.87) \tag{38}$$

$$100r^2, 99.15; F, 1251; S_{est}, 14.0; S^\circ, 0.0981; N_{\text{dp}}, 17.$$

Again, in order to include compounds in which not all the X groups are the same, it is necessary to introduce a term $M\mu$, so that the correlation equation becomes equation 39:

$$T_b = A\alpha + M\mu + a_o \tag{39}$$

Correlation of the combined data sets (sets **CP2a** and **CP2b**) with equation 38 gives the regression equation 40:

$$T_b = 831(\pm 21.5)\alpha + 30.9(\pm 4.50)\mu + 78.8(\pm 8.21) \tag{40}$$

$$100R^2, 98.32; A\ 100R^2, 98.26; F, 788.8; S_{est}, 16.4; S^{\text{o}}, 0.137; N_{dp}, 30;$$

$$r_{\alpha\mu}, 0.401; C_\alpha, 84.3; C_\mu, 15.7.$$

Clearly, polarizability is the major factor in determining the boiling point in these compounds. There is no significant difference in the values of A or in the intercepts of equations 39 and 40.

2. Hydrogen bonding

Egorochkin and coworkers[29] have measured the change in ν_{OH} for solutions of 1-substituted-2-trimethylsilylacetylenes and 1-substituted-2-trimethylstannylacetylenes containing phenol (sets **CP3** and **CP4**, Table 5). These data sets were correlated with the LDRA equation 41:

$$Q_X = L\sigma_{lX} + D\sigma_{dX} + R\sigma_{eX} + A\alpha + h \tag{41}$$

For set **CP3** the regression equation is equation 42:

$$\Delta\nu_{OH,X} = -145(\pm 21.3)\sigma_{lX} - 80.6(\pm 23.4)\sigma_{dX} + 109(\pm 32.6)\alpha + 100(\pm 8.46) \tag{42}$$

$$100R^2, 87.07; A\ 100R^2, 85.34; F, 31.41; S_{est}, 13.1; S^{\text{o}}, 0.408;$$

$$N_{dp}, 18; P_D, 35.8(\pm 11.5), \eta, 0; r_{ld}, 0.641; r_{l\alpha}, 0.557; r_{d\alpha}, 0.504;$$

$$C_l, 58.6; C_d, 32.6; C_\alpha, 8.80.$$

For set CP4 it is equation 43:

$$\Delta\nu_{OH,X} = -184(\pm 20.8)\sigma_{lX} - 93.0(\pm 22.5)\sigma_{dX} + 75.8(\pm 30.9)\alpha + 139(\pm 7.46) \tag{43}$$

$$100R^2, 88.49; A\ 100R^2, 87.05; F, 38.42; S_{est}, 12.6; S^{\text{o}}, 0.382;$$

$$N_{dp}, 19; P_D, 33.6(\pm 8.92), \eta, 0; r_{ld}, 0.627; r_{l\alpha}, 0.491; r_{d\alpha}, 0.483;$$

$$C_l, 63.0; C_d, 31.8; C_\alpha, 5.19.$$

The goodness of fit is in accord with the experimental error in the data. Both equations 42 and 43 are significant at the 99.9% confidence level. Though σ_l is significantly collinear in σ_d, a dependence on both parameters is fairly certain. Electrical effects are the predominant factor in the structural effect with the localized effect making the greater contribution.

In another paper, Egorochkin and coworkers[30] have reported $\Delta\nu_{CH}$ values for the interaction of substituted acetylenes with dimethylformamide and with tetrahydrofuran (sets **CP5** and **CP6** respectively, Table 5). Correlation of set **CP5** with the LDRA equation gave the regression equation 44:

$$\Delta\nu_{CH,X} = 73.0(\pm 7.61)\sigma_{lX} + 37.5(\pm 6.29)\sigma_{dX} + 72.9(\pm 1.25) \quad (44)$$

$$100R^2, 85.38; A\ 100R^2, 84.57; F, 49.65; S_{est}, 4.91; S^o, 0.415; N_{dp}, 20;$$

$$P_D, 34.0(\pm 6.45), \eta, 0; r_{ld}, 0.358; r_{l\alpha}, 0.449; r_{d\alpha}, 0.179; C_l, 66.0; C_d, 34.0.$$

For set **CP6**, the regression equation is given by equation 45:

$$\Delta\nu_{CH,X} = 81.7(\pm 10.2)\sigma_{lX} + 33.9(\pm 8.38)\sigma_{dX} + 64.8(\pm 1.41) \quad (45)$$

$$100R^2, 83.19; A\ 100R^2, 82.00; F, 32.18; S_{est}, 5.14; S^o, 0.455; N_{dp}, 16;$$

$$P_D, 29.3(\pm 7.99), \eta, 0; r_{ld}, 0.378; r_{l\alpha}, 0.633; r_{d\alpha}, 0.329; C_l, 70.7; C_d, 29.3.$$

In sets **CP3** and **CP4** the 1-substituted-2-trimethylsilylacetylenes and 1-substituted-2-trimethylstannylacetylenes are acting as hydrogen bond acceptors while in sets **CP5** and **CP6** they are acting as hydrogen bond donors. Electrical effects are similar in all four sets, however. The P_D values show no statistically significant difference and η is zero in all four data sets. The major difference between the donor and acceptor sets is that the latter show a dependence on polarizability while the former do not.

3. Conformation

Monosubstituted cyclohexanes, *c*-HxX, exist in two conformations, axial and equatorial. In the axial conformation the substituent X is in close proximity to the *cis* hydrogen atoms at positions 3 and 5 (H^{3c} and H^{5c}). It has been shown[16] that when X is a tetrahedral group of the type $MZ^LZS^MZ^S$ (L, M and S are large, medium and small, respectively), the *A* values for *c*-HxX ($A_X \equiv -\Delta G_X$) are a function of electrical and steric effects when M is a second period element (X_{t2} group) but solely a function of steric effects when M is a third period element (X_{t3} group). The electrical effect is the result of a weak hydrogen bond between Z^S and H^{3c} and H^{5c}. This effect may occur only when these atoms are in contact. When M is a third or higher period element, H^{3c} and H^{5c} are in contact only with M. It will not be observed in tetrahedral groups with M of any period greater than second.

The predominant effect on *A* in tetrahedral groups of a given period is a steric effect due to Z^S. The steric effect due to M is much smaller. The magnitude of the Z^S steric effect decreases as the covalent radius of M, r_{CM}, decreases. Thus we obtain equation 46:

$$S_{Z^s(p)} = a_1/r_{CM(n)} + a_o \quad (46)$$

where $S_{Z^s(p)}$ is the coefficient of the steric parameter and *p* is the period of M, while r_{CM} is the covalent radius of M.

To test the hypothesis, we consider the set of c-HxX for which X is MMe_3 and M is C, Si, Ge, Sn or Pb, for all of which A values are available (set **CP7**, Table 5). The A values for the set of interest should obey equation 47:

$$A_{MMe_3} = S_1\vartheta_M + S_{2S}\vartheta_{ZS} + A_0 \quad (47)$$

As no value of r_{CM} for Pb was available, it was replaced by l_{CM}, the bond length of the C−M bond. This is equal to the sum of the covalent radii of C and M. The value of ϑ_M for C, Si, Ge, Sn and Pb is linear in l_{CM}. Equation 48 then exists:

$$\vartheta_M = a_2 l_{CM} + a_{20} \quad (48)$$

Then A values for MMe_3 groups should be linear in l_{CM}. Correlation with equation 49:

$$A_{MMe_3} = b_1 l_{CM} + B_o \quad (49)$$

gives the regression equation 50:

$$A_{MMe_3} = -24.5(\pm 2.76) l_{CM} + 57.4(\pm 5.40) \quad (50)$$

$$100r^2, 96.33; F, 78.70; S_{est}, 1.60, S^o, 0.247; N_{dp}, 5.$$

These results support our arguments concerning the effect of tetrahedral groups on A values.

C. Physical Properties (QSPR)

1. Dipole moments of X_nMZ_{4-n}

The dipole moments μ of X_nMZ_{4-n}[31a] were correlated with the LDR equation (equation 1), and in some cases the CR equation (equation 15) and the LD equation 51:

$$Q_X = L\sigma_{lX} + D\sigma_{dX} + h \quad (51)$$

M may be Si, Ge, Sn or Pb; Z is constant throughout a given data set and may be Me, Et or H, while X varies. The data sets studied are reported in Table 6.

Correlation of μ in benzene for $XSiMe_3$ (set **PP1**) with the LDR equation gave the regression equation 52:

$$\mu_X = 4.75(\pm 0.460)\sigma_{lX} - 0.890(\pm 0.405)\sigma_{dX} + 0.270(\pm 0.133) \quad (52)$$

$$100R^2, 88.53; A\ 100R^2, 87.82; F, 57.91; S_{est}, 0.290; S^o, 0.371;$$

$$N_{dp}, 18; P_D, 15.8(\pm 7.38), \eta, 0; r_{ld}, 0.641; r_{l\alpha}, 0.557; r_{d\alpha}, 0.460.$$

Correlation of μ for $XGeMe_3$ (set **PP2**) with the CR equation and the $\sigma_{c14.3}$ and $\sigma_{c16.7}$ constants gave the regression equations 53 and 54:

$$\mu_X = 5.56(\pm 0.513)\sigma_{c14.3,X} + 0.430(\pm 0.144) \quad (53)$$

$$100r^2, 93.63; F, 117.5; S_{est}, 0.295, S^o, 0.282; N_{dp}, 10; P_D, 14.3; \eta, 0.$$

$$\mu_X = 5.61(\pm 0.519)\sigma_{c16.7,X} + 0.474(\pm 0.142) \quad (54)$$

$$100r^2, 93.61; F, 117.1; S_{est}, 0.295, S^o, 0.283; N_{dp}, 10; P_D, 16.7; \eta, 0.$$

TABLE 6. QSPR data sets[a]

PP1. μ in PhH, $XSiMe_3$. X, μ: H, 0.58; Cl, 2.02; Br, 2.31; I, 2.46; Ph, 0.42; Vi, 0.33; 2-ViVn, 0.29; CH_2Vi, 0.82; C_2H, 0.45[b], Me, 0; Bu, 0; *t*-Bu, 0; CH_2Ph, 0.68[b]; OMe, 1.18; OEt, 1.17; OPr-*i*, 1.178; OPh, 1.24; OAc, 1.86; SMe, 1.73; NMe_2, 0.67.

PP2. μ in PhH, $XGeMe_3$. X, μ: H, 0.668; F, 2.51; Cl, 2.89; Br, 2.84; I, 2.81; Ph, 0.58; Me, 0[c]; CH_2Ph, 0.63; C_2H, 0.79[b], OMe, 1.73; OEt, 1.60.

PP3. μ in PhH, $XSnMe_3$. X, μ: Cl, 3.46; Br, 3.45; I, 3.37; Ph, 0.51; Me, 0[c]; CH_2Ph, 0.91; NMe_2, 1.09; NEt_2, 0.85; Vi, 0.45.

PP4. μ in PhH, $XSiEt_3$. X, μ: H, 0.76; F, 1.74; Cl, 2.09; Br, 2.42; Ph, 0.71[d]; 2-ViVn, 0.47; CH_2Vi, 0.2; Et, 0; OH, 1.21.

PP5. μ in PhH, $XSnEt_3$. X, μ: Cl, 3.56; Br, 3.35; Ph, 0.5[d]; Et, 0; OAc, 2.05; PhC_2, 0.99.

PP6. μ in PhH, $XPbEt_3$. X, μ: Cl, 4.42; Br, 4.49; OH. 1.93; Ph, 0.82; Et, 0.3.

PP7. μ in PhH, $XSiH_3$. X, μ: H, 0; F, 1.268; Cl, 1.292; Br, 1.31; I, 1.62[e]; N_3, 2.17; SiH_3, 0; OMe, 1.165; $OSiH_3$, 0.24; C_2H, 0.316[b]; Me, (−) 0.7351; Et, (−) 0.81; Bu, (−) 0.76; *i*-Bu, (−) 0.75.

PP8. μ in PhH, $XGeH_3$. X, μ: H, 0[c]; Cl, 2.10; Br, 2.00; I, 1.81; CN, 3.99; Me, (−) 0.644; C_2H, 0.136[b].

PP9. μ in PhH, X_2SiMe_2. X, μ: H, 0.75; Cl, 1.89; Br, 2.45; OH. 1.94[f]; OMe, 1.29; OEt, 1.39; OPh, 1.278[e]; SMe, 1.25; Ph, 0.36; CH_2Vi, 0.54; Me, 0; Pr, 0[f].

PP10. μ in PhH, X_2GeMe_2. X, μ: H, 0.616; Cl, 3.14; OMe, 1.63; OEt, 1.51; OPr, 1.49; OPh, 1.45; Me, 0.

PP11. μ in PhH, X_2SnMe_2. X, μ: Cl, 4.22; Br, 3.86; I, 3.76; OEt, 2.19; NMe_2, 1.33; Me, 0.

PP12. μ in PhH, X_3SnMe. X, μ: H, 0.68; Cl, 3.62; Br, 3.77; I, 2.64; NMe_2, 1.36, Me, 0.

PP13. Ionization potentials (eV), ViX. X, IP: Me, 9.69; Et, 9.72; Pr, 9.52; Bu, 9.48; CH_2Vi, 9.62; CH_2Ph, 9.71; CH_2CN, 10.18; CH_2OH, 10.16; CH_2Cl, 10.34; CH_2Br, 10.16; NH_2, 8.64; OMe, 9.05; OEt, 9.15; OBu, 9.07; OAc, 9.85; I, 10.08; CO_2H, 10.91; CO_2Me, 11.12; CHO, 10.95; $SiMe_3$, 9.56; SiH_3, 10.37; CH_2GeMe_3, 8.85.

PP14. Ionization potentials (eV), HC_2X. X, IP: H, 11.40; Me, 10.36; Et, 10.18; Pr, 10.09; Bu, 10.05; *t*-Bu; 9.97; *c*-Hx, 9.93; CH_2OH, 10.50; CH_2Cl, 10.76; CH_2Br, 10.65; F, 11.26; Cl, 10.58; Br, 10.31; CN, 11.60; CHO, 11.57; CF_3, 12.10; SiH_3, 10.73; $SiMe_3$, 10.18; $GeMe_3$, 10.00; $SnEt_3$, 9.00; CH_2SiMe_3, 9.04.

PP15. Vertical ionization potentials (eV), π_3 (b_1) orbital of PhX. X, IP (v): H, 9.24; Me, 8.84; *t*-Bu, 8.83; CH_2Bu-*t*, 8.77; F, 9.35; Cl, 9.10; Br, 8.99; I, 8.67; N_3, 8.72[b]; NH_2, 8.05; NHMe, 7.65; NMe_2, 7.37; OH, 8.56; OMe, 8.39; SH, 8.17; SMe, 8.07; SPh, 7.89; CH_2Cl, 9.27; CH_2Br, 9.23; CH_2OMe, 9.12; CH_2OH, 8.90; CH_2Vi, 8.65; *c*-Pr, 8.66; Vi, 8.49; Ph, 8.39; C_2H, 8.82; CF_3, 9.90; CCl_3, 9.32; Ac, 9.51; CO_2Me; 9.50; CN, 9.72; SO_2Me, 9.74; SO_2Ph, 9.37; NO_2, 9.99; SiH_3, 9.18; $SiMe_3$, 9.05; SiF_3, 10.23[b]; $SiCl_3$, 9.46; $SiBr_3$, 9.06[b]; $SiMeF_2$; 9.55; $SiMeCl_2$; 9.52; $SiMeF_2$; 9.55; $SiMeBr_2$; 9.10; $SiMe_2Cl$, 9.30; $SiMe_2Ph$, 8.98; $SiPh_3$, 8.96; $GeMe_3$, 9.00; $GePh_2Br$, 9.19; $GePh_3$, 8.95; $SnMe_3$, 8.94; $SnPh_2Cl$, 9.39; $SnPh_3$; 9.04; $PbPh_3$, 8.95; CH_2SiMe_3, 8.42; $C(SiMe_3)_3$, 8.10; CH_2SnMe_3, 8.21[b].

PP16. $A^{1/2}$ ($\mathrm{l\,mol^{1/2}\,cm^{-1}}$) of the C ≡ C stretching frequency (IR) of Me_3GeC_2X. $A^{1/2}$, X: NH_2, −131.9; OMe, −121.8; OH, −114.2; F, −99.0; Me, −38.3; Vi, −25.6; H, −13.0; CF_3, 12.3; CN, 9.8; CHO, 47.7; Ac, 42.7; NO_2, 30.3; CH_2Ph, −40.2; CH_2OMe, −24.5; CH_2Br, −11.4; Ph, −37.7.

TABLE 6. (*continued*)

PP17. δ $^{13}C^{\alpha}$ (NMR) of Me_3SiC_2X. X, δ $^{13}C^{\alpha}$: H, 89.78; Me, 83.89; *t*-Bu, 82.14; Ph, 94.05; C_6F_5, 109.40; SMe, 96.90; SEt, 95.21; Cl, 74.76; Br, 86.99; I, 104.2; CH_2OEt, 90.99; CH_2Cl, 91.72; CH_2Br, 92.24; $SiMe_3$, 113.79; $SiEt_3$, 111.20; $GeMe_3$, 113.92; $GeEt_3$, 111.20; $SnMe_3$, 117.78; $SnEt_3$, 118.97; CH_2SiMe_3, 83.87.

PP18. δ ^{29}Si (NMR) of Me_3SiC_2X. X, δ ^{29}Si: H, −17.6; Me, −19.6; *t*-Bu, −19.4; Ph, −18.0; C_6F_5, −15.3; SMe, −18.5; SEt, −18.5; Cl, −16.1; Br, −16.0; I, −16.2; CH_2Cl, −17.3; CH_2Br, −17.3; $SiMe_3$, −19.4; $GeMe_3$, −20.1; $GeEt_3$, −20.1; $SnMe_3$, −20.8; $SnEt_3$. −20.3; CH_2SiMe_3, −20.1.

PP19. δ ^{119}Sn (NMR) of Me_3SiC_2X. X, δ ^{119}Sn: H, −68.3; Me, −73.7; Et, −73.2; *t*-Bu, −72.8; Ph, −67.7; C_6F_5, −57.8; SMe, −68.1; SEt, −68.4; Cl, −58.5; Br, −58.8; CH_2OEt, −69.5; CH_2Cl, −65.5; CH_2Br, −65.6; $SiMe_3$, −77.0; $SiEt_3$, −77.1; $GeMe_3$, −77.3; $GeEt_3$, −78.9; $SnMe_3$, −80.9; CH_2SiMe_3, −74.0.

[a] For abbreviations, see Section I.A.
[b] Assumed value.
[c] In CCl_4.
[d] In *c*-Hx.
[e] In dioxane.
[f] In the liquid state.

Correlation of μ for $XSnMe_3$ (set **PP3**) with the LDR equation gave the regression equation 55:

$$\mu_X = 7.47(\pm 0.727)\sigma_{lX} - 0.0193(\pm 0.200) \tag{55}$$

$$100r^2, 93.78 F, 105.6; S_{est}, 0.381, S^o, 0.283; N_{dp}, 9; P_D, 0; \eta, 0.$$

Correlation of μ for $XSiEt_3$ (set **PP4**) with the CR equation gave the regression equation 56:

$$\mu_X = 4.11(\pm 0.494)\sigma_{c16.7,X} + 0.365(\pm 0.128) \tag{56}$$

$$100r^2, 90.83; F, 69.34; S_{est}, 0.280, S^o, 0.343; N_{dp}, 9; P_D, 16.7; \eta, 0.$$

Correlation of μ for $XSnEt_3$ (set **PP5**) and $XPbEt_3$ (set **PP6**) with the LD equation gave the regression equations 57 and 58:

$$\mu_X = 7.16(\pm 1.57)\sigma_{lX} - 0.324(\pm 0.317) \tag{57}$$

$$100r^2, 83.85; F, 20.77; S_{est}, 0.690, S^o, 0.492; N_{dp}, 6; P_D, 0; \eta, 0.$$

$$\mu_X = 10.8(\pm 1.13)\sigma_{lX} - 3.72(\pm 0.941)\sigma_{dX} + 0.343(\pm 0.317) \tag{58}$$

$$100R^2, 98.62; A\ 100R^2, 98.16; F, 71.62; S_{est}, 0.341; S^o, 0.186; N_{dp}, 5;$$

$$P_D, 25.7(\pm 8.23), \eta, 0; r_{ld}, 0.544.$$

Correlation of μ for $XSiH_3$ (set **PP7**) with the CR equation and the $\sigma_{c14.3}$ and $\sigma_{c16.7}$ constants gave the regression equations 59 and 60:

$$\mu_X = 4.85(\pm 0.441)\sigma_{c16.7,X} + 7.43(\pm 2.42)\sigma_{eX} - 0.610(\pm 0.144) \tag{59}$$

$$100R^2, 93.47; A\ 100R^2, 92.82; F, 64.45; S_{est}, 0.310, S^o, 0.295; N_{dp}, 12;$$

$$P_D, 16.7; \eta, -7.64(\pm 2.39); r_{ce}, 0.544.$$

$$\mu_X = 4.78(\pm 0.435)\sigma_{c14.3,X} + 7.69(\pm 2.42)\sigma_{eX} - 0.634(\pm 0.146) \quad (60)$$

$$100R^2, 93.45; A\ 100R^2, 92.79; F, 64.19; S_{est}, 0.311, S^o, 0.296; N_{dp}, 12;$$
$$P_D, 14.3; \eta, -8.03(\pm 2.42); r_{ce}, 0.031.$$

However correlation of μ for $XGeH_3$ (set **PP8**) with the CR equation gave the regression equation 61:

$$\mu_X = 6.42(\pm 0.685)\sigma_{c16.7,X} - 0.318(\pm 0.254) \quad (61)$$

$$100r^2, 95.64; F, 87.75; S_{est}, 0.387, S^o, 0.256; N_{dp}, 6; P_D, 16.7; \eta, 0.$$

Because substituents are either overall electron donors or electron acceptors, they can in many cases cause the molecular dipole moment to have one direction for acceptors and the opposite direction for donors. In such cases, to account for this, it is necessary to assign a negative sign to the dipole moment when the substituent is an overall electron donor. This was done for alkyl groups in sets **PP7** and **PP8**. In Table 6, when this has been done, the minus sign is in parentheses to show that it has been assigned.

Correlation of μ for X_2SiMe_2 (set **PP9**) with the LDR equation gave the regression equation 62:

$$\mu_X = 4.14(\pm 0.507)\sigma_{lX} + 0.115(\pm 0.153) \quad (62)$$

$$100r^2, 88.10; F, 66.60; S_{est}, 0.295, S^o, 0.381; N_{dp}, 11; P_D, 0; \eta, 0.$$

Correlation of μ for X_2GeMe_2 (set **PP10**) with the CR equation and the $\sigma_{c14.3}$ and $\sigma_{c16.7}$ constants gave results which show no significant difference in goodness of fit between the regression equations. They are equations 63 and 64, respectively:

$$\mu_X = 5.71(\pm 0.974)\sigma_{c14.3,X} + 0.353(\pm 0.230) \quad (63)$$

$$100r^2, 87.31; F, 34.41; S_{est}, 0.380, S^o, 0.421; N_{dp}, 7; P_D, 14.3; \eta, 0.$$

$$\mu_X = 5.79(\pm 0.989)\sigma_{c16.7,X} + 0.413(\pm 0.222) \quad (64)$$

$$100r^2, 87.25; F, 34.22; S_{est}, 0.380, S^o, 0.422; N_{dp}, 7; P_D, 16.7; \eta, 0.$$

Correlation of μ for X_2SnMe_2 (set **PP11**) with the LD equation gave the regression equation 65:

$$\mu_X = 8.75(\pm 0.570)\sigma_{lX} - 0.0345(\pm 0.196) \quad (65)$$

$$100r^2, 98.33; F, 235.5; S_{est}, 0.243, S^o, 0.158; N_{dp}, 6; P_D, 0; \eta, 0.$$

Correlation of μ for X_3SnMe (set **PP12**) with the LD equation gave the regression equation 66:

$$\mu_X = 6.81(\pm 0.657)\sigma_{lX} + 0.308(\pm 0.213) \quad (66)$$

$$100r^2, 96.42; F, 107.6; S_{est}, 0.332, S^o, 0.232; N_{dp}, 6; P_D, 0; \eta, 0.$$

In discussing these results it is necessary to consider the relationship between the molecular dipole moment μ and the individual bond moments μ_b (equation 67):

$$\mu = \sum \mu_b \quad (67)$$

Then for $XMeZ_3$, equations 68 and 69 apply:

$$\mu = \mu_{b,MX} - 3\cos 70.52^o \mu_{b,MZ} \quad (68)$$

$$= \mu_{b,MX} - \mu_{b,MZ} \quad (69)$$

for X_2MZ_2, equations 70 and 71 apply:

$$\mu = 2\cos 54.76^o (\mu_{b,MX} - \mu_{b,MZ}) \quad (70)$$

$$= 1.154(\mu_{b,MX} - \mu_{b,MZ}) \quad (71)$$

and for X_3MZ, equations 72 and 73 apply:

$$\mu = 3\cos 70.52^o \mu_{b,MX} - \mu_{b,MZ} \quad (72)$$

$$= \mu_{b,MX} - \mu_{b,MZ} \quad (73)$$

Assuming that μ_b is a function of $\Delta\chi$, the electronegativity difference between X and M or Z and M is given by equations 74a and 74b:

$$\mu_b = \chi_X - \chi_M \quad (74a)$$

$$\mu_b = \chi_Z - \chi_M \quad (74b)$$

and then equations 75 and 76 apply:

$$\mu = f[(\chi_X - \chi_M) - (\chi_Z - \chi_M)] \quad (75)$$

$$= f(\chi_X - \chi_Z) \quad (76)$$

As Z is constant throughout a data set, it follows that within a data set μ should be a function of χ_X.

L and *C* are equivalent to each other insofar as the magnitude of the localized electrical effect is concerned[1]. The *L* values obtained in the correlations are the same for a given M, as they are all less than two standard deviations from each other. The mean values of *L* for Si, Ge and Sn are 4.46(±0.392), 5.90(±0.459) and 7.88(±0.846), respectively. Although only one *L* value is available for Pb, it seems that *L* increases with the atomic number for group 14 elements other than carbon.

There seems to be a linear relationship between the mean value of *L* and the first ionization potential (IP) of MMe_4 when M is Si, Ge, Sn and probably Pb as well. As the IP of $PbMe_4$ is regarded as uncertain[31b] and only one data set for lead derivatives was available, making the *L* value for lead uncertain, no definite conclusion regarding the fit of lead compounds in this relationship can be reached. The only *L* value available for carbon at this time is reliable but does not fit the *L*–IP relationship.

The value of P_D for Si and Ge data sets is generally about 16, that for Sn is generally about 0. For the only carbon data set studied it is 28, but this is a large set with a wide range of substituent types and the value is reliable. For the only lead data set available, the range of substituent types is small as is the number of data points; we therefore regard the P_D value for this set as unreliable. We conclude that for the group 14 elements the order of P_D values is C > Si ≈ Ge > Sn. In view of this sequence we suspect that the correct P_D value for Pb is probably zero and that it follows Sn in the sequence. The value of η generally obtained is 0.

Overall, the goodness of fit obtained for these correlations is about what can be expected for the quality of the data, particularly when we take into account the fact that in most cases the dipole moments were determined in benzene, a Lewis base, while the compounds studied are all Lewis acids.

2. *Ionization potentials*

First ionization potentials of substituted ethylenes[32] (set **PP13**, Table 6) were correlated with the LDRA equation (equation 38). The regression equation is equation 77:

$$IP_X = 1.58(\pm 0.218)\sigma_{lX} - 2.45(\pm 0.123)\sigma_{dX} - 1.09(\pm 0.455)\alpha_X + 10.11(\pm 0.0819) \quad (77)$$

$$100R^2, 96.24; A\ 100R^2, 95.85; F, 153.9; S_{est}, 0.140; S^o, 0.214;$$
$$N_{dp}, 22; P_D, 60.7(\pm 4.83), \eta, 0; r_{ld}, 0.157; r_{l\alpha}, 0.394; r_{d\alpha}, 0.058;$$
$$C_l, 37.3; C_d, 57.6; C_\alpha, 5.11.$$

Correlation of first ionization potentials of substituted acetylenes[33] (set **PP14**, Table 6) with the LDRA equation gave the regression equation 78:

$$IP_X = 1.13(\pm 0.324)\sigma_{lX} - 2.63(\pm 0.472)\sigma_{dX} + 7.72(\pm 3.57)\sigma_{eX} - 4.67(\pm 0.752)\alpha + 11.36(\pm 0.166) \quad (78)$$

$$100R^2, 91.94; A\ 100R^2, 90.52; F, 45.65; S_{est}, 0.249; S^o, 0.325; N_{dp}, 21;$$
$$P_D, 70.0(\pm 16.4), \eta, 2.35(\pm 1.23); r_{ld}, 0.211; r_{le}, 0.276; r_{l\alpha}, 0.634;$$
$$r_{de}, 0.748; r_{d\alpha}, 0.142; r_{e\alpha}, 0.370; C_l, 20.7; C_d, 48.1; C_e, 14.1; C_\alpha, 17.1.$$

Correlation of vertical ionization potentials of the $\pi_S(\pi_3)$ orbital of substituted benzenes[33,34] (set **PP15**, Table 6) gave, on exclusion of the values for N_3, $SiBr_3$, SiF_3 and CH_2SnMe_3, the regression equation 79:

$$IP_X = 0.895(\pm 0.117)\sigma_{lX} - 1.61(\pm 0.0976)\sigma_{dX} + 4.12(\pm 0.480)\sigma_{eX} - 0.628(\pm 0.113)\alpha + 9.28(\pm 0.0515) \quad (79)$$

$$100R^2, 93.18; A\ 100R^2, 92.75; F, 157.2; S_{est}, 0.151; S^o, 0.275; N_{dp}, 51;$$
$$P_D, 64.3(\pm 5.52), \eta, 2.56(+-0.254); r_{ld}, 0.166; r_{le}, 0.037; r_{l\alpha}, 0.464;$$
$$r_{de}, 0.346; r_{d\alpha}, 0.497; r_{e\alpha}, 0.470; C_l, 29.4; C_d, 52.1; C_e, 13.5; C_\alpha, 4.13.$$

The lack of fit of the CH_2SnMe_3 group is certainly due to an error in the value of σ_d,. The poor fit of the $SiBr_3$ and SiF_3 groups is probably also due to errors in their substituent constants. The cause for the lack of fit of the N_3 group is unknown.

The P_D values are in good agreement for all three data sets. The lack of dependence on σ_e for the substituted ethylenes is surprising; we are unable to account for it. The dependence on α is much smaller for the substituted benzenes and ethylenes than for the acetylenes. We are unable to account for this at the present time.

3. *Infrared A values*

Egorochkin and coworkers[35] have reported $A^{1/2}$ values for 1-substituted-2-trimethylgermylacetylenes (set **PP16**, Table 6), where A is the stretching frequency of the triple bond. We have correlated the data set with the LDR equation (equation 1); the regression equation is given by equation 80:

$$A_X^{1/2} = 191(\pm 5.11)\sigma_d - 8.50(\pm 1.61) \quad (80)$$

$$100r^2, 99.01; F, 1398; S_{est}, 5.94, S^o, 0.106; N_{dp}, 16; P_D, 100; \eta, 0.$$

Here, $A^{1/2}$ is a function solely of σ_d. This is surprising as it has been reported that $A^{1/2}$ is a function solely of σ_R^o, which is itself a composite parameter that is dependent on both σ_d and σ_e[36]. We have made use of equation 80 in the form of equation 81:

$$\sigma_{dX} = 0.00524A_X^{1/2} + 0.0445 \quad (81)$$

to calculate new σ_d values for a number of substituents and regard these values as very reliable.

4. NMR chemical shifts

Correlation of ^{13}C chemical shifts for 1-substituted-2-trimethylsilylacetylenes[29] (set **PP17**, Table 6) with the LDRA equation (equation 41) gave the regression equation 82:

$$\delta_X = 12.3(\pm 8.43)\sigma_{lX} - 62.4(\pm 9.81)\sigma_{dX} - 201(\pm 44.4)\sigma_{eX} - 29.1(\pm 14.3)\alpha + 87.0(\pm 3.23) \quad (82)$$

$$100R^2, 89.30; A\ 100R^2, 87.16; F, 29.20; S_{est}, 5.07; S^o, 0.381; N_{dp}, 19;$$

$$P_D, 83.2(\pm 16.4), \eta, -3.22(\pm 0.501); r_{ld}, 0.667; r_{le}, 0.144; r_{l\alpha}, 0.596;$$

$$r_{de}, 0.266; r_{d\alpha}, 0.558; r_{e\alpha}, 0.242; C_l, 12.5; C_d, 61.8; C_e, 19.9; C_\alpha, 5.78.$$

The term in σ_1 was retained because of the highly significant collinearity between σ_1 and σ_d. Correlation of chemical shifts for ^{29}Si (set **PP18**, Table 6) with the LDRA equation gave the regression equation 83:

$$\delta_X = 8.45(\pm 1.00)\sigma_{lX} - 4.37(\pm 1.09)\sigma_{dX} - 3.94(\pm 1.51)\alpha + 18.2(\pm 0.392) \quad (83)$$

$$100R^2, 89.64; A\ 100R^2, 88.26; F, 40.40; S_{est}, 0.609; S^o, 0.365; N_{dp}, 18;$$

$$P_D, 34.1(\pm 9.34), \eta, 0; r_{ld}, 0.641; r_{le}, 0.129; r_{l\alpha}, 0.557; r_{de}, 0.261;$$

$$r_{d\alpha}, 0.504; r_{e\alpha}, 0.280; C_l, 62.1; C_d, 32.1; C_\alpha, 5.79.$$

However, correlation of ^{119}Sn chemical shifts of 1-substituted-2-trimethylstannyl acetylenes (set **PP19**, Table 6) gave the regression equation 84:

$$\delta_X = 34.9(\pm 4.10)\sigma_{lX} - 9.42(\pm 4.43)\sigma_{dX} - 12.6(\pm 6.08)\alpha - 70.5(\pm 1.47) \quad (84)$$

$$100R^2, 89.17; A\ 100R^2, 87.81; F, 41.15; S_{est}, 2.49; S^o, 0.370; N_{dp}, 19;$$

$$P_D, 21.3(\pm 10.4), \eta, 0; r_{ld}, 0.627; r_{le}, 0.140; r_{l\alpha}, 0.491; r_{de}, 0.260;$$

$$r_{d\alpha}, 0.483; r_{e\alpha}, 0.366; C_l, 74.5; C_d, 20.1; C_\alpha, 5.38.$$

Structural effects on the silicon and tin chemical shifts are very similar. The P_D values do not differ significantly and η is zero in both data sets. These results are very different from those observed for ^{13}C chemical shifts.

D. Bioactivities (QSAR)

Little in the way of data sets involving the effect of structural variation on bioactivity is available for compounds of germanium and lead. Some data are available for alkylstannanes[37] although the data sets are very limited in both the number of data points and in the extent of variation in the alkyl group structure. The toxicity of trialkylstannanes to *Botrytis* and of dialkylstannanes to *B. subtilis* (sets **BA1** and **BA2**, Table 7) were modeled using the simple branching equation (equation 4) in the form of equation 85:

$$BA_{Ak} = a_1 n_1 + a_C n_C + a_o \quad (85)$$

where n_C is the number of carbon atoms in the alkyl group and a_C is its coefficient, and in the form of equation 86:

$$BA_{Ak} = a_1 n_1 + a_2 n_2 + a_o \quad (86)$$

Correlation of the data sets with equations 85 and 86 gave the regression equation 87:

$$\log BA_{Ak} = -2.48(\pm 0.478)n_1 - 0.0458(\pm 0.427) \quad (87)$$

$$100r^2, 84.97; F, 26.92; S_{est}, 0.427; S^o, 0.409; N_{dp}, 5.$$

With the exclusion of the data point for hexyl, it gave equation 88:

$$BA_{Ak} = -0.653(\pm 0.0547)n_1 - 0.478(\pm 0.0446)n_2 - 0.0458(\pm 0.0387) \quad (88)$$

$$100R^2, 99.70; A\ 100R^2, 99.60; F, 331.5; S_{est}, 0.0387; S^o, 0.0867;$$

$$N_{dp}, 5; r_{12}, 0.612; C_{n1}, 57.7; C_{n2}, 42.3; B_1/B_2, 1.37.$$

LD_{50}s for the toxicity of trialkylstannanes (set **BA3**, Table 7) to the rat were correlated with equation 89:

$$LD_{50,Ak} = S\upsilon + A\alpha + B^o \quad (89)$$

TABLE 7. QSAR data sets[a]

BA1. Toxicities of trialkylstannanes to *Botrytis* (mM). Ak, BA: Me, 0.9; Et, 0.004; Pr, 0.002; Bu, 0.001; Pe, 0.01.

BA2. Toxicities of dialkylstannanes to *B. subtilis* (mM). Ak, BA: Me, 0.9; Et, 0.2; Pr, 0.07; Bu, 0.07, Pe, 0.06, Hx, 0.14.

BA3. Toxicities of trialkylstannanes to the rat (mM). Ak, BA: Me, 0.07; Et, 0.04; Pr, 0.3; Bu, 0.7; Hx, 2; *c*-Hx, 1.

BA4. Toxicities of trialkylstannanes to crab larvae (mM). Ak, BA: Me, 0.56; Et, 0.39; Pr, 0.19; Bu, 0.055; *c*-Hx, 0.02.

BA5. Toxicities of dialkylstannanes to crab larvae (mM). Ak, BA: Me, 82; Et, 15; Pr, 2.8; *c*-Hx, 0.37.

BA6. LD_{50}'s ($mM\,kg^{-1}$) of substituted germatranes to white mice (ip). X, LD_{50}: 2-Tp, 0.0546; 3-Tp, 0.295; H, 1.46; CH_2Br, 1.14; CH_2NEt_2, 1.17; 3-Fr, 5.70; 2-Fr, 7.17; $CH_2CH_2CO_2Et$, 7.50; CH_2Cl, 11.0; $OSiMe_3$, 11.7; CH_2CH_2CN, 15.8; Vi, 22.8. OH, 35.6; CH_2Tp-2, 1.03; CH_2Fr-2, 9.87; $PnNEt_2$-4, 8.86; $PnNMe_2$-4, 10.9; $CH_2CHMeCO_2Me$, 21.3; CH_2CH_2Py-4, 7.91; CH_2CH_2Py-2, 8.65.

[a]For abbreviations, see Section I.A.

The regression equation is equation 90:

$$LD_{50,Ak} = 6.88(\pm 1.28)\alpha - 1.63(\pm 0.238) \quad (90)$$

$$100r^2, 87.90; F, 29.06; S_{est}, 0.261; S^o, 0.426; N_{dp}, 6 : r_{nl,\alpha}, 0.890.$$

As α and υ are highly collinear, a possible dependence on the latter cannot be excluded.

LD_{50}s for the toxicity of di- and trialkylstannanes to crab larvae (sets **BA4** and **BA5**, Table 7) were correlated with equation 91:

$$LD_{50,Ak} = S\upsilon + B^o \quad (91)$$

giving the regression equations 92 and 93:

$$\log LD_{50/Ak} = -6.20(\pm 1.07)\upsilon_{Ak} + 4.86(\pm 0.125) \quad (92)$$

$$100r^2, 94.34; F, 33.33; S_{est}, 0.292; S^o, 0.336; N_{dp}, 4.$$

$$\log LD_{50/Ak} = -4.18(\pm 0.826)\upsilon_{Ak} + 1.90(\pm 0.556) \quad (93)$$

$$100r^2, 89.53; F, 25.65; S_{est}, 0.225; S^o, 0.418; N_{dp}, 5.$$

As expected, the bioactivities of alkylstannanes are a function only of steric effects and polarizability.

Lukevics and Ignatovich[38] have reported intraperitoneal LD_{50} values ($mg\,kg^{-1}$) in white mice for substituted germatranes, $XGe(OCH_2CH_2)_3N$. After conversion to $mmol\,kg^{-1}$, these values (set **BA6**, Table 7) were correlated with the IMF equation in the form of equation 94:

$$Q_X = C\sigma_{c50,X} + M\mu_X + A\alpha_X + H_1 n_{HX} + H_2 n_{nX} + S_1\upsilon_{1X} + S_2\upsilon_{2X} + B^o \quad (94)$$

After excluding the values for 2- and 3-thienyl and for vinyl, regression equation 94 was obtained:

$$\log LD_{50,X} = -1.91(\pm 0.954)\sigma_{c50,X} + 0.247(\pm 0.103)\mu_X + 0.298(\pm 0.194) \quad (95)$$

$$100R^2, 41.61; A\ 100R^2, 37.72; F, 4.989; S_{est}, 0.376; S^o, 0.842; N_{dp}, 17;$$

$$P_D, 50 : r_{\sigma\mu}, 0.003; C_\sigma, 79.4; C_\mu, 20.6.$$

This equation is significant at the 97.5% confidence level, though it accounts for only about 40% of the variance of the data. There is a dependence on μ and perhaps on σ as well. It is surprising to find no dependence on polarizability. The poor fit may be due to the existence of more than one mode of activity in the data set.

VI. THE VALIDITY OF THE ESTIMATED SUBSTITUENT CONSTANTS

The values of Q calculated for various Group 14 substituents from the applications reported in the previous section provide the only evidence for the validity of the parameter estimates in Reference 1. Table 8 presents values of Q_{obs}, Q_{calc} and ΔQ. The data set from which the calculated value was obtained, and the parameter types used in the calculation are also reported. The agreement between observed and calculated values is described in terms of the number of standard deviations, N_{SD}, defined in equation 96:

$$N_{SD} = \frac{\|\Delta Q\|}{S_{est}} \quad (96)$$

TABLE 8. Values of $Q_{X,obs}$, $Q_{X,calc}$ and ΔQ^a

X	Q type	$Q_{X,obs}$	Q_{calc}	ΔQ	N_{sd}	Parameters	Set
SiH_3	IP	10.37	10.32	0.0497	0.355	σ_l, σ_d, α	**PP13**
	IP	10.73	10.94	−0.211	−0.846	σ_l, σ_d, α	**PP14**
	IP	9.18	9.27	−0.0907	−0.599	σ_l, σ_d, σ_e, α	**PP15**
$SiBr_3$	IP	9.06	9.44	−0.38	−2.52	σ_l, σ_d, σ_e, α	**PP15**
$SiCl_3$	IP	9.46	9.45	0.0125	0.0825	σ_l, σ_d, σ_e, α	**PP15**
SiF_3	IP	10.23	9.79	0.44	2.91	σ_l, σ_d, σ_e, α	**PP15**
$SiBr_2Me$	IP	9.10	9.28	−0.179	−1.18	σ_l, σ_d, σ_e, α	**PP15**
$SiCl_2Me$	IP	9.52	9.39	0.132	0.874	σ_l, σ_d, σ_e, α	**PP15**
SiF_2Me	IP	9.55	9.29	0.263	1.74	σ_l, σ_d, σ_e, α	**PP15**
$SiClMe_2$	IP	9.30	9.22	0.0801	0.529	σ_l, σ_d, σ_e, α	**PP15**
$SiMe_3$	$\log k_{rel}$	−0.824	−0.743	−0.081	−0.577	σ_l	**CR3**
	$\log k_{rel}$	−0.310	−0.400	0.090	−1.40	σ_l	**CR4**
	$\Delta\nu_{OH}$	121	132	−10.5	0.801	σ_l, σ_d, α	**CP3**
	$\Delta\nu_{OH}$	158	164	−7.64	−0.604	σ_l, σ_d, α	**CP4**
	$\Delta\nu_{OH}$	71	69.7	1.88	0.260	σ_l, σ_d	**CP5**
	$\Delta\nu_{OH}$	66	60.2	5.75	1.12	σ_l, σ_d	**CP6**
	IP	9.86	10.00	−0.136	−0.969	σ_l, σ_d, α	**PP13**
	IP	10.18	10.11	0.0695	0.279	σ_l, σ_d, α	**PP14**
	IP	9.05	9.05	−0.00201	−0.0174	σ_l, σ_d, σ_e, α	**PP15**
	δ ^{13}C	113.79	109.96	3.83	0.754	σ_l, σ_d, σ_e, α	**PP17**
	δ ^{29}Si	−19.4	−19.6	0.151	0.248	σ_l, σ_d, α	**PP18**
	δ ^{119}Sn	−77.0	−76.2	−0.849	−0.341	σ_l, σ_d, α	**PP19**
$SiMe_2Et$	$\log k_{rel}$	−0.959	−0.743	−0.216	−1.53	σ_l	**CR3**
$SiEt_3$	$\Delta\nu_{OH}$	163	176	−13.3	−1.03	σ_l, σ_d, α	**CP4**
	$\Delta\nu_{OH}$	70	69.7	0.278	0.0565	σ_l, σ_d	**CP5**
	$\Delta\nu_{OH}$	63	60.2	2.75	0.535	σ_l, σ_d	**CP6**
	δ ^{13}C	111.20	114.07	−2.87	−0.566	σ_l, σ_d, σ_e, α	**PP17**
	δ ^{119}Sn	−77.1	−77.9	2.75	0.535	σ_l, σ_d, α	**PP19**
$SiMe_2Ph$	IP	8.98	9.03	−0.0533	−0.352	σ_l, σ_d, σ_e, α	**PP15**
$SiPh_3$	IP	8.96	9.00	−0.0403	0.267	σ_l, σ_d, σ_e, α	**PP15**
$GeMe_3$	$\log k_{rel}$	−0.523	−0.597	0.074	0.525	σ_l	**CR3**
	$\log k_{rel}$	−0.268	−0.354	0.086	1.34	σ_l	**CR4**
	$\Delta\nu_{OH}$	145	130	14.6	1.11	σ_l, σ_d, α	**CP3**
	$\Delta\nu_{OH}$	181	163	17.9	1.42	σ_l, σ_d, α	**CP4**
	$\Delta\nu_{OH}$	67	71.2	−4.16	−0.848	σ_l, σ_d	**CP5**
	$\Delta\nu_{OH}$	61	62.0	−1.02	−0.199	σ_l, σ_d	**CP6**
	IP	10.00	9.99	0.00923	0.0370	σ_l, σ_d, α	**PP14**
	IP	9.00	9.02	−0.0214	−0.141	σ_l, σ_d, σ_e, α	**PP15**
	δ ^{13}C	113.92	110.33	3.59	0.707	σ_l, σ_d, σ_e, α	**PP17**
	δ ^{29}Si	−20.1	−19.4	−0.656	−1.08	σ_l, σ_d, α	**PP18**
	δ ^{119}Sn	−77.3	−75.5	−1.82	−0.730	σ_l, σ_d, α	**PP19**
$GeEt_3$	$\Delta\nu_{OH}$	146	145	0.577	0.0439	σ_l, σ_d, α	**CP3**
	$\Delta\nu_{OH}$	183	173	9.42	0.745	σ_l, σ_d, α	**CP4**

TABLE 8. (*continued*)

X	Q type	$Q_{X,obs}$	Q_{calc}	ΔQ	N_{sd}	Parameters	Set
	$\Delta\nu_{OH}$	66	71.2	−5.16	−1.05	σ_l, σ_d	**CP5**
	δ ^{13}C	111.20	114.35	−3.15	−0.622	$\sigma_l, \sigma_d, \sigma_e, \alpha$	**PP17**
	δ ^{29}Si	−20.1	−20.0	−0.111	−0.183	$\sigma_l, \sigma_d, \alpha$	**PP18**
	δ ^{119}Sn	−75.9	−77.2	−1.68	−0.674	$\sigma_l, \sigma_d, \alpha$	**PP19**
$GeBrPh_2$	IP	9.17	9.14	0.0347	0.230	$\sigma_l, \sigma_d, \sigma_e, \alpha$	**PP15**
$GePh_3$	IP	8.95	8.88	0.0661	0.437	$\sigma_l, \sigma_d, \sigma_e, \alpha$	**PP15**
$SnMe_3$	$\Delta\nu_{OH}$	158	137	21.2	1.61	$\sigma_l, \sigma_d, \alpha$	**CP3**
	$\Delta\nu_{OH}$	194	168	26.0	2.05	$\sigma_l, \sigma_d, \alpha$	**CP4**
	$\Delta\nu_{OH}$	69	70.8	−1.81	−0.368	σ_l, σ_d	**CP5**
	$\Delta\nu_{OH}$	56	61.5	−5.54	−1.08	σ_l, σ_d	**CP6**
	IP	8.94	9.97	−0.0264	−0.175	$\sigma_l, \sigma_d, \sigma_e, \alpha$	**PP15**
	δ ^{13}C	117.78	112.58	5.20	1.03	$\sigma_l, \sigma_d, \sigma_e, \alpha$	**PP17**
	δ ^{29}Si	−20.8	−19.7	−1.11	−1.81	$\sigma_l, \sigma_d, \alpha$	**PP18**
	δ ^{119}Sn	−80.9	−76.4	−4.50	−1.81	$\sigma_l, \sigma_d, \alpha$	**PP19**
$SnEt_3$	$\Delta\nu_{OH}$	162	152	10.2	0.774	$\sigma_l, \sigma_d, \alpha$	**CP3**
	$\Delta\nu_{OH}$	54	61.5	−7.54	−1.47	σ_l, σ_d	**CP6**
	IP	9.00	9.41	−0.410	−1.65	$\sigma_l, \sigma_d, \alpha$	**PP14**
	δ ^{29}Si	118.97	116.59	2.38	0.468	$\sigma_l, \sigma_d, \alpha$	**PP18**
	δ ^{119}Sn	−20.3	−20.2	−0.0616	−0.101	$\sigma_l, \sigma_d, \alpha$	**PP19**
$SnPr_3$	$\Delta\nu_{OH}$	54	61.5	−7.5	−1.46	σ_l, σ_d	**CP6**
$SnMe_2Bu$-*t*	$\Delta\nu_{OH}$	64	70.8	−6.8	−1.38	σ_l, σ_d	**CP5**
	$\Delta\nu_{OH}$	54	61.5	−7.5	−7.46	σ_l, σ_d	**CP6**
SnMe(Bu-*t*)$_2$	$\Delta\nu_{OH}$	65	70.8	−5.8	−1.18	σ_l, σ_d	**CP5**
	$\Delta\nu_{OH}$	59	61.5	−2.5	−0.486	σ_l, σ_d	**CP6**
Sn(Bu-*t*)$_3$	$\Delta\nu_{OH}$	63	70.8	−7.8	−1.59	σ_l, σ_d	**CP5**
	$\Delta\nu_{OH}$	54	61.5	−7.5	−1.46	σ_l, σ_d	**CP6**
$SnBu_3$	$\Delta\nu_{OH}$	64	70.8	−6.8	−1.38	σ_l, σ_d	**CP5**
	$\Delta\nu_{OH}$	50	61.5	−11.2	−2.24	σ_l, σ_d	**CP6**
$SnClPh_2$	IP	9.29	9.14	0.153	1.01	$\sigma_l, \sigma_d, \sigma_e, \alpha$	**PP15**
$SnPh_3$	IP	9.04	8.93	0.114	0.752	$\sigma_l, \sigma_d, \sigma_e, \alpha$	**PP15**
$PbPh_3$	IP	8.95	8.89	0.0629	0.416	$\sigma_l, \sigma_d, \sigma_e, \alpha$	**PP15**
CH_2SiMe_3	$\log k_{rel}$	−0.310	−0.354	0.045	0.316	σ_l	**CR3**
	$\log k_{rel}$	−0.301	−0.277	−0.024	−0.380	σ_l	**CR4**
	$\Delta\nu_{OH}$	145	160	14.6	1.17	$\sigma_l, \sigma_d, \alpha$	**CP3**
	$\Delta\nu_{OH}$	183	195	−12.3	−0.974	$\sigma_l, \sigma_d, \alpha$	**CP4**
	$\Delta\nu_{OH}$	68	60.2	7.84	1.60	σ_l, σ_d	**CP5**
	$\Delta\nu_{OH}$	58	51.9	9.14	1.20	σ_l, σ_d	**CP6**
	IP	9.04	8.96	0.0807	0.324	$\sigma_l, \sigma_d, \alpha$	**PP14**
	IP	8.42	8.46	−0.0372	−0.246	$\sigma_l, \sigma_d, \sigma_e, \alpha$	**PP15**
	δ ^{13}C	83.87	81.46	2.41	0.475	$\sigma_l, \sigma_d, \sigma_e, \alpha$	**PP17**
	δ ^{29}Si	−20.1	−21.0	0.880	1.44	$\sigma_l, \sigma_d, \alpha$	**PP18**
	δ ^{119}Sn	−74.0	−75.1	4.09	1.64	$\sigma_l, \sigma_d, \alpha$	**PP19**
$C(SiMe_3)_3$	IP	8.10	8.27	−0.170	−1.13	$\sigma_l, \sigma_d, \sigma_e, \alpha$	**PP15**

TABLE 8. (*continued*)

X	Q type	$Q_{X,obs}$	Q_{calc}	ΔQ	N_{sd}	Parameters	Set
CH_2GeMe_3	$\log k_{rel}$	−0.268	−0.306	0.038	0.271	σ_l	**CR3**
	$\log k_{rel}$	−0.284	−0.261	−0.022	−0.354	σ_l	**CR4**
	$\Delta\nu_{OH}$	68	59.1	8.95	1.82	σ_l, σ_d	**CP5**
	$\Delta\nu_{OH}$	59	53.0	5.99	1.17	σ_l, σ_d	**CP6**
	IP	8.85	9.02	−0.168	−1.20	σ_l, σ_d, α	**PP13**
CH_2SnMe_3	IP	8.21	8.66	−0.45	−2.98[b]	σ_l, σ_d, σ_e, α	**PP15**
	IP	8.21	8.47	−0.21	−1.39[c]	σ_l, σ_d, σ_e, α	**PP15**
$CH_2Sn(Bu\text{-}t)_3$	$\Delta\nu_{OH}$	68	59.1	8.9	1.81	σ_l, σ_d	**CP5**
	$\Delta\nu_{OH}$	55	51.8	3.2	0.623	σ_l, σ_d	**CP6**

[a]For abbreviations, see Section I. A.
[b]Using the σ_d value given in Reference 1.
[c]Using the mean of the σ_d values given in Table 1 for CH_2SiMe_3 and CH_2GeMe_3.

When $N_{SD} \leqslant 1$ the agreement is considered excellent, for $1 < N_{SD} \leqslant 2$ agreement is considered fair, for $2 < N_{SD} \leqslant 3$ poor, and for $N_{SD} > 3$ unacceptable. Values of N_{SD} are also given in Table 8.

$MZ^1Z^2Z^3$ groups. The agreement between calculated and observed values for substituents in which Z groups are H, alkyl or aryl is good. For MEt_3 with M = Si, Ge and Sn, the electrical effect substituent constants were assumed equal to those for the MMe_3 groups. The same assumption was made for several $SnAk_3$ groups. The N_{SD} values given in Table 8 show that the assumption is justified, in agreement with previous results for substituents of the type WAk_n in which W is an atom or group of atoms[39]. The electrical, steric and intermolecular force substituent constants for these substituents are probably reliable. The electrical effect substituent constants for groups in which Z is halogen gave poorer results, though the data available to test them are scanty. There continues to be a lack of data on $M(OAk)_3$ and $M(SAk)_3$ groups.

$C(MZ^1Z^2Z^3)_nH_{3-n}$ groups. We have calculated σ_d values for the CH_2SiMe_3 and CH_2GeMe_3 groups directly from the $A^{1/2}$ values of set **PP16**. These values seem to be very reliable. Contrary to the statement in Reference 1, there appears to be a special capability for electron donation in these groups. Our results for the CH_2SnMe_3 group are in agreement with this conclusion, as also are our results for the $CH_2Sn(Bu\text{-}t)_3$ group.

There is still very little data available for Pb substituents. Much more experimental work is required before we can arrive at a reliable description of substituent effects of Group 14 elements other than carbon.

VII. APPENDIX. SUPPLEMENTARY GLOSSARY

Electrical effect parameterization

CR equation: A diparametric electrical effect model using the composite $\sigma_{Ck'}$ and pure σ_e parameters. A value of P_D is assumed and a value of η calculated.

LD equation: A diparametric electrical effect model using the pure σ_l and composite σ_D parameters. A value of P_D is calculated and a value of η assumed.

Steric effect parameterization

$\acute{\upsilon}$ A steric effect parameter based on van der Waals radii that has been corrected for the difference in bond length between the X-G and Y-G bonds.

Other parameterization

τ A parameter that accounts for the effect of temperature on the reaction velocity. It is defined as $100/T$ where T is the absolute temperature.

φ A parameter that accounts for the effect of concentration of a component in a mixed solvent. It is the mole fraction of that component.

ζ A parameter used to combine two or more data sets into a single set. It is determined by choosing a substituent present in each of the data sets to be combined and defining Q for that substituent as ζ. Thus, if the substituent chosen is Z and there are sets 1, 2 and 3 to be combined, the ζ values are Q_{Z1}, Q_{Z2} and Q_{Z3} respectively.

The glossary to which this is a supplement will be found in Appendix I of Reference 1.

VIII. REFERENCES

1. M. Charton, in *The Chemistry of Organic Germanium, Tin and Lead Compounds* (Ed. S. Patai), Wiley, Chichester, 1995, pp. 603–664.
2. C. G. Derick, *J. Am. Chem. Soc.*, **33**, 1152 (1911).
3. G. K. Branch and M. Calvin, *The Theory of Organic Chemistry*, Prentice-Hall, New York, 1941, p. 211; R. W. Taft, in *Steric Effects in Organic Chemistry* (Ed. M. S. Newman), Wiley, New York, 1956, p. 556; H. B. Watson, *Modern Theories of Organic Chemistry*, Oxford Univ. Press, Oxford, 1937, p. 40; A. E. Remick, *Electronic Interpretations of Organic Chemistry*, 2nd edn., Wiley, New York, 1949, p. 93.
4. J. G. Kirkwood, *J. Chem. Phys.*, **2**, 351 (1934); J. G. Kirkwood and F. H. Westheimer, *J. Chem. Phys.*, **6**, 506 (1938); F. H. Westheimer and J. G. Kirkwood, *J. Chem. Phys.*, **6**, 513 (1938); F. H. Westheimer and M. W. Shookhoff, *J. Am. Chem. Soc.*, **61**, 555 (1939); F. H. Westheimer, *J. Am. Chem. Soc.*, **61**, 1977 (1939); J. N. Sarmousakis, *J. Chem. Phys.*, **12**, 277 (1944).
5. O. Exner and P. Fiedler, *Collect. Czech. Chem. Commun.*, **45**, 1251 (1980).
6. K. Bowden and E. J. Grubbs, *Prog. Phys. Org. Chem.*, **19**, 183 (1992); K. Bowden and E. J. Grubbs, *Chem. Soc. Rev.*, **25**, 171 (1996).
7. M. Charton, *J. Phys. Org. Chem.*, **12**, 275 (1999).
8. M. Charton and B. I. Charton, *J. Chem. Soc., Perkin Trans. 2*, 2203 (1999).
9. M. Charton, *Top. Curr. Chem.*, **114**, 107 (1983).
10. (a) M. Charton, in *Rational Approaches to the Synthesis of Pesticides* (Eds. P. S. Magee, J. J. Menn and G. K. Koan), American Chemical Society, Washington, D.C., 1984, pp. 247–278. (b) M. Charton, *Stud. Org. Chem.*, **42**, 629 (1992).
11. L. Pauling, *The Nature of the Chemical Bond*, 3rd edn., Cornell Univ. Press, Ithaca, 1960, pp. 257–264.
12. M. Charton, *J. Am. Chem. Soc.*, **91**, 615 (1969).
13. (a) M. Charton, *Prog. Phys. Org. Chem.*, **8**, 235 (1971). (b) M. Charton, *Prog. Phys. Org. Chem.*, **10**, 81 (1973).
14. M. Charton and C. Sirovich, Abstr. 27th M. A. R. M. Am. Chem. Soc., 1993, p. 129.
15. A. Verloop, W. Hoogenstraaten and J. Tipker, *Drug Design*, **7**, 165 (1976).
16. M. Charton, *Adv. Mol. Struct. Res.*, **5**, 25 (1999).
17. M. Charton, in *Classical and 3-D QSAR in Agrochemistry and Toxicology* (Eds. C. Hansch and T. Fujita), American Chemical Society, Washington, D.C., 1995, pp. 75–95.
18. M. Charton and B. I. Charton, *J. Phys. Org. Chem.*, **7**, 196 (1994).
19. M. Charton and B. I. Charton, Abstr. Int. Symp. Lipophilicity in Drug Research and Toxicology., Lausanne, 1995, p. O–3.
20. M. Charton and B. I. Charton, *J. Org. Chem.*, **44**, 2284 (1979).
21. M. Charton, *Top. Curr. Chem.*, **114**, 107 (1983).

22. C. Eaborn and B. Singh, *J. Organomet. Chem.*, **177**, 333 (1979).
23. S. Kozuka, S. Yamaguchi and W. Tagaki, *Bull. Chem. Soc. Jpn.*, **26**, 473 (1983).
24. J. Pola, J. M. Bellama and V. Chvalovsky, *Coll. Czech. Chem. Commun.*, **39**, 2705 (1974).
25. S. Boué, M. Gielen and J. Nasielski, *J. Organomet. Chem.*, **9**, 443 (1967).
26. M. H. Abraham and G. F. Johnston, *J. Chem. Soc., A*, 193 (1970).
27. G. C. Robinson, *J. Org. Chem.*, **28**, 843 (1963).
28. J. Dean (Ed.), *Lange's Handbook of Chemistry*, 13th edn., McGraw-Hill, New York, 1985; R. C. Weast (Ed.), *Handbook of Chemistry and Physics*, 67th edn., CRC Press, Boca Raton, 1986.
29. A. N. Egorochkin, S. E. Skobeleva, V. L. Tsvetkova and E. T. Bogoradovskii, *Russ. Chem. Bull.*, **42**, 1982 (1993).
30. A. N. Egorochkin, S. E. Skobeleva, E. T. Bogoradovskii and T. P. Zubova, *Russ. Chem. Bull.*, **43**, 975 (1994).
31. (a) A. L. McClellan, *Tables of Experimental Dipole Moments*, W. H. Freeman, San Francisco, 1963; A. L. McClellan, *Tables of Experimental Dipole Moments*, Vol. 2, Rahara Enterprises, El Cerrito, Cal., 1974.
    (b) G. Distefano, S. Pignataro, A. Ricci, F. P. Colonna and D. Pietropaolo, *Ann. Chim.*, **64**, 153 (1974).
32. A. N. Egorochkin, S. E. Skobeleva and T. G. Mustina, *Russ. Chem. Bull.*, **46**, 1549 (1997).
33. A. N. Egorochkin, S. E. Skobeleva and T. G. Mustina, *Russ. Chem. Bull.*, **47**, 1436 (1998).
34. A. N. Egorochkin, S. E. Skobeleva and T. G. Mustina, *Russ. Chem. Bull.*, **46**, 65 (1997); T. Kobayashi and S. Nagakura, *Bull. Chem. Soc. Jpn.*, **47**, 2563 (1974); J.-F. Gal, S. Geribaldi, G. Pfister-Guillouzo and D. G. Morris, *J. Chem. Soc., Perkin Trans. 2*, 103 (1985); J. Bastide, J. P. Maier and T. Kubota, *J. Electron Spectrosc. Relat. Phenom.*, **9**, 307 (1976).
35. A. N. Egorochkin, S. E. Skobeleva, T. G. Mustina and E. T. Bogoradovskii, *Russ. Chem. Bull.*, **47**, 1526 (1998).
36. M. Charton, *Prog. Phys. Org. Chem.*, **16**, 287 (1987).
37. M. J. Selwyn, in *The Chemistry of Tin* (Ed. P. G. Harrisn), Blackie, Glasgow, 1989, pp. 359–396.
38. E. Lukevics and L. M. Ignatovich, in *The Chemistry of Organic Germanium, Tin and Lead Compounds* (Ed. S. Patai), Wiley, Chichester, 1995, pp. 857–863.
39. M. Charton, *Prog. Phys. Org. Chem.*, **13**, 119 (1981); M. Charton, *Prog. Phys. Org. Chem.*, **16**, 287 (1987); M. Charton, *Advances in Quantitative Structure Property Relationships*, **1**, 171 (1996).

CHAPTER 9

Radical reaction mechanisms of and at organic germanium, tin and lead

MARC B. TARABAN, OLGA S. VOLKOVA, ALEXANDER I. KRUPPA and TATYANA V. LESHINA

Institute of Chemical Kinetics and Combustion, Novosibirsk-90, 630090 Russia
Fax: +7(3832) 342350; e-mail: taraban@ns.kinetics.nsc.ru

The chemistry of organic germanium, tin and lead compounds — Vol. 2
Edited by Z. Rappoport

I. INTRODUCTION

The mechanisms of organometallic reactions and, in particular the processes involving group 14 elements, in principle, do deserve an extensive survey. Interest in the derivatives of the elements of this group is so great and their reactions are so diverse that sometimes they are capable of puzzling any experienced investigator. Having no intention, however, to startle the reader with the mass of available data on reaction mechanisms, we have tried to restrict ourselves to demonstrating the wide potential of spin chemistry methods in investigations of the elementary mechanisms of organic reactions involving radical species with group 14 elements. More than 20 years have passed since the publication of the first review[1] on the applications of chemically induced dynamic nuclear polarization to study organometallic reactions. Since then, the majority of published reviews concerned either directly or indirectly with the application of spin chemistry methods to organometallic reactions were too specialized and therefore could hardly attract the attention of the organometallic community[2]. However, the methods of spin chemistry open new prospects for deeper understanding of the mechanistic features of the broad variety of organometallic reactions.

Despite the spectacular achievements in the organic syntheses of germanium, tin and lead organoelement compounds[3,4] their reactivity still requires more thorough investigation. As for reliably established reaction mechanisms, one literally could count them on the fingers of one hand. At present the investigations of the structure and properties of short-lived intermediates of organometallic reactions are acknowledged to be among the most topical lines of chemical research. At present, the overwhelming majority of the reaction mechanisms established by means of physical methods involves paramagnetic species such as free element-centered radicals, and biradicals including the heavy carbene analogs, i.e. germylenes. The present chapter is devoted to a description of the elementary mechanisms of the reactions involving these species.

There are several reasons for the interest in homolytic processes. Recent years have witnessed a strong tendency of organic syntheses to employ organometallic reagents in routes to previously unknown or otherwise hardly accessible compounds. From this viewpoint, the reactions of group 14 organoelement compounds are of special interest, in particular, photoinduced reactions of Ge and Sn compounds involving homolytic stages. It is essential that these radical reactions will occur under mild conditions, thus allowing one to obtain a wide range of products with retention of the functional element-containing group — the process characteristic for heterolytic reactions involving organometallic derivatives. These features have stimulated intensive applications of laser pulse photolysis techniques, ESR and spin chemistry methods for the investigation of the reaction mechanisms of homolytic processes involving organic Ge, Sn and Pb compounds. Due to the high informativity of spin chemistry methods the data obtained by means of these methods literally hold a central position among papers devoted to the investigation of homolytic processes involving group 14 elements. Below, we will specifically emphasize the potential of spin chemistry methods.

Another goal is to attract the attention of the organometallic community to the options rendered by spin chemistry methods. A major problem is posed by the fact that, at present, these capabilities are mainly exploited by the spin chemistry experts themselves. However, the application of spin chemistry methods would allow organic chemists to obtain more interpretable and unambiguous data on the structure and properties of short-lived paramagnetic intermediates as compared to other physical methods. When comparing different methods, it is necessary to take into account that the method of chemically induced dynamic nuclear polarization combines the simplicity and reliability of the product identification characteristic by NMR spectroscopy with an extremely high sensitivity. Moreover, the rules of qualitative analysis of the polarization effects are straightforward and do not require special training. For those interested, Section II includes an introduction to spin chemistry techniques and their broad potentialities.

II. SPIN CHEMISTRY TOOLS – GENERAL BACKGROUND

Spin chemistry is a comparatively young field of science — about 30 years old — related to the chemistry of radical reactions where the rate and direction of the process depend on the interaction of the electron and nuclear spin of the paramagnetic species which are the precursors of the reaction products. Three phenomena form the foundation of spin chemistry: (1) chemically induced dynamic nuclear (electron) polarization (CIDNP or CIDEP); (2) the magnetic field effect (MFE), which is the influence of the external magnetic field on the product yield and the reaction rates; and (3) the magnetic isotope effect (MIE), dealing with the influence of the external and internal (generated by the magnetic nuclei of the radical) magnetic fields on the distributions of magnetic isotopes (with nonzero nuclear spin) over the reaction products.

One of the most important phenomenon, chemically induced dynamic nuclear polarization (CIDNP), deserves more detailed consideration, since it forms the basis of one of the most powerful modern methods for the investigation of the structure and reactivity of short-lived (from nano- to microseconds) paramagnetic precursors of the reaction products. CIDNP manifests itself in the form of unusual line intensities and/or phases of NMR signals observed when the radical reaction takes place directly in the probe of the spectrometer. These anomalous NMR signals — enhanced absorption or emission — are observed within the time of nuclear relaxation of the diamagnetic molecule (from several seconds to several minutes). Later on, the NMR spectrum re-acquires its equilibrium form.

Theory suggests that the nonequilibrium population leading to the unusual NMR lines is generated as a result of electron–nuclear interactions in the so-called radical pair. Such a pair of paramagnetic particles may originate through the homolysis of a molecule under the action of heating, light or ionizing radiation as well as from single electron transfer between donor and acceptor molecules and occasional radical encounters in the bulk preceding the recombination.

The analysis of CIDNP effects formed in such a radical pair allows one to obtain information on the structure and reactivity of active short-lived paramagnetic species (free neutral and charged radicals), on molecular dynamics in the radical pair and on the geminate ('in-cage') and homogeneous ('escape') processes of complex chemical reactions, of great importance when studying their mechanisms. CIDNP data are informative of the multiplicity of reacting states, necessary for better understanding the nature of photochemical processes. The observation of CIDNP effects is unambiguous evidence that the relevant product had a radical precursor. One might distinguish two types of CIDNP effects: net effects (enhanced absorption or emission) and multiplet effects which take the form of intensity redistribution between individual components of multiplet signals in the

NMR spectrum. The analysis of CIDNP effects is usually carried out using the existing rules[5]. The sign of net CIDNP effect (Γ_N) observed in a high magnetic field is defined by the product of multiplication of the following parameters: $\Gamma_N = \mu \cdot \varepsilon \cdot \Delta g \cdot \mathrm{A}$, where μ is the multiplicity of the precursor radical pair ['+' for a triplet (T) and uncorrelated (F) pair, and '−' for a singlet (S) precursor], ε is '+' for 'in-cage' and '−' for 'escape' recombination products, Δg is the sign of the difference in g-factors of the radical with polarized nucleus and radical partner in the RP (radical pair), while A is the sign of the hyperfine interaction (HFI) constant of the nucleus under study in the radical. The sign of Γ_N reflects the phase of the NMR signal of the nucleus under study: '+' for enhanced absorption (A) and '−' for emission (E). For example, if one considers some group of, say, protons in the product resulting from the recombination (ε is '+') of the uncorrelated radical pair or F-pair (μ is '+'), and if this group belonged to a radical with a g-factor smaller than that of the partner radical of the radical pair (Δg is '−'), and if the sign of the hyperfine interaction for this particular group in the radical is negative (A is '−'), then the multiplication gives

$$\frac{\mu \cdot \varepsilon \cdot \Delta g \cdot A}{+ \cdot + \cdot - \cdot -} = +(A)$$

One should then observe an enhanced absorption of the NMR signal of this group. The qualitative rules for analysis of the multiplet effect (Γ_M) could be written as follows: $\Gamma_M = \mu \cdot \varepsilon \cdot J_1 \cdot J_2 \cdot A_1 \cdot A_2 \cdot \gamma$, where μ and ε are the same as for Γ_N, A_1 and A_2 are the signs of hyperfine constants of nuclei 1 and 2 in the radicals, J is the sign of the spin–spin coupling constant of these nuclei in the molecule, while γ is '+' if these nuclei belong to the same radical and '−' if they belong to different radicals constituting the pair. The sign of Γ_M defines two types of multiplet effects, '+' for E/A and '−' for A/E.

The mere fact of CIDNP observation provides no data on the contribution of the radical pathway to the product formation. To obtain this information one should employ another technique, the magnetic field effect (MFE). Basic manifestations of MFE include: (1) the variations of the ratio of geminate ('in-cage') and homogeneous ('escape') products of radical reaction as a function of the applied external magnetic field; (2) the dependence of the reaction rates (effective rate constants) on the external magnetic field. In itself, the observation of MFE is decisive evidence of the prevalence of the radical pathway of the reaction under study. Theory suggests several model mechanisms of MFE formation[6,7] and the modeling of the MFE and comparison of theoretical and experimental findings make it possible to reveal the features of the molecular dynamics (in particular, lifetimes) and the structure of reacting states, to discover new reactive intermediates and to get better insight into magnetic properties (g-factors, hyperfine constants) of the radicals involved. In certain cases, where the MFE are particularly high, one might consider the use of this method to govern the chemical reaction.

III. REACTIONS OF GERMANIUM AND TIN DERIVATIVES CONTAINING THE ELEMENT–ALKALI METAL BOND

At the very beginning of spin chemistry, the reactions of alkyl lithiums with organic halides were the systems where CIDNP effects[8] and the influence of the external magnetic field[9] (magnetic field effects) were discovered. From the mechanistic viewpoint, the reactions of triorganogermyl derivatives of alkali metals R_3GeLi(Na,K) with organic halides should follow a similar sequence of radical stages involving germanium-centered radicals. The reactions of R_3GeLi(Na,K) are of special interest not only as a method for the introduction of an organometallic function resulting in the formation of the carbon–metal bond. It will be demonstrated below how important is the role of supramolecular factors (associated states) for the mechanisms of these reactions.

Similar to the reaction with *n*-BuLi[9] the main reaction products of the interaction between benzyl chloride $PhCH_2Cl$ and Et_3GeLi, Et_3GeNa and Et_3GeK in benzene are the corresponding unsymmetrical (**1**) and symmetrical (**2** and **3**) products (equation 1).

$$PhCH_2Cl + Et_3GeLi(Na,K) \longrightarrow \underset{(\mathbf{1})}{PhCH_2GeEt_3} + \underset{(\mathbf{2})}{PhCH_2CH_2Ph} + \underset{(\mathbf{3})}{Et_3GeGeEt_3} + Li(Na,K)Cl \quad (1)$$

Table 1 lists ^{1}H CIDNP effects detected during mixing of the initial reagents directly in the probe of an NMR spectrometer[10]. The analysis of the observed chemical polarization effects in accordance with the existing rules[11] (see Section II) allows one to propose the radical pathway for the formation of the main reaction products (Scheme 1).

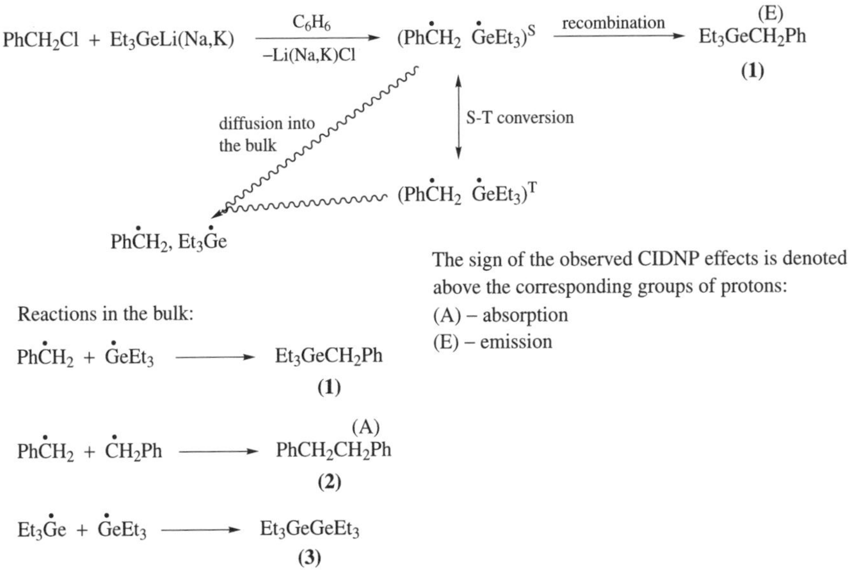

$$PhCH_2Cl + Et_3GeLi(Na,K) \xrightarrow[-Li(Na,K)Cl]{C_6H_6} (Ph\dot{C}H_2\ \dot{G}eEt_3)^S \xrightarrow{\text{recombination}} \underset{(\mathbf{1})}{\overset{(E)}{Et_3GeCH_2Ph}}$$

S-T conversion: $(Ph\dot{C}H_2\ \dot{G}eEt_3)^S \longleftrightarrow (Ph\dot{C}H_2\ \dot{G}eEt_3)^T$

diffusion into the bulk: $(Ph\dot{C}H_2\ \dot{G}eEt_3)^{S,T} \rightsquigarrow Ph\dot{C}H_2,\ Et_3\dot{G}e$

Reactions in the bulk:

$$Ph\dot{C}H_2 + \dot{G}eEt_3 \longrightarrow \underset{(\mathbf{1})}{Et_3GeCH_2Ph}$$

$$Ph\dot{C}H_2 + \dot{C}H_2Ph \longrightarrow \underset{(\mathbf{2})}{\overset{(A)}{PhCH_2CH_2Ph}}$$

$$Et_3\dot{G}e + \dot{G}eEt_3 \longrightarrow \underset{(\mathbf{3})}{Et_3GeGeEt_3}$$

The sign of the observed CIDNP effects is denoted above the corresponding groups of protons:
(A) – absorption
(E) – emission

SCHEME 1

The significant difference in the *g*-factors[12] ($\Delta g = 6.3 \times 10^{-3}$) of the radicals comprising the initial singlet radical pair of $^{\bullet}GeEt_3$ and $^{\bullet}CH_2Ph$ radicals defines the net character of the observed CIDNP effects (see Table 1). To elucidate the role of the radical pathway, it was necessary to study the dependence of the ratio of the yields of products **1** and **2** on the external magnetic field strength. The difference in *g*-factors and hyperfine interaction in the radical pair of the benzyl and triethylgermyl free radicals allows one to expect a marked influence of the external magnetic field on the recombination probability of this pair. Indeed, the decrease in the external magnetic field from 1.88 T to the geomagnetic value results in the noticeable variation in the ratio **1/2**; for the reaction of $PhCH_2Cl$ with

Et_3GeNa this effect amounts to $28 \pm 8\%$, and in the case of Et_3GeK the influence is even greater and reaches $38 \pm 12\%$ (Table 1)[10].

The observation of a magnetic field effect implies that the radical pathway of the formation of the main reaction products is prevalent. Despite the decrease in the magnitude of the magnetic field effect in the reaction of benzyl chloride with Et_3GeLi ($11 \pm 5\%$), the reaction mechanism should be similar for all organogermanium derivatives under study, Et_3GeLi(Na,K). As shown for the case of *n*-BuLi reactions with benzyl halides[9], the influence of the external magnetic field has been reliably detected only for rather significant ratios of unsymmetrical and symmetrical (cf **1** and **2**) products. This is due to the limited influence of the external magnetic field on the recombination probability of the radical pairs in nonviscous liquids in accordance with the predictions of the radical pair theory[13]. In the case under study, it is quite reasonable to assume that the increase of the **1/2** ratio in the series Et_3GeLi, Et_3GeNa and Et_3GeK leads to the growth of the observed magnetic field effect (Table 1).

High values of cage effects (Table 1) close to those detected in the reactions of alkyllithium compounds[9] suggest that, similarly to *n*-BuLi, organogermanium derivatives of alkali metals enter the reaction in the associated state. It is necessary to note that without this assumption it would be difficult to explain the large cage effect values generally uncharacteristic for the reactions of free radicals in solution[14].

The suggested involvement of Et_3GeLi associates has been supported by the results of X-ray analysis in solution[15]. Angular dependence of X-ray scattering intensity (including small angle scattering) as well as analysis of the radical distribution function has allowed one to detect the formation of triethylgermyllithium associates in benzene, cyclohexane and THF. The resulting spatial characteristics — in benzene the diameter of the associate is 12–14 Å and the Ge–Ge distance is about 4.5 Å–allowed one to propose a hexameric structure of the associated units of Et_3GeLi. Similar to the findings of the X-ray structural analysis of the single crystals of Me_3SiLi[16], one might propose a distorted octahedron structure of these associates of Et_3GeLi. Only two faces of such an octahedron are accessible for the attacking benzyl chloride molecule, since the others are blocked by the bulky Et_3Ge- substituents (Figure 1). These steric hindrances justify high values of the experimentally observed cage effects, since under these conditions $^{\bullet}GeEt_3$ radicals enter the reaction in the complex with parent associate.

TABLE 1. ^1H CIDNP effects and main products of the reactions of $PhCH_2Cl$ with 0.5 M solutions of Et_3GeLi(Na,K) in various solvents and in different magnetic fields

Reagent/Solvent	Viscosity (cP)	Products ratio, **1/2**		CIDNP sign of α-CH_2 protons[a]		
		Geomagnetic field	1.88 T	**1**	**2**	Cage effect[b]
Et_3GeLi/Hexane	0.31	—	4.1 ± 0.4	E	A	0.80
Et_3GeLi/Benzene	0.65	6.0 ± 0.4	5.4 ± 0.3	E	A	0.84
Et_3GeLi/Dodecane	1.49	—	7.0 ± 0.4	E	A	0.88
Et_3GeLi/THF	—	—	6.5 ± 0.4	E	A	0.87
Et_3GeNa/Benzene	0.65	10.0 ± 0.8	7.8 ± 0.4	E	A	0.89
Et_3GeK/Benzene	0.65	13.6 ± 1.0	9.8 ± 0.8	E	A	0.91

[a]E–emission, A–absorption.

[b]The following relationship was used to estimate the cage effect, $e = 1 - P$, where $P = \mathbf{2}/(\mathbf{1} + \mathbf{2})$. P is calculated from the intensities of the corresponding NMR signals (of α-CH_2 protons) observed during the reaction in the magnetic field 1.88 T.

This hypothesis also allows one to explain the differences in the **1/2** products ratio for the series Et_3GeLi, Et_3GeNa and Et_3GeK (Table 1). Indeed, in these different reactions the precursor of the main products is the same singlet radical pair of benzyl and triethylgermyl radicals (Scheme 1), and without the assumption of a reactive complex of $^{\bullet}GeEt_3$ radical with Et_3GeLi, Et_3GeNa and Et_3GeK associate, it would be difficult to explain the observed trend in the products ratio. Apparently, this is due to the scale of the steric hindrances, dependent on the alkaline metal forming the backbone of the associate (Figure 1, see Plate 1).

Figure 2 shows the magnetic field dependence of the ratio of the yields of symmetrical (**2**) and unsymmetrical (**1**) products for the reaction of benzyl chloride with Et_3GeNa in benzene[17]. As earlier observed in the reactions of alkyllithiums with dichlorodiphenylmethane[9], the field dependence pattern qualitatively reflects two basic mechanisms of radical pair theory—HFI and Δg mechanisms. In this particular case, the cage effects in nonviscous media (benzene) create the necessary prerequisites for the

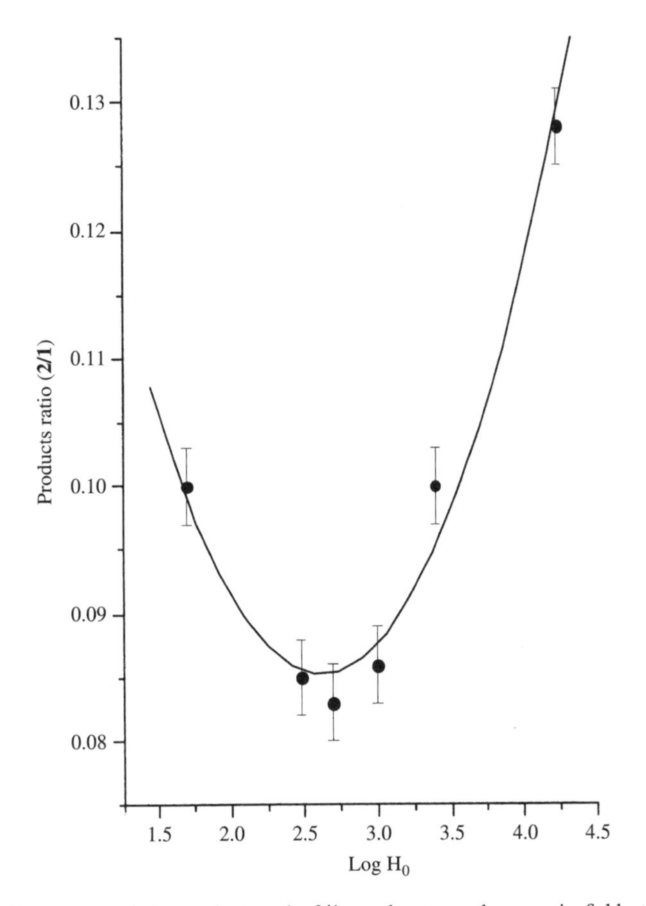

FIGURE 2. Dependence of the products ratio **2/1** on the external magnetic field strength (H_0) observed for the reaction of benzyl chloride with Et_3GeNa in benzene. The solid line shows the second order polynomial fit of the experimental data

observation of the magnetic field effect, which is greater than 30%. It is necessary to stress that in the reaction under study, high cage effects are due to the unusual features of the recombination of the radical pair of $^\bullet GeEt_3$ and $^\bullet CH_2Ph$ radicals, i.e. to the reaction in the associated state, and is therefore *sine qua non* for the observation of the magnetic field effect.

Let us demonstrate how the theoretical modeling of the experimental magnetic field dependence could support the proposed reaction mechanism and prove the involvement of the suggested radical pair–precursor of the main reaction products. According to Scheme 1, the experimentally measured products ratio **2/1** could be expressed as the following ratio of their formation rates:

$$\mathbf{2/1} = \frac{V_2}{V_1 + V_1{}^{\mathrm{DIFF}}} \tag{2}$$

where V_1 is the in-cage formation rate of **1**, $V_1{}^{\mathrm{DIFF}}$ is the formation rate of **1** in the bulk and V_2 is the formation rate of **2**; $V_1 = P_S\nu$, where ν is the generation rate of free $^\bullet GeEt_3$ and $^\bullet CH_2Ph$ radicals and P_S is the recombination probability of the initial radical pair comprised of $^\bullet GeEt_3$ and $^\bullet CH_2Ph$ radicals. The formation rate of **1** in the bulk is $V_1{}^{\mathrm{DIFF}} = k_1[^\bullet GeEt_3][CH_2Ph]$ and the formation rate of **2** is $V_2 = k_2[^\bullet CH_2Ph]^2$. Thus, the product ratio **2/1** could be rewritten in the form:

$$\mathbf{2/1} = \frac{k_2[\dot{C}H_2Ph]}{P_S\nu + k_1[\dot{G}e\,Et_3][\dot{C}H_2Ph]} \tag{3}$$

Equations 4 and 5 define the time variation of the concentrations of $^\bullet GeEt_3$ and $^\bullet CH_2Ph$ radicals:

$$\frac{d[\dot{C}H_2Ph]}{dt} = (1 - P_S)\nu - k_2[\dot{C}H_2Ph]^2 - k_1[\dot{G}eEt_3][\dot{C}H_2Ph] \tag{4}$$

$$\frac{d[\dot{G}eEt_3]}{dt} = (1 - P_S)\nu - k_3[\dot{G}eEt_3]^2 - k_1[\dot{G}eEt_3][\dot{C}H_2Ph] \tag{5}$$

Assuming that in the first approximation, $k_2 = k_3 = 1/2k_1$, under stationary conditions, we obtain equation 6.

$$\mathbf{2/1} = \frac{1 - P_S}{2(1 + P_S)} \tag{6}$$

This simple kinetic reasoning allows us to draw the interrelation between the experimentally measured products ratio **2/1** and the recombination probability of the initial singlet radical pair P_S (Scheme 1). Theoretical calculations of P_S in the frame of the semiclassical approximation[18] (which considers the precession of the electron spin of a radical around the vector sum of the external magnetic field vector and the averaged vector of the HFIs of all the magnetic nuclei of this radical) for the radical pair comprised of $^\bullet GeEt_3$ and $^\bullet CH_2Ph$ radicals which take into account the magnetic resonance parameters of these radicals known from the literature[12] demonstrate fairly good agreement between theory and experiment (cf Figures 2 and 3). These conclusions first demonstrate that the radical pathway of this reaction is prevalent, and second, they unambiguously confirm the proposed reaction mechanism and the structure of the radical pair, precursor of the main reaction products. One remarkable fact is noteworthy. The comparison of experimental (Figure 2) and calculated (Figure 3) magnetic field effects shows that, despite excellent reproduction of the field dependence pattern, theory fails to explain the magnitude of the

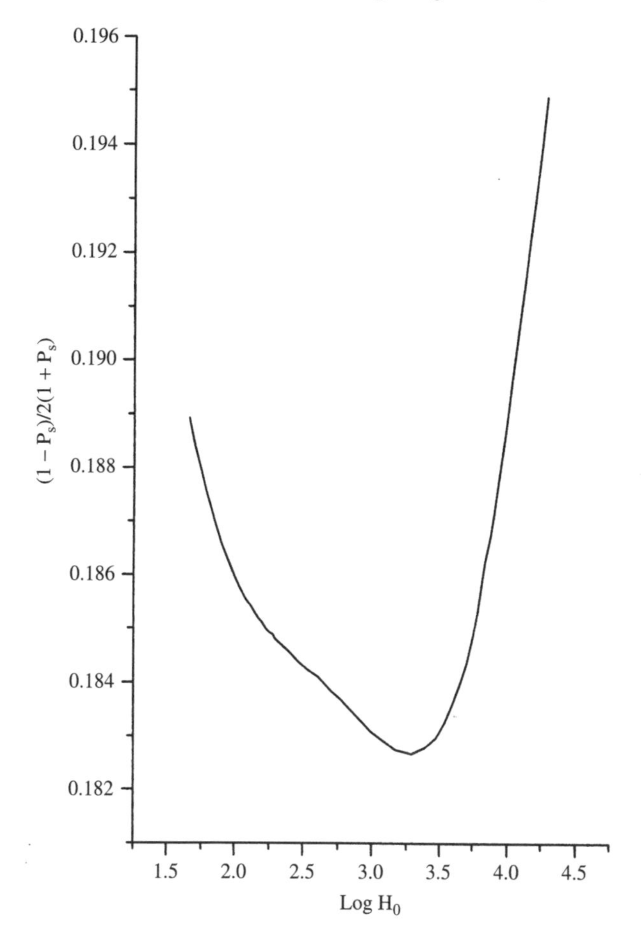

FIGURE 3. Results of theoretical modeling of the products ratio **2/1**. Calculated dependence of $\frac{(1 - P_S)}{2(1 + P_S)}$ on the external magnetic field strength H_0. P_S is the recombination probability of the radical pair comprised of $^{\bullet}GeEt_3$ and $^{\bullet}CH_2Ph$ radicals

observed magnetic effect. Later efforts[19] have shown that if the model calculations would account for the associated state (Figure 4, see Plate 2) of the $^{\bullet}GeEt_3$ radical entering the reaction with the $^{\bullet}CH_2Ph$ radical (the association factor could be included in the model as a variable electron exchange interaction between these radicals in the radical pair), it would be possible to reproduce the magnitude of the experimental effect. Thus, the above analysis demonstrates conclusively the potentialities of spin chemistry techniques in investigations of the elementary mechanisms of chemical reactions and the molecular dynamics of the elementary act (the role of electron exchange interaction and steric effects).

Despite the above self-evident demonstration of the radical nature of the intermediate species formed in the reactions of triorganogermyl derivatives of alkali metals, in a number of papers one yet might find a discussion of other mechanisms, sometimes with speculations concerning quite unusual intermediates. For instance, similar processes were observed in the reaction of $Ph_3GeLi(Na,K)$ with aryl halides. However, the radical processes are postulated to be prevalent only for aryl fluorides, iodides and bromides, while in the case of chlorides the discussion considers not only the radical mechanism, but also a pathway involving an aryne intermediate **4**[20].

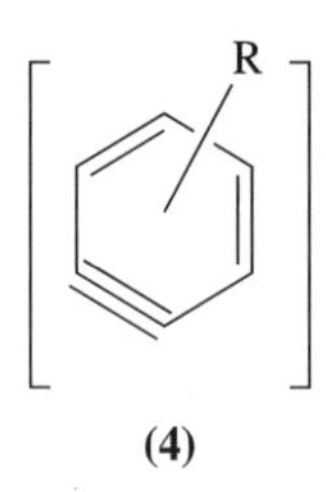

(4)

While it is difficult to ensure the reliable observation of the radical processes during the mixing of reagents, these are easily detected in the photoinduced reactions of aryl-substituted compounds of Ge and Sn. Laser pulse photolysis experiments show that direct photoionization of $Ph_3Ge(Sn)^-$ anion results in the neutral radical[21,22]. The application of the Chemically Induced Dynamic Electron Polarization (CIDEP) method has allowed the detection of polarized emission signal of the radicals, thus leading to a conclusion that direct photoionization of $Ph_3Ge(Sn)^-$ anion occurs from the triplet state (equation 7)[22].

$$R_3MLi \xrightarrow{h\nu} (R_3MLi)^S \longrightarrow (R_3MLi)^T \longrightarrow [R_3\dot{M} \cdots e^-]^T + Li^+ \qquad (7)$$

$$M = Ge, Sn$$

Similar single electron transfer processes were also observed in thermal reactions if other electron acceptors were used instead of alkyl halides. The main products of the interaction of R_3GeLi with electron acceptors such as 3,5-di-*t*-butyl-*o*-quinone, fluorenone, tetracyanoquinodimethane and 2,4,6-tri-*t*-butylnitrobenzene include the corresponding digermane and N(or O)-germyl adducts. ESR spectra of the reaction mixture demonstrate the formation of germyl radicals as well as the radical anions of the organic substrates[23]. The electron transfer reaction has also been shown to be preferable for the interaction of trialkylgermyl lithium with a paramagnetic quinoid ($R'O^\bullet$) which is completely transformed into the diamagnetic anion (equation 8)[23].

$$R_3GeLi + R'\dot{O} \longrightarrow R_3\dot{Ge} + R'O^- + Li^+ \qquad (8)$$

The interaction of R_3GeLi (R = Ph, Mes) with conjugated aldo- and keto-forms of electron acceptors (such as 2-furaldehyde, 2-thiophenecarboxaldehyde and their corresponding nitro derivatives) leads only to germylation of the initial organic compounds. The formation of the organic radical anions in these reactions has been confirmed by ESR spectroscopy, and this speaks in favor of an electron transfer process. Further addition of germyl radicals to the initial organic substrate results in the germylcarbinol formed through a C-germylation mechanism. In the presence of an excess of the aldehyde, the germyl ketone is formed; the corresponding nitro compounds mainly lead to O-germyl derivatives[24]. However, as opposed to the investigations employing spin chemistry techniques, none of the studies concerns the supramolecular chemistry aspects and considers the influence of the associated state upon the reaction mechanism.

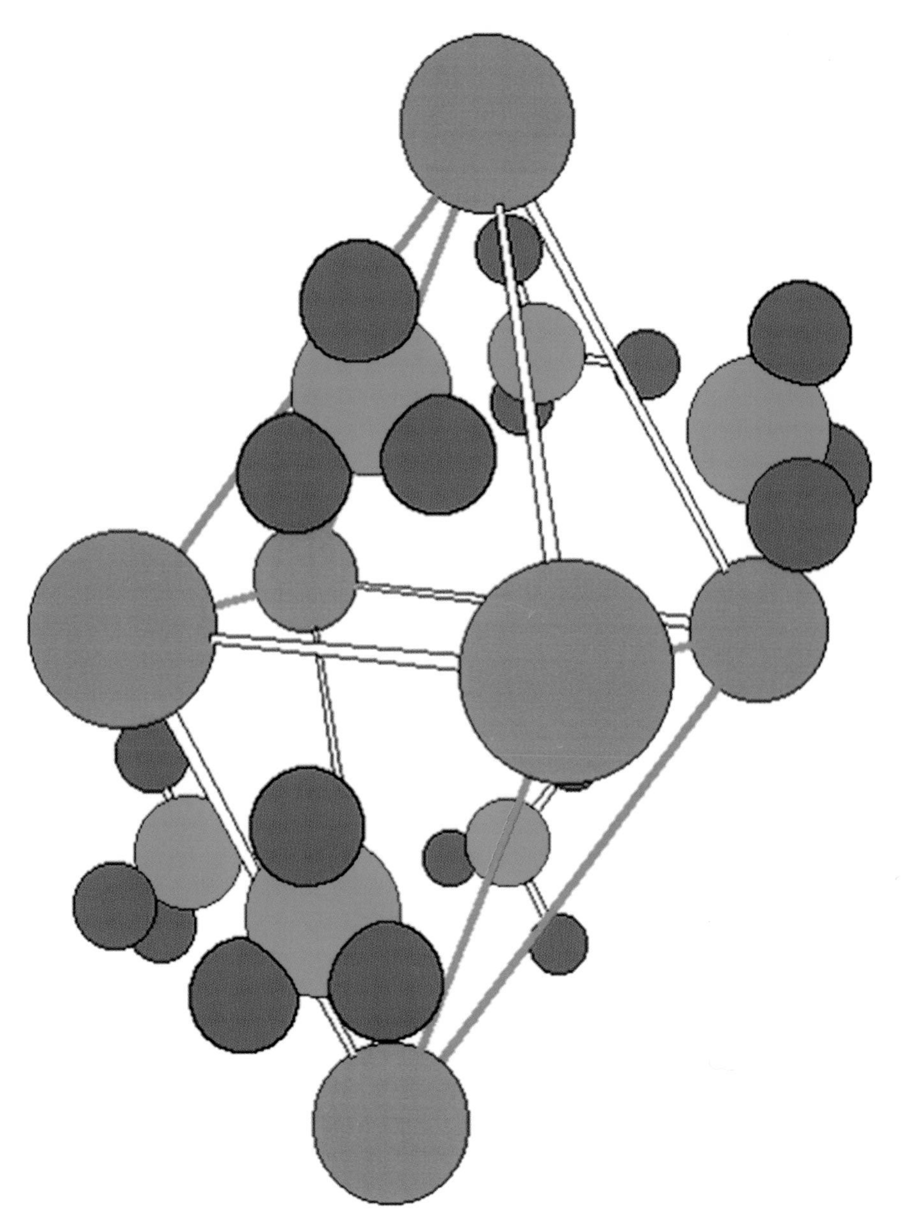

FIGURE 1. General view of the hexameric units of triethylgermyllithuim associate $(Et_3GeLi)_6$ in solution: ● (Li), ● (Ge), ● (Et), — (uncoordinative bond)

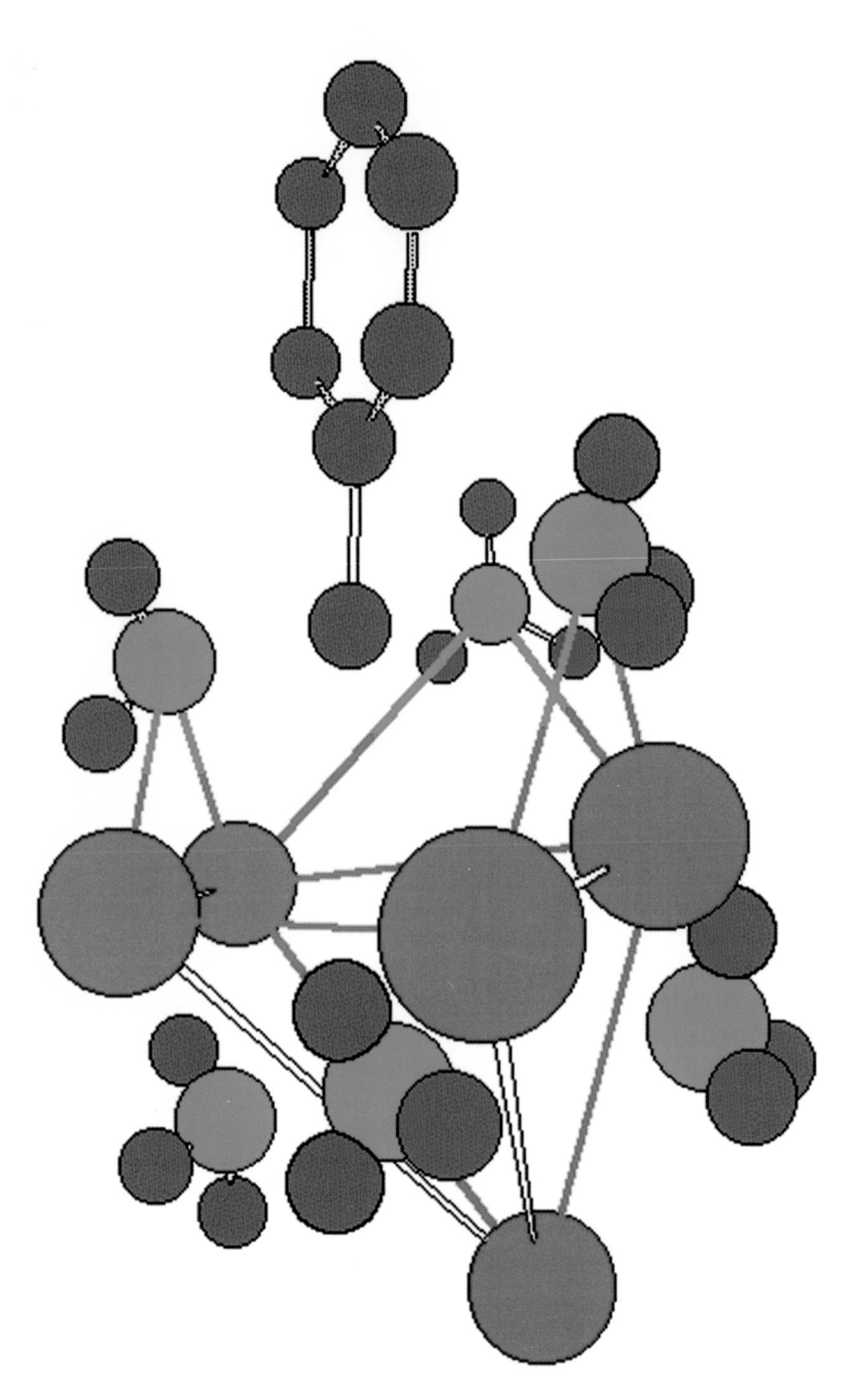

FIGURE 4. Schematic presentation of the structure of the radical pair of $^{\bullet}CH_2Ph$ radical and associated $^{\bullet}GeEt_3$ radical: ● (Li), ● (Ge), ● (Et or C), ▬ and ▬▬ (uncoordinative bonds)

In conclusion, let us highlight the capabilities of the spin chemistry methods and the necessity to take into account the structural features when applying the magnetic and spin effects to the investigations of the reaction mechanisms. Recent publications provide an illustrative example of the unsuccessful attempt to reproduce the pioneering results[9] of the observation of magnetic field influence on the reactions of n-BuLi with organic halides where the authors[25] have neglected the role of association and the long-explored fact[26–28] that the interactions of alkyllithiums with organic halides could follow several different mechanisms. In addition to the radical process, depending on the reaction conditions (solvent, temperature and the concentration of the reagents) these mechanisms might involve nucleophilic substitution and an ion exchange reaction. All the mechanisms above lead to identical products, and therefore to ensure the involvement of comparatively long-lived radical pairs which are responsible for the magnetic field effect, it is necessary to check their formation, e.g. by means of CIDNP. These pairs, in turn, appear only in the presence of the associated states of n-BuLi. However, the authors[25] have neglected all these aspects of the mechanism under study, and if such a superficial approach were used to study a previously unknown process, this would definitely lead to erroneous conclusions about the reaction mechanism.

IV. PHOTOTRANSFORMATIONS OF THE ORGANOELEMENT α-KETONES

High reactivity, the possibility to fundamentally change the nature of main reaction products through variation of the solvent polarity, excellent radical acceptor properties, unusual photochemical characteristics — all these features of organoelement α-ketones R_3MCOR' (M = Ge, Sn) define the broad prospects of using these compounds in various synthetic applications. The keen interest in organoelement α-ketones is also stimulated by recent developments in synthetic radical chemistry. For instance, the possibility to use R_3MCOR' as an excellent equivalent of carbonyl radical acceptor synthon has been demonstrated by the example of acyl germanes[29] in radical cyclization reactions. For using the organoelement α-ketones in organic syntheses it is necessary to clarify the relevant elementary reaction mechanisms. From the synthetic viewpoint, of special interest are the homolytic processes which ensure less rigid reaction conditions as compared with heterolytic ones.

A. Reaction Mechanisms of the Photolyses of α-Germyl Ketones in Various Media

Free radical mechanisms of the photolytic decomposition of R_3MCOR' (M = Ge, Sn) were considered in a number of earlier fundamental studies[30,31]. The nature of the resulting reaction products identified in nonpolar (alkanes, benzene, either in the presence or in the absence of a radical trap) or polar (alcohols, pyridine) media has allowed the proposal of a generalized reaction mechanism of the photolysis of R_3GeCOR' compounds (Scheme 2).

In the frame of the proposed mechanism, the primary act of the photolysis of α-germyl ketones in nonpolar media is a C−Ge bond cleavage (Norrish Type I) leading to element-centered and acyl (aroyl) free radicals. The observed main reaction products result from the interaction of the radicals in the bulk and from the halogen abstraction from the corresponding radical trap (e.g. organic halides), if added. In contrast, in polar media (alcohol) the formation of the final reaction products could be explained only by invoking the hypothesis of a photoinduced isomerization of R_3GeCOR' to the unusual reactive germoxycarbene intermediate, $R_3GeO\ddot{C}R'$, **5**. The insertion of the germoxycarbene into

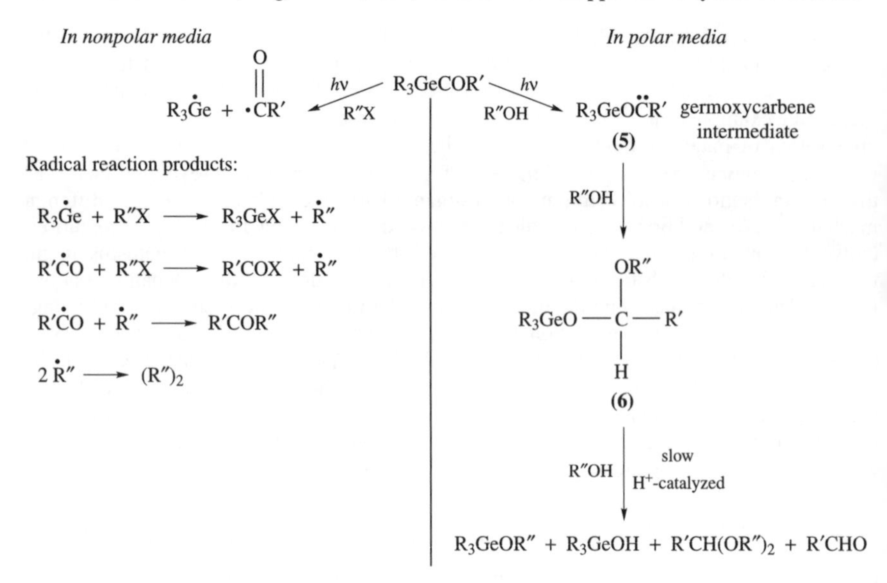

SCHEME 2

the polar O−H bond of the alcohol results in the unstable semiacetal **6** which decomposes to the main reaction products shown in Scheme 2.

What will happen during the photolysis of R_3GeCOR' in nonpolar, nonhalogenated solvents? Early assumptions[32] concerning the mechanism of photodecomposition of α-germyl ketones in nonpolar solvents in the absence of halogenated radical traps include the addition reaction of photochemically generated $R_3M^{\bullet}$ radicals to the unsaturated C=O bond of the unreacted precursor ketone. Indeed, the application of ESR spectroscopy has allowed one to observe rather stable radical adducts (**7**) formed even through the thermolysis of the aroyl derivatives R_3MCOAr (M = Ge, Sn) (equations 9 and 10)[33].

$$R_3MCOAr \xrightarrow{h\nu,\,\Delta} R_3\dot{M} + \dot{C}OAr \quad (9)$$

$$R_3\dot{M} + R_3MCOAr \longrightarrow R_3M{-}\underset{\displaystyle OMR_3}{\underset{|}{\dot{C}}}{-}Ar \quad (10)$$

(**7**)

These radical adducts have characteristic resolved ESR spectra which allow the accurate identification of all magnetic properties (g-factors and hyperfine splitting of all magnetic nuclei of the system including ^{13}C, ^{73}Ge and ^{117}Sn and ^{119}Sn of the observed species)[33].

One of the most important implications of these experiments is the conclusion that, due to the high polarity of the C=O bond, organoelement α-ketones R_3MCOR' (M = Ge, Sn) are extraordinary effective radical traps. The electronegative oxygen atom of their carbonyl group could be attacked not only by element-centered radicals $R_3M^{\bullet}$ (M = Ge, Sn), but also by thiyl radicals $^{\bullet}SR$[34] and phosphorus-centered[35] radicals. Indeed, the experimental estimates of the absolute reaction rate constants of the element-centered radicals with various radical traps show that the rate constants of the addition of $R_3M^{\bullet}$ radicals to the oxygen atom of the carbonyl bond are nearly two orders of magnitude higher than the reaction rate constants of halogen abstraction from the halogenated traps resulting in the corresponding halides[36].

Earlier suggestions of the involvement of free radical intermediates have stimulated the application of spin chemistry methods to the investigation of the detailed mechanism of the photolysis of α-germyl ketones in either polar or nonpolar solvents, in the presence or in the absence of traps of element-centered free radicals. Of special interest are problems of the multiplicity of the reactive state, and the transformations of the element-centered free radicals in the bulk.

Table 2 lists ^{1}H and ^{13}C CIDNP effects detected during the photolysis of benzoyltriethylgermane $Et_3GeCOPh$ in nonpolar (C_6D_6 or c-C_6D_{12}) and polar (CD_3OD) solvents in the absence and in the presence of the radical traps (benzyl chloride $PhCH_2Cl$ and bromide $PhCH_2Br$). Both the initial ketone and its decomposition products demonstrate the effects of chemical polarization[37,38].

Note that in all the cases considered in Table 2 the ethyl protons of the initial $Et_3GeCOPh$ demonstrate positive polarization (A). Therefore, the analysis of these effects in accordance with the existing rules[11] allows us to conclude that partially reversible photodecomposition of the ketone both in the presence and in the absence of the radical traps occurs from the triplet excited state with the formation of the triplet radical pair comprised of $Et_3Ge^{\bullet}$ and $^{\bullet}COPh$ radicals. The analysis employed the following g-factor and hyperfine interaction values of the radicals: $g(Et_3Ge^{\bullet}) = 2.0089$, $g(^{\bullet}COPh) = 2.0008$ and $A_H(CH_2) \leqslant 0.5$ mT (for $Et_3Ge^{\bullet}$)[12].

The question of the multiplicity of the reactive state in the photolysis of α-germyl ketones is rather controversial. An attempt to apply laser pulse photolysis techniques

TABLE 2. ^{1}H and ^{13}C CIDNP effects detected in the photolysis of $Et_3GeCOPh$ under various conditions[a]

	Reagents	Reaction products	Chemical shift δ (ppm)	CIDNP sign[b]
	$Et_3GeC(O)Ph$ in c-C_6D_{12}	$\underline{\underline{Et_3}}GeC(O)Ph$	1.07	A
		$\underline{\underline{Et_3}}GeCD(OGeEt_3)Ph$	0.85–1.15	E
^{1}H	$Et_3GeC(O)Ph$+ $PhCH_2Cl$ in c-C_6D_{12}	$\underline{\underline{Et_3}}GeC(O)Ph$	1.07	A
		$\underline{\underline{Et_3}}GeCl$	0.95	E
		$PhC\underline{\underline{H_2}}COPh$	3.80	E
	$Et_3GeC(O)Ph$ in CD_3OD	$\underline{\underline{Et_3}}GeC(O)Ph$	1.07	A
		$\underline{\underline{Et_3}}GeOCD(OCD_3)Ph$	1.10	E
	$Et_3GeC(O)Ph$+ $PhCH_2Br$ in C_6D_6	1) $Et_3Ge\underline{\underline{C}}(O)Ph$	239.4	E
		$Et_3GeC(O)\underline{\underline{Ph}}$	136.1	E
^{13}C		2) $(CH_3\underline{\underline{C}}H_2)_3GeBr$	5.8	A
		3) $\underline{\underline{Ph}}COBr$	134.0	A
		$Ph\underline{\underline{C}}OBr$	165.1	A
		4) $\underline{\underline{Ph}}CH_2COPh$	136.9	E
		$Ph\underline{\underline{C}}H_2COPh$	45.2	A
		$PhCH_2\underline{\underline{C}}OPh$	196.1	E

[a]Double underline denotes the polarized groups of nuclei.
[b]E—emission, A—absorption.

to investigate the photochemical transformations of α-germyl ketones[39] failed to detect the reactions of germoxycarbenes, and it was also impossible to observe the addition of germanium-centered radicals to the oxygen atom of the carbonyl group of the unreacted R_3GeCOR' (equation 10). However, transient spectra detected during the photolysis of $Ph_3GeCOPh$ and $PhMe_2GeCOPh$ in cyclohexane show the absorption signals of short-lived intermediates attributed to $Ph_3Ge^{\bullet}$ and $PhMe_2Ge^{\bullet}$ radicals (λ_{max} = 325–330 nm and λ_{max} = 315–320 nm, respectively)[40]. These experimental findings do not enable us to deduce the multiplicity of the reactive state of the photolyzed α-germyl ketones. Nevertheless, speculations based on the assumption that the lowest n,π^* state of $Ph_3GeCOPh$ and $PhMe_2GeCOPh$ (about 440 nm) lies much below the lowest π,π^* state (about 295 nm) lead to the wrong conclusion that Norrish Type I photocleavage of these ketones occur from the singlet state[39].

The decisive evidence of the reaction mechanism is furnished by ^{13}C CIDNP observations. The analysis of nuclear polarization effects detected during the photolysis of $Et_3GeCOPh$ in the presence of benzyl bromide (Table 2, Figure 5) unambiguously confirms the formation of an initial triplet radical pair of $Et_3Ge^{\bullet}$ and $^{\bullet}COPh$[38]. The following magnetic properties of free radicals were employed for the analysis of ^{13}C CIDNP effects: $Et_3Ge^{\bullet}[g = 2.0089, A_C(CH_2) < 0]$; $^{\bullet}COPh$ $[g = 2.0008, A_C(CO) > 0, A_C(ipso - Ph) > 0]$; $^{\bullet}CH_2Ph[g = 2.0025, A_C(CH_2) > 0, A_C(ipso - Ph) < 0]$[12].

When $Et_3GeCOPh$ is photolyzed in the presence of alcohols, the positive polarization of the unreacted ketone is also observed, and this points to the formation of the triplet radical pair $(Et_3Ge^{\bullet}\ ^{\bullet}COPh)^T$ in polar solvents, too. The negative polarization (emission) at δ 1.10 ppm is attributed to the ethyl groups of the semiacetal $Et_3GeOCD(OCD_3)Ph$ formed through germoxycarbene $Et_3GeO\ddot{C}Ph$ insertion into the O–D (O–H) bond of methanol. The sign of these effects, as well as the mere fact of their formation, imply that the insertion of germoxycarbene includes a stage of H(D) atom abstraction from methanol molecule resulting in the singlet radical pair $(Et_3GeO\dot{C}DPh\ ^{\bullet}OCD_3)^S$. Note that an earlier proposed mechanism of the photolysis of R_3MCOR' in polar media (alcohols) has not involved any radical stages (Scheme 2). Since the latter radical pair is incapable of regenerating the

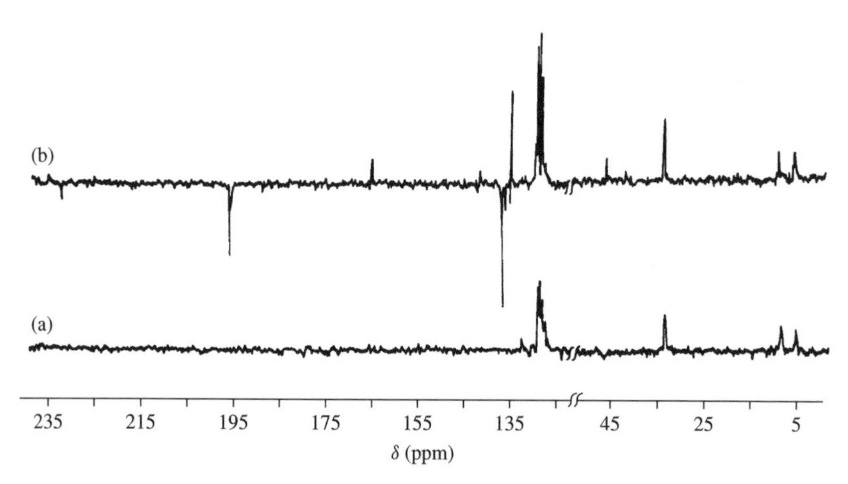

FIGURE 5. ^{13}C CIDNP spectra detected in the photolysis of $Et_3GeCOPh$ in C_6D_6 in the presence of $PhCH_2Br$: (a) initial spectrum, (b) under UV irradiation. (For line assignment, see Table 2)

initial benzoyltriethylgermane, while the polarized $Et_3GeCOPh$ is nevertheless observed, we are led to conclude that photodecomposition of $Et_3GeCOPh$ in alcohols involves both reaction mechanisms, i.e. Norrish Type I cleavage and the formation of germoxycarbene intermediate.

As already mentioned, the element-centered radicals formed through the photolysis of organoelement α-ketones are capable of attacking the most electronegative carbonyl group of the initial ketone[33]. The resulting radical adducts $R_3M\dot{C}(OMR_3)R'$ disproportionate and/or recombine to the main reaction products. Two polarized signals were observed in the NMR spectra taken during the photolysis of $Et_3GeCOPh$ in c-C_6D_{12} in the absence of any radical traps (Table 2): the absorption of the initial ketone which is again a manifestation of the starting triplet radical pair, and the emission of $Et_3GeCD(OGeEt_3)Ph$ formed due to the escape of $^{\bullet}GeEt_3$ radicals into the bulk. Similar to the earlier studied aroyl derivatives[33], the ESR spectra recorded during the photolysis of $Et_3GeCOPh$ in toluene-d_8 indicate the formation of a rather stable radical with g-factor equal to 2.0033 and hyperfine splitting characteristic for structure **8** (Figure 6). Thus, the detected radical adduct **8** appears to be the product of a sequential addition of two $Et_3Ge^{\bullet}$ radicals to the molecule of the initial ketone.

$OGeEt_3$

Et_3Ge

$GeEt_3$

(8)

$A(ortho) = 0.426$ mT; $A(meta) = 0.16$ mT

All these experimental findings allowed us to propose a comprehensive scheme of the phototransformations of benzoyltriethylgermane in various media (Scheme 3).

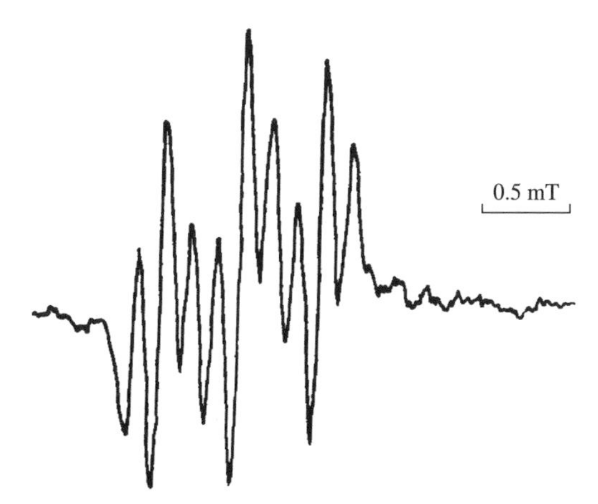

FIGURE 6. ESR spectrum of the radical formed in the photolysis of $Et_3GeCOPh$ in toluene-d_8

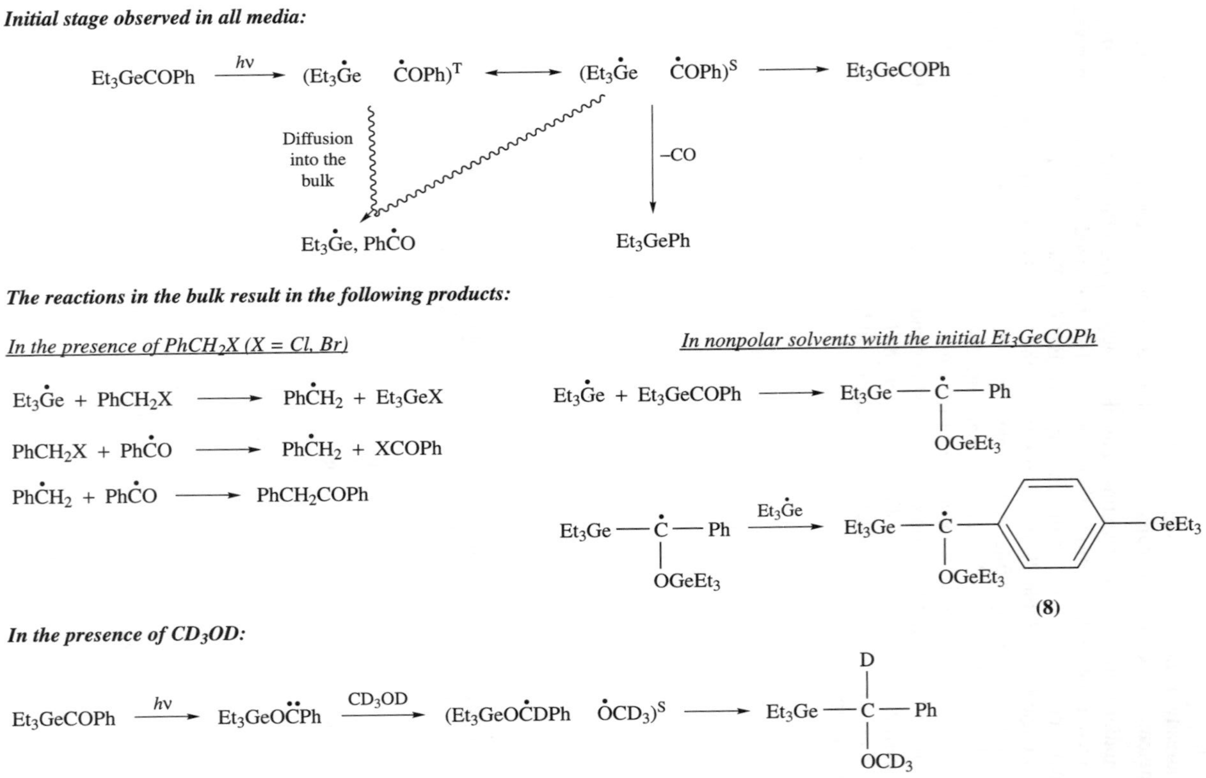

Initial stage observed in all media:

$$Et_3GeCOPh \xrightarrow{h\nu} (Et_3\dot{G}e\quad \dot{C}OPh)^T \longleftrightarrow (Et_3\dot{G}e\quad \dot{C}OPh)^S \longrightarrow Et_3GeCOPh$$

$$(Et_3\dot{G}e\quad \dot{C}OPh)^S \xrightarrow{-CO} Et_3GePh$$

Diffusion into the bulk: $Et_3\dot{G}e,\ Ph\dot{C}O$

The reactions in the bulk result in the following products:

In the presence of $PhCH_2X$ (X = Cl, Br)

$$Et_3\dot{G}e + PhCH_2X \longrightarrow Ph\dot{C}H_2 + Et_3GeX$$

$$PhCH_2X + Ph\dot{C}O \longrightarrow Ph\dot{C}H_2 + XCOPh$$

$$Ph\dot{C}H_2 + Ph\dot{C}O \longrightarrow PhCH_2COPh$$

In nonpolar solvents with the initial $Et_3GeCOPh$

$$Et_3\dot{G}e + Et_3GeCOPh \longrightarrow Et_3Ge{-}\dot{C}(OGeEt_3){-}Ph$$

$$Et_3Ge{-}\dot{C}(OGeEt_3){-}Ph \xrightarrow{Et_3\dot{G}e} Et_3Ge{-}\dot{C}(OGeEt_3){-}C_6H_4{-}GeEt_3$$

(8)

In the presence of CD_3OD:

$$Et_3GeCOPh \xrightarrow{h\nu} Et_3GeO\ddot{C}Ph \xrightarrow{CD_3OD} (Et_3GeO\dot{C}DPh\quad \dot{O}CD_3)^S \longrightarrow Et_3Ge{-}C(D)(OCD_3){-}Ph$$

SCHEME 3

B. Photolysis of α-Stannyl Ketones

From the viewpoint of classical organometallic chemistry, α-stannyl ketones were always seen like something exotic. Indeed, one could list a dozen papers concerning the failed attempts to identify and trace the reaction mechanisms of R_3SnCOR'. Some α-stannyl ketones could be very unstable and decompose *in statu nascendi.* Similar to R_3GeCOR', the anomalous bathochromic shift of n-π^* absorption of α-stannyl ketones was explained by the inductive effect of the metal which led to an energy increase in the n-orbital with simultaneous preservation of the energy of the π^*-orbital. This results in a sharp decrease in the energy of the $n \rightarrow \pi^*$ transition[41]. The first suggestions of free radical mechanisms of the reactions involving acylstannanes were based on the experimentally observed acceleration of the autooxidation of n-Bu_3SnCOR' in the presence of azobisisobutyronitile[42]. Unfortunately, the current literature lacks data on the reaction mechanisms of α-stannyl ketones. However, this class of compounds could be very attractive from the viewpoint of spin chemistry techniques due to the high natural abundance of the magnetic isotopes of tin (^{117}Sn and ^{119}Sn) and the extreme magnetic resonance parameters of tin-centered radicals. For instance, the ^{119}Sn hyperfine interaction constant in the corresponding tin-centered radicals which is of the order of 150 mT (nuclear spin $I = 1/2$), and the g-factors which differ significantly from the pure spin value (about 2.0160)[12] allow one to expect notable enhancement coefficients for ^{1}H CIDNP and provide the possibility to observe ^{119}Sn CIDNP.

However, the instability of the simplest α-tin ketones precludes the application of CIDNP methods in studying their reaction mechanisms. $Me_3SnCOMe$ decomposes on attempted isolation if exposed to daylight. Therefore, to study the mechanisms of the photodecomposition of R_3SnCOR' it is reasonable to choose more steric hindered derivatives which are relatively stable under ambient conditions. Consequently, the regularities of the reaction mechanisms of α-stannyl ketones R_3SnCOR' have been studied with 2-methylpropanoyltripropylstannane[43] $Pr_3SnCOCHMe_2$ (**9**).

Since one might expect that Norrish Type I cleavage of **9** should result in isopropanoyl $^{\bullet}COCHMe_2$, and — after decarbonylation — isopropyl $^{\bullet}CHMe_2$ radicals, it was reasonable to expect a similarity between the ^{1}H CIDNP spectra taken during the photolysis of **9** and those observed in the photolysis of diisopropyl ketone $(Me_2CH)_2CO$ (**10**). Indeed, the experiment (Figure 7) demonstrates a striking similarity between these two reactions. Moreover, analysis of the reaction mixture show the presence of the same reaction products also detected[44] in the photolysis of diisopropyl ketone. Table 3 lists the chemical shifts of the compounds under study and the line assignments of the detected nuclear polarization effects.

This evident similarity of proton polarizations observed for **9** and **10** shows that, in the case of **9**, ^{1}H CIDNP effects could only be employed to trace the fate of $^{\bullet}COCHMe_2$ and $^{\bullet}CHMe_2$ radicals and the pathways of formation of the reaction products that do not contain organotin function (Table 3). To facilitate analysis of the structure and multiplicity of the initial radical pair and the reaction pathways of tin-centered radicals, it is much more convenient to employ ^{13}C and ^{119}Sn CIDNP techniques.

^{13}C CIDNP effects (Figure 8) were detected for the carbonyl carbon of the initial **9** (emission, δ_C 249.8) and carbon monoxide (absorption, δ_C 183). Polarization of the carbonyl carbon is evidence for the formation of acyl type radicals $^{\bullet}COR$ and, in principle, proves that photodecomposition of **9** follows Norrish Type I cleavage resulting in $^{\bullet}SnPr_3$ and $^{\bullet}COCHMe_2$ radicals. This conclusion is also supported by ^{119}Sn CIDNP effects of the initial α-tin ketone and the organotin products of its photodecomposition (Table 4 and Figure 9).

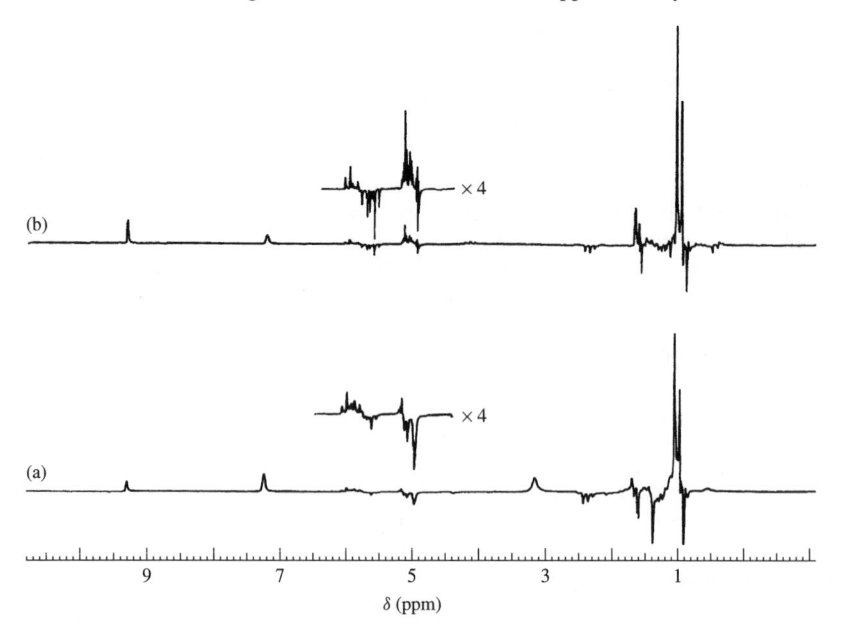

FIGURE 7. [1]H CIDNP effects detected in the photolysis of $Pr_3SnCOCHMe_2$ and diisopropyl ketone $(Me_2CH)_2CO$ in C_6D_6 (only the spectra under irradiation are shown): (a) $Pr_3SnCOCHMe_2$, (b) $(Me_2CH)_2CO$

TABLE 3. [1]H CIDNP effects observed in the photolysis of $Pr_3SnCoCHMe_2$ in C_6D_6[a]

Reaction product	Group of nuclei	Chemical shift δ_H, ppm (multiplicity)	CIDNP sign[b]
$MeCH{=}CH_2$	$\underline{\underline{Me}}-$	1.65 (dd)	A/E + E
	$-C\underline{\underline{H}}=$	5.90 (m)	A/E + A
	$C\underline{\underline{H_2}}=$	5.10 (m)	A/E + E
$Me_2CHC(O)H$	$\underline{\underline{Me_2}}-$	1.12 (d)	—
	$-C\underline{\underline{H}}-$	2.38 (sp)	—
	$-C(O)\underline{\underline{H}}$	9.33 (d)	A
$(Me_2CH)_2CO$ **(10)**	$\underline{\underline{Me_2}}-$	1.02 (d)	A/E + A
	$-C\underline{\underline{H}}-$	2.44 (sp)	A/E + E
$CH_3CH_2CH_3$	$C\underline{\underline{H_3}}-$	0.93 (t)	A/E + E
$Pr_3SnC(O)CHMe_2$ **(9)**	$C\underline{\underline{H_3}}-CH_2-CH_2-$	0.90–1.25	c
	$CH_3-C\underline{\underline{H_2}}-CH_2-$	1.67 (sp)	c
	$CH_3-CH_2-C\underline{\underline{H_2}}-$	0.90–1.25	c
	$Me_2C\underline{\underline{H}}-$	2.45 (sp)	c
	$\underline{\underline{Me_2}}CH-$	1.35 (d)	c

[a]Double underline denotes the polarized groups of nuclei.
[b]E — emission, A — absorption.
[c]The analysis of ^{1}H CIDNP effects is hampered by the strong overlap of the signals of the initial $Pr_3SnCOCHMe_2$ with those of the reaction products.

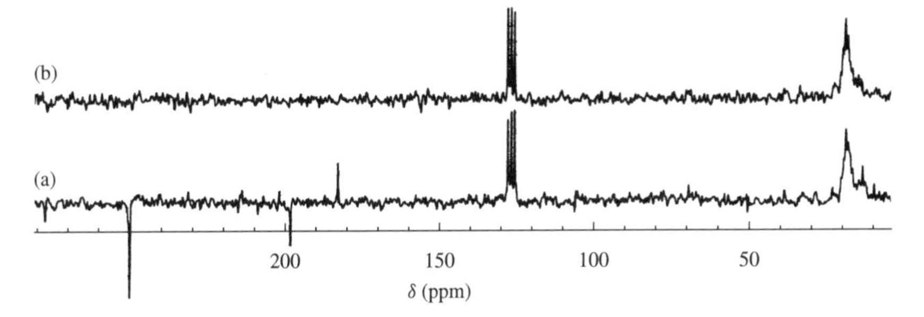

FIGURE 8. ^{13}C CIDNP spectra detected in the photolysis of $Pr_3SnCOCHMe_2$ in C_6D_6: (a) under UV irradiation, (b) after photolysis

TABLE 4. ^{119}Sn CIDNP effects observed in the photolysis of $Pr_3SnCOCHMe_2$ in C_6D_6

Reaction product	Group of nuclei	Chemical shift δ_{Sn} (ppm)[a]	CIDNP sign[b]
$Pr_3SnC(O)Pr$-*i*	–Sn–	−98.0	A
Pr_3SnPr-*i*	–Sn–	−8.0	A
	–Sn–	−82.0	A
$Pr_3SnSnPr_3$		−44.0[c]	A[c]
		−121.0[c]	E[c]
$Pr_3SnSn(Pr_2)SnPr_3$	Pr_3Sn–	−75.0[d]	A[d]
	–$Sn(Pr_2)$–	−225.0[d]	A[d]

[a]Chemical shifts relative to Me_4Sn.
[b]E — emission, A — absorption.
[c]J(^{119}Sn–^{117}Sn) satellites.
[d]Observed only after prolonged irradiation.

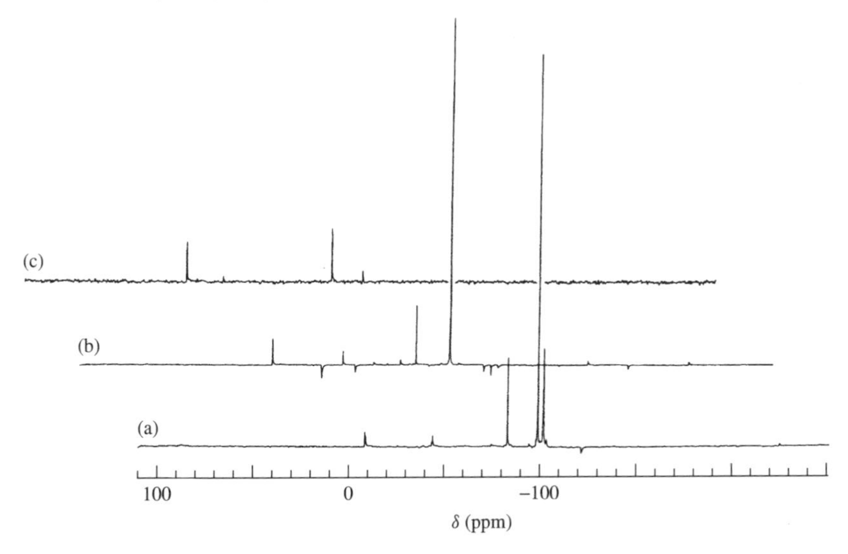

FIGURE 9. ^{119}Sn CIDNP effects detected in the photolysis of $Pr_3SnCOCHMe_2$ in C_6D_6: (a) 60 scans under UV irradiation, (b) additional 60 scans under light, (c) 2000 scans after photolysis

In accordance with the existing rules[11] of CIDNP analysis, the observed emission of the ^{13}C carbonyl group of the initial **9** (δ_C 249.8, Figure 8) shows that this polarization is generated in the triplet initial radical pair. An opposite sign of the polarization of CO (δ_C 183, Figure 8) suggests that the disproportionation of 2-methylpropanoyl radical $^{\bullet}COCHMe_2$ with CO elimination (decarbonylation) occurs mainly after the separation of the initial radical pair, followed by the diffusion of the partner radicals into the bulk[44], $K_{CO} \sim 10^7$ s^{-1}. An identical conclusion about the multiplicity of the initial radical pair could be made from the analysis of the ^{119}Sn CIDNP effects of the initial α-tin ketone (absorption, Table 4 and Figure 9). The positive ^{119}Sn polarization of the tripropylisopropylstannane $Pr_3SnCHMe_2$ also points to its formation from the triplet radical pair. However, since the above-mentioned decarbonylation takes place after the separation of the partners in the bulk, one should conclude that $Pr_3SnCHMe_2$ is a product of the homogeneous recombination of $^{\bullet}SnPr_3$ and $^{\bullet}CHMe_2$ radicals (F-pair). The absorption of the recombination product of two $^{\bullet}SnPr_3$ radicals, i.e. hexapropyldistannane, is easily assigned by means of the observed satellites (Figure 9 and Table 4) from the ^{119}Sn–^{117}Sn spin–spin interaction with a characteristic constant $J(^{119}Sn{-}^{117}Sn) = 2574$ Hz[45].

Analysis of the ^{1}H CIDNP effects could be most conveniently carried out by comparing the polarizations observed in the photolysis of **9** with those detected in phototransformations of **10** (Figure 7). It is essential to point out the similarity of the ^{1}H CIDNP effects observed in the photolysis of **9** (Figure 7a), **10** (Figure 7b) and **9** in the presence of benzyl chloride $PhCH_2Cl$ (Figure 10). In the latter case, there is no doubt that $PhCH_2Cl$ enters

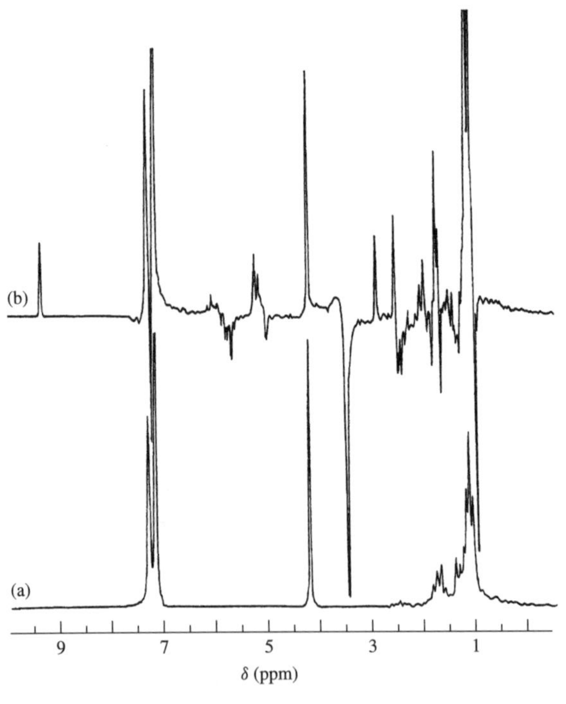

FIGURE 10. ^{1}H CIDNP effects observed in the photolysis of $Pr_3SnCOCHMe_2$ in C_6D_6 in the presence of benzyl chloride $PhCH_2Cl$: (a) initial, (b) under UV irradiation

the reaction as a trap of tin-centered $^{\bullet}SnPr_3$ radicals. The above-mentioned similarity of nuclear polarizations means that ^1H CIDNP effects in the photolysis of **9** are displayed by the radical pair which does not include an $^{\bullet}SnPr_3$ radical. Thus, the integrated analysis of all multinuclear CIDNP effects (^1H, ^{13}C and ^{119}Sn) allows one to propose a detailed scheme of the photolysis of **9** (Scheme 4).

$Pr_3SnCOCHMe_2 \longrightarrow (Pr_3\dot{S}n \quad \dot{C}OCHMe_2)^T \longleftrightarrow (Pr_3\dot{S}n \quad \dot{C}OCHMe_2)^S \longrightarrow Pr_3SnCOCHMe_2$

Diffusion into the bulk

$Pr_3\dot{S}n, \dot{C}OCHMe_2$

Reactions in the bulk:

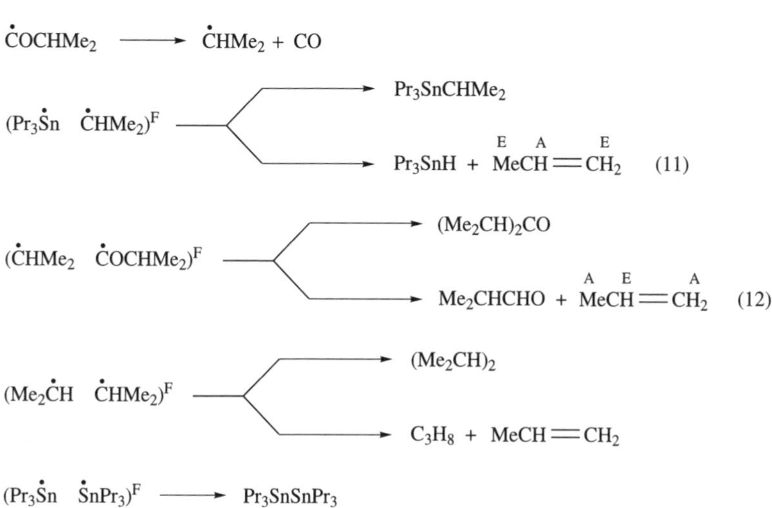

In the presence of $PhCH_2Cl$

E – emission, A – absorption denote proton polarizations observed in equations 11 and 12

$Pr_3\dot{S}n + PhCH_2Cl \longrightarrow Pr_3SnCl + Ph\dot{C}H_2$

$Ph\dot{C}H_2 + \dot{C}OCHMe_2 \longrightarrow PhCH_2COCHMe_2$

$Ph\dot{C}H_2 + \dot{C}H_2Ph \longrightarrow PhCH_2CH_2Ph$

$Ph\dot{C}H_2 + \dot{C}HMe_2 \longrightarrow PhCH_2CHMe_2$

SCHEME 4

TABLE 5. Additional ^{1}H CIDNP effects (see Table 3) observed in the photolysis of $Pr_3SnCOCHMe_2$ in C_6D_6 in the presence of $PhCH_2Cl$[a]

Reaction product	Group of nuclei	Chemical shift δ_H, ppm (multiplicity)	CIDNP sign[b]
	$\underline{\underline{Me}}-$	1.65 (dd)	A/E + A
$MeCH{=}CH_2$	$-\underline{\underline{CH}}{=}$	5.90 (m)	A/E + E
	$\underline{\underline{CH_2}}{=}$	5.10 (m)	A/E + A
$PhCH_2)_2$	$-\underline{\underline{CH_2}}-$	2.77 (s)	A
$PhCH_2C(O)CHMe_2$	$-\underline{\underline{CH_2}}-$	3.47 (s)	E
	$\underline{\underline{o\text{-}H}}$	7.20 (s)	E

[a]Double underline denotes the polarized groups of nuclei.
[b]E—emission, A—absorption.

Special attention should be paid to the opposite signs of the net polarization of the propene $MeCH{=}CH_2$ observed in the photolysis of **9** and **10** (cf Figure 7a and Figure 7b). The analysis of the net proton polarizations of $MeCH{=}CH_2$ formed in the photolysis of **9** in the absence of the radical trap shows that the propene is generated from the radical pair involving the tin-centered $^{\bullet}SnPr_3$ radical (equation 11, Scheme 4). Indeed, the net polarization effects of propene become identical with those observed in the photodecomposition of **10**, when **9** is photolyzed in the presence of benzyl chloride—the radical trap of tin-centered radicals, which precludes the participation of the $^{\bullet}SnPr_3$ radical in the reactions in the bulk (cf Table 3 and Table 5, Figure 7 and Figure 10). Evidently, in the absence of $PhCH_2Cl$, proton polarization of propene in the photolysis of **9** is formed at the stage of equation 11 (Scheme 4), while in the presence of benzyl chloride ^{1}H CIDNP effects of $MeCH{=}CH_2$ are generated at the stage of equation 12 (Scheme 4). Theoretical modeling of the polarization kinetics[43] has allowed us to define the contributions of the stages of equations 11 and 12 to the observed net polarization of propene in the absence and in the presence of $PhCH_2Cl$. It has been shown that in the absence of benzyl chloride the contribution of the stage of equation 11 to the observed net polarization of $MeCH{=}CH_2$ is 4 orders of magnitude higher than that from the stage of equation 12. However, with $PhCH_2Cl$ added, a dramatic drop in the contribution from the stage of equation 11 is observed and the stage of equation 12 becomes prevalent.

Thus, the application of multinuclear CIDNP techniques (^{1}H, ^{13}C and ^{119}Sn) has allowed us to obtain detailed information on the elementary mechanisms of the reactions of α-germanium and α-tin ketones. It has been demonstrated that photodecomposition of all the ketones under study follows the mechanism of Norrish Type I cleavage from the triplet excited state. The CIDNP results confirm the literature data[33,36] that organoelement α-ketones are the most effective traps of element-centered radicals, i.e. that the introduction of the element atom at an α-position to the carbonyl group increases its vulnerability toward free radical attack. An important distinction of the organoelement α-ketones from their carbon analogs is the tendency of α-germanium ketones to form oxycarbene intermediates in polar media, rather than ketyl-type radicals which are characteristic for carbon analogs.

V. REACTIONS OF UNSATURATED ORGANIC DERIVATIVES OF GERMANIUM, TIN AND LEAD

Despite their structural simplicity, allylic derivatives of germanium and tin $R_3MCH_2CH{=}CH_2$ (M = Ge, Sn) are, perhaps, among the most intriguing topics for mechanistic research. One peculiar fact is their capability to enter photoinduced reactions of both homolytic addition and substitution, depending on the nature of the chosen reagent. The reaction mechanisms of the homolytic reactions of $R_3MCH_2CH{=}CH_2$ are of special interest to organic chemists, since their applications to organic synthesis open the way to hitherto unknown or otherwise almost inaccessible compounds.

A. Reactions of Homolytic Addition

Photoinduced reactions of homolytic addition of bromotrichloromethane CCl_3Br to allylic derivatives of germanium and tin could be an illustrative example of the potentialities of CIDNP application to study processes where the polarization effects have not been generated at the initiation stage. An earlier proposed[46] overall scheme of CCl_3Br addition to the allylic double bond in the $R_3MCH_2CH{=}CH_2$ molecule was based on the analysis of the reaction products (Scheme 5).

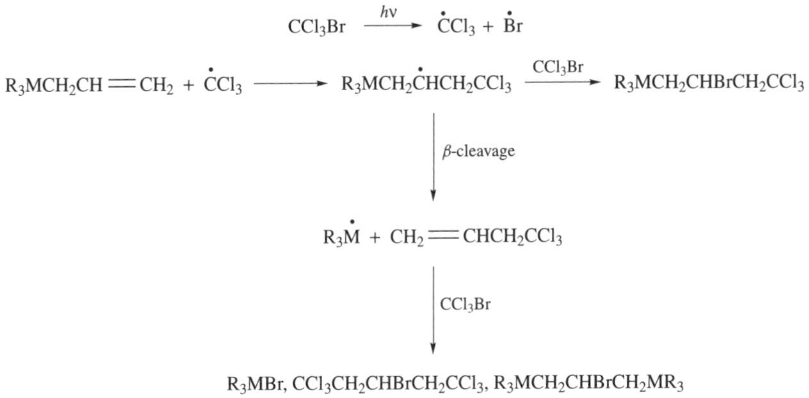

$$CCl_3Br \xrightarrow{h\nu} \dot{C}Cl_3 + \dot{B}r$$

$$R_3MCH_2CH{=}CH_2 + \dot{C}Cl_3 \longrightarrow R_3MCH_2\dot{C}HCH_2CCl_3 \xrightarrow{CCl_3Br} R_3MCH_2CHBrCH_2CCl_3$$

$$R_3MCH_2\dot{C}HCH_2CCl_3 \xrightarrow{\beta\text{-cleavage}} R_3\dot{M} + CH_2{=}CHCH_2CCl_3$$

$$\xrightarrow{CCl_3Br} R_3MBr,\ CCl_3CH_2CHBrCH_2CCl_3,\ R_3MCH_2CHBrCH_2MR_3$$

SCHEME 5

From the viewpoint of the reaction mechanism, the emphasis in Scheme 5 is focused on products with the general formula $R_3MCH_2CHBrCH_2CCl_3$ (M = Ge, Sn) with a halogen atom in a β-position to the element M (the so-called normal addition product). These compounds are believed to be unstable and to decompose with the elimination of an $R_3M^\bullet$ radical. The phenomenon is referred to as β-decomposition or β-cleavage (Scheme 5). The mechanism presented in this scheme lacks the radical pair stages, while the experimental results[47,48] demonstrate CIDNP effects observed for the initial compounds and the main reaction products of the interaction of $R_3MCH_2CH{=}CH_2$ with CCl_3Br (Table 6). Thus Scheme 5, which is based on the analysis of the reaction products, needs to be refined.

The reaction mechanism of the photoinduced interaction between $Et_3SnCH_2CH{=}CH_2$ (**11**) and CCl_3Br was also studied by means of another physical method—the so-called radiofrequency (RF) probing technique[49]. To facilitate the interpretation of CIDNP data and to identify the primary reaction stage which does not involve radical pairs, it is convenient to start from the RF probing technique.

1. The radiofrequency (RF) probing technique — general background

The groundwork for the RF probing technique was laid back in the 1960s when Forsén and Hoffman[50] proposed using the method of RF saturation to study the kinetics of fast

TABLE 6. ^{1}H CIDNP effects detected in the photolysis of $R_3MCH_2CH{=}CH_2$ (M = Ge, Sn) in c-C_6D_{12}

Reaction products	^{1}H CIDNP sign (protons in corresponding positions)[a]		
	1	**2**	**3**
$R_3M\overset{1}{C}H_2-\overset{2}{C}H{=}\overset{3}{C}H_2$	— (1.8, M = Ge)[b] (1.7, M = Sn)	A + A/E (5.8)	E (4.8, M = Ge) (4.6, M = Sn)
$CCl_3\overset{1}{C}H_2\overset{2}{C}H{=}\overset{3}{C}H_2$	A (3.3)	—	A (5.4, M = Ge) (5.1, M = Sn)
$R_3M\overset{1}{C}H_2\overset{2}{C}HBr\overset{3}{C}H_2CCl_3$	— (1.9, M = Ge)	E (4.8, M = Ge)	A (3.3, M = Ge)
$R_3M\overset{1}{C}H_2\overset{2}{C}HBr\overset{3}{C}H_2CCl_3$	E (2.4, M = Sn)	—	E (3.8, M = Sn)
$\overset{1}{C}HCl_3$	E (7.3)		

[a] E — emission, A — absorption.
[b] Chemical shifts (δ ppm) of the polarized signals are given in parentheses.

exchange processes. The procedure involves a radiofrequency saturation of the nuclei at one of the positions of the molecule with subsequent tracing of the fate of such a 'label' in the NMR spectra. In the version of this technique used in the present investigations of chemical reactions, the additional RF field is applied to a certain group of nuclei in the precursor molecule, and after the reaction the resulting NMR spectrum allows one to identify the product and location of this group of nuclei in the product molecule. In the case of homonuclear decoupling one should expect to observe a decrease in the intensity of the 'labelled' nuclei in the product molecule, while the heteronuclear decoupling leads to an increase in this intensity due to the nuclear Overhauser effect. The conditions described below should be met in order to observe the migration of the 'labelled' nuclei in the precursor to the reaction products. Let us consider a generalized reaction scheme (equation 13):

$$A \longrightarrow [B] \longrightarrow C \tag{13}$$

where A is the initial molecule, C is the reaction product and B is an intermediate species, free or charged radical, or any other reactive intermediate. The first condition relates to the lifetime τ_B of the intermediate species B that should be shorter than the relaxation time, i.e. $\tau_B < T_{1B}$, where T_{1B} is the spin–lattice relaxation time of the nuclei B. The second and third conditions relate to the reaction time and observation time, i.e. the yield of 'labelled' molecules during a time period comparable to the relaxation time of the diamagnetic product molecule should be sufficient for their detection in the NMR spectrum: $t \leqslant T_1{}^C$, where $T_1{}^C$ is the spin–lattice relaxation time of the product C. The recording time also should not exceed the relaxation time of the diamagnetic molecule, otherwise one will observe only equilibrium NMR signals: $t \sim T_1{}^C$. All the above possibilities and restrictions of the RF probing technique allow us to use this method to investigate the photoreactions of allylic derivatives of tin and germanium.

2. Homolytic addition of bromotrichloromethane to allyltriorganostannanes

The need to invoke two methods (CIDNP and RF probing) to study the mechanism of the photoinduced addition of CCl_3Br to $Et_3SnCH_2CH{=}CH_2$ (**11**) is dictated by the impossibility of analyzing the polarization effects under the assumption of CIDNP formation in a single radical pair. However, application of the RF probing technique has allowed us to 'label' the protons in different positions of the initial compounds and then to trace their fate, by identifying their locations in the reaction product molecule. This approach allows us to clarify the structure of at least one of the partner radicals comprising the radical pair responsible for the CIDNP formation.

The photoinduced reaction is given by equation 14:

$$\underset{\textbf{(11)}}{Et_3SnCH_2CH{=}CH_2} + CCl_3Br \xrightarrow{h\nu} Et_3SnBr + CH_2{=}CHCH_2CCl_3 \qquad (14)$$

When a radiofrequency field is applied to the individual groups of nuclei of allyltriethylstannane ($Et_3Sn\overset{1}{C}H_2\overset{2}{C}H{=}\overset{3}{C}H_2$) under simultaneous UV irradiation of the reaction mixture in c-C_6D_{12}, the NMR spectra demonstrate a peculiar pattern shown in Figure 11. The saturation of the proton signals of the precursor in position 3 leads to a 30–40% decrease in the intensity of the proton signal attributed to position 1′ of the product 4,4,4-trichlorobutene-1 ($CCl_3\overset{1'}{C}H_2\overset{2'}{C}H{=}\overset{3'}{C}H_2$). The saturation of the protons in position 1 of the precursor **11** results

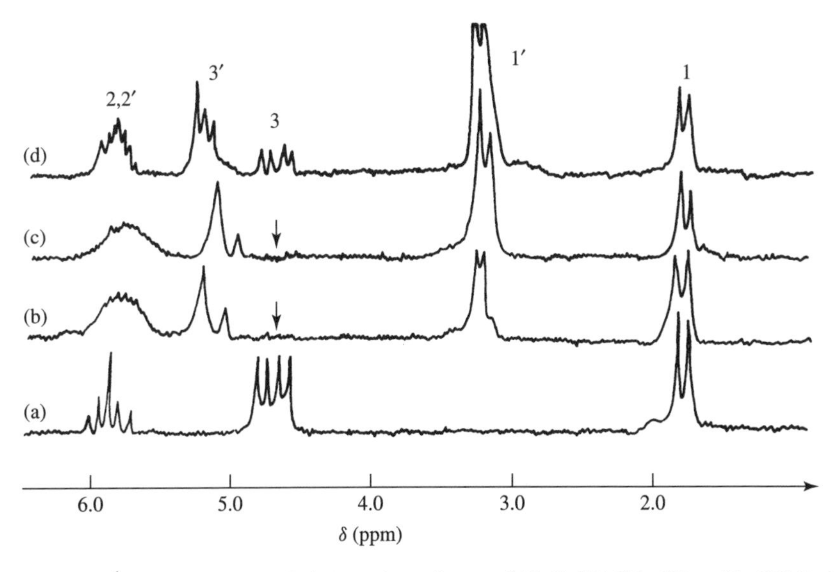

FIGURE 11. ^{1}H NMR spectra of the reaction mixture of $Et_3SnCH_2CH{=}CH_2$ with CCl_3Br in c-C_6D_{12} under a sequential saturation (application of an RF field) to the proton groups: (a) initial spectrum; (b) under irradiation with saturation of the protons of the precursor at position 3, (c) dark spectrum with saturation of the protons of the precursor at position 3, (d) the resulting equilibrium spectrum. Numerical indices denote the protons of the precursor ($Et_3Sn\overset{1}{C}H_2\overset{2}{C}H{=}\overset{3}{C}H_2$), and primed indices denote the protons of the reaction product ($CCl_3\overset{1'}{C}H_2\overset{2'}{C}H{=}\overset{3'}{C}H_2$). The part of the spectrum associated with the ethyl groups is omitted

in a decrease in the intensity of the NMR signal of the protons in position 3′ of the reaction product (Figure 11). The observed migration of the saturation 'label' $1 \rightarrow 3'$ and $3 \rightarrow 1'$ demonstrates that in full accordance with the mechanism of Scheme 5 the initial stages of the reaction involve the addition of a trichloromethyl radical $^{\bullet}CCl_3$ to the terminal carbon atom (at position 3) of the double bond of **11**. The formation of the radical adduct suggests that it is this radical which plays a role of one of the partners in the radical pair responsible for the formation of CIDNP effects. Moreover, it is quite reasonable to assume that the $^{\bullet}CCl_3$ radical will be the other partner of this radical pair.

The analysis of the net proton polarization effects (Table 6 and Figure 12) of the main reaction product trichlorobutene $Cl_3CCH_2CH{=}CH_2$ (^1H NMR: $\delta = 3.3$ and 5.4 ppm in $c\text{-}C_6H_{12}$) supports the suggestion[47] that this compound might originate from the disproportionation of the diffusion (F-pair) radical pair $(Et_3SnCH_2\dot{C}HCH_2CCl_3\ ^{\bullet}CCl_3)^F$. ^1H CIDNP spectra of the reaction mixture also demonstrate two negatively polarized signals ($\delta = 2.4$ and 3.8 ppm) which are not present in the equilibrium spectrum (Figure 12). Chemical shifts and splitting parameters (doublets, $J = 6$–7 Hz) allow us to attribute these lines to the so-called normal addition product $Et_3SnCH_2CHBrCH_2CCl_3$ (**12**). The analysis of CIDNP shows that the observed effects could not originate in the above-mentioned F-pair. Therefore, it is reasonable to conclude that the observed polarizations are formed in the act of radical β-cleavage of the normal addition product **12** which is extremely unstable and could be detected only in its polarized state (Scheme 6).

The protons of **12** are polarized due to partial back recombination of the singlet radical pair $(Et_3Sn^{\bullet}\ ^{\bullet}CH_2CHBrCH_2CCl_3)^S$. The analysis of CIDNP effects of trichlorobutene does not preclude the assumption that the polarization of $CH_2{=}CHCH_2CCl_3$ is also

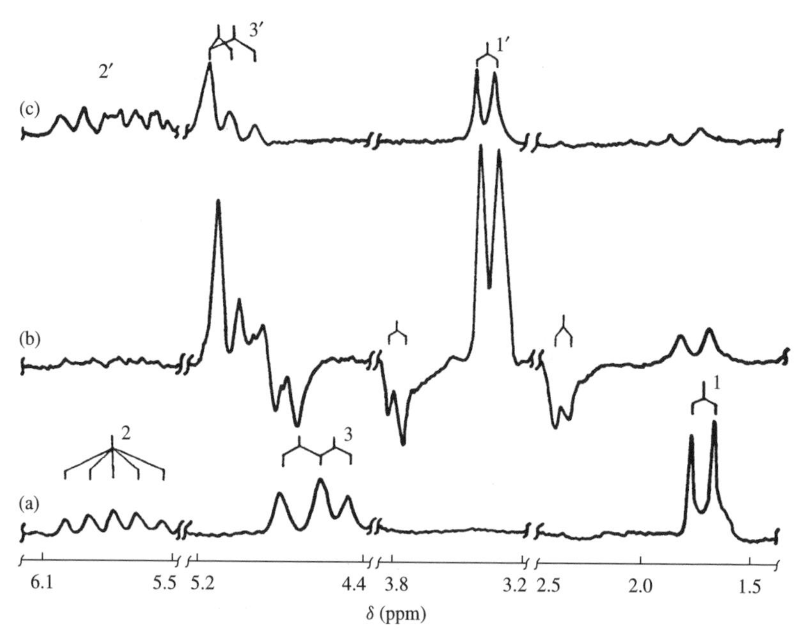

FIGURE 12. ^1H CIDNP effects detected in the reaction of photoinitiated addition of CCl_3Br to $Et_3SnCH_2CH{=}CH_2$ in $c\text{-}C_6D_{12}$: (a) initial spectrum, (b) under UV irradiation, (c) after the reaction. Numerical indices are the same as in Figure 11

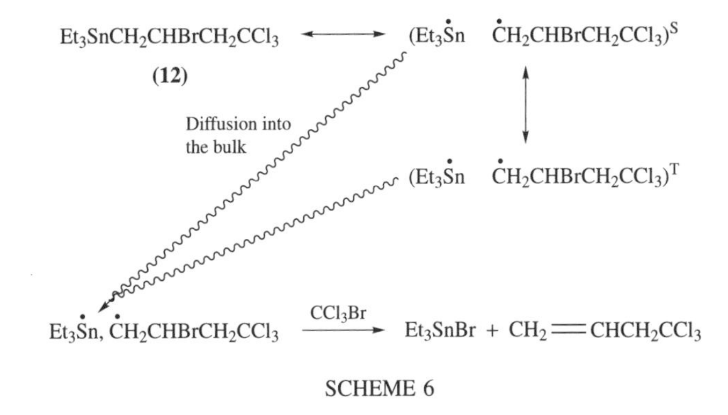

SCHEME 6

formed in the above singlet radical pair. Thus, there exist two possible pathways for the trichlorobutene formation: (i) β-cleavage of **12** and (ii) β-cleavage of the radical adduct $Et_3SnCH_2\dot{C}HCH_2CCl_3$ escaped from the F-pair $(Et_3SnCH_2\dot{C}HCH_2CCl_3\dot{C}Cl_3)^F$. However, the assumption that β-cleavage of **12** is the main source of trichlorobutene is unacceptable, since it was shown earlier[46] that the yield of **12** in the homolytic addition reactions of various reagents to **11** is determined by their reaction rate constants with radical adducts similar to $Et_3SnCH_2\dot{C}HCH_2CCl_3$. For instance, in the case of the reactions with EtSeH, EtSH and EtSD, the yield of type **12** species decreases from 100% to 33% when passing from ethylselenol to ethyldeuteriothiol in line with their relative reactivity in the chain propagation step of radical reactions. Bromotrichloromethane is a far less reactive reagent than thiols[46], and therefore one could hardly expect the formation of significant amounts of **12** in the process under study. An additional argument that **12** is not the main precursor of trichlorobutene are the RF-saturation experiments applied to the NMR spectral range of the protons of **12**. The RF field applied during the chemical reaction within the spectral ranges $\delta = 2.3$–2.5 ppm and 3.6–3.8 ppm (cf Figure 12) under conditions of full saturation has not lead to a decrease in the intensities of the corresponding protons of trichlorobutene. Thus, the main source of trichlorobutene and the other principal reaction product is the process of β-cleavage of the radical adduct $Et_3SnCH_2\dot{C}HCH_2CCl_3$.

3. Photolysis of allyltriorganogermanes in the presence of CCl_3Br and other polyhalogenated alkanes

Photoinitiated reaction of allyltriorganogermanes with polyhalogenated alkanes is much slower than the reaction of **11** with CCl_3Br. The main reaction products with CCl_3Br include the normal addition adducts $R_3GeCH_2CHBrCH_2CCl_3$ ($R_3 = Me_3, Me_2Cl, MeCl_2, Cl_3$) and 4,4,4-trichlorobutene-1, which is present in trace amounts. The normal addition products are stable in solution up to 120°C, but their attempt at isolation by fractional distillation results in a fast cleavage to R_3GeBr and trichlorobutene. It should be noted that the rate of decomposition of normal addition products depends on the electronic structure of the R_3Ge group. A sequential substitution of methyl groups at the germanium with more electronegative chlorine atoms stabilizes the resulting adducts[48].

Table 6 lists the polarization effects of the initial compounds, normal addition products and trichlorobutene observed in the photolysis of allyltriorganogermanes in the presence of CCl_3Br. The analysis of the detected ^1H CIDNP shows that, similarly to the reaction of allyltriethylstannane described above, the polarization is formed in the diffusion F-pair of the radical adduct $R_3GeCH_2\dot{C}HCH_2CCl_3$ and $^{\bullet}CCl_3$ radical (Scheme 7). The initial compound $R_3GeCH_2CH{=}CH_2$ is polarized as the 'in-cage' product of the same pair, while the polarizations of trichlorobutene $CH_2{=}CHCH_2CCl_3$ and the normal addition product $R_3GeCH_2CHBrCH_2CCl_3$ are formed in the escape of the radicals into the bulk.

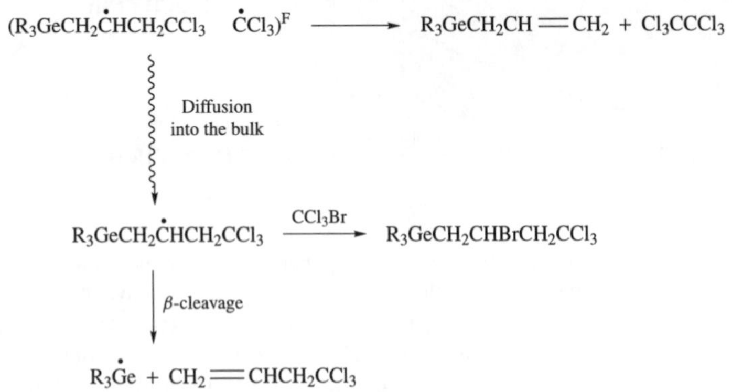

$$(R_3GeCH_2\dot{C}HCH_2CCl_3 \quad \dot{C}Cl_3)^F \longrightarrow R_3GeCH_2CH{=}CH_2 + Cl_3CCCl_3$$

Diffusion into the bulk

$$R_3GeCH_2\dot{C}HCH_2CCl_3 \xrightarrow{CCl_3Br} R_3GeCH_2CHBrCH_2CCl_3$$

β-cleavage

$$R_3\dot{G}e + CH_2{=}CHCH_2CCl_3$$

SCHEME 7

The photochemical interaction of $Et_3GeCH_2CH{=}CH_2$ (**13**) with CCl_3Br is also accompanied by the formation of trace amounts of polarized chloroform $CHCl_3$ (Table 6 and Figure 13). The sign of chloroform polarization (emission) allows one to suggest that $CHCl_3$ is a product of the 'in-cage' disproportionation of the initial

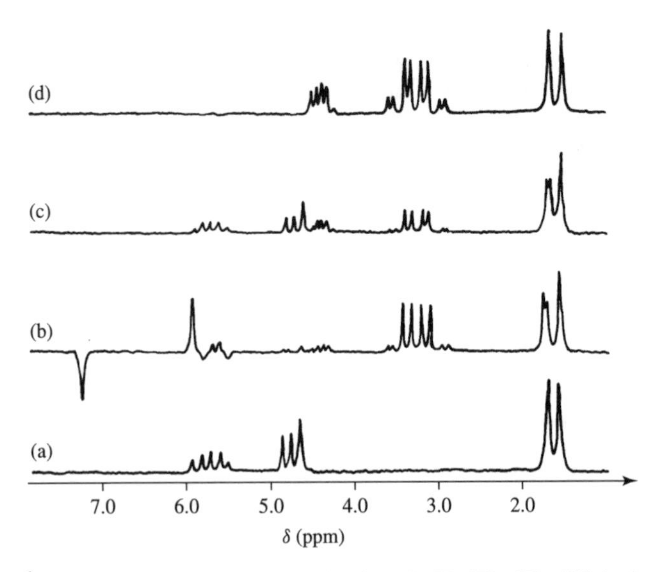

FIGURE 13. ^{1}H CIDNP effects in the photolysis of $Et_3GeCH_2CH{=}CH_2$ (**13**) in the presence of CCl_3Br in c-C_6D_{12}: (a) initial spectrum, (b) under UV irradiation, (c) dark spectrum, (d) after the photolysis. The part of the spectrum associated with the ethyl groups is omitted

diffusion pair (Scheme 7). Another product of this disproportionation is definitely $Et_3GeCH_2CH{=}CHCCl_3$, which is observed in trace amounts in the NMR spectra after completion of the reaction (equation 15).

$$(Et_3GeCH_2\dot{C}HCH_2CCl_3\ \dot{C}Cl_3)^F \longrightarrow Et_3GeCH_2CH{=}CHCCl_3 + CHCl_3 \qquad (15)$$

Thus, the detailed investigations by means of spin chemistry techniques and RF probing demonstrate that the homolytic addition reactions of bromotrichloromethane to allylic derivatives of germanium and tin $R_3MCH_2CH{=}CH_2$ proceed via mechanisms that somewhat differ for $M = Sn$ and $M = Ge$. Comparison of all the experimental results leads to the conclusion that all these differences are determined by both the nature of the element and the electron-donating properties of the R_3M group. However, the general feature of these processes is the involvement of free radical stages in the β-cleavage reaction as well as the radical addition to the terminal carbon to give the radical adducts $R_3MCH_2\dot{C}HCH_2CCl_3$.

B. Homolytic Substitution Reactions

Photoinduced homolytic addition reactions of various reagents to multiple bonds of unsaturated organic derivatives of group 14 elements, particularly to the corresponding tin derivatives, are of extreme importance as synthetic routes to a number of rather complex alicyclic and heterocyclic compounds[51]. Analysis of the basic features of these processes allows us to conclude that otherwise almost inaccessible unusual products could be formed in the homolytic substitution reactions of the hydrogen atom in the organic moiety of the organotin species, in particular, if a hydrogen atom is substituted by halogen. A major challenge is to select a proper reagent capable of entering a regiospecific reaction with the substrate which would proceed solely via a free-radical mechanism, since the ionic reactions usually lead to the elimination of an organic moiety. One example of these unusual transformations is the photoinitiated reaction of allyltriethylstannane $Et_3SnCH_2CH{=}CH_2$ (**11**) with *N*-bromohexamethyldisilazane $(Me_3Si)_2NBr$ (**14**), leading to a high yield of allene $CH_2{=}C{=}CH_2$ (*ca* 92%). The literature[52] shows that the reaction of **14** with olefins results in allylic bromination products. It has been suggested that the reaction with olefins proceeds via a free-radical mechanism, and this provides grounds to assume that the reaction of $(Me_3Si)_2NBr$ with **11** will also follow the allylic bromination mechanism.

However, analysis of the main reaction products has unexpectedly revealed 92% allene and equimolar amounts of bromotriethylstannane Et_3SnBr and hexamethyldisilazane $(Me_3Si)_2NH$ (equation 16).

$$\underset{\mathbf{(11)}}{Et_3SnCH_2CH{=}CH_2} + \underset{\mathbf{(14)}}{(Me_3Si)_2NBr} \xrightarrow{h\nu} Et_3SnBr + (Me_3Si)_2NH + CH_2{=}C{=}CH_2 \qquad (16)$$

The direct photolysis of the reaction mixture of **11** and **14** in the probe of the NMR spectrometer has allowed one to detect ^1H CIDNP effects (Figure 14) of the methyl protons of $(Me_3Si)_2NBr$ and hexamethyldisilazane, and the methylene protons of allene. The spectrum also demonstrates polarization effects in the region of olefin protons of precursor **14**.

According to earlier published data[52] the primary act of the process is the homolytic decomposition of *N*-bromodisilazane **14** (Scheme 8, equations 17 and 18). Further reaction stages include the interactions of **11** with the $(Me_3Si)_2N^\bullet$ radical and bromine atom. The above-discussed CIDNP studies of the photoinduced interaction of **11** with CCl_3Br

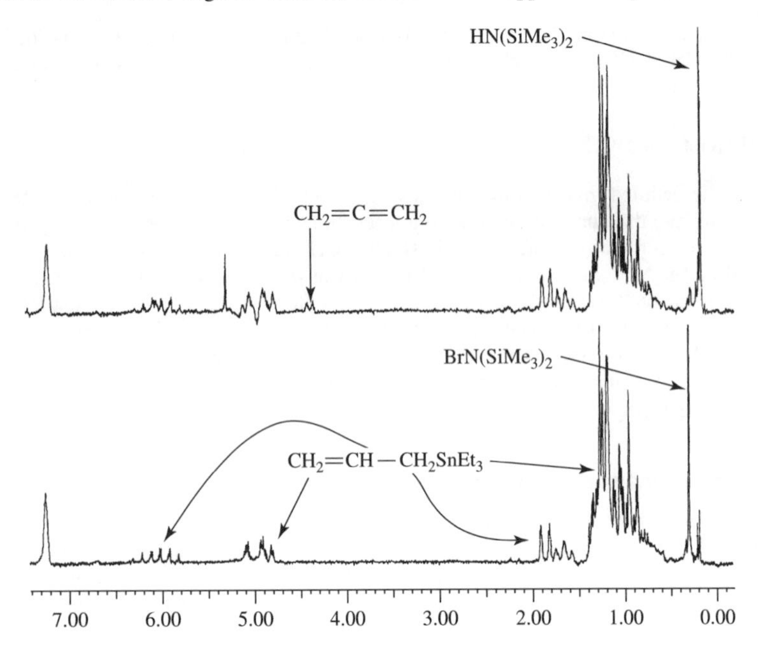

FIGURE 14. ^{1}H NMR spectra of the reaction mixture of $Et_3SnCH_2CH{=}CH_2$ with $(Me_3Si)_2NBr$ in deuteriobenzene. Bottom spectrum: the initial reaction mixture; upper spectrum: CIDNP effects detected 20 s after the UV irradiation. The spectra show polarized signals corresponding to the main reaction products

have shown that bromine adds to the least hydrogenated carbon atom of **11**. It has been shown[47] that the brominated product with the bromine atom in a β-position to the triethyltin substituent (**12**) undergoes a fast β-cleavage—only polarized signals of this product were detected, and no traces could be seen in the NMR spectrum after photolysis. In the present case, CIDNP effects of the allene protons suggest that, similarly to the main product of the photolysis of **11** with CCl_3Br, i.e. $CH_2{=}CHCH_2CCl_3$, $CH_2{=}C{=}CH_2$ results from a β-cleavage of the brominated product $Et_3SnCH_2CBr{=}CH_2$ (**15**, Scheme 8). Compound **15** originates from the disproportionation of the radical pair of $^{\bullet}CH_2CHBrCH_2SnEt_3$ and $(Me_3Si)_2N^{\bullet}$; another product of this pair is a polarized hexamethyldisilazane $(Me_3Si)_2NH$, since the $(Me_3Si)_2N^{\bullet}$ radical readily abstracts hydrogen[52] from the partner radical in the pair and it does not enter the recombination reaction.

$$(Me_3Si)_2N{-}Br \xrightarrow{h\nu} (Me_3Si)_2N^{\bullet} + Br^{\bullet} \qquad (17)$$

(14)

$$Et_3SnCH_2CH{=}CH_2 + Br^{\bullet} \longrightarrow Et_3SnCH_2CHBrCH_2^{\bullet} \qquad (18)$$

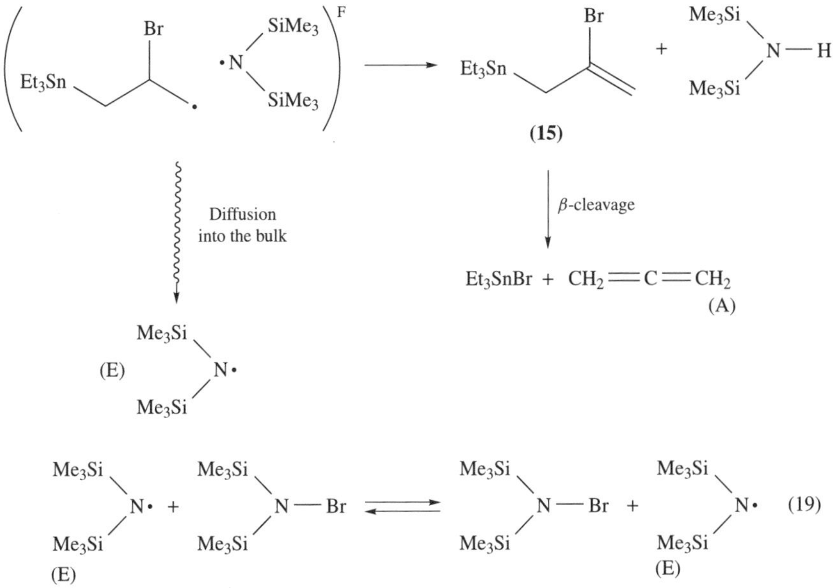

$\overline{(Et_3SnCH_2CHBr\dot{C}H_2 \;\; \cdot N(SiMe_3)_2)}^F \longrightarrow Et_3SnCH_2C(Br)=CH_2$ (15) + $(Me_3Si)_2N{-}H$

(15) $\xrightarrow{\beta\text{-cleavage}}$ $Et_3SnBr + CH_2{=}C{=}CH_2$ (A)

Diffusion into the bulk $\rightsquigarrow$ $(Me_3Si)_2N\cdot$ (E)

$$(Me_3Si)_2N\cdot \text{ (E)} + (Me_3Si)_2N{-}Br \rightleftharpoons (Me_3Si)_2N{-}Br + (Me_3Si)_2N\cdot \text{ (E)} \quad (19)$$

A, absorption; E, emission

SCHEME 8

Another polarized product, the precursor *N*-bromohexamethyldisilazane (**14**), could result from a chemical exchange reaction between the polarized $(Me_3Si)_2N^{\bullet}$ radical and the initial $(Me_3Si)_2NBr$ (Scheme 8, equation 19).

It is necessary to emphasize the unusual character of the observed results. It is common belief that radicals usually attack the terminal γ-carbon atom of an allylic moiety. Indeed, the CIDNP effects detected in the photolysis of **11** in the presence of bromotrichloromethane[47] unambiguously point to the formation of the radical adduct $Et_3SnCH_2\dot{C}HCH_2CCl_3$ resulting from the addition of the $^{\bullet}CCl_3$ radical to the terminal γ-carbon. However, in the present case we observe neither the products that could stem from the analogous addition of $(Me_3Si)_2N^{\bullet}$ radical to allylic γ-carbon, nor CIDNP effects that could be ascribed to certain unstable para- or diamagnetic intermediates formed through this addition. Instead, the $^{\bullet}Br$ attacks the least hydrogenated β-carbon atom of the allylstannane and the resulting $^{\bullet}CH_2CHBrCH_2SnEt_3$ radical becomes the precursor of the unusual products of the homolytic substitution reaction.

As demonstrated earlier[48], compounds with a β-bromine atom to R_3Ge groups (in contrast to their tin analogs) are quite stable in solution, and the β-cleavage reaction is observed only during an attempt at their separation. If the above assumptions on the mechanism of $CH_2{=}C{=}CH_2$ formation are correct (Scheme 8), a high yield of allene should not be observed in the photoreactions of **11** and $Et_3GeCH_2CH{=}CH_2$ with **14**. The experiments have indeed shown that the mixture obtained after the reaction does not contain $CH_2{=}C{=}CH_2$ at all. These results suggest the following regularity observed when analyzing the mechanisms of formation of the products of the photoinduced reaction of **14** with allylic tin (**11**) and germanium derivatives. The definitive stage of formation of the products are the interactions in the uncorrelated radical F-pair of $^{\bullet}CH_2CHBrCH_2MEt_3$ and $(Me_3Si)_2N^{\bullet}$ radicals (cf Schemes 8 and 9).

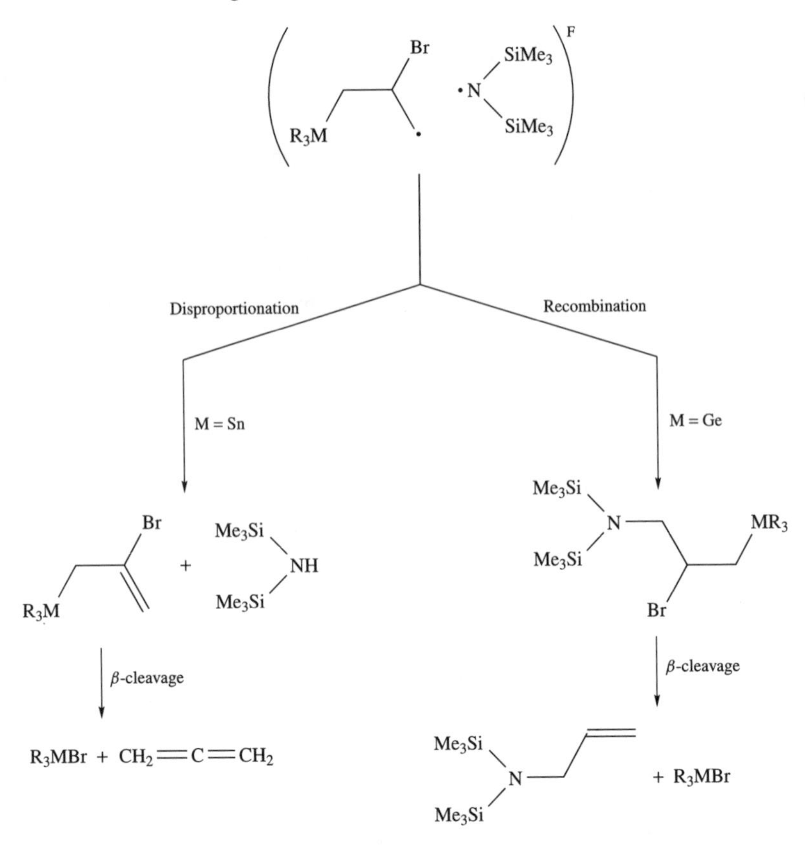

SCHEME 9

When M = Sn the disproportionation of the radical pair becomes the main reaction pathway, while for M = Ge a recombination of the pair is also possible. Undoubtedly, this difference is caused by the influence of the element atom, and this is in agreement with the evidences obtained earlier that the effect of germanium-containing substituents on the radical addition reactions is distinctly different from that of tin-containing groups[47,48].

Thus, there are two instances of the influence of the organoelement on the radical reactivity. Despite the apparent triviality, these results are unusual and the nature of the influence of organoelement function on the hydrogen atom abstraction by another radical or a recombination process is of special interest. As of now, one might only suggest that the organotin substituent, having more pronounced donor properties as compared to an organogermanium substituent, facilitates the homolytic C−H bond cleavage.

C. Other Examples of Reactions of Homolytic Addition and Substitution

Allylic derivatives $R_3MCH_2CH{=}CH_2$ (M = Ge, Sn, Pb) react with a number of element-centered radicals $^{\bullet}YPh$ (Y = S, Se, Te) formed in the decomposition of the corresponding dichalcogen compounds PhYYPh[53]. $R_3MCH_2CH{=}CH_2$ also reacts with the isopropyl free radical generated by the decomposition of *i*-PrHgCl. It has been found that allylic derivatives of lead and, under certain conditions, the similar germanium derivatives, undergo S_{2H} or $S_{2H'}$ radical substitution reactions which follow a chain mechanism similar to that described above for allyltriethylstannane (equation 20).

$$CH_2{=}CHCH_2MR_3 + {}^{\bullet}SPh \longrightarrow CH_2{=}CHCH_2SPh \qquad (20)$$

Analysis of the yield of the product reveals the following trend — the reactivity of the allylic derivatives of group 14 elements increases in the sequence: Ge < Sn < Pb. Comparison with the reactivity of the vinyl derivatives, in particular, with the β-metallostyrenes, shows that $R_3MCH_2CH{=}CH_2$ have higher reactivity in the substitution reactions of the R_3M group by alkyl or chalcogen moiety. The reactivity of the chalcogen-centered radicals toward the allylic derivatives is given by the following sequence: S > Se > Te[53].

Radical substitution reactions involving allylic tin derivatives could be accompanied by a photoinduced 1,3-rearrangement[54,55]. A photostationary mixture of cinnamyl(triphenyl)stannane with its regioisomer 1-phenylprop-2-enyl(triphenyl)stannane has been shown to form in the photolysis of (E)-cinnamyl(triphenyl)stannane in benzene under aerobic conditions, or in the presence of halogenated organic compounds or radical-trapping reagents (equation 21).

$$PhCH{=}CHCH_2SnPh_3 \underset{C_6H_6,\ air}{\overset{h\nu}{\rightleftharpoons}} PhCH(SnPh_3)CH{=}CH_2 \qquad (21)$$

It is noteworthy that this rearrangement was not observed in pure benzene under anaerobic conditions. It is suggested[54] that the rearrangement is intramolecular and occurs via a π-π^* excitation of the cinnamyl group. The simultaneous homolytic cleavage of the C−Sn bond resulting in $Ph_3Sn^{\bullet}$ and cinnamyl radicals was also observed. In the case of crotyl- and phenyl(tributyl)stannanes, this rearrangement appears to be inefficient, while the triphenyl and dibutylphenyl substituted derivatives undergo this 1,3-rearrangement via photoexcitation of the phenyl substituent at the tin atom resulting in a regioisomeric mixture of the initial linear and isomerized stannane. In both the above cases, the content of the isomerized compound in the photostationary mixture is greater than that of the precursor allylstannane.

Of special interest are also the reactions of allyl-substituted element-centered radicals. For instance, $AllBu_2Ge^{\bullet}$ and dibutyl(2-methylallyl)germyl radicals undergo disproportionation reactions leading to derivatives of tetra- and divalent germanium[56].

$$CH_2{=}CHCH_2\dot{G}eBu_2 + CH_2{=}CHCH_2\dot{G}eBu_2 \longrightarrow (CH_2{=}CHCH_2)_2GeBu_2 + {:}GeBu_2 \qquad (22)$$

Note that while homolysis of the element−carbon bond is as a rule not characteristic for allylic derivatives, similar cyclopentadienyl tin and lead derivatives undergo a direct photolysis with $Cp^{\bullet}$ radical abstraction[57]. ESR spectra taken in the photolysis of

$CpMe_3Pb$, $CpPh_3Pb$ and Cp_2Ph_2Pb have shown the formation of the cyclopentadienyl radical. Interestingly, under the same conditions the photolysis of $CpEt_3Pb$ results in the $Cp^•$ radical only at temperatures above −50 °C. Below −100 °C only the formation of the ethyl radical has been observed. The properties of the resulting lead-centered radicals are similar to those of the tin-centered species, though the former demonstrate lower reactivity toward alkyl bromides and alkenes. It has been found that lead-centered radicals are effectively trapped by α,β-diketones, in particular by biacetyl, resulting in the formation of radical adducts which could be observed by means of ESR spectroscopy. The interaction of lead-centered radicals with 2-methyl-2-nitropropane leads to the corresponding radical adduct $Me_3C(R_3PbO)NO^•$, which is observed in the ESR spectrum.

From the viewpoint of the electronic structure, the benzyl-substituted derivatives could demonstrate a certain similarity with the allylic species. However, their properties are close to that of the above-mentioned cyclopentadienyl derivatives, i.e. their photolysis results in the homolysis of the element–carbon bond. Both stationary and time-resolved ^1H CIDNP techniques were used to study the mechanism of photolysis of $Me(PhCH_2)_3Sn$ and $(PhCH_2)_3SnCl$[58]. The analysis of the detected nuclear polarization effects demonstrates that the decomposition of $Me(PhCH_2)_3Sn$ occurs from a triplet excited state. The reaction mechanism and hence the observed polarizations are independent of the solvent. On the contrary, CIDNP detected in the photolysis of $(PhCH_2)_3SnCl$ shows a marked dependence on the nature of the solvent. The analysis of polarization effects observed in benzene favors the conclusion that photodecomposition of $(PhCH_2)_3SnCl$ occurs from the singlet excited state. When $CDCl_3$ is used as a solvent, the detected polarizations correspond to the simultaneous formation of both singlet and triplet radical pairs. The formation of singlet radical pairs in the photolysis of $(PhCH_2)_3SnCl$ was taken as compelling evidence for the formation of stannylene[58]. Similar results[59] have stimulated discussions on the possibility of the involvement of germylenes in the photolyses of benzyl-substituted digermanes.

VI. REACTIONS INVOLVING GERMYLENES AND DIGERMENES

In the last decade, reactions involving short-lived para -and diamagnetic organogermanium derivatives, i.e. germylenes and unsaturated germanium derivatives containing a multiple bond Ge=X (X = Ge, N, P, O), have attracted considerable attention. Of special interest are the elementary reaction mechanisms involving these intermediates which pose a challenging problem for modern physical chemistry[60]. For the most part, the solution amounts to a choice between radical and ionic mechanisms, since the experimental data often cannot distinguish between the two options. It is clear that the most reliable way to find a solution is to identify these short-lived intermediates, the precursors of the end reaction products.

When analyzing the literature data on reactive intermediates in organometallic reactions, two basic approaches to solve this fundamental problem are used. In the first approach, which is characteristic for classical organic chemistry, the conclusion is reached on the structure of the short-lived intermediate species and on their involvement in the process under study on the basis of analysis of the end reaction products. Another approach, more typical for physical chemistry, is based on time-resolved techniques, which allow one to measure the rate constants of the reactions of intermediates. However, in this case, one usually refrains from analysis of the reaction products. Unfortunately, it should be noted that inconsistency is often observed between the spectroscopic and kinetic data on the intermediates in reactions involving short-lived derivatives of group 14 elements. Table 7 exemplifies the discrepancies of spectral data for the simplest alkyl-substituted short-lived carbenoid, dimethylgermylene Me_2Ge: (**16**).

At the same time, spin chemistry techniques are capable of providing reliable information on the nature of the generated intermediates and their consequent transformations[13].

TABLE 7. Spectroscopic parameters of dimethylgermylene Me_2Ge:

λ_{max} (nm)	Precursor	Conditions[a]	References
420	$Me_2Ge(SePh)_2$	21 K, Ar	61
506	$(Me_2Ge)_5$	77 K, 3-MP	62
430	$(Me_2Ge)_6$	77 K, 3-MP	63
420	$Me_2Ge(SePh)_2$	77 K, 3-MP	61
490	$(Me_2Ge)_5$	293 K, c-C_6H_{12}	62
450	$(Me_2Ge)_6$	293 K, c-C_6H_{12}	63
425	$PhGeMe_2SiMe_3$	293 K, c-C_6H_{12}	64
420	$Me_2Ge(SePh)_2$	293 K, c-C_6H_{12}	61
420	$Me_2Ge(SePh)_2$	293 K, CCl_4	61
380	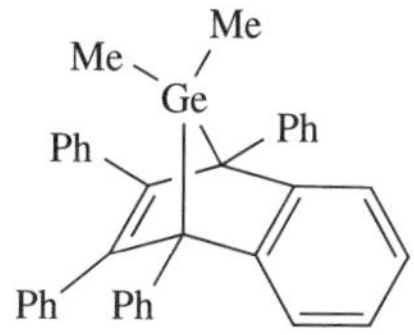 Me Me Ge Ph Ph Ph Ph (structure)	293 K, C_7H_{16}	65
480	$Me_3GeGeHMe_2$	293 K, gas phase	66

[a] 3-MP, 3-methylpentane.

From this perspective, the combination of spin chemistry techniques and laser pulse photolysis should allow one to obtain the most valuable information on the formation and decay reactions of such active short-lived derivatives as alkyl-substituted germylenes and digermenes, as well as on germanium-centered free radicals[67,68].

A. Generation of Dimethylgermylene

7-Germanorbornadienes are among the most convenient and commonly used germylene precursors[67–71]. Under mild thermal conditions or UV irradiation these compounds decompose to the inert aromatic molecule and a short-lived germylene. Equation 23 shows the generation of Me_2Ge: from 7,7′-dimethyl-1,4,5,6-tetraphenyl-2,3-benzo-7-germanorbornadiene (**17**).

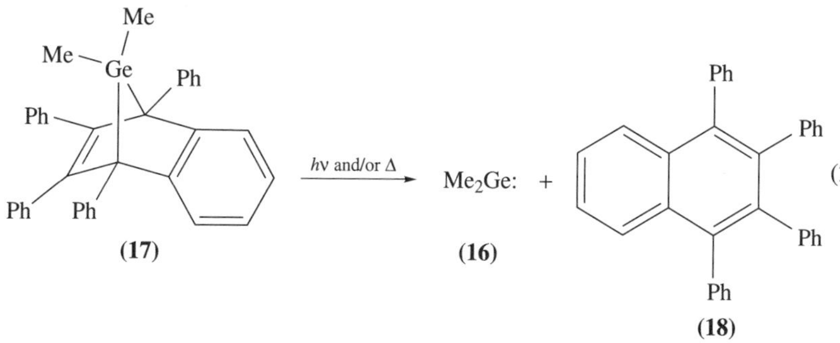

Me Me Ge Ph Ph Ph Ph **(17)** $\xrightarrow{h\nu \text{ and/or } \Delta}$ Me_2Ge: **(16)** + Ph Ph Ph Ph **(18)** (23)

It is necessary to mention the ongoing discussion about the hypothesis of the involvement of a biradical species resulting from breaking of one of the Ge–C bonds in the process described by equation 23. It has been found that the thermal decomposition of **17** monitored by the NMR spectra[67] or by the absorption spectra of 1,2,3,4-tetraphenylnaphthalene (**18**) obeys a first order kinetics and is characterized by a reaction rate constant $k = 1.1 \times 10^{-3}\ s^{-1}$ (78 °C, in toluene)[69]. The order of magnitude of the activation parameters of this reaction ($\Delta H^{\neq} = 116.5\ kJ\,mol^{-1}$, $\Delta S^{\neq} = 28.5\ J\,mol^{-1}\,K^{-1}$ in the range 65–85 °C) are in agreement with the literature data on reactions involving biradicals[69]. Photogeneration of **16** from **17** was also studied by pulse photolysis and matrix isolation[65]. In addition to **16**, an absorption band at $\lambda_{max} = 420$ nm is observed in the photolysis of **17** at 77 K. The band was attributed to a germanium-centered 1,5-biradical. This conclusion is based on the fact that the annealing of the species with a λ_{max} at 420 nm did not result in germanium-containing products and it reverted to the initial **17**. Thus, the mechanism of the generation of **16** in equation 24 has been proposed.

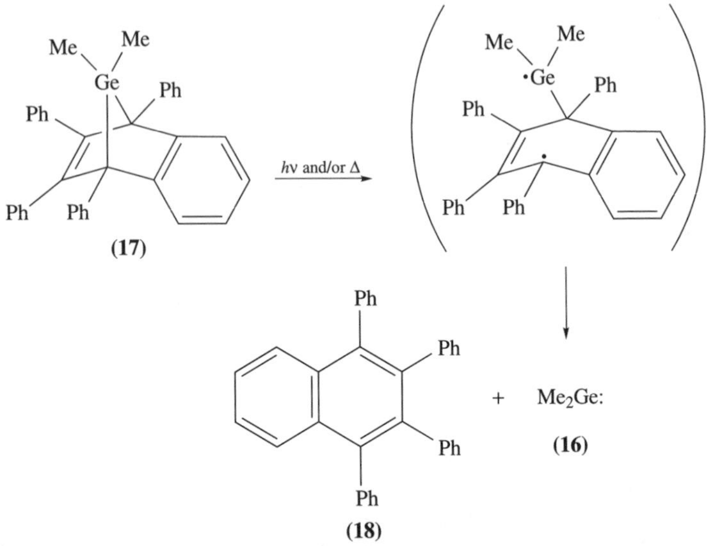

Me Me Ge Ph Ph Ph Ph (**17**) $h\nu$ and/or Δ → [Me Me •Ge Ph Ph Ph Ph (biradical)] → Ph Ph Ph Ph (**18**) + Me_2Ge: (**16**) (24)

Application of the ^1H CIDNP technique has allowed us to confirm the hypothesis of the biradical formation and to propose a detailed scheme of the photolysis of **17**[68].

It is also known[72] that **16** might react with the initial **17** at a rather high reaction rate constant, $k = 1.2 \times 10^7\ mol^{-1}\,s^{-1}$. However, the formation of the addition product of **16** to **17**, 7,7′,8,8′-tetramethyl-1,4,5,6-tetraphenyl-2,3-benzo-7,8-digermabicyclo[2.2.2]octadiene (**19**), has been observed in the NMR spectra only in the photolysis of **17** in the presence of CCl_4.

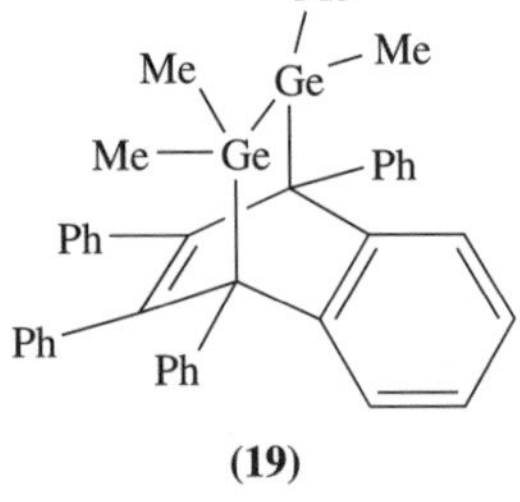

Me Me Me Ge Me Ge Ph Ph Ph Ph

(**19**)

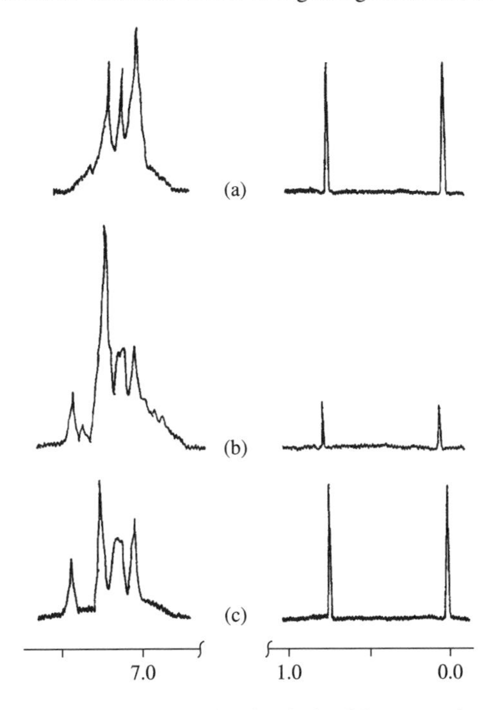

FIGURE 15. ^{1}H CIDNP effects observed in the photolysis of 7-germanorbornadiene (**17**) in C_6D_6: (a) initial spectrum, (b) under UV irradiation, (c) after the photolysis

Figure 15 shows the nuclear polarization effects detected in the photolysis of **17** in deuteriobenzene, emissions of the Me groups of the initial **17** ($\delta = 0.22$ and 0.94 ppm) and the absorption of tetraphenylnaphthalene **18** ($\delta = 7.43$ ppm). CIDNP effects of the initial **17** provide direct evidence of the reversibility of the photodecomposition of **17** and confirm the hypothesis of the formation of a biradical proposed on the basis of the above-mentioned annealing experiments. Indeed, the formation of nuclear polarization effects of the initial **17** and the reversibility of the process could be explained only through the generation of the 1,5-biradical resulting from the cleavage of one of the endocyclic Ge−C bonds. The analysis of ^{1}H CIDNP effects detected in the photolysis of **17** points to the formation of a germanium-centered 1,5-biradical in the singlet state. Its recombination results in the regeneration of the initial compound **17**, and the methyl protons demonstrate negative (emission) polarization (Figure 15). In this case, the observed positive (absorption) polarization of the protons of the main stable reaction product **18** suggests that tetraphenylnaphthalene originates from the triplet state of this 1,5-biradical. Regarding the formation of the 1,5-biradical, one should take into account that in accordance with the requirement of the retention of total spin of the system, a certain product resulting from the decomposition of the triplet biradical (equation 24) should be formed in the triplet excited state. It is quite reasonable to assume that this triplet excited product is germylene **16**, resulting from the cleavage of the second Ge−C bond in the biradical accompanied by simultaneous formation of **18**. Laser pulse photolysis studies of the analogous 7-silanorbornadiene[73] confirm that **18** is formed in the *singlet* state, thus another product of the cleavage of the *triplet* biradical should be generated in a *triplet* excited state. The combination of CIDNP and laser pulse photolysis data allows us to propose the mechanism of the formation of **16** shown in Scheme 10.

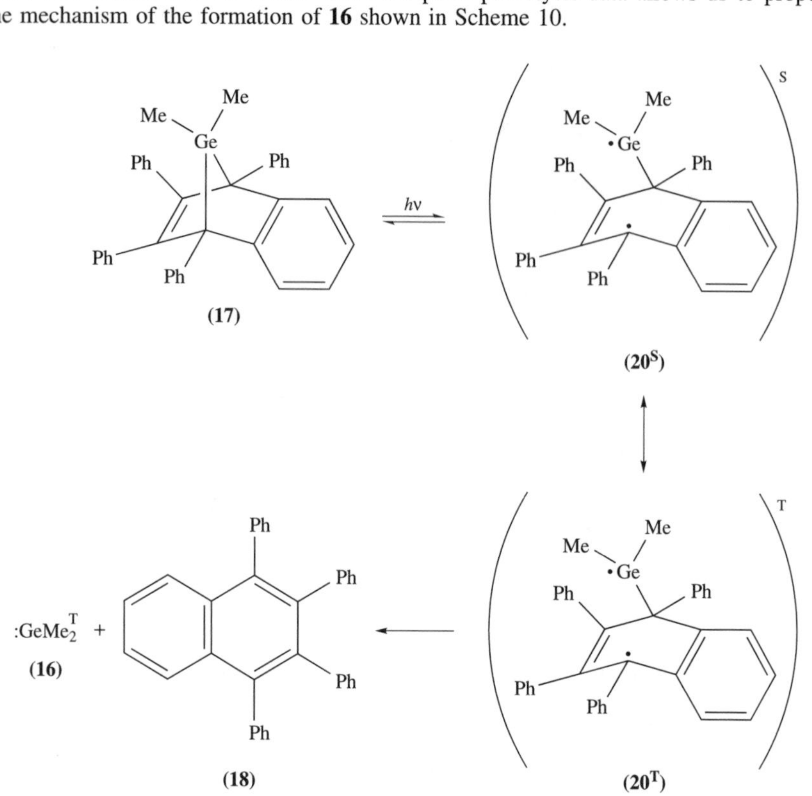

Me Me Ge Ph Ph Ph Ph **(17)** $h\nu$ Me Me •Ge Ph Ph Ph Ph S **(20^S)** Me Me •Ge Ph Ph Ph Ph T **(20^T)** :$GeMe_2^T$ + **(16)** Ph Ph Ph Ph **(18)**

SCHEME 10

However, attempts to detect CIDNP effects of the initial compound in the thermal decomposition of **17** have been unsuccessful[67,74]. Only the polarizations formed in the radical pairs involving germyl free radicals resulting from the reaction of singlet **16** with the germylene trapping agents were observed. Possible reasons for this inconsistency might include both lower concentrations of the paramagnetic species formed in the reactions of thermal generation as compared to photoinduced decomposition, and CIDNP methodology[67,74] which employed greater delays prior to the registration pulse of the NMR spectrometer. Certainly, the possibility of changes in the reaction mechanism of the thermolysis of **17** must not be ruled out, e.g. a simultaneous cleavage of both Ge−C bonds (synchronous mechanism). In this case, polarization effects will be generated only in the reactions of **16** with the trapping agents.

In concluding of this section, it is necessary to discuss some additional aspects of the process under study. In principle, as an alternative to Scheme 10 one might suggest an additional pathway for the formation of polarization effects of the initial compound **17**, namely, the regeneration of **17** via the reaction of **16** with the final reaction product **18**. However, since the reaction occurs in the bulk, the presence of the additional trapping agents of germylene should affect the observed CIDNP efficiency. However, it has been reliably established that polarization effects of the initial **17** appear to be independent of the presence of the scavengers of **16**. Similar reaction of **16** with substituted naphthalenes was also not observed in the thermal decomposition of **17**[60,72].

The nature of polarization effects of the initial **17** and of the tetraphenylnaphthalene **18** points to the realization of the so-called S-T_0 mechanism of CIDNP formation[13], characteristic for rigid biradicals. In the present case, one might expect a relatively rigid fixation of the unpaired electron orbital at the germanium atom with respect to the carbon skeleton of the 1,5-biradical **20**. These structural features of the biradical **20** resulting from 7-heteronorbornadienes have been additionally confirmed by theoretical calculations of the geometry and electron exchange interaction parameters[75]. Due to the lack of literature data on the structure and lifetimes of element-centered biradicals, the information obtained in CIDNP experiments is of obvious mechanistic interest. For instance, the observation of CIDNP effects allows one to estimate the lifetime of the biradical **20**. In accordance with radical pair theory[13], to generate the nonequilibrium population (CIDNP) in the radicals with HFI constants not greater than 0.5 mT (typical for Me protons in the biradical **20**) the lifetime of the intermediate should be longer than several nanoseconds.

It is reliably established that the singlet state is the ground state of the germylenes[65]. Therefore, the consequent stages of the process under study (Scheme 10) will include triplet to singlet conversion of germylene **16** ($:GeMe_2{}^T$ in the triplet excited state) as well as the reactions of its triplet and singlet states with the trapping agents. Note that a wide variety of approaches has been used to study the reactions of singlet germylene, while only the application of CIDNP techniques has allowed us to identify the processes involving triplet germylene. We now discuss the reactions of **16** in various spin states.

B. Reaction of Dimethylgermylene with Various Trapping Agents

Insertion processes into a C−X bond (X = OH, Hal or element) are known to be characteristic of alkyl-substituted germylenes[60]. Application of the laser pulse photolysis technique has allowed one to determine the reaction rate constants of these processes. However, all these results characterize the reactions of germylenes in the ground singlet state, and there are no literature data on the reactions of triplet germylenes identified by conventional methods of physical chemistry. The insertion reactions of germylenes in nonpolar solvents are thought[60] to follow two alternative mechanisms: (i) formation of a three-centered intermediate state and its synchronous decomposition, and (ii) a free radical pathway involving abstraction and recombination steps. It is quite difficult to distinguish these alternatives only on the basis of analysis of the end products, while the application of CIDNP techniques often allows us to elucidate the reaction mechanism and to determine the multiplicity of the reacting state of the germylene. CIDNP methods were used to study the reactions of both thermally[67,74] and photochemically[68,76] generated germylene **16**. First, it is noteworthy that the addition of the trapping agent does not affect the rate of thermolysis of **17**[67,69]. The above-described CIDNP effects of the initial compound[68] detected in the photolysis of **17** also remain essentially unchanged. Hence, these trapping agents do not participate in the reaction with the element-centered 1,5-biradical **20**.

The insertion reactions of **16** into the C-Hal bond of alkyl halides were found to be the most convenient model for CIDNP studies. Since in accordance with Scheme 10 the photolysis of **17** leads to the formation of triplet germylene **16**, we start with the first evidence of the reactions of triplet excited **16**[76].

1. Reactions with thiacycloheptyne

3,3,6,6-Tetramethyl-1-thiacyclohept-4-yne (**21**) is known to be one of the most effective trapping agents of **16**, since this germylene readily inserts into its triple bond with the formation of the corresponding germacyclopropene (**22**) (equation 25). The rate constant of this process is $k = 5 \times 10^8$ mol^{-1} s^{-1}[76]. Figure 16 and Table 8 show the polarization effects of the initial compound and **21** (emissions of Me protons) and the positively polarized protons of the reaction product **22**.

Me_2Ge: + **21** (thiacycloheptyne; Me, Me / Me, Me) ⟶ **22** (germacyclopropene; Ge with Me, Me) (25)

(16) **(21)** **(22)**

Observation of the polarization effects of the reaction product, germacyclopropene **22**, confirms its formation from paramagnetic precursors. However, in the case under study, one could hardly imagine the generation of a radical pair, and therefore it is reasonable to suggest that the intermediate step of the insertion of **16** into the triple bond of thiacycloheptyne **21** involves the formation of a 1,3-biradical. Analysis of the observed proton polarization effects (Table 8 and Figure 16) in accordance with the existing rules[11] allows us to propose the reaction mechanism in Scheme 11.

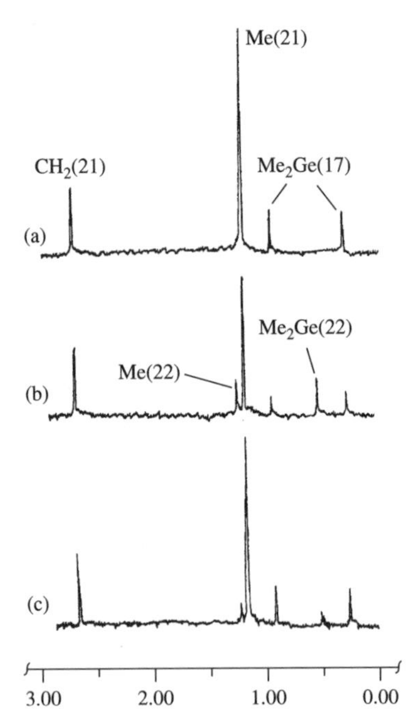

FIGURE 16. ^{1}H CIDNP effects detected in the photolysis of 7-germanorbornadiene (**17**) in the presence of thiacycloheptyne (**21**) in C_6D_6 (only the aliphatic part of the NMR spectra is present): (a) initial spectrum, (b) under UV irradiation, (c) after the photolysis

TABLE 8. ^{1}H CIDNP effects observed in the photolysis of 7-germanorbornadiene **17** in the presence of thiacycloheptyne **21** in C_6D_6

Reaction product	Chemical shift δ (ppm)	CIDNP effects[a]
21 (Me)	1.15 (s, 12 H)	E
22 (Me)	1.22 (s, 12 H)	A
22 (Me_2Ge<)	0.48 (s, 6H)	A

[a]E, emission; A, absorption.

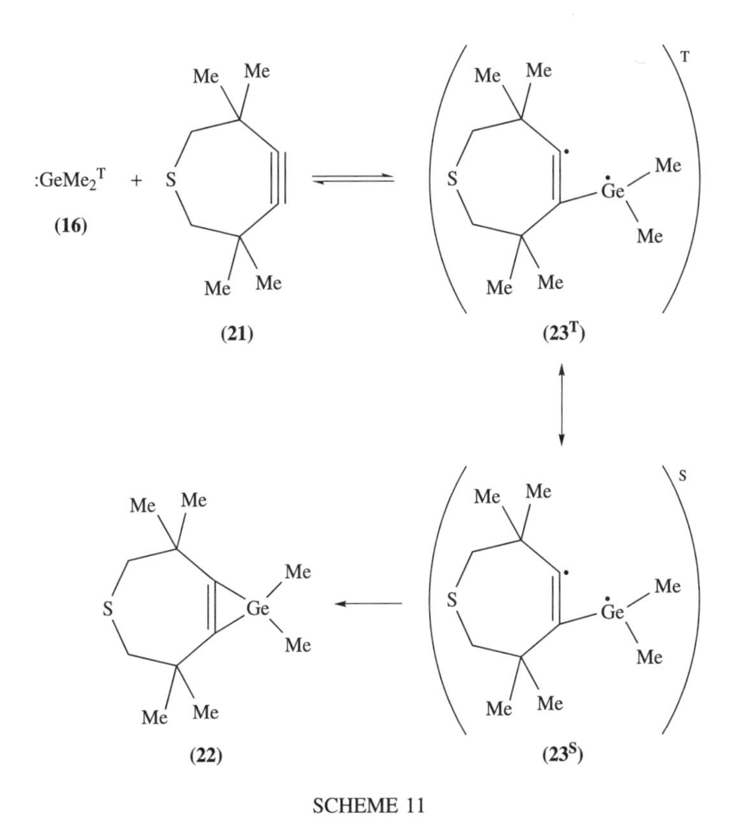

SCHEME 11

Thus, the sign of the observed polarization of the Me protons of the product **22** (absorption) is unambiguous evidence of the formation of an intermediate triplet 1,3-biradical $\mathbf{23^T}$, and this fact, in turn, means that **16** enters the reaction with **21** in the excited triplet state. The end product **22** is formed after the triplet–singlet conversion of the triplet biradical $\mathbf{23^T}$ followed by cyclization of the singlet biradical $\mathbf{23^S}$ (Scheme 11), while the triplet biradical $\mathbf{23^T}$ also reverts to the initial reagents. Only this reverse reaction could explain the negative polarization (emission) of the Me protons of the initial thiacycloheptyne **21**. It should be noted that the process described in Scheme 11 is the first example of a reaction of the excited triple state of germylene **16**[76].

2. Reactions with carbon tetrachloride

The main products of the reaction between germylene **16** and CCl_4 include the insertion product $ClMe_2GeCCl_3$ (**24**), Me_2GeCl_2 (**25**) and hexachloroethane C_2Cl_6. The yield of the insertion product **24** in the reaction of photochemically generated **16** is not greater than 20–30%, and the yield of **25** is 70–80%[65]. In the case of thermally generated **16**, the yield of **25** increases up to 95%, and this is believed to be due to thermal decomposition of **24**[67]. The kinetic parameters of this reaction were also studied by means of laser pulse photolysis, and Table 9 summarizes the reaction rate constants for different precursors of **16**.

Both thermal[67] and photochemical[77] decomposition of **17** demonstrate CIDNP effects of the methyl protons of **24** and **25** (Table 10). In addition, the above-mentioned polarization effects of the initial **17** have been also observed in the photodecomposition of **17**.

TABLE 9. Rate constants of reactions of Me_2Ge: with CCl_4 at an ambient temperature

Reaction	Precursor/Solvent	k (mol^{-1} s^{-1})	Reference
	17/C_7H_{16}	1.2×10^7	65
16 + CCl_4	$PhMe_2GeSiMe_3$/c-C_6H_{12}	3.2×10^8	64
	$(Me_2Ge)_6$/c-C_6H_{12}	4.9×10^8	63

TABLE 10. ^{1}H CIDNP effects observed in the photolysis and thermolysis of **17** in the presence of halogenated traps

Reaction mixture/solvent	Reaction products	Chemical shift δ (ppm)	CIDNP effects[a]
17 + CCl_4/MePh (1:3) (thermolysis at 80 °C)	$ClMe_2GeCCl_3$	0.80	A
	Me_2GeCl_2	0.90	E
17 [10^{-2} M] + CCl_4/C_6D_6 (1:1) (photolysis at 20 °C	$ClMe_2GeCCl_3$	0.98[b]	A
	Me_2GeCl_2	1.00[b]	E
	19 (>$GeMe_2$)	0.86, 0.88	—
17 [10^{-3} M] + CCl_4/C_6D_6 (1:3) (photolysis at 20 °C)	$Me_2ClGeCCl_3$	0.68[c]	E
	Me_2GeCl_2	0.74[c]	A
	$CHCl_3$	6.50	E
17 + Me_3SnCl/C_6D_6 (photolysis at 20 °C)	$Cl\underline{Me}_2GeSnMe_3$		A
	$ClMe_2GeSn\underline{Me}_3$		A
17 + $PhCH_2Br$/PhCl (thermolysis at 80 °C)	$Br\underline{Me}_2GeCH_2Ph$	0.55	E
	$BrMe_2Ge\underline{CH}_2Ph$	2.57	E
	Me_2GeBr_2	1.06	A
	$Ph\underline{Me}$	2.22	—
	$(Ph\underline{CH}_2)_2$	2.84	—
17+ $PhCH_2Br$/c-C_6D_{12} (photolysis at 20 °C)	Me_2GeBr_2	0.95	A
	$Ph\underline{Me}$	2.30	—
	$(Ph\underline{H}_2)_2$	2.80	—

[a] E, emission; A, absorption.

[b] NMR spectra after the photolysis show a single line at $\delta = 1.00$ ppm; the addition of **25** to the reaction mixture allows us to assign the signal to this compound.

[c] In accordance with the literature data, the sole NMR signal at $\delta = 0.72$ ppm observed after the reaction corresponds to the main reaction product **25**.

CIDNP effects of the main reaction products **24** and **25** suggest that the first stage of the interaction of **16** with the trapping agent (CCl_4) is the abstraction of a halogen atom. The resulting radical pair of $^{\bullet}GeMe_2Cl$ and $^{\bullet}CCl_3$ radicals recombines to give the product **24** of insertion of **16** into the C−Cl bond. The escape of the radicals into the bulk followed by the abstraction of a second halogen atom leads to **25**. The analysis of CIDNP effects observed in the presence of excess amounts of CCl_4 (Table 10) shows that the initial radical pair has triplet multiplicity. Thus, **16** generated in the photolysis of **17** enters the reaction with CCl_4 in the excited triplet state (Scheme 12).

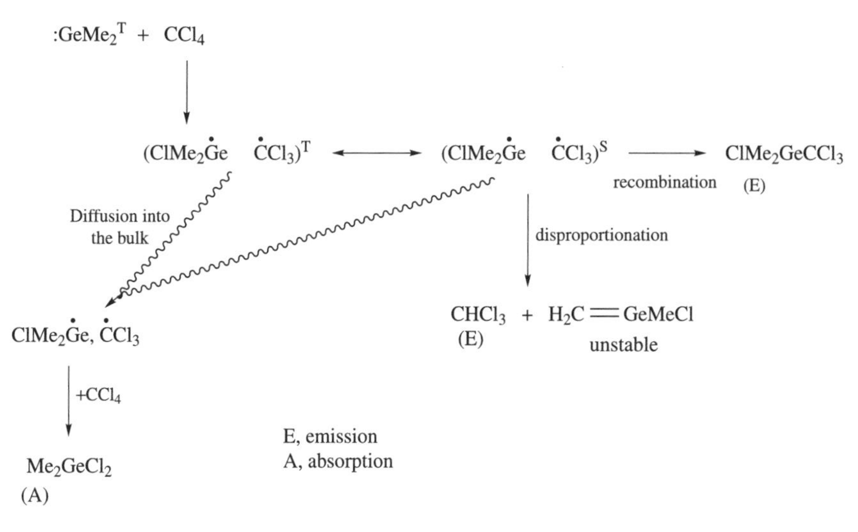

SCHEME 12

An interesting feature of these reactions observed only in CIDNP experiments[77] is the formation of chloroform (Table 10). It is reasonable to assume that a methyl group of the $^{\bullet}GeMe_2Cl$ radical is the source of the polarized proton of $CHCl_3$, and chloroform is the second cage product of the disproportionation of the singlet radical pair (Scheme 12). It is seen from Table 10 that the increase in concentration of the initial **17** results in alteration of the CIDNP signs of the 'in-cage' and 'escape' products. This effect is most likely due to the reactions of both singlet and triplet states of **16**. It could be also explained by the competitive reactions of **16** with CCl_4 and the initial **17**. As already mentioned, this reaction yields 7,8-digermabicyclo[2.2.2]octadiene **19**.

3. Reactions with chlorotrimethylstannane

According to the literature data[65], the rate constant of the reaction of **16** with Me_3SnCl is high enough ($k = 3.5 \times 10^8$ mol^{-1} s^{-1}) to expect that this scavenger will be also capable of trapping triplet germylene **16**. Indeed, the polarization effects (Table 10) observed in the reaction of photogenerated **16** with chlorotrimethylstannane[68] in the absorption of both methyl groups at Sn and Ge atoms of the insertion product $Me_3SnGeMe_2Cl$ could not be ascribed only to the reaction of singlet **16**, since in this case the Me_3Sn protons would demonstrate emission. On the other hand, if the reaction of triplet **16** would prevail, one should expect to observe opposite signs of the $ClMe_2Ge$ and Me_3Sn protons of the insertion product (emission and absorption, respectively). Taking into account the sufficiently high rate constant of the reaction of **16** with Me_3SnCl, one might suggest that chlorotrimethylstannane reacted with both spin states of **16**, and that the observed CIDNP is essentially the superposition of the polarization effects formed in the triplet and singlet radical pairs. This hypothesis could be corroborated or discounted only by analysis of the dependence of the efficiency of the CIDNP of $Me_3SnGeMe_2Cl$ on the Me_3SnCl concentration. However, the literature lacks this kind of data.

4. Reaction with benzyl bromide

CIDNP techniques have been successfully applied to investigations of the detailed mechanism of the reaction of **16** with benzyl bromide[67,74]. The yield of the end products depends on the mode of generation of **16**. In the case of thermal reaction, the reaction mainly results in the insertion product, $PhCH_2GeMe_2Br$, together with amounts of Me_2GeBr_2 (5%), PhMe (5%) and $(PhCH_2)_2$ (15%)[67]. The analysis of ^1H CIDNP effects of $PhCH_2GeMe_2Br$ and Me_2GeBr_2 (Table 10) allows one to propose the mechanism involving the reactions of ground singlet state of **16** shown in Scheme 13.

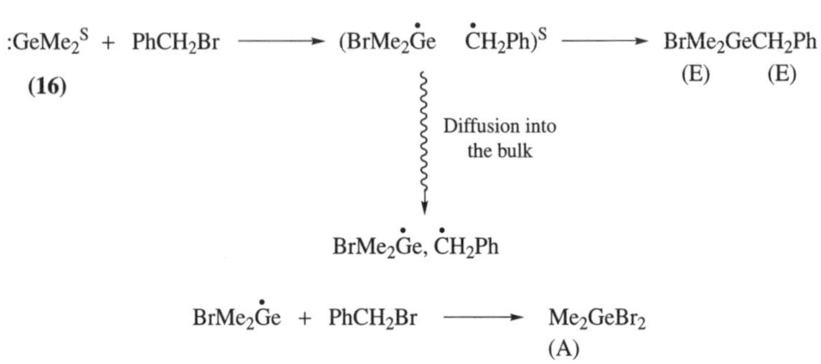

$$:GeMe_2{}^S + PhCH_2Br \longrightarrow (BrMe_2\dot{Ge}\ \ \dot{C}H_2Ph)^S \longrightarrow \underset{(E)\quad\ (E)}{BrMe_2GeCH_2Ph}$$

(16)

Diffusion into the bulk

$$BrMe_2\dot{Ge},\ \dot{C}H_2Ph$$

$$BrMe_2\dot{Ge} + PhCH_2Br \longrightarrow \underset{(A)}{Me_2GeBr_2}$$

SCHEME 13

However, in the reaction involving photogenerated **16**, only one polarized product (Me_2GeBr_2, Table 10) is observed. In accordance with Scheme 13 it is formed in the bulk and demonstrates polarization effects identical to those detected in the thermolysis. It could be suggested that in both cases the reaction follows the same mechanism, and the 'in-cage' product $PhCH_2GeMe_2Br$ appears to be unstable under the UV irradiation. Analogous singlet radical pairs are also shown to form in the reactions of singlet **16** with bromotrichloromethane, benzyl iodide and chlorodiphenylmethane[67].

C. Generation of a Digermene from a 7,8-Digermabicyclo[2.2.2]octadiene

Bicyclic molecules with a Ge−Ge bond are considered[3,4,72] to be potential precursors of another family of active short-lived germanium derivatives, the digermenes. Photochemical decomposition of 1,4-diphenyl-7,7′,8,8′-tetramethyl-2,3-benzo-7,8-digermabicyclo[2.2.2]octadiene (**26**) results in 1,4-diphenylnaphthalene (**27**) and supposedly in tetramethyldigermene[78] (**28**) identified on the basis of the products derived by its reaction with trapping agents (equation 26). The existing views of the reaction mechanisms of the decomposition of such bicyclic compounds are based on analogies with the schemes proposed for 7-heteronorbornadienes.

$$Me_2Ge{=}GeMe_2 \quad (28) \qquad (26)$$

Me Me Me Me Ge Ge Ph Ph **(26)** $\xrightarrow{h\nu}$ $Me_2Ge{=}GeMe_2$ **(28)** + Ph Ph **(27)** (26)

Similar to the case of 7-heteronorbornadienes, the application of the ^1H CIDNP method has allowed us to identify the elementary stages of the photolytic decomposition of **26**[79]. In addition to **27**, the main reaction products also include oligogermanes and the product of photorearrangement of the initial **26**, namely 6,6′,7,7′-tetramethyl-2,5-diphenyl-3,4-benzo-6,7-digermatricyclo[3.3.0.0]octane[80]. Table 11 and Figure 17 show the polarization effects of methyl and aromatic 5,6-protons of the initial **26** observed in the photolysis in C_6D_6; similar effects were also observed in *c*-C_6D_{12} and in CCl_4–C_6D_6 mixtures.

It is apparent that the polarization effects observed in this reaction (Table 11) are similar to those detected in the photolysis of 7-germanorbornadiene **17** (see Section IV.A). In this case, the opposite sign of the effects of the initial 7,8-digermabicyclooctadiene **26** and 1,4-diphenylnaphthalene **27** most likely indicate an S-T_0 mechanism of CIDNP formation. Thus, CIDNP data confirm the suggestion that photodecomposition of digermabicyclo derivatives follows a mechanism analogous to that proposed for 7-germanorbornadiene, i.e. via a cleavage of one endocyclic C–Ge bond with formation of the intermediate 1,6-biradical. The mechanism in Scheme 14 has been proposed on the basis of the analysis of the observed CIDNP effects[79].

Analysis of the observed polarization effects (Table 11 and Figure 17) shows that **26** is regenerated through the recombination of the singlet 1,6-biradical **29**, while the cleavage of the second C–Ge bond in triplet biradical **29** results in a triplet excited diphenylnaphthalene **27** and tetramethyldigermene **28**. Note that the formation of a triplet excited state of a diamagnetic molecule from the triplet biradical **29** follows from the requirement of retention of the total spin of the system. The formation of the excited triplet state of diphenylnaphthalene **27** has been confirmed in laser pulse photolysis experiments[80].

An intensive absorption band is detected immediately after the first laser pulse in the photolysis of a solution of **26**. Its characteristic maximum band ($\lambda_{max} = 420$ nm

TABLE 11. ^1H CIDNP effects observed in the photolysis of 7,8-digermabicyclooctadiene **(26)** in C_6D_6

Reaction product	Chemical shift δ (ppm)	CIDNP effects[a]
26 ($GeMe_2$)	0.35, 0.51 (both s)	E
26 (H(5),H(6))	6.15 (s)	E
27 (H(2),H(3))	7.43 (s)	A
27 (H(6-9))	7.20–8.20 (m)	—
27 (Ph)	7.25 (s, 10 H)	—

[a] E, emission; A, absorption.

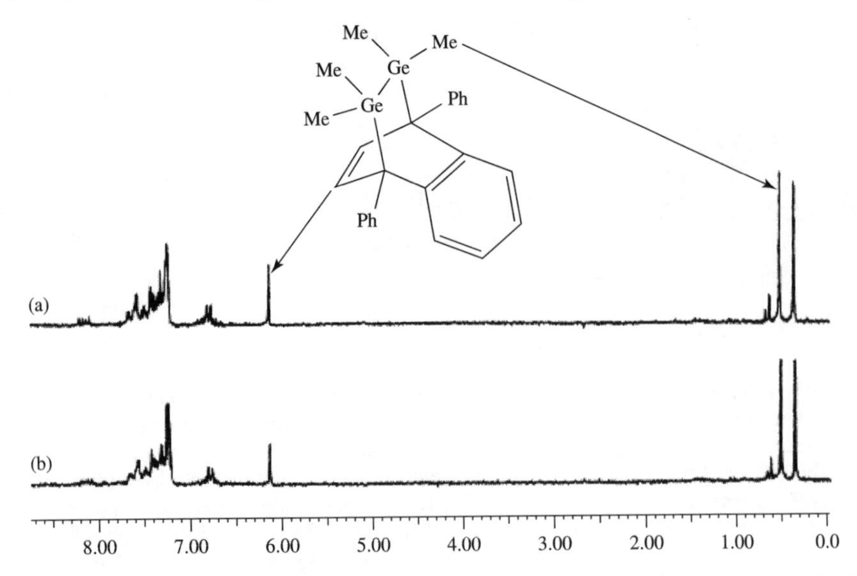

FIGURE 17. [1]H CIDNP effects detected in the photolysis of 7,8-digermabicyclooctadiene (**26**) in C_6D_6: (a) initial spectrum, (b) under UV irradiation

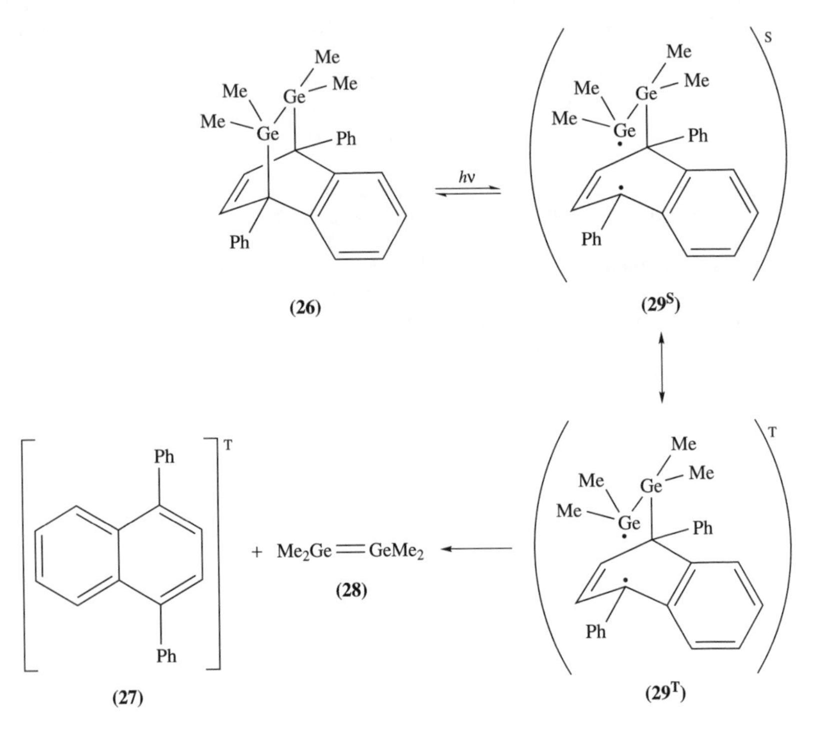

SCHEME 14

in hexane[79] and $\lambda_{max} = 430$ nm in cyclohexane[80]) is effectively quenched in the presence of oxygen ($k = 3.5 \times 10^9$ mol^{-1} s^{-1}). These observations are in good agreement with the literature data on the T-T absorption of 1,4-diphenylnaphthalene[81], and this has enabled the experimentally detected absorption band to be ascribed to the triplet excited **27** (Scheme 14).

At the same time, the use of an excimer laser as a light source ($\lambda = 308$ nm) precludes the detection of the absorption signal at $\lambda = 380$ nm which is observed with an excitation light at $\lambda = 266$ nm and ascribed to tetramethyldigermene[80]. This discrepancy could be also explained by the relatively low yield of **28** when $\lambda = 308$ nm is used for the excitation due to the significant difference in the molar extinctions of **26** in these two spectral regions.

Unfortunately, the above results do not allow one to determine unequivocally the mechanism presented in Scheme 14 as the sole possible pathway for the photodecomposition of **26**, or if there exists a possibility of a parallel reaction with sequential generation of two germylenes **17** (Scheme 15). In this case, one should expect a Ge−Ge bond cleavage in the 1,6-biradical **29** and the resulting 1,5-biradical might display two alternative pathways: (i) a recombination resulting in the corresponding 7-germanorbornadiene, or (ii) a cleavage of a second C−Ge bond generating one more germylene **16** and a triplet excited diphenylnaphthalene **27** (Scheme 15). If the latter possibility is realized, one might expect to

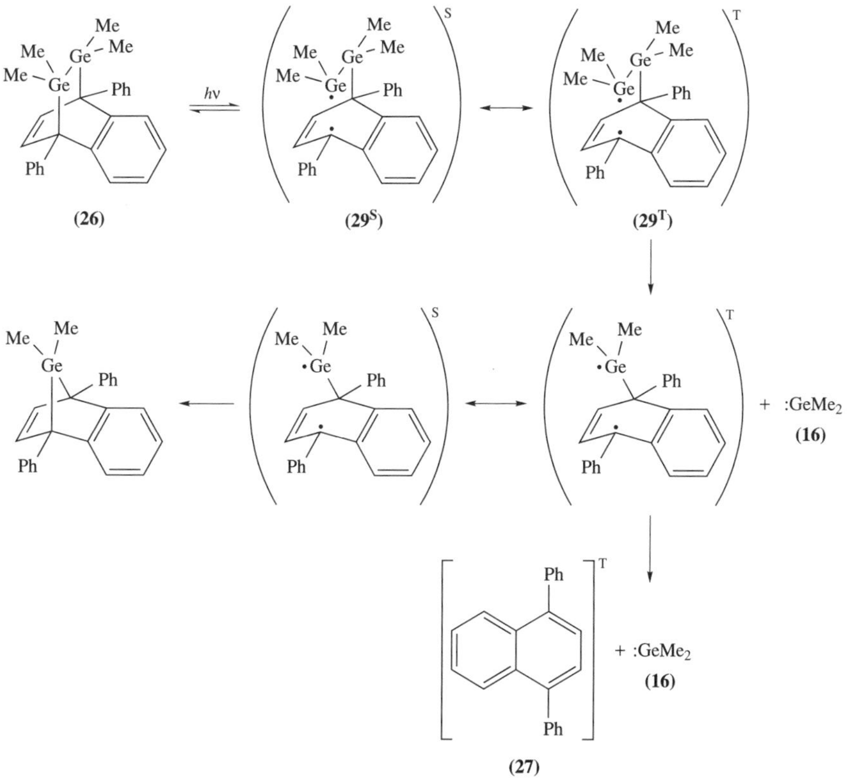

SCHEME 15

observe the polarized signals of the corresponding 7-germanorbornadiene. However, these have not been detected experimentally[79]. This does not necessarily mean that dimethylgermylene **16** is not formed in the process under study, since 7-germanorbornadiene most likely will not accumulate under the stationary UV irradiation. Moreover, the possibility of mutual transformations of germylene and digermene should not be excluded. Similar reactions are well-explored for the so-called Lappert's digermenes with $(Me_3Si)_2CH$ substituents. These compounds dissociate spontaneously to the corresponding germylenes[3,4] and, according to theoretical predictions, analogous reactions are also possible for the simplest digermenes.

In order to gain a deeper understanding of the reaction mechanism of the photolysis of **26**, it is reasonable to study this process in the presence of trapping agents capable of reacting with the expected intermediate, i.e. with germylene and/or digermene. Thus, comparison of the CIDNP effect detected in the photolysis of **17** and **26** in the presence of trapping agents will allow us to study the consecutive formation of digermene **28** and germylene **16** generated from **26** and to define the spin of the generated intermediates and their reactive states involved in the reactions with the trapping agents.

D. Reactions of the Intermediates Formed in the Photolysis of a 7,8-Digermabicyclooctadiene with Various Trapping Agents

Because of the known capability of digermenes to insert into multiple bonds[3,4], it is reasonable to start the present discussion with the reactions which occur in the photolysis of **26** in the presence of unsaturated trapping agents. Similarly to germylene **16**, digermene **28** could insert into the triple bond of thiacycloheptyne **21** with the formation of digermacyclobutene **30**.

Me Me
Me
Me
Ge
S
Ge
Me
Me
Me Me

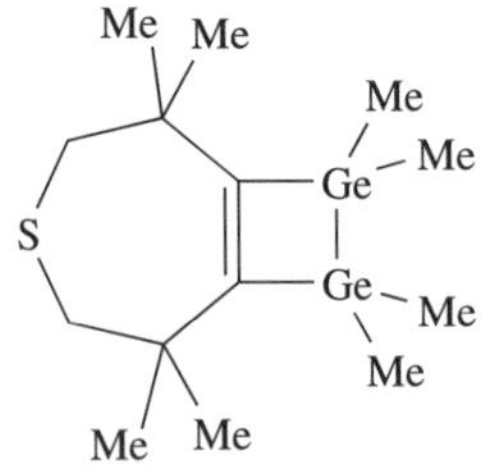

(30)

Table 12 and Figure 18 show the polarization effects detected in the photolysis of **26** in the presence of thiacycloheptyne **21** which could be compared with the effects observed under analogous conditions in the photodecomposition of 7-germanorbornadiene **17** (Table 8).

Two polarized lines ($\delta = 1.22$ and 0.48 ppm) strictly coincide with the signals earlier assigned to germacyclopropene **22** resulting from the insertion of germylene **16** into the triple bond of thiacycloheptyne **21** (Table 8 and Table 12). According to the literature[82], signals at $\delta = 1.13$ and 0.65 ppm are attributed to the digermacyclobutene **30**. Signals from **30** could be observed only at the initial stages of the photolysis and, after prolonged UV irradiation, germacyclopropene **22** remains the major reaction product (Figure 18). This points to the low stability of **30** under conditions of stationary photolysis. Photodecomposition of digermacyclobutene **30** might result in germylene **16** and germacyclopropene **22**. Another reaction resulting in the polarized germacyclopropene **22** involves the insertion of triplet excited germylene **16** into the triple bond of thiacycloheptyne **21** via the biradical intermediates (Scheme 11). A major question is to decide

TABLE 12. [1]H CIDNP effects observed in the photolysis of 7,8-digermabicyclooctadiene **26** in the presence of thiacycloheptyne **21**

Reaction product	Chemical shift δ (ppm)	CIDNP effects[a]
21 (Me)	1.15 (s, 12H)	E
22 (Me)	1.22 (s, 12H)	A
22 ($GeMe_2$)	0.48 (s, 6H)	A
30 (Me)	1.13 (s, 12H)	A
30 ($GeMe_2$)	0.65 (s, 12H)	—

[a]E, emission; A, absorption.

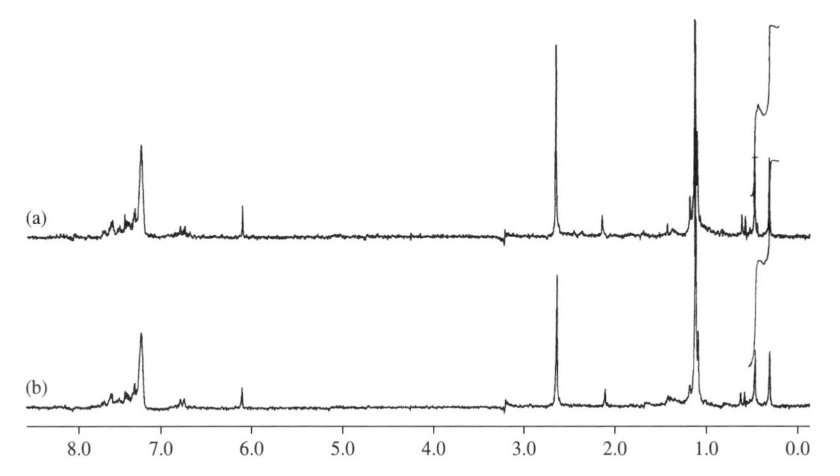

FIGURE 18. [1]H CIDNP effects observed in the photolysis of 7,8-digermabicyclooctadiene **26** in the presence of thiacycloheptyne **21** in C_6D_6: (a) under UV irradiation, (b) after the photolysis

between two possible sources of triplet germylene in the system under study: the decomposition of digermacyclobutene **30** or of tetramethyldigermene **28**? The possibility of forming both triplet and singlet germylene in the decomposition of $Me_2Ge{=}GeMe_2$ is determined by certain features of the double bond in digermene which is characterized by weak side overlap of the p-electron orbitals of germanium atoms (see equation 27)[3,4].

$$\underset{\textbf{(28)}}{Me_2Ge{=}GeMe_2} \longrightarrow \underset{(\textbf{16}^S)}{Me_2Ge{:}^S} + \underset{(\textbf{16}^T)}{{:}GeMe_2T} \qquad (27)$$

In any case, the identical polarization effects of germacyclopropene **22** observed in the photolysis of **17** and **26** in the presence of thiacycloheptyne **21** confirm the involvement of triplet excited germylene **16**. CIDNP effects of the methyl groups of digermacyclobutene **30** suggest that its formation follows a mechanism analogous to that presented in Scheme 11, though involving the corresponding 1,3-biradical. The comparison with the polarization effects detected in the presence of halogenated trapping agents could be helpful for a better understanding of the possibility of generating germylene **16** from digermene **28**.

It has been found that 7,8-digermabicyclooctadiene **26** reacts spontaneously with CCl_4 and benzyl bromide at ambient temperatures. However, these reactions are in fact much slower than the photolysis and do not preclude the observation of CIDNP. Moreover, the presence of these halogenated trapping agents does not hamper the detection of CIDNP effects formed in the photolysis of the initial **26**. Thus, the trapping agents do not enter the reaction with the intermediate 1,6-biradical **29** which is responsible for the CIDNP formation (Scheme 14). CIDNP effects detected in these reactions are summarized in Table 13 and the experimental spectra are shown in Figure 19.

Photoinitiated reaction of 7,8-digermabicyclooctadiene **26** with carbon tetrachloride yields diphenylnaphthalene **27** (95%), hexachloroethane C_2Cl_6 (37%), dichlorotetramethyldigermane $ClMe_2GeGeMe_2Cl$ (71%) and dichlorodimethylgermane Me_2GeCl_2 (27%)[80]. The formation of a significant amount of $ClMe_2GeGeMe_2Cl$ is explained by the

TABLE 13. ^{1}H CIDNP effects detected in the photolysis of 7,8-digermabicyclooctadiene **26** in the presence of halogenated trapping agents

Reaction mixture/solvent	Reaction products	Chemical shift δ (ppm)	CIDNP effects[a]
	$Me_2ClGeCCl_3$	0.67	E (weak)
26 + CCl_4/C_6D_6 (1:3)	Me_2GeCl_2	0.70	A (weak)
	$CHCl_3$	6.45	E
	Me_2GeBr_2	1.10	E
		4.30	E
26 + $PhCH_2Br/C_6D_6$	[intermediates]	3.50, (dd $J = 7.0$ Hz; $J = 4.0$ Hz)	A
		5.20, (t, $J = 7.0$ Hz)	E

[a]E, emission; A, absorption.

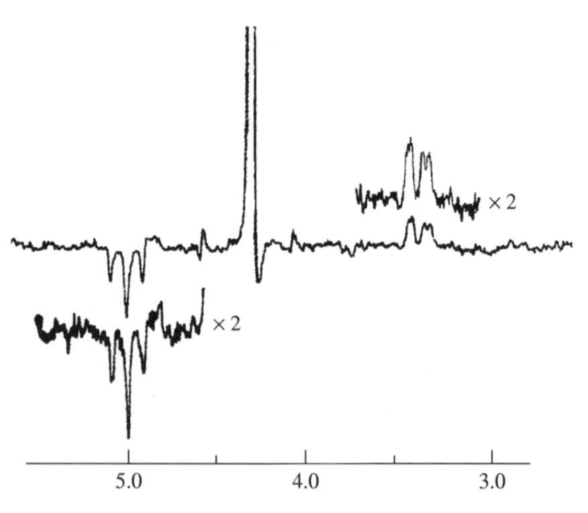

FIGURE 19. ^{1}H CIDNP effects observed in the photolysis of 7,8-digermabicyclooctadiene **26** in C_6D_6 in the presence of benzyl bromide

reaction of digermene **28** with CCl_4, resulting in a sequential abstraction of two chlorine atoms[80]. However, it has been shown that this product might also result from the insertion of germylene **16** into the Ge−Cl bond of dichlorodimethylgermane[67]. On the other hand, the presence of monogermanium derivatives as well as polarized $CHCl_3$ points to the involvement of dimethylgermylene **16** (cf Tables 10 and 13). Note that chloroform demonstrates noticeable CIDNP effects with a sign identical to that observed in the photolysis of 7-germanorbornadiene **17** against the background of a rather weak polarization of the Me protons of $ClMe_2GeCCl_3$ and Me_2GeCl_2.

Thus, the formation of polarized chloroform and monogermanium compounds in the photolysis of 7,8-digermabicyclooctadiene **26** in the presence of CCl_4 shows that the generated germylene **16** enters the reaction with CCl_4 in a triplet excited state (Scheme 12). However, judging from the weak polarizations of $ClMe_2GeCCl_3$ and Me_2GeCl_2 one should not exclude also reactions of the ground singlet state of **16**. The possibility of generating both singlet and triplet excited germylenes **16** has already been mentioned (equation 27). In fact, the CIDNP technique allows one to detect only the reaction of **16** with CCl_4 and no signs of interaction between the digermene **28** and carbon tetrachloride were observed despite the known fact that the rate constant of this process measured in laser pulse photolysis experiments[80] is rather high ($k = 1.2 \times 10^7$ mol^{-1} s^{-1}).

The reverse situation has been observed when the photolysis of 7,8-digermabicyclooctadiene **26** is performed in the presence of benzyl bromide. These observations are markedly different from the CIDNP pattern detected in the photolysis of 7-germanorbornadiene **17** with $PhCH_2Br$ (Tables 10 and 13). Dibromodimethylgermane is a characteristic product of the photolysis of both **17** and **26** in the presence of $PhCH_2Br$ and, similarly to the reactions with CCl_4, the sign of the polarization effects of Me_2GeBr_2 (Table 13) suggests the involvement of triplet excited germylene **16**. However, the major distinction between these two processes is the appearance of polarized signals close to the methylene protons of $PhCH_2Br$ ($\delta = 4.30$ ppm) and in the range characteristic for the double bond protons (Figure 19). The chemical shift of the polarized protons suggests that the reactive intermediates generated by the photolysis of 7,8-digermabicyclooctadiene **26** do not enter the C−Br bond cleavage reaction of benzyl bromide which is characteristic for germylenes, but rather attack the benzene ring, leading to the loss of aromaticity. It is known[3,4] that digermenes could insert into the double bonds of aromatic molecules (anthracene, benzene derivatives, heterocycles, etc.) to form the corresponding bicyclic compounds, but these reactions have not been detected for germylenes[60]. Therefore, it is quite reasonable to assume that the observed CIDNP effects are generated in the reaction of digermene **28** and $PhCH_2Br$. However, the resulting products are unstable, and it was impossible to accumulate detectable amounts under UV irradiation. The mass spectra of the reaction mixture demonstrate only the presence of Me_2GeBr_2, $(PhCH_2)_2$ and oligogermanes[82]. Unfortunately, only polarized signals lead to certain information on the nature of these intermediate products (Figure 19) and, at present, it is impossible to propose a reliable structure for these species. The absence of analogous polarization effects in the photolysis of **17** where **28** could be generated only through the dimerization of germylene **16** definitely supports the hypothesis that in the photolysis of **26**, germylene **16** is formed through the decomposition of digermene **28** (equation 27).

VII. CONCLUSION

The achievements of mechanistic studies employing spin chemistry methods convincingly demonstrate the potential of these techniques in revealing the detailed reaction mechanisms of a number of homolytic processes involving organic Ge, Sn and Pb compounds. The role of short-lived paramagnetic (and sometimes diamagnetic) intermediates, such as free radicals, biradicals, germylenes and digermenes, has been conclusively elucidated. The application of spin chemistry techniques, in particular, in combination with laser pulse photolysis methods allows us to attain much deeper insight into the organometallic chemistry of highly reactive species. From our viewpoint, progress in the investigation of the elementary mechanisms could be achieved only when applying a combination of physical and chemical analytical techniques. This is particularly true since the processes described in the present chapter do not compare with the rapidly growing number of newly synthesized organometallic compounds and with reactions in which they participate. If our efforts to attract the attention of the organometallic community to the spectacular potential of spin chemistry techniques turn out to be successful, we will consider our goal to have been achieved.

VIII. REFERENCES

1. R. Benn, *Rev. Chem. Intermediates*, **3**, 45 (1979).
2. I. V. Khudyakov, Yu. A. Serebrennikov and N. J. Turro, *Chem. Rev.*, **93**, 537 (1993).
3. J. Barrau, J. Escudie and J. Satgé, *Chem. Rev.*, **90**, 283 (1990).
4. T. Tsumuraya, S. A. Batcheler and S. Masamune, *Angew. Chem., Int. Ed. Engl.*, **30**, 902 (1991).
5. R. Kaptein, Ph.D. Thesis, Leiden, 1971.
6. B. Brocklehurst, *Nature,* **221**, 921 (1969).
7. Yu. N. Molin, R. Z. Sagdeev and K. M. Salikhov, *Rev. Soviet Authors, Chem. Ser.*, **1**, 1 (1979).
8. H. R. Ward and R. G. Lawler, *J. Am. Chem. Soc.*, **89**, 5518 (1967).
9. R. Z. Sagdeev, Yu. N. Molin, K. M. Salikhov, T. V. Leshina, M. A. Kamkha and S. M. Shein, *Org. Magn. Reson.*, **5**, 603 (1973).
10. T. V. Leshina, V. I. Maryasova, R. Z. Sagdeev, O. I. Margorskaya, D. A. Bravo-Zhivotovskii, O. A. Kruglaya and N. S. Vyazankin, *React. Kinet. Catal. Lett.*, **12**, 491 (1979).
11. R. Kaptein, *Chem. Commun.*, 732 (1971).
12. H. Fischer and K.-H. Hellwege (Eds.), *Landolt-Börnstein New Series. Numerical Data and Functional Relationship in Science and Technology. Magnetic Properties of Free Radicals*, Group II, Vol. 9, Parts A–C, Springer-Verlag, Berlin, 1977–1980.
13. K. M. Salikhov, Yu. N. Molin, R. Z. Sagdeev and A. L. Buchachenko, *Spin Polarization and Magnetic Effects in Radical Reactions*, Akadémiai Kiadó, Budapest, 1984.
14. J. K. Kochi, *Free Radicals*, Vol. 1, Wiley, New York, 1973.
15. V. I. Korsunsky, M. B. Taraban, T. V. Leshina, O. I. Margorskaya and N. S. Vyazankin, *J. Organomet. Chem.*, **215**, 179 (1981).
16. W. H. Isley, T. F. Schaaf, M. D. Glick and J. P. Oliver, *J. Am. Chem. Soc.*, **102**, 3769 (1980).
17. M. B. Taraban, T. V. Leshina, R. Z. Sagdeev, K. M. Salikhiv, Yu. N. Molin, O. I. Margorskaya and N. S. Vyazankin, *J. Organomet. Chem.*, **256**, 31 (1983).
18. K. Schulten and P. G. Wolynes, *J. Chem. Phys.*, **68**, 3292 (1978).
19. N. V. Shokhirev, E. C. Korolenko, M. B. Taraban and T. V. Leshina, *Chem. Phys.*, **154**, 237 (1991).
20. K. Mochida and N. Matsushige, *J. Organomet. Chem.*, **229**, 1 (1982).
21. K. Mochida, M. Wakasa, Y. Sakaguchi and H. Hayashi, *Chem. Lett.*, 773 (1986).
22. K. Mochida, M. Wakasa, Y. Sakaguchi and H. Hayashi, *J. Am. Chem. Soc.*, **109**, 7942 (1987).
23. P. Rivière, A. Castel, D. Desor and C. Abdennadher, *J. Organomet. Chem.*, **443**, 51 (1993).
24. P. Rivière, A. Castel and F. Cosledan, *Phosphorus, Sulfur, Silicon Relat. Elem.*, **104**, 169 (1995).
25. M. Wakasa and H. Hayashi, *Chem. Phys. Lett.*, **340**, 493 (2001).
26. D. Bryce-Smith, *J. Chem. Soc.*, 1603 (1955).
27. D. E. Applequist and D. F. O'Brien, *J. Am. Chem. Soc.*, **85**, 743 (1963).
28. K. West and W. Glase, *J. Chem. Phys.*, **34**, 685 (1961).
29. D. P. Curran, U. Diederichsen and M. Palovich, *J. Am. Chem. Soc.*, **119**, 4797 (1997).
30. A. G. Brook, P. J. Dillon and R. Pearce, *Can. J. Chem.*, **49**, 133 (1971).
31. M. Kosugi, H. Naka, H. Sano and T. Migita, *Bull. Chem. Soc. Jpn.*, **60**, 3462 (1987).
32. A. G. Brook, in *Adv. Organomet. Chem.*, Vol. 7 (Eds. F. G. A. Stone and R. West), Academic Press, New York, 1968, pp. 95–155.

33. A. Alberti, G. Seconi, G. F. Pedulli and A. Degl'Innocenti, *J. Organomet. Chem.*, **253**, 291 (1983).
34. A. Alberti, A. Degl'Innocenti, L. Grossi and L. Lunazzi, *J. Org. Chem.*, **49**, 4613 (1984).
35. A. Alberti, A. Degl'Innocenti, G. F. Pedulli and A. Ricci, *J. Am. Chem. Soc.*, **107**, 2316 (1985).
36. K. U. Ingold, J. Lusztyk and J. C. Scaiano, *J. Am. Chem. Soc.*, **106**, 343 (1984).
37. M. B. Taraban, V. I. Maryasova, T. V. Leshina, L. I. Rybin, D. V. Gendin and N. S. Vyazankin, *J. Organomet. Chem.*, **326**, 347 (1987).
38. M. B. Taraban, V. I. Maryasova, T. V. Leshina and D. Pfeifer, *Main Group Metal Chem.*, **14**, 33 (1991).
39. K. Mochida, K. Ichikawa, S. Okui, Y. Sakaguchi and H. Hayashi, *Chem. Lett.,* 1433 (1985).
40. H. Hayashi and K. Mochida, *Chem. Phys. Lett.*, **101**, 307 (1983).
41. G. J. D. Peddle, *J. Organomet. Chem.*, **14**, 139 (1968).
42. M. Kosugi, H. Naka, H. Sano and T. Migita, *Bull. Chem. Soc. Jpn.*, **60**, 3462 (1987).
43. A. I. Kruppa, M. B. Taraban, S. A. Svarovsky and T. V. Leshina, *J. Chem. Soc., Perkin Trans. 2*, 2151 (1996).
44. J. Lipsher and H. Fisher, *J. Phys. Chem.*, **88**, 2555 (1984).
45. B. Wrackmeyer, *Annu. Rep. NMR Spectrosc.*, **16**, 73 (1985).
46. M. G. Voronkov, V. I. Rakhlin, S. Kh. Khangazheev, R. G. Mirskov and A. S. Dneprovskii, *Dokl. Chem.*, **259**, 6386 (1981).
47. T. V. Leshina, R. Z. Sagdeev, N. E. Polyakov, M. B. Taraban, V. I. Valyaev, V. I. Rakhlin, R. G. Mirskov, S. Kh. Khangazheev and M. G. Voronkov, *J. Organomet. Chem.,* **259**, 295 (1983).
48. T. V. Leshina, V. I. Valyaev, M. B. Taraban, V. I. Maryasova, V. I. Rakhlin, S. Kh. Khangazheev, R. G. Mirskov and M. G. Voronkov, *J. Organomet. Chem.*, **299**, 271 (1986).
49. T. V. Leshina, R. Z. Sagdev, N. E. Polyakov, A. V. Yurkovskaya, A. A. Obynochny and V. I. Maryasova, *Chem. Phys. Lett.*, **96**, 108 (1983).
50. S. Forsén and R. Hoffman, *J. Chem. Phys.*, **39**, 2892 (1963).
51. V. I. Rakhlin, R. G. Mirskov and M. G. Voronkov, *Russ. J. Org. Chem.*, **32**, 6771 (1996).
52. B. P. Roberts and C. Wilson, *J. Chem. Soc., Chem. Commun.*, **17**, 752 (1978).
53. J. P. Light, M. Ridenour, L. Beard and J. W. Hershberger, *J. Organomet. Chem.*, **326**, 17 (1987).
54. A. Takuwa, T. Kanaue, K. Yamashita and Y. Nishigaichi, *J. Chem. Soc., Perkin Trans. 1*, 1309 (1998).
55. A. Takuwa, T. Kanaue, Y. Nishigaichi and H. Iwamoto, *Tetrahedron Lett.*, **36**, 575 (1995).
56. K. Mochida and I. Miyagawa, *Bull. Chem. Soc. Jpn.*, **56**, 1875 (1983).
57. A. G. Davies, J. A.-A. Hawari, Ch. Gaffney and P. G. Harrison, *J. Chem. Soc., Perkin Trans. 2*, 631 (1982).
58. A. Standt and H. Dreeskamp, *J. Organomet. Chem.*, **322**, 49 (1987).
59. K. Mochida, H. Kikkawa and Y. Nakadaira, *Bull. Chem. Soc. Jpn.*, **64**, 2772 (1991).
60. W. P. Neumann, *Chem. Rev.*, **91**, 311 (1991).
61. Sh. Tomoda, M. Shimoda, Y. Takeuchi, Y. Kajii, K. Obi, I. Tanaka and K. Honda, *J. Chem. Soc., Chem. Commun.*, 910 (1988).
62. K. Mochida and S. Tokura, *Bull. Chem. Soc. Jpn.*, **65**, 1642 (1992).
63. K. Mochida, N. Kanno, R. Kato, M. Kotani, S. Yamauchi, M. Wakasa and H. Hayashi, *J. Organomet. Chem.*, **415**, 191 (1991).
64. K. L. Bobbitt, V. M. Maloney and P. P. Gaspar, *Organometallics,* **10**, 2772 (1991).
65. O. M. Nefedov, M. P. Egorov, A. I. Ioffe, L. G. Menchikov, P. S. Zuev, V. I. Minkin, B. Yu. Simkin and M. N. Glukhovtsev, *Pure Appl. Chem.*, **64**, 265 (1992).
66. R. Becerra, S. E. Bogdanov, M. P. Egorov, V. Ya. Lee, O. M. Nefedov and R. Walsh, *Chem. Phys. Lett.*, **250**, 111 (1996).
67. J. Koecher, M. Lehnig and W. P. Neumann, *Organometallics,* **7**, 1201 (1988).
68. S. P. Kolesnikov, M. P. Egorov, A. M. Galminas, O. M. Nefedov, T. V. Leshina, M. B. Taraban, A. I. Kruppa and V. I. Maryasova, *J. Organomet. Chem.*, **391**, C1 (1990).
69. A. J. Shusterman, B. E. Landrum and R. L. Miller, *Organometallics,* **8**, 1851 (1989).
70. W. Ando, T. Tsumuraya and A. Sekiguchi, *Chem. Lett.*, 317 (1987).
71. W. Ando, H. Itoh and T. Tsumuraya, *Organometallics,* **8**, 2759 (1989).
72. P. Bleckman, R. Mincwitz, W. P. Neumann, M. Schriever, M. Thibud and B. Watta, *Tetrahedron Lett.*, **25**, 2467 (1984).

73. M. B. Taraban, V. F. Plyusnin, O. S. Volkova, V. P. Grivin, T. V. Leshina, V. Ya. Lee, V. I. Faustov, M. P. Egorov and O. M. Nefedov, *J. Phys. Chem.*, **99**, 14719 (1995).
74. J. Koecher and M. Lehnig, *Organometallics,* **3**, 937 (1984).
75. M. B. Taraban, A. I. Kruppa, O. S. Volkova, I. V. Ovcharenko, R. N. Musin and T. V. Leshina, *J. Phys. Chem. A,* **104**, 1811 (2000).
76. M. P. Egorov, M. B. Ezhova, S. P. Kolesnikov, O. M. Nefedov, M. B. Taraban, A. I. Kruppa and T. V. Leshina, *Mendeleev Commun.*, 143 (1991).
77. M. B. Taraban, V. I. Maryasova, T. V. Leshina, V. Ya. Lee, M. P. Egorov and O. M. Nefedov, *Abstracts of II International Conference on Modern Trends in Chemical Kinetics and Catalysis* (Ed. V. N. Parmon), Novosibirsk, 1995, **III**, p. 577.
78. H. Sakurai, Y. Nakadaira and H. Tobita, *Chem. Lett.*, 1855 (1982).
79. M. B. Taraban, O. S. Volkova, V. F. Plyusnin, Yu. V. Ivanov, T. V. Leshina, M. P. Egorov, O. M. Nefedov, T. Kayamori and K. Mochida, *J. Organomet. Chem.*, **601**, 324 (2000).
80. K. Mochida, T. Kayamori, M. Wakasa, H. Hayashi and M. P. Egorov, *Organometallics*, **19**, 3379 (2000).
81. I. Carmichel and G. L. Hug, *J. Phys. Chem. Ref. Data,* **15**, 1 (1986).
82. O. M. Nefedov, M. P. Egorov, A. M. Galminas, S. P. Kolesnikov, A. Krebs and J. Berndt, *J. Organomet. Chem.*, **301**, C21 (1986).

CHAPTER **10**

Free and complexed R_3M^+ cations (M = Ge, Sn, Pb)

ILYA ZHAROV[†] and JOSEF MICHL

Department of Chemistry and Biochemistry, University of Colorado, Boulder, CO 80309-0215, USA
Fax: (303) 492 0799; e-mail: michl@eefus.colorado.edu

I. INTRODUCTION

The heavier congeners of carbenium ions, R_3E^+ (E = Si, Ge, Sn, Pb), have been of long-standing interest. Their preparation and structure in condensed media have been

[†] Present address: Beckman Institute for Advanced Science and Technology, University of Illinois at Urbana-Champaign, 405 North Mathews Ave, Urbana, IL 61801.

The chemistry of organic germanium, tin and lead compounds — Vol. 2
Edited by Z. Rappoport

a subject of extensive studies and controversy for decades because of a fundamental interest in understanding similarities and differences between carbon and heavier group 14 elements. The silyl cations, R_3Si^+, have received the most attention[1–6] and their heavier analogues relatively little. Several reasons could be suggested to explain this. For example, increased metallic character of these elements places their compounds into an inorganic category, and their toxicity makes them less attractive to deal with. The latest book on the chemistry of organic germanium, tin and lead compounds[7] does not contain a chapter discussing their cations. We are unaware of any other full review on the subject. It is only mentioned briefly when the related cyclotrigermenium cation is discussed in reviews by Belzner[8] and Schleyer[9]. Therefore, we tried to include older literature relevant to the subject.

A few words about the use of the word 'cation' and the 'R_3M^+' notation in our review are in order. 'Cation' means a species with a positive charge of +1 and 'R_3M^+' might be understood as a three-coordinate metal cation free of any significant interactions. However, with a couple of possible exceptions, in the case of group 14 elements such free species are only known in the gas phase, and their structures have not been determined experimentally. In bulk, these ions generally are complexed to one or more solvent molecules or a counterion. In solution, group 14 cations would therefore be more accurately represented by '$[R_3ML_n]^+$', where n is typically 1 or 2 and M is only partially positively charged. In our review we concentrated on cations in which the three R groups are covalently bound to M and interactions with L are weak. Throughout the text we used the word 'cation' and the 'R_3M^+' notation for such complexed cations, both for simplicity and because in many cases neither n nor the exact nature of L (a solvent molecule or a counterion) are known. The existence of such complexation was not always recognized explicitly in the earlier literature.

II. PREPARATION

A. Gas Phase

The formation of triorganogermyl and triorganostannyl cations in the gas phase was first reported half a century ago[10–13]. There are several reviews of older literature discussing the preparation of gaseous group 14 cations[14–16]. In a comprehensive study Lappert and coworkers[17] studied the mass spectra and measured the ionization potentials of the species $Me_3M{-}M'Me_3$ and Me_4M (M, M′ = C, Si, Ge, Sn, Pb) and the appearance potentials of the Me_3M^+ cations (Table 1). They found that in all spectra the Me_3M^+ cations were the most abundant and that their abundance decreased with increasing atomic weight of M, while that of the Me_2M^+, MeM^+ and M^+ cations increased. Both the ionization potentials and the appearance potentials decreased with increasing atomic weight of M as would be expected for weaker M−M′ and M−C bonds. Using these measurements the authors also calculated enthalpies of formation for the Me_3M^+ cations (Table 2) which showed that the formation of Me_3Si^+ and Me_3Ge^+ cations is thermodynamically more favorable than that of the Me_3C^+ cation. The appearance potentials for Me_3Sn^+ and Me_2RSn^+ cations produced by electron impact from Me_3SnR (R = Me, Et, n-Pr, i-Pr, n-Bu, i-Bu, t-Bu and Me_3Sn) were also measured[18] and the results are listed in Table 3. Based on these measurements it was concluded that the Me_3Sn^+ cation is energetically favored over Me_3C^+ unless unusually strong bonds to the Sn atom must be broken to form the former. There appeared to be no clear relationship between the length or branching of the alkyl substituent and appearance potentials.

TABLE 1. Ionization (IP) and appearance (AP) potentials for $Me_3M-M'Me_3$ and Me_3M-Me compounds[17]

Compound	IP (eV)	AP (eV)	
		Me_3M^+	$^+M'Me_3$
$Me_3C-GeMe_3$	8.98	10.19	9.91
$Me_3C-SnMe_3$	8.34	10.03	9.32
$Me_3C-PbMe_3$	7.99	9.45	8.67
$Me_3Si-SiMe_3$	8.35	10.22	—
$Me_3Ge-GeMe_3$	8.18	9.96	—
$Me_3Sn-SnMe_3$	8.02	9.51	—
$Me_3Pb-PbMe_3$	7.41	9.02	—
$Me_3Si-GeMe_3$	8.31	10.19	9.99
$Me_3Si-SnMe_3$	8.18	10.18	9.80
$Me_3Ge-SnMe_3$	8.20	10.01	9.85
Me_3Si-Me	9.85	10.53	—
Me_3Ge-Me	9.29	10.05	—
Me_3Sn-Me	8.76	9.58	—
Me_3Pb-Me	8.26	8.77	—

TABLE 2. Standard enthalpies of formation for Me_3M^+ cations[17]

Cation	ΔH_f^0
Me_3C^+	178.2
Me_3Si^+	158.56
Me_3Ge^+	164.99
Me_3Sn^+	184.25
Me_3Pb^+	200.07

TABLE 3. Ionization (IP) and appearance (AP) potentials for Me_3Sn-R compounds[18]

R	IP (eV)	AP (eV)	
		Me_3Sn^+	Me_2RSn^+
CH_3	8.76	9.72	—
C_2H_5	—	9.49	9.88
$H_2C=CH$	—	10.44	9.56
n-C_3H_7	8.54	9.50	9.59
i-C_3H_7	8.28	9.17	10.03
$H_2C=CHCH_2$	—	8.68	9.43
n-C_4H_9	—	9.80	9.67
s-C_4H_9	8.27	9.20	9.76
i-C_4H_9	8.34	9.79	9.62
t-C_4H_9	—	9.50	10.95
$SnMe_3$	8.08	9.85	8.17[a]

[a] $Me_3Sn-SnMe_2{}^+$.

B. Strongly Acidic Media

It was found early on that, unlike silyl or germyl cations, stannyl cations can be prepared in strong acids from both triorganotin halides and tetraorganotin compounds. For instance, Robinson and coworkers[19,20] studied solutions of tetramethyltin in sulfuric acid using cryoscopy and conductometry and concluded that 'Me_3Sn^+' HSO_4^- is present. These authors also showed that in the case of Ph_4Sn no organostannyl cation is formed. Cryoscopy and conductometry suggested that 'R_3Sn^+' cations are formed from various R_4Sn and R_3SnCl precursors (R = Et, *n*-Pr, Me, but not *n*-Bu) dissolved in sulfuric acid[21]. Cationic tin species are also formed from methyltin hydrides in fluorosulfuric acid[22]. A downfield ^{119}Sn NMR signal found in the reaction of Me_4Sn with $SbCl_5$ was attributed to the 'Me_3Sn^+' cation[23].

C. Aqueous Solutions

While germyl cations are not known in aqueous solutions, reports on the formation of stannyl cations appeared as early as 1923[24]. Since then, numerous investigations[25–27] established that ions of the type $R_3M(OH_2)_n{}^+$ (M = Sn, Pb) can be prepared in aqueous solutions. The early literature on this subject was reviewed in 1966 by Tobias[28]. Among earlier studies one should mention a series of publications by the group of Rabenstein[29,30] who studied plumbyl cation complexes in aqueous solutions by ^1H NMR.

Presently, this subject received a renewed interest in terms of environmental effects of these cations. Two recent comprehensive reports[31,32] describe the hydrolysis of trimethyltin compounds in aqueous and salt media at various temperatures and ionic strengths and provide an overview of the relevant literature.

D. Organic Solvents

Perhaps the most popular method of generating group 14 'R_3M^+' cationic species in organic media is the hydride abstraction reaction[33]. For instance, in one of the early works Lambert and Schilf[34] prepared germyl cations using hydride abstraction from R_3GeH (R = Me, Ph) by $Ph_3C^+ClO_4^-$ in sulfolane and dichloromethane. Tri-*n*-butylstannnyl cations with less coordinating anions were generated similarly by Lambert and Kuhlmann[35] in benzene (Scheme 1) and by Kira and coworkers[36] in dichloromethane (Scheme 2). Halogen (and hydride) abstraction was used in the preparation of acetonitrile complexed $R_3Sn^+SbF_6^-$ salts (R = *t*-Bu, *t*-$BuCH_2$, cyclohexyl) from the corresponding triorganotin bromides (or hydrides) and SbF_5[37].

$$R_3SnH + Ph_3C^+\,ClO_4^- \longrightarrow R_3Sn^+ClO_4^- + Ph_3CH$$

$$R_3SnCl + Ag^+\,ClO_4^- \longrightarrow R_3Sn^+ClO_4^- + AgCl$$

$$R_3SnH + B(C_6F_5)_3 \longrightarrow R_3Sn^+BH(C_6F_5)_3^-$$

SCHEME 1

$$Bu_3SnH + Ph_3C^+\,B(3,5\text{-}(CF_3)_2C_6H_3)_4^- \longrightarrow Bu_3Sn^+\,B(3,5\text{-}(CF_3)_2C_6H_3)_4^- + Ph_3CH$$

SCHEME 2

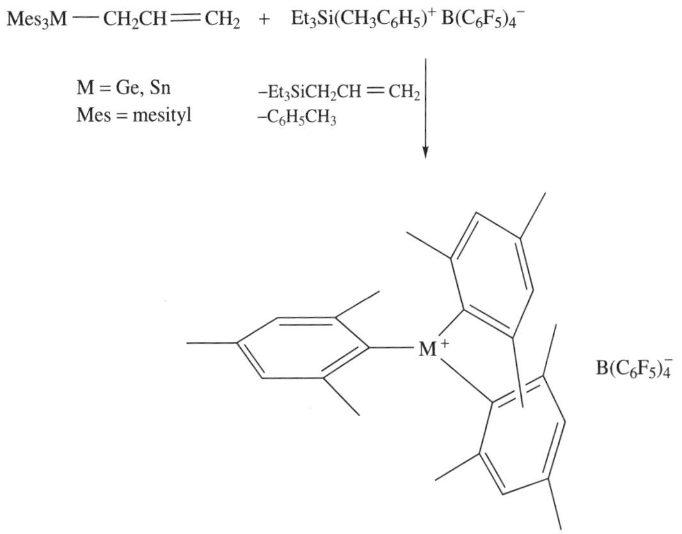

$$Mes_3M—CH_2CH{=}CH_2 + Et_3Si(CH_3C_6H_5)^+ B(C_6F_5)_4^- \xrightarrow[-C_6H_5CH_3]{-Et_3SiCH_2CH=CH_2} Mes_3M^+ \; B(C_6F_5)_4^-$$

M = Ge, Sn
Mes = mesityl

SCHEME 3

An alternative allyl leaving group approach was used by Lambert and coworkers[38] in the preparation of the sterically hindered trimesitylgermyl and trimesitylstannyl cations (Scheme 3), analogous to their earlier preparation of the trimesitylsilyl ion[39].

Several examples of oxidative generation of 'R_3Sn^+' cations have been reported. One-electron oxidation of Me_6Sn_2, $Me_3SnGeMe_3$ and $Me_3SnSiMe_3$ by the 10-methacridinium cation in acetonitrile was shown to lead to the (presumably solvent-complexed) trimethylstannyl cation[40]. Oxidation of R_4Sn, Me_3SnR and R_6Sn_2 (R = Me, Et, *n*-Bu, Ph, Vi) by the thianthrene cation radical in acetonitrile also resulted in the formation of the corresponding presumably solvent-complexed 'R_3Sn^+' cations[41]. Recently, a series of nitrile complexes of t-Bu_3M^+ cations (M = Si, Ge, Sn) has been prepared by Sekiguchi and coworkers[42] by oxidation of the corresponding t-Bu_6M_2 dimetallanes with Ph_3C^+ $B(3,5\text{-}(CF_3)_2C_6H_3)_4^-$ (Scheme 4). The solvent-complexed Ph_3Pb^+ cation was prepared by an oxidation of Ph_6Pb_2 with $AgNO_3$ in acetonitrile[43]. A crystalline solvent-free $CB_{11}Me_{12}^-$ salt of the n-Bu_3Sn^+ cation with significant anion–cation interaction was prepared by Michl and coworkers[44] in hexane by oxidizing n-Bu_6Sn_2 with $CB_{11}Me_{12}^{\bullet}$ (Scheme 5).

$$t\text{-}Bu_3M—M(Bu\text{-}t)_3 \xrightarrow[R-C\equiv N]{2\,Ph_3C^+B[3,5\text{-}(CF_3)_2C_6H_3]_4^-} 2[t\text{-}Bu_3M^+N{\equiv}C—R][B(3,5\text{-}(CF_3)_2C_6H_3)_4]^- + 2\,Ph_3C^{\bullet}$$

M = Si, Ge, Sn   R = Me, *t*-Bu

SCHEME 4

Numerous studies of the electrooxidation of various organotin and organolead compounds have been conducted in the group of Kochi[45–48] who showed electron transfer

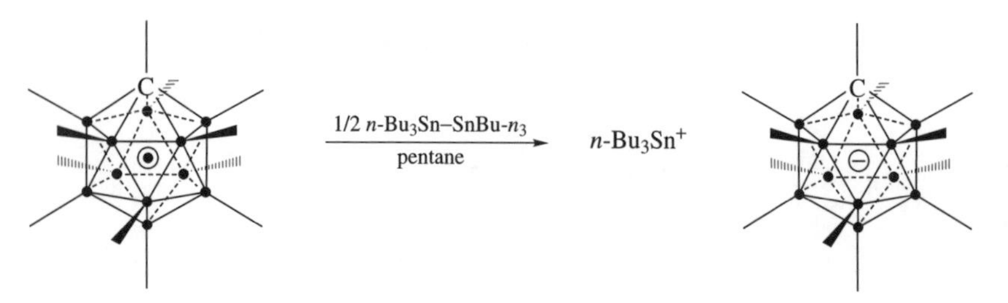

SCHEME 5

to be the limiting step of the oxidation. These oxidations are believed to first produce the radical cations $[R_3M{-}MR_3]^{\bullet +}$, which then decompose to give 'R_3M^+' and $R_3M^\bullet$, and the radical is oxidized further. A number of oxidation potentials determined for group 14 dimetallanes $R_3M{-}M'R_3$ (M and M$' =$ Si, Ge, Sn, R = Me, Et) in acetonitrile[49] and for triphenylstannyl derivatives in THF and DMF[50] are summarized in Table 4. The oxidation potentials decrease with increasing atomic weight of M as expected. Substitution of Me by Et appears to have little effect, except in the case of Me_6Ge_2 and Et_6Ge_2, where the reported oxidation potential is significantly lower for the ethyl derivative, perhaps due to a different electrode mechanism or to an experimental problem. Out of the three oxidation potentials observed for Ph_3SnH, the two lower ones were attributed to (i) $Ph_3SnH \rightarrow Ph_3Sn^\bullet + H^+ + 1e^-$ and (ii) $Ph_3Sn^\bullet \rightarrow$ 'Ph_3Sn^{+}' $+ 1e^-$, and the third higher potential to the oxidation of Ph_6Sn_2 formed by the dimerization of the $Ph_3Sn^\bullet$ radical. Oxidations of Ph_3SnI, Ph_3SnSPh and $Ph_3SnOCHO$ led directly to the 'Ph_3Sn^+' cation, while the corresponding chloride and triflate were not oxidizable.

TABLE 4. Oxidation potentials (E_p) of group 14 derivatives

Compound	E_p, (V)
	In MeCN, against Ag/AgCl/MeCN[49]
$Me_3Si{-}SiMe_3$	1.76
$Me_3Ge{-}GeMe_3$	1.70
$Me_3Sn{-}SnMe_3$	1.28
$Me_3Si{-}GeMe_3$	1.76
$Me_3Si{-}SnMe_3$	1.60
$Me_3Ge{-}SnMe_3$	1.44
$Et_3Si{-}SiEt_3$	1.76
$Et_3Ge{-}GeEt_3$	1.48
$Et_3Sn{-}SnEt_3$	1.24
$Et_3Si{-}GeEt_3$	1.70
$Et_3Si{-}SnEt_3$	1.56
$Et_3Ge{-}SnEt_3$	1.40
	In THF, against satd. calomel[50]
Ph_3SnH	0.80, 1.15, 1.50
Ph_3SnI	1.03
$Ph_3Sn{-}SnPh_3$	1.50
$Ph_3Sn{-}SiMe_2(Bu\text{-}t)$	1.63
$Ph_3Sn{-}SPh$	1.22 (in DMF)
$Ph_3Sn{-}OCHO$	1.57 (in DMF)

III. STRUCTURE AND PROPERTIES

A. Gas Phase

The structure of isolated R_3M^+ cations in the gas phase was the subject of several computational studies. Significant difficulties are associated with calculations for molecules containing the heavier elements, particularly Sn and Pb. NMR chemical shift calculations require consideration of relativistic effects (spin–orbit coupling). A discussion of these difficulties and of effective core potentials developed for these calculations is beyond the scope of this review.

Earliest studies concentrated on the H_3M^+ cations (M = C, Si, Ge, Sn, Pb) with planar D_{3h} symmetry[51]. However, CASSCF and MRSDCI calculations[52] of the ground state geometries and energies of various germanium hydrides and hydride cations found a pyramidal C_{3v} ground state structure for the H_3Ge^+ cation 4.6 kcal mol^{-1} (CASSCF) or 6 kcal mol^{-1} (MRSDCI) below the planar D_{3h} structure, and predicted a rapid umbrella inversion at room temperature.

More recently, Schleyer and coworkers[53] compared H_3C^+ with its heavier congeners using the B3LYP/6-311++G(2d,2p) method for C, Si, Ge and the B3LYP/TZ+ZP method with quasirelativistic effective core potentials for Sn and Pb, and found that at this level of calculation the D_{3h} structures are favored for C, Si and Ge. For Sn and Pb the D_{3h} cations were predicted to be metastable due to reductive elimination to $HM^+{-}H_2$, which are energetically favored. Another investigation of bonding in the D_{3h} H_3Ge^+ cation showed that the HOMO(H_2) and NLUMO(HGe^+) orbitals are indeed appreciably populated and that this cation has a $HGe^+{-}H_2$ complex character[54]. However, recently Schwarz and coworkers[55] found using the B3LYP/6-311++G(2d,2p) method that the classical D_{3h} H_3Ge^+ cation is by *ca* 10 kcal mol^{-1} more stable than the dihydrogen complex $HGe(H_2)^+$.

According to MP2/VDZ+P calculations[56], the π-donor ability of halogen substituents in Hal_3M^+ and $HalH_2M^+$ cations (Hal = F, Cl, Br, I; M = C, Si, Ge, Sn, Pb) increases from F to I for all of these cations, the thermodynamic stabilization of the cations by halogen substituents increases in the same order, and for the heavier congeners this stabilization is diminished compared to that in the carbocations.

B. Solution

1. Strong acids as solvents

Several important contributions to the understanding of the nature of 'R_3Sn^+' cations in superacids were made by Birchall and coworkers. They reported ^{1}H NMR and Mössbauer spectra for 'Me_3Sn^+' cations in sulfuric and fluorosulfuric acids[57] and showed that cationic tin species formed under these conditions have coordination numbers of 5 or 6. The 'Me_3Sn^+' cation was not very stable in these solutions and was oxidized to 'Me_2Sn^{+2}'. The ^{119}Sn NMR chemical shift (Table 5) for 'Me_3Sn^+' in 92% sulfuric acid at 0 °C is 194 ppm and, for 'Et_3Sn^+' in fluorosulfuric acid at −20 °C, it is 288 ppm[58]. Methyltin hydrides Me_nSnH_{4-n} ($n = 1$–3) also gave cationic species in fluorosulfuric acid[22]. Based on ^{119}Sn NMR and Mössbauer spectroscopy, it was concluded that the '$Me_nSnH_{3-n}{}^+$' cations have two fluorosulfates occupying the positions above and below the plane of three covalently attached substituents, producing a trigonal–bipyramidal arrangement around the tin atom. A ^{119}Sn NMR chemical shift for Me_3Sn^+ in fluorosulfuric acid at −60 °C was 322 ppm[22].

TABLE 5. ^{119}Sn NMR chemical shifts for cationic tin species

Compound	Solvent	Chemical shift vs. Me_4Sn	Reference
Me_4Sn	92% H_2SO_4	194	58
Et_4Sn	HSO_3F	288	58
Me_3SnH	HSO_3F	322	22
Me_2SnH_2	HSO_3F	156	22
$MeSnH_3$	HSO_3F	−29	22
SnH_4	HSO_3F	−194	22
Me_4Sn	$SbCl_5$	208	23
$Bu_3SnB(C_6F_5)_3H$	C_6D_6	360	35
Bu_3SnClO_4	CD_2Cl_2	245	35
Bu_3SnClO_4	C_6D_6	231	35
Bu_3SnClO_4	sulfolane	150	35
Me_3SnClO_4	CD_2Cl_2	249	35
Me_3SnClO_4	C_6D_6	234	35
$Bu_3SnB(3,5(CF_3)_2C_6H_3)_4$	CD_2Cl_2	356	36
$Bu_3SnOEt_2B(3,5(CF_3)_2C_6H_3)_4$	CD_2Cl_2	165	36
$Bu_3SnCB_{11}Me_{12}$	C_6D_{12}	454	44
$Bu_3SnCB_{11}Me_{12}$	solid	461	44
$Bu_3SnOEt_2CB_{11}Me_{12}$	C_6D_{12}	168	44
$Mes_3SnB(C_6F_5)_4$	C_6D_6	806	38

Also known are the deuterium-induced isotope effect on the ^{119}Sn shielding of −0.05 ppm/D and the primary isotope effect of -11.6 ± 7 Hz for Sn−H spin–spin coupling for the '$SnD_nH_{3-n}{}^+$' cations generated in fluorosulfuric acid at low temperature from SnD_nH_{4-n}[59]. The chemical shift of 208 ppm found[23] in the ^{119}Sn NMR of the reaction mixture of Me_4Sn and $SbCl_5$ was attributed to the 'Me_3Sn^+' cation.

A ^{209}Pb NMR chemical shift of 980 ppm has been reported[60] for the 'Me_3Pb^+' cation in fluorosulfuric acid at low temperature.

2. Organic solvents

The nature of 'Me_3M^+' cations in organic solvents has been a subject of controversy similar to that for silyl cations[1], albeit not such a heated one.

Various monohalides of triorganotin derivatives have been shown by NMR spectroscopy to ionize in polar solvents, providing the corresponding coordinated cations[61]. Solutions of tributyltin or triphenyltin chloride, perchlorate and tetrafluoroborate have been studied by Edlund and coworkers[62] in dichloromethane, sulfolane, acetonitrile, pyridine, DMPU, DMSO and HMPA (Table 6). They reported the most downfield ^{119}Sn NMR chemical shift of 220 ppm for 'n-Bu_3Sn^+' ClO_4^- in dichloromethane and showed that the ^{119}Sn NMR

TABLE 6. ^{119}Sn NMR chemical shifts vs. Me_4Sn for n-Bu_3Sn and Ph_3Sn derivatives[64]

	CH_2Cl_2	Sulfolane	MeCN	Pyridine	DMPU	DMSO	HMPA
Bu_3SnCl	156	130	119	14	18	2	−47
Bu_3SnClO_4	220	139	54	−24	0	12	−43
Bu_3SnBF_4	160	168	44	1	4	11	−44
Ph_3SnCl	−44	−93	−98	−213	−223	−227	−242
Ph_3SnClO_4	—	−157	−211	−232	−263	−236	−275
Ph_3SnBF_4	—	−152	−216	−229	−261	−236	−276

chemical shift moves upfield as the solvation power is increased from dichloromethane to HMPA for both series [a similar observation was made in the cases of 'Bu_3Sn^+' $B(3,5\text{-}(CF_3)_2C_6H_3)_4^-$[36] and '$Bu_3Sn^+$' $CB_{11}Me_{12}^-$[44] whose upfield ^{119}Sn NMR chemical shifts of 165 ppm in CD_2Cl_2 and 168 ppm in C_6D_{12}, respectively, in the presence of ether, were attributed to the formation of the solvated $Bu_3SnOEt_2^+$ cations]. Based on their observation and on the ^{35}Cl NMR line width, Sn−C scalar coupling and $^{119}Sn-^{31}P$ coupling measurements, Edlund and coworkers concluded that in these solutions neutral tetrahedral and cationic trigonal bipyramidal species are in equilibrium, with bipyramidal coordination favored in solvents of higher donicity. ESI/MS of aqueous acetonitrile solutions of R_3SnHal (R = Me, *n*-Bu, Ph, Hal = Cl, Br) showed R_3Sn^+ cations together with solvated cations such as $[(R_3Sn)_n(OH_2)_x]^+$ and $[R_3Sn(NCMe)]^+$ or $[R_3Sn(Py)]^+$ in the presence of pyridine[63].

In another early report, Lambert and Schlif[34] claimed germyl cations in sulfolane and dichloromethane to be free based on conductivity, cryoscopic measurements and ^{35}Cl NMR spectroscopy (^{35}Cl line widths and chemical shifts). Later, Lambert and Kuhlmann[35] reported a ^{119}Sn NMR chemical shift of *ca* 360 ppm for '$n\text{-}Bu_3Sn^+$' $BH(C_6F_5)_3^-$ in benzene at room temperature (but in a later paper[64] a ^{119}Sn NMR chemical shift for the similar 'Et_3Sn^+' $B(C_6F_5)_4^-$ salt in toluene was reported to be 251 ppm), while Kira and coworkers[36] reported a ^{119}Sn NMR chemical shift of 356 ppm for '$n\text{-}Bu_3Sn^+$' $B(3,5\text{-}(CF_3)_2C_6H_3)_4^-$ in dichloromethane at $-20\,^{\circ}C$ and also showed that there is no interaction with the anion, based on the ^{19}F NMR measurements. Both groups claimed that they prepared tricoordinate stannyl cations with no significant coordination to the solvent, based on the downfield ^{119}Sn NMR chemical shifts.

This conclusion was disputed later by Edlund and coworkers[60] who suggested that a solvent molecule occupies the fourth coordination site in both cases. Indeed, the ^{119}Sn NMR chemical shifts observed by Lambert and collaborators and Kira and coworkers are similar to that reported by Birchall and Manivannan[58] for 'Me_3Sn^+' in fluorosulfuric acid (322 ppm), where there is little doubt about the coordinated nature of the cationic tin species. Since an accurate computational prediction of ^{119}Sn NMR chemical shifts is hard[65], Edlund and coworkers used empirical correlations to show that the ^{119}Sn NMR chemical shifts observed lie in the range corresponding to covalent arene complexes of stannyl cations and predicted a ^{119}Sn NMR chemical shift of 1500–2000 ppm for a truly free stannyl cation[60]. They also used scalar Sn−C coupling to support their conclusion about coordination and pyramidalization in the cationic species reported by Lambert and collaborators and Kira and coworkers. Another proof of the pyramidalization of triorganostannyl cations even by weak nucleophiles was obtained recently by Michl and coworkers[44] who found a ^{119}Sn NMR chemical shift of 454 ppm for '$n\text{-}Bu_3Sn^+$' $CB_{11}Me_{12}^-$ in cyclohexane. The single crystal structure of this salt (whose solid sample had a CP-MAS ^{119}Sn NMR chemical shift of 461 ppm) showed no solvent coordination, but a significant interaction of the methyl groups of two carboranyl anions with the distinctly pyramidalized stannyl cation. The ^{119}Sn NMR signal of $n\text{-}Bu_3Sn^+CB_{11}Me_{12}^-$ lies downfield relative to the signals reported by Lambert and collaborators and Kira and coworkers, but it is far from the value predicted for a free stannyl cation.

Optimized structures of Me_nSnH_{4-n}, H_3SnX (X = OH_2, Cl), H_3Sn^+ solvated by one and two water molecules, and H_3SnOH solvated by one molecule of water have been calculated by Cremer and coworkers[65] at the HF/DZ+P and MP2/DZ+P level. They calculated ^{119}Sn chemical shifts using the IGLO/DZ and IGLO/DZ+P methods and showed that only strongly coordinated pyramidal and trigonal bipyramidal stannyl cations exist in solution. They predicted a ^{119}Sn NMR chemical shift of *ca* 1000 ppm for a free triorganostannyl cation.

The degree of cationic character of Mes_3Ge^+ in aromatic solvents could be only roughly estimated[38] from the ^{13}C NMR data due to the lack of a suitable germanium nuclide for direct NMR observations. The conclusion was reached that the positive charge on germanium in Mes_3Ge^+ is comparable to that on silicon in the Mes_3Si^+ cation[39], which is believed to be the first free silyl cation in condensed media. The trimesitylstannyl cation prepared by Lambert and coworkers[38] has a record high ^{119}Sn NMR chemical shift of 806 ppm, which still falls somewhat short of the 1000 ppm chemical shift predicted by Cremer and coworkers[65] and considerably short of the 1500–2000 ppm value estimated by Edlund and coworkers[60] for a free stannyl cation. The fact that no solvent dependence was found for the Mes_3Sn^+ cation led to the conclusion that in this case an interaction with the $B(C_6F_5)_4^-$ counterion is significant[38]. Attempts to increase the steric hindrance around the cationic tin atom sufficiently have failed[38].

The interaction between R_3M^+ cations (R = H, Me, Cl, M = C, Si, Ge, Sn, Pb) and toluene was examined computationally by Basch[66] using MP2/CEP and MP2/RCEP methods. He found that the R_3M group in R_3M^+-toluene complexes for M = Si, Ge, Sn and Pb lies almost directly above the *ipso* carbon atom with the hydrogen atom almost in the plane of the aromatic ring, and also that the binding energy increases from Si to Pb, which he attributes to an additional non-bonding stabilizing interaction between one of the M–C bonds in the R_3M^+ cation and the aromatic π system, which is especially strong when M = Pb.

C. Solid State

Several single crystal structures have been determined for solvated stannyl cations. Thus, a hydrated tributyltin cation n-$Bu_3Sn(OH_2)_2^+$ has a symmetry close to D_{3h}, with two water molecules occupying the axial positions[67]. In $[(c\text{-}C_6H_{11})_3Sn(NCMe)_2]^+SbF_6^-$ the cationic tin atom is trigonal bipyramidal, with two acetonitrile ligands occupying the axial sites[37]. The structure of the pivalonitrile complexes of the t-Bu_3Ge^+ and t-Bu_3Si^+ cations has also been established by X-ray crystallography[42].

A single crystal structure was determined recently[44] for solvent-free n-Bu_3Sn^+ $CB_{11}Me_{12}^-$, crystallized from hexane (Figure 1). The methyl groups of two carboranyl anions interact with the somewhat pyramidalized stannyl cation in an approximately trigonal bipyramidal geometry (the average Sn–C distance is 2.81 Å and the sum of valence angles around the tin atom is 353°). As noted above, the ^{119}Sn NMR chemical shift values of 454 ppm in cyclohexane and of 461 ppm in the solid are closer to those reported for solvated stannyl cations than to the chemical shift of the presumably largely free Mes_3Sn^+ cation. The authors used the B3LYP/SSD method to calculate the ion-pairing energies for four optimized isomers of an isolated Me_3Sn^+ $CB_{11}Me_{12}^-$ ion pair, with Me_3Sn^+ next to a methyl in position 1, 2, 7 or 12 of the CB_{11} cluster. The 12-isomer was found to be the most stable and to have the shortest $Sn{-}CH_3$ and the longest $B{-}CH_3$ bond length, the flattest coordinated CH_3 group, the most pyramidalized Me_3Sn^+ cation and the largest degree of inter-ion charge transfer. The calculated stability of the ion pairs decreases in the order $12 > 7 > 2$, with the Sn atom always located almost exactly on the $B{-}CH_3$ bond axis, and the stabilizing interaction not merely electrostatic but also of the donor–acceptor type. Me_3Sn^+ interacts much less with the carbon-attached 1-methyl group and the Sn atom is then tilted by 36° off the axial position toward the 2-methyl group.

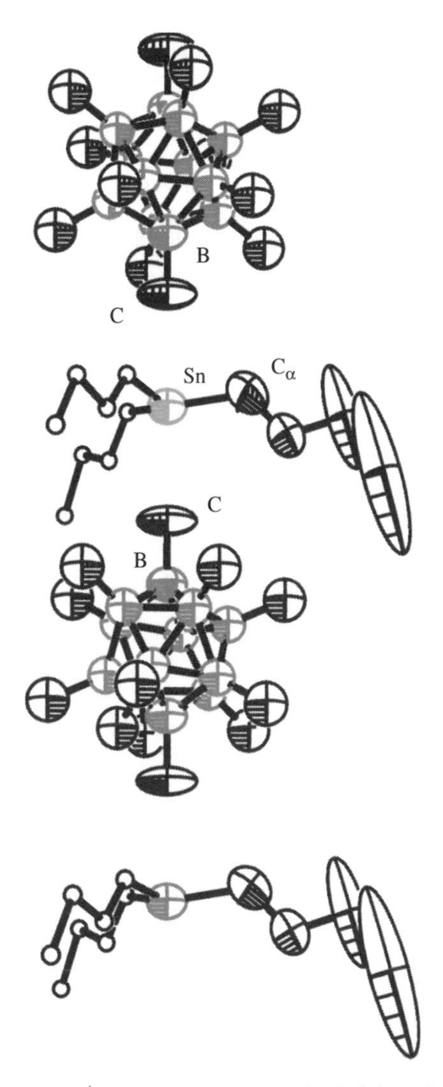

FIGURE 1. Structure of n-Bu_3Sn^+ $CB_{11}Me_{12}^-$: segment of an infinite column of alternating cations and anions. Thermal ellipsoids are at 50% (not shown on two of the butyl groups in each cation). Hydrogen atoms and one component of the disordered butyl groups are omitted, Sn atoms are dark grey, C atoms are black and B atoms are light grey. Within each cage, one of the vertices is a carbon atom (disordered, not shown). Reprinted with permission from Reference 44. Copyright 2000 American Chemical Society

D. Intramolecular Stabilization

Many compounds are known in which a cationic Ge, Sn or Pb atom is intramolecularly stabilized by a strong interaction with lone pairs on donor atoms such as N or O. For instance, a series of germyl cations **1** stabilized by an intramolecular nitrogen–germanium bond have been prepared recently in the group of Jutzi[68,69]. In these compounds there is no cation–anion interaction but the GeC_3 group is somewhat pyramidal (the sum of valence angles around Ge is *ca* 352°). Several germyl cations of type **2** stabilized by an intramolecular bond to oxygen have been reported by the group of Rivière[70,71]. In this case the germanium atom has a trigonal pyramidal geometry and a partially cationic character.

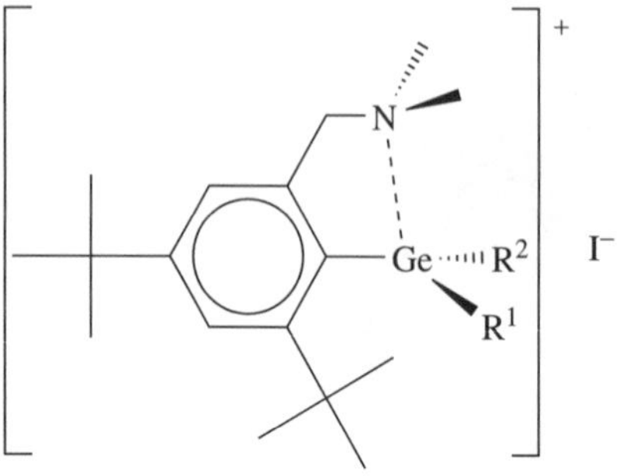

(1) 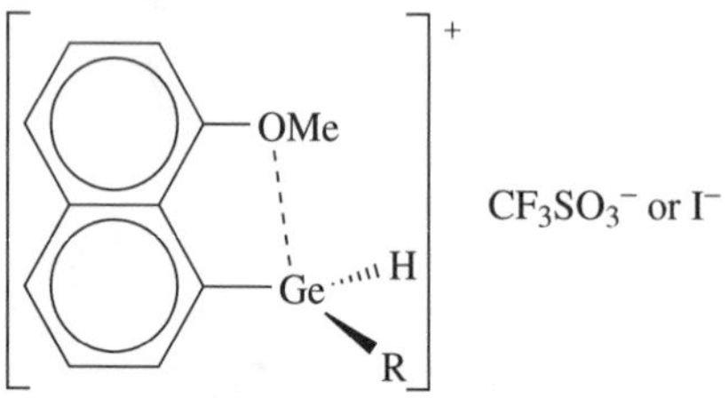 **(2)**

Protonated digermyl and distannyl ethers **3** were prepared by hydrolysis of the corresponding $R_3M^+B(C_6F_5)_4^-$ salts[64]. Their germanium and tin atoms possess a partial cationic character judging by the flattened tetrahedra around the metal atoms and elongated bonds to oxygen atoms. In stannyl cations **4**[72,73] the cationic pentacoordinate trigonal bipyramidal tin atom is stabilized by two intramolecular tin–nitrogen bonds.

Tetrastannylammonium and -phosphonium cations **5** were prepared by the reaction of $(Me_3Sn)_3E$ (E = N, P) with Me_3SnOTf or $Me_3SnF/NaBPh_4$ in toluene[74]. These cations dissociate slightly in THF and can be viewed as base-stabilized stannyl cations.

Wrackmeyer and coworkers[75,76] characterized by NMR spectroscopy intermediate stannyl cations **6** and **7**, in which cationic tin is stabilized by interactions with amino and alkynyl groups. In both cases the ^{119}Sn chemical shifts (*ca* −35 ppm for **6** and *ca* 110 ppm for **7**) indicate a strong coordination of the cationic tin atom. A stannyl cation **8** (M = Sn) stabilized only by π coordination to a carbon–carbon triple bond was also prepared[77]. Its crystal structure revealed a somewhat pyramidalized cationic tin atom with a sum of valence angles equal to 351°. The ^{119}Sn NMR chemical shift of 215 ppm also suggests a quite strong coordination of the cationic Sn atom. ^{119}Sn and ^{11}B NMR analysis of **8**, M = Sn, showed that there is an equilibrium between two structures in which the alkynyl group is attached to either the tin or the boron atom. In its lead analogue (**8**, M = Pb)[78] the cationic lead atom is less pyramidalized (a sum of valence angles is 356°), but the ^{209}Pb NMR chemical shift of 723 ppm suggests a strong coordination of the cationic lead atom (*cf* the ^{209}Pb NMR chemical shift of 980 ppm reported for the 'Me_3Pb^+' cation in fluorosulfuric acid[60]).

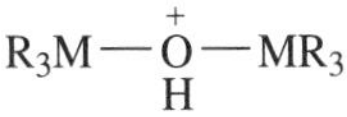

$R_3M-\overset{+}{O}(H)-MR_3$

M = Ge, R = Me

M = Sn, R = Et

(3)

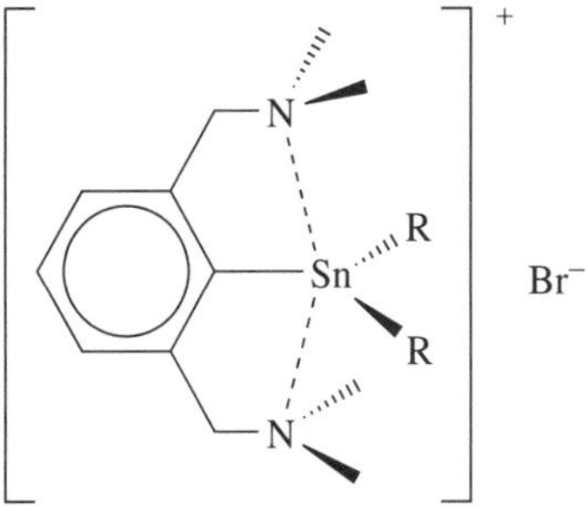

Br^-

(4)

$CF_3SO_3^-$ or BF_4^-

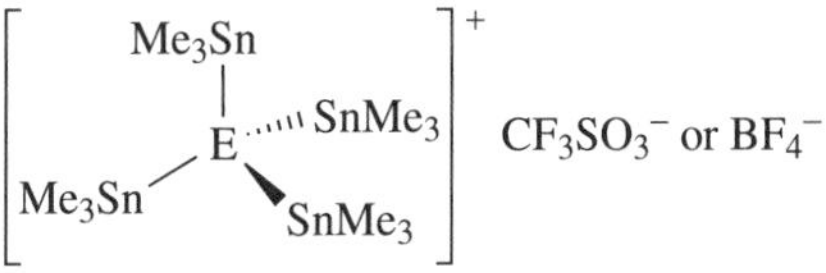

E = N, P

(5)

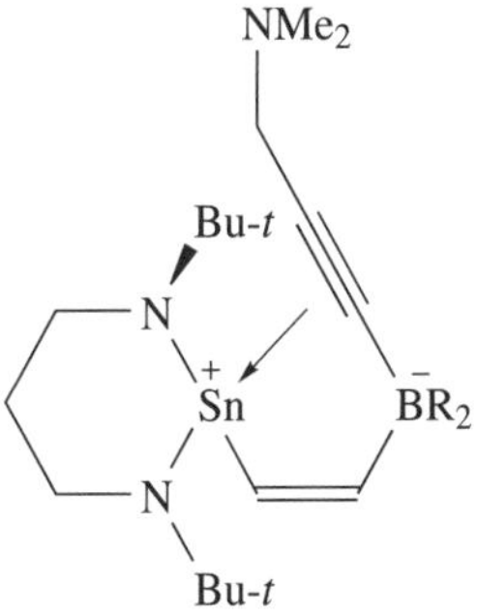

(6)

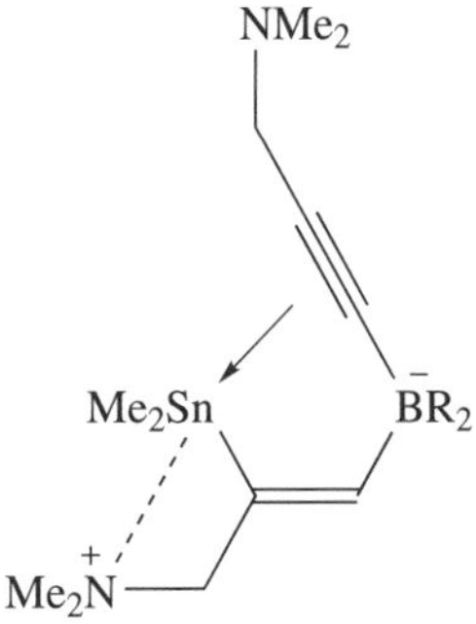

(7)

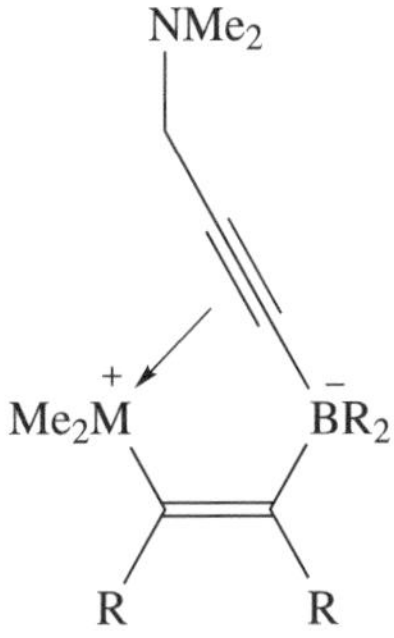

M = Sn, Pb

(8)

IV. REACTIONS

A. Gas Phase

Me_3M^+ cations (M = Si, Ge, Sn, Pb) produced by electron impact ionization of Me_4M (M = C, Si, Ge, Sn, Pb) have the highest intensity in the spectra and their fragmentation to give MeM^+ cations under higher electron impact energy increases for metals with a higher atomic number[79]. Unimolecular and collision-induced fragmentations of Me_3Ge^+, Me_3Sn^+, and Me_3Pb^+ cations show the following results[80,81]: Me_3Ge^+ loses ethylene, methyl and ethane; Me_3Sn^+ loses one, two and three methyl groups, but no ethylene; Me_3Pb^+ loses methyl groups and ethane. The differences were explained in terms of the decreasing element–carbon bond strength and the increasing preference for the +2 oxidation state in the order Ge, Sn, Pb. Gas-phase fragmentation of a labeled trineopentylstannyl cation under electrospray ionization conditions proceeds via β-methide transfer to yield labeled isobutene and the dineopentylmethylstannyl cation (Scheme 6)[82].

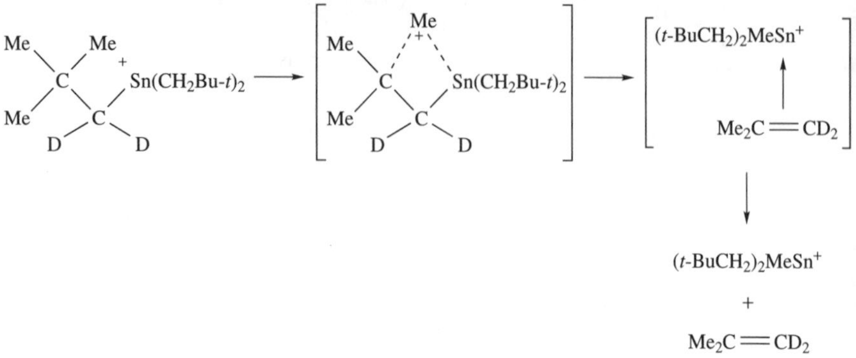

$$Me_2C(CD_2\text{-}Sn^+(CH_2Bu\text{-}t)_2) \longrightarrow [\text{Me-bridged cation}] \longrightarrow [(t\text{-}BuCH_2)_2MeSn^+ \leftarrow Me_2C{=}CD_2] \longrightarrow (t\text{-}BuCH_2)_2MeSn^+ + Me_2C{=}CD_2$$

SCHEME 6

The n-Bu_3Ge^+ cation produced from n-Bu_4Ge by electron impact, as well as Me_3Sn^+ and i-Pr_3Sn^+ cations, was found[83] to be more reactive than the corresponding carbocation. Adducts of Me_3Ge^+ to various N, O and S containing bases (such as alcohols, amines and esters) under ion cyclotron resonance conditions were found to be stable while the analogous Me_3Si^+ adducts were not[84]. The binding energy of R_3M^+ cations (R = Me, Et, n-Pr, n-Bu) to water, studied by high pressure mass spectrometry[85], decreases in the series R_3Si^+, R_3Ge^+, R_3Sn^+, in contrast to the earlier observations[84]. ΔH° decreases with the increasing size of R, and ΔS° is nearly constant.

The reaction of H_3Ge^+ cation with ethylene was found[86] by tandem mass spectrometry to yield stable adduct ions $[C_2H_7Ge]^+$ which first decompose to the ions $[C_2H_5Ge]^+$ with the proposed structure of a protonated germacyclopropane and then are proposed to rearrange to the ethylgermyl cation, which loses a hydrogen molecule and forms the vinylgermyl cation (Scheme 7).

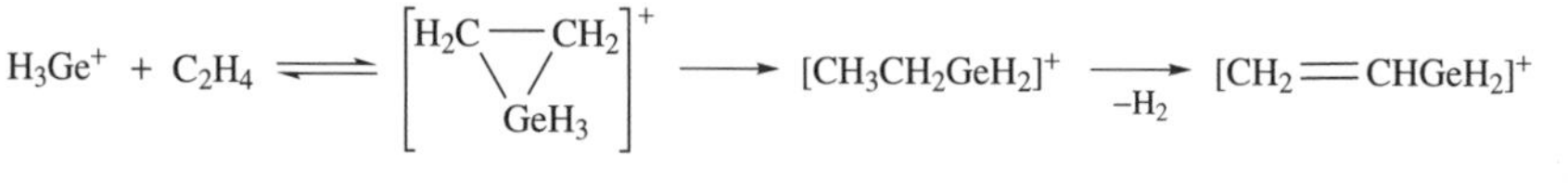

$$H_3Ge^+ + C_2H_4 \rightleftharpoons \left[\begin{matrix} H_2C\text{—}CH_2 \\ \backslash\ / \\ GeH_3 \end{matrix}\right]^+ \longrightarrow [CH_3CH_2GeH_2]^+ \xrightarrow[-H_2]{} [CH_2{=}CHGeH_2]^+$$

SCHEME 7

Ion–molecule reactions in H_4Ge/hydrocarbons[87,88], H_3GeMe/hydrocarbons[89], H_4Ge/H_4Si[90,91] and H_3GeMe/H_4Si[92–94] mixtures have been studied by the group of Vaglio using ion trap mass spectrometry. They observed a variety of cations, such as H_3M^+, H_2M^+, HM^+, M^+, $H_5M_2{}^+$, $H_3M_2{}^+$, $H_7M_3{}^+$, $H_6M_3{}^+$, $H_5M_3{}^+$, $H_4M_3{}^+$ and $H_7M_4{}^+$, and described their interconversions. Recently, Schwarz and coworkers reported a study of fragmentation patterns for the H_3Ge^+ cation (dominated by the loss of one and two hydrogens) and the H_2Ge^+ cation[55].

Gas-phase reactions of the methyl cation with Me_4M (M = Si, Ge, Sn) were studied using a radiochemically generated T_3C^+ (Scheme 8)[95]. In this case the methide anion is readily abstracted from all three tetramethyl derivatives, tetramethyltin being the most reactive. An analogous reaction of methyl cations with Et_4M (M = C, Si, Ge, Sn) resulted mostly in abstraction of the ethyl anion when M was silicon, germanium or tin[96]. As shown in Scheme 8, the authors assumed that the abstraction occurs by electrophilic substitution on the M−C bond, with front-side attack on the methyl group.

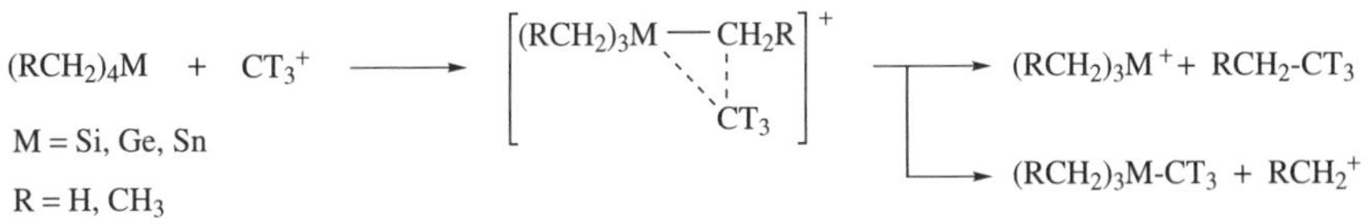

$(RCH_2)_4M + CT_3^+ \longrightarrow [(RCH_2)_3M—CH_2R \cdots CT_3]^+ \longrightarrow (RCH_2)_3M^+ + RCH_2\text{-}CT_3$ or $(RCH_2)_3M\text{-}CT_3 + RCH_2^+$

M = Si, Ge, Sn

R = H, CH_3

SCHEME 8

The following thermodynamic data have been reported[97] for a methide anion transfer in the gas-phase equilibria $Me_3M^+ + Me_4M' \rightleftharpoons Me_4M + Me_3M'^+$ (M = Si, Ge, Sn): for $Me_3Si^+ + Me_4Ge \rightleftharpoons Me_4Si + Me_3Ge^+$, $\Delta H^0 = -10.2 \pm 1.2$ kcal mol^{-1}, $\Delta S^0 = -3.7 \pm 2.4$ cal K^{-1} mol^{-1}; for $Me_3Ge^+ + Me_4Sn \rightleftharpoons Me_4Ge + Me_3Sn^+$, $\Delta H^0 = -8.1 \pm 0.9$ kcal mol^{-1}, $\Delta S^0 = -0.9 \pm 1.6$ cal K^{-1} mol^{-1}.

Me_3M^+ cations (M = C, Si, Ge, Sn) were used as a stereochemical probe in gas-phase reactions with 1,2-cyclopentanediol isomers[98]. The decomposition pattern of the [1,2-cyclopentanediol + $Me_3M]^+$ adducts depended on the stereochemistry of the diol. For *cis*-diol the decomposition led readily to hydrated $[Me_3M(OH_2)]^+$ cations, while for *trans*-diol the adduct was significantly more stable; the Me_3Ge^+ cation was a more sensitive and selective reactant than other group 14 cations.

B. Solution

Little is known about the reactivity of the group 14 cations in solution. Clearly, one would expect these strong electrophiles to react with common nucleophiles, but such trivial reactions were not documented except for the reaction of $Bu_3Sn^+CB_{11}Me_{12}^-$ with PhLi which produced Bu_3SnPh[44]. $[(c\text{-}C_6H_{11})_3Sn(NCMe)_2]^+SbF_6^-$ and compounds of this series have been shown to be effective promoters of Diels–Alder additions to furan[37].

V. RELATED SPECIES

Sekiguchi and coworkers[99,100] reported the preparation and the crystal structure of a free cyclotrigermenium cation (**9**) as a tetraphenylborate salt. The stability of this cation stems from its aromatic 2π-electron system, and from the steric bulk of tri-*tert*-butylsilyl groups. They later expanded this work to similar cyclotrigermenium cations with various substituents and counterions[101–103].

Two transition metal stabilized germyl cations, **10**[104] and **11**[105], have been reported. In the first one the strongly pyramidalized cationic germanium center is bonded to two iron atoms and one *t*-Bu group, and is stabilized by the coordination to DMAP. The second one has a planar cationic germanium bonded to two tungsten atoms and one methyl group.

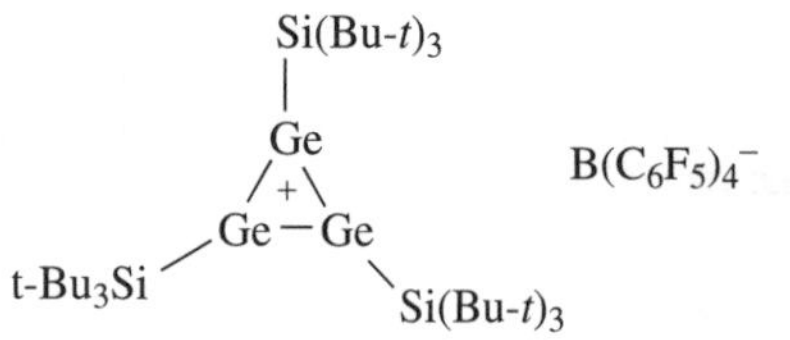

(9)

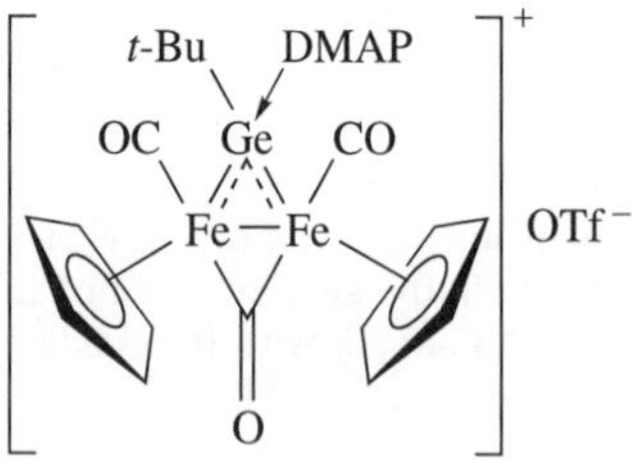

(10)

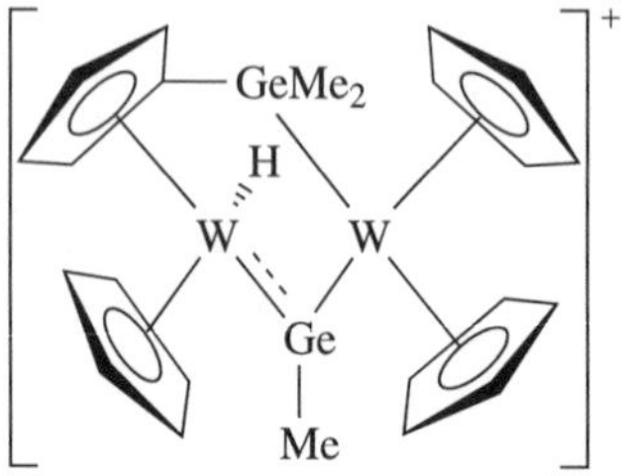

(11)

Several related cations have also been studied computationally. In the case of $C_2GeH_7^+$ cations four local minima were found, with the planar Me_2HGe^+ cation representing the global one[106]. For the cyclic cations $Si_3H_3^+$ and $Ge_3H_3^+$ the planar D_{3h} structure was found to be the global minimum[107]. Another study of the cyclic $H_3M_3^+$ cations (M = C, Si, Ge, Sn, Pb) found non-planar C_{3v} hydrogen-bridged structures for Ge, Sn and Pb to be more stable than the planar D_{3h} structures[108]. Kudo and Nagase[109] published a short review discussing calculational results for the cations of strained polycyclic Si, Ge, Sn and Pb compounds.

1,3-Migrations of a methyl group in compounds of the type $(Me_3Si)_2C(SiMe_2X)(MMe_3)$ between silicon and a heavier group 14 element, germanium[110] or tin[111], were postulated by Eaborn and coworkers to proceed via a bridged cationic intermediate **12**. Intermediacy of stannyl cations has also been postulated[112] in aromatic electrophilic substitution reactions.

Pentamethylcyclopentadienylgermyl and pentamethylcyclopentadienylstannyl cations **13** have been prepared by Jutzi and coworkers[113] as tetrafluoroborate salts in ether. These cations, of course, are of the RM^+ rather than the 'R_3M^+' kind. The structure of the tin derivative has been established by X-ray crystallography[113]. It revealed an almost C_{5v} symmetrical molecule with the average tin to ring carbon atom distance of 2.46 Å and with a somewhat short Sn−F distance of 2.97 Å. The reactivity of the pentamethylcyclopentadienylgermyl cation and its complexes with nitrogen nucleophiles has also been studied by Jutzi and coworkers[114,115]. Recently, the reaction of $Sn(C_5Me_5)_2$ with $Ga(C_6F_5)_3$ was reported to produce the first group 14 triple-decker cation $[(C_5Me_5)Sn(C_5Me_5)Sn(C_5Me_5)]^+[Ga(C_6F_5)_4]^-$ (**14**). Its structure was determined by X-ray crystallography[116].

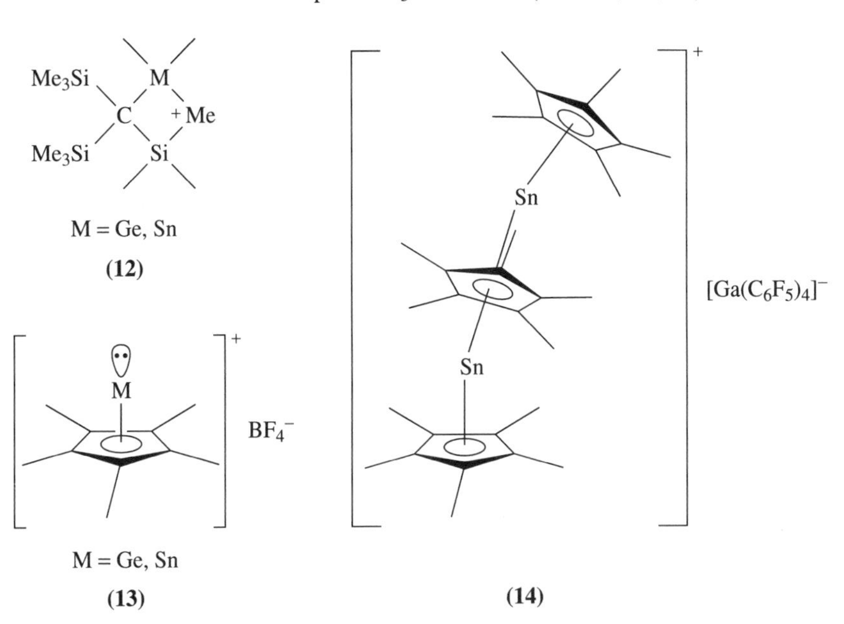

The structure of and bonding in the $CpGe^+$ cation have been analyzed computationally[117]. The calculated Ge-to-ring distance was 1.99 Å for the C_{5v} cation, in which the Ge atom is surrounded by eight valence electrons and its lone pair occupies an sp hybrid pointing away from the Cp ring.

VI. CONCLUSIONS

Several major advances have been made in the field of 'R_3M^+' cations (M = Ge, Sn, Pb) in recent years, including the preparation of free cyclotrigermenium cations and possibly close to free trimesitylgermyl and trimesitylstannyl cations. However, many challenges remain, for example, a structural characterization of free R_3M^+ cations (M = Ge, Sn, Pb). In many respects these cations appear to be even more challenging than the silyl cations, due to their weaker and longer bonds to carbon. Presently inadequate tools which need to be improved are computational methods for the NMR chemical shifts of the Sn and Pb nuclei.

VII. REFERENCES

1. P. D. Lickiss, in *The Chemistry of Organic Silicon Compounds* (Eds. Z. Rappoport and Y. Apeloig), 2nd edn., Wiley, Chichester, p. 557, 1998.
2. C. A. Reed, *Acc. Chem. Res.*, **31**, 325 (1998).
3. J. B. Lambert and Y. Zhao, *Angew. Chem., Int. Ed. Engl.*, **36**, 400 (1997).
4. J. B. Lambert, S. Zhang, C. L. Stern and J. C. Huffman, *Science*, **260**, 1917 (1993).
5. J. B. Lambert, S. Zhang and S. M. Ciro, *Organometallics*, **13**, 2430 (1994).
6. Z. Xie, J. Manning, R. W. Reed, R. Mathur, P. D. Boyd, A. Benesi and C. A. Reed, *J. Am. Chem. Soc.*, **118**, 2922 (1996).
7. S. Patai (Ed.), *The Chemistry of Organic Germanium, Tin and Lead Compounds*, Wiley, Chichester, 1995.
8. J. Belzner, *Angew. Chem., Int. Ed. Engl.*, **36**, 1277 (1997).

9. P. v. R. Schleyer, *Science*, **275**, 39 (1997).
10. B. G. Hobrock and R. W. Kiser, *J. Phys. Chem.*, **66**, 155 (1962).
11. D. B. Chambers, F. Gloking, J. R. C. Light and M. Weston, *J. Chem. Soc., Chem. Commun.*, 282 (1966).
12. B. G. Hobrock and R. W. Kiser, *J. Phys. Chem.*, **65**, 2186 (1961).
13. D. B. Chambers, F. Gloking and M. Weston, *J. Am. Chem. Soc.*, **89**, 1759 (1967).
14. M. R. Litzow and T. R. Spalding, *Mass Spectrometry of Inorganic and Organic Compounds*, Elsevier, Amsterdam, 1973.
15. J. Charalambous (Ed.), *Mass Spectrometry of Metal Compounds*, Butterworths, London, 1975.
16. J. T. Bursey, M. M. Bursey and D. G. I. Kingston, *Chem. Rev.*, **73**, 191 (1973).
17. M. F. Lappert, J. B. Pedley, J. Simpson and T. R. Spadling, *J. Organomet. Chem.*, **29**, 195 (1971).
18. A. L. Yergey and F. W. Lampe, *J. Organomet. Chem.*, **15**, 339 (1968).
19. R. J. Gillespie and E. A. Robinson, *Proc. Chem. Soc.*, 147 (1957).
20. R. J. Gillespie, R. Kapoor and E. A. Robinson, *Can. J. Chem.*, **44**, 1197 (1966).
21. R. C. Paul, J. K. Puri and K. C. Malhotra, *J. Inorg. Nucl. Chem.*, **35**, 403 (1973).
22. T. Birchall and V. Manivannan, *J. Chem. Soc., Dalton Trans.*, 2671 (1985).
23. K. B. Dillon and G. F. Hewitson, *Polyhedron*, **3**, 957 (1984).
24. C. A. Kraus and C. C. Callis, *J. Am. Chem. Soc.*, **45**, 2624 (1923).
25. J. R. Webster and W. L. Jolly, *Inorg. Chem.*, **10**, 877 (1971).
26. V. Peruzzo, S. Faleschini and G. Plazzogna, *Gazz. Chim. Ital.*, **99**, 993 (1969).
27. P. Zanella and G. Plazzogna, *Ann. Chim. (Rome)*, **59**, 1160 (1969).
28. R. S. Tobias, *Organomet. Chem. Rev.*, **1**, 93 (1966).
29. T. L. Sayer, S. Backs, C. A. Evans, E. K. Millar and D. L. Rabenstein, *Can. J. Chem.*, **55**, 3255 (1977).
30. E. K. Millar, C. A. Evans and D. L. Rabenstein, *Can. J. Chem.*, **56**, 3104 (1978).
31. V. Cannizzaro, C. Foti, A. Gianguzza and F. Marrone, *Ann. Chim. (Rome)*, **88**, 45 (1998).
32. C. De Stefano, C. Foti, A. Gianguzza, F. J. Millero and S. Sammartano, *J. Solution Chem.*, **28**, 959 (1999).
33. J. Y. Corey, *J. Am. Chem. Soc.*, **97**, 3237 (1975).
34. J. B. Lambert and W. Schilf, *Organometallics*, **7**, 1659 (1988).
35. J. B. Lambert and B. Kuhlmann, *J. Chem. Soc., Chem. Commun.*, 931 (1992).
36. M. Kira, T. Oyamada and H. Sakurai, *J. Organomet. Chem.*, **471**, C4 (1994).
37. W. A. Nugent, R. J. McKinney and R. L. Harlow, *Organometallics*, **3**, 1315 (1984).
38. J. B. Lambert, Y. Zhao, H. Wu, W. C. Tse and B. Kuhlmann, *J. Am. Chem. Soc.*, **121**, 5001 (1999).
39. J. B. Lambert and Y. Zhao, *Angew. Chem., Int. Ed. Engl.*, **36**, 400 (1997).
40. S. Fukuzumi, T. Kitano and K. Mochida, *J. Am. Chem. Soc.*, **112**, 3246 (1990).
41. S. Loczynski, B. Boduszek and H. J. Shine, *J. Org. Chem.*, **56**, 914 (1991).
42. M. Ichinohe, H. Fukui and A. Sekiguchi, *Chem. Lett.*, 600 (2000).
43. L. Doretti and S. Faleschini, *Gazz. Chim. Ital.*, **100**, 819 (1970).
44. I. Zharov, B. T. King, Z. Havlas, A. Pardi and J. Michl, *J. Am. Chem. Soc.*, **122**, 10253 (2000).
45. C. L. Wong and J. K. Kochi, *J. Am. Chem. Soc.*, **101**, 5593 (1979).
46. S. Fukuzumi, C. L. Wong and J. K. Kochi, *J. Am. Chem. Soc.*, **102**, 2928 (1980).
47. B. W. Walthner, F. Williams, W. Lau and J. K. Kochi, *Organometallics*, **2**, 688 (1983).
48. J. K. Kochi, *Angew. Chem., Int. Ed. Engl.*, **27**, 1227 (1988).
49. K. Mochida, A. Itani, M. Yokoyama, T. Tsuchiya, S. Worley and J. K. Kochi, *Bull. Chem. Soc. Jpn.*, **58**, 2149 (1985).
50. H. Tanaka, H. Ogawa, H. Suga, S. Torii, A. Jutand, S. Aziz, A. G. Suarez and C. Amatore, *J. Org. Chem.*, **61**, 9402 (1996).
51. G. Trinquier, *J. Am. Chem. Soc.*, **114**, 6807 (1992).
52. K. K. Das and K. Balasubramanian, *J. Chem. Phys.*, **93**, 5883 (1990).
53. J. Kapp, P. R. Schreiner and P. v. R Schleyer, *J. Am. Chem. Soc.*, **118**, 12154 (1996).
54. E. del Rio, M. I. Menendez, R. Lopez and T. L. Sordo, *Chem. Commun.*, **18**, 1779 (1997).
55. P. Jackson, N. Sändig, M. Diefenbach, D. Schröder, H. Schwarz and R. Srinivas, *Chem., Eur. J.*, **7**, 151 (2001).

56. G. Frenking, S. Fau, C. M. Marchand and H. Gruetzmacher, *J. Am. Chem. Soc.*, **119**, 6648 (1997).
57. T. Birchall, P. K. H. Chan and A. R Pereira, *J. Chem. Soc., Dalton Trans.*, 2157 (1974).
58. T. Birchall and V. Manivannan, *Can. J. Chem.*, **63**, 2211 (1985).
59. K. L. Leighton and R. E. Wasylishen, *Can. J. Chem.*, **65**, 1469 (1987).
60. M. Ashradi, D. Johnels and U. Edlund, *Chem. Commun.*, 1279 (1996).
61. A. G. Davies, *Organotin Chemistry*, VCH, Weinheim, 1997.
62. U. Edlund, M. Ashradi and D. Johnels, *J. Organomet. Chem.*, **456**, 57 (1993).
63. W. Henderson and M. J. Taylor, *Polyhedron*, **15**, 1957 (1996).
64. J. B. Lambert, S. M. Ciro and C. L. Stern, *J. Organomet. Chem.*, **499**, 49 (1995).
65. D. Cremer, L. Olsson, F. Reichel and E. Kraka, *Isr. J. Chem.*, **33**, 369 (1994).
66. H. Basch, *Inorg. Chim. Acta*, **242**, 191 (1996).
67. A. G. Davies, J. P. Goddard, M. B. Hursthouse and N. P. C. Walker, *J. Chem. Soc., Chem. Commun.*, 597 (1983).
68. H. Schmidt, S. Keitemeyer, B. Neumann, H.-G. Stammler, W. W. Schoeller and P. Jutzi, *Organometallics*, **17**, 2149 (1998).
69. P. Jutzi, S. Keitemeyer, B. Neumann and H.-G. Stammler, *Organometallics*, **18**, 4778 (1999).
70. F. Cosledan, A. Castel and P. Rivière, *Main Group Met. Chem.*, **20**, 7 (1997).
71. F. Cosledan, A. Castel, P. Rivière, J. Satge, M. Veith and V. Huch, *Organometallics*, **17**, 2222 (1998).
72. G. van Koten, J. T. B. H. Jastrzebski, J. G. Noltes, A. L. Spek and J. C. Schoone, *J. Organomet. Chem.*, **148**, 233 (1978).
73. A. J. Crowe, P. J. Smith and P. G. Harrison, *J. Organomet. Chem.*, **204**, 327 (1981).
74. M. Driess, C. Monse, K. Merz and C. van Wullen, *Angew. Chem., Int. Ed. Engl.*, **39**, 3684 (2000).
75. B. Wrackmeyer, G. Kehr and S. Ali, *Inorg. Chim. Acta*, **216**, 51 (1994).
76. B. Wrackmeyer, K. H. von Locquenghien and S. Kundler, *J. Organomet. Chem.*, **503**, 289 (1995).
77. B. Wrackmeyer, S. Kundler and R. Boese, *Chem. Ber.*, **126**, 1361 (1993).
78. B. Wrackmeyer, K. Horchler and R. Boese, *Angew. Chem., Int. Ed. Engl.*, **28**, 1500 (1989).
79. K. G. Heumann, K. Bachman, E. Kubassek and K. H. Leiser, *Z. Naturforsch., B*, **28**, 107 (1973).
80. G. S. Groenewold, M. L. Gross, M. M. Bursey and P. R. Jones, *J. Organomet. Chem.*, **235**, 165 (1982).
81. J. J. de Ridder and G. Dijkstra, *Recl. Trav. Chim. Pays-Bas*, **86**, 737 (1967).
82. D. Dakternieks, A. E. K. Lim and K. F. Lim, *Chem. Commun.*, **15**, 1425 (1999).
83. D. J. Harvey, M. G. Horning and P. Vouros, *Org. Mass Spectrom.*, **5**, 599 (1971).
84. V. C. Trenerry and J. H. Bowie, *Org. Mass Spectrom.*, **16**, 344 (1981).
85. J. A. Stone and W. J. Wytenburg, *Can. J. Chem.*, **65**, 2146 (1987).
86. K. P. Lim and F. W. Lampe, *Org. Mass Spectrom.*, **28**, 349 (1993).
87. P. Benzi, L. Operti, G. A. Vaglio, P. Volpe, M. Speranza and R. Gabrieli, *J. Organomet. Chem.*, **373**, 289 (1989).
88. P. Benzi, L. Operti, G. A. Vaglio, P. Volpe, M. Speranza and R. Gabrieli, *Int. J. Mass Spectrom. Ion Proc.*, **100**, 647 (1990).
89. L. Operti, M. Splendore, G. A. Vaglio, P. Volpe, M. Speranza and G. Occhiucci, *J. Organomet. Chem.*, **433**, 35 (1992).
90. L. Operti, M. Splendore, G. A. Vaglio and P. Volpe, *Spectrochim. Acta, Part A*, **49A**, 1213 (1993).
91. L. Operti, M. Spendori, G. A. Vaglio, P. Volpe, A. M. Franklin and J. F. J. Todd, *Int. J. Mass Spectrom. Ion Proc.*, **136**, 25 (1994).
92. M. Castiglioni, L. Operti, R. Rabezzana, G. A. Vaglio and P. Volpe, *Int. J. Mass Spectrom.*, **179/180**, 277 (1998).
93. L. Operti, M. Spendori, G. A. Vaglio and P. Volpe, *Organometallics*, **12**, 4509 (1993).
94. P. Antonioni, C. Canepa, A. Maranzana, L. Operti, R. Rabezzana, G. Tonachini and G. A. Vaglio, *Organometallics*, **20**, 382 (2001).
95. N. A. Gomzina, T. A. Kochina, V. D. Nefedov, E. N. Sinotova and D. V. Vrazhnov, *Zh. Obshch. Khim.*, **64**, 443 (1994); *Engl. Transl., Russ. J. Gen. Chem.*, **64**, 403 (1994).

96. N. A. Gomzina, T. A. Kochina, V. D. Nefedov and E. N. Sinotova, *Zh. Obshch. Khim.*, **64**, 630 (1994); *Engl. Transl., Russ. J. Gen. Chem.*, **64**, 574 (1994).
97. A. C. M. Wojtyniak, X. Li and J. A. Stone, *Can. J. Chem.*, **65**, 2849 (1987).
98. W. J. Meyerhaffer and M. M. Bursey, *J. Organomet. Chem.*, **373**, 143 (1989).
99. A. Sekiguchi, M. Tsukamoto and M. Ichinohe, *Science*, **275**, 60 (1997).
100. A. Sekiguchi, M. Tsukamoto, M. Ichinohe and N. Fukaya, *Phosphorus, Sulfur, Silicon Relat. Elem.*, **124/125**, 323 (1997).
101. M. Ichinohe, N. Fukaya and A. Sekiguchi, *Chem. Lett.*, 1045 (1998).
102. A. Sekiguchi, N. Fukaya and M. Ichinohe, *Phosphorus, Sulfur, Silicon Relat. Elem.*, **150/151**, 59 (1999).
103. A. Sekiguchi, N. Fukaya, M. Ichinohe and Y. Ishida, *Eur. J. Inorg. Chem.*, **6**, 1155 (2000).
104. J. Fujita, Y. Kawano, H. Tbito, M. Simoi and H. Ogino, *Chem. Lett.*, 1353 (1994).
105. L. K. Figge, P. J. Carroll and D. H. Berry, *Angew. Chem., Int. Ed. Engl.*, **35**, 435 (1996).
106. P. Antoniotti, F. Grandinetti and P. Volpe, *J. Phys. Chem.*, **99**, 17724 (1995).
107. S. P. So, *Chem. Phys. Lett.*, **313**, 587 (1999).
108. E. D. Jemmis, G. N. Srinivas, J. Leszcynski, J. Kapp, A. A. Korkin and P. v. R. Schleyer, *J. Am. Chem. Soc.*, **117**, 11361 (1995).
109. T. Kudo and S. Nagase, *Rev. Heteroat. Chem.*, **8**, 122 (1993).
110. C. Eaborn and A. K. Saxena, *J. Chem. Soc., Chem. Commun.*, 1482 (1984).
111. S. M. Dhaler, C. Eaborn and J. D. Smith, *J. Chem. Soc., Chem. Commun.*, 1183 (1987).
112. W. P. Neumann, H. Hillgärtner, K. M. Bains, R. Dicke, K. Vorspohl, U. Kobs and U. Nussbeutel, *Tetrahedron*, **45**, 951 (1989).
113. P. Jutzi, F. Kohl, P. Hofmann, C. Krueger and Y.-H. Tsay, *Chem. Ber.*, **113**, 757 (1980).
114. P. Jutzi, B. Hampel, M. B. Hursthouse and A. J. Howes, *Organometallics*, **5**, 1944 (1986).
115. F. X. Kohl, E. Schlueter, P. Jutzi, C. Krueger, G. Wolmershaeuser, P. Hofmann and P. Stauffert, *Chem. Ber.*, **117**, 1178 (1984).
116. A. H. Cowley, C. L. Macdonald, J. S. Silverman, J. D. Gorden and A. Voigt, *Chem. Commun.*, 175 (2001).
117. A. Haaland and B. E. R. Schilling, *Acta Chem. Scand., Ser. A*, **A38**, 217 (1984).

CHAPTER **11**

Alkaline and alkaline earth metal-14 compounds: Preparation, spectroscopy, structure and reactivity

PIERRE RIVIERE, ANNIE CASTEL AND MONIQUE RIVIERE-BAUDET

Laboratoire d'Hétérochimie Fondamentale et Appliquée, UMR 5069 du CNRS, Université Paul Sabatier, 31062 Toulouse cedex, France
Fax: 00 335 61 55 82 04; E-mail: riviere@chimie.ups-tlse.fr

The chemistry of organic germanium, tin and lead compounds — Vol. 2
Edited by Z. Rappoport

I. LIST OF ABBREVIATIONS

CIDEP	chemically induced dynamic electron polarization
CIP	contact ion pair
CVD	chemical vapor deposition
DBU	1,8-diazabicyclo[5.4.0]undec-7-ene
DCPH	dicyclohexylphosphine
DME	1,2-dimethoxyethane
DMF	*N,N*-dimethylformamide
DMPU	dimethylpropyleneurea
DMSO	dimethyl sulfoxide
ERO	electron rich olefin
ESR	electron spin resonance
HME	halogen metal exchange
HMPA	hexamethylphosphoramide
HMPT	hexamethylphosphorous triamide
LDA	lithium diisopropylamide
Men	mentyl
Mes	mesityl (2,4,6-trimethylphenyl)
MIMIRC	Michael–Michael ring closure
Np = Naph	Naphthyl
Nph	Neophyl: $PhMe_2CCH_2$
PMDETA	pentamethyldiethylenetriamine
SET	single electron transfer
SSIP	solvent separated ion pair
Tbt	2,4,6-tris[bis(trimethylsilyl)methyl]phenyl
TFPB	tetrakis[3,5-bis(trifluoromethyl)phenyl]borate
THF	tetrahydrofuran
Tip	2,4,6-triisopropylphenyl
TMEDA	tetramethylethylenediamine
TMS	trimethylsilyl or tetramethylsilane
Tol	tolyl: MeC_6H_4

II. INTRODUCTION

This chapter will concentrate on the chemistry of metal-14-centered anions (Ge, Sn, Pb). These compounds and their silyl analogues are ionic or polarized alkaline and alkaline earth metal-14 compounds, as well as delocalized molecules such as metalloles. Ammonium metallates $M_{14}{}^{-}R_4N^{+}$ or metal-14-centered anion radicals are also considered. The subject was explored during the 1960s and 1970s[1,2] and thoroughly reviewed in 1982 and 1995 in *Comprehensive Organometallic Chemistry*, Vols. I and II[3,4], and for silicon species in a previous volume of this series[5]. By that time the main routes to metal-14 anions were known. Since then, the subject has been developed in the topics of particular syntheses, stabilization using steric hindrance, electronic effects and complexation, spectroscopic and structural analyses[6–10].

This chapter will emphasize the synthesis, stabilization, spectroscopic and structural aspects, reactivity and synthetic use of these compounds. For completeness, a summary of the subject previously reviewed[3–10] will be included together with recent examples from the literature. The material will be divided into three sections: preparations, spectroscopic and structural studies (with some characteristic spectroscopic and structural data) and reactivity (with some applications in organic synthesis). Each part will be organized into sections dealing with a particular element in the order: germanium, tin and then lead.

In the next section dealing with preparations we shall begin with mono-anions, and continue with *gem* and vicinal metal-centered di-anions, including the metalloles and polymetallated mono- and poly-anions.

III. PREPARATIONS

A. M_{14}–Alkali Metal Compounds*

1. From M_{14}–H compounds

Because of their use in Chemical Vapor Deposition (CVD), monometallated germanes and polygermanes were synthesized directly by the reaction of the germane itself with alkali metals (equation 1)[3a].

$$GeH_4 \xrightarrow{M} H_3GeM \qquad (1)$$

$$M = Li, Na, K$$

Similarly, when GeH_4 was added to freshly prepared solutions of sodium silylsilanides $[NaSiH_n(SiH_3)_{3-n}, n = 0–3]$ in diglyme at 100 °C, a vigorous evolution of hydrogen and a loss of SiH_4 was observed[11]. The reaction led to sodium silylgermanides $[NaGeH_n(SiH_3)_{3-n}, n = 0–2]$. The silylation of sodium silylgermanide with silyl nonafluorobutanesulfonate proved to be particularly effective in the synthesis of Si–Ge containing process gases suitable for CVD (Scheme 1)[11].

Solutions of diaryldialkaligermanes were prepared by the reaction of alkali metals with the corresponding arylgermanes in HMPA/THF (equation 2)[12]. Aryldialkaligermanes have never been isolated, but their formation was chemically confirmed by the Ar_2GeD_2 formed by deuteriolysis. Their reaction with organic halides was not taken as sufficient evidence of their formation. In ^{13}C NMR they show a deshielded *ipso* aromatic carbon compared

* M_{14} denotes an element belonging to Group 14 of the periodic Table, such as Si, Ge, Sn, Pb.

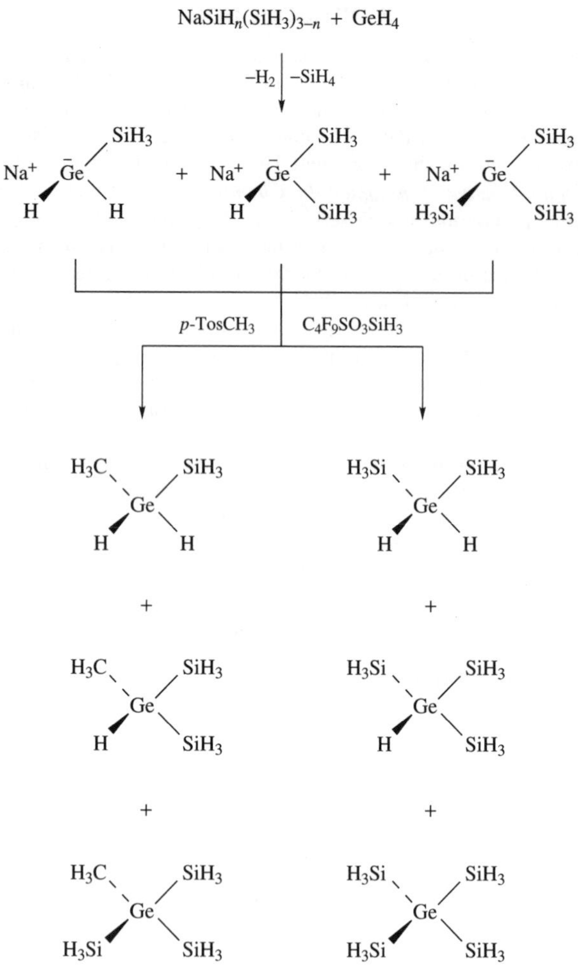

$$NaSiH_n(SiH_3)_{3-n} + GeH_4$$

$-H_2$ | $-SiH_4$

$Na^+\ \bar{Ge}(H)(H)(SiH_3)$ + $Na^+\ \bar{Ge}(H)(SiH_3)(SiH_3)$ + $Na^+\ \bar{Ge}(SiH_3)(SiH_3)(SiH_3)$

p-TosCH$_3$ | $C_4F_9SO_3SiH_3$

$H_3CGe(H)(H)(SiH_3)$ + $H_3CGe(H)(SiH_3)_2$ + $H_3CGe(SiH_3)_3$

$H_3SiGe(H)(H)(SiH_3)$ + $H_3SiGe(H)(SiH_3)_2$ + $H_3SiGe(SiH_3)_3$

SCHEME 1

to the starting arylgermane.

$$Ar_2GeH_2 + 4\ M \xrightarrow[\text{or THF}]{\text{HMPA/THF}} Ar_2GeM_2 + 2\ M\ H \qquad (2)$$

$$Ar = C_6H_5,\ p\text{-}CH_3C_6H_4,\ M = Li, Na, K$$

Hydrogermolysis of alkyllithiums in ether or THF is one of the most general ways to synthesize organogermyllithiums and organohydrogermyllithiums (equation 3)[3a,4a,9]. Highly basic solvents such as trialkylamines or DBU increase the yield of alkyl or arylhydrogermyllithiums, but also favor their dimerization through Li–H elimination.

$$R^1R^2R^3GeH + R^4Li \xrightarrow[\text{or THF}]{\text{ether}} R^1R^2R^3GeLi + R^4H \quad (3)$$

$R^1 = R^2 = R^3 = Et$, $R^4 = Bu$ (ether, 10% yield; THF + Et_3N, 78% yield)

$R^1 = R^2 = Ph$, $R^3 = H$, $R^4 = Bu$ (THF + Et_3N, −40 °C, 72% yield)

A retentive stereochemistry in these reactions has been established (equation 4)[1,13,14].

$$\alpha\text{-Naph(Ph)(R)GeH} + n\text{-BuLi} \xrightarrow{\text{retention}} \alpha\text{-Naph(Ph)(R)GeLi} + \text{BuH} \quad (4)$$

Various organogermyllithiums $R_nH_{3-n}GeLi$ (R = Ph, Mes etc., $n = 1-3$) were prepared easily following this general procedure[3a,4a,9]. Their stability depends on the nature of the R groups linked to germanium, on the possibility of complexation of the two metal centers and on the possibility of electron-pair delocalization.

Trimesitylgermyllithium was isolated as a yellow solid in the form of a THF complex in 62% yield (equation 5)[15]. The polarity of the germanium–lithium bond was evident from the high field shift of the Mes signals in ^1H NMR spectra and the low field shift of the *ipso*-aromatic carbons in ^{13}C NMR spectra.

$$Mes_3GeH + t\text{-BuLi} \xrightarrow[20\,^\circ C]{\text{THF}} Mes_3GeLi + i\text{-BuH} \quad (5)$$

In the same way, bulky organohydrogermyllithiums were prepared and stabilized in complexes with a crown ether (equation 6)[16]. They were characterized by deuteriolysis, alkylation (MeI, Me_2SO_4) and reactions with acyl chlorides.

$$R_2GeH_2 + t\text{-BuLi} \xrightarrow[-i\text{-BuH}]{\text{THF}} R_2HGeLi \xrightarrow[R = \text{Mes}]{\text{12-crown-4}} R_2HGe^-(\text{Li(12-crown-4)}_2)^+ \quad (6)$$

R = Ph(−40 °C, 95%)

R = Mes(−20 °C, 84%)

Instead of solvent complexation, intramolecular chelation can also stabilize germanium-centered anions. Aryl(8-methoxy-1-naphthyl)hydrogermyllithiums, prepared from the corresponding chelated organogermane in 50–90% yield, are stable and their spectroscopic characteristics suggest a double-metal complexed structure (equation 7)[17].

$$\text{8-(MeO)-1-(RGeH}_2\text{)-naphthalene} \xrightarrow[(-40\,^\circ C \text{ to } 20\,^\circ C)]{t\text{-BuLi/THF}} \text{8-(MeO)-1-(RGeHLi)-naphthalene (O}\cdots\text{Li}\cdots\text{Ge)} + i\text{-BuH} \quad (7)$$

R = Mes, Ph, 8-MeO-1-Np

When the same reaction was conducted with two equivalents of n-BuLi followed by alkylation with MeI, it gave the expected dimethylated germanium derivative almost quantitatively. However, the reaction mechanism might be more complex than the expected nucleophilic substitution by a germanium-centered dianion. A more complex radical process is probably involved[17] (see also Scheme 5). Similar stabilization of germanium-centered anions using 2-(dimethylamino)phenyl as a chelating ligand allowed the isolation of crystalline germylanion as a monochelated monomer with the lithium atom coordinated to one of the amino groups and to two THF molecules as established by X-ray diffraction. The compound was prepared by deprotonation of the corresponding hydrogermane with *t*-BuLi in THF at −40 °C, in 93% yield (equation 8). The yield was estimated by quenching with D_2O. It was isolated as pale yellow monocrystals after recrystallization from toluene at −20 °C[18].

$$R_2Ge(H)\text{-}C_6H_4\text{-}2\text{-}NMe_2 \xrightarrow[\text{THF, } -40\,°C,\ 1\ h]{t\text{-BuLi } (\times 3.3)} [2\text{-}(Me_2N\cdots Li^+(THF)_2)C_6H_4]GeR_2^- \quad (93\%) \tag{8}$$

$R = 2\text{-}Me_2NC_6H_4$

This compound, which is monomeric in the solid state, exists in toluene solution in equilibrium with its dimer (Scheme 2)[18] as evident from analyses of variable temperature ^1H, ^7Li and ^{13}C NMR. Whereas the monomer is favored at higher temperatures in terms of the entropy effect, the dimer is favored at lower temperatures in terms of electrostatic interactions through which electrons of the germanium negative centers are efficiently stabilized by the two lithium cations. The thermodynamic parameters for this equilibrium were estimated.

Germylpotassium reagents stabilized by chelating amino or alkoxy ligands were prepared by hydrogermolysis of benzylpotassium in THF. They were isolated as bright yellow powders in 74 to 93% yield (equation 9)[19].

$$Ar_3GeH + C_6H_5CH_2K \xrightarrow[-C_7H_8]{\text{THF}} Ar_3Ge^- \ K^+ \tag{9}$$

$Ar = (2\text{-}CH_2NMe_2)C_6H_4,\ (2\text{-}MeO)(5\text{-}Me)C_6H_3$

Another way to stabilize germanium-centered anions is to induce a partial delocalization of the lone pair. Germolyl anions are easily available from the hydrogermolysis of BuLi, $PhCH_2K$ or $(SiMe_3)_2NK$ with 12-crown-4 or 18-crown-6 complexing lithium or potassium (Scheme 3)[20–22]. They are usually sufficiently stable to be isolated as yellow crystals in good yields. They display a pyramidal germanium center with weak bond localization in the diene portion of the ring[21]. This conclusion emerges from structural studies of some of them[22] (see Section IV).

A *gem* germanium-centered dilithium has never been clearly observed, although it was postulated several times or characterized transiently in solution in reactions of alkali metals with arylhydrogermanes (equation 2), lithiosilolysis of a silylgermyllithium (equation 10)[23] and hydrogermolysis of an excess of *t*-BuLi (Scheme 4)[16]. In the last case, the pseudogermyldilithium intermediate reacted with alkyl or acyl halides to give the expected dialkylated or diacylated germanium compound in good yields, but its deuteriolysis gave

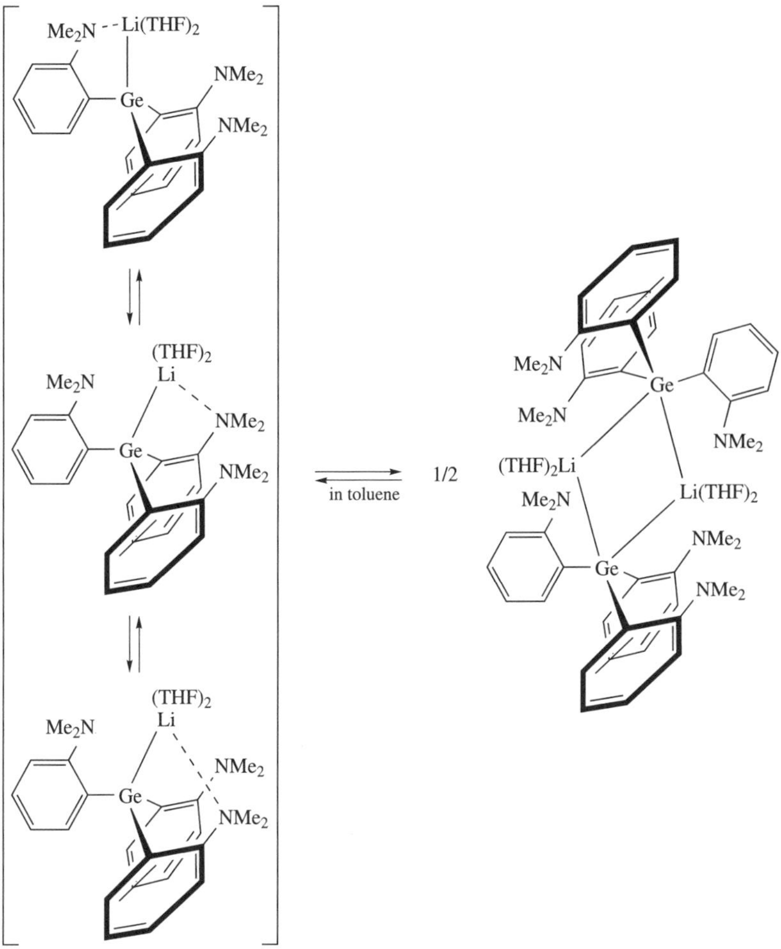

SCHEME 2

only the monodeuteriated organogermane.

$$Me_3SiGeEt_2Li + Me_3SiLi \xrightarrow{\text{HMPT}} Me_3SiSiMe_3 + \underset{(70\%)}{Et_2GeLi_2} \quad (10)$$

Analysis of pseudogermyldilithiums of this kind, obtained according to Scheme 4 and stabilized by steric hindrance, shows an aggregate between germyl and alkyllithiums. In an examination of the reaction pathway leading to dialkylation of the germanium center (Scheme 5)[24], it was shown that the reaction with alkyl halides is not an unequivocal probe for the characterization of metal-14-centered dianions. In the same way, germolyldianions have to be considered as delocalized dimeric sandwich structures with both alkali metals coordinated by the germolyldianions in an η^5 fashion (see Section IV, Figure 1)[21]. Their aromaticity has been studied (see Section IV)[22].

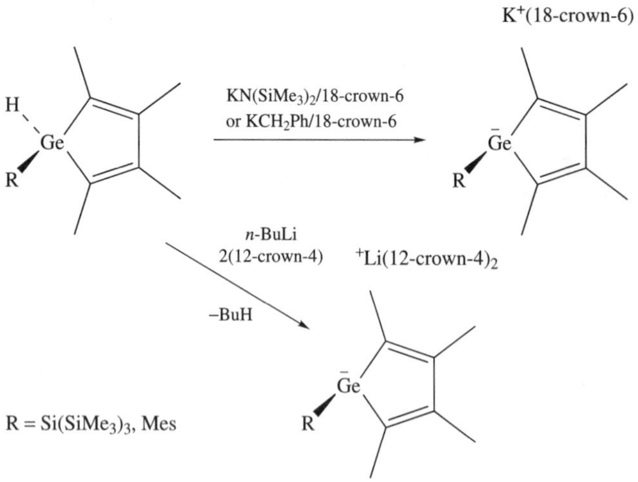

K^+(18-crown-6)

H, R–Ge (ring) $\xrightarrow{\text{KN(SiMe}_3)_2\text{/18-crown-6 or KCH}_2\text{Ph/18-crown-6}}$ R–$\bar{\text{Ge}}$ (ring)

$\xrightarrow{n\text{-BuLi, 2(12-crown-4), –BuH}}$ $^+$Li(12-crown-4)$_2$ R–$\bar{\text{Ge}}$ (ring)

R = Si(SiMe$_3$)$_3$, Mes

SCHEME 3

$Ph_2GeH_2 \xrightarrow[\text{excess}]{t\text{-BuLi}}$ (Then) —

MeI → Ph_2HGeMe (37%) + Ph_2GeMe_2 (56%)

MesCOCl → $Ph_2HGeCOMes$ (15%) + $Ph_2Ge(COMes)_2$ (19%)

D_2O → Ph_2HGeD (~100%)

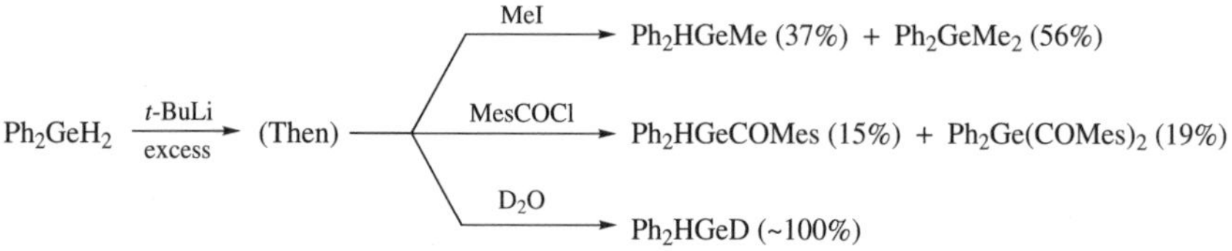

SCHEME 4

More metal centered are the vicinal organo 1,2-digermyldianions, easily prepared by metallation of the corresponding organo 1,2-dihydrodigermanes with *t*-BuLi in THF. Perfectly stable at low temperature (−40 °C), 1,2-digermyldilithium compounds offer a convenient synthesis of cyclodigermanes by their *in situ* reaction with 1,3-dibromopropane (equation 11) or with *trans*-$(Et_3P)_2PtCl_2$, as well as of cyclopolygermanes when reacted with $R_2Ge(OMe)_2$ or $R_2(MeO)GeGe(MeO)R_2$[25].

$$Ph_2HGeGeHPh_2 \xrightarrow[-40\,^{\circ}C]{t\text{-BuLi}} Ph_2LiGeGeLiPh_2 \xrightarrow{Br(CH_2)_3Br} \text{cyclo-}(Ph_2Ge\text{–}GePh_2\text{–}(CH_2)_3) \quad (11)$$

In the reaction between (chlorodimesitylsilyl)diarylgermanes and *t*-BuLi in THF, new germyllithium compounds, resulting from an intramolecular hydrogermolysis of an intermediate silyllithium (Scheme 6), were characterized[26]. The silylated germylanion was characterized by deuteriolysis and alkylation.

Equimolar amounts of lithium diisopropylamide (LDA) and tin hydrides reacted in THF to form diisopropylamine and the corresponding stannyllithium (equation 12)[27,28]. In diethyl ether or hexane, an excess of tin hydride was required for complete reaction which leads to the ditin compound and lithium hydride (equation 13).

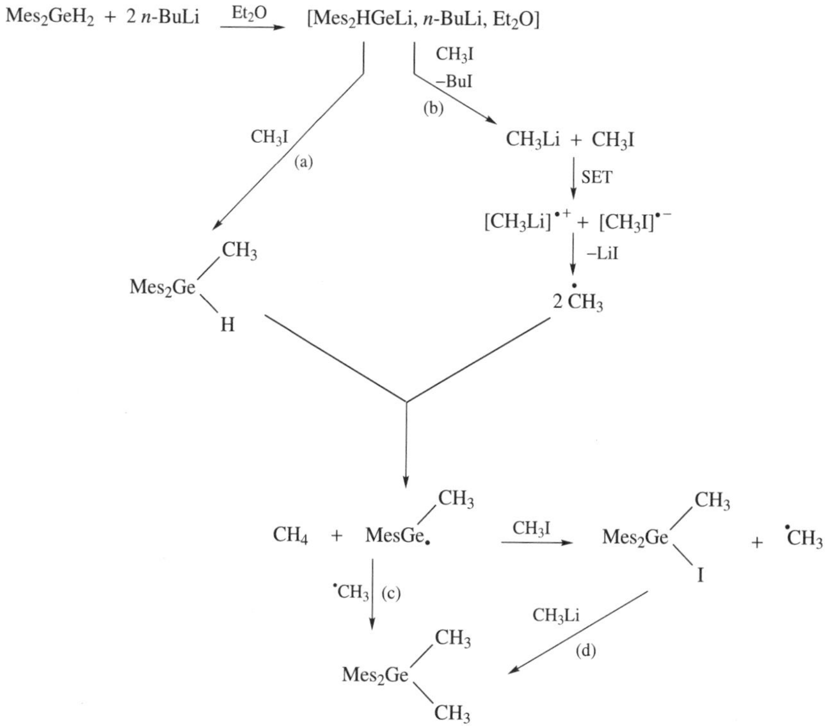

Mes_2GeH_2 + 2 n-BuLi $\xrightarrow{Et_2O}$ [Mes_2HGeLi, n-BuLi, Et_2O]

(a) CH_3I → $Mes_2Ge(CH_3)H$

(b) CH_3I, −BuI → CH_3Li + CH_3I $\xrightarrow{SET}$ $[CH_3Li]^{\bullet+}$ + $[CH_3I]^{\bullet-}$ $\xrightarrow{-LiI}$ 2 $\dot{C}H_3$

→ CH_4 + $MesGe^{\bullet}(CH_3)$ $\xrightarrow{CH_3I}$ $Mes_2Ge(CH_3)I$ + $^{\bullet}CH_3$

(c) $MesGe^{\bullet}(CH_3)$ $\xrightarrow{^{\bullet}CH_3}$ $Mes_2Ge(CH_3)_2$

(d) $Mes_2Ge(CH_3)I$ $\xrightarrow{CH_3Li}$ $Mes_2Ge(CH_3)_2$

SCHEME 5

$Ar_2Ge(H){-}SiMes_2(Cl)$ $\xrightarrow[THF]{t\text{-}BuLi}$ $Ar_2Ge(H){-}SiMes_2(Li)$ + t-BuCl

$Ar_2Ge(H){-}SiMes_2(Li)$ → $Ar_2Ge(Li){-}SiMes_2(H)$

$Ar_2Ge(Li){-}SiMes_2(H)$ $\xrightarrow{MeOH\text{-}d_4}$ $Ar_2Ge(D){-}SiMes_2(H)$

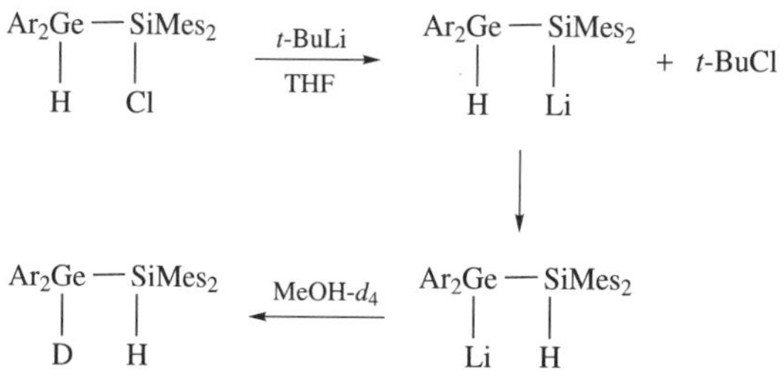

Ar = Mes, Tip

SCHEME 6

$$R_3SnH \xrightarrow[THF,\ 0\,^{\circ}C]{LDA} R_3SnLi + i\text{-}Pr_2NH \qquad (12)$$

R = Me, Bu

$$2\ R_3SnH \xrightarrow[ether]{LDA} R_3SnSnR_3 + LiH + i\text{-}Pr_2NH \qquad (13)$$

R = Me

It was shown that in equation 13 the formation of ditin proceeds from an intermediate hydrostannyllithium (equation 14)[27].

$$\mathrm{Me_3SnLi + Me_3SnH \longrightarrow [Me_3SnSn(H)Me_3]^- \; Li^+ \longrightarrow (Me_3Sn)_2 + LiH} \tag{14}$$

Starting from diorganostannanes, the same reaction gave high yields of stannylanions (equation 15), but did not afford the *gem*-dilithium metal derivatives when an excess of LDA was used[29,30].

$$\mathrm{Bu_2SnH_2 \xrightarrow{LDA^*} Bu_2SnHLi \xrightarrow{D_2O} Bu_2SnHD} \tag{15}$$

(*) : stoichiometric or excess

However, as also observed in germanium chemistry, a mixture of dihydride with an excess of lithiated reagent behaves as a stannyldianion in reactions with organic halides, polyformaldehyde or epoxides (Scheme 7)[29], and it thus appears as an interesting synthetic reagent.

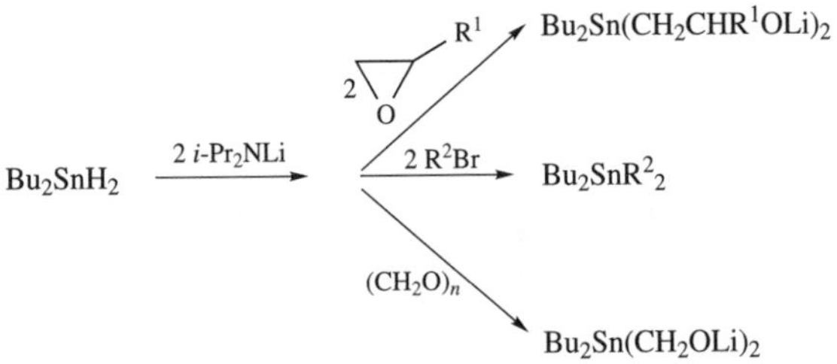

$$\mathrm{Bu_2SnH_2 \xrightarrow{2\,\mathit{i}\text{-}Pr_2NLi}} \begin{cases} \xrightarrow{2\ \text{epoxide}(R^1)} \mathrm{Bu_2Sn(CH_2CHR^1OLi)_2} \\ \xrightarrow{2\,R^2Br} \mathrm{Bu_2SnR^2{}_2} \\ \xrightarrow{(CH_2O)_n} \mathrm{Bu_2Sn(CH_2OLi)_2} \end{cases}$$

R^1 = H, Me; R^2 = allyl, benzyl, $Cl(CH_2)_3$

SCHEME 7

Whereas mixed methylneophyltin hydrides reacted with NaH in DMSO to give the corresponding organotin sodium (60–70% yield) (equation 16), the more bulky trineophyltin hydride gave hexaneophylditin as the only product (equation 17)[31].

$$\mathrm{MeNph_2SnH + NaH \xrightarrow[-H_2]{DMSO} MeNph_2SnNa} \tag{16}$$

$$\mathrm{Nph_3SnH + NaH \xrightarrow{DMSO} Nph_3SnSnNph_3} \tag{17}$$

The same neophyltin hydrides did not react with LDA, possibly because of steric hindrance.

2. *Substitution halogen/metal*

The formation of a metal anion from metal halide and lithium can be rationalized by a double and successive monoelectronic transfer, with (or without when weak $M_{14}-M_{14}$ bond is involved) formation of a transient digermane (Scheme 8).

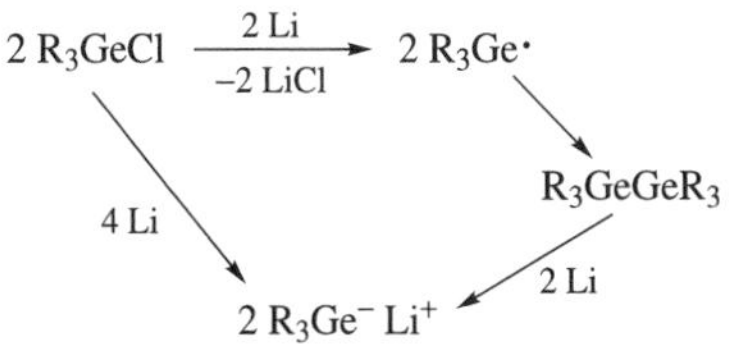

$$\begin{array}{ccc} 2\,R_3GeCl & \xrightarrow[-2\,LiCl]{2\,Li} & 2\,R_3Ge^{\bullet} \\ \searrow^{4\,Li} & & \downarrow \\ & & R_3GeGeR_3 \\ & 2\,R_3Ge^-\,Li^+ & \xleftarrow{2\,Li} \end{array}$$

SCHEME 8

Following this scheme, Me_3GeLi was prepared from Me_3GeCl and lithium in HMPA/Et_2O (equation 18)[32] and then used in the preparation of germasilanes, which are useful in the synthesis of stable silyllithiums (equation 19)[32].

$$4\,Me_3GeCl + 8\,Li \xrightarrow[-4\,LiCl]{HMPA/Et_2O(-78\,^{\circ}C)} 4\,Me_3GeLi \xrightarrow[-4\,LiCl]{SiCl_4} (Me_3Ge)_4Si \quad (18)$$

$$(Me_3Ge)_4Si + MeLi \xrightarrow[20\,^{\circ}C]{THF} Me_4Ge + (Me_3Ge)_3SiLi \quad (19)$$

A series of aryl and alkylarylgermyllithiums and stannyllithiums were synthesized in the same way with the purpose of transforming them by laser photolysis into the corresponding metal-centered radicals[33] (equation 20).

$$(Ph_nMe_{3-n}E)_2 \xrightarrow{Li} 2\,Ph_nMe_{3-n}E^- \xrightarrow{h\nu} 2\,Ph_nMe_{3-n}E^{\bullet} \quad (20)$$

$$E = Si, Ge, Sn; n = 1{-}3$$

The substitution of halogen by metal was also used to prepare metalloyl anions and, more specifically, the dianions of germoles (equation 21)[34–37].

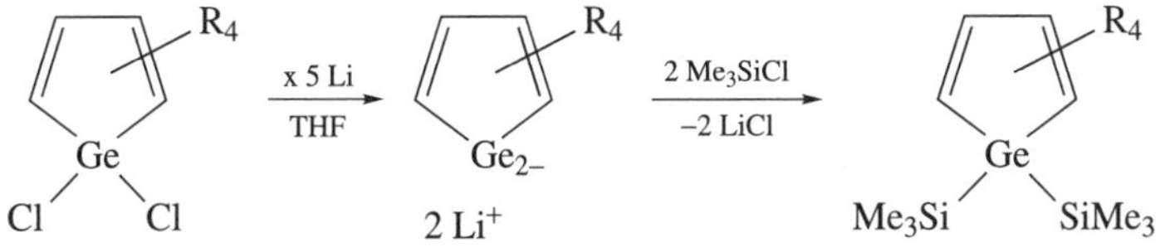

$$R_4C_4GeCl_2 \xrightarrow[THF]{x\,5\,Li} R_4C_4Ge^{2-}\;2\,Li^+ \xrightarrow[-2\,LiCl]{2\,Me_3SiCl} R_4C_4Ge(SiMe_3)_2 \quad (21)$$

R = Et, Ph

The compound with R = Ph was prepared by stirring a THF solution of the corresponding dichloride with lithium for 12 h at room temperature, followed by extraction with dioxane and recrystallization at $-20\,^{\circ}C$. The crystallographic study showed two structurally distinct forms depending on the crystallization temperature (see Section IV, Figure 8). The two forms can be considered as highly aromatic[35].

Germolyl dianion having R = Et was synthesized following a similar process but using a mixture of THF and TMEDA. An X-ray crystal structure of the resulting (2,3,4,5-tetraethyl-1-germacyclopentadiene)$^{2-}$ 2 Li^+ showed three lithium cations around one germole ring in η^1, η^5 and η^5 sites, giving the composition $[\eta^5\text{-Li.TMEDA}][\eta^5\text{-Li}]_{1/2}$ $[\eta^1\text{-Li}]_{1/2}.[Et_4C_4Ge]$. Nearly equal C−C bond distances in the Et_4C_4Ge ring suggest a delocalized π-system[36] (see Section IV).

Conversion of 1,1-dichloro-2,3-diphenyl-1-germaindene to the lithium (or sodium) dianion according to reaction 21 in THF−TMEDA led to an unusual phenomenon of aromatization of the GeC_4 portion of 1-germaindene at the expense of the aromatic C_6 ring[37].

The reaction of divalent M_{14} halides (Ge, Sn) with an excess of alkali metal (Na, K) in benzene led to vicinal-centered digermyl and distannyl dianions (equation 22)[38].

$$2\ \text{Cl-E-2,6-Tip}_2\text{C}_6\text{H}_3 \xrightarrow{\text{M}} [\text{Tip}\cdots\text{M}\cdots\text{Tip}]\ (2,6\text{-Tip}_2\text{C}_6\text{H}_3)\text{E}{-}\text{E}(\text{C}_6\text{H}_3\text{-2,6-Tip}_2)\ [\text{Tip}\cdots\text{M}\cdots\text{Tip}] \qquad (22)$$

E = Ge, M = Na
E = Sn, M = K

These compounds were isolated as red crystals (Ge: 23%, Sn: 30% yield). Their X-ray structures are similar (Figure 1), but the crystals are not isomorphous. The structure of the digermyl disodium compound is characterized by an inversion center at the midpoint of the Ge−Ge bond, whereas the distannyldipotassium has a twofold axis of symmetry along the potassium−potassium vector. In each structure the alkali metal counterions are associated with each dianion by coordination in a π sandwich fashion, and the alkali−metal$_{14}$ distances are longer than the normal single bond. By treating $ClGeC_6H_3\text{-2,6-Mes}_2$ with KC_8, the same authors obtained a crystallized trigermylallyl anion formed from the corresponding cyclotrigermyl radical (equation 23)[39]. The structure of the trigermylallyl anion, isolated as dark green crystals, indicates a planar geometry at the central germanium and the K counterion complexed in a π fashion with two mesityl ligands. The complexing mesityl rings form a bent sandwich structure at the K^+ ion.

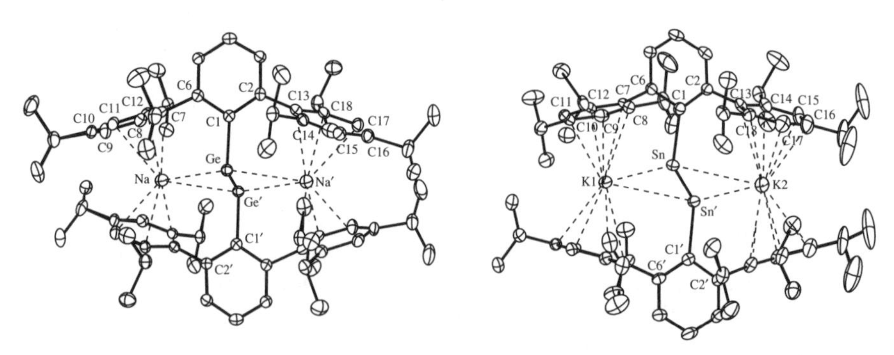

FIGURE 1. Solid state structure of Na_2 or K_2 $[E(2,6\text{-Tip}_2C_6H_3)]_2$ (E = Ge, Sn) with hydrogen atoms omitted. Reprinted with permission from Reference 38. Copyright 1998 American Chemical Society

$$R(Cl)Ge \xrightarrow[3\,h]{KC_8/THF} R_3Ge_3^{\bullet} \xrightarrow{KC_8/THF} [R_3Ge_3]^- K^+ \qquad (23)$$

R = 2,6-$Mes_2C_6H_3$ (Mes, Mes)

Bulky stannyl anions which are not available from stannyl hydrides (see Section I.A.1) were prepared by using the reaction of alkali metals with tin halides (equations 24–26)[31,40,41].

$$(Me)_n(Nph)_{3-n}SnBr + Li \xrightarrow{THF} (Me)_n(Nph)_{3-n}SnLi \qquad (24)$$

$n = 0, 1, 2$(70–75% yields)

$$Bu_3SnX \xrightarrow[THF]{[K^+/K^-]18\text{-crown-6}} Bu_3SnK \qquad (25)$$

$$Ph_3SnCl + Na \xrightarrow[-NaCl]{THF} Ph_3SnNa \qquad (26)$$

In a similar way, but in a one-pot experiment (equation 27)[42], a stable silylated stannyl anion was prepared in 44% yield and isolated as white crystals. X-ray crystallography showed a distorted tetrahedral tin atom with a Sn–Li distance (2.87 Å) which is shorter than those published for $[Li(THF)_3.Sn\{(4\text{-}MeC_6H_4)NSiMe_2\}_3CH]$ (2.89 Å) and $[Li(PMDETA).SnPh_3]$ (2.87 Å), but longer than the Li–M_{14} distances in the Ge and Si analogues: (2.67 Å) and (2.64 Å), respectively.

$$Me_3SiCl \xrightarrow[2.\ SnCl_4(0.25\ eq.)]{1.\ Li,\ THF} \underset{(44\%)}{(Me_3Si)_3SnLi(THF)_3} \qquad (27)$$

Similar reactions appear in an industrial patent using oxides instead of halides (equation 28)[43].

$$(R_3Sn)_2O \xrightarrow[-Li_2O]{Li\ (excess)} 2\ R_3SnLi \qquad (28)$$

Trialkyllead–lithium compounds were prepared by the reaction of trialkyllead bromide with an excess of lithium metal in THF at −78 °C (equation 29)[44].

$$Me_3PbBr \xrightarrow[-78\,^{\circ}C]{Li/THF} Me_3PbLi \qquad (29)$$

3. Nucleophilic cleavage of M_{14}–M_{14} bonds and transmetallation reactions

Nucleophilic cleavage by potassium *t*-butoxide of digermanes, distannanes and diplumbanes in *N*, *N′*-dimethylpropyleneurea (DMPU) as solvent provided a facile and general method for preparations of germyl-, stannyl- and plumbyl-centered anions in good yields (60–90%) (equation 30)[45]. Cleavage of mixed $M_{14}-M'_{14}$ bonds have also been investigated[46].

$$Ph_3E\text{-}EPh_3 + Me_3COK \xrightarrow{\text{DMPU}} Ph_3EK + Ph_3EOCMe_3 \quad (30)$$

$$E = Ge, Sn, Pb$$

Other nucleophilic reagents have been used, such as MeLi[32,47] (equation 31), $PhCH_2K$ (equation 32)[21] and Bu_4NF (equation 33)[48].

$$(Me_3Si)_4Ge + MeLi \xrightarrow{\text{THF}} Me_4Si + (Me_3Si)_3GeLi(THF)_3 \quad (31)$$

$$\text{(tetramethylgermole)}Ge(SiMe_3)_2 \xrightarrow[\text{18-crown-6}]{KCH_2Ph} \text{(tetramethylgermole)}Ge^{-}SiMe_3 \; K^{+}\text{(18-crown-6)} \quad (32)$$

$$Me_3Ge{-}SiMe_3 \xrightarrow[\text{HMPA, RT, 5 h}]{Bu_4NF/5\ \text{mol\%}} Me_3SiF + Me_3Ge^{-}\ {}^{+}NBu_4 \quad (33)$$

Tris(trimethylsilyl)germyl lithium obtained according to equation 31 as a THF complex[32,47] was isolated as colorless needles (88% yield). By using PMDETA in hexane, a new complexed $(Me_3Si)_3GeLi(pmdeta)$ was also obtained as colorless crystals. Both germyllithium derivatives have a similar Ge–Li distance (2.666 Å and 2.653 Å) and a tetrahedral germanium. The ammonium germanate obtained in equation 33 is highly ionic, as revealed by a bathochromic effect ($\Delta\lambda = +125$nm) observed by comparison with the starting silagermane. It gives classical nucleophilic reactions with alkyl halides and nucleophilic additions to carbonyl compounds[48].

Cleavage with cesium fluoride was used in the case of stannylsilanes (equation 34)[49]. The generated stannyl anions are very effective in synthetic applications, mainly in abstraction of halogen to initiate organic $4 + 2$ cycloadditions (equation 35)[49], a reaction which constitutes one of its chemical characterizations.

$$Bu_3SnSiMe_3 + CsF \longrightarrow [Bu_3SnSi(F)Me_3]^{-}Cs^{+} \longrightarrow Bu_3Sn^{-}Cs^{+} + Me_3SiF \quad (34)$$

Treatment of $Sn_2(CH_2Bu\text{-}t)_6$ with potassium naphthalenide in THF at 25 °C afforded crystalline $K[Sn(CH_2Bu\text{-}t)_3](THF)_2$ which in toluene gave $K[Sn(CH_2Bu\text{-}t)_3](\eta^6\text{-}C_6H_5Me)_3$ (equation 36)[50]. The X-ray structure of the latter revealed that potassium is in a distorted tetrahedral environment with a K–Sn bond length of 3.55 Å (See Section IV).

Br

Br

$Me_3SiSnBu_3$ (2.0 eq.)
CsF (2.0 eq.)
MeOOC–CH=CH–COOMe

COOMe

COOMe

93%

+

trace

(35)

$$(t\text{-BuCH}_2)_3\text{SnSn}(\text{CH}_2\text{Bu-}t)_3 \xrightarrow[\text{THF, 25 °C}]{\text{K/Naphthalene}} 2\ (t\text{-BuCH}_2)_3\text{SnK(THF)}_2 \xrightarrow{\text{PhMe, 25 °C}} 2\ (t\text{-BuCH}_2)_3\text{SnK}(\eta^6\text{-C}_6\text{H}_5\text{Me})_3 \qquad (36)$$

Nucleophilic cleavage of the lead–lead bond is one of the few ways to form plumbyl anions (equations 30, 37 and 38)[45,51,52], but in the case of a sterically hindered and thus weak lead–lead bond, a metal–metal cleavage is required (equation 39)[45,51–53].

$$\text{Bu}_3\text{PbPbBu}_3 \xrightarrow[\text{THF/0 °C}]{n\text{-BuLi}} \text{Bu}_4\text{Pb} + \text{Bu}_3\text{PbLi} \qquad (37)$$

$$\text{Ph}_3\text{PbPbPh}_3 + \text{PhLi} \xrightarrow{\text{THF}} \text{Ph}_4\text{Pb} + \text{Ph}_3\text{PbLi} \qquad (38)$$

$$\text{Mes}_3\text{PbPbMes}_3 + \text{Li} \longrightarrow 2[\text{Mes}_3\text{Pb}]^-\text{Li}^+ \qquad (39)$$

When they can be used, transmetallation reactions are also interesting synthetic routes to M_{14} anions. In trialkylgermyl alkali metal compounds, the alkali metal is easily displaced by a less electropositive alkali metal (equation 40)[54].

$$\text{Et}_3\text{GeLi} \xrightarrow[-\text{Li}]{+\text{Na}} \text{Et}_3\text{GeNa} \xrightarrow[-\text{Na}]{+\text{K}} \text{Et}_3\text{GeK} \xrightarrow[-\text{K}]{+\text{Cs}} \text{Et}_3\text{GeCs} \qquad (40)$$

In the same way, organodigermylmercury compounds led to germyllithium compounds (equations 41 and 42)[1,3a,55]

$$(R_3Ge)_2Hg + 2\ M \xrightarrow[\text{or } n\text{-}C_6H_{14}]{C_6H_6} Hg + 2\ R_3GeM \quad (41)$$

M = Li, Na, K, Rb, Cs

$$[GeR_2{-}GeR_2{-}Hg]_n \xrightarrow[\text{HMPT}]{\text{Li}} nR_2LiGe\text{-}GeLiR_2 \xrightarrow{\text{MeI}} nR_2MeGeGeMeR_2 \quad (42)$$

R = Ph, Et

4. Other syntheses and other metal-14 anions

There are other examples of reactions leading to or involving the formation of metal-centered anions, sometimes of a new type.

For example, treatment of a digermene with an equivalent of lithium naphthalenide in DME provided a digermenyl lithium species isolated as red microcrystals and identified by its reaction with methanol (equation 43)[56].

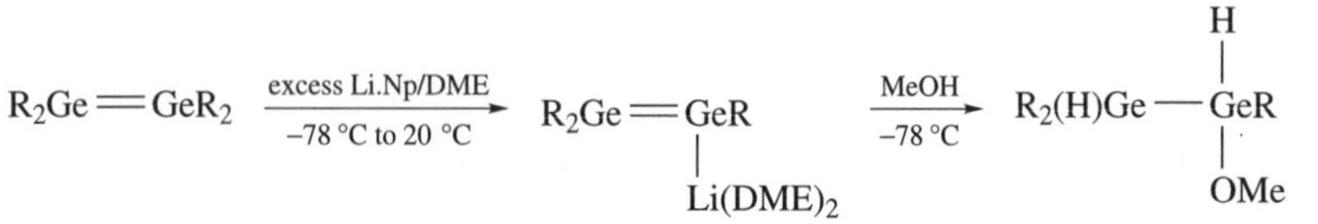

$$R_2Ge{=}GeR_2 \xrightarrow[-78\,^{\circ}C \text{ to } 20\,^{\circ}C]{\text{excess Li.Np/DME}} R_2Ge{=}GeR(Li(DME)_2) \xrightarrow[-78\,^{\circ}C]{\text{MeOH}} R_2(H)Ge{-}GeR(H)(OMe) \quad (43)$$

Oxidative addition of an organometallic compound to a halogermylene also leads to a germylanion when the reaction is carried out at low temperature to prevent the nucleophilic substitution of the germanium–halogen bond. The transient halogermylsodium thus formed (Scheme 9) undergoes a fast 'condensation' to a more stable digermanylsodium characterized *in situ* by NMR from the two different $\delta^{29}Si$ signals: 21.9 and 23.7 ppm, and by chemical means from its reaction with alkyl halides[57].

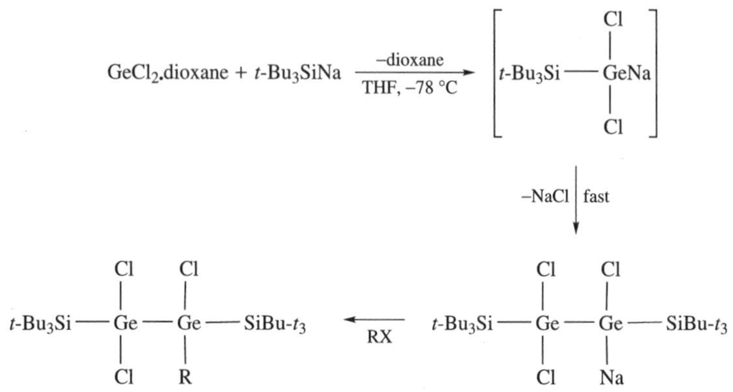

$$GeCl_2.dioxane + t\text{-}Bu_3SiNa \xrightarrow[\text{THF}, -78\,^{\circ}C]{-\text{dioxane}} [t\text{-}Bu_3Si{-}GeCl_2Na]$$

$$[t\text{-}Bu_3Si{-}GeCl_2Na] \xrightarrow[\text{fast}]{-\text{NaCl}} t\text{-}Bu_3Si{-}GeCl_2{-}GeCl(Na){-}SiBu\text{-}t_3 \xrightarrow{\text{RX}} t\text{-}Bu_3Si{-}GeCl_2{-}GeCl(R){-}SiBu\text{-}t_3$$

SCHEME 9

Synthesis of a cyclotetragermyl dianion from a vicinal digermyl dianion was achieved through nucleophilic attack of *t*-BuLi (equation 44)[58]. This compound, isolated as yellow crystals, was characterized by X-ray diffractometry. It contains a planar four-membered ring of germanium atoms.

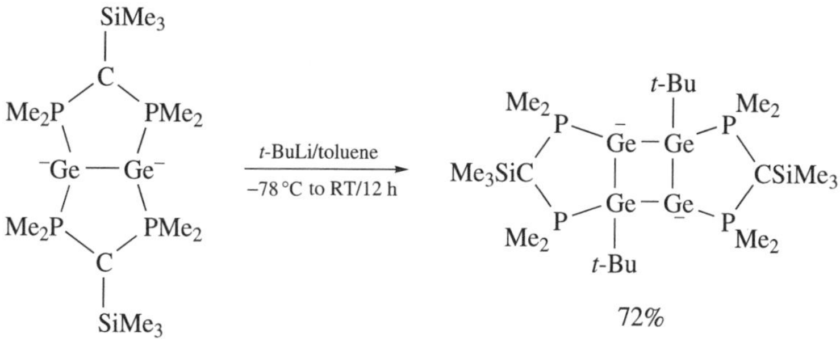

$$\text{cyclic } [(Me_2P)_2C(SiMe_3)]_2Ge_2^{2-} \xrightarrow[-78\,^\circ C \text{ to RT/12 h}]{t\text{-BuLi/toluene}} [t\text{-Bu}Ge_2\{(Me_2P)_2CSiMe_3\}]_2^{2-} \quad 72\% \qquad (44)$$

Heterobinuclear complexes featuring the bridging germole dianion ligand have been described (equation 45)[59,60].

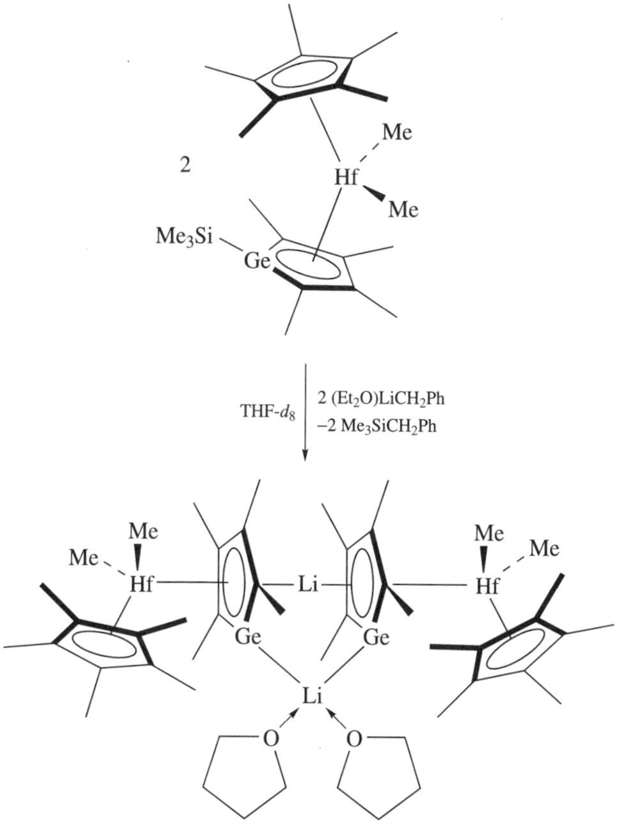

$$2\ (C_5Me_5)(C_4Me_4GeSiMe_3)HfMe_2 \xrightarrow[\text{THF-}d_8]{2\ (Et_2O)LiCH_2Ph,\ -2\ Me_3SiCH_2Ph} \qquad (45)$$

Germylanions have been generated electrochemically (equation 46)[61].

$$Ph_nMe_{3-n}GeH \xrightarrow[Bu_4N^+BF_4^-]{e} Ph_nMe_{3-n}Ge^- \ {}^+NBu_4 \qquad (46)$$

The reaction of these electrochemically generated anions with phenylacetylene (i) was compared with the same reaction with the corresponding germyllithiums (ii). The stereochemistry of the resulting vinyl intermediates was different. Reaction (ii) produced a mixture of *Z* (33%) and *E* (57%) 2-organogermylstyrene, while reaction (i) led to *Z* (75%) and *E* (25%) 2-organogermylstyrene. These differences were interpreted in terms of the dependence of the activation energy for the isomerization of the intermediate on the nature of the counterion and its state of solvation[61].

Sonication of 1,1-dichloro-2,3,4,5-tetraethylgermole with an excess of lithium in THF and TMEDA gave a red solution of trigermole dianion which was crystallized and characterized by X-ray analysis. It showed that one lithium cation is engaged in a lithocene structure, while the second one is in an environment similar to those in common organolithium compounds complexed by THF and TMEDA (equation 47)[62].

Et4 Ge Cl Cl —Li/THF, TMEDA→ Et4 Ge Et4 Ge Li Ge Et4 Li (THF)(TMEDA) (47)

Anionic pentacoordinated 1,2-oxagermetanide was synthesized quantitatively by deprotonation of the corresponding β-hydroxygermane (Scheme 10)[63]. Upon heating at 150 °C for 30 days, this compound equilibrated with another diastereoisomer and underwent a Peterson-type reaction with elimination of olefins (Scheme 10).

In a formal oxidative coupling reaction, a dimeric deltahedral Zintl ion was obtained in high yield as green-brown crystals (equation 48)[64]. The X-ray structural determination revealed that the distance between the two Ge_9 clusters is about 2.49 Å, which corresponds to a simple two-center, two-electron localized bond. There are two different cesium sites in the structure: one caps the open square face of each Ge_9 cluster and one the Ge−Ge edge of its partner in the dimer. A more interesting role is played by the second cesium cation, since it not only caps faces and edges, but does that for two neighboring clusters and thus connects the clusters into chains. The positioning of the ethylene diamine molecules is such that the chains are enveloped by them and the cryptated potassiums are located between the chains[64].

$$KCs_2Ge_9 \xrightarrow[\text{cryptand}]{H_2NCH_2CH_2NH_2} Cs_4(\text{K-cryptand})_2[(Ge)_9{-}(Ge)_9].6H_2NCH_2CH_2NH_2 \quad (48)$$

cryptand: 4,7,13,16,21,24-hexaoxa-1,10-diazabicyclo[8.8.8]hexacosane

Germanates are ammonium salts in which the counterion is a germylanion (equation 49)[65]. They are generally prepared from 'acidic' hydrogermanes by reaction with a Lewis base N. Sometimes they decompose reversibly to germylene in the presence of

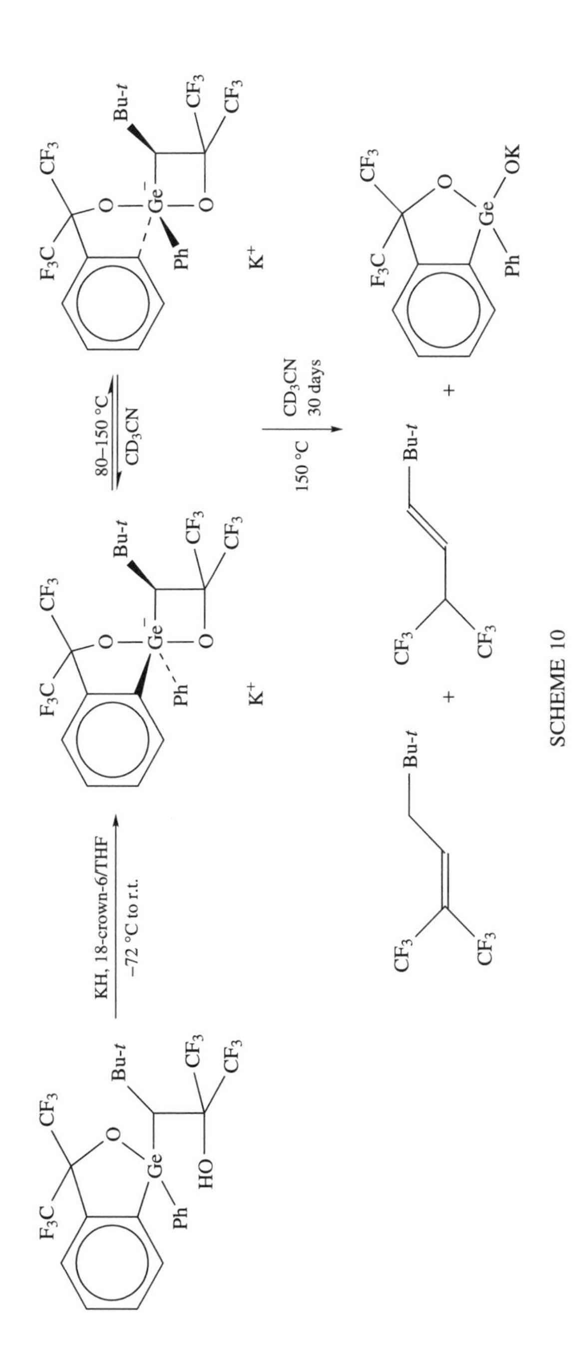

KH, 18-crown-6/THF
−72 °C to r.t.
80–150 °C
CD3CN
150 °C
CD3CN
30 days
F3C
CF3
O
Ge
Ph
HO
Bu-t
CF3
CF3
K+
OK
+

SCHEME 10

an excess of base N. When the base is DBU, the germanate can be isolated as an oil which precipitates in benzene.

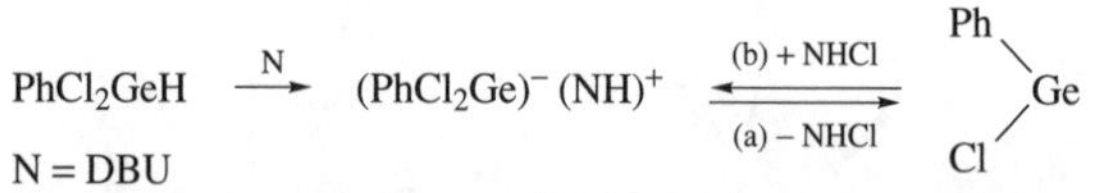

$$PhCl_2GeH \xrightarrow{N} (PhCl_2Ge)^- (NH)^+ \underset{(a)\,-\,NHCl}{\overset{(b)\,+\,NHCl}{\rightleftarrows}} PhGeCl \qquad N = DBU \tag{49}$$

Another way to germanates consists in the protonation of an amino ligand linked to germanium. Using HF, the germanium center was hexacoordinated by fluorine anions (Scheme 11)[66]. The germanate has a structure, obtained by single crystal X-ray diffractometry, in which the germanium center is a slightly distorted octahedron with Ge−F (axial) bond lengths of 1.84 Å, while the Ge−F (equatorial) bond lengths are different (1.79 Å, 1.87 Å)[66].

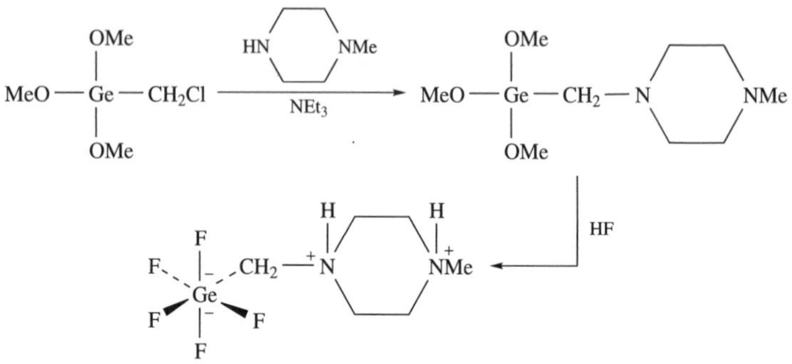

$$(MeO)_3Ge{-}CH_2Cl \xrightarrow[NEt_3]{HN(CH_2CH_2)_2NMe} (MeO)_3Ge{-}CH_2{-}N(CH_2CH_2)_2NMe \xrightarrow{HF} F_5Ge^{-}{-}CH_2{-}{}^{+}NH(CH_2CH_2)_2{}^{+}NHMe$$

SCHEME 11

The reaction of a lithium tripodal silylamide with dichlorogermylene resulted in a diamination of the germylene center coupled with an oxidative addition of one of the lithium amides (equation 50)[67]. The resulting triaminogermyllithium was isolated as colorless parallelipedic crystals in 80% yield. An X-ray analysis revealed relatively large Ge−N bond lengths (1.98 Å) along with a large Ge−Li bond length (2.90 Å), suggesting a high localization of negative charge on the germanium center, which forms a close ion-pair with the THF complexed lithium counterion.

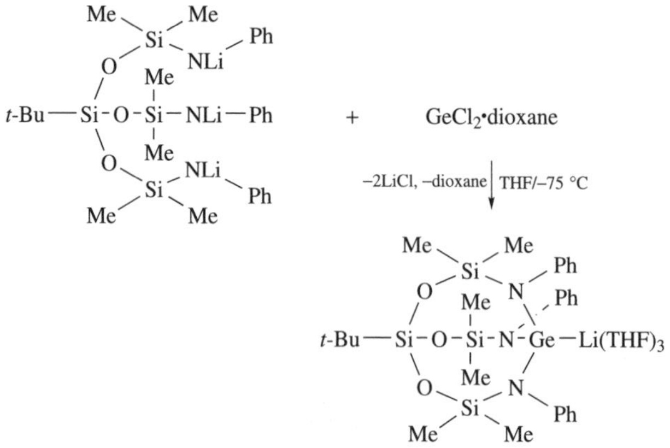

$$t\text{-}Bu{-}Si(OSiMe_2NLiPh)_3 + GeCl_2{\cdot}dioxane \xrightarrow[-2LiCl,\ -dioxane]{THF/-75\,^{\circ}C} t\text{-}Bu{-}Si(OSiMe_2NPh)_3Ge{-}Li(THF)_3 \tag{50}$$

Metal-centered radical anions have recently received increased interest. Usually formed by SET reactions from derivatives of tetravalent or divalent metal-14 (equation 51)[68,69], they have been characterized by ESR and radiofluorescence.

$$[(Me_3Si)_2CH]_2E + \text{Na mirror} \xrightarrow[\text{THF}]{20\,^{\circ}\text{C}/10\text{–}15\ \text{s}} [(Me_3Si)_2CH]_2E^{\bullet -}Na^+ \qquad (51)$$

$E = \text{Ge, Sn}$

Further contact between solutions of radical anions and sodium resulted in the complete disappearance of their ESR signals because of their transformation to a diamagnetic dianion through a second single electron transfer (equation 52)[68].

$$[(Me_3Si)_2CH]_2Ge^{\bullet -}Na^+ + Na \xrightarrow[\text{THF}]{20\,^{\circ}\text{C}} [(Me_3Si)_2CH]_2Ge\ :^{2(\bullet -)}\ 2Na^+ \qquad (52)$$

Several publications about oxidative addition of metal or organometal derivatives to stannylenes describe a new and efficient way to stannyl anions[38,70–74]. The reaction of CpLi with Cl_2Sn resulted in a mononuclear complex as colorless cubic crystals obtained in 48% yield (equation 53)[70]. Its structure was resolved by X-ray diffraction and shows a complete separation of the ion-pair[70].

$$n\ Cp_2Sn + CpLi \xrightarrow{\text{12-crown-4}} Cp_3Sn^-[Li(\text{12-crown-4})_2]^+ \qquad (53)$$

The reduction of bivalent aryltin chloride by sodium anthracenide in THF gave a new way to a bulky distannylanion in the form of a stable very close ion-pair (equation 54)[71]. The X-ray crystal structure reveals a normal Sn–Sn distance (2.81 Å) in the distannylanion and a Sn–Na bond length of 3.24 Å.

$$ArSnCl + Na(\text{anthracenide}) \xrightarrow{\text{THF}} (THF)_3Na^+\ {}^-[SnAr]_2 \qquad (54)$$

$Ar = 2,6\text{-}Tip_2C_6H_3$

New compounds having a trigonal-planar ‘paddle wheel’ triorganostannate ion were obtained from the reaction of sodium cyclopentadienide with bis(cyclopentadienyl)tin(II) and PMDETA, in a molar ratio of 1 : 1 : 1 in THF (equation 55)[72]. The reaction product, isolated as yellow crystals, was investigated by X-ray diffractometry. The structure is different from that of the cyclopentadienyl stannyl lithium obtained in equation 53. It displays a $(\eta^5\text{-Cp})_3Sn$ unit as a ‘paddle wheel’ triorganostannate anion in which one Cp ligand is additionally involved in a $Sn(\mu\text{-}\eta^5\text{-Cp})Na$ bridge. The Sn center is nearly trigonal planar, being separated from the complexed (PMDETA) sodium center by a Cp ring, with Sn–Cp and Cp–Na bond lengths of 2.73 Å and 2.55 Å.

$$Cp_2Sn + CpNa \xrightarrow[\text{THF}]{\text{PMDETA}} (\eta^5\text{-Cp})_2Sn — (\mu\text{-}\eta^5\text{-Cp}) — Na \cdot \text{PMDETA} \qquad (55)$$

A triaminostannyl lithium, prepared by oxidative addition of the corresponding amide to the stannylene bearing the same substituents (equation 56)[73], was characterized by methylation (MeI, 71% yield) and halogenation (I_2, 62% yield).

$$[Sn(NRAr)_2] \xrightarrow{Li(NRAr)(OEt_2)} [LiSn(NRAr)_3] \qquad (56)$$

$R = C(CD_3)_2Me;\ Ar = 3,5\text{-}Me_2C_6H_3$

Similarly, the reaction of dichlorostannylene with $[LiR^{N}(tmeda)]_2$ [R^{N} = CH($SiMe_2$Bu-*t*)C_5H_4N-2] afforded a lithium trialkylstannate zwitterionic cage molecule having Li−Cl chloro-bridges according to X-ray structural determination (equation 57)[74].

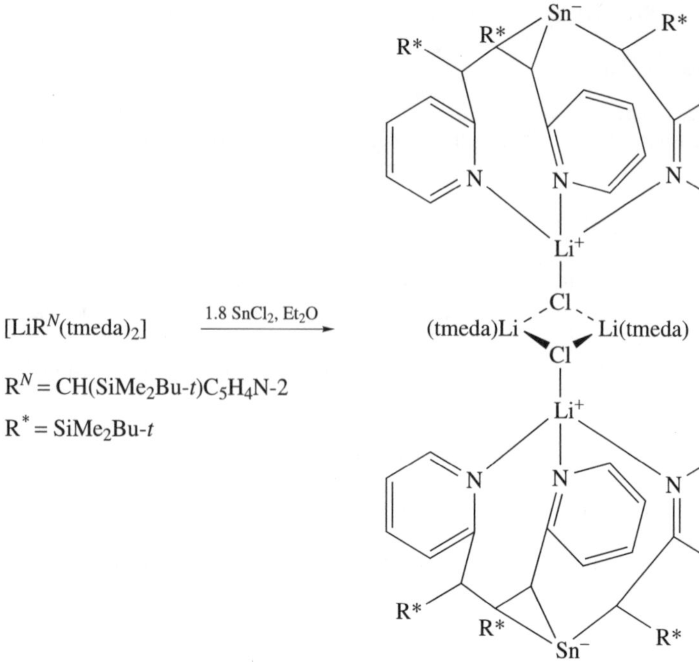

$$[LiR^{N}(tmeda)_2] \xrightarrow{1.8\ SnCl_2,\ Et_2O} \text{(tmeda)Li}(\mu\text{-Cl})_2\text{Li(tmeda)} \quad (57)$$

R^{N} = CH($SiMe_2$Bu-*t*)C_5H_4N-2

R^{*} = $SiMe_2$Bu-*t*

Stannyl anions with a highly coordinated tin center are also known. A hydridostannyl anion in the shape of a trigonal bipyramid in which two iodine atoms occupy the apical positions was obtained by oxidative addition of lithium iodide to the corresponding tin hydride (equation 58)[75]. It was characterized by ^{119}Sn NMR. Since apical iodines are more nucleophilic than the hydrogen, in its reactivity with α-ethylenic carbonyl compounds, attack by iodine precedes reduction by hydrogen, achieving regioselective 1,4 reductions.

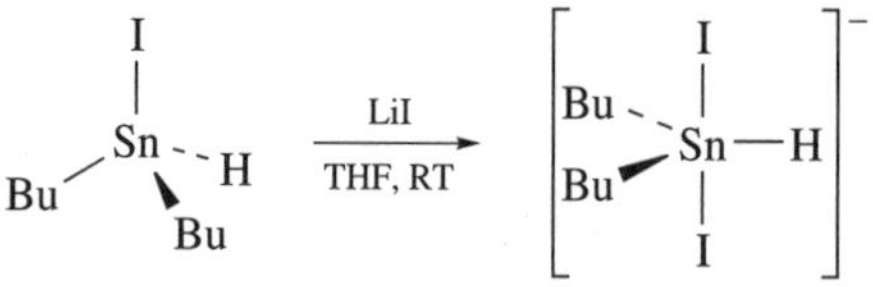

$$Bu_2Sn(I)H \xrightarrow[\text{THF, RT}]{\text{LiI}} [Bu_2Sn(I)_2H]^{-}\ Li^{+} \quad (58)$$

A 1,2-oxastannetanide, which is an example of an anionic pentacoordinate tin compound with a four-membered ring oxygen and one organic substituent in apical position, was prepared at room temperature by deprotonation of the corresponding β-hydroxystannane using KH in THF and 18-crown-6 (equation 59)[76]. The presence of a stannyl anion was evident by the appearance of a double quartet with δ_F centers at −75.09 ppm

($^{4}J_{FF}$: 8.8 Hz) and −71.77 ($^{4}J_{FF}$: 8.8 Hz), and a singlet (δ_{Sn} : −229.65 ppm). The large upfield shift (105 ppm) of δ_{Sn} compared with the precursor from β-hydroxystannane (−125.09 ppm) to the pentacoordinate stannyl anion (−229.65 ppm) strongly supports the structure of a pentacoordinate stannate complex.

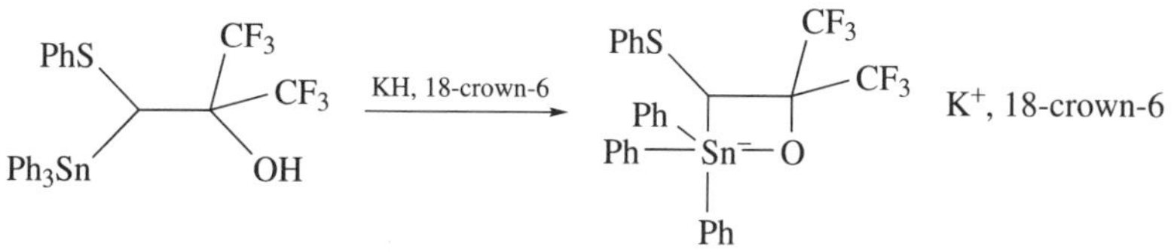

$$\text{PhS(Ph}_3\text{Sn)CH–C(CF}_3)_2\text{OH} \xrightarrow{\text{KH, 18-crown-6}} [\text{cyclic Ph}_3\text{Sn}^-\text{–O (PhS, CF}_3, \text{CF}_3)]\ \text{K}^+\text{, 18-crown-6} \qquad (59)$$

Other stannates having four tin–carbon bonds were prepared by the reaction of tetracoordinate 1,1-dialkyl-3,3′-bis(trifluoromethyl)-3*H*-2,1-benzoxastannoles with RLi at 0 °C in THF (equation 60)[77].

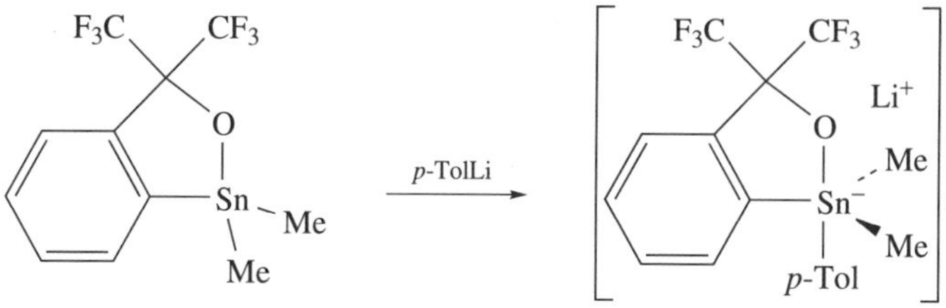

$$\text{benzoxastannole (F}_3\text{C, CF}_3\text{, O, SnMe}_2) \xrightarrow{p\text{-TolLi}} [\text{benzoxastannole (F}_3\text{C, CF}_3\text{, O, Sn}^-\text{Me}_2(p\text{-Tol}))\ \text{Li}^+] \qquad (60)$$

Similar oxidative additions of Bu_4NF to tetracoordinate monoorganostannanes synthesized from a diaminostannylene also afforded pentacoordinate stannyl anions (equation 61)[78].

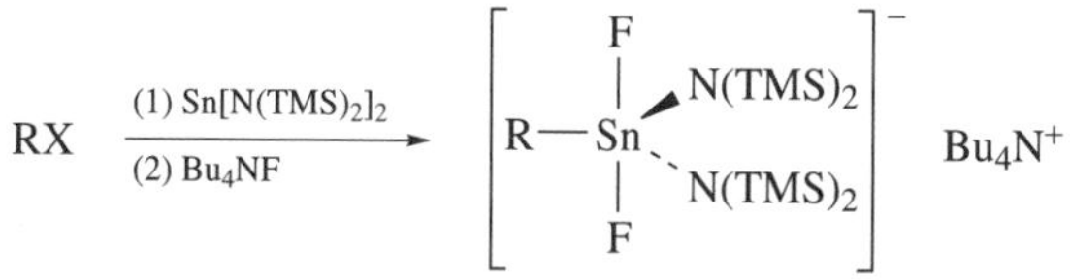

$$\text{RX} \xrightarrow[\text{(2) Bu}_4\text{NF}]{\text{(1) Sn[N(TMS)}_2]_2} [\text{R–SnF}_2(\text{N(TMS)}_2)_2]^-\ \text{Bu}_4\text{N}^+ \qquad (61)$$

R = Alkyl, Ph, PhCH = CH etc., X = I

Stannylene radical anions were generated in a way similar to that used for germylene radical anions (cf. equation 51).

Since plumbylene is a stable state in lead chemistry, oxidative addition of organometallics to plumbylene is an attractive way to lead-centered anions. Following this route, multidecker anions can be prepared for lead, if crown or cryptand ligands coordinate the alkali metal cations (equation 62)[70].

$$2[\text{Cp}_2\text{Pb}] + [\text{CpTl}] \xrightarrow[\text{2 (12-crown-4)}]{\text{2 [CpLi]}} [\text{Cp}_5\text{Pb}_2]^-[\text{Li(12-crown-4)}_2]^+ + [\text{Cp}_2\text{TlLi}] \qquad (62)$$

Cyclic pentacoordinate plumbyl anions were synthesized by the oxidative addition of tetraalkylammonium halides to a dithiolatoplumbole (equation 63)[79]. An X-ray structure analysis of the adduct revealed a monocyclic anionic geometry.

$$Ph_2Pb(S_2C_6H_4) + Et_4N^+ X^- \xrightarrow{CH_3CN} [Ph_2Pb(X)(S_2C_6H_4)]^- Et_4N^+ \quad (63)$$

X = Br, Cl, F

A stable zwitterionic carbene–plumbylene adduct has been reported (equation 64)[80]. According to its X-ray crystal structure, the central carbon atom is almost planar while the lead atom is pyramidal. As expected, the $P^-{-}C^+$ bond length (2.54 Å) is longer than a covalent P–C (about 2.38 Å) and a P=C bond, which is expected to be about 2.05 Å[80].

$$\text{carbene}(i\text{-Pr}) \xrightarrow[\text{Toluene}]{R_2Pb} C{=}PbR_2 \longleftrightarrow {}^+C{-}\overset{-}{Pb}R_2 \quad (64)$$

R = Tip

A dianionic compound with trigonal-planar coordinated lead was obtained from the reaction of disodium decacarbonyldichromate with lead nitrate (equation 65)[81]. The authors found that its ^{207}Pb NMR signal (δ: 7885 ppm) supports an unsaturated character. The π system was also chemically evident. Below 213 K, in the presence of PMe_3, a pyramidal adduct is formed quantitatively, the structure of which confirms coordination of the Lewis base PMe_3 to the coordinatively unsaturated lead center (Pb–P: 2.84 Å).

$$[Cr_2(CO)_{10}]^{2-} \xrightarrow[\text{THF}]{Pb(NO_3)_2} [Pb\{Cr(CO)_5\}_3]^{2-} \underset{-PMe_3\ (THF)}{\overset{+PMe_3}{\rightleftharpoons}} [Me_3P{-}Pb\{Cr(CO)_5\}_3]^{2-} \quad (65)$$

B. M_{14}–Group (II) Metal Compounds

For several years many reactions were rationalized in terms of transient germyl-Grignard reagents[3a,4a]. Then, a few of these compounds, along with symmetrical digermylmagnesium, were isolated mainly in transmetallation reactions[82–84]. Their structures were confirmed by chemical characterization, spectroscopic analysis and sometimes by X-ray structural studies. They possess the reactivity of nucleophilic germylanions (Scheme 12)[82].

$$R_2HGeLi + MgBr_2 \xrightarrow[-LiBr]{-10\,^{\circ}C} [R_2HGeMgBr] \xrightarrow{PhClGeH_2} R_2HGeGeH_2Ph$$

R = Ph, 56%; R = Mes, 67%

SCHEME 12

Bis(trimethylgermyl)magnesium was the first symmetrical bis(organogermyl) magnesium isolated. It was obtained from the reaction of bis(trimethylgermyl)mercury with magnesium in DME and isolated as colorless crystals complexed with DME (equation 66)[83]. The X-ray structure analysis shows a germanium–magnesium bond length of 2.7 Å (Figure 2)[84].

$$(Me_3Ge)_2Hg + Mg \xrightarrow{DME} (Me_3Ge)_2Mg \cdot 2\ DME \qquad (66)$$

Addition of a Grignard reagent to a germasilene or a digermene also leads to germyl Grignard compounds. The addition to unsymmetrical germasilene is regioselective with

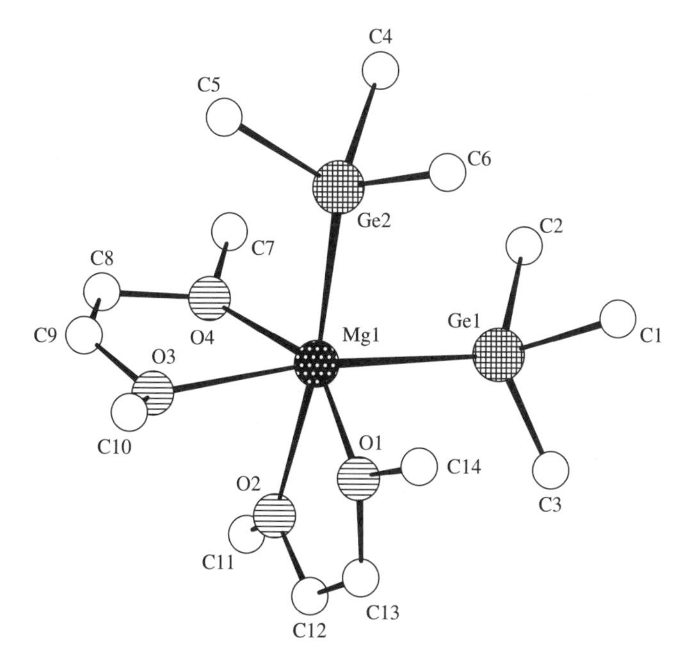

FIGURE 2. Solid state structure of $(Me_3Ge)_2Mg \cdot 2$ DME with hydrogen atoms omitted. Reproduced by permission of Verlag der Zeitschrift für Naturforschung from Reference 84

the alkyl group adding to silicon to produce a germyl magnesium halide (equation 67)[85]. As expected, upon hydrolysis, the germyl Grignard leads to the corresponding germanes.

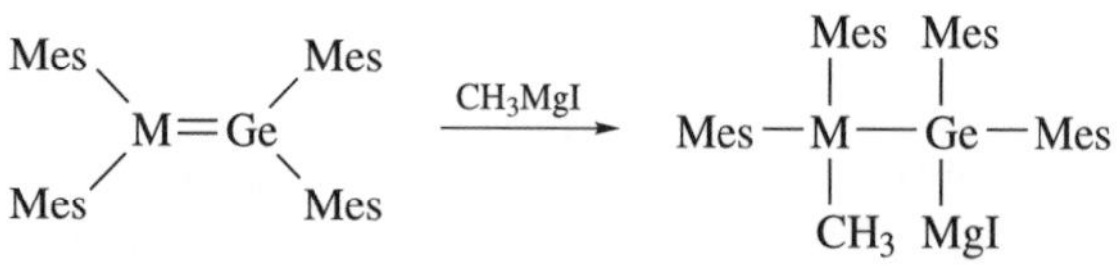

$$\mathrm{Mes_2M{=}GeMes_2} \xrightarrow{\mathrm{CH_3MgI}} \mathrm{Mes{-}M(Mes)(CH_3){-}Ge(Mes)(MgI){-}Mes} \qquad (67)$$

M = Si, Ge

Stannyl Grignard reagents are more easily available from tin hydrides (equation 68)[86,87].

$$\mathrm{Bu_3SnH} + i\text{-}\mathrm{PrMgCl} \xrightarrow[-\mathrm{C_3H_8}]{\text{ether}} \mathrm{Bu_3SnMgCl} \qquad (68)$$

Triphenyllead Grignard reagent, prepared by the reaction of a Grignard reagent with $PbCl_2$ in THF, is always used *in situ*, like a classical Grignard. It reacts as a plumbyl anion leading to substitution reactions, for example with allylic or propargylic halides (Scheme 13)[88].

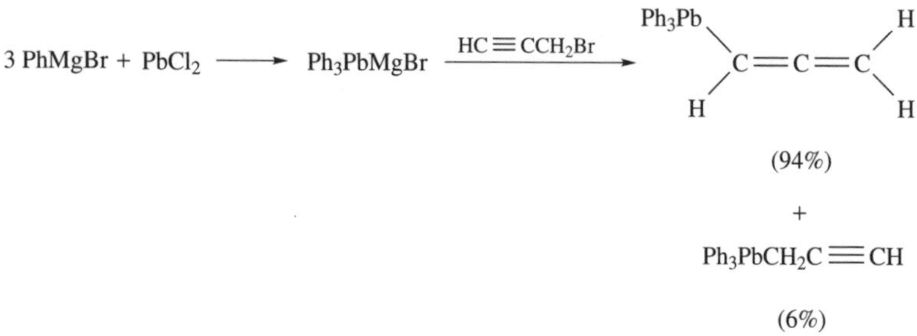

$$3\,\mathrm{PhMgBr} + \mathrm{PbCl_2} \longrightarrow \mathrm{Ph_3PbMgBr} \xrightarrow{\mathrm{HC{\equiv}CCH_2Br}} \mathrm{Ph_3Pb(H)C{=}C{=}CH_2}\ (94\%) + \mathrm{Ph_3PbCH_2C{\equiv}CH}\ (6\%)$$

SCHEME 13

$$\mathrm{R_3GeCl} + \mathrm{Ca} \xrightarrow[0.01\ \text{torr}]{900\,^{\circ}\mathrm{C}} \mathrm{R_3GeCaCl}$$

$$\mathrm{R_3GeGeR_3} + \mathrm{Ca} \xrightarrow[0.01\ \text{torr}]{900\,^{\circ}\mathrm{C}} \mathrm{(R_3Ge)_2Ca}$$

$$\xrightarrow{\mathrm{H_2O}} \mathrm{R_3GeH} \qquad \xrightarrow{\mathrm{PhBr}} \mathrm{R_3GePh}$$

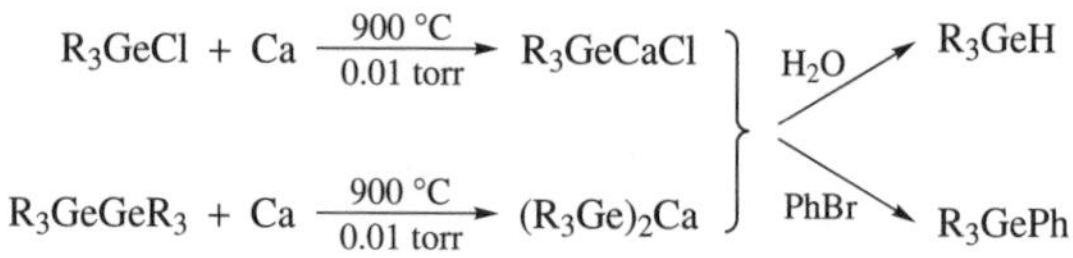

R = alkyl, aryl

SCHEME 14

Insertion reactions of calcium atoms into $M_{14}{-}M_{14}$ bonds yield symmetrical or unsymmetrical $M_{14}{-}Ca$ compounds according to Scheme 14[4] and equation 69[89]. A trimethylsilyl trimethylstannyl calcium was also characterized chemically in the cocondensation of calcium with trimethylsilyl trimethylstannane[90]. Calcium bis(stannide) (equation 69) crystallizes in the form of colorless cuboids in a centrosymmetric space group P1. The calcium atom lies on the crystallographic center of inversion in the middle of the linear Sn−Ca−Sn chain. The calcium atom is coordinated in a distorted octahedral fashion by two tin atoms

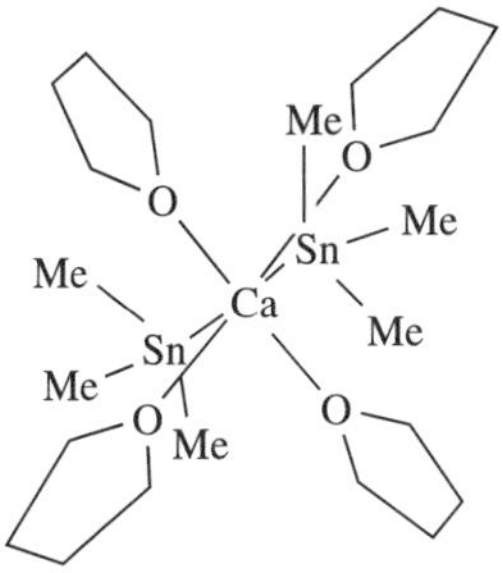

FIGURE 3. Stereoscopic projection along Sn–Ca–Sn axis in $[(Me_3Sn)_2Ca]$. 4 THF

and four oxygen atoms of the THF ligands in a *trans* configuration (Figure 3)[89].

$$Me_3SnSnMe_3 + Ca \xrightarrow{THF} [(Me_3Sn)_2Ca] \cdot 4\ THF \qquad (69)$$

When the bis(trimethylstannyl)calcium was reacted with an excess of hexamethyldistannane, a new polystannylcalcium was formed quantitatively according to equation 70[89].

$$[(Me_3Sn)_2Ca] \cdot 4\ THF + 6\ Me_3SnSnMe_3 \longrightarrow [\{(Me_3Sn)_3Sn\}_2Ca] \cdot 4\ THF + 6\ SnMe_4 \qquad (70)$$

Similar insertions of strontium or barium atoms into M_{14}–M_{14} bonds also afforded M_{14}–group (II) compounds (equation 71)[3a].

$$Ph_3GeGePh_3 + M \xrightarrow[-40\,^{\circ}C]{NH_3} (Ph_3Ge)_2M \xrightarrow{THF} (Ph_3Ge)_2M \cdot THF \qquad (71)$$

$M = Sr, Ba$

Other compounds displaying a M_{14}–metal bond with a nucleophilic center can be considered as reacting like M_{14} anions (Scheme 15, equation 72)[3a,4a], but they usually behave in a different way (Schemes 16 and 17, equation 73)[3a,4a,91] and therefore will not be discussed here.

$$[(F_3C)_3Ge]_2Zn + Cp_2TiCl_2 \xrightarrow[20\,^{\circ}C,\ 80\%]{Et_2O} Cp_2ClTiGe(CF_3)_3 \qquad (72)$$

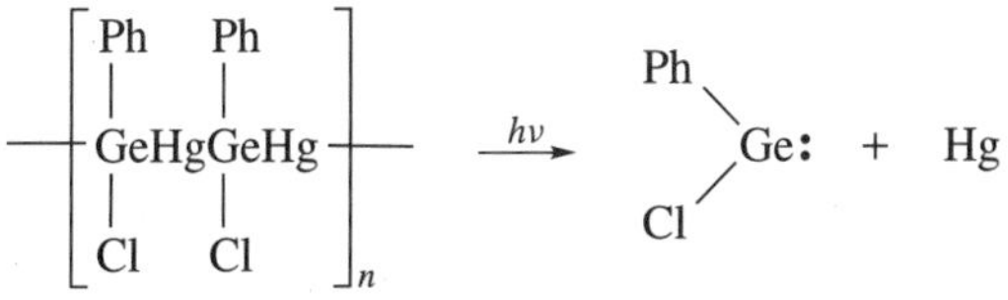

$$-\!\!\left[\begin{array}{cc} Ph & Ph \\ | & | \\ \text{GeHg} & \text{GeHg} \\ | & | \\ Cl & Cl \end{array}\right]_n\!\!- \xrightarrow{h\nu} \begin{array}{l} Ph \\ \quad \diagdown \\ \quad\ \ Ge\!: \\ \quad \diagup \\ Cl \end{array} + \ Hg \qquad (73)$$

Some representative metal-14 anions are listed in Tables 1–4.

$$(R_3Ge)_2M + ROH \longrightarrow R_3GeMOR + R_3GeH$$

$$M = Hg, Cd;\ R = \text{alkyl, aryl, acyl} \qquad R_3GeMOR \xrightarrow[-M]{+ROH} R_3GeOR$$

SCHEME 15

$$(R_3Ge)_2Hg \xrightarrow[-Hg]{h\nu \text{ or } T\,^\circ C} 2\ R_3Ge^\bullet$$

$$R_3Ge^\bullet \xrightarrow[i\text{-}Pr_2O]{CH_2{=}CHBu} R_3GeC_6H_{13}\text{-}n$$

$$R_3Ge^\bullet \longrightarrow R_3GeGeR_3$$

$$R_3Ge^\bullet \xrightarrow{PhCH{=}NOBu\text{-}t} R_3GeCH(Ph)N(O^\bullet)Bu\text{-}t$$

ESR characterized

SCHEME 16

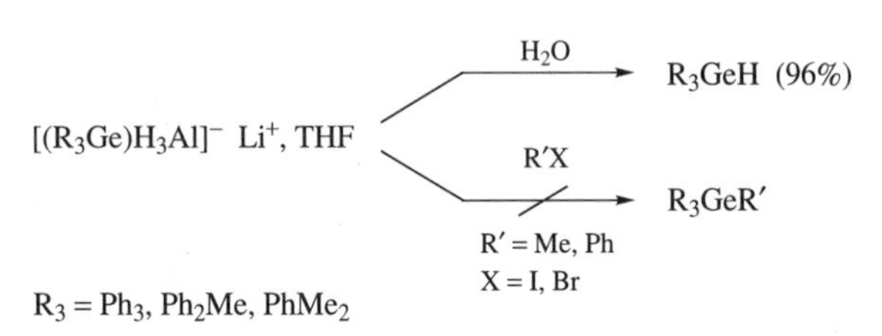

SCHEME 17

TABLE 1. Representative metallated germyl compounds

Compound	Starting reagents	References
Me_3GeLi	Me_3GeCl or $(Me_3Ge)_2/Li$	92–94, 32
Et_3GeLi, Et_3GeK, Et_3GeNa	$(Et_3Ge)_2Hg/Li$	95, 96
	$(Et_3Ge)_2/Li$, K, Na	
Ph_3GeLi	$Ph_3GeH/BuLi$, $(Ph_3Ge)_2Ge/Li$	33, 97–99
	Ph_3GeCl/Li	
Ph_3GeK	$(Ph_3Ge)_2/Me_3COK$	45
$Ph_2MeGeLi$	$Ph_2MeGeCl/Li$	33
$PhMe_2GeLi$	$PhMe_2GeCl/Li$	33
Mes_3GeLi	$Mes_3GeH/t\text{-}BuLi$	15
$(2\text{-}Me_2NC_6H_4)_3GeLi$	$(2\text{-}Me_2NC_6H_4)_3GeH/t\text{-}BuLi$	18
$(2\text{-}Me_2NCH_2C_6H_4)_3GeK$	$(2\text{-}Me_2NCH_2C_6H_4)_3GeH/PhCH_2K$	19
$(PhCH_2)_3GeLi$	$(PhCH_2)_4Ge/Li$	100
$(Me_3Si)_3GeLi$	$(Me_3Si)_4Ge/MeLi$	47, 101
Mes_2HGeLi	$Mes_2GeH_2/t\text{-}BuLi$	16
Ph_2HGeLi	$Ph_2GeH_2/t\text{-}BuLi$	16
Tip_2HGeLi	$(Tip)_2GeH_2/t\text{-}BuLi$	26

TABLE 1. (*continued*)

Compound	Starting reagents	References
$(8\text{-MeONp})_2HGeLi$	$(8\text{-MeONp})_2GeH_2$/*t*-BuLi	17
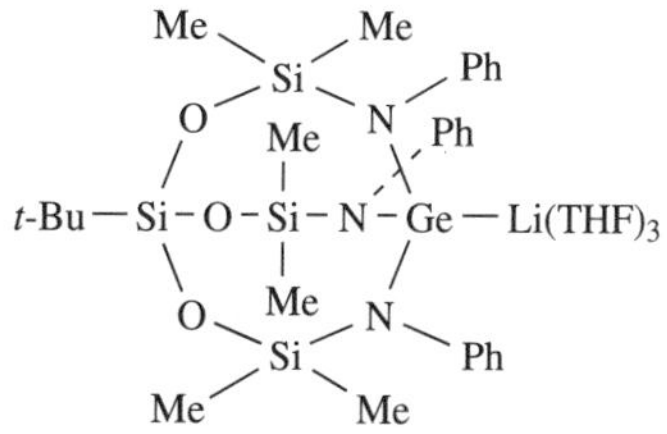 Me Me Si O Me N Ph Ph *t*-Bu—Si—O—Si—N—Ge—Li$(THF)_3$ O Me N Si Ph Me Me	*t*-$BuSi(OSiMe_2NLiPh)_3$/ Cl_2Ge.dioxane	67
$MesH_2GeLi$	$MesGeH_3$/*t*-BuLi	102
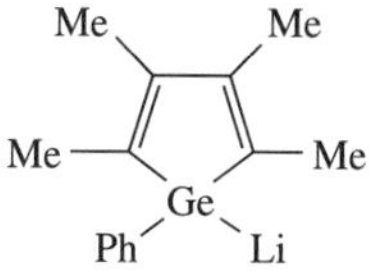 Me Me Me Me Ge Ph Li	$Ph(C_4Me_4)GeH$/*n*-BuLi	20
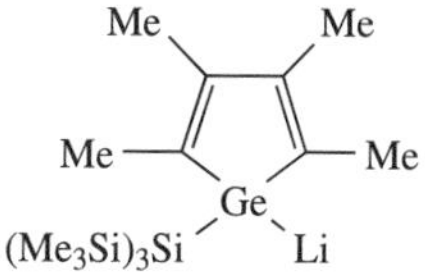 Me Me Me Me Ge $(Me_3Si)_3Si$ Li	$(Me_3Si)_3Si(C_4Me_4)GeH$/*n*-BuLi, 2 (12-crown-4)	21
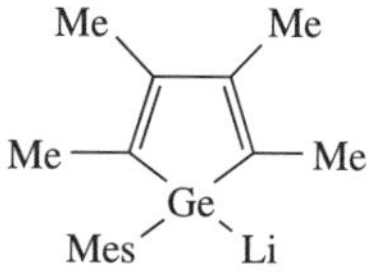 Me Me Me Me Ge Mes Li	$Mes(C_4Me_4)GeH$/*n*-BuLi, 2 (12-crown-4)	21
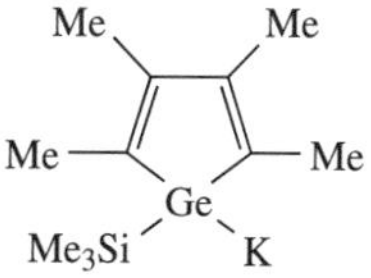 Me Me Me Me Ge Me_3Si K	$(Me_3Si)_2(C_4Me_4)Ge/PhCH_2K$, 18-crown-6	21
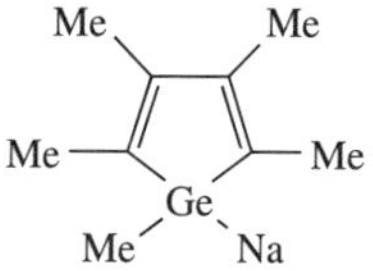 Me Me Me Me Ge Me Na	$[Me(C_4Me_4)Ge]_2$/2 Na, 15-crown-5	21
[Li(THF)(TMEDA)] [2,3,4,5-Et_4-*Ge,Ge*-{Li(2,3,4,5-$Et_4C_4Ge)_2$}C_4Ge]	$(C_4Et_4Ge)Cl_2$/Li, THF, TMEDA	62

TABLE 2. Representative dimetallated germyl and digermyl compounds

Compound	Starting reagents	References
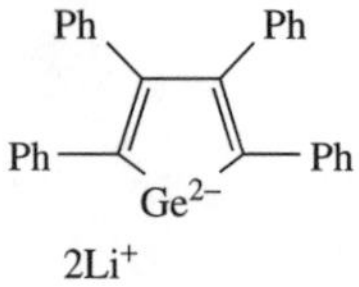 Ph, Ph, Ph, Ph, Ge^{2-}, $2Li^+$	$(C_4Ph_4)GeCl_2/Li$	34, 35
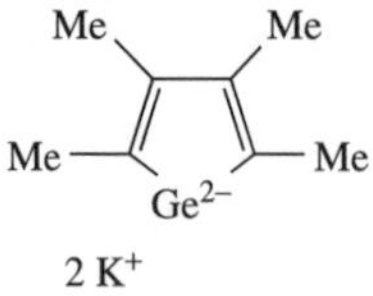 Me, Me, Me, Me, Ge^{2-}, $2\,K^+$	$(C_4Me_4)GeCl_2/K$	21
Ph_2GeK_2	Ph_2GeH_2/K	12
Et_2GeLi_2	$Me_3SiGeEt_2Li/Me_3SiLi$	103
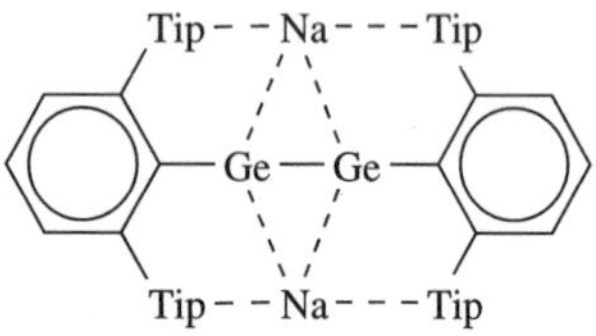 Tip- -Na- - -Tip, Ge—Ge, Tip- -Na- - -Tip	$ClGeC_6H_3Tip_2$-2,6/Na	38
$Ph_2LiGeGeLiPh_2$	$Ph_2HGeGeHPh_2/t$-BuLi	25

TABLE 3. Representative metallated stannyl compounds

Compound	Starting reagents	References
Me_3SnLi	$(Me_3Sn)_2/MeLi$ or BuLi, Me_3SnH/LDA, Me_3SnCl/Li	27, 104–106
Bu_3SnLi	Bu_3SnH/LDA, $(Bu_3Sn)_2/MeLi$ or BuLi,	28, 104, 106
Ph_3SnLi	Ph_3SnCl/Li	105
$Me_2NphSnLi$	$Me_2NphSnBr/Li$	31
$MeNph_2SnLi$	$MeNph_2SnBr/Li$	31
Nph_3SnLi	Nph_3SnBr/Li	31
$(Me_3Si)_3SnLi$	$(Me_3Si)_4Sn/MeLi$, $Me_3SiLi/SnCl_4$	42, 107
Bu_2HSnLi	Bu_2SnH_2/LDA	29, 30
Ph_2HSnLi	Ph_2SnH_2/LDA	29, 30
$(c\text{-}C_6H_{11})_2HSnLi$	$(c\text{-}C_6H_{11})_2SnH_2/LDA$	29, 30
Me_3SnNa	Me_3SnCl/Na, $(Me_3Sn)_2/Na$	41, 108
Bu_3SnNa	Bu_3SnCl/Na	41
Ph_3SnNa	Ph_3SnCl/Na	41
$Me_2NphSnNa$	$Me_2NphSnH/NaH$	31
$MeNph_2SnNa$	$MeNph_2SnH/NaH$	31
Me_3SnK	Me_3SnH/t-BuOK	109
Bu_3SnK	$Bu_3SnCl/[K^+/K^-]$	40
Ph_3SnK	$(Ph_3Sn)_2/t$-BuOK, Ph_3SnH/t-BuOK	45, 109
Bu_3SnCs	$(Bu_3Sn)_2/CsF$	49

TABLE 4. Representative metallated plumbyl compounds

Compound	Starting reagents	References
Me_3PbLi	Me_3PbBr/Li	44
Bu_3PbLi	$(Bu_3Pb)_2/BuLi$,	51
$(t\text{-}Bu)_3PbLi$	$((t\text{-}Bu)_3Pb)_2/Li$	110
Ph_3PbLi	$(Ph_3Pb)_2/PhLi$, Li	53, 111
Mes_3PbLi	$(Mes_3Pb)_2/Li$	52
$o\text{-}Tol_3PbLi$	$(o\text{-}Tol_3Pb)_2/Li$	111
$p\text{-}Tol_3PbLi$	$(p\text{-}Tol_3Pb)_2/Li$	111
$(2,4\text{-}Xyl)_3PbLi$	$((2,4\text{-}Xyl)_3Pb)_2/Li$	111
Ph_3PbK	$(Ph_3Pb)_2/t\text{-}BuOK$	45

IV. SPECTROSCOPIC AND STRUCTURAL STUDIES

UV and ^1H NMR, and more particularly ^{13}C NMR, spectroscopies are excellent tools for the analysis of molecular and electronic structures of metal-14 anions. Calculations are also very useful for evaluating the charge delocalization, particularly in the metalole series, in connection with UV studies. X-ray structural analyses are conclusive when single crystals are isolated.

A. UV-visible Spectroscopy

In UV-visible studies of metal-14 anions showed bathochromic shifts of their absorption maxima in comparison with those of the precursor organogermanes. This absorption band (Table 5) can be explained in terms of a transition from the non-bonding orbital of the metal (HOMO) to the lowest anti-bonding orbital (LUMO) of the metal–carbon bonds for alkyl metal-14 anions[93] or of the phenyl groups in the aryl series[17,33,112]. Moreover, the bathochromic shifts on going from lithium to potassium are indicative of CIP (Contact Ion Pair) formation for the aryl group-14 anions[112] with an electron localization at the metal.

Under laser photolysis, these metal-14 anions easily gave the corresponding radicals by direct photo-ejection from the group-14 element[33] (equation 74).

$$Ph_nMe_{3-n}E^- \xrightarrow{h\nu} Ph_nMe_{3-n}E^\bullet + e^- \longrightarrow 1/2\ (Ph_nMe_{3-n}E)_2 \qquad E = Si, Ge, Sn;\ n = 1\text{–}3 \qquad (74)$$

CIDEP studies indicated that photo-ejection reactions probably occurred from triplet anions[33]. Oxidation potentials (−0.29 to −0.90 V, versus SCE)[113] confirmed the electron-donor properties of the anions.

B. NMR Spectroscopy

NMR spectroscopy (^1H, ^{13}C, ^7Li, ...) is widely used to clarify the nature of the metal-14–alkalimetal bond and the possible interactions between the metal-14 anion center and its substituents. The upfield shifts of the proton NMR signals of Ge–H correlate with the negative charge on germanium[16], but the most interesting conclusions about charge delocalization in these compounds were obtained from ^{13}C studies of arylgermylanions (Table 6). A comparison of their chemical shifts with those of the starting arylgermanes

TABLE 5. UV spectra λ_{max} for group-14 centered anions

Anion	λ_{max} (nm) for M =			Reference
	Li	Na	K	
Me_3Ge	280	280	300	93
n-Bu_3Ge	<280	~280	290	93
Ph_3Ge	308		352	112
$PhMe_2Ge$	290			33
(8-MeONp)PhHGe	348			17
Ph_3Sn	298(sh)		350(sh)	112
Ph_3Pb	298(sh)		348	112

TABLE 6. ^{13}C chemical shifts (ppm) of metallated arylanions

Compound	*ipso*	*ortho*	*meta*	*para*	$\Delta\delta(ipso)^a$	$\Delta\delta(ortho)^a$	$\Delta\delta(meta)^a$	$\Delta\delta(para)^a$
PhH_2GeNa^b	163.3	139.6	127.7	124.1	+31.7	+3.6	−1.4	−5.6
$MesH_2GeLi^c$	153.7	144.0	128.7	130.9	+26.4	−0.1	0	−8.3
Ph_2HGeNa^b	164.4	137.8	126.8	123.5	+29.7	+2.0	−2.3	−6.3
Ph_2HGeLi^c	159.4	137.6	127.0	124.1	+24.4	+1.4	−2.4	−6.0
Mes_2HGeLi^c	153.9	144.4	127.2	132.6	+22.0	0	−2.3	−6.8
8-MeO−Np PhHGeLi^c	161.7	137.4	126.6	123.8	+23.5	+2.5	−2.1	−5.2
Ph_3GeNa^b	165.3	137.3	126.9	123.8	+28.9	+1.5	−2.1	−6.0
$PhEt_2GeLi^d$	173.1	135.0	124.8	120.7	+35.4	+2.0	−3.3	−7.9
Ph_3GeK^e	166.3	137.3	126.6	123.5	—	—	—	—
Mes_3GeLi^c	152.5	142.8	129.3	139.3	+17.3	−0.9	0	−0.8
$Ph_2LiGeGeLiPh_2{}^c$	157.0	137.7	127.0	124.5	+20.4	+1.5	−2.6	−1.6
$Ph_2GeK_2{}^c$	164.0	138.3	127.6	124.3	+28.3	+0.6	−2.6	−6.6
$(p$-$Tol)_2GeK_2{}^c$	160.0	138.7	128.8	133.0	+19.3	+1.7	−2.2	−0.6
Ph_3SnK^e	167.7	139.2	126.9	124.3	—	—	—	—
Ph_3PbK^e	191.1	140.2	128.1	123.7	—	—	—	—

$^a\Delta\delta = \delta(Ar_3MLi) - \delta(Ar_3MH)$.
^{b}Solutions in NH_3; benzene resonance occurs at 129.04 ppm.
^{c}In THF-d_8.
^{d}In HMPA.
^{e}In DMPU; cyclohexane used as an internal reference at 27.7 ppm.

showed a strong downfield shift of the *ipso* carbon and a moderate high field shift of the *para* carbon which was observed in all the M_{14} series[6,9,16,45,114–116]. Chemical shifts of the *meta* and *ortho* carbons were less affected by the metalation of the arylgermane. These results can be attributed to a polarization of the phenyl ring, resulting in decreased electron density at the *ipso* carbon. Such a polarization can be induced by a localized negative charge on the germanium center which is consistent with predominant inductive π-polarization effects and negligible (or absent) mesomeric effects.

In the particular case of germoles (C_4Ge rings), the corresponding anions (MLi, MK) had either a localized non-aromatic structure with a negative charge localized on germanium[20,21], or delocalized aromatic structure[34,35], depending on the nature of the metal and the substituents.

The ^{119}Sn NMR chemical shifts (Table 7) seem not to be correlated to the negative charge on the metal. Moreover, the ^{207}Pb resonance of Ph_3Pb^- appeared at an extremely low field shift (+1040 to +1060 ppm)[116] (Table 7).

TABLE 7. ^{119}Sn and ^{207}Pb chemical shifts and J couplings of group-14 anions

Compound	δM_{14}(ppm)	$J(M_{14}-M)$	Solvent
Ph_3SnLi	−106.7		THF
Ph_3SnLi		$J(^7Li-^{119}Sn) = 412$ Hz	Toluene-d_8
Ph_3SnK	−108.4		THF
Me_3SnLi	−179		THF
Et_3SnLi	−99		THF
Bu_3SnLi	−155	$J(^7Li-^{119}Sn) = 402.5$ Hz	Et_2O
$[(t\text{-}BuCH_2)_3SnK(\eta^6\text{-}C_6H_5Me)_3]$	−221		Toluene
	−211	$J(^{39}K-^{119}Sn) = 289$ Hz	Solid state
Ph_3PbLi	1062.6		THF
Me_3PbLi	512		THF
$(t\text{-}Bu)_3PbLi$	1573.8		THF

The role of covalency in group 14 atom–alkali metal interactions has been widely discussed from proton and lithium NMR chemical shifts. Cox and coworkers[117] suggested that the degree of association between the group-14 atom and lithium increased in the order Pb < Sn < Ge, and that the germanium–lithium bond had a considerable degree of covalent character. Other spectroscopic studies (^{7}Li, ^{119}Sn, ^{207}Pb NMR) of phenyl-substituted group-14 anions showed that the structure in solution can be described by a classical ion-pair model[116] with variations from CIP to SSIP (Solvent Separated Ion Pair) depending on the solvent (ether to tetrahydrofuran).

Within the alkyl series[118], the nature of the Ge–metal bond in Et_3GeM (M = Li, Na, K and Cs) was studied by proton NMR methods by measuring variations of δCH_2 of the ethyl groups ($\Delta = \delta CH_3 - \delta CH_2$) according to the metal or the solvent. It was concluded that under these conditions, the Ge–M grouping is a contact ion-pair which becomes a solvent separated ion-pair when HMPT is added.

The weakness of these bonds was also demonstrated by NMR studies of (trimethylstannyl) and (tributylstannyl) lithiums[119] in solution. The addition of more than two equivalents of HMPA produced the ion-separated complex $Bu_3Sn^-//Li(HMPA)_2^+$ both in ether and in THF. By contrast, the observation of a Sn–Li coupling at low temperature seemed to indicate a significant covalent interaction. For example, (tributylstannyl)lithium in ether at −119 °C showed Li–Sn coupling ($J_{Li-Sn} = 402.5$ Hz) in both the ^{119}Sn and ^{7}Li spectra[119]. In the ^{119}Sn NMR spectra, both the 1 : 1 : 1 : 1 quartet from coupling of ^{119}Sn to ^{7}Li and the 1 : 1 : 1 triplet derived from the natural abundance of ^{6}Li were observed. The expected ^{117}Sn and ^{119}Sn satellites were also well resolved in the ^{7}Li spectra. The same large coupling constant ($J_{Li-Sn} = 412$ Hz) was previously observed for Ph_3SnLi–PMDETA[120]. According to the authors, large coupling constants could imply that there is a significant amount of covalent character in these Sn–Li bonds. Another reason might be the use of a predominantly 5s orbital in Sn–Li bonds[6]. In the potassium series, the presence of a Sn–K bond in $[K\{Sn(CH_2Bu\text{-}t)_3\}(THF)_2]$ was confirmed by a solid state ^{119}Sn cross-polarization magic angle spinning (CP-MAS) NMR spectral study in which the coupling $^{119}Sn-^{39}K$ (with J of 289 Hz) was observed for the first time[50].

The configurational stability of germyllithium compounds has been studied by temperature variation of the proton NMR spectra[121]. The selected system $Ph(i\text{-}Pr)_2GeLi$ possesses groups (methyl of the i-Pr groups) which are diastereotopic. The methyl non-equivalence was observed in diglyme up to 185 °C, which is the upper experimental limit. Assuming that rotation around the Ge–C bond was fast on the NMR time scale, the non-equivalence of these groups was taken as evidence for slow inversion about germanium. A lower limit to inversion about trivalent germanium could therefore be set at about 24 kcal mol^{-1}.

C. X-ray Diffraction Studies

The sterically hindered compound $(Me_3Si)_3GeLi$ (donor) (donor = THF or PMDETA) was the first germanium–metal bonded complex to be characterized by X-ray structure analysis at low temperature (153 K)[47]. The germanium–lithium distances of 2.666(6) Å (THF) and 2.653(9) Å (PMDETA) are slightly greater than the sum of the covalent radii of Ge and Li (2.56 Å). The reduction of angles around the central germanium was explained by the polar nature of the Ge–Li bond (Figure 4).

A shorter Ge–Li bond of 2.598(9) Å[18] was observed for tris(2-dimethylaminophenyl) germyllithium. In this case, the distortion of the geometry around the germanium center was explained by the interaction between the lithium atom and the amino group.

X-ray structural studies of the anions of germoles have recently stimulated a great deal of interest[21,22,35,36]. Crown ethers (12-crown-4 for Li and 16-crown-6 for K) were used in monometallation reactions, giving free germacyclopentanienide ions[21,22]. They have a non-aromatic ring with a pyramidal germanium center (Figure 5).

Several types of metal coordination to germole dianions have been described[21,35,36]: η^5, η^5 (a), η^1, η^5 (b) and η^1, η^5 and η^5 (c) (Figure 6).

An example of η^5 bonding type is evident in the case of the bis(germole dianion) complex $[K_4(18\text{-crown-6})]_3[C_4Me_4Ge]_2$ (Figure 7)[21].

The dilithium salt of the tetraphenylgermole dianion has the very interesting property of crystallizing from dioxane in two structurally distinct forms (a and b) depending upon the crystallization temperature. The crystals obtained from dioxane at −20 °C have a reverse sandwich structure (a), while crystals obtained at 25 °C have one lithium atom η^5-coordinated to the ring atoms and the other η^1-coordinated to the germanium atom (b)[35] (Figure 8).

The X-ray structure determination of the germole dianion $[\eta^5\text{-Li.TMEDA}][\eta^5\text{-Li}]_{1/2}[\eta^1\text{-Li}]_{1/2}[Et_4C_4Ge]$[36] showed three lithium cations around one germole ring in the η^1, η^5 and η^5 sites (Figure 9).

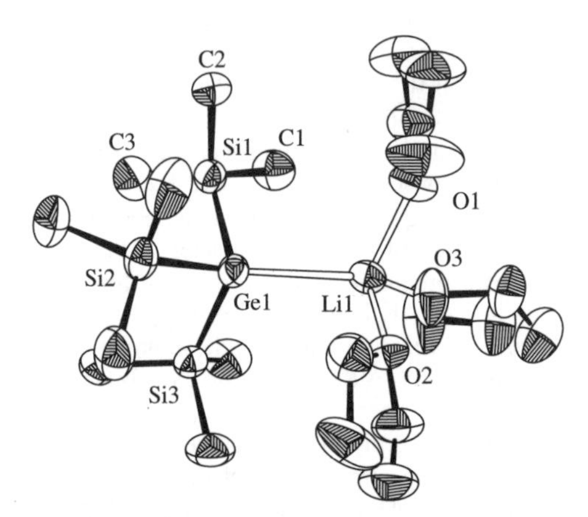

FIGURE 4. Solid state structure of $(Me_3Si)_3GeLi(THF)_3$ with hydrogen atoms omitted. Reprinted with permission from Reference 47. Copyright 1996 American Chemical Society

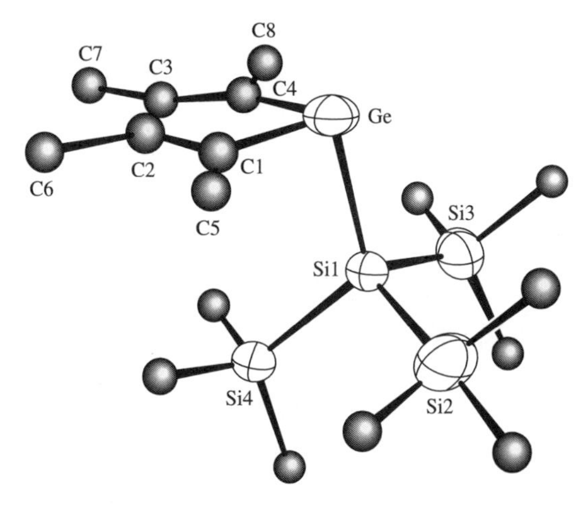

FIGURE 5. Solid state structure of $[Me_4C_4GeSi(SiMe_3)_3]$ with hydrogen atoms omitted. Reproduced by permission of Wiley-VCH from Reference 22

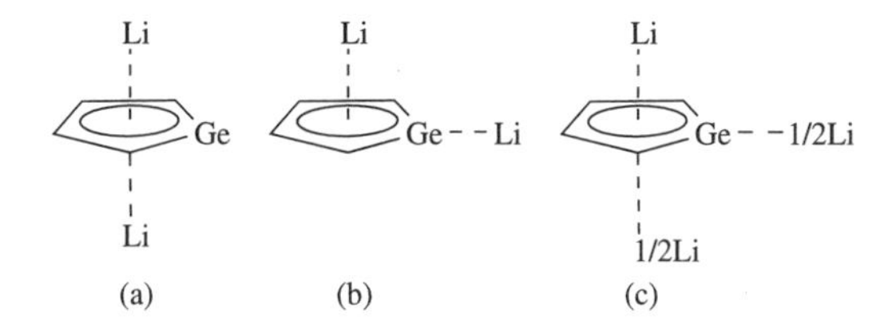

FIGURE 6. Coordination states of germolyl anions

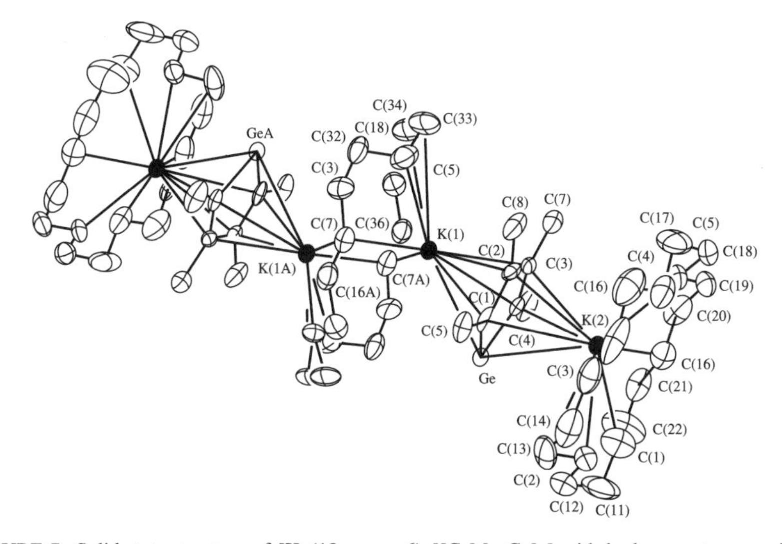

FIGURE 7. Solid state structure of $[K_4(18\text{-crown-}6)_3][C_4Me_4Ge]_2]$ with hydrogen atoms omitted. Reprinted with permission from Reference 21. Copyright 1996 American Chemical Society

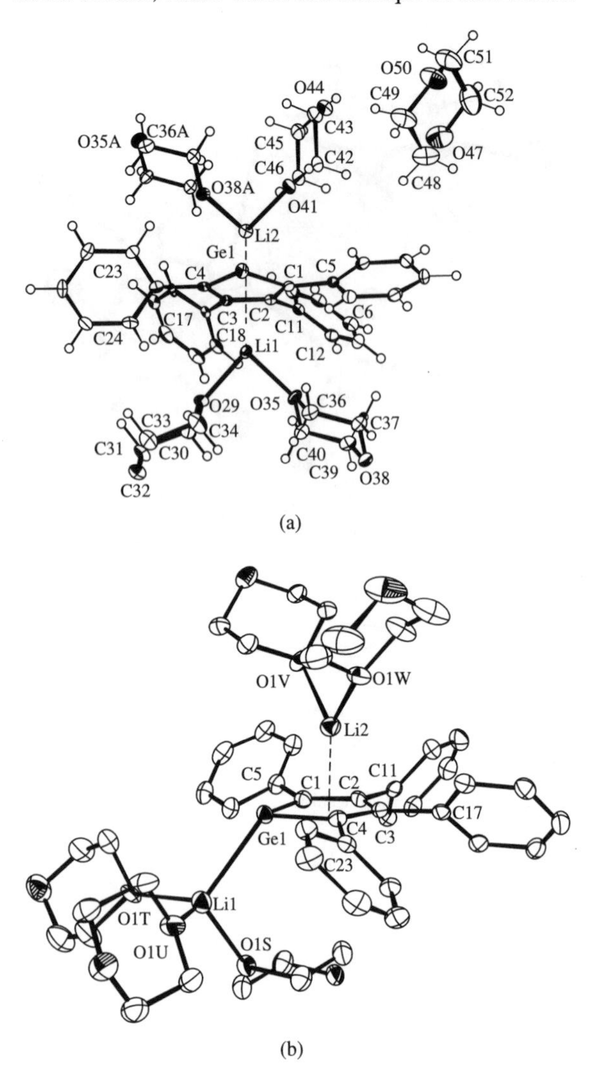

FIGURE 8. Solid state structure of $Li_2[(PhC)_4Ge]$.dioxane (forms (a) and (b)) with hydrogen atoms omitted. Reproduced by permission of Wiley-VCH from Reference 35

The germole dianions in these structures appear to possess delocalized π-systems, as evident by nearly equivalent C−C bond lengths in the five-membered rings. In the sandwich structure, the two metal atoms (K or Li) lie above and below the C_4Ge ring within bonding distance of all five ring atoms. Other structures of germylanions have also been described[38,39]. These compounds do not have metal−Ge contact, but are ion-separated species.

Although many alkali and alkaline earth stannate complexes have been structurally characterized, there are only a few reports of stannyl compounds containing true metal−Sn

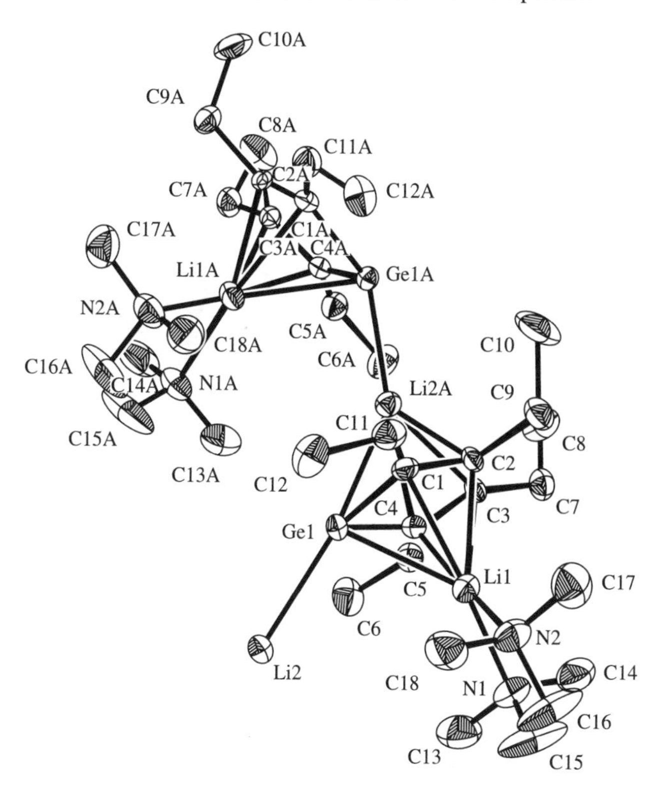

FIGURE 9. Solid state structure of the dilithium salts of the 2,3,4,5-tetraethylgermole dianion with hydrogen atoms omitted. Reprinted with permission from Reference 36. Copyright 1999 American Chemical Society

bonds. [Ph_3SnLi(PMDETA)] was the first metal–Sn bonded complex to be characterized in the solid state[120] (Figure 10). The Sn–Li bond of 2.817(7) Å (average of two independent molecules) is a little greater than the sum of the covalent radii (2.74 Å). The Sn center shows a pyramidal geometry which was also observed for the Ph_3Sn^- ion in the solid state structure of [(18-crown-6) $K^{+-}SnPh_3$][122].

The Pb analogue complex Ph_3PbLi(PMDETA)[123] shows very similar structural features. The Pb–Li distance is slightly greater than the sum of the covalent radii of Pb and Li (average over both independent molecules: 2.858 Å; sum of covalent radii: 2.81 Å). The Ph_3Pb moiety is, however, more pyramidal than the Ph_3Sn unit (average C–Sn–C angle: 96.1°; average C–Pb–C angle: 94.3°). The significant distortion in the Ph_3E unit, away from a tetrahedral geometry, is consistent with the expected increase in the effective energetic separation of s and p orbitals on descending group 14.

Depending upon the coordinated solvent around the metal, several interesting features have been described. The complex $[Li(dioxane)_4]^+[Sn(furyl)_3.Li(furyl)_3Sn]^-$.2dioxane is an ion-pair consisting of lithium ion coordinated by four dioxanes and a complex anion. The latter consists of two pyramidal $(furyl)_3Sn^-$ ions linked by their furyl O-atoms to a central 6-coordinated Li center[124]. The stannyl potassium compound [K{$Sn(CH_2Bu$-$t)_3$} (η^6-$C_6H_5Me)_3$] is the first example of a complex in which the alkali metal ion is

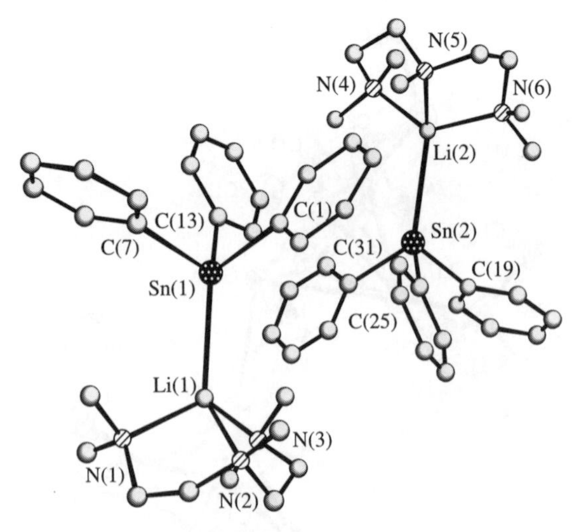

FIGURE 10. Solid state structure of $Ph_3SnLi(PMDETA)$ with hydrogen atoms omitted. Reproduced by permission of Wiley-VCH from Reference 119

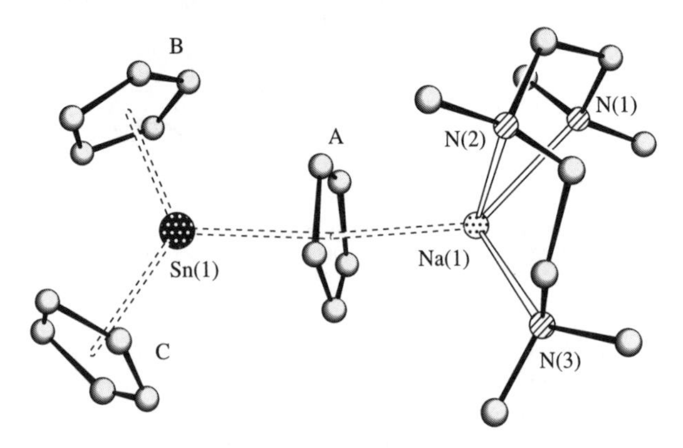

FIGURE 11. Solid state structure of $[(\eta^5\text{-Cp})_2Sn(\mu\text{-Cp})Na$ (PMDETA] with hydrogen atoms omitted. Reprinted with permission from Reference 127. Copyright 1997 American Chemical Society

not stabilized by heteroatom donors (O, N, etc.)[50]. Three π-bonded toluene molecules solvate the K^+ cation. The tin environment is pyramidal and the Sn−K distance is 3.548(3) Å.

More recently, complexes containing 'paddle wheel' $[(\eta^5\text{-}C_5H_5)_3E^-]$(E = Sn, Pb) anions were prepared and structurally characterized[126,127] (Figure 11).

These monomeric complexes are essentially isostructural and contain trigonal-planar $(\eta^5\text{-}C_5H_5)_3E^-$(E = Sn, Pb) units (Sn and Pb: 0.14 Å out of the plane of the three Cp centers), linked through a μ-Cp bridge to a $[Na(PMDETA)]^+$ cation. When the cation was

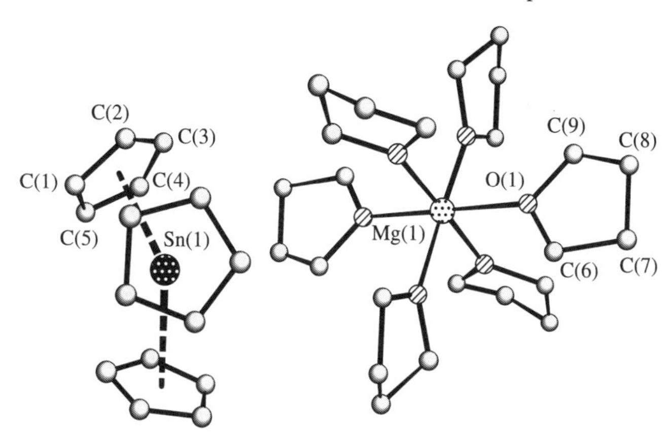

FIGURE 12. Solid state structure of $2[(\eta^3\text{-Cp})_3Sn]^-[Mg(THF)_6]^{2+}$ (only one of the $[(\eta^3\text{-Cp})_3Sn]^-$ anions is shown) with hydrogen atoms omitted. Reprinted with permission from Reference 127. Copyright 1997 American Chemical Society

TABLE 8. Bond distances of metallated group-14 compounds

Compound	M_{14}–M(Å)[a]	Reference
$(Me_3Si)_3Ge.Li(THF)_3$	2.666	47
$(Me_3Si)_3Ge.Li(PMDETA)$	2.653	47
$(o\text{-}Me_2NC_6H_4)_3GeLi$	2.598	18
$Li_2[Ph_4C_4Ge]$.5dioxane	2.730, 2.681	35
$[\eta^5\text{-Li.TMEDA}][\eta^5\text{-Li}]_{1/2}[\eta^1\text{-Li}]_{1/2}[Et_4C_4Ge]$	2.583, 2.684, 2.675	36
Me Me Si Ph O Me N Ph *t*-Bu—Si–O–Si–N–Ge—Li(THF)$_3$ O Me N Si Ph Me Me	2.904	67
$Ph_3SnLi.PMDETA$	2.817	120
$[Li(OC_6H_3Ph_2\text{-}2,6)_3Sn]$	2.784	125
$[K\{Sn(CH_2(Bu\text{-}t))_3\}(\eta^6\text{-}C_6H_5Me)_3]$	3.548	50
$Ph_3PbLi.PMDETA$	2.858	123

[a] M_{14} = Ge, Sn, Pb; M = Li, K.

exchanged for magnesium, a separate$[(\eta^3\text{-Cp})_3Sn]^-$ ion was observed (Figure 12). The Cp ligands of the anion are bonded equivalently to the Sn center. The Cp (centroid)–Sn contacts were also significantly shorter than those found in the Na complex and a more pyramidal geometry was observed for this anion.

Table 8 gives some examples of metal 14–metal bond distances.

V. REACTIVITIES

A. Hydrolysis

Hydrolysis and more particularly deuteriolysis of the $M_{14}-M_1$ or M_2 bond is one of the best ways to characterize M_{14} anions, mainly used in germanium, tin and lead chemistry[3,4,8] (equations 75 and 76)[20,86]. This characterization reaction becomes very typical in the case of hydrometal-14 anions from the coupling constant between hydrogen and deuterium observed in ^{1}H NMR spectra (equation 77)[16,29,30]. The reaction of plumbyllithium or sodium with water was reported to be more complicated and lead to PbO[8].

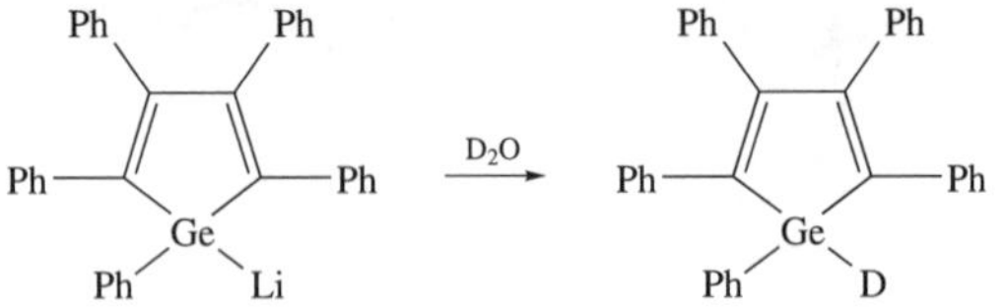

(75)

$$Bu_3SnMgBr \xrightarrow{D_2O} Bu_3SnD \qquad (76)$$

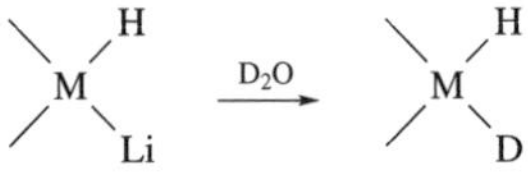

(77)

M = Ge, Sn
M = Ge, J(H–D) = 0.9 Hz

The stereochemistry of the hydrolysis of the germanium–lithium bond (retention) was established[8].

B. Oxidation

Oxygen, sulfur or selenium insert in the germanium alkali–metal bond[8] (equation 78[67] and equation 79[128]).

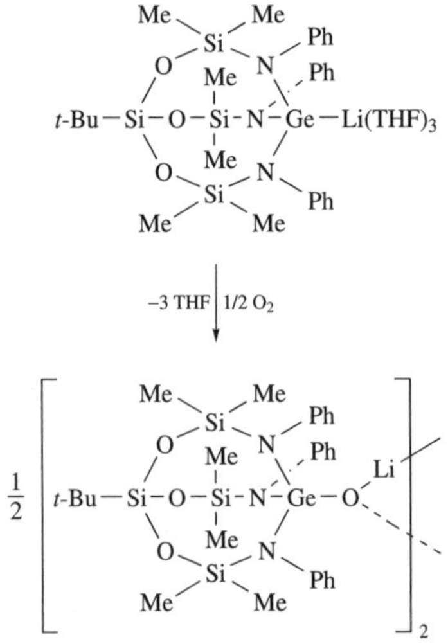

(78)

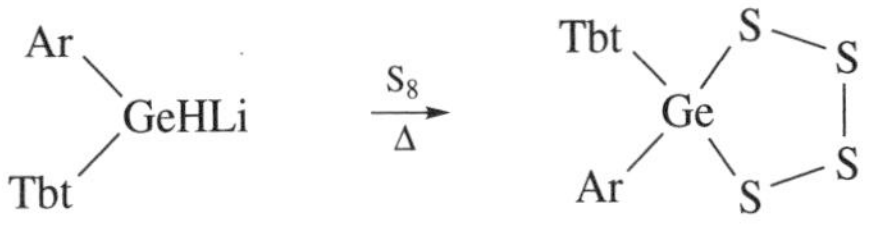

$$\text{Ar(Tbt)GeHLi} \xrightarrow[\Delta]{S_8} \text{Tbt(Ar)Ge}(\text{S}-\text{S}-\text{S}-\text{S}) \quad (79)$$

Ar = Mes, Tip

The treatment of arylhydrogermyllithium with elemental selenium produced tetraselena germolanes (equation 80)[9].

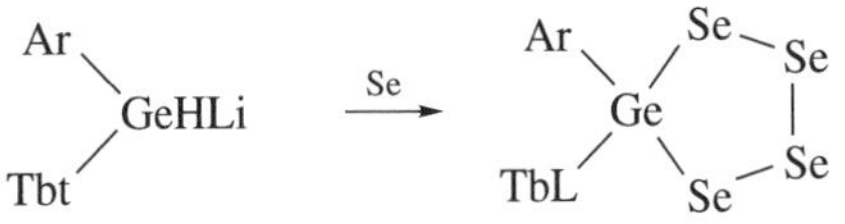

$$\text{Ar(Tbt)GeHLi} \xrightarrow{\text{Se}} \text{Ar(TbL)Ge}(\text{Se}-\text{Se}-\text{Se}-\text{Se}) \quad (80)$$

Ar: Mes, Tip

The reaction of Ph_3SnLi with selenium in THF at room temperature led to the lithium triphenyltin selenide, which had been trapped by reaction with metal-14 halides (Scheme 18)[2].

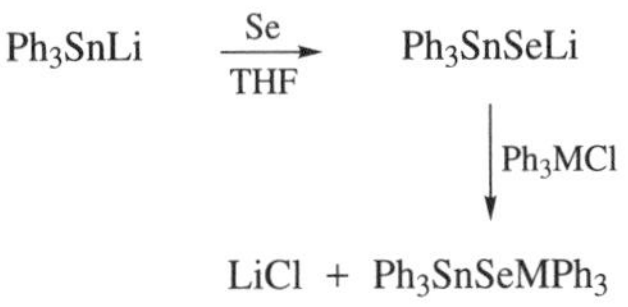

$$Ph_3SnLi \xrightarrow[\text{THF}]{\text{Se}} Ph_3SnSeLi$$

$$Ph_3SnSeLi \xrightarrow{Ph_3MCl} LiCl + Ph_3SnSeMPh_3$$

M = Ge, Sn, Pb

SCHEME 18

C. Substitution

Metal-14 anions react with alkyl halides (RX) mostly by nucleophilic substitution (S_N2), the stereochemistry of which is dependent on the structure of R and X, the solvent and the nature of the counterion. Other reactions were also observed: nucleophilic substitution at halogen [also called halogen/metal exchange (HME)] and single electron processes. In some cases steric hindrance around the reactant results in elimination.

1. Substitutions at carbon

The reaction of organogermylmetal compounds with organic halides is an effective route to form germanium–carbon bonds. The stereochemistry of these reactions was established as predominantly retention (equation 81)[129,130,131]; see Section V.C.2.

$$R^1R^2R^3Ge^{*}Li + R^4Cl \xrightarrow{\text{Retention}} R^1R^2R^3R^4Ge^{*} + LiCl \quad (81)$$

Methyl iodide was widely used for the characterization of metal-14 centered anions. Generally, the reaction occurs at room temperature and leads to almost quantitative yields (equation 82). However, it was shown in germanium chemistry that one has to be careful in the interpretation of the results because complexes such as [GeLi, RLi, ether] lead to the same dialkylation on metal-14 (see Section III. A. 1, Scheme 5).

$$\geq M_{14}{-}M + MeI \longrightarrow \geq M_{14}{-}Me + MI \tag{82}$$

M = group I or group II metal

Because of steric hindrance and high coordination of the germanium center, in some cases methyl iodide undergoes halogen/metal exchange (equation 83)[17].

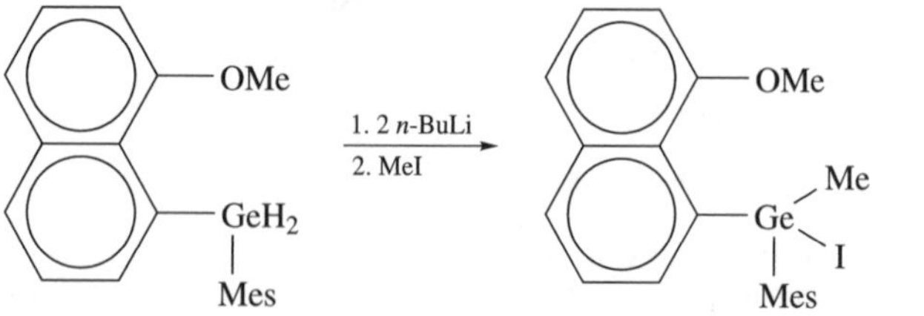

(83)

Within the tin series, the trimethylstannyl anion has to be considered as one of the most powerful simple nucleophiles available; thus an S_N2 reaction at carbon with inversion of configuration is often observed (equation 84)[4b].

$$R^1{}_3Sn^- + R^2X \longrightarrow R^1{}_3SnR^2 + X^- \tag{84}$$

R^1 = Me, Bu, Ph

S_N2 substitutions at the halogen were also observed (equation 85)[4b].

$$R^1{}_3Sn^- + R^2X \longrightarrow R^1{}_3SnX + R^{2-} \tag{85}$$

R^1 = Me, Ph

$S_{RN}1$ reactions with aryl halides were also observed. Their general mechanism is presented in Scheme 19[4b] and illustrated in the subsection on SET reactions.

$$R_3Sn^- + R'X \longrightarrow (R'X)^{\bullet -} + R_3Sn^{\bullet}$$

$$(R'X)^{\bullet -} \longrightarrow R'^{\bullet} + X^-$$

$$R'^{\bullet} + R_3Sn^- \longrightarrow (R_3SnR')^{\bullet -}$$

$$(R_3SnR')^{\bullet -} + R'X \longrightarrow R_3SnR' + (R'X)^{\bullet -}$$

SCHEME 19

Formal nucleophilic substitutions have been studied by simple trapping techniques designed to separate and estimate contributions of reactions proceeding by way of free radicals, by way of anions, geminate or synchronous processes[108]. Reactions of trimethyltin sodium with organic halides in THF at 0 °C were examined using dicyclohexylphosphine for trapping free radicals and *t*-butylamine for trapping free anionoids. Among the twenty-two halides included in this study, nine were shown to involve two or all three of the mechanistic pathways.

Primary chlorides reacted predominantly by a direct mechanism (S_N2 or a multicentered process). Isobutyl or neopentyl halides led to contributions from electron transfer (free radicals) and halogen–metal exchange (anionoid) mechanisms.

Secondary bromides reacted predominantly by an electron transfer and competitive but minor halogen–metal exchange, while the relative contribution from these were reversed in the case of iodides.

Triethylcarbinyl chlorides reacted exclusively by elimination while the bromide reacted by electron transfer in competition with elimination.

1- and 2-bromoadamantanes reacted, because of high steric hindrance, by electron transfer, and 1-chloroadamantane, which is less reactive, gave no reaction under the same conditions.

The results obtained in the case of primary halides were confirmed by kinetic studies of their reactions with stannylanions using a stopped flow technique. The resulting rate constants were much greater than those calculated for an electron transfer according to the Hush–Marcus theory which supports a nucleophilic reactivity rather than a single electron transfer pathway[132].

On the contrary, in the case of 1-iodonorbornane (a tertiary halide), the result of the reaction with trimethylstannyl reagents (Me_3SnM, M = Li, Na), both in the absence and in the presence of trapping agents, confirmed that the nucleophilic substitution process is governed by competition between polar and radical mechanisms[133].

As in germanium chemistry, MeI was also mainly used to characterize stannylanions (equation 86)[134,135]. The structure of the methylated compound was determined by X-ray analysis.

$$HC(SiMe_2NR)_3Sn^- \; [Li(THF)_n]^+ \xrightarrow{MeI} HC(SiMe_2NR)_3SnMe \qquad (86)$$

R = Tol, *t*-Bu

Substitution of alkyl halides by triphenylplumbyl anions also easily gave alkylation of the metal-14 center through a process which proceeds with inversion of configuration at the carbon center (equation 87)[3c].

$$Ph_3Pb^-Na^+ + (S)\text{-}(+)\text{-}s\text{-}BuBr \longrightarrow (R)\text{-}(-)\text{-}Ph_3Pb\text{-}Bu\text{-}s + NaBr \qquad (87)$$

Non-classical metal-14 anions often react in the same way. Thus, ammonium germanates were alkylated with MeI (equation 88)[65].

$$(PhCl_2Ge)^-(DBUH)^+ \xrightarrow[-DBU,HI]{MeI} PhCl_2GeMe \qquad (88)$$

Metal dianions and polymetal anions were also alkylated by MeI or other alkyl halides RX (equations 89 and 90) in high yields (70–80%)[35,25].

$$\text{2,3,4,5-Ph}_4\text{C}_4\text{Ge}^{2-}\ 2\,\text{Li}^+ \xrightarrow[-2\,\text{LiX}]{2\,\text{RX}} \text{2,3,4,5-Ph}_4\text{C}_4\text{GeR}_2 \quad (89)$$

$$Ph_2LiGeGeLiPh_2 + 2\,MeI \longrightarrow Ph_2MeGeGeMePh_2 + 2\,LiI \quad (90)$$

All of these reactions which lead to M_{14}–carbon bonds allow the synthesis of various functional metal-14 organometallic compounds.

The reaction of germyllithiums with chloromethyl methyl ether gave the germylmethyl methyl ether in good yield (equation 91)[9,136].

$$R_3GeLi + ClCH_2OCH_3 \xrightarrow{-LiCl} R_3GeCH_2OCH_3 \quad (91)$$

R_3Ge: PhH_2Ge(73% yield)

R_3Ge: $TMS_2(Me_3Ge)Ge$(89% yield)

The reaction of Ph_3MLi (M = Ge, Sn) or Me_3SnNa with 6-bromo-1-heptene gave the expected 6-M_{14}-1-heptene as the major product, but also (2-methylcyclopentyl)methyl derived metal-14 compound (equation 92)[137]. An intermediate 1-methyl-5-hexenyl radical was proposed, but its participation was not clearly established. Distannylation of haloalkylpropene was also described (equation 93)[138].

$$CH_2{=}CH(CH_2)_3CH(CH_3)Br + Ph_3MLi \xrightarrow[15-25\,^\circ C]{THF} CH_2{=}CH(CH_2)_3CH(CH_3)MPh_3 + (2\text{-methylcyclopentyl})CH_2MPh_3 \quad (92)$$

M = Ge, Sn

$$ClCH_2C(=CH_2)CH_2Cl \xrightarrow[THF]{2\,R_3SnLi} R_3SnCH_2C(=CH_2)CH_2SnR_3 \quad (93)$$

The arylstannylation of the aniline skeleton using *o*-, *m*- or *p*-bromo-*N,N*-dimethylanilines has been reported (equation 94)[139]. In the case of the three bromoanilines, a modified process was used (Scheme 20)[139].

$$o\text{-}BrC_6H_4NMe_2 + Me_3SnLi \longrightarrow o\text{-}Me_3SnC_6H_4NMe_2 + LiBr \quad (94)$$

Organic *gem*-di- and tri-halides, or carbon tetrachloride gave complete stannylation because α-stannylalkyl halides are much more reactive than α-stannylalkyl *gem*-dihalides

$$x\text{-}BrC_6H_4NH_2 + 3\,BuLi \longrightarrow x\text{-}LiC_6H_4NLi_2 \xrightarrow{MgBr_2} x\text{-}BrMgC_6H_4NLi_2 \xrightarrow{Ph_3SnI} x\text{-}Ph_3SnC_6H_4NLi_2 \xrightarrow{H_2O} x\text{-}Ph_3SnC_6H_4NH_2$$

$x = o, m, p$

SCHEME 20

(equations 95 and 96)[4b].

$$3\ Me_3SnLi + HCCl_3 \xrightarrow{-3\ LiCl} HC(SnMe_3)_3 \tag{95}$$

$$2\ Me_3SnLi + (TMS)_2CCl_2 \xrightarrow{-2\ LiCl} (TMS)_2C(SnMe_3)_2 \tag{96}$$

In lead chemistry, the level of metallation was temperature-dependent (equation 97)[3c].

$$Ph_3PbLi + CCl_4 \xrightarrow{-60\,^{\circ}C} Ph_3PbCCl_3; \quad Ph_3PbLi + CCl_4 \xrightarrow{r.t.} (Ph_3Pb)_2CCl_2 \tag{97}$$

In the case of 1,2-dihaloethanes and 1,3-dihalopropanes, an elimination was observed (equation 98)[3c]; see Scheme 21[140].

$$2\ Ar_3PbMgX + XCH_2CH_2X \longrightarrow Ar_6Pb_2 + CH_2 = CH_2 + 2\ MgX_2 \tag{98}$$

$X = Cl, Br$

$$R_3SnNa + XCH_2CH_2CH_2X \longrightarrow R_3SnH + XCH_2CH{=}CH_2 + NaX$$

$$R_3SnNa + XCH_2CH{=}CH_2 \longrightarrow R_3SnCH_2CH{=}CH_2 + NaX$$

$$2R_3SnH \xrightarrow{R_3SnNa} R_3SnSnR_3 + H_2$$

SCHEME 21

When the chain becomes longer, dimetallation occurs preferentially (equation 99)[140].

$$2\ Me_3SnNa + Cl(CH_2)_nCl \xrightarrow{-2\ NaCl} Me_3Sn(CH_2)_nSnMe_3 \tag{99}$$

$n = 4\text{–}6$

2-stannylpyrimidines were synthesized by stannylanion substitution of 2-chloro- or 2-bromopyrimidines (equation 100)[141].

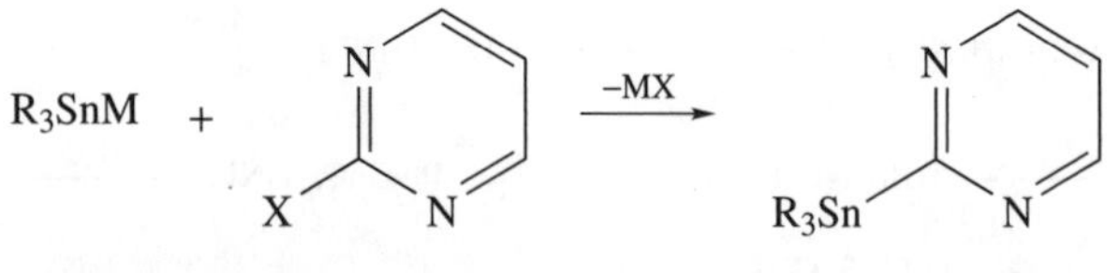

$$R_3SnM + \text{2-X-pyrimidine} \xrightarrow{-MX} \text{2-}R_3Sn\text{-pyrimidine} \quad (100)$$

R = Me, Bu, Ph; M = Na, Li;
X = Cl, Br

Another application of the direct alkylation of metal-14 anions is the synthesis of polymer-supported organotin hydrides. These were prepared by the reaction of ω-haloalkylpolystyrenes with hydridobutylstannyllithium. The stannyl group was separated from the phenyl ring of polystyrene by two, three or even four carbon spacers. These polymers were found to contain 0.8–1.4 mmol of Sn−H per gram. The reducing ability of the polymer-supported organotin hydrides was monitored by reactions with haloalkanes (Scheme 22)[142].

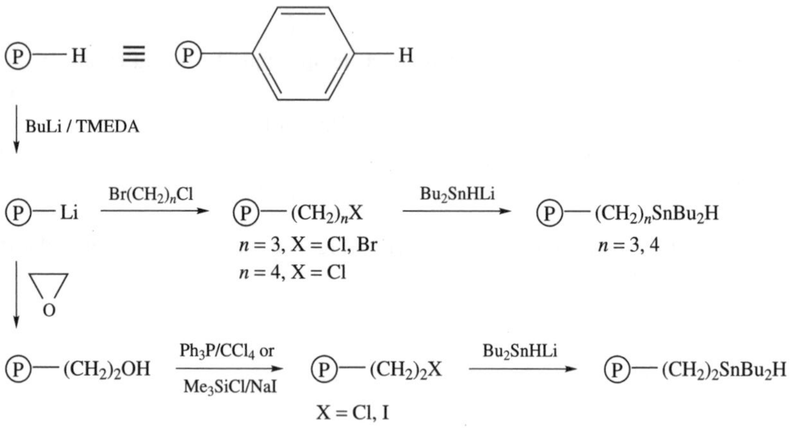

Ⓟ—H ≡ Ⓟ—C_6H_4—H

↓ BuLi / TMEDA

Ⓟ—Li $\xrightarrow{Br(CH_2)_nCl}$ Ⓟ—$(CH_2)_nX$ $\xrightarrow{Bu_2SnHLi}$ Ⓟ—$(CH_2)_nSnBu_2H$

$n = 3$, X = Cl, Br; $n = 4$, X = Cl → $n = 3, 4$

↓ (ethylene oxide)

Ⓟ—$(CH_2)_2OH$ $\xrightarrow[Me_3SiCl/NaI]{Ph_3P/CCl_4 \text{ or}}$ Ⓟ—$(CH_2)_2X$ $\xrightarrow{Bu_2SnHLi}$ Ⓟ—$(CH_2)_2SnBu_2H$

X = Cl, I

SCHEME 22

A convenient, general and efficient (96% yield) synthesis of primary α-alkoxyorganostannanes from stannylanions and α-haloethers has been reported (equation 101)[28].

$$Bu_3SnLi \xrightarrow[-78\,^{\circ}C \to r.t.]{ROCH_2Cl} Bu_3SnCH_2OR \quad (101)$$

The substitution of aromatic acyl chloride by organogermyl groups occurs undoubtedly by an addition–elimination process (S_N acyl), but under particular conditions SET processes were also involved[143] (see Section V.E). These reactions yield as major products either α-germylketones (equations 102 and 103)[15,16,144] or bis(organogermyl) carbinols, depending on reagents and conditions.[8]

$$Ph_3GeLi + ArCOCl \xrightarrow{-LiCl} Ph_3GeCOAr \quad (102)$$

Ar = Ph, p-$MeOC_6H_4$, p-FC_6H_4, p-$CF_3C_6H_4$

$$Ar_2HGeLi + PhCOCl \xrightarrow{-LiCl} Ar_2HGeCOPh \quad (103)$$

$Ar = Ph, Mes$

With the more sterically hindered 2,4,6-trimethylbenzoyl chloride, the expected reaction occurred but it also gave an unexpected germa-β-diketone which might be formed through a benzoylgermyllithium (Scheme 23)[16]. Digermyl diketones were synthesized in the same way (equation 104)[9].

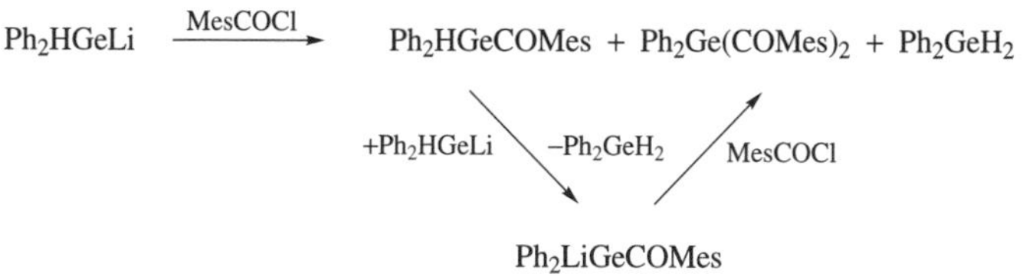

$$Ph_2HGeLi \xrightarrow{MesCOCl} Ph_2HGeCOMes + Ph_2Ge(COMes)_2 + Ph_2GeH_2$$

$Ph_2HGeCOMes \xrightarrow{+Ph_2HGeLi,\ -Ph_2GeH_2} Ph_2LiGeCOMes \xrightarrow{MesCOCl} Ph_2Ge(COMes)_2$

SCHEME 23

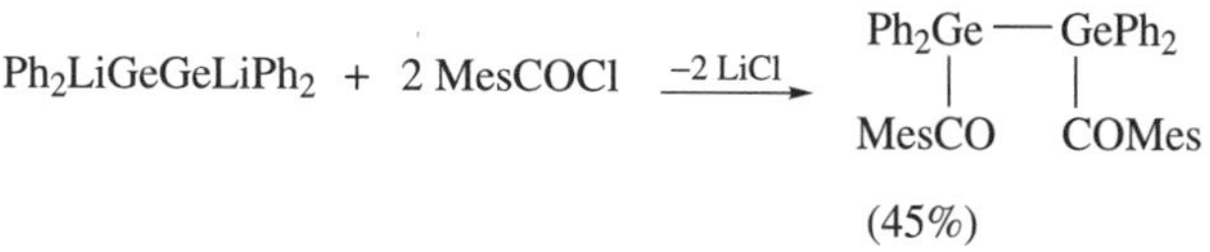

$$Ph_2LiGeGeLiPh_2 + 2\ MesCOCl \xrightarrow{-2\ LiCl} (MesCO)Ph_2Ge{-}GePh_2(COMes)\ (45\%) \quad (104)$$

The reaction of PhH_2GeLi with MesCOCl gave an unexpected triacylgermane, no doubt by successive *trans*-lithiation of a transient hydrogermyl ketone (equation 105)[102].

$$3\ PhH_2GeLi + 3\ MesCOCl \xrightarrow[\text{hydrolysis}]{-3\ LiCl} \underset{(81\%)}{PhGe(COMes)_3} + 2\ PhGeH_3 \quad (105)$$

When the steric effect around germanium and the carbonyl did not prevent subsequent addition of the germyllithium to the germyl ketone, the reaction gave mainly the α-digermyl alcohol (equation 106)[102].

$$MesH_2GeLi + PhCOCl \xrightarrow[2.\ H_2O]{1.\ -40\,^\circ C,\ -LiCl} \underset{(29\%)}{(MesH_2Ge)_2C(OH)Ph} \quad (106)$$

The reaction of a stannylanion with an acyl chloride also constitutes a general access to acylstannanes, but in low yields (equation 107)[145]. These can be improved by using other functions derived from carboxylic acids[8] (see equation 118 below).

$$R'_3SnLi + RCOCl \xrightarrow{-LiCl} R'_3SnCOR \quad (107)$$

$R = aryl; R' = allyl, aryl$

In lead chemistry, the reaction between the trimesitylplumbyllithium and acyl chlorides led to acylplumbanes in high yield and to the first isolable acylplumbane as a yellow crystalline compound, whose structure was confirmed by single crystal X-ray analysis (equation 108)[52].

$$\mathrm{Mes_3PbLi + RCOCl \xrightarrow{-LiCl} Mes_3PbCOR} \tag{108}$$

$$\mathrm{R = Me, Ph}$$

Other less stable acylplumbanes were not isolated but characterized *in situ* (equation 109)[3c].

$$\mathrm{Ph_3PbLi \xrightarrow[-LiCl]{PhCOCl} Ph_3PbCOR} \tag{109}$$

$$\mathrm{R = Me, Ph, OEt, NEt_2}$$

The reaction of vicinal M_{14} dianions with organic dihalides gave heterocyclization (equation 110)[9,25].

$$\mathrm{Ph_2LiGeGeLiPh_2 + Br(CH_2)_3Br \xrightarrow{-2\,LiBr} \underset{(26\%)}{\text{cyclo-}[Ph_2Ge{-}GePh_2{-}(CH_2)_3]}} \tag{110}$$

In the case of benzyl halides, a lithium halogen exchange prevents the cyclization and the reaction gives only oligomers (equation 111)[9,25].

$$\mathrm{Ph_2LiGeGeLiPh_2 \xrightarrow{o\text{-}C_6H_4(CH_2Br)_2} Ph_2(Br)GeGe(Li)Ph_2 \xrightarrow{-LiBr} \frac{1}{n}\,(Ph_2Ge)_n} \tag{111}$$

In tin chemistry, this type of reaction was used to prepare a distannacyclooctane, but in low yield (11%) (equation 112)[140].

$$\mathrm{Me_2(Na)SnSn(Na)Me_2 + Br(CH_2)_3Br \xrightarrow{-2\,NaBr} \text{cyclo-}[Me_2Sn(CH_2)_3]_2\text{ (bicyclic, }Me_2Sn\ \ SnMe_2)} \tag{112}$$

Other carbon–heteroelement bonds have also been used to obtain substitution at carbon by metal-14 anions.

Methylation of germylanions was achieved with Me_2SO_4 (equation 113)[16].

$$\mathrm{R_2HGeLi \xrightarrow[2.\ H_2O]{1.\ Me_2SO_4} R_2HGeMe} \tag{113}$$

$$\mathrm{R = Ph\ (95\%);\ R = Mes\ (95\%)}$$

Tosylate and other alkoxy groups have been used as leaving groups (equations 114[4b], 115[146] and 116[147]).

$$\mathrm{Me_3SnLi + R\text{-}C_6H_{10}\text{-}OTs \xrightarrow{-TsOLi} R\text{-}C_6H_{10}\text{-}SnMe_3} \tag{114}$$

$$\mathrm{Bu_3SnMgCl + AcOCH(OR)_2 \xrightarrow{-AcOMgCl} Bu_3SnCH(OR)_2} \tag{115}$$

$$\mathrm{Bu_3SnLi} + \text{3-ethoxycyclopent-2-enone} \xrightarrow[-78\,^\circ\mathrm{C}]{\mathrm{THF}} \text{3-(tributylstannyl)cyclopent-2-enone} + \mathrm{EtOLi} \tag{116}$$

Amino groups are also suitable leaving groups, as shown in equation 117[148,149].

$$\mathrm{RR^1N{-}CH(R^2){-}Bt} \xrightarrow[\mathrm{THF,\ 0\,^\circ C;\ 38\text{–}85\%}]{\mathrm{Bu_3SnLi\ (2\ equiv)}} \mathrm{RR^1N{-}CH(R^2){-}SnBu_3} \tag{117}$$

(Bt = benzotriazol-1-yl)

α-Metal-14 ketones were obtained by the reaction of metal-14 anions with esters or amides (equation 118)[143,145,150]. The yields were often better than those obtained using acyl chlorides[8].

$$\mathrm{R'_3MLi + RCOX \xrightarrow{-LiX} R'_3MCOR} \tag{118}$$

$$\mathrm{M = Ge,\ Sn,\ Pb}$$

$$\mathrm{X = OR',\ SR',\ NR'_2}$$

$$\mathrm{R = alkyl,\ aryl,\ R''_2N}$$

2. Substitutions at metal

Transmetallation occurs between metal-14 anions and various halometal compounds to yield a variety of organometallic compounds.

Organogermyllithiums have been used to prepare germanium–magnesium compounds which could not be isolated but led to selective germylation (Scheme 24)[9,82]. Similarly, germylmercury compounds were obtained (equation 119)[9,15].

$$\mathrm{Mes_3GeLi + HgCl_2 \xrightarrow{-LiCl} Mes_3GeHgCl \xrightarrow[-LiCl]{Mes_3GeLi} (Mes_3Ge)_2Hg} \tag{119}$$

The reaction of dimesitylgermyllithium etherate with diethylchloroaluminium gave an etherate of the corresponding germyl–aluminium compound in 66% yield (equation 120)[151].

$$R_2HGeLi + MgBr_2 \xrightarrow[-LiBr]{-10\,^\circ C} [R_2HGeMgBr] \xrightarrow{PhClGeH_2} R_2HGeGeH_2Ph$$

R = Ph, 56%
R = Mes, 67%

SCHEME 24

$$Mes_2(H)GeLi, Et_2O + Et_2AlCl \xrightarrow[-78\,^\circ C]{hexane} Mes_2(H)Ge{-}AlEt_2 \leftarrow Et_2O \quad (120)$$

The same reaction was observed in tin chemistry (equation 121)[4b] and applied also to the synthesis of a tin–zinc compound (equation 122)[4b].

$$Bu_3SnLi + Et_2AlCl \xrightarrow{-LiCl} Bu_3SnAlEt_2 \quad (121)$$

$$2\ Bu_3SnLi + ZnBr_2 \xrightarrow{-LiBr} (Bu_3Sn)_2Zn \quad (122)$$

The reaction of trimethylgermyllithium with silicon tetrachloride, because of steric hindrance, gave a low yield of tetra(trimethylgermyl)silane, and hexamethyldigermane was obtained as the major product. The digermane resulted from a lithium/halogen exchange reaction (equation 123)[32].

$$4\ Me_3GeLi + SiCl_4 \xrightarrow{pentane} \underset{(19\%)}{(Me_3Ge)_4Si} + \underset{(30\%)}{Me_3GeGeMe_3} \quad (123)$$

Less bulky germyllithiums gave higher yields of germasilanes (equation 124)[105].

$$PhMe_2GeLi + t\text{-}BuMe_2SiCl \xrightarrow[THF]{-25\,^\circ C} \underset{(71\%)}{PhMe_2GeSiMe_2Bu\text{-}t} + LiCl \quad (124)$$

In the same way catenated stannyl, germyl silanes were prepared (equation 125)[105]. The structure of an aryl compound was resolved by X-ray analysis[105].

$$t\text{-}BuMe_2SiGeMe_2Cl + R_3SnLi \xrightarrow{-LiCl} t\text{-}BuMe_2SiGeMe_2SnR_3 \quad (125)$$

R = Me, Ph

Transmetallations of germoles have also been studied starting from the localized monoanion and the germolyldianion (equations 126 and 127)[20,35].

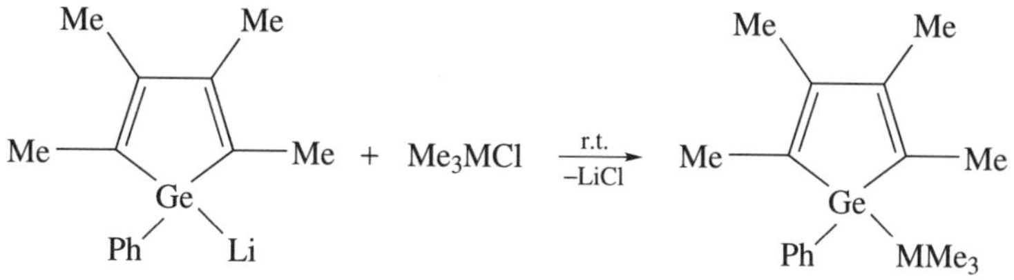

$$\text{(1-Ph-2,3,4,5-Me}_4\text{C}_4\text{Ge)Li} + \text{Me}_3\text{MCl} \xrightarrow[-\text{LiCl}]{\text{r.t.}} \text{1-Ph-1-MMe}_3\text{-2,3,4,5-Me}_4\text{C}_4\text{Ge} \qquad (126)$$

M = Si (75%); Sn (60%)

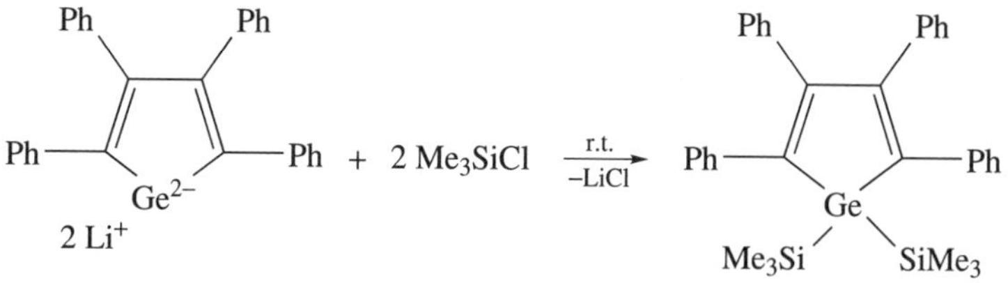

$$\text{(2,3,4,5-Ph}_4\text{C}_4\text{Ge}^{2-})\ 2\,\text{Li}^+ + 2\,\text{Me}_3\text{SiCl} \xrightarrow[-\text{LiCl}]{\text{r.t.}} \text{1,1-(Me}_3\text{Si)}_2\text{-2,3,4,5-Ph}_4\text{C}_4\text{Ge} \qquad (127)$$

Starting from a metal-14 dihalide, selective monogermylation or complete germylation can be obtained by using germylpotassium or germyllithium compounds (equations 128 and 129)[152].

$$Ph_2SiCl_2 + Ph_3GeK \xrightarrow[-78\,^{\circ}C,\,5\ h]{Et_2O} Ph_3GeSiClPh_2 + KCl \qquad (128)$$

$$Ph_2SiCl_2 + 2\ Ph_3GeLi \xrightarrow[-78\,^{\circ}C]{DME} Ph_3GeSiPh_2GePh_3 + 2\ LiCl \qquad (129)$$

Non-symmetrical organohydropolygermanes are usually difficult to obtain. In some cases, the reaction of a germylanion with an organohalohydrogermane allows their synthesis (Scheme 25, a), but when the organohalohydrogermane is too 'acidic', a competitive lithiation reaction (Scheme 25, b) gives polygermanes through α-elimination[16].

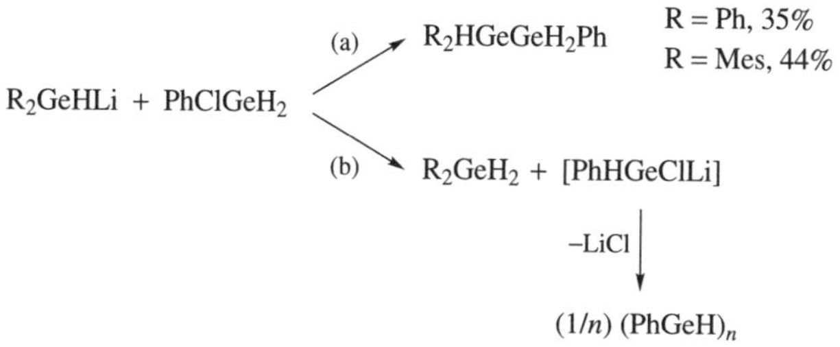

$$R_2GeHLi + PhClGeH_2 \xrightarrow{(a)} R_2HGeGeH_2Ph \quad R = Ph,\ 35\%;\ R = Mes,\ 44\%$$

$$R_2GeHLi + PhClGeH_2 \xrightarrow{(b)} R_2GeH_2 + [PhHGeClLi] \xrightarrow{-LiCl} (1/n)\,(PhGeH)_n$$

SCHEME 25

Novel half-sandwich complexes with divalent Ge, Sn and Pb were obtained from the reaction of CpSnCl with the corresponding metal-14 anion in more than 80% yield (equation 130)[153]. The single crystal X-ray crystallographic analysis of the compound with

M = Ge revealed an average effect of η^3 and η^1 bonding modes of the cyclopentadienyl ring to the tin atom[153].

$$\mathrm{CpSnCl} + \mathrm{KM(OBu\text{-}}t)_3 \xrightarrow[25\,^\circ\mathrm{C}]{\text{Toluene}} \mathrm{CpSn}(\mu\text{-}\mathrm{O}t\text{-}\mathrm{Bu})_2\mathrm{MOBu\text{-}}t + \mathrm{KCl} \qquad (130)$$

M = Ge, Sn, Pb

The X-ray analysis of diphenylbis[tris(trimethylsilyl)germyl]plumbane, obtained from the reaction of diphenyldichlorolead with the appropriate germyllithium (equation 131), showed a staggering of methyl groups around the lead center[154].

$$(\mathrm{C_6H_5})_2\mathrm{PbCl_2} + 2(\mathrm{THF})_{2.5}\mathrm{LiGe(SiMe_3)_3} \xrightarrow[-78\,^\circ\mathrm{C\ to\ r.t.}]{(\mathrm{C_2H_5})_2\mathrm{O}} (\mathrm{C_6H_5})_2\mathrm{Pb(Ge(SiMe_3)_3)_2} + 2\ \mathrm{LiCl} \qquad (131)$$

Metal-14-metal-14 compounds were also prepared from tin- or lead-centered anions as shown in equations 132 and 133[155].

$$(o\text{-Tol})_3\mathrm{M\ Li} + (o\text{-Tol})_3\mathrm{SnI} \xrightarrow[-\mathrm{LiI}]{\mathrm{THF}} (o\text{-Tol})_3\mathrm{SnM(Tol\text{-}}o)_3 \qquad (132)$$

M = Sn, 2 h, r.t., 66% yield (X-ray: Sn−Sn bond length : 2.883 Å)

M = Pb, −78 °C, 57% yield

$$\mathrm{R_3PbLi} + \mathrm{Ph_3SnCl} \xrightarrow[-60\,^\circ\mathrm{C}]{\mathrm{THF}} \mathrm{R_3PbSnPh_3} \qquad (133)$$

R = o-Tol, 67% yield (X-ray: Sn−Pb bond length : 2.845 Å)

R = Mes, 44% yield

The reaction of one equivalent of $[\mathrm{Li(THF)_3.Sn(SiMe_3)_3}]$ with $[\mathrm{Sn(2\{\text{-}(Me_3Si)_2C\}\text{-}C_5H_4N)Cl}]$ in $\mathrm{Et_2O}$ at ambient temperature gave the corresponding divalent Sn−tetravalent Sn−compound in high yield (equation 134)[156]. A single crystal X-ray diffraction study revealed the distannyl compound to be monomeric and the divalent Sn−tetravalent Sn−bond to be of length 2.869 Å. This was the first measurement of such a bond length. The ^{119}SnNMR spectrum exhibits two singlets at $\delta = 897$ ppm and 502 ppm, each with well resolved isotropically shifted ^{119}Sn and ^{117}Sn satellites (J: 6700 and 6400 Hz, respectively). This was also the first measurement of a ^1J coupling between Sn atoms of different valences.

Polygermanes $\mathrm{R_3Ge(GeEt_2)_{\mathit{n}}GeR_3}$ (R = alkyl, aryl) and polystannanes $\mathrm{Ph_3Sn\text{-}}(t\text{-}\mathrm{Bu_2Sn})_n\mathrm{SnPh_3}$ ($n = 1$ to 4) were prepared (equation 135) and studied by ^{119}Sn NMR and UV. A relationship between the electronic excitation and ^{119}Sn NMR chemical shifts of the central Sn atom was obtained, as well as a linear relationship between the coupling constant $^2J(^{119}\mathrm{Sn}/^{119}\mathrm{Sn})$ and the 'non-bonding' distance $d(\mathrm{Sn}\cdots\mathrm{Sn})$. These correlations point to a smooth transition between the covalently bonded polystannanes and metallic tin[157,158].

$$\text{(2-pyridyl)C(SiMe}_3)_2\text{SnCl} \xrightarrow[\text{Et}_2\text{O}]{\text{Li(THF)}_3\text{•Sn(SiMe}_3)_3} \text{(2-pyridyl)C(SiMe}_3)_2\text{Sn–Sn(SiMe}_3)_3 \quad (134)$$

$$2\ R_3MLi + X(t\text{-}Bu_2Sn)_nX \xrightarrow[-2\ LiX]{-78\,^\circ C} R_3M(t\text{-}Bu_2Sn)_nMR_3 \quad (135)$$

$$X = Cl, Br \qquad M = Ge, R = Et, Ph$$
$$M = Sn, R = Ph$$

Treatment of a solution of tri(t-butyl)plumbyllithium or triarylplumbyllithiums in THF with a variety of group-14 electrophiles gave a number of dimetalla derivatives of lead (equations 136 and 137)[110,111]. The values and signs of the coupling constants $^1J(^{207}Pb-^{29}Si)$, $^1J(^{207}Pb-^{119}Sn)$ and $^1J(^{207}Pb-^{207}Pb)$ were determined.

$$t\text{-}Bu_3PbLi \xrightarrow[\substack{\text{or } RR'_2MBr(M{=}Ge,\ Pb)\\(THF, -30\,^\circ C)}]{RR'_2MCl(M{=}Si,\ Sn)} t\text{-}Bu_3PbMRR'_2 \quad (136)$$

$$R,R' = \text{allyl, aryl}$$

$$Ar_3PbLi + Ar'_3MX \xrightarrow[-60\,^\circ C]{THF} Ar_3PbMAr'_3 \quad (137)$$

$$M = Ge, X = Cl, Br$$

$$M = Pb, X = Br, I$$

$$Ar, Ar' = Ph,\ p\text{-tolyl, 2,4-xylyl, } p\text{-anisyl, 2-Np}$$

A tin-containing silylferrocenes was prepared according to equation 138[159].

$$Fe[C_5H_4SiMe_2CH_2Cl]_2 \xrightarrow[-2\ NaCl]{2\ Me_3SnNa,\ NH_3,\ -78\,^\circ C} Fe[C_5H_4SiMe_2CH_2SnMe_3]_2 \quad (138)$$

Various other compounds bearing another metal (usually a transition metal or a lanthanide) linked to the M_{14} (Ge, Sn, Pb) have been reported. Some of them were prepared from metal-14 anions (equations 139[160], 140[161], 141[4b], 142[162], 143[3c], 144[19] and 145[163]).

$$\text{(cyclic } t\text{-BuN(CH}_2\text{)}_3\text{N}t\text{-Bu)PCl} + \text{Et}_3\text{GeLi} \xrightarrow{-\text{LiCl}} \text{(cyclic } t\text{-BuN(CH}_2\text{)}_3\text{N}t\text{-Bu)PGeEt}_3 \quad (71\%) \tag{139}$$

$$2\ \text{Me}_3\text{GeLi} \xrightarrow[\text{THF, }-78\,^\circ\text{C}]{\text{CuBr}\cdot\text{Me}_2\text{S}} (\text{Me}_3\text{Ge})_2\text{CuLi} \tag{140}$$

$$2\ \text{Me}_3\text{SnLi} \xrightarrow{\text{CuBr}} (\text{Me}_3\text{Sn})_2\text{CuLi} \tag{141}$$

$$\text{R}^1\text{R}^2\text{MCl}_2 + \text{Ph}_3\text{GeLi} \xrightarrow[-\text{LiCl}]{\text{THF}} \text{R}^1\text{R}^2\text{M(Cl)(GePh}_3) \tag{142}$$

$R^1, R^2 = Cp, Cp^*; M = Zr, Hf$

$$\text{Ph}_3\text{PbLi} + [\text{Et}_4\text{N}][\text{ClM(CO)}_5] \xrightarrow{-\text{LiCl}} [\text{Et}_4\text{N}][\text{Ph}_3\text{PbM(CO)}_5] \tag{143}$$

$$M = Cr, Mo, W$$

$$2\ [(2\text{-Me}_2\text{NCH}_2\text{C}_6\text{H}_4)_3\text{Ge}^-\text{K}^+] + \text{LnI}_2 \xrightarrow{-2\ \text{KI}} ((2\text{-Me}_2\text{NCH}_2\text{C}_6\text{H}_4)_3\text{Ge})_2\text{Ln} \tag{144}$$

Ln = Yb (14%); Sm (13%)

$$\text{Ph}_3\text{GeK} + \text{Cp}_3\text{UCl} \xrightarrow{\text{THF, }-20\,^\circ\text{C}} \underset{(80\%)}{\text{Ph}_3\text{GeUCp}_3} + \text{KCl} \tag{145}$$

An interesting application of M_{14} anions in transition metal chemistry was the synthesis of optically active anions containing a transition metal–germanium bond. Thus, reaction of the optically active methyl (α-naphthyl) phenylgermyllithium R_3GeLi with $Mo(CO)_6$, $W(CO)_6$, $Fe(CO)_5$, $(\eta^5\text{-MeC}_5H_4)Mn(CO)_3$ or $(\eta^5\text{-C}_5H_5)(Ph_3P)NiCl$ led to anionic complexes (equation 146)[164] isolated for the four first ones as the Et_4N^+ salts.

$$\text{R}_3\text{GeLi} + \text{MCO} \xrightarrow{\text{Et}_2\text{O}} \text{R}_3\text{GeM}^-\text{Li}^+ \tag{146}$$

$$M = W(CO)_5, Mo(CO)_5, (\eta^5\text{-MeC}_5H_4)Mn(CO)_2, Fe(CO)_4$$

The reaction of metal-14 anions with the metal–halogen bond gave heterocyclization, as shown in equation 147[9,165].

$$Ph_2LiGeGeLiPh_2 + trans\text{-}(Et_3P)_2PtCl_2 \xrightarrow[-40\,^{\circ}C]{-LiCl} (Et_3P)_2Pt\langle(GePh_2)(GePh_2)\rangle \quad (147)$$

In some cases a competition between the reaction of metal-14 anions with the carbon–halogen and the metal–halogen bond was observed. For example, in the reaction of chloro(chloromethyl)dimethyl silane or germane with group-14 element nucleophiles $R_3M'Li$ ($M' = Si$, Ge, Sn), the expected monometallated $R_3M'-MMe_2-CH_2Cl$ was obtained in a very low yield, while disubstituted compounds $R_3M'-MMe_2-CH_2-M'R_3$ were mainly produced because, in the monometallated M' compound, the carbon–halogen bond is activated by the β-effect of the R_3M' group (equation 148)[165].

$$Me_2M(Cl)CH_2Cl \xrightarrow[2.\ H_2O]{1.\ R_3M'Li} Me_2M(M'R_3)CH_2Cl + Me_2M(M'R_3)CH_2M'R_3 \quad (148)$$

$M = Si, Ge$

$R_3M' = Me_3Ge, PhMe_2Ge, Me_3Sn$

Nucleophilic substitutions at metal can also involve a leaving group other than halogen. Digermanes were obtained in high yields (80–85%) by the lithiogermolysis of germyl triflates (equation 149)[166]. The stereochemistry of such nucleophilic substitution of an alkoxygermane by a germyllithium reagent was studied. The germyllithium reagents retain their configuration whereas inversion occurs for the alkoxygermane (equation 150 and Scheme 26)[167]. Organohydrodigermanes (equation 151)[9] and cyclopolygermanes (equations 152 and 153)[9,25] were obtained in the same way.

$$R_3GeLi + R'_3GeOSO_2CF_3 \longrightarrow R_3GeGeR'_3 + CF_3SO_2OLi \quad (149)$$

$$(-)\text{-}MePhNpGe^*OMen \xrightarrow[\text{Inversion}]{Ph_3GeLi} (+)\text{-}MePhNpGe^*GePh_3 \quad (150)$$

$$|\alpha|_D^{25} : -49 \qquad |\alpha|_D^{25} : +7.5$$

$$MesH_2GeLi + MesH_2GeOMe \xrightarrow{-MeOLi} \underset{(48\%)}{MesH_2GeGeH_2Mes} \quad (151)$$

$$Ph_2LiGeGeLiPh_2 + Et_2(MeO)GeGe(OMe)Et_2 \xrightarrow{-MeOLi} \underset{(49\%)}{\text{cyclo-}(Ph_2Ge-GePh_2-GeEt_2-Et_2Ge)} \quad (152)$$

$$Ph_2LiGeGeLiPh_2 + (MeO)_2Ge(\text{3,4-dimethylgermacyclopent-3-ene}) \xrightarrow{-2\,MeOLi} \underset{(56\%)}{(Ph_2Ge-Ph_2Ge)Ge(\text{3,4-dimethylgermacyclopent-3-ene})} \quad (153)$$

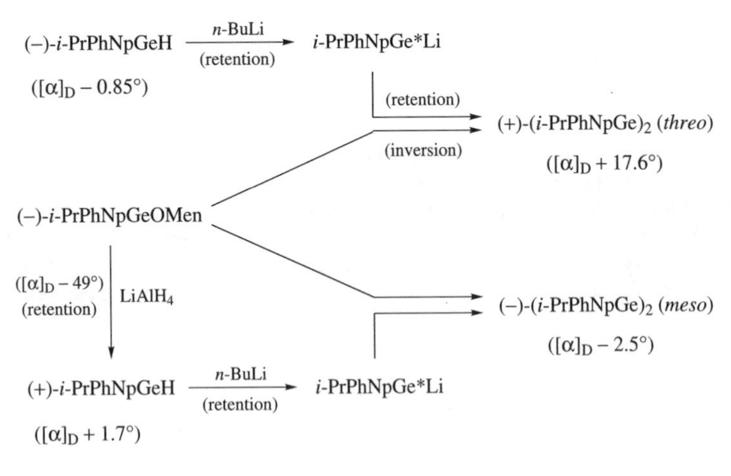

SCHEME 26

The reaction of Ph_3SnNa with $(Bu_2SnS)_3$ gave the symmetric tin sulfide $(Ph_3Sn)_2S$ (equation 154)[41].

$$Ph_3SnNa \xrightarrow{(Bu_2SnS)_3} (Ph_3Sn)_2S + Ph_3SnSnPh_3 \qquad (154)$$

An important application of this method is in the preparation of (triphenylstannyl)diphenylphosphine (equation 155)[41].

$$Ph_3SnNa \xrightarrow{Ph_2PX} Ph_3SnPPh_2 + NaX \qquad (155)$$

In transition metal complexes, the metal–carbonyl bond is also easily cleaved by M_{14} anions (Scheme 27)[4a].

Cyclotrigermenes were obtained from the reaction of metal-14 anions with tris(tri-*t*-butylsilyl)cyclotrigermenium tetrakis[3,5-bis(trifluoromethyl)phenyl]borate, the TFPB behaving as the leaving group (equation 156)[168]. The cyclotrigermene structure was established by X-ray determination.

$$[\{(t\text{-}Bu)_3Si\}_3Ge_3]^{+}\ TFPB^{-} \xrightarrow[Et_2O]{RM} \text{cyclo-}Ge(R)(Si(Bu\text{-}t)_3)\text{–}Ge(Si(Bu\text{-}t)_3){=}Ge(Si(Bu\text{-}t)_3) \qquad (156)$$

RM = $(t\text{-}Bu)_3SiNa$, $(t\text{-}Bu)_3GeNa$, $(Me_3Si)_3SiLi{\cdot}3THF$

$(Me_3Si)_3GeLi{\cdot}3THF$, MesLi

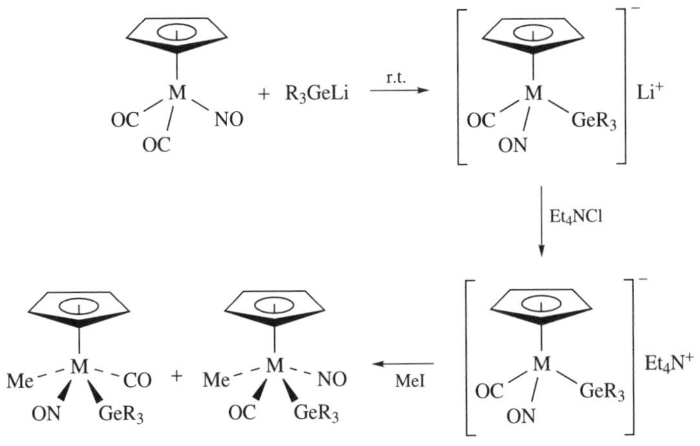

$R_3 = MePh(1\text{-}C_{10}H_7)$;
M = W, Mo

SCHEME 27

D. Nucleophilic Additions

M_{14} anions were characterized by carbonation when the M_{14}-methanoic acids formed were sufficiently stable[8]. The stereochemistry of this reaction was determined as retention (Scheme 28)[169,170].

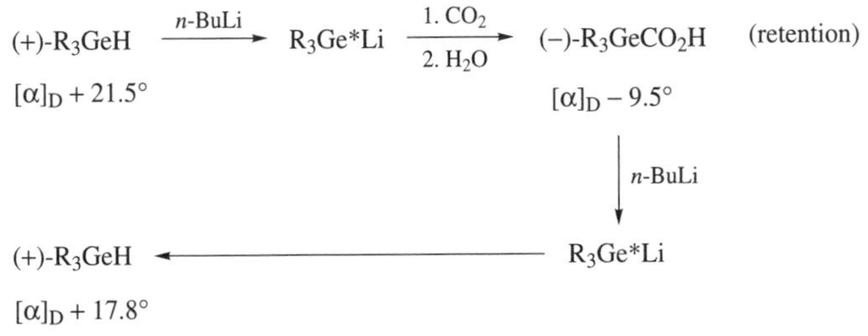

$$(+)\text{-}R_3GeH \xrightarrow{n\text{-BuLi}} R_3Ge^{*}Li \xrightarrow[2.\ H_2O]{1.\ CO_2} (-)\text{-}R_3GeCO_2H \quad \text{(retention)}$$

$[\alpha]_D + 21.5°$ $\qquad$ $[\alpha]_D - 9.5°$

$(-)\text{-}R_3GeCO_2H \xrightarrow{n\text{-BuLi}} R_3Ge^{*}Li \longrightarrow (+)\text{-}R_3GeH$

$[\alpha]_D + 17.8°$

R_3 = Me, Ph, α-Np

SCHEME 28

Reactions of M_{14}-anions with aldehydes resulted in facile 1,2-additions leading to α-metallated alcohols (equations 157[16,24] and 158[9]). When the aldehyde was highly conjugated (equation 158), the nucleophilic addition did not occur and a SET reaction took place (see Section V.E). In the case of tin, the reaction was used for the synthesis of α-stannylalcohol (equation 159)[4b] and then applied to the preparation of α-alkoxystannanes (equations 160 and 161)[104,171] and α-iodostannanes[102,172].

$$R^1R^2HGeLi + R^3CHO \xrightarrow[2.\ H_2O]{1.\ -30\,^\circ C} R^1R^2HGeCH(OH)R^3 \quad (157)$$

$R^1 = R^2 = Ph$, $R^3 = Ph$ (87% yield)

$R^1 = R^2 = Mes$, $R^3 = Ph$ (62% yield)

$R^1 = Mes$, $R^2 = H$, $R^3 = Ph$ (23% yield)

$R^1 = R^3 = Mes$ (53% yield)

$$R_3GeLi + R'CH{=}CHCHO \xrightarrow[2.\ HCl]{1.\ r.t} R'CH{=}CHCH(OH)GeR_3 \quad (158)$$

$R = Ph$, $R' = Ph$ (69% yield)

$R = Ph$, $R' = p\text{-}O_2NC_6H_4$ (traces)

$$Bu_3SnLi + RCHO \xrightarrow[hydrolysis]{after} Bu_3SnCH(OH)R \quad (159)$$

$$Bu_3SnLi + RCHO \longrightarrow RCH(O^-Li^+)SnBu_3 \xrightarrow{R'X} RCH(OR')SnBu_3 \quad (160)$$

Aldehyde (OBu-*t*, TBSO, OMOM substituted cyclohexane) $\xrightarrow[\text{then MOMCl}/i\text{-}Pr_2NEt]{Bu_3SnLi,\ THF,\ -70\,^\circ C}$ MOMO/Bu_3Sn product (OBu-*t*, TBSO, OMOM) (161)

$MOMCl = MeOCH_2Cl$

Trimethylgermyllithium reacted with aliphatic ketones to give the corresponding trimethylgermylcarbinols stereoselectively (equation 162)[173].

t-Bu-cyclohexanone $\xrightarrow[2.\ H_2O]{1.\ Me_3GeLi,\ Et_2O\text{–}HMPA,\ -30\,^\circ C}$ *t*-Bu-cyclohexane-C(OH)($GeMe_3$)

(*cis*/*trans* = 15/85)

(162)

In the reaction of acylgermanes with triethylgermyllithium (equation 163)[174], *gem*-(bisgermyl)alcohols were isolated.

$$R_3E\overset{O}{\overset{\|}{C}}R' + LiGeEt_3 \xrightarrow[(H^+)]{hexane} R_3E\underset{R'}{\overset{OH}{C}}GeEt_3 \quad (163)$$

$R_3E = Et_3Ge, Et_3GeGeEt_2, Me_3SiGeEt_2,$
$R' = CMe_3, Ph$

Thermostable silaenolate anions were obtained by the reaction of tris(trimethylsilyl)acylsilanes with triethylgermyllithium in THF (equation 164)[174].

$$RC(=O)—E(SiMe_3)_2(R') \xrightarrow[-Me_3SiGeEt_3]{Et_3GeLi} RC(O^-\,Li^+)\!=\!ESiMe_3(R') \quad (164)$$

$R = CMe_3, R' = SiMe_3, E = Si, Ge$
$R = \text{adamantyl}, R' = SiMe_3, E = Si$
$R = R' = CMe_3, E = Si$

Ammonium germanates gave the same 1,2-addition to a carbonyl group (equation 165)[48].

$$Me_3Ge^-Bu_4N^+ \xrightarrow[2.\ H_2O]{1.\ PhCOMe} \underset{(75\%)}{MePh(Me_3Ge)COH} + PhMeCHOH \quad (165)$$

In the case of α,β-unsaturated ketones, enolates of 3-M_{14} ketones were obtained regioselectively and characterized chemically (Scheme 29)[173]. The reaction with α,β-unsaturated amides is shown in equation 166[175].

$$Et_3GeLi + PhCH{=}CHCONMe_2 \xrightarrow[hydrolysis]{after} Et_3Ge\overset{Ph}{C}HCH_2CONMe_2 \quad (166)$$

These 1,4-additions were also observed in tin chemistry (equations 167 and 168)[176–179].

$$R^1C(=O)CH{=}CHR^2 \xrightarrow[2.\ Ac_2O]{1.\ Me_3SnLi} R^1C(OAc){=}CHCH(R^2)SnMe_3 \quad (167)$$

$R^1 = Me, R^2 = H$

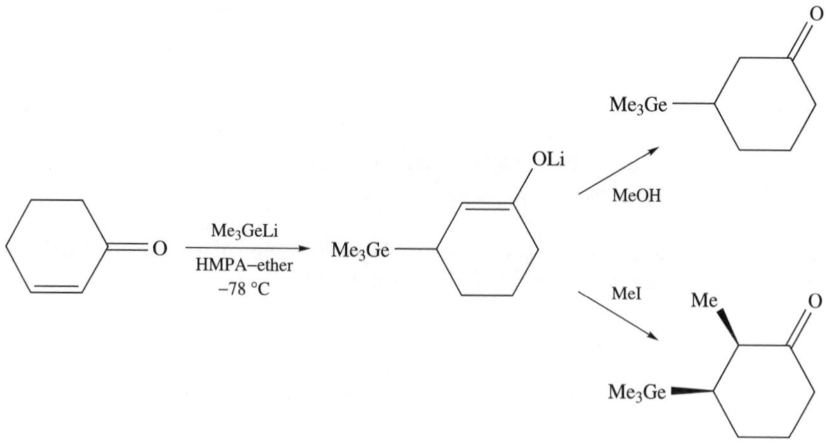
O
Me3Ge
OLi
MeOH
Me3GeLi
HMPA–ether
−78 °C
Me3Ge
MeI
Me
O
Me3Ge

SCHEME 29

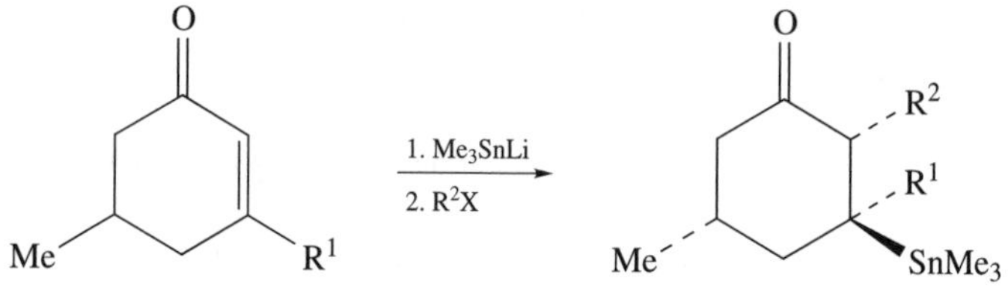
O
Me
R1
1. Me3SnLi
2. R2X
O
R2
R1
Me
SnMe3

(168)

$R^1 = Me$, $R^2 =$ allyl, propargyl, benzyl

These nucleophilic additions to a carbonyl group were also used to synthesize a germene by Peterson's reaction (equation 169)[180].

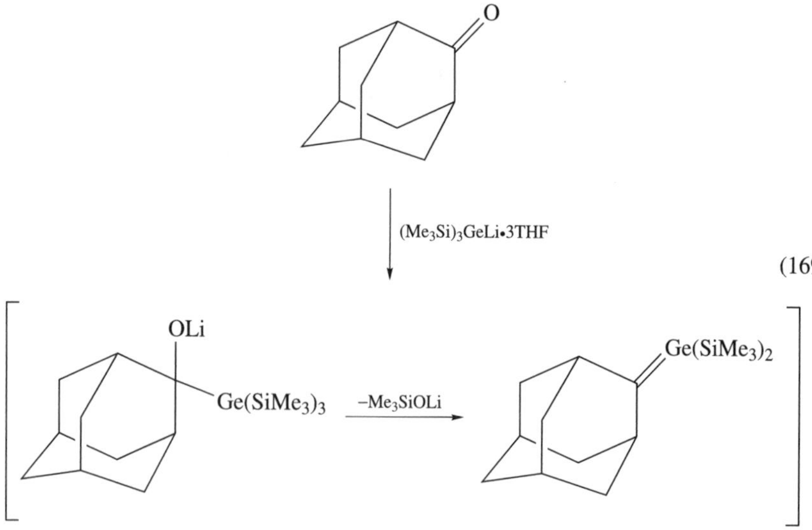
O
(Me3Si)3GeLi•3THF
OLi
Ge(SiMe3)3
−Me3SiOLi
Ge(SiMe3)2

(169)

The complex $Et_3GeNa{-}YCl_3$ reacted as a strong base and abstracted the acidic hydrogen α to carbonyl, but did not lead to the expected nucleophilic addition (equation 170)[181].

$$RC(O)CH_2R' \xrightarrow[\text{2. MeI/}-78\,^\circ\text{C}]{\text{1. }Et_3GeNa\text{-}YCl_3\text{ (3/1) THF/HMPA/}-78\,^\circ\text{C}} RC(O)CH(Me)R' \quad (97\text{–}98\%) \tag{170}$$

The cleavage of epoxides by metal-14 anions afforded β-metallated alcohols (Scheme 7, equations 171–173)[3c,4b,8,182]. Thiiranes and aziridines gave similar reactions (equation 171).

$$Ph_3PbLi + \underset{Y}{\triangledown} \xrightarrow[\text{hydrolysis}]{\text{after}} Ph_3PbCH_2CH_2YH \tag{171}$$

Y = O, S, NR

$$Me_3SnLi + \text{cyclohexene oxide} \longrightarrow \textit{trans}\text{-2-}(Me_3Sn)\text{cyclohexanol} \tag{172}$$

$$\text{(BnO)}_3\text{-glycal 1,2-epoxide} \xrightarrow[\text{2. }NH_4Cl]{\text{1. }Bu_3SnLi} \text{(BnO)}_3\text{-2-hydroxy-1-}SnBu_3\text{ pyranose} \tag{173}$$

3-germa-β-diketiminates were obtained from the reaction of metal-14 anions with ArCN (equation 174)[183]. The X-ray crystal structures of two representatives of this new class of compounds were determined and shown to have, in contrast to related C-analogues, the anionic charge localized at M_{14}.

$$R_3GeLi{\cdot}(THF)_3 + PhC{\equiv}N \longrightarrow [R_2Ge(PhC{=}NR)_2Li]_2 \quad (90\%) \tag{174}$$

Metal-14 anions can add to activated ethylenic bonds[8]. Thus, the reaction of Ph_3ELi (E = Ge, Sn) with cobaltocenium or decamethylcobaltocenium salts resulted in a nucleophilic addition forming cyclopentadiene–cyclopentadienyl–cobalt complexes and a competitive single electron reduction giving a cobaltocene. The proportion of the nucleophilic addition decreased from germanium to tin and also when Cp was changed to Cp* (Scheme 30)[184].

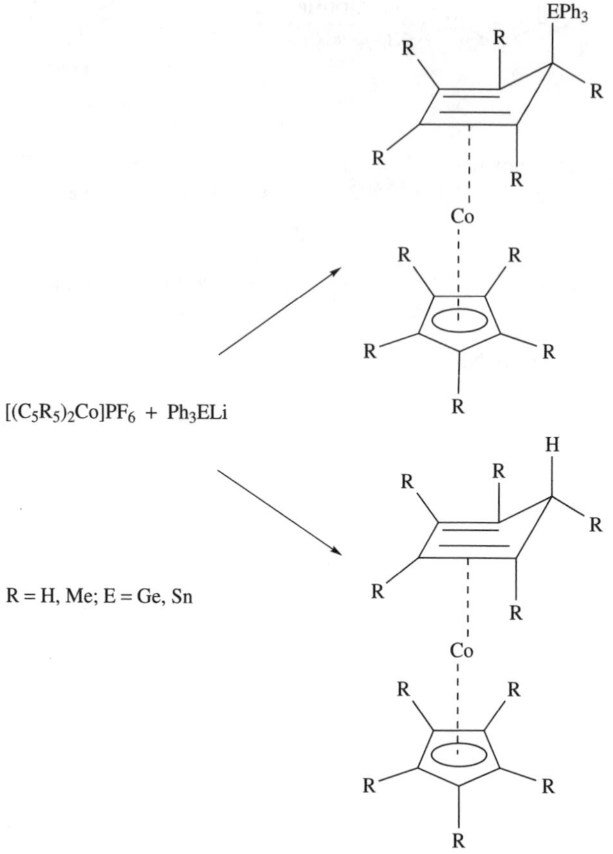

$[(C_5R_5)_2Co]PF_6$ + Ph_3ELi

R = H, Me; E = Ge, Sn

SCHEME 30

The reaction of germyllithiums with C_{60} gave different 1,2-monoadducts (equation 175)[185], the structures of which were resolved by X-ray analysis.

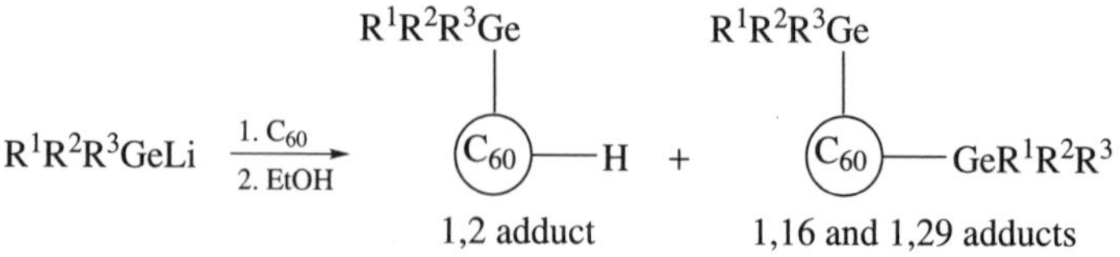

$$R^1R^2R^3GeLi \xrightarrow[2.\ EtOH]{1.\ C_{60}} R^1R^2R^3Ge{-}C_{60}{-}H + R^1R^2R^3Ge{-}C_{60}{-}GeR^1R^2R^3 \quad (175)$$

1,2 adduct; 1,16 and 1,29 adducts

Examples of additions of M_{14} anions to the acetylenic bond (equation 176)[186] and to dienes (equation 177)[187] have been described[8].

$$\text{R-C}\equiv\text{C-CH}_2\text{CH}_2\text{OPMB} \xrightarrow[-78\ ^\circ\text{C, then MeOH}]{Bu_3SnLi{\cdot}CuBr{\cdot}Me_2S} Bu_3Sn(\text{C}=\text{CHR})\text{CH}_2\text{CH}_2\text{OPMB} \qquad (176)$$

R = H, COOMe (84–98%)

OPMB = *p*-methoxybenzyloxy

$$\text{CH}_2{=}\text{CH-CH}{=}\text{CH-CO}_2\text{Me} \xrightarrow[2.\ R^1R^2CO]{1.\ Bu_3SnLi} Bu_3Sn\text{CH}_2\text{CH}{=}\text{CHCH(CO}_2\text{Me)C(OH)}R^1R^2 \qquad (177)$$

R^1R^2CO = MeCOMe, PhCHO, cyclohexanone, *p*-$MeOC_6H_4CHO$

E. SET Reactions

The notion that group-14 organometallic molecules react at ambient or moderate temperatures, preferentially by electron pair mechanisms involving concerted or polar (ionic) bond breaking, has prevailed for a long time. Homolytic cleavage, on the other hand, was thought to be a mechanism typical of high temperature or radical initiated (photo or chemically) reactions. During the last decade, new reactions involving the exchange of a single electron between closed-shell diamagnetic molecules, so-called SET reactions, have been largely regarded as a less and less exotic phenomenon[188]. Recent investigations in the field of the chemistry of organometal-14 compounds have shown that various functional compounds (A), strongly conjugated and, because of this, having a very low LUMO, react with organometal-14 compounds by inducing a single electron transfer. These reactions yield, depending on the polarity of the metal center, either a metal-centered intermediate radical (equation 178) or a metal-centered cation (equation 179)[188].

$$R_3\overset{\delta-}{M}-\overset{\delta+}{Y} + A \longrightarrow (R_3M-Y)^{\bullet+} + (A)^{\bullet-} \longrightarrow R_3Ge^{\bullet} + (A)^{\bullet-}\,Y^{+} \qquad (178)$$

Y = H, Li

$$\overset{\delta-}{R_3M}-\overset{\delta+}{Z} + A \longrightarrow (R_3M-Z)^{+\bullet} + (A)^{\bullet-} \longrightarrow Z^{\bullet} + R_3M^{+}\,(A)^{\bullet-} \qquad (179)$$

M = metal 14

A : electron acceptor Z = Cl, NMe_2

These reactions can be rationalized on the basis that the organometal-14 compound reacts as a prometal-centered radical (equation 178) and as a prometal-centered cation (equation 179)[188]. Metal-14 anions react in SET reactions, according to equation 178.

1. SET at carbon and metal

The possibility of a single electron transfer process in substitution reactions at carbon and metal was at first gradually and now widely accepted.

In germanium chemistry the importance of free radical pathways in substitution reactions of secondary bromides with R_3GeLi ($R = CH_3, C_6H_5$) reagents is strongly indicated by product stereochemistry in cyclohexyl systems and by cyclization of the *cis*-heptene-2-yl moiety to yield [(2-methylcyclopentenyl)methyl]germanes, with the appropriate *cis/trans* ratio, as shown in equation 180, Table 9 and equations 181 and 182[189].

$$\text{CH}_2{=}\text{CH(CH}_2)_3\text{CHBrCH}_3 + R_3MLi \xrightarrow[15\,^{\circ}C]{\text{solvent}} \text{CH}_2{=}\text{CH(CH}_2)_3\text{CH(MR}_3)\text{CH}_3 + cis\text{-}(2\text{-methylcyclopentyl})CH_2MR_3 + trans\text{-}(2\text{-methylcyclopentyl})CH_2MR_3 \qquad (180)$$

$R_3 = (CH_3)_{3-x}(C_6H_5)_x$

$$4\text{-}CH_3C_6H_{10}Br + (C_6H_5)_3GeLi \xrightarrow{\text{THF}} 4\text{-}CH_3C_6H_{10}Ge(C_6H_5)_3\ (34\%) + 4\text{-}CH_3C_6H_{10}Ge(C_6H_5)_3\ (66\%) \qquad (181)$$

TABLE 9. Reaction profile of R_3MLi with 6-bromo-1-heptene

Entry	M	*x*	Solvent	Product ratio[a]			Yield %
				MR_3 (6-substituted 1-heptene)	CH_2MR_3 / CH_3 (cyclopentane)	CH_2MR_3 / Me (cyclopentane)	
1	Sn	0	THF	21	58	21	60
2	Sn	0	HMPA	82	13	5	16
3	Sn	1	THF	15	61	24	86
4	Sn	2	THF	86	10	4	93
5	Sn	3	THF	100			83
6	Ge	0	HMPA	83[b]	13	4.4	52
7	Ge	3	THF	68	22.4	9.6	27

[a] Product ratios established by ^1H and ^{13}C NMR spectroscopy.
[b] The (83%) non-cyclized product germane consisted of 35% 6-germyl-1-heptene, 40% *cis*-6-germyl-2-heptene and 8% *trans*-6-germyl-2-heptene.

Br

CH_3 + $(C_6H_5)_3GeLi$ $\xrightarrow{THF}$ CH_3 $Ge(C_6H_5)_3$ 45% + $Ge(C_6H_5)_3$ CH_3 55% (182)

Upon steady light illumination of aromatic compounds (A) in the presence of the triphenylstannylanion metal cation pair (Ph_3Sn^-, M^+) in tetrahydrofuran, radical anions of the aromatic compounds ($A^{\bullet -}, M^+$) were produced by electron transfer from Ph_3Sn^-, M^+ to the excited aromatic compounds. After the light was cut off, the radical anions of perylene and tetracene persisted for a long time; for anthracene and pyrene, the radical anions formed transiently and decayed rapidly. The decay rate depended on the reduction potentials of (A). The decay processes were attributed to back electron transfer from ($A^{\bullet -}$, M^+) to the distannane ($Ph_3SnSnPh_3$) which was produced by the coupling of $Ph_3Sn^\bullet$ radicals, since the rate constants calculated on the basis of the above reactions were in good agreement with the observed rate constants. Slow decay of ($A^{\bullet -}, M^+$) can be realized when the reduction potentials of (A) are less negative than that of the distannane (Scheme 31, Figure 13)[190].

The reaction of trimethylstannyl sodium with primary halides has been studied in detail with emphasis on the effect of solvents and added radical and carbanion traps. By lowering

$$A \xrightarrow{h\nu} A^*$$

$$A^* + Ph_3Sn^- \longrightarrow A^{\bullet -} + Ph_3Sn^{\bullet}$$

$$2\,Ph_3Sn^{\bullet} \longrightarrow Ph_3SnSnPh_3$$

$$A^{\bullet -} + Ph_3SnSnPh_3 \underset{k_b}{\overset{k_a}{\rightleftharpoons}} A + Ph_3SnSnPh_3^{\bullet -}$$

$$Ph_3SnSnPh_3^{\bullet -} \longrightarrow Ph_3Sn^- + Ph_3Sn^{\bullet}$$

$$Ph_3Sn^{\bullet} + A^{\bullet -} \longrightarrow Ph_3Sn^- + A$$

SCHEME 31

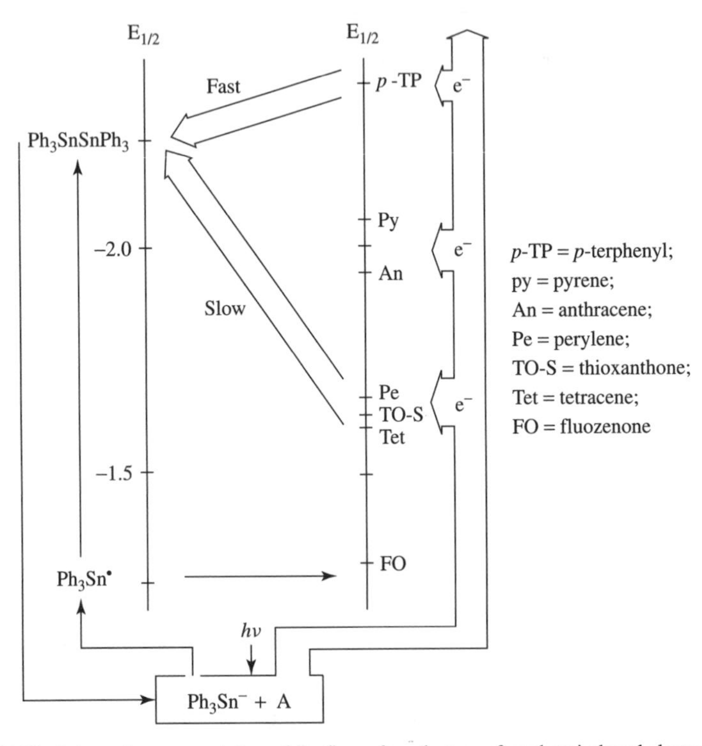

FIGURE 13. Schematic representation of the flow of an electron after photoinduced electron transfer; (⇒) flow of the electron and (→) flow of the chemical reaction. Reprinted with permission from Reference 190. Copyright 1982 American Chemical Society

the viscosity of the solvent, lowering the cation coordinating ability of the solvent or running the reactions in the presence of a radical trap, it was shown that radical intermediates were involved (Scheme 32, Table 10)[191].

Furthermore, the reaction of a primary alkyl iodide having a cyclizable radical probe with Me_3SnNa did not occur exclusively by S_N2 and HME pathways, as previously

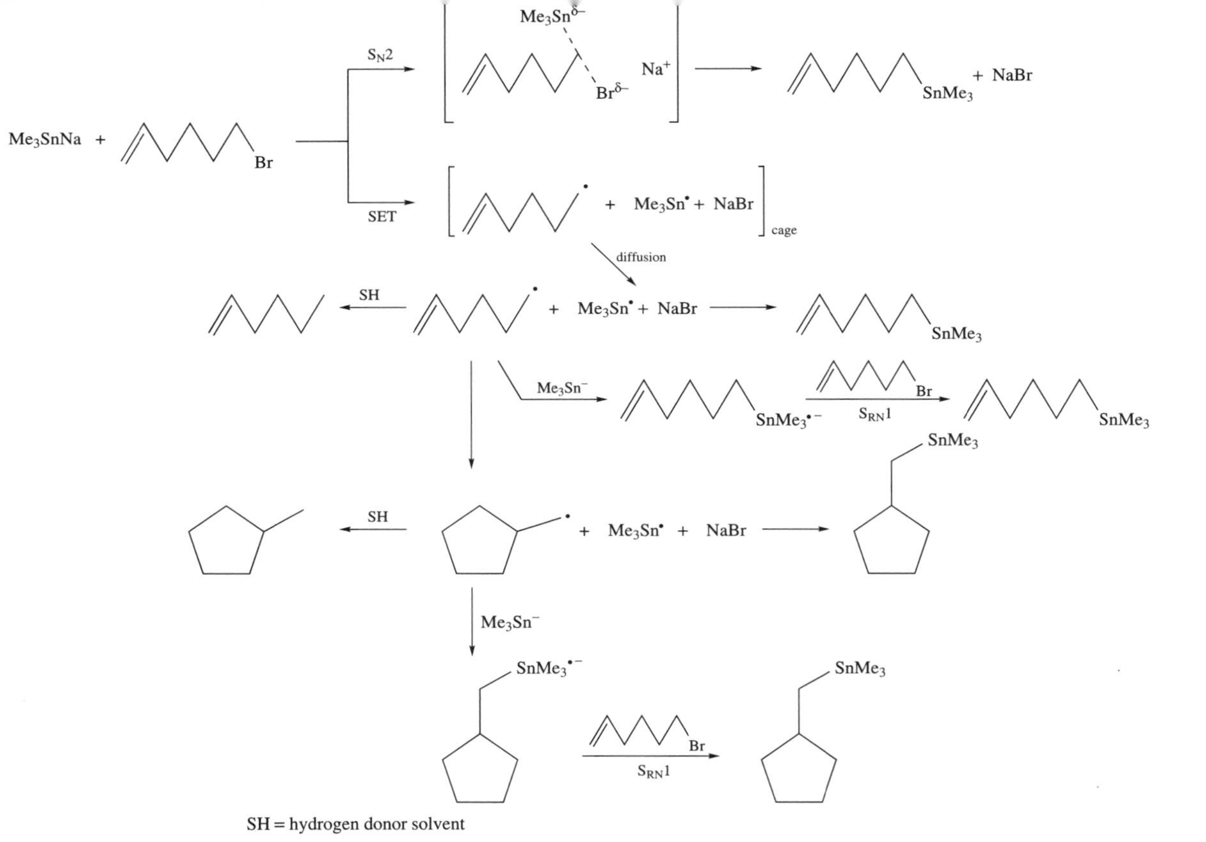

Me3SnNa +
Br
SN2
SET
Me3Snδ–
Brδ–
Na+
SnMe3
+ NaBr
+ Me3Sn• + NaBr
cage
diffusion
SH
+ Me3Sn• + NaBr
SnMe3
Me3Sn–
SnMe3•–
Br
SRN1
SnMe3
SnMe3
SH
+ Me3Sn• + NaBr
Me3Sn–
SnMe3•–
Br
SRN1
SnMe3
SH = hydrogen donor solvent

SCHEME 32

TABLE 10. Reaction profile of Me_3SnNa with 6-bromo-1-hexene[a]

Entry	Solvents (ratio)	Yield of product[b] (%)			
		 5-hexenyl-$SnMe_3$	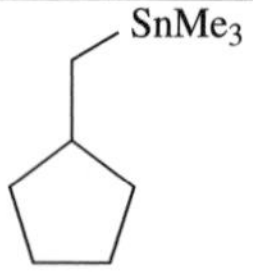 cyclopentylmethyl-$SnMe_3$	 1-hexene	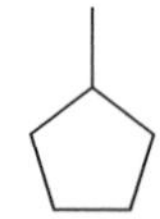 methylcyclopentane
1	THF-$C_{12}H_{26}$ (2 : 8)	92.2	6.7	...	...
2	THF-$C_{12}H_{26}$ (3 : 7)	94.1	4.6	...	...
3	THF-$C_{12}H_{26}$ (5 : 5)	95.5	2.3	...	...
4	THF-$C_{12}H_{26}$ (7 : 3)	98.0	trace	...	...
5	THF-$C_{12}H_{26}$ (9 : 1)	98.8	0	...	...
6	THF-Et_2O (2 : 8)	86.4	12.6	trace	...
7	THF-Et_2O (3 : 7)	93.5	5.2	...	...
8	THF-Et_2O (5 : 5)	98.8	trace	...	...
9	THF-Et_2O (7 : 3)	98.5	0	...	...
10	THF-Et_2O (9 : 1)	99.1	0	...	...
11	THF-C_5H_{12} (2 : 8)	83.4	15.1	trace	...
12	THF-C_5H_{12} (3 : 7)	89.9	9.0	trace	...
13	THF-C_5H_{12} (5 : 5)	95.0	4.0	...	...
14	THF-C_5H_{12} (7 : 3)	98.7	trace	...	...
15	THF-C_5H_{12} (9 : 1)	98.5	0	...	...

[a] Reactions were conducted by using 0.024 M concentrations of RB_2 and 0.048 M concentrations of Me_3SnNa at 0 °C for 15 min.
[b] Yields are based on the RB_2 consumed.

TABLE 11. Reaction of *endo*-5-(2-iodoethyl)-2-norbornene with Me_3SnNa[a]

Entry	Additive[b] (molar equiv)	Temp (°C)	Yield of product[b] (%)		
			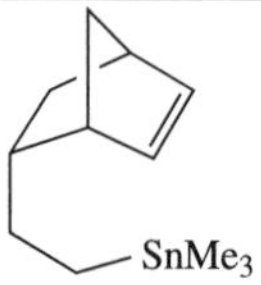 5-(2-trimethylstannylethyl)-2-norbornene ($SnMe_3$)	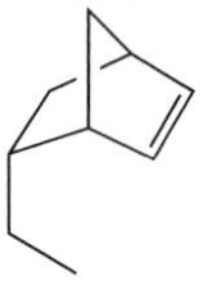 5-ethyl-2-norbornene	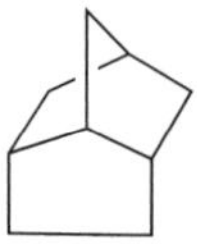 tricyclic product
1	none	0	90.5	7.3	2.6
2	DCPH, 1	0	51.5	20.1	30.0
3	DCPH, 10	0	60.0	14.1	25.3
4	TBA[c], 1	0	75.3	15.0	5.2
5	TBA, 10	0	75.0	22.3	0
6	DCPH, 10; TBA, 10	0	48.6	33.0	19.0
7	none	−23	96.0	3.0	3.5
8	DCPH, 10	−23	80.0	7.4	14.2
9	TBA, 10	−23	95.0	3.0	0.5
10	none	−78	97.0	0	0
11	DCPH, 10	−78	90.0	2.7	2.0
12	TBA, 10	−78	96.0	trace	trace

[a] Reactions were conducted by using 0.05 M concentration of RI and 0.1 M concentration of Me_3SnNa in THF for 15 min.
[b] Yields are based on the RI consumed.
[c] TBA = *t*-butylamine.

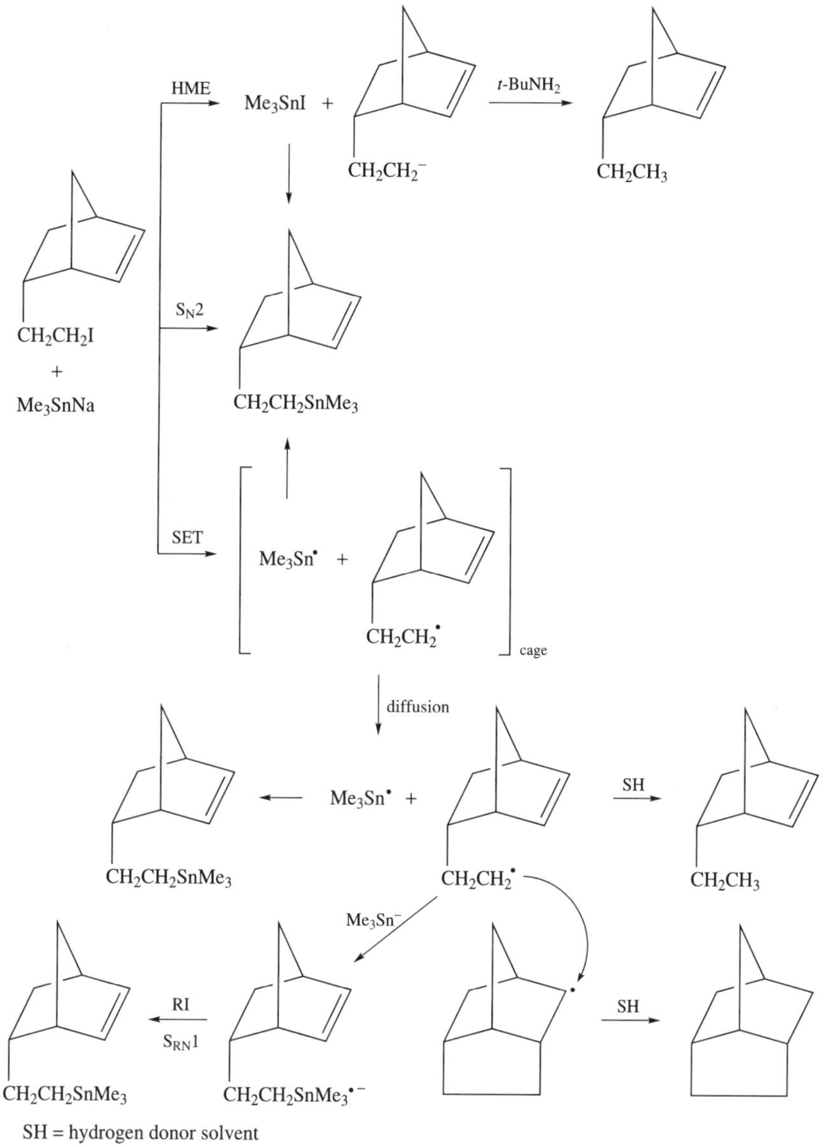

SH = hydrogen donor solvent

SCHEME 33

reported, but also by an electron transfer pathway to a significant extent (Scheme 33, Table 11)[191].

Evidence for an intermediate stannyl radical implication in such SET processes was obtained directly by a stopped flow technique in the reaction of tributylstannyl anion with *s*- and *t*-butyl bromides and iodides (Scheme 34, Figure 14)[192].

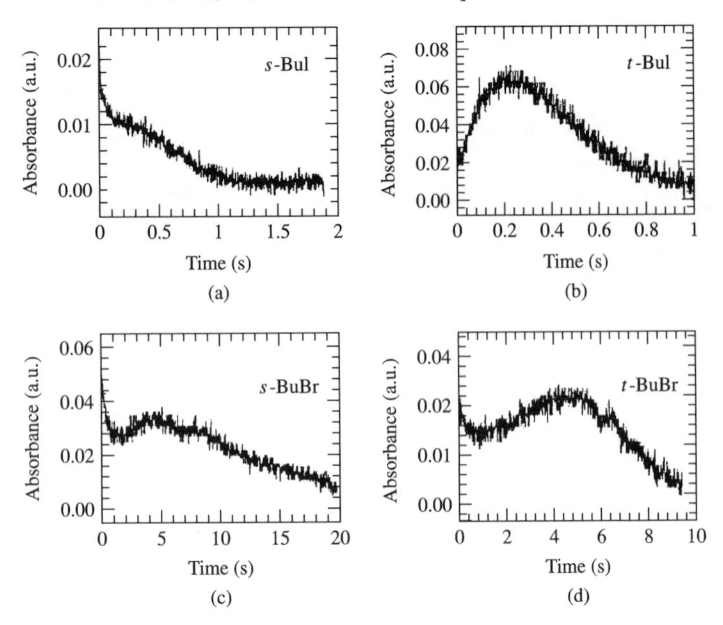

FIGURE 14. Time profiles observed at 400 nm for the reaction of tributylstannyl anion with (a) *s*-BuI (5.0×10^{-3} mol dm^{-3}), (b) *t*-BuI (1.25×10^{-3} mol dm^{-3}), (c) *s*-BuBr (2.5×10^{-3} mol dm^{-3}) and (d) *t*-BuBr (1.25×10^{-2} mol dm^{-3}). Reprinted with permission from Reference 192. Copyright 1998 American Chemical Society

Another example of SET substitution at carbon and metal by metal-14 anions was evident in the reaction of Me_3SnNa with 2-chloropyridine, *p*-chlorobenzonitrile, *o*- and *m*-dichlorobenzene, 1,3,5-trichlorobenzene, 2,5-,2,6- and 3,5-dichloropyridine, which gave good yields of substitution products through a suggested $S_{RN}1$ mechanism (equation 183)[193].

Cl, Cl (ring with X) $\xrightarrow[2\ Me_3SnNa]{h\nu}$ $SnMe_3$, Me_3Sn (ring with X) (183)

X = CH, N; 11 examples, yield = 4–90%

When *p*-bis(trimethylstannyl)benzene was treated with sodium metal in liquid ammonia, a dianion was formed which upon photostimulation with C_6H_5Cl afforded the disubstitution product in 70% yield (Scheme 35)[193].

The reaction between triorganostannyl ions and haloarenes in liquid ammonia can lead to substitution and reduction products. It was found that in some cases the reaction can follow an $S_{RN}1$ mechanism exclusively. With triphenylstannyl ions, good yields of products of the $S_{RN}1$ mechanism were obtained when reactions were conducted with chloroarenes

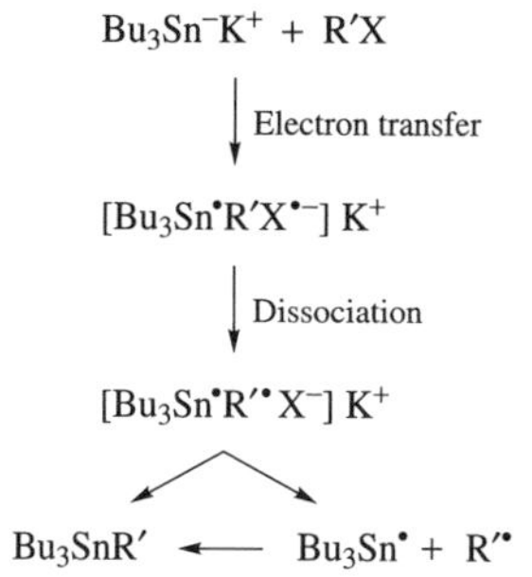

$Bu_3Sn^-K^+ + R'X$

↓ Electron transfer

$[Bu_3Sn^{\bullet}R'X^{\bullet-}]\ K^+$

↓ Dissociation

$[Bu_3Sn^{\bullet}R'^{\bullet}X^-]\ K^+$

$Bu_3SnR' \leftarrow Bu_3Sn^{\bullet} + R'^{\bullet}$

SCHEME 34

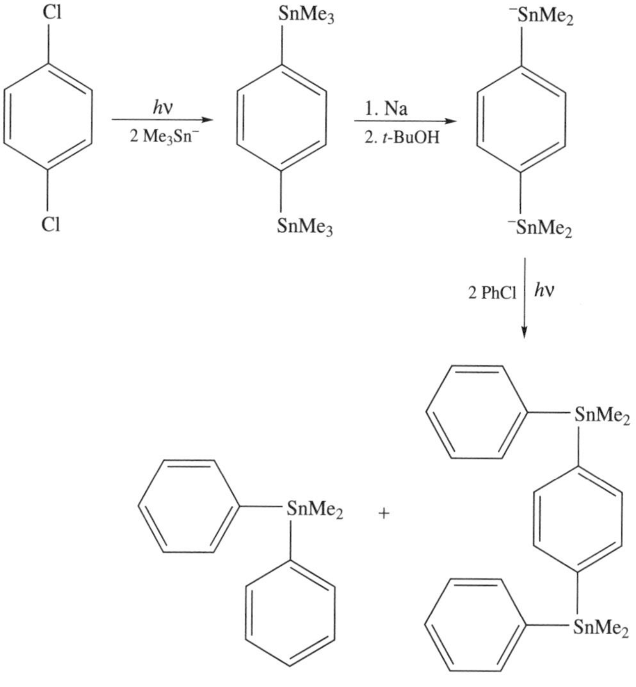

Cl, Cl; $h\nu$, 2 Me_3Sn^-; $SnMe_3$, $SnMe_3$; 1. Na, 2. *t*-BuOH; $^-SnMe_2$, $^-SnMe_2$; 2 PhCl, $h\nu$; $SnMe_2$ +; $SnMe_2$, $SnMe_2$

SCHEME 35

(*p*-chlorotoluene, *p*-dichlorobenzene, 1-chloronaphthalene and 1-chloroquinoline) and with some bromoarenes as shown in Scheme 36[194].

The reaction of trimethylstannylsodium with two geminal dihalides, 6,6-dichloro-5,5-dimethyl-1-hexene and 6,6-diiodo-5,5-dimethyl-1-hexene, gave evidence of a single electron transfer pathway. An initial electron transfer from Me_3Sn^- to the geminal dihalides leads to the haloradical ($X^{\bullet}$), which then serves as the precursor to all the reactions and products detailed in Scheme 37[195].

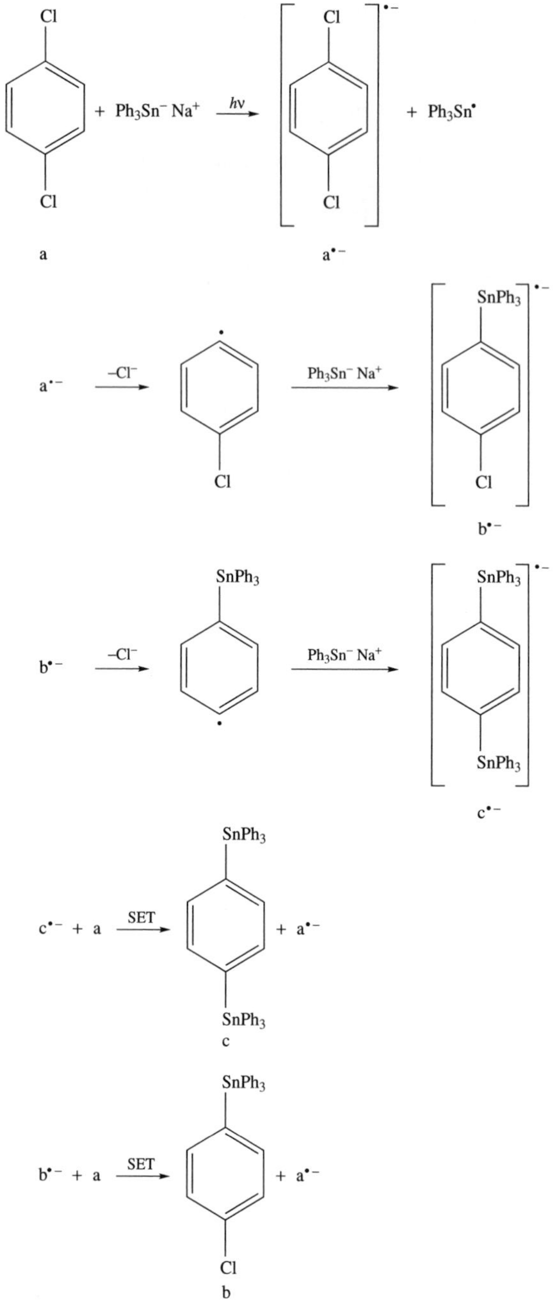

Cl–C₆H₄–Cl (a) + $Ph_3Sn^- Na^+$ $\xrightarrow{h\nu}$ [Cl–C₆H₄–Cl]$^{\bullet-}$ ($a^{\bullet-}$) + $Ph_3Sn^{\bullet}$

$a^{\bullet-}$ $\xrightarrow{-Cl^-}$ •C₆H₄–Cl $\xrightarrow{Ph_3Sn^- Na^+}$ [Ph_3Sn–C₆H₄–Cl]$^{\bullet-}$ ($b^{\bullet-}$)

$b^{\bullet-}$ $\xrightarrow{-Cl^-}$ Ph_3Sn–C₆H₄• $\xrightarrow{Ph_3Sn^- Na^+}$ [Ph_3Sn–C₆H₄–$SnPh_3$]$^{\bullet-}$ ($c^{\bullet-}$)

$c^{\bullet-}$ + a $\xrightarrow{SET}$ Ph_3Sn–C₆H₄–$SnPh_3$ (c) + $a^{\bullet-}$

$b^{\bullet-}$ + a $\xrightarrow{SET}$ Ph_3Sn–C₆H₄–Cl (b) + $a^{\bullet-}$

SCHEME 36

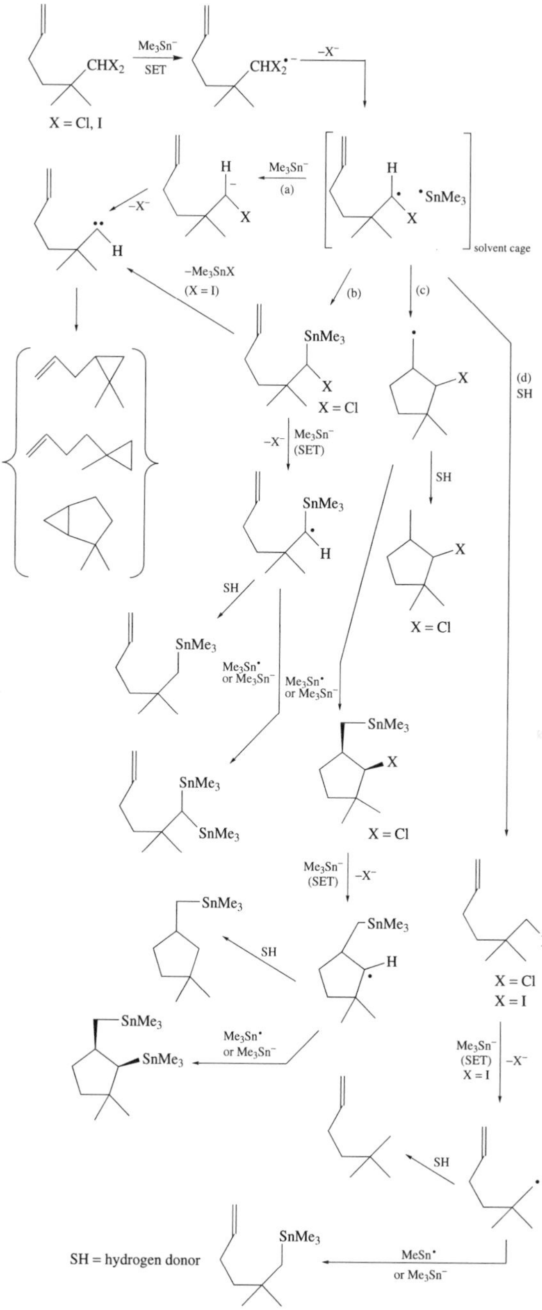

SH = hydrogen donor

SCHEME 37

When steric hindrance makes nucleophilic substitution difficult, germyllithiums reacted with acyl chlorides to give a competitive SET reaction (equation 184 and Scheme 38)[15].

$$n\ \mathrm{Mes_3GeLi} + m\ \mathrm{PhCOCl} \longrightarrow \mathrm{Mes_3GeCOPh} + \mathrm{Mes_3GeCl} + \mathrm{Mes_3GeH} + \mathrm{PhCOCOPh} \quad (184)$$

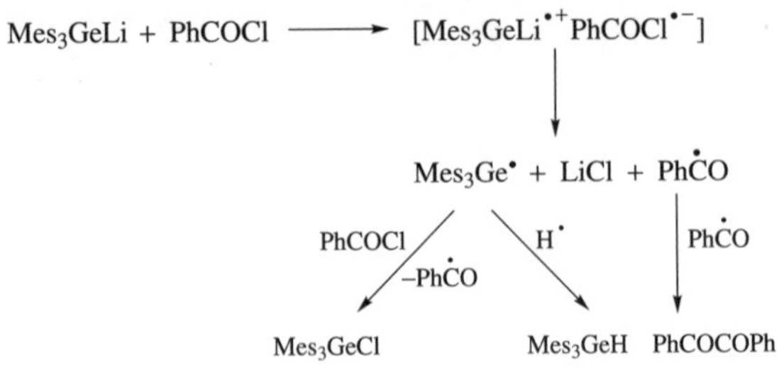

$$\mathrm{Mes_3GeLi} + \mathrm{PhCOCl} \longrightarrow [\mathrm{Mes_3GeLi}^{\bullet+}\mathrm{PhCOCl}^{\bullet-}]$$

$$\downarrow$$

$$\mathrm{Mes_3Ge^{\bullet}} + \mathrm{LiCl} + \mathrm{Ph\dot{C}O}$$

$$\xrightarrow[-\mathrm{Ph\dot{C}O}]{\mathrm{PhCOCl}} \mathrm{Mes_3GeCl} \qquad \xrightarrow{\mathrm{H^{\bullet}}} \mathrm{Mes_3GeH} \qquad \xrightarrow{\mathrm{Ph\dot{C}O}} \mathrm{PhCOCOPh}$$

SCHEME 38

Diethyl arylphosphates have also been shown to react with alkali metaltriorganostannides through a SET mechanism involving stannyl radicals and affording arylstannane in excellent yields (equation 185)[196].

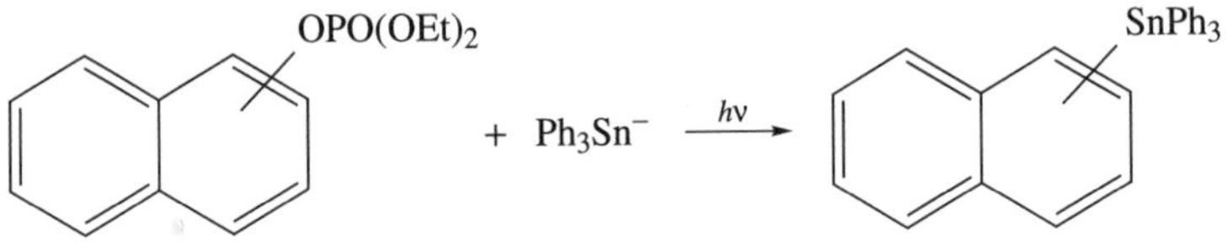

$$\mathrm{C_{10}H_7OPO(OEt)_2} + \mathrm{Ph_3Sn^-} \xrightarrow{h\nu} \mathrm{C_{10}H_7SnPh_3} \quad (185)$$

$$\text{1-OPO(OEt)}_2 \longrightarrow \text{1-SnPh}_3$$

$$\text{2-OPO(OEt)}_2 \longrightarrow \text{2-SnPh}_3$$

Aluminum alkoxides have been shown to influence the selectivity of reactions involving a single electron transfer stage. Triethylgermyllithium reacts with amides R_2NCO-X ($X = Cl$, OMe, NMe_2, Ph) by a mechanism which includes a free radical stage. The radical anion salt of *N,N*-diethylbenzamide, which is thermally stable in hydrocarbons, was detected in the course of these reactions which gave (*N,N*-dialkylcarbamoyl)germanes when they were performed in the presence of an equimolar amount of $(s\text{-BuO})_3Al$ (Scheme 39)[143].

2. SET additions

Single electron processes were also evident in addition reactions of metal-14 anions to conjugated carbonyl compounds and other conjugated molecules having a low LUMO. When organogermyllithiums R_3GeLi (R = Ph, Mes) were reacted with several carbonyl conjugated substrates (2-furan carboxaldehyde, 2-thiophene carboxaldehyde and their corresponding nitro derivatives), only germylation of the carbonyl groups was observed with a regioselectivity depending on the nature of the unsaturated ring. With 2-furancarboxaldehyde and 2-thiophenecarboxaldehyde, the germylcarbinols were obtained by mainly nucleophilic C-germylation (equation 186)[197]. In the case of the nitro compounds, O-germylation was dominant, and a single electron transfer mechanism was corroborated by

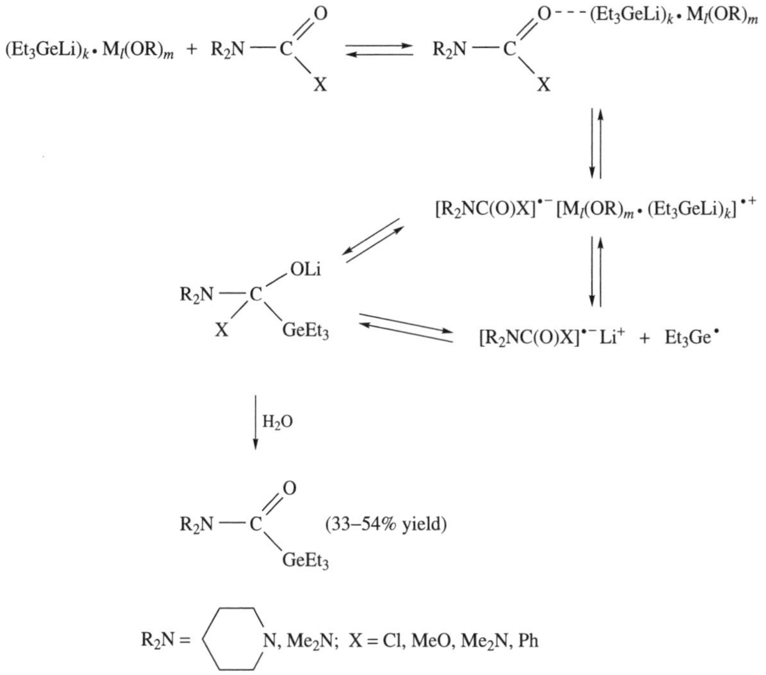

$(Et_3GeLi)_k \cdot M_l(OR)_m + R_2N{-}C(=O)X \rightleftharpoons R_2N{-}C(=O\cdots(Et_3GeLi)_k \cdot M_l(OR)_m)X$

$\rightleftharpoons [R_2NC(O)X]^{\bullet -}[M_l(OR)_m \cdot (Et_3GeLi)_k]^{\bullet +}$

$\rightleftharpoons [R_2NC(O)X]^{\bullet -}Li^+ + Et_3Ge^\bullet$

$[R_2NC(O)X]^{\bullet -}[M_l(OR)_m \cdot (Et_3GeLi)_k]^{\bullet +} \rightleftharpoons R_2N{-}C(OLi)(X)(GeEt_3) \rightleftharpoons [R_2NC(O)X]^{\bullet -}Li^+ + Et_3Ge^\bullet$

$R_2N{-}C(OLi)(X)(GeEt_3) \xrightarrow{H_2O} R_2N{-}C(=O){-}GeEt_3$ (33–54% yield)

R_2N = piperidino N, Me_2N; X = Cl, MeO, Me_2N, Ph

SCHEME 39

ESR measurements. The presence of the intermediate organic radical anion was confirmed by comparison with a similar intermediate obtained from the SET of the nitro-substituted aldehyde with an electron-rich olefin (Scheme 40[197], equation 187[197], Figures 15 and 16).

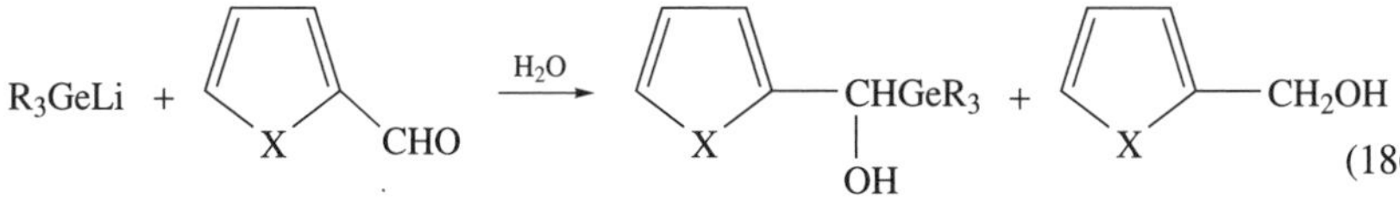

$$R_3GeLi + \text{2-furyl/2-thienyl-CHO (ring X)} \xrightarrow{H_2O} \text{(ring X)-CH(OH)GeR}_3 + \text{(ring X)-CH}_2\text{OH} \quad (186)$$

X = O, R_3 = Ph_3, Ph_2H, Mes_2H

X = S, R_3 = Ph_3, Mes_2H

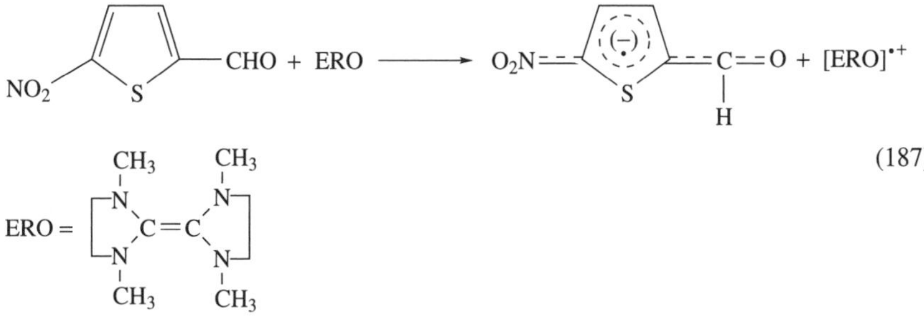

$$NO_2\text{-(thiophene, S)-CHO} + ERO \longrightarrow [O_2N{=}\text{(thiophene}^{(-)}\text{, S)}{=}C(H){=}O] + [ERO]^{\bullet +} \quad (187)$$

ERO = $(CH_3)N$ / $(CH_3)N$ imidazolidinylidene C=C imidazolidinylidene $N(CH_3)$ / $N(CH_3)$ (tetramethyl-biimidazolidinylidene)

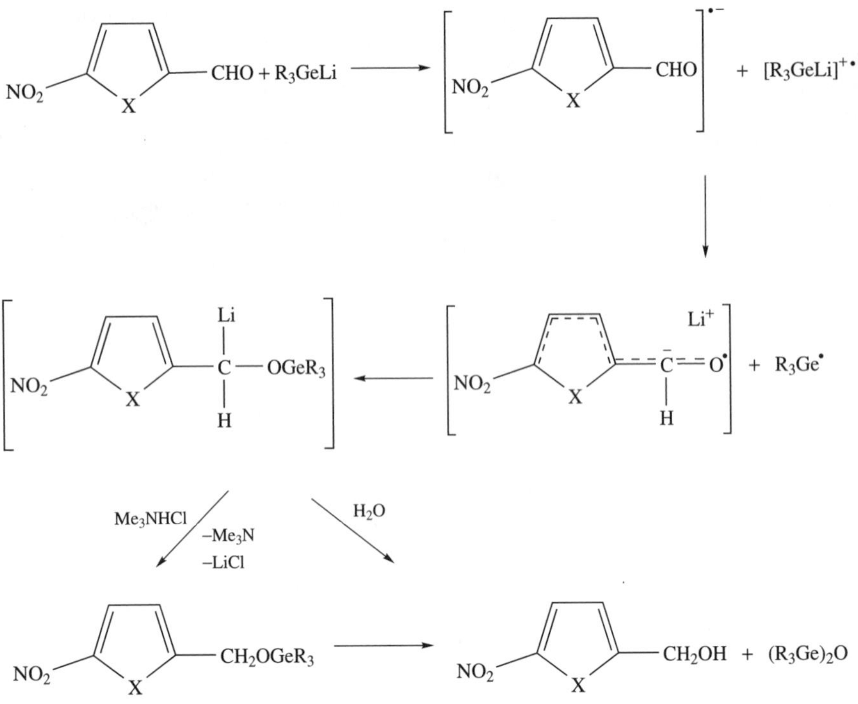

$R_3 = Ph_3, Ph_2H, Mes_2H$
$X = O, S$

SCHEME 40

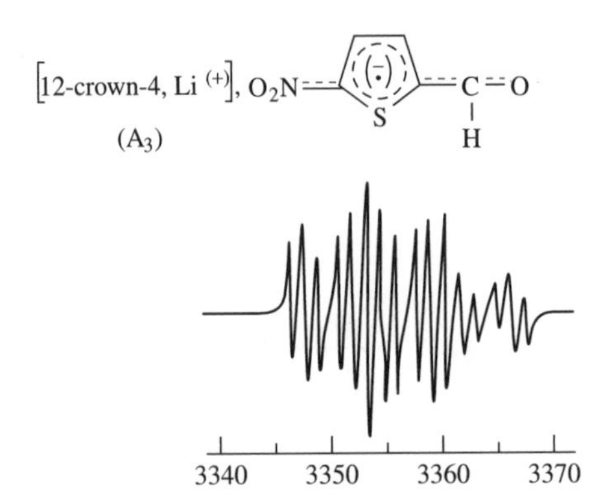

FIGURE 15. ESR spectrum of the radical anion generated from the reaction of 5-nitrothiophene-2-carboxaldehyde with germyllithiums.12-crown-4. From Reference 197 with permission from Gordon and Breach Publishing, copyright Taylor & Francis

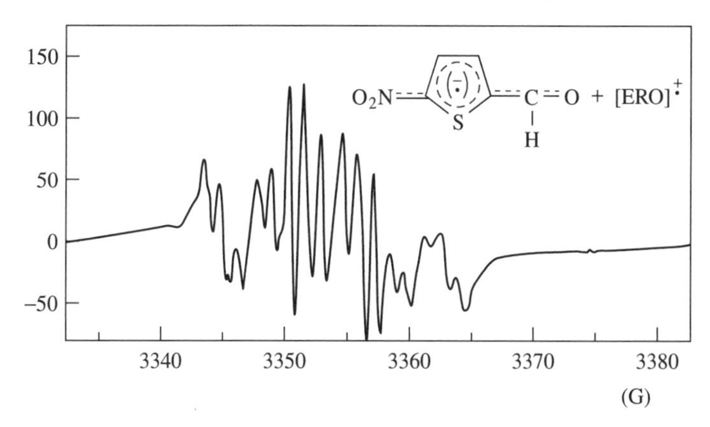

FIGURE 16. ESR spectrum of the radical anion generated from the reaction of 5-nitrothiophene carboxaldehyde with ERO. From Reference 197 with permission from Gordon and Breach Publishing, copyright Taylor and Francis

The reaction of trimethyl and triphenylstannyl potassium with mono- and di-substituted enones in acetonitrile as solvent led in nearly quantitative yields either to a mixture of diastereomers or to a pure diastereomer of β-stannylketones (Scheme 41)[109]. There was experimental support for the existence of a SET mechanism, i.e. partial or total inhibition of the reaction by addition of a free radical scavenger (galvinoxyl) or a radical anion scavenger (*p*-dinitrobenzene). The possibility of a SET depends on the one-electron donor ability of the nucleophile and the electron acceptor ability of the ketone. These reactions are stereoselective.

Reactions between organogermyllithiums R_3GeLi and several substrates favoring SET reactions (3,5-di-*t*-butyl-*o*-quinone, fluorenone, tetracyanoquinodimethane, 2,4,6-tri-*t*-butylnitrosobenzene, etc.) led mainly to the digermanes and O- or N-germyl adducts. These reactions mainly proceed by a SET. An ESR study of the reaction showed transient organic radical anions and germanium-centered radicals $R_3Ge^\bullet$. The digermanes were formed by recombination of $R_3Ge^\bullet$ radicals, as well as by lithiogermolysis of the reaction adducts.

In the case of SET addition of R_3GeLi to 3,5-di-*t*-butyl-*o*-quinone (Scheme 42)[198,199], the two radical intermediates were observed by ESR when R = Mes. The dominance of this mechanism was demonstrated by the reaction between R_3GeLi and a paramagnetic quinonic species, the galvinoxyl radical, which was almost completely transformed into a diamagnetic anion (Scheme 43)[199].

A study by ESR of the reaction between R_3GeLi (R = Ph, Mes) with fluorenone showed the transient formation of the radical anion derived from fluorenone. The reaction results were, however, influenced by the experimental procedure. When fluorenone was added slowly to an excess of germyllithium, digermane was mainly formed (equation 188), whereas when the germyllithium was added to fluorenone, only traces of digermanes were observed (equation 189)[199]. These results were explained by the fact that fluorenone also behaved as a spin trap for the transient germyl radical leading to an O-germylated adduct, the lithiogermolysis of which in the presence of excess germyllithium gave digermane (Scheme 44)[199].

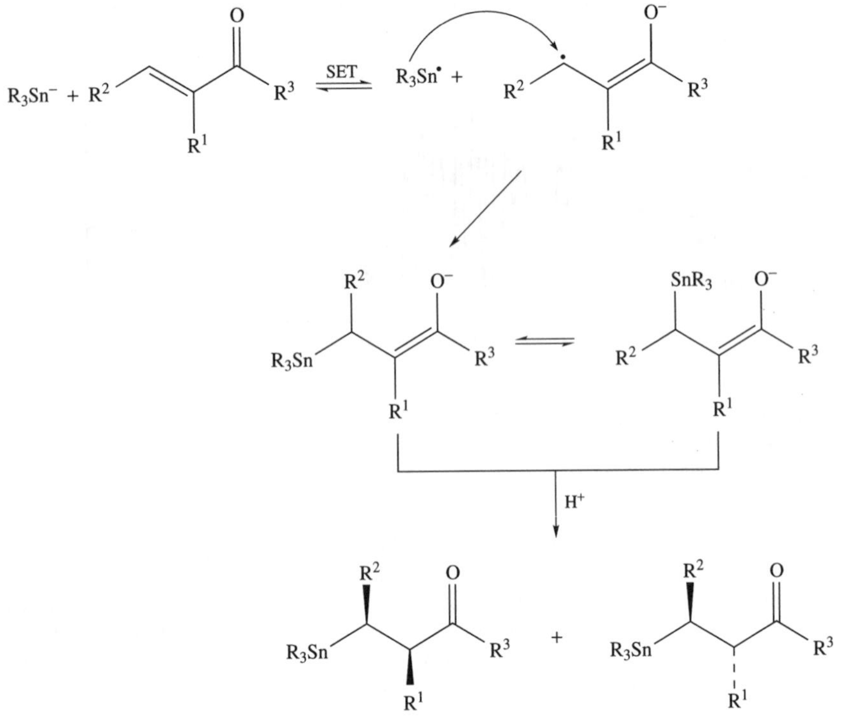

$R_3Sn^- + R^2CH{=}C(R^1)C(O)R^3 \xrightleftharpoons{SET} R_3Sn^\bullet + R^2\dot{C}H{-}C(R^1){=}C(O^-)R^3$

$\rightarrow R_3Sn{-}CH(R^2){-}C(R^1){=}C(O^-)R^3 \rightleftharpoons R^2{-}CH(SnR_3){-}C(R^1){=}C(O^-)R^3$

$\xrightarrow{H^+}$ $R_3Sn{-}CH(R^2){-}CH(R^1){-}C(O)R^3$ (*threo*) + $R_3Sn{-}CH(R^2){-}CH(R^1){-}C(O)R^3$ (*erythro*)

R, R^1, R^2, R^3 = Me, Ph

SCHEME 41

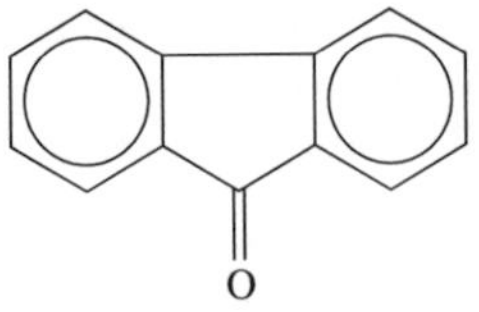

$$\text{fluorenone} + 2\,Ph_3GeLi \xrightarrow[10\%]{HCl} \underset{(33\%)}{Ph_3GeH} + \underset{(67\%)}{Ph_3GeGePh_3} + \underset{(traces)}{Ph_3GeCl} + \text{fluorenone} + \text{fluorenol} \qquad (188)$$

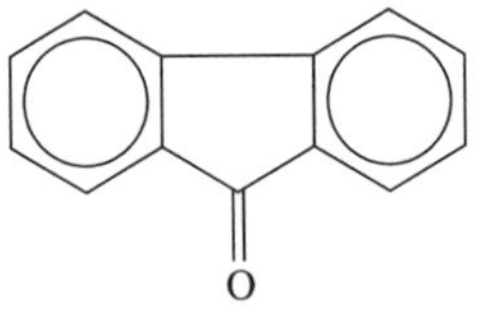

$$\text{fluorenone} + Ph_3GeLi \xrightarrow[10\%]{HCl} \underset{(traces)}{Ph_3GeH} + \underset{(traces)}{Ph_3GeGePh_3} + \underset{(98\%)}{Ph_3GeCl} + \text{fluorenone} + \text{fluorenol} \qquad (189)$$

The possibility of an internal spin trapping of the transient germanium-centered radical by the starting single electron acceptor was evident in the reaction of germyllithium with tri-*t*-butylnitrosobenzene (BNB) (Scheme 45)[199]. The transient germylated anilino radicals [B] were observed (R = Ph, Mes), but the formation of digermane was hindered by steric effects in the case of R = Mes. Steric effects inhibited coupling of the radical (e) and the lithiogermolysis of the O-germylated adduct (f).

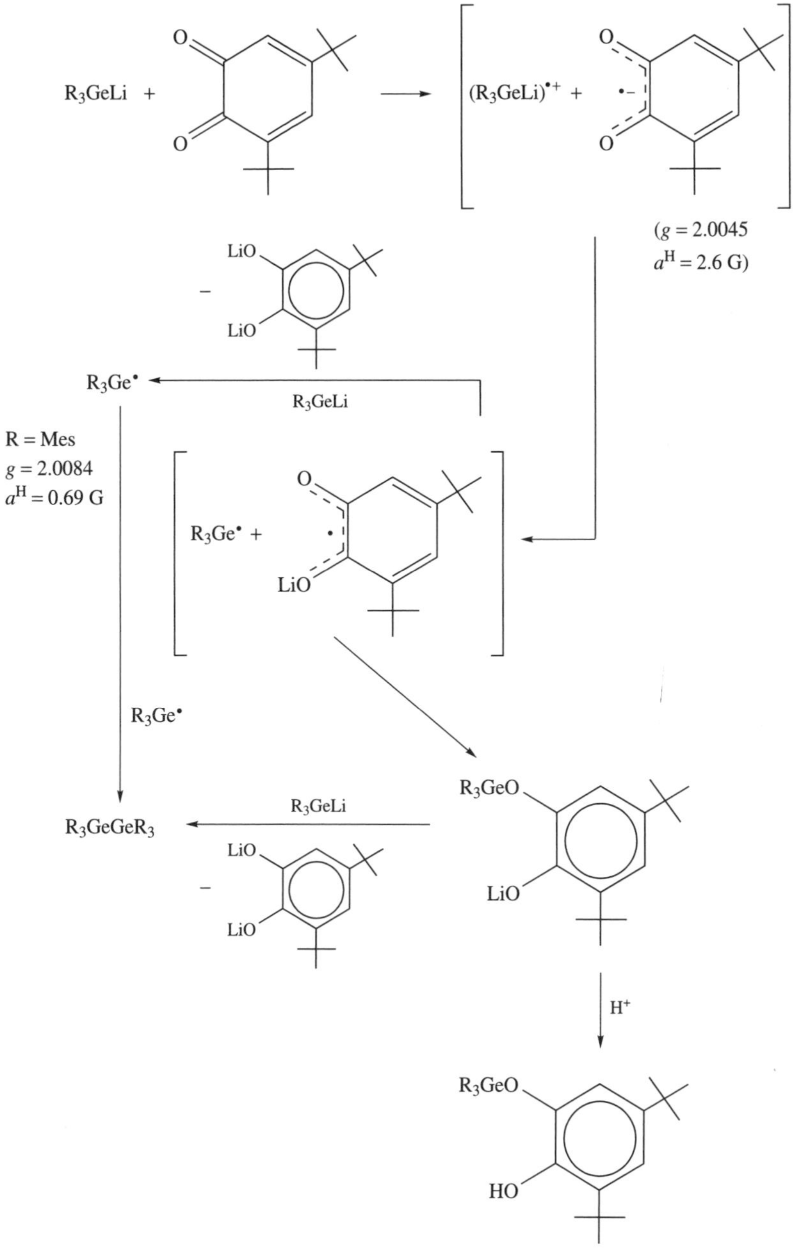

R_3GeLi + O, O → [$(R_3GeLi)^{\bullet+}$ + O, O]

($g = 2.0045$
$a^H = 2.6$ G)

− LiO, LiO

$R_3Ge^\bullet$ ← R_3GeLi

R = Mes
$g = 2.0084$
$a^H = 0.69$ G

[$R_3Ge^\bullet$ + O, LiO]

$R_3Ge^\bullet$

R_3GeGeR_3 ← R_3GeLi — R_3GeO, LiO

− LiO, LiO

H^+

R_3GeO, HO

SCHEME 42

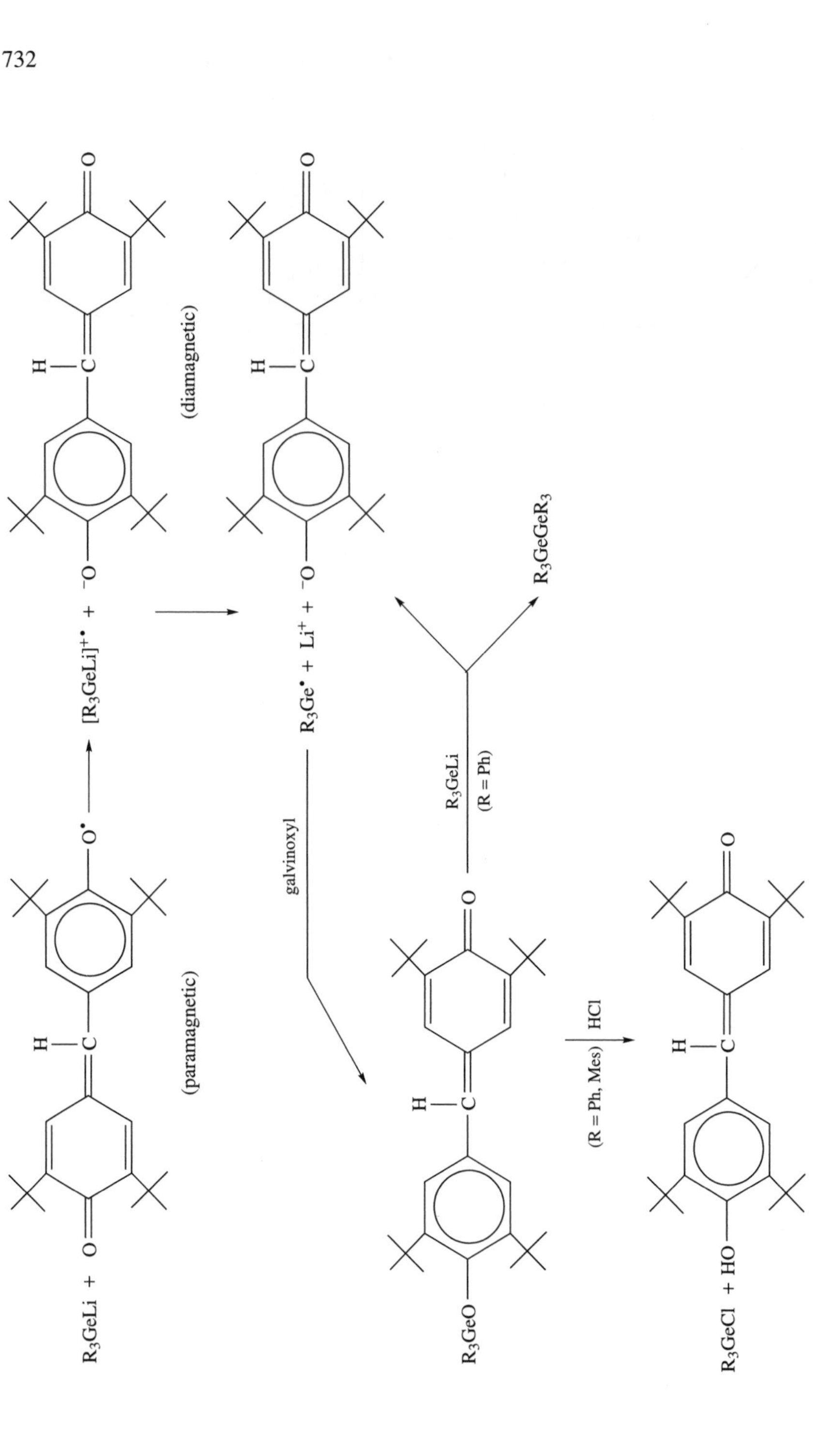

R_3GeLi + O= ... (paramagnetic) → $[R_3GeLi]^{+\bullet}$ + ^{-}O ... (diamagnetic)

↓

$R_3Ge^{\bullet}$ + Li^+ + ^{-}O ...

galvinoxyl

R_3GeO ...

R_3GeLi

(R = Ph)

R_3GeGeR_3

(R = Ph, Mes) HCl

R_3GeCl + HO ...

SCHEME 43

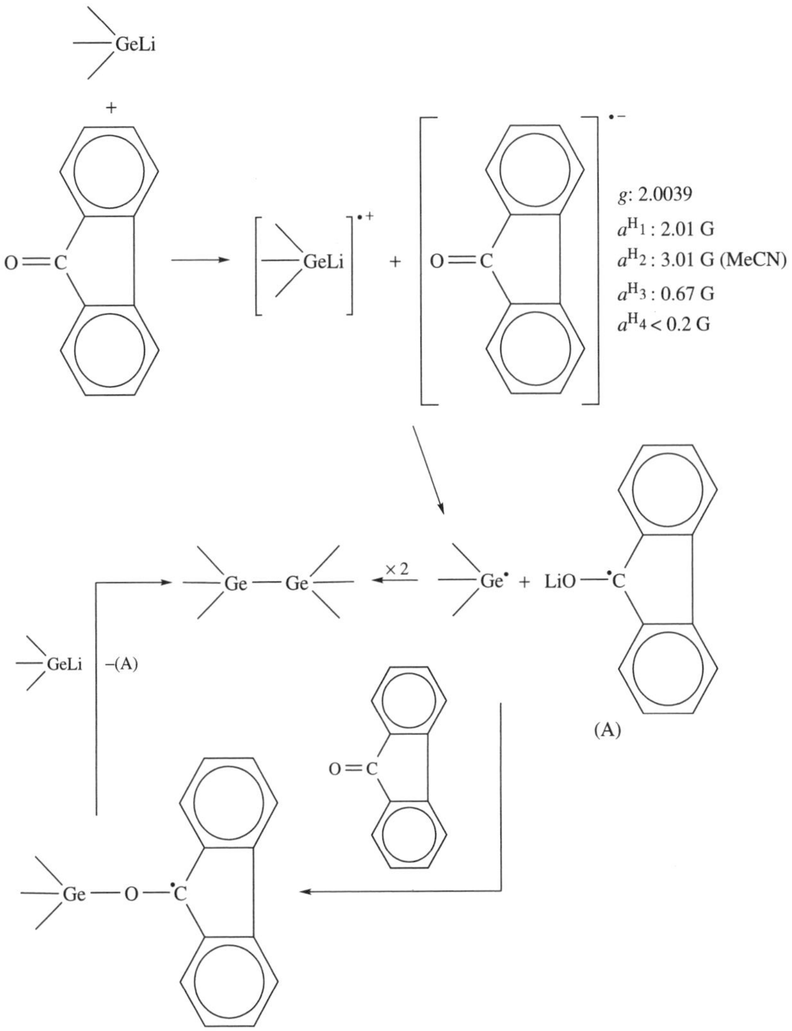
GeLi
+
O=C
GeLi
O=C
g: 2.0039
a^H1 : 2.01 G
a^H2 : 3.01 G (MeCN)
a^H3 : 0.67 G
a^H4 < 0.2 G
Ge — Ge
×2
Ge• + LiO—•C
GeLi
−(A)
(A)
O=C
Ge—O—•C

SCHEME 44

The change in the addition mechanism from nucleophilic to single electron transfer is often initiated by high conjugation of the substrate and steric effects which are able to prevent the nucleophilic attack by the M_{14} anion. For example, in the case of benzophenone, Ph_2HGeLi gave a nucleophilic addition, while Mes_2HGeLi gave a SET addition (Scheme 46)[16].

3. SET cleavages

Triethylgermyllithium, which added easily to benzoyl triethylgermane in hexane to give, after hydrolysis, bis(triethylgermyl)phenylcarbinol in 70% yield (equation 190)[200],

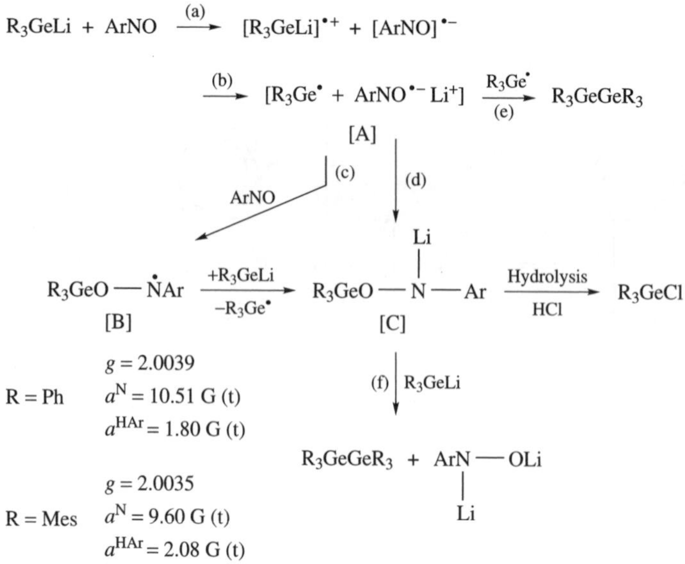

$$R_3GeLi + ArNO \xrightarrow{(a)} [R_3GeLi]^{\bullet +} + [ArNO]^{\bullet -}$$

$$\xrightarrow{(b)} [R_3Ge^{\bullet} + ArNO^{\bullet -}Li^+] \;[A] \xrightarrow[(e)]{R_3Ge^{\bullet}} R_3GeGeR_3$$

[A] $\xrightarrow[(c)]{ArNO}$ $R_3GeO-\dot{N}Ar$ [B]

[A] $\xrightarrow{(d)}$ $R_3GeO-N(Li)-Ar$ [C]

$$R_3GeO-\dot{N}Ar \;[B] \xrightarrow[-R_3Ge^{\bullet}]{+R_3GeLi} R_3GeO-N(Li)-Ar \;[C] \xrightarrow[HCl]{Hydrolysis} R_3GeCl$$

[C] $\xrightarrow{(f)\; R_3GeLi}$ $R_3GeGeR_3 + ArN(Li)-OLi$

R = Ph: $g = 2.0039$, $a^N = 10.51$ G (t), $a^{HAr} = 1.80$ G (t)

R = Mes: $g = 2.0035$, $a^N = 9.60$ G (t), $a^{HAr} = 2.08$ G (t)

SCHEME 45

underwent in HMPA a completely different reaction which led to digermane and phenyllithium. This reaction proceeded through a SET mechanism as shown by detection of the transient radical anion of benzoylgermane using ESR (Scheme 47)[200].

$$Et_3GeLi + Et_3GeCOPh \xrightarrow[\text{after hydrolysis}]{\text{hexane}} (Et_3Ge)_2C(OH)Ph \qquad (190)$$

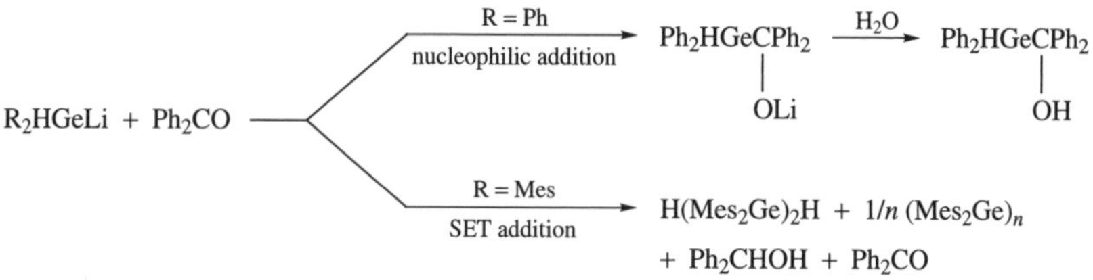

$R_2HGeLi + Ph_2CO$ $\xrightarrow[\text{nucleophilic addition}]{R = Ph}$ $Ph_2HGeCPh_2(OLi)$ $\xrightarrow{H_2O}$ $Ph_2HGeCPh_2(OH)$

$R_2HGeLi + Ph_2CO$ $\xrightarrow[\text{SET addition}]{R = Mes}$ $H(Mes_2Ge)_2H + 1/n\,(Mes_2Ge)_n + Ph_2CHOH + Ph_2CO$

SCHEME 46

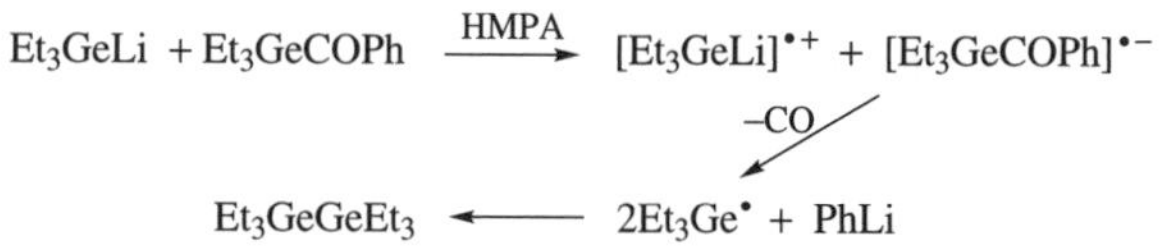

$$Et_3GeLi + Et_3GeCOPh \xrightarrow{HMPA} [Et_3GeLi]^{\bullet +} + [Et_3GeCOPh]^{\bullet -}$$

$$[Et_3GeCOPh]^{\bullet -} \xrightarrow{-CO} 2Et_3Ge^{\bullet} + PhLi$$

$$2Et_3Ge^{\bullet} \longrightarrow Et_3GeGeEt_3$$

SCHEME 47

F. Miscellaneous Reactions

1. Decomposition

Organohydrogermyllithiums R_2HGeLi have a stability in solution which depends on the solvent used and the nature of the R group linked to the metal. For example, dimesitylhydrogermyllithium is stable in solvents such as pentane, THF and amines, but diphenylhydrogermyllithium decomposed slowly in THF at 20 °C within 24 h. This decomposition was faster in the presence of an amine (Et_3N or Et_2NMe), and gave di-, tri- and tetra-germyllithiums as confirmed by hydrolysis (equation 191)[16]. The nature of these polygermanes depends mainly on the reaction time. The selective synthesis of di-, tri- or tetra-germanes can be achieved and monitored by GC analysis.

$$Ph_2GeH_2 + t\text{-}BuLi \xrightarrow[-BuH,\ -LiH]{Et_3N} Ph_2HGe(GePh_2)_nGeLiPh_2 \xrightarrow{H_2O} Ph_2HGe(GePh_2)_nGeHPh_2 \quad n = 0\text{–}2 \qquad (191)$$

2. Rearrangement reactions

Treatment of hydrogermanium cyclopentadiene transition metal complexes with LDA can lead initially to a competition between the deprotonation of the hydrogen linked to germanium or to the cyclopentadienyl ring, but a migration of the germyl group to cyclopentadiene was actually observed (equation 192)[9].

$$(\eta^5\text{-}C_5H_5)Fe(CO)_2GeHPh_2 \xrightarrow{LDA} (\eta^5\text{-}C_5H_4GeHPh_2)Fe(CO)_2Li \xrightarrow[-LiI]{MeI} (\eta^5\text{-}C_5H_4GeHPh_2)Fe(CO)_2Me \qquad (192)$$

3. Insertion of a bivalent metal-14

The insertion of a bivalent metal-14 derivatives into a germanium–metal bond led to polymetal anions which gave, after alkylation, polygermanes (equations 193 and 194)[201,202].

$$3\ Et_3GeLi + GeI_2 \longrightarrow (Et_3Ge)_3GeLi \xrightarrow{MeI} (Et_3Ge)_3GeMe \qquad (193)$$

$$R_3SnLi + R_2Sn \longrightarrow R_3SnSnR_2Li \qquad (194)$$

R = Me, Et, Bu

4. Insertion of a transition metal complex

Insertion of copper(I) cyanide into the Sn–Li bond resulted in an oxidation of Cu to cuprates (equation 195)[4b].

$$R_3SnLi + CuY \longrightarrow [R_3SnCuY]Li \quad (195)$$

$$R = Me;\ Y = CN$$

$$R = Bu;\ Y = SPh$$

Bis(naphthalene)titanium complexes were prepared by insertion of freshly prepared $Ti(C_{10}H_8)_2$ into the Sn−K bond (Scheme 48)[203].

$$Me_3SnK + [Ti(C_{10}H_8)_2] \xrightarrow[\text{THF, }-78\,^{\circ}\text{C}]{15.C.5} [K(15.C.5)_2]\,[Ti(C_{10}H_8)_2(SnMe_3)]\ (75\text{–}90\%)$$

15.C.5 = 15-crown-5
$C_{10}H_8$ = naphthalene

$$\downarrow Me_3SnK,\ THF,\ 15.C.5$$

$$[K(15.C.5)_2]_2\,[Ti(C_{10}H_8)_2(SnMe_3)_2]\ (55\%)$$

SCHEME 48

5. Ligand exchange

The reaction of tri(substituted allyl) stannyllithium with (substituted allyl)lithium formed an equilibrium mixture of tri(substituted allyl) stannyllithiums having all possible combinations of substituents (equation 196)[204].

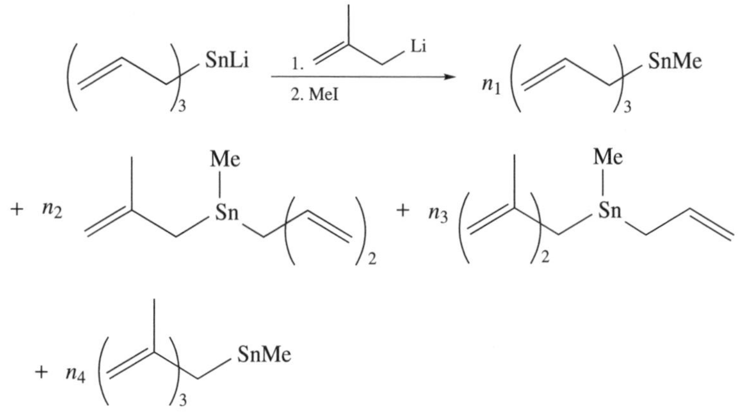

(CH2=CHCH2)3SnLi —[1. CH2=C(Me)CH2Li; 2. MeI]→ n_1 (CH2=CHCH2)3SnMe + n_2 CH2=C(Me)CH2Sn(Me)(CH2CH=CH2)2 + n_3 (CH2=C(Me)CH2)2Sn(Me)CH2CH=CH2 + n_4 (CH2=C(Me)CH2)3SnMe (196)

6. Heterocyclic rearrangements initiated by nucleophilic addition to carbonyl

The reaction of Brook's ketone with an excess of Et_3GeLi followed by hydrolysis gave a 1 : 2 mixture of a trisilacyclobutane and (adamantoyl) adamantyl carbinol. It was suggested that this reaction involves the formation of a transient disilene. The structure of the trisilacyclobutane was established by X-ray analysis (equation 197)[205].

$$(Me_3Si)_3SiC(=O)Ad \xrightarrow[\text{THF/r.t./48 h}]{Et_3GeLi} \underset{54\%}{Me_3SiSi(OGeEt_3)-Si(SiMe_3)_2-Si(SiMe_3)_2-C(H)Ad} + \underset{87.5\%}{AdC(=O)-CH(OH)Ad} \quad (197)$$

7. Elimination reactions

Metal-14 anion centers have been used to initiate elimination reactions for the synthesis of doubly bonded metal-14 compounds, as shown in equation 198[206].

$$Tip_2Sn(F)-GeMes_2(H) \xrightarrow{t\text{-BuLi}} Tip_2Sn(F)-GeMes_2(Li) \xrightarrow{-LiF} Tip_2Sn{=}GeMes_2 \quad (198)$$

8. Base activity

It was shown, using NMR analysis, that gradual introduction of HMPT into a benzene solution of Et_3GeM (M = Li, Na, K and Cs) caused increased solvation of the M^+ cation and the formation of a real ion-pair separated by solvent. Comparison of the reactivity of Et_3GeM in benzene solution and (Et_3Ge^-) (M^+, HMPT) with methyl *t*-butyl ketone showed a drastic difference. As expected, Et_3GeM gave the germylcarbinol while the ion-pair (Et_3Ge^-) (M^+, HMPT) gave almost quantitative proton abstraction (90%) with formation of Et_3GeH (equations 199 and 200)[118].

$$Et_3GeLi + MeCOBu\text{-}t \xrightarrow[\text{hydrolysis}]{\text{after}} Et_3Ge-C(Me)(OH)-Bu\text{-}t \quad (199)$$

$$Et_3Ge^- (Li,HMPT)^+ + MeCOBu\text{-}t \longrightarrow Et_3GeH + t\text{-}BuC(OLi){=}CH_2 \quad (200)$$

9. Photoreduction

Direct photo-ejection of an electron from the corresponding metal-14 anion reduced it to a metal-14 centered radical (equation 74)[4a].

G. Synthetic Applications

Because of the high cost of germanium and the high toxicity of lead, these metal-14 elements have been little used in organic synthesis compared with tin. The use of tin compounds in syntheses has been reviewed[4b,207]. We shall here illustrate the particular application of metal-14 anions.

Metal-14 anions have been used due to their high basic reactivity. They are able to abstract acidic hydrogen (equation 201)[181], induce elimination (equation 202)[49] and lead to stereospecific cyclization (equations 203 and 204)[49].

1. $Et_3GeNa{\cdot}YCl_3$ 3:1 THF/HMPA −78 °C
2. MeI/−78 °C

97–98%

(201)

R, R′ = Ph, cyclohexyl, C_5H_{11}, Me_2CHCH_2, Me, Et

Br
Br
$^{-}SnBu_3$

(202)

MeCOO COOMe

CO_2Me
CO_2Me

93%

Br
Br
COOMe

$Me_3SiSnBu_3$ (3.0 eq)
CsF (3.0 eq), DMF

H
H
COOMe

83%

(203)

Br
Br
COOMe

$Me_3SiSnBu_3$ (3.0 eq)
CsF (3.0 eq), DMF

H
H
COOMe

63%

(204)

A few key steps in the synthesis of optically active forskolin were achieved using allylation of aldehydes by a tin route (equation 205)[208].

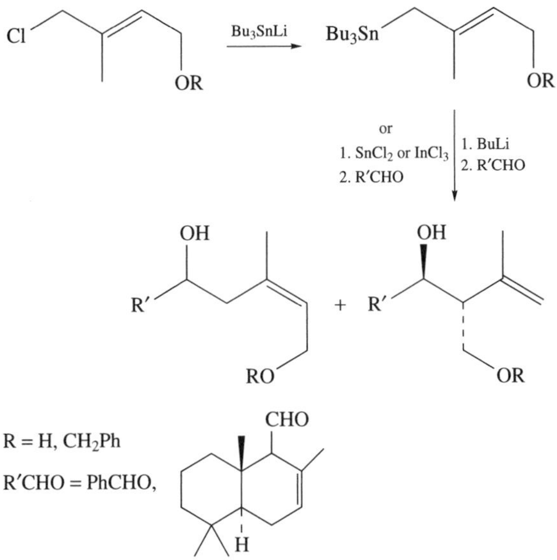

Cl, OR, Bu_3SnLi, Bu_3Sn, OR

or
1. $SnCl_2$ or $InCl_3$
2. R′CHO
1. BuLi
2. R′CHO

OH, R′, RO, +, OH, R′, OR

(205)

R = H, CH_2Ph

R′CHO = PhCHO, CHO, H

R_3SiO, O, O

2. Bu_3SnLi, THF, −78 °C
3. Separation
1. AcCl

R_3SiO, O, $SnBu_3$ and R_3SiO, O, $SnBu_3$

inversion BuLi, −78 °C inversion

R_3SiO, OH, R_3SiO, OH

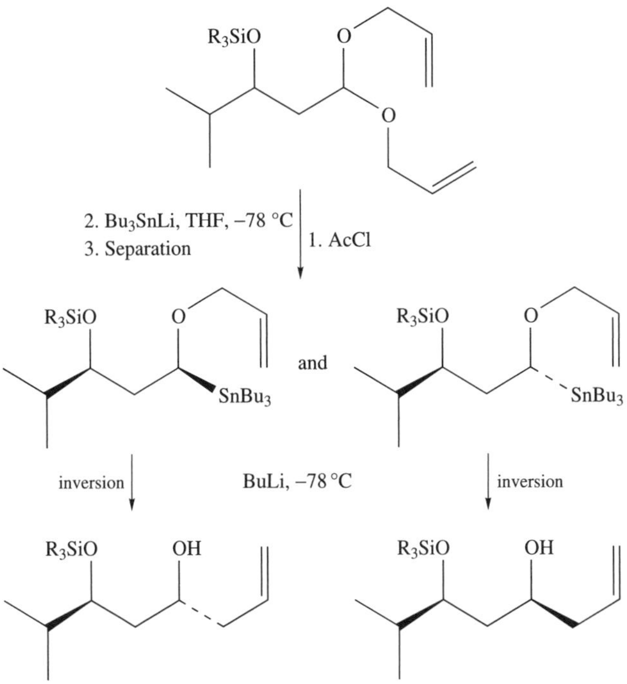

SCHEME 49

α-(Trialkylstannyl) ether, obtained from the stannylation of α-chloroallyl ether by Bu_3SnLi, allowed the stereoselective synthesis of lithioether and then, after Wittig rearrangement, the corresponding alcohol (Scheme 49)[209].

A one-pot sequential Michael–Michael ring closure (MIMIRC) reaction using 2-cyclohexenone as the initial Michael acceptor allowed an effective construction of various polyfunctionalized polycyclic compounds (Scheme 50)[210].

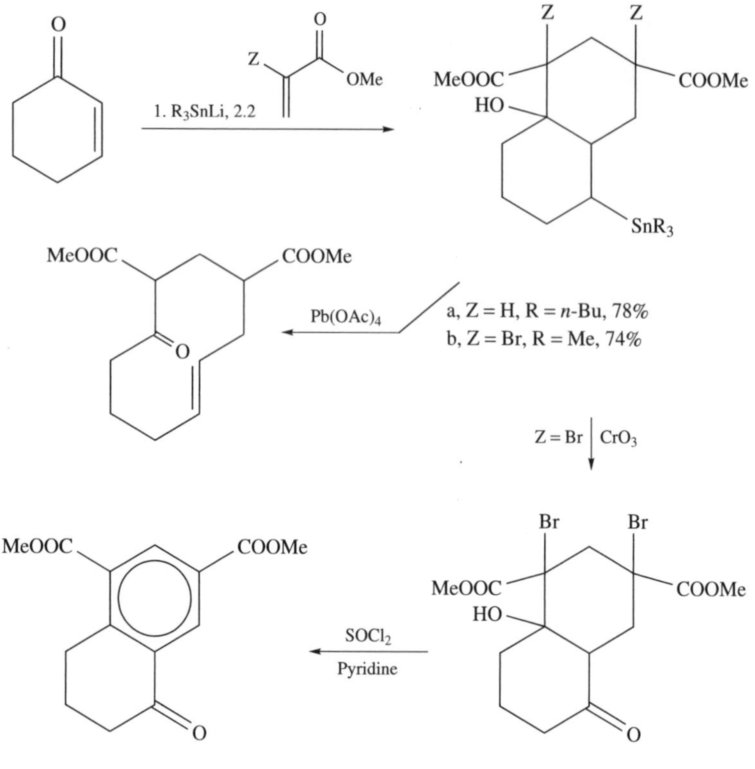

1. R_3SnLi, 2.2

a, Z = H, R = *n*-Bu, 78%
b, Z = Br, R = Me, 74%

$Pb(OAc)_4$

Z = Br | CrO_3

$SOCl_2$
Pyridine

SCHEME 50

Sulfuration of hypervalent anionic tin complexes led to a practical synthesis of disulfides involving non-odorous reagents and avoiding the use of H_2S (equation 206)[211].

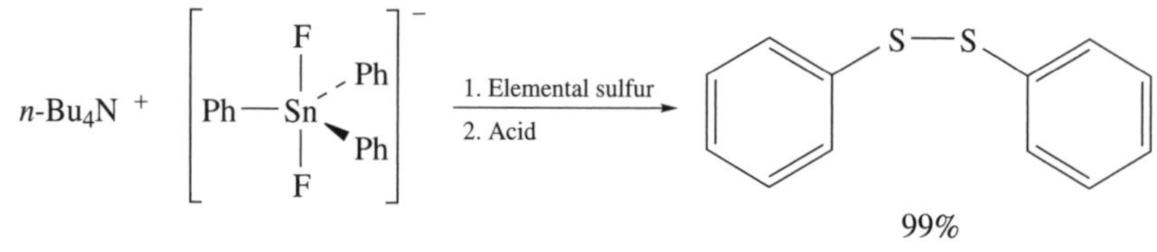

n-Bu_4N^+ $[Ph_3SnF_2]^-$ → (1. Elemental sulfur; 2. Acid) PhS–SPh 99%

(206)

The stannylation of conjugated ethylenic ketones coupled with an oxidative destannylation provided the synthetic means to various new carbonyl compounds (equations 207 and 208)[106].

1. Me_3SnLi
2. MeLi

Me_3Sn OH

$CrO_3 \cdot 2C_5H_5N$ (207)

O OH

63%

O

THF-NH_3
$C_5H_{11}I$
Me_3SnLi

(208)

O

1. MeLi
2. $CrO_3 \cdot 2C_5H_5N$
3. NaOH

$SnMe_3$ O

Oxyfunctional organolithium reagents which are very useful synthons are easy to make by stannylation of aldehydes followed by a destannylation with an organolithium (equations 209 and 210)[104].

RCHO $\xrightarrow[\text{2. R'X}]{\text{1. } Bu_3SnLi}$ R–CH(OR′)($SnBu_3$) $\xrightarrow{\text{BuLi}}$ R–CH(OR′)(Li) (209)

RCHO = furan-3-carboxaldehyde, R′X = 2-chloroethyl ether

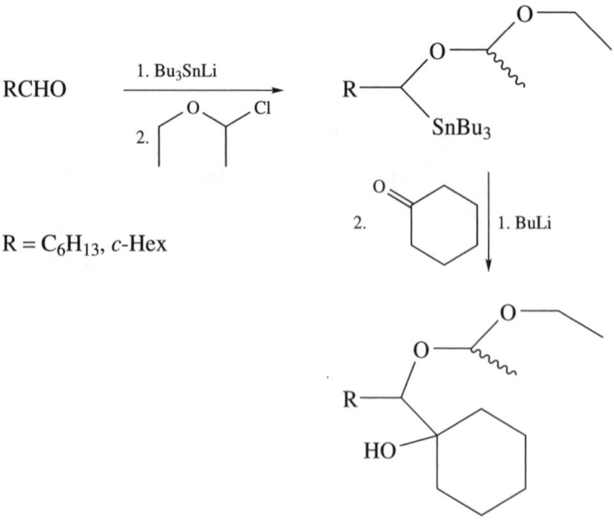

RCHO —(1. Bu_3SnLi; 2. ClCH(CH_3)OCH_2CH_3)→ R–CH(SnBu_3)–O–CH(CH_3)–O–Et —(1. BuLi; 2. cyclohexanone)→ R–CH(O–CH(CH_3)–O–Et)–C(OH)(cyclohexyl)

$R = C_6H_{13}$, *c*-Hex

(210)

Tributylstannylmagnesium chloride and tributylstannylalkalis reacted with immonium salts to form non-substituted, α-substituted or α,α-disubstituted aminomethyltributyltins. Transmetallation of the aminomethyltributyl tins with butyllithium followed by condensation with carbonyls provided a regiospecific route to β-aminoalcohols (Scheme 51)[87].

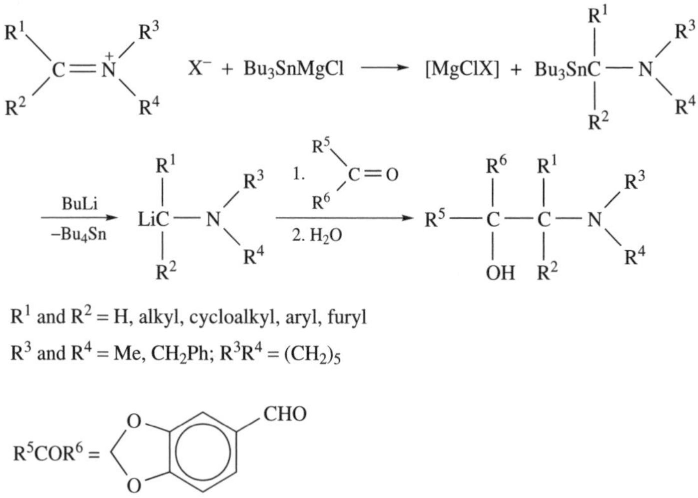

$$R^1R^2C{=}\overset{+}{N}R^3R^4\ X^- + Bu_3SnMgCl \longrightarrow [MgClX] + Bu_3SnCR^1R^2{-}NR^3R^4$$

$$\xrightarrow[-Bu_4Sn]{BuLi} LiCR^1R^2{-}NR^3R^4 \xrightarrow[2.\ H_2O]{1.\ R^5R^6C{=}O} R^5{-}CR^6(OH){-}CR^1R^2{-}NR^3R^4$$

R^1 and R^2 = H, alkyl, cycloalkyl, aryl, furyl

R^3 and R^4 = Me, CH_2Ph; $R^3R^4 = (CH_2)_5$

R^5COR^6 = 3,4-methylenedioxybenzaldehyde (CHO)

SCHEME 51

Trimethylstannyllithium can catalyze the rearrangement of 1,4-bis(trimethylstannyl)-2-butyne to 2,3-bis(trimethylstannyl)-1,3-butadiene, which is a useful reagent in the preparation of their silicon analogues (Scheme 52)[212].

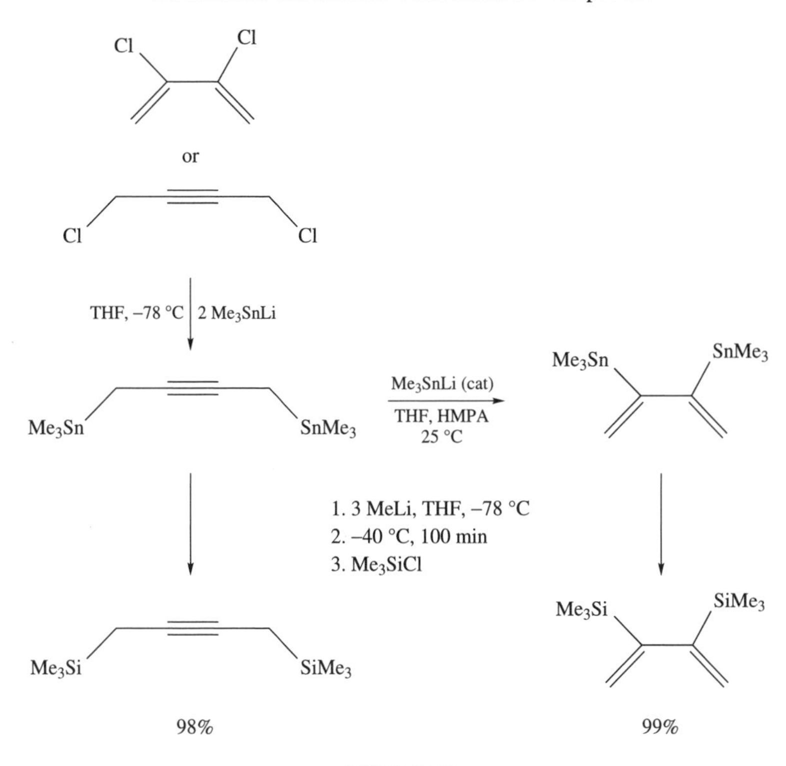

SCHEME 52

VI. REFERENCES

1. M. Lesbre, P. Mazerolles and J. Satgé, *The Organic Compounds of Germanium*, Wiley, London, 1971.
2. M. J. Newlands, Chap. 11 in *Organotin Compounds with Tin—Other Metal Bonds in Organotin Compounds* (Ed. A. K. Sawyer), Marcel Dekker Inc., New York, 1972, p. 881.
3. G. Wilkinson (Ed.), *Comprehensive Organometallic Chemistry* (*COMC I*, Vol. 2), Pergamon Press, Oxford, 1982.
   (a) P. Rivière, M. Rivière-Baudet and J. Satgé, Germanium, Chap 10.
   (b) A. G. Davies and P. J. Smith, Tin, Chap. 11.
   (c) P. G. Harrison, Lead, Chap. 12.
4. G. Wilkinson (Ed.), *Comprehensive Organometallic Chemistry* (*COMC II*, Vol. 2), Pergamon Press, Oxford, 1995.
   (a) P. Rivière, M. Rivière-Baudet and J. Satgé, Germanium, Chap 5.
   (b) A. G. Davies and P. J. Smith, Tin, Chap. 6.
   (c) P. G. Harrison, Lead, Chap. 7.
5. J. P. Belzner and U. Dehnert, Chap. 14 in *The Chemistry of Organic Silicon Compounds*, Vol. 2 (Eds. Z. Rappoport and Y. Apeloig), Wiley, Chichester, 1998.
6. M. A. Paver, C. A. Russell and D. S. Wright, *Angew. Chem., Int. Ed. Engl.*, **34**, 1545 (1995).
7. B. Goldfuss and P. v. R. Schleyer, *Organometallics*, **16**, 1543 (1997).
8. N. S. Vyazankin, G. A. Razuvaev and O. A. Kruglaya, in *Organometallic Reactions* (Eds. I. Becker and M. Tsutsui), Vol. 5, Wiley-Interscience, New York, 1970, p. 101.
9. A. Castel, P. Rivière and J. Satgé, *J. Organomet. Chem.*, **462**, 97 (1993).
10. H. K. Sharma, R. J. Villazana, F. Cervantes-Lee, L. Parkanyi and K. H. Pannel, *Phosphorus, Sulfur, Silicon, Relat. Elem.*, **87**, 257 (1994).

11. T. Lobreyer, H. Oberhammer and W. Sundermeyer, *Angew. Chem., Int. Ed. Engl.*, **32**, 586 (1993).
12. K. Mochida, N. Matsushige and M. Hamashima, *Bull. Chem. Soc. Jpn.*, **58**, 1443 (1985).
13. G. Bahr, H. O. Kalinowski and S. Pawlenko, in *Methoden der Organische Chemie (Houben-Weyl), Met. Org. Verbindungen (Ge, Sn)*, Thieme Verlag, Stuttgart, 1978.
14. E. Colomer and R. J. P. Corriu, *Top. Curr. Chem.*, **96**, 79 (1981).
15. A. Castel, P. Rivière, J. Satgé, H. Y. Ko and D. Desor, *J. Organomet. Chem.*, **397**, 7 (1990).
16. A. Castel, P. Rivière, J. Satgé and H. Y. Ko, *Organometallics*, **9**, 205 (1990).
17. A. Castel, P. Rivière, F. Cosledan, J. Satgé, M. Onyszchuk and A. M. Lebuis, *Organometallics*, **15**, 4488 (1996).
18. A. Kawachi, Y. Tanaka and K. Tamao, *Eur. J. Inorg. Chem.*, 461 (1999).
19. D. J. Berg, C. K. Lee, L. Walker and G. W. Bushnell, *J. Organomet. Chem.*, **493**, 47 (1995).
20. P. Dufour, J. Dubac, M. Dartiguenave and Y. Dartiguenave, *Organometallics*, **9**, 3001 (1990).
21. W. P. Freeman, T. D. Tilley, L. M. Liable-Sands and A. L. Rheingold, *J. Am. Chem. Soc.*, **118**, 10457 (1996).
22. W. P. Freeman, T. D. Tilley, F. P. Arnold, A. L. Rheingold and P. K. Gantzel, *Angew. Chem., Int. Ed. Engl.*, **34**, 1887 (1995).
23. D. A. Bravo-Zhivotovskii, S. D. Pigarev, O. A. Vyazankina and N. S. Vyazankin, *Z. Obshch, Khim.*, **57**, 2735 (1987); *J. Gen. Chem. USSR (Engl. Transl.)*, **57**, 2644 (1987).
24. F. Cosledan, A. Castel and P. Rivière, *Phosphorus, Sulfur, Silicon, Relat. Elem.*, **129**, 1 (1997).
25. A. Castel, P. Rivière, J. Satgé, D. Desor, M. Ahbala and C. Abdenadher, *Inorg. Chim. Acta*, **212**, 51 (1993).
26. K. M. Baines, R. J. Groh, B. Joseph and U. R. Parshotam, *Organometallics*, **11**, 2176 (1992).
27. W. Reimann, H. G. Kuivila, D. Farah and T. Apoussidis, *Organometallics*, **6**, 557 (1987).
28. T. S. Kaufman, *Synlett*, 1377 (1997).
29. M. F. Connil, B. Jousseaume, N. Noiret and M. Pereyre, *Organometallics*, **13**, 24 (1994).
30. M. F. Connil, B. Jousseaume and M. Pereyre, *Organometallics*, **15**, 4469 (1996).
31. J. C. Podesta, A. B. Chopa, N. N. Giagante and A. E. Zuniga, *J. Organomet. Chem.*, **494**, 5 (1995).
32. K. M. Baines, K. A. Mueller and T. K. Sham, *Can. J. Chem.*, **70**, 2884 (1992).
33. K. Mochida, M. Wakasa, Y. Sakaguchi and H. Hayashi, *J. Am. Chem. Soc.*, **109**, 7942 (1987).
34. J. H. Hong and P. Boudjouk, *Bull. Soc. Chim. Fr.*, **132**, 495 (1995).
35. R. West, H. Sohn, D. R. Powell, T. Müller and Y. Apeloig, *Angew. Chem., Int. Ed. Engl.*, **35**, 1002 (1996).
36. S. B. Choi, P. Boudjouk and J. H. Hong, *Organometallics*, **18**, 2919 (1999).
37. S. B. Choi, P. Boudjouk and K. Qin, *Organometallics*, **19**, 1806 (2000).
38. L. Pu, M. O. Senge, M. M. Olmstead and P. P. Power, *J. Am. Chem. Soc.*, **120**, 12682 (1998).
39. M. M. Olmstead, L. Pu, R. S. Simons and P. P. Power, *J. Chem. Soc., Chem. Commun.*, 1595 (1997).
40. L. Hevesi and B. Lacave-Goffin, *Synlett*, 1047 (1995).
41. L. R. Allain, C. A. L. Filgueiras and A. Abras, *J. Braz. Chem. Soc.*, **7**, 119 (1996).
42. C. J. Cardin, D. J. Cardin, W. Clegg, S. J. Coles, S. P. Constantine, J. R. Rowe and S. J. Teat, *J. Organomet. Chem.*, **573**, 96 (1999).
43. W. Storch, M. Vosteen, U. Emmel and U. Wietelmann, *PCT Int. Appl. WO 97 47, 630, (CI.C07F7/22, 18)* (1997).
44. B. Wrackmeyer and K. Horchler, *Z. Naturforsch. B*, **44**, 1195 (1989).
45. E. Buncel and T. K. Venkatachalam, *Heteroatom Chem.*, **5**, 201 (1994).
46. T. Kohei and K. Atsushi, *Angew. Chem., Int. Ed. Engl.*, **34**, 818 (1995).
47. S. Freitag, R. Herbst-Irmer, L. Lameyer and D. Stalke, *Organometallics*, **15**, 2839 (1996).
48. K. Mochida, H. Suzuki, M. Namba, T. Kugita and Y. Yokoyama, *J. Organomet. Chem.*, **499**, 83 (1995).
49. H. Sato, N. Isono, K. Okamura, T. Date and M. Mori, *Tetrahedron Lett.*, **35**, 2035 (1994).
50. P. B. Hitchcock, M. F. Lappert, G. A. Lawless and B. Royo, *J. Chem. Soc., Chem. Commun.*, 554 (1993).
51. I. Suzuki, T. Furuta and Y. Yamamoto, *J. Organomet. Chem.*, **443**, C6 (1993).
52. R. Villazana, H. Sharma, F. Cervantes-Lee and K. H. Pannell, *Organometallics*, **12**, 4278 (1993).
53. L. C. Willemsens and G. J. M. Van der Kerk, *J. Organomet. Chem.*, **15**, 117 (1968).

54. D. A. Bravo-Zhivotovskii, I. D. Kalikhman, O. A. Kruglaya and N. S. Vyazankin, *Izv. Akad. Nauk SSSR, Ser. Khim.*, 508; *Bull. Acad. Sci. USSR*, 444 (1978).
55. P. Riviere, A. Castel and J. Satgé, *J. Organomet. Chem.*, **212**, 351 (1981).
56. J. Park, S. A. Batcheller and S. Masamune, *J. Organomet. Chem.*, **367**, 39 (1989).
57. M. Ichinohe, H. Sekiyama, N. Fukaya and A. Sekiguchi, *J. Am. Chem. Soc.*, **122**, 6781 (2000).
58. H. H. Karsch, G. Baumgartner and S. Gamper, *J. Organomet. Chem.*, **462**, C3 (1993).
59. J. M. Dysard and T. D. Tilley, *Organometallics*, **19**, 2671 (2000).
60. J. M. Dysard and T. D. Tilley, *J. Am. Chem. Soc.*, **122**, 3097 (2000).
61. T. Kugita, Y. Tanoguchi, M. Okano and H. Hamano, *Main Group Met. Chem.*, **20**, 321 (1997).
62. J. H. Hong, Y. Pan and P. Boudjouk, *Angew. Chem., Int. Ed. Engl.*, **35**, 186 (1996).
63. T. Kawashima, Y. Nishiwaki and R. Okazaki, *J. Organomet. Chem.*, **499**, 143 (1995).
64. L. Xu and S. C. Sevov, *J. Am. Chem. Soc.*, **121**, 9245 (1999).
65. P. Riviere, A. Castel, D. Guyot and J. Satgé, *J. Organomet. Chem.*, **290**, C15 (1985).
66. R. Tacke, J. Heermann and M. Puelm, *Z. Naturforsch. B*, **53**, 535 (1998).
67. M. Veith, O. Schutt and V. Huch, *Angew. Chem., Int. Ed. Engl.*, **39**, 601 (2000).
68. M. P. Egorov, O. M. Nefedov, T. S. Lin and P. P. Gaspar, *Organometallics*, **14**, 1539 (1995).
69. V. A. Bagryansky, V. I. Borovkov, Yu. N. Molin, M. P. Egorov and O. M. Nefedov, *Chem. Phys. Lett.*, **295**, 230 (1998).
70. M. A. Beswick, H. Gornitzka, J. Kaercher, M. E. G. Mosquera, J. S. Palmer, P. R. Raithby, C. A. Russell, D. Stalke, A. Steiner and D. S. Wright, *Organometallics*, **18**, 1148 (1999).
71. L. Pu, S. T. Haubrich and P. P. Power, *J. Organomet. Chem.*, **582**, 100 (1999).
72. M. G. Davidson, D. Stalke and D. S. Wright, *Angew. Chem., Int. Ed. Engl.*, **31**, 1226 (1992).
73. E. Catalina, W. M. Davis and C. C. Cummins, *Organometallics*, **14**, 577 (1995).
74. W. P. Leung, L. H. Weng, W. H. Kwok, Z. Y. Zhou, Z. Y. Zhang and T. C. W. Mak, *Organometallics*, **18**, 1482 (1999).
75. T. Suwa, I. Shibata and A. Baba, *Organometallics*, **18**, 3965 (1999).
76. T. Kawashima, N. Ywama and R. Okazaki, *J. Am. Chem. Soc.*, **115**, 2506 (1993).
77. Y. Yamamoto, A. Sakaguchi, N. Ohashi and K. Y. Akiba, *J. Organomet. Chem.*, **506**, 259 (1996).
78. E. Fouquet, M. Pereyre and A. L. Rodriguez, *J. Org. Chem.*, **62**, 5242 (1997).
79. V. Chandrasekhar, A. Chandresekaran, R. O. Day, J. M. Holmes and R. R. Holmes, *Phosphorus, Sulfur, Silicon, Relat. Elem.*, **115**, 125 (1996).
80. F. Stabenow, W. Saak and M. Weidenbruch, *Chem. Commun.*, 1131 (1999).
81. P. Rutsch and G. Huttner, *Angew. Chem., Int. Ed.*, **39**, 2118 (2000).
82. A. Castel, P. Rivière, J. Satgé and Y. H. Ko, *J. Organomet. Chem.*, **342**, C1 (1988).
83. L. Rösch, *Angew. Chem., Int. Ed. Engl.*, **20**, 872 (1981).
84. L. Rösch, C. Krueger and A. P. Chiang, *Z. Naturforsch. B*, **39**, 855 (1984).
85. C. E. Dixon, M. R. Netherton and K. M. Baines, *J. Am. Chem. Soc.*, **120**, 10365 (1988).
86. J. C. Lahournère and J. Valade, *J. Organomet. Chem.*, **22**, C3 (1970).
87. B. Elissondo, J. B. Verlhac, J. P. Quintard and M. Pereyre, *J. Organomet. Chem.*, **339**, 267 (1988).
88. D. Seyferth, D. Y. Son and S. Shah, *Organometallics*, **13**, 2105 (1994).
89. M. Westerhausen, *Angew. Chem., Int. Ed. Engl.*, **33**, 1493 (1994).
90. K. Mochida and T. Yamanishi, *Bull. Chem. Soc. Jpn.*, **60**, 3429 (1987).
91. N. Duffaut, J. Dunogues, R. Calas, P. Rivière, J. Satgé and A. Cazes, *J. Organomet. Chem.*, **149**, 57 (1978).
92. E. J. Bulten and J. G. Noltes, *J. Organomet. Chem.*, **29**, 397 (1971).
93. K. Mochida, T. Kugita and Y. Nakadaira, *Polyhedron*, **9**, 2263 (1990).
94. K. Mochida and N. Matsushige, *J. Organomet. Chem.*, **229**, 11 (1982).
95. A. N. Egorochkin, S. Y. Khorshev, N. S. Vyazankin, E. N. Gladishev, V. T. Bychkov and O. A. Kruglaya, *Z. Obshch. Khim.*, **38**, 276 (1968); *Chem. Abstr.*, **69**, 6600s (1968).
96. R. J. Cross and F. Glockling, *J. Organomet. Chem.*, **3**, 253 (1965).
97. M. A. Bush and P. Woodward, *J. Chem. Soc.*, 1833 (1967).
98. E. Amberger, W. Stoeger and H. R. Honigschmid-Grossich, *Angew. Chem.*, **78**, 549 (1966).
99. R. J. Cross and F. Glockling, *J. Organomet. Chem.*, **3**, 146 (1965).
100. R. J. Cross and F. Glockling, *J. Chem. Soc.*, 4125 (1964).
101. A. G. Brook, F. Abdesaken and H. Söllradl, *J. Organomet. Chem.*, **299**, 9 (1986).
102. A. Castel, P. Rivière, J. Satgé and D. Desor, *J. Organomet. Chem.*, **433**, 49 (1992).

103. D. A. Bravo-Zhivotovskii, S. D. Pigarev, M. G. Voronkov and N. S. Vyazankin, *Z. Obshch. Khim.*, **59**, 863 (1989); *J. Gen. Chem. (Engl. Transl.)*, **59**, 761 (1989).
104. W. C. Still, *J. Am. Chem. Soc.*, **100**, 1481 (1978).
105. H. K. Sharma, F. Cervantes-Lee, L. Parkanyi and K. H. Pannell, *Organometallics*, **15**, 429 (1996).
106. W. C. Still, *J. Am. Chem. Soc.*, **99**, 4836 (1977).
107. F. Preuss, T. Wieland, J. Perner and G. Heckmann, *Z. Naturforsch., Teil B*, **47**, 1355 (1992).
108. G. F. Smith, H. G. Kuivila, R. Simon and L. Sultan, *J. Am. Chem. Soc.*, **103**, 833 (1981).
109. A. B. Chopa, A. P. Murray and M. T. Lockhart, *J. Organomet. Chem.*, **585**, 35 (1999).
110. M. Herberhold, V. Tröbs and B. Wrackmeyer, *J. Organomet. Chem.*, **541**, 391 (1997).
111. H. J. Koglin, K. Behrends and M. Dräger, *Organometallics*, **13**, 2733 (1994).
112. E. Buncel, R. D. Gordon and T. K. Venkatachalam, *J. Organomet. Chem.*, **507**, 81 (1996).
113. K. Mochida and T. Kugita, *Main Group Met. Chem.*, **11**, 215 (1988).
114. R. J. Batchelor and T. Birchall, *J. Am. Chem. Soc.*, **105**, 3848 (1983).
115. E. Buncel, T. K. Venkatachalam, B. Eliasson and U. Edlund, *J. Am. Chem. Soc.*, **107**, 303 (1985).
116. U. Edlund, T. Lejon, P. Pyykko, T. K. Venkatachalam and E. Buncel, *J. Am. Chem. Soc.*, **109**, 5982 (1987).
117. R. H. Cox, E. G. Janzen and W. B. Harrison, *J. Magn. Reson.*, **4**, 274 (1971).
118. D. A. Bravo-Zhivotovskii, I. D. Kalikhman, O. A. Kruglaya and N. S. Vyazankin, *Izv. Akad. Nauk SSSR, Ser. Khim.*, 1934; *Bull. Acad. Sci. USSR*, 1800 (1977).
119. H. J. Reich, J. P. Borst and R. R. Dykstra, *Organometallics*, **13**, 1 (1994).
120. D. Reed, D. Stalke and D. S. Wright, *Angew. Chem., Int. Ed. Engl.*, **30**, 1459 (1991).
121. J. B. Lambert and M. Urdaneta-Perez, *J. Am. Chem. Soc.*, **100**, 157 (1978).
122. T. Birchall and J. A. Vétrone, *J. Chem. Soc., Chem. Commun.*, 877 (1988).
123. D. R. Armstrong, M. G. Davidson, D. Moncrieff, D. Stalke and D. S. Wright, *J. Chem. Soc., Chem. Commun.*, 1413 (1992).
124. M. Veith, C. Ruloff, V. Huch and F. Töllner, *Angew. Chem., Int. Ed. Engl.*, **27**, 1381 (1988).
125. D. Smith, P. E. Fanwick and I. P. Rothwell, *Inorg. Chem.*, **28**, 618 (1989).
126. M. A. Paver, C. A. Russel, D. Stalke and D. S. Wright, *J. Chem. Soc., Chem. Commun.*, 1349 (1993).
127. D. R. Armstrong, M. J. Duer, M. G. Davidson, D. Moncrieff, C. A. Russel, C. Stourton, A. Steiner, D. Stalke and D. S. Wright, *Organometallics*, **16**, 3340 (1997).
128. T. Matsumoto, N. Tokitoh, R. Okazaki and M. Goto, *Organometallics*, **14**, 1008 (1995).
129. C. Eaborn, R. E. E. Hill and P. Simpson, *J. Organomet. Chem.*, **37**, 275 (1972).
130. J. P. Quintard, S. Hauvette-Frey, M. Pereyre, C. Couret and J. Satgé, *C.R. Acad. Sci. Paris, Ser. C*, **287**, 247 (1978).
131. E. Matarasso-Tchiroukhine and P. Cadiot, *J. Organomet. Chem.*, **121**, 155 (1976).
132. T. Kugita, M. Wakasa, J. Tamura and H. Hayashi, *Inorg. Chem. Commun.*, **1**, 386 (1998).
133. W. Adcock and C. I. Clark, *J. Org. Chem.*, **60**, 723 (1995).
134. K. W. Hellmann, C. Bott, L. H. Gade, I. J. Scowen and M. McPartlin, *Polyhedron*, **17**, 737 (1998).
135. H. M. J. C. Creemers, J. G. Noltes and G. J. M. Van der Kerk, *J. Organomet. Chem.*, **14**, 217 (1968).
136. G. Märkl and R. Wagner, *Tetrahedron Lett.*, **27**, 4015 (1986).
137. K. W. Lee and J. San Filippo Jr., *Organometallics*, **2**, 906 (1983).
138. S. Chandrasekhar, S. Latour, J. D. Wuest and B. Zacharie, *J. Org. Chem.*, **48**, 3810 (1983).
139. V. S. Zavgorodnii, L. E. Mikhailov and A. V. Novoselov, *Zh. Org. Khim.*, **32**, 879 (1996); *Chem. Abstr.*, **126**, 212204q (1997).
140. D. Farah, K. Swami and H. G. Kuivila, *J. Organomet. Chem.*, **429**, 311 (1992).
141. J. Saudosham, *Tetrahedron*, **50**, 275 (1994).
142. G. Dumartin, G. Ruel, J. Kharboutli, B. Delmond, M. F. Connil, B. Jousseaume and M. Pereyre, *Synlett*, 952 (1994).
143. A. V. Seleznev, D. A. Bravo-Zhivotovskii, T. I. Vakul'skaya and M. G. Voronkov, *Polyhedron*, **9**, 227 (1990).
144. D. A. Nicholson and A. L. Allred, *Inorg. Chem.*, **4**, 1747 (1965).
145. A. Capperucci, A. Degl'innocenti, C. Faggi, G. Reginato and A. Ricci, *J. Org. Chem.*, **54**, 2966 (1989).

146. I. Beaudet, A. Duchene, J. L. Parrain and J. P. Quintard, *J. Organomet. Chem.*, **427**, 201 (1992).
147. E. Laborde, L. E. Lesheski and J. S. Kiely, *Tetrahedron Lett.*, **31**, 1837 (1990).
148. A. R. Katritzky, H. X. Chang and J. Wu, *Synthesis*, **9**, 907 (1994).
149. W. H. Pearson and E. P. Stevens, *Synthesis*, **9**, 904 (1994).
150. S. I. Kiyooka and A. Miyauchi, *Chem. Lett.*, 1829 (1985).
151. F. Cosledan, thèse n° 2724, Toulouse, France, 1997.
152. M. Charisse, M. Mathes, D. Simon and M. Dräger, *J. Organomet. Chem.*, **445**, 39 (1993).
153. M. Veith, C. Mathur and V. Huch, *Organometallics*, **15**, 2858 (1996).
154. S. P. Mallela and R. A. Geanangel, *Inorg. Chem.*, **33**, 6357 (1994).
155. C. Schneider-Koglin, K. Behrends and M. Dräger, *J. Organomet. Chem.*, **448**, 29 (1993).
156. C. J. Cardin, D. J. Cardin, S. P. Constantine, A. K. Todd, S. J. Teat and S. Coles, *Organometallics*, **17**, 2144 (1998).
157. S. Adams and M. Dräger, *Angew. Chem., Int. Ed. Engl.*, **26**, 1255 (1987).
158. A. Castel, P. Rivière, B. St. Roch, J. Satgé and J. P. Malrieu, *J. Organomet. Chem.*, **247**, 149 (1983).
159. R. Altmann, O. Gausset, D. Horn, K. Jurkschat and M. Schürmann, *Organometallics*, **19**, 430 (2000).
160. E. E. Nifantiev, N. S. Vyazankin, S. F. Sorokina, L. A. Vorobieva, O. A. Vyazankina, D. A. Bravo-Zhivotovskii and A. R. Bekker, *J. Organomet. Chem.*, **277**, 211 (1984).
161. E. Piers and R. M. Lemieux, *Organometallics*, **17**, 4213 (1998).
162. H. G. Woo, W. P. Freeman and T. D. Tilley, *Organometallics*, **11**, 2198 (1992).
163. M. Porchia, F. Ossola, G. Rossetto, P. Zanella and N. Brianese, *J. Chem. Soc., Chem. Commun.*, 550 (1987).
164. E. Colomer and R. J. P. Corriu, *J. Chem. Soc., Chem. Commun.*, 435 (1978).
165. S. Inoue and Y. Sato, *Organometallics*, **8**, 1237 (1989).
166. W. Uhlig, *J. Organomet. Chem.*, **409**, 377 and **421**, 189 (1991).
167. F. Carré and R. Corriu, *J. Organomet. Chem.*, **65**, 349 (1974).
168. A. Sekiguchi, N. Fukaya and M. Ichinohe, *J. Am. Chem. Soc.*, **121**, 11587 (1999).
169. A. G. Brook and C. J. D. Peddle, *J. Am. Chem. Soc.*, **85**, 2338 (1963).
170. C. Eaborn, R. E. E. Hill and P. Simpson, *J. Organomet. Chem.*, **37**, 267 (1972).
171. S. Arseniyadis, J. I. M. Hernando, J. Q. Del Moral, D. V. Yashunsky and P. Potier, *Synlett*, 1010 (1998).
172. J. M. Chong and S. B. Park, *J. Org. Chem.*, **58**, 523 (1993).
173. K. Mochida and M. Nanba, *Polyhedron*, **13**, 915 (1994).
174. I. S. Biltueva, D. A. Bravo-Zhivotovskii, I. D. Kalikhman, V. Yu. Vitkovskii, S. G. Shevchenko, N. S. Vyazankin and M. G. Voronkov, *J. Organomet. Chem.*, **368**, 163 (1989).
175. D. A. Bravo-Zhivotovskii, S. D. Pigarev, I. D. Kalikhman, O. A. Vyazankina and N. S. Vyazankin, *Zh. Obshch. Khim.*, **52**, 1935 (1982); *Chem. Abstr.*, **97**, 198301t (1982).
176. J. Fujiwara, M. Watanabe and T. Sato, *J. Chem. Soc., Chem. Commun.*, 349 (1994).
177. T. Sato, T. Watanabe, T. Hayata and T. Tsukui, *J. Chem. Soc., Chem. Commun.*, 153 (1989).
178. S. Kusuda, Y. Watanabe, Y. Ueno and T. Toru, *J. Org. Chem.*, **57**, 3145 (1992).
179. A. Krief and L. Provins, *Tetrahedron Lett.*, **39**, 2017 (1998).
180. D. A. Bravo-Zhivotovskii, I. Zharov, M. Kapon and Y. Apeloig, *J. Chem. Soc., Chem. Commun.*, 1625 (1995).
181. Y. Yokoyama and K. Mochida, *J. Organomet. Chem.*, **499**, C4 (1995).
182. F. Burkhart, M. Hoffmann and H. Kessler, *Tetrahedron Lett.*, **39**, 7699 (1998).
183. P. B. Hitchcock, M. F. Lappert and M. Layh, *Chem. Commun.*, 2179 (1998).
184. V. P. Mar'in, P. V. Petrovskii and E. V. Krasil'nikova, *Izv. Akad. Nauk SSSR, Ser. Khim.*, 2557 (1996); *Chem. Abstr.*, **126**, 157621w (1997).
185. T. Kusukawa and W. Ando, *J. Organomet. Chem.*, **561**, 109 (1998).
186. J. Baxter, E. G. Mata and E. J. Thomas, *Tetrahedron*, **54**, 14359 (1998).
187. A. Krief and L. Provins, *Synlett*, 505 (1997).
188. P. Rivière, M. Rivière-Baudet and A. Castel, *Main Group Met. Chem.*, **17**, 679 (1994).
189. W. Kitching, H. Olszowy and K. Harvey, *J. Org. Chem.*, **46**, 2423 (1981).
190. T. Aruga, O. Ito and M. Matsuda, *J. Phys. Chem.*, **86**, 2950 (1982).
191. E. C. Ashby, W. Y. Su and T. N. Pham, *Organometallics*, **4**, 1493 (1985).
192. M. Wakasa and T. Kugita, *Organometallics*, **17**, 1913 (1998).

193. E. F. Corsico and R. A. Rossi, *Synlett.*, 227 (2000).
194. C. C. Yammal, J. C. Podesta and R. A. Rossi, *J. Org. Chem.*, **57**, 5720 (1992).
195. E. C. Ashby and A. K. Deshpande, *J. Org. Chem.*, **59**, 7358 (1994).
196. A. B. Chopa, M. T. Lockhart and G. Silbestri, *Organometallics*, **19**, 2249 (2000).
197. P. Rivière, A. Castel and F. Cosledan, *Phosphorus, Sulfur, Silicon, Relat. Elem.*, **104**, 169 (1995).
198. P. Rivière, A. Castel, Y. H. Ko and D. Desor, *J. Organomet. Chem.*, **386**, 147 (1990).
199. P. Rivière, A. Castel, D. Desor and C. Abdennadher, *J. Organomet. Chem.*, **443**, 51 (1993).
200. D. A. Bravo-Zhivotovskii, I. S. Biltueva, T. I. Vakul'skaya, N. S. Vyazankin and M. G. Voronkov, *Izv. Akad. Nauk SSSR, Ser. Khim.*, 2872; *Bull. Acad. Sci. USSR*, 2671 (1987).
201. G. A. Razuvaev, *J. Organomet. Chem.*, **200**, 243 (1980).
202. K. Kobayashi, M. Kawanisi, T. Hitomi and S. Kozima, *J. Organomet. Chem.*, **233**, 299 (1982).
203. J. E. Ellis, D. W. Blackburn, P. Yuen and M. Jang, *J. Am. Chem. Soc.*, **115**, 11616 (1993).
204. Y. Horikawa and T. Takeda, *J. Organomet. Chem.*, **523**, 99 (1996).
205. D. A. Bravo-Zhivotovskii, Y. Apeloig, Y. Ovchinnikov, V. Igonin and Y. T. Struchkov, *J. Organomet. Chem.*, **446**, 123 (1993).
206. M. A. Chaubon, J. Escudié, H. Ranaivonjatovo and J. Satgé, *J. Chem. Soc., Chem. Commun.*, 2621 (1996).
207. T. Sato, *Synthesis*, 259 (1990).
208. D. Behnke, L. Hennig, M. Findeisen, P. Welzel, D. Muller, M. Thormann and H. J. Hofmann, *Tetrahedron*, **56**, 1081 (2000).
209. R. Hoffmann and R. Bruchner, *Angew. Chem., Int. Ed. Engl.*, **31**, 647 (1992).
210. G. H. Posner and E. Asirvatham, *Tetrahedron Lett.*, **27**, 663 (1986).
211. S. S. Kerverdo, X. Fernandez, S. Poulain and M. Gingras, *Tetrahedron Lett.*, **41**, 5841 (2000).
212. H. J. Reich, I. L. Reich, K. E. Yelm, J. E. Holladay and D. Gschneidner, *J. Am. Chem. Soc.*, **115**, 6625 (1993).

CHAPTER **12**

Spectroscopic studies and quantum-chemical calculations of short-lived germylenes, stannylenes and plumbylenes

SERGEY E. BOGANOV, MIKHAIL P. EGOROV, VALERY I. FAUSTOV and OLEG M. NEFEDOV

N. D. Zelinsky Institute of Organic Chemistry of the Russian Academy of Sciences, Leninsky prospect, 47, 119991 Moscow, Russian Federation
Fax: 7-095-1356390; e-mail: mpe@mail.ioc.ac.ru

The chemistry of organic germanium, tin and lead compounds — Vol. 2
Edited by Z. Rappoport

I. LIST OF ABBREVIATIONS

AM1 semiempirical method based on the modified neglect of diatomic overlap approximation using radial gaussian functions to modify the core–core repulsion term

ANO atomic natural orbital

AO atomic orbital
B3LYP Becke's three-parameter hybrid functional using the Lee, Yang and Parr correlation functional
BLYP Becke's exchange functional and the Lee, Yang and Parr correlation functional
CA carbene analog
CAS SCF complete active space self-consistent field
CCSD(T) coupled cluster calculations, using single and double substitutions from the Hartree–Fock determinant. Include triple excitations non-iteratively
CI configuration interaction
CIS configuration interaction with single excitations
CNDO semiempirical method based on the complete neglect of differential overlap approximation
CVD chemical vapor deposition
DFT density functional theory
DHF Dirac–Hartree–Fock
DZ double-zeta
DZ + d double-zeta + polarization functions on heavy atoms
DZP double-zeta + polarization functions on all atoms
E_a activation energy
EA electron affinity
ΔE_{ST} energy difference between the first singlet and triplet states ($\Delta E_{ST} = E_{triplet} - E_{singlet}$)
ECP effective core potential
EHMO extended Hückel MO method
EIMS electron impact mass spectrometry
ESR electron spin resonance
G2 Gaussian-2 methods that correspond effectively to energy calculations at the QCISD(T)/6–311+G(3df,2p)//MP2(full)/6–31G(d) level with ZPE from HF/6–31G(d) level and higher level corrections
HF Hartree–Fock
HOMO highest occupied molecular orbital
ICLAS intracavity laser absorption spectroscopy
IE ionization energy
IR infrared spectroscopy
IRDLKS infrared diode laser kinetic spectroscopy
IRDLS infrared diode laser spectroscopy
ISEELS inner-shell electron energy loss spectroscopy
LIF laser-induced fluorescence
LIP laser-induced phosphorescence
LRAFKS laser resonance absorption flash kinetic spectroscopy
LUMO lowest unoccupied molecular orbital
MNDO semiempirical method based on the modified neglect of diatomic overlap-approximation
MO molecular orbital
MP_n ($n = 2–4$) n-th order of Møller–Plesset correlation energy correction
MRSDCI multireference single + double configuration interaction
MRSDCI + Q multireference single + double configuration interaction plus a Davidson-correction for uncoupled quadruple clusters
MW microwave spectroscopy

PBE	generalized gradient functional of Perdew, Burke and Ernzerhof
PES	potential energy surface
PM3	semiempirical AM1-type method using an alternative set of parameters
QCISD(T)	quadratic configuration interaction calculation, including single and double-substitutions and non-iteratively triple excitations
R^2	correlation coefficient
RECP	relativistic effective core potential
r.t.	room temperature
SCF	self-consistent field
SO	spin–orbital coupling term
SOCI	second-order configuration interaction
SR	synchrotron radiation
TS	transition state
TZP	triple-zeta + a set of polarization functions on all atoms
TZ2P	triple-zeta + two sets of polarization functions on all atoms
UV	ultraviolet
VSEPR	valence-shell electron pair repulsion
VUV	vacuum ultraviolet

II. INTRODUCTION

Carbene analogs are electrically neutral species, which are characterized by the presence of an atom possessing either at least one lone pair of s or p electrons and at least one unoccupied p orbital or two unpaired electrons in s and p or in two p orbitals, but having no unoccupied d level with a quantum number lower than that of the valent p electrons[1,2]. Carbene analogs of the Group 14 elements (CAs), silylenes (SiR_2), germylenes (GeR_2), stannylenes (SnR_2) and plumbylenes (PbR_2), represent an important class of chemical derivatives of these elements. Short-lived representatives of CAs play a significant role in chemical transformations of organogermanium, -tin and -lead compounds, including those of practical interest. Since the 1970s the synthetic approaches to the preparation of stable CAs have also been developed and a lot of stable CAs have been synthesized. The chemistry of CAs has been surveyed in many reviews and monographs in detail, the most recent being those on silylenes[3,4], germylenes[5–7] and stannylenes[5].

The first spectroscopic studies of CAs were performed in the 1930s[8–10]. A great number of CAs have since been characterized by different spectroscopic methods. Application of molecular spectroscopy to studies of short-lived CAs has enabled us to obtain the most unambiguous experimental evidence of their existence as kinetically independent species. It also provided very important information on the nature of bonding in these molecules, their electronic structures and geometries, necessary for understanding their chemical reactivity. Information on spectral characteristics of some CAs has already found analytic use for their detection in complex reaction mixtures (for example, in CVD processes), for studies of their further transformations and for identification of the simplest CAs in interstellar clouds. In spite of extensive data on molecular spectroscopy of CAs and the fast growing number of publications in this field, there is still no comprehensive review devoted to spectroscopic studies of CAs, although their spectral characteristics were partially listed in different reviews. The most complete consideration of spectroscopic studies of silylenes is presented in a recent review by Gaspar and West[3]. Spectral characteristics of many CAs are collected in a compilation by Nefedov and coworkers[11], some data on vibrational and electronic spectroscopy can be found in a serial compilation of Jacox[12–15] and those on ionization energies and electron affinities in the compilation of Lias and coworkers[16]. Results of matrix IR studies of CAs have been presented in a review by Korolev and Nefedov[17]. Different physicochemical properties of CAs are collected in the NIST Chemistry WebBook[18].

The present review is devoted to spectroscopic studies of short-lived germylenes, stannylenes and plumbylenes, and their complexes with Lewis bases. Some data on silylenes are also given for comparison, though detailed consideration of all available data on silylenes is beyond the scope of this review. The spectroscopic studies of monoxides, EO (E = Si, Ge, Sn, Pb), and related diatomic molecules with formal divalent E, as well as those of isonitrile analogs, ENR (E = Si, Ge, Sn, Pb), are not considered here, as the parent carbon monoxide and C≡NH are not classified as carbenes. Since the electronic structure of carbonylgermylene, GeCO, and carbonylstannylene, SnCO, differ significantly from the electronic structure of carbenes and CAs[19], their spectra too are not discussed here. At the same time spectroscopic data for complexes of germylenes, stannylenes and plumbylenes with Lewis bases are included, because such complexes are closely related to free germylenes, stannylenes and plumbylenes and, as is obvious now, they play an important role in chemical transformations involving these reactive intermediates.

The results of electronic, vibrational, microwave and photoelectron spectroscopy studies will be discussed below. The data obtained by these methods for stable germylenes, stannylenes and plumbylenes (except those for cyclopentadienyl derivatives because of their quite specific electronic structure) and their complexes are also discussed in this review owing to their importance for understanding the general trends. At the same time, nuclear magnetic resonance data available for most stable CAs and Mössbauer spectroscopy data available for a huge number of stable Sn(II) compounds are not included, as these methods were solely applied to study the stable CAs (with only one exception[20] where Mössbauer spectra of $SnMe_2$ and SnMeH were measured in low-temperature Ar matrices). Consideration of the microwave spectroscopy studies of short-lived germylenes, stannylenes and plumbylenes is supplemented with analysis of their geometries obtained by different experimental methods. The structures of stable germylenes, stannylenes and plumbylenes are not discussed here. Discussion of the structures of stable germylenes and their stable complexes can be found in earlier reviews[6,21]. Structural data for stable plumbylenes and their stable complexes are available in a recent compilation[22]. Section IX devoted to photoelectron spectroscopy studies is supplemented with data on ionization energies and electron affinities measured by mass spectrometry and also with a short overview of the electrochemistry of germylenes and stannylenes.

Quantum-chemical calculations are employed extensively for the interpretation of experimentally observed spectra. Increasing accuracy of modern quantum-chemical methods in the prediction of spectral characteristics, geometries and energies allows one to a certain extent to fill the gaps in experimental data available for germylenes, stannylenes and plumbylenes. Therefore we review in Section X the results of quantum-chemical studies of germylenes, stannylenes and plumbylenes.

III. ELECTRONIC TRANSITIONS IN TRIATOMIC AND TETRAATOMIC GERMYLENES, STANNYLENES AND PLUMBYLENES

This and subsequent sections are devoted to the electronic spectra of germylenes, stannylenes and plumbylenes. All experimentally known CAs have a singlet ground state. Typically, in this state the HOMO of the CA represents the nonbonding lone-pair orbital of n type and σ symmetry with significant localization on the divalent silicon, germanium, tin or lead atom and the LUMO is a π-type MO with the main contribution from the empty p orbital of the same atom (see Section IX). Therefore, the first allowed electronic transition between HOMO and LUMO in CAs belongs to the n$-\pi$ type and the corresponding bands lie in the visible or in the near-UV region. Thus, the UV-vis absorption spectroscopy is a very convenient tool for a direct detection of CAs. Indeed, a great number of CAs have been characterized by low resolution UV-vis absorption spectra in the condensed phase, mainly in low-temperature matrices with the standard absorption technique and also in solutions using a flash photolysis technique. The spectra obtained by these methods usually consist of one or more broad bands, each being actually an unresolved superposition of a series of overlapping vibronic lines broadened due to interactions with the environment, and it will be considered in the next section. Generation of CAs in the gas phase and use of a high resolution probe technique allow one to resolve vibronic and in some cases even rotational structure in the electronic spectra, provided that the molecule has not too many degrees of freedom. Currently it has been done only for some triatomic germylenes, stannylenes and plumbylenes of two types: EX_2 and EXY, E = Ge, Sn, Pb, and for germylidene, $H_2C{=}Ge$. The spectra of the tri- and tetraatomic germylenes, stannylenes and plumbylenes will be considered in this section.

The assignment of the observed bands to the particular CA is typically based on the following data: the stoichiometry of the precursor decomposition, exclusion of alternative possible products with known spectral characteristics, correspondence of the observed rovibronic band structure with that expected on the basis of selection rules, agreement of some frequencies obtained from analysis of vibronic structure with the ground state frequencies determined from vibrational spectra, use of different precursors for generation of the same CA, and also on some other criteria, depending on the method of generation and on the probing technique. Lately, high level quantum-chemical calculations have found wide application in interpretation of the main features of the observed spectra and in assignment of the bands to the particular electronic transition. At present reliable and self-consistent data have been obtained for most of the triatomic germylenes, stannylenes and plumbylenes.

The singlet ground state (S_0) of the triatomic CAs is designated as $\tilde{X}^1A_1$ for EX_2 and $\tilde{X}^1A'$ for EXH in accordance with the symmetry of their HOMO. The first excited electronic state of these molecules is the low-lying triplet (T_1) state: $\tilde{a}^3B_1$ for EX_2 and $\tilde{a}^3A''$ for EXH, followed by the excited singlet (S_1) state $\tilde{A}^1B_1$ for EX_2 and $\tilde{A}^1A''$ for EXH. Transition T_1-S_0 is forbidden by the selection rules. However, this type transitions could have nonzero probability and thus appear in the spectra as relatively weak bands. Due to its low intensity the detection of this transition is quite complicated. The lifetime of the T_1 state is usually quite long (*ca* 10^{-5}s). Transition S_1-S_0 is allowed by the selection rules; the typical lifetime of the S_1 state is $10^{-9}-10^{-6}$s. The bands corresponding to this transition are usually intensive. The lifetime of the excited state can be used as a criterion for the assignment of the observed transition to a T_1-S_0 or to an S_1-S_0 band. With rare exceptions (see below) all the observed and thoroughly studied electronic bands of triatomic germylenes, stannylenes and plumbylenes are attributed to these two transitions. A particular vibrational level of an electronic state of a triatomic CA will be designated as $T(\nu_1, \nu_2, \nu_3)$, where T is the corresponding electronic term (like $^1A''$ or $^1B_1$), and v_i are the quantum numbers for vibrational levels of i-th fundamental modes.

The following important data can be obtained from analysis of the rovibronic structure of the electronic transition band: the difference between zero vibrational levels of the upper and lower electronic states, T_{00}, fundamental frequencies for the fully symmetric vibrational modes (for symmetric stretching, ν_1, and bending, ν_2, modes of the EX_2, or stretching, ν_1 and ν_3, and bending ν_2 modes of EHX) in both states (designated as ν_i" for the lower electronic state and as ν_i' for the upper state), rotational constants for each of the vibrational levels of the electronic states involved (if rotational structure is resolved), and, from the rotational constants, two geometrical parameters (because only two rotational constants are independent for the planar molecules) for the corresponding electronic state.

Some conclusions on the excited state dynamics (such as the existence of nonradiative and, particularly, predissociative processes, the estimation of dissociation limit, etc.) can be derived from a detailed consideration of the rovibronic structure and measurements of the lifetimes. It is noteworthy that the observation of the fine rovibronic structure can be hampered and its analysis can be complicated by the presence of several isotopes of the constituent atoms (e.g. Ge, Sn, Pb, Cl and Br) and also by a significant increase in the principal moments of inertia due to the presence of heavy elements in the molecule.

Some numerical characteristics of the observed electronic transitions in triatomic germylenes, stannylenes and plumbylenes of the EX_2 and EXY types are collected in Table 1 and Table 2, respectively. Similar data on the triatomic silylenes are included for comparison in these tables. Some molecular constants of germylidene and its silicon analog are compared in Table 3.

A. Prototype EH_2 Molecules

The laser-induced fluorescence (LIF) excitation and dispersed fluorescence spectra of the $^1B_1-^1A_1$ transition in the prototype germylene, GeH_2, and its deuteriated derivative, GeD_2, were first measured by Saito and Obi under supersonic jet conditions[23–25]. GeH_2 and GeD_2 were produced by 193-nm laser photolysis of $PhGeH_3$ and $PhGeD_3$, respectively. The LIF spectra consisted of progressions of the upper state bending vibrations (ν_2') up to vibrational quantum number 4 for GeH_2 and 6 for GeD_2 in the 650–470 nm region. The term value, T_e (the energy of the vibrational zero point of the upper state relative to that of the ground state), of the $^1B_1$ state has been determined to be 16330 cm^{-1}[23]. The observed predominant excitation of the bending mode is compatible with the large change in bond angle between the ground and the excited states. From the absence of transitions from higher vibrational levels of the $^1B_1$ state it has been suggested that a predissociation channel to Ge(^{1}D) + H_2 (or D_2) opens at *ca* 20000 cm^{-1}[24,25]. Based on these findings the heat of formation of GeH_2 is estimated to be within 19053–19178 cm^{-1}(228–229 kJ mol^{-1})[25]. All vibronic bands had simple rotational structure due to the cooling of the internal degrees of freedom under the jet conditions[23–25]. Isotopic splitting of the rotational lines due to germanium isotopomers was well resolved. The missing rotational r-subbranch in the LIF excitation spectra has been noted, suggesting a heterogeneous (rotational dependent) K_a' dependent predissociation to Ge(^{3}P) + H_2 (or D_2) in the $^1B_1$ state of germylene. The presence of *J*-dependent predissociation has been noted too[25]. Mechanisms for all the predissociation processes have been proposed. The fluorescence lifetimes of single rovibronic levels of the upper state have been measured[25]. It has been found that the lifetimes decrease from *ca* 2.5 to *ca* 0.5 μs with increasing vibrational quantum number. This trend could be related to a nonradiative process. However, the decrease in lifetimes calculated in the framework of this hypothesis deviated from that observed experimentally. From the dispersed fluorescence spectra the bending frequencies of the ground state have been obtained for both GeH_2 and GeD_2[23].

Later[26], in the LIF spectra of jet-cooled GeH_2 (produced from GeH_4 by an electric discharge) a set of additional vibronic bands of the $^1B_1-^1A_1$ transition was recorded, including $^1B_1(1, v, 0)-^1A_1(0, 0, 0)$ progression, with $v = 0$–3. The germanium isotopic splitting was observed for most of the bands. A number of new rotational lines were revealed in the $^1B_1(0, 0, 0)-^1A_1(0, 0, 0)$ band, including a few very weak lines in the $^rR_0$ and $^rQ_0$ branches, terminating on upper state levels involving $K_a' = 1$. It allowed one to determine the ground and the excited state rotational constants and effective (r_0) molecular structure of GeH_2 in both states for the first time. The measured fluorescence lifetimes[26] have been found to be in good agreement with the previous data[25] for lower vibronic levels and noticeably longer for higher levels. The reason of this discrepancy is related to a revealed sensitivity of the observed fluorescence lifetimes to the experimental conditions[26], and the recently measured lifetimes[26] are in much better agreement with the theoretically predicted ones than those reported previously[25]. The mechanisms of predissociation processes have been discussed in detail on the basis of the experimental results and CAS SCF calculations[26].

A small part of the rotationally resolved absorption spectrum of GeH_2 at room temperature in the region of the $^1B_1(0,1,0)–^1A_1(0,0,0)$ transition has been recorded by a laser resonance absorption flash kinetic spectroscopy (LRAFKS)[27]. The most intensive (not assigned) lines (17118.67 and 17111.31 cm^{-1}) have been used further for kinetic studies[27,28]. Later, the room temperature absorption spectrum of GeH_2 (produced in a continuous flow discharge of GeH_4 diluted with argon) was recorded by the intracavity laser absorption spectroscopy (ICLAS) in a wider region (17090–17135 cm^{-1}), corresponding to the central part of the same $^1B_1(0,1,0)–^1A_1(0,0,0)$ transition[29]. The rotational constants for all five germanium isotopomers in the excited state have been obtained from analysis of the observed rotational structure, and the equilibrium geometry, r_e, of GeH_2 in the $^1B_1$ state has been estimated[29]. Recently, laser optogalvanic spectroscopy was used to investigate the central part of the $^1B_1(0,0,0)–^1A_1(0,0,0)$ band of GeH_2 generated from GeH_4 by rf discharge at room temperature[30]. Analysis of the rotational structure of these spectra as well as in LIF spectra of jet-cooled GeD_2 has yielded improved ground and excited state rotational constants for a number of germanium and hydrogen isotopomers of germylene. It allowed us to obtain more accurate r_0 geometries for the ground and the excited states and approximate equilibrium structures for both states of the germylene[30]. Additional $^1B_1(1,v,0)–^1A_1(0,0,0)$ progression, with $v = 0–3$, was observed in LIF spectra of jet-cooled GeD_2, which gave the ν_1' fundamental frequency for the $^1B_1$ state of this molecule. The ground state ν_1'' and ν_2'' vibrational frequencies have also been obtained from dispersed fluorescence spectra of GeH_2 and GeD_2[30].

Electronic spectra of the prototype stannylene, SnH_2, and plumbylene, PbH_2, have not been reported so far.

B. $EHal_2$ Molecules

The first low-resolution ultraviolet absorption spectrum of GeF_2 was reported by Hauge and coworkers[31]. GeF_2 was generated by interaction of GeF_4 with germanium metal at 250 °C and by evaporation of germanium difluoride $(GeF_2)_x$ at 150 °C. The spectrum consisted of $(0,v',0)–(0,v'',0)$ progressions in the region 240–220 nm and was attributed to $^1B_1–^1A_1$ transition. LIF spectrum of jet-cooled GeF_2 produced by reaction of Ge with F_2 at 450 °C was reported later[32]. It represents a poorly resolved band system extending from 231 to 224 nm, corresponding to the same electronic transition[32]. Progressions including both bending and symmetric stretching vibrational modes of the upper state, terminating on some lowest ground state bending mode levels, have been revealed. The vibronic structure was accompanied by a background due to a $^1B_1$ state dissociation process with onset near 225 nm[32]. An unresolved emission band of $^1B_1–^1A_1$ transition in GeF_2, peaking at 235 nm and spanning 265–215 nm, has been observed by irradiation of GeF_4 with synchrotron radiation (SR) with energy above 14 eV[33]. The lifetime of the $GeF_2\,^1B_1$ state measured at 235 nm has been found to be 9.3 ± 0.1 ns[33].

A structureless absorption band with maximum at 146.3 nm, assigned to $^1B_2–^1A_1$ transition in GeF_2 (produced by pyrolysis of GeF_4) was reported by Cole and coworkers[34].

The $^3B_1–^1A_1$ transition in GeF_2 in the 370–325 nm region was first observed using emission spectroscopy[35]. GeF_2 was generated by microwave discharge in GeF_4 vapor. The vibronic structure of this band was analyzed. Emission from the $^3B_1$ state of GeF_2 has also been reported by other groups for GeF_2 produced by rf glow discharge in GeH_4, H_2 and CF_4 or C_2F_6 mixtures[36,37] and in GeF_4 vapor[38]. All the results[35–38] agree well with one another. An unresolved broad chemiluminescence band, spanning 490–270 nm and peaking at 407 nm, was detected in the reaction of Ge with ClF_3 or SF_4 and has also been ascribed to this transition in GeF_2[39]. In the course of studying vacuum-UV fluorescence spectroscopy of GeF_4, an unresolved emission band was observed in the 380–300 nm region with a maximum at 340 nm, corresponding to $^3B_1-^1A_1$ transition in GeF_2 (produced by GeF_4 dissociation at a radiation energy of *ca* 14 eV)[33]. The lifetime of the $^3B_1$ state has been estimated from decay measurements at the band maximum; it was found to be greater than 500 ns[33]. Analysis of the high resolution laser-induced phosphorescence (LIP) spectra of jet-cooled GeF_2 (produced by reaction of Ge with F_2) in the 331–305 nm region[32] has resulted in the correction of the previous[35] assignments of the bands in the vibronic structure of this transition and in a more precise definition of the T_{00} value. Besides, in addition to progressions involving activities of bending modes of both states, two new weaker progressions have been revealed[32]. The first corresponds to $^3B_1(1, v', 0)-^1A_1(0, 0, 0)$ transitions. The second progression should involve the ground state symmetric stretching frequency, but such an assignment should imply an increase in the previous[40] value of this frequency from 692 cm^{-1} to 721 cm^{-1}[32]. At the same time no alternative way of assignment of all the observed progressions has been found[32].

Vibrationally resolved chemiluminescent emission spectrum of $GeCl_2$ in the 490–410 nm region has been obtained by burning $GeCl_4$ in potassium vapor[41]. From analysis of the vibrational structure, bending and symmetric stretching frequencies were obtained for both the ground and excited states of $GeCl_2$[41]. The energy T_{00} was also measured, but the nature of the excited state was left unknown[41]. Later[42], this emission was attributed to the $^3B_1-^1A_1$ transition. The absorption spectrum of $GeCl_2$, generated by evaporation of polymeric germanium dichloride or by interaction of $GeCl_4$ with Ge, with resolved vibrational structure due to bending and symmetric stretching modes in the upper state, was recorded in the 330–300 nm region and assigned to the $^1B_1-^1A_1$ transition using extended Hückel calculations[42]. The continuous absorption starting at about 310 nm was suggested to be due to predissociation of the Ge−Cl bond. In the corresponding emission spectrum, obtained by microwave discharge in $GeCl_4$ vapor, the vibrational structure was not resolved[42]. Unresolved emission bands corresponding to both $^3B_1-^1A_1$ and $^1B_1-^1A_1$ transitions in $GeCl_2$ were observed during the reaction of a high temperature Ge beam with ICl[43] and in the vacuum ultraviolet (VUV) photolysis of Me_nGeCl_{4-n} $(n = 0–2)$[44] and $GeCl_4$[45,46]. In the case of Me_nGeCl_{4-n} photolysis the band in the 560–390 nm region ($^3B_1-^1A_1$) was more intensive than the band in the 370–310 nm region ($^1B_1-^1A_1$), which forced the authors to reassign the longer wavelength band to $^1B_1-^1A_1$ transition in $GeCl_2$, while the shorter wavelength band was attributed to $^1A_2-^1A_1$ transition[44]. However, the proposed[44] assignment did not find further support later[46,47]. The measured[44] lifetime of the upper state responsible for the shorter wavelength band was found to equal *ca* 90 ns, while the longer wavelength band showed a pressure-dependent biexponential decay with zero-pressure components of 17.4 and 101 μs. Such values of lifetimes are typical for the excited singlet and triplet states, respectively.

High resolution laser-induced emission excitation spectra in both regions of $GeCl_2$ transitions (450–400 and 320–300 nm) have been recorded using the supersonic jet technique[47]. $GeCl_2$ was produced by pyrolysis of $HGeCl_3$ at 200 °C. This study represents the first direct LIP detection of the excited triplet state of any carbene or carbene analogs. More accurate definitions have been given to the band origins (T_{00}) and fully symmetric vibration frequencies in all the involved electronic states of $GeCl_2$. In spite of small rotational constants of the molecule under consideration and overlapping of the lines due to different germanium and chlorine isotopomers, the authors have succeeded in observing poorly resolved rotational band contours for a number of bands of shorter-wavelength band systems. It has also been shown that the part of the spectrum below 316 nm was mainly caused by emission of GeCl, a product of a predissociative process for $GeCl_2$, rather than by $GeCl_2$ itself. All the assignments made have been supported by *ab initio* calculations[47].

Interaction of K or Na vapor with $GeCl_4$ is accompanied by a very weak chemiluminescence in the even longer wavelength region (666–560 nm)[48]. Its emitter has been proposed to be $GeCl_2$, because two frequencies obtained from analysis of the vibrational structure of the band coincided with fundamental frequencies of the full-symmetric (stretching and bending) modes of the ground state of $GeCl_2$. However, further attempts to observe this emission system failed[43,47], and the nature of this band remains unknown.

Reactions of Ge atoms (in both the ground ^3P_J and metastable $^1D_2$ states) with Br_2 and I_2 produced structureless emission bands with maxima at *ca* 480, 380 and 600, 500 nm respectively, which have been assigned to $^3B_1{-}^1A_1$ (longer wavelength bands) and $^1B_1{-}^1A_1$ (shorter wavelength bands) transitions in $GeBr_2$ and GeI_2, respectively[43]. The low resolution absorption spectrum of GeI_2 vapors contains bands with maxima at 575, 475, 360 and 225 nm[49,50].

Only low resolution absorption spectra have been recorded for SnF_2 (in the 246–237 nm region) and PbF_2 (band maximum at 243.5 nm) in the region of $^1B_1{-}^1A_1$ transitions[51]. Monomeric SnF_2 and PbF_2 were produced by evaporation of the corresponding difluorides $(EF_2)_x$ at *ca* 700 and *ca* 800 K, respectively. An analysis of the vibronic structure observed only on the longer wavelength side of the SnF_2 band gave values for the bending frequencies in the ground and in the excited states and the T_{00} value. The band of PbF_2 was not resolved vibrationally, but the T_{00} value has been estimated, assuming an analogy in relative positions of the band maxima and the (0,0,0)–(0,0,0) transitions for EF_2, E = Ge, Sn, Pb. A broad emission band with maximum at *ca* 400 nm observed in beam reactions of $Sn(^3P)$ atoms with ClF_3, SF_4 and SF_6 was attributed to the $^3B_1{-}^1A_1$ transition in SnF_2[39].

$^3B_1{-}^1A_1$ transition in $SnCl_2$ (generated by electric discharge in $SnCl_4$ vapor) has been detected in the 500–400 nm region as an emission band with clear vibronic structure[8] and as chemiluminescence (produced in reactions of $SnCl_4$ with nitrogen or hydrogen atoms[52] or by burning of $SnCl_4$ in potassium vapor[53]). An unresolved absorption band of $SnCl_2$ (as a vapor over the molten salt at about 570 K) with maximum at 322 nm has been assigned to the $^1B_1{-}^1A_1$ transition[42]. A weak emission band due to this transition was recorded in the course of the reaction of Sn atoms in the $^1D_2$ state with Cl_2[43]. Beam-gas reactions of Sn atoms in the ground ^3P_J and in the metastable $^1D_2$ states with ICl result in the appearance of two intensive chemiluminescence bands peaking at 350 and 425 nm, which correspond to $^1B_1{-}^1A_1$ and $^3B_1{-}^1A_1$ transitions, respectively[43]. Structureless absorption bands with maxima at 245 and 195 nm were also observed for $SnCl_2$ vapors[42,54]. These absorptions and an absorption below 320 nm within the region of the $^1B_1{-}^1A_1$ transition lead to dissociation of $SnCl_2$ with the formation of SnCl and Cl in the ground and excited states[42,54–56].

Beam-gas reaction of Sn atoms (in the ^3P_J state) with Br_2 and I_2 results in chemiluminescence continuous bands peaking at 470 and 553 nm respectively, attributed to $^3B_1{-}^1A_1$ transitions in $SnHal_2$(Hal = Br, I) molecules[39]. Excitation of $SnBr_2$ vapors with $N_2(A^3\Sigma_u{}^+)$ also produces luminescence in the visible region (in the range of 550–440 nm) with a maximum at 505 nm[57]. Low resolution absorption spectra recorded for $SnBr_2$ (at 608–749 K) and SnI_2 (at 593–920 K) vapors revealed bands peaking at *ca* 480, 365, 285, 245 and 200 nm for $SnBr_2$[49,54] and at 550, 458, 358, 225 and below

TABLE 1. Energies, geometries and fundamental frequencies of triatomic carbene analogs of EX_2 type in ground and excited states[a,b]

	Electronic state	Region of transition from or to the ground state (nm)	T_{00} (cm^{-1})	ν_1 (cm^{-1})	ν_2 (cm^{-1})	Bond length, r_0 (Å)	Bond angle, θ_0 (deg)	Reference
SiH_2	$^{1}A_1$	—	0		1 009			62
						1.51402[c]	91.9830[c]	63
	$^{3}B_1$	—	4 800					64
			6 300 or 7 300					65
	$^{1}B_1$	650–460	15 547.7730		856.53	1.48532[c]	122.4416[c]	63
				1 990				66
SiD_2	$^{1}A_1$	—	0		731			62
						1.515	92.12	67
	$^{1}B_1$	640–460	15 539.8751			1.483	123.2	67
					616			62
GeH_2	$^{1}A_1$	—	0	1 856	916	1.5934	91.28	30
						1.5883[c]	91.22[c]	30
	$^{1}B_1$	650–470	16 325.544	1 798	783			26
						1.5564	123.02	30
			16 312					23–25
						1.5422[c]	122.82[c]	29
						1.5471[c]	123.44[c]	30
GeD_2	$^{1}A_1$	—	0	1 335	657			30
	$^{1}B_1$	650–470	16 323					23–25
			16 324.3387	1 304	561			30
SiF_2	$^{1}A_1$	—	0	853	344			68
	$^{3}B_1$	420–360	26 319.478		278.2	1.586	113.1	68
	$^{1}B_1$	280–210	44 113.9		250	1.601	115.9	69
	$^{1}B_2$	165–155	62 278–62 281		320			34
				795				70
$SiCl_2$	$^{1}A_1$	—	0	521.6	200.6	2.068	101.5	71
	$^{3}B_1$	640–500	18 943		157.4			72
			18 473					73
	$^{1}B_1$	360–290	30 013.5	428.9	149.8	2.032	123.4	71
$SiBr_2$	$^{1}A_1$	—	0					
	$^{3}B_1$	550–450	~18 000[d]					43
	$^{1}B_1$	400–340	~25 000[d]					43, 74
SiI_2	$^{1}A_1$	—	0					
	$^{1}B_1$	550–450	~18 000[d]					43
GeF_2	$^{1}A_1$	—	0	721[e]	263			32
	$^{3}B_1$	370–300	30 582.7	673.3	191.3			32
							112	35
	$^{1}B_1$	245–220	43 860.9	625.2	160.6			32
	$^{1}B_2$	156–136	~68 000					34
$GeCl_2$	$^{1}A_1$	—	0	391	159			47
	$^{3}B_1$	500–400	22 315.0	393	118			47
	$^{1}B_1$	370–300	30 622.0	354	104			47
$GeBr_2$	$^{1}A_1$	—	0					
	$^{3}B_1$	530–430	~19 000[d]					43
	$^{1}B_1$	440–340	~23 000[d]					43

(*continued overleaf*)

TABLE 1. (*continued*)

	Electronic state	Region of transition from or to the ground state (nm)	T_{00} (cm^{-1})	ν_1 (cm^{-1})	ν_2 (cm^{-1})	Bond length, r_0 (Å)	Bond angle, θ_0 (deg)	Reference
GeI_2	$^1A_1$	—	0					
	$^3B_1$	650–550	$\sim$15 000^d					43, 49
	$^1B_1$	550–450	$\sim$18 000^d					43, 49
SnF_2	$^1A_1$	—	0		180			51
	$^3B_1$	500–300	$\sim$20 000^d					39
	$^1B_1$	246–237	40 741		120			51
$SnCl_2$	$^1A_1$	—	0	350	120			53
	$^3B_1$	500–400	22 249	240	80			53
	$^1B_1$	370–270	$\sim$27 000^d					43
$SnBr_2$	$^1A_1$	—	0					
	$^3B_1$	550–370	$\sim$18 000^d					39
SnI_2	$^1A_1$	—	0					
	$^3B_1$	600–520	$\sim$17 000^d					39
PbF_2	$^1A_1$	—	0					
	$^1B_1$	$\sim$243.5	$\sim$40 560					51
PbI_2	$^1A_1$	—	0	168	44			60
	$^3B_1$	525–475	20 200	149				60

^{a}The data correspond to isotopomers, containing the most abundant isotopes, or represent the average values if the isotopic structure has not been observed.
^{b}The number of the significant digits in the values exhibited corresponds to that presented in the original publications.
^{c}Equilibrium geometry (r_e, θ_e).
^{d}A rough estimate from the longer wavelength limit of the band.
^{e}Tentative value.

200 nm for SnI_2[49,54,58,59]. Thermal luminescence for SnI_2 (in the 1125–1423 K range) consisted of bands with maxima at 615, 570 and 500 nm[59]. Tentative interpretations of the observed spectra have been proposed[49,59]; in particular the longest wavelength absorption bands have been assigned to the $^1B_1-^1A_1$ transition. Interpretation of the absorption spectra of $SnHal_2$ (and $PbHal_2$) taking into account dissociation processes was first conducted by Samuel[56]. The ultraviolet absorption cross sections of $SnHal_2$(Hal = Cl, Br, I) in the 400–200 nm region were measured and ways of photodissociation of these molecules depending on the absorbed light wavelength in this region were considered[54].

In the absorption spectra of $PbCl_2$ vapors three continuous bands peaking at 360, 320 and below 291.6 nm have been observed[9,10,42]. The first band has been attributed to $^1B_1-^1A_1$ transition and the others to dissociation continuum[9,10,42,56]. A number of structureless bands with maxima at about 450, 330 and 230 nm and at about 530, 430, 300 and 200 nm have been found in the low resolution absorption spectra of $PbBr_2$ and PbI_2 vapors, respectively[49]; the first bands were tentatively (and obviously erroneously) attributed to $^1B_1-^1A_1$ transitions.

Vibrationally resolved LIP and dispersed fluorescence spectra in the 525–475 nm region have been recorded for the heaviest triatomic plumbylene PbI_2 in the 670–1170 K temperature interval[60]. The observed band system was assigned to the $^3B_1-^1A_1$ transition, based on SCF-Xα-SW calculations. The frequencies of the bending mode of the ground state and the symmetric stretching modes of both the ground and the excited states of

PbI_2 have been obtained from vibronic structure analysis together with the T_{00} value[60]. Eight well-resolved emission bands in the 710–395 nm region have been observed under photoexcitation of PbI_2[61]. Some of them can belong to vibrational progressions of the $^3B_1-^1A_1$ transition in PbI_2.

C. EHHal Molecules

Electronic transitions in triatomic CAs of the EXY type have been reported only for monohalogermylenes and monohalosilylenes. The latter are beyond the scope of the current review. The $^1A''-^1A'$ transition in GeHCl and GeDCl was first detected in the 520–445 nm region as chemiluminescence, produced by interaction of GeH_4 or GeD_4 with Cl_2[75]. An analysis of the vibronic structure of the band has been carried out and the bending and Ge–Cl stretching frequencies in the ground and in the excited states have been determined. The electronic spectra of HGeCl[76] and HGeBr[77] produced at ambient temperatures by the reaction of GeH_4 with chlorine and bromine atoms have been reported in LIF studies. The spectra represented progressions of the bending modes in the $^1A''$ excited state and included activity of the bending mode in the ground $^1A'$ state in the case of HGeBr. From the analysis of the observed rotational structure the bond angles in both states have been estimated for HGeCl as well as HGeBr.

The LIF spectra of jet-cooled HGeCl, HGeBr, HGeI and their deuterium isotopomers in the region of the $^1A''-^1A'$ transition have been recorded using the pulsed discharge of the corresponding monohalogermanes[78,79]. All three excited state vibrational frequencies have been measured for each molecule. Analysis of the rotational structure of the (0,0,0)–(0,0,0) transition gave the rotational constants for both the states of all the molecules. However, direct determination of the geometrical parameters from the rotational constants appeared to be impossible due to the large correlation found between the Ge–H bond lengths and the bond angles. Finally, the approximate r_0 structures for both states of HGeHal molecules have been obtained when the bond angles were constrained to previous *ab initio* values[80]. The lifetimes of the $^1A''$ states of HGeHal and DGeHal were found to equal 548 ns for GeHCl, 527 ns for GeDCl, 736 ns for GeHBr and 733 ns for GeDBr[78] and to be dependent on the rovibronic state for HGeI and DGeI and lying within 1.5–2.3 μs for the vibrationally unexcited and monoexcited levels of the $^1A''$ state[79], which implies existence of some nonradiative processes for the latter germylenes. Comparison of the lifetimes of $^1A''$ states of the analogous silylenes[81–83] and germylenes shows that the lifetimes increase with increasing mass of both the central atom and the halogen atom[78,79].

Attempts to detect HGeF in the reaction of GeH_4 with fluorine atoms or in a pulsed discharge of H_3GeF using the LIF technique were unsuccessful[78] and it has been concluded that HGeF either cannot be obtained by these methods or does not fluoresce in the excited state[78].

D. Comparison of the Molecular Constants of the Triatomic Carbene Analogs

In accordance with the data of Tables 1 and 2 the parent dihydrides EH_2(E = Si, Ge) have the lowest energy of transition to the first excited singlet state among the triatomic CAs. Consecutive introduction of halogen substituents into a molecule of CA results in the gradual increase of this energy, which is obviously associated with lowering the energy of the HOMO in the ground state due to admixing halogen p atomic orbitals. Apparently, this conclusion is also valid for transitions to the first excited triplet state. The same mechanism is responsible for variations of the singlet–triplet splitting in triatomic carbenes[88]. In both series of mono- and dihalogenides the S_1-S_0 and T_1-S_0 energy gaps decrease with increasing mass of the halogen atoms, i.e., on decreasing the substituent electronegativity,

TABLE 2. Energies, geometries and fundamental frequencies of triatomic carbene analogs of EXY type in ground and excited states[a,b]

	Electronic state	Region of the transition (nm)	T_{00} (cm^{-1})	ν_1 (cm^{-1})	ν_2 (cm^{-1})	ν_3 (cm^{-1})	r_0(E-H) (Å)	r_0(E-X) (Å)	Bond angle θ_0 (deg)[c]	Reference
SiHF	$^1A'$		0		859					84
							1.548	1.606	97.0	85
							1.528[d]	1.603[d]	96.9[d]	85
	$^1A''$	430–390	23 260.021	1 546.9	558.4	856.9				81
							1.557	1.602	114.4	85
							1.526[d]	1.597[d]	115.0[d]	85
SiDF	$^1A'$		0		643					85
	$^1A''$	455–380	23 338.723	1 174.3	424.8	854.4				85
SiHCl	$^1A'$		0	1 968.8	805.9	522.8	1.5214	2.0729	95.0	82
							1.525[d]	2.067[d]	96.9[d]	82
	$^1A''$	600–410	20 717.769	1 747.1	563.9	532.3	1.505	2.047	116.5	82
							1.532[d]	2.040[d]	118.1[d]	82
SiDCl	$^1A'$		0	1 434.4	592.3	518.1				82
	$^1A''$	600–410	20 773.431	1 300.8	408.6	543.2				82
SiHBr	$^1A'$		0	1 976.2	771.9	412.4	1.518	2.237	93.4	83
							1.522[d]	2.231[d]	95.9[d]	83
	$^1A''$	600–410	19 902.851	1 787.0	535.3	416.5	1.497	2.208	116.4	83
							1.546[d]	2.199[d]	119.4[d]	83
SiDBr	$^1A'$		0	1 439.5		408.0				83
	$^1A''$	600–410	19 953.677	1 325.6	376.4	434.3				83

SiHI	$^{1}A'$		0			350.0	1.534	2.463	92.4	86
					727					87
	$^{1}A''$	550–470	18 259.020	1 852.5	485.0	335.7	1.515	2.436	114.9	86
SiDI	$^{1}A'$		0							
	$^{1}A''$	560–460	18 302.96	1 367	367.9	324.8				86
GeHCl	$^{1}A'$		0			441.9	1.592	2.171	(94.3)	78
					689.2					75
	$^{1}A''$	520–430	21 514.68	1 262.9	431.2	398.8	1.613	2.146	(114.5)	78
GeDCl	$^{1}A'$		0							
	$^{1}A''$	470–430	21 614.48	979.6	321.7	398.9				78
GeHBr	$^{1}A'$		0		695					77
						288.7	1.598	2.329	(93.9)	78
	$^{1}A''$	500–440	20 660.30	1 380.8	419.3	280.3	1.615	2.308	(116.3)	78
GeDBr	$^{1}A'$		0							
	$^{1}A''$	485–440	20 746.43	1 047.6	312.1	278.5				78
GeHI	$^{1}A'$		0				1.593	2.525	(93.5)	79
	$^{1}A''$	530–490	18 929.294	1 542.5	337.1	203.4	1.618	2.515	(116.2)	79
GeDI	$^{1}A'$		0							
	$^{1}A''$	530–490	19 001.759	1 144.2	259.7	205.0				79

[a] The data correspond to isotopomers, containing the most abundant isotopes, or represent the average values if the isotopic structure has not been observed.
[b] The number of the significant digits in the values exhibited corresponds to that presented in the original publications.
[c] The bond angles given in parentheses were taken from quantum-chemical calculations and were used in calculations of the bond lengths from the rotational constants.
[d] Equilibrium geometry (r_e, θ_e).

which is in accord with a qualitative theory[89]. The effect of the central atom on the energy separation is not obvious, perhaps partly due to the lack of reliable data.

Geometries of some triatomic CAs in both the ground and excited states have been obtained from the rotational constants determined from analysis of the rotational structure of the electronic bands (see Tables 1 and 2). The ground state geometries will be considered in Section VIII. Due to lack of experimental data it is difficult to find any trend in the central atom or substituent effects on the bond angles or bond lengths in the triatomic CAs in the excited states. However, it is clear that these effects are less pronounced in the excited states than in the ground state. Comparison of structural parameters for the triatomic CAs in the ground and excited states shows that, in general, electronic excitation results in slight contraction of valence bonds (except for the E−H bonds in EHHal) and in significant increase of bond angle, with the latter trend being in agreement with electrostatic force theory[90,91].

In accordance with the selection rules, only excitation of full-symmetric vibrational modes can accompany the electronic transitions. One can conclude that the vibrational progressions for electronic transitions in the triatomic CAs more often result from excitation of bending modes. It is caused by significant change in the bond angle at the excitation.

As can be seen from Tables 1 and 2, electronic excitation is usually accompanied by lowering of the fundamental frequencies, with the fundamental frequencies in the first excited singlet state being lower than those in the first triplet state. Apparently, diminution of the bending frequencies is related to the increase of bond angle in the excited states, whereas lowering the stretching frequencies reflects weakening of the bonds in spite of their shortening.

E. Intrinsic Defects in Solid Silicon and Germanium Dioxides

In the final part of the treatment of the electronic spectra of triatomic germylenes, stannylenes and plumbylenes it is worth mentioning studies of intrinsic defects on the surface and in the bulk of solid SiO_2 and GeO_2, which have been discussed in detail in a review[92]. Among the different types of defects, active centers, containing two coordinated silicon and germanium, have been revealed. The reactivity of these centers closely resembles that of dihalosilylenes and dihalogermylenes. The optical properties of such unusual silylenes and germylenes have also been found to be similar to those of difluorosilylene and difluorogermylene. Indeed, the absorption maxima of silylene- and germylene-type defects on the surface lie at 243–234 nm and at *ca* 230 nm respectively, whereas the emission maxima are at *ca* 460 nm (T_1-S_0 transition, the lifetime of the upper state is 19 ms) and 285 nm (S_1-S_0 transition, the lifetime of the upper state is 5.1 ns) for the silylene[93–97] and at 390 nm (T_1-S_0 transition, the lifetime of the upper state is 145 μs) and at 290 nm (T_1-S_0 transition) for the germylene[92]. The absorption maxima of the silylene and germylene defects in the bulk are at 248 and 243 nm, respectively, and the emission maxima are at 459 (T_1-S_0 transition, the lifetime of the upper state is *ca* 10 ms) and 282 nm (S_1-S_0 transition) for the silylene and at 400 (T_1-S_0 transition, the lifetime of the upper state is *ca* 100 μs) and 288 nm for the germylene[98]. Both emission bands of these silylenes and germylenes arise under their excitation into the S_1 state, which suggests the existence of nonradiative S_1-T_1 transition. This transition has an activation barrier because, depending on the temperature, the major luminescence channel can be either S_1-S_0 or T_1-S_0 transition. A plausible mechanism for the S_1-T_1 conversion has been suggested[92]; it involves the second triplet state and explains the existence of the activation barrier by the necessity of linearization of the carbene analogs in the course of the conversion.

TABLE 3. Energies, geometries and fundamental frequencies for the $X_2C = E$ (X = H, D; E = Si, Ge) molecules in ground and excited states

	Electronic state	Region of the transition (nm)	T_{00} (cm^{-1})	ν_1 (cm^{-1})[a]	ν_2 (cm^{-1})[b]	ν_3 (cm^{-1})[c]	ν_4 (cm^{-1})[d]	ν_6 (cm^{-1})[e]	r_0(E–C), (Å)	r_0(C–H), (Å)	θ_0(HCH), (deg)	Reference
$H_2C = Si$	$^1A_1$	—	0	—	—	—	—	—	1.706	1.105	114.4	101, 102
	$^1B_2$	342–300	29 319.875	2 997	1 102	702	697	731	1.814	1.087	134.0	101, 102
$D_2C = Si$	$^1A_1$	—	0	—	—	—	—	—				
	$^1B_2$	342–300	29 237.348	2 180	831	681	547	549				101, 102
$H_2C = Ge$	$^1A_1$	—	0	—	—	—	—	—	1.7908	1.1022	115.05	101
	$^1B_2$	367–345	27 330.423	—	1 024	548	613	692	1.914	1.082	139.3	101
$D_2C = Ge$	$^1A_1$	—	0	—	—	—	—	—				
	$^1B_2$	367–345	27 278.450	—	782	528	465	513				101

[a] C–H stretching; [b] CH_2 scissoring; [c] E–C stretching; [d] Out-of-plane bending; [e] CH_2 rocking.

F. Tetraatomic Carbene Analogs: Germylidene and Silylidene

The first observation of 1-germavinylidene or germylidene, $H_2C{=}Ge$, and its deuteriated derivative, $D_2C{=}Ge$, have been reported in 1997[99]. Both were produced in a pulsed electric discharge jet using $GeMe_4$ and $Ge(CD_3)_2$ as precursors. An attempt to obtain stannylidene by using the same method was unsuccessful[99]. The identification of the germylidenes was based on both the detailed analysis of their LIF spectra, recorded in the 367–345 nm region, and comparison of the observed spectral features with those predicted by quantum-chemical calculations[99]. The observed band system has been assigned to the allowed transition from the second excited singlet electronic state $\tilde{B}^1B_2$ [with the $\ldots(11a_1)^1(4b_1)^2(5b_2)^1$ electron configuration, where $11a_1$ MO is the lone electron pair on the Ge atom, $4b_1$ MO is the Ge−C π-bonding orbital and $5b_2$ MO is essentially the in-plane p_y orbital on Ge[100]] to the ground $\tilde{X}^1A_1$ singlet state [with the $\ldots(11a_1)^2(4b_1)^2$ electron configuration]. The $^1A_2{-}^1A_1$ transition from the first excited singlet state $\tilde{A}^1A_2[\ldots(11a_1)^2(4b_1)^1(5b_2)^1]$ is forbidden by the selection rules. Further study of germylidenes enriched with the ^{74}Ge isotope[101] allowed one to analyze in detail the rotational structure of the vibronic subband, corresponding to the transition between vibrationally unexcited levels of the electronic states, and to obtain geometries of the germylidene in the ground and in the excited states. The molecule has been found to be planar and to have C_{2v} symmetry, as expected. Some characteristics of the germylidene and its silicon analog obtained from analysis of their LIF spectra are compared in Table 3. As can be seen from Table 3, the T_{00} energy for germylidene is lower than that for silylidene, which is in agreement with the results of quantum-chemical calculations[99,100]. The structures of germylidene and silylidene closely resemble each other in both the ground and excited states. Electronic excitation results in elongation of the E=C double bond and in a slight contraction of the C−H bonds and opening of the CH_2 angle. The lifetime of the vibrationally unexcited level of the upper state of germylidene has been measured to be *ca* 2 μs and depends on the rotational level[101]. Many rovibronic levels of germylidene reveal quantum beats in their fluorescence decays, originating from interaction of the excited state with some other states, mainly with the ground state. The same effect has been found for almost all rovibronic levels of the upper state of silylidene[102].

IV. ELECTRONIC SPECTRA OF POLYATOMIC GERMYLENES, STANNYLENES AND PLUMBYLENES

This section is devoted to the electronic spectra of germylenes, stannylenes and plumbylenes carrying polyatomic substituents. The available data on the absorption maxima of these species are collected in Table 4, from which it can be seen that only stable polyatomic stannylenes and plumbylenes have been characterized by UV spectroscopy.

A. Detection and Identification

Most of the short-lived germylenes were generated by photolysis of suitable precursors in low-temperature (77 K) hydrocarbon glasses or by flash photolysis in solutions at room temperature and characterized by low-resolution absorption UV spectroscopy, using a fast response probe technique in the last case. With rare exception, their UV spectra consisted of a single broad absorption band. Because the position of this band is the only measured quantity which can be obtained from the low-resolution UV spectrum, and it is not sufficient for unequivocal identification of the absorbing species, these studies are usually accompanied by experiments on chemical trapping of the germylenes. Chemical trapping experiments were carried out under photolysis of the same precursor in solutions in the presence of suitable scavengers to prove predominant formation of the germylene upon photodecomposition of the chosen precursor, and also after melting the hydrocarbon glasses doped with a trapping agent to prove the presence of the germylene in the glass. Thus, the assignment of the observed band to the particular species was mostly based on the stoichiometry of the precursor photodecomposition (from analysis of stable end products) and a chemical proof of the intermediacy of the germylene. Besides that, in some studies several different precursors were used for generation of the same germylene.

UV spectra of the two germylenes, dimethylgermylene, $GeMe_2$[103,104] and 1-germacyclopent-3-en-1-ylidene[105], were recorded in low-temperature (11–21 K) Ar matrices. Their UV detection was supplemented by studying IR bands of these germylenes in the same matrices[104,105] (see Section VI), which significantly enhanced the reliability of the UV identification. Reversible phototransformations of the 1-germacyclopent-3-enylidene into the isomeric 1-germacyclopent-2,4-diene (germole) and 1-germacyclopent-1,3- and 1,4-dienes have also been observed[105]. This allowed one to establish unambiguously the correspondence of the UV and IR bands assigned to the germylene by observing simultaneous changes in their intensities.

The gas-phase UV spectrum of $GeMe_2$ generated from two different precursors has been recorded by laser resonance absorption flash kinetic spectroscopy[106]. The identity of the germylene was supported by chemical trapping experiment and by analysis of the stable end products of the decomposition of the chosen precursors.

Most of the direct UV detection of germylenes in the liquid phase as well as the detection of $GeMe_2$ in the gas phase were followed by kinetic studies of the germylene reactions. Unfortunately, the obtained liquid-phase rate constants show a large scatter and differ appreciably from those obtained in the gas phase. Thus, the rate constants of the germylene reactions can hardly be used to confirm the identity of the germylene in the liquid phase.

UV spectra of stable germylenes, stannylenes and plumbylenes were recorded in solutions or in the solid phase under inert atmosphere in order to prevent access of oxygen and moisture. Thermochromic transitions have been revealed for many stable solid germylenes, stannylenes and plumbylenes. Their nature has not been studied.

B. Complications

Generation of short-lived germylenes is often accompanied by formation of other short-lived species. Germyl radicals with absorption maxima in the 300–350 nm region were identified upon photodecomposition of trigermanes[107,108], linear oligogermanes $Me(Me_2Ge)_5Me$[109] and polygermanes $(R_2Ge)_n, n = 10{-}100$[110], phenyl-substituted digermanes[111] and germatrisilacyclobutanes[112]. A short wavelength band detected besides the $GePh_2$ band[113] upon photolysis of $(Me_3Si)_2GePh_2$ was also attributed to a radical species[107]. In general, the presence of aromatic substituents in the precursor molecule seems to favor the formation of germanium centered radicals[114,115]. A similar effect of aromatic substituents has been also noted for silylene precursors[116,117]. A check of the presence of radical species in hydrocarbon glasses by ESR spectroscopy has been used in some studies[103,109,118–120]. Germylene generation from arylgermanes can be accompanied by formation of germaethenes (germenes)[114,115,121]. In the course of photolysis of 7,7-dimethyl-1,2,3,4-tetraphenyl-6,6-benzo-7-germanorbornadiene, a known photochemical precursor of $GeMe_2$, another labile product has been detected and tentatively identified as the corresponding germanorcaradiene[122], a product of isomerization of the precursor. Dimerization of germylenes leading to formation of digermenes is a process often observed in solutions and upon annealing of hydrocarbon glasses. The list of labile by-products which accompany formation of short-lived germylenes, stannylenes and plumbylenes will probably be expanded in the future. The presence of such species in the systems under investigation represents the most seriously interfering factor in detection and identification of UV absorptions of polyatomic germylenes, stannylenes and plumbylenes.

The majority of the studies cited here have produced forcible arguments for the performed assignment of the detected absorption to the particular short-lived germylene. However, the absorption maxima observed for the same germylene formed from different precursors differ remarkably from each other (Table 4). The largest scatter in the reported absorption maxima (from 380 to 506 nm) exists for $GeMe_2$, which UV spectrum has been studied in the greatest number of studies. Such discrepancies can partly be explained by low precision in determination of the maximum for typically broad bands of low intensity belonging to short lived germylenes (for stable germylenes, stannylenes and plumbylenes the extinction coefficient is usually $<1000–2000\ M^{-1}\ cm^{-1}$) and also by medium effects. In the case of low-temperature hydrocarbon glasses, the influence of the by-products of precursor photodecomposition, which remain in the same cage of the hydrocarbon matrix, can be the main factor determining the band shift[119]. A special type of such influence, which can affect the UV band position in both liquid and solid phases, is the formation of a CA complex with by-products (and precursors) possessing Lewis basicity. The effect of complexation on the UV absorptions of germylenes, stannylenes and plumbylenes will be considered in detail in Section V. A series of labile complexes of germylenes, stannylenes and plumbylenes has also been detected by IR spectroscopy and they will be considered in Section VII. The important conclusions which can be drawn from these considerations, are that (i) UV absorptions of germylenes, stannylenes and plumbylenes seem to be very sensitive to complexation, (ii) complexation usually results in a blue shift of the UV absorption, and (iii) many classes of organic compounds (aromatics, unsaturated organic compounds, organic derivatives of the Group 15, 16 and 17 elements etc.) could form complexes with CAs. Thus, complex formation is expected to be a very common reaction accompanying generation of short-lived polyatomic germylenes. The reasons mentioned above can explain qualitatively most disagreements in the reported UV band positions of short-lived germylenes, but not all of them. For example, it can hardly be understood in the framework of these reasons why the absorption of $GeMe_2$, generated from cyclo-$(GeMe_2)_5$ should be strongly red-shifted relative to its absorption upon its generation from cyclo-$(GeMe_2)_6$ or $Me(GeMe_2)_5Me$ under the same conditions (Table 4). Obviously, the reasons for the discrepancy in the UV band positions of short-lived polyatomic germylenes and the differences in their reaction kinetics, which were mentioned above, are of the same nature, and these reasons are not yet completely understood.

C. Nature of Electronic Transitions in Polyatomic Germylenes, Stannylenes and Plumbylenes

It follows from numerous theoretical studies that the ground electronic state of all CAs (except those with very peculiar substituents, like Li atoms[123,124] or extra-bulky t-Bu_3Si groups[125]) is the singlet state, in which the HOMO is typically a lone pair of the divalent atom E (E = Si, Ge, Sn, Pb) and the LUMO is essentially a p orbital of the same atom. The promotion of the lone electron pair to the empty p orbital corresponds to the first allowed electronic transition. The absorption maxima (presented in Table 4) of germylenes, stannylenes and plumbylenes (the longest wavelength absorption maxima for those possessing more than one absorption band in the near UV-vis region) with alkyl, aryl and silyl substituents lie in the usual region for such $n \rightarrow p$ transitions and therefore can surely be assigned to this transition. For series of polyatomic silylenes[126–128] and

$GeMe_2$[129] such an assignment has been supported by *ab initio* calculations of the corresponding transition energies, which were found to be in reasonable agreement with the experimentally observed band positions.

It will be seen in Section IX that in the case of germylenes, stannylenes and plumbylenes bearing two amino-substituents[130–133] there are three high-lying occupied MOs which are close in energy. The lone pair of the divalent germanium, tin or lead atom lies in energy between two MOs, which represent a nitrogen lone-pair orbital antibonding (HOMO) and bonding combinations. A similar order of the highest occupied MOs has also been noted for $GeCl(N(SiMe_3)_2)$[134]. Thus, one can expect that for such germylenes, stannylenes and plumbylenes (Table 4) the second band from the red edge of the spectrum corresponds to the excitation of an electron from the lone pair of the divalent atom of the Group 14 element. In the case of 1,3-dineopentylpyrido[*b*]-1,3,2λ^2-diazasilole, -germole and -stannole only the third MO corresponds to the divalent atom lone pair[135]. Two higher-lying MOs represent combinations of nitrogen atom lone pairs and the π orbital of the aromatic ring. Calculations at the CIS level with ECP DZ basis of the MO structure of diphosphanyl- and diarsanyl-substituted germylenes, stannylenes and plumbylenes has shown that the three highest occupied MOs in these molecules represent combinations of lone-pair orbitals of the divalent germanium, tin or lead atom and phosphorus or arsenic atoms with some admixture of p-AO of the divalent atom of the Group 14 element, while the LUMO corresponds to the p orbital of the divalent atom[136]. Therefore, two or three observed UV bands for experimentally studied diphosphanyl- and diarsanyl-substituted germylenes, stannylenes and plumbylenes (Table 4) have been assigned to electronic transitions from these MOs of mixed character to the p orbital[136].

To the best of our knowledge there are no published studies of the MO structure of CAs of $E(XR)_2$ type, where X = O, S, Se, Te. Consequently, the assignment of UV bands of these molecules is not clear, but it is usually assumed that the HOMO of such CAs is the lone pair of the divalent atom of the Group 14 element. By analogy with silylenes[128], one can expect that the HOMO of germylenes, stannylenes and plumbylenes containing only one substituent with a lone pair at the α-atom (which belongs to the Group 15, 16 or 17 elements) is the lone pair of divalent germanium, tin or lead atom, and the longest wavelength electronic transition in these species is the $n \rightarrow p$ transition.

D. Effects of Substituents on the Position of the Electronic Transition Band

The nature of substituent effects on the position of absorption maxima corresponding to $n \rightarrow p$ transition in polyatomic silylenes have been considered in detail in theoretical studies of Apeloig and coworkers[126,128]. Unfortunately, there is no similar study for germylenes, stannylenes or plumbylenes. However, the available experimental data show that the main conclusions obtained by Apeloig and coworkers are also applicable to polyatomic germylenes, stannylenes and plumbylenes.

Substituent effects can be divided into two types: electronic and steric. The steric effect results from steric repulsion of bulky substituents, and destabilizes the ground electronic state and simultaneously results in an increasing bond angle at the divalent atom of the Group 14 element. Because the bond angle in the first excited singlet state is much larger for CAs (see experimental data for triatomic systems in Section III and the theoretical data for $GeMe_2$ by Barthelat and coworkers[129]), the repulsion of bulky substituents will destabilize this state to a much lower extent. Therefore, the energy of the $n \rightarrow p$ electronic transition is expected to decrease on increasing the ground state bond angle for a series of CAs bearing substituents with similar electronic properties at the same divalent atom E (E = Si, Ge, Sn, Pb). Assuming essential resemblance of the potential energy curves for the first excited singlet states for such CAs, one can expect that the red shift of the experimentally observed vertical electronic transitions will be even more pronounced. Simple MO consideration leads to the same conclusion. Increasing the bond angle in the ground electronic state of a CA results in an increase of the degree of hybridization of the divalent atom of the Group 14 element and therefore in a rise of the HOMO (the divalent atom lone pair) energy and lowering of the LUMO (p orbital of this divalent atom) energy in accordance with Bent's rule[137] and thereby lowering the energy of the corresponding electronic transition.

Dependence of the absorption maximum on the bond angle can be illustrated by data on stable germylenes, stannylenes and plumbylenes containing alkyl and aryl substituents (Table 5). It is noteworthy that according to the X-ray analysis, the aromatic rings in the aryl-substituted germylenes, stannylenes and plumbylenes presented in Table 5 are rotated out of the plane of the carbene center. This excludes conjugation effects and allows one to consider alkyl- and aryl-substituted compounds together. It can be seen that there is a clear parallel between the values of the bond angles and the positions of the UV bands for these stable species. Variations in band positions depending on the bond angle are most prominent in the case of germylenes. Observation of the absorption bands of $(2,4,6\text{-}(CF_3)_3C_6H_2)_2E$ (E = Ge, Sn) at unexpectedly low wavelengths is explained by intramolecular coordination of the germanium and tin atoms to fluorine atoms of the *o*-CF_3 groups, which has been established by X-ray analysis[138,139]. The extra-high value of the absorption maximum of $((Me_3Si)_3Si)_2Pb$ reflects the strong σ-donor effect of the $(Me_3Si)_3Si$ group.

Besides the absorption spectra, the fluorescence spectra were recorded for $GeMe_2$[118,119], GeMePh[119] and $GePh_2$[119] (Table 4). The slightly larger Stokes shift (difference between maxima of the fluorescence and absorption bands) found for $GeMe_2$ can also be attributed to a slightly larger change in the C−Ge−C bond angle upon excitation for this germylene compared to GeMePh and $GePh_2$. This suggests some increase in the ground state bond angle at the Ge atom upon introduction of the bulkier phenyl group.

The nature of the electronic effects of a number of substituents has been elucidated in the course of theoretical studies of the n $\rightarrow$ p electronic transitions in silylenes[126,128]. σ-Acceptor substituents increase and σ-donor substituents (like Me_3Si) reduce the singlet–singlet energy gap in accordance with Bent's rule[137] due to change in the hybridization of the divalent atom of the Group 14 element. Thus, the steric effect of bulky substituents is equivalent to the weak σ-donor effect. π-Donors affect the p orbital of the divalent atom of the Group 14 element, raising its energy, which increases the energy of the n $\rightarrow$ p transition. Substituents like Hal, NR_2, OR or SR show both σ-acceptor and π-donor properties, affecting the transition energy in the same direction. α-Unsaturated organic substituents (aryl, vinyl, ethynyl) have two orbitals which can interact with the p orbital of the divalent atom of the Group 14 element. Those are occupied π and empty low-lying π^* orbitals. Interaction with the π orbital results in raising the p orbital energy, while interaction with the π^* orbital results in lowering its energy. It has been found[126] that in the case of silylenes, the latter interaction prevails over the former for aryl and vinyl groups, whereas for ethynyl group the situation is the reverse. Substitution of an H atom by a Me group slightly increases the transition energy. Thus, the Me group acts as a weak σ-acceptor. In the case of silylenes, increasing the transition energy by an Me group is stronger than that by the ethynyl group. Experimental data on absorption maxima of the corresponding silylenes[140,141] are in excellent agreement with the transition energies calculated in the course of these studies[126,128]. The available data on absorption maxima of germylenes, stannylenes and plumbylenes (Table 4) are also in qualitative agreement with these predictions[142].

The energy of the 0–0 transition in GeH_2 is 16320 cm^{-1} (613 nm, see Section III). The corresponding energy for $GeMe_2$ can be roughly estimated as an average of the

TABLE 4. Absorption maxima of polyatomic germylenes, stannylenes and plumbylenes

Carbene analog[a]	Precursor	Conditions of generation and spectrum recording[b]	λ^c_{max}	Reference
Me_2Ge	$Me_2Ge(N_3)_2$	$h\nu$, Ar matrix, 12–18 K	405	104
	$Me_2Ge(SePh)_2$	$h\nu$, Ar matrix, 21 K	420	103
Me_2Ge	$Me_2Ge(N_3)_2$	$h\nu$, 3-MP, 77 K	*ca* 405	104
	7-germanorbornadiene[d]	$h\nu$, 3-MP, 77 K	416 (620)	119
	$Me_2Ge(SePh)_2$	$h\nu$, 3-MP, 77 K	420	103
	7-germanorbornadiene[d]	$h\nu$, 3-MP, 77 K	420	142, 144
	7-germanorbornadiene[d]	$h\nu$, 3-MP/IP (3 : 7), 77 K	420	145
	$(PhMe_2Ge)_2GeMe_2$	$h\nu$, 3-MP, 77 K	422	107, 108
	$(PhMe_2Ge)_2GeMe_2$	$h\nu$, 3-MP, 77 K	422 (623)	119
	c-$(Me_2Ge)_6$	$h\nu$, 3-MP, 77 K	430 (650)	118
	c-$(Me_2Ge)_6$	$h\nu$, 3-MP, 77 K	430 (630)	119
	$(GeMe_2)_3$ (biphenyl-bridged)	$h\nu$, 3-MP, 77 K	430	146
	$Me(Me_2Ge)_5Me$	$h\nu$, 3-MP, 77 K	436 (628)	119
	$Me(Me_2Ge)_5Me$	$h\nu$, 3-MP, 77 K	437	109
	c-$(Me_2Ge)_5$	$h\nu$, 3-MP, 77 K	506	120
Me_2Ge	$(PhMe_2Ge)_2$	$h\nu$, THF, r.t.	440	111, 114
	7-germanorbornadiene[d]	$h\nu$, C_6H_6, r.t.	380	147
	7-germanorbornadiene[d]	$h\nu$, n-C_7H_{16}, r.t.	380	122, 147
	$(PhMe_2Ge)_2GeMe_2$	$h\nu$, c-C_6H_{12}, r.t.	420	107, 108
	$Me_2Ge(SePh)_2$	$h\nu$, c-C_6H_{12}, r.t.	420	103
	$PhMe_2GeSiMe_3$	$h\nu$, c-C_6H_{12}, r.t.	425	121
	$PhMe_2GeSiMe_3$	$h\nu$, c-C_6H_{12}, r.t.	430	115
	$PhMe_2GeGeMe_3$	$h\nu$, c-C_6H_{12}, r.t.	430	115
	c-$(Me_2Ge)_6$	$h\nu$, c-C_6H_{12}, r.t.	450	118
	c-$(Me_2Ge)_5$	$h\nu$, c-C_6H_{12}, r.t.	490	120
Me_2Ge	Me_5Ge_2H or Me_2Ge (cyclic)	$h\nu$, gas phase	480	106

(continued overleaf)

TABLE 4. (*continued*)

Carbene analog[a]	Precursor	Conditions of generation and spectrum recording[b]	λ^{c}_{max}	Reference
Et_2Ge	$Et_2Ge(SePh)_2$	$h\nu$, 3-MP, 77 K	425	148
	7-germanorbornadiene[d]	$h\nu$, 3-MP, 77 K	440	142, 144
	$(Et_2Ge)_n$, $n = 10–100$	$h\nu$, c-C_6H_{12}, r.t.	430	110
Pr_2Ge	$Pr_2Ge(SePh)_2$	$h\nu$, 3-MP, 77 K	425	148
i-Pr_2Ge	c-$(i$-$Pr_2Ge)_4$	$h\nu$, 3-MP, 77 K	542	149
	c-$(i$-$Pr_2Ge)_4$	$h\nu$, c-C_6H_{12}, r.t.	560	149
Bu_2Ge	$Bu_2Ge(SePh)_2$	$h\nu$, 3-MP, 77 K	425	148
	7-germanorbornadiene[d]	$h\nu$, 3-MP, 77 K	440	142
	$(Bu_2Ge)_n$, $n = 10–100$	$h\nu$, c-C_6H_{12}, r.t.	450	110
Hex_2Ge	$(Hex_2Ge)_n$, $n = 10–100$	$h\nu$, c-C_6H_{12}, r.t.	460	110
Ge	$Ge(N_3)_2$	$h\nu$, Ar matrix	410, 248	105
Me_3Si $SiMe_3$ Ge Me_3Si $SiMe_3$	CA is stable	i-C_6H_{14}, 77–293 K, THF, r.t.	450, 280	150
$(Me_3SiCH_2)_2Ge$	c-$(R_2Si)_3Ge(CH_2SiMe_3)_2$,	$h\nu$, c-C_6H_{12}, r.t.	470	112
	R = i-Pr, t-$BuCH_2$	$h\nu$, 3-MP, 77 K	470	112
	$(Me_3SiCH_2)_2Ge(SiMe_3)_2$	$h\nu$, 3-MP, 77 K	470	112
	c-$((t$-$BuCH_2)_2Si)_2Ge(CH_2SiMe_3)_2$	$h\nu$, 3-MP, 77 K	460	151
	c-$O((t$-$BuCH_2)_2Si)_2Ge(CH_2SiMe_3)_2$	$h\nu$, 3-MP, 77 K	445	151
$((Me_3Si)_2CH)_2Ge$	CA is stable	C_6H_{14}, r.t.	414, 302, 227	132, 152, 153
PhMeGe	7-germanorbornadiene[d]	$h\nu$, 3-MP, 77 K	440	142
	$(Me_3Ge)_2GeMePh$	$h\nu$, 3-MP, 77 K	456	107, 108
	$(Me_3Ge)_2GeMePh$	$h\nu$, 3-MP, 77 K	456 (645)	119
	$(Ph_2MeGe)_2$	$h\nu$, THF, r.t.	450	111, 114
	$(Me_3Ge)_2GeMePh$	$h\nu$, c-C_6H_{12}, r.t.	440	107, 108
	$(PhMeGe)_n$, $n = 10–100$	$h\nu$, c-C_6H_{12}, r.t.	440	110

Mes(*t*-Bu)Ge	Mes(*t*-Bu)Ge$(SiMe_3)_2$	$h\nu$, 3-MP, 77 K	508	142, 144
Ph_2Ge	$Ph_2Ge(GeMe_3)_2$	$h\nu$, 3-MP, 77 K	462 (651)	107, 108, 119
	$Ph_2Ge(SiMe_3)_2$	$h\nu$, 3-MP, 77 K	463	145
	$Ph_2Ge(SiMe_3)_2$	$h\nu$, 3-MP, 77 K	466	142, 144
	7-germanorbornadiene[d]	$h\nu$, 3-MP, 77 K	466	142, 144
	$Ph_2Ge(SiMe_3)_2$	$h\nu$, *c*-C_6H_{12}, r.t.	445	113
	$Ph_2Ge(GeMe_3)_2$	$h\nu$, *c*-C_6H_{12}, r.t.	450	107, 108
	$(Ph_3Ge)_2$	$h\nu$, THF, r.t.	470	111, 114
Mes_2Ge	$Mes_2Ge(SiMe_3)_2$	$h\nu$, 3-MP or 3-MP/IP, 77 K	550	142, 144, 145
	$Mes_2Ge(SiMe_3)_2$	$h\nu$, C_6H_{14}, r.t.	550, 325	155
	c-$(Mes_2Ge)_3$	$h\nu$, C_6H_{14}, r.t.	550, 325	155
$(4\text{-}MeC_6H_4)_2Ge$	$(4\text{-}MeC_6H_4)_2Ge(SiMe_3)_2$	$h\nu$, 3-MP, 77 K	471	142
$(2,6\text{-}Me_2C_6H_3)_2Ge$	$(2,6\text{-}Me_2C_6H_3)_2Ge(SiMe_3)_2$	$h\nu$, 3-MP, 77 K	543	142
$(2,6\text{-}Et_2C_6H_3)_2Ge$	$(2,6\text{-}Et_2C_6H_3)_2Ge(SiMe_3)_2$	$h\nu$, 3-MP or 3-MP/IP, 77 K	544	142, 145
Tip_2Ge	$Tip_2Ge(SiMe_3)_2$	$h\nu$, 3-MP or 3-MP/IP, 77 K	558	142, 145
(2,4,6-*t*-$Bu_3C_6H_2)_2Ge$	CA is stable	C_6H_{14} or THF, r.t.	430	156
	CA is stable	solid, r.t.	405	156
TbtMesGe	CA is stable	C_6H_{14}, r.t.	575	157
$(2,6\text{-}Mes_2C_6H_3)_2Ge$	CA is stable	Et_2O, r.t.	578	158
TbtTipGe	CA is stable	C_6H_{14}, −73–60 °C	580	159, 160
	CA is stable	C_6H_{14}, r.t.	581	161
$(2,4,6\text{-}(CF_3)_3C_6H_2)_2Ge$	CA is stable	C_6H_{14}, r.t.	374	138
$(2,4,6\text{-}(Me_2NCH_2)_3C_6H_2O)_2Ge$	CA is stable	*c*-C_6H_{12}, r.t.	385	143
(*t*-Bu)N, N(Bu-*t*), Si, (*i*-Pr)N, Ge, N(Pr-*i*)	CA is stable	solvent is not reported, r.t.	557	162
(*t*-Bu)N, N(Bu-*t*), Si, RN, Ge, NR, R = $2,6\text{-}Me_2C_6H_3$	CA is stable	solvent is not reported, r.t.	460	162

(*continued overleaf*)

TABLE 4. (*continued*)

Carbene analog[a]	Precursor	Conditions of generation and spectrum recording[b]	λ^{c}_{max}	Reference
$((t\text{-}Bu)_2N)_2Ge$	CA is stable	$c\text{-}C_6H_{12}$, r.t.	445, 310, 227	163
$(N\text{—}Ge)_2$	CA is stable	C_6H_{14}, r.t.	426, 250, 217	164
$(t\text{-}Bu(Me_3Si)N)_2Ge$	CA is stable	C_6H_{14}, r.t.	392, 325, 230	132
$((Me_3Si)_2N)_2Ge$	CA is stable	C_6H_{14}, r.t.	364, 300, 228	132
N, NNp, Ge, NNp	CA is stable	$i\text{-}C_6H_{14}$, r.t.	360, 307, 241	135
$((Tip_2FSi)(i\text{-}Pr_3Si)P)_2Ge$	CA is stable	C_6H_{14}, r.t.	626, 396	136
$(2,6\text{-}Tip_2C_6H_3)ClGe$	CA is stable	C_6H_{14}, r.t.	393	165
$C(SiMe_3)_2$, Sn, $C(SiMe_3)_2$	CA is stable	C_6H_{14}, r.t.	484, 370, 280, 247	166
$((Me_3Si)_2CH)_2Sn$	CA is stable	C_6H_{14}, r.t.	495, 332, 239	132, 152
$((Me_3Si)_3Si)_2Sn$	CA is stable	not reported	838, 559	167
$(2,4,6\text{-}(CF_3)_3C_6H_2)((Me_3Si)_3Si)Sn$	CA is stable	C_5H_{12}, r.t.	540	168
$(2\text{-}t\text{-}Bu\text{-}4,5,6\text{-}Me_3C_6H)((Me_3Si)_3Si)Sn$	CA is stable	C_6H_{14}, r.t.	643, 368	169
TbtTipSn	CA is stable	C_6H_{14}, r.t.	561	170
TbtTcpSn	CA is stable	C_6H_{14}, r.t.	586	171
TbtTppSn	CA is stable	C_6H_{14}, r.t.	563	171
$(2,6\text{-}Mes_2C_6H_3)_2Sn$	CA is stable	Et_2O, r.t.	553	158
$(2,4,6\text{-}t\text{-}Bu_3C_6H_2)_2Sn$	CA is stable	C_5H_{12}, r.t.	476	172
$(2\text{-}t\text{-}Bu\text{-}4,5,6\text{-}Me_3C_6H)_2Sn$	CA is stable	solvent is not reported, r.t.	479	173
$(2,4,6\text{-}(CF_3)_3C_6H_2)_2Sn$	CA is stable	toluene, r.t.	345	139
$(2,4,6\text{-}(Me_2NCH_2)_3C_6H_2O)_2Sn$	CA is stable	$c\text{-}C_6H_{12}$, r.t.	372	143

$(2,4,6\text{-}t\text{-}Bu_3C_6H_2C(S)S)_2Sn$	CA is stable	solid, r.t.	464, 364	174
$(t\text{-}Bu(Me_3Si)N)_2Sn$	CA is stable	C_6H_{14}, r.t.	433, 330, 305, 236	132
$((Me_3Si)_2N)_2Sn$	CA is stable	C_6H_{14}, r.t.	487, 389, 287, 230	131, 132
		C_6H_{14}, r.t.	387	175
	CA is stable	C_6H_{14}, r.t.	475, 222	164
$((Tip(t\text{-}Bu)FSi)(i\text{-}Pr_3Si)P)_2Sn$	CA is stable	C_6H_{14}, r.t.	579, 438	136
$((Tip_2FSi)(i\text{-}Pr_3Si)P)_2Sn$	CA is stable	C_6H_{14}, r.t.	644, 433, 384	136
$((Tip(t\text{-}Bu)FSi)(i\text{-}Pr_3Si)As)_2Sn$	CA is stable	C_6H_{14}, r.t.	641, 459	136
$(2,6\text{-}Tip_2C_6H_3)ISn$	CA is stable	C_6H_{14}, r.t.	428	165
	CA is stable	C_6H_{14}, r.t.	610, 385	176
$((Me_3Si)_3Si)_2Pb$	CA is stable	not reported	1 056, 578	167
$(2,6\text{-}Tip_2C_6H_3)MePb$	CA is stable	C_6H_{14}, r.t.	466, 332	177
$(2,6\text{-}Tip_2C_6H_3)(t\text{-}Bu)Pb$	CA is stable	C_6H_{14}, r.t.	470	177
$(3,5\text{-}(t\text{-}Bu)_2C_6H_3CMe_2CH_2)(2,4,6\text{-}(t\text{-}Bu)_3C_6H_2)Pb$	CA is stable	solid, r.t.	406	178
$((Me_3Si)_2CH)TbtPb$	CA is stable	C_6H_{14}, r.t.	531	179, 180
$((Me_3Si)_3Si)(2,3,4\text{-}Me_3\text{-}6\text{-}t\text{-}BuC_6H)Pb$	CA is stable	solvent is not reported, r.t.	610, 341, 303	178
$((Me_3Si)_3Si)(2,4,6\text{-}(CF_3)_3C_6H_2)Pb$	CA is stable	C_5H_{12}, r.t.	1 025, 586	168
Tbt_2Pb	CA is stable	C_6H_{14}, r.t.	610	179, 180
TtmTbtPb	CA is stable	C_6H_{14}, r.t.	560	179, 180
TipTbtPb	CA is stable	C_6H_{14}, r.t.	550	179, 180

(continued overleaf)

TABLE 4. (*continued*)

Carbene analog[a]	Precursor	Conditions of generation and spectrum recording[b]	λ^{c}_{max}	Reference
Tip_2Pb	CA is stable	C_6H_{14}, r.t.	541, 385, 321	181
$(2,3,4\text{-}Me_3\text{-}6\text{-}t\text{-}BuC_6H)_2Pb$	CA is stable	solid, r.t.	490	178
$(2,6\text{-}Mes_2C_6H_3)_2Pb$	CA is stable	Et_2O, r.t.	526	158
$(2,6\text{-}Tip_2C_6H_3)PhPb$	CA is stable	C_6H_{14}, r.t.	460	177
$(2,4,6\text{-}(Me_2NCH_2)_3C_6H_2O)_2Pb$	CA is stable	$c\text{-}C_6H_{12}$, r.t.	360	143
Tbt(TbtS)Pb	CA is stable	toluene, r.t.	540	182
$(TbtS)_2Pb$	CA is stable	C_6H_{14}, r.t.	538	183
$((Tip(t\text{-}Bu)FSi)(i\text{-}Pr_3Si)P)_2Pb$	CA is stable	C_6H_{14}, r.t.	645, 465, 346	136
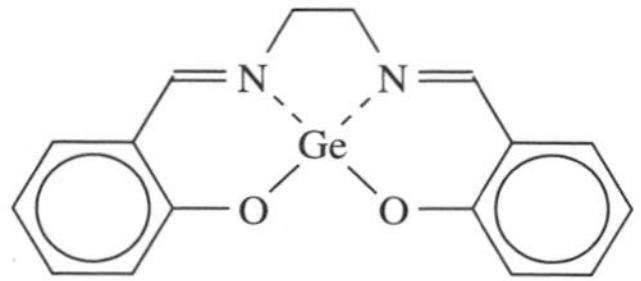 N, N, Ge, O, O	CA is stable	DMF, r.t.	354	184
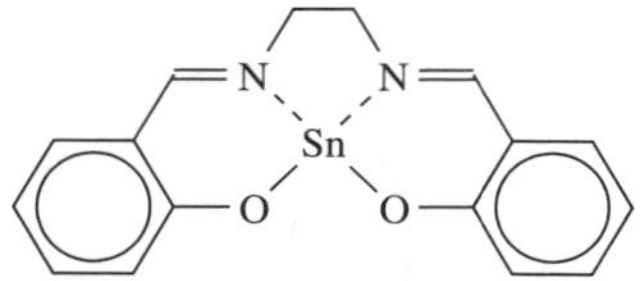 N, N, Sn, O, O	CA is stable	DMF, r.t.	366	184
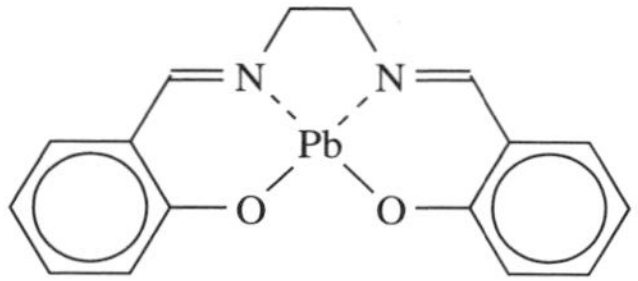 N, N, Pb, O, O	CA is stable	DMF, r.t.	379	184

[a] $Tbt = 2,4,6\text{-}((Me_3Si)_2CH)_3C_6H_2$; $Tip = 2,4,6\text{-}(i\text{-}Pr)_3C_6H_2$; $Ttm = 2,4,6\text{-}((Me_3SiCH_2)_3C_6H_2$; Np = neopentyl; $Tcp = 2,4,6\text{-}(c\text{-}C_6H_{11})_3C_6H_2$; Tpp = 2,4,6-tris(1-ethylpropyl)phenyl.

[b] 3-MP = 3-methylpentane, IP = isopentane.

[c] The fluorescence maxima are shown in parentheses.

[d] 7-germanorbornadiene = 7,7-disubstituted 1,2,3,4-tetraphenyl-5,6-benzo-7-germanorbornadiene; substituents at Ge atom correspond to those in the germylene generated from

TABLE 5. The long-wavelength absorption maxima and bond angles at divalent germanium, tin and lead atoms for stable carbene analogs

Carbene analog	λ_{max} (nm)	Reference	Bond angle (deg) (method[a])	Reference
$(2,6\text{-}Mes_2C_6H_3)_2Ge$	578	158	114.2 (X-ray)	158
$(2,4,6\text{-}t\text{-}Bu_3C_6H_2)_2Ge$	430, 405[b]	156	108.0 (X-ray)	156
$((Me_3Si)_2CH)_2Ge$	414	152, 153	107 (ED)	185
Me_3Si $SiMe_3$ Ge Me_3Si $SiMe_3$	450	150	90.97 (X-ray)	150
$(2,4,6\text{-}(CF_3)_3C_6H_2)_2Ge$	374	138	99.95 (X-ray)	138
$(2,6\text{-}Mes_2C_6H_3)_2Sn$	553	158	114.7 (X-ray)	158
$(2,4,6\text{-}t\text{-}Bu_3C_6H_2)_2Sn$	476	172	103.6 (X-ray)	172
$((Me_3Si)_2CH)_2Sn$	495	152	97 (ED)	185
$C(SiMe_3)_2$ Sn $C(SiMe_3)_2$	484	166	86.7 (X-ray)	166
$(2,4,6\text{-}(CF_3)_3C_6H_2)_2Sn$	345	139	98.3 (X-ray)	139
$((Me_3Si)_3Si)_2Pb$	1 056	167	113.56 (X-ray)	167
$SiMe_3$ Me_2Si $SiMe_3$ Pb Me_2Si $SiMe_3$ $SiMe_3$	610	176	117.1 (X-ray)	176
Tbt_2Pb	610	179, 180	116.3 (X-ray)	179, 180
$(2,6\text{-}Mes_2C_6H_3)_2Pb$	526	158	114.5 (X-ray)	158
$(2,3,4\text{-}Me_3\text{-}6\text{-}t\text{-}BuC_6H)_2Pb$	490[b]	178	103.04 (X-ray)	178
$(2,6\text{-}Tip_2C_6H_3)MePb$	466	177	101.4 (X-ray)	177
$(2,6\text{-}Tip_2C_6H_3)(t\text{-}Bu)Pb$	470	177	100.5 (X-ray)	177
$(2,6\text{-}Tip_2C_6H_3)PhPb$[c]	460	177	95.64 (X-ray)	177
$(3,5\text{-}(t\text{-}Bu)_2C_6H_3CMe_2CH_2)PbR$, $R = 2,4,6\text{-}(t\text{-}Bu)_3C_6H_2$	406[b]	178	94.8 (X-ray)	178
$(t\text{-}Bu)N$ $N(Bu\text{-}t)$ Si RN Ge NR, $R = 2,6\text{-}Me_2C_6H_3$	460	162	97.5 (X-ray)	162
$(2,6\text{-}Tip_2C_6H_3)ClGe$	393	165	101.31 (X-ray)	165
$(2,6\text{-}Tip_2C_6H_3)ISn$	428	165	102.6 (X-ray)	165
Tbt(TbtS)Pb	540[d]	182	100.2 (X-ray)	182

(*continued overleaf*)

TABLE 5. (*continued*)

Carbene analog	λ_{max}(nm)	Reference	Bond angle (deg) (method[a])	Reference
(2,2,6,6-tetramethylpiperidino)$_2$Ge [structure]	426, 250, 217	164	111.4 (X-ray)	163
$((Me_3Si)_2N)_2Ge$	364, 300, 228	132	101 (ED)	186
			107.1 (X-ray)	187
$((Me_3Si)_2N)_2Sn$	487, 389, 287	131, 132	96.0 (ED)	164
			104.7 (X-ray)	186
$((Tip(t\text{-}Bu)FSi)(i\text{-}Pr_3Si)P)_2Sn$	579, 438	136	98.78 (X-ray)	136
$((Tip(t\text{-}Bu)FSi)(i\text{-}Pr_3Si)As)_2Sn$	641, 459	136	94.64 (X-ray)	136
$((Tip(t\text{-}Bu)FSi)(i\text{-}Pr_3Si)P)_2Pb$	645, 465, 346	136	97.88 (X-ray)	136

[a] X-ray = X-ray analysis, ED = electron diffraction.
[b] absorption in solid phase.
[c] The plane of the Ph ring almost coincides with the plane of the plumbylene center.
[d] This absorption probably belongs to a complex of this plumbylene with solvent (toluene).

absorption and the fluorescence band maxima. This gives *ca* 19000 cm^{-1} (520 nm) using the data of Mochida and coworkers[119]. Thus, Me groups seem to reduce the energy of the $n \rightarrow p$ transition in the germylene series as they did in the silylene series. Lengthening of the alkyl chain does not affect much the absorption maximum position for germylenes, which can be seen from comparison of the data for $GeMe_2$, $GeEt_2$, and $GeBu_2$ (Table 4). A small bathochromic shift of the absorption band (relative to the band of $GeMe_2$) upon introduction of isopropyl or hexyl substituents is caused by steric rather than by electronic factors. Consecutive substitution of the Me groups in $GeMe_2$ by Ph groups results in red shift of the absorption bands. This shift can be caused by both steric and electronic effects of the phenyl groups. A rough estimation of the 0–0 transition energy in $GePh_2$ using the data of Mochida and coworkers[119] gives *ca* 16500 cm^{-1} (605 nm), which is close to that for GeH_2. However, because the latter value is too approximate, it is not clear whether the $n-\pi^*$ interaction is stronger than the $n-\pi$ interaction for phenyl-substituted germylenes or not. The lowest energies of the vertical electronic transition which have been measured for germylenes, stannylenes and plumbylenes are 610 nm ($Ge(SiMe_3)Ph$), 838 nm ($Sn(Si(SiMe_3)_3)_2$) and 1056 nm for ($Pb(Si(SiMe_3)_3)_2$) (Table 4). Obviously, the σ-donor effect of the $SiMe_3$ group is responsible for the long-wavelength absorption of these compounds. Introduction of Cl, OR, NR_2 or PR_2 groups shifts the bands of the $n \rightarrow p$ transition in germylenes, stannylenes and plumbylenes to lower wavelengths (Table 4) in accordance with the electronic effects of these substituents. In the case of $(2,4,6\text{-}(Me_2NCH_2)_3C_6H_2O)_2E$ (E = Ge, Sn, Pb)[143] the additional hypsochromic shift of the absorption band is due to intramolecular coordination of the germanium, tin or lead atom by the nitrogen atoms of $o\text{-}Me_2NCH_2$ groups.

E. Effect of the Nature of the Element E in R_2E Species on the Position of their Electronic Transition Band

Similarly to the case of the triatomic species (Section III), comparison of absorption maxima of stable germylenes, stannylenes and plumbylenes bearing the same substituents at different divalent atoms of the Group 14 elements (Table 4) does not reveal any firm trends. In the series of ETbtTip, $E(2,6\text{-}Mes_2C_6H_3)_2$, $ETip_2$, $E(Si(Me_3Si)_3)(2\text{-}t\text{-}Bu\text{-}4,5,6\text{-}Me_3C_6H)$ the absorption bands exhibit slight hypsochromic shift on going from Ge to Pb, although the complete series are available only for the first two types of compounds. In the series of the $(2,4,6\text{-}(Me_2NCH_2)_3C_6H_2O)_2E$, $(2,4,6\text{-}(CF_3)_3C_6H_2)_2E$ compounds, additionally stabilized by intramolecular coordination of the divalent atom E with the N and F atoms of the *ortho*-substituents, the absorption bands shift to the blue region on passing from Ge to Pb too. The slight bathochromic shift of the UV bands along the same row of divalent atoms of the Group 14 elements is observed for the following series: $E(CH(SiMe_3)_2)_2$, $E(Si(SiMe_3)_3)_2$, $E(2,4,6\text{-}t\text{-}Bu_3C_6H_2)_2$, $E(2\text{-}t\text{-}Bu\text{-}4,5,6\text{-}Me_3C_6H)_2$, $E(Si(SiMe_3)_3)(2,4,6\text{-}(CF_3)_3C_6H_2)$, $E(NR_2)_2$, $E(PR_2)_2$. None of these series is complete. Obviously, the same substituents reveal their effects to a different extent, depending on the nature of the divalent atom. Unfortunately, there are no UV data on labile stannylenes and plumbylenes with relatively simple substituents.

V. ELECTRONIC SPECTRA OF INTERMOLECULAR COMPLEXES OF GERMYLENES, STANNYLENES AND PLUMBYLENES WITH LEWIS BASES

Owing to the presence of an empty p-MO, the carbene analogs can form donor–acceptor complexes with Lewis bases. Formation of such molecular complexes has been repeatedly suggested as the first step of many reactions of CAs (see, e.g. References[3–5]). Complexation stabilizes the CAs and leads to changes in their reactivity and spectral properties. In this section the electronic spectra of intermolecular complexes of germylenes, stannylenes and plumbylenes with n-donor agents are considered. The available data are collected in Table 6. As one can see, they include, with a single exception, only germylene complexes. The data on spectral properties of silylene complexes have recently been reviewed by Gaspar and West[3].

The stronger the interaction of the lone electron pair of the n-donor agent with an empty p-MO of CA, the more stable is the complex, the higher is the energy of the LUMO of a complex, and the larger is the hypsochromic shift of the absorption maximum of a CA. Such a qualitative assessment of the strength of donor–acceptor complexes of CAs is widely used[5]. Based on this approach it was suggested[188] that the strength of silylene complexes decreases in the following series of n-donor agents: amines > phosphines > ethers > disulfides > halogenides. Experimentally observed shifts in absorption maxima of complexes of germylenes agree in general with this tendency (Table 6). However, the nature of substituents in CA and in an n-donor agent affects the strength of the complex formed. Typically, the hypsochromic shift has a magnitude of 100–150 nm. However, it is still unclear why the absorption maxima of Me_2Ge, MePhGe and Ph_2Ge recorded in hydrocarbon solutions and in coordinating solvents (THF) at room temperature (Table 4) are the same[111,114]. The absorption maxima of the stable germylene $(2,4,6\text{-}t\text{-}Bu_3C_6H_2)_2Ge$ in cyclohexane and THF solutions coincide[156]. In this case a steric hindrance at the germylene center[156] hampers a complexation. For the same reason the spectra of $(2,6\text{-}Mes_2C_6H_3)_2E$ (E = Ge, Sn, Pb) recorded in diethyl ether[158] correspond to free CAs, but not to their complexes with the solvent molecules.

Quantum-chemical studies showed that the ability to form complexes with Lewis bases decreases on going from silylenes to stannylenes and increases in the following series of n-donor agents: amines < phosphines < arsines < stibines[189]. This series somewhat differs from that proposed based on experimental data[188]. Calculations were successfully used to predict the absorption maxima shifts on complexation of silylenes with amines[190].

Whereas coordination of a CA to a lone electron pair of a heteroatom containing only single bonds typically results in a hypsochromic shift of the CA absorption maxima, upon complexation of germylenes with S atoms of thiocarbonyl compounds (and also

TABLE 6. Absorption maxima of germylene and stannylene complexes with n-donor agents

Complex	Conditions of observation of the complex[a]	λ_{max} of the complex	λ_{max} of the free CA[b]	Reference
$Cl_2Ge \cdot PPh_3$	C_6H_{14}, (−8) − +62 °C	233.8	310	198
$Me_2Ge \cdot PPh_3$	C_7H_{16}, r.t.	370	380	199
$Me_2Ge \cdot ClPh$	3-MP/IP (3 : 7), 77 K	392	420	142, 145
$Me_2Ge \cdot (C_6H_{11}Cl\text{-}c)$	3-MP/IP (3 : 7), 77 K	341	420	142, 145
$Me_2Ge \cdot S{=}C{=}C(Bu\text{-}t)_2$	3-MP/IP (1 : 4), 77 K	595	420	192
$Me_2Ge \cdot PhH$	3-MP/IP (4 : 1), 77 K	436–423	436	119
$Ph_2Ge \cdot (2\text{-}MeTHF)$	3-MP/IP (3 : 7), 77 K	325	463	142, 145
$Ph_2Ge \cdot OH(Et)$	3-MP/IP (3 : 7), 77 K	320	466	142
$Ph_2Ge \cdot OH(i\text{-}Pr)$	3-MP/IP (3 : 7), 77 K	324	466	142
$Ph_2Ge \cdot OH(Bu\text{-}n)$	3-MP/IP (3 : 7), 77 K	325	466	142
$Ph_2Ge \cdot OH(Bu\text{-}t)$	3-MP/IP (3 : 7), 77 K	332	466	142
$Ph_2Ge \cdot S$ 	3-MP/IP (3 : 7), 77 K	332	463	142, 145
$Ph_2Ge \cdot SMe_2$	3-MP/IP (3 : 7), 77 K	326	463	142, 145
$Ph_2Ge \cdot N$ 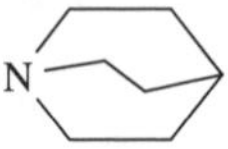	3-MP/IP (3 : 7), 77 K	334	463	142, 145
$Ph_2Ge \cdot (C_6H_{11}Cl\text{-}c)$	3-MP/IP (3 : 7), 77 K	374	463	142, 145
$Ph_2Ge \cdot ClPh$	3-MP/IP (3 : 7), 77 K	403	463	142, 145
$Ph_2Ge \cdot S{=}C{=}C(Bu\text{-}t)_2$	3-MP/IP (1 : 4), 77 K	565	463	192
$Mes_2Ge \cdot (2\text{-}MeTHF)$	3-MP/IP (3 : 7), 77 K	360	550	145
	2-MeTHF, 77 K	373	550	145
$Mes_2Ge \cdot OH(Et)$	3-MP/IP (3 : 7), 77 K	333	550	142
$Mes_2Ge \cdot OH(i\text{-}Pr)$	3-MP/IP (3 : 7), 77 K	339	550	142
$Mes_2Ge \cdot OH(Bu\text{-}n)$	3-MP/IP (3 : 7), 77 K	359	550	142
$Mes_2Ge \cdot OH(Bu\text{-}t)$	3-MP/IP (3 : 7), 77 K	362	550	142
$Mes_2Ge \cdot S$ 	3-MP/IP (3 : 7), 77 K	352	550	142, 145
$Mes_2Ge \cdot SMe_2$	3-MP/IP (3 : 7), 77 K	348	550	142, 145
$Mes_2Ge \cdot S(Et)CH_2CH{=}CH_2$	3-MP/IP (4 : 6), 77 K	380	550	142, 145
$Mes_2Ge \cdot N$ 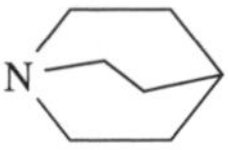	3-MP/IP (3 : 7), 77 K	349	550	142, 145
$Mes_2Ge \cdot NEt_3$	3-MP/IP (3 : 7), 77 K	414	550	142
$Mes_2Ge \cdot PBu_3$	3-MP/IP (3 : 7), 77 K	306	550	142, 145
$Mes_2Ge \cdot (C_6H_{11}Cl\text{-}c)$	3-MP/IP (3 : 7), 77 K	495	550	142, 145
$Mes_2Ge \cdot ClPh$	3-MP/IP (3 : 7), 77 K	538	550	142, 145
$Mes_2Ge \cdot ClCH_2CH{=}CH_2$	3-MP/IP (3 : 7), 77 K	530	550	142, 145
$Mes_2Ge \cdot ClCH_2CH{=}CHCH_3$	3-MP/IP (3 : 7), 77 K	515	550	142, 145
$Mes_2Ge \cdot S{=}C{=}C(Bu\text{-}t)_2$	3-MP, 77 K	580	550	192
	3-MP/IP (1 : 4), 77 K	582	550	192
$Mes_2Ge \cdot S{=}(Ad\text{-}2)$	3-MP, 77 K	690	550	193
$(2,6\text{-}Et_2C_6H_3)_2Ge \cdot (2\text{-}MeTHF)$	3-MP/IP (3 : 7), 77 K	369	544	145
$(2,6\text{-}Et_2C_6H_3)_2Ge \cdot OH(Et)$	3-MP/IP (3 : 7), 77 K	332	544	142
$(2,6\text{-}Et_2C_6H_3)_2Ge \cdot OH(i\text{-}Pr)$	3-MP/IP (3 : 7), 77 K	341	544	142
$(2,6\text{-}Et_2C_6H_3)_2Ge \cdot OH(Bu\text{-}n)$	3-MP/IP (3 : 7), 77 K	343	544	142

TABLE 6. (*continued*)

Complex	Conditions of observation of the complex[a]	λ_{max} of the complex	λ_{max} of the free CA[b]	Reference
$(2,6-Et_2C_6H_3)_2Ge \cdot OH(Bu\text{-}t)$	3-MP/IP (3 : 7), 77 K	367	544	142
$(2,6-Et_2C_6H_3)_2Ge \cdot SMe_2$	3-MP/IP (3 : 7), 77 K	357	544	142, 145
$(2,6-Et_2C_6H_3)_2Ge \cdot S$(cyclo-$C_4H_8$)	3-MP/IP (3 : 7), 77 K	359	544	142, 145
$(2,6-Et_2C_6H_3)_2Ge \cdot N$(quinuclidine)	3-MP/IP (3 : 7), 77 K	356	544	142, 145
$(2,6-Et_2C_6H_3)_2Ge \cdot PBu_3$	3-MP/IP (3 : 7), 77 K	314	544	142, 145
$(2,6-Et_2C_6H_3)_2Ge \cdot (C_6H_{11}Cl\text{-}c)$	3-MP/IP (3 : 7), 77 K	508	544	142, 145
$(2,6-Et_2C_6H_3)_2Ge \cdot ClPh$	3-MP/IP (3 : 7), 77 K	532	544	142, 145
$(2,4,6\text{-}i\text{-}Pr_3C_6H_2)_2Ge \cdot (2\text{-MeTHF})$	3-MP/IP (3 : 7), 77 K	376	558	142, 145
$(2,4,6\text{-}i\text{-}Pr_3C_6H_2)_2Ge \cdot SMe_2$	3-MP/IP (3 : 7), 77 K	357	558	142, 145
$(2,4,6\text{-}i\text{-}Pr_3C_6H_2)_2Ge \cdot S$(cyclo-$C_4H_8$)	3-MP/IP (3 : 7), 77 K	366	558	142, 145
$(2,4,6\text{-}i\text{-}Pr_3C_6H_2)_2Ge \cdot NEt_3$	3-MP/IP (3 : 7), 77 K	445	558	142
$(2,4,6\text{-}i\text{-}Pr_3C_6H_2)_2Ge \cdot N$(quinuclidine)	3-MP/IP (3 : 7), 77 K	363	558	142, 145
$(2,4,6\text{-}i\text{-}Pr_3C_6H_2)_2Ge \cdot PBu_3$	3-MP/IP (3 : 7), 77 K	334	558	142, 145
$(2,4,6\text{-}i\text{-}Pr_3C_6H_2)_2Ge \cdot (C_6H_{11}Cl\text{-}c)$	3-MP/IP (3 : 7), 77 K	544	558	142, 145
$(2,4,6\text{-}i\text{-}Pr_3C_6H_2)_2Ge \cdot ClPh$	3-MP/IP (3 : 7), 77 K	553	558	142, 145
TbtTipGe· THF	THF, r.t.	430	580	159
$((Me_3Si)_2CH)_2Sn \cdot S{=}C{=}C(Bu\text{-}t)_2$	3-MP, 77 K	600	495	194

[a] 3-MP = 3-methylpentane, IP = isopentane.
[b] Absorption maximum of the corresponding CA recorded under the same conditions without donor substrate.

of silylenes with both carbonyl and thiocarbonyl compounds[191]) a bathochromic shift has been observed[192–194]. The nature of this effect was not discussed. Complexation of Me_2Si with the N atom of MeCN has been shown to result in a hypsochromic shift[195].

There are several examples when molecular complexes of CAs with Lewis bases obtained in hydrocarbon glasses transformed into insertion or cycloaddition products upon annealing or melting the matrices. Namely, the transformation of complexes of allyl chloride and allyl mercaptan with Mes_2Ge into insertion products of Mes_2Ge into the C−Cl or S−H bond has been described[145]. Similarly, intermediate complexes of germylenes[142] and silylenes[196] with alcohols were found to isomerize into O−H bond insertion products. Formation of thia(oxa)siliranes was observed upon annealing (and also upon irradiation at a suitable wavelength) of matrices containing complexes of silylenes with ketones and thioketones[191]. These facts represent a compelling argument for intermediate formation of such CA complexes in these and similar reactions under other conditions.

Complexation of germylenes with such weak Lewis bases as aromatics was suggested[119]. The authors observed hypsochromic shifts of the absorption maximum of Me_2Ge if its precursors contained aromatic substituents. Moreover, the absorption maximum of

Me_2Ge generated in the presence of benzene from precursors containing no aromatic moieties was found to shift hypsochromically, with the shift increasing with an increase in benzene concentration (up to *ca* 20 nm)[119]. At the same time, the existence of a complex of Me_2Ge with benzene as a discrete compound was not established. Complexation of Me_2Ge with aromatics can be a reason for the large scatter of its reported absorption maxima (see Table 4). Similarly, the absorption spectrum of the stable Tbt(TbtS)Pb in toluene solution could in fact be due to its complex with the solvent[182]. At the same time, the position of the absorption band of $(2,4,6\text{-}(CF_3)_3C_6H_2)_2Sn$ observed in toluene at very low wavelengths is determined by intramolecular coordination of the tin atom by the fluorine atoms of *o*-CF_3 groups but not by interaction with the solvent[139].

Complexation of CAs with dinitrogen will be discussed in Section VII. Here, we just note that such complexation is probably responsible for large shifts in the UV absorption maxima of MeClSi, MeHSi and Me_2Si on going from noble gas matrices to nitrogen matrices[197]. The UV spectra of germylenes, stannylenes or plumbylenes in nitrogen matrices were not reported.

VI. VIBRATIONAL SPECTRA OF GERMYLENES, STANNYLENES AND PLUMBYLENES

Vibrational spectra of labile CAs were recorded using matrix IR and Raman spectroscopy and, in the case of some triatomic CAs, using IR and Raman spectroscopy in the gas phase. The gas phase spectra are often complicated by the decomposition of CAs at the temperatures required to obtain adequate vapor pressures, besides that the gas phase Raman spectra are often complicated by laser-induced resonance fluorescence, which is the greatest interfering factor in the Raman spectroscopy of gases[200]. The full-symmetric fundamental frequencies for the ground electronic state of a molecule can also be obtained from high resolution electronic spectra. The reported frequencies of CAs measured in the course of electronic spectroscopy studies are also included and discussed in this section. IR spectra of stable germylenes, stannylenes and plumbylenes considered below have been recorded in Nujol mull, in the liquid phase or in thin films. The vibrational spectra of stable dihalides of germanium, tin and lead, $(EHal_2)_x$, in the solid phase are not considered in this section, because the crystal structure of solid $(EHal_2)_x$ contains no $EHal_2$ units, as is well known.

The general approaches used in the studies considered below for assignment of the observed vibrational bands to the short-lived molecules are analogous to those described in Sections III and IV. The assignment of the revealed bands to normal, or fundamental, vibrational modes has been based on taking into account selection rules, observations of the bands in characteristic regions, observations of isotopic shifts, results of depolarization measurements in the Raman spectra and results of normal coordinate analysis. (It is noteworthy that Raman depolarization measurements can be conducted for matrix isolated species as well; see Reference[200] and references cited therein.) Lately, quantum-chemical calculations of vibrational spectra have become an important tool for both identification of CAs and assignment of their vibrational spectra.

Frequencies assigned to the fundamental vibrations of the triatomic germylenes, stannylenes and plumbylenes are collected in Table 7 together with corresponding data on triatomic silylenes for comparison. The vibrational frequencies are sensitive to the environment of the molecule. Therefore, the most precise frequency values measured under each type of condition used (in the gas phase and in different low-temperature inert matrices) by different spectroscopic methods are shown in the table for each of the CAs. Nonfundamental frequencies of triatomic and observed frequencies of polyatomic germylenes, stannylenes and plumbylenes are listed in the text.

A. Prototype EH_2 Molecules

The data on vibrational frequencies of the prototype germylene, GeH_2, are ambiguous. Several reactive species were produced by the vacuum-ultraviolet (VUV) photolysis (by H_2 and Xe microwave discharge lamps) of GeH_nD_{1-n}, $n = 0$–4, isolated in an Ar matrix at 4–23 K[201]. Two series of IR absorptions at 1839, 1813, 928, 850 cm^{-1} and at 1887, 1864, 920 cm^{-1} were tentatively attributed to germyl radical, GeH_3, and germylene, GeH_2, respectively on the basis of a normal coordinate analysis, taking into account the observed deuterium shifts. A similar mixture of the reactive products was formed from multipole dc discharge of GeH_4 and detected in low temperature Ar matrices by IR spectroscopy[202]. In disagreement with previous conclusions[201] it has clearly been observed that the intensity of the 928 cm^{-1} band correlated with that of the 1870 cm^{-1} (1864 cm^{-1} in Reference[201]) band upon annealing of the matrices, whereas the intensity of the 913 cm^{-1} band correlated with the intensity of the 1814 cm^{-1} (1813 cm^{-1} in Reference[201]) band. Thus the bands at 1890, 1870 and 928 cm^{-1} have been attributed to the symmetric stretching (ν_1), antisymmetric stretching (ν_3) and bending (ν_2) modes of GeH_2, respectively. (This numbering of the fundamental frequencies of symmetric triatomic CAs will be used below.) Doping the matrices with hydrogen, passed through discharge plasma, resulted in a faster decrease of the intensities of the GeH_2 bands in comparison with those of GeH_3 bands upon annealing, indicating that the reaction of GeH_2 with hydrogen atoms is faster than the reaction of GeH_3[202].

Another assignment of the vibrational bands observed in the 1800–1900 cm^{-1} region upon the VUV photolysis of GeH_4[201] was proposed later[203]. It was suggested that the bands attributed[201] to GeH_2 actually belong to GeH_3 and vice versa, and also that $\nu_3 > \nu_1$ for GeH_2. This suggestion was based on the similarity of the germanium and silicon analogs. For the latter it had been shown theoretically[203,204] that the Si–H stretching frequencies decrease in the series SiH_4, SiH_3, SiH_2, SiH, and the corresponding order of stretching frequencies, $\nu_3 > \nu_1$, had been found both experimentally[205] and theoretically[203] for SiH_2. The proposed[203] order of stretching frequencies of GeH_2 was later supported by *ab initio* calculations[206]. Positions of IR bands attributed to matrix isolated complexes of GeH_2 and GeD_2 with one and two molecules of HF[203] were well understood based on this new assignment of the GeH_2 bands, keeping the initial[201] assignment of GeD_2 bands without change.

At the same time the value 1856 cm^{-1} obtained for the ν_1 frequency of GeH_2 in the ground electronic state from the analysis of the vibronic structure of the ${}^1B_1 - {}^1A_1$ transition[30] is closer to the 1864 cm^{-1} frequency, which was initially[201] attributed to the matrix isolated GeH_2, but assigned to the ν_3 mode. Thus, the initial[201] identification of the GeH_2 bands can be correct, but requires changes in assignment of the Ge–H stretching frequencies to symmetric and antisymmetric modes in accordance with the requirement that $\nu_3 > \nu_1$.

The studies of vibrational spectra of GeH_2 paralleled those of SiH_2. The VUV photolysis of SiH_4 in Ar matrices also resulted in the formation of several reactive species, detected by IR spectroscopy[207]. Two of them were identified as SiH_2 and SiH_3[207]. However, it has recently been argued that the observed sets of bands attributed[207] to SiH_2 and SiH_3 can belong to SiH_2 molecules occupying different matrix sites[208]. Using the same arguments one can conclude that both sets of bands observed upon the VUV photolysis of GeH_4 in Ar matrices[201], except the band at 850 cm^{-1}, can also be attributed to GeH_2 occupying different matrix sites.

The vibrational spectra of SnH_2 and PbH_2 have not been reported so far.

B. $EHal_2$ Molecules

IR spectra of GeF_2 in the region of Ge−F stretching frequencies have been recorded both in the gas phase[40] and in solid Ne[40] and Ar matrices[40,209]. Monomeric GeF_2 was produced by evaporation of a $(GeF_2)_x$ sample at temperatures up to *ca* 425 K[40] or by reaction of GeF_4 vapor with metallic Ge at 570–620 K[209]. The first process was accompanied by production of dimeric species in a significant amount, but the dimer absorption, fortunately, did not overlap the bands of the monomer[40]. Two bands of monomeric GeF_2 were observed in all the cases. The germanium isotopic structure of these bands (at 685 and 655 cm^{-1}) was well resolved in the Ne matrix. Using the known relation[210] between bond angle and antisymmetric stretching frequencies (υ_3) of isotopomers, the bond angle in GeF_2 was calculated for two possible assignments of the observed bands to symmetric and antisymmetric stretching modes[40]. Assuming that the lower frequency was υ_3, the bond angle was found to be $94 \pm 2°$, whereas assignment of the higher frequency to υ_3 gave an angle of only $82 \pm 3°$. The latter value is unreasonably small. Thus, the higher of the observed frequencies has been assigned to the υ_1 mode, while the lower frequency has been assigned to υ_3. This assignment was confirmed later by observing the fluorine spin weight effects on the intensities of the rotational lines in the microwave spectra of GeF_2 in the first excited vibrational states[211]. The value 97.148° for the GeF_2 bond angle determined from the microwave spectrum[211] has also turned out to be very close to that calculated from the isotopic structure of the υ_3 band.

Raman spectra of GeF_2 (generated by reaction of GeF_4 vapor with Ge metal) isolated in N_2 and Ar matrices were also recorded only in the Ge−F stretching vibration region[209]. The Raman spectra in N_2 matrices were of a higher quality than those in Ar matrices. Among the number of bands observed in N_2 matrices only the bands at 702 and 653 cm^{-1} could be attributed to monomeric GeF_2; the other bands were assigned to GeF_2 oligomers on the basis of warm-up experiments. Raman depolarization measurements showed that the strong band at 653 cm^{-1} was clearly polarized and therefore corresponded to a ν_1 mode, in disagreement with the previous[40] assignment. The polarization of the band at 702 cm^{-1} was not measured because of its low intensity, it is noteworthy that the use of dinitrogen as a matrix gas is not favorable for the matrix studies of CAs, because N_2 is a weak Lewis base whereas CAs exhibit Lewis acid properties (see Sections V and VII). Although possible complexation with dinitrogen can hardly be expected to significantly affect the depolarization measurements, nevertheless it would be of interest to obtain results of such measurements for GeF_2 in Ar matrices. Unfortunately, the depolarization measurements in the Ar matrices were not carried out[209]. The Raman spectra of GeF_2 in Ar matrices contained only three bands at 705, 689 and 659 cm^{-1}. The strong band at 659 cm^{-1} was assigned to a ν_1 vibration of monomeric GeF_2 by analogy with the assignment in the N_2 matrix spectrum. The assignment of the 689 cm^{-1} band, whose position is closest to that of one of the IR bands of GeF_2 observed in Ar matrices, has not been reported[209].

Ab initio calculations of fundamental frequencies of GeF_2 at different levels of theory predict the ν_1 frequency to be higher than the ν_3 frequency[32], in support of the initial assignment[40] of the GeF_2 stretching vibrations. In the LIF spectrum of GeF_2 in the region of the $^1B_1 - ^1A_1$ transition (see Section III) a minor progression was revealed, which can be assigned only on the assumption that the gas-phase ν_1 frequency of GeF_2 in the ground state is equal to *ca* 721 cm^{-1} [32]. This value is much higher than that obtained from the gas-phase IR spectrum[40]. However, the GeF_2 ground state fundamental frequencies obtained from *ab initio* calculations[32] rather agree with the IR data, so the question as to the assignment of these progressions remains open. Thus, there is apparent disagreement in the assignments of the stretching frequencies of GeF_2 and experimental reexamination of its IR and Raman spectra both in the gas phase and in inert low-temperature matrices is desired. The frequency of the bending vibration (ν_2) of GeF_2 has only been obtained from analysis of the vibronic structure of the $^1B_1-^1A_1$[31,32] and $^3B_1-^1A_1$[32,35–38] electronic transitions in this molecule.

IR spectra of $GeCl_2$ were recorded in the region of the Ge−Cl stretching vibrations in Ar matrices[212–215]. Monomeric $GeCl_2$ was produced by VUV photolysis of GeH_nCl_{4-n}, $n =$ 0–2[212,215] or by evaporation of $(GeCl_2)_x$ polymers[213,214]. The stretching frequencies observed in all the studies are in good agreement. Both germanium and chlorine isotopic patterns of the stretching vibration bands were almost completely resolved in the spectra obtained by Maltsev and coworkers[214]. The bond angle equal to $99 \pm 4°$ has been computed using the measured ν_3 frequencies for different isotopomers[214]. This value is in excellent agreement with the values obtained by other methods (see Section VIII). Raman spectra of $GeCl_2$ were recorded in the gas phase[216,217] and in N_2 matrices[200]. $GeCl_2$ was generated by evaporation of polymeric $(GeCl_2)_x$ at 600–800 K[216], by reaction of $GeCl_4$ with metallic Ge[200] or by reaction of gaseous $GeCl_4$ with solid GeAs at temperatures above 700 K[217]. All three bands due to fundamental vibrations have been observed[200,217]. Depolarization measurements performed in the gas phase and in the N_2 matrices in the region of stretching vibrations[200,216,217] indicate that the ν_1 frequency is higher than ν_3, in agreement with the tentative assignment made earlier[213].

Monomeric $GeBr_2$ was produced by UV photolysis of H_2GeBr_2[204,218], by evaporation of polymeric $(GeBr_2)_x$[218], or by reaction of gaseous $GeBr_4$ with solid GeAs at temperatures above 700 K[217] and detected by IR spectroscopy in Ar matrices[204,218] or by Raman spectroscopy in the gas phase[217]. The initial tentative assignment of stretching frequencies to symmetric and antisymmetric modes was based on the fact that in IR spectra the ν_3 bands of dihalides of the Group 14 elements are usually more intensive than the ν_1 bands, and also on normal coordinate analysis performed taking into account isotopic splitting pattern of these bands[218]. This assignment was supported later by the Raman depolarization measurements[217].

Raman spectra of GeI_2 were recorded in the gas phase[217,219,220] and its IR spectra were obtained in the gas phase and in Ar matrices[50]. Assignment of the GeI_2 stretching frequencies to ν_1 and ν_3 modes was based on the Raman depolarization measurements[217,219]. The bond angle calculated from the observed ν_3 frequency isotopic shifts has been found to be equal to *ca* 105°[50]. This value is close to that of 102° obtained from electron diffraction measurements:[221]. Besides the fundamental frequencies, a series of overtones and differential and combinational frequencies were observed in the gas-phase Raman spectra of GeI_2 at elevated temperatures[219].

In the studies considered below dihalostannylenes and dihaloplumbylenes were typically generated by evaporation of the corresponding salts at appropriate temperatures.

IR spectra of SnF_2[222] and PbF_2[222,223] were recorded in Ar[222,223] and Ne[222] matrices, both in the E−F (E = Sn, Pb) stretching and bending vibration regions. Interaction of SnF_4 with Sn metal was also used to produce SnF_2[222]. Besides monomeric species, their dimers $(EF_2)_2$[222,223] and products of interaction of the monomers or dimers with the metal atoms were detected in the matrices[222]. Assignment of the observed bands of EF_2 to fundamental modes in the stretching vibration region has been based on assumptions that the intensity of the band of the antisymmetric vibration is higher than that of the symmetric one[222] and that the order of stretching frequencies of EF_2 is the same as for the corresponding dichlorides (see Table 7)[223]. The F−Sn−F bond angle calculated from the isotopic structure of the ν_3 band is equal to $94 \pm 5°$; calculation of the bond angle assuming an alternative assignment of the SnF_2 bands in the stretching vibration region also gave a reasonable value, $90 \pm 5°$[222]. No other measurement of the bond angle in SnF_2 has so far been reported. Raman spectra of SnF_2 and PbF_2 have not been reported.

Three bands of $SnCl_2$ were identified in the gas-phase Raman spectra[216,224,225]. Their assignment to the fundamental modes was performed based on the depolarization measurements. In the Raman spectra of $SnCl_2$ isolated in N_2 matrix all $SnCl_2$ fundamental frequencies were also observed, whereas in Ar matrix the bending vibration region was not recorded[200]. Depolarization measurements were not carried out for matrix isolated $SnCl_2$. IR spectra of $SnCl_2$ isolated in Ar matrices were recorded in the stretching vibration region only[213,215,223]. Both observed bands showed splitting due to chlorine isotopes. Assignment of these bands to fundamental modes ν_1 and ν_3 was initially conducted[213] by taking into account the gas-phase value for ν_1 known from the analysis of the vibronic structure of the $^3B_1 - ^1A_1$ electronic transition[53]. This assignment is in agreement with the Raman data discussed above. VUV photolysis of $SnCl_4$ was also used to generate $SnCl_2$[215]. This process is accompanied by formation of an $SnCl_3$ radical and a number of ionic species[215].

The Raman spectrum of $PbCl_2$ in the gas phase has been obtained at 1270 K in the presence of Cl_2 to suppress its decomposition to PbCl[216]. Only two bands assigned to ν_1 and ν_2 modes on the basis of depolarization measurements have been observed[216]. The band corresponding to the ν_3 mode has not been observed, apparently due to its low intensity. Three fundamental frequencies of $PbCl_2$ have been obtained from the Raman spectra in N_2 and Ar matrices[200]. In the Ar matrices, the chlorine isotopic splitting of the $PbCl_2$ bands was well resolved. Depolarization measurements were only carried for the species isolated in the Ar matrices. IR spectra of $PbCl_2$ isolated in Ar[213,223,226] and N_2[226] matrices were recorded in the stretching vibration region. The chlorine isotopic structure of the bands was much better resolved in Ar matrices[226]. From the measured ν_3 frequency values for different isotopomers, the ClPbCl bond angle was computed to be $96 \pm 3°$[226].

Only Raman spectra have been reported for $SnBr_2$ and $PbBr_2$. The gas-phase Raman spectrum of $PbBr_2$ has been recorded in the presence of Br_2[216]. Only two bands corresponding to ν_1 and ν_2 modes according to the Raman depolarization measurements were observed. The spectrum was complicated by laser-induced resonance fluorescence processes. The strong laser-induced (514.5 nm) resonance fluorescence precluded one from recording the gas-phase $SnBr_2$ Raman spectrum[216]. However, the bending frequency of $SnBr_2$ was obtained from the separation of the vibronic bands in the observed resonance fluorescence spectrum[216]. Raman spectra of $SnBr_2$ isolated in both Ar and N_2 matrices and $PbBr_2$ isolated in N_2 matrices have been recorded[200]. The bending frequency for $PbBr_2$ has not been observed because of its appearance in a region difficult for detection. Depolarization measurements carried out in the E−Br stretching vibration region allowed one to distinguish ν_1 and ν_3 frequencies of $SnBr_2$ and $PbBr_2$.

Attempts to record the gas-phase Raman spectra of SnI_2 and PbI_2 failed[216]. It was impossible to obtain any information on vibrational frequencies from the PbI_2 spectrum due to resonance fluorescence and emissions by products of PbI_2 decomposition[216]. In the case of SnI_2, only strong resonance fluorescence has been observed. From the separation of vibronic bands in this fluorescence spectrum, the ν_2 frequency of SnI_2 has been determined[216]. IR spectra of monomeric SnI_2 and PbI_2 were recorded in Ar and Xe matrices in the stretching vibration region and in the gas phase at elevated temperatures in the bending vibration region[227]. Thus, three bands corresponding to three fundamental modes were detected for each of the diiodides. In addition to monomeric SnI_2, an oligomeric species, probably $(SnI_2)_2$, has also been detected in the matrices. Its single band was distinguished from the bands of the monomer by warm-up experiments. The matrix IR spectra of PbI_2 were not complicated by the presence of oligomers. The same bands of PbI_2 in Ar matrices were obtained upon matrix reaction of Pb atoms with I_2. Assignment of the SnI_2 and PbI_2 bands observed in the stretching region to the symmetric and antisymmetric modes was based on the assumption that the ν_3 bands are usually more intensive than the ν_1 bands in the IR spectra of dihalides of the Group 14 elements. The ν_1 frequency of PbI_2 obtained from the IR spectrum is close to that obtained from the vibronic structure of the $^3B_1$–$^1A_1$ electronic transition in this molecule[60].

C. Mixed EXY Molecules

VUV and UV photolysis of GeH_3Cl in Ar matrices resulted in formation of a GeH_2Cl radical and a minor neutral labile product, containing only one hydrogen atom and characterized by a single IR band in the Ge–H stretching vibration region, which shows typical deuterium shift by use of GeD_3Cl as precursor[228]. This product was tentatively identified as monochlorogermylene, GeHCl, which can be formed by secondary photolysis of the GeH_2Cl radical. Similarly, the minor products of UV photolysis of GeH_3Br and GeD_3Br in Ar matrices were GeHBr and GeDBr, respectively[229]. Each of these germylenes was characterized by three bands, corresponding to the three fundamental vibrations. The bands corresponding to Ge–H(D) stretching and bending modes were split due to different trapping sites. The bending and Ge–Br stretching frequencies of GeHBr obtained from the IR spectra are in excellent agreement with those determined from the vibronic structure of the $^1A''$–$^1A'$ electronic transition in this molecule[77,78].

Evaporation of the mixtures of $SnCl_2$ with $SnBr_2$ (2 : 1) and $PbCl_2$ with $PbBr_2$ (2 : 1) at 500 K and 740 K, respectively, resulted in formation of SnClBr and PbClBr species detected by Raman spectroscopy in N_2 matrices[200]. The spectra were recorded only in the E–Hal stretching vibration region. It was noted that the E–Hal stretching frequencies of the mixed halides lay between the symmetric and antisymmetric E–Hal stretching frequencies of the corresponding $EHal_2$, isolated in N_2 matrices. This fact has been explained by simple force field analysis. In the gas-phase Raman spectrum of SnClBr produced by the same method, two of three bands corresponding to the bending and Sn–Br stretching fundamental vibrations were found[216]. The third band corresponding to the Sn–Cl stretching vibration of SnClBr was believed to coincide with the strong ν_1 band of $SnCl_2$, which was also present in the vapor phase. The values of the Sn–Br vibration frequency of SnClBr measured in both studies[200,216] are in good agreement. Strong resonance fluorescence was observed during an attempt to obtain the Raman spectrum of SnClI generated by the evaporation of a mixture of $SnCl_2$ and SnI_2(20 : 1)[216]. From the separation of vibronic bands in this fluorescence spectrum, the bending frequency of SnClI was determined.

D. Some Conclusive Remarks on the Vibrational Spectra of the Triatomic Carbene Analogs

As discussed above, some discrepancies still remain in the identification and assignment of the bands of GeH_2 and GeF_2 (and also of SiH_2[205,207,208,230], $SiCl_2$ and $SiBr_2$[71,217,231,232]; see Table 7). The complete sets of the fundamental frequencies have been established for other germylenes, stannylenes and plumbylenes. Although different fundamental frequencies were often measured under different conditions, the frequency shifts are usually not large on going from the gas phase to inert matrices. It can be seen from the data of Table 7 that with a rare exception, the frequencies of triatomic CAs decrease on going from the gas phase to matrices and decrease in matrices formed by

TABLE 7. Fundamental frequencies of triatomic carbene analogs in the ground electronic state[a]

	$\nu_1 (cm^{-1})^b$	$\nu_2 (cm^{-1})^b$	$\nu_3 (cm^{-1})^b$	Conditions and detection method[c]	Reference
SiH_2	2032–1967[d]	996	2032–1967[d]	Ar matrix, 4–14 K, IR	208
	1964.4	994.8	1973.3	Ar matrix, 10 K, IR	205, 207[i]
	2022–1985[d]	1001–996[d]	2022–1985[d]	Kr matrix, 6 K, IR	208
	1995.9280	998.6229	1992.816	gas phase, 300 K, IRDLS	230
	—	1009	—	gas phase, *ca* 20 K, ES	62
SiHD	1973.3	854.3	1436.9	Ar matrix, 10 K, IR	205
SiD_2	1426.9	719.8	1439.1	Ar matrix, 10 K, IR	205
	1461–1439[d]	721	1461–1439[d]	Kr matrix, 6 K, IR	208
	—	731	—	gas phase, *ca* 20 K, ES	62
GeH_2	1887	928	1864	Ar matrix, 4–23 K, IR	201, 202
	1813	913	1839	Ar matrix, 4–23 K, IR	201[i], 202[i], 203
	1856	916	—	gas phase, *ca* 20 K, ES	30
GeHD	1884	806	1322	Ar matrix, 4–23 K, IR	201
GeD_2	1327	658	1338	Ar matrix, 4–23 K, IR	201
	1335	657	—	gas phase, *ca* 20 K, ES	30
SiF_2	851.0	—	864.6	Ne matrix, 5 K, IR	234
	842.8	—	852.9	Ar matrix, 15 K, IR	234
	855.010	—	870.405	gas phase, 1400 K, IR	237
	—	343.6	—	gas phase, 1400 K, MW	238
	853	344	—	gas phase, *ca* 20 K, ES	68
$SiCl_2$	518.7	—	509.4	Ne matrix, 5 K, IR	232
	512.5	202.2	501.4	Ar matrix, 15 K, IR	232
	512.0	—	501.2	Ar matrix, 15–20 K, IR	236
	509.9	—	496.3	N_2 matrix, 15 K, IR	232
	—	155	—	gas phase, >700 K, Raman	217
	521.6	200.6	—	gas phase, *ca* 20 K, ES	71
	513	195	502	force field calculations	231
$SiBr_2$	402.6	—	399.5	Ar matrix, 15 K, IR	232
	399.9	—	394.1	N_2 matrix, 15 K, IR	232
	312	—	—	gas phase, >670 K, Raman	217
	404	130	400	force field calculations	231
SiI_2	—	88	—	gas phase, >670 K, Raman	217
GeF_2	685.0	—	655.0	Ne matrix, 5 K, IR	40
	676	—	648	Ar matrix, 5–11 K, IR	40
	643.0	—	673.5	Ar matrix, 4 K, IR	209
	659	—	705	Ar matrix, 4 K, Raman	209
	653	—	702	N_2 matrix, 4 K, Raman	209
	692	—	663	gas phase, 420 K, IR	40
	721	263	—	gas phase, *ca* 20 K, ES	32
$GeCl_2$	398.6	—	373.5	Ar matrix, 15 K, IR	214
	390	163	362	N_2 matrix, 4 K, Raman	200
	399	159		gas phase, 570–770 K, Raman	216
	392	157	372	gas phase, >670 K, Raman	217
	391	159	—	gas phase, *ca* 20 K, ES	47
$GeBr_2$	286	110	276	Ar matrix, 20 K, IR	218
	288	102	267	gas phase, >670 K, Raman	217
GeI_2	228	75	242	gas phase, 670–1170 K, Raman	217, 219
	220	—	226.3	Ar matrix, 14 K, IR	50
	—	78	—	gas phase, 670–1170 K, IR	50
SnF_2	605.4	201	584.4	Ne matrix, 5 K, IR	222
	592.7	197	570.9	Ar matrix, 15 K, IR	222
	—	180	—	gas phase, *ca* 20 K, ES	51
$SnCl_2$	353	—	332	Ar matrix, 4 K, Raman	200
	354.8	—	334.6	Ar matrix, 15 K, IR	213
	341	124	320	N_2 matrix, 4 K, Raman	200
	352	120	—	gas phase, 920 K, Raman	216, 224

TABLE 7. (*continued*)

	$\nu_1(cm^{-1})^b$	$\nu_2(cm^{-1})^b$	$\nu_3(cm^{-1})^b$	Conditions and detection method[c]	Reference
	358	121	340	gas phase, 666–1 047 K, Raman[f]	225
	355	121	347	gas phase, 666–1 047 K, Raman[g]	225
	362	127	344	gas phase, 666–1 047 K, Raman[h]	225
	350	120	—	gas phase, *ca* 20 K, ES	53
$SnBr_2$	244	82	231	Ar matrix, 4 K, Raman	200
	237	84	223	N_2 matrix, 4 K, Raman	200
	—	80	—	gas phase, 900 K, Raman	216
SnI_2	196	—	187	Ar matrix, 14 K, IR	227
	188	—	181	Xe matrix, 14 K, IR	227
	—	60	—	gas phase, 770–1 120 K, IR	227
	—	61	—	gas phase, 1100 K, Raman	216
PbF_2	545.7	170	522.5	Ne matrix, 5 K, IR	222
	531.2	165	507.2	Ar matrix, 10–15 K, IR	222, 223
$PbCl_2$	322.3	103	300.7	Ar matrix, 4 K, Raman	200
	321.0	—	299.0	Ar matrix, 15 K, IR	226
	305	104	281	N_2 matrix, 4 K, Raman	200
	306	—	282	N_2 matrix, 15 K, IR	226
	314	99	—	gas phase, 1300 K, Raman	216
$PbBr_2$	208	—	189	N_2 matrix, 4 K, Raman	200
	200	64	—	gas phase, elevated temp., Raman	216
PbI_2	163	—	158	Ar matrix, 14 K, IR	227
	158	—	153	Xe matrix, 14 K, IR	227
	—	43	—	gas phase, 970–1 270 K, IR	227
	168	44	—	gas phase, *ca* 20 K, ES	60
SiHF	1 913.1[e]	859.0[e]	833.7[e]	Ar matrix, 15 K, IR	239
	—	859	—	gas phase, *ca* 20 K, ES	84
SiDF	1 387.4[e]	638.4[e]	833.3[e]	Ar matrix, 15 K, IR	239
		643		gas phase, *ca* 20 K, ES	85
SiHCl	1 968.8	805.9	522.8	gas phase, *ca* 20 K, ES	82
SiDCl	1 434.4	592.3	518.1	gas phase, *ca* 20 K, ES	82
SiHBr	1 976.2	771.9	412.4	gas phase, *ca* 20 K, ES	83
SiDBr	1 439.5	—	408.0	gas phase, *ca* 20 K, ES	83
SiHI	—	727	350.0	gas phase, *ca* 20 K, ES	87, 86
GeHCl	1 862	—	—	Ar matrix, 6–23 K, IR	228
	—	689.2	441.9	gas phase, *ca* 20 K, ES	75, 78
GeDCl	1 343	—	—	Ar matrix, 6–23 K, IR	228
GeHBr	1 858	701	283	Ar matrix, 8–24 K, IR	229
	—	695	288.7	gas phase, *ca* 20 K, ES	77, 78
GeDBr	1 336	502	281	Ar matrix, 8–24 K, IR	229
SnClBr	328	—	228	N_2 matrix, 4 K, Raman	200
	352	100	240	gas phase, 940 K, Raman	216
SnClI	—	91	—	gas phase, 1050 K, Raman	216
PbClBr	295	—	200	N_2 matrix, 4 K, Raman	200

[a]The fundamental frequencies correspond to isotopomers, containing the most abundant isotopes, or represent the effective values if the isotopic structure has not been observed.

[b]ν_1, ν_3 are stretching frequencies, symmetric and antisymmetric, respectively, in the case of symmetric EX_2 molecules; ν_2 are bending frequencies.

[c]IRDLS = infrared diode laser spectroscopy; ES = electronic spectroscopy, spectra with resolved vibrational structure; MW = microwave spectroscopy; force field calculations denote harmonic frequencies obtained on the basis of combined analysis of electron diffraction and vibrational spectroscopy data.

[d]The bands are split due to different trapping sites in the matrix.

[e]The frequencies correspond to the more stable trapping site in the matrix.

[f]Excitation at 457.9 nm.

[g]Excitation at 480 nm.

[h]Excitation at 514.5 nm.

[i]Values calculated from data in this reference.

different matrix gases in the following order: Ne > Ar > Xe and N_2. The bending frequency is the least sensitive to the environment. Such a matrix effect is quite usual for different types of compounds; however, it is worth emphasizing here that in nitrogen matrices the matrix shift of CA frequencies can be determined predominantly by specific donor–acceptor interaction (by complexation) with dinitrogen molecules (see below). The order of the fundamental frequencies remains the same in both the gas phase and matrices for all CAs, with the possible exception of SiH_2[230]. The observation of two IR active stretching vibrations for some of the symmetric germylenes, stannylenes and plumbylenes was historically the first firm experimental evidence of their bent structure. Besides the fundamental frequencies, those of other types have been observed for GeI_2 (a set of overtones, differential and combinational frequencies)[219] and for SiH_2 ($\nu_1/2\nu_2$ Fermi and $2\nu_1/2\nu_3$ Darling-Dennison resonances)[205,230,233]. In some studies the recorded matrix spectra of CAs were complicated by band splittings due to trapping the molecules in different matrix sites. Such matrix splittings can be a source of contradictions in the identification of the bands of matrix isolated SiH_2[205,207,208] and GeH_2[201–203].

The relative intensities of the stretching vibration bands of the symmetric triatomic CAs seem to be very characteristic: in all cases when the symmetric and antisymmetric frequencies were identified unambiguously (SiF_2, $GeCl_2$, GeI_2, $SnCl_2$, $SnBr_2$, $PbCl_2$, $PbBr_2$) the ν_1 Raman band was much more intensive than the ν_3 one, while the intensity of the ν_1 IR band was lower than that of the ν_3 IR band. This observation has been used for assignments of stretching vibration bands of some other triatomic CAs ($GeBr_2$, SnF_2, SnI_2, PbF_2, PbI_2). Based on the observed isotopic splitting of the ν_3 bands of SiF_2[234], $SiCl_2$[232,235,236], $SiBr_2$[232], GeF_2[40], $GeCl_2$[214], GeI_2[50], SnF_2[222] and $PbCl_2$[226], bond angles were computed for these molecules. For SnF_2 there is no other experimental measurement of the bond angle. The obtained values for other molecules are in good agreement with more precise values determined in microwave and electron diffraction studies (presented in Section VIII).

The stretching (and also bending) frequencies of EX_2 and EXY decrease in the series in the order Si > Ge > Sn > Pb and F > Cl > Br > I (except for the Si–H stretching frequencies of HSiHal, which increase with increasing halogen weight). This reflects not only the increase in the weights of the composing atoms, but also a real weakening of the bonds, seen by comparing the stretching force constants reported in most of the studies performed. Initially, this conclusion was reached by Andrews and Frederick who compared stretching frequencies of dichlorides[213]. It has also been noted that dichlorides have smaller stretching force constants than the tetrachlorides, due to more p character of the E–Hal bonds in the former compounds[213]. This is also valid for other pairs of EX_2 and EX_4 molecules[11].

E. Polyatomic Germylenes, Stannylenes and Plumbylenes

The number of polyatomic germylenes, stannylenes and plumbylenes characterized by their vibrational spectra is still very limited. Only IR spectroscopy was used for this purpose. Unstable molecules were studied in low-temperature inert matrices. The stable germylenes, stannylenes and plumbylenes were treated by standard means.

Hydroxygermylene, HGeOH, was first produced in Ar matrix at 15 K upon photoinduced (340–300 nm) intramolecular insertion of Ge atom into the OH bond of H_2O submolecule in a $Ge \cdot OH_2$ complex, formed by co-deposition of Ge atoms and water with excess Ar[240]. Three observed IR bands at 1741.3, 661.3 and 566.2 cm^{-1} were assigned to $Ge \cdot OH$ stretching, Ge–O stretching and torsion vibrational modes. Later, HGeOH was identified as one of the products of the photochemical reaction of GeH_4 with O_3 in Ar matrices[241]. All the fundamental frequencies of this molecule [$\nu_1(OH) = 3652.0$,

ν_2(HGe) = 1741.1 (being in Fermi resonance with the overtone $2\nu_3$ observed as a weak band at 1757.6 cm^{-1}), ν_3(GeOH) = 885.2, ν_4(HGeO) = 708.7, ν_5(GeO) = 661.0 (shows characteristic Ge isotope splitting) and ν_6(torsion) = 566.0 cm^{-1}] were observed in its IR spectrum. These frequencies were assigned to the normal vibrational modes by observing isotopic shifts, when deuterium-substituted germane and ozone containing ^{16}O and ^{18}O isotopes were used in this reaction. The complete set of fundamental frequencies was also obtained for $HGe^{18}OH$, $DGe^{16}OD$ and $DGe^{18}OD$, whereas $HGe^{16}OD$ and $HGe^{18}OD$ were characterized by ν(HGe) frequency only. Similarly to $HGe^{16}OH$, Fermi resonance between ν_2 and $2\nu_3$ was revealed in the case of $HGe^{18}OH$. Unlike hydroxysilylene[242,243] the bands of only one conformer were present in the IR spectrum of hydroxygermylene[241]. Taking into account results of *ab initio* calculations[244], which showed that the *s-cis* conformer of HGeOH is lower in energy than the *s-trans* conformer, the authors[241] concluded that the observed conformer is the *s-cis* one. Quantum-chemical calculations at a higher level of theory have confirmed that the *s-cis* conformer is the more stable of the two conformers, but the difference in their energies is very small (<0.4 kcal mol^{-1})[245]. Thus, quantum-chemical calculations do not allow one to identify the observed[240,241] conformer of HGeOH unequivocally.

Similarly to Ge atoms (and Si atoms[240]) co-condensation of Sn atoms with water molecules results in the formation of the $Sn \cdot OH_2$ complex, stabilized and observed in Ar matrices[240]. Upon UV irradiation (340–300 nm) this complex transforms into HSnOH. The hydroxystannylene molecule has been characterized by five IR bands: ν_2 (HSn) as doublet at 1608.0 and 1597.7 cm^{-1} (the source of this splitting is not clear), ν_3 (SnOH) at 782.6, ν_5 (SnO) at 569.3 and ν_6 (torsion) at 475.5 cm^{-1} [240].

Methylgermylene, GeHMe, has been detected in the course of a matrix (Ar matrices, 12 K) FTIR spectroscopy study of vacuum pyrolysis of 1,1-dimethyl-1-germa-3-thietane and 1,1,3,3-tetramethyl-1-germacyclobutane (equation 1)[246].

1,1-dimethyl-1-germa-3-thietane (Me_2Ge ring with S) —800–1200 K, $-H_2C{=}S$→ $Me_2Ge{=}CH_2$

1,1,3,3-tetramethyl-1-germacyclobutane (Me_2Ge ring with Me, Me) —950–1200 K, $-H_2C{=}CMe_2$→ $Me_2Ge{=}CH_2$

$Me_2Ge{=}CH_2$ → GeMeEt → [cyclopropane-type ring–GeHMe] —$-H_2C{=}CH_2$→ GeHMe

(1)

Methylgermylene has been shown to result from thermal decomposition of an intermediate 1,1-dimethyl-1-germene, $Me_2Ge{=}CH_2$. Five of the twelve IR bands of this germylene have been observed and assigned to normal vibrational modes, based on results of B3LYP calculations. The most intensive band at 1798.6 cm^{-1}, which undoubtedly corresponds to stretching vibration of the Ge–H bond, is slightly lower than the Ge–H stretching frequencies of GeH_2, GeHHal and slightly higher than that of GeHOH (see above). The band at 535.6 cm^{-1}, which exhibited a clear quadruplet structure due to natural Ge isotope content, was identified as the Ge–C stretching vibration band. Other observed frequencies at 2891.6, 1201.2 and 868.8 cm^{-1} were attributed to C–H stretching mode and rocking modes of the methyl group.

Three other weak bands at 528.0, 554.8 and 783.6 cm^{-1} revealed in these experiments[246] were tentatively attributed to ethylmethylgermylene, GeMeEt, which is believed to be an intermediate product of decomposition of $Me_2Ge{=}CH_2$ to GeHMe (equation 1).

In accordance with the B3LYP calculations, these frequencies have been assigned to Ge−C stretching vibrations of the Ge−CH_2 and Ge−CH_3 groups and to CH_3 rocking mode, respectively.

Dimethylgermylene, $GeMe_2$, dimethylstannylene, $SnMe_2$, and its perdeuteriated derivative, $Sn(CD_3)_2$, were produced by the reactions shown in equation 2[104] and equations 3–5[247] and stabilized in low-temperature Ar matrices.

$$\mathrm{Me_2Ge(N_3)_2} \xrightarrow[\text{Ar, 12–18 K}]{h\nu,\ 248 \text{ or } 254 \text{ nm}} \mathrm{GeMe_2 + 3N_2} \tag{2}$$

$$\text{Ar, 5 K}$$

$$c\text{-}(\mathrm{Me_2Sn})_6 \xrightarrow{400\text{ K}} \mathrm{SnMe_2} + c\text{-}(\mathrm{Me_2Sn})_5 \tag{3}$$

$$\mathrm{Me_2SnH_2 + Ar^*} \longrightarrow \mathrm{SnMe_2 + H_2} \tag{4}$$

$$c\text{-}((\mathrm{CD_3})_2\mathrm{Sn})_6 \xrightarrow{400\text{ K}} \mathrm{Sn(CD_3)_2} + c\text{-}((\mathrm{CD_3})_2\mathrm{Sn})_5 \tag{5}$$

Frequencies obtained from matrix IR spectra of $GeMe_2$, $SnMe_2$ and $SiMe_2$ and their assignments are shown in Table 8. It is noteworthy that the close similarity of the IR spectra observed for these species is an additional argument for correct identification of these CAs. *Ab initio* calculated fundamental frequencies of $SiMe_2$ and $GeMe_2$, and the results of normal coordinate analysis of the $SnMe_2$ spectrum are also presented in Table 8. The theoretically predicted and experimentally observed frequencies are in good agreement.

Matrix IR spectra of 1-germacyclopent-3-enylidene and its d_6 analogue were obtained during a study of the photochemical interconversions shown in equation 6[105].

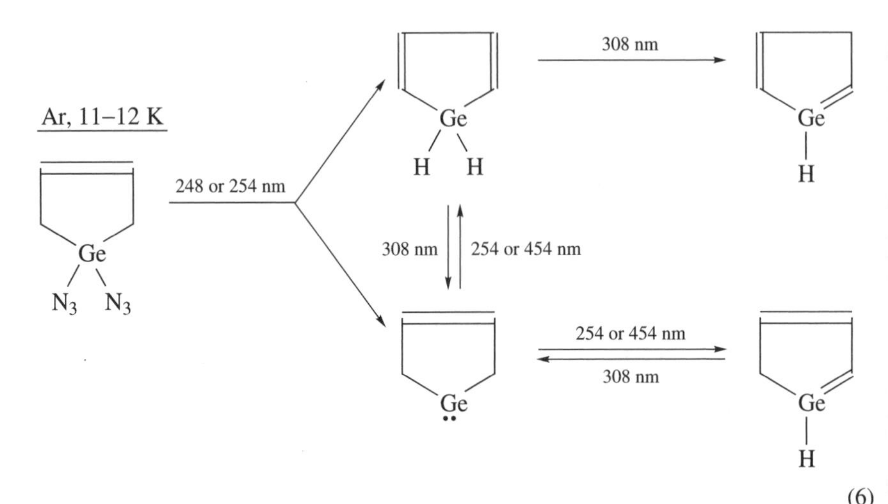

(6)

Each of the species has been characterized by a large number of IR bands and by its UV absorption. Full vibrational assignment in the IR spectra of these molecules has been performed on the basis of RHF/DZ+d calculations for both non-deuteriated and deuteriated analogs. The bands of 1-germacyclopent-3-enylidene at 508 and 478 (438 for

TABLE 8. Experimentally observed and calculated fundamental frequencies (cm^{-1}) of Me_2Si, Me_2Ge and Me_2Sn[a]

Type of vibration[247]	Me_2Si exp.[250,251]	Me_2Si calc.[b,c 127]	Me_2Ge exp.[104]	Me_2Ge calc.[d 129]	Me_2Sn exp.[247]	Me_2Sn calc.[e 247]
ν(C−H)		3261 (15.7)	2987 w.		2990 m.	3008
ν(C−H)		3261 (39.0)	2974 s.		2924 m.	2923
ν(C−H)		3227 (56.9)	2957 s.			
ν(C−H)		3219 (1.8)	2897 w.			
ν(C−H)		3160 (20.8)				
ν(C−H)		3156 (19.4)				
$\delta(CH_3)$	1435 m.	1600 (26.2)				
$\delta(CH_3)$		1586 (1.1)				
$\delta(CH_3)$		1580 (3.0)	1234 m.			
$\delta(CH_3)$		1570 (10.0)	1217 w.			
$\delta(CH_3)$	1220 s.	1436 (38.2)	1205 m.		1187 w.	1181
$\delta(CH_3)$	1210 m.	1426 (23.8)	1195 w.		1182 sh.	1179
$\rho(CH_3)$	850 s.	958 (49.1)	882 m.		774 s.	755
$\rho(CH_3)$	806 v.s.	803 (14.8)	817 m.		745 sh.	752
$\rho(CH_3)$		670 (14.8)			739 v.s.	751
$\rho(CH_3)$		635 (8.4)				
ν(E−C), sym.	690 m.	697 (12.6)	541 w.	560	504 s.	509
ν(E−C), asym.	735 m.	690 (49.4)	527 v.s.	497	518 s.	522
δ (CEC)		266 (3.5)		288		
$\tau(E-CH_3)$		124 (0.0)				
$\tau(E-CH_3)$		55 (0.0)				

[a] w. = weak, m. = medium, s. = strong, v.s. = very strong, sh. = shoulder.
[b] SCF/DZP calculations.
[c] Calculated intensities (km/mol^{-1}) are presented in parentheses.
[d] SCF/ECP DZP calculations.
[e] Force field calculations.
[f] Tentative assignment to $SiMe_2$.

1-germacyclopent-3-enylidene-d_6) cm^{-1} were attributed to the symmetric and antisymmetric Ge−C stretching vibrations, respectively. Analogous study of the similar silicon-containing systems was published earlier[248,249]. The symmetric and antisymmetric Si−C stretching frequencies of 1-silacyclopent-3-enylidene are at 616 and 741 cm^{-1}, whereas those of 3,4-dimethyl-1-silacyclopent-3-enylidene are at 626 and 775 cm^{-1}, respectively.

The IR spectrum of the stable $Sn(CH(SiMe_3)_2)_2$ has been recorded in hexane solution[152]. Assignment of the observed bands was carried out by analogy with the spectra of related compounds; the bands corresponding to the stretching vibrations of Sn−C bonds have not been revealed.

The IR spectra of stable cyclic $Me_2Si(t\text{-BuN})_2E$(E = Ge, Sn) have been obtained in films[252]. The IR spectrum of their labile silicon analog, $Me_2Si(t\text{-BuN})_2Si$, generated by photolysis of the diazide $Me_2Si(t\text{-BuN})_2Si(N_3)_2$ in Ar matrix, has been recorded too[252]. All spectra are similar to each other. The bands of the silylene, the germylene and the stannylene, at 781, 771 and 764 cm^{-1} respectively, were tentatively assigned to ν(E−N), those at 830, 814 and 811 cm^{-1} to ν(Si−C) and those at 883, 853 and 845 cm^{-1} to $\rho(Me_2Si)$. Assignment of the other bands has not been reported.

The IR spectra of a series of stable symmetric diaminogermylenes, diaminostannylenes and diaminoplumbylenes [$(t\text{-Bu}(Me_3Si)N)_2E$, $((Me_3Si)_2N)_2E$, E = Ge, Sn, Pb; $(t\text{-Bu}_2N)_2E$, $((Et_3Si)_2N)_2E$, $((Me_3Ge)_2N)_2E$, $((Et_3Ge)_2N)_2E$, $((Ph_3Ge)_2N)_2E$, E = Ge,

Sn] have been obtained for pure liquids or for Nujol mull[163,132]. The frequencies of antisymmetric stretching vibrations of the E–N bonds in these CAs have been identified in the 430–380 cm^{-1} region. Some decrease in the frequencies on passing from germylenes through to stannylenes with the same substituents has been observed. IR frequencies of the stretching Sn–N vibrations have also been reported for a series of cyclic diaminostannylenes $(Me_3Si)N\text{-}(CH_2)_n\text{-}N(SiMe_3)\text{-}Sn$, $n = 2–4$[175]. The frequencies were found in the 400–355 cm^{-1} region. Their position clearly depends on the strain in the cycle, decreasing with its increase.

The apparent disagreement in the identification of the ν(E–N) vibrations in the cited studies[132,163,175,252] clearly indicates that more careful analysis of the vibrational spectra of these CAs is needed.

The stretching frequency of the Ge–O bond in $Ge(OC_6H_2(CH_2NMe_2)_3\text{-}2,4,6)_2$ recorded in benzene solution, has been reported to lie at 1040 cm^{-1}[143]. However, this germylene is additionally stabilized by intramolecular coordination of the Ge center by the N atoms of two dimethylaminomethyl groups attached to different benzene rings in the *ortho*-positions and in fact can be considered as an intramolecular donor–acceptor complex[143].

Several other stable germylenes[138,156,158,184,253–260], stannylenes[158,184,254,261–271] and plumbylenes[158,184,271,272] were characterized by their IR spectra. The assignments of the observed bands in these studies were restricted to identification of a number of characteristic frequencies of the substituents only.

VII. VIBRATIONAL SPECTRA OF COMPLEXES OF GERMYLENES, STANNYLENES AND PLUMBYLENES WITH LEWIS BASES

A number of complexes of CAs with Lewis bases ($X_2E \cdot B_n$) were studied by matrix IR spectroscopy. Their absorptions are collected in Table 9. Data for complexes of silylenes are included in the table for comparison.

Weak complexes $H_2Si \cdot HF$ and $H_2Si \cdot (HF)_2$ and also cyclic $SiHF \cdot HF$ with coordination of both the H and F atoms of HF to the F and H atoms of SiHF, respectively, were produced in Ar matrices (at 13 K) by interaction of silane (and its deuteriated analog) with F_2 upon codeposition, followed by UV photolysis[203]. Identification of these complexes was based on results of HF/DZP calculations[203].

Similar reaction of GeH_4 and GeD_4 with F_2 were used to generate the corresponding complexes $H_2Ge \cdot HF$, $H_2Ge \cdot (HF)_2$, their deuteriated analogs as well as complexes $GeHF(GeDF) \cdots HF(DF)$ in Ar matrices[203]. In this case both cyclic and open [with coordination of the H(D) atom of HF(DF) to the F atom of HGeF] complexes between GeHF(GeDF) and HF(DF) have been observed.

The donor–acceptor complex $H_2Ge \cdot OH_2$ was obtained by photochemical reaction of germane and ozone in solid argon at 14–18 K[273]. Isotopic substitution provided a basis for assignment of the IR absorptions observed and the suggestion that in the $H_2Ge \cdot OH_2$ complex a H_2Ge submolecule has inequivalent hydrogen atoms, while the hydrogen atoms of the H_2O submolecule are equivalent. Later, quantum-chemical calculations at different levels of theory supported this suggestion[274]. Complexes $H(OH)Si \cdot OH_2$ and $H(OH)Ge \cdot OH_2$ were detected by IR spectroscopy in the reaction of Si and Ge (in this case under UV photolysis) atoms with excess of water in Ar matrices[240]. Under similar conditions complex $H(OH)Sn \cdot OH_2$ was not observed[240].

1 : 1 Complex formation of $SnCl_2$, PbF_2, $PbCl_2$, $PbBr_2$ and PbI_2 with CO, as well as $SnCl_2$ and PbF_2 with NO and N_2 in Ar matrices was studied with IR spectroscopy by

TABLE 9. Infrared absorptions (cm^{-1}) of complexes ($X_2E \cdot B_n$) of carbene analogs (EX_2) with Lewis bases (B)

$X_2E \cdot B_n$	Frequencies (cm^{-1}) of EX_2 moiety and assignments	Frequencies (cm^{-1}) of B_n moieties and assignments	Reference
$H_2Si \cdot HF$	1 985.7 ν_{as}(Si–H)	3 828.1 ν(H–F)	203
$D_2Si \cdot DF$	1 448.3 ν_{as}(Si–D)	2 798 ν(D–F)	203
$H_2Si \cdot (HF)_2$	1 942.8 ν_{as}(Si–H)		203
$D_2Si \cdot (DF)_2$	1 424.7 ν_{as}(Si–D)		203
cyclic-SiHF · HF	1 908.0 ν(Si–H)	3 796 ν(H–F)	203
	865.5 ν(Si–F)		
	751.6 δ(F–Si–H)		
cyclic-SiDF · DF	1 385.4 ν(Si–D)	2 784 ν(D–F)	203
$H(OH)Si \cdot OH_2$	1 929.4 ν(Si–H)		240
	778.5 ν(Si–O)		
$H_2Ge \cdot HF$	1 870.7 ν_{as}(Ge–H)	3 730.6 ν(H–F)	203
$D_2Ge \cdot DF$	1 846.8 ν_{as}(Ge–D)	2 739 ν(D–F)	203
$H_2Ge \cdot (HF)_2$	1 819.0 ν_{as}(Ge–H)		203
$D_2Ge \cdot (DF)_2$	1 312.5 ν_{as}(Ge–D)		203
$H_2Ge \cdot {}^{16}OH_2$	1 813.6 ν(Ge–H)	3 686.0 ν_{as}(O–H)	273
	1 777.2 ν(Ge–H)	3 597.4 ν_s(O–H)	
	1 794.4 2δ(H–Ge–H)	1 586.1 δ(H–O–H)	
	897.8 δ(H–Ge–H)		
$H_2Ge \cdot {}^{18}OH_2$	1 813.6 ν(Ge–H)	3 672.5 ν_{as}(O–H)	273
	1 777.2 ν(Ge–H)	3 590.3 ν_s(O–H)	
	1 794.4 2δ(H–Ge–H)	1 580.1 δ(H–O–H)	
	897.8 δ(H–Ge–H)		
$HDGe \cdot {}^{16}OHD$	1 811.6 ν(Ge–H)	3 637.3 ν(O–H)	273
	1 782.0 ν(Ge–H)	2 682.3 ν(O–D)	
	1 307.4 ν(Ge–D)	1 398.4 δ(H–O–D)	
	1 287.0 ν(Ge–D)		
$HDGe \cdot {}^{18}OHD$	1 811.6 ν(Ge–H)	3 626.3 ν(O–H)	273
	1 782.0 ν(Ge–H)	2 667.6 ν(O–D)	
	1 307.4 ν(Ge–D)	1 390.9 δ(H–O–D)	
	1 287.0 ν(Ge–D)		
$D_2Ge \cdot {}^{16}OD_2$	1 308.9 ν(Ge–D)	2 738.5 ν_{as}(O–D)	273
	1 281.6 ν(Ge–D)	2 627.8 ν_s(O–D)	
	1 293.3 2δ(D–Ge–D)	1 173.6 δ(D–O–D)	
	646.4 δ(D–Ge–D)		
$D_2Ge \cdot {}^{18}OD_2$	1 308.9 ν(Ge–D)	2 718.3 ν_{as}(O–D)	273
	1 281.6 ν(Ge–D)	2 617.2 ν_s(O–D)	
	1 293.3 2δ(D–Ge–D)	1 164.8 δ(D–O–D)	
	646.4 δ(D–Ge–D)		
cyclic-GeHF · HF	1 784.6 ν(Ge–H)	3 662.1 ν(H–F)	203
cyclic-GeDF · DF	1 287.8 ν(Ge–D)	2 693.2 ν(D–F)	203
linear-HGeF · HF	1 795.2 ν(Ge–H)	3 717.2 ν(H–F)	203
linear-DGeF · DF	1 296.2 ν(Ge–D)	2 732.1 ν(D–F)	203
$H(OH)Ge \cdot OH_2$	1 763.1 ν(Ge–H)		240
	609.1 ν(Ge–O)		
$Cl_2Ge \cdot PPh_3$	328 ν(P–$GeCl_2$)		282
	313 ν(P–$GeCl_2$)		
	300 ν(P–$GeCl_2$)		
$Cl_2Ge \cdot P(Bu\text{-}t)_3$	322 ν(P–$GeCl_2$)		282
	298 ν(P–$GeCl_2$)		

(*continued overleaf*)

TABLE 9. (*continued*)

$X_2E \cdot B_n$	Frequencies (cm^{-1}) of EX_2 moiety and assignments	Frequencies (cm^{-1}) of B_n moieties and assignments	Reference
$Br_2Ge \cdot PPh_3$	242 ν(P−GeBr$_2$) 227 ν(P−GeBr$_2$) 205 ν(P−GeBr$_2$)		282
$F_2Sn \cdot N_2$	588 ν_s(Sn−F) 565 ν_{as}(Sn−F)		275
$F_2Sn \cdot (N_2)_2$	583 ν_s(Sn−F) 557 ν_{as}(Sn−F)		275
$F_2Sn \cdot C_2H_4$	551.5 ν_s(Sn−F) 537.5 ν_{as}(Sn−F)		277
$F_2Sn \cdot HC{\equiv}CC_5H_{11}$	565 ν_s(Sn−F) 540 ν_{as}(Sn−F)	3 256 ν(≡C−H) 2 088 ν(C≡C) 1 011 δ(H−C≡C)	278, 279
$F_2Sn \cdot PhH$	564 ν_s(Sn−F) 542 ν_{as}(Sn−F)	696 δ(C−H) 688 δ(C−H)	280
$F_2Sn \cdot (PhH)_2$	562 ν_s(Sn−F) 538 ν_{as}(Sn−F)		280
$F_2Sn \cdot PhCl$	566 ν_s(Sn−F) 549 ν_{as}(Sn−F)	764 δ(C−H) 753 δ(C−H)	280
$F_2Sn \cdot PhMe$	563 ν_s(Sn−F) 536 ν_{as}(Sn−F)	740 δ(C−H)	280
$F_2Sn \cdot ClMe$	567 ν_s(Sn−F) 543 ν_{as}(Sn−F)		281
$Cl_2Sn \cdot {}^{12}CO$	324.3 ν_{as}(Sn−Cl in Sn^{35}Cl$_2$) 319.9 ν_{as}(Sn−Cl in Sn^{35}Cl^{37}Cl)	2 175.5 ν(C−O)	223
$Cl_2Sn \cdot {}^{13}CO$	324.3 ν_{as}(Sn−Cl in Sn^{35}Cl$_2$) 319.9 ν_{as}(Sn−Cl in Sn^{35}Cl^{37}Cl)	2 127.8 ν(C−O)	223
$Cl_2Sn \cdot NO$	326.9 ν_{as}(Sn−Cl in Sn^{35}Cl$_2$) 323 ν_{as}(Sn−Cl in Sn^{35}Cl^{37}Cl)	1 891.7 ν(N−O)	223
$Cl_2Sn \cdot N_2$	329.8 ν_{as}(Sn−Cl in Sn^{35}Cl$_2$) 326.7 ν_{as}(Sn−Cl in Sn^{35}Cl^{37}Cl)		223
$Cl_2Sn \cdot (N_2)_2$	326.1 ν_{as}(Sn−Cl in Sn^{35}Cl$_2$) 322.3 ν_{as}(Sn−Cl in Sn^{35}Cl^{37}Cl)		223
$Cl_2Sn \cdot P(Bu\text{-}t)_3$	290 ν(P−SnCl$_2$) 250 ν(P−SnCl$_2$)		282
$Br_2Sn \cdot P(Bu\text{-}t)_3$	204 ν(P−SnBr$_2$) 192 ν(P−SnBr$_2$)		282
$F_2Pb \cdot CO$	520.6 ν_s(Pb−F) 496.3 ν_{as}(Pb−F)	2 176.4 ν(C−O)	223
$F_2Pb \cdot NO$	522.6 ν_s(Pb−F) 498.7 ν_{as}(Pb−F)	1 891.4 ν(N−O)	223
$F_2Pb \cdot N_2$	526.6 ν_s(Pb−F) 502.2 ν_{as}(Pb−F)		223
$Cl_2Pb \cdot {}^{12}CO$	315.2 ν_s(Pb−Cl) 292.6 ν_{as}(Pb−Cl)	2 174.5 ν(C−O)	223
$Cl_2Pb \cdot {}^{13}CO$	315.2 ν_s(Pb−Cl) 292.6 ν_{as}(Pb−Cl)	2 126 ν(C−O)	223
$Br_2Pb \cdot CO$		2 161.2 ν(C−O)	223

Tevault and Nakamoto[223]. It was found that the bands of CO and NO shift to higher (up to 2%) frequencies whereas CA bands shift to lower (up to 10%) frequencies by complex formation. No band assignable to the N_2 stretching mode was observed, evidently due to weak interaction between $SnCl_2(PbF_2)$ and N_2. When the N_2 concentration reached 4% in Ar the new bands corresponding to the $SnCl_2 \cdot (N_2)_2$ complex were detected (Table 9).

The magnitude of the observed shifts upon complexation was used as a measure of the extent of σ donation from the ligand to the metal center of the carbene analogs, i.e. as a measure of the strength of the complexes[223]. The effect of changing the ligand has been elucidated in the $SnCl_2 \cdot B$ and $PbF_2 \cdot B$ series. The magnitude of the negative shifts of the EX_2 stretching bands follows the order CO > NO > N_2. The effect of changing the halogen was studied in the $PbX_2 \cdot CO$ series. The CO stretching frequencies of these complexes follow the order: $PbF_2 > PbCl_2 > PbBr_2 > PbI_2$. Thus, the strength of the complexes decreases from CO to N_2 and from fluorides to iodides.

Recently, complex formation between SnF_2 and N_2 was studied by matrix IR spectroscopy[275]. Complexes of SnF_2 with one and two molecules of N_2 were detected. Based on the magnitudes of the shifts of the EX_2 stretching bands upon complexation, it has been demonstrated that the strength of complexes of these molecules with N_2 of the same composition [1 : 1: $N_2 \cdot SnF_2$[275], $N_2 \cdot SnCl_2$, $N_2 \cdot PbF_2$[223], and 1 : 2: $(N_2)_2 \cdot SnF_2$[275], $(N_2)_2 \cdot SnCl_2$[223]] is nearly identical. The structure and stability of $N_2 \cdot SnF_2$ and $(N_2)_2 \cdot SnF_2$ were studied by *ab initio* methods (see Section X)[275].

The undesirability of using dinitrogen as a matrix gas in studies of CAs is one of the important conclusions of these works[223,275]. The vibrational bands recorded in N_2 matrices for $SiCl_2$, $SiBr_2$[232], GeF_2[209], $GeCl_2$, $SnCl_2$, $SnBr_2$, SnBrCl, $PbBr_2$, PbBrCl[200], $PbCl_2$[200,226], MeSiCl, MeSiH and Me_2Si[197,276] which were assigned to the corresponding CAs can in fact belong to complexes of CAs with N_2 (probably of 1 : 2 composition). Complexation with N_2 results in only small shifts in the IR bands of CAs, but probably affects significantly the position of the absorption maximum in the UV-VIS spectra of the CAs (see Sections IV and V, and compare UV-spectral data of MeClSi, MeHSi and Me_2Si in different matrices[197]).

Complexes between SnF_2 and very weak donors of electron density like ethylene[277], heptyne-1[278,279] and aromatics (PhH, PhCl and PhMe)[280] of 1 : 1 composition were formed in Ar matrices at 12 K and studied by IR spectroscopy. In addition, the complex $F_2Sn \cdot (PhH)_2$ was detected[280] (Table 9). The magnitudes of shifts of the SnF_2 stretching bands upon complexation with unsaturated compounds testify to the similar strength of these complexes. In this series a C=C double bond is found to be a slightly stronger electron density donor than a triple one. In the series of monosubstituted benzenes, an electron donor substituent (Me) increases while an electron-withdrawing substituent (Cl) decreases the strength of complexes formed.

The only example of a complex of CA with alkyl halide detected by matrix IR spectroscopy is the complex $MeCl \cdot SnF_2$[281]. Its structure and stability were studied by *ab initio* methods (see Section X).

The ability of SnF_2 to form labile complexes with varied electron density donors provides a better understanding of the mechanisms of action of Sn(II) salts as co-catalysts in important industrial processes.

To conclude this section we note that IR absorptions corresponding to the vibrations of $P \cdots EHal_2 (E = Ge, Sn)$ fragments were described for the room-temperature stable complexes $Ph_3P \cdot GeCl_2$, $(t\text{-}Bu)_3P \cdot GeCl_2$, $Ph_3P \cdot GeBr_2$, $(t\text{-}Bu)_3P \cdot SnCl_2$ and $(t\text{-}Bu)_3P \cdot SnBr_2$[282].

VIII. MICROWAVE SPECTRA OF GERMYLENES. STRUCTURES OF SHORT-LIVED CARBENE ANALOGS

A. Rotational Transitions

Rotational transitions in the ground electronic state have been studied for only two germylenes: GeF_2 and $GeCl_2$. There are no microwave studies of stannylenes or plumbylenes. GeF_2 was produced by evaporation of germanium difluoride at 363 K[283] or by electric discharge in GeF_4[284]. $GeCl_2$ was generated in a glow discharge of $GeCl_4$ or by interaction of $GeCl_4$ with metallic Ge at 770 K[285].

The microwave spectrum of GeF_2 in the region 8–35 GHz[283] and the millimeter-wave spectrum of $GeCl_2$ in the frequency range 108–160 GHz[285] have been recorded for the ground (0,0,0) and vibrationally excited (1,0,0), (0,1,0), (0,0,1) and (0,2,0) states. In the case of GeF_2 only lines originating from molecules containing germanium isotopes with zero nuclear spin (^{70}Ge, ^{72}Ge, ^{74}Ge, ^{76}Ge) were assigned initially[283], using characteristic Stark patterns and isotope shifts. Later[211], the lines of $^{73}GeF_2$ were also assigned and the quadrupole coupling constants obtained from the observed hyperfine structure due to ^{73}Ge nuclear quadrupole. No hyperfine structure due to chlorine nuclear quadrupoles was resolved in the spectrum of $GeCl_2$[285].

Rotational constants obtained for both the ground and the three first excited vibrational states allowed one to derive the equilibrium molecular structures of GeF_2 (r_e = 1.7321 Å, $\theta_e = 97.148°$[211]) and $GeCl_2$ (r_e = 2.169452 Å, $\theta_e = 99.8825°$[285]). From measurements of the Stark effect the dipole moment of GeF_2 has been determined to be 2.61 Debye[283]. The harmonic and anharmonic force constants up to the third order have been obtained for both molecules and reported too[283,285].

A further study of rotational transitions for four isotopomers of GeF_2 with germanium isotopes 70, 72, 74 and 76 (and also for $^{28}SiF_2$) using a cavity pulsed microwave Fourier transform spectrometer[284] has allowed us to observe the ^{19}F hyperfine structure owing to the higher resolution achieved in the spectra. Spin–spin and spin–rotation coupling constants have been obtained from the analysis of this hyperfine structure[284]. The F–F internuclear distance has been determined from these constants. It turned out to be only slightly smaller than the distance, which can be derived from the equilibrium geometry, reported previously[211].

B. The Ground State Geometries of Short-lived Carbene Analogs

The ground state geometries of labile CAs obtained by different methods are collected in Table 10. There is good agreement in the results reported for each of the CAs. The bond lengths in triatomic CAs increase with increasing atomic mass of the central atom and the substituents, reflecting the increase in the covalent radii of the composing atoms. It has been noted repeatedly that the E–X bond lengths in EX_2 are always longer than those in EX_4 (E = Si, Ge, Sn, Pb; X = H, Hal)[11,283,286–288]. This was ascribed to a higher ionic character of the E–X bonds in EX_2 than in EX_4[283,288] and to the presence of occupied antibonding MOs in EX_2, which are absent in EX_4, resulting in lengthening and weakening of the E–X bonds in EX_2[286,287] (a detailed description of the valence shell MOs in triatomic germylenes, stannylenes and plumbylenes is presented in Section IX).

The bond angle in triatomic CAs with the same substituents decreases upon increasing the atomic number of the central atom, although it is worth noting that uncertainties in the determination of bond angles for plumbylenes are extremely large[289]. In the case

TABLE 10. Geometries of labile carbene analogs in the ground electronic state

	Method[a]	Bond length(s) (Å)	Bond angle(s) (deg)	Reference
SiH_2	IRDLKS	1.525^b; 1.514^c	91.8^b, 92.08^c	291
	ES	1.51402^c	91.9830^c	63
SiHF	ES	1.548^b, 1.606^b	97.0^b	85
	ES	1.528^c, 1.603^c	96.9^d	85
SiF_2	MW	1.5901^c	100.77^c	292
SiFCl	MW	1.5960^d, 2.0714^d	100.85^d	293
SiHCl	ES	1.5214^b, 2.0729^b	95.0^b	82
	ES	1.525^c, 2.067^c	96.9^c	82
$SiCl_2$	MW	2.065310^c	101.3240^c	294
	ED	2.083^e, 2.089^f	102.8^e	295
	ED	2.076^c	104.2^c	231
	ES	2.068^b	101.5^b	71
SiHBr	ES	1.518^b, 2.237^b	93.4^b	83
	ES	1.522^c, 2.231^c	95.9^c	83
$SiBr_2$	ED	2.243^e; 2.249^f	102.7^e	295
	ED	2.227^c	103.1^c	231
SiHI	ES	1.534^b, 2.463^b	92.4^b	86
GeH_2	ES	1.5934^b	91.28^b	30
	ES	1.5883^c	91.22^c	30
GeF_2	MW	1.7321^c	97.148^c	211
GeHCl	ES	1.592^b, 2.171^b	$(94.3)^{b,h}$	78
$GeCl_2$	MW	2.169452^c	99.8825^c	285
	ED	2.183^e; 2.186^f	100.3^e	288
GeHBr	ES	1.598^b, 2.329^b	$(93.9)^{b,h}$	78
$GeBr_2$	ED	2.359^f	101.0^f	296
GeHI	ES	1.593^b, 2.525^b	$(93.5)^{b,h}$	79
GeI_2	ED	2.540^g	102.1^g	221
SnF_2	ED	2.06^f	—	297
$SnCl_2$	ED	2.335^c	99.1^c	298
$SnBr_2$	ED	2.501^c	100.0^c	298
	ED	2.512^f	100.0^f	299
SnI_2	ED	2.688^c	105.3^c	298
	ED	2.706^f	103.8^f	299
PbF_2	ED	2.041^f	97^f	300
$PbCl_2$	ED	2.444^f	97^f	300
$PbBr_2$	ED	2.598^f	97^f	300
PbI_2	ED	2.807^f	97^f	300
$H_2C=Si$	ES	1.105^b, 1.706^b	114.4^b (HCH)	101, 102
$H_2C=Ge$	ES	1.1022^b, 1.7908^b	115.05^b (HCH)	101

[a]IRDLKS = infrared diode laser kinetic spectroscopy, investigation of the rotational structure of the ν_2 band; MW = microwave spectroscopy; ED = electron diffraction; ES = electronic spectroscopy, investigation of the rovibronic structure of electronic transitions; bond angles calculated from isotopic shifts of the ν_3 bands of triatomic CAs are not presented here due to their large experimental error; these values are reported in Section VI.

[b]r_0, θ_0. [c]r_e, θ_e. [d]r_z, θ_z. [e]r_a, θ_a. [f]r_g, θ_g. [g]r_α, θ_α.

[h] The bond angles presented in parentheses have been transferred from quantum-chemical calculations and used to obtain the bond lengths presented.

of symmetric triatomic CAs the bond angle increases upon changing the substituents from H to F and then down the Group 17 elements. This can be explained in that the increasing ratio of the substituent covalent radius to the central atom covalent radius brings about increasing spatial repulsion between the two substituents[288]. This trend is also in agreement with predictions of valence-shell electron pair repulsion (VSEPR) theory[290]. In accordance with Bent's rule[137] the central atom s AO contribution to bonding decreases and the s character of the lone electron pair increases with the increasing difference in electronegativities of the central atom and the substituents, which leads to a decrease in the bond angle. However, a decrease in the bond angle can be seen for asymmetric triatomic CAs in the series F > Cl > Br > I. Apparently, lengthening the E−Hal bonds on passing successively from F to Br results in diminution of the spatial interaction between the halogen and hydrogen atoms in these CAs.

Besides the triatomic CAs, the molecular structures have been determined experimentally for only labile silylidene and germylidene, as shown in Table 10. At the same time most of the stable CAs have been characterized by X-ray analysis or by electron diffraction. The available structural data for some stable germylenes, stannylenes and plumbylenes have partly been presented in Section IV. The comprehensive consideration of the geometries of stable CAs is beyond the scope of the present review.

IX. PHOTOELECTRON SPECTRA, IONIZATION ENERGIES, ELECTRON AFFINITIES AND REDOX POTENTIALS OF GERMYLENES, STANNYLENES AND PLUMBYLENES

Photoelectron spectroscopy is an important tool for studying the molecular orbital (MO) structure. Photoelectron spectroscopy probes the occupied molecular orbitals by measuring the ionization energies of electrons. The angle distribution of the photoelectrons gives information on the symmetry of the MOs. In accordance with Koopmans' theorem[301], ionization energy (IE), or binding energy, of an electron is the negative of the energy of the corresponding molecular orbital. Thus photoelectron spectroscopy gives a set of MO energies for a molecule. However, it is noteworthy that there are some restrictions in the application of Koopmans' theorem. At the same time no significant deviations from predictions made on the basis of Koopmans' theorem have been revealed in photoelectron spectroscopy studies of carbene analogs (CAs). Most studies of germylenes, stannylenes and plumbylenes were performed by means of ultraviolet photoelectron spectroscopy. Assignment of bands in photoelectron *(PE)* spectra is based on quantum-chemical calculations at different levels, which have been carried out in parallel with the photoelectron spectroscopy studies. Correlations with previous assignments of bands of other CAs have been taken into account too. For some germylenes, stannylenes and plumbylenes the first adiabatic ionization energy (IE) has also been evaluated by means of mass spectrometry.

A. Prototype EH_2 Molecules

The upper limit for the adiabatic first IE of GeH_2 has been obtained by means of photoionization mass spectrometry in the course of photoionization studies of GeH_n molecules, generated by abstraction of hydrogen atoms from GeH_4 by fluorine atoms[302]. It was found to be 9.25 eV[302]. On the basis of simple MO considerations it can be concluded that this ionization comes from HOMO, which is of the germanium atom lone-pair character. This is the only study of molecules of this simplest type.

B. $EHal_2$ Molecules

The PE spectra of $EHal_2$ were recorded in a molecular effusive beam. Monomeric $SnHal_2$ and $PbHal_2$ were produced by evaporating the corresponding salts $(EHal_2)_x$ at temperatures above the melting point. The formation of SnF_2 in this process was accompanied by the appearance of another species (a band at 10.63 eV), which was tentatively identified as the dimer $(SnF_2)_2$[303]. This is the only complication noted for the evaporation processes used. For the production of vapors of sufficiently pure monomeric $GeHal_2$

(Hal = F, Cl, Br, I), a solid state reaction between germanium sulfide and suitable lead dihalides at *ca* 570 K was used[304–306]. The observed IEs, corresponding to ionizations from the valence shell MOs, are presented in Table 11.

$EHal_2$ molecules belong to the C_{2v} symmetry group and have nine occupied valence MOs, four of which are of a_1 symmetry, three of b_2 symmetry, one of b_1 symmetry and one of a_2 symmetry, taking the YZ plane as the molecular plane with the Z axis being the C_2 axis.

Quantum-chemical calculations [relativistic HF/TZ, using geometrical parameters derived from experimental data for $GeHal_2$[304–306], pseudo-potential LCAO-MO-SCF[307] and CI[308] for $SnCl_2$, as well as CNDO for $SnCl_2$ and $SnBr_2$[130], and extended Hückel MO (EHMO) for a series of $EHal_2$[309]] predict the following sequence of the valence shell MOs.

The HOMO, $4a_1$ (only valence shell orbitals are numbered), represents a combination of halogen p AOs destabilized by an antibonding interaction with a central atom sp hybrid orbital. Thus this MO is antibonding E–Hal MO in nature and has quite a large contribution from a central atom valence s orbital. This explains the quite low IEs observed for ionizations from $4a_1$ MO, incompatible with a lone-pair orbital mainly localized on the central atom. This MO determines the Lewis base properties of the central atom in $EHal_2$.

The following four MOs have a predominantly halogen lone-pair character. The $3b_2$ and $1a_2$ MOs are almost completely localized on the halogen atoms. The $3b_2$ MO is formed by in-plane halogen p AOs and corresponds to antibonding halogen–halogen through space interaction. Therefore, it is slightly destabilized relative to the $1a_2$ MO, which arises from two spatially remote, slightly overlapping, out-of-plane halogen p AOs. The $1b_1$ and $3a_1$ orbitals, being also localized mainly on the halogen atoms, have some stabilizing contribution from the central atom p orbitals. The $1b_1$ MO is constructed from the halogen out-of-plane p AOs and the central atom unoccupied p_x orbital, while the $3a_1$ MO corresponds to a combination of the in-plane halogen p AOs with some contribution (depending on the halogen attached) from the central atom p_z orbital.

The $2b_2$ bonding MO represents an in-phase combination of the central atom p_y and the radial halogen p AOs. Hence this MO is of a bonding E–Hal nature and can be identified as σ(E–Hal). However, CNDO calculations predict that the energy of the $3a_1$ MO is lower than that of the $2b_2$ MO[130], in disagreement with *ab initio* calculations[307,308].

The $2a_1$ MO has mainly a central atom s AO character, but is stabilized through interaction with a radial halogen p orbital combination (Hal = F, Cl, Br), and can be identified as σ(E–Hal), or it is destabilized through a halogen s orbital combination of a_1 symmetry (Hal = I). Some halogen s AO contribution is present in this MO, even in the case of F, Cl and Br substituents.

$1b_2$ and $1a_1$ MOs represent halogen s orbital combinations of corresponding symmetries, with $1a_1$ being stabilized through interaction with the central atom s AO in the case of iodine owing to quite high energy of the 5s AO of this atom.

The PE spectra of all $EHal_2$ molecules conform to the described sequence of MOs, although it is worth mentioning that the initial assignment[310,311] of the bands of $SnHal_2$ and $PbHal_2$, Hal = Cl, Br, based on correlations with PE bands of linear $HgHal_2$ and diatomic TlHal and InHal, differed notably from the assignment presented in Table 11.

Relative band intensity changes in the PE spectra recorded using He I and He II radiation[303,307,309] or synchrotron radiation with similar photon energies (24 and 50 eV)[308,312] provide experimental support to the predominant central atom orbital character of the $4a_1$ and $2a_1$ MOs, assuming that MO photoionization cross-sections are linked with those of atomic orbitals. It comes from losses in intensity of the $4a_1^{-1}$ and $2a_1^{-1}$ bands (the superscript "–1" means an electron detachment from this MO) relative to the

TABLE 11. Ionization energies (eV) for ionizations from the valence molecular orbitals of $EHal_2$ and their assignments[a,b]

	$4a_1$ 	$3b_2$ 	$1a_2$ 	$1b_1$ 	$3a_1$ 	$2b_2$ 	$2a_1$ 	$1a_1 + 1b_2$ 	Exciting irradiation	Reference
GeF_2	11.98	14.4	14.4	15.9	15.55	16.2	18.70		He I	304
$GeCl_2$	10.55	11.44	11.70	12.58	12.69	13.41	16.73		He I	305
$GeBr_2$	10.02	10.54	10.86	11.67	11.82	12.61	16.42		He I	305
GeI_2	9.08	9.50	9.83	10.62	10.62	11.57			He I	306
SnF_2	11.48	13.61	13.61	14.37	14.37	14.37	17.04		He I, He II	303
$SnCl_2$	10.31	11.31	11.31	12.07	12.07	12.72	15.9		He I	310
	10.31	11.27	11.27	12.02	12.02	12.69	15.81		He I, He II	309
	10.38	11.39	11.39	12.39	12.39	12.78	15.94		He I, He II	307
	10.37	11.0 sh	11.33	12.12	12.12	12.77	15.90		He I	130
								22.61	SR (24 eV)	308, 312
$SnBr_2$	9.85	10.63	10.63	11.39	11.39	12.10	15.6		He I	310
	9.83	10.58	10.58	11.33	11.33	12.05	15.56		He I, He II	309
	9.87	10.2 sh	10.65	11.35	11.35	12.05	15.24		He I	130
SnI_2	9.05	9.23	9.55	10.35	10.35	11.19	15.50		He I, He II	309
PbF_2	11.84	12.89	12.89	13.59	13.59	13.59			He I, He II	303
$PbCl_2$	10.34	10.86	10.86	11.58	11.58	12.19	16.47		He I, He II	309
	10.1	10.6	10.6	11.5	11.5	12.0			He I	310
	10.11	10.82	10.82	11.56	11.56	11.97			He I	311
$PbBr_2$	9.81	10.25	10.25	10.94	11.13	11.63	16.3		He I	310
	9.85	10.29	10.29	11.07	11.07	11.71	16.19		He I, He II	309
PbI_2	8.90	9.20	9.49	10.20	10.32	10.91	16.32		He I, He II	309[c]

[a] Schematic MO representations are based on those shown in References 304–306 for $GeHal_2$ (Hal = F, Cl, Br, I). Only the most important contributing AOs are depicted leaving out variations, which differ for different molecules.

[b] sh = shoulder.

[c] The He I PE spectrum of PbI_2 is also shown in Reference 311.

halogen lone-pair originating bands ($3b_2^{-1}$, $1a_2^{-1}$, $1b_1^{-1}$ and $3a_1^{-1}$) in the He I PE spectra of $SnHal_2$ and $PbHal_2$ compared to the He II spectra. It is consistent with the lower photoionization cross-section of the halogen valence p orbitals relative to the tin and lead p and s valence orbitals with respect to the He II radiation. In spite of the predominant halogen p AOs character of the $2b_2$ MO, the $2b_2^{-1}$ band of $SnCl_2$ was found not to decrease in intensity in the He II spectrum[307], which implies an appreciable contribution from the central atom AOs.

It has been noted repeatedly that the bands originating from ionizations from the $4a_1$ and $2a_1$ MOs are weaker than the other five observed for $EHal_2$ in the He I PE spectra, with the $2a_1^{-1}$ band not being observed in the PE spectrum of GeI_2 and PbF_2 (see Table 11). This conclusion should be treated with care, because $3b_2^{-1}$ and $1a_2^{-1}$ ionizations as well as those from $1b_1$ and $3a_1$ often form a single unresolved band, while $4a_1^{-1}$ and $2a_1^{-1}$ bands usually have no contribution from the other ionizations. Nevertheless, this fact can again be rationalized by a considerable central atom AO character of the $4a_1$ and $2a_1$ MOs and, particularly, by the predominant central atom valence s character of the $2a_1$ orbital (and predominant Ge 4s and I 5s character of $2a_1$ in GeI_2[306]). The more intensive bands arise from MOs of mainly halogen lone-pair p character.

The predominant halogen valence-shell s character of $1b_2$ and $1a_1$ MOs, and therefore their low cross-section to He II radiation, results in extremely low intensity of the $1b_2^{-1}$ and $1a_1^{-1}$ bands. Actually, the band corresponding to ionizations from both MOs has been observed only for $SnCl_2$, when intensive synchrotron radiation was used[308,312].

IEs originating from MOs with considerable central atom character ($4a_1, 2a_1$) show smaller shifts due to exchange of halogen atoms than IEs originating from MOs with predominant halogen character ($3b_2, 1a_2, 1b_1, 3a_1$)[305]. This trend can be seen from the data of Table 11. It is interesting to note that such shifts for ionizations from $2b_2$ have an intermediate value. The general trend in decreasing IEs of $EHal_2$ on passing from F through to I substituents parallels the IEs of the halogen atoms[130,304].

Photoelectron spectroscopy of solid Sn(II) and Pb(II) halides, $(EHal_2)_x$, have also been recorded[303,309]. A remarkable resemblance of the solid-phase and gas-phase spectra of $EHal_2$ (E = Sn, Pb, Hal = F, Cl, Br, I) was found, although the solid-phase bands were naturally shifted to a lower energy region and were broadened compared to the gas-phase bands. This suggests that much of the molecular orbital character of stannylenes and plumbylenes is carried over to the orbital structure of the solids, in spite of the fact that the crystal structure of solid $(EHal_2)_x$ does not include $EHal_2$ units[313,314].

The elements Sn and Pb have outer d-shell electrons, the ionization energy of which fall in the range normally regarded as the valence region of ionization energies ($\leqslant$ *ca* 30 eV), although the d orbitals are essentially atomic in nature. Due to spin–orbit interaction in the final d^{-1} ion it can be formed in the $^2D_{5/2}$ and $^2D_{3/2}$ states, with the $^2D_{5/2}$ state being of lower energy. Because the energy separation between these states is quite large, the features associated with d-shell ionization show characteristic $^2D_{5/2}/^2D_{3/2}$ splittings. The d-shell orbital IEs for $EHal_2$ are shown in Table 12. Central atom outer d-shell IEs of $EHal_2$ show a shift to higher values with increasing halogen electronegativity[307,315], similar to the valence-shell IEs. It has been shown that these IEs are consistent with the atomic description for the central atom outer d orbitals[315,316]. It has been found also that these IEs can be reasonably described by a simple electrostatic model[315,316]. The asymmetry parameters β for ionization from the 5d subshell of the metal atom in $PbHal_2$ (as well as in $HgHal_2$ and TlHal), Hal = F, Cl, Br, I, have been determined from He II_α photoelectron spectra recorded at two angles to an unpolarized photon beam[317].

TABLE 12. Outer d-shell ionization energies (eV) for $EHal_2$(E = Sn, Pb)

	$^2D_{5/2}$	$^2D_{3/2}$	Exciting irradiation	Reference
SnF_2	33.7	34.5	He II_α	303
$SnCl_2$	33.48	34.53	He II_α, He II_β	309, 315, 316
	33.34	34.69	He II	307
	33.48	34.53	SR	308, 312
$SnBr_2$	33.15	34.21	He II_α, He II_β	309, 315, 316
SnI_2	33.6	34.9	He II	309
PbF_2	27.63	30.34	He II_α	303, 317
$PbCl_2$	27.34	29.92	He II_α, He II_β	309, 315–317
$PbBr_2$	27.02	29.58	He II_α, He II_β	309, 315–317
PbI_2	26.48	29.20	He II_α, He II_β	309, 315–317

The most detailed photoelectron spectroscopy study has been carried out for $SnCl_2$[308] using synchrotron radiation (SR). The use of dispersed SR as an ionizing source for photoelectron spectroscopy studies has great benefit due to its features of high intensity of this radiation, continuous tunability over a wide spectral range and some others. The use of SR (21–52 eV) first allowed one to investigate the complete outer- and- inner-valence shells, as well as Sn 4d subshell photoionization of $SnCl_2$. Some new features in the PE spectrum of $SnCl_2$ were revealed. On the basis of configuration interaction (CI) calculations it has been concluded that the Koopmans' theorem assignment for the outer-valence ionizations $4a_1^{-1}, 3b_2^{-1}, 1a_2^{-1}, 1b_1^{-1}, 3a_1^{-1}$ and $2b_2^{-1}$ has been satisfactory, whereas the spectral strength of the $2a_1^{-1}$ main line has been considerably reduced and three new weak broad features, observed around the $2a_1^{-1}$ band, have been mainly due to satellites of the $2a_1^{-1}$ band. A band at 22.61 eV, corresponding to an ionization from $1a_1 + 1b_2$ MOs of mainly Cl 3s character, has been detected for the first time. The quite low IE value for this ionization has been explained by high electron density on the chlorine atoms in $SnCl_2$ due to tin donation. For the $4d^{-1}$ doublet the branching ratio ($^2D_{5/2}$: $^2D_{3/2}$) has been measured. The value obtained (1.39) is not far from the statistical value of 1.5. An Auger widespread band (independent of whether the radiation energy was 50 or 52 eV) due to the 4d hole decay has been observed in addition to the 4d ionization. This band corresponds to $SnCl_2^{++}$ states lying approximately within the 26.5–28.7 eV range. Records of constant ionic state spectra revealed two resonances at 25.03 and 26.11 eV, assigned to 4d $\rightarrow 8b_1$ (LUMO) transition with spin–orbit splitting. The excitation of $SnCl_2$ at these energies induces intensive autoionization processes following the 4d $\rightarrow$ $8b_1$ electron promotion, resulting in strongly resonant behavior of most of the bands observed.

The adiabatic first IEs for some of $EHal_2$ have also been determined by means of electron impact mass spectrometry (EIMS) from ionization efficiency curves. In these studies the vapor of monomeric GeF_2 was produced by reaction of CaF_2 with metallic germanium at *ca* 1500 K[318] or by evaporation of melted germanium difluoride at *ca* 400 K[319]. Monomeric $GeCl_2$ and $GeBr_2$ were obtained by interaction of the vapor of $GeCl_4$ (at 520–660 K[320,321]) or $GeBr_4$ (at 623 K[320]) with metallic germanium. Other monomeric species were produced by evaporation of the corresponding melted salts. The EIMS results are compared with the photoelectron spectroscopy data in Table 13. As can be seen, the EIMS and photoelectron spectroscopy data (except those for $SnBr_2$, $PbBr_2$ and some data for SnI_2) are in reasonable agreement.

C. EHalHal′ Molecules

Three molecules of this type have been studied by EIMS and their adiabatic first IEs determined[322–324]. The values are given in Table 13. The EHalHal′ molecules were produced by evaporation of 1 : 1 mixtures of $(EHal_2)_x$ and $(EHal'_2)_x$ at Knudsen conditions[322–324]. From comparison of the IE data in Table 13 it may be deduced that the available first IEs for EHalHal′ are somewhat overestimated.

D. Acyclic Organyl and Aminogermylenes, -stannylenes and -plumbylenes

The compounds ERR′, where E = Ge, Sn, Pb; R, R′ = $(Me_3Si)_2CH$, t-Bu$(Me_3Si)N$ or $(Me_3Si)_2N$, belong to the first known representatives of stable CAs. Their gas-phase PE spectra (evaporation temperature <390 K) were reported in a series of publications[130–132,153,330,331]. The IEs obtained are collected in Table 14. The assignments of PE bands of ER_2(R = $(Me_3Si)_2CH$, t-Bu$(Me_3Si)N$ and $(Me_3Si)_2N$) have been made on the basis of a comparison of the observed spectra of ERR′ with those of the parent molecules RH, as well as of ER_4 and also of HgR_2 and ZnR_2 (which differ from ER_2 by the absence of a lone-pair at the central atom), assuming a local C_{2v} symmetry at the E atom[130].

The first bands in the PE spectra of $E(CH(SiMe_3)_2)_2$ were assigned to the ionization from the a_1 HOMO, which represents a combination of sp hybrid orbitals of the divalent atom E and the carbon 2p orbitals. This MO is mainly localized on the atom E and has its lone-pair character due to a predominant contribution from s AO. Interaction of the E atom p AO with the E atom s AO lowers the energy of the resulting MO, while the interaction with α-carbon p AOs raises it relative to the energy of the p AO of the free atom E. Both effects approximately balance out, and the resulting IE for ionization from the HOMO of ER_2 has been noted to be close to the atomic first IE due to ionization from p AO.

TABLE 13. First ionization energies (eV) of $EHal_2$

	EIMS	Reference	Photoelectron spectroscopy[a]		Reference
	adiabatic		adiabatic	vertical	
GeF_2	11.6 ± 0.3	318	11.65	11.98 ± 0.03	304
	11.8 ± 0.1	319			
$GeCl_2$	10.4 ± 0.3	320	10.20 ± 0.05	10.55 ± 0.03	305
	10.2 ± 0.1	321			
$GeBr_2$	9.5 ± 0.3	320	9.60 ± 0.05	10.02 ± 0.03	305
$SnCl_2$	10.1 ± 0.4	325	—	10.38	307
$SnBr_2$	10.0 ± 0.4	326	—	9.83	309
	10.6 ± 0.2	322			
SnI_2	9.8 ± 0.2	322	—	9.05	309
	9.3 ± 0.5	327			
	8.83 ± 0.10	328			
$PbCl_2$	10.3 ± 0.1	324	—	10.34	309
$PbBr_2$	10.2 ± 0.1	324	—	9.85	309
PbI_2	8.86 ± 0.03[b]	329	—	8.90	309
SnBrCl	10.3 ± 0.3	323	—	—	
SnBrI	10.0 ± 0.2	322	—	—	
PbBrCl	10.4 ± 0.1	324	—	—	

[a]Typical error for values obtained by photoelectron spectroscopy is of a few hundredths of eV.
[b]Photoionization MS.

The second band in the PE spectra of $E(CH(SiMe_3)_2)_2$ has been assigned to ionization from the b_2 MO, which is antisymmetric bonding MO, localized mainly on the C_2E fragment. The next band originating from the a_1 symmetric bonding MO ionization has not been observed since it is obscured by the broad unresolved bands due to ionizations from the substituent's orbitals.

In the case of $E(N(SiMe_3)_2)_2$ and $E(N(Me_3Si)(t\text{-Bu}))_2$ the first band in the PE spectra has been assigned to the ionization from the b_2 MO, corresponding to the nitrogen lone-pair orbital antibonding combination, whereas only the second band arises from ionization from the a_1 MO of the divalent atom E lone-pair character. The third band is due to ionization from nitrogen lone-pair bonding MO of a_1 symmetry. The fourth band, observed only for stannylenes and plumbylenes, but obscured for germylenes by broad bands due to ionizations from the substituent's MOs, corresponds to ionization from the b_2 bonding MO, $\sigma_{asym}(E{-}N)$.

The energy of the MO corresponding to the lone-pair of the atom E and $\sigma_{asym}(E{-}N)$ MO in $E(N(SiMe_3)_2)_2$ and $E(N(SiMe_3)(t\text{-Bu}))_2$ rises with increasing mass of the atom E, following the trend in the atomic first IEs of the Group 14 elements, in accordance with considerable localization of these MOs at the E atoms. The slight increase in the first IEs and the slight decrease in the third IEs on going from the germylenes to the plumbylenes reflect the increasing spatial separation between nitrogen atoms in this series. It therefore reflects the weakening of 'through space' interaction between the nitrogen lone-pair AOs, resulting in that antibonding combination (b_2) becoming 'less' antibonding, whereas the bonding combination (a_1) becomes 'less' bonding.

Comparison of the IEs corresponding to ionization from lone-pairs of the atoms E in $E(N(SiMe_3)_2)_2$ and $E(N(SiMe_3)(t\text{-Bu}))_2$ shows that the IE values are lower in the latter case in spite of the fact that the more electronegative substituents should stabilize the HOMO, increasing its IE, and that the $N(SiMe_3)_2$ group is expected to be less electronegative than the $N(SiMe_3)(t\text{-Bu})$ group due to the presence of two electropositive trimethylsilyl substituents. This fact can be explained by a significant $p_\pi{-}d_\pi$ interaction between the nitrogen lone-pairs and the vacant silicon 3d orbitals, which increases the overall electronegativity of the $N(SiMe_3)_2$ group compared to the $N(SiMe_3)(t\text{-Bu})$ group[130]. Such $p_\pi{-}d_\pi$ interaction lowers not only the energy level of the MO with the central atom lone-pair character, but also the energy levels of the MO representing nitrogen lone-pairs.

The proposed explanation is consistent with the reported data on the first IEs of stable $E(NR_2)_2(E = Ge, Sn; R = t\text{-Bu}$ or $NR_2 = \text{cyclo-}NCMe_2(CH_2)_3CMe_2)$, bearing only β-carbon atoms in the substituents[163,164]. The first IEs of these CAs are much lower than those of $E(N(SiMe_3)_2)_2$ and $E(N(SiMe_3(t\text{-Bu}))_2$ (Table 14) due to the absence of $p_\pi(N){-}d_\pi(Si)$ stabilization of their HOMO. It can be expected that the second IE, corresponding to the ionization from the divalent atom E lone-pair, is also lowered for the CAs bearing only β-carbon atoms in the substituents. These IEs have not been published, but there are indirect arguments in favor of this expectation based on easier oxidizing addition of MeI to $E(NR_2)_2(E = Ge, Sn; NR_2 = \text{cyclo-}NCMe_2(CH_2)_3CMe_2)$[164].

On the basis of the IE values for ionizations from the divalent atom E lone-pair, one can expect a decrease in the basicity of germylenes, stannylenes and plumbylenes in the following series: $E(CH(Me_3Si)_2)_2 > E(N(Me_3Si)(t\text{-Bu}))_2 > E(N(Me_3Si)_2)_2$[130]. This ordering is in agreement with available data on the chemical behavior of these molecules[130,131]. At the same time it is necessary to note that $EHal_2$ have higher first IE (corresponding to the divalent atom E lone-pair MO), but form complexes with Lewis acids more readily than $E(N(SiMe_3)_2)_2$[130]. This indicates that there are other factors which play an important role in the formation of complexes by these CAs[130].

TABLE 14. Valence-shell ionization energies (eV) for polyatomic ER_2, E = Ge, Sn, Pb

								Exciting irradiation	Reference
		a_1 (E:)		$b_2(\sigma_{asym}$(E-C))					
$((Me_3Si)_2CH)_2Ge$		7.75		8.87				He I	130, 153, 331
$((Me_3Si)_2CH)_2Sn$		7.42		8.33				He I	130, 330, 331
$((Me_3Si)_2CH)_2Pb$		7.25		7.98				He I	130, 330, 331
		b_2 (N:)		a_1 (E:)	a_1 (N:)	$b_2(\sigma_{asym}$(E-N))			
$(t\text{-Bu}(Me_3Si)N)_2Ge$		7.24		8.27	8.61			He I	130–132
$(t\text{-Bu}(Me_3Si)N)_2Sn$		7.26		7.90	8.47	9.33		He I	130–132
$(t\text{-Bu}(Me_3Si)N)_2Pb$		7.26		7.69	8.49	9.00		He I	130–132
$((Me_3Si)_2N)_2Ge$		7.71		8.68	8.99			He I	130–132
$((Me_3Si)_2N)_2Sn$		7.75		8.38	8.85	9.50		He I	130–132
$((Me_3Si)_2N)_2Pb$		7.92		8.16	8.81	9.39		He I	130–132
$((t\text{-}Bu_2N)_2Ge$		6.78						He I	163
$((t\text{-}Bu_2N)_2Sn$		6.74						He I	163
$((CH_3)_2\overline{C(CH_2)_3C(CH_3)_2N})_2Ge$		6.90						He I	164
$((CH_3)_2\overline{C(CH_2)_3C(CH_3)_2N})_2Sn$		6.80						He I	164
		(N: + Cl:)		(E:)					
$GeCl(N(SiMe_3)_2)$[a]		9.2		10.0				He I	134
	$\pi-3$	σ-lp	$\pi-2$	σ(Ge-N)	$\pi-1^* + \sigma(t\text{-Bu})$	σ	σ		
t-Bu, N, Ge[b], N, t-Bu	6.65, 6.85, 6.97	8.60	8.80	10.55	11.12	12.68	14.96	He I, He II	133

[a]This molecule is assumed to have a local C_s symmetry. For description of the MOs, see the text.
[b]This molecule has C_s symmetry. Designation of the MOs is shown in the authors' notation; for a more detailed description of the MOs, see the text.

Core levels in $Sn(N(SiMe_3)_2)_2$ have been studied by means of X-ray photoelectron spectroscopy[132]. The following binding energies are observed: Sn ($3d_{5/2}$) 491.93, N (1s) 402.01, Cl (1s) 289.36 and Si ($2p_{3/2}$) 105.97 eV.

The He I PE spectrum of the labile germylene, $GeCl(N(SiMe_3)_2)$, has been obtained recently[134]. This germylene was generated by pyrolysis of the corresponding germacyclopentene in the gas phase at 713 K, as shown in equation 7.

$$\text{(3,4-dimethyl-1-chloro-1-}N(SiMe_3)_2\text{-germacyclopent-3-ene)} \xrightarrow[-\ \text{2,3-dimethylbutadiene}]{713\ K} GeCl(N(SiMe_3)_2) \xrightarrow[-Me_3SiCl]{1023\ K} Ge{=}NSiMe_3 \qquad (7)$$

At higher temperatures $GeCl(N(SiMe_3)_2)$ underwent further decomposition to trimethylsilylgermaisonitrile. The He I PE spectra of a series of germaimines and germaisonitriles, generated from similar precursors, have also been reported in this paper[134].

The assignment of the bands attributed to $GeCl(N(SiMe_3)_2)$ has been done with the aid of B3LYP calculations on model structures. The first band at 9.2 eV belongs to ionization from MO which represents antibonding combination of the nitrogen and chlorine lone-pair. The second band at 10.0 eV corresponds to ionization from MO of germanium atom lone-pair character. The bands around 10.7 eV are associated with a chlorine atom lone-pair and σ(Si−C) MO. Unresolved bands around 13.5 eV have also been observed.

E. Cyclic Molecules

The PE spectrum of the stable aromatic 1,3-di-*tert*-butyl-1,3,2-diazagermol-2-ylidene (Table 14), as well as of its carbon and silicon analogs, has been obtained using He I and He II radiation[133]. Assignment of the observed bands has been carried out on the basis of DFT calculations of the compounds under consideration and Koopmans' theorem[133]. The first band of the germylene corresponds to ionization from the π-type HOMO, arising from out-of-phase mixing of the C=C π-orbital of the imidazole ring with combination of the nitrogen lone-pairs and the germanium atom p orbital. It clearly shows vibrational structure, tentatively assigned to a stretching frequency of the molecular ion in the ground state. Similarly to the stable $E(NR_2)_2$, considered above, the ionization from the MO of the germanium atom lone-pair character is responsible for the second band. The third band is due to ionization from MO, which is essentially the out-of-phase combination of the nitrogen lone-pairs. The fourth band corresponds to ionization from σ-MO, which mainly comprises Ge−N bonds with some contribution from *tert*-butyl C−C bonds. The next three bands are due to ionization from MOs with a large contribution from the *tert*-butyl groups.

The 1,3-di-*tert*-butyl-1,3,2-diazasilol-2-ylidene has been found to have similar molecular orbital structure[133]. At the same time, in the case of the carbon analog, 1,3-di-*tert*-butylimidazol-2-ylidene, the first IE corresponds to ionization from MO of the carbene center atom lone-pair character, which suggests in the framework of Koopmans' theorem that this compound has reverse ordering of the two highest occupied MOs[133]. However, the subsequent *ab initio* calculations with account of electron correlation have shown that the reason for the low IE in the ionization from the carbene center atom lone-pair orbital of 1,3-di-*tert*-butylimidazol-2-ylidene is the large relaxation of the corresponding wave function in the cation[332]. In this case Koopmans' theorem should not be valid. The degree of aromatic stabilization in this type of carbene and carbene analogs has also been discussed in detail[133,332,333] based on the established electronic structure. The influence of the electronic structure of these CAs on their chemical and physicochemical properties has also been considered[133,332,333].

The unoccupied molecular orbital structure of 1,3-di-*tert*-butylimidazol-2-ylidene and its silicon and germanium analogs, as well as of their saturated analogs, 2,5-di-*tert*-butyl-1-E-2,5-diazacyclopentaylidenes (E = C, Si, Ge), has been studied by means of inner-shell electron energy loss spectroscopy (ISEELS)[334]. ISEELS consists in excitation of electrons from core-shell levels to the virtual levels and provides information on both the energies of the unoccupied MOs and the spatial distribution of excited electron density in those orbitals. Each of the carbenes and carbene analogs as well as a number of model compounds have been characterized by C 1s and N 1s spectra. The Ge 3p spectra have also been recorded for germylenes. The analysis of these spectra with the aid of *ab initio* calculations resulted in some interesting conclusions concerning the nature of bonding in these molecules. Particularly, it has been shown that there is considerable π-allyl delocalization over the N−E−N fragment in all molecules and additional aromatic delocalization in the unsaturated molecules[334].

F. Electron Affinities

Electron affinity (EA) is a measure of the ability of a molecule to attach an electron. This value is equal to an adiabatic IE of the corresponding negative ion. EAs for some $EHal_2$ have been derived from appearance energies of the corresponding anions formed by low-energy electron dissociative resonance capture from $EHal_4$[335–337] in the course of mass spectrometric studies. For the purpose of evaluation of the EAs, many possible processes following the electron capture were considered, but most of them were found not to fit the experiments. The EAs corresponding to appropriate processes are presented in Table 15. In the case of $GeCl_2$ and $GeBr_2$ the authors[337] were unable to decide between two possible processes, but inclined to favor the process leading to electronically excited Hal_2.

Relatively low values of IP and significant values of EA obtained experimentally for a number of CAs suggest that these species can participate in electron transfer interactions with a variety of electron acceptors/donors.

G. Electrochemistry

One-electron electrochemical oxidation (E_{ox}) and reduction (E_{red}) potentials are quantities closely related to the first IEs and EAs, respectively. One of the main differences between the two series of quantities is that the former embrace effects of solvation of both the initial neutral precursers and the final ions. Unfortunately, the solvation energies cannot be evaluated easily, therefore it is usually difficult to correlate electrochemical

TABLE 15. Electron affinities (eV)

	Assumed type of dissociative resonance capture		Reference
	$EHal_4 + e \rightarrow EHal_2^- + 2Hal$	$EHal_4 + e \rightarrow EHal_2^- + Hal_2^*$	
GeF_2	1.3	—	335
$GeCl_2$	2.90	2.56	337
$GeBr_2$	1.80	1.61	337
$SnCl_2$	—	1.04	336
$SnBr_2$	—	1.33	336
SnI_2	—	1.74	336

potentials with the IEs and EAs. Since many reactions occur in the liquid phase, the oxidation and reduction potentials are of great importance for chemists.

Data on the redox potentials of germylenes, stannylenes, plumbylenes and their complexes are scarce. In fact, only the electrochemistry of dihalogermylenes, dihalostannylenes and their complexes with Lewis bases[338] as well as with chromium, molybdenum and tungsten pentacarbonyles[339] has been studied.

Cyclic voltammetry of dihalogermylenes, dihalostannylenes and their complexes with Lewis bases revealed one reduction and one oxidation peak (both are one-electron)[338]. The $E_{1/2}$ values are given in Table 16. Most of the reduction and oxidation waves of EX_2 and $EX_2 \cdot B$ were found to be irreversible, suggesting that the corresponding radical ions are very unstable. Quasi-reversible oxidation waves were observed for $GeX_2 \cdot$ dioxane ($X = Cl$, Br) and $GeCl_2 \cdot Py$ complexes[338]. Quasi-reversible reductions were found for the $GeCl_2 \cdot$ dioxane complex and GeI_2. Lifetimes of the $GeCl_2 \cdot$ dioxane and GeI_2 radical anions were estimated to be *ca* 4 and 2.5 s at 20 °C, respectively[338].

The nature of ligand B affects the redox properties of $GeCl_2 \cdot B$ complexes. The oxidation potentials increase in the order: bpy $<$ $AsPh_3$ $<$ Py $\sim$ PPh_3 $<$ dioxane. The reduction potentials tend to become more negative in the opposite sequence: dioxane - PPh_3 - Py - $AsPh_3$ - bpy (Table 16). A linear correlation was found between the oxidation and reduction potentials of $GeCl_2 \cdot B$, suggesting that the molecular orbitals involved in the electrochemical oxidation and reduction processes are located on the germanium moiety[338]. Using a standard electrochemical method the equilibrium formation constants, K, for $GeCl_2 \cdot PPh_3$ ($K = 7 \times 10^3$ mol l^{-1}, MeCN, 20 °C) and $GeCl_2 \cdot AsPh_3$ ($K = 2 \times 10^4$ mol l^{-1}, MeCN, 20 °C) were determined. The values obtained are close to that measured by using UV spectroscopy (for $GeCl_2 \cdot PPh_3$, $K = 2 \times 10^3$ mol l^{-1}, 23 °C, in *n*-Bu_2O[198].

TABLE 16. Redox potentials and electrochemical gaps of dihalogermylenes, dihalostannylenes and their complexes with Lewis bases in MeCN at 20 °C (platinum electrode, Bu_4NBF_4 as supporting electrolyte, vs. Ag/AgCl/KCl (sat.))[338]

EX_2	B	$E_{1/2}$ (ox)(V)	$-E_{1/2}$(red)(V)	G(V)^a
$GeCl_2$	dioxane	1.46^b	0.41^b, 0.44^c	1.87
$GeCl_2$	PPh_3	1.14(2e)^d	0.58	1.72
$GeCl_2$	$AsPh_3$	1.05	0.59	1.64
$GeCl_2$	Py	1.12^b	0.56	1.68
$2GeCl_2$	bpy^f	0.91 (2e)^e	0.74	1.65
$GeBr_2$	dioxane	1.08^b	0.45	1.53
$GeBr_2$	PPh_3	0.72	0.38	1.10
GeI_2	—	—	0.99^b	>3.59
GeI_2	PPh_3	1.44	—	>3.15
SnF_2	—	—	0.94	>3.54
$SnCl_2$	—	1.88	0.21, 1.20	2.09
$SnCl_2$	dioxane	1.67	0.78	2.45
$SnBr_2$	—	1.82	0.40, 1.31	2.22
SnI_2	—	1.16	0.02, 0.51	1.14

^{a}Electrochemical gap, $G = E_{ox} - E_{red}$.
^{b}Quasi-reversible.
^{c}In DMF[339].
^{d}A two-electron wave is due to simultaneous oxidation of free PPh_3 at this potential as well.
^{e}A two-electron wave is due to coordination of two $GeCl_2$ moieties.
^{f}bpy = α, α-bipyridinyl.

Taking GeI_2 and $SnCl_2$ as examples, one can conclude that in general the complexation with n-donors results in a decrease of the oxidation potentials and in shifts of the reduction potentials of the CAs to the more cathodic region, as expected.

The values of the oxidation potentials of dihalogermylenes (stannylenes) and their complexes with the Lewis bases indicate that these compounds should react as typical reducing agents. However, the low reduction potentials of these compounds suggest that they should be quite strong oxidizing agents too. Indeed, the reduction potentials of $GeX_2 \cdot B$(X = Cl, Br; B = dioxane, PPh_3, $AsPh_3$, Py) lie in the region from −0.4 to −0.60 V, which is typical of the convenient organic [e.g. *p*-benzoquinone, $E_{1/2}$(red) = −0.52 V] or inorganic [e.g. O_2, $E_{1/2}$(red) = −0.82 V] oxidizing agents. Of the compounds studied, SnI_2 was found to be the most powerful oxidizing agent. Its reduction potential (−0.02 V) is close to the reduction potential of such a strong oxidant as TCNQ (+0.12 V). Several reactions illustrating the oxidizing properties of dihalogermylenes, dihalostannylenes and their complexes were found[338].

An electrochemical gap ($G = E_{ox} - E_{red}$) characterizes the energy gap between the HOMO and LUMO. The $GeBr_2 \cdot PPh_3$ complex and SnI_2 have the smallest G value while GeI_2 and $GeI_2 \cdot PPh_3$ have the largest (Table 16). Thus, one may expect that, of the compounds studied, SnI_2 and the $GeBr_2 \cdot PPh_3$ complex, possessing both low oxidation and reduction potentials, have a tendency to give a radical or a single electron transfer radical ion reaction, while ionic processes would be more effective with GeI_2, $GeI_2 \cdot PPh_3$ and SnF_2.

X. QUANTUM-CHEMICAL CALCULATIONS

A wide range of quantum-chemical methods have been used for studying properties of carbene analogs R_2E (E = Ge, Sn, Pb) and various aspects of their chemistry. They include semiempirical methods (MNDO, AM1 and PM3), *ab initio* calculations at various levels and approaches based on density functional theory (DFT) which, in the last decade, emerged as a reliable and economic tool for modeling ground state properties and reaction dynamics of intermediates.

Most of the R_2E are transient species. Even their formation typically occurs via sequences of complex chemical reactions. *A priori* several reaction channels involving these intermediates could be envisaged. Quantum-chemical calculations have been used to explore and compare possible pathways.

The chemistry of carbenes and silylenes was developed extensively in the last decades. Parallels in the chemistry of singlet carbenes and silylenes with that of their heavier analogs are widely recognized heuristic tools to characterize reactivity of these less studied species. Calculations allow one to study in detail the trends in the prototype series of model chemical systems involving the derivatives of two-coordinated C, Si, Ge, Sn, Pb and thus get a deeper insight into the chemistry of these species. Moreover, quantum-chemical calculations are an important tool used in the interpretation of photoelectron, IR and UV spectra of CAs.

It is often difficult to deduce from the restricted experimental data the main trends in reactivity. Theoretical calculations could give the needed information to fill these gaps. In other words, they could be used to sometimes 'interpolate' scarce experimental data, e.g. in studying substituent effects. To examine general trends in the series of Group 14 elements carbene analogs, data for silylenes and even for carbenes are presented in this section for some cases.

A. Methods

Relativistic effects are known to be important for molecules with heavy elements. The influence of these effects on the ground state properties of hydrides EH_2 (E = Si,

Ge, Sn, Pb) has been studied by Dyall at the SCF level using all-electron molecular DHF calculations[340]. The known relativistic shortening of bond lengths involving a heavy atom in the series of EH_2 was estimated to be 0.00004 Å (E = Si), 0.003 Å (E = Ge), 0.011 Å (E = Sn) and 0.044 Å (E = Pb). The same effect on the H−E−H bond angle in EH_2 is very small ($<0.4°$). More significant changes were induced on the dipole moments of EH_2 (0.02, 0.10, 0.25, and 0.69 D for E = Si, Ge, Sn, and Pb, respectively). An analysis of the electron density in EH_2 shows that relativistic effects reduce the electronegativity of the heavy atom. Relativistic effects decrease the calculated stretching frequencies.

Relativistic effects in heavy atoms are most important for inner-shell electrons. In *ab initio* and DFT calculations these electrons are often treated through relativistic effective core potentials (RECP), also known as pseudopotentials. This approach is sometimes called quasirelativistic, because it accounts for relativity effects in a rather simplified scalar way. The use of pseudopotentials not only takes into account a significant part of the relativistic corrections, but also diminishes the computational cost.

Geometry optimization followed by vibrational frequency calculations is now common practice in quantum-chemical studies. Thus structural information as well as harmonic IR band positions are produced in most computational studies of CAs. A recent survey[289] on the structure of symmetric dihalides $EHal_2$ (E = Si, Ge, Sn, Pb, Hal = F, Cl, Br, I) shows that *ab initio* calculations reproduce nicely both bond lengths and bond angles[289]. The observed variations in the structural parameters both down Group 14 and from fluoride to iodide are also well reproduced. The only notable exception is the bond angle in SnF_2, where experiment and theory are in obvious disagreement[289]. Arguments based on the trend established for other EX_2 dihalides favor the *ab initio* value[289].

DFT-based methods are also proved to be quite reliable for calculations of both geometry and vibrational frequencies of CAs (see, for example, references[341–343]).

In recent years IR frequency calculations of CAs have been carried out routinely not only for IR spectra interpretations itself, but also for reaction studies in order to verify the nature of stationary points found and for predicting vibrational band shifts upon complexation. Several papers[342,344] were devoted to comparative study of the accuracy of various DFT functionals for calculating various properties of germylenes, stannylenes and plumbylenes. A complete set of EX_2 and EX_4 (E = C, Si, Ge, Sn, Pb; X = F, Cl, Br, I) molecules was studied by DFT methods using RECP[342]. Semilocal and hybrid functionals and B3LYP especially were found to be superior to the Hartree–Fock level for calculating reaction energies. It should be noted that functionals which do not rely on empirical adjustments (like the currently most popular B3LYP) but on universal physical constraints (so-called nonempirical functionals) are expected to give more accurate predictions for all systems, including elements from the whole Periodic Table. One of these nonempirical density functionals, PBE[345], was tested extensively[346,347] and compared with *ab initio* G2 calculations in a study of germylene cycloaddition reactions[348].

B. Calculations of Electronic Transition Energies. Singlet–Triplet Energy Separations ΔE_{ST}

According to calculations all experimentally known silylenes[3], germylenes, stannylenes and plumbylenes have a singlet ground state. Theory predicts a triplet ground state for SiLiH, $SiLi_2$[3], GeHLi and $GeLi_2$[349,123]. These lithiated species are of interest from a purely theoretical point of view because they have little chance of being experimentally observed. Recent DFT calculations by Apeloig and coworkers[125] predicted a triplet ground state for several organic silylenes bearing bulky substituents with α-electropositive atoms with $(t\text{-}Bu_3Si)_2Si$ being the most promising candidate for experimental verification (calculated singlet–triplet splitting $\Delta E_{ST} = -0.31$ eV). No such predictions were reported for germylenes as well as for stannylenes and plumbylenes. Spectroscopic studies on singlet–triplet transitions are very limited compared to the wealth of data on singlet–singlet transitions. Quantum-chemical calculations were used to fill this gap. Most of these were performed for tri- and tetraatomic germylenes, stannylenes and plumbylenes. Below we briefly consider the results obtained using the most advanced methods. In discussing trends within the ER_2 series we restrict ourselves to data obtained at the same computational level. Values of ΔE_{ST} calculated at *ab initio* and DFT levels are presented and compared with available experimental data in Table 17.

TABLE 17. Calculated singlet–triplet splitting ΔE_{ST} (eV) in germylenes, stannylenes and plumbylenes

	ΔE_{ST}	Method	Reference
GeH_2	1.05	MCSCF/ECP-DZP + SO	350
GeH_2	1.00	CAS + MCSCF RCI	351
GeH_2	0.99	MRD CI	352
GeH_2	0.93	CASSCF/TZ(2df,2p)	26
GeH_2	1.18	B3LYP/6–311G*	123
GeF_2	3.59 (3.79)[a]	CCSD(T)/DZP	32
GeF_2	3.54	MRSDCI + Q	353
GeF_2	3.64	B3LYP/6–311G*	123
$GeCl_2$	2.61 (2.77)[a]	CAS-MCSCF + MRCI	80
$GeCl_2$	2.76	B3LYP/6–311G*	123
$GeBr_2$	2.41 (2.36)[a]	CAS-MCSCF + MRCI	80
$GeBr_2$	2.46	B3LYP/6–311G*	123
GeI_2	1.84 (1.86)[a]	CAS-MCSCF + MRCI	80
GeHF	2.01	B3LYP/6–311G*	123
GeHCl	1.73	CAS-MCSCF + MRCI	80
GeHCl	1.83	B3LYP/6–311G*	123
GeHBr	1.64	CAS-MCSCF + MRCI	80
GeHBr	1.75	B3LYP/6–311G*	123
GeHI	1.51	CAS-MCSCF + MRCI	80
GeHLi	−0.20	B3LYP/6–311G*	123
$GeLi_2$	−0.26	B3LYP/6–311G*	123
GeHMe	1.24	B3LYP/6–311G*	123
$GeMe_2$	1.33	B3LYP/6–311G*	123
$GeMe_2$	0.61	SCF/DZP	129
$Ge(NH_2)_2$	2.42	B3LYP/6–31G*	349
$Ge(OH)_2$	2.92	B3LYP/6–31G*	349
$Ge=CH_2$	2.50	CISD + Q//TZ(2df,2pd)	100
$Ge=CH_2$	2.56	B3LYP/6–311G*	123
SnH_2	1.03	MCSCF/ECP-DZP + SO	350
SnH_2	1.03	CAS + MCSCF RCI	351
SnF_2	3.40 (*ca* 2.48)[a]	MRSDCI + Q	353
$SnCl_2$	2.60 (2.76)[a]	CASSCF + MRSDCI	354
$SnBr_2$	2.41 (2.23)[a]	CASSCF + MRSDCI	354
SnI_2	2.04 (2.11)[a]	CASSCF + MRSDCI	354
PbH_2	1.70	MCSCF/ECP-DZP + SO	350
PbH_2	1.76	CAS + MCSCF RCI	351
PbF_2	4.08 (5.03)[a]	MRSDCI + Q	353
$PbCl_2$	3.02	CASSCF + MRSDCI	354
$PbBr_2$	2.82	CASSCF + MRSDCI	354
PbI_2	2.33 (2.50)[a]	CASSCF + MRSDCI	354

[a]Experimental ΔE_{ST} values in parentheses are taken from Table 1.

The differences between the calculated and experimental ΔE_{ST} values are usually less than 10%, the only exceptions being SnF_2 with its large experimental uncertainty in ΔE_{ST} (see above) and PbF_2. It is worth noting the good accuracy achieved in DFT calculations for ΔE_{ST} of germylenes[123] and dihalostannylenes[355].

Halogen substitution in germylenes, stannylenes and plumbylenes increases the singlet–triplet gap in the same way as in carbenes[11] and silylenes[3]. It is often implicitly assumed that ΔE_{ST} as well as many other characteristics should change down the Periodic Table in a monotonous way. Thus the expected order of ΔE_{ST} in the CA series is $SiR_2 < GeR_2 < SnR_2 < PbR_2$. The data of Table 17 show an interesting irregularity on going from germylenes to stannylenes. Calculations by Gordon and coworkers[350] on the EH_2 series predict a small increase (0.02 eV) in ΔE_{ST} on going from GeH_2 to SnH_2. Previous CAS+MCSCF RCI calculations by Balasubramanian[351] gave a 'normal' ordering: $\Delta E_{ST}(GeH_2) < \Delta E_{ST}(SnH_2)$, but with a very small (0.03 eV) difference between both values. For most dihalides $EHal_2$ (Hal = F, Cl, Br) calculations at the MRSDCI level and available experimental data (Table 17) show an 'inverted' order $\Delta E_{ST}(GeR_2) > \Delta E_{ST}(SnR_2)$. On going from SnR_2 to PbR_2 the ΔE_{ST} values grow sharply due to strong relativistic effects in PbR_2 (see below). The 'special' position of stannylenes in the series GeF_2, SnF_2 and PbF_2 was also mentioned while comparing some calculated ground state characteristics like IP, F–E–F bond angle, E–F overlap and E–F bond strengths (E = Ge, Sn, Pb)[353].

The role of relativistic effects in low-lying states of GeH_2, SnH_2 and PbH_2 has been analyzed by Balasubramanian[351] using CAS MCSCF computations followed by large-scale relativistic CI, which include the spin–orbit integrals. The spin–orbit mixings of the ${}^1A_1$ and ${}^3B_1(A_1)$ states in the relativistic CI wave functions of PbH_2 and SnH_2 were found to be quite significant, especially for PbH_2. This mixing lowers the ${}^1A_1$ state of PbH_2 by 1308 cm^{-1} while the ${}^3B_1(A_1)$ state is raised by 1371 cm^{-1} with respect to the ${}^3B_1$ state without the spin–orbit splitting. Thus relativistic effects increase ΔE_{ST} in PbH_2 by 0.33 eV, which accounts for half the difference between the respective values for SnH_2 and PbH_2 (Table 17). A similar conclusion was reached by Gordon and coworkers[350], who explored the coupling of the ${}^1A_1$ and ${}^3B_1$ states in EH_2 (E = C, Si, Ge, Sn and Pb) with a different approach. Relativistic potential energy surfaces were constructed using a spin–orbit coupling term. The relativistic effect does not affect the singlet-triplet gap in CH_2, SiH_2 and GeH_2 since the spin-orbit coupling in these molecules is relatively small. The relativistic ΔE_{ST} values in SnH_2 and PbH_2 are about 0.04 eV and 0.25 eV larger than the corresponding adiabatic ones.

Use of *ab initio* calculations to assign the singlet–singlet transitions in spectroscopic studies of small germylenes, stannylenes and plumbylenes has now become standard practice[47,80,100,351,354]. For unsymmetric triatomics HGeR (R = Cl, Br, I), *ab initio* calculated geometric parameters were used to resolve some problems in determining molecular structures for both the ground and excited states[78,79].

C. Reactions of R_2E

Quantum-chemical calculations have been used to probe all the characteristic chemical reactions of CAs (at least in the case of silylenes and germylenes). The theoretical studies cover intramolecular rearrangements, insertions into σ-bonds, additions to double and triple bonds and dimerizations. Note that experimental data on the mechanisms of these reactions are still scarce and the results of theoretical studies are needed to understand the main trends in the reactivity of germylenes, stannylenes and plumbylenes.

1. Intrinsic stability of R_2E. Intramolecular isomerizations

The reactions that eventually determine the existence of any species are intramolecular isomerizations and fragmentations. CAs are usually considered to be stable toward fragmentations, though numerical data on bond energies are scarce.

Calculated[353,354,356] and experimental energies of dissociation reactions $EX_2 \rightarrow EX + X$ (E = Ge, Sn, Pb; R = H, F, Cl, Br, I) are collected in Table 18. Dissociation could produce EX and X in various low-lying electronic states. The bond energies ΔE in Table 18 correspond to the lowest electronic state of the products. The trends in bond energies for dihalides EX_2 are similar to those in tetrahalides EX_4 : ΔE decreases in the series Ge > Sn > Pb and, with the same E, in the series F > Cl > Br > I. The high values of ΔE (Table 18) support the general assumption of stability of the two coordinated Ge, Sn and Pb species toward dissociation. However, ionization could dramatically reduce (to 37.9 kcal mol^{-1} for GeH_2^+) the strength of the E–H bond[356].

Intramolecular rearrangement of germylenes, stannylenes and plumbylenes into a doubly bonded isomer can be another cause of their intrinsic instability. For example, in the matrix IR study of 1-germacyclopent-3-en-1,1-ylidene, direct experimental evidence was obtained for a photochemical germylene–germene isomerization[105,357].

Isomerizations in the ground state were studied theoretically only in the case of germylenes. Early *ab initio* calculations show methylgermylene to be more stable compared to germaethene by 23[358], 15[359], and 17.6[360] kcal mol^{-1}. The most recent theoretical estimates of isomerization energies and activation barriers[361] for reactions $HGeEH_3 \rightarrow H_2Ge=EH_2$ (E = C, Si, Ge) are given in Table 19.

TABLE 18. Dissociation energies ΔE(kcal mol^{-1}) for reactions: $EX_2 \rightarrow EX + X$ (E = Ge, Sn, Pb; X = H, F, Cl, Br, I)

EX_2	Calculated[a]	Exp.	Reference
$GeH_2(^1A_1)$	69.2	66	356
$GeF_2(^1A_1)$	116.2	116	353
$SnF_2(^1A_1)$	113	112.8	353
$SnCl_2(^1A_1)$	78.6		354
$SnBr_2(^1A_1)$	70.3		354
$SnI_2(^1A_1)$	57.4		354
$PbF_2(^1A_1)$	86	88.7	353
$PbCl_2(^1A_1)$	73.6	76.1	354
$PbBr_2(^1A_1)$	70.6		354
$PbI_2(^1A_1)$	57.9		354

[a] At the CASSCF/MRSDCI + RECP level.

TABLE 19. Relative energies (kcal mol^{-1}) of isomers and transition states (TS) in $HGeEH_3 \rightarrow H_2Ge{=}EH_2$ isomerization reactions calculated at the CCSD(T)/ANO level[361]

E	$HGeEH_3$	TS	$H_2Ge{=}EH_2$
C	0.0	21.7	11.4
Si	0.0	9.8	3.2
Ge	0.0	14	−2.0

Calculations show that the nature of the substituent EH_3 has a dramatic effect on the germylene stability. Thus, both methyl- and silylgermylene are more stable than germene and germasilene, while germylgermylene is slightly less stable than digermene. The isomerization barriers in systems with E = Si, Ge are not high, implying a facile thermal interconversion.

The relative stabilities of divalent and tetravalent EH_2O_2 (E = C, Si, Ge, Sn, Pb) isomers were calculated using the BLYP and B3LYP density functionals with DZP and TZ2P basis sets, as well as CCSD and CCSD(T) single-point energies at the BLYP/TZ2P optimized geometries[362]. Of four structures considered, $E(OH)_2$ was found to be a global minimum with E = Si, Ge, Sn, Pb. The difference in energies between the $E(OH)_2$ and HE(= O)OH isomers increases in the series C, Si, Ge, Sn, Pb: −42, 8, 30, 50 and 74 kcal mol^{-1}, respectively.

2. Dimerization

Experimental aspects of the dimerization of germylenes, stannylenes and plumbylenes were discussed in a review[363]. Relations between the characteristics of ER_2 (E = C, Si, Ge, Sn, Pb; R = H, F) and the structure of their dimers E_2R_4 were studied theoretically by Trinquier and coworkers[364–367].

A total of six E_2R_4 (E = C, Si, Ge, Sn, Pb; R = H, F) structures **1–6** were considered on the singlet potential energy surface (PES) at the CI + MP2/DZP//SCF/DZP level[364,365]. The most important trend observed in this series was a decrease in stability of the olefin-type structure **1** manifested in *trans*-bent structure **2** of Ge_2H_4, and a dramatic increase in the stability of the *trans*-bridged structures **4**, which become a global minimum for Sn_2R_2 and Pb_2R_2 (Table 20). For every E and R, the *cis*-structures **3** and **5** were found to be less stable compared to the corresponding *trans*-isomers **2** and **4**.

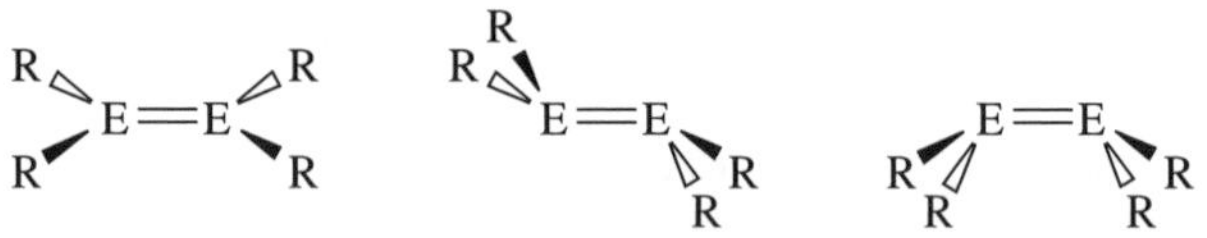

(1) **(2)** **(3)**

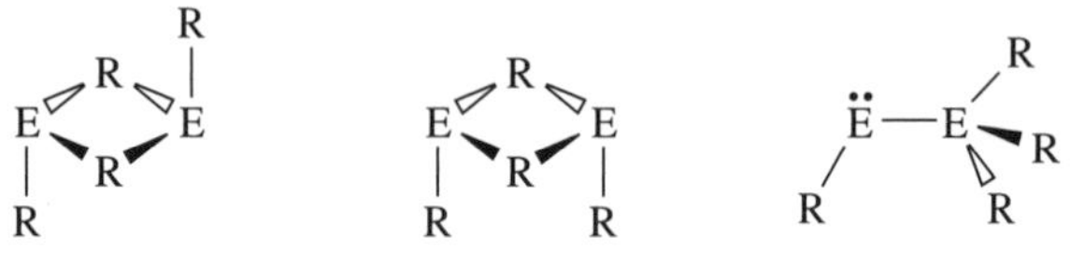

(4) **(5)** **(6)**

E = C, Si, Ge, Sn, Pb; R = H, F

A simple rule for the occurrence of *trans*-bent distorted structures **2** at homopolar double bonds was derived from an elementary molecular orbital model treating $\sigma-\pi$ mixing[364] and a valence bond treatment[367]. The relation between the singlet–triplet separation (ΔE_{ST}) of the constituent ER_2 and $\sigma+\pi$ bond energy $E_{\sigma+\pi}$ was used as a criterion for determining the expected structure of $R_2E{=}ER_2$. The *trans*-bent geometry **2** occurs when $1/4E_{\sigma+\pi} \leqslant \Delta E_{ST} < 1/2E_{\sigma+\pi}$. The first part of the inequality determines the *trans*-bending distortion of the double bond, while the second part determines the existence of a direct E=E link.

Singlet potential surfaces for E_2H_4 (E = C, Si, Ge, Sn) systems were explored through *ab initio* SCF + CI calculations with DZP (E = C, Si, Ge) and DZP-ECP (E = Sn, Pb) bases[365]. In all cases except E = C, the bridged structures were found to be true minima. Planar **1** or *trans*-bent **2** $HE_2{=}EH_2$ species were found to be true minima in all cases except E = Pb, where it is only a saddle point. The most stable structures of Si_2H_4 and Ge_2H_4 are **2**, while the most stable structures of Sn_2H_4 and Pb_2H_4 are the *trans*-bridged forms **4** (Table 20). The *cis*-bridged form **5** with E = Si, Ge, Sn, Pb is less stable relative to the *trans*-bridged **4** by *ca* 2 kcal mol^{-1}. The H_3EEH isomers **6** lie between these two symmetrical forms and are never found to be at the absolute minimum on the PES. The *trans*-bridged structures have a rather constant binding energy (Table 20) with respect to $2EH_2(^1A_1)$ whatever the nature of the E atom (30 ± 3 kcal mol^{-1}). The stability of the bridged structures may be due to the significant ionicity of the bridges $E^+{-}H^-{-}E^+$ in the planar four-membered rings.

The effect of fluorine substituents in ER_2 on the structures and stability of **1–6** dimers was studied theoretically at the same level[366]. The planar π-bonded structure **1** was found to be a true minimum on the C_2F_4 PES, but a saddle point on the Si_2F_4, Ge_2F_4 and Sn_2F_4 surfaces. The isomer **6** was found to be a true minimum in all cases except when E = Pb. Two nearly degenerate doubly bridged structures (*cis*-**5** and *trans*-**4**) were found to be true minima in all cases except when E = C. The preferred isomers are tetrafluoroethylene for C_2F_4, F_3SiSiF for Si_2F_4, and the *trans*-bridged structures for Ge_2F_4, Sn_2F_4 and Pb_2F_4.

With respect to two singlet EF_2 fragments, the bridged structures have binding energies that increase regularly from 3 kcal mol^{-1} (in Si_2F_4) to 62 kcal mol^{-1} (in Pb_2F_4), whereas bridged C_2F_4 is largely unbound (Table 20). The potential wells corresponding to the bridged structures were found to be rather flat, possibly inducing small distortions associated with very slight energy changes. The in-plane $C_{2h} \rightarrow C_i$ deformation found for the planar four-membered ring of **4** (E = Ge, R = F) is in agreement with its solid state geometry. A structural and energetic comparison was made for fluorine bridges in

TABLE 20. Energies (kcal mol^{-1}) of singlet isomers E_2R_4 relative to $2ER_2(^1A_1)$ (E = C, Si, Ge, Sn; R = H, F) calculated at the CI + MP2/DZP//SCF/DZP level

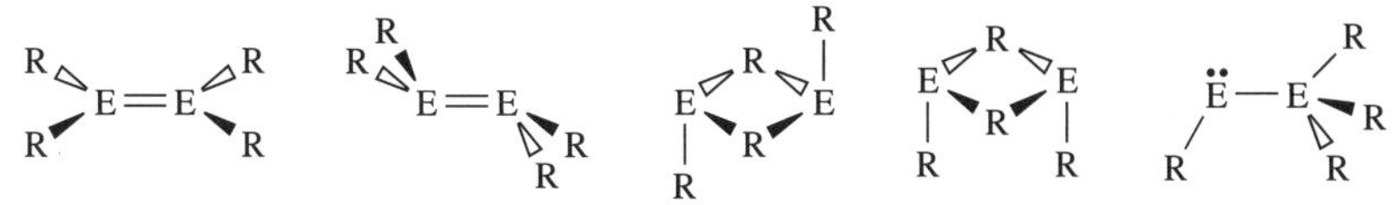

E	$R_2E{=}ER_2$ (**1**)	$R_2E{=}ER_2$ (**2**)	(**4**)	(**5**)	$\ddot{E}{-}ER_3$ (**6**)
		R = H[a]			
C	−192	—	−51.7	−27.3	−112.9
Si	−53.7	—	−31.2	−28.5	−43.9
Ge	−32.7	−35.9	−26.9	−24.3	−33.5
Sn	−14.7	−24.1	−33.2	−30.9	−26.2
Pb	15	−4.8	−28.7	−26.7	−11.2
		R = F[b]			
C	−53.6	—	65.3	65.3	−14.7
Si	44.2	—	−2.6	−2.1	−6.3
Ge	73.5	—	−23.8	−21.9	8.3
Sn	82.2	—	−50.9	−48.3	9.1
Pb	149.9	—	−62.3	−57.5	26.4

[a] Reference 365; [b] Reference 366.

E_2F_4 and the hydrogen bridges in E_2H_4. Some results were compared with available spectroscopic data for the monomers and dimers of SnF_2 and PbF_2.

Being very important, these results still raise some questions and provide an incentive for future studies. One is the influence of electronic correlation on the geometry of **1–6**. Use of more advanced theoretical methods is clearly needed. The conspicuous differences between the hydrogen and fluorine bridges need to be rationalized. The reaction path for interconversion of E_2R_4 isomers should also be explored. The stability of olefin-type structures $H_2E{=}EH_2$ (E = Ge, Sn) was studied computationally at the SOCI/3-21G(d)//MCSCF/3-21G(d) level using a four-electron four-orbital full optimized reaction space[368]. The ability of germanium to form π bonds was found to be higher than that of tin.

3. Insertions of CAs into H–H bond and their reverse reactions

Reaction with molecular hydrogen (equation 8) is the simplest example of CA insertion:

$$EH_2 + H_2 \longrightarrow EH_4 \; (E = Si, Ge, Sn, Pb) \qquad (8)$$

Reactions 8 of silylene, SiH_2[3] and germylene, GeH_2[369] are potentially of great significance in CVD systems where H_2 is present. The kinetics of SiH_2 reactions with H_2 were investigated in detail[3,370]. For the same reaction of GeH_2 experimental estimates of the reaction rate were also reported[369], whereas for SnH_2 and PbH_2 no kinetic studies are available. From the theoretical point of view reaction 8 is not only a prototype of all EH_2 insertions into single bonds but also provides a good test system for assessing the accuracy of various quantum-chemical schemes used for studying mechanisms of CA reactions.

The PES of the SiH_2 insertion reaction 8 was studied extensively at various theoretical levels[3]. It was established that account for electron correlation has a dramatic effect on both the shape of the PES and the barrier height[371]. The MP2 calculations[371] reveal the presence of a weak pre-reaction complex on the PES of the $SiH_2 + H_2$ system. In a recent paper[369] calculations at the QCSD(T) and DFT B3LYP levels have shown that GeH_2 also forms a pre-reaction complex with H_2. Complexes and transition states for SiH_2 and GeH_2 insertions have similar structures (Figure 1). Some geometrical parameters of these structures are shown in Table 21. Results from the B3LYP and QCISD calculations are in good agreement with each other, not only for the stable species (reagents and products) but also for the transition states and complexes. The differences between silylene and germylene complexes (Table 21) are of the degree to be expected from the characteristic differences in Ge–H and Si–H bond lengths. The calculated energies are shown in Table 22.

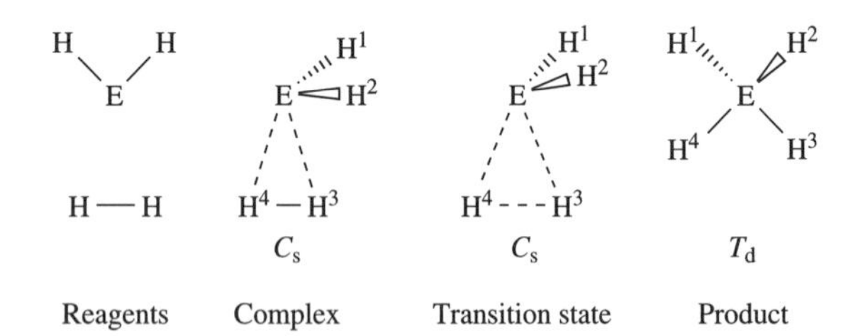

FIGURE 1. The stationary points of reactions $EH_2 + H_2 \rightarrow EH_4$; E = Si[369,371], Ge[369], calculated with B3LYP, MP2, and QCISD methods. The symmetry group is given below each structure

TABLE 21. Geometric parameters R (Å), A (deg) of stationary points of reactions $EH_2 + H_2 \rightarrow EH_4$ (E = Si, Ge) calculated by various methods. For atom numbering see Figure 1

Structure	Method	$R(E-H^1)$	$R(E-H^3)$	$R(E-H^4)$	$R(H^3-H^4)$	$A(H^1-E-H^2)$	Reference
	E = Si						
Complex	MP2[a]	1.505	1.858	1.787	0.792	94	371
Complex	QCISD[b]	1.509	1.936	1.859	0.783	95	369
Complex	B3LYP[c]	1.513	1.870	1.778	0.805	95	369
Transition State	MP2[a]	1.477	1.636	1.515	1.135	110	371
Transition State	QCISD[b]	1.479	1.661	1.522	1.117	110	369
Transition State	B3LYP[c]	1.484	1.647	1.522	1.140	110	369
	E = Ge						
Complex	QCISD[b]	1.589	2.276	2.225	0.758	92	369
Complex	B3LYP[c]	1.590	2.108	2.033	0.773	92	369
Transition State	QCISD[b]	1.539	1.742	1.573	1.215	111	369
Transition State	B3LYP[c]	1.539	1.721	1.568	1.250	111	369

[a]MP2/6-311G(2d,2p); [b]QCISD/6-311G(d,p); [c]B3LYP/6-311++G(3df,2pd).

TABLE 22. Energies (ΔE + ZPE) (kcal mol^{-1}) of stationary points on the PES of reaction 8

EH_2	$EH_2 + H_2$	Complex	TS[a]	XH_4	Method	Reference
SiH_2	0.0	—	5.0	−57.7	CCSD(T)[b]	372
SiH_2	0.0	−3.1	1.7	−55.2	MP4[c]	371
SiH_2	0.0	−3.3	1.4 (0.1)	−53.9	B3LYP[d]	369
SiH_2	0.0	−2.1	3.1 (1.7)	−54.0	QCISD(T)[e]	369
SiH_2	0.0	—	−0.5	−57.1	Exp.	369
GeH_2	0.0	—	17.6	−38.8	CCSD(T)[b]	372
GeH_2	0.0	−1.3	13.4 (12.1)	−35.7	B3LYP[d]	369
GeH_2	0.0	−0.9	14.0 (12.6)	−38.2	QCISD(T)[d]	369
GeH_2	0.0	—	15–20	−35.1 to −40.2	Exp.	369
SnH_2	0.0	—	37.9	−15.4	CCSD(T)[b]	372
PbH_2	0.0	—	53.0	7.7	CCSD(T)[b]	372

[a]Transition state. ΔE values given in parentheses are calculated at 298 K.
[b]CCSD(T)/ECP DZP.
[c]MP4(SDTQ)/6-311++G(3df,3pd)//MP2/6-311G(2d,2p).
[d]B3LYP/6-311++G(3df,2pd)//B3LYP/6-311++G(3df,2pd).
[e]QCISD(T)/6-311++G(3df,2pd)//QCISD(T)/6-311G(d,p).

The pre-reaction complex in the reaction of germylene with H_2 is only about half as strong (−0.9 to −1.3 kcal mol^{-1}) as that found for silylene (−2.1 to −3.3 kcal mol^{-1}) (Table 22). Both calculations and experiment[369] show that the insertion of GeH_2 into the H−H bond require overcoming a high activation barrier (calculation *ca* 12.1 kcal mol^{-1}, experiment *ca* 15–20 kcal mol^{-1}) in contrast to the reaction of SiH_2 with H_2, which in fact has no barrier. The calculations revealed no mechanistic differences in silylene and germylene insertions.

For heavier carbene analogs EH_2 theoretical results on reaction 8 are more scarce and experimental data are absent. The thermochemistry of reaction 8 for E = Si, Ge, Sn, Pb at the DHF level was investigated by Dyall[340]. Activation barriers for the whole series of elements (Si, Ge, Sn, Pb) were computed by Hein, Thiel and Lee[372] at the CCSD(T)/ECP DZP level (Table 22).

No complexes between EH_2 and H_2 were reported in Reference[372], probably because a search for them was not attempted. The question of complex intermediacy in reactions of SnH_2 and PbH_2 with H_2 remains open. One could expect their formation, but it is reasonable to assume that the complexes should be weaker compared to those formed by SiH_2 and GeH_2.

The reaction energies (Figure 2) increase significantly on going from E = Si to Pb and even become positive for E = Pb (Table 22). This reflects the increased stability of E(II) relative to E(IV) compounds for heavier Group 14 elements. Although PbH_4 is thermodynamically unstable with regard to PbH_2 elimination, it should be long-lived under normal conditions because the corresponding barrier is as high as 45.3 kcal mol^{-1}[372].

4. *Insertions into other σ bonds*

The first and so far only study on the whole set of EH_2 insertions into the H−E bond of the methane analogs EH_4 (E = C, Si, Ge, Sn, Pb) (equation 9) has been reported by Trinquier[373].

$$EH_2 + EH_4 \longrightarrow E_2H_6 (E = \text{Si, Ge, Sn, Pb}) \quad (9)$$

Calculations were conducted at the MP4/DZP//HF/DZP level (ECP DZP for X = Sn, Pb). Some conclusions of this work[373] should be considered with caution due to restrictions of the SCF approach used in the geometry optimization. The most important conclusion, supported for E = Si[374] and Ge[375] by recent calculations using more sophisticated

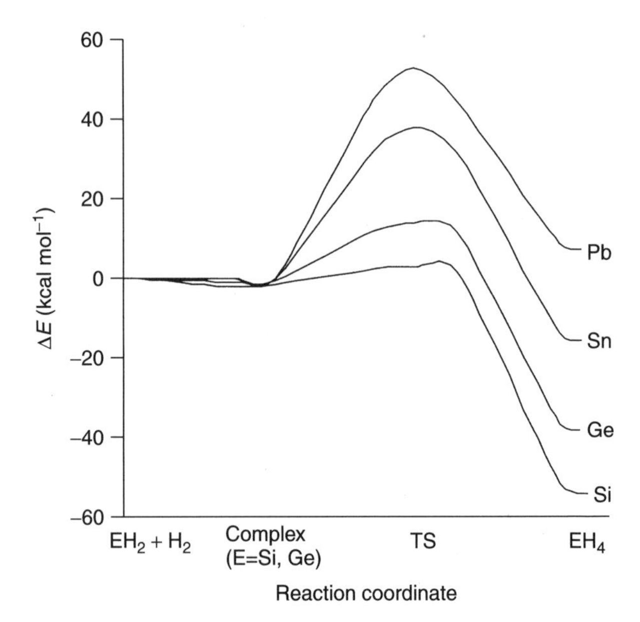

FIGURE 2. Schematic energy profiles for the reaction $EH_2 + H_2 \rightarrow EH_4$ (E = Si - Pb) based on the results of QCISD (E = Si, Ge[369]) and CCSD(T) (E = Sn, Pb[372]) calculations

TABLE 23. Relative energies[a] (kcal mol^{-1}) for reactions $EH_4 + EH_2 \rightarrow E_2H_6$

Structure	Si[b]	Si[c]	Ge[b]	Ge[d]	Sn[b,e]	Pb[b,e]
$EH_4 + EH_2$	0.0		0.0		0.0	0.0
Complex	−4.8	−12.3	−5.1	−6.0	−9.8	−9.1
TS	−5.0	−11.6 (0.8[f])	1.0	−4.0 (−1.2[f])	2.6	11.0
E_2H_6	−53.3	−56.4 (−54.4[f])	−41.4	−42.2 (−39.7[f])	−33.6	−17.9

[a] MP4/DZP//HF/DZP level without ZPE corrections. [b] Reference[373]. [c] Reference[374]. [d] Reference[375]. [e] RECP + DZP used. [f] Experimental value.

methods (see below), is that reactions 9 involve formation of H-bridged intermediate complexes with *syn* orientation of the EH_2 group. A complex with *anti* orientation does not exist at the HF level, although later it was found in the course of MP2/6-311G(d,p) calculations on reactions with E = Si[374] and Ge[375]. The reaction exothermicity falls steeply and E_a rises on going from carbenes to their heavier analogs (Table 23).

syn *anti*

Detailed kinetic and theoretical studies have been performed on the prototype GeH_2 insertions into Si−H[376] and Ge−H[375] bonds (equations 10 and 11).

$$GeH_2 + SiH_4 \longrightarrow H_3SiGeH_3 \quad (10)$$

$$GeH_2 + GeH_4 \longrightarrow Ge_2H_6 \quad (11)$$

Both reactions show the characteristic pressure dependence of a third-body assisted association reaction. The high pressure rate constants, obtained by extrapolation, gave the Arrhenius equations 12 and 13.

$$\log(k^{\infty}/\text{cm}^3\ \text{molecule}^{-1}\text{s}^{-1}) = (-11.73 \pm 0.06) + (1.10 \pm 0.10\ \text{kcal mol}^{-1})/RT \ln 10\ (\text{reaction 10})^{376} \quad (12)$$

$$\log(k^{\infty}/\text{cm}^3\ \text{molecule}^{-1}\text{s}^{-1}) = (-11.17 \pm 0.10) + (1.24 \pm 0.17\ \text{kcal mol}^{-1})/RT \ln 10\ (\text{reaction 11})^{375} \quad (13)$$

The Arrhenius parameters are consistent with a moderately fast reaction occurring at approximately one-thirtieth (one-fifth with GeH_4) of the collision rate. Both reactions are somewhat slower compared to SiH_2 insertion in silane (equation 14) which is known to be a fast, nearly collisionally controlled process[340] (equation 15).

$$SiH_2 + SiH_4 \longrightarrow Si_2H_6 \quad (14)$$

$$\log(k^{\infty}/\text{cm}^3\ \text{molecule}^{-1}\ \text{s}^{-1}) = (-9.91 \pm 0.04) + (0.79 \pm 0.17\ \text{kcal mol}^{-1})/RT \ln 10 \quad (15)$$

Ab initio MP2/6-311G(d,p) and DFT B3LYP/6-311++G(3df,2pd) calculations have shown that these reactions proceed with intermediate formation of weak H-bridged complexes C1 and C2 which rearrange into the final products via TS1 and TS2, respectively. The topology of the PES of these systems and the linking of pathways revealed in the calculations[374–376] are shown in Figure 3.

The potential energy surfaces of the reactions in equations 10 and 11 are more complex than that of equation 14 and involve reaction path bifurcation and low symmetry structures in each channel. The salient features of structures of these complexes and TSs are shown in Figure 4.

Variations in interatomic distances for the complexes and transition states on going from E = Si to Ge are rather small and of the same magnitude as those found in the products. Energies of stationary points for reactions 9 and 10 are presented in Table 24. An additional feature of the complex C1 is its C_1 symmetry, and the existence as left (C1l) or right (C1r) handed forms, which are separated by the very low ($E_a = 0.4$ Kcal mol^{-1}) rotational transition state TS0. A similar situation was found for transition state TS2. It also has a C_1 symmetry, and possesses left (TS2l) and right (TS2r) handed forms divided by a low rotational maximum.

For reactions 10 and 11, calculations show the C1 + TS1 pathway is favored energetically over the C2 + TS2 route. However, the energy differences between the two pathways are less than 1 kcal mol^{-1} for reaction 10 and *ca* 4.2 for reaction 11. It suggests that both channels are probably operative for reaction 10 and only one (C1 + TS1) for reaction 11. The energies and E_a for the reactions calculated by G2 (reactions 10 and 11) and B3LYP/6-311++G(3df,2pd) (reaction 10) methods are in good agreement with experimental estimates (Table 24). The main difference between these reactions is that the transition states for reactions 10 and 11 are tighter than that for reaction 14, and the high pressure limiting *A* factors of reactions 10 and 11 are smaller by a factor of approximately 10 than that of reaction 14.

Other insertion reactions of germylenes, stannylenes and plumbylenes studied theoretically so far are mostly limited to germylenes. Su and Chu[123,349] have reported DFT B3LYP/6-311G* calculations on reactions of GeH_2, Ge=CH_2, GeHMe, $GeMe_2$, GeHF, GeF_2, GeHCl, $GeCl_2$, GeHBr, $GeBr_2$, GeHLi and $GeLi_2$ with methane. All the germylenes react with initial formation of a loose donor–acceptor complex, followed by a high-energy three-membered-ring TS and an insertion product. Complexation energies ΔE_{cp} are less

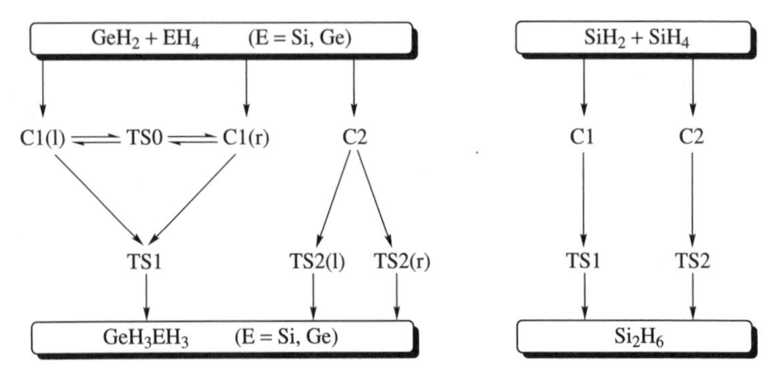

FIGURE 3. Topology of the potential energy surfaces for the reactions $GeH_2 + EH_4 \rightarrow H_3GeEH_3$ (E = Si, Ge) and $SiH_2 + SiH_4 \rightarrow Si_2H_6$, from MP2/6-311G(d,p) and B3LYP/6311++G(3df,2pd) (E = Si) calculations[374–376]. See text for definitions of left(l) and right(r) handed forms and Figure 4 for their structures

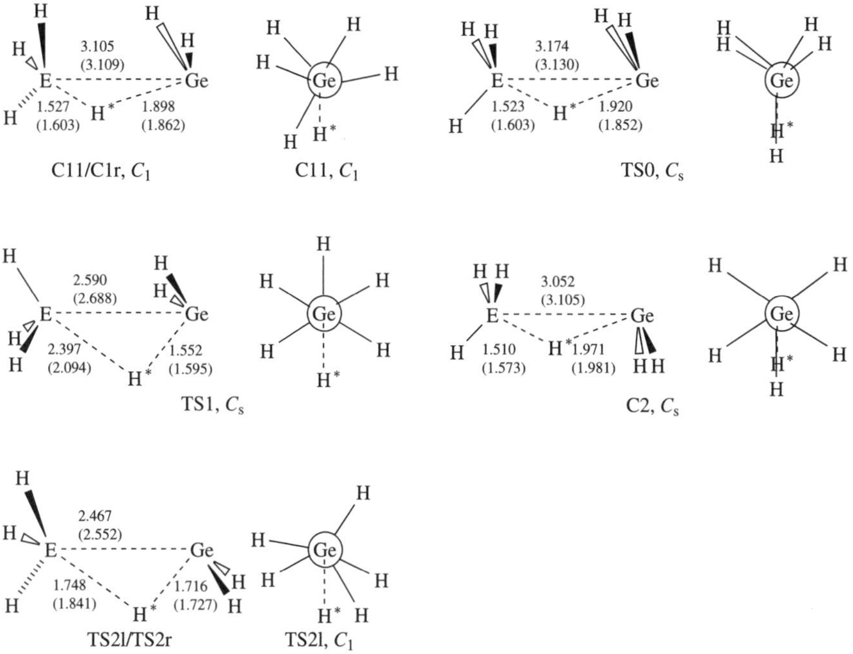

C1l/C1r, C_1 C1l, C_1 TS0, C_s

TS1, C_s C2, C_s

TS2l/TS2r TS2l, C_1

FIGURE 4. *Ab initio* MP2/6-311G(d,p) calculated geometries of the local minima and TSs on the $GeH_2 + EH_4 \rightarrow H_3EGeH_3$ (E = Si, Ge) potential energy surfaces. Point groups are given beside the structure label. Interatomic distances are in Å with values for E = Ge given in parentheses. The migrating H atom is marked by an asterisk. See text for definitions of left(l) and right(r) handed forms

than 1 kcal mol^{-1} (Table 25), the only exception being GeF_2($\Delta E_{cp} = 1.4$ kcal mol^{-1}). Activation barriers E_a are high. Even for the very reactive germylenes GeH_2[27] and $GeMe_2$[5], the calculations gave E_a values of 33.2 and 39.1 kcal mol^{-1} (Table 25). These high E_a values explain why germylenes usually do not insert into C−H bonds[5]. A configuration mixing model based on the theory of Pross[377] and Shaik[378] has been used to interpret the barrier origin[349]. It suggests that the singlet–triplet splitting ΔE_{ST} in the GeXY species can be used as a guide to predict its activity in insertion reactions. For the series GeH_2, $Ge{=}CH_2$, GeHMe, $GeMe_2$, GeHF, GeF_2, GeHCl, $GeCl_2$, GeHBr and $GeBr_2$, a linear correlation between the calculated E_a and ΔE_{ST} (equation 16) was found[349].

$$E_a = 0.818\Delta E_{ST} + 13.5\,(10 \text{ points}, R^2 = 0.94) \tag{16}$$

Likewise, a linear correlation between E_a and the reaction enthalpy ΔH, also obtained at the same level of theory (Table 25), is given in equation 17.

$$E_a = 1.29\Delta H + 69.6\,(10 \text{ points}, R^2 = 0.95) \tag{17}$$

It was concluded[349] that electronic rather than steric factors play a decisive role in determining the chemical reactivity of the germylenes.

TABLE 24. *Ab initio* G2 and DFT[a] calculated and experimental relative energies (kcal mol^{-1}) of stationary points of the PES for $GeH_2 + EH_4 \rightarrow H_3EGeH_3$ (E = Si, Ge) reactions

	$GeH_2 + EH_4$	Complex C1	TS0(Rot)	TS1	Complex C2	TS2	H_3EGeH_3	Reference
E = Si								
ΔE(0 K)	0.0	−6.0 (−4.6)	−5.7 (−4.5)	−1.2 (1.6)	−5.1 (−3.6)	−0.2 (2.1)	−35.7 (−34.8)	376
ΔE(298 K)	0.0	−6.2 (−4.8)	−6.3 (−5.1)	−1.6 (1.3)	−5.1 (−3.5)	−1.0 (1.4)	−36.2 (−35.2)	376
ΔH(298 K)	0.0	−6.8 (−5.4)	−6.9 (−5.7)	−2.2 (0.7)	−5.7 (−4.1)	−1.6 (0.8)	−36.8 (−35.8)	376
Experimental				−1.1			−35.3	376
E = Ge								
ΔE(0 K)	0.0	−6.0	−5.9	−4.0	−4.7	0.2	−42.2	375
Experimental	0.0			−1.2			(−39.7 ± 3)	375

[a] B3LYP/6-311++G(3df,2pd) values are in parentheses.

TABLE 25. B3LYP/6-311G* calculated relative energies (kcal mol^{-1}) of stationary points in the reaction GeXY + CH_4 → HGeMeXY[349].

GeXY	ΔE^a_{comp}	$\Delta E^{\ddagger b}$	ΔH^c
GeH_2	−0.94	+33.2	−28.0
$Ge{=}CH_2$	−0.05	+58.0	−5.3
GeHLi	−0.45	+24.6	−28.6
$GeLi_2$	+0.37	+29.3	−17.9
GeHMe	−0.19	+35.8	−26.8
$GeMe_2$	−0.02	+39.1	−25.1
GeHF	−0.98	+48.6	−17.1
GeF_2	−1.40	+77.7	+0.4
GeHCl	−0.23	+48.1	−15.6
$GeCl_2$	−0.09	+70.0	−0.1
GeHBr	−0.17	+48.2	−14.5
$GeBr_2$	+0.02	+68.3	+1.3

^{a}The stabilization energy of the precursor complex relative to GeXY + CH_4.
^{b}Energy of the TS relative to GeXY + CH_4.
^{c}The reaction enthalpy.

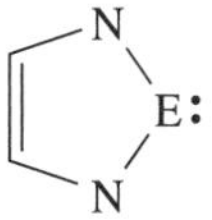

(**7**) E = C, (**8**) E = Si, (**9**) E = Ge

A comparison of Arduengo-type carbene, silylene and germylene **7–9** insertion into the C−H bond of methane has been carried out using calculations at the B3LYP/6-31G* and CCSD(T)/6-31G**//B3LYP/6-31G* levels[379]. These reactions also involve formation of pre-reaction complexes. The main energetic characteristics of the stationary points found are given in Table 26. The calculated singlet–triplet gap ΔE_{ST} in **7–9** is very high (3.6–2.2 eV) and shows an opposite trend compared to the ΔE_{ST} of the parent species (CH_2, SiH_2 and GeH_2), as well as to those of CAs with π-donor substituents [$C(NH_2)_2$, $Si(NH_2)_2$ and $Ge(NH_2)_2$]. This peculiarity is related to the unusual nature of the highest occupied MO of **7–9** detected by photoelectron spectroscopy[332] (see Section IX). The activation barriers E_a of insertions are also very high and obey the expected order C < Si < Ge. There is good agreement between the B3LYP and CCSD(T) values.

Insertion of $GeMe_2$ into the C−H, Si−H, N−H, P−H, O−H, S−H, F−H and Cl−H bonds was studied by DFT B3LYP/6-311G* and *ab initio* MP2/6-311G* methods[343]. Results of CCSD(T) calculations for the same set of reactions were reported[380]. All the reactions include an initial formation of a donor–acceptor complex, followed by the TS leading to the insertion product. The agreement between the geometries of the stationary points calculated at the MP2 and B3LYP levels is reasonably good for most structures. Energies of stationary points along the reaction paths, which are given in Table 27, show that the complexation energies range from 25 to *ca* 1 kcal mol^{-1} and decrease in the order NH_3 > H_2O > PH_3 > H_2S ~ HF > HCl > SiH_4 ~ CH_4. It is noteworthy that the DFT B3LYP calculations systematically give lower (by *ca* 2.5 kcal mol^{-1}) complexation

TABLE 26. Singlet–triplet splittings ΔE_{ST}/eV in **7–9**, and relative energies[a] (kcal mol^{-1}) for the process **7** (**8** and **9**) $+CH_4 \rightarrow$ Complex $\rightarrow$ TS $\rightarrow$ Product[379]

(cyclic N,N E: structure)	ΔE_{ST}	ΔE^{b}_{comp}	$\Delta E^{\ddagger c}$	ΔH^{d}
7	3.66 (3.58)	−2.0 (−2.0)	62.2 (56.4)	−10.7 (−8.7)
8	2.83 (2.59)	−0.5 (−0.1)	75.8 (77.8)	−1.2 (−1.0)
9	2.34 (2.19)	−2.9 (−1.6)	82.9 (86.5)	13.0 (19.4)

[a]At the CCSD(T)/6-31G**//B3LYP/6-31G* and B3LYP/6-31G* (in parentheses) levels.
[b]The stabilization energy of the precursor complex, relative to the corresponding reactants.
[c]The energy of the TS, relative to the corresponding reactants.
[d]The energy of the product, relative to the corresponding reactants.

TABLE 27. Relative energies (kcal mol^{-1}) for the process $GeMe_2 + H\text{-}XH_{n-1} \rightarrow$ Complex $\rightarrow$ TS $\rightarrow$ $Me_2Ge(H)XH_{n-1}$ (X = C, N, O, F, Si, P, S, and Cl; $n = 1\text{–}4$) calculated by using the MP2/6-311G* and B3LYP/6-311G* (in parentheses) methods[343]

XH_n	$GeMe_2 + XH_n$	ΔE^{a}_{comp}	$\Delta E^{\ddagger b}$	ΔH^{c}
CH_4	0.0 (0.0)	−1.1 (−0.02)	35.6 (39.1)	−32.6 (−25.1)
NH_3	0.0 (0.0)	−25.0 (−20.8)	22.7 (25.1)	−40.4 (−33.3)
H_2O	0.0 (0.0)	−16.2 (−13.9)	15.1 (14.8)	−50.2 (−45.1)
HF	0.0 (0.0)	−7.1 (−7.2)	9.6 (4.74)	−61.2 (−59.1)
SiH_4	0.0 (0.0)	−2.2 (−0.6)	11.4 (15.7)	−41.3 (−33.8)
PH_3	0.0 (0.0)	−14.1 (−9.0)	8.1 (11.7)	−46.1 (−37.4)
H_2S	0.0 (0.0)	−9.7 (−6.1)	3.9 (6.0)	−54.9 (−45.8)
HCl	0.0 (0.0)	−3.7 (−1.6)	2.5 (1.2)	−64.2 (−56.4)

[a]The stabilization energy of the precursor complex, relative to reactants.
[b]The energy of the TS, relative to the corresponding reactants.
[c]The energy of the product, relative to the corresponding reactants.

energies compared to MP2. All germylene insertions into X−H bonds occur in a concerted manner via a three-membered-ring TS, and the stereochemistry at the heteroatom X center is preserved. Differences in E_a calculated by both methods are of various signs with an average absolute deviation of 2.8 kcal mol^{-1}. The MP2 calculated E_a increase in the order of substrates: HCl < H_2S < PH_3 < HF < SiH_4 < H_2O < NH_3 < CH_4. For all the substrates B3LYP calculations give lower exothermicity of insertion compared to the MP2 values (by *ca* 6.9 kcal mol^{-1}). In spite of some differences in the absolute values the general trend in E_a and ΔH obtained by both methods is essentially the same. The larger the atomic number of the heteroatom X in a given row, the easier the insertion reaction of XH_n hydrides occurs and the larger its exothermicity (Table 27). These results indicate that the B3LYP method could be recommended for investigations of molecular geometries, electronic structures and kinetic features of the germylene reactions.

MNDO calculations on insertion reactions of Me_2Sn into Cl−Sn and I−C bonds were reported by Dewar and coworkers[381]. Two alternative pathways including concerted carbene-like insertion (a) and nonconcerted radical two-step insertion (b) were investigated.

$$Me_2Sn + ClSnRMe_2 \xrightarrow{} \left[Me_2Sn \cdots SnMe_2R \ (\cdots Cl) \right]^{\ddagger} \ \text{(a)} \rightarrow Me_2ClSnSnMe_2R$$
$$Me_2Sn + ClSnRMe_2 \rightarrow Me_2\dot{S}nR + Me_2\dot{S}nCl \ \text{(b)} \rightarrow Me_2ClSnSnMe_2R \quad (18)$$

R = Cl, Me

$$Me_2Sn + MeI \rightarrow \left[Me_2Sn \cdots Me \ (\cdots I) \right]^{\ddagger} \ \text{(a)} \rightarrow Me_3SnI$$
$$Me_2Sn + MeI \rightarrow Me_2\dot{S}nI + \dot{M}e \ \text{(b)} \rightarrow Me_3SnI \quad (19)$$

In reactions of Me_2Sn with $ClSnRMe_2$ a concerted mechanism (equation 18a) is favored ($E_a = 10.1$ and 14.2 kcal mol^{-1} for R = Cl and Me, respectively) over a radical route (equation 18b), $E_a = 20.0$ kcal mol^{-1} (R = Cl) and 28.4 kcal mol^{-1} (R = Me). In the case of MeI the situation is reversed: a radical mechanism (equation 19b) ($E_a = 33.0$ kcal mol^{-1}) is preferred over a carbene-like insertion (equation 19a, $E_a = 42.3$ kcal mol^{-1}).

5. Cycloadditions

The classical barrier heights and the thermodynamics of cycloaddition of EX_2 (E = C, Si, Ge, Sn; X = H, F) to acetylene were calculated by Boatz, Gordon and Sita[382] using MP2/3-21G(d)//HF/3-21G(d) energies. The nature of the ring bonding in cyclo-($EX_2C_2H_2$) was investigated via analysis of the total electron density and was found to have little or no π-complex character. The exothermicity of the cycloaddition falls dramatically in the order C > Si > Ge > Sn for both EH_2 and EF_2 species. For GeF_2 and SnF_2 the reactions even become endothermic ($\Delta E = 14.4$ and 16.5 kcal mol^{-1}). Cycloaddition of all hydrides EH_2 proceeds without a barrier. Fluorine substitution induces substantial barriers to the formation of all the corresponding heterocyclopropenes.

Horner, Grev and Schaefer[383] compared the energies of the decomposition reactions cyclo-$EH_2XH_2YH_2 \rightarrow EH_2 + H_2X{=}YH_2$ for E, X, Y = C, Si, Ge using results of CCSD/DZP calculations. Of the ten rings studied, germirane (cyclo-$GeH_2CH_2CH_2$) was by far the least stable with respect to dissociation, being only about 20 kcal mol^{-1} more stable than the isolated $GeH_2 + H_2C{=}CH_2$. It was concluded that the known difficulties in germirane synthesis have a thermochemical origin. It agrees with the fact that the only known example of successful synthesis of germirane[384] involved a special type of olefin and a Lappert-type germylene. The three-membered ring decomposition enthalpy can be predicted semiquantitatively from a simple model using the strain energies along with the single bond dissociation energies, π-bond energies and divalent state stabilization energies[383].

DFT B3LYP/6-31G* calculations have been performed on the potential energy surfaces for cycloaddition of germylenes GeH_2, $GeMe_2$, $Ge(NH_2)_2$, $Ge(OH)_2$, GeF_2, $GeCl_2$, $GeBr_2$ and $Ge{=}CH_2$ to the C=C double bond of ethylene[385]. Unlike the case of silylene (SiH_2[386], SiF_2 and $SiCl_2$[387]) additions, a π-complex intermediate is formed between

TABLE 28. Relative energies ($kcal\,mol^{-1}$) of stationary points for the $GeR_2 + H_2C=CH_2 \rightarrow$ cyclo-$GeR_2CH_2CH_2$ reactions from DFT B3LYP/6-31G(d) calculations[385]

GeR_2	Complex	TS	cyclo-$GeR_2CH_2CH_2$
GeH_2	−23.5	−21.5	−27.4
$GeMe_2$	−15.6	−14.2	−27.3
$Ge(NH_2)_2$	−2.3	16.8	8.7
$Ge(OH)_2$	−3.9	23.3	11.4
GeF_2	−8.2	27.9	13.8
$GeCl_2$	−9.8	18.2	8.6
$GeBr_2$	−11.8	13.5	5.5
$Ge{=}CH_2$	−12.6	5.4	3.2

germylenes and ethylene. Of the germylenes studied, only reactions of GeH_2 and $GeMe_2$ are feasible from both a kinetic and a thermodynamic point of view (Table 28). Formation of three-membered rings by other germylenes is an endothermic process. The origin of barrier heights was discussed using the aforementioned configuration mixing model of Pross and Shaik. A linear correlation was found between calculated E_a and the singlet–triplet splitting ΔE_{ST} (equation 20 and Table 28).

$$E_a = 0.906\Delta E_{ST} - 40.7 \text{ (8 points, } R^2 = 0.923) \tag{20}$$

Calculations of the PES of GeH_2 cycloaddition to ethylene at the MP2/6-31G(d)//RHF/6–31G(d) level were followed by computation of kinetic properties at different temperatures using statistical thermodynamics and transition state theory[388]. The reaction was shown to proceed without formation of an intermediate complex, which is in agreement with the results of Anwari and Gordon[386], but in clear disagreement with DFT B3LYP/6–31G(d) calculations[385]. Calculations on this prototype reaction using more rigorous methods are needed.

Dihalogermylenes and dihalostannylenes are supposed to be rather inert species, yet MNDO calculations predict a low activation barrier ($E_a = 19.5\ kcal\,mol^{-1}$) for the cheletropic addition of Br_2Sn to butadiene with formation of cyclo-$Br_2SnC_4H_6$ (equation 21)[381].

Br_2Sn + [butadiene] ⟶ [cyclo-C_4H_6]$SnBr_2$ (21)

6. Miscellaneous

Ab initio MP2 calculations with DZ quality basis sets were performed on the reactions of C_2H_4 with $GeH_n (n = 0–3)$[389]. Single-point calculations at the QCISD(T)/6-311G(3df,2p) level were performed. The results were used to speculate on the mechanisms of reactions occurring during radiolysis of germane/ethylene mixtures.

Formation of GeF_2 in reactions between GeF_4 and Si_2H_6 was studied by CCSD(T)//B3LYP calculations using the basis set of the DZP quality[390]. These reactions are related to the mechanism of silane activation in the low-temperature thermal CVD deposition of Ge films from a GeF_4 source. Several reactions between GeF_4 and SiH_4/Si_2H_6 were investigated. The most important are those leading to SiH_3GeF_3, which could easily decompose ($E_a = 31.9$ kcal mol^{-1}) into SiH_3F and GeF_2. Disilane was suggested to be an efficient activator because it more easily produces SiH_3GeF_3 compared to silane.

High level *ab initio* calculations have been reported on the PES of singlet SiH_2 and GeH_2 reactions with water, methanol, ethanol, dimethyl ether and trifluoromethanol[391]. Besides the classical route for EH_2 (E = Si, Ge) insertion into X−O (X = H, C) bonds via 1,2 hydrogen atom shift reaction (equation 22), two new reaction channels (equations 23 and 24) were identified on each PES, except for reactions involving dimethyl ether. Equations 23 and 24 display routes for H_2 elimination, following the initial formation of an association complex.

$H_2E + HOR$ → (10) → [(11)]‡ → (14) (22)

(10) → [(12)]‡ $\xrightarrow{-H_2}$ (15) (23)

(10) → [(13)]‡ $\xrightarrow{-H_2}$ (16) (24)

E = Si, Ge; R = H, CH_3, CF_3

The processes via transition states **12** and **13** have activation energies comparable with that of the classical route via TS **11** (Table 29). For reactions involving SiH_2 and water, a simple activated complex theory analysis predicts that these newly identified reaction channels (equations 23 and 24) are equally likely to be accessed as that in equation 22. For reactions involving GeH_2 and water, a similar analysis predicts that equations 23 and 24 will occur in preference to the 1,2 hydrogen shift in **10**. Indeed, the room-temperature rate constant for H_2 elimination from the germanium complex **10** was predicted to be approximately 5 orders of magnitude larger than for the H atom migration channel.

TABLE 29. Relative energies (kcal mol^{-1}) of stationary points[a] on the PES of the reaction H_2E + ROH calculated at the MP2/6-311++G(d,p)//MP2/6-311++G(d,p) level[391]

	$SiH_2 + H_2O$[b]		SiH_2 + CH_3OH	SiH_2 + C_2H_5OH	$SiH_2 + CF_3OH$	$GeH_2 + H_2O$	GeH_2 + CH_3OH
Reagents	0.0	(0.0)	0.0	0.0	0.0	0.0	0.0
10	−12.7	(−11.6)	−18.1	−18.8	−6.6	−11.0	−15.1
11	9.2	(12.2)	3.0	2.1	9.3	20.2	15.2
12	9.2	(11.4)	3.8	3.1	7.0	13.5	9.0
13	8.8	(11.1)	3.6	2.7	6.8	13.5	9.1
14	−70.3	(−66.0)	−72.7	−73.1		−40.9	−42.8
15 + H_2	−26.6	(−25.2)	−30.6	−31.3		−16.7	−19.6
16 + H_2	−26.8	(−25.4)	−30.0	−30.8		−16.5	−19.0

[a]ZPE correction included.
[b]Values given in parentheses are calculated at the QCISD(T)/6-311++G(d,p)//QCISD(T)/6-311++G(d,p) level.

D. Complexes with Lewis Bases

In this section we consider the results of calculations devoted to studying properties of donor–acceptor complexes between CAs and Lewis bases.

Complexes of EH_2 (E = Si, Ge, Sn) with donor molecules, AH_3 (A = N, P, As, Sb, Bi) and AH_2 (A = O, S, Se, Te), were studied by Schöller and Schneider[189] with *ab initio* MP2 calculations using RECP and basis sets of DZP quality. Association energies in the range of 15–30 kcal mol^{-1} were found. They decrease in the order $SiH_2 > GeH_2 > SnH_2$. The population analysis indicates for NH_3 and BiH_3 only a weak bonding toward the EH_2 fragment while the higher homologs with A = P, As, Sb form 1,2-dipolar ylide structures. A dual parameter relationship between (a) the HOMO energies of the donor (n-orbital of the AH_3 unit, n, p orbitals for AH_2) and (b) the known covalent bond energies versus the binding energies of the donor–acceptor complexes was examined and found to describe satisfactorily the essential features of the stabilities of the donor–acceptor structures.

Nowek and Leszczynski used *ab initio* post-Hartree–Fock and DFT B3LYP calculations to resolve the problem of the structure of the $H_2Ge\cdots OH_2$ complex[274]. The molecular geometries of the nonplanar **17** (C_1) and planar **18** (C_s) conformers were optimized at the DFT and MP2 levels of theory using TZP and TZ2P basis sets. The nonplanar complex **17** with a Ge···O distance of 2.214 Å (MP2), 2.268 Å (B3LYP) corresponds to a global minimum while complex **18** was found to be very weakly bound (if bound at all).

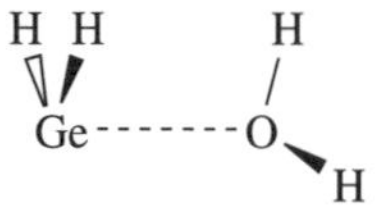

(17) 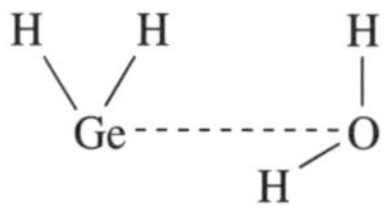 **(18)**

Calculated interaction energies (corrected for the basis set superposition error and ZPE) are relatively large and amount to 9.8 kcal mol^{-1} by B3LYP and 8.9 kcal mol^{-1} by CCSD(T) methods. Harmonic vibration frequencies calculated for monomers and the complex are in reasonable agreement with experimental data[241]. Agreement is good for IR band shifts due to complexation and isotopic substitution.

Formation of complexes with electron donors (Lewis bases) seems to be an ubiquitous feature of all CAs. Dihalostannylenes and dihaloplumbylenes which are inert in most of the characteristic carbene analog reactions form complexes with such weak electron donors as heptyne, methyl chloride and even dinitrogen (see above). Quantum-chemical calculations in Nefedov's group were used to assign the IR bands of these complexes recorded in low-temperature Ar matrices and to get information on their structure and stability[275,278,279,281].

Semiempirical AM1 and PM3 calculations on the reaction of SnF_2 with hept-1-yne[278,279] show the formation of π-complex **19** (Figure 5). Its stability was estimated as 7.4 (AM1) and 9.1 (PM3) kcal mol^{-1}. In agreement with the study of the SnF_2 reaction with acetylene[382], the cycloaddition of SnF_2 to the triple bond of hept-1-yne was calculated to be highly endothermic. PM3 calculations of the dimer Sn_2F_4 give the bridged structure **4**[279], which agrees with spectral data.

The PES of the system $SnF_2 + CH_3Cl$ was studied by *ab initio* MP2/3-21G(d)//HF/3-21G(d) and PM3 methods[281]. Calculations have shown that the reaction between SnF_2 and CH_3Cl results in the formation of a donor–acceptor complex **20** (Figure 5). The orientation of SnF_2 and CH_3Cl in **20** is determined by the dipole–dipole interaction. The complexation energy is 14.2 kcal mol^{-1} (MP2) and 15.7 kcal mol^{-1} (PM3). Rearrangement of **20** into the insertion product $MeSnF_2Cl$ is favorable from an energetic point of view [$\Delta E = -47.4$ kcal mol^{-1} (MP2) and -15.8 kcal mol^{-1} (PM3)] but the activation energy is very high: 47.0 (MP2) and 34.6 kcal mol^{-1} (PM3), so that this reaction does not occur under the experimental conditions[281]. Quantum-chemical calculations were used to interpret the IR spectrum of the complex recorded in Ar matrix.

Ab initio MP2/3-21G(d2)//HF/3-21G(d2) calculations on the $SnF_2 + N_2$ and $SnF_2 + 2N_2$ systems[275] have revealed the presence of the local minima corresponding to complexes **21** and **22** (Figure 6). Both are stabilized by interaction of the lowest unoccupied p-MO (LUMO) of SnF_2 with the lone pair of N_2 (p,n interaction). Alternative structures of complexes stabilized by interaction of the p-LUMO of stannylene with the occupied π-MO of dinitrogen (p,π interaction), an analog of **19** in reaction of SnF_2 with hept-1-yne, or structures resulting from orbital σ,p and σ,n interactions employing σ-MO of the SnF_2 stannylene center were not found.

Calculated MP2 interaction energies with one and two N_2 molecules which include corrections for ZPE and basis set superposition error are 4.6 and 8.9 kcal mol^{-1}, respectively. The calculations well reproduce the experimentally observed shifts of the valence vibrational bands of SnF_2 upon complexation with N_2. They also indicate a small polarization of the N_2 ligands in the complexes, resulting in their nonzero intensities in the IR spectrum. However, these polarization effects are too small to be observed under the experimental conditions[275].

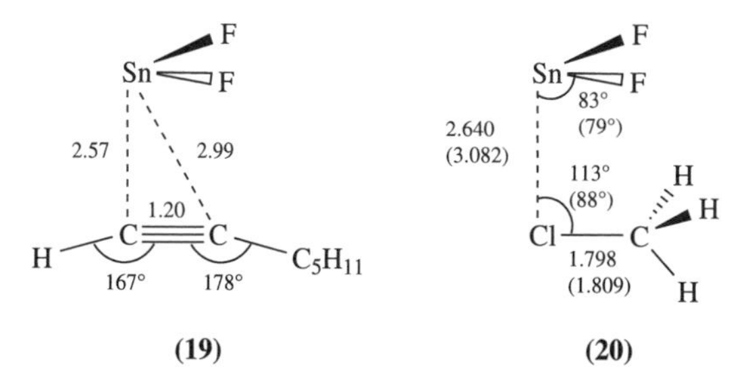

FIGURE 5. Main structural parameters of complexes **19** and **20** calculated by the PM3 and RHF/3-21G(d) (in parentheses) methods. Interatomic distances are in Å

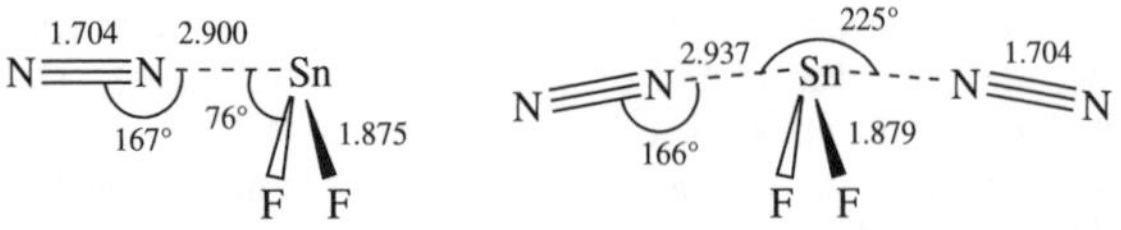

(21) C_s **(22)** C_{2v}

FIGURE 6. Structural parameters of complexes $SnF_2 + N_2$ and $SnF_2 + 2N_2$ calculated by *ab initio* MP2/3-21G(d2)//HF/3-21G(d2). Point groups are given beside the structure label. Interatomic distances are in Å

The binding energies, changes in Sn−F bond lengths and SnF_2 frequency shifts in complex **22** with two N_2 moieties are approximately twice as big as in complex **21** with one N_2 ligand, i.e. complexation of SnF_2 with one N_2 molecule does not affect its interaction with the second N_2. This means that despite the relatively high stability of **21**, its formation does not change the reactivity of SnF_2 in a dramatic way. Cycloaddition of SnF_2 to the triple $N \equiv N$ bond is energetically extremely unfavorable, and the three-membered cycle SnF_2N_2 was found to be unstable[275].

XI. CONCLUSIONS

This review shows that despite the huge amount of information accumulated on fundamental structural, electronic and spectral characteristics of germylenes, stannylenes and plumbylenes, many gaps still exist in this area. Modern quantum-chemical calculations partially help to fill these gaps, but reliable experimental data are needed.

Very little is known about the nature of the weak interactions of CAs in solutions where a vast majority of their chemical reactions has been studied. Particularly, the study of donor–acceptor complexes of CAs by modern physical-chemical methods is still of great interest. Besides, complexation of CAs with donors or acceptors of electron density is a useful tool for modifying the stability, reactivity and spectral properties of CAs. Systematic investigations of the redox properties of CAs are needed in order to elucidate the role of electron transfer in the transformations of CAs.

The quest for CAs with a triplet ground state, as well as experimental determination of their structure in excited states, the values of singlet–triplet gaps and the reactivity of triplet CAs appear to be important venues in the chemistry of CAs in forthcoming years. The possibility of solving the aforementioned problems in the chemistry of CAs depends significantly on the availability of suitable precursors of CAs. In fact, only a few good precursors of germylenes, and especially of stannylenes and plumbylenes, are currently known. Therefore, the development of new precursors and new approaches to the generation of these species remain an urgent problem in the chemistry of germylenes, stannylenes and plumbylenes.

XII. ACKNOWLEDGMENTS

Our own work on germylenes and stannylenes has been supported by the Russian Foundation for Basic Research, INTAS and NATO grants.

XIII. REFERENCES

1. M. E. Volpin, Yu. D. Koreshkov, V. G. Dulova and D. N. Kursanov, *Tetrahedron*, **18**, 107 (1962).
2. O. M. Nefedov and M. N. Manakov, *Angew. Chem., Int. Ed. Engl.*, **5**, 1021 (1966).

3. P. P. Gaspar and R. West, in *The Chemistry of Organic Silicon Compounds*, Vol. 2 (Eds. Z. Rappoport and Y. Apeloig), Wiley, New York, 1998, pp. 2463–2568.
4. R. Becerra and R. Walsh, in *Research in Chemical Kinetics*, Vol. 3 (Eds. R. G. Compton and G. Hancock), Elsevier, Amsterdam, 1995, pp. 263–326.
5. W. P. Neumann, *Chem. Rev.*, **91**, 311 (1991).
6. J. Barrau and G. Rima, *Coord. Chem. Rev.*, **178–180**, 593 (1998).
7. M. P. Egorov and P. P. Gaspar, in *Encyclopedia of Inorganic Chemistry*, Vol. 3 (Ed. R. B. King), Wiley, New York, 1995, pp. 1229–1319.
8. R. K. Asundi, S. M. Karim and R. Samuel, *Proc. Phys. Soc.*, **50**, 581 (1938).
9. K. Butkow, *Phys. Ziets. Sowjetunion, A*, **4**, 577 (1933); *Chem. Abstr.*, **28**, 1276^7 (1934).
10. K. Butkow, *Phys. Ziets. Sowjetunion, A*, **5**, 906 (1934); *Chem. Abstr.*, **29**, 405^1 (1935).
11. O. M. Nefedov, M. P. Egorov, A. I. Ioffe, L. G. Menchikov, P. S. Zuev, V. I. Minkin, B. Ya. Simkin and M. N. Glukhovtsev, *Pure Appl. Chem.*, **64**, 265 (1992).
12. M. E. Jacox, *J. Phys. Chem. Ref. Data*, **13**, 945 (1984).
13. M. E. Jacox, *J. Phys. Chem. Ref. Data*, **17**, 269 (1988).
14. M. E. Jacox, *J. Phys. Chem. Ref. Data*, **19**, 1387 (1990).
15. M. E. Jacox, *J. Phys. Chem. Ref. Data*, **27**, 115 (1998).
16. S. G. Lias, J. E. Bartmess, J. F. Liebman, J. L. Holmes, R. D. Levin and W. G. Mallard, *J. Phys. Chem. Ref. Data*, **17**, Suppl. 1, 1 (1988).
17. V. A. Korolev and O. M. Nefedov, *Adv. Phys. Org. Chem.*, **30**, 1 (1995).
18. Internet site: http://webbook.nist.gov
19. L. Zhang, J. Dong and M. Zhou, *J. Chem. Phys.*, **113**, 8700 (2000).
20. Y. Yamada, T. Kumagawa, Y. T. Yamada and T. Tominaga, *J. Radioanal. Nucl. Chem. Lett.*, **201**, 417 (1995).
21. S. N. Tandura, S. N. Gurkova and A. I. Gusev, *Zh. Strukt. Khim.*, **31**, 154 (1990) (in Russian); *Chem. Abstr.*, **113**, 65418 (1990).
22. C. E. Holloway and M. Melnic, *Main Group Met. Chem.*, **20**, 399 (1997).
23. K. Saito and K. Obi, *Chem. Phys. Lett.*, **215**, 193 (1993).
24. K. Obi, M. Fukushima and K. Saito, *Appl. Surf. Sci.*, **79–80**, 465 (1994).
25. K. Saito and K. Obi, *Chem. Phys.*, **187**, 381 (1994).
26. J. Karolczak, W. W. Harper, R. S. Grev and D. J. Clouthier, *J. Chem. Phys.*, **103**, 2839 (1995).
27. R. Becerra, S. E. Boganov, M. P. Egorov, O. M. Nefedov and R. Walsh, *Chem. Phys. Lett.*, **260**, 433 (1996).
28. U. N. Alexander, N. A. Trout, K. D. King and W. D. Lawrance, *Chem. Phys. Lett.*, **299**, 291 (1999).
29. A. Campargue and R. Escribano, *Chem. Phys. Lett.*, **315**, 397 (1999).
30. T. C. Smith, D. J. Clouthier, W. Sha and A. G. Adam, *J. Chem. Phys.*, **113**, 9567 (2000).
31. R. Hauge, V. M. Khanna and J. L. Margrave, *J. Mol. Spectrosc.*, **27**, 143 (1968).
32. J. Karolczak, R. S. Grev and D. J. Clouthier, *J. Chem. Phys.*, **101**, 891 (1994).
33. K. J. Boyle, D. P. Seccombe, R. P. Tuckett, H. Baumgaertel and H. W. Jochims, *Chem. Phys. Lett.*, **294**, 507 (1998).
34. J. L. Cole, R. H. Hauge, J. L. Margrave and J. W. Hastie, *J. Mol. Spectrosc.*, **43**, 441 (1972).
35. R. W. Martin and A. J. Merer, *Can. J. Phys.*, **51**, 727 (1973).
36. S. Yagi and N. Takahashi, *Appl. Phys. Lett.*, **61**, 2677 (1992).
37. S. Yagi, T. Ohta, N. Takahashi, K. Saito and K. Obi, *Nippon Kagaku Kaishi*, 231 (1994); *Chem. Abstr.*, **120**, 258950 (1994).
38. S. Yagi, T. Ohta, K. Saito and K. Obi, *J. Appl. Phys.*, **74**, 1480 (1993).
39. W. J. Rosano and J. M. Parson, *J. Chem. Phys.*, **84**, 6250 (1986).
40. J. W. Hastie, R. Hauge and J. L. Margrave, *J. Phys. Chem.*, **72**, 4492 (1968).
41. A. Tewarson and H. B. Palmer, *J. Mol. Spectrosc.*, **22**, 117 (1967).
42. J. W. Hastie, R. H. Hauge and J. L. Margrave, *J. Mol. Spectrosc.*, **29**, 152 (1969).
43. J. H. Wang, B. S. Cheong and J. M. Parson, *J. Chem. Phys.*, **91**, 2834 (1989).
44. T. Ibuki, *Chem. Phys. Lett.*, **169**, 64 (1990).
45. H. Biehl, K. J. Boyle, D. P. Seccombe, D. M. Smith, R. P. Tuckett, H. Baumgaertel and H. W. Jochims, *J. Electron Spectrosc. Relat. Phenom.*, **97**, 89 (1998).
46. T. Ibuki and A. Kamamoto, *Chem. Phys. Lett.*, **260**, 314 (1996).

47. J. Karolczak, Q. Zhuo, D. J. Clouthier, W. M. Davis and J. D. Goddard, *J. Chem. Phys.*, **98**, 60 (1993).
48. C. M. Pathak and H. B. Palmer, *J. Mol. Spectrosc.*, **31**, 170 (1969).
49. I. A. Topol and S. A. Zaitsev, *Vestn. Mosk. Univ., Ser. 2: Khim.*, **32**, 564 (1991) (in Russian); *Chem. Abstr.*, **116**, 139050 (1992).
50. S. A. Zaitsev, S. B. Osin, V. A. Koryazhkin and V. F. Shevelkov, *Vestn. Mosk. Univ., Ser. 2: Khim.*, **31**, 128 (1990) (in Russian); *Chem. Abstr.*, **114**, 14135 (1991).
51. R. H. Hauge, J. W. Hastie and J. L. Margrave, *J. Phys. Chem.*, **72**, 3510 (1968).
52. P. Deschamps and G. Pannetier, *J. Chim. Phys.*, **61**, 1547 (1964).
53. D. Naegeli and H. B. Palmer, *J. Mol. Spectrosc.*, **21**, 325 (1966).
54. J. Maya, *J. Chem. Phys.*, **67**, 4976 (1977).
55. G. A. Oldershaw and K. Robinson, *J. Chem. Soc., A*, 2963 (1971).
56. R. Samuel, *Rev. Mod. Phys.*, **18**, 103 (1946).
57. T. Kobayashi, E. Ikehara, M. Tsukada and N. Fujii, *J. Photochem. Photobiol., A*, **87**, 1 (1995).
58. I. G. Murgulesen, E. Ivana and E. Popa, *Rev. Roum. Chim.*, **27**, 695 (1982).
59. R. J. Zollweg and L. S. Frost, *J. Chem. Phys.*, **50**, 3280 (1969).
60. S. A. Zaitsev, A. P. Monyakin, A. V. Buchkin and V. A. Koryazhkin, *Vestn. Mosk. Univ., Ser. 2: Khim.*, **32**, 329 (1991) (in Russian); *Chem. Abstr.*, **116**, 71151 (1992).
61. G. Rodrigues, C. M. Herring, R. D. Fraser and J. G. Eden, *J. Opt. Soc. Am., B*, **13**, 1362 (1996).
62. M. Fukushima, S. Mayama and K. Obi, *J. Chem. Phys.*, **96**, 44 (1992).
63. R. Escribano and A. Campargue, *J. Chem. Phys.*, **108**, 6249 (1998).
64. A. Kasdan, E. Herbst and W. C. Lineberger, *J. Chem. Phys.*, **62**, 541 (1975).
65. J. Berkowitz, J. P. Green, H. Cho and B. Ruscic, *J. Chem. Phys.*, **86**, 1235 (1987).
66. H. Ishikawa and O. Kajimoto, *J. Mol. Spectrosc.*, **160**, 1 (1993).
67. M. Fukushima and K. Obi, *J. Chem. Phys.*, **100**, 6221 (1994).
68. J. Karolczak, R. H. Judge and D. J. Clouthier, *J. Am. Chem. Soc.*, **117**, 9523 (1995).
69. R. N. Dixon and M. Halle, *J. Mol. Spectrosc.*, **36**, 192 (1970).
70. R. D. Johnson, J. W. Hudgens and M. N. R. Ashfold, *Chem. Phys. Lett.*, **261**, 474 (1996).
71. J. Karolczak and D. J. Clouthier, *Chem. Phys. Lett.*, **201**, 409 (1993).
72. K. Du, X. Chen and D. W. Setser, *Chem. Phys. Lett.*, **181**, 344 (1991).
73. H. Sekiya, Y. Nishimura and M. Tsuji, *Chem. Phys. Lett.*, **176**, 477 (1991).
74. B. P. Ruzsicska, A. Jodhan, I. Safarik, O. P. Strausz and T. N. Bell, *Chem. Phys. Lett.*, **139**, 72 (1987).
75. R. I. Patel and G. W. Stewart, *Can. J. Phys.*, **55**, 1518 (1977).
76. H. Ito, E. Hirota and K. Kuchitsu, *Chem. Phys. Lett.*, **175**, 384 (1990).
77. H. Ito, E. Hirota and K. Kuchitsu, *Chem. Phys. Lett.*, **177**, 235 (1991).
78. W. W. Harper and D. J. Clouthier, *J. Chem. Phys.*, **108**, 416 (1998).
79. W. W. Harper, C. M. Clusek and D. J. Clouthier, *J. Chem. Phys.*, **109**, 9300 (1998).
80. M. Benavides-Garcia and K. Balasubramanian, *J. Chem. Phys.*, **97**, 7537 (1992).
81. W. W. Harper, J. Karolczak, D. J. Clouthier and S. C. Ross, *J. Chem. Phys.*, **103**, 883 (1995).
82. W. W. Harper and D. J. Clouthier, *J. Chem. Phys.*, **106**, 9461 (1997).
83. H. Harjanto, W. W. Harper and D. J. Clouthier, *J. Chem. Phys.*, **105**, 10189 (1996).
84. H. U. Lee and J. P. Deneufville, *Chem. Phys. Lett.*, **99**, 394 (1983).
85. W. W. Harper, D. A. Hasutler and D. J. Clouthier, *J. Chem. Phys.*, **106**, 4367 (1997).
86. D. J. Clouthier, W. W. Harper, C. M. Klusek and T. C. Smith, *J. Chem. Phys.*, **109**, 7827 (1998).
87. J. Billingsley, *Can. J. Phys.*, **50**, 531 (1972).
88. V. I. Minkin, B. Ya. Simkin and M. N. Glukhovtsev, *Usp. Khim.*, **58**, 1067 (1989) (in Russian); *Chem. Abstr.*, **112**, 75977 (1990).
89. J. F. Harrison, R. C. Liedtke and J. F. Liebman, *J. Am. Chem. Soc.*, **101**, 7162 (1979).
90. H. Nakatsuji, *J. Am. Chem. Soc.*, **95**, 345 (1973).
91. H. Nakatsuji, *J. Am. Chem. Soc.*, **95**, 354 (1973).
92. V. A. Radtsig, *Chem. Phys. Rep.*, **14**, 1206 (1995).
93. L. N. Skuja, A. N. Streletsky and A. B. Pakovich, *Solid State Commun.*, **50**, 1069 (1984).
94. A. A. Bobyshev and V. A. Radtsig, *Kinet. Katal.*, **29**, 638 (1988) (in Russian); *Chem. Abstr.*, **109**, 156983 (1988).
95. V. A. Radtsig, *Khim. Fiz.*, **10**, 1262 (1991) (in Russian); *Chem. Abstr.*, **115**, 240633 (1991).

96. V. A. Radzig, *Colloids Surf., A: Physicochem. Eng. Aspects*, **74**, 91 (1993).
97. V. N. Bagratashvili, S. I. Tsypina and V. A. Radtsig, *J. Non-Cryst. Solids.*, **180**, 221 (1995).
98. L. N. Skuja, *J. Non-Cryst. Solids.*, **149**, 77 (1992).
99. W. W. Harper, E. A. Ferrall, R. K. Hillard, S. M. Stogner, R. S. Grev and D. J. Clouthier, *J. Am. Chem. Soc.*, **119**, 8361 (1997).
100. S. M. Stogner and R. S. Grev, *J. Chem. Phys.*, **108**, 5458 (1998).
101. D. A. Hostutler, T. C. Smith, H. Li and D. J. Clouthier, *J. Chem. Phys.*, **111**, 950 (1999).
102. W. W. Harper, K. W. Waddell and D. J. Clouthier, *J. Chem. Phys.*, **107**, 8829 (1997).
103. S. Tomoda, M. Shimoda, Y. Takeudu, Y. Kajii, K. Obi, I. Tanaka and K. Honda, *J. Chem. Soc., Chem. Commun.*, 910 (1988).
104. J. Barrau, D. L. Bean, K. M. Welsh, R. West and J. Michl, *Organometallics*, **8**, 2606 (1989).
105. V. N. Khabashesku, S. E. Boganov, D. Antic, O. M. Nefedov and J. Michl, *Organometallics*, **15**, 4714 (1996).
106. R. Becerra, S. E. Boganov, M. P. Egorov, V. Ya. Lee, O. M. Nefedov and R. Walsh, *Chem. Phys. Lett.*, **250**, 111 (1996).
107. K. Mochida, I. Yoneda and M. Wakasa, *J. Organometal. Chem.*, **399**, 53 (1990).
108. M. Wakasa, I. Yoneda and K. Mochida, *J. Organometal. Chem.*, **366**, C1 (1989).
109. K. Mochida, H. Chiba and M. Okano, *Chem. Lett.*, 109 (1991).
110. K. Mochida, K. Kimijima, H. Chiba, M. Wakasa and H. Hiyashi, *Organometallics*, **13**, 404 (1994).
111. K. Mochida, M. Wakasa, Y. Nakadira, Y. Sakaguchi and H. Hayashi, *Organometallics*, **7**, 1869 (1988).
112. H. Suzuki, K. Okabe, R. Kato, N. Sato, Y. Fukuda and H. Watanabe, *Organometallics*, **12**, 4833 (1993).
113. S. Konieczny, S. J. Jacobs, J. K. Wilking and P. P. Gaspar, *J. Organometal. Chem.*, **341**, C17 (1988).
114. K. Mochida, M. Wakasa, Y. Sakaguchi and H. Hayashi, *Bull. Chem. Soc. Jpn.*, **64**, 1889 (1991).
115. K. Mochida, H. Kikkawa and Y. Nakadira, *J. Organometal. Chem.*, **412**, 9 (1991).
116. P. P. Gaspar, D. Holten, S. Konieczny and J. Y. Corey, *Acc. Chem. Res.*, **20**, 329 (1987).
117. P. P. Gaspar, B. H. Boo, S. Chari, A. K. Ghosh, D. Holden, C. Kirmaier and S. Konieczny, *Chem. Phys. Lett.*, **105**, 153 (1984).
118. K. Mochida, N. Kanno, R. Kato, M. Kotani, S. Yamauchi, M. Wakasa and H. Hayasi, *J. Organometal. Chem.*, **415**, 191 (1991).
119. K. Mochida, S. Tokura and S. Murata, *J. Chem. Soc., Chem. Commun.*, 250 (1992).
120. K. Mochida and S. Tokura, *Bull. Chem. Soc. Jpn.*, **65**, 1642 (1992).
121. K. L. Bobbitt, V. M. Maloney and P. P. Gaspar, *Organometallics*, **10**, 2772 (1991).
122. M. P. Egorov, A. S. Dvornikov, V. A. Kuzmin, S. P. Kolesnikov and O. M. Nefedov, *Izv. Akad. Nauk SSSR, Ser. Khim.*, 1200 (1987) (in Russian); *Bull. Acad. Sci. USSR, Div. Chem. Sci.*, **36**, 1114 (1987) (in English).
123. M.-D. Su and S.-Y. Chu, *Tetrahedron Lett.*, **40**, 4371 (1999).
124. J. Kalcher and A. F. Sax, *J. Mol. Struct. (Theochem)*, **253**, 287 (1992).
125. M. C. Holthausen, W. Koch and Y. Apeloig, *J. Am. Chem. Soc.*, **121**, 2623 (1999).
126. Y. Apeloig, M. Karni, R. West and K. Welsh, *J. Am. Chem. Soc.*, **116**, 9719 (1994).
127. R. S. Grev and H. F. Schaefer, *J. Am. Chem. Soc.*, **108**, 5804 (1986).
128. Y. Apeloig and M. Karni, *J. Chem. Soc., Chem. Commun.*, 1048 (1985).
129. J. C. Barthelat, B. S. Roch, G. Trinquier and J. Satgé, *J. Am. Chem. Soc.*, **102**, 4080 (1980).
130. D. H. Harris, M. F. Lappert, J. B. Pedley and G. J. Sharp, *J. Chem. Soc., Dalton Trans.*, 945 (1976).
131. D. H. Harris and M. F. Lappert, *J. Chem. Soc., Chem. Commun.*, 895 (1974).
132. M. J. S. Gynane, D. H. Harris, M. F. Lappert, P. P. Power, P. Rivière and M. Rivière-Baudet, *J. Chem. Soc., Dalton Trans.*, 2004 (1977).
133. A. J. Arduengo, H. Bock, H. Chen, M. Denk, D. A. Dixon, J. C. Green, W. A. Herrmann, N. L. Jones, M. Wagner and R. West, *J. Am. Chem. Soc.*, **116**, 6641 (1994).
134. S. Foucat, T. Pigot, G. Pfister-Guillouzo, H. Lavayssiere and S. Mazieres, *Organometallics*, **18**, 5322 (1999).
135. J. Heinicke, A. Opera, M. K. Kindermann, T. Karpati, L. Nyulaszi and T. Veszpremi, *Chem. Eur. J.*, **4**, 541 (1998).

136. M. Driess, R. Janoschek, H. Pritzkow, S. Rell and U. Winkler, *Angew. Chem., Int. Ed. Engl.*, **34**, 1614 (1995).
137. H. A. Bent, *Chem. Rev.*, **68**, 587 (1968).
138. J. E. Bender, M. M. B. Holl and J. W. Kampf, *Organometallics*, **16**, 2743 (1997).
139. H. Grützmacher, H. Pritzkow and F. T. Edelmann, *Organometallics*, **10**, 23 (1991).
140. M. J. Michalczyk, M. J. Fink, D. J. De Young, C. W. Carlson, K. M. Welsh, R. West and J. Michl, *Silicon, Germanium, Tin and Lead Comp.*, **9**, 75 (1986).
141. R. West, *Pure Appl. Chem.*, **56**, 163 (1984).
142. W. Ando, H. Itoh and T. Tsumuraya, *Organometallics*, **8**, 2759 (1989).
143. J. Barrau, G. Rima and T. El Amraoui, *Organometallics*, **17**, 607 (1998).
144. W. Ando, T. Tsumuraya and A. Sekiguchi, *Chem. Lett.*, 317 (1987).
145. W. Ando, H. Itoh, T. Tsumuraya and H. Yoshida, *Organometallics*, **7**, 1880 (1988).
146. H. Sakurai, K. Sakamota and M. Kira, *Chem. Lett.*, 1379 (1984).
147. S. P. Kolesnikov, M. P. Egorov, A. S. Dvornikov, V. A. Kuzmin and O. M. Nefedov, *Metalloorg. Khim.*, **2**, 799 (1989) (in Russian); *Chem. Abstr.*, **112**, 179255 (1990).
148. S. Tomoda, M. Shimoda and Y. Takeuchi, *Nippon Kagaku Kaishi*, 1466 (1989); *Chem. Abstr.*, **112**, 158413 (1990).
149. K. Mochida and S. Tokura, *Organometallics*, **11**, 2752 (1992).
150. M. Kira, S. Ishida, T. Iwamoto, M. Ichinoche, C. Kabuto, L. Ignatovich and H. Sakurai, *Chem. Lett.*, 263 (1999).
151. H. Suzuki, K. Okabe, S. Uchida, H. Watanabe and M. Goto, *J. Organomet. Chem.*, **509**, 177 (1996).
152. P. J. Davidson, D. H. Harris and M. F. Lappert, *J. Chem. Soc., Dalton. Trans.*, 2268 (1976).
153. D. E. Goldberg, D. H. Harris, M. F. Lappert and K. M. Thomas, *J. Chem. Soc., Chem. Commun.*, 261 (1976).
154. M. Kira, T. Maruyama and H. Sakurai, *Chem. Lett.*, 1345 (1993).
155. N. P. Toltl, W. J. Leigh, G. M. Kollegger, W. G. Stibbs and K. M. Baines, *Organometallics*, **15**, 3732 (1996).
156. P. Jutzi, H. Schmidt, B. Neumann and H.-G. Stammler, *Organometallics*, **15**, 741 (1996).
157. K. Kishikawa, N. Tokitoh and R. Okazaki, *Chem. Lett.*, 239 (1998).
158. R. S. Simons, L. Pu, M. M. Olmstead and P. P. Power, *Organometallics*, **16**, 1920 (1997).
159. N. Tokitoh, K. Manmaru and R. Okazaki, *Nippon Kagaku Kaishi*, 240 (1994); *Chem. Abstr.*, **121**, 134283 (1994).
160. N. Tokitoh, K. Manmaru and R. Okazaki, *Organometallics*, **13**, 167 (1994).
161. N. Tokitoh, K. Kishikawa, T. Matsumoto and R. Okazaki, *Chem. Lett.*, 827 (1995).
162. A. Schaefer, W. Saak and M. Weidenbruch, *Z. Anorg. Allg. Chem.*, **624**, 1405 (1998).
163. M. F. Lappert, M. J. Slade, J. L. Atwood and M. J. Zaworotko, *J. Chem. Soc., Chem. Commun.*, 621 (1980).
164. M. F. Lappert, P. P. Power, M. J. Slade, L. Hedberg, K. Hedberg and V. Schomaker, *J. Chem. Soc., Chem. Commun.*, 369 (1979).
165. L. Pu, M. M. Olmstead, P. P. Power and B. Schiemenz, *Organometallics*, **17**, 5602 (1998).
166. M. Kira, R. Yanchibara, R. Hirano, C. Kabuto and H. Sakurai, *J. Am. Chem. Soc.*, **113**, 7785 (1991).
167. K. W. Klinkhammer and W. Schwarz, *Angew. Chem., Int. Ed. Engl.*, **34**, 1334 (1995).
168. K. W. Klinkhammer, T. F. Fässler and H. Grützmacher, *Angew. Chem., Int. Ed. Engl.*, **37**, 124 (1998).
169. M. Stürmann, W. Saak, K. W. Klinkhammer and M. Weidenbruch, *Z. Anorg. Allg. Chem.*, **625**, 1955 (1999).
170. N. Tokitoh, M. Saito and R. Okazaki, *J. Am. Chem. Soc.*, **115**, 2065 (1993).
171. M. Saito, N. Tokitoh and R. Okazaki, *Chem. Lett.*, 265 (1996).
172. M. Weidenbruch, J. Schlaefke, A. Schaefer, K. Peters, H. G. von Schnering and H. Marsmann, *Angew. Chem., Int. Ed. Engl.*, **33**, 1846 (1994).
173. M. Weidenbruch, H. Kilian, K. Peters, H. G. von Schnering and H. Marsmann, *Chem. Ber.*, **128**, 983 (1995).
174. M. Weidenbruch, U. Grobecker, W. Saak, E.-M. Peters and K. Peters, *Organometallics*, **17**, 5206 (1998).
175. C. D. Schaefer and J. J. Zuckerman, *J. Am. Chem. Soc.*, **96**, 7160 (1974).

176. C. Eaborn, T. Ganicz, P. B. Hitchcock, J. D. Smith and S. E. Sözerli, *Organometallics*, **16**, 5621 (1997).
177. L. H. Pu, B. Twamley and P. P. Power, *Organometallics*, **19**, 2874 (2000).
178. M. Stürmann, M. Weidenbruch, K. W. Klinkhammer, F. Lissner and H. Marsmann, *Organometallics*, **17**, 4425 (1998).
179. N. Kano, K. Shibata, N. Tokitoh and R. Okazaki, *Organometallics*, **18**, 2999 (1999).
180. N. Kano, N. Tokitoh and R. Okazaki, *J. Synth. Org. Chem. Japan*, **56**, 919 (1998).
181. M. Stürmann, W. Saak, H. Marsmann and M. Weidenbruch, *Angew. Chem., Int. Ed. Engl.*, **38**, 187 (1999).
182. N. Kano, N. Tokitoh and R. Okazaki, *Organometallics*, **16**, 4237 (1997).
183. N. Kano, N. Tokitoh and R. Okazaki, *Organometallics*, **17**, 1241 (1998).
184. D. Agustin, G. Rima, H. Gornitzka and J. Barrau, *J. Organomet. Chem.*, **592**, 1 (2000).
185. T. Fjeldberg, A. Haaland, B. E. R. Schilling, M. F. Lappert and A. J. Thorne, *J. Chem. Soc., Dalton. Trans.*, 1551 (1986).
186. T. Fjeldberg, H. Hope, M. F. Lappert, P. P. Power and A. J. Thorne, *J. Chem. Soc., Chem. Commun.*, 639 (1983).
187. R. W. Chorley, P. B. Hitchcock, M. F. Lappert, W. P. Leung, P. P. Power and M. M. Olmstead, *Inorg. Chim. Acta*, **198–200**, 203 (1992).
188. W. Ando, A. Sekiguchi, K. Hagiwara, A. Sakakibara and H. Yoshida, *Organometallics*, **7**, 558 (1988).
189. W. W. Schöller and R. Schneider, *Chem. Ber. Recl.*, **130**, 1013 (1997).
190. R. T. Conlin, D. Laakso and P. Marshall, *Organometallics*, **13**, 838 (1994).
191. W. Ando, K. Hagiwara and A. Sekiguchi, *Organometallics*, **6**, 2270 (1987).
192. W. Ando and T. Tsumuraya, *Organometallics*, **8**, 1467 (1989).
193. T. Tsumuraya, S. Sato and W. Ando, *Organometallics*, **8**, 161 (1989).
194. T. Ohtaki, Y. Kabe and W. Ando, *Organometallics*, **12**, 4 (1993).
195. G. Levin, P. K. Das, C. Bilgrien and C. L. Lee, *Organometallics*, **8**, 1206 (1989).
196. G. R. Gillette, G. H. Noren and R. West, *Organometallics*, **8**, 487 (1989).
197. G. Maier, G. Mihm, H. P. Reisenauer and D. Littmann, *Chem. Ber.*, **117**, 2369 (1984).
198. S. P. Kolesnikov, I. S. Rogozhin, A. Ya. Shteinshneider and O. M. Nefedov, *Izv. Akad. Nauk SSSR, Ser. Khim.*, 799 (1980) (in Russian); *Bull. Acad. Sci. USSR, Div. Chem. Sci.*, **29**, 554 (1980) (in English).
199. S. P. Kolesnikov, M. P. Egorov, A. S. Dvornikov, V. A. Kuzmin and O. M. Nefedov, *Izv. Akad. Nauk SSSR, Ser. Khim.*, 2654 (1988) (in Russian); *Bull. Acad. Sci. USSR, Div. Chem. Sci.*, **37**, 2397 (1988) (in English).
200. G. A. Ozin and A. V. Voet, *J. Chem. Phys.*, **56**, 4768 (1972).
201. G. R. Smith and W. A. Guillory, *J. Chem. Phys.*, **56**, 1423 (1972).
202. A. Lloret, M. Oria, B. Seondi and L. Abouaf-Marguin, *Chem. Phys. Lett.*, **179**, 329 (1991).
203. T. C. McInnis and L. Andrews, *J. Phys. Chem.*, **96**, 5276 (1992).
204. W. D. Allen and H. F. Schaefer, *Chem. Phys.*, **108**, 243 (1986).
205. L. Fredin, R. H. Hauge, Z. H. Kafafi and J. L. Margrave, *J. Chem. Phys.*, **82**, 3542 (1985).
206. P. R. Bunker, R. A. Phillips and R. J. Buenker, *Chem. Phys. Lett.*, **110**, 351 (1984).
207. D. E. Milligan and M. E. Jacox, *J. Chem. Phys.*, **52**, 2594 (1970).
208. N. Legay-Sommaire and F. Legay, *J. Phys. Chem., A*, **102**, 8759 (1998).
209. H. Huber, E. P. Kundig, G. A. Ozin and A. V. Voet, *Can. J. Chem.*, **52**, 95 (1974).
210. G. Herzberg, *Infrared and Raman Spectra*, D. Van Nostrand Co., Princeton, New York, 1945.
211. H. Takeo and R. F. Curl, *J. Mol. Spectrosc.*, **43**, 21 (1972).
212. W. A. Guillory and G. R. Smith, *J. Chem. Phys.*, **53**, 1661 (1970).
213. L. Andrews and D. L. Frederick, *J. Am. Chem. Soc.*, **92**, 775 (1970).
214. A. K. Maltsev, V. A. Svyatkin and O. M. Nefedov, *Dokl. Akad. Nauk SSSR*, **227**, 1151 (1976); *Chem. Abstr.*, **85**, 113990 (1976).
215. J. H. Miller and L. Andrews, *J. Mol. Struct.*, **77**, 65 (1981).
216. I. R. Beattie and R. O. Perry, *J. Chem. Soc., A*, 2429 (1970).
217. J. Bouix, R. Hillel and A. Michaelides, *J. Raman Spectrosc.*, **7**, 346 (1978).
218. R. J. Isabel, G. R. Smith, R. K. McGraw and W. A. Guillory, *J. Chem. Phys.*, **58**, 818 (1973).
219. S. Choukroun, J. C. Launay, M. Pouchard, P. Hagenmuller, J. Bouix and R. Hillel, *J. Cryst. Growth*, **43**, 597 (1978).
220. R. C. McNutt, *Sci. Tech. Aerosp. Rep.*, **14**, Abstr. 76–11260 (1976).

221. N. I. Giricheva, G. V. Girichev, S. A. Shlykov, V. A. Titov and T. P. Chusova, *J. Mol. Struct.*, **344**, 127 (1995).
222. R. H. Hauge, J. W. Hastie and J. L. Margrave, *J. Mol. Spectrosc.*, **45**, 420 (1973).
223. D. Tevault and K. Nakamoto, *Inorg. Chem.*, **15**, 1282 (1976).
224. R. O. Perry, *J. Chem. Soc., Chem. Commun.*, 886 (1969).
225. M. Fields, R. Devonshire, H. G. M. Edwards and V. Fawcett, *Spectrochim. Acta, A*, **51**, 2249 (1995).
226. J. W. Hastie, R. H. Hauge and J. L. Margrave, *High Temp. Sci.*, **3**, 56 (1971).
227. S. A. Zaitsev, S. B. Osin and V. F. Shevelkov, *Vestn. Mosk. Univ., Ser. 2: Khim.*, **29**, 564 (1988) (in Russian); *Chem. Abstr.*, **110**, 124150 (1989).
228. R. J. Isabel and W. A. Guillory, *J. Chem. Phys.*, **55**, 1197 (1971).
229. R. J. Isabel and W. A. Guillory, *J. Chem. Phys.*, **57**, 1116 (1972).
230. E. Hirota and H. Ishikawa, *J. Chem. Phys.*, **110**, 4254 (1999).
231. A. G. Gershikov, N. Yu. Subbotina and M. Hargittai, *J. Mol. Spectrosc.*, **143**, 293 (1990).
232. G. Maass, R. H. Hauge and J. L. Margrave, *Z. Anorg. Allg. Chem.*, **392**, 295 (1972).
233. H. Ishikawa and O. Kajimoto, *J. Mol. Spectrosc.*, **174**, 270 (1995).
234. J. W. Hastie, R. H. Hauge and J. L. Margrave, *J. Am. Chem. Soc.*, **91**, 2536 (1969).
235. D. E. Milligan and M. E. Jacox, *J. Chem. Phys.*, **49**, 1938 (1968).
236. V. A. Svyatkin, A. K. Maltsev and O. M. Nefedov, *Izv. Akad. Nauk, Ser. Khim.*, 2236 (1977) (in Russian); *Chem. Abstr.*, **88**, 13888 (1978).
237. G. L. Caldow, C. M. Deeley, P. H. Turner and I. M. Mills, *Chem. Phys. Lett.*, **82**, 434 (1981).
238. V. M. Rao and R. F. Curl, *J. Chem. Phys.*, **45**, 2032 (1966).
239. Z. K. Ismail, L. Fredin, R. H. Hauge and J. L. Margrave, *J. Chem. Phys.*, **77**, 1626 (1982).
240. J. W. Kauffman, R. H. Hauge and J. L. Margrave, *ACS, Symp. Ser.*, **179**, 355 (1982).
241. R. Withnall and L. Andrews, *J. Phys. Chem.*, **94**, 2351 (1990).
242. Z. K. Ismail, R. H. Hauge, L. Fredin, J. W. Kauffman and J. L. Margrave, *J. Chem. Phys.*, **77**, 1617 (1982).
243. R. Withnall and L. Andrews, *J. Phys. Chem.*, **89**, 3261 (1985).
244. G. Trinquier, M. Pelissier, B. Saint-Roch and H. Lavassiére, *J. Organomet. Chem.*, **214**, 169 (1981).
245. J. Kapp, M. Remko and P. v. R. Schleyer, *J. Am. Chem. Soc.*, **118**, 5745 (1996).
246. V. N. Khabashesku, K. N. Kudin, J. Tamas, S. E. Boganov, J. L. Margrave and O. M. Nefedov, *J. Am. Chem. Soc.*, **120**, 5005 (1998).
247. P. Bleckmann, H. Maly, R. Minkwitz, W. P. Neumann, B. Watta and G. Olbrich, *Tetrahedron Lett.*, **23**, 4655 (1982).
248. V. N. Khabashesku, V. Balaji, S. E. Boganov, S. A. Bashkirova, P. M. Matveichev, E. A. Chernyshev, O. M. Nefedov and J. Michl, *Mendeleev Commun.*, 38 (1992).
249. V. N. Khabashesku, V. Balaji, S. E. Boganov, O. M. Nefedov and J. Michl, *J. Am. Chem. Soc.*, **116**, 320 (1994).
250. C. A. Arrington, K. A. Klingensmith, R. West and J. Michl, *J. Am. Chem. Soc.*, **106**, 525 (1984).
251. G. Raabe, H. Vancik, R. West and J. Michl, *J. Am. Chem. Soc.*, **108**, 671 (1986).
252. M. Veith, E. Werle, R. Lisowsky, R. Koeppe and H. Schnöckel, *Chem. Ber.*, **125**, 1375 (1992).
253. H. Schmidt, S. Keitemeyer, B. Neumann, H.-G. Stammler, W. W. Schoeller and P. Jutzi, *Organometallics*, **17**, 2149 (1998).
254. P. B. Hitchcock, M. F. Lappert and A. J. Thorne, *J. Chem. Soc., Chem. Commun.*, 1587 (1990).
255. L. D. Silverman and M. Zeldin, *Inorg. Chem.*, **19**, 272 (1980).
256. P. Jutzi and W. Steiner, *Chem. Ber.*, **109**, 1575 (1976).
257. J. T. Ahlemann, H. W. Roesky, R. Murugavel, E. Parisiny, M. Noltemeyer, H. G. Schmidt, O. Mueller, R. Herbst-Irmer, L. N. Markovskii and Y. G. Shermolovich, *Chem. Ber.*, **130**, 1113 (1997).
258. S. Mazieres, H. Lavayssiere, G. Dousse and J. Satgé, *Inorg. Chim. Acta*, **252**, 25 (1996).
259. M. Rivière-Baudet, M. Dahrouch and H. Gornitzka, *J. Organomet. Chem.*, **595**, 153 (2000).
260. G. L. Wegner, R. J. F. Berger, A. Schier and H. Schmidbaur, *Organometallics*, **20**, 418 (2001).
261. W. W. du Mont and M. Grenz, *Chem. Ber.*, **118**, 1045 (1985).
262. M. A. Matchett, M. Y. Chiang and W. E. Buhro, *Inorg. Chem.*, **33**, 1109 (1994).
263. H. H. Karsch, A. Appelt and G. Müller, *Organometallics*, **5**, 1664 (1986).

264. B. Cetinkaya, P. B. Hitchcock, M. F. Lappert, M. C. Misra and A. J. Thorne, *J. Chem. Soc., Chem. Commun.*, 148 (1984).
265. D. Haensgen, J. Kuna and B. Ross, *Chem. Ber.*, **109**, 1797 (1976).
266. M. Westerhausen, M. M. Enzelberger and W. Schwarz, *J. Organomet. Chem.*, **491**, 83 (1995).
267. M. Mehring, C. Löw, M. Schürmann, F. Uhlig, K. Jurkschat and B. Mahien, *Organometallics*, **19**, 4613 (2000).
268. A. J. Edwards, M. A. Paver, P. R. Raithky, M.-A. Rennie, C. A. Russell and D. S. Wright, *J. Chem. Soc., Dalton Trans.*, 1587 (1995).
269. R. Cea-Olivares, J. Novosad, J. D. Woollins, A. M. Z. Slawin, V. Garcia-Montalvo, G. Espinoza-Perez and P. Garcia y Garcia, *J. Chem. Soc., Chem. Commun.*, 519 (1996).
270. M. Westerhausen, J. Greul, H.-D. Hausen and W. Schwarz, *Z. Anorg. Allg. Chem.*, **622**, 1295 (1996).
271. S. Wingerter, H. Gornitzka, R. Bertermann, S. K. Pandey, J. Rocha and D. Stalke, *Organometallics*, **19**, 3890 (2000).
272. S. Brooker, J.-K. Buijink and F. T. Edelmann, *Organometallics*, **10**, 25 (1991).
273. R. Withnall and L. Andrews, *J. Phys. Chem.*, **94**, 2351 (1990).
274. A. Nowek and J. Leszczynski, *J. Phys. Chem., A*, **101**, 3784 (1997).
275. S. E. Boganov, V. I. Faustov, M. P. Egorov and O. M. Nefedov, *Izv. Akad. Nauk, Ser. Khim.*, 1087 (1998) (in Russian); *Russ. Chem. Bull. (Engl. Transl.)*, **47**, 1054 (1998).
276. H. P. Reisenauer, G. Mihm and G. Maier, *Angew. Chem.*, **94**, 864 (1982).
277. P. F. Meier, D. L. Perry, R. H. Hauge and J. L. Margrave, *Inorg. Chem.*, **18**, 2051 (1979).
278. S. E. Boganov, V. I. Faustov, M. P. Egorov and O. M. Nefedov, *Izv. Akad. Nauk, Ser. Khim.*, 54 (1994) (in Russian); *Russ. Chem. Bull. (Engl. Transl.)*, **43**, 47 (1994).
279. S. E. Boganov, V. I. Faustov, M. P. Egorov and O. M. Nefedov, *High Temp. Mater. Sci.*, **33**, 107 (1995).
280. S. E. Boganov, M. P. Egorov and O. M. Nefedov, *Izv. Akad. Nauk, Ser. Khim.*, 97 (1999) (in Russian); *Russ. Chem. Bull. (Engl. Transl.)*, **48**, 98 (1999).
281. S. E. Boganov, V. I. Faustov, S. G. Rudyak, M. P. Egorov and O. M. Nefedov, *Izv. Akad. Nauk, Ser. Khim.*, 1121 (1996) (in Russian); *Russ. Chem. Bull. (Engl. Transl.)*, **45**, 1061 (1996).
282. W.-W. Du Mont, B. Neudert, G. Rudolph and H. Schumann, *Angew. Chem.*, **88**, 303 (1976).
283. H. Takeo, R. F. Curl and P. W. Wilson, *J. Mol. Spectrosc.*, **38**, 464 (1971).
284. C. Styger and M. C. L. Gerry, *J. Mol. Spectrosc.*, **158**, 328 (1993).
285. M. J. Tsuchiya, H. Honjou, K. Tanaka and T. Tanaka, *J. Mol. Struct.*, **352–353**, 407 (1995).
286. O. M. Nefedov, S. P. Kolesnikov and A. I. Ioffe, *Organomet. Chem. Library*, **5**, 181 (1977).
287. A. I. Ioffe and O. M. Nefedov, *Zh. Vses. Khim. O-va*, **24**, 475 (1979) (in Russian); *Chem. Abstr.*, **92**, 57650 (1980).
288. Gy. Shultz, J. Tremmel, I. Hargittai, I. Berecz, S. Bohatka, N. D. Kagramanov, A. K. Maltsev and O. M. Nefedov, *J. Mol. Struct.*, **55**, 207 (1979).
289. M. Hargittai, *Chem. Rev.*, **100**, 2233 (2000).
290. R. J. Gillespie and I. Hargittai, *The VSEPR Model of Molecular Geometry*, Allyn & Bacon, Boston, 1991.
291. C. Yamada, H. Kanamori, E. Hirota, N. Nishiwaki, N. Itabashi, K. Kato and T. Goto, *J. Chem. Phys.*, **91**, 4582 (1989).
292. H. Shoji, T. Tanaka and E. Hirota, *J. Mol. Spectrosc.*, **47**, 268 (1973).
293. M. Fujitake and E. Hirota, *J. Mol. Struct.*, **413–414**, 21 (1997).
294. M. Fujitake and E. Hirota, *Spectrochim. Acta, A*, **50**, 1345 (1994).
295. I. Hargittai, Gy. Schultz, J. Tremmel, N. D. Kagramanov, A. K. Maltsev and O. M. Nefedov, *J. Am. Chem. Soc.*, **105**, 2895 (1983).
296. Gy. Schultz, M. Colonits and M. Hargittai, *Struct. Chem.*, **11**, 161 (2000).
297. P. A. Akishin, V. P. Spiridonov and A. N. Khodchenkov, *Zh. Fis. Khim.*, **32**, 1679 (1958) (in Russian); *Chem. Abstr.*, **53**, 849 (1959).
298. K. V. Ermakov, B. S. Butayev and V. P. Spiridonov, *J. Mol. Struct.*, **248**, 143 (1991).
299. A. V. Demidov, A. G. Gershikov, E. Z. Zasorin, V. P. Spiridonov and A. A. Ivanov, *Zh. Strukt. Khim.*, **24**, 9 (1983) (in Russian); *J. Struct. Chem. (Engl. Transl.)*, **24**, 7 (1983).
300. V. I. Bazhanov, *Zh. Strukt. Khim.*, **32**, 54 (1991) (in Russian); *J. Struct. Chem. (Engl. Transl.)*, **32**, 44 (1991).
301. T. Koopmans, *Physica*, **1**, 104 (1934).

302. B. Ruscic, M. Schwarz and J. Berkowitz, *J. Chem. Phys.*, **92**, 1865 (1990); erratum: *J. Chem. Phys.*, **92**, 6338 (1990).
303. I. Novak and A. W. Potts, *J. Chem. Soc., Dalton Trans.*, 2211 (1983).
304. G. Jonkers, S. M. van der Kerk, R. Mooyman and C. A. de Lange, *Chem. Phys. Lett.*, **90**, 252 (1982).
305. G. Jonkers, S. M. van der Kerk and C. A. de Lange, *Chem. Phys.*, **70**, 69 (1982).
306. G. Jonkers, S. M. van der Kerk, R. Mooyman, C. A. de Lange and J. G. Snijders, *Chem. Phys. Lett.*, **94**, 585 (1983).
307. C. Cauletti, M. de Simone and S. Stranges, *J. Electron Spectrosc. Relat. Phenom.*, **57**, R1 (1991).
308. S. Stranges, M.-Y. Adam, C. Cauletti, M. de Simone, C. Furlani, M. N. Piancastelli, P. Decleva and A. Lisini, *J. Chem. Phys.*, **97**, 4764 (1992).
309. I. Novak and A. W. Potts, *J. Electron Spectrosc. Relat. Phenom.*, **33**, 1 (1984).
310. S. Evans and A. F. Orchard, *J. Electron Spectrosc. Relat. Phenom.*, **6**, 207 (1975).
311. J. Berkowitz, in *Electron Spectroscopy*, (Ed. D. A. Shirley), North-Holland Publ. Co., Amsterdam, 1972, 391.
312. S. Stranges, M.-Y. Adam, C. Cauletti, M. de Simone and M. N. Piancastelli, *AIP Conf. Proc.*, **258**, 60 (1992).
313. R. C. McDonald, H. Ho-Kuen Han and K. Eriks, *Inorg. Chem.*, **15**, 762 (1976).
314. G. M. Bancroft, T. K. Sham, D. E. Estman and W. Gudat, *J. Am. Chem. Soc.*, **99**, 1752 (1977).
315. A. W. Potts and M. L. Lyus, *J. Electron Spectrosc. Relat. Phenom.*, **13**, 327 (1978).
316. A. W. Potts and W. C. Price, *Phys. Scr.*, **16**, 191 (1977).
317. I. Novak and A. W. Potts, *J. Phys., B*, **17**, 3713 (1984).
318. T. C. Ehlert and J. L. Margrave, *J. Chem. Phys.*, **41**, 1066 (1964).
319. K. F. Zmbow, J. W. Hastie, R. Hauge and J. L. Margrave, *Inorg. Chem.*, **7**, 608 (1968).
320. O. M. Uy, D. W. Muenov and J. L. Margrave, *Trans. Faraday Soc.*, **65**, 1296 (1969).
321. E. Vajda, I. Hargittai, M. Colonits, K. Ujszaszy, J. Tamas, A. K. Maltsev, R. G. Mikaelian and O. M. Nefedov, *J. Organomet. Chem.*, **105**, 33 (1976).
322. C. Hirayama and R. D. Straw, *Thermochim. Acta*, **80**, 297 (1984).
323. S. Ciach, D. J. Knowles, A. J. C. Nicholson and D. L. Swingler, *Inorg. Chem.*, **12**, 1443 (1973).
324. J. W. Hastie, H. Bloom and J. D. Morrison, *J. Chem. Phys.*, **47**, 1580 (1967).
325. A. S. Buchanan, D. J. Knowles and D. L. Swingler, *J. Phys. Chem.*, **73**, 4394 (1969).
326. D. J. Knowles, A. J. C. Nicholson and D. L. Swingler, *J. Phys. Chem.*, **74**, 3642 (1970).
327. C. Hirayama and R. L. Kleinosky, *Thermochim. Acta*, **47**, 355 (1981).
328. K. Hilpert and K. A. Gingerich, *Int. J. Mass Spectrom. Ion Phys.*, **47**, 247 (1983).
329. J. Berkowitz, *Adv. High Temp. Chem.*, **3**, 123 (1971).
330. P. J. Davidson and M. F. Lappert, *J. Chem. Soc., Chem. Commun.*, 317 (1973).
331. P. J. Davidson, D. H. Harris and M. F. Lappert, *J. Chem. Soc., Chem. Commun.*, 2268 (1976).
332. C. Heinemann, T. Müller, Y. Apeloig and H. Schwarz, *J. Am. Chem. Soc.*, **118**, 2023 (1996).
333. C. Boehme and G. Frenking, *J. Am. Chem. Soc.*, **118**, 2039 (1996).
334. J. F. Lehmann, S. G. Urquhart, L. E. Ennis, A. P. Hitchcock, K. Hatano, S. Gupta and M. K. Denk, *Organometallics*, **18**, 1862 (1999).
335. P. W. Harland, S. Cradock and J. C. J. Thynne, *Int. J. Mass Spectrom. Ion Phys.*, **10**, 169 (1972).
336. R. E. Pabst, D. L. Perry, J. L. Margrave and J. L. Franklin, *Int. J. Mass Spectrom. Ion Phys.*, **24**, 323 (1977).
337. R. E. Pabst, J. L. Margrave and J. L. Franklin, *Int. J. Mass Spectrom. Ion Phys.*, **25**, 361 (1977).
338. V. Ya. Lee, A. A. Basova, I. A. Matchkarovskaya, V. I. Faustov, M. P. Egorov, O. M. Nefedov, R. D. Rakhimov and K. P. Butin, *J. Organomet. Chem.*, **499**, 27 (1995).
339. M. P. Egorov, A. A. Basova, A. M. Gal'minas, O. M. Nefedov, A. A. Moiseeva, R. D. Rakhimov and K. P. Butin, *J. Organomet. Chem.*, **574**, 279 (1999).
340. K. G. Dyall, *J. Chem. Phys.*, **96**, 1210 (1992).
341. B. Delley and G. Solt, *J. Mol. Struct. (Theochem)*, **139**, 159 (1986).
342. S. Escalante, R. Vargas and A. Vela, *J. Phys. Chem., A*, **103**, 5590 (1999).
343. M.-D. Su and S.-Y. Chu, *J. Phys. Chem. A*, **103**, 11011 (1999).
344. T. Mineva, N. Russo, E. Sicilia and M. Toscano, *Int. J. Quantum Chem.*, **56**, 669 (1995).

345. J. P. Perdew, K. Burke and M. Ernzerhof, *Phys. Rev. Lett.*, **77**, 3865 (1996).
346. M. Ernzerhof and G. E. Scuseria, *J. Chem. Phys.*, **110**, 5029 (1999).
347. C. Adamo and V. Barone, *J. Chem. Phys.*, **110**, 6158 (1999).
348. V. I. Faustov, B. G. Kimel, M. P. Egorov and O. M. Nefedov, unpublished results.
349. M.-D. Su and S.-Y. Chu, *J. Am. Chem. Soc.*, **121**, 4229 (1999).
350. N. Matsunaga, S. Koseki and M. S. Gordon, *J. Chem. Phys.*, **104**, 7988 (1996).
351. K. Balasubramanian, *J. Chem. Phys.*, **89**, 5731 (1988).
352. R. A. Philips, R. J. Bunker, R. Bearsworth, P. R. Bunker, P. Jensen and W. P. Kraemer, *Chem. Phys. Lett.*, **118**, 60 (1985).
353. D. G. Dai, M. M. Al-zahrani and K. Balasubramanian, *J. Phys. Chem.*, **98**, 9233 (1994).
354. M. Benavides-Garcia and K. Balasubramanian, *J. Chem. Phys.*, **100**, 2821 (1994).
355. E. Sicilia, M. Toscano, T. Mineva and N. Russo, *Int. J. Quantum Chem.*, **61**, 571 (1997).
356. K. K. Das and K. Balasubramanian, *J. Chem. Phys.*, **93**, 5883 (1990).
357. V. N. Khabashesku, S. E. Boganov, K. N. Kudin, J. L. Margrave, J. Michl and O. M. Nefedov, *Izv. Akad. Nauk SSSR, Ser. Khim.*, 2027 (1999) (in Russian); *Russ. Chem. Bull. (Engl. Transl.)*, **48**, 2003 (1999).
358. T. Kudo and S. Nagase, *Chem. Phys. Lett.*, **84**, 375 (1981).
359. G. Trinquier, J. C. Barthelat and J. Satgé, *J. Am. Chem. Soc.*, **104**, 5931 (1982).
360. S. Nagase and T. Kudo, *Organometallics*, **3**, 324 (1984).
361. R. Grev and H. F. Schaefer III, *Organometallics*, **11**, 3489 (1992).
362. N. A. Richardson, J. C. Rienstra-Kiracofe and H. F. Schaefer III, *Inorg. Chem.*, **38**, 6271 (1999).
363. M. Weidenbruch, *Eur. J. Inorg. Chem.*, **3**, 373 (1999).
364. J.-P. Malrieu and G. Trinquier, *J. Am. Chem. Soc.*, **111**, 5916 (1989).
365. G. Trinquier, *J. Am. Chem. Soc.*, **112**, 2130 (1990).
366. G. Trinquier and J.-C. Barthelat, *J. Am. Chem. Soc.*, **112**, 9121 (1990).
367. G. Trinquier and J.-P. Malrieu, *J. Phys. Chem.*, **94**, 6184 (1990).
368. T. S. Windus and M. S. Gordon, *J. Am. Chem. Soc.*, **114**, 9559 (1992).
369. R. Becerra, S. E. Boganov, M. P. Egorov, V. I. Faustov, O. M. Nefedov and R. Walsh, *Can. J. Chem.*, **78**, 1428 (2000).
370. J. M. Jasinski, R. Becerra and R. Walsh, *Chem. Rev.*, **95**, 1203 (1995).
371. M. S. Gordon, D. R. Gano, J. S. Binkley and M. J. Frisch, *J. Am. Chem. Soc.*, **108**, 2191 (1986).
372. T. A. Hein, W. Thiel and T. J. Lee, *J. Phys. Chem.*, **97**, 4381 (1993).
373. G. Trinquier, *J. Chem. Soc., Faraday Trans.*, **89**, 775 (1993).
374. R. Becerra, H. M. Frey, B. P. Mason, R. Walsh and M. S. Gordon, *J. Chem. Soc., Faraday Trans.*, **91**, 2723 (1995).
375. R. Becerra, S. E. Boganov, M. P. Egorov, V. I. Faustov, O. M. Nefedov and R. Walsh, *J. Am. Chem. Soc.*, **120**, 12657 (1998).
376. R. Becerra, S. E. Boganov, M. P. Egorov, V. I. Faustov, O. M. Nefedov and R. Walsh, *Phys. Chem. Chem. Phys.*, **3**, 184 (2001).
377. A. Pross, *Theoretical and Physical Principles of Organic Reactivity*, John Wiley & Sons Inc., New York, 1995.
378. S. Shaik, H. B. Schlegel and S. Wolfe, *Theoretical Aspects of Physical Organic Chemistry*, John Wiley & Sons Inc., New York, 1992.
379. M.-D. Su and S.-Y. Chu, *Inorg. Chem.*, **38**, 4819 (1999).
380. M.-D. Su and S.-Y. Chu, *J. Chin. Chem. Soc.*, **47**, 135 (2000).
381. M. S. Dewar, J. E. Friedheim and G. L. Grady, *Organometallics*, **4**, 1784 (1985).
382. J. A. Boatz, M. S. Gordon and L. R. Sita, *J. Phys. Chem.*, **94**, 5488 (1990).
383. D. A. Horner, R. S. Grev and H. F. Schaefer, *J. Am. Chem. Soc.*, **114**, 2093 (1992).
384. W. Ando, H. Ohgaki and Y. Kabe, *Angew. Chem., Int. Ed. Engl.*, **33**, 659 (1994).
385. M.-D. Su and S.-Y. Chu, *J. Am. Chem. Soc.*, **121**, 11478 (1999).
386. F. Anwari and M. S. Gordon, *Isr. J. Chem.*, **23**, 129 (1983).
387. M. S. Gordon and W. Nelson, *Organometallics*, **14**, 1067 (1995).
388. X. H. Lu, Y. X. Wang and C. B. Liu, *Acta Chim. Sinica*, **57**, 1343 (1999).
389. P. Antoniotti, P. Benzi, M. Castiglioni and P. Volpe, *Eur. J. Inorg. Chem.*, **2**, 323 (1999).
390. K. Sakata and A. Tachibana, *Chem. Phys. Lett.*, **320**, 527 (2000).
391. M. W. Heaven, G. F. Metha and M. A. Buntine, *J. Phys. Chem. A*, **105**, 1185 (2001).

CHAPTER **13**

Multiply bonded germanium, tin and lead compounds

NORIHIRO TOKITOH

Institute for Chemical Research, Kyoto University, Gokasho, Uji, Kyoto 611-0011, Japan
Fax: 81-774-38-3209; E-mail: tokitoh@boc.kuicr.kyoto-u.ac.jp

and

RENJI OKAZAKI

Department of Chemical and Biological Sciences, Faculty of Science, Japan Women's University, 2-8-1 Mejirodai, Bunkyo-ku, Tokyo 112–8681, Japan
Fax: 81-3-5981-3664; E-mail: okazaki@jwu.ac.jp

The chemistry of organic germanium, tin and lead compounds — Vol. 2
Edited by Z. Rappoport

I. INTRODUCTION

In the last two decades, remarkable progress has been made in the chemistry of low-coordinate compounds of heavier group 14 elements[1]. Following the successful synthesis and isolation of the first stable silene[2a] and disilene[2b] in 1981, a variety of low-coordinated silicon compounds such as Si=Pn (Pn = N[3], P[4], As[5]) and Si=Ch (Ch = S[6], Se[7]), 1-silaallenes[8], silabenzene[9], 2-silanaphthalene[10], and tetrasila-1,3-butadiene[11] have been synthesized as stable compounds by taking advantage of kinetic stabilization with bulky substituents (so-called steric protection) and most of them are structurally well-characterized These successful results in the chemistry of doubly bonded silicon compounds naturally provoked the challenge to extend this chemistry to that of their heavier congeners, i.e. the corresponding low-coordinated germanium, tin and lead compounds.

Although some review articles are now available on the syntheses and properties of low-coordinate species of heavier group 14 elements, most of them are restricted to those dealing with the most thoroughly investigated elements, silicon and germanium[1]. In view of the present situation of low-coordinated compounds of heavier group 14 elements, it should be timely to survey the recent progress in this field and to make a systematic comparison of the multiply bonded systems with full periodic range covering germanium, tin and lead.

This review is divided into several sections according to the type of compounds, and in each section the similarity and/or difference among the germanium, tin and lead analogues will be delineated. Some comparisons with the related carbon and/or silicon analogues are added when necessary.

II. HEAVIER CONGENERS OF OLEFINS

Although a number of excellent review articles have appeared on disilenes and digermenes[1], the chemistry of dimetallenes of heavier group 14 elements, i.e. the heavier congeners of olefins, is summarized here again with the addition of updated examples. In this field it is very interesting to elucidate whether all the group 14 elements including the heaviest case of lead can generate a doubly bonded system and also how the character of such a heavy double bond differs from those of olefins and disilenes. In other words, a systematic investigation of these doubly bonded compounds is very important to determine whether or not the common concepts in the organic chemistry of elements in the second row, e.g. hybridization and conjugation, are acceptable to the whole group 14 elements.

A. Digermenes

Ge–Ge double-bond compounds, i.e. digermenes, do not exist in a monomer form under normal conditions, because they undergo ready oligomerization or polymerization as in the case of disilenes. The bulkiness of the substituents on the germanium atoms has a very large effect on the stability of digermenes. Thus, digermenes bearing small substituents are not isolable but transient, giving oligomers or polymers, while the substitution by too bulky ligands results in the formation of the corresponding germylenes. Only the digermenes bearing moderately bulky substituents such as 2,6-diethylphenyl[12], 2,6-diisopropylphenyl[13] or bis(trimethylsilyl)methyl[14] can exist as stable compounds. Some digermenes such as 1,2,3,4-tetrakis[bis(trimethylsilyl)methyl]digermene (**1**)[14] retain their digermene structure but only in the solid state, while in solution they exist as equilibrium mixtures with the corresponding germylene (**2**) (Scheme 1).

$$[(Me_3Si)_2CH]_2Ge{=}Ge[CH(SiMe_3)_2]_2 \rightleftharpoons 2\,[(Me_3Si)_2CH]_2Ge{:}$$

(**1**) (**2**)

SCHEME 1

The molecular structures of some digermenes have been determined by X-ray structural analysis[12,15–20] and it was found that most of the digermenes have *trans*-bent geometry except in a few cases. The pyramidal geometry of the germanium atoms in digermenes is in sharp contrast to the trigonal planar carbons of olefins. Even the sterically hindered digermenes are extremely sensitive to oxygen and moisture and exist as stable compounds only under an inert atmosphere.

1. Synthesis of digermenes

a. Synthesis of digermenes from germylenes. Tetraalkyldigermene **1** is synthesized by the reaction of a dichlorogermylene–dioxane complex with a Grignard reagent or by the reaction of a stable diaminogermylene with an organolithium reagent (Scheme 2)[14]. Tetraaryldigermenes are also prepared by the treatment of the corresponding dihalogermylenes with an appropriate organometallic reagent (Scheme 2)[13,20].

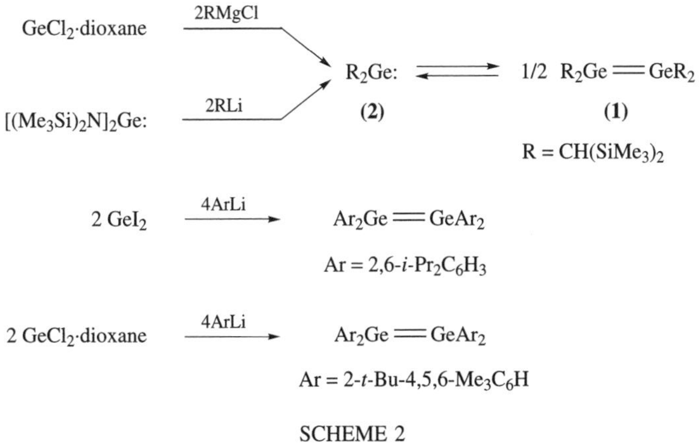

$$GeCl_2{\cdot}dioxane \xrightarrow{2RMgCl} R_2Ge{:}\ (\mathbf{2})$$

$$[(Me_3Si)_2N]_2Ge{:} \xrightarrow{2RLi} R_2Ge{:}\ (\mathbf{2})$$

$$R_2Ge{:}\ (\mathbf{2}) \rightleftharpoons 1/2\ R_2Ge{=}GeR_2\ (\mathbf{1})$$

$R = CH(SiMe_3)_2$

$$2\,GeI_2 \xrightarrow{4ArLi} Ar_2Ge{=}GeAr_2$$

$Ar = 2,6\text{-}i\text{-}Pr_2C_6H_3$

$$2\,GeCl_2{\cdot}dioxane \xrightarrow{4ArLi} Ar_2Ge{=}GeAr_2$$

$Ar = 2\text{-}t\text{-}Bu\text{-}4,5,6\text{-}Me_3C_6H$

SCHEME 2

b. Synthesis of digermenes by photolysis. Photolysis of cyclotrigermanes[12] or bis(trialkylsilyl)germanes[21,22] having bulky substituents gives the corresponding digermenes. For examples, digermenes **3** and **4** have been synthesized by these methods (Scheme 3). These methods can be utilized for the synthesis of both stable digermenes and transient ones, but they cannot be applied to the synthesis of extremely bulky digermenes. In such congested systems, the corresponding cyclotrigermanes or bis(trialkylsilyl)germanes are not available due to the steric repulsion between the substituents introduced.

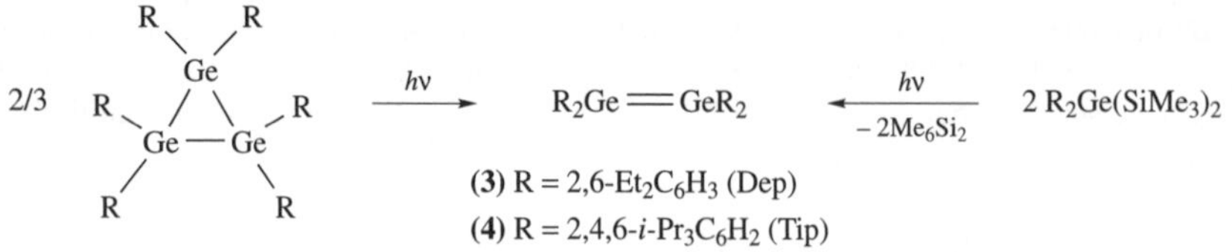

$$2/3\ (R_2Ge)_3 \xrightarrow{h\nu} R_2Ge{=}GeR_2 \xleftarrow[-\,2Me_6Si_2]{h\nu} 2\ R_2Ge(SiMe_3)_2$$

(3) $R = 2,6\text{-}Et_2C_6H_3$ (Dep)
(4) $R = 2,4,6\text{-}i\text{-}Pr_3C_6H_2$ (Tip)

SCHEME 3

c. Synthesis of digermenes by the reduction of dihalogermanes. Reactions of overcrowded diaryldihalogermanes with lithium naphthalenide give the corresponding digermenes **5**, **6** and **10** (Scheme 4)[13,19,23]. The tetrakis(trialkylsilyl)digermenes **7**–**9** can be prepared by a similar method[17]. This route is particularly useful for the synthesis of sterically hindered digermenes, which cannot be produced from cyclotrigermanes due to their steric hindrance.

$$2\ R^1R^2GeX_2 \xrightarrow{LiNaph} R^1R^2Ge{=}GeR^1R^2$$

(5–10)

(5) $R^1 = R^2 = 2,6\text{-}i\text{-}Pr_2C_6H_3$ (Dip)
(6) $R^1 =$ mesityl (Mes); $R^2 =$ Dip
(7) $R^1 = R^2 = i\text{-}Pr_2MeSi$
(8) $R^1 = R^2 = t\text{-}BuMe_2Si$
(9) $R^1 = R^2 = i\text{-}Pr_3Si$
(10) $R^1 =$ Mes; $R^2 = 2,4,6\text{-}[(Me_3Si)_2CH]_3C_6H_2$ (Tbt)

SCHEME 4

There have already been several reports on the reduction of dichlorogermanes bearing bulky aryl groups with lithium naphthalenide (Scheme 5). Upon treatment with lithium naphthalenide, dichloro[bis(2,6-dimethylphenyl)]germane **11a** and dichloro[bis(2,6-diethylphenyl)]germane **11b** afforded the corresponding cyclotrigermanes **12a,b**[12,24]. On the other hand, digermene **5** was obtained as a major product from dichloro[bis(2,6-diisopropylphenyl)]germane **11c** under similar conditions (Scheme 5)[13].

In general, the reduction of a dichlorogermane (**13**) is considered to proceed as illustrated via **14**–**17** and **19**, **20** to give **18** and **21** in Scheme 6[1j,13]. This mechanism involves a linear chain elongation mechanism. Dihalogermanes with small ligands gave oligomeric products by the chain elongation reaction.

As for dihalogermanes having relatively bulky ligands (2,6-dimethylphenyl or 2,6-diethylphenyl) on the germanium atom, the chain elongation may proceed to give a trimeric compound **19**. The steric congestion due to the substituents used forces the

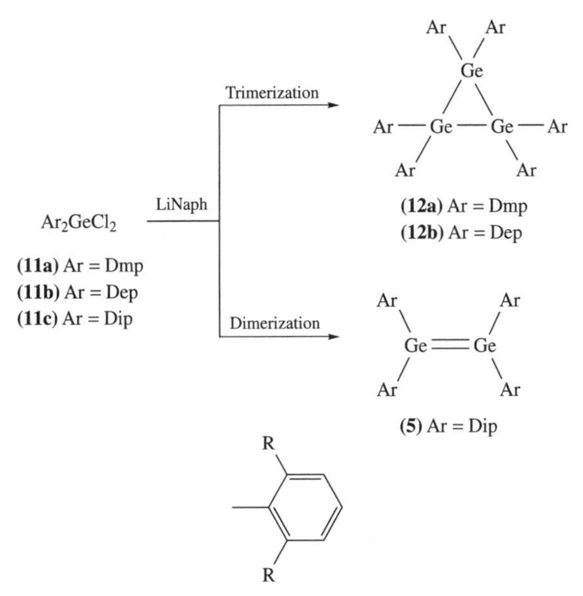

SCHEME 5

Ar_2GeCl_2 (13) $\xrightarrow{\text{LiNaph}}$ $Ar_2Ge(Li)Cl$ (14) $\xrightarrow[\alpha \text{ elimination}]{-\text{LiCl}}$ $Ar_2Ge:$ (15)

(14) $\xrightarrow{Ar_2GeCl_2}$ $ClAr_2Ge{-}GeAr_2Cl$ (16) $\xrightarrow{\text{LiNaph}}$ $LiAr_2Ge{-}GeAr_2Cl$ (17) $\xrightarrow[\beta \text{ elimination}]{-\text{LiCl}}$ $Ar_2Ge{=}GeAr_2$ (18)

(16) $\xrightarrow{Ar_2GeCl_2}$ $ClAr_2Ge{-}GeAr_2{-}GeAr_2Cl$ (19) $\xrightarrow{\text{LiNaph}}$ $LiAr_2Ge{-}GeAr_2{-}GeAr_2Cl$ (20) $\xrightarrow[\text{ring closure}]{-\text{LiCl}}$ $cyclo\text{-}(Ar_2Ge)_3$ (21)

(20) $\xrightarrow{\text{chain elongation}}$ oligomeric, or polymeric products

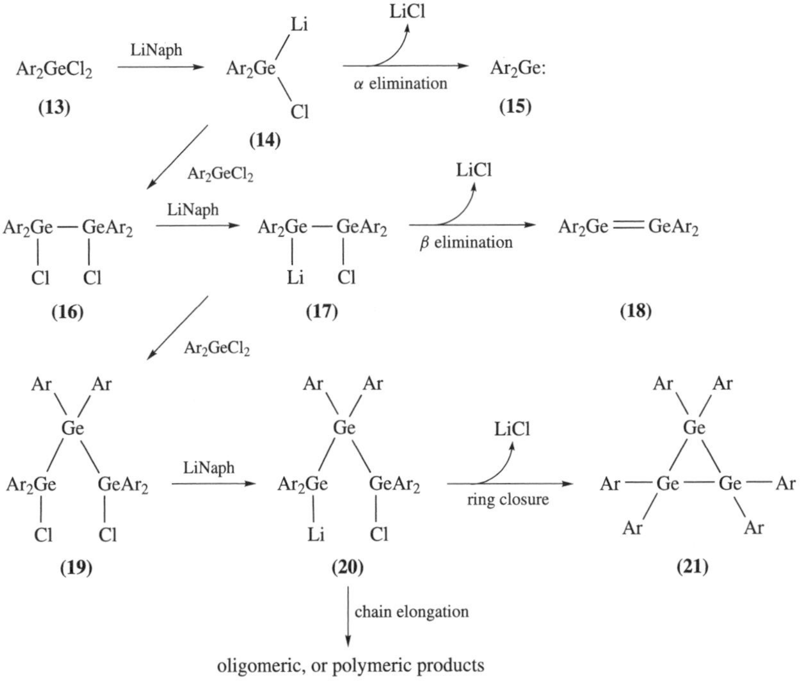

SCHEME 6

ends of the chain to get close to one another so that the chain cyclotrimerizes. When the substituents on the germanium atom become bulkier than those in **11a** or **11b** (i.e. 2,6-diisopropylphenyl in **11c**), the intermediate **17** becomes too congested to permit the chain elongation and it rather prefers to undergo β-elimination leading to the digermene **18**. In the case of an extremely hindered dichlorogermane such as Tbt(Mes)$GeCl_2$, the attempted reductive coupling of the initial intermediate **14** may be prevented by the steric repulsion between the bulky ligands, and hence the intermediate undergoes an α-elimination to give the germylene **15**. This takes place in the case of digermene **10** in Scheme 4[23].

Actually, an equilibrium between digermene **10** and the corresponding germylene **22** was observed in solution and the thermodynamic parameters were determined (Scheme 7)[23]. Crystallographic analysis of digermene **10** revealed that it has a considerably elongated Ge=Ge double bond[25] which is in good agreement with the lability of this bond in solution (*vide infra*).

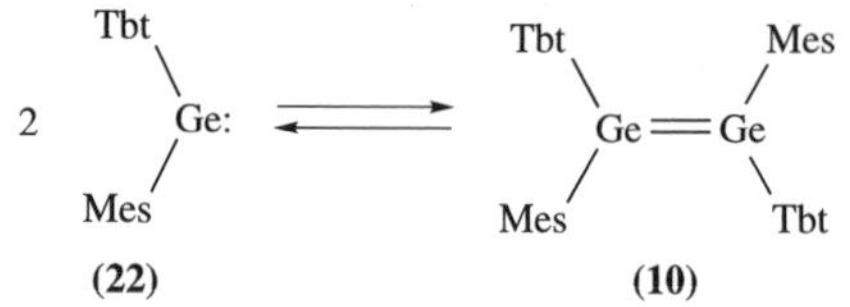

2 Tbt(Mes)Ge: **(22)** ⇌ Tbt(Mes)Ge=Ge(Mes)Tbt **(10)**

$\Delta H = 14.7 \pm 0.2$ kcal mol^{-1}

$\Delta S = 42.4 \pm 0.8$ cal mol^{-1} K^{-1}

SCHEME 7

d. Equilibrium between the Digermene and Germylene. Tbt- and Mes-substituted germylene **22** was readily generated by the reduction of Tbt(Mes)GeX_2 (X = Cl or Br) with 2 equivalents of lithium naphthalenide in THF[23]. The color of **22** in solution is blue and **22** shows an absorption maximum at $\lambda_{max} = 575$ nm, which is assignable to an n–p transition. As for the electronic spectra of germylenes, the steric effect of substituents on n–p transitions is well investigated[26]. The electronic absorptions of some germylenes reported so far are listed in Table 1[26,27].

It can be seen in Table 1 that the bulkier the substituents on the germanium atom become, the longer λ_{max} is observed for the germylenes [Ph_2Ge: (466 nm) < Dmp_2Ge:

TABLE 1. Electronic absorptions of germylenes

Germylenes[a]	λ_{max} (nm)	Color	Conditions
Me_2Ge:	420	yellow	in 3-MP at 77 K
Et_2Ge:	440	yellow	in 3-MP at 77 K
n-Bu_2Ge:	440	yellow	in 3-MP at 77 K
MePhGe:	440	yellow	in 3-MP at 77 K
Ph_2Ge:	466	yellow-orange	in 3-MP at 77 K
p-Tol_2Ge:	471	yellow-orange	in 3-MP at 77 K
Mes(*t*-Bu)Ge:	508	red	in 3-MP at 77 K
Dmp_2Ge:	543	purple	in 3-MP at 77 K
Dep_2Ge:	544	purple	in 3-MP at 77 K
Mes_2Ge:	550	purple	in 3-MP at 77 K
Tip_2Ge:	558	purple	in 3-MP at 77 K
Mes^*_2Ge:	430	orange	in hexane or THF
$Dmtp_2Ge$:	578	purple	in ethyl ether
Tbt(Mes)Ge:	575	blue	in hexane at r.t.

[a] *p*-Tol = 4-methylphenyl, Dmp = 2,6-dimethylphenyl, Dep = 2,6-diethylphenyl, Tip = 2,4,6-triisopropylphenyl, Mes* = 2,4,6-tri-*t*-butylphenyl, Dmtp = 2,6-dimesitylphenyl

(543 nm) < Dep_2Ge: (544 nm); Mes_2Ge: (550 nm) < Tip_2Ge: (558 nm)]. In the theoretical CI calculation of H_2Ge: the equilibrium value of the H-Ge-H angle in the excited state is larger than that in the ground state (123.2 and 92.6°, respectively)[28].

When the R-Ge-R angle of a germylene becomes large, its ground state is destabilized, while the excited state is conversely stabilized. Thus, the energy difference between the ground state ($^1A_1$) and excited state ($^1B_1$) becomes small and hence a red shift of λ_{max} is observed (Scheme 8). The λ_{max} value (575 nm) of **22**[23] indicates that the bulkiness of the combination of Tbt and Mes group in **22** is similar to that of two Dmtp groups in the Power's germylene **23** ($Dmtp_2Ge$:), which is isolated as stable crystals[27b].

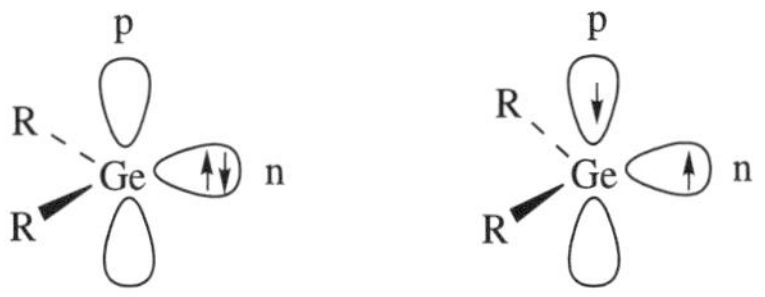

ground state ($^1A_1$) excited state ($^1B_1$)

SCHEME 8

Interestingly, a hexane solution of **22** showed a unique thermochromic character. It is blue (λ_{max} = 575 nm) at room temperature, but it turns orange-yellow (λ_{max} = 439 nm) at a low temperature[23]. The same change in color was observed in the process of concentration of the solution of **22**. Although a dilute solution of **22** is blue, its concentrated solution is orange-yellow[23].

X-ray crystallographic analysis of the orange single crystals, which were obtained after the removal of inorganic salts and naphthalene, has revealed that the structure of the orange crystal is that of the digermene (*E*)-Tbt(Mes)Ge=Ge(Mes)Tbt **10**, the dimer of the germylene Tbt(Mes)Ge: **22**[25]. The details of the structural analysis of **10** will be discussed later.

e. Synthesis of cyclic digermenes. In 1995, Sekiguchi and his coworkers reported the first cyclic digermenes, i.e. cyclotrigermenes **24a,b**, which were synthesized by the reaction of $GeCl_2$–dioxane complex with t-Bu_3SiNa or t-Bu_3GeLi (Scheme 9)[16].

The successful isolation of these unique cyclic digermenes is of particular note not only for the extension to the synthesis of other cyclic digermenes **25**[18] and germasilenes **26–28**[29] but also for the application of this ring system to the chemistry of unprecedented germaaromatic systems, i.e. cyclotrigermenium ions **29**[30] (Scheme 9). The structures and properties of these novel cyclic digermenes **24** and the related low-coordinated germanium compounds are not described here in detail, since they are fully accounted for in Chapter 14 of this volume.

2. Structure of digermenes

a. Acyclic digermenes. Digermenes have the following structural characteristics: (1) shortened Ge=Ge bond distances of 2.21–2.35 Å relative to the known Ge–Ge single bonds (2.457–2.463 Å)[1p,17,20], and (2) *trans*-bent double bonds and bent angles. The angle δ formed by the R-Ge-R plane and Ge–Ge axis is (7–36°). Theoretical studies using various basis sets predict a *trans*-bent geometry for the parent system (H_2Ge=GeH_2) with relatively short Ge=Ge bond lengths of 2.27–2.33 Å and a significant bent angle of 34–40°[15,31]. The *trans*-bent conformation for digermenes is explained in terms of

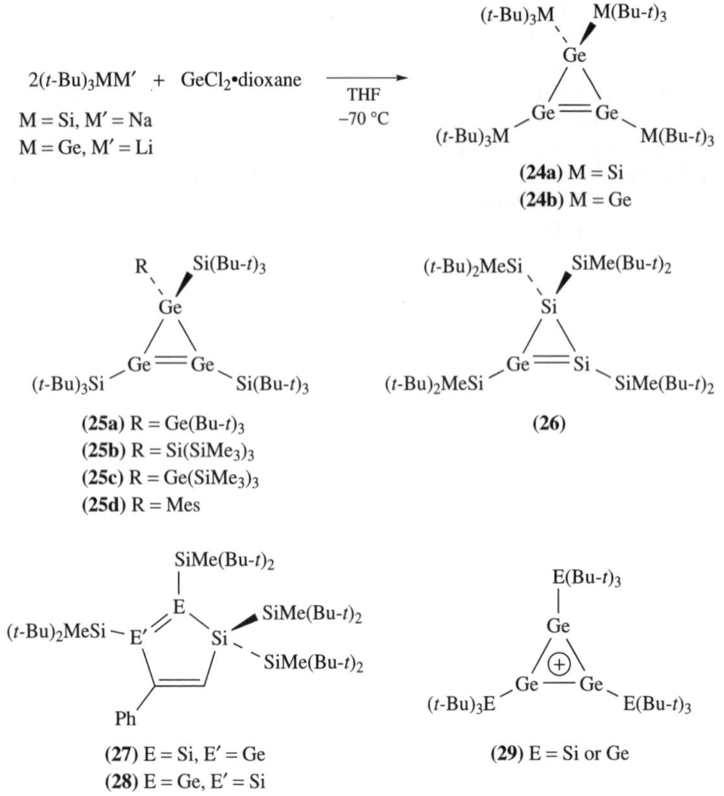

$2(t\text{-Bu})_3MM' + GeCl_2\text{•dioxane} \xrightarrow[-70\,^{\circ}C]{THF}$

M = Si, M′ = Na
M = Ge, M′ = Li

(t-Bu)$_3$M M(Bu-t)$_3$ Ge Ge=Ge (t-Bu)$_3$M M(Bu-t)$_3$

(**24a**) M = Si
(**24b**) M = Ge

R Si(Bu-t)$_3$ Ge Ge=Ge (t-Bu)$_3$Si Si(Bu-t)$_3$

(**25a**) R = Ge(Bu-t)$_3$
(**25b**) R = Si(SiMe$_3$)$_3$
(**25c**) R = Ge(SiMe$_3$)$_3$
(**25d**) R = Mes

(t-Bu)$_2$MeSi SiMe(Bu-t)$_2$ Si Ge=Si (t-Bu)$_2$MeSi SiMe(Bu-t)$_2$

(**26**)

SiMe(Bu-t)$_2$ E (t-Bu)$_2$MeSi—E′ Si SiMe(Bu-t)$_2$ SiMe(Bu-t)$_2$ Ph

(**27**) E = Si, E′ = Ge
(**28**) E = Ge, E′ = Si

E(Bu-t)$_3$ Ge ⊕ Ge—Ge (t-Bu)$_3$E E(Bu-t)$_3$

(**29**) E = Si or Ge

SCHEME 9

the stabilization of the HOMO orbital by mixing with the Ge–Ge σ^* orbital which predominates over destabilization of the Ge-Ge σ bonding. As a result, the *trans*-bent form becomes more stable than the planar form. As to the *cis*-bent form, the mixing of the HOMO orbital with the antibonding σ^* orbital is forbidden by symmetry and the energy of the *cis*-bent form increases with increasing the bent angle δ.

As mentioned in the previous section, the unique equilibrium of the highly congested digermene (E)-**10** with the corresponding germylene **22** implies a weakness of its Ge–Ge double bond[23]. Therefore, it should be important to make a systematic comparison of the structural features of (E)-**10** with other digermenes, the results of which are summarized in Table 2[15,17,19,20,32]. Although the twist angle (γ) along the Ge–Ge axis and the bent angles (the angle δ formed by the C–Ge–C plane and Ge–Ge axis) of (E)-**10** are 12, 16 and 18°, respectively, being in the range of those for the previously reported digermenes, the Ge(1)-Ge(2) bond length [2.416(4) Å] of digermene (E)-**10** is remarkably longer than those for the previously formed digermenes [2.21–2.35 Å] and close to the germanium–germanium single bond lengths [e.g. 2.465 Å in $(Ph_2Ge)_4^{33}$ or 2.463 and 2.457 Å in $(Ph_2Ge)_6^{34}$].

These results show that the large steric repulsion between the Tbt and Mes groups facing each other might be released by lengthening the germanium–germanium double

TABLE 2. Structural comparison of isolated digermenes

Digermene	r (Å)	α (deg)	β (deg)	γ (deg)	δ (deg)	Reference
$Dep_2Ge{=}GeDep_2$ (**3**)[a]	2.213(2)	115.4(2)	118.7(1) 124.3(1)	10	12	12
$Ar_2Ge{=}GeAr_2$ (**41**)[b]	2.2521(8)	128.0(2) 128.5(2)	116.1(2) 116.0(2) 115.0(2) 116.6(2)	20.4	7.9 10.4	20
(Z)-Dip(Mes)Ge=Ge(Mes)Dip (**6**)[c]	2.301(1)	109.9(2)	111.6(2) 124.2(2)	7	36	19
$Dis_2Ge{=}GeDis_2$ (**1**)[d]	2.347(2)	112.5(2)	113.7(3) 122.3(2)	0	32	15, 32
$[(i\text{-}Pr)_2MeSi]_2Ge{=}Ge[SiMe(Pr\text{-}i)_2]_2$ (**7**)	(i) 2.266(1)	117.0(0)	120.9(0) 121.4(0)	0	7.1	17
	(ii) 2.268(1)	118.2(0)	123.5(0) 117.9(0)	0	5.9	
$[(i\text{-}Pr)_3Si]_2Ge{=}Ge[Si(Pr\text{-}i)_3]_2$ (**9**)	2.298(1)	115.3(0)	125.2(4) 116.5(5)	0	16.4	17
(E)-Tbt(Mes)Ge=Ge(Mes)Tbt (**10**)	2.416(4)	109.3(8) 108.5(9)	130.3(6) 133.4(7) 117.0(6) 113.9(6)	12	16 18	25

[a] Dep = 2,6-diethylphenyl.
[b] Ar = 2-t-Bu-4,5,6-Me_3C_6H.
[c] Dip = 2,6-diisopropylphenyl.
[d] Dis = bis(trimethylsilyl)methyl.

bond. This may be the longest germanium–germanium double bond reported so far. The sums of the bond angles around Ge(1) and Ge(2) are 356.6 and 355.8°, respectively. A closely related doubly bonded system of silicon, an extremely hindered disilene bearing the same substituents as (E)-**10**, i.e. (E)-Tbt(Mes)Si=Si(Mes)Tbt (**30**), has already been synthesized and characterized by X-ray crystallographic analysis[35]. The twist angle and the bent angles of disilene **30** are 8.7, 14.6 and 9.4°, respectively. The Si=Si bond for disilene (E)-**30** is 3.8% longer than the mean value of the Si=Si bond lengths (2.147 Å) in other carbon-substituted disilenes. In the case of digermene (E)-**10**, the Ge=Ge bond is lengthened by 6.0% as compared to the mean value of the Ge=Ge bond lengths (2.278 Å) in other carbon-substituted digermenes. The elongation of the Ge=Ge bond length of (E)-**10** shows that the Ge–Ge double bond is softer than the Si–Si double bond, which is known to be again softer than the C–C double bond.

b. Cyclic digermenes. Some of the isolated cyclotrigermenes have been characterized by X-ray crystallographic analysis. Although they are embedded in such a strained three-membered ring systems, the bond lengths of their Ge–Ge double bonds were found to lie in the range of previously reported acyclic digermenes. However, it should be noted that the planarity of the Ge–Ge double bond in cyclotrigermenes is highly dependent on their substituents. Thus, cyclotrigermene **24a** has a completely planar geometry around its Ge–Ge double bond[16], while unsymmetrically substituted cyclotrigermene **25b** showed an unusual *cis*-bent geometry[18]. The detailed description is given in Chapter 14.

3. Reactions of digermenes

Tetraaryldigermenes react with various reagents as shown in Scheme 10[1p]. For example, addition of methanol gives methoxydigermanes[12,24]. In a reaction with an appropriate chalcogen source, chalcogenadigermiranes, which are [2 + 1]cycloadducts of the digermenes with chalcogen atoms, are obtained[36,37]. Reactions with ketone[36,38] or alkyne derivatives[39] afford [2 + 2]cycloadducts. Reactions with diazomethane and phenyl azide give three-membered ring compounds[36a,40]. Thus, digermenes are useful building blocks for the synthesis of small-ring compounds containing a germanium–germanium bond.

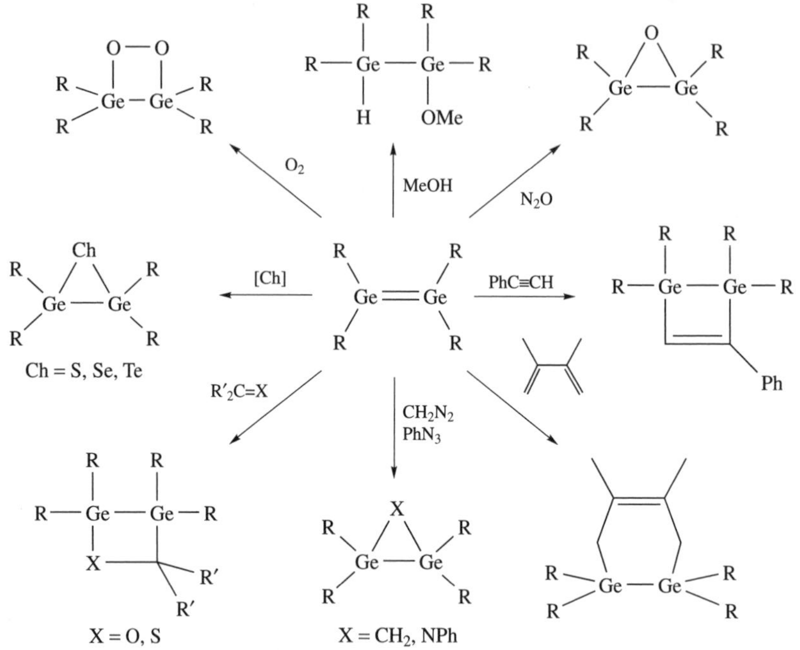

SCHEME 10

a. Estimation of π-bond energies. Masamune and coworkers synthesized stable geometric isomers of digermenes **31a,b** and experimentally determined the π-bond strength from their isomerization (Scheme 11). The enthalpy of activation for **31** has been determined

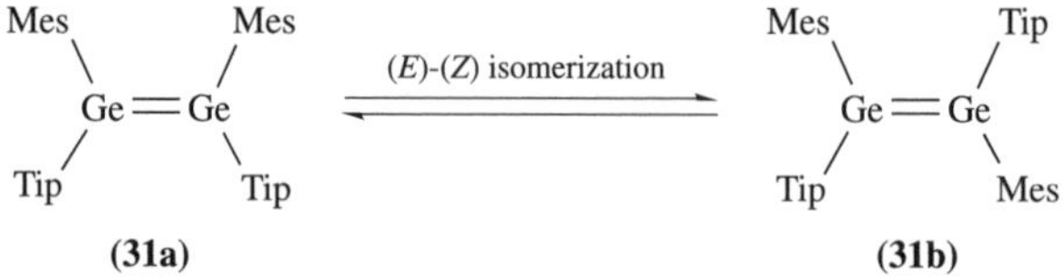

(31a) **(31b)**

SCHEME 11

from kinetic studies (for $Z-E$ conversion $\Delta H^{\ddagger} = 22.2 \pm 0.3$ kcal mol^{-1}, and for $E-Z$ conversion it is 20.0 ± 0.3 kcal mol^{-1})[19]. These values are in good agreement with the calculated value[41].

b. Thermal behavior of digermenes. There have been some interesting reports on the thermolysis of digermenes. Thermolysis of hexamesitylcyclotrigermane **32** in the presence of triethylsilane or 2,3-dimethyl-1,3-butadiene gave **33** and **34** or **35** and **36**, respectively[42]. The most reasonable explanation for the generation of these products is as follows. Thermolysis of cyclotrigermane **32** affords digermene **37** and germylene **38** and the latter reacts with the silane or diene to afford **34** or **36**, respectively. On the other hand, **37** may undergo 1,2-mesityl shift to give germylgermylene **39**, which is then trapped with the silane or diene to afford **33** or **35**, respectively (Scheme 12)[42].

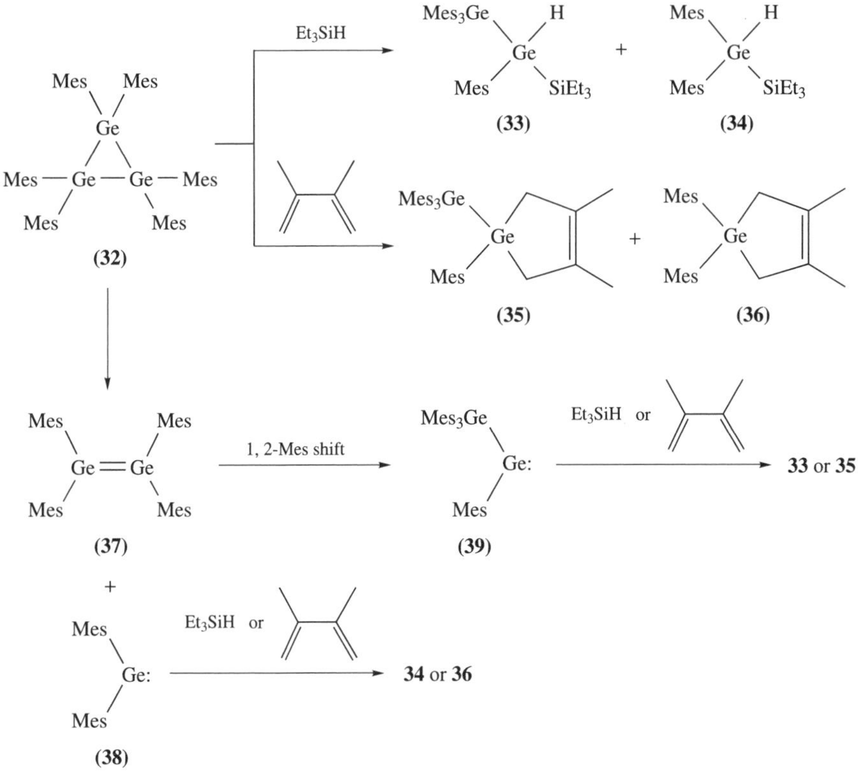

SCHEME 12

Tetrakis(2,6-diethylphenyl)digermene **3** is reportedly converted into hexakis(2,6-diethylphenyl)cyclotrigermane **12b** on heating in solution[1j]. The proposed mechanism involves the dissociation of digermene **3** into two germylenes Dep_2Ge: **40**, followed by their reaction with the second digermene **3** to form the cyclotrigermane **12b** (Scheme 13). In this process the dissociation energy of tetrakis(2,6-diethylphenyl)digermene was estimated to be <30 kcal mol^{-1} [1j].

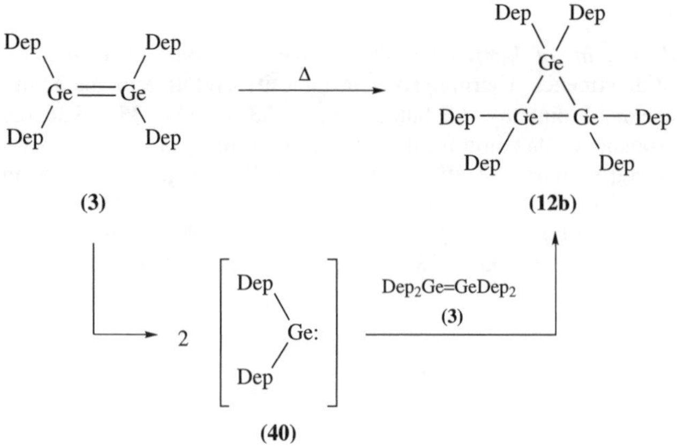

$Dep_2Ge{=}GeDep_2$ **(3)** —Δ→ cyclo-$(GeDep_2)_3$ **(12b)**

(3) → 2 [Dep_2Ge:] **(40)** —$Dep_2Ge{=}GeDep_2$ **(3)**→ **(12b)**

SCHEME 13

c. Dissociation of digermenes into the corresponding germylenes. There have been three reports on the dissociation of a digermene into the corresponding germylenes.

The tetraalkyldigermene $[(Me_3Si)_2CH]_2Ge{=}Ge[CH(SiMe_3)_2]_2$ **1** was synthesized by dimerization of the corresponding germylene **2** as the first example of a stable digermene (Scheme 2)[32]. X-ray crystallographic analysis of digermene **1** revealed that the Ge−Ge bond length of **1** [2.347(2) Å] is shorter than the typical Ge−Ge single bond length (2.457–2.463 Å). Interestingly, however, the chemical properties of **1** in solution suggest that it dissociates into the monomer (germylene) **2**. Thus, it undergoes oxidative addition with some reagents to form tetravalent germanium compound (Scheme 14)[15,32].

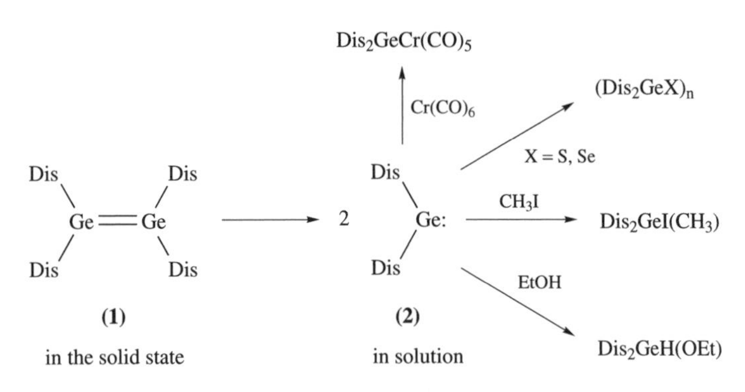

SCHEME 14

On the other hand, $(2\text{-}t\text{-Bu-}4,5,6\text{-}Me_3C_6H)_2Ge{=}Ge(2\text{-}t\text{-Bu-}4,5,6\text{-}Me_3C_6H)_2$ **41** was also synthesized as a stable digermene and was characterized by X-ray crystallography[20]. In solution digermene **41** reacted with benzil to afford **42**, which is a [4 + 1]cycloadduct of the germylene **43** with benzil (Scheme 15). The digermene **41** was found by cryoscopic molecular weight measurement to exist as the corresponding monomer, germylene **43**, in solution.

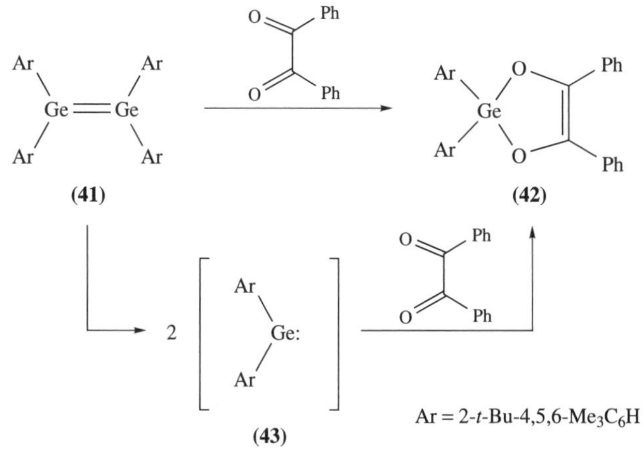

Ar = 2-*t*-Bu-4,5,6-Me_3C_6H

SCHEME 15

Furthermore, as described in the previous section, the spectroscopic observation of an equilibrium between a digermene and germylenes was first achieved in the case of highly crowded digermene (*E*)-**10** by UV-vis spectroscopy (see Scheme 7)[23].

B. Germenes

Germenes[43] are germanium analogues of alkenes, which play a very important role in organic chemistry. Their synthesis and isolation have been reported only in few papers, because of their tendency to undergo dimerization. The first stable germene was synthesized in 1987 by taking advantage of steric protection[44]. The structural analysis of some germenes has been reported and they are known to have the trigonal planar geometry on the Ge and C atoms in sharp contrast to digermenes, which have the pyramidal geometry on the Ge atoms.

Germenes are highly reactive and readily undergo 1,2-addition with various single-bond compounds or $[2 + n]$cycloaddition with multiple-bond compounds to give germanium–carbon singly bonded compounds or germacycles, respectively.

1. Synthesis of germenes

The first stable germenes **46** were synthesized by Berndt and coworkers via the coupling reactions of stable nitrogen-substituted germylenes **45a,b** with the electrophilic cryptocarbene **44** (**44′**) (Scheme 16).

The boron atoms of **46a** and **46b** ($\delta_B = 66$ and 65 ppm) are more strongly shielded than that of the 1,3-diboretane **47a** ($\delta_B = 82$ ppm)[44b,45]. Although the germanium–carbon double bond distance of **46a** (1.827 Å) is shorter by 8% than that of a Ge–C single bond (1.98 Å)[46], it is longer than that of the calculated value for the parent germene $H_2Ge{=}CH_2$ (1.71–1.81 Å)[47]. The sums of the bond angles of Ge and C atoms of a germene unit of **46a** are 359.9 and 359.7°, respectively. These facts show that a large amount of negative

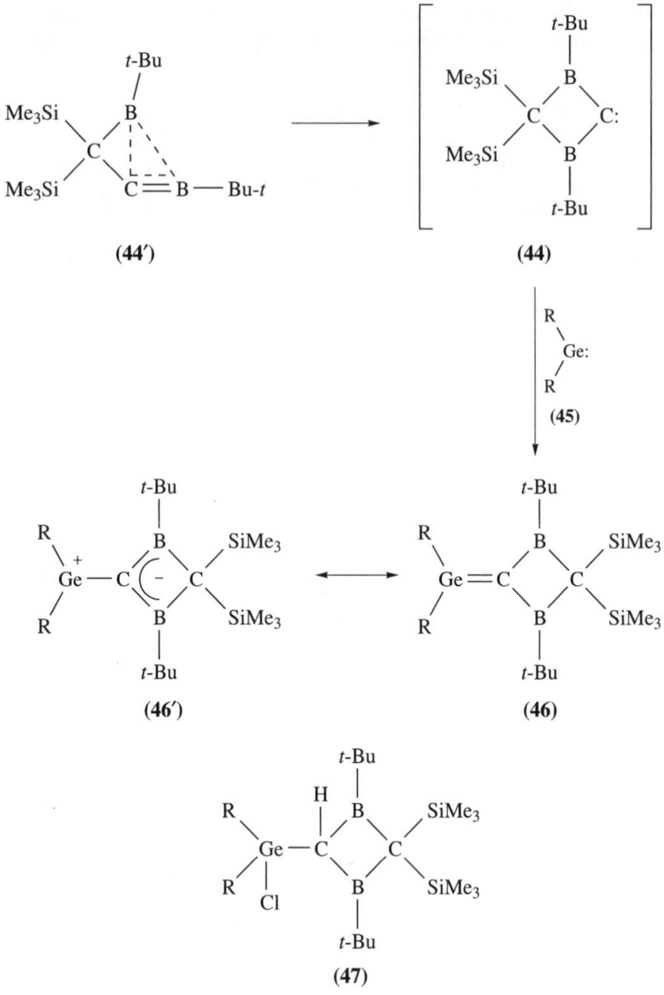

(44′) (44) (45) (46′) (46) (47)

(a) R = $N(SiMe_3)_2$, **(b)** RR = *t*-BuN($Si(Me_3)_2$)NBu-*t*

SCHEME 16

charge is located on the boron atoms and germenes **46** are stabilized by resonance with an ylide form **46′**.

Escudié, Satgé and coworkers have reported the synthesis of the first germene **49** bearing only carbon substituents by dehydrofluorination of the corresponding fluorogermane **48**[44a,c]. X-ray crystallographic analysis revealed that the length of the germanium–carbon double bond of **49** is 1.80 Å, which is shorter by 9% than that of the germanium–carbon single bond and consistent with the calculated value[47]. The sums of the bond angles of Ge and C atoms of the germene unit of **49** are 360.0 and 360.0°, respectively. This synthetic route has been used extensively for other germenes[1p].

In connection with this method, the treatment of fluorovinylgermane **50** with *t*-BuLi resulted in the formation of dimesitylneopentylgermene **51** via an addition–elimination reaction (Scheme 17)[48].

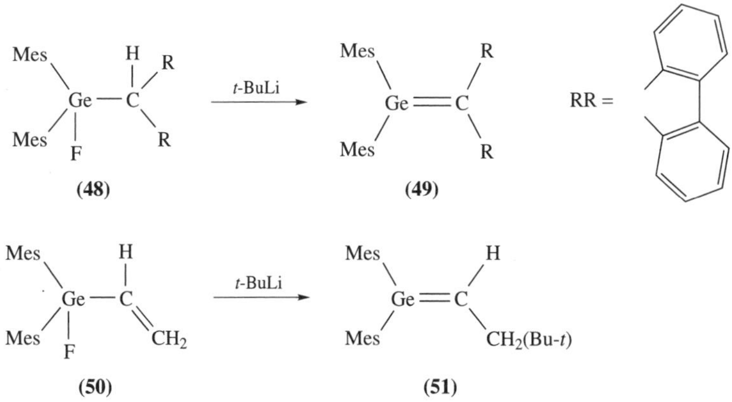

SCHEME 17

Tokitoh, Okazaki and coworkers have reported the synthesis of another unique germene **53** by the reaction of overcrowded diarylgermylene **52** with carbon disulfide (Scheme 18)[49]. The structure of **53** was confirmed by X-ray crystallographic analysis[49]. The distance of the germanium-carbon double bond of **53** is 1.77 Å, which is shorter by 11% than that of a typical Ge-C single bond length (1.98 Å). The sums of the bond angles of Ge and C atoms of the germene unit of **53** are 359.7 and 360.0°, respectively.

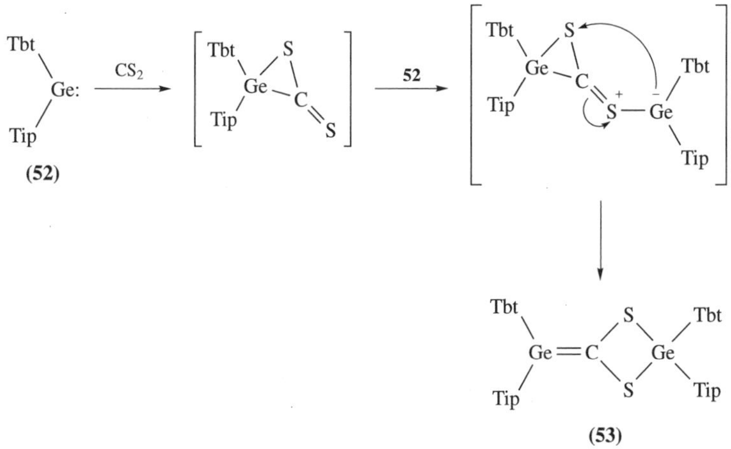

SCHEME 18

In addition to the stable germenes mentioned above, an ylide-type compound **56** was synthesized by the reaction of the nitrogen-substituted germylene **45a** with bis(dialkylamino)cyclopropenylidene **55** prepared *in situ* by the treatment of cyclopropenium cation

54 with *n*-BuLi (Scheme 19)[50]. The structure of **56** was confirmed by X-ray crystallographic analysis, where the distance (2.08 Å) between the germanium and the carbene-*C* atoms is appreciably longer than that of other germenes (1.77–1.83 Å) and close to that of a single bond (1.98 Å). The sum of the bond angles around the Ge atom (ΣGe) of **56** is 303.0°, indicating a trigonal–pyramidal structure. This structural feature is different from that of other germenes, which have a trigonal–planar geometry for the germene unit (ΣGe = 359.7–360.0° and ΣC = 359.7–360.0°)[44,45,49].

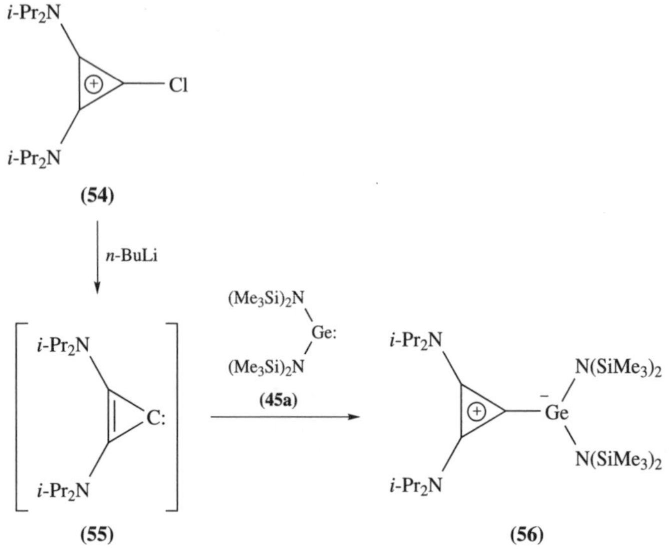

i-Pr2N
Cl
i-Pr2N
(54)
n-BuLi
i-Pr2N
C:
i-Pr2N
(55)
(Me3Si)2N
Ge:
(Me3Si)2N
(45a)
i-Pr2N
N(SiMe3)2
Ge
N(SiMe3)2
i-Pr2N
(56)

SCHEME 19

2. Reactions of germenes

Germenes react with various reagents as shown in Scheme 20[1e,1p,51]. For example, addition of methanol affords a methoxygermane. In the reactions with ketones and aldehydes oxagermetane derivatives are obtained. The reactions of α,β-unsaturated aldehydes and ketones afford [4+2]cycloadducts. These reactions proceed regiospecifically, according to the $Ge^{\delta+}{=}C^{\delta-}$ polarity.

In the case of germaketenedithioacetal **53**, unique reaction with molecular oxygen has been reported (Scheme 21)[52]. On exposure of **53** to air, dihydroxygermane **57** and 1,3,2-dithiagermeta-4-one **58** were obtained as the reaction products. On the other hand, the reaction of **53** with oxygen in the presence of methanol afforded **57** and **58** together with hydroxymethoxygermane **59**, i.e. the methanolysis product of germanone **61**. The formation of **57**, **58** and **59** can be reasonably explained as shown in Scheme 21. Germene **53** reacts with oxygen to form an intermediary [2+2]cycloadduct, 1,2,3-dioxagermetane **60**, the cycloreversion of which may give **58** and germanone **61**. Since germanones are known to be highly reactive and unstable species[44], the intermediate **61** might quickly react with H_2O or methanol to give **57** or **59**.

Brook and coworkers have already reported a similar reaction mode in the reaction of silene with oxygen[53]. In that case the intermediary silanone which arises from the

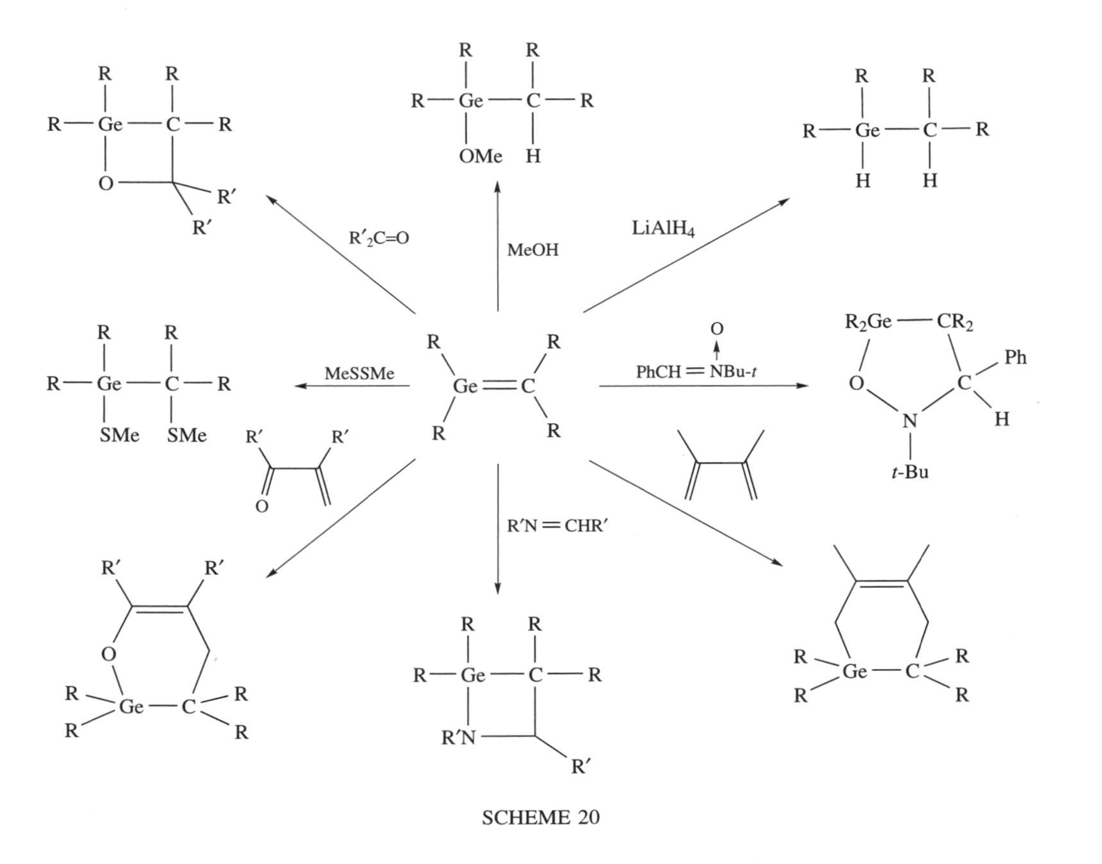

SCHEME 20

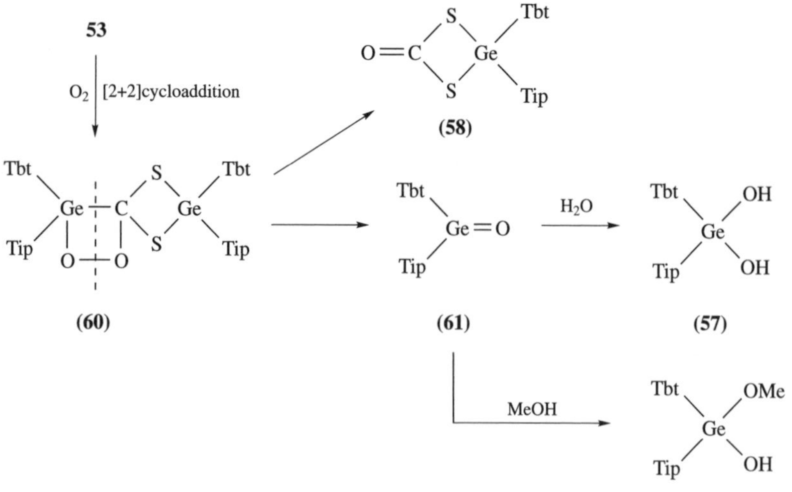

SCHEME 21

breakdown of 1,2,3-dioxasiletane undergoes a ready trimerization to afford the cyclic siloxane[53]. The lack of such a polymerization product of germanone **61** suggests the effectiveness of the steric protection system. Indeed, Tokitoh and coworkers have reported the formation and reaction of the first stable germanone in solution derived from a kinetically stabilized germylene[54]. It is interesting and noteworthy that the same type of reactions are observed in the oxidation of a silene and a germene.

The germanium–carbon double bond in germene **53** was found to undergo an interesting thermal dissociation (Scheme 22). When the benzene-d_6 solution of germene **53** was

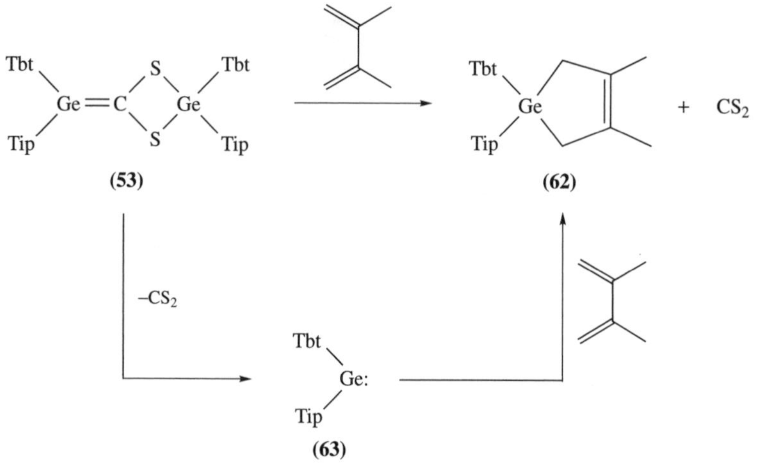

SCHEME 22

heated at 95 °C in the presence of 2,3-dimethyl-1,3-butadiene in a sealed tube, germacyclopentene **62**, a [1+4]cycloadduct of germylene **63** with butadiene, was obtained[55]. The formation of **62** indicates the generation of germylene **63** in the thermolysis of germene **53**. This mechanism was supported by the fact that carbon disulfide, the counterpart of the thermal dissociation of **53**, was detected by ^{13}C NMR spectroscopy[52].

Raasch has reported that the thermolysis of a ketenedithioacetal resulted in the dissociation into the corresponding thioketene and thioketone[56]. It should be noted that it is not the retro[2+2]cycloaddition similar to the carbon analogue, but the cleavage of the germanium–carbon double bond that has taken place in the thermolysis of germene **53**.

C. Distannenes and Stannenes

In contrast to the extensively studied doubly bonded systems of silicon and germanium, the chemistry of the corresponding tin compounds, i.e. distannenes and stannenes, has not been fully disclosed yet probably due to the much higher reactivity and instability of such low-coordinated organotin compounds. In the following sections we briefly describe the synthesis and properties of stable distannenes and stannenes.

1. Distannenes

In 1976, the first stable distannene **64** bearing two bis(trimethylsilyl)methyl groups on each tin atom was isolated as brick red crystals by Lappert and coworkers (Scheme 23)[57]. Crystallographic analysis of **64** revealed that it is nonplanar and centrosymmetric with each tin atom in a pyramidal environment (sum of the angles at tin is 342°, compared with 360° expected for pure sp^2 hybridization at Sn or 327° for sp^3 hybridization) having a fold angle of 41°. (The fold angle is defined as the angle between the Sn–Sn vector and the SnC_2 plane of each monomer.) The Sn–Sn bond distance is 2.768 Å, which is comparable to an average Sn–Sn single bond length.

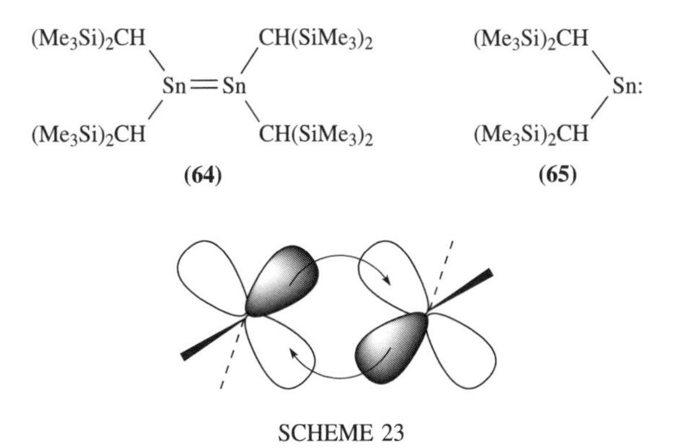

SCHEME 23

According to the molecular orbital calculations of a tin–tin double bond, a calculated equilibrium structure is predicted to be *trans*-bent, which is consistent with the experimental findings. This structural distortion from a planar form to a *trans*-bent form is achieved through mixing an M–M p-orbital (b_u) (HOMO) with an M–M σ^* orbital. On the other hand, the energy of an M–M σ orbital (a_g) increases upon *trans*-folding, mainly due to the loss of M–M σ bonding. For distannene, the energy drop of the b_u orbital predominates over destabilization of the a_g orbital. When calculations are performed on a *cis*-folded form, mixing of a b_u(HOMO) with an antibonding σ^* is forbidden by symmetry and the energy increases with the fold angle. In solution, distannene **64** was found to exist as a distannene–stannylene **65** equilibrium mixture. Attempts to observe its ^{119}Sn NMR in solution at room temperature were not successful, but two signals at 165 K in either ether or toluene at 740 and 725 ppm assignable to distannene **64** were observed[58]. At 375 K, only the signal of the stannylene **65** was observed at 2315 ppm. The NMR studies show quite a small dissociation energy of 12.8 kcal mol^{-1} and a low $^1J(^{119}\text{Sn}-^{117}\text{Sn})$ value of 1340 ± 10 Hz. Both theoretical and experimental findings suggest that the Sn=Sn bond in **64** is exceptionally weak and not a covalent bond in the usual sense. The bonding mode in **64** is described as a double dative bond, in which the lone pair on each monomer interacts with the empty p_z orbital on the other as shown in Scheme 23.

The first aryl-substituted distannene **66** was synthesized by Masamune and Sita through photolysis of the corresponding cyclotristannane **67** in solution at $-78\,^\circ$C (Scheme 24)[59]. Although ^{119}Sn NMR of **66** at $-68\,^\circ$C showed a singlet at 427 ppm with considerably larger tin–tin coupling satellites [$^1J(^{119}\text{Sn}-^{117}\text{Sn}) = 2930$ Hz] than that of **64**, there is a thermal equilibrium between **66** and **67**. Cyclotristannane **67** is stable in an inert solvent at $0\,^\circ$C or lower temperature to $-78\,^\circ$C, but a rapid equilibrium between **66** and **67** occurs at room temperature or above with **66** being favored at higher temperature.

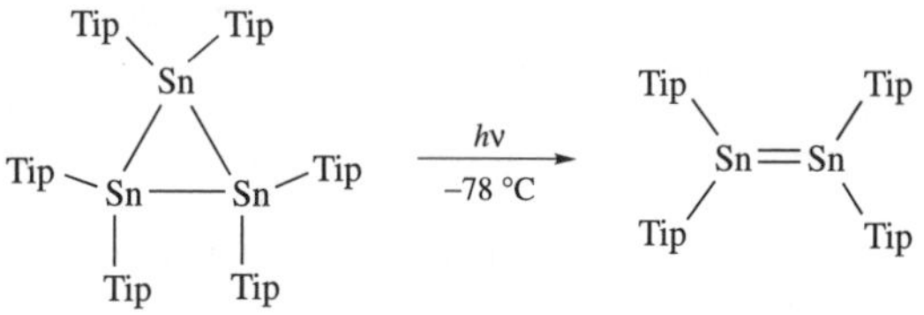

(67) —$h\nu$, $-78\,^\circ$C→ (66)

SCHEME 24

Weidenbruch and coworkers reported that the ^{119}Sn NMR signal of the corresponding stannylene **68** appeared at 1420 ppm ($40\,^\circ$C) in the equilibrated mixture of **66** and **67** (Scheme 25)[60].

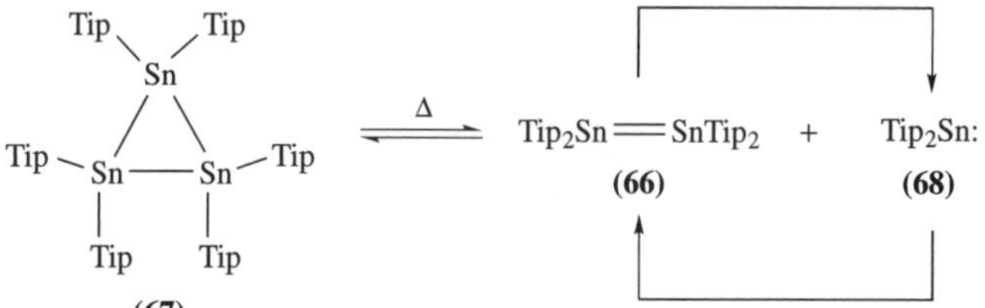

(67) ⇌(Δ) $Tip_2Sn{=}SnTip_2$ (66) + Tip_2Sn: (68)

SCHEME 25

Later, they also reported the first X-ray structural analysis of the aryl-substituted distannene by introduction of 2-*t*-butyl-4,5,6-trimethylphenyl groups[61]. A conspicuous feature of its structure is the long tin–tin bond length of 2.91 Å, which is markedly longer than that the typical tin–tin single bond. In its ^{119}Sn NMR at 373 K a sharp signal at 1401 ppm is observed in the same region as that of **68** assignable to a monomeric stannylene. At room temperature no ^{119}Sn NMR signals can be observed as in the case of **64**. This temperature dependence in ^{119}Sn NMR shows that it also exists as a monomer–dimer equilibrium mixture. A distannene which does not dissociate into stannylenes in solution is still unknown.

2. Stannenes

The chemistry of stannenes, which are also analogous to ethylene, has been less developed than that of silenes and germenes. In 1987, two stannenes **69** and **70** stabilized by not only steric protection but also by electronic perturbation were synthesized and characterized by X-ray crystallographic analysis by Berndt and coworkers (Scheme 26)[45,62].

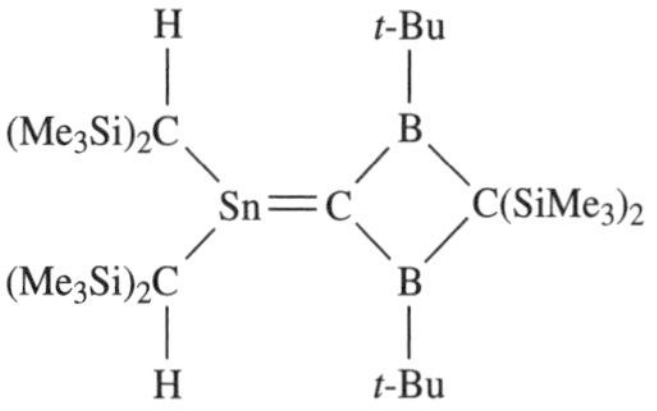

(69) (70)

SCHEME 26

A decisive proof for the presence of tricoordinated tin in **69** is provided by the very low-field chemical shift of the ^{119}Sn NMR signal at $\delta = 835$. The Sn=C bond length is 2.025 Å in **69** which is comparable to calculated values for $H_2C{=}SnH_2$ [2.063 Å at MCSCF/3-21 GG(d)[31e] and 1.98 Å at SCF/3-21 GG(d)[63]]. The average twist angle around the Sn=C bond is 61° and the Sn atom is slightly pyramidalized. In 1992, Satgé and coworkers reported the synthesis of bis(2,4,6-triisopropylphenyl)(fluorenylidene)stannene **71** (Scheme 27) by dehydrofluorination of the corresponding fluorostannane[64].

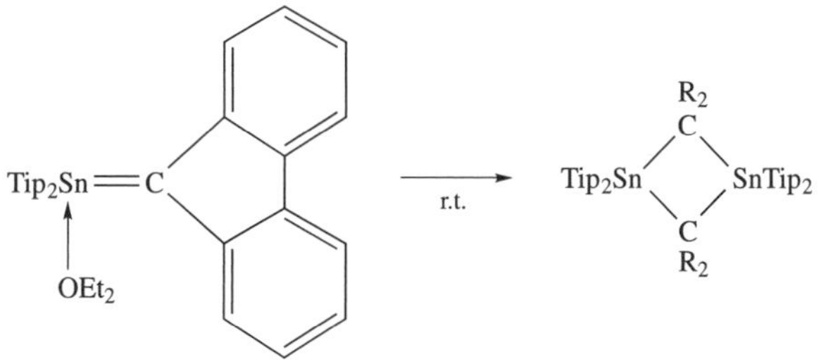

(71) r.t. (72)

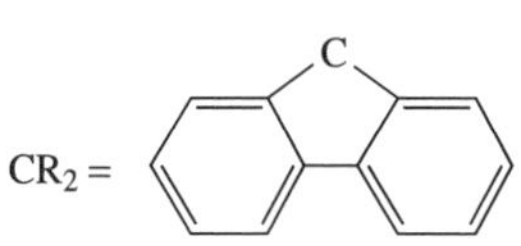

SCHEME 27

Although the X-ray structural analysis of this stannene **71** was not carried out, its formation was inferred from its low ^{119}Sn NMR resonance (288 ppm). In comparison with the doubly bonded tin derivatives mentioned above, however, it shows a higher field chemical shift. This high chemical shift can be rationalized in terms of complexation of tin with ether used in the synthesis, as evidenced by the broad signal for the OCH_2 hydrogens in the ^1H NMR spectrum. At room temperature, **71** slowly converts to the head-to-tail dimer **72** (Scheme 27). When two bis(trimethylsilyl)methyl groups were used as substituents on the tin atom instead of two Tip groups in **71**, attempts to stabilize a stannene were unsuccessful[65]. In 1995, it was found that bis(2,4,6-triisopropylphenyl)stannylene **74** reacted with 4,5-dimethyl-1,3-diisopropylimidazol-2-ylidene **73** to furnish an adduct **75** (Scheme 28)[66]. An X-ray structural analysis of **75** revealed the presence of a long tin–carbon double bond with a length of 2.379 Å. The geometry of the molecule observed here clearly indicates that the adduct **75** is better described as the limiting formula **75b** rather than as the stannene form **75a**.

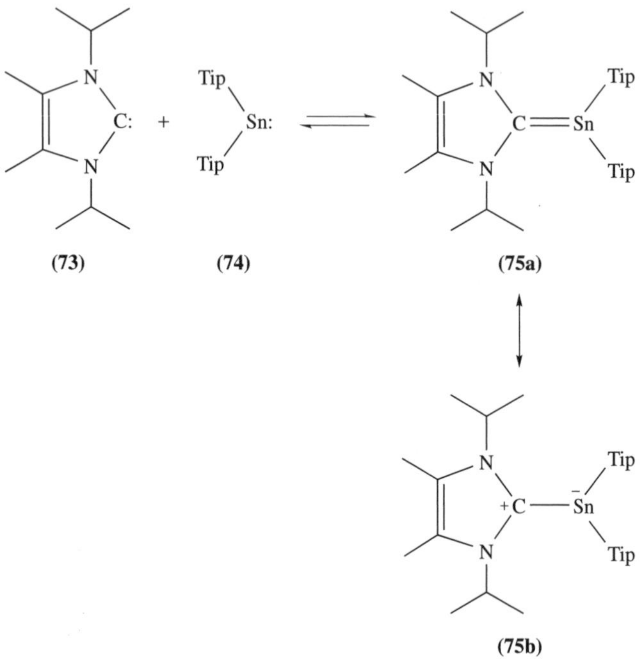

(73) (74) (75a)

(75b)

SCHEME 28

D. Diplumbenes

In contrast to the remarkable progress in the chemistry of divalent organic compounds of silicon, germanium and tin, the heaviest congeners of this series, i.e. divalent organolead compounds (plumbylenes), are less investigated. They usually occur as reactive intermediates in the preparation of plumbanes R_4Pb and undergo polymerization and/or disproportionation in the absence of suitable stabilizing groups on the lead atom[67]. However, there has been very little information on the dimers of plumbylenes, i.e. diplumbenes. Furthermore, so far no experimental information is available on the chemistry of lead–carbon doubly bonded systems, i.e. plumbenes.

1. Stable plumbylenes

In 1974 the first stable diaminoplumbylene (**76**) was synthesized by Lappert and coworkers[68] and since then many other stable plumbylenes with heteroatom substituents have been reported. Recently, the synthesis and characterization of a stable aryl(arylthio)-plumbylene (**77**), which is one of the rare examples of heteroleptic plumbylenes, have also been reported (Scheme 29)[69].

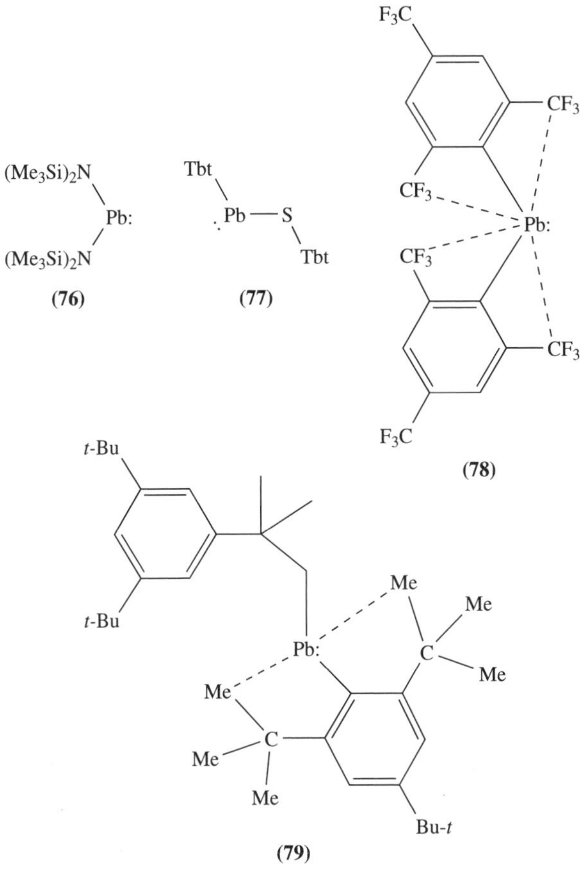

(76) **(77)** **(78)** **(79)**

SCHEME 29

In contrast to the heteroatom-substituted plumbylenes, a few plumbylenes bearing only carbon substituents have been reported. Some of them are stabilized by intramolecular coordination of the lone pair of a donor group in the organic substituent, thus giving the lead a coordination number greater than 2 as can be seen in the diarylplumbylene R^f_2Pb

(**78**; R^f = bis[2,4,6-tris(trifluoromethyl)phenyl]) and the alkylarylplumbylene (**79**)[70]. On the other hand, kinetically stabilized plumbylenes bearing only organic substituents that do not contain donor groups are scarce, and their structures and reactivities are almost unexplored. The first dialkylplumbylene, Dis_2Pb [**80**; Dis = bis(trimethylsilyl)methyl)], was obtained in only 3% yield,[57b,c,71] and its structure was crystallographically determined recently[72]. Another dialkylplumbylene **81** with a lead atom in a seven-membered ring system was synthesized and characterized by X-ray diffraction (Scheme 30)[73]. Two stable diarylplumbylenes, **82**[27b] and **83**[70], were also structurally characterized (Scheme 30). Furthermore, an extremely hindered diarylplumbylene, Tbt_2Pb: (**84**), has recently been obtained by the nucleophilic substitution reaction of $[(Me_3Si)_2N]_2Pb$: (**76**) with two molar amounts of TbtLi as stable blue crystals (Scheme 30)[74].

$[(Me_3Si)_2CH]_2Pb$:

(80)

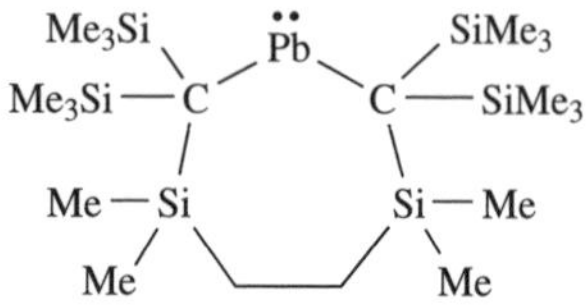

(81)

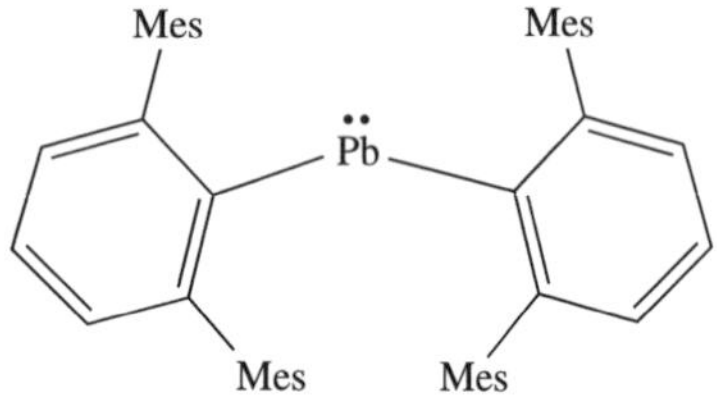

(82)

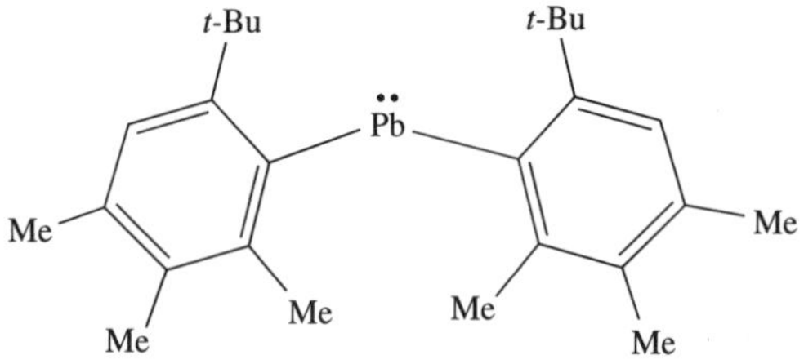

(83)

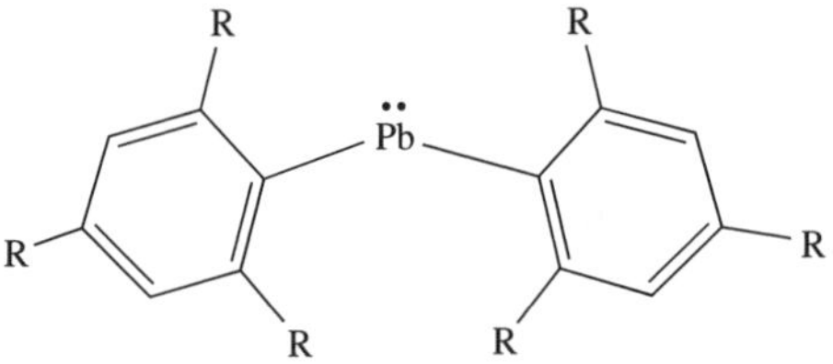

$[R = CH(SiMe_3)_2]$

(84)

SCHEME 30

In all cases mentioned above, the isolated plumbylenes exist as V-shaped monomers, and no evident bonding interaction between the lead centers was observed for the plumbylenes either in solution or in the solid state. In other words, until quite recently there has been no experimental information for the existence of a lead–lead double bond (diplumbene), the dimer of plumbylenes.

2. Lead–lead double bonds in the solid state

In 1998, Klinkhammer and coworkers found a quite interesting feature for their new heteroleptic plumbylene $R^f[(Me_3Si)_3Si]Pb$: (**86**)[75]. Although plumbylene **86** was synthesized by a novel ligand disproportionation between the corresponding stannylene R^f_2Sn: (**85**) and plumbylene $[(Me_3Si)_3Si]_2Pb$:, the crystallographic analysis of the isolated lead product revealed that **86** has a dimeric form **86′** in the solid state with a considerably short Pb···Pb separation [3.537(1) Å] and a *trans*-bent angle of 40.8° (Scheme 31). Weidenbruch and coworkers also reported the synthesis of another heteroleptic plumbylene (**87**), the structural analysis of which showed a dimeric structure **87′** in the solid state with a similar short Pb···Pb separation [3.370(1) Å] and a *trans*-bent angle of 46.5° (Scheme 32)[70].

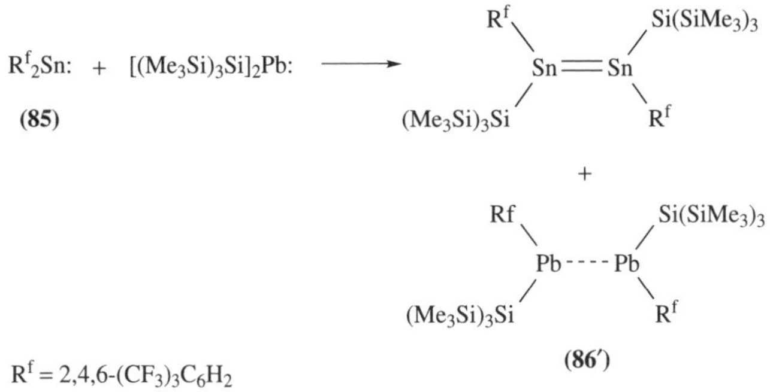

R^f_2Sn: + $[(Me_3Si)_3Si]_2Pb$: ⟶ $R^f[(Me_3Si)_3Si]Sn{=}Sn[Si(SiMe_3)_3]R^f$ + $R^f[(Me_3Si)_3Si]Pb\cdots Pb[Si(SiMe_3)_3]R^f$

(**85**) (**86′**)

$R^f = 2,4,6\text{-}(CF_3)_3C_6H_2$

SCHEME 31

Although these two plumbylenes **86** and **87** were found to have a close intramolecular contact and exist as a dimer in the solid state, the lead–lead distances in their dimeric form are still much longer than the theoretically predicted values (2.95–3.00 Å) for the parent diplumbene, $H_2Pb{=}PbH_2$[31d,f,76]. In 1999, however, Weidenbruch and coworkers succeeded in the synthesis and isolation of the dimer of less hindered diarylplumbylene Tip_2Pb: (**88**), i.e. $Tip_2Pb{=}PbTip_2$ (**89**) (Scheme 33)[77]. Compound **89** showed a rather shorter Pb–Pb length [3.0515(3) Å] and much larger *trans*-bent angles (43.9° and 51.2°) than those observed for **86′** and **87′**, strongly indicating that **89** is the first molecule with a lead–lead double bond in the solid state, although **89** was found to dissociate into the monomeric plumbylene **88** in solution. Furthermore, they examined the synthesis of a heteroleptic plumbylene, $Tip[(Me_3Si)_3Si]Pb$: (**90**), by the treatment of diarylplumbylene **88** and disilylplumbylene $[(Me_3Si)_3Si]_2Pb$:. The X-ray structure analysis revealed that the product has the centrosymmetrical diplumbene structure **91** in the solid state (Scheme 33) with a *trans*-bent angle of 42.7° and a Pb–Pb bond length of 2.9899(5) Å[78], which is even shorter than that of **89** and very close to the theoretically predicted value for the parent diplumbene.

In order to elucidate the relationship between the structure of plumbylene dimers and the bulkiness of substituents, Weidenbruch and coworkers synthesized and characterized

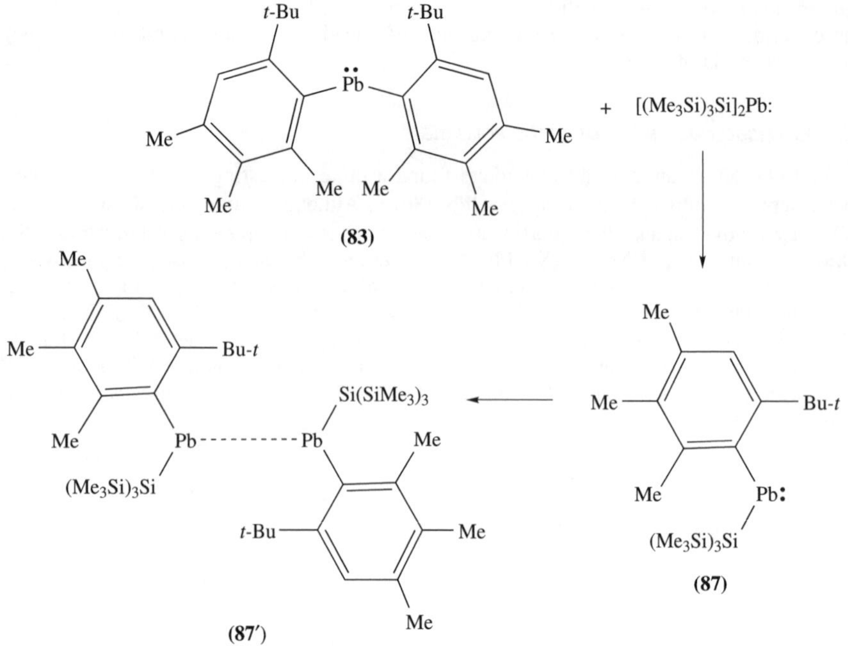

t-Bu t-Bu
Pb
Me Me
Me Me Me Me
(**83**)
+ $[(Me_3Si)_3Si]_2Pb$:

Me
Me Bu-t
Me Pb
$(Me_3Si)_3Si$
(**87**)

Me
Me Bu-t
$Si(SiMe_3)_3$
Me Pb- - - - - - - - - -Pb Me
$(Me_3Si)_3Si$
t-Bu Me
Me
(**87′**)

SCHEME 32

4 TipMgBr + 2 $PbCl_2$ → (THF, −110 °C; −2 $MgBr_2$; −2 $MgCl_2$) 2 Tip_2Pb: ⇌ Tip Tip Pb=Pb Tip Tip
(**88**) solution (**89**) solid state

Tip_2Pb: + $[(Me_3Si)_3Si]_2Pb$: → (n-pentane, r.t.) 2 $(Me_3Si)_3Si$ Pb: Tip ⇌ $Si(SiMe_3)_3$ Tip Pb=Pb Tip $(Me_3Si)_3Si$
(**88**) (**90**) solution (**91**) solid state

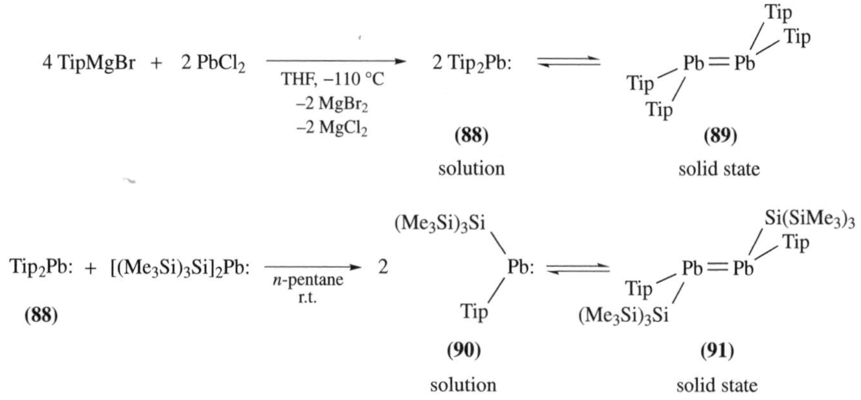

SCHEME 33

much less hindered diarylplumbylene Mes_2Pb: (**92**), which was isolated as a plumbylene dimer (**93**) stabilized with coexisting magnesium salt $[MgBr_2(THF)_4]$ (Scheme 34)[78]. The large Pb···Pb separation [3.3549(6) Å] and *trans*-bent angle (71.2°) of **93** suggest that the character of the lead–lead bonding interaction in plumbylene dimers is delicate and changeable.

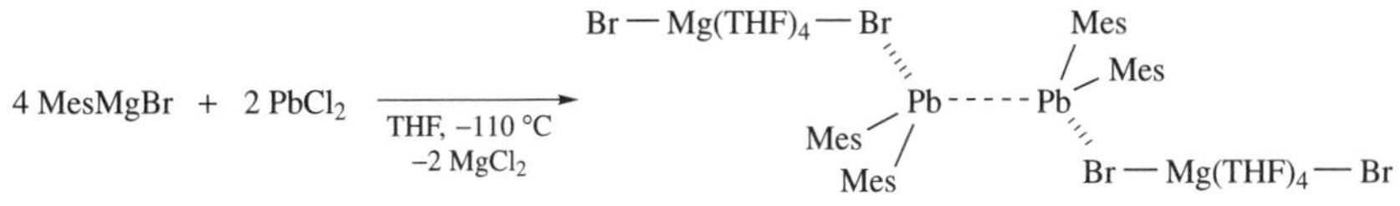

$$4\ MesMgBr + 2\ PbCl_2 \xrightarrow[-2\ MgCl_2]{THF,\ -110\ ^\circ C} [Br{-}Mg(THF)_4{-}Br]\cdot Mes_2Pb\text{-----}PbMes_2\cdot [Br{-}Mg(THF)_4{-}Br]\ \mathbf{(93)}$$

SCHEME 34

E. Conjugated Doubly Bonded Systems

In the last few years, further progress has been made in the chemistry of doubly bonded compounds of heavier group 14 elements. The successful isolation and characterization naturally prompted the chemists to examine the synthesis of the more sophisticated systems such as conjugated doubly bonded systems.

Thus, in 1997 Weidenbruch and coworkers have reported the synthesis and isolation of the first stable conjugated Si–Si double-bond compound, i.e. hexatipyltetrasilabuta-1,3-diene **94**[79]. Tetrasilabutadiene **94** was prepared through a rather unique synthetic route starting from the corresponding tetraaryl-substituted disilene **95** via the mono-lithiated disilene **96** as shown in Scheme 35.

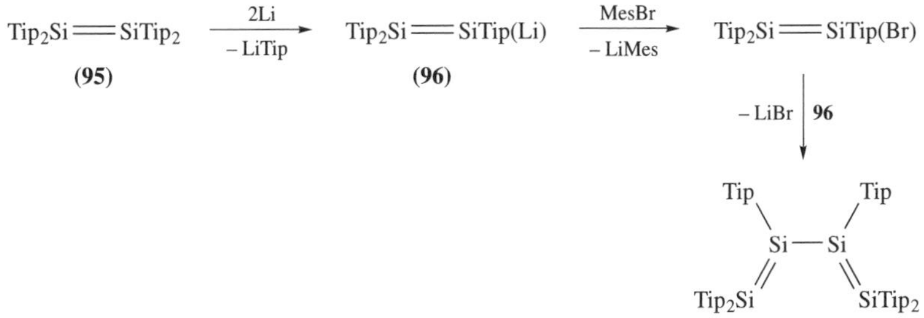

$$Tip_2Si{=}SiTip_2\ \mathbf{(95)} \xrightarrow[-\,LiTip]{2Li} Tip_2Si{=}SiTip(Li)\ \mathbf{(96)} \xrightarrow[-\,LiMes]{MesBr} Tip_2Si{=}SiTip(Br) \xrightarrow[-\,LiBr]{\mathbf{96}} Tip_2Si{=}Si(Tip){-}Si(Tip){=}SiTip_2\ \mathbf{(94)}$$

SCHEME 35

In the same year, Tokitoh and coworkers have succeeded in the synthesis and isolation of the first stable 2-silanaphthalene **97** by taking advantage of the steric protection afforded by the Tbt group (Scheme 36)[10]. These two compounds **94** and **97** should be noted as the first examples which showed that double bonds containing a silicon can make a conjugated system as well as the parent carbon analogues. In 2000, a stable silabenzene **98**, a much simpler silaaromatic system than **97**, was also synthesized and isolated using a similar kinetic stabilization method (Scheme 36)[9].

1. Conjugated double-bond compounds containing germanium

Weidenbruch and coworkers have recently succeeded in extending their chemistry of tetrasilabutadiene (*vide supra*) to its heavier congener, i.e. hexaaryltetragermabuta-1,3-diene **100**[80]. In this case, tetragermabutadiene **100** was prepared from digermene **4** via **99** by the synthetic method similar to that of its silicon analogue (Scheme 37)[79].

Although the crystallographic analysis of the structure of **100** revealed the existence of two Ge–Ge double bonds, no information has been given concerning the possible conjugation between the two digermene units. A more convincing insight was obtained from the electronic spectrum of **100**. The longest wavelength absorption at 560 nm for

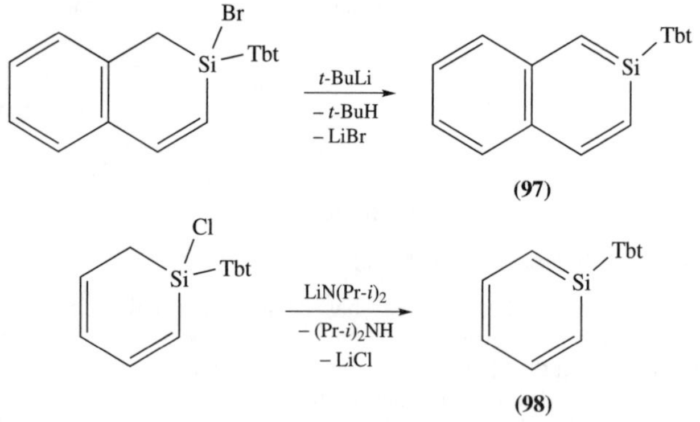

SCHEME 36

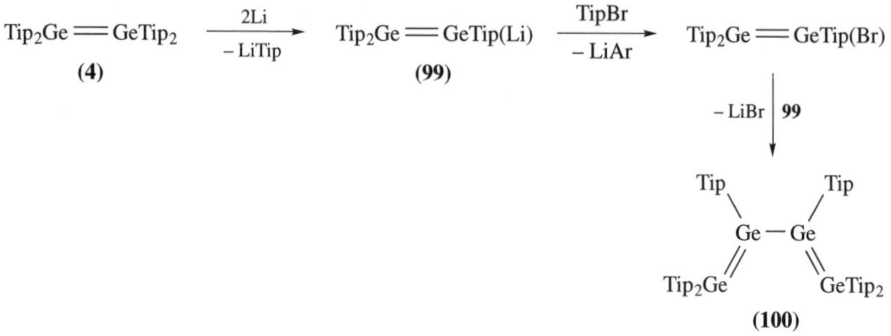

SCHEME 37

100 in hexane is reasonably interpreted in terms of the bathochromic shifts of those observed for digermenes, which showed yellow or orange color in solution (λ_{max} = 408–440 nm)[1p,20,23].

In view of the successful isolation of tetragermabutadiene **100** and its conjugated electronic properties, it may be possible to construct other types of conjugated systems containing germanium atom(s). Actually, Tokitoh and coworkers have recently succeeded in isolating a stable 2-germanaphthalene **101** bearing a Tbt group on the Ge atom[81]. The synthesis of germanaphthalene **101** should be noted not only as giving a new example of a cyclic conjugated germene, but also as the first stable example of a neutral germaaromatic compound (Scheme 38).

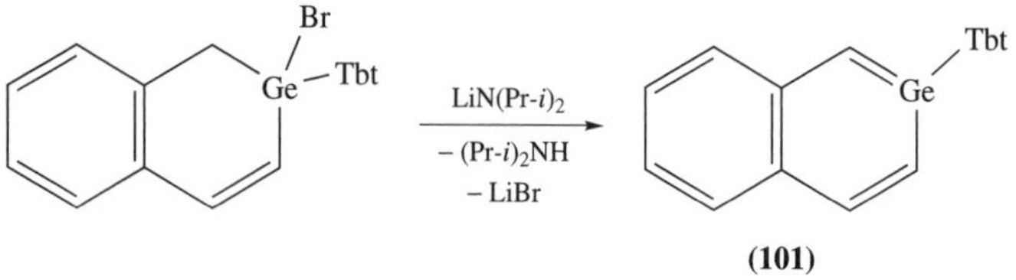

SCHEME 38

III. HEAVIER CONGENERS OF KETONES

Carbonyl compounds such as ketones and aldehydes are another important class of doubly bonded systems in organic chemistry. However, their heavier element congeners, 'heavy ketones', are much less explored because of the extremely high reactivities.

In the past few decades, almost all of the heavier chalcogen analogues of ketones, i.e. thioketones[82], selenoketones[83], and telluroketones[84], have been synthesized and characterized. Both thermodynamic and kinetic stabilization methods have been applied to stabilize these unstable double-bond species. In contrast to the doubly bonded systems between carbon and heavier chalcogens, heavier group 14 element analogues of ketones are much more reactive and unstable and hence their structures and properties have not been fully disclosed until recently[85].

In the series of silicon-containing heavy ketones, however, some suggestive theoretical calculations by Kudo and Nagase[86] and the first isolation of the thermodynamically stabilized silanethione **102a,b** by Corriu and coworkers[87] have strongly stimulated the chemistry of this field (Scheme 39). Then, in 1994 Okazaki, Tokitoh and coworkers reported the synthesis and isolation of the first example of a kinetically stabilized silanethione **103** by taking advantage of steric protection using the extremely bulky aryl group, Tbt (Scheme 39)[6,88].

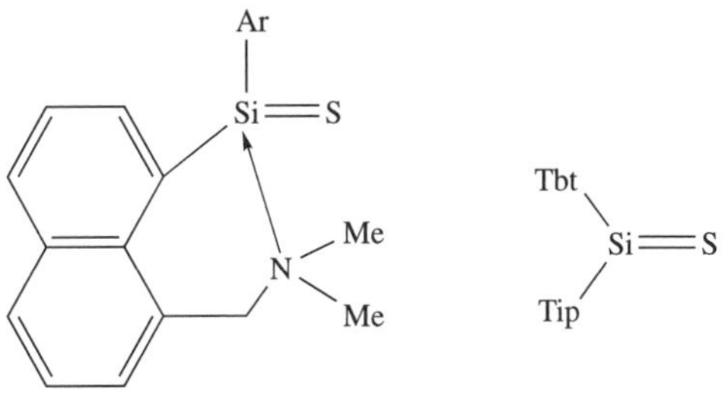

Ar
Si=S
N
Me
Me
Tbt
Si=S
Tip

(102a) Ar = Ph
(102b) Ar = α-Naph
(103)

SCHEME 39

The detailed background of the chemistry of heavy ketones and the recent progress in the field of silicon-containing heavy ketones has already appeared as a chapter in the previous volume of this series[85a] and also in other reviews[88]. Therefore, in the following sections we will discuss the chemistry of the heavier congeners containing germanium, tin and lead. Systematic comparisons for silicon through lead compounds reveal interesting differences in their properties depending on the elements. At first, it may be useful to compile the calculated σ and π bond energies of all the combinations of doubly bonded systems between group 14 and 16 elements (Table 3)[6,88]. As can be seen in Table 3 that all doubly bonded systems have an energy minimum, suggesting the possibility of their isolation if an appropriate synthetic method is available.

A. Germanium-containing Heavy Ketones

Until recently, there were only two examples of stable Ge–S double-bond compounds and one each for Ge–Se and Ge–Te double-bond compounds, but both of them are stabilized by the intramolecular coordination of a nitrogen ligand to the germanium center.

TABLE 3. Bond energies (kcal mol^{-1}) and lengths (Å) for $H_2M{=}X$ systems calculated at the B3LYP/TZ(d,p) level

$H_2M{=}X$		X			
		O	S	Se	Te
$H_2C{=}X$	σ^a	93.6	73.0	65.1	57.5
	π^b	95.3	54.6	43.2	32.0
	d^c	1.200	1.617	1.758	1.949
	Δ^d	15.5	11.9	11.1	10.1
$H_2Si{=}X$	σ^a	119.7	81.6	73.7	63.2
	π^b	58.5	47.0	40.7	32.9
	d^c	1.514	1.945	2.082	2.288
	Δ^d	8.1	9.4	9.3	8.7
$H_2Ge{=}X$	σ^a	101.5	74.1	67.8	59.1
	π^b	45.9	41.1	36.3	30.3
	d^c	1.634	2.042	2.174	2.373
	Δ^d	8.6	9.5	9.2	8.6
$H_2Sn{=}X$	σ^a	94.8	69.3	64.3	56.4
	π^b	32.8	33.5	30.6	26.3
	d^c	1.802	2.222	2.346	2.543
	Δ^d	7.6	8.9	8.5	8.1
$H_2Pb{=}X$	σ^a	80.9	60.9	57.0	50.3
	π^b	29.0	30.0	27.8	24.4
	d^c	1.853	2.273	2.394	2.590
	Δ^d	8.5	9.2	8.9	8.1

[a] σ bond energy.
[b] π bond energy.
[c] Length of the M=X double bond.
[d] Value of % reduction in a bond length defined as [(single bond length – double bond length)/single bond length] × 100.

In 1989, Veith and coworkers reported the synthesis of a base-stabilized Ge–S double-bond species **104** (Scheme 40)[89]. The X-ray structural analysis[40] shows that the sum of the bond angles around Ge atom was 355°, which indicates that the geometry for the Ge atom can be described as distorted tetrahedral, or better still as trigonal planar with an additional bond (N→Ge). The Ge–S bond distance of 2.063(3) Å was about 0.2 Å shorter than the value for a Ge–S single bond. The ^1H NMR spectrum showed three signals assigned to nonequivalent *t*-butyl groups which also indicate this coordination of the nitrogen atom. The synthesis of a germanone bearing the same substituent from the corresponding germylene **105** was also studied (Scheme 40)[89a], but the trial was unsuccessful, since it resulted in the formation of **106**, a dimer of the corresponding germanium–oxygen double-bond species. This is most likely due to the high polarity of the Ge=O bond[90] in spite of the thermodynamic stabilization.

As another synthetic approach to the doubly bonded systems containing a germanium, Kuchta and Parkin reported the synthesis of a series of terminal chalcogenido complexes of germanium **107**, **108** and **109** (Scheme 41)[91]. X-ray structural analyses of **107**, **108** and **109** revealed that they have unique germachalcogenourea structures stabilized by the intramolecular coordination of nitrogen atoms but the central Ge–X (X = S, Se, Te) bond of **107**, **108** and **109** should be represented by a resonance structure, $Ge^+{-}X^- \leftrightarrow Ge{=}X$.

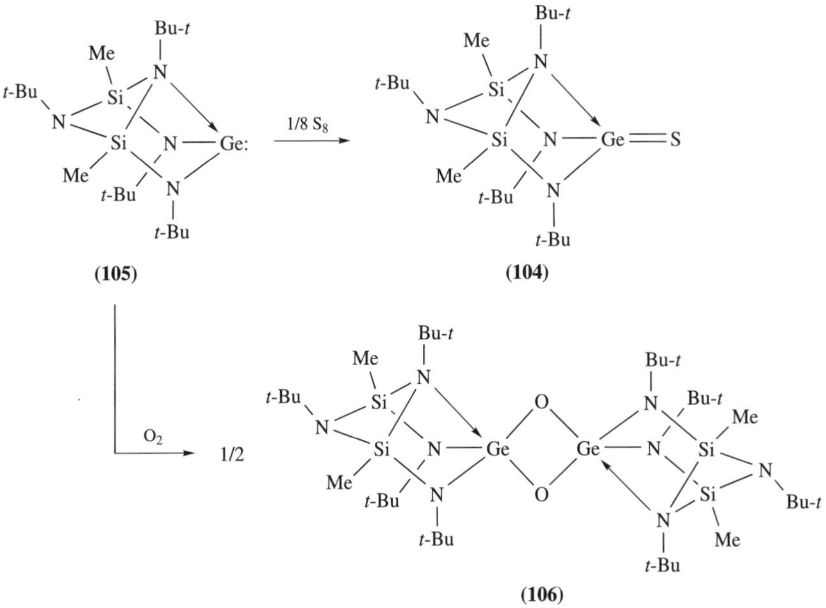

(105)

$1/8\ S_8$

(104)

O_2

1/2

(106)

SCHEME 40

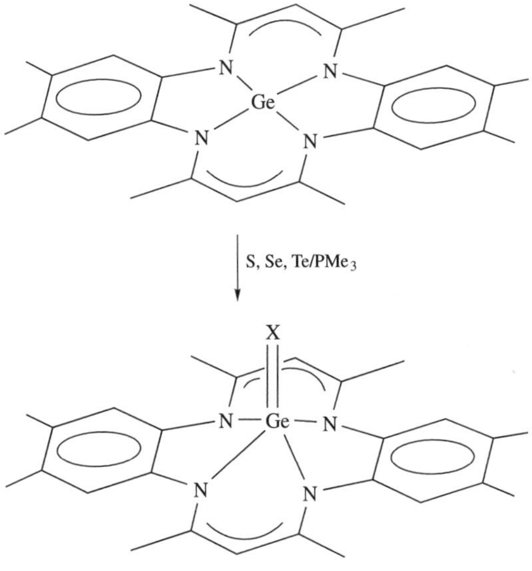

S, Se, Te/PMe_3

(107) X = S
(108) X = Se
(109) X = Te

SCHEME 41

Their bond lengths are somewhat longer compared to the sums of theoretically predicted double-bond covalent radii: 2.110(2) Å for Ge=S (**107**), 2.247(1) Å for Ge=Se (**108**) and 2.446(1) Å for Ge=Te (**109**).

1. Synthetic strategies for stable germanium-containing heavy ketones

A variety of preparation methods are known for transient germanium–chalcogen double-bond species; some of them seem to be also useful for the synthesis of kinetically stabilized systems. Indeed, Tokitoh, Okazaki, and coworkers found that the reaction of a germylene with an appropriate chalcogen source is one of the most versatile and general methods for the synthesis of stable germanium-containing heavy ketones (Scheme 42)[85b].

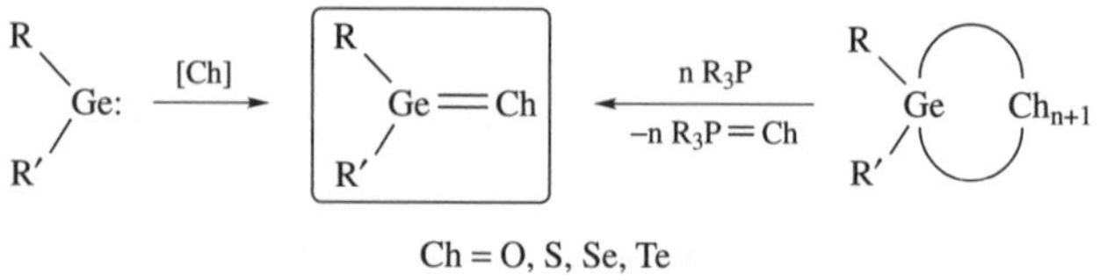

$$RR'Ge: \xrightarrow{[Ch]} RR'Ge{=}Ch \xleftarrow[-n\,R_3P{=}Ch]{n\,R_3P} RR'Ge(Ch_{n+1})$$

Ch = O, S, Se, Te

SCHEME 42

As in the case of silanethione **103**, there has been developed an efficient synthetic method for stable germanethiones and germaneselones, i.e. germanium–sulfur and germanium–selenium double-bond compounds, via dechalcogenation reactions of the corresponding overcrowded germanium-containing cyclic polychalcogenides with a phosphine reagent (Scheme 42)[85b]. This method is superior to the direct chalcogenation of germylenes in view of the easy separation and isolation of the heavy ketones by simple filtration of the phosphine chalcogenides formed. However, it cannot be applied to the synthesis of germanones and germanetellones due to the lack of stable precursors, i.e. cyclic polyoxides and polytellurides.

a. Synthesis of a stable diarylgermanethione. A series of overcrowded cyclic polysulfides bearing two bulky aryl groups, i.e. 1,2,3,4,5-tetrathiagermolanes **110a–c**, have been synthesized as the precursors for kinetically stabilized germanethiones[92]. The desulfurization of tetrathiolane **110a** resulted in the formation of 1,3,2,4-dithiadigermetane **111a**, a dimer of germanethione **112a**, suggesting that the combination of Tbt and Mes groups is not sufficient to stabilize the reactive Ge=S system (Scheme 43)[93].

In contrast, desulfurization of **110b** bearing a bulkier Tip group gave germanethione Tbt(Tip)Ge=S **112b** without forming any dimer (Scheme 43)[94]. Although **112b** is highly reactive toward water and oxygen, it can be isolated quantitatively under argon atmosphere as orange-yellow crystals. It should be noted that **112b** is the first kinetically stabilized isolable germanethione[94]. It melted at 163–165 °C without decomposition, and no change was observed even after heating its hexane solution at 160 °C for 3 days in a sealed tube. The absorption maximum at 450 nm observed for the orange-yellow hexane solution of **112b** was attributable to the n–π^* transition of the Ge=S double bond.

Desulfurization of tetrathiagermolane **110c** bearing a Dep group, having a bulk between those of Mes and Tip, also gave a dimer of the corresponding germanethione **111c** as in the case of **110a** (Scheme 43)[95]. At the beginning of the reaction, however, the electronic spectrum of the hexane solution is reported to show the appearance of a transient absorption at 450 nm attributable to the intermediary germanethione **112c**.

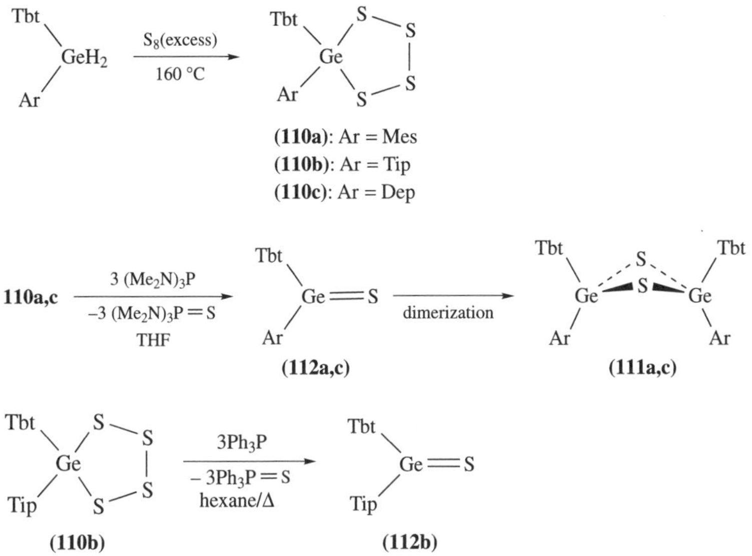

Tbt(Ar)GeH_2 —[S_8(excess), 160 °C]→ Tbt(Ar)GeS_4

(**110a**): Ar = Mes
(**110b**): Ar = Tip
(**110c**): Ar = Dep

110a,c —[3 $(Me_2N)_3P$, −3 $(Me_2N)_3P{=}S$, THF]→ Tbt(Ar)Ge=S (**112a,c**) —[dimerization]→ (**111a,c**)

(**110b**) —[3Ph_3P, − 3$Ph_3P{=}S$, hexane/Δ]→ Tbt(Tip)Ge=S (**112b**)

SCHEME 43

These results can be reasonably interpreted in terms of the bulkiness of the protecting groups on the germanium atom, indicating that a combination of Tbt and Tip groups is necessary in order to isolate a germanethione[95].

b. Synthesis of a stable diarylgermaneselone. As in the case of germanethiones, it is known that the treatment of less hindered tetraselenagermolane Tbt(Mes)$GeSe_4$ **113a**[95,96] with triphenylphosphine gives only 1,3,2,4-diselenadigermetane **115**, a dimer of the corresponding germaneselone Tbt(Mes)Ge=Se (**114**), even in the presence of an excess amount of 2,3-dimethyl-1,3-butadiene. This result clearly shows the high reactivity of a germaneselone and the insufficient steric protection by the combination of Tbt and Mes groups (Scheme 44)[95].

By contrast, Tokitoh, Okazaki and coworkers have reported that deselenation of the bulkier precursor Tbt(Tip)$GeSe_4$ **113b** with 3 molar equivalents of triphenylphosphine in refluxing hexane under argon resulted in the quantitative isolation of the first stable germaneselone Tbt(Tip)Ge=Se **116** as red crystals (Scheme 44)[95,97]. Dimerization of **116** was not observed even in refluxing hexane, in spite of the longer Ge=Se bond distance than that of Ge=S. Germaneselone **116** was extremely sensitive to moisture but thermally quite stable under inert atmosphere. One can see that the combination of Tbt and Tip groups is sufficiently effective to stabilize the reactive germaselenocarbonyl unit of **116**, as is the case of the germanethione **112b**.

c. Stable diaryl-substituted germanetellone. In contrast to the extensive studies on thiocarbonyl and selenocarbonyl compounds[7,8], the chemistry of tellurocarbonyl compounds has been much less studied owing to their instabilities[83c,98]. The chemistry of a germanetellone, the germanium analogue of a tellone, has also been very little explored. Theoretical calculations for $H_2Ge{=}Te$ at B3LYP/TZ(d,p) level have predicted that it has even smaller σ (59.1 kcal mol^{-1}) and π (30.3 kcal mol^{-1}) bond energies than those of the

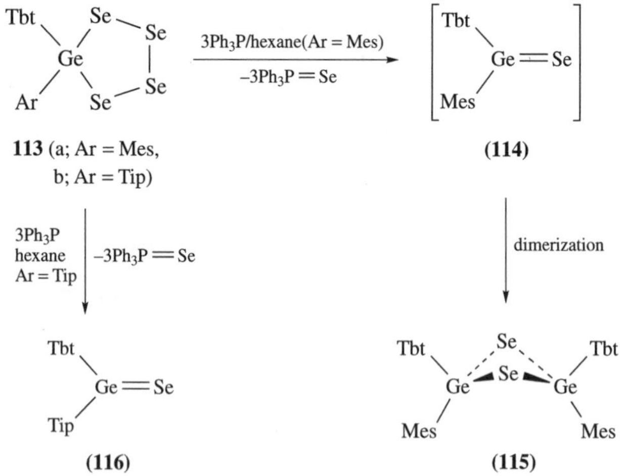

SCHEME 44

corresponding germanethione and germaneselone, but it still exists at an energy minimum, suggesting the possibility of its isolation[6b]. Kuchta and Parkin have already reported the synthesis and crystallographic structure of germatellurourea **109** (Scheme 41), which is stabilized by intramolecular coordination of nitrogen atoms onto the Ge atom[91], as the only report on the chemistry of germanium–tellurium double-bond species. In view of these facts, the synthesis and isolation of a kinetically stabilized germanetellone are significant not only in order to clarify the character of the Ge–Te double bond by itself, but also to elucidate systematically the properties of germanium-containing heavy ketones.

The successful isolation of stable germanethione **112b** and germaneselone **116** suggests that an overcrowded cyclic polytelluride might be a useful precursor for a germanetellone, if it is available. However, no isolable cyclic polytelluride has been obtained, probably owing to the instability of polytellurides. Hence, another synthetic approach was necessary to generate and isolate the germanetellones.

Tokitoh, Okazaki and coworkers have reported that when the stable diarylgermylene **117**, obtained by reduction of the corresponding dibromide with lithium naphthalenide, was allowed to react with an equimolar amount of elemental tellurium in THF, a germanetellone **118** was obtained directly (Scheme 45)[99].

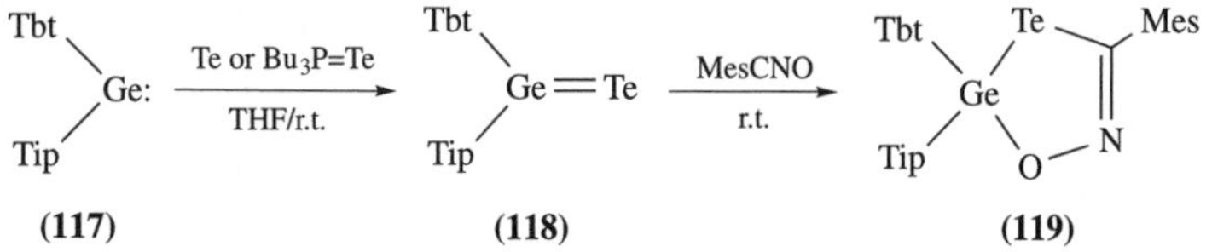

SCHEME 45

The color change of the solution from blue (λ_{max} = 581 nm) due to **117** to green (λ_{max} = 623 nm) was indicative of the generation of a germanetellone **118**[99]. The trapping experiment with mesitonitrile oxide leading to the formation of 37% oxatellurazagermole **119**, the [3+2]cycloadduct of **118**, also suggested the generation of **118** as a stable species in solution (Scheme 45)[99,100]. Although this is the first generation and direct observation of a kinetically stabilized germanetellone, an alternative synthetic method for **118** was considered to be necessary for its isolation in view of the low efficiency of its generation and the practical purification procedures.

d. Isolation of a stable germanetellone. Tokitoh, Okazaki and coworkers have reported that when germylene **117**, generated from dibromogermane Tbt(Tip)$GeBr_2$ and lithium naphthalenide, was allowed to react with diphenylacetylene it afforded germirene **120** in good yield as white crystals (Scheme 46)[101]. This germirene is kinetically stable owing to the bulky groups, in contrast to previously reported germirenes[102] which are known to be hydrolyzed rapidly in air. They found that germirene **120** was thermally labile and, on heating at 70 °C in the presence of 2,3-dimethyl-1,3-butadiene, it gave germacyclopentene **121** (95%) and diphenylacetylene (100%) with complete consumption of the starting material (Scheme 46). This cheletropic reaction is reversible; in the absence of the trapping reagent, the colorless solution at room temperature turns pale blue at 50 °C showing the regeneration of germylene **117**, and becomes colorless again on cooling. These results indicate that germirene **120** is a useful precursor for diarylgermylene **117** under neutral conditions without forming any reactive byproducts (Scheme 46)[101].

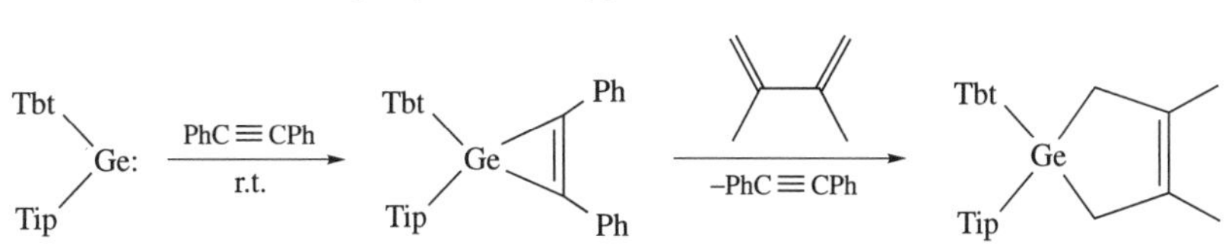

Tbt(Tip)Ge: **(117)** —PhC≡CPh, r.t.→ **(120)** —(2,3-dimethyl-1,3-butadiene), −PhC≡CPh→ **(121)**

120 $\rightleftharpoons$ (70 °C / C_6D_6) **117** + PhC≡CPh

SCHEME 46

For the synthesis of germanetellone **118**, germirene **120** and an equimolar amount of elemental tellurium were allowed to react in benzene-d_6 at 80 °C. On heating the mixture for 9 days and monitoring by ^{1}H NMR, the appearance of new signals along with those of diphenylacetylene at the expense of those assigned to **120** was observed and the solution turned green. The almost quantitative generation of germanetellone **118** was confirmed by the trapping experiment with mesitonitrile oxide giving the corresponding [3+2]cycloadduct **119** in 94% yield (Scheme 47)[99,100]. Removal of the solvent from the green solution without the addition of mesitonitrile oxide gave quantitatively germanetellone **118** as green crystals (Scheme 47)[99,100]. This is the first isolation of a kinetically stabilized germanetellone. Germanetellone **118** was sensitive toward moisture, especially in solution, but thermally quite stable; it melted at 205–210 °C without decomposition.

e. Synthesis of alkyl,aryl-substituted germanium-containing heavy ketones[85a]. Furthermore, Tokitoh, Okazaki, and coworkers have examined the chalcogenation of an alkyl,aryl-disubstituted germylene, **122**[101], for a systematic synthesis and isolation of alkyl, aryl-disubstituted germanium-containing heavy ketones (Scheme 48). Germylene **122** was generated by the loss of diphenylacetylene from the corresponding germirene **123** under the conditions similar to those for diarylgermylene **117**.

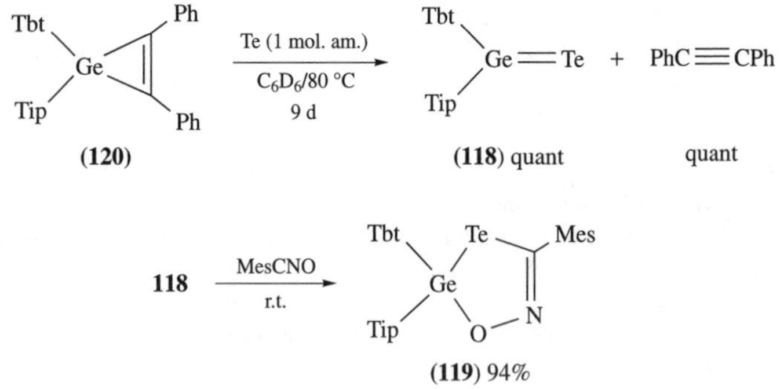

Te (1 mol. am.), C_6D_6/80 °C, 9 d: **(120)** → **(118)** quant + PhC≡CPh quant

118 + MesCNO, r.t. → **(119)** 94%

SCHEME 47

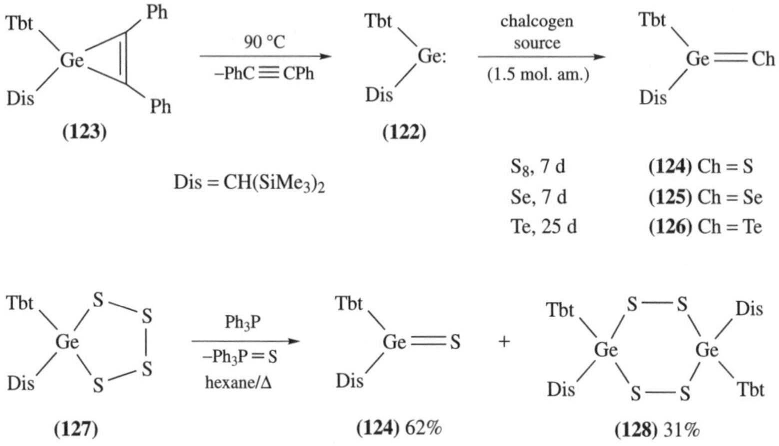

(123) → 90 °C, −PhC≡CPh → **(122)** → chalcogen source (1.5 mol. am.) → Tbt(Dis)Ge=Ch

Dis = $CH(SiMe_3)_2$

S_8, 7 d — **(124)** Ch = S
Se, 7 d — **(125)** Ch = Se
Te, 25 d — **(126)** Ch = Te

(127) → Ph_3P, $-Ph_3P{=}S$, hexane/Δ → **(124)** 62% + **(128)** 31%

SCHEME 48

Thermal reaction of germirene **123** was performed in the presence of 1/8 molar amount of S_8 to give germanethione Tbt(Dis)Ge=S (**124**) quantitatively; it was isolated as yellow crystals in a glove box under pure argon (Scheme 48). Similarly, germaneselone Tbt(Dis)Ge=Se (**125**) and germanetellone Tbt(Dis)Ge=Te (**126**) were also synthesized quantitatively and isolated as orange-red and blue-green crystals, respectively (Scheme 48)[99,103]. In case of germanethione **124**, an alternative synthetic route starting from the corresponding tetrathiagermolane Tbt(Dis)GeS_4 (**127**) was examined. The desulfurization of **127** with three molar equivalent amounts of triphenylphosphine in hexane resulted in the formation of the expected germanethione **124** as a major product together with a tetrathiadigermacyclohexane derivative **128** as a minor product (Scheme 48).

f. Synthesis of stable dialkyl-substituted germanium-containing heavy ketones. Although there have so far been no reports on the synthesis of kinetically stabilized germanium-containing heavy ketones bearing only alkyl substituents, pentacoordinate germanechalcogenones **130–132** bearing two alkyl ligands were recently synthesized by the chalcogenation of the corresponding base-stabilized germylenes **129** (Scheme 49)[104].

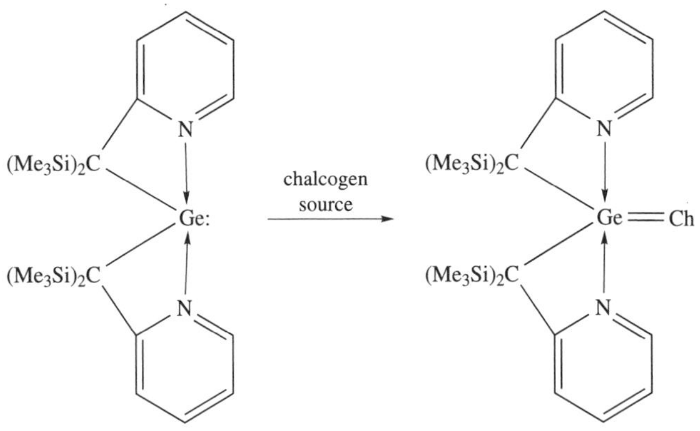

$(Me_3Si)_2C$ ... Ge: ... chalcogen source → $(Me_3Si)_2C$ ... Ge=Ch

(129)

(130) Ch = S

(131) Ch = Se

(132) Ch = Te

SCHEME 49

Their crystallographic analysis revealed that these pentacoordinate Ge complexes have pseudo-trigonal bipyramidal geometry with a trigonal planar arrangement of the chalcogen and the two carbon atoms around the germanium center. The Ge–chalcogen bond lengths of **130**–**132** were found to be intermediate between typical single and double bond lengths (*vide infra*).

g. Heteroatom-substituted germanium–chalcogen double-bond compounds. In addition to the extensive studies on Ge-containing heavy ketones, there have been recently reported several examples of stable germanium–chalcogen doubly bonded systems having two heteroatom substituents. These include **134**[105], **136**[106], **137**,[106a] **139a,b**[107] and **140a–d** (Scheme 50)[107] which were mostly prepared by the direct chalcogenation of the corresponding germylenes **133**, **135** and **138** which are thermodynamically stabilized by the heteroatom substituents.

B. Tin-containing Heavy Ketones

a. Synthesis of stable stannanethiones and stannaneselones. The successful isolation of Si- and Ge-containing heavy ketones naturally provoked the challenge for the synthesis and isolation of the much heavier congeners, i.e. tin–chalcogen doubly bonded systems. However, the combination of Tbt and Tip groups is not bulky enough to stabilize the stannanethione Tbt(Tip)Sn=S **141**[108], in contrast to the lighter analogues such as silanethione **103** and germanethione **112b**.

Although the stability of stannanethione **141**, which was generated in the sulfurization of the corresponding stannylene Tbt(Tip)Sn: (**142**) with elemental sulfur, was evidenced by the characteristic absorption maxima (λ_{max} = 465 nm) in its electronic spectra, concentration of the reaction mixture resulted in the dimerization of **141** giving 1,3,2,4-dithiadistannetane derivative **143** (Scheme 51)[108a,c]. Stannanethione **141** can also be formed in the desulfurization of the corresponding tetrathiastannolane **144** with triphenylphosphine, but the formation of **141** has been confirmed only by trapping experiments (*vide infra*)[108b] and the ^{119}Sn NMR chemical shift for the central tin atom of **141**

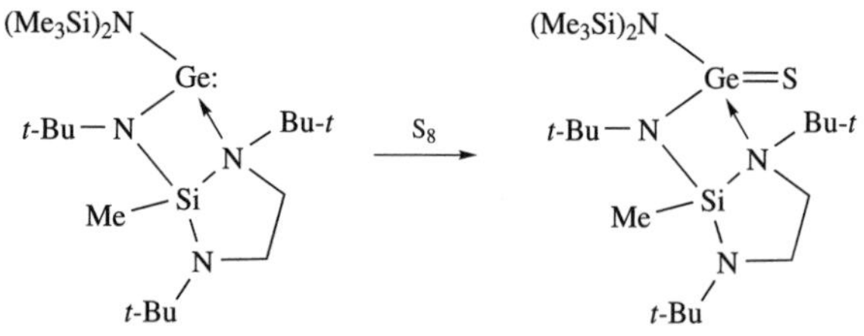

$(Me_3Si)_2N$
Ge:
t-Bu — N
Bu-*t*
N
Me — Si
N
t-Bu
(133)

S_8

$(Me_3Si)_2N$
Ge═S
t-Bu — N
Bu-*t*
N
Me — Si
N
t-Bu
(134)

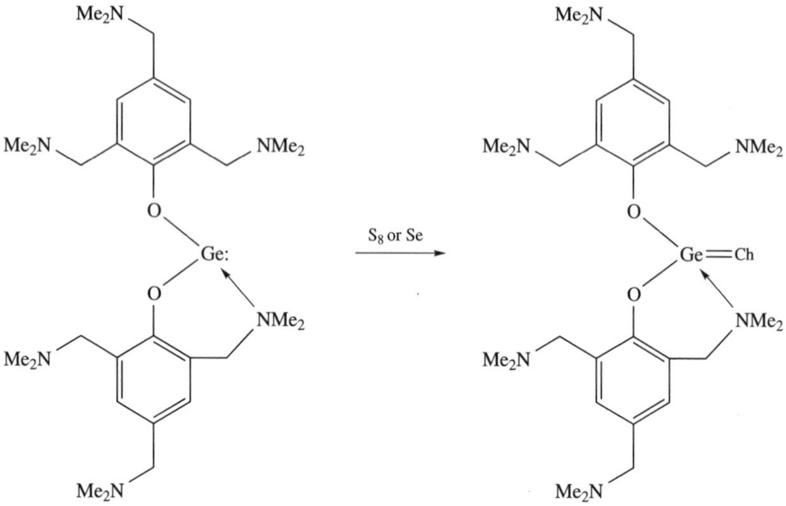

Me_2N
Me_2N
NMe_2
O
Ge:
O
NMe_2
Me_2N
Me_2N
(135)

S_8 or Se

Me_2N
Me_2N
NMe_2
O
Ge═Ch
O
NMe_2
Me_2N
Me_2N
(136) Ch = S
(137) Ch = Se

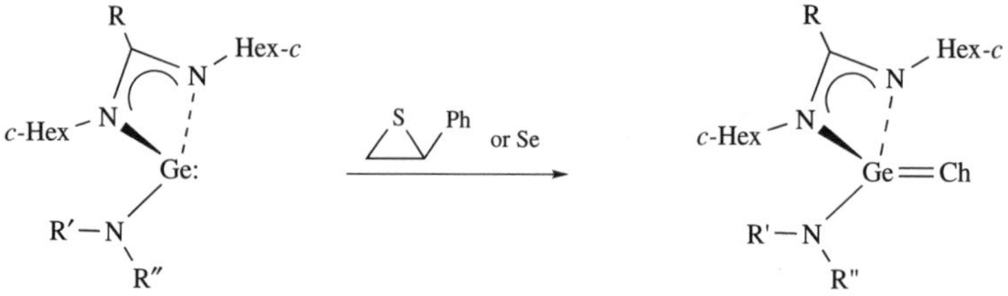

R
Hex-*c*
N
c-Hex — N
Ge:
R′ — N
R″
(138)

S
Ph
or Se

R
Hex-*c*
N
c-Hex — N
Ge═Ch
R′ — N
R"
(139a,b) Ch = S
(140a–d) Ch = Se

(a) R = Me, R′ = *c*-Hex, R″ = C(Me)=N(Hex-*c*)
(b) R = *t*-Bu, R′ = *c*-Hex, R″ = C(Me)=N(Hex-*c*)
(c) R = Me, R′ = R″ = $SiMe_3$
(d) R = *t*-Bu, R′ = R″ = $SiMe_3$

SCHEME 50

was not measured. The formation of similarly substituted stannaneselone **146** was also demonstrated by the intramolecular trapping experiments (*vide infra*), but no spectroscopic evidence was obtained (Scheme 51).[108b,109]

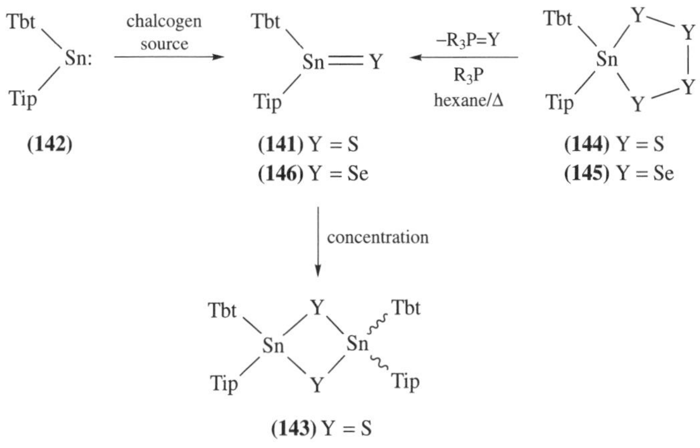

Tbt(Tip)Sn: **(142)** —chalcogen source→ Tbt(Tip)Sn=Y **(141)** Y = S; **(146)** Y = Se ←$-R_3P{=}Y$ / R_3P, hexane/Δ— Tbt(Tip)Sn(Y_4) **(144)** Y = S; **(145)** Y = Se

Tbt(Tip)Sn=Y —concentration→ **(143)** Y = S

SCHEME 51

The high reactivity of tin–chalcogen double bonds was somewhat suppressed by the modification of the steric protection group combined with the Tbt group from the Tip group to more hindered ones such as 2,4,6-tricyclohexylphenyl (Tcp), 2,4,6-tris(1-ethylpropyl)-phenyl (Tpp) and 2,2″-dimethyl-*m*-terphenyl (Dmtp) groups[110,111]. Thus, in these bulkier systems, the corresponding stannanethiones (**148**, **151** and **154**) and stannaneselones (**157**, **160** and **163**), formed respectively from **147**, **150** and **153** and from **156**, **159** and **162**, showed characteristic orange and red colors, respectively, and all of them showed considerably deshielded ^{119}Sn NMR chemical shifts ($\delta_{Sn} = 467$ to 643 ppm) at ambient temperature, suggesting the sp^2 character of their tin centers and also their stability in solution (Scheme 52). On concentration of the samples used for the NMR measurements, however, these tin-containing heavy ketones underwent dimerization to give the corresponding dimers (**149**, **152**, **155**, **158**, **161** and **164**) although it was unsuccessful in isolating them as a stable solid or crystallines.

As the final goal of the synthesis and isolation of tin–chalcogen double-bond species, further modification of Dmtp group was examined[111]. Thus, the fine tuning of the side chains of the terphenyl unit of Dmtp, i.e. the use of 2,2″-diisopropyl-*m*-terphenyl (Ditp) group which bears two isopropyl groups instead of methyl groups in 2 and 2″ positions, allowed isolation of the corresponding stannanethione **166** and stannaneselone **168** as stable orange and red crystals, respectively. Both **166** and **168** were synthesized by the dechalcogenation of the corresponding Tbt- and Ditp-substituted tetrachalcogenastannolanes **165** and **167** with three equivalents of phosphine reagents (Scheme 53)[111]. No dimerization of **166** and **168** was observed even after concentration to the solid state, and the molecular geometry of stannaneselone **169** was definitively determined by X-ray crystallographic analysis.

In addition to the successful isolation of the stable stannaneselone **168**, it should be noted that deselenation of **167** with two molar equivalents of triphenylphosphine resulted in the isolation of a novel tin-containing cyclic diselenide, i.e. diselenastannirane **169**, as a stable crystalline compound[111].

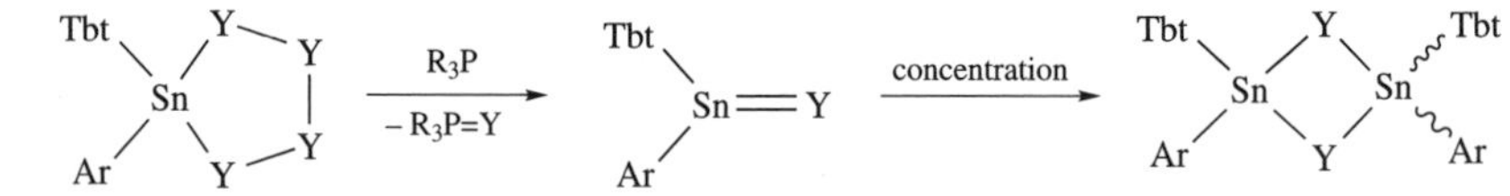

(**147**) Ar = Tcp, Y = S
(**150**) Ar = Tpp, Y = S
(**153**) Ar = Dmtp, Y = S
(**156**) Ar = Tcp, Y = Se
(**159**) Ar = Tpp, Y = Se
(**162**) Ar = Dmtp, Y = Se

(**148**) Ar = Tcp, Y = S
(**151**) Ar = Tpp, Y = S
(**154**) Ar = Dmtp, Y = S
(**157**) Ar = Tcp, Y = Se
(**160**) Ar = Tpp, Y = Se
(**163**) Ar = Dmtp, Y = Se

(**149**) Ar = Tcp, Y = S
(**152**) Ar = Tpp, Y = S
(**155**) Ar = Dmtp, Y = S
(**158**) Ar = Tcp, Y = Se
(**161**) Ar = Tpp, Y = Se
(**164**) Ar = Dmtp, Y = Se

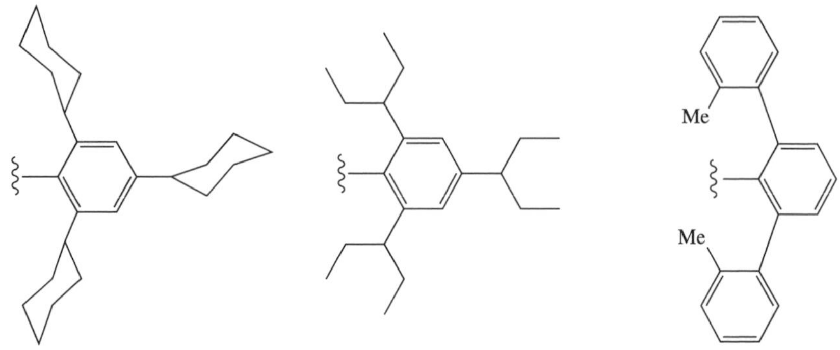

Tcp group

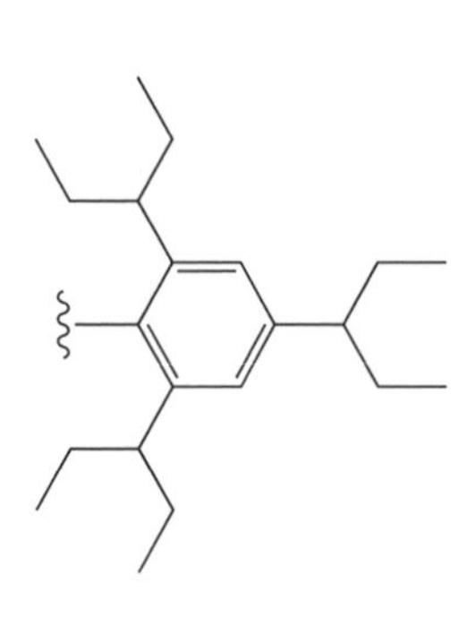

Tpp group

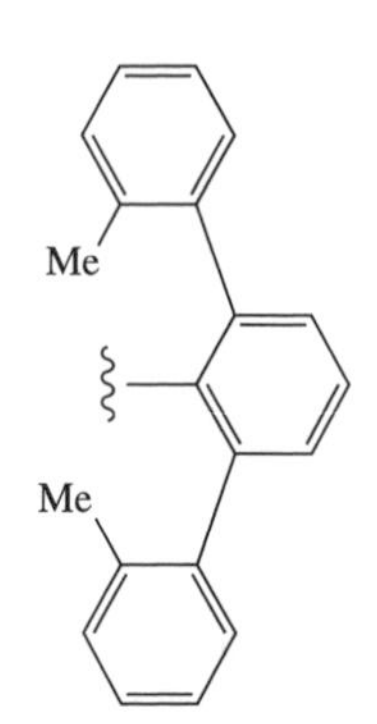

Dmtp group

SCHEME 52

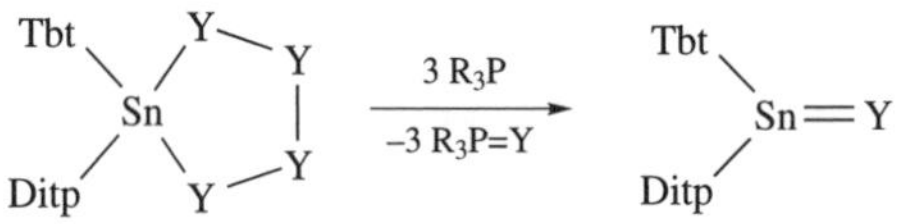

(**165**) Y = S　R = Bu　(**166**) Y = S
(**167**) Y = Se　R = Ph　(**168**) Y = Se

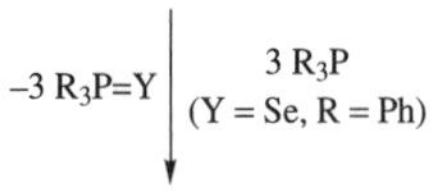

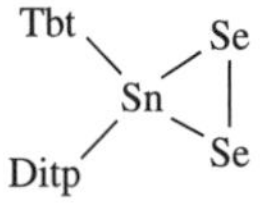

(**169**) Y = S

Ditp group = 2,6-(2′-*i*-$PrC_6H_4)_2C_6H_3$

SCHEME 53

C. Lead-containing Heavy Ketones[74b]

a. Synthesis of plumbanethiones by desulfurization of tetrathiaplumbolanes. In anticipation of kinetic stabilization of a plumbanethione, the heaviest congener of metallanethiones of group 14 elements, a series of hindered tetrathiaplumbolanes **170–172**, were desulfurized with 3 equivalents of hexamethylphosphorous triamide at low temperature (−78 °C) in THF. The color of the reaction solution turned red for **170** and **171** or orange for **172**, indicating the generation of plumbanethiones **173**, **174** or **175**, respectively[74b,112]. Subsequent addition of mesitonitrile oxide to these solutions at −78 °C gave the corresponding oxathiazaplumboles **176–178**, the [3+2]cycloadducts of plumbanethione **173–175**, in moderate yields in each case (Scheme 54)[74b,112]. Plumbanethiones **173** and **175** were also trapped with phenyl isothiocyanate to give the [2+2]cycloaddition products **179** and **181**, respectively, though **174** did not afford such an adduct **180**. The formation of new Pb-containing heterocycles **176–178**, **179** and **181** is noteworthy as the first examples of the trapping of plumbanethiones. Observation of the color changes of the reaction solution suggests that plumbanethiones **173–175** thus formed are stable in solution, at least below −20 °C.

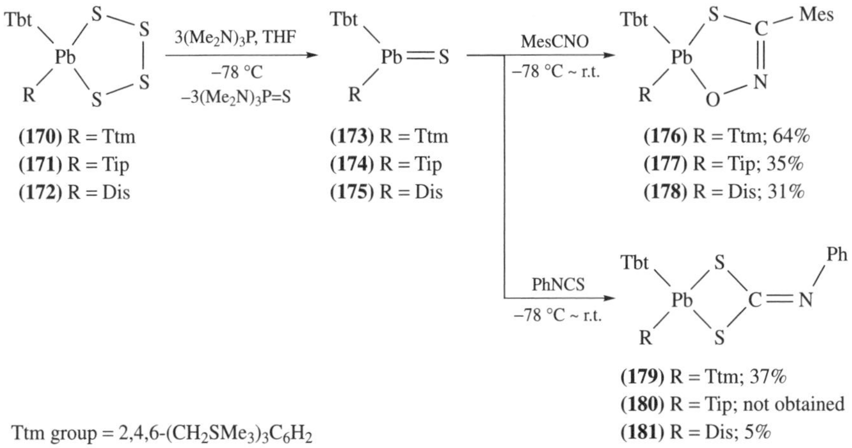

$3(Me_2N)_3P$, THF, −78 °C, $-3(Me_2N)_3P{=}S$; MesCNO, −78 °C ~ r.t.; PhNCS, −78 °C ~ r.t.

(170) R = Ttm
(171) R = Tip
(172) R = Dis

(173) R = Ttm
(174) R = Tip
(175) R = Dis

(176) R = Ttm; 64%
(177) R = Tip; 35%
(178) R = Dis; 31%

(179) R = Ttm; 37%
(180) R = Tip; not obtained
(181) R = Dis; 5%

Ttm group = 2,4,6-$(CH_2SMe_3)_3C_6H_2$

SCHEME 54

In order to elucidate the thermal stability of the plumbanethione **174**, tetrathiaplumbolane **171** was desulfurized at higher temperature. Reaction of **171** with 3 equivalents of triphenylphosphine in toluene at 50 °C gave a deep red solution, from which plumbylene **182** was precipitated as pure deep red crystals (Scheme 55)[113]. Besides **182**, 1,3,2,4-dithiadiplumbetane **183** was obtained as another lead-containing major product. It should

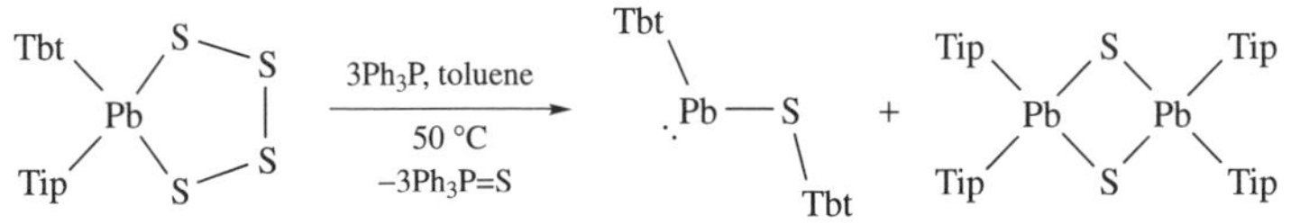

$3Ph_3P$, toluene, 50 °C, $-3Ph_3P{=}S$

(171) **(182)** 26% **(183)** 32%

SCHEME 55

be noted that the final product **182** bears only Tbt groups, while **183** has only Tip groups, though the starting material **171** bears both Tbt and Tip groups on the lead atom.

The fact that the final products, **182** and **183**, bear only Tbt and Tip groups, respectively, indicates the presence of some comproportionation process in their formation from **174**. A plausible mechanism is shown in Scheme 56 for the formation of **182** and **183**, though a detailed reaction mechanism is not clear at present[113]. The tetrathiaplumbolane **171** is first desulfurized to provide plumbanethione **174**, as has been seen in the desulfurization of **171** at low temperature. Plumbanethione **174** might undergo 1,2-aryl migration to give plumbylenes **184** and **185**, which subsequently react to afford an arylsulfido bridged organolead heterocycle **186**, since a less hindered arylthioplumbylene is known to have a tendency to oligomerize, forming a cyclic compound[114a]. The retro [2+2]cycloaddition reaction of **186** affords **182** and **187**. Dithiadiplumbetane **183** may be obtained from the less crowded plumbylene **187**.

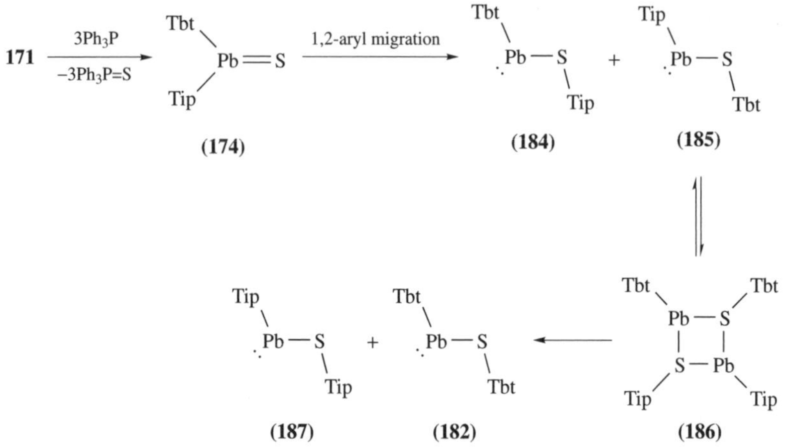

SCHEME 56

X-ray crystallographic analysis reveals that plumbylene **182** exists as a monomer and there is no intermolecular interaction between the lead and sulfur atoms[113a], while other heteroatom-substituted plumbylenes so far known exist as the heteroatom-bridged cyclic oligomers[114].

b. Synthesis of plumbanethione by sulfurization of plumbylene. In order to prove the 1,2-aryl migration process proposed for plumbanethione **174**, the synthesis of plumbanethione **189** by sulfurization of a diarylplumbylene **188** with one atom equivalent of elemental sulfur at low temperature and the successive trapping reaction of **189** with mesitonitrile oxide were carried out (Scheme 57)[74b]. The formation of cycloadduct **190** indicates the generation of intermediary plumbanethione **189**. Furthermore, the fact that heating of a toluene solution of isolated plumbylene **188** with one atom equivalent of elemental sulfur at 50 °C gave **182** clearly demonstrates the occurrence of 1,2-aryl migration in **189** (Scheme 57)[74b]. The formation of **182** in this experiment is the first experimental demonstration that R(RS)Pb is more stable than $R_2Pb{=}S$.

This interpretation of the experimental results was corroborated by *ab initio* calculations on the relative stabilities of a series of double-bond compounds, $[H_2Pb{=}X]$, and

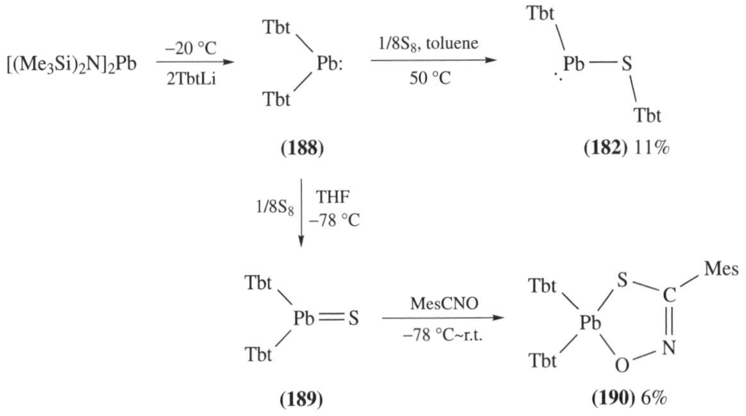

$[(Me_3Si)_2N]_2Pb$ —(−20 °C, 2TbtLi)→ Tbt_2Pb: (188) —($1/8S_8$, toluene, 50 °C)→ Tbt–Pb–S–Tbt (182) 11%

(188) —($1/8S_8$, THF, −78 °C)→ $Tbt_2Pb{=}S$ (189) —(MesCNO, −78 °C~r.t.)→ (190) 6%

SCHEME 57

their plumbylene-type isomers, [*trans*-H-Pb-X-H] and [*cis*-H-Pb-X-H] (X=O, S, Se, and Te)[74b]. The results of the theoretical calculations and experiments for lead–chalcogen double-bond species are in sharp contrast to those of the other group 14 element analogues which do not isomerize to the divalent compounds[88].

D. Structures and Properties of Heavy Ketones

It is very important to reveal the structural features of heavy ketones and to make a systematic comparison with features of the carbonyl analogue such as the bond shortening and the trigonal planar geometry which result from the sp^2 hybridization between the carbon and oxygen atoms. In the following sections, the experimentally and theoretically obtained features of heavy ketones, including the silicon-containing heavy ketones, are systematically compared.

1. X-ray crystallographic analysis

First, a crystallographic analysis of the thermodynamically stabilized silanethione was established for the bulky silanethione **102b** (Scheme 39). Although the Si–S bond [2.013(3) Å] in **102b** is shorter than a typical Si–S single bond (2.13–2.16 Å)[115] suggesting that it has a double-bond character to some extent, it is still 0.07 Å longer than the calculated value for the parent silanethione $H_2Si{=}S$. The Si–N distance (1.964 Å) in **102b**, which is slightly longer than a Si–N σ bond (1.79 Å), supports a very strong coordination of the nitrogen atom of the dimethylaminomethyl group in **102b** to the central silicon atom. Such intramolecular coordination in turn makes the silathiocarbonyl unit of **102b** considerably deviant from the ideal trigonal planar geometry; the sum of the angles around the central silicon atom is 344.9°. The authors concluded that a resonance betaine structure contributes strongly to the electronic distribution of the internally coordinated silanethiones **102a,b**.

Therefore, the elucidation of the intrinsic structural parameters of heavy ketones has to be done with kinetically stabilized systems. Most of the heavy ketones synthesized by taking advantage of the steric protection with the Tbt group have provided single crystals suitable for X-ray structural analysis. The results for silanethione **103**, germanethione

TABLE 4. Structural parameters of heavy ketones Tbt(R)M=X

R	Tip	Tip	Tip	Dis	Tip	Dis	Ditp
M	Si	Ge	Ge	Ge	Ge	Ge	Sn
X	S	S	Se	Se	Te	Te	Se
Compound	**103**	**112b**	**116**	**125**	**118**	**126**	**168**
M–X (Å)	1.948(4)	2.049(3)	2.180(2)	2.173(3)	2.398(1)	2.384(2)	2.375(3)
Δ_{obs} (%)[a]	9	9	9	8	9	8	9
$\Sigma\angle$M (deg)[b]	359.9	359.4	359.3	360.0	359.5	360.0	359.9

[a] The bond shortening (%) compared to the corresponding single bonds.
[b] Summation of the bond angles around the M atom.

112b, germaneselones **116**, **125**, germanetellones **118**, **126** and stannaneselone **168** are summarized in Table 4.

The structural parameters of all heavy ketones examined show that they have an almost completely trigonal planar geometry and a distinct double-bond nature. The observed double-bond lengths and Δ_{obs} [bond shortening (%) compared to the corresponding single bonds] values are in good agreement with calculated values for $H_2M{=}X$[6b,88]. These findings clearly indicate that heavy ketones have structural features similar to those of a ketone, although their double-bond character is lower than that of the corresponding carbon analogues as judged by their Δ_{obs} values.

It should be noted that all the M=X bond lengths observed are significantly shorter than those reported for the corresponding double-bond compounds stabilized by intramolecular coordination of heteroatoms. In other words, thermodynamically stabilized systems suffer from considerable electronic perturbation by heteroatom substituents.

Although some examples of thermodynamically stabilized double-bond systems between group 14 and 16 elements showed trigonal planar geometry due to their structural restriction, almost all of their bond lengths are longer than those of kinetically stabilized systems and of the values theoretically predicted[116]. These results clearly show that considerable electronic perturbation is inevitably involved in the thermodynamically stabilized systems.

2. NMR spectra

As in the case of ^{13}C NMR, low-field chemical shifts are characteristic of sp^2-hybridized nuclei also in the heavy ketones. For example, the ^{29}Si chemical shifts of silanethione **103** and silaneselone [Tbt(Dip)Si=Se (**191**)][88,117] are 167 and 174 ppm, respectively. Silanetellone [Tbt(Dip)Si=Te (**192**)][88,117] also shows its ^{29}Si NMR signal at 171 ppm. In contrast, the ^{29}Si NMR signal of Corriu's compound **102a** (X=S) appears at 22.3 ppm[87], indicating the high sp^3 nature of the silicon centers in **102a**. In the cases of the above-mentioned silaneselone and silanetellone, the characteristic low-field ^{77}Se and ^{125}Te chemical shifts of **191** ($\delta_{Se} = 635$) and **192** ($\delta_{Te} = 731$) are also indicative of their sp^2-hybridized chalcogen atoms.

Similarly, stannanethione **166** and stannaneselone **168** kinetically stabilized by Tbt and Ditp groups have ^{119}Sn chemical shifts of 531 and 440 ppm, respectively. Kuchta and Parkin have reported the synthesis of stable terminal chalcogenido complexes of tin **193** (Scheme 58), and the chemical shifts for the central tin atom appears at a much higher field, i.e. −303 ppm (**193a**, X=S) and −444 ppm (**193b**, X=Se)[91]. This clearly shows that the thermodynamically stabilized tin–chalcogen double bonds in **193** (X=S, Se) are electronically perturbed to a great extent.

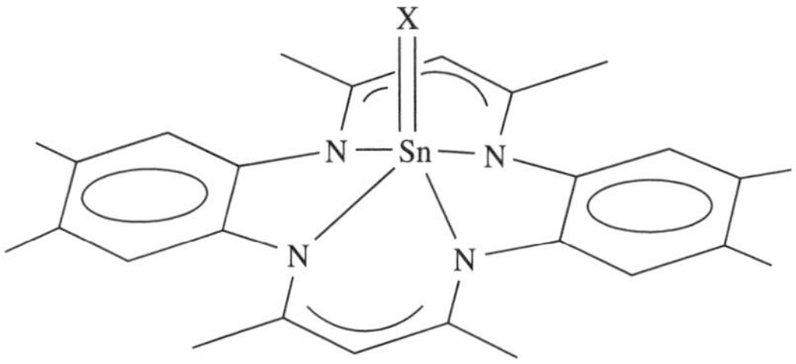

(193a) X = S
(193b) X = Se

SCHEME 58

On the other hand, such information is not available for germanium-containing heavy ketones because of the difficulty in observing Ge NMR spectra, but the low-field shifted ^{77}Se and ^{125}Te NMR signals of germaneselones [$\delta_{Se} = 941$ (for **116**) and 872 (for **125**)][95,97] and germanetellones [$\delta_{Te} = 1143$ (for **118**) and 1009 (for **126**)][99,103] are in agreement with the sp^2-hybridization of these elements as in the cases of silaneselone **191** and silanetellone **192**.

3. UV-vis spectra

Table 5 lists characteristic visible absorptions observed for the heavy ketones kinetically stabilized by the Tbt group.

It is known that the n–π^* absorptions of a series of $R_2C{=}X$ (X = O, S, Se, Te) compounds undergo a systematic red shift on going down the Periodic Table[84]. A similar tendency is observed also for the Si, Ge and Sn series of Tbt(R)Si=X (X = S, Se), Tbt(R)Ge=X (X = S, Se, Te) and Tbt(R)Sn=X (X = S, Se). In contrast, one can see a very interesting trend in the absorption maxima of two series Tbt(R)M=S and

TABLE 5. Electronic spectra (n→ π^*) of doubly bonded compounds between group 14 elements and chalcogens

Observed λ_{max} (nm)^a			Calcd^j		
	X = S	X = Se		λ_{max} (nm)	$\Delta\varepsilon^*_{n\pi}$ (eV)^k
TbtCH=X	587^b	792^c	$H_2C{=}S$	460	10.81
Tbt(R)Si=X	396^d	456^e	$H_2Si{=}S$	352	10.27
Tbt(R)Ge=X^f	450^g	519^g	$H_2Ge{=}S$	367	9.87
Tbt(R)Sn=X	473^h	531^i	$H_2Sn{=}S$	381	9.22
			$H_2Pb{=}S$	373	9.11

^{a}In hexane.
^{b}Reference 127.
^{c}Reference 128.
^{d}R = Tip. Reference 6.
^{e}In THF. R = Dip.
^{f}X = Te: R = Tip, λ_{max} 640 nm. Reference 99.
^{g}R = Tip. Reference 85.
^{h}R = Tip. Reference 108c.
^{i}R = Ditp. Reference 111.
^{j}CIS/TZ(d,p)//B3LYP/TZ(d,p).
$^k\varepsilon_{LUMO(\pi^*)} - \varepsilon_{HOMO(n)}$.

Tbt(R)M=Se (M = C, Si, Ge, Sn). In both series the λ_{max} values are greatly blue-shifted on going from carbon to silicon congeners, whereas the λ_{max} values of silicon, germanium and tin congeners are red-shifted with increasing atomic number of the group 14 elements. This trend is also found in the calculated values for $H_2M{=}S$ (M = C, Si, Ge, Sn)[6,88]. Since the calculated $\Delta\varepsilon_{n\pi^*}$ values increase continuously from $H_2Sn{=}S$ to $H_2C{=}S$, a long wavelength absorption for $H_2C{=}S$ (and hence for TbtCH=S[69]) most likely results from a large repulsion integral ($J^*_{n\pi}$) for the carbon–sulfur double bond, as in the case of $H_2C{=}O$ vs $H_2Si{=}O$[118].

4. Raman spectra

The stretching vibrations of the M=X bond were measured by Raman spectra for silanethione **103** (724 cm^{-1}), germanethiones **112b** (521 cm^{-1}) and **124** (512 cm^{-1}), and germanetellones **118** (381 cm^{-1}) and **126** (386 cm^{-1}). These values are in good agreement with those calculated for $H_2M{=}X$ compounds [723 (Si=S), 553 (Ge=S) and 387 (Ge=Te) cm^{-1}, respectively][6,88]. It is noteworthy that the observed value for Tbt(Tip)Ge=S **112b** is very close to that observed by IR spectroscopy for $Me_2Ge{=}S$ (518 cm^{-1}) in an argon matrix at 17–18 K[119], indicating similarity in the nature of the bond of both germanethiones in spite of the great difference in the size and nature of the substituents.

E. Reactivities of Heavy Ketones

As mentioned in the previous sections, heavy ketones undergo ready head-to-tail dimerization (or oligomerization) when the steric protecting groups on the group 14 element are not bulky enough to suppress their high reactivity. Indeed, Tbt(Mes)M=X (M = Si, Ge; X = S, Se) cannot be isolated at an ambient temperature and instead the dimerization products are obtained. In the cases of tin-containing heavy ketones, the dimerization reactions are much easier than in the lighter group 14 element congeners, and only the combination of Tbt and Ditp groups can stabilize the long and reactive tin–chalcogen double bonds (*vide supra*).

Although the high reactivity of metal–chalcogen double bonds of isolated heavy ketones is somewhat suppressed by the steric protecting groups, Tbt-substituted heavy ketones allow the examination of their intermolecular reactions with relatively small substrates. The most important feature in the reactivity of a carbonyl functionality is the reversibility in the reactions across its carbon–oxygen double bond (via the addition–elimination mechanism via a tetracoordinate intermediate) as is observed, for example, in reactions with water and alcohols. The energetic basis of this reversibility is that there is very little difference in the σ and π bond energies of the C=O bond (Table 3). In contrast, an addition reaction involving a heavy ketone is highly exothermic and hence essentially irreversible because of the much smaller π bond energy than the corresponding σ bond energy of these species.

All heavy ketones kinetically stabilized by a Tbt group react with water and methanol almost instantaneously to give tetracoordinate adducts (Scheme 59). They also undergo cycloadditions with unsaturated systems such as phenyl isothiocyanate, mesitonitrile oxide and 2,3-dimethyl-1,3-butadiene to give the corresponding [2+2]-, [2+3]- and [2+4]cycloadducts, respectively (Scheme 59). The former two reactions proceed at room temperature, while the reaction with the diene takes place at higher temperature, with the lighter homologues requiring more severe conditions.

Reaction of **112b** with methyllithium followed by alkylation with methyl iodide gives a germophilic product **194**. The [2+4]cycloaddition of **112b** with 2-methyl-1,3-pentadiene affords **195** regioselectively. When a hexane solution of **195** is heated at 140 °C in a sealed tube in the presence of excess 2,3-dimethyl-1,3-butadiene, a dimethylbutadiene adduct **196**

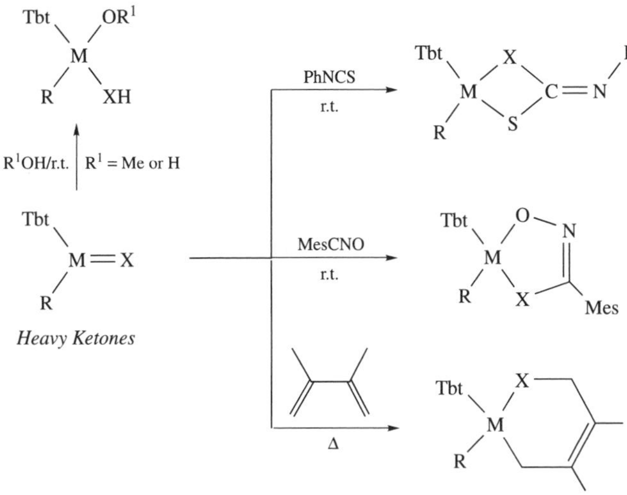

SCHEME 59

is obtained in a high yield, indicating that the Diels–Alder reaction of germanethione **112b** with a diene is reversible, and hence that a diene adduct such as **195** or **196** can be a good precursor of germanethione **112b** (Scheme 60)[95]. Similar reactivity was observed for germaneselone **116**, and the retrocycloaddition of **197** (shown by the formation of **198**) takes place at a much lower temperature (50 °C) than that for **195**, suggesting a weaker C–Se bond than a C–S bond (Scheme 60).

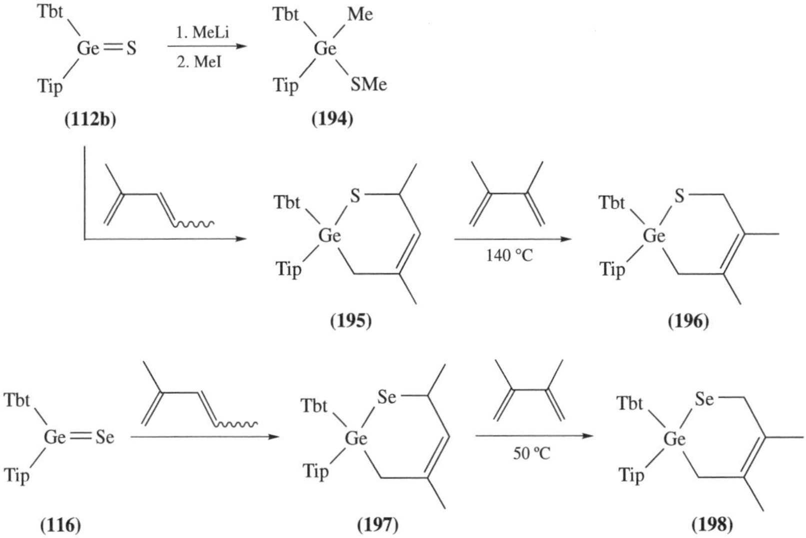

SCHEME 60

In contrast to the considerable thermal stability of the isolated heavy ketones of silicon, germanium and tin, a plumbanethione behaves differently. When stable plumbylene Tbt_2Pb **188**[74a,120] was sulfurized by 1 molar equivalent of elemental sulfur at 50 °C, the heteroleptic plumbylene TbtPbSTbt (**182**) was obtained (Scheme 57)[74b] instead of plumbanethione $Tbt_2Pb{=}S$ **189**, which is an expected product in view of the reactivity observed for divalent species of silicon, germanium and tin (*vide supra*). The formation of **182** is most reasonably explained in terms of 1,2-migration of the Tbt group in the intermediate plumbanethione **189**, and this observation is supporting evidence for the 1,2-aryl migration in plumbanethione **174** proposed in the reaction shown in Scheme 56. This unique 1,2-aryl migration in a plumbanethione is in keeping with a theoretical calculation which reveals that plumbylene HPb(SH) is about 39 $kcal\,mol^{-1}$ more stable than plumbanethione $H_2Pb{=}S$[74b].

IV. HEAVIER CONGENERS OF ALLENES

a. Silaallene. As can be seen in the previous sections, a variety of heavier element analogues of alkenes and ketones have been isolated by taking advantage of kinetic stabilization using bulky substituents and their structures have been characterized by X-ray crystallographic analysis. Also, there have been reported some stable examples of heavier congeners of imines such as silaneimines, $R_2Si{=}NR$[1m,3] and germaneimines, $R_2Ge{=}NR$[121]. On the other hand, the chemistry of cumulative double-bond compounds of heavier group 14 elements is less explored. In 1993, the first stable silacumulene, 1-silaallene **200**, was synthesized by the reaction of well-designed alkynyl fluorosilane **199** with *t*-butyllithium by West and coworkers (Scheme 61)[8a]. Silaallene **200** was isolated as stable crystals and characterized by X-ray structural analysis, but very little is known about its reactivity[8].

b. Germaallenes. Regarding the cumulene-type compounds containing a germanium atom, germaphosphaallene **202** was synthesized by Escudié and coworkers in 1996 using the reduction of (fluorogermyl)bromophosphaalkene **201** (Scheme 62)[122]. Although the generation of **202** was confirmed by ^{13}C and ^{31}P NMR at −40 °C, **202** dimerized at room temperature to afford two types of dimers, i.e. **203** and **204** (Scheme 62)[122]. This result shows that the combinations of the substituents on the germanium and phosphorous atoms were not bulky enough to prevent dimerization of **202**.

In 1997, Tokitoh, Okazaki and coworkers reported evidence for the generation of a kinetically stabilized 1-germaallene **207**[25,123]. It was synthesized by two different synthetic approaches as shown in Scheme 63. One is the dechalcogenation reaction of the corresponding alkylidenetelluragermirane **205** with a phosphine reagent, and the other is the reduction of (1-chlorovinyl)chlorogermane **206** with *t*-butyllithium.

Both reactions afforded an identical product as judged by its NMR spectrum, and 1-germaallene **207** was found to be marginally stable in solution, showing the characteristic low-field ^{13}C NMR chemical shift for its central *sp* carbon atom ($\delta_c = 243.6$ in C_6D_6). The down-field shift for the central *sp* carbon has been also observed for 1-silaallene **200** ($\delta_c = 226$)[8] and germaphosphaallene **202** ($\delta_c = 281$)[122]. Although **207** can be trapped with methanol, mesitonitrile oxide and elemental sulfur to give the corresponding adducts **208**–**210**, it slowly undergoes an intramolecular cyclization in solution at room temperature to give compound **211** (Scheme 64)[25,123]. Concentration and isolation of **207** as crystals at low temperature has failed, and no crystallographic information has been obtained for **207** so far.

In 1998, West and coworkers succeeded in the synthesis and isolation of 1-germaallene **212**, the first example of a stable 1-germaallene either in solution or in the solid state

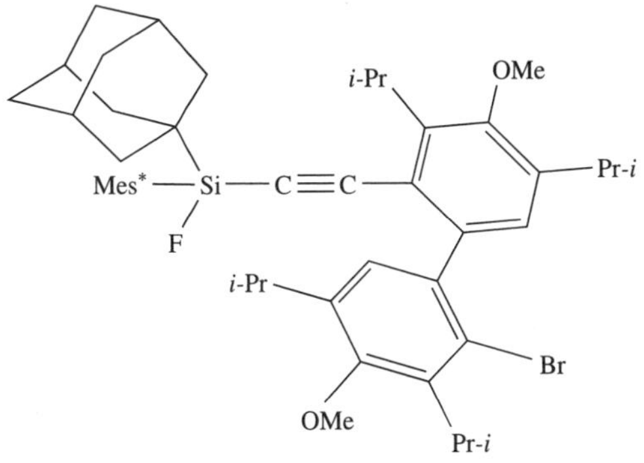

(199)

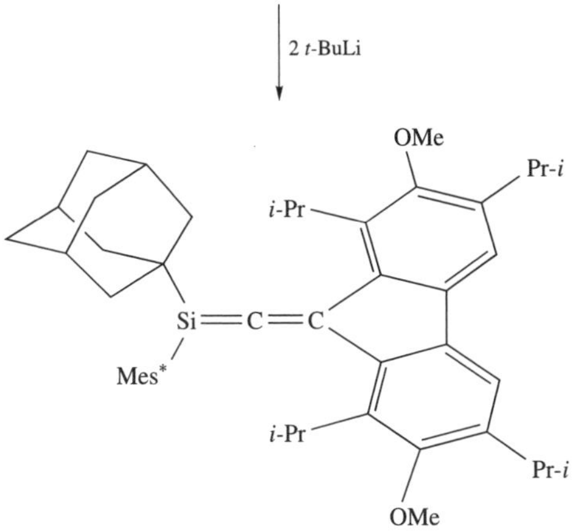

2 t-BuLi

(200)

Mes* = 2,4,6-(t-Bu)$_3$C$_6$H$_2$

SCHEME 61

(Scheme 65)[124]. **212** was stable in ether solution up to 0 °C but is completely decomposed after 15 h at 25 °C, whereas it remained unchanged in toluene up to 135 °C. Gemaallene **212** was isolated by crystallization from ether at −20 °C as colorless crystals, which showed a low-field ^{13}C NMR signal at 235.1 ppm in toluene-d_8 as in the case of **207**[124].

A crystallographic analysis of **212** had revealed the structural parameters for this unique cumulative bonding[124]. The G=C bond length is 1.783(2) Å and the Ge=C=C unit is not linear, having a bending angle of 159.2°. The sum of the bond angles at Ge (348.4°) indicates a strong pyramidalization around the Ge center.

c. Tristannaallene. Quite recently, Wiberg and coworkers reported the first example of 1,2,3-tristannaallene **214** by taking advantage of the tri-t-butylsilyl (supersilyl) group[125]. It was synthesized by the reaction of diaminostannylene **213** with supersilyl sodium (Scheme 66).

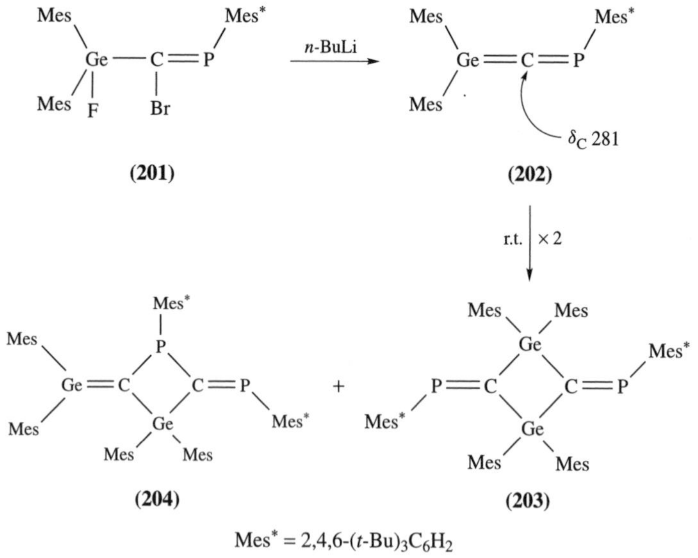

Mes
Mes*
Ge—C=P
Mes F Br
n-BuLi
Mes
Mes*
Ge=C=P
Mes
δ_C 281

(201) (202)

r.t. ×2

Mes*
P
Mes
Ge=C
C=P
Mes
Ge
Mes*
Mes Mes
+
Mes Mes
Ge
Mes*
P=C
C=P
Mes*
Ge
Mes Mes

(204) (203)

Mes* = 2,4,6-(t-Bu)$_3C_6H_2$

SCHEME 62

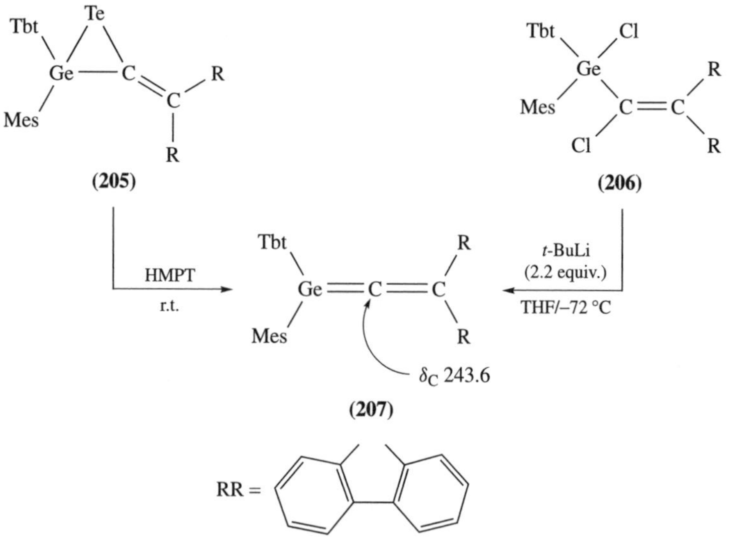

Tbt
Te
Ge—C
R
C
Mes
R

Tbt
Cl
Ge
R
Mes
C=C
Cl
R

(205) (206)

HMPT
r.t.
Tbt
R
Ge=C=C
Mes
R
δ_C 243.6
t-BuLi
(2.2 equiv.)
THF/−72 °C

(207)

RR =

SCHEME 63

At the first stage below −25 °C (in pentane/C_6D_6) this reaction yields the tristannaallene **214**, but it rearranges at 25 °C to the isomeric cyclotristannene **215**. The formation of **214** can be interpreted in terms of an initial formation of disupersilylstannylene **216** (Scheme 67)[125].

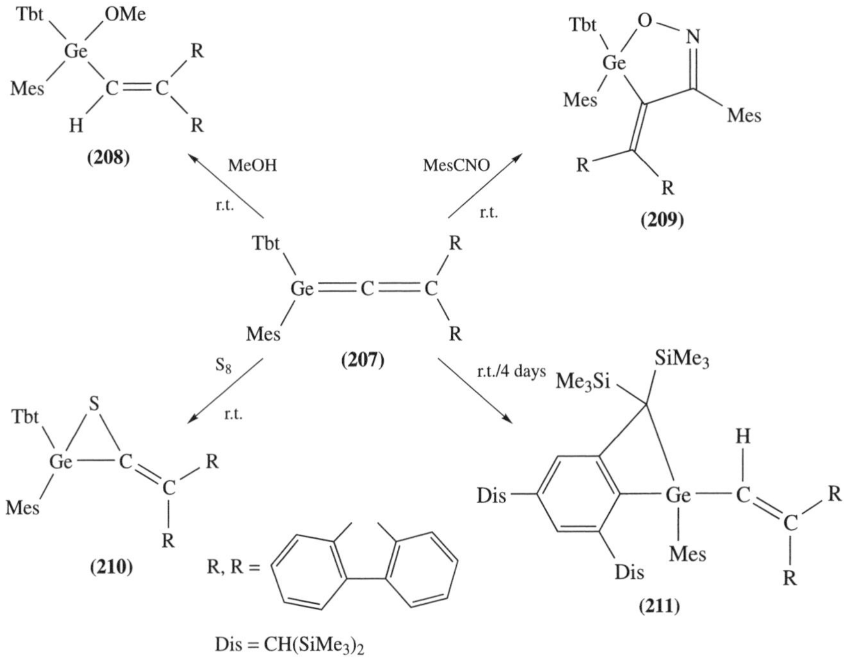

Dis = $CH(SiMe_3)_2$

SCHEME 64

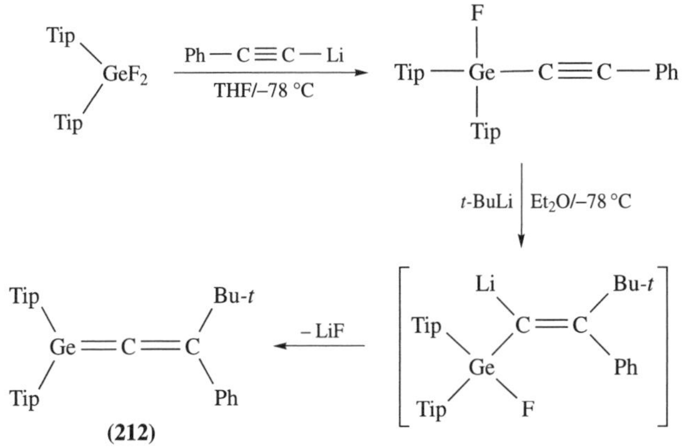

SCHEME 65

Although several examples of heavy allenes have been synthesized as stable compounds[126], most of them are isolated as marginally stable species at room temperature. Further systematic progress is necessary in order to elucidate the intrinsic properties of these interesting doubly bonded systems of heavier group 14 elements.

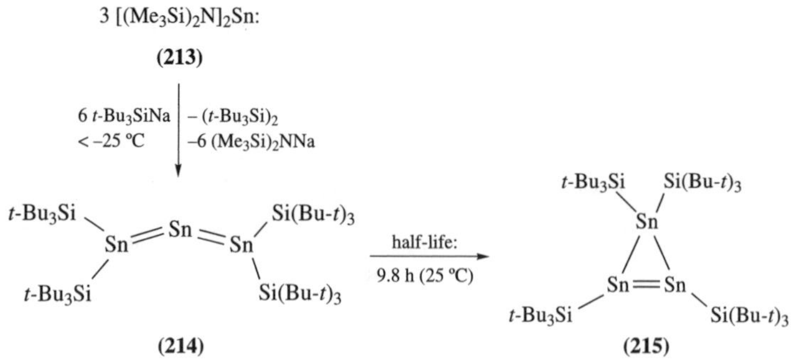

3 $[(Me_3Si)_2N]_2Sn$:

(213)

6 *t*-Bu_3SiNa, < −25 °C; − $(t\text{-}Bu_3Si)_2$, −6 $(Me_3Si)_2NNa$

$t\text{-}Bu_3Si$, $t\text{-}Bu_3Si$ Sn=Sn=Sn $Si(Bu\text{-}t)_3$, $Si(Bu\text{-}t)_3$

(214)

half-life: 9.8 h (25 °C)

(215)

SCHEME 66

$[(Me_3Si)_2N]_2Sn$: **(213)** —(2 *t*-Bu_3SiNa, −2 $(Me_3Si)_2NNa$)→ $(t\text{-}Bu_3Si)_2Sn$: **(216)** —(**216**, − *t*-$Bu_3Si^{\bullet}$)→ [$(t\text{-}Bu_3Si)_2Sn=Sn^{\bullet}Si(Bu\text{-}t)_3$] —(− *t*-$Bu_3Si^{\bullet}$, **216**)→ **(214)**

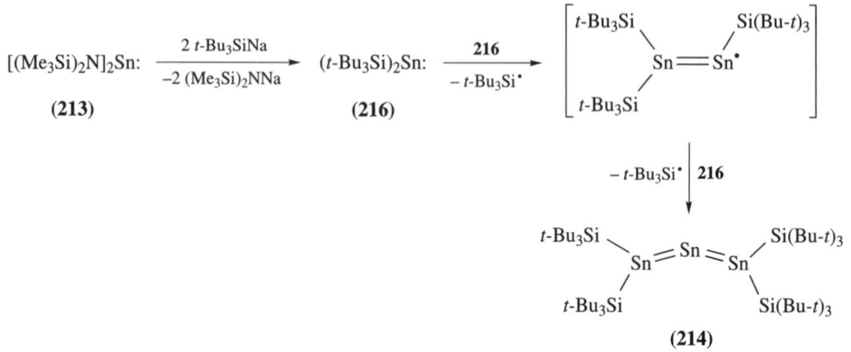

SCHEME 67

V. OUTLOOK AND FUTURE

This chapter outlines the recent progress in the chemistry of multiply bonded species of Ge, Sn, and Pb. As can be concluded from this chapter and related recent reviews which deal with the chemistry of this family of compounds which are stable at room temperature, the field of low-coordinated species of heavier group 14 elements has matured considerably in recent years.

For example, we can now make a systematic comparison for heavy ketones through silicon to lead. However, heavy ketones containing an oxygen atom are still elusive species and neither their isolation nor spectroscopic detection has been achieved so far probably due to their extremely high reactivity caused by their highly polarized structure. The most fascinating and challenging target molecules in this area should be the stable oxygen-containing heavy ketones **217** (Scheme 68).

As for the multiply bonded systems of heavier group 14 elements, all the homonuclear double bonds were successfully isolated and characterized by the end of the last century. However, the chemistry of heteronuclear double bonds and those of conjugated systems are still in growth. The synthesis and isolation of unprecedented cumulative systems such as **217** and aromatic systems containing Ge, Sn and Pb atoms, e.g. **219–222** (Scheme 68), will also be within the range of this future chemistry. Likewise, the synthesis and isolation of the triply bonded systems **223** (Scheme 68) should be one of the most challenging projects in this field.

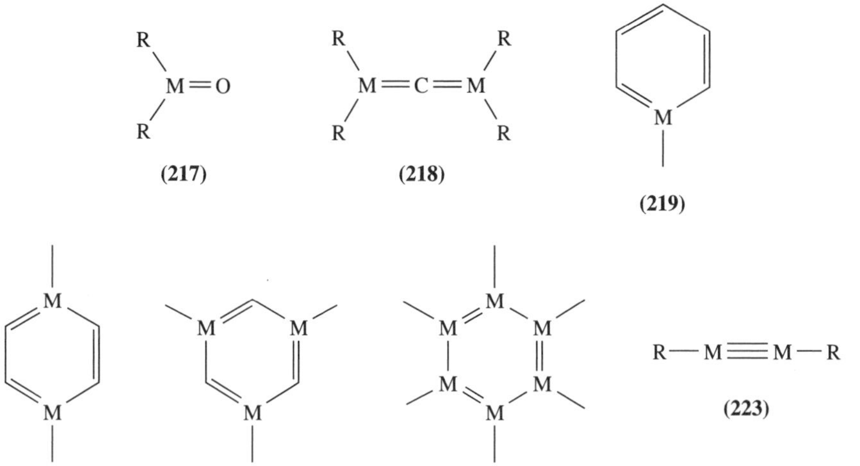

(217) (218) (219)

(220) (221) (222) (223)

M = Ge, Sn, and Pb

SCHEME 68

As can be seen from this chapter, kinetic stabilization has been used as a much superior method to thermodynamic stabilization in order to elucidate the intrinsic nature of the chemical bonding containing low-coordinated heavier group 14 elements. However, it might be useful to combine the two types of different stabilization methods for the synthesis and isolation of much more reactive systems as mentioned above.

Elucidation of the intrinsic properties of unprecedented chemical bondings of heavier group 14 elements and their systematic comparison will be of great importance in efforts to extend the conventional organic chemistry to that of the whole main group elements.

VI. REFERENCES

1. For reviews, see:
(a) R. West, *Pure Appl. Chem.*, **56**, 163 (1984).
(b) J. Satgé, *Pure Appl. Chem.*, **56**, 137 (1984).
(c) G. Raabe and J. Michl, *Chem. Rev.*, **85**, 419 (1985).
(d) R. West, *Angew. Chem., Int. Ed. Engl.*, **26**, 1201 (1987).
(e) J. Barrau, J. Escudié and J. Satgé, *Chem. Rev.*, **90**, 283 (1990).
(f) P. Jutzi, *J. Organomet. Chem.*, **400**, 1 (1990).
(g) M. F. Lappert and R. S. Rowe, *Coord. Chem. Rev.*, **100**, 267 (1990).
(h) J. Satgé, *J. Organomet. Chem.*, **400**, 121 (1990).
(i) W. P. Neumann, *Chem. Rev.*, **91**, 311 (1991).
(j) T. Tsumuraya, S. A. Batcheller and S. Masamune, *Angew. Chem., Int. Ed. Engl.*, **30**, 902 (1991).
(k) M. Weidenbruch, *Coord. Chem. Rev.*, **130**, 275 (1994).
(l) A. G. Brook and M. Brook, *Adv. Organomet. Chem.*, **39**, 71 (1996).
(m) I. Hemme and U. Klingebiel, *Adv. Organomet. Chem.*, **39**, 159 (1996).
(n) M. Driess, *Adv. Organomet. Chem.*, **39**, 193 (1996).
(o) R. Okazaki and R. West, *Adv. Organomet. Chem.*, **39**, 232 (1996).
(p) K. M. Baines and W. G. Stibbs, *Adv. Organomet. Chem.*, **39**, 275 (1996).
(q) M. Driess and H. Grützmacher, *Angew. Chem., Int. Ed. Engl.*, **35**, 827 (1996).

(r) P. P. Power, *J. Chem. Soc., Dalton Trans.*, 2939 (1998).
(s) M. Weidenbruch, *Eur. J. Inorg. Chem.*, 373 (1999).
(t) N. Tokitoh and R. Okazaki, *Coord. Chem. Rev.*, **210**, 251 (2000).
(u) See also, the reviews in *The Chemistry of Organic Silicon Compounds,* Vol. 2 (Eds. Z. Rappoport and Y. Apeloig), Wiley, Chichester, 1998.

2. (a) A. G. Brook, F. Abdesaken, B. Gutekunst, G. Gutekunst and R. K. Kallury, *J. Chem. Soc., Chem. Commun.*, 191 (1981).
(b) R. West, M. J. Fink and J. Michl, *Science*, **214**, 1343 (1981).
3. N. Wiberg, K. Schurz and G. Fischer, *Angew. Chem., Int. Ed. Engl.*, **24**, 1053 (1985).
4. C. N. Smit, M. F. Lock and F. Bickelhaupt, *Tetrahedron Lett.*, **25**, 3011 (1984).
5. M. Driess and H. Pritzkow, *Angew. Chem., Int. Ed. Engl.*, **31**, 316 (1992).
6. (a) H. Suzuki, N. Tokitoh, S. Nagase and R. Okazaki, *J. Am. Chem. Soc.*, **116**, 11578 (1994).
(b) H. Suzuki, N. Tokitoh, R. Okazaki, S. Nagase and M. Goto, *J. Am. Chem. Soc.*, **120**, 11096 (1998).
7. N. Tokitoh, *Phosphorus, Sulfur Silicon Relat. Elem.*, **136–138**, 123 (1998).
8. (a) G. E. Miracle, J. L. Ball, D. R. Powell and R. West, *J. Am. Chem. Soc.*, **115**, 11598 (1993).
(b) M. Trommer, G. E. Miracle, B. E. Eichler, D. R. Powell and R. West, *Organometallics*, **16**, 5737 (1997).
9. (a) K. Wakita, N. Tokitoh, R. Okazaki and S. Nagase, *Angew. Chem. Int. Ed.*, **39**, 634 (2000).
(b) K. Wakita, N. Tokitoh, R. Okazaki, N. Takagi and S. Nagase, *J. Am. Chem. Soc.*, **122**, 5648 (2000).
10. (a) N. Tokitoh, K. Wakita, R. Okazaki, S. Nagase, P. v. R. Schleyer and H. Jiao, *J. Am. Chem. Soc.*, **119**, 6951 (1997).
(b) K. Wakita, N. Tokitoh, R. Okazaki, S. Nagase, P. v. R. Schleyer and H. Jiao, *J. Am. Chem. Soc.*, **121**, 11336 (1999).
11. M. Weidenbruch, S. Willms, W. Saak and G. Henkel, *Angew. Chem., Int. Ed. Engl.*, **36**, 2503 (1997).
12. J. T. Snow, S. Murakami S. Masamune and D. J. Williams, *Tetrahedron Lett.*, **25**, 4191 (1984).
13. J. Park, S. A. Batcheller and S. Masamune, *J. Organomet. Chem.*, **367**, 39 (1989).
14. T. Fjeldberg, A. Haaland, B. E. R. Schilling, M. F. Lappert and A. J. Thorne, *J. Chem. Soc., Dalton Trans.*, 1551 (1986).
15. D. E. Goldberg, P. B. Hitchcock, M. F. Lappert, K. M. Thomas, A. J. Thorne, T. Fjeldberg, A. Haaland and B. E. R. Schilling, *J. Chem. Soc., Dalton Trans.*, 2387 (1986).
16. (a) A. Sekiguchi, H. Yamazaki, C. Kabuto and H. Sakurai, *J. Am. Chem. Soc.*, **117**, 8025 (1995).
(b) M. Ichinohe, H. Sekiyama, N. Fukaya and A. Sekiguchi, *J. Am. Chem. Soc.*, **122**, 6781 (2000).
17. M. Kira, T. Iwamoto, T. Maruyama, C. Kabuto and H. Sakurai, *Organometallics*, **15**, 3767 (1996).
18. A. Sekiguchi, N. Fukaya, M. Ichinohe, N. Takagi and S. Nagase, *J. Am. Chem. Soc.*, **121**, 11587 (1999).
19. S. A. Batcheller, T. Tsumuraya, O. Tempkin, W. M. Davis and S. Masamune, *J. Am. Chem. Soc.*, **112**, 9394 (1990).
20. M. Weidenbruch, M. Stürmann, H. Kilian, S. Pohl and W. Saak, *Chem. Ber.*, **130**, 735 (1997).
21. S. Collins, S. Murakami, J. T. Snow and S. Masamune, *Tetrahedron Lett.*, **26**, 1281 (1985).
22. T. Tsumuraya, S. Sato and W. Ando, *Organometallics*, **9**, 2061 (1990).
23. K. Kishikawa, T. Tokitoh and R. Okazaki, *Chem. Lett.*, 239 (1998).
24. S. Masamune, Y. Hanzawa and D. J. Williams, *J. Am. Chem. Soc.*, **104**, 6136 (1982).
25. K. Kishikawa, Ph.D. Thesis, The University of Tokyo, 1997.
26. (a) W. Ando, T. Tsumuraya and A. Sekiguchi, *Chem. Lett.*, 317 (1987).
(b) W. Ando, H. Itoh and T. Tsumuraya, *Organometallics*, **8**, 2759 (1989).
27. (a) P. Jutzi, H. Schmit, B. Neumann and H. Stammler, *Organometallics*, **15**, 741 (1996).
(b) R. S. Simons, L. Pu, M. M. Olmstead and P. P. Power, *Organometallics*, **16**, 1920 (1997).
28. J. -C. Barthelat, B. S. Roch, G. Trinquier and J. Satgé, *J. Am. Chem. Soc.*, **102**, 4080 (1980).
29. (a) V. Y. Lee, M. Ichinohe, A. Sekiguchi, N. Takagi and S. Nagase, *J. Am. Chem. Soc.*, **122**, 9034 (2000).
(b) V. Y. Lee, M. Ichinohe and A. Sekiguchi, *J. Am. Chem. Soc.*, **122**, 12604 (2000).

30. (a) A. Sekiguchi, M. Tsukamoto and M. Ichinohe, *Science*, **275**, 60 (1997).
(b) M. Ichinohe, N. Fukaya and A. Sekiguchi, *Chem. Lett.*, 1045 (1998).
31. (a) G. Trinquier, J. -P. Malrieu and P. Rivière, *J. Am. Chem. Soc.*, **104**, 4529 (1982).
(b) S. Nagase and T. Kudo, *J. Mol. Struct. THEOCHEM*, **103**, 35 (1983).
(c) C. Liang and L. C. Allen, *J. Am. Chem. Soc.*, **112**, 1039 (1990).
(d) G. Trinquier, *J. Am. Chem. Soc.*, **112**, 2130 (1990).
(e) R. S. Grev, H. F. Schaefer III and K. M. Baines, *J. Am. Chem. Soc.*, **112**, 9458 (1990).
(f) T. L. Windus and M. S. Gordon, *J. Am. Chem. Soc.*, **114**, 9559 (1992).
(g) H. Jacobsen and T. Ziegler, *J. Am. Chem. Soc.*, **116**, 3667 (1994).
32. P. B. Hitchcock, M. F. Lappert, S. J. Miles and A. J. Thorne, *J. Chem. Soc., Chem. Commun.*, 480 (1984).
33. L. Ross and M. Dräger, *J. Organomet. Chem.*, **199**, 195 (1980).
34. M. Dräger and L. Ross, *Z. Anorg. Allg. Chem.*, **476**, 95 (1981).
35. (a) N. Tokitoh, H. Suzuki, R. Okazaki and K. Ogawa, *J. Am. Chem. Soc.*, **115**, 10428 (1993).
(b) H. Suzuki, N. Tokitoh, R. Okazaki, J. Harada, K. Ogawa, S. Tomoda and M. Goto, *Organometallics*, **14**, 1016 (1995).
36. (a) S. A. Batcheller and S. Masamune, *Tetrahedron Lett.*, **29**, 3383 (1988).
(b) T. Tsumuraya, Y. Kabe and W. Ando, *J. Chem. Soc., Chem. Commun.*, 1159 (1990).
37. T. Tsumuraya, S. Sato and W. Ando, *Organometallics*, **7**, 2015 (1988).
38. (a) W. Ando and T. Tsumuraya, *J. Chem. Soc., Chem. Commun.*, 770 (1989).
(b) W. Ando and T. Tsumuraya, *Organometallics*, **8**, 1467 (1989).
39. T. Tsumuraya, Y. Kabe and W. Ando, *J. Organomet. Chem.*, **482**, 131 (1994).
40. W. Ando and T. Tsumuraya, *Organometallics*, **7**, 1882 (1988).
41. R. S. Grev, *Adv. Organomet. Chem.*, **33**, 125 (1991).
42. (a) K. M. Baines, J. A. Cooke, C. E. Dixon, H. W. Liu and M. R. Netherton *Organometallics*, **13**, 631 (1994).
(b) K. M. Baines, J. A. Cooke and J. J. Vittal, *J. Chem. Soc., Chem. Commun.*, 1484 (1992).
43. (a) J. Escudié and H. Ranaivonjatovo, *Adv. Organomet. Chem.*, **44**, 113 (1999).
(b) P. P. Power, *Chem. Rev.*, **99**, 3463 (1999).
(c) J. Escudié, C. Couret, H. Ranaivonjatovo and J. Satgé, *Coord. Chem. Rev.*, **130**, 427 (1994).
(d) J. Escudié, C. Couret and H. Ranaivonjatovo, *J. Organomet. Chem.*, **178–180**, 565 (1998).
44. (a) C. Couret, J. Escudié, J. Satgé and M. Lazraq, *J. Am. Chem. Soc.*, **109**, 4411 (1987).
(b) H. Meyer, G. Baum, W. Massa and A. Berndt, *Angew. Chem., Int. Ed. Engl.*, **26**, 798 (1987).
(c) M. Lazraq, J. Escudié, C. Couret, J. Satgé, M. Dräger and R. Dammel, *Angew. Chem., Int. Ed. Engl.*, **27**, 828 (1988).
45. A. Berndt, H. Meyer, G. Baum, W. Massa and S. Berger, *Pure Appl. Chem.*, **59**, 1011 (1987).
46. E. G. Rochow and E. W. Abel, *The Chemistry of Germanium, Tin and Lead*, Pergamon, Oxford, 1975.
47. (a) J. Barrau, G. Rima and J. Satgé, *J. Organomet. Chem.*, **252**, C73 (1983).
(b) K. D. Dobbs and W. J. Hehre, *Organometallics*, **5**, 186 (1986).
48. C. Couret, J. Escudié, G. Delpon-Lacaze and J. Satgé, *Organometallics*, **11**, 3176 (1992).
49. N. Tokitoh, K. Kishikawa and R. Okazaki, *J. Chem. Soc., Chem. Commun.*, 1425 (1995).
50. H. Schumann, M. Glanz, F. Grigsdies, F. E. Hahn, M. Tamm and A. Grzegorzewski, *Angew. Chem., Int. Ed. Engl.*, **36**, 2232 (1997).
51. M. Larzraq, C. Couret, J. Escudié, J. Satgé and M. Dräger, *Organometallics*, **10**, 1771 (1991).
52. K. Kishikawa, N. Tokitoh and R. Okazaki, *Chem. Lett.*, 695 (1996).
53. A. G. Brook, S. C. Nyburg, F. Abdesaken, B. Gutekunst, G. Gutekunst, R. K. Kallury, Y. C. Poon, Y-M. Chang and W-N. Winnie, *J. Am. Chem. Soc.*, **104**, 5667 (1982).
54. N. Tokitoh, T. Matsumoto and R. Okazaki, *Chem. Lett.*, 1087 (1995).
55. For the reactivity of germylene **63**. see:
(a) N. Tokitoh, K. Manmaru and R. Okazaki, *Organometallics*, **13**, 167 (1994).
(b) N. Tokitoh, K. Kishikawa, T. Matsumoto and R. Okazaki, *Chem. Lett.*, 827 (1995).
56. M. S. Raasch, *J. Org. Chem.*, **35**, 3470 (1970).
57. (a) D. E. Goldberg, D. H. Harris, M. F. Lappert and K. M. Thomas, *J. Chem. Soc., Chem. Commun.*, 261(1976).
(b) P. J. Davidson, D. H. Harris and M. F. Lappert, *J. Chem. Soc., Dalton Trans.*, 2268 (1976).

(c) J. D. Cotton, P. J. Davidson and M. F. Lappert, *J. Chem. Soc., Dalton Trans.*, 2275 (1976).
(d) J. D. Cotton, P. J. Davidson, M. F. Lappert and J. D. Donaldson, *J. Chem. Soc., Dalton Trans.*, 2286 (1976).

58. K. W. Zilm, G. A. Lawless, R. M. Merril, J. M. Millar and G. G. Webb, *J. Am. Chem. Soc.*, **109**, 7236 (1987).
59. S. Masamune and L. R. Sita, *J. Am. Chem. Soc.*, **107**, 6390 (1985).
60. M. Weidenbruch, A. Schäfer, H. Kilian, S. Pohl and W. Saak, *Chem. Ber.*, **125**, 563 (1992).
61. M. Weidenbruch, H. Kilian, K. Peters and H. G. Schnering, *Chem. Ber.*, **128**, 983 (1995).
62. H. Meyer, G. Baum, W. Massa, S. Berger and A. Berndt, *Angew. Chem., Int. Ed. Engl.*, **26**, 546 (1987).
63. K. Dobbs and W. J. Hehre, *Organometallics*, **5**, 2057 (1986).
64. G. Anselme, H. Ranaivonjatovo, J. Escudié, C. Couret and J. Satgé, *Organometallics*, **11**, 2747 (1992).
65. G. Anselme, C. Couret, J. Escudié, S. Richelme and J. Satgé, *J. Organomet. Chem.*, **418**, 321 (1991).
66. A. Schäfer, M. Weidenbruch, W. Saak and S. Pohl, *J. Chem. Soc., Chem. Commun.*, 1157 (1995).
67. For reviews, see:
(a) P. G. Harrison, in *Comprehensive Organometallic Chemistry* (Eds. G. Wilkinson, F. G. A. Stone and E. A. Abel), Vol. 2. Pergamon, New York, 1982, p. 670.
(b) P. G. Harrison, in *Comprehensive Organometallic Chemistry II*, (Eds. G. Wilkinson, F. G. A. Stone and E. A. Abel) Vol. 2 Pergamon, New York, 1995, p. 305.
(c) P. G. Harrison, in *Comprehensive Coordination Chemistry*, (Ed. G. Wilkinson, Vol. Eds. R. D. Gillard and J. A. McCleverty), Vol. 3, Pergamon, Oxford, 1987, p. 185.
(d) E. W. Abel, in *Comprehensive Inorganic Chemistry*, (Eds. J. C. Bailar Jr., H. J. Emeleus, R. Nyholm and A. F. Trotman-Dickenson), Vol. 2, Pergamon, Oxford, 1973, p. 105.
68. (a) D. H. Harris and M. F. Lappert, *J. Chem. Soc., Chem. Commun.*, 895 (1974).
(b) M. J. S. Gyane, D. H. Harris, M. F. Lappert, P. P. Power, P. Rivière and M. Rivière-Baudet, *J. Chem. Soc., Dalton Trans.*, 2004 (1977).
(c) T. Fjeldberg, H. Hope, M. F. Lappert, P. P. Power and A. J. Thorne, *J. Chem. Soc., Chem. Commun.*, 639 (1983).
69. (a) N. Kano, N. Tokitoh and R. Okazaki, *Organometallics*, **16**, 4237 (1997).
(b) N. Kano, N. Tokitoh and R. Okazaki, *Phosphorus, Sulfur and Silicon*, **124–125**, 517 (1997).
70. M. Stürumann, M. Weidenbruch, K. W. Klinkhammer, F. Lissner and H. Marsmann, *Organometallics*, **17**, 4425 (1998).
71. (a) P. J. Davidson and M. F. Lappert, *J. Chem. Soc., Chem. Commun.*, 317 (1973).
(b) J. D. Cotton, P. J. Davidson, D. E. Goldberg, M. F. Lappert and K. M. Thomas, *J. Chem. Soc., Chem. Commun.*, 893 (1974).
72. K. W. Klinkhammer, unpublished results. See also, reference 1s.
73. C. Eaborn, T. Ganicz, P. B. Hitchcock, J. D. Smith and S. E. Sözerli, *Organometallics*, **16**, 5621 (1997).
74. (a) N. Kano, K. Shibata, N. Tokitoh and R. Okazaki, *Organometallics*, **18**, 2999 (1999).
(b) N. Kano, N. Tokitoh and R. Okazaki, *J. Synth. Org. Chem., Jpn. (Yuki Gosei Kagaku Kyokai Shi)*, **56**, 919 (1998) (In English).
75. K. W. Klinkhammer, T. F. Fässler and H. Grützmacher, *Angew. Chem. Int. Ed.*, **37**, 124 (1998).
76. G. Trinquier and J.-P. Marliew, *J. Am. Chem. Soc.*, **109**, 5303 (1987).
77. M. Stürumann, W. Saak, H. Marsmann and M. Weidenbruch, *Angew. Chem. Int. Ed.*, **38**, 187 (1999).
78. M. Strümann, W. Saak and M. Weidenbruch, *Z. Anorg. Alleg. Chem.*, **625**, 705 (1999).
79. M. Weidenbruch, S. Willms, W. Saak and G. Henkel, *Angew. Chem., Int. Ed. Engl.*, **36**, 2503 (1997).
80. H. Schäfer, W. Saak and M. Weidenbruch, *Angew. Chem. Int. Ed.*, **39**, 3703 (2000).
81. N. Tokitoh, N. Nakata and N. Takeda, unpublished results.
82. For reviews, see:
(a) F. Duus, in *Comprehensive Organic Chemistry* (Eds. D. H. R. Barton and W. D. Ollis), Vol. 3, Pergamon Press, Oxford, 1979, p. 373.

(b) V. A. Usov, L. V. Timokhina and M. G. Voronkov, *Sulfur Rep.*, **12**, 95 (1992).
(c) J. Voss, in *Houben-Weyl Methoden der Organischen Chemie* (Ed. D. Klamann), Band 11, George Thieme Verlag, Stuttgart 1985, p. 188.
(d) R. Okazaki, *Yuki Gosei Kagaku Kyokai Shi,* **46**, 1149 (1988).
(e) W. M. McGregor and D. C. Sherrington, *Chem. Soc. Rev.*, 199 (1993).
(f) W. G. Whittingham, in *Comprehensive Organic Functional Group Transformations* (Eds. A. R. Katritzky, O. Meth-Cohn and C. W. Rees), Vol. 3, Pergamon Press, Oxford, 1995, p. 329.

83. For reviews, see:
(a) P. D. Magnus, in *Comprehensive Organic Chemistry* (Eds. D. H. R. Barton and W. D. Ollis), Vol. 3, Pergamon Press, Oxford, 1979, p. 491.
(b) C. Paulmier, in *Selenium Reagents and Intermediates in Organic Synthesis* (Ed. C. Paulmier), Pergamon Press, Oxford, 1986, p. 58.
(c) F. S. Guziec, Jr., in *The Chemistry of Organic Selenium and Tellurium Compounds* (Ed. S. Patai), Vol. 2, Wiley, Chichester, 1987, p. 215.
(d) F. S. Guziec, Jr. and L. J. Guziec, in *Comprehensive Organic Functional Group Transformations*, (Eds. A. Katritzky, O. Meth-Cohn and C. W. Rees), Vol. 3, Pergamon Press, Oxford, 1995, p. 381.
84. M. Minoura, T. Kawashima and R. Okazaki, *J. Am. Chem. Soc.*, **115**, 7019 (1993).
85. (a) For silicon–chalcogen double bonds, see: N. Tokitoh and R. Okazaki, chap. 17 in *The Chemistry of Organosilicon Compounds Vol. 2*, Part 2 (Eds. Z. Rappoport and Y. Apeloig, Wiley, Chichester, 1998. p. 1063.
(b) For germanium–chalcogen doubly bonds, see: N. Tokitoh, T. Matsumoto and R. Okazaki, *Bull. Chem. Soc. Jpn.*, **72**, 1665 (1999).
(c) For review on stable double bonded compounds of germanium and tin, see: K. M. Baines and W. G. Stibbs, in *Advances in Organometallic Chemistry* (Eds. F. G. Stone and R. West), Vol. 39, Academic Press, San Diego, 1996, p. 275.
(d) G. Raabe and J. Michl, in *The Chemistry of Organic Silicon Compounds, Part 2*, (Eds. S. Patai and Z. Rappoport), Wiley, Chichester, 1989, p. 1015.
(e) R. West, *Angew. Chem., Int. Ed. Engl.*, **26**, 1201 (1987).
(f) L. E. Gusel'nikov and N. S. Nametkin, *Chem. Rev.*, **79**, 529 (1979).
86. (a) T. Kudo and S. Nagase, *J. Phys. Chem.*, **88**, 2833 (1984).
(b) T. Kudo and S. Nagase, *Organometallics*, **5**, 1207 (1986).
87. P. Arya, J. Boyer, F. Carré, R. Corriu, G. Lanneau, J. Lapasset, M. Perrot and C. Priou, *Angew. Chem., Int. Ed. Engl.*, **28**, 1016 (1989).
88. R. Okazaki and N. Tokitoh, *Acc. Chem. Res.*, **33**, 625 (2000).
89. (a) M. Veith, S. Becker and V. Huch, *Angew. Chem., Int. Ed. Engl.*, **28**, 1237 (1989).
(b) M. Veith, A. Detemple and V. Huch, *Chem. Ber.*, **124**, 1135 (1991).
(c) M. Veith and A. Detemple, *Phosphorus, Sulfur and Silicon*, **65**, 17 (1992).
90. (a) G. Trinquier, J. C. Barthelat and J. Satgé, *J. Am. Chem. Soc.*, **104**, 5931 (1982).
(b) G. Trinquier, M. Pelissier, B. Saint-Roch and H. Lavayssiere, *J. Organomet. Chem.*, **214**, 169 (1981).
91. M. C. Kuchta and G. Parkin, *J. Chem. Soc., Chem. Commun.*, 1351 (1994).
92. (a) N. Tokitoh, H. Suzuki, T. Matsumoto, Y. Matsuhashi, R. Okazaki and M. Goto, *J. Am. Chem. Soc.*, **113**, 7047 (1991).
(b) T. Matsumoto, N. Tokitoh, R. Okazaki and M. Goto, *Organometallics*, **14**, 1008 (1995).
93. N. Tokitoh, T. Matsumoto, H. Ichida and R. Okazaki, *Tetrahedron Lett.*, **32**, 6877 (1991).
94. N. Tokitoh, T. Matsumoto, K. Manmaru and R. Okazaki, *J. Am. Chem. Soc.*, **115**, 8855 (1993).
95. T. Matsumoto, N. Tokitoh and R. Okazaki, *J. Am. Chem. Soc.*, **121**, 8811 (1999).
96. N. Tokitoh, T. Matsumoto and R. Okazaki, *Tetrahedron Lett.*, **33**, 2531 (1992).
97. T. Matsumoto, N. Tokitoh and R. Okazaki, *Angew. Chem., Int. Ed. Engl.*, **33**, 2316 (1994).
98. (a) G. Erker and R. Hock, *Angew. Chem., Int. Ed. Engl.*, **28**, 179 (1989).
(b) M. Segi, T. Koyama, Y. Tanaka, T. Nakajima and S. Suga, *J. Am. Chem. Soc.*, **111**, 8749 (1989).
(c) R. Boese, A. Haas and C. Limberg, *J. Chem. Soc., Chem. Commun.*, 1378 (1991).
(d) A. Haas and C. Limberg, *Chimia*, **46**, 78 (1992).
(e) A. G. M. Barrett, D. H. R. Barton and R. W. Read, *J. Chem. Soc., Chem. Commun.*, 645 (1979).

(f) A. G. M. Barrett, R. W. Read and D. H. R. Barton, *J. Chem. Soc., Perkin Trans. 1*, 2191 (1980).
(g) T. Severengiz and W.-W. du Mont, *J. Chem. Soc., Chem. Commun.*, 820 (1987).
(h) K. A. Lerstrup and L. Henriksen, *J. Chem. Soc., Chem. Commun.*, 1102 (1979).
(i) M. F. Lappert, T. R. Martin and G. M. McLaughlin, *J. Chem. Soc., Chem. Commun.*, 635 (1980).
(j) M. Segi, A. Kojima, T. Nakajima and S. Suga, *Synlett*, **2**, 105 (1991).
(k) M. Minoura, T. Kawashima and R. Okazaki, *J. Am. Chem. Soc.*, **115**, 7019 (1993).
99. N. Tokitoh, T. Matsumoto and R. Okazaki, *J. Am. Chem. Soc.*, **119**, 2337 (1997).
100. T. Matsumoto, Ph.D. Thesis, University of Tokyo, 1994.
101. N. Tokitoh, K. Kishikawa, T. Matsumoto and R. Okazaki, *Chem. Lett.*, 827 (1995).
102. (a) A. Krebs and J. Berndt, *Tetrahedron Lett.*, **24**, 4083 (1983).
(b) M. P. Egorov, S. P. Kolesnikov, Yu. T. Struchkov, M. Yu Antipin, S. V. Sereda and O. M. Nefedov, *J. Organomet. Chem.*, **290**, C27 (1985).
103. N. Tokitoh and R. Okazaki, *Main Group Chemistry News*, **3**, 4 (1995).
104. G. Ossig, A. Meller, C. Brönneke, O. Müller, M. Schäfer and R. Herbst-Irmer, *Organometallics*, **16**, 2116 (1997).
105. M. Veith and A. Z. Rammo, *Z. Anorg. Allg. Chem.*, **623**, 861 (1997).
106. (a) J. Barrau, G. Rima and T. El Amraoui, *J. Organomet. Chem.*, **570**, 163 (1998).
(b) J. Barrau, G. Rima and T. El Amraoui, *Inorg. Chim. Acta*, **241**, 9 (1996).
107. S. R. Foley, C. Bensimon and D. S. Richeson, *J. Am. Chem. Soc.*, **119**, 10359 (1997).
108. (a) N. Tokitoh, M. Saito and R. Okazaki, *J. Am. Chem. Soc.*, **115**, 2065 (1993).
(b) Y. Matsuhashi, N. Tokitoh and R. Okazaki, *Organometallics*, **12**, 2573 (1993).
(c) M. Saito, N. Tokitoh and R. Okazaki, *Organometallics*, **15**, 4531 (1996).
109. M. Saito, N. Tokitoh and R. Okazaki, *J. Organomet. Chem.*, **499**, 43 (1995).
110. M. Saito, Ph.D. Thesis, University of Tokyo, 1997.
111. M. Saito, N. Tokitoh and R. Okazaki, *J. Am. Chem. Soc.*, **119**, 11124 (1997).
112. N. Kano, N. Tokitoh and R. Okazaki, *Chem. Lett.*, 277 (1997).
113. (a) N. Kano, N. Tokitoh and R. Okazaki, *Organometallics*, **16**, 4237 (1997).
(b) N. Kano, N. Tokitoh and R. Okazaki, *Phosphorus, Sulfur and Silicon*, **124–125**, 517 (1997).
114. (a) P. B. Hitchcock, M. F. Lappert, B. J. Samways and E. L. Weinburg, *J. Chem. Soc., Chem. Commun.*, 1492 (1983).
(b) S. C. Goel, M. Y. Chiang and W. E. Buhro, *Inorg. Chem.*, **29**, 4640 (1990).
(c) C. Eaborn, K. Izod, P. B. Hitchcock, S. E. Sözerli and J. D. Smith, *J. Chem. Soc., Chem. Commun.*, 1829 (1995).
115. (a) W. S. Sheldrick, in *The Chemistry of Organic Silicon Compounds* (Eds. S. Patai and Z. Rappoport), Part 1, Wiley, Chichester, 1989. pp. 227–304.
(b) See also R. K. Sibao, N. L. Keder and H. Eckert, *Inorg. Chem.*, **29**, 4163 (1990).
116. For a recent review, see Reference 43a.
117. (a) N. Tokitoh, T. Sadahiro, N. Takeda and R. Okazaki, *The 12th International Symposium on Organosilicon Chemistry, 4A23, Sendai, Japan (1999).*
(b) K. Hatano, T. Sadahiro, N. Tokitoh and R. Okazaki, *The 12th International Symposium on Organosilicon Chemistry, P-85 Sendai, Japan (1999).*
118. T. Kudo and S. Nagase, *Chem. Phys. Lett.*, **128**, 507 (1986).
119. V. N. Khabashesku, S. E. Boganov, P. S. Zuev and O. M. Nefedov, *J. Organomet. Chem.*, **402**, 161 (1991).
120. N. Kano, Ph.D. Thesis, The University of Tokyo, 1997.
121. A. Meller, G. Ossig, W. Maringgele, D. Stalk, R. Herbst-Irmer, S. Freitag and G. M. Sheldrick, *J. Chem. Soc., Chem. Commun.*, 1123 (1991).
(b) For a recent review, see Reference 1p.
122. H. Ramdane, H. Ranaivonjatovo and J. Escudié, *Organometallics*, **15**, 3070 (1996).
123. N. Tokitoh, K. Kishikawa and R. Okazaki, *Chem. Lett.*, 811 (1998).
124. (a) B. E. Eichler, D. R. Powell and R. West, *Organometallics*, **17**, 2147 (1998).
(b) B. E. Eichler, D. R. Powell and R. West, *Organometallics*, **18**, 540 (1999).
125. N. Wiberg, H.-W. Lerner, S.-K. Vasisht, S. Wagner, K. Karaghiosoff, H. Nöth and W. Ponikwar, *Eur. J. Inorg. Chem.*, 1211 (1999).
126. For recent reviews, see:

(a) J. Escudié, H. Ranaivonjatovo and L. Rigon, *Chem. Rev.*, **100**, 3639 (2000).
(b) B. E. Eichler and R. West, *Adv. Organomet. Chem.*, **46**, 1 (2001).
127. (a) N. Tokitoh, N. Takeda and R. Okazaki, *J. Am. Chem. Soc.*, **116**, 7907 (1994).
(b) N. Takeda, N. Tokitoh and R. Okazaki, *Chem. Eur. J.*, **3**, 62 (1997).
128. N. Takeda, N. Tokitoh and R. Okazaki, *Angew. Chem., Int. Ed. Engl.*, **35**, 660 (1996).

CHAPTER **14**

Unsaturated three-membered rings of heavier Group 14 elements

VLADIMIR YA. LEE and AKIRA SEKIGUCHI

Department of Chemistry, University of Tsukuba, Tsukuba, Ibaraki 305-8571, Japan
Fax: (+81)-298-53-4314; e-mail: sekiguch@staff.chem.tsukuba.ac.jp

I. INTRODUCTION

The chemistry of stable, small cyclic compounds consisting of heavy Group 14 elements (i.e. Si, Ge and Sn) has a relatively short but very impressive history[1–5]. In contrast to their transient congeners, which were proposed a long time ago, the stable three-membered

The chemistry of organic germanium, tin and lead compounds — Vol. 2
Edited by Z. Rappoport

ring compounds were reported for the first time only in 1982 by Masamune's group[6]. After the synthesis of the first representatives of cyclotrisilanes[6], cyclotrigermenes[7] and cyclotristannanes[8], the chemistry of such molecules has been greatly developed during the past two decades. These compounds, similar to their carbon analogue — cyclopropane, were found to be very reactive because of their significant ring strain and the weakness of the endocyclic metal–metal bonds. The most fundamental discovery of their reactivity was the cycloelimination reaction to produce two kinds of key reactive species — heavy carbene analogues and dimetallaalkenes (dimetallenes)[1]. This is now a well-established way to generate such reactive intermediates. Heterocyclotrimetallanes consisting of different heavier Group 14 elements can be easily imagined as a convenient source for the preparation of heterodimetallenes, which can possess unusual structural and chemical properties. Nevertheless, until now there have been only a few examples of such 'mixed' cyclotrimetallanes[9–12], which can be explained by the difficulties in their preparation relative to their 'homo' analogues.

Cyclotrimetallenes, compounds that combine the cyclotrimetallane skeleton and an endocyclic metal–metal double bond in one molecule, have not been synthesized until quite recently because of the great ring strain and high reactivity of the metal–metal double bond. Nevertheless, employment of bulky silyl substituents, which sterically protect the molecule and decrease the ring strain, makes possible the successful preparation of the cyclotrimetallenes. After the first report of the synthesis of cyclotrimetallenes[13], there was an explosive growth in the development of their chemistry[14–20]. However, even now there is a very limited number of methods for their synthesis, which are not general and usually more complicated than in the case of cyclotrimetallanes. The reactivity of cyclotrimetallenes, which combine the chemical properties of both cyclotrimetallanes and dimetallenes, was found to be very rich. Various addition and cycloaddition reactions across the metal–metal double bond give access to new cyclic and bicyclic compounds, whereas insertion reactions into the three-membered ring produce ring enlargement products[21–23]. In the present review, we will concentrate only on the very recent developments in the chemistry of cyclotrimetallenes (unsaturated three-membered ring systems) of heavier Group 14 elements. Special attention will be paid to the chemistry of the cyclotrigermenium cation — the free germyl cation with a 2π-electron system[24]. The chemistry of cyclotrimetallanes, that is, their saturated analogues, will not be considered in this article, since it has already been described in previous reviews[1–5].

II. CYCLOTRIMETALLENES – UNSATURATED THREE-MEMBERED RINGS OF HEAVIER GROUP 14 ELEMENTS

A. Cyclotrisilenes

The synthesis of the first cyclotrisilenes has required a longer time than for the cyclotrigermenes due to the lack of suitable stable silylenes, in contrast, for example, to the well-known dichlorogermylene–dioxane complex[25]. Therefore, the preliminary preparation of the silylene precursors, which can generate silylenes *in situ*, was necessary for the successful synthesis of cyclotrisilenes. Until now, only two examples of cyclotrisilenes have been reported in the literature, of which only one was structurally characterized.

These first reports on the preparation of the stable cyclotrisilenes appeared in 1999, when two groups independently published consecutive papers on the cyclotrisilenes bearing different substituents. Cyclotrisilene **1**, which was prepared in Kira's group, has three *tert*-butyldimethylsilyl substituents and a bulky tris(*tert*-butyldimethylsilyl)silyl group attached to one unsaturated silicon atom[14]. This compound was obtained by the reaction

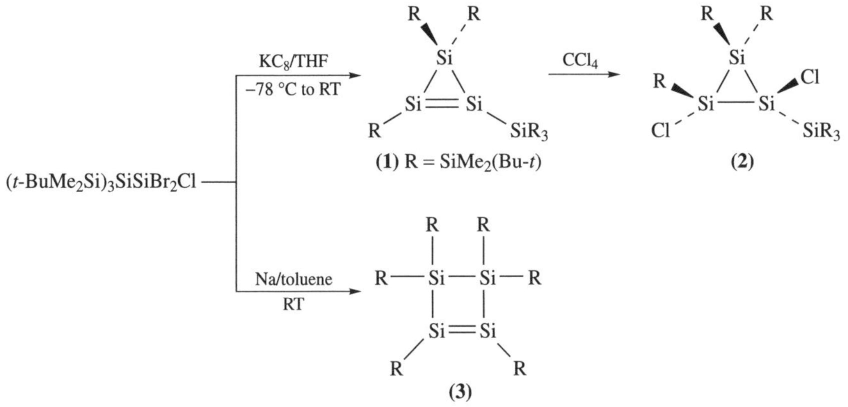

$(t\text{-BuMe}_2\text{Si})_3\text{SiSiBr}_2\text{Cl}$ $\xrightarrow{\text{KC}_8\text{/THF, } -78\,^\circ\text{C to RT}}$ (1) R = $\text{SiMe}_2(\text{Bu-}t)$ $\xrightarrow{\text{CCl}_4}$ (2)

$(t\text{-BuMe}_2\text{Si})_3\text{SiSiBr}_2\text{Cl}$ $\xrightarrow{\text{Na/toluene, RT}}$ (3)

SCHEME 1

of $R_3SiSiBr_2Cl$ ($R = SiMe_2Bu$-t) with potassium graphite in THF at $-78\,^\circ C$ in 11% yield as dark red crystals (Scheme 1). The existence of the doubly-bonded silicon atoms was determined from the ^{29}Si NMR spectrum, which showed two down-field signals at 81.9 and 99.8 ppm. These are significantly shifted up-field relative to those for the acyclic tetrasilyldisilenes (142–154 ppm)[26]. The structure of **1** was proved by reaction with CCl_4, for which the structure of the product **2** was established by X-ray crystallography. It is interesting that the final product of the reduction of $R_3SiSiBr_2Cl$ depends strongly on the reducing reagent. Thus, treatment of $R_3SiSiBr_2Cl$ with sodium in toluene at room temperature gave the cyclotetrasilene **3**[27] in 64% yield without any formation of **1**.

The second example of a cyclotrisilene was reported by Sekiguchi's group[15]. This compound was prepared by the reductive coupling of R_2SiBr_2 and $RSiBr_3$ [$R = SiMe(Bu$-$t)_2$] with sodium in toluene at room temperature (Scheme 2). Cyclotrisilene **4** was isolated as red-orange crystals in 9% yield. The ^{29}Si NMR spectrum of **4** showed a down-field signal at 97.7 ppm, which is attributable to the unsaturated silicon atoms, and an up-field signal at -127.3 ppm, which is typical for a saturated silicon atom in a three-membered ring system.

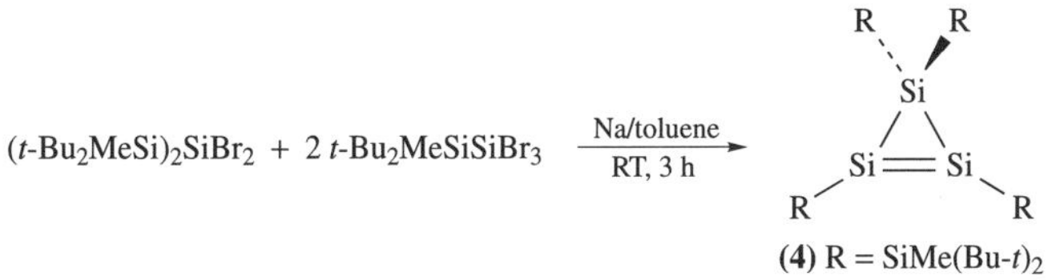

$(t\text{-Bu}_2\text{MeSi})_2\text{SiBr}_2 + 2\ t\text{-Bu}_2\text{MeSiSiBr}_3 \xrightarrow{\text{Na/toluene, RT, 3 h}}$ (4) R = $\text{SiMe}(\text{Bu-}t)_2$

SCHEME 2

Cyclotrisilene **4** has a symmetrical structure (C_{2v} symmetry), which allowed the growth of a single crystal and the determination of its crystal structure (Figure 1). X-ray analysis has revealed an almost isosceles triangle with bond angles of 62.8(1), 63.3(1) and 53.9(1)°. The geometry around the Si=Si double bond is not planar, but *trans*-bent with a torsion angle Si4–Si1–Si2–Si5 of 31.9(2)°. The Si=Si double bond length in **4** is 2.138(2) Å, which was recognized as one of the shortest distances among the Si=Si double bond

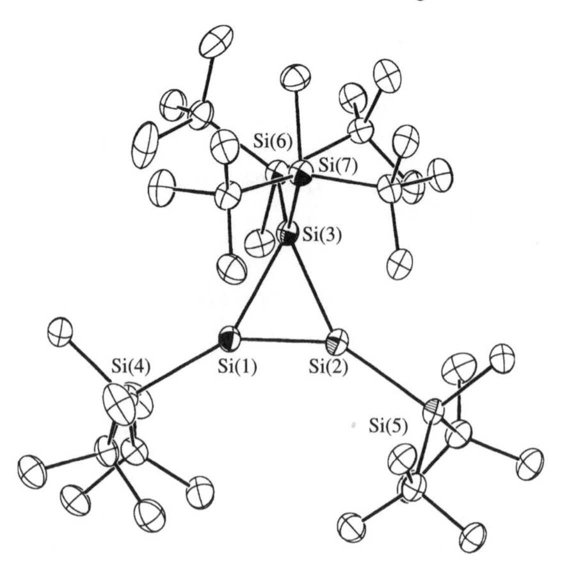

FIGURE 1. ORTEP drawing of cyclotrisilene **4**. Reproduced by permission of Wiley-VCH from Reference 15

lengths reported thus far[28]. The UV-Vis spectrum of **4** showed four bands with maxima at 223 ($\varepsilon = 7490$), 259 (3610), 297 (1490) and 466 nm (440).

B. Cyclotrigermenes

Historically, cyclotrigermenes were the first cyclotrimetallenes of Group 14 elements to be prepared. At present, six cyclotrigermenes have been described in the scientific literature, and four of them have been structurally characterized. In 1995, Sekiguchi and coworkers reported the unexpected formation of an unsaturated three-membered ring system consisting of Ge atoms by the reaction of dichlorogermylene–dioxane complex with tris(*tert*-butyl)silyl sodium or tris(*tert*-butyl)germyl lithium at $-70\,^{\circ}$C in THF (Scheme 3)[13]. Cyclotrigermenes **5** and **6** were isolated as dark-red crystals by gel-permeation chromatography. The NMR spectra of cyclotrigermenes are very simple, because of their symmetrical structures, showing only two sets of signals in ^{1}H, ^{13}C and ^{29}Si NMR spectra. The structures of both **5** and **6** were confirmed by X-ray crystallography, which showed the completely planar geometry around the Ge=Ge double

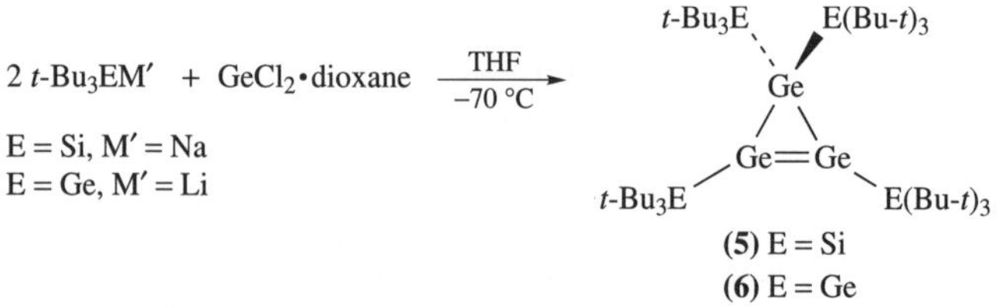

$$2\ t\text{-Bu}_3\text{EM}' + \text{GeCl}_2\cdot\text{dioxane} \xrightarrow[-70\,^{\circ}\text{C}]{\text{THF}}$$

E = Si, M′ = Na
E = Ge, M′ = Li

t-Bu$_3$E, E(Bu-*t*)$_3$, Ge, Ge=Ge, *t*-Bu$_3$E, E(Bu-*t*)$_3$

(**5**) E = Si
(**6**) E = Ge

SCHEME 3

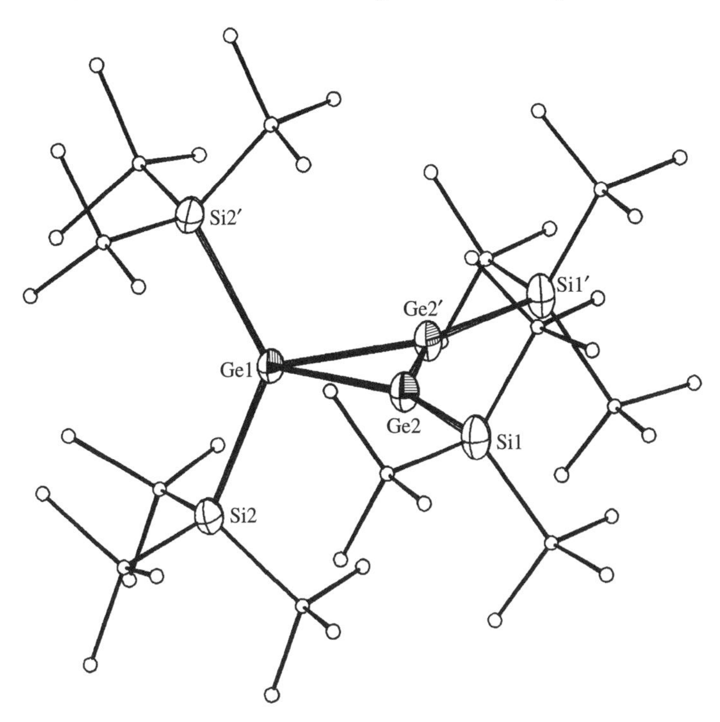

FIGURE 2. ORTEP drawing of cyclotrigermene **5**. Reprinted with permission from Reference 13. Copyright 1995 American Chemical Society

bond with a Ge=Ge double bond length of 2.239(4) Å (for **5** see Figure 2). Such planarity is somewhat unusual, since all digermenes reported before have a *trans*-bent configuration of the Ge=Ge double bond with folding angles of 12–36°[29–32].

The mechanism of the formation of a three-membered unsaturated ring was clarified later, when this reaction was reexamined by the same authors in detail[18]. They found that the reaction of dichlorogermylene–dioxane complex with one equivalent of t-Bu_3SiNa in THF at $-78\,^{\circ}C$ led to the formation of *cis,trans*-1,2,3-trichloro-1,2,3-tris(tri-*tert*-butylsilyl)cyclotrigermene **7** in 98% yield (Scheme 4). The *cis,trans* conformation of **7** was established by NMR spectroscopy and X-ray analysis (Figure 3). Treatment of **7** with two equivalents of t-Bu_3SiNa in THF at $-78\,^{\circ}C$ cleanly produced cyclotrigermene **5** (Scheme 4), which gives evidence that **7** is a precursor for **5**.

Monitoring of the reaction by ^{29}Si NMR allowed the detection of the reaction intermediates, which cannot be isolated, but were evidenced by trapping reactions (Scheme 5 and Figure 4). The first intermediate—digermenoid **8** (two signals at 21.9 and 23.7 ppm in the ^{29}Si NMR spectrum)—was quenched with hydrochloric acid at $-78\,^{\circ}C$ with the formation of a protonated product **9**. With iodomethane, a methylated product **10** was quantitatively obtained. Above $-8\,^{\circ}C$ the digermenoid **8** undergoes selective β-elimination to give another intermediate—digermene t-$Bu_3Si(Cl)Ge{=}Ge(Cl)Si(Bu$-$t)_3$ **11**, which cannot be seen in the ^{29}Si NMR spectrum, but can be trapped with dienes (Scheme 5). With both isoprene and 2,3-dimethyl-1,3-butadiene the corresponding cyclohexene derivatives **12** and **13** were obtained, whereas germacyclopentene derivatives were not found, which

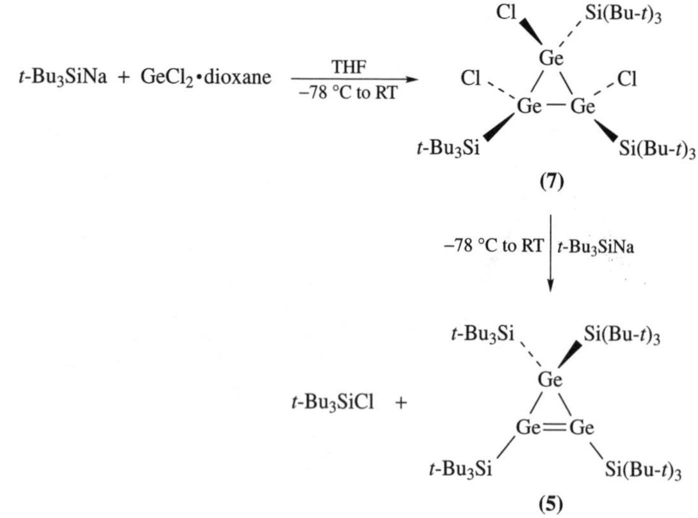

t-Bu_3SiNa + $GeCl_2$•dioxane $\xrightarrow[-78\ °C\ to\ RT]{THF}$ (7)

(7) $\xrightarrow[-78\ °C\ to\ RT]{t\text{-}Bu_3SiNa}$ t-Bu_3SiCl + (5)

(7): Ge, Ge—Ge, Cl, Cl, Cl, $Si(Bu$-$t)_3$, $Si(Bu$-$t)_3$, t-Bu_3Si

(5): Ge, Ge=Ge, t-Bu_3Si, $Si(Bu$-$t)_3$, t-Bu_3Si, $Si(Bu$-$t)_3$

SCHEME 4

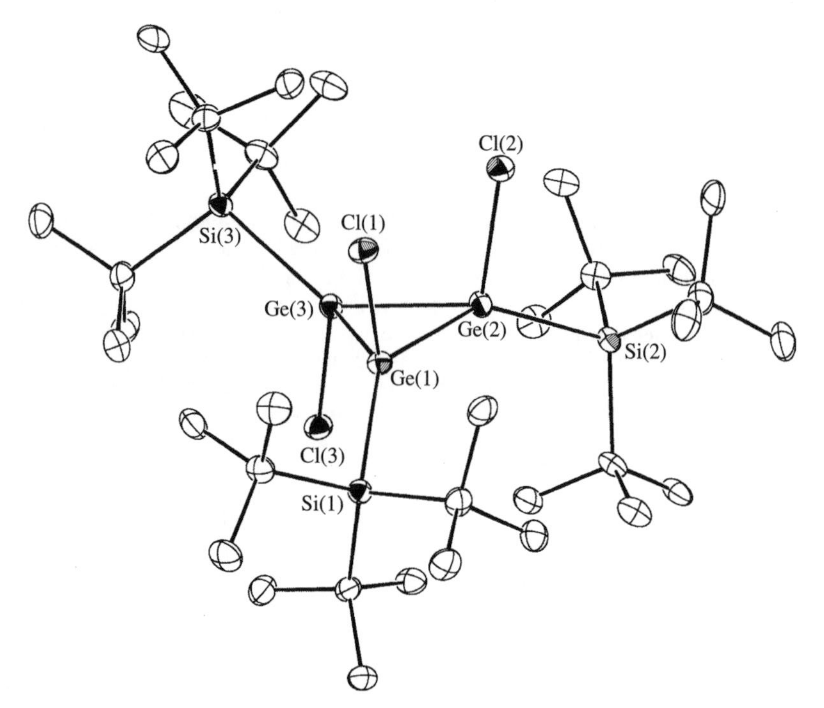

FIGURE 3. ORTEP drawing of cyclotrigermene **7**. Reprinted with permission from Reference 18. Copyright 2000 American Chemical Society

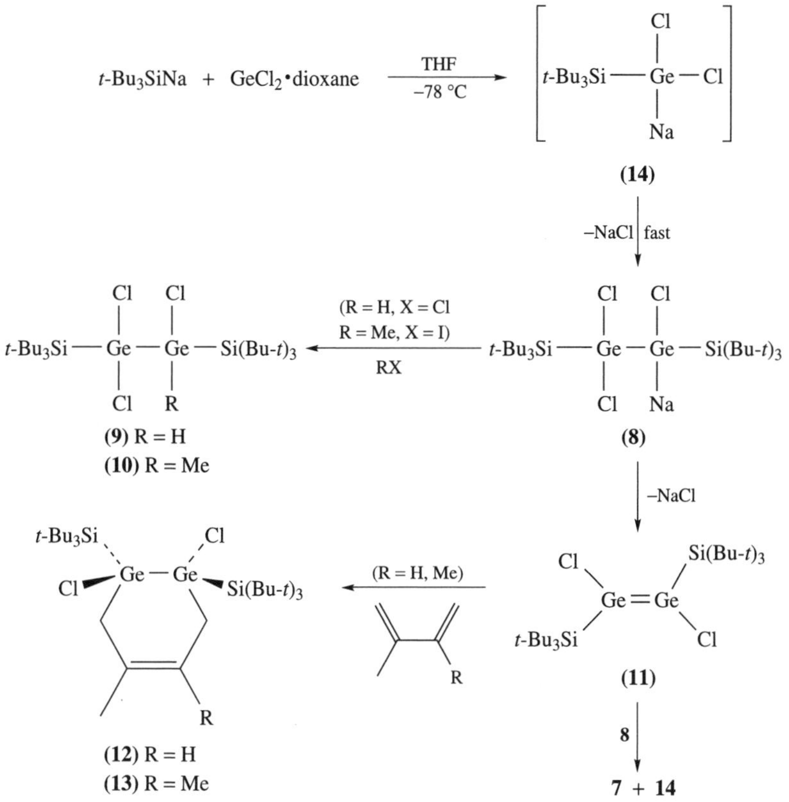

t-Bu$_3$SiNa + GeCl$_2$•dioxane —THF, −78 °C→ [t-Bu$_3$Si—GeCl$_2$—Na] (14)

(14) —−NaCl, fast→ t-Bu$_3$Si—GeCl$_2$—GeCl(Na)—Si(Bu-t)$_3$ (8)

(8) —RX (R = H, X = Cl; R = Me, X = I)→ t-Bu$_3$Si—GeCl$_2$—GeClR—Si(Bu-t)$_3$ (9) R = H; (10) R = Me

(8) —−NaCl→ (Cl)(t-Bu$_3$Si)Ge=Ge(Si(Bu-t)$_3$)(Cl) (11)

(11) + diene (R = H, Me) → (12) R = H; (13) R = Me

(11) —8→ 7 + 14

SCHEME 5

indicates that α-elimination to generate a germylene species is not involved in the reaction pathway. Thus, the reaction of $GeCl_2$•dioxane with t-Bu$_3$SiNa can be explained by the following mechanism. First, insertion of dichlorogermylene into the Si–Na bond occurred to form the germylenoid t-Bu$_3$SiGeCl$_2$Na **14**, which undergoes self-condensation to afford digermenoid **8**, stable at low temperature. Second, above −8 °C thermal decomposition of **8** gives digermene **11** as an intermediate. And finally, this digermene reacts with digermenoid **8**, followed by cyclization with formation of trichlorocyclotrigermene **7** and germylenoid **14**. The resulting **7** then transforms into the final cyclotrigermene **5**.

Other examples of cyclotrigermenes were synthesized in a different way by Sekiguchi and coworkers, taking advantage of the previously prepared cyclotrigermenium cation, whose synthesis will be described later (see Section II.E.3). Thus, unsymmetrically substituted cyclotrigermenes were prepared by the reactions of tris(tri-*tert*-butylsilyl)cyclotrigermenium tetrakis[3,5-bis(trifluoromethyl)phenyl]borate (TFPB$^-$) **15** with the appropriate nucleophiles (Scheme 6)[17]. This method seems to be a convenient route for the preparation of new cyclotrigermenes. Thus, reaction of **15** with t-Bu$_3$SiNa, t-Bu$_3$GeNa, (Me$_3$Si)$_3$SiLi•3THF, (Me$_3$Si)$_3$GeLi•3THF or MesLi at −78 °C quickly produced the corresponding unsymmetrically substituted cyclotrigermenes **5**, **16–19** in high yields. It was quite interesting to know that the geometry around the Ge=Ge

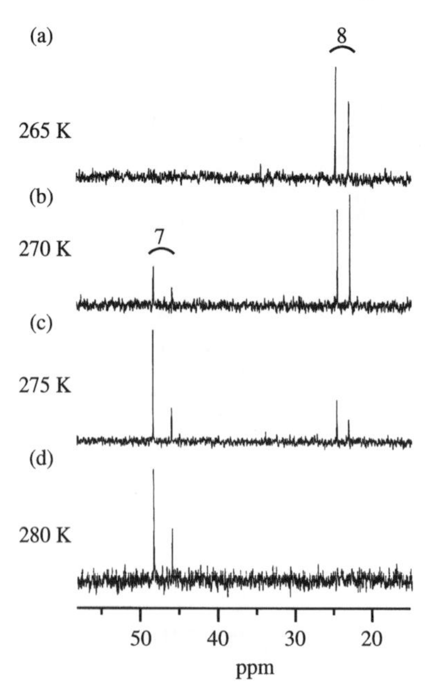

FIGURE 4. Reaction intermediate in the reaction of $GeCl_2$·dioxane with t-Bu_3SiNa in THF-d_8 monitored by ^{29}Si NMR spectroscopy: formation of digermenoid **8** and its thermal transformation to a cyclotrigermene **7**. Reprinted with permission from Reference 18. Copyright 2000 American Chemical Society

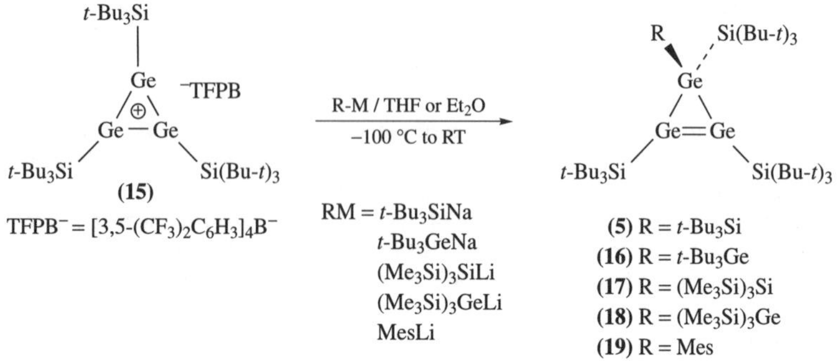

t-Bu_3Si Ge ⊕ Ge—Ge $^{-}$TFPB t-Bu_3Si Si(Bu-t)$_3$

(15)

$TFPB^- = [3,5\text{-}(CF_3)_2C_6H_3]_4B^-$

R-M / THF or Et_2O

−100 °C to RT

RM = t-Bu_3SiNa
t-Bu_3GeNa
$(Me_3Si)_3SiLi$
$(Me_3Si)_3GeLi$
MesLi

R Si(Bu-t)$_3$ Ge Ge=Ge t-Bu_3Si Si(Bu-t)$_3$

(5) R = t-Bu_3Si
(16) R = t-Bu_3Ge
(17) R = $(Me_3Si)_3Si$
(18) R = $(Me_3Si)_3Ge$
(19) R = Mes

SCHEME 6

double bond in **17** is not *trans*-bent, which is the general case for the dimetallenes of Group 14[29–32], but *cis*-bent with folding angles of 12.5° for the Ge3 atom and 4.4° for the Ge2 atom (Figure 5). The Ge=Ge double bond length is 2.264 Å, which lies in the normal region. Such an unusual *cis*-bent configuration was well reproduced by *ab initio* calculations: the *cis* folding angles are 8.8° and 5.8° for Ge3 and Ge2 atoms, respectively. No energy minimum was found for the *trans*-bent form. Therefore, the *cis*-bent geometry is caused by both steric and electronic effects of the substituents at the endocyclic saturated

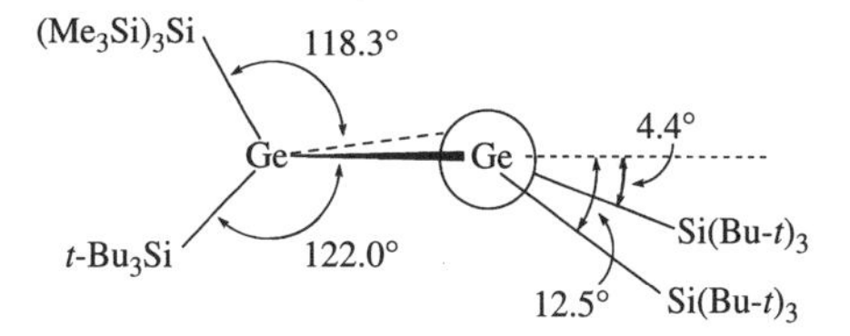

FIGURE 5. *cis*-bent geometry around Ge=Ge double bond of cyclotrigermene **17**. Reprinted with permission from Reference 17. Copyright 1999 American Chemical Society

germanium atom, but not by a crystal packing force. As was mentioned above, usually the digermenes exhibit a *trans*-bent configuration, which becomes less pronounced upon the introduction of electropositive substituents, such as R_3Si groups[33]. In the case of cyclotrigermene **17**, using such electropositive and bulky silyl substituents caused the formation of a Ge=Ge double bond with an unprecedented *cis*-bent geometry.

C. Cyclotristannene

Until now, there was only one example of a cyclotristannene — tetrakis(tri-*tert*-butylsilyl)cyclotristannene **20** reported by Wiberg and coworkers in 1999[16]. The title compound **20** was prepared by the reaction of t-Bu_3SiNa with stable stannylenes $:Sn[N(SiMe_3)_2]_2$ or $:Sn(OBu$-$t)_2$ in pentane at room temperature (Scheme 7). It was isolated as dark, red-brown crystals in 27% yield. In agreement with the symmetrical structure of the molecule, **20** exhibits only two sets of signals in the ^{1}H, ^{13}C and ^{29}Si NMR spectra. The most important and informative was the ^{119}Sn NMR spectrum, which showed both up-field (−694 ppm) and down-field (412 ppm) resonances. The last one is typical for the three-coordinated doubly-bonded tin atoms[34,35], whereas the first one can be assigned to an endocyclic Sn atom in a three-membered ring system[8,34].

The structure of **20** was determined by X-ray analysis, which showed an almost planar environment around the Sn=Sn double bond. All the previously reported distannenes have a *trans*-bent configuration around the Sn=Sn double bond[34–38]. The Sn=Sn double bond

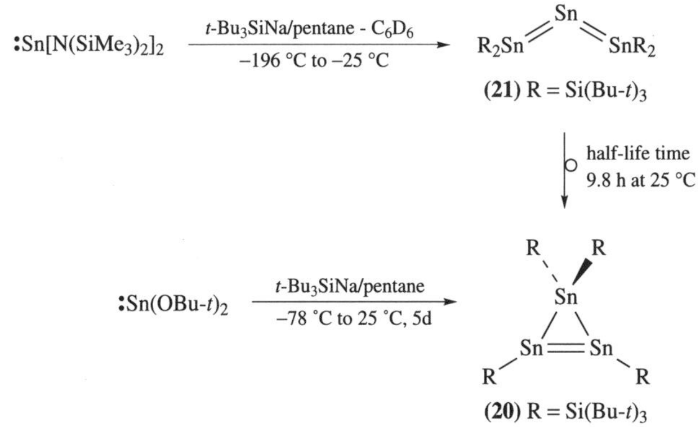

$:Sn[N(SiMe_3)_2]_2$ —(t-Bu_3SiNa/pentane - C_6D_6; −196 °C to −25 °C)→ $R_2Sn=Sn=SnR_2$

(21) R = Si(Bu-*t*)$_3$

half-life time 9.8 h at 25 °C

$:Sn(OBu$-$t)_2$ —(t-Bu_3SiNa/pentane; −78 °C to 25 °C, 5d)→

(20) R = Si(Bu-*t*)$_3$

SCHEME 7

length was only 2.59 Å — the shortest among all distannenes structurally characterized so far[34–38].

One of the most important findings of the reaction was a discovery of the intermediately formed tristannaallene **21**, which is not very thermally stable and at room temperature rearranges to the isomeric cyclotristannene **20** (Scheme 7). Such rearrangement apparently implies the migration of one silyl substituent followed by cyclization. Nevertheless, it was possible to isolate compound **21** in 20% yield as dark-blue crystals, which were highly air and moisture sensitive and isomerized slowly at room temperature to form cyclotristannene **20**. Such isomerization takes place by a first-order reaction with a half-life of 9.8 h at 25 °C. The structure of the tristannaallene **21** was supported by ^{119}Sn and ^{29}Si NMR spectra. In the ^{119}Sn NMR spectrum of **21**, two down-field signals at 503 and 2233 ppm were observed with an intensity ratio 2 : 1. The first one was attributed to the terminal Sn atoms in an allene unit, which is in agreement with the only reported distannene compound that is stable at room temperature in solution, (2,4,6-(*i*-Pr)$_3$C$_6$H$_2$)$_2$Sn=Sn(2,4,6-(Pr-*i*)$_3$C$_6$H$_2$)$_2$ ($\delta = 427.3$)[34]. The central allenic Sn atom, which formally has sp type hybridization, resonates at a much lower field (2233 ppm), which is the usual shift for the stannylenes[39–41]. Such similarity suggests that the central tin atom in **21** has a considerable stannylene character, therefore the bonding situation in **21** is best described by the resonance formula shown in Scheme 8. The ^{29}Si NMR spectrum of **21** showed only one signal at 77.3 ppm.

The crystal structure of **21** was finally confirmed by X-ray analysis, which showed an unexpectedly short Sn=Sn double bond length of 2.68 Å (Figure 6). This bond length is the shortest among all distannenes reported thus far (2.77–2.91 Å)[34–38], although it is longer than that in **20** (2.59 Å). Another important feature of the allene **21** is a bent, rather

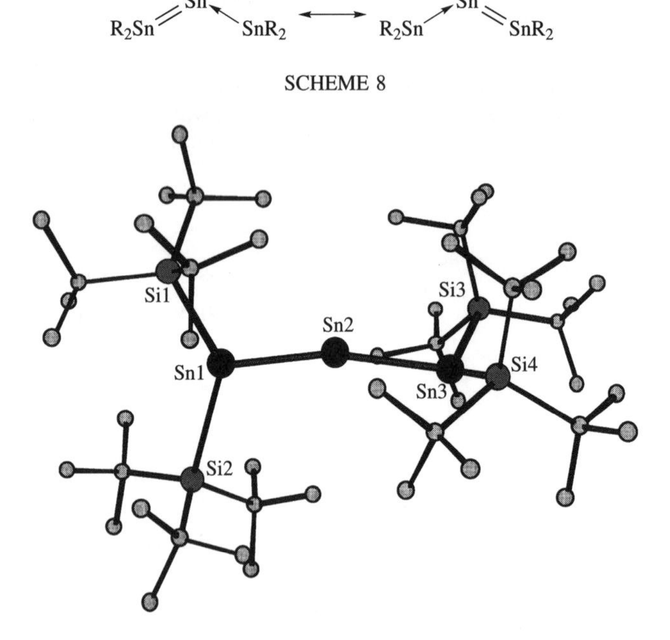

SCHEME 8

FIGURE 6. Molecular structure of tristannaallene **21**

than linear, Sn_3 chain. Such a phenomenon was explained by the large contribution of the resonance formulae [$R_2Sn \rightarrow SnR_2 \leftrightarrow R_2Sn \leftarrow SnR_2$] in the structure of **21** rather than the real allenic structure. The geometry around the terminal tin atoms in the Sn_3 chain in **21** is not planar, as in the case of the corresponding carbon atoms in allenes. The $(t\text{-}Bu_3Si)_2Sn$ groups in **21** adopt, similar to the R_2C groups in allenes, a *gauche* configuration; thus the four $t\text{-}Bu_3Si$ groups mask the Sn_3 chain almost completely to prevent intermolecular reactions between the Sn_3 units.

D. 'Mixed' Cyclotrimetallenes

It was expected that 'mixed' cyclotrimetallenes — that is, the three-membered ring compounds consisting of different Group 14 elements — would possess some interesting and unusual properties that may distinguish them from their homonuclear analogues (cyclotrisilenes, cyclotrigermenes and cyclotristannenes). First, such interest concerned the structural characteristics and the specific reactivity of the 'mixed' cyclotrimetallenes. The synthesis of the title compounds was not obvious, and until now only two examples of such molecules have been reported. In 2000, Sekiguchi and coworkers reported the first representatives of such unsaturated molecules: 1- and 2-disilagermirenes **22** and **23**[19]. 1-Disilagermirene **22** was prepared by the Würtz-type reductive coupling reaction of R_2GeCl_2 and $RSiBr_3$ [R = SiMe(*t*-Bu)$_2$] with sodium in toluene (Scheme 9)[19,20]. Compound **22** was isolated as hexagonal ruby crystals in 40% yield and appeared to be highly air and moisture sensitive. The ^1H and ^{13}C NMR spectra corresponded well with a symmetrical structure for **22**, showing only two sets of signals for methyl- and *tert*-butyl groups, whereas the ^{29}Si NMR spectrum displayed three resonances at 18.7, 25.6 and 107.8 ppm, of which the first two belong to the silyl substituents, and the last one is characteristic of the Si=Si double bond.

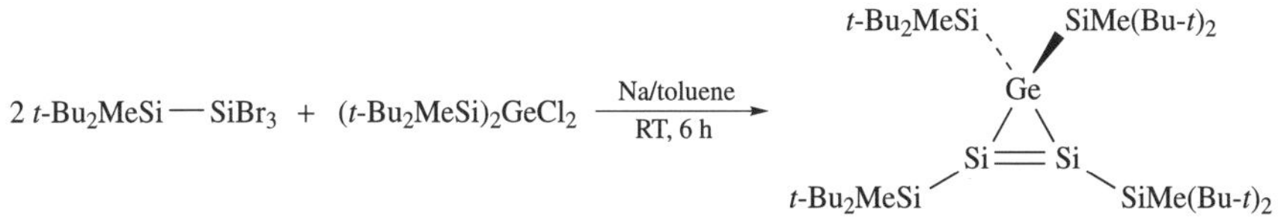

2 *t*-Bu$_2$MeSi—SiBr$_3$ + (*t*-Bu$_2$MeSi)$_2$GeCl$_2$ —(Na/toluene, RT, 6 h)→

t-Bu2MeSi SiMe(Bu-t)2
Ge
Si=Si
t-Bu2MeSi SiMe(Bu-t)2
(22)

SCHEME 9

The molecular structure of 1-disilagermirene **22** was determined by X-ray crystallography (Figure 7). The three-membered ring represents an almost isosceles triangle with bond angles of 52.71(3), 63.76(3) and 63.53(3)°. The silicon–silicon double bond length of **22** is 2.146(1) Å, which is rather short compared with other Si=Si double bond lengths reported so far (2.138–2.289 Å)[28,42]. The average bond length between Ge and the two Si atoms in the ring is 2.417 Å, which is intermediate between the endocyclic Ge–Ge bond length of 2.522 (4) in cyclotrigermene[13] and the Si–Si bond length of 2.358 (3) Å in cyclotrisilene[15]. The geometry around the Si=Si bond is not planar, but *trans*-bent with a torsional angle Si3–Si1–Si2–Si4 of 37.0(2)°. One of the possible reasons for such twisting of the Si=Si double bond may be the eclipsed conformation of the two $t\text{-}Bu_2MeSi$ substituents connected to the unsaturated silicon atoms.

Under photolysis of the benzene solution of **22** with a high pressure Hg lamp ($\lambda >$ 300 nm) for 4 h, a migration of the silyl substituent with the formation of an endocyclic Si=Ge double bond system takes place (Scheme 10). The reaction proceeds quite cleanly

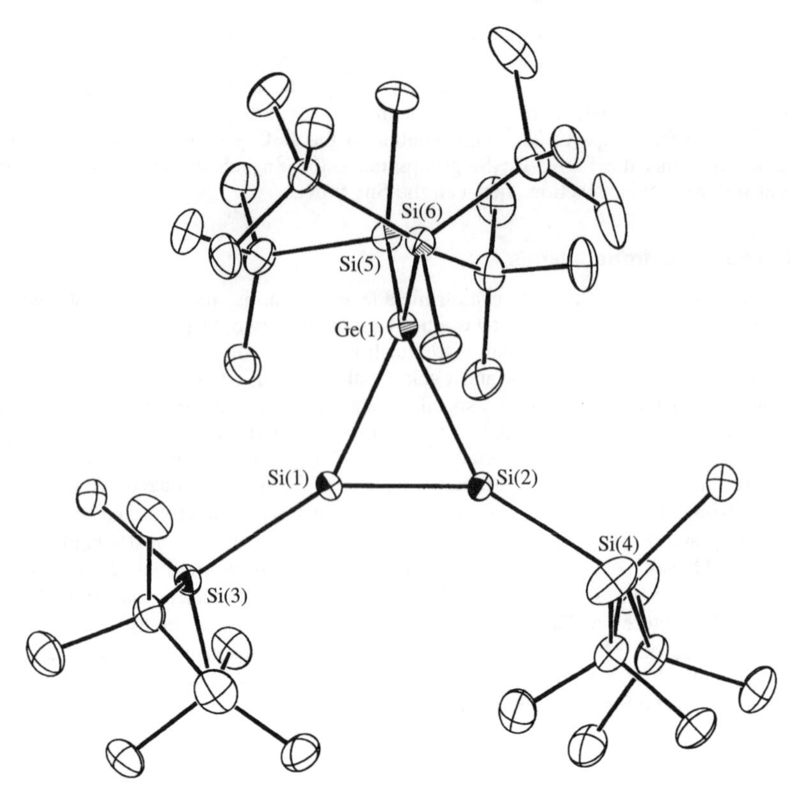

FIGURE 7. ORTEP drawing of 1-disilagermirene **22**. Reprinted with permission from Reference 19. Copyright 2000 American Chemical Society

and the 2-disilagermirene **23** was formed almost quantitatively. Compound **23** represents the first example of a stable germasilene reported to date, since the only previously reported example by Baines and Cooke, tetramesitylgermasilene, is unstable and can survive only at a low temperature[43].

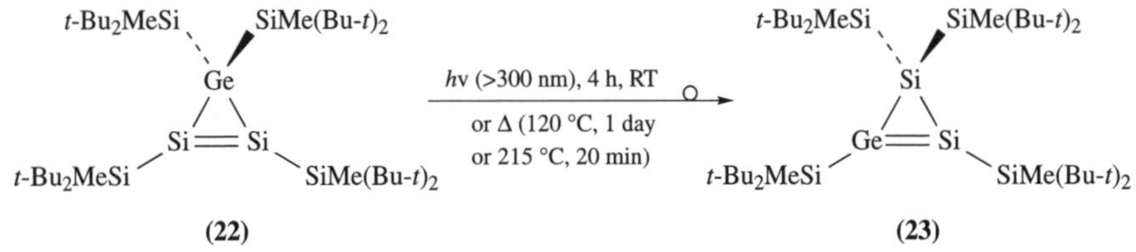

t-Bu$_2$MeSi, SiMe(Bu-*t*)$_2$ — Ge — Si=Si — *t*-Bu$_2$MeSi, SiMe(Bu-*t*)$_2$ **(22)**

$h\nu$ (>300 nm), 4 h, RT
or Δ (120 °C, 1 day
or 215 °C, 20 min)

t-Bu$_2$MeSi, SiMe(Bu-*t*)$_2$ — Si — Ge=Si — *t*-Bu$_2$MeSi, SiMe(Bu-*t*)$_2$ **(23)**

SCHEME 10

The 2-disilagermirene **23** was isolated from hexane solution as scarlet plate crystals and appeared to be extremely thermally stable with a melting point of 194–196 °C. The ^{1}H and ^{13}C NMR spectra of **23** are more complicated than those of **22**, because the 2-disilagermirene **23** has lost the C_{2v} symmetry of the 1-disilagermirene **22**. Thus, the ^{1}H NMR spectrum of **23** showed three resonances for three types of methyl groups and four resonances for non-equivalent *tert*-butyl groups, whereas the ^{13}C NMR spectrum showed three sets of signals both for methyl and *tert*-butyl groups. The ^{29}Si NMR spectrum showed five signals, of which three belong to the silyl substituents, 39.5, 27.8 and 6.9; the endocyclic doubly-bonded Si atom exhibits a down-field resonance at 106.7 and the endocyclic saturated Si atom has an up-field resonance at −120.1 ppm.

The molecular structure of 2-disilagermirene **23** was established by X-ray crystallography, which revealed a triangular structure composed of one saturated silicon atom, one unsaturated silicon atom and one unsaturated germanium atom. Although the accurate determination of bond lengths and angles in the three-membered ring was impossible because of significant disorder in the positions of doubly bonded Si and Ge atoms, it was possible to determine the geometry around the Si=Ge double bond, which also has a *trans*-bent configuration with a torsional angle of 40.3(5)°.

The isomerization of **22** to **23** can also be performed under thermal conditions. Thus, thermolysis of a solution of **22** in mesitylene at 120 °C cleanly produced 2-disilagermirene **23** in one day. Thermolysis can also be performed in benzene solution in a temperature interval from 80 to 100 °C, but it requires a longer reaction time, about 4 days. Thermal reaction of **22** produced an equilibrium mixture of **22** (2%) and **23** (98%), from which it was estimated that **23** is more stable than **22** by *ca* 3 kcal mol^{-1}. Thermolysis of **22** without solvent at 215 °C cleanly produces 2-disilagermirene **23** quantitatively in 20 minutes without any side products.

Ab initio calculations on the model H_3Si-substituted 1-disilagermirene **24** and 2-disilagermirene **25** at the MP2/DZd and B3LYP/DZd levels show the Si=Si double bond length in **24** to be 2.105 Å (MP2) and 2.107 Å (B3LYP), which agree well with the experimental value of 2.146 Å. The Si=Ge double bond length in **25** was predicted to be 2.180 Å (MP2) and 2.178 Å (B3LYP). It was also found that **25** is more stable than **24** by 3.9 (MP2) and 2.3 (B3LYP) kcal mol^{-1}. These values are in good agreement with the experimentally estimated value of *ca* 3 kcal mol^{-1}.

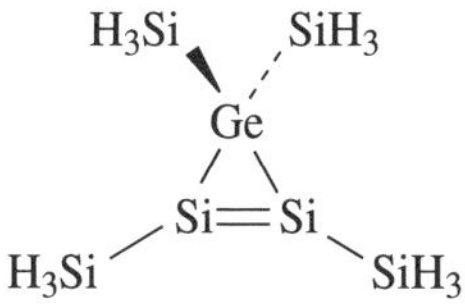

H_3Si SiH_3 Ge Si=Si H_3Si SiH_3

(24)

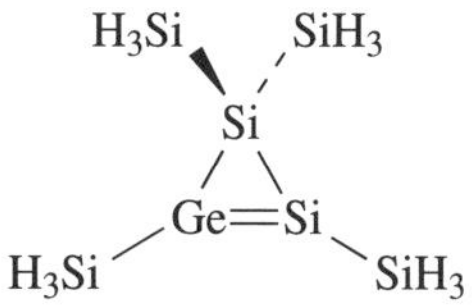

H_3Si SiH_3 Si Ge=Si H_3Si SiH_3

(25)

E. Reactivity of Cyclotrimetallenes

The reactivity of the three-membered unsaturated rings of heavier Group 14 elements is still largely unknown, although one can easily expect them to have very interesting properties arising from their unusual structures, which combine both a highly reactive metal-metal double bond and highly strained three-membered skeleton in one molecule. From all the cyclotrimetallenes described above, the reactivity has been studied for cyclotrisilenes, cyclotrigermenes and "mixed" compounds. The chemistry of the only reported cyclotristannene is still open for investigation.

1. Addition reactions

a. Reactions with CCl_4. The reactivity of cyclotrimetallenes with CCl_4 was studied in the case of cyclotrisilenes and 'mixed' cyclotrimetallenes. Thus, cyclotrisilene **4**, 1-disilagermirene **22** and 2-disilagermirene **23** were reacted with an excess of CCl_4 to form the corresponding *trans*-dichloro derivatives **26–28** even at low temperature in nearly quantitative yield (Scheme 11)[23]. As was mentioned before, the cyclotrisilene **1** also reacts with CCl_4 to produce the *trans*-1,2-dichloro derivative **2**[14]. The reaction proceeds selectively to produce only one isomer — *trans*, which can be explained by the steric requirements. The crystal structure of compound **27** is shown in Figure 8. It is interesting that the dichloro derivatives **26–28** can be quantitatively converted back to the corresponding starting cyclotrimetallenes **4**, **22** and **23** by treatment with *t*-Bu_3SiNa (Scheme 11)[23].

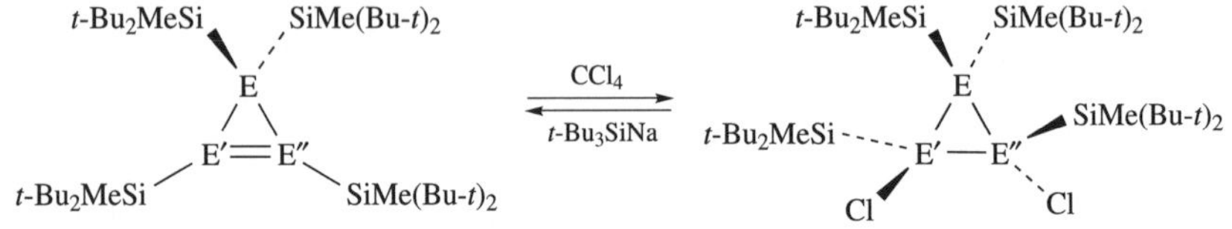

t-Bu_2MeSi, $SiMe(Bu\text{-}t)_2$, E, E′=E″, *t*-Bu_2MeSi, $SiMe(Bu\text{-}t)_2$

CCl_4 (forward) / *t*-Bu_3SiNa (reverse)

t-Bu_2MeSi, $SiMe(Bu\text{-}t)_2$, E, *t*-Bu_2MeSi, E′—E″, $SiMe(Bu\text{-}t)_2$, Cl, Cl

(4) E = E′ = E″ = Si
(22) E = Ge, E′ = E″ = Si
(23) E = E″ = Si, E′ = Ge

(26) E = E′ = E″ = Si
(27) E = Ge, E′ = E″ = Si
(28) E = E″ = Si, E′ = Ge

SCHEME 11

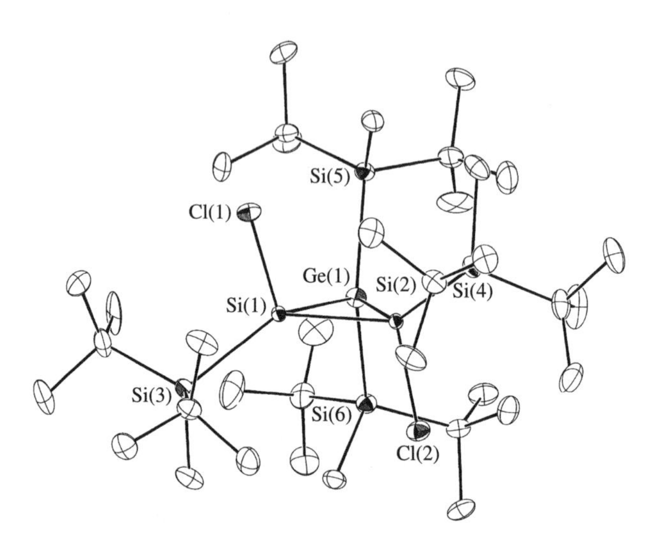

FIGURE 8. ORTEP drawing of 1,2-dichloro-1,2,3,3-tetrakis[di-*tert*-butyl(methyl)silyl]-1,2-disilagermirane **27**

2. Cycloaddition reactions

a. [2 + 2] *Cycloaddition reactions.*

i. Reactions with phenylacetylene. The behavior of cyclotrimetallenes toward phenylacetylene was surprisingly different, showing the different nature of the three-membered ring compounds composed of the Si, the Ge, or both Si and Ge atoms. Thus, reaction of phenylacetylene with mesityl-substituted cyclotrigermene **19** proceeds as a [2 + 2] cycloaddition reaction with the formation of the resulting bicyclic three- and four-membered ring compound **29** as orange crystals in the form of two stereoisomers **29a** and **29b** (Scheme 12)[21]. The crystal structure analysis of **29a** showed a highly folded bicyclic skeleton with a dihedral angle between the planes of the three- and four-membered rings of 97.4° (Figure 9).

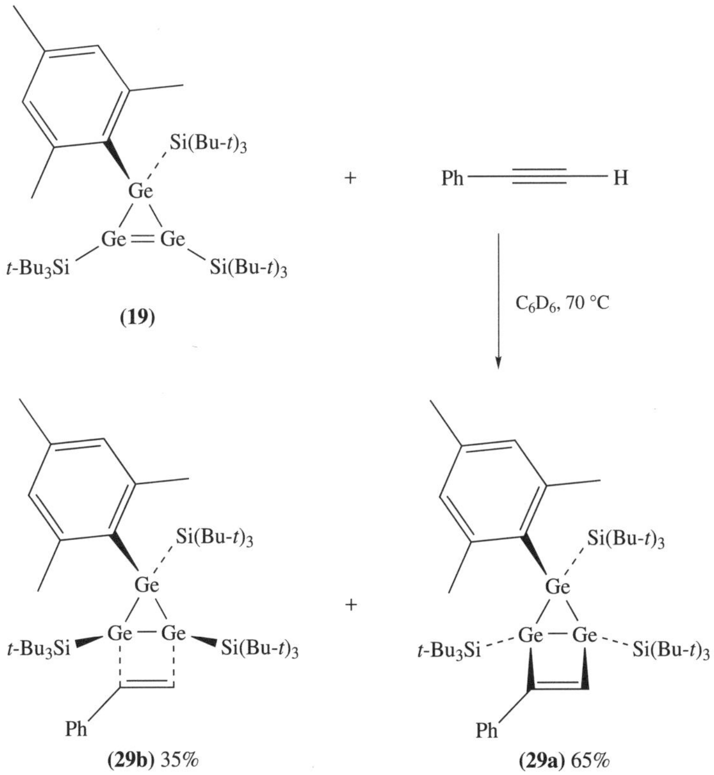

SCHEME 12

The reaction of the 'mixed' cyclotrimetallenes **22** and **23** with phenylacetylene also proceeds through the initial [2 + 2] cycloaddition of the first molecule of phenylacetylene across the E=Si (E = Si, Ge) double bond with the formation of the three- and four-membered bicyclic compounds **30** and **31**[22]. But the reaction does not stop at this stage: valence isomerization of the bicyclic compound takes place to form the silole type structures **32** and **33** with one Si=C and one Ge=C double bond, which in turn

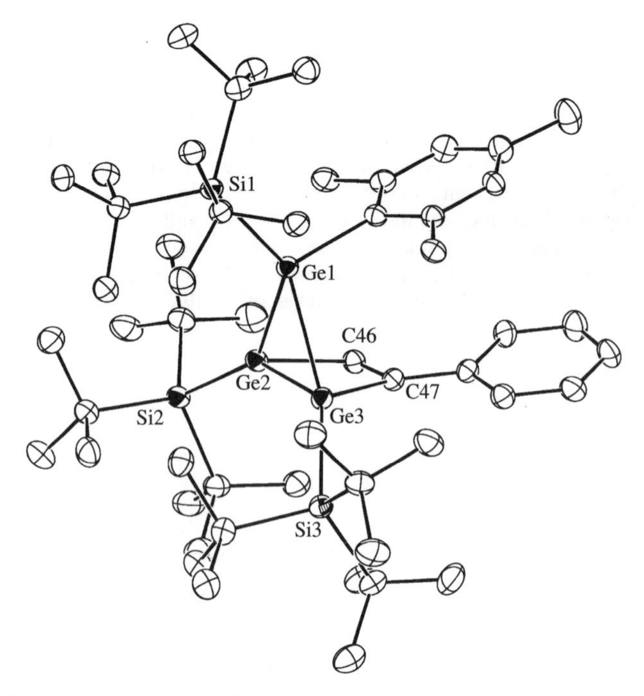

FIGURE 9. ORTEP drawing of a [2 + 2] cycloadduct **29a**. Reproduced by permission of Wiley-VCH from Reference 21

quickly isomerizes to give the thermodynamically more stable siloles **34** and **35** with one Si=Ge and one C=C double bond (Scheme 13). The silole-type compounds **34** and **35** react with a second molecule of phenylacetylene in a [2 + 2] cycloaddition manner to give the final four- and five-membered bicyclic compounds **36** and **37**. In the case of 2-disilagermirene **23**, it was possible to isolate the intermediate silole **35**, representing the first metalladiene of the type E=E′-C=C (E,E′ — heavier Group 14 elements). Two examples of the isolable metalladienes of Group 14 elements have been recently reported by Weidenbruch's group: hexakis(2,4,6-tri-isopropylphenyl)tetrasila-1,3-butadiene[44] and hexakis(2,4,6-tri-isopropylphenyl)tetragerma-1,3-butadiene[45].

The structure of compound **35** was determined by means of all spectral data. Thus, the ^{29}Si NMR spectrum showed five resonances, of which three belong to the silyl substituents: 19.4, 26.6 and 30.1 ppm, the down-field signal at 124.2 ppm was attributable to the doubly bonded silicon atom and the up-field signal at −45.6 ppm corresponds to the endocyclic sp^3 Si atom. X-ray analysis of **35** revealed an almost planar five-membered ring, although the Si=Ge double bond has a twisted (*trans*-bent) configuration with a torsional angle Si(6)−Ge(1)−Si(2)−Si(5) of 38.6(1)° (Figure 10). The Si=Ge double bond length, which was determined experimentally for the first time, is 2.250(1) Å, which is intermediate between the typical values for Si=Si and Ge=Ge double bond lengths. From the experimental data, i.e. X-ray crystallography, the UV-Vis spectrum and reactivity of the silole **35**, it was found that there is almost no conjugation between the two double bonds in the cyclopentadiene ring of **35**. This seems to be curious since all the known cyclopentadiene compounds were described as fully conjugated systems, for which

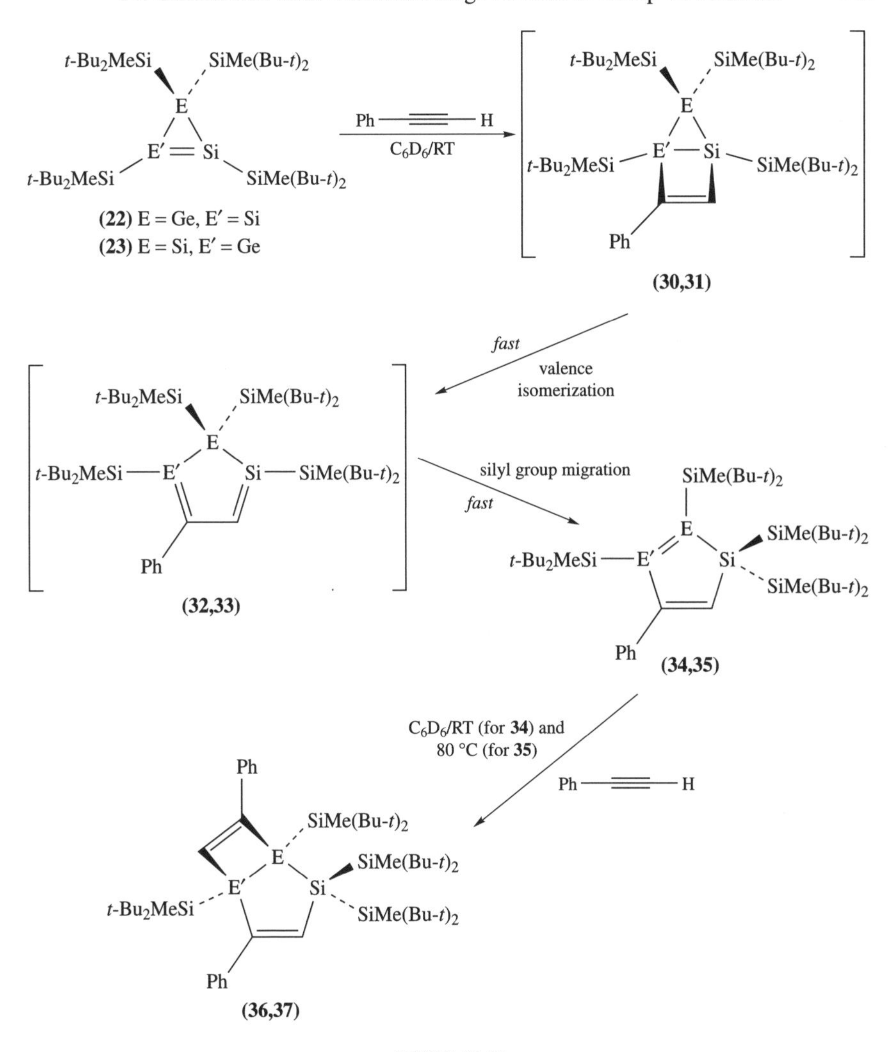

SCHEME 13

Diels–Alder cycloaddition reactions are typical[46]. Apparently, such unusual behavior is caused by both the great energy difference and the difference in the size of the atoms of the Si=Ge and C=C double bonds, which prevent an effective overlapping of the molecular orbitals of the two π-bonds necessary for real conjugation. In contrast, for the symmetrical heavier Group 14 elements containing 1,3-diene systems with two equal double bonds, such as 2,3-digerma-1,3-butadiene $H_2C{=}GeH{-}GeH{=}CH_2$, theoretical calculations have predicted about half the degree of conjugation compared with that of the parent 1,3-butadiene[47]. The cyclotrisilene **4** also readily reacts with phenylacetylene to form finally a bicyclic compound similar to that of **36** and **37**[48]. In this case the isolation of the cyclopentadiene-type compound was impossible due to its very short life time.

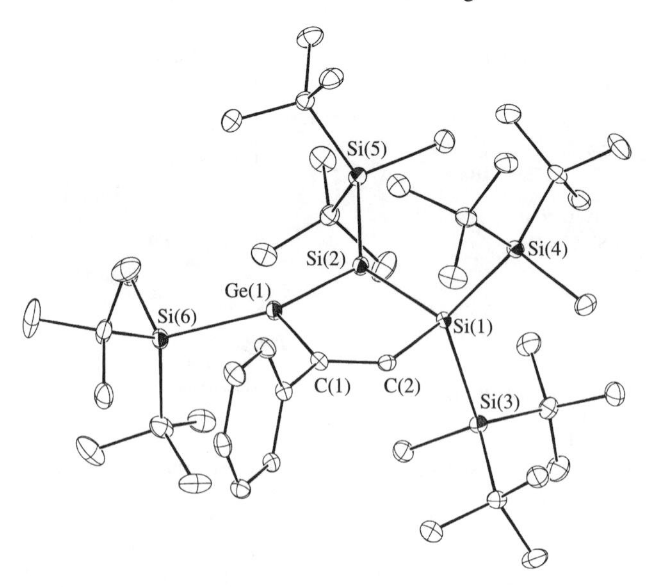

FIGURE 10. ORTEP drawing of silole derivative **35**. Reprinted with permission from Reference 22. Copyright 2000 American Chemical Society

ii. Reactions with aldehydes and ketones. Reaction with carbonyl compounds is also very sensitive to the steric requirements. Thus, highly sterically protected tetrakis(tri-*tert*-butylsilyl)cyclotrigermene **5** does not react with benzaldehyde[49], whereas the reaction of 1-disilagermirene **22** with benzaldehyde proceeds almost immediately to give a set of products depending on the reaction conditions and the ratio of the reagents (Scheme 14)[50]. Thus, at room temperature 1-disilagermirene **22** reacts with one molecule of benzaldehyde to give the bicyclic three- and four-membered ring compound **38**, similar to the above case of phenylacetylene[22]. Compound **38** can be considered as a kinetically controlled product, which presumably is stabilized by an interaction of the electron-rich phenyl group and the empty σ^*-orbital of the exocyclic Ge−Si bond. Nevertheless, such an arrangement of the phenyl group is not favorable due to the steric repulsion with silyl substituents on the Ge atom; therefore, upon heating at mild conditions compound **38** was isomerized quantitatively to a thermodynamically more stable compound **39**, whose structure was established by X-ray analysis (Figure 11). Compound **39** has a highly folded skeleton with a dihedral angle between the two planes of 107.8°.

The initially formed bicyclic compound **38** has a highly strained and very reactive bridgehead endocyclic Si−Si bond. It can easily react with a second molecule of benzaldehyde by the insertion pathway to form a new bicyclic compound **40** with a norbornane type skeleton (Scheme 14 and Figure 12). Although this last reaction closely resembles the previous case of phenylacetylene[22], the mechanism is evidently different: in the case of phenylacetylene the final product **36** is a result of [2 + 2] *cycloaddition* of the second molecule of phenylacetylene across the new Si=Ge double bond, whereas in the case of benzaldehyde the final norbornane **40** is a result of the *insertion* of the second molecule of benzaldehyde into the strained Si−Si single bond. Apparently, the reactions of disilagermirenes with phenylacetylene and benzaldehyde have the same initial steps to form bicyclic compounds, but then the reaction pathways become different due to the different nature of these intermediate bicyclic compounds.

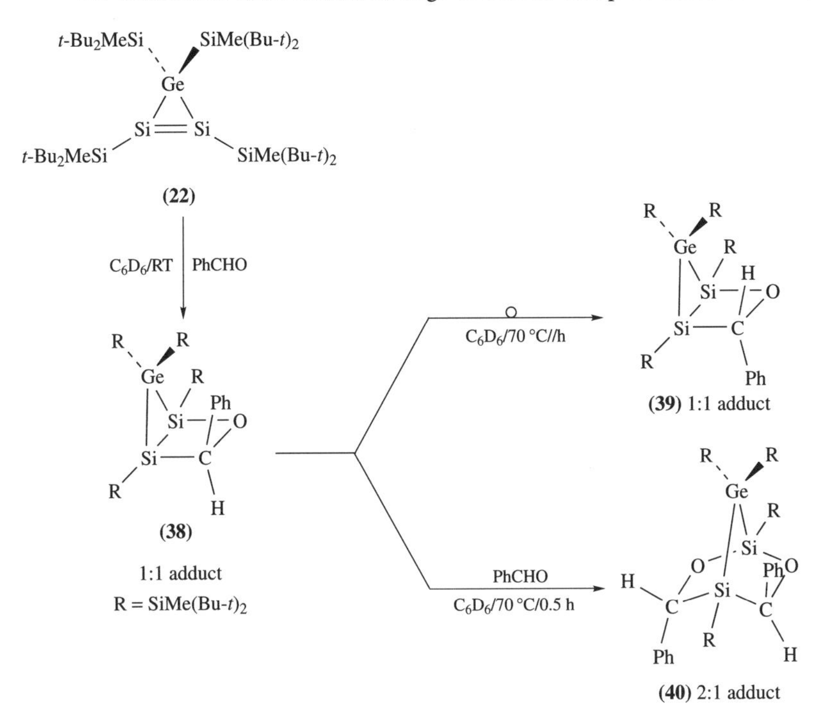

SCHEME 14

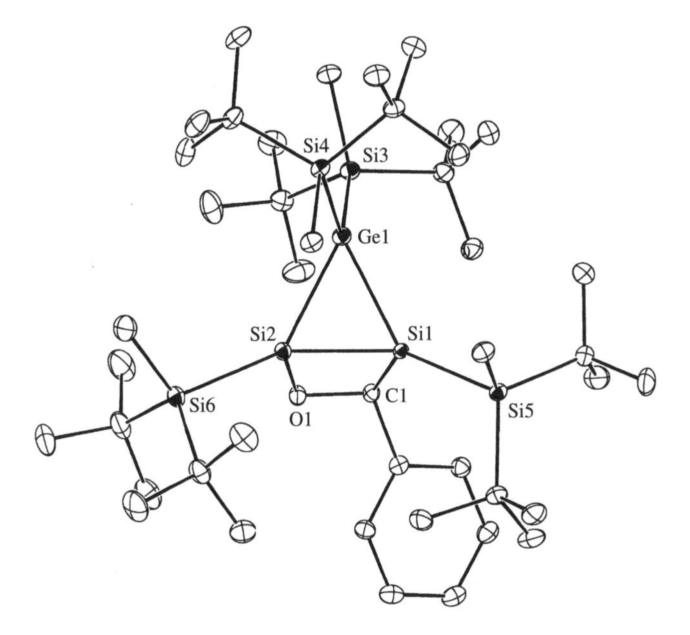

FIGURE 11. ORTEP drawing of bicyclic compound **39**

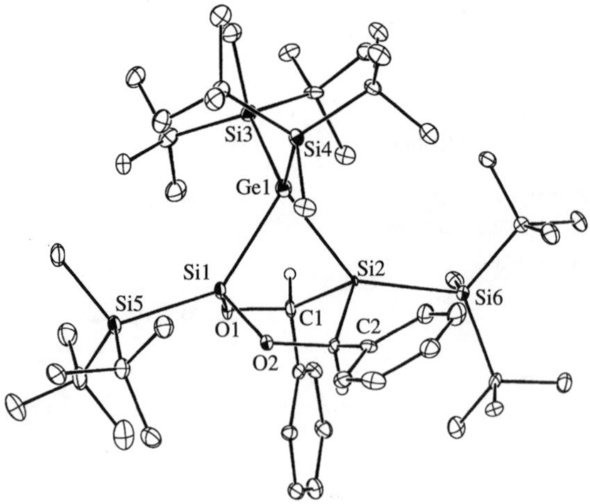

FIGURE 12. ORTEP drawing of bicyclic compound **40** with a norbornane type skeleton

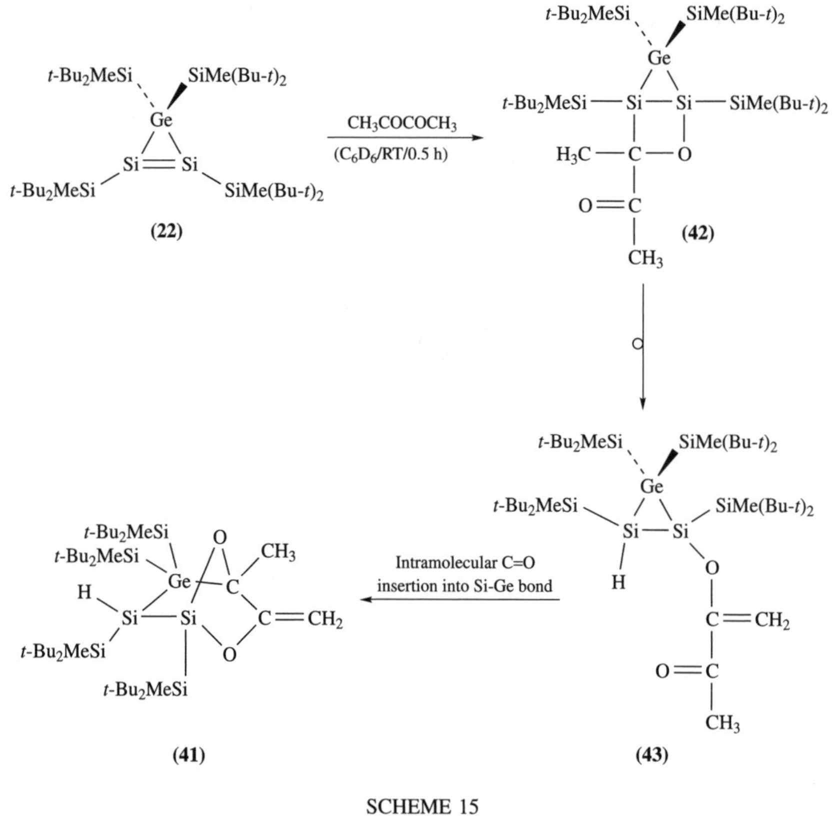

SCHEME 15

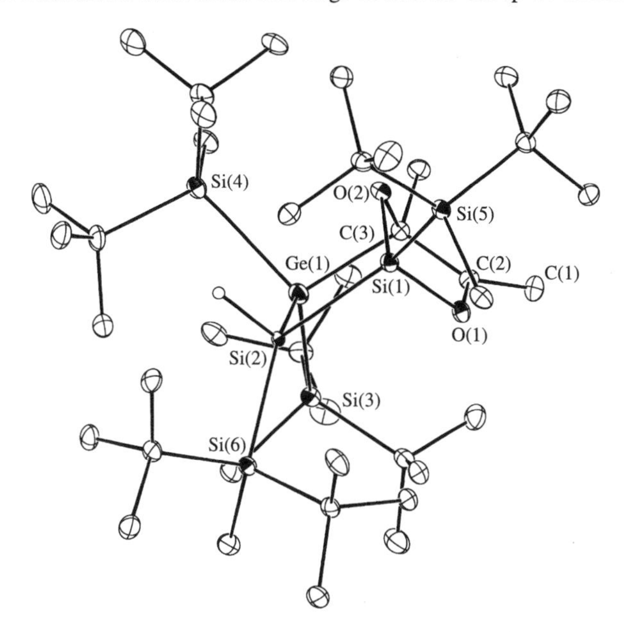

FIGURE 13. ORTEP drawing of bicyclic compound **41**

The reaction of 1-disilagermirene **22** with ketones is similar to the benzaldehyde case. Thus, reaction with butane-2,3-dione gives a final bicyclic product **41**, which also has a norbornane type skeleton (Scheme 15, Figure 13)[50]. Formation of this compound can be reasonably explained by the initial [2 + 2] cycloaddition of one carbonyl group across the Si=Si bond to form the three- and four-membered ring bicyclic compound **42**, followed by the isomerization of disilaoxetane **42** to an enol ether derivative **43**. The intramolecular insertion of the second carbonyl group into the endocyclic Si−Ge single bond in **43** completes this reaction sequence to produce the final norbornane **41**. In this case, C=O insertion occurred into the Si−Ge bond rather than the Si−Si bond, which is reasonable due to the weakness of Si−Ge bond.

b. [4 + 2] *Cycloaddition reactions.*

i. Reactions with 1,3-dienes. Conjugated dienes, such as 2,3-dimethyl-1,3-butadiene and isoprene, are traditionally widely used as trapping reagents, for both transient and stable dimetallenes of Group 14 elements to form the corresponding Diels−Alder adducts. While reacting with the cyclotrimetallenes of Group 14 elements, conjugated dienes have produced the corresponding [4 + 2] adducts in the form of fused three- and six-membered ring compounds. For example, the reaction of mesityl-substituted cyclotrigermene **19** with both 2,3-dimethyl-1,3-butadiene and isoprene yields bicyclic adducts **44** and **45** (Scheme 16)[21].

X-ray analysis confirmed the structure of compound **44** (Figure 14). Only one of the two possible stereoisomers was formed, which corresponds to the attack of the isoprene on the Ge=Ge double bond from the mesityl side. This probably can be explained by the lower steric bulkiness of the mesityl group compared with the t-Bu_3Si group. Due to

Si(Bu-t)$_3$
Ge
Ge=Ge
t-Bu$_3$Si Si(Bu-t)$_3$
(**19**)
+
R Me
(R = H, Me)
hexane, RT
Si(Bu-t)$_3$
Ge
t-Bu$_3$Si Ge — Ge Si(Bu-t)$_3$
R Me
(**44**) R = H
(**45**) R = Me

SCHEME 16

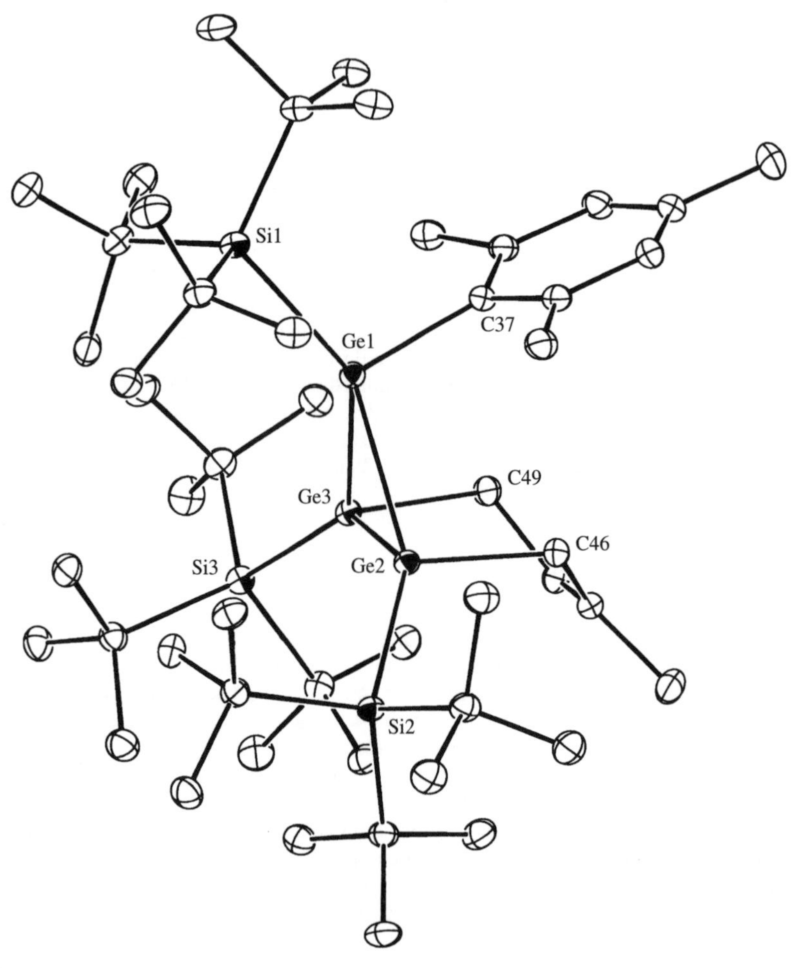

FIGURE 14. ORTEP drawing of a [4 + 2] cycloadduct **44**. Reproduced by permission of Wiley-VCH from Reference 21

steric reasons, the three t-Bu_3Si groups occupy the less hindered pseudoequatorial positions, whereas the mesityl group and $CH_2C(Me)=CHCH_2$ moiety occupy the pseudoaxial positions (Figure 14).

It is noteworthy that the bicyclic three- and six-membered compounds **44** and **45** can serve as precursors for both germylene and digermene at 70 °C[21]. Such species can be effectively trapped by 2,3-dimethyl-1,3-butadiene to form germacyclopentene **48** and bicyclic compounds **49** and **50** (Scheme 17).

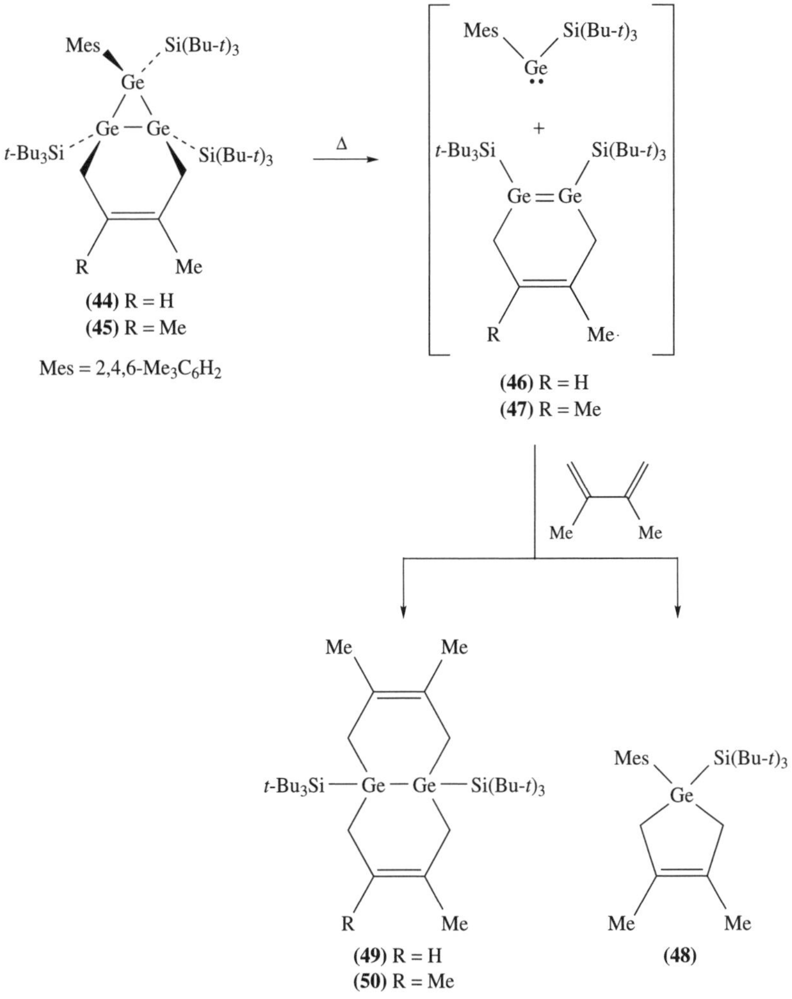

(**44**) R = H
(**45**) R = Me
Mes = 2,4,6-$Me_3C_6H_2$

(**46**) R = H
(**47**) R = Me

(**49**) R = H
(**50**) R = Me

(**48**)

SCHEME 17

The most interesting point in these reactions is the ring contraction, which takes place in the intermediate digermacyclohexadienes **46** and **47** during reaction with diphenylacetylene (Scheme 18)[51]. Thus, during thermolysis of the bicyclic

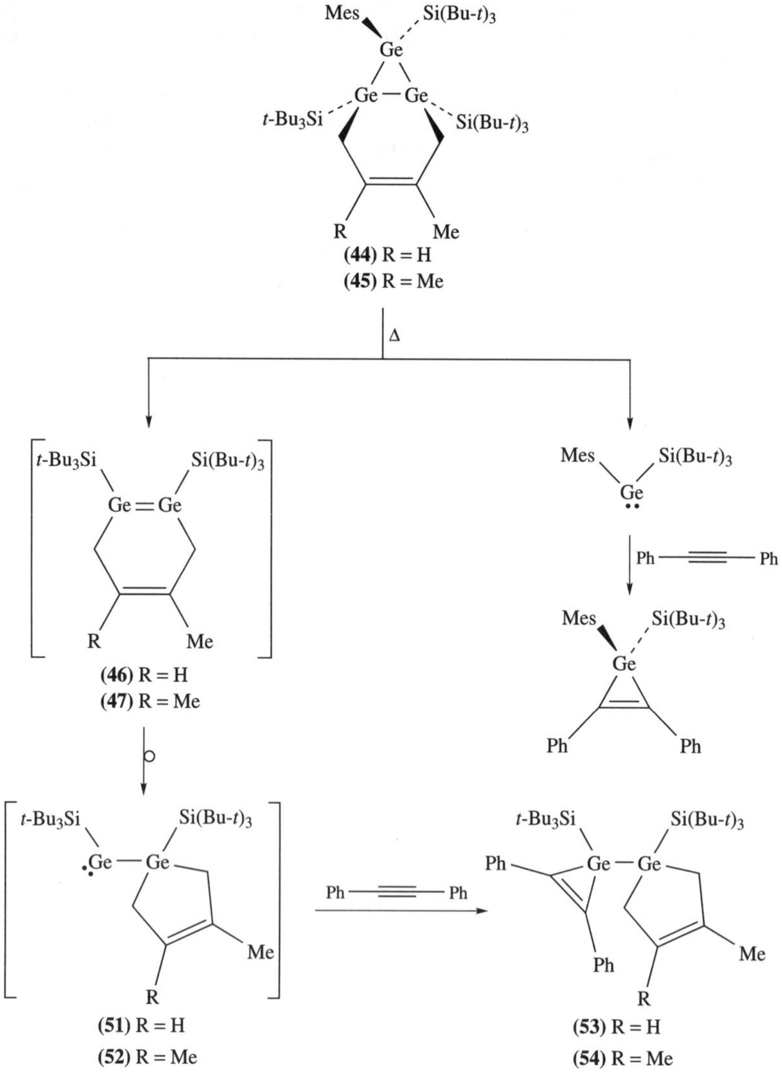

SCHEME 18

compounds **44** and **45**, a formal migration of the CH_2 group from one Ge atom to another one in the intermediate digermenes **46** and **47** takes place, resulting in ring contraction to form germylgermylene species **51** and **52**. These last germylenes can be trapped by diphenylacetylene to produce the corresponding germacyclopropenyl-substituted germacyclopentenes **53** and **54** (Scheme 18 and Figure 15). Similar digermene–germylgermylene rearrangement of tetramesityldigermene[52,53] and germasilene–silylgermylene rearrangement of tetramesitylgermasilene[43] were previously reported by Baines and Cooke.

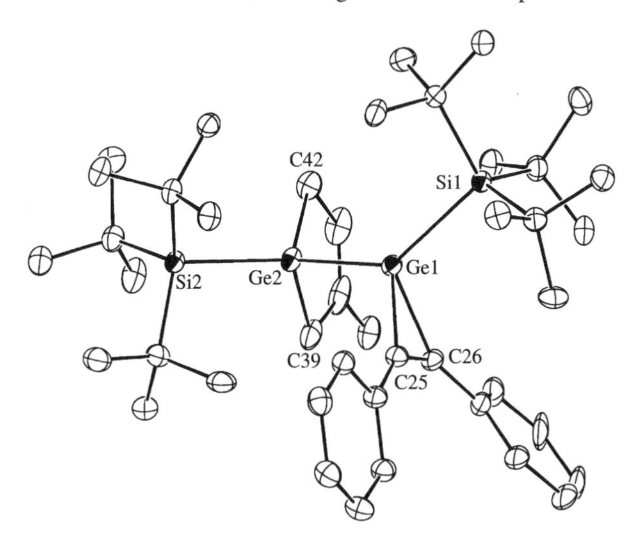

FIGURE 15. ORTEP drawing of germacyclopropenyl-substituted germacyclopentene **53**

3. Oxidation of cyclotrigermenes — formation of the 'free' germyl cation

It is well known that the cyclopropenium cation with a Hückel-type aromatic 2 π-electron system is the simplest and smallest aromatic compound, which is relatively stable due to its resonance stabilization despite the very large ring strain[54–56]. Although the chemistry of the cyclopropenium cation is well established now, the analogues of this compound consisting of heavier Group 14 elements were unknown until recently. Theoretical calculations on the stability of $A_3H_3^+$ cations (A = C, Si, Ge, Sn, Pb)[57] predicted a preference for the classical cyclopropenium cation structures with D_{3h} symmetry for the carbon and silicon cases, whereas C_{3v} hydrogen-bridged forms were expected to be favored for germanium, tin and lead. In contrast to these calculations, the first free germyl cation with a 2 π-electron system was reported by Sekiguchi and coworkers in 1997 as a classical cyclopropenium-type cation[58,59]. Thus, tris(tri-*tert*-butylsilyl)cyclotrigermenium tetraphenylborate $[(t\text{-Bu}_3\text{SiGe})_3{}^+ \bullet \text{TPB}^-]$ (TPB^- = tetraphenylborate) **55** was prepared by the treatment of tetrakis(tri-*tert*-butylsilyl)cyclotrigermene **5**[13] with trityl tetraphenylborate in dry benzene and was isolated as air and moisture sensitive yellow crystals (Scheme 19).

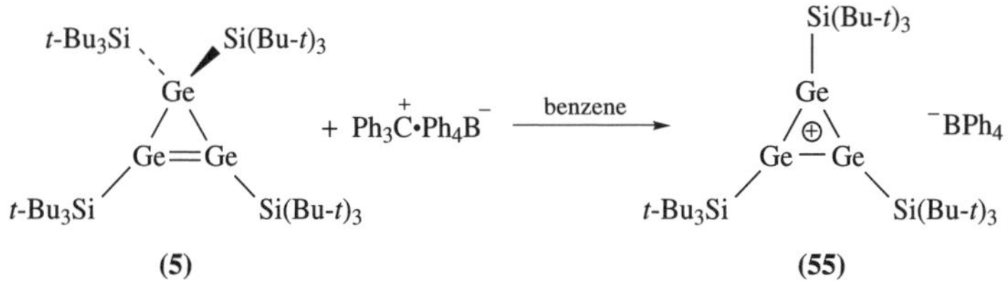

t-Bu$_3$Si, Si(Bu-t)$_3$ / Ge / Ge=Ge / t-Bu$_3$Si, Si(Bu-t)$_3$ **(5)** + $\text{Ph}_3\overset{+}{\text{C}}\bullet\text{Ph}_4\overset{-}{\text{B}}$ —benzene→ Si(Bu-t)$_3$ / Ge / $\oplus$ / Ge—Ge / t-Bu$_3$Si, Si(Bu-t)$_3$ $^-$BPh$_4$ **(55)**

SCHEME 19

The structure of **55** was determined on the basis of NMR spectral data and finally confirmed by X-ray crystallographic analysis (Figure 16). The three germanium atoms

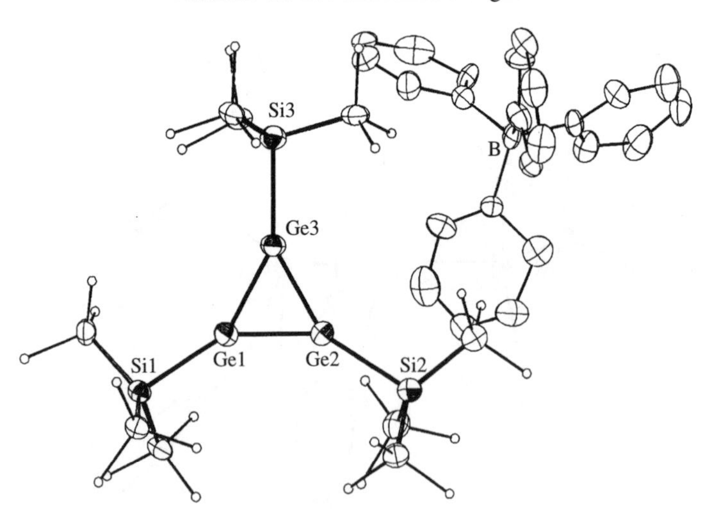

FIGURE 16. ORTEP drawing of $[(t\text{-Bu}_3\text{SiGe})_3{}^+ \bullet \text{TPB}^-]$ **55**. Reprinted with permission from Reference 58. Copyright 1997, American Association for the Advancement of Science

form an equilateral triangle [Ge−Ge bond lengths 2.321(4)–2.333(4) Å and Ge−Ge−Ge bond angles 59.8(1)–60.3(1)°]. The Ge−Ge bond lengths observed in **55** are intermediate between the Ge=Ge double bond [2.239(4) Å] and the Ge−Ge single bond [2.522(4) Å] of the precursor **5**[13]. The closest distance between germanium and the aromatic carbon atoms of TPB^- is greater than 4 Å, well beyond the range of any significant interaction. These structural features indicate that **55** has a cyclotrigermenium skeleton and it is a free germyl cation with a 2 π-electron system. The aromatic stabilization of the cyclotrigermenium ion and the charge delocalization explain the observed lack of any close interaction with the counter anion.

However, the problem of TPB^- was its chemical instability[60], because **55** can survive in a solution of dichloromethane only at temperatures below −78 °C. $[3,5\text{-}(CF_3)_2C_6H_3]_4B^-$ ($TFPB^-$, tetrakis{3,5-bis(trifluoromethyl)phenyl}borate)[61], $(C_6F_5)_4B^-$ ($TPFPB^-$, tetrakis (pentafluorophenyl)borate)[62] and $[4\text{-}(t\text{-BuMe}_2\text{Si})C_6F_4]_4B^-$ ($TSFPB^-$, tetrakis{4-[*tert*-butyl(dimethyl)silyl]-2,3,5,6-tetrafluorophenyl}borate)[63,64] are known to be stable borate anions, which can increase the stability of the resulting cyclotrigermenium ion. Therefore, the reactions of cyclotrigermenes **5** and **6** with $Ph_3C^+ \bullet TFPB^-$, $Ph_3C^+ \bullet TPFPB^-$ and $Ph_3C^+ \bullet TSFPB^-$ were studied with the hope of obtaining the stable cyclotrigermenium salts[65–67]. In fact, the reaction of $(t\text{-Bu}_3\text{Si})_4\text{Ge}_3$ **5** and $Ph_3C^+ \bullet TFPB^-$ in benzene at room temperature produced the salt $(t\text{-Bu}_3\text{SiGe})_3{}^+ \bullet \text{TFPB}^-$ **15**, which was isolated as a yellow powder in 81% yield (Scheme 20). The reaction of **6** with $Ph_3C^+ \bullet TFPB^-$ in benzene also proceeded smoothly to give $(t\text{-Bu}_3\text{GeGe})_3{}^+ \bullet \text{TFPB}^-$ **56** in 76% yield. In a similar way, the reaction of **5** and **6** with $Ph_3C^+ \bullet TPFPB^-$ in benzene produced $(t\text{-Bu}_3\text{SiGe})_3{}^+ \bullet \text{TPFPB}^-$ **57** (80%) and $(t\text{-Bu}_3\text{GeGe})_3{}^+ \bullet \text{TPFPB}^-$ **58** (80%), respectively. The reaction of **5** with $Ph_3C^+ \bullet TSFPB^-$ produced $(t\text{-Bu}_3\text{SiGe})_3{}^+ \bullet \text{TSFPB}^-$ **59** as orange crystals in 88% yield. The resulting germyl cations can survive for extensive periods without decomposition both in solution and in the solid state (Scheme 20).

The molecular structure of **15** is shown in Figure 17. The three-membered ring consisting of germanium atoms is almost an equilateral triangle with the Ge−Ge

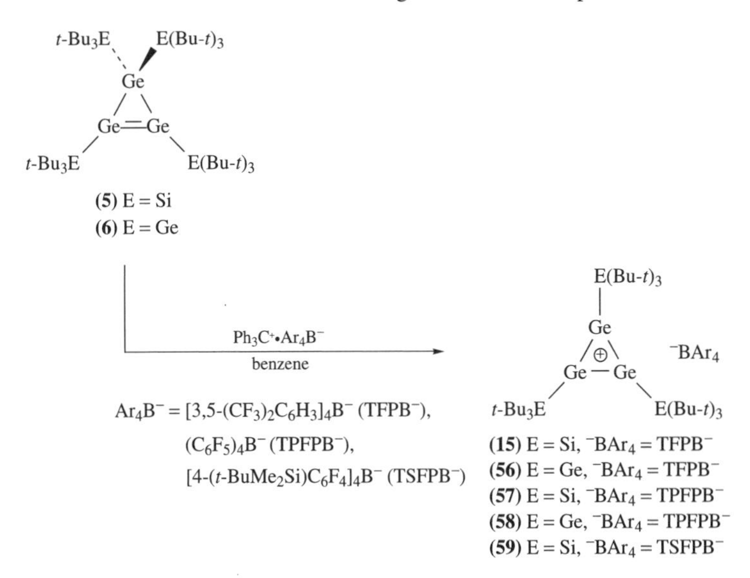

SCHEME 20

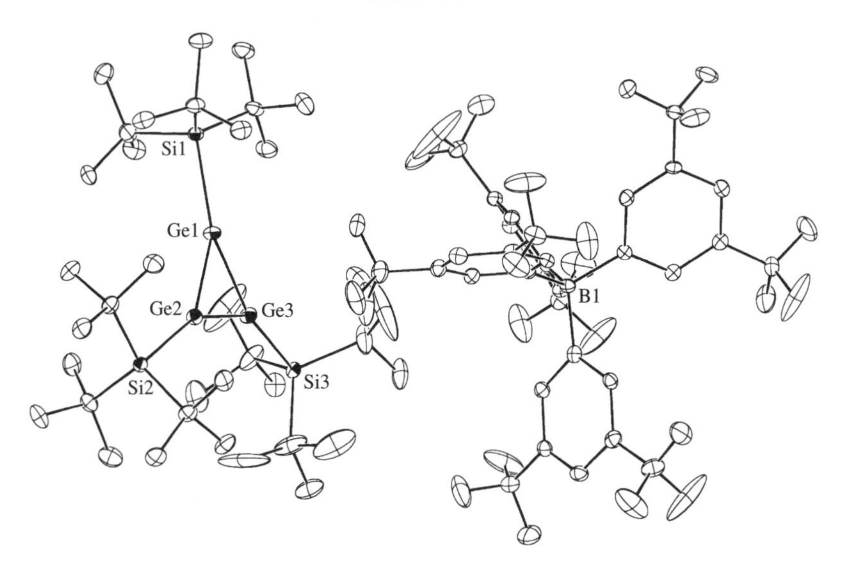

FIGURE 17. ORTEP drawing of $[(t\text{-Bu}_3\text{SiGe})_3{}^{+}\bullet\text{TFPB}^{-}]$ **15**. Reproduced by permission of Wiley-VCH from Reference 67

distances of the three-membered ring almost equal, ranging from 2.3284(8) to 2.3398(8) Å [av. 2.3333(8) Å]. The Ge−Si bond lengths [Ge1−Si1, 2.425(1); Ge2−Si2, 2.442(1); Ge3−Si3, 2.444(1) Å] of **15** are shortened compared with those of **5**[13] [2.629(7) Å for the Ge−Si bond length of sp^3 Ge atom and 2.448(7) Å for the Ge−Si bond length of sp^2 Ge atoms]. The perspective view appears to show a weak electrostatic interaction between the germanium and fluorine atoms. The three closest distances range from 3.823 to 5.097 Å; however, these distances are longer than the sum (3.57 Å) of the van der Waals radii for germanium and fluorine atoms[68].

X-ray diffraction data were also obtained for **59**. Due to the steric bulkiness of the t-BuMe$_2$Si group attached to the *para* positions of the phenyl rings of the borate anion, no interaction between the cation moiety and the counter anion can be found (Figure 18). As a consequence, the skeleton of the three-membered framework forms an equilateral triangle; the Ge−Ge bond lengths are 2.3310(8) for Ge1−Ge2, 2.3315(7) for Ge1−Ge3 and 2.3349(8) Å for Ge2−Ge3, and the Ge−Ge−Ge bond angles are 60.10(2)° for Ge2−Ge1−Ge3, 59.96(2)° for Ge1−Ge2−Ge3 and 59.94(2)° for Ge1−Ge3−Ge2. The structural features for **15** and **59** are practically the same as those of **55**[58].

The evidence for the existence of the free cyclotrigermenium ion in solution was supported by the NMR spectroscopic data. The ^{1}H, ^{13}C and ^{29}Si NMR chemical shifts for the cyclotrigermenium moiety of **15**, **57** and **59** in CD_2Cl_2 are practically the same. For example, the ^{29}Si NMR chemical shifts of **15**, **57** and **59** are independent of the counter anion. The ^{29}Si NMR chemical shift of **15** is also essentially the same in different solvents, appearing at $\delta = 64.0$ in CD_2Cl_2, $\delta = 64.2$ in $CDCl_3$, $\delta = 64.4$ in toluene-d$_8$ and $\delta = 64.5$ in Et$_2$O. This independence from both the counter anion and solvent clearly indicates that $(t\text{-Bu}_3\text{SiGe})_3{}^+$ is a free germyl cation in solution. The large down-field shifted ^{29}Si NMR resonance of $(t\text{-Bu}_3\text{SiGe})_3{}^+$, relative to that of the precursor **5** ($\delta = 37.2$ for the t-Bu$_3$Si substituent attached to the saturated Ge atom and 50.1 for the t-Bu$_3$Si group attached to the Ge=Ge double bond)[13], is due to the positive charge of the cyclotrigermenium ion.

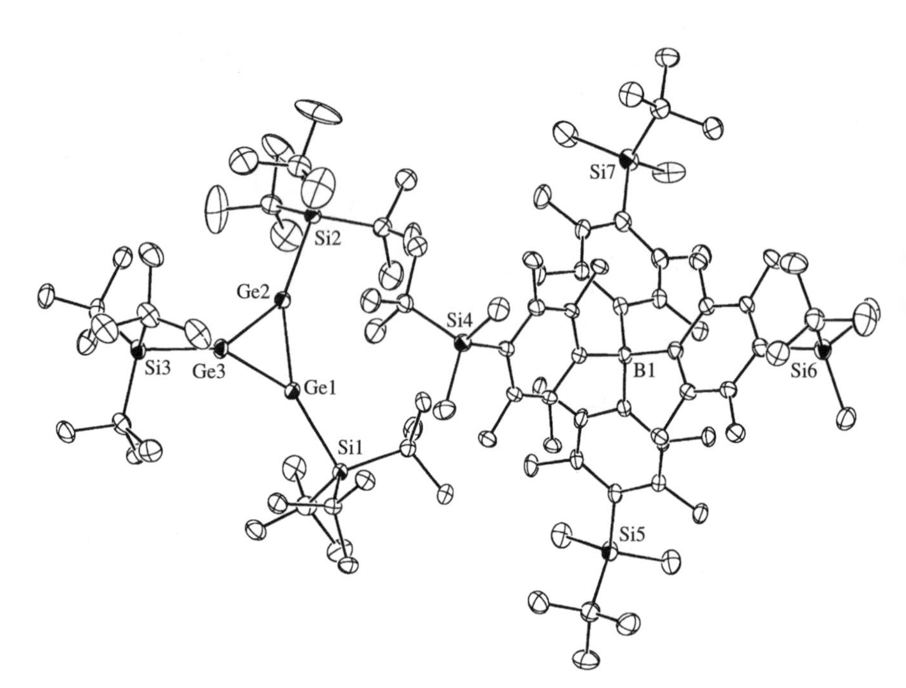

FIGURE 18. ORTEP drawing of [$(t\text{-Bu}_3\text{SiGe})_3{}^+$•TSFPB$^-$] **59**. Reproduced by Permission of Wiley-VCH from Reference 67

The positive charge is not localized on the germanium atoms, but is significantly transferred to the silicon centers. The atomic (Mulliken) charges on $(H_3SiGe)_3{}^+$ according to HF/6-31G* level calculations indicate a delocalization of the positive charge: −0.07 for the ring germanium atoms and +0.64 for the silicon atoms of the SiH_3 substituents.

III. EPILOGUE AND OUTLOOK

The present review has no pretensions to be a comprehensive one and to cover all the chemistry of small ring systems containing heavier Group 14 elements: this field is quite vast and greatly exceeds the framework of this article. Here we have concentrated on the relatively narrow and rather new field of the unsaturated three-membered rings consisting of heavier Group 14 elements. In organic chemistry, cyclopropene, which is the smallest unsaturated ring system, and its derivatives are among the most important classes of organic compounds due to their enhanced reactivity caused by a great ring strain and existence of the endocyclic C=C double bond. Apparently, the heavy cyclopropene analogues, that is cyclotrimetallenes composed of Si, Ge, Sn and Pb atoms, would occupy a similar important position in the chemistry of Group 14 elements, since their high reactivity is even more pronounced than that of cyclopropene because of the extremely reactive endocyclic metal–metal double bond and the weakness of endocyclic metal–metal single bonds. Even now, cyclotrimetallenes, which quite recently were considered to be synthetically inaccessible, can be considered as unusual molecules, and their number is still very limited. The chemistry of such compounds, which often exhibit great differences in the structures, properties and reactivity from their carbon analogues, has started quite recently, and there are still many questions and problems to be solved. Thus, the Pb-containing representative—i.e. cyclotriplumbene—has not yet been prepared. Several possible combinations for the 'mixed' cyclotrimetallenes can also be imagined as the next target molecules. The reactivity of cyclotrimetallenes also needs to be studied in detail; moreover, the first preliminary investigations showed the extremely high synthetic possibilities of such compounds, since very rich chemistry can be developed from them. Thus, one can expect much progress in this 'hot' field of Group 14 elements chemistry in the near future, which will have an important impact on our understanding of the nature of bonding and reactivity in systems with elements heavier than carbon.

IV. ACKNOWLEDGMENT

We highly appreciate Dr. Masaaki Ichinohe and Mrs. Norihisa Fukaya, Tadahiro Matsuno, Hiroshi Sekiyama and Yutaka Ishida for their invaluable experimental contributions. We wish to thank Professor Shigeru Nagase for the theoretical calculations. This work was supported by a Grant-in-Aid for Scientific Research (Nos. 13029015, 13440185, 12042213) from the Ministry of Education, Science and Culture of Japan, and TARA (Tsukuba Advanced Research Alliance) fund.

V. REFERENCES

1. T. Tsumuraya, S. A. Batcheller and S. Masamune, *Angew. Chem., Int. Ed. Engl.*, **30**, 902 (1991).
2. M. Weidenbruch, *Chem. Rev.*, **95**, 1479 (1995).
3. K. M. Baines and W. G. Stibbs, *Adv. Organomet. Chem.*, **39**, 275–324 (1996).
4. M. Driess and H. Grützmacher, *Angew. Chem., Int. Ed. Engl.*, **35**, 828 (1996).
5. M. Weidenbruch, *Eur. J. Inorg. Chem.*, 373 (1999).
6. S. Masamune, W. Hanzawa, S. Murakami, T. Bally and J. F. Blount, *J. Am. Chem. Soc.*, **104**, 1150 (1982).

7. S. Masamune, W. Hanzawa and D. J. Williams, *J. Am. Chem. Soc.*, **104**, 6137 (1982).
8. S. Masamune, L. R. Sita and D. J. Williams, *J. Am. Chem. Soc.*, **105**, 630 (1983).
9. K. M. Baines and J. A. Cooke, *Organometallics*, **10**, 3419 (1991).
10. A. Heine and D. Stalke, *Angew. Chem., Int. Ed. Engl.*, **33**, 113 (1994).
11. H. Suzuki, K. Okabe, S. Uchida, H. Watanabe and M. Goto, *J. Organomet. Chem.*, **509**, 177 (1996).
12. M.-A. Chaubon, J. Escudié, H. Ranaivonjatovo and J. Satgé, *J. Chem. Soc., Chem. Commun.*, 2621 (1996).
13. A. Sekiguchi, H. Yamazaki, C. Kabuto, H. Sakurai and S. Nagase, *J. Am. Chem. Soc.*, **117**, 8025 (1995).
14. T. Iwamoto, C. Kabuto and M. Kira, *J. Am. Chem. Soc.*, **121**, 886 (1999).
15. M. Ichinohe, T. Matsuno and A. Sekiguchi, *Angew. Chem., Int. Ed.*, **38**, 2194 (1999).
16. N. Wiberg, H.-W. Lerner, S.-K. Vasisht, S. Wagner, K. Karaghiosoff, H. Nöth and W. Ponikwar, *Eur. J. Inorg. Chem.*, 1211 (1999).
17. A. Sekiguchi, N. Fukaya, M. Ichinohe, N. Takagi and S. Nagase, *J. Am. Chem. Soc.*, **121**, 11587 (1999).
18. M. Ichinohe, H. Sekiyama, N. Fukaya and A. Sekiguchi, *J. Am. Chem. Soc.*, **122**, 6781 (2000).
19. V. Ya. Lee, M. Ichinohe, A. Sekiguchi, N. Takagi and S. Nagase, *J. Am. Chem. Soc.*, **122**, 9034 (2000).
20. V. Ya. Lee, M. Ichinohe and A. Sekiguchi, *Phosphorus Sulfur Silicon Relat. Elem.*, in press.
21. N. Fukaya, M. Ichinohe and A. Sekiguchi, *Angew. Chem., Int. Ed.*, **39**, 3881 (2000).
22. V. Ya. Lee, M. Ichinohe and A. Sekiguchi, *J. Am. Chem. Soc.*, **122**, 12604 (2000).
23. V. Ya. Lee, T. Matsuno, M. Ichinohe and A. Sekiguchi, *Heteroatom Chem.*, in press.
24. For a review on the cyclotrigermenium cation systems, see: V. Ya. Lee, A. Sekiguchi, M. Ichinohe and N. Fukaya, *J. Organomet. Chem.*, **611**, 228 (2000).
25. O. M. Nefedov, S. P. Kolesnikov and I. S. Rogozhin, *Izv. Akad. Nauk SSSR, Ser. Khim.*, 170 (1980); *Chem. Abstr.*, **95**, 18242k (1980).
26. M. Kira, T. Maruyama, C. Kabuto, K. Ebata and H. Sakurai, *Angew. Chem., Int. Ed. Engl.*, **33**, 1489 (1994).
27. M. Kira, T. Iwamoto and C. Kabuto, *J. Am. Chem. Soc.*, **118**, 10303 (1996).
28. M. Kaftory, M. Kapon and M. Botoshansky, in *The Chemistry of Organic Silicon Compounds*, (Eds. Z. Rappoport and Y. Apeloig), Vol. 2, Part 1, Chap. 5, Wiley, Chichester (1998).
29. J. Escudié, C. Couret, H. Ranaivonjatovo and J. Satgé, *Coord. Chem. Rev.*, **130**, 427 (1994).
30. P. P. Power, *J. Chem. Soc., Dalton Trans.*, 2939 (1998).
31. P. P. Power, *Chem. Rev.*, **99**, 3463 (1999).
32. J. Escudié and H. Ranaivonjatovo, *Adv. Organomet. Chem.*, **44**, 113 (1999).
33. M. Kira, T. Iwamoto, T. Maruyama, C. Kabuto and H. Sakurai, *Organometallics*, **15**, 3767 (1996).
34. S. Masamune and L. R. Sita, *J. Am. Chem. Soc.*, **107**, 6390 (1985).
35. M. A. D. Bona, M. C. Cassani, J. M. Keates, G. A. Lawless, M. F. Lappert, M. Stürmann and M. Weidenbruch, *J. Chem. Soc., Dalton Trans.*, 1187 (1998).
36. (a) D. E. Goldberg, D. H. Harris, M. F. Lappert and K. M. Thomas, *J. Chem. Soc., Chem. Commun.*, 261 (1976).
(b) D. E. Goldberg, P. B. Hitchcock, M. F. Lappert, K. M. Thomas, A. J. Thorne, T. Fjeldberg, A. Haaland and B. E. R. Schilling, *J. Chem. Soc., Dalton Trans.*, 2387 (1986).
37. K. W. Klinkhammer and W. Schwarz, *Angew. Chem., Int. Ed. Engl.*, **34**, 1334 (1995).
38. K. W. Klinkhammer, T. F. Fässler and H. Grützmacher, *Angew. Chem., Int. Ed. Engl.*, **37**, 124 (1998).
39. M. Kira, R. Yauchibara, R. Hirano, C. Kabuto and H. Sakurai, *J. Am. Chem. Soc.*, **113**, 7785 (1991).
40. K. W. Zilm, G. A. Lawless, R. M. Merrill, J. M. Millar and G. G. Webb, *J. Am. Chem. Soc.*, **109**, 7236 (1987).
41. N. Tokitoh, M. Saito and R. Okazaki, *J. Am. Chem. Soc.*, **115**, 2065 (1993).
42. T. A. Schmedake, M. Haaf, Y. Apeloig, T. Müller, S. Bukalov and R. West, *J. Am. Chem. Soc.*, **121**, 9479 (1999).
43. K. M. Baines and J. A. Cooke, *Organometallics*, **11**, 3487 (1992).
44. M. Weidenbruch, S. Willms, W. Saak and G. Henkel, *Angew. Chem., Int. Ed. Engl.*, **36**, 2503 (1997).

45. H. Schäfer, W. Saak and M. Weidenbruch, *Angew. Chem., Int. Ed.*, **39**, 3703 (2000).
46. J. March, *Advanced Organic Chemistry*, Wiley, 4th Edn. New York, 1992.
47. C. Jouany, S. Mathieu, M.-A. Chaubon-Deredempt and G. Trinquier, *J. Am. Chem. Soc.*, **116**, 3973 (1994).
48. M. Ichinohe, T. Matsuno and A. Sekiguchi, *Chem. Commun.*, 183 (2001).
49. N. Fukaya and A. Sekiguchi, unpublished results.
50. V. Ya. Lee, M. Ichinohe and A. Sekiguchi, submitted.
51. N. Fukaya, M. Ichinohe, Y. Kabe and A. Sekiguchi, submitted.
52. K. M. Baines, J. A. Cooke, C. E. Dixon, H. W. Liu and M. R. Netherton, *Organometallics*, **13**, 631 (1994).
53. K. M. Baines, J. A. Cooke and J. J. Vittal, *J. Chem. Soc., Chem. Commun.*, 1484 (1992).
54. R. Breslow, *J. Am. Chem. Soc.*, **79**, 5318 (1957).
55. R. Breslow and C. Yuan, *J. Am. Chem. Soc.*, **80**, 5991 (1958).
56. R. Breslow, *Pure Appl. Chem.*, **28**, 111 (1971).
57. E. D. Jemmis, G. N. Srinivas, J. Leszczynski, J. Kapp, A. A. Korkin and P. v. R. Schleyer, *J. Am. Chem. Soc.*, **117**, 11361 (1995).
58. A. Sekiguchi, M. Tsukamoto and M. Ichinohe, *Science*, **275**, 60 (1997).
59. A. Sekiguchi, M. Tsukamoto, M. Ichinohe and N. Fukaya, *Phosphorus Sulfur Silicon Relat. Elem.*, **124–125**, 323 (1997).
60. S. H. Strauss, *Chem. Rev.*, **93**, 927 (1993).
61. S. R. Bahr and P. Boudjouk, *J. Org. Chem.*, **57**, 5545 (1992).
62. J. C. W. Chien, W.-M. Tsai and M. D. Raush, *J. Am. Chem. Soc.*, **113**, 8570 (1991).
63. L. Jia, X. Yang, A. Ishihara and T. J. Marks, *Organometallics*, **14**, 3135 (1995).
64. L. Jia, X. Yang, C. L. Stern and T. J. Marks, *Organometallics*, **16**, 842 (1997).
65. M. Ichinohe, N. Fukaya and A. Sekiguchi, *Chem. Lett.*, 1045 (1998).
66. A. Sekiguchi, N. Fukaya and M. Ichinohe, *Phosphorus Sulfur Silicon Relat. Elem.*, **150–151**, 59 (1999).
67. A. Sekiguchi, N. Fukaya, M. Ichinohe and Y. Ishida, *Eur. J. Inorg. Chem.*, 1155 (2000).
68. A. Bondi, *J. Phys. Chem.*, **68**, 443 (1964).

Contents of Volume 1